S0-AXW-594

MasteringBIOLOGY

Your gateway to the most widely-used online biology self-study materials and more...

With your purchase of a new copy of Freeman's *Biological Science*, Third Edition, you should have received a **Student Access Kit for MasteringBiology™**. This kit contains instructions and a code for accessing this dynamic website. Your Student Access Kit looks like this:

DON'T THROW YOUR ACCESS KIT AWAY!

If you did not purchase a new textbook or cannot locate the Student Access Kit, you can purchase your subscription to MasteringBiology™ online with a major credit card by selecting Buy Now. To do so, go to www.masteringbio.com. Identify the exact title and edition of your textbook when prompted.

INSTRUCTORS:

Look for your "Instructor Access Kit" that you received with your book. Contact your Pearson Benjamin Cummings Sales Representative if you need assistance.

MasteringBIOLOGY What is MasteringBiology™?

MasteringBiology™ offers two valuable learning systems:

1. The **Study Area** for studying on your own or in a study group
2. **MasteringBiology™ Assignments** that your instructor may require

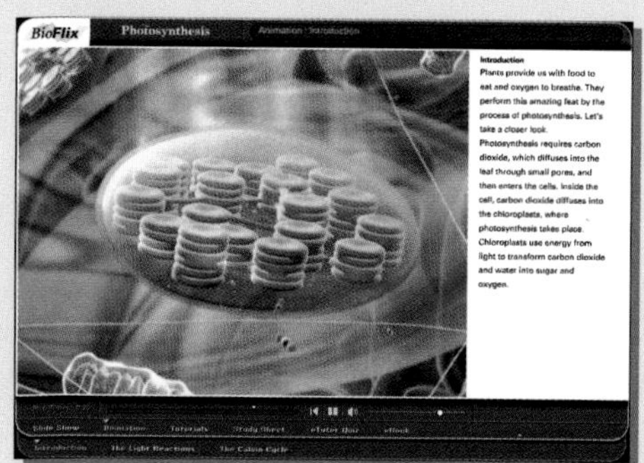

The **Study Area** provides state-of-the-art BioFlix™ 3-D Animations, Discovery videos, Web Animations, Cell Biology videos, interactive eBook, answers to textbook questions and much more (see other side).

MasteringBiology™ Assignments may include Tutorials and other Multiple Choice Questions required by your instructor. MasteringBiology™ tutorials coach you with feedback specific to your needs and offer hints when you get stuck, optimizing your study time and maximizing your learning.

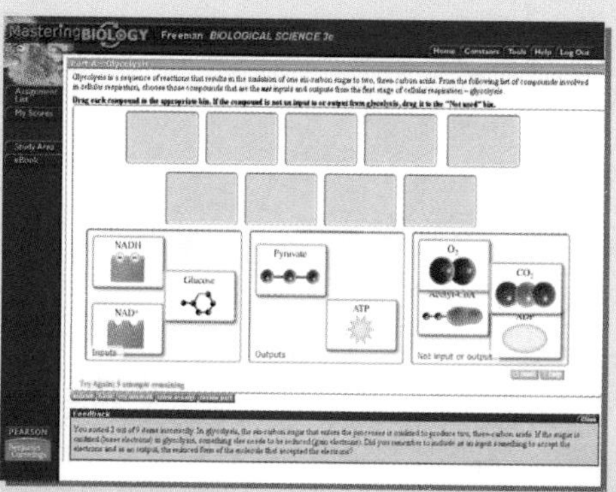

Features of the MasteringBiology™ Study Area for Freeman's *Biological Science,* Third Edition

CHAPTER GUIDE

Prepare for the test by reviewing important information outlined in the Chapter Guide. The Chapter Guide provides a quick overview of key concepts, practice quizzes, and review questions to help students study for the exam. Link directly to the BioFlix™ 3-D animations, Discovery Videos, BioSkills, Cell Biology Animations, and interactive eBook to help students master biology's most difficult topics.

BIOFLIX™ 3-D ANIMATIONS

BioFlix

BioFlix™ are 3-D movie-quality animations with carefully constructed student tutorials, labeled slide shows, study sheets, and quizzes which help bring biology to life. Topics include tour of an animal cell, tour of a plant cell, cellular respiration, photosynthesis, mitosis, meiosis, protein synthesis, water transport in plants, how neurons work, and muscle contraction.

VIDEOS

Watch brief Discovery Channel™ video clips on 29 different topics including antibiotic resistance, fighting cancer, and introduced species.

WEB ANIMATIONS

Web Animation Learn about biological concepts in a visual format with online web animations. Animations include pre-quizzes and post-quizzes to test students' understanding of dynamic concepts and processes. Additional Cell Biology Animations provide vivid images of the functions and processes of the cell.

GRAPHIT!

Graphing tutorials show students how to plot, interpret, and critically evaluate real data.

CHAPTER QUIZZES

Students can assess their comprehension of material with 20 multiple-choice quiz questions for each chapter.

ANSWER KEY

Find answers to the book's figure and table caption questions and exercises, "you should be able to..." activities, and end-of-chapter questions.

ART

View and print artwork from the textbook to follow along in class or use as a study tool in review sessions.

AUDIO GLOSSARY

Build your biology vocabulary with our audio glossary. Hear the correct pronunciation and learn the meaning of all key terms.

E-BOOK

Refer to a convenient online version of the book while you study.

BIOLOGICAL SCIENCE

ABOUT THE PHOTO:
He'e, *Octopus* sp. (juvenile), photographed at Midway Atoll National Wildlife Refuge, located in the Northwestern Hawaiian Islands, 29 March 2003, by David Liittschwager and Susan Middleton, authors of *Archipelago: Portraits of Life in the World's Most Remote Island Sanctuary* (National Geographic Society, 2005).

BIOLOGICAL SCIENCE

THIRD EDITION

SCOTT FREEMAN
University of Washington

San Francisco Boston New York
Cape Town Hong Kong London Madrid Mexico City Montreal Munich Paris Singapore Sydney Tokyo Toronto

Editor-in-Chief, Biology: Beth Wilbur
Sponsoring Editors: Andrew Gilfillan, Susan Winslow, Becky Ruden
Editorial Project Managers: Ann Heath, Sonia DiVittorio
Development Editors: Erin Mulligan, Mary Catherine Hager, Susan Weisberg
Senior Production Supervisor: Shannon Tozier
Associate Editor: Mercedes Grandin
Assistant Editor: Anna Amato
Editorial Assistant: Lisa Tarabokjia
Media Producer: Ericka O'Benar
Executive Directors of Development: Carol Trueheart, Deborah Gale
Executive Marketing Manager: Lauren Harp
Director of Marketing: Christy Lawrence
Managing Editor: Mike Early
Executive Managing Editor: Erin Gregg
Production Service and Composition: Pre-Press PMG
Illustrations: Kim Quillin, Imagineering Media Services, Inc., Greg Williams
Interior Design: Marilyn Perry
Cover Design: Kim Quillin, Marilyn Perry
Cover Production: Laura Wieglab and Side by Side Studios
Manufacturing Buyer: Mike Early and Michael Pene
Director, Image Resource Center: Melinda Patelli
Manager, Rights and Permissions: Zina Arabia
Visual Research Coordinator: Elaine Soares
Image Permission Coordinator: Debbie Latronica
Photo Researcher: Yvonne Gerin
Cover Printer: Phoenix Color
Printer and Binder: R.R. Donnelley, Willard

Cover Image: Octopus–he'e, *Octopus* sp. (juvenile): David Liittschwager/Susan Middleton

Library of Congress Cataloging-in-Publication Data

Freeman, Scott
Biological science / Scott Freeman.—3rd ed.
p. cm.
Includes index.
ISBN 0-13-224950-2 (student edition)—ISBN 0-13-224985-5 (professional copy)—ISBN 0-13-225308-9 (volume 1)—ISBN 0-13-232543-8 (volume 2)—ISBN 0-13-156816-7 (volume 3)
1. Biology. I. Title.
QH308. 2. F73 2007
570—dc22

2007036866

ISBN 10-digit 0-13-224950-2; 13-digit 978-0-13-224950-8 (Student edition)
ISBN 10-digit 0-13-224985-5; 13-digit 978-0-13-224985-0 (Professional copy)
ISBN 10-digit 0-13-225308-9; 13-digit 978-0-13-225308-6 (Volume 1)
ISBN 10-digit 0-13-232543-8; 13-digit 978-0-13-232543-1 (Volume 2)
ISBN 10-digit 0-13-156816-7; 13-digit 978-0-13-156816-7 (Volume 3)

1 2 3 4 5 6 7 8 9 10—DOW—11 10 09 08 07
www.aw-bc.com

Brief Contents

About the Author

Scott Freeman received his PhD. in Zoology from the University of Washington and was subsequently awarded an Alfred P. Sloan Postdoctoral Fellowship in Molecular Evolution at Princeton University. His current research focuses on the scholarship of teaching and learning—specifically, (1) how active learning and peer teaching techniques increase student learning and improve performance in introductory biology, and (2) how the levels of exam questions vary among introductory biology courses, standardized post-graduate entrance exams, and professional school courses. He has also done research in evolutionary biology on topics ranging from nest parasitism to the molecular systematics of the blackbird family. Scott teaches introductory biology for majors at the University of Washington and is co-author, with Jon Herron, of the standard-setting undergraduate text *Evolutionary Analysis*.

UNIT ADVISORS

An elite group of eleven content experts and star teachers worked with Scott and Kim on every aspect of the third edition. These advisors came to be treasured as they read and interpreted reviews, provided recommendations on outstanding recent papers to check, answered questions, and provided advice on an array of specific issues. The quality and accuracy of this book are a tribute to their efforts and skills.

Ross Feldberg, *Tufts University* (Unit 1)

David Wilson, *Parkland College* (Unit 1)

Paula Lemons, *Duke University* (Unit 2)

Greg Podgorski, *Utah State University* (Units 3 and 4)

George Gilchrist, *College of William and Mary* (Unit 5)

Brianna Timmerman, *University of South Carolina* (Unit 6)

Marc Perkins, *Orange Coast College* (Unit 6)

Michael Black, *California Polytechnic State University* (Units 6 and 8)

Diane Marshall, *University of New Mexico* (Unit 7)

James M. Ryan, *Hobart and William Smith Colleges* (Unit 8)

Alan Molumby, *University of Illinois, Chicago* (Unit 9)

ILLUSTRATOR

Kim Quillin combines expertise in biology and information design to create pedagogically effective and scientifically accurate visual representations of biological principles. She received her B.A. in Biology at Oberlin College and her Ph.D. in Integrative Biology (as a National Science Foundation Graduate Fellow) from the University of California, Berkeley, and has taught undergraduate biology at both schools. Students and instructors alike have praised Kim's illustration programs for *Biological Science*, as well as *Biology: A Guide to the Natural World*, by David Krogh, and *Biology: Science for Life*, by Colleen Belk and Virginia Borden, for their success at applying core principles of information design to convey complex biological ideas in a visually appealing manner.

Preface to Instructors

☞ STUDENTS, There is also a preface for you, located right before Chapter 1. It's called "Using this Book as a Tool for Learning." Please read it—it should help you get organized and be successful in this course. ***Scott Freeman***

This book is for instructors who want to help their students learn how to think like a biologist. The content knowledge, problem-solving ability, and analytical skills that this requires can help students become better human beings, in addition to preparing them for success in clinical medicine, scientific research, conservation, law, teaching, journalism, and other careers.

A course goal that focuses on thinking and learning, rather than simply memorizing, is important because today's biology students are going to be tomorrow's problem solvers. This is biology's century, not only because of the breathtaking pace of research but also because many of the most profound challenges we face today—resource shortages, overpopulation, species extinctions, drug resistance, global warming—are biological in nature. The world needs our students.

A Student-Centered Textbook

The first edition of *Biological Science* focused on offering a new approach to teaching biology—one that emphasized higher-order thinking skills over an encyclopedic grasp of what is known about biology. The second edition stayed true to this vision, but added topics and features that made the book easier for professors to use. The third edition also went through several rounds of revision driven by expert reviewers and advisors. Based on input from hundreds of professors around the world, I made thousands of changes to make the book even more accurate, current, and easy to use. But fundamentally, this edition is all about students. For the past three years, the book team and I have used insights from research on student learning—as well as direct feedback from students—as a way to create a better teaching tool.

I plunged into the literature on student learning, and members of the book team and I conducted two dozen focus groups with over 130 students. We asked students the same questions, over and over: What topics are most difficult in this chapter? What helped you "get it"? What tripped you up? How can this figure, or table, or passage of text teach better? Our goal was to create a text that is innovative, engaging, filled with the excitement that drives research, and inspired by data on how students learn.

Fairly quickly, our reading and research identified two fundamental problems:

1. *Novice learners have trouble picking out important information.* This is one of the most striking findings emerging from research on how people learn. It's also an issue that we hear about all the time as instructors. "Do we have to know X?" "Will Y be on the test?" We also see it when students come to office hours and open up their text—they've highlighted everything. Students need help figuring out which material is really important and which offers supporting details. This is crucial to their success, because if we're doing our job right, we're going to test them on the important stuff—not details that they'll forget five minutes after the exam.
2. *Novice learners have a terrible time with self-assessment.* How many times have students told you, "I understood the concepts *so well*, but just did badly on the exam"? Novice learners struggle to understand when they don't understand something. They'll sit in class or read the text and tell themselves, "Yeah, I get this, I get this," but then they crash on the exam. Experts are much more skeptical—they make themselves *use* information and ideas before they're confident they understand what's going on.

Fundamentally, our task is simple: We need to help our students become better students. Textbooks should help learners acquire the skills they need to make the novice-to-expert transition.

Supporting Novice Learners

The Gold Thread—"Learn It"

Our response to the "can't pick out the important points" problem is a battery of tools, highlighted in gold throughout the text.

- **Key Concepts** are listed at the start of the chapter. When material related to these key concepts is presented in the chapter itself, it is flagged with a gold bullet.
- **Chapter Summaries** revisit each of the key concepts. In this way, the big ideas in each chapter are laid out at the start, developed in detail, and then summarized.
- **Check Your Understanding** boxes appear at the ends of key sections within each chapter. Each box briefly summarizes

one or two fundamental points—the key ideas that students ought to have mastered before they move on.

- **Highlighted Passages** help students focus on particularly important information throughout the text.
- **Summary Tables** pull information together into a compact format that is easy to review.

In effect, the gold thread offers students expert guidance on picking out and focusing on the important, unifying points in an information-laden discipline. In addition, and in the same spirit,

- **"Pointing Hands"** in illustrations act like your hand at the whiteboard, guiding students' attention to a figure's central teaching points.

The Blue Thread—"Practice It"

Our response to the self-assessment problem is an array of questions and exercises, highlighted in blue throughout the text.

- **In-text "You should be able to's"** offer exercises on topics that professors and students have identified as the most difficult concepts in each chapter.
- **Caption Questions and Exercises** challenge students to critically examine the information in a figure or table—not just absorb it.
- **Check Your Understanding** boxes present two to three tasks that students should be able to complete in order to demonstrate a mastery of specified key ideas.
- **Chapter Summaries** include "You should be able to" problems or exercises related to each of the key concepts declared in the gold, "Learn It" thread.
- **End-of-Chapter Questions** are organized around Bloom's taxonomy of learning, so students can test their understanding at the knowledge, comprehension, and application levels.

The fundamental idea is that if students really understand a piece of information or a concept, they should be able to do something with it.

Supporting Visual Learners

Kim Quillin—this book's illustrator—is a 1-in-6.6-billion talent. She combines superb academic training in biology (a Ph.D. from the University of California, Berkeley), insights from Edward Tufte's research on information architecture, impressive artistic talent, and a teacher's sensitivity to students.

Kim's goals in the first edition were to build a visual narrative that was (1) a direct extension of the text narrative and (2) guided by rigorous attention to the latest advances in information design. For example, figures that illustrate stepwise processes in biology were presented in a standardized, linear, and intuitive format, to spare students the "speed bumps" of convoluted compositions. In all of the figures, the drawings, colors, and labels were placed with surgical precision to maximize cleanliness and present a clear hierarchy of information. And to help bring a narrative voice to the figures, Kim integrated the "pointing hand" feature—inspired by National Academy of Sciences member M.A.R. Koehl—as a quiet and friendly way of highlighting important or challenging information such as data trends in graphs, or implications of phylogenetic trees.

The second edition stayed faithful to the original art philosophy, adding details requested by professors, formalizing experiments into boxes that made the experimental process consistent and explicit, and increasing the vibrancy and color of the drawings. Refinements in the third edition have focused on bringing the content of many individual figures up-to-date and honing the clarity, impact, and student-friendliness of every figure.

Supporting Skill Building

When instructors write learning objectives for introductory biology courses, they are paying close attention to skill building, in addition to the traditional emphasis on mastering content and concepts and the newer focus on developing higher-order thinking skills.

To aid skill building in introductory biology, I've added a set of nine new appendixes, called BioSkills, to the end matter of the book—just before the glossary. The BioSkills are meant to provide background on skills and techniques that are used throughout biological science. In addition, some of the BioSkills should help shore up problems with student preparation in mathematics and chemistry. They are called out at relevant points in the text. Please point them out to students who are having trouble with specific skills.

1. Reading Graphs
2. Reading a Phylogenetic Tree
3. Using Statistical Tests and Interpreting Standard Error Bars
4. Reading Chemical Structures
5. Using Logarithms
6. Making Concept Maps
7. Using Electrophoresis to Separate Molecules
8. Observing Microscopic Structures and Processes
9. Combining Probabilities

Serving a Community of Teachers

There is nothing that inspires me more than watching a passionate teacher work with motivated students. Teaching and learning is the essence of humanity, of who we are.

I write *Biological Science* because it's my way of being part of that work. As a textbook author, my greatest reward comes from interacting with inspired instructors and students from around the world—through e-mails, phone calls, focus groups, reviews, seminars, and teaching workshops.

Introductory biology courses are undergoing a slow and steady, yet remarkable, change: from a memorization-driven, largely passive exercise to a dynamic, active interchange that emphasizes higher-order thinking skills. The change is good. Instructors are stimulated and having more fun. They're approaching teaching problems in a hypothesis-testing framework, collecting data, and changing course designs based on empirical evidence. Students are performing at a higher level and should be better prepared for graduate school, professional school, and careers related to biology. My hope is that *Biological Science* can support this change.

Thank you for considering this text, for your passion about biology, and for your work on behalf of your students. What you do is *important*.

Scott Freeman
University of Washington

This book is dedicated to the world's greatest profession: teaching. Teachers are like Johnny Appleseed—they sow seeds, but seldom get to see the trees or fruit. This edition reflects in particular four teachers who had an exceptional impact on me: Vern Bailey, Owen Jenkins, Sievert Rohwer, and Barbara Wakimoto. Thank you.

Detailed Contents

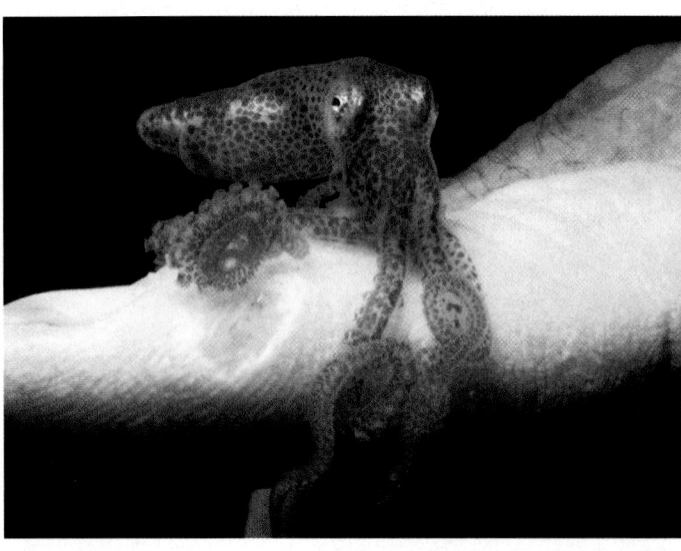

Acknowledgments

Expert Feedback

Work on this edition was organized around two pillars: feedback from our advisory board and peer reviewers, and feedback from students.

Advisory Board

A cadre of highly select advisors analyzed reviews, provided citations for particularly important recent papers, and worked with me and Kim Quillin to brainstorm ideas for improving each chapter and unit. It was a joy to work with them, and their expertise and perspective were vital to this edition.

Unit 1 Ross Feldberg, *Tufts University*
David Wilson, *Parkland College*
Unit 2 Paula Lemons, *Duke University*
Unit 3 Greg Podgorski, *Utah State University*
Unit 4 Greg Podgorski, *Utah State University*
Unit 5 George Gilchrist, *William & Mary College*
Unit 6 Brianna Timmerman, *University of South Carolina*
Marc Perkins, *Orange Coast College*
Michael Black, *California Polytechnic State University, San Luis Obispo*
Unit 7 Diane Marshall, *University of New Mexico*
Unit 8 Jim Ryan, *Hobart and William Smith Colleges*
Michael Black, *California Polytechnic State University, San Luis Obispo*
Unit 9 Alan Molumby, *University of Illinois, Chicago*

Reviewers

Peer review is the backbone of scientific publication. Acting as a reviewer is a fundamental component of our service as professional biologists, and reviewing chapters from an introductory text is one of the most important things we can do to improve the training of the next generation of biologists. I am deeply grateful for the insights offered by the following reviewers, who drew on their extensive content expertise, teaching experience, and research practice.

Marc Albrecht, *University of Nebraska, Kearney*
David Asch, *Youngstown State University*
Mariette Baxendale, *University of Missouri, St. Louis*
Greg Beaulieu, *University of Victoria*
Christopher Beck, *Emory University*
Peter Berget, *Carnegie Mellon University*
Janet Bester-Meredith, *Seattle Pacific University*
Cynthia Bishop, *Seattle Pacific University*
Michael Black, *California Polytechnic State University, San Luis Obispo*
Anthony Bledsoe, *University of Pittsburgh*
Patrice Boily, *University of New Orleans*
Scott Bowling, *Auburn University*
Maureen Brandon, *Idaho State University*
John Briggs, *Arizona State University*
Art Buikema, *Virginia Tech University*
Kim Caldwell, *University of Alabama*
Jeff Carmichael, *University of North Dakota*
Patrick Carter, *Washington State University*
John Caruso, *University of New Orleans*
Mary Lynn Casem, *California State University, Fullerton*
Cynthia Church, *Metropolitan State College*
Alison Cleveland, *University of South Florida*
Anita Davelos Baines, *University of Texas, Pan-American*
Jeff Demuth, *Indiana University*
Todd Duncan, *University of Colorado, Denver*
Johnny El Rady, *University of South Florida*
Peter Facchini, *University of Calgary*
Zen Faulkes, *University of Texas, Pan-American*
Ross Feldberg, *Tufts University*
Lewis Feldman, *University of California, Berkeley*
Jonathan Fisher, *St. Louis University*
Steve Frankel, *Northeastern University*
Amy Frary, *Mount Holyoke College*
Jed Fuhrman, *University of Southern California*
Caitlin Gabor, *Texas State University, San Marcos*
Michael Gaines, *University of Miami*
John R. Geiser, *Western Washington University*
D. Timothy Gerber, *University of Wisconsin, La Crosse*
Lisa Gerheart, *University of California, Davis*
Kathy Gillen, *Kenyon College*
Florence K. Gleason, *University of Minnesota, Twin Cities*
John Godwin, *North Carolina State University*
Reuben Goforth, *Michigan State University*
Linda Green, *University of Virginia*
Joe Harsh, *University of North Carolina, Charlotte*
Clare Hays, *Metropolitan State University*
Kerry Heafner, *University of Louisiana, Monroe*
Harold Heatwole, *North Carolina State University*
Brian Helmuth, *University of Southern California*
Susan Hengeveld, *Indiana University*
Mark Hens, *University of North Carolina, Greensboro*
Albert Herrera, *University of Southern California*
Malcolm Hill, *University of Richmond*
Ron Hoham, *Colgate University*
Kelly Howe, *University of New Mexico*
Cindy Johnson-Groh, *Gustavus Adolphus College*

Walter Judd, *University of Florida*
Nancy Kaufmann, *University of Pittsburgh*
Loren Knapp, *University of South Carolina*
Scott Knight, *Montclair State University*
Paul Lagos, *University of Mississippi*
Paula Lemons, *Duke University*
Vicky Lentz, *SUNY, College at Oneonta*
Georgia Lind, *Kingsborough Community College*
Chris Little, *University of Texas, Pan-American*
Andrea Lloyd, *Middlebury College*
Christopher Loretz, *University of Buffalo*
Cindy Martinez Wedig, *University of Texas, Pan-American*
Andrew McCubbin, *Washington State University*
Kelly McLaughlin, *Tufts University*
Victoria McMillan, *Colgate University*
Jennifer Miskowski, *University of Wisconsin, La Crosse*
Alan Molumby, *University of Illinois, Chicago*
Daniel Moon, *University of North Florida*
Mike Muller, *University of Illinois, Chicago*
Dana Nayduch, *Georgia Southern University*
Jacalyn S. Newman, *University of Pittsburgh*
Harry Nickla, *Creighton University*
Mary Jane Niles, *University of San Francisco*
Shawn Nordell, *St. Louis University*
Celia Norman, *Arapahoe Community College*
Nicole Obert, *University of Illinois, Urbana-Champaign*
John Osterman, *University of Nebraska, Lincoln*
John Palisano, *University of the South, Sewanee*
Glenn Parsons, *University of Mississippi, Oxford*
Andrew Pease, *Villa Julie College*
Deborah Pelli, *University of North Carolina, Greensboro*
Shelley A. Phelan, *Fairfield University*
Debra Pires, *University of California, Los Angeles*
Peggy Pollak, *Northern Arizona University*
Harvey Pough, *Rochester Institute of Technology*
Colin Purrington, *Swarthmore College*
Margaret Qazi, *Gustavus Adolphus College*
Rajinder Ranu, *Colorado State University*
Pamela C. Rasmussen, *Michigan State University*
Ann E. Rushing, *Baylor University*
James Ryan, *Hobart and William Smith Colleges*
Adam Ryburn, *SUNY, College at Oneonta*
Margaret Saha, *College of William and Mary*
Mark Sandheinrich, *University of Wisconsin, La Crosse*
Glenn Sauer, *Fairfield University*
Stephen G. Saupe, *St. John's University*
Andrew Scala, *Dutchess Community College*
Richard Showman, *University of South Carolina, Columbia*
Walter Shriner, *Mt. Hood Community College*
Sue Simon-Westendorf, *Ohio University*
Mark Spiro, *Bucknell University*
Paul Stapp, *California State University, Fullerton*
Scott Steinmaus, *California Polytechnic State University, San Luis Obispo*
John Stiller, *Eastern Carolina University*
John Stolz, *Duquesne University*
Kirk A. Stowe, *University of South Carolina*
Brianna Timmerman, *University of South Carolina*
Martin Tracey, *Florida International University*
Ashok Updhyaya, *University of South Florida*
Ann Vogel, *Illinois State University*
Fred Wasserman, *Boston University*
Elizabeth Weiss, *University of Texas, Austin*
Susan Whittemore, *Keene State College*
Ted Zerucha, *Appalachian State University*

Accuracy Reviewers

Once several rounds of peer review are completed and revised chapters and figures are produced, we rely on accuracy reviewers to check every page for errors. This work is demanding intellectually and has to be done under time pressure. The accuracy reviewers for this edition were exceptionally talented and timely.

Unit 1 Wayne Becker, *University of Wisconsin, Madison*
Unit 2 James Manser, *Harvey Mudd College (formerly)*
Unit 3 Peter Berget, *Carnegie Mellon University*
Mary Rose Lamb, *University of Puget Sound*
Unit 4 James Manser, *Harvey Mudd College (formerly)*
Unit 5 Jeffrey Feder, *University of Notre Dame*
Andrew Forbes, *University of Notre Dame*
Andrew Michel, *University of Notre Dame*
Tom Powell, *University of Notre Dame*
Unit 6 Laura Baumgartner, *University of Colorado, Boulder*
Michael Black, *California Polytechnic State University, San Luis Obispo*
Kimberly Erickson, *University of Colorado, Boulder*
Steve Trudell, *University of Washington, Seattle*
Unit 7 Susan Waaland, *University of Washington, Seattle*
Unit 8 Warren Burggren, *University of North Texas*
Susan Whittemore, *Keene State College*
Unit 9 Mark Johnston, *Dalhousie University*

BioSkills Julie Aires, *Florida Community College at Jacksonville;* Ross Feldberg, *Tufts University;* George Gilchrist, *William and Mary College;* Doug Luckie, *Michigan State University;* Greg Podgorski, *Utah State University*

Correspondents

I am grateful to colleagues who take the initiative to contact me directly or through my publisher to make suggestions on how to improve the text and figures. Please never hesitate to do this—I take your comments to heart, in the spirit of a shared commitment to improved student learning. This list also includes friends and colleagues who were kind enough to respond to emails or calls from me, asking for ideas on how to clarify specific topics.

Julie Aires, *Florida Community College, Jacksonville*
Gerald Borgia, *University of Maryland*
Scott Bowling, *Auburn University*
Elizabeth Cowles, *Rice University*
Fred Delcomyn, *University of Illinois, Urbana-Champaign*
Leslie Dendy, *University of New Mexico, Los Alamos*
John Dudley, *University of Illinois, Urbana-Champaign*
Larry Forney, *University of Idaho*
Arthur Gibson, *University of California, Los Angeles*
Matt Gilg, *University of Northern Florida*
Jean Heitz, *University of Wisconsin, Madison*
Jack Hogg, *University of Montana*
Johnathan Kupferer, *University of Illinois, Chicago*
Hans Landel, *Edmonds Community College*
Frederick Lanni, *Carnegie Mellon University*
Andi Lloyd, *Middlebury College*
Carmen Mannella, *Wadsworth Center, SUNY Albany*
Andrew McCubbin, *Washington State University*
Tim Nelson, *Seattle Pacific University*
Shawn Nordell and students, *University of St. Louis*
Carol Pollock, *University of British Columbia*
Joelle Presson, *University of Maryland*
William Saunders, *LaGuardia Community College*
David Senseman, *University of Texas, San Antonio*
Bryan Spohn, *Florida Community College, Jacksonville*
Scott Steinmaus, *California Polytechnic State University, San Luis Obispo*
Judy Stone, *Colby College*
Dean Wendt, *California Polytechnic State University, San Luis Obispo*

Student Feedback

The second pillar of this edition—in addition to the role played by advisors and reviewers—was an extensive series of focus groups with students who were currently taking introductory biology or who had just completed the course.

Student Focus Group Coordinators

Planning and implementing the student focus groups would have been impossible without the support of key faculty members, who went out of their way to provide opportunities for their students to be heard.

Julie Aires, *Florida Community College, Jacksonville*
Frank Cantelmo, *St. Johns University, New York*
Matt Gilg, *University of Northern Florida*
Bill Hoese, *California State University, Fullerton*
John Nagey, *Scottsdale Community College*
Debra Pires, *University of California, Los Angeles*
Emily Taylor, *California Polytechnic State University, San Luis Obispo*
John Weser, *Scottsdale Community College*

Student Focus Group Participants

The students who attended focus groups were asked three questions about chapters they were assigned to read: (1) what were the most difficult concepts, (2) why were they hard, and (3) what helped you finally get it? We usually had them work in groups, and when they reported back to us, we never failed to be deeply impressed by the quality of their ideas and their ability to articulate them. It is not possible to overstate how important student feedback was to this edition. Combined with the superb advice I was getting from advisors, reviewers, and other colleagues, I had a wealth of ideas on how to make each chapter work better for both instructors and learners. These students were *inspiring*.

California Polytechnic State University, San Luis Obispo
Jenna Arruda, Katie Camfield, Benjamin Capper, Mandsa Chandra, Rebekah Clarke, Annalisa Conn, Marisa Crawford, Katie Duffield, T.J. Eames, Megan Fay, Margaret Hackney, Steffani Hall, Gemma Hill, John Kong, Taylor Lindblom, Adam Marre, Vik Mediratta, Serena Moen, Sunil Patel, Corinne Ross, Teresa Sais, Jessie Singer, Stephanie Szeto, Kelsey Tallon, Gregory Thurston, Greg Vidovic, Melody Wilkinson, Taiga Young

California State University, Fullerton
Redieat Assefa, Josemari Feliciano, Civon Gewelber, Sarah Harpst, Jeff Kuhnlein, Linda Ong, T. Richard Parenteau, Robert Tran, Nicole Bournival

Fairfield University
Sally Casper, Tamika Dickens, Pryce Gaynor, Cindi Munden

Florida Community College, Jacksonville
Algen Albritten III, Danielle Boss, Chantel Callier, Eugenia Cruz, Lauren Faulkner, Jonathan Hopkins, Chantae Knight, David Lambert, Amber McCurdy, Tara Pladsen, Lauren Spruiell, Courtney Torgeon, Theresa Tran

LaGuardia Community College
Kristine Azzoli, Felicita Gonzalez, Pedro Granados, Mike Levine, Kris Ragoonath, Maria Reyes

Scottsdale Community College
Tatum Arthur, Shadi Asayesh, Angela Bikl, Abrey Britt, Drew Bryck, Jason Butler, Cindy Clifton, Dean Doty, Tannaz Farahani, Bethany Garcia, Jeff Godfrey, Troy Graziadei, Dina Habhab, Loreley Hall, Crista Jackson, Paul Krueger, David Levine, Chad Massena, Jessica Massena, Brian Martinez, Sam Mohmand, Esther Morantz, Jill Patel, Staci Puckett, Rebecca Rees, David Rosenbaum, Samantha Schrepel, Chris Schroeder, Kelsey Thomsen, Chris Volpe, Jamie Wagner, Lianne Wharton

St. Johns University, New York
Diana Carroccia, Milea Emmons, Tunc Ersoy, Blayre Linker, Zain Mirza, Mohammed Sheikh, Richardson Talarera, Michael Weinberg, Win Aung Yeni

University of California, Los Angeles
Farhan Banani, Stephanie Davis, Krystal De La Rosa, Samantha Hammer, Jennifer Okuda Hein, Neha Jashi, Marissa Lee, Calvin Leung, Venkat Mocherla, David Nguyen, Isabella Niu, Aya Obara

University of Maryland
Megan Berg, Lauren Fitzgerald, Megan Janssen, Avita Jones, Deidre Robinson

University of Northern Florida
Elysia Brennan, Christopher Ferrara, Lindsay Googe, Samantha Grogan, Marie Haagensen, Crystal Harris, Madeline Parhalo, Sherline Pierre, Stacy Pohlman, Nichole Polito, Sarah Lynn Redding, Megan Richardson, Megan Smart, Frank Snyder

Supplements Contributors

Our goal for the supplements package to accompany the Third Edition was to create learning tools that incorporate principles of active learning. Research shows that students do better in class when they are asked to use the material they are learning about. From a new workbook that encourages students to practice biology, to interactive web animations that test their knowledge, our supplements ask students to work with information, not just memorize it. My sincere thanks to the following people for their important contribution to the book's core teaching values.

Media Supplements

Marc Albrecht, *University of Nebraska, Kearney*
John Bell, *Brigham Young University*
Michael Black, *California Polytechnic State University, San Luis Obispo*
Warren Burggren, *University of North Texas*
Fannie Chen
Carol Chihara, *University of San Francisco*
Clarissa Dirks, *University of Washington, Seattle*
Kimberly Erickson, *University of Colorado, Boulder*
Zen Faulkes, *University of Texas, Pan-American*
Kathy Gillen, *Kenyon College*
Mary Catherine Hager
Susan Hengeveld, *Indiana University, Bloomington*
Loren Knapp, *University of South Carolina*
Jonathan Lochamy, *Georgia Perimeter College*
James Manser, *Harvey Mudd College (formerly)*
Cynthia Martinez-Wedig, *University of Texas, Pan-American*
Victoria McMillan, *Colgate University*
Andrew Pease, *Villa Julie College*
Debra Pires, *University of California, Los Angeles*
Pamela Rasmussen, *Michigan State University*
Susan Rouse, *Brenau University*
Christina Russin, *Northwestern University*
William Russin, *Northwestern University*
Cheryl Ingram Smith, *Clemson University*
Ellen M. Smith
Mark Spiro, *Bucknell University*
Eric Stavney, *DeVry University*
Michael Wenzel, *California State University, Sacramento*

Print Supplements

Marc Albrecht, *University of Nebraska, Kearney*
Charles Austerberry, *Creighton University*
Brian Bagatto, *University of Akron*
Jay Brewster, *Pepperdine University*
Warren Burggren, *University of North Texas*
Cynthia Giffen, *University of Wisconsin, Madison*
Jean Heitz, *University of Wisconsin, Madison*
Laurel Hester, *Cornell University*
Cynthia Martinez-Wedig, *University of Texas, Pan-American*
Jenny McFarland, *Edmonds Community College*
Greg Podgorski, *Utah State University*
Carol Pollock, *University of British Columbia*
Susan Rouse, *Southern Wesleyan University*
Elena Shpak, *University of Tennessee*
Sally Sommers Smith, *Boston University*
Briana Timmerman, *University of South Carolina*
David Wilson, *Parkland College*

Book Team

Finally, this edition would not have been published without the encouragement and support of our publishing partners at Pearson Arts & Science. I would like to acknowledge those individuals in the Pearson Science Group who helped make the Third Edition possible.

Prentice Hall

This edition was launched by Prentice Hall and then transferred—along with all other Prentice Hall biology titles—to their sister company, Benjamin Cummings. The editorial team at Prentice Hall was responsible for establishing the vision that directed this edition. In addition, they recruited the first advisors and media and supplement contributors, and implemented the initial set of student focus groups, prior to turning the project over to their Benjamin Cummings colleagues. I am grateful for their talent, energy, and friendship, and for the extraordinary efforts they made to make the management transition as smooth as possible. These people are Andrew Gilfillan (Sponsoring Editor), Ann Heath (Executive Project Manager), Erin Mulligan (Development Editor), Lisa Tarabokjia (Editorial Assistant), and Carol Trueheart (VP, Executive Director of Development).

Benjamin Cummings

The Benjamin Cummings editorial and production team welcomed *Biological Science* into their publishing house and moved the project forward through the final critical stages of development and into production. The team was initially led by Sponsoring Editor Susan Winslow, who brought a fresh perspective to the project. Then, Market Development Manager Becky Ruden took over the Sponsoring Editor reins with focused energy and verve. She is a bright, young star. Special thanks go to Project Manager Sonia DiVittorio, who has proven herself one of the sharpest talents in textbook publishing. Sonia's tireless pursuit of quality is evident on every page. In tribute, I have given her a new title at the company: Goddess of Bookmaking.

Thanks also go to Senior Production Supervisor Shannon Tozier, who led the production team with enormous skill and perseverance, and to Design Manager Marilyn Perry, who helped create the striking interior and cover designs. Others on the book team deserving acknowledgement include: Mary Catherine Hager and Susan Weisberg (Development Editors), Anna Amato (Assistant Editor), Mercedes Grandin (Associate Editor), Ericka O'Benar (Media Producer), Yvonne Gerin, Elaine Soares, and Debbie Latronica (photo research team), Christy Lawrence, Lauren Harp, Lillian Carr, and Mansour Bethany (marketing team), Josh Frost (market development), and Deborah Gale (Executive Director of Development).

Additionally, I'd like to thank Chris Thillen (copyeditor); Frank Purcell, Ellen Sanders, Pete Shanks (proofreaders); production editor Katy Faria and her production colleagues at Pre-Press PMG; and the illustration team at Imagineering for all the hours and energy put into making this edition the best it can be.

Finally, I'd like to extend my appreciation to the people whose vantage point allows them to assess the bigger picture. I am grateful to Linda Davis, President of Pearson Arts and Sciences, for helping me understand Benjamin Cummings' overarching publishing goals and how *Biological Science* fits into them. Thank you, Beth Wilbur, Editor-in-Chief of Biology, for providing editorial leadership and helping the book take root in its new home and allowing room for its personality to flourish. A very special thank-you goes to Paul Corey, President of the Pearson Science Group. Paul has been an advocate for the core values of *Biological Science* from its beginning over a decade ago. I am grateful for his continuing friendship and professional guidance.

Supplements

The Freeman *Biological Science* supplements package offers a robust suite of print and electronic tools designed to help instructors make the most of their limited time and to help students study efficiently.

INSTRUCTOR RESOURCES

- The entire textbook illustration program is available in JPEG format. All tables, photos, and line drawings with (and without) labels are individually enhanced for optimal in-class projection and are pre-loaded into chapter-correlated PowerPoint presentations.
- A second set of PowerPoint presentations consists of lecture outlines for each chapter, augmented by key text illustrations and hyperlinks to Web Animations.
- A third set of PowerPoint presentations allows select key figures to be presented in a step-by-step manner. In-text figure caption exercises with illustrated answers are included in this step-edited set.
- In-class active lecture questions correlated by chapter can be used with any classroom response system and are available in PowerPoint format.
- The Instructor Guide includes lecture outlines, active learning lecture activities, answers to end-of-chapter questions, and innovative material to help motivate and engage students.
- The Printed Test Bank and Computerized Test Bank have been peer reviewed and student tested. Test questions are ranked according to Bloom's taxonomy and the improved TestGen® software makes assembling tests much easier. The Test Bank is also available in Course Management systems and in Word® format on the Instructor Resources DVD.
- Four-color Transparency Acetates with every illustration from the text are available. Labels and images have been enlarged and modified to ensure optimal projection in a large lecture hall.

STUDENT RESOURCES

- NEW! *Practicing Biology: A Student Workbook* offers a variety of activities such as modeling, mapping, and graphing to help students with different learning styles visualize and understand biological processes.
- NEW! The eBook addresses the changing needs of students and instructors in the majors biology course by offering an electronic version of the text that links directly to animations, quizzes, and videos.
- NEW! Complimentary access to Pearson Tutor Services provides highly interactive one-on-one biology tutoring by qualified instructors seven nights a week during peak study hours. Students will be able to "drop in" for live online help, submit questions to an e-structor anytime, or pre-schedule a tutoring session with an e-structor to receive help at their convenience.
- The Study Guide presents a breakdown of key biological concepts, and helps students focus on the fundamentals of each chapter. It is designed in two parts to help students study more effectively. Part I is intended as a "survival guide," and Part II explores the material in the textbook, chapter by chapter.

MULTIMEDIA RESOURCES

- NEW! MasteringBIOLOGY MasteringBiology™ offers in-depth online tutorials on biology's toughest topics. These tutorials provide hints and feedback specific to each student's misconceptions. MasteringBiology also includes diagnostic test questions and is useful for "just-in-time" teaching.
- NEW! **BioFlix** BioFlix cover the most difficult biology topics with 3-D, movie-quality animations, labeled slide shows, carefully constructed student tutorials, study sheets, and quizzes that support all types of learners. Topics include Tour of an Animal Cell, Tour of a Plant Cell, Membrane Transport, Cellular Respiration, Photosynthesis, Mitosis, Meiosis, DNA Replication, Protein Synthesis, Water Transport in Plants, How Neurons Work, Synapses, and Muscle Contraction.
- Web Animation Web Animations add depth and visual clarity to the most important topics and processes described in the text. Animations include pre-quizzes and post-quizzes to help students prepare for exams.
- NEW! Discovery Channel video clips on 29 different topics include antibiotic resistance, fighting cancer, and introduced species. Additional Cell Biology animations provide vivid images of the functions and processes of the cell.
- NEW! Access to BioForum, a new online community forum created just for biology educators. Learn from your peers about how they're teaching difficult topics or what resources they use to help them teach. Post activities, handouts, Web links, or other tools you would like to share with forum participants.
- NEW! Video demonstration shows biology educators using active learning techniques you can incorporate in your classroom immediately.
- Course Management content for *Biological Science,* 3rd Edition, is available for institutions using **WebCT** or **Blackboard** and is also available in our nationally hosted **Course Compass** course management system. If desired, WebCT and Blackboard cartridges containing only the Test Bank are available for download.

FOCUS ON THE GOLD THREAD ...

The gold thread helps you pick out important information.

Start with *Key Concepts* on the first page of every chapter. Read these gold bullet points first to familiarize yourself with the chapter's big ideas.

Watch for material related to Key Concepts inside chapters; it will also be flagged with a gold bullet. Slow down and pay close attention to these highlighted passages.

Other particularly important information is highlighted in gold. Gold highlighting is always a signal to slow down and pay special attention.

The gold half of ***Check Your Understanding*** boxes summarizes important information from the section you just read. Stop and ask yourself, Do I *really* understand every bullet point?

Summary of Key Concepts at the ends of chapters is a good place to start reviewing when it's time to study for an exam. Key Concepts are revisited here in detail.

... PRACTICE WITH THE BLUE THREAD.

The blue thread helps you practice what you've learned.

The best way to succeed on exams is to *practice*. If you really understand a piece of information or a concept, **you should be able to** *do something with it.* If you can't, you haven't mastered the material.

Practice It activities are always flagged with blue bullets and blue type. Answers to all ***Practice It*** activities are available at **www.masteringbio.com** under the Study Area button.

Text passages flagged with a blue bullet, blue type, and the words *"you should be able to"* offer exercises on topics that professors and students have identified as most difficult. These are the topics most students struggle with on exams.

Many figures and some tables include blue ***Questions*** or ***Exercises*** to help you check your understanding of the material they present.

The blue half of ***Check Your Understanding*** boxes asks you to do something with the information in the gold half. If you can't complete these exercises, go back and re-read that section of the chapter.

Blue *"you should be able to"* exercises also let you test your understanding of each item in the ***Summary of Key Concepts*** at the ends of chapters.

PAY SPECIAL ATTENTION TO FIGURES AND SUMMARY TABLES

Half of this book is text. The other half is figures and tables. Figures always focus on important concepts. Tables present the raw material of science; they are data rich.

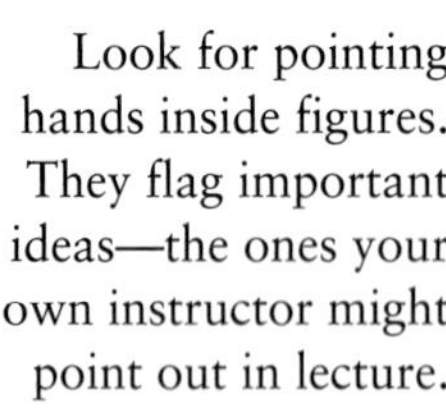

Look for pointing hands inside figures. They flag important ideas—the ones your own instructor might point out in lecture.

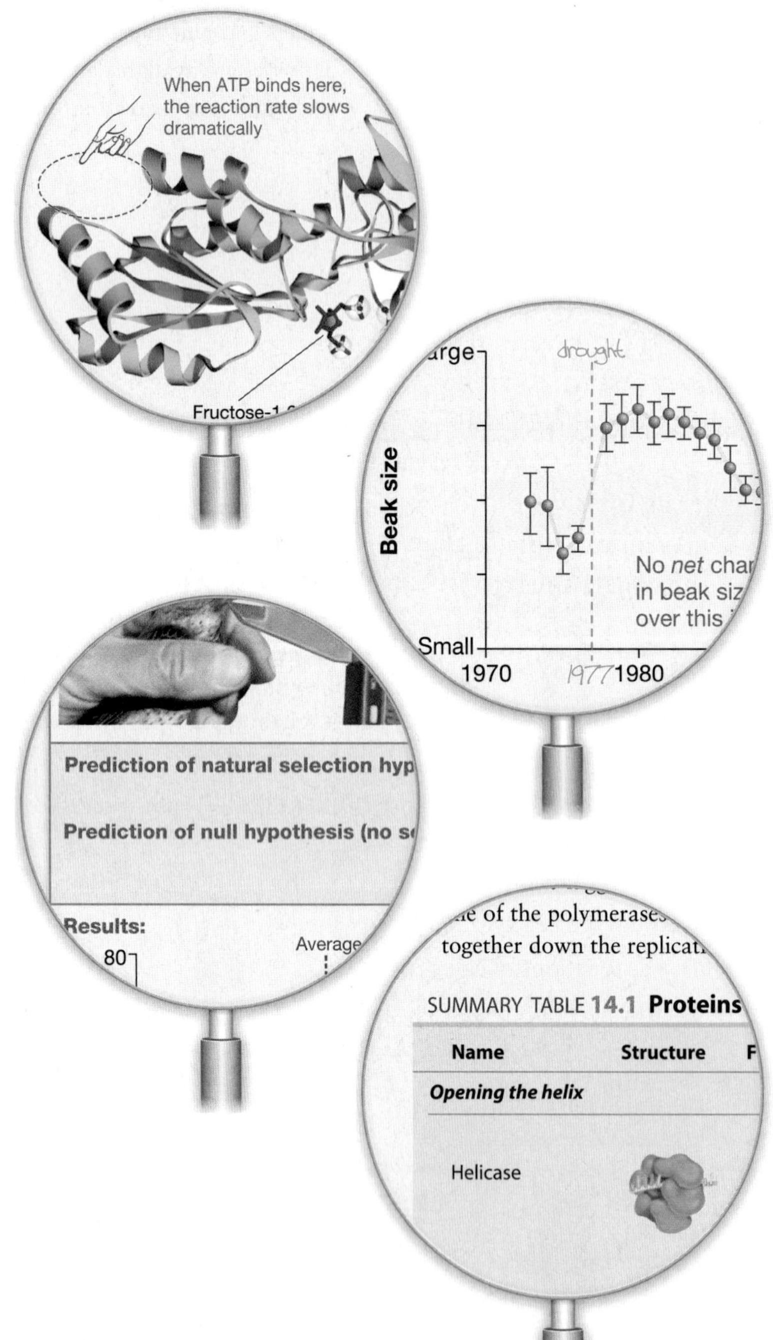

Some figures and tables ask you to to analyze data and fill in information. Take the time to do these exercises; they will help you master important concepts.

Experiment Boxes will help you understand how experiments are designed and give you practice interpreting data. Some leave space open for you to fill in null hypotheses, predicted outcomes, or the conclusion.

Summary Tables pull important information together in a format that's easy to review. A complete list of Summary Tables is provided inside the back cover.

Part I: ***The Study Area*** includes Web Animations and BioFlix.

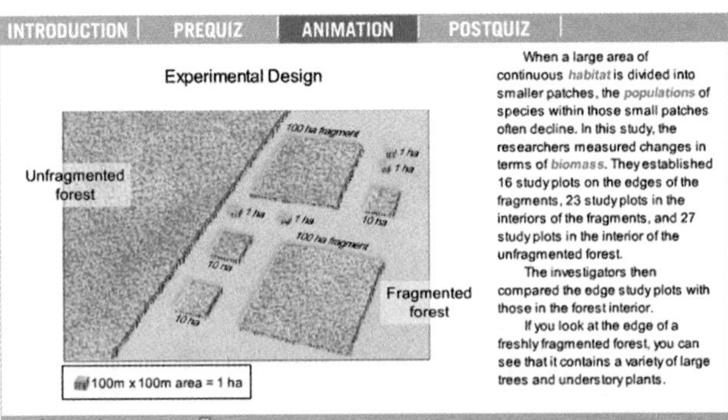

Web Animations Web Animation help you review important textbook topics. Pre- and post-quizzes let you test your proficiency.

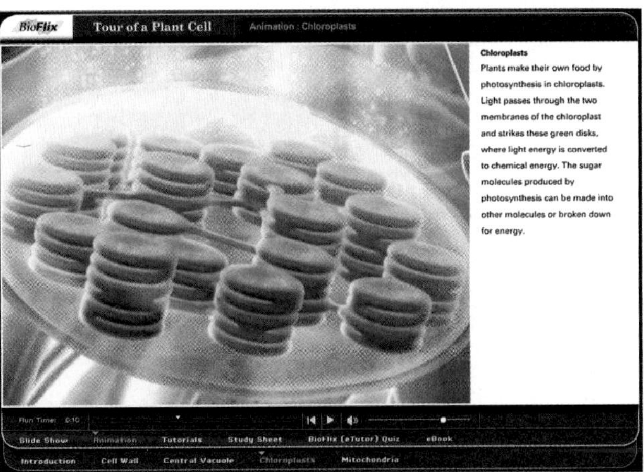

BioFlix BioFlix cover the most difficult biology topics with 3-D, movie-quality animations, labeled slide shows, carefully constructed tutorials, study sheets, and quizzes that support all types of learners. Topics include: Tour of an Animal Cell, Tour of a Plant Cell, Membrane Transport, Cellular Respiration, Photosynthesis, Mitosis, Meiosis, DNA Replication, Protein Synthesis, Water Transport in Plants, How Neurons Work, Synapses, and Muscle Contraction.

Part 2: ***MasteringBiology Instructor Assignments*** include in-depth tutorials on biology's toughest topics.

If you are working on a tutorial and get stuck, . . .

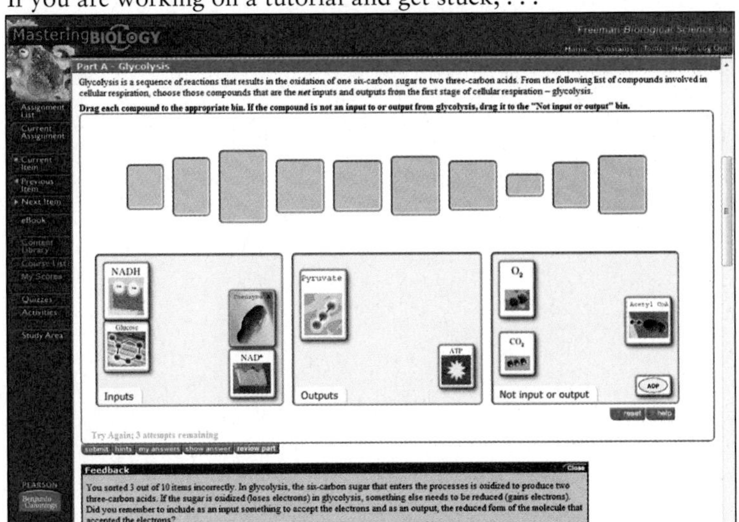

. . . you will receive instant feedback specific to your error.

You can also ask for hints . . .

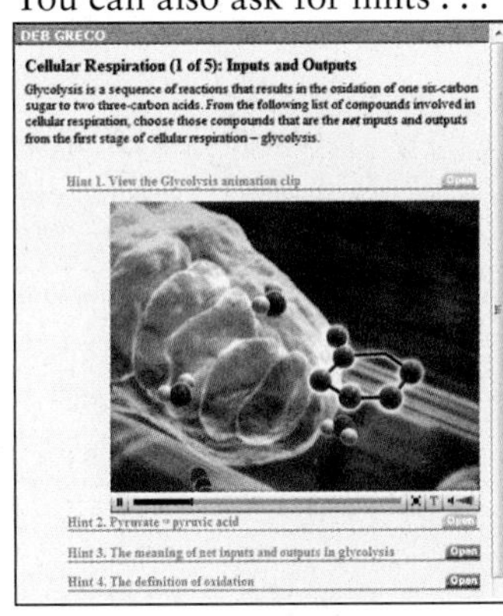

. . . to receive background information and Socratic questions and exercises designed to guide you to mastery of the topic.

Biology and the Tree of Life

1

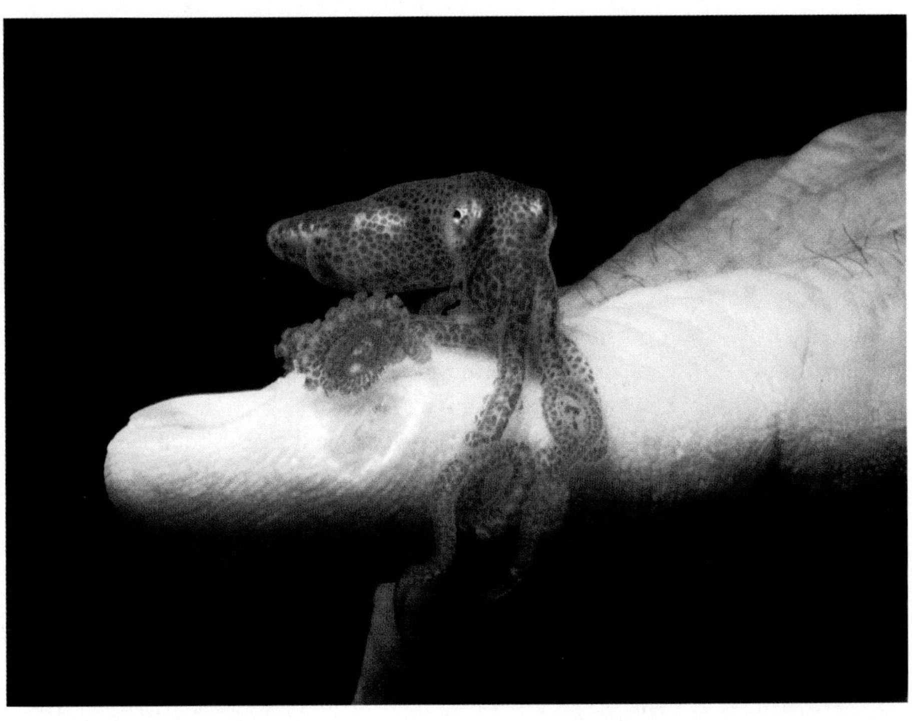

A young reef octopus, from Hawaii, also appears on the cover of this book. It is one of perhaps 10 million species alive today.

KEY CONCEPTS

- Biological science was founded with the development of (1) the cell theory, which proposes that all organisms are made of cells and that all cells come from preexisting cells, and (2) the theory of evolution by natural selection, which maintains that species change through time because individuals with certain heritable traits produce more offspring than other individuals do.
- A phylogenetic tree is a graphical representation of the evolutionary relationships among species. These relationships can be estimated by analyzing similarities and differences in traits. Species that share distinctive traits are closely related and are placed close to each other on the tree of life.
- Biologists ask questions, generate hypotheses to answer them, and design experiments that test the predictions made by competing hypotheses.

In essence, biology is a search for ideas and observations that unify our understanding of the diversity of life, from bacteria living in rocks a mile underground to octopuses and humans. Chapter 1 is an introduction to this search.

The goals of this chapter are to introduce the amazing variety of life-forms alive today, consider some fundamental traits shared by all organisms, and explore how biologists go about answering questions about life. The chapter also introduces themes that will resonate throughout this book: (1) analyzing how organisms work at the molecular level, (2) understanding why organisms have the traits they do in terms of their evolutionary history, and (3) helping you learn how to think like a biologist.

We begin by examining two of the greatest unifying ideas in all of science: the cell theory and the theory of evolution by natural selection. When these concepts emerged in the mid-1800s, they revolutionized the way that biologists understand the world. The cell theory proposed that all organisms are made of cells and that all cells come from preexisting cells. The theory of evolution by natural selection maintained that species have changed through time and that all species are related to one another through common ancestry. The theory of evolution by natural selection established that bacteria, mushrooms, roses, robins, and humans are all part of a family tree, similar to the genealogies or family trees that connect individual people.

A **theory** is an explanation for a very general class of phenomena or observations. The cell theory and the theory of evolution provide a foundation for the development of modern biology because they focus on two of the most general questions possible: What are organisms made of? Where did they come from? Let's begin by tackling the first of these two questions.

1.1 The Cell Theory

The initial conceptual breakthrough in biology—the cell theory—emerged after some 200 years of work. In 1665 Robert Hooke used a crude microscope to examine the structure of cork (a bark tissue) from an oak tree. The instrument magnified objects to just 30 times (303) their normal size, but it allowed Hooke to see something extraordinary. In the cork he observed small, pore-like compartments that were invisible to the naked eye (**Figure 1.1a**). These structures came to be called cells.

Soon after Hooke published his results, Anton van Leeuwenhoek succeeded in developing much more powerful microscopes, some capable of magnifications up to 3003 . With these instruments, Leeuwenhoek inspected samples of pond water and made the first observations of single-celled organisms like the *Paramecium* in **Figure 1.1b**. He also observed and described the structure of human blood cells and sperm cells.

In the 1670s a researcher who was studying the leaves and stems of plants with a microscope concluded that these large, complex structures are composed of many individual cells. By the early 1800s enough data had accumulated for a biologist to claim that *all* organisms consist of cells. But between then and now, biologists have developed microscopes that are tens of thousands of times more powerful than Leeuwenhoek's and have described well over a million new species. Did the biologist's claim hold up?

(a) The first view of cells: Robert Hooke's drawing from 1665

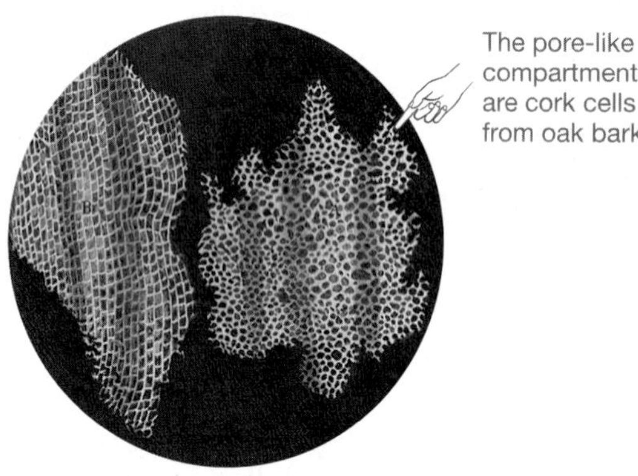

(b) Anton van Leeuwenhoek was the first to view single-celled "animalcules" in pond water.

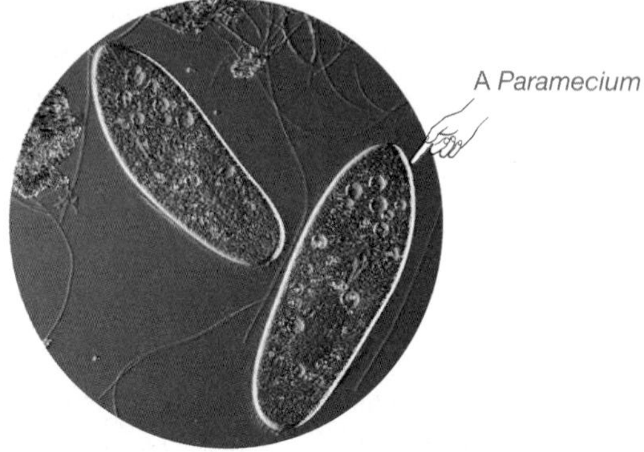

FIGURE 1.1 The Discovery of Cells.

Are *All* Organisms Made of Cells?

The smallest organisms known today are bacteria that are barely 200 nanometers wide, or 200 *billionths* of a meter. (See the endpapers of this book to review the metric system and its prefixes.) It would take 5000 of these organisms lined up end to end to span a millimeter. This is the distance between the smallest hash marks on a metric ruler. In contrast, sequoia trees can be over 100 meters tall. This is the equivalent of a 20-story building. Bacteria and sequoias are composed of the same fundamental building block, however—the cell. Bacteria consist of a single cell; sequoias are made up of many cells.

Biologists have become increasingly dazzled by the diversity and complexity of cells as advances in microscopy have made it possible to examine cells at higher magnifications. The basic conclusion made in the 1800s is intact, however: As far as is known, all organisms are made of cells. Today, a **cell** is defined as a highly organized compartment that is bounded by a thin, flexible structure called a plasma membrane and that contains concentrated chemicals in an aqueous (watery) solution. The chemical reactions that sustain life take place inside cells. Most cells are also capable of reproducing by dividing—in effect, by making a copy of themselves.

The realization that all organisms are made of cells was fundamentally important, but it formed only the first part of the cell theory. In addition to understanding what organisms are made of, scientists wanted to understand how cells come to be.

Where Do Cells Come From?

Most scientific theories have two components: The first describes a pattern in the natural world, while the second identifies a mechanism or process that is responsible for creating that pattern. Hooke and his fellow scientists had articulated the pattern component of the cell theory. In 1858 Rudolph Virchow added the process component by stating that all cells arise from preexisting cells. The complete cell theory, then, can be stated as follows: All organisms are made of cells, and all cells come from preexisting cells.

This claim was a direct challenge to the prevailing explanation, called spontaneous generation. At the time, most biologists believed that organisms arise spontaneously under certain conditions. For example, the bacteria and fungi that spoil foods such as milk and wine were thought to appear in these nutrient-rich media of their own accord—they spring to life from nonliving materials. Spontaneous generation was a **hypothesis**: a proposed explanation. The all-cells-from-cells hypothesis, in contrast, maintained that cells do not spring to life spontaneously but are produced only when preexisting cells grow and divide. Biologists usually use *theory* to refer to proposed explanations for broad patterns in nature and *hypothesis* to refer to explanations for more tightly focused questions.

Soon after the all-cells-from-cells hypothesis appeared in print, Louis Pasteur set out to test its predictions experimentally. A **prediction** is something that can be measured and that must be correct if a hypothesis is valid. Pasteur wanted to determine whether microorganisms could arise spontaneously in a nutrient broth or whether they appear only when a broth is exposed to a source of preexisting cells. To address the question, he created two treatment groups: a broth that was not exposed to a source of preexisting cells and a broth that was. The spontaneous generation hypothesis predicted that cells would appear in both treatments. The all-cells-from-cells hypothesis predicted that cells would appear only in the treatment exposed to a source of preexisting cells.

Figure 1.2 shows Pasteur's experimental setup. Note that the two treatments are identical in every respect but one. Both used glass flasks filled with the same amount of the same nutrient broth. Both were boiled for the same amount of time to kill any existing organisms such as bacteria or fungi. But because the flask pictured in **Figure 1.2a** had a straight neck, it was exposed to preexisting cells after sterilization by the heat treatment. These preexisting cells are the bacteria and fungi that cling to dust particles in the air. They could drop into the nutrient broth because the neck of the flask was straight. In contrast, the flask

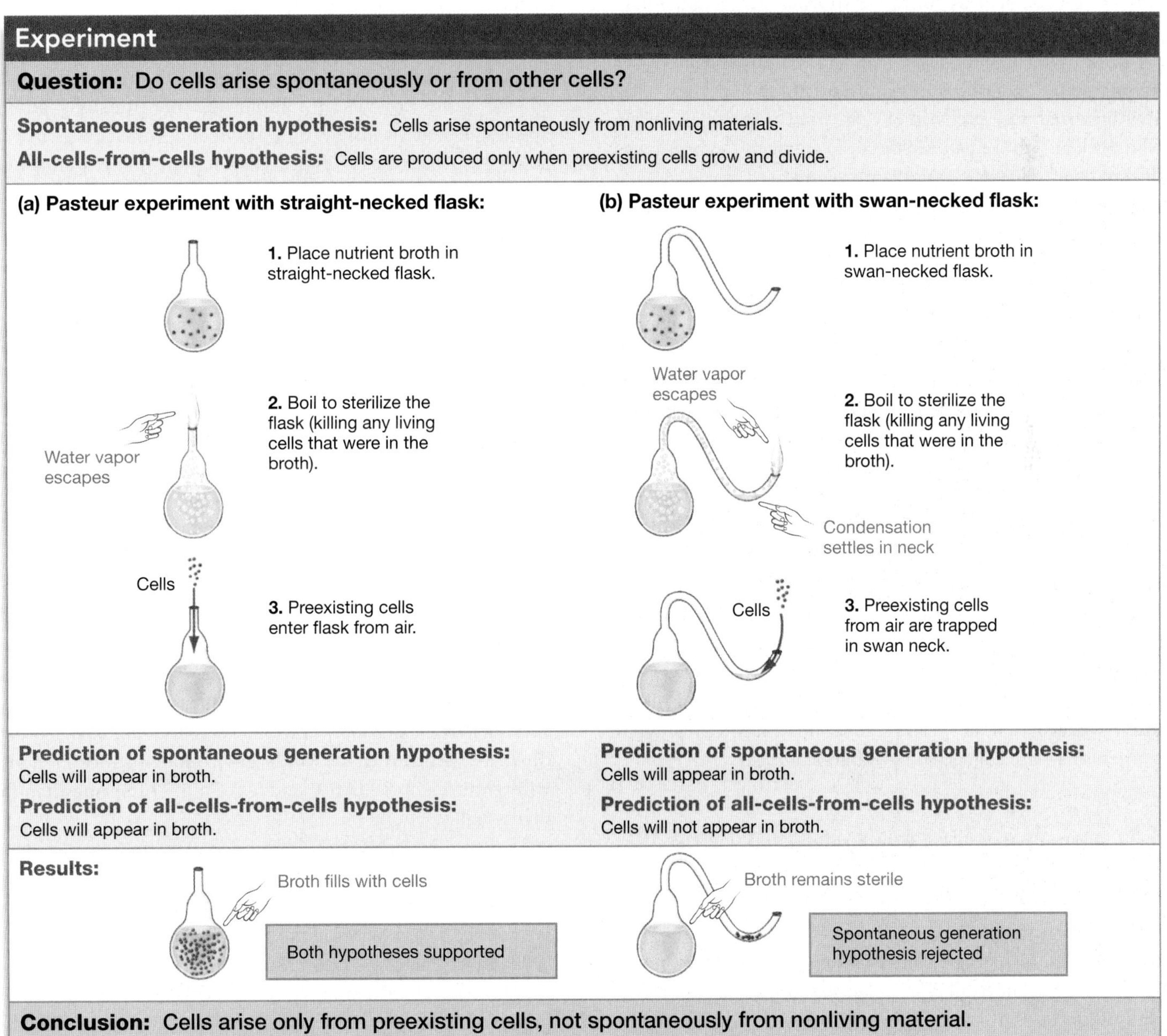

FIGURE 1.2 The Spontaneous Generation Hypothesis Was Tested Experimentally.

drawn in **Figure 1.2b** had a long swan neck. Pasteur knew that water would condense in the crook of the swan neck after the boiling treatment and that this pool of water would trap any bacteria or fungi that entered on dust particles. Thus, the swan-necked flask was isolated from any source of preexisting cells even though it was still open to the air.

Pasteur's experimental setup was effective because there was only one difference between the two treatments and because that difference was the factor being tested—in this case, a broth's exposure to preexisting cells. ● If you understand this concept, you should be able to identify problems that would arise if he had put different types of broth in the two treatments, heated them for different lengths of time, or used a ceramic flask for one treatment and a glass flask for the other.

And Pasteur's results? As Figure 1.2 shows, the treatment exposed to preexisting cells quickly filled with bacteria and fungi. This observation was important because it showed that the heat sterilization step had not altered the nutrient broth's capacity to support growth, and because it supported the hypothesis that growth started with preexisting cells. But the treatment in the swan-necked flask remained sterile. Even when the broth was left standing for months, no organisms appeared in it.

Because Pasteur's data were in direct opposition to the predictions made by the spontaneous generation hypothesis, the results persuaded most biologists that the all-cells-from-cells hypothesis was correct.

The success of the cell theory's process component had an important implication: If all cells come from preexisting cells, it follows that all individuals in a population of single-celled organisms are related by common ancestry. Similarly, in a multicellular individual such as you, all of the cells present are descended from preexisting cells, tracing back to a fertilized egg. A fertilized egg is a cell created by the fusion of sperm and egg—cells that formed in individuals of the previous generation. In this way, all of the cells in a multicellular organism are connected by common ancestry.

The second great founding idea in biology is similar, in spirit, to the cell theory. It also happened to be published the same year as the all-cells-from-cells hypothesis. This was the realization, made independently by Charles Darwin and Alfred Russel Wallace, that all *species*—all distinct, identifiable types of organisms—are connected by common ancestry.

1.2 The Theory of Evolution by Natural Selection

In 1858 short papers written separately by Darwin and Wallace were read to a small group of scientists attending a meeting of the Linnean Society of London. A year later, Darwin published a book that expanded on the idea summarized in those brief papers. The book was called *The Origin of Species*. The first edition sold out in a day.

What Is Evolution?

Like the cell theory, the theory of evolution by natural selection has a pattern and a process component. Darwin and Wallace's theory made two important claims concerning patterns that exist in the natural world. The first claim was that species are related by common ancestry. This contrasted with the prevailing view in science at the time, which was that species represent independent entities created separately by a divine being. The second claim was equally novel. Instead of accepting the popular hypothesis that species remain unchanged through time, Darwin and Wallace proposed that the characteristics of species can be modified from generation to generation. Darwin called this process "descent with modification."

Evolution, then, means that species are not independent and unchanging entities, but are related to one another and can change through time. This part of the theory of evolution—the pattern component—was actually not original to Darwin and Wallace. Several scientists had already come to the same conclusions about the relationships among species. The great insight by Darwin and Wallace was in proposing a process, called **natural selection**, that explains *how* evolution occurs.

What Is Natural Selection?

Natural selection occurs whenever two conditions are met. The first is that individuals within a population vary in characteristics that are **heritable**—meaning, traits that can be passed on to offspring. A **population** is defined as a group of individuals of the same species living in the same area at the same time. Darwin and Wallace had studied natural populations long enough to realize that variation among individuals is almost universal. In wheat, for example, some individuals are taller than others. As a result of work by wheat breeders, Darwin and Wallace knew that short parents tend to have short offspring. Subsequent research has shown that heritable variation exists in most traits and populations. The second condition of natural selection is that in a particular environment, certain versions of these heritable traits help individuals survive better or reproduce more than do other versions. For example, if tall wheat plants are easily blown down by wind, then in windy environments shorter plants will tend to survive better and leave more offspring than tall plants will.

If certain heritable traits lead to increased success in producing offspring, then those traits become more common in the population over time. In this way, the population's characteristics change as a result of natural selection acting on individuals. This is a key insight: Natural selection acts on individuals, but evolutionary change affects only populations. In this example, populations of wheat that grow in windy environments tend to become shorter from generation to generation. But in any given generation, none of the individual wheat plants get taller or shorter as a result of natural selection. This sort of change in the characteristics of a population, over time, is evolution. Evolution occurs when heritable variation leads to differential success in reproduction. ● If you understand this concept, you should be able to graph how average stem height will change over time in a population that occupies a windy environment versus a non-windy environment—one where tall plants have an advantage due to better access to light. (For help with reading and making graphs, see **BioSkills 1** in the back of this book.)

Darwin also introduced some new terminology to identify what is happening during natural selection. For example, in everyday English the word *fitness* means health and well-being. But in biology, **fitness** means the ability of an individual to produce offspring. Individuals with high fitness produce many offspring. Similarly, the word *adaptation* in everyday English means that an individual is adjusting and changing to function in new circumstances. But in biology, an **adaptation** is a trait that increases the fitness of an individual in a particular environment. Once again, consider wheat: In windswept habitats, wheat plants with short stalks have higher fitness than do individuals with long stalks. Short stalks are an adaptation to windy environments.

To clarify further how natural selection works, consider the origin of the vegetables called the cabbage family plants. Broccoli, cauliflower, Brussels sprouts, cabbage, kale, savoy, and collard greens descended from the same species—the wild plant in the mustard family pictured in **Figure 1.3a**. To create the plant called broccoli, horticulturists selected individuals of the wild mustard species with particularly large and compact flowering stalks. In mustards, the size and shape of the flowering stalk is a heritable trait. When the selected individuals were mated with one another, their offspring turned out to have larger and more compact flowering stalks, on average, than the original population (**Figure 1.3b**). By repeating this process over many generations, horticulturists produced a population with extraordinarily large and compact flowering stalks. The derived population has been artificially selected for the size and shape of the flowering stalk; as **Figure 1.3c** shows, it barely resembles the ancestral form. Note that during this process, the size and shape of the flowering stalk in each individual plant did not change within its lifetime—the change occurred in the characteristics of the population over time. Darwin's great insight was that natural selection changes the characteristics of a wild population over time, just as the deliberate manipulation of "artificial selection" changes the characteristics of a domesticated population over time.

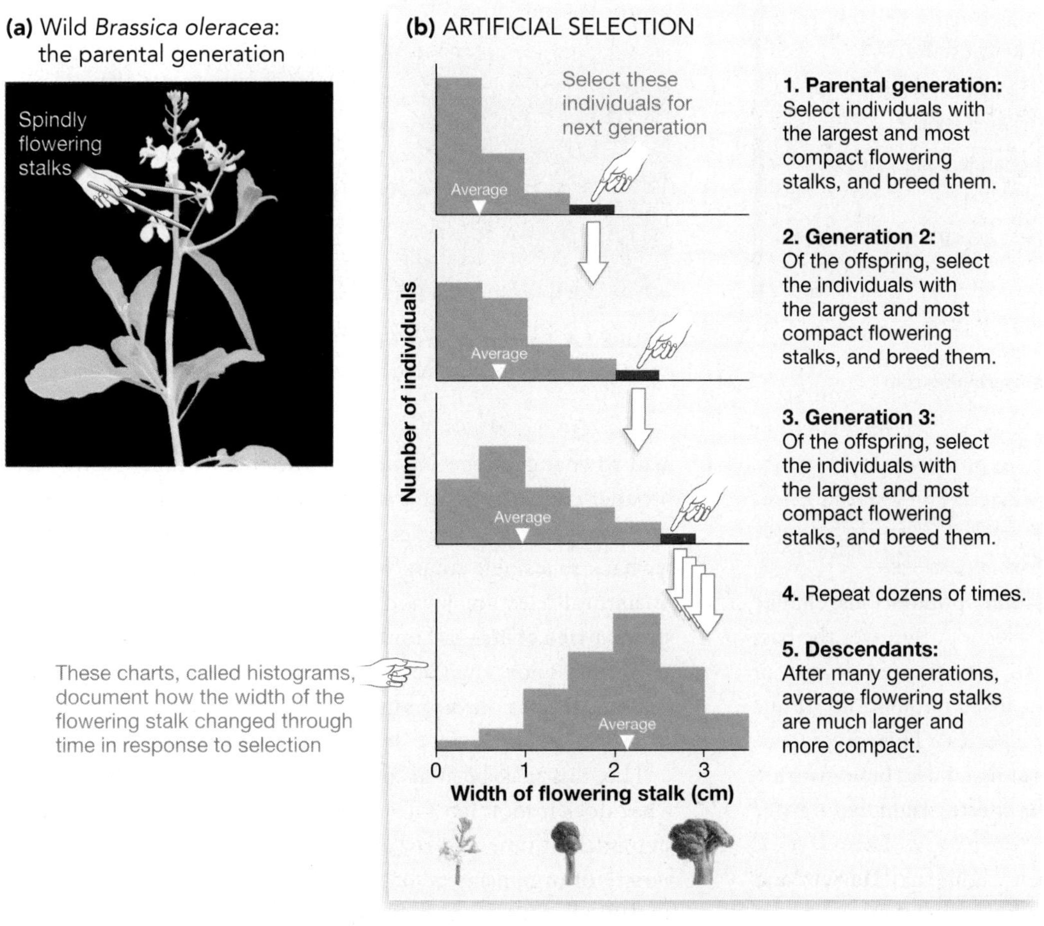

FIGURE 1.3 Artificial Selection Can Produce Dramatic Changes in Organisms.

Since Darwin and Wallace published their work, biologists have succeeded in measuring hundreds of examples of natural selection in wild populations. They have accumulated a massive body of evidence documenting that species have changed through time.

Together, the cell theory and the theory of evolution provided the young science of biology with two central, unifying ideas:

1. The cell is the fundamental structural unit in all organisms.
2. All species are related by common ancestry and have changed over time in response to natural selection.

Check Your Understanding

If you understand that...

- Natural selection occurs when heritable variation in certain traits leads to improved success in reproduction. Because individuals with these traits produce many offspring with the same traits, the traits increase in frequency and evolution occurs.
- Evolution is simply a change in the characteristics of a population over time.

You should be able to...

Explain why each of the following common misconceptions about evolution by natural selection is incorrect, using the example of selection on the height of wheat stalks.

1) evolution is progressive, meaning that species always get larger, more complex, or "better" in some sense;
2) individuals as well as populations change when natural selection occurs; or
3) individuals with high levels of fitness are stronger or bigger or "more dominant."

at www.masteringbio.com

Artificial Selection

(a) *Tragopogon mirus* (left) evolved from *Tragopogon dubius* (right).

(b) Different soapberry bug species feed on native (left) or introduced (right) plants.

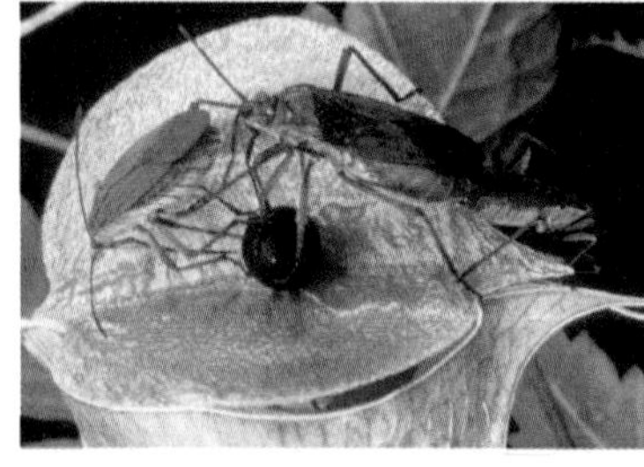

(c) Different maggot fly species feed on hawthorn (left) versus apple (right) fruits.

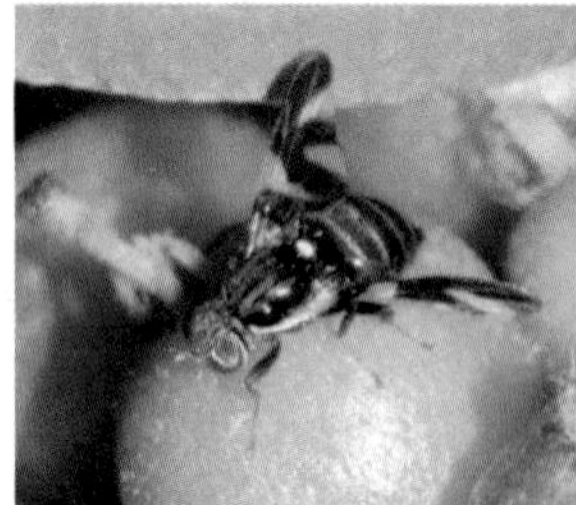

FIGURE 1.4 Speciation in Action. The pairs of organisms shown here are in the process of becoming independent species (see Chapter 26).

1.3 The Tree of Life

Section 1.2 focused on how individual populations change through time in response to natural selection. But over the past several decades, biologists have also documented dozens of cases in which natural selection has caused populations of one species to diverge and form new species. This divergence process is called **speciation**. In several instances, biologists are documenting the formation of new species right before our eyes (**Figure 1.4**).

Research on speciation supports a claim that Darwin and Wallace made over a century ago—that natural selection can lead to change *between* species as well as within species. The broader conclusions are that all species come from preexisting species and that all species, past and present, trace their ancestry back to a single common ancestor. If the theory of evolution by natural selection is valid, biologists should be able to reconstruct a **tree of life**—a family tree of organisms. If life on Earth arose just once, then such a diagram would describe the genealogical relationships among species with a single, ancestral species at its base.

Has this task been accomplished? If the tree of life exists, what does it look like? To answer these questions, we need to step back in time and review how biologists organized the diversity of organisms *before* the development of the cell theory and the theory of evolution.

Linnaean Taxonomy

In science, the effort to name and classify organisms is called **taxonomy**. This branch of biology began to flourish in 1735 when a botanist named Carolus Linnaeus set out to bring order to the bewildering diversity of organisms that were then being discovered.

The building block of Linnaeus' system is a two-part name unique to each type of organism. The first part indicates the organism's **genus** (plural: **genera**). A genus is made up of a closely related group of species. For example, Linnaeus put humans in the genus *Homo*. Although humans are the only living species in this genus, several extinct organisms, all of which walked upright and made extensive use of tools, were later also assigned to *Homo*. The second term in the two-part name identifies the organism's species. Section 1.1 defined a species as a distinct, identifiable type of organism. More formally, a **species** is made up of individuals that regularly breed together or have characteristics that are distinct from those of other species. Linnaeus gave humans the specific name *sapiens*.

An organism's genus and species designation is called its scientific name or Latin name. Scientific names are always italicized. Genus names are always capitalized, but species names are not—for instance, *Homo sapiens*. Scientific names are based on Latin or Greek word roots or on words "Latinized" from other languages (see **Box 1.1**). Linnaeus gave a scientific name to every species then known. (He also Latinized his own name—from Karl von Linné to Carolus Linnaeus.)

Linnaeus maintained that different types of organisms should not be given the same genus and species names. Other species may be assigned to the genus *Homo*, and members of other genera may be named *sapiens*, but only humans are named *Homo sapiens*. Each scientific name is unique.

Linnaeus' system has stood the test of time. His two-part naming system, or **binomial nomenclature**, is still the standard in biological science.

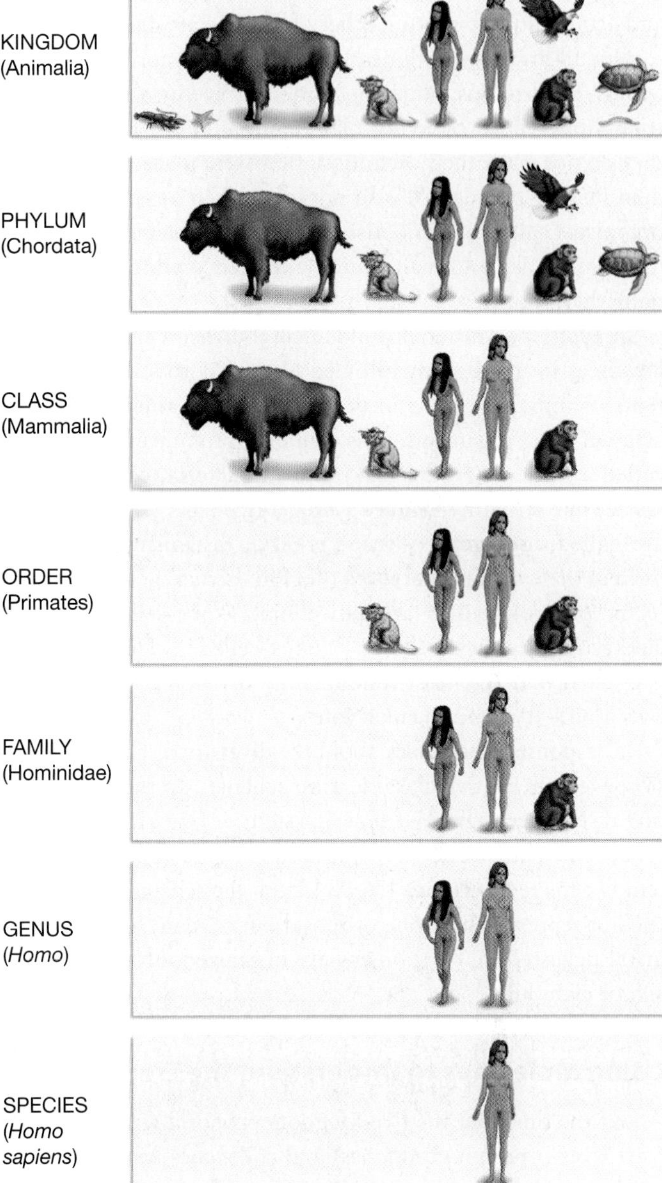

FIGURE 1.5 Linnaeus Defined Taxonomic Levels. In the Linnaean system, each animal species is placed in a taxonomic hierarchy with seven levels. Lower levels are nested within higher levels.

Taxonomic Levels To organize and classify the tremendous diversity of species being discovered in the 1700s, Linnaeus created a hierarchy of taxonomic groups: From the most specific grouping to the least specific, the levels are **species**, **genus**, **family**, **order**, **class**, **phylum** (plural: **phyla**), and **kingdom**. **Figure 1.5** shows how this nested, or hierarchical, classification scheme works, using humans as an example. Although our species is the sole living member of the genus *Homo*, humans are now grouped with the orangutan, gorilla, common chimpanzee, and bonobo in a family called Hominidae. Linnaeus grouped members of this family with gibbons, monkeys, and lemurs in an order called Primates. The Primates are grouped in the class Mammalia with rodents, bison, and other organisms that have fur and produce milk. Mammals, in turn, join other animals with structures called notochords in the phylum Chordata, and they join all other animals in the kingdom Animalia. Each of these named groups—primates, mammals, or *Homo sapiens*—can be referred to as a **taxon** (plural: **taxa**). The essence of Linnaeus's system is that lower-level taxa are nested within higher-level taxa.

Aspects of this hierarchical scheme are still in use. As biological science matured, however, several problems with Linnaeus's original proposal emerged.

How Many Kingdoms Are There? Linnaeus proposed that species could be organized into two kingdoms—plants and animals. According to Linnaeus, organisms that do not move and that produce their own food are plants; organisms that move and acquire food by eating other organisms are animals.

Not all organisms fall neatly into these categories, however. Molds, mushrooms, and other fungi survive by absorbing nutrients from dead or living plants and animals. Even though they do not make their own food, they were placed in the kingdom Plantae because they do not move. The tiny, single-celled organisms called bacteria also presented problems. Some bacteria can move, and many can make their own food. But initially they, too, were thought to be plants.

In addition, a different fundamental division emerged when advances in microscopy allowed biologists to study the contents of individual cells in detail. In plants, animals, and an array of other organisms, cells contain a prominent component called a nucleus (**Figure 1.6a**). But in bacteria, cells lack this kernel-like structure (**Figure 1.6b**). Organisms with a nucleus are called **eukaryotes** ("true-kernel"); organisms without a nucleus are called **prokaryotes** ("before-kernel"). The vast majority of prokaryotes are unicellular ("one-celled"); many eukaryotes are multicellular ("many-celled"). These findings suggested that the most fundamental division in life was between prokaryotes and eukaryotes.

In response to new data about the diversity of life, biologists proposed alternative classification schemes. In the late 1960s one researcher suggested that a system of five kingdoms best reflects the patterns observed in nature. This five-kingdom system is depicted in **Figure 1.7**. Although the scheme was widely used, it represents just one proposal out of many. Other biologists suggested that organisms are organized into three, four, six, or eight kingdoms.

(a) Eukaryotic cells have a membrane-bound nucleus.

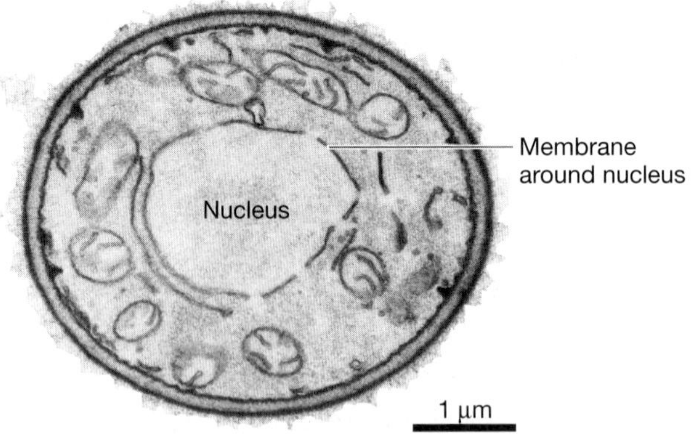

(b) Prokaryotic cells do *not* have a membrane-bound nucleus.

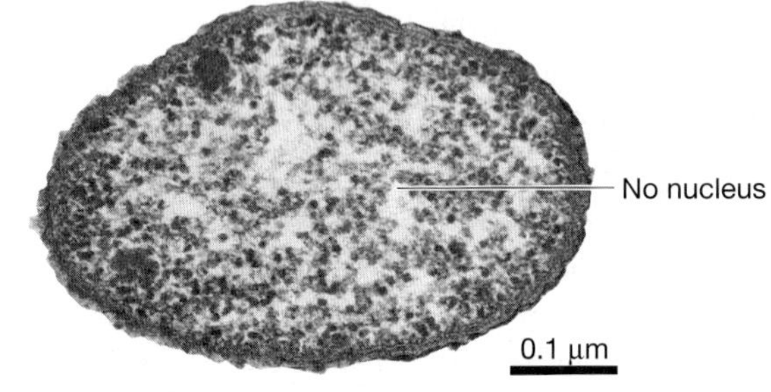

FIGURE 1.6 Eukaryotes and Prokaryotes.

EXERCISE Study the scale bars; then draw two ovals that accurately represent the relative sizes of a eukaryotic cell and a prokaryotic cell.

Using Molecules to Understand the Tree of Life

About the time that the five-kingdom proposal was published, Carl Woese (pronounced *woes*) and colleagues began working on the problem from a radically different angle. Instead of assigning organisms to kingdoms based on characteristics such as the presence of a nucleus or the ability to move or to manufacture food, they attempted to understand the relationships among organisms by analyzing their chemical components. Their goal was to understand the **phylogeny** (meaning "tribe-source") of all organisms—their actual genealogical relationships.

To understand which organisms are closely versus distantly related, Woese and co-workers needed to study a molecule that is found in all organisms. The molecule they selected is called small subunit rRNA. It is an essential part of the machinery that all cells use to grow and reproduce.

Although rRNA is a large and complex molecule, its underlying structure is simple. The rRNA molecule is made up of sequences of four smaller chemical components called ribonucleotides. These ribonucleotides are symbolized by the letters

BOX 1.1 Scientific Names and Terms

Scientific names and terms are often based on Latin or Greek word roots that are descriptive. For example, *Homo sapiens* is derived from the Latin *homo* for "man" and *sapiens* for "wise" or "knowing." The yeast that bakers use to produce bread and that brewers use to brew beer is called *Saccharomyces cerevisiae*. The Greek root *saccharo* means "sugar," and *myces* refers to a fungus. *Saccharomyces* is aptly named "sugar fungus" because yeast is a fungus and because the domesticated strains of yeast used in commercial baking and brewing are often fed sugar. The specific name of this organism, *cerevisiae*, is Latin for "beer." Loosely translated, then, the scientific name of brewer's yeast means "sugar fungus for beer."

Most biologists find it extremely helpful to memorize some of the common Latin and Greek roots. To aid you in this process, new terms in this text are often accompanied by a reference to their Latin or Greek word roots in parentheses.

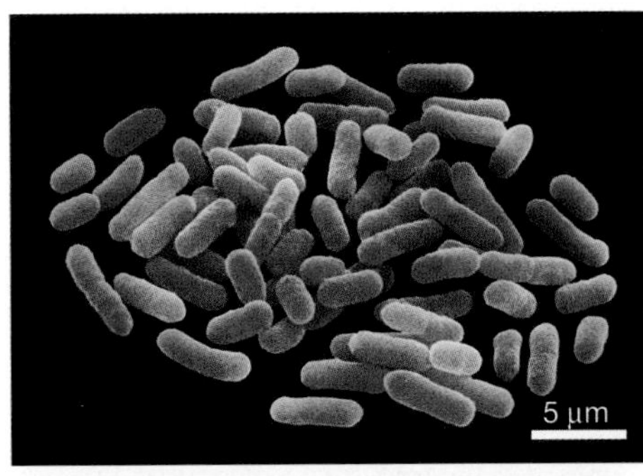

KINGDOM MONERA (includes all prokaryotes)

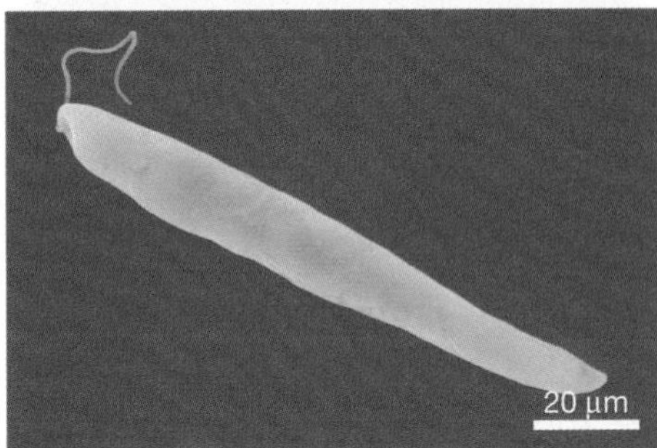

KINGDOM PROTISTA (includes several groups of unicellular eukaryotes)

KINGDOM PLANTAE

KINGDOM FUNGI

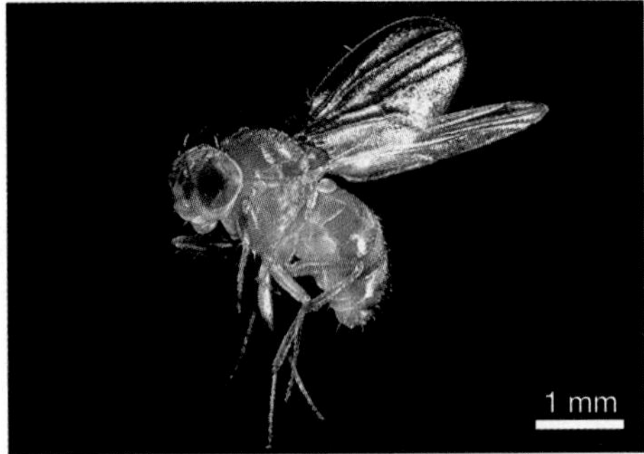

KINGDOM ANIMALIA

FIGURE 1.7 The Five-Kingdom Scheme. For decades, most biologists accepted the hypothesis that organisms naturally fall into the five kingdoms depicted here.

QUESTION How many times bigger is a fruit fly than one of the prokaryotic cells pictured here?

A, U, C, and G. In rRNA, ribonucleotides are connected to one another linearly, like boxcars of a freight train (**Figure 1.8**).

Why might rRNA be useful for understanding the relationships among organisms? The answer is that the ribonucleotide sequence in rRNA is a trait, similar to the height of wheat stalks or the size of flowering stalks of broccoli, that can change during the course of evolution. Although rRNA performs the same function in all organisms, the sequence of ribonucleotide building blocks in this molecule is not identical among species. In land plants, for example, the molecule might start with the sequence A-U-A-U-C-G-A-G. In green algae, which are closely related to land plants, the same section of the molecule might contain A-U-A-U-G-G-A-G. But in brown algae, which are not closely related to green algae or to land plants, the same part of the molecule might consist of A-A-A-U-G-G-A-G.

The research program that Woese and co-workers pursued was based on a simple premise: If the theory of evolution is correct, then rRNA sequences should be very similar in closely related organisms but less similar in organisms that are less closely related. Groups that are closely related, like the plants, should share certain changes in rRNA that no other species have.

To test this premise, the researchers determined the sequence of ribonucleotides in the rRNA of a wide array of species. Then they considered what the similarities and differences in the sequences implied about relationships among the species. The goal was to produce a diagram that described the phylogeny of the organisms in the study. A diagram that depicts evolutionary history in this way is called a **phylogenetic tree**. Just as a family tree shows relationships among individuals, a phylogenetic tree shows relationships among species. On a phylogenetic tree, branches that are close to one another represent species that are closely related; branches that are farther apart represent species that are more distantly related.

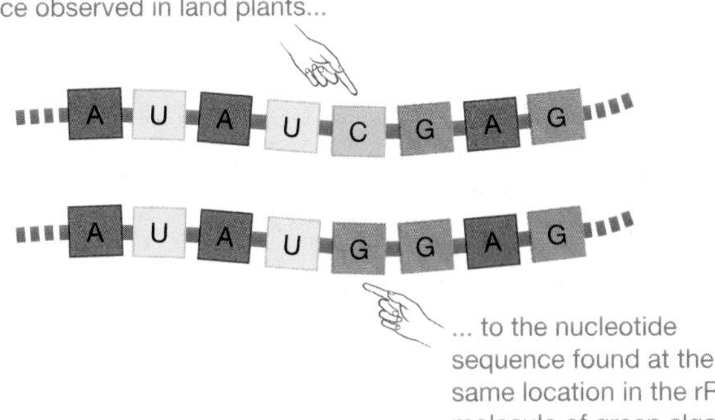

FIGURE 1.8 RNA Molecules Are Made Up of Smaller Molecules. The complete small subunit rRNA molecule contains about 2000 ribonucleotides; just 8 are shown in this comparison.

QUESTION Suppose that in the same portion of rRNA, molds and other fungi have the sequence A-U-A-U-G-G-A-C. According to these data, are fungi more closely related to green algae or to land plants? Explain your logic.

The Tree of Life Estimated from an Array of Genes To construct a phylogenetic tree, researchers use a computer to find the arrangement of branches that is most consistent with the similarities and differences observed in the data. Although the initial work was based only on the sequences of ribonucleotides observed in rRNA, biologists now use data sets that include sequences from a wide array of genes. A recent tree produced by comparing these sequences is shown in **Figure 1.9**. Because this tree includes species from many different kingdoms and phyla, it is often called the universal tree, or the tree of life. For help in learning how to read a phylogenetic tree, see **BioSkills 2**.

The tree of life implied by rRNA and other genetic data astonished biologists. For example:

- The fundamental division in organisms is not between plants and animals or even between prokaryotes and eukaryotes. Rather, *three* major groups occur: (1) the Bacteria; (2) another group of prokaryotic, single-celled organisms called the Archaea; and (3) the eukaryotes. To accommodate this new perspective on the diversity of organisms, Woese created a new taxonomic level called the **domain**. As Figure 1.9 indicates, the three domains of life are now called the Bacteria, Archaea, and Eukarya.
- Some of the kingdoms that had been defined earlier do not reflect how evolution actually occurred. For example, recall that Linnaeus grouped the multicellular eukaryotes known as fungi with plants. But the genetic data indicate that fungi are much more closely related to animals than they are to plants.
- Bacteria and Archaea are much more diverse than anyone had imagined. If the differences among animals, fungi, and plants warrant placing them in separate kingdoms, then dozens of kingdoms exist among the prokaryotes.

The Tree of Life Is a Work in Progress Just as researching your family tree can help you understand who you are and where you came from, so the tree of life helps biologists understand the relationships among organisms and the history of species. For example, the discovery of the Archaea and the placement of lineages such as the fungi qualify as exciting breakthroughs in our understanding of life's diversity. Work on the tree of life continues at a furious pace, however, and the location of certain branches on the tree is hotly debated. As databases expand and as techniques for analyzing data improve, the shape of the tree of life presented in **Figure 1.9** will undoubtedly change.

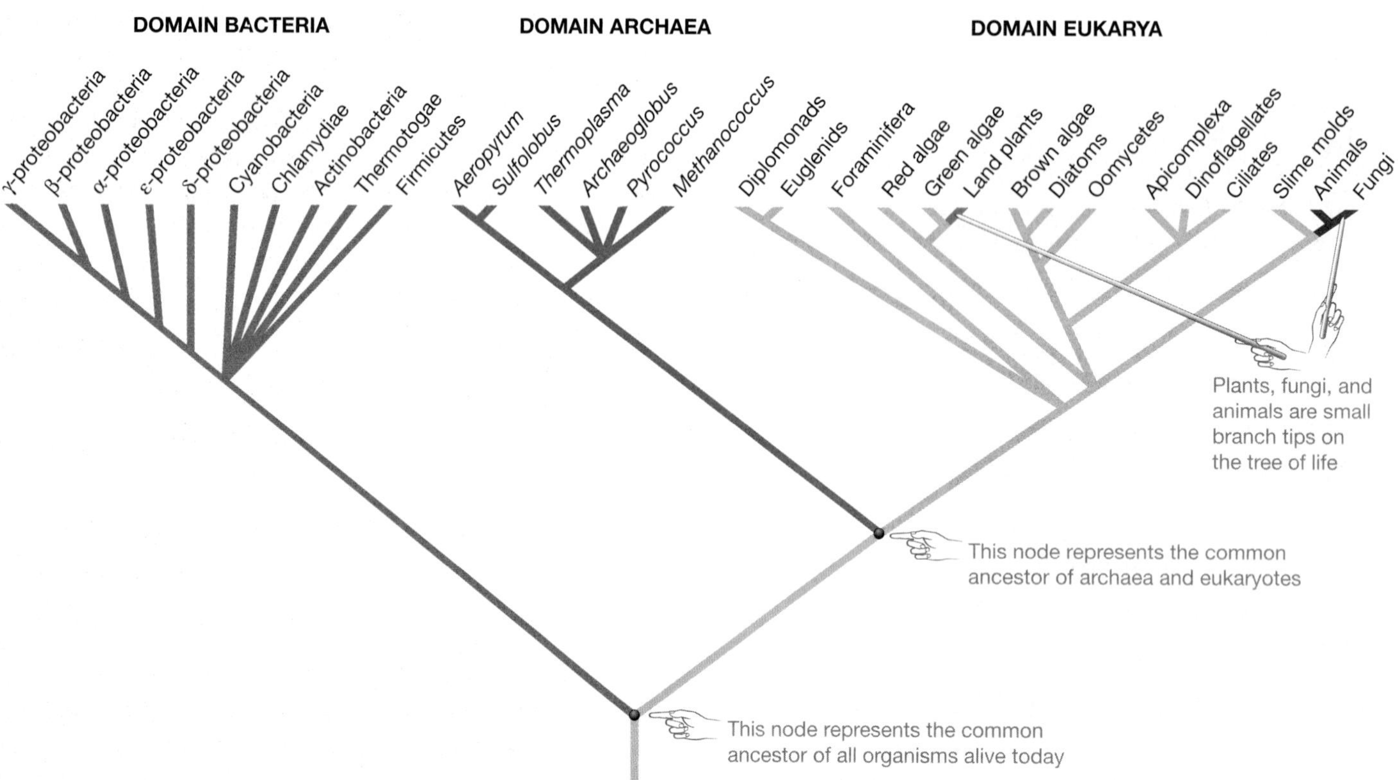

FIGURE 1.9 The Tree of Life. "Universal tree" estimated from a large amount of gene sequence data. The three domains of life revealed by the analysis are labeled. Common names are given for most lineages in the domains Bacteria and Eukarya. Genus names are given for members of the domain Archaea, because most of these organisms have no common names.

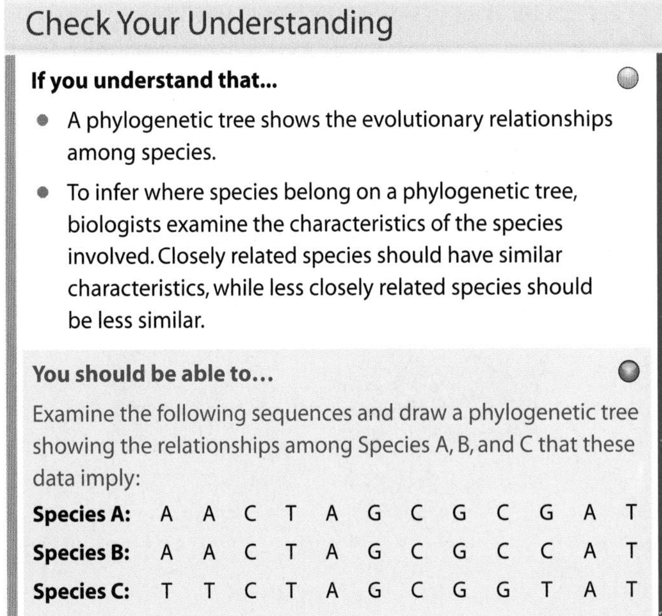

Check Your Understanding

If you understand that...

- A phylogenetic tree shows the evolutionary relationships among species.
- To infer where species belong on a phylogenetic tree, biologists examine the characteristics of the species involved. Closely related species should have similar characteristics, while less closely related species should be less similar.

You should be able to...

Examine the following sequences and draw a phylogenetic tree showing the relationships among Species A, B, and C that these data imply:

Species A:	A	A	C	T	A	G	C	G	C	G	A	T
Species B:	A	A	C	T	A	G	C	G	C	C	A	T
Species C:	T	T	C	T	A	G	C	G	G	T	A	T

1.4 Doing Biology

This chapter has introduced some of the great ideas in biology. The development of the cell theory and the theory of evolution by natural selection provided cornerstones when the science was young; the tree of life is a relatively recent insight that has revolutionized the way researchers understand the diversity of life on Earth.

These theories are considered great because they explain fundamental aspects of nature, and because they have consistently been shown to be correct. They are considered correct because they have withstood extensive testing. How do biologists test ideas about the way the natural world works? The answer is that they test the predictions made by alternative hypotheses, often by setting up carefully designed experiments. To illustrate how this approach works, let's consider two questions currently being addressed by researchers.

Why Do Giraffes Have Long Necks? An Introduction to Hypothesis Testing

If you were asked why giraffes have long necks, you might say that long necks enable giraffes to reach food that is unavailable to other mammals. This hypothesis is expressed in African folktales and has traditionally been accepted by many biologists. The food competition hypothesis is so plausible, in fact, that for decades no one thought to test it. Recently, however, Robert Simmons and Lue Scheepers assembled data suggesting that the food competition hypothesis is only part of the story. Their analysis supports an alternative hypothesis—that long necks allow giraffes to use their heads as effective weapons for battering their opponents.

How did biologists test the food competition hypothesis? What data support their alternative explanation? Before attempting to answer these questions, it's important to recognize that hypothesis testing is a two-step process. The first step is to state the hypothesis as precisely as possible and list the predictions it makes. The second step is to design an observational or experimental study that is capable of testing those predictions. If the predictions are accurate, then the hypothesis is supported. If the predictions are not met, then researchers do further tests, modify the original hypothesis, or search for alternative explanations.

The Food Competition Hypothesis: Predictions and Tests Stated precisely, the food competition hypothesis claims that giraffes compete for food with other species of mammals. When food is scarce, as it is during the dry season, giraffes with longer necks can reach food that is unavailable to other species and to giraffes with shorter necks. As a result, the longest-necked individuals in a giraffe population survive better and produce more young than do shorter-necked individuals, and average neck length of the population increases with each generation. To use the terms introduced earlier, long necks are adaptations that increase the fitness of individual giraffes during competition for food. This type of natural selection has gone on so long that the population has become extremely long necked.

The food competition hypothesis makes several explicit predictions. For example, the food competition hypothesis predicts that (1) neck length is variable among giraffes; (2) neck length in giraffes is heritable; and (3) giraffes feed high in trees, especially during the dry season, when food is scarce and the threat of starvation is high.

The first prediction is correct. Studies in zoos and natural populations confirm that neck length is variable among individuals.

The researchers were unable to test the second prediction, however, because they studied giraffes in a natural population and were unable to do breeding experiments. As a result, they simply had to accept this prediction as an assumption. In general, though, biologists prefer to test every assumption behind a hypothesis.

What about the prediction regarding feeding high in trees? According to Simmons and Scheepers, this is where the food competition hypothesis breaks down. Consider, for example, data collected by a different research team about the amount of time that giraffes spend feeding in vegetation of different heights. **Figure 1.10a** shows that in a population from Kenya, both male and female giraffes spend most of their feeding time eating vegetation that averages just 60 percent of their full height. Studies on other populations of giraffes, during

(a) Most feeding is done below neck height.

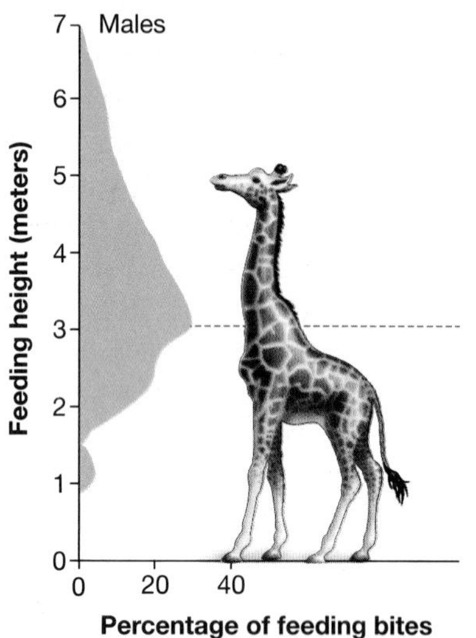

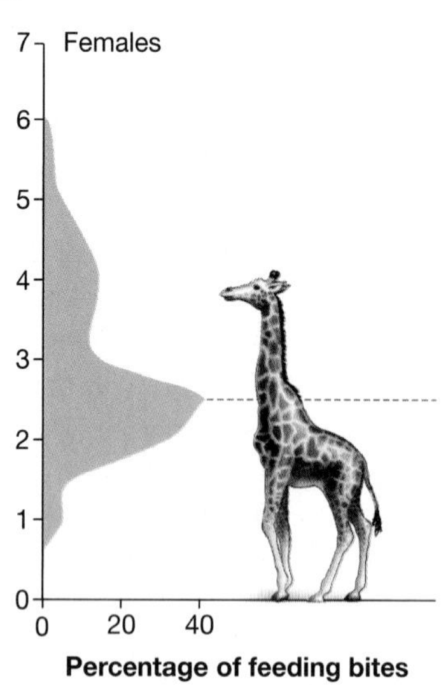

(b) Typical feeding posture in giraffes

FIGURE 1.10 Giraffes Do Not Usually Extend Their Necks to Feed.

both the wet and dry seasons, are consistent with these data. Giraffes usually feed with their necks bent (Figure 1.10b).

These data cast doubt on the food competition hypothesis, because one of its predictions does not appear to hold. Biologists have not abandoned this hypothesis completely, though, because feeding high in trees may be particularly valuable during extreme droughts, when a giraffe's ability to reach leaves far above the ground could mean the difference between life and death. Still, Simmons and Scheepers have offered an alternative explanation for why giraffes have long necks. The new hypothesis is based on the mating system of giraffes.

The Sexual Competition Hypothesis: Predictions and Tests Giraffes have an unusual mating system. Breeding occurs year round rather than seasonally. To determine when females are coming into estrus (or "heat") and are thus receptive to mating, the males nuzzle the rumps of females. In response, the females urinate into the males' mouths. The males then tip their heads back and pull their lips to and fro, as if tasting the liquid. Biologists who have witnessed this behavior have proposed that the males taste the females' urine to detect whether estrus has begun.

Once a female giraffe enters estrus, males fight among themselves for the opportunity to mate. Combat is spectacular. The bulls stand next to one another, swing their necks, and strike thunderous blows with their heads. Researchers have seen males knocked unconscious for 20 minutes after being hit and have cataloged numerous instances in which the loser died. Giraffes are the only animals known to fight in this way.

These observations inspired a new explanation for why giraffes have long necks. The sexual competition hypothesis is based on the idea that longer-necked giraffes are able to strike harder blows during combat than can shorter-necked giraffes. In engineering terms, longer necks provide a longer moment arm. A long moment arm increases the force of the impact. (Think about the type of sledge hammer you'd use to bash down a concrete wall—one with a short handle or one with a long handle?) Thus, longer-necked males should win more fights and, as a result, father more offspring than do shorter-necked males. If neck length in giraffes is inherited, then the average neck length in the population should increase over time. Under the sexual competition hypothesis, long necks are adaptations that increase the fitness of males during competition for females.

Although several studies have shown that long-necked males are more successful in fighting and that the winners of fights gain access to estrous females, the question of why giraffes have long necks is not closed. With the data collected to date, most biologists would probably concede that the food competition hypothesis needs further testing and refinement and that the sexual selection hypothesis appears promising. It could also be true that both hypotheses are correct. For our purposes, the important take-home message is that all hypotheses must be tested rigorously.

In many cases in biological science, testing hypotheses rigorously involves experimentation. Experimenting on giraffes is difficult. But in the case study considered next, biologists were able to test an interesting hypothesis experimentally.

Why Are Chili Peppers Hot? An Introduction to Experimental Design

Experiments are a powerful scientific tool because they allow researchers to test the effect of a single, well-defined factor on a particular phenomenon. Because experiments testing the effect of neck length on food and sexual competition in giraffes have yet to be done, let's consider a different question: Why do chili peppers taste so spicy?

The jalapeño, Anaheim, and cayenne peppers used in cooking descended, via artificial selection, from a wild shrub that is native to the deserts of the American Southwest. As **Figure 1.11a** shows, wild chilies produce fleshy fruits with seeds inside, just like their domesticated descendants. In both wild chilies and the cultivated varieties, the "heat" or pungent flavor of the fruit and seeds is due to a molecule called capsaicin. In humans and other mammals, capsaicin binds to heat-sensitive cells in the tongue and mouth. In response to this binding, signals are sent to the brain that produce the sensation of burning. Similar signals would be transmitted if you drank boiling water. Asking why chilies are hot, then, is the same as asking why chilies contain capsaicin.

Josh Tewksbury and Gary Nabhan proposed that the presence of capsaicin is an adaptation that protects chili fruits from being eaten by animals that destroy the seeds inside. To understand this hypothesis, it's important to realize that the seeds inside a fruit have one of two fates when the fruit is eaten. If the seeds are destroyed in the animal's mouth or digestive system, then they never germinate (sprout). In this case, "seed predation" has occurred. But if seeds can travel undamaged through the animal, then they are eventually "planted" in a new location along with a valuable supply of fertilizer. In this case, seeds are dispersed. Here's the key idea: Natural selection should favor fruits that taste bad to animal species that act as seed predators. But these same fruits should not deter species that act as seed dispersers. This proposal is called the directed dispersal hypothesis.

Does capsaicin deter seed predators, as the directed dispersal hypothesis predicts? To answer this question, the researchers captured some cactus mice (**Figure 1.11b**) and birds called curve-billed thrashers (**Figure 1.11c**). These species are among the most important fruit- and seed-eating animals in habitats where chilies grow. Based on earlier observations, the biologists predicted that cactus mice destroy chili seeds but that curve-billed thrashers disperse them effectively.

To test the directed dispersal hypothesis, the biologists offered both cactus mice and curve-billed thrashers three kinds of fruit: hackberries, fruits from a strain of chilies that can't synthesize capsaicin, and pungent chilies that have lots of capsaicin. The non-pungent chilies are about the same size and color as normal chilies and have similar nutritional value. The hackberries look like chilies, except that they are not quite as red and contain no capsaicin. The three fruits were present in equal amounts. For each animal tested, the researchers recorded the percentage of hackberry, non-pungent chili, and pungent chili that was eaten during a specific time interval. Then they calculated the average amount of each fruit that was eaten by five test individuals from each species.

The directed dispersal hypothesis predicts that seed dispersers will eat the pungent chilies readily but that seed predators won't. Recall that a prediction specifies what we should observe if a hypothesis is correct. Good scientific hypotheses make testable predictions—predictions that can be supported or rejected by collecting and analyzing data. If the directed dispersal hypothesis is wrong, however, then there shouldn't be any difference in what various animals eat. This latter possibility is called a **null hypothesis**. A null hypothesis specifies what we should observe when the hypothesis being tested

(a) Wild chilies produce fruits that contain seeds.

(b) Cactus mice are seed eaters.

(c) Curve-billed thrashers are fruit eaters.

FIGURE 1.11 Chilies... and Chili Eaters?

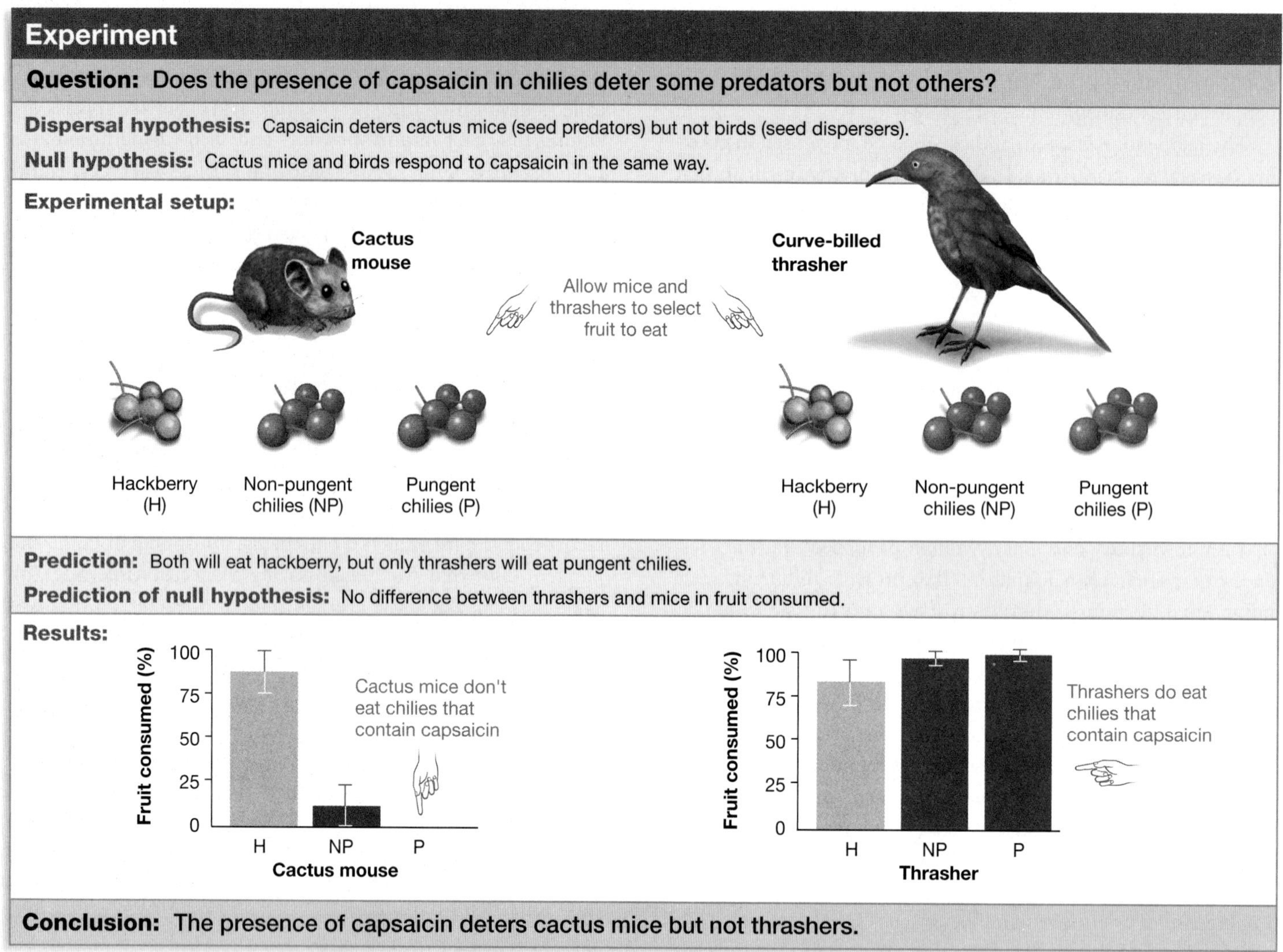

FIGURE 1.12 An Experimental Test: Does Capsaicin Deter Some Fruit Eaters? The graphs in the Results section indicate the average percentage of fruit consumed by the animals that were tested. The thin vertical lines indicate the standard error associated with each average. The standard error is a measure of variability and uncertainty in the data. For more details, see **BioSkills 3**.

doesn't hold. These predictions are listed in **Figure 1.12**. Do the predictions of the directed dispersal hypothesis hold? To answer this question, look at the results plotted in Figure 1.12. View the data for each type of fruit to see if different amounts were eaten by different types of animals, and ask yourself whether the two animals ate the same or different amounts of each type of fruit. Based on your analysis of the data, decide whether the results support the directed dispersal hypothesis or the null hypothesis. Use the conclusion stated in Figure 1.12 to check your answer.

In relation to designing effective experiments, this study illustrates several important points:

- It is critical to include control groups. A control checks for factors, other than the one being tested, that might influence the experiment's outcome. For example, if hackberries had not been included as a control, it would have been possible to claim that the cactus mice in the experiment didn't eat pungent chilies simply because they weren't hungry. But the not-hungry hypothesis can be rejected because all of the animals ate hackberries.

- The experimental conditions must be carefully controlled. The investigators used the same feeding choice setup, the same time interval, and the same definition of what "fruit consumed" meant in each test. Controlling all of the variables except one—the types of fruits presented—is crucial because it eliminates alternative explanations for the results. For example, what types of problems could arise if the cactus mice were given less time to eat than the thrashers, or if the test animals were always presented with hackberries first and pungent chilies last?

- Repeating the test is essential. It is almost universally true that larger sample sizes in experiments are better. For example, suppose that the experimenters had used just one cactus mouse instead of five, and that this mouse was unlike other cactus mice because it ate almost anything. If so, the resulting data would be badly distorted. By testing many individuals, the amount of distortion or "noise" in the data caused by unusual individuals or circumstances is reduced.

(MB) **Web Animation** at www.masteringbio.com
Introduction to Experimental Design

If you understand these concepts, you should be able to design an experiment that tests the directed dispersal hypothesis by adding capsaicin to small pieces of apple—a fruit that both mice and thrashers eat readily.

To test the assumption that cactus mice are seed predators and curve-billed thrashers are seed dispersers, the researchers did a follow-up experiment. They fed fruits of the non-pungent chili to each type of predator. When the seeds had passed through the animals' digestive systems and were excreted, the researchers collected and planted the seeds—along with 14 uneaten seeds. Planting uneaten seeds served as a control treatment, because it tested the hypothesis that the seeds were viable and would germinate if they were not eaten. About 50 percent of the uneaten seeds germinated, and almost 60 percent of the seeds eaten by a thrasher germinated. But none of the seeds eaten by cactus mice germinated. The data indicate that seeds pass through curve-billed thrashers unharmed but are destroyed when eaten by cactus mice.

Based on the outcomes of these two experiments, the researchers concluded that curve-billed thrashers are efficient seed dispersers and are not deterred by capsaicin. The cactus mice, in contrast, refuse to eat chilies. If they ate chilies, the mice would kill the seeds. These are exactly the results predicted by the directed dispersal hypothesis. The biologists concluded that the presence of capsaicin in chilies is an adaptation that keeps their seeds from being destroyed by mice. In habitats that contain cactus mice, the production of capsaicin increases the fitness of individual chili plants.

These experiments are a taste of things to come. In this text you will encounter hypotheses and experiments on questions ranging from how water gets to the top of 100-meter-tall sequoia trees to why the bacterium that causes tuberculosis has become resistant to antibiotics. A commitment to tough-minded hypothesis testing and sound experimental design is a hallmark of biological science. Understanding their value is an important first step in becoming a biologist.

Check Your Understanding

If you understand that...

- Hypotheses are proposed explanations that make testable predictions.
- Predictions are observable outcomes of particular conditions.
- Well-designed experiments alter just one condition—a condition relevant to the hypothesis being tested.

You should be able to...

1) Design an experiment to test the hypothesis that the use of capsaicin as cooking spice is an adaptation—specifically, that the presence of capsaicin in food kills disease-causing bacteria.
2) State the predictions of the adaptation hypothesis and the null hypothesis in your experiment.
3) Answer the following questions about your experimental design:

   How does the presence of a control group in your experiment allow you to test the null hypothesis?

   Why isn't the experiment valid without the presence of a control group?

   How are experimental conditions controlled or standardized in a way that precludes alternative explanations of the data?

   Why do you propose to repeat the experiment many times?

Chapter Review

SUMMARY OF KEY CONCEPTS

For over two hundred years, biologists have been discovering traits that unify the spectacular diversity of living organisms.

Biological science was founded with the development of (1) the cell theory, which proposes that all organisms are made of cells and that all cells come from preexisting cells, and (2) the theory of evolution by natural selection, which maintains that the characteristics of species change through time—primarily because individuals with certain heritable traits produce more offspring than do individuals without those traits.

The cell theory is an important unifying principle in biology, because it identified the fundamental structural unit common to all life. The theory of evolution by natural selection is another key unifying principle, because it states that all organisms are related by common ancestry. It also offered a robust explanation for why species change through time and why they are so well adapted to their habitats.

You should be able to describe the evidence that supported the cell theory. You should also be able to explain why a population of wild *Brassica oleracea* will evolve by natural selection, if in response to global warming individuals with large leaves begin producing the most offspring.

MB Web Animation at www.masteringbio.com
Artificial Selection

A phylogenetic tree is a graphical representation of the evolutionary relationships among species. Phylogenies can be established by analyzing similarities and differences in traits. Species that share many traits are closely related and are placed close to each other on the tree of life.

The cell theory and the theory of evolution predict that all organisms are part of a genealogy of species, and that all species trace their ancestry back to a single common ancestor. To reconstruct this phylogeny, biologists have analyzed the sequence of components in rRNA and other molecules found in all cells. A tree of life, based on similarities and differences in these molecules has three major lineages: the Bacteria, Archaea, and Eukarya.

You should be able to explain why biologists can determine whether newly discovered species are members of the Bacteria, Archaea, or Eukarya by analyzing their rRNA or other molecules.

Biologists ask questions, generate hypotheses to answer them, and design experiments that test the predictions made by competing hypotheses.

Another unifying theme in biology is a commitment to hypothesis testing and to sound experimental design. Analyses of neck length in giraffes and the capsaicin found in chilies are case studies in the value of testing alternative hypotheses and conducting experiments. Biology is a hypothesis-driven, experimental science.

You should be able to explain (1) the relationship between a hypothesis and a prediction, and (2) why experiments are convincing ways to test predictions.

MB Web Animation at www.masteringbio.com
Introduction to Experimental Design

QUESTIONS

Test Your Knowledge

1. Anton van Leeuwenhoek made an important contribution to the development of the cell theory. How?
   a. He articulated the pattern component of the theory—that all organisms are made of cells.
   b. He articulated the process component of the theory—that all cells come from preexisting cells.
   c. He invented the first microscope and saw the first cell.
   d. He invented more powerful microscopes and was the first to describe the diversity of cells.

2. Suppose that a proponent of the spontaneous generation hypothesis claimed that cells would appear in Pasteur's swan-necked flask eventually. According to this view, Pasteur did not allow enough time to pass before concluding that life does not originate spontaneously. Which of the following is the best response?
   a. The spontaneous generation proponent is correct: Spontaneous generation would probably happen eventually.
   b. Both the all-cells-from-cells hypothesis and the spontaneous generation hypothesis could be correct.
   c. If spontaneous generation happens only rarely, it is not important.
   d. If spontaneous generation did not occur after weeks or months, it is not reasonable to claim that it would occur later.

3. What does the term *evolution* mean?
   a. The strongest individuals produce the most offspring.
   b. The characteristics of an individual change through the course of its life, in response to natural selection.
   c. The characteristics of populations change through time.
   d. The characteristics of species become more complex over time.

4. What does it mean to say that a characteristic of an organism is heritable?
   a. The characteristic evolves.
   b. The characteristic can be passed on to offspring.
   c. The characteristic is advantageous to the organism.
   d. The characteristic does not vary in the population.

5. In biology, what does the term *fitness* mean?
   a. how well trained and muscular an individual is, relative to others in the same population
   b. how slim an individual is, relative to others in the same population
   c. how long a particular individual lives
   d. the ability to survive and reproduce

6. Could *both* the food competition hypothesis and the sexual selection hypothesis explain why giraffes have long necks? Why or why not?
   a. No. In science, only one hypothesis can be correct.
   b. No. Observations have shown that the food competition hypothesis cannot be correct.
   c. Yes. Long necks could be advantageous for more than one reason.
   d. Yes. All giraffes have been shown to feed at the highest possible height and fight for mates.

Test Your Knowledge answers: 1. d; 2. d; 3. c; 4. b; 5. d; 6. c

Test Your Understanding

Answers are available at www.masteringbio.com

1. The Greek roots of the term taxonomy can be translated as "arranging rules." Explain why these roots were an appropriate choice for this term.

2. It was once thought that the deepest split among life-forms was between two groups: prokaryotes and eukaryotes. Draw and label a phylogenetic tree that represents this hypothesis. Then draw and

label a phylogenetic tree that shows the actual relationships among the three domains of organisms.

3. Why was it important for Linnaeus to establish the rule that only one type of organism can have a particular genus and species name?
4. What does it mean to say that an organism is adapted to a particular habitat?
5. Compare and contrast natural selection with the process that led to the divergence of a wild mustard plant into cabbage, broccoli, and Brussels sprouts.
6. The following two statements explain the logic behind the use of molecular sequence data to estimate evolutionary relationships:

   "If the theory of evolution is true, then rRNA sequences should be very similar in closely related organisms but less similar in organisms that are less closely related."

   "On a phylogenetic tree, branches that are close to one another represent species that are closely related; branches that are farther apart represent species that are more distantly related."

   Is the logic of these statements sound? Why or why not?

Applying Concepts to New Situations

Answers are available at www.masteringbio.com

1. A scientific theory is a set of propositions that defines and explains some aspect of the world. This definition contrasts sharply with the everyday usage of the word theory, which often carries meanings such as "speculation" or "guess." Explain the difference between the two definitions, using the cell theory and the theory of evolution by natural selection as examples.
2. Turn back to the tree of life shown in Figure 1.9. Note that Bacteria and Archaea are prokaryotes, while Eukarya are eukaryotes. On the simplified tree below, draw an arrow that points to the branch where the structure called the nucleus originated. Explain your reasoning.

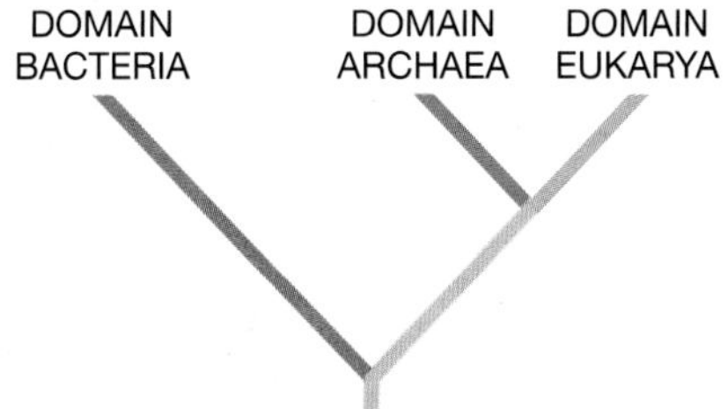

3. The proponents of the cell theory could not "prove" that it was correct in the sense of providing incontrovertible evidence that all organisms are made up of cells. They could state only that all organisms examined to date were made of cells. Why was it reasonable for them to conclude that the theory was valid?
4. How do the tree of life and the taxonomic categories created by Linnaeus (kingdom, phylum, class, order, family, genus, and species) relate to one another?

www.masteringbio.com is also your resource for • Answers to text, table, and figure caption questions and exercises • Answers to *Check Your Understanding* boxes • Online study guides and quizzes • Additional study tools including the *E-Book for Biological Science* 3rd ed., textbook art, animations, and videos.

2

Water and Carbon: The Chemical Basis of Life

KEY CONCEPTS

- Molecules form when atoms bond to each other. Chemical bonds are based on electron sharing. The degree of electron sharing varies from nonpolar covalent bonds, to polar covalent bonds, to ionic bonds.
- Of all small molecules, water is the most important for life. Water is highly polar and readily forms hydrogen bonds. Hydrogen bonding makes water an extremely efficient solvent with a high capacity for absorbing energy.
- Chemical reactions tend to be spontaneous if they lead to lower potential energy and higher entropy (more disorder). An input of energy is required for nonspontaneous reactions to occur.
- Most of the important compounds in organisms contain carbon. Early in Earth's history, the energy in sunlight triggered nonspontaneous reactions that led to the formation of key carbon-containing molecules.

Life began in the ocean, over 3 billion years ago.

Chapter 1 introduced experiments on the hypothesis of spontaneous generation, which tested the idea that life arises from nonliving materials. This work helped build a consensus that spontaneous generation does not occur. But for life to exist now, spontaneous generation must have occurred at least once, early in Earth's history. How did life begin? This simple query has been called "the mother of all questions."

This chapter examines a theory, called chemical evolution, that attempts to answer this question. Like the theories introduced in Chapter 1, the theory of chemical evolution has a pattern component that makes a claim about the natural world and a process component that explains that pattern. **Chemical evolution** is the proposition that early in Earth's history, simple chemical compounds in the atmosphere and ocean combined to form larger, more complex substances. As a result, the chemistry of the oceans and atmosphere changed over time. The theory's name is appropriate because in its simplest sense, the term *evolution* means change through time. According to the theory, the process responsible for this pattern was the conversion of the kinetic energy in sunlight and heat to chemical energy in the form of bonds between atoms that formed large, complex compounds.

The theory also maintains that larger, more complex substances began to accumulate and then reacted with one another to produce even more complicated compounds, and that continued chemical evolution eventually led to the origin of life. More specifically, the hypothesis is that one of these complex compounds was able to make a copy of itself, or self-replicate. As this molecule multiplied, the process of evolution by natural selection took over—there was a switch from chemical evolution to biological evolution. Eventually, a self-replicating molecule became surrounded by a membrane, and cellular life began.

At first glance, the theory of chemical evolution seems wildly implausible. But is it? What evidence do biologists have that chemical evolution occurred? Let's start with the fundamentals—the atoms and molecules that would have combined to get chemical evolution started.

2.1 The Building Blocks of Chemical Evolution

Just four types of atoms—hydrogen, carbon, nitrogen, and oxygen—make up 96 percent of all matter found in organisms today. Many of the molecules found in living cells contain thousands, or even millions, of these atoms bonded together. But early in Earth's history, it is likely that these elements existed only in simple substances such as water and carbon dioxide, which contain just three atoms apiece.

The theory of chemical evolution maintains that simple compounds in the ancient atmosphere and ocean combined to form the larger, more complex substances found in living cells. To understand how this process could have begun, we need to consider the following questions:

- What is the physical structure of hydrogen, carbon, nitrogen, oxygen, and other atoms found in living cells?
- What is the structure of water, carbon dioxide, and the other simple molecules that acted as the building blocks of chemical evolution?

These questions focus on one of the most central themes in biology: Function follows from structure. To understand how a molecule affects your body or the role it played in chemical evolution, you have to understand how it is put together.

What Atoms Are Found in Organisms?

Figure 2.1 shows a simple way of depicting the structure of an atom, using hydrogen and carbon as examples. Extremely small particles called electrons orbit an atomic nucleus made up of large particles called protons and neutrons. Protons have a positive electric charge, neutrons are electrically neutral, and electrons have a negative electric charge. Opposite charges attract; like charges repel. When the number of protons and the number of electrons in an atom (or molecule) are the same, the charges balance and the atom is electrically neutral.

Figure 2.2 shows a segment of the periodic table of the elements. Notice that each of the **elements** contains a characteristic number of protons, called its **atomic number**. The atomic number is given as a subscript of each element symbol in the table. The number of neutrons present in an element can vary, however. Forms of an element with different numbers of neutrons are known as **isotopes** ("equal-places"). For example, all

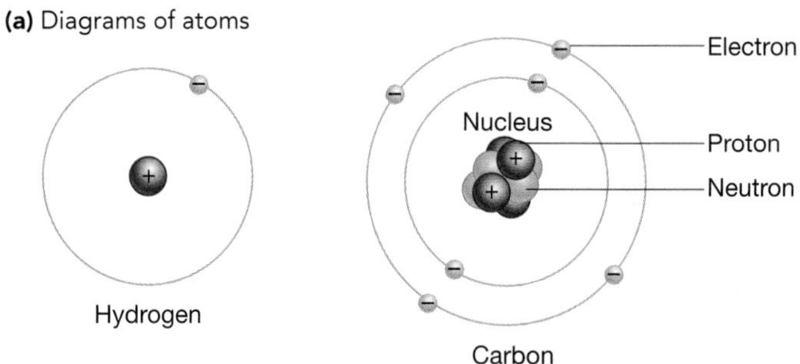

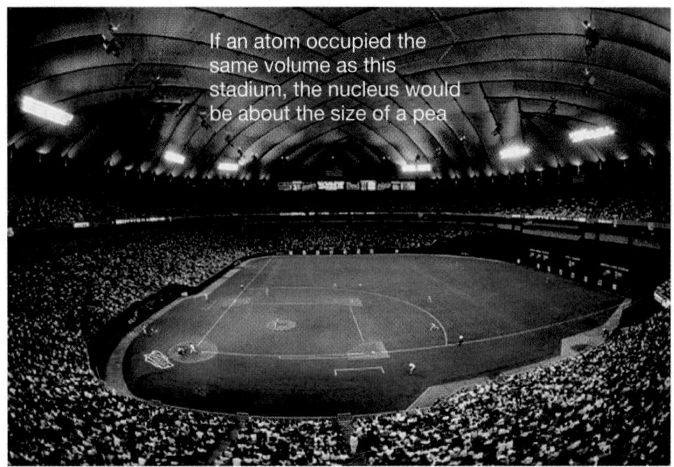

FIGURE 2.1 Parts of an Atom. These diagrams of the hydrogen and carbon atoms show how the nucleus, made up of protons and neutrons, is surrounded by orbiting electrons. In reality, electrons do not orbit the nucleus in circles; their actual orbits are complex. As the note on this photograph shows, the diagrams are not to scale.

atoms of the element carbon have 6 protons. But naturally occurring isotopes of carbon can have either 6 or 7 neutrons, giving them a total of 12 or 13 protons and neutrons, respectively. The sum of the protons and neutrons in an atom is called its **mass number**, given as a superscript of each symbol in Figure 2.2. Although the masses of protons, neutrons, and

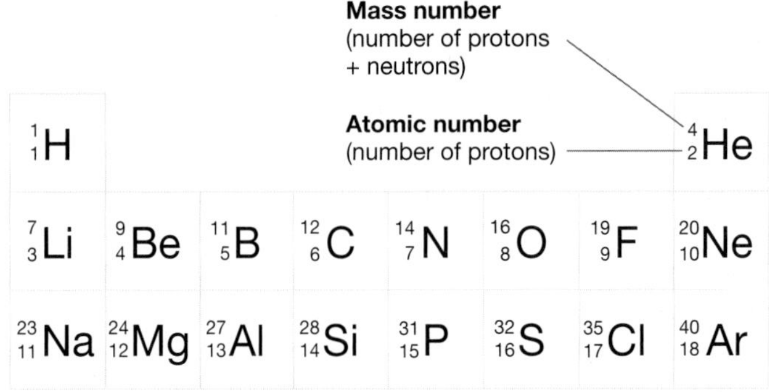

FIGURE 2.2 A Portion of the Periodic Table. Each element has a unique atomic number and is represented by a unique one- or two-letter symbol.

electrons can be measured in grams, the numbers involved are so small that chemists and physicists prefer to use a special unit called the **atomic mass unit (amu)**, or the dalton. The masses of protons and neutrons are virtually identical and are routinely rounded to 1 amu. A carbon atom that contains 6 protons and 6 neutrons has a mass of 12 amu and a mass number of 12, while a carbon atom with 7 neutrons would have a mass number of 13. These isotopes would be written as ^{12}C and ^{13}C, respectively.

To understand how the atoms involved in chemical evolution behave, though, focus on how electrons are arranged around the nucleus. Electrons move around atomic nuclei in specific regions called **orbitals**. Each electron orbital has a distinctive shape, and each orbital can hold up to two electrons. Orbitals, in turn, are grouped into levels called **electron shells**. These are numbered 1, 2, 3, and so on, to indicate their relative distance from the nucleus. The electrons of an atom fill the innermost shells first, before filling outer shells.

To understand how the structures of atoms differ, take a moment to study **Figure 2.3**. This chart highlights the atoms that are most abundant in living cells. The diagrams in each box of the table show how electrons are distributed in the shells of carbon and other key elements. The outermost shell of an atom is called the atom's valence shell, and the electrons found in that shell are referred to as **valence electrons**. Two observations are important:

1. In each of the highlighted elements, the outermost electron shell is not full. The highlighted elements have at least one unpaired valence electron.

2. The number of unpaired electrons in the valence shell varies among elements. Carbon has four unpaired electrons in its outermost shell; hydrogen has one. The number of unpaired electrons found in an atom is called its **valence**. Carbon's valence is four; hydrogen's valence is one.

These observations are significant because an atom is most stable when its valence shell is filled. One way that shells can be filled is through the formation of **chemical bonds**—strong attractions that bind atoms together.

How Does Covalent Bonding Hold Molecules Together?

To understand how atoms can become more stable by sharing electrons, consider hydrogen. The hydrogen atom has just one electron, which resides in a shell that can hold two electrons. When two atoms of hydrogen approach each other, the two electrons present become shared by the two nuclei (**Figure 2.4**).

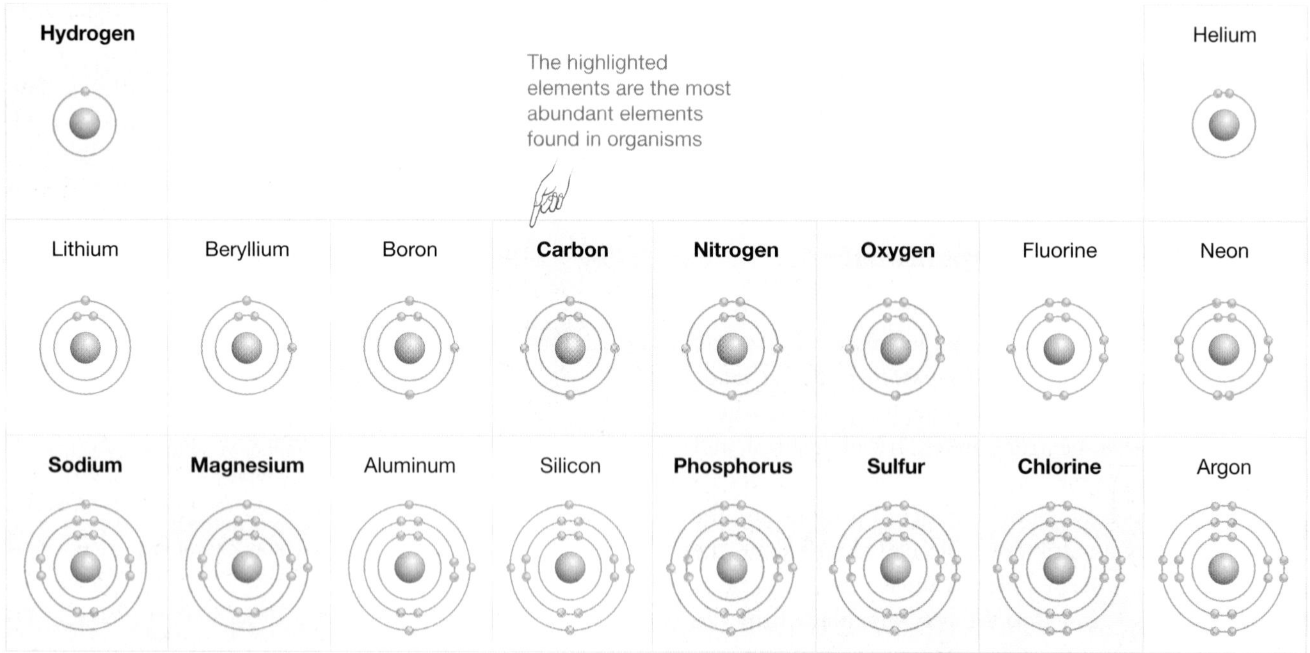

FIGURE 2.3 The Structure of Atoms Found in Organisms. This diagram represents the atomic nucleus as a closed sphere. The first (inner), second, and third electron shells are shown as rings. The dots on the rings represent electrons. Electrons are drawn in pairs if they occupy filled orbitals within the same electron shell; they are drawn singly if they occupy unfilled orbitals.

EXERCISE Unpaired electrons in an unfilled shell can form chemical bonds. Write down the number of bonds each highlighted atom is capable of forming.

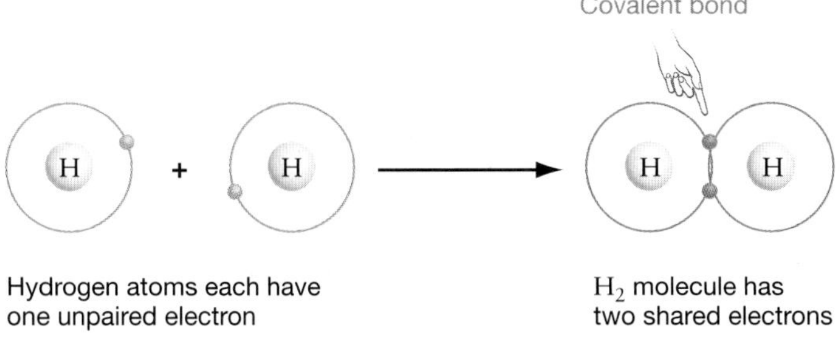

FIGURE 2.4 Covalent Bonds Result from Electron Sharing. When two hydrogen atoms come into contact, their electrons are attracted to the positive charge in each nucleus. As a result, their orbitals overlap, the electrons are shared by each nucleus, and a covalent bond forms.

(a) Nonpolar covalent bond in hydrogen molecule

(b) Polar covalent bonds in water molecule

FIGURE 2.5 Electron Sharing and Bond Polarity. (a) The electrons in bonds are shared equally if the two atoms involved have equal electronegativity, but **(b)** unequally if the atoms have unequal electronegativity. Delta (δ) symbols in polar covalent bonds refer to partial positive and negative charges that arise due to unequal electron sharing.

Both atoms now have a completely filled shell, and they have formed a chemical bond. Together, the bonded hydrogen atoms are more stable than the two individual hydrogen atoms. The shared electrons "glue" the atoms together in a **covalent bond**. Substances that are held together by covalent bonds are called **molecules**. In this case of two hydrogen atoms, the bonded atoms form a single molecule of hydrogen, written as H H or H_2.

It's often helpful to think about covalent bonding as electrical attraction and repulsion. As two hydrogen atoms move closer together, their positively charged nuclei repel each other and their negatively charged electrons repel each other. But each proton attracts both electrons, and each electron attracts both protons. Covalent bonds form when the attractive forces overcome the repulsive forces. This is the case when hydrogen atoms interact to form the hydrogen molecule (H_2).

To visualize what's happening when a covalent bond forms, imagine two children who want the same toy. If both children grasp the toy tightly, through it they become attached to each other. For as long as they hold onto the toy, neither child can leave the other's side. ● If you understand this concept, you should be able to explain what the children, the toy, and the children's hold on the toy actually represent, and why the "children" want the toy.

To summarize, covalent bonding is based on electron sharing. In a hydrogen molecule, the electrons involved in the covalent bond are shared between the two nuclei. This situation is illustrated in **Figure 2.5a**. The covalent bond between hydrogen atoms is represented by a dash, and the electrons are drawn as dots between the two nuclei. An equal sharing of electrons, as illustrated in Figure 2.5a, results in a **nonpolar covalent bond**—a covalent bond that is symmetrical.

It's important to note, though, that the electrons participating in covalent bonds are not always shared equally between the atoms involved. Asymmetric sharing of electrons results in polar covalent bonds. Some atoms hold electrons in covalent bonds much more tightly than do other atoms, so the degree of sharing varies by element. Chemists call this property an atom's **electronegativity**. If you understand this concept, you should be able to extend the analogy of a covalent bond being represented by two children clutching the same toy. What would this situation look like if one of the children wanted the toy more than the other? In this case, which of the two bonds would be weaker—meaning, easier to break?

Oxygen is among the most electronegative of all elements: It attracts covalently bonded electrons more strongly than does any other atom commonly found in organisms. Nitrogen's electronegativity is somewhat lower than oxygen's. Carbon and hydrogen, in turn, have relatively low and approximately equal electronegativities. The electronegativities of the four most abundant elements in organisms are related as follows: $O >> N > C \cong H$.

To put these ideas into practice, consider the water molecule. Water consists of oxygen bonded to two hydrogen atoms and is written H_2O. As **Figure 2.5b** illustrates, the electrons involved in the covalent bonds in water are not shared equally but are held much more tightly by the oxygen nucleus than by the hydrogen nuclei. Hence, water has two polar covalent bonds. Because electrons are shared unequally, the oxygen atom has a partial negative charge and the hydrogen atoms have a partial positive charge. Notice that the partial charges on the molecule are symbolized by the lowercase Greek letter delta, δ. As Section 2.2 will show, the partial charges on water molecules—due simply to the difference in electronegativity between oxygen and hydrogen—are one of the primary reasons life exists.

How Does Ionic Bonding Hold Ionic Compounds Together?

Ionic bonds are similar in principle to covalent bonds, but instead of sharing electrons between two atoms, the electrons in

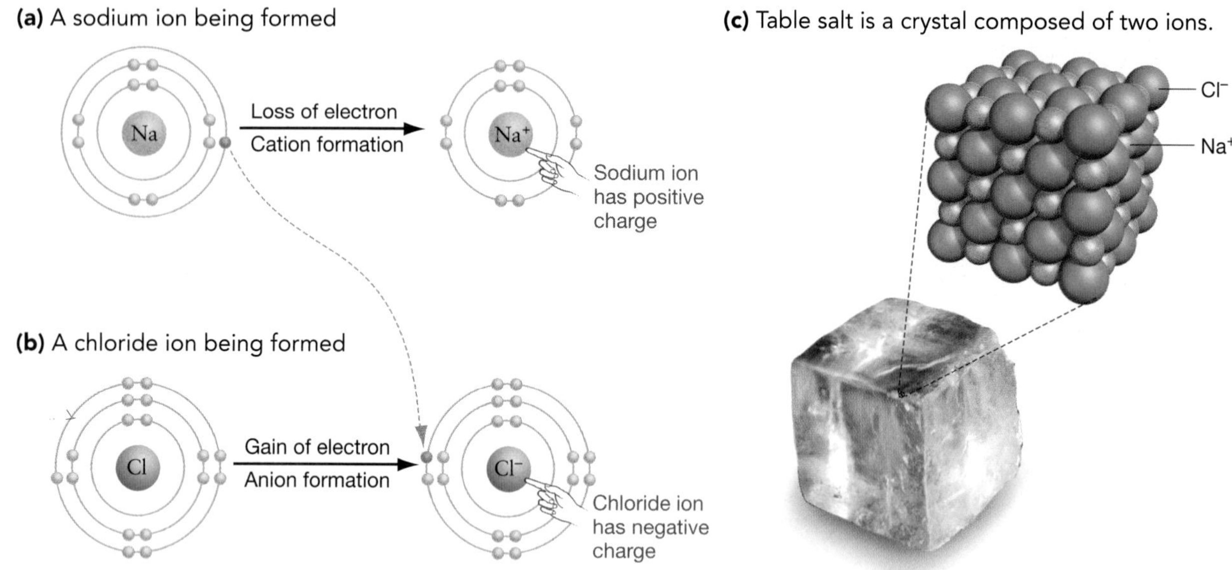

FIGURE 2.6 Ion Formation and Ionic Bonding. (a) When an atom of sodium (Na) loses an electron, it forms a sodium ion (Na^+). The sodium ion is stable because it has a full valence shell. **(b)** When an atom of chlorine (Cl) gains an electron, it forms a chloride ion (Cl^-), which also has a full valence shell. **(c)** In table salt (NaCl), sodium and chloride ions pack into a crystal structure held together by electrical attraction.

ionic bonds are completely transferred from one atom to the other. The electron transfer occurs because it gives the resulting atoms a full outermost shell. Sodium atoms (Na), for example, tend to lose an electron, leaving them with a full second shell. This is a much more stable arrangement, energetically, than having a lone electron in their third shell (**Figure 2.6a**). The atom that results has a net electric charge of +1, because it has one more proton than it has electrons. An atom or molecule that carries a charge is called an **ion**. The sodium ion is written Na^+ and, like other positively charged ions, is called a **cation**. Chlorine atoms (Cl), in contrast, tend to gain an electron, filling their outermost shell (**Figure 2.6b**). The atom has a net charge of −1, because it has one more electron than protons. This negatively charged ion, or **anion**, is written Cl^- and is called chlori*de*. When sodium and chlorine combine to form table salt (sodium chloride, NaCl), the atoms pack into a crystal structure consisting of sodium cations and chloride anions (**Figure 2.6c**). The electrical attraction between the ions is so strong that salt crystals are difficult to break apart.

This discussion of covalent and ionic bonding supports an important general observation: The degree to which electrons are shared in chemical bonds forms a continuum. As the left-hand side of **Figure 2.7** shows, covalent bonds between atoms with exactly the same electronegativity—for example, between the atoms of hydrogen in H_2—represent one end of the continuum. The electrons in these nonpolar bonds are shared equally. In the middle of the continuum are bonds where one atom is much more electronegative than the other. In these asymmetric bonds, substantial partial charges exist on each of the atoms. These types of polar covalent bonds occur

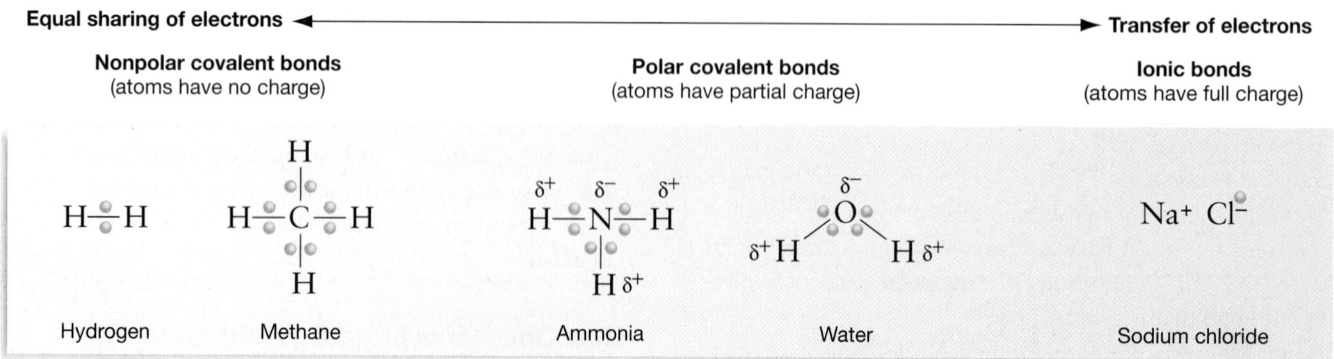

FIGURE 2.7 The Electron-Sharing Continuum. The degree of electron sharing in chemical bonds can be thought of as a continuum, from equal sharing in nonpolar covalent bonds to no sharing in ionic bonds.

QUESTION Why do most polar covalent bonds involve nitrogen or oxygen?

when a highly electronegative atom such as oxygen or nitrogen is bound to an atom with a lower affinity for electrons, such as carbon or hydrogen. Water (H_2O) and ammonia (NH_3) contain polar covalent bonds. At the right-hand side of the continuum are molecules made up of atoms with extreme differences in their electronegativities. In this case, electrons are transferred rather than shared, the atoms have full charges, and the bonding is ionic. Common table salt, NaCl, is a familiar example. In biology, ionic bonding is rare. The chemical bonds that occur in, biological molecules are on the left-hand side and the middle of this continuum.

Some Simple Molecules Formed from H, C, N, and O

Look again at Figure 2.3 and count the number of unpaired electrons in the valence shells of carbon, nitrogen, and oxygen atoms: Carbon has four, nitrogen has three, and oxygen has two. Each unpaired electron can make up half of a covalent bond. As a result, a carbon atom can form a total of four covalent bonds; nitrogen can form three; and oxygen, two. When each of the four unpaired electrons of a carbon atom covalently bonds with a hydrogen atom, the molecule that results is written CH_4 and is called methane (**Figure 2.8a**). This is the most common molecule found in natural gas. When a nitrogen atom's three unpaired electrons bond with three hydrogen atoms, the result is NH_3, or ammonia. Similarly, an atom of oxygen can form covalent bonds with two atoms of hydrogen, resulting in a water molecule (H_2O).

Double and Triple Bonds Atoms with more than one unpaired electron in the valence shell can form multiple single bonds and sometimes double or triple bonds. **Figure 2.8b** shows how carbon forms double bonds with oxygen atoms to produce carbon dioxide (CO_2). Triple bonds result when three pairs of electrons are shared. **Figure 2.8c** shows the structure of molecular nitrogen (N_2), which forms when two nitrogen atoms establish a triple bond.

The Geometry of Some Simple Molecules Proponents of the theory of chemical evolution argue that simple molecules such as CH_4, NH_3, H_2O, CO_2, and N_2 were important components of Earth's ancient atmosphere and ocean. This reasoning is based on the observation that these molecules are found in volcanic gases on Earth and in the atmospheres of nearby planets. If so, then such molecules provided the building blocks for chemical evolution. Before analyzing how these molecules could have combined to form more complex compounds, however, we need to consider their shape and how scientists represent their composition and geometry. Recall that this is a crucial point—one that will resonate throughout this text. Function follows from structure. For example, the overall shape of a molecule often dictates its behavior and function.

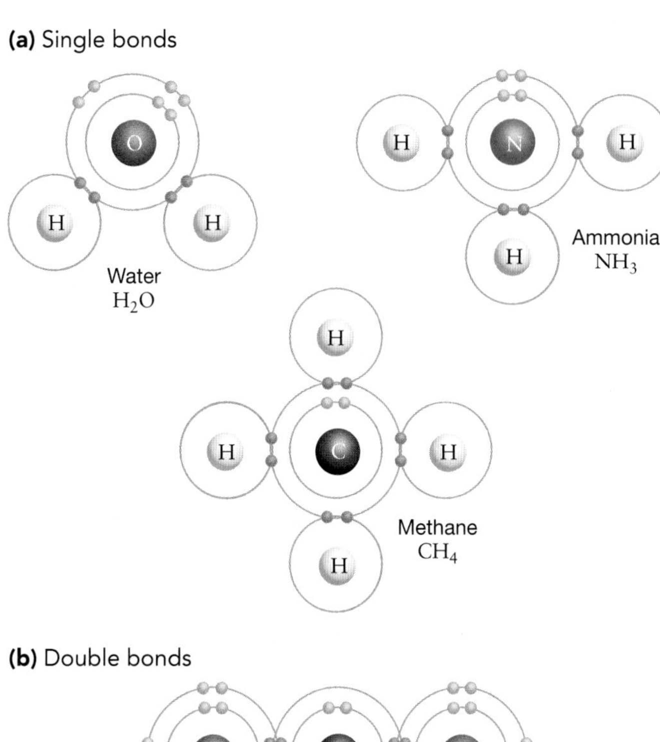

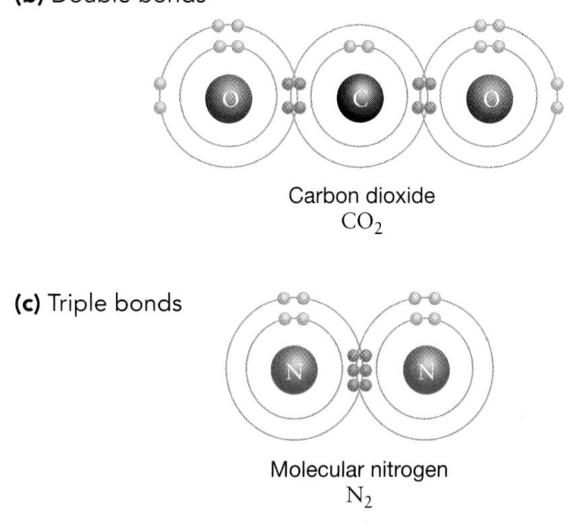

FIGURE 2.8 Unpaired Electrons in the Valence Shell Participate in Covalent Bonds. Covalent bonding is based on sharing of electrons in the outermost shell. Covalent bonds can be **(a)** single, **(b)** double, or **(c)** triple.

QUESTION Why can carbon participate in four single bonds, while oxygen can participate in only two and hydrogen in only one?

The shape of a simple molecule is governed by the geometry of its bonds. Bond angles, in turn, are determined by electrical repulsion that occurs between pairs of electrons. For example, **Figure 2.9a** shows the shape of methane (CH_4).

- The four pairs of electrons in carbon's outer shell form a tetrahedron when bonded to hydrogen because repulsion among the shared electron pairs orients them at an angle of 109.5° to one another.
- The same outer shells and bond geometry occur in ammonia (NH_3), but one of the four orbitals in the nitrogen atom's outer shell is already filled with a pair of electrons—so just three orbitals with unpaired electrons bond with hydrogen.

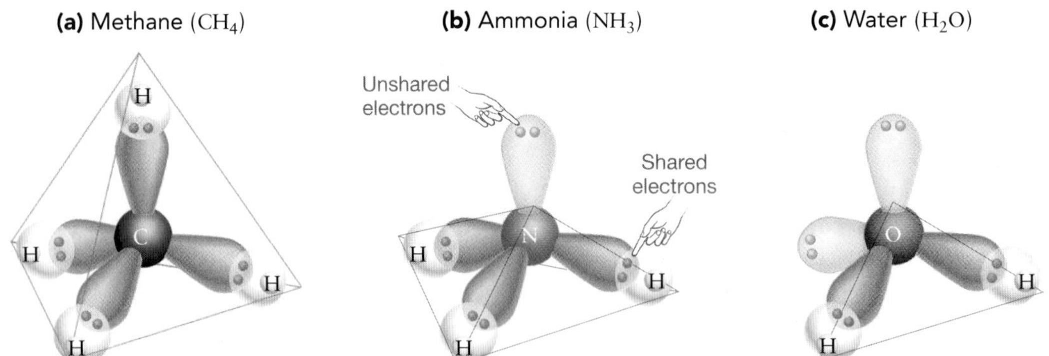

FIGURE 2.9 The Geometry of Methane, Ammonia, and Water. Diagrams showing the shapes of the orbitals in the outermost shells of **(a)** methane, **(b)** ammonia, and **(c)** water. Though some of the same orbitals are involved in bonding, the overall shape of each molecule is radically different.

● **EXERCISE** Label which molecule is bent and planar, which forms a tetrahedron (having four identical faces), and which forms a pyramid.

In this case, the geometry of the orbitals and the geometry of the molecule are different (**Figure 2.9b**).

- In the oxygen molecule, two of the four orbitals present in the outermost shell are filled with electron pairs. When hydrogen atoms bond to the remaining two electrons to form water (H_2O), the resulting molecule is bent and two-dimensional, or planar (**Figure 2.9c**). Section 2.2 will explore how water's shape, in combination with the differing electronegativities between oxygen and hydrogen, make it the most important molecule on Earth.

Representing Molecules Molecules can be represented in a variety of ways. The simplest representation is a **molecular formula**, which indicates the numbers and types of atoms in a molecule. Water has the molecular formula H_2O; methane has the molecular formula CH_4 (**Figure 2.10a**).

Molecular formulas are a compact way of stating a molecule's composition, but they contain no information about how the molecule is put together. **Structural formulas** indicate which atoms are bonded together. In a structural formula, single, double, and triple bonds are represented by single, double, and triple dashes, respectively. **Figure 2.10b** gives structural formulas for molecules that were important components of ancient Earth's atmosphere and ocean. Structural formulas also provide some information on the molecule's geometry. Water's structural formula indicates that it is bent, while the oxygen molecule is drawn as a linear molecule.

The limitation of structural formulas is that they are two-dimensional, while molecules are three-dimensional. More complex diagrams, such as ball-and-stick models and space-filling models, are based on the three-dimensional structure of molecules. **Figures 2.10c** and **2.10d** show how ball-and-stick and space-filling models give a more sophisticated view of a molecule's geometry as well as the relative sizes of the atoms involved. For more information on interpreting chemical structures, see **BioSkills 4** in the back of the book.

Quantifying the Concentration of Key Molecules When a **chemical reaction** occurs, one substance is combined with others or broken down into another substance. Chemicals react in simple whole-number combinations. For example, 1 molecule of acetic acid reacts with 1 molecule of ethanol to form 1 molecule of ethyl acetate. But if you wanted to set up this reaction in an experiment, how would you know how much acetic acid and ethanol to add? The problem is that there is no simple way of counting the numbers of molecules present in a solution. Researchers solve this problem using the mole concept.

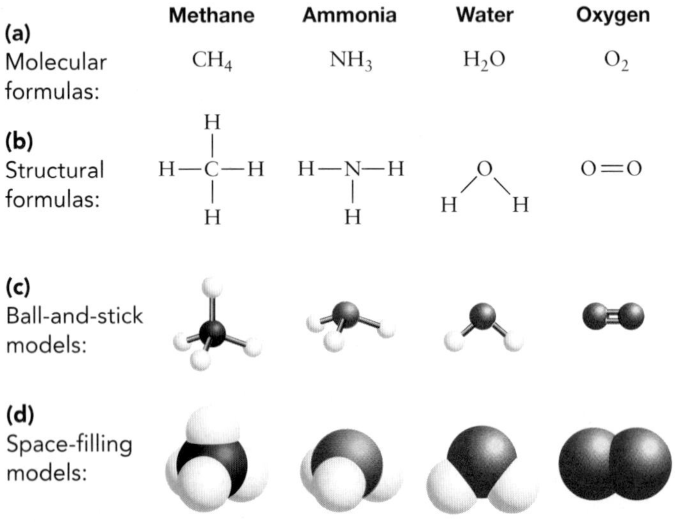

	Methane	Ammonia	Water	Oxygen
(a) Molecular formulas:	CH_4	NH_3	H_2O	O_2
(b) Structural formulas:	H above, H—C—H, H below	H—N—H, H below	O bonded to H and H (bent)	O=O
(c) Ball-and-stick models:				
(d) Space-filling models:				

FIGURE 2.10 Molecules Can Be Represented Several Ways. Each method of representing a molecule has particular advantages. **(a)** Molecular formulas are compact and indicate the number and identity of atoms involved. **(b)** Structural formulas show which atoms are bonded to one another and indicate geometry in two dimensions. **(c)** Ball-and-stick models take up more space than do structural formulas but include information about three-dimensional shape. **(d)** Space-filling models are not as easy to read as ball-and-stick models but more accurately depict the spatial relationships between atoms.

A **mole** refers to the number 6.022×10^{23}—just as the unit called the dozen refers to the number 12 or the unit million refers to the number 1×10^6. The mole is a useful unit because the mass of one mole of any molecule is the same as its molecular weight expressed in grams. **Molecular weight** is the sum of the mass numbers of all the atoms in a molecule. For example, summing the mass numbers of 2 atoms of hydrogen and 1 atom of oxygen gives 1 + 1 + 16, or 18, which is water's molecular weight. It follows that if you weighed a sample of 18 grams of water, it would contain 6.022×10^{23} water molecules, or 1 mole of water molecules.

Now let's get back to your experiment with acetic acid and ethanol. To obtain an equal number of each type of molecule, you use a balance to measure out the two substances and add an equal number of moles of each. For example, if you wanted to react one mole of acetic acid and one mole of ethanol, you'd weigh out a number of grams of acetic acid and ethanol equal to their respective molecular weights and make up a **solution**—a homogenous (uniform) mixture of one or more substances dissolved in a liquid. When substances are dissolved in a liquid, their concentration is expressed in terms of molarity (symbolized by "M"). **Molarity** is the number of moles of the substance present per liter of solution. A 1-molar solution of acetic acid in water, for instance, means that 1 mole of acetic acid—that is, 6.022×10^{23} acetic acid molecules—is contained in 1 liter of water.

Check Your Understanding

If you understand that...

- Covalent bonds are based on electron sharing. Ionic bonds are based on electrical attraction between ions with opposite charges.
- Covalent bonds can be polar or nonpolar, depending on whether the electronegativities of the two atoms involved are the same or different.

You should be able to...

1) Draw the structural formulas of methane (CH_4) and ammonia (NH_3) and add dots to indicate the relative locations of the covalently bonded electrons, based on the relative electronegativities of C, H, and N.
2) Draw the electron shells around sodium ions (Na^+) and chloride ions (Cl^-) and explain why table salt (NaCl) exists.

2.2 The Early Oceans and the Properties of Water

Life is based on water. In a typical living cell, over 75 percent of the volume consists of this molecule. Nearly 70 percent of your body, by weight, is water (**Figure 2.11**). You can survive for weeks without eating, but you aren't likely to live more than 3–4 days without drinking. In humans, a water loss of just 2 percent leads to weakness, dizziness, irritability, headache, trouble urinating, and a general feeling of being unwell.

FIGURE 2.11 Water Is the Most Abundant Molecule in Organisms. Fruits shrink when they are dried because they consist primarily of water.

Water is essential for life today, and virtually all researchers agree that most of the important steps in chemical evolution, including the origin of life itself, occurred in water. Water is vital for a simple reason: As a **solvent**—an agent for getting substances into solution—water can dissolve more types of substances than any other molecule known. The chemical reactions occurring inside your body right now and the reactions that caused chemical evolution some 3.5 billion years ago depend on direct, physical interaction between the reactants. Substances are most likely to collide and react when they are dissolved. Life is based on water primarily because of water's solvent properties. The formation of Earth's first ocean, about 3.8 billion years ago, was a turning point in chemical evolution, because it gave chemical evolution a place to happen.

Why Is Water Such an Efficient Solvent?

To understand why water is such an effective solvent, recall from Figure 2.5b that the molecule contains two hydrogen atoms bonded to an oxygen atom, and that oxygen and hydrogen differ in their electronegativity—meaning, their ability to attract the electrons involved in covalent bonds. Because oxygen is among the most electronegative of all elements, it attracts covalently bonded electrons much more strongly than does the hydrogen nucleus. As a result, the electrons that participate in the bonds are not shared equally. The covalent bonds in water are polar.

If water were a linear molecule like carbon dioxide, the polarity of its covalent bonds would not matter very much. But because the water molecule is bent, the molecule as a whole is **polar,** meaning that the overall distribution of charge is

(a) Water is polar.

Electrons are pulled toward oxygen

δ^-

O

H H

δ^+ δ^+

(b) Hydrogen bonds form between water molecules.

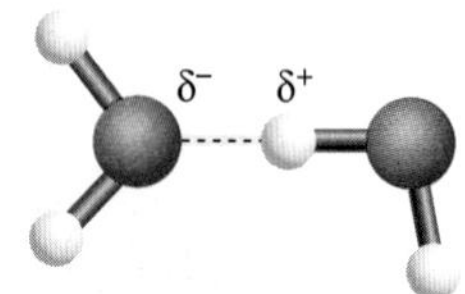

FIGURE 2.12 Water Is Polar and Participates in Hydrogen Bonds.
(a) Because of oxygen's high electronegativity, the electrons that are shared between hydrogen and oxygen spend more time close to the oxygen nucleus—giving the oxygen atom a partial negative charge and the hydrogen atom a partial positive charge. **(b)** The electrical attraction that occurs between the partial positive and negative charges on water molecules forms a hydrogen bond.

● **EXERCISE** Label the hydrogen bond in part (b).

● **EXERCISE** Water molecules routinely form multiple hydrogen bonds. Add 2 hydrogen bonds to the H_2O molecule on the right side of part (b).

asymmetrical. As **Figure 2.12a** shows, the side of the molecule containing the oxygen atom is slightly more negative, and the side with the hydrogen atoms is slightly more positive. As before, the partial charges on the molecule are symbolized by the lowercase Greek letter delta, δ.

Figure 2.12b illustrates how the polarity of water affects its interactions with other water molecules. When two liquid water molecules approach one another, the partial positive charge on hydrogen attracts the partial negative charge on oxygen. This weak electrical attraction forms a hydrogen bond between the molecules. ● If you understand how water's polarity makes hydrogen bonding possible, you should be able to (1) draw a version of Figure 2.12b that shows water as a linear (not bent) molecule, with partial charges on the oxygen and hydrogen atoms; and (2) explain why electrostatic attractions between water molecules would be much weaker or even nonexistent, as a result.

In a water-based, or aqueous, solution, hydrogen bonds also form between water molecules and other polar molecules. Similar interactions occur between water and ions. Ions and polar molecules stay in solution because of their interactions with water's partial charges (**Figure 2.13a**). Substances that interact with water in this way are said to be **hydrophilic** ("water-loving").

In contrast, compounds that are uncharged and nonpolar do not interact with water through hydrogen bonding and do not dissolve in water. Instead, they interact with each other (**Figure 2.13b**). Substances that do not interact with water are said to be **hydrophobic** ("water-hating"). The interactions between hydrophobic atoms and molecules are much weaker than interactions based on hydrogen bonding, however.

● To summarize, hydrogen bonding makes it possible for almost any charged or polar molecule to dissolve in water. Although individual hydrogen bonds are not nearly as strong as covalent or ionic bonds, they are much more common. Hydrogen bonding is extremely important in biology due to the sheer number of hydrogen bonds that form between water and other molecules that are polar or carry a charge.

How Does Water's Structure Correlate with Its Properties?

Water's structure is unusual. Its small size, bent shape, highly polar covalent bonds, and overall polarity are unique among molecules. Because the structure of molecules routinely correlates with their function, it's not surprising that water has some remarkable properties. We've already reviewed the most important of water's chemical attributes—its ability to form hydrogen bonds with ions and polar compounds and its capacity to act as a solvent.

(a) Polar molecules and ions dissolve readily in water.

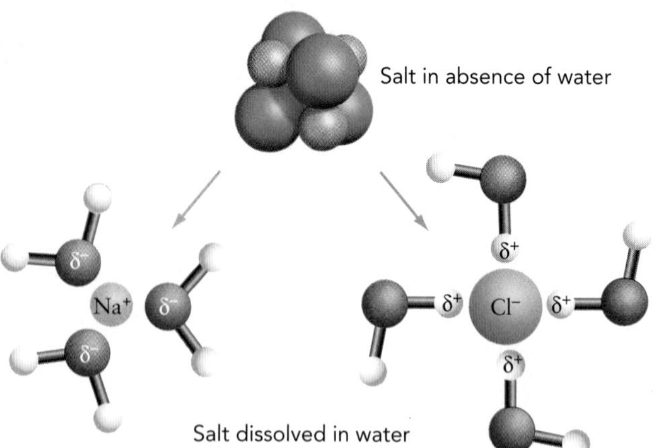

(b) Nonpolar molecules do not dissolve readily in water.

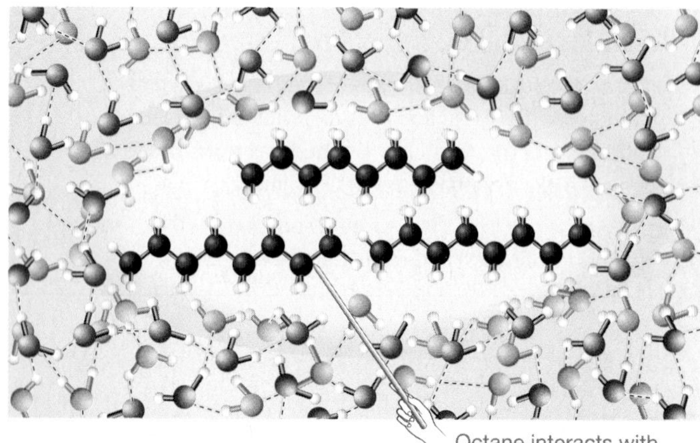

FIGURE 2.13 Polar Molecules and Ions Dissolve Readily in Water; Nonpolar Molecules Do Not.
(a) Water's polarity makes it a superb solvent for polar molecules and ions. **(b)** In aqueous solution, nonpolar molecules and compounds interact with each other but not with water.

● **QUESTION** Explain the physical basis of the expression, "Oil and water don't mix."

Water has several other striking physical properties. Each of these is the direct result of hydrogen bonding. In relation to chemical evolution, the most important of these properties are that water expands as it changes from a liquid to a solid and that it has an extraordinarily large capacity for absorbing heat. Let's analyze why hydrogen bonding explains these properties, along with the property called surface tension.

Cohesion, Adhesion, and Surface Tension Binding between like molecules is called **cohesion**. Water is cohesive—meaning that it stays together—because of the hydrogen bonds that form between individual molecules. Binding between unlike molecules, in contrast, is called **adhesion**. Adhesion is usually analyzed in regard to interactions between a liquid and a solid surface. Water adheres to surfaces that have any polar or charged components.

Cohesion and adhesion are important in explaining how water moves from the roots of plants to their leaves, but you can also see them in action in the concave surface, or **meniscus**, that forms in a glass of water (**Figure 2.14a**). Because water molecules at the surface hydrogen-bond with water molecules below them, they experience a net downward pull. But water molecules at the surface also adhere to the glass, allowing them to resist the downward pull. The result of these two forces is the formation of a meniscus.

Cohesion is also instrumental in the phenomenon known as surface tension. Because hydrogen bonding exerts a pulling force, or tension, at the surface of any body of water, water molecules are not stable at the surface—they are constantly being pulled away. It follows that a body of water as a whole is most stable when its total surface area is minimized. This fact has an important consequence: Water resists any force that increases its surface area. More specifically, any force that depresses a water surface meets with resistance. This resistance makes a water surface act as if it had an elastic membrane (**Figure 2.14b**)—a property called **surface tension**.

All liquids have a surface tension. Water's surface tension is extraordinarily high because of the extensive hydrogen bonding that occurs between molecules. In water, the "elastic membrane" at the surface is stronger than it is in other liquids.

Water Is Denser as a Liquid than as a Solid When factory workers pour molten metal or plastic into a mold and allow it to cool to the solid state, the material shrinks. When molten lava pours out of a volcano and cools to solid rock, it shrinks. But when you fill an ice tray with water and put it in the freezer to make ice, the water expands.

Unlike most substances, water is denser as a liquid than it is as a solid. In other words, there are more molecules of water in a given volume of liquid water than there are in the same

(a) Cohesion and adhesion explain why a meniscus forms where water meets a solid surface.

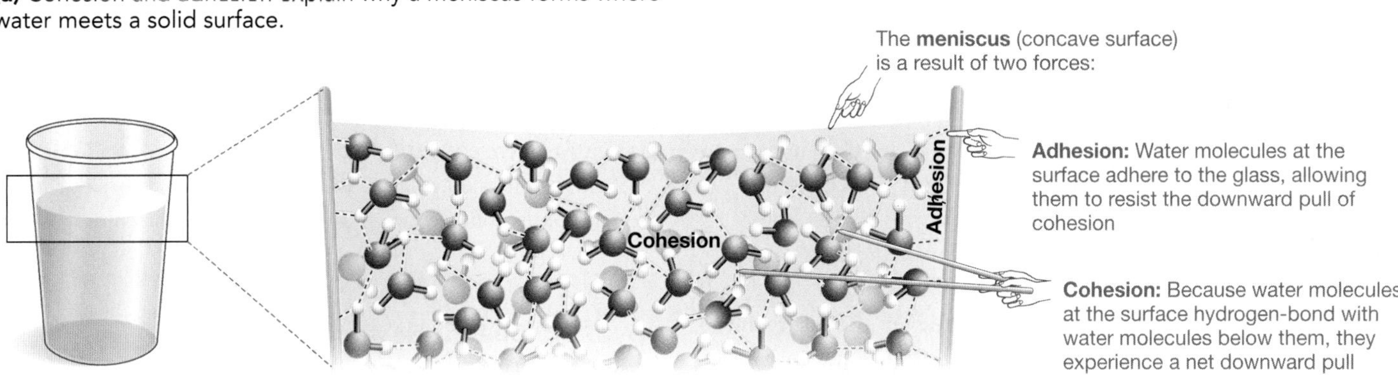

(b) Water has high suface tension.

FIGURE 2.14 Cohesion, Adhesion, and Surface Tension. (a) A combination of adhesion and cohesion causes a meniscus to form when water is bounded by a surface. **(b)** Water resists forces—like the weight of an insect—that increase its surface area. The resistance is great enough that light objects do not break through the surface.

(a) In ice, water molecules form a crystal lattice.

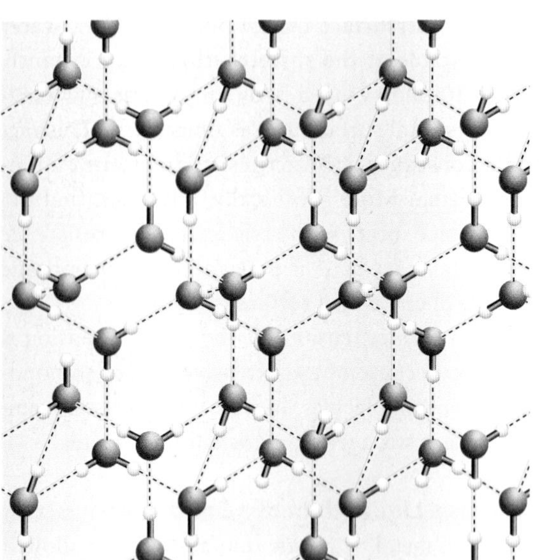

(b) In liquid water, no crystal lattice forms.

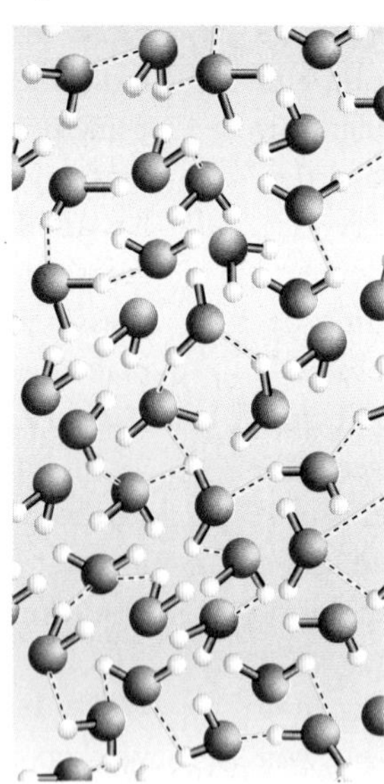

(c) Liquid water is denser than ice. As a result, ice floats.

FIGURE 2.15 Hydrogen Bonding Forms the Crystal Structure of Ice.

EXERCISE In ice, each molecule can form four hydrogen bonds at one time. (Each oxygen atom can form two; each hydrogen atom can form one.) Choose two molecules in part (a) and circle or draw in all four H bonds.

volume of solid water. **Figure 2.15a** illustrates why this is so. Note that in ice, each water molecule participates in four hydrogen bonds. These hydrogen bonds cause the water molecules to form a regular and repeating structure, or crystal. The crystal structure of ice is fairly open, meaning that there is a relatively large amount of space between molecules.

Now compare the extent of hydrogen bonding and the density of ice with that of liquid water, illustrated in **Figure 2.15b**. Although the exact structure of liquid water is not fully understood, it is clearly dynamic. Hydrogen bonds are constantly being made and broken. Overall, the extent of hydrogen bonding in liquid water is much less than that found in ice. As a result, molecules in the liquid phase are packed much more closely together than are molecules of solid water, even though their temperature is higher. Normally, heating a substance causes it to expand because molecules begin moving faster and colliding more often and with greater force. But heating ice causes hydrogen bonds to break and the open crystal structure to collapse to a less-dense configuration in liquid water. In this way, hydrogen bonding explains why water is denser as a liquid than as a solid.

This property of water has an important result: Ice floats (**Figure 2.15c**). If it did not, ice would sink to the bottom of lakes, ponds, and oceans soon after it formed. The ice would stay frozen in the cold depths. If water weren't so unusual, it is almost certain that Earth's oceans would have frozen solid before life had a chance to start.

Water Has a High Capacity for Absorbing Energy

Hydrogen bonding is also responsible for another of water's remarkable physical properties: Water has a high capacity for absorbing energy. For example, water has an extraordinarily high specific heat. **Specific heat** is the amount of energy required to raise the temperature of 1 gram of a substance by 1°C. Water's specific heat is high because, when a source of energy such as sunlight or a flame strikes it, hydrogen bonds must be broken before heat can be transferred and the water molecules begin moving faster. As a result, it takes an extraordinarily large amount of energy to change the temperature of water and other molecules where extensive hydrogen bonding occurs (**Table 2.1**).

Similarly, it takes a large amount of energy to break the hydrogen bonds in liquid water and change the molecules from the

TABLE **2.1 Specific Heats of Some Liquids**

The specific heats reported in this table were measured at 25°C and are given in units of joules per gram of substance per degree Celsius. (The **joule** is a unit of energy.)

Liquids with extensive hydrogen bonding	**Specific Heat**
Ammonia	4.70
Water	4.18
Liquids with some hydrogen bonding	
Ethanol (CH_3CH_2OH)	2.44
Ethylene glycol ($HOCH_2CH_2OH$; used in antifreeze)	2.22
Liquids with little or no hydrogen bonding	
Benzene (C_6H_6)	1.80
Xylene (C_8H_{10})	1.72
Sulfuric acid (H_2SO_4)	1.40

liquid phase to the gas phase. Water's **heat of vaporization**—the energy required to change 1 gram of it from a liquid to gas—is higher than that of most molecules that are liquid at room temperature. As a result, water has to absorb a great deal of energy to evaporate. Water's high heat of vaporization is the reason that sweating or dousing yourself with water is an effective way to cool off on a hot day. Because water molecules have to absorb a great deal of energy from your body in order to evaporate, you lose heat.

These properties of water are important to the theory of chemical evolution. As Section 2.3 will show, the energy in sunlight can drive the formation of complex molecules from simple building-block compounds. But energy inputs can also break complex molecules apart. Water's high specific heat insulates dissolved substances from sources of energy like intense sunlight. Because compounds that were important in chemical evolution dissolve readily in water, they would have formed in the ocean or rained out of the atmosphere into the ocean. Once these compounds were in an environment where they were better protected, they would have persisted and slowly increased in concentration over time. Similarly, if chemical evolution took place at the edges of ponds or on beaches, the evaporation of water would have kept the surfaces relatively cool. Hydrogen bonding has a moderating influence on water temperature.

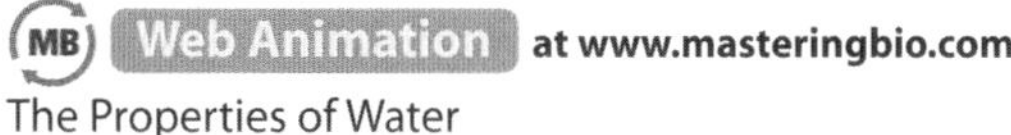

at www.masteringbio.com
The Properties of Water

Acid–Base Reactions and pH

One other aspect of water's chemistry is important to understanding chemical evolution and how organisms work today. Water is not a completely stable molecule. In reality, water molecules continually undergo a chemical reaction with themselves. This "dissociation" reaction can be written as follows:

$$H_2O \rightleftharpoons H^+ + OH^-$$

The double arrow indicates that the reaction proceeds in both directions.

The molecules on the right-hand side of the expression are the **hydrogen ion** (H^+) and the **hydroxide ion** (OH^-). A hydrogen ion is simply a proton. In reality, however, protons never exist by themselves. In water, for example, protons associate with water molecules to form the hydronium ion, H_3O^+. Thus, the dissociation of water is more accurately written

$$H_2O + H_2O \rightleftharpoons H_3O^+ + OH^-$$

One of the water molecules on the left-hand side of the expression has given up a proton, while the other water molecule has accepted a proton. Substances that give up protons during chemical reactions and raise the hydrogen ion concentration of water are called **acids**; molecules or ions that acquire protons during chemical reactions and lower the hydrogen ion concentration of water are called **bases**. A chemical reaction that involves a transfer of protons is called an acid–base reaction. Every acid–base reaction requires a proton donor and a proton acceptor—an acid and a base. Most acids act only as acids, and most bases act only as bases; but water can act as both an acid and a base.

Water is an extremely weak acid, though—very few water molecules dissociate to form hydronium ions and hydroxide ions. In contrast, strong acids like the hydrochloric acid in your stomach readily give up a proton when they react with water.

$$HCl + H_2O \rightleftharpoons H_3O^+ + Cl^-$$

Strong bases readily acquire a proton when they react with water. For example, sodium hydroxide (NaOH, commonly called lye), dissociates completely in water to form Na^+ and OH^-. The hydroxide ions then accept a proton from hydronium ions in the water, forming more water molecules.

$$NaOH(aq) \rightarrow Na^+ + OH^-$$

$$OH^- + H_3O^+ \rightleftharpoons 2H_2O$$

To summarize, adding an acid to a solution increases the concentration of protons; adding a base to a solution lowers the concentration of protons.

These points are important because in a solution, the tendency for acid–base reactions to occur is largely a function of the number of protons present. When an acid–base reaction occurs, the transfer of a proton changes the charge of the proton donor and the proton acceptor. When a compound's charge changes, its tendency to participate in hydrogen bonding and other types of electrostatic interactions also changes. This is critical. Even if the concentration of protons in your blood doubles, it is still extremely low. Yet the acid–base reactions triggered by the change in proton concentration are likely to kill you within a few minutes (**Box 2.1** introduces important molecules called buffers, which prevent dramatic pH changes in blood and elsewhere.)

Chemists can measure the concentration of protons in a solution directly. In a sample of pure water at 25°C, the concentration of H^+ is 1.0×10^{-7} M. (Recall that M represents molarity, or moles per liter.) Because this is such a small number (1 ten-millionth), the exponential notation is cumbersome. So chemists and biologists prefer to express the concentration of protons in a solution with a logarithmic notation called **pH**. (The term *pH* is derived from the French *puissance d'hydrogène*, or "power of hydrogen.") By definition, the pH of a solution is the negative of the base-10 logarithm, or log, of the hydrogen ion concentration:

$$pH = -\log[H^+]$$

(The square brackets are a standard notation for indicating "concentration of" a substance in solution.) Taking antilogs gives

$$[H^+] = \text{antilog}(-pH) = 10^{-pH}$$

BOX 2.1 What Is a Buffer?

Acid–base reactions add or remove a proton from ions or molecules. The addition or removal of a proton changes the charge on a compound and thus how it interacts with other polar or charged molecules. Cells are extremely sensitive to changes in pH because they contain many polar or charged substances. In addition, many of the key reactions that occur in cells involve the addition or removal of a proton, and pH affects how likely these reactions are to occur.

Based on these observations, it shouldn't be surprising to learn that minimizing changes in pH inside cells is an important part of maintaining optimal conditions for life. Compounds that minimize variation in pH are called **buffers** because they buffer a solution against the damaging effects of pH change.

Most buffers are weak acids, meaning that they are somewhat likely to give up a proton in solution. To see how they work, consider the disassociation of acetic acid in water to form acetate ions and protons.

$$CH_3COOH \rightleftharpoons CH_3COO^- + H^+$$

When acetic acid and acetate are present in about equal concentrations in a solution, they function as a buffering system. If the concentration of protons increases slightly, the protons react with acetate ions to form acetic acid and pH does not change. If the concentration of protons decreases slightly, acetic acid gives up protons and pH does not change.

In cells, a wide array of naturally occurring molecules act as buffers. Biologists have also developed an array of synthetic compounds to buffer experimental solutions that contain pH-sensitive molecules.

pH, then, is a convenient way to indicate the concentration of protons in a solution. To review logarithms, see **BioSkills 5**.

Figure 2.16 shows the pH scale and reports the pH of some common solutions. The pH of pure water at 25°C is 7. Pure water is used as a standard, or point of reference, on the pH scale. Solutions that contain acidic molecules (molecules that act as acids) have a proton concentration larger than 1×10^{-7} M and thus a pH < 7, because acidic molecules tend to release protons into solution. In contrast, solutions that contain basic molecules (molecules that act as bases) have a proton concentration less than 1×10^{-7} M and thus a pH > 7, because basic molecules tend to accept protons from solution. Solutions with a pH of 7 are considered neutral solutions, neither acidic nor basic. The solution inside living cells is about 7. ● If you understand this concept, you should be able to explain the difference in hydrogen ion concentration between a slightly basic solution with pH 8 and a slightly acidic solution with pH 6.

Although the pH of today's oceans is about 8, the early oceans probably were much like today's lakes in their pH and chemical composition—simply because it would have taken hundreds of millions of years for rain and wind to slowly erode rocks, and for rivers to deliver dissolved ions that increased the ocean's salinity and increased its pH.

As chemical evolution began, then, water provided the physical environment for key reactions to take place. In some cases, water also acted as an important reactant. Although acid–base reactions were not critical to the initial stages of chemical evolution, they became extremely important once the process was under way.

Table 2.2 summarizes some of the key properties of water for you, as a reference. Now, before looking more closely at the role of carbon in life, let's move on to consider the reactions that created the building blocks of chemical evolution.

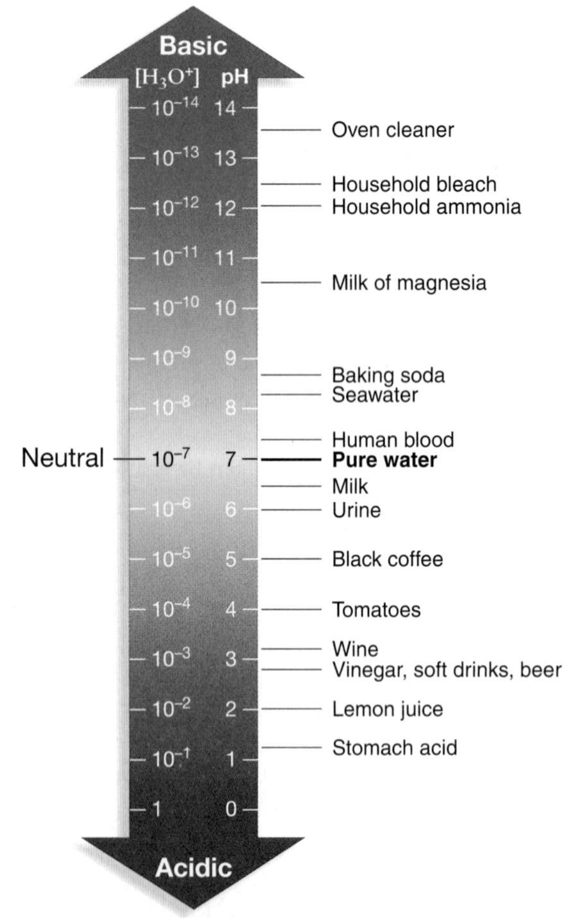

FIGURE 2.16 The pH Scale. Because the pH scale is logarithmic, a change in one unit of pH represents a change in the concentration of hydrogen ions equal to a factor of 10. Coffee has a hundred times more H^+ than pure water has.

● **QUESTION** How is the pH of black coffee affected if you add milk?

SUMMARY TABLE 2.2 **Properties of Water**

Property	Cause	Biological consequences
Solvent for charged or polar compounds		Most chemical reactions important for life take place in aqueous solution.
High **cohesion**	Hydrogen bonds form between water molecules.	Creates surface tension; also important in water transport in plants.
High **adhesion**		Plays a role in water transport in plants.
High **surface tension**	High cohesion makes a water surface resist forces that increase surface area.	Important in water transport in plants; small organisms can walk on water.
Denser as a liquid than a solid	Hydrogen bonding leads to formation of low-density crystal structure in ice.	
High **specific heat**	Water molecules must absorb lots of heat energy to break hydrogen bonds and experience increased movement (and thus temperature).	Oceans absorb and release heat slowly, moderating coastal climates.
High **heat of vaporization**		Evaporation of water from an organism cools the body.

● **EXERCISE** You should be able to fill in the missing cells in this table. You should also be able to make a concept map relating water's structure to the properties listed here. (For an introduction to concept mapping, see **BioSkills 6**.)

2.3 Chemical Reactions, Chemical Evolution, and Chemical Energy

Proponents of the theory of chemical evolution contend that simple molecules present in the atmosphere and ocean of ancient Earth participated in chemical reactions that produced larger, more complex molecules. According to one hypothesis, the early atmosphere was made up of gases ejected from volcanoes. Carbon dioxide, water vapor, and nitrogen are the dominant gases ejected from volcanoes today; a small amount of molecular hydrogen (H_2) and methane (CH_4) may also be present. But when these molecules are placed in a glass tube together and allowed to interact, very little happens. Water vapor condenses to liquid water as the mixture cools, but the simple molecules do not suddenly link together to create large, complex substances like those found in living cells. Instead, their bonds remain intact. How, then, did chemical evolution occur?

To answer this question we need to explore two topics: (1) how chemical reactions occur, and (2) how conditions on ancient Earth made certain reactions possible.

How Do Chemical Reactions Happen?

Chemical reactions are written in a format similar to mathematical equations, with the initial, or **reactant**, atoms or molecules shown on the left and the resulting reaction **product**(s) shown on the right. For example, the most common reaction in the mix of gases and water that emerge from volcanoes is

$$CO_2(g) + H_2O(l) \rightleftharpoons H_2CO_3(aq)$$

This expression indicates that carbon dioxide (CO_2) reacts with water (H_2O), forming carbonic acid (H_2CO_3). The state of each reactant and product is indicated as gas (g), liquid (l), in aqueous solution (aq), or solid (s).

Note that the expression is balanced; that is, 1 carbon, 3 oxygen, and 2 hydrogen atoms are present on each side of the expression. Note also that the expression contains a double arrow, meaning that the reaction is reversible. When the forward and reverse reactions proceed at the same rate, the quantities of reactants and products remain constant, although not necessarily equal. A dynamic but stable state such as this is termed a **chemical equilibrium**. A chemical equilibrium can be disturbed by changing the concentration of reactants or products. For example, adding CO_2 to the mixture would drive the reaction to the right, creating more H_2CO_3 until the equilibrium proportions of reactants and products are reestablished. Adding H_2CO_3 instead would drive the reaction to the left. Removing CO_2 would also drive the reaction to the left; removing H_2CO_3 would drive it to the right.

A chemical equilibrium can also be altered by changes in temperature. For example, the water molecules in this set of interacting elements, or **system**, would be present as a combination of liquid water and water vapor:

$$H_2O(l) \rightleftharpoons H_2O(g)$$

If liquid water molecules absorb enough heat, they transform to the gaseous state. This is called an **endothermic** ("within heating") process because heat is absorbed during the process. In contrast, the transformation of water vapor to liquid water releases heat and is called **exothermic** ("outside heating"). Raising the temperature of this system drives the equilibrium to the right; cooling the system drives it to the left.

In relation to chemical evolution, though, these reactions and changes of state are not particularly interesting. Carbonic acid is not an important intermediate in the formation of more complex molecules. According to models developed by a series of researchers, however, interesting things do begin to happen when the system being studied is the ions and molecules present in the oceans and atmosphere of early Earth, and when large amounts of energy are added.

What Is Energy?

Energy can be defined as the capacity to do work or to supply heat. This capacity exists in one of two ways—as a stored potential or as an active motion.

Stored energy is called **potential energy**. An object gains or loses its ability to store energy because of its position. An electron that resides in an outer electron shell will, if the opportunity arises, fall into a lower electron shell closer to the positive charges on the protons in the nucleus. Because of its position farther from the positive charges in the nucleus, an electron in an outer electron shell has more potential energy than does an electron in an inner shell (**Figure 2.17**).

Kinetic energy is the energy of motion. Molecules have kinetic energy because they are constantly in motion. This form of kinetic energy—the kinetic energy of molecular motion—is called **thermal energy**. The **temperature** of an object is a measure of how much thermal energy its molecules possess. If an object has a low temperature, its molecules are moving slowly. (We perceive this as "cold".) If an object has a high temperature, its molecules are moving rapidly. (We perceive this as "hot".) When two objects with different temperatures come into contact, thermal energy is transferred between them. We call this transferred energy **heat**.

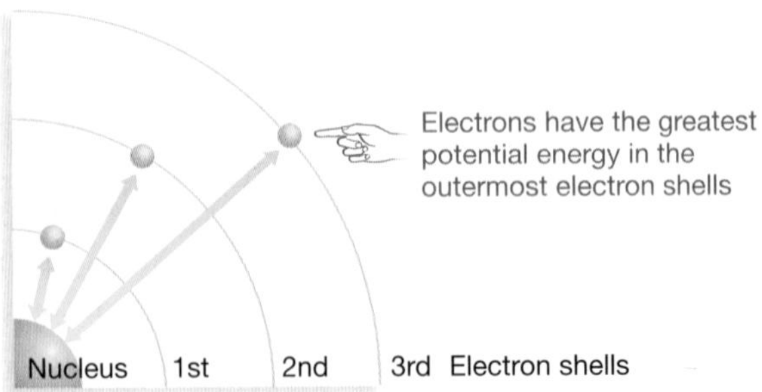

FIGURE 2.17 Potential Energy as a Function of Electron Shells. Electrons in outer shells have more potential energy than do electrons in inner shells, because negative charges in outer shells are farther from the positive charges in the nucleus. Each shell represents a distinct level of potential energy.

There are many forms of potential and kinetic energies, and energy can change from one form into another. To drive this point home, consider a water molecule sitting at the top of a waterfall, as in **Figure 2.18a**. This molecule has potential energy because of its position. If the molecule passes over the waterfall, its potential energy is converted to the kinetic energy of motion. When the molecule has reached the rocks below, it has experienced a change in potential energy because it has changed position. The change in potential energy is transformed into an equal amount of energy in other forms: mechanical energy, which tends to break up the rocks; heat (thermal energy), which raises the temperature of the rocks and the water itself; and sound.

An electron in an outer electron shell is analogous to the water molecule at the top of a waterfall (**Figure 2.18b**). If the electron falls to a lower shell, its potential energy is converted to the kinetic energy of motion. After the electron occupies the lower electron shell, it experiences a change in potential energy. As panel 3 in Figure 2.18b shows, the change in potential energy is transformed into an equal amount of energy in other forms—usually thermal energy, but sometimes light.

These examples illustrate the **first law of thermodynamics**, which states that energy is conserved. Energy cannot be created or destroyed, but only transferred and transformed.

Energy transformation is the heart of chemical evolution. According to the best data available, molecules that were part of the young Earth were exposed to massive inputs of energy. Part of this energy was in the form of heat, from the molten mass that initially formed the planet. Part of the energy was in the form of high-energy radiation from the Sun, which today is screened out by molecules in the atmosphere. How would the application of large amounts of heat and radiation affect the course of chemical evolution?

Chemical Evolution: A Model System

To assess the impact of energy inputs on the simple molecules present in the early oceans and atmosphere, researchers have constructed computer models to simulate the reactions that can occur among carbon dioxide, water, ammonia, and hydrogen molecules. The goal of one such study was quite specific: The researchers wanted to determine whether a molecule called formaldehyde (H_2CO) could be produced. Along with hydrogen cyanide (HCN), formaldehyde is a key intermediate in the creation of the larger, more complex molecules found in cells. Forming formaldehyde and hydrogen cyanide is the critical first step in chemical evolution—a trigger that could set the process in motion.

The research group began by proposing that the following reaction could take place:

$$CO_2(g) + 2H_2(g) \rightarrow H_2CO(g) + H_2O(g)$$

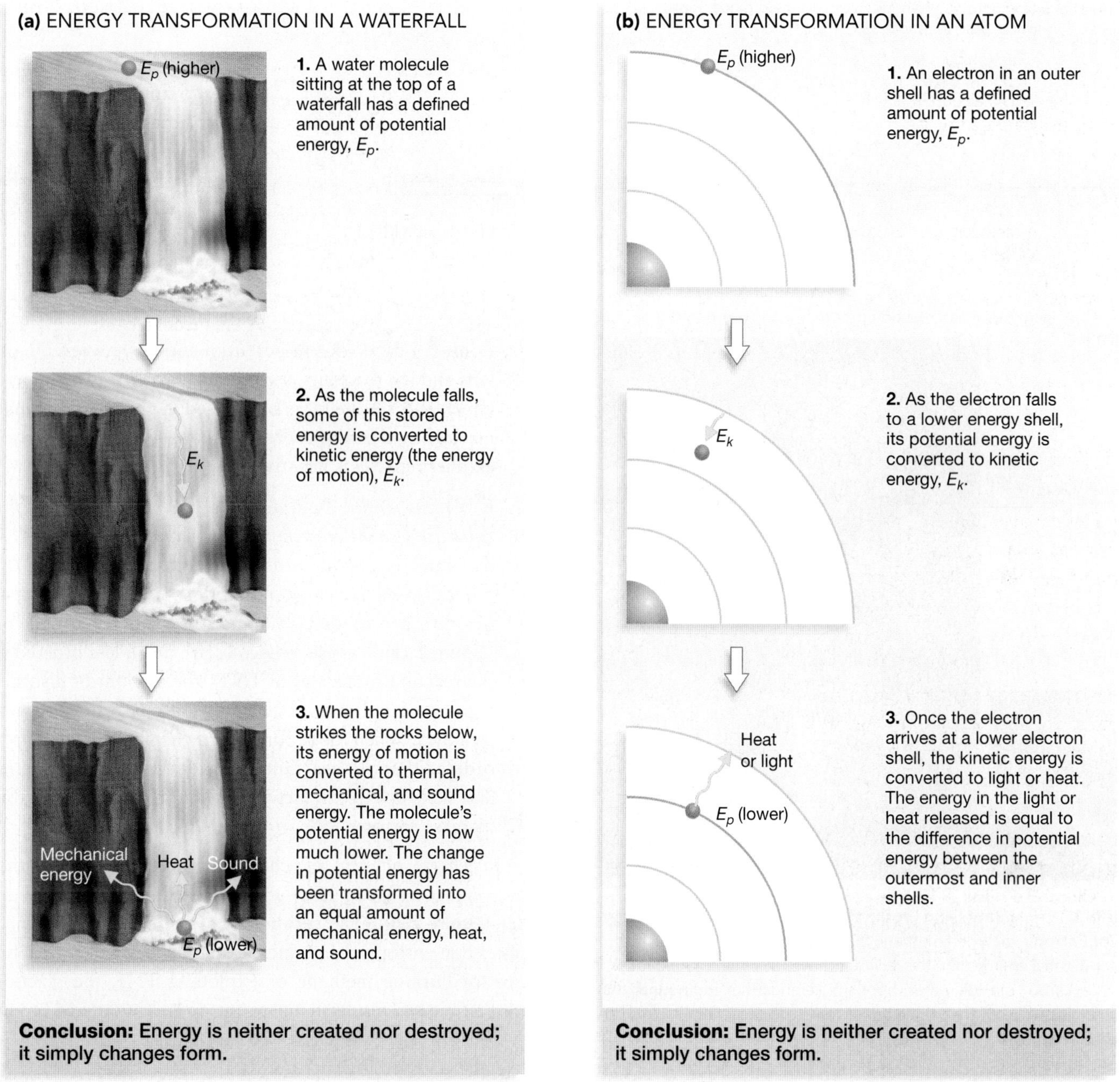

(a) ENERGY TRANSFORMATION IN A WATERFALL

E_p (higher)

1. A water molecule sitting at the top of a waterfall has a defined amount of potential energy, E_p.

E_k

2. As the molecule falls, some of this stored energy is converted to kinetic energy (the energy of motion), E_k.

Mechanical energy
Heat
Sound
E_p (lower)

3. When the molecule strikes the rocks below, its energy of motion is converted to thermal, mechanical, and sound energy. The molecule's potential energy is now much lower. The change in potential energy has been transformed into an equal amount of mechanical energy, heat, and sound.

Conclusion: Energy is neither created nor destroyed; it simply changes form.

(b) ENERGY TRANSFORMATION IN AN ATOM

E_p (higher)

1. An electron in an outer shell has a defined amount of potential energy, E_p.

E_k

2. As the electron falls to a lower energy shell, its potential energy is converted to kinetic energy, E_k.

Heat or light
E_p (lower)

3. Once the electron arrives at a lower electron shell, the kinetic energy is converted to light or heat. The energy in the light or heat released is equal to the difference in potential energy between the outermost and inner shells.

Conclusion: Energy is neither created nor destroyed; it simply changes form.

FIGURE 2.18 Energy Transformations. During an energy transformation, the total amount of energy in the system remains constant.

Before we explore how they tested this hypothesis, however, it will be helpful to know why this reaction doesn't occur spontaneously—that is, why doesn't it occur *without* an input of energy?

What Makes a Chemical Reaction Spontaneous? When chemists say that a reaction is spontaneous, they have a precise meaning in mind: ● Chemical reactions are spontaneous if they proceed on their own, without any continuous external influence such as added energy. Two factors determine whether a reaction is spontaneous or nonspontaneous:

1. *Reactions tend to be spontaneous if the products have lower potential energy than the reactants.* Reaction products have lower potential energy if their electrons are held more tightly than are the electrons of the reactants. Recall that highly electronegative atoms such as oxygen and nitrogen hold electrons much more tightly than do atoms with a lower electronegativity, such as carbon and hydrogen. For

(a) Potential energy drops when electrons are held more tightly in reaction products than in reactants.

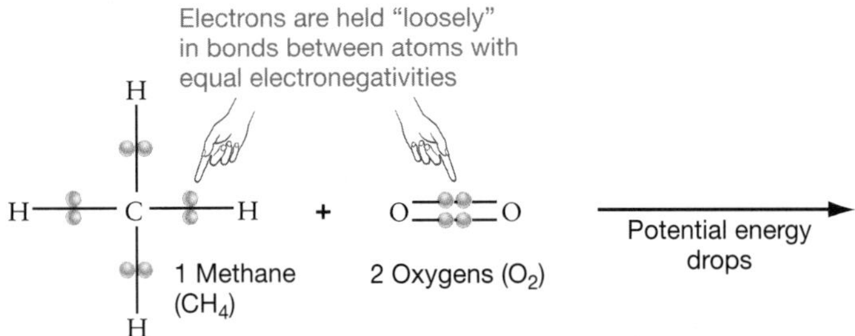

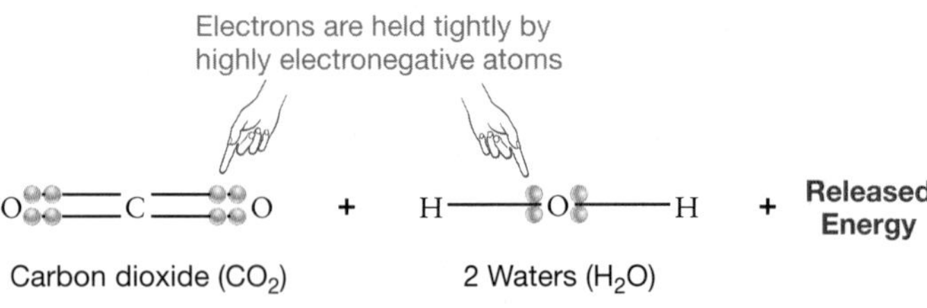

(b) Entropy increases when reaction products are less ordered than reactants.

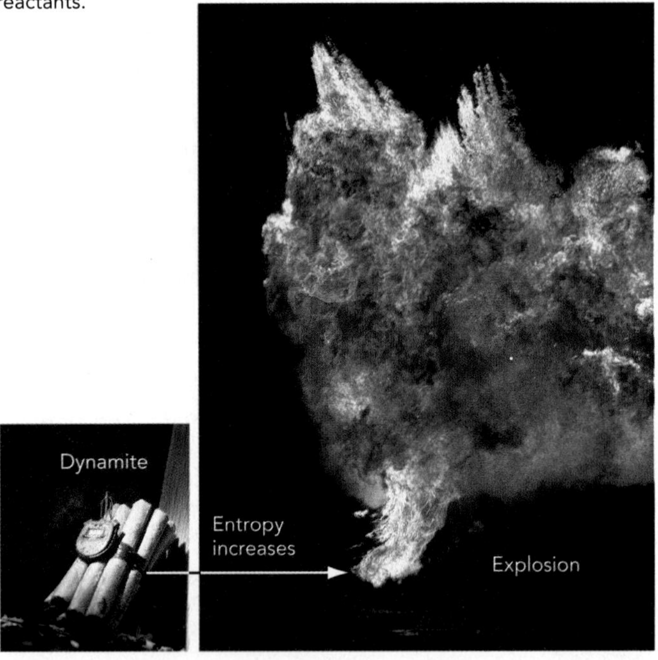

FIGURE 2.19 Reactants and Products May Differ in Potential Energy and/or Entropy. (a) When methane burns, the products have much lower potential energy than the reactants have. **(b)** When TNT explodes, it produces carbon dioxide, water vapor, smoke, and other compounds that are much less ordered than the original system. The reaction results in an increase in entropy.

EXERCISE In part (a), label which electrons have relatively low potential energy and which have relatively high potential energy.

example, when natural gas burns, methane reacts with oxygen gas to produce carbon dioxide and water:

$$CH_4(g) + 2O_2(g) \rightarrow CO_2(g) + 2H_2O(g)$$

The electrons involved in the C=O and H–O bonds of carbon dioxide and water are held much more tightly than they were in the C–H and O=O bonds of methane and oxygen (**Figure 2.19a**). As a result, the products have much lower potential energy than the reactants. The difference in potential energy between reactants and products is given off as heat, so the reaction is exothermic. In chemical reactions, the difference in potential energy between the products and the reactants is symbolized by ΔH. (The uppercase Greek letter Δ, delta, is often used in chemical and mathematical notation to represent change.) When a reaction is exothermic, ΔH is negative.

2. *Reactions tend to be spontaneous when the product molecules are less ordered than the reactant molecules.* TNT, or dynamite, is a highly ordered structure. But when TNT explodes, gases like carbon dioxide, carbon monoxide, various nitrogen oxides, and small particulates are given off (**Figure 2.19b**). These molecules are much less ordered than the reactant molecules in TNT. The amount of disorder in a group of molecules is called its **entropy**, which is symbolized by S. When the products of a chemical reaction are less ordered than the reactant molecules are, entropy increases and ΔS is positive. Reactions tend to be spontaneous if they increase entropy.

In general, physical and chemical processes proceed in the direction that results in lower potential energy and increased disorder (**Figure 2.20**). The **second law of thermodynamics**, in fact, states that entropy always increases in an isolated system. In the case of burning methane or exploding TNT, the reaction is exothermic *and* results in higher entropy—less-ordered products.

Because reactions tend to be spontaneous when ΔH is negative and ΔS is positive, it's necessary to assess the *combined* contributions of changes in heat and disorder to determine whether a chemical reaction is spontaneous. To do this, chemists define a quantity called the **Gibbs free-energy change**, symbolized by ΔG.

$$\Delta G = \Delta H - T\Delta S$$

Here, T stands for temperature measured on the Kelvin scale. The $T\Delta S$ term simply means that entropy becomes more important in determining free-energy change as the temperature of the molecules increases. The faster molecules are moving, the more important entropy becomes in determining the overall free-energy change.

Chemical reactions are spontaneous when ΔG is less than zero. Such reactions are said to be **exergonic**. Reactions are nonspontaneous when ΔG is greater than zero. Such reactions are termed **endergonic**. When ΔG is zero, reactions are at equilibrium. ● If you understand these concepts, you should be able to explain why the same reaction can be nonspontaneous at low temperature but spontaneous at high temperature. You should also be able to explain why some exothermic reactions are nonspontaneous.

Here are the important things to remember about free energy: Free energy changes when the potential energy and/or entropy of substances changes. Chemical reactions run in the direction that lowers the free energy of the system. Exergonic reactions are spontaneous and release energy; endergonic reactions are nonspontaneous and require an input of energy to proceed.

Energy Inputs and the Start of Chemical Evolution The reaction between carbon dioxide (CO_2) and hydrogen gas (H_2) that forms formaldehyde (H_2CO) and water is endergonic. Formaldehyde and water have more potential energy and are more highly ordered than CO_2 and H_2. The reaction is nonspontaneous because ΔG is positive. For the reaction to occur, a large input of energy is required.

To explore how formaldehyde formation and chemical evolution got started, a research group constructed a computer model of the ancient atmosphere. The model consisted of a list of all possible chemical reactions that can occur among CO_2, H_2O, N_2, NH_3, CH_4, and H_2 molecules. In addition to the spontaneous reactions, they included reactions that occur when these molecules are struck by sunlight. This was crucial because sunlight represents a source of energy.

The sunlight that strikes Earth is made up of packets of light energy called photons. The amount of light energy contained in a photon can vary widely. Today, most of the higher-energy photons in sunlight never reach Earth's lower atmosphere. Instead, they are absorbed by a molecule called ozone (O_3) in the upper atmosphere. But if Earth's early atmosphere was filled with volcanic gases released as the molten planet cooled, it is extremely unlikely that appreciable quantities of ozone existed. Based on this logic, researchers infer that when chemical evolution was occurring, large quantities of high-energy photons bombarded the planet.

To understand why this energy source was so important, recall that the atoms in hydrogen and carbon dioxide have full

(a) Water has a higher potential energy at the top of a waterfall than at the bottom.

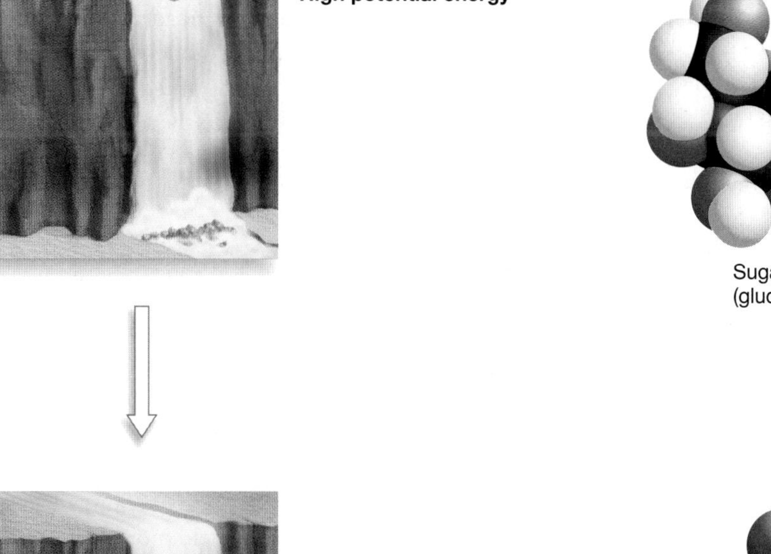

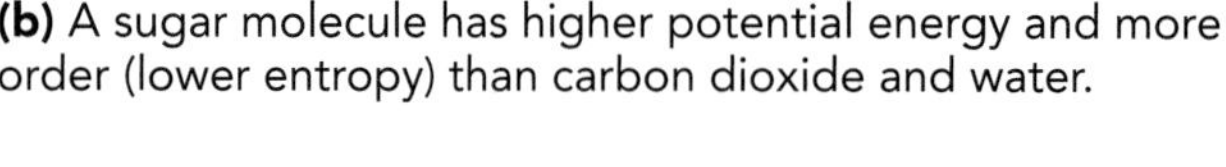
(b) A sugar molecule has higher potential energy and more order (lower entropy) than carbon dioxide and water.

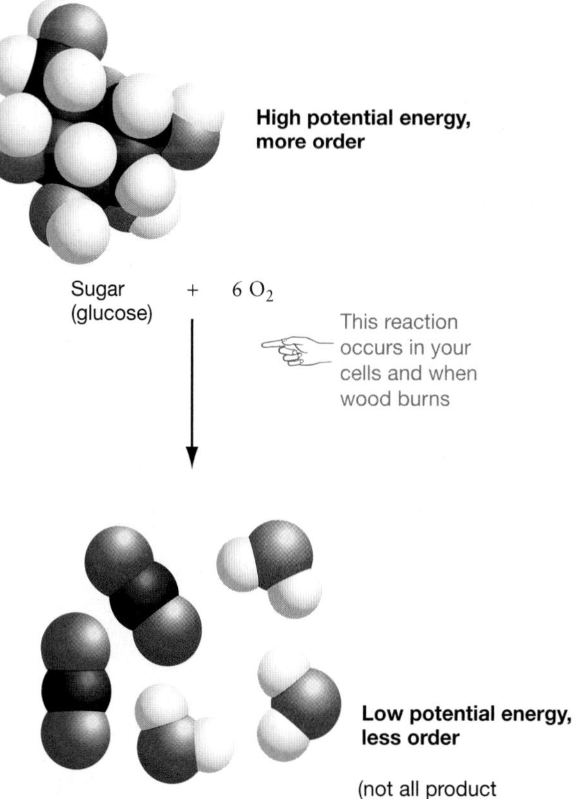

FIGURE 2.20 Spontaneous Processes Result in Lower Potential Energy or Increased Disorder, or Both.

outermost shells. This arrangement makes these molecules largely unreactive. But energy from photons can break molecules apart by knocking electrons away from the outer shells of atoms. The atoms that result, called **free radicals**, have unpaired electrons and are extremely unstable (**Figure 2.21**). To mimic the conditions on early Earth more accurately, the computer model included several reactions that produce highly reactive free radicals.

To understand which of the long list of possible reactions would actually occur, and to estimate how much formaldehyde could have been produced in the ancient atmosphere, the researchers needed to consider the effects of two additional factors: temperature and concentration.

The Roles of Temperature and Concentration in Chemical Reactions Even if a chemical reaction occurs spontaneously, it may not happen quickly. The reactions that convert iron to rust or sugar molecules to carbon dioxide and water are spontaneous, but at room temperature they occur very slowly, if at all. For most reactions to proceed, one chemical bond has to break and another one has to form. For this to happen, the substances involved must collide in a specific orientation that brings the electrons involved near each other.

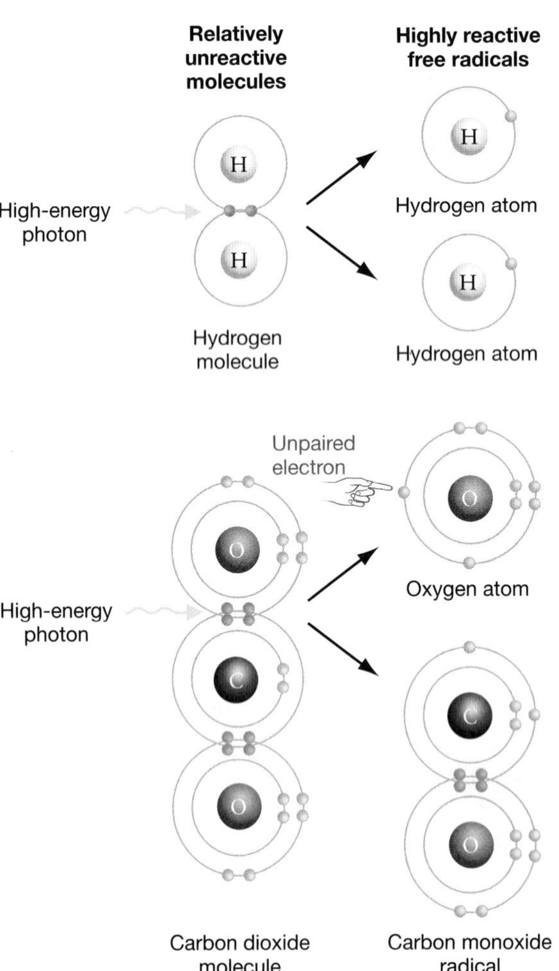

FIGURE 2.21 Free Radicals Are Extremely Reactive. When a high-energy photon strikes a hydrogen or carbon dioxide molecule, free radicals can be created. Formation of free radicals is thought to be responsible for some key reactions in chemical evolution.

The number of collisions occurring among the substances in a mixture depends on the temperature and the concentrations of the reactants. When the concentration of reactants is high, more collisions should occur and reactions should proceed more quickly. When their temperature is high, reactants should move faster and collide more frequently. Higher concentrations and higher temperatures should tend to speed up chemical reactions. **Figure 2.22** provides data from a set of experiments performed by students from Parkland College in Champaign, Illinois, to test these predictions. As the data in the Results section show, the reaction rate speeded up significantly when the concentration of reactants was increased and when the temperature of the reaction mixture was higher.

To model the behavior of simple molecules in the ancient atmosphere, then, the researchers who were working on the formaldehyde synthesis reaction needed to specify the concentration of each molecule as well as the temperature. Then they were able to assign a rate to each of the reactions listed in their model based on the actual reaction rates observed in experiments conducted at controlled temperatures and concentrations, like those in Figure 2.22.

Their result? They calculated that, under temperature and concentration conditions accepted as reasonable approximations of early Earth conditions by most atmospheric scientists, appreciable quantities of formaldehyde would have been produced. Using a similar model, other researchers have shown that significant amounts of hydrogen cyanide (HCN) could also have been produced in the ancient atmosphere. According to this research, large quantities of the critical intermediates in chemical evolution would have formed in the ancient atmosphere.

How Did Chemical Energy Change during Chemical Evolution?

The initial products of chemical evolution are important, for a simple reason: They have more potential energy than the reactant molecules have. When formaldehyde is produced, an increase in potential energy occurs because the electrons that bond CO_2 and H_2 together are held more tightly than they are in H_2CO or H_2O. This form of potential energy—the potential energy stored in chemical bonds—is called **chemical energy**.

This observation gets right to the heart of chemical evolution: The energy in sunlight was converted to chemical energy—potential energy in chemical bonds. This energy transformation explains how chemical evolution was possible. When small, simple molecules absorb energy, chemical reactions can occur that transform the external energy into potential energy stored in chemical bonds. More specifically, the energy in sunlight was converted to chemical energy in the form of formalde-

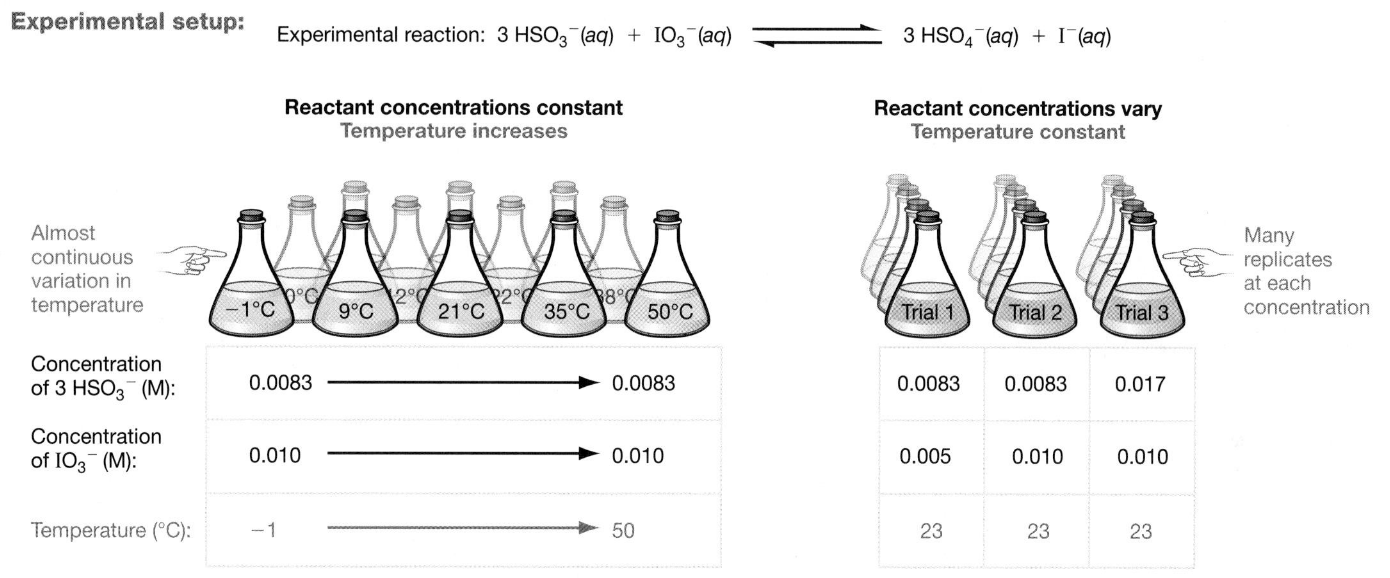

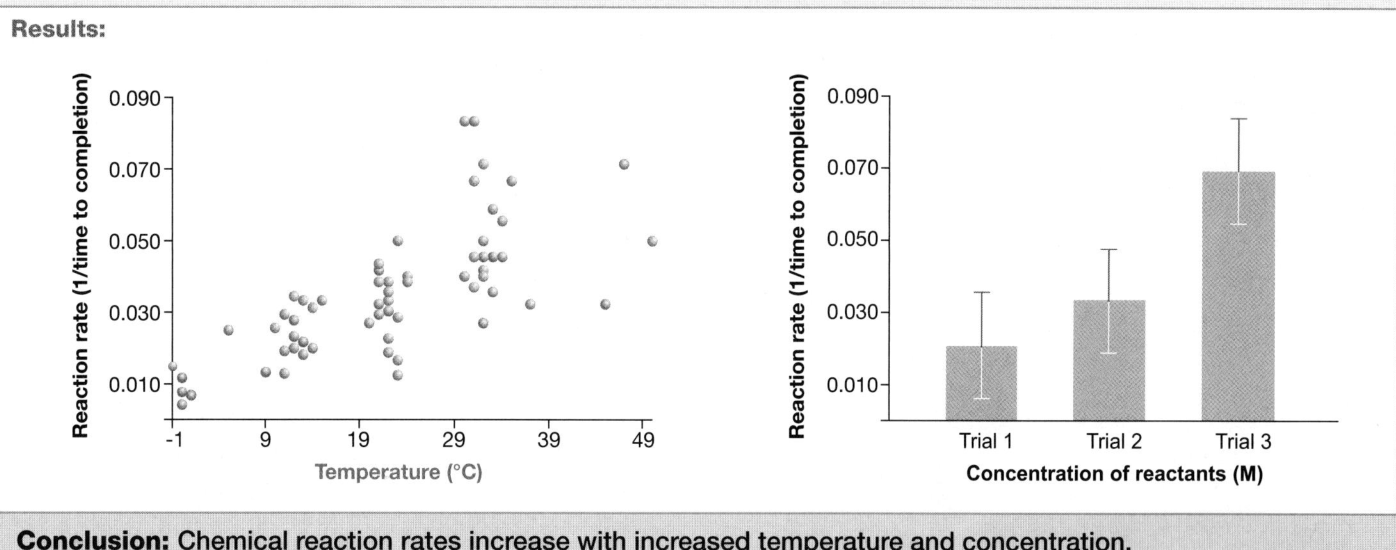

FIGURE 2.22 Testing the Hypothesis That Reaction Rates Are Sensitive to Changes in Concentration and Temperature.

hyde and hydrogen cyanide. The complete reaction that results in the formation of formaldehyde is written as

$$CO_2(g) + 2H_2(g) + \text{sunlight} \rightarrow H_2CO(g) + H_2O(g)$$

Notice that the reaction is balanced in terms of the atoms *and* the energy involved. An increase in chemical energy occurred that made the production of larger, more complex molecules possible.

This result makes sense intuitively. Energy is the capacity to do work, and it seems logical that building larger, more complex molecules requires work to be done. More specifically, the reactions involved in chemical evolution are endergonic, so inputs of energy were required. Now the question is, what happened to these first building blocks of chemical evolution?

Check Your Understanding

If you understand that...

- Chemical reactions tend to be spontaneous if they lead to lower potential energy and higher entropy (more disorder).
- The combined effects of potential energy and entropy changes are summarized in the equation for the Gibbs free-energy change.

You should be able to...

1) Write out the Gibbs equation and define each of the components.
2) Explain why potential energy might decrease as the result of a chemical reaction and why entropy might increase.

2.4 The Importance of Carbon

Life has been called a carbon-based phenomenon, and with good reason. Except for water, almost all of the molecules found in organisms contain this atom.

Carbon has great importance in biology because it is the most versatile atom on Earth. Due to its four valence electrons, it can form many covalent bonds. With different combinations of single and double bonds, an almost limitless array of molecular shapes are possible. You have already examined the tetrahedral structure of methane and the linear shape of carbon dioxide. When molecules contain more than one carbon atom, these atoms can be bonded to one another in long chains, as in the component of the gasoline called octane (C_8H_{18}; **Figure 2.23a**), or in a ring, as in the sugar glucose ($C_6H_{12}O_6$; **Figure 2.23b**). Molecules that contain carbon are called **organic** molecules. (Other types of molecules are referred to as *inorganic* compounds.) Carbon atoms provide the structural framework for virtually all of the important compounds associated with life.

Linking Carbon Atoms Together

The formation of carbon-carbon bonds was an important event in chemical evolution. It represented a crucial step toward the production of the types of molecules found in living organisms. Once organic compounds such as formaldehyde and hydrogen cyanide had formed, continued chemical evolution could occur by the addition of heat alone. For example, when molecules of formaldehyde are heated, they react with one another to form a molecule called acetaldehyde. Acetaldehyde contains a carbon-carbon bond. With continued heating, reactions between formaldehyde and acetaldehyde molecules can produce the larger carbon-containing compounds called sugars.

Figure 2.24 summarizes these steps. In studying the figure, be sure to note two key messages: (1) Chemical evolution got under way because abundant sources of outside energy were available to trigger endergonic reactions; and (2) some of these molecules produced by chemical evolution are found in organisms living today. Once chemical evolution was under way, a wide variety of small organic molecules would have accumulated in the ancient Earth.

Functional Groups

In general, the carbon atoms in an organic molecule furnish a skeleton that gives the molecule its overall shape. But the chemical behavior of the compound—meaning the types of reactions

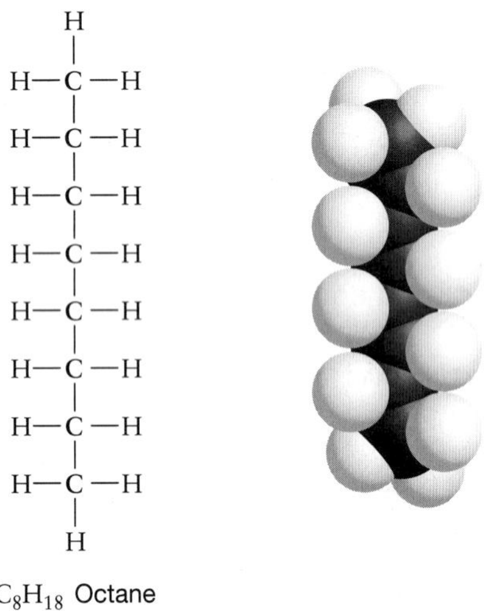

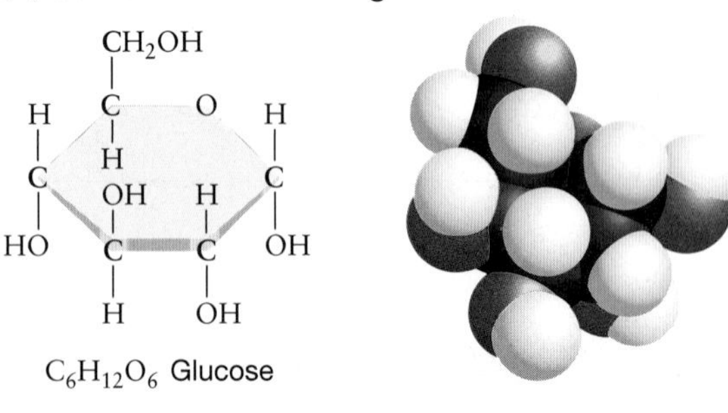

FIGURE 2.23 The Shapes of Carbon-Containing Molecules. (a) Octane is one of the primary ingredients in gasoline. It is a linear molecule. **(b)** Glucose is a sugar that can form the ring-like structure illustrated here.

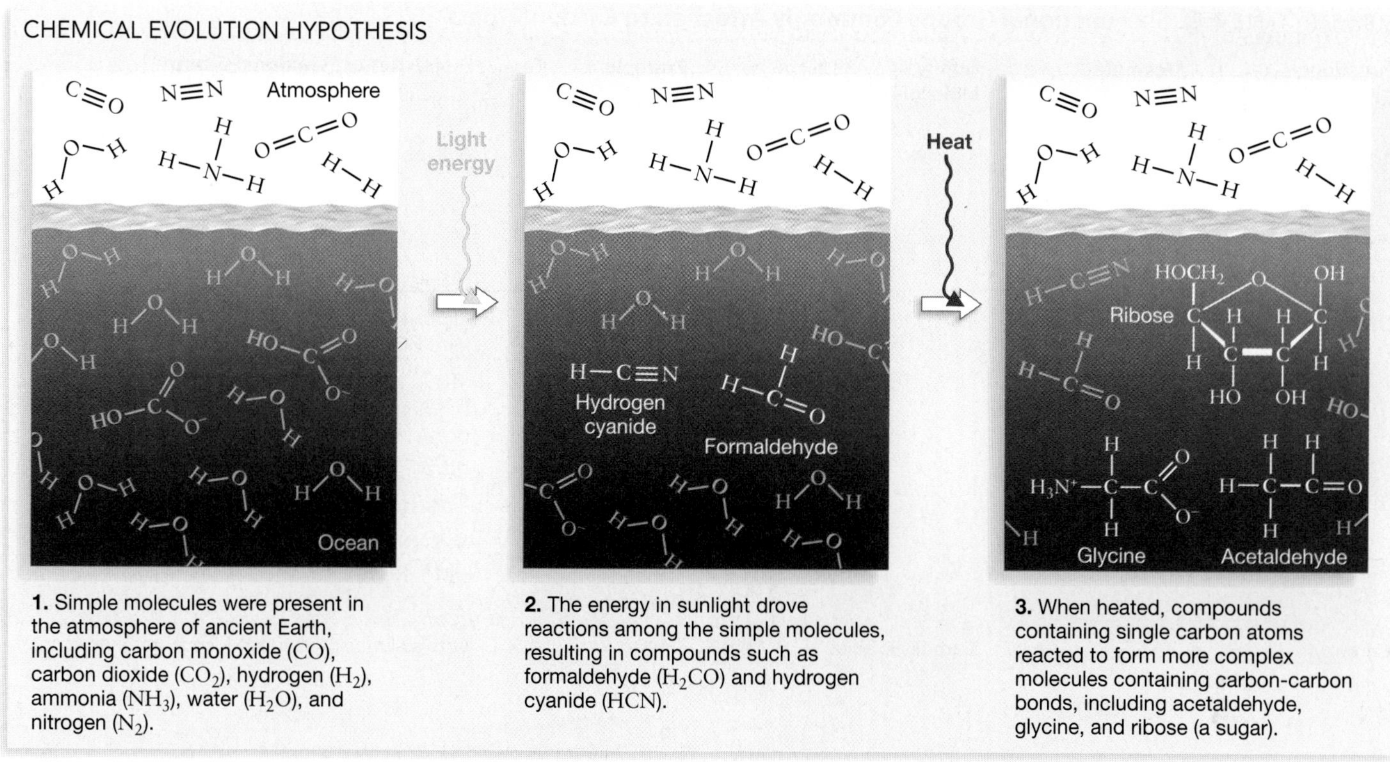

FIGURE 2.24 The Start of Chemical Evolution—An Overview. Chemical evolution is a process whereby simple molecules containing C, H, O, and N react to form organic compounds with higher potential energy in the form of carbon-carbon bonds. The process is triggered by an energy source such as sunlight or the heat released in a volcanic eruption.

that it participates in—is dictated by groups of H, N, or O atoms that are bonded to one of the carbon atoms in a specific way. These critically important H-, N-, and O-containing groups found in organic compounds are called **functional groups**. The composition and properties of six prominent functional groups recognized by organic chemists are summarized in **Table 2.3**. To understand the role that organic compounds play in organisms, it is important to analyze how these functional groups behave.

- In solution, the amino and carboxyl functional groups tend to attract or drop a proton, respectively. Amino groups function as bases, while carboxyl groups act as acids. During chemical evolution and in organisms today, the most important types of amino- and carboxyl-containing molecules are the amino acids, which Chapter 3 analyzes in detail. Amino acids contain both an amino group and a carboxyl group. These molecules can be linked together by covalent bonds that form between amino- and carboxyl-containing groups on different amino acids.

- The carbonyl group is found on aldehyde and ketone molecules such as formaldehyde, acetaldehyde, and acetone. This functional group is the site of reactions that link these molecules into larger, more complex compounds, like the sugar ribose illustrated in Figure 2.24.

- Hydroxyl groups are important because they act as weak acids. In many cases, the protons involved in acid–base reactions that occur in cells come from hydroxyl groups on organic compounds. Because hydroxyl groups are polar, molecules containing a number of hydroxyl groups will form hydrogen bonds and be highly soluble in water.

- Phosphate groups carry two negative charges. When phosphate groups are transferred from one organic compound to another, the change in charge often dramatically affects the recipient molecule.

- Sulfhydryl groups consist of a sulfur atom bonded to a hydrogen atom. They are important because sulfhydryl groups on different molecules can link together via disulfide (S–S) bonds.

To summarize, functional groups make things happen. The number and type of functional groups attached to a framework of carbon atoms imply a great deal about how that molecule is going to behave. When you encounter an organic compound that is new to you, it's important to do two things: Examine its overall size and shape, and locate any functional groups. Understanding these two features will help you understand the molecule's role in chemical evolution and today's cells.

SUMMARY TABLE **2.3** **Six Functional Groups Commonly Attached to Carbon Atoms**

Functional Group	*Formula	Family of Molecules	Example	Properties of Functional Group
Amino	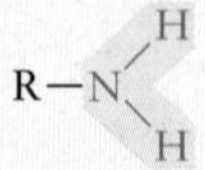 $R-NH_2$	Amines	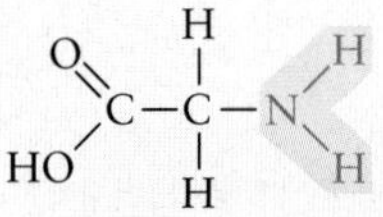 Glycine (an amino acid): $HO(O=)C-CH_2-NH_2$	Acts as a base—tends to attract a proton to form: 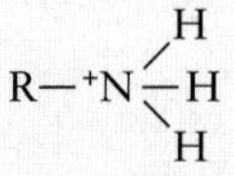 $R-{}^{+}NH_3$
Carbonyl	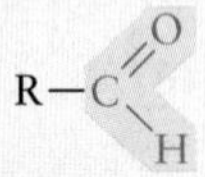 $R-C(=O)H$	Aldehydes	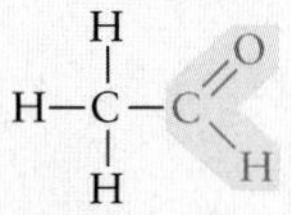 Acetaldehyde: $CH_3-C(=O)H$	Aldehydes, especially, react with compounds of form HR_2 to produce larger molecules with form: 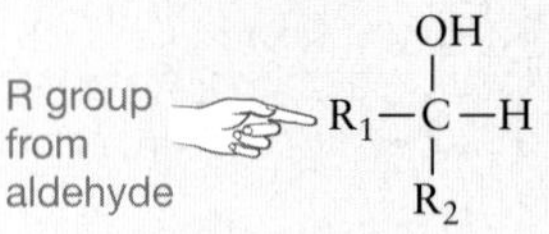 $R_1-C(OH)(H)-R_2$ (R group from aldehyde: R_1)
	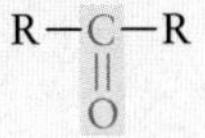 $R-C(=O)-R$	Ketones	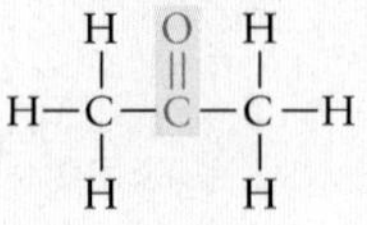 Acetone: $CH_3-C(=O)-CH_3$	
Carboxyl	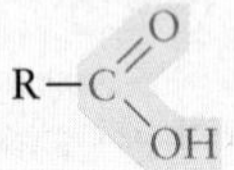 $R-C(=O)OH$	Carboxylic acids	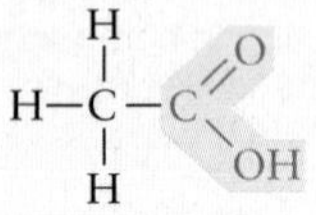 Acetic acid: $CH_3-C(=O)OH$	Acts as an acid—tends to lose a proton to form: 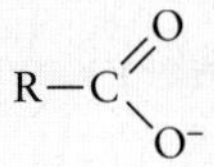 $R-C(=O)O^-$
Hydroxyl	R—OH	Alcohols	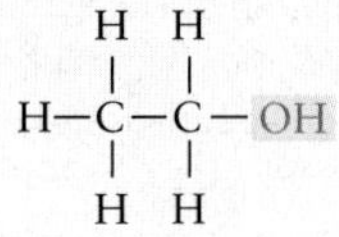 Ethanol: CH_3-CH_2-OH	Highly polar, so makes compounds more soluble through hydrogen bonding with water; may also act as a weak acid and drop a proton
Phosphate	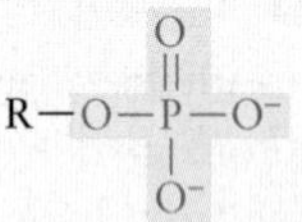 $R-O-P(=O)(O^-)-O^-$	Organic phosphates	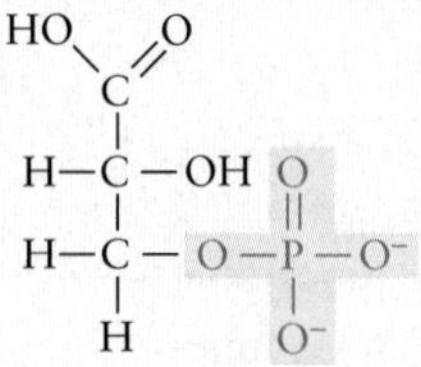 3-Phosphoglyceric acid: $HO(O=)C-CH(OH)-CH_2-O-P(=O)(O^-)-O^-$	When several phosphate groups are linked together, breaking O–P bonds between them releases large amounts of energy
Sulfhydryl	R—SH	Thiols	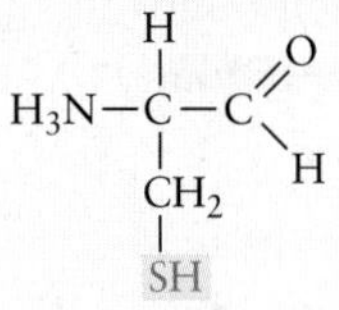 Cysteine: $H_3N-CH(CH_2-SH)-C(=O)H$	When present in proteins, can form disulfide (S–S) bonds that contribute to protein structure

EXERCISE Predict whether each functional group is polar or nonpolar, based on the electronegativities of the atoms involved.

Once carbon-containing molecules with functional groups had appeared early in Earth history, what happened next? For chemical evolution to continue, two things had to happen. First, reactions among relatively small and simple organic compounds had to produce the building blocks of the large molecules found in living cells. Second, these building blocks had to link together to form proteins, nucleic acids, and carbohydrates—the large, complex compounds found in organisms. Our task in the next three chapters is to analyze how these events occurred and how proteins, nucleic acids, and carbohydrates function in organisms today.

As far as we know, the jump from nonlife to life has occurred only once in the history of the universe. According to the theory of chemical evolution, it happened as molecules featured in the next three chapters began to accumulate in the waters of the ancient Earth.

Chapter Review

SUMMARY OF KEY CONCEPTS

The theory of chemical evolution maintains that early in Earth's history, complex molecules formed from simple precursor compounds in the ancient atmosphere and ocean, as chemical reactions converted the energy in sunlight and other sources into chemical energy. Chemical energy is a form of potential energy and can be stored in bonds between atoms.

Molecules form when atoms bond to each other. Chemical bonds are based on electron sharing. The degree of electron sharing varies from nonpolar covalent bonds, to polar covalent bonds, to ionic bonds.

When atoms participate in chemical bonds to form molecules, the shared electrons give the atoms full valence shells and thus contribute to the atoms' stability. The electrons may be shared equally or unequally, depending on the relative electronegativities of the two atoms involved. Nonpolar covalent bonds result from equal sharing; polar covalent bonds are due to unequal sharing. Ionic bonds form when an electron is completely transferred from one atom to another.

You should be able to draw the electron-sharing continuum and place molecular oxygen (O_2), carbon dioxide (CO_2), and calcium chloride ($CaCl_2$) on it.

Of all small molecules, water is the most important for life. Water is highly polar and readily forms hydrogen bonds. Hydrogen bonding makes water an extremely efficient solvent with a high capacity for absorbing energy.

Water is the most effective solvent known because it is polar, meaning that it has partial positive and negative charges. Water is polar because it is bent and has two polar covalent bonds. Polar molecules and charged substances, including ions, interact with water and stay in solution via hydrogen bonding and electrostatic attraction. Water's ability to participate in hydrogen bonding also gives it an extraordinarily high capacity to absorb heat and cohere to molecules.

You should be able to draw how water interacts with ammonia in solution.

 Web Animation at www.masteringbio.com

The Properties of Water

Chemical reactions tend to be spontaneous if they lead to lower potential energy and higher entropy (more disorder). An input of energy is required for nonspontaneous reactions to occur.

The reactions involved in chemical evolution resulted in products that had higher potential energy and lower entropy than the reactants had. As a result, these reactions were nonspontaneous. They could occur only because there was an input of kinetic energy in the form of sunlight and heat.

You should be able to explain why cells cannot stay alive without constant inputs of energy.

Most of the important compounds in organisms contain carbon. Early in Earth history, the energy in sunlight triggered nonspontaneous reactions that led to the formation of key carbon-containing molecules.

Organic molecules are critical to life because they have complex shapes provided by a framework of carbon atoms, along with complex chemical behavior due to the presence of functional groups. The first step in chemical evolution was the formation of the small organic compounds called formaldehyde and hydrogen cyanide from molecules such as ammonia (NH_3), methane (CH_4), molecular hydrogen (H_2), and carbon dioxide (CO_2.) These reactions occur readily when a source of intense energy, such as the radiation in sunlight, is present.

You should be able to explain why molecules with carbon-carbon bonds have more potential energy and lower entropy than carbon dioxide.

QUESTIONS

Test Your Knowledge

1. Which of the following occurs when a covalent bond forms?
   a. The potential energy of electrons drops.
   b. Electrons in valence shells are shared between nuclei.
   c. Ions of opposite charge interact.
   d. Polar molecules interact.

2. If a reaction is exothermic, then which of the following statements is true?
   a. The products have lower potential energy than the reactants.
   b. Energy must be added for the reaction to proceed.
   c. The products have lower entropy (are more ordered) than the reactants.
   d. It occurs extremely quickly.

3. If a reaction is exergonic, then which of the following statements is true?
   a. The products have lower free energy than the reactants.
   b. Energy must be added for the reaction to proceed.
   c. The products have lower entropy (are more ordered) than the reactants.
   d. It occurs extremely quickly.

4. What is thermal energy?
   a. a form of potential energy
   b. the temperature increase that occurs when any form of energy is added to a system
   c. mechanical energy
   d. the kinetic energy of molecular motion, measured as heat

5. What determines whether a chemical reaction is spontaneous?
   a. if it increases the disorder, or entropy, of the substances involved
   b. if it decreases the potential energy of the substances involved
   c. the temperature only—reactions are spontaneous at high temperatures and nonspontaneous at low temperatures
   d. the combined effect of changes in potential energy and entropy

6. Which of the following is *not* an example of an energy transformation?
   a. A shoe drops, converting potential energy to kinetic energy.
   b. A chemical reaction converts the energy in sunlight into the chemical energy in formaldehyde.
   c. The electrical energy flowing through a light bulb's filament is converted into light and heat.
   d. Sunlight strikes a prism and separates into distinct wavelengths.

Test Your Knowledge answers: 1. b; 2. a; 3. a; 4. d; 5. d; 6. d

Test Your Understanding

Answers are available at www.masteringbio.com

1. Consider the reaction between carbon dioxide and water, which forms carbonic acid:

   $$CO_2(g) + H_2O(l) \rightleftharpoons H_2CO(aq)$$

   In aqueous solution, carbonic acid immediately dissociates to form a proton and the bicarbonate ion, as follows:

   $$H_2CO_3(aq) \rightleftharpoons H^+(aq) + HCO_3^-(aq)$$

   Does this second reaction raise or lower the pH of the solution? Does the bicarbonate ion act as an acid or a base? If an underwater volcano bubbled additional CO_2 into the ocean, would this sequence of reactions be driven to the left or the right? How would this affect the pH of the ocean?

2. When chemistry texts introduce the concept of electron shells, they emphasize that shells represent distinct potential energy levels. In introducing electron shells, this chapter also emphasized that they represent distinct distances from the positive charges in the nucleus. Are these two points of view in conflict? Why or why not?

3. Draw a ball-and-stick model of the water molecule, and explain why this molecule is bent. Indicate the location of the partial electric charges on it. Why do these partial charges exist?

4. Hydrogen bonds form because the opposite, partial electric charges on polar molecules attract. Covalent bonds form as a result of the electrical attraction between electrons and protons. Covalent bonds are much stronger than hydrogen bonds. Explain why, in terms of the electrical attractions involved.

5. Explain why extensive hydrogen bonding gives water an extraordinarily high specific heat.

6. Explain the relationship between the carbon atoms in an organic molecule and the functional groups on the same molecule.

Applying Concepts to New Situations

Answers are available at www.masteringbio.com

1. Why isn't CO_2 bent and polar, like H_2O? Why is H_2O much more likely to participate in chemical reactions than CO_2?

2. Oxygen is extremely electronegative, meaning that its nucleus pulls in electrons shared in covalent bonds. Because these electrons are close to the oxygen nucleus, they have lower potential energy. Explain the changes in electron position that are illustrated in Figure 2.19a based on oxygen's electronegativity.

3. When nuclear reactions take place, some of the mass in the atoms involved is converted to energy. The energy in sunlight is created during nuclear fusion reactions on the Sun. Explain what astronomers mean when they say that the Sun is burning down and that it will eventually burn out.

4. Why do coastal regions tend to have climates with moderate temperatures and lower annual variation in temperature than do inland areas at the same latitude?

www.masteringbio.com is also your resource for • Answers to text, table, and figure caption questions and exercises • Answers to *Check Your Understanding* boxes • Online study guides and quizzes • Additional study tools including the *E-Book for Biological Science* 3rd ed., textbook art, animations, and videos.

Protein Structure and Function

3

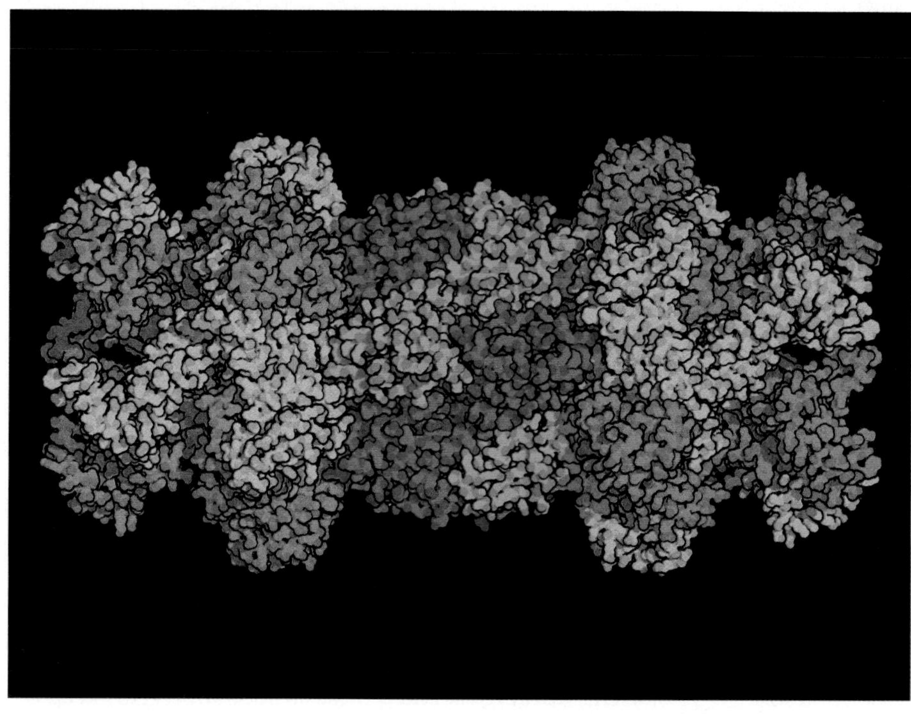

A space-filling model of an AAA+ protease, a type of protein whose function is to chop up and dispose of old or damaged proteins in cells.

KEY CONCEPTS

- Proteins are made of amino acids. Amino acids vary in structure and function because their side chains vary in composition.
- Proteins vary widely in structure. The structure of a protein can be analyzed at four levels that form a hierarchy—the amino acid sequence, substructures called α-helices and β-pleated sheets, interactions between amino acids that dictate a protein's overall shape, and combinations of individual proteins that make up larger, multiunit molecules.
- Proteins vary widely in function. In cells, most proteins are enzymes that function as catalysts. Chemical reactions occur much faster when they are catalyzed by enzymes. During enzyme catalysis, the reactants bind to an enzyme's active site in a way that allows the reaction to proceed efficiently.

Chapter 2 introduced the hypothesis that chemical reactions in the atmosphere and ocean of ancient Earth led to the formation of the first complex carbon-containing compounds. This idea, called chemical evolution, was first proposed in 1923 by Alexander I. Oparin. The hypothesis was published again—independently and six years later—by J. B. S. Haldane. Today, the Oparin–Haldane proposal can best be understood as a formal scientific theory. As Chapter 1 noted, scientific theories typically have two components: a statement about a pattern that exists in the natural world and a proposed mechanism or process that explains the pattern. In the case of chemical evolution, the pattern is that increasingly complex carbon-containing molecules formed in the atmosphere and ocean of ancient Earth. The process responsible for this pattern was the conversion of energy, from sunlight and other sources, into chemical energy in the bonds of large, complex molecules.

Scientific theories are continuously refined as new information comes to light, and many of Oparin and Haldane's original ideas about how chemical evolution occurred have been extensively revised. In its current form, the theory can be broken into four steps, each requiring an input of energy:

1. Chemical evolution began with the production of small organic compounds such as formaldehyde (H_2CO) and hydrogen cyanide (HCN), from reactants such as H_2, CO_2, and CH_4, and NH_3.
2. Formaldehyde, hydrogen cyanide, and other simple organic compounds reacted to form the mid-sized molecules called

amino acids, sugars, and nitrogenous bases. These building-block molecules accumulated in the shallow waters of the ancient ocean, forming a complex solution called the **prebiotic soup**.

3. Mid-sized, building-block molecules linked to form the types of large molecules found in cells today, including proteins, nucleic acids, and complex carbohydrates. These large molecules are each made up of distinctive chemical subunits that join together: Proteins are composed of amino acids; nucleic acids are composed of nucleotides; and complex carbohydrates are composed of sugars.
4. Life became possible when one of these large, complex molecules acquired the ability to make a copy of itself. This self-replicating molecule began to multiply by means of chemical reactions that it controlled. At that point, life had begun. Chemical evolution gave way to biological evolution.

Analyzing how the final three steps in chemical evolution occurred is the subject of this chapter and the next three chapters. Each of these chapters focuses on one of the four primary classes of biological molecules found in cells living today: proteins, nucleic acids, carbohydrates, and lipids. What do the subunits of these large molecules look like, and how could they have been created by chemical evolution? How are the subunits linked to form a larger molecule—a protein, nucleic acid, a complex carbohydrate, or lipid? What do these large molecules look like, and what do they do in living cells? Finally, which type of molecule was responsible for the origin of life? This last question is particularly intriguing, because researchers around the world are racing to synthesize a self-replicating molecule—to produce life in a test tube. Which types of molecule are they working with, and why? Let's begin our analysis with the workhorse molecules called proteins.

3.1 What Do Proteins Do?

As a group, proteins serve more types of functions in cells than any other type of molecule does. Consider the red blood cells that are moving through your arteries, capillaries, and veins right now. Each of these cells contains about 300 million copies of the protein called hemoglobin. Hemoglobin carries oxygen from your lungs to cells throughout the body. But every red blood cell also has thousands of copies of a protein called carbonic anhydrase, which is important for moving carbon dioxide from cells back to the lungs, where it can be breathed out. Other proteins protect the body by attacking disease-causing bacteria and viruses; still others make movement possible or form structures that give cells their shape.

Proteins are crucial to most tasks required for cells to exist. The diverse functions of proteins include the following:

- *Defense.* Proteins called antibodies and complement proteins attack and destroy viruses and bacteria that cause disease.
- *Movement.* Motor proteins and contractile proteins are responsible for moving the cell itself, or for moving large molecules and other types of cargo inside the cell. As you turn this page, for example, specialized proteins called actin and myosin will slide past one another as they work to flex or extend muscle cells in your fingers and arm.
- *Catalysis.* Many proteins are specialized to **catalyze**, or speed up, chemical reactions, most of which would not otherwise be able to proceed. A protein that functions as a catalyst is called an **enzyme**. The carbonic anhydrase molecules in red blood cells are an example, and so is the protein called salivary amylase that is present in your mouth. Salivary amylase helps begin the digestion of starch and other complex carbohydrates into simple sugars.
- *Signaling.* Proteins called peptide hormones bind to receptor proteins on particular cells. In response, the activity of the cell with the receptor protein changes. In this way, proteins are involved in carrying and receiving signals from cell to cell inside the body.
- *Structure.* Structural proteins give cells mechanical support. They also create structures such as fingernails and hair. The membranes of red blood cells have structural proteins that connect to structural proteins inside the cell itself. Working together, these structural proteins keep the cells flexible and in their normal disc-like shape.

SUMMARY TABLE **3.1** **Protein Functions**

Protein Type	Role in Cell or Organism
Antibodies and complement proteins	Defense—destruction of disease-causing viruses and bacteria
Contractile proteins and motor proteins	Movement
Enzymes	Catalyze chemical reactions
Hormones	Act as signals that help coordinate the activities of many cells
Receptor proteins	Receive chemical signals from outside cell and initiate response
Structural proteins	Provide support for cells and tissues; form structures such as hair, feathers, cocoons, and spider webs
Transport proteins	Move substances across cell membrane; substances throughout body

- *Transport.* Proteins are responsible for allowing particular molecules to enter or exit cells, and for carrying specific compounds throughout the body. Hemoglobin is a particularly well-studied transport protein, but virtually every cell is studded with membrane proteins that control the passage of specific molecules and ions.

These aspects of protein function are summarized in **Table 3.1**; each will be explored in much greater detail later in the text. In many if not all cases, the function of the protein involved is closely correlated with its structure. Before analyzing how proteins are put together, let's briefly consider how the building blocks of proteins could have been synthesized before life began.

3.2 Early Origin-of-Life Experiments

In 1953 a graduate student named Stanley Miller performed a breakthrough experiment in the study of chemical evolution. Miller wanted to answer a simple question: Can complex organic compounds be synthesized from the simple molecules present in Earth's early atmosphere and ocean? In other words, is it possible to recreate the first steps in chemical evolution by simulating ancient Earth conditions in the laboratory?

Miller's experimental setup (**Figure 3.1**) was designed to produce a microcosm of ancient Earth. The large glass flask represented the atmosphere and contained the gases methane (CH_4), ammonia (NH_3), and hydrogen (H_2), all of which have high free energy. This large flask was connected to a smaller flask by glass tubing. The small flask held a tiny ocean—200 milliliters (mL) of liquid water. Miller boiled this water constantly so that water vapor was added to the mix of gases in the large flask. As the vapor cooled and condensed, it flowed back into the smaller flask, where it boiled again. In this way, water vapor circulated continuously through the system. This was important: If the molecules in the simulated atmosphere reacted with one another, the "rain" would carry them into the simulated ocean, forming a simulated version of the prebiotic soup.

Had Miller stopped there, however, little or nothing would have happened. Even at the boiling point of water (100°C), the molecules involved in the experiment are stable. They do not undergo spontaneous chemical reactions, even at high temperatures.

Something did start to happen in the apparatus, however, when Miller sent electrical discharges across the electrodes he'd inserted into the atmosphere. These miniature lightning bolts added a crucial element to the reaction mix—pulses of intense electrical energy. After a day of continuous boiling and sparking, the solution in the boiling flask began to turn pink. After a week, it was deep red and cloudy. When Miller analyzed the molecules dissolved in the solution, he found that several complex carbon-containing compounds were present, including several with carbon-carbon bonds. The experiment, driven by

Experiment

Question: Can simple molecules and kinetic energy lead to chemical evolution?

Hypothesis: If kinetic energy is added to a mix of simple molecules with high free energy, reactions will occur that produce more complex molecules, perhaps including some with carbon-carbon bonds.

Null hypothesis: Chemical evolution will not occur, even with an input of energy.

Experimental setup:

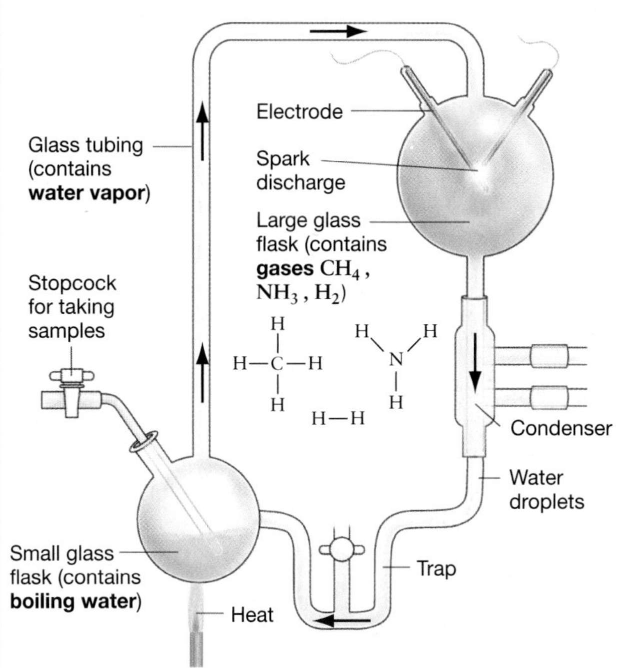

Prediction: Complex organic compounds will be found in the liquid water.

Prediction of null hypothesis: Only the starting molecules will be found in the liquid water.

Results:

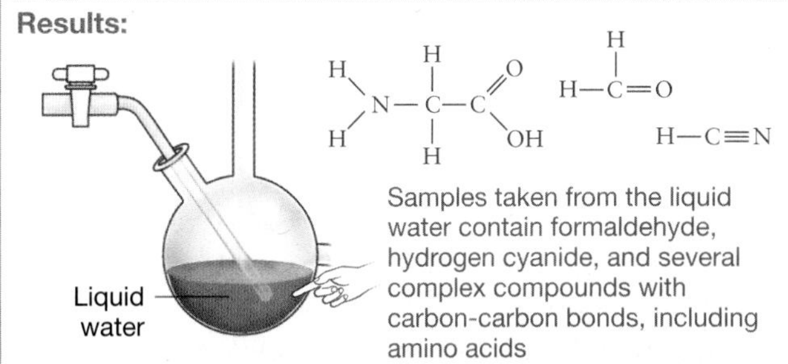

Conclusion: Chemical evolution occurs readily if simple molecules with high free energy are exposed to a source of kinetic energy.

FIGURE 3.1 Miller's Spark-Discharge Experiment. The arrows in the "Experimental setup" diagram indicate the flow of water vapor or liquid. The large glass flask can contain any mixture of gases desired. The condenser consists of a jacket with cold water flowing through it.

EXERCISE Label the parts of the apparatus that mimic the ocean, the atmosphere, rain, and lightning.

heat energy and the energy in the electrical discharges, had recreated the start of chemical evolution.

To find out exactly which products resulted from the initial reactions in the simulated atmosphere, Miller drew samples from the apparatus at intervals. In these samples he found large quantities of hydrogen cyanide (HCN) and formaldehyde (H_2CO). The data were exciting because HCN and (H_2CO) are required for reactions that lead to the synthesis of more complex organic molecules. Indeed, some of these more complex compounds were actually present in the miniature ocean. The sparks and heating had led to the synthesis of compounds that are fundamental to life: amino acids.

3.3 Amino Acids and Polymerization

Based on the presence of amino acids, Miller claimed that his experiment simulated the second stage in chemical evolution—the formation of a prebiotic soup. Although the assumptions behind his experiments, and hence the results, eventually came under fire, follow-up studies have confirmed that the second step in chemical evolution occurred early in Earth's history. For example, organic compounds like methane (CH_2), formaldehyde, and acetaldehyde form when water vapor, carbon monoxide, carbon dioxide, hydrogen (H_2), and other volcanic gases are put into a glass flask and exposed to the types of high-energy radiation found in sunlight. These are realistic simulations of conditions on early Earth (see Chapter 2). Based on results like these, there is now a strong consensus that amino acids and other components of the prebiotic soup are readily produced under conditions that accurately simulate the atmosphere and oceans of ancient Earth.

Now let's look at the molecules themselves. What are amino acids, and how are they linked to form proteins?

The Structure of Amino Acids

The bacterial cells that live on your skin contain several thousand different proteins; as a group, the cells in your body produce tens of thousands of distinct proteins. But most of these proteins are composed of just 21 different building blocks, called **amino acids**. All 21 amino acids have a common backbone structure.

To understand the structure of an amino acid, recall that carbon atoms have four unpaired valence electrons and form four covalent bonds. Every amino acid has a carbon that makes three similar bonds. The first bond attaches this carbon to NH_2—the amino functional group (**Figure 3.2a**). The second bond links this carbon to COOH—the carboxyl functional group. You might recall from Chapter 2 that the carboxyl group is acidic because its two oxygen atoms are highly electronegative. They pull electrons away from the hydrogen atom, which means that it is relatively easy for this group to lose a proton. The combination of the amino and carboxyl groups on these

(a) Non-ionized form of amino acid

Amino group — H_2N—C(H)(R)—C(=O)OH — Carboxyl group

Non-ionized | Side chain | Non-ionized

(b) Ionized form of amino acid

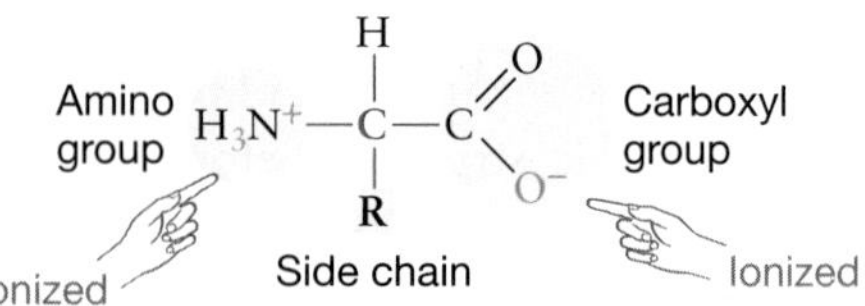

FIGURE 3.2 Amino Acid Structure. All amino acids have the same general structure: a central carbon, shown in red, bonded to an amino functional group, a carboxyl functional group, a hydrogen atom, and a side chain, or R-group.

molecules inspired the name amino acid. A third bond links the highlighted carbon to a hydrogen atom. In all amino acids, then, a carbon atom is bonded to an amino group, a carboxyl group, and hydrogen. As we'll see in a moment, the nature of the fourth attachment is what makes each amino acid unique.

The presence of an amino group and a carboxyl group in amino acids is important. **Figure 3.2b** shows what happens to these functional groups in solution. In water at pH 7, the concentration of protons causes the amino group to act as a base. It attracts a proton to form NH_3^+. The carboxyl group, in contrast, loses a proton to form COO^-. The charges on these functional groups (1) help amino acids stay in solution where they can interact with each other and with other solutes, and (2) add to their chemical reactivity.

The Nature of Side Chains In Figure 3.2, the highlighted carbon atom in the amino acid forms a fourth bond with an atom or a group of atoms, abbreviated as "R." Chemists use this symbol to indicate additional atoms called a side chain. In every amino acid, a carbon atom is linked to a hydrogen atom, an amino group, a carboxyl group, and an R-group. The 21 amino acids found in organisms are different because their R-groups are different. These R-groups vary from a single hydrogen atom to large structures containing carbon atoms linked into rings. Several of the side chains found in amino acids contain functional groups highlighted in Chapter 2, Table 2.3. Specifically, some amino acids have R-groups with carboxyl, sulfhydryl, hydroxyl, or amino functional groups.

Figure 3.3 shows the most common 20 amino acids found in cells and sorts them based on whether their side chain is nonpolar, polar, or electrically charged. (The amino acid that is not shown—selenocysteine—is similar in structure to cysteine but has a selenium atom in place of sulfur.) This property of side chains is important because amino acids with nonpolar side chains do not have charged or electronegative

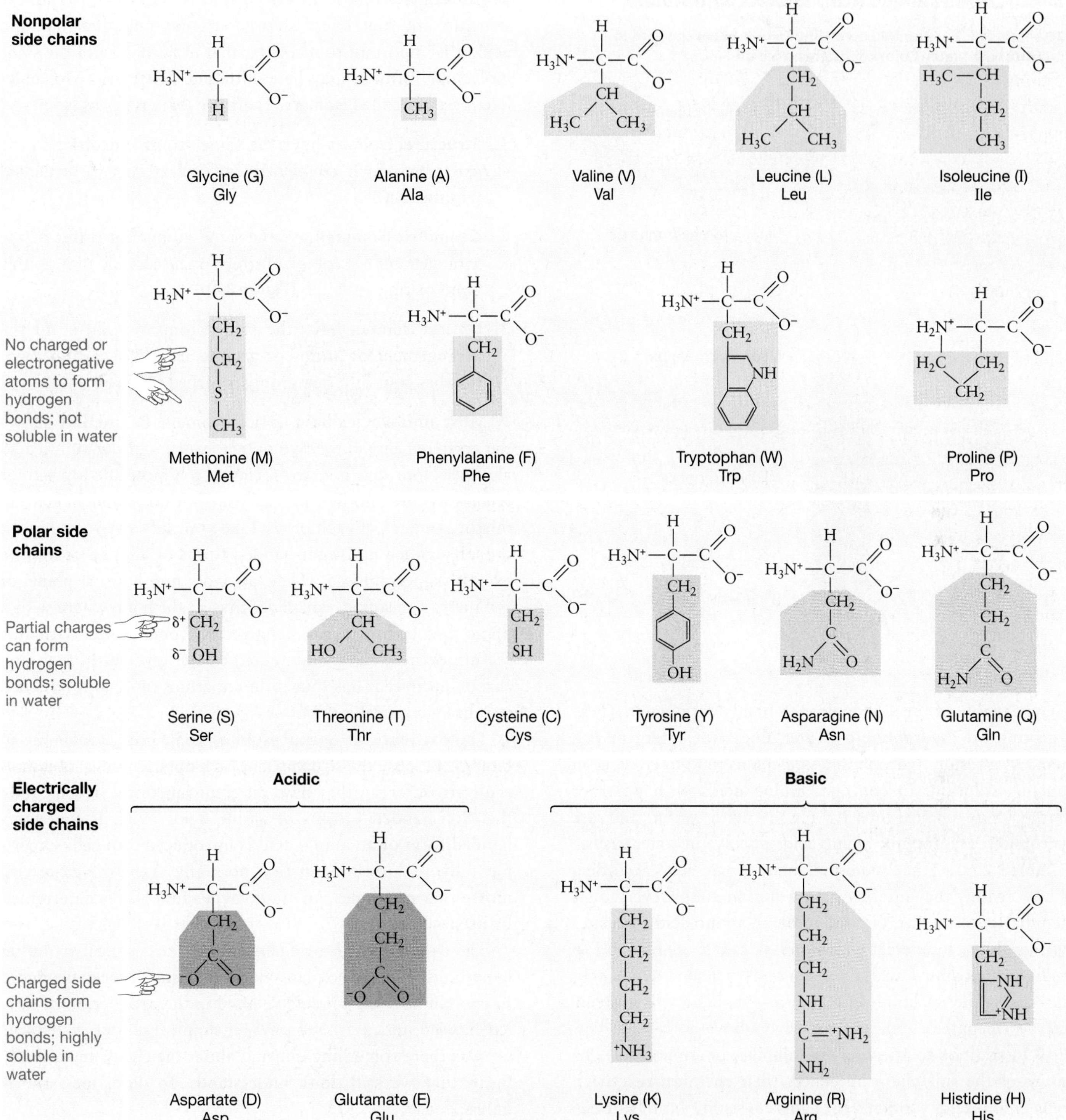

FIGURE 3.3 The 20 Major Amino Acids Found in Organisms. At the pH (about 7.0) found in cells, the 20 major amino acids found in organisms have the structural formulas shown here. The side chains are highlighted, and standard single-letter and three-letter abbreviations for each amino acid are given. For clarity, the carbon atoms in the ring structures of phenylalanine, tyrosine, tryptophan, and histidine are not shown; each bend in a ring is the site of a carbon atom. The hydrogen atoms in these structures are also not shown. A double line inside a ring indicates a double bond.

● **EXERCISE** Explain why the green R-groups are nonpolar and why the pink R-groups are polar, based on the relative electronegativities of O, N, C, and H. Note that sulfur (S) has an electronegativity almost equal that of carbon, and slightly higher than that of hydrogen.

TABLE 3.2 **How Amino Acids Interact with Water**

20 amino acids are ranked according to how likely they are to interact with water. Color codes are based on Figure 3.3.

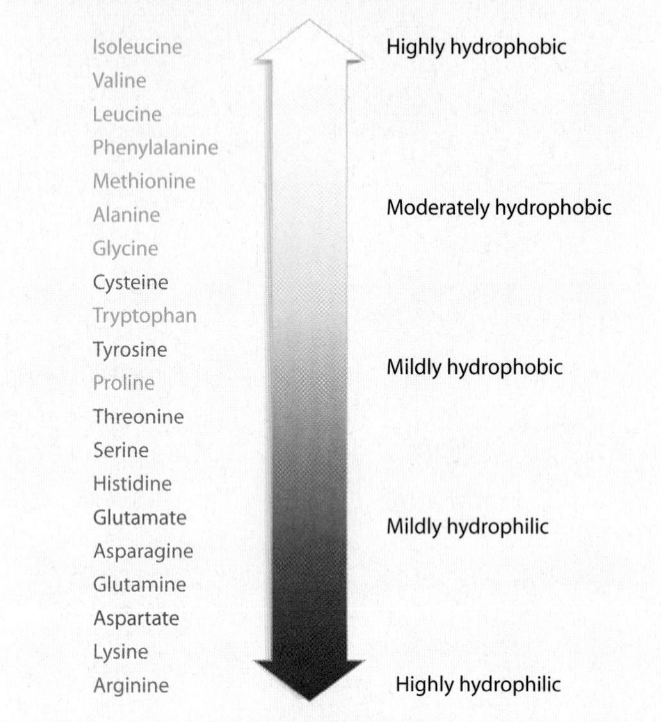

Isoleucine	Highly hydrophobic
Valine	
Leucine	
Phenylalanine	
Methionine	
Alanine	Moderately hydrophobic
Glycine	
Cysteine	
Tryptophan	
Tyrosine	
Proline	Mildly hydrophobic
Threonine	
Serine	
Histidine	
Glutamate	Mildly hydrophilic
Asparagine	
Glutamine	
Aspartate	
Lysine	
Arginine	Highly hydrophilic

atoms capable of forming hydrogen bonds with water. These R-groups are **hydrophobic**—water does not interact with them. As a result, hydrophobic side chains tend to coalesce in aqueous solution. In contrast, amino acids with polar or charged side chains interact readily with water and are **hydrophilic**. Hydrophilic amino acids dissolve in water easily.

Table 3.2 ranks the 20 most common amino acids according to how readily they interact with water. In almost every case, the polarity of the R-group found on an amino acid correlates with its ability to interact with water. You should be able to predict which amino acids are most and least soluble in water, and—using the structures in Figure 3.3—explain why cysteine and tyrosine interact relatively poorly with water.

In addition to affecting the solubility of amino acids, the nature of the side chain influences their chemical reactivity. Some amino acids contain side chains consisting entirely of carbon and hydrogen atoms. These R-groups rarely participate in chemical reactions. As a result, the chemical behavior of these amino acids depends primarily on their size and shape rather than reactivity. In contrast, amino acids that have hydroxyl, amino, carboxyl, or sulfhydryl functional groups in their side chains are more reactive. For example, amino acids with sulfur atoms (S) in their side chains can form bonds that help link different parts of large proteins. The key point here is that variation in the structure of the R-groups in amino acids explains variation in their properties and functions.

What Are Optical Isomers? When you are analyzing the structure and function of an organic compound like an amino acid, it is important to recognize that molecules with the same molecular formula may have different structures. Molecules like these are called **isomers**. There are three types of isomers:

1. **Structural isomers** have the same atoms but differ in the order in which covalently bonded atoms are attached (**Figure 3.4a**).
2. **Geometric isomers** have the same atoms but differ in the arrangement of atoms or groups on either side of a double bond or ring structure (**Figure 3.4b**).
3. **Optical isomers** have the same atoms but differ in the arrangement of atoms or groups around a carbon atom that has four different groups attached (**Figure 3.4c**).

Most amino acids have optical isomers. Figure 3.4c shows the arrangements in the two optical isomers of the amino acid alanine. Note that the two forms of the molecule are mirror images of one another, just as your left and right hands are mirror i+mages of each other. Like your left and right hands, the left-handed and right-handed forms of alanine cannot be exactly superimposed. They also do not have a plane of symmetry—meaning, you cannot bisect them anywhere with a plane and create identical halves. Carbon atoms with this feature exist in every amino acid except glycine. In fact, any carbon atom that has four different atoms or groups attached to it has an optical isomer.

The existence of optical isomers is an important issue in biology. Because the structures of each optical isomer of a molecule are different, they have different functions. In cells, only the "left-handed" forms of amino acids exist. If the right-handed form of an amino acid is introduced into cells experimentally, it does not function normally. As is the case with most other molecules, an amino acid's function is determined by its structure.

These observations are a challenge to the chemical evolution hypothesis, because no plausible mechanism has been proposed to explain how the process resulted in the production of only left-handed optical isomers. Was it simply a matter of chance? Or was there something unusual about the chemistry of early Earth that we still don't understand? To date, the issue is unresolved.

How Do Amino Acids Link to Form Proteins?

Amino acids link to each other to form proteins. Similarly, the molecular building blocks called nucleotides attach to one another to form nucleic acids, and simple sugars connect to form complex carbohydrates. In general, a molecular subunit such as an amino acid, a nucleotide, or a sugar is called a **monomer** ("one-part"). When monomers bond together, the resulting structure is called a **polymer** ("many-parts"). The process of

(a) Structural isomers differ in the order which their atoms are attached.

Ethanol (C_2H_6O) — CH_3CH_2OH

Dimethyl ether (C_2H_6O) — CH_3OCH_3

(b) Geometric isomers differ in the arrangement of atoms around a double bond.

trans-2-butene (C_4H_8)

cis-2-butene (C_4H_8)

(c) Optical isomers are mirror images of one another—they cannot be exactly superimposed.

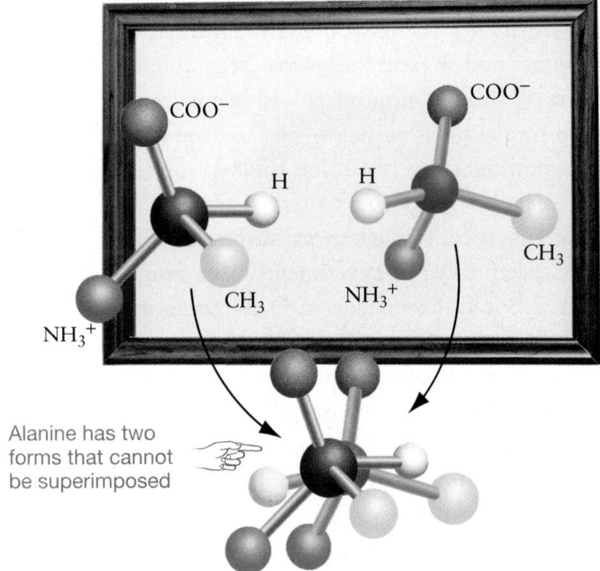

Hands are mirror images, just as optical isomers are.

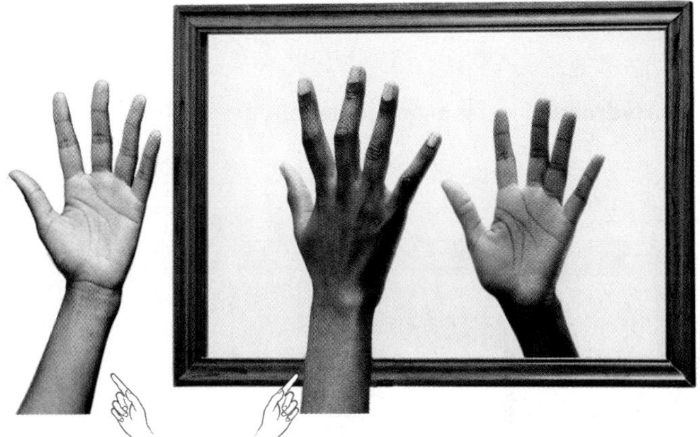

Your left and right hands cannot be superimposed (that is, with both thumbs facing the same direction and both palms facing the same direction)

FIGURE 3.4 Structural, Geometric, and Optical Isomers. (a) Ethanol is the active ingredient in alcoholic beverages; dimethyl ether, a gas at room temperature, is used in refrigeration. **(b)** The molecules *trans*-2-butene and *cis*-2-butene are used in the production of gasoline, synthetic rubber, and solvents. **(c)** All amino acids except glycine have optical isomers.

linking monomers is called **polymerization** (**Figure 3.5**). Thus, amino acids polymerize to form proteins. Biologists also use the word **macromolecule** to denote a very large molecule that is made up of smaller molecules joined together. A **protein** is a macromolecule—a polymer—that consists of linked amino acid monomers.

The theory of chemical evolution states that monomers in the prebiotic soup polymerized to form proteins and other types of macromolecules found in organisms. This is a difficult step, because monomers such as amino acids do not spontaneously self-assemble into macromolecules such as proteins. According to the second law of thermodynamics reviewed in Chapter 2, this is not surprising. Complex and highly organized molecules are not expected to form spontaneously from simpler constituents, because polymerization organizes the molecules involved into a more complex, ordered structure. Stated another way, polymerization decreases the disorder, or entropy, of the molecules involved. In addition, polymers are energetically much less stable than their component monomers. Because the ΔH term of the Gibbs free-energy equation is positive and the $T\Delta S$ term is negative, ΔG is positive at all temperatures. Polymerization reactions are endergonic and nonspontaneous. Monomers must absorb energy in order to link together. How could this have happened during chemical evolution?

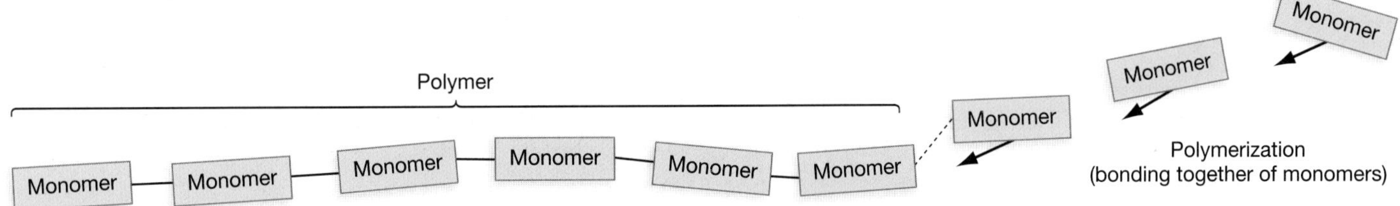

FIGURE 3.5 Monomers Are the Building Blocks of Polymers. Monomers polymerize to form polymers.

(a) Condensation reaction: monomer in, water out

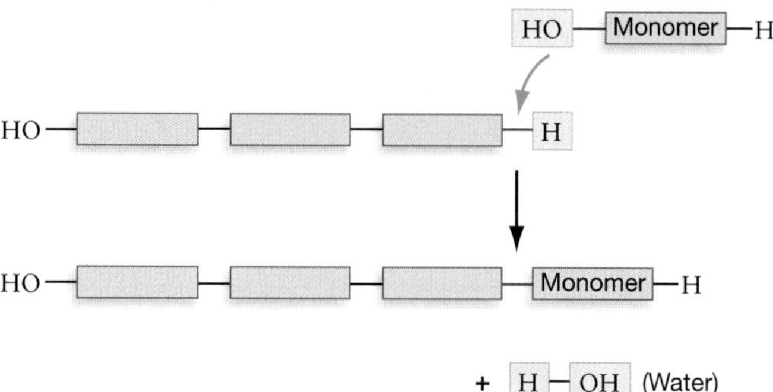

(b) Hydrolysis: water in, monomer out

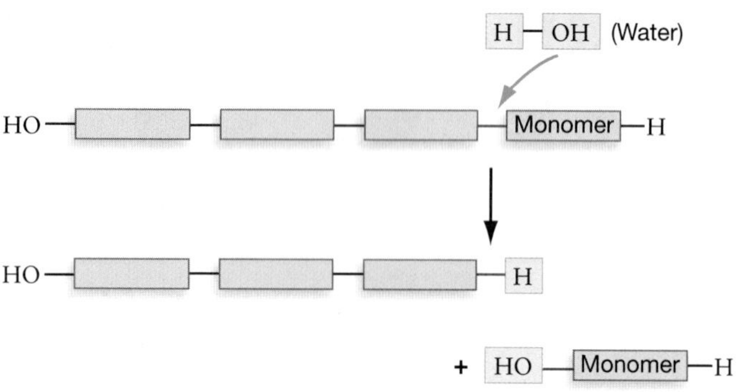

FIGURE 3.6 Polymers Can Be Extended or Broken Apart. (a) In a condensation reaction, a monomer is added to a polymer to make a longer polymer. The new bond that forms results in the formation of a water molecule. **(b)** In hydrolysis, a water molecule reacts with the bond linking the monomers. A monomer is broken off the chain, resulting in a shorter polymer.

Could Polymerization Occur in the Energy-Rich Environment of Early Earth? Researchers who added heat or electrical discharges to solutions of either amino acids or nucleic acids found that polymerization reactions proceed slowly, if at all. This is because monomers polymerize through **condensation reactions**, also known as **dehydration reactions**. These reactions are aptly named because the newly formed bond results in the loss of a water molecule (**Figure 3.6a**). The reverse reaction, called **hydrolysis**, breaks polymers apart by adding a water molecule (**Figure 3.6b**). The water molecule reacts with the bond linking the monomers, separating one monomer from the polymer chain.

In a solution, such as the prebiotic soup, condensation and hydrolysis represent the forward and reverse reactions of a chemical equilibrium. Hydrolysis dominates because it increases entropy and because it is energetically favorable—it lowers the potential energy of the electrons involved. This is why polymerization reactions are expected to proceed slowly or not at all.

According to recent experiments, the key to overcoming hydrolysis during chemical evolution was, quite literally, as common as mud. Researchers have been able to create stable polymers by mixing monomers with a source of chemical energy and tiny mineral particles—the size found in clay or mud. These experiments are based on the hypothesis that growing macromolecules would be protected from hydrolysis if they clung, or adsorbed, to the mineral surfaces.

More specifically, the experiments were designed to simulate events that could have occurred in the prebiotic soup. In one experiment, researchers put amino acids in a solution with a source of chemical energy and tiny mineral particles and allowed them to react. After a day, the researchers separated the mineral particles from the solution. They then put the particles into a fresh solution containing amino acids and a source of chemical energy. After repeating this procedure for several days, they analyzed the mineral particles and found polymers up to 55 amino acids long. These results support the hypothesis that adsorption to mineral particles protects polymers from hydrolysis. Because the experimental procedure was designed to mimic coastal environments where beaches are repeatedly washed with fresh waves or tidal flows, the results make it reasonable to claim that at least some muddy tide pools and beaches became covered with small proteins early in Earth's history.

The Peptide Bond Exactly how do amino acids polymerize? As **Figure 3.7** shows, amino acids polymerize when a bond forms between the carboxyl group of one amino acid and the amino group of another. The C–N bond that results from this condensation reaction is called a **peptide bond.** Peptide bonds are particularly stable because electrons are partially shared between the neighboring carbonyl functional group and the

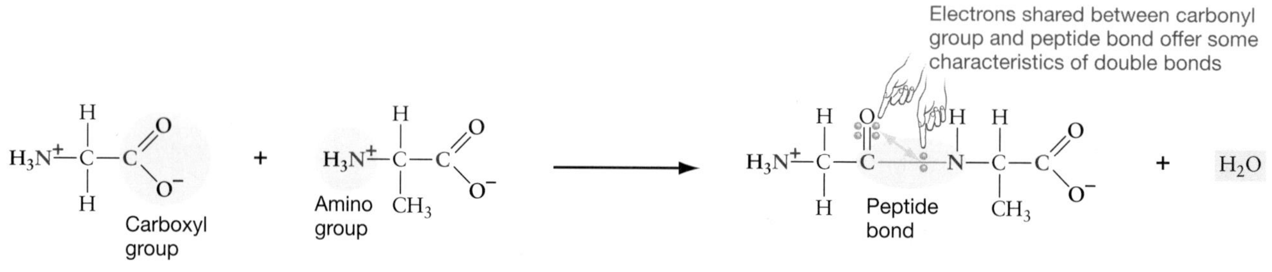

Electrons shared between carbonyl group and peptide bond offer some characteristics of double bonds

FIGURE 3.7 Peptide Bond Formation. When the carboxyl group of one amino acid reacts with the amino group of a second amino acid, a peptide bond forms.

QUESTION Is a peptide bond a hydrogen bond, a nonpolar covalent bond, a polar covalent bond, or an ionic bond?

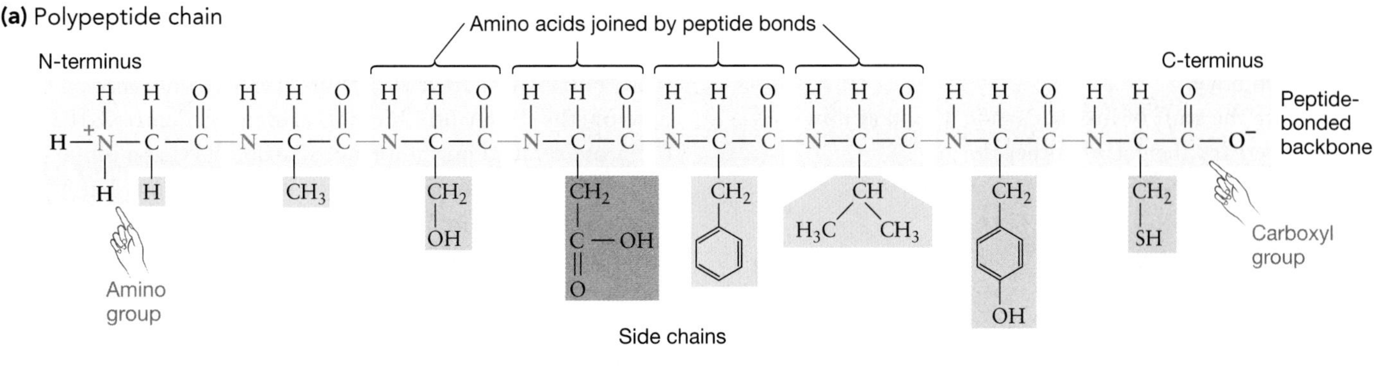

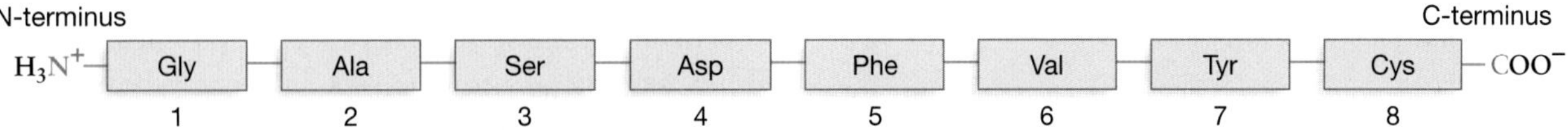

FIGURE 3.8 Amino Acids Polymerize to Form Polypeptides. (a) Amino acids can be linked into long chains, called polypeptides, by peptide bonds. **(b)** By convention, the sequence of amino acids in a polypeptide chain is numbered from the N-terminus to the C-terminus.

peptide bond. The degree of electron sharing is great enough that peptide bonds actually have some of the characteristics of a double bond. For example, the peptide bond is planar.

When amino acids are linked by peptide bonds into a chain, the amino acids are referred to as **residues** and the resulting molecule is called a **polypeptide. Figure 3.8a** shows how the chain of peptide bonds in a polypeptide gives the molecule a structural framework, or a "backbone." Three points are important to note about this backbone: (1) The side chains present in each residue extend out from it, (2) it has directionality, and (3) it is flexible. The backbone is directional because there is an amino group ($-NH_3^+$) on one end of every polypeptide chain and a carboxyl group ($-COO^-$) on the other. By convention, biologists always write amino acid sequences in the same direction. The end of the sequence that has the free amino group is placed on the left and is called the N-terminus, or amino-terminus, and the end with the free carboxyl group appears on the right-hand side of the sequence and is called the C-terminus, or carboxy-terminus. The amino acids in the chain are always numbered starting from the N-terminus (**Figure 3.8b**), because the N-terminus is the start of the chain when proteins are synthesized in cells. Although the peptide bond itself cannot rotate because of its double-bond nature, the single bonds on either side of the peptide bond can rotate. As a result, the structure as a whole is flexible (**Figure 3.9**).

When fewer than 50 amino acids are linked together in this way, the resulting polypeptide is called an **oligopeptide** ("few peptides") or simply a **peptide**. Polypeptides that contain 50 or

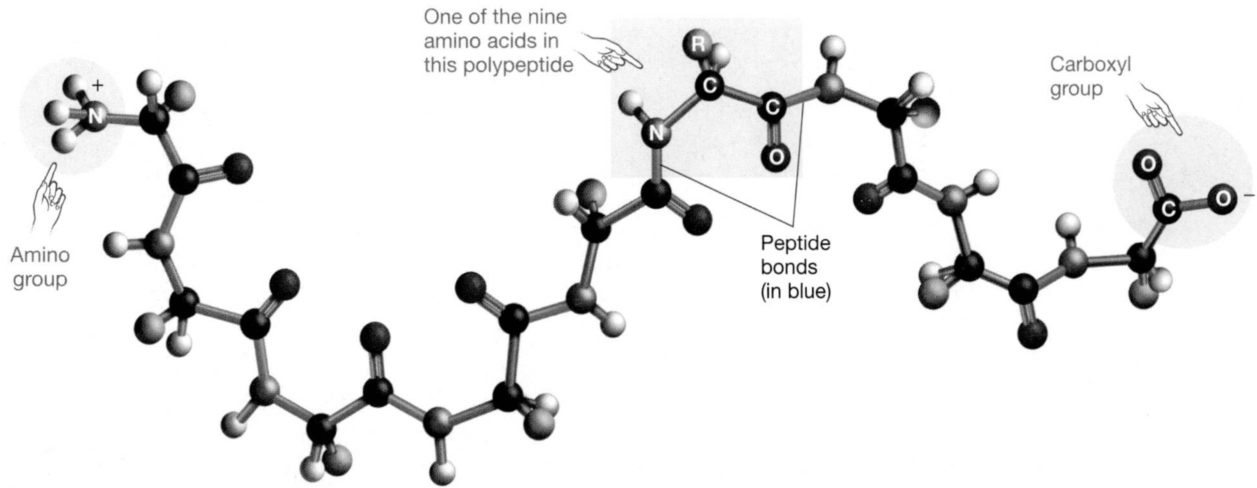

FIGURE 3.9 Proteins Are Flexible. Polypeptide chains are flexible because the groups on either side of each peptide bond can rotate about their single bonds.

● **EXERCISE** Draw an arrow to each peptide bond in this molecule. Circle each amino acid.

more amino acids are formally called **proteins**. Proteins may consist of single polypeptides or multiple polypeptides that are bonded to each other.

Proteins are the stuff of life. Let's take a look at how they are put together, and then at what they do.

Check Your Understanding

If you understand that...

- Amino acids are small molecules with a carbon atom bonded to a carboxyl group, an amino group, a hydrogen atom, and a side chain called an R-group.
- Each amino acid has distinctive chemical properties because each has a unique R-group.
- When the carboxyl group of one amino acid reacts with the amino group of another amino acid, a strong covalent bond called a peptide bond forms.
- Polypeptides are polymers made up of peptide-bonded amino acids. Small polypeptides are called oligopeptides, and large polypeptides are called proteins.

You should be able to ...

1) Draw the general form of an amino acid.
2) Draw and label two amino acids linked by a peptide bond.

MB Web Animation at www.masteringbio.com
Condensation and Hydrolysis Reactions

3.4 What Do Proteins Look Like?

With respect to their structure, proteins may be the most diverse class of molecules known. This is important, because diversity in structure and in the chemical reactivity of R-groups explains the incredible functional diversity of proteins.

Figure 3.10 illustrates some of the diverse sizes, shapes, and functions observed in proteins. Note that these proteins range from butterfly-shaped to donut-shaped to globular. In the case of the TATA-box binding protein in Figure 3.10a and the protein called porin in Figure 3.10b, the shape of the molecule has an obvious correlation with its function. The TATA-box binding protein has a groove where DNA molecules fit; porin has a hole that forms a pore. The groove in the TATA-box binding protein interacts with specific regions of DNA inside cells, while porin fits in cell membranes and allows certain hydrophilic molecules to pass through. But most of the proteins found in cells function as enzymes and are globular (Figure 3.10c).

The unparalleled diversity of proteins—in size, shape, and other aspects of structure—is important because function follows from structure. Proteins can serve diverse functions in cells because they are diverse in size and shape as well as in the chemical properties of their amino acids.

How can biologists make sense of this diversity of protein size and shape? Initially, the amount of variation seems overwhelming. Fortunately, it is not. No matter how large or complex a protein may be, its underlying structure can be broken down into just four basic levels of organization.

Primary Structure

Every protein has a unique sequence of amino acids. That simple revelation was the culmination of 12 years of study by Frederick Sanger and co-workers during the 1940s and 1950s. Sanger's group worked out the first techniques for determining the amino acid sequence of a protein and published the completed sequence of the hormone insulin, a protein that helps regulate sugar concentrations in the blood of humans and other mammals. When other proteins were analyzed, it rapidly became clear that each protein has a definite and distinct amino acid sequence.

(a) TATA-box binding protein

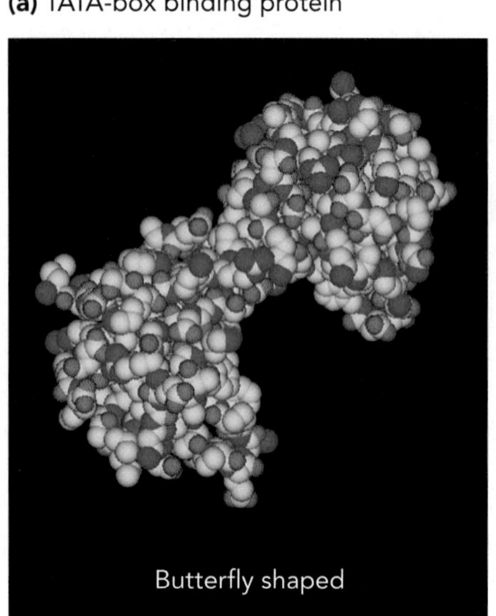

(b) Porin

(c) Pyrophosphatase

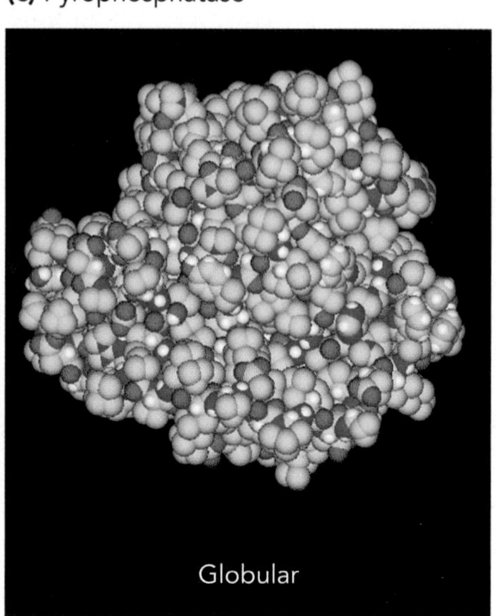

FIGURE 3.10 In Size and Shape, Proteins Are the Most Diverse Class of Molecules Known.

Biochemists call the unique sequence of amino acids in a protein the **primary structure** of that protein. The sequence of amino acids in Figure 3.8, for example, defines that polypeptide's primary structure. With 21 types of amino acids available and size ranging from two amino acid residues to tens of thousands, the number of primary structures that are possible is practically limitless. There may, in fact, be 21^n different polypeptides of length n. For a polypeptide that is just 10 amino acids long, 21^{10} primary sequences are possible. This is over 16,500 billion.

Recall that the R-groups present on each amino acid affect its solubility and chemical reactivity. Based on this observation, it is reasonable to predict that the R-groups present in a polypeptide will affect the polypeptide's properties and function. In some cases, even a single change in the sequence of amino acids can cause radical changes in the way the macromolecule as a whole behaves. As an example, consider the hemoglobin protein of humans. In some individuals, hemoglobin has a valine instead of a glutamate at the amino acid numbered 6 in a strand of 146 amino acids (**Figure 3.11a**). Valine's side chain is very different from the R-group in glutamate. The change results in a protein that tends to crystallize instead of staying in solution when oxygen concentrations in the blood are low. When hemoglobin crystallizes, the red blood cells that carry the protein adopt a sickled shape (**Figure 3.11b**) and can no longer efficiently pass through the blood vessels called capillaries. In people whose hemoglobin contains this single amino acid change, cells downstream of blocked capillaries become starved for oxygen. A debilitating illness called sickle-cell disease results. A protein's primary structure is fundamental to its function and to determining higher levels of protein structure: secondary, tertiary, and quaternary.

Secondary Structure

Even though variation in the amino acid sequence of a protein is virtually limitless, it is only the tip of the iceberg in terms of generating structural diversity. The next level of organization in proteins is known as **secondary structure**, which is created in part by hydrogen bonding. Secondary structures are distinctively shaped sections of proteins that are stabilized largely by hydrogen bonding that occurs between the carboxyl oxygen of one amino acid residue and the hydrogen on the amino group of another. The oxygen atom in the carboxyl group has a partial negative charge due to its high electronegativity, while the hydrogen atom in the amino group has a partial positive charge because it is bonded to nitrogen, which has high electronegativity (**Figure 3.12a**).

Note that secondary structure is not created by interactions among side chains, but instead by interactions between atoms that are part of a protein's peptide-bonded backbone. This is a key point, because hydrogen bonding between sections of the backbone is possible only when different parts of the same polypeptide bend in a way that puts carboxyl and amino groups close together. The bending that aligns parts of the backbone and allows these bonds to form occurs in several distinct ways. **Figure 3.12b** shows two of the most important configurations that allow formation of hydrogen bonds: (1) an

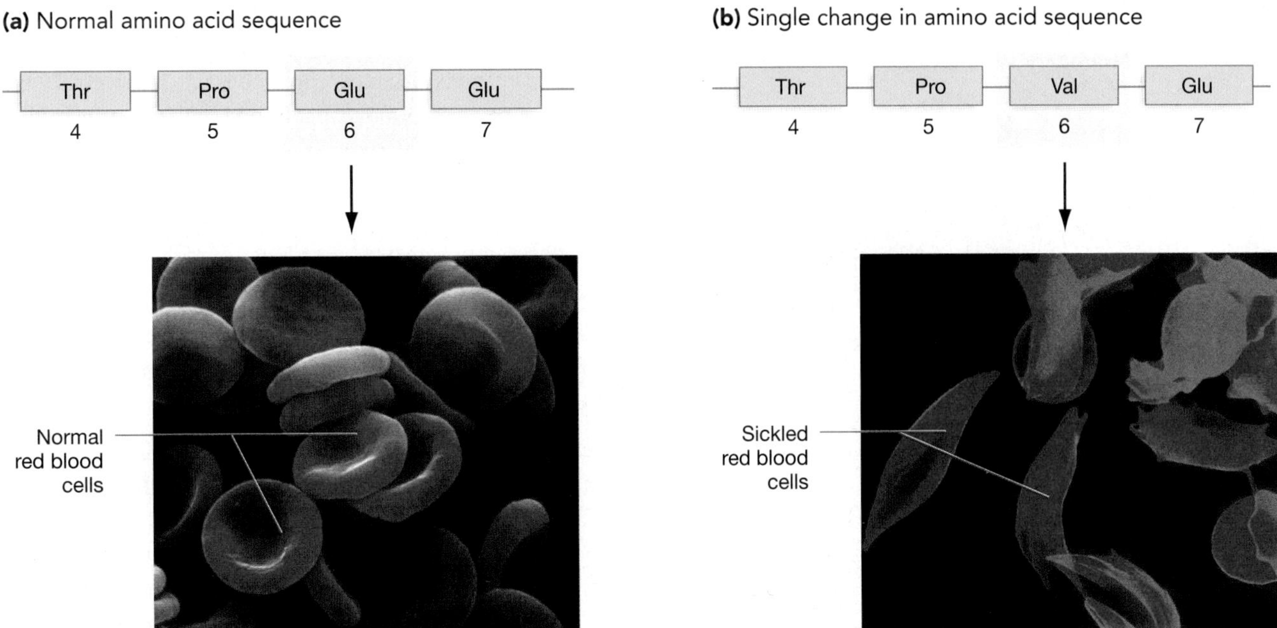

FIGURE 3.11 Changes in Primary Structure Affect Protein Function. Compare the primary structure of **(a)** normal hemoglobin with that of **(b)** hemoglobin molecules of people with sickle-cell disease. The single amino acid change causes red blood cells to change from their normal disc shape in (a) to a sickled shape in (b) when oxygen concentrations are low. Each red blood cell contains about 300 million hemoglobin molecules.

(a) Hydrogen bonds form between peptide chains.

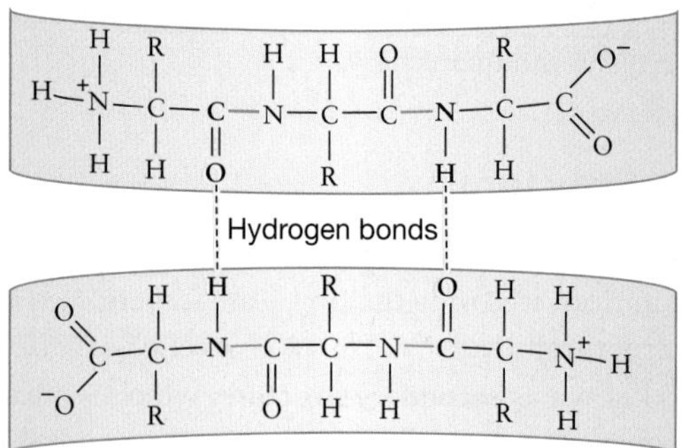

(b) Secondary structures of proteins result.

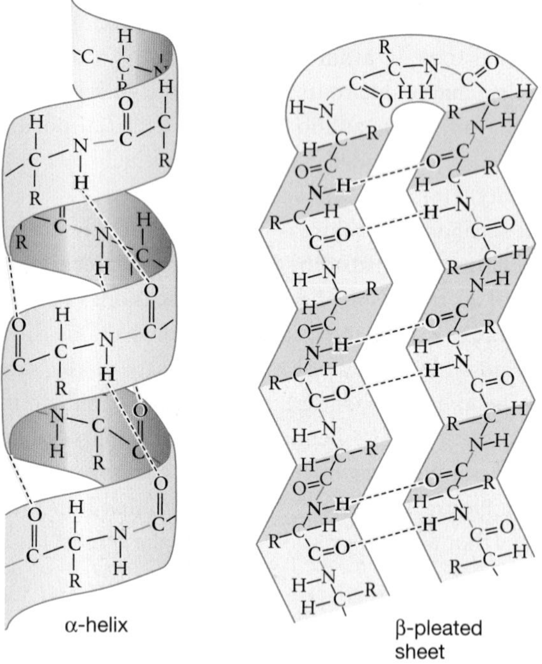

(c) Ribbon diagrams of secondary structure.

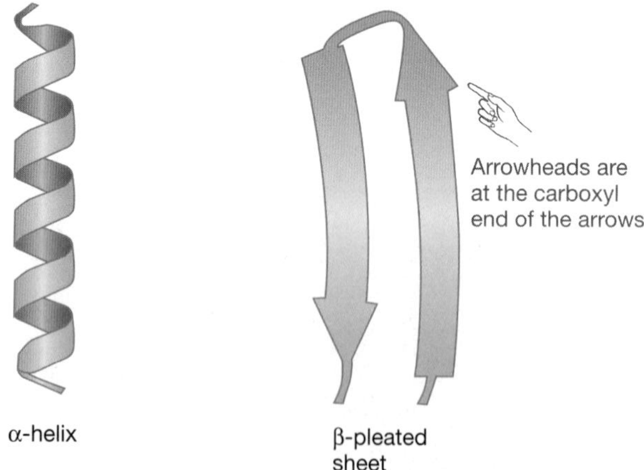

FIGURE 3.12 Secondary Structures of Proteins. The peptide-bonded backbone of a protein can coil or fold in on itself when hydrogen bonds form between amino groups and carboxyl groups.

EXERCISE β-pleated sheets often involve three or more strands. Add a third strand to the β-pleated sheets in part (c).

α-helix (alpha helix), in which the polypeptide's backbone is coiled, and (2) a **β-pleated sheet** (beta-pleated sheet), in which segments of a peptide chain bend 180° and then fold in the same plane. When biologists use illustrations called ribbon diagrams to represent the shape of a protein, α-helice are shown as coils while β-pleated sheets are shown by groups of arrows in a plane (**Figure 3.12c**).

In many cases, secondary structure consists exclusively of α-helices and β-pleated sheets. Which one forms, if either, depends on the molecule's primary structure—specifically, the geometry of the amino acids in the sequence. Methionine and glutamic acid, for example, are much more likely to be involved in α-helices than in β-pleated sheets. The opposite is true for valine and isoleucine. Proline, in contrast, is unlikely to be involved in either type of secondary structure. This is a key point: Which secondary structures form depends on the protein's primary structure.

Although each of the hydrogen bonds in an α-helix or a β-pleated sheet is very weak relative to a covalent bond, the large number of hydrogen bonds in these structures makes them highly stable. As a result, they increase the stability of the molecule as a whole and help define its shape. For overall shape and stability, though, a protein's tertiary structure is even more important.

Tertiary Structure

Alpha helices and β-pleated sheets form due to interactions between components of a protein's peptide-bonded backbone. In contrast, most of the overall shape, or **tertiary structure**, of a polypeptide results from interactions between R-groups or between R-groups and the peptide backbone. As **Figure 3.13a** shows, side chains can be involved in a wide variety of bonds and interactions. Because each contact between R-groups causes the peptide-bonded backbone to bend and fold, each contributes to the distinctive three-dimensional shape of a polypeptide.

Four types of interactions that involve side chains are particularly important:

1. Hydrogen bonds form in several ways: between hydrogen atoms and the carboxyl group in the peptide-bonded backbone, and between hydrogen and atoms with partial negative charges in side chains.
2. In an aqueous solution, water molecules interact with hydrophilic side chains and force hydrophobic side chains to coalesce. Once the hydrophobic side chains are close to each other, they are stabilized by electrical attractions known as **van der Waals interactions**. These weak attractions occur because the constant motion of electrons gives molecules a tiny asymmetry in charge that changes with time. If molecules get extremely close to each other, the minute partial charge on one molecule induces an opposite

(a) Interactions that determine the tertiary structure of proteins

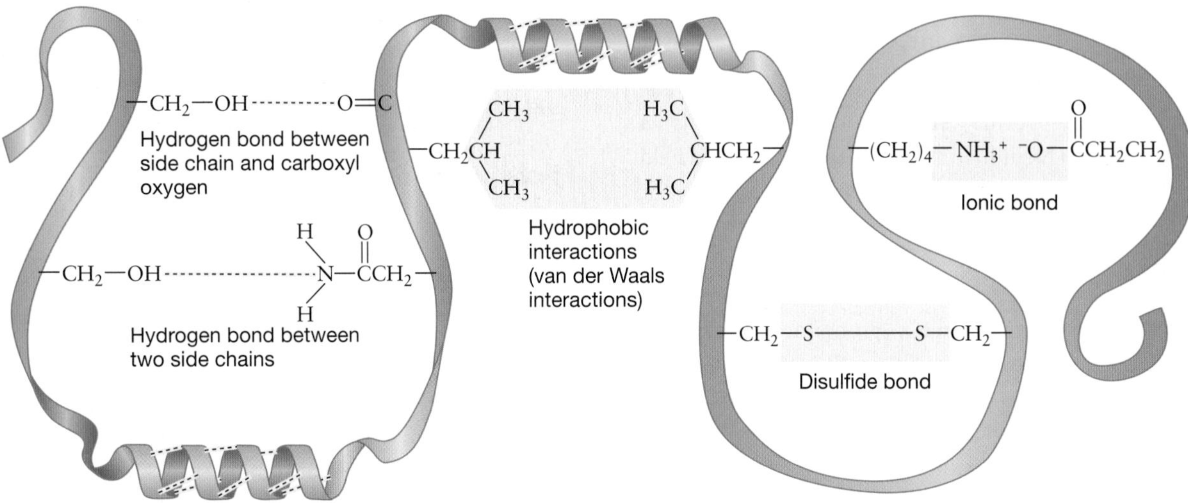

(b) Tertiary structures are diverse.

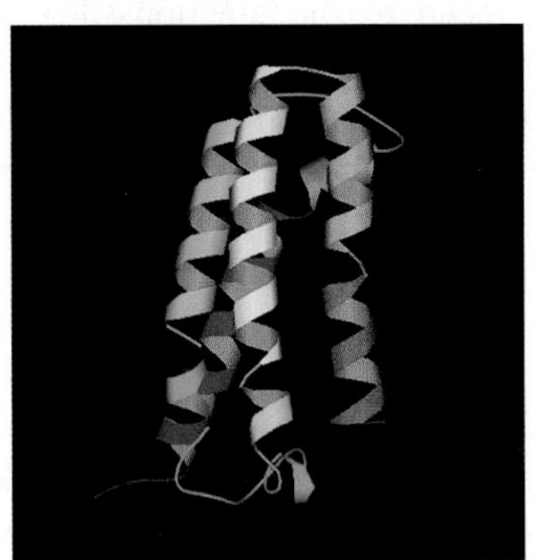

A tertiary structure composed mostly of α-helices

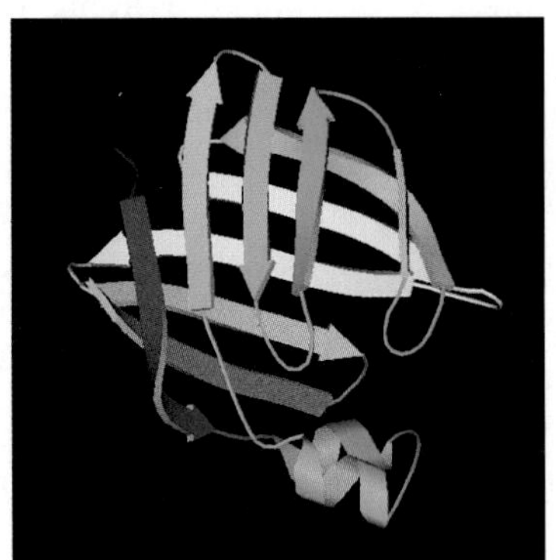

A tertiary structure composed mostly of β-pleated sheets

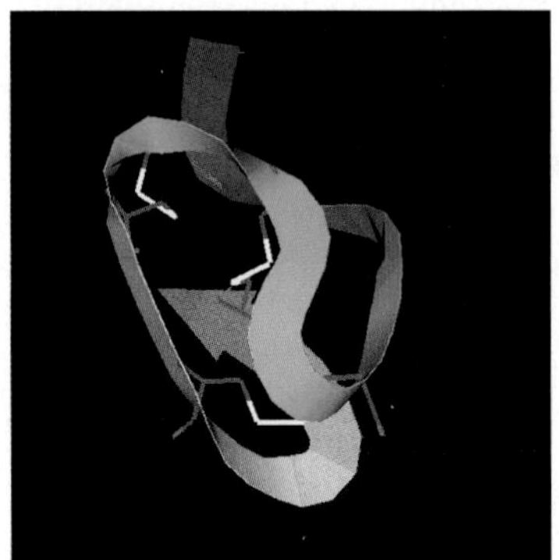

A tertiary structure rich in disulfide bonds

FIGURE 3.13 Tertiary Structure of Proteins Results from Interactions Involving R-Groups (a) Each protein has a unique overall shape called its tertiary structure. Tertiary structure is created by bonds and other interactions that cause proteins to fold in a precise way. **(b)** The tertiary structure of these proteins includes interactions between α-helices and β-pleated sheets. The polypeptide chains are color-coded so that you can follow the chain from the amino-end (red) to the carboxy-end (dark blue).

EXERCISE In part (b): Label the four a-helices in the left-hand protein, the α-helix and one of the β-pleated sheets in the center protein, and the yellow disulfide bonds in the right-hand protein.

partial charge in the nearby molecule and causes an attraction. Although the attraction is very weak relative to covalent bonds or even hydrogen bonds, a large number of van der Waals interactions can occur in a polypeptide when many hydrophobic residues congregate. The result is a significant increase in stability. It is important to recognize how important this is. Hydrophobic interactions are responsible for folding that results in the globular shape of many proteins.

3. Covalent bonds can form between sulfur atoms when a reaction occurs between sulfur-containing R-groups, such as those found in cysteine. These **disulfide** ("two-sulfur") **bonds** are frequently referred to as bridges because they create strong links between distinct regions of the same polypeptide.

4. Ionic bonds form between groups that have full and opposing charges, such as the ionized amino and carboxyl functional groups highlighted on the right in Figure 3.13a.

It's also important to note that the overall shape of many proteins depends in part on the presence of secondary structures like α-helices and β-pleated sheets. Tertiary structure depends on primary structure and secondary structure.

With so many interactions possible between side chains and peptide-bonded backbones, it's not surprising that polypeptides vary in shape from rod-like filaments to ball-like masses. (See **Figure 3.13b,** and look again at Figure 3.10.)

Quaternary Structure

The first three levels of protein structure involve individual polypeptides. But many proteins contain several distinct polypeptides that interact to form a single structure. The combination of polypeptide subunits gives proteins a **quaternary structure**. The individual polypeptides may be held together by bonds or other interactions among R-groups or sections of their peptide backbones.

In the simplest case, a protein with quaternary structure can consist of just two subunits that are identical. The Cro protein found in a virus called bacteriophage λ (pronounced *LAMB-da*) is an example (**Figure 3.14a**). Proteins with two polypeptide subunits are called dimers ("two-parts"). More than two polypeptides can be linked into a single protein, however, and the polypeptides involved may be distinct in primary, secondary, and tertiary structure. Hemoglobin, for example, is a tetramer ("four-parts"). As **Figure 3.14b** shows, the completed protein consists of two copies of each of two different polypeptides. Proteins that consist of a single polypeptide lack quaternary structure.

Table 3.3 summarizes the four levels of protein structure, using hemoglobin as an example. The key thing to note is that protein structure is hierarchical. Quaternary structure is based on tertiary structure, which is based in part on secondary structure. All three of the higher-level structures are based on primary structure. You should be able to describe and/or label elements of the primary, secondary, tertiary, and quaternary structure of the proteins in Figure 3.14b. (Check your work by using the information about hemoglobin given in Table 3.3.)

The summary table and preceding discussion have two important messages. First, the combination of primary, secondary, tertiary, and quaternary levels of structure is responsible for the fantastic diversity of sizes and shapes observed in proteins. Second, most elements of protein structure are based on folding of polypeptide chains. Does this folding occur spontaneously? What happens if normal folding is disrupted?

Folding and Function

If you were able to synthesize one of the polypeptides in hemoglobin from individual amino acids, and if you placed the resulting chain in water, it would spontaneously fold into the shape of the tertiary structure shown in Table 3.3. This result seems counterintuitive. Because an unfolded protein has many more ways to move about, it has much higher entropy than the folded version. Folding is spontaneous in some cases, however, because the bonds and van der Waals interactions that occur make the folded molecule more stable energetically than the unfolded molecule. Thus, folding may release enough free energy to be exergonic and occur spontaneously.

Folding is also crucial to the function of a completed protein. This point was hammered home in a set of classic experiments by Christian Anfinson and colleagues during the 1950s. Anfinson studied a protein called ribonuclease that is found in many organisms. Ribonuclease is an enzyme that breaks ribonucleic acid polymers apart. Anfinson found that the ribonuclease could be unfolded, or **denatured**, by treating it with compounds that break hydrogen bonds and disulfide bonds (**Figure 3.15**). The denatured ribonuclease was unable to function normally—it could no longer break apart nucleic acids. This is not surprising, given that the function of a protein depends on its structure.

When Anfinson removed the denaturing agents, however, the molecule refolded and began to function normally again. These

(a) Cro protein, a dimer

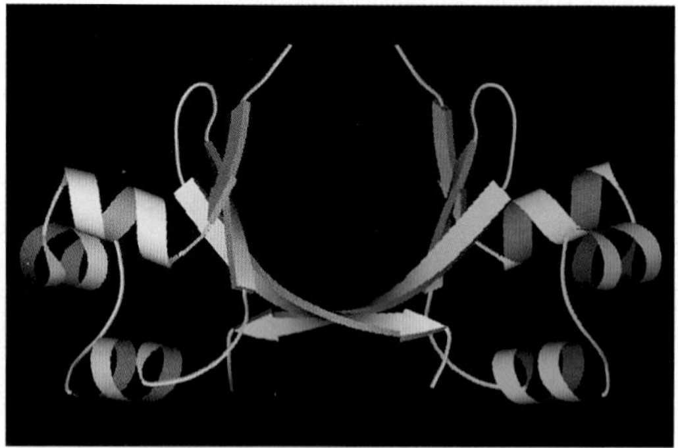

(b) Hemoglobin, a tetramer

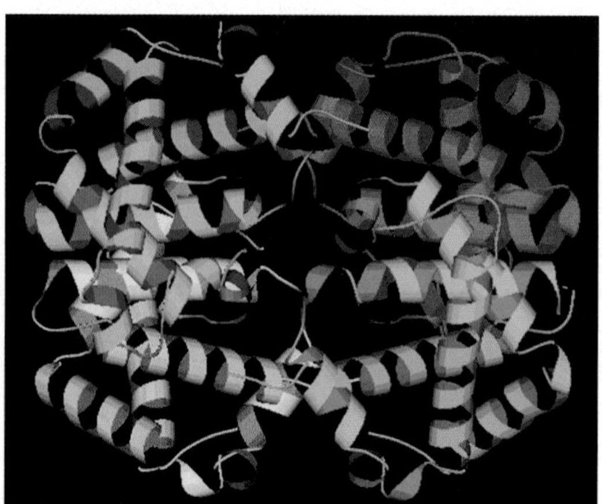

FIGURE 3.14 Quaternary Structures of Proteins Are Created by Multiple Polypeptides. These are ribbon diagrams, which represent the primary sequence as a ribbon. **(a)** The Cro protein is a dimer—it consists of two polypeptides. The polypeptides are identical in this case but are colored yellow and green here. **(b)** Hemoglobin is a tetramer—it consists of four polypeptides. Hemoglobin is composed of two copies of two polypeptides. The yellow and purple polypeptides are identical; so are the blue and green polypeptides.

EXERCISE Label the β-pleated sheet that holds the Cro dimer together.

SUMMARY TABLE **3.3 Protein Structure**

Level	Description	Stabilized by	Example: Hemoglobin
Primary	The sequence of amino acids in a polypeptide	Peptide bonds	Gly – Ser – Asp – Cys
Secondary	Formation of α-helices and β-pleated sheets in a polypeptide	Hydrogen bonding between groups along the peptide-bonded backbone; thus, depends on primary structure.	
Tertiary	Overall three-dimensional shape of a polypeptide (includes contribution from secondary structures)	Bonds and other interactions between R-groups, or between R-groups and the peptide-bonded backbone; thus, depends on primary structure.	One of hemoglobin's subunits
Quaternary	Shape produced by combinations of polypeptides (thus, combinations of tertiary structures)	Bonds and other interactions between R-groups, and between peptide backbones of different polypeptides; thus, depends on primary structure.	Hemoglobin, which consists of four polypeptide subunits

EXERCISE Label the secondary structures within the tertiary structure, and the tertiary structures within the quaternary structure.

experiments confirmed that ribonuclease folds spontaneously and that folding is essential for normal function.

More recent work has shown that in cells, folding is often facilitated by specific proteins called **molecular chaperones**. Many molecular chaperones belong to a family of molecules called the heat-shock proteins. These compounds are produced in large quantities after cells experience high temperatures or other treatments that make proteins lose their tertiary structure. Heat-shock proteins speed the refolding of other proteins into their normal shape after denaturation has occurred.

To summarize, a protein's function depends on its shape. In most cases, the final shape of a protein depends on folding. To drive this point home, **Box 3.1** investigates the way a change in the shape of the prion (pronounced *PREE-on*) protein of mammals converts normal molecules into abnormally folded proteins. Misfolded prions cause the brains of humans and other mammals to disintegrate.

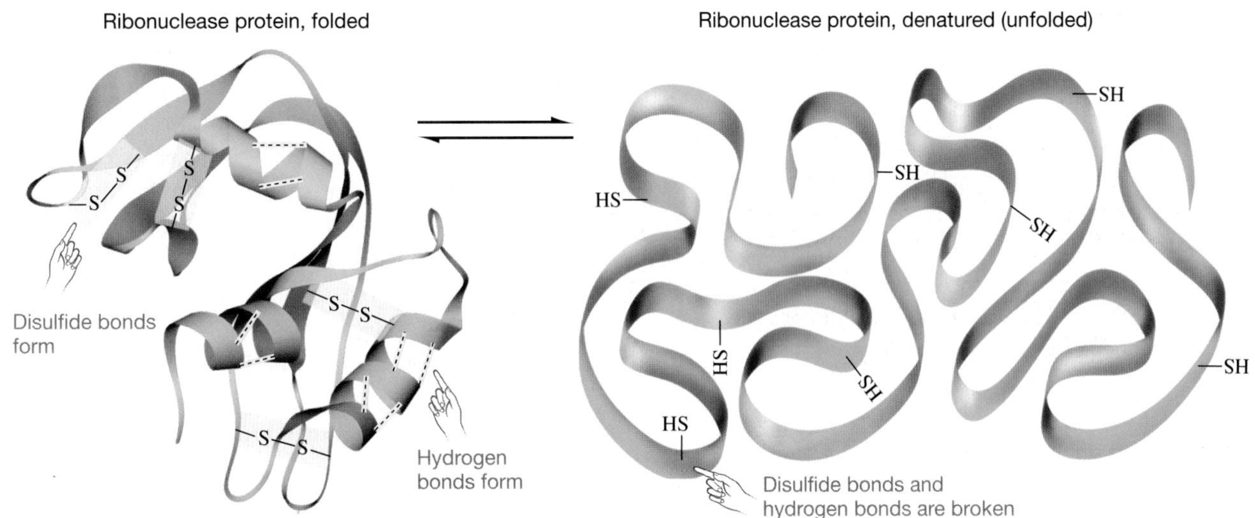

FIGURE 3.15 Proteins Fold into Their Normal, Active Shape. (a) The tertiary structure of ribonuclease is defined primarily by four disulfide bonds, shown in yellow. The protein's primary structure is represented by the green ribbon. **(b)** When the disulfide bonds and various non-covalent bonds are broken, the protein denatures (unfolds).

BOX 3.1 Prions

Over the past several decades, evidence has accumulated that certain proteins can act as infectious, disease-causing agents. The proteins involved are called **prions**, or proteinaceous infectious particles. Where do these molecules come from, and how do they work?

According to the prion hypothesis developed by Stanley Prusiner, infectious proteins are improperly folded forms of normal proteins that are present in healthy individuals. The infectious and normal forms do not necessarily differ in amino acid sequence, however. Instead, their shapes are radically different. Further, the infectious form of a protein can induce other normal protein molecules to change their shape to the altered form.

Figure 3.16 illustrates the shape differences observed in the normal and infectious forms of the first prion ever described. The molecule in Figure 3.16a is called the prion protein (PrP) and is a normal component of mammalian cells. Mutant versions of this protein, like the one in Figure 3.16b, are found in a wide variety of species and cause a family of diseases known as the spongiform encephalopathies—literally, "sponge-brain-illnesses." Hamsters, cows, goats, and humans afflicted with these diseases undergo massive degeneration of the brain. Cattle suffer from "mad cow disease"; sheep and goats acquire scrapie (so called because the animals itch so badly that they scratch off their wool or hair); humans develop kuru or Creutzfeldt-Jakob disease. Although some spongiform encephalopathies can be inherited, in many cases the disease is transmitted when individuals eat tissues containing the infectious form of PrP. All of the prion illnesses are fatal.

(a) Normal prion protein

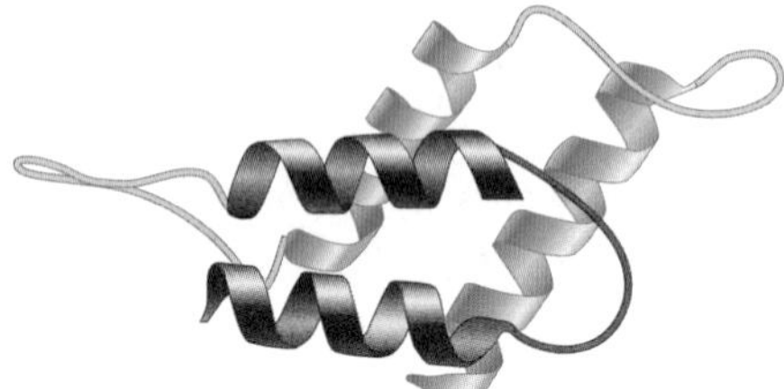

(b) Misfolded prion protein

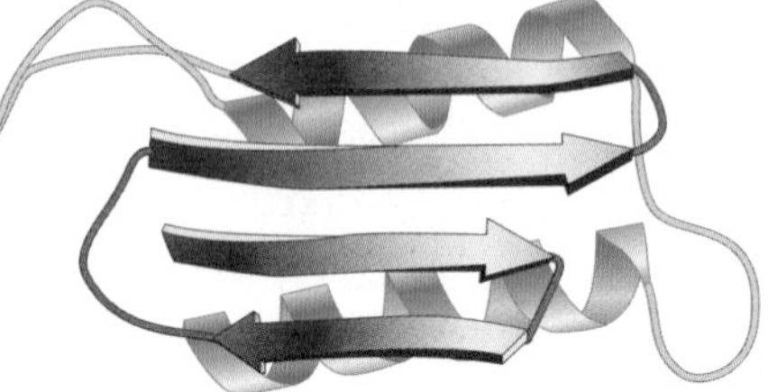

FIGURE 3.16 Prions Are Improperly Folded Proteins. Ribbon model of **(a)** a normal prion protein and **(b)** the misfolded form that causes mad cow disease in cattle. Secondary structure is represented by coils α-helices and arrows (β-pleated sheets).

Given how proteins are put together and how they are folded, let's take a detailed look at how they function as catalysts. Diversity in structure and chemical reactivity explain why proteins are easily the most diverse catalysts known.

Check Your Understanding

If you understand that...

- Proteins have up to four levels of structure.
- Primary structure consists of the sequence of amino acids.
- Secondary structure results from interactions between atoms in the peptide-bonded backbone of the same polypeptide, and yields structures such as α-helices and β-pleated sheets.
- Tertiary structure is a consequence of bonds or other interactions between R-groups or between R-groups and the peptide-bonded backbone of a polypeptide, and produces distinctive folds.
- Quaternary structure occurs when multiple polypeptides interact to form a single protein.

You should be able to ...

1) Draw a protein that you make up.
2) Describe elements of its primary through quaternary structure.

3.5 Enzymes: An Introduction to Catalysis

Of all the functions that proteins perform in cells, catalysis may be the most important. To understand why, recall from Chapter 2 that changes in potential energy and entropy determine whether a reaction is spontaneous. Spontaneous chemical reactions are not necessarily fast, however. The reaction that makes iron rust is exothermic and spontaneous, but it occurs slowly. Most of the chemical reactions in cells don't occur fast enough to support life unless a catalyst is present.

To appreciate how enzymes work, it is essential to get a deeper understanding of the factors that limit the speed at which chemical reactions occur. Reaction rates depend on how the chemical bonds involved are broken and re-formed. Recall from Chapter 2 that reactions take place when reactants (1) collide in a precise orientation and (2) have enough kinetic energy to overcome repulsion between electrons that come into contact as a bond forms. Part of the reason enzymes are such effective catalysts is that they bring reactant molecules—called **substrates**—together in a precise orientation, so that the electrons involved in the reaction can interact. But enzymes can also affect the amount of kinetic energy reactants must have for a reaction to proceed. To understand why, it's critical to realize that a collision between reactants creates a combination of old and new bonds called a **transition state**. The amount of free

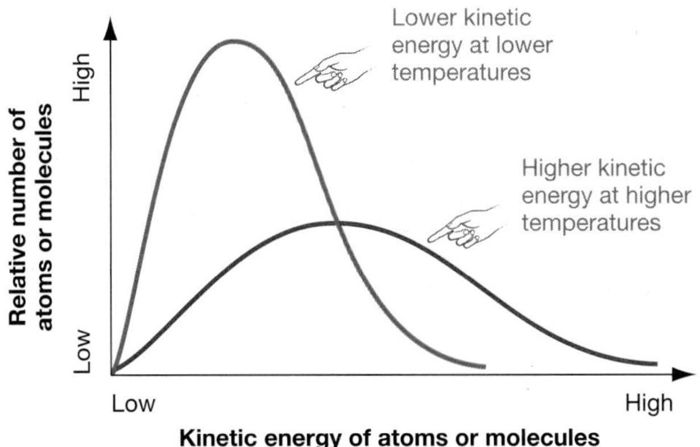

FIGURE 3.17 Atoms and Molecules Have More Kinetic Energy at High Temperatures than at Low Temperatures.

EXERCISE Add a vertical line two-thirds of the way along the horizontal axis. Label this line "Activation energy for reaction." At low and high temperatures, which molecules have enough kinetic energy for this reaction to proceed?

energy required to reach the transition state is called the **activation energy** of the reaction. Reactions happen when reactants have enough kinetic energy to reach the transition state. The kinetic energy of molecules, in turn, is a function of their temperature (**Figure 3.17**). This is why chemical reactions tend to proceed faster at higher temperatures, as noted in Chapter 2.

Figure 3.18 graphs the changes in free energy that take place during the course of a chemical reaction. In this graph, ΔG indicates the overall change in free energy in the reaction—that is, the energy of the products minus the energy of the reactants. In this case the products have lower potential energy than the reactants, meaning that the reaction is exothermic. But because the activation energy for this reaction, symbolized by E_a is high, the reaction would proceed slowly. The more unstable the transition state, the higher the activation energy and the less likely a reaction is to proceed quickly.

Reaction rates, then, depend on both the kinetic energy of the reactants and the activation energy of the particular reaction—meaning the free energy of the transition state. If the kinetic energy of the participating molecules is high, then molecular collisions are likely to result in completed reactions. But if the activation energy of a particular reaction is also high, then collisions are less likely to result in completed reactions.

In many cases, the electrons in the transition-state molecule can be stabilized when they interact with another ion, atom, or molecule. When this occurs, the activation energy required for the reaction drops and the reaction rate increases. A substance that lowers the activation energy of a reaction and increases the rate of the reaction is called a **catalyst**. It's important to note that a catalyst is not consumed in a chemical reaction, even though it participates in the reaction. The composition of a catalyst is exactly the same after the reaction as it was before.

Figure 3.19 diagrams how catalysts lower the activation energy for a reaction by lowering the free energy of the transition state. Note that the presence of a catalyst does not affect the overall energy change, ΔG, or change the energy of the reactants or the products. A catalyst changes only the free energy of the transition state.

Most enzymes are quite specific in their activity—they catalyze just a single reaction by lowering the activation energy that is required—and many are astonishingly efficient. Most of the important reactions in biology do not occur at all, or else

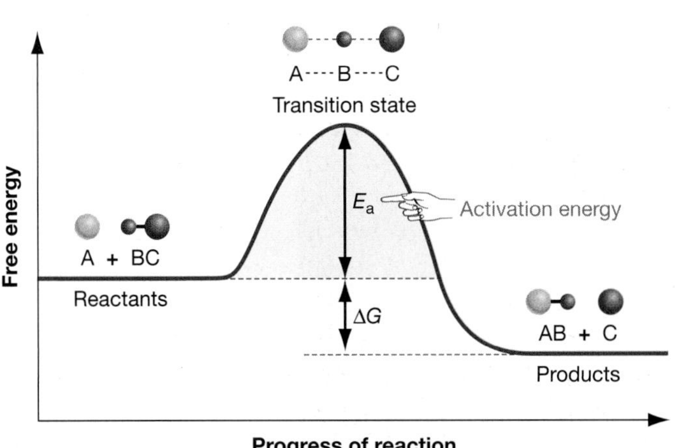

FIGURE 3.18 Changes in Free Energy during a Chemical Reaction. The changes in free energy that occur over the course of a hypothetical reaction between an atom A and a molecule containing atoms B and C. The overall reaction would be written as $A + BC \rightarrow AB + C$. E_a is the activation energy of the reaction, and ΔG is the overall change in free energy.

EXERCISE This graph illustrates an exergonic reaction. Draw the same type of graph for an endergonic reaction.

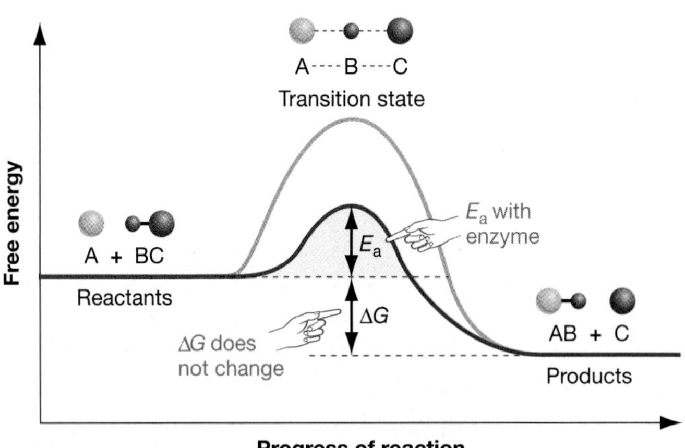

FIGURE 3.19 A Catalyst Changes the Activation Energy of a Reaction. The energy profile for the same reaction diagrammed in Figure 3.18, but with a catalyst present. Even though the energy barrier to the reaction, E_a is much lower, ΔG does not change.

QUESTION Can a catalyst make a nonspontaneous reaction occur spontaneously?

EXERCISE Go back to Figure 3.17 and add another vertical line one-third of the way along the horizontal axis. Label this line "Activation energy for reaction with catalyst." Label the molecules that have enough energy to undergo a reaction with a catalyst but not without a catalyst.

proceed at imperceptible rates, without a catalyst. In contrast, a single molecule of the enzyme carbonic anhydrase can catalyze over 1,000,000 reactions *per second*. It's not unusual for enzymes to speed up reactions by a factor of a million; some enzymes make reactions go many *trillions* of times faster than they would without a catalyst.

To summarize, enzymes bring reactants together in precise orientations and stabilize transition states. Protein catalysts are important because they speed up the chemical reactions that are required for life. Now the question is, how do enzymes do what they do?

How Do Enzymes Work?

The initial hypothesis for how enzymes work was proposed by Emil Fischer in 1894. According to Fischer's lock-and-key model, enzymes are rigid structures analogous to a lock. The keys are substrates that fit into the lock and then react.

Several important ideas in this model have stood the test of time. For example, Fischer was correct in proposing that enzymes bring substrates together in a precise orientation that makes reactions more likely. His model also accurately explained why most enzymes can catalyze only one specific reaction. Enzyme specificity is a product of the geometry and chemical properties of the sites where substrates bind.

As researchers began to test and extend Fischer's model, the location where substrates bind and react became known as the enzyme's **active site**. The active site is where catalysis actually occurs. When techniques for solving the three-dimensional structure of enzymes became available, it turned out that enzymes tend to be very large relative to substrates and roughly globular, and that the active site is in a cleft or cavity within the globular shape. The enzyme hexokinase, which is at work in most cells of your body now, is a good example. (Many enzymes have names that end with *–ase*.) As the left-hand side of **Figure 3.20** shows, the active site in hexokinase is a small notch in an otherwise large, crescent-shaped enzyme.

As knowledge of enzyme action grew, however, Fischer's model was modified. Perhaps the most important change was based on the realization that enzymes are not rigid and static, but flexible and dynamic. In fact, many enzymes undergo a significant change in shape, or conformation, when reactant molecules bind to the active site. This conformational change, called an **induced fit**, is noticeable in the hexokinase molecule on the right of Figure 3.20. As hexokinase binds its substrate—the sugar glucose—the enzyme rocks forward over the active site.

In addition, recent research has clarified the nature of Fischer's key. When one or more substrate molecules enter the active site, they are held in place through hydrogen bonding or other electrical interactions with amino acids in the active site. Once the substrate is bound, one or more R-groups in the active site come into play. The degree of interaction between the substrate and enzyme increases and reaches a maximum when the transition state is formed. Thus, Fischer's key is actually the transition state.

Interactions with R-groups at the active site stabilize the transition state and thus lower the activation energy required for the reaction to proceed. At the atomic level, R-groups that line the active site may form short-lived covalent bonds that assist with the transfer of atoms or groups of atoms from one reactant to another. More commonly, the presence of acidic or basic R-groups allows the reactants to lose or gain a proton more readily.

Figure 3.21 provides a specific example of how substrates interact with an active site. The panels show a substrate fitting into the active site of the enzyme called ribonuclease (the same enzyme that denatured in Figure 3.15). Notice that once the substrate has bound, R-groups in the active site interact with the substrate to form the transition state and get the reaction under

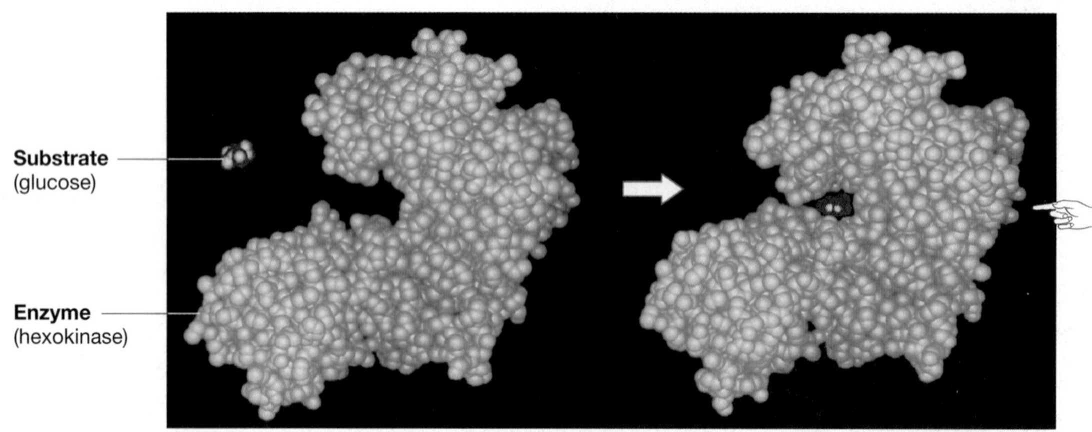

FIGURE 3.20 Reactant Molecules Bind to Specific Locations in an Enzyme. The reactant molecule, shown in red, fits into a precise location, called the active site, in the green enzyme. In the enzyme shown here and in many others, the binding event causes the protein to change its shape.

EXERCISE Label the active site in the enzyme on the left.

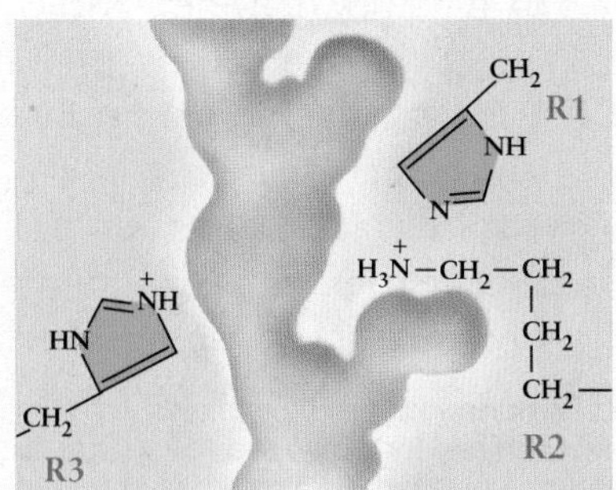

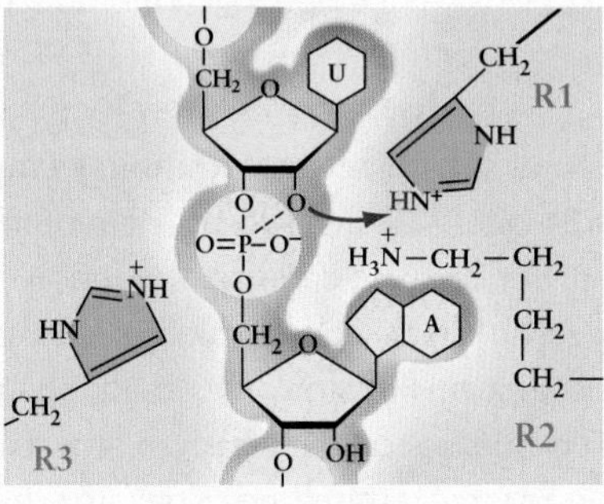

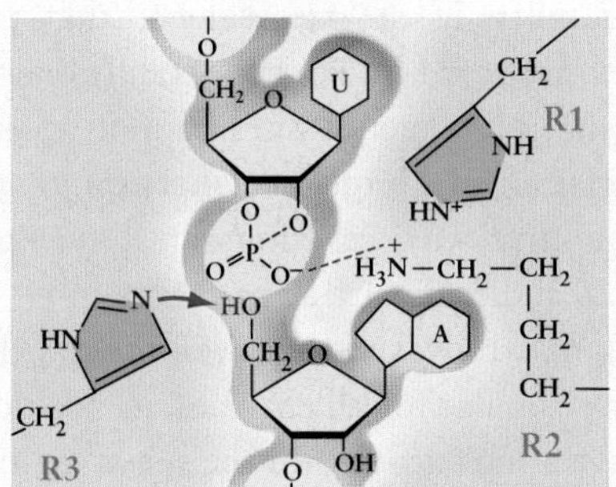

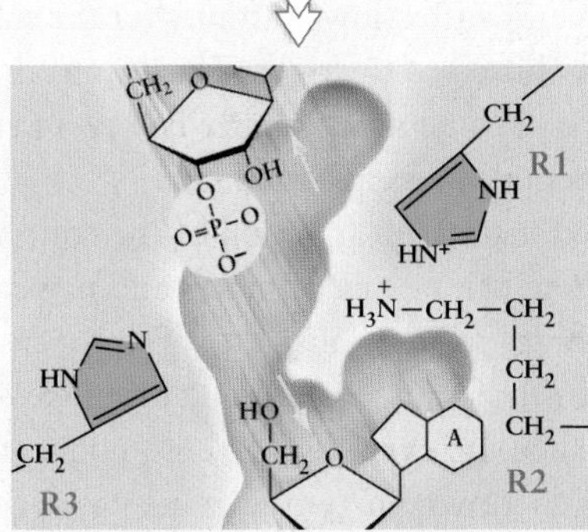

FIGURE 3.21 R-groups in an Enzyme's Active Site Stabilize the Transition State. Part of the reaction sequence that occurs when the enzyme ribonuclease catalyzes a reaction that cuts a substrate in two. Note how precisely the substrate fits into the active site and how a few R-groups in the enzyme are involved in key steps in the reaction.

way. ● You should be able to analyze this figure and identify two acid–base reactions that occur inside the active site and stabilize the transition state.

Whatever the mechanism, enzymes increase reaction rates by bringing substrates together and lowering the energy required to form the transition state. The products of the reaction have a much lower affinity for the active site than do either the reactants or the transition state, however. Therefore, they are released from the enzyme once they form.

Figure 3.22 summarizes these principles. Enzyme catalysis can be analyzed as a three-step process:

1. **Initiation.** Instead of reactants occasionally colliding in a random fashion, enzymes orient reactants precisely as they bind at specific locations within the active site.
2. **Transition state facilitation.** The act of binding induces the formation of the transition state. In some cases the transition state is stabilized by a change in the enzyme's shape. The interaction between the substrate and R-groups in the enzyme's active site lowers the activation energy required for the reaction. As a result, inside a catalyst's active site, more reactant molecules have sufficient kinetic energy to reach this lowered activation energy. Thus, the catalyzed reaction proceeds much more rapidly than the uncatalyzed reaction.
3. **Termination.** The reaction products have considerably less affinity for the active site than does the transition state. Thus binding ends, the enzyme returns to its original conformation, and the products are released.

Fischer's model inspired intensive and fruitful research into the mechanism of enzyme action.

● You should be able to complete the following sentences:

1. Enzymes speed reaction rates by ________________ and lowering activation energy.
2. Activation energies drop because enzymes destabilize bonds in the reactant, stabilize the ________________, make acid–base reactions more favorable, and/or change the reaction mechanism through a covalent bonding interaction.
3. Enzyme specificity is a function of the active site's shape and the chemical properties of the ________________ at the active site.
4. In enzymes, as in many molecules, function follows from ________________.

Do Enzymes Act Alone? Atoms or molecules that are not part of the enzyme's primary structure are often required for the enzyme to function normally. These enzyme **cofactors** can be either (1) metal ions such as Zn^{2+} (zinc) or Mg^{2+} (magnesium), or (2) small organic **molecules** called **coenzymes**. In

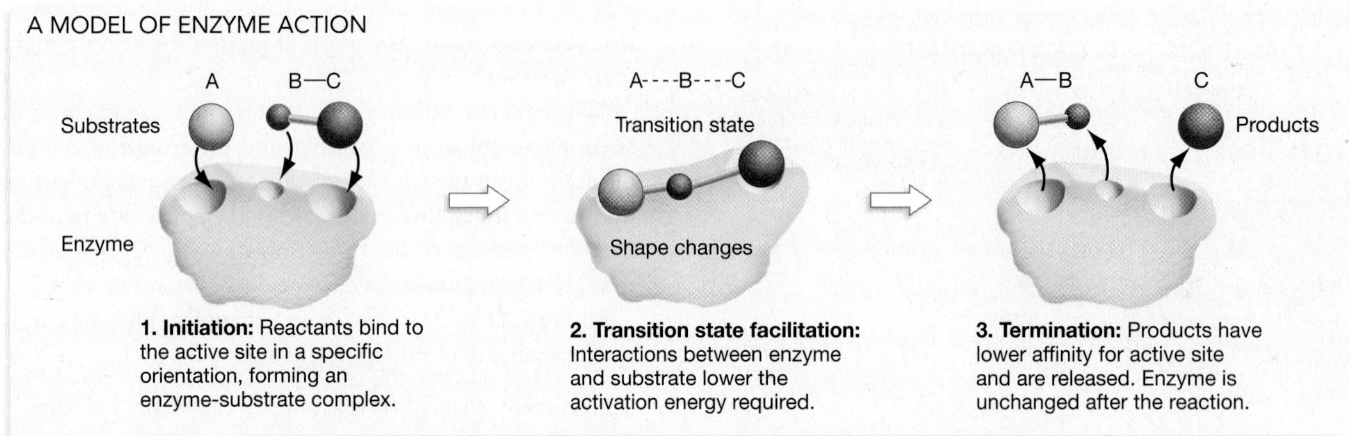

FIGURE 3.22 Enzyme Action Can Be Analyzed as a Three-Step Process.

many cases, the cofactor binds to the active site and is thought to play a key role in stabilizing the transition state during the reaction. The presence of the cofactor is therefore essential for catalysis. To drive this point home, consider that many of the vitamins in your diet are required for the production of enzyme cofactors. Vitamin deficiencies cause enzyme-cofactor deficiencies. Lack of cofactors, in turn, disrupts normal enzyme function and causes disease. For example, thiamine (vitamin B_1) is required for the production of an enzyme cofactor called thiamine pyrophosphate, which is required by three different enzymes. Lack of thiamine in the diet dramatically reduces the activity of these enzymes and causes an array of nervous system and heart disorders collectively known as beriberi.

In addition, most enzymes are regulated by molecules that are not part of the enzyme itself. An enzyme is active or inactive, depending on the presence or absence of molecules that change the protein's structure in some way. In some cases, catalysis is inhibited when a molecule that is similar in size and shape to the substrate(s) binds to the active site. This event is called **competitive inhibition,** because the molecule involved competes with the substrate(s) for access to the enzyme's active site (Figure 3.23a). In other cases, a regulatory molecule binds at a location other than the active site. This type of regulation is called **allosteric** ("different-structure") **regulation,** because the molecule involved does not affect the active site directly. Instead, the binding event changes the shape of the enzyme in a way that makes the active site accessible or inaccessible (**Figure 3.23b**). Later chapters will provide detailed examples of how regulatory molecules change the activity of specific enzymes.

MB Web Animation **at www.masteringbio.com**
Activation Energy and Enzymes

What Limits the Rate of Catalysis? For several decades after Fischer's model was published, most research on enzymes focused on rates of enzyme action, or what biologists call enzyme kinetics. Researchers observed that, when the amount of product produced per second—indicating the speed of the reaction—is plotted as a function of substrate concentration, a graph like that shown in **Figure 3.24** results. When substrate concentrations are low, the speed of an enzyme-catalyzed reaction increases in a linear fashion. As substrate concentrations increase, however, the increase in speed begins to slow. Eventually the reaction rate plateaus at a maximum speed. This pattern is in striking contrast to the situation for uncatalyzed reactions, in which reaction speed tends to show a continuing linear increase with substrate concentration. The "saturation kinetics" of enzyme-catalyzed reactions were taken as strong evidence that the enzyme-substrate complex proposed by Fischer actually exists. The idea was that at some point, active sites cannot accept substrates any faster, no matter how large the concentration of substrates gets. Stated another way, reaction rates level off because all available enzyme molecules are being used.

It's important to recognize, however, that even though the general shape of the curve is similar for all enzymes, its position varies. In enzymes that have a high affinity for the substrate, the curve is shifted to the left. This means that substrate concentrations do not have to be very high for these enzymes to function at maximum speed. In enzymes with a lower affinity for the substrate, though, the curve is shifted to the right. This means that substrate concentrations have to be much higher before an enzyme is able to catalyze reactions at its maximal rate.

Thus, an enzyme's speed is partly a function of the concentration of its substrates (there can be one or more) and partly a function of its intrinsic affinity for the substrates. The processing of ethanol—the active ingredient in alcoholic beverages—in the human liver furnishes an example of these concepts. Ethanol is a poison broken down in two enzyme-catalyzed steps. First ethanol is converted to acetaldehyde (CH_3CHO). Then acetaldehyde is converted to acetic acid (CH_3COOH)—a harmless molecule that gives vinegar its bite. Humans have two versions of the enzyme that catalyzes the second reaction. One version of the enzyme works quickly and the other much more slowly. In some people, the fast-working version of the enzyme has a single amino acid change in its primary sequence that

(a) Competitive inhibition directly blocks the active site.

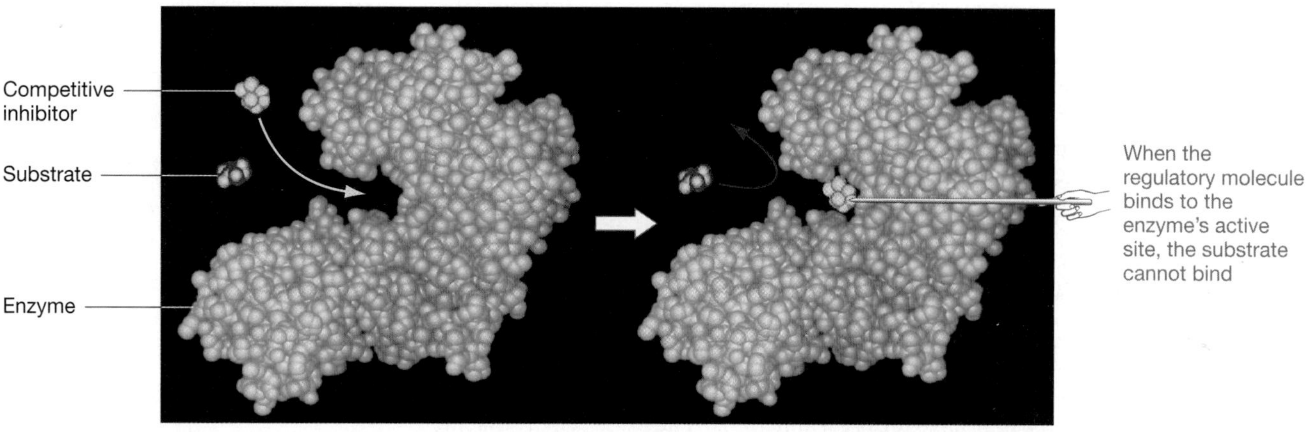

(b) Allosteric regulation occurs when a regulatory molecule binds somewhere other than the active site.

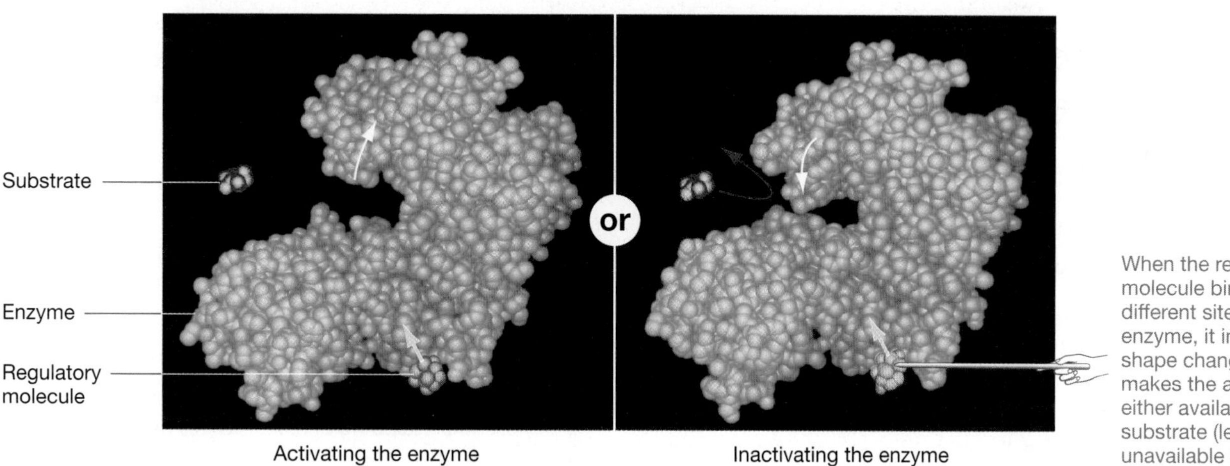

FIGURE 3.23 An Enzyme's Activity Is Precisely Regulated. Enzymes are turned on or off when specific molecules bind to them.

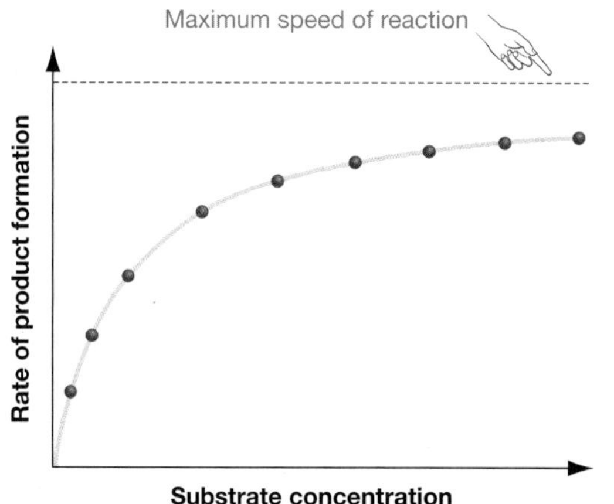

FIGURE 3.24 Kinetics of an Enzyme-Catalyzed Reaction. The general shape of this curve is characteristic of enzyme-catalyzed reactions.

● **EXERCISE** Label the part of the graph where (1) reaction is most sensitive to changes in substrate concentration, and (2) most or all of the active sites present are occupied.

affects the enzyme's active site. In these individuals, the fast-working version of the enzyme is completely inactivated, so only the slow-working version functions normally. Acetaldehyde must be at a high concentration before the slow-working version of the enzyme works at maximum speed. In these individuals, acetaldehyde builds up to high enough concentrations in the bloodstream that uncomfortable symptoms result, including a rapid heartbeat and pronounced flushing of the skin. People who lack the fast-acting version of the enzyme have virtually zero tolerance for alcoholic beverages. Imbibing even a small amount of ethanol makes them feel ill.

How Do Physical Conditions Affect Enzyme Function?

Given that an enzyme's structure is critical to its function, it's not surprising that an enzyme's activity is sensitive to conditions that alter its structure. In particular, the activity of an enzyme often changes drastically as a function of temperature and pH. Temperature affects the movement of the enzyme as well as the kinetic energy of the substrates; pH affects the makeup and charge of amino acid side chains with carboxyl or amino groups, and the active site's ability to participate in proton-transfer or electron-transfer reactions.

Do data support these assertions? **Figure 3.25a** shows how the activity of an enzyme changes as a function of temperature. Data for the enzyme glucose-6-phosphatase, which is helping to produce usable energy in your cells right now, are shown for two species of bacteria. Note that in both of the bacterial species illustrated, the enzyme has a distinct optimum—a temperature at which it functions best. One of the bacterial species lives inside your gut, where the temperature is about 40°C, while the other lives in hot springs, where temperatures can be close to 100°C. The temperature optimum for the enzyme that reflects these environments. The two types of bacteria have different versions of the enzyme that differ in primary structure. Natural selection—the process introduced in Chapter 1—has favored different structures that have different functions. The enzymes are adaptations that allow each species to thrive at different temperatures.

Figure 3.25b makes the same point for pH. The enzyme in this graph, called chitinase, protects bacterial cells by digesting a molecule found in the cell walls of fungi that use bacteria as food. The data come from a species of bacterium that lives in acidic pools and a bacterial species that lives in the soil under palm trees. The organism that thrives in an acidic environment has a version of the enzyme that performs best at low pH; the organism that lives near palms has a version of the enzyme that functions best near neutral pH. Each enzyme is sensitive to changes in pH, but each species' version of the enzyme has a structure that allows it to function best at the pH of its environment.

To summarize, the rate of an enzyme-catalyzed reaction depends not only on substrate concentration and the enzyme's intrinsic affinity for the substrate but also on temperature and pH. Temperature affects the movement of the substrates and enzyme; pH affects the enzyme's shape and reactivity.

(a) Enzymes from different organisms may function best at different temperatures.

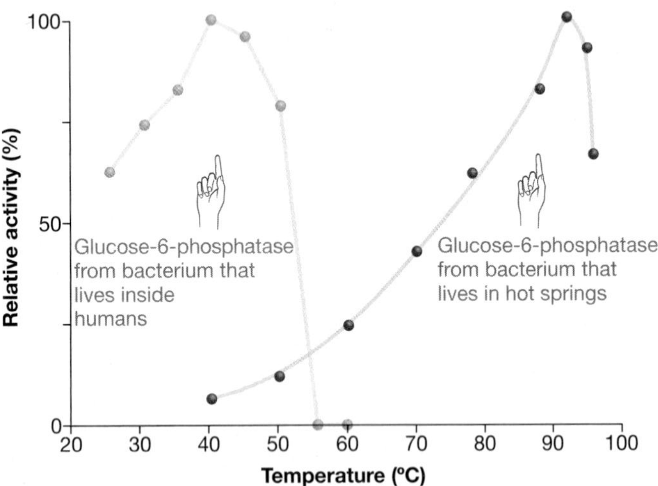

(b) Enzymes from different organisms may function best at different pHs.

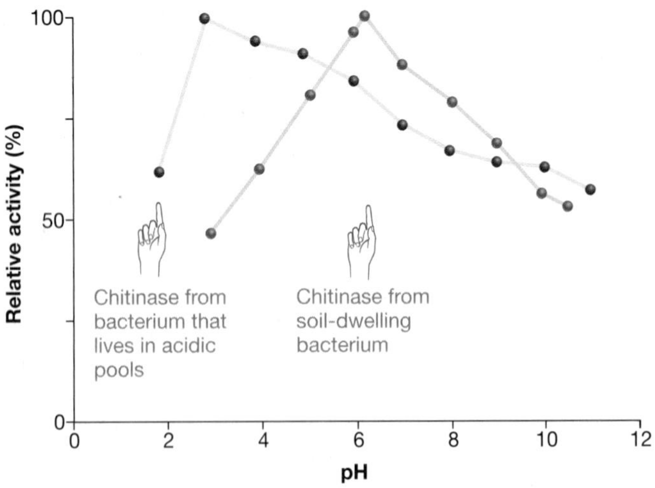

FIGURE 3.25 Enzymes Have an Optimal Temperature and pH. Enzymes are sensitive to changes in temperature and pH. Further, the structures of the enzymes found in a particular organism allow that organism to function well at the **(a)** temperature and **(b)** pH of its environment.

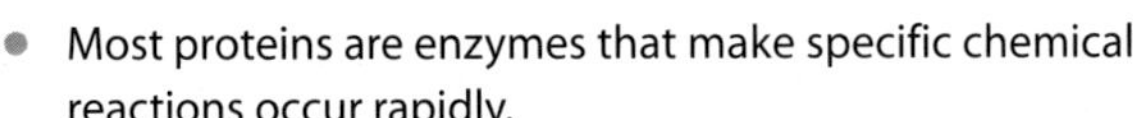

If you understand that...

- Most proteins are enzymes that make specific chemical reactions occur rapidly.
- Enzymes catalyze reactions by means of a three-step mechanism:

  Step 1: Binding of reactants in a precise orientation inside the active site.

  Step 2: Facilitation of the transition state, thus lowering the activation energy required for the reaction. This step often involves a shape change in the enzyme, resulting in an "induced fit" between active site and substrate. Enzyme cofactors are often required at this step.

  Step 3: Release of products, which do not bind tightly to the active site.
- The activity of enzymes is controlled through allosteric regulation or competitive inhibition. Different enzymes work at different rates, and enzymes are sensitive to changes in temperature and pH.

You should be able to...

1) Diagram the three steps in enzyme catalysis.

2) Add drawings and notes indicating how competitive inhibition and allosteric regulation affect these steps.

Was the First Living Entity a Protein?

The theory of chemical evolution holds that life began with a molecule that could make a copy of itself. This self-replicating entity increased in number and formed a population of individuals. The population then began to evolve by natural selection.

For such a molecule to make a copy of itself quickly, it must have been able to catalyze the reactions required for the copying process. This chapter's examination of protein structure and function raises a key question: Was the first self-replicating molecule on Earth a protein?

Several observations argue that the answer to this question is yes. Experimental studies have shown that amino acids were likely to be abundant in the prebiotic soup and that they could have polymerized to form small proteins. In addition, proteins are the most efficient catalysts known, and a self-replicating molecule had to act as a catalyst during the as-sembly and polymerization of its copy. These observations support the hypothesis that the self-replicating molecule was a polypeptide. Indeed, several laboratories that are currently working to create life have focused on synthesizing a self-replicating protein.

To date, however, attempts to simulate the origin of life with proteins have not been successful. Most origin-of-life researchers are increasingly skeptical about the hypothesis that life began with a protein. Their reasoning is that to make a copy of something, a mold or template is required. Proteins cannot furnish this information. Nucleic acids, in contrast, can. How they do so is the subject of Chapter 4.

Chapter Review

SUMMARY OF KEY CONCEPTS

Proteins are made of amino acids. Amino acids vary in structure and function because their side chains vary in composition.

Once monomers such as amino acids have formed, the next key event in chemical evolution is their polymerization into macromolecules. Researchers have observed polypeptide formation from the polymerization of amino acids on clay particles. In proteins, amino acids are joined by a peptide bond between the carboxyl group of one amino acid and the amino group of a second amino acid.

You should be able to draw a generalized amino acid and a peptide bond. You should also be able to explain why some R-groups on amino acids are soluble versus insoluble in water, and why some R-groups are reactive versus nonreactive.

MB Web Animation at www.masteringbio.com
Condensation and Hydrolysis Reactions

Proteins vary widely in structure. The structure of a protein can be analyzed at four levels that form a hierarchy. These are the amino acid sequence, substructures called α-helices and β-pleated sheets, interactions between amino acids that dictate a protein's overall shape, and combinations of individual proteins that make up larger, multiunit molecules.

A protein's primary structure, or sequence of amino acids, is responsible for most of its chemical properties. Interactions that take place between carboxyl and amino groups in the same peptide-bonded backbone create secondary structures, which are stabilized primarily by hydrogen bonding. Interactions between R-groups found in the same polypeptide and between R-groups and the peptide-bonded backbone allow the protein to fold into a characteristic overall shape. A complete protein frequently consists of several different polypeptides, bonded together.

You should be able to make a concept map (see BioSkills 6) that relates the four levels of protein structure to each other and to how proteins function as catalysts. You should be able to describe how variation in each of these four levels of protein structure can affect a protein's function.

Proteins vary widely in function. In cells, most proteins are enzymes that function as catalysts. Chemical reactions occur much faster when they are catalyzed by enzymes. During enzyme catalysis, the reactants bind to an enzyme's active site in a way that allows the reaction to proceed efficiently.

In organisms, proteins function in defense, movement, signaling, structural support, and transport of materials. In addition, they are far and away the most diverse and efficient class of catalysts known. A catalyst speeds the rate of that reaction by lowering the activation energy required, even though the catalyst itself is unchanged by the reaction. Enzymes are protein catalysts that lower activation energy by stabilizing the transition state of the reaction.

Enzyme-catalyzed reactions take place at a location called the active site of the enzyme. The active site in each type of enzyme has unique chemical properties and a distinctive size and shape. As a result, most enzymes are able to catalyze only one specific reaction. As a group, enzymes are able to catalyze many types of reactions because the chemical and physical structures of their active sites are so diverse. This diversity is due to the variety of amino acids and the four levels of protein structure.

Many enzymes function only with the help of cofactors. In addition, virtually all enzyme activity in cells is regulated by molecules that bind at the active site or at locations on the protein that induce a change in the size or shape of the active site. The rate at which enzymes work depends on substrate concentration, their affinity for the substrate, pH, and temperature.

You should be able to draw an energy diagram for a chemical reaction with and without a catalyst, and relate the change in activation energy to events at an enzyme's active site.

MB Web Animation at www.masteringbio.com
Activation Energy and Enzymes

QUESTIONS

Test Your Knowledge

1. What two functional groups are present on every amino acid?
   a. a carbonyl group and a carboxyl group
   b. an amino group and a carbonyl group
   c. an amino group and a hydroxyl group
   d. an amino group and a carboxyl group

2. Twenty-one different amino acids are found in the proteins of cells. What distinguishes these molecules?
   a. the location of their carboxyl group
   b. the location of their amino group
   c. the composition of their side chains, or R-groups
   d. their ability to form peptide bonds

3. What determines the primary structure of a polypeptide?
   a. its sequence of amino acids
   b. hydrogen bonds that form between carboxyl and amino groups on different residues
   c. hydrogen bonds and other interactions between side chains
   d. the number, identity, and arrangement of polypeptides that make up a protein

4. In a polypeptide, what is most responsible for the secondary structure called an α-helix?
   a. the sequence of amino acids
   b. hydrogen bonds that form between carboxyl and amino groups on different residues
   c. hydrogen bonds and other interactions between side chains
   d. the number, identity, and arrangement of polypeptides that make up a protein

5. What is a transition state?
   a. the complex formed as covalent bonds are being broken and reformed during a reaction
   b. the place where an allosteric regulatory molecule binds to an enzyme
   c. an interaction between reactants with high kinetic energy, due to high temperature
   d. the shape adopted by an enzyme that has an inhibitory molecule bound at its active site

6. By convention, biologists write the sequence of amino acids in a polypeptide in which direction?
   a. carboxy- to amino-terminus
   b. amino- to carboxy-terminus
   c. polar residues to nonpolar residues
   d. charged residues to uncharged residues

Test Your Knowledge answers: 1. d; 2. c; 3. a; 4. b; 5. a; 6. b

Test Your Understanding

Answers are available at www.masteringbio.com

1. Explain the lock-and-key model of enzyme activity. Be sure to comment on what role the active site of an enzyme plays.

2. Isoleucine, valine, leucine, phenylalanine, and methionine are amino acids with highly hydrophobic side chains. Suppose a section of a protein contains a long series of these hydrophobic residues. How would you expect this portion of the protein to behave when the molecule is in aqueous solution?

3. Compare and contrast competitive inhibition and allosteric regulation.

4. Does it take energy for polymerization reactions to proceed, or do they occur spontaneously? Why or why not?

5. A major theme in this chapter is that the structure of molecules correlates with their function. Use this theme to explain why proteins can perform so many different functions in organisms and why proteins as a group are such effective catalysts.

6. Explain why temperature and pH affect enzyme function.

Applying Concepts to New Situations

Answers are available at www.masteringbio.com

1. Researchers are searching for life on other planets. Is it reasonable to expect that protein-based life-forms could exist in extraterrestrial environments that are very hot or highly acidic? Why or why not?

2. Recently, researchers were able to measure movement that occurred in a single amino acid in an enzyme as reactions were taking place in its active site. The amino acid that moved was located in the active site, and the rate of movement correlated closely with the rate at which the reaction was taking place. Discuss the significance of these findings, using the information in Figures 3.20 and 3.22.

3. Researchers can analyze the atomic structure of enzymes during catalysis. In one recent study, investigators found that the transition state included the formation of a free radical, and that a coenzyme bound to the active site donated an electron to help stabilize the free radical. How would the reaction rate and the stability of the transition state change if the coenzyme were not available?

4. In a recent experiment, researchers found that if they took a mixture containing two optical isomers of a monomer and put a thin layer of the mixture on water, a crystal-like structure formed. Only one of the optical isomers—not both—was included in the crystal. The researchers suggested that a similar phenomenon might occur if both optical isomers of a monomer formed a thin coating on a clay mineral particle. How might this observation relate to the fact that proteins in cells contain only one isomer of each amino acid?

www.masteringbio.com is also your resource for • Answers to text, table, and figure caption questions and exercises • Answers to *Check Your Understanding* boxes • Online study guides and quizzes • Additional study tools including the *E-Book for Biological Science* 3rd ed., textbook art, animations, and videos.

Nucleic Acids and the RNA World

4

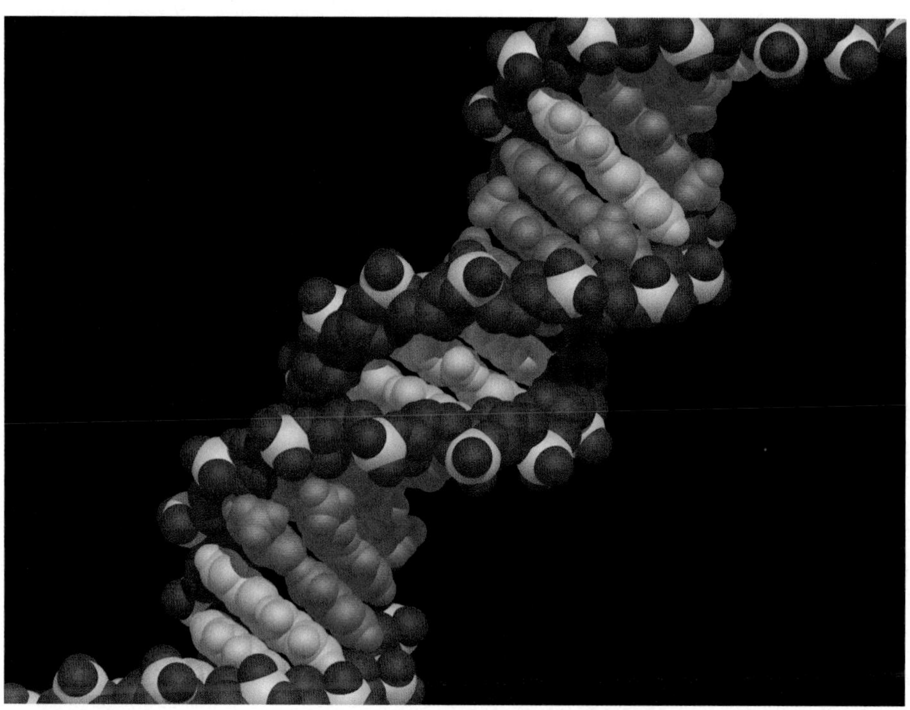

A space-filling model of deoxyribonucleic acid, or DNA. In today's cells, DNA functions as an information repository. But the nucleic acid called ribonucleic acid, or RNA, was probably responsible for the origin of life.

KEY CONCEPTS

- Nucleotides are monomers that consist of a sugar, a phosphate group, and a nitrogen-containing base. Ribonucleotides polymerize to form RNA. Deoxyribonucleotides polymerize to form DNA.
- DNA's primary structure consists of a sequence of nitrogen-containing bases, which contain information in the form of a molecular code. DNA's secondary structure consists of two DNA strands running in opposite directions. The strands are held together by complementary base pairing and are twisted into a double helix.
- RNA's primary structure also consists of a sequence of nitrogen-containing bases that contain information in the form of a molecular code. Its secondary structure includes short double helices and structures called hairpins. RNA molecules called ribozymes catalyze important chemical reactions.

Chapter 3 began with experimental evidence that chemical evolution produced the monomers called amino acids and their polymers called proteins. But the chapter ended by stating that even though proteins are the workhorse molecules of today's cells, few researchers still favor the hypothesis that life began as a protein molecule. Instead, the vast majority of biologists now contend that life began as a polymer called a nucleic acid—specifically, a molecule of ribonucleic acid (RNA). This proposal is called the RNA world hypothesis. Just what is a nucleic acid? How could these types of molecules have been produced by chemical evolution? What do the nucleic acids RNA and deoxyribonucleic acid (DNA) do in cells today, and why isn't it plausible to propose that life began with DNA? Before attempting to answer these questions, let's consider an even more basic issue: How would we know when a molecule became alive? Stated another way, What *is* life?

Like many simple questions, the issue of what defines life is not easy to answer. Although there is no precise definition of what constitutes life, most biologists point to two attributes to distinguish life from nonlife. The first attribute is **reproduction,** the ability of something to make a copy of itself. (In organisms that reproduce asexually, the copy is exact. In the case of sexually reproducing organisms, the copy is not exact. Instead, traits from male and female individuals are combined to produce an offspring.) In essence, this chapter is about the first molecule that could make a copy of itself.

For something to be alive, however, most biologists insist that it have a second attribute called **metabolism**: the ability to acquire particular molecules and use them in controlled chemical reactions that maintain conditions suitable for life and that contribute to growth. In today's organisms, chemical reactions are precisely controlled because enzymes and reactants are contained inside a cell by a plasma membrane. To many biologists, then, the presence of a plasma membrane is also required for life.

According to the theory of chemical evolution, these two attributes of life did not emerge simultaneously. Instead, the theory predicts that chemical evolution first led to the existence of a molecule that could make copies of itself. Some researchers argue that this self-replicating molecule was also capable of metabolism, because it had to be able to catalyze the polymerization reactions that allowed copies of itself to form. Later, the theory holds, a descendant of this molecule became enclosed in a membrane. This event created the first cell. The formation of the cell allowed for reactants to be enclosed. As a result, it made precise control over reactions possible.

The key idea is that life started with replication and later became cellular. To distinguish a naked (membrane-less) self-replicator from later, cellular forms of "true" life, this chapter refers to the self-replicator as the first living entity or the first life-form—but *not* as the first organism. The term **organism** is reserved for cellular-based life. This distinction isn't trivial. Naked, self-replicating molecules undoubtedly existed early in Earth's history, and researchers are almost certain to produce one in the laboratory within your lifetime. But it is much less likely that biologists will be able to create cellular life in the foreseeable future. Will humans be able to create life in a test tube? The answer depends, in part, on how you define life.

To understand how life began, it's helpful to delve into more of the chemistry that underlies biology. Let's begin with an analysis of nucleic acids as monomers and end with current research efforts to create a self-replicating molecule.

4.1 What Is a Nucleic Acid?

Nucleic acids are polymers, just as proteins are polymers. But instead of being made up of monomers called amino acids, **nucleic acids** are made up of monomers called **nucleotides. Figure 4.1a** diagrams the three components of a nucleotide: (1) a phosphate group, (2) a sugar, and (3) a nitrogenous (nitrogen-containing) base. The phosphate is bonded to the sugar molecule, which in turn is bonded to the nitrogenous base. A **sugar** is an organic compound with a carbonyl group (>C=O) and several hydroxyl (OH) groups. Notice that the "prime" symbols (′) in

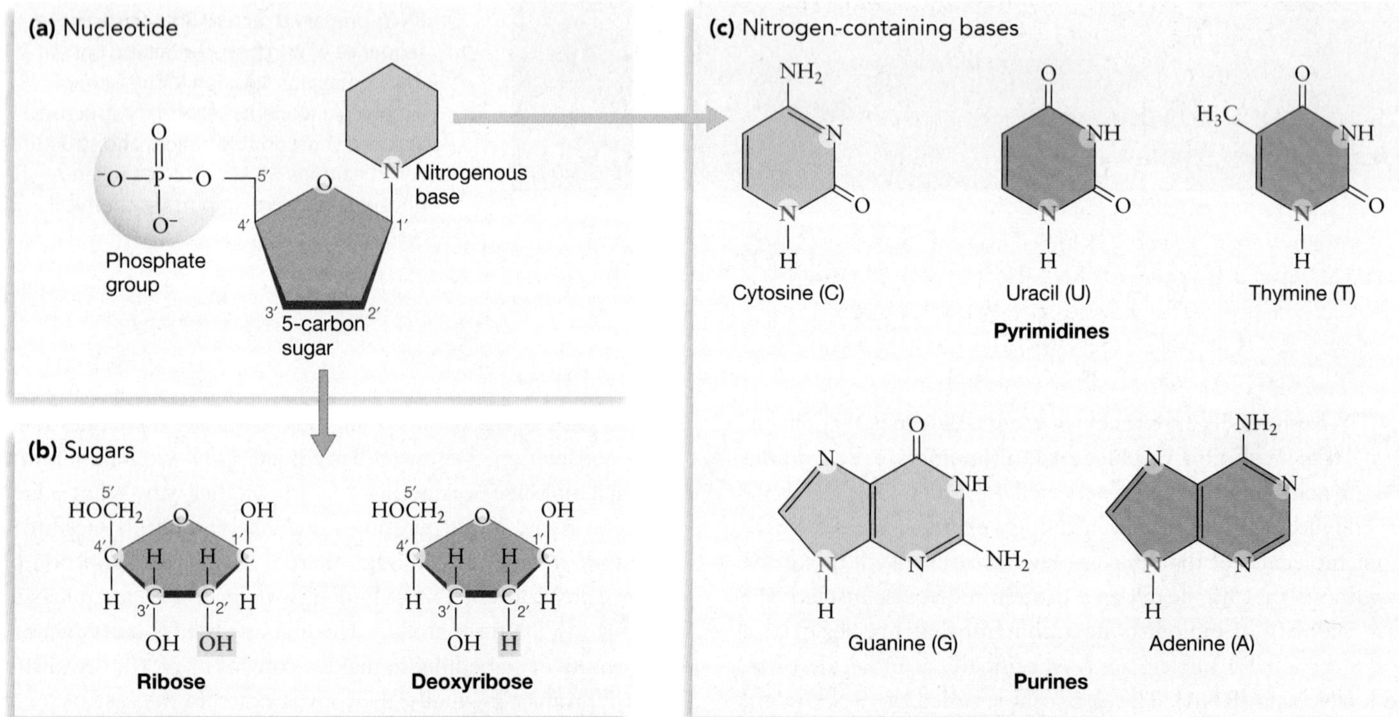

FIGURE 4.1 The General Structure of a Nucleotide. (a) The relationship between the phosphate group, the sugar, and the nitrogenous base of a nucleotide. The numbers indicate the position of each carbon in the sugar's ring. The nitrogenous base is bonded to carbon number 1 in the ring, while the phosphate is bonded to carbon number 5. The bond between the phosphate group and the sugar is called a 5′ linkage. Although hydrogen atoms are bonded to the carbon atoms in the ring (see part b), biologists routinely omit them in representations, for clarity. **(b)** Ribose and deoxyribose are similar sugars that are found in nucleotides. **(c)** Purines and pyrimidines are nitrogenous bases. A C–N bond links them to the sugar in a nucleotide. This bond forms at the nitrogen atom that is highlighted in each base. Purines are substantially larger than pyrimidines.

Figure 4.1 indicate that the carbon being referred to is part of the sugar and not of the attached nitrogenous base.

Although a wide variety of nucleotides are found in living cells, origin-of-life researchers concentrate on two types: **ribonucleotides** and **deoxyribonucleotides.** In ribonucleotides, the sugar is ribose; in deoxyribonucleotides, it is deoxyribose. As **Figure 4.1b** shows, these two sugars differ by a single atom. Ribose has an –OH group bonded to the second carbon in the ring. Deoxyribose has an H instead at the same location. Note that *deoxy* means "lacking oxygen."

Cells today have four different ribonucleotides, each of which contains a different nitrogenous base. These bases, diagrammed in **Figure 4.1c**, belong to structural groups called **purines** and **pyrimidines**. Ribonucleotides include the purines adenine (A) and guanine (G), and the pyrimidines cytosine (C) and uracil (U).

Similarly, four different deoxyribonucleotides are found in cells today and are distinguished by the structure of their nitrogenous base. Like ribonucleotides, deoxyribonucleotides include adenine, guanine, and cytosine. But instead of uracil, a closely related pyrimidine called thymine (T) occurs in deoxyribonucleotides (**Figure 4.1c**). You should be able to diagram a ribonucleotide and a deoxyribonucleotide, using a ball for the phosphate group, a pentagon to represent the sugar subunit, and a hexagon to represent the nitrogenous base. Label the 2′, 3′, and 5′ carbons on the sugar molecule, and add the atoms that are bonded to the 2′ carbon.

Could Chemical Evolution Result in the Production of Nucleotides?

Based on data presented in Chapter 3, most researchers contend that amino acids were abundant early in Earth's history. As yet, however, no one has observed the formation of a nucleotide via chemical evolution. The problem lies with mechanisms for synthesizing the sugar and nitrogenous base components of these molecules. Let's consider each issue in turn.

Laboratory simulations have shown that many sugars can be synthesized readily under conditions that mimic the prebiotic soup. Specifically, when formaldehyde (H_2CO) molecules are heated in solution, they react with one another to form almost all of the sugars that have five or six carbons. (These are called pentoses and hexoses, respectively.) Ironically, the ease of forming these sugars creates a problem. In modern experiments the various pentoses and hexoses are produced in approximately equal amounts, but it seems logical to predict that ribose would have had to be particularly abundant for RNA or DNA to form in the prebiotic soup. How ribose came to be the dominant sugar during chemical evolution is still a mystery. Origin-of-life researchers refer to this issue as "the ribose problem."

The origin of the pyrimidines is equally challenging. Simply put, origin-of-life researchers have yet to discover a plausible mechanism for the synthesis of cytosine, uracil, and thymine molecules prior to the origin of life. Purines, in contrast, are readily synthesized by reactions among hydrogen cyanide (HCN) molecules. For example, both adenine and guanine have been found in the solutions recovered after spark-discharge experiments.

The ribose problem and the questions about the origin of pyrimidine bases are two of the most serious challenges that remain for the theory of chemical evolution. Research on how large quantities of ribose could have formed in the prebiotic soup, along with the pyrimidines, continues. But once formed, how did nucleic acids polymerize to form RNA and DNA? This question has an answer.

How Do Nucleotides Polymerize to Form Nucleic Acids?

Nucleic acids form when nucleotides polymerize. As **Figure 4.2** shows, the polymerization reaction involves the formation of a bond between the phosphate group of one nucleotide and the hydroxyl group of the sugar component of another nucleotide. The result of this condensation reaction is called a **phosphodiester linkage**, or a phosphodiester bond. In Figure 4.2, a phosphodiester linkage joins the 5′ carbon on the ribose of one nucleotide to the 3′ carbon on the ribose of the other. When the nucleotides involved contain the sugar ribose, the polymer that is produced is called **ribonucleic acid**, or simply **RNA**. If the nucleotides contain the sugar deoxyribose instead, then the resulting polymer is **deoxyribonucleic acid**, or **DNA**.

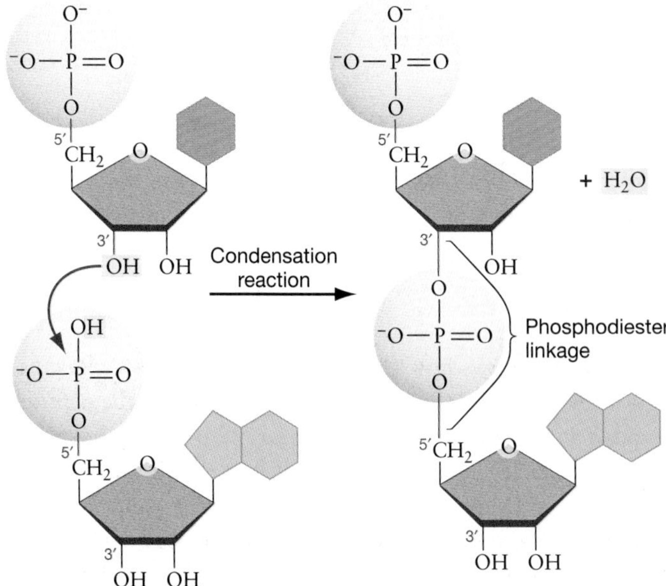

FIGURE 4.2 Nucleotides Polymerize via Phosphodiester Linkages. Ribonucleotides can polymerize via condensation reactions. The linkage that results, between the 3′ carbon of one ribonucleotide and the 5′ carbon of another ribonucleotide, is called a phosphodiester linkage.

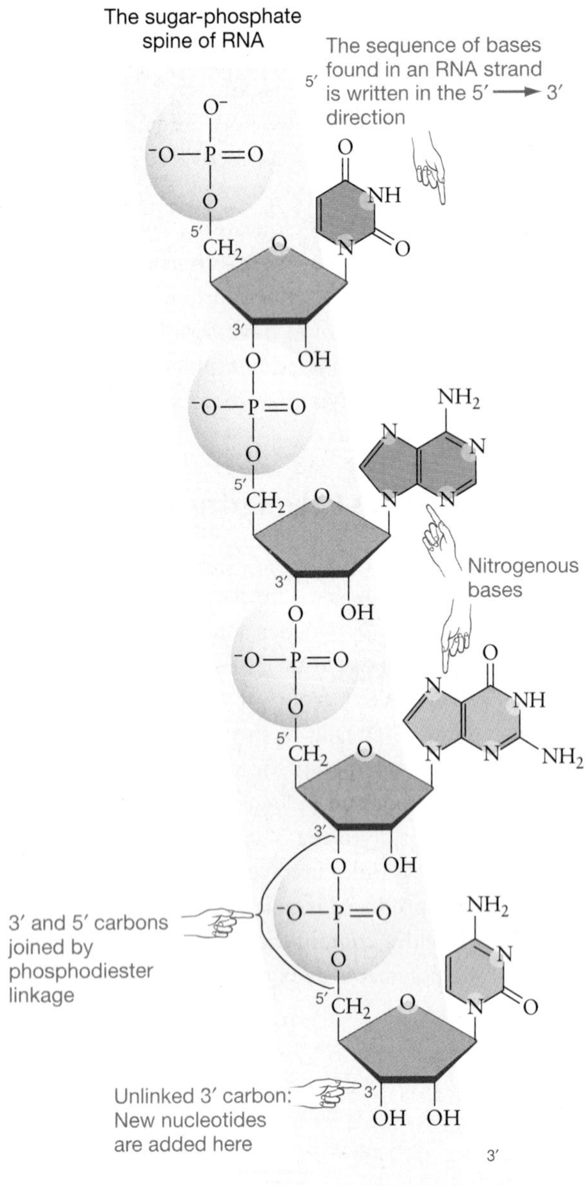

FIGURE 4.3 RNA Has a Sugar-Phosphate Backbone.

● **EXERCISE** Identify the four bases in this RNA strand, using Figure 4.1c as a key. Then write down the base sequence, starting at the 5′ end.

Figure 4.3 shows how the chain of phosphodiester linkages in a nucleic acid acts as a backbone, analogous to the peptide-bonded backbone found in proteins. The sugar-phosphate spine of a nucleic acid is directional, as is the peptide-bonded spine of a polypeptide. In a strand of RNA or DNA, one end has an unlinked 5′ carbon while the other end has an unlinked 3′ carbon—meaning a carbon that is not linked to another nucleotide. By convention, the sequence of bases found in an RNA or DNA strand is always written in the 5′ → 3′ direction. (The system is logical because in cells, RNA and DNA are always synthesized in this direction. Bases are added at the 3′ end of the growing molecule.) The sequence of nitrogenous bases forms the primary structure of the molecule, analogous to the sequence of amino acids in a polypeptide. When biologists write the primary structure of a stretch of DNA, they simply list the sequence of nucleotides using their single-letter abbreviations. For example, a six-base-long DNA sequence might be ATTAGC.

In cells, the polymerization reactions that form nucleotides are catalyzed by enzymes. Like other polymerization reactions, the process is endergonic. Polymerization can take place in cells because the free energy of the nucleotide monomers is first raised by reactions that add two phosphate groups to the ribonucleotides or deoxyribonucleotides, creating nucleotide triphosphates (**Figure 4.4**). Molecules that have phosphate groups attached in this way are said to be **phosphorylated**. This is a key point, and one that you will encounter again and again in this text: The addition of one or more phosphate groups raises the potential energy of the substrate molecules enough to make an endergonic reaction possible. (Chapter 9 explains how this happens.) In the case of nucleic acid polymerization, researchers refer to the phosphorylated nucleotides as "activated."

During chemical evolution, activated nucleotides probably polymerized on the surfaces of clay-sized (very fine-grained) mineral particles. In a suite of experiments analogous to those discussed in Chapter 3 for protein synthesis, researchers have produced RNA molecules by incubating activated ribonucleotides with tiny mineral particles. The hypothesis was that polymerization could occur without an enzyme if the ribonucleotides and growing RNA strands adhered to the clay particles.

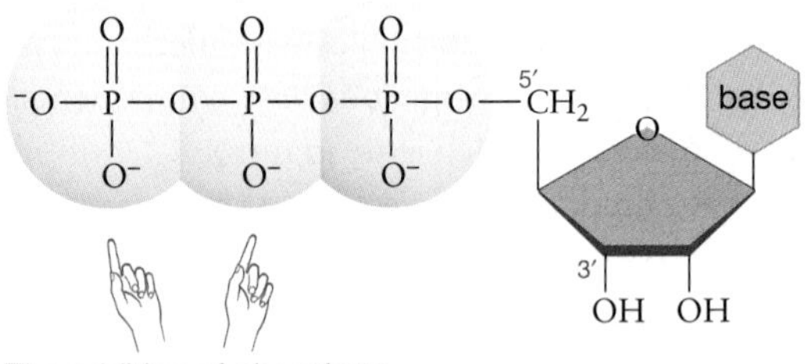

FIGURE 4.4 Activated Monomers Drive Endergonic Polymerization Reactions. Polymerization reactions involving ribonucleotides are endergonic. But polymerization reactions involving ribonucleotide triphosphates are exergonic.

● **EXERCISE** The reaction shown in Figure 4.2 does not actually occur—only nucleotide triphosphates like the one shown here can form phosphodiester bonds with other nucleotides or with nucleic acids. Add two phosphate groups to the nucleotide in Figure 4.2 to make the reaction realistic. On the right side of the reaction, the products should include a hydroxl group and two phosphate groups.

In one experiment, researchers isolated the clay particles after a day of incubation and then added a fresh batch of activated nucleotides. They repeated this reaction-isolation-reaction sequence for a total of 14 days, or 14 additions of fresh ribonucleotides. At the end of the two-week experiment, they analyzed the mineral particles, using techniques called **gel electrophoresis** and **autoradiography** (see **BioSkills 7** at the end of this book), and found RNA molecules up to 40 nucleotides long. Based on these results, there is a strong consensus that if ribonucleotides and deoxyribonucleotides were able to form during chemical evolution, they would be able to polymerize and form RNA and DNA. Now, what would these nucleic acids look like, and what could they do?

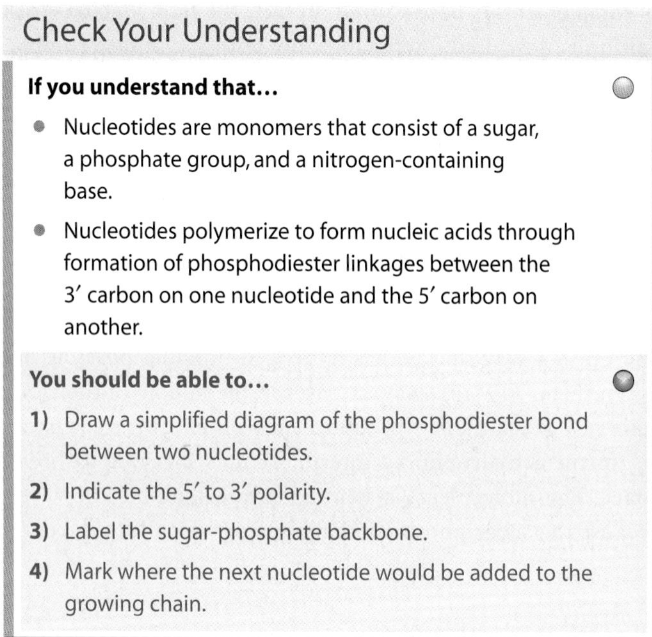

Check Your Understanding

If you understand that...

- Nucleotides are monomers that consist of a sugar, a phosphate group, and a nitrogen-containing base.
- Nucleotides polymerize to form nucleic acids through formation of phosphodiester linkages between the 3′ carbon on one nucleotide and the 5′ carbon on another.

You should be able to...

1) Draw a simplified diagram of the phosphodiester bond between two nucleotides.
2) Indicate the 5′ to 3′ polarity.
3) Label the sugar-phosphate backbone.
4) Mark where the next nucleotide would be added to the growing chain.

4.2 DNA Structure and Function

Nucleic acids have a primary structure that is somewhat similar to the primary structure of proteins. Proteins have a peptide-bonded backbone with a series of R-groups that extend from it; DNA and RNA molecules have a sugar-phosphate backbone, created by phosphodiester linkages, and a sequence of four nitrogenous bases that extend from it. Recall that the primary structures of DNA and RNA contrast in two fundamental aspects: The sugar in DNA's backbone is deoxyribose instead of the ribose in RNA, and RNA contains the nitrogenous base uracil instead of the thymine found in DNA.

DNA and RNA also have secondary structure, analogous to the secondary structure of proteins. But the secondary structures that form in DNA and RNA differ from each other and contrast with those observed in proteins. Distinctive secondary structure, combined with differences in primary structure, dictate why DNA and RNA have different functions in organisms. Let's analyze the secondary structure and function of DNA first, and then dig into the secondary structure and function of RNA.

What Is the Nature of DNA's Secondary Structure?

The solution to DNA's secondary structure, announced in 1953, ranks among the great scientific breakthroughs of the twentieth century. James Watson and Francis Crick presented a model for the secondary structure of DNA in a one-page paper published in *Nature*. At the time, Watson was a 25-year-old postdoctoral fellow and Crick was a 37-year-old graduate student.

Watson and Crick's finding was a hypothesis based on a series of results from other laboratories. They were trying to propose a secondary structure that could explain several important observations about the DNA found in cells:

- Chemists had worked out the structure of nucleotides and knew that DNA polymerized through the formation of phosphodiester linkages. Thus, Watson and Crick knew that the molecule had a sugar-phosphate backbone.
- By analyzing the nitrogenous bases in DNA samples from different organisms, Erwin Chargaff had established two empirical rules: (1) The total number of purines and pyrimidines in DNA is the same; and (2) the numbers of T's and A's in DNA are equal, and the numbers of C's and G's in DNA are equal.
- By bombarding DNA with X-rays and analyzing how it scattered the radiation, Rosalind Franklin and Maurice Wilkins had calculated the distances between groups of atoms in the molecule (see **Bioskills 8** for an introduction to this techique, called **X-ray crystallography**). The scattering patterns showed that three distances were repeated many times: 0.34 nanometer (nm), 2.0 nm, and 3.4 nm. Because the measurements repeated, the researchers inferred that DNA molecules had a regular and repeating structure. The pattern of X-ray scattering suggested that the molecule was helical or spiral in nature.

Based on this work, understanding DNA's structure boiled down to understanding the nature of the helix involved. What type of helix would have a sugar-phosphate backbone and explain both Chargaff's rules and the Franklin–Wilkins measurements?

To answer this question, Watson and Crick began by analyzing the size and geometry of deoxyribose, phosphate groups, and the nitrogenous bases. The bond angles and measurements suggested that the distance of 2.0 nm probably represented the width of the helix and that 0.34 nm was likely to be the distance between bases stacked in a spiral. Now they needed to make sense of Chargaff's rules and the 3.4-nm distance, which

FIGURE 4.5 Building a Physical Model of DNA Structure. Watson (left) and Crick (right) represented the arrangement of the four deoxyribonucleotides in a double-helical arrangement, using wires with precise lengths and geometries.

appeared to be exactly 10 times the distance between a single pair of bases.

To solve this problem, Watson and Crick constructed a series of physical models such as the one pictured in **Figure 4.5**. Building these models allowed them to tinker with different types of helical configurations. After many false starts, they hit on an idea that looked promising. They arranged two strands of DNA side by side and running in opposite directions—meaning that one strand ran in the $5' \rightarrow 3'$ direction while the other was oriented $3' \rightarrow 5'$. Strands with this orientation are said to be antiparallel. Watson and Crick found that if the antiparallel strands were twisted together to form a double helix, the coiled sugar-phosphate backbones ended up on the outside of the spiral and the nitrogenous bases on the inside. But for the bases from each backbone to fit in the interior of the 2.0-nm-wide structure, they had to form purine-pyrimidine pairs (see **Figure 4.6a**). With that came a fundamental insight: Inside the double helix, the bases lined up in a way that allowed hydrogen bonds to form between certain purines and pyrimidines. More specifically, adenine could form hydrogen bonds with thymine, and guanine could form hydrogen bonds with cytosine (**Figure 4.6b**). Because of this specificity, the A-T and G-C bases were said to be complementary. Two hydrogen bonds form when A and T pair, but three hydrogen bonds form when G and C pair. As a result, the G-C interaction is slightly stronger than the A-T bond.

Watson and Crick had discovered the phenomenon known as **complementary base pairing**. In fact, the term **Watson–Crick pairing** is now used interchangeably with the phrase complementary base pairing.

Figure 4.7 shows how antiparallel strands of DNA form when complementary bases line up and form hydrogen bonds. As you study the figure, notice that DNA is put together like a ladder whose ends have been twisted in opposite directions. The sugar-phosphate backbone forms the supports of the ladder; the base pairs represent the rungs of the ladder. The twisting occurs because it allows the nitrogenous bases to line up in a way that makes hydrogen bonding possible between them. The nitrogenous bases in the middle of the DNA helix also stack tightly on top of each other. This tight packing forms a hydrophobic interior that is difficult to break apart. The molecule as a whole is water soluble, however, because the sugar-phosphate backbones, which face the exte-

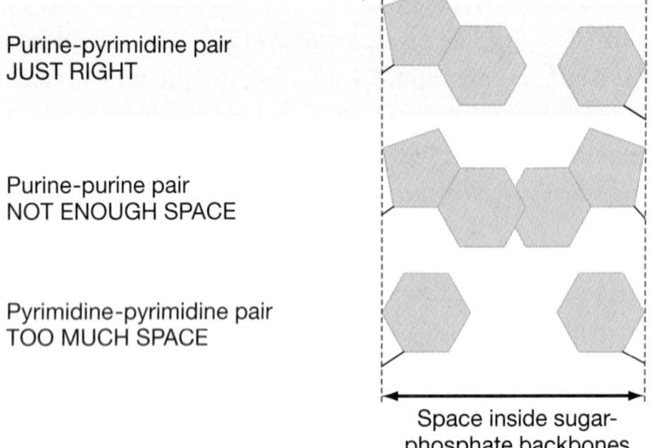

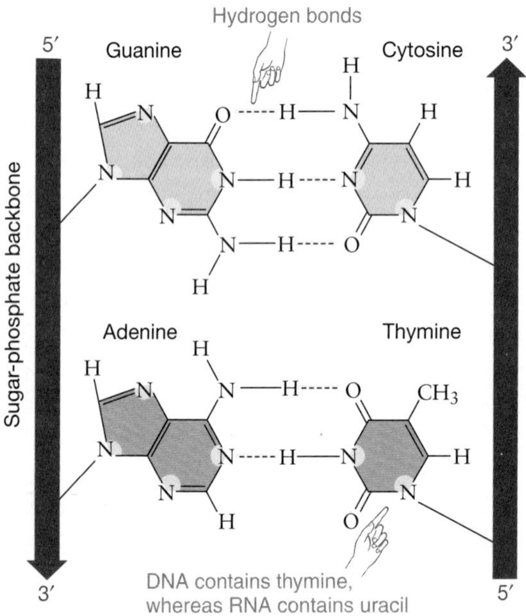

FIGURE 4.6 Complementary Base Pairing Is Based on Hydrogen Bonding. (a) Only purine-pyrimidine pairs fit inside the double helix effectively. **(b)** Hydrogen bonds form when guanine and cytosine pair, and when adenine and thymine pair.

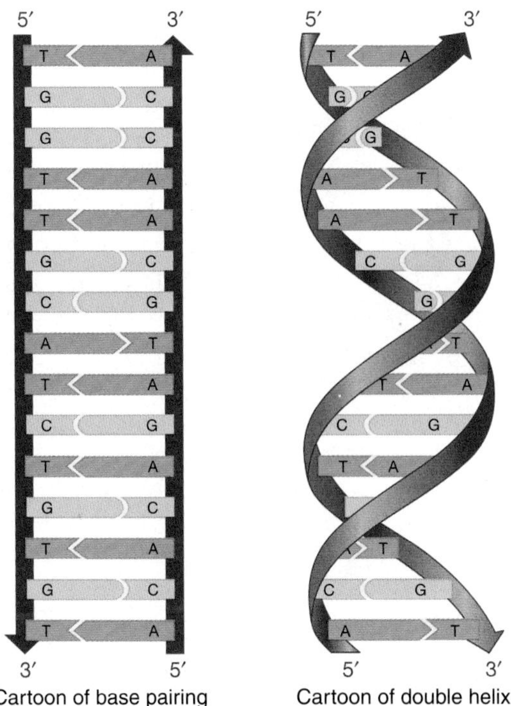

FIGURE 4.7 The Secondary Structure of DNA Is a Double Helix. Complementary base pairing is responsible for twisting DNA into a double helix.

rior of the molecule, are negatively charged and thus are hydrophilic.

It is also important to note that the outside of the helical DNA molecule forms two types of grooves and that the types differ in size. The larger of the two is known as the major groove, and the smaller one is known as the minor groove. **Figure 4.8** highlights these grooves and illustrates how DNA's secondary structure explains the measurements observed by Franklin and Wilkins (see **Box 4.1**).

Since the model of the double helix was published, experimental tests have shown that the hypothesis is correct in almost every detail. DNA's secondary structure consists of two antiparallel strands twisted into a **double helix**. The molecule is stabilized by hydrophobic interactions in its interior and by hydrogen bonding between the complementary base pairs A-T and G-C. ● If you understand DNA's secondary structure, you should be able to explain why the major and minor grooves exist, and why the strands are antiparallel to each other rather than parallel. Now the question is, How does this secondary structure affect the molecule's function?

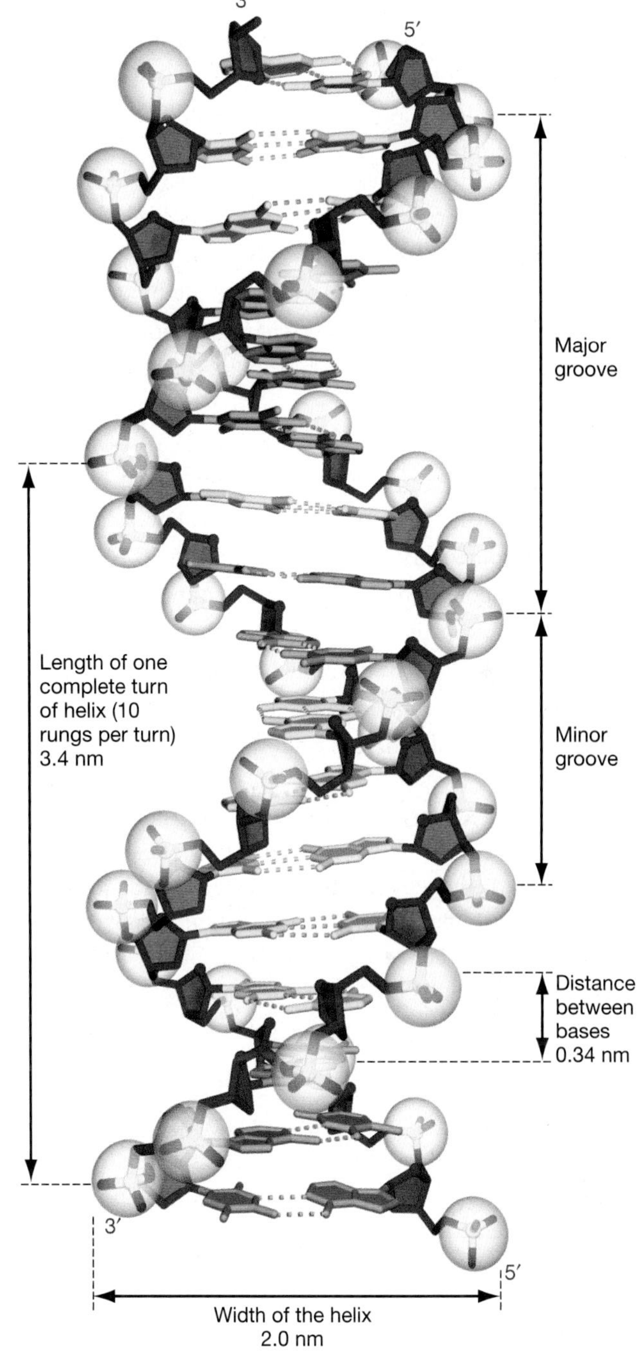

FIGURE 4.8 Dimensions of DNA Secondary Structure. The double-helix hypothesis explains the measurements inferred from X-ray analysis of DNA molecules.

● **QUESTION** What do the red pentagons and the yellow balls represent? What do the lines of green and purple dots represent?

DNA Is an Information-Containing Molecule

Watson and Crick's model created a sensation because it revealed how DNA could store and transmit biological information. In literature, information consists of letters on a page. In music, information is composed of the notes on a staff. But inside cells, information consists of a sequence of monomers in a nucleic acid. The four nitrogenous bases function like letters of the alphabet. A particular sequence of bases is like the sequence of letters in a word—it has meaning. In all cells that have been examined to date, from tiny bacteria to gigantic redwood trees, DNA carries the information required for the organism's growth and reproduction. Exploring how this information is encoded and translated into action is the heart of Chapters 15 through 18.

Here, however, our focus is on how life began. The theory of chemical evolution holds that life began as a self-replicating molecule—a molecule that could make a copy of itself. **Figure 4.9** shows how a copy of DNA can be made by complementary base pairing. In step 1 of this diagram, heating or enzyme-catalyzed reactions cause the double helix to separate. In step 2, free deoxyribonucleotides form hydrogen bonds with complementary bases on the original strand of DNA—also called a **template strand**. As they do, their sugar-phosphate groups form phosphodiester linkages to create a new strand—also called a **complementary strand**. Note that the $5' \rightarrow 3'$ directionality of the complementary strand is opposite that of the template strand. In this way, complementary base pairing allows each strand of a DNA double helix to be copied exactly, producing the two daughter molecules shown in step 3. Thus, DNA contains the information required for a copy of itself to be made.

Watson and Crick ended their paper on the double helix with one of the classic understatements in the scientific literature: "It has not escaped our notice that the specific pairing we have postulated immediately suggests a possible copying mechanism." The central idea is that DNA's primary structure serves as a mold or template for the synthesis of a complementary strand.

at www.masteringbio.com

Nucleic Acid Structure

In today's cells, however, DNA does not self-replicate spontaneously. Instead, the molecule is copied through a complicated series of reactions catalyzed by a large suite of enzymes. Consistent with these data, researchers consider it extremely unlikely that a DNA molecule began copying itself early in Earth's history. Let's explore why.

Is DNA a Catalytic Molecule?

The DNA double helix is a highly structured molecule that is much more stable than RNA and most proteins. DNA is regular and symmetric, with few chemical groups exposed that can participate in chemical reactions. For example, the lack of a 2′ hydroxl group on deoxyribonucleotides (see Figure 4.1b) makes the polymer much less reactive than RNA and highly resistant to chemical degradation. All of these features increase DNA's stability and thus its effectiveness as a reliable information-bearing molecule. Intact stretches of DNA have been recovered from fossils that are tens of thousands of years old. Despite death and exposure to a wide array of pH,

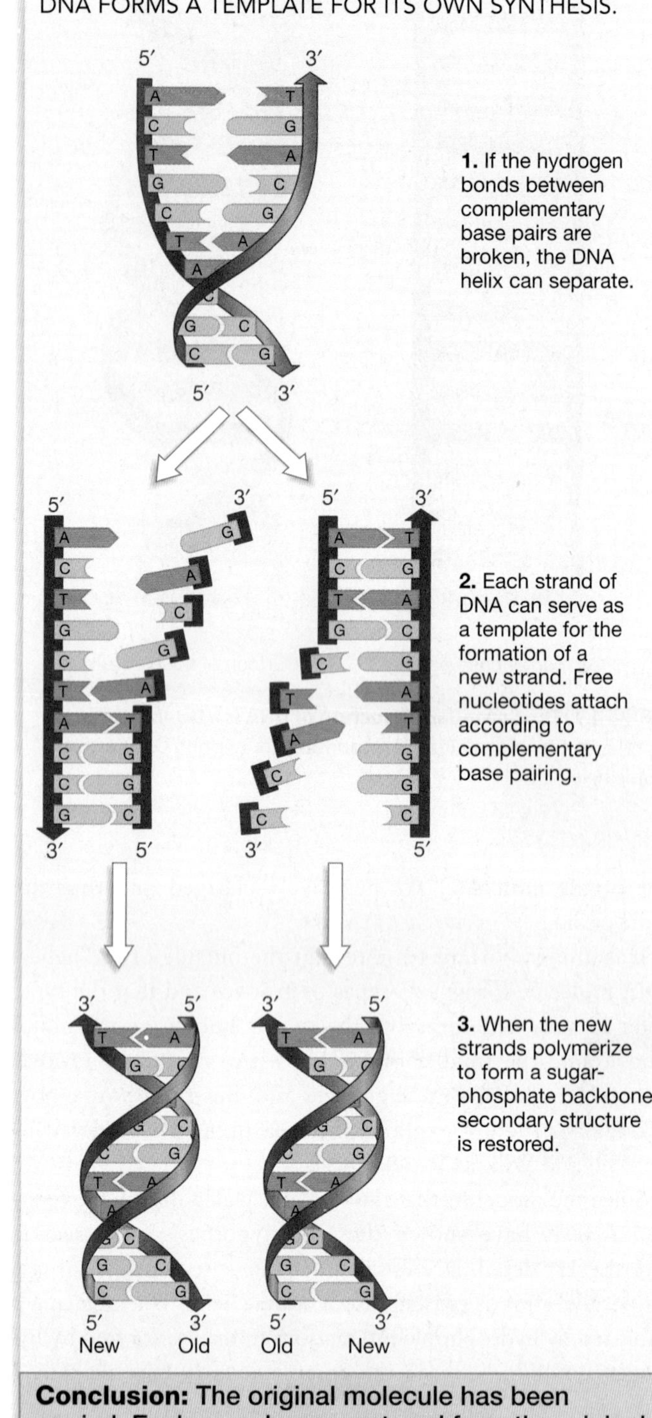

FIGURE 4.9 Making a Copy of DNA. If new bases are added to each of the two strands of DNA via complementary base pairing, a copy of the DNA molecule can be produced.

QUESTION When double-stranded DNA is heated to 95°C, the bonds between complementary base pairs break and single-stranded DNA results. Based on this observation, is the reaction shown in step 1 endergonic or exergonic?

BOX 4.1 The Human Side of Research

The discovery of the double helix was a critically important advance in the early 1950s, because experiments conducted in the 1940s—and described in detail in Chapter 14—had established that DNA is the hereditary material. Stated another way, scientists had recently discovered that genes are made of DNA. Immediately thereafter, understanding DNA's structure became *the* question in biology. Researchers who were working on the problem knew that the greatest honor in science—the Nobel Prize—would almost certainly be awarded to the person or persons who were first to discover the correct structure. The race was on: Many of the brightest and most famous scientists of the day were involved, working in different laboratories around the world.

Arguably the most important event in the research was Wilkins's and Franklin's ability to produce X-ray pictures of DNA. Recall that the photographs made it possible to make crucial measurements and inferences about the molecule's structure. Although Wilkins and Franklin are justifiably given joint credit for this work, they actually did not cooperate closely. It was Franklin who took the photographs that led to the realization that DNA is helical, but it was Wilkins who showed her photographs to Watson and Crick. Wilkins did so without Franklin's knowledge or permission. Today, such an act would be considered a serious form of professional misconduct. Another interesting wrinkle to this story is that Linus Pauling—undoubtedly the century's greatest biochemist—was also working on the problem at the time, but he was denied permission to travel from the United States to England to see the data. The United States was then in the throes of the Cold War, and the U.S. State Department had revoked Pauling's passport because his political views were considered too liberal.

As expected, Watson and Crick, along with Wilkins, received the Nobel Prize. The award was made in 1962, after experimental work had confirmed that the double-helix model was correct. Tragically, Franklin died of ovarian cancer in 1958, when she was just 37. She never received the Nobel, because the honor is not awarded posthumously.

The moral of the story, if there is one, may be simply that scientific research is a very human enterprise. In addition, times have changed. Although competition can still be intense, biology is an increasingly cooperative and international enterprise; misconduct is usually detected and addressed promptly. Rosalind Franklin's professional life was made difficult at times because she was a woman. Today most of the graduate students in molecular biology are female.

temperature, and chemical conditions, the molecules have the same sequence of bases as the organisms had when they were alive.

The orderliness and stability that make DNA such a dependable information repository make it extraordinarily inept at catalysis, however. Recall from Chapter 3 that enzyme function is based on a specific binding event between a substrate and a protein catalyst. Thanks to the enormous diversity of primary through quaternary structures found in proteins, a wide array of binding events can occur. In comparison, DNA's primary and secondary structures are simple. It is not surprising, then, that DNA has never been observed to catalyze any reaction in any organism. Even though researchers have been able to construct single-stranded DNA molecules that can catalyze a few simple reactions in the laboratory, the number and diversity of reactions involved is a minute fraction of the activity catalyzed by proteins.

In short, DNA furnishes an extraordinarily stable template for copying itself. But due to its inability to act as an effective catalyst, virtually no researchers support the hypothesis that the first life-form consisted of DNA. Instead, most biologists who are working on the origin of life support the hypothesis that life began with RNA.

Check Your Understanding

If you understand that...

- DNA's primary structure consists of a sequence of deoxyribonucleotides.
- DNA's secondary structure consists of two DNA molecules that run in opposite $5' \rightarrow 3'$ orientations to each other. The two strands are twisted into a double helix, and they are held together through hydrogen bonds between A-T and G-C pairs and hydrophobic interactions among atoms inside the helix.
- The sequence of deoxyribonucleotides in DNA contains information, much like the sequence of letters and punctuation on this page. Due to complementary base pairing, each DNA strand also contains the information required to form the complementary strand.

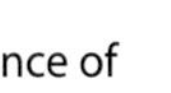

You should be able to...

Make a sketch of a DNA molecule that

1) Indicates at least four complementary base pairs.
2) Labels the molecule's sugar-phosphate backbone, the hydrogen bonds between complementary bases, and the location of hydrophobic interactions that stabilize the helix.
3) Indicates the $5' \rightarrow 3'$ orientation of each strand.

4.3 RNA Structure and Function

Like DNA, RNA has a primary structure consisting of a sugar-phosphate backbone formed by phosphodiester linkages and, extending from that backbone, a sequence of four types of nitrogenous bases. But it's important to recall two significant differences between these nucleic acids:

1. The pyrimidine base thymine does not exist in RNA. Instead, RNA contains the closely related pyrimidine base uracil.
2. The sugar in the sugar-phosphate backbone of RNA is ribose, not deoxyribose as in DNA.

The second point is critical because the hydroxyl (–OH) group on the 2′-carbonof ribose is much more reactive than the hydrogen atom on the 2′-carbonof deoxyribose. This functional group can participate in reactions that tear the polymer apart. The presence of this –OH group makes RNA much less stable than DNA.

A comparison of secondary structure in RNA and DNA is just as instructive. Like DNA molecules, most RNA molecules have secondary structure that results from complementary base pairing between purine and pyrimidine bases. In RNA, adenine forms hydrogen bonds only with uracil, and guanine again forms hydrogen bonds with cytosine. (Guanine can also bond with uracil, but it does so much less effectively than with cytosine.) Thus, the complementary base pairs in RNA are A-U and G-C. Three hydrogen bonds form between guanine and cytosine, but only two form between adenine and uracil.

How do the secondary structures of RNA and DNA differ? In the vast majority of cases, the purine and pyrimidine bases in RNA undergo hydrogen bonding with complementary bases on the same strand, rather than forming hydrogen bonds with complementary bases on a different strand, as in DNA. **Figure 4.10** shows how this works. The key is that when bases on one part of an RNA strand fold over and align with ribonucleotides on another segment of the same strand, the two sugar-phosphate strands are antiparallel. In this orientation, hydrogen bonding between complementary bases results in a stable double helix.

If the section where the fold occurs includes a large number of unbonded bases, then the stem-and-loop configuration shown in Figure 4.10 results. This type of secondary structure is called a **hairpin**. Several other types of RNA secondary structures are possible, each involving a different length and arrangement of base-paired segments.

Like the α-helicesand β-pleatedsheets observed in many proteins, RNA secondary structures are stabilized by hydrogen bonding and occur spontaneously. Even though hairpins and other types of secondary structures reduce the entropy of RNA molecules, they form spontaneously because the energy released in formation of H-bonds makes the overall process favorable. Hydrogen bond formation is exothermic; and overall, the formation of hairpins is exergonic.

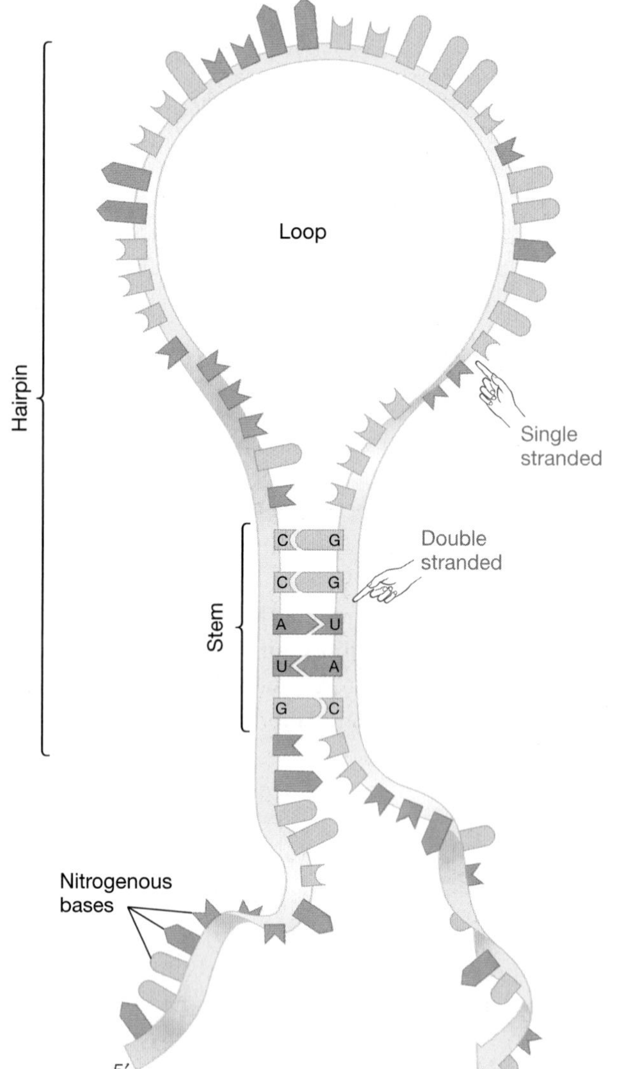

FIGURE 4.10 Complementary Base Pairing and Secondary Structure in RNA: Stem-and-Loop Structures. This RNA molecule has secondary structure. The double-stranded "stem" and single-stranded "loop" form a hairpin. The bonded bases in the stem are antiparallel, meaning that they are oriented in opposite directions.

RNA molecules can also have tertiary structure and quaternary structure, owing to interactions that fold secondary structures into complex shapes or that hold different RNA strands together. As a result, RNA molecules with different base sequences can have very different overall shapes and chemical properties. Although RNA molecules do not begin to rival proteins in overall structural diversity or the presence of various functional groups, RNA molecules are much more diverse in size, shape, and reactivity than DNA molecules are. Structurally and chemically, RNA is intermediate between the complexity of proteins and the simplicity of DNA.

Based on these observations, it is logical to note that RNA is also intermediate in terms of its function in cells. RNA molecules cannot archive information nearly as efficiently as DNA molecules do; but RNAs do perform key functions in information processing. Likewise, they cannot catalyze as many reactions as proteins do, but certain RNAs catalyze particularly important reactions. RNA molecules are also involved in defending cells, providing structure, and regulating the cell's response to changes in the environment. In cells, RNA molecules function like a jackknife or a pocket tool with an array of attachments: They perform a wide variety of tasks reasonably well.

The diversity of roles that RNA plays in cells will be a prominent theme in later chapters. Here let's focus on the roles that RNA could have played in the origin of life—as an information-containing entity and as a catalyst.

RNA as an Information-Containing Molecule

Because RNA contains a sequence of bases analogous to the letters in a word, it can function as an information-containing molecule. And because hydrogen bonding occurs specifically between A-U pairs and G-C pairs in RNA, it is possible for RNA to furnish the information required to make a copy of itself. The process occurs as shown for DNA in Figure 4.9, except that the template RNA exists as a single strand instead of a double strand. When RNA is copied, free ribonucleotides form hydrogen bonds with complementary bases on the original strand of RNA—the template strand. As they do, their sugar-phosphate groups form phosphodiester linkages to produce the complementary strand. Finally, the hydrogen bonds between the strands are broken by heating or by a catalyzed reaction. The new RNA molecule now exists independently of the original strand. If these steps were repeated with the new strand used as a template, the resulting molecule would be a copy of the original. In this way, an RNA's primary sequence serves as a mold. ● You should be able to draw the steps involved in synthesizing a complementary strand of RNA, analogous to the steps described for DNA in Figure 4.9.

Although complementary base pairing allows RNA to carry the information required for the molecule to be copied, RNA is not nearly as stable a repository for particular sequences as is DNA. In the prebiotic soup, could an RNA molecule have made a copy of itself before it would have degraded?

Is RNA a Catalytic Molecule?

In terms of diversity in chemical reactivity and overall shape, RNA molecules do not begin to match proteins. The primary structure of RNA molecules is much more restricted because there are only four types of nitrogenous bases in RNA versus the 21 types of amino acids found in proteins, and RNA molecules cannot form the variety of bonds that give proteins their extensive tertiary structure. But because RNA has a degree of structural and chemical complexity, it should be capable of stabilizing a few transition states and catalyzing at least a limited number of chemical reactions. Indeed, Sidney Altman and Thomas Cech shared the 1989 Nobel Prize in chemistry for showing that catalytic, enzyme-like RNAs, or **ribozymes**, exist in organisms. The ribozymes they isolated, from a single-celled organism called *Tetrahymena*, catalyzed both the hydrolysis and condensation of phosphodiester linkages. Building on that accomplishment, researchers have discovered ribozymes that catalyze dozens of different reactions in cells. For example, ribozymes catalyze the formation of peptide bonds when amino acids polymerize to form polypeptides. Ribozymes are at work in your cells right now.

The discovery of ribozymes was a watershed event in origin-of-life research. Before Altman and Cech published their discovery, biologists thought that proteins were the only type of molecule capable of catalyzing chemical reactions in organisms. But if a ribozyme in *Tetrahymena* catalyzes polymerization reactions similar to those diagrammed in steps 1 and 2 of Figure 4.9, it raises the possibility that an RNA molecule could make a copy of itself. Such a molecule could copy itself and qualify as the first living entity. Does any experimental evidence support this hypothesis?

4.4 The First Life-Form

The theory of chemical evolution maintains that life began as a naked self-replicator—a molecule that existed by itself in solution, without being enclosed in a membrane. To make a copy of itself, that first living molecule had to provide a template that could be copied. It also had to catalyze polymerization reactions that would link monomers into a copy of that template. Because RNA is capable of both processes, most origin-of-life researchers propose that the first life-form was made of RNA. As noted in the introduction to this chapter, that proposal is called the RNA world hypothesis.

Because no self-replicating RNA molecules exist today, researchers test the hypothesis by trying to simulate the RNA world in the laboratory. The eventual goal is to create an RNA molecule that can catalyze its own replication. When this goal is achieved, a life-form will have been created in the laboratory.

To understand how researchers go about this work, consider a recent set of experiments by Wendy Johnston and others working in David Bartel's laboratory. This research group's goal was ambitious; they wanted to create, from scratch, a ribozyme that was capable of catalyzing the addition of ribonucleotide triphosphates to an existing strand, via complementary base pairing. Such a ribozyme would be called RNA replicase. This research program is creating considerable excitement among biologists interested in the origin of life, because adding ribonucleotides to a growing strand is a key attribute of an RNA replicase.

The strategy Bartel's group pursued is outlined in **Figure 4.11**. As you study this figure, note that their procedure essentially mimics the process of natural selection introduced in

Experiment

Question: Can a ribozyme follow a template and catalyze the addition of bases to a growing RNA strand?

Hypothesis: Repeated selection among randomly generated RNA sequences can result in an efficient enzyme.

Null hypothesis: Selection cannot produce an efficient enzyme.

Experimental setup:

1. Randomly generate many billions of large RNA molecules, each with a unique base sequence.

2. Incubate the large RNAs from step 1 with small RNA templates that have a short primer. (The template + primer is shown magnified here to clarify base pairing.)

3. Isolate RNA molecules that are best at catalyzing addition of nucleotides to primer, creating complementary strand. These are ribozymes.

4. Make many copies of the ribozymes, in such a way that each new RNA has a small number of randomly generated changes in base sequence.

5. Repeat steps 2–4 seventeen times.

Prediction: After many rounds, the ribozyme will be able to catalyze addition of all 14 ribonucleotides.

Prediction of null hypothesis: None of the large RNA molecules produced will be able to catalyze addition of ribonucleotides.

Results:

An autoradiograph of the RNA molecules synthesized by the round-18 ribozyme

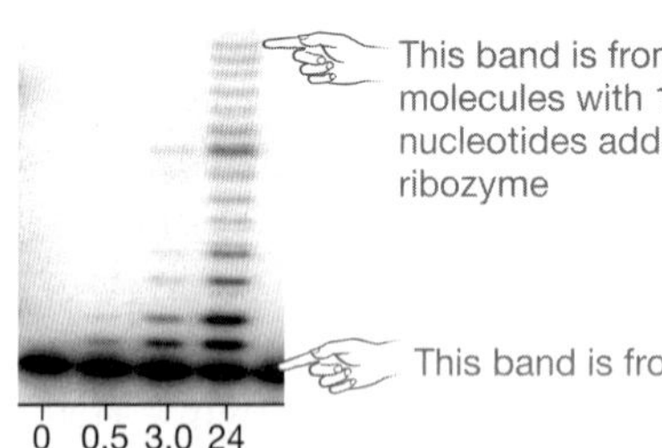

Conclusion: If allowed to react for 24 hours, the round-18 ribozyme can add all 14 bases to the primer, producing a strand complementary to the template.

FIGURE 4.11 Selection of a Novel Ribozyme. Researchers use variations of this protocol to produce ribozymes that can catalyze a wide variety of reactions.

Chapter 1. For example, the researchers began their experiments by establishing a large population of RNA molecules that were variable in structure and function, analogous to a large population of organisms with a variety of features. In this case, the researchers created millions of large RNA molecules that contained randomly generated primary sequences. They incubated these newly synthesized, large RNAs with a small RNA "primer," similar to the short RNAs that bind to template strands when RNA is being synthesized in cells. When the large RNAs and the primers were allowed to react with a large supply of free ribonucleotides, the researchers found that some of the large molecules could catalyze the addition of a few ribonucleotides that were complementary to the bases in the template. In this "round-1" experiment, the polymerization didn't happen very quickly or accurately, but it did occur. From their initial pool of millions of randomly generated molecules, Bartel's team had discovered a ribozyme that could catalyze the synthesis of RNA.

To continue the experiment, the researchers isolated the round-1 ribozymes. This step was analogous to natural selection, because only certain variants survived to produce "offspring." To create these offspring, which would represent the next generation of ribozymes, Bartel's team copied the selected molecules in a way that introduced a few random changes to their sequence of bases. These changes were analogous to the types of mutations that occur each generation in natural populations. Mutation is a random process that produces new traits. When the researchers incubated the modified ribozymes with a new template RNA, they found that most of the modified ribozymes worked worse than the original ones. Some worked better, however.

The researchers selected the best of these round-2 ribozymes, copied them in a way that introduced a few additional, random changes in their base sequences, and allowed them to react with yet another template. The best ribozymes in this round 3 were isolated, and then the copying, reaction, and selection process was repeated . . . and repeated . . . and repeated. By round 18, the group had found a ribozyme that was a much better catalyst than the original molecule. Evolution had occurred. They had created a ribozyme that was reasonably proficient at adding ribonucleotides to a growing strand.

As this book goes to press, the round-18 ribozyme is the closest biologists have come to creating life. Thanks to similar efforts at other laboratories around the world, researchers have produced an increasingly impressive set of ribozymes—an array of molecules capable of catalyzing many of the key reactions responsible for replication and metabolism. Each result provides support for the RNA world hypothesis. Each result also brings research teams closer to the creation of an RNA replicase. If this goal is met, human beings will have created a living entity in a test tube.

Chapter Review

SUMMARY OF KEY CONCEPTS

Nucleotides are monomers that consist of a sugar, a phosphate group, and a nitrogen-containing base. Ribonucleotides polymerize to form RNA. Deoxyribonucleotides polymerize to form DNA.

Ribonucleotides have a hydroxyl (–OH) group on their 2′ carbon, while deoxyribonucleotides do not. Both types of nucleotides polymerize via phosphodiester bonds to form nucleic acids, which have a sugar-phosphate backbone with nitrogenous bases attached.

You should be able to diagram the difference between a ribonucleotide and a deoxyribonucleotide. You should also be able to draw a generalized nucleic acid and label the sugar-phosphate backbone, phosphodiester linkages, 5′ to 3′ polarity, and nitrogenous bases.

DNA's primary structure consists of a sequence of nitrogen-containing bases, which contain information in the form of a molecular code. DNA's secondary structure consists of two DNA strands running in opposite directions. The strands are held together by complementary base pairing and are twisted into a double helix.

DNA is an extremely stable molecule that serves as a superb archive for information in the form of base sequences. DNA is stable because deoxyribonucleotides lack a reactive 2′ hydroxyl group and because antiparallel DNA strands form a secondary structure called a double helix. The DNA double helix is stabilized by hydrogen bonds that form between complementary purine and pyrimidine bases and by hydrophobic interactions between bases stacked on the inside of the spiral. This same structural stability and regularity make DNA ineffective at catalysis, however.

In addition to being stable, DNA is readily copied via complementary base pairing. Complementary base pairing occurs between A-T and G-C pairs in DNA.

You should be able to explain why DNA molecules with a high percentage of guanine and cytosine are particularly stable. You should also be able to draw a DNA double helix where there is a purine–purine pair or a pyrimidine–pyrimidine pair.

MB Web Animation at www.masteringbio.com
Nucleic Acid Structure

RNA's primary structure also consists of a sequence of nitrogen-containing bases that contain information in the form of a molecular code. Its secondary structure includes short double helices and structures called hairpins. RNA molecules called ribozymes catalyze important chemical reactions.

RNA molecules can have secondary structure because complementary base pairing occurs between A-U and G-C pairs on the same strand. Some tertiary and quarternary structure also occurs, because RNA molecules fold in precise ways and may interact with each other.

Compared with proteins and DNA, RNA is exceptionally versatile. The primary function of proteins is to catalyze chemical reactions, and the primary function of DNA is to carry information. But RNA is an "all-purpose" macromolecule that can do both.

You should be able to explain why RNA molecules can have tertiary and quaternary structure, while DNA molecules do not.

QUESTIONS

Test Your Knowledge

1. What are the four nitrogenous bases found in RNA?
   a. uracil, guanine, cytosine, thymine (U, G, C, T)
   b. adenine, guanine, cytosine, thymine (A, G, C, T)
   c. adenine, uracil, guanine, cytosine (A, U, G, C)
   d. alanine, threonine, glycine, cysteine (A, T, G, C)

2. What determines the primary structure of an RNA molecule?
   a. the sugar-phosphate backbone
   b. complementary base pairing and the formation of hairpins
   c. the sequence of deoxyribonucleotides
   d. the sequence of ribonucleotides

3. DNA attains a secondary structure when hydrogen bonds form between the nitrogenous bases called purines and pyrimidines. What are the complementary base pairs that form in DNA?
   a. A-T and G-C
   b. A-U and G-C
   c. A-G and T-C
   d. A-C and T-G

4. By convention, biologists write the sequence of bases in RNA and DNA in which direction?
   a. $3' \rightarrow 5'$
   b. $5' \rightarrow 3'$
   c. N-terminal to C-terminal
   d. C-terminal to N-terminal

5. The secondary structure of DNA is called a double helix. Why?
   a. Two strands wind around one another in a helical, or spiral, arrangement.
   b. A single strand winds around itself in a helical, or spiral, arrangement.
   c. It is shaped like a ladder.
   d. It stabilizes the molecule.

6. In RNA, when does the secondary structure called a hairpin form?
   a. when hydrophobic residues coalesce
   b. when hydrophilic residues interact with water
   c. when complementary base pairing between ribonucleotides on the same strand creates a stem-and-loop structure
   d. when complementary base pairing forms a double helix

Test Your Knowledge answers: 1. c; 2. d; 3. a; 4. b; 5. a; 6. c

Test Your Understanding

Answers are available at www.masteringbio.com

1. Make a concept map (see **BioSkills 6**) that relates DNA's primary structure to its secondary structure. Your diagram should include deoxyribonucleotides, purines, pyrimidines, phosphodiester bonds, DNA, complementary base pairing, and antiparallel strands.

2. Make a generalized diagram of monomers undergoing condensation reactions to form a polymer. Label the type of bond that forms when nucleotides polymerize into RNA or DNA. Does it take energy for polymerization reactions to proceed, or do they occur spontaneously? Why or why not?

3. Growing strands of nucleic acids are always extended in the $5' \rightarrow 3'$ direction. What do the $5'$ and $3'$ refer to? Now recall that only activated nucleotides (nucleotide triphosphates) can be added to a growing strand. Draw an activated nucleotide being added to an RNA strand.

4. Why is DNA such a stable molecule compared with proteins or RNA?

5. Explain how the secondary structures called hairpins form in RNA. Include a labeled diagram.

6. A major theme in this chapter is that the structure of molecules correlates with their function. Explain why DNA's secondary structure limits its catalytic abilities compared with that of RNA. Why is it logical that RNA molecules can catalyze a modest but significant array of reactions? Why are proteins the most effective catalysts?

Applying Concepts to New Situations

Answers are available at www.masteringbio.com

1. Do you agree with the "definition" of life given at the start of this chapter? Why or why not? If you were looking for life on Jupiter's moon Europa or on Mars, how would you know when you found it?
2. Suppose that experiments like those reviewed in Section 4.4 succeeded in producing a molecule that could make a copy of itself. Outline a one-page opinion piece for your local newspaper that explains the nature of the research and discusses the ethical and philosophical implications of the discovery.
3. Before Watson and Crick published their model of the DNA double helix, Linus Pauling offered a model based on a triple helix. Draw your conception of what a DNA triple helix might look like. What interactions would keep such a secondary structure together? How could such a molecule be copied?
4. Origin-of-life researcher Robert Crabtree maintains that experiments simulating early Earth conditions are a valid way to test the theory of chemical evolution. Crabtree claims that if scientists working in the field agree that an experiment is a plausible reproduction of early Earth conditions, it is valid to infer that its results are probably correct—that the simulation effectively represents events that occurred some 3.5 billion years ago. Do you agree? Do you find the models and experiments presented in this chapter and previous chapters to be convincing tests of the theory? Explain your answers.

5 An Introduction to Carbohydrates

KEY CONCEPTS

- Sugars and other carbohydrates are highly variable in structure.
- Monosaccharides are monomers that polymerize to form polymers called polysaccharides, via different types of glycosidic linkages.
- Carbohydrates perform a wide variety of functions in cells, ranging from energy storage to formation of tough structural fibers.

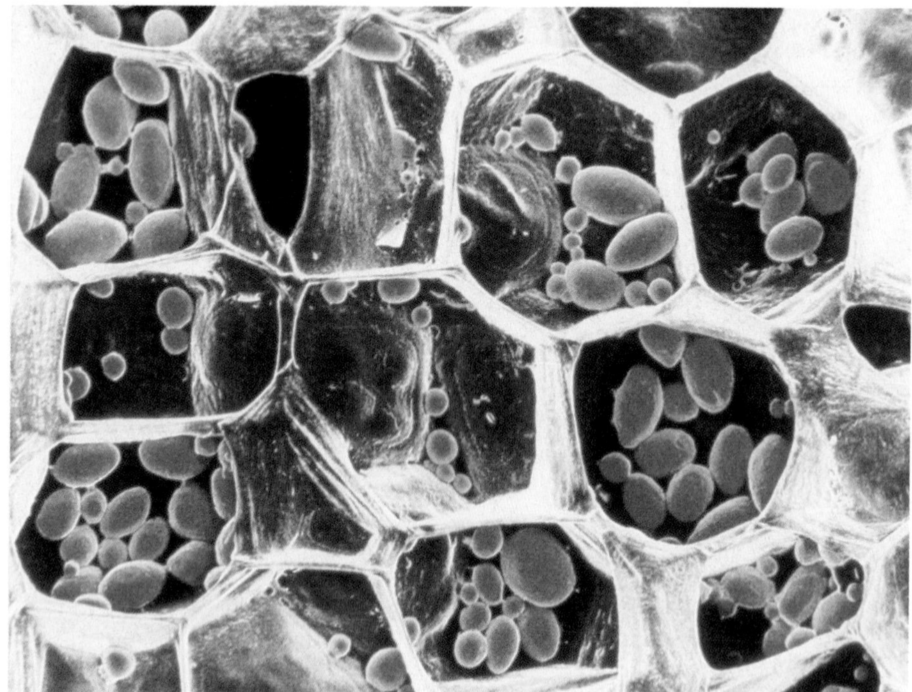

This scanning electron micrograph shows plant cell walls (yellow honeycombed structures), which consist primarily of cellulose. The orange granules are stored starch. Both cellulose and starch are carbohydrates. Cellulose provides structural support for these cells. Starch supplies chemical energy.

This unit highlights the four types of macromolecules that are prominent in today's cells: proteins, nucleic acids, carbohydrates, and lipids. Understanding the structure and function of each of these macromolecules is a basic requirement for exploring how life began and how organisms work. Chapters 3 and 4 analyzed the way proteins and nucleic acids are put together and what they do. This chapter focuses on carbohydrates; Chapter 6 will introduce lipids.

The term **carbohydrate** encompasses the monomers called monosaccharides ("one-sugar"), small polymers called oligosaccharides ("few-sugars"), and the large polymers called polysaccharides ("many-sugars"). The name carbohydrate is logical because the chemical formula of many of these molecules is $(CH_2O)_n$, where the *n* refers to the number of "carbon-hydrate" groups. The name can also be misleading, though, because carbohydrates do not consist of carbon atoms bonded to water molecules. Instead, they are molecules with a carbonyl ($\rangle C{=}O$) and several hydroxyl (–OH) functional groups, along with several to many carbon-hydrogen (C–H) bonds.

Let's begin by considering the structure of monosaccharides and continue by analyzing how these building blocks polymerize to form polysaccharides. The chapter closes with an analysis of the role of carbohydrates in the origin of life and a look at what carbohydrates do in cells today.

5.1 Sugars as Monomers

Sugars are fundamental to life. They provide chemical energy in cells and furnish some of the molecular building blocks required for the synthesis of larger, more complex compounds. Monosaccharides were also important during chemical evolution, early in Earth's history. For example, the sugar called ribose is required for the formation of nucleotides. Laboratory simulations have shown that ribose and many other monosaccharides could have been produced in the prebiotic soup. What are these compounds, and how do they differ from one another?

Figure 5.1 illustrates the structure of the monomer called a **monosaccharide**, or simple sugar. The carbonyl group that serves as one of monosaccharides' distinguishing features can be found either at the end of the molecule, forming an aldehyde sugar (an aldose), or within the carbon chain, forming a ketone sugar (a ketose). The presence of a carbonyl group along with multiple hydroxyl groups provides an array of functional groups in sugars. Based on this observation, it's not surprising that sugars are able to participate in a large number of chemical reactions.

The number of carbon atoms present also varies in monosaccharides. By convention, the carbons in a monosaccharide are numbered consecutively, starting with the end nearest the carbonyl group. Figure 5.1 features three-carbon sugars, or **trioses**. Ribose, which acts as a building block for nucleotides, has five carbons and is called **pentose**; the glucose that is coursing through your bloodstream and being used by your cells right now is a six-carbon sugar, or **hexose**.

Besides varying in the location of the carbonyl group and the total number of carbon atoms present, monosaccharides can vary in the spatial arrangement of their atoms. There is, for example, a wide array of pentoses and hexoses. Each is distinguished by the configuration of its hydroxyl functional groups. **Figure 5.2** illustrates glucose and galactose, which are six-carbon sugars that are optical isomers—they have the same chemical formula ($C_6H_{12}O_6$) but not the same structure. Although both are aldose sugars with six carbons, they differ in the spatial arrangement of the hydroxyl group at the carbon highlighted in Figure 5.2. Because their structures differ, their functions differ. In cells, glucose is used as a source of chemical energy that sustains life. But for galactose to be used as a source of energy, it first has to be converted to glucose via an enzyme-catalyzed reaction. A total of eight different hexoses exist, due to the different ways of configuring the hydroxyl groups in space. In addition, each hexose comes in two forms, each a mirror image of the other. Thus, 16 distinct structures with the molecular formula $C_6H_{12}O_6$ are possible.

Sugars do not usually exist in the form of linear chains as illustrated in Figure 5.1 and Figure 5.2, however. Instead, in aqueous solution they tend to form ring structures. The chain and ring forms exist in equilibrium when simple sugars are in solution, although almost all are in the ring shape. Glucose serves as the example in **Figure 5.3**. When the cyclic structure forms in glucose, the carbon that is numbered 1 in the linear chain forms a bond with an oxygen atom and with a hydroxyl group that can be oriented in two distinct ways. As the right-hand side of Figure 5.3 shows, the different configurations result in the molecules α-glucose and β-glucose.

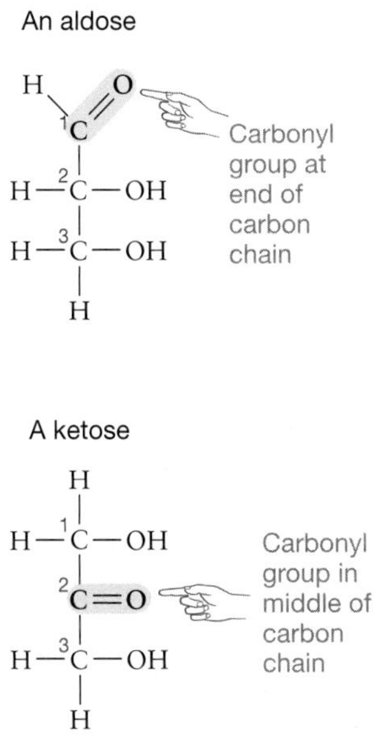

FIGURE 5.1 The Carbonyl Group in a Sugar Occurs in One of Two Configurations.

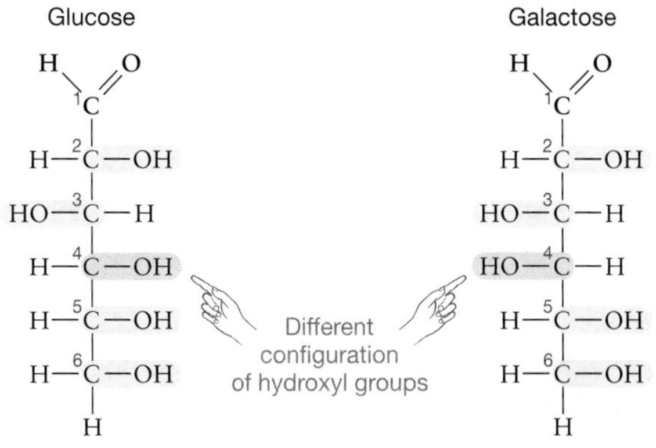

FIGURE 5.2 Sugars May Vary in the Configuration of Their Hydroxyl Groups. The two six-carbon sugars shown here vary only in the spatial orientation of their hydroxyl groups. These molecules are optical isomers, but they are not mirror images of each other.

EXERCISE Mannose is a six-carbon sugar that is identical to glucose, except that the hydroxyl (OH) group on carbon number 2 is switched in orientation. Next to these structures, draw the structural formula of mannose.

QUESTION What kind of isomers are mannose and glucose?

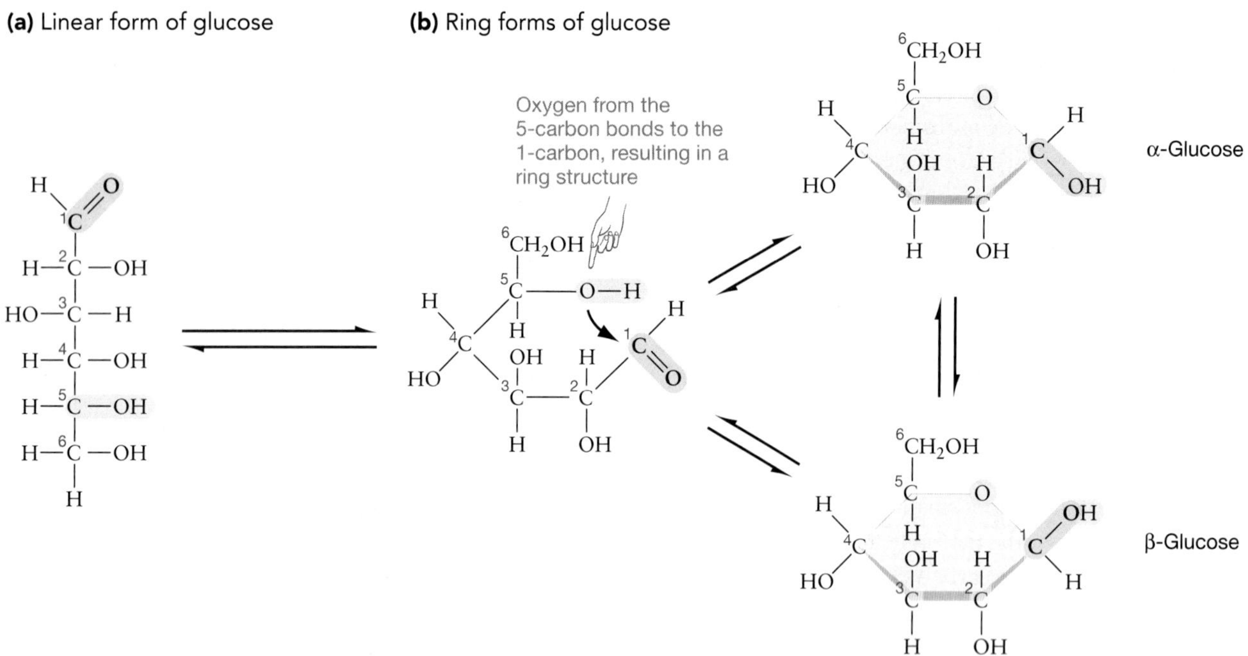

FIGURE 5.3 Sugars Exist in Linear and Ring Forms. (a) The linear form of glucose is rare. **(b)** In solution, almost all glucose molecules spontaneously bend into one of two ring structures, called the α and β forms of glucose. The difference between the two forms lies in whether the hydroxyl group on carbon number 1 is above or below the plane of the ring. The two forms exist in equilibrium, but the β form is more common because it is slightly more stable than the α form.

To summarize, many distinct monosaccharides exist because so many aspects of their structure are variable: alternative ring forms, optical isomers with different arrangements of hydroxyl groups in space and different mirror-image forms, variation in carbon number, and aldose or ketose placement of the carbonyl group. Each monosaccharide has a unique structure and function.

Laboratory simulations have shown that most monosaccharides are readily synthesized under conditions that mimic the prebiotic soup. For example, when formaldehyde (H_2CO) molecules are heated in solution, they react with one another to form almost all of the pentoses and hexoses, as well as some seven-carbon sugars. In addition, researchers recently announced the discovery of the three-carbon ketose illustrated in Figure 5.1, along with a wide array of compounds closely related to sugars, on the Murchison meteorite that struck Australia in 1969. Based on these observations, investigators suspect that sugars are synthesized on dust particles and other debris in interstellar space and could have rained down onto Earth as the planet was forming. Most researchers interested in chemical evolution maintain that a wide diversity of monosaccharides existed in the prebiotic soup. But it remains a mystery why ribose might have predominated and made the synthesis of nucleotides possible. It also appears highly unlikely that monosaccharides were able to polymerize to form the polysaccharides found in today's cells. Let's explore why.

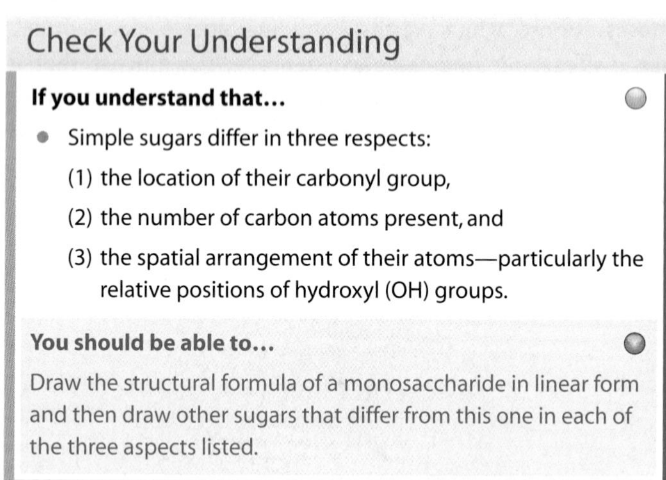
Check Your Understanding

If you understand that...

- Simple sugars differ in three respects:
  (1) the location of their carbonyl group,
  (2) the number of carbon atoms present, and
  (3) the spatial arrangement of their atoms—particularly the relative positions of hydroxyl (OH) groups.

You should be able to...

Draw the structural formula of a monosaccharide in linear form and then draw other sugars that differ from this one in each of the three aspects listed.

5.2 The Structure of Polysaccharides

Polysaccharides are polymers that form when monosaccharides are linked together. They are also known as complex carbohydrates. The simplest polysaccharides consist of two sugars and are known as **disaccharides**. The two monomers involved may be identical, as in the two α-glucose molecules that link to form maltose. Or they may be different, as in the combination of a glucose molecule and a galactose molecule

(a) Monosaccharides polymerize when hydroxyl groups react to form glycosidic linkages...

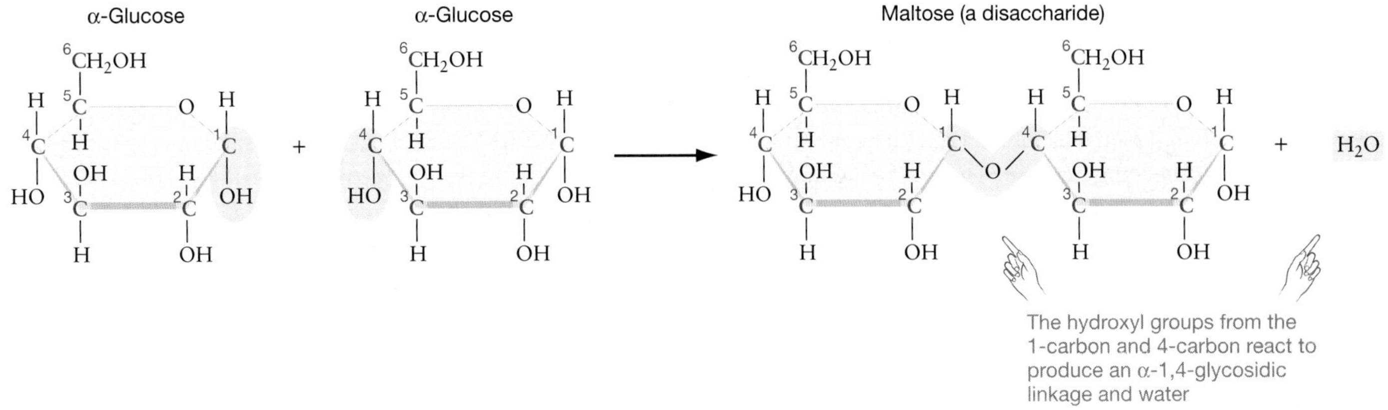

α-Glucose

α-Glucose

Maltose (a disaccharide)

+ H_2O

The hydroxyl groups from the 1-carbon and 4-carbon react to produce an α-1,4-glycosidic linkage and water

(b) ...between various carbons and with various geometries.

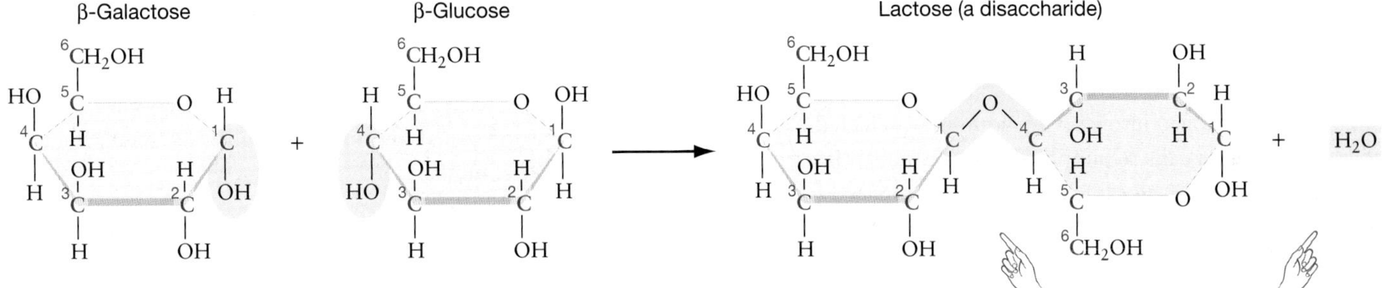

In this case, the hydroxyl groups from the 1-carbon and 4-carbon react to produce a β-1,4-glycosidic linkage and water

FIGURE 5.4 Monosaccharides Polymerize through Formation of Glycosidic Linkages. A glycosidic linkage occurs when hydroxyl groups on two monosaccharides undergo a condensation reaction to form a bond.

that forms lactose—the most important sugar in milk. (**Box 5.1** on page 88 explains how problems with metabolizing lactose or galactose can lead to illness in humans.)

Simple sugars polymerize when a condensation reaction occurs between two hydroxyl groups, resulting in a covalent bond called a **glycosidic linkage** (**Figure 5.4a**). Glycosidic linkages are analogous to the peptide bonds that hold proteins together and to the phosphodiester bonds that connect the nucleotides in nucleic acids. There is an important difference, however. Peptide and phosphodiester bonds always form at the same location in their monomers. But because glycosidic linkages form between hydroxyl groups, and because every monosaccharide contains at least two hydroxyl groups, the location and geometry of glycosidic linkages can vary widely among polysaccharides (**Figure 5.4b**). To drive this point home, consider the structures of the most common polysaccharides found in organisms today: starch, glycogen, cellulose, and chitin, along with a modified polysaccharide called peptidoglycan. Each of these macromolecules can consist of a few hundred to many thousands of monomers (**Figure 5.5**), joined by glycosidic linkages at different locations.

Starch: A Storage Polysaccharide in Plants

In plant cells, monosaccharides are stored for later use in the form of starch. **Starch** consists entirely of α-glucose monomers

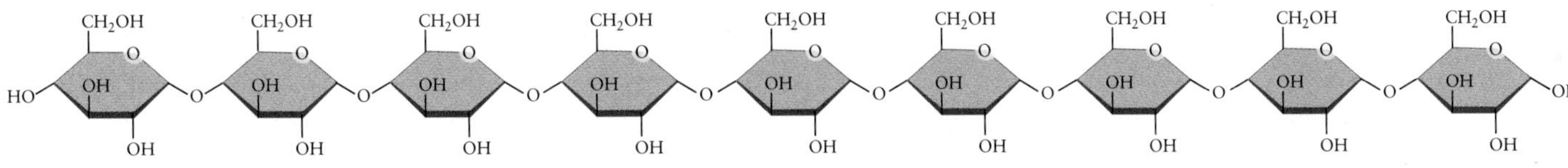

FIGURE 5.5 Polysaccharides Are Polymers of Monosaccharides. Note that sugars are often drawn in simplified form for legibility.

EXERCISE Circle the glycosidic linkages in this polysaccharide.

that are joined by glycosidic linkages. As the top panel in **Table 5.1** shows, the angle of the linkages between carbons 1 and 4 causes the chain of glucose subunits to coil into a helix. Starch is actually a mixture of two such polysaccharides, however. One is an unbranched molecule called amylose, which contains only α-1,4-glycosidic linkages. The other is a branched molecule called amylopectin. The branching in amylopectin occurs when glycosidic linkages form between carbon 1 of a glucose monomer on one strand and carbon 6 of a glucose monomer on another strand. In amylopectin, branches occur in about one out of every 30 monomers.

Glycogen: A Highly Branched Storage Polysaccharide in Animals

Glycogen performs the same storage role in animals that starch performs in plants. In humans, for example, glycogen is stored in the liver and in muscles. When you start exercising, enzymes begin breaking glycogen into glucose monomers, which are then processed in muscle cells to supply energy. Glycogen is a polymer of α-glucose and is nearly identical to the branched form of starch. But instead of an α-1,6-glycosidic linkage occurring in about one out of every 30 monomers, a branch occurs in about one out of every 10 glucose subunits (see Table 5.1).

Cellulose: A Structural Polysaccharide in Plants

As Chapter 1 noted, all cells are enclosed by a membrane. In most unicellular and most multicellular organisms living today, the cell is also surrounded by a layer of material called a wall. A **cell wall** is a protective sheet that occurs outside the membrane. In algae, plants, bacteria, fungi, and many other groups, the cell wall is composed primarily of one or more polysaccharides.

In plants, cellulose is the major component of the cell wall. **Cellulose** is a polymer of β-glucose monomers, joined by β-1,4-glycosidic linkages. As Table 5.1 shows, the geometry of the bond is such that each glucose monomer in the chain is flipped in relation to the adjacent monomer. This arrangement increases the stability of cellulose strands, because the flipped orientation makes it possible for multiple hydrogen bonds to form between adjacent and parallel strands of cellulose. As Table 5.1 also shows, cellulose often occurs in long, parallel strands that are joined by these hydrogen bonds. The linked cellulose fibers are strong and provide the cell with structural support.

Chitin: A Structural Polysaccharide in Fungi and Animals

Chitin is a polysaccharide that stiffens the cell walls of fungi. It is also found in a few algae and in many animals; it is, for example, the most important component of the external skeletons of insects and crustaceans.

Chitin is similar to cellulose, but instead of consisting of glucose monomers, the monosaccharide involved is one called *N*-acetylglucosamine. These monomers are abbreviated "NAc." NAc are joined by β-1,4-glycosidic linkages (see Table 5.1). As in cellulose, the geometry of these bonds results in every other residue being flipped in orientation.

Like the glucose monomers in cellulose, the subunits of *N*-acetylglucosamine in chitin form hydrogen bonds between adjacent strands. The result is a tough sheet that provides stiffness and protection.

Peptidoglycan: A Structural Polysaccharide in Bacteria

Bacteria, like plants, have cell walls. But unlike plants, in bacteria the ability to produce cellulose is extremely rare. Instead, a polysaccharide called **peptidoglycan** gives bacterial cell walls strength and firmness.

Peptidoglycan is the most complex of the polysaccharides discussed thus far. It has a long backbone formed by two types of monosaccharides that alternate with each other and are linked by β-1,4-glycosidic linkages. In addition, a chain of amino acids is attached to one of the two sugar types. When molecules of peptidoglycan align, peptide bonds link the amino acid chains on adjacent strands. **Box 5.2** on page 88 describes how certain antibiotics kill bacteria by disrupting the enzymes that catalyze the formation of the peptide-bonded cross-links.

Although the presence of amino acids gives peptidoglycan a more complex structure than either cellulose or chitin, it is important to note that all three of these polysaccharides have an important feature in common. Structural polysaccharides usually exist as sets of long, parallel strands that are linked to each other. This design gives materials that are made of these molecules the ability to withstand forces that pull or push on them—what an engineer would call tension and compression. In this way, the structure and function of structural polysaccharides are correlated.

MB **Web Animation** at www.masteringbio.com
Carbohydrate Structure and Function

Polysaccharides and Chemical Evolution

Cellulose is the most abundant organic compound on Earth today, and chitin is probably the second most abundant by weight. Virtually all organisms manufacture glycogen or starch. But despite their current importance to organisms, polysaccharides probably played little to no role in the origin of life. This conclusion is supported by several observations:

SUMMARY TABLE 5.1 **Polysaccharides Differ in Structure**

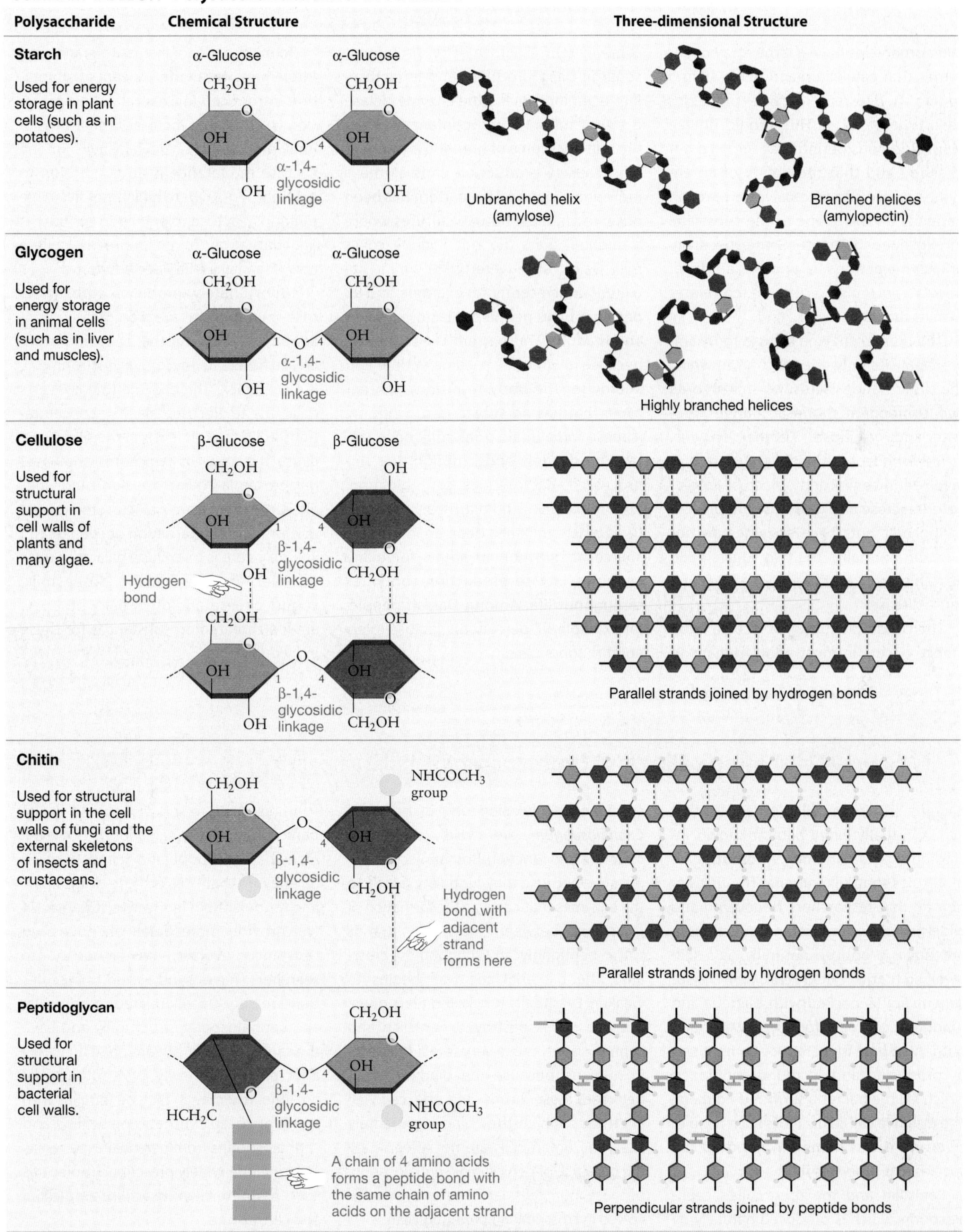

Polysaccharide	Chemical Structure	Three-dimensional Structure
Starch Used for energy storage in plant cells (such as in potatoes).	α-Glucose; α-Glucose; CH_2OH; O; OH; 1; O; 4; α-1,4-glycosidic linkage	Unbranched helix (amylose); Branched helices (amylopectin)
Glycogen Used for energy storage in animal cells (such as in liver and muscles).	α-Glucose; α-Glucose; CH_2OH; O; OH; 1; O; 4; α-1,4-glycosidic linkage	Highly branched helices
Cellulose Used for structural support in cell walls of plants and many algae.	β-Glucose; β-Glucose; CH_2OH; OH; O; 1; 4; β-1,4-glycosidic linkage; Hydrogen bond	Parallel strands joined by hydrogen bonds
Chitin Used for structural support in the cell walls of fungi and the external skeletons of insects and crustaceans.	CH_2OH; O; OH; 1; 4; β-1,4-glycosidic linkage; $NHCOCH_3$ group; Hydrogen bond with adjacent strand forms here	Parallel strands joined by hydrogen bonds
Peptidoglycan Used for structural support in bacterial cell walls.	CH_2OH; O; OH; 1; O; 4; β-1,4-glycosidic linkage; HCH_2C; $NHCOCH_3$ group; A chain of 4 amino acids forms a peptide bond with the same chain of amino acids on the adjacent strand	Perpendicular strands joined by peptide bonds

BOX 5.1 Lactose Intolerance and Galactosemia

Most organisms have a large array of enzymes that catalyze reactions involving sugars. A different enzyme processes each type of sugar. This should not be surprising, considering that sugars differ in shape and that the ability of an enzyme's active site to catalyze a reaction depends on the shape of the substrate. As explained in Chapter 3, the specificity of enzyme action is a prime example of how a molecule's structure correlates with its function.

The human enzyme lactase, for example, catalyzes only one reaction, in which the disaccharide lactose is split into its two component monosaccharides, glucose and galactose. Another enzyme, called 1-phosphate uridyl transferase, is involved in converting galactose to glucose. If lactose and galactose are present in the diet but these enzymes do not function normally, the two sugars build up to high concentration in the body and cause disease.

The condition known as lactose intolerance occurs in most adult humans to some extent. Lactose is a prominent sugar in dairy products; and in lactose-intolerant individuals, the enzyme lactase is produced only during infancy—when the primary source of nutrition is mother's milk. Lactase production ends as these individuals mature. This pattern has been observed in most human cultures where traditional diets did not include cow's milk. Dairy cattle were unknown in Thai culture until recently, for example, and 97 percent of Thai people are lactose intolerant as adults. This pattern is also logical, because it would be a waste of time and energy for the body to produce an enzyme that has no substrate to work on. If these individuals drink milk once lactase production ends, though, the unprocessed lactose causes bloating, abdominal pain, and diarrhea. But in cultures where milk has been an important source of nutrition in adults for many centuries, lactase production continues throughout life. Among Danish people, for example, 97 percent are lactose tolerant throughout life.

To make sense of lactose tolerance and intolerance, recall the concept of adaptation introduced in Chapter 1. An adaptation is a trait that increases the fitness of individuals in a particular environment. Lactose intolerance is an adaptation to environments where milk is not available to adults. Lactose tolerance, in contrast, is an adaptation to environments where milk is readily available to adults.

Unfortunately, problems with galactose metabolism are completely nonadaptive—meaning that they do not increase fitness in any known environment. If a child lacks the enzyme that converts galactose to glucose, galactose concentrations build to high levels in the blood and cause the array of symptoms known as galactosemia ("galactose-sign"). Individuals who suffer from galactosemia are at risk for mental retardation or even death. The only cure is to exclude galactose from the diet. In large enough doses, milk, yogurt, cheese, and other dairy products are life-threatening substances for these individuals.

BOX 5.2 How Do the Penicillins and Cephalosporins Kill Bacteria?

Bacteria are single-celled organisms that live in virtually every habitat known, including the human body. **Antibiotics** are molecules that kill bacteria. The first antibiotic that researchers discovered and widely used to cure bacterial infections in humans is produced naturally, by a soil-dwelling fungus called *Penicillium chrysogenum*. Soil is packed with bacteria and fungi that compete for space, water, and nutrients. Some fungi reduce competition by producing and secreting antibiotics, which kill the bacteria surrounding them. The drug we call penicillin is such a molecule, and it was named in honor of the species that produces it.

Penicillin and the drug called cephalosporin, which is produced naturally by species of soil-dwelling fungi in the genus *Cephalosporium*, are closely related in structure and function. Both molecules are effective because they bind very tightly to the enzymes that catalyze the formation of cross-links between individual strands within peptidoglycan. Without these cross-links, the bacterial cell wall begins to weaken. Eventually it fails and tears open, and the cell is destroyed. Penicillin and cephalosporin cause very few side effects in humans, because the binding event between these drugs and the cell-wall enzyme to which they bind is extremely specific. Human cells are not affected by the presence of penicillin or cephalosporin, because our cells lack the enzymes involved in cross-linking peptidoglycan.

In response to the widespread use of penicillin and cephalosporin, however, many bacterial populations have evolved an enzyme that breaks these drugs apart and renders them ineffective. (Chapter 24 will provide more detail on how drug resistance evolves.) In response, researchers have synthesized molecules that are very closely related to penicillin and cephalosporin structurally and have the same function, but are not as severely affected by the newly evolved bacterial enzyme. You may, in fact, be familiar with some of these synthetic penicillins and cephalosporins from personal experience. They include widely prescribed molecules such as methicillin, oxacillin, ampicillin, and ceftriaxone.

- No plausible mechanism exists for the polymerization of monosaccharides under conditions that prevailed early in Earth's history. In cells and in laboratory experiments, the glycosidic linkages illustrated in Figure 5.4 and Table 5.1 form only with the aid of specialized enzymes.
- It is highly likely that life began in the form of an RNA molecule. However, researchers have yet to discover a ribozyme that can join simple sugars by catalyzing the formation of glycosidic linkages. Thus, it appears extremely unlikely that polysaccharides were present in significant quantities during the RNA world.
- Even though monosaccharides contain large numbers of hydroxyl and carbonyl groups, they lack the diversity of functional groups found in amino acids. Polysaccharides also have simple secondary structures, consisting of linkages between adjacent strands. Thus, they lack the structural and chemical complexity that makes proteins, and to a lesser extent RNA, effective catalysts. To date, no reactions have been discovered that are catalyzed by polysaccharides.
- The monomers in polysaccharides are not capable of complementary base pairing. Like proteins, but unlike DNA and RNA, polysaccharides cannot provide the information required for themselves to be copied. As far as is known, no polysaccharides store information in cells. Thus, no one has proposed that the first living entity might have been a polysaccharide.

Even though polysaccharides probably did not play a significant role in the earliest forms of life, they became enormously important once cellular life evolved. Let's take a detailed look at how they function in today's cells.

Check Your Understanding

If you understand that...

- Polysaccharides form when enzymes catalyze the formation of glycosidic linkages between monosaccharides that are in the α or β form.
- Most polysaccharides are long, linear molecules, but some branch extensively. Among linear forms, it is common for adjacent strands to be linked by hydrogen bonding or other types of linkages.

You should be able to...

1) Make a rough sketch that shows the structures of glycogen and cellulose.
2) Label the glycosidic linkages in each structure.
3) Indicate where hydrogen bonds form between adjacent strands of cellulose.
4) Check your work by using Table 5.1.

5.3 What Do Carbohydrates Do?

Chapter 4 introduced one of the four basic functions that carbohydrates perform in organisms: furnishing the building blocks of molecules needed by cells. Recall that both RNA and DNA contain sugars—the five-carbon sugars ribose and deoxyribose, respectively. In nucleotides, which consist of a sugar, a phosphate group, and a nitrogenous base, the sugar itself acts as a subunit of the larger molecule. But sugars frequently furnish only the raw "carbon skeletons" that are used in the synthesis of important molecules. Amino acids are being synthesized by your cells right now, for example, using sugars as a starting point.

Although the details of how sugars are used in synthesizing amino acids and other complex molecules are beyond the scope of this book, it will be productive to delve into the other three major roles of carbohydrates. Carbohydrates have diverse functions in cells: They provide fibrous structural materials, indicate cell identity, and store chemical energy.

Carbohydrates as Structural Molecules

Cellulose and chitin, along with the modified polysaccharide peptidoglycan, are key structural compounds. They form fibers that give cells and organisms strength and elasticity.

To appreciate why cellulose, chitin, and peptidoglycan are effective as structural molecules, recall that they form long strands and that bonds can form between adjacent strands. Table 5.1 detailed how these bonds form. In the cell walls of plants, a collection of about 80 cellulose molecules are cross-linked by hydrogen bonding to create a tough fiber. These cellulose fibers, in turn, crisscross to form a tough sheet (**Figure 5.6a**). Groups of chitin molecules are also organized into fibers held together by hydrogen bonding. Although each individual hydrogen bond in cellulose and chitin is relatively weak, the combination of many weak bonds allows these structural polysaccharides to form large fibers that are strong, yet flexible. Chitin fibers stiffen the cell walls of fungi, and they overlap in the external skeletons of insects to form a dense, waterproof sheet (**Figure 5.6b**). In the cell walls of bacteria, the presence of amino acid residues in peptidoglycan allows adjacent fibers to cross-link, via peptide bonds, into durable sheets (**Figure 5.6c**).

Besides being stiff and strong, the structural carbohydrates are durable. Almost all organisms have the enzymes required to break the α-1,4- and α-1,6-glycosidic linkages that hold starch and glycogen molecules together, but only a few organisms have enzymes capable of hydrolyzing the β-1,4-glycosidic linkages in cellulose, chitin, and peptidoglycan. The shape and orientation of β-1,4-glycosidic linkages make them difficult to break, and few enzymes have active sites with the correct geometry and reactive groups to do so. As a result, the structural polysaccharides are resistant to degradation and decay.

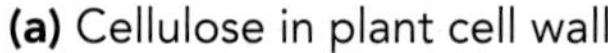

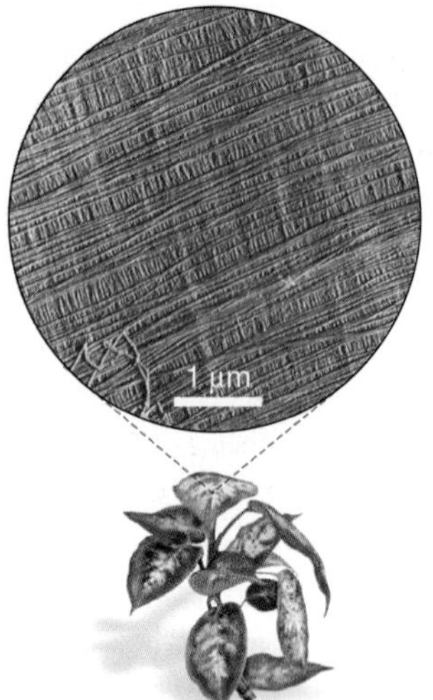

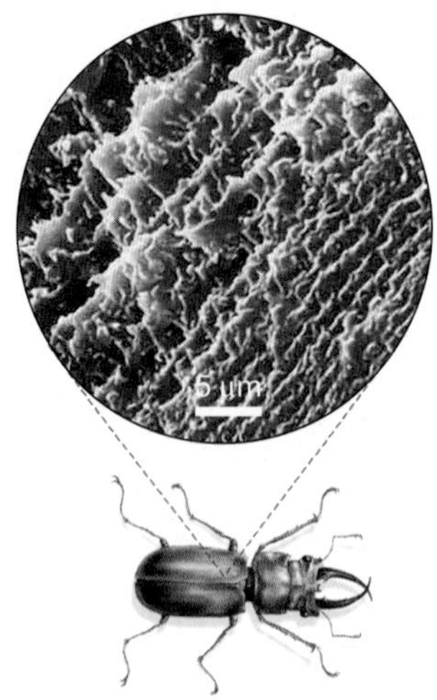

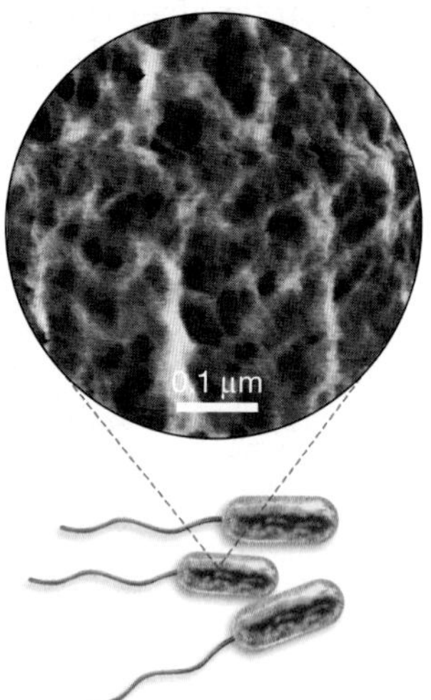

FIGURE 5.6 Cellulose, Chitin, and Peptidoglycan Form Tough Fibers or Sheets.

Ironically, the durability of cellulose is important for digestion—even though the cellulose that you ingest when you eat plant cells passes through your gut undigested. The cellulose that you eat is referred to as dietary fiber. The cellulose in dietary fiber, like the cellulose in a paper towel, absorbs water. The presence of dietary fiber adds moisture to the feces. In addition, cellulose adds bulk that helps fecal material move through the intestinal tract more quickly, preventing colon cancer, constipation, and other problems.

The Role of Carbohydrates in Cell Identity

Polysaccharides do not store information in cells, but they do *display* important information—specifically, on the outer surface of a cell. **Figure 5.7** shows how this information display happens. Certain molecules project outward from the cell surface into the surrounding environment. These molecules are called glycoproteins. A **glycoprotein** is a protein that is covalently bonded to a carbohydrate—usually the relatively short chain of sugars called oligosaccharides.

Glycoproteins are key molecules in what biologists call cell–cell recognition and cell–cell signaling. Each cell in your body has glycoproteins on its surface that identify it as part of your body. Immune system cells use these glycoproteins to distinguish your body's cells from foreign cells, such as bacteria. In addition, each distinct type of cell in a multicellular organism—for example, the nerve cells and muscle cells in your body—displays a different set of glycoproteins on its surface. In cells, form a "sugar coating" that acts like the magnetic stripe on the back of a credit card or the personal identification number (PIN) that you use to access a bank account—it immediately identifies the individual that bears it.

● If you understand this concept, you should be able to draw two cells, and then add and label (1) glycoproteins that identify each of them as distinct cell types and (2) glycoproteins that identify them as part of your body.

The identification information displayed by glycoproteins helps cells recognize and communicate with each other. The key point here is to recognize that the enormous number of structurally distinct monosaccharides makes it possible for an enormous number of unique oligosaccharides to exist. As a result, each cell type and each species can display a unique identity.

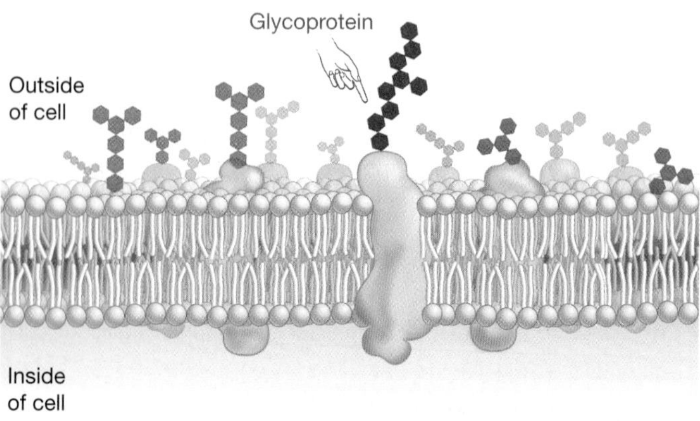

FIGURE 5.7 Carbohydrates Are an Identification Badge for Cells. Glycoproteins contain sugar groups that project from the surface of the plasma membrane. These sugar groups have distinctive structures that identify the type or species of the cell.

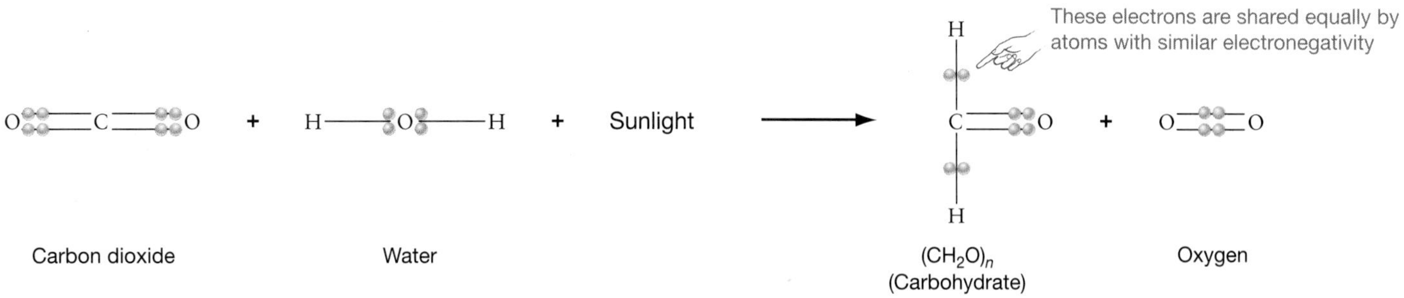

FIGURE 5.8 Carbohydrates Have High Free Energy. In these diagrams, the horizontal lines indicate covalent bonds. The dots represent the relative positions of electrons in those bonds.

EXERCISE Circle the bonds in this diagram that have the highest free energy.

The Role of Carbohydrates in Energy Storage

Candy-bar wrappers promise a quick energy boost, and ads for sports drinks claim that their products provide the "carbs" needed for peak activity. If you were to ask friends or family members what carbohydrates do in your body, they would probably say something like "They give you energy." And after pointing out that carbohydrates are also used to establish cell identity, as a structural material, and as a source of carbon skeletons for the synthesis of other complex molecules, you'd have to agree. Carbohydrates store and provide chemical energy in cells. What is it about carbohydrates that makes this possible?

Carbohydrates and Free Energy Recall from earlier chapters that the essence of chemical evolution was the conversion of kinetic energy in sunlight and heat into chemical energy stored in the bonds of molecules such as formaldehyde (H_2CO) and hydrogen cyanide (HCN). Today, the kinetic energy in sunlight is converted to chemical energy stored in the bonds of carbohydrates, via the process known as **photosynthesis**. Photosynthesis is a complex set of reactions that can be summarized most simply as follows:

$$CO_2 + H_2O + \text{sunlight} \rightarrow (CH_2O)_n + O_2$$

where $(CH_2O)_n$ represents a carbohydrate. **Figure 5.8** shows the structural formulas of the molecules involved and represents the relative positions of their covalently bonded electrons. The keys to understanding this figure are to compare the positions of the electrons in the reactants and the products, and note that the covalently bonded electrons in the C–H bonds of carbohydrates are shared more equally, and thus held less tightly, than they are in the C–O bonds of carbon dioxide. The same is true for the carbon-carbon (C–C) bonds of carbohydrates: Electrons are shared more equally than they are in the C–O bonds of carbon dioxide.

Now, recall from Chapter 2 that when electrons are held tightly by a highly electronegative atom like oxygen, they have low potential energy. But when they are bonded to a less-electronegative atom like carbon or hydrogen, they are farther from the nucleus and have high potential energy. C–C and C–H bonds have much higher free energy than C–O bonds have. As a result, carbohydrates have much more free energy than carbon dioxide has.

The essence of photosynthesis, then, is that energy in sunlight is transformed into chemical energy that is stored in the C–H and C–C bonds of carbohydrates. The C–C and C–H bonds have high free energy. When a cell needs to tap this stored energy, carbohydrates participate in exergonic reactions that synthesize a molecule called **adenosine triphosphate (ATP).** More specifically, the energy that is released when sugars are processed is used to synthesize ATP from a precursor called adenosine diphosphate (ADP). The overall reaction can be written as follows:

$$CH_2O + O_2 + ADP + P_i \rightarrow CO_2 + H_2O + ATP$$

To put this in words, the chemical energy stored in the C–H and C–C bonds of carbohydrate is transferred to chemical energy in the form of the third phosphate group in ATP. The free energy in ATP drives what would otherwise be endergonic reactions like polymerization, moves your muscles, and performs other types of work in cells. Carbohydrates are like the water that piles up behind a dam; ATP is like the electricity, generated at a dam, that lights up your home. Carbohydrates store chemical energy; ATP "spends" it.

Later chapters will analyze in detail how sugars and other carbohydrates are made in organisms, and how these carbohydrates are then broken down to provide cells with usable chemical energy in the form of ATP. For now, the important thing is to recognize that carbohydrates store and provide chemical energy to cells because they contain a large number of C–H bonds and because C–H bonds have high free energy.

Figure 5.9 summarizes and extends these points. For example, compare the structures of carbon dioxide and the carbohydrate shown in Figure 5.9a and Figure 5.9b, and you'll see that hydrogen atoms—an electron plus a proton—have been added to the

(a) Carbon dioxide

O=C=O

(b) A carbohydrate

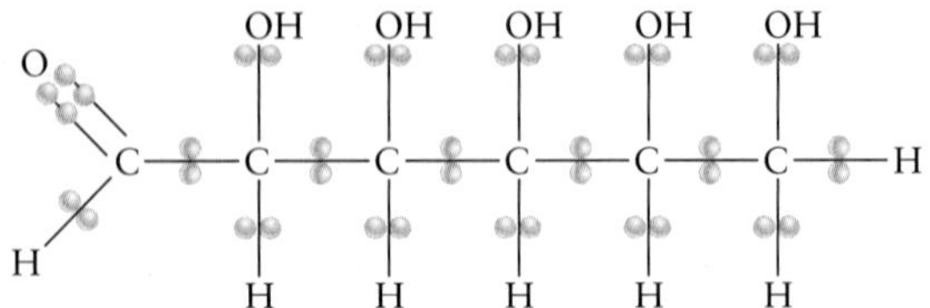

(c) A fatty acid (a component of fat molecules)

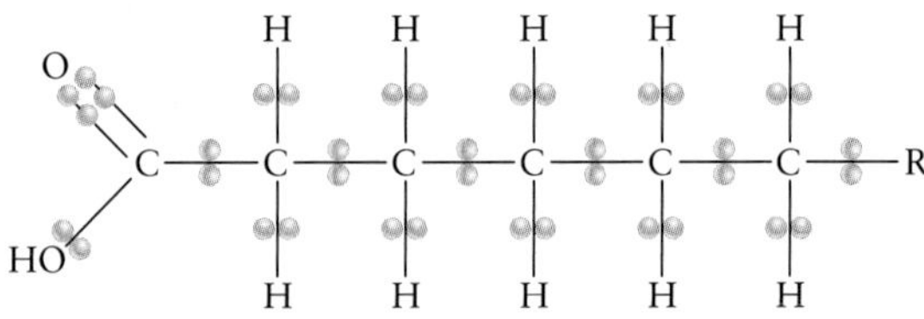

FIGURE 5.9 In Organisms, Free Energy is Stored in C–H and C–C Bonds. (a) In carbon dioxide, the electrons involved in covalent bonds are held tightly by oxygen atoms. **(b)** In carbohydrates such as the sugar shown here, many of the covalently bonded electrons are held loosely between C and H atoms. **(c)** The fatty acids found in fat molecules have more C–H bonds and fewer C–O bonds than carbohydrates do.

EXERCISE Label the molecule with the highest amount of free energy, and the molecule with the lowest amount.

QUESTION Which molecule has the highest amount of potential energy? Of chemical energy?

carbohydrate. As Figure 5.9c shows, the fatty-acid subunits found in fats have even more free energy than carbohydrates have. Fatty acids are made up largely of long chains of carbon atoms bonded to hydrogen atoms. Compared with carbohydrates, fats contain many more C–C and C–H bonds and many fewer C–O bonds. To restate briefly, C–C and C–H bonds have high free energy because the electrons are shared equally by atoms with low electronegativities, while C–O bonds have low free energy because the highly electronegative oxygen atom holds the electrons so tightly. Both carbohydrates and fats are used as fuel in cells. Fats will be discussed in more detail in Chapter 6.

How Do Carbohydrates Store Energy? Starch and glycogen are efficient energy-storage molecules because they polymerize via α-glycosidic linkages instead of the β-glycosidic linkages observed in the structural polysaccharides. The α-linkages in storage polysaccharides are readily hydrolyzed to release sugars, while the β-linkages in structural polysaccharides resist enzymatic degradation. The glucose subunits that are hydrolyzed from starch and glycogen are then processed in reactions that result in the production of ATP. Starch and glycogen are like a candy bar that has segments, so you can break off chunks whenever you need a boost.

The most important enzyme involved in catalyzing the hydrolysis of α-glycosidic linkages in glycogen is called **phosphorylase**. The enzymes involved in breaking these linkages in starch are called **amylases**. Most of your cells contain phosphorylase, so they can break down glycogen to provide glucose on demand. Your salivary glands and pancreas also produce amylases that are secreted into your mouth and small intestine, respectively. These amylases are responsible for digesting the starch that you eat.

Check Your Understanding

If you understand that...

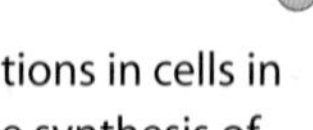

- Carbohydrates have several important functions in cells in addition to providing building blocks for the synthesis of more complex compounds.
- Polysaccharides such as cellulose, chitin, and peptidoglycan form cell walls, which give cells structural strength.
- Glycoproteins project from the surface of cells. They provide a molecular PIN that identifies the cell's type or species.
- Starch and glycogen store sugars for later use in reactions that produce ATP. Sugars contain large amounts of chemical energy because they contain carbon atoms that are bonded to hydrogen atoms or other carbon atoms. These bonds have high free energy because the electrons are shared equally by atoms with low electronegativity.

You should be able to...

1) Identify two aspects of the structures of cellulose, chitin, and peptidoglycan that correlate with their function as structural molecules.
2) Describe how the carbohydrates you ate during breakfast today are functioning in your body right now.

Chapter Review

SUMMARY OF KEY CONCEPTS

Sugars and other carbohydrates are highly variable in structure.

Carbohydrates are organic compounds that have a carbonyl group and several to many hydroxyl groups. They can occur as the simple sugars called monosaccharides or as complex polysaccharides. Many types of monosaccharide exist, each distinguished by one or more of the following features: (1) the location of their carbonyl group—either at the end of the molecule or within it; (2) the number of carbon atoms they contain—from three to seven, usually; and (3) the orientation of their hydroxyl groups in the linear chain and/or the ring form. In addition, most monosaccharides have optical isomers.

You should be able to draw two monosaccharides and explain at least two aspects of their structure that differ.

Monosaccharides are monomers that polymerize to form polymers called polysaccharides, via different types of glycosidic linkages.

Monosaccharides can be linked together by covalent bonds, called glycosidic linkages, that join hydroxyl groups on adjacent molecules. Unlike the peptide bonds that form proteins and the phosphodiester bonds of nucleic acids, glycosidic linkages may form at several locations on a monosaccharide. The different disaccharides and polysaccharides are distinguished by the type of monomers involved and the location of the glycosidic linkages between them. The most common polysaccharides in organisms today are starch, glycogen, cellulose, and chitin; peptidoglycan is an abundant polysaccharide that has short chains of amino acids attached.

You should be able to draw two disaccharides and explain at least two aspects of their structure that differ.

Carbohydrates perform a wide variety of functions in cells, ranging from energy storage to formation of tough structural fibers.

Starch and glycogen are made up of glucose molecules that are in the α ring form and that are joined by glycosidic linkages between their first and fourth carbons. In some forms of starch, individual chains are occasionally joined to other chains; in glycogen these links are abundant enough to form a highly branched molecule. Both starch and glycogen function as energy-storage molecules. Because sugars contain many C–C and C–H bonds, they contain a significant amount of chemical energy. When cells need energy, enzymes hydrolyze the α-1,4-glycosidic linkages in starch or glycogen. The reaction releases individual glucose molecules, which then are processed in reactions that lead to the production ATP. Chemical energy in the form of ATP is readily usable by cells.

Cellulose, chitin, and peptidoglycan are made up of monosaccharides joined by β-1,4-glycosidic linkages. When individual molecules of these polysaccharides align side by side, bonds form between them. In cellulose and chitin, the individual molecules are joined by hydrogen bonds; in peptidoglycan, the intermolecular links consist of peptide bonds between amino acid chains that extend from the polysaccharide. Hydrogen bonding produces bundles of cellulose and chitin molecules that form strong, elastic fibers; when peptide bonding occurs between many peptidoglycan strands, the result is a tough fibrous or sheetlike substance. Cellulose, chitin, and peptidoglycan provide cells and organisms with support. In addition, few organisms have enzymes that can degrade β-1,4-glycosidic linkages, which makes cell walls made of cellulose, chitin, and peptidoglycan resistant to attack. In carbohydrates, as in proteins and nucleic acids, structure correlates with function.

You should be able to describe two key differences in the structure of polysaccharides that function in energy storage versus structural support.

(MB) Web Animation at www.masteringbio.com
Carbohydrate Structure and Function

QUESTIONS

Test Your Knowledge

1. What is the difference between a monosaccharide, a disaccharide, and a polysaccharide?
   a. the number of carbon atoms in the molecule
   b. the type of glycosidic linkage between monomers
   c. the spatial arrangement of the various hydroxyl residues in the molecule
   d. the number of monomers in the molecule
2. What type of bond allows sugars to polymerize?
   a. glycosidic linkage
   b. phosphodiester bond
   c. peptide bond
   d. hydrogen bonds
3. What holds cellulose molecules together in bundles large enough to form fibers?
   a. the cell wall
   b. peptide bonds
   c. hydrogen bonds
   d. hydrophobic interactions between different residues in the cellulose helix
4. What are the primary functions of carbohydrates in cells?
   a. energy storage, cell identity, structure, and building blocks for synthesis
   b. catalysis, structure, and transport
   c. information storage and catalysis
   d. signal reception, signal transport, and signal response

5. Why is it unlikely that carbohydrates played a large role in the origin of life?
   a. They cannot be produced by chemical evolution.
   b. They have optical isomers.
   c. More types of glycosidic linkages are possible than are actually observed in organisms.
   d. They do not polymerize without the aid of enzymes.

6. What is a "quick and dirty" way to assess how much free energy an organic molecule has?
   a. Count the number of oxygen atoms it contains.
   b. Count the number of C–H bonds it contains.
   c. Count the number of C–O bonds it contains.
   d. Determine whether it contains a carbonyl group.

Test Your Knowledge answers: 1. d; 2. a; 3. c; 4. a; 5. d; 6. b

Test Your Understanding

Answers are available at www.masteringbio.com

1. Explain why the structure of carbohydrates supports their function in signaling the identity of a cell.
2. Draw the ring structure of glucose in the α form. Now add another glucose molecule with an α-1,4-glycosidic linkage, and then a third glucose molecule with an α-1,6-glycosidic linkage.
3. Compare and contrast the structures and functions of starch and glycogen. How are these molecules similar? How are they different?
4. Why do the bonds in a carbohydrate store a large amount of chemical energy compared to the chemical energy stored in the bonds of carbon dioxide?
5. What aspects of the structure of cellulose and chitin supports their function in protecting and stiffening cells and organisms?
6. Both glycogen and cellulose consist of glucose monomers that are linked end to end. How do the structures of these polysaccharides differ? How do their functions differ?

Applying Concepts to New Situations

Answers are available at www.masteringbio.com

1. A weight-loss program for humans that emphasizes minimal consumption of carbohydrates has become popular in recent years. What is the logic behind this diet? (Note: This diet plan has caused controversy and is not endorsed by some physicians and researchers.)
2. To treat galactosemia, physicians exclude the monosaccharide galactose from the diet. Why does the disaccharide lactose also have to be excluded from the diet?
3. Amylase, an enzmye found in human saliva, catalyzes the hydrolysis of the α-1,4-glycosidic linkages in starch. If you hold a salty cracker in your mouth long enough, it will begin to taste sweet. Why?
4. Lysozyme, an enzyme found in human saliva, tears, and other secretions, catalyzes the hydrolysis of the β-1,4-glycosidic linkages in peptidoglycan. What effect does contact with this enzyme have on bacteria?

www.masteringbio.com is also your resource for • Answers to text, table, and figure caption questions and exercises • Answers to *Check Your Understanding* boxes • Online study guides and quizzes • Additional study tools including the *E-Book for Biological Science* 3rd ed., textbook art, animations, and videos.

Lipids, Membranes, and the First Cells

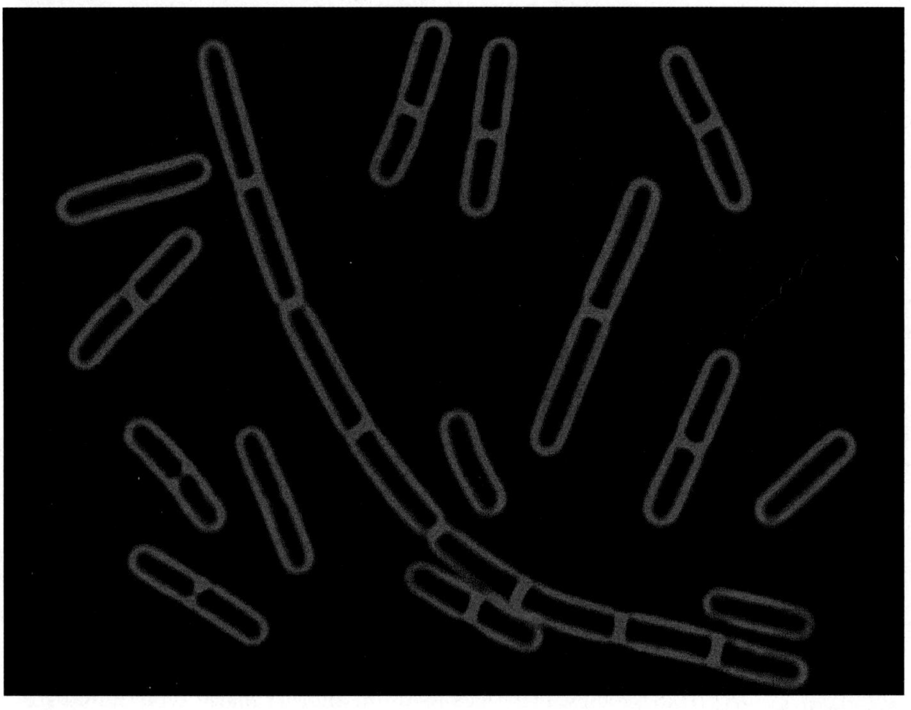

These bacterial cells have been stained with a red compound that inserts itself into the plasma membrane. The plasma membrane defines the cell—the basic unit of life. In single-celled organisms like those shown here, the membrane creates a physical separation between life on the inside and nonlife on the outside.

KEY CONCEPTS

- Phospholipids are amphipathic molecules—they have a hydrophilic region and a hydrophobic region. In solution, they spontaneously form bilayers that are selectively permeable—meaning that only certain substances cross them readily.
- Ions and molecules diffuse spontaneously from regions of high concentration to regions of low concentration. Water moves across lipid bilayers from regions of high concentration to regions of low concentration via osmosis—a special case of diffusion.
- In cells, membrane proteins are responsible for the passage of ions, polar molecules, and large molecules that can't cross the membrane on their own because they are not soluble in lipids.

The research discussed in previous chapters suggests that biological evolution began with an RNA molecule that could make a copy of itself. As the offspring of this molecule multiplied in the prebiotic soup, natural selection would have favored versions of the molecule that were particularly stable and efficient at catalysis. Another great milestone in the history of life occurred when a descendant of this replicator became enclosed within a membrane. This event created the first cell and thus the first organism.

The **cell membrane**, or **plasma membrane**, is a layer of molecules that surrounds the cell, separating it from the external environment and selectively regulating the passage of molecules and ions into or out of the cell. The evolution of the plasma membrane was a momentous development because it separated life from nonlife. Before plasma membranes existed, self-replicating molecules probably clung to clay-sized mineral particles, building copies of themselves as they randomly encountered the appropriate nucleotides in the prebiotic soup that washed over them. But the membrane made an internal environment possible—one that could have a chemical composition different from that of the external environment. This was important for two reasons. First, the chemical reactions necessary for life could occur much more efficiently in an enclosed area, because reactants could collide more frequently. Second, the membrane could serve as a selective barrier. That is, it could keep compounds out of the cell that might damage the replicator, but it might allow the entry of compounds required by the replicator. The membrane not only created the cell but also made it into an efficient and dynamic reaction vessel.

The goal of this chapter is to investigate how membranes behave, with an emphasis on how they differentiate the internal environment from the external environment. Let's begin by examining the structure and properties of the most abundant molecules in plasma membranes: the "oily" or "fatty" compounds called lipids. Then we can delve into analyzing the way lipids behave when they form membranes. Which ions and molecules can pass through a membrane that consists of lipids? Which cannot, and why? The chapter ends by exploring how proteins that become incorporated into a lipid membrane can control the flow of materials across the membrane.

6.1 Lipids

Most biochemists are convinced that the building blocks of membranes, called lipids, existed in the prebiotic soup. This conclusion is based on the observation that several types of lipids have been produced in experiments designed to mimic the chemical and energetic conditions that prevailed early in Earth's history. For example, the spark-discharge experiments reviewed in Chapter 3 succeeded in producing at least two types of lipid.

An observation made by A. D. Bangham illustrates why this result is interesting. In the late 1950s, Bangham performed experiments to determine how lipids behave when they are immersed in water. But until the electron microscope was invented, he had no idea what these lipid-water mixtures looked like. Once **transmission electron microscopes** became available, Bangham was able to produce high magnification, high resolution images of his lipid-water mixtures. (Transmission electron microscopy is introduced in **BioSkills 8.**) The images that resulted, called **micrographs**, were astonishing. As **Figure 6.1a** shows, the lipids had spontaneously formed enclosed compartments filled with water. Bangham called these membrane-bound structures vesicles and noted that they resembled cells (**Figure 6.1b**). Bangham had not done anything special to the lipid-water mixtures; he had merely shaken them by hand.

The experiment raises a series of questions: How could these structures have formed? Is it possible that vesicles like these existed in the prebiotic soup? If so, could they have surrounded a self-replicating molecule and become the first plasma membrane? Let's begin answering these questions by investigating what lipids are and how they behave.

What Is a Lipid?

Earlier chapters analyzed the structures of the organic molecules called amino acids, nucleotides, and monosaccharides and explored how these monomers polymerize to form macromolecules. Here let's focus on another major type of mid-sized molecule found in living organisms: lipids.

(a) In solution, lipids form water-filled vesicles.

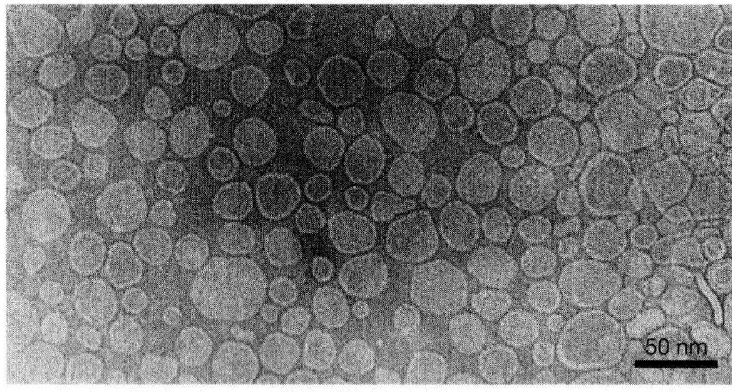

(b) Red blood cells resemble vesicles.

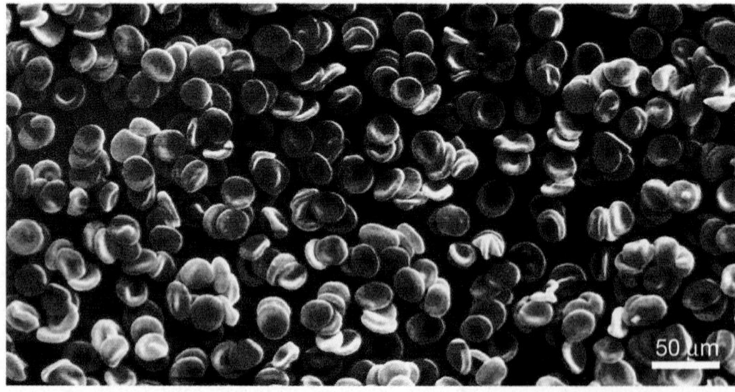

FIGURE 6.1 Lipids Can Form Cell-like Vesicles When in Water. **(a)** Transmission electron micrograph showing a cross section through the tiny, bag-like compartments that formed when a researcher shook a mixture of lipids and water. **(b)** Scanning electron micrograph showing red blood cells from humans. Note the scale bars.

Lipid is a catch-all term for carbon-containing compounds that are found in organisms and are largely nonpolar and hydrophobic—meaning that they do not dissolve readily in water. (Recall from Chapter 2 that water is a polar solvent.) Lipids do dissolve, however, in liquids consisting of nonpolar organic compounds.

To understand why lipids do not dissolve in water, examine the five-carbon compound called isoprene illustrated in **Figure 6.2a**; notice that it consists of a group of carbon atoms bonded to hydrogen atoms. Molecules that contain only carbon and hydrogen, such as isoprene or octane (see Chapter 2) are known as **hydrocarbons**. Hydrocarbons are nonpolar, because electrons are shared equally in carbon-hydrogen bonds. This property makes hydrocarbons hydrophobic. Thus, the reason lipids do not dissolve in water is that they have a significant hydrocarbon component. **Figure 6.2b** is a type of compound called a **fatty acid**, which consists of a hydrocarbon chain bonded to a carboxyl (COOH) functional group. Isoprene and fatty acids are key building blocks of the lipids found in organisms.

(a) Isoprene

$H_2C{=}C(CH_3){-}C(H){=}CH_2$

(b) Fatty acid

HO–C(=O)– Carboxyl group

H_2C–CH_2–H_2C–CH_2–H_2C–CH_2–H_2C–CH_2–H_2C–CH_2–H_2C–CH_2–H_2C–CH_2–H_3C Hydrocarbon chain

FIGURE 6.2 Hydrocarbon Groups Make Lipids Hydrophobic. **(a)** Isoprenes are hydrocarbons. Isoprene subunits can be linked end to end to form long hydrocarbon chains. **(b)** Fatty acids typically contain a total of 14-20 carbon atoms, most found in their long hydrocarbon tails.

● **EXERCISE** Circle the hydrophobic portion of a fatty acid.

A Look at Three Types of Lipids Found in Cells

Unlike amino acids, nucleotides, and carbohydrates, lipids are defined by a physical property—their solubility—instead of their chemical structure. As a result, the structure of lipids varies widely. To drive this point home, consider the structures of the most important types of lipids found in cells: fats, steroids, and phospholipids.

- **Fats** are composed of three fatty acids that are linked to a three-carbon molecule called **glycerol**. Because of this structure, fats are also called triacylglycerols or triglycerides. As **Figure 6.3a** shows, fats form when a dehydration reaction occurs between a hydroxyl group of glycerol and the carboxyl group of a fatty acid. The glycerol and fatty-acid molecules become joined by an **ester linkage**, which is analogous to the peptide bonds, phosphodiester bonds, and glycosidic linkages in proteins, nucleic acids, and carbohydrates, respectively. Fats are not polymers, however, and fatty acids are not monomers. As **Figure 6.3b** shows, fatty acids are not linked together to form a macromolecule in the way that amino acids, nucleotides, and monosaccharides are. ● After studying the structure in Figure 6.3b, you should be able to explain why fats store a great deal of chemical energy, and why they are hydrophobic.

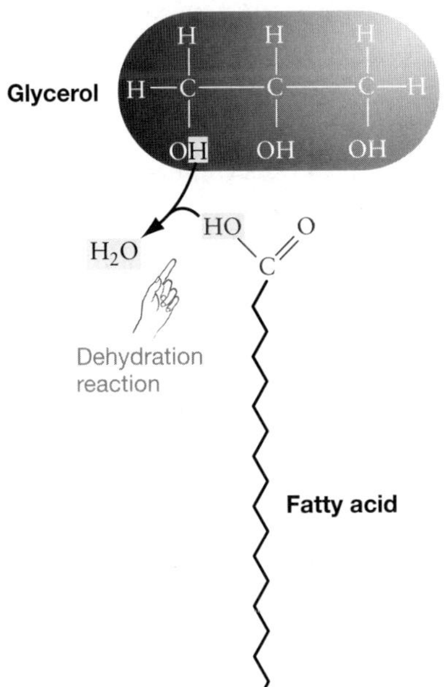

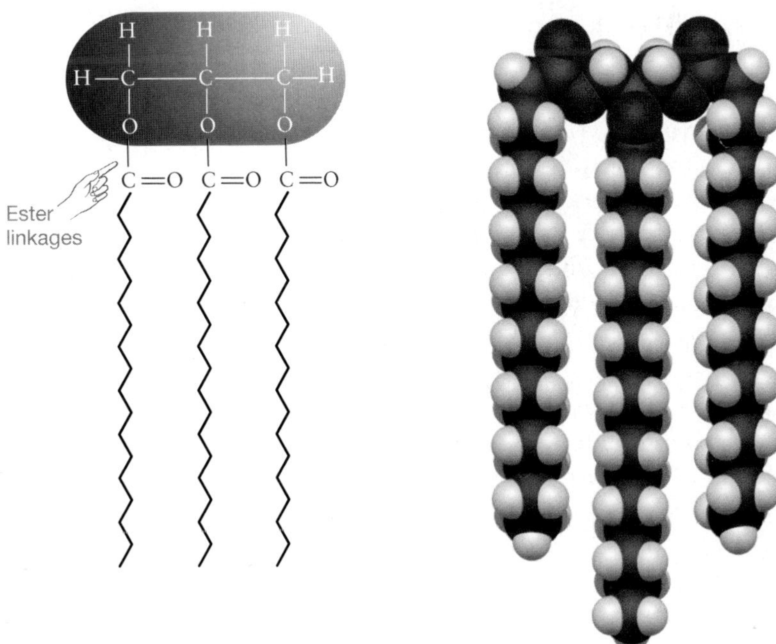

FIGURE 6.3 Fats Are One Type of Lipid Found in Cells. (a) When glycerol and a fatty acid react, a water molecule leaves. (b) The covalent bonds that result from this reaction are termed an ester linkage. The fat shown here as a structural formula and a space-filling model is tristearin, the most common type of fat in beef.

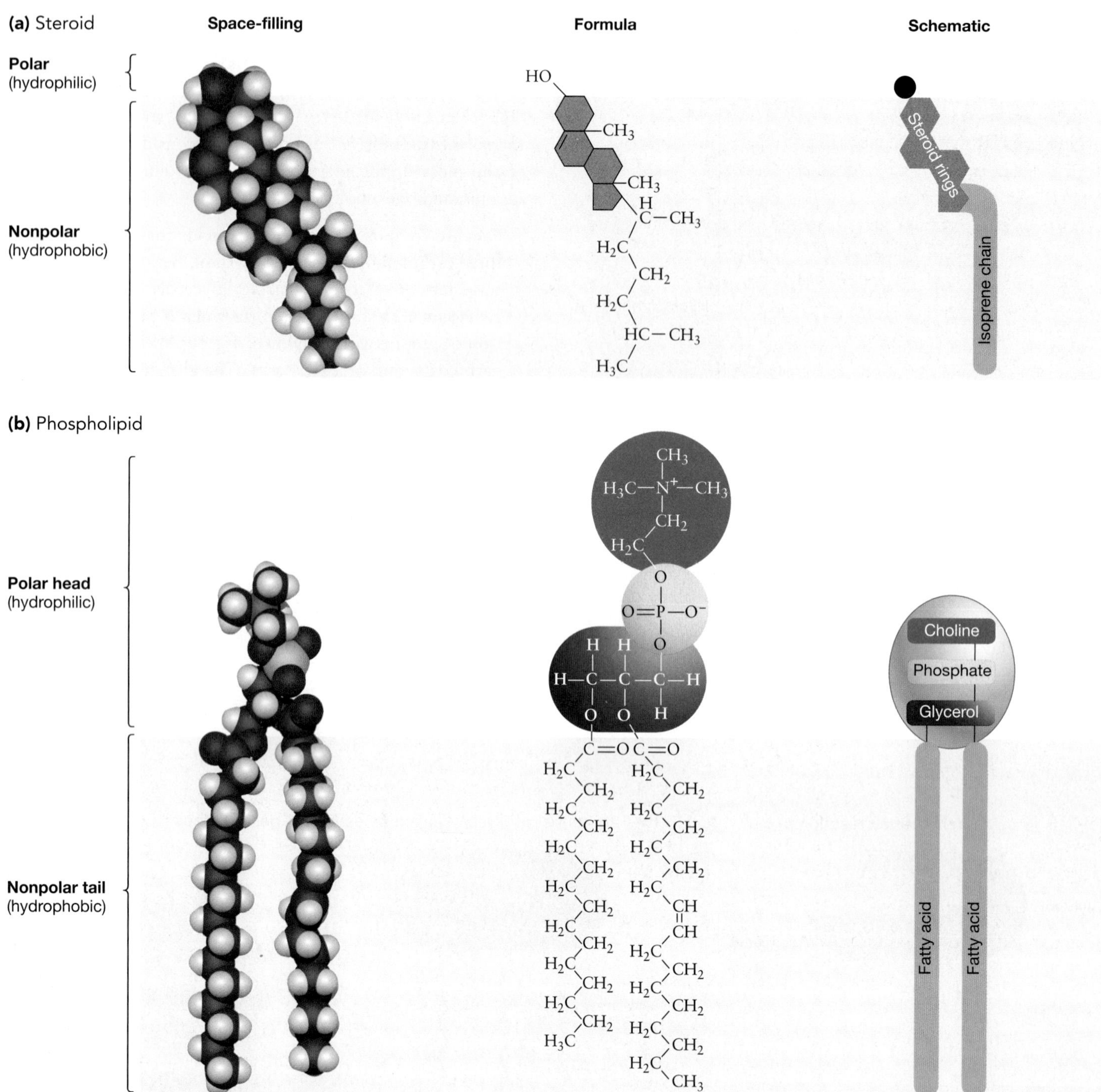

FIGURE 6.4 Amphipathic Lipids Contain Hydrophilic and Hydrophobic Elements. (a) All steroids have a distinctive four-ring structure. **(b)** All phospholipids consist of a glycerol that is linked to a phosphate group and to either two chains of isoprene or two fatty acids.

QUESTION What makes cholesterol—the steroid shown in part (a)—different from other steroids?

QUESTION If these molecules were in solution, where would water molecules interact with them?

- **Steroids** are a family of lipids distinguished by the four-ring structure shown in solid orange in **Figure 6.4a**. The various steroids differ from one another by the functional groups or side groups attached to those rings. The molecule pictured in Figure 6.4a is cholesterol, which is distinguished by a hydrocarbon "tail" formed of isoprene subunits. Cholesterol is an important component of plasma membranes in many organisms. In mammals, it is also used as the starting point for the synthesis of several of the signaling molecules called hormones. Estrogen, progesterone, and testosterone are examples of hormones derived from cholesterol. These molecules are responsible for regulating sexual development and activity in humans.
- **Phospholipids** consist of a glycerol that is linked to a phosphate group (PO_4^{2-}) and to either two chains of isoprene or two fatty acids. In some cases, the phosphate group is bonded to another small organic molecule, such as the choline shown on the phospholipid in **Figure 6.4b**. Phospholipids with isoprene tails are found in the domain Archaea introduced in Chapter 1; phospholipids composed of fatty acids are found in the domains Bacteria and Eukarya. In all three domains of life, phospholipids are critically important components of the plasma membrane.

To summarize, the lipids found in organisms have a wide array of structures and functions. In addition to storing chemical energy and serving as signals between cells, lipids act as pigments that capture or respond to sunlight, form waterproof coatings on leaves and skin, and act as vitamins used in an array of cellular processes. The most important lipid function, however, is their role in the plasma membrane. Let's take a closer look at the specific types of lipids found in membranes.

The Structures of Membrane Lipids

Not all lipids can form the artificial membranes that Bangham and his colleagues observed. In fact, just two types of lipids are usually found in plasma membranes. Membrane-forming lipids have a polar, hydrophilic region in addition to the nonpolar, hydrophobic region found in all lipids. To better understand this structure, take another look at the phospholipid illustrated in Figure 6.4b. Notice that the molecule has a "head" region containing highly polar covalent bonds as well as positive and negative charges. The charges and polar bonds in the head region interact with water molecules when a phospholipid is placed in solution. In contrast, the long isoprene or fatty-acid tails of a phospholipid are nonpolar. Water molecules cannot form hydrogen bonds with the hydrocarbon tail, so they do not interact with this part of the molecule.

Compounds that contain both hydrophilic and hydrophobic elements are **amphipathic** ("dual-sympathy"). Phospholipids are amphipathic. As Figure 6.4a shows, cholesterol is also amphipathic. It has both hydrophilic and hydrophobic regions.

The amphipathic nature of phospholipids is far and away their most important feature biologically. It is responsible for their presence in plasma membranes.

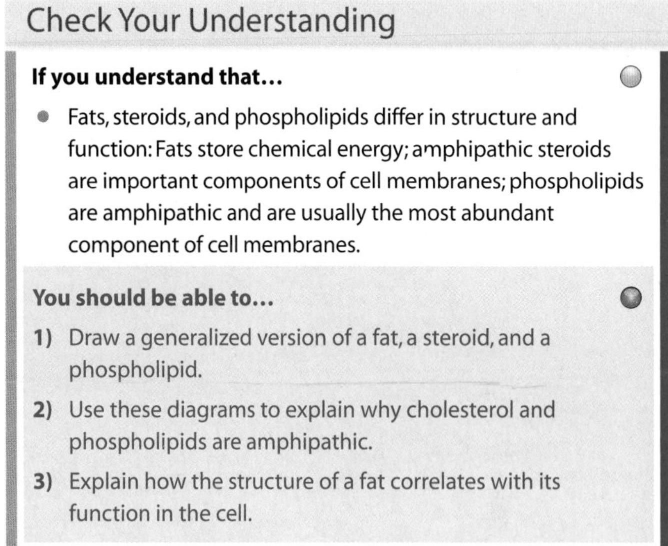
Check Your Understanding

If you understand that...

- Fats, steroids, and phospholipids differ in structure and function: Fats store chemical energy; amphipathic steroids are important components of cell membranes; phospholipids are amphipathic and are usually the most abundant component of cell membranes.

You should be able to...

1) Draw a generalized version of a fat, a steroid, and a phospholipid.
2) Use these diagrams to explain why cholesterol and phospholipids are amphipathic.
3) Explain how the structure of a fat correlates with its function in the cell.

6.2 Phospholipid Bilayers

Phospholipids do not dissolve when they are placed in water. Water molecules interact with the hydrophilic heads of the phospholipids, but not with their hydrophobic tails. Instead of dissolving in water, then, phospholipids form one of two types of structures: micelles or lipid bilayers.

Micelles (**Figure 6.5a**) are tiny droplets created when the hydrophilic heads of phospholipids face the water and the hydrophobic tails are forced together, away from the water. Phospholipid bilayers, or simply, **lipid bilayers**, are created when two sheets of phospholipid molecules align. As **Figure 6.5b** shows, the hydrophilic heads in each layer face a surrounding solution while the hydrophobic tails face one another inside the bilayer. In this way, the hydrophilic heads interact with water while the hydrophobic tails interact with each other. Micelles tend to form from phospholipids with relatively short tails; bilayers tend to form from phospholipids with longer tails.

Once you understand the structure of micelles and phospholipid bilayers, the most important point to recognize about them is that they form spontaneously. No input of energy is required. This concept can be difficult to grasp, because the formation of these structures clearly decreases entropy. Micelles and lipid bilayers are much more highly organized than phospholipids floating free in the solution. The key is to recognize that micelles and lipid bilayers are much more stable energetically than are independent molecules in solution.

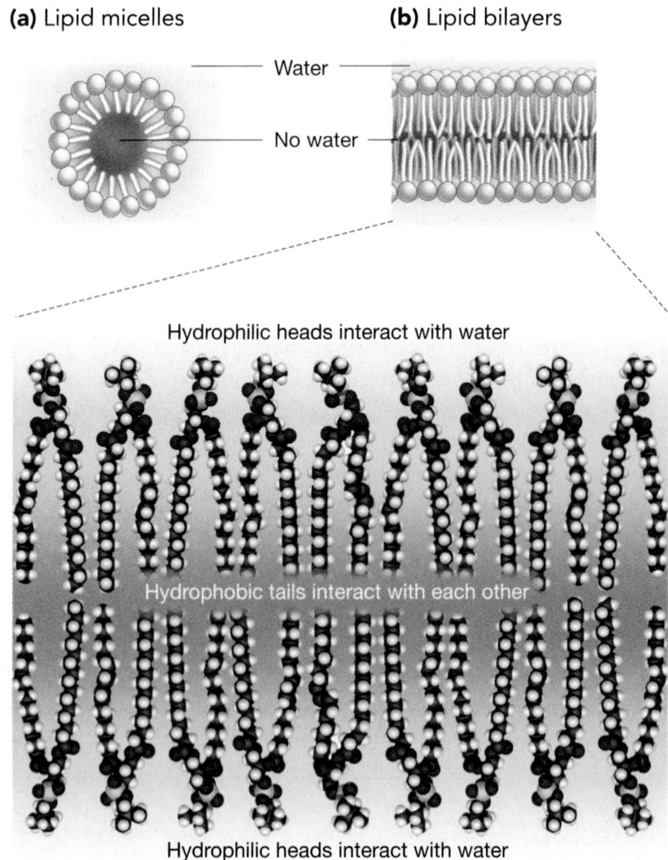

FIGURE 6.5 Phospholipids Form Bilayers in Solution. In **(a)** a micelle or **(b)** a lipid bilayer, the hydrophilic heads of phospholipids face out, toward water; the hydrophobic tails face in, away from water. Plasma membranes consist in part of lipid bilayers.

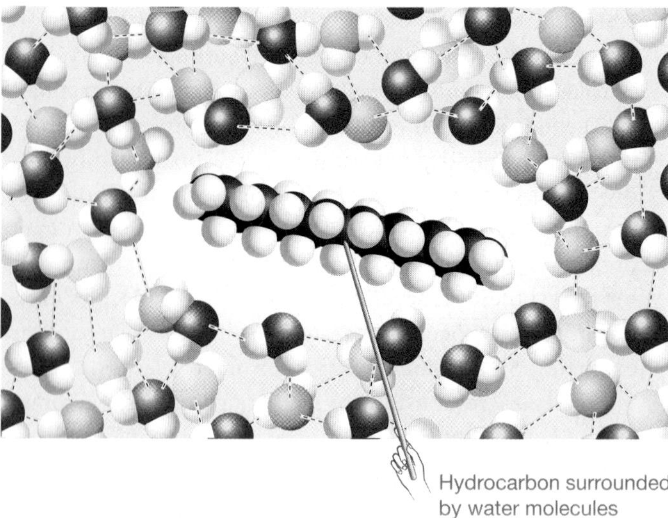

FIGURE 6.6 Hydrocarbons Disrupt Hydrogen Bonds between Water Molecules. Hydrocarbons are unstable in water because they disrupt hydrogen bonding between water molecules.

● **EXERCISE** Label the area where no hydrogen bonding is occurring between water molecules.

● **QUESTION** Hydrogen bonds pull water molecules closer together. Which way are the water molecules in this figure being pulled, relative to the hydrocarbon?

Stated another way, micelles and lipid bilayers have much lower potential energy than do independent phospholipids in solution. Independent phospholipids are unstable in water because their hydrophobic tails disrupt hydrogen bonds that otherwise would form between water molecules (**Figure 6.6**; see also Figure 2.13b). As a result, amphipathic molecules are much more stable in aqueous solution when their hydrophobic tails avoid water and instead participate in the hydrophobic (van der Waals) interactions introduced in Chapter 3. In this case, the loss of potential energy outweighs the decrease in entropy. Overall, the free energy of the system decreases. Lipid bilayer formation is endergonic and spontaneous.

● If you understand this reasoning, you should be able to add water molecules that are hydrogen-bonded to each to Figure 6.5, and explain the logic behind your drawing.

Artificial Membranes as an Experimental System

When lipid bilayers are agitated by shaking, the layers break and re-form as small, spherical structures. This is what happened in Bangham's experiment. The resulting vesicles had water on the inside as well as the outside because the hydrophilic heads of the lipids faced outward on each side of the bilayer.

Researchers have produced these types of vesicles by using dozens of different types of phospholipids. Artificial membrane-bound vesicles like these are called liposomes. The ability to create them supports an important conclusion: If phospholipid molecules accumulated during chemical evolution early in Earth's history, they almost certainly formed water-filled vesicles.

To better understand the properties of vesicles and plasma membranes, researchers began creating and experimenting with liposomes and other types of artificial bilayers. Some of the first questions they posed concerned the permeability of lipid bilayers. The **permeability** of a structure is its tendency to allow a given substance to pass across it. Once a membrane forms a water-filled vesicle, can other molecules or ions pass in or out? If so, is this permeability selective in any way? The permeability of membranes is a critical issue, because if certain molecules or ions pass through a lipid bilayer more readily than others, the internal environment of a vesicle can become different from the outside. This difference between exterior and interior environments is a key characteristic of cells.

Figure 6.7 shows the two types of artificial membranes that are used to study the permeability of lipid bilayers. Figure 6.7a shows liposomes, roughly spherical vesicles. Figure 6.7b illustrates planar bilayers, which are lipid bilayers constructed across a hole in a glass or plastic wall separating two aqueous (watery) solutions.

(a) Liposomes: Artificial membrane-bound vesicles

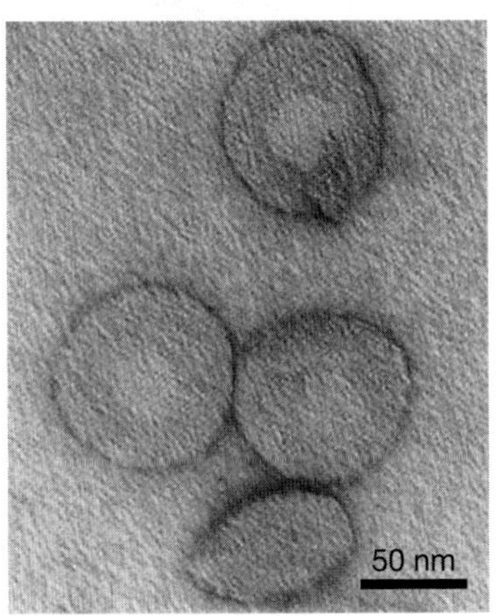

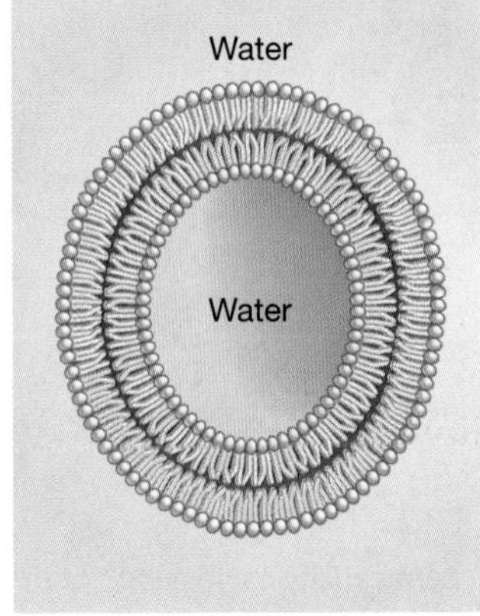

(b) Planar bilayers: Artificial membranes

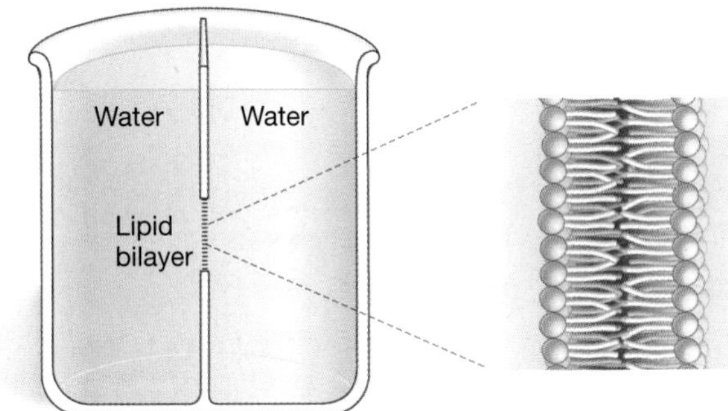

(c) Artificial-membrane experiments

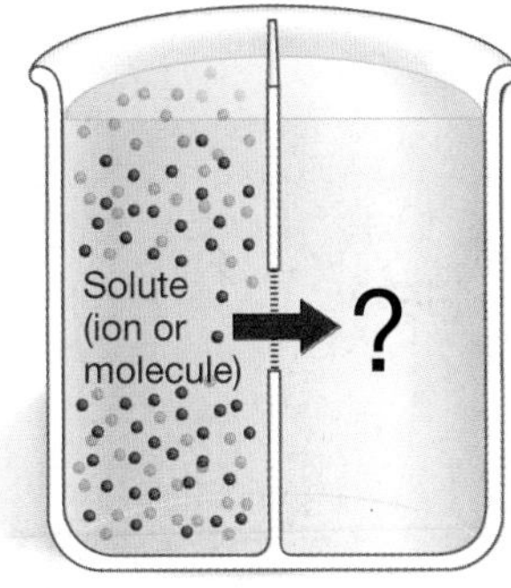

How rapidly can different solutes cross the membrane (if at all) when ...

1. Different types of phospholipids are used to make the membrane?

2. Proteins or other molecules are added to the membrane?

FIGURE 6.7 Liposomes and Planar Bilayers Are Important Experimental Systems. (a) Electron micrograph of liposomes in cross section (left) and a cross-sectional diagram of the lipid bilayer in a liposome. **(b)** The construction of planar bilayers across a hole in a glass wall separating two water-filled compartments (left), and a close-up sketch of the bilayer. **(c)** A wide variety of experiments are possible with liposomes and planar bilayers; a few are suggested here.

Using liposomes and planar bilayers, researchers can study what happens when a known ion or molecule is added to one side of a lipid bilayer (Figure 6.7c). Does the ion or molecule cross the membrane and show up on the other side? If so, how rapidly does the movement take place? What happens when a different type of phospholipid is used to make the artificial membrane? Does the membrane's permeability change when proteins or other types of molecules become part of it?

Biologists describe such an experimental system as elegant and powerful because it gives them precise control over which factor changes from one experimental treatment to the next. Control, in turn, is why experiments are such an effective way to explore scientific questions. You might recall from Chapter 1 that good experimental design allows researchers to alter one factor at a time and determine what effect, if any, each has on the process being studied.

Equally important for experimental purposes, liposomes and planar bilayers provide a clear way to determine whether a given change in conditions has an effect. By sampling the solutions on both sides of the membrane before and after the treatment and then analyzing the concentration of ions and molecules in the samples, researchers have an effective way to determine whether the treatment had any consequences.

Using such systems, what have biologists learned about membrane permeability?

Selective Permeability of Lipid Bilayers

When researchers put molecules or ions on one side of a liposome or planar bilayer and measure the rate at which the molecules arrive on the other side, a clear pattern emerges: Lipid bilayers are *highly* selective. **Selective permeability** means that some substances cross a membrane more easily than other substances can. Small, nonpolar molecules move across bilayers quickly. In contrast, large molecules and charged substances cross the membrane slowly, if at all. According to the data in **Figure 6.8**, small, nonpolar molecules such as oxygen (O_2) move across selectively permeable membranes more than a billion times faster than do chloride ions (Cl^-). Very small and uncharged molecules such as water (H_2O) can also cross membranes relatively rapidly, even if they are polar. Small, polar molecules such as glycerol and urea have intermediate permeability.

The leading hypothesis to explain this pattern is that charged compounds and large, polar molecules can't pass through the nonpolar, hydrophobic tails of a lipid bilayer. Because of their electrical charge, ions are more stable in solution where they form hydrogen bonds with water than they are in the interior of membranes, which is electrically neutral. ● If you understand this hypothesis, you should be able to predict whether amino acids and nucleotides will cross a membrane readily. To test the hypothesis, researchers have manipulated the size and structure of the tails in liposomes or planar bilayers.

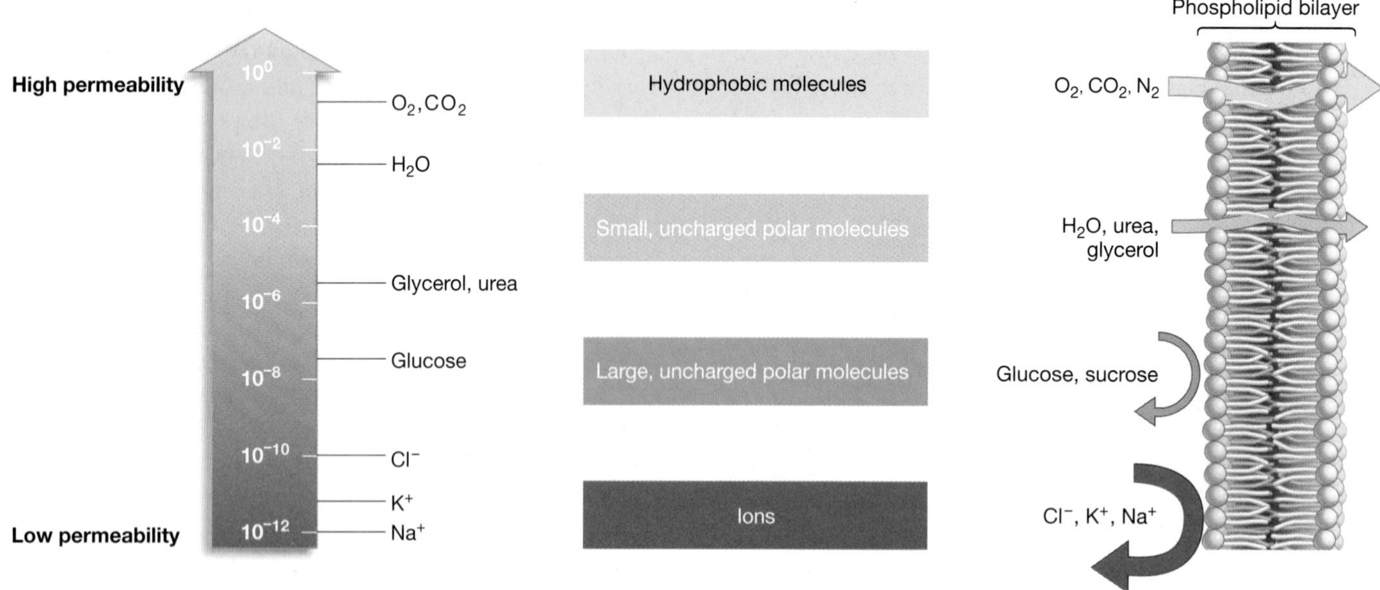

FIGURE 6.8 Selective Permeability of Lipid Bilayers. (a) The numbers represent "permeability coefficients," or the rate (cm/sec) at which an ion or molecule crosses a lipid bilayer. **(b)** The relative permeabilities of various molecules and ions, based on data like those presented in part (a).

QUESTION About how fast does water cross the lipid bilayer?

Does the Type of Lipid in a Membrane Affect Its Permeability?

Theoretically, two aspects of a hydrocarbon chain could affect the way the chain behaves in a lipid bilayer: (1) the number of double bonds it contains and (2) its length. Recall from Chapter 2 that when carbon atoms form a double bond, the attached atoms are found in a plane instead of a (three-dimensional) tetrahedron. The carbon atoms involved are also locked into place. They cannot rotate freely, as they do in carbon-carbon single bonds. As a result, a double bond between carbon atoms produces a "kink" in an otherwise straight hydrocarbon chain (**Figure 6.9**).

When a double bond exists between two carbon atoms in a hydrocarbon chain, the chain is said to be **unsaturated**. Conversely, hydrocarbon chains without double bonds are said to be **saturated**. This choice of terms is logical, because if a hydrocarbon chain does not contain a double bond, it is saturated with the maximum number of hydrogen atoms that can attach to the carbon skeleton. If it is unsaturated, then fewer than the maximum number of hydrogen atoms are attached. Because they contain more C–H bonds, which have much more free energy than C=C bonds, saturated fats have much more chemical energy than unsaturated fats do. People who are dieting are

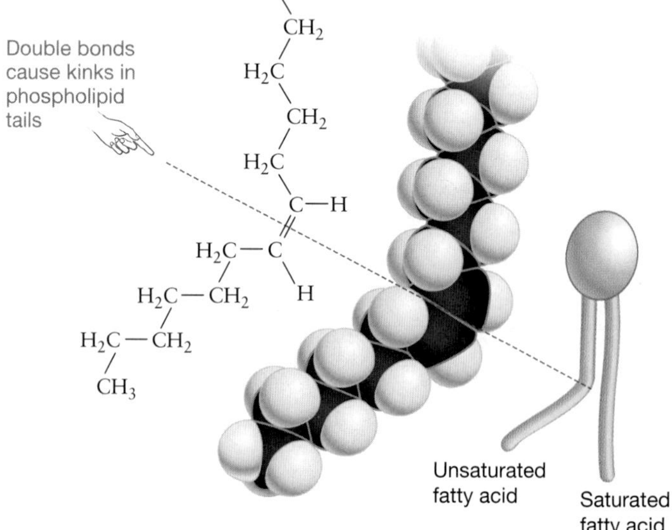

FIGURE 6.9 Unsaturated Hydrocarbons Contain Carbon-Carbon Double Bonds. A double bond in a hydrocarbon chain produces a "kink." The icon on the right indicates that one of the hydrocarbon tails in a phospholipid is unsaturated and therefore kinked.

EXERCISE Draw the structural formula and a schematic diagram for an unsaturated fatty acid containing two double bonds.

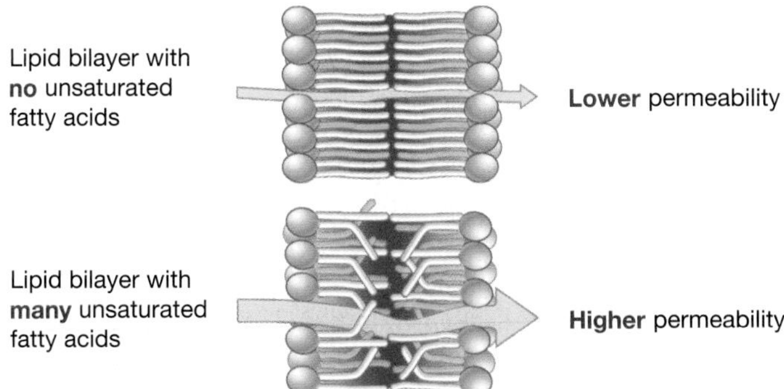

FIGURE 6.10 Fatty-Acid Structure Changes the Permeability of Membranes. Lipid bilayers containing many unsaturated fatty acids have more gaps and should be more permeable than are bilayers with few unsaturated fatty acids.

often encouraged to eat fewer saturated fats. Foods that contain lipids with many double bonds are said to be polyunsaturated and are advertised as healthier than foods with more-saturated fats.

Why do double bonds affect the permeability of membranes? When hydrophobic tails are packed into a lipid bilayer, the kinks created by double bonds produce spaces among the tightly packed tails. These spaces reduce the strength of hydrophobic interactions among the tails. Because the interior of the membrane is "glued together" less tightly, the structure should become more fluid and more permeable (**Figure 6.10**).

Hydrophobic interactions also become stronger as saturated hydrocarbon tails increase in length. Membranes dominated by phospholipids with long, saturated hydrocarbon tails should be stiffer and less permeable because the interactions among the tails are stronger.

A biologist would predict, then, that bilayers made of lipids with long, straight, saturated fatty-acid tails should be much less permeable than membranes made of lipids with short, kinked, unsaturated fatty-acid tails. Experiments on liposomes have shown exactly this pattern. Phospholipids with long, saturated tails form membranes that are much less permeable than membranes consisting of phospholipids with shorter, unsaturated tails.

The central point here is that the degree of hydrophobic interactions dictates the behavior of these molecules. This is another example in which the structure of a molecule—specifically, the number of double bonds in the hydrocarbon chain and its overall length—correlates with its properties and function.

These data are also consistent with the basic observation that highly saturated fats are solid at room temperature (**Figure 6.11a**). Lipids that have extremely long hydrocarbon tails, as **waxes** do, form stiff solids at room temperature due to the extensive hydrophobic interactions that occur (**Figure 6.11b**). Birds, sea otters, and many other organisms synthesize waxes and spread them on their exterior surface as a waterproofing; plant cells secrete a waxy layer that covers the surface of leaves and stems and keeps water from evaporating. In contrast, highly unsaturated fats are liquid at room temperature (**Figure 6.11c**). Liquid triacylglycerides are called **oils.**

Besides exploring the role of hydrocarbon chain length and degree of saturation on membrane permeability, biologists have investigated the effect of adding cholesterol molecules. Because the steroid rings in cholesterol are bulky, adding cholesterol to

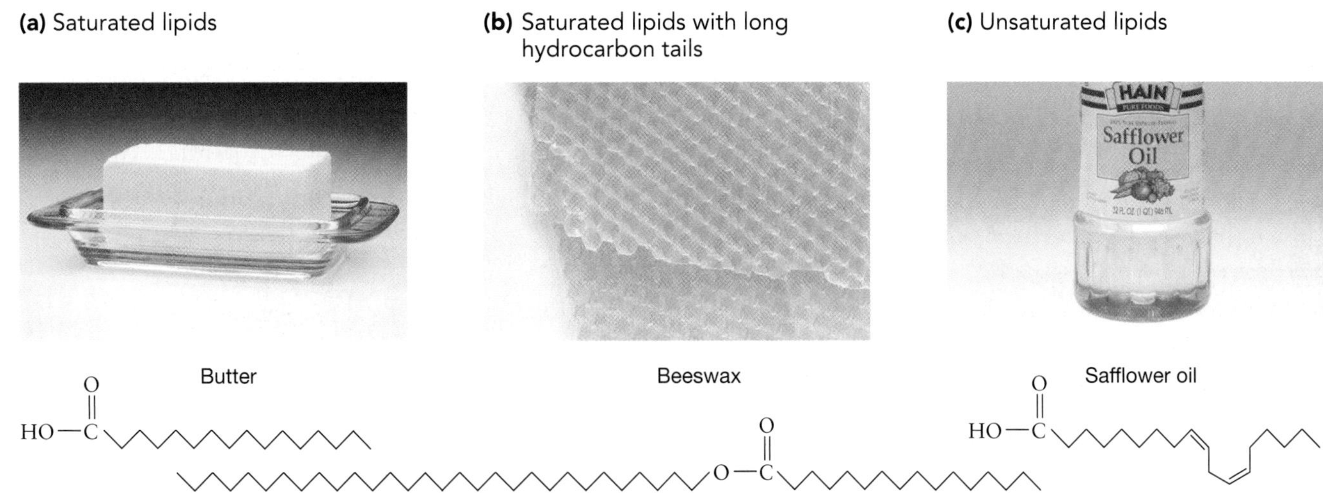

FIGURE 6.11 The Fluidity of Lipids Depends on the Characteristics of Their Hydrocarbon Chains. The fluidity of a lipid depends on the length and saturation of its hydrocarbon chain. **(a)** Butter consists primarily of saturated lipids. **(b)** Waxes are lipids with extremely long hydrocarbon chains. **(c)** Oils are dominated by "polyunsaturates"—lipids with hydrocarbon chains that contain multiple double bonds.

QUESTION Why are waxes so effective for waterproofing floors?

a membrane should increase the density of the hydrophobic section. As predicted, researchers found that adding cholesterol molecules to liposomes dramatically reduced the permeability of the liposomes. The data behind this claim are presented in **Figure 6.12**. The graph in this figure makes another important point, however: Temperature has a strong influence on the behavior of lipid bilayers.

Experiment

Question: Does adding cholesterol to a membrane affect its permeability?

Hypothesis: Cholesterol reduces permeability because it fills spaces in phospholipid bilayers.

Null hypothesis: Cholesterol has no effect on permeability.

Experimental setup:

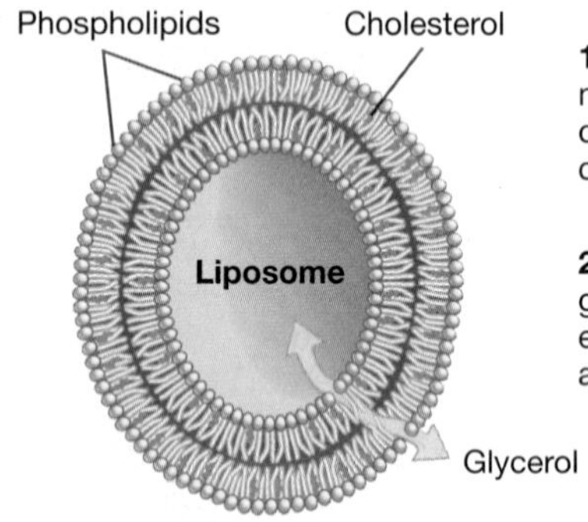

Phospholipids

Cholesterol

Liposome

Glycerol

1. Create liposomes with no cholesterol, 20% cholesterol, and 50% cholesterol.

2. Record how quickly glycerol moves across each type of membrane at different temperatures.

Prediction: Liposomes with higher cholesterol levels will have reduced permeability.

Prediction of null hypothesis: All liposomes will have the same permeability.

Results:

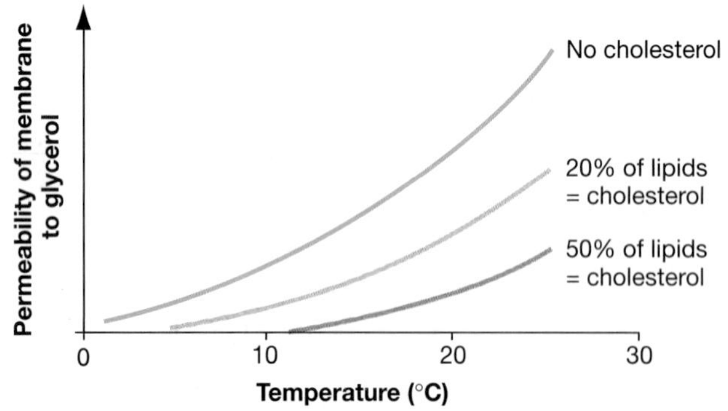

Permeability of membrane to glycerol

No cholesterol

20% of lipids = cholesterol

50% of lipids = cholesterol

0 10 20 30

Temperature (°C)

Conclusion: Adding cholesterol to membranes decreases their permeability to glycerol. The permeability of all membranes analyzed in this experiment increases with increasing temperature.

FIGURE 6.12 The Permeability of a Membrane Depends on Its Composition.

Why Does Temperature Affect the Fluidity and Permeability of Membranes?

At about 25°C—or "room temperature"—the phospholipids found in plasma membranes are liquid, and bilayers have the consistency of olive oil. This fluidity, as well as the membrane's permeability, decreases as temperature decreases. As temperatures drop, individual molecules in the bilayer move more slowly. As a result, the hydrophobic tails in the interior of membranes pack together more tightly. At very low temperatures, lipid bilayers begin to solidify. As the graph in Figure 6.12 indicates, low temperatures can make membranes impervious to molecules that would normally cross them readily.

The fluid nature of membranes also allows individual lipid molecules to move laterally within each layer, a little like a person moving about in a dense crowd (**Figure 6.13**). By tagging individual phospholipids and following their movement, researchers have clocked average speeds of 2 micrometers (μm)/second at room temperature. At these speeds, phospholipids could travel the length of a small bacterial cell in a second.

These experiments on lipid and ion movement demonstrate that membranes are dynamic. Phospholipid molecules whiz around each layer while water and small, nonpolar molecules shoot in and out of the membrane. How quickly molecules move within and across membranes is a function of temperature and the structure of the hydrocarbon tails in the bilayer.

Given these insights into the permeability and fluidity of lipid bilayers, an important question remains: *Why* do certain molecules move across membranes spontaneously?

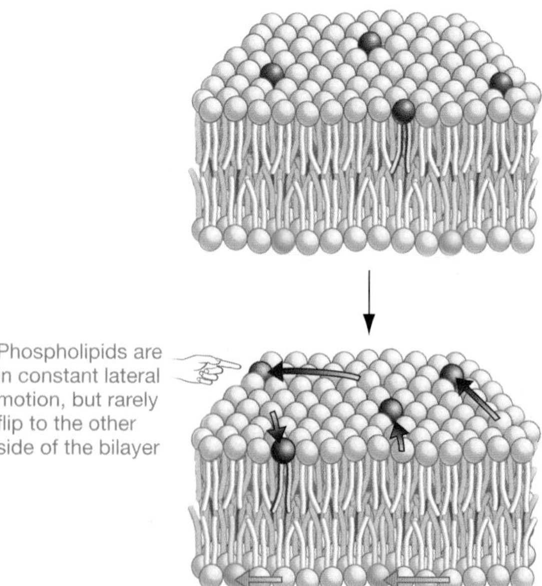

FIGURE 6.13 Phospholipids Move within Membranes. Membranes are dynamic—in part because phospholipid molecules move within each layer in the structure.

Check Your Understanding

If you understand that...

- In solution, phospholipids form bilayers that are selectively permeable—meaning that some substances cross them much more readily than others do.
- Permeability is a function of temperature, the amount of cholesterol in the membrane, and the length and degree of saturation of the hydrocarbon tails in membrane phospholipids.

You should be able to...

Fill in a chart with rows called "Temperature," "Cholesterol," "Length of hydrocarbon tails," and "Saturation of hydrocarbon tails" and columns named "Factor," "Effect on permeability," and "Reason."

6.3 Why Molecules Move across Lipid Bilayers: Diffusion and Osmosis

A thought experiment can help explain why molecules and ions are able to move across membranes spontaneously. Suppose you rack up a set of blue billiard balls on a pool table containing many white balls and then begin to vibrate the table. Because of the vibration, the balls will move about randomly. They will also bump into one another. After these collisions, some blue balls will move outward—away from their original position. In fact, the overall (or net) movement of blue balls will be outward. This occurs because the random motion of the blue balls disrupts their original, nonrandom position—as they move at random, they are more likely to move away from each other than to stay together. Eventually, the blue billiard balls will be distributed randomly across the table. The entropy of the blue billiard balls has increased. Recall from Chapter 2 that entropy is a measure of the randomness or disorder in a system. The second law of thermodynamics states that in a closed system, entropy always increases.

This hypothetical example illustrates why molecules or ions located on one side of a lipid bilayer move to the other side spontaneously. The dissolved molecules and ions, or **solutes**, have thermal energy and are in constant, random motion. Movement of molecules and ions that results from their kinetic energy is known as **diffusion**. Because solutes change position randomly due to diffusion, they tend to move from a region of high concentration to a region of low concentration. A difference in solute concentrations creates a **concentration gradient**. Molecules and ions still move randomly in all directions when a concentration gradient exists, but there is a net movement from regions of high concentration to regions of low concentration. Diffusion along a concentration gradient is a spontaneous process because it results in an increase in entropy.

Once the molecules or ions are randomly distributed throughout a solution, equilibrium is established. For example, consider two aqueous solutions separated by a lipid bilayer. **Figure 6.14** shows how molecules that pass through the bilayer diffuse to the other side. At equilibrium, molecules continue to move back and forth across the membrane, but at equal rates—simply because each molecule or ion is equally likely to move in any direction. This means that there is no longer a net movement of molecules across the membrane.

What about water itself? As the data in Figure 6.8 showed, water moves across lipid bilayers relatively quickly. Like

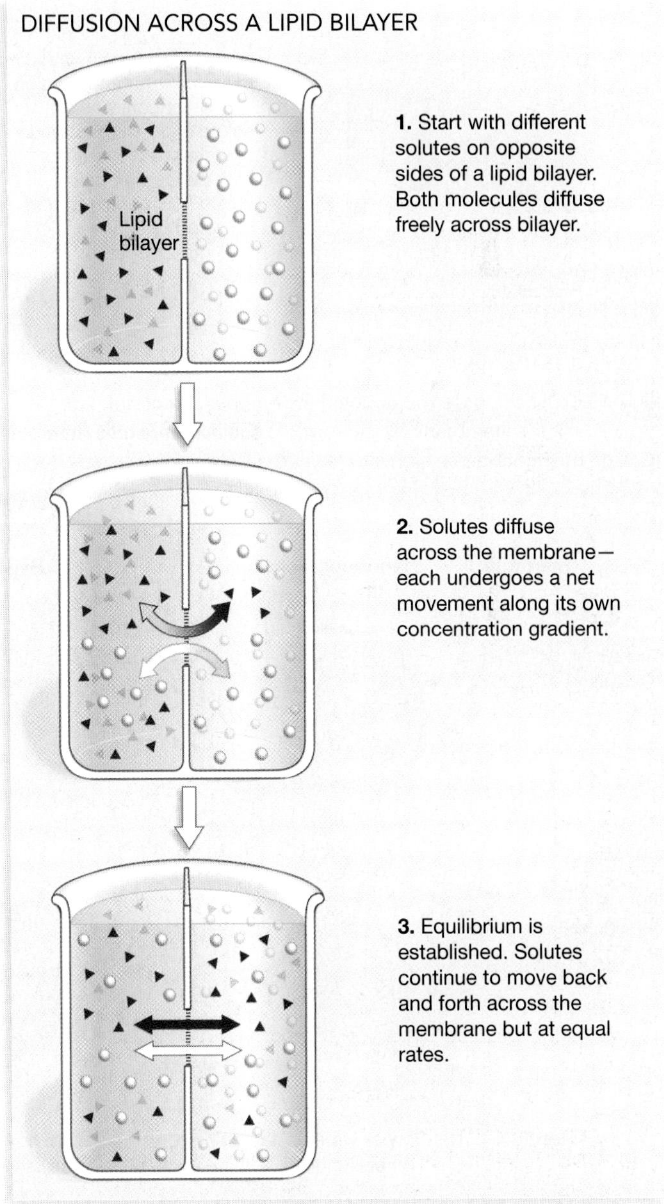

DIFFUSION ACROSS A LIPID BILAYER

Lipid bilayer

1. Start with different solutes on opposite sides of a lipid bilayer. Both molecules diffuse freely across bilayer.

2. Solutes diffuse across the membrane—each undergoes a net movement along its own concentration gradient.

3. Equilibrium is established. Solutes continue to move back and forth across the membrane but at equal rates.

FIGURE 6.14 Diffusion across a Selectively Permeable Membrane.

EXERCISE If a solute's rate of diffusion increases linearly with its concentration difference across the membrane, write an equation for the rate of diffusion across a membrane.

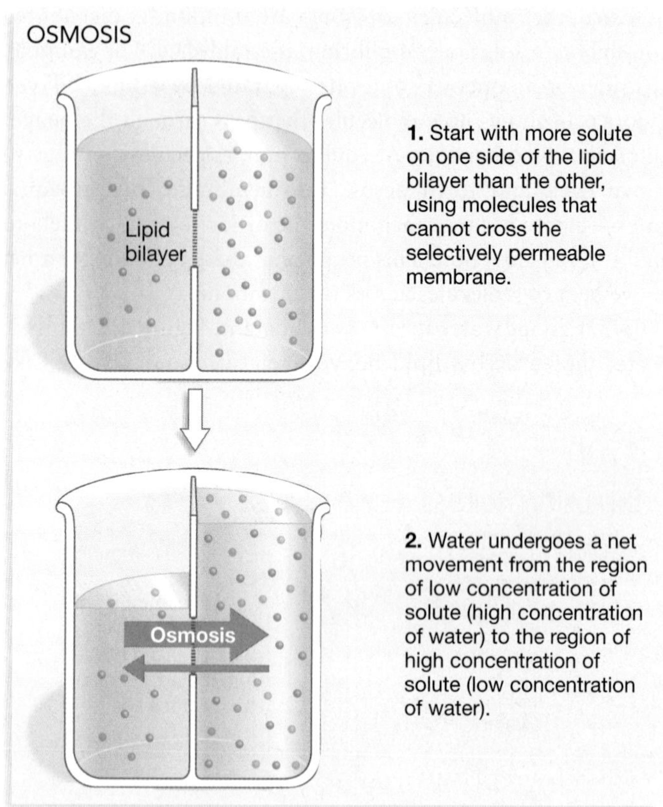

FIGURE 6.15 Osmosis.

● **QUESTION** Suppose you doubled the number of molecules on the right side of the membrane (at the start). At equilibrium, would the water level on the right side be higher or lower than what is shown here?

other substances that diffuse, water moves along its concentration gradient—from higher to lower concentration. The movement of water is a special case of diffusion that is given its own name: **osmosis**. Osmosis occurs only when solutions are separated by a membrane that is permeable to some molecules but not others—that is, a selectively permeable membrane.

The best way to think about water moving in response to a concentration gradient is to focus on the concentration of solutes in the solution. Let's suppose the concentration of a particular solute is higher on one side of a selectively permeable membrane than it is on the other side (**Figure 6.15**, step 1). Further, suppose that this solute cannot diffuse through the membrane to establish equilibrium. What happens? Water will move from the side with a lower concentration of solute to the side with a higher concentration of solute (step 2). It dilutes the higher concentration and equalizes the concentrations on both sides. This movement is spontaneous. It is driven by the increase in entropy achieved when solute concentrations are equal on both sides of the membrane.

Another way to think about osmosis is to realize that water is at higher concentration on the left side of the beaker in Figure 6.15 than it is on the right side of the beaker. As water diffuses, then, there will be net movement of water molecules from the left side to the right side: from a region of high concentration to a region of low concentration.

The movement of water by osmosis is important because it can swell or shrink a membrane-bound vesicle. Consider the liposomes illustrated in **Figure 6.16**. If the solution outside the

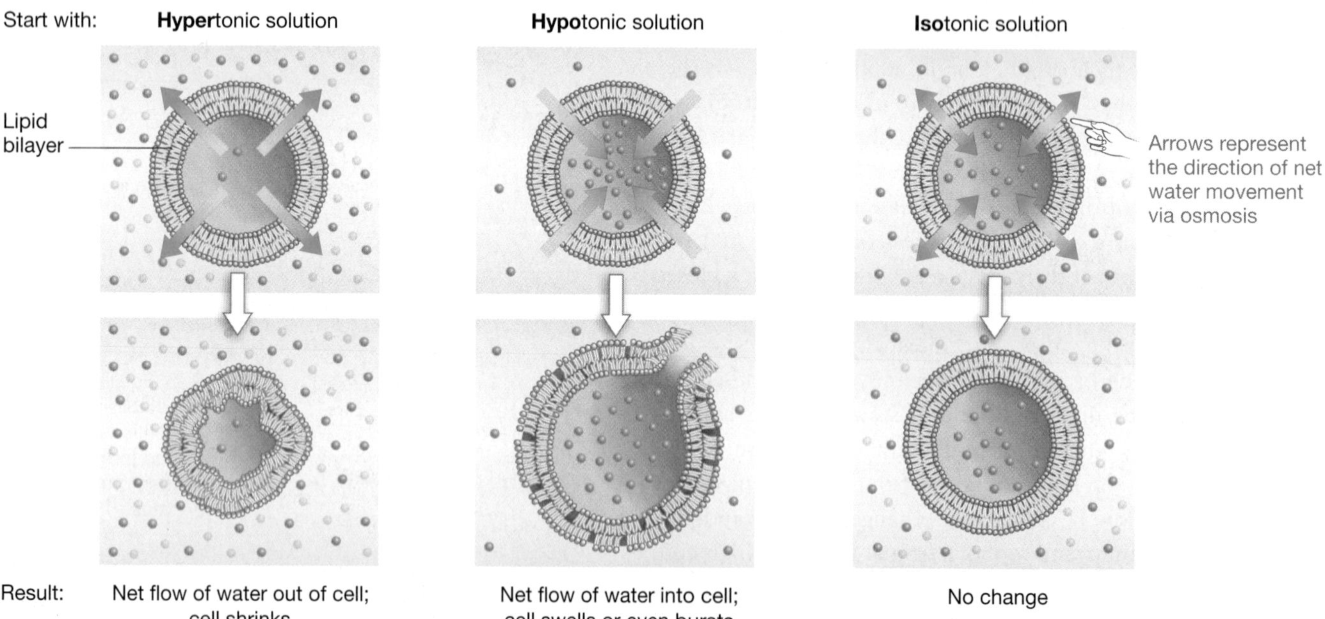

FIGURE 6.16 Osmosis Can Shrink or Burst Membrane-Bound Vesicles.

● **QUESTION** Some species of bacteria can live in extremely salty environments, such as saltwater-evaporation ponds. Is this habitat likely to be hypertonic, hypotonic, or isotonic relative to the interior of the cells?

membrane has a higher concentration of solutes than the interior has, and the solutes are not able to pass through the lipid bilayer, then water will move out of the vesicle into the solution outside. As a result, the vesicle will shrink and the membrane shrivel. Such a solution is said to be **hypertonic** ("excess-tone") relative to the inside of the vesicle. The word root *hyper* refers to the outside solution containing more solutes than the solution on the other side of the membrane. Conversely, if the solution outside the membrane has a lower concentration of solutes than the interior has, water will move into the vesicle via osmosis. The incoming water will cause the vesicle to swell or even burst. Such a solution is termed **hypotonic** ("lower-tone") relative to the inside of the vesicle. Here the word root *hypo* refers to the outside solution containing fewer solutes than the inside solution has. If solute concentrations are equal on either side of the membrane, the liposome will maintain its size. When the outside solution does not affect the membrane's shape, that solution is called **isotonic** ("equal-tone").

Note that the terms hypertonic, hypotonic, and isotonic are relative—they can be used only to express the relationship between a given solution and another solution. ● If you understand this concept, you should be able to draw liposomes in Figure 6.16 that change the relative "tonicity" of the surrounding solution. Specifically, draw (1) a liposome on the left such that the surrounding solution is hypotonic relative to the solution inside the liposome, and (2) a liposome in the center where the surrounding solution is hypertonic relative to the solution inside the liposome.

(MB) **Web Animation** at www.masteringbio.com
Diffusion and Osmosis

To summarize, diffusion and osmosis move solutes and water across lipid bilayers. What does all this have to do with the first membranes floating in the prebiotic soup? Osmosis and diffusion tend to *reduce* differences in chemical composition between the inside and outside of membrane-bound structures. If liposome-like structures were present in the prebiotic soup, it's unlikely that their interiors offered a radically different environment from the surrounding solution. In all likelihood, the primary importance of the first lipid bilayers was simply to provide a container for self-replicating molecules. Experiments have shown that ribonucleotides can diffuse across lipid bilayers. Further, it is clear that cell-like vesicles grow as additional lipids are added and then divide if sheared shaking, bubbling, or wave action. Based on these observations, it is reasonable to hypothesize that once a self-replicating ribozyme had become surrounded by a lipid bilayer, this simple life-form and its descendants would continue to occupy cell-like structures that grew and divided.

Now let's investigate the next great event in the evolution of life: the formation of a true cell. How can lipid bilayers become a barrier capable of creating and maintaining a specialized internal environment that is conducive to life? How could an effective plasma membrane—one that admits ions and molecules needed by the replicator while excluding ions and molecules that might damage it—evolve in the first cell?

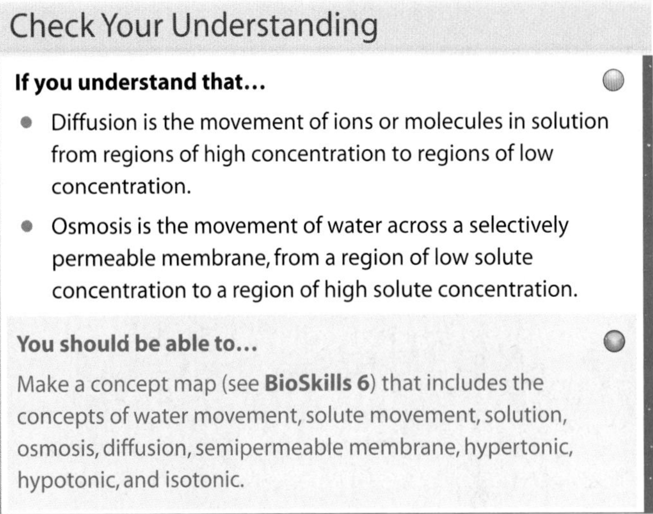
Check Your Understanding

If you understand that...

- Diffusion is the movement of ions or molecules in solution from regions of high concentration to regions of low concentration.
- Osmosis is the movement of water across a selectively permeable membrane, from a region of low solute concentration to a region of high solute concentration.

You should be able to...

Make a concept map (see **BioSkills 6**) that includes the concepts of water movement, solute movement, solution, osmosis, diffusion, semipermeable membrane, hypertonic, hypotonic, and isotonic.

6.4 Membrane Proteins

What sort of molecule could become incorporated into a lipid bilayer and affect the bilayer's permeability? The title of this section gives the answer away. Proteins that are amphipathic can be inserted into lipid bilayers.

Proteins can be amphipathic because they are made up of amino acids and because amino acids have side chains, or R-groups, that range from highly nonpolar to highly polar. (Some are even charged; see Figure 3.3 and Table 3.2.) It's conceivable, then, that a protein could have a series of nonpolar amino acids in the middle of its primary structure, but polar or charged amino acids on both ends of its primary structure, as illustrated in **Figure 6.17a**. The nonpolar amino acids would be stable in the interior of a lipid bilayer, while the polar or charged amino acids would be stable alongside the polar heads and surrounding water (**Figure 6.17b**). Further, because the secondary and tertiary structures of proteins are almost limitless in their variety and complexity, it is possible for proteins to form tubes and thus function as some sort of channel or pore across a lipid bilayer.

Based on these considerations, it is not surprising that when researchers began analyzing the chemical composition of plasma membranes in eukaryotes they found that proteins were just as common, in terms of mass, as phospholipids. How were these two types of molecules arranged? In 1935 Hugh Davson and James Danielli proposed that plasma membranes were struc-

(a) Proteins can be amphipathic.

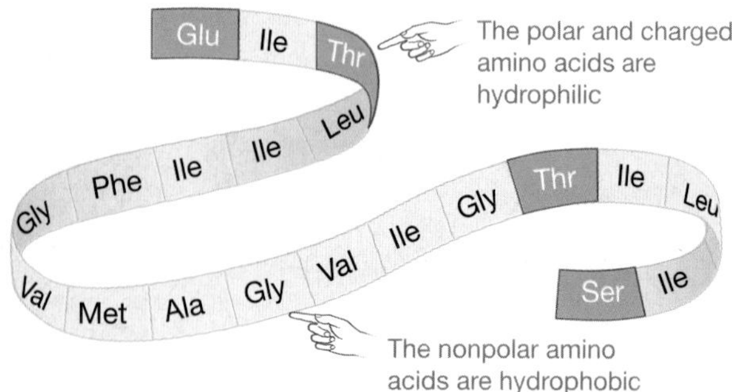

(b) Amphipathic proteins can integrate into lipid bilayers.

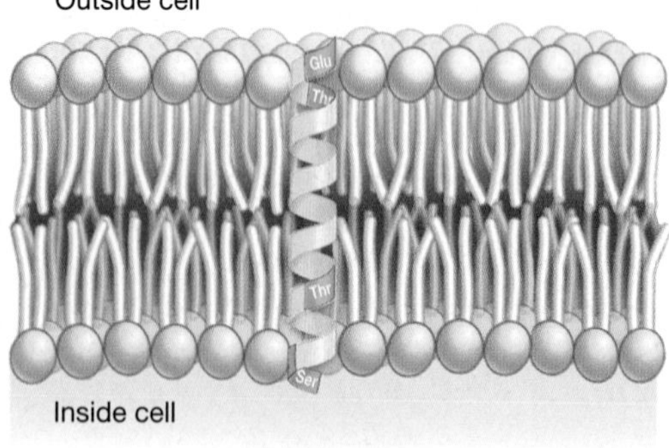

FIGURE 6.17 Proteins Can Be Amphipathic.

● **QUESTION** Researchers can analyze the primary structure of a membrane protein and predict which portions are embedded in the membrane and which are exposed to the cell's interior or exterior. How is this possible?

● **QUESTION** What type of secondary structure is shown in part (b)?

(a) Sandwich model

Membrane proteins on cell exterior

Phospholipid bilayer

Membrane proteins on cell interior

(b) Fluid-mosaic model

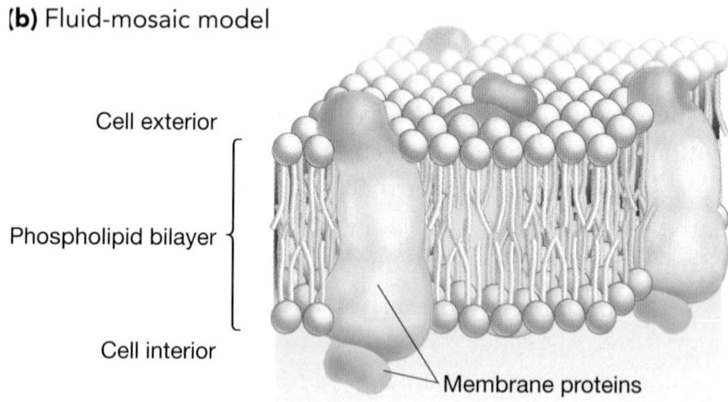

FIGURE 6.18 Past and Current Models of Membrane Structure. **(a)** The protein-lipid-lipid-protein sandwich model was the first hypothesis for the arrangement of lipids and proteins in plasma membranes. **(b)** The fluid-mosaic model was a radical departure from the sandwich hypothesis.

tured like a sandwich, with hydrophilic proteins coating both sides of a pure lipid bilayer (**Figure 6.18a**). Early electron micrographs of plasma membranes seemed to be consistent with the sandwich model, and for decades it was widely accepted.

The realization that membrane proteins could be amphipathic led S. Jon Singer and Garth Nicolson to suggest an alternative hypothesis, however. In 1972, they proposed that at least some proteins span the membrane instead of being found only outside the lipid bilayer. Their hypothesis was called the **fluid-mosaic model**. As **Figure 6.18b** shows, Singer and Nicolson suggested that membranes are a mosaic of phospholipids and different types of proteins. The overall structure was proposed to be dynamic and fluid.

The controversy over the nature of the plasma membrane was resolved in the early 1970s with the development of an innovative technique for visualizing the surface of plasma membranes. The method is called freeze-fracture electron microscopy, because the steps involve freezing and fracturing the membrane before examining it with a **scanning electron microscope,** which produces images of an object's surface (see **BioSkills 8**). As **Figure 6.19** shows, the technique allows researchers to split plasma membranes and view the middle of the structure. The scanning electron micrographs that result show pits and mounds studding the inner surfaces of the lipid bilayer. Researchers interpret these structures as the locations of membrane proteins. As step 4 in Figure 6.19 shows, the pits and mounds are hypothesized to represent proteins that span the lipid bilayer.

These observations conflicted with the sandwich model but were consistent with the fluid-mosaic model. Based on these and subsequent observations, the fluid-mosaic model is now widely accepted.

Figure 6.20 summarizes the current hypothesis for where proteins and lipids are found in a plasma membrane. Note that some proteins span the membrane and have segments facing both the interior and exterior surfaces. Proteins such as these are called

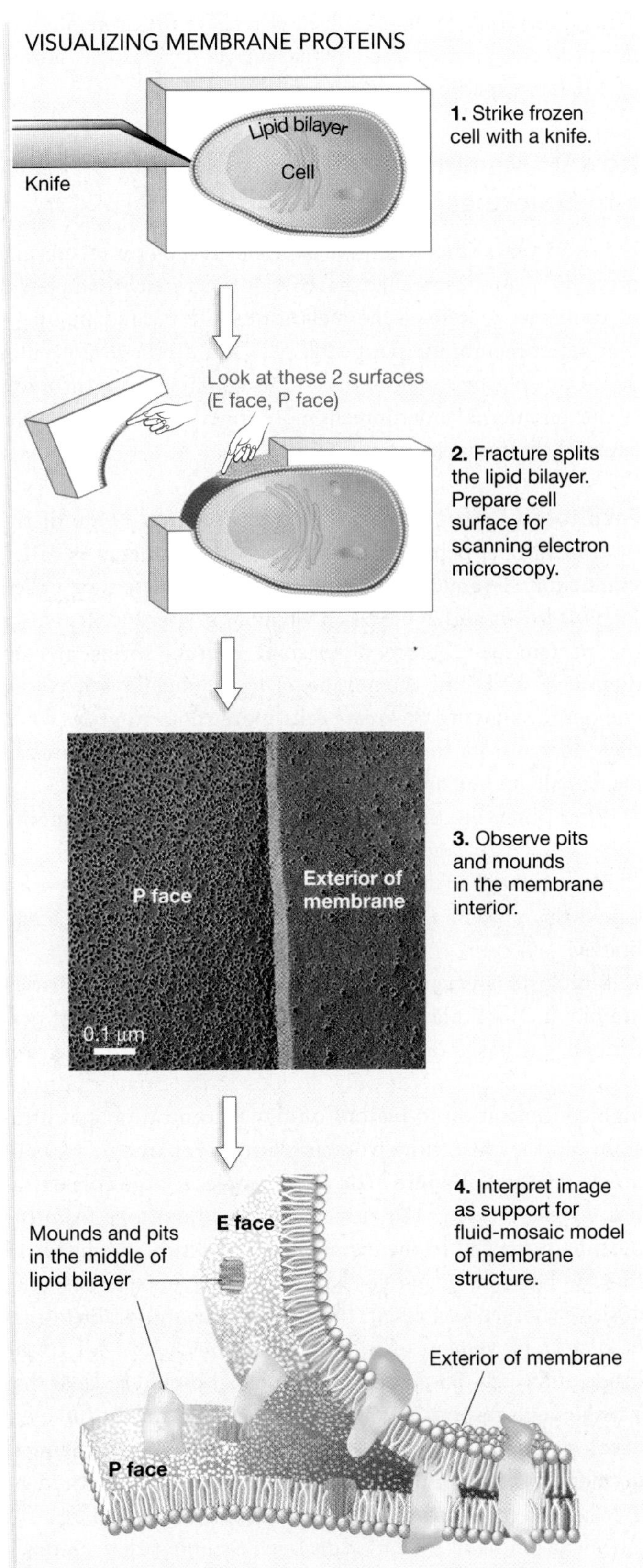

FIGURE 6.19 Freeze-Fracture Preparations Allow Biologists to View Membrane Proteins.

EXERCISE Draw what the micrograph in step 3 would look like if the sandwich model of membrane structure were correct.

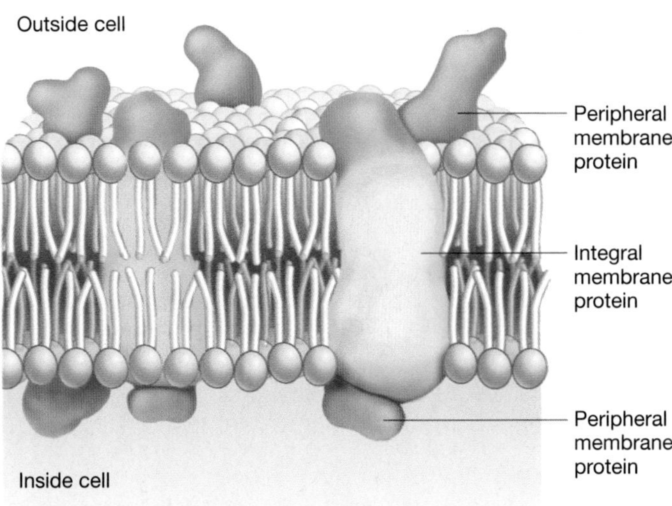

FIGURE 6.20 Integral and Peripheral Membrane Proteins. Integral membrane proteins are also called transmembrane proteins because they span the membrane. Peripheral membrane proteins are often attached to integral membrane proteins.

QUESTION Are the external and internal faces of a plasma membrane the same or different? Explain.

integral membrane proteins, or **transmembrane proteins**. Other proteins, called **peripheral membrane proteins**, are found only on one side of the membrane. Often, peripheral membrane proteins are attached to an integral membrane protein. In most cases, specific peripheral proteins are found only in the inside of the plasma membrane and thus inside the cell, while others are found only on the outside of the plasma membrane and thus facing the surrounding environment. The location of peripheral proteins is one of several reasons that the exterior surface of the plasma membrane is very different from the interior surface. It's also important to realize that the position of these proteins is not static. Like the phospholipids in the bilayer, membrane proteins are in constant motion, diffusing through the oily film.

What do all these proteins do? Later chapters will explore how certain membrane proteins act as enzymes or are involved in cell-to-cell signaling or making physical connections between cells. Here, let's focus on how integral membrane proteins are involved in the transport of selected ions and molecules across the plasma membrane.

Systems for Studying Membrane Proteins

The discovery of integral membrane proteins was consistent with the hypothesis that proteins affect membrane permeability. The evidence was not considered conclusive enough, though, because it was also plausible to claim that integral membrane proteins were structural components that influenced membrane strength or flexibility. To test whether proteins actually do affect

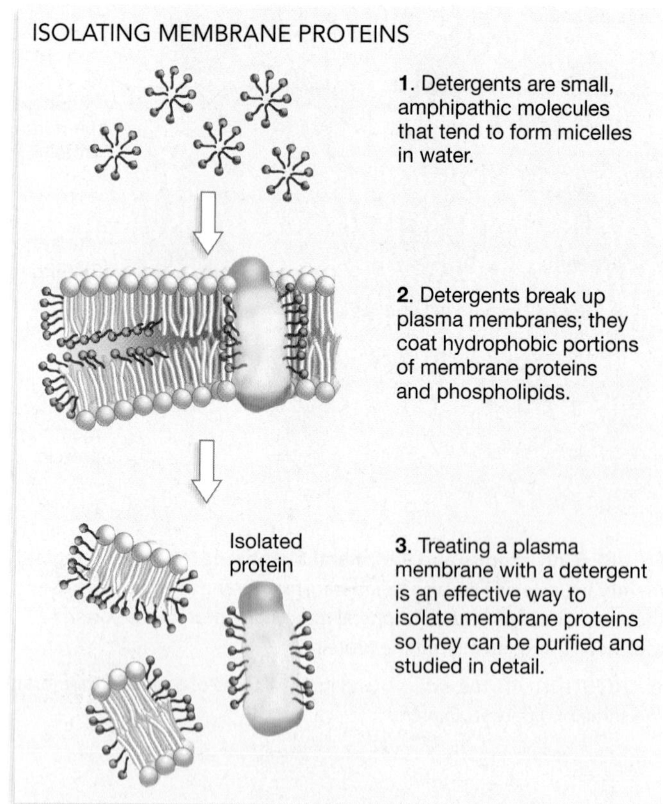

FIGURE 6.21 Detergents Can Be Used to Get Membrane Proteins into Solution.

membrane permeability, researchers needed some way to isolate and purify membrane proteins.

Figure 6.21 outlines one method that researchers developed to separate proteins from membranes. The key to the technique is the use of detergents. A detergent is a small, amphipathic molecule. As step 1 of Figure 6.21 shows, the hydrophobic tails of detergents clump in solution, forming micelles. When detergents are added to the solution surrounding a lipid bilayer, the hydrophobic tails of the detergent molecule interact with the hydrophobic tails of the lipids. In doing so, the detergent tends to disrupt the bilayer and break it apart (step 2). If the membrane contains proteins, the hydrophobic tails of the detergent molecules also interact with the hydrophobic parts of the membrane proteins. The detergent molecules displace the membrane phospholipids and end up forming water-soluble, detergent-protein complexes (step 3).

To isolate and purify these membrane proteins once they are in solution, researchers use the technique called gel electrophoresis, introduced in **BioSkills 6**. When detergent-protein complexes are loaded into a gel and a voltage is applied, the larger protein complexes migrate more slowly than smaller proteins. As a result, the various proteins isolated from a plasma membrane separate from each other. To obtain a pure sample of a particular protein, the appropriate band is cut out of the gel. The gel material is then dissolved to retrieve the protein. Once this protein is inserted into a planar bilayer or liposome, dozens of different experiments are possible.

How Do Membrane Proteins Affect Ions and Molecules?

In the 55 years since intensive experimentation on membrane proteins began, researchers have identified three broad classes of **transport proteins**—channels, transporters, and pumps—that affect membrane permeability. What do these molecules do? Can plasma membranes that contain these proteins create an internal environment more conducive to life than the external environment is?

Facilitated Diffusion via Channel Proteins One of the first membrane peptides to be investigated in detail is called gramicidin. Gramicidin is produced by a bacterium called *Bacillus brevis* and is used as a weapon: *B. brevis* cells release the protein just before a resistant coating forms around their cell wall and membrane. The gramicidin wipes out competitors, giving *B. brevis* cells more room to grow when they emerge from the resistant phase. Gramicidin is also used medicinally in humans as an antibiotic.

After observing that experimental cells treated with gramicidin seemed to lose large numbers of ions, researchers became interested in understanding how the molecule works. Could this protein alter the flow of ions across plasma membranes?

Biologists answered this question by inserting purified gramicidin into planar bilayers. The experiment they performed was based on an important fact about ion movement across membranes: Not only do ions move from regions of high concentration to regions of low concentration via diffusion, but they also flow from areas of like charge to areas of unlike charge. In **Figure 6.22**, for example, a large concentration gradient favors the movement of sodium ions from the outside of the cell to the inside. But in addition, the inside of this cell has a net negative charge while the outside has a net positive charge. As a result, there is also a charge gradient that favors the movement of sodium ions from the outside to the inside of the cell. Based on this example, it should be clear that ions move in response to a combined concentration and electrical gradient, or what biologists call an **electrochemical gradient.** ● If you understand this concept, you should be able to add an arrow to Figure 6.22, indicating the electrochemical gradient for chloride ions assuming that chloride concentrations are equal on both sides of the membrane.

To determine whether gramicidin affected the membrane's permeability to ions, the researchers measured the flow of electric current across the membrane. Because ions carry a charge, the movement of ions produces an electric current. This prop-

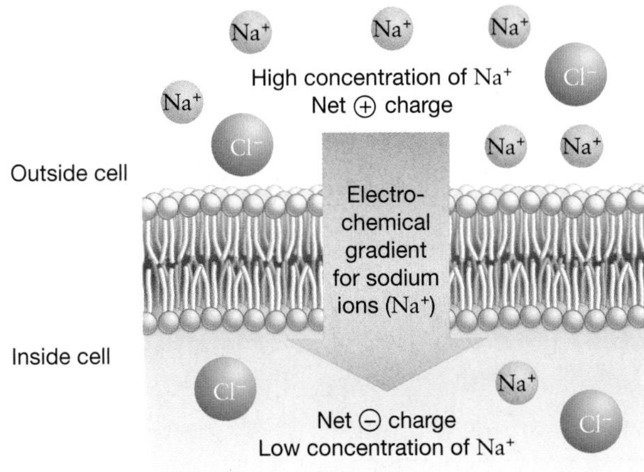

FIGURE 6.22 Electrochemical Gradients. When ions build up on one side of a membrane, they establish a combined concentration and electrical gradient.

● **EXERCISE** By adding sodium ions to this figure, illustrate a situation where there is no electrochemical gradient favoring the movement of either Na^+ or Cl^-.

erty provides an elegant and accurate test for assessing the bilayer's permeability to ions—one that is simpler and more sensitive than taking samples from either side of the membrane and determining the concentrations of solutes present. If gramicidin facilitates ion movement, then an investigator should be able to detect an electric current across planar bilayers that contain gramicidin.

The result? The graph in **Figure 6.23** shows that when gramicidin was absent, no electric current passed through the membrane. But when gramicidin was inserted into the membrane, current began to flow. Based on this observation, biologists proposed that gramicidin is an **ion channel**. An ion channel is a peptide or protein that makes lipid bilayers permeable to ions. (Recall from Chapter 3 that peptides are proteins containing fewer than 50 amino acids.) Follow-up work corroborated that gramicidin is selective. It allows only positively charged ions, or cations, to pass. Gramicidin does not allow negatively charged ions, or anions, to pass through the membrane. It was also established that gramicidin is most permeable to hydrogen ions (or protons, (H^+) and somewhat less permeable to other cations, such as potassium (K^+) and sodium (Na^+).

Researchers gained additional insight into the way gramicidin works when they determined its amino acid sequence (that is, primary structure) and tertiary structure. **Figure 6.24** provides a view from the outside of a cell to the inside through gramicidin. The key observation is that the molecule forms a hole. The portions of amino acids that line this hole are hydrophilic, while regions on the exterior (in contact with the membrane phospholipids) are hydrophobic. The molecule's structure correlates with its function.

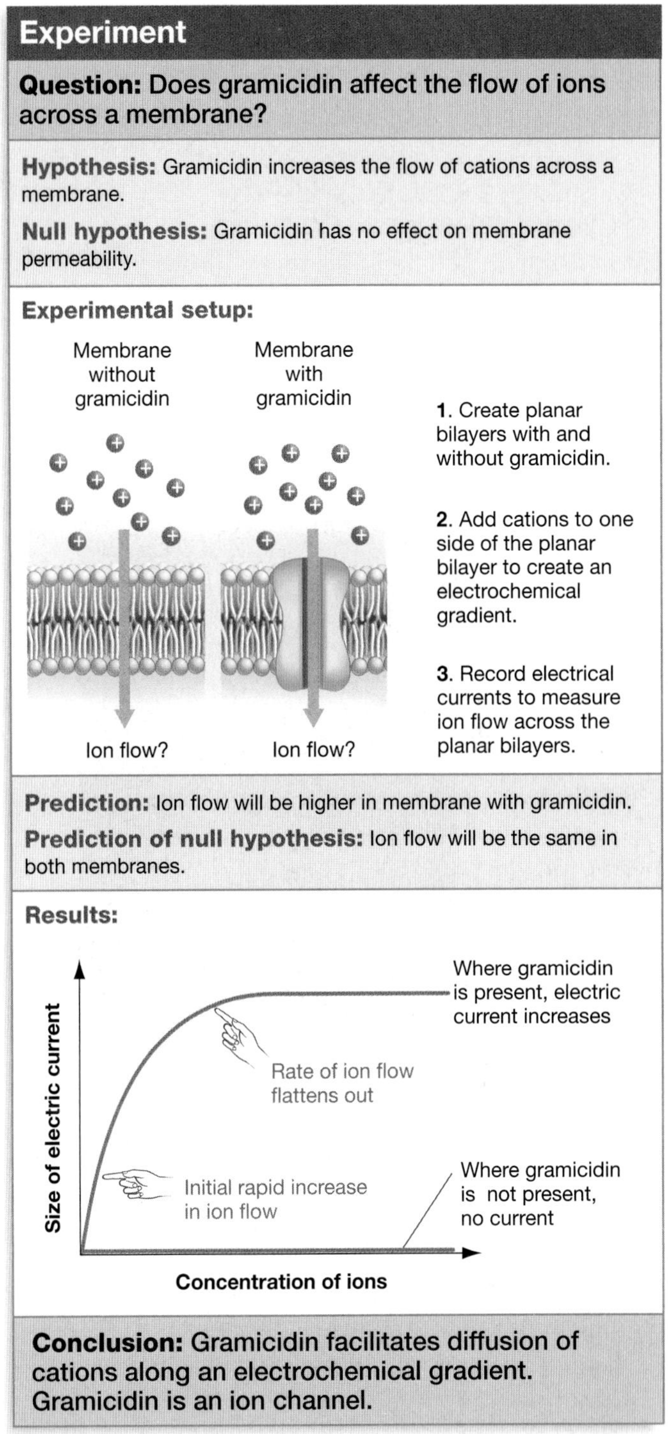

Experiment

Question: Does gramicidin affect the flow of ions across a membrane?

Hypothesis: Gramicidin increases the flow of cations across a membrane.

Null hypothesis: Gramicidin has no effect on membrane permeability.

Experimental setup:

Membrane without gramicidin

Membrane with gramicidin

Ion flow?

Ion flow?

1. Create planar bilayers with and without gramicidin.

2. Add cations to one side of the planar bilayer to create an electrochemical gradient.

3. Record electrical currents to measure ion flow across the planar bilayers.

Prediction: Ion flow will be higher in membrane with gramicidin.

Prediction of null hypothesis: Ion flow will be the same in both membranes.

Results:

Size of electric current

Concentration of ions

Where gramicidin is present, electric current increases

Rate of ion flow flattens out

Initial rapid increase in ion flow

Where gramicidin is not present, no current

Conclusion: Gramicidin facilitates diffusion of cations along an electrochemical gradient. Gramicidin is an ion channel.

FIGURE 6.23 Measuring Ion Flow through the Channel Gramicidin. Experiment for testing the hypothesis that gramicidin is an ion channel.

● **QUESTION** Why does the curve in the Results section flatten out?

Subsequent research has shown that cells have many different types of **channel proteins** in their membranes, each featuring a structure that allows it to admit a particular type of ion or small molecule. For example, Peter Agre and co-workers recently discovered channels called **aquaporins** ("water-pores") that allow

(a) Top view of gramicidin

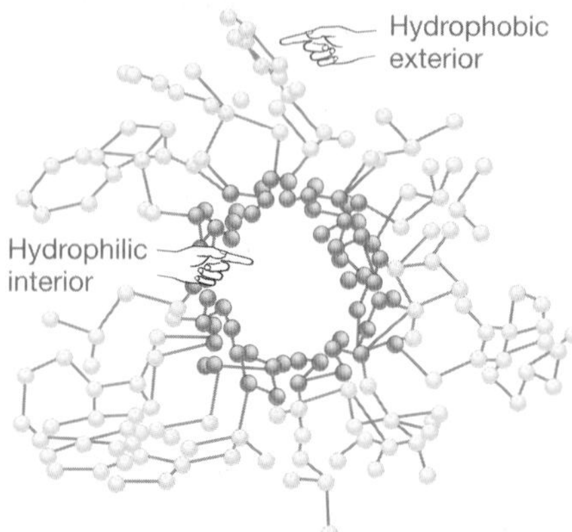

(b) Side view of gramicidin

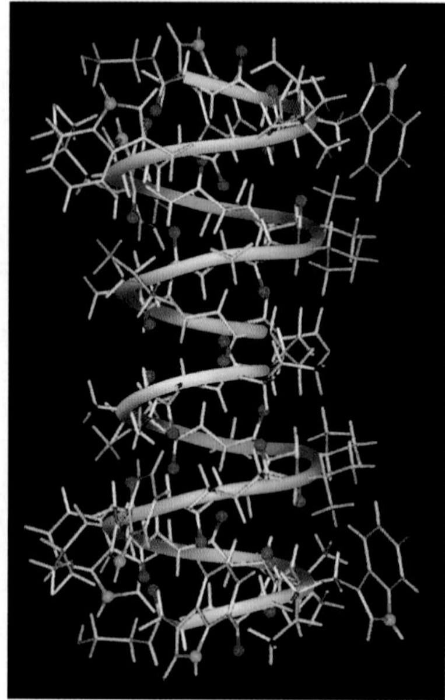

FIGURE 6.24 The Structure of a Channel Protein. Gramicidin is an α-helix consisting of only 15 amino acids. **(a)** In top view, the molecule forms a hole or pore. **(b)** In side view, a green helix traces the peptide-bonded backbone of the polypeptide. R-groups hang off the backbone to the outside. The interior of the channel is hydrophilic; the exterior is hydrophobic.

EXERCISE In (a) and (b), add symbols indicating the locations of phospholipids relative to gramicidin in a plasma membrane.

(a) Water pores allow only water to pass through.

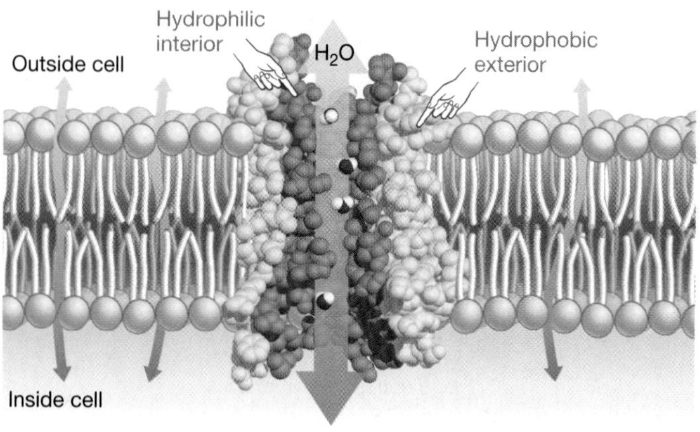

(b) Potassium channels allow only potassium ions to pass through.

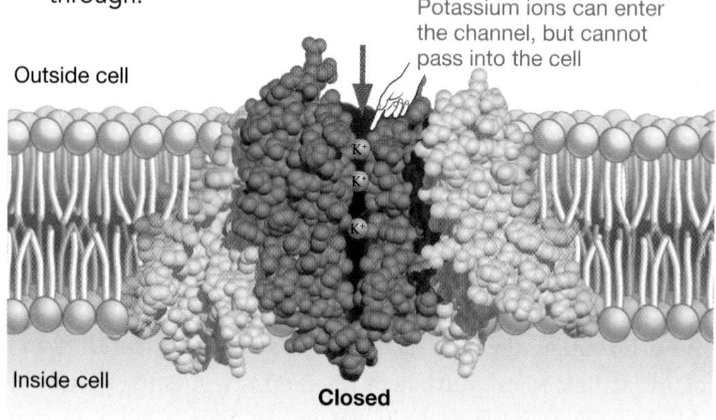

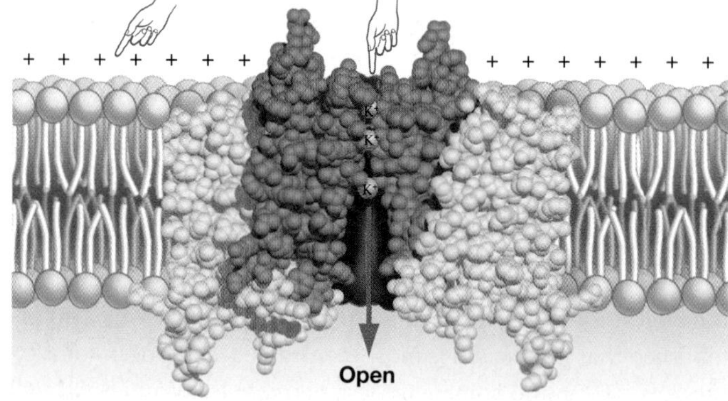

FIGURE 6.25 Most Membrane Channels Are Highly Selective and Highly Regulated (a) A cutaway view looking at the side of an aquaporin—a membrane channel that admits only water. Water moves through its pore via osmosis over 10 times faster than it can move through the lipid bilayer. **(b)** A model of a K^+ channel in the open and closed configurations.

water to cross the plasma membrane over 10 times faster than it does in the absence of aquaporins. **Figure 6.25a** shows a cutaway view from the side of an aquaporin, indicating how it fits in a plasma membrane. Like gramicidin, the channel has a pore that is lined with polar regions of amino acids—in this case, functional groups that interact with water. Hydrophobic groups form the outside of the structure and interact with the lipid bilayer. Unlike gramicidin, aquaporins are extremely selective. They admit water but not other small molecules or ions.

Selectivity turns out to be a prominent characteristic of most channel proteins. The vast majority of these proteins admit only a single type of ion. In many cases, researchers are now able to identify exactly which amino acids are responsible for making the pore selective.

Recent research has also shown that the aquaporins and ion channels are **gated channels**—meaning that they open or close in response to the binding of a particular molecule or to a change in the electrical charge on the outside of the membrane. For example, **Figure 6.25b** shows a potassium channel in the open and closed configuration. When the electrical charge on the membrane becomes positive on the outside relative to the inside, the protein's structure changes in a way that opens the channel and allows potassium ions to cross. The important point here is that in almost all cases, the flow of ions and small molecules through membrane channels is carefully controlled.

In all cases, the movement of substances through channels is passive—meaning it does not require an expenditure of energy. **Passive transport** is powered by diffusion along an electrochemical gradient. Channel proteins enable ions or polar molecules to move across lipid bilayers efficiently. ● If you understand the nature of membrane channels, you should be able to (1) draw the structure of a channel that admits calcium ions (Ca^{2+}) when a signaling molecule binds to it, (2) label hydrophilic and hydrophobic portions of the channel, (3) add ions to the outside and inside of a membrane containing the channel to explain why an electrochemical gradient favors entry of Ca^{2+}, (4) sketch the channel in the open versus closed configuration, and (5) suggest a hypothesis to explain why it might be important for the channel to be selective.

To summarize, membrane proteins such as gramicidin, aquaporins, and potassium channels circumvent the lipid bilayer's impermeability to small, charged compounds. They are responsible for **facilitated diffusion**: the passive transport of substances that otherwise would not cross a membrane readily. The presence of channels *reduces* differences between the interior and exterior. Water molecules and ions are not the only substances that move across membranes through membrane proteins, however. Larger molecules can, too.

Facilitated Diffusion via Carrier Proteins Even though facilitated diffusion does not require an expenditure of energy, it is facilitated—aided—by the presence of a specialized membrane protein. Facilitated diffusion can occur through channels or through **carrier proteins**, also called **transporters**, that change shape during the process. Perhaps the best-studied transporter is specialized for moving glucose into cells.

Next to ribose, the six-carbon sugar glucose is the most important sugar found in organisms. Virtually all cells alive today use glucose as a building block for important macromolecules and as a source of stored chemical energy. But as Figure 6.8 showed, lipid bilayers are only moderately permeable to glucose. It is reasonable to expect, then, that plasma membranes have some mechanism for increasing their permeability to this sugar.

This prediction was supported when researchers compared the permeability of glucose across planar bilayers with its permeability across membranes from cells. The plasma membrane in this study came from human red blood cells, which are among the simplest cells known. A mature red blood cell consists of a membrane, about 300 million hemoglobin molecules, and not much else (**Figure 6.26**, step 1). When these cells are placed in a hypotonic solution (step 2), water rushes into them by osmosis. As water flows inward, the cells swell. Eventually they burst, releasing the hemoglobin molecules and other cell contents. This leaves researchers with pure preparations of plasma membranes called red blood cell "ghosts" (step 3). Experiments have shown that these membranes are much more permeable to glucose than are pure lipid bilayers. Why?

After isolating and analyzing many proteins from red blood cell ghosts, researchers found one protein that specifically increases membrane permeability to glucose. When this purified protein was added to liposomes, the artificial membrane transported glucose at the same rate as a membrane from a living cell. This experiment convinced biologists that a membrane protein was indeed responsible for transporting glucose across plasma membranes. Follow-up work showed that this glucose-

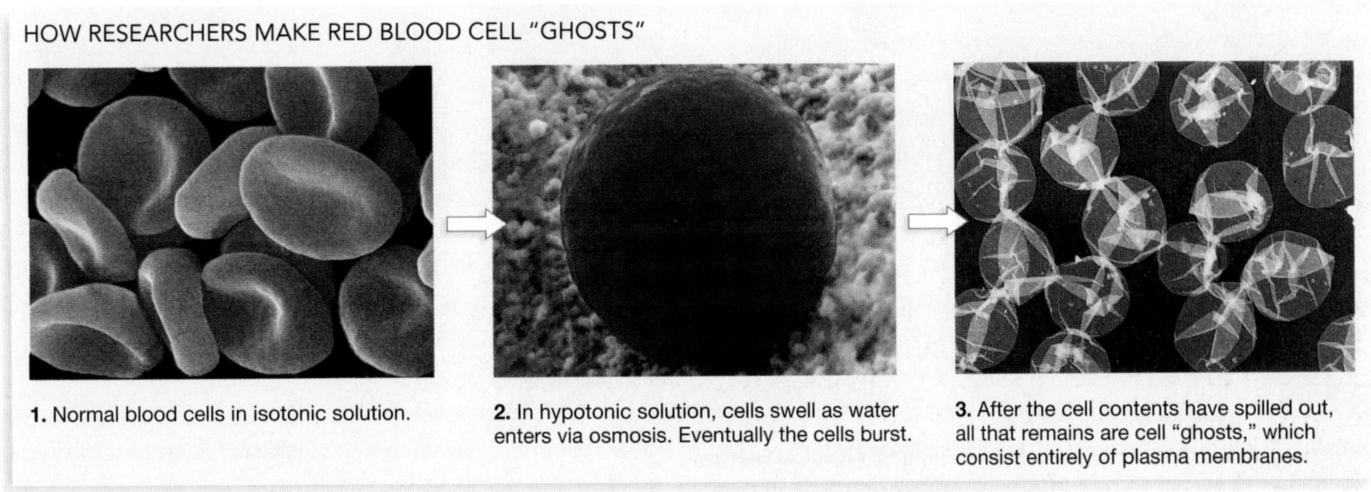

HOW RESEARCHERS MAKE RED BLOOD CELL "GHOSTS"

1. Normal blood cells in isotonic solution.

2. In hypotonic solution, cells swell as water enters via osmosis. Eventually the cells burst.

3. After the cell contents have spilled out, all that remains are cell "ghosts," which consist entirely of plasma membranes.

FIGURE 6.26 Red Blood Cell "Ghosts." Red blood cell ghosts are simple membranes that can be purified and studied in detail.

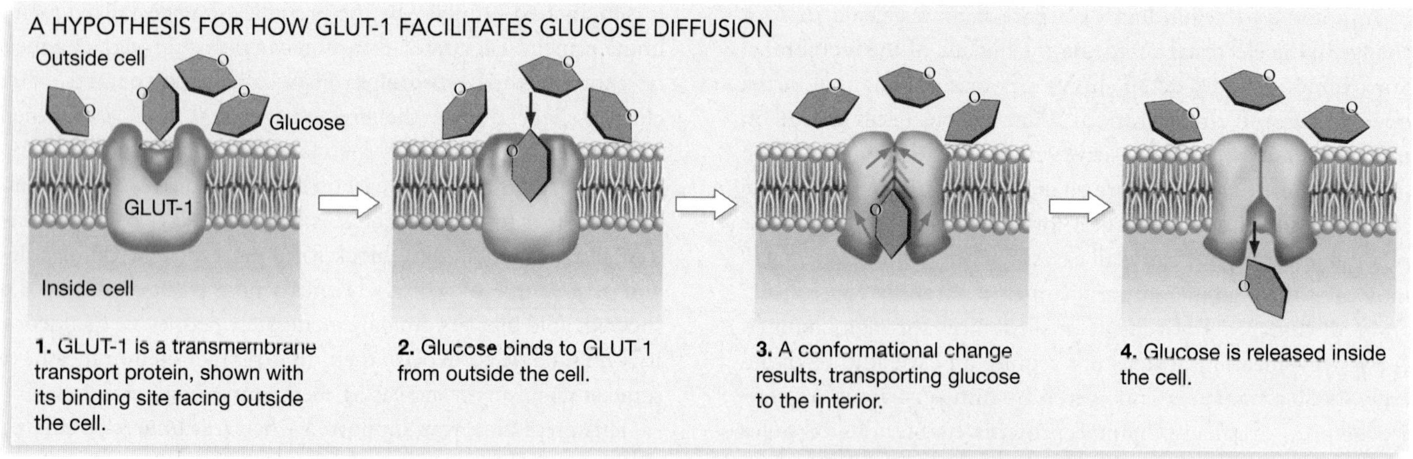

FIGURE 6.27 A Hypothesis to Explain How Membrane Transport Proteins Work. This model suggests that the GLUT-1 transporter acts like an enzyme. It binds a substrate (in this case, a glucose molecule), undergoes a conformation change, and releases the substrate.

QUESTION GLUT's binding site has the same affinity for glucose in both of its conformations. Explain how this trait allows glucose to diffuse along its concentration gradient.

transporting protein, a carrier that is now called GLUT-1, facilitates transport of the "right-handed" optical isomer of glucose but not the left-handed form. Cells use only the right-handed form of glucose, and GLUT-1's binding site is specific for the right-handed form. To make sense of these observations, biologists hypothesize that GLUT proteins with binding sites that interact with the right-handed form of glucose were favored by natural selection. Stated another way, cells with proteins like GLUT-1 thrived much better than cells without a glucose-transport protein or with proteins that transported the left-handed form.

Exactly how GLUT-1 works is a focus of ongoing research. Biologists who are working on the problem have noted that because glucose transport by GLUT-1 is so specific, it is logical to predict that the mechanism resembles the action of enzymes. One hypothesis is illustrated in **Figure 6.27**. The idea is that glucose binds to GLUT-1 on the exterior of the membrane and that this binding induces a conformational change in the protein which transports glucose to the interior of the cell. Recall from Chapter 3 that enzymes frequently change shape when they bind substrates and that such conformational changes are often a critical step in the catalysis of chemical reactions.

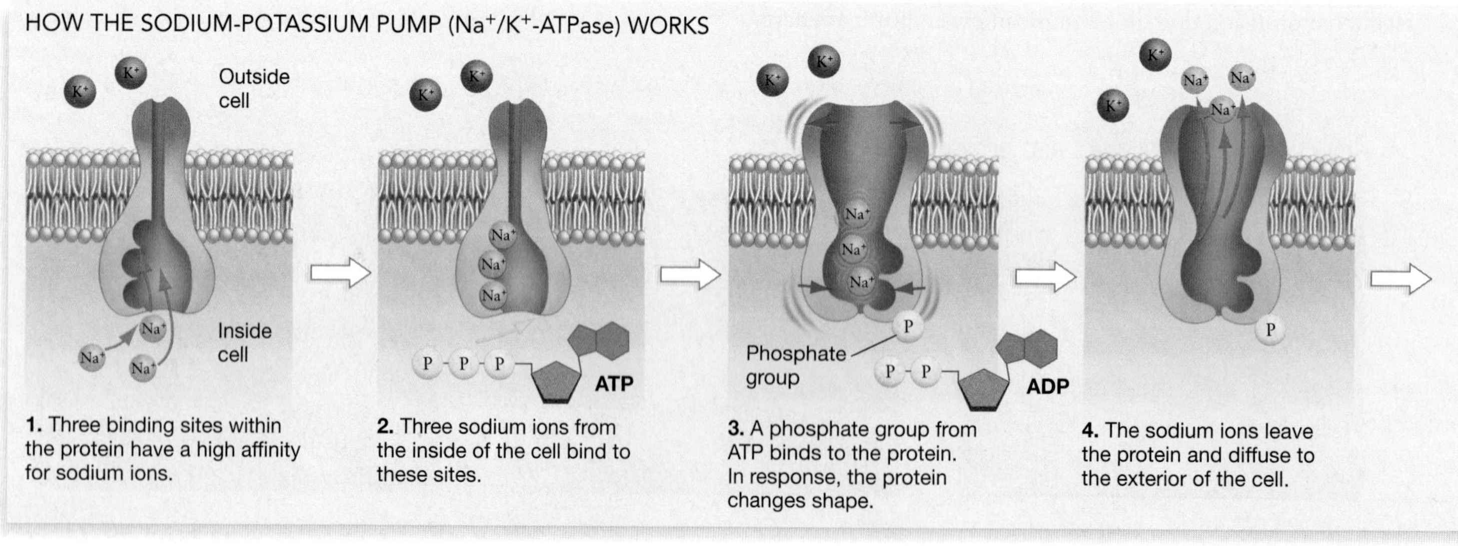

FIGURE 6.28 Active Transport Depends on an Input of Chemical Energy.

EXERCISE Circle the two steps where addition or deletion of a phosphate group causes the protein to change conformation. Label each "Shape change."

Importing molecules into cells via carrier proteins is still powered by diffusion, however. When glucose enters a cell via GLUT-1, it does so because it is following its concentration gradient. If the concentration of glucose is the same on both sides of the plasma membrane, then no net movement of glucose occurs even if the membrane contains GLUT-1. A large array of molecules moves across plasma membranes via facilitated diffusion through specific carrier proteins.

Active Transport by Pumps Whether diffusion is facilitated by channel proteins or by carrier proteins, it is a passive process that makes the cell interior and exterior more similar. But it is also possible for today's cells to import molecules or ions against their electrochemical gradient. Accomplishing this task requires energy, however, because the cell must counteract the entropy loss that occurs when molecules or ions are concentrated. It makes sense, then, that transport against an electrochemical gradient is called **active transport.**

In cells, the energy required to move substances against their electrochemical gradient is provided by a phosphate group (HPO_4^{2-}) from adenosine triphosphate, or ATP. ATP contains three phosphate groups. When one of these phosphate groups leaves ATP and binds to a protein, two negative charges are added to the protein. These charges repel other charges on the protein's amino acids. The protein's potential energy increases in response, and its conformation (shape) usually changes. As Chapter 9 will detail, proteins usually move when a phosphate group binds to them or when a phosphate group drops off. When a phosphate group leaves ATP, the resulting molecule is adenosine diphosphate (ADP), which has two phosphate groups.

Figure 6.28 shows how ions or molecules can move against an electrochemical gradient when membrane proteins called **pumps** change shape. The figure highlights the first pump that was discovered and characterized: a protein called the **sodium-potassium pump**, or more formally, Na^+/K^+-ATPase. The Na^+/K^+ part of this expression refers to the ions that are transported; ATP indicates that adenosine triphosphate is used; and –ase implies that the molecule acts like an enzyme. Unlike the situation with GLUT-1, the mechanism of action in Na^+/K^+-ATPase is now well known. When the protein is in the conformation shown in step 1 of Figure 6.28, binding sites with a high affinity for sodium ions are available. As step 2 shows, three sodium ions from the inside of the cell bind to these sites. A phosphate group from ATP then binds to the pump (step 3). When the phosphate group attaches, the pump's shape changes in a way that reduces its affinity for sodium ions. As a result, the ions leave the protein and diffuse to the exterior of the cell (step 4). In this conformation, though, the protein has binding sites with a high affinity for potassium ions (step 5). As step 6 shows, two potassium ions from outside the cell bind to the pump. When they do, the phosphate group drops off the protein and its shape changes in response—back to the original shape (step 7). In this conformation, the pump has low affinity for potassium ions. As step 8 shows, the potassium ions leave the protein and diffuse to the interior of the cell. The cycle then repeats.

This movement of ions can occur even if an electrochemical gradient exists that favors the outflow of potassium and the inflow of sodium. By exchanging three sodium ions for every two potassium ions, the outside of the membrane becomes positively charged relative to the inside. In this way, the

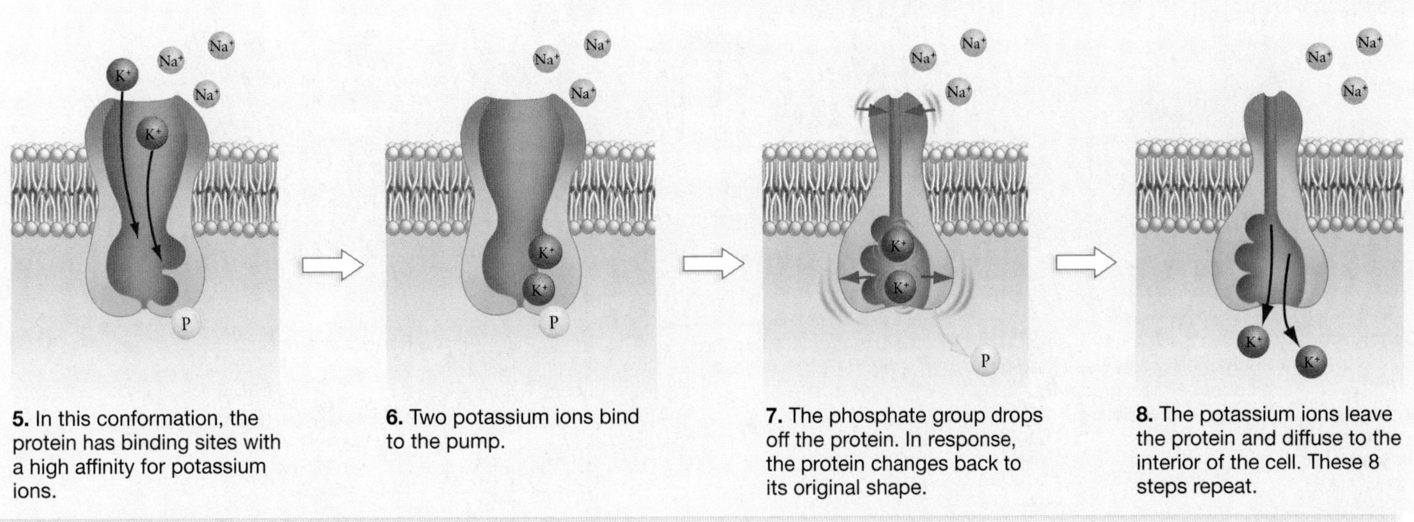

5. In this conformation, the protein has binding sites with a high affinity for potassium ions.

6. Two potassium ions bind to the pump.

7. The phosphate group drops off the protein. In response, the protein changes back to its original shape.

8. The potassium ions leave the protein and diffuse to the interior of the cell. These 8 steps repeat.

sodium-potassium pump sets up an electrical gradient as well as a chemical gradient across the membrane.

Similar pumps are specialized for moving protons (H^+), calcium ions (Ca^{2+}), or other ions or molecules. In this way, cells are capable of concentrating certain substances or setting up electrochemical gradients. It is difficult to overemphasize the importance of these gradients. For example, the electrochemical gradients produced by proton pumps allow plants to take up nutrients from the soil; the gradients established by the Na^+/K^+-ATPase and calcium pumps allow your nerve cells to transmit electrical signals throughout your body. You will encounter active transport, membrane pumps, and electrochemical gradients throughout this text.

(MB) **Web Animation** at www.masteringbio.com
Membrane Transport Proteins

Taken together, the lipid bilayer and the proteins involved in passive transport and active transport enable cells to create an internal environment that is much different from the external one. Membrane proteins allow ions and molecules to cross the plasma membrane, even though they are not lipid soluble. (**Figure 6.29**). When membrane proteins first evolved, then, the early cells acquired the ability to create an internal environment that was conducive to life—meaning that such an environment contained the substances required for manufacturing ATP and copying ribozymes. Cells with particularly efficient and selective membrane proteins would be favored by natural selection and would come to dominate the population. Cellular life had begun.

Some 3.5 billion years later, cells continue to evolve. What do today's cells look like, and how do they produce and store the chemical energy that makes life possible? How do they use ATP to move pumps and channels and other molecules and machines where they're needed? Answering these and related questions is the focus of Unit 2.

Check Your Understanding

If you understand that...

- Membrane proteins allow ions and molecules that ordinarily do not readily cross lipid bilayers to enter or exit cells.
- Substances may move along an electrochemical gradient via facilitated diffusion through channel proteins or transport proteins, or they may move against an electrochemical gradient in response to work done by pumps.

You should be able to...

1) Sketch a phospholipid bilayer.
2) Indicate how ions and large molecules cross it via each major type of membrane transport protein.

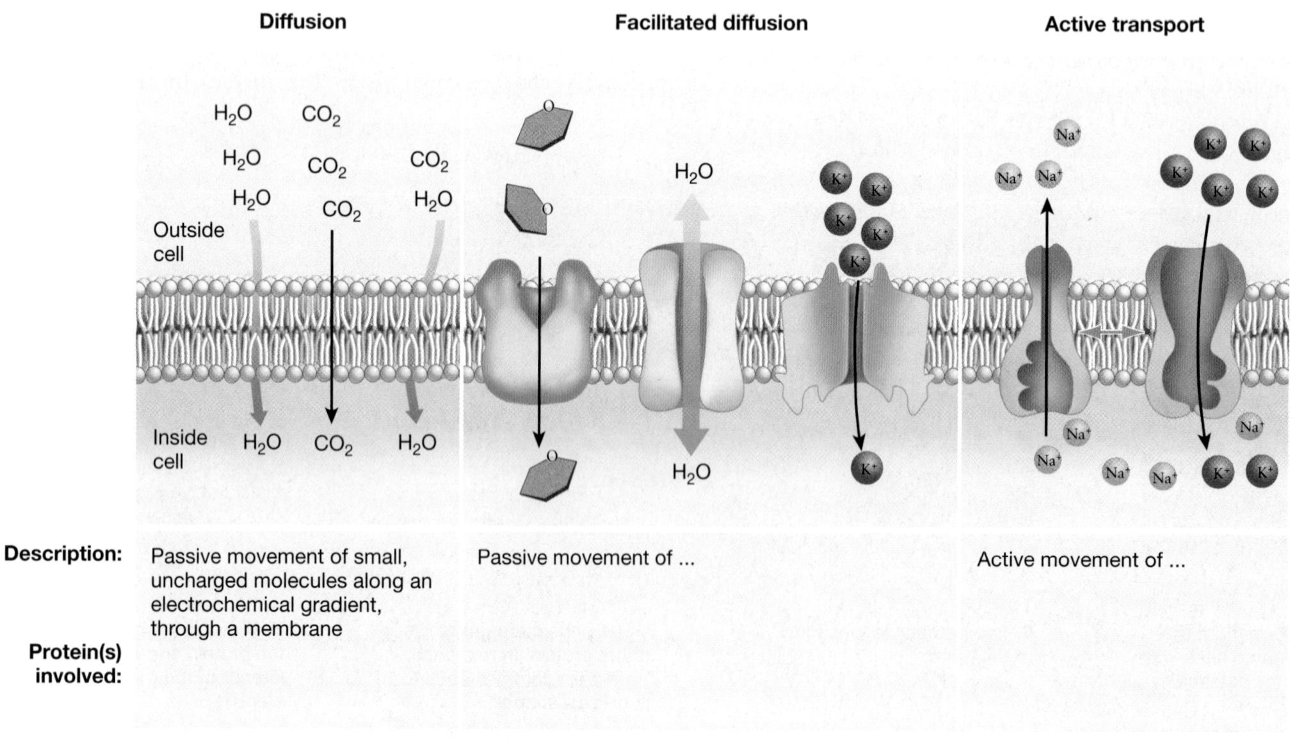

FIGURE 6.29 Mechanisms of Membrane Transport: A Summary.

EXERCISE Complete the chart.

Chapter Review

SUMMARY OF KEY CONCEPTS

Phospholipids are amphipathic molecules—they have a hydrophilic region and a hydrophobic region. In solution, they spontaneously form bilayers that are selectively permeable—meaning that only certain substances cross them readily.

The plasma membrane is a structure that forms a physical barrier between life and nonlife. The basic structure of plasma membranes is created by a phospholipid bilayer. Phospholipids have a polar head and a nonpolar tail. The nonpolar tail consists of a lipid—usually a fatty acid or an isoprene. Lipids do not dissolve in water.

Small, nonpolar molecules tend to move across membranes readily; ions and other charged compounds cross rarely, if at all. The permeability and fluidity of lipid bilayers depend on temperature and on the types of phospholipids present. For example, because phospholipids that contain long, saturated fatty acids form a dense and highly hydrophobic membrane interior, they tend to be less permeable than phospholipids containing shorter, unsaturated fatty acids.

You should be able to draw phospholipid bilayers that are highly permeable and fluid versus highly impermeable and lacking in fluidity.

Ions and molecules diffuse spontaneously from regions of high concentration to regions of low concentration. Water moves across lipid bilayers from regions of high concentration to regions of low concentration via osmosis—a special case of diffusion.

Diffusion is movement of ions or molecules due to their kinetic energy. Solutes move via diffusion from a region of high concentration to a region of low concentration. This is a spontaneous process driven by an increase in entropy. Water also moves across membranes spontaneously if a molecule or an ion that cannot cross the membrane is found in different concentrations on the two sides. In osmosis, water moves from the region with a lower concentration of solutes to the region of higher solute concentration. Osmosis is a passive process driven by an increase in entropy.

You should be able to draw a beaker with solutions on either side separated by a plasma membrane, and then predict what will happen after addition of a solute to one side if the solute (1) crosses the membrane readily, versus (2) is incapable of crossing the membrane.

MB Web Animation at www.masteringbio.com
Diffusion and Osmosis

In cells, membrane proteins are responsible for the passage of ions, polar molecules, and large molecules that can't cross the membrane on their own because they are not soluble in lipids.

The permeability of lipid bilayers can be altered significantly by membrane transport proteins, which are scattered throughout the plasma membrane. Channel proteins, for example, are molecules that provide holes in the membrane and facilitate the diffusion of specific ions into or out of the cell. Transport proteins are enzyme-like proteins that allow specific molecules to diffuse into the cell. In addition to these forms of facilitated diffusion, membrane proteins that act as energy-demanding pumps actively move ions or molecules against their electrochemical gradient. In combination, the selective permeability of phospholipid bilayers and the specificity of transport proteins make it possible to create an environment inside a cell that is radically different from the exterior.

You should be able to draw and label the membrane of a cell that pumps hydrogen ions to the exterior, has channels that admit calcium ions along an electrochemical gradient, and has carriers that admit lactose (a sugar) molecules along a concentration gradient. Your drawing should include arrows and labels indicating the direction of solute movement and the direction of the appropriate electrochemical gradients.

Membrane Transport Proteins

QUESTIONS

Test Your Knowledge

1. What does the term *hydrophilic* mean when it is translated literally?
   a. "oil loving"
   b. "water loving"
   c. "oil fearing"
   d. "water fearing"

2. If a solution surrounding a cell is hypotonic relative to the inside of the cell, how will water move?
   a. It will move into the cell via osmosis.
   b. It will move out of the cell via osmosis.
   c. It will not move, because equilibrium exists.
   d. It will evaporate from the cell surface more rapidly.

3. If a solution surrounding a cell is hypertonic relative to the inside of the cell, how will water move?
   a. It will move into the cell via osmosis.
   b. It will move out of the cell via osmosis.
   c. It will not move, because equilibrium exists.
   d. It will evaporate from the cell surface more rapidly.

4. When does a concentration gradient exist?
   a. when membranes rupture
   b. when solute concentrations are high
   c. when solute concentrations are low
   d. when solute concentrations differ on the two sides of a membrane

5. Which of the following must be true for osmosis to occur?
   a. Water must be at room temperature or above.
   b. Solutions with the same concentration of solutes must be separated by a selectively permeable membrane.
   c. Solutions with different concentrations of solutes must be separated by a selectively permeable membrane.
   d. Water must be under pressure.

6. Why are the lipid bilayers in cells called "selectively permeable"?
   a. They are not all that permeable.
   b. Their permeability changes with their molecular composition.
   c. Their permeability is temperature dependent.
   d. They are permeable to some substances but not others.

Test Your Knowledge answers: 1. b; 2. a; 3. b; 4. d; 5. c; 6. d

Test Your Understanding

Answers are available at www.masteringbio.com

1. Cooking oil is composed of lipids that consist of long hydrocarbon chains. Would you expect these lipids to form membranes spontaneously? Why or why not? Describe, on a molecular level, how you would expect these lipids to interact with water.
2. Explain why phospholipids form a bilayer in solution.
3. Ethanol, the active ingredient in alcoholic beverages, is a small, polar, uncharged molecule. Would you predict that this molecule crosses plasma membranes quickly or slowly? Explain your reasoning.
4. Why can osmosis occur only if solutions are separated by a selectively permeable membrane? What happens in solutions that are *not* separated by a selectively permeable membrane?
5. The text claims that the portion of membrane proteins that spans the hydrophobic tails of phospholipids is itself hydrophobic (see Figure 6.17b). Why is this logical? Look back at Figure 3.3 and Table 3.2, and make a list of amino acids you would expect to find in these regions of transmembrane proteins.
6. Examine the membrane in the accompanying figure. Label the molecules and ions that will pass through the membrane as a result of osmosis, diffusion, and facilitated diffusion. Draw arrows to indicate where each of the molecules and ions will travel.

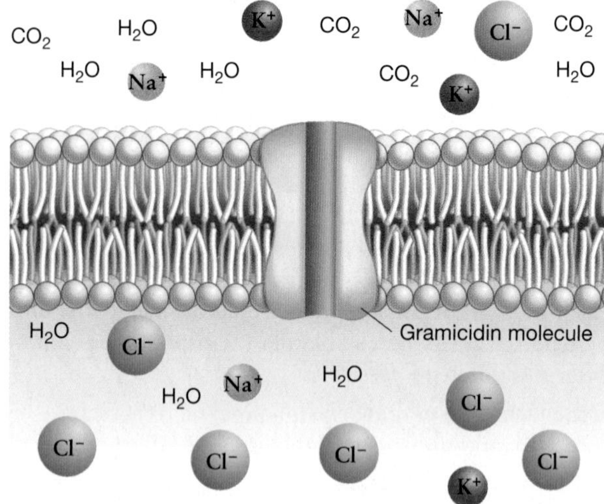

Applying Concepts to New Situations

Answers are available at www.masteringbio.com

1. When phospholipids are arranged in a bilayer, it is theoretically possible for individual molecules in the bilayer to flip-flop. That is, a phospholipid could turn 180° and become part of the membrane's other surface. Sketch this process. From what you know about the behavior of polar heads and nonpolar tails, predict whether flip-flops are frequent or rare. Then design an experiment, using a planar bilayer made up partly of fatty acids that contain a dye molecule on their hydrophilic head, to test your prediction.
2. Unicellular organisms that live in extremely cold habitats have an unusually high proportion of unsaturated fatty acids in their plasma membranes. Some of these membranes even contain polyunsaturated fatty acids, which have more than one double bond in each hydrocarbon chain. Draw a picture of this type of membrane and comment on the hypothesis that membranes with unsaturated fatty-acid tails function better at cold temperatures than do membranes with saturated fatty-acid tails. Make a prediction about the structure of fatty acids found in organisms that live in extremely hot environments.
3. When biomedical researchers design drugs that must enter cells to be effective, they sometimes add methyl (CH_3) groups to make the drug molecules more likely to pass through plasma membranes. Conversely, when researchers design drugs that act on the exterior of plasma membranes, they sometimes add a charged group to decrease the likelihood that the drugs will pass through membranes and enter cells. Explain why these strategies are effective.
4. Advertisements frequently claim that laundry and dishwashing detergents "cut grease." What the ad writers mean is that the detergents surround oil droplets on clothing or dishes, making the droplets water soluble. When this happens, the oil droplets can be washed away. Explain how this happens on a molecular level.

www.masteringbio.com is also your resource for • Answers to text, table, and figure caption questions and exercises • Answers to *Check Your Understanding* boxes • Online study guides and quizzes • Additional study tools including the *E-Book for Biological Science* 3rd ed., textbook art, animations, and videos.

Inside the Cell

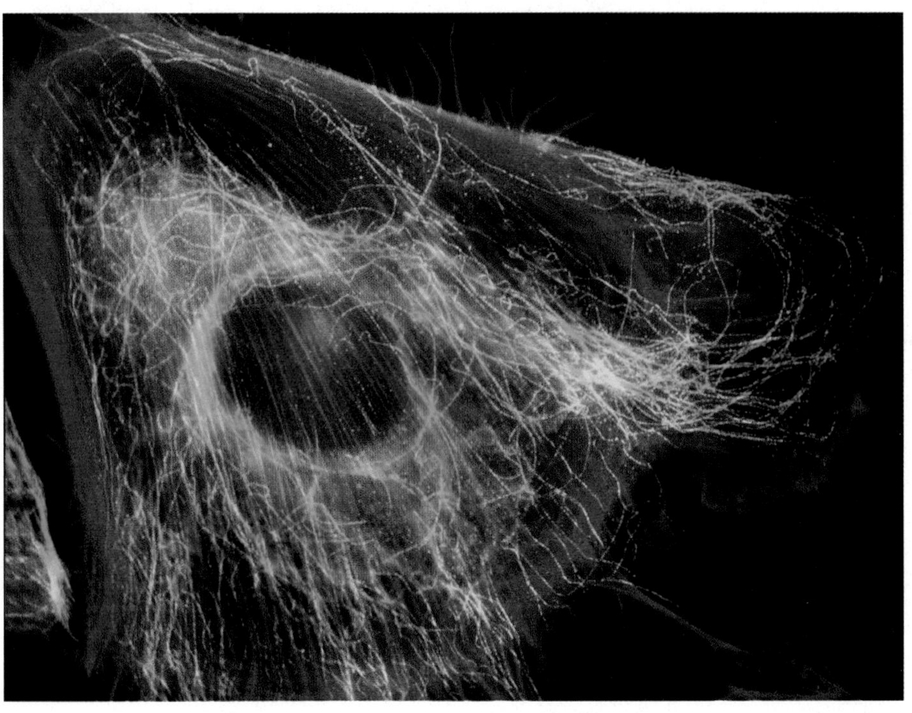

This cell has been treated with fluorescing molecules that bind to its fibrous skeleton. Microtubules (large protein fibers) are green; microfilaments (smaller fibers) are red. The cell's nucleus has been stained blue.

KEY CONCEPTS

- The structure of cell components is closely related to their function.
- Inside cells, materials are transported to their destinations with the help of molecular "zip codes."
- Cells are dynamic. Thousands of chemical reactions occur each second within cells; molecules constantly enter and exit across the plasma membrane; cell products are shipped along protein fibers; and elements of the cell's internal skeleton grow and shrink.

In Chapter 1 you were introduced to the cell theory, which states that all organisms consist of cells and that all cells are derived from preexisting cells. Since this theory was initially developed and tested in the 1850s and 1860s, an enormous body of research has confirmed that the cell is the fundamental structural and functional unit of life. Life on Earth is cellular.

In a very real sense, then, understanding how an organism works is a matter of understanding how cells are structured and how they function. To drive this point home, recall from Chapter 1 that many eukaryotic organisms and virtually all bacteria and archaea are unicellular. In terms of sheer numbers, unicellular organisms dominate life on Earth. For researchers who study these unicellular species, understanding the cell is synonymous with understanding the organism as a whole. Even in plants, animals, and other multicellular eukaryotes, complex behavior originates at the level of the cell. For example, your ability to read this page begins with changes in light-sensitive molecules located in cells at the back of your eyes. When these molecules change shape, they trigger changes in the membranes of nerve cells that connect your eyes to your brain. To understand complex processes such as vision, then, researchers often begin by studying the structure and function of the individual cells involved—the parts that make up the whole.

Chapter 6 introduced an essential part of the cell: the **plasma membrane**. Thanks to the selective permeability of phospholipid bilayers and the activity of membrane transport proteins, this structure creates an internal environment that differs from conditions outside the cell. Our task now is to explore the structures that are found inside the membrane and analyze what they do. The chapter focuses on several particularly dynamic structures and processes and introduces some of the experimental approaches that biologists use to understand them. Let's begin by surveying the structures that make up a cell, and then go on to analyze three of the most important cell systems: the nucleus, the protein manufacturing and shipping complex called the endomembrane system, and the network of protein fibers called the cytoskeleton.

7.1 What's Inside the Cell?

In Chapter 1 you read about the two fundamental types of cells observed in nature—eukaryotes and prokaryotes. Eukaryotic cells have a membrane-bound compartment called a nucleus, while prokaryotic cells do not. According to **morphology** ("form-science"), then, species fall into the two broad categories: (1) prokaryotes and (2) eukaryotes. But according to **phylogeny** ("tribe-source"), or evolutionary history, organisms fall into the three broad groups called (1) Bacteria, (2) Archaea, and (3) Eukarya. Members of the Bacteria and Archaea are prokaryotic; members of the Eukarya—including algae, fungi, plants, and animals—are eukaryotic.

In the late seventeenth century, biologists began using microscopes to study the structure of cells. Over time, improvements in optics and cell preparation techniques allowed researchers to catalogue the structures reviewed in this section. When electron microscopes became widely available in the 1950s, investigators described the internal anatomy of these structures in more detail. More recent advances in microscopy have allowed investigators to videotape certain types of cell processes in living cells (see **BioSkills 8**).

What have anatomical studies based on microscopy revealed? Let's look first at the general anatomy of prokaryotic cells and eukaryotic cells, and then consider how the structures that have been identified help cells function.

Prokaryotic Cells

Figure 7.1 shows the general structure of a prokaryotic cell. For most bacterial and archaeal species, the plasma membrane encloses a single compartment—meaning that the cell has few or no substructures separated from the rest of the cell by internal membranes. Closer examination reveals a series of important internal structures, however. Let's take a look at a typical prokaryotic cell, starting at the outside and working in.

As Chapter 6 pointed out, the cell membrane, or plasma membrane, consists of a phospholipid bilayer and proteins that either span the bilayer or attach to one side. Inside the membrane, the contents of a cell are collectively termed the **cytoplasm** ("cell-formed"). Because the cytoplasm contains a high concentration of solutes, in most habitats it is hypertonic relative to the surrounding environment. When this is the case,

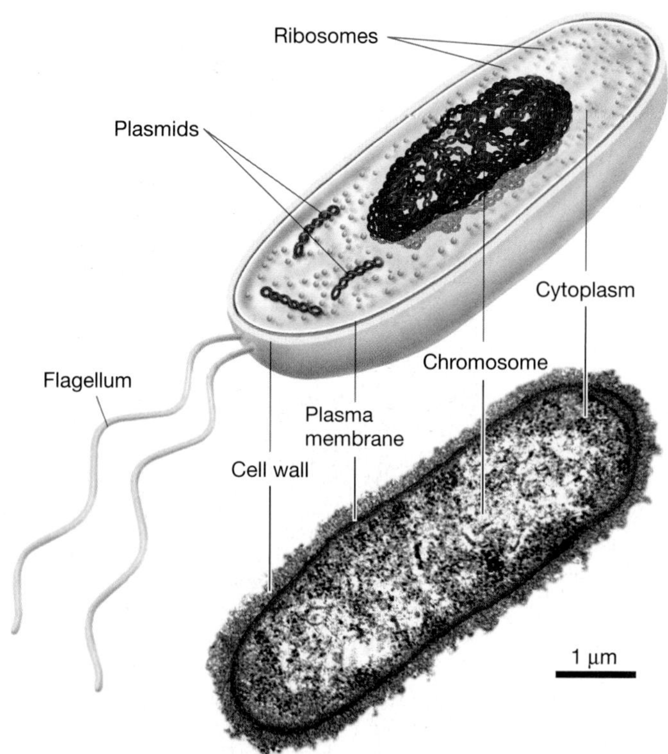

FIGURE 7.1 A Prokaryotic Cell. Prokaryotic cells are identified by a negative trait—the absence of a membrane-bound nucleus. Although there is wide variation in the size and shape of bacterial and archaeal cells, they all contain a plasma membrane, chromosomes, and protein-synthesizing ribosomes; almost all have a stiff cell wall. Some prokaryotes have flagella that power swimming and/or inner membranes where photosynthesis takes place.

EXERCISE Label the nucleoid in the drawing.

water enters the cell via osmosis and makes the cell's volume expand. In virtually all bacteria and archaea, this pressure is resisted by a stiff **cell wall**. Bacterial and archaeal cell walls are a tough, fibrous layer that surrounds the plasma membrane (**Figure 7.2**). The pressure of the plasma membrane against the cell wall is about the same as the pressure in an automobile tire. The cell wall protects the organism and gives it shape and rigidity. In addition, many bacteria have another protective layer outside the cell wall that consists of lipids with polysaccharides attached. Lipids that contain carbohydrate groups are termed **glycolipids**.

In the cytoplasm of a prokaryotic cell, the most prominent structure is the chromosome. Most prokaryotic species have a single, circular **chromosome** that consists of a large DNA molecule associated with a small number of proteins. The DNA molecule contains information, while the proteins provide structural support for the DNA. You might recall from Chapter 4 that the information in DNA is encoded in the form of its sequence of nitrogenous bases. More specifically, the primary structure of DNA contains the instructions for making the proteins, RNAs, and other molecules needed by the cell. A segment of DNA that contains the information for building an RNA

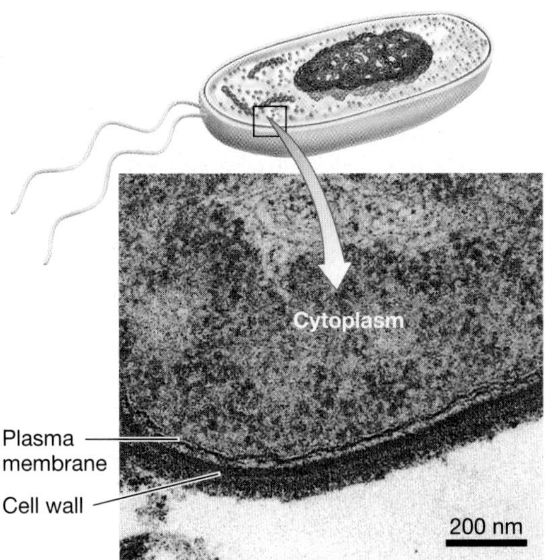

FIGURE 7.2 The Bacterial Cell Wall. In bacteria and archaea, the cell wall consists of peptidoglycan or similar polymers that are cross-linked into tough sheets. The inside of the cell wall contacts the plasma membrane, which pushes up against the wall. The outside of the cell wall makes direct contact with the outside environment, which is almost always filled with competitors and predators.

molecule or a polypeptide is called a **gene**. To summarize, chromosomes contain DNA, which contains genes.

Prokaryotic chromosomes are found in a localized area of the cell called the **nucleoid**. The nucleoid is usually found in the center of the cell and typically represents about 20 percent of the cell's total volume. It's important to note, though, that the genetic material in the nucleoid is not separated from the rest of the cytoplasm by a membrane. In the well-studied bacterium *Escherichia coli*, the circular chromosome would be over 1 mm long if it were linear—500 times longer than the cell itself. This situation is typical in prokaryotes. To fit into the cell, the DNA double helix coils on itself with the aid of enzymes to form the highly compact, "supercoiled" structure shown in **Figure 7.3**.

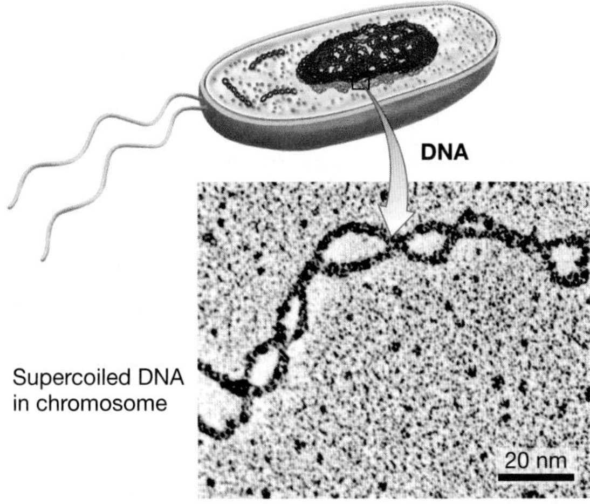

FIGURE 7.3 Bacterial DNA Is Supercoiled. The circular chromosomes of bacteria and archaea must be coiled extensively, into "supercoils," to fit in the cell.

Supercoiled regions of DNA resemble a string that has been held at either end and then twisted until it coils back upon itself.

Depending on the species and population being considered, prokaryotic cells may also contain one to about a hundred small, usually circular, supercoiled DNA molecules called **plasmids**. Plasmids contain genes but are physically independent of the main, cellular chromosome. In most cases the genes carried by plasmids are not required under normal conditions; instead they help cells adapt to unusual circumstances, such as the sudden presence of a poison in the environment. As a result, plasmids can be considered auxiliary genetic elements.

Two other prominent cell structures found in prokaryotes are **ribosomes**, which manufacture proteins, and **flagella** (singular: **flagellum**), which rotate to power swimming in aquatic species. Ribosomes are observed in all prokaryotic cells and are found throughout the cytoplasm. Bacterial ribosomes are complex structures consisting of a total of three distinct RNA molecules and over 50 different proteins. These molecular components are organized into two major structural elements, called the large subunit and small subunit (**Figure 7.4**). It is not unusual for a single cell to contain 10,000 ribosomes. Both ribosomes and prokaryotic flagella lack a membrane. Not all bacterial species have flagella, however. When present they are usually few in number and are located on the surface of the cell. Over 40 different proteins are involved in building and controlling bacterial flagella. At top speed, flagellar movement can drive a bacterial cell through water at 60 cell lengths per

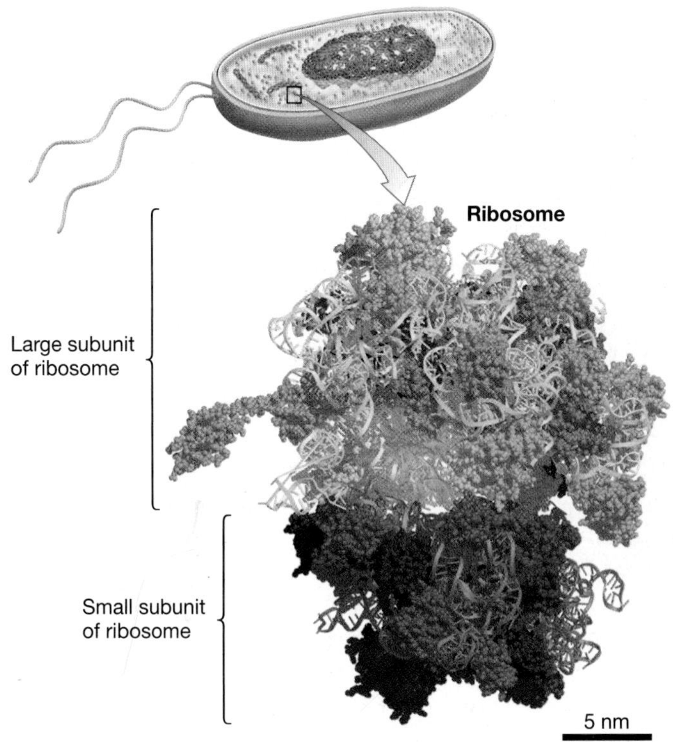

FIGURE 7.4 The Bacterial Ribosome. Bacterial ribosomes are made of RNA and protein molecules that are organized into large and small subunits.

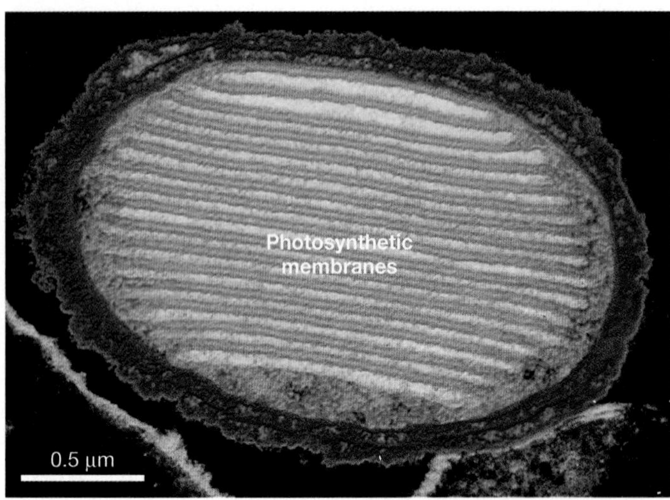

FIGURE 7.5 Photosynthetic Membranes in Bacteria. The green stripes in this photosynthetic bacterium are infoldings of the plasma membrane. They are green because they contain the pigments and enzymes required for photosynthesis.

second. In contrast, the cheetah qualifies as the fastest land animal but can sprint at a mere 25 body lengths per second.

Prokaryotes lack a nucleus, but it is not correct to say that they lack membrane-bound internal structures. Many species contain membrane-bound storage containers, and extensive internal membranes occur in bacteria and archaea that perform photosynthesis. Photosynthesis is the suite of chemical reactions responsible for converting the energy in sunlight into chemical energy stored in sugars. The photosynthetic membranes observed in prokaryotes contain the enzymes and pigment molecules required for these reactions to occur and develop as infoldings of the plasma membrane. In some cases, vesicles pinch off as the plasma membrane folds in. In other cases, flattened stacks of photosynthetic membrane, like those shown in **Figure 7.5**, form from the infolded sections of the plasma membrane.

In addition, recent research indicates that several bacterial species have internal compartments that qualify as organelles ("little organs"). An **organelle** is a membrane-bound compartment in the cytoplasm that contains enzymes or structures specialized for a particular function. One type of bacterial organelle is specialized for storing calcium ions; another contains crystals of the mineral magnetite. The magnetite crystals function like a compass needle to help cells orient themselves in a magnetic field and swim in a directed way.

Recent research has also shown that bacteria and archaea contain long, thin fibers that serve a structural role inside the cell. All bacterial species, for example, contain protein fibers. These filaments are essential for cell division to take place. Some species also have protein filaments that help maintain cell shape. Protein filaments such as these form the basis of the **cytoskeleton** ("cell skeleton").

Even though internal membranes and some cytoskeletal components are found in all prokaryotic cells studied to date, their extent pales in comparison with that in eukaryotes. When typical prokaryotic and eukaryotic cells are compared side by side, four outstanding differences jump out: (1) Eukaryotic chromosomes are found inside a membrane-bound compartment called the nucleus; (2) eukaryotic cells are often much larger; (3) they contain extensive amounts of internal membrane; and (4) they feature a diverse and dynamic cyto-skeleton (**Table 7.1**).

Eukaryotic Cells

The lineage called Eukarya includes forms ranging from unicellular species to 100-meter-tall redwoods. Brown algae, red algae, fungi, amoebae, and slime molds are all eukaryotic, as are green plants and animals. The first thing that strikes biologists about eukaryotic cells is how much larger they are on average than bacteria and archaea (**Figure 7.6**). Most eukaryotic cells range from about 5 to 100 μm in diameter, while most prokaryotic cells vary between 1 and 10 μm in diameter. A micrograph of an average eukaryotic cell, at the same scale as the bacterial cell in Figure 7.5, would fill this page. This difference in size inspired the hypothesis that when eukaryotes first evolved, they made their living by ingesting bacterial and archaeal cells whole. Stated another way, the evolution of large cell size is thought to have made it possible for eukaryotic cells to act as predators—organisms that kill and consume other organisms. Thousands of unicellular eukaryotic species alive today still survive by surrounding and taking in whole bacterial and archaeal cells.

SUMMARY TABLE 7.1 **How Do the Structures of Prokaryotic and Eukaryotic Cells Differ?**

	Location of DNA	Internal Membranes and Organelles	Cytoskeleton	Overall Size
Bacteria and Archaea	In nucleoid (not membrane bound); plasmids also common	Extensive internal membranes only in photosynthetic species; limited types and numbers of organelles	Limited in extent, relative to eukaryotes	Usually small relative to eukaryotes
Eukaryotes	Inside nucleus (membrane bound); plasmids extremely rare	Large numbers of organelles; many types of organelles	Extensive—usually found throughout volume of cell	Most are larger than prokaryotes

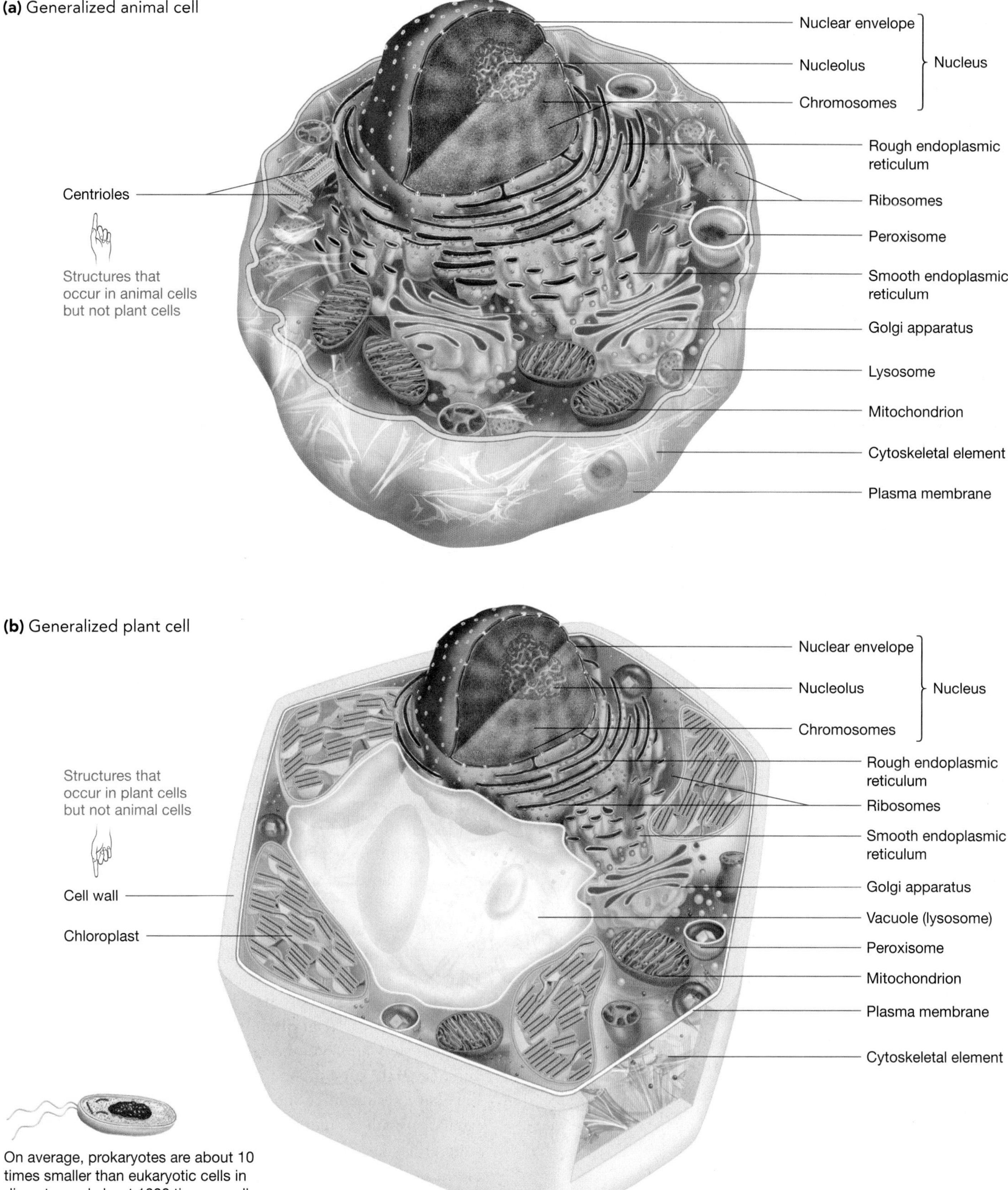

FIGURE 7.6 Animal and Plant Cells. Generalized, or "typical," **(a)** animal and **(b)** plant cells. (Compare with the prokaryotic cell, shown at true relative size at bottom left.)

The evolution of large cells has a downside, however. Ions and small molecules such as adenosine triphosphate (ATP), amino acids, and nucleotides cannot diffuse across a large volume quickly. If the ATP that supplies chemical energy in cells is used up on one side of a large cell, ATP from the other side of the cell would take a long time to diffuse to that location. Prokaryotic cells are small enough that ions and small molecules arrive where they are needed via diffusion. In fact, the size of prokaryotic cells is probably limited by the distance that molecules must diffuse or be transported inside the cell.

How do eukaryotic cells solve the problems that size can engender? The answer lies in the numerous organelles observed in eukaryotic cells. In effect, the huge volume inside a eukaryotic cell is compartmentalized into a large number of bacterium-sized parts. Because eukaryotic cells are subdivided, the molecules required for specific chemical reactions are often located within a given compartment or organelle. Compartmentalization offers two key advantages:

- Incompatible chemical reactions can be separated. For example, new fatty acids can be synthesized in one organelle while excess or damaged fatty acids are degraded and recycled in a different organelle.
- The efficiency of chemical reactions is increased. First, the substrates required for particular reactions can be localized and maintained at high concentration within organelles. Second, if substrates are used up in a particular part of the organelle, they can be replaced by substrates that have only a short distance to diffuse. Third, groups of enzymes that work together can be clustered on internal membranes instead of floating free in the cytoplasm. When the product of a reaction is the substrate for a reaction catalyzed by another enzyme, clustering the enzymes increases the speed and efficiency of a reaction sequence.

If bacteria and archaea can be compared to small machine shops, then eukaryotic cells resemble sprawling industrial complexes. The organelles and other structures found in eukaryotes are analogous to highly specialized buildings that act as factories, power stations, warehouses, transportation corridors, and administrative centers. Figure 7.6 shows how organelles are arranged in a typical animal cell and plant cell. What are these structures, and what do they do? As you read about each of these cell components in the pages that follow, focus on identifying how their structure correlates with their function. Then use the summary table on page 132 as a study guide (**Table 7.2**).

The Nucleus The **nucleus** is among the largest organelles and is highly organized (**Figure 7.7**). It is enclosed by a unique structure—a complex double membrane called the **nuclear envelope**. As section 7.2 will detail, the nuclear envelope is studded with pore-like openings, and its inside surface is associated with fibrous proteins that form a lattice-like sheet called the **nuclear lamina**. The nuclear lamina stiffens the structure and maintains its shape.

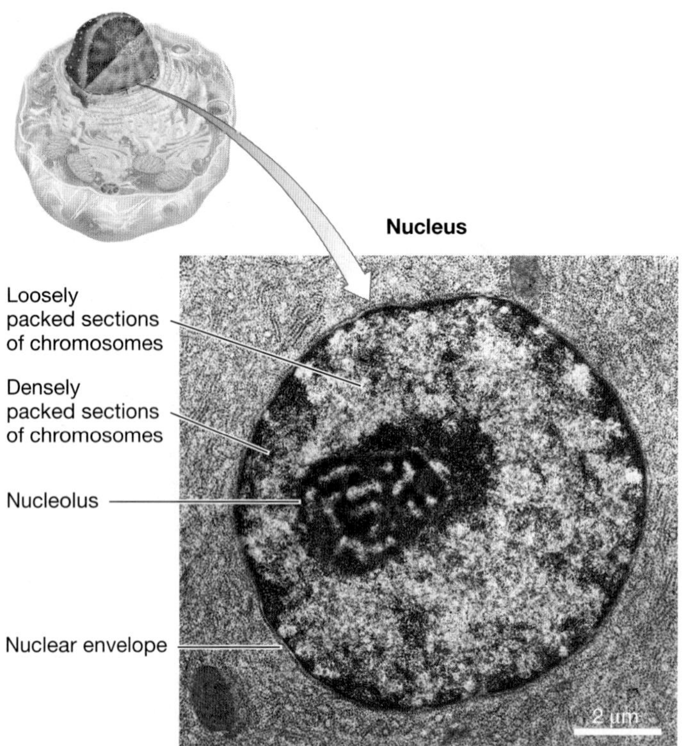

FIGURE 7.7 The Nucleus Is the Eukaryotic Cell's Information Storage and Retrieval Center. The genetic, or hereditary, information is encoded in DNA, which is a component of the chromosomes inside the nucleus.

Chromosomes do not float free inside the nucleus—instead, each chromosome occupies a distinct area and is attached to the nuclear lamina in at least one location. The nucleus also includes a distinctive region called the **nucleolus**, where the RNA molecules found in ribosomes are manufactured and the large and small ribosomal subunits are assembled.

Ribosomes In eukaryotes, the cytoplasm consists of everything inside the plasma membrane excluding the nucleus; the fluid portion of the cytoplasm is called the **cytosol**. Many of the cell's millions of ribosomes are scattered throughout the cytosol. Like the bacterial ribosome pictured in Figure 7.4, the ribosomes shown in **Figure 7.8** comprise two subunits, one large and one small. Each subunit is composed of proteins and RNA. In eukaryotes the large subunit contains three RNA molecules while the small subunit contains one. (In prokaryotes the large subunit has just two RNA molecules.) Neither ribosomal subunit is enclosed by a membrane. When the two subunits come together, they form a complex molecular machine that synthesizes proteins.

Rough Endoplasmic Reticulum In addition to the ribosomes found free in the cytosol, many ribosomes are

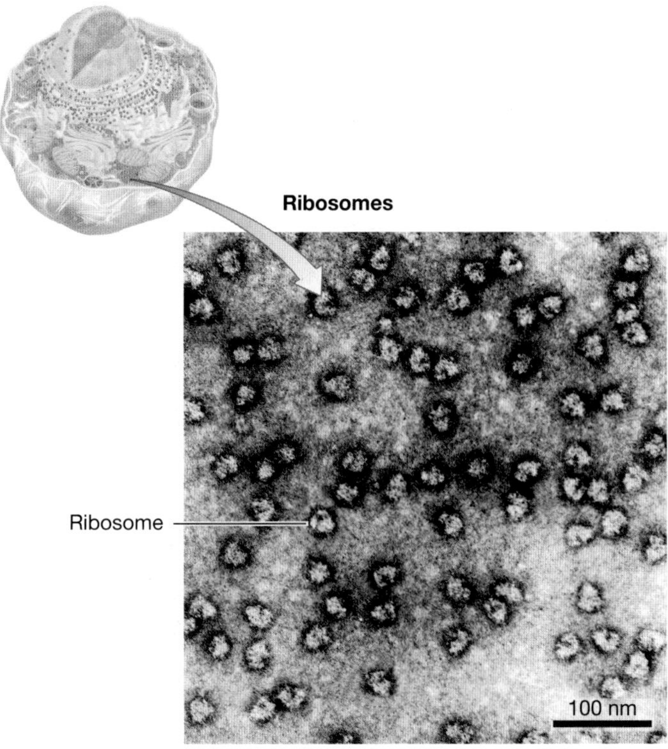

FIGURE 7.8 Ribosomes Are the Site of Protein Synthesis. Eukaryotic ribosomes are similar in structure to bacterial and archaeal ribosomes—though not identical. They are composed of large and small subunits, each of which contains both RNA molecules and proteins.

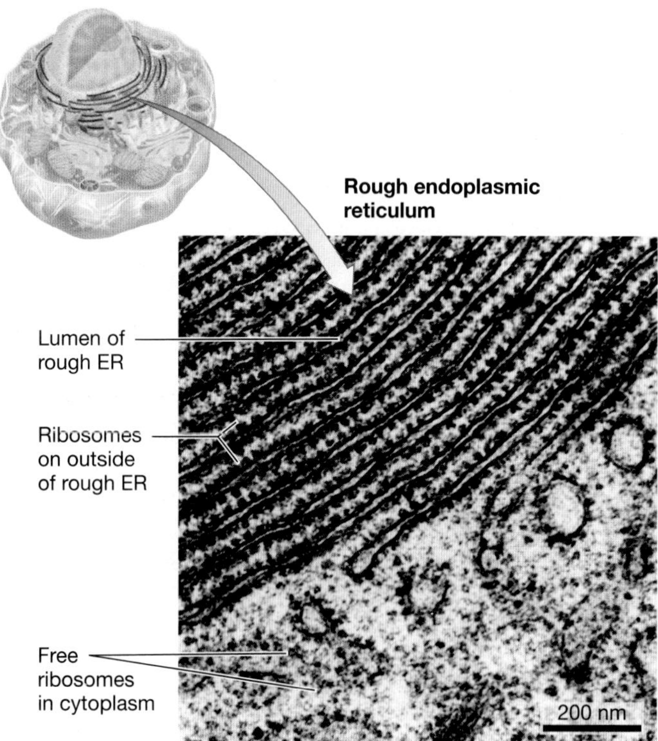

FIGURE 7.9 Rough ER Is a Protein Synthesis and Processing Complex. Rough ER is a system of membrane-bound sacs and tubules with ribosomes attached. It is continuous with the nuclear envelope and with smooth ER.

associated with membranes. More specifically, hundreds of thousands of ribosomes are attached to a network of membrane-bound sacs and tubules called the **rough endoplasmic reticulum**, or **rough ER**. Translated literally, *endoplasmic reticulum* means "inside-formed network." Look back at Figure 7.6 and notice that the ER is continuous with the outer membrane of the nuclear envelope. From the nuclear envelope, the layers of sacs that make up the ER extend into the cytoplasm.

The ribosomes associated with the rough ER are responsible for synthesizing proteins that will be inserted into the plasma membrane, secreted to the cell exterior, or shipped to an organelle called the lysosome. As they are being manufactured by ribosomes, these proteins move to the interior of the sac-like component of the rough ER (**Figure 7.9**). The interior of the rough ER, like the interior of any sac-like structure in a cell or body, is called the **lumen**. In the lumen of the rough ER, newly manufactured proteins undergo folding and other types of processing.

The proteins produced in the rough ER have a variety of functions. Some carry messages to other cells; some act as membrane transport proteins or pumps; others are enzymes. The common theme is that rough ER products are destined for transport to various distant destinations—often to the surface of the cell or beyond.

Golgi Apparatus In many cases, the products of the rough ER pass through the Golgi apparatus before they reach their final destination. The **Golgi apparatus** consists of flattened, membranous sacs called **cisternae** (singular: **cisterna**), which are stacked on top of one another (**Figure 7.10**). The organelle also has a distinct polarity, or sidedness. The *cis* ("this side") surface is closest to the rough ER and nucleus, and the *trans* ("across") surface is oriented toward the plasma membrane. The *cis* side of a Golgi apparatus receives products from the rough ER, and the *trans* side ships them out toward the cell surface. In between, within the cisternae, the rough ER's products are processed and packaged for delivery. Micrographs often show "bubbles" on either side of the Golgi stack. These are membrane-bound vesicles that carry proteins or other products to and from the organelle. Section 7.3 analyzes the intracellular movement of molecules from the rough ER to the Golgi apparatus and beyond in more detail.

Smooth Endoplasmic Reticulum Although the ER forms one continuous structure, not all of it is associated with transport of material to the Golgi apparatus and not all of it has ribosomes attached. In electron micrographs, parts of the ER that contain ribosomes look dotted and rough while portions that are free of ribosomes appear smooth and even. Appropriately enough, these parts of the ER are called **smooth endoplasmic**

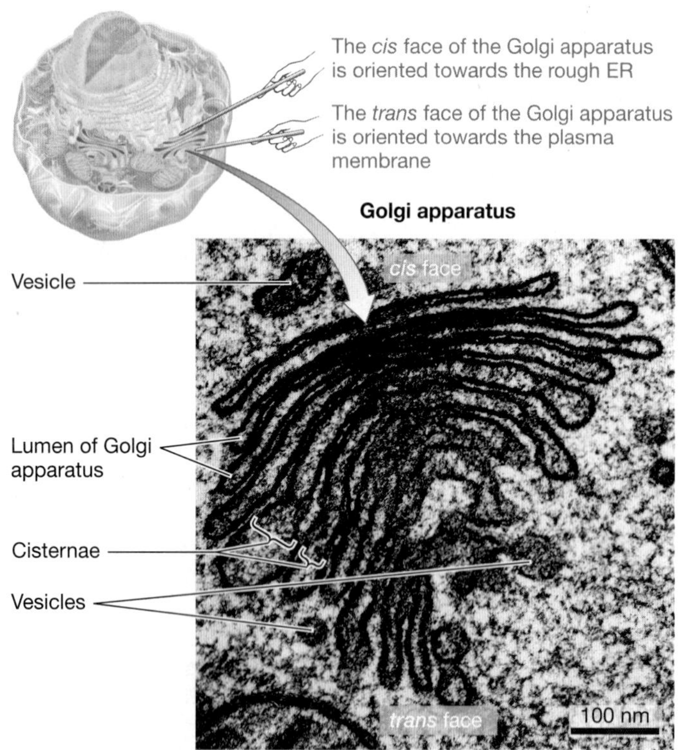

FIGURE 7.10 The Golgi Apparatus Is a Site of Protein Processing, Sorting, and Shipping. The Golgi apparatus is a collection of flattened vesicles called cisternae.

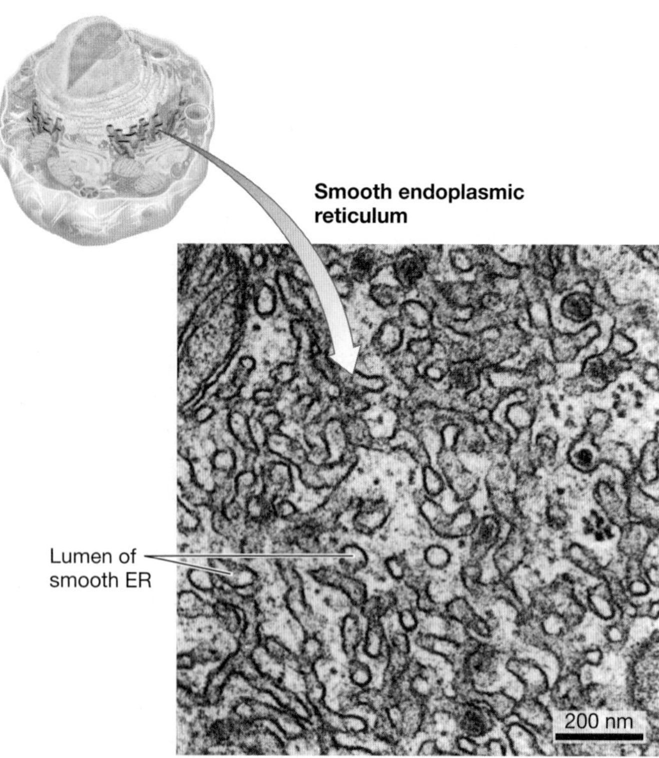

FIGURE 7.11 Smooth ER Is a Lipid-Handling Center and a Storage Facility. Smooth ER is a system of membrane-bound sacs and tubules that lacks ribosomes.

reticulum, or **smooth ER** (**Figure 7.11**). The smooth ER membrane contains enzymes that catalyze reactions involving lipids. Depending on the type of cell, these enzymes may be involved in synthesizing lipids needed by the organism, or in breaking down lipids that are poisonous. Smooth ER is also the manufacturing site for phospholipids used in cell membranes. In addition to processing lipids, smooth ER functions as a reservoir for calcium ions (Ca^{2+}) that act as a signal triggering a wide array of activities inside the cell.

The structure of endoplasmic reticulum correlates closely with its function. Rough ER has ribosomes and functions primarily as a protein-manufacturing center; smooth ER lacks ribosomes and functions primarily as a lipid-processing center. Together with the Golgi apparatus and lysosomes, the endoplasmic reticulum forms the **endomembrane system**. The endomembrane ("inner-membrane") system is the primary center for protein and lipid synthesis in eukaryotic cells.

Peroxisomes Virtually all eukaryotic cells contain globular organelles called **peroxisomes** (**Figure 7.12**). These organelles have a single membrane and grow and divide independently of other organelles. Although different types of cells from the same individual may have distinct types of peroxisomes, these organelles all share a common function: Peroxisomes are centers for oxidation reactions. As Chapter 9 will explain in detail, oxidation reactions remove electrons from atoms and molecules. In many cases the products of these reactions include hydrogen peroxide (H_2O_2), which is highly corrosive. If hydrogen peroxide escaped from the peroxisome, the H_2O_2 would quickly damage organelle membranes and the plasma membrane. This is rare, however. Inside the peroxisome, the enzyme catalase quickly "detoxifies" hydrogen peroxide by converting it to water and oxygen.

Different types of peroxisomes contain different suites of oxidative enzymes. As a result, each is specialized for oxidizing particular compounds. For example, the peroxisomes in your liver cells contain enzymes that oxidize an array of toxins, including the ethanol in alcoholic beverages. The products of these oxidation reactions are usually harmless and are either excreted from the body or used in other reactions. Other peroxisomes contain enzymes that catalyze the oxidation of fatty acids. These reactions produce molecules that include acetyl CoA, which is used for the synthesis of important molecules elsewhere in the cell. In plant leaves, specialized peroxisomes called **glyoxisomes** are packed with enzymes that convert one of the products of photosynthesis into a sugar that can be used to store energy for the cell. In contrast, seeds do not perform photosynthesis and lack this type of peroxisome. Instead, they have peroxisomes with enzymes that oxidize fatty acids to yield glucose. The glucose is then used by the young plant as it begins to grow. In each case, there is a clear connection between structure and function: The enzymes found inside the peroxisome make a specialized set of oxidation reactions possible.

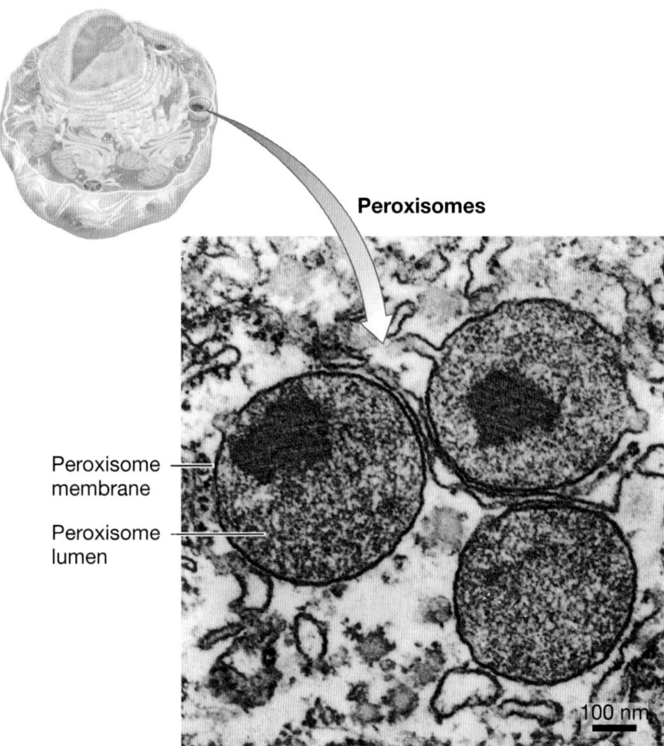

FIGURE 7.12 Peroxisomes Are the Site of Oxidation Reactions. Peroxisomes are globular organelles with a single membrane.

Lysosomes The major structures involved in solid-waste processing and materials storage in the cell are called **lysosomes.** The size and shape of these organelles vary widely, and in the cells of plants, fungi, and certain other groups they are referred to as vacuoles. In animal cells, lysosomes function as digestive centers (**Figure 7.13**). The organelle's interior, or lumen, is acidic because proton pumps in the lysosome membrane import enough hydrogen ions to maintain a pH of 5.0. This organelle also contains about 40 different enzymes. Each of these proteins is specialized for breaking up a different type of macromolecule—proteins, nucleic acids, lipids, or carbohydrates—into its component monomers. The monomers are then excreted or recycled into new polymers. The digestive enzymes are collectively called acid hydrolases because they catalyze hydrolysis reactions that break monomers from macromolecules most efficiently at pH 5.0. In the cytosol, where the pH is about 7.2, these enzymes are less active.

Figure 7.14 illustrates two ways that materials are delivered to lysosomes in animal cells. When **phagocytosis** (literally, "eat-cell-act") occurs, the plasma membrane of a cell surrounds a

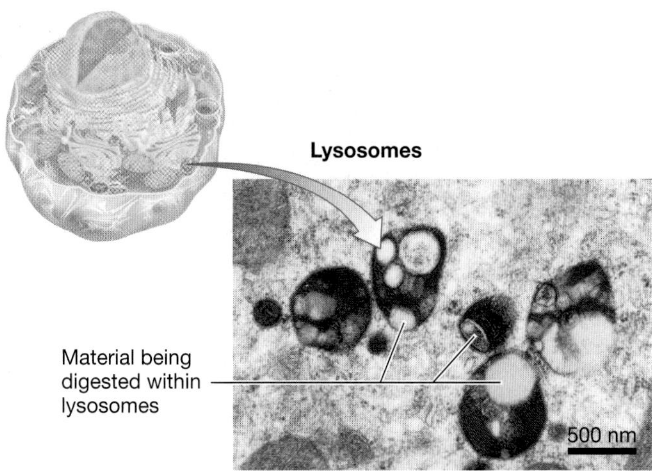

FIGURE 7.13 Lysosomes Are Recycling Centers. Lysosomes are usually oval or globular and have a single membrane.

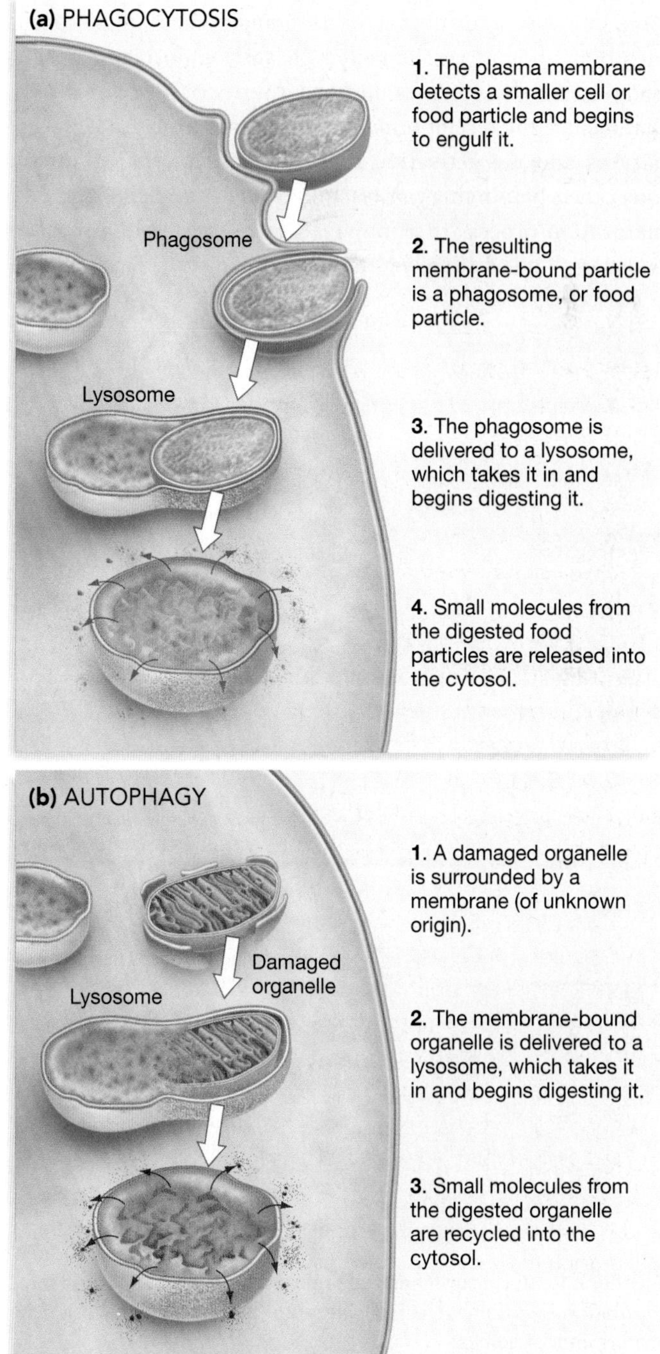

FIGURE 7.14 Two Ways to Deliver Materials to Lysosomes. Materials can be transported to lysosomes after phagocytosis or via autophagy.

smaller cell or a food particle and engulfs it, forming a structure called a phagosome or food vacuole. This structure is delivered to a lysosome, where it is taken in and digested. During **autophagy** ("same-eating"), damaged organelles are surrounded by a membrane and delivered to a lysosome. There the components are digested and recycled.

Materials can also be processed by lysosomes as a result of **receptor-mediated endocytosis**. As **Figure 7.15** shows, this sequence of events begins when macromolecules outside the cell bind to membrane proteins that act as receptors. More than 25 distinct receptors have now been characterized, each specialized for responding to a different macromolecule. Once binding occurs, the plasma membrane folds in and pinches off to form a membrane-bound vesicle called an **early endosome** ("inside-body"). Early endosomes undergo a series of processing steps that include the receipt of digestive enzymes from the Golgi apparatus and the activation of proton pumps that gradually lower their pH. In this way, early endosomes undergo a gradual maturation process that may lead to the formation of a **late endosome** and eventually a fully functioning lysosome.

Regardless of whether the materials in lysosomes originate via phagocytosis, autophagy, or receptor-mediated endocytosis, the result is similar: Molecules are hydrolyzed. The amino acids, nucleotides, sugars, and other molecules that result from acid hydrolysis leave the lysosome via transport proteins in the organelle's membrane. Once in the cytoplasm, they can be reused.

It is important to note, however, that not all of the materials that are surrounded by membrane and taken into a cell end up in lysosomes. **Endocytosis** ("inside-cell-act") refers to any pinching off of the plasma membrane that results in the uptake of material from outside the cell. Endocytosis can occur in three ways: (1) phagocytosis, (2) receptor-mediated endocytosis, and (3) **pinocytosis** ("drink-cell-act"). Pinocytosis brings fluid into the cytoplasm via tiny vesicles that form from infoldings of the plasma membrane. The fluid inside these vesicles is not transported to lysosomes, but is used elsewhere in the cell. In addition, most of the macromolecules that collect in early endosomes are selectively removed and used long before the structure becomes a lysosome.

Compared with the lysosomes of animal cells, the **vacuoles** of plant and fungal cells are large—sometimes taking up as much as 80 percent of a plant cell's volume (**Figure 7.16**). Although

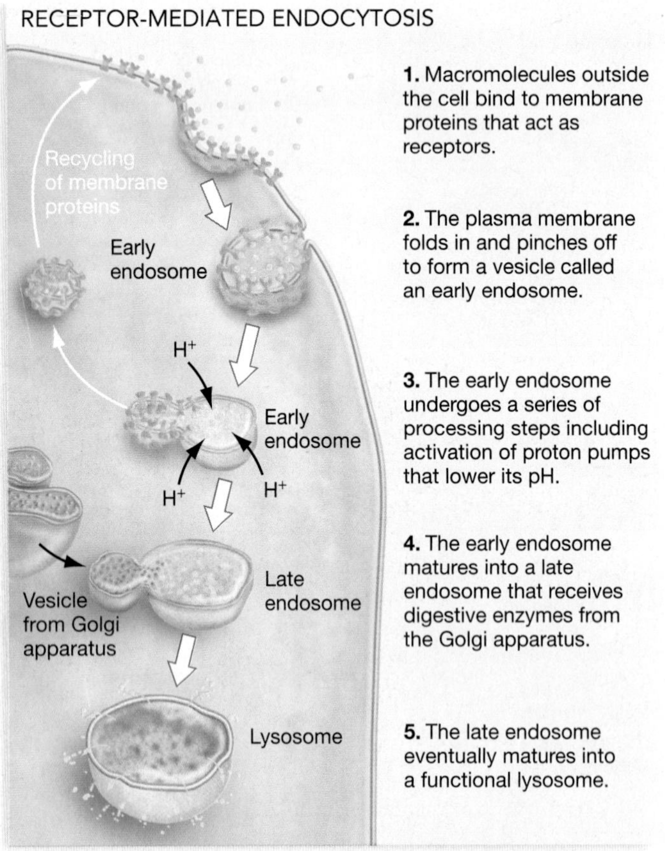

RECEPTOR-MEDIATED ENDOCYTOSIS

Recycling of membrane proteins

Early endosome

H^+

Early endosome

H^+ H^+

Vesicle from Golgi apparatus

Late endosome

Lysosome

1. Macromolecules outside the cell bind to membrane proteins that act as receptors.

2. The plasma membrane folds in and pinches off to form a vesicle called an early endosome.

3. The early endosome undergoes a series of processing steps including activation of proton pumps that lower its pH.

4. The early endosome matures into a late endosome that receives digestive enzymes from the Golgi apparatus.

5. The late endosome eventually matures into a functional lysosome.

FIGURE 7.15 Receptor-Mediated Endocytosis May Lead to Lysosome Formation. Endosomes created by receptor-mediated endocytosis may mature into lysosomes.

QUESTION Why is it significant that vesicles from the Golgi apparatus fuse with early endosomes?

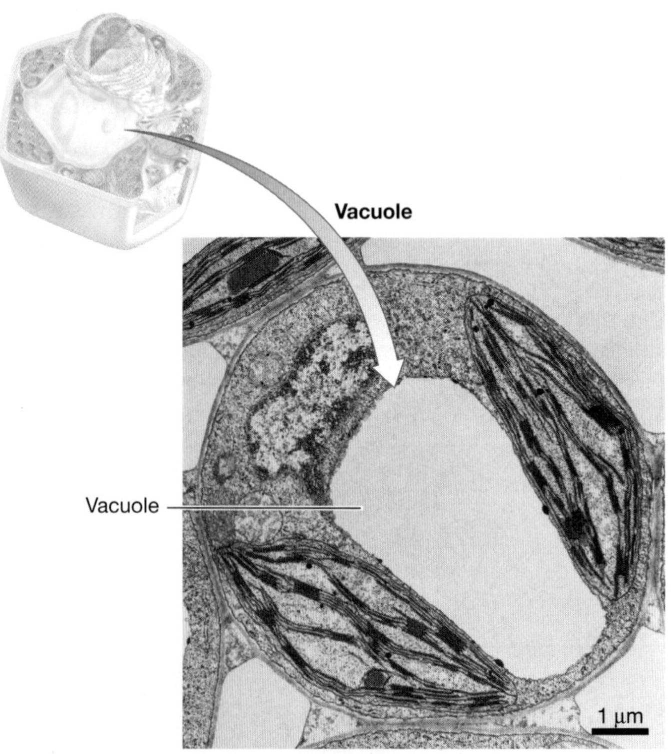

FIGURE 7.16 Vacuoles Are Storage Centers. Vacuoles are variable in size and function. Some contain digestive enzymes and serve as recycling centers; most are large storage containers.

QUESTION Why are toxins like nicotine, cocaine, and caffeine stored in vacuoles instead of the cytosol?

QUESTION About what percentage of this cell's volume is taken up by the vacuole?

some vacuoles contain enzymes that are specialized for digestion, most of the vacuoles observed in plant and fungal cells act as storage depots. In many cases, the stored material is water, which maintains the cell's normal volume, or ions such as potassium (K^+) and chloride (Cl^-). Inside seeds, cells may contain a large vacuole filled with proteins. When the embryonic plant inside the seed begins to grow, enzymes begin digesting these proteins to provide amino acids for the growing individual. In cells that make up flower petals or fruits, vacuoles are filled with colorful pigments. Elsewhere in the plant, vacuoles may be packed with noxious compounds that protect leaves and stems from being eaten by predators. The type of chemical involved varies by species, ranging from bitter-tasting tannins to toxins such as nicotine, morphine, caffeine, or cocaine.

Mitochondria The chemical energy required to build all of these organelles and do other types of work comes from adenosine triphosphate (ATP), most of which is produced in the cell's **mitochondria** (singular: mitochondrion). As **Figure 7.17** shows, each mitochondrion has two membranes. The outer membrane defines the organelle's surface, while the inner membrane is connected to a series of sac-like **cristae**. The solution inside the inner membrane is called the **mitochondrial matrix**. In eukaryotes, most of the enzymes and molecular machines responsible for synthesizing ATP are embedded in the membranes of the cristae or suspended in the matrix. Depending on the type of cell, from 50 to more than a million mitochondria may be present.

Each mitochondrion possesses a small chromosome that contains genes, independent of the main chromosomes in the nucleus. This mitochondrial DNA is a component of a circular and supercoiled chromosome that is similar in structure to bacterial chromosomes. Mitochondria also manufacture their own ribosomes. Like most organelles, mitochondria can grow and divide independently of nuclear division and cell division.

Chloroplasts Most algal and plant cells possess an organelle called the **chloroplast**, in which sunlight is converted to chemical energy during photosynthesis. The chloroplast has a double membrane around its exterior, analogous to the structure of a mitochondrion (**Figure 7.18**). Instead of featuring sac-like cristae that connect to the inner membrane, though, the interior of the chloroplast is dominated by hundreds of membrane-bound, flattened vesicles called **thylakoids**, which are independent of the inner membrane. Thylakoids are stacked like pancakes into piles called **grana** (singular: *granum*). Many of the pigments, enzymes, and molecular machines responsible for converting light energy into carbohydrates are embedded in the thylakoid membranes. Certain critical enzymes and substrates, however, are found outside the thylakoids in the region called the **stroma**.

The number of chloroplasts per cell varies from none to several dozen. Like mitochondria, each chloroplast contains a

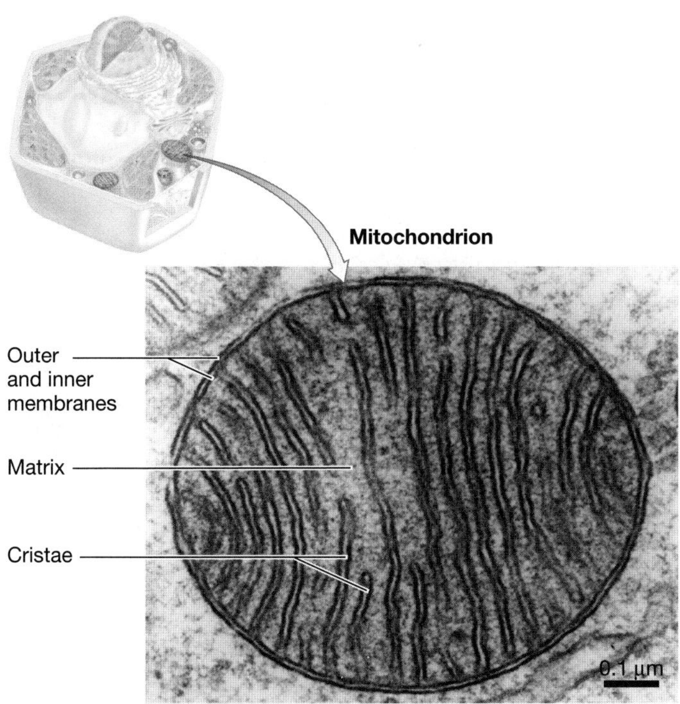

FIGURE 7.17 Mitochondria Are Power-Generating Stations. Mitochondria are variable in size and shape, but all have a double membrane with sac-like cristae inside.

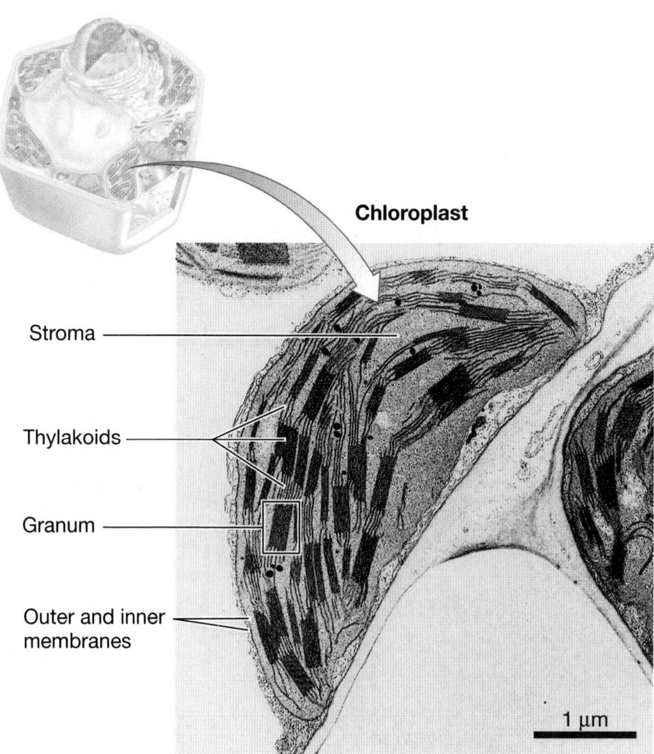

FIGURE 7.18 Chloroplasts Are Sugar-Manufacturing Centers. Many of the enzymes and other molecules required for photosynthesis are located in membranes inside the chloroplast. These membranes are folded into thylakoids and stacked into grana.

circular chromosome. Chloroplast DNA is independent of the main genetic material inside the nucleus. Chloroplasts also grow and divide independently of nuclear division and cell division.

Cytoskeleton The final major structural feature that is common to all eukaryotic cells is an extensive system of protein fibers called the cytoskeleton. The cytoskeleton contains several distinct types of proteins and fibers and has an array of functions which are analyzed in detail in Section 7.4. In addition to giving the cell its shape and structural stability, cytoskeletal proteins are involved in moving the cell itself and in moving materials within the cell.

The Cell Wall In fungi, algae, and plants, cells possess an outer cell wall in addition to their plasma membrane (**Figure 7.19**). The cells of animals, amoebae, and other groups lack this feature. Although the composition of the cell wall varies among species and even between types of cells in the same individual, the general plan is similar: Rods or fibers composed of a carbohydrate run through a stiff matrix made of other polysaccharides and proteins (see Chapter 8 for details on this). In addition, some plant cells produce a secondary cell wall that features a particularly tough molecule called lignin. Lignin forms a branching, cagelike network that is almost impossible for enzymes to attack. The combination of cellulose fibers and lignin in secondary cell walls makes up most of the material we call wood.

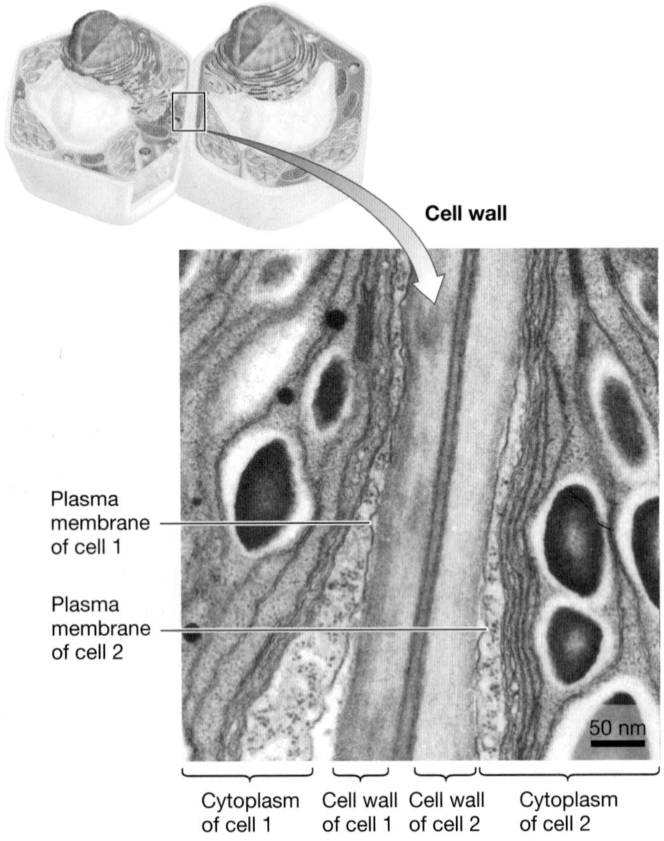

FIGURE 7.19 Cell Walls Protect Plants and Fungi. Plants have cell walls that contain cellulose; in fungi the major structural component of the cell wall is chitin.

QUESTION Is the cell wall inside or outside of the plasma membrane?

MB Web Animation *BioFlix* at www.masteringbio.com
Tour of an Animal Cell; Tour of a Plant Cell

How Does Cell Structure Correlate with Function?

The preceding discussion emphasized how the structure of each organelle fits with its role in the cell. As **Table 7.2** (page 132) indicates, an organelle's membrane and its complement of enzymes correlate closely with its function. If you understand this concept, you should be able to draw and label a generalized eukaryotic cell and indicate which components serve as the administrative/information hub, protein manufacturing and shipping system, recycling center, support beams, roads, storage depot, power station, food manufacturing facility, fatty-acid processing center, and perimeter fencing with secured gates.

The same connection between structure and function occurs at the level of the entire cell. Inside an individual plant or animal, cells are specialized for certain tasks and have a structure that correlates with those tasks. For example, the muscle cells in your upper leg are extremely long, tube-shaped structures. They are filled with protein fibers that slide past one another as the entire muscle flexes or extends. It is this sliding motion that allows your muscles to contract or extend as you run. Muscle cells are also packed with mitochondria, which produce the ATP required for the sliding motion to occur. In contrast, nearby fat cells are rounded, globular structures that store fatty acids. They consist of little more than a plasma membrane, a nucleus, and a fat droplet. Neither cell bears a close resemblance to the generalized animal cell pictured in Figure 7.6a

To drive home the correlation between the overall structure and function of a cell, examine the transmission electron micrographs in **Figure 7.20**. The animal cell in Figure 7.20a is from the pancreas and is specialized for the manufacture and export of digestive enzymes. It is packed with rough ER and Golgi, which make this function possible. The animal cell in Figure 7.20b is from the testis and synthesizes the steroid hormone called testosterone. This cell is dominated by smooth ER, where processing of steroids and other lipids takes place. The plant cell in Figure 7.20c is from the leaf of a potato and is specialized for absorbing light and manufacturing sugar; the cell in Figure 7.20d is from a potato tuber (part of an underground stem) and functions as a starch storage container. The leaf cell contains hundreds of chloroplasts, while the tuber cell has a prominent storage vacuole filled with carbohydrate. In each case, the type of organelles in each cell and their size and number correlate with the cell's specialized function.

(a) Animal pancreatic cell: Exports digestive enzymes.

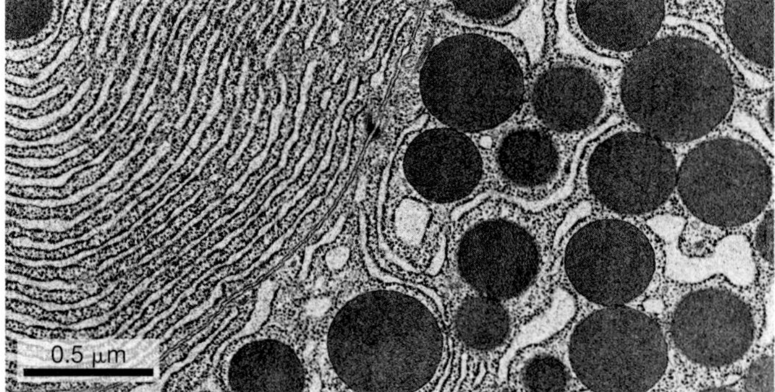

(b) Animal testis cell: Exports lipid-soluble signals.

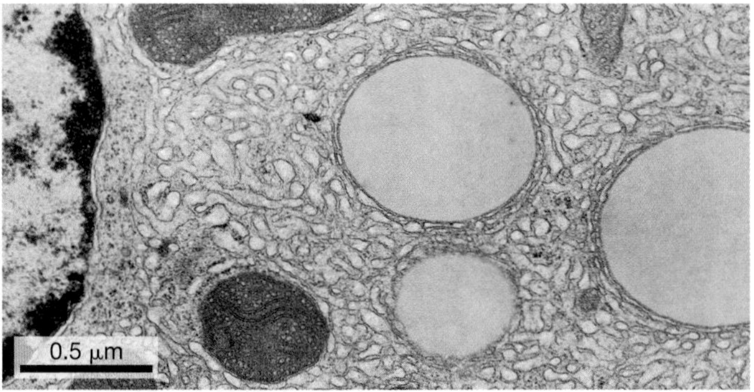

(c) Plant leaf cell: Manufactures ATP and sugar.

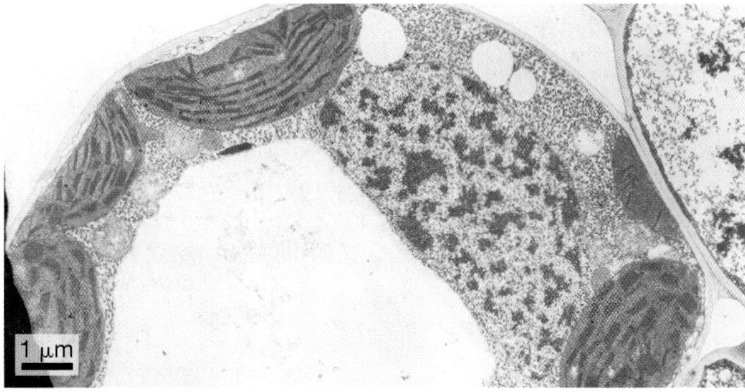

(d) Plant root cell: Stores starch.

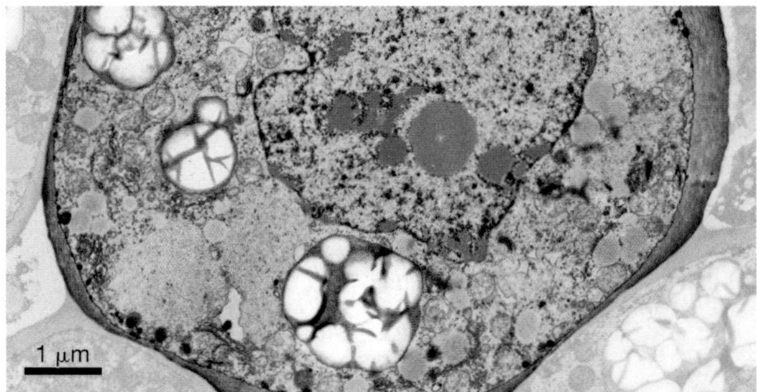

FIGURE 7.20 Cell Structure Correlates with Function.

EXERCISE In part (a), label the rough ER and secretory vesicles. (They are dark and round.) In (b), label the smooth ER. In (c), label the chloroplasts, vacuole, and nucleus. In (d), label the starch granules (they are white in this micrograph).

The Dynamic Cell

Biologists study the structure and function of organelles and cells with a combination of tools and approaches. Light microscopes and transmission electron microscopes have allowed researchers to see cells at increasingly high magnification and increasingly better resolution. These types of microscopy allowed biologists to characterize the basic size and shape of organelles and where they occurred in the cell; a technique called **differential centrifugation** made it possible to isolate particular cell components and analyze their chemical composition. As **Box 7.1** explains, differential centrifugation is based on breaking cells apart to create a complex mixture and then separating components in a centrifuge. The individual parts of the cell can then be purified and studied in detail, in isolation from other parts of the cell.

Although these techniques have led to an increasingly sophisticated understanding of how cells work, they have a limitation. Transmission electron microscopy is based on a fixed "snapshot" of the cell that is to be observed, and differential centrifugation is based on splitting cells into parts that are analyzed independently. Neither technique allows investigators to directly explore how things move from place to place in the cell or how parts interact. The information gleaned from these techniques can make cells seem static. In reality, cells are dynamic.

The amount of chemical activity and the speed of molecular movement inside cells is nothing short of fantastic. Bacterial ribosomes add up to 20 amino acids per second to a growing polypeptide, and eukaryotic ribosomes typically add two per second. Given that there are about 15,000 ribosomes in each bacterium and possibly a million in an average eukaryotic cell, hundreds or even thousands of new protein molecules can be finished each second in every cell. In the same amount of time, a typical cell in your body uses an average of 10 million ATP molecules and synthesizes just as many. It's not unusual for a cellular enzyme to catalyze 25,000 or more reactions per second; most cells contain hundreds or thousands of enzymes. A minute is more than enough time for each membrane phospholipid in your body to travel the breadth of the organelle or cell where it resides. The hundreds of trillions of mitochondria inside you are completely replaced about every 10 days, for as long as you live. The plasma membrane is fluid, and its composition is constantly changing.

Because humans are such large organisms, it's impossible for us to imagine what life is really like inside a cell. At the scale of a ribosome or an organelle or a cell, gravity is inconsequential. Instead, the dominant forces are the charge- or polarity-based electrostatic attractions between molecules and the kinetic energy of motion. At this level, events take nanoseconds, and speeds are measured in micrometers per second. Contemporary methods for studying cells, including some of the imaging techniques featured in **BioSkills 8**, capture

SUMMARY TABLE **7.2 Eukaryotic Cell Components**

	Structure		Function
	Membrane	**Components**	
Nucleus	Double ("envelope"); openings called nuclear pores	Chromosomes Nucleolus Nuclear lamina	Genetic information Assembly of ribosome subunits Structural support
Ribosomes	None	Large/small subunits—complex of RNA and proteins	Protein synthesis
Endomembrane system			
Rough ER	Single; contains receptors for entry of selected proteins	Network of branching sacs Ribosomes associated	Protein synthesis and processing
Golgi apparatus	Single; contains receptors for products of rough ER	Stack of flattened cisternae	Protein processing (e.g., glycosylation)
Smooth ER	Single; contains enzymes for synthesizing phospholipids	Network of branching sacs Enzymes for synthesizing lipids	Lipid synthesis
Peroxisomes	Single; contains transporters for selected macromolecules	Enzymes that catalyze oxidation reactions Catalase (processes peroxide)	Processing of fatty acids
Lysosomes	Single; contains proton pumps	Acid hydrolases (catalyze hydrolysis reactions)	Digestion and recycling
Vacuoles	Single; contains transporters for selected molecules	Varies—pigments, oils, carbohydrates, water, or toxins	Varies—coloration, storage of oils, carbohydrates, water, or toxins
Mitochondria	Double; outer contains enzymes for processing pyruvate; inner contains enzymes for ATP production	Enzymes that catalyze oxidation-reduction reactions, ATP synthesis	ATP production
Chloroplasts	Double; plus membrane-bound sacs in interior	Pigments Enzymes that catalyze oxidation-reduction reactions	Production of ATP and sugars via photosynthesis
Cytoskeleton	None	Actin filaments Intermediate filaments Microtubules	Structural support; movement of materials; in some species, movement of whole cell
Plasma membrane	Single; contains transport and receptor proteins	Phospholipid bilayer with transport and receptor proteins	Selective permeability—maintains intracellular environment
Cell wall	None	Carbohydrate fibers running through carbohydrate or protein matrix	Protection, structural support

this dynamism by tracking how organelles and molecules move and interact over time.

The rest of this chapter focuses on this theme of cellular dynamism and movement. To begin, let's look at how molecules move into and out of the cell's control center—the nucleus. Then we'll consider how proteins move from ribosomes into the lumen of the rough ER and then to the Golgi apparatus and beyond. The chapter closes by analyzing how cytoskeletal elements help transport cargo inside the cell or move the cell itself.

BOX 7.1 How Does a Centrifuge Work?

For decades, the centrifuge was among the most common tools used by biologists who study life at the level of molecules and cells. It was vital to early studies of organelles and other cell structures because it can separate cell components efficiently. A centrifuge accomplishes this by spinning cells in a solution that allows molecules and other cell components to separate according to their density or size and shape.

The first step in preparing a cell sample for centrifugation is to release the organelles and cell components by breaking the cells apart. This can be done by putting them in a hypotonic solution, by exposing them to high frequency vibration, by treating cells with a detergent, or by grinding them up. Each of these methods breaks apart plasma membranes and releases the contents of the cells.

The pieces of plasma membrane broken up by these techniques quickly reseal to form small vesicles, often trapping cell components inside. The solution that results from the homogenization step is a mixture of these vesicles, free-floating macromolecules released from the cells, and organelles. A solution such as this is called a cell extract or cell homogenate.

When a cell homogenate is placed in a centrifuge tube and spun at high speed, the components that are in solution tend to move outward, along the red arrow in **Figure 7.21a**. The effect is similar to a merry-go-round, which seems to push you outward in a straight line away from the spinning platform. In response to this outward-directed force, the solution containing the cell homogenate exerts a centripetal ("center-seeking") force that pushes the homogenate away from the bottom of the tube. Larger, denser molecules or particles resist this inward force more readily than do smaller, less dense ones and so reach the bottom of the centrifuge tube faster.

To separate the components of a cell extract, researchers often perform a series of centrifuge runs. Steps 1 and 2 of **Figure 7.21b** illustrate how an initial treatment at low speed causes larger, heavier parts of the homogenate to move below smaller, lighter parts. The material that collects at the bottom of the tube is called the **pellet**, and the solution and solutes left behind form the **supernatant** ("above

(Continued on next page)

(a) How a centrifuge works

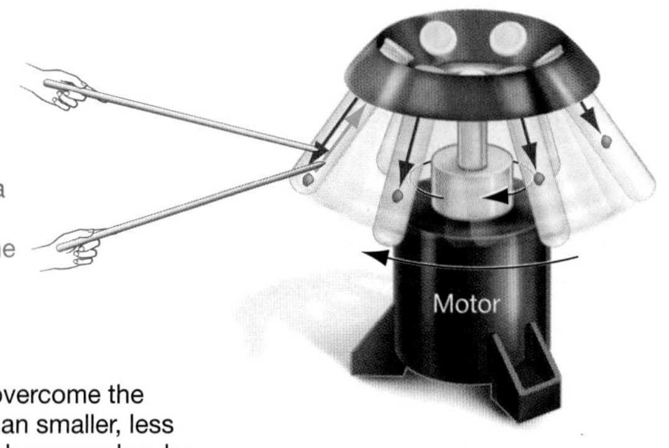

(b) DIFFERENTIAL CENTRIFUGATION

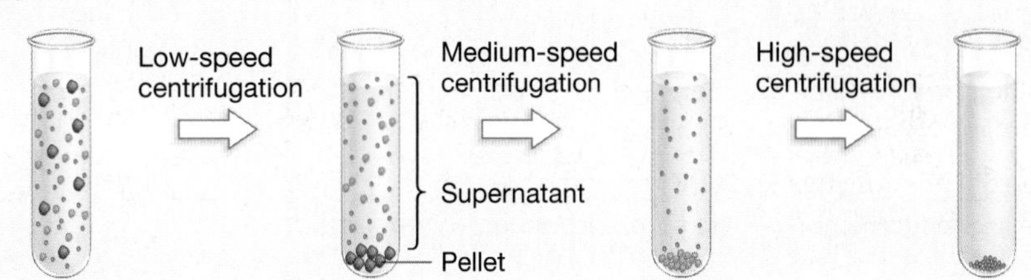

(c) DENSITY GRADIENT CENTRIFUGATION

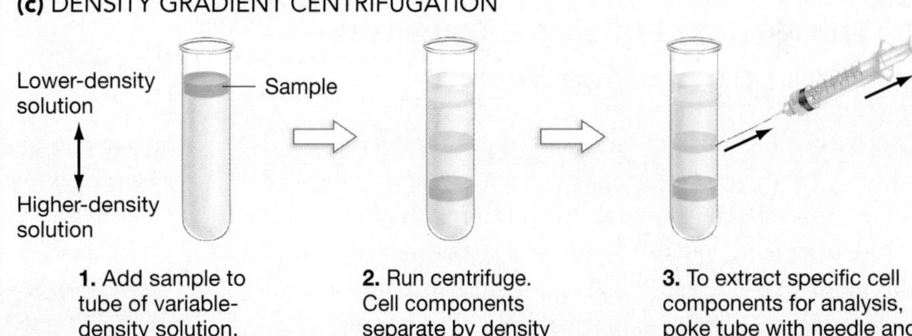

FIGURE 7.21 Cell Components Can Be Separated by Centrifugation. (a) The forces inside a centrifuge tube allow cell components to be separated. **(b)** Through a series of centrifuge runs made at increasingly higher speeds, an investigator can separate fractions of a cell homogenate by size via differential centrifugation. **(c)** A high-speed centrifuge run can achieve extremely fine separation among cell components by density gradient centrifugation.

(continued)

swimming"). The supernatant is placed in a fresh tube and centrifuged at increasingly higher speeds and longer durations. Each centrifuge run continues to separate cell components based on their size and density.

To accomplish even finer separation of macromolecules or organelles, researchers frequently follow up with centrifugation at extremely high speeds. One strategy is based on filling the centrifuge tube with a series of sucrose solutions of increasing density. The density gradient allows cell components to separate on the basis of small differences in size, shape, and density. When the centrifuge run is complete, each cell component occupies a distinct band of material in the tube, based on how quickly each moves through the increasingly dense gradient of sucrose solution during the centrifuge run. A researcher can then collect the material in each band for further study.

Check Your Understanding

If you understand that...

- Each structure found in prokaryotic and eukaryotic cells performs a function vital to the cell's survival.

You should be able to...

1) Summarize the differences between the structure of a prokaryotic cell and a eukaryotic cell.
2) Describe the structure and function of the eukaryotic nucleus, endomembrane system, mitochondria, chloroplast, and cytoskeleton.
3) Explain how the structure of peroxisomes and lysosomes correlate with their function.

7.2 The Nuclear Envelope: Transport into and out of the Nucleus

The nucleus is the information center of eukaryotic cells. It is a corporate headquarters, design center, and library all rolled into one. Appropriately enough, its interior is highly organized. The organelle's overall shape and structure are defined by the mesh-like nuclear lamina, which also helps anchor each chromosome. The remainder of each chromosome occupies a well-defined region in the nucleus, and specific centers exist where the genetic information in DNA is decoded and processed. At these locations, large suites of enzymes interact to produce RNA messages from specific genes at specific times. Meanwhile, the nucleolus functions as the site of ribosome synthesis.

Consistent with its role as information repository and processing center, the nucleus is separated from the rest of the cell by the nuclear envelope. Biologists began to understand exactly how the nuclear envelope is structured when transmission electron microscopy became available in the 1950s. Micrographs of cross sections though the envelope showed that the structure is supported by the fibrous nuclear lamina and is bounded by two membranes, each consisting of a lipid bilayer. As the diagram in **Figure 7.22** indicates, the inner membrane and the outer membrane are separated by a space that is continuous with the lumen of the endoplasmic reticulum.

Later, scanning electron microscopes allowed biologists to view the surface of the envelope. As the micrograph in **Figure 7.23** shows, the envelope contains thousands of openings called **nuclear pores**. Because these pores extend through both inner

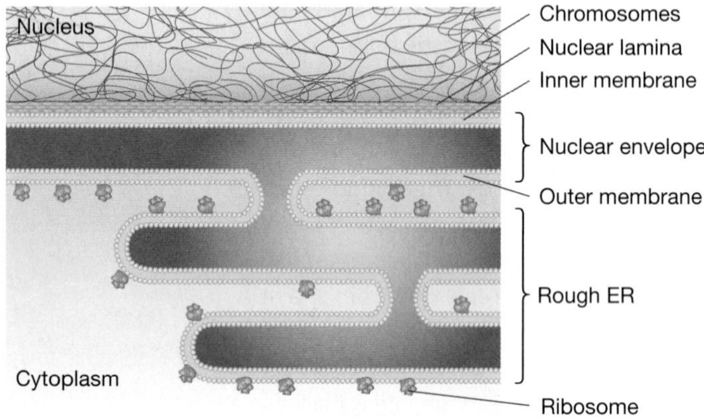

FIGURE 7.22 The Structure of the Nuclear Envelope. The nuclear envelope has a double membrane and is continuous with the endoplasmic reticulum.

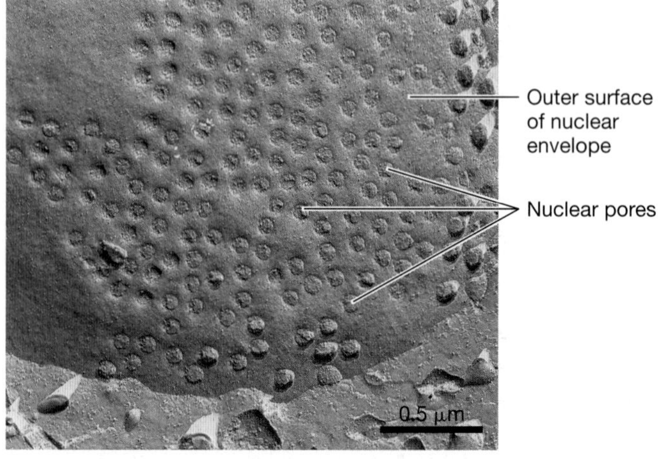

FIGURE 7.23 The Nuclear Envelope Is Studded with Pores. Scanning electron micrograph of the surface of the nuclear envelope.

and outer nuclear membranes, they connect the inside of the nucleus with the cytoplasm. Follow-up research showed that each pore consists of over 50 different proteins. These protein molecules form an elaborate structure called the **nuclear pore complex** (**Figure 7.24a**).

A series of experiments in the early 1960s showed that molecules travel into and out of the nucleus through the nuclear pore complexes. The initial studies were based on injecting tiny gold particles into cells and then preparing the cells for electron microscopy. In electron micrographs, gold particles show up as black dots. One or two minutes after injection, the micrographs showed that most of the gold particles were in the cytoplasm. A few, however, were closely associated with nuclear pores. Ten minutes after injection, particles were inside the nucleus as well as in the cytoplasm. These data supported the hypothesis that the pores function as the doors to the nucleus. Follow-up work confirmed that the nuclear pore complex is the only gate between the cytoplasm and the nucleus and that only certain molecules go in and out. Passage through the nuclear pore is selective.

(a) Cross-sectional view of nuclear pore

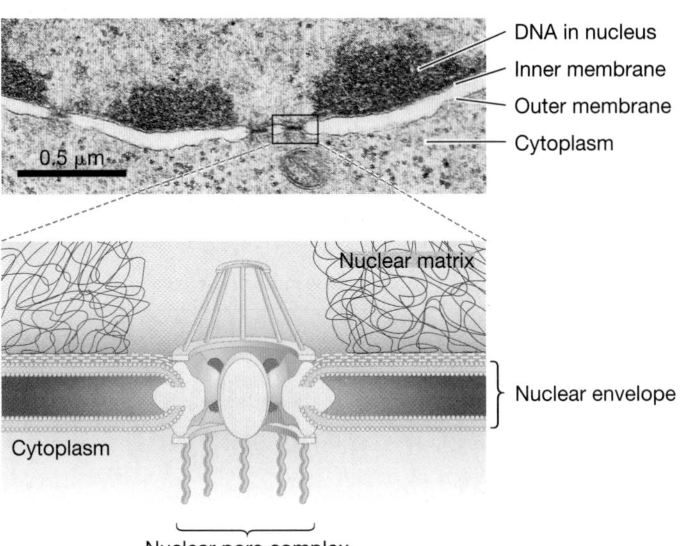

(b) Molecules move into and out of the nucleus through the nuclear pore

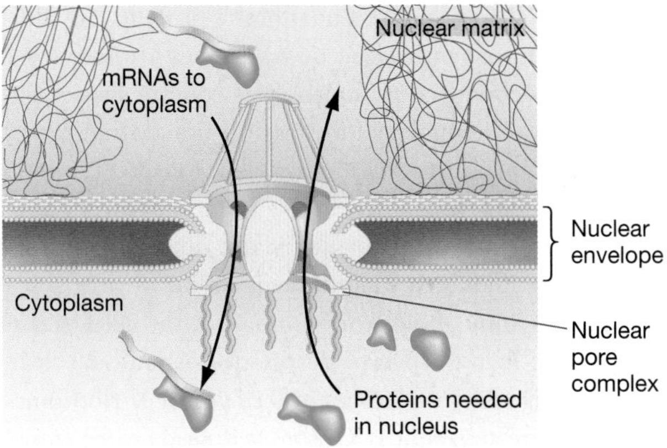

FIGURE 7.24 Structure and Function of the Nuclear Pore.
(a) The drawing is based on electron micrographs of the nuclear pore. **(b)** Messenger RNAs are synthesized in the nucleus and must be exported to the cytoplasm. Proteins needed in the nucleus are synthesized in the cytoplasm and have to be imported to the nucleus.

What substances traverse nuclear pores? DNA clearly does not—it never leaves the nucleus. But information coded in DNA is used to synthesize RNA inside the nucleus. Several types of RNA molecules are produced, each distinguished by size and function. For example, most ribosomal RNAs are manufactured in the nucleolus, where they bind to proteins to form completed ribosomal subunits. The functional ribosomes are then exported to the cytoplasm. Similarly, molecules called messenger RNAs carry the information required to manufacture proteins out to the cytoplasm, where protein synthesis takes place. Traffic in the other direction is also impressive. Nucleotide triphosphates that act as building blocks for DNA and RNA enter the nucleus, as do the proteins responsible for copying DNA, synthesizing RNAs, extending the nuclear lamina, assembling ribosomes, or building chromosomes (**Figure 7.24b**).

To summarize, ribosomal subunits and various types of RNAs exit the nucleus; proteins that are needed inside enter it. In a typical cell, over 500 molecules pass through each of the 3000–4000 nuclear pores every second. The traffic is intense. How is it regulated and directed?

How Are Molecules Imported into the Nucleus?

The first experiments on how molecules move through the nuclear pore focused on proteins that are produced by viruses. **Viruses** are parasites that use the cell's machinery to make copies of themselves. When a virus infects a cell, certain of its proteins enter the nucleus. Investigators noticed that if a particular amino acid in one of these proteins happens to be altered, the viral protein was no longer able to pass through the nuclear pore. This simple-sounding observation led to a key hypothesis: Proteins that are synthesized by ribosomes in the cytosol but are headed for the nucleus contain a "zip code"—a molecular address tag that marks them for transport through the nuclear pore complex. The idea was that viral proteins were able to enter the nucleus only if they have the same address tag as normal cellular proteins have. Thus, the proteins with the altered amino acid were thwarted. This zip code came to be called the **nuclear localization signal (NLS)**.

A series of experiments on a protein called nucleoplasmin helped researchers better understand the nature of this signal. Nucleoplasmin plays an important role in the assembly of chromosomes and happens to have a distinctive structure: It consists of a globular protein core surrounded by a series of extended protein "tails." When researchers labeled nucleoplasmin with a radioactive atom and injected it into the cytoplasm of living cells, they found that the radioactive signal quickly ended up in the nucleus.

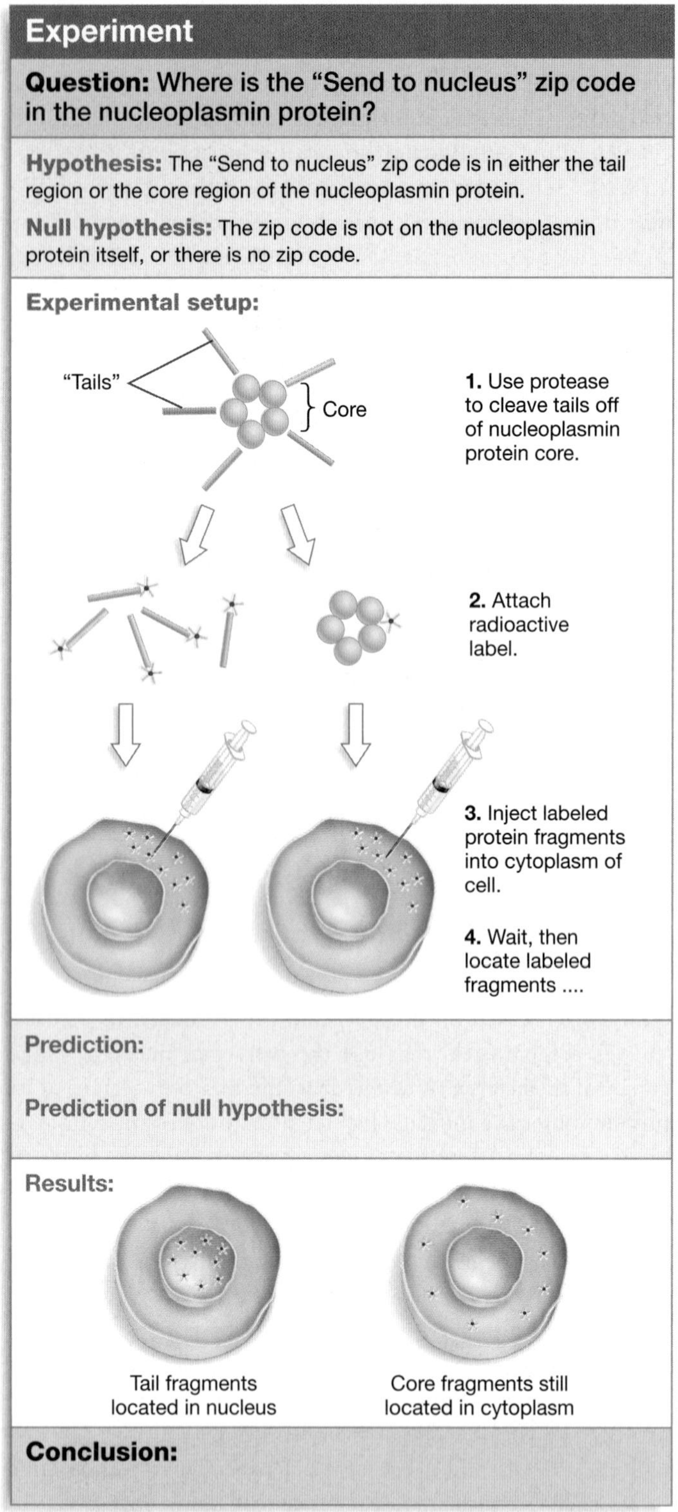

Experiment

Question: Where is the "Send to nucleus" zip code in the nucleoplasmin protein?

Hypothesis: The "Send to nucleus" zip code is in either the tail region or the core region of the nucleoplasmin protein.

Null hypothesis: The zip code is not on the nucleoplasmin protein itself, or there is no zip code.

Experimental setup:

"Tails"

Core

1. Use protease to cleave tails off of nucleoplasmin protein core.

2. Attach radioactive label.

3. Inject labeled protein fragments into cytoplasm of cell.

4. Wait, then locate labeled fragments ....

Prediction:

Prediction of null hypothesis:

Results:

Tail fragments located in nucleus

Core fragments still located in cytoplasm

Conclusion:

FIGURE 7.25 Where Is the "Send to Nucleus" Zip Code in the Nucleoplasmin Protein?

● **EXERCISE** Without looking at the text, fill in the predictions and conclusion(s) in this experiment.

Figure 7.25 outlines how the nuclear localization signal in nucleoplasmin was found. Researchers began by using enzymes called proteases to separate the core sections of nucleoplasmin from the tails. After separating the two components, the researchers labeled each part with radioactive atoms and injected them into the cytoplasm of different cells. When they examined the experimental cells with the electron microscope, they found that tail fragments were transported to the nucleus. Core fragments, in contrast, remained in the cytoplasm. These data suggested that the zip code must be somewhere in the tail part of the protein.

By analyzing different stretches of the tail section, the biologists eventually found a 17-amino-acid-long section that had to be present to direct proteins to the nucleus. The biologists therefore concluded that the NLS zip code consisted of 17 specific amino acids in the tail.

Follow-up work confirmed that other proteins bound for the nucleus have similar localization signals. More recent research has shown that the movement of proteins and other large molecules into and out of the nucleus is an energy-demanding process that involves transport proteins called importins and exportins. Importins and exportins function like trucks that haul cargo in and out of the nucleus, through the nuclear pore complex.

Currently, biologists are trying to unravel how all this traffic in and out of the nucleus is regulated to avoid back-ups and head-on collisions. The goal is to understand the precise mechanisms responsible for moving cargo into and out of the cell's headquarters.

(MB) Web Animation at www.masteringbio.com
Transport into the Nucleus

7.3 The Endomembrane System: Manufacturing and Shipping Proteins

The nuclear membrane is not the only place in cells where cargo moves in a regulated and energy-demanding fashion. For example, Chapter 6 highlighted how specific ions and molecules are pumped into and out of cells or transported across the plasma membrane by specialized membrane proteins. In addition, proteins that are synthesized by ribosomes in the cytosol for use inside mitochondria or chloroplasts contain special signal sequences, like the nuclear localization signal, that target the proteins for transport to these organelles. Ions, ATP, amino acids, and other small molecules diffuse randomly throughout the cell, but the movement of proteins and other large molecules is energy demanding and tightly regulated.

If you pause to think about it for a moment, the need to sort proteins and ship them to specific destinations should be clear. Proteins are produced by ribosomes in the cytosol or ribosomes on the ER. Each protein that is synthesized needs to be transported to one of the many compartments inside the eukaryotic cell. Acid hydrolases need to end up in lysosomes, catalase needs to shipped to peroxisomes, and ribosomal proteins need to be transported to the nucleolus. To get to the right location, each protein has to have an address tag and a transport and delivery system.

To get a better understanding of how protein sorting and transport occur in eukaryotic cells, let's consider perhaps the most intricate of all manufacturing and shipping complexes: the endomembrane system. In this system, proteins that are synthesized in the rough ER move to the Golgi apparatus for processing, and from there travel to the cell surface or other destinations. The idea that materials might move through the endomembrane system in an orderly way was inspired by a simple observation. According to electron micrographs, cells that secrete digestive enzymes, hormones, or other types of products have particularly large amounts of rough ER and Golgi. This correlation led to the idea that these cells have a "secretory pathway" that starts in the rough ER and ends with products leaving the cell (**Figure 7.26**). How does this hypothesized pathway work?

George Palade and colleagues did pioneering research on the secretory pathway with an experimental approach known as a **pulse-chase experiment**. The strategy is based on providing experimental cells with a large concentration of a radioactively labeled molecule for a short time. For example, if a cell receives a large amount of labeled amino acid for a short time, virtually all of the proteins synthesized during that interval will be labeled. This "pulse" of labeled molecule is followed by a chase—large amounts of an unlabeled version of the same molecule, provided for a long time. If the chase consists of unlabeled amino acid, then the proteins synthesized during the chase period will *not* be labeled. The general idea is to mark a population of molecules at a particular interval and then follow their fate over time. This approach is analogous to adding a small amount of dye to a stream and then following the movement of the dye molecules.

In testing the secretory pathway hypothesis, Palade's team focused on pancreatic cells that were growing in culture, or in vitro.[1] These cells are specialized for secreting digestive enzymes into the small intestine and are packed with rough ER and Golgi. The basic experimental approach was to supply the cells with a 3-minute pulse of the amino acid leucine, labeled with a radioactive atom, followed by a long chase with nonradioactive leucine. Because the radioactive leucine was incorporated into all proteins being produced during the pulse, those proteins were labeled. Then the researchers prepared a sample of the cells for electron microscopy and autoradiography (see **BioSkills 7 and 8**). **Figure 7.27a** shows part of a single cell that was examined immediately after the pulse. A close look at this micrograph should convince you that most of the newly synthesized

[1]The term *in vitro* is Latin for "in glass." Experiments that are performed outside living organisms are done in vitro. The term *in vivo,* in contrast, is Latin for "in life." Experiments performed with living organisms are done in vivo.

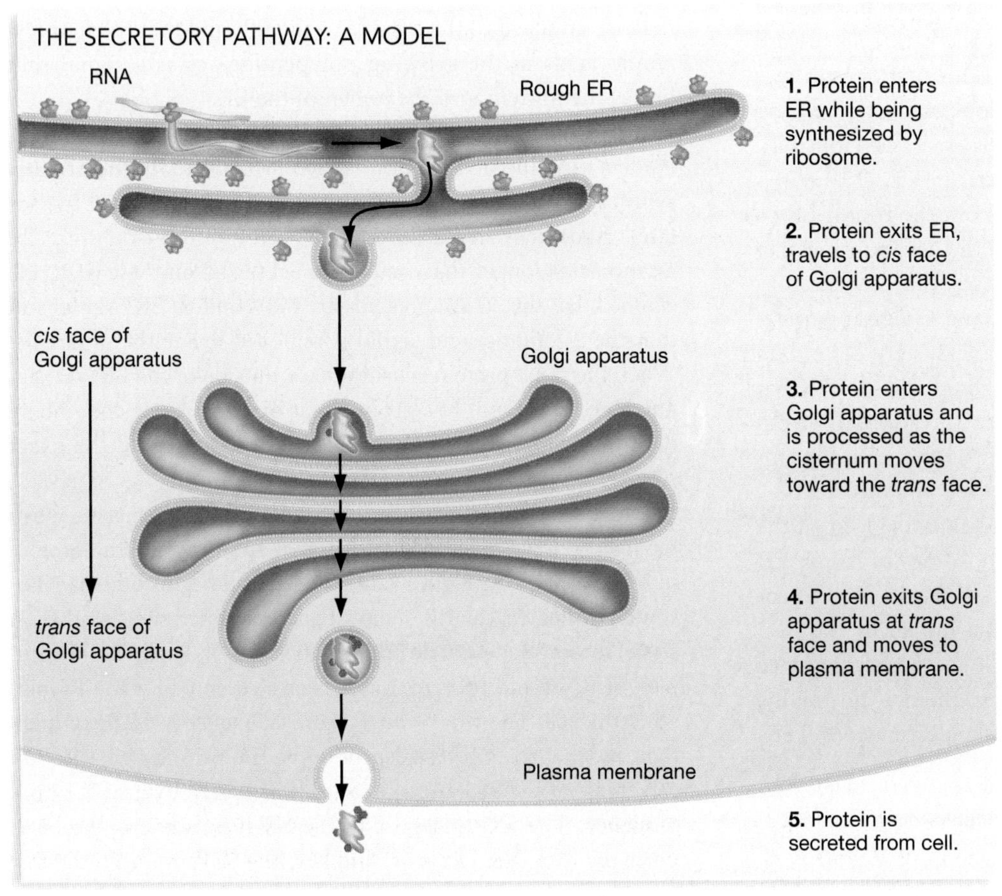

FIGURE 7.26 The Secretory Pathway Hypothesis. The secretory pathway hypothesis proposes that proteins intended for secretion from the cell are synthesized and processed in a highly prescribed set of steps.

(a) Immediately after labeling.

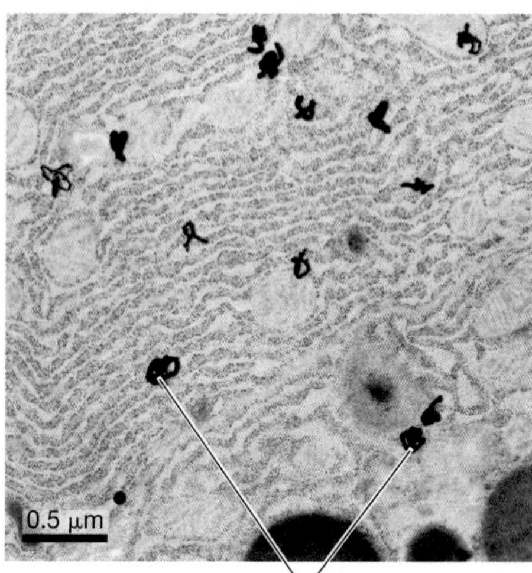

(b) 7 minutes after end of labeling.

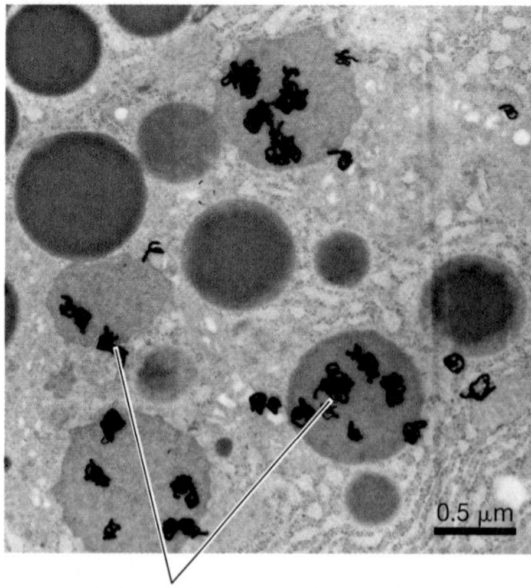

(c) 80 minutes after end of labeling.

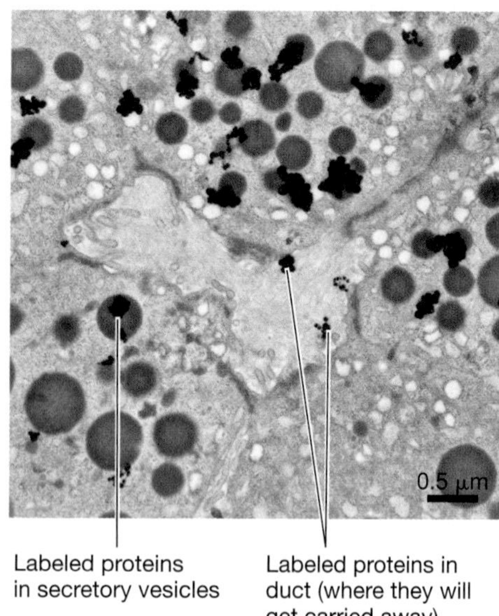

FIGURE 7.27 Results of a Pulse-Chase Experiment.

proteins are inside this cell's rough ER. **Figure 7.27b**, in contrast, shows part of a single cell 7 minutes later, after the pulse ended. Now the situation has changed. Few of the labeled proteins were in the rough ER. Instead, most of the labeled proteins are inside structures called secretory vesicles on the *trans* side of a Golgi apparatus (some are found inside the Golgi apparatus itself). The micrograph in **Figure 7.27c** is at lower magnification and shows parts of five cells. The structure in the middle is a duct that carries digestive enzymes from pancreatic cells toward their destination in the small intestine. This micrograph was taken 80 minutes after the pulse. If you look at the figure carefully, you'll note that most labeled proteins are in secretory vesicles or actually outside the cell, in the duct.

Because the labeled proteins move from the rough ER to Golgi apparatus to secretory vesicles to the cell exterior over time, the results support the hypotheses that a secretory pathway exists and that the rough ER and Golgi apparatus function as an integrated endomembrane system. The data suggest that proteins produced in the rough ER don't float around the cytoplasm aimlessly or drift randomly from organelle to organelle. Instead, traffic through the endomembrane system appears to be highly organized and directed.

Now let's break the system down and examine four of the steps in more detail. Recall that the ribosomes in rough ER are bound to the outside of the membrane. How do the proteins that they manufacture get into the lumen of the ER? How do they move from the ER to the Golgi apparatus? Once they're inside the Golgi, what happens to them? And finally, how do the finished proteins get to their destination? Let's consider each question in turn.

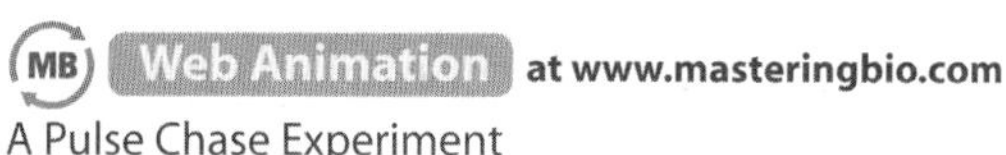

at www.masteringbio.com
A Pulse Chase Experiment

Entering the Endomembrane System: The Signal Hypothesis

How do proteins enter the endomembrane system? The signal hypothesis predicted that proteins bound for the endomembrane system have a zip code analogous to the nuclear localization signal. The idea, proposed by Günter Blobel and colleagues, was that these proteins are synthesized by ribosomes that are attached to the outside of the ER, and that the first few amino acids in the growing polypeptide act as a signal that brings the protein into the lumen of the ER.

This hypothesis received important support when researchers made a puzzling observation: When proteins that are normally synthesized in the rough ER are instead manufactured by isolated ribosomes in vitro—with *no* ER present—they are 20 amino acids longer than usual. Blobel seized on these data. He claimed that the 20 amino acids are the "send-to-ER" signal and that the 20-amino-acid signal is removed inside the organelle. When the same protein is synthesized outside of the ER, the signal is not removed. His group went on to produce convincing data that supported the hypothesis: They identified the exact series of amino acids in the **ER signal sequence**.

More recent work has documented the mechanisms responsible for receiving the send-to-ER signal and inserting the protein into the rough ER (**Figure 7.28**). The action begins when a ribosome synthesizes the ER signal sequence, which then binds to a **signal recognition particle (SRP)** in the cytosol. An SRP is a complex of RNA and protein that acts as a receptor for the ER signal sequence. The ribosome + signal sequence + SRP complex then attaches to an SRP receptor in the ER membrane itself. You can think of the SRP as a key that is activated by an ER signal sequence. The SRP receptor in the ER membrane is the lock. Once the lock (the receptor) and key (the SRP) connect, the rest

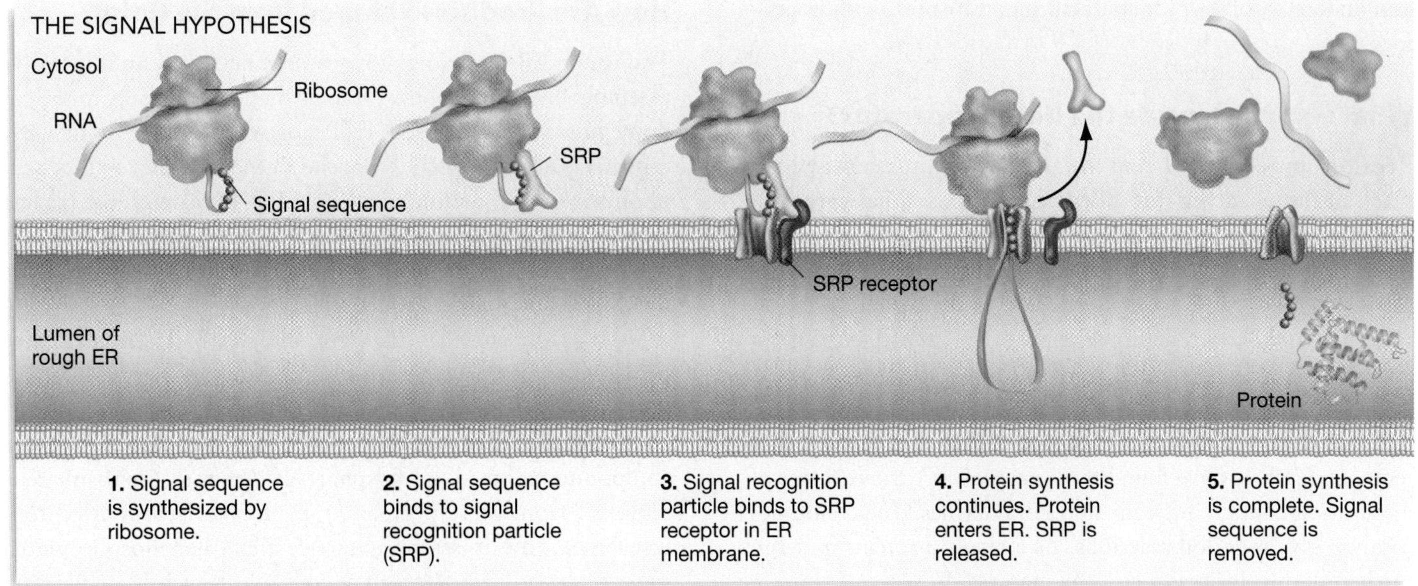

FIGURE 7.28 The Signal Hypothesis Explains How Proteins Destined for Secretion Enter the Endomembrane System. According to the signal hypothesis, proteins destined for secretion contain a short stretch of amino acids that interact with a signal recognition particle (SRP) in the cytoplasm. This interaction allows the protein to enter the ER.

of the protein is synthesized, and then the signal sequence is removed. If the finished polypeptide will eventually be shipped to an organelle or secreted from the cell, then it enters the lumen of the rough ER. If the finished polypeptide is a membrane protein, it remains in the rough ER membrane while it is being processed.

Once proteins are inside the rough ER or inserted into its membrane, they fold into their three-dimensional shape with the help of chaperone proteins introduced in Chapter 3. In addition, proteins that enter the lumen interact with enzymes that catalyze the addition of carbohydrate side chains. Because carbohydrates are polymers of sugar monomers, the addition of one or more carbohydrate groups is called **glycosylation** ("sugar-together"). The resulting molecule is called a **glycoprotein** ("sugar-protein"; see Chapter 5). As **Figure 7.29** shows, proteins that enter the ER often gain a specific carbohydrate that consists of 14 sugar subunits. Thus, proteins that are synthesized in the rough ER are also folded and modified by glycosylation there. The completed glycoproteins are ready for shipment to the Golgi apparatus.

Getting from the ER to the Golgi

How do proteins travel from the ER to the Golgi apparatus? Palade's group thought they had the answer, based on data from the pulse-chase experiments that first confirmed the existence of the endomembrane system. When labeled proteins appeared in a region between the rough ER and the Golgi apparatus, they appeared to be inside small membrane-bound structures. Based on these observations, the biologists suggested that proteins are transported between the two organelles in vesicles. The idea was that vesicles bud off of the ER, move away, fuse with the membrane on the *cis* face of the Golgi apparatus, and dump their contents inside.

This hypothesis was supported when other researchers used differential centrifugation to isolate and characterize the vesicles that contained labeled proteins. Using this approach, investigators have established that a distinctive type of vesicle carries proteins from the rough ER to the Golgi apparatus.

Taken together, these results support a general conclusion about the endomembrane system: It is a sophisticated, highly organized complex. Proteins move through it in a directed way

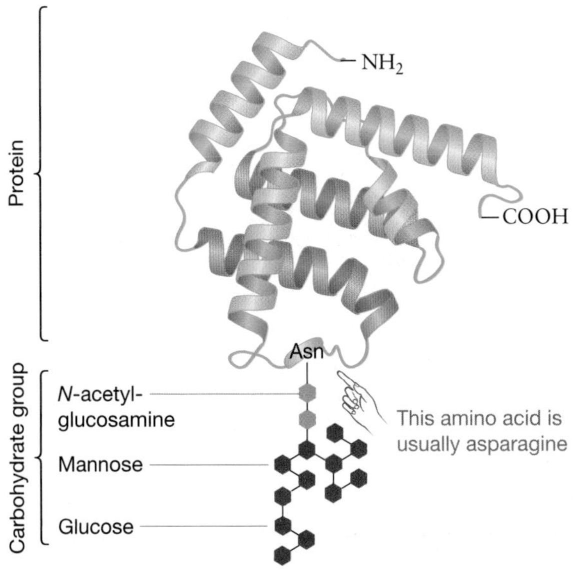

FIGURE 7.29 Glycosylation Adds Carbohydrate Groups to Proteins. When proteins enter the ER, most acquire the 14 sugar residues shown here. Some of these sugars may be removed or others added as proteins pass through the Golgi apparatus.

and undergo a series of manufacturing, transport, and processing steps.

What Happens inside the Golgi Apparatus?

Recall from Section 7.1 that the Golgi apparatus consists of a stack of flattened vesicles called cisternae, and that cargo enters one side of the structure and exits the other. Recent research has shown that the composition of the Golgi apparatus is dynamic. New cisternae are constantly created at the *cis* face, while old cisternae break apart at the *trans* face. Once a cisternum forms, it gradually moves toward the *trans* face. As it does, it changes in composition and activity.

By separating individual cisternae and analyzing their contents, researchers have found that cisternae at various stages of maturation contain different suites of enzymes. These enzymes catalyze glycosylation reactions. As a result, proteins undergo further modification as a cisterna matures. While cisternae are still near the *cis* face, some of the proteins inside have sugar-phosphate groups added. Later, the carbohydrate group that was added in the rough ER is removed. Near the *trans* face, various types of carbohydrate chains are attached that may protect the protein or help it attach to surfaces.

If the rough ER is like a foundry and stamping plant where rough parts are manufactured, then the Golgi is like a finishing area where products are polished and painted and readied for shipping.

How Are Products Shipped from the Golgi?

The rough ER and Golgi apparatus function as an impressive assembly line. Some of the proteins they produce stay in the endomembrane system itself, replacing worn-out molecules. But if proteins are processed to the end of the line, they will be sent to one of several destinations, including lysosomes, the plasma membrane, or the outside of the cell. How are these finished products put into the right shipping containers, and how are the different containers addressed?

Studies on enzymes that are shipped to lysosomes have provided some answers to both questions. A key finding was that lysosome-bound proteins have a phosphate group attached to a specific sugar subunit on their surface, forming the compound mannose-6-phosphate. If mannose-6-phosphate is removed from these proteins, they are not transported to a lysosome. This is strong evidence that the phosphorylated sugar serves as a zip code, analogous to the nuclear localization and rough ER signals discussed earlier. More specifically, data indicate that mannose-6-phosphate binds to a protein in the membranes of certain vesicles. These vesicles, in turn, have proteins on their surface that interact specifically with proteins in the lysosomal membranes. In this way, the presence of mannose-6-phosphate targets proteins for vesicles that deliver their contents to lysosomes.

Figure 7.30 pulls these observations together into a comprehensive model explaining how the products of the endomem-

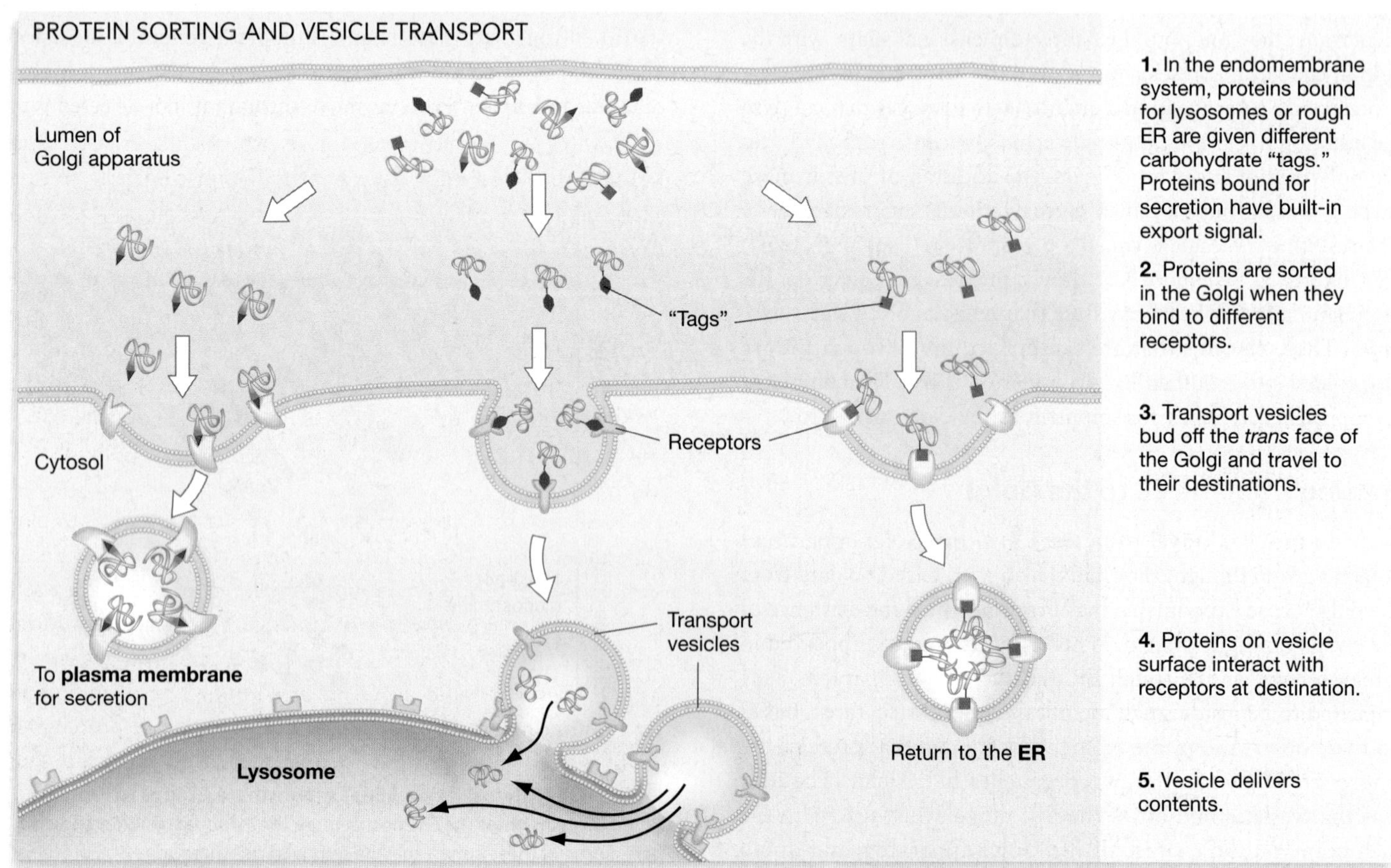

FIGURE 7.30 In the Golgi Apparatus, Proteins Are Sorted into Vesicles That Are Targeted to a Destination.

brane system are loaded into specific vesicles and shipped to their correct destination. The key observation is that each protein that comes out of the Golgi apparatus has a molecular tag that places it in a particular type of transport vesicle. Each type of transport vesicle, in turn, has a tag that allows it to be transported to the correct destination—the plasma membrane, a lysosome, or the ER.

In particular, notice that the transport vesicle shown on the left of Figure 7.30 is bound for the plasma membrane, where it will secrete its contents to the outside. This process is called **exocytosis** ("outside-cell-act"). When exocytosis occurs, the vesicle membrane and plasma membrane make contact and fuse, analogous to the way that the vesicle membrane and lysosome membrane are shown fusing in the center, bottom of the figure. At the plasma membrane, the two sets of lipid bilayers rearrange in a way that exposes the interior of the vesicle to the outside of the cell. The vesicle's contents then diffuse away from the cell into the space outside the cell. In this way, cells in your pancreas deliver digestive enzymes to the duct that leads to the small intestine—where digestion actually takes place.

Proteins that are synthesized in the cytoplasm also have zip codes directing them to mitochondria, chloroplasts, or other destinations. To summarize, then, the array of proteins produced in a cell are sorted via distinctive zip codes. These molecular addresses allow proteins to be shipped to the compartment where they function.

If vesicles function like shipping containers for products that move between organelles, do they travel along some sort of road or track? What molecule or molecules function as the delivery truck, and does ATP supply the gas? In general, what physical mechanisms are responsible for moving vesicles to their destination? Let's delve into these questions in Section 7.4.

Check Your Understanding

If you understand that...

- In cells, the transport of proteins and other large molecules is energy demanding and tightly regulated.
- Proteins must have the appropriate molecular zip code to enter or leave the nucleus, enter the lumen of the rough ER, or become incorporated into vesicles destined for lysosomes or the plasma membrane.
- In many cases, proteins and other types of cargo are shipped in vesicles that contain molecular zip codes on their surface.

You should be able to...

1) Predict what happens to proteins that lack an ER signal sequence.
2) Predict the outcome of an experiment where secreted proteins are placed inside vesicles with a zip code associated with shipment to lysosomes.

7.4 The Dynamic Cytoskeleton

Based on early observations with light microscopes, biologists concluded that the cytoplasm of eukaryotic cells was a fluid-filled space devoid of structure. As microscopy improved, however, researchers realized that cell cytoplasm contains an extremely dense and complex network of fibers. This cytoskeleton helps maintain cell shape by providing structural support. It's important to recognize, though, that the cytoskeleton is not a static structure like the scaffolding used at construction sites. The fibrous proteins that make up the cytoskeleton move and change to alter the cell's shape, to move materials from place to place, and to move the entire structure. Like the rest of the cell, the cytoskeleton is dynamic.

As **Summary Table 7.3** shows, there are several distinct types of cytoskeletal elements: actin filaments (also known as microfilaments), intermediate filaments, and microtubules. Each of these elements has a distinct size, structure, and function. Let's look at each one in turn.

Actin Filaments

Actin filaments are sometimes referred to as **microfilaments** because they are the cytoskeletal element with the smallest diameter. As Summary Table 7.3 indicates, **actin filaments** are fibrous structures made of a globular protein called actin. In animal cells, actin is often the most abundant of all proteins—typically it represents 5–10 percent of the total protein in the cell. Each of your liver cells contains about half a billion of these molecules.

Actin filaments form when individual actin molecules polymerize. The completed structure resembles two long strands that coil around each other. Because each actin subunit in the strand is asymmetrical, the structure as a whole has a distinct polarity. The two ends of an actin filament are different and are referred to as plus and minus ends. Each filament grows and shrinks as actin subunits are added or subtracted from each end of the structure. This phenomenon is called treadmilling, because the dynamics of the fibers resemble those of a treadmill. Actin filaments tend to grow at the plus end, however, because polymerization occurs fastest there.

In animal cells, actin filaments are particularly abundant just under the plasma membrane. **Figure 7.31a** shows a fluorescence micrograph of the actin filaments in a mammalian cell. Notice that groups of actin filaments are organized into long bundles or dense networks. Whether they are arranged in parallel as part of bundles or crisscrossed in networks, individual actin filaments are linked to one another by other proteins. In combination, bundles and networks of actin filaments help stiffen the cell and define its shape.

In addition to providing structural support, actin filaments are involved in movement when they interact with the specialized protein **myosin.** Myosin is a **motor protein:** a protein that

SUMMARY TABLE 7.3 **Cytoskeletal Filaments**

	Actin Filaments (Microfilaments)	Intermediate Filaments	Microtubules
	The three types of filaments found in the cytoskeleton are distinguished by their size and structure, and by the protein subunit of which they are made.		
Protein subunits	Actin	Keratin, vimentin, lamin, others	a-Tubulin and b-tubulin dimers
Structure	Strands in double helix 2 end 1 end 7 nm Actin subunit	Fibers wound into thicker cables 10 nm Keratin subunits	Hollow tube 2 end 1 end 25 nm a-Tubulin and b-tubulin dimer
Functions	• maintain cell shape by resisting tension (pull) • motility via muscle contraction or cell crawling • cell division in animals • movement of organelles and cytoplasm in plants, fungi, and animals	• maintain cell shape by resisting tension (pull) • anchor nucleus and some other organelles	• maintain cell shape by resisting compression (push) • motility via flagella or cilia • move chromosomes during cell division • formation of cell plate during plant cell division • move organelles • growth of plant cell walls

converts the chemical energy in ATP into the mechanical work of movement, just as a car's motor converts the chemical energy in gasoline into movement. The interaction between actin and myosin that produces movement is discussed in detail in Chapter 46. For now, it's enough to recognize that when ATP binds to myosin and is then hydrolyzed to ADP, the "head" region of the myosin molecule binds to actin and moves. The movement of this protein causes the actin filament to slide (**Figure 7.32a**). This type of movement is analogous to a line of people who are passing along a long log or pole. The people are myosin molecules; the log or pole is actin.

As **Figure 7.32b** shows, the (ATP-powered) interaction between actin and myosin is the basis for an array of cell movements:

- **Cell crawling** occurs in a wide array of organisms and cell types, including amoebae, slime molds, and certain types of human cells. Cell crawling is based on three processes: a directional extension of actin filaments that pushes the plasma membrane into bulges called **pseudopodia** ("false-feet"), adherence to a solid substrate, and a myosin-driven contraction of actin filaments at the cell's other end. In combination, the three events result in directed movement by whole cells.

(a) Actin filaments in mammalian cell

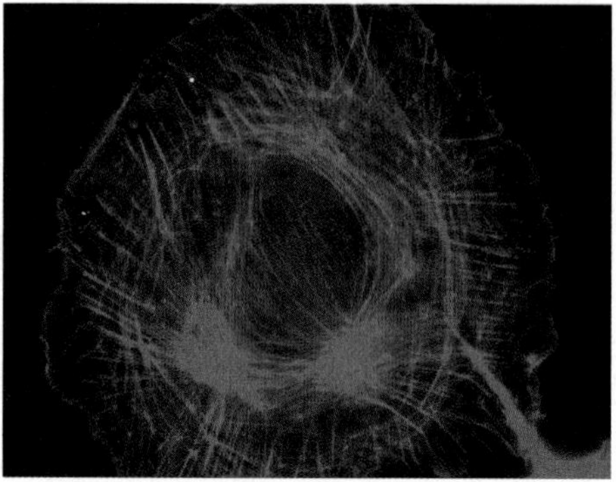

(b) Intermediate filaments in mammalian cell

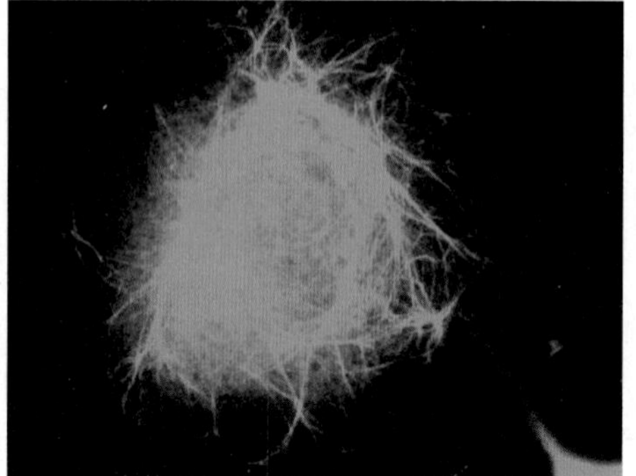

(c) Microtubules in mammalian cell

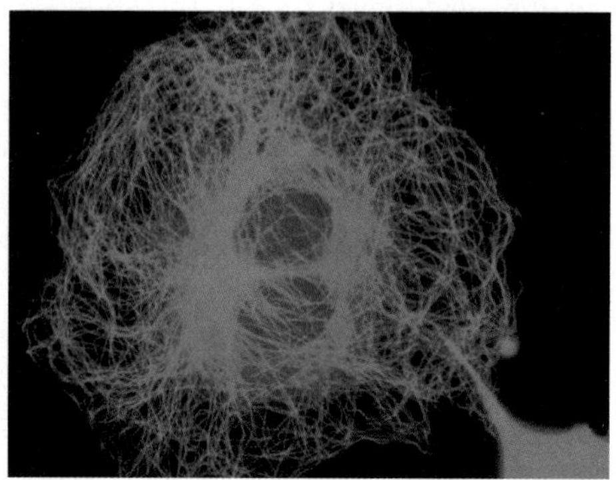

FIGURE 7.31 How Are Cytoskeletal Elements Distributed in the Cell? To make these fluorescence micrographs, researchers attached a fluorescent compound to **(a)** actin, the protein subunit of actin filaments, to **(b)** a protein found in intermediate filaments, and to **(c)** tubulin dimers.

(a) Actin and myosin interact to cause movement.

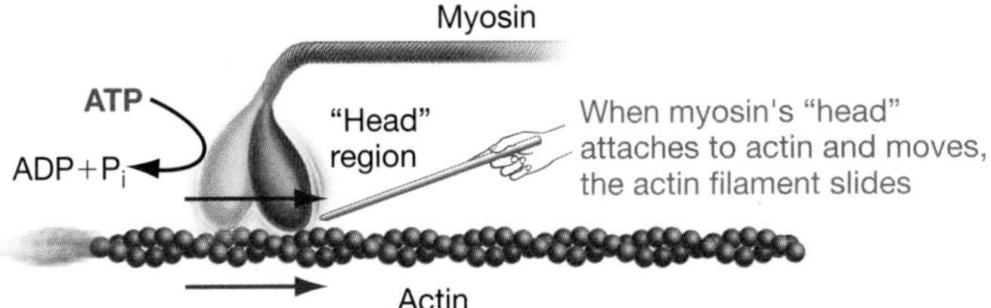

(b) Actin-myosin interactions produce several types of movement.

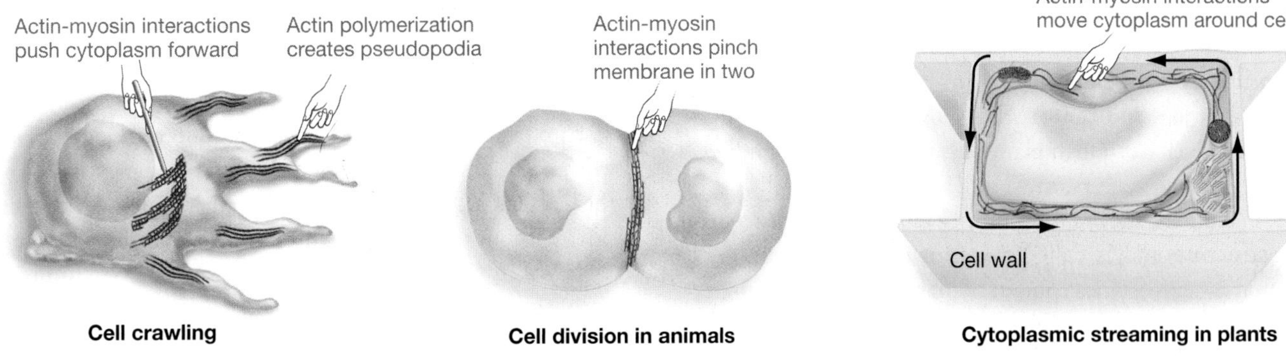

FIGURE 7.32 Many Cellular Movements Are Based on Actin-Myosin Interactions. (a) When the "head" region of the myosin protein interacts with ATP, myosin attaches to actin and changes shape. The movement causes the actin filament to slide. **(b)** Actin-myosin interactions can move cells, divide cells, and move organelles and cytoplasm.

- **Cytokinesis** ("cell-moving") is the process of cell division in animals. For these cells to divide in two, actin filaments that are arranged in a ring under the plasma membrane must slide past one another. Because they are connected to the plasma membrane, the movement of the actin fibers pinches the cell in two (see Chapter 11).
- **Cytoplasmic streaming** is the directed flow of cytosol and organelles around plant and fungal cells. The movement occurs along actin filaments and is powered by myosin. It is especially common in large cells, where the circulation of cytoplasm facilitates material transport.

Extension of actin filaments is also responsible for the expansion of long, thin fungal cells into soil or rotting wood, toward fresh supplies of nutrients. The same mechanism causes structures called pollen tubes to grow toward the egg cells of plants so sperm can be delivered prior to fertilization. You'll see actin and myosin moving cells and organisms around throughout this text.

Intermediate Filaments

Unlike actin filaments and microtubules, **intermediate filaments** (**Figure 7.31b**) are defined by size rather than composition. Many types of intermediate filaments exist, each consisting of a different protein. In many cases, different types of cells in the same organism contain different types of intermediate filaments. This is in stark contrast to actin filaments and microtubules, which are made from the same protein subunits in all eukaryotic cells. In addition, intermediate filaments are not polar; instead, each end of these filaments is identical. As a result, intermediate filaments do not treadmill, and they are not involved in directed movement driven by myosin or related proteins. Intermediate filaments serve a purely structural role in eukaryotic cells.

The intermediate filaments that you are most familiar with belong to a family of molecules called the keratins. The cells that make up your skin and that line surfaces inside your body contain about 20 types of keratin. The presence of these intermediate filaments provides the mechanical strength required for these cells to resist pressure and abrasion. Skin cells manufacture another 10 distinct forms of keratin. Depending on the location of the skin cell and keratins involved, the secreted filaments form fingernails, toenails, or hair.

Nuclear lamins, which make up the nuclear lamina layer introduced in Section 7.1, also qualify as intermediate filaments. These fibers form a dense mesh under the nuclear envelope. Recall that in addition to giving the nucleus its shape, they anchor the chromosomes. They are also involved in the breakup and reassembly of the nuclear envelope when cells divide. Some intermediate filaments project from the nucleus through the cytoplasm to the plasma membrane, where they are linked to intermediate filaments that run parallel to the cell surface. In this way, intermediate filaments form a flexible skeleton that helps shape the cell surface and hold the nucleus in place.

Microtubules

Microtubules are composed of the proteins α-tubulin and β-tubulin and are the largest cytoskeletal components in terms of diameter (**Figure 7.31c**). Molecules of α-tubulin and β-tubulin bind to form **dimers** ("two-parts"), compounds formed by the joining of two monomers. Tubulin dimers then polymerize to form the large, hollow tube called a microtubule. Because each end of a tubulin dimer is different, each end of a microtubule has a distinct polarity. Like actin filaments, microtubules are dynamic and usually grow at their plus end. Microtubules grow and shrink in length as tubulin dimers are added or subtracted.

Microtubules originate from a structure called the **microtubule organizing center** and grow outward, radiating throughout the cell. Although plant cells typically have hundreds of these organizing centers, most animal and fungal cells have just one. In animals, the microtubule organizing center has a distinctive structure and is called a **centrosome**. As **Figure 7.33** shows, animal centrosomes contain two bundles of microtubules called **centrioles**. The centrioles are not required for microtubules to form and begin growing from the centrosome, however. Although biologists have been aware for decades that they exist, their role in the cell is still unknown.

In function, microtubules are similar to actin filaments: They provide stability and are involved in movement. The microtubules that radiate out from an organizing center function like steel girders in a skyscraper. They stiffen the cell by resisting compression forces. Microtubules may also provide a structural framework for organelles. If microtubules are prevented from forming, the ER no longer assembles in its normal network-like configuration.

During cell division, microtubules from the organizing center are responsible for moving chromosomes from the original cell to each of the two resulting cells (see Chapters 11 and 12). But microtubules are involved in many other types of cellular movement as well. The remainder of this chapter focuses on how microtubules function in moving materials inside cells and in moving the entire cell.

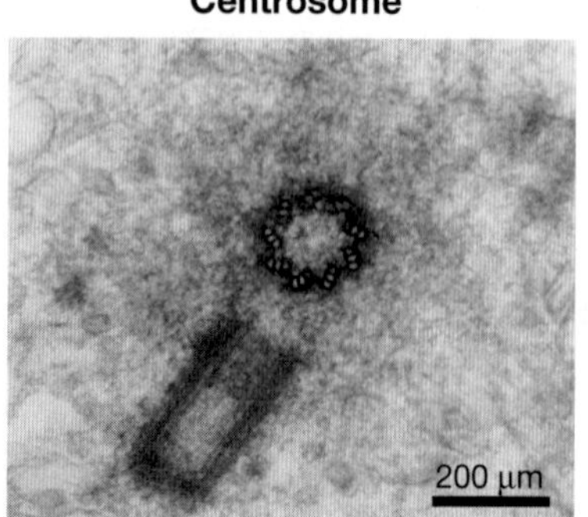

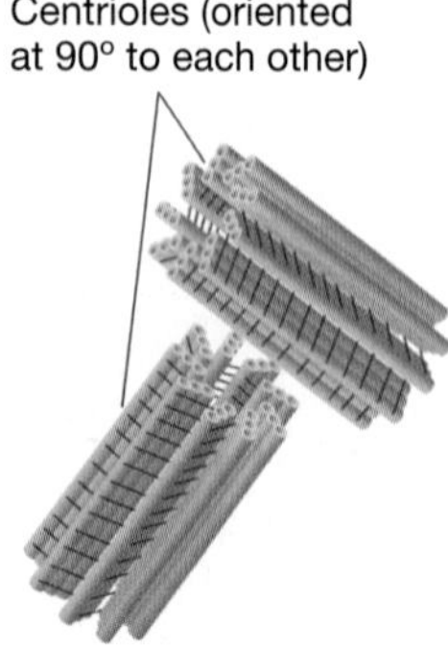

FIGURE 7.33 Centrosomes Are a Type of Microtubule Organizing Center. Microtubules emanate from microtubule organizing centers, which in animals are called centrosomes. The centrioles inside a centrosome are made of microtubules.

Studying Vesicle Transport Materials are transported to a wide array of destinations inside cells via vesicles. To study how this movement happens, Ronald Vale and colleagues focused on a cell in squid called the giant axon. The giant axon is an extremely large nerve cell that runs the length of a squid's body. If the animal is disturbed, the cell signals muscles to contract so the individual can jet away to safety.

The researchers decided to study this particular cell for three reasons. First, the giant axon is so large that it is relatively easy to see and manipulate. Second, large numbers of molecules are synthesized in the cell's ER and then transported in vesicles down the length of the cell, where they are released. As a result, a large amount of cargo moves a long distance. Third, the researchers found that if they gently squeeze the cytoplasm out of the cell, vesicle transport still occurs in the cytoplasmic material. This allowed them to do experiments on vesicle transport without the plasma membrane being in the way.

In short, the squid giant axon provided a system that could be observed and manipulated efficiently in the lab. What did the biologists find out?

Microtubules Act as "Railroad Tracks" To watch vesicle transport in action, the researchers mounted a video camera to a microscope. As **Figure 7.34** shows, this technique allowed them to document that vesicle transport was occurring along a filamentous track. A simple experiment convinced the group that this movement is an energy-dependent process. If they depleted the amount of ATP in the cytoplasm, vesicle transport stopped.

(a) Electron micrograph

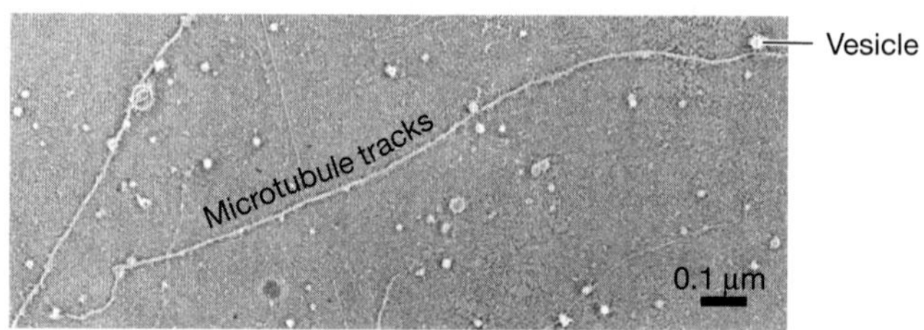

(b) Video image

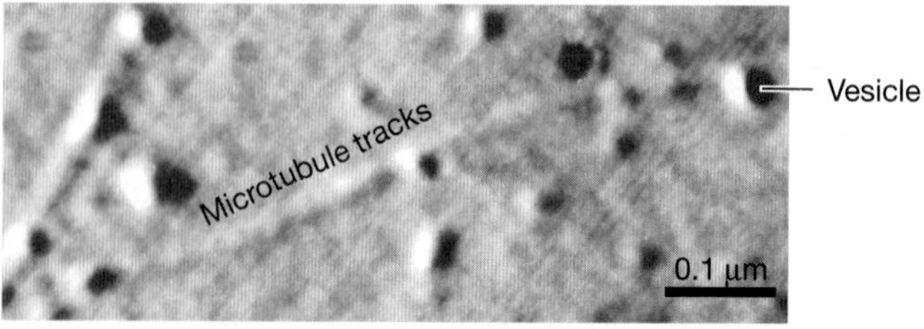

FIGURE 7.34 Transport Vesicles Move along Microtubule Tracks. The images are of extruded cytoplasm from a squid giant axon. **(a)** An electron micrograph that allowed researchers to measure the diameter of the filaments and confirm that they are microtubules. In the lower part of this image, you can see a vesicle on a "track." **(b)** A slightly fuzzy but higher-magnification videomicroscope image, in which researchers actually watched vesicles move.

To identify the type of filament involved, the biologists measured the diameter of the tracks and analyzed their chemical composition. Both types of data indicated that the tracks consist of microtubules. Microtubules also appear to be required for movement of materials elsewhere in the cell. If experimental cells are treated with a drug that disrupts microtubules, the movement of vesicles from the rough ER to the Golgi apparatus is impaired.

The general message of these experiments is that transport vesicles move through the cell along microtubules. How? Do the tracks themselves move, like a conveyer belt, or are vesicles carried along on some sort of molecular truck?

A Motor Protein Generates Motile Forces To study the way vesicles move along microtubules, Vale's group set out to take the squid axon's transport system apart and then put it back together. To begin, they assembled microtubule fibers from purified α-tubulin and β-tubulin. Then they used differential centrifugation to isolate transport vesicles. But when they mixed purified microtubules and vesicles with ATP, no transport occurred.

Something had been left out—but what?

To find the missing element or elements, the researchers purified one subcellular part after another, using differential centrifugation, and added it to the microtubule + vesicle + ATP system. Through trial and error, they found something that triggered movement. After further purification steps, the researchers finally succeeded in isolating a protein that generated vesicle movement. They named the molecule **kinesin**, from the Greek word *kinein* ("to move"). Like myosin, kinesin is a **motor protein**. Kinesin converts chemical energy in ATP into mechanical energy in the form of movement. More specifically, when ATP is added to kinesin or drops off, the protein moves.

Biologists began to understand how kinesin works when X-ray diffraction studies—similar to those that revealed the helical nature of DNA—revealed the three-dimensional structure of kinesin. As **Figure 7.35a** shows, the protein consists of two large, intertwined polypeptide chains associated with two small polypeptides. It has three major regions: a head section with two globular pieces, a tail associated with the small polypeptides, and a stalk that connects the head and tail. Follow-up studies confirmed that the two globular components of the head bind to the microtubule. The tail region binds to the transport vesicle. A kinesin molecule is like a delivery truck that carries transport vesicles along microtubule tracks. Cells contain several different kinesin proteins, each specialized for carrying a different type of vesicle.

How does kinesin move? More detailed studies of this protein's structure indicated that each of the globular components of the molecule's head has a site for binding ATP as well as a site that binds to the microtubule. To pull these observations together, biologists propose that kinesin transports vesicles by "walking" along a microtubule. The idea is that each part of the head region changes shape when it binds ATP. As Chapter 3 showed, these types of shape changes often alter the activity of a protein. As **Figure 7.35b** shows, the ATP-dependent conformational change in kinesin results in a step forward. As each head alternately binds and hydrolyzes ATP, the protein and its cargo move down the microtubule track.

In short, kinesins move molecular cargo to destinations throughout the cell. They are not the only type of motor protein active inside cells, however. Recall that myosin causes actin filaments to slide, resulting in the movement of cells or cytoplasm. Myosin is also involved in the movement of organelles along tracks made of actin. And a third motor protein, called dynein, powers the transport of certain organelles as well as swimming movements that move the entire cell. Let's take a closer look at how cells swim.

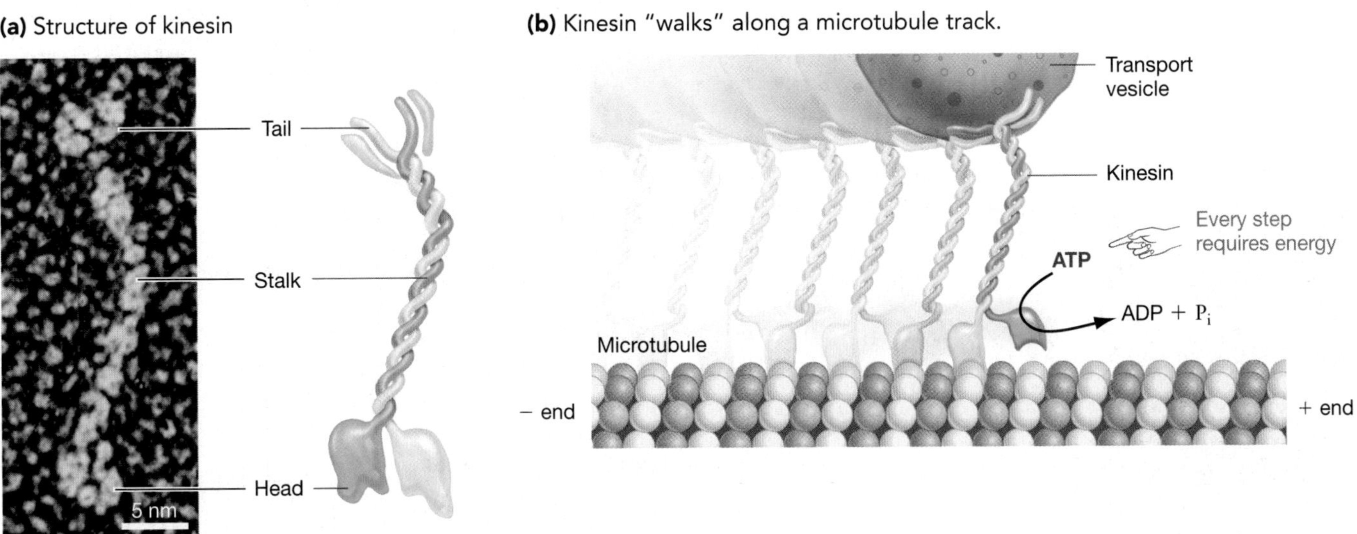

FIGURE 7.35 A Motor Protein Moves Vesicles along Microtubules. (a) Kinesin has three major segments. **(b)** The current model depicting how kinesin "walks" along a microtubule track to transport vesicles. The two head segments act like feet that alternately attach and release in response to the gain or loss of a phosphate group.

Cilia and Flagella: Moving the Entire Cell

Flagella are long, hairlike projections from the cell surface that function in movement. Flagella are found in many bacteria and eukaryotes. The structure of flagella is completely different in the two groups, however. Bacterial flagella are made of a protein called flagellin; eukaryotic flagella are constructed from microtubules (tubulin). Bacterial flagella move the cell by rotating like a ship's propeller; eukaryotic flagella move the cell by undulating—they whip back and forth. Eukaryotic flagella are surrounded by plasma membrane; bacterial flagella are not. Based on these observations, biologists conclude that the two structures evolved independently, even though their function is similar.

To understand how cells move, let's focus on eukaryotic flagella. Eukaryotic flagella are closely related to structures called **cilia** (singular: cilium), which are short, filament-like projections that are also found in some eukaryotic cells. Unicellular eukaryotes may have either flagella or cilia, while some multicellular organisms have both. In humans, for example, the cells that line the respiratory tract have cilia; sperm cells have flagella.

Flagella are generally longer than cilia, and a cell will typically have just one or two flagella but many cilia (**Figure 7.36**). But when researchers examined the two structures with the electron microscope, they found that their underlying organization is identical.

How Are Cilia and Flagella Constructed? In the 1950s, anatomical studies established that both cilia and flagella have a characteristic "9 + 2" arrangement of microtubules. As **Figure 7.37a** shows, nine microtubule pairs, or doublets, surround two central microtubules. The doublets consist of one complete and one incomplete microtubule and are arranged around the periphery of the structure. The entire 9 + 2 structure is called the **axoneme** ("axle-thread"). The axoneme attaches to the cell at a structure called the **basal body**. The basal body is identical in structure to a centriole. But unlike centrioles, the function of the basal body is well understood: It plays a central role in the growth of the axoneme.

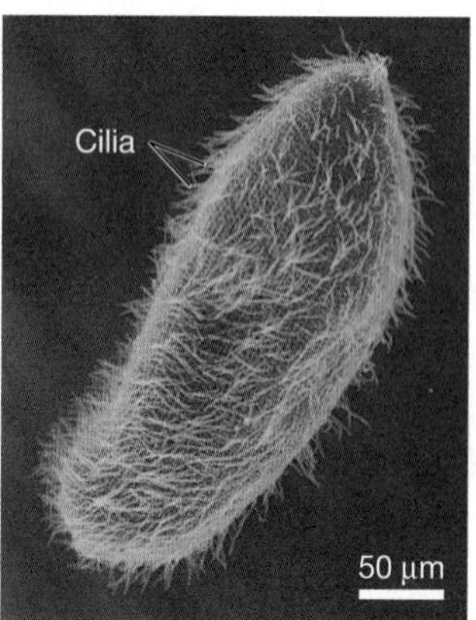

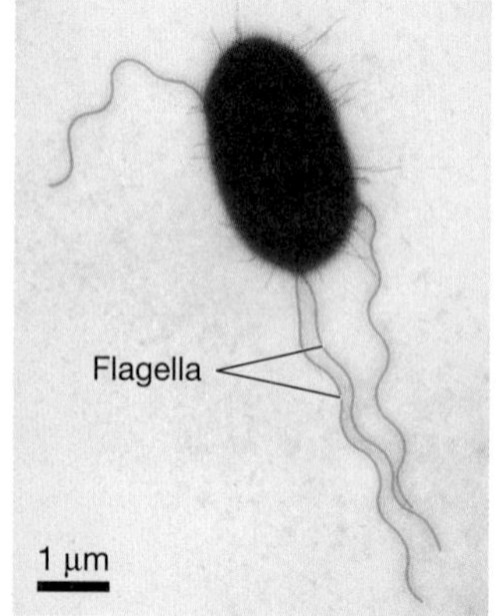

FIGURE 7.36 Cilia and Flagella Differ in Length and Number. Cilia are relatively short and large in number; flagella are relatively long and few in number.

As electron microscopy improved, biologists gained a more detailed view of the structure. As the sketch in **Figure 7.37b** illustrates, spoke-like structures connect each doublet to the central pair of microtubules. In addition, molecular links connect the nine doublets to one another. Finally, each of the doublets has a set of arms that project toward an adjacent doublet.

Microtubules are complex. How do their components interact to generate motion?

A Motor Protein in the Axoneme In the 1960s Ian Gibbons began studying the cilia of a common unicellular eukaryote called *Tetrahymena*, which lives in pond water. Gibbons found that by using a detergent to remove the plasma membrane that surrounds cilia and then subjecting the resulting solution to differential centrifugation, he could isolate axonemes. Further, the isolated structures would beat if Gibbons supplied them with ATP. These results confirmed that the beating of cilia is an energy-demanding process. They also provided Gibbons with a cell-free system for exploring the molecular mechanism of movement. Cell-free systems are elegant to study because they are isolated—they are relatively easy to manipulate, and no other cell components are present that might confuse experimental results.

(a) Transmission electron micrograph of axoneme

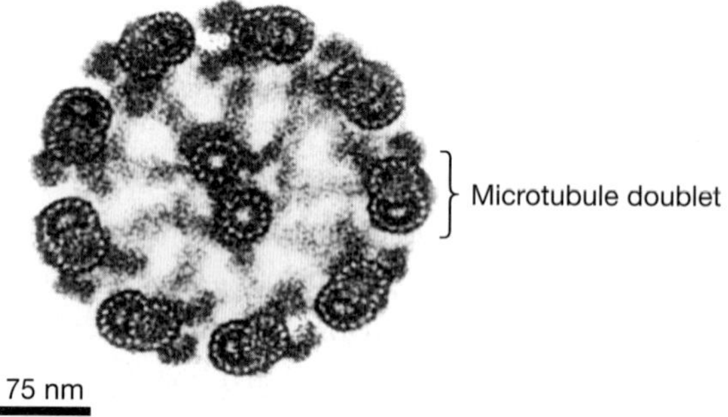

(b) Diagram of axoneme

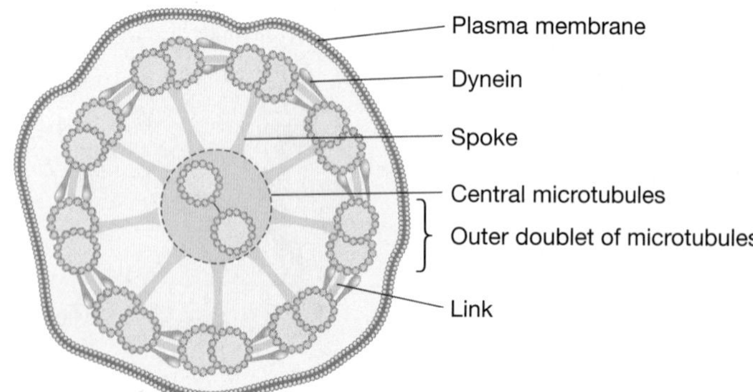

FIGURE 7.37 The Structure of Cilia and Flagella. (a) Transmission electron micrograph of a cross section through an axoneme. **(b)** The major structural elements in cilia and flagella. The microtubules are connected by links and spokes, and the entire structure is surrounded by the plasma membrane.

EXERCISE In part (a), write the numbers 1–9 next to the microtubule doublets around the outside of the axoneme.

In an early experiment with isolated axonemes, Gibbons treated the structures with a molecule that affects the ability of proteins to bind to one another. The axonemes that resulted from this treatment could not bend or use ATP. When Gibbons examined them in the electron microscope, he found that the arms had fallen off. This observation led to the hypothesis that the arms are required for movement. Follow-up work showed that the arms are made of a large protein that Gibbons named **dynein** (from the Greek word *dyne,* meaning "force").

Like myosin and kinesin, dynein is a motor protein. Structural and chemical studies have shown that dynein changes shape when a phosphate group from ATP attaches to it. More specifically, the end of a dynein molecule changes shape when it is phosphorylated. This shape change moves the molecule along the nearby microtubule. When the dynein molecule reattaches, it has succeeded in walking up the microtubule. This walking motion allows the microtubule doublets to slide past one another.

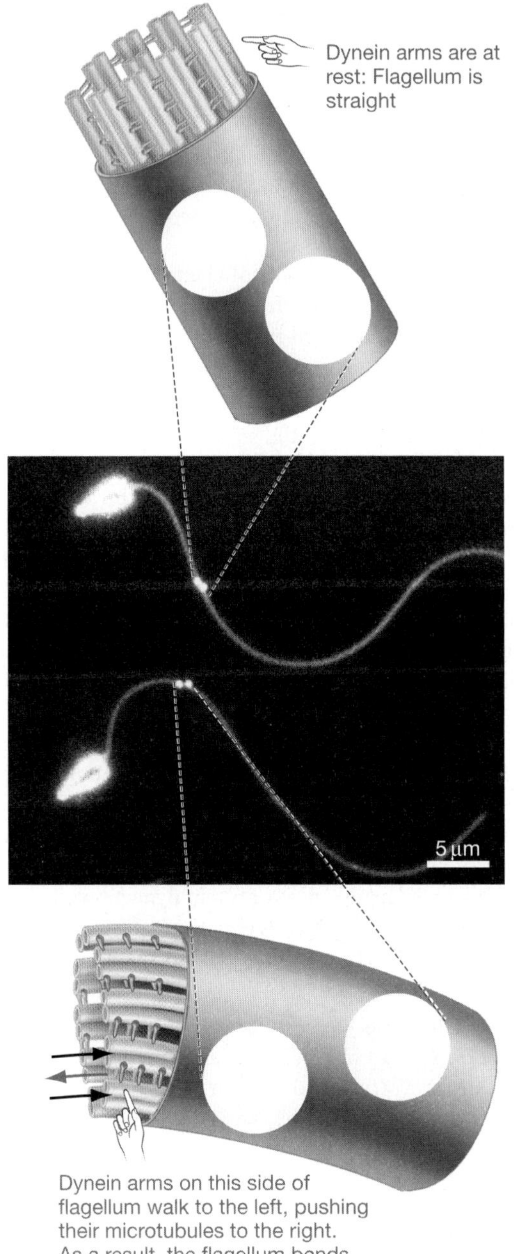

FIGURE 7.38 How Do Flagella Bend? Researchers attached a pair of gold beads to a flagellum and photographed its movement over a short time sequence. When dynein arms walk along the microtubule doublets on one side of a flagellum, the structure bends.

The outcome of dynein walking is very different from the outcome of a myosin-actin interaction, however. To understand why, remember that each of the nine doublets in the axoneme is connected to the central pair of microtubules by a spoke, and that all of the doublets are connected to each other by molecular links. As a result, the sliding motion that results from dynein walking is constrained—if one doublet slides, it transmits force to the rest of the axoneme via the links and spokes (**Figures 7.38**).

If you understand this concept, you should be able to study the top drawing in Figure 7.38 and describe what happens when movement by dynein arms causes the doublet at the bottom of the axoneme to move toward the bottom right-hand corner of the page. If the dynein arms on just one side of the axoneme walk and cause some of the doublets to slide while those on the other side are at rest, the result of the constrained, localized movement is bending (see Figure 7.38, bottom). The result of the bending of cilia or flagella is a swimming motion.

Scaled for size, flagellar-powered swimming can be rapid. In terms of the number of body or cell lengths traveled per second, a sperm cell from a bull moves faster than a human world-record-holder does when swimming freestyle. At the level of the cell, life is fast paced.

Taken together, the data reviewed in this chapter can be summed up in three words: Cells are dynamic. Chemical reactions take place at mind-boggling speeds. Microfilaments and microtubules grow and shrink. The endomembrane system synthesizes, sorts, and ships an array of products in a highly regulated manner. How does all this activity inside the cell relate to what is going on outside? This is the issue we'll take up in Chapter 8.

Check Your Understanding

If you understand that...

- Each component of the cytoskeleton has a unique structure and set of functions. In addition to providing structural support, actin filaments and microtubules work in conjunction with motor proteins to move the cell or materials inside the cell. Intermediate filaments provide structural support.
- Most elements of the cytoskeleton are dynamic—they grow and shrink over time.

You should be able to...

Predict what will happen when experimental cells are treated with drugs that inhibit formation of actin filaments or microtubules.

Chapter Review

SUMMARY OF KEY CONCEPTS

The structure of cell components is closely correlated with their function.

Because all organisms consist of cells, many questions in biology can be answered by understanding the structure and function of cells and cell components. There are two basic cellular designs: prokaryotic and eukaryotic. Eukaryotic cells are usually much larger and more structurally complex than prokaryotic cells. Most prokaryotic cells consist of a single membrane-bound compartment in which nearly all cellular functions occur. Eukaryotic cells contain numerous membrane-bound compartments called organelles. Organelles allow eukaryotic cells to compartmentalize functions and grow to a large size.

Eukaryotic organelles are specialized for carrying out different functions, and their structure is often correlated closely with their function. The defining organelle of eukaryotic cells is the nucleus, which contains the cell's chromosomes and serves as its control center. Mitochondria and chloroplasts have extensive internal membrane systems where the enzymes responsible for ATP generation and photosynthesis reside. Rough ER is named for the ribosomes that attach to it. Ribosomes are protein-making machines, and rough ER is a site for protein synthesis and processing. Smooth ER lacks ribosomes because it is a center for lipid synthesis and processing.

You should be able to predict what would happen to cells that are exposed to a drug that poisons mitochondria, an enzyme that degrades the cell wall, or a drug that prevents ribosomes from functioning.

MB Web Animation *BioFlix* at www.masteringbio.com
Tour of an Animal Cell; Tour of a Plant Cell

Inside cells, materials are transported to their destinations with the help of molecular "zip codes."

Cells have sophisticated systems for making sure that proteins and other products end up in the right place. For example, traffic across the nuclear envelope occurs through nuclear pores, which contain a multiprotein nuclear pore complex that serves as gatekeeper. Both passive and active transports of materials occur through these nuclear pore complexes. Movement of proteins and RNAs is highly regulated, allowing proteins into the nucleus only if they contain a specific molecular signal.

The endomembrane system is an extensive, interconnected system of membranes and membrane-bound compartments that can extend from the nucleus to the plasma membrane. Two principal organelles in the endomembrane system are the endoplasmic reticulum (ER) and the Golgi apparatus. The ER is the site of synthesis for a wide array of proteins and lipids. Most ER products are shipped to the Golgi apparatus, which serves as a processing and dispatching station. In many proteins the major processing step is glycosylation, or the addition of carbohydrate groups.

The movement of materials through the endomembrane system is highly organized and takes place inside membrane-bound transport organelles called vesicles. Before products leave the endomembrane system, they are sorted with molecular zip codes that direct them to vesicles headed for their final destination. The vesicles contain proteins that interact with receptor proteins on the surface of a target organelle or the plasma membrane so that the contents can be delivered.

You should be able to explain, based on information in Chapter 3, why proteins are the molecules responsible for "reading" the array of molecular zip codes in cells.

MB Web Animation at www.masteringbio.com
Transport into the Nucleus; A Pulse-Chase Experiment

Cells are dynamic. Thousands of chemical reactions occur each second within cells; molecules constantly enter and exit across the plasma membrane; cell products are shipped along protein fibers; and elements of the cell's internal skeleton grow and shrink.

The cell is a membrane-bound structure with a highly organized, dynamic interior. Inside the cell, thousands of different chemical reactions take place at incredible speeds. The products of these chemical reactions allow the cell to acquire resources from the environment, synthesize additional molecules, dispose of wastes, and reproduce.

The cytoskeleton is an extensive system of fibers that serves as a structural support for eukaryotic cells. Elements of the cytoskeleton also provide paths for moving vesicles inside cells and the machinery for moving the cell as a whole through the beating of flagella or cilia, or cell crawling. Both the movement of vesicles inside cells and cell motility depend on motor proteins, which can convert chemical energy stored in ATP into movement. For example, movement of transport vesicles occurs as the motor protein kinesin "walks" along microtubule tracks. Cilia and flagella bend as the motor protein dynein "walks" along microtubule doublets. The bending motion allows these structures to beat back and forth, enabling cells to swim or generate water currents.

You should be able to compare and contrast the structure and function of actin filaments, intermediate filaments, and microtubules. You should also be able to explain why actin filaments and microtubules have directionality and change length , and why it is misleading to refer to the cytoskeleton as "scaffolding?"

QUESTIONS

Test Your Knowledge

1. Which of the following best describe the nuclear envelope?
   a. It is continuous with the endomembrane system.
   b. It is continuous with the nucleolus.
   c. It is continuous with the plasma membrane.
   d. It contains a single membrane and nuclear pores.

2. What is a nuclear localization signal?
   a. a stretch of amino acids that directs proteins from the nucleus to the ER
   b. a molecule that is attached to nuclear proteins so that they are retained inside the nucleus
   c. a signal built into a protein that directs it to the nucleus
   d. a component of the multiprotein nuclear pore complex

3. Which of the following is *not* true of secreted proteins?
   a. They are synthesized in ribosomes.
   b. They are transported through the endomembrane system in membrane-bound transport organelles.
   c. They are transported from the Golgi apparatus to the ER.
   d. They contain a signal sequence that directs them into the ER.

4. To find the nuclear localization signal in the protein nucleoplasmin, researchers separated the molecule's core and tail segments, labeled both with a radioactive atom, and injected them into the cytoplasm. Why did the researchers conclude that the signal is in the tail region of the protein?
   a. The protein reassembled and folded into its normal shape spontaneously.
   b. Only the tail segments appeared in the nucleus.
   c. With a confocal microscope, tail segments were clearly visible in the nucleus.
   d. The tail and head segments both appeared in the nucleus.

5. Molecular zip codes direct molecules to particular destinations in the cell. How are these signals read?
   a. They bind to receptor proteins.
   b. They enter transport vesicles.
   c. They bind to motor proteins.
   d. They are glycosylated by enzymes in the Golgi apparatus.

6. The number and size of organelles in a cell correlates with that cell's function. Propose a function for cells that contain extensive rough ER.
   a. rapid cell division in growing bones or muscle tissues
   b. production and processing of fatty acids and other lipids
   c. movement via cell crawling
   d. acquiring ions and other nutrients via specialized membrane proteins

Test Your Knowledge answers: 1. a; 2. c; 3. c; 4. b; 5. a; 6. d

Test Your Understanding

Answers are available at www.masteringbio.com

1. Compare and contrast the structure of a generalized plant cell, animal cell, and prokaryotic cell. Which features are common to all cells? Which are specific to just prokaryotes, or just plants, or just animals?

2. Make a concept map that traces the movement of a secreted protein from its site of synthesis to the outside of a eukaryotic cell (see **BioSkills 6**). Identify all of the organelles that the protein passes through. Your linking verbs should indicate what is happening to the protein at each step.

3. Describe how a motor protein such as kinesin can move a transport vesicle down a microtubule track. Include all necessary steps and components.

4. Describe the logic of a pulse-chase experiment. How was this approach used to document the pattern of protein transport through the endomembrane system?

5. Briefly describe how researchers use centrifugation to isolate particular cell components for further study.

6. Structurally, what is the difference between microtubules involved in structural support, vesicle movement, chromosome movement during cell division, and flagella?

Applying Concepts to New Situations

Answers are available at www.masteringbio.com

1. In addition to delivering cellular products to specific organelles, eukaryotic cells can take up material from the outside and route it to specific organelles. For example, specialized cells of the human immune system ingest bacteria and viruses and then deliver them to lysosomes for degradation. Suggest a hypothesis for how this material is tagged and directed to lysosomes. How would you test this hypothesis?

2. The leading hypothesis to explain the origin of the nuclear envelope is that a deep infolding of the plasma membrane occurred in an ancient prokaryote. Draw a diagram that illustrates this infolding hypothesis. Does your model explain the existence of the structure's inner and outer membranes? Explain.

3. Propose a function for cells that contain (a) a large number of lysosomes, (b) a particularly extensive cell wall, and (c) many peroxisomes.

4. Suggest a hypothesis or a series of hypotheses to explain why bacteria, archaea, algae, and plants have cell walls. Suppose that mutant individuals from each group lacked a cell wall. How could you use these individuals to test your idea(s)?

www.masteringbio.com is also your resource for • Answers to text, table, and figure caption questions and exercises • Answers to *Check Your Understanding* boxes • Online study guides and quizzes • Additional study tools including the *E-Book for Biological Science* 3rd ed., textbook art, animations, and videos.

8 Cell-Cell Interactions

KEY CONCEPTS

- Extracellular material strengthens cells and helps bind them together.
- Cell-cell connections help adjacent cells adhere. Cell-cell gaps allow adjacent cells to communicate.
- Intercellular signals are responsible for creating an integrated whole from many thousands of independent parts.

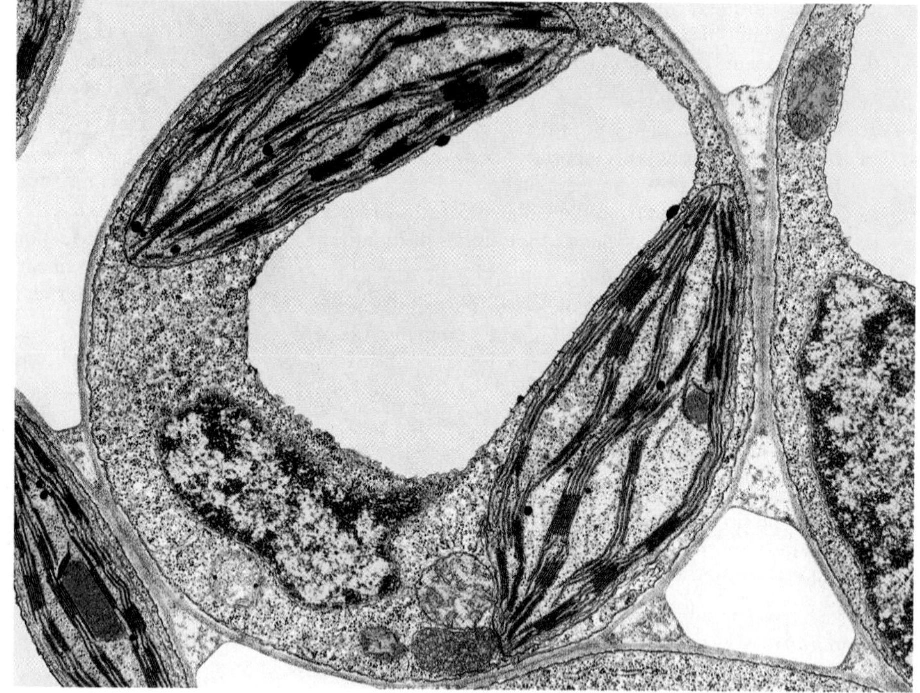

Transmission electron micrograph highlighting the cell walls (colored yellow) of several plant cells. The cell surface is where adjacent cells join and where signals from distant cells arrive.

Chapter 6 introduced the structure and function of the plasma membrane, which is the defining feature of the cell. Chapter 7 surveyed the organelles, molecular machines, and cytoskeletal elements that fill the space inside that membrane and explored how cargo moves from sources to destinations within the cell. Both chapters highlighted the breathtaking speed and diversity of events that take place at the cellular level.

The cell is clearly a bustling enterprise. But it would be a mistake to think that cells are self-contained—that they are worlds in and of themselves. Instead, cells constantly interact with other cells and the surrounding environment. They continuously adjust their activities in response to stimulation from other cells and to changes in environmental conditions.

To understand the life of a cell thoroughly, it is critical to analyze how the cell interacts with the world outside its membrane. How do cells obtain information about the outside world and respond to that information? In particular, how do cells interact with nearby cells?

For most unicellular species, the outside environment is made up of soil or water that is teeming with other organisms, either of the same or different species. In your mouth and gut and on your skin, hundreds of billions of bacterial cells are jostling for space and resources. Similar numbers of single-celled organisms are found in every tablespoon of good-quality soil. In addition to interacting with these individuals, each unicellular organism must contend with constantly shifting aspects of the physical environment, such as heat, light, ion concentrations, and food supplies. If unicellular organisms are unable to sense these conditions and respond appropriately, they die.

In multicellular species, the environment outside the cell is made up of other cells, both neighboring and distant. The cells that make up a redwood tree, an *Amanita* mushroom, or your body are intensely social. Although biologists often study cells in isolation, an individual tree, fungus, or person is actually an interdependent community of cells. If those cells do not communicate and cooperate, the whole will break into dysfunctional parts and die.

To introduce the ways that cells interact with each other, this chapter focuses on how cells in a multicellular organism communicate. Let's begin with the cell surface—with the molecules that separate the cell from its environment.

8.1 The Cell Surface

Chapter 6 introduced the currently accepted model for the structure of the plasma membrane. The fluid-mosaic hypothesis contends that the plasma membrane is a phospholipid bilayer with interspersed proteins. Many of these phospholipids and proteins have carbohydrate groups attached, forming glycoproteins and glycolipids (**Figure 8.1**). Data from electron micrographs and chemical treatments suggest that membrane proteins can be either integral, meaning that they are embedded in the bilayer, or peripheral, meaning that they are attached to one surface. Recall that if an integral protein spans the entire membrane, it is termed a transmembrane protein. Membrane proteins may float independently in the lipid bilayer or be anchored by connections to the cytoskeleton or materials outside the cell.

Chapter 6 also analyzed the primary function of the plasma membrane: to create an environment inside the cell that is different from conditions outside. Ions and molecules move across plasma membranes by direct diffusion through the phospholipid bilayer or via several types of membrane proteins. Transport of materials across the membrane can be energy demanding or passive, but it is always selective.

This picture—of a dynamic, complex plasma membrane that selectively admits or blocks passage of specific substances—is accurate but not complete. The plasma membrane does not exist in isolation. Cytoskeletal elements introduced in Chapter 7 attach to the interior face of the bilayer, and a complex array of extracellular structures exists outside. Let's consider the nature of the material outside the cell first and then analyze how the cell interacts with other cells.

The Structure and Function of an Extracellular Layer

In species from across the tree of life, it is extremely rare for cells to be bounded simply by a plasma membrane. Most cells possess a layer or wall that forms just beyond the membrane. The extracellular material helps define the cell's shape and either attaches it to another cell or acts as a first line of defense against the outside world. This observation holds for both unicellular organisms and the various types of cells in multicellular species.

Virtually all types of extracellular structures, in turn—from the cell walls of bacteria, algae, fungi, and plants to the extracellular matrix that surrounds most animal cells—follow the same fundamental design principle. Like reinforced concrete and fiberglass, they are "fiber composites": They consist of a cross-linked network of long filaments embedded in a stiff surrounding material, or ground substance (**Figure 8.2**). The molecules that make up the rods and the encasing material vary from group to group, but the engineering principle is the same.

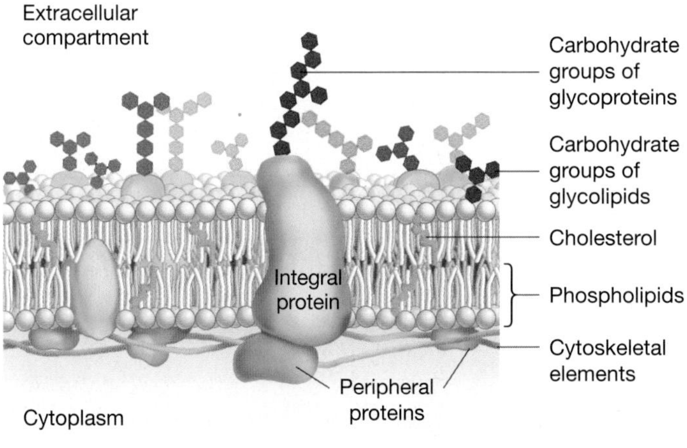

Extracellular compartment
Carbohydrate groups of glycoproteins
Carbohydrate groups of glycolipids
Cholesterol
Integral protein
Phospholipids
Cytoskeletal elements
Peripheral proteins
Cytoplasm

FIGURE 8.1 The Plasma Membrane Is a Mosaic of Phospholipids and Proteins. According to the fluid-mosaic model, the plasma membrane consists of integral and peripheral proteins scattered throughout a phospholipid bilayer.

FIGURE 8.2 Fiber Composites Resist Tension and Compression. Fiber composites such as reinforced concrete consist of a massive ground substance that fills spaces between cross-linked rods.

Fiber composites are a successful design because rods and filaments are extremely effective at withstanding stretching and straining forces, or tension. The steel rods in reinforced concrete and the cellulose fibers in a plant cell wall are unlikely to break as a result of being pulled or pushed lengthwise. Scaled for size, in fact, steel and cellulose are *equally* unlikely to break. Cellulose fibers have the same tensile strength—or resistance to tearing—as steel. In addition, the stiff surrounding substance is effective at withstanding the pressing forces called compression. Concrete performs this function in highways, and a semisolid, gelatinous mixture of polysaccharides achieves the same end in plant cell walls.

Thanks to the combination of tension- and compression-resisting elements, fiber composites are particularly rugged. And in many living cells, the fiber and composite elements are flexible as well as sturdy. When this is the case, the extracellular material is both supple and strong.

What molecules make up the rods and the ground substance found on the surface of plant and animal cells? How are these extracellular layers synthesized, and what do they do?

The Plant Cell Wall

Most plant cells are surrounded by a layer of extracellular material called a cell wall. When new cells first form, they secrete an initial fiber composite designated a **primary cell wall.** As Chapter 5 indicated, the fibrous component of the cell wall consists of long strands of the polysaccharide cellulose, which is cross-linked by polysaccharide filaments and bundled into stout, cable-like structures termed microfibrils. Microfibrils form a crisscrossed network that becomes filled with gelatinous polysaccharides (**Figure 8.3**). Chief among these gelatin-forming carbohydrates are **pectins**—the molecules that are used to thicken jams and jellies. Pectins are not as strong as cellulose, because the polysaccharides that make up their structure do not form tightly packed filaments strengthened with cross-linkages. Instead, the polysaccharides in pectin are hydrophilic. They attract and hold large amounts of water to help keep the cell wall moist.

Pectins and other gelatinous polysaccharides that form the ground substance of the cell wall are synthesized in the rough endoplasmic reticulum (ER) and Golgi apparatus and secreted to the extracellular space. The cellulose microfibrils, in contrast, are synthesized by a complex of enzymes in the plasma membrane itself.

The primary cell wall defines the shape of a plant cell. Under normal conditions, the nucleus and cytoplasm fill the entire volume of the cell and push the plasma membrane up against the wall. Because the concentration of solutes is higher inside the cell than outside, water tends to enter the cell via osmosis. The incoming water inflates the plasma membrane, exerting a force against the wall that is known as **turgor pressure**. Although plant cells experience turgor pressure throughout their lives, it is particularly important in young cells that are actively growing. Young plant cells secrete enzymes named expansins into their cell-wall matrix. Expansins catalyze reactions that allow the cellulose microfibrils in the matrix to slide past one another. Turgor pressure then forces the wall to elongate and expand. The result is cell growth.

As certain plant cells mature and stop growing, they secrete a layer of material—a **secondary cell wall**—inside the primary cell wall. The structure of the secondary cell wall varies from cell to cell in the plant and correlates with that cell's function. For example, cells on the surface of a leaf have cell walls that are impregnated with waxes that form a waterproof coating. Cells that furnish structural support for the plant's stem have secondary

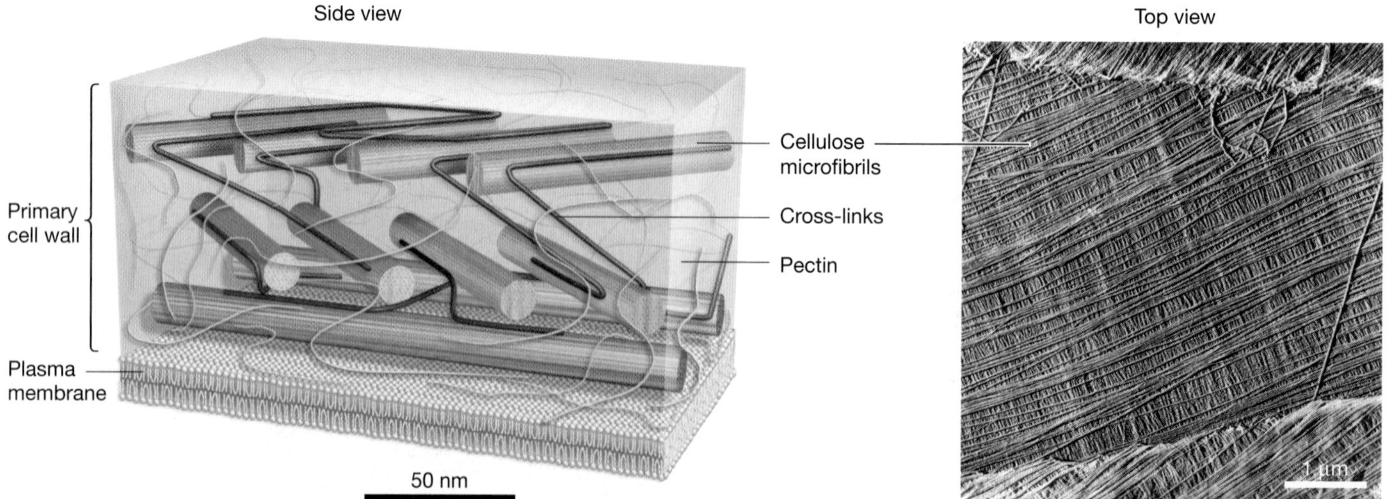

FIGURE 8.3 Primary Cell Walls of Plants Are Fiber Composites. In the primary plant cell wall, cellulose microfibrils are cross-linked by polysaccharide chains. The spaces between the microfibrils are filled with pectin molecules, which form a gelatinous solid.

cell walls that contain a great deal of cellulose. In some species, the secondary cell walls of support cells include a complex and extremely tough substance named **lignin**. Lignin forms a rigid network that is exceptionally stiff and strong. Cells that form wood secrete a secondary cell wall that contains large amounts of lignin. Cells that have thick cell walls of cellulose and lignin help plants withstand the forces of gravity and wind.

Plant cell walls are also dynamic. If they are damaged by insect attack, they may release signaling molecules that trigger the reinforcement of walls in nearby cells. When a seed germinates (sprouts), the walls of cells that store oils or starch are actively broken down, releasing these nutrients to the growing plant. Cell walls are also degraded in a controlled way as fruits ripen, making the fruits softer and more digestible for the animals that disperse the seeds inside.

Although animal cells do not make a cell wall, they do form a fiber composite outside their plasma membrane. What is this substance, and what does it do?

The Extracellular Matrix in Animals

Virtually all animal cells secrete a fiber composite called the **extracellular matrix (ECM)**. The design of this substance follows the same principles observed in the cell walls of bacteria, archaea, algae, fungi, and plants. The fibrous component of the animal ECM consists of protein fibers instead of polysaccharide filaments, however. **Figure 8.4a** illustrates the most abundant protein found in the ECM—a fibrous, cable-like molecule termed **collagen**. As in cell walls, the matrix that surrounds collagen and the other fibrous components consists of gel-forming polysaccharides—most of which are attached to a protein core. But because collagen and the other common ECM proteins are much more elastic and bendable than cellulose or lignin, the structure as a whole is relatively pliable. Most ECM components are synthesized in the rough ER, processed in the Golgi apparatus, and secreted from the cell via exocytosis. Some of the protein-bound polysaccharides that form the composite material are synthesized by membrane proteins, however.

The amount of ECM varies among different types of cells in the same organism. In bone and cartilage, for example, the number of cells is relatively small and the amount of ECM very large (**Figure 8.4b**). Skin cells, in contrast, are packed together with a minimal amount of ECM between them. The composition of the ECM also varies among cell types. For example, the ECM of lung cells contains large amounts of a rubber-like protein called elastin, which allows the ECM to expand and contract during breathing movements. The structure of a cell's ECM correlates with that cell's function.

Wherever it is found in the body, one of the ECM's most important functions is structural support—a common theme in the extracellular materials found in other organisms. ECM provides support because the combination of protein filaments and surrounding polysaccharide gel creates a strong external layer, and because transmembrane proteins "grab" proteins in the ECM. As **Figure 8.5** shows, actin filaments in the cytoskeleton are connected to transmembrane proteins called **integrins**. The integrins bind to nearby proteins in the ECM named **fibronectins**, which in turn bind to collagen fibers. This direct linkage between the cytoskeleton and ECM is critical. In addition to keeping individual cells in place, it helps adjacent cells adhere to each other via their common connection to the ECM.

The cytoskeleton-ECM linkage is also important in the development of cancer. When the cell-ECM connection is lost—through mechanisms that are not well understood—cells that are growing in an uncontrolled fashion can leave the site

(a) Collagen molecules are made of three chains that wind around each other.

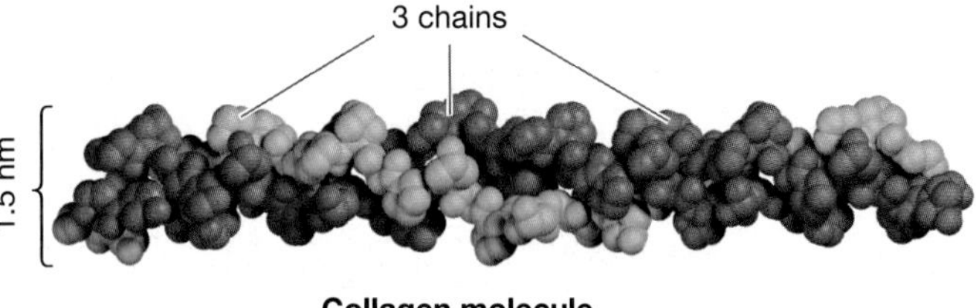

(b) Collagen fibrils in the extracellular matrix

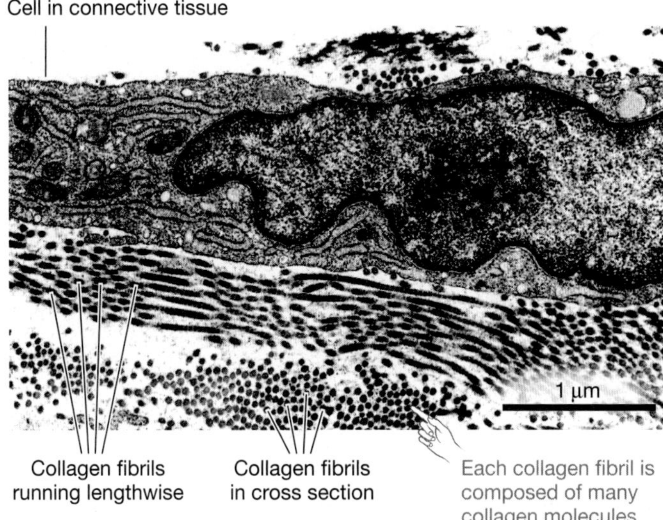

FIGURE 8.4 The Extracellular Matrix Is a Fiber Composite. (a) Although several types of fibrous proteins are found in the ECM of animal cells, the most abundant is collagen. Groups of collagen molecules coalesce to form collagen microfibrils, and bundles of microfibrils link up to form collagen fibers. **(b)** A cross section of ECM from monkey cartilage tissue, showing a cell surrounded by abundant ECM. The spaces between the collagen fibers are filled with gelatinous polysaccharides.

QUESTION In part (b), collagen fibrils appear lengthwise and in cross section. How does this observation relate to the diagram in part (a) of Figure 8.3?

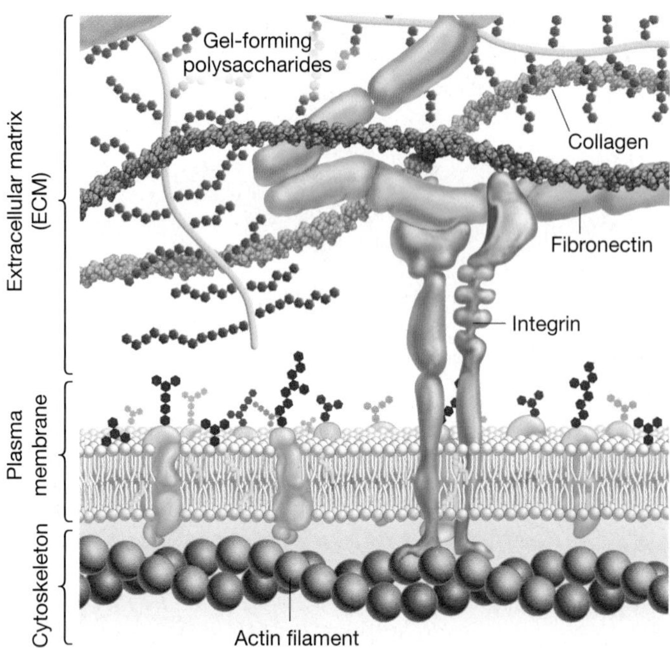

FIGURE 8.5 The Extracellular Matrix Connects to the Cytoskeleton.

EXERCISE Circle the direct physical connection among actin filaments, integrins, fibronectin, and collagen fibers.

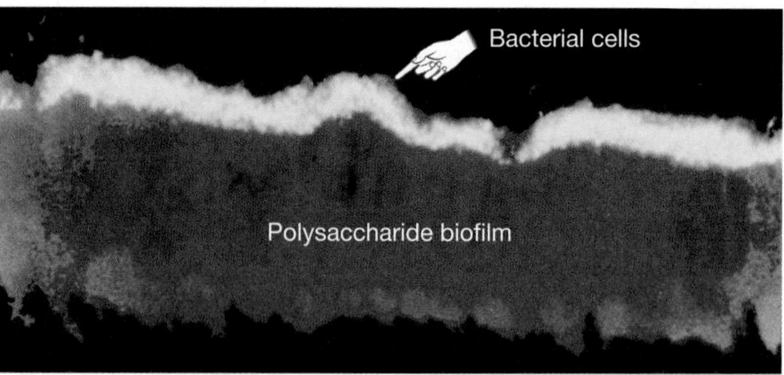

FIGURE 8.6 Groups of Unicellular Organisms May Secrete Biofilms. The yellow band consists of *Pseudomonas aeruginosa* cells—a common soil bacterium that also causes infections in cystic fibrosis patients. The red layer is a biofilm secreted by the cells that helps them stick to the substrate.

of a tumor and begin migrating throughout the body. Eventually these cells settle in a new region of the body and seed the formation of new tumors. This increases the severity of disease, because cancerous cells begin growing at many sites instead of just one. **Box 8.1** details other types of maladies that result from breakdown of the ECM.

Check Your Understanding

If you understand that...

- Most cells secrete a layer of structural material that stiffens the cell and helps define its shape. The extracellular material is usually a fiber composite—a combination of cross-linked filaments surrounded by a ground substance.

You should be able to...

1) Diagram the general structure of a fiber composite.
2) Compare and contrast the molecular composition of a plant cell wall and the ECM of animal cells.

8.2 How Do Adjacent Cells Connect and Communicate?

To understand a cell fully, it is important to appreciate it as a social entity. Cells continually interact with other cells. This is true even for unicellular organisms, which usually live in habitats packed with other unicellular species. For example, the 500 species of bacteria and archaea that inhabit your mouth compete with each other for space and nutrients. Like many other prokaryotes that are adapted for life on a solid surface, certain oral bacteria secrete a hard, polysaccharide-rich substance that accumulates outside their cell walls. This substance creates a biofilm that encases the cells and attaches them to surfaces. Biofilms that form in the mouth are termed dental plaque, and they contain an array of species. In some cases, biofilm formation is actually a product of cooperation among members of the same species. When cells of the bacterium *Pseudomonas aeruginosa* settle on a lung cell, for instance, they secrete a signaling molecule that recruits other *P. aeruginosa* cells to the site. The cells then collectively secrete a biofilm that attaches them to the surface of the lung and protects them from attack by your immune system cells or antibiotics (**Figure 8.6**).

Although unicellular species may live in close proximity and even communicate with each other, they do not make physical connections regularly. Physical connections between cells are the basis of **multicellularity** (**Figure 8.7**). The billions of cells that make up a rose bush, a mushroom-forming fungus, or a mountain gorilla must be held together somehow. And because cells in multicellular species are specialized for particular tasks, the cells must be able to communicate with each other to function as an integrated whole. It is logical, then, that rose and fungus and gorilla cells are organized into functional units, called tissues. **Tissues** consist of groups of similar cells that perform a similar function. In your body, individual muscle cells are grouped into muscle tissue that contracts and relaxes to make movement possible. Several tissues, in turn, combine to make up the integrated structures called **organs**. Organs perform specialized functions such as reproduction, digestion, and support. Flowers, roots, and stems are organs; your gonads, small intestine, and bones also are organs.

A look at the phylogeny in **Figure 8.8** should convince you that multicellularity has evolved numerous times, in an array of independent groups. (For help interpreting this tree, see

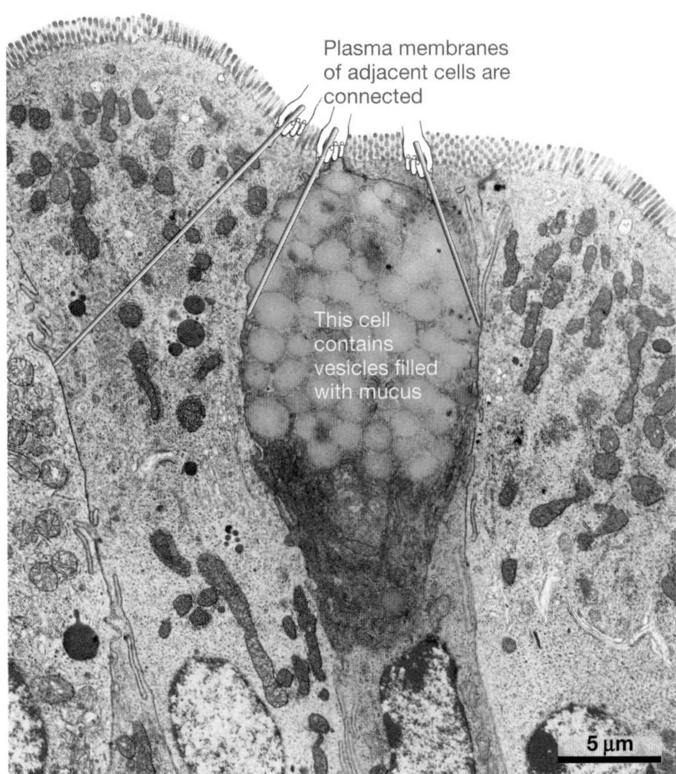

FIGURE 8.7 In Multicellular Organisms, Cells Are Connected. A cross section through cells that line the small intestine. Notice that cells are physically connected and that even adjacent cells can have different structures and functions. The cell in the middle secretes mucus that neutralizes stomach acid. Secretions from the cells to the left and right help in the digestive process.

BioSkills 2.) Based on this observation, it is logical to predict that brown algae, land plants, animals, and other multicellular groups do not share the same types of intercellular (literally, "between-cell") connections and communication mechanisms. Instead of inheriting the same structures and processes for cell-cell interactions from a common ancestor that was multicellular, each multicellular group independently evolved ways of organizing their tissues and organs.

Many of the structures responsible for cell-cell adhesion and communication in these groups were first described in the 1950s, when transmission electron microscopy made it possible to examine tissues at high magnification. In the decades that followed, researchers made exciting progress in understanding the molecular nature of cell-cell attachments and openings. Let's look first at the structures that physically attach adjacent cells to each other, then at the openings that allow nearby cells to exchange materials and information.

Cell-Cell Attachments

The structures that hold cells together vary among multicellular organisms. To illustrate this diversity, consider the intercellular connections observed in the best-studied groups of organisms: plants and animals. Transmission electron microscopy showed that a distinctive layer of material joins plant cells together. (For an introduction to electron microscopy, see **BioSkills 8.**) The same tool allowed scientists to examine the structure of cell-cell attachment points in animal cells.

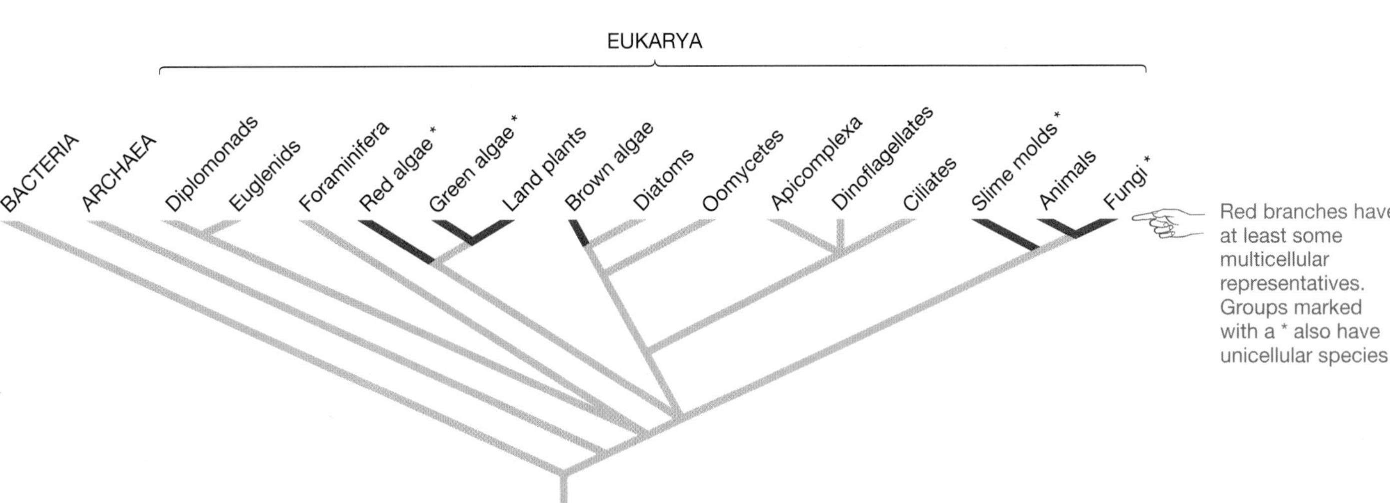

FIGURE 8.8 Multicellularity Has Evolved Numerous Times. A phylogeny of eukaryotes, indicating lineages where multicellular forms exist. Based on these data, biologists conclude that multicellularity evolved many times independently.

EXERCISE Suppose that multicellularity arose just once. If this hypothesis is correct, put a mark on the tree to indicate where multicellularity evolved.

QUESTION Why is it more logical to infer that multicellularity arose many times, instead of just once?

The extracellular space between adjacent plant cells comprises three layers (**Figure 8.9**). The primary cell walls of adjacent cells sandwich a central layer designated the middle lamella, which consists primarily of gelatinous pectins. Because this gel layer is continuous with the primary cell walls of the adjacent cells, it serves to glue them together. The two cell walls are like slices of bread; the middle lamella is like a layer of peanut butter. If enzymes degrade the middle lamella, as they do when flower petals and leaves detach and fall, the surrounding cells separate.

In many animal tissues, integrins connect the cytoskeleton of each cell to the extracellular matrix (see Section 8.1). A middle-lamella-like layer of gelatinous polysaccharides runs between adjacent animal cells, so cytoskeleton-ECM connections help hold individual animal cells together. In addition, in certain animal tissues the polysaccharide glue is reinforced by cable-like proteins that span the ECM to connect adjacent cells. These structures are particularly important in **epithelia** (singular: **epithelium**)—tissues that form external and internal surfaces. Epithelial cells form layers that separate organs and other structures. These epithelial layers must be sealed to prevent mixing of solutions from adjacent organs or structures. A variety of cell-cell attachment structures exist in epithelia and other tissues. Here, however, we'll analyze just two: tight junctions and desmosomes.

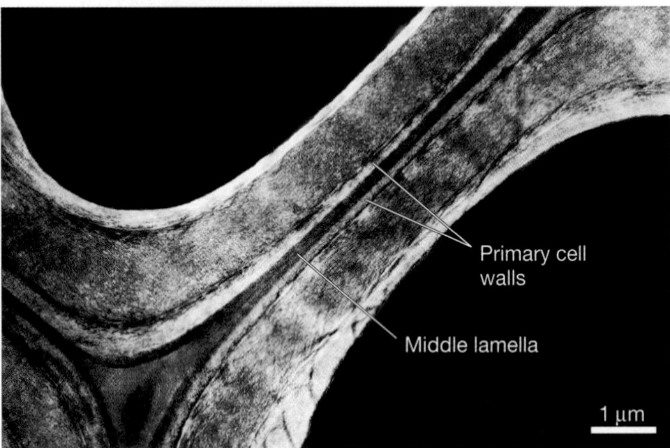

FIGURE 8.9 The Middle Lamella Connects Adjacent Plant Cells. The middle lamella contains gelatinous polysaccharides called pectins.

Tight Junctions The electron micrograph in **Figure 8.10a** shows a cross section through epithelial cells that are linked by tight junctions. A **tight junction** is a cell-cell attachment composed of specialized proteins in the plasma membranes of adjacent animal cells. As the drawing in **Figure 8.10b** indicates, these proteins line up and bind to one another. The resulting structure resembles quilting, with the proteins acting as stitches.

Because tight junctions form a watertight seal, this type of junction is common in cells that form a barrier, such as the cells lining your stomach and intestines. There, they prevent the ions and molecules in your gut contents from leaking between stomach cells or intestinal cells and from diffusing into your body. Instead, only selected nutrients enter the cells. These ions and molecules are admitted via specialized transport proteins and channels in the plasma membrane.

Although tight junctions are tight, they are also variable and dynamic. For example, the tight junctions in the cells lining your bladder are much tighter than tight junctions in the cells lining your small intestine, because they consist of different proteins. As a result, small ions can pass through the surface of the small intestine much more easily than they can

BOX 8.1 What Happens When the Extracellular Matrix Is Defective?

Defects in ECM have serious consequences. Consider, for example, the effects of collagen breakdown. One of collagen's most common constituents is the amino acid proline. When collagen fibers first assemble in the ECM, many of the proline residues participate in a reaction that forms hydroxyproline. This reaction depends on ascorbic acid, also called vitamin C. If vitamin C is lacking in the diet, hydroxyproline formation slows and collagen fibers begin to weaken or disintegrate. The ECM surrounding capillaries—the smallest blood vessels in the body—is particularly sensitive to this problem. When collagen fibers in the ECM of capillary cells weaken, the structure as a whole begins to lose its integrity. Blood cells begin leaking out, forming bruises. If vitamin C deficiency continues, capillaries weaken to the point where extensive internal bleeding occurs. The potentially fatal disease called scurvy results. Scurvy is readily cured by eating citrus fruits or other foods rich in vitamin C.

Skin is also sensitive to collagen damage. Skin cells are layered over a thick, collagen-rich ECM called the **basal lamina**. In response to intense sunlight, skin cells produce and secrete the enzyme metalloproteinase-1, which causes collagen fibers to break down. As a result, people who work for long periods in bright sunlight or who cultivate a tanned skin are almost certain to experience photoaging—changes in skin caused by the loss of collagen.

Defects in other ECM components can also cause problems. Marfan syndrome develops when a genetic defect prevents people from manufacturing the normal form of the protein fibrillin. Along with elastin, fibrillin is a key component of the elastic fibers in the ECM. People with Marfan syndrome tend to have exceptionally long fingers and be extremely tall and thin. They are also susceptible to severe heart problems. Biomedical researchers have hypothesized that Abraham Lincoln may have had Marfan syndrome.

(a) Electron micrograph of a tight junction

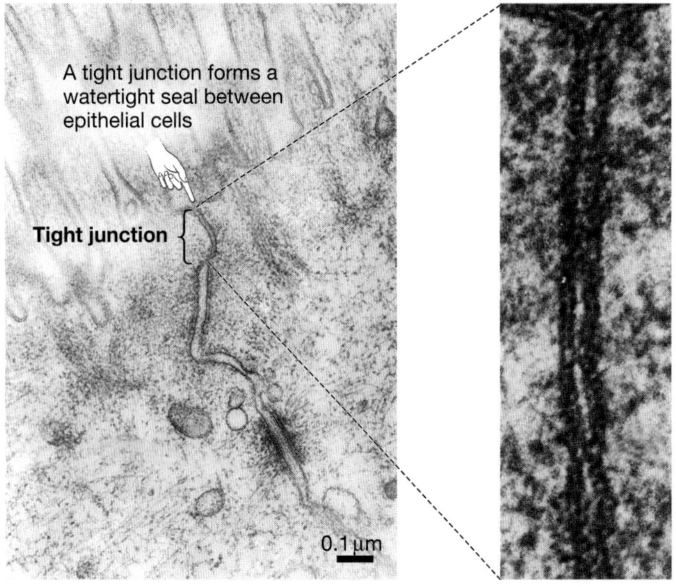

(b) Three-dimensional view of a tight junction

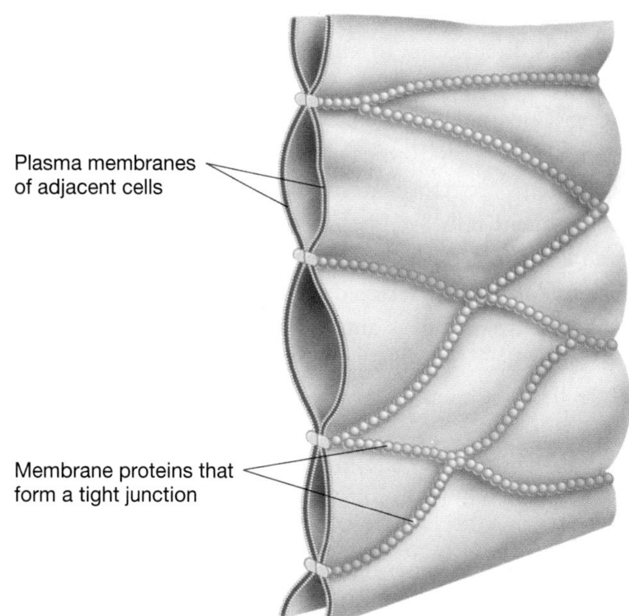

FIGURE 8.10 In Animals, Tight Junctions May Form a Seal between Adjacent Cells. (a) A cross section through adjacent cells that are attached by a tight junction. **(b)** Proteins in the membranes of each cell are arranged in a quilted pattern and bind to one another to form the junction.

EXERCISE Add lines connecting the structures in part (a) to the labels in part (b).

pass through the surface of the bladder—helping you absorb ions in your food and eliminate them in your waste. The structures are dynamic because they can be loosened in some cases to permit more transport between epithelial cells—for example, in the small intestine after a meal—and then "retightened" later.

Desmosomes **Figure 8.11a** illustrates another type of cell-cell connection found in animals: desmosomes. **Desmosomes** are particularly common cell-cell attachments in epithelial cells and in certain types of muscle cells. The structure and function of a desmosome are analogous to the rivets that hold pieces of sheet metal together. As **Figure 8.11b** indicates, though, desmosomes

(a) Micrograph of desmosome

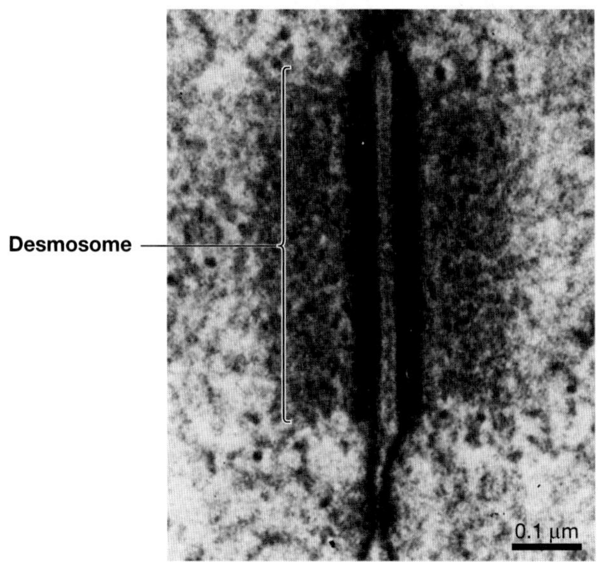

(b) Three-dimensional view of desmosome

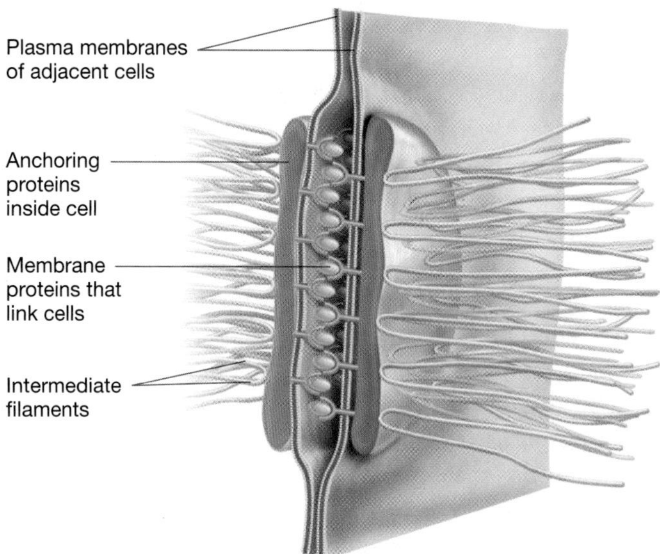

FIGURE 8.11 Adjacent Animal Cells Are Linked by Desmosomes, Which Bind Cytoskeletons Together.
(a) A cross section through a desmosome, showing the connections between adjacent cells and the intermediate filaments that link the desmosome to the cytoskeleton of each cell. **(b)** The major components of a desmosome.

EXERCISE Add lines connecting the structures in part (a) to the labels in part (b).

EXERCISE Label the desmosome in Figure 8.10a, left micrograph.

are extremely sophisticated connections. At their heart are proteins that form a physical link between the cytoskeletons of the adjacent cells. In addition to binding to each other, these proteins bind to larger proteins that anchor intermediate filaments in the cytoskeletons of the two cells. In this way, desmosomes bind together the cytoskeletons of the adjacent cells.

The proteins that form desmosomes may be involved in attaching animal cells long before a desmosome is complete, however. Early in the development of an animal, an embryo's first tissues form when selected cells begin to adhere to each other. Mature desmosomes and adult tissues and organs form much later. Which proteins are involved in these most basic types of cell-cell adhesion, early in development?

Selective Adhesion Long before electron micrographs revealed the presence of desmosomes, biologists realized that some sort of molecule must bind animal cells to each other. This insight grew out of experiments conducted on the cells of sponges in the early 1900s. Sponges are aquatic animals, and the sponge species used in this study consists of just two basic types of cells. When a biologist treated adult sponges with chemicals that made the cells dissociate from each other and separate, the result was a jumbled mass of individual and unconnected cells. But when normal chemical conditions were restored, the cells gradually began to move and stick to other cells. With time, cells of each type began to combine and aggregate, because they adhered to cells of the same tissue type. This phenomenon came to be called **selective adhesion**. In this experiment, the cells eventually re-formed complete and functional adult sponges.

In an even more dramatic experiment, the researcher dissociated the cells of adult sponges from two differently pigmented sponge species and randomly mixed them together in a culture dish. As **Figure 8.12** shows, the cells eventually sorted themselves out into distinct aggregates containing cells from only one species and from only one cell type. These results implied that the cells had some way of physically linking to each other and that the linkage was specific to the species and cell type involved. Based on these observations, researchers hypothesized that there must be some sort of molecule on the surface of animal cells that attaches only to cells of the same species and cell or tissue type.

Experiment

Question: Do animal cells adhere selectively?

Hypothesis: Cells of the same type and from the same species have a mechanism for selectively adhering to each other.

Null hypothesis: Cells do not adhere or they adhere to each other randomly, not selectively.

Experimental setup:

1. Start with two adult sponges of different species.

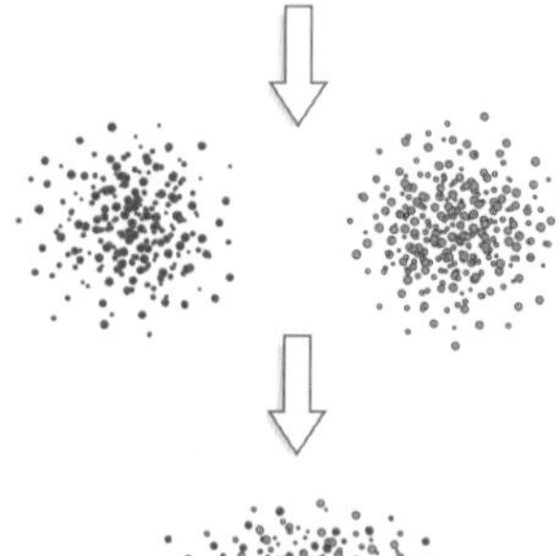

2. Use a chemical treatment to dissociate the cells.

3. Mix cells from the two species.

Prediction: Cells of the same cell type will adhere to one another, and cells from the same species will adhere to one another.

Prediction of null hypothesis:

Results:

Cells spontaneously reaggregate by cell type and species.

Conclusion: Cells of the same cell type and species have specific adhesion molecules or mechanisms.

FIGURE 8.12 Do Animal Cells Adhere Selectively?

● **EXERCISE** In the space provided, fill in the prediction made by the null hypothesis.

The Discovery of Cadherins What is the molecular nature of selective adhesion? The initial hypothesis, proposed in the 1970s, was that certain membrane proteins might be involved. The idea was that if different types of cells produce different types of adhesion proteins in their membranes, the molecules might interact in a way that anchors cells of the same type to each other.

This hypothesis was tested through experiments that relied on molecules called antibodies. An **antibody** is a protein that binds specifically to a section of another protein. Because they stick to certain proteins so specifically, antibodies can be used to block certain portions of proteins or mark proteins so they can be seen. To test the hypothesis that cell-cell adhesion takes place via interactions between membrane proteins, researchers pursued the following strategy (**Figure 8.13**):

1. Isolate the membrane proteins from a certain cell type. Create pure preparations of each protein.

2. Inject one of the membrane proteins into a rabbit. The rabbit's immune system cells respond by creating antibodies to the membrane protein. Purify antibodies that bind tightly to the injected membrane protein. Repeat this procedure for the other membrane proteins that were isolated. In this way,

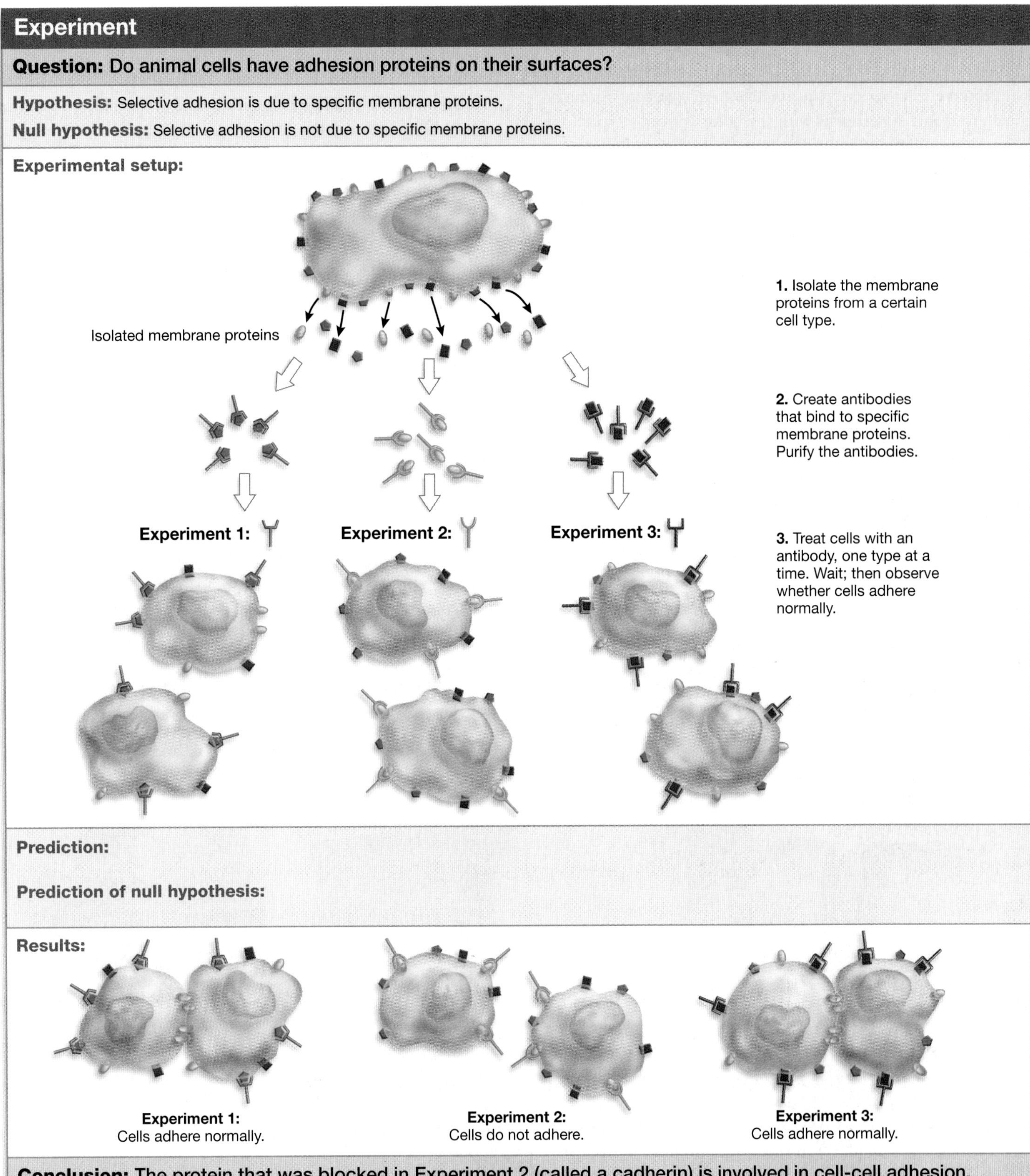

FIGURE 8.13 Do Animal Cells Have Adhesion Proteins on Their Surfaces?

EXERCISE Fill in the prediction made by each hypothesis.

obtain a large collection of antibodies—each of which binds specifically to one (and only one) type of membrane protein.

3. Add one antibody type to the mixture of dissociated cells. Observe whether the cells are able to reaggregate normally. Repeat this experiment with the other antibody types, one type at a time. (Three different antibodies are shown in Figure 8.13, so three experimental treatments are required.)
4. If treatment with a particular antibody prevents the cells from attaching to each other, the antibody is probably bound to an adhesion protein. The hypothesis is that the interaction between antibody and adhesion protein blocks the adhesion protein and keeps it from functioning properly. In effect, the antibody "shakes hands" with the adhesion protein, which prevents the adhesion protein from "shaking hands" with other adhesion proteins to attach cells.

Based on studies such as this, researchers have identified several major classes of cell adhesion proteins. The molecules in desmosomes belong to a group that came to be named **cadherins**. Further research has shown that each major type of cell in the body has a different type of cadherin in its plasma membrane, and each type of cadherin can bind only to cadherins of the same type. In this way, cells of the same tissue type attach specifically to one another.

To summarize: Animal cells attach to each other in a selective manner because different types of cell adhesion proteins can bind and rivet certain cells together. Cadherins provide the physical basis for selective adhesion in many cells and are

(a) Plasmodesmata create gaps that connect plant cells.

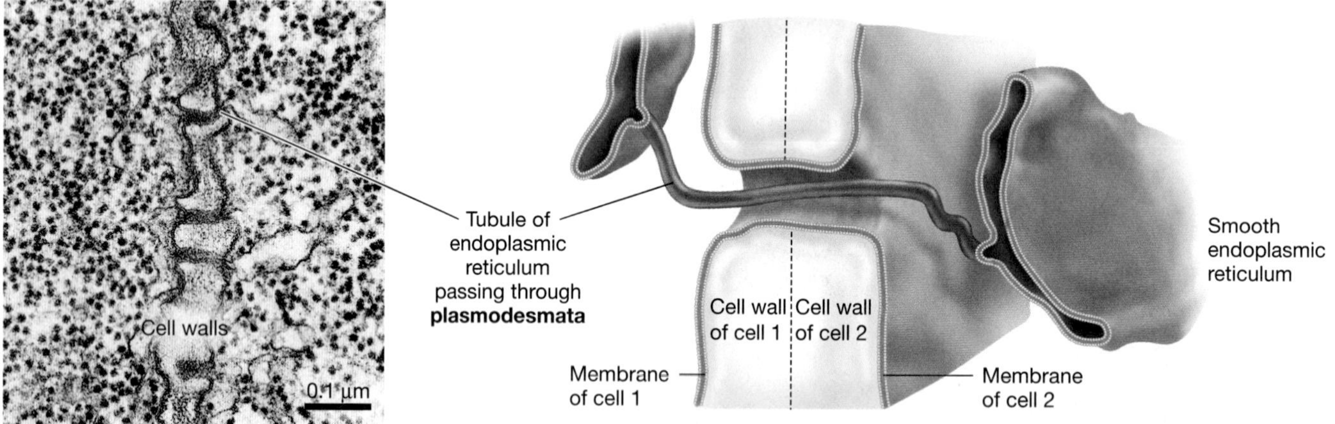

(b) Gap junctions create gaps that connect animal cells.

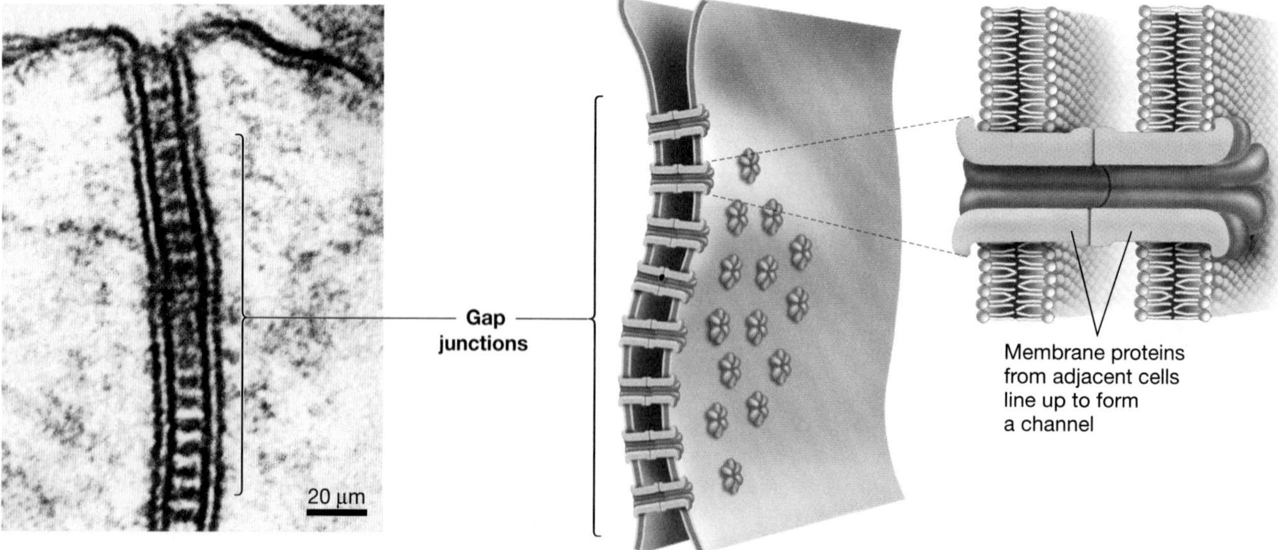

FIGURE 8.14 Adjacent Plant Cells and Adjacent Animal Cells Communicate Directly. (a) Plasmodesmata are holes in plasma membranes and cell walls of adjacent plant cells. A tubule of smooth ER runs through the hole. **(b)** Gap junctions are collections of protein-lined holes in the membranes of adjacent animal cells.

a critical component of the desmosomes that join mature cells. If you understand cell-cell attachments, you should be able to predict (1) whether molecules can pass between adjacent cells through middle lamellae and tight junctions, and (2) what would happen if you treated cells in a developing frog embryo with a molecule that blocked a cadherin observed in muscle tissue.

Cell-Cell Gaps

The presence of a middle lamella, continuous ECM, tight junctions, desmosomes, and cadherins bind adjacent cells to each other. How do they communicate? In both plants and animals, direct connections between cells in the same tissue help the cells work in a coordinated fashion. The result emerged from careful analyses of electron micrographs.

In plants, gaps in cell walls create direct connections between the cytoplasm of adjacent cells. At these connections, named **plasmodesmata** (singular: **plasmodesma**), the plasma membrane and the cytoplasm of the two cells are continuous. Smooth endoplasmic reticulum (ER) runs through these holes (**Figure 8.14a**). Growing evidence suggests that plasmodesmata also contain proteins that regulate the passage of specific proteins, making the connections similar in function to the nuclear pore complex introduced in Chapter 7. At least some of the proteins that are transported through plasmodesmata are involved in coordinating the activity of adjacent cells. Plasmodesmata are communication portals.

In most animal tissues, adjacent cells are connected by structures called **gap junctions**. The key feature of gap junctions is the specialized proteins that create channels between cells (**Figure 8.14b**). These channels allow water, ions, and small molecules such as amino acids, sugars, and nucleotides to move between adjacent cells. The flow of small molecules can help adjacent cells communicate and coordinate their activities by allowing the rapid passage of regulatory ions or molecules. In the muscle cells of your heart, for example, a flow of ions through gap junctions acts as a signal that coordinates contractions. Without this cell-cell communication, a normal heartbeat would be impossible.

In effect, adjacent plant cells and adjacent animal cells share most or all of the ions and small molecules in their cytoplasm but retain their own organelles, proteins, and nucleic acids. The existence of cell-cell gaps makes it possible for cells that are immediate neighbors to communicate with each other efficiently. A tissue can act as an integrated whole if its component cells are connected by gaps in their extracellular material and in their plasma membranes (**Figure 8.15**).

How do more distant cells in a multicellular organism communicate? For example, suppose that leaf cells in a maple tree are being attacked by caterpillars or that the muscle cells in your arm are exercising so hard that they begin to run low on sugar. How do these cells signal tissues or organs elsewhere in the body to release materials that might be needed to fend off caterpillars or exhaustion? Distant cell communication is the subject of Section 8.3.

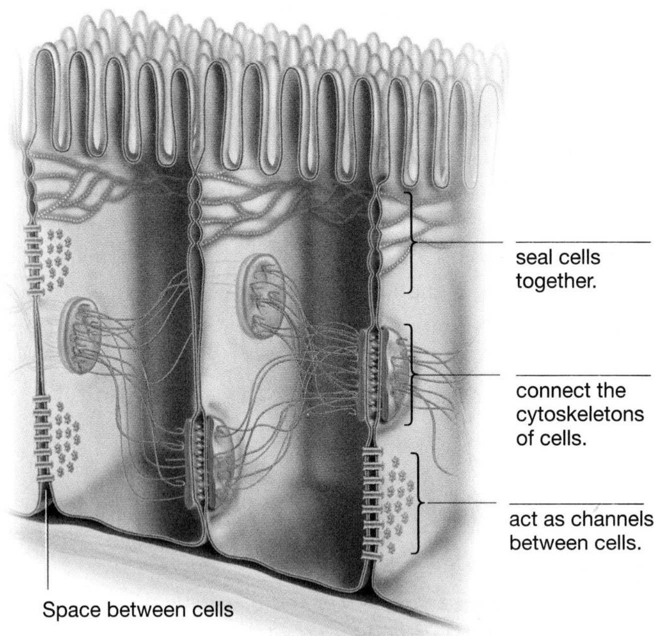

FIGURE 8.15 An Array of Structures Is Involved in Cell-Cell Adhesion and Communication.

EXERCISE In the blanks on the right, write the name of each type of cell-cell attachment.

Check Your Understanding

If you understand that...

- In plants and animals, adjacent cells are physically connected and communicate with each other through openings in their plasma membranes.

You should be able to...

1) Compare and contrast the structure and function of the middle lamella of plants and the tight junctions and desmosomes of animals.
2) Describe the structure and function of plasmodesmata and gap junctions.
3) Explain the molecular mechanism of selective adhesion.

8.3 How Do Distant Cells Communicate?

Humans, sequoia trees, and other large multicellular organisms typically contain trillions of cells, hundreds of different organs, and dozens of different tissue types. For all of these parts to work together as a unified whole, information has to be exchanged. As an example, suppose that cells in your brain sense that you are becoming dehydrated. Brain cells can't do much about the water you lose during urination, but kidney cells can. In response to dehydration, certain brain cells release a signaling molecule that travels to the kidneys. The arrival of the signal changes the activity of cells that control the amount of water being excreted in urine. The response to the signal leads to less water being lost and more water being retained—an important aspect of fighting dehydration.

Thanks to long-distance signals, the activities of cells, tissues, and organs in different parts of a multicellular body are coordinated. Even as you read this, in fact, molecules that carry information about conditions throughout your body are traveling through your bloodstream and arriving at cells from your head to your toes.

How do cells receive and respond to signals that arrive from distant cells? Let's take a step-by-step approach to answering this question, starting with a quick look at the nature of long-distance chemical signals.

Hormones Are Long-Distance Messengers

Intercellular (or "cell-cell") signals are chemical messengers. Several distinctive types of cell-cell signals exist, but here we'll focus on hormones—the molecules involved in long-distance signaling. A **hormone** is an information-carrying molecule that is secreted from a cell, circulates in the body, and acts on target cells far from the original cell that sent the signal. Hormones are usually small molecules and are typically present in minute concentrations. Even so, they have a large impact on the activity of target cells and the condition of the body as a whole. Hormones are like a fleeting scent or whispered phrase from someone you are attracted to—a tiny signal, but one that makes your cheeks flush and your heart pound.

As **Table 8.1** indicates, hormones have a wide array of effects. In humans they are responsible for regulating most aspects of body chemistry and triggering sexual maturation and sexual activity. In plants they direct most aspects of growth and development and mobilize defenses against attacks by insects and other threats.

Table 8.1 should also convince you that hormones vary widely in chemical structure. The important point about a hormone's structure, though, is not whether it is a gas or peptide or glycoprotein, but whether it is lipid soluble. Steroids and other types of lipid-soluble hormones diffuse across the plasma membrane and enter the cytoplasm of their target cells. Hydrophilic hormones, in contrast, do not cross the plasma membrane. This is important because a hormone has to be recognized to have an effect on a target cell. Where does this recognition and reception step occur—inside the cell or outside?

TABLE **8.1 Hormones Have Diverse Structures and Functions**

Hormone Name	Chemical Structure	Function of Signal
Ethylene	C_2H_4 (a gas)	Stimulates fruit ripening, regulates aging
Insulin	Protein, 51 amino acids	Stimulates glucose uptake in animal bloodstream
Systemin	Peptide, 18 amino acids	Stimulates plant defenses against herbivores
Estrogen	Steroid	Stimulates development of female characteristics in animals
Brassinosteroids	Steroid	Stimulates plant cell elongation
Prostaglandins	Modified fatty acid	Perform a variety of functions in animal cells
Thyroxine (T4)	Modified amino acid	Regulates metabolism in animals
FSH	Glycoprotein	Stimulates egg maturation, sperm production in animals
Auxin	Small organic compound	Signals changes in long axis of plant body

Signal Reception

Hormones and other types of cell-cell signals deliver their message by binding to receptor molecules. When the molecule that carries the message: "We're getting dehydrated—conserve water" reaches your kidneys, only certain cells respond because only certain cells have the appropriate receptor. The presence of an appropriate receptor dictates which cells will be able to respond to a particular hormone. Bone and muscle cells don't respond to the "conserve water" message, because they don't have a receptor for it.

The association between signal and receptor brings up another important point about long-distance signaling between cells. In many cases, cells in a wide array of tissues and organs respond to the same signal. This is possible because they all have the appropriate receptor. For example, if you are startled by a loud noise, cells in your adrenal glands secrete a hormone that carries the message: "Get ready to fight or run." In response, your heart rate increases, your breathing rate increases, and cells in your liver release sugars that your muscles can use to power rapid movement. This is the basis of an "adrenalin rush." Heart, lung, and liver cells respond to adrenalin because they have an adrenalin receptor. The general message here is that long-distance signaling coordinates the activities of cells throughout a multicellular organism.

If receptors are the key to a cell's response, how do they work? **Signal receptors** are proteins that change conformation—meaning, overall shape—or activity after binding to a signaling molecule. The signal receptors that respond to steroid hormones are located inside the cell, because steroid hormones are lipid soluble and thus readily diffuse through the plasma membrane. But most signal receptors are located in the plasma membrane because most hormones are not lipid soluble. No matter where they are located, it's important to recognize two additional points about signal receptors:

- Receptors are dynamic. The number of receptors in a particular cell may decline if hormonal stimulation occurs at high levels over a long period of time. The ability of a receptor molecule to bind tightly to a hormone may also decline in response to intensive stimulation. As a result, the sensitivity of a cell to a particular hormone may change over time.

- Receptors can be blocked. The drugs called beta-blockers, for example, bind to certain receptors for the hormone adrenalin (adrenalin is also called epinephrine). When adrenalin binds to receptors in heart cells, the cells are stimulated to contract. So if a physician wants to reduce heart cell contraction as a way to lower heart rate and thus blood pressure, she is likely to prescribe a beta-blocker.

The most important and most general characteristic of signal receptors, though, is that their physical conformation changes when a hormone binds to them. This is a critical event in cell-cell signaling. The change in receptor structure means that the signal has been received. It's like throwing an "on" switch. What happens next?

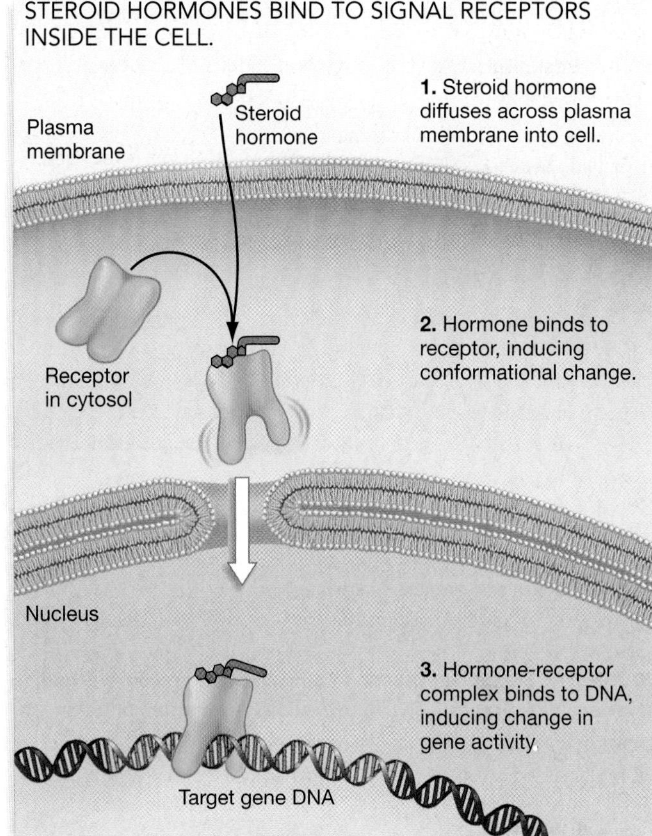

FIGURE 8.16 Some Cell-Cell Signals Enter the Cell and Bind to Receptors in the Cytoplasm. Because they are lipids, steroid hormones can diffuse across the plasma membrane and bind to signal receptors inside the cell. The hormone-receptor complex is transported to the nucleus and binds to genes, changing their activity.

Signal Processing

Once a cell receives a signal, something has to happen to initiate the cell's response. In some cases the "something" occurs directly. For example, steroid hormones such as testosterone and estrogen diffuse through the plasma membrane and enter the cytoplasm, where they bind to a receptor protein. The hormone-receptor complex is then transported to the nucleus, where it triggers changes in the genes being expressed in the cell (**Figure 8.16**). Early in human development, the arrival of testosterone in cells leads to the expression of genes that trigger the development of male reproductive organs. (In the absence of testosterone, female reproductive organs develop.) Later in life, a surge of testosterone triggers the changes associated with puberty in boys, while the arrival of estrogen molecules spurs the development of adult female characteristics in girls.

The general point here is that when lipid-soluble hormones bind to receptors located inside the cell, they usually trigger a change in the cell's activity directly. The change in activity might involve certain types of genes being activated or shut down, or certain pumps in the plasma membrane being stimulated. In each case, the hormone-receptor complex directly triggers the change by binding to the genes or to the pumps.

The effects of hormones that *cannot* diffuse across the plasma membrane and enter the cytoplasm are not nearly so direct. It is not possible for hormones that bind to receptors on the plasma membrane to directly change the activity of pumps or genes or other molecules and structures inside the cell. Instead, the signal that arrives at the surface of the cell has to be changed to an intracellular signal. The knock at the door has to be answered from inside. Thus, a hormone's arrival at the cell surface triggers **signal transduction**—the conversion of the signal from one form to another. A long and often complex series of events ensues, collectively called a signal transduction pathway.

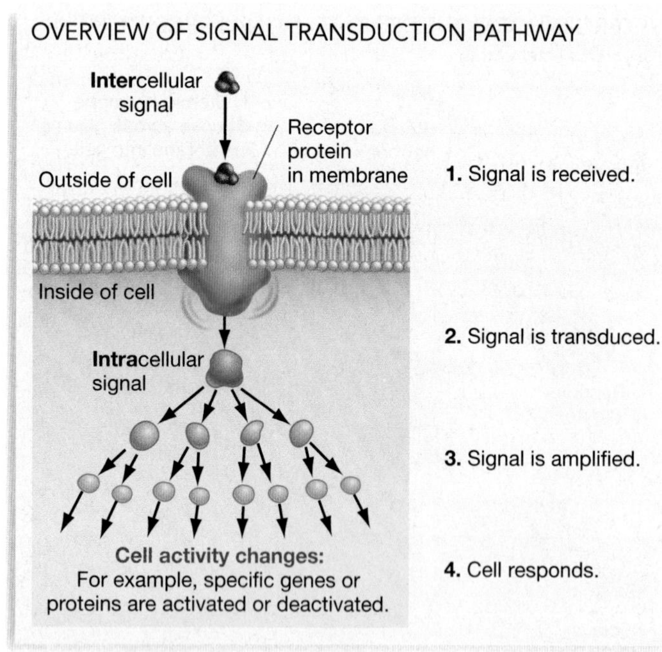

FIGURE 8.17 Signal Transduction Converts an Intercellular Signal to an Intracellular Signal. Signal transduction is a multistep process.

QUESTION What do the word roots inter– and intra– mean?

Signal transduction is a common occurrence in everyday life. For example, the e-mail messages you receive are transmitted from one computer to another over cables or wireless transmissions. These types of electronic or electromagnetic signals can be transmitted efficiently over long distances. But to you, they are meaningless. Software in your computer has to transduce the signals to letters and words on the screen—into a form that you can understand and respond to. If you own a portable music player, songs that are encoded as 1s and 0s in a digital file are transduced to sound waves that you listen to. Reading the 1s and 0s directly would be meaningless to you.

Signal transduction pathways work the same way. In a cell, signal transduction converts an extracellular signal to an intracellular signal (**Figure 8.17**). As in an e-mail transmission, a signal that is easy to transmit is converted to a signal that is easily understood and that triggers a response. There is another important aspect of signal transduction, though, in addition to this change in form. When a hormone arrives at the cell surface, the message it transmits is amplified as it changes form. Recall that hormones are present in miniscule concentrations but trigger a large response from cells. Signal amplification is one reason this is possible. In everyday life, the amplifier in a portable music player performs an analogous function. Once it is amplified, a tiny sound signal can get a whole roomful of people dancing.

In cells, signal transduction occurs at the plasma membrane; amplification occurs inside. One major type of signal transduction system consists of membrane channels that open to allow a flow of ions into the cell, changing the electrical properties of the membrane. This type of signal transduction is analyzed in detail in Chapter 45. Here let's focus on the other major types of signal transduction systems. Each involves a distinctive type of membrane protein: (1) G proteins and (2) enzyme-linked receptors. G proteins work by triggering the production of an intracellular or "second" messenger. Enzyme-linked receptors work by triggering the activation of a series of proteins inside the cell, through the addition of phosphate groups. There are many types of G proteins and enzyme-linked receptors, but each of these protein families works in the same general way. Let's look at these two signal transduction systems in turn.

G Proteins Many signal receptors span the plasma membrane and are closely associated with peripheral membrane proteins inside the cell called **G proteins**. The *G* is appropriate because these proteins bind guanosine triphosphate (GTP) and guanosine diphosphate (GDP).

GTP is a nucleoside triphosphate that is similar in structure to adenosine triphosphate (ATP). As Chapter 4 noted, nucleoside triphosphates have high potential energy because their three phosphate groups have four negative charges close together. The cluster of charges results in electrons repulsing each other and moving farther from nearby nuclei. And as Chapter 6 noted when discussing the sodium-potassium pump, ATP's high potential energy allows it to make things happen in cells—such as pumping ions across an electrochemical gradient. When ATP binds to a protein, or when a single phosphate group from ATP binds to a protein, the addition of the negative charges makes the protein change shape. Changes in shape produce changes in activity. GTP works the same way. When GTP binds to a protein, the protein changes its structure and function. G proteins are turned on or activated when they bind GTP; they are off or inactivated when a phosphate group drops off to form GDP. An activated G protein triggers the production of a second messenger inside the cell.

To understand how G proteins work, follow the events diagrammed in **Figure 8.18**. As step 1 in the figure shows, the sequence begins when a hormone arrives and binds to a receptor in the plasma membrane. Notice that the receptor is a transmembrane protein, and that it is coupled to a peripheral G protein on the membrane's inner surface. In response to hormone binding, the receptor changes shape. The shape change is a switch that activates its G protein. More specifically, the G protein releases the GDP molecule that was keeping it in an inactive state and binds GTP instead. Compared to GDP, GTP has an additional phosphate group with two negative charges. When GTP is attached, the G protein responds by undergoing

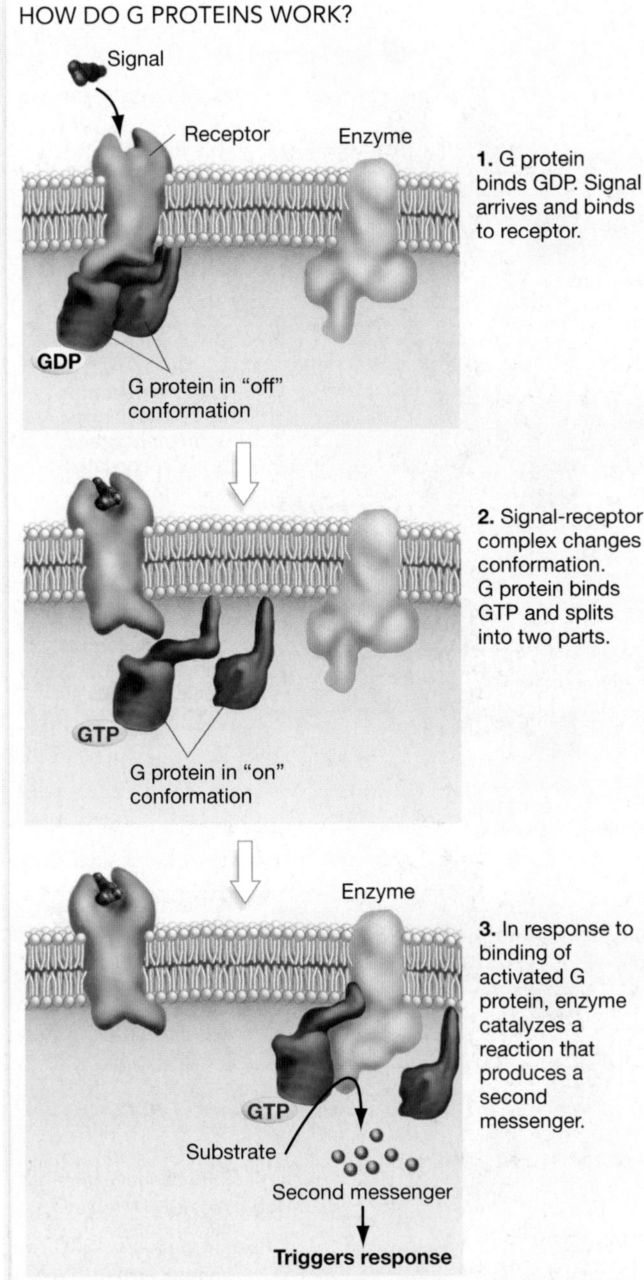

FIGURE 8.18 G Proteins Trigger the Production of a Second Messenger. If a receptor that is coupled to a G protein is activated by a signal, the G protein binds guanosine triphosphate (GTP) and splits into two parts. The change in the G protein activates a nearby enzyme and results in the production of a second messenger.

a radical shape change: It splits into two parts (see step 2). One of these parts activates a nearby enzyme that is embedded in the plasma membrane.

The enzymes that are activated by G proteins catalyze the production of small molecules called second messengers. A "first messenger" is a hormone that arrives at the cell surface; **second messengers** are nonprotein signaling molecules that increase in concentration inside a cell and elicit a response to the first messenger. They are intracellular signals that spread the message carried by the extracellular signal—the hormone.

Second messengers are effective because they are small and because they diffuse rapidly to spread the signal throughout the cell. In addition, they can be produced in large quantities in a short time. This characteristic is important. Because the arrival of a single hormone molecule can stimulate the production of many second messenger molecules, the signal transduction event amplifies the original signal.

Several types of small molecules act as second messengers in cells. **Table 8.2** lists some of the best-studied second messengers and provides an example of how specific types of cells respond to each second messenger. Note that several second messengers activate **protein kinases**—enzymes that activate or inactivate other proteins by adding a phosphate group to them. If you read the table closely, you'll see evidence for two important conclusions: (1) Second messengers aren't restricted to a single role or single cell type—the same second messenger can initiate dramatically different events in different cell types; and (2) it is common for more than one second messenger to be involved in triggering a cell's response to the same extracellular signal.

To make sure that you understand how G proteins and second messengers work, imagine the following movie scene: A spy arrives at a castle gate. The guard receives a note from the spy, but he cannot read the coded message. Instead, the guard turns to the queen. She reads the note and summons a nearby teacher, telling him to send children from the school throughout the castle to warn everyone of approaching danger. ● You should be able to identify which characters in the scene correspond to the second messenger, G protein, hormone, receptor, and enzyme activated by the G protein.

It's also important to appreciate what happens when a signal transduction system like this doesn't work. For example,

TABLE 8.2 **Examples of Second Messengers**

Name	Type of response
Cyclic guanosine monophosphate (cGMP)	Opens ion channels; activates certain protein kinases
Diacylglycerol (DAG)	Activates certain protein kinases
Inositol triphosphate (IP_3)	Opens calcium channels—mobilizes stored calcium ions
Cyclic adenosine monophosphate (cAMP)	Activates certain protein kinases
Calcium ions (Ca^{2+})	Binds to a receptor called calmodulin; Ca^{2+}/calmodulin complex then activates proteins

suppose the guard in the movie scene refuses to acknowledge the arrival of the spy. Would the inhabitants of the castle respond to the advancing army? An analogous defect can cause the condition in humans known as type II diabetes mellitus—a disease that Chapter 43 discusses in detail. In many cases, Type II diabetes develops when the cell-surface receptors for the hormone insulin do not function normally. As a result, cells do not receive the message carried by insulin and do not initiate their normal response—importing glucose from the bloodstream. Thus, glucose concentrations in the blood remain high, causing tissue damage and disease.

Enzyme-Linked Receptors Another major class of signal transduction systems is based on what biologists call enzyme-linked receptors. Instead of activating a nearby G protein, enzyme-linked receptors transduce the signal from a hormone by directly catalyzing a reaction inside the cell. To see how this works and understand the consequences, follow the events illustrated in **Figure 8.19**. The figure focuses on what is perhaps the best-studied group of enzyme-linked receptors: the **receptor tyrosine kinases (RTKs)**.

When a hormone binds to an RTK, the protein responds by forming a dimer (step 1). In this conformation, the receptor has a binding site for a phosphate group from ATP inside the cell (see step 2). Once it is phosphorylated, an RTK becomes an active enzyme. Proteins inside the cell then form a bridge between the activated RTK and a peripheral membrane protein called **Ras,** which functions like a G protein (step 3). The formation of the RTK bridge activates Ras. Specifically, GTP binds to Ras. Once Ras binds GTP, it triggers the phosphorylation of another protein, activating it (step 4). The phosphorylated protein then catalyzes the phosphorylation of still other proteins, which phosphorylate yet another population of proteins (step 5).

This sequence of events is termed a **phosphorylation cascade**. Because each enzyme in the cascade catalyzes the phosphorylation of numerous "downstream" enzymes, the original signal is amplified many times over. One activated receptor tyrosine kinase can trigger the activation of thousands of proteins at the end of the cascade. In most cases, the proteins that take part in a phosphorylation cascade are held in close physical proximity by scaffolding proteins. The presence of this structure increases the speed and efficiency of the reaction sequence.

To make sure that you understand how RTKs and phosphorylation cascades work, imagine that you had one red domino, one black domino, and a large supply of green, blue, yellow, pink, and orange dominos. The red domino represents an RTK, and the black domino represents Ras. Each of the other colors represents a type of protein in a phosphorylation cascade. ● You should be able to explain (1) how you would set up the dominos to simulate a phosphorylation cascade and (2) why tipping the red domino simulates what happens when RTK becomes phosphorylated in response to hormone binding.

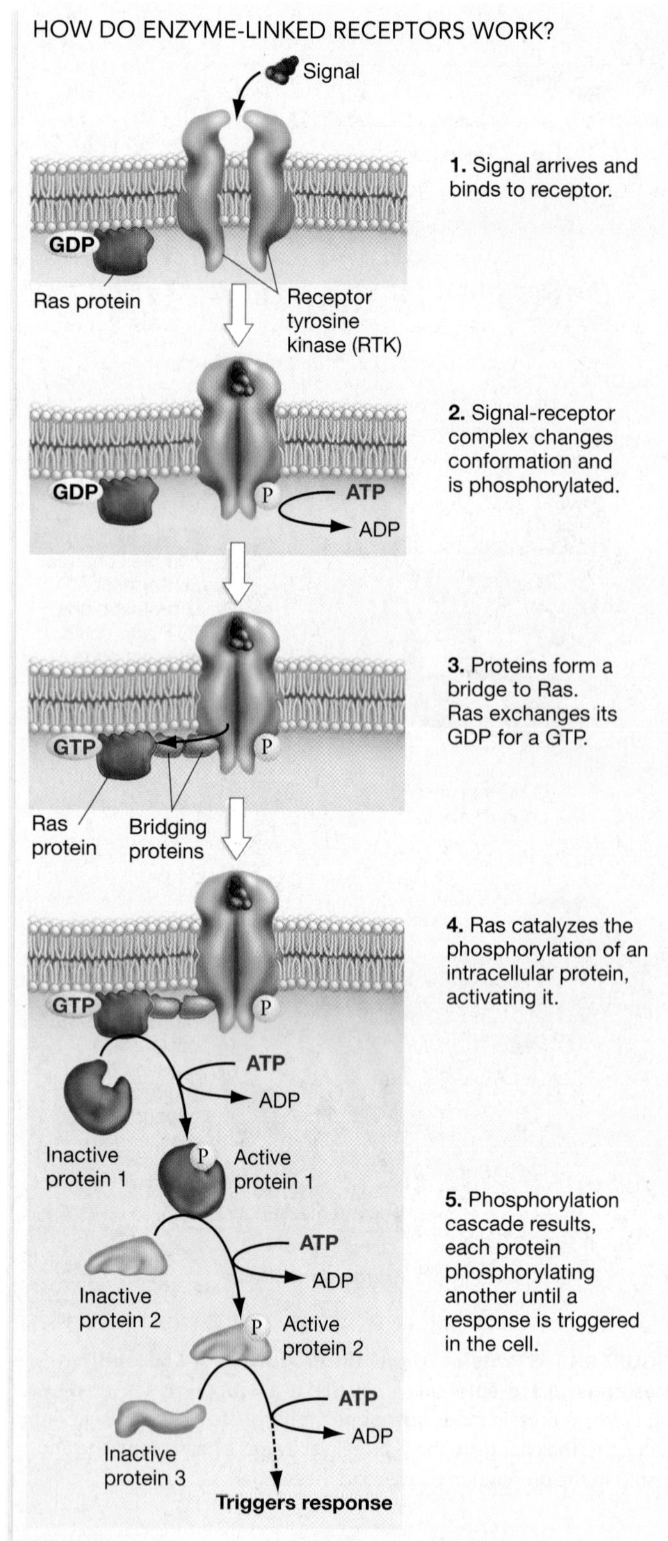

FIGURE 8.19 Enzyme-Linked Receptors Trigger a Phosphorylation Cascade.

Although this discussion linked G proteins with the production of second messengers and enzyme-linked receptors with phosphorylation cascades, it's important to recognize that some G proteins trigger phosphorylation cascades and that some phosphorylation cascades result in the production of second messengers. In addition, it's common for many signaling pathways to exist in the same cell, and for intracellular signals that are part of one pathway to participate in other pathways as well. As a result, signal transduction pathways intersect and connect. Instead of being strictly linear like the pathways illustrated in Figures 8.17 through 8.19, in reality signal transduction pathways form more of a network. This complexity is important: It allows cells to respond to an array of extracellular signals in an integrated way.

To summarize: Many of the key signal transduction events observed in cells occur via G proteins or enzyme-linked receptors. The signal transduction event has two results: (1) It converts an easily transmitted extracellular message into an intracellular message, and (2) it amplifies the original message many times over. Although the hormone at the surface of the cell eventually disengages from its receptor and diffuses away or disintegrates, a "louder" intracellular signal—whether it is a large number of second messenger molecules or a large population of phosphorylated proteins—carries the information throughout the cell and triggers a response.

Signal Response

What is the ultimate response to the messages carried by hormones? The answer varies from signal to signal and from cell to cell. Recall that steroid hormones bind to intracellular receptors and directly change which genes are active in the cell. As a result, the cell's activity changes. Similar types of changes in gene expression can occur in response to second messengers or a cascade of protein phosphorylation events.

In wheat seeds, for example, embryos secrete a hormone called GA_1 when they start to grow. The hormone binds to receptors in cells near the starch-storing part of the seed. Once hormone-receptor binding occurs, the concentration of a second messenger called cGMP rises inside the cells. The second messenger triggers the production of a protein that activates the gene for a starch-digesting enzyme. Once the gene is turned on, large quantities of the starch-digesting enzyme α-amylase begin to enter the cell's endomembrane system. But the G protein activated by GA_1 also triggers a rise in intracellular Ca^{2+}. High Ca^{2+} concentrations allow vesicles that are packed with α-amylase to fuse with the plasma membrane and release their contents into the starch storage area. Thus, target cells begin producing *and* secreting an enzyme that digests stored starch. When the enzymes begin to work on the starch, sugars are released. By releasing a hormone, the germinating embryo has signaled for the release of the nutrients it needs to grow.

Second messengers and phosphorylation cascades result in the activation or deactivation of a particular target protein that already exists in the cell—an enzyme, a membrane channel, or a protein that activates certain genes. Whatever the mechanism, the activity of the target cell changes dramatically in response to the arrival of the signal.

We've analyzed long-distance cell-cell communication as a three-step system: signal reception, signal transduction and amplification, and response. Now the question is, how is the signal turned off? Cells near the starch storage area of wheat seeds do not need to secrete unlimited quantities of α-amylase. Abnormalities would result if the morphological changes induced by testosterone and estrogen continued past your early teenage years into adulthood. What keeps signals from being amplified indefinitely?

Signal Deactivation

Cells have built-in systems for turning off intracellular signals. For example, activated G proteins convert GTP to GDP in a reaction that releases a phosphate group. The G protein's conformation changes as a result. Activation of its associated enzyme stops, and production of the second messenger ceases. The messengers in the cytosol are also short lived. For example, pumps in the membrane of the smooth ER return calcium ions to storage, and enzymes called phosphodiesterases convert active cAMP (see Table 8.2) and cGMP to inactive AMP and GMP.

Phosphorylation cascades wind down in a similar way. Enzymes called phosphatases are always present in cells, where they catalyze reactions that remove phosphate groups from proteins. If hormone stimulation of a receptor tyrosine kinase ends, phosphatases are able to dephosphorylate enough components of the phosphorylation cascade that the response begins to slow. Eventually it stops.

Although an array of specific mechanisms are involved, here is the general observation: Signal transduction systems trigger a rapid response and can be shut down quickly. As a result, they are exquisitely sensitive to small changes in the concentration of hormones or in the number and activity of signal receptors.

It is critical, though, to appreciate what happens when a signal transduction system does not shut down properly. For example, recall that Ras activates a phosphorylation cascade when it binds GTP but is deactivated when it binds GDP. With surprising frequency, human cells begin producing Ras proteins that are defective and no longer convert GTP to GDP once they are activated. As a result, GTP stays bound and the defective Ras proteins stay in the "on" position. They continue stimulating a phosphorylation cascade even when no growth hormones are present. Cells with defective Ras are likely to keep dividing as a result, which may lead to the development of cancer. It is estimated that 25–30 percent of all human cancers involve cells with this type of defective Ras protein. The family of diseases called cancer is explored in much more detail in Chapter 11 and Chapter 18.

The four steps of cell-cell signaling—reception, processing, response, and deactivation—allow hormone-secreting cells to elicit a specific response from cells in nearby or distant tissues. Despite this diversity in structure and function, all hormones have the same general role in the organism: They coordinate the activities of cells in response to information from outside or inside the body. In multicellular organisms ranging from wheat to humans, cell-cell signaling helps millions of individual cells function as an integrated whole. As a result, multicellular organisms can respond to changing conditions in an appropriate way.

Cell-Cell Communication

Check Your Understanding

If you understand that...

- Intercellular signals coordinate the activities of cells throughout the body in response to changes in internal or external conditions.
- If intercellular signals do not enter the cell, they bind to a receptor on the plasma membrane. In response, the intercellular signal is transduced to an intracellular signal that the cell responds to.

You should be able to...

1) Explain what happens during each of the four steps involved in signal transduction (reception, processing, response, and deactivation).
2) Diagram a phosphorylation cascade versus production of a second messenger.
3) Explain why only certain cells respond to particular signals.
4) Explain how signals are amplified.

Chapter Review

SUMMARY OF KEY CONCEPTS

Extracellular material strengthens cells and helps bind them together.

The vast majority of cells secrete an extracellular layer. In bacteria, archaea, algae, and plants, the extracellular material is stiff and is called a cell wall. In animals, the secreted layer is flexible and is called the extracellular matrix (ECM). Although the types of molecules present in the external coating vary widely, the basic structure and function of the extracellular layer are the same: It is a fiber composite that defines the cell's shape and helps protect it from damage. Fiber composites consist of cross-linked filaments that provide tensile strength and a ground substance that fills space and resists compression. In plants the extracellular filaments are cellulose microfibrils; in animals the most abundant filaments are made of the protein collagen. In both plants and animals, the ground substance is composed of gel-forming polysaccharides.

You should be able to predict what happens to animal cells when (1) they are treated with an enzyme that cuts integrin molecules, or (2) collagen fibers degrade.

Cell-cell connections help adjacent cells adhere. Cell-cell gaps allow adjacent cells to communicate.

In both unicellular and multicellular organisms, cell-cell (intercellular) interactions are mediated by molecules in the extracellular layer and plasma membrane. Most interactions among unicellular species are competitive in nature, but the cells of multicellular organisms are intensely social and cooperate with each other. Many cells of multicellular organisms are physically bound to one another via the glue-like middle lamella that forms between plant cells or the tight junctions and desmosomes observed between animal cells. The cytoplasm of adjacent cells is in direct communication, via openings called plasmodesmata in plants and gap junctions in animals.

You should be able to explain why the hypothesis that multicellularity evolved independently in plants and animals is supported by the observation that they have different structures serving similar functions in cell-cell adhesion and communication.

Intercellular signals are responsible for creating an integrated whole from many thousands of independent parts.

Distant cells in multicellular organisms communicate through signaling molecules that bind to receptors found on or in specific target cells. Once these signals are received, they are often transduced to a new type of intracellular signal, which is amplified. This internal signal triggers a sequence of events that leads to the activation of certain enzymes, the release or uptake of specific ions or molecules, or a change in the activity of target genes. Because enzymes inside the cell quickly deactivate the signal, the cell's response is tightly regulated. A continued response usually depends on continued stimulation by the signaling molecule. Thus, cell-cell communication occurs by a four-step process: (1) signal reception, (2) signal processing, (3) signal response, and (4) signal deactivation. As a result, cells and tissues throughout the body can alter their activity in response to changing conditions, and do so in a coordinated way.

You should be able to explain why the hormone adrenalin can stimulate cells in both the heart and liver, yet trigger different responses (increasing heart rate versus release of glucose).

Cell-Cell Communication

QUESTIONS

Test Your Knowledge

1. Which of the following statements represents a fundamental difference between the fibers found in the extracellular layers of plants and those of animals?
   a. Plant fibers are thicker; they are also stronger because they have more cross-linkages.
   b. Animal fibers consist of proteins; plant fibers consist of polysaccharides instead.
   c. Plant extracellular fibers never move; animal fibers can slide past one another.
   d. Cellulose microfibrils run parallel to each other; collagen filaments crisscross.
2. In animals, where are most components of the extracellular material synthesized?
   a. smooth ER
   b. the rough ER and Golgi apparatus
   c. in the extracellular layer itself
   d. in the plasma membrane
3. Treating dissociated cells with certain antibodies makes the cells unable to reaggregate. Why?
   a. The antibodies bind to cell adhesion proteins called cadherins.
   b. The antibodies bind to the fiber component of the extracellular matrix.
   c. The antibodies bind to receptors on the cell surface.
   d. The antibodies act as enzymes that break down desmosomes.
4. What does it mean to say that a signal is transduced?
   a. The signal enters the cell directly and binds to a receptor inside.
   b. The physical form of the signal changes between the outside of the cell and the inside.
   c. The signal is amplified, such that even a single molecule evokes a large response.
   d. The signal triggers a sequence of phosphorylation events inside the cell.
5. Why are tight junctions found in only certain types of tissues, while desmosomes are found in a wide array of cells?
   a. Tight junctions are required only in cells where communication between adjacent cells is particularly important.
   b. Tight junctions are not as strong as desmosomes.
   c. Tight junctions have different structures but the same functions.
   d. Tight junctions are found only in epithelial cells that must be watertight.
6. What physical event represents the receipt of an intercellular signal?
   a. the passage of ions through a desmosome
   b. the activation of the first protein in a phosphorylation cascade
   c. the binding of a hormone to a signal receptor, which changes conformation in response
   d. the activation of a G protein associated with a signal receptor

Test Your Knowledge answers: 1. b; 2. b; 3. a; 4. b; 5. d; 6. c

Test Your Understanding

Answers are available www.masteringbio.com

1. Why is it difficult to damage a fiber composite?
2. Why does a phosphorylation cascade amplify an intercellular signal?
3. Compare and contrast a tight junction and a desmosome. How are they similar, and how are they different? Then compare and contrast desmosomes and plasmodesmata.
4. Animal cells adhere to each other selectively. Summarize experimental evidence that supports this statement. Explain the molecular basis of selective adhesion.
5. The cell-cell signal called insulin acts via the receptor tyrosine kinase. Make a concept map showing the reception, processing, and response steps in the signal transduction pathway for insulin. Your diagram should also include the concepts of amplification, phosphorylation cascade, activated transport proteins, and importing glucose from the bloodstream.
6. Why do researchers use the term *pathway* in reference to signal transduction? What is the significance of the observation that many signal transduction pathways intersect or overlap, creating a network?

Applying Concepts to New Situations

Answers are available at www.masteringbio.com

1. Suppose that an animal species and a plant species each lacked the ability to secrete an extracellular matrix. What would these organisms look like, and how would they live?
2. Suppose that a signal that is released from one cell triggered a response in another cell in the *absence* of signal transduction. Diagram the sequence of events that would occur between the arrival of the signal and the cell's response. Compare and contrast the response to a signal that does involve signal transduction.
3. In most species of fungi, chitin is a major polysaccharide found in cell walls. Review the structure of chitin as described in Chapter 5, and then make a sketch predicting the structure of the fungal cell wall.
4. Suppose you created an antibody that bound to the receptor illustrated in Figure 8.18. How would the signal transduction pathway be affected? How would the signal transduction pathway be affected by a drug that bound permanently to the receptor?

9 Cellular Respiration and Fermentation

KEY CONCEPTS

- In cells, the endergonic reactions required for life occur in conjunction with an exergonic reaction involving ATP.
- Cellular respiration produces ATP from molecules with high potential energy—often glucose.
- Glucose processing has four components: (1) glycolysis, (2) pyruvate processing, (3) the Krebs cycle, and (4) electron transport coupled with oxidative phosphorylation.
- Fermentation pathways allow glycolysis to continue when the lack of an electron acceptor shuts down electron transport chains.
- Respiration and fermentation are carefully regulated.

When table sugar is heated, it undergoes the uncontrolled oxidation reaction known as burning. Burning gives off heat. In cells, the simple sugar called glucose is oxidized through a long series of carefully regulated reactions. Instead of being given off as heat, some of the energy produced by these reactions is used to synthesize ATP.

Cells are dynamic. Vesicles move cargo from the Golgi apparatus to the plasma membrane and other destinations, enzymes synthesize a complex array of macromolecules, and millions of membrane proteins pump ions and molecules to create an environment conducive to life. These cell activities change almost constantly in response to signals from other cells or the environment. What fuels all this action? The answer lies in the molecule **adenosine triphosphate (ATP)**. ATP has high potential energy and acts as the major energy currency in cells.

Earlier chapters explored the role that ATP plays in moving materials through cells, in moving cells themselves through the environment, and in responding to signals. Recall that phosphate groups contain a negative charge. When a phosphate group from ATP is transferred to a protein, the addition of the negative charge causes the newly phosphorylated protein to change shape. When a protein pump is phosphorylated, its shape changes in a way that transports ions or molecules across the plasma membrane against an electrochemical gradient (see Chapter 6). In some cases, an entire ATP molecule—which contains four tightly clustered negative charges—is added to a protein. When ATP binds to a motor protein such as kinesin or dynein or myosin, the protein moves in a way that transports materials or changes the cell's shape (Chapter 7). When ATP or a phosphate group from ATP binds to a specific site on an enzyme, the enzyme's shape and activity can change in a way that triggers synthesis of a second messenger or contributes to a phosphorylation cascade (Chapter 8).

Key Concept | Important Information | Practice It

In short, ATP allows cells to do work. Because staying alive takes work, there is no life without ATP.

The goal of this chapter is to investigate how cells make adenosine triphosphate, starting from sugars and other compounds that have high potential energy. The energy that is released as sugar is processed in cells and is used to transfer a phosphate group to **adenosine diphosphate (ADP)**, generating ATP. Section 9.1 introduces the events involved by reviewing some fundamental points about free energy, analyzing how most ATP is used in cells, and exploring the principles of reduction-oxidation reactions, which are responsible for producing ATP. Section 9.2 follows up with an overview of the reactions involved in processing glucose, the most common fuel used by organisms. Sections 9.3 through 9.6 delve into the glucose-processing reactions in detail. Section 9.7 introduces an alternative route for ATP production—called fermentation—that occurs in many bacteria, archaea, and eukaryotes. Section 9.8, which closes the chapter, examines how cells divert certain carbon-containing compounds away from ATP production and into the synthesis of DNA, RNA, amino acids, and other molecules. This chapter is your introduction to **metabolism**—the chemical reactions that occur in living cells.

9.1 The Nature of Chemical Energy and Redox Reactions

Recall from Chapter 2 that chemical energy is a form of potential energy. Potential energy is energy that is associated with position or configuration. In cells, electrons are the most important source of chemical energy. The amount of potential energy that an electron has is based on its position relative to other electrons and to the protons in the nuclei of nearby atoms. If an electron is close to negative charges on other electrons and far from the positive charges in nuclei, it has high potential energy. In general, the potential energy of a molecule is a function of the way its electrons are configured or positioned.

ATP makes things happen in cells because it has a great deal of potential energy. To understand why this is so, study the structural formula in **Figure 9.1a**. Notice that the three phosphate groups in ATP contain a total of four negative charges, and that these charges are confined to a small area. In part because these negative charges repel each other, the potential energy of the electrons in the phosphate groups is extraordinarily high.

When ATP reacts with water during a hydrolysis reaction, the bond between ATP's outermost phosphate group and its neighbor is broken, resulting in the formation of ADP and inorganic phosphate, P_i, which has the formula $H_2PO_4^-$ (**Figure 9.1b**). This reaction is highly exergonic. Recall that **exergonic** reactions release energy, while **endergonic** reactions require an input of energy. Under standard conditions of temperature and pressure in the laboratory, a total of 7.3 kilocalories (kcal) of energy, per mole of ATP, is released during the reaction.

You might remember from Chapter 2 that changes in free energy dictate whether a reaction is exergonic or endergonic. Changes in free energy, in turn, depend on the relationship between reactants and products in terms of potential energy and entropy (disorder). The hydrolysis of ATP is exergonic because the entropy of the product molecules is much higher than that of the reactants, and because there is a large drop in potential energy when ADP and P_i are formed from ATP. The change in

(a) ATP consists of three phosphate groups, ribose, and adenine.

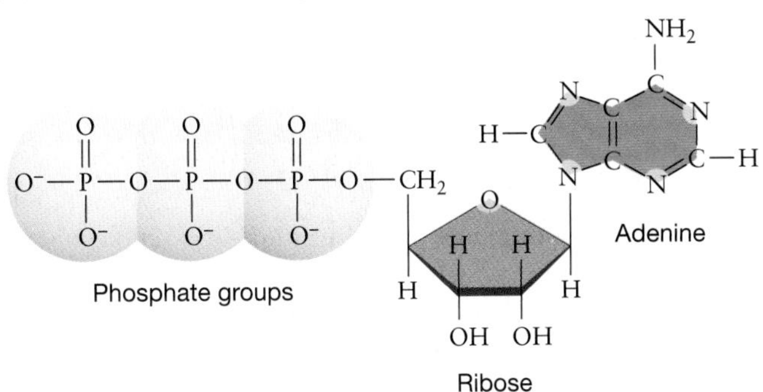

(b) Energy is released when ATP is hydrolyzed.

FIGURE 9.1 Adenosine Triphosphate (ATP) Has High Potential Energy. **(a)** ATP has high potential energy, in part because four negative charges are clustered in its three phosphate groups. The negative charges repel each other, raising the potential energy of the electrons. **(b)** When ATP is hydrolyzed to ADP and inorganic phosphate, a large free-energy change occurs.

potential energy occurs in part because the electrons from ATP's phosphate groups are now spread between two molecules instead of being clustered on one molecule, meaning that there is less electrical repulsion. In addition, the negative charges on ADP and P_i are stabilized much more efficiently by interactions with the partial positive charges on surrounding water molecules than are the charges on ATP.

If the reaction diagrammed in Figure 9.1b occurred in a test tube, the energy released would be lost as heat. But cells don't lose that 7.3 kcal/mole as heat. Instead, they use it to make things happen. More specifically, things start to happen when ATP is hydrolyzed and a phosphate group is transferred to a protein. The addition of a phosphate group to a substrate is called **phosphorylation**. Phosphorylation of proteins is exergonic because the electrons in ADP and the phosphate group have much less potential energy than they did in ATP.

When phosphorylation adds a negative charge to a protein, the electrons in the protein change configuration in response. The molecule's overall shape, or conformation, is usually altered as well (**Figure 9.2**). Part of the protein moves. Protein movement—either in response to phosphorylation or to binding of an entire ATP molecule—is what transports materials inside cells, powers flagella or cilia, and pumps ions across membranes. It is also responsible for driving the endergonic reactions required for life.

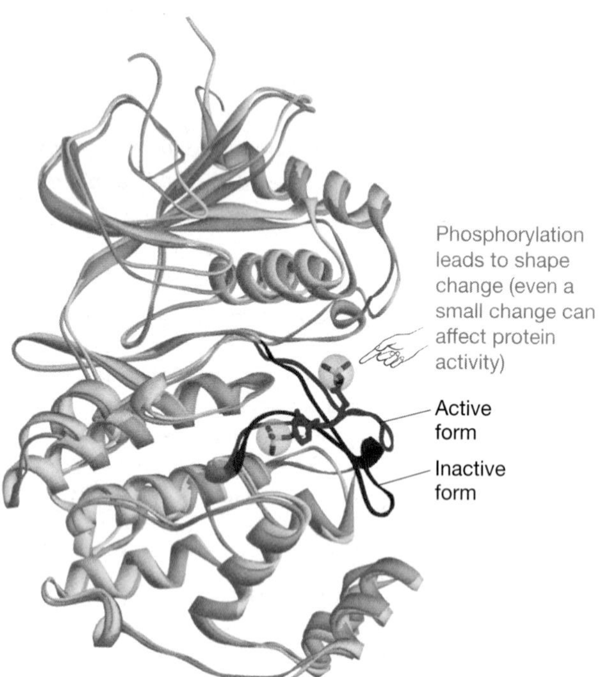

FIGURE 9.2 Phosphorylation Changes the Shape and Activity of Proteins. When proteins are phosphorylated or an ATP molecule binds to them, they often change shape in a way that alters their activity. The shape changes might seem subtle to you, but at the atomic scale, where reactions occur, they are huge. The figure shows an enzyme called MAP kinase. Note that two phosphate groups (yellow) activate this molecule.

How Does ATP Drive Endergonic Reactions?

In the time it takes to read this sentence, millions of endergonic reactions will have occurred in your cells. This chemical activity is possible because entire ATP molecules or phosphate groups from ATP are being added to reactant molecules or enzymes. To see how this process works, consider an endergonic reaction between a compound A and compound B that results in a product AB needed by your cells. This reaction can happen only when ATP reacts with the substrate to produce a phosphorylated molecule and ADP. If the reactant that is phosphorylated is compound B, it is referred to as an activated substrate. Activated substrates contain a phosphate group and have high free energy—high enough that the reaction between compound A and the activated form of compound B is exergonic. The two compounds then go on to react and form the product molecule AB.

In some cases, the enzyme that catalyzes the reaction is phosphorylated instead of a reactant. When either a substrate or an enzyme is phosphorylated, the exergonic phosphorylation reaction is said to be coupled to an endergonic reaction. In cells, endergonic reactions become exergonic when the substrates or enzymes involved are phosphorylated.

Figure 9.3 graphs how **energetic coupling** between an exergonic and endergonic reaction works. To begin reading this graph, note that the reaction between A and B is endergonic—ΔG is positive. But when the exergonic reaction occurs that results in a phosphate group leaving ATP and binding to B, the free energy of the reactants A and B is now high enough to make the reaction that forms AB exergonic. This is what coupling between phosphorylation and endergonic reactions does.

If you understand the principles of energetic coupling, consider the following endergonic reaction: ribulose + $CO_2 \rightarrow$ 2 glycerate. You should be able to:

(1) Draw an enzyme and add ribulose and CO_2 nearby. (Don't worry about what ribulose actually looks like.) Explain why the reaction can't occur.

(2) Show 2 ATPs adding 2 phosphate groups to the ribulose. Explain why the resulting molecule, called ribulose bisphosphate, is called an activated substrate. Draw the activated substrate and CO_2 binding to the enzyme.

(3) Then draw the outcome of the reaction—two molecules of glycerate, each of which has a phosphate group attached, leaving the enzyme. Summarize how the coupling of an exergonic reaction and endergonic reaction occurred.

It is hard to overstate the importance of energetic coupling. Without it, life is impossible. If you run out of ATP, then enzymes and reactants can no longer be phosphorylated. If this happens, you will die within minutes. The energetic coupling going on in your cells right now is keeping you alive.

Now the issue is, if a great deal of energy is released when ATP loses a phosphate group and a protein is phosphorylated,

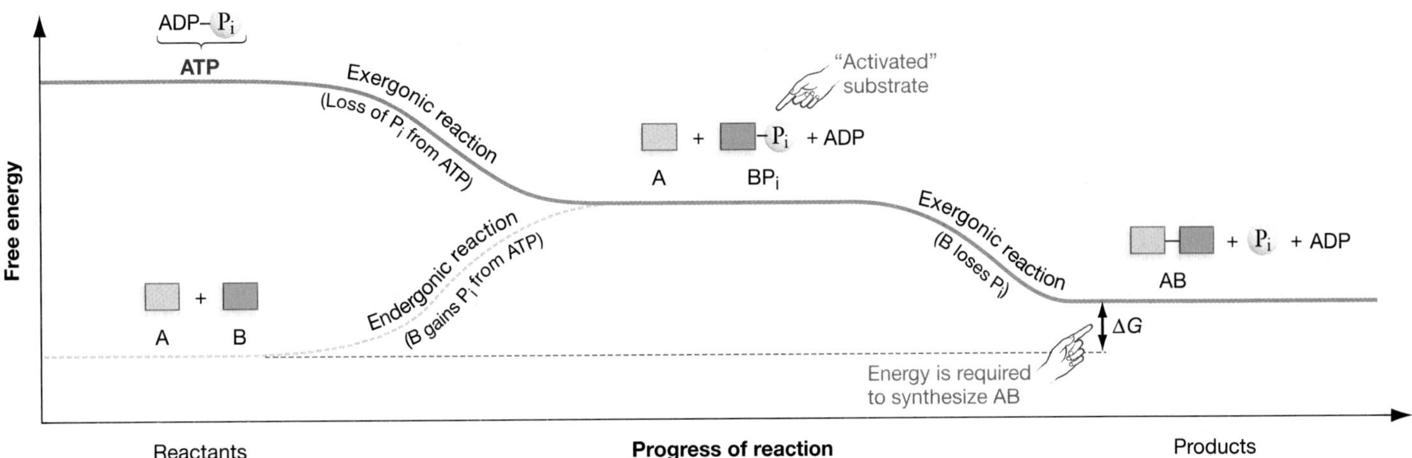

FIGURE 9.3 Exergonic Phosphorylation Reactions Are Coupled to Endergonic Reactions. In cells, endergonic reactions occur if one reactant (or an enzyme) undergoes phosphorylation—the addition of a phosphate group from ATP. The phosphorylated reactant molecule (or enzyme) has high enough free energy that the subsequent reaction is exergonic.

then a great deal of energy must be required to synthesize ATP from ADP by adding P_i. Where does this energy come from? The answer is redox reactions.

What Is a Redox Reaction?

Reduction-oxidation reactions, or **redox reactions**, are a class of chemical reactions that involve the loss or gain of one or more electrons. Many reactions, like the acid-base reactions introduced in Chapter 2, do not involve the gain or loss of an electron. Redox reactions are central in biology, though, because they drive the formation of ATP.

In a redox reaction, the atom that loses one or more electrons is said to be oxidized, and the atom that gains one or more electrons is said to be reduced. To help keep these terms straight, chemists use the mnemonic "LEO the lion goes GER"—Loss of Electrons is **Oxidation**; Gain of Electrons is **Reduction**. (An alternative is OIL RIG—Oxidation Is Loss; Reduction Is Gain.) Oxidation events are always coupled with a reduction; if one atom loses an electron, another has to gain it. Stated another way, a reactant that acts as an **electron donor** is always paired with a reactant that acts as an **electron acceptor**.

The gain or loss of an electron can be relative, however. During a redox reaction, electrons can be transferred completely from one atom to another, or the electrons can simply shift their positions in covalent bonds. For example, consider the overall reaction for photosynthesis diagrammed in **Figure 9.4**. Plants take in carbon dioxide (CO_2) and water (H_2O); and

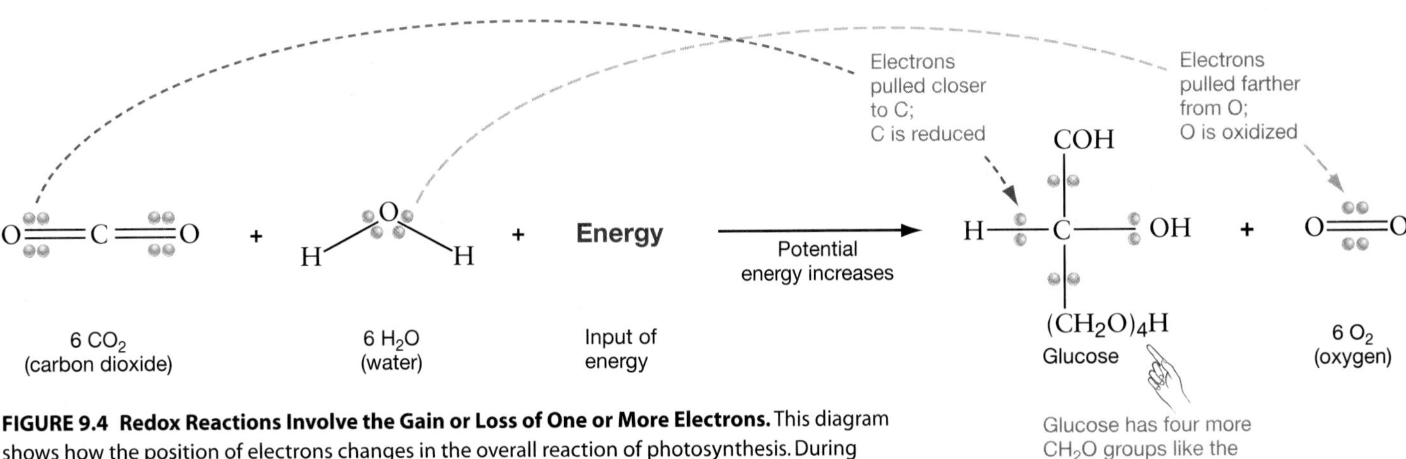

FIGURE 9.4 Redox Reactions Involve the Gain or Loss of One or More Electrons. This diagram shows how the position of electrons changes in the overall reaction of photosynthesis. During photosynthesis, carbon atoms in CO_2 are reduced to form glucose and other sugars. The process is endergonic and requires an input of energy. Glucose has much higher potential energy than carbon dioxide does.

● **EXERCISE** Draw the rest of the structure of glucose (the four CH_2O groups, with the last carbon also bonded to an H), and add the positions of the electrons. Circle and label an atom that is reduced (has "gained" electrons); circle and label an atom that is oxidized (has "lost" electrons).

with the aid of sunlight, they synthesize carbohydrate—in this example, the sugar glucose ($C_6H_{12}O_6$)—and release molecular oxygen (O_2) and water. The orange dots in the illustration represent the positions of the electrons involved in covalent bonds. Now compare the position of the electrons in the first reactant, carbon dioxide, with their position in the first product, glucose. Notice that many of the electrons have moved closer to the carbon nucleus in glucose. This means that carbon has been reduced. It has "gained" electrons. The change occurred because the carbon and oxygen atoms in CO_2 do not share electrons equally, while the carbon and hydrogen atoms in glucose do. In CO_2, the high electronegativity of the oxygen atoms pulled electrons away from the carbon atom. Now compare the position of the electrons in the reactant water molecules with their position in the O_2 molecules that are produced. In O_2, the electrons have moved farther from the oxygen atoms than they were in the water molecules, meaning that the oxygen atoms have been oxidized. Oxygen has "lost" electrons. In the overall reaction for photosynthesis, carbon atoms are reduced while oxygen atoms are oxidized.

These shifts in electron position change the amount of chemical energy in the reactants and products. When photosynthesis occurs, electrons are held much more loosely in the product molecules than in the reactant molecules. This means their potential energy has increased. The entropy of the products is also much lower than the reactants. As a result, the reaction is endergonic. It can take place only with an input of energy from sunlight.

Some Other Approaches to Understanding Redox

Redox reactions are absolutely fundamental to life, so it's critical to gain a solid understanding of redox chemistry. To give you a deeper understanding of how these reactions work, here are some other descriptions of redox dynamics:

- The electrons and bonds involved in redox reactions can be likened to a seesaw. In this analogy, an electron pair that is being shared equally by two atoms is like a seesaw balanced on its fulcrum.

Because this state is inherently unstable, it has high potential energy. The electrons involved in a covalent bond between carbon and hydrogen correspond to this situation—they are shared equally, because carbon and hydrogen are equally electronegative. Now suppose that a redox reaction occurs, and electrons are transferred toward a more electronegative atom such as oxygen. In this case, the bond is like a seesaw with one end lowered to the ground.

In this state, the seesaw is stable—meaning that it has much lower potential energy. This is what happens in redox reactions when electrons that were formerly involved in covalent bonds between carbon and hydrogen instead participate in bonds between carbon and oxygen. In a reaction like this, the difference in potential energy between the two states is released as heat or transferred to another atom.

- During the redox reactions that occur in cells, electrons (e^-) are often transferred from an atom in one molecule to an atom in a different molecule. When this occurs, the electron is usually accompanied by H^+. The proton transfer is not a reduction reaction, but the upshot is that the molecule that contains the reduced atom gains a hydrogen (H) atom. Such molecules tend to have high potential energy, because the electrons in C–H bonds are relatively far from the positive charges in a nucleus. This observation should sound familiar, from the introduction to carbohydrates in Chapter 5. Molecules that have a large number of C–H bonds, such as carbohydrates and fats, store a great deal of potential energy.

Conversely, molecules that are oxidized in cells often lose a proton along with an electron. Instead of having many C–H bonds, oxidized molecules in cells tend to have many C–O bonds. Such molecules also have lower potential energy. To understand why, remember that oxygen atoms have extremely high electronegativity. Because oxygen atoms hold electrons so tightly, the electrons involved in bonds with oxygen atoms have low potential energy.

In many redox reactions in biology, understanding where oxidation and reduction have occurred becomes a matter of following hydrogen atoms. In many biological systems, reduction means "adding H's" and oxidation means "removing H's."

What Happens When Glucose Is Oxidized? To test your understanding of redox reactions, consider what happens when glucose undergoes the uncontrolled oxidation reaction called burning:

$$C_6H_{12}O_6 + 6\ O_2 \rightarrow 6\ CO_2 + 6\ H_2O + \text{energy}$$

The photograph at the start of the chapter shows this reaction occurring, and **Figure 9.5** provides a diagram. ● If you understand the fundamental principles of reduction-oxidation, you should be able to fill in the electron positions in each bond in the figure and answer the following questions:

1. Are the carbon atoms in glucose oxidized or reduced?
2. Are the oxygen atoms in the oxygen molecule (O_2) oxidized or reduced?
3. Glucose is the molecule that acts as an electron donor in this reaction. Which molecule acts as the electron acceptor?
4. Which has higher potential energy: the reactants or the products? Based on your answer, add "Energy" to the appropriate side of the equation; then add a label indicating "Input of energy" or "Release of energy."

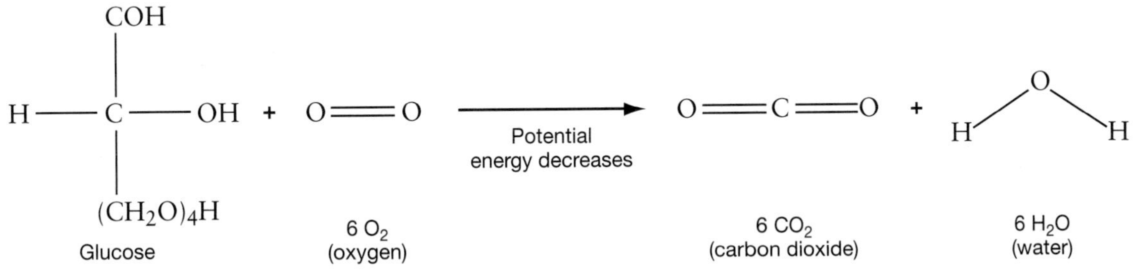

COH | H—C—OH | $(CH_2O)_4H$ + O=O → (Potential energy decreases) O=C=O + H–O–H

Glucose + $6\ O_2$ (oxygen) → $6\ CO_2$ (carbon dioxide) + $6\ H_2O$ (water)

FIGURE 9.5 Tracking Electron Transfer during the Oxidation of Glucose.

EXERCISE Fill in the electron positions in each bond shown; then use these data to explain why the reaction is exergonic. (Check your work, using the electron positions diagrammed in Figure 9.4).

When glucose burns, the change in potential energy is converted to kinetic energy in the form of heat. More specifically, a total of 686 kcal of heat is released when one mole of this sugar is oxidized.

Glucose does not burn in cells, however. Instead, the glucose in cells is oxidized through a long series of carefully controlled redox reactions. These reactions are occurring, millions of times per minute, in your cells right now. Instead of being given off as heat, much of the energy released by the drop in potential energy is being used to make the ATP you need to read, think, move, and stay alive. In cells, the change in free energy that occurs during the oxidation of glucose is used to synthesize ATP from ADP and P_i.

9.2 An Overview of Cellular Respiration

The first thing to appreciate about glucose processing and ATP production in cells is just how dynamic it is. In general, a cell contains only enough ATP to last from 30 seconds to a few minutes. Like many other cellular processes, the production and use of ATP is fast.

A second point to consider is where glucose comes from. Most glucose is produced by plants and other photosynthetic species, which use the energy in sunlight to reduce carbon dioxide (CO_2) to glucose and other carbohydrates. When photosynthetic species decompose or are eaten, they provide glucose to animals, fungi, and many bacteria and archaea. All organisms use glucose as a building block in the synthesis of fats, carbohydrates such as starch and glycogen, and other energy-storage compounds. The chemical energy stored in fats and carbohydrates acts like a savings account. ATP, in contrast, is like cash. To make ATP and get cash, carbohydrates must first be converted to glucose (**Figure 9.6**). When a cell needs energy, glucose is used to produce ATP through processes called cellular respiration and fermentation.

Chapter 10 focuses on how plants and other photosynthetic organisms produce glucose from carbon dioxide, water, and sunlight. This chapter focuses on cellular respiration and fermentation. Some of the chemistry involved in processing glucose actually predates the evolution of photosynthesis.

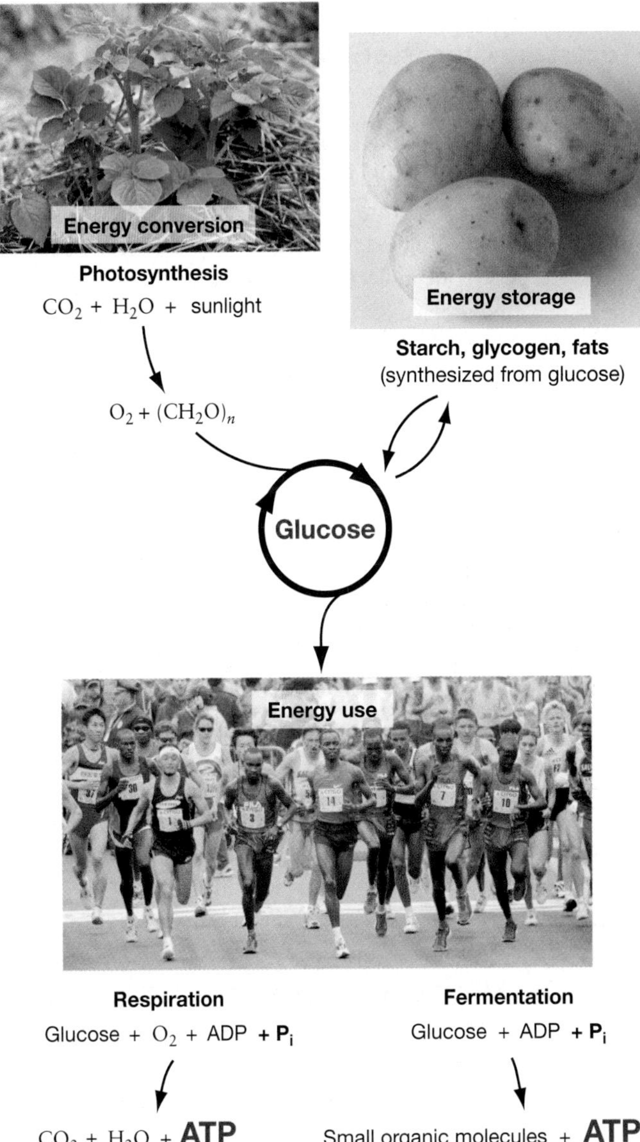

FIGURE 9.6 Glucose Is the Hub of Energy Processing in Cells. Glucose is the end-product of photosynthesis. Both plants and animals store glucose and oxidize it to provide chemical energy in the form of ATP.

EXERCISE The CO_2 produced by cellular respiration is used by plants during photosynthesis. The O_2 produced by photosynthesis is used during cellular respiration. Add arrows to this diagram to indicate these relationships.

Indeed, the reactions analyzed early in this chapter were probably occurring as the first cell sprang to life. The ancestry of some of the enzymes introduced in this chapter dates back about 3.5 billion years.

How does the oxidation of glucose take place in a way that supports the production of ATP? In many organisms, cellular respiration is a four-step process: (1) Glucose is broken down to a three-carbon compound called pyruvate; (2) pyruvate is processed to form a compound called acetyl-CoA; (3) acetyl-CoA is oxidized to CO_2; and (4) compounds that were reduced in steps 1–3 are oxidized in reactions that lead to the production of ATP. Let's get an overview of the way these steps work, and then pursue a more detailed analysis in Sections 9.3 through 9.6.

Processing Glucose: Glycolysis

An enormous diversity of organisms use the monosaccharide **glucose** as their primary fuel. In virtually all species examined to date, the first step in the oxidation of glucose is a sequence of 10 chemical reactions that are collectively called **glycolysis** (literally, "sugar breakdown"). These reactions occur in the cytosol of cells. During glycolysis, one molecule of the six-carbon sugar glucose is broken into two molecules of the three-carbon compound pyruvate. Some of the potential energy released by this sequence of reactions is used to phosphorylate ADP molecules, forming ATP. In addition, one of the reactions in the sequence reduces a molecule called **nicotinamide adenine dinucleotide**, symbolized **NAD^+**. As **Figure 9.7** shows, the addition of two electrons and a proton to NAD^+ produces **NADH**. We'll soon see that NADH readily donates electrons to other molecules. As a result, it is called an **electron carrier** and is said to have "reducing power."

Figure 9.8a summarizes the results of the reactions of glycolysis. Notice that glucose goes in and that ATP, NADH, and pyruvate come out.

Pyruvate Processing

What happens to the pyruvate produced by glycolysis? If an electron acceptor such as oxygen is present in the cell, pyruvate undergoes a sequence of reactions that results in the product molecule acetyl-CoA. These redox reactions have two results: (1) another molecule of NADH is synthesized, and (2) one of the three carbon atoms in pyruvate is oxidized to carbon dioxide. This CO_2 molecule contributes to the carbon dioxide that you exhale with every breath.

In eukaryotes, pyruvate processing occurs in the matrix inside mitochondria. During this step in the glucose oxidation system, pyruvate goes in and NADH, CO_2, and acetyl-CoA come out (**Figure 9.8b**).

The Krebs Cycle

The acetyl CoA produced by pyruvate processing enters a key sequence of reactions called the **Krebs cycle**. In eukaryotic cells, the enzymes involved in the Krebs cycle are located inside mitochondria. Some of these enzymes are found in the matrix inside the organelle, while others are associated with the surface of the mitochondrion's inner membrane.

During the Krebs cycle, each acetyl-CoA molecule is oxidized to two molecules of carbon dioxide. Some of the potential energy released by these reactions is used to (1) reduce NAD^+ to NADH; (2) reduce another electron carrier, called **flavin adenine dinucleotide** (**FAD**), to **$FADH_2$**; and (3) phosphorylate ADP, yielding ATP (**Figure 9.8c**). Acetyl-CoA enters this sequence of reactions; carbon dioxide, NADH, $FADH_2$, and ATP come out of the sequence.

With the completion of the Krebs cycle, glucose has been completely oxidized to CO_2. This observation raises several questions, though. According to the overall reaction for the oxidation of glucose ($C_6H_{12}O_6 + 6\ O_2 \rightarrow 6\ CO_2 + 6\ H_2O +$ energy), molecular oxygen (O_2) is a reactant. Where does oxygen come into play? And what happens to all of the electrons that have been transferred from glucose to NADH and $FADH_2$?

Electron Transport

In cells, the high potential energy of electrons carried by NADH and $FADH_2$ is gradually "stepped down" by molecules that participate in a series of redox reactions. Stated another way, the potential energy in the electrons is gradually decreased via a series of step-by-step reactions. The proteins involved in the redox reactions make up what is known as an **electron transport chain (ETC)**. As electrons are passed along the electron transport chain, they gradually fall from a higher to a lower potential energy. In eukaryotes, the ETC components are located in the inner membrane of mitochondria; in bacteria and archaea, they are found in the plasma membrane.

The transfer of an electron to each molecule in the electron transport chain changes the molecule's shape and activity. In several cases, the molecule that receives the electron responds by changing conformation in a way that transfers a proton across the mitochondrion's inner membrane. This is crucial. In effect, the movement of electrons through the chain results in the pumping of protons to the outside of the membrane. It's a little bit like using electricity to pump water up into a storage tank above a water wheel. The subsequent buildup of protons results in a strong electrochemical gradient across the mitochondrial membrane. The gradient drives the protons through a membrane protein called ATP synthase. In eukaryotes, ATP synthase is located in the inner membrane of mitochondria. The force generated by the flow of protons through ATP synthase makes part of the protein spin. This change in conformation drives the phosphorylation of ADP. Indirectly, then, the sequence of redox reactions in the electron transport chain results in the production of ATP (**Figure 9.8d**).

To summarize: A series of reduction-oxidation reactions allows the proteins in the electron transport chain to pump

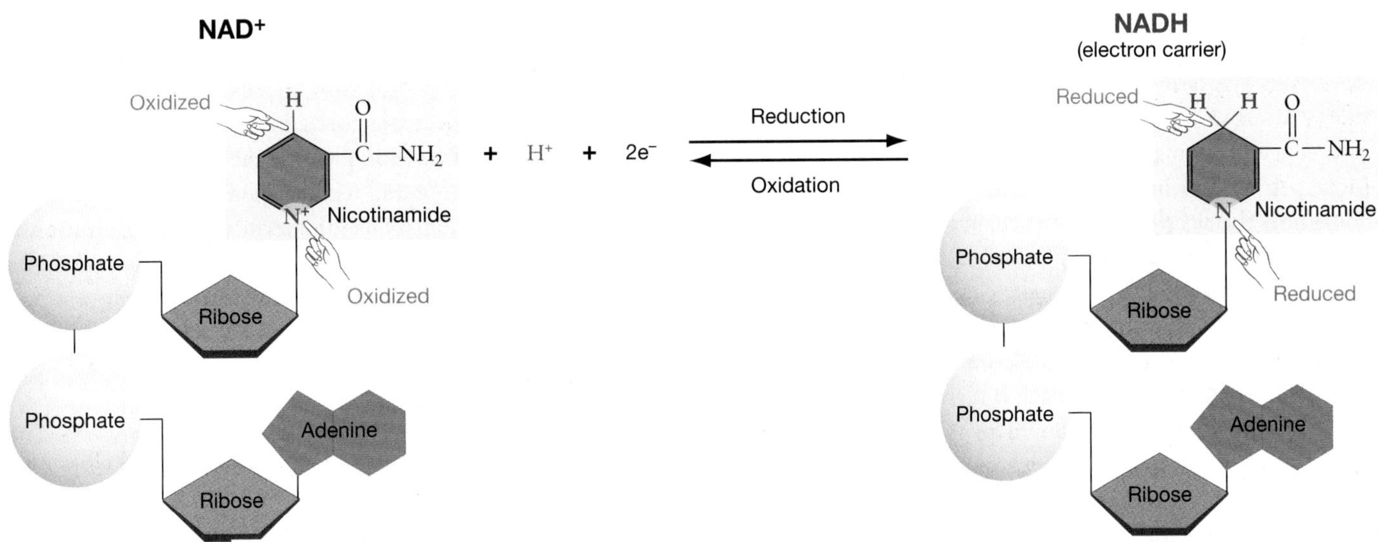

FIGURE 9.7 NAD$^+$ and NADH and Oxidized and Reduced Forms. NADH is the reduced form of NAD$^+$.

QUESTION Compare the structure of NAD$^+$ with the structure of ATP in Figure 9.1. Which portions of the two molecules are identical?

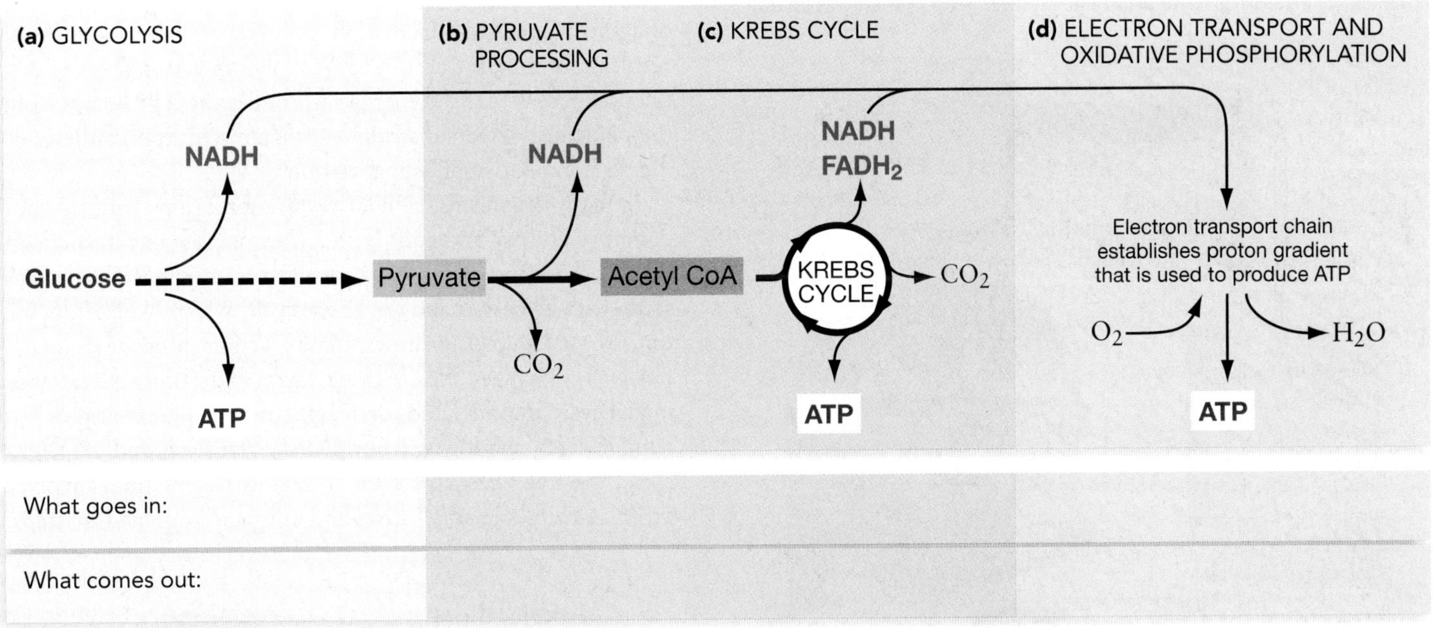

FIGURE 9.8 An Overview of Glucose Oxidation. Cells produce ATP from glucose via a series of processes: **(a)** glycolysis, **(b)** pyruvate processing, **(c)** the Krebs cycle, and **(d)** electron transport combined with oxidative phosphorylation. Each component produces at least some ATP or NADH. Because the four components are connected, glucose oxidation is an integrated metabolic pathway. Glycolysis, pyruvate processing, and the Krebs cycle complete the oxidation of glucose. The NADH and $FADH_2$. they produce then feed the electron transport chain.

EXERCISE Fill in the chart along the bottom.

protons across the inner membrane of the mitochondria. When the protons flow back across the membrane through ATP synthase, ATP is synthesized from ADP. ADP diffuses into mitochondria along a concentration gradient; ATP diffuses out along a concentration gradient.

Most people (me included!) initially have trouble understanding how the electron transport chain works. To make sure that you understand what is happening, consider this analogy: The proteins in the electron transport chain are like an electric pump—they use a flow of electrons to pump water up to a storage tank above a water wheel (**Figure 9.9**). When the water falls down through the water wheel, it turns and drives a millstone that grinds flour. ● You should be able to identify which parts of the water-pumping analogy correspond to the electron transport chain, ATP production, the pumping of protons, ATP synthase, and the proton gradient.

Once the electrons donated by NADH and $FADH_2$ have passed through the electron transport chain, they are transferred to a final electron acceptor, which in many organisms is oxygen. A transfer of these electrons to oxygen, along with protons, results in the formation of water as an end-product. The four molecules in the overall reaction for the oxidation of glucose—glucose, oxygen, carbon dioxide, and water—are now accounted for. Chemical energy has been transferred from glucose to ATP, via NADH and $FADH_2$. These reactions are keeping you alive right now.

Most of the ATP production in cells results from the reactions in the electron transport chain. Hence, glycolysis and the Krebs cycle can be thought of as mechanisms for stripping electrons from glucose and feeding them to the electron transport chain. ● Biologists use the term **cellular respiration** for any process of ATP production that involves each of the following: a compound that acts as an electron donor, an electron transport chain, and an electron acceptor.

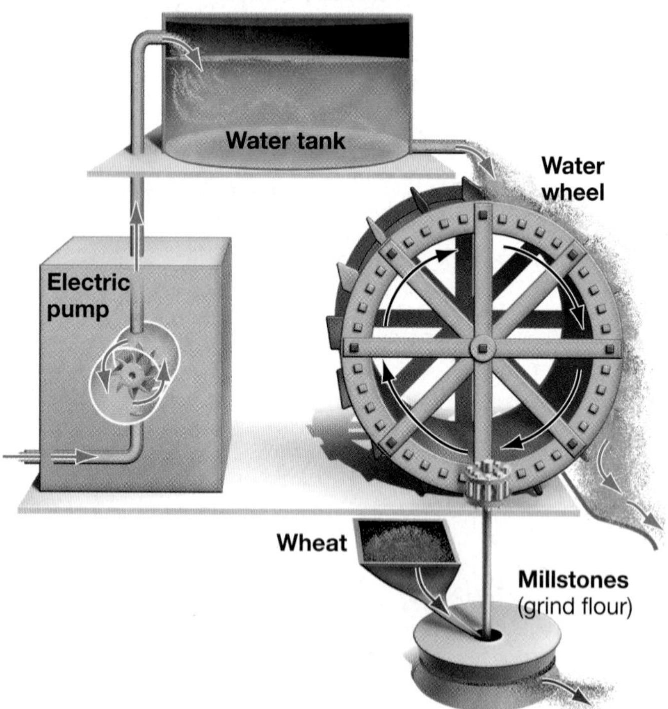

FIGURE 9.9 An Analogy for Oxidative Phosphorylation.

● **QUESTION** Which parts of the diagram represent endergonic reactions? Which parts represent exergonic reactions?

● **EXERCISE** Once you understand which components of this drawing correspond to the ETC, protons, proton gradient, and ATP synthase, sketch in the inner mitochondrial membrane where you think it exists.

Methods of Producing ATP

The phosphorylation events that are catalyzed by ATP synthase are powered by a stream of protons. The gradient that drives these protons is set up by an electron transport chain that uses a molecule with low free energy—usually oxygen—as the final electron acceptor. Because this mode of ATP production links the phosphorylation of ADP with the oxidation of NADH and $FADH_2$, it is called **oxidative phosphorylation.**

During glycolysis and the Krebs cycle, however, ATP is produced by enzyme-catalyzed phosphorylation reactions. These reactions are called **substrate-level phosphorylation**. In substrate-level phosphorylation, an enzyme catalyzes the transfer of a phosphate group from a phosphorylated substrate to ADP, resulting in the production of ATP (**Figure 9.10**). The key distinction here is that the energy to produce the ATP comes from the phosphorylated substrate—not from a proton gradient, as it does in oxidative phosphorylation.

In many organisms, cellular respiration begins with the oxidation of glucose and ends with a flow of electrons through an electron transport chain to a final electron acceptor. The process starts with glycolysis and the Krebs cycle, which oxidize glucose to CO_2. As **Figure 9.11** shows, these reactions produce dramatic free-energy changes. Notice that relative to the other compounds in glycoysis and the Krebs cycle, glucose has a great deal of potential energy and low entropy. During glycolysis and the Krebs cycle, the free energy of each subsequent compound involves drops—at times in small steps and at other times in large steps.

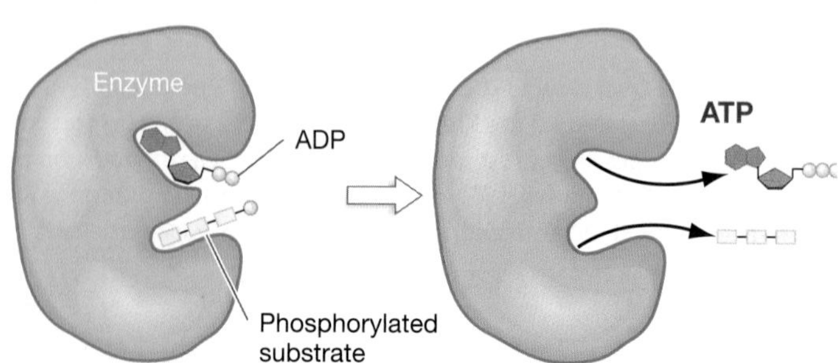

FIGURE 9.10 Substrate-Level Phosphorylation Involves an Enzyme and a Phosphorylated Substrate. Substrate-level phosphorylation occurs when an enzyme catalyzes the transfer of a phosphate group from a phosphorylated substrate to ADP, forming ATP.

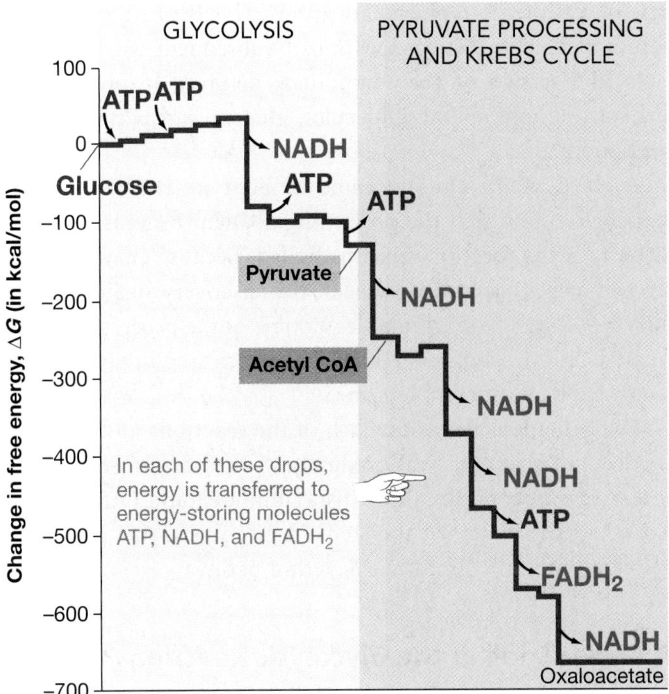

FIGURE 9.11 Free Energy Changes as Glucose Is Oxidized. Graph of free-energy changes during glycolysis and the Krebs cycle. Drops in free energy are associated with the production of ATP, NADH, and $FADH_2$. Oxaloacetate is the endpoint of the Krebs cycle.

QUESTION Which is associated with larger changes in free energy, production of ATP or production of NADH and $FADH_2$? Add labels indicating where in the cell glycolysis and the Krebs cycle take place.

The small drops in free energy are coupled with the production of ATP; the large drops are associated with the synthesis of NADH and $FADH_2$. In effect, then, the free energy present in glucose is transferred to ATP, NADH, and $FADH_2$. In the third (and final) stage of cellular respiration, NADH and $FADH_2$ carry these electrons to the electron transport chain.

If the molecule that functions as the final electron acceptor is unavailable, however, then cellular respiration cannot occur. In this case, alternative pathways called fermentation reactions take over (**Figure 9.12**). In eukaryotes, for example, fermentation occurs when oxygen is lacking. As we'll see in Section 9.5, fermentation consists of reactions that allow glycolysis to continue even if pyruvate is not drawn off to be processed into acetyl-CoA. By-products such as ethanol (the active ingredient in alcoholic beverages) and lactic acid result from the fermentation pathways present in various species of eukaryotes. In bacteria and archaea, other fermentation pathways and products are possible.

This overview of cellular respiration is intended simply to introduce the processes responsible for most ATP production in cells. Now we need to explore glycolysis, pyruvate processing, the Krebs cycle, and electron transport chains in more detail. Let's take a closer look at each process in turn.

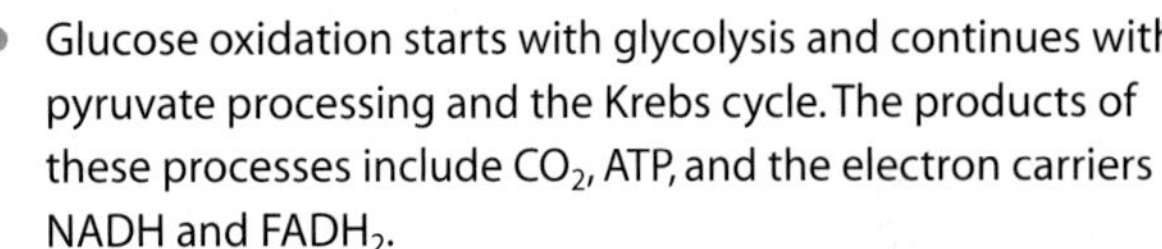

If you understand that...

- Glucose oxidation starts with glycolysis and continues with pyruvate processing and the Krebs cycle. The products of these processes include CO_2, ATP, and the electron carriers NADH and $FADH_2$.
- NADH and $FADH_2$ donate electrons to an electron transport chain, which pumps protons to create a proton gradient. Oxygen or another compound with low free energy serves as the final electron acceptor.
- Protons flowing down the gradient established by the electron transport chain allow ATP synthase to make ATP.

You should be able to...

Explain how oxidative phosphorylation and substrate-level phosphorylation differ in terms of

1) the nature of the enzyme or physical mechanism involved,
2) where they occur during the processing of glucose, and
3) the source of the energy to drive the endergonic reaction forming ATP.

BioFlix at www.masteringbio.com
Cellular Respiration

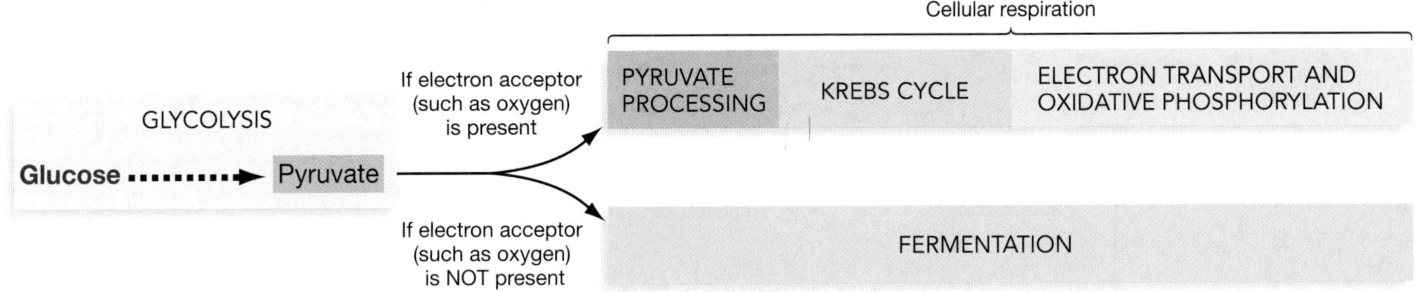

FIGURE 9.12 Cellular Respiration and Fermentation Are Alternative Pathways for Producing Energy. When oxygen or another electron acceptor used by the ETC is present in a cell, the pyruvate produced by glycolysis enters the Krebs cycle and the electron transport system is active. But if no electron acceptor is available to keep the ETC running, the pyruvate undergoes reactions known as fermentation.

9.3 Glycolysis

Glycolysis may be the most fundamental of all metabolic pathways, but it was discovered by accident. In the late 1890s Hans and Edward Buchner were working out techniques for manufacturing extracts of baker's yeast for commercial or medicinal use. (Yeast extracts are still added to some foods as a flavor enhancer or nutritional supplement.) In one set of experiments the Buchners added sucrose, or table sugar, to their extracts. Sucrose is a disaccharide consisting of glucose linked to fructose, another six-carbon sugar. At the time, sucrose was commonly used as a preservative—a substance used to preserve food from decay. But the Buchners found that instead of preserving the yeast extracts, the sucrose was quickly broken down and fermented, with alcohol appearing as a by-product. This was a key finding because it showed that fermentation and other types of cellular metabolism could be studied in vitro—outside the organism. Until then, researchers thought that metabolism could take place only in intact organisms.

The Buchners and other researchers followed up on this observation by trying to determine how the sugar was being processed. An important early observation was that the reactions could go on much longer than normal if inorganic phosphate was added to the mixture. This finding implied that some of the compounds involved were being phosphorylated. Soon after, a molecule called fructose bisphosphate was isolated. (The prefix *bis*– means that two phosphate groups are attached to the fructose molecule at distinct locations.) Subsequent work showed that all but two of the compounds involved in glycolysis—the starting and ending molecules, glucose and pyruvate—are phosphorylated.

A third major finding came to light in 1905, when researchers found that the processing of sugar by yeast extracts stopped if the reaction mix was boiled. Because enzymes were known to be inactivated by heat, this discovery suggested that enzymes were involved in at least some of the processing steps. Years later, investigators realized that each step in glycolysis is catalyzed by a different enzyme.

Over the next 35 years, each of the reactions and enzymes involved in glycolysis was gradually worked out by several different researchers. Because Gustav Embden and Otto Meyerhof identified many of the steps in glycolysis, this process is also known as the Embden–Meyerhof pathway.

A Closer Look at the Glycolytic Reactions

By breaking open cells, separating the cell components via differential centrifugation (introduced in Chapter 7), and testing which components could sustain glycolysis, biologists discovered something interesting: All 10 reactions of glycolysis occur in the cytosol. **Figure 9.13** details the 10 reactions involved in glycolysis. There are three key points to note about the reaction sequence:

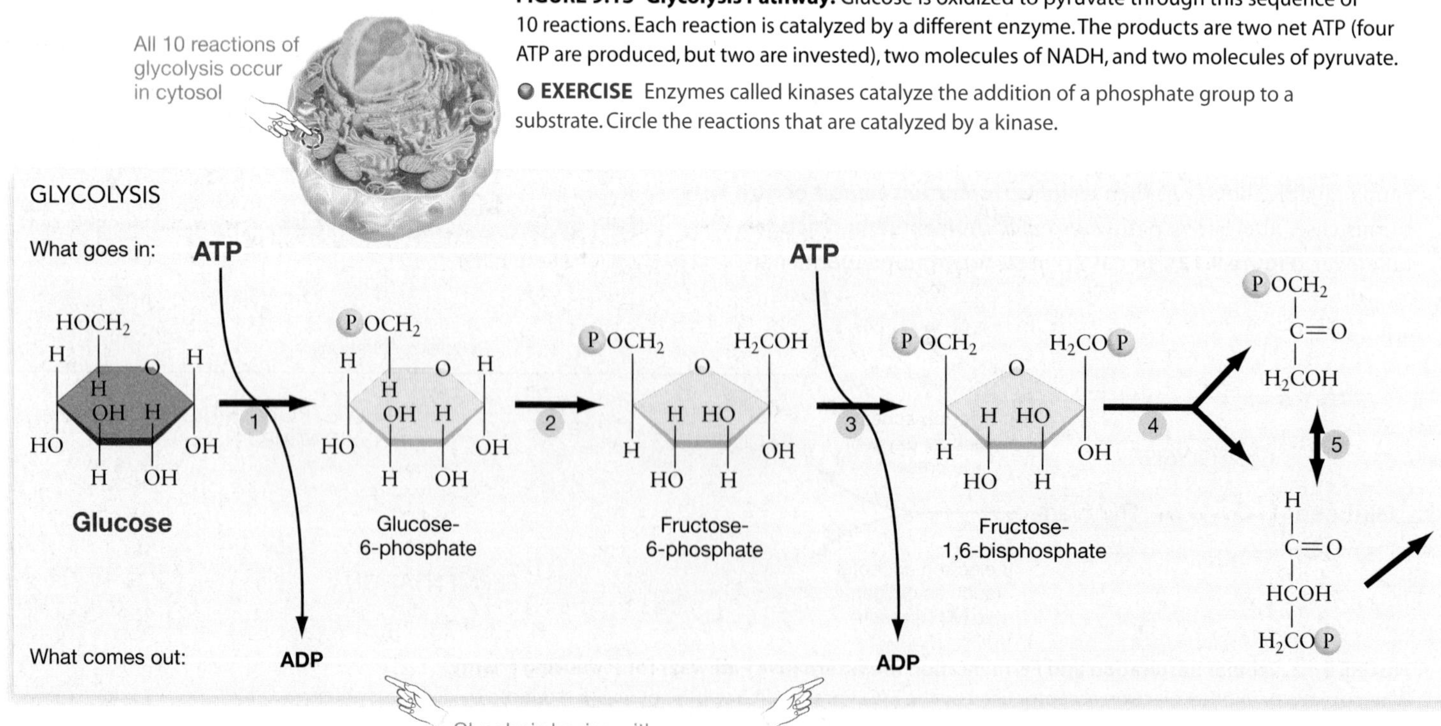

FIGURE 9.13 Glycolysis Pathway. Glucose is oxidized to pyruvate through this sequence of 10 reactions. Each reaction is catalyzed by a different enzyme. The products are two net ATP (four ATP are produced, but two are invested), two molecules of NADH, and two molecules of pyruvate.

● **EXERCISE** Enzymes called kinases catalyze the addition of a phosphate group to a substrate. Circle the reactions that are catalyzed by a kinase.

1. Contrary to what most researchers expected, glycolysis starts by *using* ATP, not producing it. In the initial step, glucose is phosphorylated to form glucose-6-phosphate. After an enzyme rearranges this molecule to fructose-6-phosphate in the second step, the third step in the reaction sequence adds a second phosphate group, forming the fructose-1,6-bisphosphate observed by early researchers. Thus, two ATP molecules are used up before any ATP is produced by glycolysis.

2. Once this energy-investment phase of glycolysis is complete, the subsequent reactions represent an energy payoff phase. The sixth reaction in the sequence results in the reduction of two molecules of NAD^+; the seventh produces two molecules of ATP and erases the energy "debt" of two molecules of ATP invested early in glycolysis. The final reaction in the sequence produces another two ATPs. For each molecule of glucose processed, the net yield is two molecules of NADH, two of ATP, and two of pyruvate.

3. The production of ATP during glycolysis occurs by substrate-level phosphorylation. In reactions 7 and 10 of Figure 9.13, an enzyme catalyzes the transfer of a phosphate group from a phosphorylated intermediate in glycolysis directly to ADP.

The discovery and elucidation of the glycolytic pathway ranks as one of the great achievements in the history of biochemistry. Because the enzymes involved have been observed in nearly every bacterium, archaean, and eukaryote examined, it is logical to infer that the ancestor of all organisms living today made ATP by glycolysis. The reactions outlined in Figure 9.13 are among the most ancient and fundamental of life processes.

How Is Glycolysis Regulated?

Once the glycolytic pathway was worked out, researchers focused on the structures and functions of the enzymes involved and the way the sequence is regulated. It seemed logical to predict that glycolysis does not occur at the maximum rate at all times. Instead, researchers predicted that glycolysis would proceed only when cells needed fresh supplies of ATP. An important advance occurred when biologists observed that high levels of ATP inhibit a key glycolytic enzyme called phosphofructokinase. **Phosphofructokinase** catalyzes step 3 in Figure 9.13—the synthesis of fructose-1,6-bisphosphate from fructose-6-phosphate.

The discovery that ATP inhibits phosphofructokinase was important because this enzyme catalyzes a key step in the reaction sequence. After steps 1 and 2 occur, an array of enzymes that aren't involved in glycolysis can convert the products of the reactions to reactants used in other metabolic pathways. Before step 3, then, the sequence can be interrupted and the intermediates used efficiently elsewhere in the cell. This is not true of later steps in glycolysis. Once fructose-1,6-bisphosphate is synthesized, then, there is no point in stopping the process. Based on these observations, it makes sense that the pathway is turned on or off at step 3.

The observation that ATP inhibits phosphofructokinase raises an important question, however: Why would ATP—a substrate that is required for the reaction in step 3—also inhibit the reaction? In the vast majority of cases, the addition of a substrate *speeds* the rate of a chemical reaction instead of slowing it.

To explain this unusual situation, biologists hypothesized that high levels of ATP are a signal that the cell does not need to produce more ATP. When an enzyme in a pathway is

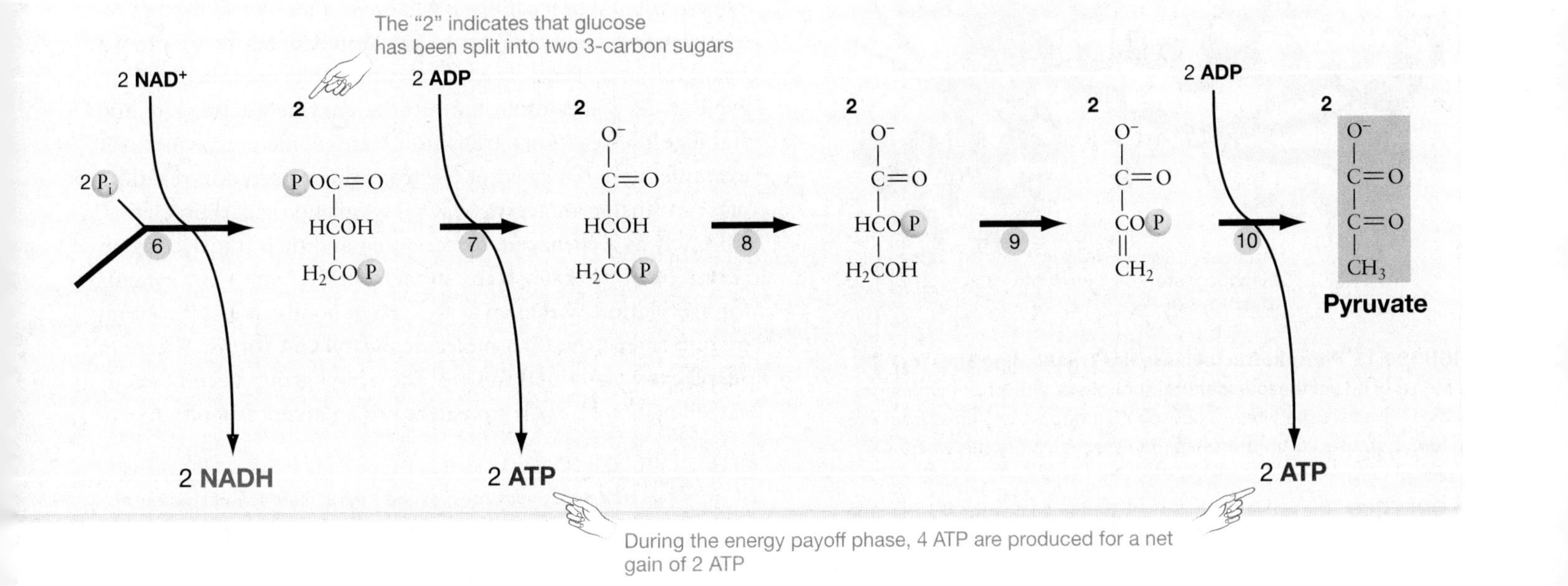

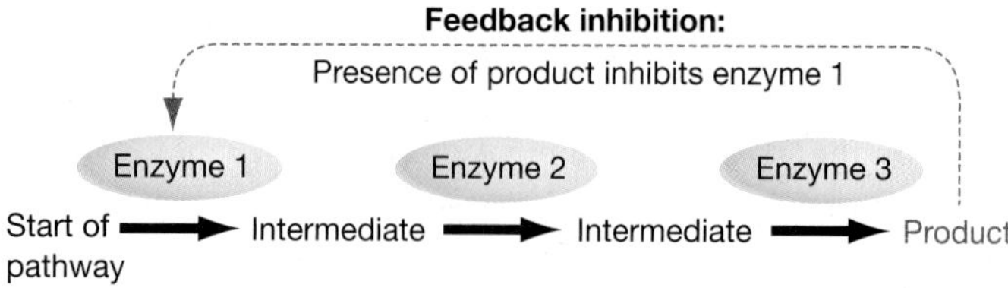

FIGURE 9.14 Metabolic Pathways May Be Regulated by Feedback Inhibition. Feedback inhibition occurs when the product of a metabolic pathway inhibits an enzyme that is active early in the pathway.

QUESTION Why would regulation be less effective if the product inhibited enzyme 3 instead of enzyme 1?

inhibited by the product of the reaction sequence, **feedback inhibition** is said to occur (**Figure 9.14**). The product molecule "feeds back" to stop the reaction sequence when the product is abundant. In glycolysis, cells that are able to stop the glycolytic reactions when ATP is abundant can conserve their stores of glucose for times when ATP is scarce. As a result, natural selection should favor individuals who have phosphofructokinase molecules that are inhibited by high concentrations of ATP.

This hypothesis provided a satisfying explanation for why ATP inhibits the key step of glycolysis, but it did not explain *how* high levels of the substrate inhibit the enzyme. The answer came later, when researchers were able to determine the three-dimensional structure of phosphofructokinase. As **Figure 9.15** shows, phosphofructokinase has two distinct binding sites for ATP. ATP can bind at the enzyme's active site or at a site that changes the enzyme's activity—a **regulatory site**. At the active site, ATP is converted to ADP and the phosphate group is transferred to fructose-6-phosphate. This reaction results in the synthesis of fructose-1,6-bisphosphate. If ATP concentrations are high, however, the molecule also binds at a regulatory site on phosphofructokinase. When ATP binds at this second location, the enzyme's conformation changes in a way that makes the reaction rate at the active site drop dramatically. In this case, ATP acts as an allosteric regulator (see Chapter 3).

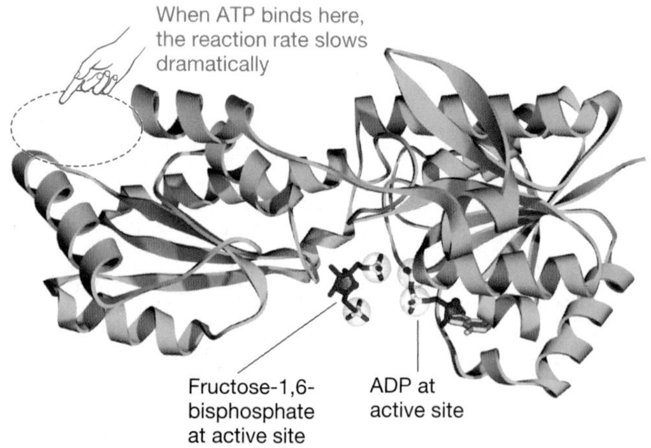

FIGURE 9.15 Phosphofructokinase Has Two Binding Sites for ATP. A model of one of the four identical subunits of phosphofructokinase. Notice the active site, where a phosphate group is transferred from ATP to form fructose-1,6-bisphosphate, and the regulatory site, where ATP binds.

QUESTION The active site has much higher affinity for ATP than the regulatory site does. What would be the consequences if the regulatory site had higher affinity for ATP than the active site did?

Thanks to advances such as these, glycolysis is among the best understood of all metabolic pathways. Now the question is, what happens to its product? How is pyruvate processed?

9.4 Pyruvate Processing

In eukaryotes, the pyruvate produced by glycolysis is transported from the cytosol to mitochondria. As noted in Chapter 7 noted, mitochondria are organelles found in virtually all eukaryotes. **Figure 9.16** shows a diagram and an image of this organelle generated with a technique called cryo-electron tomography.[1] Notice that a mitochondrion has two membranes and that the interior is filled with layers of sac-like structures called **cristae**. Cristae are connected to the main part of the inner membrane by short tubes. The region inside the inner membrane but outside the cristae is called the **mitochondrial matrix**.

The pyruvate produced by glycolysis moves across the mitochondrion's outer membrane and into the intermembrane space through small pores. Entry into the matrix occurs via active transport, through a membrane protein called the pyruvate carrier, located in the inner membrane. Because the transport of pyruvate into the mitochondrion is an active process, it represents an energy-consuming step in glucose oxidation.

What happens to pyruvate once it enters the mitochondrion? This issue remained unanswered until it was discovered that an organic compound called **coenzyme A (CoA)** serves as a cofactor in a wide variety of reactions catalyzed by cellular enzymes. You might recall from Chapter 3 that many enzymes require some type of cofactor to function. Cofactors are often a metal ion or a relatively small organic molecule called a coenzyme. Coenzymes often bind to the enzyme's active site and stabilize the reaction's transition state. Some coenzymes, for example, donate or accept electrons during a redox reaction; others transfer an acetyl ($-COCH_3$) group or carbon chains. CoA acts as a coenzyme by accepting and then transferring an acetyl group to a substrate. In fact, the "A" in "CoA" stands for acetylation. When an acetyl group bonds to a sulfur atom on one end of the CoA molecule, **acetyl CoA** forms. When an acetyl group is bound to CoA, the acetyl group becomes activated and can be easily transferred to an acceptor molecule.

[1]Cryo-electron tomography has provided a much more accurate picture of mitochondrial morphology compared to inferences based on transmission electron microscopy. For a recent review, see C. Mannella, *Biochemica et Biophysica Acta* 1762 (2006):140–147.

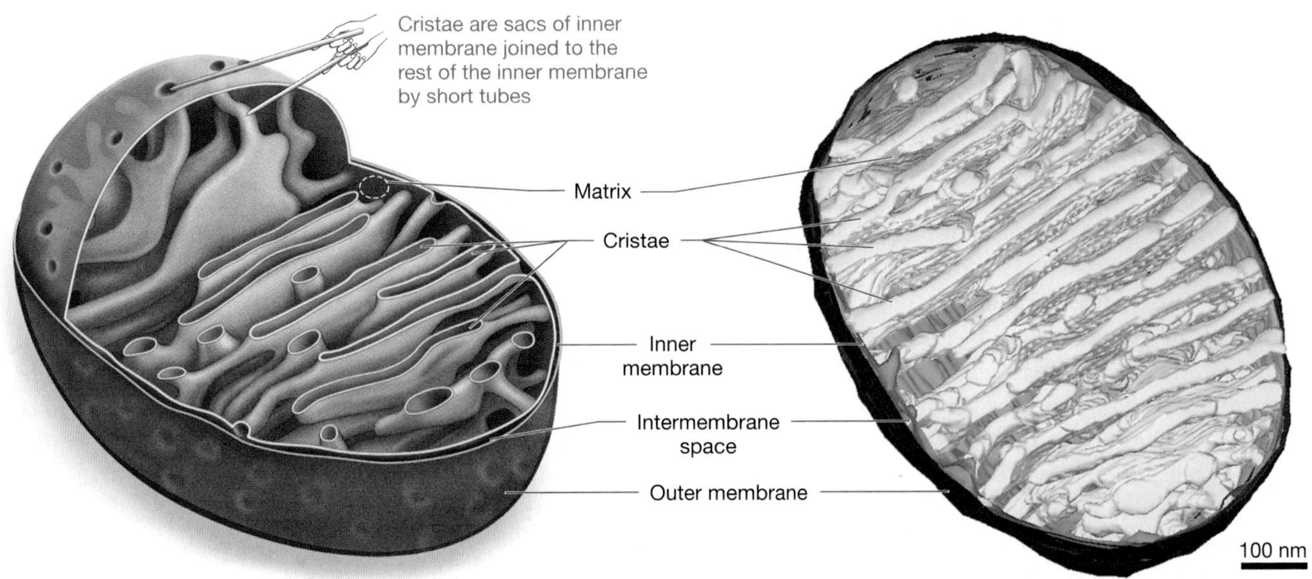

FIGURE 9.16 The Structure of the Mitochondrion. These images are based on recent research using an imaging technique called cryo-electron tomography. Notice that the mitochondria have outer and inner membranes, and that the inner membrane is connected by short tubes to sac-like structures called cristae. Pyruvate processing occurs within the mitochondrial matrix.

Soon after CoA's role as a coenzyme was revealed, other investigators found that pyruvate reacts with CoA to produce **acetyl CoA**. Follow-up work showed that the conversion of pyruvate to acetyl CoA occurs in a series of steps inside an enormous and intricate enzyme complex called **pyruvate dehydrogenase.** In eukaryotes, pyruvate dehydrogenase is located in the mitochondrial matrix. In bacteria and archaea, pyruvate dehydrogenase is located in the cytosol.

As pyruvate is being processed, NAD^+ is reduced to NADH and one of the carbons in the pyruvate is oxidized to CO_2. The remaining two-carbon acetyl unit is transferred to CoA (**Figure 9.17**). Acetyl CoA is the final product of the pyruvate processing step in glucose oxidation.

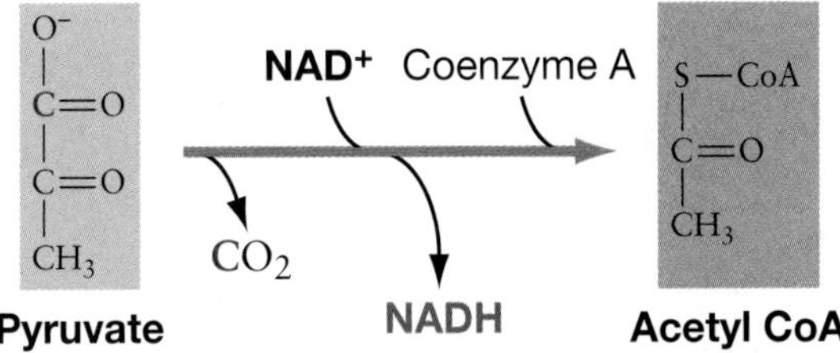

FIGURE 9.17 Pyruvate Is Oxidized to Acetyl CoA. The reaction shown here is catalyzed by pyruvate dehydrogenase.

EXERCISE Above the reaction arrow, list 3 molecules whose presence speeds up the reaction. Label them "Positive control." Below the reaction arrow, list 3 molecules whose presence slows down the reaction. Label them "Feedback inhibition."

EXERCISE Circle the acetyl group ($-C_2H_3O$) on acetyl CoA. Label the sulfur atom on CoA where the acetyl group bonds.

When supplies of ATP are abundant, however, the events just described fail to occur. Pyruvate processing stops when the pyruvate dehydrogenase complex becomes phosphorylated. In this case, phosphorylation changes the shape of pyruvate dehydrogenase in a way that inhibits catalytic activity. High concentrations of other products—specifically acetyl CoA and NADH—also increase the rate of phosphorylation of the enzyme complex. Like the control of phosphofructokinase activity in glycolysis, these are examples of feedback inhibition. Reaction products feed back to stop or slow down the pathway.

On the contrary, the presence of high concentrations of NAD^+, CoA, or adenosine monophosphate (AMP)—which indicates low ATP supplies—speeds up the reactions catalyzed by the pyruvate dehydrogenase complex. To capture this point, biologists say that pyruvate processing is under both positive and negative control. Large supplies of products inhibit the enzyme complex; large supplies of reactants and low supplies of products stimulate it. Pyruvate processing is a key regulatory point in glucose oxidation.

If ATP supplies are low in a cell, though, the complex actively churns out acetyl CoA. This compound participates in a reaction that kicks off the next major step in glucose oxidation: the Krebs cycle.

9.5 The Krebs Cycle

While Embden, Meyerhof, and others were working out the sequence of reactions in glycolysis, biologists in several laboratories were focusing on a different set of redox reactions that take place in actively respiring cells. These reactions involve small organic acids such as citrate, malate, and succinate. Because they have the form R-COOH, these molecules are called carboxylic acids. In some cases, the redox reactions that produce carboxylic acids also result in the production of carbon dioxide. Recall from Section 9.1 that carbon dioxide is a highly oxidized form of carbon and that it is the endpoint of glucose metabolism. Thus, it was logical for researchers to propose that the oxidation of small carboxylic acids could be an important component of glucose oxidation.

Early researchers made several key observations about these reactions. First, a total of eight small carboxylic acids are oxidized rapidly enough to imply that they are involved in glucose metabolism—the most rapid set of oxidation reactions predicted to occur in cells. Second, when one of these carboxylic acids is added to cells, the rate of glucose oxidation increases. The added molecules did not appear to be used up, however. Instead, virtually all of the carboxylic acid added seemed to be recovered later. Why this should be so was a puzzle. Even while this observation remained unexplained, however, the biochemists working on the reactions were able to determine the order in which the eight acids were oxidized. The result is the reaction pathway shown in **Figure 9.18a**.

The mystery of why intermediates in the pathway were not used up was resolved when Hans Krebs realized that the reaction sequence might occur in a cyclical fashion instead of a linear pathway. Krebs had another crucial insight when he suggested that the reaction sequence was directly tied to the processing of pyruvate produced by glycolysis. To test these hypotheses, Krebs and a colleague set out to determine whether pyruvate—the endpoint of the glycolytic pathway—could react with oxaloacetate—the most oxidixed carboxylic acid. Because oxaloacetate was the endpoint of the pathway in Figure 9.18a, the cycle hypothesis predicted that a reaction of oxaloacetate with pyruvate should produce the starting point of the pathway—the six-carbon carboxylic acid called citrate (**Figure 9.18b**).

The result? As predicted, citrate formed when the biologists added oxaloacetate and pyruvate to cells. Based on this result, Krebs proposed that pyruvate is oxidized to carbon dioxide through a cycle of reactions. In honor of this insight, the pathway became known as the Krebs cycle.[2]

When radioactive isotopes of carbon became available in the early 1940s, researchers in several labs used them to confirm the cyclical nature of the Krebs cycle. For example, by adding radioactively labeled citrate or pyruvate to cells and analyzing the radioactive compounds that resulted, it was possible to show that carbon atoms cycle through the sequence of reactions just as Krebs had proposed. Later, the discovery of the pyruvate dehydrogenase complex showed that acetyl CoA is the key compound that feeds the Krebs cycle, not pyruvate.

To summarize Sections 9.4 and 9.5, the reactions catalyzed by pyruvate dehydrogenase act as a preparatory step for the Krebs cycle. The acetyl CoA that is produced by the pyruvate processing goes on to react with oxaloacetate, forming citrate. When processing of pyruvate in the mitochondrion is complete, all three carbons in pyruvate are oxidized to carbon dioxide. One

[2]The Krebs cycle is also known as the *citric acid cycle* because it starts with citrate, which becomes citric acid when protonated. Because citric acid has three carboxyl groups, the reaction sequence is also called the *tricarboxylic acid (TCA) cycle*.

(a) The carboxylic acids that are oxidized during cellular respiration

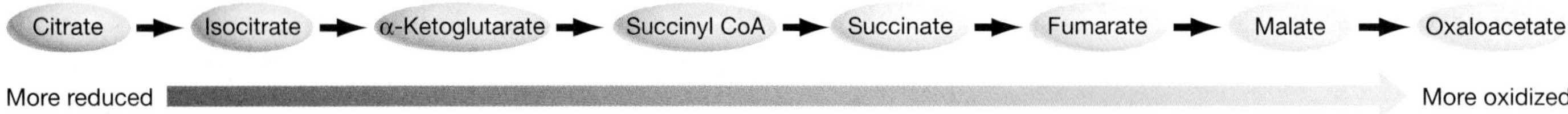

(b) The Krebs cycle hypothesis

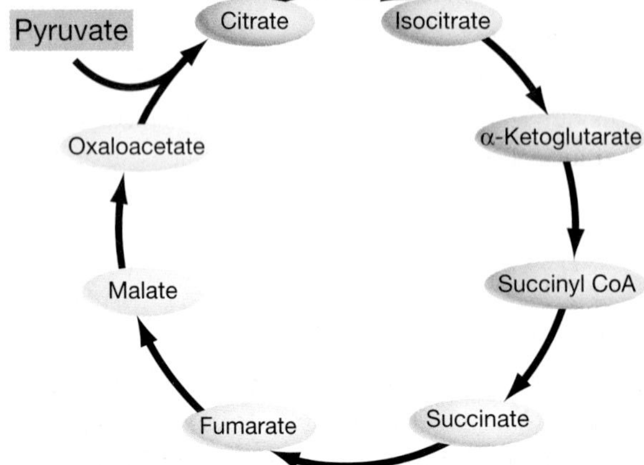

FIGURE 9.18 A Series of Carboxylic Acids Is Oxidized in the Krebs Cycle. (a) The eight molecules that are oxidized in actively respiring cells are pictured in order of their tendency to donate or accept electrons in redox reactions. **(b)** Hans Krebs proposed that if the pyruvate from glycolysis reacted with oxaloacetate to form citrate, then the eight reactions in part (a) would form a cycle.

EXERCISE Krebs turned out to be wrong about pyruvate starting the cycle. In part (b), add the correct molecule that links pyruvate to the cycle.

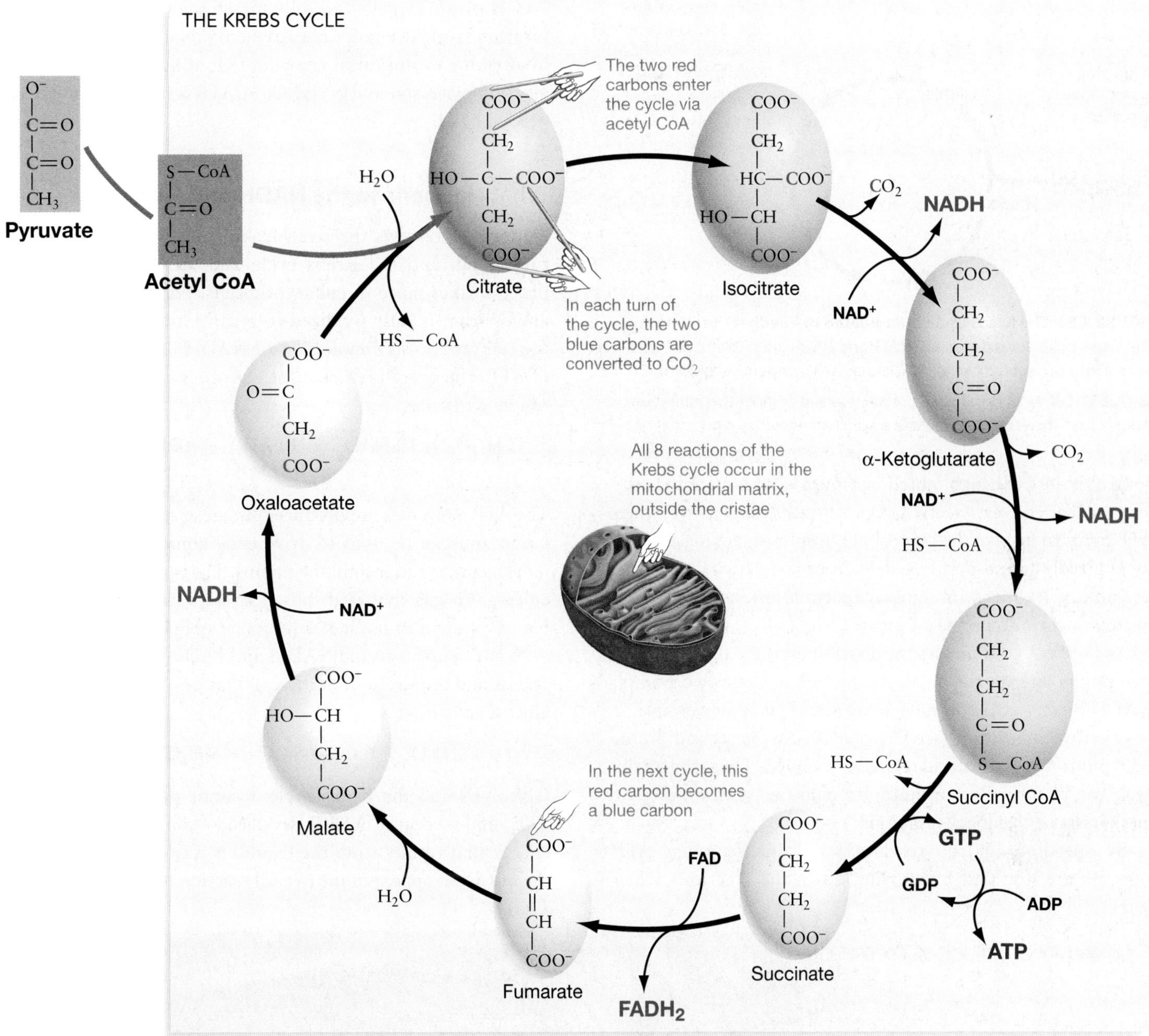

FIGURE 9.19 The Krebs Cycle Completes the Oxidation of Glucose. Acetyl CoA goes into the Krebs cycle, and carbon dioxide, NADH, $FADH_2$, and GTP come out. The GTP is produced by substrate-level phosphorylation and then converted to ATP. If you follow individual carbon atoms around the cycle several times, you'll come to an important conclusion: All the carbons in the cycle are eventually positioned as "blue carbons" that are released as CO_2. This occurs because the "red carbons" in fumarate—which is symmetric—can flip position when malate forms.

of the carbons is oxidized in the pyruvate dehydrogenase complex, and two are oxidized in the Krebs cycle itself. As **Figure 9.19** indicates, the energy released by the oxidation of one molecule of acetyl CoA is used to produce three molecules of NADH, one of $FADH_2$, and one of **guanosine triphosphate (GTP)** through substrate-level phosphorylation. GTP is then converted to ATP. Pyruvate and CoA enter the sequence of reactions; CoA, carbon dioxide, ATP, NADH, and $FADH_2$ come out. In bacteria and archaea, the enzymes responsible for pyruvate processing and the Krebs cycle are located in the cytosol. In eukaryotes, most of the enzymes responsible for the Krebs cycle are located in the mitochondrial matrix.

How Is the Krebs Cycle Regulated?

Like glycolysis and pyruvate processing, the Krebs cycle is carefully regulated. Reaction rates are high when ATP is scarce; reaction rates are low when ATP is abundant. The major

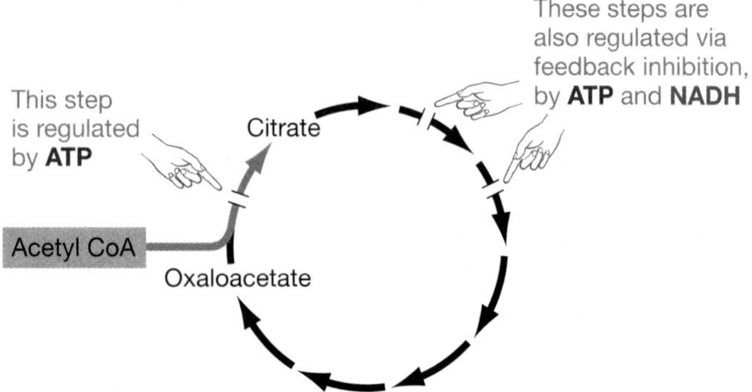

FIGURE 9.20 The Krebs Cycle Is Regulated by Feedback Inhibition. The Krebs cycle slows down when ATP and NADH are plentiful. ATP acts as an allosteric regulator, while NADH acts as a competitive inhibitor.

● **QUESTION** How do allosteric regulation and competitive inhibition differ? (Your answer should include a sketch of how they work.)

control points are highlighted in **Figure 9.20**. Notice that the enzyme that converts acetyl CoA to citrate is shut down when ATP binds to it. Like the control of phosphofructokinase activity in glycolysis and pyruvate dehydrogenase activity, this is an example of feedback inhibition. Reaction products feed back to stop or slow down the pathway.

As Figure 9.20 indicates, feedback inhibition also occurs at two points later in the cycle. At the first of these two points, NADH binds to the enzyme's active site. This is an example of competitive inhibition (see Chapter 3). At the second point, ATP binds to an allosteric regulatory site. ● Thus, the Krebs cycle can be turned off at multiple points, via several different mechanisms of feedback inhibition.

By speeding up the oxidation of glucose when ATP supplies are low and by slowing down the reactions when ATP is plentiful, cells carefully match the rate of cellular respiration to their energy requirements. Natural selection has favored the evolution of enzymes that allow cells to conserve glucose, pyruvate, and acetyl CoA. ATP is produced on an as-needed basis.

What Happens to the NADH and $FADH_2$?

Figure 9.21 reviews the relationships of glycolysis, pyruvate processing, and the Krebs cycle and identifies where each process takes place in eukaryotic cells. For each molecule of glucose that is fully oxidized to 6 carbon dioxide molecules, the cell produces 10 molecules of NADH, 2 of $FADH_2$, and 4 of ATP. The overall reaction for glycolysis and the Krebs cycle can be written as

$$C_6H_{12}O_6 + 10\ NAD^+ + 2\ FAD + 4\ ADP + 4\ P_i \rightarrow 6\ CO_2 + 10\ NADH + 2\ FADH_2 + 4\ ATP$$

The ATP molecules are produced by substrate-level phosphorylation and can be used to drive endergonic reactions, power movement, or run membrane pumps. The carbon dioxide molecules are a gas that is disposed of as waste—you exhale it; plants release it or use it as a reactant in photosynthesis.

What happens to the NADH and $FADH_2$ produced by glycolysis and the Krebs cycle? Recall that the overall reaction for glucose oxidation is

$$C_6H_{12}O_6 + 6\ O_2 \rightarrow 6\ CO_2 + 6\ H_2O + \text{energy}$$

Glycolysis and the Krebs cycle account for the glucose, the CO_2, and—because ATP is produced—some of the chemical energy that results from the overall reaction. But the O_2 and the H_2O that appear in the overall reaction for the oxidation of

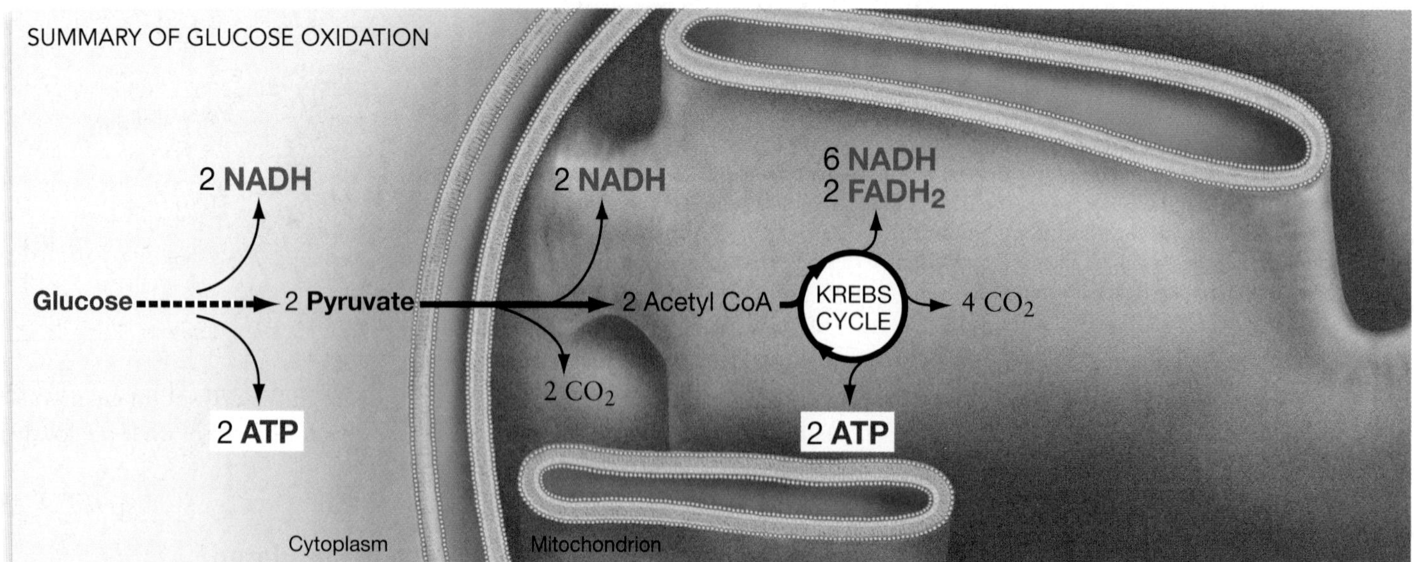

FIGURE 9.21 A Summary of Glucose Oxidation. Glucose is completely oxidized to carbon dioxide via glycolysis, the subsequent oxidation of pyruvate, and then the Krebs cycle. In eukaryotes, glycolysis occurs in the cytosol; pyruvate oxidation and the Krebs cycle take place in the mitochondrial matrix.

● **EXERCISE** Label the outer membrane, the inner membrane, and the sac-like cristae of the mitochondrion.

glucose are still unaccounted for. As it turns out, so is much of the chemical energy. The reaction that has yet to occur is

$$\text{NADH} + \text{FADH}_2 + \text{O}_2 + \text{ADP} + \text{P}_i \rightarrow \text{NAD}^+ + \text{FAD} + \text{H}_2\text{O} + \text{ATP}$$

In this reaction, oxygen is reduced to form water. The electrons that drive the redox reaction come from NADH and $FADH_2$. These molecules are oxidized to NAD^+ + FAD.

In effect, glycolysis, pyruvate processing, and the Krebs cycle transfer electrons from glucose to NAD^+ and FAD, creating NADH and $FADH_2$. These molecules then carry the electrons to oxygen, which serves as the final electron acceptor in eukaryotic cells. When oxygen accepts electrons, water is produced. All the components of the overall reaction for glucose oxidation are accounted for.

How does this final part of the process occur? Specifically, how is ATP generated as electrons are transferred from NADH or $FADH_2$ to O_2? In the 1960s—decades after the details of glycolysis and the Krebs cycle had been worked out—a startling and highly controversial answer to these questions emerged.

Check Your Understanding

If you understand that...

- During glycolysis, pyruvate processing, and the Krebs cycle, one molecule of glucose ($C_6H_{12}O_6$) is completely oxidized to six molecules of carbon dioxide (CO2).

You should be able to...

Model the following components of cellular respiration by pretending that a large piece of paper is a cell. Draw a large mitochondrion inside it. Cut out small squares of paper and label them as glucose, glycolytic reactions, Krebs cycle reactions, pyruvate dehydrogenase complex, pyruvate, acetyl CoA, CO_2, ADP → ATP, NAD → NADH, and FAD → $FADH_2$.

1) Put each of the squares in the appropriate location in the cell.
2) Draw arrows to connect the appropriate molecules and reactions.
3) Using dimes for electrons, demonstrate how glucose is oxidized to CO_2.
4) Label points where regulation occurs.

Once your model is working, you'll be ready to consider what happens to the NADH and $FADH_2$ you've produced.

9.6 Electron Transport and Chemiosmosis

The answer to one fundamental question about the oxidation of NADH and $FADH_2$ turned out to be relatively straightforward. To determine where the oxidation reactions take place in eukaryotes, researchers isolated mitochondria by using differential centrifugation techniques (introduced in Chapter 7). Then they broke open the organelles and separated the inner and outer membranes from the mitochondrial matrix and the solution in the intermembrane space. They found that NADH could be oxidized in the isolated membranes, but not in the matrix or in the intermembrane fluid. Follow-up work showed that the oxidation process takes place on the inner membrane of the mitochondria and the membranes of cristae. In prokaryotes, the oxidation of NADH occurs in the plasma membrane.

Biologists who analyzed the components of these membranes made a key discovery when they isolated molecules that switch between a reduced and an oxidized state during respiration. The molecules were hypothesized to be the key to processing NADH and $FADH_2$. What are these molecules, and how do they work?

Components of the Electron Transport Chain

Collectively, the molecules responsible for the oxidation of NADH and $FADH_2$ are designated the electron transport chain (ETC). As electrons are passed from one protein to another in the chain, the energy released by the redox reactions is used to pump protons across the inner membrane of mitochondria. Once this proton gradient is established, a stream of protons through the enzyme ATP synthase drives the production of ATP from ADP and Pi. Once the electrons at the bottom of the chain are accepted by oxygen to form water, the oxidation of glucose is complete.

Several points are fundamental to understanding how the ETC works:

- Most of these molecules are proteins that contain distinctive chemical groups where the redox events take place. The active groups include ring-containing structures called flavins or iron-sulfur complexes or iron-containing heme groups. Each of these subunits is readily reduced or oxidized.
- The inner membrane of the mitochondrion also contains a molecule called **ubiquinone**, which is not a protein. Ubiquinone got its name because it belongs to a family of compounds called quinones and because it is nearly ubiquitous in organisms. It is also called **coenzyme Q** or simply **Q**. Ubiquinone consists of a carbon-containing ring attached to a long tail made up of isoprene subunits. The structure of Q determines the molecule's function. The long, isoprene-rich tail is hydrophobic. As a result, Q is lipid soluble and can move throughout the mitochondrial membrane efficiently. In contrast, all but one of the proteins in the ETC are embedded in the membrane.
- The molecules involved in processing NADH and $FADH_2$ differ in electronegativity, or their tendency to hold electrons.

Because Q and the ETC proteins can cycle between a reduced state and an oxidized state, and because they differ in electronegativity, investigators realized that it should be possible to arrange these molecules into a logical sequence. The idea

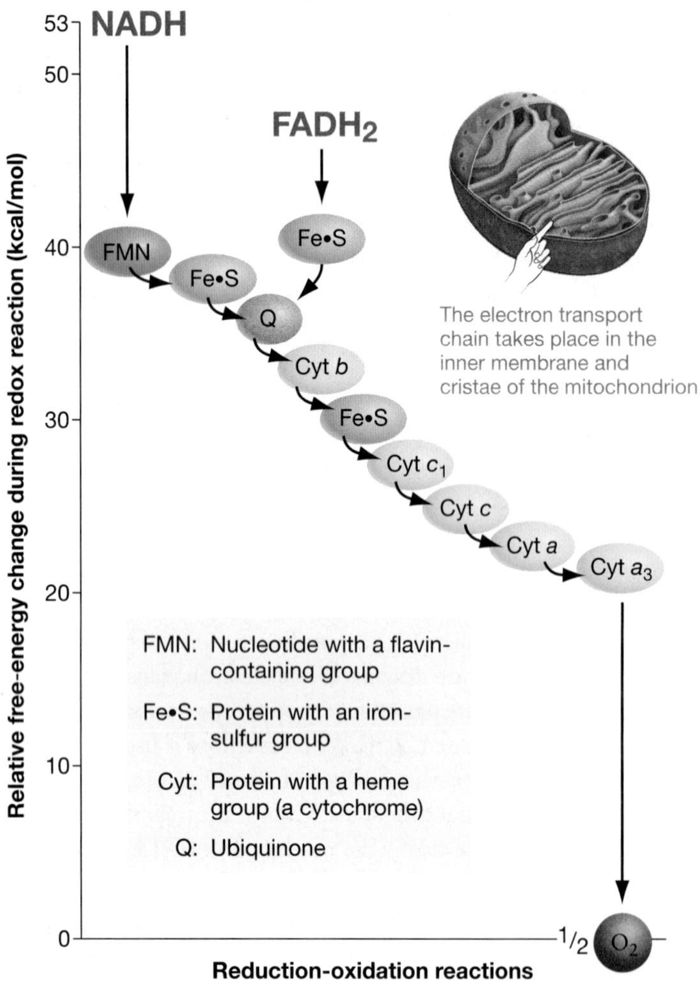

FIGURE 9.22 A Series of Reduction-Oxidation Reactions Occurs in an Electron Transport Chain. Electrons step down in potential energy from the electron carriers NADH and $FADH_2$ through an electron transport chain to a final electron acceptor. When oxygen is the final electron acceptor, water is formed. The overall free-energy change of 53 kcal/mol (from NADH to oxygen) is broken into small steps.

EXERCISE Label the most and least electronegative molecules in the figure.

was that electrons would pass from a molecule with lower electronegativity to one with higher electronegativity, via a redox reaction. As electrons moved through the chain, they would be held in covalent bonds more and more tightly. A small amount of energy would be released in each reaction, and the potential energy in each successive bond would lessen.

Figure 9.22 shows how electrons step down in potential energy from the electron carriers NADH and $FADH_2$ to O_2. Researchers worked out the sequence of compounds in the chain by experimenting with poisons that inhibit particular proteins in the inner membrane. For example, when an electron transport chain is treated with the drug antimycin A, cytochrome *b* and Q are reduced; but all of the other elements in the chain remain oxidized. This pattern makes sense only if electrons flow from NADH and $FADH_2$ to cytochrome *b* and Q before being passed on to other components.

Experiments with other poisons showed that NADH donates an electron to a flavin-containing protein at the top of the chain, while $FADH_2$ donates electrons to an iron-sulfur-containing protein that then passes them directly to Q. After passing through each of the remaining components in the chain, the electrons are finally accepted by oxygen. Under standard conditions of temperature and pressure in the laboratory, the total potential energy difference from NADH to oxygen is a whopping 53 kilocalories/mole (abbreviated kcal/mol).

To summarize, electrons from NADH and $FADH_2$ pass through an electron transport chain consisting of a series of proteins and Q. Each successive molecule in the chain has slightly higher electronegativity than the previous one. As the electrons move from one link in the chain to the next, they are held tighter and tighter and their potential energy lessens. A molecule with particularly high electronegativity—oxygen, in plants, animals, fungi, and many bacteria and archaea—acts as the final electron acceptor. Electron transport takes place within the inner mitochondrial membrane.

Once the nature of the electron transport chain became clear, biologists understood the fate of the electrons carried by NADH and $FADH_2$ and how oxygen acts as the final electron acceptor. All of the electrons that were originally present in glucose were now accounted for. This is satisfying, except for one crucial question: How do these redox reactions generate ATP? Does substrate-level phosphorylation take place in the electron transport chain, just as it does in glycolysis and the Krebs cycle?

The Chemiosmotic Hypothesis

Throughout the 1950s most biologists working on cellular respiration assumed that electron transport chains include enzymes that catalyze substrate-level phosphorylation. Despite intense efforts, however, no one was able to find a component of the ETC that phosphorylated ADP to produce ATP.

In 1961 Peter Mitchell made a radical break with prevailing ideas by proposing that the connection between electron transport and ATP production is indirect. His hypothesis: The real job of the electron transport chain is to pump protons from the matrix of the mitochondrion through the inner membrane and out to the intermembrane space or the interior of cristae. According to Mitchell, the pumping activity of the electron transport chain would lead to a buildup of protons in these areas. In this way, the intermembrane space or inside of cristae would become positively charged relative to the matrix and would have a much higher concentration of protons. The result would be a strong electrochemical gradient favoring the movement of protons back into the matrix. This **proton-motive force**, he hypothesized, is used by an enzyme in the inner membrane to synthesize ATP.

Mitchell called the production of ATP via a proton gradient **chemiosmosis**, and his proposal of an indirect link between electron transport and ATP production is the chemiosmotic hypothesis. Although proponents of a direct link between

Experiment

Question: How are the electron transport chain and ATP production linked?

Chemiosmotic hypothesis: The linkage is indirect. The ETC creates a proton-motive force that drives ATP synthesis by a mitochondrial protein.

Alternative hypothesis: The linkage is direct. The ETC is associated with enzymes that perform substrate-level phosphorylation.

Experimental setup:

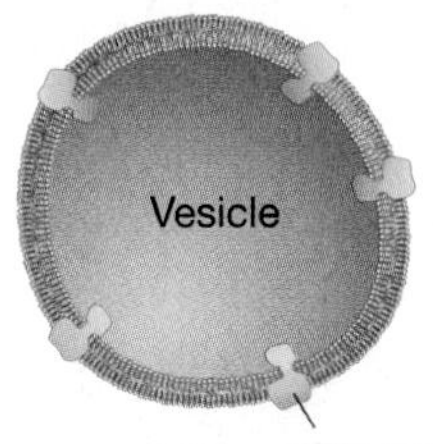

1. Create vesicles from artificial membranes; add ATP-synthesizing enzyme found in mitochondria.

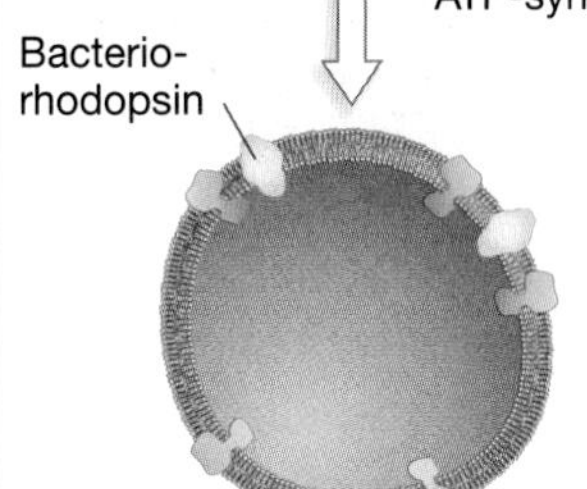

2. Add bacteriorhodopsin, a protein that acts as a light-activated proton pump.

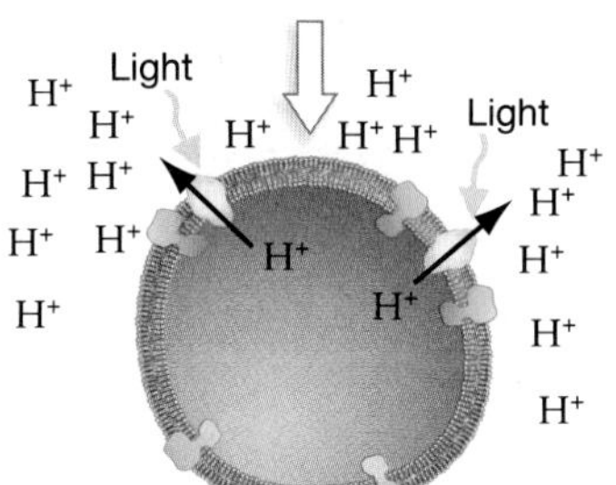

3. Illuminate vesicle so that bacteriorhodopsin pumps protons out of vesicle, creating a proton gradient.

Prediction of chemiosmotic hypothesis: ATP will be produced within the vesicle.

Prediction of alternative hypothesis: No ATP will be produced.

Results:

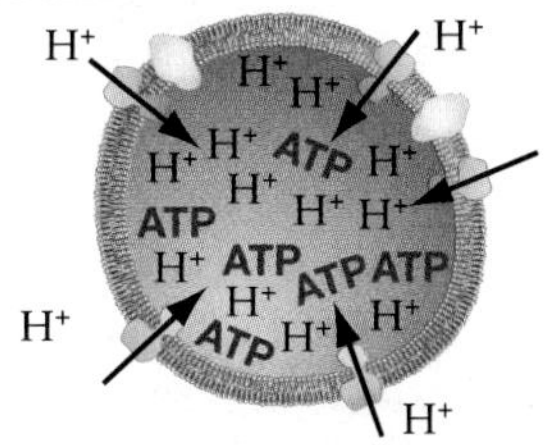

ATP is produced within the vesicle, in the absence of the electron transport chain.

Conclusion: The linkage between electron transport and ATP synthesis is indirect; the movement of protons drives the synthesis of ATP.

FIGURE 9.23 Evidence for the Chemiosmotic Hypothesis.

QUESTION Do you regard this as a convincing test of the chemiosmotic hypothesis? Why or why not?

electron transport and substrate-level phosphorylation objected vigorously to Mitchell's idea, several key experiments supported it.

One of these experiments demonstrated the existence of a key element in Mitchell's scheme (**Figure 9.23**): A mitochondrial enzyme can use a proton gradient to synthesize ATP. The biologists who did this experiment began by making vesicles from artificial membranes that contained an ATP-synthesizing enzyme found in mitochondria. Along with this enzyme, they inserted bacteriorhodopsin. Bacteriorhodopsin is a well-studied membrane protein that acts as a light-activated proton pump. When light strikes this protein, the protein absorbs some of the light energy and changes conformation in a way that results in the pumping of protons from the interior of a cell or membrane-bound vesicle to the outside. As a result, the experimental vesicles established a strong electrochemical gradient favoring proton movement to the interior. When the vesicles were illuminated to initiate proton pumping, ATP began to be produced from ADP inside the vesicles. This result provided strong support for the chemiosmosis hypothesis. Mitchell's prediction was correct: In this situation, ATP production depended solely on the existence of a proton-motive force. It could occur in the *absence* of an electron transport chain.

Based on this experiment and others, most biologists accepted the chemiosmotic hypothesis as valid. Instead of being produced by enzymes inside the mitochondrion, ATP is produced by a flow of protons. In Section 9.1 you considered an analogy between the ETC system and a waterwheel. You can also liken chemiosmosis to a hydroelectric dam. The electron transport chain is analogous to a series of gigantic pumps that force water up and behind a dam. The inner mitochondrial membrane functions as the dam, and ATP synthase is like the turbines inside the dam. At a hydroelectric dam, the movement of water makes turbines spin and generate electricity. In a mitochondrion, protons are pumped instead of water. When protons move through ATP synthase, the protein spins and generates ATP. If you understand chemiosmosis, you should be able to explain why ATP production during cellular respiration is characterized as indirect. More specifically, you should be able to explain the relationship between glucose oxidation, the proton gradient, and ATP synthase.

Electron transport chains and ATP synthase occur in organisms throughout the tree of life. They are humming away in your cells now. Let's look in more detail at how they function.

How Is the Electron Transport Chain Organized?

Once the predictions of the chemiosmotic hypothesis were verified, researchers focused on understanding the three-dimensional structure of the components of the electron transport chain and on determining how electron transport is coupled to proton pumping. This work has shown that the

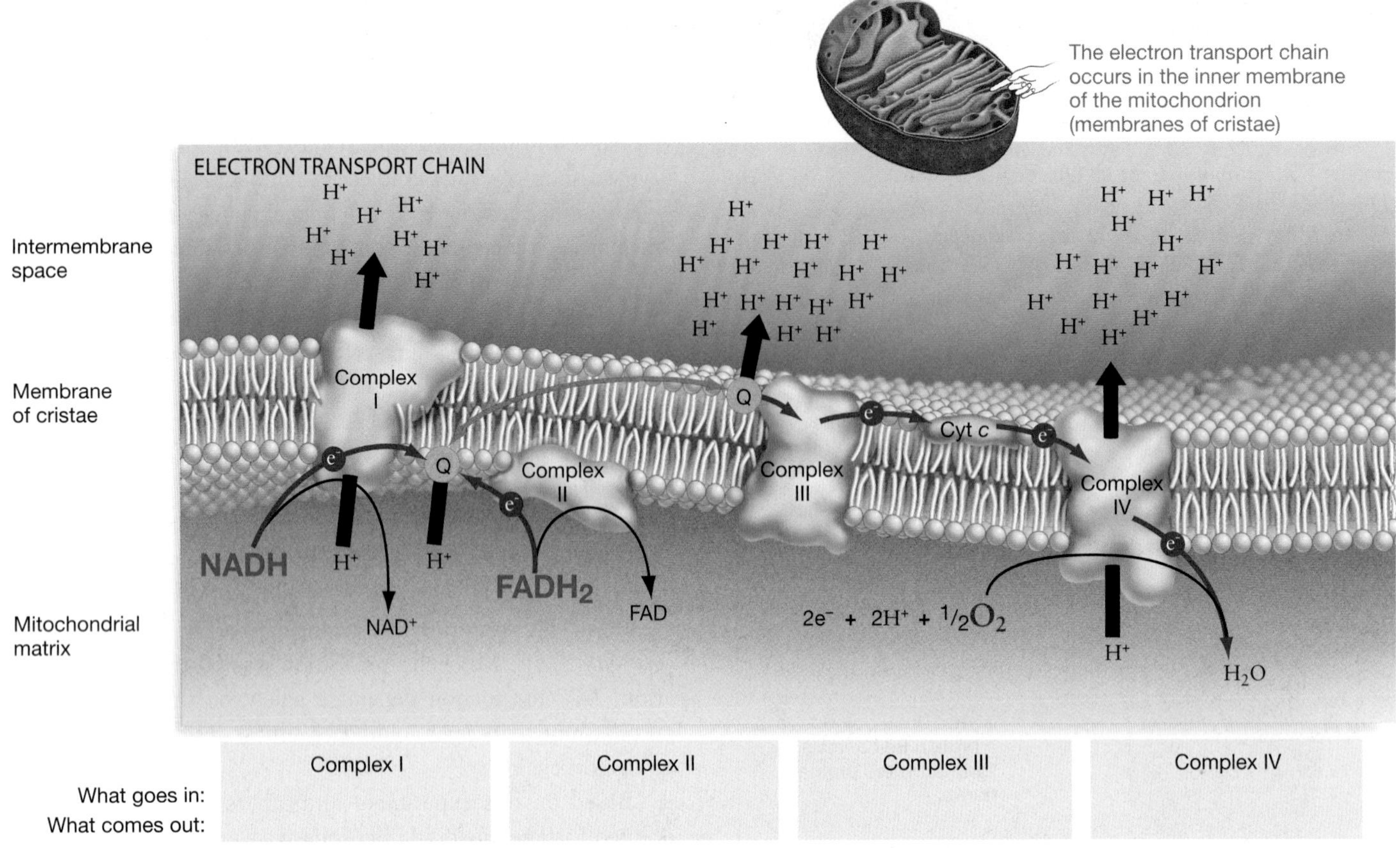

FIGURE 9.24 How Is the Electron Transport Chain Organized? The individual components of the electron transport chain diagrammed in Figure 9.22 are grouped into large multiprotein complexes. Electrons are carried from one complex to another by Q and by cytochrome *c*; Q also shuttles protons across the membrane. The orange arrow indicates Q moving back and forth. Complexes I and IV use the potential energy released by the redox reactions to pump protons from the mitochondrial matrix to the intermembrane space.

EXERCISE Add an arrow across the membrane and label it "Proton gradient." In the boxes at the bottom, list "What goes in" and "What comes out" for each complex.

ETC components are organized into four large complexes of proteins and cofactors (**Figure 9.24**). Protons are pumped by two of the complexes. Q and the protein **cytochrome *c*** act as shuttles that transfer electrons between complexes. Q also carries a proton across the membrane along with an electron.

The structural studies completed to date confirm that in complexes I and IV, protons actually pass directly through a sequence of electron carriers. The exact route taken by the protons is still being worked out. It is also not clear how the redox reactions taking place inside each complex—as electrons step down in potential energy—make proton movement possible.

The best-understood interaction between electron transport and proton pumping takes place in complex III. Research has shown that when Q accepts electrons from complex I or complex II, it also gains protons. The reduced form of Q then diffuses to the outer side of the inner membrane, where its electrons are used to reduce a component of complex III near the intermembrane space. The protons held by Q are released to the intermembrane space. In this way, Q shuttles electrons and protons from one side of the membrane to the other. The electrons proceed down the transport chain, and the protons released to the intermembrane space contribute to the proton-motive force.

Now, how does this proton gradient make the production of ATP possible?

The Discovery of ATP Synthase

In 1960 Efraim Racker made several key observations about how ATP is synthesized in mitochondrial membranes. When he used mitochondrial membranes to make vesicles, Racker noticed that some happened to form with their membrane inside out. Electron microscopy revealed that the inside-out membranes had numerous large proteins studded along their surfaces. Each of these proteins appeared to have a base in the membrane with a lollipop-shaped stalk and a knob that projected out from the membrane (**Figure 9.25a**). If the solution was vibrated or treated with a compound called urea, the stalks and knobs fell off.

Racker seized on this technique to isolate the stalks and knobs and do experiments with them. For example, he found that isolated stalks and knobs could hydrolyze ATP, forming

(a) A vesicle formed from an "inside-out" mitochondrial membrane

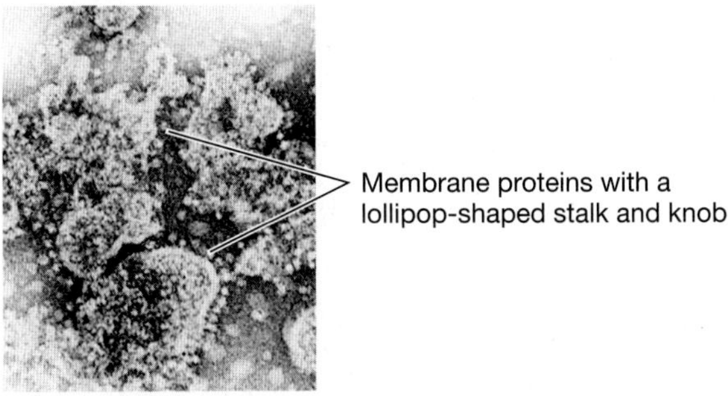

(b) The F_o unit is the base; the F_1 unit is the knob.

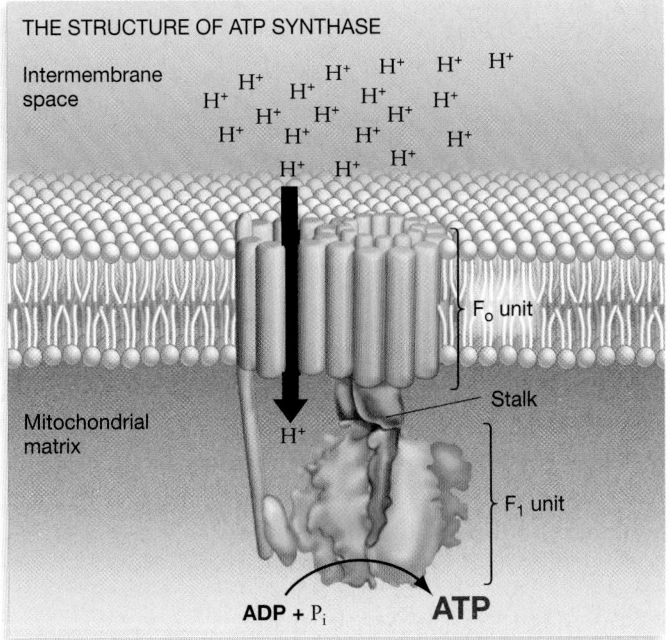

FIGURE 9.25 The Structure of ATP Synthase. (a) When patches of mitochondrial membrane turn inside out and form vesicles, proteins that have a lollipop-shaped stalk-and-knob structure face outward. Normally, the stalk and knob face inward, toward the mitochondrial matrix. **(b)** ATP synthase has two major components, designated F_o and F_1, connected by a stalk.

ADP and inorganic phosphate. The vesicles that contained just the base component, without the stalks and knobs, could not process ATP. The base components were, however, capable of transporting protons across the membrane.

Based on these observations, Racker proposed that the stalk-and-knob component of the protein was an enzyme that both hydrolyzes and synthesizes ATP. To test this idea, he added the stalk-and-knob components back to vesicles that had been stripped of them and confirmed that the vesicles were then capable of synthesizing ATP. Follow-up work also confirmed his hypothesis that the membrane-bound base component is a proton channel.

As **Figure 9.25b** shows, the structure of this protein complex is now reasonably well understood. The ATPase "knob" component is called the F_1 unit; the membrane-bound, proton-transporting base component is the F_o unit. The F_1 and F_o units are connected by the stalk. The entire complex is known as **ATP synthase**. According to the current model, here is how this enzyme functions: A flow of protons through the F_o unit causes the stalk connecting the two subunits to spin. By attaching long actin filaments to the stalk and examining them with a videomicroscope, researchers have been able to see the rotation, which can reach speeds of 350 revolutions per second. As the F_1 unit rotates along with the stalk, its subunits are thought to change conformation in a way that catalyzes the phosphorylation of ADP to ATP. Understanding exactly how this reaction occurs is currently the focus of intense research. ATP synthase makes most of the ATP that keeps you alive.

Oxidative Phosphorylation

As discussed earlier, formation of ATP through the combination of proton pumping by electron transport chains and the action of ATP synthase is called oxidative phosphorylation. The term is appropriate because the phosphorylation of ADP is based on the oxidation of NADH and $FADH_2$.

Figure 9.26 summarizes glucose oxidation and cellular respiration by tracing the fate of the carbon atoms and electrons in glucose. Notice that electrons from glucose are transferred to NADH and $FADH_2$, passed through the electron transport chain, and accepted by oxygen. The pumping of protons during electron transport creates the proton-motive force that drives ATP synthesis. The diagram also indicates the approximate yield of ATP from each component of the process. Recent research has shown that about 30 ATP molecules are produced from each molecule of glucose.[3] Of these, 26 ATP molecules are produced by ATP synthase. The fundamental message here? The vast majority of the "payoff" from the oxidation of glucose occurs via oxidative phosphorylation.

It is important to recognize, though, that cellular respiration can occur without oxygen. Oxygen is the electron acceptor used by all eukaryotes and a wide diversity of bacteria and archaea. Species that depend on oxygen as an electron acceptor for the ETC are said to use **aerobic** respiration and are called aerobic organisms. (The Latin root *aero–* means "air.") But many thousands of bacterial and archaeal species rely on electron acceptors other than oxygen and electron donors other than glucose. As Chapter 28 will show, some bacteria and archaea use H_2, H_2S, CH_4, or other inorganic compounds as electron donors. For species in oxygen-poor environments, nitrate (NO_3^{2-}) and sulfate (SO_4^{2-}) are particularly common electron acceptors. Cells that depend on

[3]Traditionally, biologists thought that 36 ATP would be synthesized for every mole of glucose oxidized. More recent work has shown that actual yield is about 30 ATP (see P.C. Hinkle et al. *Biochemistry* 30 (1991): 3576–3582). Also, it's important to note that yield varies with conditions in the cell.

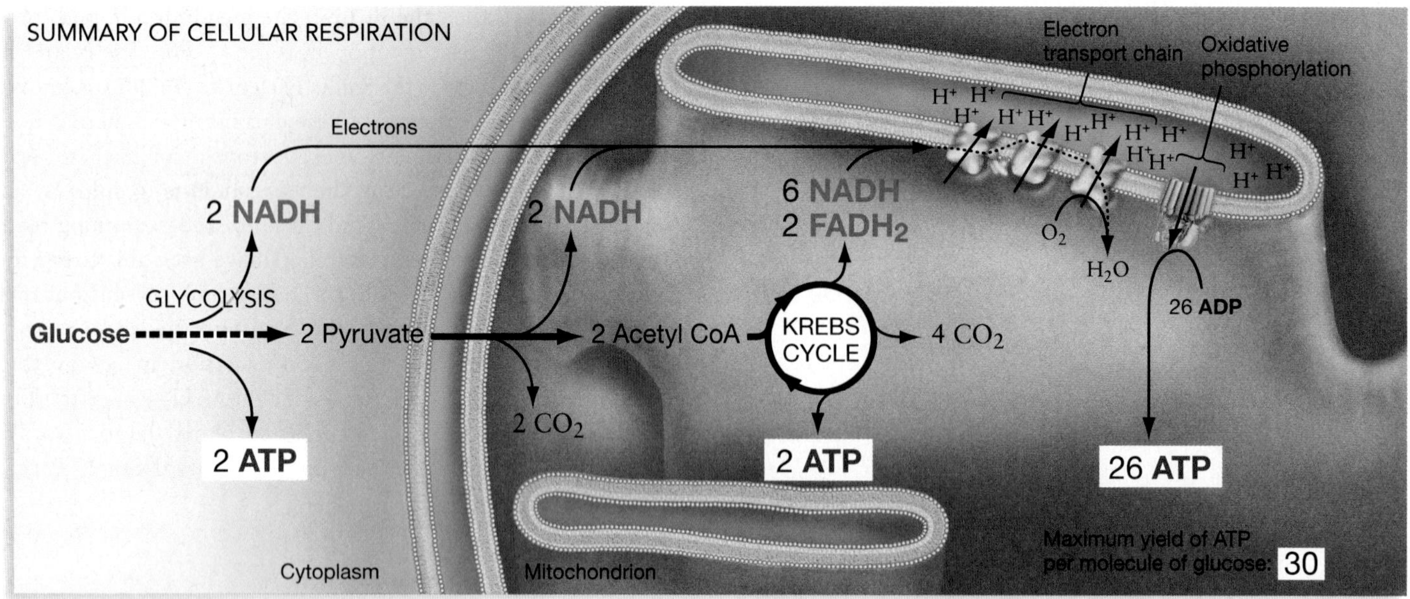

FIGURE 9.26 A Summary of Cellular Respiration

EXERCISE Across the top of the figure, write the overall reaction for the oxidation of glucose. Explain what happens to each reactant. Identify the source of each product.

electron acceptors other than oxygen are said to use **anaerobic** ("no air") respiration. Even though the starting and ending points of cellular respiration are different, these cells are still able to use electron transport chains to create a proton-motive force that drives the synthesis of ATP. In bacteria and archaea, the ETC and ATP synthase are located in the plasma membrane.

Oxygen is the most effective of all electron acceptors, however, because of its high electronegativity. Because oxygen holds electrons so tightly, the potential energy of electrons in a bond between an oxygen atom and a non-oxygen atom is low. As a result, there is a large difference between the potential energy of electrons in NADH and the potential energy of electrons bonded to an oxygen atom (see Figure 9.22). The large differential in potential energy means that the electron transport chain can generate a large proton-motive force. Cells that do not use oxygen as an electron acceptor cannot generate such a large potential energy difference. As a result, they cannot make as much ATP as do cells that use aerobic respiration. This is important: It means that anaerobic organisms tend to grow much more slowly than aerobic organisms. If cells that use anaerobic respiration compete with cells using aerobic respiration, the cells that use oxygen as an electron acceptor almost always grow faster and reproduce more. Because of this, species that rely on anaerobic respiration usually live in environments where oxygen is not present. In oxygen-free habitats, they do not have to compete with aerobic species.

What happens when oxygen or other electron acceptors are temporarily used up and are unavailable? Without oxygen or another electron acceptor in place, the electrons carried by NADH have no place to go. When this happens, the electron transport chain stops. All of the NAD^+ in the cell quickly becomes NADH. This situation is life threatening. When there is no longer any NAD^+ to supply the reactions of glycolysis, no ATP can be produced. If NAD^+ cannot be regenerated somehow, the cell will die. How do cells cope?

Check Your Understanding

If you understand that...

- As electrons from NADH and $FADH_2$ move through the electron transport chain, protons are pumped into the intermembrane space of mitochondria.
- The electrochemical gradient across the inner mitochondrial membrane drives protons through ATP synthase, resulting in the production of ATP from ADP.

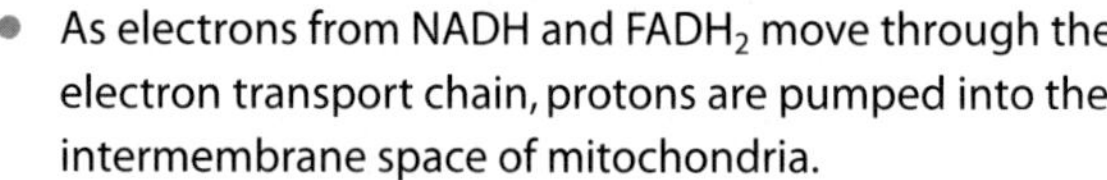

You should be able to...

Add paper cutouts labeled ETC and ATP synthase to the model you made in Section 9.3, and then explain the steps in electron transport and chemiosmosis, using dimes to represent electrons and pennies to represent protons.

9.7 Fermentation

Fermentation is a metabolic pathway that regenerates NAD+ from stockpiles of NADH and allows glycolysis to continue producing ATP in the absence of the electron acceptor required by the ETC. It occurs when pyruvate or a molecule derived from pyruvate—*not* oxygen—accepts electrons from NADH. When NADH gets rid of electrons in this way, NAD^+ is produced. With NAD^+ present, glycolysis can continue to produce ATP via substrate-level phosphorylation and the cell can stay alive, even when electron transport chains are shut down for lack of an electron acceptor. In many cases, the molecule that is formed by the addition of an electron to pyruvate (or another electron acceptor) cannot be used by the cell. In some cases, this by-product is toxic and is excreted from the cell as waste (**Figure 9.27a**).

In organisms that usually use oxygen as an electron acceptor, fermentation is an alternative mode of energy production when oxygen supplies temporarily run out. For example, if you sprint a long distance, your muscles begin metabolizing glucose so fast that your lungs and circulatory system cannot supply oxygen rapidly enough to keep electron transport chains active. When oxygen is absent, the electron transport chains shut down and NADH cannot donate its electrons there. The pyruvate produced by glycolysis then begins to accept electrons from NADH, and fermentation takes place (**Figure 9.27b**). The result of this process, called **lactic acid fermentation**, is the formation of a product molecule called lactate and regeneration of NAD^+.

Figure 9.27c illustrates a different type of fermentation pathway. These reactions, called **alcohol fermentation**, occur in the fungus *Saccharomyces cerevisiae*—baker's and brewer's yeast. When these fungal cells are placed in an environment such as bread dough or a bottle of grape juice and begin growing there, they quickly use up all the available oxygen. They continue to use glycolysis to metabolize sugar, however, by enzymatically converting pyruvate to the two-carbon compound acetaldehyde. This reaction gives off carbon dioxide, which causes bread to rise and creates the bubbles in champagne and beer. Acetaldehyde then accepts electrons from NADH, forming the NAD^+ required to keep glycolysis going. The addition of electrons to acetaldehyde results in the formation of ethanol as a waste product. Ethanol is excreted from yeast cells as waste but functions as the active ingredient in alcoholic beverages.

Bacteria and archaea that exist exclusively through fermentation are present in phenomenal numbers in the oxygen-free environment of your small intestine and in the rumen (first stomach) of cows. The rumen is a specialized digestive organ that contains over 10^{10} (10 billion) bacterial and archaeal cells per *milliliter* of fluid. The fermentations that occur in these cells result in the production of an array of fatty acids. Cows use these fermentation by-products as a source of energy. Cattle don't actually live off grass directly—they only eat it to feed the bacteria and archaea that they do live off. Cells that employ other types of fermentations are used commercially in the production of soy sauce, tofu, yogurt, cheese, vinegar, and other products.

Even though fermentation is a widespread and commercially important type of metabolism, it is extremely inefficient compared with cellular respiration. Fermentation produces just 2 molecules of ATP for each molecule of glucose metabolized, while cellular respiration produces about 30—approximately 15 times more energy per glucose molecule than fermentation. The reason for this disparity is that oxygen has much higher electronegativity than electron acceptors such as pyruvate and

(a) Fermentation pathways allow cells to regenerate NAD^+ for glycolysis.

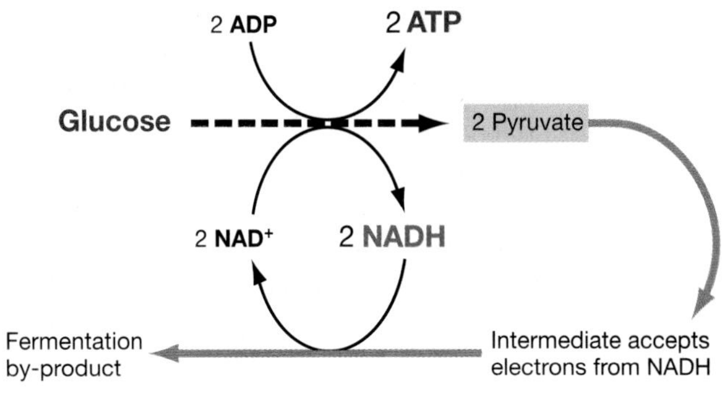

(b) Lactic acid fermentation occurs in humans.

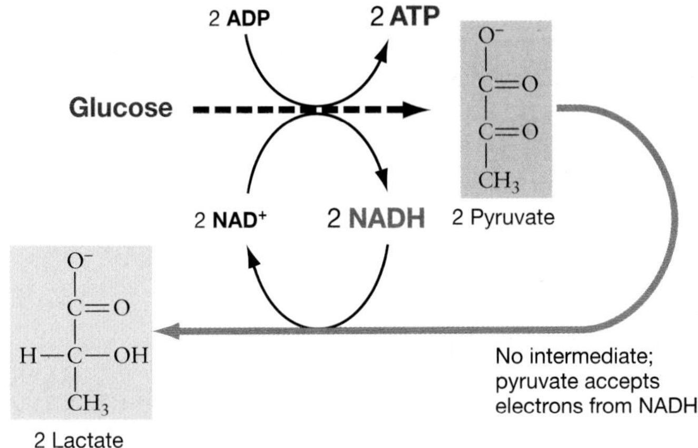

(c) Alcohol fermentation occurs in yeast.

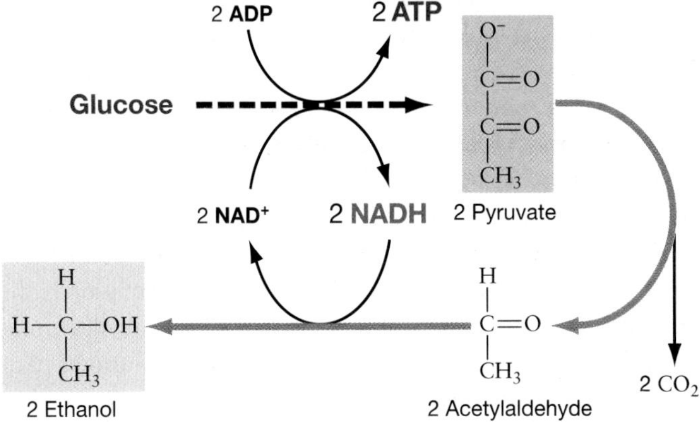

FIGURE 9.27 Fermentation Regenerates NAD^+ So that Glycolysis Can Continue. Among the bacteria, archaea, and eukaryotes, many types of fermentation occur.

acetaldehyde. As a result, the potential energy drop between the start and end of fermentation is a tiny fraction of the potential energy change that occurs during cellular respiration.

Based on these observations, it is not surprising that organisms capable of both processes never use fermentation when an appropriate electron acceptor is available for cellular respiration. Organisms that can switch between fermentation and cellular respiration that uses oxygen as an electron acceptor are called **facultative aerobes**. The term aerobe refers to using oxygen, while the adjective facultative reflects the ability to use cellular respiration when oxygen is present and fermentation when it is absent.

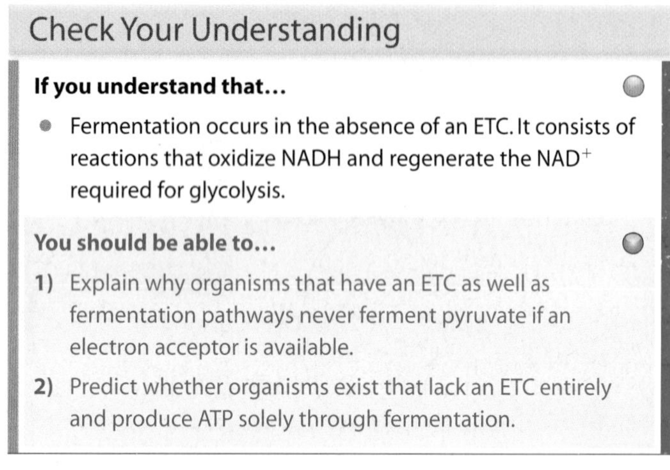

Check Your Understanding

If you understand that...

- Fermentation occurs in the absence of an ETC. It consists of reactions that oxidize NADH and regenerate the NAD^+ required for glycolysis.

You should be able to...

1) Explain why organisms that have an ETC as well as fermentation pathways never ferment pyruvate if an electron acceptor is available.
2) Predict whether organisms exist that lack an ETC entirely and produce ATP solely through fermentation.

(MB) Web Animation at www.masteringbio.com
Glucose Metabolism

9.8 How Does Cellular Respiration Interact with Other Metabolic Pathways?

The enzymes, products, and intermediates involved in cellular respiration and fermentation do not exist in isolation. Instead, they are part of a huge and dynamic inventory of chemicals inside the cell. Because metabolism includes thousands of different chemical reactions, the amounts and identities of molecules inside cells are constantly in flux (**Figure 9.28**). Fermentation pathways, electron transport, and other aspects of carbohydrate metabolism may be crucial to the life of a cell, but they also have to be seen as parts of a whole.

To make sense of the chemical inventory inside cells and the full scope of metabolism, it is critical to recognize that cells have two fundamental requirements to stay alive, grow, and reproduce: energy and carbon. More formally, cells need a source of high-energy electrons for generating chemical energy in the form of ATP, and a source of carbon-containing molecules that can be used to synthesize DNA, RNA, proteins, fatty acids, and other molecules. Reactions that result in the breakdown of molecules and the production of ATP are called **catabolic pathways**; reactions that result in the synthesis of larger molecules from smaller components are called **anabolic pathways**.

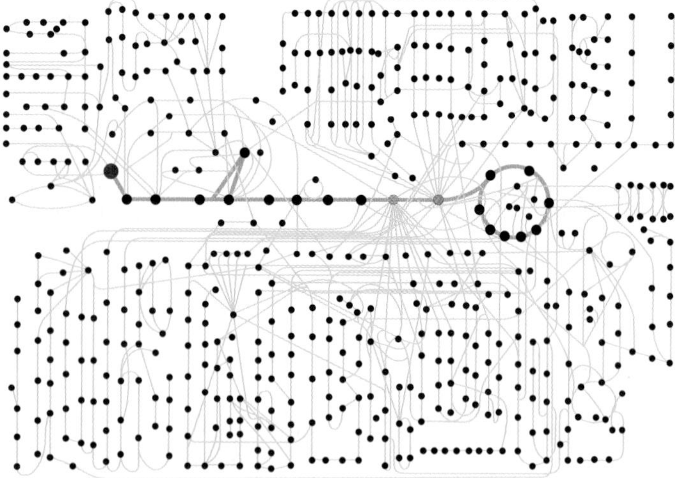

FIGURE 9.28 Metabolic Pathways Interact. A representation of a few of the thousands of chemical reactions that occur in cells. The dots represent molecules, and the lines represent enzyme-catalyzed reactions.

EXERCISE Label the red dot, and circle the 10 reactions of glycolysis. Draw a box around the Krebs cycle.

How do glycolysis and the Krebs cycle interact with other catabolic pathways and with anabolic pathways? Let's consider how molecules other than carbohydrates are used as fuel in eukaryotes, and then examine how molecules involved in glycolysis and the Krebs cycle are sometimes used as building blocks in the synthesis of cell components.

Processing Proteins and Fats as Fuel

Most organisms ingest, synthesize, or absorb a wide variety of carbohydrates. These molecules range from sucrose, maltose, and other simple sugars to large polymers such as glycogen and starch. As Chapter 5 noted, glycogen is the major form of stored carbohydrate in animals, while starch is the major form of stored carbohydrate in plants. Recall that both glycogen and starch are polymers of glucose, but they differ in the way their long chains of glucose branch. Using enzyme-catalyzed reactions, cells can produce glucose from glycogen, starch, and most simple sugars. Glucose and fructose can then be processed by the enzymes of the glycolytic pathway.

Carbohydrates are not the only important source of carbon compounds used in the catabolic pathways, however. As Chapter 6 pointed out, fats are highly reduced macromolecules consisting of glycerol bonded to chains of fatty acids. In cells, fats are routinely broken down by enzymes to form glycerol and acetyl CoA. Glycerol enters the glycolytic pathway once it has been oxidized and phosphorylated to form glyceraldehyde-3-phosphate—one of the intermediates in the 10-reaction sequence. Acetyl CoA enters the Krebs cycle.

Proteins can also be catabolized, meaning that they can be broken down and used to produce ATP. Once they are broken

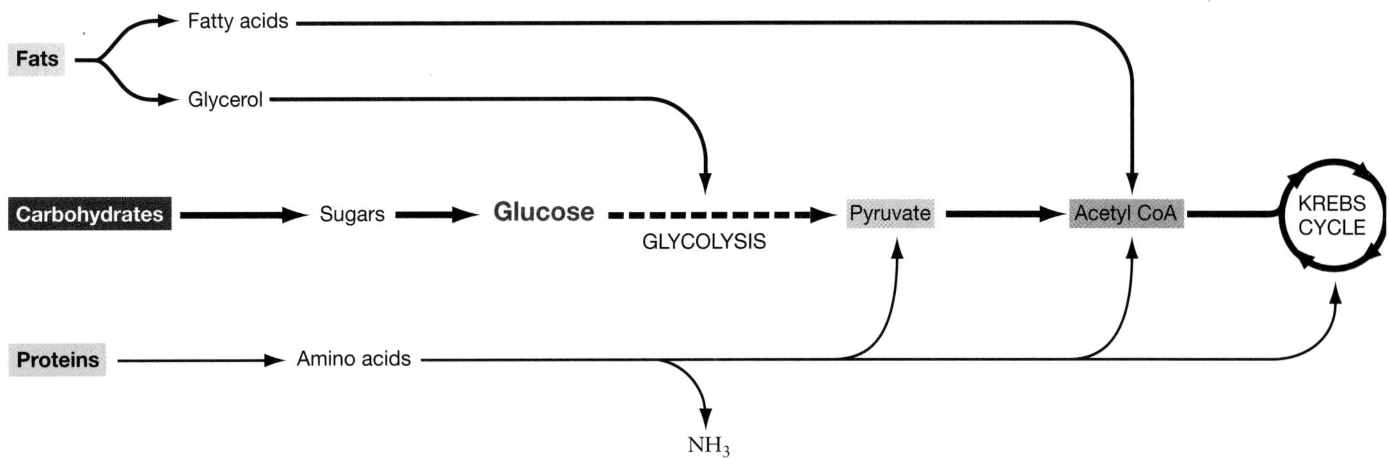

FIGURE 9.29 Proteins, Carbohydrates, and Fats Can All Furnish Substrates for Cellular Respiration. A variety of carbohydrates can be converted to glucose and processed by glycolysis. If carbohydrates are scarce, cells can obtain high-energy compounds from fats or even proteins for ATP production. These are catabolic reactions.

down to their constituent amino acids, the amino ($-NH_2$) groups are removed in enzyme-catalyzed reactions. These amino groups are excreted in urine as waste. The carbon compounds that remain after this catabolic step are converted to pyruvate, acetyl CoA, and other intermediates in glycolysis and the Krebs cycle.

Figure 9.29 summarizes the catabolic pathways of carbohydrates, fats, and proteins and shows how their breakdown products feed an array of steps in glucose oxidation and cellular respiration. When all three types of molecules are available in the cell to generate ATP, carbohydrates are used up first, then fats, and finally proteins.

Anabolic Pathways Synthesize Key Molecules

Where do cells get the precursor molecules required to synthesize amino acids, RNA, DNA, phospholipids, and other cell components? Not surprisingly, the answer often involves intermediates in carbohydrate metabolism. For example,

- In humans, about half of the 21 required amino acids can be synthesized from molecules siphoned off from the Krebs cycle.
- Acetyl CoA is the starting point for anabolic pathways that result in the synthesis of fatty acids.
- The molecule that is produced by the first reaction in glycolysis can be oxidized to start the synthesis of ribose-5-phosphate—a key intermediate in the production of ribonucleotides and deoxyribonucleotides. These nucleotides, in turn, are required for manufacturing RNA and DNA.
- If ATP is abundant, pyruvate and lactate (from fermentation) can be used as a substrate in the synthesis of glucose. Excess glucose is converted to glycogen and stored.

Figure 9.30 summarizes how intermediates in carbohydrate metabolism are drawn off to synthesize macromolecules. The take-home message is that the same molecule can serve many different functions in the cell. As a result, catabolic and anabolic pathways are closely intertwined.

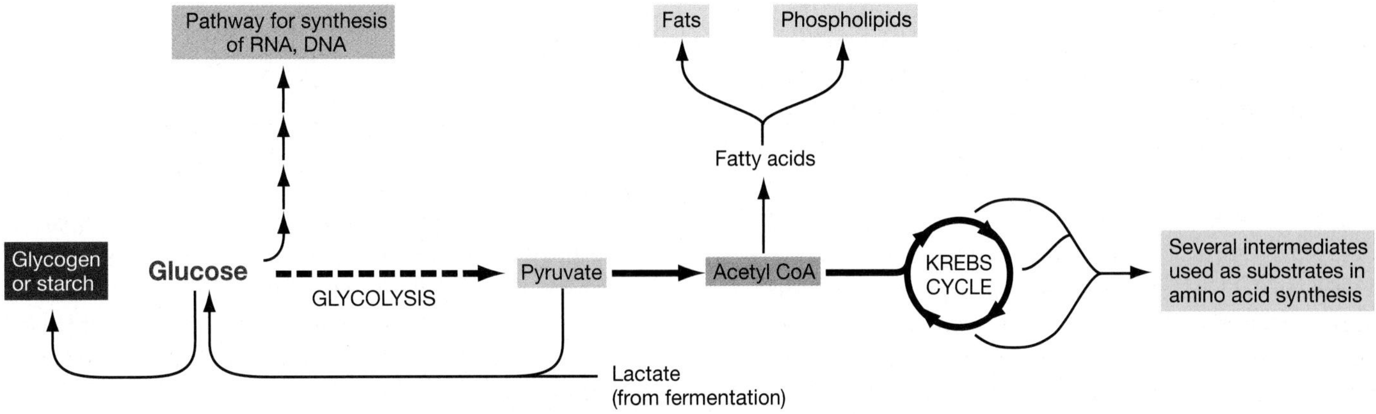

FIGURE 9.30 Intermediates in Carbohydrate Metabolism Can Be Drawn Off to Synthesize Cell Components. Several of the intermediates in carbohydrate metabolism act as precursor molecules in anabolic reactions leading to the synthesis of RNA, DNA, glycogen or starch, amino acids, and fatty acids.

Chapter Review

SUMMARY OF KEY CONCEPTS

In cells, the endergonic reactions required for life occur in conjunction with an exergonic reaction involving ATP.

ATP is the currency that cells use to pump ions, drive endergonic reactions, move cargo, and perform other types of work. When proteins bind ATP or are phosphorylated, they respond by changing conformation, or shape, in a way that allows them to act as enzymes, pumps, or motors.

Almost all of the reactions that occur inside cells are endergonic. When ATP or a phosphate group from ATP is added to a substrate or enzyme that participates in an endergonic reaction, the potential energy of the substrate or enzyme is raised enough to make the reaction exergonic and thus spontaneous. In this way, ATP drives reactions that otherwise would not occur.

You should be able to explain why the addition of phosphate groups raises the potential energy of proteins, and why phosphorylation often causes proteins to change shape.

Cellular respiration produces ATP from molecules with high potential energy—often glucose.

Cells produce ATP from sugars or other compounds with high free energy by using one of two general pathways: (1) cellular respiration or (2) fermentation. Cellular respiration is based on redox reactions that transfer electrons from a compound with high free energy, such as glucose, to a molecule with lower free energy, such as oxygen, through an electron transport chain. Fermentation involves the transfer of electrons from one organic compound to another without participation by an electron transport chain.

You should be able to explain how cellular respiration occurs in organisms that use electron donors other than glucose and electron acceptors other than oxygen. Which of the four steps in glucose oxidation would also occur in these organisms? Could fermentation also occur?

Glucose oxidation has four components: (1) glycolysis, (2) pyruvate processing, (3) the Krebs cycle, and (4) electron transport coupled with oxidative phosphorylation.

In eukaryotes, glycolysis takes place in the cytosol. The Krebs cycle occurs in the mitochondrial matrix, and electron transport and oxidative phosphorylation proceed in the inner membranes of mitochondria.

Glycolysis is a 10-step reaction sequence in which glucose is broken down into two molecules of pyruvate. For each molecule of glucose processed during glycolysis, two molecules of ATP are produced by substrate-level phosphorylation and two molecules of NAD^+ are reduced to two molecules of NADH.

Prior to the Krebs cycle, each pyruvate is oxidized to acetyl CoA by a series of reactions that results in the production of one molecule of carbon dioxide and the synthesis of one molecule of NADH. The Krebs cycle begins when acetyl CoA reacts with oxaloacetate to form citrate. The ensuing series of reactions results in the regeneration of oxaloacetate, the production of 2 molecules of carbon dioxide, the synthesis of 1 molecule of ATP by substrate-level phosphorylation, the reduction of 1 molecule of FAD to $FADH_2$, and the reduction of 3 molecules of NAD^+ to 3 of NADH.

In essence, glycolysis, pyruvate processing, and the Krebs cycle strip electrons from glucose and use them to reduce NAD^+ and FAD. The resulting electron carriers—NADH and $FADH_2$—then donate electrons to an electron transport chain, which gradually steps the electrons down in potential energy until they are finally accepted by oxygen or by another final electron acceptor, resulting in the formation of water in aerobic organisms. The components of the electron transport chain use the energy released by the oxidation of NADH and $FADH_2$ to pump protons across the inner mitochondrial membrane. The pumping activity creates an electrochemical gradient, or proton-motive force, that ATP synthase uses to produce ATP. Oxidative phosphorylation is the production of ATP by electron transport and the generation of a proton-motive force.

You should be able to describe what would happen to NADH levels in a cell in the first few seconds after a drug has poisoned phosphofructokinase, the enzyme that converts acetyl CoA to citrate, cytochrome *c*, or ATP synthase.

MB Web Animation at www.masteringbio.com
Cellular Respiration

Fermentation pathways allow glycolysis to continue when the lack of an electron acceptor shuts down electron transport chains.

If no electron acceptor such as oxygen is available, cellular respiration stops, because electron transport chains cannot continue. As a result, all NAD^+ is converted to NADH and glycolysis must stop. Fermentation pathways regenerate NAD^+ when an organic molecule such as pyruvate accepts electrons from NADH. Depending on the molecule that acts as an electron acceptor, fermentation pathways produce lactate, ethanol, or other reduced organic compounds as a by-product.

You should be able to explain why organisms that are producing ATP only via fermentation grow much more slowly than organisms that are producing ATP via cellular respiration.

MB Web Animation at www.masteringbio.com
Glucose Metabolism

Cellular respiration and fermentation are carefully regulated.

When supplies of ATP, NADH, and $FADH_2$ are high in the cell, feedback regulation occurs—that is, product molecules bind to and inhibit enzymes involved in ATP production. The glycolytic pathway slows when ATP binds to phosphofructokinase, and the pyruvate dehydrogenase complex is inhibited when it is phosphorylated by ATP. The enzyme that converts acetyl CoA to citrate slows when ATP binds to it, and certain enzymes in the Krebs cycle are inhibited when NADH or ATP bind to them. As a result, ATP is produced only when needed.

You should be able to draw a graph predicting how the rate of ATP production changes as a function of ATP concentration. (Write "ATP concentration" on the *x*-axis and "ATP production" on the *y*-axis.)

QUESTIONS

Test Your Knowledge

1. When does feedback inhibition occur?
   a. when lack of an appropriate electron acceptor makes an electron transport chain stop
   b. when an enzyme that is active early in a metabolic pathway is inhibited by a product of the pathway
   c. when ATP synthase reverses and begins pumping protons out of the mitochondrial matrix
   d. when cellular respiration is inhibited and fermentation begins

2. Where does the Krebs cycle occur in eukaryotes?
   a. in the cytosol
   b. in the matrix of mitochondria
   c. in the inner membrane of mitochondria
   d. in the intermembrane space of mitochondria

3. What does the chemiosmotic hypothesis claim?
   a. Substrate-level phosphorylation occurs in the electron transport chain.
   b. Substrate-level phosophorylation occurs in glycolysis and the Krebs cycle.
   c. The electron transport chain is located in the inner membrane of mitochondria.
   d. Electron transport chains generate ATP indirectly, by the creation of a proton-motive force.

4. What is the function of the reactions in a fermentation pathway?
   a. to generate NADH from NAD^+, so electrons can be donated to the electron transport chain
   b. to synthesize pyruvate from lactate
   c. to generate NAD^+ from NADH, so glycolysis can continue
   d. to synthesize electron acceptors, so that cellular respiration can continue

5. When do cells switch from cellular respiration to fermentation?
   a. when electron acceptors required by the ETC are not available
   b. when the proton-motive force runs down
   c. when NADH and $FADH_2$ supplies are low
   d. when pyruvate is not available

6. Why are NADH and $FADH_2$ said to have "reducing power"?
   a. They are the reduced forms of NAD^+ and FAD.
   b. They donate electrons to components of the ETC, reducing those components.
   c. They travel between the cytosol and the mitochondrion.
   d. They have the power to reduce carbon dioxide to glucose.

Test Your Knowledge answers: 1. b; 2. b; 3. d; 4. c; 5. a; 6. b

Test Your Understanding

Answers are available at www.masteringbio.com

1. Explain why NADH and $FADH_2$ are called electron carriers. Where do these molecules get electrons, and where do they deliver them? In eukaryotes, what molecule do these electrons reduce?
2. Compare and contrast substrate-level phosphorylation and oxidative phosphorylation.
3. What is the relationship between cellular respiration and fermentation? Why does cellular respiration produce so much more ATP than fermentation does? Why does aerobic respiration produce much more ATP than anaerobic respiration?
4. Diagram the relationship among the four components of cellular respiration: glycolysis, pyruvate processing, the Krebs cycle, and electron transport. What molecules connect these three processes? Where does each process occur in a eukaryotic cell?
5. Explain the relationship between electron transport and oxidative phosphorylation. What does ATP synthase look like, and how does it work?
6. Describe the relationship among carbohydrate metabolism, the catabolism of proteins and fats, and anabolic pathways.

Applying Concepts to New Situations

Answers are available at www.masteringbio.com

1. Cyanide ($C{\equiv}N^-$) blocks complex IV of the electron transport chain. Suggest a hypothesis for what happens to the ETC when complex IV stops working. Your hypothesis should explain why cyanide poisoning is fatal.
2. The presence of many sac-like cristae results in a large amount of membrane inside mitochondria. Suppose that some mitochondria had few cristae. How would their output of ATP compare with that of mitochondria with many cristae? Explain your answer.
3. When yeast cells are placed into low-oxygen environments, the mitochondria in the cells become reduced in size and number. Suggest an explanation for this observation.
4. Most agricultural societies have come up with ways to ferment the sugars in barley, wheat, rice, corn, or grapes to produce alcoholic beverages. Historians argue that this was an effective way for farmers to preserve the chemical energy in grains and fruits in a form that would not be eaten by rats or spoiled by bacteria or fungi. Why does a great deal of chemical energy remain in the products of fermentation pathways?

www.masteringbio.com is also your resource for • Answers to text, table, and figure caption questions and exercises • Answers to *Check Your Understanding* boxes • Online study guides and quizzes • Additional study tools including the *E-Book for Biological Science* 3rd ed., textbook art, animations, and videos.

10 Photosynthesis

KEY CONCEPTS

- Photosynthesis consists of two linked sets of reactions. In reactions driven directly by light, ATP and the electron carrier NADPH are produced. In subsequent reactions called the Calvin cycle, the ATP and NADPH are used to reduce carbon dioxide (CO_2) to carbohydrate $(CH_2O)_n$. In eukaryotic cells, both processes take place in chloroplasts.
- The light-dependent reactions transform the energy in sunlight to chemical energy in the form of electrons with high potential energy. Excited electrons either are used to produce NADPH or are donated to an electron transport chain, which results in the production of ATP via chemiosmosis.
- The Calvin cycle starts with the enzyme rubisco, which catalyzes the addition of CO_2 to a five-carbon molecule. The resulting compound undergoes a series of reactions that use ATP and NADPH and lead to the production of sugar.

Plants and other photosynthetic organisms convert the energy in sunlight to chemical energy in the bonds of sugar. The sugar produced by photosynthetic organisms fuels cellular respiration and growth. Photosynthetic organisms, in turn, are consumed by animals, fungi, and a host of other organisms. Directly or indirectly, most organisms on Earth get their energy from photosynthesis.

About three billion years ago, a novel combination of light-absorbing molecules and enzymes gave a bacterial cell the capacity to convert light energy into chemical energy in the carbon-carbon and carbon-hydrogen bonds of sugar. When sunlight is used to manufacture carbohydrate, **photosynthesis** is said to occur. The origin of photosynthesis ranks as one of the great events in the history of life. Since this process evolved, photosynthetic organisms have dominated the Earth in terms of abundance and mass.

The vast majority of organisms alive today rely on photosynthesis, either directly or indirectly, to stay alive. Photosynthetic organisms such as trees and mosses and ferns are termed **autotrophs** (literally, "self-feeders"), because they make all of their own food from ions and simple molecules. Non-photosynthetic organisms such as humans and fungi and bacteria like *Escherichia coli* are called **heterotrophs** ("different-feeders") because they have to obtain the sugars and many of the other macromolecules they need from other organisms. Because there could be no heterotrophs without autotrophs, photosynthesis is fundamental to almost all life. Glycolysis may qualify as the most ancient set of energy-related chemical reactions from an evolutionary viewpoint; but ecologically—meaning, in terms of how organisms interact with each other—photosynthesis is easily the most important.

All organisms perform respiration or fermentation, but only selected groups are also capable of photosynthesis. This chapter presents a step-by-step analysis of how photosynthetic species manufacture sugar from sunlight, carbon dioxide, and water. After studying how algae, plants, and other photosynthetic organisms do this remarkable chemistry, you should have a deeper appreciation for the food you eat and the oxygen you breathe.

Key Concept · Important Information · Practice It

10.1 An Overview of Photosynthesis

Research on photosynthesis began early in the history of biological science. Starting in the 1770s, a series of experiments showed that photosynthesis takes place only in the green parts of plants; that sunlight, carbon dioxide (CO_2), and water (H_2O) are required; and that oxygen (O_2) is produced as a by-product. By the early 1840s enough was known about this process for biologists to propose that photosynthesis allows plants to convert electromagnetic energy in the form of sunlight into chemical energy in the bonds of carbohydrates. Eventually the overall reaction was understood to be

$$CO_2 + 2\ H_2O + \text{light energy} \rightarrow (CH_2O)_n + H_2O + O_2$$

The $(CH_2O)_n$ is a generic carbohydrate. (The subscript "n" indicates that different carbohydrates have different multiples of CH_2O.) In essence, energy from light is transformed to chemical energy in the C–H and C–C bonds of carbohydrates. When glucose is the carbohydrate that is eventually produced, the reaction can be written as

$$6\ CO_2 + 12\ H_2O + \text{light energy} \rightarrow C_6H_{12}O_6 + 6\ O_2 + 6\ H_2O$$

Take a moment to look at this summary reaction carefully, and make note of a key point: It is the reverse of cellular respiration. Photosynthesis is an endergonic suite of reactions that reduces carbon dioxide to glucose or other sugars. Cellular respiration is an exergonic suite of reactions that oxidizes glucose to carbon dioxide and results in the production of ATP.

How does photosynthesis happen? Based on the overall reaction, early investigators assumed that CO_2 and H_2O react directly to form CH_2O, and that the oxygen atoms in carbon dioxide are given off as oxygen gas (O_2). Both hypotheses proved to be incorrect. Let's see why.

Photosynthesis: Two Linked Sets of Reactions

During the 1930s two independent lines of research on photosynthesis converged, leading to a major advance. The first research program, led by Cornelius van Niel, focused on how photosynthesis occurs in purple sulfur bacteria. Van Niel and his group found that these cells can grow in the laboratory on a food source that lacks sugars. Based on this observation, he concluded they must be autotrophs that manufacture their own carbohydrates. But to grow, the cells had to be exposed to sunlight and hydrogen sulfide (H_2S). Van Niel also showed that these cells did not produce oxygen as a by-product of photosynthesis. Instead, elemental sulfur (S) accumulated in their medium. In these organisms, the overall reaction for photosynthesis was

$$CO_2 + 2\ H_2S + \text{light energy} \rightarrow (CH_2O)_n + H_2O + 2\ S$$

Van Niel's work was crucial for two reasons. First, it showed that CO_2 and H_2O do *not* combine directly during photosynthesis. Instead of acting as a reactant in photosynthesis in these species, H_2O is a product of the process. Second, van Niel's data showed that the oxygen atoms in CO_2 are not released as oxygen gas (O_2). This conclusion was logical because no oxygen was produced by the purple sulfur bacteria, even though carbon dioxide participated in the reaction—just as it did in plants.

Based on these findings, biologists hypothesized that the oxygen atoms released during plant photosynthesis must come from H_2O. This proposal was supported by experiments with isolated chloroplasts, which produced oxygen in the presence of sunlight even if no CO_2 was present. The hypothesis was confirmed when heavy isotopes of oxygen—^{18}O compared with the normal isotope, ^{16}O—became available to researchers. Biologists then exposed algae or plants to H_2O that contained ^{18}O, collected the oxygen gas that was given off as a by-product of photosynthesis, and confirmed that the released oxygen gas contained the heavy isotope. As predicted, the reaction that produced this oxygen occurred only in the presence of sunlight. The light-dependent reactions of photosynthesis result in the production of oxygen from water.

A second major line of research helped support these discoveries. Between 1945 and 1955, a team led by Melvin Calvin began feeding labeled carbon dioxide ($^{14}CO_2$) to algae and identifying the molecules that subsequently became labeled with the radioisotope. These experiments allowed researchers to identify the exact sequence of reactions involved in reducing CO_2 to sugars. Because Calvin played an important role in this research, the reactions that reduce carbon dioxide and result in the production of sugar came to be known as the **Calvin cycle**. Later research showed that the Calvin cycle can function only if the light-dependent reactions are occurring.

To summarize: Early research on photosynthesis showed that the process consists of two linked sets of reactions. One set is triggered by light; the other set—the Calvin cycle—requires the products of the light-dependent reactions. The light-dependent reactions result in the production of oxygen from water; the Calvin cycle results in the production of sugar from carbon dioxide.

How are the two sets of reactions linked? The short answer is by electrons. More specifically, electrons are released when water is split to form oxygen gas. During the light-dependent reactions, these electrons are promoted to a high energy state by light and then transferred to a phosphorylated version of NAD^+, called $NADP^+$, that is abundant in photosynthesizing cells. This reaction forms NADPH. Like NADH, NADPH is an electron carrier. ATP is also produced in the light-dependent reactions (see the middle portion of **Figure 10.1**). During the Calvin cycle, the electrons in NADPH and the potential energy in ATP are used to reduce

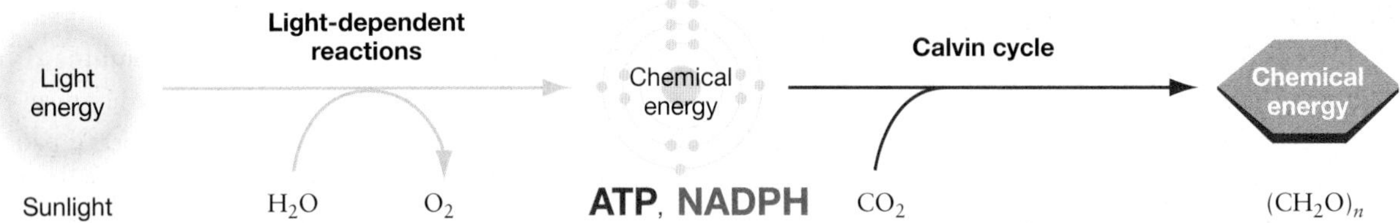

FIGURE 10.1 Photosynthesis Has Two Linked Components. In the light-dependent reactions of photosynthesis, light energy is transformed to chemical energy in the form of ATP and NADPH. During the Calvin cycle, the ATP and NADPH produced in the light-dependent reactions are used to reduce carbon dioxide to carbohydrate.

CO_2 to carbohydrate (the right-hand side of Figure 10.1). The resulting sugars are used in cellular respiration to produce ATP for the cell. Plants oxidize sugars in their mitochondria and consume O_2 in the process, just like animals and other eukaryotes do.

Where does all this activity take place?

The Structure of the Chloroplast

Experiments with various plant tissues established that photosynthesis takes place only in the green portions of plants. Follow-up work with light microscopes suggested that the reactions occur inside the bright green organelles called **chloroplasts** ("green-formed"). Leaf cells typically contain from 40 to 50 chloroplasts, and a square millimeter of leaf averages about 500,000 chloroplasts (**Figure 10.2a**). When membranes derived from chloroplasts were found to release oxygen after exposure to sunlight, the hypothesis that chloroplasts are the site of photosynthesis became widely accepted.

When electron microscopy became available in the 1950s, researchers observed that chloroplasts are extremely membrane rich. As **Figure 10.2b** shows, the organelle is enclosed by an

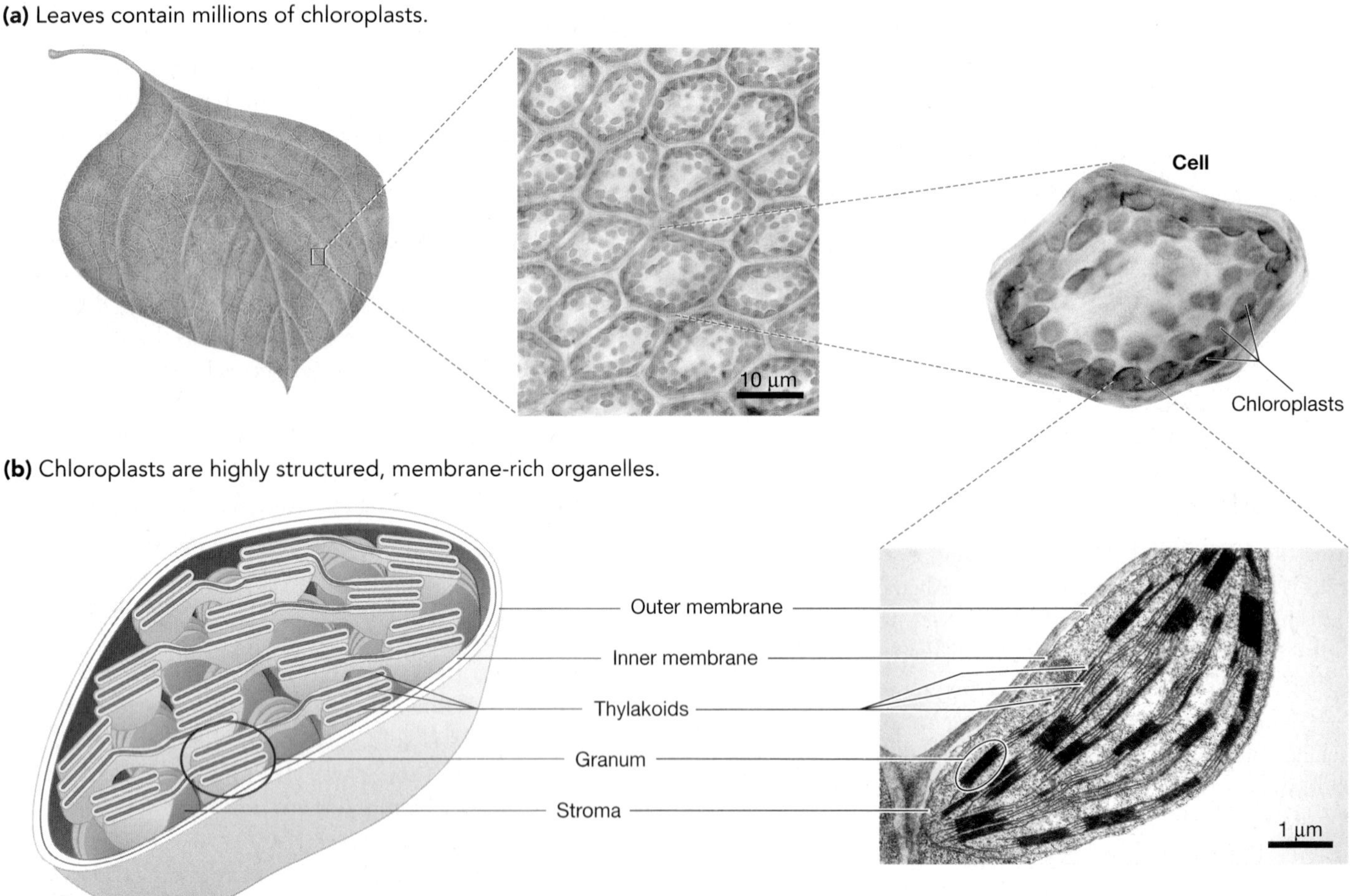

(a) Leaves contain millions of chloroplasts.

Cell

Chloroplasts

10 μm

(b) Chloroplasts are highly structured, membrane-rich organelles.

Outer membrane

Inner membrane

Thylakoids

Granum

Stroma

1 μm

FIGURE 10.2 Photosynthesis Takes Place in Chloroplasts. (a) In plants, photosynthesis takes place in organelles called chloroplasts. **(b)** The internal membranes of chloroplasts form flattened, vesicle-like structures called thylakoids, some of which form stacks called grana.

outer membrane and an inner membrane. The interior is dominated by vesicle-like structures called **thylakoids**, which often occur in interconnected stacks called **grana** (singular: **granum**). The space inside a thylakoid is its **lumen**. (Recall that *lumen* is a general term for the interior of any sac-like structure. Your stomach and intestines have a lumen.) The fluid-filled space between the thylakoids and the inner membrane is the **stroma**.

When researchers analyzed the chemical composition of thylakoid membranes, they found huge quantities of pigments. **Pigments** are molecules that absorb only certain wavelengths of light—other wavelengths are either transmitted or reflected. Pigments have colors because we see the wavelengths that pass through or bounce off them. The most abundant pigment found in the thylakoid membranes turned out to be chlorophyll. Chlorophyll ("green-leaf") reflects or transmits green light. As a result, it is responsible for the green color of plants, some algae, and many photosynthetic bacteria.

Developmental studies showed that chloroplasts are derived from colorless organelles called proplastids ("before-plastids"). Proplastids are found in the cells of embryonic plants and in the rapidly dividing tissues of mature plants. As cells mature, proplastids develop into chloroplasts or other types of plastids specialized for that cell's particular task (see **Box 10.1**).

Before plunging into the details of how photosynthesis occurs inside a chloroplast, take a moment to consider just how astonishing the process is. Chemists have synthesized an amazing diversity of compounds from relatively simple starting materials, but their achievements pale in comparison to cells that synthesize sugar from just carbon dioxide, water, and sunlight. If it is not *the* most sophisticated chemistry on Earth, photosynthesis is certainly a contender. What's more, photosynthetic cells accomplish this feat in environments ranging from mountaintop snowfields to the open ocean, tropical rain forests, and polar ice caps. Photosynthesis can occur in virtually any habitat where light is available.

10.2 How Does Chlorophyll Capture Light Energy?

Photosynthesis begins with the light-dependent reactions. The light-dependent reactions, in turn, begin with the simple act of sunlight striking chlorophyll. To understand the consequences of this event, it's helpful to review the nature of light. Light is a type of electromagnetic radiation, which is a form of energy. The essence of photosynthesis is converting electromagnetic

BOX 10.1 Types of Plastids

Plastids are a family of double-membrane-bound organelles found in plants. As **Figure 10.3** shows, they develop from small, unspecialized organelles called proplastids. As a cell matures and takes on a specialized function in the plant, its proplastids also develop a distinctive form and function. If a developing plant cell in a stem or leaf is exposed to light, for example, its proplastids are usually stimulated to develop into chloroplasts.

There are three major types of plastids: (1) chloroplasts, (2) leucoplasts, and (3) chromoplasts. Figure 10.2 detailed the structure of chloroplasts. Leucoplasts ("white-formed") often function as energy storehouses. More specifically, leucoplasts may store chemical energy in the bonds of molecules with high potential energy. Some leucoplasts store oils; others synthesize and sequester starch; still others store proteins. Chromoplasts ("color-formed") are brightly colored because they synthesize and hoard large amounts of orange, yellow, or red pigments in their vacuoles. High concentrations of chromoplasts are responsible for many of the bright colors of fruits and some flowers.

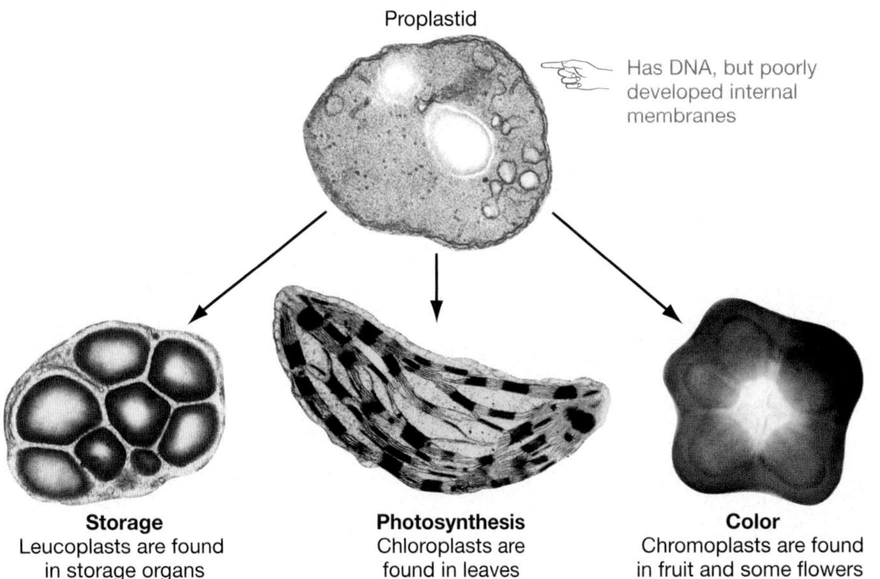

FIGURE 10.3 Chloroplasts Are a Specialized Type of Plastid. Each type of plastid has a distinctive structure and function, but all plastids develop from organelles called proplastids.

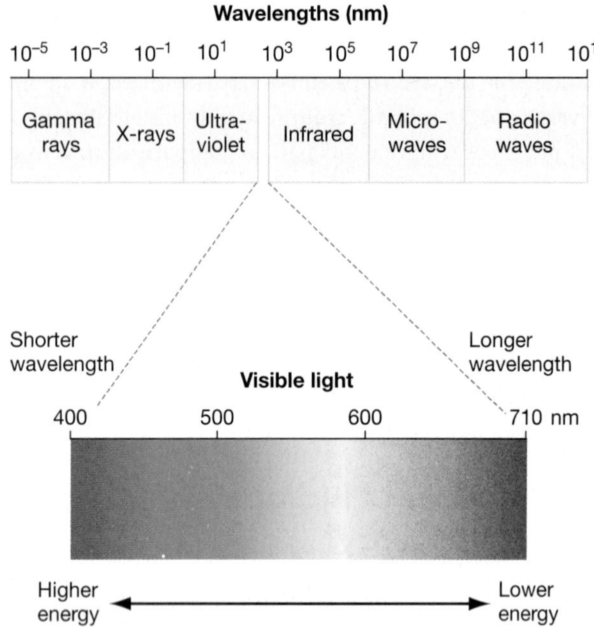

FIGURE 10.4 The Electromagnetic Spectrum. Electromagnetic energy radiates through space in the form of waves. Humans can see radiation at wavelengths between about 400 nm and about 710 nm. The shorter the wavelength of electromagnetic radiation, the higher its energy.

energy in the form of sunlight into chemical energy in the C–C and C–H bonds of sugar.

Physicists describe light's behavior as both wavelike and particle-like. As is true of all waves, including waves of water or air, electromagnetic radiation is characterized by its **wavelength**—the distance between two successive wave crests (or wave troughs). The wavelength determines the type of electromagnetic radiation.

Figure 10.4 illustrates the **electromagnetic spectrum**—the range of wavelengths of electromagnetic radiation. Humans cannot see all these wavelengths, however. Electromagnetic radiation that humans can see is called **visible light**; it ranges in wavelength from about 400 to about 710 nanometers (nm, or 10^{-9} m). Shorter wavelengths of electromagnetic radiation contain more energy than longer wavelengths do, so blue light and ultraviolet light contain much more energy than red light and infrared light do.

To emphasize the particle-like nature of light, physicists point out that it exists in discrete packets called **photons**. In understanding photosynthesis, the important point to consider is that each photon and each wavelength of light have a characteristic amount of energy. Pigment molecules absorb this energy. How?

Photosynthetic Pigments Absorb Light

When a photon strikes an object, the photon may be absorbed, transmitted, or reflected. A pigment molecule absorbs particular wavelengths of light. Sunlight is white light, which consists of all wavelengths in the visible portion of the electromagnetic spectrum at once. If a pigment absorbs all of the visible wavelengths, no visible wavelength of light is reflected back to your eye, and the pigment appears black. If a pigment absorbs many or most of the wavelengths in the blue and green parts of the spectrum but transmits or reflects red wavelengths, it appears red.

What wavelengths do various plant pigments absorb? In one approach to answering this question, researchers grind up leaves and add a liquid that acts as a solvent for lipids. The solvent extracts pigment molecules from the leaf mixture. As **Figure 10.5a** shows, the pigments in the extract can then be separated from each other using a technique called thin layer chromatography. To begin, spots of a raw leaf extract are placed

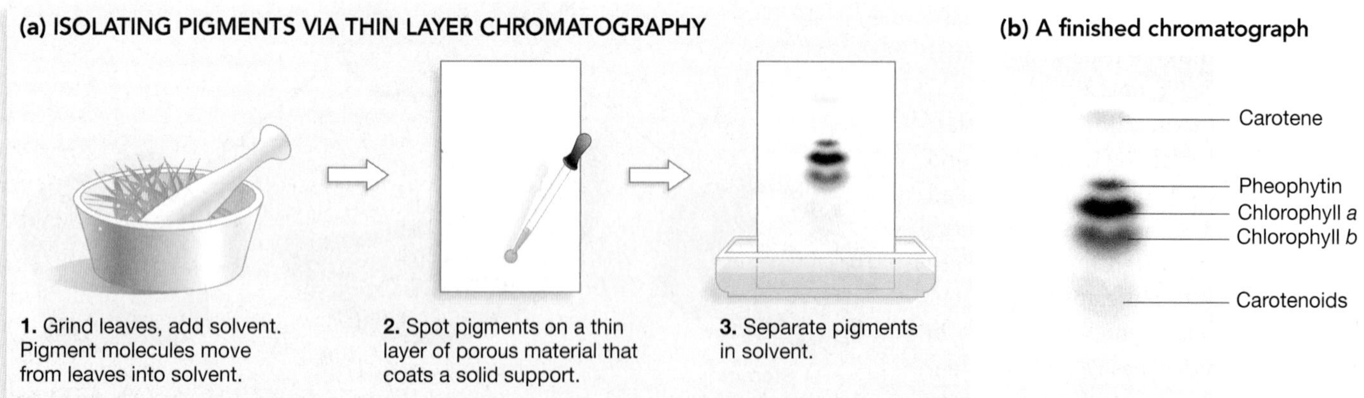

FIGURE 10.5 Pigments Can Be Isolated via Chromatography. (a) Thin layer chromatography is an effective way to isolate the various pigments in photosynthetic tissue. **(b)** Photosynthetic tissues, including those from the grass leaves used to make this chromatograph, typically contain several pigments. Different species of photosynthetic organisms may contain different types and quantities of pigments.

near the bottom of a stiff support that is coated with a thin layer of silica gel, cellulose, or similar porous material. The coated support is then placed in a solvent solution. As the solvent wicks upward through the coating, the pigment molecules in the mixture are carried along. Because the pigment molecules in the extract vary in size, solubility, or both, they are carried along with the solvent at different rates.

Figure 10.5b shows a chromatograph from a grass-leaf extract. Notice that this leaf contains an array of pigments. To find out which wavelengths are absorbed by each of these molecules, researchers cut out a single region (color band) of the filter paper, extract the pigment, and use an instrument called a spectrophotometer to record the wavelengths absorbed (**Box 10.2**).

Using this approach, biologists have produced data like those shown in **Figure 10.6a**. This graph is a plot of light absorbed versus wavelength and is called an **absorption spectrum**. Research based on these techniques has confirmed that there are two major classes of pigment in plant leaves: chlorophylls and carotenoids. The **chlorophylls**, designated chlorophyll *a* and chlorophyll *b*, absorb strongly in the blue and red regions of the visible spectrum. The presence of chlorophylls makes plants look green because they reflect and transmit green light, which they do not absorb. Beta-carotene (β-carotene)and other **carotenoids** constitute a different family of pigments, which absorb in the blue and green parts of the visible spectrum. Thus, carotenoids appear yellow, orange, or red.

Which of these wavelengths drive photosynthesis? T. W. Engelmann answered this question by laying a filamentous alga across a glass slide that was illuminated with a spectrum of colors. The idea was that the alga would begin performing photosynthesis in response to the various wavelengths of light and produce oxygen as a by-product. To determine exactly where oxygen was being produced, Engelmann added bacterial cells from a species that is attracted to oxygen. As **Figure 10.6b** shows, most of the bacteria congregated in the blue and red regions of the slide. Because wavelengths in these parts of the spectrum were associated with high oxygen concentrations, Engelmann concluded that they defined the **action spectrum** for photosynthesis. The data suggested that blue and red photons are the most effective at driving photosynthesis. Because the chlorophylls absorb these wavelengths, the data also suggested that chlorophylls are the main photosynthetic pigments.

Before analyzing what happens during the absorption event itself, let's look quickly at the structure and function of the chlorophylls and carotenoids.

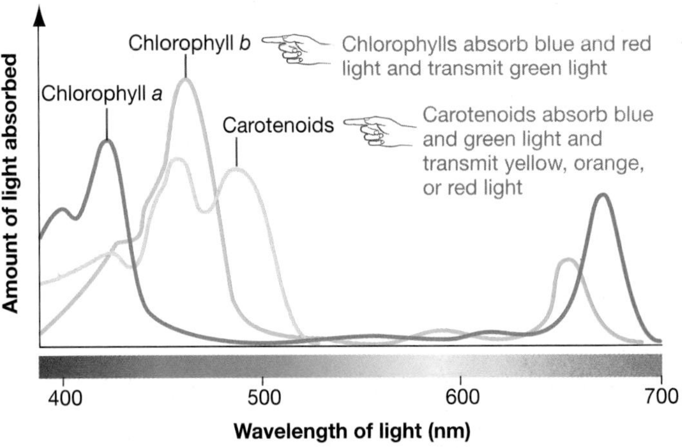

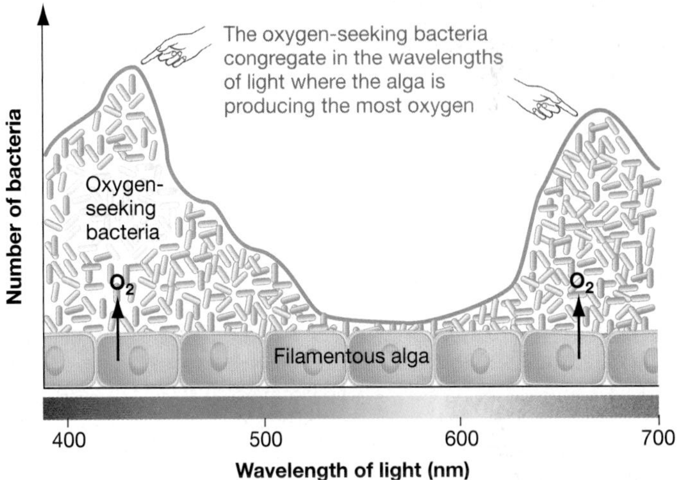

FIGURE 10.6 There Is a Strong Correlation between the Absorption Spectrum of Pigments and the Action Spectrum for Photosynthesis.

EXERCISE Some photosynthetic species have versions of chlorophyll pigments that absorb most strongly at 390 nm and 880 nm. Draw what part (b) would look like if, instead of the alga shown here, these organisms were placed along the bottom.

What Is the Role of Carotenoids and Other Accessory Pigments? Carotenoids are called accessory pigments because they absorb light and pass the energy on to chlorophyll. The carotenoids found in plants belong to two classes, called carotenes and xanthophylls. **Figure 10.8a** shows the structure of β-carotene, which gives carrots their orange color. A xanthophyll called zeaxanthin, which gives corn kernels their bright yellow color, is nearly identical to β-carotene, except that the ring structures on either end of the molecule contain a hydroxyl (–OH) group.

Both xanthophylls and carotenes are found in chloroplasts. In autumn, when the leaves of deciduous trees begin to die, their chlorophyll degrades first. The wavelengths scattered by the carotenoids that remain turn northern forests into spectacular displays of yellow, orange, and red.

Carotenoids absorb wavelengths of light that are not absorbed by chlorophyll. As a result, they extend the range of wavelengths that can drive photosynthesis. But researchers

BOX 10.2 How Do Researchers Measure Absorption Spectra?

Researchers use an instrument called a **spectrophotometer** to measure the wavelengths of light absorbed by a particular pigment. As **Figure 10.7** shows, a solution of purified pigment molecules is exposed to a specific wavelength of light. If the pigment absorbs a great deal of this wavelength, then little light will be transmitted through the sample. But if the pigment absorbs the wavelength poorly, then most of the incoming light will be transmitted. The light that passes through the sample strikes a photoelectric tube, which converts it to an electric current. The amount of the electric current that is generated, as measured by a meter, is proportional to the intensity of the incoming light. High current readings indicate low absorption at a particular wavelength, while low current readings signal high absorption. By testing one wavelength after another, investigators can measure the full absorption spectrum of a pigment.

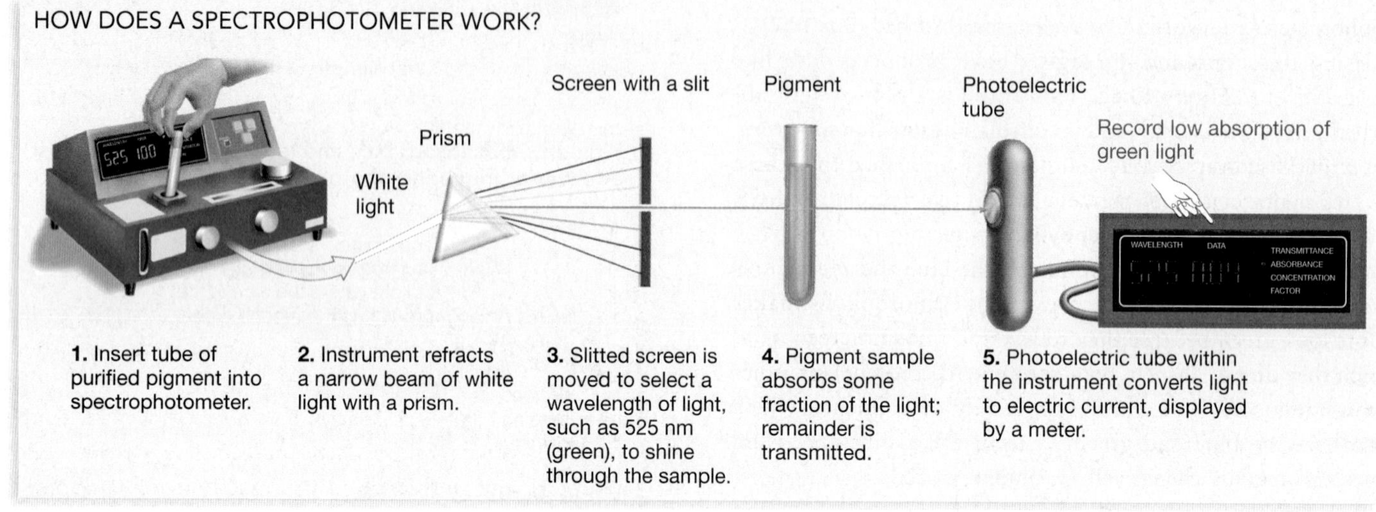

FIGURE 10.7 Spectrophotometers Can Measure the Light Wavelengths Absorbed by Pigments

QUESTION Suppose that you were analyzing a sample of the green pigment shown here and that you changed the wavelength selector in the spectrophotometer to 480 nm (blue light). How would the amount of absorption change?

discovered an even more important function by analyzing what happens to leaves when carotenoids are destroyed. Many herbicides, for example, work by inhibiting enzymes that are involved in carotenoid synthesis. Plants lacking carotenoids rapidly lose their chlorophyll, turn white, and die. Based on these results, researchers have concluded that carotenoids also serve a protective function.

To understand the molecular basis of carotenoid function, recall from Chapter 2 that energy from photons—especially the high-energy, short-wavelength photons in the ultraviolet part of the electromagnetic spectrum—contain enough energy to knock electrons out of atoms and create free radicals. Free radicals, in turn, trigger reactions that degrade molecules. Fortunately, carotenoids can "quench" free radicals by accepting or stabilizing unpaired electrons. As a result, they protect

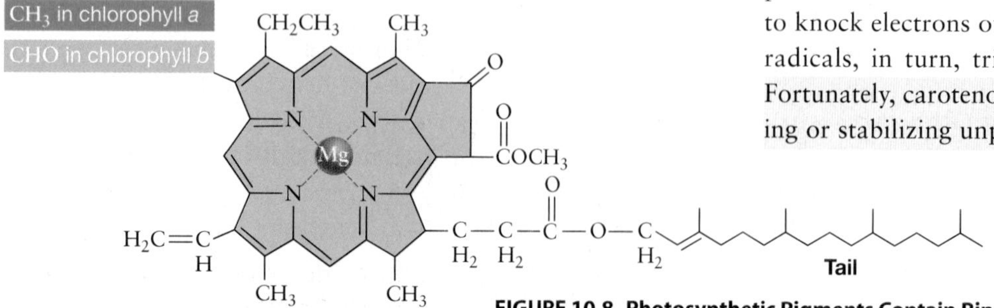

FIGURE 10.8 Photosynthetic Pigments Contain Ring Structures. (a) *β*-Carotene is an orange pigment found in carrot roots and other plant tissues. **(b)** Although chlorophylls *a* and *b* are very similar structurally, they have the distinct absorption spectra shown in Figure 10.6a.

chlorophyll molecules from harm. When carotenoids are absent, chlorophyll molecules are destroyed. As a result, photosynthesis stops. Starvation and death follow.

Carotenoids are not the only molecules that protect plants from the damaging effects of sunlight, however. Researchers have recently analyzed individuals of the mustard plant *Arabidopsis thaliana* that are unable to synthesize pigments called flavonoids, which are normally stored in the vacuoles of leaf cells. Because flavonoids absorb ultraviolet radiation, individuals that lack flavonoids are subject to damage from the free radicals resulting from exposure to UV light. In effect, these pigments function as a sunscreen for leaves and stems. Without them, chlorophyll molecules would be broken apart by high-energy radiation.

The general message here is that the energy in sunlight is a double-edged sword. It makes photosynthesis possible, but it can also lead to the formation of free radicals that damage cells. The role of carotenoids and flavonoids as protective pigments is crucial.

The Structure of Chlorophyll As **Figure 10.8b** shows, chlorophyll *a* and chlorophyll *b* are very similar in structure as well as in absorption spectra. Both have two fundamental parts: a long tail made up of isoprene subunits (introduced in Chapter 6) and a "head" consisting of a large ring structure with a magnesium atom in the middle. The tail keeps the molecule embedded in the thylakoid membrane; the head is where light is absorbed.

But just what is "absorption"? Stated another way, what happens when a photon of a particular wavelength—say, red light with a wavelength of 680 nm—strikes a chlorophyll molecule?

When Light Is Absorbed, Electrons Enter an Excited State

When a photon strikes a chlorophyll molecule, the photon's energy can be transferred to an electron in the chlorophyll molecule's head region. In response, the electron is "excited," or raised to a higher electron shell—one with greater potential energy. As **Figure 10.9** shows, the excited electron states that are possible in a particular pigment are discrete—meaning, incremental rather than continuous—and can be represented as lines on an energy scale. In the figure, the ground state, or unexcited state, is shown as 0 and the higher energy states are designated 1 and 2. If the difference between the possible energy states is the same as the energy in the photon, then the photon can be absorbed and an electron is excited to that energy state.

In chlorophyll, for example, the energy difference between the ground state and state 1 is equal to the energy in a red photon, while the energy difference between state 0 and state 2 is equal to the energy in a blue photon. Thus, chlorophyll can readily absorb red photons and blue photons. Chlorophyll does not absorb green light well, because there is no discrete step—no difference in possible energy states for its electrons—that corresponds to the amount of energy in a green photon.

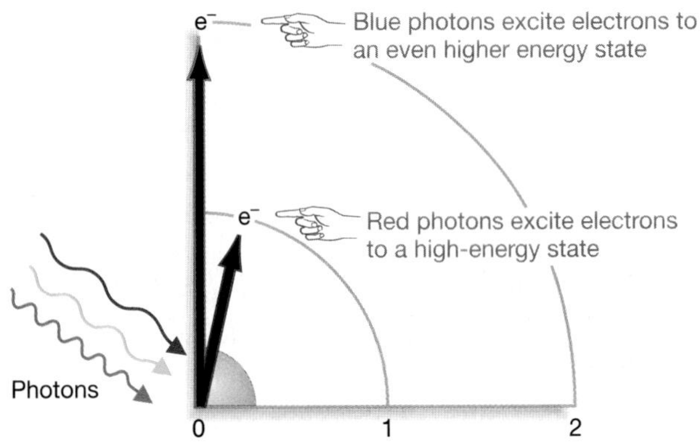

FIGURE 10.9 Electrons Are Promoted to High-Energy States When Photons Strike Chlorophyll. When a photon strikes chlorophyll, an electron can be promoted to a higher energy state, depending on the energy in the photon.

QUESTION Suppose a pigment had a discrete energy state that corresponded to the energy in green light. Where would you draw this energy state on this diagram?

Wavelengths in the ultraviolet part of the spectrum have so much energy that they may actually eject electrons from a pigment molecule. In contrast, wavelengths in the infrared regions have so little energy that in most cases they merely increase the movement of atoms in the pigment, generating heat rather than exciting electrons.

If an electron is excited by a photon, it means that energy in the form of electromagnetic radiation is transferred to an electron, which now has high potential energy. What happens to the potential energy in this excited electron? If it simply falls right back to its ground state, some of the absorbed energy is released as heat—meaning molecular movement—while the rest is released as electromagnetic radiation. This is the phenomenon known as **fluorescence** (**Figure 10.10**). Because some

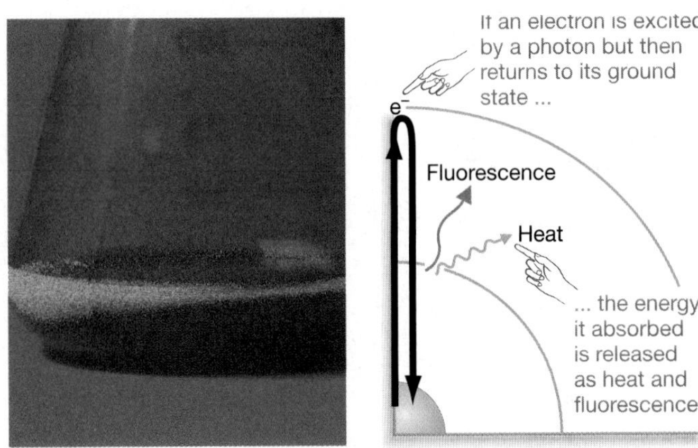

FIGURE 10.10 Fluorescence Occurs When Excited Electrons Fall Back to the Ground State. A pure solution of chlorophyll exposed to ultraviolet light. Electrons are excited to a high-energy state but immediately fall back to a low-energy state, emitting red photons and heat.

of the energy in the original photon is transformed to heat, the electromagnetic radiation that is given off during fluorescence has lower energy and a longer wavelength than the original photon does. When chlorophyll is in a chloroplast, though, only about 2 percent of the red and blue photons that it absorbs normally produce fluorescence. What happens to the other 98 percent of the excited electrons?

An answer to this question began to emerge when experiments showed that chlorophyll molecules work in groups—not individually. In the thylakoid membrane, 200–300 chlorophyll molecules and accessory pigments such as carotenoids are grouped together in an array of proteins, forming a complex called a **photosystem**. Each photosystem, in turn, has two major elements: an antenna complex and a reaction center.

The Antenna Complex When a red or blue photon strikes a pigment molecule in the antenna complex, the energy is absorbed and an electron is excited in response. This energy—but not the electron itself—is passed along to a nearby chlorophyll molecule, where another electron is excited in response. This phenomenon is known as resonance. Chlorophyll molecules in the antenna center transmit resonance energy.

Once the energy is transmitted, the original excited electron falls back to its ground state. In this way, energy is transmitted from one chlorophyll molecule to the next inside the antenna complex. The system acts like a radio antenna that receives a specific wavelength of electromagnetic radiation and transmits it to a receiver. In a photosystem, the receiver is called the reaction center.

The Reaction Center When energy from the antenna complex reaches the reaction center of a photosystem, an all-important energy transformation event occurs. At the **reaction center**, excited electrons are transferred to a molecule that acts as an electron acceptor. When this molecule becomes reduced, the energy transformation event that started with the absorption of light becomes permanent: Electromagnetic energy is transformed to chemical energy. It cannot be reemitted as fluorescence. The redox reaction that occurs in the reaction center results in the production of chemical energy from sunlight.

The key to understanding the reaction center is that, in the absence of light, chlorophyll cannot reduce the electron acceptor because the reactions are endergonic. But when light excites electrons in chlorophyll to a high-energy state, the reactions become exergonic. In this way, a light-induced, exergonic reaction is coupled to an endergonic reaction. The chemistry of the reaction center is analogous to the way that exergonic phosphorylation reactions drive other endergonic reactions in cells (see Chapter 9).

Figure 10.11 summarizes how chlorophyll interacts with light energy by illustrating the three possible fates of electrons that are excited by photons in photosynthetic pigments. They can (1) drop back down to a low energy level and cause fluorescence, (2) excite an electron in a nearby pigment and induce resonance, or (3) be transferred to an electron acceptor in a redox reaction. Fluorescence is typical of isolated pigments, resonance occurs in antenna complex pigments, and redox occurs in reaction center pigments.

Now the question is, what happens to the high-energy electrons that are transferred in the reaction center? Specifically, how are they used to manufacture sugar?

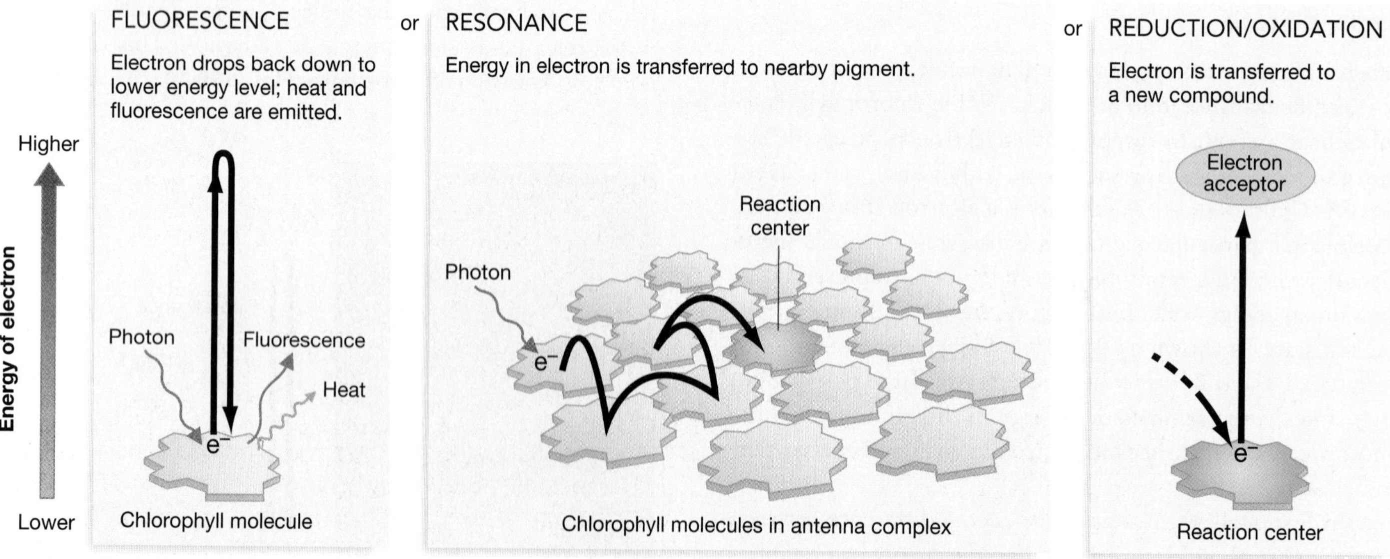

FIGURE 10.11 Three Fates for Excited Electrons in Photosynthetic Pigments. When sunlight promotes electrons in pigments to a high-energy state, three things can happen: They can fluoresce, pass energy to a nearby pigment via resonance, or transfer the electron to an electron acceptor.

Check Your Understanding

If you understand that...

- Pigments absorb certain wavelengths of light.
- When a chlorophyll molecule in the antenna complex of a chloroplast membrane absorbs red or blue light, one of its electrons is promoted to a high-energy state.
- In the antenna complex, high-energy electrons transmit their energy among chlorophyll molecules. When energy is transferred to a chlorophyll molecule in the reaction center, the electron that is excited in response reduces an electron acceptor. In this way, light energy is transformed to chemical energy.

You should be able to...

Explain which of these processes is analogous to what happens when you strike a tuning fork and then touch the vibrating tuning fork against another tuning fork.

10.3 The Discovery of Photosystems I and II

During the 1950s the fate of the high-energy electrons in photosystems was the central issue facing biologists interested in photosynthesis. Ironically, a central insight into this issue came from a simple experiment on how green algae responded to various wavelengths of light.

The experimental setup was based on the observation that photosynthetic cells respond to specific wavelengths of light. In particular, the algal cells being studied responded strongly to wavelengths of 700 nm and 680 nm, which are in the far-red and red portions of the visible spectrum, respectively. In a key experiment, Robert Emerson found that if cells were illuminated with either far-red light or red light, the photosynthetic response was moderate (**Figure 10.12**). But if cells were exposed to a combination of far-red and red light, the rate of photosynthesis increased dramatically. When both wavelengths were present, the photosynthetic rate was much more than the sum of the rates produced by each wavelength independently. This phenomenon was called the enhancement effect. Why it occurred was a complete mystery at the time.

The puzzle posed by the enhancement effect was eventually solved by Robin Hill and Faye Bendall. These biologists synthesized data emerging from an array of labs and proposed that green algae and plants have two distinct types of reaction centers rather than just one. Hill and Bendall proposed that one reaction center, which came to be called **photosystem II**, interacts with a different reaction center, now referred to as **photosystem I** (because it was discovered first). According to the two-photosystem hypothesis, the enhancement effect occurs because photosynthesis is much more efficient when both photosystems are operating together.

Experiment

Question: Light at both the red and far-red wavelengths stimulates photosynthesis. How does a combination of these wavelengths affect the rate of photosynthesis?

Hypothesis: When red light and far-red light are combined, the rate of photosynthesis will double.

Null Hypothesis: When red light and far-red light are combined, the rate of photosynthesis will not double.

Experimental setup:

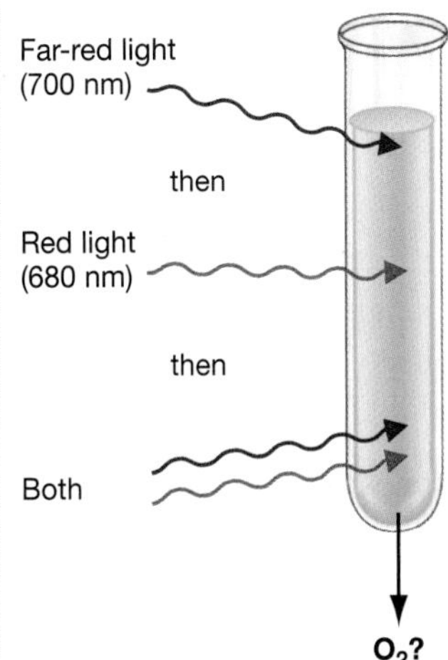

1. Expose algal cells to an intensity of far-red (700 nm) or red (680 nm) light that maximizes rate of photosynthesis at each wavelength. Then expose same algal cells to same intensities of each wavelength at same time.

2. Record the rate of photosynthesis as the amount of oxygen produced.

Prediction: When the wavelengths are combined, the rate of photosynthesis will be double the maximum rate observed for each wavelength independently.

Prediction of null hypothesis: When the wavelengths are combined, the rate of photosynthesis will not be double the maximum rate observed for each wavelength independently.

Results:

Relative rate of photosynthesis
Far-red light on
Red light on
Both lights on
Time

Conclusion: There is an enhancement effect for red and far-red light. The combination of 700 nm and 680 nm wavelengths more than doubles the rate of photosynthesis.

FIGURE 10.12 Discovery of the "Enhancement Effect" of Red and Far-Red Light.

Subsequent work has shown that the two-photosystem hypothesis is correct. In green algae and land plants, thylakoid membranes contain photosystems that differ in structure and function but complement each other.

To figure out how the two photosystems work, investigators chose not to study them together. Instead, they focused on species of photosynthetic bacteria that have photosystems similar to either photosystem I or II, but not both. Once each type of photosystem was understood in isolation, they turned to understanding how the two photosystems work in combination in green algae and land plants. Let's do the same—we'll analyze photosystem II, then photosystem I, and then how the two interact.

How Does Photosystem II Work?

To analyze photosystem II, researchers focused on studying species from the purple nonsulfur bacteria and the purple sulfur bacteria. These cells have a single photosystem that has many of the same components observed in photosystem II of cyanobacteria ("blue-green bacteria"), algae, and plants.

In photosystem II, the action begins when the antenna complex transmits energy to the reaction center and the molecule pheophytin comes into play. Structurally, **pheophytin** is identical to chlorophyll except that pheophytin lacks a magnesium atom in its head region. Functionally, though, the two molecules are different. Instead of acting as a pigment that promotes an electron when it absorbs a photon, pheophytin acts as an electron acceptor. When an electron in the reaction center chlorophyll is excited energetically, the electron binds to pheophytin and the reaction center chlorophyll is oxidized. When pheophytin is reduced in this way, the energy transformation step that started with the absorption of light is completed.

Electrons from Pheophytin Enter an Electron Transport Chain Electrons that reach pheophytin are passed to an electron transport chain in the thylakoid membrane. In both structure and function, this group of molecules is similar to the electron transport chain in the inner membrane of mitochondria (Chapter 9). For example, the electron transport chain associated with photosystem II contains several quinones and cytochromes. Electrons in both chains participate in a series of reduction-oxidation reactions and are gradually stepped down in potential energy. In mitochondria as well as chloroplasts, the redox reactions result in protons being pumped from one side of an internal membrane to the other. In both organelles, the resulting proton gradient drives ATP production via ATP synthase. ● Photosystem II triggers chemiosmosis and ATP synthesis in the chloroplast.

Figure 10.13a details the sequence of events in the electron transport chain of thylakoids. One of the key molecules involved is a quinone called **plastoquinone**, symbolized PQ.

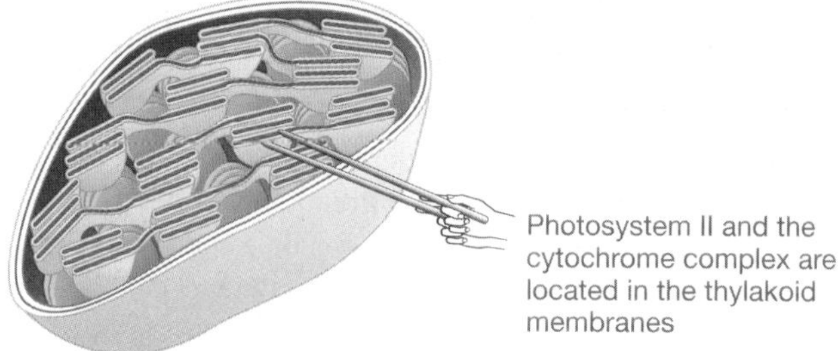

(a) In photosystem II, excited electrons feed an electron transport chain.

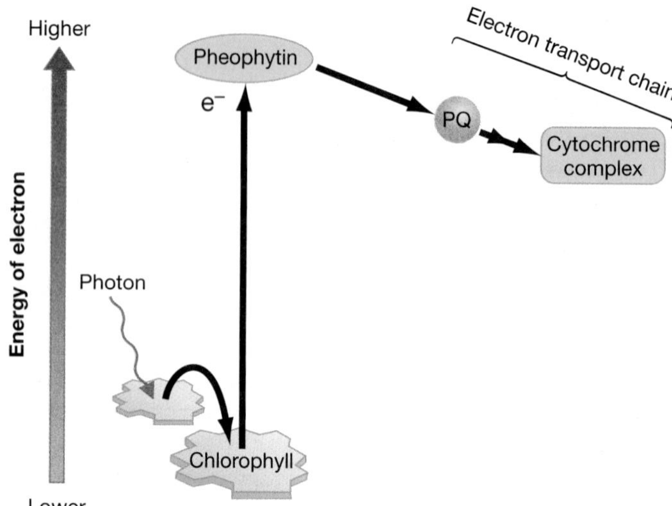

(b) Plastoquinone carries protons to the inside of thylakoids, creating a proton-motive force.

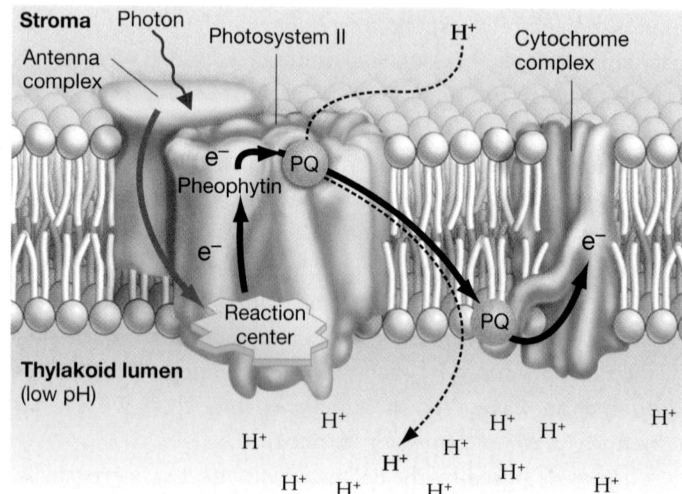

FIGURE 10.13 Photosystem II Feeds an Electron Transport Chain That Pumps Protons. **(a)** When an excited electron leaves the chlorophyll molecule in the reaction center of photosystem II, the electron is accepted by pheophytin, transferred to plastoquinone (PQ), and then stepped down in energy along an electron transport chain. **(b)** PQ carries electrons from photosystem II along with protons from the stroma. The electrons are passed to the cytochrome complex, and the protons are released in the thylakoid lumen.

● **EXERCISE** Add an arrow to part (b), indicating the direction of the proton-motive force.

Recall from Chapter 9 that quinones are small hydrophobic molecules. Because plastoquinone is lipid soluble and not anchored to a protein, it is free to move from one side of the thylakoid membrane to the other. When it receives electrons from pheophytin, plastoquinone carries them to the other side of the membrane and delivers them to more electronegative molecules in the chain. These electron acceptors are found in a complex that contains a cytochrome similar to the types of cytochromes present in mitochondrial membranes.

In this way, plastoquinone shuttles electrons from pheophytin to the cytochrome complex. The electrons are then passed through a series of iron- and copper-containing proteins in the cytochrome complex. The potential energy released by these reactions allows protons to be added to other plastoquinone molecules, which carry them to the lumen side of the thylakoid membrane. As **Figure 10.13b** shows, the protons transported by plastoquinone result in a large concentration of protons in the thylakoid lumen. When photosystem II is active, the pH of the thylakoid interior reaches 5 while the pH of the stroma hovers around 8. Because the pH scale is logarithmic, the difference of 3 units means that the concentration of H^+ is $10 \times 10 \times 10 = 1000$ times higher in the lumen than in the stroma. In addition, the stroma becomes negatively charged relative to the thylakoid lumen. The net effect of electron transport, then, is to set up a large proton gradient that will drive H^+ out of the thylakoid lumen and into the stroma. Based on your reading of Chapter 9, it should come as no surprise that this proton-motive force drives the production of ATP.

ATP Synthase Uses the Proton-Motive Force to Phosphorylate ADP In mitochondria, NADH and $FADH_2$ donate electrons to an electron transport chain, and the redox reactions that occur in the chain result in protons being pumped out of the matrix and into the intermembrane space. In chloroplasts, pheophytin molecules that are part of photosystem II donate electrons to an electron transport chain, and the redox reactions that occur result in protons being pumped out of the stroma and into the thylakoid lumen.

In both mitochondria and chloroplasts, protons diffuse down the resulting electrochemical gradient. This is an exergonic process that drives the endergonic synthesis of ATP. More specifically, the flow of protons through the enzyme ATP synthase causes conformational changes that drive the phosphorylation of ADP. In plants, chemiosmosis occurs in chloroplasts as well as mitochondria.

MB Web Animation at www.masteringbio.com
Chemiosmosis

In photosystem II, then, the light energy captured by chlorophyll is transformed to chemical energy stored in ATP. This process is called **photophosphorylation**. Substrate-level phosphorylation, oxidative phosphorylation, and photophosphorylation all result in the production of ATP.

To summarize, photosystem II starts with an electron being promoted to a high-energy state and ends with the production of ATP via chemisosmosis. ● If you understand these aspects of photosystem II, you should be able to compare and contrast photosystem II to the electron transfer chain (ETC) of mitochondria. You also should be able to explain where high-energy electrons come from in each system and what the eventual product is. The photosystem II story is not yet complete, however, because we haven't accounted for the electrons that flow through the system. The electron that was transferred from chlorophyll to pheophytin in the photosystem II reaction center needs to be replaced. In addition, the electron transport chain of photosystem II needs to donate its electrons to some final electron acceptor.

Where do the electrons required by photosystem II come from, and where do they go? The parallel between the photosystem in purple sulfur bacteria and photosystem II of plants ends here. In purple sulfur bacteria, cytochrome donates an electron back to the reaction center, so the same electron can again be promoted to a high-energy state when a photon is absorbed. In this way, electrons cycle through the system. But in photosystem II, electrons come from a different source.

Photosystem II Obtains Electrons by Oxidizing Water To understand where the electrons that feed photosystem II come from, think back to the overall reaction for photosynthesis: $CO_2 + 2\,H_2O$ + light energy $\rightarrow (CH_2O) + H_2O + O_2$. In the presence of sunlight, carbon dioxide and water are used to produce carbohydrate, water, and oxygen gas. Recall that experiments with radioisotopes of oxygen showed that the oxygen atoms in O_2 come from water, not from carbon dioxide. As it turns out, the electrons that enter photosystem II also come from water. The oxygen-generating reaction can be written as

$$2\,H_2O \rightarrow 4\,H^+ + 4\,e^- + O_2$$

Because electrons are removed from water, the molecule becomes oxidized. This reaction is referred to as "splitting" water. It supplies a steady stream of electrons for photosystem II and is catalyzed by enzymes that are physically integrated into the photosystem II complex. When excited electrons leave photosystem II and enter the ETC, the photosystem becomes so electronegative that enzymes can strip electrons away from water, leaving protons and oxygen. The oxygen-generating reaction is highly endergonic. It is possible only because the energy in sunlight drives it by removing electrons from photosystem II.

Among all life-forms, photosystem II is the only protein complex that can catalyze the splitting of water molecules.

Organisms such as cyanobacteria, algae, and plants that have this type of photosystem are said to perform **oxygenic** ("oxygen-producing") photosynthesis, because they generate oxygen as a by-product of the process. The purple sulfur and purple nonsulfur bacteria cannot oxidize water. They perform **anoxygenic** ("no oxygen-producing") photosynthesis.

It is difficult to overstate the importance of this unique reaction. The oxygen that is keeping you alive right now was produced by it. O_2 was, in fact, almost nonexistent on Earth before enzymes evolved that could catalyze the oxidation of water. According to the fossil record, oxygen levels in the atmosphere and oceans began to rise only about 2 billion years ago, as organisms that perform oxygenic photosynthesis increased in abundance. This change was a disaster for anaerobic organisms, because oxygen is toxic to them. But as oxygen became even more abundant, certain bacterial cells evolved the ability to use it as an electron acceptor during cellular respiration. This was a momentous development. O_2 is so electronegative that it creates a huge potential energy drop for the electron transport chains involved in cellular respiration. As a result, organisms that use O_2 as an electron acceptor in cellular respiration can produce much more ATP than can organisms that use other electron acceptors (see Chapter 9). Aerobic organisms grow so efficiently that they have long dominated our planet. Biologists rank the evolution of the oxygen-rich atmosphere as one of the most important events in the history of life.

Despite its fundamental importance, though, the mechanism responsible for the oxygen-generating reaction is not yet understood. Determining exactly how photosystem II splits water may be the greatest challenge currently facing researchers interested in photosynthesis. This issue has important practical applications as well, because if human chemists could replicate the reaction in an industrial setting, it might be possible to produce huge volumes of O_2 and hydrogen gas (H_2) from water. If this could be accomplished inexpensively, the H_2 produced could be used as a clean fuel for cars and trucks.

If you understand the essence of photosystem II, you should be able to make a concept map summarizing how it works. Your diagram should include a reaction center, pheophytin, an ETC, the proton gradient, ATP synthase, and the enzyme complex that splits water.

How Does Photosystem I Work?

Recall that researchers dissected photosystem II by studying similar, but simpler, photosystems in purple nonsulfur and purple sulfur bacteria. To understand the structure and function of photosystem I, they turned to heliobacteria ("sun-bacteria").

Like purple nonsulfur and purple sulfur bacteria, heliobacteria use the energy in sunlight to promote electrons to a high-energy state. But instead of being passed to an electron transport chain that pumps protons across a membrane, the high-energy electrons in heliobacteria are used to reduce NAD^+. When NAD^+ gains two electrons and a proton, NADH is produced. In the cyanobacteria and algae and land plants, a similar set of reactions reduces a phosphorylated version of NAD^+, symbolized $NADP^+$, yielding NADPH. Both NADH and NADPH function as electron carriers.

Figure 10.14 shows how photosystem I works. When pigments in the antenna complex absorb photons and pass the energy on to the reaction center in photosystem I, excited electrons are passed through an electron transport chain. Specifically, they pass along a series of iron- and sulfur-containing proteins inside the photosystem, and then to a molecule called **ferredoxin**. The electrons then move from ferredoxin to the enzyme ferredoxin/$NADP^+$ oxidoreductase—also called $NADP^+$ reductase—which transfers two electrons and a proton to $NADP^+$. This reaction forms NADPH. The photosystem itself and $NADP^+$ reductase are anchored in the thylakoid membrane; ferredoxin is closely associated with the bilayer.

To summarize: Photosystem I results in the production of NADPH. **NADPH** is similar in function to the NADH and $FADH_2$ produced by the Krebs cycle. It is an electron carrier that can donate electrons to other compounds and thus reduce them. Photosystem II, in contrast, results in the production of a proton gradient that drives the synthesis of ATP.

In combination, then, photosystems I and II produce chemical energy stored in ATP as well as reducing power in the form of NADPH. Although several groups of bacteria have just one of the two photosystems, the cyanobacteria, algae, and plants have both. In these organisms, how do the two photosystems interact?

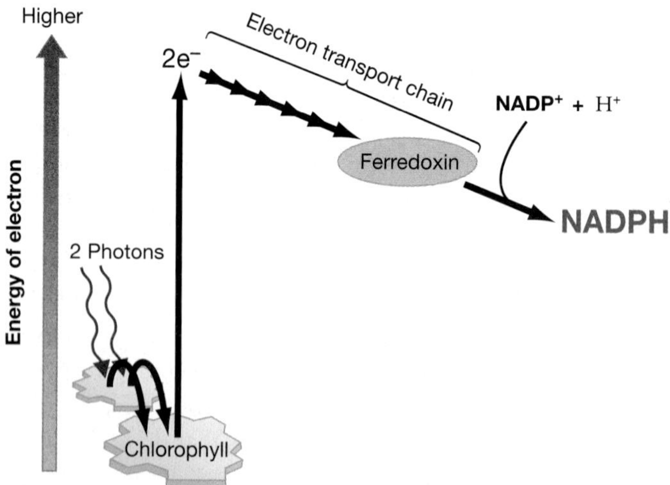

FIGURE 10.14 Photosystem I Produces NADPH. When excited electrons leave the chlorophyll molecule in the reaction center of photosystem I, they pass through a series of iron- and sulfur-containing proteins until they are accepted by ferredoxin. In an enzyme-catalyzed reaction, the reduced form of ferredoxin reacts with $NADP^+$ to produce NADPH.

QUESTION Is $NADP^+$ reduced or oxidized in photosystem I?

The Z Scheme: Photosystems I and II Work Together

When they realized that photosystems I and II have distinct but complementary functions, Robin Hill and Faye Bendall proposed that these systems interact as shown in **Figure 10.15**. The diagram illustrates a model, known as the **Z scheme**, that furnished a breakthrough in research on photosynthesis. The name was inspired by the shape of the proposed path of electrons through the two photosystems, when that path was plotted on a vertical axis representing the changes occurring in their potential energy.

Following the path of electrons through the Z scheme—by tracing the route of electrons through Figure 10.15 with your finger—will help drive home how photosynthesis works. The process starts when photons excite electrons in the chlorophyll molecules of photosystem II's antenna complex. When the energy in the excited electron is transmitted to the reaction center, a special pair of chlorophyll molecules named P680 passes excited electrons to pheophytin. From there the electron is gradually stepped down in potential energy through redox reactions among a series of quinones and cytochromes, which act as an electron transport chain. Using the energy released by the redox reactions, plastoquinone (PQ) carries protons across the thylakoid membrane. ATP synthase uses the resulting proton-motive force to phosphorylate ADP, creating ATP.

When electrons reach the end of photosystem II's electron transport chain, they are passed to a small diffusible protein called **plastocyanin** (symbolized PC in Figure 10.15). Plastocyanin picks up an electron from the cytochrome complex, diffuses through the lumen of the thylakoid, and donates the electron to photosystem I. A single plastocyanin molecule can shuttle over 1000 electrons per second between photosystems. In this way, plastocyanin forms a physical link between photosystem II and photosystem I.

The flow of electrons between photosystems, by means of plastocyanin, is important because it replaces electrons that are carried away from a chlorophyll molecule called P700 in the photosystem I reaction center. The electrons flowing from photosystem II to P700 are eventually transferred to the protein ferredoxin, which then passes electrons to an enzyme that catalyzes the reduction of $NADP^+$ to NADPH. The electrons that initially left photosystem II are replaced by electrons that are stripped away from water, producing oxygen gas as a by-product.

To summarize, the Z scheme traces the path of electrons from water through photosystem II to photosystem I to NADPH. ● You should be able to explain (1) the role of plastocyanin in linking the two photosystems, (2) where the electrons that flow through the system have their highest potential energy, and (3) why the Z scheme is sometimes referred to as **noncyclic electron flow.**

The Z scheme helps explain the enhancement effect in photosynthesis documented in Figure 10.12. When algal cells are illuminated with wavelengths at 680 nm, in the red portion of the spectrum, only photosystem II can run at a maximum rate. The overall rate of electron flow through the Z scheme is moderate because photosystem I's efficiency is reduced. Similarly, when cells receive only wavelengths at 700 nm, in the far red, only photosystem I is capable of peak efficiency; photosystem II is working at a below-maximum rate, so the overall rate of electron flow is reduced. But when both wavelengths are available

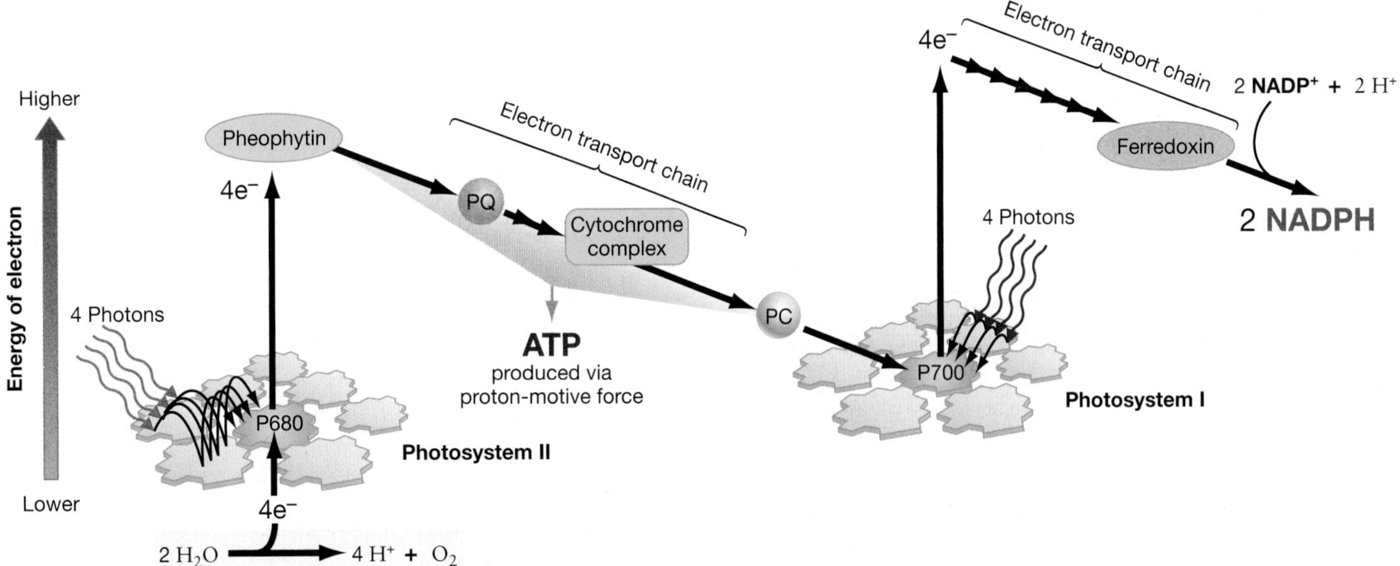

FIGURE 10.15 The Z Scheme Links Photosystems I and II. The Z scheme proposes that electrons from photosystem II enter photosystem I, where they are promoted to an energy state high enough to make the reduction of $NADP^+$ possible.

● **QUESTION** How do the electron transport chains in photosystems I and II compare and contrast?

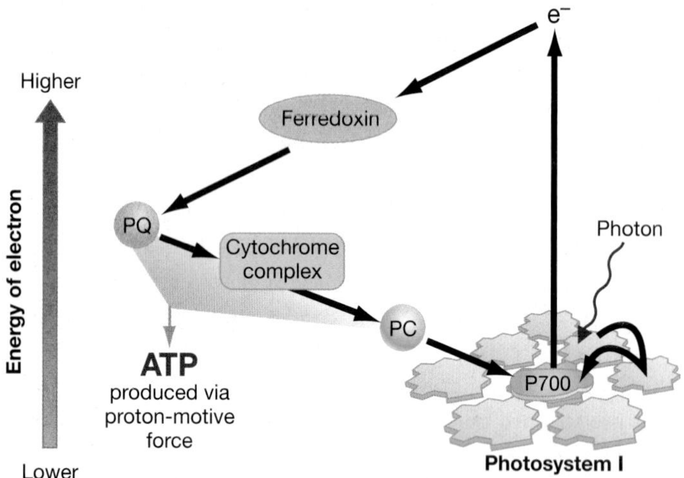

FIGURE 10.16 Cyclic Photophosphorylation Produces ATP. Cyclic electron transport is an alternative to the Z scheme. Instead of being donated to $NADP^+$, electrons cycle through the system, and this results in the production of additional ATP via photophosphorylation.

at the same time, both photosystem II and photosystem I are activated by light and work at a maximum rate, leading to enhanced efficiency.

Recent evidence indicates that a different electron path also occurs in green algae and plants. This pathway, called **cyclic photophosphorylation**, is illustrated in **Figure 10.16**. During cyclic photophosphorylation, photosystem I transfers electrons back to the electron transport chain in photosystem II, to augment ATP generation through photophosphorylation. This "extra" ATP is required for the chemical reactions that reduce carbon dioxide (CO_2) and produce sugars. In this way, cyclic photophosphorylation coexists with the Z scheme and produces additional ATP.

Although the Z-scheme model has held up well under experimental tests, a major mystery remains. As **Figure 10.17** shows, photosystem II is much more abundant in the interior, stacked membranes of grana, while photosystem I and ATP synthase are much more common in the exterior, unstacked membranes. Notice that the proton gradient established by photosystem II drives protons into the stroma, meaning that the stroma is the site of ATP production. The physical separation between the photosystems is perplexing, though, considering that their functions are so tightly integrated according to the Z scheme. Why they are found in different parts of the thylakoid is currently the focus of intense research and debate.

In contrast to the Z scheme, the fate of the ATP and NADPH produced by photosystems I and II is well documented. Chloroplasts use ATP and NADPH to reduce carbon dioxide to sugar. Your life, and the life of most other organisms, depends on this process. How does it happen?

Photosynthesis

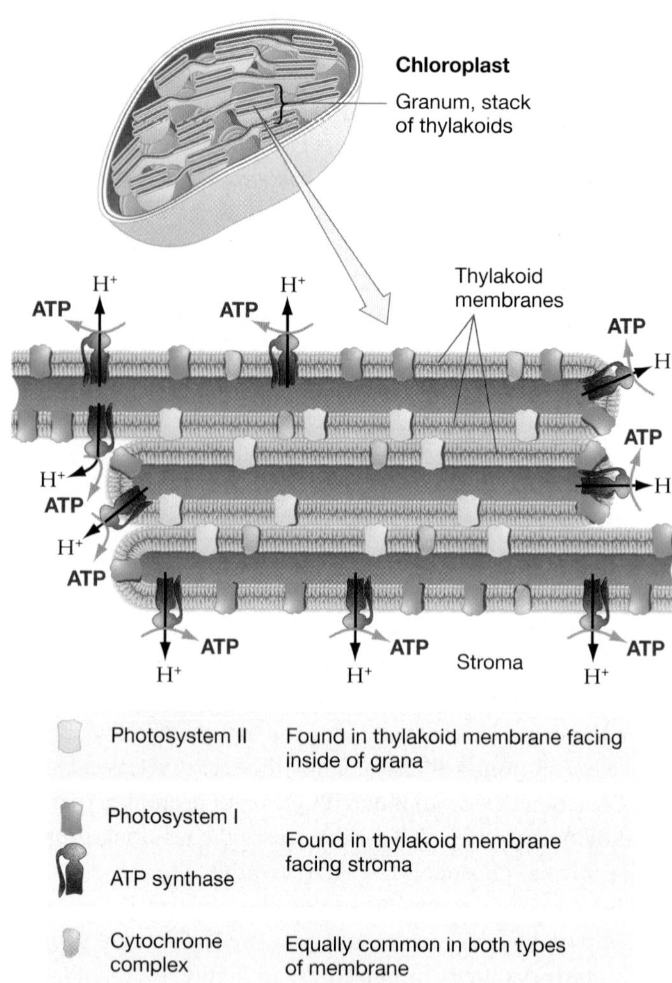

FIGURE 10.17 Photosystems I and II Occur in Separate Regions of Thylakoid Membranes within Grana.

EXERCISE On the figure, draw the path of an electron that follows the Z scheme from photosystem II to photosystem I. Then draw the path of an electron that participates in cyclic photophosphorylation.

Check Your Understanding

If you understand that...

- Photosystem II contributes high-energy electrons to an electron transport chain that pumps protons, creating a proton-motive force that drives ATP synthase.
- Photosystem I makes NADPH.

You should be able to...

1) Make a model of the Z scheme using paper cutouts. On pieces of paper, label the following: the antenna systems of photosystems II and I, the ETCs in photosystems I and II, pheophytin, plastoquinone, plastocyanin, ferredoxin, and the reaction that splits water.
2) Using dimes to represent electrons, explain how they flow through the photosystems.

10.4 How Is Carbon Dioxide Reduced to Produce Glucose?

The reactions analyzed in Section 10.3 are triggered by light. This is logical, because their entire function is focused on transforming energy—converting electromagnetic energy in the form of sunlight to chemical energy in the phosphate bonds of ATP and the electrons of NADPH. The reactions that lead to the production of sugar from carbon dioxide, in contrast, are not triggered directly by light. Instead, they depend on the ATP and NADPH produced by the light-dependent reactions of photosynthesis.

The realization that the energy transformation and carbon dioxide reduction components of photosynthesis are two separate but linked processes was a fundamental insight. Research on the CO_2-reduction reactions gained momentum just after World War II, when radioactive isotopes of carbon became available for research purposes.

The Calvin Cycle

To unravel the reaction sequence that reduces carbon dioxide, Melvin Calvin's group used the pulse-chase strategy introduced in Chapter 7. Recall that pulse-chase experiments introduce a pulse of labeled compound followed by a chase of unlabeled compound. The fate of the labeled compound is then followed through time. In this case, the researchers fed green algae a pulse of $^{14}CO_2$ followed by a large amount of unlabeled CO_2 (**Figure 10.18**). After waiting a specified amount of time, they ground up the cells to form a crude extract, separated individual molecules in the extract via chromatography, and laid X-ray film over the chromatography surface. If radioactively labeled molecules were present, the energy they emitted would expose the film and create a dark spot. The labeled compounds could then be isolated and identified.

By varying the amount of time between starting the pulse of labeled $^{14}CO_2$ and analyzing the cells, Calvin and co-workers began to piece together the sequence in which various intermediates formed. For example, when the team analyzed cells almost immediately after starting the $^{14}CO_2$ pulse, they found that the three-carbon compound 3-phosphoglycerate predominated. This result suggested that 3-phosphoglycerate was the initial product of carbon reduction. Stated another way, it appeared that carbon dioxide reacted with some unknown molecule to produce 3-phosphoglycerate.

This was an interesting result, because 3-phosphoglycerate is one of the ten intermediates in glycolysis. The finding that glycolysis and carbon reduction share intermediates was intriguing because of the relationship between the two pathways. The ATP and NADPH reactions lead to the manufacture of carbohydrate; glycolysis breaks it down. Because the two processes are related in this way, it was logical that at least some intermediates in glycolysis and CO_2 reduction are the same.

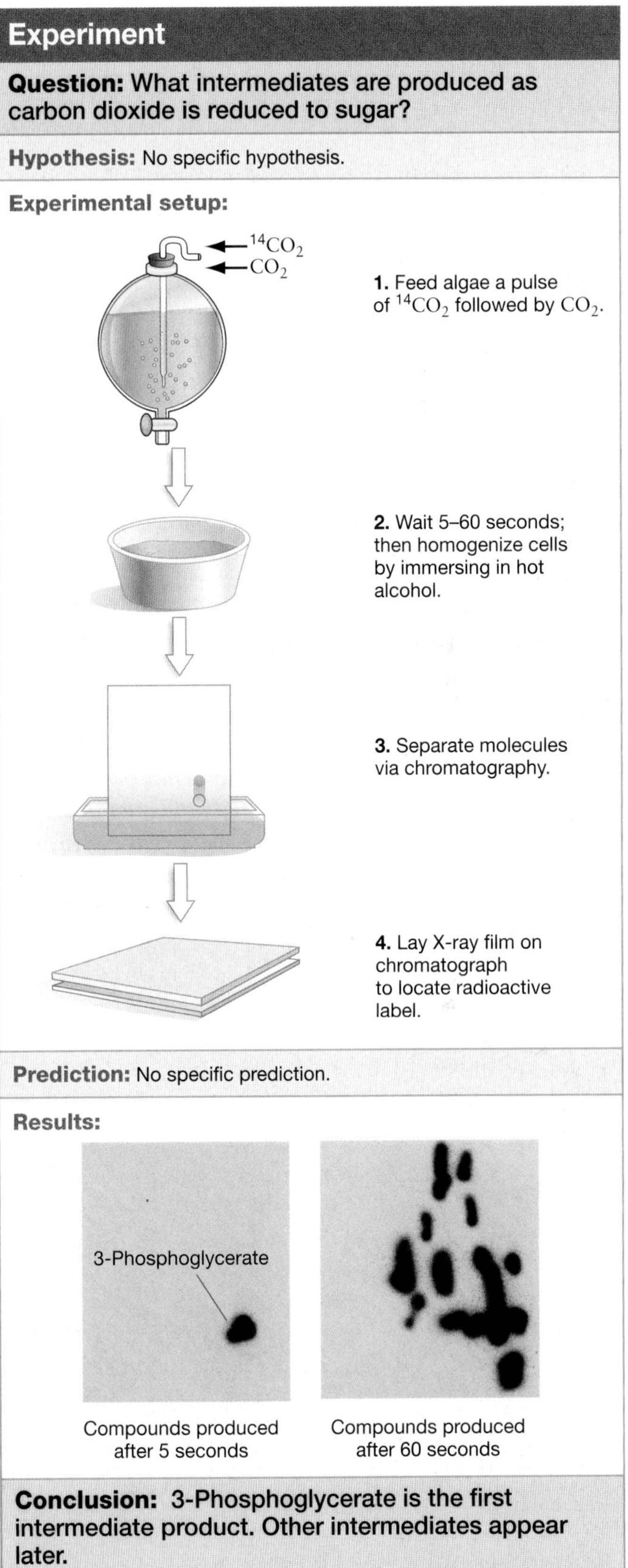

FIGURE 10.18 Experiments Revealed the Reaction Pathway Leading to Reduction of CO_2.

QUESTION Why wasn't this experiment based on a specific hypothesis and set of predictions?

As Calvin's group pieced together the sequence of events in carbon dioxide reduction, an important question remained unanswered: What compound reacts with CO_2 to produce 3-phosphoglycerate? This was the key, initial step. The group searched in vain for a two-carbon compound that might serve as the initial carbon dioxide acceptor and yield 3-phosphoglycerate. Then, while Calvin was running errands one day, it occurred to him that the molecule reacting with carbon dioxide might contain five carbons, not two. The idea was that adding CO_2 to a five-carbon molecule would produce a six-carbon compound, which could then split in half to form 2 three-carbon molecules.

Experiments to test this hypothesis confirmed that the five-carbon compound **ribulose bisphosphate (RuBP)** is the initial reactant. Eventually three phases of CO_2 reduction were worked out and became known as the Calvin cycle (**Figure 10.19**):

1. *Fixation phase.* The events begin when CO_2 reacts with RuBP. This phase "fixes" carbon dioxide by attaching it to a more complex molecule. It also leads to the production of two molecules of 3-phosphoglycerate. **Carbon fixation** is the addition of carbon dioxide to an organic compound, putting CO_2 into a biologically useful form.
2. *Reduction phase.* Next, 3-phosphoglycerate is phosphorylated by ATP and then reduced by electrons from NADPH. The product is the phosphorylated sugar **glyceraldehyde-3-phosphate (G3P)**. Some of the resulting G3P is drawn off to manufacture glucose and fructose, which are linked to form the disaccharide sucrose.
3. *Regeneration phase.* The rest of the G3P keeps the cycle going by serving as the substrate for the third phase in the cycle: reactions that result in the regeneration of RuBP.

All three phases take place in the stroma of chloroplasts. One turn of the Calvin cycle fixes one molecule of CO_2; three turns of the cycle are required to produce one molecule of G3P. If you understand the essence of the Calvin cycle, you should be able to explain the relationship between CO_2, G3P, RuBP, and 3-phosphoglycerate.

The discovery of the Calvin cycle clarified how the ATP and NADPH produced by light-dependent reactions allow cells to reduce CO_2 to carbohydrate $(CH_2O)_n$. Because sugars store a great deal of potential energy, producing them takes a great deal of chemical energy. During photosynthesis, the energy required to reduce CO2 to sugar is provided by ATP and NADPH synthesized in the light-dependent reactions.

Once the reaction sequence in the Calvin cycle was confirmed, attention focused on the initial phase—the reaction between RuBP and CO_2. The reaction between RuBP and CO_2 is one of only two reactions that are unique to the Calvin cycle. Most reactions involved in reducing CO_2 also occur during glycolysis or other metabolic pathways.

The reaction between CO_2 and RuBP starts the transformation of carbon dioxide gas from the atmosphere into sugars. Plants use sugars to fuel cellular respiration and build leaves, roots, flowers, seeds, tree trunks, and other structures. Millions of non-photosynthesizer organisms—including fish, insects, fungi, and mammals—also depend on this reaction to provide the sugars they need for cellular respiration. Ecologically, the

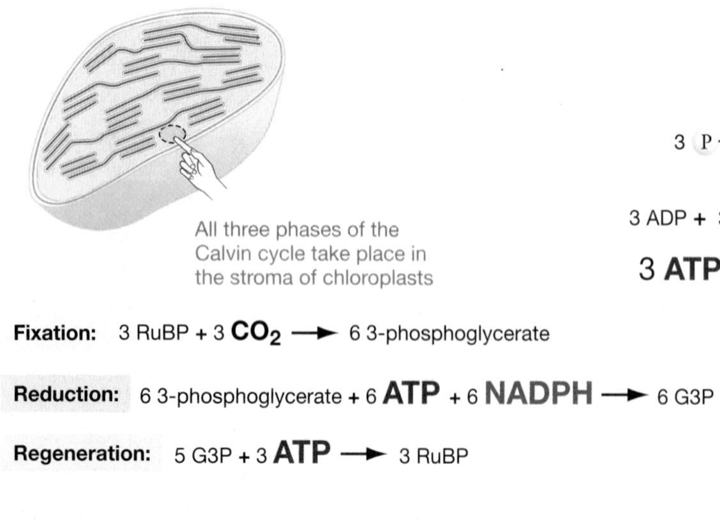

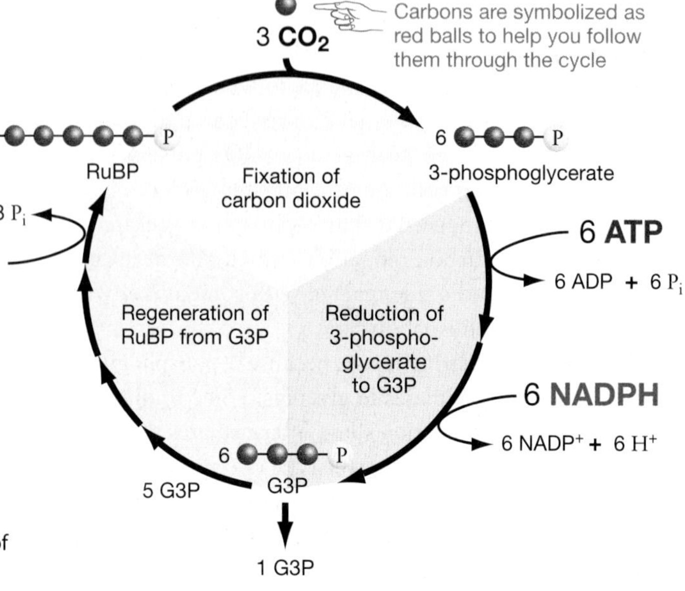

FIGURE 10.19 Carbon Dioxide Is Reduced in the Calvin Cycle. The reactions of the Calvin cycle depend on the products of the light-dependent reactions.

addition of CO_2 to RuBP may be the most important chemical reaction on Earth. The enzyme that catalyzes it is fundamental to all life. How does this molecule work?

The Discovery of Rubisco

To find the enzyme that fixes CO_2, Arthur Weissbach and colleagues ground up spinach leaves, purified a large series of proteins from the resulting cell extracts, and then tested each protein to see if it could catalyze the incorporation of $^{14}CO_2$ into RuBP to form 3-phosphoglycerate. Eventually they were able to isolate an enzyme that catalyzes the reaction. The enzyme turned out to be extremely abundant in leaf tissue. The researchers' data suggested that the enzyme constituted at least 10 percent of the total protein found in spinach leaves.

The CO_2-fixing enzyme was eventually purified and analyzed. Ribulose-1,5-bisphosphate carboxylase/oxygenase is its full name, but it is commonly referred to as **rubisco**. Rubisco is found in all photosynthetic organisms that use the Calvin cycle to fix carbon, and it is thought to be the most abundant enzyme on Earth. Its three-dimensional structure has been determined as well. The molecule is shaped like a cube and has a total of eight active sites where CO_2 is fixed.

Even though it has a large number of active sites, rubisco is a slow enzyme. Each active site catalyzes just three reactions per second; other enzymes typically catalyze thousands of reactions per second. Plants synthesize huge amounts of rubisco, possibly as an adaptation compensating for its lack of speed.

Besides being slow, rubisco is extremely inefficient. Rubisco is inefficient because it catalyzes the addition of O_2 to RuBP as well as the addition of CO_2 to RuBP. This is a key point. Oxygen and carbon dioxide compete at the enzyme's active sites, which slows the rate of CO_2 reduction. Why would an active site of rubisco accept both molecules? Given rubisco's importance in producing food for photosynthetic species, the observation is puzzling. It appears to be **maladaptive**—a trait that reduces the fitness of individuals. One hypothesis to explain the dual nature of the active site is based on the observation that rubisco was present in photosynthetic organisms long before the evolution of oxygenic photosynthesis. As a result, O_2 was extremely rare in the atmosphere when rubisco evolved. According to this hypothesis, rubisco's inefficiency is a historical artifact. The idea is that rubisco is adapted to an atmosphere that no longer exists—one that was extremely rich in CO_2 and poor in O_2.

Unfortunately, the reaction of O_2 with RuBP does more than simply compete with the reaction of CO_2 at the same active site. One of the molecules that results from the addition of oxygen to RuBP is processed in reactions that consume ATP and release CO_2. Part of this pathway occurs in chloroplasts, and part in peroxisomes and mitochondria. The reaction sequence resembles respiration, because it consumes oxygen and produces carbon dioxide. As a result, it is called **photorespiration**. Because photorespiration consumes energy and undoes carbon fixation, it can be considered a reverse photosynthesis (**Figure 10.20**). When photorespiration occurs, the overall rate of photosynthesis declines.

Reaction with carbon dioxide during photosynthesis:

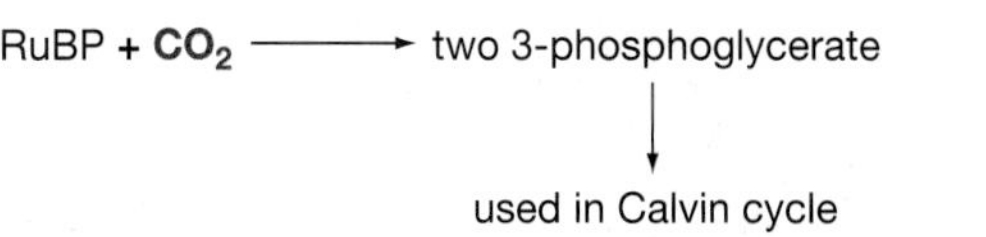

Reaction with oxygen during "photorespiration":

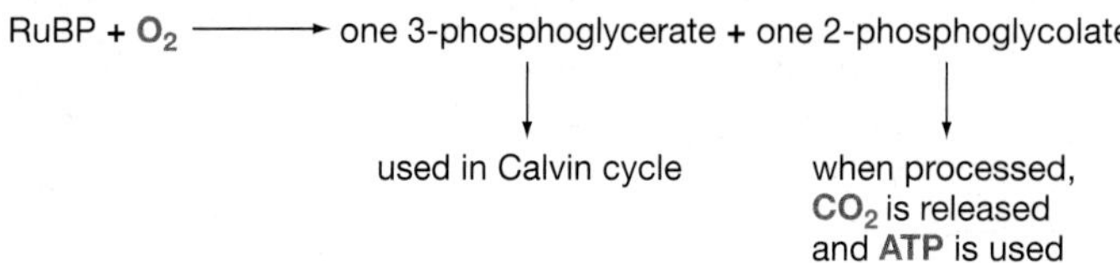

FIGURE 10.20 Photorespiration Competes with Photosynthesis. Rubisco catalyzes competing reactions with very different outcomes.

● **QUESTION** After studying these reactions, explain why biologists say that photorespiration "undoes" photosynthesis.

The oxygenation reaction that triggers photorespiration is favored when oxygen concentrations are high and CO_2 concentrations are low. But as long as carbon dioxide concentrations in leaves are high, the CO_2-fixation reaction is favored and photorespiration is relatively rare.

How Is Carbon Dioxide Delivered to Rubisco?

Carbon dioxide is present in the atmosphere and is continuously used as a reactant in photosynthesizing cells. It would seem straightforward, then, for CO_2 to diffuse directly into plants along a concentration gradient. But the situation is not this simple, because plants are covered with a waxy coating called a cuticle. This lipid layer prevents water from evaporating out of tissues, but it also prevents CO_2 from entering them.

How does CO_2 get into photosynthesizing tissues? A close look at a leaf surface, such as the one in **Figure 10.21a**, provides the answer. The leaf surface is dotted with openings bordered by two distinctively shaped cells. The paired cells are called **guard cells**, the opening is called a pore, and the entire structure is called a **stoma** (plural: **stomata**). If CO_2 concentrations inside the leaf are low as photosynthesis gets under way, chemical signals activate proton pumps in the membranes of guard cells. These pumps establish a charge gradient across the membrane. In response, potassium ions (K^+) move into the guard cells. When water follows along the newly created osmotic gradient, the cells swell and create a pore. As **Figure 10.21b** shows, an open stoma allows CO_2 from the atmosphere to diffuse into

(a) Leaf surfaces contain stomata.

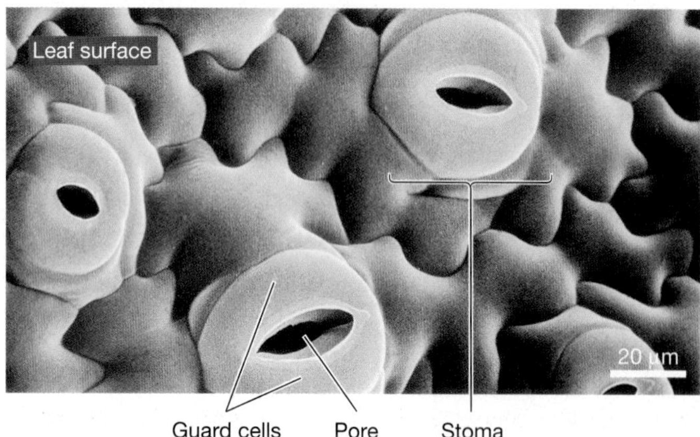

(b) Carbon dioxide diffuses into leaves through stomata.

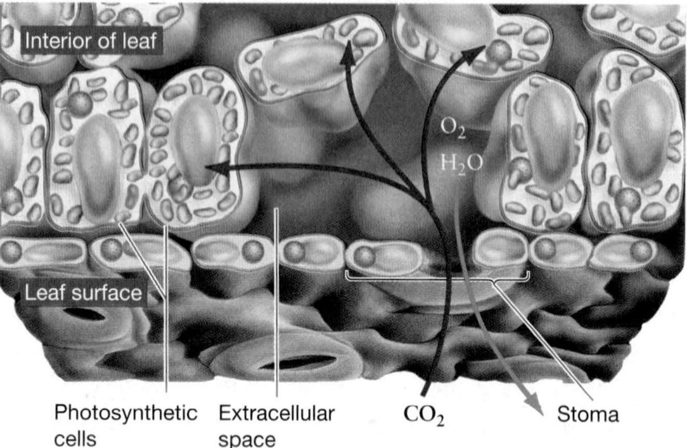

FIGURE 10.21 Leaf Cells Obtain Carbon Dioxide through Stomata. **(a)** Stomata consist of two guard cells and a pore. **(b)** When a stoma is open, CO_2 diffuses into the leaf along a concentration gradient.

filled spaces inside the leaf, and from there into the extracellular fluid surrounding photosynthesizing cells. Eventually the CO_2 diffuses along a concentration gradient into the chloroplasts of the cells. A strong concentration gradient favoring entry of CO_2 is maintained by the Calvin cycle, which constantly uses up the CO_2 in chloroplasts.

Stomata are normally open during the day, when photosynthesis is occurring, and closed at night. But if the daytime is extremely hot and dry, leaf cells may begin losing a great deal of water to evaporation through their stomata. When this occurs, they must either close the openings and halt photosynthesis or risk death from dehydration. When conditions are hot and dry, then, photosynthesis and growth stop. How do plants that live in hot, dry environments cope? An answer emerged as biologists struggled to understand a surprising experimental result.

C_4 Photosynthesis After the Calvin cycle had been worked out in algae, researchers in a variety of labs used the same pulse-chase approach to investigate how carbon fixation occurs in other species. Just as Calvin had done, Hugo Kortschak and colleagues and Y. S. Karpilov and associates exposed leaves of sugarcane and maize (corn) to radioactive carbon dioxide ($^{14}CO_2$) and sunlight, and then isolated and identified the product molecules. Both research teams expected to find the first of the radioactive carbon atoms in 3-phosphoglycerate—the normal product of carbon fixation by rubisco. Instead, they found that in their species the radioactive carbon atom ended up in four-carbon compounds such as malate and aspartate—not in three-carbon sugars.

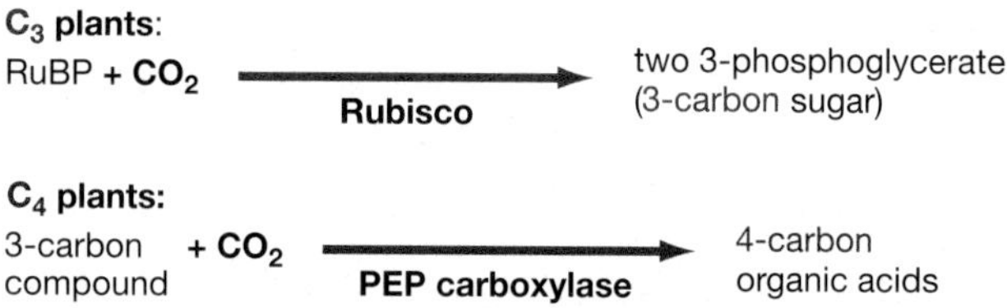

FIGURE 10.22 Initial Carbon Fixation in C_4 Plants Is Different from That in C_3 Plants.

The experiments revealed a twist on the usual pathway for carbon fixation. Instead of creating a three-carbon sugar, it appeared that in their CO_2 fixation produced four-carbon sugars. The two pathways became known as **C_3 photosynthesis** and **C_4 photosynthesis,** respectively (**Figure 10.22**).

Researchers who followed up on the initial reports found that, in some plant species, carbon dioxide can be added to RuBP by rubisco *or* to three-carbon compounds by an enzyme called **PEP carboxylase**. They also showed that the two enzymes are found in distinct cell types within the same leaf. PEP carboxylase is common in **mesophyll cells** near the surface of leaves, while rubisco is found in **bundle-sheath cells** that surround the vascular tissue in the interior of the leaf (**Figure 10.23a**). **Vascular tissue** conducts water and nutrients in plants.

Based on these observations, Hal Hatch and Roger Slack proposed a three-step model to explain how CO_2 that is fixed to a four-carbon sugar feeds the Calvin cycle (**Figure 10.23b**):

1. PEP carboxylase fixes CO_2 in mesophyll cells.
2. The four-carbon organic acids that result travel to bundle-sheath cells.
3. The four-carbon organic acids release a CO_2 molecule that rubisco uses as a substrate to form 3-phosphoglycerate. This step initiates the Calvin cycle.

In effect, then, the C_4 pathway acts as a CO_2 pump. The reactions that take place in mesophyll cells require energy in the form of ATP, but they increase CO_2 concentrations in cells where rubisco is active. Because it increases the ratio of carbon

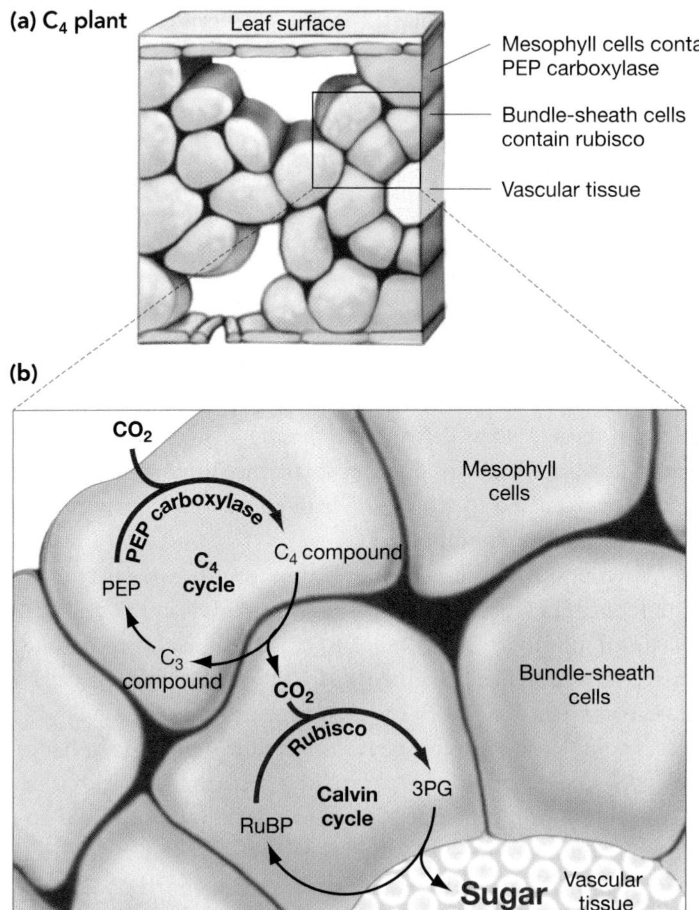

FIGURE 10.23 In C_4 plants, Carbon Fixation Occurs Independently of the Calvin Cycle. (a) The carbon-fixing enzyme PEP carboxylase is located in mesophyll cells, while rubisco is in bundle-sheath cells. **(b)** CO_2 is fixed to the three-carbon compound PEP by PEP carboxylase, forming a four-carbon organic acid. A CO_2 molecule from the four-carbon sugar then feeds the Calvin cycle.

dioxide to oxygen in photosynthesizing cells, less O_2 binds to rubisco's active sites. Stated another way, CO_2 fixation is favored over O_2 fixation when carbon dioxide concentrations in leaves are high. As a result, the C_4 pathway limits the damaging effects of photorespiration. The pathway is an adaptation that keeps CO_2 concentrations in leaves high. Later experiments supported the Hatch and Slack model in almost every detail.

Follow-up work showed that the C_4 pathway is found almost exclusively in plants that thrive in hot, dry habitats. To understand why, it is critical to realize that plants have to close their stomata on hot, dry days, to avoid catastrophic water loss. When stomata close, oxygen levels build up in leaves, CO_2 concentration declines, and photorespiration is favored. PEP carboxylase is not affected by high oxygen levels, however, and has a higher affinity for CO_2 than rubisco does. Thus, PEP carboxylase can continue to deliver CO_2 to rubisco and minimize the impact of photorespiration even when oxygen is plentiful and CO_2 is rare inside leaves. As a result, C_4 plants can continue to produce sugars on hot, dry days much more efficiently than C_3 plants can.

Sugarcane, maize (corn), and crabgrass are some familiar C_4 plants, but the pathway is actually found in several thousand species in 19 distinct lineages of flowering plants. These observations suggest that the C_4 pathway has evolved independently several times. It is not the only mechanism that plants use to continue growth under hot, dry conditions, however.

CAM Plants Some years after the discovery of C_4 photosynthesis, researchers studying a group of flowering plants called the Crassulaceae came across a second mechanism for limiting the effects of photorespiration. This photosynthetic pathway became known as **crassulacean acid metabolism**, or **CAM**. Like the C_4 pathway, CAM is a CO_2 pump that acts as an additional, preparatory step to the Calvin cycle. It has the same effect: It increases the concentration of CO_2 inside photosynthesizing cells. It also involves an organic acid with four carbons. But unlike the C_4 pathway, CAM occurs at a different time than the Calvin cycle does—not in a different place.

CAM occurs in cacti and other species living in such hot, dry environments that individuals routinely keep their stomata closed all day. When night falls and conditions become cooler and moister, CAM plants open their stomata and take in huge quantities of CO_2. These molecules are temporarily fixed to organic acids and stored in the central vacuoles of photosynthesizing cells. During the day, the molecules are processed in reactions that release the CO_2 and feed the Calvin cycle.

Figure 10.24 summarizes the similarities and differences between C_4 photosynthesis and CAM. Both C_4 photosynthesis and CAM function as CO_2 pumps to minimize the amount of photorespiration that occurs when stomata are closed and CO_2 cannot diffuse in directly from the atmosphere. Both are found in flowering plant species that live in hot, dry environments. But while C_4 plants stockpile CO_2 in cells where rubisco is not active, CAM plants store CO_2 when rubisco is inactive. In C_4 plants, the reactions catalyzed by PEP-carboxylase and rubisco are separated in space; in CAM plants, the reactions are separated in time.

Obtaining and reducing CO_2 is fundamental to photosynthesis. In a larger sense, photosynthesis is fundamental to the millions of species that depend on plants, algae, and cyanobacteria for food. What do photosynthetic organisms do with the sugar they synthesize? More specifically, what happens to the G3P that is drawn off from the Calvin cycle?

at www.masteringbio.com
Strategies for Carbon Fixation

(a) C_4 plants sequester CO_2 in certain cells.

(b) CAM plants sequester CO_2 at night.

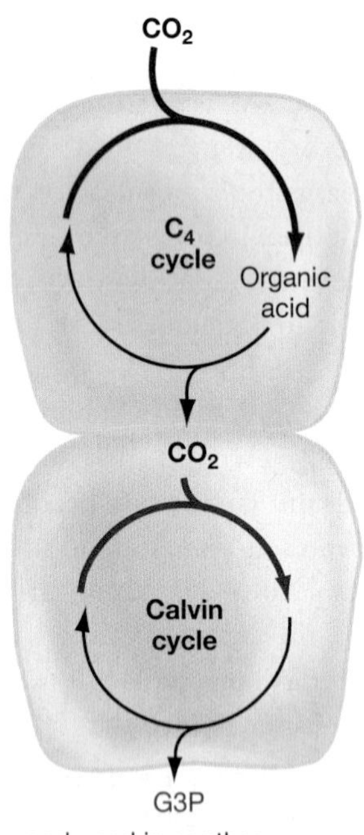

FIGURE 10.24 C_4 Photosynthesis and CAM Accomplish the Same Task in Different Ways. (a) In C_4 plants, CO_2 is fixed to organic acids in some cells and then released to other cells where the Calvin cycle enzymes are located. **(b)** CAM plants open their stomata at night and fix CO_2 to organic acids.

What Happens to the Sugar That Is Produced by Photosynthesis?

The G3P molecules that exit the Calvin cycle enter one of several reaction pathways. The most important of these pathways results in the production of the monosaccharides glucose and fructose, which in turn combine to form the disaccharide **sucrose** (**Figure 10.25**). The reaction sequence starts with G3P from the Calvin cycle, involves a series of other phosphorylated three-carbon sugars, includes the synthesis of the familiar six-carbon sugar glucose, and ends with the production of sucrose. An alternative pathway results in the production of glucose molecules that polymerize to form **starch.** Starch production occurs inside the chloroplast; sucrose synthesis takes place in the cytosol.

All of the intermediates involved in the production of glucose, as well as of G3P, also occur in glycolysis. The general observation here is that many of the enzymes and intermediates involved in glucose processing are common to both respiration and photosynthesis.

When photosynthesis is taking place slowly, almost all the glucose that is produced is used to make sucrose. As Chapter 5 noted, sucrose is a disaccharide ("two-sugar") consisting of a glucose molecule bonded to a fructose molecule. Sucrose is water soluble and is readily transported to other parts of the plant. If the sucrose is delivered to rapidly growing parts of the plant, it is broken down to fuel cellular respiration and growth. If it is transported to storage cells in roots, it is converted to starch and stored for later use.

When photosynthesis is proceeding rapidly and sucrose is abundant, glucose is used to synthesize starch in the chloroplasts of photosynthetic cells. Recall from Chapter 5 that starch is a polymer of glucose. In photosynthesizing cells, starch acts as a temporary sugar-storage product. Starch is not water soluble, so it cannot be transported from photosynthetic cells to other areas of the plant. At night, the starch that is temporarily stored in leaf cells is broken down and used to manufacture sucrose molecules. The sucrose is then used by the photosynthetic cell in respiration or transported to other parts of the plant. In this way, chloroplasts provide sugars for cells throughout the plant by day and by night.

If a mouse eats the starch that is stored in a chloroplast or root cell, however, the chemical energy in the C–C and C–H bonds of the starch is used to fuel the mouse's growth and reproduction. If the mouse is then eaten by an owl, the chemical energy in the mouse's tissues fuels the predator's growth and reproduction. In this way, virtually all cell growth and reproduction can be traced back to the chemical energy that was originally captured by photosynthesis. Photosynthesis is the staff of life.

Check Your Understanding

If you understand that...

- The Calvin cycle is a three-phase process: CO_2 fixation (synthesis of 3-phosphoglycerate), production of carbohydrate (synthesis of G3P), and regeneration of RUBP.

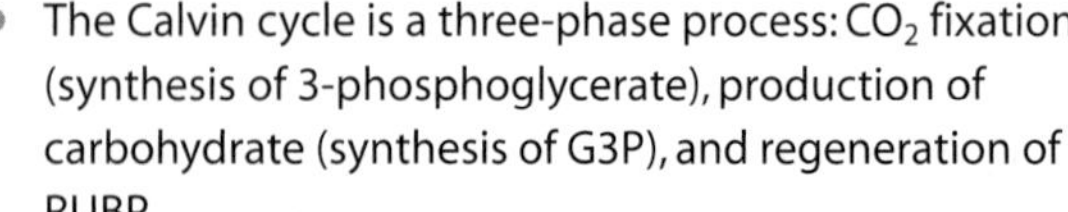

You should be able to...

1) Describe three ways that CO_2 is delivered to rubisco: (a) via organic acids in mesophyll cells, (b) via organic acids synthesized at night and stored in vacuoles, and (c) directly.
2) Explain how the solubility of starch and sucrose affect their use by plants.

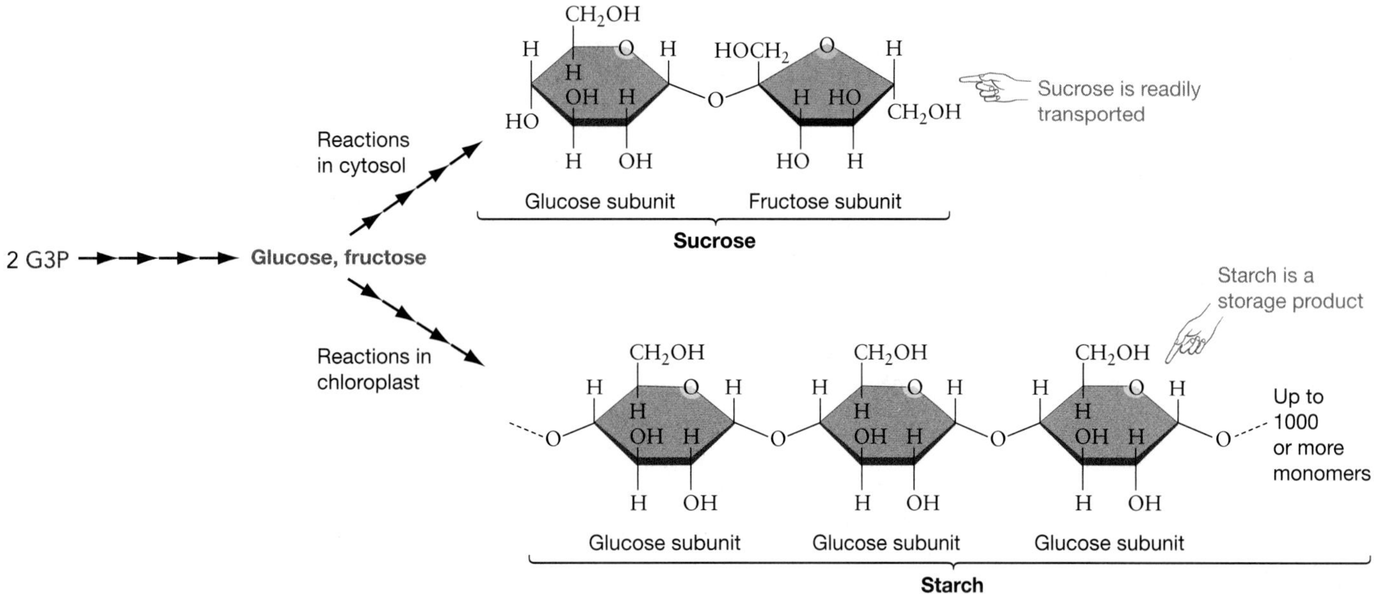

FIGURE 10.25 Sucrose and Starch Are the Main Photosynthetic Products. In plants, sugars are transported in the form of sucrose and stored in the form of starch.

EXERCISE Sucrose can be converted to starch, and starch to sucrose. Add an element to the diagram to indicate this.

Chapter Review

SUMMARY OF KEY CONCEPTS

Photosynthesis is the conversion of light energy to chemical energy, stored in the bonds of carbohydrates. The sucrose generated by photosynthesis fuels cellular respiration and is a substrate for the synthesis of complex carbohydrates, amino acids, fatty acids, and other cell components. As the primary food source for a diverse array of heterotrophs, photosynthetic organisms provide the energy that sustains most life on Earth.

Photosynthesis consists of two linked sets of reactions. In reactions driven directly by light, ATP and the electron carrier NADPH are produced. In subsequent reactions called the Calvin cycle, the ATP and NADPH are used to reduce carbon dioxide (CO_2) to carbohydrate [$(CH_2O)_n$]. In eukaryotic cells, both processes take place in chloroplasts.

The light-dependent reactions occur in internal membranes of the chloroplast that are organized into structures called thylakoids in stacks known as grana. The Calvin cycle takes place in a fluid portion of the chloroplast called the stroma.

You should be able to explain why biologists have stopped using the phrase and "light-independent reactions" to describe the Calvin cycle.

The light-dependent reactions transform the energy in sunlight to chemical energy in the form of electrons with high potential energy. Excited electrons either are used to produce NADPH or are donated to an electron transport chain, which results in the production of ATP via chemiosmosis.

The energy transformation step of photosynthesis begins when a pigment molecule in an antenna complex absorbs a photon in the blue or red part of the visible spectrum. When absorption occurs, the energy in the photon is transferred to an electron in the pigment molecule. The energy in the excited electron is eventually transferred to chlorophyll molecules that act as reaction centers. There a high-energy electron is transferred to an electron acceptor, which becomes reduced. In this way, light energy is transformed to chemical energy.

Plants and algae have two types of reaction centers, which are part of larger complexes called photosystem I and photosystem II. Each photosystem consists of an antenna complex with 200–300 chlorophyll and carotenoid molecules, a reaction center, and an electron acceptor that completes energy transformation.

In photosystem II, high-energy electrons are accepted by the electron acceptor pheophytin. Electrons are then passed along an electron transport chain, where they are gradually stepped down in potential energy. The energy released by these redox reactions is used to pump protons across the thylakoid membrane. The resulting proton gradient drives the synthesis of ATP by ATP synthase. This method of producing ATP is called photophosphorylation. Electrons donated to the electron transport chain by photosystem II are replaced by electrons taken from water, resulting in the production of oxygen as a by-product.

In photosystem I, high-energy electrons are accepted by iron- and sulfur-containing proteins and passed to ferredoxin. In an enzyme-catalyzed reaction, the reduced form of ferredoxin passes electrons to $NADP^+$ to form NADPH.

The Z scheme describes how photosystems I and II are thought to interact. The scheme begins with the movement of an electron from photosystem II to the electron transport chain. At the end of the chain, the protein plastocyanin carries electrons to photosystem I. There the electrons are promoted to a high energy state in response to the absorption of a photon, and subsequently used to reduce $NADP^+$. Electrons from photosystem I may occasionally be passed to the electron transport chain instead of being used to reduce $NADP^+$, resulting in a cyclic flow of electrons between the two photosystems to produce the additional ATP needed to reduce carbon dioxide.

MB **Web Animation** at www.masteringbio.com
Chemiosmosis; Photosynthesis

You should be able to predict what happens to NADP levels in the first few seconds after a drug poisons photosystem I or photosystem II. You should also be able to explain why the rate of photosynthesis can be estimated by measuring the rate of oxygen production in chloroplasts.

The Calvin cycle starts with the enzyme rubisco, which catalyzes the addition of CO_2 to a five-carbon molecule. The compound that results undergoes a series of reactions that use ATP and NADPH and lead to the production of sugar.

The CO_2-reduction reactions of photosynthesis depend on the products of the light-dependent reactions. The process begins when CO_2 is attached to a five-carbon compound called ribulose bisphosphate (RuBP). This reaction is catalyzed by the enzyme rubisco. The six-carbon compound that results immediately splits in half to form two molecules of 3-phosphoglycerate. Subsequently, 3-phosphoglycerate is reduced to a sugar called glyceraldehyde-3-phosphate (G3P). Some G3P is used to synthesize glucose and fructose, which combine to form sucrose; the rest participates in reactions that regenerate RuBP so the cycle can continue.

Rubisco catalyzes the addition of oxygen as well as carbon dioxide to RuBP. The reaction with oxygen leads to a loss of fixed CO_2 and ATP and is called photorespiration. Photorespiration is particularly problematic in hot, dry conditions, when stomata close to prevent excessive water loss. Because the closure of stomata reduces CO_2 levels in photosynthesizing cells, the reaction of O_2 with RuBP is favored. C_4 and CAM plants have distinct but functionally similar mechanisms for augmenting CO_2 concentrations in photosynthesizing cells and thus for limiting photorespiration.

MB **Web Animation** at www.masteringbio.com
Strategies for Carbon Fixation

You should be able to predict what would happen in a chloroplast containing an unusual form of rubisco—either a form that did not bind oxygen or a form that worked 10 times as fast as most forms of the enzyme.

QUESTIONS

Test Your Knowledge

1. What is the stroma of a chloroplast?
   a. the inner membrane
   b. the pieces of membrane that connect grana
   c. the interior of a thylakoid
   d. the fluid inside the chloroplast but outside the thylakoids

2. Why is chlorophyll green?
   a. It absorbs all wavelengths in the visible spectrum, transmitting ultraviolet and infrared light.
   b. It absorbs wavelengths only in the red and far-red portions of the spectrum (680 nm, 700 nm).
   c. It absorbs wavelengths in the blue and red parts of the visible spectrum.
   d. It absorbs wavelengths only in the blue part of the visible spectrum and transmits all other wavelengths.

3. What does it mean to say that CO_2 becomes fixed?
   a. It becomes bonded to an organic compound.
   b. It is released during cellular respiration.
   c. It acts as an electron acceptor.
   d. It acts as an electron donor.

4. What do the light-dependent reactions of photosynthesis produce?
   a. G3P
   b. RuBP
   c. ATP and NADPH
   d. sucrose or starch

5. Why do the absorption spectrum for chlorophyll and the action spectrum for photosynthesis coincide?
   a. Photosystems I and II are activated by different wavelengths of light.
   b. Wavelengths of light that are absorbed by chlorophyll trigger the light-dependent reactions.
   c. Energy from wavelengths absorbed by carotenoids is passed on to chlorophyll.
   d. The rate of photosynthesis depends on the amount of light received.

6. What happens when an excited electron is passed to an electron acceptor in a photosystem?
   a. It drops back down to its ground state, resulting in the phenomenon known as fluorescence.
   b. The chemical energy in the excited electron is released as heat.
   c. The electron acceptor is oxidized.
   d. Energy in sunlight is transformed to chemical energy.

Test Your Knowledge answers: 1. d; 2. c; 3. a; 4. c; 5. b; 6. d

Test Your Understanding

Answers are available at www.masteringbio.com

1. Explain how the energy transformation step of photosynthesis occurs. How is light energy converted to chemical energy in the form of ATP and NADPH?
2. Explain how the carbon reduction step of photosynthesis occurs. How is carbon dioxide fixed? Why are both ATP and NADPH required to produce sugar?
3. Sketch the Z scheme. Explain how photosystem I and photosystem II interact by tracing the path of an electron through the Z scheme. What molecule connects the two photosystems?
4. When does photorespiration occur? What are its consequences for the plant?
5. Make a sketch showing how C_4 photosynthesis and CAM separate CO_2 acquisition from the Calvin cycle in space and time, respectively.
6. Why do plants need both chloroplasts and mitochondria?

Applying Concepts to New Situations

Answers are available at www.masteringbio.com

1. Compare and contrast mitochondria and chloroplasts. In what ways are their structures similar and different? What molecules or systems function in both types of organelles? Which enzymes or processes are unique to each organelle?
2. Do you agree with the hypothesis that photorespiration is an evolutionary "holdover?" Why or why not?
3. In addition to providing their protective function, carotenoids absorb certain wavelengths of light and pass the energy on to the reaction centers of photosystem I and II. Based on their function, predict exactly where carotenoids are located in the chloroplast. Explain your rationale. How would you test your hypothesis?
4. Consider plants that occupy the top, middle, or ground layer of a forest, and algae that live near the surface of the ocean or in deeper water. Would you expect the same photosynthetic pigments to be found in species that live in these different habitats? Why or why not? How would you test your hypothesis?

www.masteringbio.com is also your resource for • Answers to text, table, and figure caption questions and exercises • Answers to *Check Your Understanding* boxes • Online study guides and quizzes • Additional study tools including the *E-Book for Biological Science* 3rd ed., textbook art, animations, and videos.

11 The Cell Cycle

KEY CONCEPTS

- After chromosomes are copied, mitosis distributes one chromosome copy to each of two daughter cells. Mitosis and cytokinesis produce two cells that are identical to the parent cell.
- Over their life span, eukaryotic cells go through a cycle that consists of four carefully controlled phases.
- In multicellular organisms, uncontrolled cell growth leads to cancer. Different types of cancer result from different types of defects in control over the cell cycle.

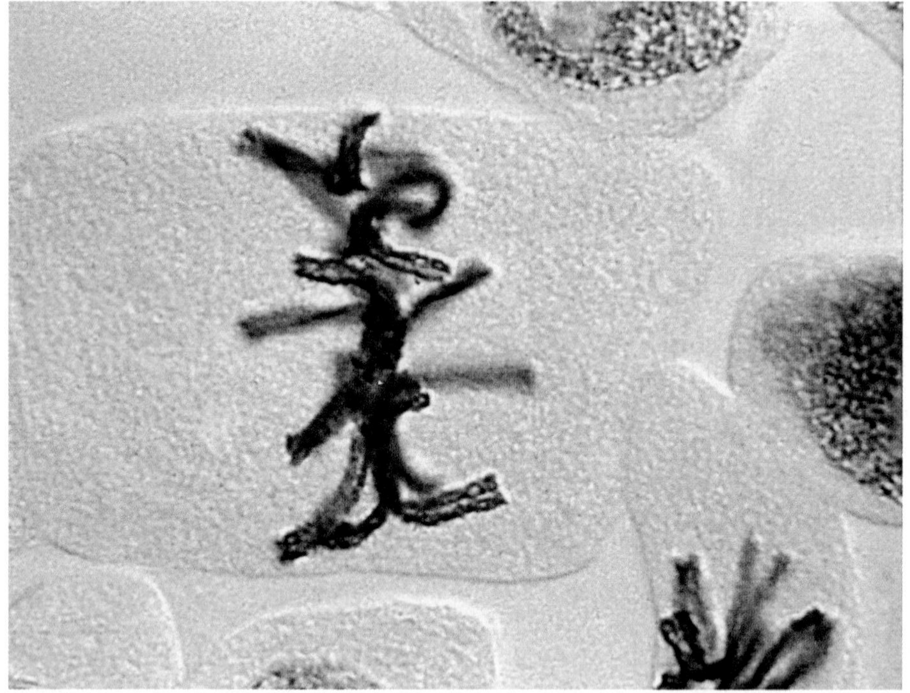

This cell, from a hyacinth, is undergoing a type of cell division called mitosis. Understanding how mitosis occurs is a major focus of this chapter.

Chapter 1 introduced the cell theory, which maintains that all organisms are made of cells and that all cells arise from preexisting cells. Although the cell theory was widely accepted among biologists by the 1860s, a great deal of confusion remained about how cells reproduced. Most proponents of the cell theory believed that new cells arose within preexisting cells by a process that resembled the growth of mineral crystals. But Rudolf Virchow proposed that new cells arise through the division of preexisting cells—that is, **cell division**.

In the late 1800s, careful microscopic observations of newly developing individuals, or **embryos**, confirmed Virchow's hypothesis. Research documented that multicellular eukaryotes start life as single-celled embryos and grow through a series of cell divisions. As better microscopes became available, biologists were able to observe and describe the division process in more detail. In particular, researchers focused their attention on the nuclei of dividing cells and the fate of the chromosomes. Chromosomes are the carriers of hereditary material—the instructions for building and operating the cell. These early studies revealed two fundamentally different ways that nuclei divide prior to cell division. In animals, one type of nuclear division leads to the production of sperm and eggs, and the other type of nuclear division leads to the production of all other cell types. Sperm and eggs are male and female reproductive cells, termed **gametes**; all other cell types are referred to as **somatic** (literally, "body-belonging") **cells**. In both kinds of cell division, a so-called parent cell is said to give rise to daughter cells.

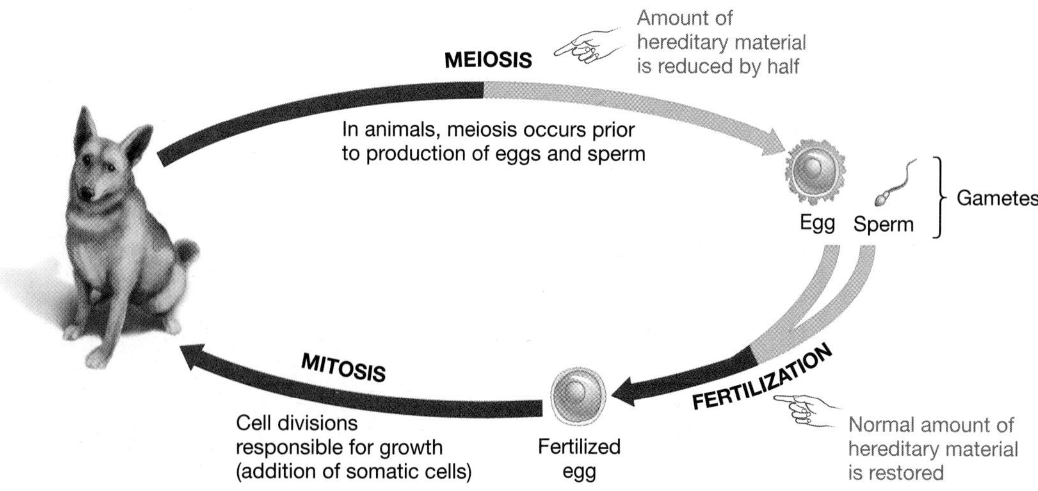

FIGURE 11.1 Two Types of Nuclear Division Occur in Many Species. In animals, meiosis leads to the production of eggs and sperm. Mitosis is responsible for producing somatic cells.

During the type of nuclear division that leads to the production of sperm and eggs, the amount of hereditary material found in the parent cell nucleus is reduced by half. As a result, the daughter cells that become sperm or eggs do not contain the same genetic material as the parent cell. This type of nuclear division is called **meiosis**. In animals, meiosis occurs only prior to the formation of gametes (**Figure 11.1**). In general, meiosis produces reproductive cells and is the basis of sexual reproduction. Chapter 12 explores the mechanism and consequences of meiosis in detail.

When new somatic cells form in eukaryotes, the amount of hereditary material in the original cell and the daughter cells remains constant. **Mitosis** is a division of the genetic material that produces daughter cells that are genetically identical to their parent cell. Mitosis is usually accompanied by **cytokinesis** ("cell movement")—the division of the cytoplasm into the two daughter cells. Together, mitosis and cytokinesis are the processes responsible for three key events in multicellular eukaryotes.

1. *Growth* The ancestry of the trillions of genetically identical cells that make up your body can be traced back through a series of mitotic divisions to a single fertilized egg—the result of the union of a sperm and an egg from your parents.

2. *Wound repair* When you suffer a scrape or cut, the cells that repair your skin and heal the wound are generated via mitosis and cytokinesis.

3. *Reproduction* When yeast cells greatly increase in number in a pile of bread dough or in a vat of beer, they are reproducing by mitosis and cytokinesis. In both unicellular and multicellular species, mitosis followed by cytokinesis is the basis of asexual reproduction. **Asexual reproduction** results in the production of offspring that are genetically identical to the parent.

This chapter's our goals are to explore how mitosis occurs and how cell division is regulated. The first section introduces the relationship between mitosis and other major stages in a cell's life cycle. The next two sections provide an in-depth look at each event in mitosis and explore how mitosis and other events in the cell's life cycle are regulated. The chapter concludes by examining why uncontrolled cell division and cancer can result when the systems that regulate cell division break down.

11.1 Mitosis and the Cell Cycle

In the course of studying cell division, nineteenth-century biologists found that certain chemical dyes made threadlike structures visible in the nuclei of dividing eukaryotic cells. In 1879 Walther Flemming followed up on this discovery by documenting how the threadlike structures in salamander embryos change as the cells divide. As **Figure 11.2a** shows, the threads that Flemming observed were paired when they first appeared, just before cell division. During cell division, each pair of threads split to produce single, unpaired threads in the daughter cells. Flemming introduced the term *mitosis*, from the Greek *mitos* ("thread"), to describe this division process.

Soon after Flemming made that discovery, similar observations were reported for the roundworm *Ascaris*. In addition to confirming that each pair of threads split in this species, investigators reported that the total number of threads in a cell remained constant during subsequent divisions. Thus, all of the cells in a roundworm's body had the same number and types of threads.

In 1888 Wilhelm Waldeyer coined the term **chromosome** ("colored-body") to refer to the threadlike structures observed in dividing cells (**Figure 11.2b**). Chromosomes are

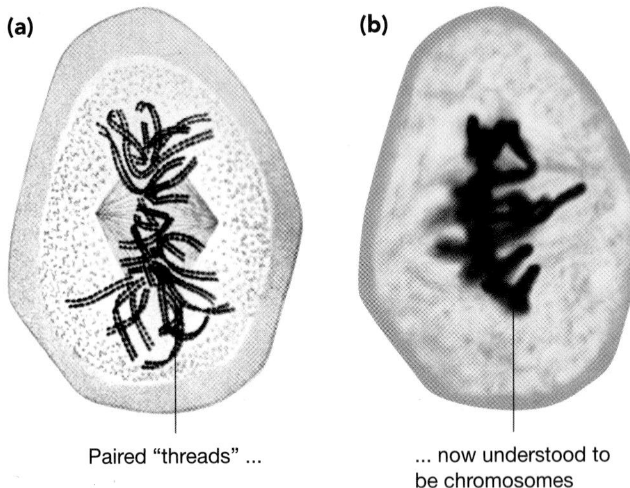

FIGURE 11.2 Chromosomes Move during Mitosis. (a) Walther Flemming's 1879 drawing of mitosis in the salamander embryo. The black threads are chromosomes. **(b)** Chromosomes can be stained with dyes and observed in the light microscope.

made up in part of deoxyribonucleic acid (DNA). More specifically, a chromosome consists of a single, long DNA double helix that is wrapped around proteins in a highly organized manner. DNA encodes the cell's hereditary information, or genetic material. A gene is a length of DNA that codes for a particular protein or ribonucleic acid (RNA) found in the cell. By observing how chromosomes moved during mitosis, biologists realized that the purpose of mitosis was to distribute the parent cell's genetic material to daughter cells during cell division. Prior to mitosis, each chromosome is copied. As mitosis starts, the chromosomes also condense from long, thin, fibrous forms into compact structures that can be moved around the cell efficiently. In the remainder of mitosis, one of the chromosome copies is distributed to each of two daughter cells. **Figure 11.3** illustrates unreplicated chromosomes, replicated chromosomes before they have condensed prior to mitosis, and replicated chromosomes that have condensed at the start of mitosis.

Before delving into the details of how mitosis occurs, though, let's examine how it fits into the other events in the life of a cell.

M Phase and Interphase

As early researchers studied the fate of chromosomes during cell division, they realized that even rapidly growing plant and animal cells do not divide continuously. Instead, growing cells cycle between a dividing phase called the **mitotic** (or **M**) **phase** and a nondividing phase called **interphase** ("between-phase"). Chromosomes can be stained and observed with a light microscope only during M phase, when they are condensed into compact structures. Cells actually spend most of their time in interphase, however. No dramatic changes are observed in the nucleus during interphase, when chromosomes are uncoiled into extremely long, thin structures. Even when stains are used,

(a) Unreplicated chromosome

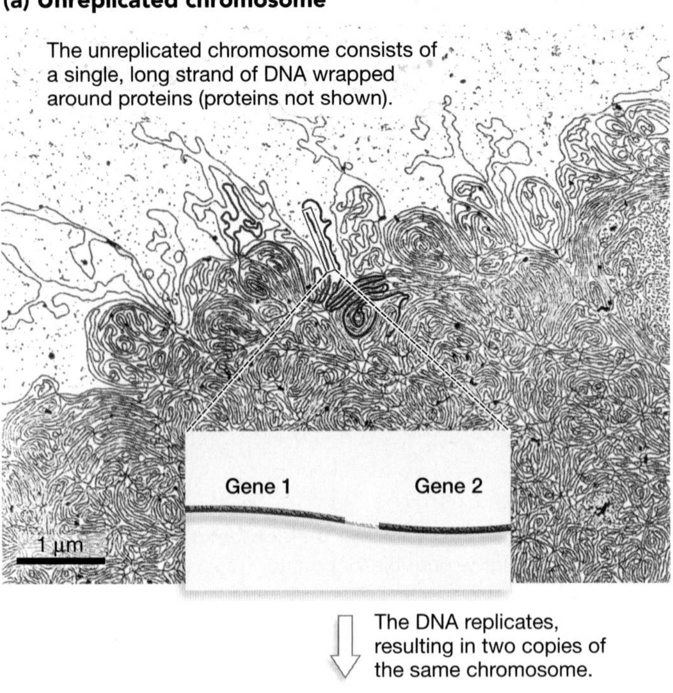

The DNA replicates, resulting in two copies of the same chromosome.

(b) Replicated chromosome

Copies of same chromosome

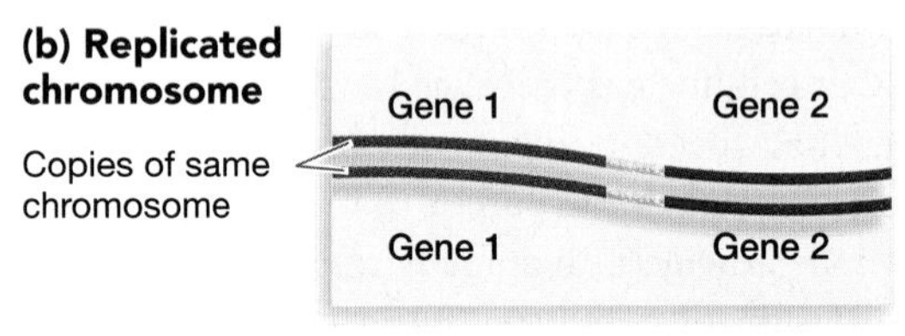

The DNA condenses around its associated proteins, resulting in a compact chromosome that is 10,000 times shorter than its original length.

(c) Condensed replicated chromosome

Copies of same chromosome, condensed

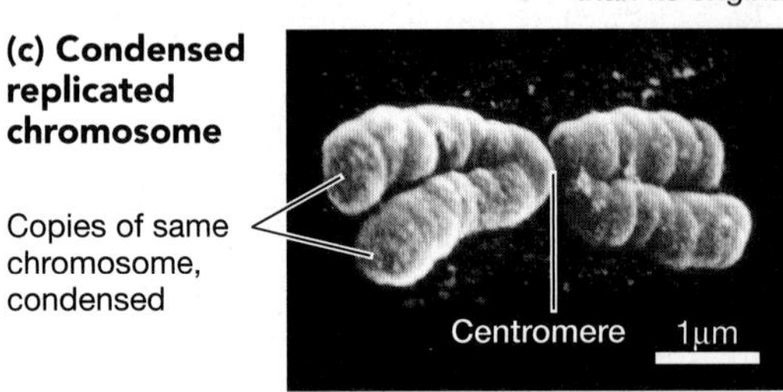

FIGURE 11.3 Chromosome Morphology Changes before and during Mitosis. (a) Chromosomes consist of long DNA molecules associated with proteins. Genes are segments of DNA. **(b)** Chromosomes are replicated prior to mitosis, forming two identical copies that stay attached to each other. **(c)** Early in mitosis, replicated chromosomes condense into highly compact structures, attached at a region called the centromere.

individual interphase chromosomes are not normally visible in the light microscope.

Nineteenth-century biologists knew that chromosomes moved to daughter cells during mitosis, and that each daughter cell ended up with the same number of chromosomes as the parent cell. Thus, it was logical to infer that the chromosomes had to be duplicated in the parent cell at some point in the cell cycle. When did this replication step occur?

The Discovery of the Cell Cycle

The question of when chromosomes were copied was not resolved until the early 1950s, when radioactive isotopes became available. To test hypotheses about the timing of chromosome replication, researchers exposed growing cells to a radioactive isotope that would be incorporated into DNA as it was being synthesized. The specific compounds used were radioactive phosphorus or radioactive thymidine. Phosphorus is a component of deoxyribonucleotides such as thymidine, and deoxyribonucleotides are components of DNA (see Chapter 4). The idea was to label DNA as chromosomes were being copied, then wash away any radioactive isotope that hadn't been incorporated. The labeled DNA could then be visualized by exposing the treated cells to X-ray film. Emissions from radioactive phosphorus or thymidine create a black dot in the film. This is the technique called autoradiography (see **BioSkills 7**).

When Alma Howard and Stephen Pelc did the first experiments using this approach, they simply looked to see if black dots—indicating active DNA synthesis—were present in the nuclei of M-phase cells or interphase cells. When they looked at cells soon after the exposure to a radioactive isotope ended, they found that black dots appeared only in interphase cells, but never in M-phase cells. This was strong evidence that DNA replication occurs during interphase.

Howard and Pelc had identified a new stage in the life of a cell; they called it **synthesis** (or **S**) **phase**, for DNA synthesis. S phase is part of interphase. Their data showed that duplication of the genetic material occurs independently of mitosis—the process that distributes chromosome copies to daughter cells. To describe the regular alternation between M phase and interphase, Howard and Pelc began referring to the cell cycle. The **cell cycle** is the orderly sequence of events that occurs starting from the formation of a eukaryotic cell through the duplication of its chromosomes to the time it undergoes division itself. During the cycle, two key events are (1) the replication, or copying, of the hereditary material in chromosomes and (2) the partitioning of copied chromosomes to the two daughter cells during M phase. The hereditary material is duplicated, with one copy going to each daughter cell during mitosis. As a result, daughter cells contain genetic information identical to that of the parent cell.

The Discovery of the Gap Phases

Howard and Pelc, along with researchers in other labs, followed up on these early results by asking how long the S phase lasted. The researchers used the following experimental strategy to answer this question: They exposed growing cells to radioactive phosphorus or thymidine and then waited various lengths of time before looking at the cells. In one experiment, researchers worked with cells that were growing in culture and stopped the labeling of DNA after 30 minutes of exposure to radioactive thymidine. Cultured cells are powerful experimental tools because they can be manipulated much more easily than cells in an intact organism can (see **Box 11.1**). To stop DNA from being labeled with radioactive thymidine, they flooded the solution surrounding the cultured cells with nonradioactive

BOX 11.1 Cell-Culture Methods

For researchers, there are important advantages to growing plant and animal cells outside the organism itself. Cell cultures provide large populations of a single type of cell and the opportunity to control experimental conditions precisely.

The first successful attempt to culture animal cells occurred in 1907, when a researcher cultivated amphibian nerve cells in a drop of fluid from the spinal cord. But it was not until the 1950s and 1960s that biologists could routinely culture plant and animal cells in the laboratory. The long lag time was due to the difficulty of recreating conditions that exist in the intact organism.

To grow in culture, animal cells must be provided with a liquid mixture containing the nutrients, vitamins, and hormones that stimulate growth. Initially, this mixture was provided through the use of serum, which is the liquid portion of blood; now serum-free media are available for certain cell types. Serum-free media are preferred because they are much more precisely defined chemically than serum. In addition, many types of animal cells will not grow in culture unless they are provided with a solid surface that mimics the types of surfaces to which cells in the intact organisms adhere. As a result, cells are typically cultured in flasks like the one shown in Figure 11.4.

Even under optimal conditions, though, normal cells display a finite life span in culture. In contrast, many cultured cancerous cells grow indefinitely. In culture, cancerous cells also do not adhere tightly to the surface of the culture flask. These characteristics correlate with two features of cancerous cells in organisms: they grow in a continuous and uncontrolled fashion and can break away from the site where they originated and infiltrate new tissues and organs.

Because of their immortality and relative ease of growth, cultured cancer cells are commonly used in research on basic aspects of cell structure and function. For example, the first human cell type to be grown in culture was isolated in 1951 from a malignant tumor of the uterine cervix. These cells are called HeLa cells in honor of their source, Henrietta Lacks, who died soon thereafter from the cervical cancer. HeLa cells continue to grow in laboratories around the world. They have been used in numerous studies of human cell function, including the experiments that documented chromosome replication occurring independently of mitosis.

thymidine. Stated another way, they used the pulse-chase approach introduced in Chapter 7 to mark a small population of cells. Then they analyzed the labeled cells 2 hours after the end of labeling, then 4 hours after labeling, 6 hours after labeling, and so on. When they graphed the percentage of marked cells that were undergoing mitosis versus the time after exposure, they got the result indicated in **Figure 11.4**.

The cells that these researchers were studying take about 24 hours to complete one cell cycle. The cells do not divide synchronously, though—at any time, the cells in a sample are in all different parts of interphase and M phase. The key to interpreting Figure 11.4, then, is to realize that when the pulse ended, some cells should have been just completing S phase. If they went into M phase immediately, some labeled M-phase cells should have been present just after the pulse ended. But instead, no labeled dividing cells appeared for about 4 hours. The roughly 4-hour time lag between the end of the pulse and the appearance of the first labeled mitotic nuclei corresponds to a time lag that occurs between the end of S phase and the beginning of M phase. Stated another way, there is a gap in the cycle. The gap represents the period when chromosome replication is complete but mitosis has not yet begun.

This "lag" in the cell cycle came to be called **G_2 phase**, for second gap. It was considered the second gap because data indicate that another gap exists. To follow the logic, note in Figure 11.4 that labeled nuclei undergoing mitosis are observed over a period of about 6 to 8 hours. Because all of these cells had to be somewhere in S phase—ranging from just beginning DNA synthesis to almost ending—when radioactive thymidine was available, it is logical to conclude that S phase lasts 6 to 8 hours. When the times to get through the S, G_2, and M phases are added up and compared with the 24 hours it takes these cells to complete one cell cycle, though, there is a discrepancy of 7 to 9 hours. This discrepancy represents the gap called **G_1 phase**, for first gap. As **Figure 11.5** shows, G_1 phase occurs after M phase but before S phase. In these cells, the G_1 phase is about twice as long as G_2

Why do the gap phases exist? Besides copying their chromosomes during S phase, dividing cells also must replicate organelles and manufacture additional cytoplasm. Before mitosis can take place, the parent cell must grow large enough and synthesize enough organelles that its daughter cells will be normal in size and function. The two gap phases provide the time required to accomplish these tasks. They allow the cell to complete all the requirements for cell division other than chromosome replication.

Given this overview of the major events in the life of a cell, let's turn now to M phase and delve into the process of mitosis. Once the genetic material has been copied, how do cells divide it up between daughter cells? Recall that a gene is a segment of DNA that contains the information for synthesizing a particular protein or RNA molecule. How do cells ensure that each daughter cell receives an identical complement of chromosomes

Experiment

Question: How long does S phase last?

Hypothesis: No explicit hypothesis—this experiment was exploratory in nature.

Experimental setup:

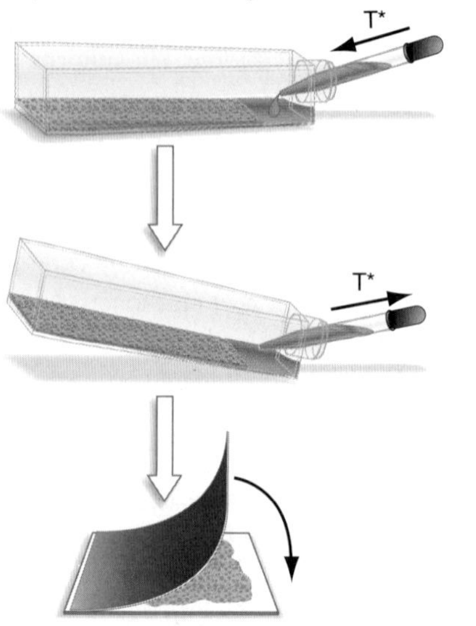

1. Feed radioactive thymidine (T*) to cells growing in culture. Only the cells synthesizing DNA (in S phase) will incorporate the T* label.

2. After 30 minutes, wash unincorporated T* out of cell culture.

3. Spread out cells and expose to photographic emulsion 2, 4, 6,...16 hours after end of labeling.

Prediction: No explicit predictions.

Results:

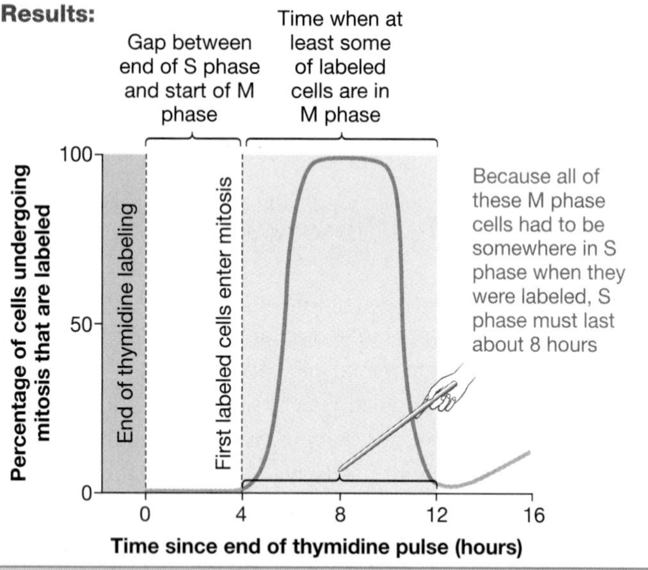

Conclusion: S phase is about 8 hours long. More important, there is a gap between S phase and M phase.

FIGURE 11.4 A Gap Occurs between Chromosome Replication and Mitosis. Because thymidine is incorporated into newly synthesized DNA, a short pulse of radioactive thymidine labels a population of cells that are in S phase. The graph shows the percentage of cells undergoing mitosis that are labeled versus the time since the pulse ended.

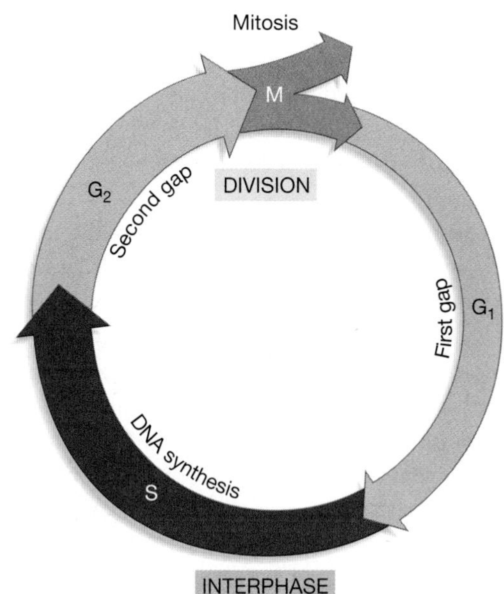

FIGURE 11.5 The Cell Cycle Has Four Phases. A representative cell cycle. The time required for the G_1 and G_2 phases varies dramatically among cells and organisms.

and thus an identical complement of genes? (The discussion that follows describes mitosis in plant cells and in animal cells; **Box 11.2** on page 230 describes cell division in bacteria.)

11.2 How Does Mitosis Take Place?

Early observations of cell division focused on the fate of the parent cell's chromosomes. Because chromosomes that are undergoing mitosis are visible under the light microscope when they are stained, investigators could watch mitosis occur. As a result, the major events in mitosis were well understood long before the cell cycle was fully described.

Mitosis results in the division of replicated chromosomes and the formation of two daughter nuclei with identical chromosomes and genes. Mitosis is usually accompanied by cytokinesis—cytoplasmic division and the formation of two daughter cells. Let's take a closer look at the events in mitosis, beginning with observations about the nature of a eukaryotic cell's chromosomes.

Events in Mitosis

The first drawing in **Figure 11.6** shows the chromosomes found in a hypothetical plant cell or animal cell. The number of chromosomes in each cell varies widely among species. Both humans and potato plants have a total of 46 chromosomes in each cell; a maize (corn) plant has 20, dogs have 66, and fruit flies have 8. In Figure 11.6 there are a total of four chromosomes in the cell shown. (Although this cell is in interphase, the chromosomes are shown partially condensed simply to make them visible.) Eukaryotic chromosomes normally exist as extremely long, threadlike strands consisting of DNA associated with globular proteins called **histones**. In eukaryotes the DNA-protein material is called **chromatin**.

The second drawing in Figure 11.6 shows chromosomes that have been copied prior to mitosis. Each of the DNA copies in a replicated chromosome is called a **chromatid**. The two chromatids are joined together along their entire length as well as at a specialized region of the chromosome called the **centromere**. Chromatids from the same chromosome are referred to as **sister chromatids**. Sister chromatids represent

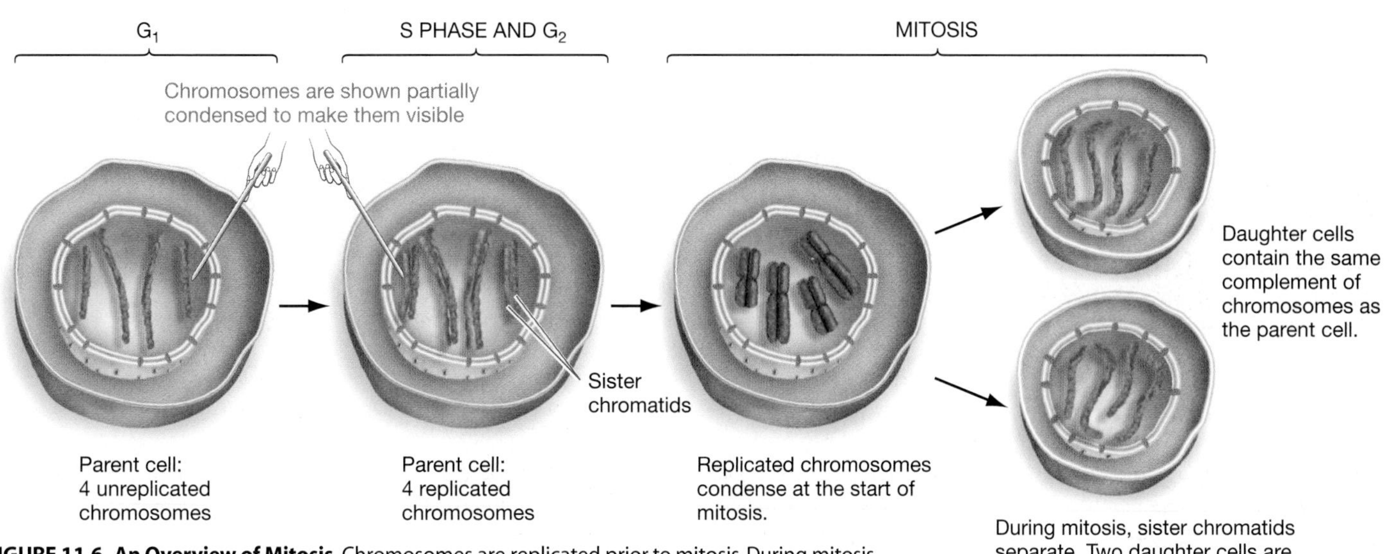

FIGURE 11.6 An Overview of Mitosis. Chromosomes are replicated prior to mitosis. During mitosis, the replicated chromosomes are partitioned to the two daughter nuclei. In most cases, mitosis is followed by cytokinesis.

QUESTION In the daughter cells of mitosis, are chromosomes replicated or unreplicated?

exact copies of the same genetic information. Each chromatid contains one long DNA double helix. At the start of M phase, then, each chromosome consists of two sister chromatids that are attached to one another at the centromere. You should be able to explain the relationship between chromosomes and (1) genes, (2) chromatin, and (3) sister chromatids.

The third drawing in Figure 11.6 shows that as mitosis begins, chromatin condenses to form a much more compact structure. Replicated, condensed chromosomes correspond to the paired threads observed in salamander cells by early biologists (see Figure 11.2a).

The final drawing in Figure 11.6 shows that during mitosis, the two sister chromatids separate to form independent chromosomes, and one copy of each chromosome goes to each of the two daughter cells. As a result, each daughter cell receives a copy of the genetic information that is contained in each chromosome. Every daughter cell ends up with exactly the same complement of chromosomes as the parent cell had prior to replication, and thus every daughter cell receives the same genetic information.

Although mitosis is a continuous process, biologists routinely identify several subphases within M phase on the basis of distinctive events that occur. These subphases of mitosis are designated prophase, prometaphase, metaphase, anaphase, and telophase. Some students use the mnemonic device IPPMAT to remind themselves that interphase is followed by the mitotic subphases prophase, prometaphase, metaphase, anaphase, and telophase.

To understand how mitosis proceeds, let's look at each subphase in turn.

Prophase Mitosis begins with the events of **prophase** ("before-phase"), as shown in **Figure 11.7**. The chromosomes have already replicated during interphase (Figure 11.7, step 1); during prophase (step 2), they condense into compact structures. Chromosomes first become visible in the light microscope during prophase.

In the cytoplasm, prophase is marked by the formation of the mitotic spindle. The **mitotic spindle** is a structure that produces mechanical forces that pull chromosomes into the daughter cells during mitosis. The mitotic spindle consists of an array of microtubules—components of the cytoskeleton that were introduced in Chapter 7. Groups of microtubules attach to the chromosomes and are called **spindle fibers**. In all eukaryotes, spindle fibers originate from a microtubule organizing center. Although the nature of this organizing region varies among plants, animals, fungi, and other eukaryotic groups, the function of the spindle itself is the same. Figure 11.7 illustrates an animal cell undergoing mitosis, so the microtubule organizing center is a **centrosome**—a structure that contains a pair of **centrioles** (see Chapter 7). During prophase in all eukaryotes, the mitotic spindles either begin moving to opposite sides of the cell or form on opposite sides.

Prometaphase Once chromosomes have condensed, the nucleolus disappears and the nuclear envelope fragments. After the nuclear envelope has disintegrated, spindle fibers from each mitotic spindle attach to one of the two sister chromatids of each chromosome. These events occur during **prometaphase** ("before middle-phase"); see Figure 11.7, step 3.

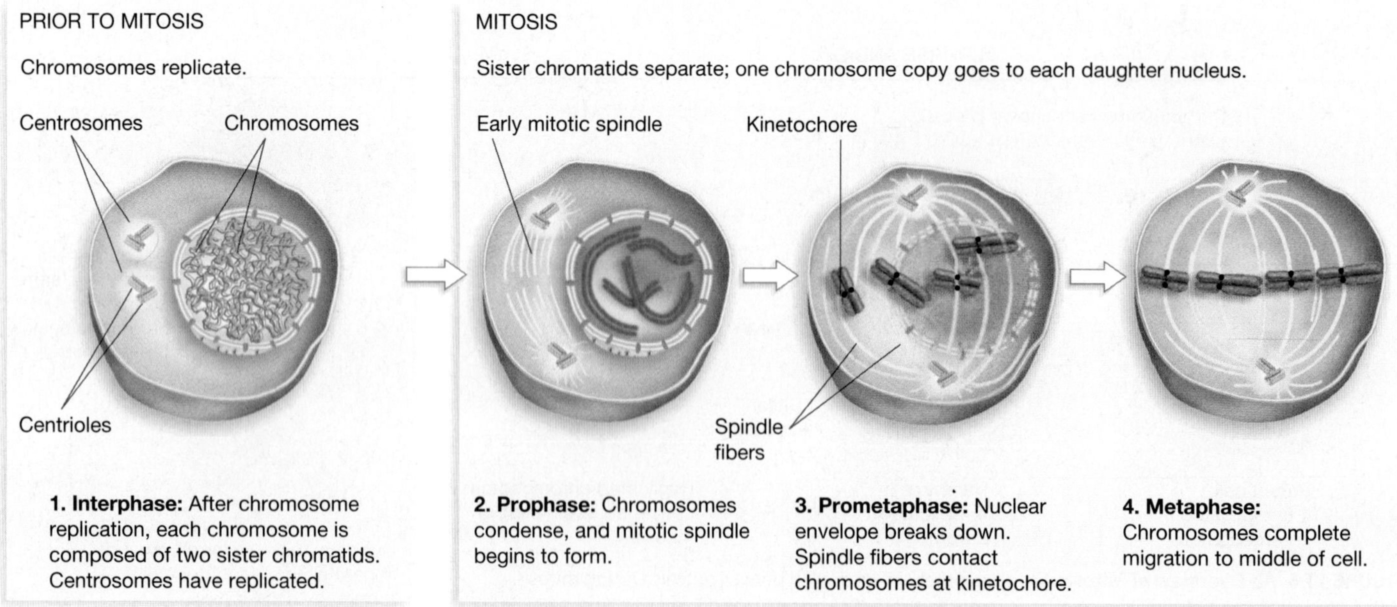

FIGURE 11.7 Mitosis and Cytokinesis.

The attachment between the spindle fibers and each chromatid is made at a structure called the **kinetochore**. Kinetochores are located at the centromere region of the chromosome, where sister chromatids are attached to each other. Each chromosome has two kinetochores where spindle fibers attach—one on each side.

During prometaphase in animals, the centrosomes continue their movement to opposite poles of the cell. In all groups, the microtubules that are attached to the kinetochores begin moving the chromosomes to the middle of the cell.

Metaphase During **metaphase** ("middle-phase"), animal centrosomes complete their migration to the opposite poles of the cell (Figure 11.7, step 4). In all groups, the kinetochore microtubules finish moving the chromosomes to the middle of the cell. When metaphase is over, the chromosomes are lined up along an imaginary plane called the **metaphase plate**. At this point, the formation of the mitotic spindle is complete. Each chromatid is attached to spindle fibers that run from its kinetochore to one of the two poles of the cell. Each chromosome is held by kinetochore spindle fibers reaching to opposite poles and exerting the same amount of tension or pull. A tug of war is occurring, with kinetochore spindle fibers pulling each chromosome in opposite directions.

Anaphase At the start of **anaphase** ("against-phase"), the centromeres that are holding sister chromatids together split (Figure 11.7, step 5). Because they are under tension, sister chromatids are pulled apart equally—with the same amount of force—to create independent chromosomes. The kinetochore spindle fibers then begin to shorten. As they do, motor proteins pull the chromosomes to opposite poles of the cell. The two poles of the cell are also pushed away from each other by motor proteins associated with microtubules that are not attached to chromosomes.

During anaphase, replicated chromosomes split into two identical sets of unreplicated chromosomes. The separation of sister chromatids to opposite poles is a critical step in mitosis, because it ensures that each daughter cell receives the same complement of chromosomes. When anaphase is complete, each pole of the cell has an equivalent and complete collection of chromosomes that are identical to those present in the parent cell prior to chromosome replication.

Telophase During **telophase** ("end-phase"), a nuclear envelope begins to form around each set of chromosomes (Figure 11.7, step 6). The mitotic spindle disintegrates, and the chromosomes begin to de-condense. Once two independent nuclei have formed, mitosis is complete.

Cytokinesis

Prior to the onset of M phase, mitochondria, lysosomes, chloroplasts, and other organelles have replicated, and the rest of the cell contents have grown. During cytokinesis (Figure 11.7, step 7), the cytoplasm divides to form two daughter cells, each with its own nucleus and complete set of organelles. Cytokinesis normally occurs immediately following mitosis.

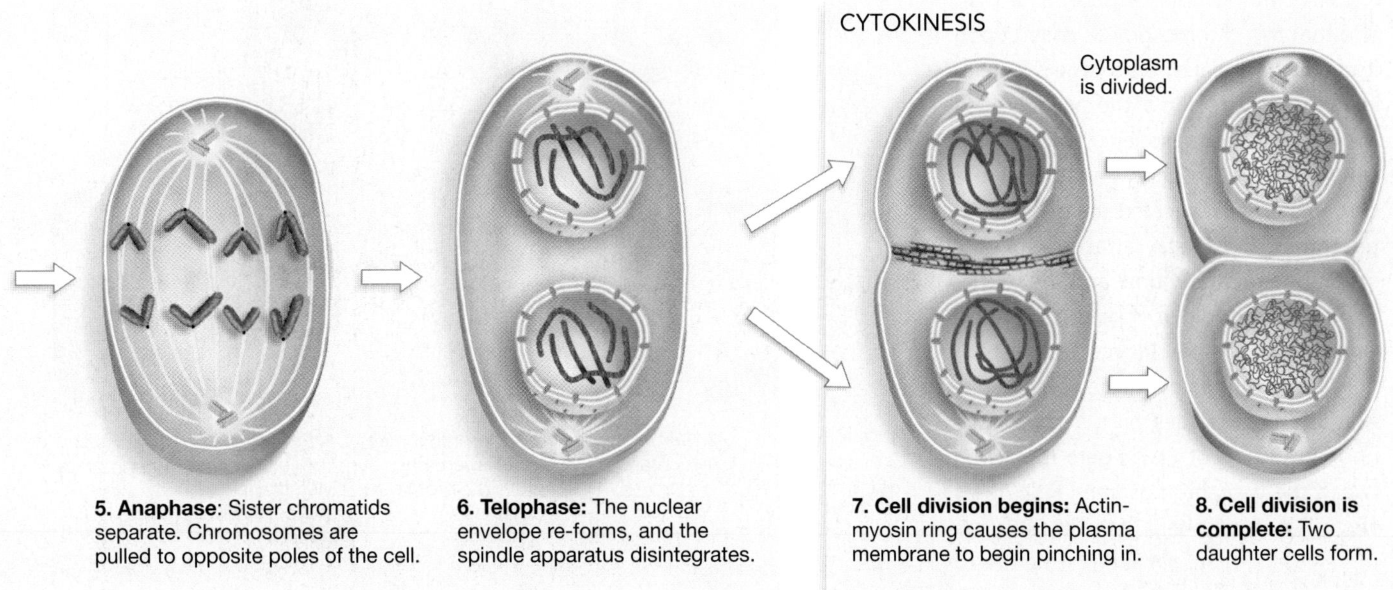

5. Anaphase: Sister chromatids separate. Chromosomes are pulled to opposite poles of the cell.

6. Telophase: The nuclear envelope re-forms, and the spindle apparatus disintegrates.

7. Cell division begins: Actin-myosin ring causes the plasma membrane to begin pinching in.

8. Cell division is complete: Two daughter cells form.

In animals, fungi, and slime molds, cytokinesis begins with the formation of a **cleavage furrow** (**Figure 11.9a**). The furrow appears because a ring of actin filaments forms just inside the plasma membrane, in a plane that bisects the cell. A motor protein called myosin binds to these actin filaments. When myosin binds to ATP or ADP, part of the protein moves in a way that causes actin filaments to slide (see Chapter 46). As myosin moves the ring of actin filaments on the inside of the plasma membrane, the ring shrinks in size and tightens. Because the ring is attached to the plasma membrane, the shrinking ring pulls the membrane with it. As a result, the plasma membrane pinches inward. The actin and myosin filaments continue to slide past each other, tightening the ring further, until the original membrane is pinched in two and cell division is complete.

In plants, the mechanism of cytokinesis is different. A series of microtubules and other proteins define and organize the region where the new plasma membranes and cell walls will form. Vesicles from the Golgi apparatus are then transported to the middle of the dividing cell, where they form a structure called the **cell plate** (**Figure 11.9b**). The vesicles carry components of the cell wall and plasma membrane that gradually build up, completing the cell plate and dividing the two daughter cells.

To help you review the major events in cell division, **Table 11.1** summarizes the key structures involved, and **Figure 11.10** shows photographs of cells in interphase and undergoing mitosis and cytokinesis. ● After you've studied this table and figure and reviewed Figure 11.7, you should be able to make a table with rows titled (1) microtubule organizing center and spindle fibers, (2) nuclear envelope, and (3) chromosomes, and columns titled with the five phases of mitosis. Fill in the table by summarizing what happens to the three structures during each phase of mitosis.

Once mitosis and cytokinesis had been described in detail, biologists turned their attention to understanding the molecular mechanisms that are responsible for them. In particular, researchers wanted to know two things: (1) How do sister chromatids separate to become independent chromosomes, and (2) how do those chromosomes move to daughter cells? The exact and equal partitioning of genetic material to the two daughter cells is the most fundamental aspect of mitosis. How does this process occur?

MB Web Animation *BioFlix* at www.masteringbio.com
The Phases of Mitosis; Mitosis

BOX 11.2 How Do Bacteria Divide?

To reproduce, bacterial cells divide into two genetically identical daughters. This process is called **binary fission**. **Figure 11.8** sketches the major events in involved. Recall that most bacteria contain a single, circular chromosome composed of DNA and that this chromosome is coiled upon itself. The bacterial chromosome is located in a distinctive region of the cytosol called the nucleoid, but a nucleus is lacking. Recent research has shown that after the chromosome is replicated, the two daughter chromosomes stay attached to each other for a period of time and then separate to opposite sides, or poles, of the cell. Understanding how this movement happens is currently an active area of research. As the chromosomes part, a contractile ring composed of FtsZ fibers forms between the two chromosomes. FtsZ fibers are a major component of the bacterial cytoskeleton and are similar in structure to microtubules. As the FtsZ ring closes, the bacterial cytoplasm is divided in two, completing cell division.

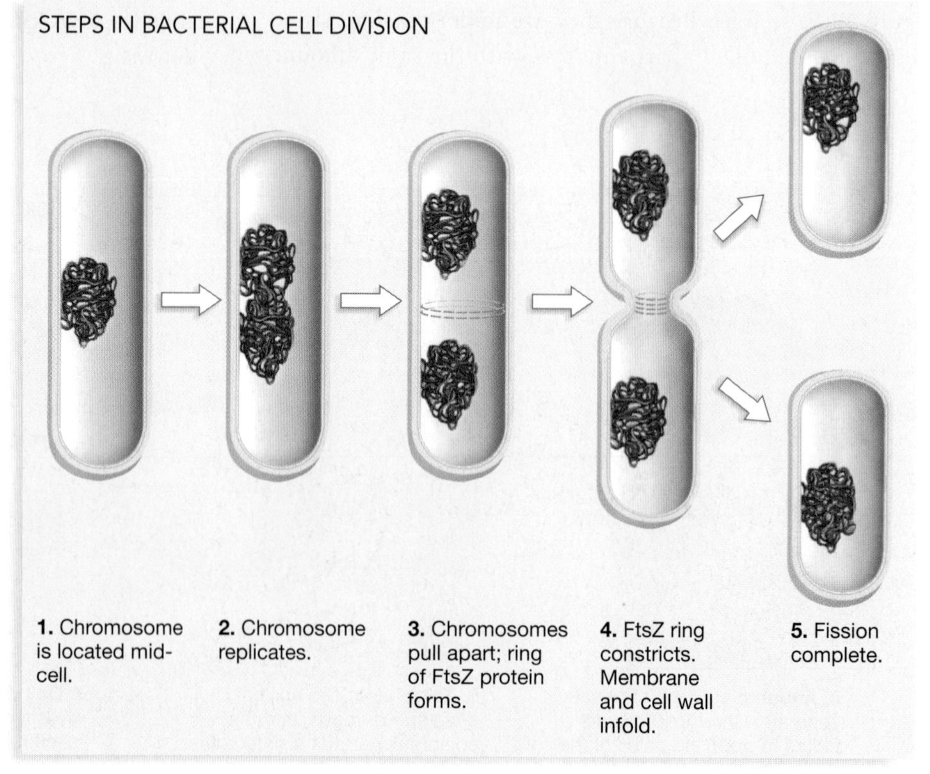

FIGURE 11.8 Steps in Bacterial Cell Division.

● **QUESTION** Are the daughter cells of bacterial cell division identical to each other in chromosomal makeup, or are they different?

(a) Cytokinesis in animals

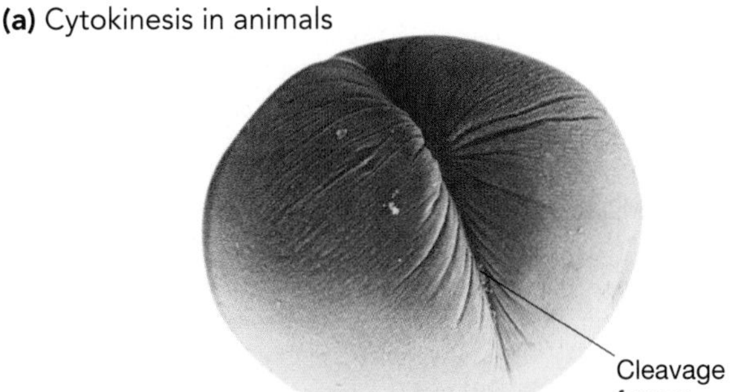

(b) Cytokinesis in plants

FIGURE 11.9 The Mechanism of Cytokinesis Varies among Eukaryote Groups. (a) In animals, the cytoplasm is divided by a cleavage furrow that pinches the parent cell in two. **(b)** In plants, the cytoplasm is divided by a cell plate that forms in the middle of the parent cell.

SUMMARY TABLE **11.1 Structures Involved in Mitosis**

Structure	Definition
Chromosome	A structure composed of a DNA molecule and associated proteins.
Chromatin	The material that makes up eukaryotic chromosomes. Consists of a DNA molecule complexed with histone proteins.
Chromatid	One strand of a replicated chromosome, with its associated proteins.
Sister chromatids	The two strands of a replicated chromosome. When chromosomes are replicated, they consist of two sister chromatids. The genetic material in sister chromatids is identical. When sister chromatids separate during mitosis, they become independent chromosomes.
Centromere	The structure that joins sister chromatids.
Kinetochore	The structure on sister chromatids where spindle fibers attach.
Microtubule organizing center	Any structure that organizes microtubules.
Centrosome	The microtubule organizing center in animals.
Centriole	Cylindrical structures that comprise micro-tubules, located inside animal centrosomes.

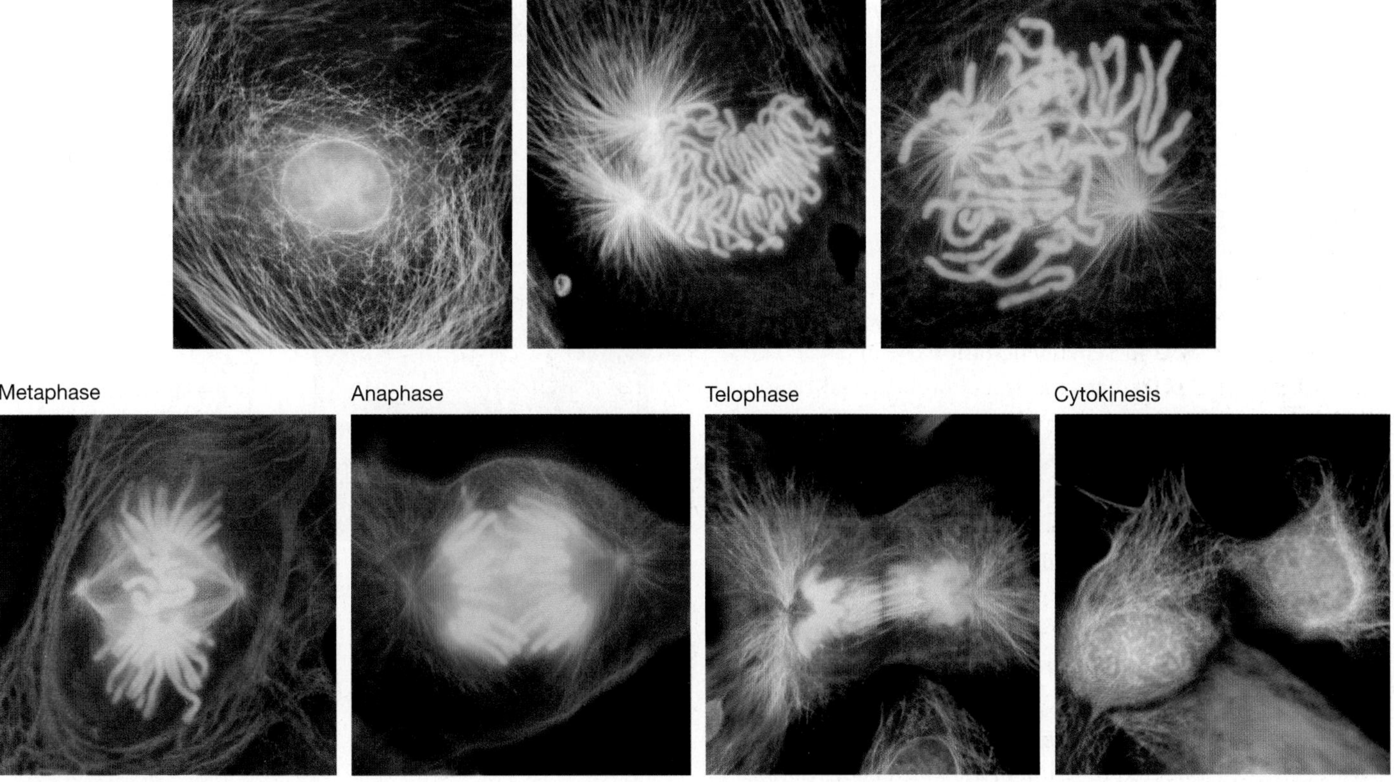

FIGURE 11.10 Mitosis and Cytokinesis in Action. Micrographs showing newt cells in interphase and undergoing mitosis and cytokinesis. Chromosomes are stained blue, microtubules are green, and actin filaments are red.

● **EXERCISE** Next to each image, describe what is happening in less than five words. In at least two of the photographs, label the chromosomes, centrosomes, and mitotic spindle. In the metaphase cell, label the metaphase plate.

How Do Chromosomes Move during Mitosis?

To understand how sister chromatids separate and move to daughter cells, biologists have focused on understanding how the mitotic spindle functions. Do spindle microtubules act as railroad tracks, the way they do in vesicle transport? Is some sort of motor protein involved? And what is the nature of the kinetochore, where the chromosome and microtubules are joined?

Mitotic Spindle Forces Spindle fibers are composed of microtubules. Recall from Chapter 7 that microtubules are composed of α-tubulin and β-tubulin dimers, that the length of a microtubule is determined by the number of tubulin dimers it contains, that microtubules are asymmetric—meaning they have a plus end and a minus end—and that microtubules grow at their plus end. During mitosis, spindle fibers grow from the microtubule organizing center until their plus end attaches to the kinetochore.

These observations suggest two possible mechanisms for the movement of chromosomes during anaphase. Is the spindle microtubule shortening due to a loss of tubulin dimers from one end? Alternatively, it is possible that microtubules slide past each other, much as actin and myosin filaments do during cytokinesis in animal cells, and that this sliding action pulls chromosomes toward the poles of the cell.

To test these contrasting hypotheses, biologists introduced fluorescently labeled tubulin subunits into prophase or metaphase cells. This treatment made the entire mitotic spindle visible. (See step 1 in the "Experimental setup" section of **Figure 11.11**.) Then, once anaphase had begun, the researchers marked a region of the spindle with a bar-shaped beam of laser light. The laser quenched the fluorescence in the exposed region, making that section of the spindle dark, or "photobleached," although it was still functional (see step 2 of the "Experimental setup"). The "Results" section of Figure 11.11 shows that as anaphase progressed, two things happened: (1) the photobleached region remained stationary, and (2) The spindle fibers got shorter between the photobleached region and the kinetochore.

To explain this result, the biologists concluded that the kinetochore microtubules must remain stationary during anaphase. The hypothesis suggested that, instead of sliding past each other as actin fibers do, microtubules shorten at the kinetochore because tubulin subunits are lost from the plus ends. As kinetochore microtubules shorten at the kinetochore, the chromosomes are pulled along. How does this happen?

A Kinetochore Motor **Figure 11.12** shows the current model for the structure and function of the kinetochore during chromosome movement. Research is continuing, however, and it is likely that the model presented in the figure will be modified as additional data become available.

Experiment

Question: How do microtubules shorten to pull sister chromatids apart at anaphase?

Hypothesis: Microtubules shorten at one end.

Alternate hypothesis: Microtubules slide past each other like actin filaments.

Experimental setup:

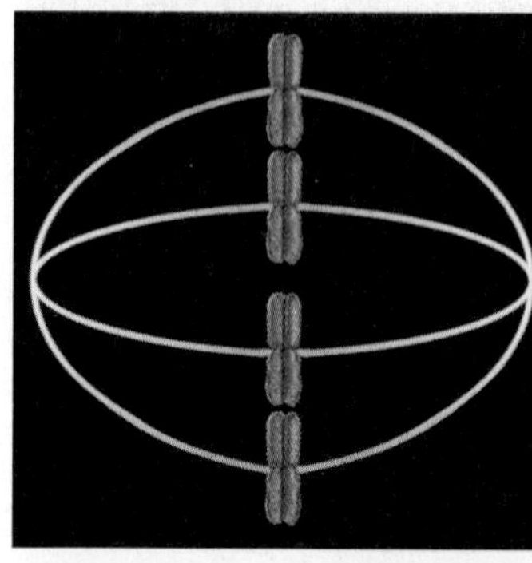

1. Use fluorescent labels to make the metaphase chromosomes fluoresce blue and the microtubules fluoresce yellow.

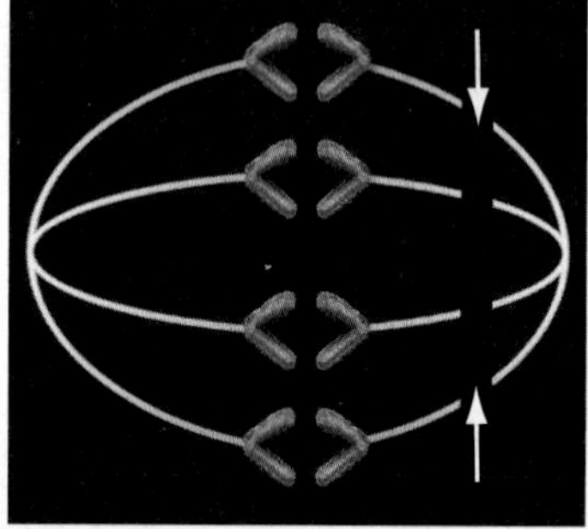

2. At the start of anaphase, photobleach a section of microtubules to mark them without changing their function.

Prediction: The photobleached section will still be visible as chromosomes begin to move.

Prediction of alternative hypothesis: The photobleached section will disappear when chromosomes begin to move.

Results:

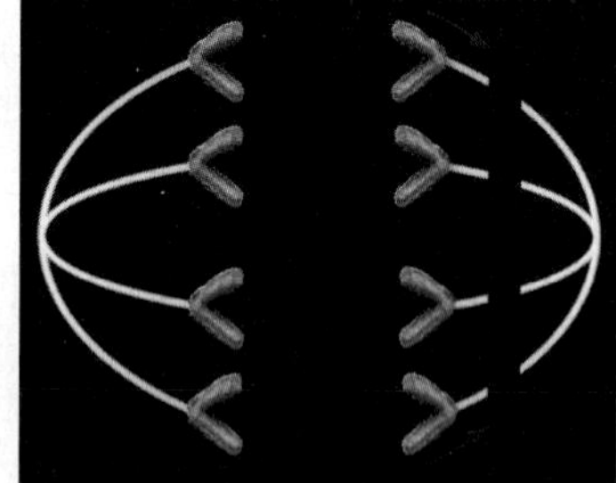

The photobleached section remained visible, but the distance between chromosomes and photobleached section lessened.

Conclusion: Microtubules shorten at one end—at the kinetochore.

FIGURE 11.11 During Anaphase, Microtubules Shorten at the Kinetochore.

EXERCISE Add drawings to show the outcomes you would expect if microtubules shorten at the end opposite the chromosome or if microtubules slide past each other.

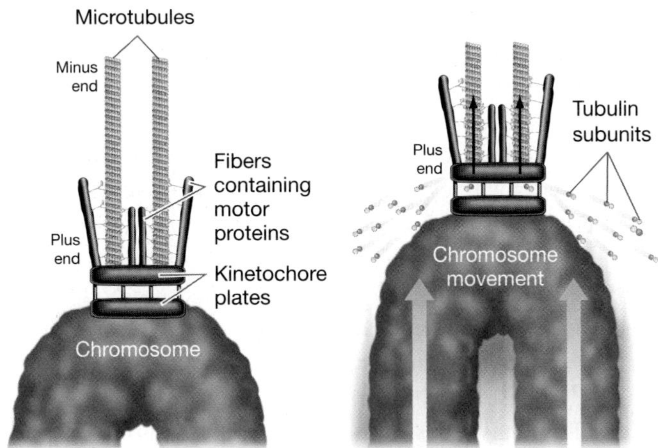

FIGURE 11.12 How Do Microtubules Move Chromosomes during Mitosis? The kinetochore consists of an inner plate, an outer plate, and associated fibers that contain motor proteins. The fibers are thought to act as "legs" that walk the chromosomes down the length of the kinetochore microtubules as the microtubules shorten, due to loss of tubulin dimers at the kinetochore.

Note that the kinetochore is thought to have a base that attaches to the centromere region of the chromosome and a "crown" of fibrous proteins projecting outward. Recent research has shown that the kinetochore contains dyneins and other motor proteins. These motor proteins are thought to be attached to the kinetochore's fibrous crown and to be capable of "walking" down microtubules—from their plus ends near the kinetochore toward their minus ends at the spindle. This observation suggests that the mechanism of chromosome movement is reminiscent of the way that kinesin walks down microtubules during vesicle transport or that dynein walks down the microtubule doublets in flagella. As Chapter 7 explained, motor proteins such as kinesin and dynein not only attach or detach from microtubules when they bind to ATP versus ADP, they also change shape. The combination of shape changes and attachment-detachment cycles causes motor proteins to walk down the length of a microtubule or other fiber.

To put all of these ideas together, biologists hypothesize that as anaphase gets under way, proteins in the kinetochore catalyze the loss of tubulin subunits at the plus end of the spindle fiber, while dyneins and other kinetochore motor proteins walk toward the minus end. As the microtubule shortens and the detach-move-reattach cycle of the motor proteins repeats, the chromosome is pulled to one end of the mitotic spindle.

Efforts to understand spindle structure and movement bring us to the frontier of research on mitosis. Having explored how the process occurs, let's focus on how it is controlled. When does a cell divide, and when does it stop dividing? How is cell division regulated? These questions are fundamental. When cell division occurs in an uncontrolled manner, cancerous tumors can form.

Check Your Understanding

If you understand that...

- After chromosomes replicate, mitosis distributes one copy of each chromosome to each daughter cell.
- Mitosis and cytokinesis lead to the production of cells with the same genetic material as that of the parent cell.

You should be able to...

1) Draw an unreplicated chromosome and a replicated chromosome, and label the sister chromatids and the centromere on the replicated chromosome.
2) Sketch what happens to the spindle apparatus, nuclear envelope, and chromosomes during prophase, prometaphase, metaphase, anaphase, and telophase.

11.3 Control of the Cell Cycle

Although the events of mitosis are virtually identical in all eukaryotes, other aspects of the cell cycle are variable. For example, the length of the cell cycle can vary enormously among different cell types, even in the same individual. In humans, intestinal cells routinely divide more than twice a day to renew tissue that is lost during digestion; mature human nerve and muscle cells do not divide at all. Most of these differences are due to variation in the length of the G_1 phase. In rapidly dividing cells, G_1 is essentially eliminated. Most nondividing cells, in contrast, are permanently stuck in G_1. Researchers refer to this arrested stage as the G_0 state, or simply "G zero." Cells that are in G_0 have effectively exited the cell cycle and are sometimes referred to as post-mitotic. Nerve cells, muscle cells, and many other cell types enter G_0 once they have matured.

A cell's division rate can also vary in response to changes in conditions. For example, human liver cells normally divide about once per year. But if part of the liver is damaged or lost, the remaining cells divide every one or two days until repair is accomplished. Cells of unicellular organisms such as yeasts, bacteria, or archaea divide rapidly only if the environment is rich in nutrients; otherwise, they enter a quiescent (inactive) state.

To explain the existence of so much variability, biologists hypothesized that the cell cycle must be regulated in some way and that regulation varies among cells and organisms. Understanding how the cell cycle is controlled is now the most prominent issue in research on cell division—partly because defects in control can lead to uncontrolled, cancerous growth.

The Discovery of Cell-Cycle Regulatory Molecules

The first solid evidence for cell-cycle control molecules came to light in 1970, when researchers published the results of experiments on fusing pairs of mammalian cells that were growing in cell culture. In the presence of certain chemicals, viruses, or an electric shock, the membranes of two cells that are growing in culture can be made to fuse, creating a single cell with two nuclei.

How did cell fusion experiments point to the existence of cell-cycle control molecules? When investigators fused cells that were in different stages of the cell cycle, certain nuclei changed phases. For example, when a cell in M phase was fused with one in interphase, the nucleus of the interphase cell initiated M phase (**Figure 11.13a**). The biologists hypothesized that the cytoplasm of M-phase cells contains a regulatory molecule that induces interphase cells to enter M phase.

This hypothesis was supported by experiments on the South African claw-toed frog, *Xenopus laevis*. As an egg of these frogs matures, it changes from a cell called an **oocyte**, which is arrested in a phase similar to G_2, to a mature egg that has entered M phase. The eggs are attractive to study, partly because they are more than 1 mm in diameter. Their large size makes it relatively easy to purify large amounts of cytoplasm and to use syringes with extremely fine needles to inject the eggs with cytoplasm from eggs in different stages of development. When biologists purified cytoplasm from M-phase frog eggs and injected it into the cytoplasm of frog oocytes arrested in the G_2 phase, the immature oocytes entered M phase (**Figure 11.13b**). But when cytoplasm from interphase cells was injected into G_2 oocytes, the cells remained in the G_2 phase. The researchers concluded that the cytoplasm of M-phase cells—but not the cytoplasm of interphase cells—contains a factor that drives immature oocytes into M phase to complete their maturation.

Experiment

Question: Do regulatory molecules control entry into specific phases of the cell cycle?

Hypothesis: A regulatory molecule triggers mitosis (M phase).

Null hypothesis: No regulatory molecule triggers mitosis.

Experimental setup:

(a) Cell fusion experimental setup:

M-phase cell

Interphase cell

Fuse M-phase cells with G_1, S, or G_2 phase cells.

(b) Microinjection experimental setup:

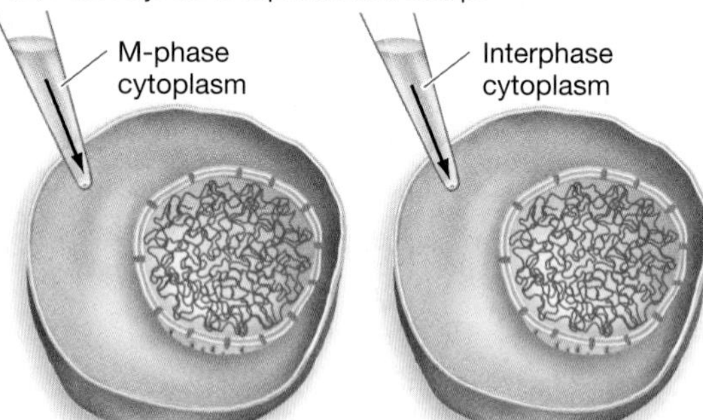

Inject cytoplasm from M-phase cell into one frog oocyte and cytoplasm from interphase cell into another frog oocyte.

Prediction: Interphase cells will begin M phase.

Prediction of null hypothesis: Interphase cells will not begin M phase.

Prediction: One or both of the frog oocytes will begin M phase.

Prediction of null hypothesis: Neither of the frog oocytes will begin M phase.

Results:

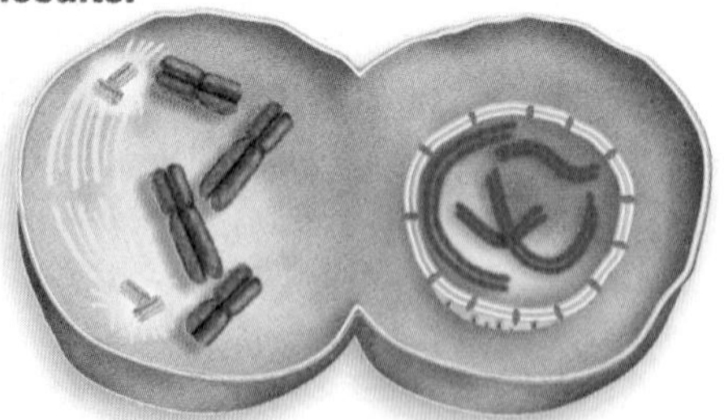

Chromosomes in interphase cell condense, signaling start of M phase.

Results:

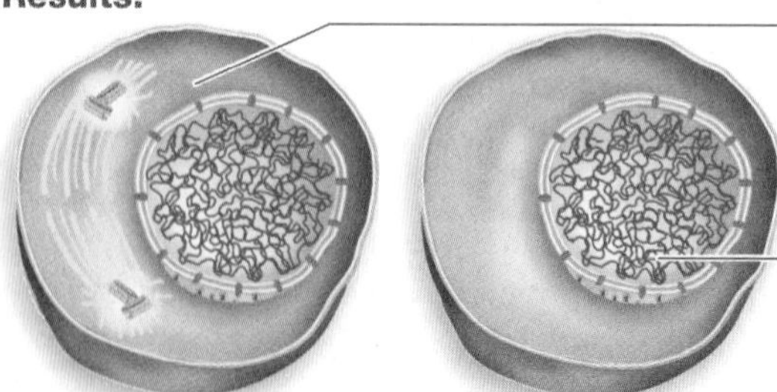

Conclusion: M-phase cytoplasm contains a regulatory molecule that induces M phase in interphase cells.

FIGURE 11.13 Experimental Evidence for Cell-Cycle Control Molecules. (a) When M-phase cells are fused with cells in G_1 S, or G_2 phase, the interphase chromosomes condense and begin M phase. **(b)** Microinjection experiments supported the hypothesis that a regulatory molecule induces M phase.

This factor was eventually purified and is now called **mitosis-promoting factor**, or **MPF**. Subsequent experiments showed that MPF induces mitosis in all eukaryotes. For example, injecting M-phase cytoplasm from mammalian cells into immature frog eggs results in egg maturation. Human MPF can trigger mitosis in yeast cells. The molecule appears to be a general signal that says "Start mitosis." How does it work?

MPF Contains a Protein Kinase and a Cyclin Once MPF had been isolated and purified, researchers found that it is made up of two distinct polypeptide subunits. One of the subunits is a **protein kinase**—an enzyme that catalyzes the transfer of a phosphate group from ATP to a target protein. Recall from Chapter 9 that proteins can be activated or inactivated by phosphorylation. Because the addition of a phosphate group changes the target protein's shape and activity, protein kinases frequently act as regulatory elements in the cell.

These observations suggested that MPF acts by phosphorylating a protein that triggers the onset of mitosis. But research showed that the concentration of MPF protein kinase is more or less constant throughout the cell cycle. How can MPF trigger mitosis if the protein kinase subunit is always present?

The answer to this question lies in the second MPF subunit, which belongs to a family of proteins called the **cyclins**. Cyclins got their name because their concentrations fluctuate throughout the cell cycle. As **Figure 11.14a** shows, the cyclin associated with MPF builds up in concentration during interphase and peaks during M phase. This increase in concentration is important because the protein kinase subunit in MPF can be active only when it is bound to the cyclin subunit. As a result, the protein kinase subunit of MPF is called a **cyclin-dependent kinase**, or **Cdk**. MPF is a dimer consisting of a cyclin subunit and a cyclin-dependent kinase subunit. The cyclin subunit functions as a regulatory protein; the kinase subunit is the part that catalyzes the phosphorylation of other proteins to start mitosis.

According to Figure 11.14a, the number of complete MPF dimers builds up steadily during interphase. Why doesn't this increasing concentration of MPF trigger the onset of M phase? The answer is that MPF's Cdk subunit becomes phosphorylated at two sites after it binds to cyclin. When Cdk is phosphorylated, its conformation changes in a way that renders the protein inactive. Late in G_2 phase, however, enzymes cause one of the phosphate groups on the Cdk subunit to drop off. This dephosphorylation reaction changes MPF's shape in a way that activates it.

Once MPF is activated, it binds to the proteins listed in **Figure 11.14b** and catalyzes their phosphorylation. Chromosomal proteins activated by MPF cause chromosomes to condense into the threads visible during M phase. In addition, microtubule-associated proteins phosphorylated by MPF

(a) Cyclin concentration regulates MPF concentration.

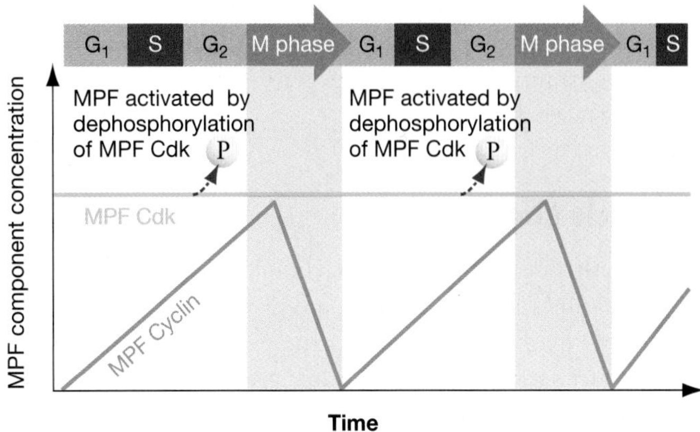

(b) Activated MPF has an array of effects.

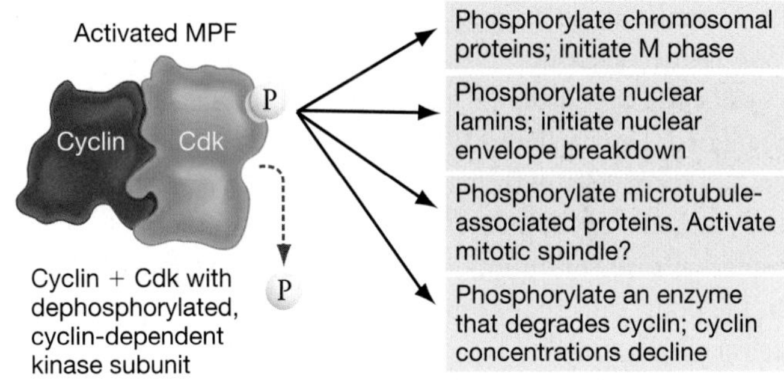

FIGURE 11.14 M-Phase Promoting Factor Is Created When a Cyclin Binds to a Protein Kinase. (a) Cyclin concentrations cycle in dividing cells, reaching a peak in M phase. **(b)** Cyclin binds to a cyclin-dependent kinase, creating MPF. When MPF is activated, it activates genes that initiate M phase. It also activates proteins that degrade cyclin.

QUESTION Why is it important that MPF activates proteins that degrade cyclin?

may be involved in assembling the mitotic spindle apparatus. In this way, MPF triggers the onset of M phase.

MPF also activates an enzyme complex that promotes the degradation of MPF's own cyclin subunit, however. By activating this enzyme complex, MPF triggers its own destruction. This is an example of **negative feedback**—similar to the mechanisms of feedback inhibition introduced in Chapter 9. In response to MPF activity, then, the concentration of cyclin declines rapidly. Slowly, it builds up again during interphase. In this way, an oscillation in cyclin concentration is set up. If you understand this aspect of cell-cycle regulation, you should be able to describe what MPF does. You should also be able to explain the relationship between MPF and (1) cyclin, (2) Cdk, and (3) the enzymes that phosphorylate MPF, dephosphorylate MPF, and degrade MPF.

The dramatic oscillation in cyclin concentration and activation acts as a clock that drives the ordered events of the cell cycle. These events are happening millions of times each day in locations throughout your body. Over a 24-hour period, you swallow millions of cheek cells. Millions of additional cells are lost from your intestinal lining each day and leave your body as waste. To replace them, cells in your cheek and intestinal tissue are constantly making and degrading cyclin and pushing themselves through the cell cycle.

Cell-Cycle Checkpoints

MPF is only one of many protein complexes involved in regulating the cell cycle. For example, a different cyclin and protein kinase are involved in triggering the passage from G_1 phase into S phase, and several regulatory proteins are involved in maintaining the G_0 state of quiescent cells. A complex array of regulatory molecules is involved in either holding cells in particular stages or in stimulating passage to the next phase.

To make sense of these observations, Leland Hartwell and Ted Weinert introduced the concept of a cell-cycle checkpoint. A **cell-cycle checkpoint** is a critical point in the cell cycle that is regulated. Hartwell and Weinert identified checkpoints by analyzing yeast cells with defects in the cell cycle. The defective cells lacked a specific checkpoint and, as a result, kept dividing under culture conditions when normal cells stopped growing. In the body, cells that keep growing in this way form a mass of cells called a **tumor**.

As **Figure 11.15** indicates, there are three distinct checkpoints during the four phases of the cell cycle. In effect, interactions among regulatory molecules at each checkpoint allow a cell to "decide" whether to proceed with division. If these regulatory molecules are defective, the checkpoint may fail. As a result, cells may start growing in an uncontrolled fashion.

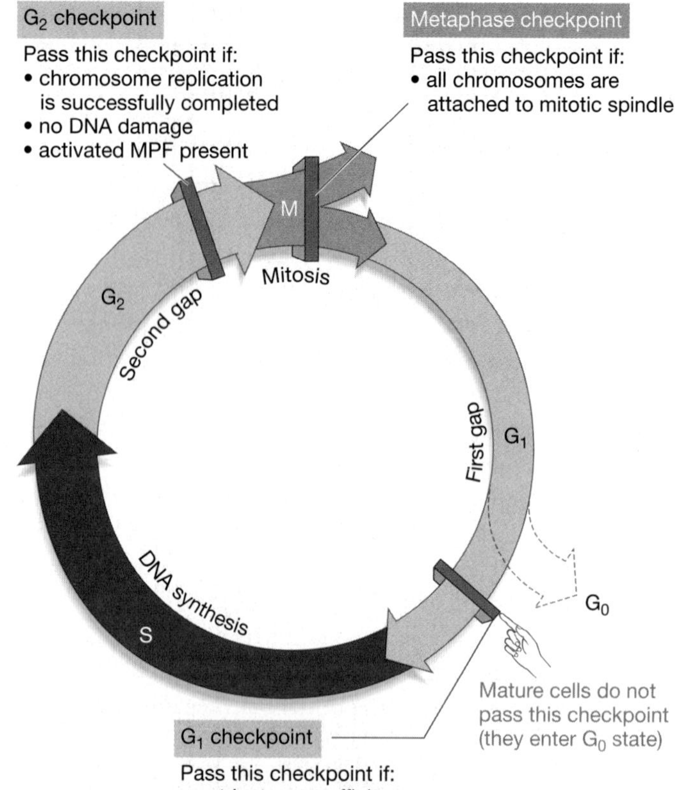

FIGURE 11.15 The Three Cell-Cycle Checkpoints

QUESTION Why is it advantageous for a cell to arrest at the M-phase checkpoint if not all chromosomes are attached to the mitotic spindle?

The first cell-cycle checkpoint occurs late in G_1. For most cells, this checkpoint is the most important in establishing whether the cell will continue through the cycle and divide or exit the cycle and enter G_0. What determines whether a cell passes the G_1 checkpoint?

- Because a cell must reach a certain size before its daughter cells will be large enough to function normally, biologists hypothesize that some mechanism exists to arrest the cell cycle if the cell is too small.
- Unicellular organisms arrest at the G_1 checkpoint if nutrient conditions are poor.
- Cells in multicellular organisms pass (or do not pass) through the G_1 checkpoint in response to signaling molecules from other cells, or what are termed social signals.
- If DNA is physically damaged, the protein **p53** activates genes that either stop the cell cycle until the damage can be repaired or lead to the cell's programmed, controlled destruction—a phenomenon known as **apoptosis**. In this way, p53 acts as a brake on the cell cycle. Cancer can develop if "brake" molecules such as p53 are defective, because the damaged DNA remains unrepaired. If the damage occurs in certain genes, cells can begin growing in an uncontrolled fashion and spread to other locations in the body. Consequently, regulatory proteins like p53 are called **tumor suppressors**.

The general message here is that the components of the G_1 checkpoint have the same function: ensuring that the cell is healthy and should replicate its DNA and divide.

The second checkpoint occurs after S phase, at the boundary between the G_2 and M phases. Specifically, cells appear to arrest at the G_2 checkpoint if chromosome replication has not been completed properly or if DNA is damaged. Because MPF is the key signal triggering the onset of M phase, investigators were not surprised to find that it is involved in the G_2 checkpoint. Although much remains to be learned, data suggest that if DNA is damaged or if chromosomes are not replicated correctly, the dephosphorylation and activation of MPF are blocked. When MPF is not activated, cells remain in G_2 phase. Some data indicate that cells at this checkpoint may also respond to signals from other cells and to internal signals relating to their size.

The final checkpoint occurs during mitosis. If not all chromosomes are properly attached to the mitotic spindle, M phase arrests at metaphase. Specifically, anaphase is delayed until all kinetochores are properly attached to mitotic spindle fibers. If this checkpoint did not exist, some chromosomes might not separate properly, and daughter cells would receive an incorrect number of chromosomes during anaphase. Because they would receive too much or too little genetic material, the effect on the daughter cells could be disastrous.

To summarize, the three cell-cycle checkpoints have the same purpose: They prevent the division of cells that are damaged or that have other problems, and they prevent the growth of mature cells that are in the G_0 state and should not grow any more. If one of the checkpoints fails, the affected cells may begin growing in an uncontrolled fashion. For the organism as a whole, the consequences of uncontrolled cell division are dire: cancer.

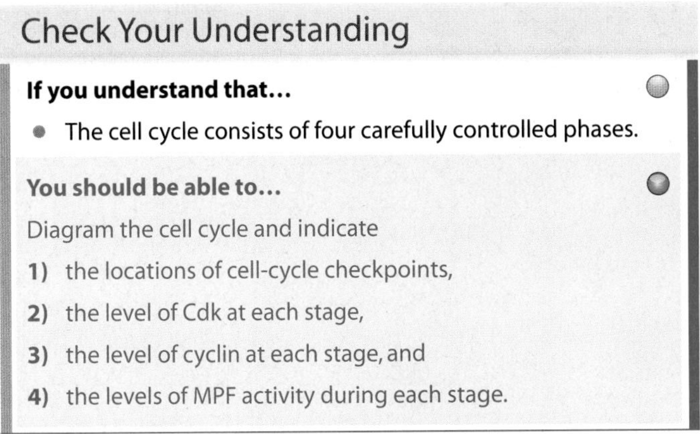

Check Your Understanding

If you understand that...

- The cell cycle consists of four carefully controlled phases.

You should be able to...

Diagram the cell cycle and indicate

1) the locations of cell-cycle checkpoints,
2) the level of Cdk at each stage,
3) the level of cyclin at each stage, and
4) the levels of MPF activity during each stage.

(MB) **Web Animation** at www.masteringbio.com
The Cell Cycle

11.4 Cancer: Out-of-Control Cell Division

Few diseases inspire more fear than cancer does. The fear springs from the difficulty of treating many forms of cancer, the potentially fatal nature of many cancers, and their frequency of occurrence. Most of us know someone who has had some form of cancer, and most of us have known someone who died from the disease. According to the American Cancer Society, 50 percent of American men and 33 percent of American women will develop cancer during their lifetime. In the United States, one in four of all deaths are from cancer. It is the second leading cause of death, exceeded only by heart disease.

Cancer is a general term for disease caused by cells that are growing in an uncontrolled fashion, that invade nearby tissues, and that spread to other sites in the body. Cancerous cells cause disease because they use nutrients and space needed by normal cells and disrupt the function of normal tissues. Humans suffer from at least 200 types of cancer. Stated another way, cancer is not a single illness but a complex family of diseases that affect an array of organs, including the breast, colon, brain, lung, and skin. In addition, several types of cancer can affect the same organ. Skin cancers, for example, come in multiple forms. Some are relatively easy to treat; others are often fatal. **Figure 11.16** provides data on how the number of deaths due to different types of cancer has changed through time in the United States.

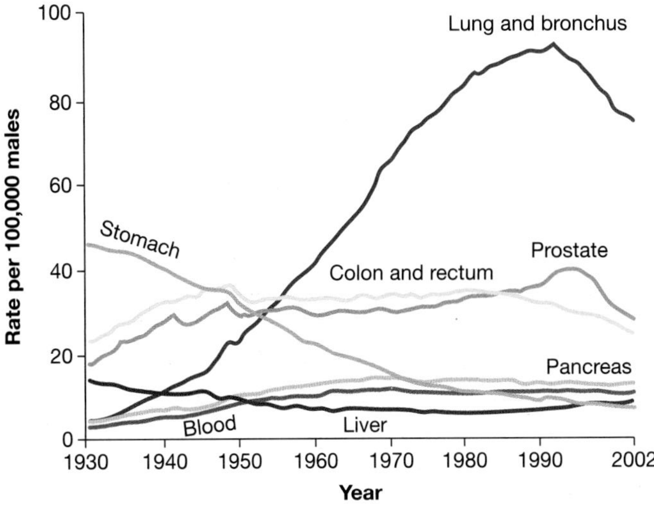

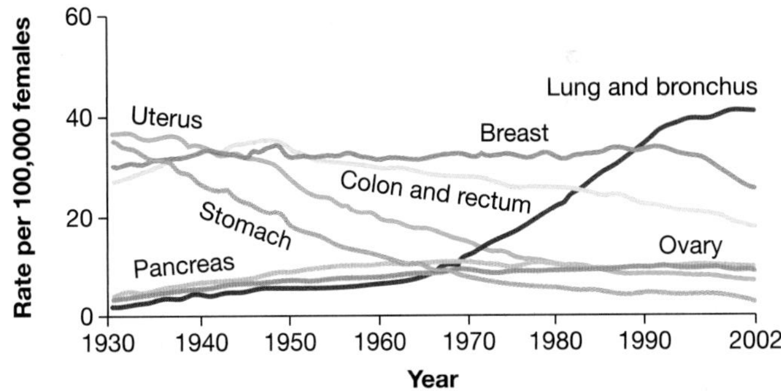

FIGURE 11.16 Changes in Cancer Death Rates over Time. Changes in the death rates from various types of cancer in **(a)** men and **(b)** women in the United States.

QUESTION In this population, how do changes over time in deaths due to lung cancer compare in males and females? Suggest a hypothesis to explain the similarities and differences.

Although cancers vary in time of onset, growth rate, seriousness, and cause, they have a unifying feature: Cancers arise from cells in which cell-cycle checkpoints have failed. More specifically, cancerous cells have two types of defects: (1) defects that make the proteins required for cell growth active at all times, and (2) defects that prevent tumor suppressor genes from shutting down the cell cycle. The protein called Ras, for example, was introduced in Chapter 8 as a key component in signal transduction systems—including phosphorylation cascades that trigger cell growth. Many cancerous cells have defective forms of Ras that do not become inactivated when they should. Instead, the defective Ras constantly send signals that trigger mitosis and cell division. Likewise, a large percentage of cancerous cells have defective forms of the tumor suppressor p53. Instead of being arrested or destroyed, cells that have damaged DNA are allowed to continue growth.

To get an understanding of why cancer occurs, let's review general characteristics of these diseases and then delve into the details of why regulatory mechanisms become defective.

Properties of Cancer Cells

When even a single cell in a multicellular organism begins to divide in an uncontrolled fashion, a mass of cells called a tumor results. For example, most cells in the adult human brain do not divide. But if a single abnormal brain cell begins unrestrained division, the growing tumor that results may disrupt the brain's function. What can be done? If the tumor can be removed without damaging the affected organ, a cure might be achieved. This is why surgical removal of the tumor is usually the first step in the treatment of a cancer. Often, though, surgery does not cure cancer. Why?

In addition to growing quickly, cancer cells are invasive—meaning that they are able to spread throughout the body via lymphatic vessels (introduced in Chapter 49), which collect excess fluid from tissues and return it to the bloodstream, or via the bloodstream itself. Invasiveness is a defining feature of a **malignant tumor**—one that is cancerous. Masses of noninvasive cells are noncancerous and form **benign tumors**. Some noninvasive tumors, like warts, grow slowly and are largely harmless. Others grow quickly and can cause problems if they are located in the brain or other sensitive parts of the body.

Cells become malignant and cancerous if they gain the ability to detach from the original tumor and invade other tissues. By spreading from the primary tumor site where uncontrolled growth originated, cancer cells can establish secondary tumors elsewhere in the body (**Figure 11.17**). This process is called **metastasis**. If metastasis has occurred by the time the original tumor is detected, then secondary tumors have begun to form and surgical removal of the primary tumor will not lead to a cure. As a result, the disease can be very difficult to treat. This is why early detection is the key to treating cancer most effectively.

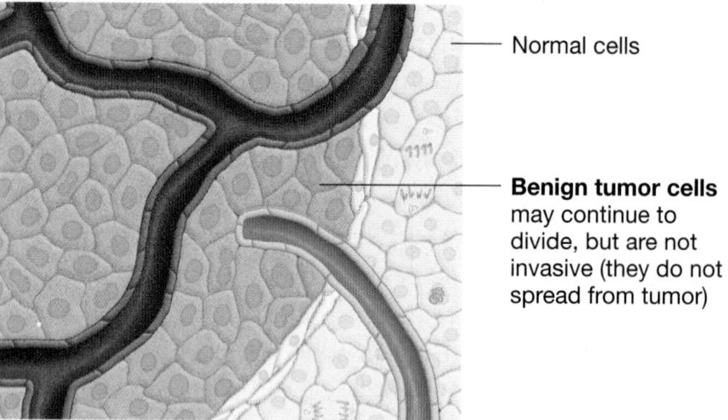

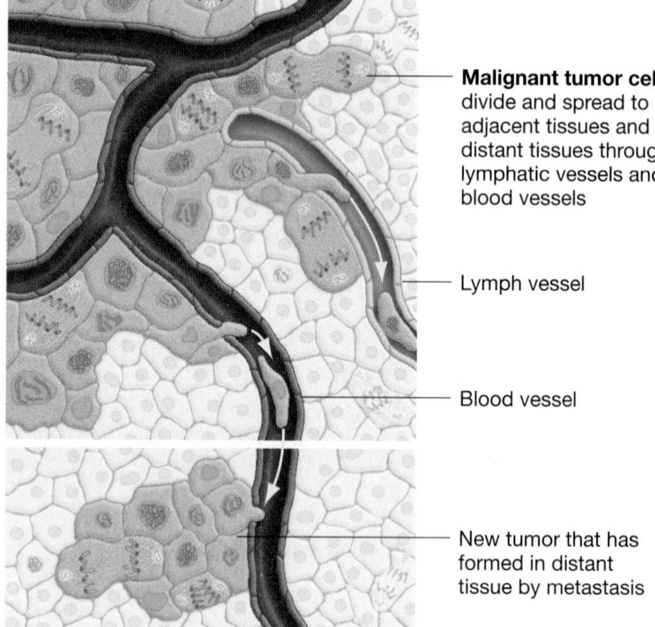

FIGURE 11.17 Cancers Spread to New Locations in the Body. **(a)** Benign tumors grow in a single location. **(b)** Malignant tumors are metastatic—meaning that their cells can spread to distant parts of the body and initiate new tumors. Malignant tumors cause cancer.

Cancer Involves Loss of Cell-Cycle Control

If cancer is caused by uncontrolled cell growth, what is the molecular nature of the disease? Recall that when many cells mature, they enter the G_0 phase—meaning their cell cycle is arrested at the G_1 checkpoint. In contrast, cells that do pass through the G_1 checkpoint are irreversibly committed to replicating their DNA and entering G_2. Based on this observation, biologists hypothesized that many or even most types of cancer involve defects in the G_1 checkpoint. To understand the molecular nature of the disease, then, researchers focused on understanding the normal mechanisms that operate at that checkpoint. In this way, cancer research and research on the normal cell cycle have become two sides of the same coin.

Social Control In unicellular organisms, passage through the G_1 checkpoint is thought to depend primarily on cell size and the availability of nutrients. If nutrients are plentiful, cells pass through the checkpoint and grow rapidly. In multicellular organisms, however, cells receive a constant supply of adequate nutrients via the bloodstream. Because nutrients generally are not limiting, most cells in multicellular organisms divide in response to some other type of signal. Because these signals arrive from other cells, biologists refer to *social control* over cell division. The general idea is that individual cells should be allowed to divide only when their growth is in the best interests of the organism as a whole.

Social control of the cell cycle is based on the types of cell-cell signals introduced in Chapter 8—more specifically, signals called growth factors. **Growth factors** are polypeptides or small proteins responsible for stimulating cell division. Many growth factors were discovered in the course of working out techniques for growing cells in culture. When researchers isolated mammalian cells in culture and provided them with adequate nutrients, the cells arrested in G_1 phase. They began to grow only when biologists added **serum**—the liquid that remains after blood clots and the blood cells have been removed. Some component of serum allowed cells to pass through the G_1 checkpoint. What was it?

In 1974 biologists succeeded in identifying one of these serum components. The component was a protein called platelet-derived growth factor (PDGF). As its name implies, PDGF is released by blood components called **platelets**, which promote blood clotting at wound sites. PDGF binds to receptor tyrosine kinases in the plasma membranes of cells in the area. When these cells receive the growth signal, they are stimulated to divide. The increased cell numbers facilitate wound healing.

Researchers subsequently found that PDGF is produced by an array of cell types. Investigators also succeeded in isolating and identifying a diverse array of other growth factors. For different types of cells to grow in culture, different combinations of growth factors must be supplied. Based on this result, biologists infer that different types of cells in an intact multicellular organism are controlled by different combinations of growth factors. Cancer cells are another story, however. Cancerous cells can often be cultured successfully without externally supplied growth factors. This observation suggests that the normal social controls on the G_1 checkpoint have broken down in cancer cells.

Social Controls and Cell-Cycle Checkpoints Why don't the social controls of the G_1 checkpoint work in cancer cells? Let's start by understanding the relationship between growth factors and the G_1 checkpoint in normal cells.

Recall from Section 11.3 that for an interphase cell to begin mitosis, a specific cyclin-Cdk complex must be activated. In cells that are in the G_0 resting state, the arrival of growth factors stimulates the production of cyclins. The arrival of growth factors also stimulates the production of a key regulatory protein called E2F. E2F is analogous to MPF, which activates genes required for M phase. When E2F is activated, it triggers the expression of genes required for S phase.

When E2F is first produced, however, it binds to a tumor suppressor protein called Rb. **Rb protein** is one of the key molecules that enforces the G_1 checkpoint. When E2F is bound to Rb, it is in the "off" position—it can't activate the genes required for S phase. In effect, a cell's commitment to dividing depends on whether Rb stays bound to the E2F protein. As long as Rb binds to E2F, the cell remains in G_0.

This situation changes dramatically if growth factors continue to arrive. With continued social stimulation, production of cyclins and thus cyclin-Cdk complexes continues. The cyclins involved are different from the cyclins involved in initiating M phase, however. They are specific to the G_1 checkpoint. (There are also cyclins specific to the G_2 checkpoint.) When G_1 cyclin-Cdk complexes are activated, they begin catalyzing the phosphorylation of Rb (**Figure 11.18**). When Rb is phosphorylated, it changes shape and no longer binds to E2F. The unbound E2F is free to activate its target genes, and

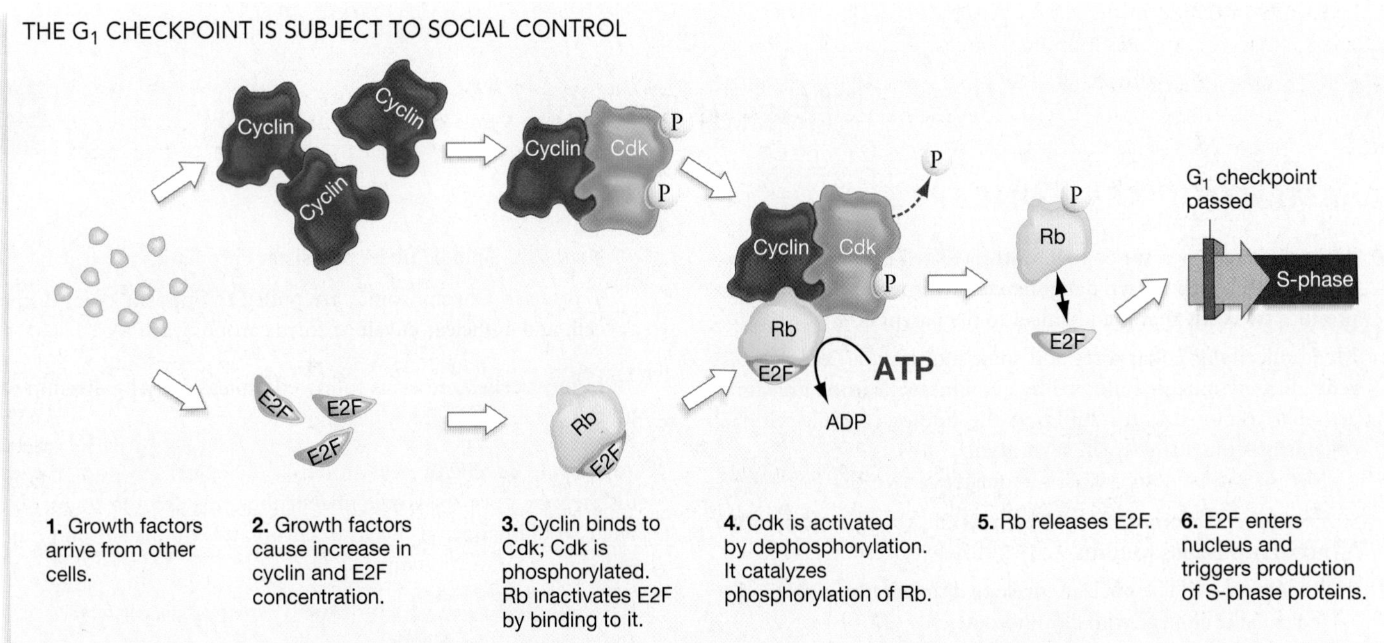

FIGURE 11.18 The G_1 Checkpoint Is Subject to Social Control. Cells that are in G_0 do not exit and begin dividing again unless they receive signals from other cells, in the form of growth factors. If enough signals arrive, the proteins that respond override the inhibitory effects of the Rb protein and begin S phase.

S phase gets under way. The growth factors thus function as a social signal that says, "It's OK to override Rb. Go ahead and pass the G_1 checkpoint and divide."

Cells can become cancerous when social controls fail—meaning, when cells begin dividing in the *absence* of a go-ahead signal from growth factors. One of two things go wrong.

- In some cancers the G_1 cyclin is overproduced. When cyclins stay at high concentrations, the Cdk that binds to cyclin phosphorylates Rb continuously, activating E2F and sending the cell into S phase. Cyclin overproduction results from the presence of excessive amounts of growth factors or from cyclin production in the absence of growth signals—usually due to a defect in a signaling pathway. The specific pathway involved starts with a receptor tyrosine kinase that responds to growth factors and includes the Ras protein highlighted in Chapter 8. This is why it is common to find defective Ras proteins in cancerous cells.

- The second possibility is that Rb is defective. If the Rb protein is missing or if it does not bind to E2F normally, any E2F that is present pushes the cell through the G_1 checkpoint and into S phase. In this way, defects in Rb lead to uncontrolled cell division.

The link between defective Rb protein and cancer was originally discovered in children with retinoblastoma, a cancer that occurs in about 1 in 20,000 children. The disease is characterized by the appearance of malignant tumors in the light-sensing tissue, or retina, of the eye. But follow-up work showed that most human cancer cells have defects in the Rb protein, not just cells that lead to retinoblastoma.

To summarize, overproduction of cyclins or defects in Rb itself can lead to the inactivation of Rb as a tumor suppressor. When Rb not longer works properly, the result can be failure of social control over the G_1 checkpoint, rampant cell growth, and tumor formation. ● If you understand the relationship between the failure of social controls and cancer, you should be able to make diagrams like Figure 11.18 showing the consequences of (1) constant cyclin production and (2) defective Rb proteins.

Cancer Is a Family of Diseases One of the broadest messages to come out of research on p53, Rb, and other molecules involved in arresting cell growth is that a wide variety of defects can lead to the failure of the G_1 checkpoint and the onset of cancer. Because many proteins are essential to the G_1 checkpoint, many different defects can cause the checkpoint to fail. In addition, researchers now realize that cancer is seldom due to a single defect. Most cancers develop only after several genes have been damaged. The combined dam-age is then enough to break cell-cycle control and induce uncontrolled growth and metastasis. Each type of cancer is due to a unique combination of errors. Stated another way, cancer can be caused by hundreds if not thousands of different defects.

Because cancer is actually a family of diseases with a complex and highly variable molecular basis, there will be no "magic bullet," or single therapy that cures all forms of the illness. Still, recent progress in understanding the cell cycle and the molecular basis of cancer has been dramatic, and cancer prevention and early detection programs are increasingly effective. The prognosis for many cancer patients is remarkably better now than it was even a few years ago.

Chapter Review

SUMMARY OF KEY CONCEPTS

● **After chromosomes are copied, mitosis distributes one chromosome copy to each of two daughter cells. Mitosis and cytokinesis produce two cells that are identical to the parent cell.**

Most unicellular eukaryotes and some multicellular eukaryotes reproduce by mitosis followed by cytokinesis. Mitotic divisions are also responsible for building the bodies of multicellular organisms from a single-celled embryo.

Mitosis can be described as a sequence of five phases:

Prophase Chromosomes condense, and the mitotic spindle apparatus begins to form.

Prometaphase The nuclear envelope disintegrates, and spindle fibers make contact with chromosomes.

Metaphase Spindle fibers move chromosomes to the metaphase plate, an imaginary plane in the middle of the cell.

Anaphase Spindle fibers pull sister chromatids apart.

Telophase Chromosomes are pulled to opposite poles of the cell, and a nuclear envelope forms around each set.

In most cells, mitosis is followed immediately by division of all cell contents, or cytokinesis.

You should be able to explain why the nuclear envelope has to disintegrate prior to spindle fibers contacting chromosomes, and what would happen if the sister chromatids of one chromosome failed to separate at anaphase. ●

The Phases of Mitosis; Mitosis

Over their life span, eukaryotic cells go through a cycle that consists of four carefully controlled phases.

Dividing cells alternate between the dividing phase, called M phase, and a nondividing phase known as interphase. Chromosome synthesis and replication, or S phase, occurs during interphase. S phase and M phase are separated by gap phases called G_1 and G_2. Cell growth and replication of nonnuclear cell components occur during the gap phases. The order of cell-cycle phases is $G_1 \rightarrow S \rightarrow G_2 \rightarrow M \rightarrow G_1$ and so forth.

Progression through the cell cycle is regulated at three checkpoints. Passage through the G_1 checkpoint represents an irreversible commitment to divide and is contingent on cell size, nutrient availability, lack of DNA damage, and/or growth signals from other cells. Mature cells do not pass this checkpoint. At the G_2 checkpoint, progression through the cycle is delayed until chromosome replication has been successfully completed. At the metaphase checkpoint during M phase, anaphase is delayed until all chromosomes are correctly attached to the mitotic spindle.

An array of different molecules regulates progression through the cell-cycle checkpoints. The most important of these regulatory molecules are cyclin-dependent kinases (Cdks) and cyclins, which are found in all eukaryotic cells. Active Cdks are enzymes that trigger progress through a checkpoint by phosphorylating important target proteins. Cyclins are proteins whose concentrations oscillate during the cell cycle, regulating the activity of Cdks. In multicellular organisms, cyclin concentrations are partially controlled by growth factors from other cells. As a result, the G_1 checkpoint is said to be under social control.

You should be able to predict what happens if an influx of growth factors leads to an increase in the production of cyclins.

MB **Web Animation** at www.masteringbio.com
The Cell Cycle

In multicellular organisms, uncontrolled cell growth leads to cancer. Different types of cancer result from different types of defects in control over the cell cycle.

Cancer is a common disease that is characterized by loss of control at the G_1 checkpoint, resulting in cells that divide in an uncontrolled fashion. Cancer cells also have the ability to spread throughout the body, which makes treatment difficult. In normal cells, growth-factor signals are required to produce G_1 cyclin-Cdk complexes and trigger division. An important target of activated G_1 cyclin-Cdk complexes is the retinoblastoma (Rb) protein, which regulates a protein responsible for initiating S phase. Defects in G_1 cyclin and Rb protein are common in human cancer cells.

You should be able to explain why cancer is considered a family of diseases, and why there is unlikely to be a single cure for cancer.

QUESTIONS

Test Your Knowledge

1. Which statement about the daughter cells of mitosis is correct?
   a. They differ genetically from one another and from the parent cell.
   b. They are genetically identical to one another and to the parent cell.
   c. They are genetically identical to one another but are different from the parent cell.
   d. Only one of the two daughter cells is genetically identical to the parent cell.
2. Progression through the cell cycle is regulated by oscillations in the concentration of which type of molecule?
   a. p53, Rb, and other tumor suppresssors
   b. receptor tyrosine kinases
   c. cyclin-dependent kinases
   d. cyclins
3. According to the data in Figure 11.4, the first labeled mitotic cells appear about 4 hours after the labeling period ends. From these data, researchers concluded that G_2 lasted about 4 hours. Why?
   a. The lengths of the total cell cycle and the G_1, S, and M phases were known, so the length of G_2 could be found by subtraction.
   b. It confirmed other data indicating that G_2 lasts about 4 hours in cultured cells of this type.
   c. Cyclins were labeled, so their concentration had to build up for 4 hours.
   d. Labeled cells are in S phase, so 4 hours passed between the end of S phase and the onset of M phase.
4. What major events occur during anaphase of mitosis?
   a. Chromosomes replicate, so each chromosome consists of two identical sister chromatids.
   b. Chromosomes condense and the nuclear envelope disappears.
   c. The chromosomes end up at opposite ends of the cell and two nuclear envelopes form.
   d. Sister chromatids separate, forming independent chromosomes.
5. What evidence suggests that during anaphase, spindle fibers shorten at the kinetochore and not at the base of the mitotic spindle?
   a. Motor proteins are located at the kinetochore.
   b. Motor proteins are located at the kinetochore *and* at the base of the mitotic spindle.
   c. When fluorescing microtubules are bleached in the middle, the bleached segment stays stationary as the fibers shorten near the kinetochore.
   d. When fluorescing microtubules are bleached in the middle, the bleached segment moves toward the base of the mitotic spindle as the fibers shorten near the microtubule organizing center.
6. The normal function of the Rb protein is to bind to E2F, inactivating it. What are the consequences of defects in Rb?
   a. It binds more tightly to E2F, so E2F cannot stimulate the production of molecules needed in S phase. The cell cannot progress through the cell cycle.
   b. It no longer responds to phosphorylation—it cannot be regulated.
   c. It does not bind to E2F properly, so E2F continuously stimulates the production of molecules needed in S phase.
   d. Rb acts as a tumor suppressor, similar in function to p53.

Test Your Knowledge answers: 1. b; 2. d; 3. d; 4. d; 5. c; 6. c

Test Your Understanding

Answers are available at www.masteringbio.com

1. Sketch the phases of mitosis in a cell with 6 chromosomes listing the major events of each phase. Identify at least two events that must be completed successfully for daughter cells to share an identical complement of chromosomes.
2. Make a concept map illustrating normal and abnormal events at the G_1 checkpoint. Your diagram should include p53, Rb, E2F, social signals, G_1 Cdk, G_1 cyclin, S-phase proteins, uncontrolled growth, phosphorylated (inactivated) cyclin-Cdk, dephosphorylated (activated) cyclin-Cdk, phosphorylated (inactivated) Rb.
3. Explain how cell fusion and microinjection experiments supported the hypothesis that specific molecules are involved in the transition from interphase to M phase.
4. Why are most protein kinases considered regulatory proteins?
5. Why are cyclins called cyclins? Explain their relationship to cyclin-dependent kinases and to growth factors.
6. Early detection is the key to surviving most cancers. Why?

Applying Concepts to New Situations

Answers are available at www.masteringbio.com

1. In multicellular organisms, nondividing cells stay in G_1 phase. For the cell, why is it better to be held in G_1 rather than S, G_2, or M phase?
2. When fruit fly embryos first begin to develop, mitosis occurs without cytokinesis. What is the result?
3. Predict the outcome of an experiment involving the fusion of a cell in G_1 phase with a cell in G_2 phase. What would happen to the G_1-phase nucleus? To the G_2-phase nucleus? Why?
4. Cancer is primarily a disease of older people. Further, a group of individuals may share a genetic predisposition to developing certain types of cancer, yet vary a great deal in time of onset—or not get the disease at all. Discuss these observations in light of the claim that several defects usually have to occur for cancer to develop.

www.masteringbio.com is also your resource for • Answers to text, table, and figure caption questions and exercises • Answers to *Check Your Understanding boxes* • Online study guides and quizzes • Additional study tools including the *E-Book for Biological Science* 3rd ed., textbook art, animations, and videos.

Meiosis

12

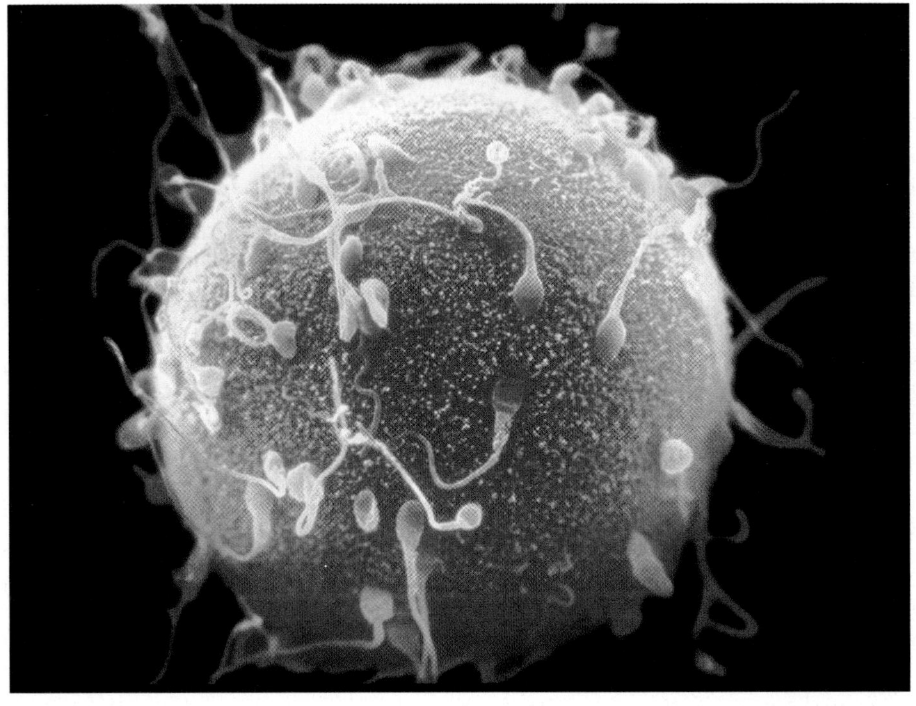

Scanning electron micrograph showing human sperm attempting to enter a cell that will become an egg. This chapter introduces the type of nuclear division called meiosis, which in animals occurs prior to the formation of sperm and eggs.

KEY CONCEPTS

- Meiosis is a type of nuclear division. It results in cells that have half as many chromosomes as the parent cell, and in animals it is involved in the formation of eggs and sperm. When an egg and a sperm cell combine to form an offspring, the original number of chromosomes is restored.
- Each cell produced by meiosis receives a different combination of chromosomes. Because genes are located on chromosomes, each cell produced by meiosis receives a different complement of genes. Meiosis leads to offspring that are genetically distinct from each other and from their parents.
- The leading hypothesis to explain meiosis is that genetically variable offspring are more likely to thrive in changing environments.
- If mistakes occur during meiosis, the resulting egg and sperm cells may contain the wrong number of chromosomes. It is rare for offspring with an incorrect number of chromosomes to develop normally.

Why sex?

Simple questions—such as why sexual reproduction exists—are sometimes the best. This chapter asks what sexual reproduction is and why some organisms employ it. Understanding sex has been one of the great mysteries of classical and contemporary biology.

Before plunging into the question of why sex exists, it is important to realize that there are two basic levels of explanation in biology: ultimate explanation and proximate explanation. An **ultimate explanation** is evolutionary in nature. It explains *why* something happens. In terms of sex, biologists want to know why it evolved. This is an important question because most organisms never reproduce sexually, or they do so only occasionally. It's also a difficult question, because asexual reproduction is much more efficient than sexual reproduction. Organisms that reproduce asexually don't have to flash colorful ornaments, exude exotic scents, or dance or sing to woo a mate and reproduce sexually. Instead, they just clone themselves. In eukaryotes that reproduce asexually, the cloning is done via mitosis.

In searching for an ultimate explanation for sex, biologists focus on a simple observation: Asexually produced offspring are genetically identical to their parents. Sexually produced offspring, in contrast, are genetically unlike their parents. Presumably, sexual reproduction is worth all the time and trouble involved because genetic variation is a good thing for offspring in certain types of environments. Why is sex advantageous for some species? This question is the focus of Sections 12.2 and 12.3.

A **proximate explanation** is mechanistic in nature. It explains *how* something happens. For example, biologists want to know how the process of sexual reproduction occurs in terms of the molecular and cellular structures and events that are involved. It has long been understood that during sexual reproduction, a male reproductive cell—a **sperm**—and a female reproductive cell—an **egg**—unite to form a new individual. The process of uniting sperm and egg is called **fertilization**. The first biologists to observe fertilization studied the large, translucent eggs of sea urchins. Due to the semi-transparency of the sea urchin egg cell, researchers were able to see the nuclei of a sperm and an egg fuse.

When these results were published in 1876, they raised an important question at the proximate level, concerning the number of chromosomes found in parents, sperm, eggs, and offspring. Cell biologists had already established that the number of chromosomes is constant from cell to cell within a multicellular organism. It was also accepted that chromosome number is the same in the parent and daughter cells of mitosis. Biologists confirmed that all of the cells in a newly growing offspring, or **embryo**, are the products of mitotic divisions, and that all cells in the body are the direct descendants of the nucleus that formed at fertilization. The question is, How can the chromosomes from a sperm cell and an egg cell combine, but form an offspring that has the same chromosome number as its mother and its father?

A hint at the answer came in 1883, when a researcher noted that cells in the body of roundworms of the genus *Ascaris* have four chromosomes, while their sperm and egg nuclei have only two chromosomes apiece. Four years later, August Weismann formally proposed a hypothesis to explain the riddle: During the formation of **gametes**—reproductive cells such as sperm and eggs—there must be a distinctive type of cell division that leads to a reduction in chromosome number. Specifically, if the sperm and egg contribute an equal number of chromosomes to the fertilized egg, Weismann reasoned, they must each contain half of the usual number of chromosomes. Then, when sperm and egg combine, the resulting cell has the same chromosome number as its mother's cells and its father's cells have.

In the decades that followed, biologists confirmed this hypothesis by observing gamete formation in a wide variety of plant and animal species. Eventually this form of cell division came to be called meiosis (literally, "lessening-act"). **Meiosis** is nuclear division that leads to a halving of chromosome number. It precedes the formation of eggs and sperm in animals, and it provides a satisfying proximate explanation for how sexual reproduction occurs. To a biologist, asking "Why sex?" is equivalent to asking "Why meiosis?" Let's delve in by looking at how meiosis occurs.

12.1 How Does Meiosis Occur?

When cell biologists began to study the cell divisions that lead to gamete formation, they made an important observation: Each organism has a characteristic number of chromosomes. Consider the drawing in **Figure 12.1**, based on a paper published by Walter Sutton in 1902. It shows the chromosomes of the lubber grasshopper during the cell divisions leading up to the formation of a sperm. In total, there are 23 chromosomes in the cell. Sutton realized, however, that there are just 12 distinct types of chromosomes based on size and shape. In this species, there were two chromosomes of each type.

Sutton designated 11 of the chromosomes by the letters *a* through *k* and the twelfth by the letter X. Some years later, Nettie Stevens established that the X chromosome is associated with the sex of the individual and is called a **sex chromosome**. Non-sex chromosomes, such as *a–k* in Sutton's grasshopper cell, are known as **autosomes**. It turns out that lubber grasshoppers have just one type of sex chromosome. In this species, females have two sex chromosomes and are designated XX; males have just one sex chromosome and are designated XO, where the O refers to the "missing" chromosome. Two types of sex chromosomes, known as X and Y, exist in humans and other mammals. Human females have two X chromosomes while males have one X and one Y chromosome.

Sutton also introduced an important term for the pairs of sex chromosomes or autosomes. He referred to the two chromosomes of each type as **homologous** ("same proportion") **chromosomes**. (They can also be called **homologs**.) The two chromosomes labeled *c*, for example, have the same size and shape and are homologous.

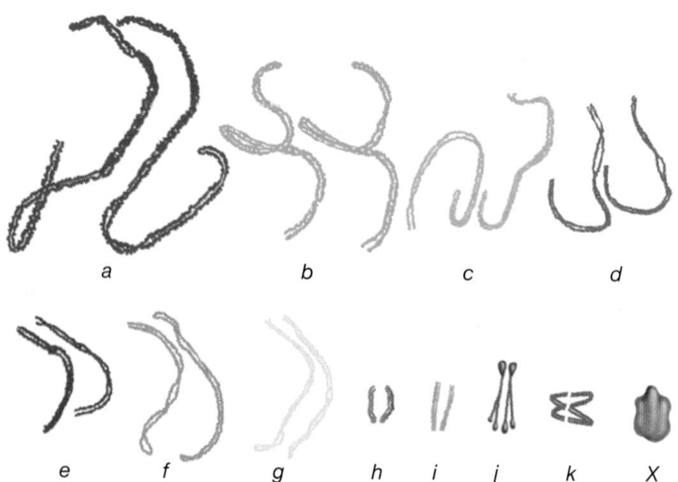

FIGURE 12.1 Cells Contain Different Types of Chromosomes, and Chromosomes Come in Pairs. Letters designate each of the 12 distinct types of chromosomes found in lubber grasshopper cells. There are two of each type of chromosome (except the X, as this is a male). The two members of a chromosome pair are called homologs.

Later work showed that homologous chromosomes are similar not only in size and shape but also in content. Homologous chromosomes carry the same genes. A **gene** is a section of DNA that influences one or more hereditary traits in an individual. A trait is a characteristic. For example, each copy of chromosome *c* found in lubber grasshoppers might carry genes that influence eye formation, body size, singing behavior, or jumping ability. The versions of a gene found on homologous chromosomes may differ, however. Biologists use the term **allele** to denote different versions of the same gene. For example, the alleles on each copy of chromosome *c* in a lubber grasshopper might contribute to rounder eyes versus narrower eyes, larger body size versus smaller body size, faster songs versus slower songs, and so on. Homologous chromosomes carry the same genes, but each homolog may contain different alleles.

At this point in his study, Sutton had succeeded in determining the lubber grasshopper's **karyotype**—meaning the number and types of chromosomes present (see **Box 12.1** on page 252). As karyotyping studies expanded, cell biologists realized that, like lubber grasshoppers, the vast majority of plants and animals have more than one of each type of chromosome. These investigators designated terms to identify the number of chromosome copies they observed. Organisms such as lubber grasshoppers, humans, and cedar trees are called **diploid** ("double-form"), because they have two versions of each type of chromosome. Diploid organisms have two alleles of each gene—one on each of the homologous pairs of chromosomes. Organisms such as bacteria, archaea, and many algae are called **haploid** ("single-form"), because their cells contain just one of each type of chromosome. Haploid organisms do not contain homologous chromosomes. They have just one allele of each gene.

Researchers also invented a compact notation to indicate the number of chromosomes and chromosome sets in a particular organism or type of cell. By convention, the letter n stands for the number of distinct types of chromosomes in a given cell and is called the **haploid number**. If sex chromosomes are present, they are counted as a single type in the haploid number. Diploid cells have a haploid number, which indicates the number of different types of chromosomes present. To indicate the number of complete chromosome sets observed, a number is placed before the n. Thus, a cell can be n, or $2n$, or $3n$, and so on. The combination of a number and n is termed the cell's **ploidy**. Diploid cells or species are designated $2n$, because two chromosomes of each type are present—one from each parent. Haploid cells or species are labeled simply n, because they have just one set of chromosomes—no homologs are observed. In haploid cells, the number 1 in front of n is implied and is not written out.

Later work revealed that it is common for species in some lineages—particularly land plants such as ferns—to contain more than two of each type of chromosome. Instead of having two homologous chromosomes per cell, as many organisms do, **polyploid** ("many-form") species may have three or more of each type of chromosome in each cell. Depending on the number of homologs present, such species are called triploid ($3n$), tetraploid ($4n$), hexaploid ($6n$), octoploid ($8n$), and so on. Why some species are haploid versus diploid or tetraploid is currently the subject of debate and research.

To summarize, the haploid number n indicates the number of distinct types of chromosomes present. Different species have different haploid numbers (see **Table 12.1**). Human cells contain 23 distinct types of chromosomes, so $n = 23$. In the grasshopper cells that Sutton studied, $n = 12$. In contrast, a cell's ploidy (n, $2n$, $3n$, etc.) indicates the number of each type of chromosome present. Stating a cell's ploidy is the same as stating the number of haploid chromosome sets present. Because most human and grasshopper cells contain two of each type of chromosome, they are diploid. In humans $2n = 46$; in Sutton's grasshoppers $2n = 24$. Ploidy refers to the number of each type of chromosome present; haploid number identifies how many different types of chromosomes occur.

Table 12.2 summarizes the vocabulary that biologists use to describe the number and types of chromosomes found in a cell.

Sutton and the other early cell biologists did more than describe the karyotypes observed in their study organisms, however. Through careful examination, they were able to track how chromosome numbers change during meiosis. These studies confirmed Weismann's hypothesis that a special type of cell division occurs during gamete formation. This result was a major advance in understanding sex at the proximate level.

TABLE **12.1** **The Number of Chromosomes Found in Some Familiar Organisms**

Organism	Number of Different Types of Chromosomes (haploid number n)	Diploid Chromosome Number ($2n$)
Humans	23	46
Domestic dog	36	72
Fruit fly	4	8
Chimpanzee	24	48
Bulldog ant	1	2
Garden pea	7	14
Corn (maize)	10	20

An Overview of Meiosis

Recall that cells replicate each of their chromosomes before undergoing meiosis. At the start of the process, then, chromo-

SUMMARY TABLE 12.2 **Vocabulary Used to Indicate the Chromosomal Makeup of a Cell**

Term	Definition	Example or Comment
Chromosome	Structure made up of DNA and proteins; carries the cell's hereditary information (genes)	Eukaryotes have linear, threadlike chromosomes; most bacteria and archaea have just one circular chromosome
Sex chromosome	Chromosome associated with an individual's sex	X and Y chromosomes of humans: Males are XY, females XX. Also, Z and W chromosomes of birds and buttterflies: Males are ZZ, females ZW.
Autosome	A non-sex chromosome	Chromosomes 1–22 in humans.
Unreplicated chromosome	A chromosome that consists of a single copy (in eukaryotes, a single "thread")	
Replicated chromosome	A chromosome that has been copied; consists of two linear structures joined at the centromere	Centromere
Sister chromatids	The chromosome copies in a replicated chromosome	Sister chromatids
Homologous chromosomes (homologs)	In a diploid organism, chromosomes that are similar in size, shape, and gene content	You have a chromosome 22 from your mother (red) and a chromosome 22 from your father (blue). Homologous chromosomes
Non-sister chromatids	The chromosome copies in homologous chromosomes	Non-sister chromatids
Tetrad	Homologous chromosomes that are joined together	Tetrad
Haploid number	The number of different types of chromosomes in a cell; symbolized n	Humans have 23 different types of chromosomes ($n = 23$)
Diploid number	The number of chromosomes present in a diploid cell; symbolized $2n$	In humans all cells except gametes are diploid and contain 46 chromosomes ($2n = 46$)
Ploidy	The number of each type of chromosome present; equivalent to the number of haploid chromosome sets present	
Haploid	Having one of each type of chromosome (n)	Bacteria and archaea are haploid, as are many algae; plant and animal gametes are haploid
Diploid	Having two of each type of chromosome ($2n$)	Most familiar plants and animals are diploid
Polyploid	Having more than two of each type of chromosome; cells may be triploid ($3n$), tetraploid ($4n$), hexaploid ($6n$), and so on	Seedless bananas are triploid; many ferns are tetraploid; bread wheat is hexaploid

somes are in the same state they are in prior to mitosis. When chromosome replication is complete, each chromosome consists of two identical **sister chromatids**. Sister chromatids contain the same genetic information and are physically joined at a portion of the chromosome called the **centromere** as well as along their entire length (**Figure 12.2a**).

To understand meiosis, it is critical to understand the relationship between chromosomes and sister chromatids. An unreplicated chromosome consists of a single DNA molecule with its associated proteins, while a replicated chromosome consists of two sister chromatids. The trick is to recognize that unreplicated and replicated chromosomes are each considered *single*

(a) Each chromosome replicates prior to undergoing meiosis.

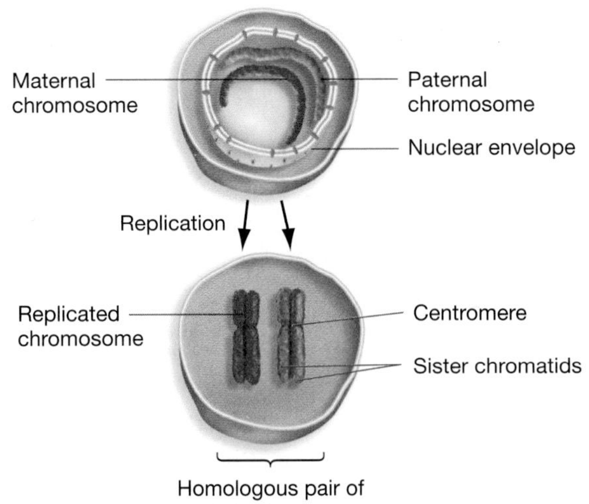

(b) During meiosis, chromosome number in each cell is reduced.

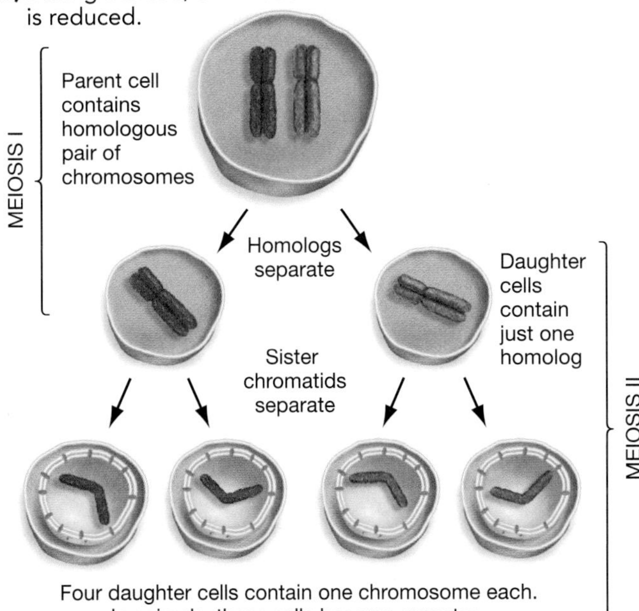

(c) A full complement of chromosomes is restored during fertilization.

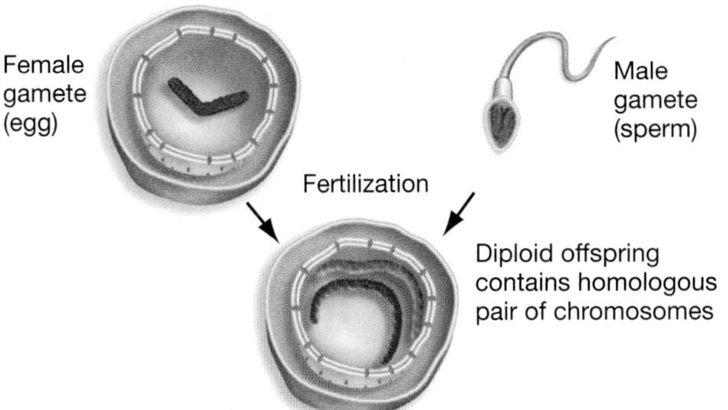

FIGURE 12.2 The Major Events in Meiosis. Meiosis reduces chromosome number by half. In diploid organisms, the products of meiosis are haploid.

EXERCISE In parts (b) and (c), write *n* or 2*n* next to each cell to indicate its ploidy.

chromosomes—even though the replicated chromosome comprises *two* sister chromatids. A chromosome is still just one chromosome whether it is unreplicated (consisting of a single strand) or replicated (consisting of two strands). This is a key point. If you understand this concept, you should be able to draw the same chromosome in the replicated and unreplicated state, explain why both structures represent a single chromosome, and then label the sister chromatids in the replicated chromosome.

Meiosis consists of two cell divisions, called **meiosis I** and **meiosis II**. As **Figure 12.2b** shows, the two divisions occur consecutively but differ sharply. During meiosis I, the homologs in each chromosome pair separate from each other. One homolog goes to one daughter cell; the other homolog goes to the other daughter cell. The homolog that came from the individual's mother is colored red in Figure 12.2; the homolog that came from the father is colored blue. It is a matter of chance which daughter cell receives which homolog. The end result is that the daughter cells from meiosis I have one of each type of chromosome instead of two, and thus half as many chromosomes as the parent cell has. During meiosis I, the diploid (2*n*) parent cell produces two haploid (*n*) daughter cells. Each chromosome still consists of two sister chromatids, however.

During meiosis II, sister chromatids from each chromosome separate. One sister chromatid goes to one daughter cell; the other sister chromatid goes to the other daughter cell. The cell that starts meiosis II has one of each type of chromosome, but each chromosome has been replicated (meaning it still consists of two sister chromatids). The cells produced by meiosis II also have one of each type of chromosome, but now the chromosomes are unreplicated.

To reiterate, sister chromatids separate during meiosis II, just as they do during mitosis. Meiosis II is actually equivalent to mitosis occurring in a haploid cell. As in mitosis, chromosome movements during meiosis I and II are caused by spindle fibers that attach at the centromere of each chromosome and pull sister chromatids apart.

Sutton and a host of other early cell biologists worked out this sequence of events through careful observation of cells with the light microscope. Based on these studies, they came to a key realization: The outcome of meiosis is a reduction in chromosome number. For this reason, meiosis is known as a reduction division. In most plants and animals, the original cell is diploid and the four daughter cells are haploid. These four haploid daughter cells, each containing one of each homologous chromosome, eventually go on to form egg cells or sperm cells via a process called **gametogenesis** ("gamete-origin"), which is described in Chapter 22, Chapter 40, and Chapter 48. When two gametes fuse during fertilization, a full complement of chromosomes is restored (**Figure 12.2c**). The cell that results from fertilization is diploid and is called a **zygote**. In this way, each diploid individual receives both a haploid chromosome set from its mother and a haploid set from its father. Homologous chromosomes are therefore referred to as either maternal or paternal

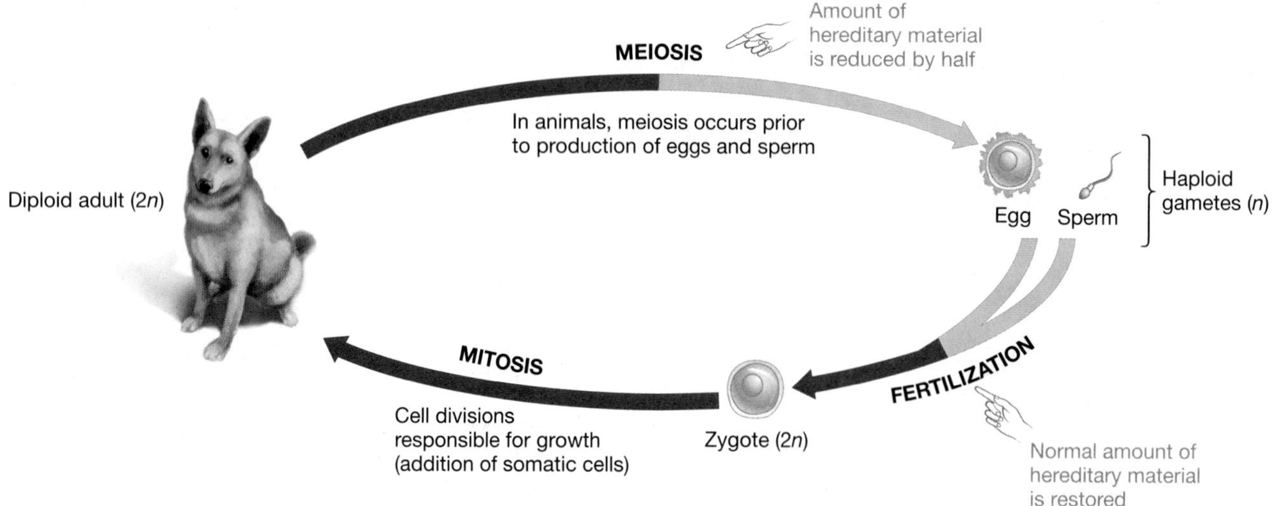

FIGURE 12.3 Ploidy Changes during the Life Cycle of a Dog. This type of life cycle is called diploid dominant, because the organism is diploid throughout most of its life cycle.

EXERCISE In many algae, the only diploid phase is the zygote. Zygotes undergo meiosis to form haploid cells, which then grow into haploid adults. Adults produce haploid sperm and eggs. Sketch this "haploid-dominant" life cycle.

in origin. **Maternal chromosomes** come from the mother; **paternal chromosomes** come from the father.

Figure 12.3 puts these events into the context of an animal's **life cycle**—the sequence of events that occurs over the life span of an individual, from fertilization to the production of offspring. As you study the figure, note how ploidy changes as the result of meiosis and fertilization. In the case of the dog illustrated here, meiosis in a diploid adult results in the formation of haploid gametes, which combine to form a diploid zygote. It's also important to realize that meiosis and gametogenesis occur in specialized organs—the testes or ovaries of a dog, for example—and that meiosis represents a tiny fraction of all cell division events that occur during the life cycle.

Once Sutton and others had published their work on the nature of meiosis and changes in ploidy, the mystery of fertilization was finally solved. To appreciate the consequences of meiosis fully, though, let's now analyze the events in more detail.

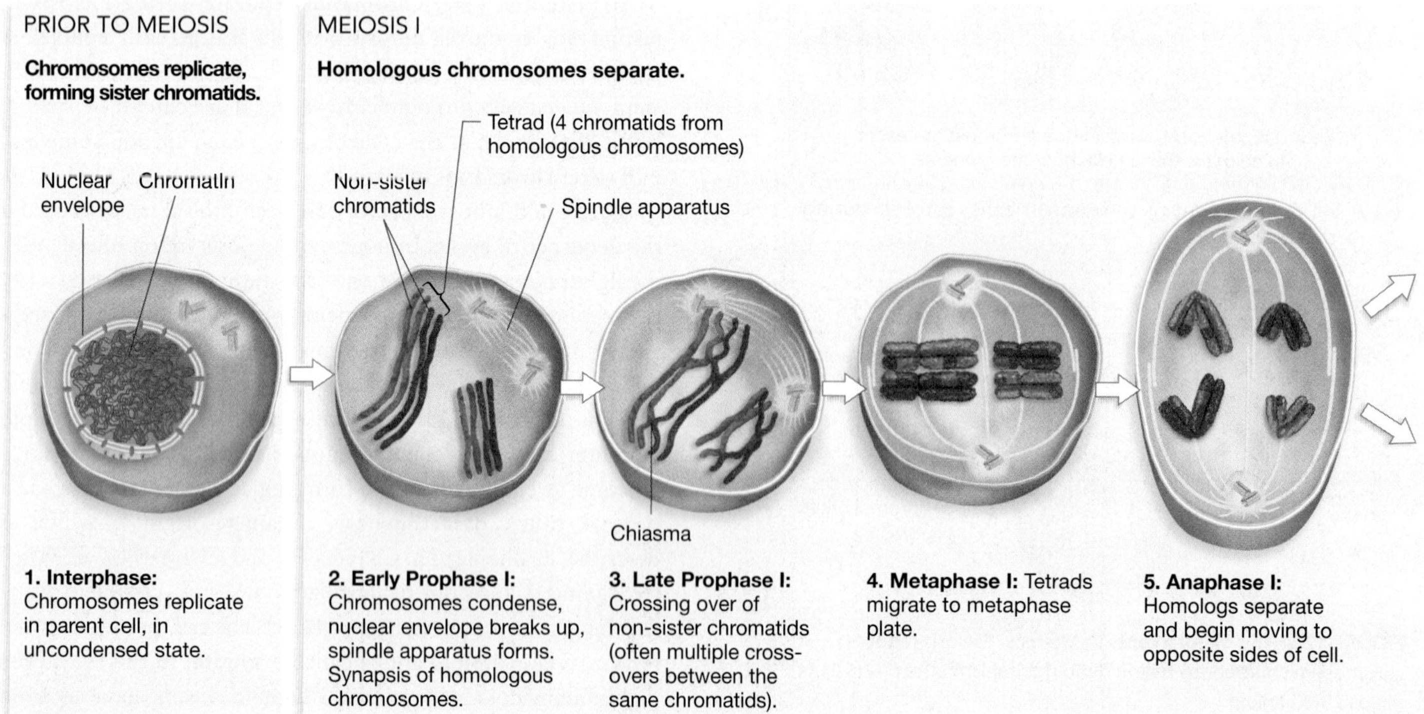

FIGURE 12.4 The Phases of Meiosis.

EXERCISE In step 3, circle all of the other chiasmata.

The Phases of Meiosis I

Meiosis begins after chromosomes have been replicated during S phase. Prior to the start of meiosis, chromosomes are extremely long structures, just as they are during interphase of the normal cell cycle. The major steps that occur once meiosis begins are shown in **Figure 12.4**, using a diploid species with a haploid number of 2 as an example ($n = 2$; $2n = 4$). As in Figure 12.3, maternal chromosomes are red and paternal chromosomes are blue.

During early prophase I the chromosomes condense, the spindle apparatus forms, and the nuclear envelope begins to disappear. The next event illustrated, still during early prophase of meiosis I, is crucial: Homologous chromosome pairs come together. This pairing process is called **synapsis** and is illustrated in step 2 of Figure 12.4. Synapsis is possible because regions of homologous chromosomes that are similar at the molecular level attract one another, via mechanisms that are currently the subject of intense research. The structure that results from synapsis is called a **tetrad** (*tetra* means four in Greek). A tetrad consists of two homologous chromosomes, with each homolog consisting of two sister chromatids. The chromatids from the homologs are referred to as **non-sister chromatids**. In the figure, the red-colored chromatids are non-sister chromatids with respect to the blue-colored chromatids.

During late prophase I, the non-sister chromatids begin to separate at many points along their length. They stay joined at certain locations, however, and look as if they cross over one another. Each crossover forms an X-shaped structure called a **chiasma** (plural: *chiasmata*). (In the Greek alphabet, the letter *X* is "chi.") Normally, at least one chiasma forms in every pair of homologous chromosomes; usually there are several chiasmata. As step 3 of Figure 12.4 shows, the chromatids involved in chiasma formation are homologous but not sisters. Consistent with this observation, Thomas Hunt Morgan proposed that a physical exchange of paternal and maternal chromosomes occurs at chiasmata. According to this hypothesis, paternal and maternal chromatids break and rejoin at each chiasma, producing chromatids that have both paternal and maternal segments. Morgan called this process of chromosome exchange **crossing over**. In step 4 of Figure 12.4, the result of crossing over is illustrated by chromosomes with a combination of red and blue segments. When crossing over occurs, the chromosomes that result have a mixture of maternal and paternal alleles.

The next major stage in meiosis I occurs during metaphase I, when pairs of homologous chromosomes (tetrads) are moved to a region called the **metaphase plate** by spindle fibers (step 4). Two points are key here: Each tetrad moves to the metaphase plate independently of the other tetrads, and the alignment of maternal and paternal homologs from each chromosome is random. During anaphase I, the homologous chromosomes in each tetrad separate and begin moving to opposite sides of the cell (step 5). Meiosis I concludes with telophase I, when the homologs finish moving to opposite sides of the cell (step 6). When meiosis I is complete, **cytokinesis** (division of cytoplasm) occurs and two haploid daughter cells form.

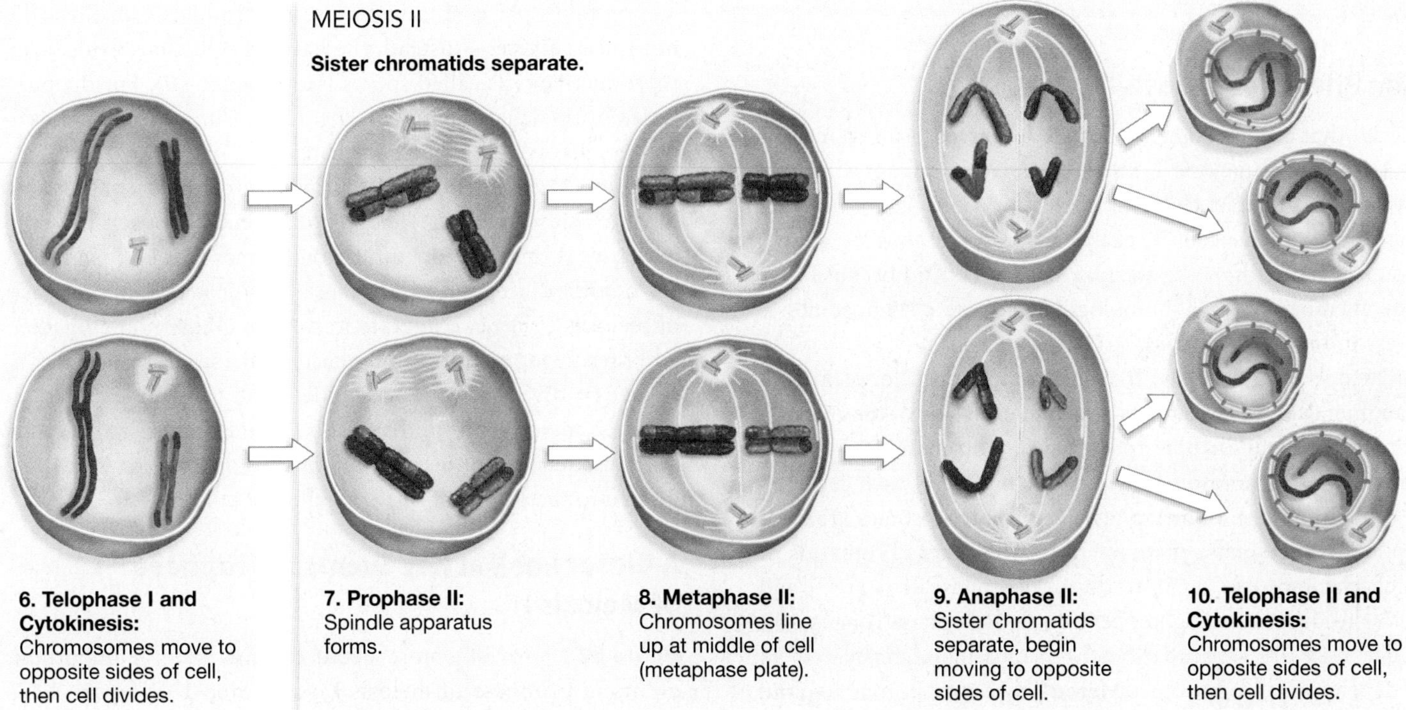

The end result of meiosis I is that one chromosome of each homologous pair is distributed to a different daughter cell. A reduction division has occurred: The daughter cells of meiosis I are haploid. The sister chromatids remain attached in each chromosome, however, meaning that the haploid daughter cells produced by meiosis I still contain replicated chromosomes. The chromosomes in each cell are a random assortment of maternal and paternal chromosomes as a result of crossing over and the random distribution of maternal and paternal homologs during metaphase.

Although meiosis I is a continuous process, biologists summarize the events by identifying these distinct phases:

- Early Prophase I—Replicated chromosomes condense, the spindle apparatus forms, and the nuclear envelope disappears. Synapsis of homologs forms pairs of homologous chromosomes (tetrads). Spindle fibers attach to the kinetochores at the centromeres of chromosomes.
- Late Prophase I—Crossing over results in a mixing of chromosome segments from maternal and paternal chromosomes.
- Metaphase I—Pairs of homologous chromosomes (tetrads) migrate to the metaphase plate and line up.
- Anaphase I—Homologs separate and begin moving to opposite ends of the cell.
- Telophase I—Homologs finish moving to opposite sides of the cell. In some species, a nuclear envelope re-forms around each set of chromosomes.

When meiosis I is complete, the cell divides.

The Phases of Meiosis II

Recall that chromosome replication occurred prior to meiosis I. Throughout meiosis I, sister chromatids remain attached. Because no further chromosome replication occurs between meiosis I and meiosis II, each chromosome consists of two sister chromatids at the start of meiosis II. And because only one member of each homologous pair of chromosomes is present, the cell is haploid.

Next, during prophase II, a spindle apparatus forms in both daughter cells. Spindle fibers attach to each side of the chromosomes—one spindle fiber to each sister chromatid—and begin moving the chromosomes toward the middle of each cell (step 7 of Figure 12.4). In metaphase II, the chromosomes are lined up at the metaphase plate (step 8). The sister chromatids of each chromosome separate during anaphase II (step 9) and move to different daughter cells during telophase II (step 10). Once they are separated, each chromatid is considered an independent chromosome. Meiosis II results in four haploid cells, each with one chromosome of each type.

As in meiosis I, biologists routinely designate distinct phases in meiosis II:

- Prophase II—The spindle apparatus forms. If a nuclear envelope formed at the end of meiosis I, it breaks apart.
- Metaphase II—Replicated chromosomes, consisting of two sister chromatids, are lined up at the metaphase plate.
- Anaphase II—Sister chromatids separate. The unreplicated chromosomes that result begin moving to opposite sides of the cell.
- Telophase II—Chromosomes finish moving to opposite sides of the cell. A nuclear envelope forms around each haploid set of chromosomes.

When meiosis II is complete, each cell divides to form two daughter cells. Because meiosis II occurs in both daughter cells of meiosis I, the process results in a total of four daughter cells from each original, parent cell. In a nutshell, one diploid cell with replicated chromosomes gives rise to four haploid cells with unreplicated chromosomes. It should make sense, after you examine the right side of Figure 12.4 in detail, that the movement of chromosomes during meiosis II is virtually identical to what happens in a mitotic division in a haploid cell.

In male animals the cells produced by meiosis go on to form sperm through a series of events that are detailed in later chapters. In females of at least some animal species, though, meiosis begins in cells called oocytes and then stops and restarts at several points as oocytes mature into eggs. In mammals, for example, meiosis is not completed until after fertilization takes place. And in plants, the products of meiosis do not form gametes—instead, the haploid cells that result form reproductive cells called spores (see Chapter 40). For the purposes of this chapter, however, you can think of meiosis as a process involved in gamete formation.

Figure 12.5 and **Table 12.3** provide a detailed comparison of mitosis and meiosis. The key difference between the two processes is that homologous chromosomes pair early in meiosis, but not during mitosis. Because homologs pair in prophase of meiosis I, they can migrate to the metaphase plate together and then separate during anaphase of meiosis I, resulting in a reduction division. ● If you understand this key distinction between meiosis and mitosis, you should be able to describe the consequences for meiosis if homologs do not pair. Now let's delve into the details of this critical event.

A Closer Look at Key Events in Prophase of Meiosis I

Figure 12.7 provides more detail on how several important events in prophase of meiosis I occur. Step 1 of Figure 12.7 shows that after chromosome replication is complete, sister

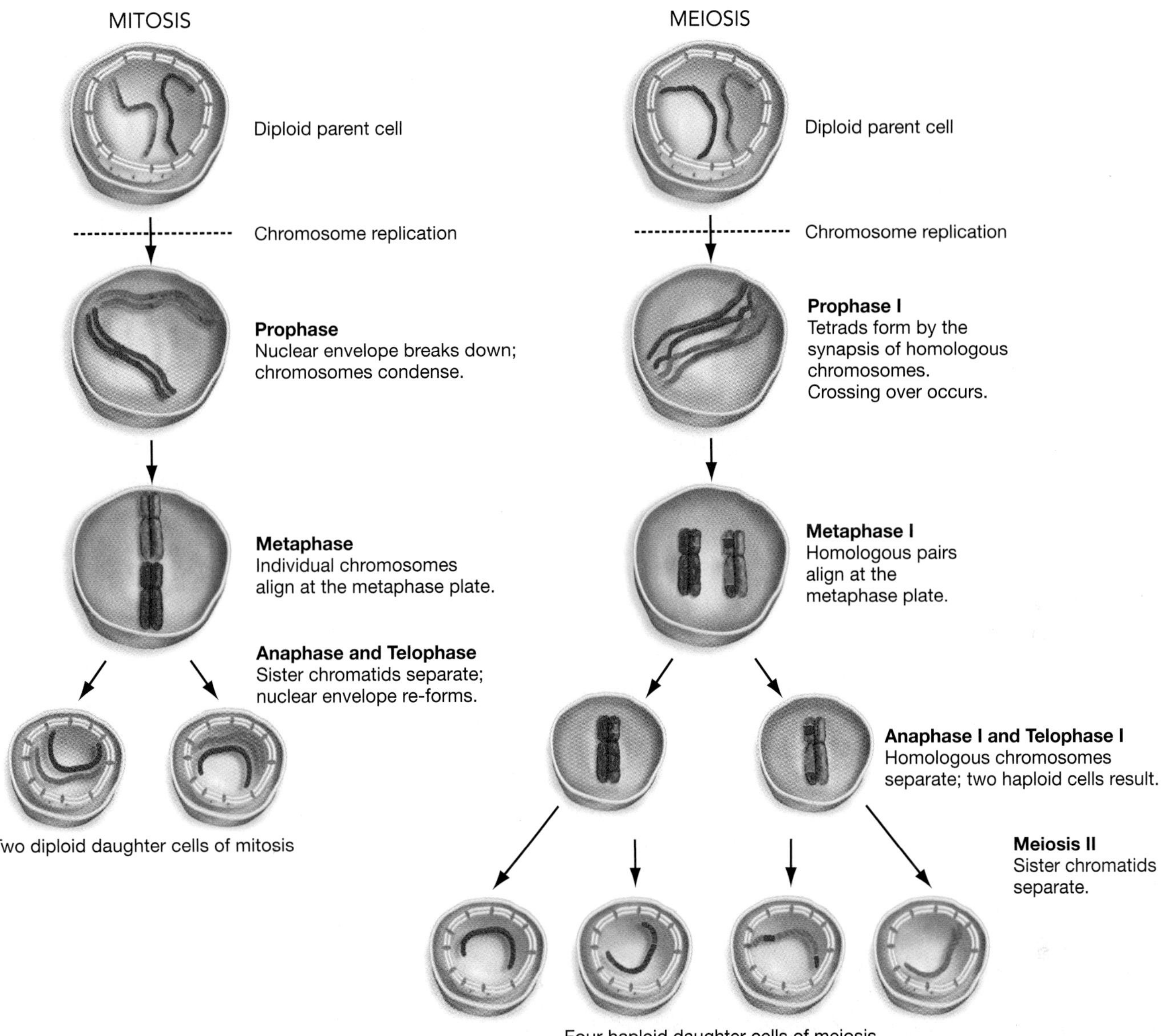

FIGURE 12.5 A Comparison of Mitosis and Meiosis. Mitosis produces two daughter cells with chromosomal complements identical to the parent cell. Meiosis produces four haploid cells with chromosomal complements unlike each other and unlike the diploid parent cell.

SUMMARY TABLE **12.3 Key Differences between Mitosis and Meiosis**

Feature	Mitosis	Meiosis
Number of cell divisions	One	Two
Number of chromosomes in daughter cells, compared with parent cell	Same	Half
Synapsis of homologs	No	Yes
Number of crossing-over events	None	One or more per pair of homologous chromosomes
Makeup of chromosomes in daughter cells	Identical	Different—only one of each chromosome type present, paternal and maternal segments mixed within chromosomes
Role in life cycle	Asexual reproduction in eukaryotes; cell division for growth of multicellular organisms	Precedes production of gametes in sexually reproducing animals

BOX 12.1 Karyotyping Techniques

Although chromosomes maintain their individuality and physical integrity throughout the cell cycle, they are readily visible only during mitosis or meiosis, when they are condensed. To describe an individual's karyotype, then, biologists must study cells undergoing cell division.

The first step in generating a karyotype is to obtain a sample of cells from the individual being studied. Cancer researchers might collect cells from a tumor; physicians who are concerned about the possibility of birth defects might obtain a few cells from the developing embryo inside the mother. The next step is to grow the cells in culture, using techniques introduced in Chapter 11. When the cultured cells are dividing rapidly, they are treated with a compound called colchicine. Colchicine stops mitosis at metaphase by disrupting the formation of the mitotic spindle. At this stage the chromosomes are relatively easy to study, because they are condensed and consist of sister chromatids. The chromosomes of colchicine-treated cells are then stained and examined with the light microscope.

Researchers can distinguish condensed chromosomes by size; by the position of the centromere, which holds sister chromatids together; and by striping or banding patterns that appear in response to some stains. Subtler differences among chromosomes are apparent when a higher-resolution technique for karyotyping called **spectral karyotyping** (**SKY**), or **chromosome painting**, is used. The "painting" is done with fluorescent dyes that are attached to short DNA molecules. The dyed pieces of DNA bind to particular regions of particular chromosomes. By using a combination of dyes, technicians can give each pair of homologous chromosomes a distinctive suite of colors (**Figure 12.6a**). The high-resolution image produced by this technique allows clinicians to diagnose an array of chromosomal abnormalities. Consider some examples:

- **Figure 12.6b** shows chromosomal changes associated with a cancer called chronic myelogenous leukemia. The defect is a **translocation**—a swapping of chromosome segments. In this case, a piece of chromosome 9 and a piece of chromosome 22 have changed places. The translocation causes a genetic change that leads to uncontrolled cell growth.
- **Figure 12.6c** shows the sex chromosomes of an individual with **Klinefelter syndrome**, which develops in people who have two X chromosomes and a Y chromosome instead of a single X and a single Y. People with this syndrome have male sex organs but are sterile. They may also develop some female-like traits, such as enlarged breasts.

Karyotyping is an important diagnostic tool. As techniques for generating karyotypes improve, our ability to detect and interpret chromosomal defects also improves.

(a) Normal human karyotype

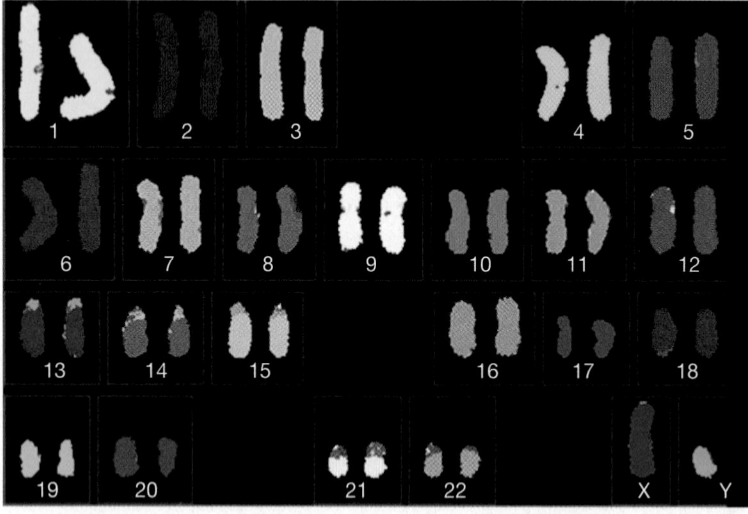

(b) Translocations are abnomalities

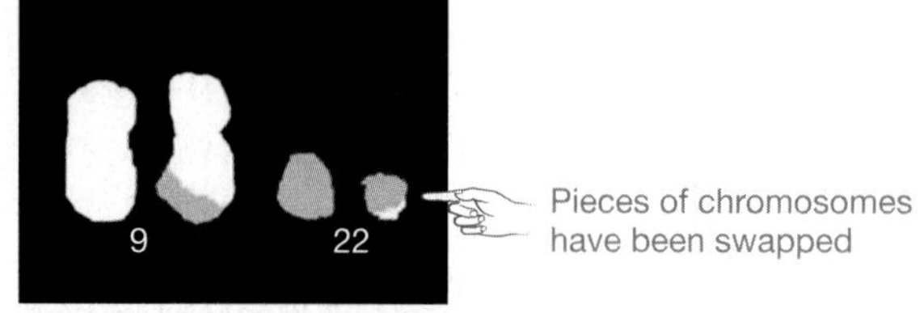

(c) Sex chromosomes in Klinefelter syndrome

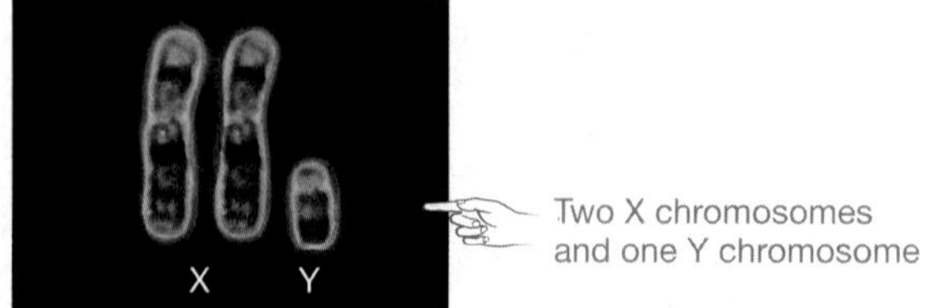

FIGURE 12.6 Human Chromosomes. (a) Unlike the orderly arrangement seen here, condensed chromosomes that are undergoing mitosis are arranged randomly when first observed with the microscope. To determine a karyotype, a technician uses a computer to separate the image of each condensed chromosome, place homologous pairs side by side, and arrange the homologs by number. Note that even though these are replicated chromosomes, the fluorescent labeling is so strong that individual chromatids can't be distinguished. **(b)** In this individual, a piece of chromosome 9 (white) and a piece of chromosome 22 (purple) have been swapped via translocation. **(c)** Individuals with Klinefelter syndrome have two X chromosomes and a Y chromosome.

QUESTION Is the karyotype in part (a) of a male or female?

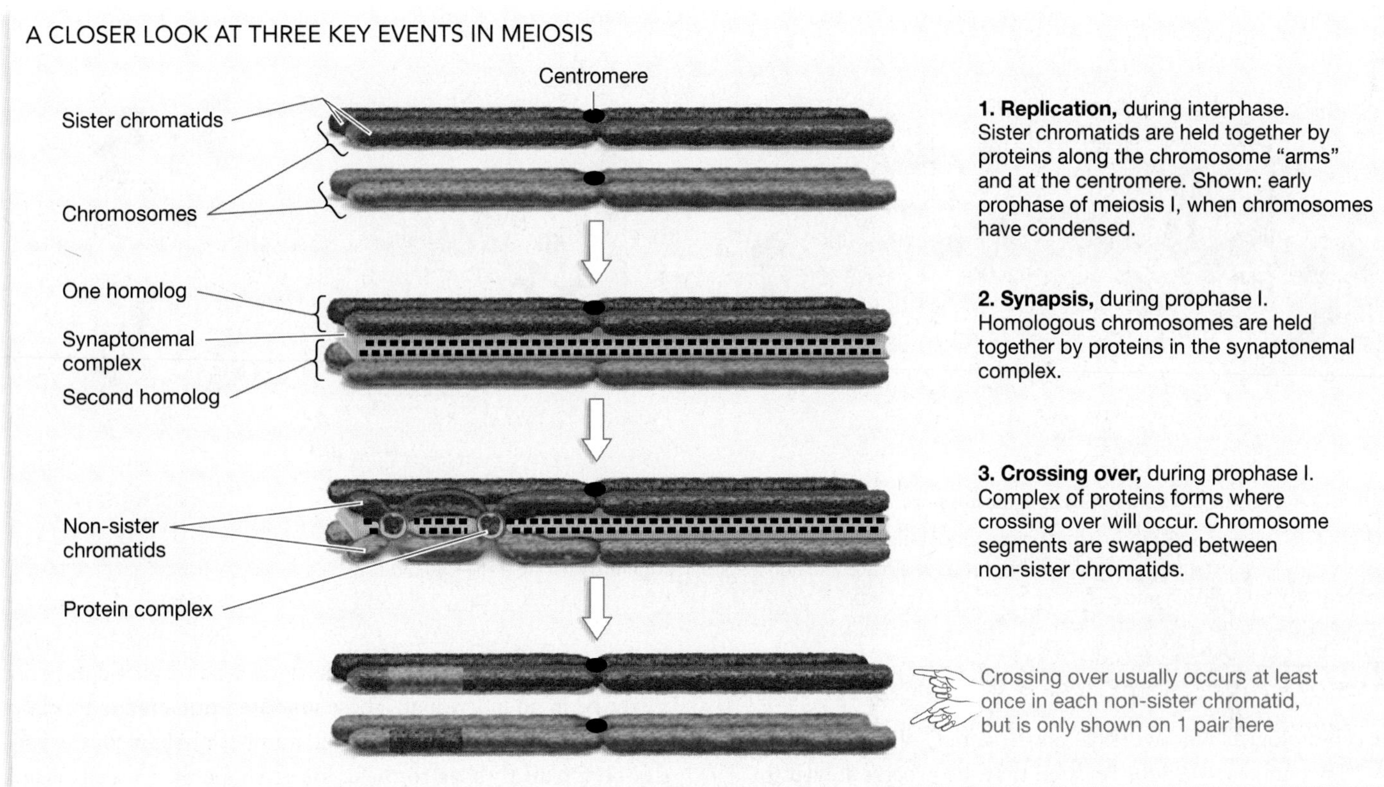

FIGURE 12.7 A Closer Look at Chromosome Replication, Synapsis, and Crossing Over.

chromatids stay very tightly joined along their entire length. When homologs synapse, two pairs of non-sister chromatids are brought close together and are held there by a network of proteins called the **synaptonemal complex** (step 2). Crossing over can occur at many locations along the length of this paired structure. In most species crossing over typically occurs at least once in each pair of homologs; in some species it occurs two or three times. During the process, a complex of proteins cuts the chromosomes and then reattaches segments from homologs. The key to understanding crossing over is to recognize that at each point where it occurs—as in the circles labeled "protein complex" in step 3 of Figure 12.7—the non-sister chromatids from each homolog get physically broken at the same point and *attached to each other*. As a result, segments of maternal and paternal chromosomes are swapped, as shown at the bottom of Figure 12.7. ● If you understand what is happening during crossing over, you should be able to make two chains of paper clips to simulate non-sister chromatids and then simulate a crossing over event between them.

This more detailed look at replication, synapsis, and crossing over should help you interpret the micrographs in **Figure 12.8**, which show grasshopper cells undergoing meiosis. Once you have a solid understanding of how meiosis occurs, you will be ready to consider ultimate explanations for why meiosis and sexual reproduction exist. Meiosis is an intricate, tightly regulated process that involves dozens if not hundreds of different proteins. Given this complexity, it is logical to hypothesize that this type of nuclear division is extremely important. Why?

Meiosis

Check Your Understanding

If you understand that...

- Meiosis is called a reduction division because the total number of chromosomes present is cut in half.

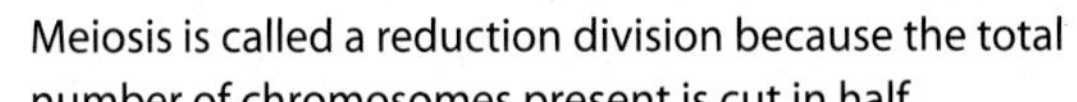

- During meiosis, a single diploid parent cell with replicated chromosomes gives rise to four haploid daughter cells.

You should be able to...

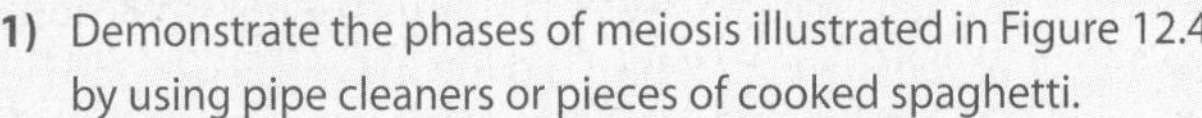

1) Demonstrate the phases of meiosis illustrated in Figure 12.4 by using pipe cleaners or pieces of cooked spaghetti.
2) Identify the event that makes meiosis a reduction division, unlike mitosis, and explain why it is responsible for reduction division.

(a) Prophase of meiosis I

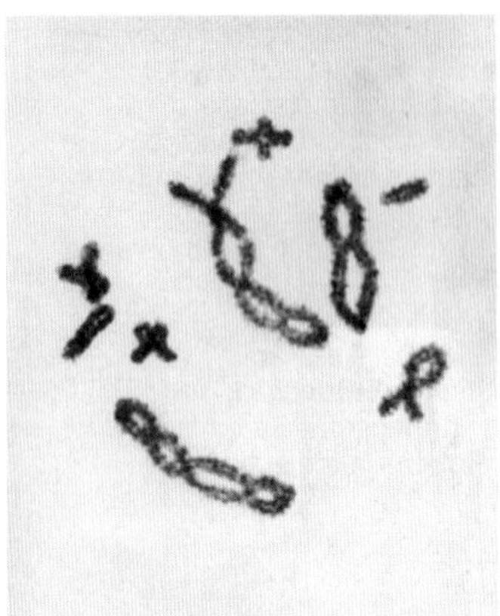

(b) Anaphase of meiosis I

(c) Metaphase of meiosis II

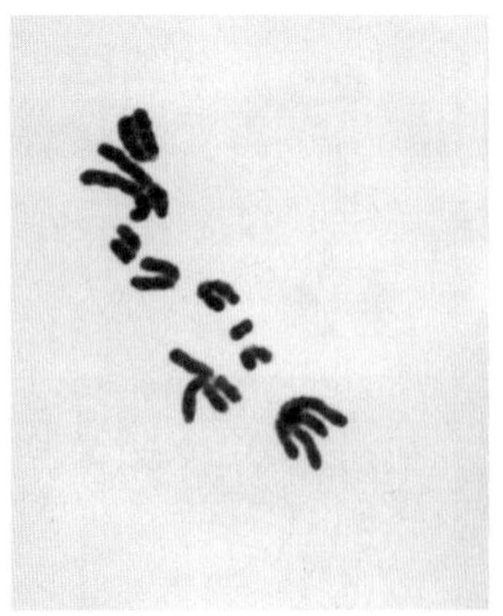

(d) Anaphase of meiosis II

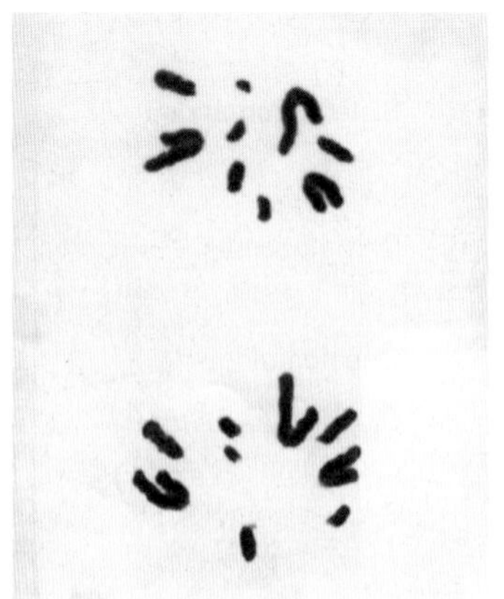

FIGURE 12.8 Meiotic Chromosomes in Grasshoppers.

EXERCISE (a) Label several chiasmata, sister chromatids, tetrads, and non-sister chromatids; (b) add the spindle apparatus; (c) label sister chromatids and the metaphase plate, and add a spindle apparatus; (d) label the haploid number and ploidy of each nucleus.

12.2 The Consequences of Meiosis

The cell biologists who worked out the details of meiosis in the late 1800s and early 1900s realized that the process solved the riddle of fertilization. Weissman's hypothesis—that a reduction division precedes gamete formation in animals—was confirmed. This was an important advance in providing a proximate explanation for sexual reproduction. But researchers also understood that meiosis has another important outcome: Thanks to the independent shuffling of maternal and paternal chromosomes and crossing over during meiosis I, the chromosomes in gametes are different from the chromosomes in parental cells. Subsequently, fertilization brings haploid sets of chromosomes from a mother and father together to form a diploid offspring. The chromosome complement of this offspring is unlike that of either parent. It is a random combination of genetic material from each parent.

This change in chromosomal complement is crucial. The critical observation is that changes in chromosome configuration occur only during sexual reproduction—*not* during asexual reproduction. **Asexual reproduction** refers to any mechanism of producing offspring that does not involve the fusion of gametes. Asexual reproduction in eukaryotes is usually based on mitosis, and the chromosomes in the daughter cells of mitosis are identical to the chromosomes in the parental cell (see Chapter 11). In contrast, **sexual reproduction** refers to the production of offspring through the fusion of gametes. Sexual reproduction results in offspring that have chromosome complements unlike each other and their parents. Why is this difference important?

Chromosomes and Heredity

The changes in chromosomes produced by meiosis and fertilization are significant because chromosomes contain the cell's hereditary material. Stated another way, chromosomes contain the instructions for specifying what a particular trait might be in an individual. These inherited traits range from eye color and height in humans to the number or shape of the bristles on a fruit fly's leg to the color or shape of the seeds found in pea plants.

In the early 1900s biologists began using the term gene to refer to the inherited instructions for a particular trait. Chapter 13 explores the experiments that confirmed that each chromosome is composed of a series of genes encoding information for different traits. In humans, for example, a single chromosome might contain genes that influence height, hair color, the spacing of teeth, and the tendency to develop colon cancer; another chromosome might include genes that affect eye color, susceptibility to allergies, and predisposition to schizophrenia. In most cases there are hundreds or thousands of genes on each chromosome. Recall from Section 12.1 that the term *allele* refers to a particular version of a gene and that homologous chromosomes may carry different alleles. You might have a chromosome containing alleles that tend to produce medium height, black hair color, and a low predisposition to colon cancer, as well as a homologous chromosome containing alleles that contribute to short height, blond hair color, and a predisposition to colon cancer. Other individuals you know might have alleles associated with extreme height, red hair color, and a moderate predisposition to colon cancer.

Chromosomes are composed of genes, and identical copies of chromosomes are distributed to daughter cells during mitosis. Thus, cells that are produced by mitosis are genetically identical to the parent cell, and offspring produced during asexual reproduction are genetically identical to one another as well as to their parent. The offspring of asexual reproduction are **clones**—or exact copies—of their parent. But the offspring produced by sexual reproduction are genetically different from one another and unlike either their mother or their father. Let's analyze three aspects of meiosis that create variation among chromosomes—and hence the genetic makeup—of sexually produced offspring: (1) separation and distribution of homologous chromosomes, (2) crossing over, and (3) fertilization.

How Does the Separation and Distribution of Homologous Chromosomes Produce Genetic Variation?

Each cell in your body contains 23 homologous pairs of chromosomes and 46 chromosomes in total. Half of these chromosomes came from your mother, and half came from your father. Each chromosome is composed of genes that influence particular traits. For example, one gene that affects your eye color might be located on one chromosome, while one of the genes that affects your hair color might be located on a different chromosome (**Figure 12.9a**).

Suppose that the chromosomes you inherited from your mother contain alleles that tend to produce brown eyes and black hair, but the chromosomes you inherited from your father include alleles that tend to specify green eyes and red hair. (This is a simplification for the purpose of explanation. In reality, several genes with various alleles interact in complex ways to produce human eye color and hair color.) Will any particular gamete you produce contain the genetic instructions inherited from your mother or the instructions inherited from your father?

To answer this question, study the diagram of meiosis in **Figure 12.9b**. It shows that when pairs of homologous chromosomes line up during meiosis I and the homologs separate, a variety of combinations of maternal and paternal chromosomes can result. Each daughter cell gets a random assortment of maternal and paternal chromosomes. As Chapter 13 will explain in detail, this phenomenon is known as the principle of independent assortment. In the example given here, meiosis results in gametes with alleles for brown eyes and black hair, like your mother, and green eyes and red hair, like your father. But two additional combinations also occur: brown eyes and red hair, or green eyes and black hair. Four different combinations of paternal and maternal chromosomes are possible when two chromosomes are distributed to daughter cells during meiosis I. ● If you understand how independent assortment produces genetic variation in the daughter cells of meiosis, you should be able to explain how genetic variation would be affected if maternal chromosomes always lined up together on one side of the metaphase plate during meiosis I and paternal chromosomes always lined up on the other side.

How many different combinations of maternal and paternal homologs are possible when more chromosomes are involved? In an organism with three chromosomes per haploid set ($n = 3$) eight types of gametes can be generated by randomly grouping maternal and paternal chromosomes. In general, a diploid organism can produce 2^n combinations of maternal and paternal chromosomes, where n is the haploid chromosome number. This means a human ($n = 23$) can produce 2^{23}, or about 8.4 million, gametes that differ in their combination of maternal and paternal chromosome sets. Clearly, the random assortment of whole chromosomes generates an impressive amount of genetic variation among gametes.

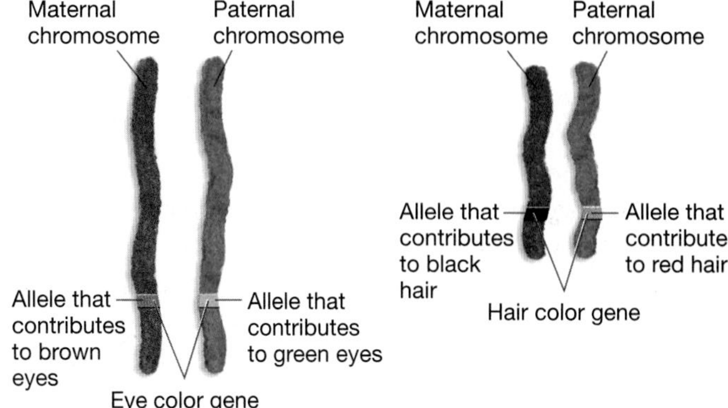

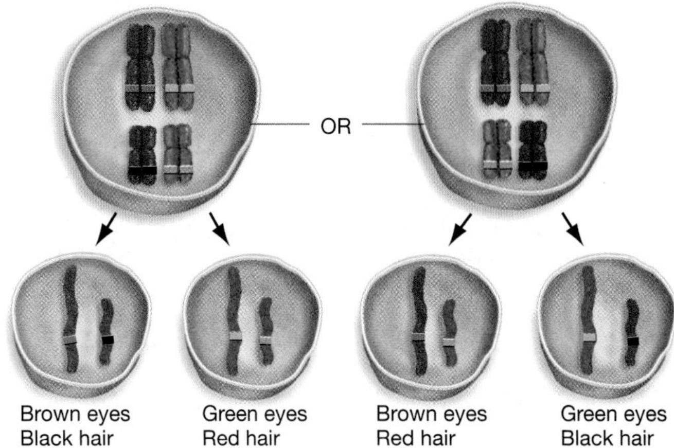

FIGURE 12.9 Separation of Homologous Chromosomes Results in Variable Combinations of Genes. (a) A hypothetical example: Genes that influence eye color and hair color in humans are on different chromosomes. **(b)** The cells along the bottom are products of meiosis. Notice that each cell has a different combination of genes, due to the separation of homologous chromosomes during meiosis I.

The Role of Crossing Over

Recall from Section 12.1 that segments of paternal and maternal chromatids exchange at each chiasma that forms during meiosis I. Thus, crossing over produces new combinations of alleles on the same chromosome—combinations that did not exist in either parent. This phenomenon is known as recombination. **Genetic recombination** is any change in the combination of alleles on a given chromosome. In species that reproduce sexually, recombination occurs via crossing over during meiosis. But genetic recombination also occurs in haploid organisms such as bacteria, which cannot undergo meiosis (see **Box 12.2**).

Crossing over and recombination are important because they dramatically increase the genetic variability of gametes

BOX 12.2 How Does Recombination Occur in Bacteria?

In many eukaryotes, sexual reproduction occurs when haploid cells that are formed by meiosis combine to form a new diploid individual. Sexual reproduction does not exist in bacteria, however. Bacteria are haploid throughout their lives and do not undergo meiosis. Genes can move from one individual to another, however, via a mechanism called conjugation.

Conjugation takes place when a physical connection, called a conjugation tube, forms between two bacterial cells. In many cases conjugation is followed by an event called plasmid transfer: As **Figure 12.10a** shows, a small loop of DNA called a plasmid is copied in one cell and transferred to the other cell. Plasmids were introduced briefly in Chapter 7 and are analyzed in more detail in Chapter 19.

As **Figure 12.10b** shows, however, conjugation can also result in genetic recombination. This occurs when a plasmid that has become integrated into the main bacterial chromosome is copied and transferred, along with genes from the main bacterial chromosome, through the conjugation tube to a recipient cell. In some cases the transferred portion of chromosome replaces the homologous portion in the recipient cell's original chromosome, resulting in a new combination of alleles on that chromosome—meaning, recombination.

The figure highlights two key points about bacterial "sex": (1) It is a one-way transfer of genetic material instead of an exchange between individuals, and (2) instead of involving all of the genes present, the transfer is limited to a plasmid or a small portion of the genes in the main chromosome.

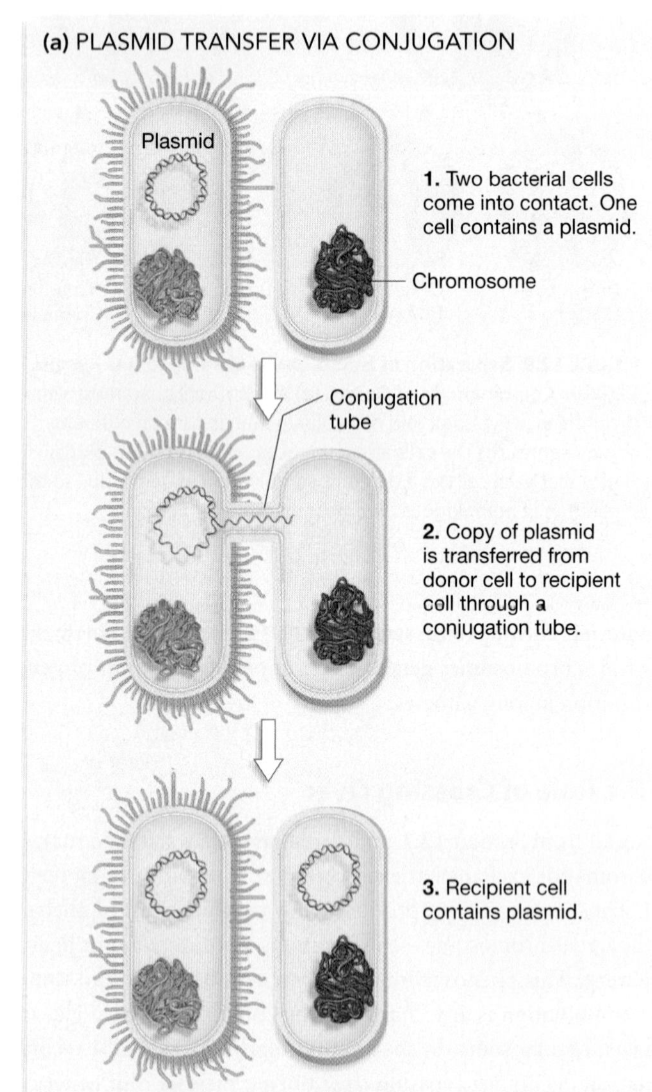

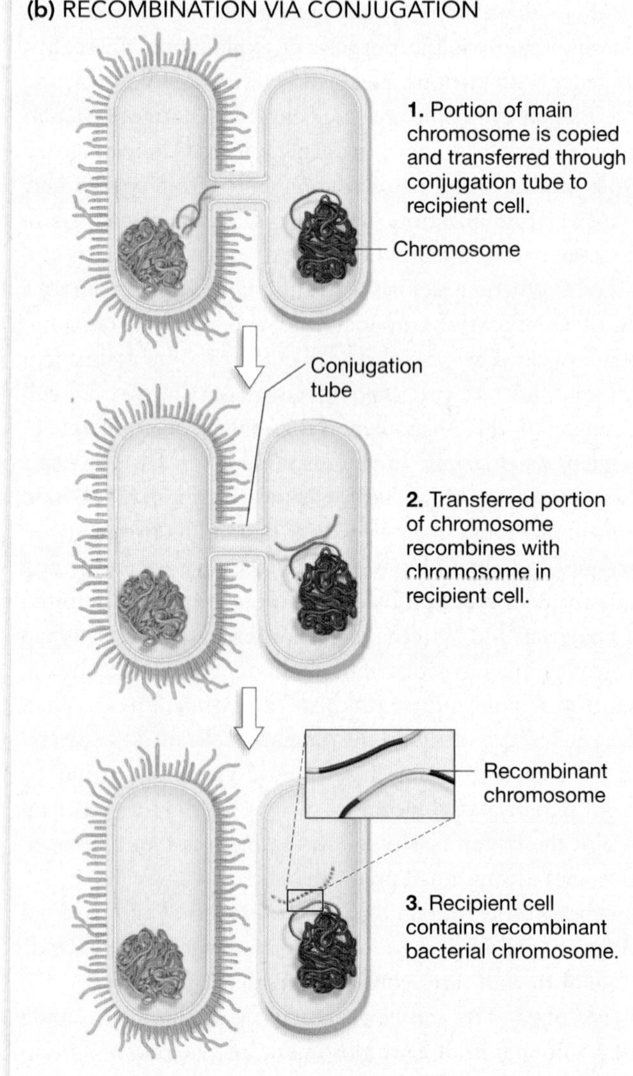

FIGURE 12.10 Bacterial "Sex" Is One-Way and Involves Only a Few Genes.

produced by meiosis. Recall that the separation and distribution of homologous chromosomes during meiosis varies the combination of chromosomes present. But in addition, crossing over varies the combinations of alleles within each chromosome. Recent data on humans, for example, indicate that an average of about 50 chiasmata occur in each cell undergoing meiosis I. As a result, the number of genetically different gametes (PE) that an individual can produce is much more than the 8.4 million produced by the separation and distribution of homologs. When crossing over occurs a total of 50 or more times throughout the entire suite of tetrads, the number of genetically distinct gametes that you can produce is virtually limitless.

Biologists say that meiosis "shuffles" alleles, because each daughter cell gets a random assortment of maternal and paternal chromosomes and because most chromosomes contain both maternal and paternal alleles. As Section 12.3 will show, the genetic variation produced by meiosis has profound consequences for the ability of offspring to survive and reproduce.

How Does Fertilization Affect Genetic Variation?

Crossing over and the random mixing of maternal and paternal chromosomes ensure that each gamete is genetically unique. Even if two gametes produced by the same individual fuse to form a diploid offspring—meaning **self-fertilization**, or "selfing," takes place—the offspring are very likely to be genetically different from the parent (**Figure 12.11**). Selfing is common in some plant species. It also occurs in the many animal species in which single individuals contain both male and female sex organs.

Self-fertilization is rare or nonexistent in many sexually reproducing species, however. Instead, gametes from different individuals combine to form offspring. This is called **outcrossing**. Outcrossing increases the genetic diversity of offspring because it combines chromosomes from different individuals, which are likely to contain different alleles.

How many genetically distinct offspring can be produced when outcrossing occurs? Let's answer this question using humans as an example. Recall that a single human can produce about 8.4 million different gametes—even in the absence of crossing over. When a person mates with a member of the opposite sex, the number of different genetic combinations that can result is equal to the product of the number of different gametes produced by each parent. In humans this means that potentially 8.4 million $\times$ 8.4 million $= 70.6 \times 10^{12}$ genetically distinct offspring can result from any one mating. This number

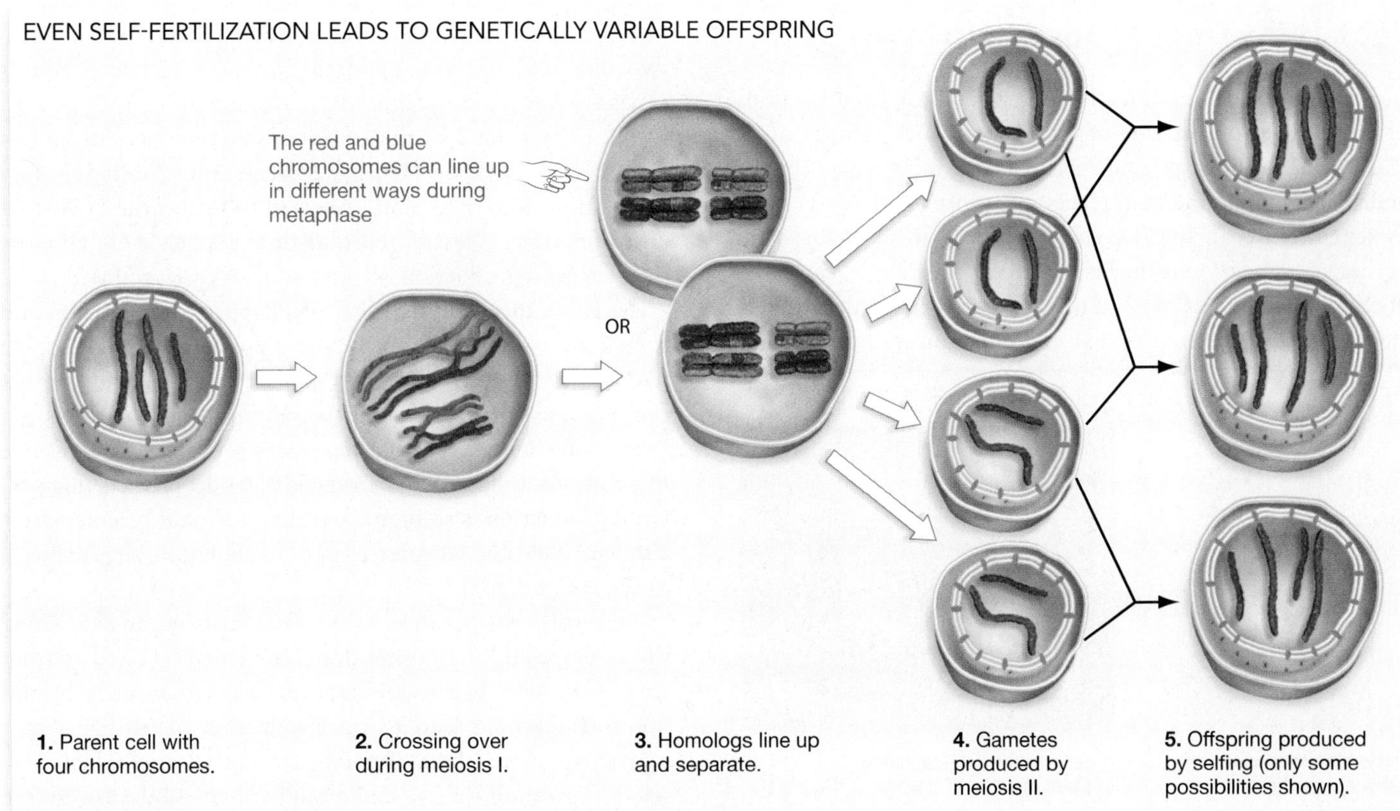

FIGURE 12.11 Even If Self-Fertilization Takes Place, Offspring Are Genetically Variable. Some possible results of self-fertilization in an organism with four chromosomes ($2n = 4$).

EXERCISE In step 5, only a few of the many types of offspring that could be produced are shown. Sketch two additional types that are different from those shown.

is far greater than the total number of people who have ever lived—and the calculation does not even take into account variation generated by crossing over, which occurs at least once along each chromosome. Sexual reproduction results in genetically diverse offspring.

Check Your Understanding

If you understand that...

- The daughter cells produced by meiosis are genetically different from the parent cell because maternal and paternal homologs align randomly at metaphase of meiosis I and because crossing over leads to recombination within chromosomes.

You should be able to...

1) Draw a diploid parent cell with $n = 3$ (three types of chromosomes), and then sketch six of the many genetically distinct types of daughter cells that may result when this parent cell undergoes meiosis.
2) Compare and contrast the degree of genetic variation that results from asexual reproduction, selfing, and outcrossing.

12.3 Why Does Meiosis Exist? Why Sex?

Meiosis and sexual reproduction occur in only a small fraction of the lineages on the tree of life. Bacteria and archaea normally undergo only asexual reproduction; most algae, fungi, and some animals and land plants reproduce asexually as well as sexually. Recall that asexual reproduction in eukaryotes occurs via mitosis. Quaking aspen trees, for example, can produce new individuals by sending up shoots from underground stems (**Figure 12.12**). Asexual reproduction is found even among the vertebrates. For example, several species of guppy in the genus *Poeciliopsis* reproduce exclusively via mitosis.

FIGURE 12.12 Asexual Reproduction in Quaking Aspen. Quaking aspen trees send out underground stems, from which new individuals sprout. Both the underground stems and the new shoots are produced by mitosis. As this process continues, a large group of genetically identical individuals forms. The photograph shows such a group in autumn, when quaking aspen leaves turn bright yellow.

Sexual reproduction is common among multicellular ("many-celled") organisms, however. It is the major mode of reproduction in the insects, which number over 43 million species, as well as in species-rich groups like the molluscs (clams, snails, squid) and vertebrates.

Even though sex plays an important role in the life of many organisms, until recently it was not clear why it occurs. On the basis of theory, biologists had good reason to think that sexual reproduction should not exist. Let's take a closer look.

The Paradox of Sex

In 1978 John Maynard Smith pointed out that the existence of sexual reproduction presents a paradox. Maynard Smith developed a mathematical model showing that because asexually reproducing individuals do not have to produce male offspring, their progeny can produce twice as many grand-offspring as can individuals that reproduce sexually. **Figure 12.13** diagrams this result by showing the number of females (♀) and males (♂) produced over several generations by asexual versus sexual reproduction. In this example, each individual produces four offspring over the course of his or her lifetime. In the asexual population, each individual is a female that produces four offspring. But in the sexual population, it takes two individuals—one male and one female—to produce four offspring. Thus, two out of every four children that each female produces sexually—the males—cannot have children of their own. As a result, generation 2 of the sexual population has just half as many child-producing offspring as generation 2 in the asexual population. Maynard Smith referred to this as the "two-fold cost of males." Asexual reproduction is much more efficient than sexual reproduction because no males are produced.

Based on this analysis, what will happen when asexual and sexual individuals exist in the same population and compete with one another? If all other things are equal, individuals that reproduce asexually should increase in frequency in the population while individuals that reproduce sexually should decline in frequency. In fact, Maynard Smith's model predicts that sexual reproduction is so inefficient that it should be completely eliminated. At the ultimate level of explanation, the existence of sex is a paradox.

To resolve the paradox of sex, biologists began examining the assumption "If all other things are equal." Stated another way, biologists began looking for ways that meiosis and outcrossing could lead to the production of offspring that reproduce more than asexually produced individuals do. After decades of debate and analysis, two solid answers are beginning to emerge.

The Purifying Selection Hypothesis

The first response to the paradox of sex hinges on a simple observation: If a gene is damaged or changed in a way that

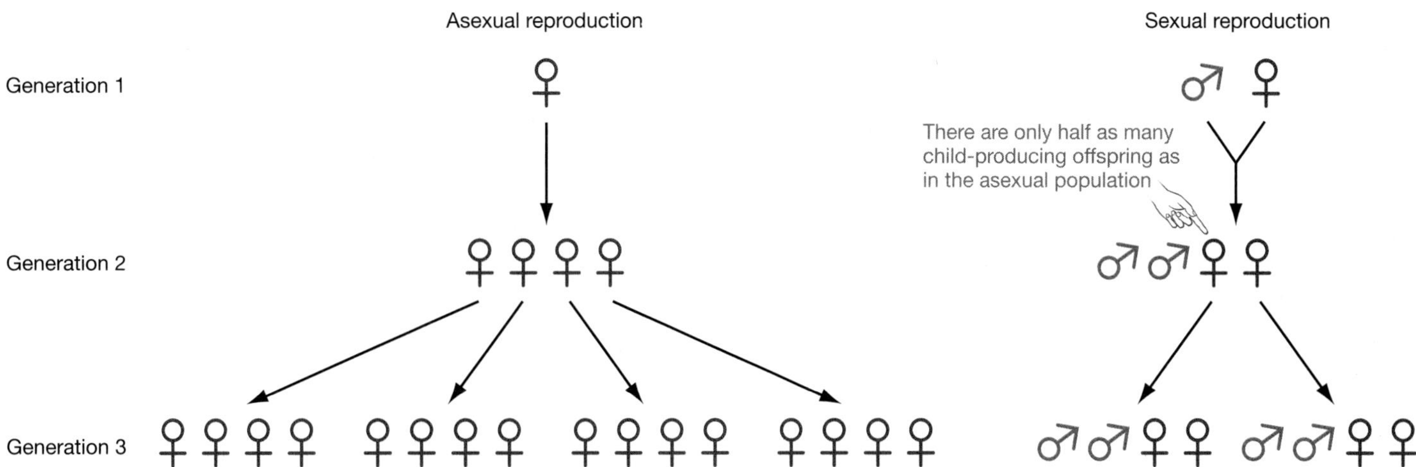

FIGURE 12.13 Asexual Reproduction Confers a Large Numerical Advantage. Each female symbol (♀) and male symbol (♂) represents an individual. In this hypothetical example, every individual produces four offspring over the course of a lifetime, sexually reproducing individuals produce half males and half females, and all offspring survive to breed.

QUESTION How many asexually produced offspring would be present in generation 4? How many sexually produced offspring?

causes it to function poorly, it will be inherited by *all* of that individual's offspring when asexual reproduction occurs. Suppose the damaged gene arose in generation 1 of Figure 12.13. If the damaged gene is important enough, it might cause the four asexual females present in generation 2 to produce fewer than four offspring apiece—perhaps because they die young. If so, then generation 3 will not have twice as many individuals in the asexual lineage compared to the sexual lineage. An allele that functions poorly and lowers the fitness of an individual is said to be deleterious. Asexual individuals are doomed to transmitting all of their deleterious alleles to all of their offspring.

Suppose, however, that the same deleterious allele arose in the sexually reproducing female in generation 1 of Figure 12.13. If the female also has a normal copy of the gene, and if she mates with a male that has normal copies of the gene, then on average half of her offspring will lack the deleterious allele. Sexual individuals are likely to have offspring that lack deleterious alleles present in the parent.

Natural selection against deleterious alleles is called purifying selection. Over time, purifying selection should steadily reduce the numerical advantage of asexual reproduction.

To test this hypothesis, researchers recently compared the same genes in closely related species of *Daphnia*—a common inhabitant of ponds and lakes—that reproduce asexually versus sexually. As predicted, they found that individuals in the asexual species contained many more deleterious alleles than individuals in the sexual species. Results like these have convinced biologists that purifying selection is an important factor limiting the success of asexual reproduction.

The Changing Environment Hypothesis

The second hypothesis to explain sexual reproduction focuses on the benefits of producing genetically diverse offspring. Here's the key idea: If the environment changes from one generation to the next in a way that makes parents poorly adapted, then offspring that are genetically different from their parents and from each other are more likely to survive and produce offspring of their own. Conversely, offspring that are genetic clones of their parents are less likely to thrive if the environment changes.

What type of environmental change would favor genetically diverse offspring? The possibilities include changes in temperature, moisture, predators, competitors, and food sources. But recently researchers have focused on one particular component of environmental change—the emergence of new strains of disease-causing agents.

The genetic characteristics of disease-causing organisms and viruses tend to change very quickly over time. In your own lifetime, for example, several new disease-causing agents have emerged that afflict humans; they include the SARS virus, new strains of HIV, the parasite that causes malaria, and the tuberculosis bacterium. In addition to employing drugs to combat these types of agents, humans have hundreds of genes that are involved in defense. At many of these genes, certain alleles help individuals fight off particular strains of bacteria, parasitic eukaryotes, or viruses. As you might predict, the presence of certain disease-fighting alleles is crucial for plants and animals that cannot rely on drug therapies for help.

What happens if all of the offspring produced by an individual are genetically identical? If a new strain of disease-causing agent evolves, then all of the asexually produced offspring are likely to be susceptible to that new strain. But if the offspring are genetically variable, then it is likely that at least some offspring will have combinations of alleles that enable them to fight off the new disease and produce offspring of their own.

The logic of the hypothesis is sound. Do any data support it?

Testing the Changing-Environment Hypothesis Curtis Lively and colleagues tested the changing-environment hypothesis recently by studying a species of snail that is native to New Zealand. This type of snail lives in ponds and other freshwater habitats and is susceptible to infection by over a dozen species of parasitic trematode worms. Snails that become infected cannot reproduce—the worms eat their reproductive organs. The parasites are rare in some habitats and common in others.

The biologists were interested in working on this snail species because some individuals reproduce only sexually while others reproduce only asexually. If the changing-environment hypothesis for the advantage of sex is correct, then the frequency of sexually reproducing individuals should be much higher in habitats where parasites are common than it is in habitats where parasites are rare (**Figure 12.14**). The logic here is that asexually reproducing individuals should have high fitness in environments where parasites are rare. Because they produce so many offspring in parasite-free environment, alleles that are associated with asexual reproduction should increase in frequency. In contrast, sexually reproducing individuals should have high fitness in habitats where parasites are common. In these environments, alleles that are associated with sexual reproduction should be frequent.

To test these predictions, the researchers collected a large number of individuals from different habitats. Lively and coworkers examined snails in habitats where parasites were more or less common and calculated the frequency of individuals that reproduce sexually versus those that reproduce asexually. The results are plotted in Figure 12.14. The data show that habitats where parasite infection rates are high have a relatively large number of sexually reproducing individuals compared with habitats that have low parasite incidence.

This result and a variety of other studies support the changing-environment hypothesis. Although the paradox of sex remains an active area of research, more biologists are becoming convinced that sexual reproduction is an adaptation that reduces the impact of purifying selection and increases the fitness of individuals in environments where disease-causing organisms are common. It's a way to rid your lineage of bad alleles and make sure your offspring have good alleles.

12.4 Mistakes in Meiosis

When homologous chromosomes separate during meiosis I, a complete set of chromosomes is transmitted to each daughter cell. But what happens if there is a mistake and the chromosomes are not properly distributed? What are the consequences for offspring if gametes contain an abnormal set of chromosomes?

In 1866 Langdon Down described a distinctive suite of co-occurring conditions observed in some humans. The syndrome was characterized by mental retardation, a high risk for heart problems and leukemia, and a degenerative brain disorder

Experiment

Question: Why does sexual reproduction occur?

Hypothesis: In habitats where parasitism is common, sexually produced offspring have higher fitness than do asexually produced individuals.

Null hypothesis: There is no relationship between the presence of parasites and reproduction method.

Experimental setup:

1. Collect snails from a wide array of habitats.

2. Document percentage of males in each population, as an index of frequency of sexual reproduction. More males means that more sexual reproduction is occurring.

3. Note two types of populations: In one, males are common; in the other, males are almost nonexistent. Infer that sexual reproduction is either common or almost nonexistent.

Parasitism rate in this population?

Parasitism rate in this population?

4. Document percentage of individuals infected with parasites in sexually versus asexually reproducing populations.

Prediction: In populations where sexual reproduction is common, parasitism rates are high. In populations with only asexual reproduction, infection rates are low.

Prediction of null hypothesis: No difference in parasitism rate between populations that reproduce sexually versus asexually.

Results:

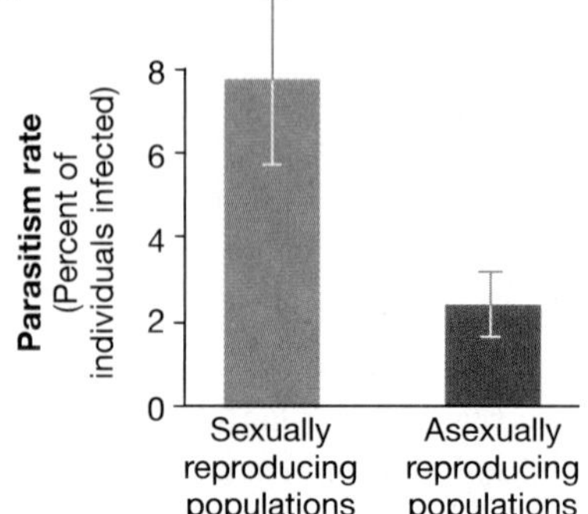

Conclusion: Sexual reproduction is common in habitats where parasitism is common. Asexual reproduction is common in habitats where parasitism is rare.

FIGURE 12.14 Is Sexual Reproduction Favored when Disease or Parasitism Rates Are High?

similar to Alzheimer's disease. **Down syndrome**, as the disorder came to be called, is observed in about 0.15 percent of live births (1 infant in every 666). For over 80 years the cause of the syndrome was unknown. Then in the late 1950s a researcher published observations on the chromosome sets of nine Down syndrome children. The data suggested that the condition is associated with the presence of an extra copy of chromosome 21. This situation is called a **trisomy** ("three-bodies")—in this case, trisomy-21—because each cell has three copies of the chromosome. To explain why trisomy occurred, the biologist proposed that the extra chromosome resulted from a mistake during meiosis in one of the parents.

How Do Mistakes Occur?

For a gamete to get one complete set of chromosomes, two steps in meiosis must be perfectly executed.

1. Each pair of homologous chromosomes must separate from each other during the first meiotic division, so that only one homolog ends up in each daughter cell. If both homologs move to the same pole of the parent cell, the products of meiosis will be abnormal. This sort of meiotic error, illustrated in **Figure 12.15**, is referred to as **nondisjunction**, because the homologs do not separate or disjoin. Notice that two daughter cells have two copies of the same chromosome (blue in Figure 12.15), while the other two lack that chromosome entirely. Gametes that contain an extra chromosome are symbolized as $n + 1$; gametes that lack one chromosome are symbolized as $n - 1$. If an $n + 1$ gamete (PE) is fertilized by a normal n gamete, the resulting zygote will be $2n + 1$. This situation is a trisomy. If the $n - 1$ gamete is fertilized by a normal n gamete, the resulting zygote will be $2n - 1$. This situation is called **monosomy**. Cells that have too many or too few chromosomes are said to be **aneuploid** ("without-form").
2. Sister chromatids must separate from each other and move to opposite poles of the dividing cell during meiosis II. If this step fails, then the resulting daughter cells will be $n + 1$ and $n - 1$. It is relatively rare, though, for abnormal $n + 1$ and $n - 1$ gametes to be produced during nondisjunction in the second meiotic division.

Taken together, these two types of meiotic mistakes occur at a relatively high frequency. In humans, for example, researchers estimate that nondisjunction events occur in as many as 10 percent of meiotic divisions. The types of mistakes vary, but the consequences are almost always severe when defective gametes participate in fertilization. In a recent study of human pregnancies that ended in early embryonic or fetal death, 38 percent of the 119 cases involved atypical chromosome complements that resulted from mistakes in meiosis. Trisomy accounted for 36 percent of the abnormal karyotypes found. It was also common to find the incorrect number of complete chromosome sets, called triploidy ($3n$). Less common were abnormally sized or shaped chromosomes and monosomy ($2n - 1$). ● Based on these data and a large number of other studies, it is clear that mistakes in meiosis are a major cause of spontaneous abortion in humans.

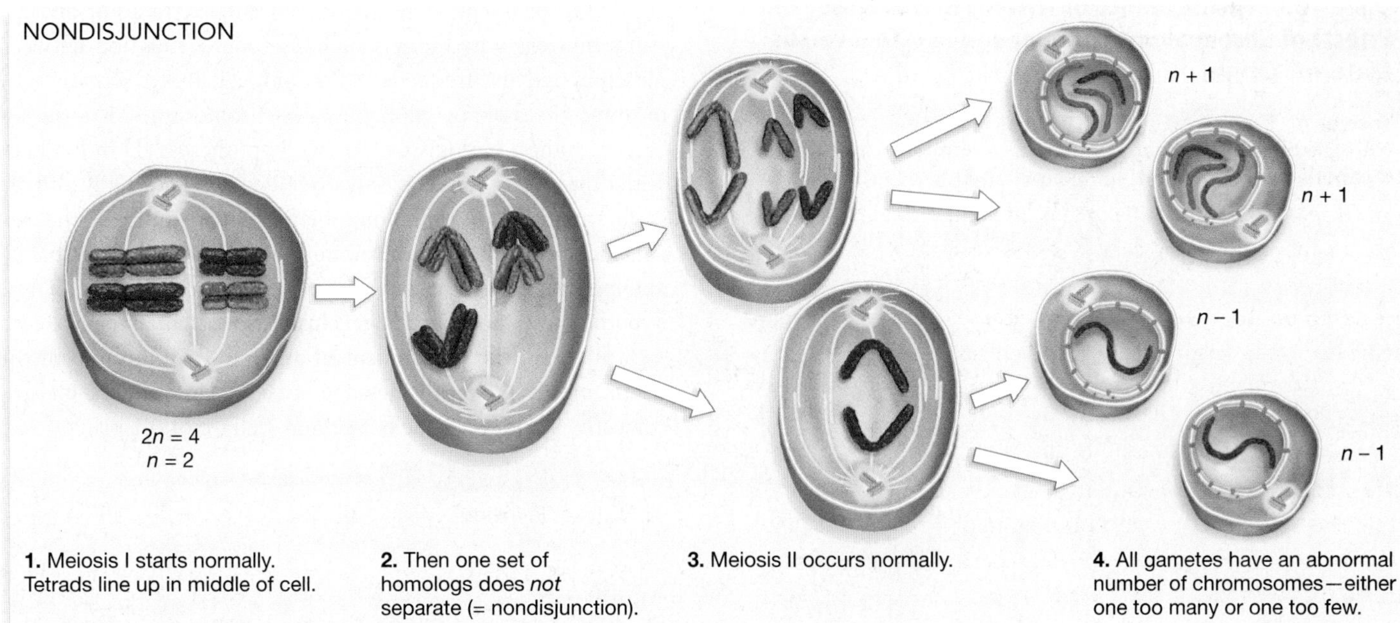

FIGURE 12.15 Nondisjunction Leads to Gametes with Abnormal Chromosome Numbers. If homologous chromosomes fail to separate during meiosis I, the gametes that result will have an extra chromosome or will lack a chromosome.

● **EXERCISE** Nondisjunction also results when meiosis I occurs normally but two of the sister chromatids illustrated in step 3 fail to separate (nondisjunction at meiosis II). Redraw steps 2–4 to show how this mechanism of nondisjunction occurs.

Why Do Mistakes Occur?

The leading hypothesis to explain the incidence of trisomy and other meiotic mistakes is that they are accidents—random errors that occur during meiosis. Consistent with this proposal, there does not seem to be any genetic or inherited predisposition to trisomy or other types of dysfunction. Most cases of Down syndrome, for example, occur in families with no history of the condition.

Even though meiotic errors may be random, there are still strong patterns in their occurrence:

- With the exception of trisomy-21, most of the trisomies and monosomies observed in humans involve the sex chromosomes. Klinefelter syndrome, which develops in XXY individuals—as described in Box 12.1—occurs in about 1 in 1000 live male births. Trisomy X (karyotype XXX) occurs in about 1 in 1000 live births and results in females who may or may not have symptoms such as impaired mental function and sterility. **Turner syndrome** develops in XO individuals, where the "O" stands for lack of a second X, and occurs in about 1 in 5000 live births. Individuals with this syndrome are female but are sterile.
- **Table 12.4** shows data collected on trisomies in the autosomes of human fetuses and infants. Three observations deserve mention: (1) Trisomy is much more common in the smaller chromosomes (numbers 13–22) than it is in the larger chromosomes (numbers 1–12); (2) trisomy-21 is far and away the most common type of trisomy observed; and (3) maternal errors account for most incidences of trisomy. For example, over 90 percent of cases of Down syndrome are due to chromosomal defects in eggs.
- Maternal age is an important factor in the occurrence of trisomy. As **Figure 12.16** shows, the incidence of Down syndrome increases dramatically in mothers over 35 years old.

TABLE 12.4 **The Incidence of Trisomy in Humans: Effects of Chromosome Number and Paternal versus Maternal Origin**

Trisomy (chromosome number)	Total Number of Cases	Due to Error in Sperm	Due to Error in Egg	Maternal Errors (%)
2–12	16	3	13	81
13	7	2	5	71
14	8	2	6	75
15	11	3	8	73
16	62	0	62	100
18	73	3	70	96
21	436	29	407	93
22	11	0	11	100

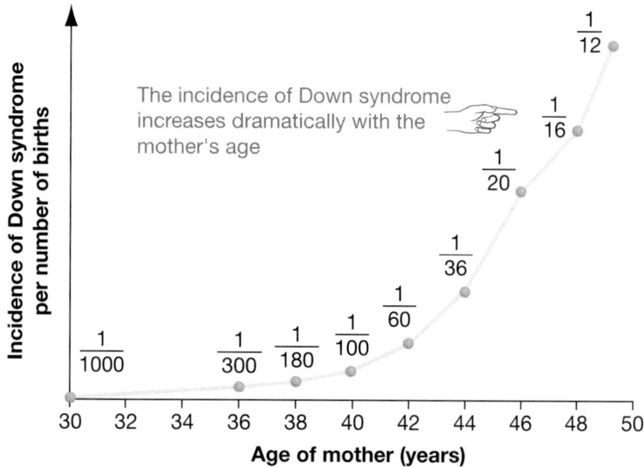

FIGURE 12.16 The Frequency of Down Syndrome Increases as a Function of a Mother's Age.

QUESTION Suppose that you are an obstetrician. Based on these data, at what age would you recommend that pregnant mothers undergo procedures to check the karyotype of the embryos they carry?

Why do these patterns occur? Biologists still do not have a good explanation for why most cases of aneuploidy in humans are due to problems with meiosis in females. In addition, it is still a mystery why there is such a strong correlation between maternal age and frequency of trisomy-21. But to explain why most instances of aneuploidy involve chromosome 21 or the sex chromosomes, biologists offer two hypotheses: (1) Individuals with other types of aneuploidy do not develop normally and are spontaneously aborted, long before birth; and (2) the sex chromosomes and chromosome 21 are more susceptible to aneuploidy than are other types of chromosomes. These hypotheses are not mutually exclusive, so both may be correct.

Studies on the mechanism of aneuploidy continue. In the meantime, one overall message is clear: Undergoing meiosis correctly is critical to the health and welfare of offspring.

MB **Web Animation** at www.masteringbio.com
Mistakes in Meiosis

Chapter Review

SUMMARY OF KEY CONCEPTS

Meiosis is a type of nuclear division. It results in cells that have half as many chromosomes as the parent cell, and in animals it is involved in the formation of eggs and sperm. When an egg and a sperm cell combine to form an offspring, the original number of chromosomes is restored.

When biologists confirmed that sperm and egg nuclei fuse during fertilization, it led to the hypothesis that a special type of cell division must precede gamete formation. Specifically, the proposal was that a sperm and egg must each have half of the normal number of chromosomes found in other cells. This hypothesis was confirmed when researchers observed meiosis and established that it results in gametes with half the normal chromosome number.

As the details of meiosis were being worked out, biologists realized that chromosomes exist in sets. In diploid organisms, individuals have two versions of each type of chromosome. One of the versions is inherited from the mother, and one from the father. The similar, paired chromosomes are called homologs. Haploid organisms, in contrast, have just one of each type of chromosome.

Each chromosome is replicated before meiosis begins. At the start of meiosis I, each chromosome consists of a pair of sister chromatids joined along their length and at a centromere. Homologous pairs of chromosomes synapse early in meiosis I, forming a tetrad—a group of two homologous chromosomes. After non-sister chromatids from the homologous chromosomes undergo crossing over, the pair of homologous chromosomes migrates to the metaphase plate. At the end of meiosis I, the homologous chromosomes separate and are distributed to two daughter cells. During meiosis II, sister chromatids separate and are distributed to two daughter cells.

You should be able to explain why meiosis does not occur in bacteria—most of which have a singular, circular chromosome.

MB Web Animation *BioFlix* at www.masteringbio.com
Meiosis

Each cell produced by meiosis receives a different combination of chromosomes. Because genes are located on chromosomes, each cell produced by meiosis receives a different complement of genes. Meiosis leads to offspring that are genetically distinct from each other and from their parents.

When meiosis and outcrossing occur, the chromosome complements of offspring differ from one another and from their parents, for three reasons: (1) Maternal and paternal homologs are distributed randomly when chromosomes separate at the end of meiosis I; (2) maternal and paternal homologs exchange segments during crossing over; and (3) outcrossing results in a combination of chromosome sets from different individuals. The consequences of these differences became clear when biologists realized that chromosomes contain the hereditary material. Meiosis leads to genetic differences among offspring and between parents and offspring.

You should be able to explain why monozygotic twins, which develop from the same fertilized egg, are much more similar than dizygotic twins, which develop from two different eggs that were fertilized by two different sperm.

The leading hypothesis to explain meiosis is that genetically variable offspring are more likely to thrive in changing environments.

Asexual reproduction is much more efficient than sexual reproduction because males do not have to be produced and no time or energy has to be spent in courtship. Sexual reproduction is favored in many groups, however, because it allows parents to produce offspring that lack deleterious alleles and because genetically diverse offspring are better able to resist parasites than are genetically uniform offspring. Some species alternate between sexual and asexual reproduction over the course of the year.

You should be able to predict whether sexual reproduction occurs during seasons when conditions are stable or seasons when conditions change rapidly.

If mistakes occur during meiosis, the resulting egg and sperm cells may contain the wrong number of chromosomes. It is rare for offspring with an incorrect number of chromosomes to develop normally.

Mistakes during meiosis lead to gametes and offspring with an unbalanced set of chromosomes. Children with Down syndrome, for example, have an extra copy of chromosome 21. The leading hypothesis to explain these mistakes is that they are random accidents resulting in a failure of homologous chromosomes or sister chromatids to separate properly during meiosis.

You should be able to explain what happens when none of the homologous chromosomes present separate at anaphase of meiosis, but their sister chromatids separate normally at meiosis II.

MB Web Animation at www.masteringbio.com
Mistakes in Meiosis

QUESTIONS

Test Your Knowledge

1. In the roundworm *Ascaris*, eggs and sperm have two chromosomes, but all other cells have four. Observations such as this inspired which important hypothesis?
   a. Before gamete formation, a special type of cell division leads to a quartering of chromosome number.
   b. Before gamete formation, a special type of cell division leads to a halving of chromosome number.
   c. After gamete formation, half of the chromosomes are destroyed.
   d. After gamete formation, either the maternal or the paternal set of chromosomes disintegrates.

2. What are homologous chromosomes?
   a. chromosomes that are similar in their size, shape, and gene content
   b. similar chromosomes that are found in different individuals of the same species
   c. the two "threads" in a replicated chromosome (they are identical copies)
   d. the products of crossing over, which contain a combination of segments from maternal chromosomes and segments from paternal chromosomes

3. What is a tetrad?
   a. the "X" that forms when chromatids from homologous chromosomes cross over
   b. a group of four chromatids produced when homologs synapse
   c. the four points where homologous chromosomes touch as they synapse
   d. the group of four genetically identical daughter cells produced by mitosis

4. What is genetic recombination?
   a. the synapsing of homologs during prophase of meiosis I
   b. the new combination of maternal and paternal chromosome segments that results when homologs cross over
   c. the new combinations of chromosome segments that result when outcrossing occurs
   d. the combination of a haploid phase *and* a diploid phase in a life cycle

5. What is meant by a paternal chromosome?
   a. the largest chromosome in a set
   b. a chromosome that does not separate correctly during meiosis I
   c. the member of a homologous pair that was inherited from the mother
   d. the member of a homologous pair that was inherited from the father

6. Meiosis II is similar to which process?
   a. mitosis in haploid cells
   b. nondisjunction
   c. outcrossing
   d. meiosis I

Test Your Knowledge answers: 1. b; 2. a; 3. b; 4. b; 5. d; 6. a

Test Your Understanding

Answers are available at www.masteringbio.com

1. Triploid ($3n$) watermelons are produced by crossing a tetraploid ($4n$) strain with a diploid ($2n$) plant. Briefly explain why this mating produces a triploid individual. Why can mitosis proceed normally in triploid cells, but meiosis cannot?

2. Meiosis is called a reduction division, but all of the reduction occurs during meiosis I—no reduction occurs during meiosis II. Explain why meiosis I is a reduction division but meiosis II is not.

3. Some plant breeders are concerned about the resistance of asexually cultivated plants, such as seedless bananas, to new strains of disease-causing bacteria, viruses, or fungi. Briefly explain their concern by discussing the differences in the genetic "outcomes" of asexual and sexual reproduction.

4. Explain why nondisjunction leads to trisomy and other types of abnormal chromosome complements. Draw an example of how an abnormal chromosome complement is produced during meiosis. In what sense are these chromosome complements "unbalanced?"

5. Explain the relationship between homologous chromosomes and the relationship between sister chromatids.

6. Lay four pens and four pencils on a tabletop, and imagine that they represent replicated chromosomes in a diploid cell with $n = 2$. Explain the phases of meiosis by moving the pens and pencils around. (If you don't have enough pens and pencils, use strips of paper or fabric.)

Applying Concepts to New Situations

Answers are available at www.masteringbio.com

1. The gibbon has 44 chromosomes per diploid set, and the siamang has 50 chromosomes per diploid set. In the 1970s a chance mating between a male gibbon and a female siamang produced an off spring. Predict how many chromosomes were observed in the somatic cells of the offspring. Do you predict that this individual would be able to form viable gametes? Why or why not?

2. Meiosis results in a reassortment of maternal and paternal chromosomes. If $n = 3$ for a given organism, there are eight different combinations of paternal and maternal chromosomes. If no crossing over occurs, what is the probability that a gamete will receive *only* paternal chromosomes?

3. Some researchers predict that spontaneous abortion should be rare in older females, because they are less likely than young females to be able to have offspring in the future. How does this claim relate to Figure 12.16, which graphs mother's age versus the incidence of Down syndrome?

4. The data on snail populations that were used to test the changing-environment hypothesis have been criticized because they are observational and not experimental in nature. As a result, they do not control for factors other than parasites that might affect the frequency of sexually reproducing individuals.
   a. Design an experimental study that would provide stronger evidence that the frequency of parasite infection causes differences in the frequency of sexually versus asexually reproduced individuals in this species of snail.
   b. In defense of the existing data, comment on the value of observing patterns like this in nature, versus under controlled conditions in the laboratory.

www.masteringbio.com is also your resource for • Answers to text, table, and figure caption questions and exercises • Answers to *Check Your Understanding* boxes • Online study guides and quizzes • Additional study tools including the *E-Book for Biological Science* 3rd ed., textbook art, animations, and videos.

Mendel and the Gene

13

Experiments on garden peas and sweet peas (shown here) helped launch the science of genetics.

KEY CONCEPTS

- Mendel discovered that in garden peas, individuals have two alleles, or versions, of each gene. Prior to the formation of eggs and sperm, the two alleles of each gene separate so that one allele is transmitted to each egg or sperm cell.
- Genes are located on chromosomes. The separation of homologous chromosomes during anaphase of meiosis I explains why alleles of the same gene segregate to different gametes.
- If genes are located on different chromosomes, then the alleles of each gene are transmitted to egg cells and sperm cells independently of each other. This occurs because chromosomes line up randomly in metaphase of meiosis I.
- There are important exceptions and extensions to the basic patterns of inheritance that Mendel discovered.

The science of biology is built on a series of great ideas. Two of these—the cell theory and the theory of evolution—were introduced in Chapter 1. The cell theory describes the basic structure of organisms; the theory of evolution by natural selection clarifies why species change through time. These theories explain fundamental features of the natural world and answer some of our most profound questions about the nature of life: What are organisms made of? Where did species come from?

A third great idea in biology addresses an equally important question: Why do offspring resemble their parents? An Austrian monk named Gregor Mendel provided part of the answer in 1865 when he announced that he had worked out the rules of inheritance through a series of experiments on garden peas. The other part of the answer was provided during the final decades of the nineteenth century by the biologists who described the details of meiosis. The chromosome theory of inheritance, formulated in 1903 by Walter Sutton and Theodor Boveri, linked these two results. This theory contends that meiosis, introduced in Chapter 12, causes the patterns of inheritance that Mendel observed. It also asserts that the hereditary factors called genes are located on chromosomes.

This chapter focuses on the evidence for the chromosome theory of inheritance. Let's begin with a basic question: What are the rules of inheritance that Mendel discovered?

13.1 Mendel's Experiments with a Single Trait

Gregor Mendel was a monk who lived and worked in the city of Brünn, located 70 miles north of Vienna. (Brünn was then part of Austria. Today the city is called Brno and is part of the

Czech Republic.) Mendel was educated in the natural sciences at the University of Vienna and also studied physics and mathematics under Christian Doppler, who discovered the Doppler effect for sound and light waves.

In Mendel's day, questions about **heredity**—meaning inheritance, or the transmission of traits from parents to offspring—were primarily the concern of animal breeders and horticulturists. A **trait** is any characteristic of an individual, ranging from overall height to the primary structure of a particular membrane protein. In Brünn, for example, there was a particular interest in how selective breeding could result in hardier and more productive varieties of sheep, fruit trees, and vines. To that end, an Agricultural Society had been formed. Its members emphasized the importance of research that would help breeding programs become more efficient. Mendel was an active member of this society; the monastery he belonged to was also devoted to scientific teaching and research.

What Questions Was Mendel Trying to Answer?

Mendel set out to address the most fundamental of all issues concerning heredity: What are the basic patterns in the transmission of traits from parents to offspring? At the time, two hypotheses had been formulated to answer this question. The first, called blending inheritance, claimed that the traits observed in a mother and father blend together to form the traits observed in their offspring. As a result, an offspring's traits are intermediate between the mother's and father's traits. For example, blending inheritance contended that black sheep have hereditary determinants for black wool and that white sheep have hereditary determinants for white wool. When these individuals mate, their hereditary determinants blend to form a new hereditary determinant for gray wool—their offspring should be gray.

The second hypothesis was called the inheritance of acquired characters, which claimed that traits present in parents are modified, through use, and passed on to their offspring in the modified form. The classical prediction of this hypothesis is that adult giraffes acquire longer necks by straining to reach leaves high in the tops of trees and that they subsequently produce longer-necked offspring. The idea here is that the genetic determinants present in an individual are modified through use, and passed on to offspring in the modified form. Jean-Baptiste Lamarck was a prominent proponent of this hypothesis in the eighteenth century, and it had many adherents until the late 1800s.

These hypotheses were being promoted by the greatest scientists of Mendel's time. But are they correct?

Garden Peas Serve as the First Model Organism in Genetics

Mendel was certainly not the first scientist interested in studying the basic mechanisms of heredity. Why was he successful where others failed? Several factors came into play. One of the most important was that Mendel chose an appropriate model organism to study. A **model organism** for genetics studies is a species where individuals are small, short-lived, inexpensive to care for, able to produce large numbers of offspring, and easy to manipulate experimentally. Such species are called models because the conclusions drawn from studying them turn out to apply to many other species as well.

After investigating and discarding several candidates, Mendel chose to study the pea plant *Pisum sativum*. His reasons were practical: Peas are inexpensive, are easy to grow from seed, have a relatively short reproductive cycle, and produce reasonably large numbers of seeds. These features made it possible for Mendel to continue experiments over several generations and to collect data from a large number of individuals. Because of his choice, garden peas became the first model organism in genetics. **Genetics** is the branch of biology that focuses on the inheritance of traits.

Two additional features of the pea made it possible for Mendel to design his experiments: He could control which parents were involved in a mating, and he could arrange matings between individuals that differed in easily recognizable traits such as flower color or seed shape. Why was this important?

How Did Mendel Arrange Matings? **Figure 13.1a** shows a garden pea flower, including its male and female reproductive organs. Sperm cells are produced in **pollen grains**, which are small sacs that mature in the male reproductive structure of the plant. Eggs are produced in the female reproductive structure. Fertilization begins when pollen grains are deposited on a tubelike section of the female reproductive organs. Sperm cells travel down this tube to the egg cells, where fertilization takes place.

Under normal conditions, garden peas pollinate themselves rather than requiring pollen from other pea plants for fertilization to occur. **Self-fertilization** (or selfing) takes place when pollen from one flower falls on the female reproductive organ of that same flower. Selfing is common because pollen from other plants rarely reaches the flower—its petals form a compartment that encloses the male and female reproductive organs and tends to exclude bees and other types of pollinating insects.

As **Figure 13.1b** shows, however, Mendel could circumvent this arrangement by removing the male reproductive organs from a flower before any pollen formed. Later he could transfer pollen from another flower to that flower's female reproductive organ with a brush. This type of mating is referred to as a **cross-pollination**, or simply a cross. Using this technique, Mendel could control the matings of his model organism.

What Traits Did Mendel Study? Mendel conducted his experiments on varieties of peas that differed in seven traits: seed shape, seed color, pod shape, pod color, flower color, flower and pod position, and stem length. As **Figure 13.2** shows, each trait

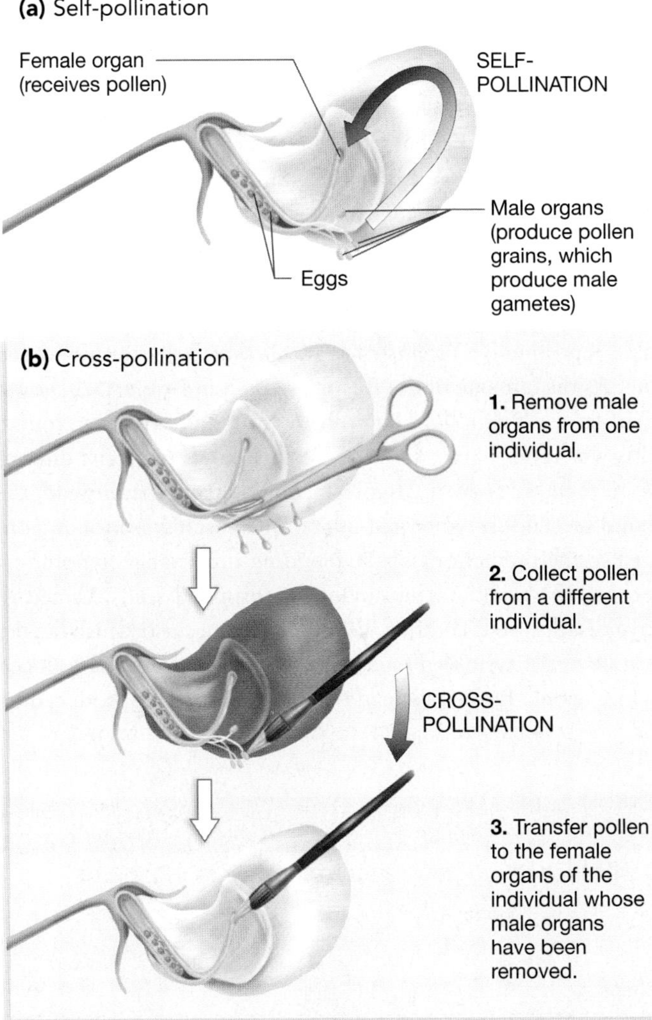

FIGURE 13.1 Peas Can Be Self-Pollinated or Cross-Pollinated. **(a)** The petals of a pea form an enclosed compartment. As a result, most fertilization takes place when pollen grains from the male reproductive organ of one flower fall on the female reproductive organ of the same flower. **(b)** Mendel arranged matings between individuals by removing the male organs from one flower and then using a paintbrush to dust its female organ with pollen collected from a different flower.

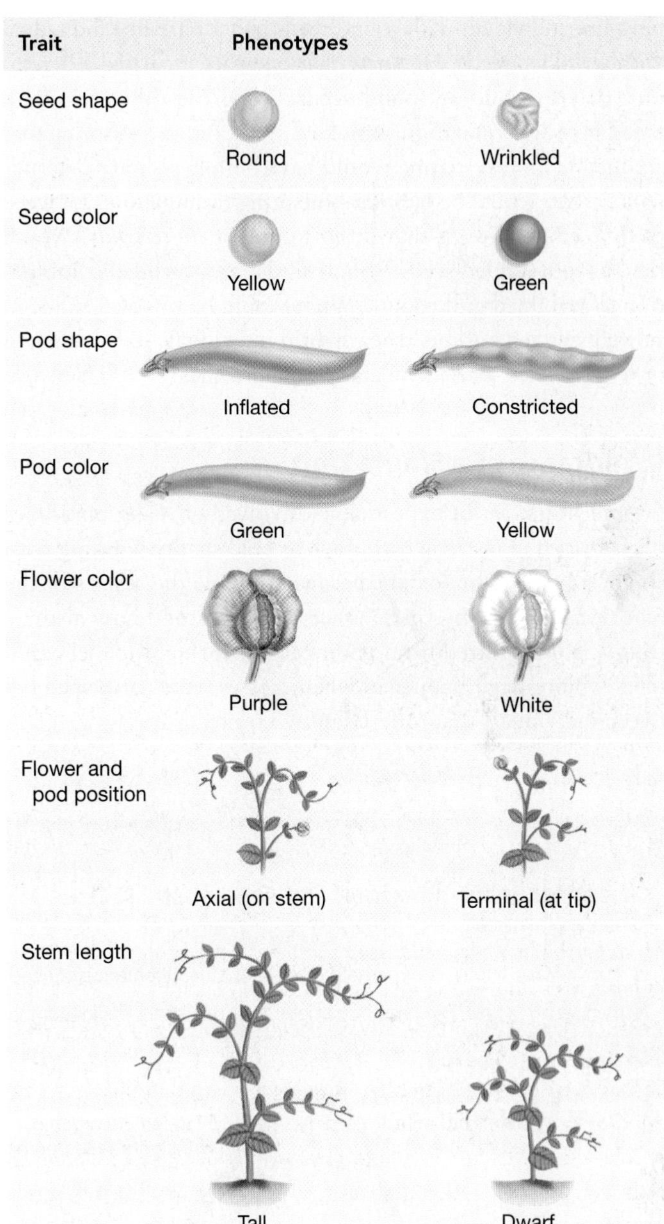

FIGURE 13.2 Mendel Studied Seven Traits That Were Variable in Garden Peas. Two distinct phenotypes existed for each of the seven traits that Mendel studied in garden peas.

exhibited one of two forms. Biologists refer to the observable features of an individual, such as the shape of a pea seed or the eye color of a human, as its **phenotype** (literally, "show-type"). In the pea populations that Mendel studied, two distinct phenotypes existed for all seven traits.

Mendel began his work by obtaining individuals from what breeders called pure lines or true-breeding lines. A **pure line** consists of individuals that produce offspring identical to themselves when they are self-pollinated or crossed to another member of the pure-line population. For example, earlier breeders had developed pure lines for wrinkled seeds and round seeds. During two years of trial experiments, Mendel confirmed that individuals that germinated from his wrinkled seeds produced only wrinkled-seeded offspring when they were mated to themselves or to another pure-line individual that germinated from a wrinkled seed; individuals from his round seeds produced only round-seeded offspring when they were mated to themselves or to another pure-line individual from a round seed.

Why is this result important? Remember that Mendel wanted to find out how traits are transmitted from parents to offspring. Once he had confirmed that he was working with pure lines, he could predict how matings within each line would turn out; in other words, he knew what the offspring from these matings would look like. He could then compare these results with the outcomes of crosses between individuals from different pure lines. For example, suppose he arranged matings between a pure-line individual with round seeds and a pure-line individual with wrinkled seeds. He knew that one parent carried a hereditary determinant for round seeds, while the other carried a hereditary determinant for wrinkled seeds. But the offspring that resulted from this mating would have both hereditary determinants. They would be **hybrids**—offspring from matings between true-breeding parents that differ in one or more traits. Would they have wrinkled seeds, round seeds, or a blended combination of wrinkled and round? What would be the seed shape in subsequent generations when hybrid individuals self-pollinated or were crossed with members of the pure lines?

Inheritance of a Single Trait

Mendel's first set of experiments involved crossing pure lines that differed in just one trait, such as seed shape. Working with single traits was important because it made the results of the matings easier to interpret. Once he understood how a single trait was transmitted from parents to offspring, Mendel could then explore what happened when crosses were performed between individuals that differed in two traits.

Mendel began his single-trait crosses by crossing individuals from round-seeded and wrinkled-seeded pure lines. The adults used in an initial experimental cross such as this represent the **parental generation**. Their progeny (that is, offspring) are called the **F_1 generation**. F_1 stands for "first filial"; the Latin roots *fili* and *filia* mean *son* and *daughter*. Subsequent generations are symbolized as the F_2 generation, F_3 generation, and so on.

Certain Traits "Recede" In his first set of crosses, Mendel took pollen from round-seeded plants and placed it on the female reproductive organs of plants from the wrinkled-seeded line. As the left side of the Results section in **Figure 13.3** shows, all of the seeds produced by progeny from this cross were round. This was remarkable, for two reasons. First, the traits did not blend together to form an intermediate phenotype. Instead, the round-seeded form appeared intact. This result was in stark contrast to the predictions of the blending-inheritance hypothesis. Second, the genetic determinant for wrinkled seeds seemed to have disappeared. Did this disappearance occur because the determinant for wrinkled seeds was located in the egg (produced by the female flower) instead of in the sperm cells produced by

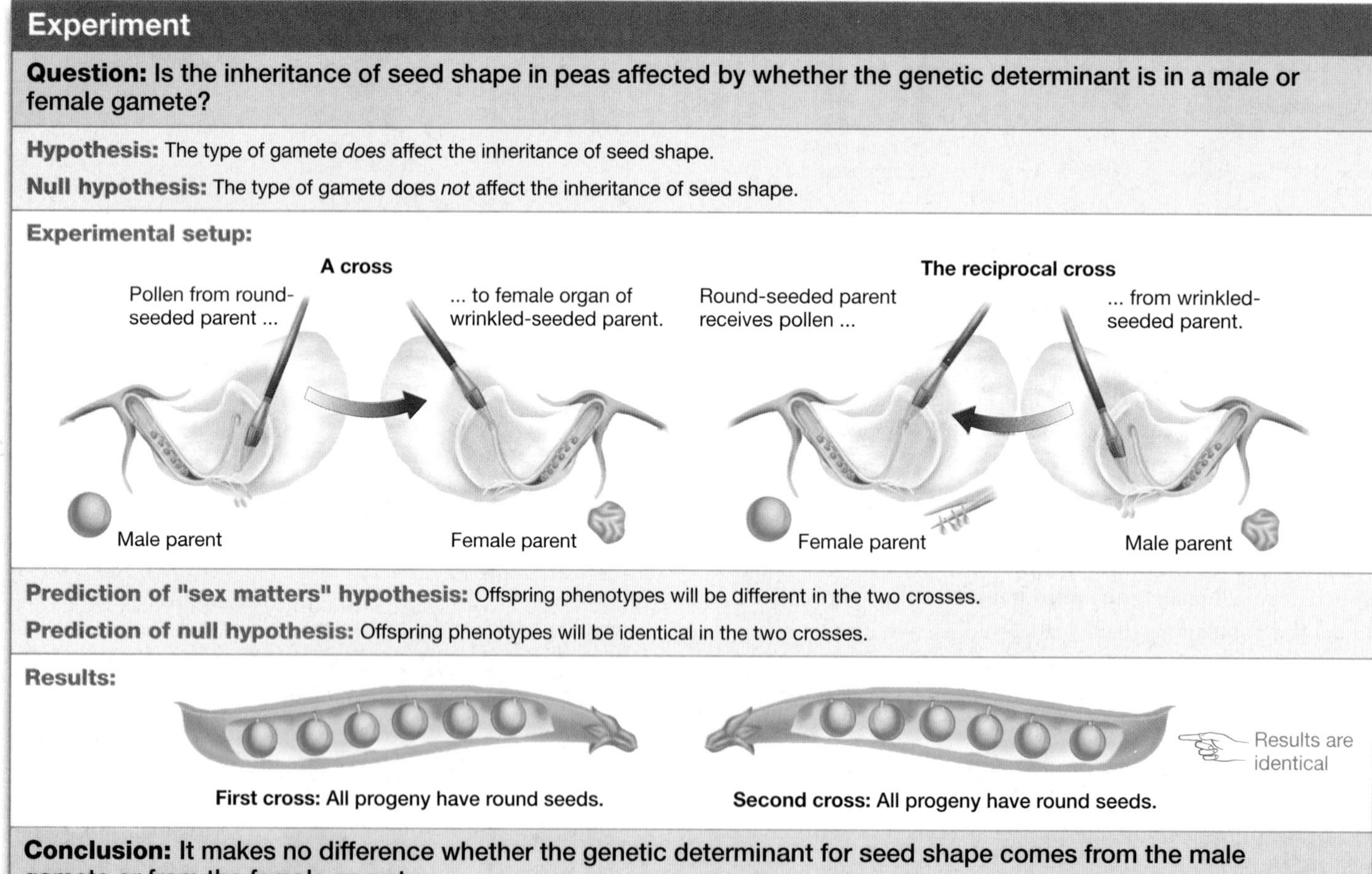

FIGURE 13.3 A Reciprocal Cross.

QUESTION What is the purpose of a reciprocal cross?

pollen from the male flower? In general, did it matter which parent and gamete type had a particular genetic determinant?

To answer these questions, Mendel performed a second set of crosses—this time with pollen taken from an individual germinated from a wrinkled-seeded pea (see Figure 13.3). These crosses completed a **reciprocal cross**—a set of matings where the mother's phenotype in the first cross is the father's phenotype in the second cross, and the father's phenotype in the first cross is the mother's phenotype in the second cross.

In this case the results of the reciprocal crosses were identical: All of the F_1 progeny in the second cross had round seeds, just as in the first cross. This second cross established that it does not matter whether the genetic determinants for seed shape are located in the male or female parent. But a question still remained: What had happened to the genetic determinant for wrinkled seeds?

Dominant and Recessive Traits Mendel planted the F_1 seeds and allowed the individuals to self-pollinate when they matured. He collected the seeds that were produced by many plants in the resulting F_2 generation and observed that 5,474 were round and 1,850 were wrinkled. This observation was striking. The wrinkled seed shape reappeared in the F_2 generation after disappearing completely in the F_1 generation. This was astonishing. No one had observed the phenomenon before, simply because it had been customary for biologists to stop their breeding experiments with F_1 offspring.

Mendel invented some important terms to describe this result. He designated the genetic determinant for the wrinkled shape as **recessive**. This was an appropriate term because none of the F_1 individuals had wrinkled seeds—meaning the determinant for wrinkled seeds appeared to recede or temporarily become latent or hidden. In contrast, Mendel referred to the genetic determinant for round seeds as **dominant**. This term was apt because the round-seed determinant appeared to dominate over the wrinkled-seed determinant when both were present. It's important to note, though, that in genetics the term *dominant* has nothing to do with the everyday English usage as powerful or superior. Subsequent research has shown that individuals with the dominant phenotype do not necessarily have higher fitness than do individuals with the recessive phenotype. Nor are dominant genetic determinants necessarily more common than recessive ones. For example, a rare, dominant allele in humans causes a fatal illness—a type of brain degeneration called Huntington's disease. In genetics, the terms *dominance* and *recessiveness* identify *only* which phenotype is observed in individuals carrying two different genetic determinants. ● If you understand this concept, you should be able to explain why the allele for polydactyly, which causes extra digits to form in humans, is considered dominant while alleles associated with formation of five digits are considered recessive.

Mendel also noticed that the round and wrinkled seeds of the F_2 generation were present in a ratio of 2.96:1, or essentially 3:1. The 3:1 ratio means that for every four individuals, on average three had the dominant phenotype and one had the recessive phenotype. In other words, about 3/4 of the F_2 seeds were round and 1/4 were wrinkled.

Before trying to interpret this pattern, however, it was important for Mendel to establish that the results were not restricted to inheritance of seed shape. So he repeated the experiments with each of the six other traits of pea plants. In each case, he obtained similar results: The products of reciprocal crosses were the same; one form of the trait was always dominant regardless of the parent it came from; the F_1 progeny showed only the dominant trait and did not exhibit an intermediate phenotype; and in the F_2 generation, the ratio of individuals with dominant and recessive phenotypes was about 3 to 1.

How could these patterns be explained? Mendel answered this question with a series of propositions about the nature and behavior of the hereditary determinants. These hypotheses rank as some of the most brilliant insights in the history of biological science.

The Nature and Behavior of the Hereditary Determinants

Mendel's results were clearly inconsistent with the hypothesis of blending inheritance. To explain the patterns that he observed, Mendel proposed a competing hypothesis called particulate inheritance. He maintained that the hereditary determinants for traits do not blend together or acquire new or modified characteristics through use. In fact, hereditary determinants maintain their integrity from generation to generation. Instead of blending together, they act like discrete entities or particles.

Mendel's hypothesis was the only way to explain the observation that phenotypes disappeared in one generation and reappeared intact in the next. It also represented a fundamental break with ideas that had prevailed for hundreds of years.

What Are Genes, Alleles, and Genotypes? Today geneticists use the word **gene** to indicate the hereditary determinant for a trait. For example, the hereditary factor that determines the difference between round and wrinkled seeds in garden peas is referred to as the gene for seed shape.

Mendel's insights were even more penetrating, however. He also proposed that each individual has two versions of each gene. Today different versions of the same gene are called **alleles**. Different alleles are responsible for the variation in the traits that Mendel studied. In the case of the gene for seed shape, one allele of this gene is responsible for the round form of the seed while another allele is responsible for the wrinkled form. The alleles that are found in a particular individual are called its **genotype**. An individual's genotype has a profound effect on its phenotype—its physical traits.

The hypothesis that alleles exist in pairs was important because it gave Mendel a framework for explaining dominance and recessiveness. He proposed that some alleles are dominant and others are recessive. Recall that dominance and recessiveness identify which phenotype actually appears in an individual when both alleles are present. In garden peas, the allele for round seeds is dominant; the allele for wrinkled seeds is recessive. Therefore, as long as one allele for round seeds is present, seeds are round. When both alleles present are for wrinkled seeds (thus no allele for round seeds is present), seeds are wrinkled.

These hypotheses explain why the phenotype for wrinkled seeds disappeared in the F_1 generation and reappeared in the F_2 generation. But why did round- and wrinkled-seeded plants exist in a 3:1 ratio in the F_2 generation?

The Principle of Segregation To explain the 3:1 ratio of phenotypes in F_2 individuals, Mendel reasoned that the two alleles of each gene must segregate—that is, separate—into different gamete cells during the formation of eggs and sperm in the parents. As a result, each gamete contains one allele of each gene. This idea is called the **principle of segregation**.

To show how this principle works, Mendel used a letter to indicate the gene for a particular trait. For example, *R* represents the gene for seed shape. He used uppercase (*R*) to symbolize a dominant allele and lowercase (*r*) to symbolize a recessive allele. (Note that the symbols for genes are always italicized.)

Using this notation, Mendel could describe the genotype of the individuals in the pure line with round seeds (dominant) as *RR*. The genotype of the pure line with wrinkled seeds (recessive) is *rr*. Because *RR* and *rr* individuals have two copies of the same allele, they are said to be **homozygous** for the seed-shape gene (*homo* is the Greek root for "same," while *zygo* means "yoked together"). Pure-line individuals always produce offspring with the same phenotype because they are homozygous—no other allele is present.

Figure 13.4a diagrams what happened to these alleles when Mendel crossed the *RR* and *rr* pure lines. According to his analysis, *RR* parents produce eggs and sperm that carry the *R* allele, while *rr* parents produce gametes with the *r* allele. When two gametes—one from each parent—are fused together, they create offspring with the *Rr* genotype. Such individuals, with two different alleles for the same gene, are said to be **heterozygous** (*hetero* is the Greek root for "different"). Because the *R* allele is dominant, all of these F_1 offspring produced round seeds.

Why do the two phenotypes appear in a 3:1 ratio in the F_2 generation? A mating between parents that are both heterozygous at the gene in question is called a **monohybrid cross**. Mendel proposed that during gamete formation in the F_1 (heterozygous) individuals, the paired *Rr* alleles separate into different gamete cells. As a result, about half of the gametes carry the *R* allele and half carry the *r* allele (**Figure 13.4b**). During self-fertilization, a given sperm has an equal chance of fertilizing either an *R*-bearing egg or an *r*-bearing egg.

(a) A cross between two homozygotes

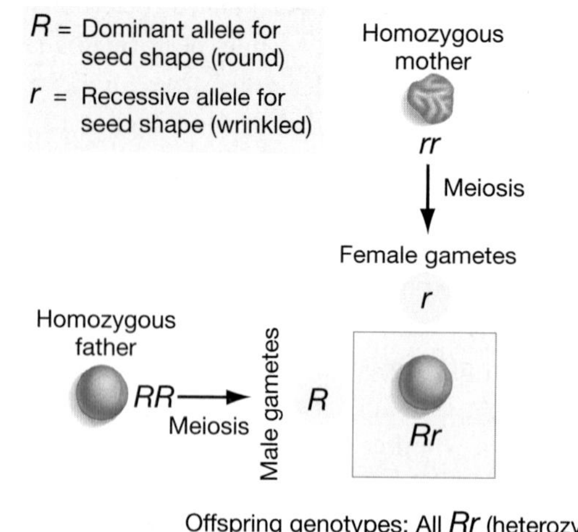

(b) A cross between two heterozygotes

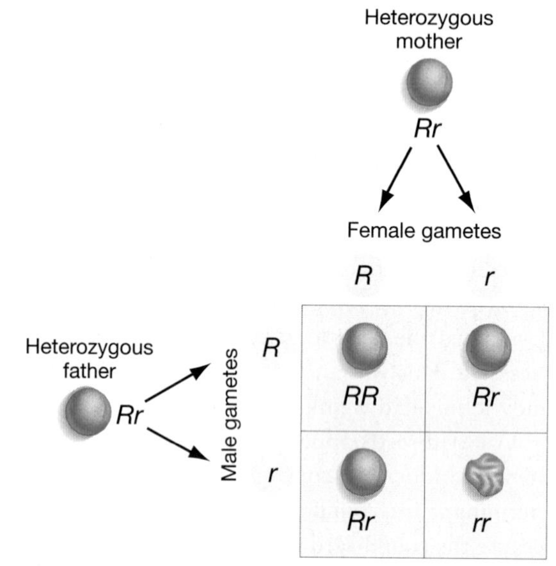

FIGURE 13.4 Mendel Analyzed the F_1 and F_2 Offspring of a Cross between Pure Lines. Notice that when you construct a Punnett square, you only need to list each unique type of gamete at the head of the rows and columns. For example, even though the *RR* alleles segregate in the male parent of part (a), you only have to list the *R* allele in the square—not *R* and *R*.

QUESTION In constructing a Punnett square, does it matter whether the male or female gametes go on the left or across the top? Why or why not?

Years after Mendel published his work, R. C. Punnett invented a straightforward technique for predicting the genotypes and phenotypes that should appear in the resulting offspring. A **Punnett square** is based on the following steps:

1. Write each unique gamete genotype produced by one parent in a row.
2. Write each unique gamete genotype produced by the other parent in a column.
3. Create boxes in the square or rectangle composed of the rows and columns.
4. Fill in the boxes with the offspring genotypes that result from fusion of the gamete genotypes in the appropriate row and column.
5. Predict the proportions of each offspring genotype and phenotype by tallying the offspring genotypes and phenotypes present in the boxes. **BioSkills 9** explains why this tallying process works.

● If you understand these concepts, you should be able to state the purpose of a Punnett square and explain why it is not necessary to create a 2 × 2 square in Figure 13.4a, with the two *r* alleles that segregated from the *rr* mother as columns and the two *R* alleles that segregated from the *RR* father as rows.

As an example of the concluding step in analyzing a cross, the Punnett square in Figure 13.4b predicts that 1/4 of the F_2 offspring will be *RR*, 1/2 will be *Rr*, and 1/4 will be *rr*. Because the *R* allele is dominant to the *r* allele, 3/4 of the offspring should be round seeded and 1/4 should be wrinkled seeded. These results are *exactly* what Mendel found in his experiments with peas. In the simplest and most elegant fashion possible, his interpretation explains the 3:1 ratio of round to wrinkled seeds observed in the F_2 offspring and the mysterious reappearance of the wrinkled seeds.

The term **genetic model** refers to a set of hypotheses that explains how a particular trait is inherited. **Table 13.1** summarizes Mendel's model for explaining the basic patterns in the transmission of traits from parents to offspring. These hypotheses are sometimes referred to as Mendel's rules. His genetic model was a radical break from the hypotheses of blending inheritance and inheritance of acquired characters that previously dominated scientific thinking about heredity.

Testing the Model

Mendel's model explained his results in a logical way. But is it correct? To answer this question, Mendel conducted a series of experiments with the F_2 progeny described in Figure 13.4b. These experiments tested two important predictions:

1. Plants with the recessive phenotype—wrinkled seeds—are *rr*. Thus, they should produce only *rr* offspring when they are self-pollinated or crossed with another individual with wrinkled seeds.
2. Plants with the dominant phenotype (round seeds) may be either *Rr* or *RR*. These two genotypes should be present in the ratio 2:1. (That is, there should be twice as many heterozygotes as homozygotes among individuals with round seeds.) Individuals with the *RR* genotype should produce only *RR* offspring when they are self-pollinated. In contrast, *Rr* individuals should produce offspring with the same 3:1 ratio of round:wrinkled phenotypes observed in the cross diagrammed in Figure 13.4b.

Mendel planted the F_2 seeds and allowed the plants to self-pollinate when they matured. He then examined the phenotypes of the F_3 seeds. He confirmed the first prediction: F_2 plants with wrinkled seeds always produced offspring with wrinkled seeds. This result was consistent with the claim that these F_2 plants have an *rr* genotype.

What about the offspring of parents with the dominant phenotype? Mendel let 565 round-seeded plants self-pollinate. Of these, 193 plants produced only round-seeded offspring. Mendel inferred that these parents had the *RR* genotype. In contrast, 372 of the round-seeded parents produced seeds that

SUMMARY TABLE **13.1 Mendel's Model to Explain the Results of a Cross between Pure Lines**

Mendel's Claim	Notes
1. Peas have two versions, or alleles, of each gene.	This also turns out to be true for many other organisms.
2. Alleles do not blend together.	The hereditary determinants maintain their integrity from generation to generation.
3. Each gamete contains one allele of each gene.	This is due to the principle of segregation—the alleles of each gene segregate during the formation of gametes.
4. Males and females contribute equally to the genotype of their offspring.	When gametes fuse, offspring acquire a total of two alleles for each gene—one from each parent.
5. Some alleles are dominant to other alleles.	When a dominant allele and a recessive allele for the same gene are found in the same individual, that individual has the dominant phenotype.

● **QUESTION** What is the difference between genes and alleles, a genotype and a phenotype, a homozygous individual and a heterozygous individual, and dominant alleles and recessive alleles?

were either round or wrinkled. Mendel inferred that in these individuals, the round-seeded parental genotype was *Rr*. In this experiment, the ratio of *Rr* to *RR* in the parents—based on the number of round and wrinkled seeds they produced—was 372:193 or 1.93:1. This is extremely close to the prediction of 2:1. Mendel observed the same patterns when he let F_2 individuals from the other six crosses (involving seed color and the other traits) self-fertilize. These results furnished overwhelming support for his model.

13.2 Mendel's Experiments with Two Traits

Working with one trait at a time allowed Mendel to establish that blending inheritance does not occur. It also allowed him to infer that each gene he was working with had two alleles and to recognize the principle of segregation. His next step was to extend these results. The most important question he addressed was whether the principle of segregation holds true if parental lines differ with respect to two traits.

To explore this issue, Mendel crossed a pure-line parent that produced round, yellow seeds with a pure-line parent that produced wrinkled, green seeds. The F_1 offspring of this cross should be heterozygous at both genes. A mating between parents that are both heterozygous for two traits is called a **dihybrid cross**.

Mendel's earlier experiments had established that the allele for yellow seeds was dominant to the allele for green seeds; these alleles were designated *Y* for yellow and *y* for green. As **Figure 13.5** indicates, two distinct possibilities existed for how the alleles of these two different genes—the gene for seed shape and the gene for seed color—would be transmitted to offspring. The first possibility was that the allele for seed shape and the allele for seed color present in each parent would separate from one another and be transmitted independently. This hypothesis is called independent assortment, because the two alleles would separate and sort themselves into gametes independently of each other (Figure 13.5a). The second possibility was that the allele for seed shape and the allele for seed color would be transmitted to gametes together. This hypothesis can be called dependent assortment, because the transmission of one allele would depend on the transmission of another (Figure 13.5b).

As Figure 13.5 shows, the F_1 offspring of Mendel's mating are expected to have the dominant round and yellow phenotypes whether the alleles were transmitted together or independently. When Mendel did the cross and observed the F_1 individuals, this is exactly what he found. All of the F_1 offspring had round, yellow seeds. All of these individuals were heterozygous at both genes.

In contrast to the situation in the F_1 generation, however, the two hypotheses make radically different predictions about what Mendel should have observed when the F_1 individuals were allowed to self-fertilize and produce an F_2 generation. If the alleles assort independently and combine randomly to form gametes, then each heterozygous parent should produce four different gamete genotypes, as illustrated in Figure 13.5a. A 4-row-by-4-column Punnett square results, and it predicts that there should be 9 different offspring genotypes and 4 phenotypes. Further, the yellow-round, green-round, yellow-wrinkled, and green-wrinkled phenotypes should be present in the frequencies 9/16, 3/16, 3/16, and 1/16, respectively. This is equivalent to a ratio of 9:3:3:1. But if the alleles from each parent stay together, then a 2-row-by-2-column Punnett square would predict only three possible offspring genotypes and just two phenotypes, as Figure 13.5b shows. The hypothesis of dependent assortment predicts that F_2 offspring should be yellow-round or green-wrinkled, present in a ratio of 3:1. Note that the Punnett squares are making explicit predictions about the outcome of an experiment, based on a specific hypothesis about which alleles are present in each parent and how they are transmitted.

When Mendel examined the phenotypes of the F_2 offspring, he found that they conformed to the predictions of the hypothesis of independent assortment. Four phenotypes were present in frequencies that closely approximated the predicted frequencies of 9/16, 3/16, 3/16, and 1/16 and the predicted ratio of 9:3:3:1 (Figure 13.5c). Based on these data, he accepted the hypothesis that alleles of different genes are transmitted independently of one another. This result became known as the **principle of independent assortment.**

● If you understand the principle of independent assortment, it should make sense to you that an individual with the genotype *AaBb* produces gametes with the genotypes *AB*, *Ab*, *aB*, and *ab*. You should be able to predict the genotypes of the gametes produced by individuals with the genotypes *AABb*, *PpRr*, and *AaPpRr*.

Using a Testcross to Confirm Predictions

Mendel did experiments with combinations of traits other than seed shape and color and obtained results similar to those in Figure 13.5c. Each paired set of traits produced a 9:3:3:1 ratio of progeny phenotypes in the F_2 generation. He even did a limited set of crosses examining three traits at a time. Although all of these data were consistent with the principle of independent assortment, his most powerful support for the hypothesis came from a different type of experiment.

In designing this study, Mendel's goal was to test the prediction that an *RrYy* plant produces four different types of gametes in equal proportions. To accomplish this, Mendel invented a technique called a testcross. A **testcross** uses a parent that contributes only recessive alleles to its offspring to help determine the unknown genotype of the second parent. Testcrosses are useful because the genetic contribution of the homozygous

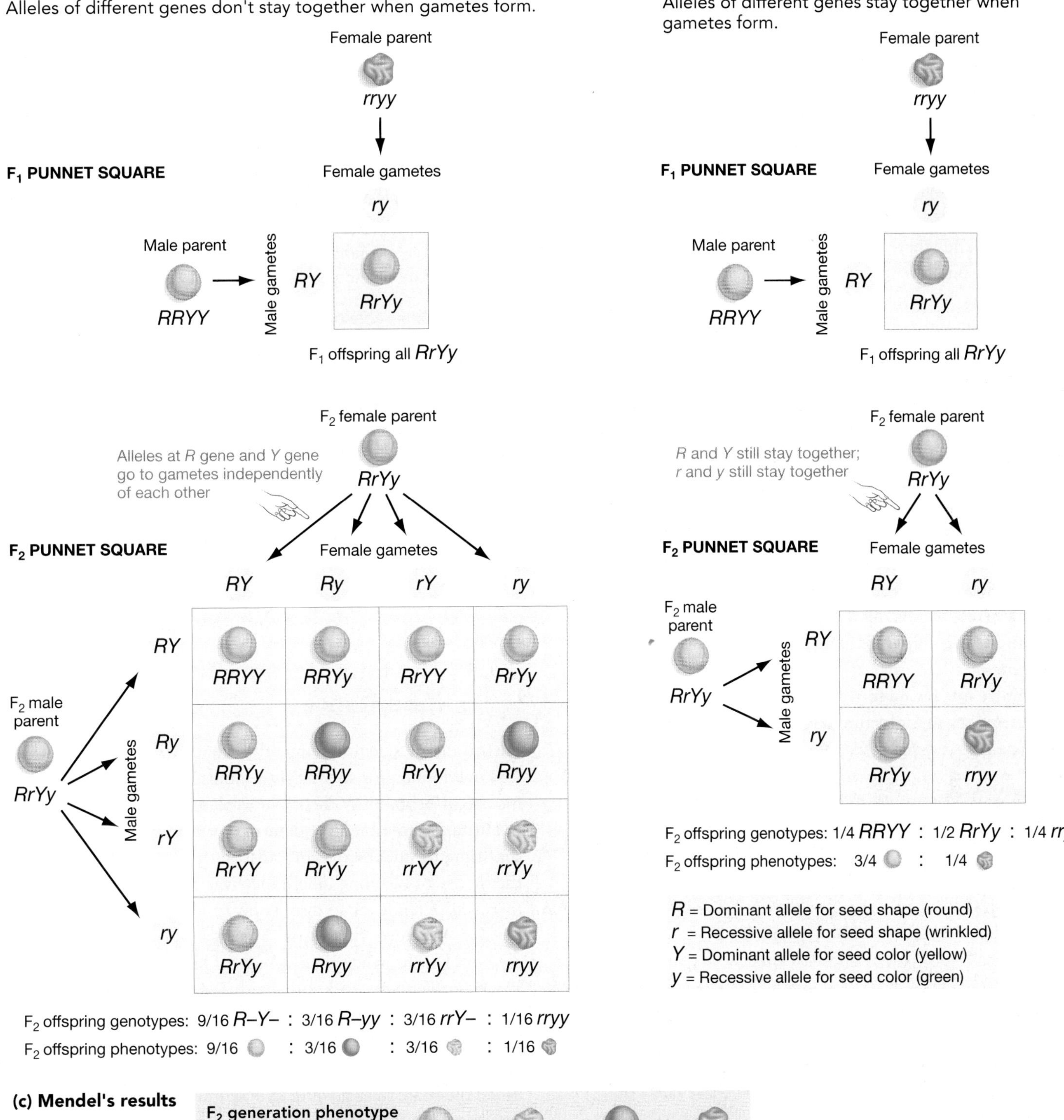

(c) Mendel's results

F_2 generation phenotype					
Number	315	101	108	32	556 total
Fraction of offspring	9/16	3/16	3/16	1/16	These data are consistent with the predictions of independent assortment

FIGURE 13.5 Mendel Analyzed the F_1 and F_2 Offspring of a Cross between Pure Lines for Two Traits. Either of two events could occur when alleles of different genes are transmitted to offspring: Alleles could be sorted into gametes independently of each other, or alleles from the same parent could be transmitted together, generation after generation.

recessive parent is easy to predict and analyze. As a result, a testcross allows experimenters to test the genetic contribution of the other parent. If the other parent has the dominant phenotype but an unknown genotype, the results of the testcross allow researchers to infer whether that parent is homozygous or heterozygous for the dominant allele.

In this case, Mendel performed a testcross between parents that were *RrYy* and *rryy*. The types and proportions of offspring that should result can be predicted with the Punnett square shown in **Figure 13.6**. If the principle of independent assortment is valid, there should be four types of offspring in equal proportions.

What were the actual proportions observed? Mendel did this experiment and examined the seeds produced by the progeny. He found that 31 were round and yellow, 26 were round and green, 27 were wrinkled and yellow, and 26 were wrinkled and green. As predicted, these numbers are nearly identical to the 27.5 individuals expected with each genotype, given the total of 110 individuals. The predicted ratio of phenotypes was 1:1:1:1, which matched the observed ratio. The testcross had confirmed the principle of independent assortment.

Mendel's work provided a powerful conceptual framework for thinking about transmission genetics—the patterns that occur as alleles pass from one generation to the next. This framework was based on (1) the segregation of discrete, paired alleles into separate gametes and (2) the independent assortment of alleles that affect different traits. The experiments you've just reviewed were brilliant in design, execution, and interpretation. Unfortunately, they were ignored for 34 years.

(MB) **Web Animation** at www.masteringbio.com
Mendel's Experiments
The Principle of Independent Assortment

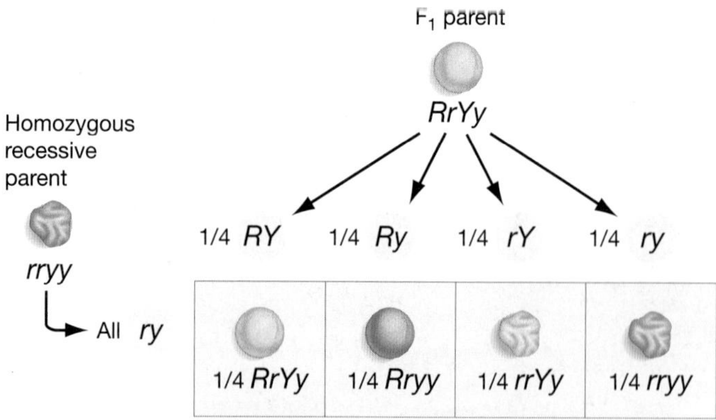

FIGURE 13.6 The Predictions Made by the Principle of Independent Assortment Can Be Evaluated in a Testcross. If the principle of independent assortment is correct and *RrYy* parents produce four types of gametes in equal proportions, then a mating between *RrYy* and *rryy* parents should produce four types of offspring in equal proportions, as this Punnett square shows.

Check Your Understanding

If you understand that...

- Mendel discovered that individuals have two alleles of each gene, and that each gamete receives one of the two alleles present in a parent. This is the principle of segregation.
- Mendel found that alleles from different genes are transmitted to gametes independently of each other. This is the principle of independent assortment.
- The alleles that Mendel analyzed were either dominant or recessive, meaning heterozygous individuals had the dominant phenotype.

You should be able to...

Use the genetic problems at the end this chapter to practice the following skills:

1) Starting with parents of known genotypes, create and analyze Punnett squares to predict the genotypes and phenotypes that will occur in their F_1 and F_2 offspring, and then calculate the expected frequency of each genotype and phenotype (Genetics Problems 12, 17, 19, 21).
2) Given the outcome of a cross, infer the genotypes and phenotypes of the parents (Genetics Problems 13, 15, 23, 24, 25).

13.3 The Chromosome Theory of Inheritance

Historians of science frequently debate why Mendel's work was overlooked for so long. It is almost undoubtedly true that his use of probability theory in explaining his results and his quantitative treatment of data were difficult for biologists of that time to understand and absorb. It may also be true that the theory of blending inheritance was so well entrenched that there was a tendency to dismiss his results as peculiar or unbelievable. Whatever the reason, Mendel's work was not appreciated until 1900, when three groups of biologists, working with a variety of plants and animals, independently "discovered" Mendel's main conclusions.

The rediscovery of Mendel's work, 16 years after his death, ignited the young field of genetics. Mendel's experiments established the basic rules governing how traits are passed from parents to offspring. They described the pattern of inheritance. But what process is responsible for these patterns? Two biologists, working independently, came up with the answer. Walter Sutton and Theodor Boveri each realized that meiosis could be responsible for Mendel's rules. When this hypothesis was published in 1903, research in genetics exploded.

Recall from Chapter 12 that meiosis is the type of cell division that precedes gamete formation. The details of the process were worked out in the final decades of the nineteenth century. What Sutton and Boveri grasped is that meiosis not only reduces

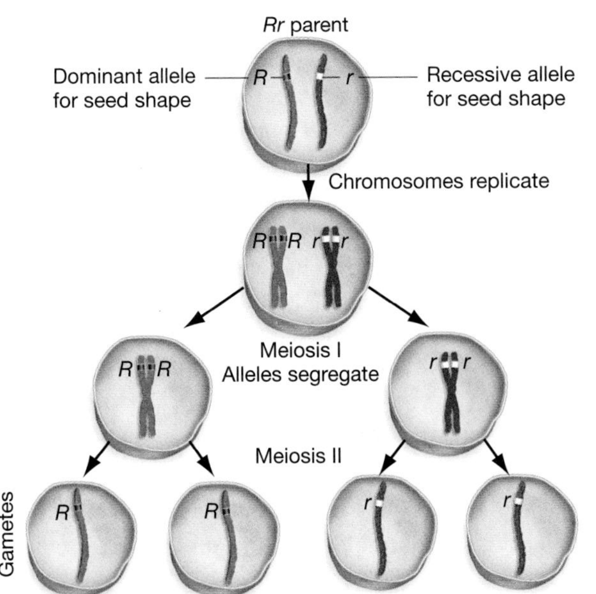

FIGURE 13.7 Meiosis Is Responsible for the Principle of Segregation. A parent's two alleles segregate into different gametes, as Mendel hypothesized, because homologous chromosomes separate during meiosis I.

chromosome number by half but also explains the principle of segregation and the principle of independent assortment.

The cell nucleus at the top of **Figure 13.7** illustrates Sutton and Boveri's central insight—the hypothesis that chromosomes are composed of Mendel's hereditary determinants, or genes. In this example, the gene for seed shape is shown at a particular position along a chromosome. This location is known as a **locus** ("place"; plural: *loci*). A genetic locus is the physical location of a gene. The paternal and maternal chromosomes shown in Figure 13.7 happen to possess different alleles of the gene for seed shape: One allele specifies round seeds (*R*), while the other specifies wrinkled seeds (*r*).

The subsequent steps in Figure 13.7 show how these alleles segregate into different daughter cells during meiosis I, when homologous chromosomes separate. The physical separation of alleles during anaphase of meiosis I is responsible for Mendel's principle of segregation.

Figure 13.8 follows the fate of the alleles for two different genes—in this case, for seed shape and seed color—as meiosis proceeds. Because these genes are located on different nonhomologous chromosomes, they assort independently of one another at meiosis I. Four types of gametes, produced in equal proportions, result. This is the physical basis of Mendel's principle of independent assortment.

Sutton and Boveri formalized these observations in the **chromosome theory of inheritance**. Like other theories in

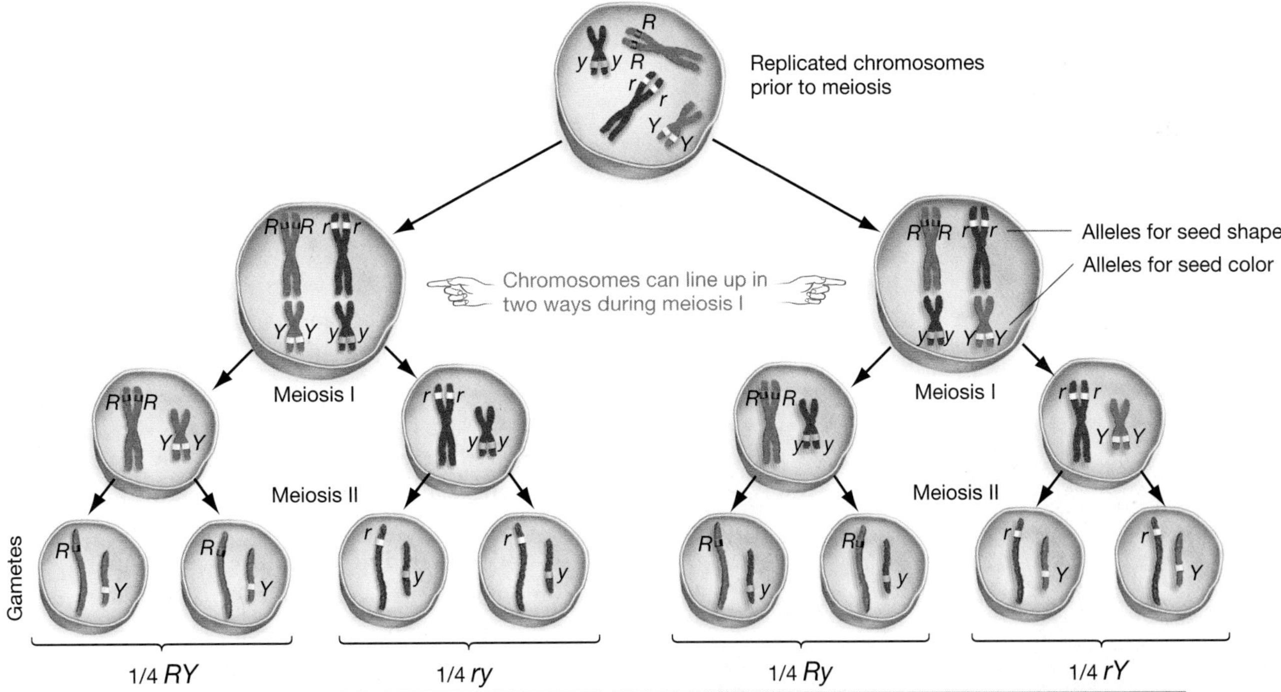

FIGURE 13.8 Meiosis Is Responsible for the Principle of Independent Assortment. The alleles for different traits assort independently, as Mendel hypothesized, because nonhomologous chromosomes assort independently during meiosis I. Maternal and paternal chromosomes are shown in different colors for clarity

biology, the chromosome theory consists of a pattern—a set of observations about the natural world—and a process that explains the pattern. The chromosome theory states that Mendel's rules can be explained by the independent alignment and separation of homologous chromosomes at meiosis I.

When Sutton and Boveri published their findings, however, the hypothesis that chromosomes consist of genes was untested. What experiments confirmed that chromosomes contain genes?

13.4 Testing and Extending the Chromosome Theory

During the first decade of the twentieth century, an unassuming insect rose to prominence as a model organism for testing the chromosome theory of inheritance. This organism—the fruit fly *Drosophila melanogaster*—has been at the center of genetic studies ever since. *Drosophila melanogaster* has all the attributes of a useful model organism for experimental studies in genetics: small size, ease of rearing in the lab, a short reproductive cycle (about 10 days), and abundant offspring (up to a few hundred per mating). The elaborate external anatomy of this insect also makes it possible to identify interesting phenotypic variation among individuals (**Figure 13.9a**).

(a) The fruit fly *Drosophila melanogaster*

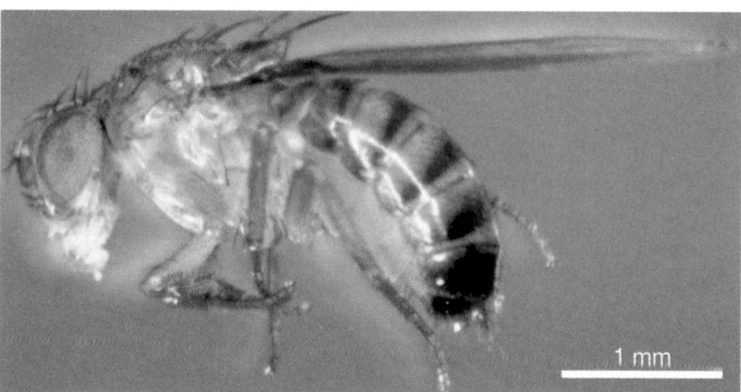

(b) Eye color is a variable trait.

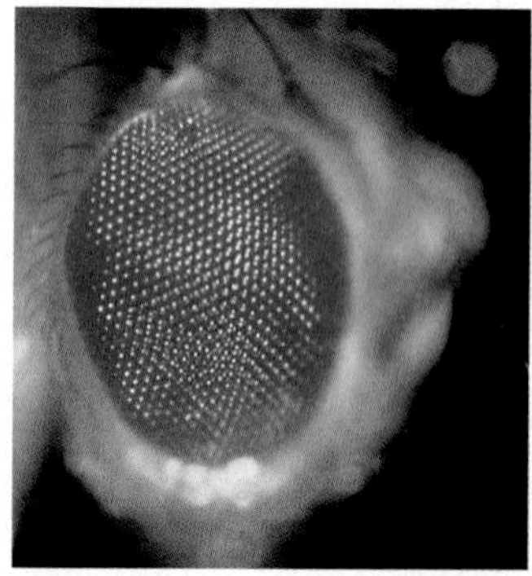

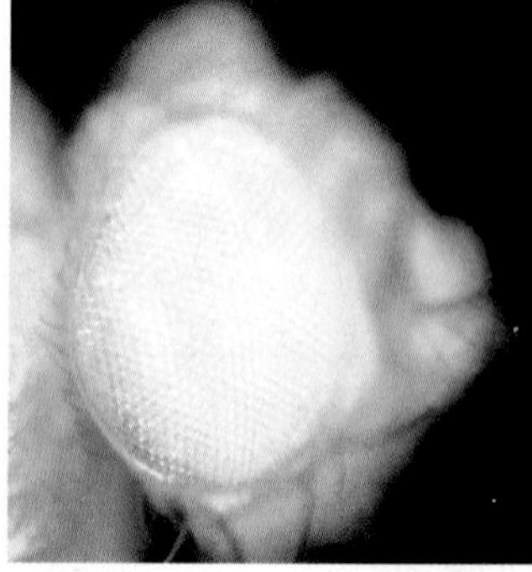

FIGURE 13.9 The Fruit Fly *Drosophila melanogaster* Is an Important Model Organism in Genetics.

● **EXERCISE** In part (b), label the phenotype that is considered wild type. Label the phenotype that is a rare mutant.

Drosophila was adopted as a model organism by Thomas Hunt Morgan and his students. But because *Drosophila* is not a domesticated species like the garden pea, Morgan had no readily available phenotypic variants such as Mendel's round and wrinkled seeds. Consequently, an early goal of Morgan's research was simply to find and characterize individuals with different phenotypes.

Morgan's group referred to individuals with the most common phenotype as **wild type**. But while examining his cultures, Morgan discovered a male fly that had white eyes rather than the wild-type red eyes (**Figure 13.9b**). This individual had a discrete and easy-to-recognize phenotype different from the normal phenotype. Morgan inferred that the white-eyed phenotype resulted from a **mutation**—a change in a gene (in this case, a gene that affects eye color). Individuals with white eyes (or other traits attributable to mutation) are referred to as **mutants**.

To explore how the white-eye trait is inherited in fruit flies, Morgan mated a red-eyed female fly with the mutant white-eyed male fly. All of the F_1 progeny had red eyes. But when Morgan did the reciprocal cross, by mating white-eyed females to red-eyed males, he got a different result: All F_1 females had red eyes, but all F_1 males had white eyes. Recall that Mendel's reciprocal crosses had always given results that were similar to each other. But Morgan's reciprocal crosses did not. The experiment suggested a definite relationship between the sex of the progeny and the inheritance of eye color. When the female parent was white-eyed, the white-eyed phenotype appeared in F_1 males. How could these observations be reconciled with Mendel's rules of inheritance?

The Discovery of Sex Chromosomes

Nettie Stevens began studying the karyotypes of insects about the time that Morgan began his work with *Drosophila*. One of her outstanding observations was of a striking difference in the chromosome complements of males and females in the beetle *Tenebrio molitor*. In females of this species, diploid cells contain 20 large chromosomes. But diploid cells in males contain 19 large chromosomes and 1 small one. Stevens called the small chromosome the Y chromosome. This Y chromosome paired with one of the large chromosomes at meiosis I, which had already been named the X chromosome. The X and Y were different in size and shape, but they acted like homologs during meiosis. Later work showed that even though X and Y chromosomes contain different genes, they have regions that are similar enough to lead to proper pairing during prophase of meiosis I. The X and Y are now called sex chromosomes.

Stevens also noticed that female beetles had two X chromosomes, while male beetles had an X and a Y. Based on this observation, she proposed that an individual's genotype at the sex chromosomes (XX in females, XY in males) determines an individual's gender.

X-Linked Inheritance and the Chromosome Theory

To explain the results of his crosses with white-eyed flies, Morgan put his genetic data together with Stevens's observations on sex chromosomes. *Drosophila* females, like *Tenebrio* females, have two X chromosomes; male fruit flies carry an X and a Y, just as *Tenebrio* males do. As a result, half the gametes produced by a male fruit fly should carry an X chromosome; the other half, a Y chromosome (**Figure 13.10**).

Morgan realized that the transmission pattern of the X chromosome in males and females could account for the results of his reciprocal crosses. Specifically, he proposed that the gene for white eye color in fruit flies is located on the X chromosome and that the Y chromosome does not carry an allele of this gene. This hypothesis is called **X-linked inheritance** or **X-linkage**. (Proposing that a gene resides on the Y chromosome is called **Y-linked inheritance** or **Y-linkage**. Proposing that a gene is on either sex chromosome is termed **sex-linked inheritance** or **sex-linkage**.) The key observation is that even though the X and Y chromosomes synapse during prophase of meiosis I, they differ in size, shape, and gene content.

According to the hypothesis of X-linkage, a female has two copies of the gene that specifies eye color because she has two X chromosomes. One of these chromosomes came from her female parent, and the other from her male parent. A male, in contrast, has only one copy of the eye-color gene because he has only one X chromosome, which comes from his mother.

The Punnett squares in **Figure 13.11** show that Morgan's experimental results can be explained if the gene for eye color is located on the X chromosome, and if the allele for red color is dominant to the allele for white color. In this figure, the allele for red eyes is denoted w^+, while the allele for white eyes is denoted w. (In fruit-fly genetics, the + symbol always indicates the wild-type trait and the gene symbol often refers to the mutant phenotype—for example, w refers to the mutant white-eye phenotype.) The Y chromosome present in males is simply designated as Y. Using this notation, the genotypes used in the experiment are written as $w^+ \, w^+$ for red-eyed females; wY for white-eyed males; ww for white-eyed females; and w^+Y for red-eyed males. If you study the offspring genotypes, you should be convinced that the results predicted by the hypothesis of X-linkage match the observed results.

When reciprocal crosses give different results, such as those illustrated in Figure 13.11, it is likely that the gene in question

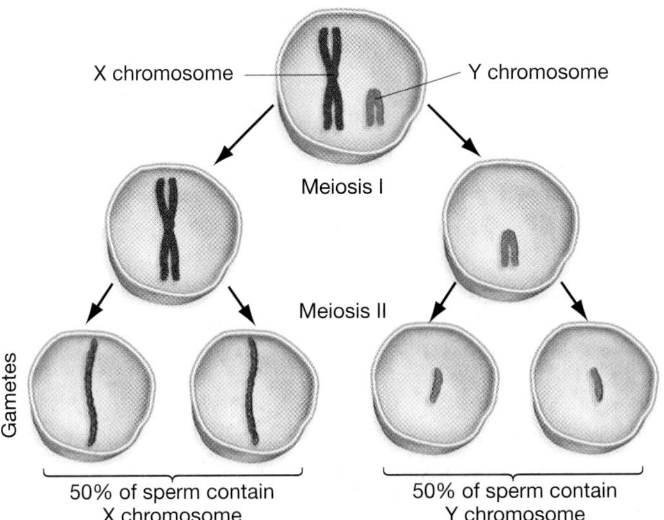

FIGURE 13.10 Sex Chromosomes Pair during Meiosis I, Then Segregate to Form X-Bearing and Y-Bearing Gametes. Sex chromosomes synapse at meiosis I in male fruit flies, even though the X and Y chromosomes differ in size and shape. (No crossing over occurs in male fruit flies.) Thus, half the sperm cells that result from meiosis bear an X chromosome; half have a Y chromosome.

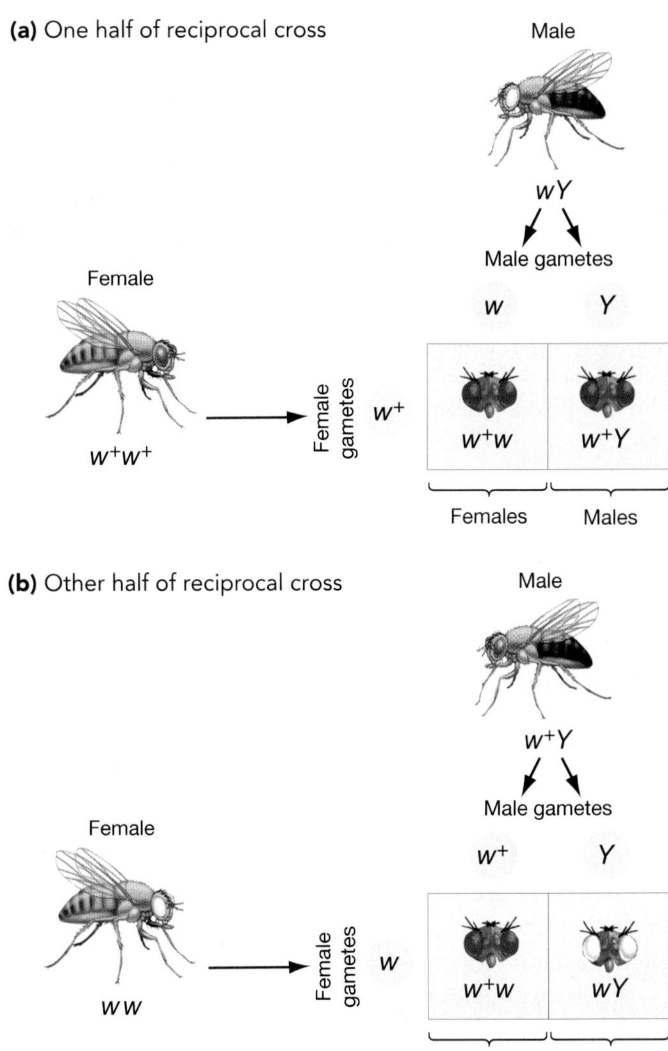

FIGURE 13.11 Reciprocal Crosses Confirm that Eye Color in *Drosophila* Is an X-Linked Trait. When Morgan crossed red-eyed females with white-eyed males and then crossed white-eyed females with red-eyed males, he observed strikingly different results. This was consistent with his hypothesis that eye color is an X-linked trait in fruit flies.

● **EXERCISE** Morgan also crossed red-eyed w^+w females with red-eyed males, w^+Y. Create a Punnett square to predict the types and proportions of offspring genotypes and phenotypes that would result.

is located on a sex chromosome. Recall from Chapter 12 that non-sex chromosomes are called **autosomes**. Genes on non-sex chromosomes are said to show **autosomal inheritance**.

Morgan's discovery of X-linked inheritance carried an even more important message, however. In *Drosophila*, the gene for white eye color is clearly correlated with inheritance of the X chromosome. This correlation was important evidence in support of the hypothesis that chromosomes contain genes. The discovery of X-linked inheritance convinced most biologists that the chromosome theory of inheritance was correct. ● If you understand this reasoning, you should be able to explain why sex-linked inheritance was consistent with predictions of the chromosome theory.

What Happens When Genes Are Located on the Same Chromosome?

When later experiments confirmed that genes are indeed the physical components of chromosomes, the result prompted Morgan and other geneticists to reevaluate Mendel's principle of independent assortment. The key issue was that genes could not undergo independent assortment if they were located on the same chromosome. The physical association of genes that are found on the same chromosome is called **linkage**. (Notice that the terms *linkage* and *sex-linkage* are different in meaning. If two or more genes are linked, it means that they are located on the same chromosome. If a single gene is sex-linked, it means that it is located on a sex chromosome.)

The first examples of linked genes involved the X chromosome of fruit flies. After Morgan established that the white-eye gene was located on *Drosophila*'s X chromosome, he and colleagues established that one of the several genes that affects body color is also located on the X. Red eyes and gray body are the wild-type phenotypes in this species; white eyes and a yellow body occur as rare mutant phenotypes. The alleles for red eyes (w^+) and gray body (y^+) are dominant to the alleles for white eyes (w) and yellow body (y). (Be sure not to confuse the notation for the Y chromosome in males, Y, with the yellow body allele, y.)

Thus, it seemed logical to predict that the linked genes would always be transmitted together during gamete formation. Stated another way, linked genes should violate the principle of independent assortment. Recall from Section 13.2 that independent assortment is observed when genes are on different chromosomes, because the alleles of unlinked genes segregate to gametes independently of one another during meiosis I. But when genes are on the same chromosome, their alleles are carried to gametes together. **Figure 13.12** shows that a female fruit fly with one X chromosome carrying the w and y^+ alleles and with a second X chromosome carrying the w^+ and y alleles should generate just two classes of gametes in equal numbers during meiosis, instead of the four classes that are predicted under the principle of independent assortment. Is this what actually occurs?

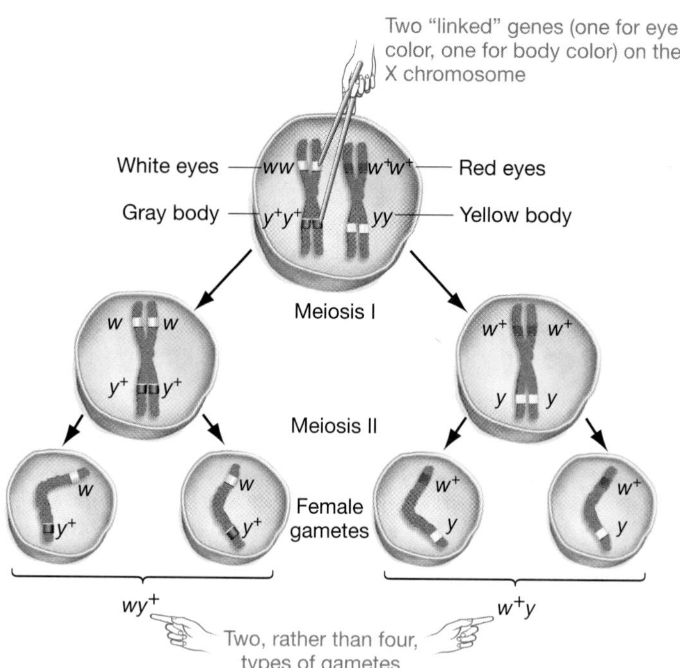

FIGURE 13.12 Linked Genes Are Inherited Together. If the *w* and *y* genes were found on different chromosomes, then this female would generate four different types of gametes instead of just two types as shown here.

● **EXERCISE** List the four genotypes that would be generated if the *w* and *y* genes were not linked.

The First Studies of Linked Genes To determine whether linked traits behave as predicted, Morgan performed crosses like the one described in the Experimental setup section of **Figure 13.13**. This figure introduces some new notation. When you express a genotype in writing, you are simply listing the relevant alleles that are present. When referring to linked genes, biologists use a slash symbol (/) to separate the alleles that are found on homologous chromosomes. For example, the female fruit flies in this experiment had wy^+ alleles on one X chromosome and w^+y alleles on the other X chromosome. This genotype is written as wy^+/w^+y.

The Results table in Figure 13.13 summarizes the phenotypes and genotypes observed in the male offspring of this experimental cross. Notice that most of these males carried an X chromosome with one of the two combinations of alleles found in their mothers: wy^+ or w^+y. Thus, *white* and *yellow* alleles do not segregate independently of each other most of the time. But a small percentage of males had novel phenotypes and genotypes: wy and w^+y^+. Morgan referred to these individuals as **recombinant** because the combination of alleles on their X chromosome was different from the combinations of alleles present in the parental generation.

To explain this result, Morgan proposed that gametes with new, recombinant genotypes were generated when crossing over occurred during prophase of meiosis I in the females. Recall from Chapter 12 that crossing over involves a physical exchange

Experiment

Question: Will genes undergo independent assortment if they are on the same chromosome?

Linkage hypothesis: Linked genes will violate the principle of independent assortment.

Null hypothesis: Linked genes will adhere to the principle of independent assortment.

Experimental setup:

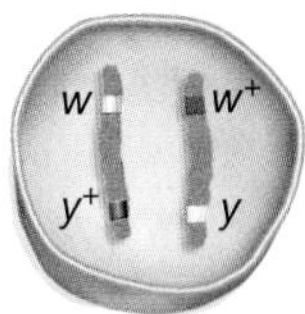

Red-eyed gray-bodied female wy^+/w^+y

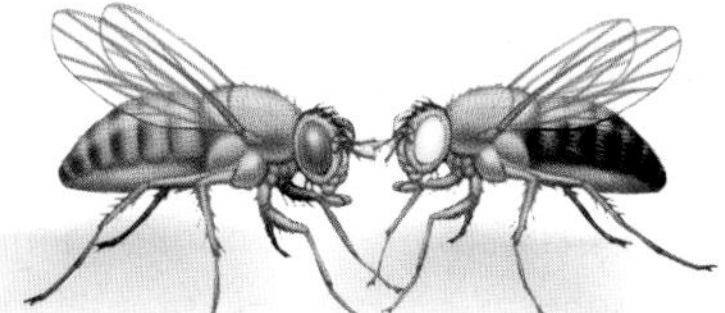

White-eyed gray-bodied male $wy^+/$ Y

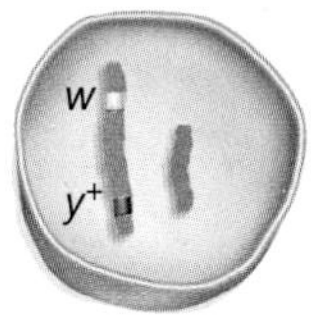

Prediction: Because these two genes are X-linked, male offspring will have only one copy of each gene, from their mother; the two possible male offspring genotypes are wy^+/Y and w^+y/Y.

Prediction of null hypothesis: Four male genotypes are possible (wy^+/Y : w^+y/Y : wy/Y : w^+y^+/Y) and will occur in equal frequency.

Results:

Male offspring

	Phenotype	Genotype	Number
		wy^+ / Y	4292
		w^+y / Y	4605
Recombinant genotypes		wy / Y	86
Recombinant genotypes		w^+y^+ / Y	44

Four male genotypes were observed (rather than two), but not in equal frequency as predicted by independent assortment

Conclusion: Neither hypothesis is fully supported. Independent assortment does not apply to linked genes—linked genes segregate together *except* when crossing over and genetic recombination have occurred.

FIGURE 13.13 Linked Genes Are Inherited Unless Recombination Occurs.

of segments from homologous chromosomes. Crossing over occurs at least once in every synapsed pair of homologous chromosomes, and usually multiple times. (Male fruit flies are an exception to this rule. For unknown reasons, no crossing over occurs in male fruit flies.) As **Figure 13.14** shows, a crossing-over event occurred somewhere between the *w* and *y* genes in the wy^+/w^+y females. The recombinant chromosomes that resulted would have the genotypes *wy* and w^+y^+ (see the middle of the bottom row of the figure). If these chromosomes ended up in a male offspring, they would produce individuals with yellow bodies and white eyes, along with individuals with gray bodies and red eyes. This is exactly what Morgan observed.

The take-home message of Morgan's experiments was that linked genes are inherited together unless crossing over occurs. When crossing over takes place, genetic recombination occurs. These results cemented the connection between the events of meiosis I and Mendel's laws. Linked alleles segregate together unless there is a physical crossover between homologous chromosomes.

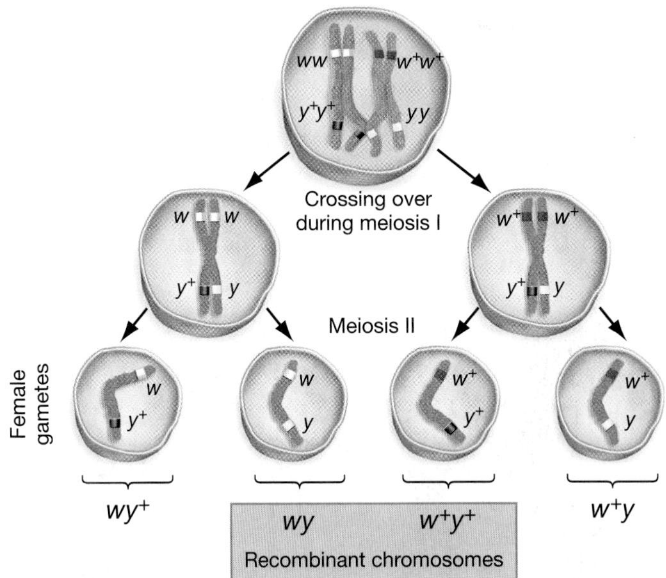

FIGURE 13.14 Genetic Recombination Results from Crossing Over. To explain the results in Figure 13.13, Morgan proposed that crossing over occurred between the *w* genes and the *y* genes in a small percentage of the F_1 females during meiosis I. The recombinant chromosomes that resulted would produce the recombinant phenotypes observed in F_2 males. (Note that the *w* and *y* genes are actually much closer than shown here.)

Linkage Mapping In an experiment like the one diagrammed in Figure 13.13, researchers calculate the recombination frequency as the number of offspring with recombinant phenotypes divided by the total number of offspring. With crosses involving white and yellow, about 1.4 percent of offspring have recombinant phenotypes, which indicates that they have genotypes. But when Morgan and his co-workers performed the same types of crosses with different pairs of X-linked traits, the team found that the fraction of recombinant gametes varied. For example, when crosses involved fruit flies with X-linked genes for white eyes and a mutant phenotype called singed bristles (*sn*), males with recombinant chromosomes for these two genes were produced about 19.6 percent of the time.

Morgan explained these observations by making a conceptual breakthrough: He proposed that genes are arranged in a linear array along a chromosome. According to Morgan's hypothesis, the physical distance between genes determines how frequently crossing over occurs between them. Crossing over occurs at random and can take place at locations all along the length of the chromosome. Based on these observations, it is logical to predict that the shorter the distance between any two genes on a chromosome, the lower the probability that crossing over will take place somewhere in between. The fundamental idea here is that greater physical distance between genes increases the chance that crossing over will take place between them (**Figure 13.15**).

A. H. Sturtevant, an undergraduate who was studying with Morgan, realized that variation in recombination frequency had an important implication: If genes are lined up along the chromosome and if the frequency of crossing over is a function of the physical distance between genes, then it should be possible to figure out where genes are in relation to each other based on the frequency of recombinants between various pairs. That is, it should be possible to create a **genetic map**. A genetic map is a diagram that shows the relative positions of genes along a particular chromosome.

Sturtevant proposed that in constructing a genetic map, the unit of distance along a chromosome should simply be the percentage of offspring that have recombinant phenotypes with respect to two genes. Sturtevant called this unit the centiMorgan (cM), in honor of his mentor. One map unit, or 1 centiMorgan, represents the physical distance that produces 1 percent recombinant offspring. For example, he proposed that the eye-color and bristle-shape genes of fruit flies are 19.6 map units apart on the X chromosome, because recombination between these genes results in 19.6 percent recombinant offspring, on average (**Figure 13.16a**). The yellow (*y*) gene for body color and the gene (*w*) for white eye color, in contrast, are just 1.4 map units apart.

Where is the *yellow* gene relative to the *singed-bristles* gene? Recombinants occurred in 21 percent of the gametes produced by females that are $y^+sn^+/y\ sn$, meaning that *y* and *sn* are 21 cM apart. Sturtevant inferred that the gene for white eyes must be located *between* the genes for yellow body and singed bristles, as shown in Figure 13.16a. Mapping genes relative to each other is like fitting puzzle pieces, and mapping *w* between *y* and *sn* is the only way to make the pairwise distances sum correctly. The key observation is that 21 cM—the distance between *y* and *sn*—is equal to 1.4 cM + 19.6 cM, or the sum of the distances between *y* and *w* and *w* and *sn*. ● If you understand this concept, you should be able to explain why *w* can't be mapped *above* *y* on Figure 3.16a.

Figure 13.16b provides a partial genetic map of the X chromosome in *Drosophila melanogaster* and the data on which the map positions are based. Using this logic and data like these, Sturtevant assembled the first genetic map. ● If you understand this concept, you should be able to add the gene called *crossveinless wings* to the chromosome in Figure 13.16b. It is 13.7 cM from *yellow*, 12.3 cM from *white*, 7.3 cM from *singed bristles*, and 22.4 cm from *miniature wings*.

Do *All* Linked Genes Violate the Principle of Independent Assortment? When crossing over occurs, the new genotypes that result match the ones that would be produced if the genes involved were located on different chromosomes. Stated another way, crossing over breaks up linkage and makes it appear as if independent assortment is occurring.

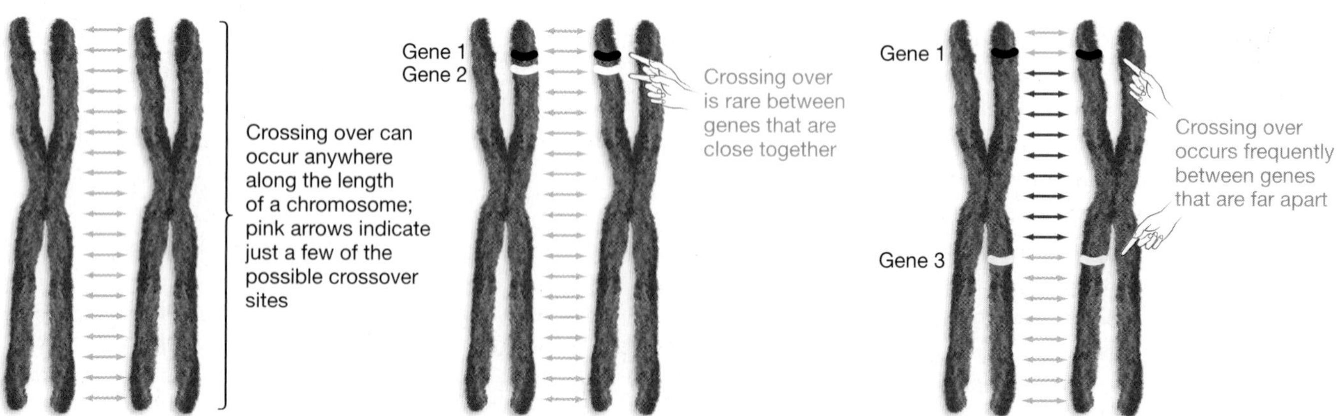

FIGURE 13.15 The Physical Distance between Genes Determines the Frequency of Crossing Over. Crossing over generally occurs at least once in every meiosis in each homologous pair of chromosomes, and often multiple times.

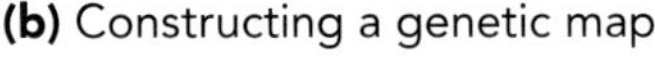

(a) Frequency of recombinant offspring can be used to map genetic distance.

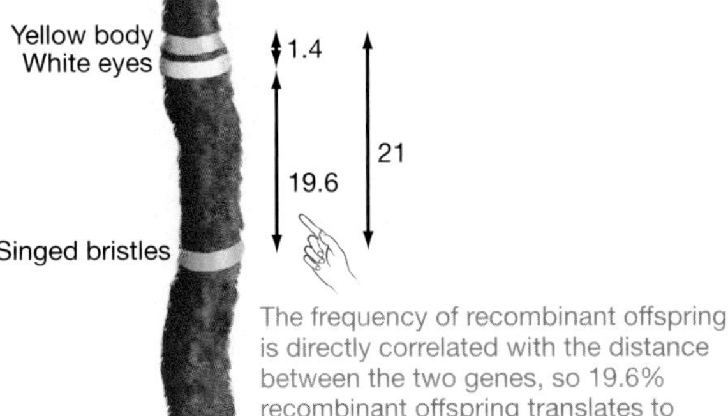

(b) Constructing a genetic map

Frequency of crossing over (%) between some genes on the X chromosome of fruit flies

	Miniature Wings	**Ruby Eyes**
Yellow body	36.1	7.5
White eyes	34.7	6.1
Singed bristles	15.1	13.5
Miniature wings	—	28.6

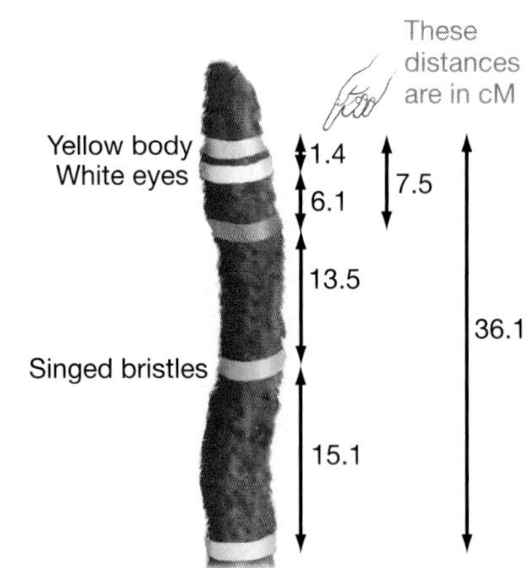

FIGURE 13.16 The Locations of Genes Can Be Mapped by Analyzing the Frequency of Recombination. **(a)** Recombinants between the yellow-body locus and the white-eye locus occur 1.4 percent of the time. Therefore, these genes are 1.4 map units apart on the chromosome. Recombinants between the yellow-body locus and the singed-bristles locus of another X-linked gene occur 21 percent of the time. But recombinants between the white-eye locus and the singed-bristles locus occur just 19.6 percent of the time. Therefore, the genes must be arranged as shown. **(b)** A partial genetic map of the X chromosome in fruit flies.

EXERCISE On the chromosome in part (b), label the orange and blue genes. (Which is *ruby* and which is *miniature wings?*)

To convince yourself that this is so, look back at the four gamete genotypes listed at the bottom of Figure 13.14. Recall that these were produced by a female with genotype wy^+/w^+y. The four gamete genotypes listed are w^+y^+, w^+y, wy^+, and wy. It should be clear that these are the same gamete genotypes that meiosis would produce if the female's genotype were w^+wy^+y, meaning the *white* and *yellow* genes were not linked.

If crossing over occurs frequently enough that all four of these gamete genotypes are produced in equal proportions, then the results of a cross involving linked genes would be indistinguishable from the predictions of independent assortment. When genes are 50 or more map units apart on the same chromosome, that is exactly what happens. Genes behave as if they assort independently when 50 percent of the gametes are recombinant with respect to the genes. As it turns out, this situation was occurring in Mendel's experiments. The genes for two of the traits he worked with are located on chromosome number 1 in peas, and at least two of the genes are located on chromosome 4. In each case, however, the genes he analyzed are located far apart. As a result, crossing over was frequent enough to produce the 9:3:3:1 ratios of F_2 phenotypes predicted by the hypothesis of independent assortment.

Before analyzing some extensions to Mendel's rules, let's step back and consider the events we've just surveyed—from the rediscovery of Mendel's work in 1900 to Sturtevant's first chromosome map, which was published in 1913. In 1899 geneticists did not understand the basic rules of heredity. But by 1913 they could map the locations of genes on chromosomes. A remarkable knowledge explosion had occurred.

Check Your Understanding

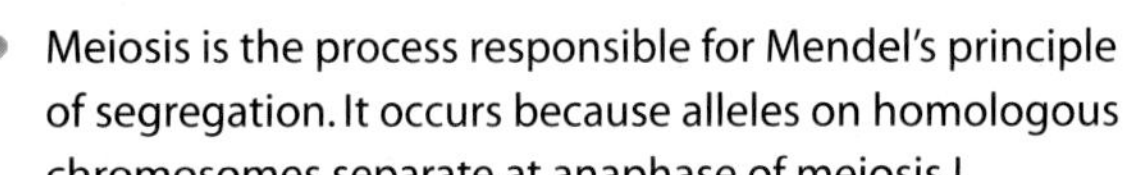

If you understand that...

- Meiosis is the process responsible for Mendel's principle of segregation. It occurs because alleles on homologous chromosomes separate at anaphase of meiosis I.
- Meiosis is the process responsible for Mendel's principle of independent assortment. Alleles of different genes go to gametes independently because pairs of homologous chromosomes line up randomly at metaphase of meiosis I.
- Because the frequency of crossing over between genes correlates with the physical distance between them, genes can be mapped on chromosomes, using the percentage of recombinant offspring observed as a unit.

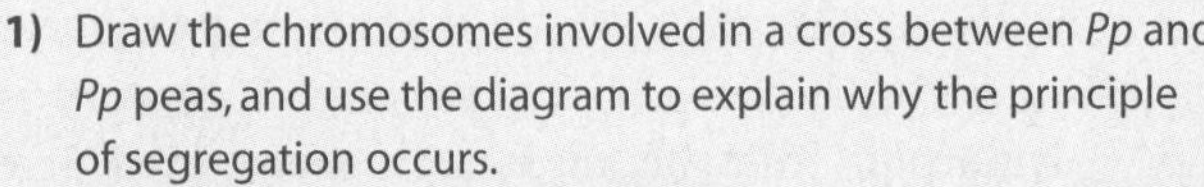

You should be able to...

1) Draw the chromosomes involved in a cross between *Pp* and *Pp* peas, and use the diagram to explain why the principle of segregation occurs.
2) Draw the chromosomes involved in a cross between *YyRr* and *YyRr* peas, and use the diagram to explain why the principle of independent assortment occurs.

13.5 Extending Mendel's Rules

Biologists point out that Mendel analyzed the simplest possible genetic system. The traits that he was studying were not sex-linked and were influenced by just two alleles of a single gene, with alleles that were completely dominant or recessive.

Even so, Mendel was able to discover the most fundamental rules of inheritance. Investigating simple model systems is an extremely important research strategy in biological science. Researchers almost always choose to analyze the simplest situation possible before going on to explore more complicated systems. Mendel probably would have failed, as so many others had done before him, had he worked with more complex patterns of inheritance.

Once Mendel's work was rediscovered, though, researchers began to analyze traits and alleles whose inheritance was more complicated. If experimental crosses produced F_2 progeny that did not conform to the expected 3:1 or 9:3:3:1 ratios in phenotypes, researchers had a strong hint that something interesting was going on. In many cases it turned out that unraveling the cause of surprising results led to new insights into how genes and heredity worked. The discovery of linkage and sex-linkage were among several important exceptions and extensions to Mendel's work. How can traits that don't appear to follow Mendel's rules contribute to a more complete understanding of heredity?

Multiple Alleles and Polymorphic Traits

Mendel worked with a total of seven traits and just 14 alleles—two for each trait. In most populations, however, dozens of alleles can be identified at each gene. The existence of more than two alleles of the same gene is known as **multiple allelism**. As an example, consider the gene for the β-globin protein in humans. β-globin makes up part of the hemoglobin protein, which carries oxygen from the lungs to tissues. Over the past few decades, biologists have identified and named over 500 different alleles of the β-globin gene. Many of these alleles are associated with distinctive phenotypes. Some alleles produce polypeptides with normal oxygen-carrying capacity, while others lead to reduced oxygen-carrying capacity and various types of anemia. Still other β-globin alleles are associated with adaptation to living at high altitudes, decreased stability at high temperatures, or resistance to the parasites that infect red blood cells and cause malaria.

When more than two distinct phenotypes are present in a population due to multiple allelism, the trait is **polymorphic** ("many-formed"). Oxygen-carrying capacity in humans is a highly polymorphic trait, because many alleles exist at the β-globin gene.

(a) Flower color is variable in four-o'clocks.

(b) Incomplete dominance in flower color

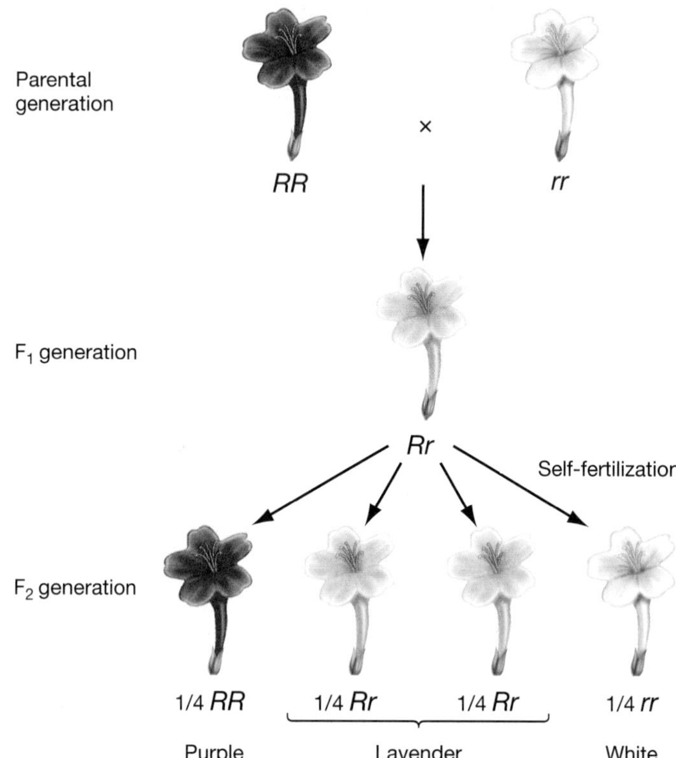

FIGURE 13.17 When Incomplete Dominance Occurs, Heterozygotes Have Intermediate Phenotypes. (a) Four-o'clocks got their name because their flowers open in the late afternoon. Flower color is highly variable. **(b)** The hypothesis illustrated is that flower color is controlled by a single gene with two alleles, symbolized *R* and *r*. These alleles exhibit incomplete dominance.

Incomplete Dominance and Codominance

The terms *dominant* and *recessive* describe which phenotype is observed when two different alleles of a gene occur in the same individual. In all seven traits that Mendel studied, only the phenotype associated with one allele—the "dominant" one—appeared in heterozygous individuals. But consider the flowers called four-o'clocks, pictured in **Figure 13.17a**. In this species, biologists have developed a pure line that has purple flowers and a pure line that has white flowers. When individuals from these strains are mated, all of their offspring are lavender (**Figure 13.17b**). In Mendel's peas, crosses between purple- and white-flowered parents produced all purple-flowered offspring. Why the difference?

Biologists answered this question by examining the phenotypes of F_2 offspring. These are the progeny of self-fertilization in lavender-flowered F_1 individuals. Of the F_2 plants, 1/4 have purple flowers, 1/2 have lavender flowers, and 1/4 have white flowers. This 1:2:1 ratio of *phenotypes* is unlike any we have seen to date, but it exactly matches the 1:2:1 ratio of *genotypes* that is produced when flower color is controlled by one gene with two alleles.

To convince yourself that this explanation is sound, study the genetic model shown in Figure 13.17b. According to the hypothesis shown in the diagram, the inheritance of flower color in four-o'clocks and peas is identical, except that the four-o'clock alleles show incomplete dominance rather than complete dominance. When **incomplete dominance** occurs, heterozygotes have an intermediate phenotype. In the case of four-o'clocks, neither purple nor white alleles dominate. Instead, the F_1 progeny—all heterozygous—show a phenotype intermediate between the two parental strains.

Incomplete dominance illustrates an important general point: Dominance is not necessarily an all-or-none phenomenon. In fact, many alleles show a relationship called **codominance**. When **codominance** occurs, heterozygotes have the phenotype associated with *both* alleles present. As an example, consider the *L* gene in humans. This gene codes for a plasma membrane glycoprotein that is found in red blood cells. Different alleles of the *L* gene lead to the appearance of membrane glycoproteins with different amino acid sequences. In most human populations, only two alleles are present. These alleles are designated L^M and L^N and code for glycoproteins called M and N. Thus, three genotypes are found: L^ML^M, L^ML^N, and L^NL^N. In L^ML^M individuals, all of the MN-type membrane glycoproteins present in red blood cells have the primary sequence associated with the L^M allele. In L^NL^N individuals, all of the MN-type proteins have the L^N allele's primary sequence. But in L^ML^N individuals, red blood cells have some membrane glycoproteins with the M phenotype and some with the N phenotype. Because both alleles are represented in the phenotype, the alleles are considered codominant.

Why Do Dominance Relationships Vary Among Alleles?

Once you understand what genes do on the molecular level, it is straightforward to see why some alleles are dominant, recessive, codominant, or intermediate in dominance. Genes contain the information required to manufacture the molecules that are found in cells. Some of these molecules are proteins that form structures in the cell or body, such as the MN proteins in the plasma membranes of red blood cells. Because the *M* and *N* alleles code for proteins with different amino acid sequences, both are present in the cells of heterozygous individuals. Both gene products are detected in the phenotype, so the alleles are considered codominant. Codominance is common in alleles that are associated with membrane proteins such as the MN protein.

Complete and incomplete dominance tend to occur in alleles that code for enzymes. The plant-height alleles that Mendel studied, for example, code for enzymes that are involved in the synthesis of a key growth hormone. A recessive allele has a defect that keeps the enzyme from functioning. Homozygous recessive individuals lack the growth hormone and are dwarfed. The dominant allele, in contrast, codes for an enzyme that functions normally. In heterozygotes, the one normal allele that is present produces enough enzyme to make a normal amount of growth hormone. Normal height results, and the allele is considered dominant.

The situation in four-o'clock flowers is different. The *R* gene analyzed in Figure 13.17b codes for an enzyme involved in producing a purple pigment in flower petals. The recessive allele codes for a dysfunctional enzyme and leads to a complete lack of pigment. Recessive homozygotes are white as a result. In heterozygotes, though, the one normal copy of the gene can produce enough enzyme to result in a small amount of purple pigment. The lavender phenotype results. Homozygous dominant individuals, in contrast, produce large amounts of normal enzyme and thus enough pigment to produce purple petals.

The message here is that alleles can interact in many different ways, depending on what their products do and how they work. In addition, recall that dominance relationships have nothing to do with fitness—the ability of an individual to produce offspring in a particular environment. Dominant alleles do not necessarily confer high fitness, and they are not necessarily more common than recessive alleles. Recall that the allele that causes Huntington's disease in humans is fatal, dominant, and rare.

Pleiotropy

As far as is known, the alleles that Mendel analyzed affect just a single trait. The gene for seed color in garden peas, for example, does not appear to affect other aspects of the individual's phenotype. In contrast, many cases have been documented in which a single allele affects a wide variety of traits. A gene that influences many traits, rather than just one trait, is said to be **pleiotropic** ("more-turning"). The gene responsible for **Marfan syndrome** in humans, called *FBN1,* is a good example. Although current research suggests that just a single gene is involved, individuals with Marfan syndrome exhibit a wide array of phenotypic effects: increased height, disproportionately long limbs and fingers, an abnormally shaped chest, and potentially severe heart problems. A large percentage of these individuals also suffer from problems with their backbone. The gene associated with Marfan syndrome is pleiotropic.

Genes Are Affected by the Physical Environment and Genetic Environment

When Mendel analyzed height in his experiments, he ensured that each plant received a similar amount of sunlight and grew in similar soil. This was important because individuals with alleles for tallness are stunted if they are deprived of nutrients, sunlight, or water—so much so that they look similar to individuals with alleles for dwarfing. For Mendel to analyze the hereditary determinants of height, he had to control the environmental determinants of height. Let's consider how two aspects of the environment affect phenotypes: (1) the individual's physical surroundings and (2) the alleles present at other genes.

The Physical Environment Has a Profound Effect on Phenotypes The phenotypes produced by most genes and alleles are strongly affected by the individual's physical environment. Consequently, an individual's phenotype is often as much a product of its physical environment as it is a product of its genotype. For example, people with the genetic disease called **phenylketonuria** (PKU) lack an enzyme that helps convert the amino acid phenylalanine to the amino acid tyrosine. As a result, phenylalanine and a related molecule, phenylpyruvic acid, accumulate in these people's bodies. The molecules interfere with the development of the nervous system and produce profound mental retardation. But if PKU individuals are identified at birth and placed on a low-phenylalanine diet, then they develop normally. In many countries, newborns are routinely tested for the defect. PKU is a genetic disease, but it is neither inevitable nor invariant. Through a simple change in their environment (their diet), individuals with a PKU genotype can have a normal phenotype.

Interactions with Other Genes Have a Profound Effect on Phenotypes In Mendel's pea plants, a single locus influenced seed shape. Further, Mendel's data showed that the seed-shape phenotype does not appear to be affected by the action of genes for seed color, seed-pod color, seed-pod shape, or other traits. The pea seeds he analyzed were round or wrinkled regardless of the types of alleles present at other loci.

In many cases, however, genes are not as independent as the gene for seed shape in peas. Consider a classic experiment published in 1905 on comb morphology in chickens. The researchers, William Bateson and R. C. Punnett, crossed parents from pure-breeding lines with rose and pea combs and found that the F_1 offspring had a different phenotype, called walnut combs. When these individuals bred, their offspring were walnut, rose, pea, and a fourth phenotype—called single—in a 9:3:3:1 ratio (**Figure 13.18a**). The genetic model in **Figure 13.18b** explains the data. If comb morphology results from interactions between two genes, symbolized *R* and *P*, if a dominant and a recessive allele exist at each gene, and if the four comb phenotypes are associated with the genotypes indicated at the bottom of the figure, then a cross between *RRpp* and *rrPP* parents would give the results that Bateson and Punnett observed.

The key point is that the alleles of different genes affect each other in a way that influences the phenotype observed in these individuals. When these types of gene interactions occur, the phenotype produced by an allele depends on the action of alleles of other genes. If a chicken has an *R* allele, its phenotype depends on the allele present at the *P* gene. A special case of this phenomenon is known as **epistasis** ("stopping" or "diminishing"). Epistasis occurs when an allele of one gene masks the effect of an allele at another gene.

Quantitative Traits

Mendel worked with **discrete traits**—characteristics that are qualitatively different. In garden peas, seed color is either yellow or green—no intermediate phenotypes exist. But many traits in peas and other organisms don't fall into discrete categories.

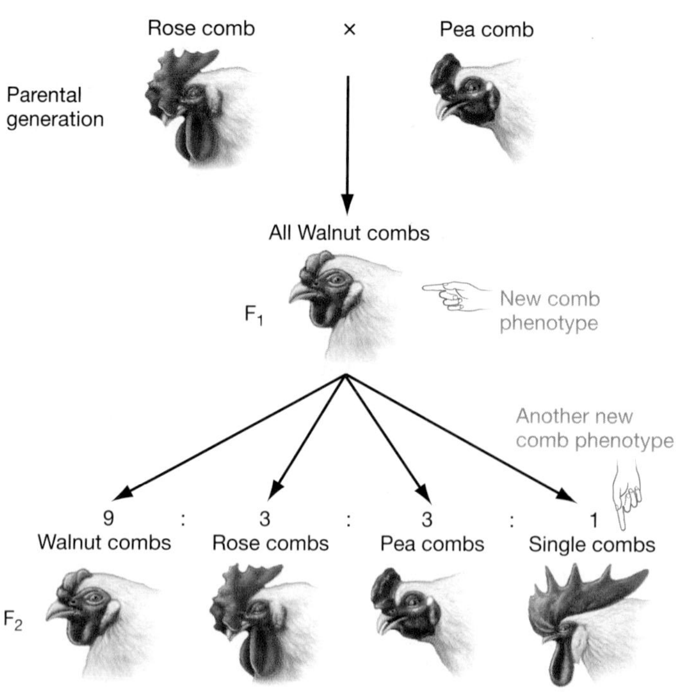

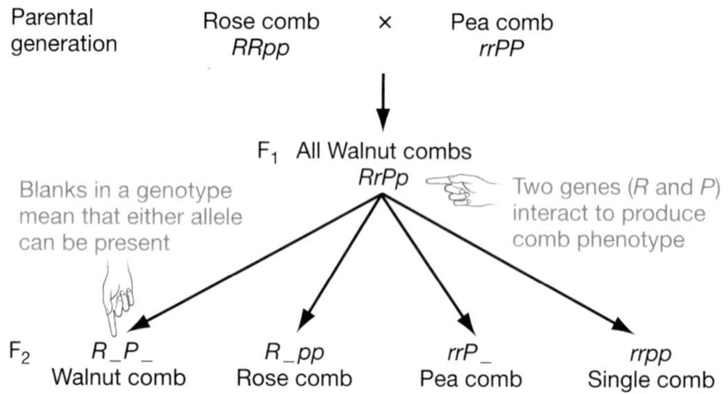

FIGURE 13.18 Different Genes Can Interact to Influence a Trait.
(a) This cross is unusual because new phenotypes show up in the F_1 and the F_2 generation. **(b)** To explain the results, researchers hypothesized that comb shape depends on two genes that interact. The phenotype associated with the alleles at each gene depends on the genetic environment—specifically, the alleles at the other gene.

In humans, for example, height, weight, and skin color fall anywhere on a continuous scale of measurement. People are not 160 cm tall or 180 cm tall, with no other heights possible. Height and many other characteristics exhibit quantitative variation—meaning that individuals differ by degree—and are called **quantitative traits**. Like discrete traits, quantitative traits are highly influenced by the physical environment. For example, the effects of nutrition on human height and intelligence have been well documented.

Quantitative traits share a common characteristic: When the frequencies of different values observed in a population are

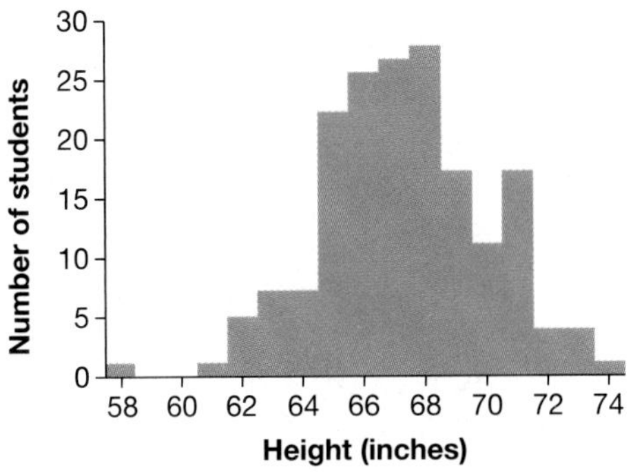

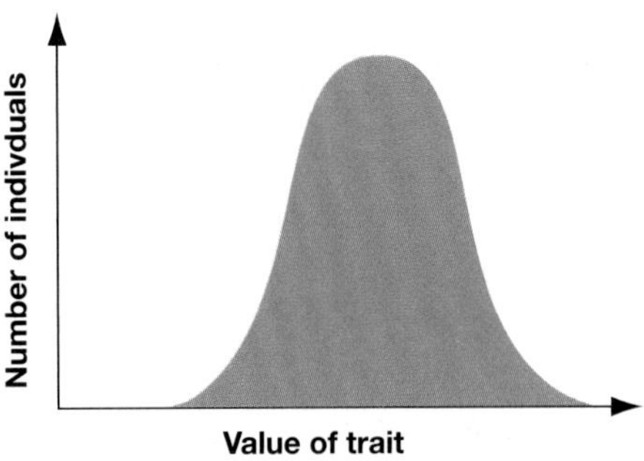

FIGURE 13.19 Quantitative Traits Are Normally Distributed. (a) A histogram plotting the heights of male undergraduates at Connecticut Agricultural College in 1914. **(b)** In most cases, the distribution of a quantitative trait in a population approximates a bell-shaped curve, called a normal distribution.

QUESTION The shortest student plotted on the histogram is 4′10″ (147 cm), and the tallest is 6′2″ (188 cm). How would a histogram of men in your class compare with the distribution in part (b)?

plotted on a histogram, or frequency distribution, they usually form a bell-shaped curve (**Figure 13.19a**). This distribution is observed so frequently that it is often called a normal distribution. In a normal distribution, high and low values occur at low frequency; intermediate values occur at high frequency (**Figure 13.19b**).

In 1909 Herman Nilsson-Ehle showed that if many genes each contribute a small amount to the value of a quantitative trait, then a continuous, bell-shaped (normal) distribution results. Nilsson-Ehle established this finding using strains of wheat that differed in kernel color. **Figure 13.20a** shows the results of a cross he performed between pure lines of white wheat and dark-red wheat. Notice that the frequency of colors in F_2 progeny forms a bell-shaped curve. To explain these results, Nilsson-Ehle proposed the set of hypotheses illustrated in **Figure 13.20b**:

- The parental strains differ with respect to three genes that control kernel color: *AABBCC* produces dark-red kernels, and *aabbcc* produces white kernels.
- The three genes assort independently. When the *AaBbCc* F_1 individuals self-fertilize, white F_2 individuals would occur at a frequency of 1/4 (*aa*) × 1/4 (*bb*) × 1/4 (*cc*) = 1/64 *aabbcc*.

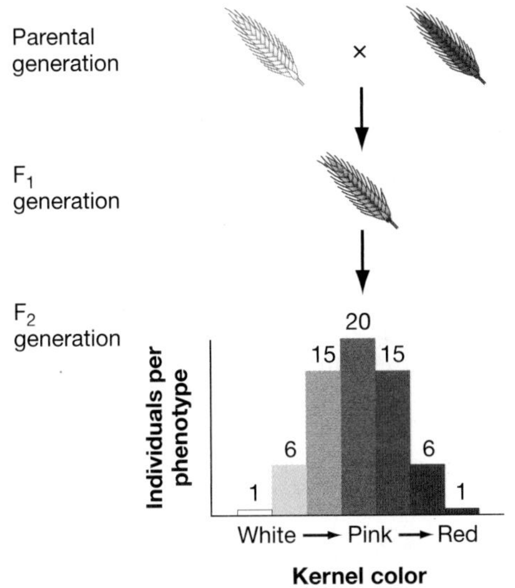

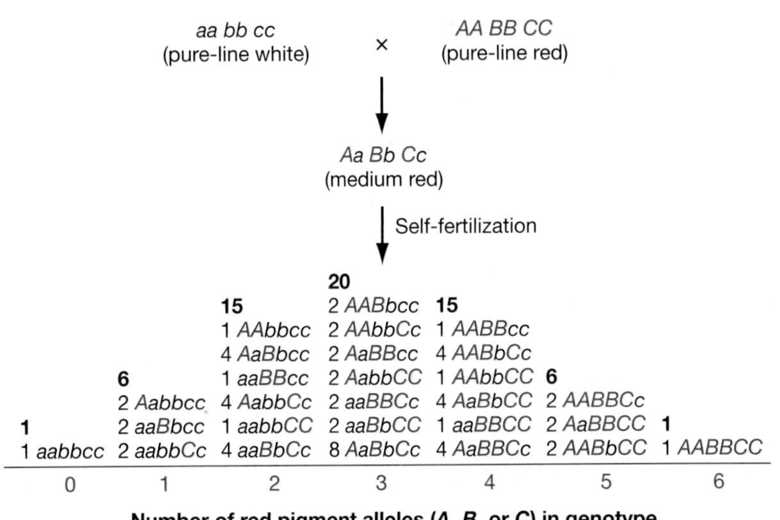

FIGURE 13.20 Quantitative Traits Result from the Action of Many Genes. (a) When wheat plants with white kernels were crossed with wheat plants with red kernels, the F_2 offspring showed a range of kernel colors. The frequency of these phenotypes approximates a normal distribution. **(b)** This model attempts to explain the results of part (a).

EXERCISE Confirm that the distribution of genotypes shown in part (b) is correct by drawing a Punnett square based on the gametes produced by the F_1 parents and then filling in the F_2 offspring genotypes that result. (The square will be 8 × 8 with 64 boxes!)

SUMMARY TABLE 13.2 **Some Extensions to Mendel's Rules**

Type of Inheritance	Definition	Consequences or Comments
Sex-linkage	Genes located on sex chromosomes.	Patterns of inheritance in males and females differ.
Linkage	Two genes found on same chromosome.	Linked genes violate principle of independent assortment.
Incomplete dominance	Heterozygotes have intermediate phenotype.	Polymorphism—heterozygotes have unique phenotype.
Codominance	Heterozygotes have phenotype of both alleles.	Polymorphism is possible—heterozygotes have unique phenotype.
Multiple allelism	In a population, more than two alleles present at a locus.	Polymorphism is possible.
Polymorphism	In a population, more than two phenotypes present.	Can result from actions of multiple alleles, incomplete dominance, and/or codominance.
Pleiotropy	A single allele affects many traits.	This is common.
Variation in the genetic environment	In discrete traits, the phenotype associated with an allele depends on which alleles are present at another gene.	One allele can be associated with different. phenotypes
Variation in the physical environment	Phenotype influenced by environment experienced by individual.	Same genotypes can be associated with different phenotypes.
Polygenic inheritance of quantitative traits	Many genes are involved in specifying traits that exhibit continuous variation.	Unlike alleles that determine discrete traits, each allele adds a small amount to phenotype.

- The *a*, *b*, and *c* alleles do not contribute to pigment production, but the *A*, *B*, and *C* alleles contribute to pigment production in an equal and additive way. As a result, the degree of red pigmentation is determined by the number of *A*, *B*, or *C* alleles present. Each uppercase (dominant) allele that is present makes a wheat kernel slightly darker red.

Check Your Understanding

If you understand that...

- Genes on the same chromosome violate the principle of independent assortment. They are not transmitted to gametes independently of each other unless crossing over occurs between them.
- Incomplete dominance, codominance, multiple allelism, pleiotropy, environmental effects, gene interactions, and polygenic inheritance are aspects of inheritance that Mendel did not study. When they occur, monohybrid and dihybrid crosses do not result in classical Mendelian ratios of offspring phenotypes.

You should be able to...

1) Define linkage, incomplete dominance, and the other phenomena listed above and give an example of each.
2) Explain why each of these phenomena makes it unlikely to observe the 3:1 and 9:3:3:1 phenotype ratios that Mendel observed in his monohybrid and dihyrid crosses.

Later work showed that Nilsson-Ehle's model hypotheses were correct in virtually every detail. Quantitative traits are produced by the independent actions of many genes, although it is now clear that some genes have much greater effects on the trait in question than other genes do. As a result, the transmission of quantitative traits is said to result from polygenic ("many-genes") inheritance. In **polygenic inheritance**, each gene adds a small amount to the value of the phenotype.

In the decades immediately after the rediscovery of Mendel's work, analyses of phenomena such as sex-linkage, multiple allelism, incomplete dominance, environmental effects, gene interactions, and polygenic inheritance provided a fairly comprehensive answer to the question of why offspring resemble their parents. **Table 13.2** summarizes some of the key exceptions and extensions to Mendel's rules and gives you a chance to compare and contrast their effects on patterns of inheritance.

13.6 Applying Mendel's Rules to Humans

When researchers set out to study how a particular gene is transmitted in wheat or fruit flies or garden peas, they begin by making a series of controlled experimental crosses. For obvious reasons, this research strategy is not possible with humans. But suppose you are concerned about an illness that runs in your family, and that you go to a genetic counselor to find out how likely your children are to have the disease. To advise you, the counselor needs to know how the trait is transmitted, including whether the gene involved is autosomal or sex-linked and what type of dominance is shown by the disease allele. To understand the transmission of human traits, investigators have to analyze human crosses that already exist. They do so by constructing a **pedigree**, or family tree, of affected individuals.

A pedigree records the genetic relationships among the individuals in a family along with each person's sex and phenotype with respect to the trait in question. If the trait is due to a single gene, then analyzing the pedigree may reveal whether the trait is due to a dominant or recessive allele and whether the gene responsible is located on a sex chromosome or on an autosome. Let's look at a series of specific case histories to see how this work is done.

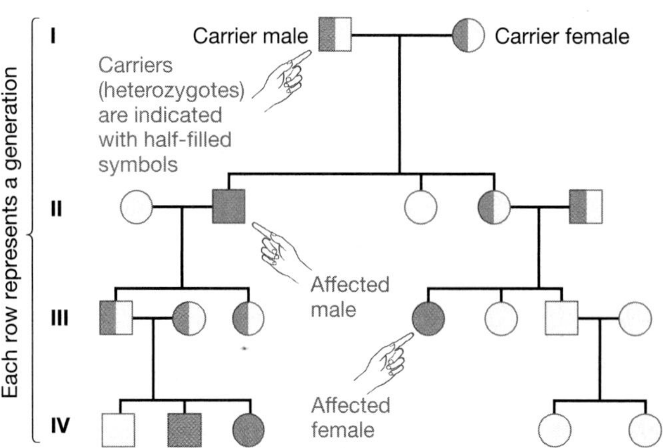

FIGURE 13.21 Pedigree of a Family with an Autosomal Recessive Disease. Diseases like sickle cell, which are inherited as autosomal recessives, appear in both males and females. For an individual to be affected, both parents must carry the allele responsible. If the recessive allele is rare, affected individuals may not appear in every generation.

QUESTION In a pedigree where an autosomal recessive disease appears, an affected parent frequently has unaffected children. But unaffected parents can have affected offspring. Why?

Are Alleles Recessive or Dominant?

To analyze the inheritance of a trait that shows discrete variation, biologists begin by assuming that a single autosomal gene is involved and that the alleles present in the population have a simple dominant-recessive relationship. This is the simplest possible situation. If the pattern of inheritance fits this model, then the assumptions—of inheritance via a single gene and simple dominance—are supported. Let's first analyze the pattern of inheritance that is typical of autosomal recessive traits and then examine patterns that emerge in pedigrees for autosomal dominant traits.

Patterns of Inheritance: Autosomal Recessive Traits In analyzing the inheritance of traits, it's helpful to distinguish conditions that *must* be met when a particular pattern of inheritance occurs versus conditions that are *likely* to be met. For example, if a phenotype is due to an autosomal recessive allele, then individuals with the trait must be homozygous. If the parents of an affected individual do not have the trait, then the parents must be heterozygous for the trait. Heterozygous individuals who carry a recessive allele for an inherited disease are referred to as **carriers** of the disease. These individuals carry the allele and transmit it even though they do not exhibit signs of the disease. When two carriers mate, they should produce offspring with the recessive phenotype about 25 percent of the time.

Figure 13.21 is the pedigree from a family in which an autosomal recessive trait, such as sickle cell disease, occurs. The key feature to notice in this pedigree is that some boys and girls exhibit the trait even though their parents do not. This is the pattern you would expect to observe under the model of an autosomal recessive trait if the parents of individuals with the trait are heterozygous. It is also logical to observe that when an affected individual has children, those children do not necessarily have the trait. This pattern is predicted if affected people marry individuals who are homozygous for the wild-type allele, which is likely to be true if the recessive allele is rare in the population.

The recessive phenotype should show up in offspring only when both parents carry that recessive allele and pass it on to their offspring. By definition, a recessive allele produces a given phenotype only when the individual is homozygous for that allele.

Patterns of Inheritance: Autosomal Dominant Traits By definition, individuals who are homozygous or heterozygous for an autosomal dominant trait will have the dominant phenotype. For autosomal dominant traits, individuals with a single copy of the allele must have the dominant phenotype. Even if one parent is heterozygous and the other is homozygous recessive, on average half of their children should show the dominant phenotype. And unless a new mutation has occurred, any child with the trait must have a parent with this trait. The latter observation is in strong contrast to the pattern seen in autosomal recessive traits.

Figure 13.22 shows the consequences of autosomal dominant inheritance, in the pedigree of a family affected by a degenerative brain disorder called **Huntington's disease**. The pedigree has two features that indicate this disease is passed to the next generation as an autosomal dominant allele. First, if a child shows the trait, then one of its parents shows the trait as well. Second, if families have a large number of children, the trait usually shows up in every generation—due to the high probability of heterozygous parents having affected children.

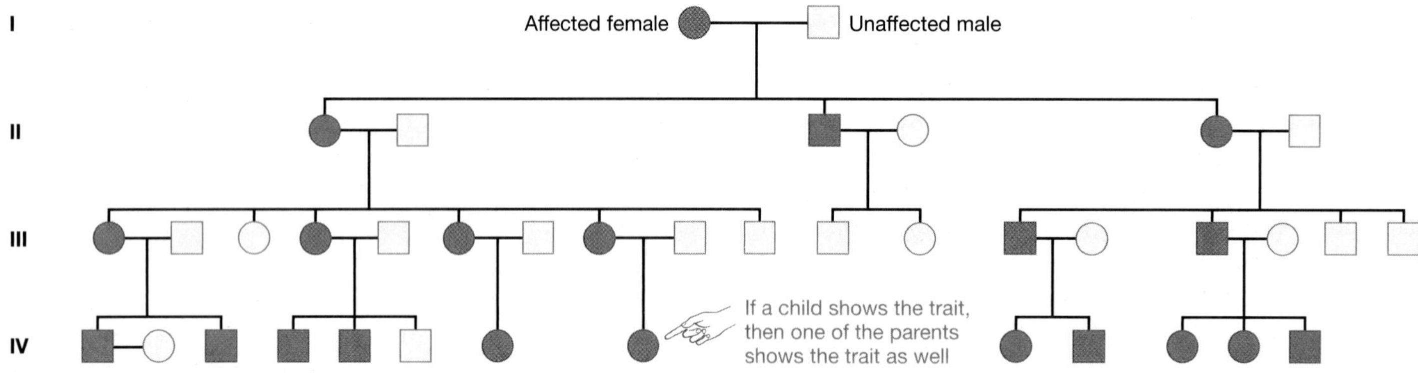

FIGURE 13.22 A Pedigree of a Family with an Autosomal Dominant Disease. Diseases like Huntington's, which are inherited as autosomal dominants, may appear in both males and females and tend to appear every generation.

Is the Trait Autosomal or Sex-Linked?

When it is not possible to arrange reciprocal crosses, can data in a pedigree indicate whether a trait is autosomal or sex-linked? The answer is based on a simple premise: If a trait appears about equally often in males and females, then it is likely to be autosomal. For example, the data in the Huntington's pedigree indicates that the disease appears in both males and females at about equal rates. But if males are much more likely to have the trait in question than females are, then the gene responsible may be found on the X chromosome. Because so few genes occur on the Y chromosome, Y-linked inheritance is rare.

To understand why a sex bias in phenotypes implicates sex-linked inheritance, recall from Section 13.4 that sex-linked genes are located on one of the sex chromosomes. Because human males have one X chromosome and one Y chromosome, they have just one copy of each X-linked gene. But because human females have two X chromosomes, they have two copies of each X-linked gene. These simple observations are critical. In humans—just as in fruit flies and in every other species that has sex chromosomes—the pattern of inheritance in sex-linked traits is different in males and females, because the complement of sex chromosomes differs in the two sexes.

What does the pedigree of an X-linked trait look like? To answer this question, let's consider the pedigree of a classic X-linked trait—the occurrence of hemophilia in the descendants of Queen Victoria and Prince Albert, who were monarchs of England in the mid-to-late 1800s. **Hemophilia** is caused by a defect in an important blood-clotting factor. Hemophiliacs are at a high risk of bleeding to death, because even minor injuries result in prolonged bleeding. The key point to notice in the pedigree of Queen Victoria's descendants is that only males developed hemophilia (**Figure 13.23**). Also note the affected male in generation II (the second square from the right in row II). His two sons were unaffected, but the trait reappeared in a grandson. Stated another way, the occurrence of hemophilia skipped a generation.

This pattern is logical because hemophilia is due to an X-linked *recessive* allele. Because males have only one X chromo-

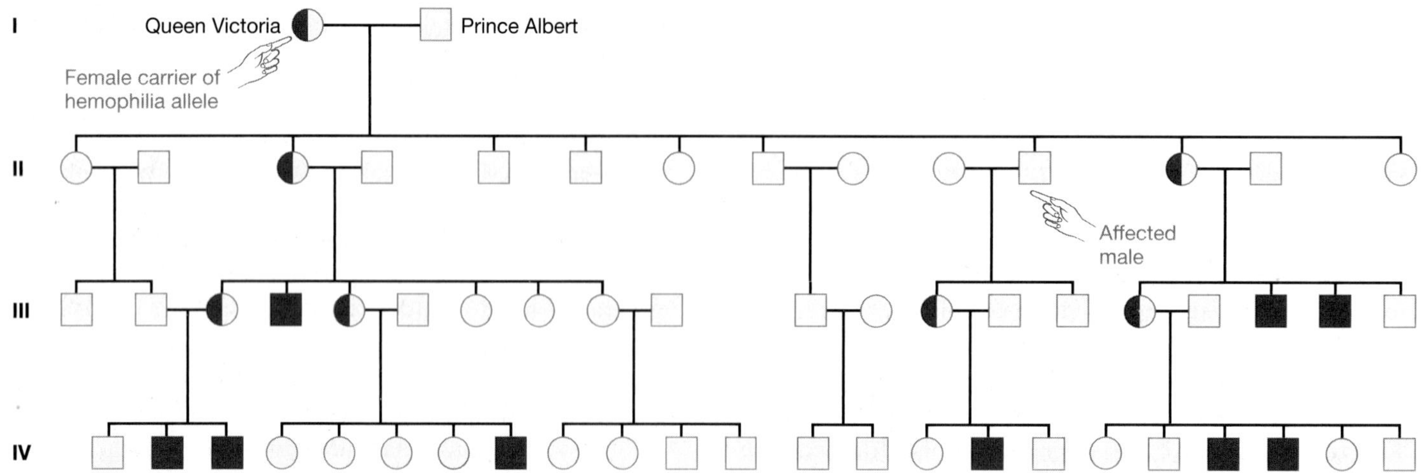

FIGURE 13.23 A Pedigree of an X-Linked Recessive Disease.

EXERCISE Redraw this pedigree, showing the pattern of inheritance if the allele for hemophilia were X-linked dominant.

some, the phenotype associated with an X-linked recessive allele appears in every male that carries it. Further, the appearance of an X-linked recessive trait skips a generation in a pedigree because the affected male passes his only X chromosome on to his daughters. But because his daughters almost always received a wild-type allele from their mother, the daughters don't show the trait. They will pass the defective allele on to about half of their sons, however.

In contrast, X-linked traits that are dominant appear in every individual who has the defective allele. A good indicator of an X-linked dominant trait is a pedigree in which an affected male has all affected daughters but no affected sons. This pattern is logical because every female offspring of an affected father gets an X chromosome from him and will herself be affected, while his sons get their only X chromosome from their unaffected mother. Besides the inherited form of a bone disease called rickets, however, very few diseases are known to be due to X-linked dominant alleles.

By analyzing pedigrees in this way, biomedical researchers have been able to characterize how most of the common genetic diseases in humans are inherited. By the 1940s, though, the burning question in genetics was no longer the nature of inheritance, but the nature of the gene itself. What are genes made of, and how are they copied so that parents pass their alleles on to their offspring? These are the questions we turn to in Chapter 14.

Chapter Review

SUMMARY OF KEY CONCEPTS

Mendel discovered that in garden peas, individuals have two alleles, or versions, of each gene. Prior to the formation of eggs and sperm, the two alleles of each gene separate so that one allele is transmitted to each egg or sperm cell.

Gregor Mendel discovered that inheritance is particulate and proposed two basic principles of transmission genetics. The first was the principle of segregation, which claimed that each trait was specified by paired hereditary determinants (alleles of genes) that separate from each other during gamete formation.

You should be able to cite data that explain why observing intermediate phenotypes in hybrid offspring, due to incomplete dominance or polygenic inheritance, is consistent with the principle of segregation and does not support the hypothesis of blending inheritance.

Genes are located on chromosomes. The separation of homologous chromosomes during anaphase of meiosis I explains why alleles of the same gene segregate to different gametes.

The chromosome theory of inheritance claimed that genes are located on chromosomes and that the movements of chromosomes during meiosis provide a physical basis for Mendel's observations. The chromosome theory provided a physical basis for the principle of segregation. Alleles of the same gene segregate from each other and end up in different gametes because homologous chromosomes segregate during meiosis I.

You should be able to draw the phases of meiosis in a diploid organism with $n = 2$, where the individual has the genotype *BbRr*, and identify where and why the *B* and *b* and the *R* and *r* each separate into different daughter cells and thus gametes.

If genes are located on different chromosomes, then the alleles of each gene are transmitted to egg cells and sperm cells independently of each other. This occurs because chromosomes line up randomly in metaphase of meiosis I.

Mendel's second basic conclusion was the principle of independent assortment, which stated that the segregation of one pair of genes—controlling a given trait—was not influenced by the segregation of other gene pairs. The chromosome theory provided a physical basis for the principle of independent assortment. Genes located on different chromosomes move to gametes independently of each other during meiosis. The principle of independent assortment holds because chromosomes line up independently of each other at the metaphase plate during meiosis I. Thus, gametes receive random combinations of chromosomes.

You should be able to add to your drawing of the phases of meiosis, indicating that the individual has the genotype *BP/bp Rr*, and explain why the *B* and *R* genes assort independently while the *B* and *P* genes do not.

MB **Web Animations** at www.masteringbio.com
Mendel's Experiments
The Principle of Independent Assortment

There are important exceptions and extensions to the basic patterns of inheritance that Mendel discovered.

Thomas Hunt Morgan and colleagues extended Mendel's work by describing X-linked inheritance and by showing that genes located on the same chromosome do not exhibit independent assortment. Studies of X-linked traits helped confirm that genes are found on chromosomes, while studies of linked traits led to the first genetic maps showing the locations of genes on chromosomes. Later studies confirmed that many traits are influenced by the interaction of several genes and that phenotypes are influenced by the environment an individual has experienced, as well as by its genotype. Analysis of pedigrees allowed researchers to work out the basic modes of inheritance for most of the common genetic diseases in humans.

You should be able to explain why crossing over is said to break up linkage between alleles, why males do not pass their X-linked traits on to their sons, why highly homozygous populations are used to control for gene interactions in experiments, and why raising individuals in similar environments is important in genetics experiments.

GENETICS QUESTIONS

1. In studies of how traits are inherited, what makes certain species candidates for model organisms?
   a. They are the first organisms to be used in a particular type of experiment, so they are a historical "model" of what researchers expect to find.
   b. They are easy to study because a great deal is already known about them.
   c. They are the best or most fit of their type.
   d. They are easy to maintain, have a short life cycle, produce many offspring, and yield data that are relevant to many other organisms.
2. Why is the allele for wrinkled seed shape in garden peas considered recessive?
   a. It "recedes" in the F_2 generation when homozygous parents are crossed.
   b. The trait associated with the allele is not expressed in heterozygotes.
   c. Individuals with the allele have lower fitness than that of individuals with the dominant allele.
   d. The allele is less common than the dominant allele. (The wrinkled allele is a rare mutant.)
3. The alleles found in haploid organisms cannot be dominant or recessive. Why?
   a. Dominance and recessiveness describe interactions between two alleles of the same gene in the same individual.
   b. Because only one allele is present, alleles in haploid organisms are always dominant.
   c. Alleles in haploid individuals are transmitted like mitochondrial DNA or chloroplast DNA.
   d. Most haploid individuals are bacteria, and bacterial genetics is completely different from eukaryotic genetics.
4. Why can you infer that individuals that are "pure line" are homozygous for the gene in question?
   a. They are highly inbred.
   b. Only two alleles are present at each gene in the populations to which these individuals belong.
   c. In a pure line, phenotypes are not affected by environmental conditions or gene interactions.
   d. No other phenotype arises in a pure-line population because no other alleles are present.
5. The genes for the traits that Mendel worked with are either located on different chromosomes or so far apart on the same chromosome that crossing over almost always occurs between them. How did this circumstance help Mendel recognize the principle of independent assortment?
   a. Otherwise, his dihybrid crosses would not have produced a 9:3:3:1 ratio of F_2 phenotypes.
   b. The occurrence of individuals with unexpected phenotypes led him to the discovery of recombination.
   c. It led him to the realization that the behavior of chromosomes during meiosis explained his results.
   d. It meant that the alleles involved were either dominant or recessive, which gave 3:1 ratios in the F_1 generation.
6. The text claims that Mendel worked with the simplest possible genetic system. Is this claim legitimate?
   a. Yes—discrete traits, two alleles, simple dominance and recessiveness, no sex chromosomes, and unlinked genes are the simplest situation known.
   b. Yes—the ability to self-fertilize or cross-pollinate made it simple for Mendel to set up controlled crosses.
   c. No—Mendel was unaware of meiosis and the chromosome theory of inheritance, so it was not easy to reach the conclusions he did.
   d. No—Mendel's experimental designs and his rules of inheritance are actually quite complex and sophisticated.
7. Mendel's rules do not correctly predict patterns of inheritance for tightly linked genes or the inheritance of alleles that show incomplete dominance or epistasis. Does this mean that his hypotheses are incorrect?
   a. Yes, because they are relevant to only a small number of organisms and traits.
   b. Yes, because not all data support his hypotheses.
   c. No, because he was not aware of meiosis or the chromosome theory of inheritance.
   d. No, it just means that his hypotheses are limited to certain conditions.
8. The artificial sweetener NutraSweet consists of a phenylalanine molecule linked to aspartic acid. The labels of diet sodas that contain NutraSweet include a warning to people with PKU. Why?
   a. NutraSweet stimulates the same taste receptors that natural sugars do.
   b. People with PKU have to avoid phenylalanine in their diet.
   c. In people with PKU, phenylalanine reacts with aspartic acid to form a toxic compound.
   d. People with PKU cannot lead normal lives, even if their environment is carefully controlled.
9. When Sutton and Boveri published the chromosome theory of inheritance, research on meiosis had not yet established that paternal and maternal homologs assort independently of each other. Then, in 1913, Elinor Carothers published a paper about a grasshopper species with an unusual karyotype: One chromosome had no homolog (meaning no pairing partner at meiosis I); another chromosome had homologs that could be distinguished under the light microscope. If chromosomes assort independently, how often should Carothers have observed each of the four products of meiosis shown in the following figure?

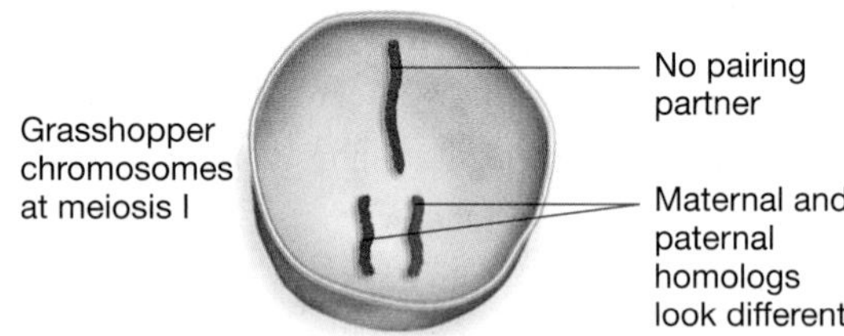

Four types of gametes possible
(each meiotic division can produce only two of the four)

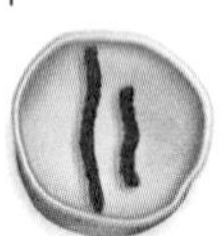  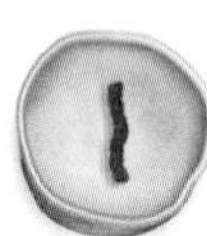 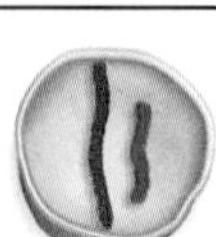

   a. Only the gametes with one of each type of chromosome would occur.
   b. The four types of gametes should be observed to occur at equal frequencies.
   c. The chromosome with no pairing partner would disintegrate, so only gametes with one copy of the other chromosome would be observed.

d. Gametes with one of each type of chromosome would occur twice as often as gametes with just one chromosome.

10. Which of the following is the strongest evidence that a trait might be influenced by polygenic inheritance?
   a. F_1 offspring of parents with different phenotypes have an intermediate phenotype.
   b. F_1 offspring of parents with different phenotypes have the dominant phenotype.
   c. The trait shows qualitative (discrete) variation.
   d. The trait shows quantitative variation.

Genetics Questions answers: 1. d; 2. b; 3. a; 4. d; 5. a; 6. a; 7. d; 8. b; 9. b; 10. d

SOLVING GENETICS PROBLEMS

The best way to test and extend your knowledge of transmission genetics is to work problems. Most genetics problems are set up as follows: You are given some information about the genotypes or phenotypes of one or both parents, along with data on the phenotypes of F_1 or F_2 offspring. Your task is to generate a set of hypotheses—or what biologists call a genetic model—to explain the results. More specifically, you will need to generate a hypothesis to address each of the following questions:

- Is the trait involved discrete or quantitative?
- Is the phenotype a product of one gene or many genes?
- For each gene involved, how many alleles are present—one, two, or many?
- Do the alleles involved show complete dominance, incomplete dominance, or codominance?
- Are the genes involved sex-linked or autosomal?
- If more than one gene is involved, are they linked or unlinked? If they are linked, does crossing over occur frequently?

It's also helpful to ask yourself whether gene interactions or pleiotropy might be occurring and whether it is safe to assume that the experimental design carefully controlled for effects of variation in other genes or the environment.

In working the problem, be sure to start with the simplest possible explanation. For example, if you are dealing with a discrete trait, you might hypothesize that the cross involves a single autosomal gene with two alleles that show complete dominance. Your next step is to infer what the parental genotypes are (if they are not already given), and then do a Punnett square to predict the offspring phenotypes and their frequencies based on your hypotheses. Next, check whether these predictions match the observed results given in the problem. If the answer is yes, you have a valid solution. But if the answer is no, you need to go back and change one of your hypotheses, redo the Punnett square, and check to see if the predictions and observations match. Keep redoing these steps until you have a model that fits the data.

Example Problems

Example 1 *Plectritis congesta* plants produce fruits that either have or do not have prominent structures called wings. The alleles involved are W^+ = winged fruit; W^- = wingless fruit. Researchers collected an array of individuals from the field and performed a series of crosses. The results are given in the following table. Complete the table by writing down the genotype of the parent or parents involved in each cross.

Parental Phenotype(s)	Number of Offspring with Winged Fruits	Number of Offspring with Wingless Fruits	Parental Genotype(s)
Wingless (self-fertilized)	0	80	
Winged (self-fertilized)	90	30	
Winged × wingless	46	0	
Winged × winged	44	0	

Solution Here you're given offspring phenotypes and you're asked to infer parental genotypes. To do this you have to hypothesize that certain parental genotypes are correct, make a Punnett square to predict the offspring genotypes, and then see if the offspring phenotypes match the data. In this case, coming up with a hypothesis for the parental genotypes is relatively straightforward, because the problem states that the trait is due to one gene and two alleles. No information on sex is given, so assume the gene is autosomal. Now look at the second entry in the chart. It shows a 3:1 ratio of offspring from a winged individual that self-fertilizes. This result is consistent with the hypothesis that W^+ is dominant and W^- recessive and that this parent's genotype is W^+W^-.

Now let's look at the first cross in the chart. If W^+ is dominant, then a wingless parent must be W^-W^-. When you do the Punnett square to predict offspring genotypes from selfing, you find that all the offspring will produce wingless fruits, consistent with the data. In the third cross, all the offspring make winged fruits even though one of the parents produces wingless fruits and thus is W^-W^-. This would happen only if the winged parent is W^+W^+. (If this reasoning isn't immediately clear to you, work the Punnett square.) In the fourth cross, you could get offspring that all make winged fruits if the parents were W^+W^+ and W^+W^+, or if the parents were W^+W^+ and W^+W^-. Both answers are correct. Again, you can write out the Punnett squares to convince yourself that this is correct.

Example 2 Two black female mice are crossed with a brown male. In several litters, female I produced 9 blacks and 7 browns; female II produced 57 blacks. What deductions can you make concerning the inheritance of black and brown coat color in mice? What are the genotypes of the parents in this case?

Solution Here you are given parental and offspring phenotypes and are asked to infer the parental genotypes. As a starting point, assume that the coat colors are due to the simplest genetic system possible: one autosomal gene with two alleles, where one allele is dominant and the other recessive. Because female II produces only black offspring, it's logical to suppose that black is dominant to brown. Let's use *B* for black and *b* for brown. Then the male parent is *bb*. To produce offspring with a 1:1 ratio of black : brown coats, female I must be *Bb*. But to produce all black offspring, female II must be *BB*. This model explains the data, so you can accept it as correct.

GENETICS PROBLEMS

11. *Tay-Sachs disease* causes nerve cells to malfunction and results in death by age 4. Two healthy parents know from blood tests that each parent carries a recessive allele responsible for Tay-Sachs. If their first three children have the disease, what is the probability that their fourth child will not? Assuming they have not yet had a child, what is the probability that, if they have four children, all four will have the disease? If their first three children are male, what is the probability that their fourth child will be male?

12. Suppose that in garden peas the genes for seed color and seed-pod shape are linked, and that Mendel crossed *YYII* parents (which produce yellow seeds in inflated pods) with *yyii* parents (which produce green seeds in constricted pods).
    - Draw the F_1 Punnett square and predict the expected F_1 phenotype(s).
    - List the genotype(s) of gametes produced by F_1 individuals if no crossing over occurs.
    - Draw the F_2 Punnett square if no crossing over occurs. Based on this Punnett square, predict the expected phenotype(s) in the F_2 generation and the expected frequency of each phenotype.
    - If crossing over occurs during gamete formation in F_1 individuals, give the genotype of the recombinant gamete(s) that result.
    - Add the recombinant gametes to the F_2 Punnett square above. Will any additional phenotypes be observed at low frequency in the F_2 generation? If so, what are they?

13. In Jimson weed, the allele that results in violet flowers (V) is dominant to one that results in white flowers (v); at another locus, the allele that makes prickly seed capsules (P) is dominant to one that results in smooth capsules (p). A plant with white flowers and prickly capsules is crossed with one that has violet flowers and smooth capsules. The F_1 consisted of 47 plants with white flowers and prickly capsules, 45 plants with white flowers and smooth capsules, 50 plants with violet flowers and prickly capsules, and 46 plants with violet flowers and smooth capsules. What are the genotypes of the parents? Are these two genes linked?

14. In cats, the *Manx* allele (M) causes a short or absent tail, while a recessive allele m confers a normal, long tail. Cats of genotype MM die as embryos. If two Manx cats mate, what is the probability that each *surviving* kitten has a long tail?

15. A plant with orange-spotted flowers was grown in the greenhouse from a seed collected in the wild. The plant was self-pollinated and gave rise to the following progeny: 88 orange with spots, 34 yellow with spots, 32 orange with no spots, and 8 yellow with no spots. What can you conclude about the dominance relationships of the alleles responsible for the spotted and unspotted phenotypes? Of orange and yellow phenotypes? What can you conclude about the genotype of the original plant that had orange, spotted flowers?

16. As a genetic counselor, you routinely advise couples about the possibility of genetic disease in their offspring based on their family histories. This morning you met with an engaged couple, both of whom are phenotypically normal. The man, however, has a brother who died of Duchenne-type muscular dystrophy, an X-linked condition that results in death before the age of 20. The allele responsible for this disease is recessive. His prospective bride, whose family has no history of the disease, is worried that the couple's sons or daughters might be afflicted.
    - How would you advise this couple?
    - The sister of this man is planning to marry his fiancé's brother. How would you advise this second couple?

17. Suppose you are heterozygous for two genes that are located on different chromosomes. You carry alleles A and a for one gene and alleles B and b for the other. Draw a diagram illustrating what happens to these genes and alleles when meiosis occurs in your reproductive tissues. Label the stages of meiosis, the homologous chromosomes, sister chromatids, nonhomologous chromosomes, genes, and alleles. Be sure to list all of the genetically different gametes that could form and indicate how frequently each type should be observed. On the diagram, identify the events responsible for the principle of segregation and the principle of independent assortment.

18. In humans, the ABO blood type is a polymorphic trait due to alleles i, I^A, and I^B. I^Ai and I^AI^A individuals have blood type A; I^Bi and I^BI^B individuals have blood type B; I^AI^B individuals have blood type AB; ii individuals have blood type O. Suppose a woman with blood type O married a man with blood type AB. What phenotypes and genotypes would you expect to observe in their offspring, and in what proportions? Answer the same question for a heterozygous mother with blood type A and a heterozygous father with blood type B.

19. Mr. Spock's father came from the planet Vulcan, and his mother was from planet Earth. A Vulcan has pointed ears and a heart located on the right side of the chest. Mr. Spock has both of these traits, which are known to be determined by two different genes, each with two alleles. Suppose that Mr. Spock married an Earth woman and that they had many children. About half of their children have pointed ears and a right-sided heart, like Spock, and about half have rounded ears and a left-sided heart, like their mother.
    - What would Mendel predict the progeny phenotypes and ratios to be? Explain your answer using formal genetic terminology.
    - How do you explain the actual results?

20. In Klingons, one gene determines hair texture; another gene determines whether the individual will have a sagittal crest (a protrusion on the forehead). The two genes are not linked.

    K = curly Klingon hair (dominant)

    k = silky earthling-like hair (recessive)

    S = large sagittal crest (dominant)

    s = smooth, flat, earthling-like forehead (recessive)

    Kayless has the genotype *KkSs*. He mates with an individual who is also heterozygous for both genes.
    - Set up a Punnett square for this dihybrid cross.
    - What are the four possible phenotypes that may result from this mating? Include a description of both hair and forehead for each phenotype.
    - What is the expected phenotypic ratio from the dihybrid cross?
    - What fraction of the progeny are expected to be heterozygous for both genes?

- What fraction are expected to be homozygous for both genes?
- Are Kayless and his mate more likely to see an actual ratio close to the predicted values if they have 16 children or 160? Why? Explain why sample size does or does not affect observed phenotypic ratios.

21. The blending-inheritance hypothesis proposed that the genetic material from parents is irreversibly mixed in the offspring. As a result, offspring should always appear intermediate in phenotype to the parents. Mendel, in contrast, proposed that genes are discrete and that their integrity is maintained in the offspring and in subsequent generations. Suppose the year is 1890. You are a horse breeder and have just read Mendel's paper. You don't believe his results, however, because you often work with cremello (very light-colored) and chestnut (reddish-brown) horses. You know that if you cross a cremello individual from a pure-breeding line with a chestnut individual from a pure-breeding line, the offspring will be palomino—meaning they have an intermediate (golden-yellow) body color. What additional crosses would you do to test whether Mendel's model is valid in the case of genes for horse color? List the crosses and the offspring genotypes and phenotypes you'd expect to obtain. Explain why these experimental crosses would provide a test of Mendel's model.

22. Two mothers give birth to sons at the same time in a busy hospital. The son of couple 1 is afflicted with hemophilia A, which is a recessive X-linked disease. Neither parent has the disease. Couple 2 has a normal son even though the father has hemophilia A. The two couples sue the hospital in court, claiming that a careless staff member swapped their babies at birth. You appear in court as an expert witness. What do you tell the jury? Make a diagram that you can submit to the jury.

23. You have crossed two *Drosophila melanogaster* individuals that have long wings and red eyes—the wild-type phenotype. In the progeny, the mutant phenotypes, called curved wings and lozenge eyes, appear as follows:

    Females
    600 long wings, red eyes
    200 curved wings, red eyes

    Males
    300 long wings, red eyes
    300 long wings, lozenge eyes
    100 curved wings, red eyes
    100 curved wings, lozenge eyes

    - According to these data, is the curved-wing allele autosomal recessive, autosomal dominant, sex-linked recessive, or sex-linked dominant?
    - Is the lozenge-eye allele autosomal recessive, autosomal dominant, sex-linked recessive, or sex-linked dominant?
    - What is the genotype of the female parent?
    - What is the genotype of the male parent?

24. In parakeets, two autosomal genes that are located on different chromosomes control the production of feather pigment. Gene *B* codes for an enzyme that is required for the synthesis of a blue pigment, and gene *Y* codes for an enzyme required for the synthesis of a yellow pigment. Recessive, loss-of-function mutations are known for both genes. Suppose that a bird breeder has two green parakeets and mates them. The offspring are green, blue, yellow, and albino (unpigmented).
    - Based on this observation, what are the genotypes of the green parents? What is the genotype of each type of offspring? What fraction of the total progeny should exhibit each type of color?
    - Suppose that the parents were the progeny of a cross between two true-breeding strains. What two types of crosses between true-breeding strains could have produced the green parents? Indicate the genotypes and phenotypes for each cross.

25. The pedigree shown below is for the human trait called osteopetrosis, which is characterized by bone fragility and dental abscesses. Is the gene that affects bone and tooth structure autosomal or sex-linked? Is the allele for osteopetrosis dominant or recessive?

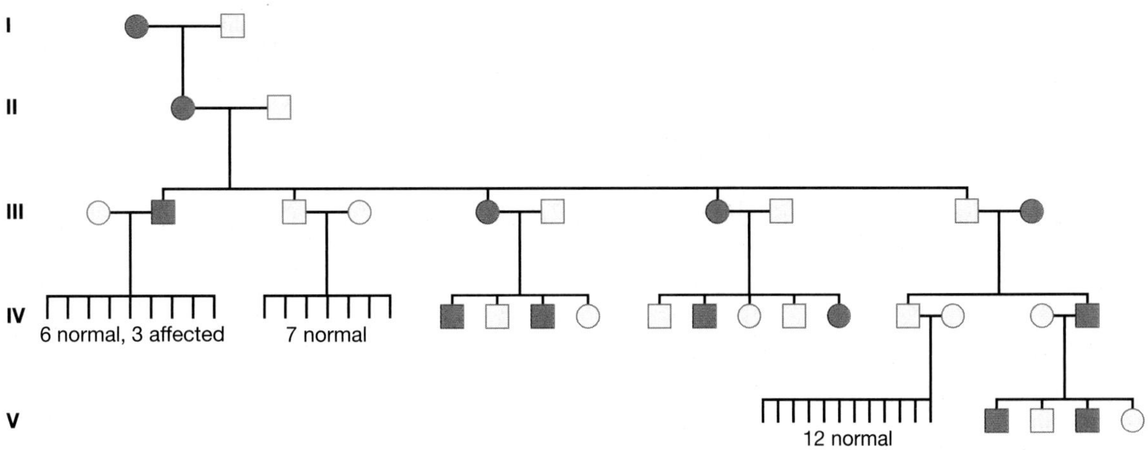

Pedigree of the trait osteopetrosis

Genetics Problems answers:

11. 3/4; 1/256 (see **BioSkills 9**); 1/2 (the probabilities of transmitting the alleles or having sons does not change over time). 12. Your answer should conform to the cross diagrammed in Figure 13.5b, except that different alleles and traits are being analyzed. 13. Parental genotypes are *vvPp* × *Vvpp*. Based on these data, we can't tell if they are linked or not. Both hypotheses make the same prediction. 14. 1/3 (the question asks about the *surviving* offspring). 15. This is a dihybrid cross that yields progeny phenotypes in a 9:3:3:1 ratio. Let *O* stand for the allele for orange petals and *o* the allele for yellow petals; let *S* stand for the allele for spotted petals and *s* the allele for unspotted petals. Start with the hypothesis that *O* is dominant to *o*, that *S* is dominant to *s*, that the two genes are found on different chromosomes so they assort independently, and that the parent individual's genotype is *OoSs*. If you do a Punnett square for the *OoSs* × *OoSs* mating, you'll find that progeny phenotypes should be in the observed 9:3:3:1 proportions. 16. Let *D* stand for the normal allele and *d* stand for the allele responsible for Duchenne-type muscular dystrophy. The woman's family has no history of the disease, so her genotype is almost certainly *DD*. The man is not afflicted, so he must be *DY*. (The trait is X-linked, so he has only one allele; the "*Y*" stands for the *Y* chromosome.) Their children are not at risk. The man's sister could be a carrier, however—meaning she has the genotype *Dd*. If so, then half of the second couple's male children are likely to be affected. 17. Your stages of meiosis should look like Figure 12.4. The *A* and *a* alleles could be on the red and blue versions of the longer chromosome, and the *B* and *b* alleles could be on the red and blue versions of the smaller chromosomes. The places you draw them are the locations of the *A* and *B* genes, but each chromosome only has one allele. Each pair of red and blue chromosomes is a homologous pair. Sister chromatids bear the same allele (e.g., both sister chromatids of the long blue chromosomes might bear the *a* allele). Chromatids from the long and short chromosomes are not homologous. To identify the events that cause the principles of segregation and independent assortment, see Figures 13.7 and 13.8, and substitute *A* and *B* for *R* and *Y*. 18. Half of their offspring should have the genotype iI^A and the type A blood phenotype. Half of their offspring should have the genotype iI^B and the type B blood phenotype. Second case: the genotype and phenotype ratios would be 1:1:1:1 I^AI^B (type AB): I^Ai (type A): I^Bi (type B); *ii* (type O). 19. Presumably, Vulcans and earthlings are homozygous for both traits because they are "pure lines." Spock, then, should be heterozygous at both loci. Because he has the Vulcan phenotype, though, we can infer that pointed ears and a right-sided heart are dominant to rounded ears and a left-sided heart. Let *P* be the allele for pointed ears and *p* be the allele for rounded ears; let *R* be the allele for a right-sided heart and *r* the allele for a left-sided heart. Spock is *PpRr;* his earthling wife is *pprr*. Mendel would predict that their offspring would have the genotypes *PpRr, Pprr, ppRr*, and *pprr* in equal proportions. Thus a quarter of their children should have pointed ears and left-sided hearts or rounded ears and right-sided hearts. The only logical explanation for the actual results is that the two genes are closely linked. If so, then Spock's gametes are either *PR* or *pr,* and his wife's gametes are all *pr*. 20. Four phenotypes should be observed in the ratio 9:3:3:1, as follows: curly and large: curly and smooth: silky and large: silky and smooth. One-quarter of the progeny should be heterozygous for both traits. One-eighth should be homozygous for both traits. They are more likely to see the predicted ratios with a large family, because there would be fewer fluctuations due to chance. 21. According to Mendel's model, palomino individuals should be heterozygous at the locus for coat color. If you mated palomino individuals, you would expect to see a combination of chestnut, palomino, and cremello offspring. If blending inheritance occurred, however, all of the offspring should be palomino. 22. Because this is an X-linked trait, the father who has hemophilia could not have passed the trait on to his son. Thus, the mother in couple 1 must be a carrier and have passed the recessive allele on to her son, who is XY and affected. To educate a jury about the situation, you should draw what happens to the X and Y during meiosis, then make a drawing showing the chromosomes in couple 1 and couple 2, with a Punnett square showing how these chromosomes get passed to the affected and unaffected children. 23. The curved-wing allele is autosomal recessive; the lozenge-eye allele is sex-linked (specifically, X-linked) recessive. Let l^+ be the allele for long wings and *l* be the allele for curved wings; let r^+ be the allele for red eyes and *r* the allele for lozenge eyes. The female parent is l^+lr^+r; the male parent is l^+lr^+Y. 24. Albinism indicates the absence of pigment, so let b stand for an allele that gives the absence of blue and y for an allele that gives the absence of yellow pigment. If blue and yellow pigment blend to give green, then both green parents are *BbYy*. The green phenotype is found in *BBYY, BBYy, BbYY*, and *BbYy* offspring. The blue phenotype is found in *BByy* or *Bbyy* offspring. The yellow phenotype is observed in *bbYY* or *bbYy* offspring. Albino offspring are *bbyy*. The phenotypes of the offspring should be in the ratio 9:3:3:1 as green:blue:yellow:albino. Two types of crosses yield *BbYy* F_1 offspring: *BByy* × *bbYY* (blue × yellow) and *BBYY* × *bbyy* (green × albino). 25. Autosomal dominant.

www.masteringbio.com is also your resource for • Answers to text, table, and figure caption questions and exercises • Answers to *Check Your Understanding* boxes • Online study guides and quizzes • Additional study tools including the *E-Book for Biological Science* 3rd ed., textbook art, animations, and videos.

DNA and the Gene: Synthesis and Repair

14

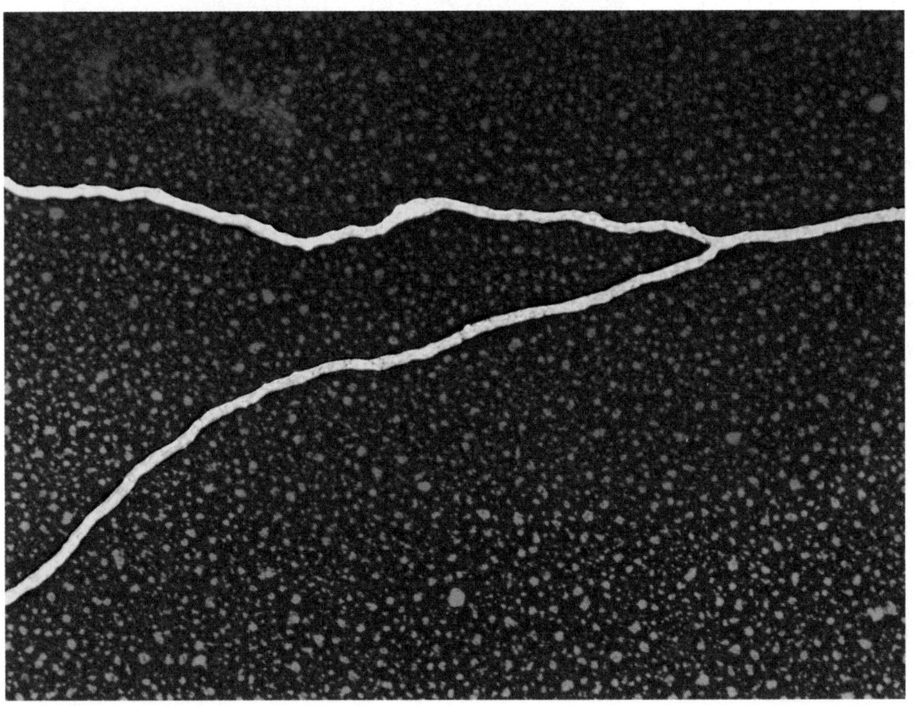

Electron micrograph showing DNA in the process of replication. The "Y" shape results from a structure called the replication fork, where DNA synthesis is taking place. The DNA double helix on the right is being replicated into two DNA double helices on the left.

KEY CONCEPTS

- Genes are made of DNA. When DNA is copied, each strand of a DNA double helix serves as the template for the synthesis of a complementary strand.
- DNA is synthesized only in the $5' \rightarrow 3'$ direction. When a DNA molecule is being copied, a large suite of specialized enzymes is involved in unwinding the double helix, continuously synthesizing the "leading strand" in the $5' \rightarrow 3'$ direction and synthesizing the "lagging strand" as a series of fragments that are then linked together.
- Specialized enzymes repair mistakes in DNA synthesis and damaged DNA. If these repair enzymes are defective, the mutation rate increases. Mutations in many types of genes can lead to cancer.

What are genes made of, and how are they copied so that they are faithfully passed on to offspring? These questions dominated biology during the middle of the twentieth century. Since Mendel's time, the predominant research strategy in genetics had been to conduct a series of experimental crosses, create a genetic model to explain the types and proportions of phenotypes that resulted, and then test the model's predictions through reciprocal crosses, testcrosses, or other techniques. This strategy was extremely productive. It led to virtually all of the discoveries analyzed in Chapter 13, including Mendel's rules, sex-linkage, linkage, genetic mapping, incomplete dominance and codominance, multiple allelism and polymorphic traits, pleiotropy, and quantitative inheritance.

But the molecular nature of Mendel's hereditary factors—what came to be called genes—remained a mystery. Further, biologists knew that genes and chromosomes were replicated during S phase of the cell cycle (see Chapter 11), with copies distributed to daughter cells during mitosis and meiosis. No one had the slightest clue about how the copying occurred, however.

The goal of this chapter is to explore how researchers solved these mysteries. Let's begin with studies that identified deoxyribonucleic acid (DNA) as the genetic material, explore how this molecule is copied during the synthesis phase of the cell cycle, and conclude by analyzing how damaged or incorrectly copied genes are repaired. Once the molecular nature of the gene was known, the nature of biological science changed forever.

14.1 DNA as the Hereditary Material

The first hint of what genes might be made of was published in 1928, when Frederick Griffith reported the discovery of a mysterious phenomenon involving hereditary traits. Griffith referred to this phenomenon as **transformation.** His transformation experiments appeared to isolate the hereditary material.

In the 1920s Griffith had been doing experiments in an attempt to develop a vaccine against the *Streptococcus pneumoniae* bacterium, which is a leading cause of pneumonia, earaches, sinusitis, and meningitis in humans. For his experiments, however, he worked with strains of *S. pneumoniae* that infect mice. A **strain** is a population of genetically identical individuals. As is the case with the strains that affect humans, the strains that affect mice vary in their **virulence**—their ability to cause disease and death. Virulent strains cause disease; avirulent (or benign) strains do not.

As **Figure 14.1** shows, the strains that Griffith happened to work with can be identified by eye when grown on a nutrient-containing medium in a petri dish. A **medium** is a liquid or solid that is suitable for growing cells. On a solid medium, cells from the nonvirulent strain form colonies that look rough; cells from the virulent strain form colonies that look smooth. Logically enough, Griffith called the nonvirulent strain R for rough and the virulent strain S for smooth.

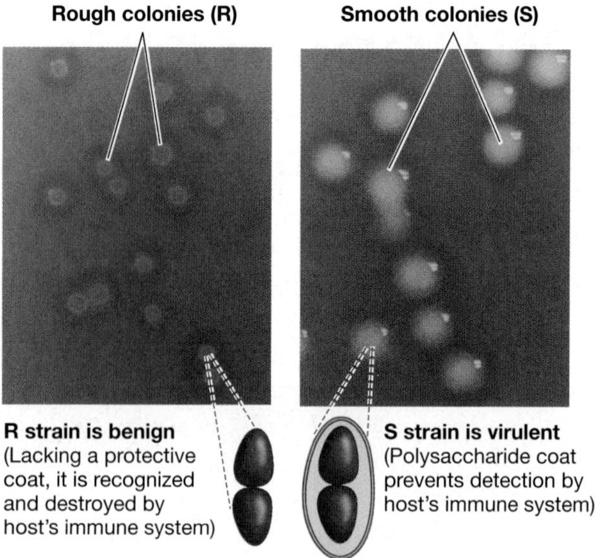

FIGURE 14.1 There Are Two Strains of *Streptococcus pneumoniae*. Nonvirulent (R) and virulent (S) strains of *Streptococcus* differ in colony appearance and at the cellular level.

To understand how the strains interact, Griffith carried out the experiment illustrated in **Figure 14.2**. This figure shows four experimental treatments that he designed. In the first treatment, he injected mice with cells of the nonvirulent R strain. As he expected, these mice lived. In the second treatment, he injected

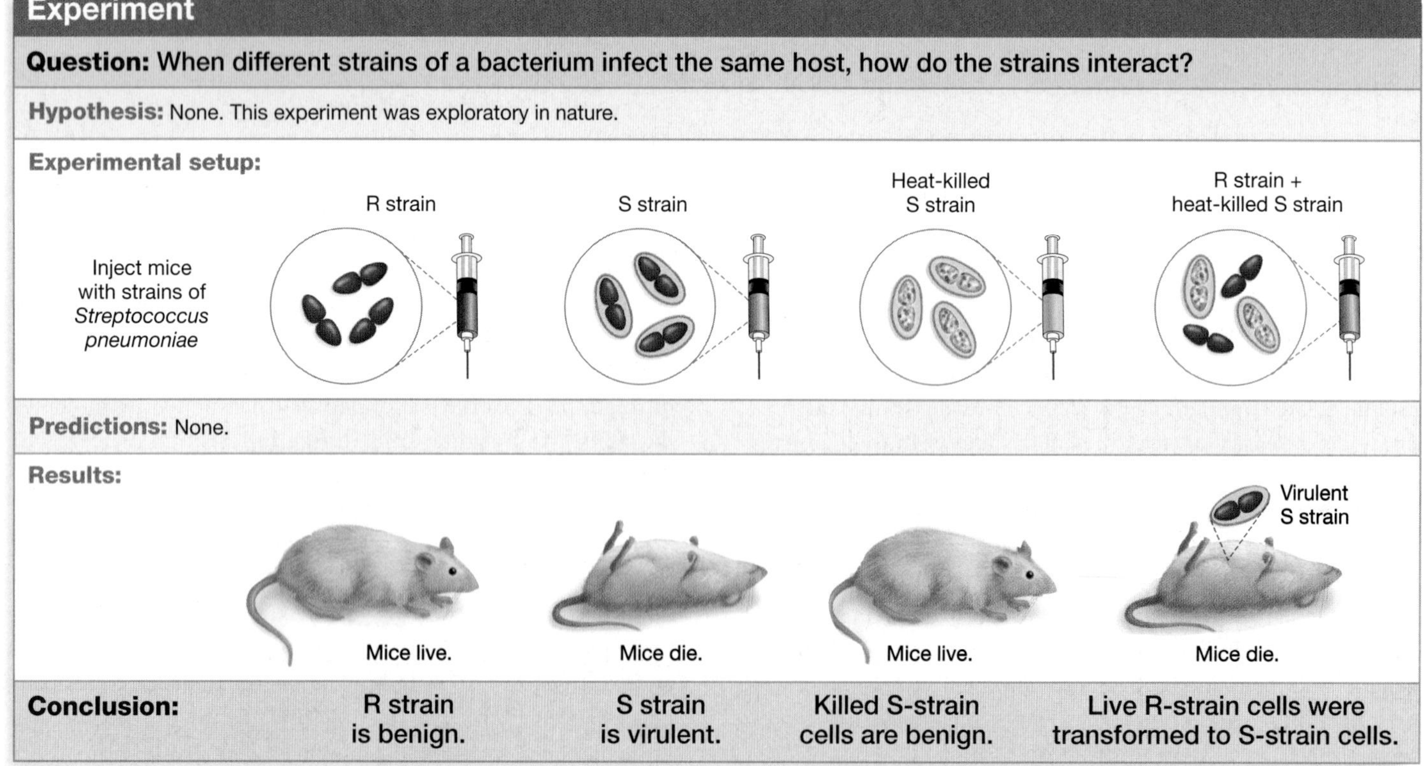

FIGURE 14.2 The Discovery of "Transformation." Mice died after being injected with a combination of live R cells and heat-killed S cells.

mice with cells of the virulent S strain. Not surprisingly, these mice died of pneumonia. So far, so good: These first two treatments were controls showing the effect that each strain of *S. pneumoniae* has on mice. In the third treatment, Griffith killed cells of the virulent S strain by heating them and then injected them into mice. These mice lived. This experimental treatment was interesting because it showed that dead S cells do not cause disease. In the final treatment, Griffith injected mice with heat-killed S cells *and* live nonvirulent R cells. Unexpectedly, these mice died. Autopsies confirmed pneumonia as the cause of death. When Griffith isolated and grew *Streptococcus* from these dead mice, he found virulent S cells, *not* nonvirulent R cells.

What was going on? Griffith proposed that something from the heat-killed S cells had "transformed" the nonvirulent R cells. Stated another way, something had changed the appearance and behavior of the R cells from R-like to S-like. Because this "something" appeared in the growing population of cells that Griffith isolated from the dead mice, it had been passed on to the offspring of the transformed cells. It was clearly some sort of hereditary, or genetic, factor. Next the question became, What is it?

Is DNA the Genetic Material?

The chromosome theory of inheritance, introduced in Chapter 13, proposed that chromosomes are composed of genes. It had been known since the late 1800s that chromosomes are a complex of DNA and proteins. Because the chromosome theory had been confirmed around 1920, it was clear that Griffith's transforming factor had to consist of either protein or DNA.

Initially, most biologists backed the hypothesis that genes are made of proteins. The arguments in favor of this hypothesis were compelling. Hundreds, if not thousands, of complex and highly regulated chemical reactions occur in even the simplest living cells. The amount of information required to specify and coordinate these reactions is almost mind-boggling. With their almost limitless variation in structure and function, proteins are complex enough to contain this much information.

DNA, in contrast, was known to be composed of just four types of deoxyribonucleotides—the monomers introduced in Chapter 4. It was also thought to be a simple molecule with some sort of repetitive and uninteresting structure. So when researchers published the first experimental evidence that DNA was the hereditary material, most biologists had the same reaction: They didn't believe it.

The Avery et al. Experiment In the early 1940s Oswald Avery, Colin M. MacLeod, and Maclyn McCarty set out to understand the molecular basis of Griffith's result. They used an elegant experimental strategy to isolate the transforming factor. To determine whether protein, ribonucleic acid (RNA—see Chapter 4), or DNA was responsible for transformation, they grew quantities of virulent S cells in culture. A **culture** is a collection of cells that grows under controlled conditions—usually suspended in a liquid medium or on the surface of a solid growth medium. Avery's group killed the cultured cells with heat, broke them open to create a cell extract, and then used chemical treatments to remove the lipids and carbohydrates from the extracts. These steps left a mixture containing protein, RNA, and DNA from the virulent S cells. The researchers divided the sample into three treatments and used different enzymes to destroy a specific macromolecule in each. One sample was treated with proteases, enzymes that destroy proteins. Another sample was treated with ribonuclease, an enzyme that breaks apart RNA. A third sample was treated with DNAase, an enzyme that cuts up DNA (**Figure 14.3**). When small quantities of the three resulting solutions were added to cultures containing nonvirulent R cells, virulent S cells appeared in all of the cultures that still contained S-cell DNA; no S cells appeared in the sample that lacked DNA. The biologists concluded that DNA—not protein or RNA—must be the transforming factor.

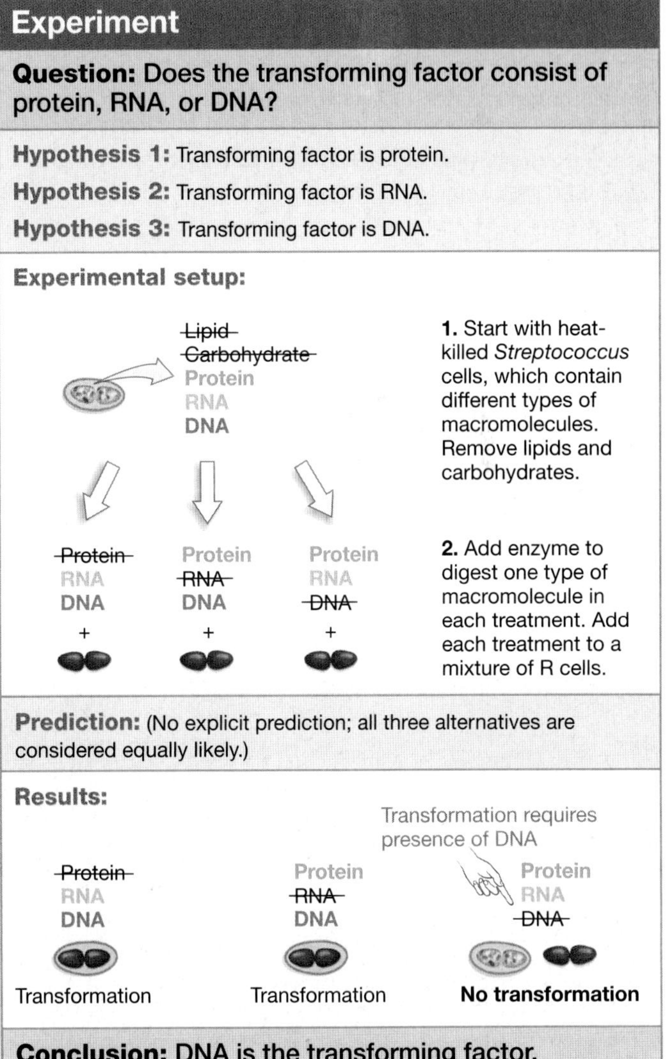

FIGURE 14.3 Experimental Evidence that DNA Is the Transforming Factor.

QUESTION If genes were made of protein, how would the results of this experiment change?

This experiment provided strong evidence that DNA is the hereditary material. But the result was not widely accepted. Researchers who advocated the protein hypothesis claimed that the enzymatic treatments were not sufficient to remove all of the proteins present and that enough protein could remain to transform R cells.

The Hershey-Chase Experiment Alfred Hershey and Martha Chase took up the question of whether genes were made of protein or DNA by studying how a virus called T2 infects the bacterium *Escherichia coli*. Hershey and Chase knew that T2 infections begin when the virus attaches to the cell wall of *E. coli* and injects its genes into the cell's interior (**Figure 14.4a**). These genes then direct the production of a new generation of virus particles inside the infected cell, which acts as a **host** for the parasitic virus. During the infection, the protein coat of the original, parent virus is left behind, still attached to the exterior of the host cell as a "ghost" (**Figure 14.4b**). Hershey and Chase also knew that T2 is made up almost exclusively of protein and DNA. But was it protein or DNA that entered the host cell and directed the production of new viruses?

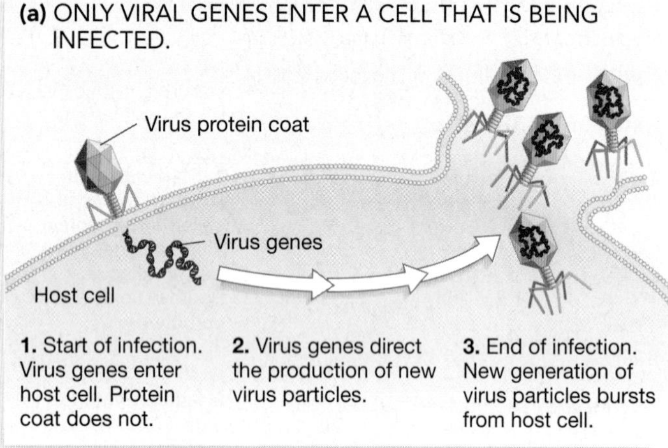

(b) The virus's protein coat stays on the outside of the host cell.

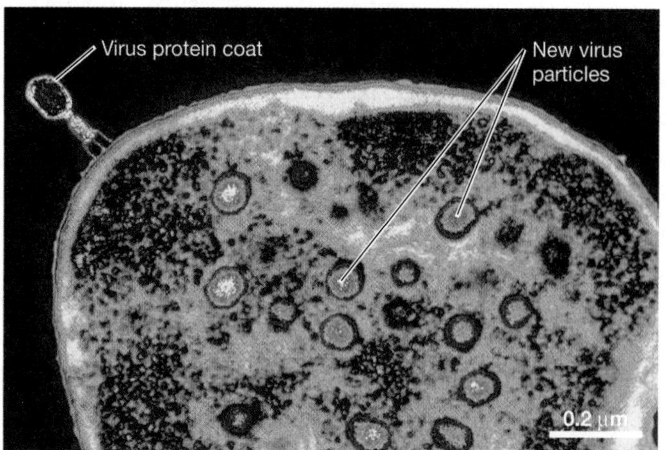

FIGURE 14.4 Viruses Inject Genes into Bacterial Cells and Leave a Protein Coat Behind. (a) Viruses that infect a bacterial cell start by injecting their genes into the cell. **(b)** The protein coat of the original virus is left behind, attached to the bacterial cell wall.

Hershey and Chase's strategy for determining which part of the virus enters the cell and acts as the hereditary material was based on two facts: (1) Proteins present in T2 contain sulfur but not phosphorus, and (2) DNA contains phosphorus but not sulfur. As **Figure 14.5** shows, the researchers began their work by growing viruses in the presence of either the radioactive isotope

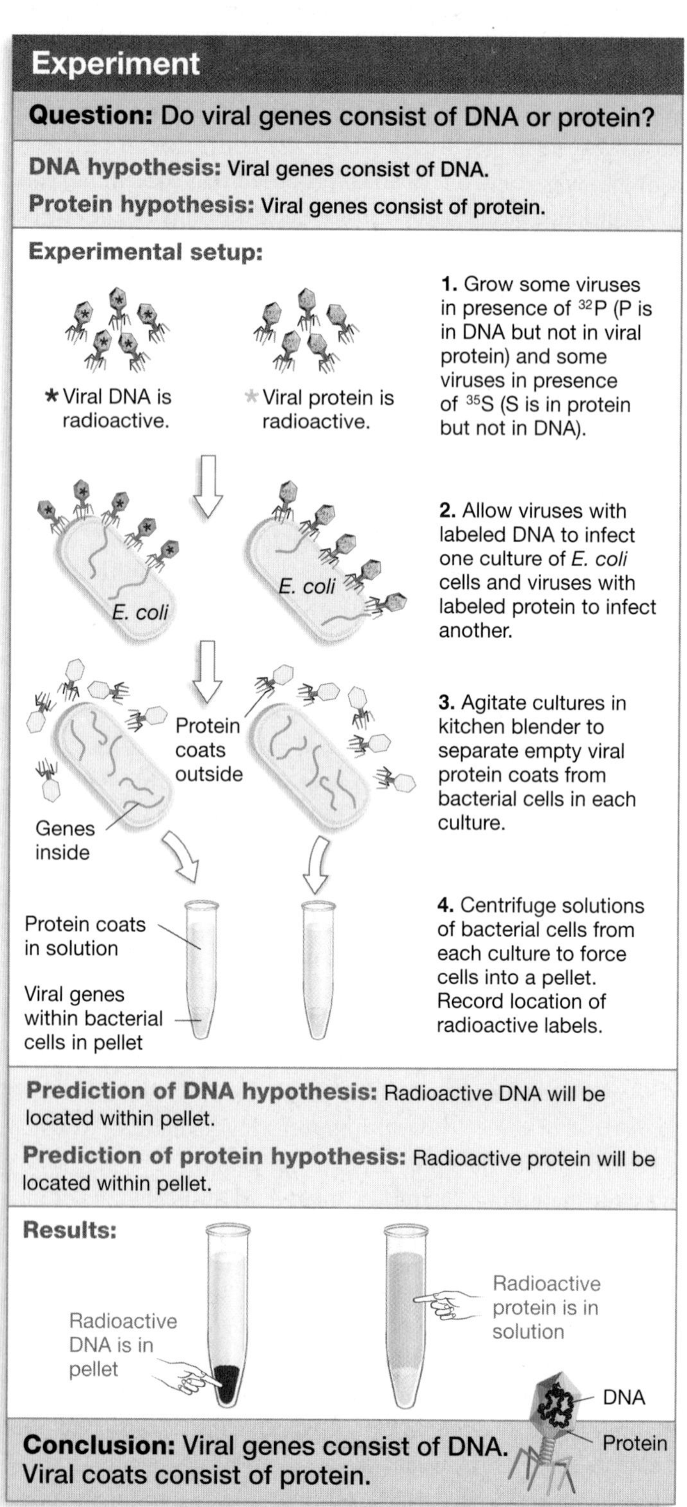

FIGURE 14.5 More Experimental Evidence that DNA Is the Hereditary Material. The experiment shown here convinced most biologists that DNA is indeed the hereditary material.

of sulfur (^{35}S) or the radioactive isotope of phosphorus (^{32}P). Because these molecules were incorporated into newly synthesized proteins and DNA, this step produced a population of viruses with radioactive proteins and a population with radioactive DNA. Then Hershey and Chase allowed each set of radioactive viruses to infect *E. coli* cells. If genes consist of DNA, then the radioactive protein should be found in the empty, or "ghost," protein coats outside the infected host cells, while the radioactive DNA should be located inside the cells. But if genes consist of proteins, the opposite should be true.

To test these predictions, Hershey and Chase sheared the ghosts off the cells by agitating each of the cultures in kitchen blenders. When the researchers spun the samples in a centrifuge, the ghosts stayed in the solution while the cells formed a pellet at the bottom of the centrifuge tube. As predicted by the DNA hypothesis, the biologists found that virtually all of the radioactive protein was in the ghosts, while virtually all of the radioactive DNA was inside the host cells. Because the injected component of the virus directs the production of a new generation of virus particles, it is this component that represents the virus's genes.

After these results were published, proponents of the protein hypothesis had to admit that DNA, not protein, must be the hereditary material. In combination, the evidence from the bacterial transformation experiments and the virus-labeling experiments was convincing. The claim that a seemingly simple molecule contained all the information for life's complexity was finally accepted.

The realization that genes are made of DNA was one of the great advances of twentieth-century biology. The insight raised two crucial questions, however: (1) How did the simple primary structure and secondary structure of DNA hold the information required to make life possible? (2) How is DNA copied, so that genetic information is faithfully passed from one cell to another during growth and from parents to offspring during reproduction? Understanding how DNA contains information is the focus of Chapter 15. The remainder of this chapter concentrates on how genes are replicated prior to mitosis and meiosis, and how damaged genes are repaired.

14.2 Testing Early Hypotheses about DNA Synthesis

The DNA inside each cell is like an ancient text that has been painstakingly copied and handed down, generation after generation. But while the most ancient of all human texts contain messages that are thousands of years old, the DNA in living cells has been copied and passed down for billions of years. And instead of being copied by monks or clerks, DNA is replicated by molecular scribes. What molecules are responsible for copying DNA, and how do they work?

Chapter 4 introduced Watson and Crick's model for the secondary structure of DNA, which was proposed in 1953. Recall that DNA is a long, linear polymer made up of monomers called deoxyribonucleotides, which consist of a deoxyribose molecule, a phosphate group, and a nitrogenous base (**Figure 14.6a**). Deoxyribonucleotides link together into a polymer when a phosphodiester bond forms between a hydroxyl group on the 3′ carbon of deoxyribose and the phosphate group attached to the 5′ carbon of deoxyribose. As **Figure 14.6b** shows, the primary structure of a DNA molecule has two major components: (1) a "backbone" made up of the sugar and phosphate groups of deoxyribonucleotides and (2) a series of nitrogen-containing bases that project from the backbone. A strand of DNA has a directionality or polarity: One end has an exposed hydroxyl group on the 3′ carbon of deoxyribose, while the other has an exposed phosphate group on a 5′ carbon. Thus, the molecule has a 3′ end and a 5′ end.

As they explored different models for the secondary structure of DNA, Watson and Crick hit on the idea of lining up two of these long strands in the opposite direction, or in what

(a) Structure of a deoxyribonucleotide

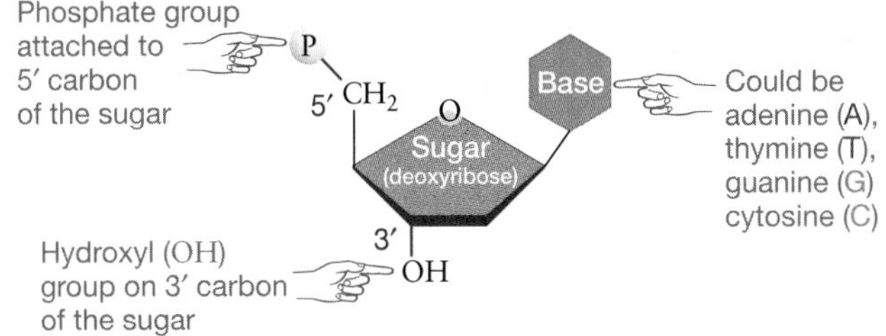

(b) Primary structure of DNA

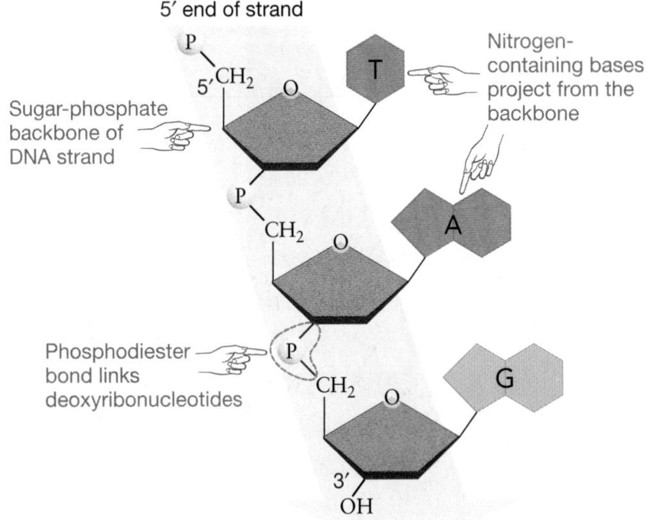

FIGURE 14.6 DNA's Primary Structure. (a) Deoxyribonucleotides are monomers that polymerize to form DNA. **(b)** DNA's primary structure is made up of a sequence of deoxyribonucleotides. Notice that the structure has a sugar-phosphate "backbone" with nitrogen-containing bases attached.

EXERCISE Write the base sequence of the DNA in part (b), in the 5′ → 3′ direction.

is called antiparallel fashion (**Figure 14.7a**). They realized that antiparallel strands will twist around each other into a spiral or helix because certain of the nitrogen-containing bases fit together in pairs inside the spiral and form hydrogen bonds. The double-stranded molecule that results is called a **double helix** (**Figure 14.7b**). This secondary structure is stabilized by hydrogen bonds that form between the nitrogenous bases adenine (A) and thymine (T) and between the bases guanine (G) and cytosine (C), along with hydrophobic interactions among the nitrogenous bases inside the helix. The specific pairing rules for hydrogen bonding of nitrogen-containing bases are called **complementary base pairing**.

Watson and Crick realized that the A-T and G-C pairing rules suggested a way for DNA to be copied when chromosomes are replicated during S phase of the cell cycle—prior to mitosis and meiosis. They suggested that the existing strands of DNA served as a template (pattern) for the production of new strands, with bases being added to the new strands according to complementary base pairing. For example, if the template strand contained a T, then an A would be added to the new strand to pair with that T. Similarly, a G on the template strand would dictate the addition of a C on the new strand.

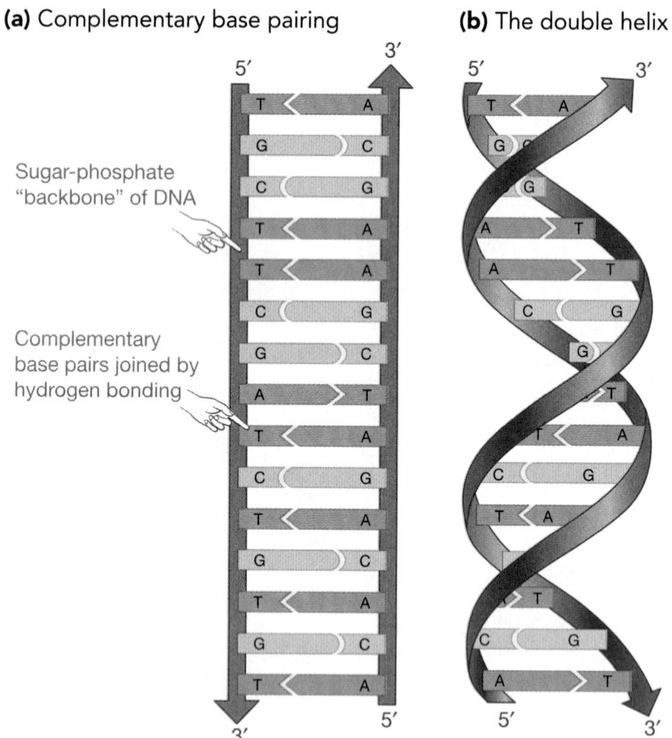

FIGURE 14.7 DNA's Secondary Structure: The Double Helix.
(a) Deoxyribonucleic acid normally has a secondary structure consisting of two strands, each with a sugar-phosphate backbone. Nitrogen-containing bases project from each strand and form hydrogen bonds. Only A-T and G-C pairs fit together in a way that allows hydrogen bonding to occur. **(b)** That bonding between complementary bases twists the molecule into a double helix.

QUESTION Why are some bases colored blue while others are green?

Biologists had three alternative hypotheses about how the old and new strands might interact during replication, however:

1. If the old strands of DNA separated, each could then be used as a template for the synthesis of a new, daughter strand. This hypothesis is called semiconservative replication, because each new daughter DNA molecule would consist of one old strand and one new strand.
2. If the bases temporarily turned outward so that complementary strands no longer faced each other, they could serve as a template for the synthesis of an entirely new double helix all at once. This hypothesis is called conservative replication. Conservative replication results in the intact parental strands and a daughter DNA molecule consisting entirely of newly synthesized strands.
3. If the parent helix was cut and unwound in short sections before being copied and put back together, then new and old strands would intermingle—old sections of DNA would be interspersed with new DNA down the length of each daughter molecule. This possibility is called **dispersive** replication.

Which of these three hypotheses is correct?

The Meselson–Stahl Experiment

Matthew Meselson and Frank Stahl realized that if they could tag parental and daughter strands of DNA in a way that would make them distinguishable from each other, they could determine whether replication was conservative, semiconservative, or dispersive. Soon after their results were published in 1958, the Meselson-Stahl work became recognized as a classic experiment in biological science.

Before they could do any tagging or publishing, however, they needed to choose an organism to study. They decided to work with a common inhabitant of the human gastrointestinal tract, the bacterium *Escherichia coli*. Because *E. coli* is small and grows quickly and readily in the laboratory, it had become a favored model organism in studies of biochemistry and molecular genetics.

Like all organisms, bacterial cells copy their entire complement of DNA, or their **genome**, before every cell division. To distinguish parental strands of DNA from daughter strands when *E. coli* replicates, Meselson and Stahl grew the cells for many generations in the presence of one of two isotopes of nitrogen: either ^{15}N or ^{14}N. Because ^{15}N contains an extra neutron, it is heavier than the normal isotope, ^{14}N.

This difference in mass, which creates a difference in density of ^{14}N-containing and ^{15}N-containing DNA, was the key to the experiment. **Figure 14.8** summarizes Meselson and Stahl's experimental strategy. The biologists reasoned that if different nitrogen isotopes were available in the growth medium when

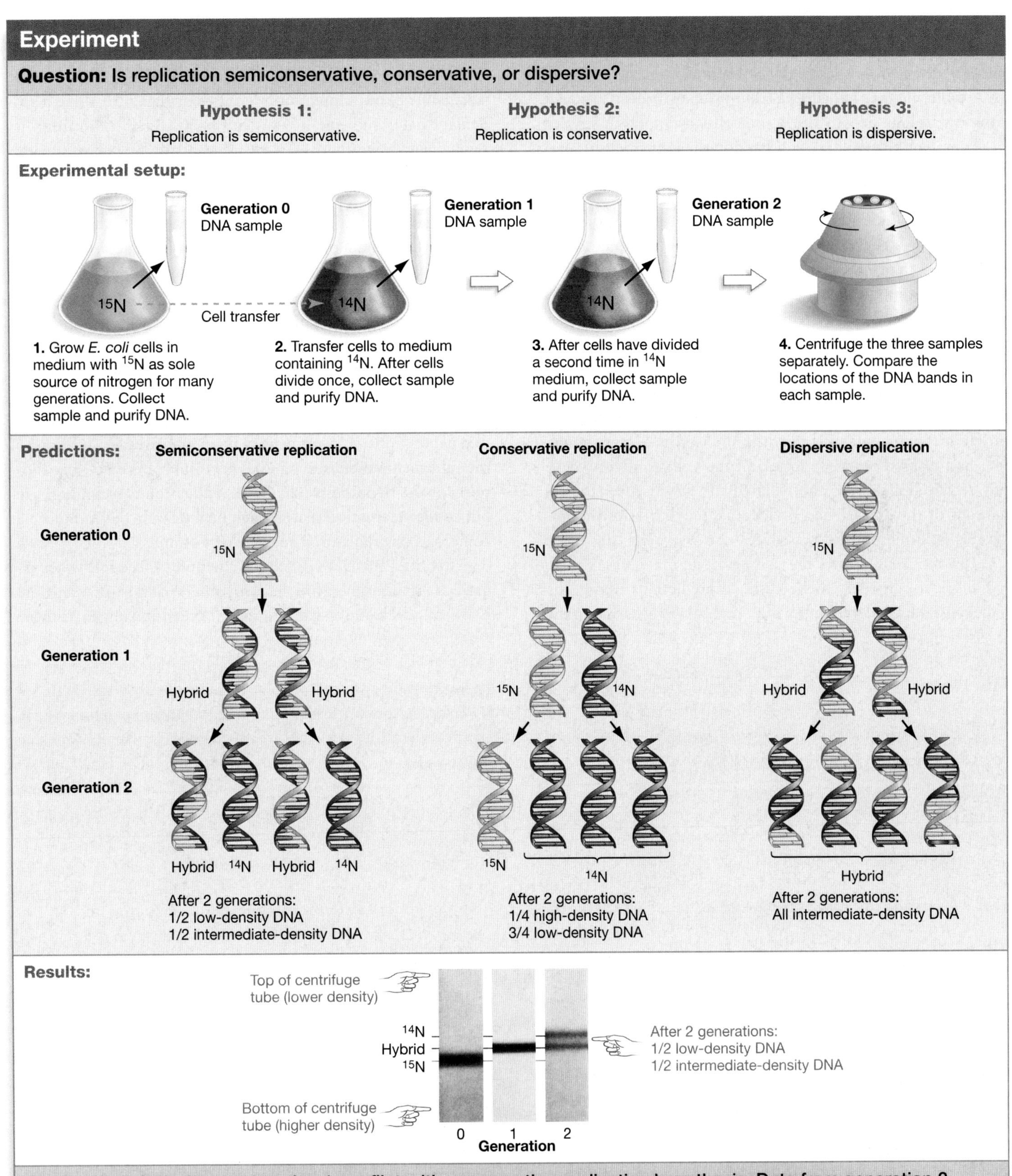

FIGURE 14.8 The Meselson–Stahl Experiment.

EXERCISE Meselson and Stahl actually let their experiment run for 4 generations, with cultures growing in the presence of ^{14}N. Sketch what data from 3rd and 4th generation DNA should look like—that is, where the DNA band(s) should be.

parental and daughter strands of DNA were produced, then the two types of strands should behave differently during centrifugation. More specifically, when intact, double-stranded DNA molecules are added to a solution that forms a gradient from low to high density during centrifugation, DNA strands that contain ^{14}N should form a band in the lower-density part of the centrifuge tube. In contrast, DNA strands that contain ^{15}N should form a band in the higher-density part of the centrifuge tube. Because the highest-density solution is at the bottom of the tube, DNA that contains ^{15}N should be found lower in the tube than DNA containing ^{14}N. In this way, DNA strands containing ^{15}N or ^{14}N should form separate bands. How could this tagging system be manipulated to test whether replication is semiconservative, conservative, or dispersive?

Meselson and Stahl began by growing *E. coli* cells with nutrients that contained only ^{15}N. They purified DNA from a sample of these cells and transferred the rest of the culture to a growth medium containing only the ^{14}N isotope. After enough time had elapsed for these experimental cells to divide once, they removed a sample and isolated the DNA. After the remainder of the culture had divided again, they removed another sample and purified the DNA.

As Figure 14.8 shows, the conservative, semiconservative, and dispersive models make distinct predictions about the makeup of the DNA molecules after replication occurred in the first and second generation. For example, if replication is conservative, then the daughter cells should have double-stranded DNA with either ^{14}N or ^{15}N, but not both. As a result, two distinct DNA bands should form in the centrifuge tube—one high-density band and one low-density band. But if replication is semiconservative or dispersive, then all of the experimental DNA should contain an equal mix of ^{14}N or ^{15}N after one generation, and one intermediate-density band should form in the centrifuge tube. But after two generations, half of the daughter cells should contain only ^{14}N if replication is semiconservative—meaning a second, lower-density band should appear in the centrifuge tube. But the dispersive model predicts that there will be just one band at intermediate density.

The photograph at the bottom of Figure 14.8 shows the experiment's results. After one generation, the density of the DNA molecules was intermediate. These data suggested that the hypothesis of conservative replication was wrong. After two generations, a lower-density band appeared in addition to the intermediate-density band. This result offered strong support for the hypothesis that DNA replication is not dispersive but **semiconservative;** that is, each newly made DNA molecule comprises one old strand and one new strand.

How does the DNA synthesis reaction proceed? Does it require an input of energy in the form of ATP, or it is spontaneous? Is it catalyzed by an enzyme, or does it occur quickly on its own?

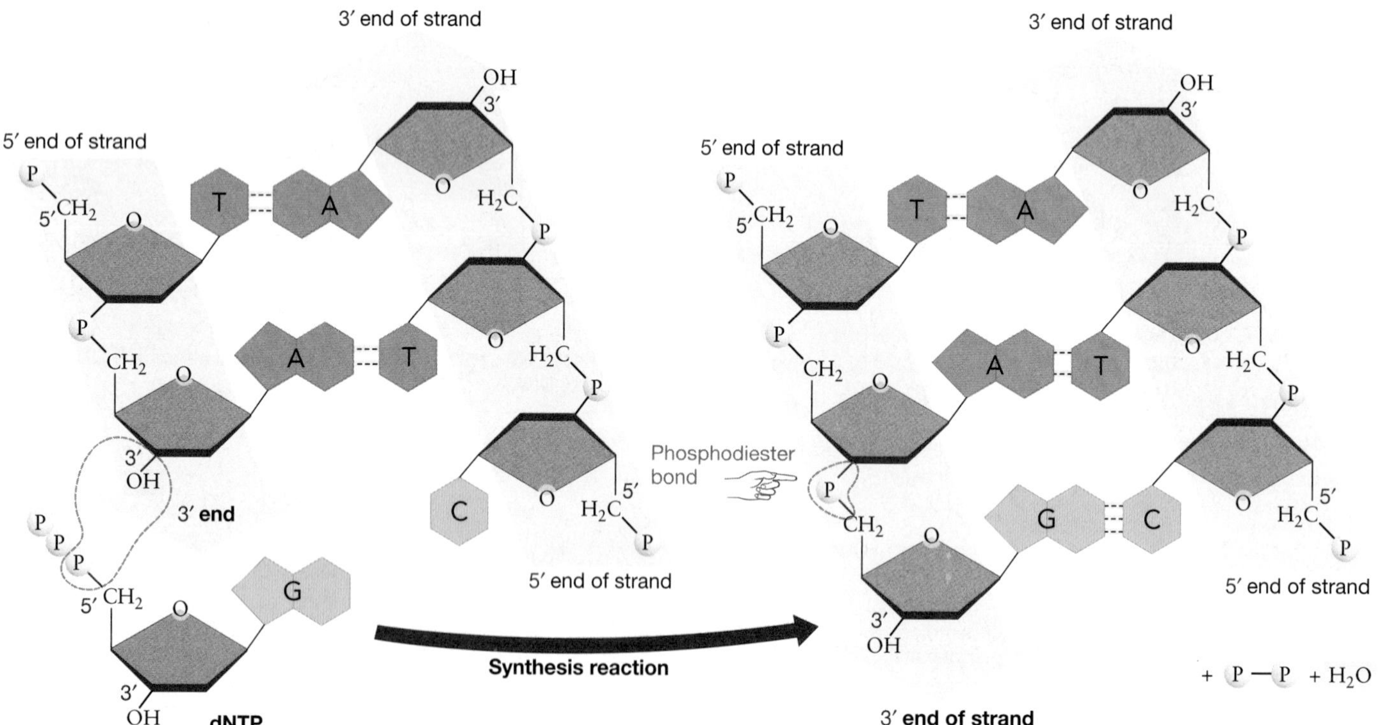

FIGURE 14.9 The DNA Synthesis Reaction. A dNTP (deoxynucleoside triphosphate) monomer is added to a DNA polymer when a phosphodiester bond forms between the 3′ carbon on the end of a DNA strand and the 5′ carbon on the dNTP in a condensation reaction.

QUESTION Does DNA synthesis proceed from the 5′ end toward the 3′ end, or in the 3′ to 5′ direction?

EXERCISE The "P–P" group reacts with H_2O, forming 2 phosphate ions (HPO_4^-) and releasing energy. Add this reaction to the figure.

14.3 A Comprehensive Model for DNA Synthesis

The initial breakthrough in research on DNA replication came with the discovery of an enzyme named **DNA polymerase**, because it polymerizes deoxyribonucleotides to DNA. This protein catalyzes DNA synthesis. Follow-up work showed that organisms contain several types of DNA polymerases. DNA polymerase III, for example, is the enzyme that is primarily responsible for copying *E. coli*'s chromosome prior to cell division.

Figure 14.9 illustrates a critical point about DNA polymerases: They can work in only one direction. More specifically, DNA polymerases can add deoxyribonucleotides to only the 3′ end of a growing DNA chain. As a result, DNA synthesis always proceeds in the 5′ → 3′ direction. ● If you understand this concept, you should be able to draw two lines indicating a DNA molecule, assign the 3′ to 5′ polarity of each strand, and then label the direction that DNA synthesis will proceed for each strand.

Figure 14.9 makes another important point about DNA synthesis. You might recall from earlier chapters that polymerization reactions are endergonic. But in cells, the reaction is exergonic because the monomers that are added to the growing strand are **deoxyribonucleoside triphosphates**, or **dNTPs**. The *N* in dNTP stands for any of the four bases found in DNA: adenine, thymine, guanine, or cytosine. Because they have three phosphate groups close together, dNTPs have high potential energy (see Chapter 9)—high enough to make the formation of phosphodiester bonds in a growing DNA strand exergonic.

Once the basic synthesis reaction was understood, researchers could delve into questions about how DNA replication gets started and proceeds.

How Does Replication Get Started?

A second major insight into the mechanism of DNA synthesis emerged when electron micrographs caught chromosome replication in action. As **Figure 14.10a** shows, a "bubble" forms in a chromosome when DNA is actively being synthesized. Bacterial chromosomes have a single location where the replication process begins, and thus a single bubble forms. Initially, the replication bubble forms at a specific sequence of bases called the **origin of replication** (**Figure 14.10b**). Replication bubbles grow as DNA replication proceeds, because synthesis is bidirectional—that is, it occurs in both directions (though always 5′ → 3′ because the strands are antiparallel) at the same time. Eukaryotes also have bidirectional replication; but they have multiple sites along each chromosome where DNA synthesis begins, and thus multiple replication bubbles (**Figure 14.10c**). ● If you understand this concept, you should be able to draw a replication bubble, assign the 3′ and 5′ polarity of each

(a) A chromosome being replicated

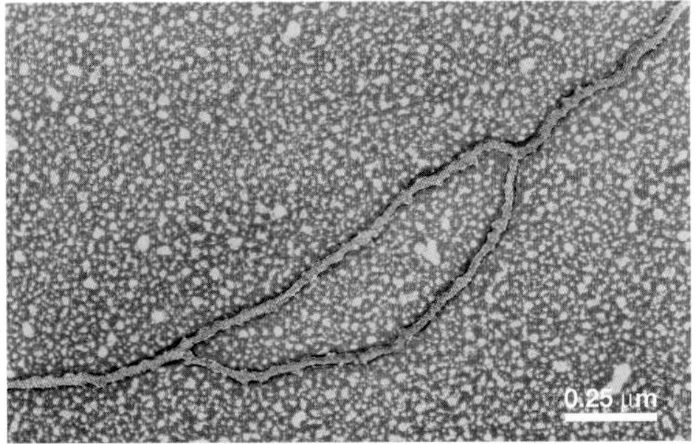

(b) Bacterial chromosomes have a single point of origin.

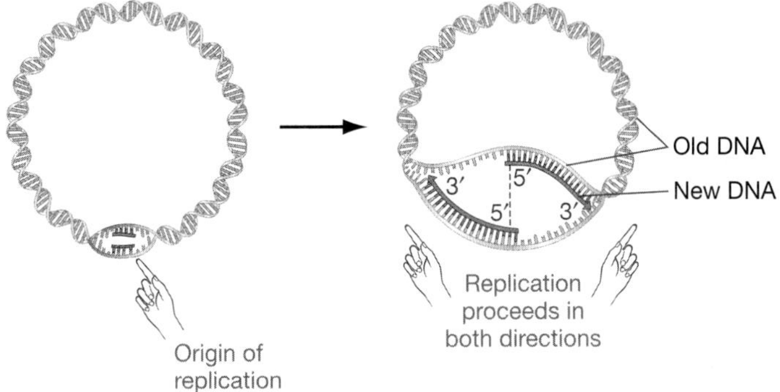

(c) Eukaryotic chromosomes have multiple points of origin.

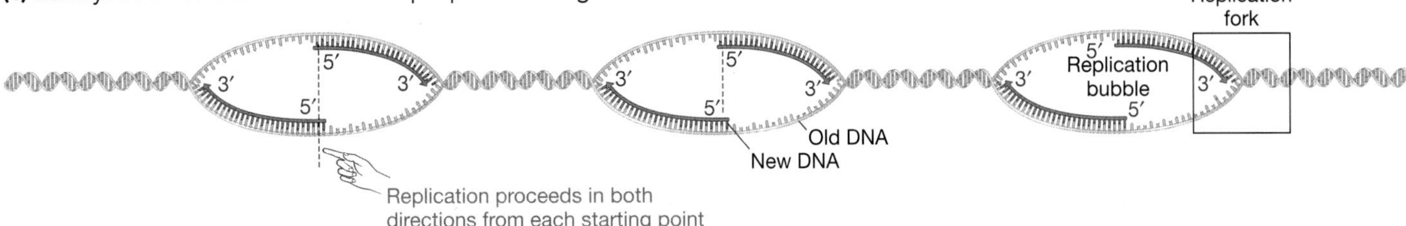

FIGURE 14.10 DNA Synthesis Proceeds in Two Directions from a Point of Origin. (a) A micrograph of a "replication bubble." **(b)** In most bacteria, chromosomes are circular and there is a single point of origin during DNA replication. **(c)** Most eukaryotes have linear chromosomes; each contains several to many points of origin for DNA synthesis.

● **EXERCISE** Circle the two replication forks in part (a).

strand of DNA present and then label the direction of DNA synthesis for each strand at both ends of the bubble.

A specific suite of proteins is responsible for recognizing sites where replication begins and opening the double helix at those points. These proteins are activated by the proteins responsible for initiating S phase in the cell cycle (see Chapter 11). Once a replication bubble opens, a different suite of enzymes takes over and initiates replication. The action takes place in the corners of each replication bubble—at a structure called the replication fork. A **replication fork** is a Y-shaped region where the parent DNA double helix is split into two single strands, which are then copied. How does the splitting event occur, and how are each of the resulting strands replicated?

How Is the Helix Opened and Stabilized?

Several key events occur at the point where the double helix opens. An enzyme called a **helicase** catalyzes the breaking of hydrogen bonds between deoxyribonucleotides. This reaction causes the two strands of DNA to separate. Proteins called **single-strand DNA-binding proteins** (**SSBPs**) attach to the separated strands and prevent them from snapping back into a double helix. In combination, then, the helicase and single-strand DNA-binding proteins open up the double helix and make both strands available for copying (**Figure 14.11**, step 1).

The "unzipping" process that occurs at the replication fork creates tension farther down the helix, however. To understand why, imagine what would happen if you started to pull apart the twisted strands of a rope. The untwisting movements at one end would force the intact section to rotate in response. If the intact end of the rope were fixed in place, though, it would eventually begin to coil on itself and kink in response to the twisting forces. This does not happen in DNA, because the twisting stress induced by helicase is relieved by proteins called topoisomerases. A **topoisomerase** is an enzyme that cuts and rejoins the DNA downstream of the replication fork. Topoisomerases do this cutting and pasting in a way that undoes twists and knots.

Now, what happens once the DNA helix is open and has stabilized?

How Is the Leading Strand Synthesized?

The keys to understanding what happens at the start of DNA synthesis are to recall that DNA polymerase III works only in the $5' \rightarrow 3'$ direction and to recognize that to start synthesis, it requires both a 3′ end *and* a single-stranded template. The single-stranded template dictates which deoxyribonucleotide should be added next, while a **primer**—which consists of a few nucleotides bonded to the template—provides a free 3′ hydroxyl (–OH) group that can combine with an incoming dNTP to form a phosphodiester bond.

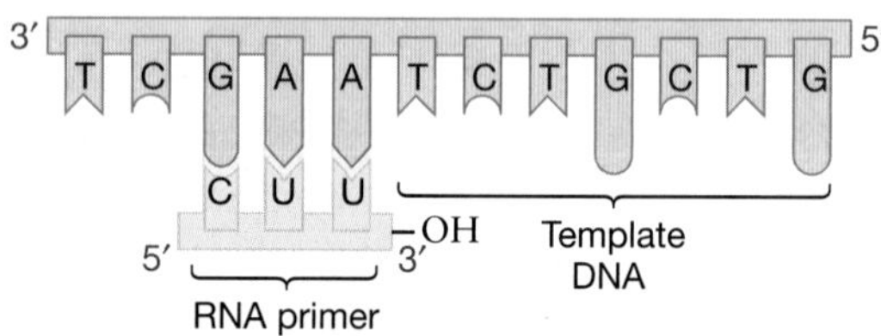

Once a primer is added to a single-stranded template, DNA polymerase III begins working in the $5' \rightarrow 3'$ direction and adds deoxyribonucleotides to complete the complementary strand.

Before DNA synthesis can get under way, then, an enzyme called **primase** has to synthesize a short stretch of ribonucleic acid (RNA) that acts as a primer for DNA polymerase. Primase is a type of **RNA polymerase**—an enzyme that catalyzes the polymerization of ribonucleotides to RNA (see Chapter 4 to review RNA's structure). Unlike DNA polymerases, primase

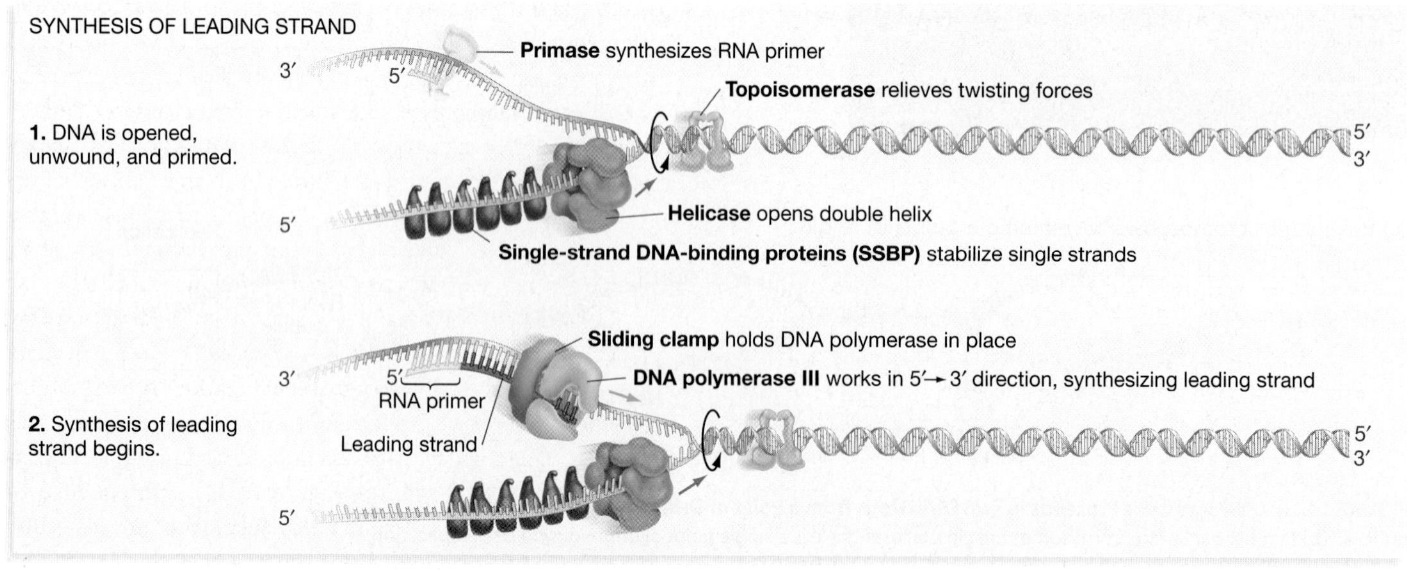

FIGURE 14.11 Synthesizing the Leading Strand during DNA Replication.

and other RNA polymerases do not require a primer. These enzymes can simply match ribonucleotides directly by complementary base pairing on single-stranded DNA. In this way, primase creates a primer for DNA synthesis.

Once the primer is in place, DNA polymerase III begins adding deoxyribonucleotides to the 3′ end of the new strand, in a sequence that is complementary to the template strand. As Figure 14.11 step 2 shows, the shape of DNA polymerase III grips the DNA strand during synthesis, a little like your hand clasping a rope. Catalysis takes place in the groove inside the enzyme, at an active site between the enzyme's "thumb" and "fingers." As DNA polymerase moves along the DNA molecule, a doughnut-shaped structure behind it, called the sliding clamp, holds the enzyme in place. The enzyme's product is called the **leading strand**, or **continuous strand**, because it leads into the replication fork and is synthesized continuously. ● If you understand leading-strand synthesis, you should be able to list the enzymes involved and for each one, predict the consequences if it were defective.

How Is the Lagging Strand Synthesized?

Synthesis of the leading strand is straightforward after an RNA primer is in place—DNA polymerase III chugs along, adding bases to the 3′ end of that strand. The enzyme moves into the replication fork, which "unzips" ahead of it. By comparison, events on the opposite strand are much more involved. To understand why, recall that the two strands in the DNA double helix are *antiparallel*—meaning they are parallel to one another but oriented in opposite directions. DNA polymerase works in only one direction, however, so if the DNA polymerase that is synthesizing the leading strand works into the replication fork, then a DNA polymerase must work away from the replication fork to synthesize the other strand in the 5′ → 3′ direction. The strand that is synthesized in the opposite direction of the replication fork is called the **lagging strand**, because it lags behind the fork. ● If you understand this concept, you should be able to draw a replication bubble in DNA, assign the 3′ and 5′ polarity of each strand, label the direction of DNA synthesis for each strand at both ends of the bubble, and indicate which of the four newly synthesized strands—two at each end of the bubble—are the leading strands and lagging strands.

The synthesis of the lagging strand starts when primase synthesizes a short stretch of RNA that acts as a primer. DNA polymerase III then adds bases to the 3′ end of the lagging strand. The key observation is that the enzyme moves away from the replication fork. But behind it, helicase continues to open the replication fork and expose new single-stranded DNA on the lagging strand.

These events create a paradox. New single-stranded template DNA is constantly appearing behind DNA polymerase III—in the direction *opposite* the direction of lagging-strand synthesis—as the helix continues to unzip during leading-strand synthesis.

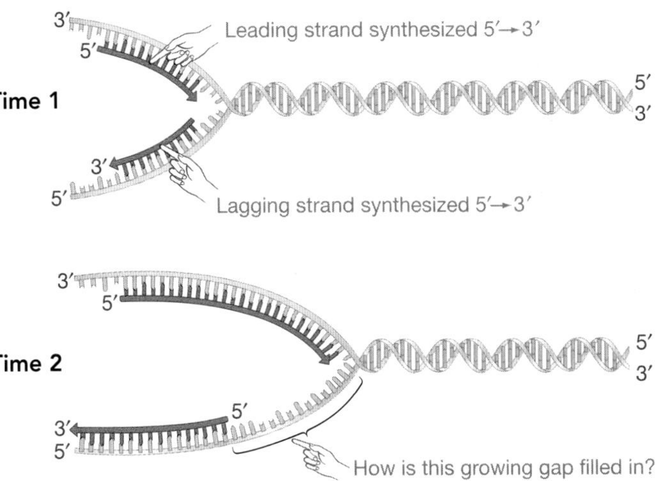

FIGURE 14.12 Gaps Form in the Lagging Strand Gaps form in the lagging strand as helicase opens the double helix ahead of the DNA polymerase synthesizing the leading strand. The gaps occur because DNA polymerase can work only away from the replication fork—in the 5′ → 3′ direction—on the lagging strand.

To see this point, consider the state of the lagging strand at time 1 and time 2 in **Figure 14.12**. As DNA polymerase works into the replication fork in the 5′ → 3′ direction and away from the replication fork in the 5′ → 3′ direction, gaps appear in the lagging strand.

The puzzle posed by lagging-strand synthesis was resolved when Reiji Okazaki and colleagues tested a hypothesis called discontinuous replication. This hypothesis stated that once primase synthesizes an RNA primer on the lagging strand (step 1 in **Figure 14.13**), DNA polymerase III might synthesize short fragments of DNA along the lagging strand and that these fragments would later be linked together to form a continuous whole. The idea was that primase would add a primer to the newly exposed single-stranded DNA at intervals and that DNA polymerase III would then synthesize the lagging strand until it reached the fragment produced earlier.

To test this hypothesis, Okazaki's group set out to test a key prediction: Could they document the existence of short DNA fragments produced during replication? Their critical experiment was based on the pulse-chase strategy introduced in Chapter 7. Specifically, they added a short "pulse" of a radioactive deoxyribonucleotide to *E. coli* cells, followed by a large "chase" of nonradioactive deoxyribonucleotide. According to the discontinuous replication model, some of this radioactive deoxyribonucleotide should end up in short, single-stranded fragments of DNA.

As predicted, the researchers succeeded in finding these fragments when they purified DNA from the experimental cells and separated the molecules by centrifugation. A small number of labeled pieces of DNA, about 1000 base pairs long, were present. These short sections came to be known as **Okazaki fragments** (steps 2 and 3 of Figure 14.13). Subsequent work showed that Okazaki fragments in eukaryotes are just 100 to 200 base pairs long, though a chromosome typically contains millions of base pairs.

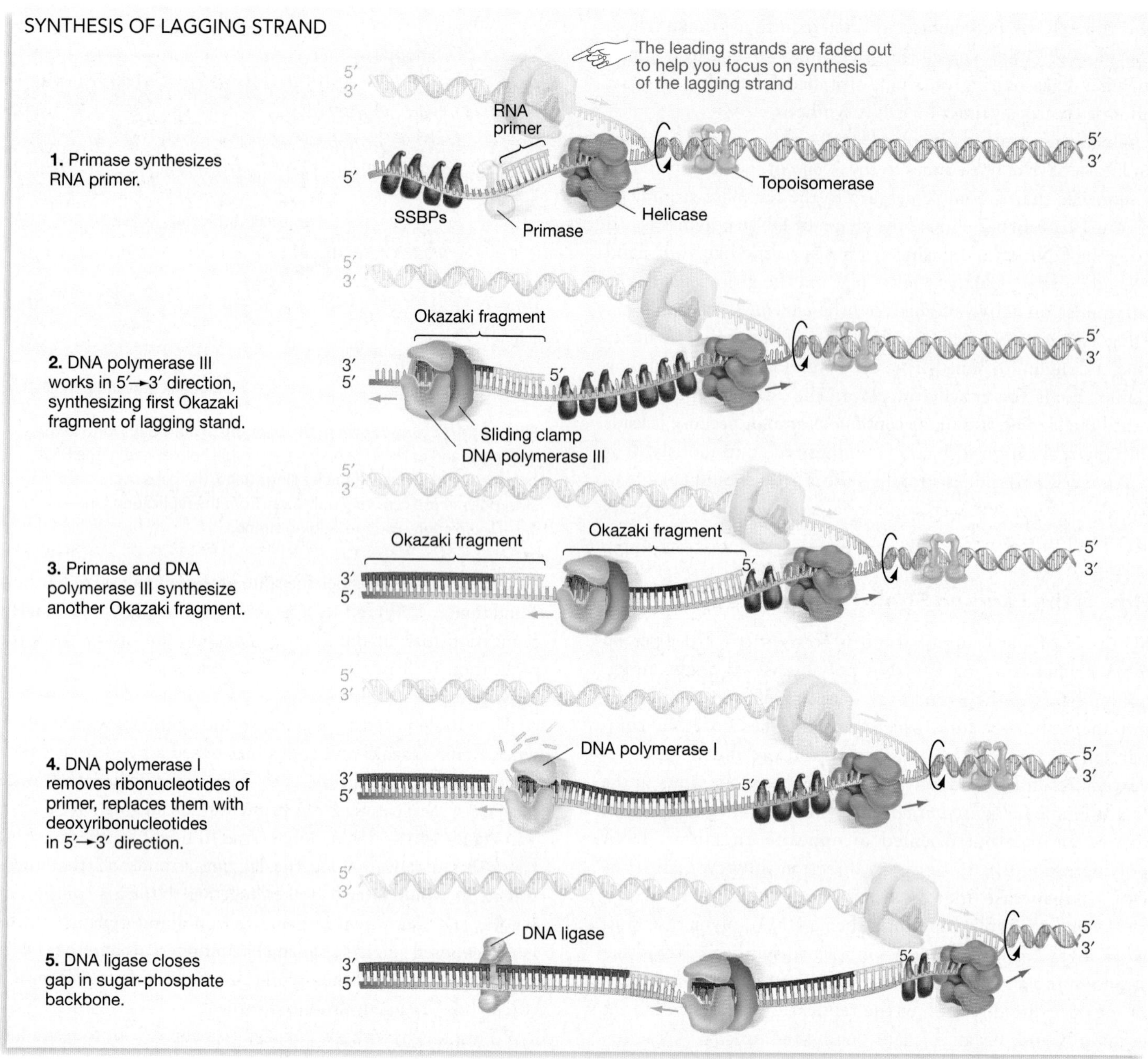

FIGURE 14.13 The Completion of DNA Replication.

How are Okazaki fragments connected into a continuous whole? As step 4 of Figure 14.13 shows, DNA polymerase I removes the RNA primer at the start of each fragment and fills in the appropriate deoxyribonucleotides. Finally, an enzyme called **DNA ligase** catalyzes the formation of a phosphodiester bond between the adjacent fragments (Figure 14.13, step 5). Because Okazaki fragments are synthesized independently and joined together later, the lagging strand is also called the **discontinuous strand.** ● If you understand lagging-strand synthesis, you should be able to draw what the two molecules resulting from DNA synthesis—starting from a single origin of replication—would look like if DNA ligase were defective.

In combination, then, the enzymes that open the replication fork and manage the synthesis of the leading and lagging strands succeed in producing a faithful copy of the original DNA molecule prior to mitosis or meiosis. Given the number of enzymes and proteins involved in the synthesis machinery, it is not surprising that it took 25 years for biologists to assemble the results summarized in **Table 14.1.**

It is also worth mentioning the following points:

- Although Figures 14.12 and 14.13 show the enzymes involved in DNA synthesis at different locations around the replication fork, in reality most are joined into one large multi-enzyme machine at the replication fork. This complex of proteins opens the double helix and synthesizes both the leading strand and the Okazaki fragments on the lagging strand. The lagging strand actually loops out and around one of the polymerases so that the entire machine can move together down the replication fork.

SUMMARY TABLE **14.1 Proteins Involved in DNA Synthesis**

Name	Structure	Function
Opening the helix		
Helicase		Catalyzes the breaking of hydrogen bonds between base pairs and the opening of the double helix
Single-strand DNA-binding proteins		Stabilize single-stranded DNA
Topoisomerase		Breaks and rejoins the DNA double helix to relieve twisting forces caused by the opening of the helix
Leading strand		
Primase		Catalyzes the synthesis of the RNA primer
DNA polymerase III		Extends the leading strand
Sliding clamp		Holds DNA polymerase in place during strand extension
Lagging strand		
Primase		Catalyzes the synthesis of the RNA primer on an Okazaki fragment
DNA polymerase III		Extends an Okazaki fragment
Sliding clamp		Holds DNA polymerase in place during strand extension
DNA polymerase I		Removes the RNA primer and replaces it with DNA
DNA ligase		Catalyzes the joining of Okazaki fragments into a continuous strand

- The basic elements of DNA synthesis were worked out in experiments on *E. coli*. Follow-up research has shown that the fundamental elements of this process are almost identical in eukaryotes. The exceptions are that (1) some of the enzymes at the replication fork in eukaryotes are larger and more complex than similar enzymes in bacteria; and (2) on the lagging strand, two different DNA polymerases do the work done by DNA polymerase III in bacteria. In almost every other detail, however, the critical components of DNA synthesis have been highly conserved over the course of evolution.

Most aspects of the model summarized in Figures 14.11 and 14.13 had emerged in the early 1980s, and subsequent decades saw major advances in our understanding of the detailed molecular structures of the major enzymes involved. But in addition to studying the replication process itself, biologists have been exploring other questions about DNA synthesis. This chapter closes with a look at two of these research areas: How are the ends of linear chromosomes replicated? And how can incorrect or damaged bases be repaired? As it turns out, both questions have practical implications—particularly for the origin of certain types of cancer.

MB Web Animation at www.masteringbio.com
DNA Synthesis

Check Your Understanding

If you understand that...

- DNA synthesis begins at specific points of origin on the chromosome and then proceeds in both directions.
- A three-step process occurs at the replication fork.
  (1) Helicase opens the double helix, SSBPs stabilize the exposed single strands, and topoisomerase prevents kinks downstream of the fork;
  (2) DNA polymerase synthesizes the leading strand after primase has added an RNA primer; and
  (3) a series of enzymes synthesizes the lagging strand.
- Lagging-strand synthesis cannot be continuous, because it moves away from the replication fork. In bacteria, enzymes called primase, DNA polymerase III, DNA polymerase I, and ligase work in sequence to synthesize Okazaki fragments and link them into a continuous whole.

You should be able to...

1) Draw a replication fork.
2) Label the leading and lagging strands and indicate their polarity.
3) Label primase, topoisomerase, single-stranded DNA binding proteins, the sliding clamp and DNA polymerase on each strand, and DNA ligase.
4) Explain the function of each structure involved.

14.4 Replicating the Ends of Linear Chromosomes

The circular DNA molecules in bacteria and archaea can be synthesized via the sequence of events diagrammed in Figures 14.11 and 14.13. Similarly, the leading and lagging strands of the linear DNA molecules found in eukaryotes are copied efficiently by the enzymes illustrated in those figures. But copying the ends of linear chromosomes is another story.

The region at the end of a linear chromosome is called a **telomere** (literally, "end-part"). **Figure 14.14** illustrates the problem that arises during the replication of telomeres. Steps 1 and 2 show that as the replication fork nears the end of a linear chromosome, DNA polymerase synthesizes the leading strand all the way to the end of the parent DNA template. As a result, leading-strand synthesis results in a normal copy of the DNA molecule. But the situation is different on the lagging strand. As step 2 of Figure 14.14 illustrates, primase adds an RNA primer close to the tip of the chromosome. DNA polymerase III synthesizes the final Okazaki fragment on the lagging strand (step 3), and DNA polymerase I removes the primer. The DNA polymerase III is unable to add DNA near the tip of the chromosome, however, because there is not enough room for primase to add a new RNA primer (step 4). As a result, the single-stranded DNA that is left once the primer has been removed must stay single stranded.

The single-stranded DNA that remains after telomeres are copied is eventually degraded, which results in the shortening of the chromosome. If this process were to continue, every chromosome would shorten by 50–100 deoxyribonucleotides on average each time that DNA replication occurred prior to mitosis or meiosis. Over time, linear chromosomes would be expected to disappear completely.

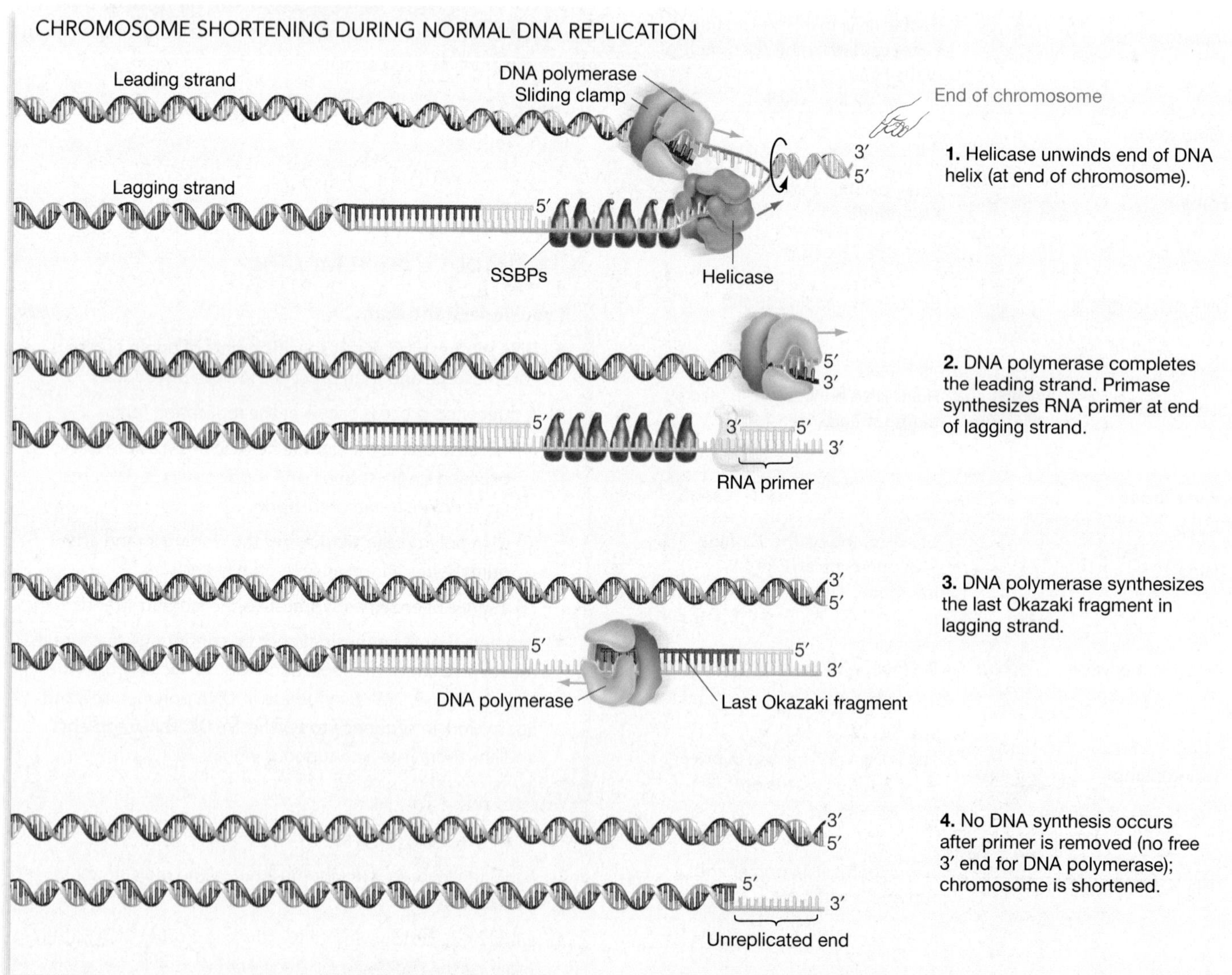

FIGURE 14.14 Telomeres Shorten during Normal DNA Replication. An RNA primer is added to the lagging strand near the end of the chromosome. Once the primer is removed, it cannot be replaced with DNA. As a result, the chromosome shortens.

Bacteria and archaea do not have this end-replication problem, because virtually all of these cells have a single, circular chromosome. How do eukaryotes maintain the integrity of their linear chromosomes? An answer emerged after two important discoveries were made:

1. Telomeres do not contain genes that code for products needed in the cell. Instead, they consist of short stretches of bases that are repeated over and over. In humans, for example, the base sequence TTAGGG is repeated thousands of times. (Because DNA and RNA sequences are always written $5' \rightarrow 3'$, the base sequence on the complementary strand is CCCTAA rather than AATCCC.) Human telomeres consist of a total of about 10,000 deoxyribonucleotides with the sequence TTAGGGTTAGGGTTAGGG....
2. An interesting enzyme called telomerase is involved in replicating telomeres. **Telomerase** is remarkable because it catalyzes the synthesis of DNA from an RNA template. Further, the enzyme carries an RNA molecule with it that acts as a built-in template. Telomerase catalyzes the addition of deoxyribonucleotides to the ends of chromosomes. The deoxyribonucleotides that are added are complementary to the ribonucleotides in its built-in RNA template As a result, telomerase can add DNA onto the end of a chromosome and prevent it from getting shorter.

Figure 14.15 shows how this process works. Step 1 shows the unreplicated segment at the 3′ end of the lagging strand at the telomere. It forms a single-strand "overhang." In step 2, telomerase binds to this overhanging section of single-stranded parent DNA. Once it has bound, it begins catalyzing the addition of deoxyribonucleotides, in the $5' \rightarrow 3'$ direction, to the overhang. In step 3, telomerase begins moving in the $5' \rightarrow 3'$ direction and continues to catalyze the addition of deoxyribonucleotides. In this way, telomerase lengthens the overhang on the lagging strand. The normal machinery of DNA synthesis—primase, DNA polymerase, and ligase—then synthesizes the lagging strand in the $5' \rightarrow 3'$ direction (see step 4). The result is that the lagging strand becomes slightly longer than it was originally.

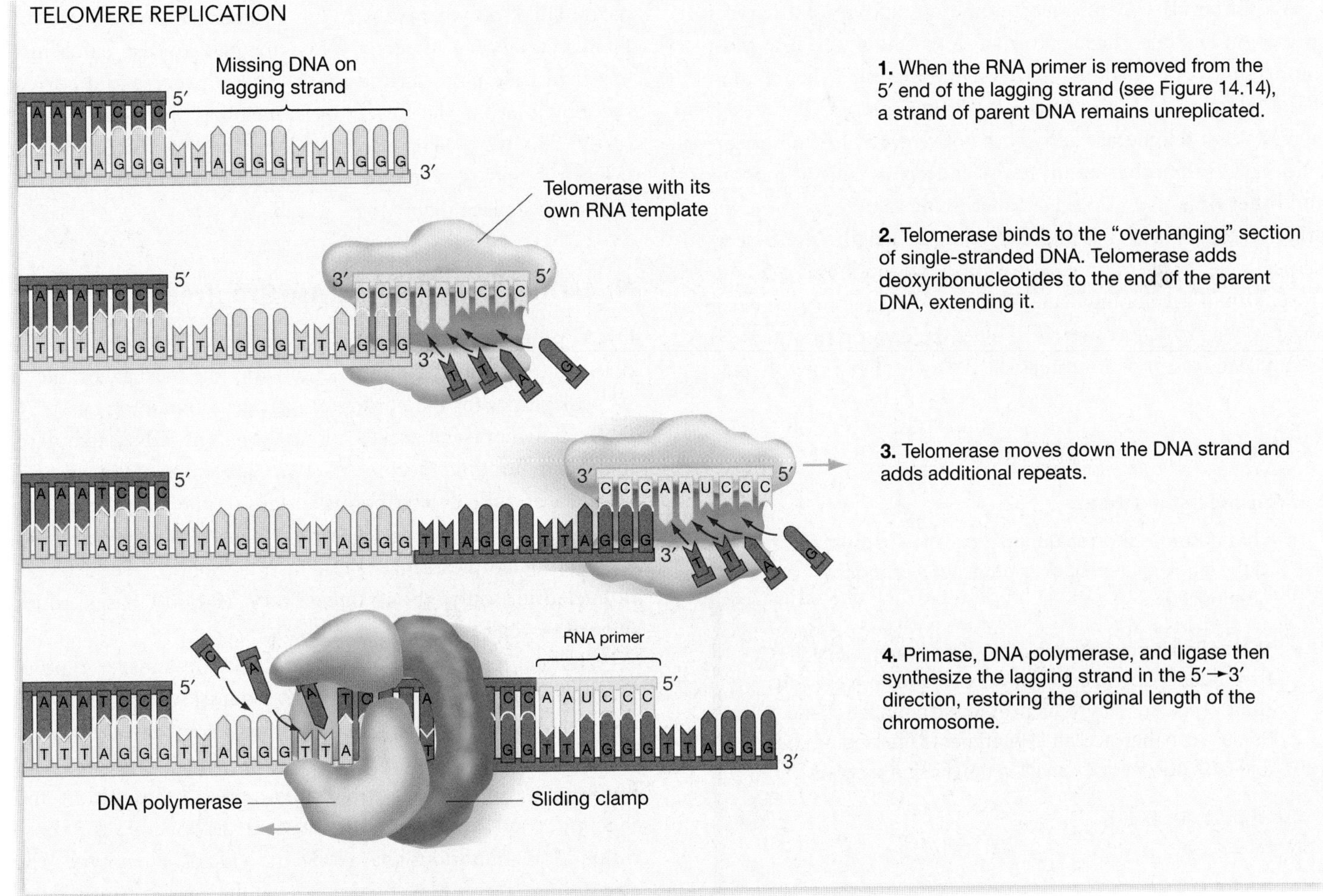

FIGURE 14.15 Telomerase Prevents Shortening of Telomeres during Replication. By extending the number of repeated sequences in the $5' \rightarrow 3'$ direction, telomerase provides room for enzymes to add an RNA primer to the lagging strand. DNA polymerase can then fill in the missing section of the lagging strand.

QUESTION Would this telomerase work as well if its RNA template had a different sequence?

It is important to recognize, though, that telomerase is not active in most types of cells. In humans, for example, active telomerase is found primarily in the cells of reproductive organs—specifically, in the cells that eventually undergo meiosis and produce gametes. Cells that are not involved in gamete formation, or what biologists call **somatic cells**, normally lack telomerase. As predicted, the chromosomes of somatic cells gradually shorten with each mitotic division, getting progressively smaller as an individual grows and ages.

These observations inspired a pair of important hypotheses. The first was that telomere shortening causes cells to stop dividing and enter the non-dividing state called G_0 (see Chapter 11). The second was that if telomerase were mistakenly activated in a somatic cell, telomeres would fail to shorten. This would allow the cell to keep dividing and possibly contribute to uncontrolled growth and cancer.

To test the first hypothesis, biologists added functioning telomerase to human cells growing in vitro. The treated cells continued dividing long past the age when otherwise identical cells stop growing. These results have convinced most biologists that telomere shortening is involved in limiting how long cells remain in an actively growing state.

A link between continued telomerase activity and cancer formation has been harder to nail down, however. The most important work to date involves mutant mice that cannot make telomerase. Many of these develop tumors. This result suggests that telomerase activity is not required for tumor formation. On the other hand, many cancerous cells in humans and other organisms do have functioning telomerase or some other mechanism for maintaining telomere length. Noncancerous cells in these organisms do not have functioning telomerase. Would developing drugs that knock out telomerase be an effective way to fight cancer? To date, the data on this question are unclear. Research continues.

Check Your Understanding

If you understand that...

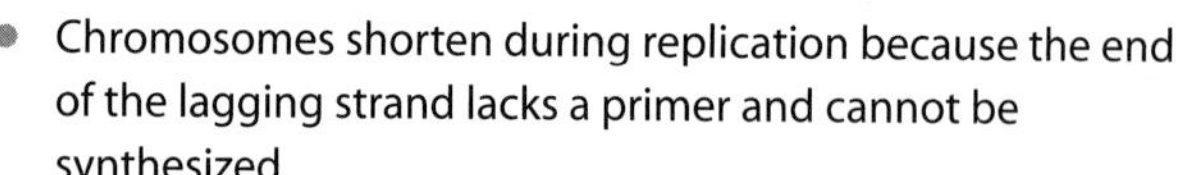

- Chromosomes shorten during replication because the end of the lagging strand lacks a primer and cannot be synthesized.
- Shortening is prevented in many cells—particularly those that produce sperm and egg—because telomerase adds short, repeated DNA sequences to the template strand. Primase can then add an RNA primer to the lagging strand, and DNA polymerase can fill in the missing sections.

You should be able to...

1) Diagram the sequence of events involved in telomere replication.
2) Add labels to indicate the enzymes involved and the polarity of each DNA strand.

14.5 Repairing Mistakes and Damage

DNA polymerases work fast. In yeast, for instance, each replication fork is estimated to move at a rate of about 50 bases per second. But the replication process is also astonishingly accurate. In organisms ranging from *E. coli* to animals, the error rate during DNA replication averages less than one mistake per *billion* deoxyribonucleotides. This level of accuracy is critical. Humans, for example, develop from a fertilized egg that has DNA containing over 6 billion deoxyribonucleotides. The DNA inside the fertilized egg is replicated over and over to create the trillions of cells that eventually make up the adult body. If more than one or two mutations occurred during each cell division cycle as a human grew, genes would be riddled with errors by the time the individual reached maturity. Genes that contain errors are often defective. Based on these observations, it is no exaggeration to claim that the accurate replication of DNA is a matter of life and death. Natural selection favors individuals with enzymes that copy DNA as exactly as possible.

How can the enzymes involved in replicating DNA be as precise as they are? The answer to this question has several parts. DNA polymerase is highly selective in matching complementary bases correctly. This enzyme can also catch mismatched base pairs during the synthesis process and correct them by inserting the correct base, meaning that it can "proofread." Finally, if mistakes remain after synthesis is complete or if DNA is damaged, repair enzymes can cut out the defective bases and replace them.

How Does DNA Polymerase Proofread?

DNA polymerases are selective because the correct base pairings (A-T and G-C) are energetically the most favorable of all possibilities for the pairing of nitrogen-containing bases. As DNA polymerase marches along a parent DNA template, hydrogen bonding occurs between incoming deoxyribonucleotides and the deoxyribonucleotides on the template strand. Because DNA polymerase checks the match between paired deoxyribonucleotides, the enzyme inserts the incorrect deoxyribonucleotide only about once every 100,000 bases added (**Figure 14.16a**).

What happens when a G-T, an A-C, or another type of mismatch occurs? An answer to this question emerged when researchers found mutants in *E. coli* with error rates that were 100 times greater than normal. Recall from Chapter 13 that mutants are individuals with traits caused by mutation and that mutation is a change in the gene responsible for those traits. Many mutations change the individual's phenotype. The change may result in a trait such as white eyes in fruit flies or an elevated mutation rate in *E. coli*. At the molecular level, a mutant phenotype usually results from a change in an enzyme or other type of protein.

(a) Mismatched bases

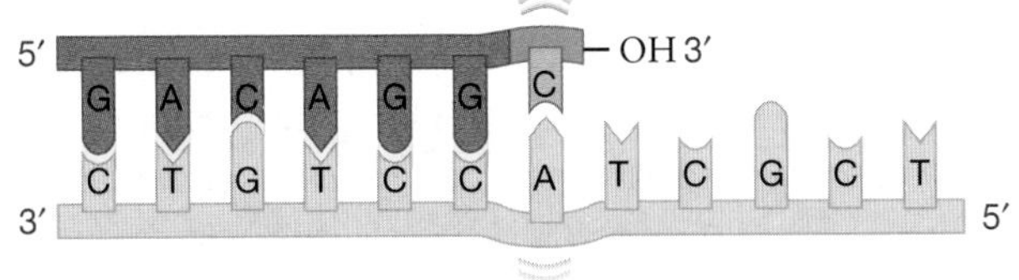

(b) DNA polymerase III can repair mismatches.

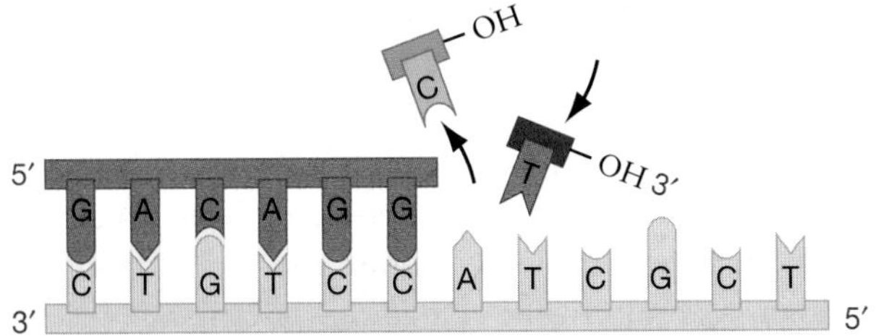

FIGURE 14.16 DNA Polymerase Can Proofread. (a) In bacteria, DNA polymerase adds an incorrect base to a growing strand of DNA about once in every 100,000 bases added. The result is a mismatch such as the pairing of A with C. **(b)** DNA polymerase can act as a 3′ → 5′ exonuclease, meaning that it can remove bases in that direction. The enzyme then adds the correct base.

In the case of *E. coli* cells with high mutation rates, biologists found that the mutation was localized to a particular portion of the the ends of DNA strands polymerase III enzyme, called the ε (epsilon) subunit. Further analyses showed that this subunit of the enzyme acts as an exonuclease—meaning an enzyme that removes deoxyribonucleotides from the ends of DNA strands (**Figure 14.16b**). The DNA polymerase III exonuclease activity removes deoxyribonucleotides in a 3′ to 5′ direction, and only if they are not hydrogen bonded to a base on the complementary strand. These results led to the conclusion that DNA polymerase III can **proofread**. If the wrong base is added during DNA synthesis, the enzyme pauses, removes the mismatched base that was just added, and then proceeds with synthesis.

Eukaryotic DNA polymerases have the same type of proofreading ability. Typically, proofreading reduces a DNA polymerase's error rate to about 1×10^{-7} (one mistake per 10 million bases). Is this accurate enough? The answer is no. If DNA polymerase leaves a mismatched pair behind in the DNA sequence by mistake, a battery of enzymes springs into action to correct the problem. **Mismatch repair** occurs when mismatched bases are corrected after DNA synthesis is complete.

The proteins responsible for mismatch repair were discovered in the same way that the proofreading capability of DNA polymerase III was uncovered—by analyzing *E. coli* mutants. In this case, the mutants had normal DNA polymerase III but abnormally high mutation rates. The first mutant gene that caused a deficiency in mismatch repair was identified in the late 1960s and was called *mutS*. (The *mut* is short for "mutator.") By the late 1980s researchers had identified 10 proteins involved in the identification and repair of base-pair mismatches in *E. coli*. These proteins recognize the mismatched pair, remove a section of the newly synthesized strand that contains the incorrect base, and fill in the correct bases. The mismatch repair enzymes are like a copy editor who corrects typos that the author—DNA polymerase, in this case—did not catch.

Nucleotide Excision Repair

Even after DNA is synthesized and proofread and mismatches repaired, all is not well. Genes are under constant assault. Nucleotides are damaged by chemicals like the hydroxyl (OH) radicals produced during aerobic metabolism, the aflatoxin B1 found in moldy peanuts and corn, and the benzo[α]pyrene in cigarette smoke. Radiation is also dangerous. Ultraviolet (UV) light, for example, can cause a covalent bond to form between adjacent pyrimidine bases. The thymine-thymine pair illustrated in **Figure 14.17** is an example. This defect is called a thymine dimer. Thymine dimers create kinks in the secondary structure of DNA that stall the enzymes responsible for transcription and DNA synthesis. The kinks also stall the movement of the replication fork during DNA replication. If the damage is not repaired, the cell may die.

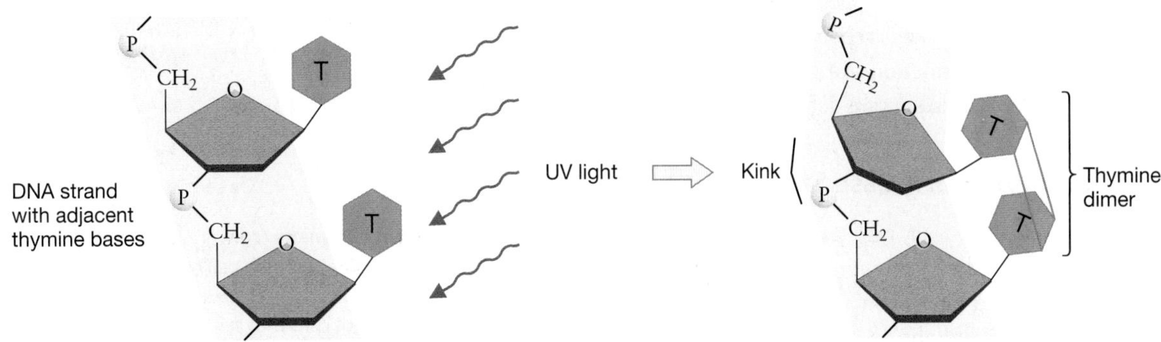

FIGURE 14.17 UV Light Damages DNA. When UV light strikes a section of DNA with adjacent thymines, the energy can break bonds within each base and result in the formation of bonds *between* them. The thymine dimer that is produced causes a kink in DNA.

QUESTION Why are infrared wavelengths much less likely to damage DNA rather than UV?

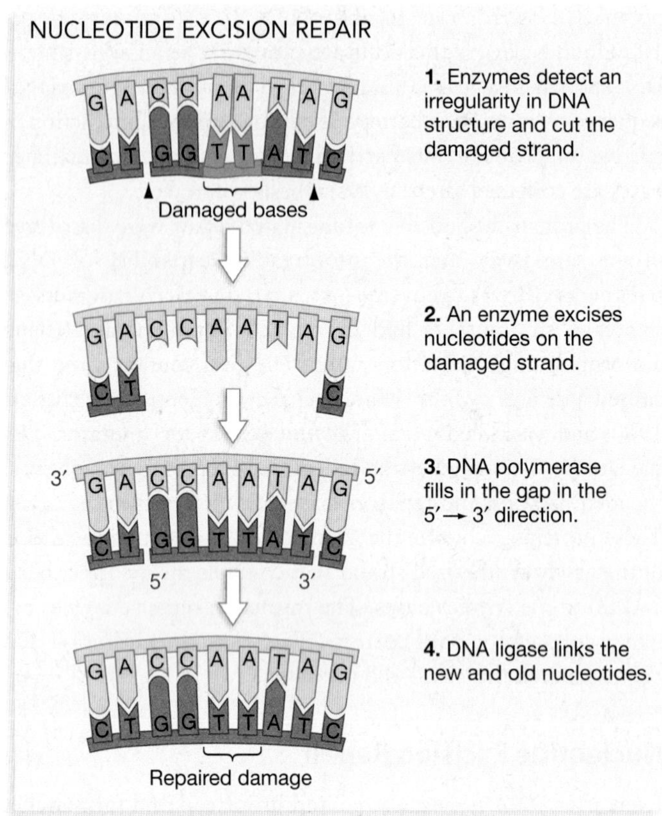

FIGURE 14.18 In Excision Repair, Defective Bases Are Removed and Replaced.

Cells have an efficient system to fix these types of problems, called **nucleotide excision repair** (**Figure 14.18**). As step 1 in the figure shows, the symmetry and regularity of DNA's secondary structure makes it possible for repair proteins to recognize thymine dimers and other types of damaged bases that produce an irregularity in the molecule. Once a damaged region is recognized, enzymes remove the single-stranded DNA around the damaged site. The presence of a DNA strand complementary to the damaged strand provides a template for resynthesis of the defective sequences. In this way, DNA's structure makes accurate repair possible. Because the molecule is repairable, DNA's structure supports its function in information storage and processing. But what happens when repair systems are defective?

Xeroderma Pigmentosum: A Case Study

Xeroderma pigmentosum (XP) is a rare autosomal recessive disease in humans. Individuals with this condition are extremely sensitive to ultraviolet (UV) light. Their skin develops lesions after even slight exposure to sunlight. In unaffected individuals, these kinds of lesions develop only after extensive exposure to UV light, X-rays, or other forms of high-energy radiation.

In 1968 James Cleaver proposed a connection between XP and DNA excision repair systems. He knew that in *E. coli*, mutations in certain genes cause DNA nucleotide excision repair proteins to fail. Cells with these mutations have an increased sensitivity to radiation. Cleaver's hypothesis was that people with XP have similar mutations. He claimed that they are extremely sensitive to sunlight because they are unable to repair the damage that occurs when the nitrogen-containing bases of DNA absorb UV light.

Cleaver's hypothesized connection between DNA damage, faulty nucleotide excision repair, and XP turned out to be correct. Much of the work that he and other investigators did relied on the use of cell cultures (see Chapter 11). In this case, the researchers collected skin cells from people with XP and from people with a normal phenotype for excision repair. When these cell populations were grown in culture and exposed to increasing amounts of ultraviolet radiation, a striking difference emerged: Cells from xeroderma pigmentosum individuals died off much more rapidly than did unaffected cells (**Figure 14.19a**).

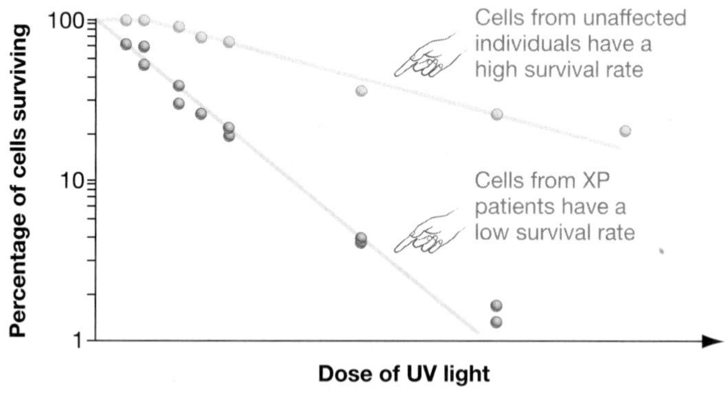

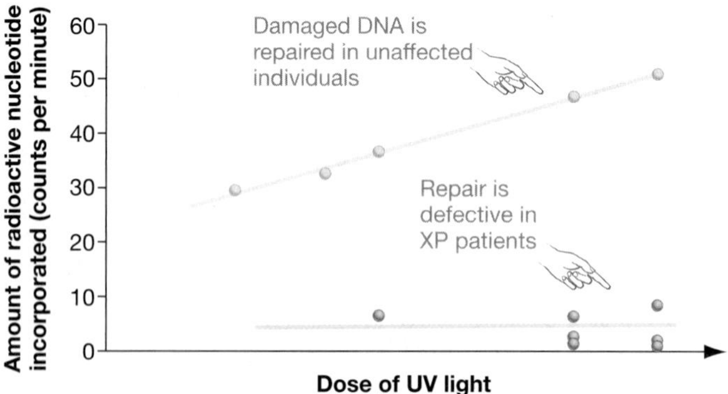

FIGURE 14.19 DNA Damage from UV Light Is Not Repaired Properly in Individuals with XP. (a) When cell cultures from unaffected individuals and from XP patients are irradiated with various doses of UV light, the percentage of cells that survive is strikingly different. **(b)** When cell cultures from unaffected individuals and from XP patients are irradiated with various doses of UV light and then fed radioactive thymidine, only unaffected individuals incorporate the labeled base.

● **QUESTION** Why are people who cultivate a skin tan likely to develop cancer? (Hint: Tanning is a response to UV light.)

The connection to excision repair systems was confirmed when Cleaver exposed cells from unaffected individuals and cells from XP individuals to various amounts of UV light, then fed the cells a radioactive deoxyribonucleotide to label DNA synthesized during the repair period. If repair is defective in the XP individuals, then their cells should incorporate virtually no radioactive deoxyribonucleotide into their DNA. Cells from unaffected individuals, in contrast, should incorporate large amounts of labeled deoxyribonucleotide into their DNA. As **Figure 14.19b** shows, this is exactly what happens. These data are consistent with the hypothesis that excision repair is virtually nonexistent in XP individuals.

More recently, genetic analyses of XP patients have shown that the condition can result from mutations in any one of seven genes. This result is not surprising in light of the large number of enzymes involved in repairing damaged DNA.

Finally, defects in the genes required for DNA repair are frequently associated with cancer. Individuals with xeroderma pigmentosum, for example, are 1000 to 2000 times more likely to get skin cancer than are individuals with intact excision repair systems. To explain this pattern, biologists suggest that if mutations in the genes involved in the cell cycle (see Chapter 11) go unrepaired, the cell may begin to grow in an uncontrolled manner. Tumor formation could result. Stated another way, if the overall mutation rate in a cell is elevated because of defects in DNA repair genes, then the mutations that trigger cancer become more likely. Research on "simple" model organisms can be extraordinarily fruitful. In this case, research on fundamental aspects of DNA replication and repair, using *Escherichia coli* as a model system, led directly to a major advance in understanding a form of cancer in humans.

Chapter Review

SUMMARY OF KEY CONCEPTS

Genes are made of DNA. When DNA is copied, each strand of a DNA double helix serves as the template for the synthesis of a complementary strand.

By labeling DNA with ^{15}N or ^{14}N, Meselson and Stahl were able to validate the hypothesis that DNA replication is semiconservative. Each strand of a parent DNA molecule provides a template for the synthesis of a daughter strand, resulting in two complete DNA double helices.

You should be able to write a sequence of double-stranded DNA that is 10 bases long, separate the sequence, and write in the bases that are added during DNA replication.

DNA is synthesized only in the $5' \rightarrow 3'$ direction. When a DNA molecule is being copied, a large suite of specialized enzymes is involved in unwinding the double helix, continuously synthesizing the "leading strand" in the $5' \rightarrow 3'$ direction and synthesizing the "lagging strand" as a series of fragments that are then linked together.

DNA synthesis is an enzyme-catalyzed reaction that takes place in one direction. DNA synthesis requires both a template and a primer sequence, and it takes place at the replication fork where the double helix is being opened.

Synthesis of the leading strand in the $5' \rightarrow 3'$ direction is straightforward, but synthesis of the lagging strand is more complex because DNA polymerase has to work away from the replication fork. By feeding *E. coli* cells a short pulse of radioactive deoxyribonucleotide, Okazaki and co-workers confirmed that short DNA fragments form on the lagging strand. These Okazaki fragments are primed by a short strand of RNA and are linked together after synthesis.

At the ends of linear chromosomes in eukaryotes, the enzyme telomerase adds short, repeated sections of DNA so that the lagging strand can be synthesized without shortening the chromosome. Telomerase is active in reproductive cells that eventually undergo meiosis. As a result, gametes contain chromosomes of normal length.

You should be able to compare and contrast the functions of the three polymerases introduced in this chapter: DNA polymerase, primase, and telomerase.

MB Web Animation at www.masteringbio.com
DNA Synthesis

Most mistakes that occur in DNA synthesis are repaired by specialized enzymes. If these repair enzymes are defective, the mutation rate increases. Mutations in many types of genes can lead to cancer.

DNA replication is remarkably accurate because DNA polymerase proofreads and because mismatch repair enzymes remove incorrect bases once synthesis is complete and replace them with the correct sequence. In addition, DNA repair occurs after bases have been damaged by chemicals or radiation. Nucleotide excision repair systems cut out damaged portions of genes and replace them with correct sequences. Several types of human cancers are associated with defects in the genes responsible for DNA repair.

You should be able to explain the logical connections between failure of repair systems, increases in mutation rate, and high likelihood of cancer developing.

QUESTIONS

Test Your Knowledge

1. The experiment that removed proteins, RNA, or DNA from *Streptococcus* extracts purported to show that DNA is the hereditary material. Why did it fail to convince many skeptics?
   a. It had no control treatments, so it was poorly designed.
   b. Genes had already been shown to be made of protein.
   c. The trait that was studied—virulence—is not genetic.
   d. Critics argued that the chemical treatments had not removed all of the protein present.
2. Which of the following is *not* a property of DNA polymerase?
   a. It catalyzes the addition of dNTPs only in the $5' \rightarrow 3'$ direction.
   b. It requires a primer to work.
   c. It is associated with a sliding clamp only on the leading strand.
   d. It can proofread because it has an exonuclease activity.
3. What is the function of topoisomerase?
   a. holding DNA polymerase steady as it moves down the leading or lagging strand
   b. opening the DNA helix at the replication fork
   c. stabilizing single strands of DNA, once the replication fork is open
   d. preventing kinks in DNA as the replication fork opens and unwinds
4. What is the function of primase?
   a. synthesis of the short section of double-stranded DNA required by DNA polymerase
   b. synthesis of a short section of RNA, complementary to single-stranded DNA
   c. closing the gap at the $3'$ end of DNA after excision repair
   d. removing primers and synthesizing a short section of DNA to replace them
5. Where and how are Okazaki fragments synthesized?
   a. on the leading strand, in a $5' \rightarrow 3'$ direction
   b. on the leading strand, in a $3' \rightarrow 5'$ direction
   c. on the lagging strand, in a $5' \rightarrow 3'$ direction
   d. on the lagging strand, in a $3' \rightarrow 5'$ direction
6. What does telomerase do?
   a. It adds a protein primer to the ends of linear chromosomes.
   b. It adds double-stranded DNA to the "blunt end" of a linear chromosome.
   c. It adds double-stranded DNA to the lagging strand at the end of a linear chromosome.
   d. It adds single-stranded DNA to the lagging strand at the end of a linear chromosome.

Test Your Knowledge answers: 1. d; 2. c; 3. d; 4. b; 5. c; 6. d

Test Your Understanding

Answers are available at www.masteringbio.com

1. Researchers try to design experiments so that an experimental treatment shows the effect that one, and only one, condition or agent has on the phenomenon being studied. Examine Figure 14.2, and decide which treatments in the transformation study acted as experimental treatments and which acted as controls. Write a general statement explaining the role of control treatments in experimental design.
2. What does it mean to say that DNA replication is "bidirectional?"
3. Why is the synthesis of the lagging strand of DNA discontinuous? How is it possible for the synthesis of the leading strand to be continuous?
4. Explain how telomerase prevents linear chromosomes from shortening during replication.
5. List in chronological order the events that increase the accuracy of DNA replication. Indicate which events involve DNA polymerase III and which involve specialized repair enzymes.
6. Explain how the structure of DNA makes it relatively easy for proteins to recognize base-pair mismatches or damaged bases. How does DNA's secondary structure make it possible for damaged sections or incorrect bases to be removed and repaired?

Applying Concepts to New Situations

Answers are available at www.masteringbio.com

1. If DNA polymerase III did not require a primer, which steps in DNA synthesis would differ from what is observed? Would any special enzymes be required to replicate telomeres? Explain your answers.
2. In the late 1950s Herbert Taylor grew bean root-tip cells in a solution of radioactive thymidine and allowed them to undergo one round of DNA replication. He then transferred the cells to a solution without the radioactive deoxyribonucleotide, allowed them to replicate again, and examined their chromosomes for the presence of radioactivity. His results are shown in the following figure, where red indicates a radioactive chromatid.
   a. Draw diagrams explaining the pattern of radioactivity observed in the sister chromatids after the first and second rounds of replication.
   b. What would the results of Taylor's experiment be if eukaryotes used a conservative mode of DNA replication?

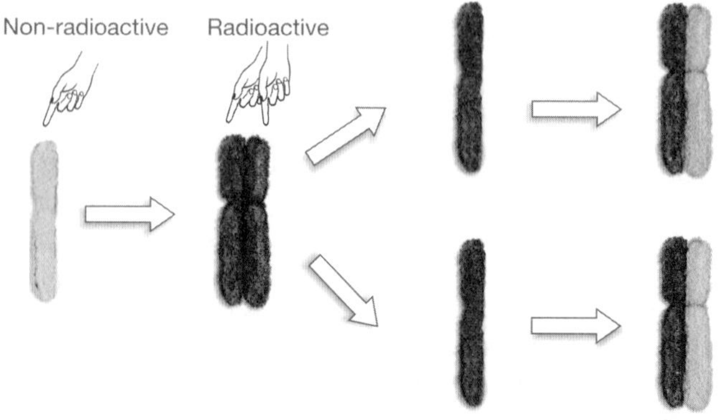

1. One round of DNA replication in radioactive solution **2.** Mitosis **3.** One round of DNA replication in non-radioactive solution

3. The graph that follows shows the survival of four different *E. coli* strains after exposure to increasing doses of ultraviolet light. The wild-type strain is normal, but the other strains have a mutation in genes called *uvrA*, *recA*, or both *uvrA* and *recA*.
   a. Which strains are most sensitive to UV light? Which strains are least sensitive?
   b. What are the relative contributions of these genes to the repair of UV damage?

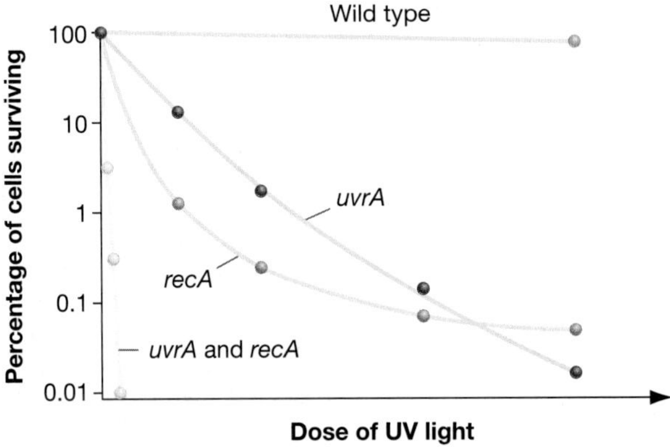

4. One widely used test to identify whether certain chemicals, such as pesticides or herbicides, might be carcinogenic (cancer causing) involves exposing bacterial cells to a chemical and recording whether the exposure leads to an increased mutation rate. In effect, this test equates cancer-causing chemicals with mutation-causing chemicals. Why is this an informative test?

www.masteringbio.com is also your resource for • Answers to text, table, and figure caption questions and exercises • Answers to *Check Your Understanding* boxes • Online study guides and quizzes • Additional study tools including the *E-Book for Biological Science* 3rd ed., textbook art, animations, and videos.

15 How Genes Work

KEY CONCEPTS

- Most genes code for proteins.
- In cells, information flows from DNA to RNA to proteins. DNA is transcribed to messenger RNA by RNA polymerase, and then messenger RNA is translated to proteins by ribosomes.
- Each amino acid in a protein is specified by a group of three bases in RNA.

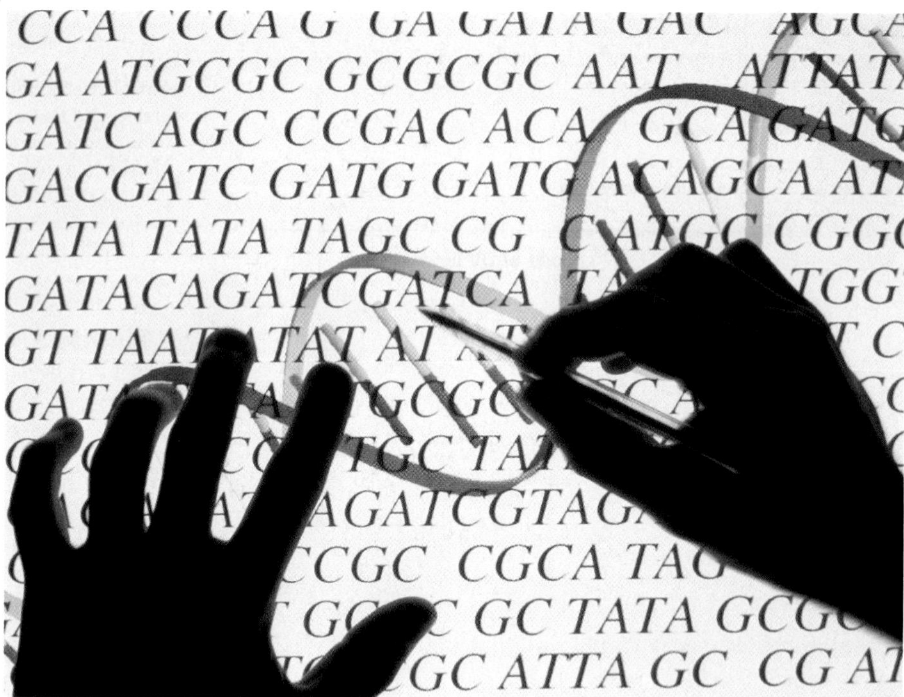

Each A, T, G, and C here represents a nitrogen-containing base in DNA. This chapter explores how biologists determined that genes are made of DNA and how a base sequence in DNA is translated into an amino acid sequence in a protein.

DNA has been called the blueprint of life. If an organism's DNA is like a set of blueprints, then its cells are like a construction site. How does the DNA inside each cell specify the types and amounts of lumber, nails, and cement that are needed as the cell grows and as its structure and function change? If the enzymes inside a cell are like construction workers, how does DNA organize them into a team that can build and maintain the cell, and perhaps remodel it when conditions change?

Mendel provided insights that made it possible to ask these questions. He discovered that particular alleles are associated with certain phenotypes and that alleles are transmitted faithfully from parent to offspring. The chromosome theory of inheritance established that genes are found in chromosomes, and the theory detailed how the movement of chromosomes during meiosis explains Mendel's results. The molecular era in biology began with the discovery that DNA is the hereditary material, and that DNA is a double helical structure containing sequences of four bases. Based on these early advances, it was clear that genes are made of DNA and that genes carry information—the instructions for making and maintaining an individual. But biologists still didn't know how the information in DNA is translated into action. How does an organism's genotype specify its phenotype?

This chapter introduces some of the most central ideas in all of biology—ideas that connected genotypes and phenotypes by revealing how genes work at the molecular level. Understanding how genes work, in turn, triggered a major transition in the history of biological science. Instead of thinking about genes solely in relation to their effects on eye color in fruit flies or on seed shape in garden peas, biologists could begin analyzing the molecular composition of both genes and their products. The molecular revolution in biology took flight.

15.1 What Do Genes Do?

Although in the early twentieth century biologists made tremendous progress in understanding how genes are inherited, an explicit hypothesis explaining what genes do did not appear until 1941. That year George Beadle and Edward Tatum published a series of breakthrough experiments on a bread mold called *Neurospora crassa*. Their research was inspired by an idea that was brilliant in its simplicity. As Beadle said: "One ought to be able to discover what genes do by making them defective." The idea was to knock out a gene by damaging it and then infer what the gene does by observing the phenotype of the mutant individual.

Today, alleles that do not function at all are called **knock-out mutants, null mutants**, or **loss-of-function mutants**. Creating and analyzing the effects of knock-out mutant alleles is still one of the most common research strategies in studies of gene function. But Beadle and Tatum were the pioneers.

To start their work, the researchers exposed a large number of *N. crassa* individuals to radiation. As Chapter 14 indicated, high-energy radiation damages the double-helical structure of DNA—often in a way that makes the affected gene nonfunctional. Then the researchers examined the mutant individuals. Eventually they succeeded in finding mutant *N. crassa* individuals that could not make specific compounds. For example, one of the mutants could not make pyridoxine, also called Vitamin B6, even though normal individuals can. Further, Beadle and Tatum showed that the inability to synthesize pyridoxine was due to a defect in a single gene, and that the inability to synthesize other molecules was due to defects in other genes.

These results inspired their **one-gene, one-enzyme hypothesis**. Beadle and Tatum proposed that the mutant *N. crassa* individual could not make pyridoxine because it lacked an enzyme required to synthesize the compound and that the lack of the enzyme was due to a genetic defect. Based on analyses of knock-out mutants, the one-gene, one-enzyme hypothesis claimed that genes contain the information needed to make proteins, many of which function as enzymes.

Three years later, Adrian Srb and Norman Horowitz published a rigorous test of the one-gene, one-enzyme hypothesis. These biologists focused on the ability of *N. crassa* individuals to synthesize the amino acid arginine. In the lab, normal cells of this bread mold grow well on a laboratory culture medium that lacks arginine. This is possible because *N. crassa* cells are able to synthesize their own arginine.

Previous work had shown that organisms synthesize arginine via a series of steps, or what biologists call a **metabolic pathway**. As **Figure 15.1** shows, compounds called ornithine and citrulline act as intermediates in the metabolic pathway leading to arginine. Further, specific enzymes are required to synthesize ornithine, convert ornithine to citrulline, and change citrulline to arginine. Srb and Horowitz hypothesized that specific genes in *N. crassa* cells are responsible for producing each of the three enzymes involved.

To test this idea, Srb and Horowitz used radiation to create a large number of mutant individuals. High-energy radiation is equally likely to damage DNA in any part of the organism's genome, however, and most organisms have thousands of genes. Of the many mutants that the biologists created, then, it was likely that only a handful contained knock-out mutations in the pathway for arginine synthesis.

To find the mutants they were looking for, the researchers performed what is now known as a genetic screen. A **genetic screen** is a technique for picking certain types of mutants out of many thousands of randomly generated mutants.

Srb and Horowitz began their screen by raising the irradiated cells on a medium that included arginine. Then they grew a sample of each type of cell on a medium that *lacked* arginine. If an individual could grow in the presence of arginine but failed to grow without arginine, they concluded that it couldn't make its own arginine. The biologists followed up by confirming that the offspring of these cells also had this defect. Based on these data, they were confident that they had isolated individuals with mutations in the genes for one or more of the enzymes shown in Figure 15.1.

To follow up on this result, the biologists grew the mutants on normal media that lacked arginine and was supplemented with nothing, or ornithine, or citrulline, or arginine. In each case, the media were solid and had a slanted surface to provide

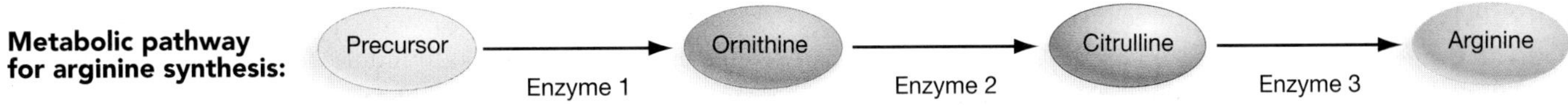

FIGURE 15.1 There Are Several Steps in the Metabolic Pathway for Arginine Synthesis. Most organisms synthesize the amino acid arginine via the series of steps shown here. Each step is catalyzed by a different enzyme.

QUESTION If a cell lacked enzyme 2 but got ornithine in its diet, could it still grow? Could it still grow if it got citrulline instead?

Experiment

Question: What do genes do?

Hypothesis: Each gene contains the information required to make a different enzyme.

Null hypothesis: Genes have nothing to do with making enzymes.

Experimental strategy: Knock out specific genes; see if the enzymes required for different steps in the pathway for synthesizing arginine are missing.

Experimental setup: Isolate mutant *N. crassa* that cannot synthesize arginine. Grow each type of mutant on normal medium with:

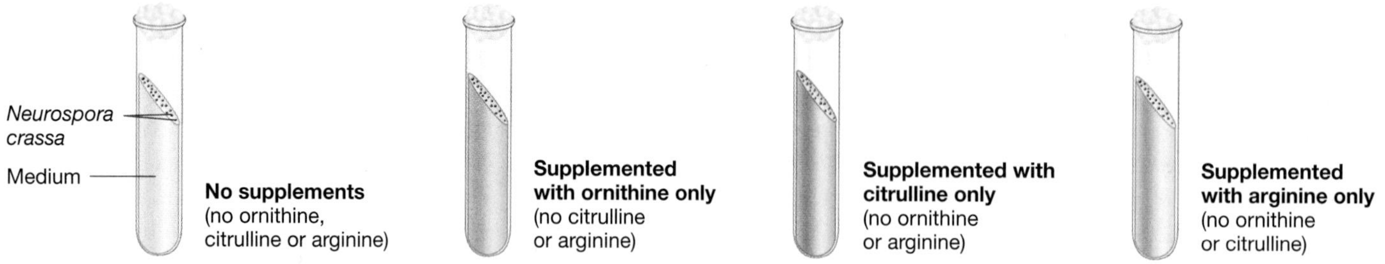

Neurospora crassa

Medium

No supplements (no ornithine, citrulline or arginine)

Supplemented with ornithine only (no citrulline or arginine)

Supplemented with citrulline only (no ornithine or arginine)

Supplemented with arginine only (no ornithine or citrulline)

Prediction: There will be three distinctive types of mutants, corresponding to defects in enzyme 1, enzyme 2, and enzyme 3 in the pathway for synthesizing arginine. Each type of mutant will be able to grow on different combinations of the four types of media.

Prediction of null hypothesis: There will not be distinctive types of mutants.

Results: There are three distinctive types of mutants, called *arg 1*, *arg 2*, and *arg 3*.

Mutant type	Supplement type: None	Ornithine only	Citrulline only	Arginine only
arg 1	no growth	GROWTH	GROWTH	GROWTH
arg 2	no growth	no growth	GROWTH	GROWTH
arg 3	no growth	no growth	no growth	GROWTH

Interpretation:

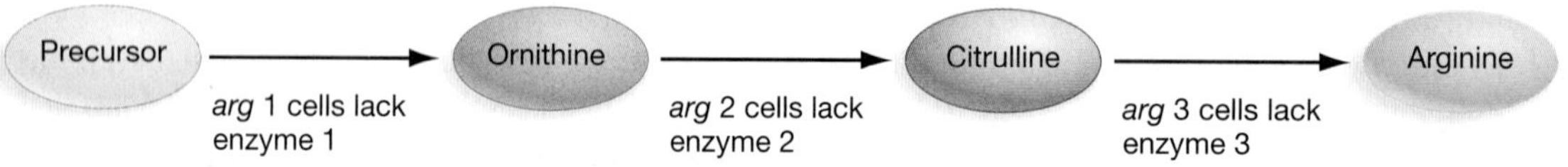

Precursor → Ornithine → Citrulline → Arginine

arg 1 cells lack enzyme 1

arg 2 cells lack enzyme 2

arg 3 cells lack enzyme 3

Conclusion: The one-gene, one-enzyme hypothesis is supported.

FIGURE 15.2 Experimental Support for the One-Gene, One-Enzyme Hypothesis. The association between specific genetic defects in *N. crassa* and specific deficits in the metabolic pathway for arginine synthesis convinced biologists that the one-gene, one-enzyme hypothesis was correct.

QUESTION Why did they bother to do the "No supplements" treatment?

adequate room for cells to grow. As **Figure 15.2** shows, the results from these growth experiments were dramatic. Some of the mutant cells were able to grow in the presence of some of these compounds but not in the presence of others. More specifically, the mutants fell into three distinct classes, which the researchers called *arg1*, *arg2*, and *arg3*. As the Interpretation section of the figure shows, the data make sense only if each type of mutant lacked a different, specific step in a metabolic pathway because of a defect in a particular gene. Srb and Horowitz had documented a correlation between a specific genetic defect and a defect at a specific point in a metabolic pathway. Their results convinced most investigators that the one-gene, one-enzyme hypothesis was correct.

Follow-up work showed that genes are responsible for all of the different types of proteins produced by cells—not just enzymes. Biologists finally understood what most genes do: They contain the instructions for making proteins.

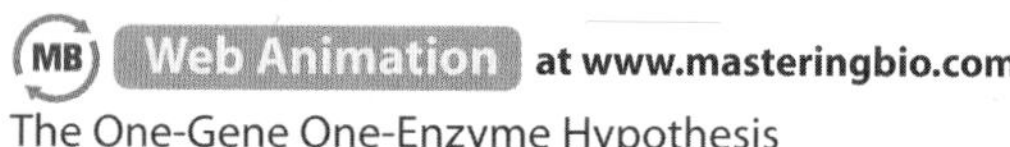

The One-Gene One-Enzyme Hypothesis

15.2 The Central Dogma of Molecular Biology

How does a gene specify the production of a protein? After Beadle and Tatum's hypothesis had been supported in *N. crassa* and a variety of other organisms, this question became a central one. Part of the answer lay in understanding the molecular nature of the gene. Biochemists knew that the primary components of DNA were four nitrogen-containing bases: the pyrimidines thymine (abbreviated T) and cytosine (C), and the purines adenine (A) and guanine (G). They also knew that these bases were connected in a linear sequence by a sugar-phosphate backbone. Watson and Crick's model for the secondary structure of the DNA molecule, introduced in Chapter 4 and reviewed in Chapter 14, revealed that two strands of DNA are wound into a double helix, held together by hydrogen bonds between the complementary base pairs A-T and G-C.

Given DNA's structure, it appeared extremely unlikely that DNA directly catalyzed the reactions that produce proteins. Its shape was too regular to suggest that it could bind a wide variety of substrate molecules and lower the activation energy for chemical reactions. Instead, Crick proposed that the sequence of bases in DNA might act as a code. The idea was that DNA was *only* an information-storage molecule. The instructions it contained would have to be read and then translated into proteins.

Crick offered an analogy with Morse code. Morse code is a message-transmission system using dots and dashes to represent the letters of the alphabet. Crick's idea was that different combinations of bases could specify the 20 amino acids, just as different combinations of dots and dashes specify the 26 letters of the alphabet. In this way, simple dots and dashes can code for the complexity of information in a human language.

A particular stretch of DNA, then, could contain the information needed to specify the amino acid sequence of a particular enzyme. In code form, the tremendous quantity of information needed to build and run a cell could be stored compactly. This information could also be copied through complementary base pairing and transmitted efficiently from one generation to the next.

It soon became apparent, however, that the information encoded in the base sequence of DNA is not translated into the amino acid sequence of proteins directly. Instead, the link between DNA as information repository and proteins as cellular machines is indirect.

RNA as the Intermediary between Genes and Proteins

The first clue that biological information does not flow directly from DNA to proteins came from data on the structure of cells. In eukaryotic cells, DNA is enclosed within a membrane-bound organelle called the nucleus (see Chapter 7). But the structures called ribosomes, where protein synthesis takes place, are outside the nucleus, in the cytoplasm. This observation began to make sense after François Jacob and Jacques Monod suggested that RNA molecules act as a link between genes and the protein-manufacturing centers. Jacob and Monod's hypothesis is illustrated in **Figure 15.3**. They predicted that short-lived molecules of RNA called **messenger RNA**, or **mRNA**, carry information from DNA to the site of protein synthesis. Messenger RNA is one of several distinct types of RNA in cells.

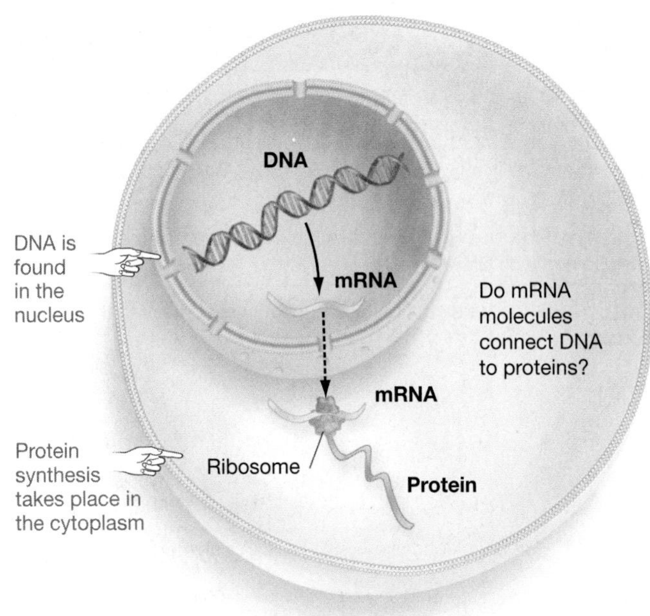

FIGURE 15.3 The Messenger RNA Hypothesis. In the cells of plants, animals, fungi, and other eukaryotes, most DNA is found only in the nucleus, but proteins are manufactured outside the nucleus, at ribosomes. Biologists proposed that the information coded in DNA is carried from inside the nucleus out to the ribosomes by messenger RNA (mRNA).

● **QUESTION** Bacteria and archaea do not have a nucleus. Based on the logic presented here, would they need to make an mRNA? Explain why or why not.

Follow-up research confirmed that the messenger RNA hypothesis is correct. One particularly important piece of evidence was the discovery of an enzyme that catalyzes the synthesis of RNA. This protein is called **RNA polymerase**, because it polymerizes ribonucleotides into strands of RNA. The key observation was that RNA polymerase synthesizes RNA molecules according to the information provided by the sequence of bases in a particular stretch of DNA. Unlike DNA polymerase, RNA polymerase does not require a primer to begin adding ribonucleotides to a growing strand of RNA.

Figure 15.4 provides experimental evidence on the transfer of information from DNA to mRNA. The experiment high-

Experiment

Question: How is information transferred from DNA to RNA?

Hypothesis: RNA is synthesized by complementary base pairing with DNA.

Null hypothesis: RNA is not synthesized by complementary base pairing with DNA.

Experimental setup: Create a reaction mix containing three elements:

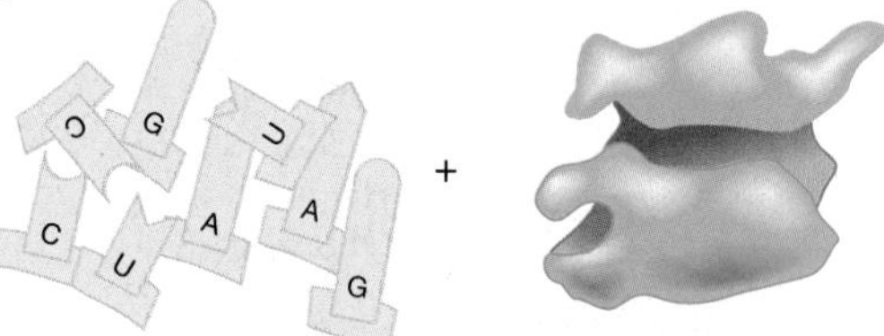

Prediction: An RNA strand will be produced containing only adenine (A).

Prediction of null hypothesis: No RNA strand will be produced, or an RNA strand with a random assortment of bases will be produced.

Results:

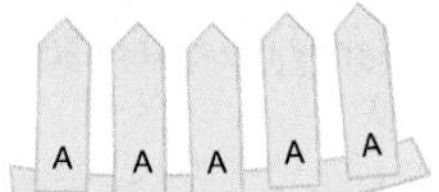

Conclusion: Information is transferred from DNA to RNA via complementary base-pairing.

FIGURE 15.4 Complementary Base Pairing Transfers Information from DNA to RNA. This experiment supported the hypothesis that RNA molecules are synthesized with a base sequence that is complementary to the corresponding DNA sequence.

lighted here began with a reaction mix containing three critical elements: (1) the enzyme RNA polymerase; (2) ribonucleotides containing the bases adenine (A), uracil (U), guanine (G), and cytosine (C); and (3) copies of a strand of synthetic DNA that contained deoxyribonucleotides in which the only base was thymine (T). (Recall from Chapter 4 that RNA contains the base uracil instead of thymine. If an RNA molecule binds to a DNA or RNA molecule, uracil forms complementary base pairs with adenine.) When researchers allowed the polymerization reaction to proceed, the only base in the RNA molecules that resulted was adenine. This result strongly supported the hypothesis that RNA polymerase synthesizes RNA according to the rules of complementary base pairing introduced in Chapter 4, because thymine pairs with adenine. In this case, a sequence of DNA containing thymine as the only base directed the polymerization of a complementary RNA sequence containing adenine as the only base. Similar experiments showed that synthetic DNAs containing no bases other than cytosine result in the production of RNA molecules containing no bases other than guanine.

The Central Dogma

Once the mRNA hypothesis was accepted, Francis Crick articulated what became known as the central dogma of molecular biology. The **central dogma** summarizes the flow of information in cells. It simply states that DNA codes for RNA, which codes for proteins:

$$\text{DNA} \rightarrow \text{RNA} \rightarrow \text{proteins}$$

Crick's simple statement encapsulates much of the research reviewed in this chapter. DNA is the hereditary material. Genes consist of specific stretches of DNA that code for products used in the cell. The sequence of bases in DNA specifies the sequence of bases in an RNA molecule, which specifies the sequence of amino acids in a protein. In this way, genes ultimately code for proteins.

Many proteins function as enzymes that catalyze chemical reactions in the cell. Other proteins perform the types of roles introduced in earlier chapters: Motor proteins and contractile proteins move the cell itself or cellular cargo, structural proteins provide support for the cell or tracks for transporting cargo, peptide hormones carry signals from cell to cell, membrane transport proteins conduct specific ions or molecules across the plasma membrane, and antibodies and other immune system proteins provide defense by recognizing and destroying invading viruses and bacteria.

Biologists use specialized vocabulary to summarize the sequence of events encapsulated in the central dogma. For example, biologists say that DNA is transcribed to RNA. In everyday English, the word *transcription* simply means making a copy of information. The scientific use of the term is appropriate because DNA acts as a permanent record—an archive or blueprint containing the information needed to build and run the cell. This permanent record is copied, via **transcription**, to the short-lived form called mRNA. The information is then transferred to a new molecular form—a sequence of amino acids. In everyday English, the word *translation* refers to the transferring of information from one language to another. In biology, the synthesis of protein from mRNA is called **translation**. Translation is the transferring of information from one type of molecule to another—from the "language" of nucleic acids to the "language" of proteins.

Translation is also often referred to simply as protein synthesis. **Figure 15.5a** summarizes the relationship between transcription and translation, as well as the relationship among

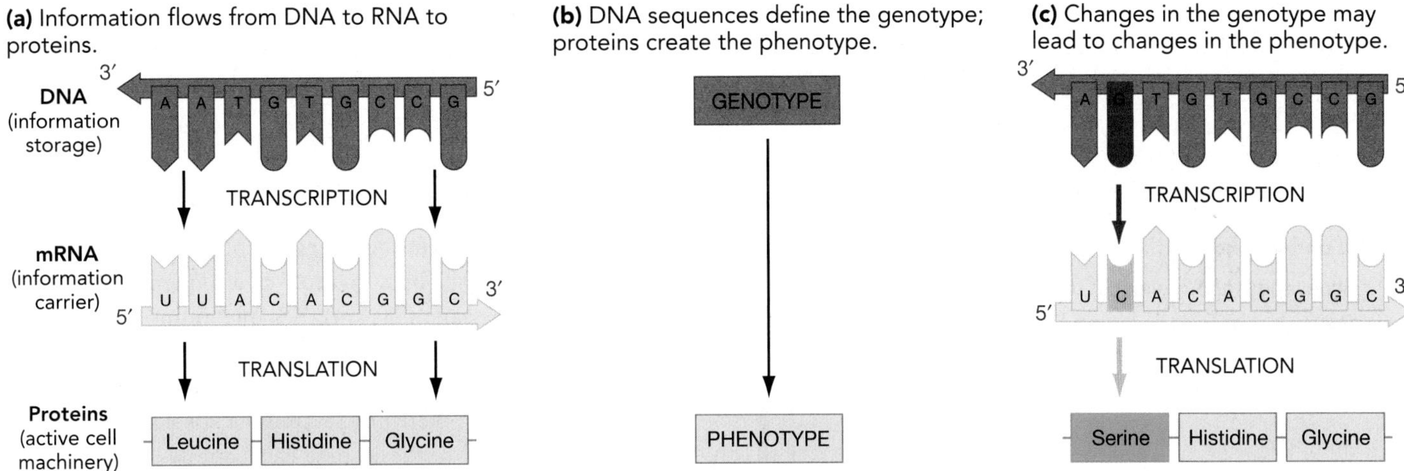

FIGURE 15.5 The Flow of Information in the Cell. (a) DNA is the information archive in the cell. mRNA is a short-lived copy of the information in DNA, used to manufacture proteins. Most of the machinery and many of the structures in cells consist of proteins. **(b)** The central dogma revealed the connection between genotype and phenotype. **(c)** Alleles of genes differ in their DNA base sequence. As a result, they may produce proteins with different primary structures and thus different functions. In this way, different alleles may be associated with different phenotypes.

DNA, RNA, and proteins. The following diagram presents these terms and relationships in a condensed form:

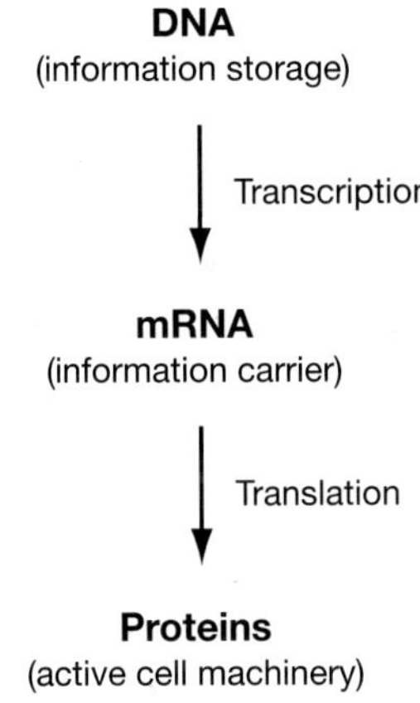

According to the central dogma, an organism's genotype is determined by the sequence of bases in its DNA while its phenotype is a product of the proteins it produces (**Figure 15.5b**). To drive this point home, consider that the enzymes and other proteins encoded by genes are what make the "stuff" of the cell and dictate which chemical reactions occur inside. In garden peas, for example, the "gene for height" that Mendel studied actually codes for an enzyme involved in synthesizing a growth hormone. An important aspect of a pea plant's phenotype—its height—is determined in part by the DNA sequence at the gene for this hormone-producing enzyme.

Later work revealed that alleles of the same gene differ in their DNA sequence. As a result, the proteins produced by different alleles of the same gene frequently differ in their amino acid sequence (**Figure 15.5c**). If the primary structures of proteins vary, their functions are likely to vary as well. The dwarfed pea plants that Mendel studied, for example, have a DNA sequence that is different from the normal allele. The altered sequence is dysfunctional, meaning that the mutants have a knock-out allele at the gene for height. These individuals do not produce the enzyme required to synthesize growth hormone. As a result, their cells do not receive the message to continue growth and the plants are short. (Dwarfing varieties like these are actually preferred in many types of crop plants and fruit trees.) The variation in phenotypes that we observe depends in part on variations in DNA sequences.

The central dogma provided an important conceptual framework for the burgeoning field called molecular genetics, and it inspired a series of fundamental questions about how genes and cells work. But important modifications to the central dogma have occurred in the decades since Frances Crick first proposed it.

- Many genes code for RNA molecules that do not function as mRNAs and are not translated into proteins. Stated another way, there are many types of RNA in addition to mRNA that perform important functions in cells. Besides carrying information from DNA so translation can occur, there are RNAs that (1) help regulate which genes are transcribed at various times, (2) process mRNAs prior to translation, (3) transport the amino acids required to build proteins, and (4) catalyze the formation of peptide bonds during translation. For these genes, information flow would be diagrammed as simply DNA → RNA.
- The central dogma suggests that information flow occurs in only one direction—from DNA to RNA to proteins. RNA does not code for the production of other RNA or DNA,

and proteins do not code for the production of RNA or DNA or other proteins. But there are important exceptions to these rules. In some cases, information flows from RNA back to DNA.

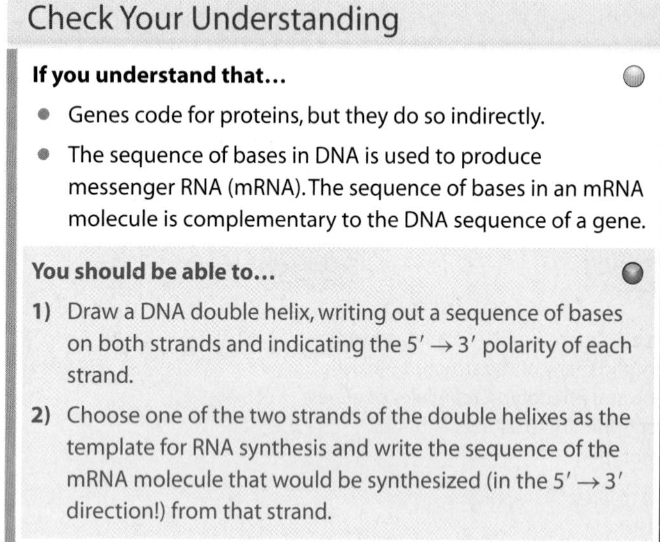

Check Your Understanding

If you understand that...

- Genes code for proteins, but they do so indirectly.
- The sequence of bases in DNA is used to produce messenger RNA (mRNA). The sequence of bases in an mRNA molecule is complementary to the DNA sequence of a gene.

You should be able to...

1) Draw a DNA double helix, writing out a sequence of bases on both strands and indicating the 5′ → 3′ polarity of each strand.
2) Choose one of the two strands of the double helixes as the template for RNA synthesis and write the sequence of the mRNA molecule that would be synthesized (in the 5′ → 3′ direction!) from that strand.

15.3 The Genetic Code

Once biologists understood the basic nature of information flow in the cell, the next challenge was to understand the final link between DNA and proteins. Exactly how does the sequence of bases in a strand of mRNA code for the sequence of amino acids in a protein? If this question could be answered, biologists would have cracked the **genetic code**—the rules that specify the relationship between a sequence of nucleotides in DNA or RNA and the sequence of amino acids in a protein. Researchers from all over the world took up the challenge. A race was on.

How Long Is a Word in the Genetic Code?

The first step in cracking the genetic code was to determine how many bases make up a "word." In a sequence of mRNA, how long is a message that specifies one amino acid? Based on some simple logic, George Gamow suggested that each code word contains three bases. His reasoning derived from the observation that there are 20 amino acids commonly used in cells, and the hypothesis that each amino acid must be specified by a particular sequence of mRNA. Because there are only four different bases in ribonucleotides (A, U, G, and C), a one-base code could specify only four amino acids. Similarly, a two-base code could represent just 4 × 4, or 16, amino acids. A three-base code, though, could specify 4 × 4 × 4, or 64, different amino acids. As a result, a three-base code would provide more than enough messages to code for all 20 amino acids (**Figure 15.6**). This three-base code is known as a **triplet code.**

These simple calculations highlight an important point: Gamow's hypothesis predicted that the genetic code is redundant. That is, more than one triplet of bases might specify the same amino acid. The idea was that different three-base sequences in an mRNA—say, AAA and AAG—might code for the same amino acid—say, lysine. The group of three bases that specifies a particular amino acid is called a **codon.** According to the triplet code hypothesis, many of the 64

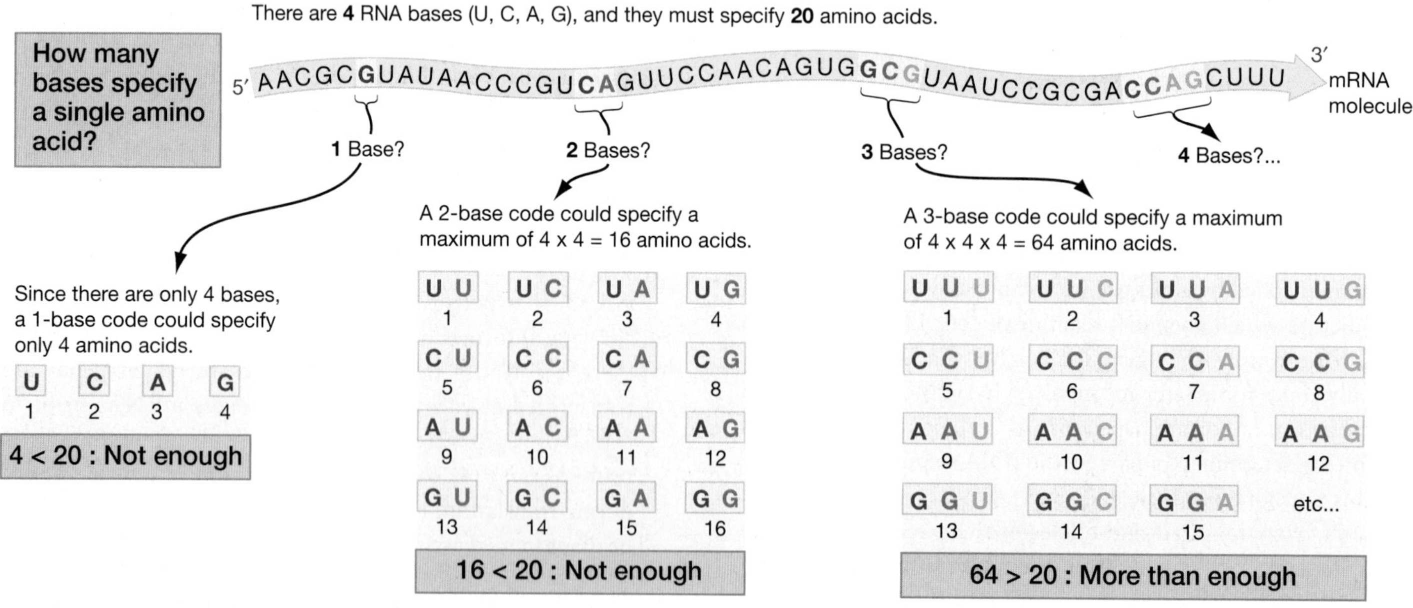

FIGURE 15.6 In the Genetic Code, How Many Bases Form a "Word"?

QUESTION How many amino acids could be specified by a four-base code? Why did biologists conclude that a four-base code was extremely unlikely? Your answer should include an explanation for why a four-base code is unlikely to evolve.

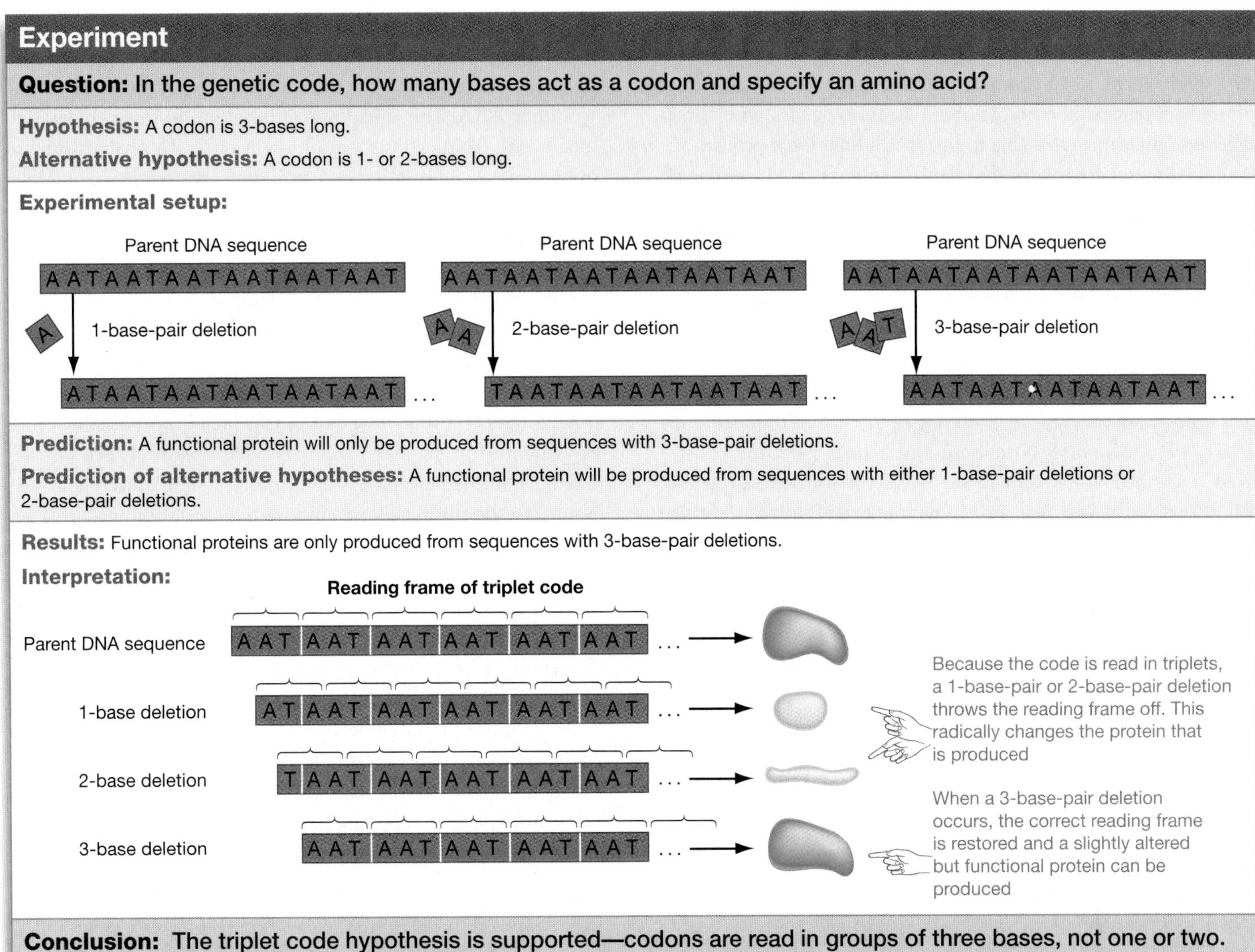

FIGURE 15.7 The Experiments That Confirmed the Triplet Code.

EXERCISE Write a sequence of bases in a strand of DNA. Make vertical marks between every third base to separate codons and indicate the reading frame. At random, select a base in the sequence and delete it. Write the new sequence and indicate its reading frame. Circle the codons that have changed after the deletion. Do the same exercise when you delete two randomly chosen bases in the sequence, and when you delete three.

codons that are possible actually specify the same amino acids.

Work by Francis Crick and Sydney Brenner confirmed that codons are three bases long. Their experiments used chemicals that caused an occasional addition or deletion of a base in DNA. As predicted for a triplet code, a one-base addition or deletion in the base sequence led to a loss of function in the gene being studied. This is because a single addition or deletion mutation throws the sequence of codons, or the **reading frame**, out of register. To understand how this works, consider the sentence "The fat cat ate the rat." The reading frame of this sentence is a three-letter word and a space. If the fourth letter in this sentence—the *f* in *fat*—were deleted but the reading frame stayed intact, the sentence would be transformed into "The atc ata tet her at." This is gibberish. When the reading frame in a DNA sequence is thrown out of register by the addition or deletion of a base, the composition of each codon changes just like the letters in each word of the example sentence. The protein produced from the altered DNA sequence has a completely different sequence of amino acids. In terms of its normal function, this protein is gibberish.

Figure 15.7 summarizes some of the experiments showing that codons are three bases long. In the Experimental setup portion of the figure, notice that the parent DNA sequence (top line) shows the reading frame for a series of AAT codons. In the sequences just below this, you can see what happened when Crick and co-workers used chemicals to remove one, two, or three bases from the parent sequence. Now look at the Results section of the figure to see how each type of mutation affected the protein. The only time functional proteins were produced was when three bases were removed. In the sentence "The fat cat ate the rat," the combination of removing one letter from each of the first three words might result in "Tha tca ate the rat." Just as the altered sentence still conveys some meaning, genes with three deletion mutations were able to produce a functional protein.

The researchers interpreted these results as strong evidence in favor of the triplet code hypothesis. Most other biologists agreed. The results also launched a long, laborious, and ultimately successful effort to determine which amino acid is specified by each of the 64 codons.

How Did Researchers Crack the Code?

The initial advance in deciphering the genetic code came in 1961, when Marshall Nirenberg and Heinrich Matthaei created a method for synthesizing RNAs of known sequence. Their method was based on an enzyme called polynucleotide phosphorylase, which catalyzes the formation of phosphodiester bonds between any ribonucleotides available, in random order. By providing ribonucleotides whose only base was uracil (U) to a reaction mix containing this enzyme, Nirenberg and Matthaei were able to create a long polymer of uracil-containing ribonucleotides. These synthetic RNAs were added to an in vitro system for synthesizing proteins. The researchers analyzed the resulting amino acid chain and determined that it was polyphenylalanine—a polymer consisting of the amino acid phenylalanine.

This result could mean only one thing: The RNA triplet UUU codes for the amino acid phenylalanine. By complementary base pairing, it was clear that the corresponding DNA sequence would be AAA. This initial observation was followed by experiments with RNAs consisting of only A or C. RNAs consisting of only AAAAA.... produced polypeptides consisting of only lysine; poly-C RNAs (RNAs consisting of only CCCCC...) produced polypeptides composed entirely of proline.

Nirenberg and Philip Leder later devised a system for synthesizing specific codons. Once they had copies of specific codons, they performed a series of experiments in which they added each of the codons to a cell extract that included the 20 different amino acids, ribosomes, and other molecules required for protein synthesis. As Chapter 7 noted, ribosomes are the multimolecular machines where proteins are synthesized. Then the researchers determined which amino acid was bound to the ribosomes when a particular codon was present. For example, when the codon CAC was in the reaction mix, the amino acid histidine would bind to the ribosomes. This result confirmed that CAC codes for histidine. These ribosome-binding experiments allowed Nirenberg and Leder to determine which of the 64 codons coded for each of the 20 amino acids (**Figure 15.8**).

In addition to matching codons to amino acids, researchers discovered that certain codons are punctuation marks signaling "start of message" or "end of message." These codons relay information that the protein chain is complete or that

SECOND BASE

FIRST BASE	U	C	A	G	THIRD BASE
U	UUU, UUC – Phenylalanine (Phe) UUA, UUG – Leucine (Leu)	UCU, UCC, UCA, UCG – Serine (Ser)	UAU, UAC – Tyrosine (Tyr) UAA – **Stop codon** UAG – **Stop codon**	UGU, UGC – Cysteine (Cys) UGA – **Stop codon** UGG – Tryptophan (Trp)	U C A G
C	CUU, CUC, CUA, CUG – Leucine (Leu)	CCU, CCC, CCA, CCG – Proline (Pro)	CAU, CAC – Histidine (His) CAA, CAG – Glutamine (Glu)	CGU, CGC, CGA, CGG – Arginine (Arg)	U C A G
A	AUU, AUC, AUA – Isoleucine (Ile) AUG – Methionine (Met) **Start codon**	ACU, ACC, ACA, ACG – Threonine (Thr)	AAU, AAC – Asparagine (Asn) AAA, AAG – Lysine (Lys)	AGU, AGC – Serine (Ser) AGA, AGG – Arginine (Arg)	U C A G
G	GUU, GUC, GUA, GUG – Valine (Val)	GCU, GCC, GCA, GCG – Alanine (Ala)	GAU, GAC – Aspartic acid (Asp) GAA, GAG – Glutamic acid (Glu)	GGU, GGC, GGA, GGG – Glycine (Gly)	U C A G

FIGURE 15.8 The Genetic Code. To read the code, match the first base in an mRNA codon, in the red band on the left side, with the second base in a codon, in the blue band along the top, and the third base in a codon, in the green band on the right side. The 64 codons, along with the amino acid or stop (termination) signal that they specify, are given in the boxes. By convention, codons are always written in the 5′ → 3′ direction.

● **EXERCISE** Pick four codons at random. Next to each of them, write the three bases in DNA that specify the codon. (Remember to write the DNA sequence in the 5′ → 3′ direction.)

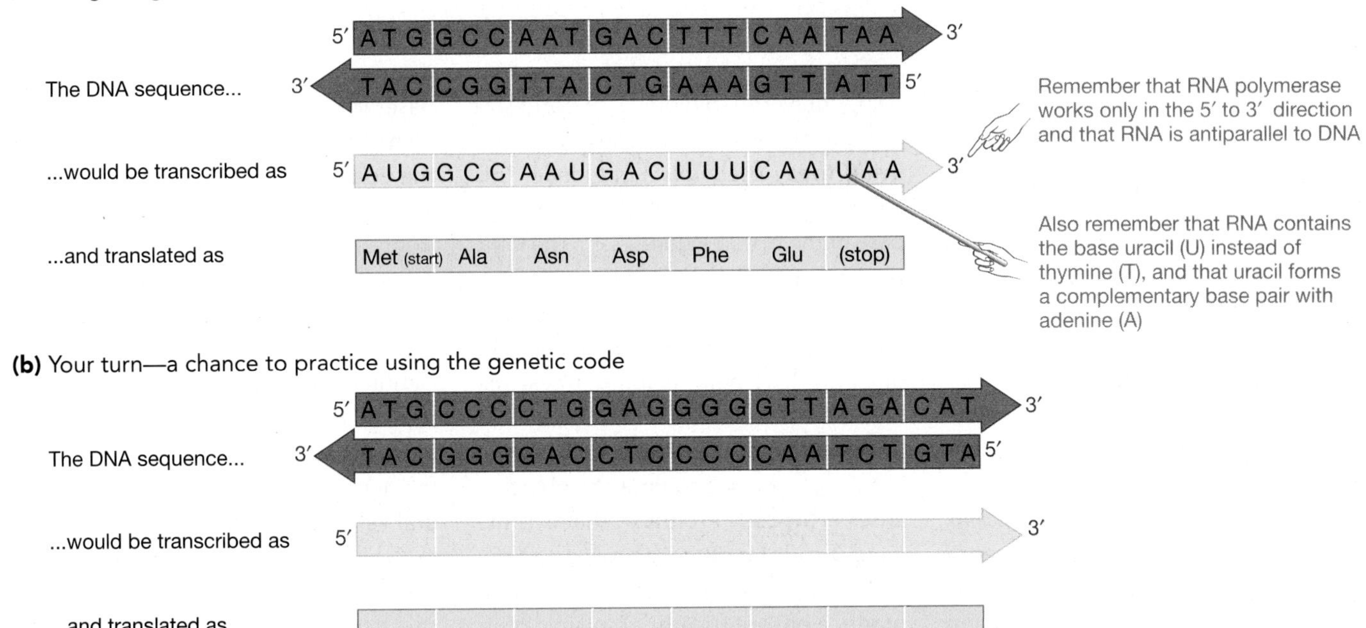

FIGURE 15.9 Using the Genetic Code to Predict Sequences. Biologists often have to predict RNA, DNA, or amino acid sequences from the types of data given here.

EXERCISE Write an mRNA that codes for the amino acid sequence Ala-Asn-Asp-Phe-Glu that is different from the one given here. Indicate the mRNA's 5′ → 3′ polarity. Then write the double-stranded DNA that corresponds to this mRNA. Indicate the 5′ → 3′ polarity of both strands.

EXERCISE Write the RNA sequence that is transcribed from the top DNA strand in part (a). Then write the amino acid sequence translated from that RNA. Do the same for the top strand in part (b).

protein synthesis should start at a given codon. There is one **start codon** (AUG), which codes for the amino acid methionine, and which also signals that protein synthesis should start at that point on the mRNA molecule. There are three **stop codons** (UAA, UAG, and UGA). The stop codons are also known as termination codons, because they signal that the protein is complete and end the translation process.

Using the genetic code, biologists are able to predict the amino acid sequences encoded by a particular DNA sequence (**Figure 15.9a**) and predict the mRNA sequence implied by a particular sequence of amino acids. If you understand how to read the genetic code, you should be able to identify the codons in Figure 15.5 and check to make sure that they are translated correctly. You should also be able to complete the exercises in **Figure 15.9b.**

The deciphering of the full genetic code, presented in Figure 15.8, is a tremendous achievement. It represents more than five years of work by several teams of researchers. As predicted, the genetic code is redundant. All amino acids except methionine and tryptophan are coded by more than one codon. The code is also unambiguous, however, because one codon never codes for more than one amino acid. Later work showed that the genetic code is also nearly universal: With a few minor exceptions, all codons specify the same amino acids in all organisms. **Box 15.1** explores the evolutionary implications of the universal code.

Check Your Understanding

If you understand that...

- The sequence of bases in mRNA forms a code in which particular combinations of three bases lead to the addition of a specific amino acid to the protein encoded by the gene.
- The genetic code is redundant because 64 combinations of bases occur, but only 20 amino acids and a stop "punctuation mark"—indicating the end of the coding sequence—need to be specified.

You should be able to...

1) Explain the consequences of mutations that occur in DNA and that change the codon ATA to one of the following codons: GTA, TTA, or GCA.
2) In each case, describe the effect on the resulting protein.

 **Web Animation** at www.masteringbio.com

The Triplet Nature of the Genetic Code

BOX 15.1 The Evolution of the Genetic Code

Figure 15.10 is a simplified version of the tree of life that was introduced in Chapter 1. Recall that the three largest branches on this tree represent what biologists call the three domains of life: the Bacteria, the Archaea, and the Eukarya. Smaller branches on the tree represent groups of species within each of the three domains, such as the cyanobacteria, α-proteobacteria, land plants, fungi, and animals.

Virtually all organisms on this tree use the genetic code in Figure 15.8. Exceptions are few. For example, in the single-celled eukaryotes *Tetrahymena* and *Paramecium*, UAA and UAG code for glutamine instead of stop; and in the yeast *Candida cylindracea*, CUG codes for serine instead of leucine. In a few species other codons in addition to AUG initiate protein synthesis.

To explain why the genetic code is nearly universal, biologists hypothesize that the common ancestor of all species living today used this same code. This common ancestor is indicated at the base of the tree in Figure 15.10. The logic here is that if all species living today use the same genetic code, then it is reasonable to infer that their common ancestor did also. (For help reading trees, see **BioSkills 2.**)

Given that the genetic code has been in existence for a long time, the question arises, Is the code arbitrary? That is, is the genetic code random with respect to the ability of organisms to survive and produce offspring? A brief examination of Figure 15.8 suggests that the answer is no. Note that when several codons specify the same amino acid, the first two bases in those codons are almost always identical. This observation suggests that the code does not represent a random assemblage of bases, like letters drawn from a hat. In fact, extensive analysis has shown that, compared with randomly generated codes, the existing genetic code is structured in a way that efficiently minimizes the phenotypic effects of small changes in DNA and errors during translation. This result suggests that the genetic code has been honed by natural selection. Stated another way, it is likely that other genetic codes may have existed early in the history of life. But because the code in Figure 15.8 is particularly efficient, it emerged as the most successful version.

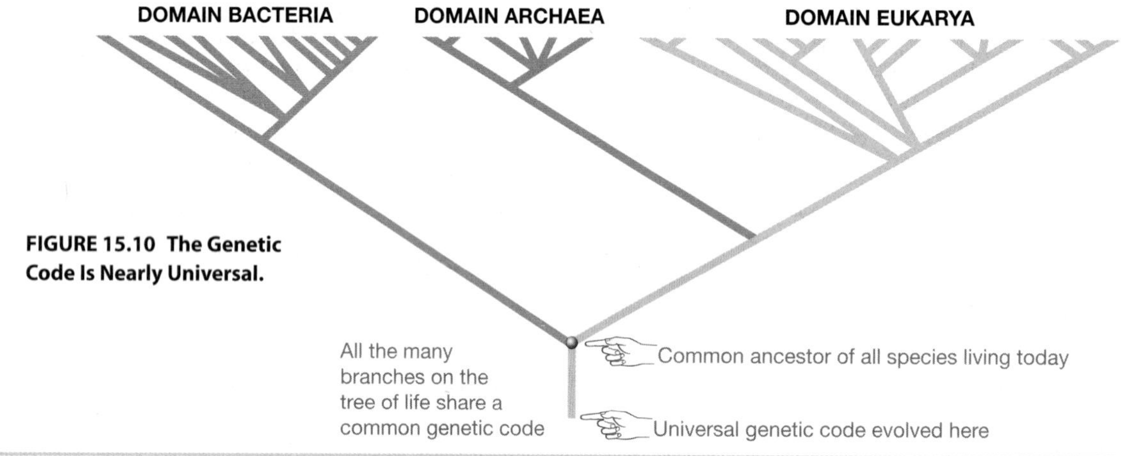

FIGURE 15.10 The Genetic Code Is Nearly Universal.

Chapter Review

SUMMARY OF KEY CONCEPTS

- **Most genes code for proteins.**

The question "What do genes do?" was solved in a series of experiments on the bread mold *Neurospora crassa*. Researchers isolated mutants that cannot grow unless they are supplied with a specific amino acid and then showed that different mutants were unable to make different chemical precursors of the amino acid. The one-gene, one-enzyme hypothesis explains these results by proposing that most genes code for proteins and that most proteins act as enzymes that catalyze specific reactions in metabolic pathways.

You should be able to explain the molecular basis of an organism's genotype and phenotype—including why alleles of the same gene exist and why different alleles produce different phenotypes.

(MB) Web Animation at www.masteringbio.com
The One-Gene, One-Enzyme Hypothesis

- **In cells, information flows from DNA to RNA to proteins. DNA is transcribed to messenger RNA by RNA polymerase, and then messenger RNA is translated to proteins by ribosomes.**

Experiments confirmed that DNA does not code for proteins directly. Instead, mRNA molecules are transcribed from DNA and then translated into proteins. One-way flow of information, from DNA to RNA to proteins, is called the central dogma of molecular biology.

You should be able to explain why the structure of DNA supports its function as an information archive, why the structure of RNA supports its array of functions—including information carrier—and why the structure of proteins supports their function as active cell machinery.

Each amino acid in a protein is specified by a group of three bases in RNA.

By synthesizing RNAs of known base composition and then observing how they functioned during the process of translation, researchers were able to unravel the genetic code. It is now established that the code is read in triplets and that the code is redundant—meaning that most of the 20 amino acids are specified by more than one codon.

You should be able to explain why some changes in DNA sequences do not change the corresponding protein—meaning, why certain changes in an organism's genotype do not change its phenotype. 

MB **Web Animation** at www.masteringbio.com
The Triplet Nature of the Genetic Code

QUESTIONS

Test Your Knowledge

1. What does the one-gene, one-enzyme hypothesis state?
   a. Genes are composed of stretches of DNA.
   b. Genes are made of protein.
   c. Genes code for ribozymes.
   d. A single gene codes for a single protein.
2. Which of the following is not an important exception to the one-gene, one-enzyme hypothesis?
   a. Many genes code for messenger RNAs.
   b. Many proteins are not enzymes.
   c. Many genes code for RNAs that function directly in the cell.
   d. Many enzymes are composed of multiple polypeptides, encoded by different genes.
3. DNA's primary structure is made up of just four different bases, and its secondary structure is regular and highly stable. How can a molecule with these characteristics hold all of the information required to build and maintain a cell?
   a. The information is first transcribed, then translated.
   b. The messenger RNA produced from DNA has much more complex secondary structures, and thus holds much more information.
   c. A protein produced (indirectly) from DNA has much more complex primary and secondary structures, and thus holds much more information.
   d. The information in DNA is in code form.
4. Why did researchers suspect that DNA does not code for proteins directly?
   a. In eukaryotes, DNA is found inside the nucleus, but proteins are produced outside the nucleus.
   b. In prokaryotes, DNA and proteins are never found together.
   c. When DNA was damaged by ultraviolet radiation or other sources of energy, the proteins in the cell did not change accordingly.
   d. There are several distinct types of RNA, of which only one functions as messenger RNA.
5. Which of the following describes an important experimental strategy in deciphering the genetic code?
   a. comparing the amino acid sequences of proteins with the base sequence of their genes
   b. analyzing the sequence of RNAs produced from known genes
   c. analyzing mutants that changed the code
   d. examining the proteins produced when RNAs of known sequence were translated
6. Knock-out mutations often produce recessive alleles. When a normal allele and a knock-out allele are paired in a heterozygous individual, the normal copy of the gene is frequently able to produce enough functional protein to give a normal phenotype. Why are these facts important?
   a. They explain why some mutants are still able to produce functional proteins.
   b. They explain the phenomenon of dominance and recessiveness discovered by Mendel.
   c. They offer a serious challenge to the one-gene, one-enzyme hypothesis.
   d. They illustrate why the central dogma is considered central.

Test Your Knowledge answers: 1. d; 2. a; 3. d; 4. a; 5. d; 6. b

Test Your Understanding

Answers are available at www.masteringbio.com

1. DNA is referred to as an "information-storage molecule." How can the base sequence of DNA store information?
2. Draw a hypothetical metabolic pathway involving a sequence of five substrates, five enzymes, and a product called BiolSciazine. Number the substrates 1–5, and label the enzymes A–E, in order. (For instance, enzyme A applies to the reaction between substrates 1 and 2.)
   - Suppose a mutation made the gene for enzyme C nonfunctional. What molecule would accumulate in the affected cells?
   - Suppose some individuals with a mutation affecting this metabolic pathway can survive if given substrate 5 in the diet. But they die even if given substrates 1, 2, 3, and 4. State a hypothesis for which enzyme in the pathway is affected by this mutation.
3. Why did experiments with *Neurospora crassa* mutants support the one-gene, one-enzyme hypothesis?
4. When researchers discovered that a combination of three deletion mutations or three addition mutations would restore the function of a gene, most biologists were convinced that the genetic code was read in triplets. Explain the logic behind this conclusion.
5. Why is the genetic code redundant?
6. Explain how a single-base deletion disrupts the reading frame of a gene.

Applying Concepts to New Situations

Answers are available at www.masteringbio.com

1. Recall that DNA and RNA are synthesized only in the 5′ → 3′ direction and that DNA and RNA sequences are always written in the 5′ → 3′ direction. Consider the following DNA sequence:

   5′ T T G A A A T G C C C G T T T G G A G A T C G G G T T A C A G C T A G T C A A A G 3′

   3′ A A C T T T A C G G G C A A A C C A C T A G C C C A A T G T C G A T C A G T T T C 5′

   - Identify the bases in the bottom strand that code for start and stop codons. Write the mRNA sequence that would be transcribed between them if the bottom strand served as the template.
   - Write the amino acid sequence that would be translated from the mRNA sequence you just wrote.

2. What problems would arise if the genetic code contained only 22 codons—one for each amino acid, a start signal, and a stop signal? (Hint: When DNA is copied prior to mitosis or meiosis, random errors occur that change its primary base sequence.)

3. Scientists say that a phenomenon is a "black box" if they can describe it and study its effects but don't yet know the underlying mechanism that causes it. In what sense was genetics—meaning the transmission of heritable traits—a black box before the central dogma of molecular biology was understood?

4. One of the possibilities that researchers interested in the genetic code had to consider was that the code was overlapping, meaning that a single base could be part of more than one codon. Make a diagram showing how an overlapping code would work, assuming that each codon is three bases long.

www.masteringbio.com is also your resource for • Answers to text, table, and figure caption questions and exercises • Answers to *Check Your Understanding* boxes • Online study guides and quizzes • Additional study tools including the *E-Book for Biological Science* 3rd ed., textbook art, animations, and videos.

Transcription and Translation

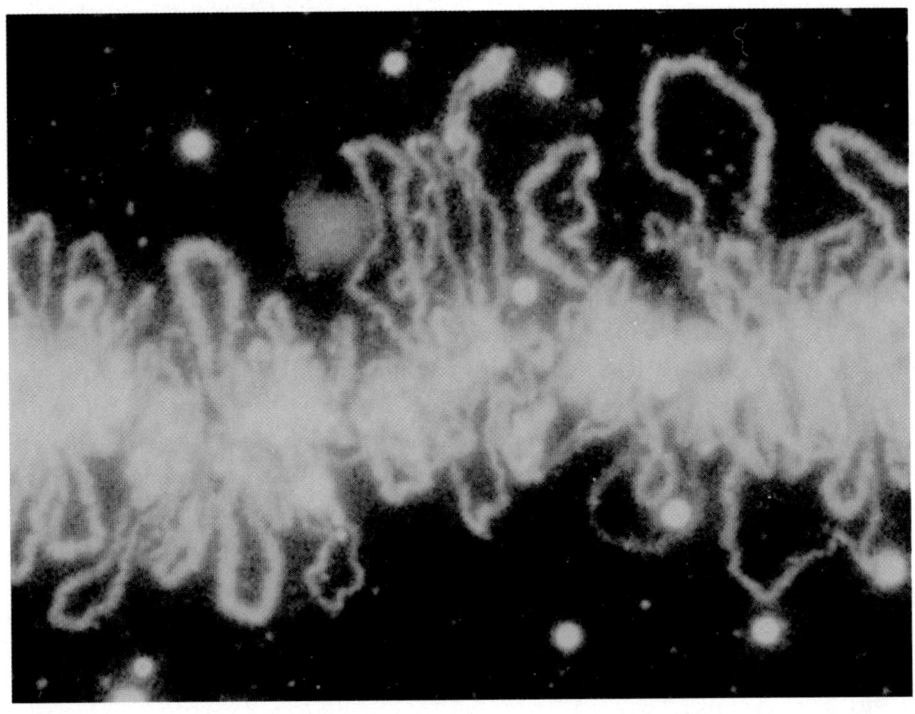

A section of a chromosome inside a frog egg, where extensive transcription is taking place. Proteins involved in transcription are stained a fluorescent green.

KEY CONCEPTS

- After RNA polymerase binds DNA with the help of other proteins, it catalyzes the production of an RNA molecule whose base sequence is complementary to the base sequence of the DNA template strand.
- Some sections of an RNA are encoded by gene regions called exons, while others are encoded by gene regions called introns. During RNA processing, introns are removed and the ends of the RNA receive a cap and tail.
- Inside ribosomes, mRNAs are translated to proteins via intermediary molecules called transfer RNAs. Transfer RNAs carry an amino acid and have a three-base-pair anticodon, which binds to a three-base-long mRNA codon. The amino acid carried by the transfer RNA is then added to the growing protein via formation of a peptide bond.
- Mutations are random changes in DNA that may or may not produce changes in the phenotype.

Proteins are the stuff of life. They give cells their shape, control the chemical reactions that go on inside them, and regulate the import, export, and movement of materials. No one knows exactly how many different proteins can be made in your body's cells, but 100,000 is a reasonable guess. While some of those 100,000 different proteins may not be manufactured at all in certain types of cells, other proteins may be present in quantities ranging from millions of copies to fewer than a dozen. If even one of those types of protein is defective, the cell may sicken or even die.

A cell builds the proteins it needs from instructions encoded in its genome. Chapter 15 explored how the information archived in DNA is retrieved and converted into action in the form of proteins. The first step in the process is the transcription of a gene and the production of a messenger RNA, or mRNA. The RNA message is a short-lived copy of the archived instructions in DNA. The sequence of ribonucleotides in the mRNA is then translated into a sequence of amino acids. This series of events defines the flow of information in the cell:

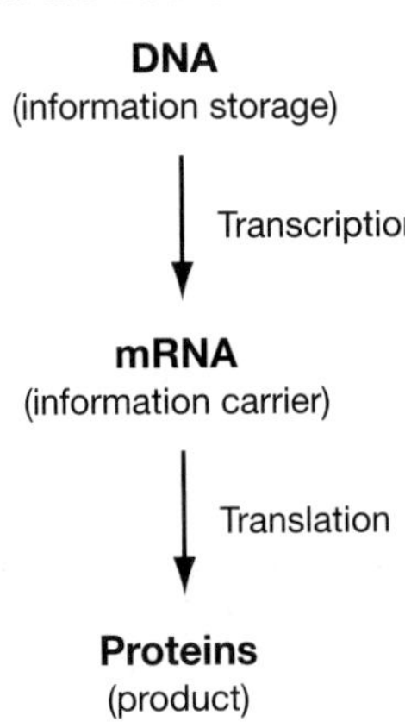

The discovery of the relationship between DNA and proteins was one of the great scientific advances of the twentieth century. But once the relationship between transcription and translation had been firmly established, biologists puzzled over how cells actually accomplish the feat of converting genetic information into protein. How do transcription and translation work? For example, how does RNA polymerase know where to start transcribing a gene and where to end? Once an RNA message is produced, how is the linear sequence of ribonucleotides translated into the linear sequence of amino acids in a protein?

This chapter explores the experiments that answered these questions. Besides providing a deeper understanding of how cells work, the results summarized here have had a significant impact on biomedicine and drug development. Many of the antibiotics prescribed today work by disrupting translation in bacteria that cause disease. In addition, research on transcription and translation set the stage for an explosion of work, introduced in Chapters 17 and 18, on how the expression of specific genes is regulated. Most genes are transcribed and translated only at certain times or in certain types of cells, and in carefully regulated amounts.

Let's begin by examining research on the mechanisms of transcription in bacteria and eukaryotes, and then delve into the mechanics of translation. The chapter concludes by examining how changes in DNA can lead to changes in protein structure and function.

16.1 Transcription in Bacteria

The first step in converting genetic information into proteins is the synthesis of a messenger RNA version of the instructions archived in DNA. Enzymes called RNA polymerases, introduced in Chapters 14 and 15, are responsible for synthesizing mRNA. Like the DNA polymerases introduced in Chapter 14, an RNA polymerase performs a template-directed synthesis in the $5' \rightarrow 3'$ direction. (Recall that 5′ and 3′ refer to carbons on the sugar subunits of ribonucleotides.) But unlike DNA polymerases, RNA polymerases do not require a primer to begin transcription.

Transcription occurs when RNA polymerase matches the base in a ribonucleotide triphosphate with the complementary base in a gene—a section of DNA that codes for a protein or RNA. Once a matching ribonucleotide is in place, RNA polymerase catalyzes the formation of a phosphodiester bond between the 3′ end of the growing mRNA chain and the new ribonucleotide. As this matching-and-catalysis process continues, an RNA that is complementary to the gene is synthesized (**Figure 16.1**). Notice that only one of the two DNA strands is used as a template and transcribed, or "read," by RNA polymerase. The strand that is read by the enzyme is called the **template strand**; the other strand is called the **non-template strand.** The non-template strand is also known as the coding strand because its sequence matches the sequence of the RNA that is transcribed from the template strand and codes for a

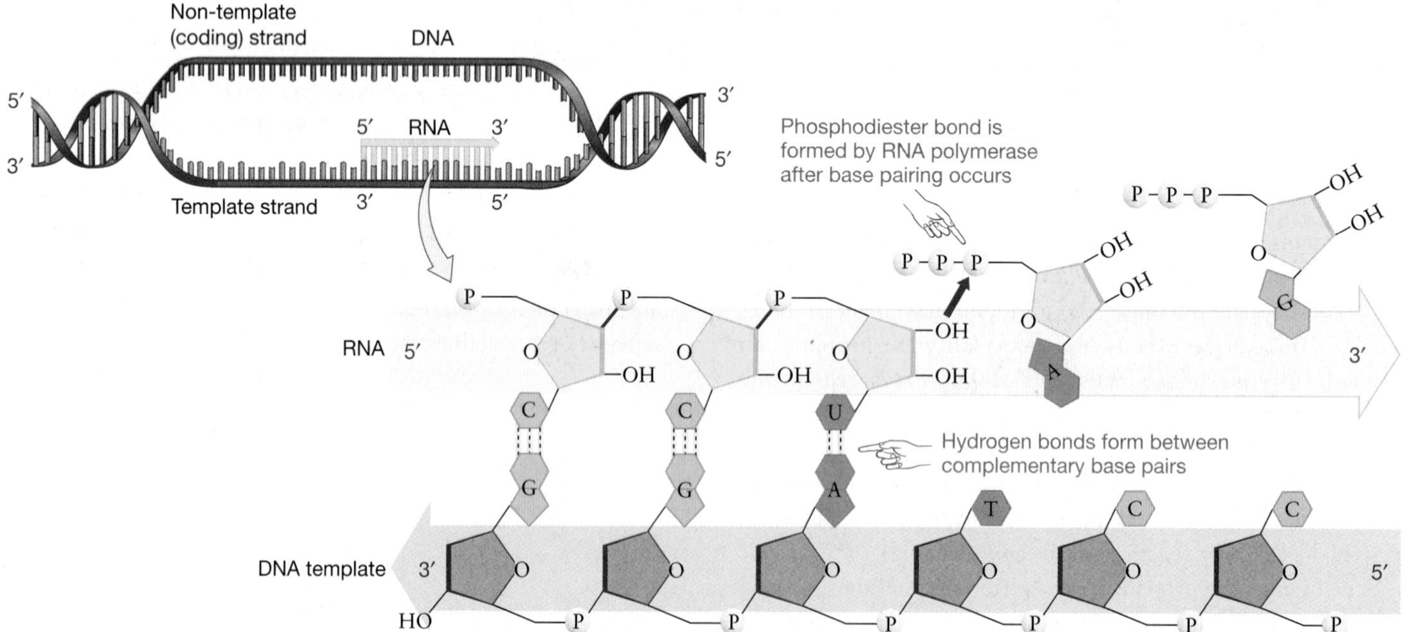

FIGURE 16.1 Transcription Is the Synthesis of RNA from a DNA Template. The reaction catalyzed by RNA polymerase results in the formation of a phosphodiester bond between ribonucleotides. RNA polymerase produces an RNA strand whose sequence is complementary to the bases in the DNA template.

● **QUESTION** In which direction is RNA synthesized, $5' \rightarrow 3'$ or $3' \rightarrow 5'$? In which direction is the DNA template "read"?

polypeptide. A key difference is that RNA has uracil (U) rather than the thymine (T) found in the non-template or coding strand. An adenine (A) in the DNA template strand specifies a U in the complementary RNA strand.

Biologists reached these conclusions by studying RNA synthesis in cell-free, or in vitro, systems. Once the basic chemical reaction was understood, an entirely new set of questions arose:

- What does RNA polymerase look like?
- Are any other proteins or factors involved in transcription, or does RNA polymerase act alone?
- How does the enzyme know where to start transcription on the DNA template?

RNA Polymerase Structure and Function

To understand the structure of RNA polymerase, researchers employed the technique called X-ray crystallography introduced in **BioSkills 8**. This procedure allows biologists to obtain information about the three-dimensional structure of large, complex molecules.

The most recent models of bacterial RNA polymerase estimated from X-ray crystallographic data indicate that the enzyme is large and globular and has several prominent channels running through the interior. The enzyme's active site, where phosphodiester bonds form, is located where several of these channels intersect.

Because the structure of molecules correlates with their function, it is logical to predict that DNA fits into one of the enzyme's channels and that the two strands in the double helix separate inside the enzyme to expose a single-stranded template at the active site. But how do the enzyme and DNA come together in the first place?

Initiation: How Does Transcription Begin?

Soon after the discovery of RNA polymerase, researchers realized that the enzyme cannot initiate transcription on its own. Instead, a detachable protein subunit called sigma must bind to RNA polymerase before transcription can begin. RNA polymerase and sigma form what biologists call a **holoenzyme** ("whole enzyme"). A holoenzyme consists of a **core enzyme**, which contains the active site for catalysis, and other required proteins (**Figure 16.2**).

If RNA polymerase is the core enzyme of this holoenzyme, what does sigma do? When researchers mixed RNA polymerase, sigma, and DNA together, they found that the holoenzyme bound tightly to specific sections of DNA. These binding sites were named **promoters**, because they are sections of DNA where transcription begins. The discovery of promoters suggested that sigma's function is regulatory in nature. Sigma appeared to be responsible for guiding RNA polymerase to specific locations

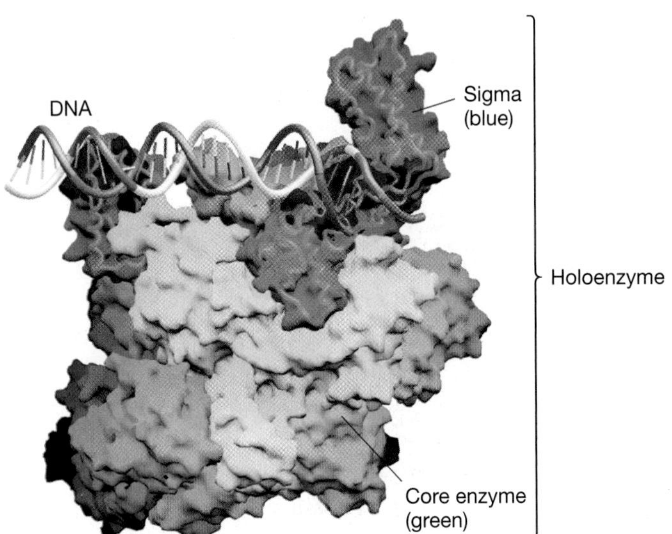

FIGURE 16.2 The Three-Dimensional Structure of RNA Polymerase. Bacterial RNA polymerase is a globular enzyme. Sigma is a regulatory component, and the core enzyme contains the active site.

where transcription should begin. What is the nature of these specific locations? What do promoters look like, and what do they do?

David Pribnow offered an initial answer to these questions in the mid-1970s. When Pribnow analyzed the base sequence of promoters from a variety of bacteria and from viruses that infect bacteria, he found that the promoters were located on the non-template strand, were 40–50 base pairs long, and had a particular section that looked similar. This similar segment of DNA had a series of bases identical or similar to TATAAT. This six-base-pair sequence is now known as the −10 box, because it is centered about 10 bases from the point where RNA polymerase starts transcription. (DNA that is located in the direction RNA polymerase moves during transcription is said to be **downstream**; DNA in the opposite direction is said to be **upstream**. Thus, the −10 box is centered about 10 bases upstream from the transcription start site.) The place where transcription begins is called the +1 site. Soon after, researchers recognized that the sequence TTGACA occurred in these same promoters, centered about 35 bases upstream from the +1 site. This second key sequence is called the **−35 box**. Sequences that are inside the promoter but outside the −10 and −35 boxes vary widely among genes.

Follow-up work showed that transcription begins when sigma, as part of the holoenzyme complex, binds to the −35 and −10 boxes (**Figure 16.3**, step 1). This is a key point. Sigma, and not RNA polymerase, makes the initial contact with DNA that starts transcription in bacteria. This finding supports the hypothesis that sigma is a regulatory protein. Sigma tells RNA polymerase where and when to start synthesizing RNA.

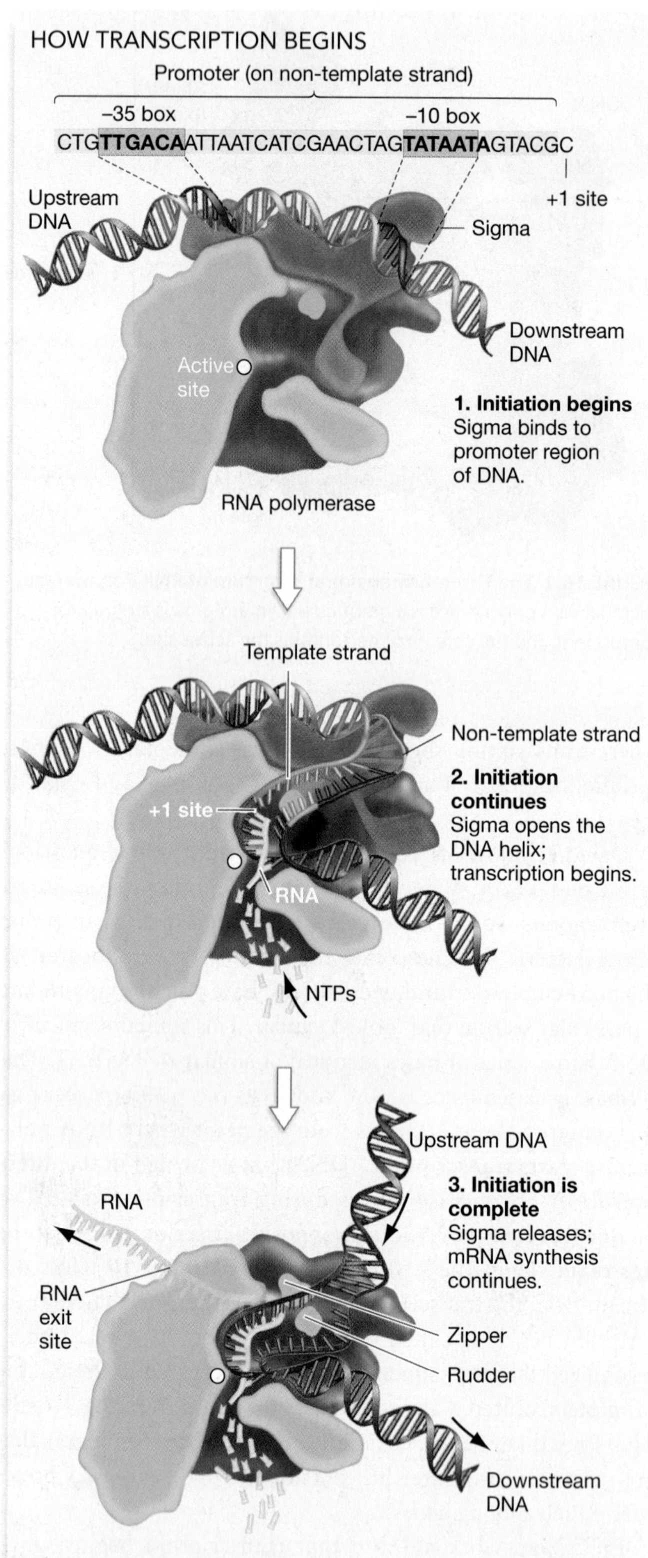

FIGURE 16.3 In Bacteria, Sigma Plays a Key Role in Initiating Transcription. When sigma attaches to RNA polymerase, the holoenzyme is able to bind to the –35 box and the –10 box of a promoter. Once it has bound to a promoter, sigma opens the DNA helix and threads the template strand through the enzyme's active site. When transcription is under way, sigma disengages from RNA polymerase.

● **EXERCISE** Circle and label grooves in the enzyme that form exit sites for RNA and DNA.

Once sigma binds to a promoter, the DNA helix opens and creates two strands of single-stranded DNA. As shown in step 2 of Figure 16.3, the template strand is threaded through a channel that leads to the active site inside RNA polymerase. Monomers known as ribonucleoside triphosphates, or NTPs, enter a channel at the bottom of the enzyme and diffuse to the active site. NTPs are like the dNTPs introduced in Chapter 14's discussion of DNA synthesis.

When an incoming NTP pairs with a complementary base on the template strand of DNA, RNA polymerization begins. The reaction catalyzed by RNA polymerase is exergonic and spontaneous because NTPs have so much potential energy, owing to their three phosphate groups. As step 3 of Figure 16.3 shows, sigma is released once RNA synthesis is under way. The **initiation** phase of transcription is complete.

Recent work has shown that most bacteria have several types of sigma proteins, each with a distinctive structure and function. *Escherichia coli* has seven different sigma proteins, for example, while *Streptomyces coelicolor* has more than 60. Each of these proteins binds to promoters with slightly different DNA base sequences. Although all promoters have a –10 box and a –35 box, the sequences within each box vary somewhat. These variations in DNA sequences affect the ability of different sigma proteins to bind. As a result, each type of sigma protein allows RNA polymerase to bind to a different type of promoter and thus different kind of gene. The identity of the sigma protein in the RNA polymerase holoenzyme determines which types of genes will be transcribed. Controlling which sigma proteins are active is one of the ways that bacterial cells control which genes are expressed.

Elongation and Termination

During the **elongation** phase of transcription, RNA polymerase moves along the DNA template in the $3' \rightarrow 5'$ direction of the template strand, synthesizing RNA in the $5' \rightarrow 3'$ direction. In the interior of the enzyme, a group of projecting amino acids called the enzyme's zipper helps open the double helix at the upstream end, and a nearby group of amino acids called the rudder helps steer the template and non-template strands through channels inside the enzyme (see Figure 16.3, step 3). Meanwhile, the enzyme's active site catalyzes the addition of nucleotides to the $3'$ end of the growing RNA molecule at the rate of about 50 nucleotides per second.

Note that during the elongation phase of transcription, all of the prominent channels or grooves in the enzyme are filled (Figure 16.3, step 3). Double-stranded DNA goes into and out of one groove; ribonucleoside triphosphates enter another; and the growing RNA strand exits to the rear. In this way, the enzyme's structure correlates closely with its function.

Transcription ends with a **termination** phase. In most cases, transcription stops when RNA polymerase reaches a stretch of DNA sequence that functions as a transcription termination signal. The bases that make up the termination signal code for

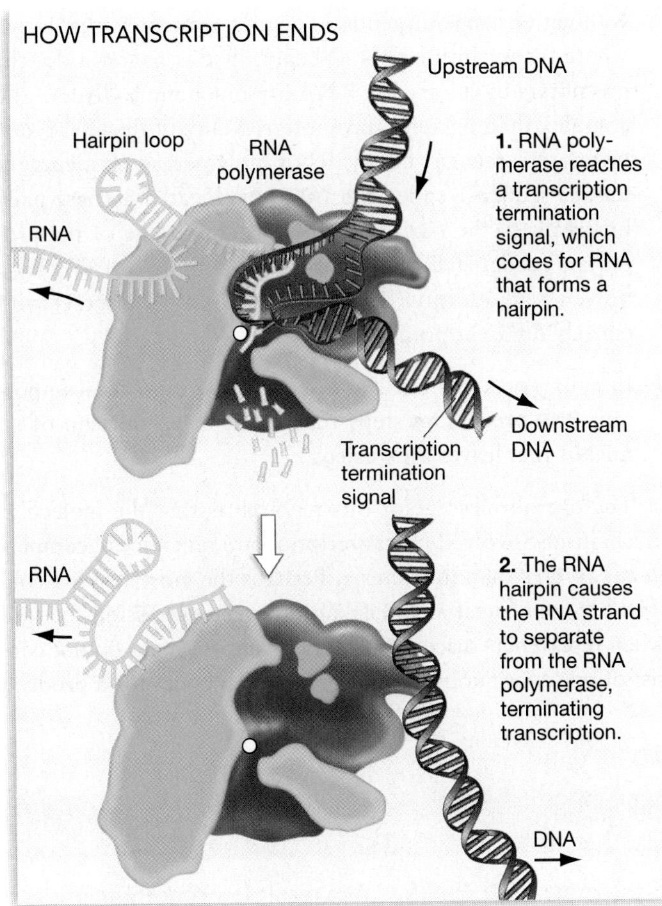

FIGURE 16.4 Transcription Terminates When a Hairpin Forms

a stretch of RNA with an unusual property: As soon as it is synthesized, the RNA sequence folds back on itself and forms a short double helix that is held together by complementary base pairing. The secondary structure that results is called a hairpin (**Figure 16.4**). The formation of the hairpin structure is thought to disrupt the interaction between RNA polymerase and the RNA transcript, resulting in the physical separation of the enzyme and its product.

● To summarize, transcription begins when sigma, as part of the holoenzyme complex, binds to the promoter at the start of a gene. Once binding occurs, RNA polymerase begins to synthesize mRNA by adding ribonucleotides that are complementary to the template strand in DNA. Transcription ends when a termination signal at the end of the gene leads to the formation of a hairpin in the mRNA, disrupting the transcription complex. ● If you understand these concepts, you should be able to (1) draw the RNA polymerase holoenzyme, (2) label sigma and the RNA polymerase core enzyme and describe their function, and (3) diagram the initiation, elongation, and termination phases of transcription in bacteria.

(MB) **Web Animation** at www.masteringbio.com
Transcription

16.2 Transcription and RNA Processing in Eukaryotes

The data reviewed in Section 16.1 established how transcription occurs in bacteria. How do these results compare with research on the mechanism of transcription in eukaryotes? What similarities and differences have come to light?

One of the first conclusions about transcription in eukaryotes is that it is similar to bacterial transcription in an important respect: RNA polymerase does not bind directly to promoter sequences by itself. Instead, proteins that came to be called **basal transcription factors** initiate eukaryotic transcription by matching the enzyme with the appropriate promoter region in DNA. In eukaryotes, the function of basal transcription factors is analogous to the function of the sigma proteins in bacteria, except that basal transcription factors interact with DNA independent of RNA polymerase.

The discovery of basal transcription factors emerged from studies with cell-free, or in vitro, systems. The basic approach was to purify RNA polymerase from human cells, add template DNA, and analyze the RNAs that were produced. But in early experiments, researchers found that RNA polymerase began copying at random locations on the template DNA instead of specifically at promoter regions. In addition, when RNA polymerase was alone, both strands of the DNA were transcribed instead of just one. This observation inspired the hypothesis that eukaryotic cells must contain sigma-like proteins that are required for normal transcription. To test this hypothesis, biologists added proteins isolated from human cells to the cell-free reaction system. By adding one or a few proteins at a time and recording which ones enabled RNA polymerase to bind to promoters and correctly transcribe the template strand, researchers were gradually able to characterize a suite of proteins required for proper initiation.

Follow-up work confirmed several important distinctions about how transcription works in bacteria and eukaryotes:

- In bacteria a single sigma protein binds to a promoter and initiates transcription, but in eukaryotes many basal transcription factors are required to initiate transcription. In eukaryotes, the machinery required to start transcription is complex.
- Eukaryotes have three distinct types of RNA polymerase (abbreviated RNA pol) instead of just one. As **Table 16.1** shows, RNA pol I, pol II, and pol III each transcribe a discrete class of RNA. Only RNA pol II transcribes the genes that code for proteins—meaning that it produces mRNA. RNA pol I makes the large RNA molecules that are found in ribosomes. RNA pol III manufactures one of the small RNAs that are found in ribosomes and the molecules called transfer RNAs, which are required for translation. **Box 16.1** explains how research on toxic mushrooms led to the discovery that eukaryotes have three different RNA polymerases.

TABLE **16.1** Eukaryotic RNA Polymerases

Name of Enzyme	**Type of Gene Transcribed**
RNA polymerase I (RNA pol I)	Genes that code for most of the large RNA molecules (rRNAs) found in ribosomes (see Section 16.3)
RNA polymerase II (RNA pol II)	Protein-coding genes (produce mRNAs)
RNA polymerase III (RNA pol III)	Genes that code for transfer RNAs (see Section 16.4), and genes that code for one of the small RNA molecules (rRNAs) found in ribosomes
RNA pol II and RNA pol III	RNA molecules found in snRNPs (see Section 16.2)

- Although eukaryotic genomes contain promoters that signal where transcription should begin, just as bacteria do, the promoters in eukaryotic DNA are much more diverse and complex than bacterial promoters. Many of the eukaryotic promoters recognized by RNA polymerase II include a unique sequence called the **TATA box**, located 30 base pairs upstream of the transcription start site. Some of the promoters recognized by pol II do not contain a TATA box, however. In addition, RNA pol I and pol III interact with entirely different promoters.
- In eukaryotes, transcription is followed by several important **RNA processing** steps that result in production of an mRNA that leaves the nucleus.

The overall message of this research is that the molecular mechanisms involved in transcription are much more complex in eukaryotes than in bacteria. Perhaps the most striking contrast between bacterial and eukaryotic gene expression emerged when researchers discovered that eukaryotic genes do not consist of one continuous DNA sequence that codes for a product,

BOX 16.1 Toxins and Transcription

Although wild mushrooms can be delicious, dining on certain species may be lethal. For example, in late summer and fall the reproductive structures of the death cap mushroom (*Amanita phalloides*) appear in forests, fields, and backyards in North America and Europe (**Figure 16.5**).

FIGURE 16.5 The Death Cap Mushroom. These mushrooms contain the toxin α-amanitin, which at low concentrations poisons the eukaryotic enzyme RNA polymerase II.

QUESTION State a hypothesis to explain why the presence of α-amanitin increases the fitness of death cap mushrooms—meaning, their ability to survive and produce offspring.

The death cap is among the most poisonous of all mushrooms. Worldwide, it is thought to account for 90 percent of mushroom poisoning deaths. Eating a single death cap can kill a healthy adult.

The active agent in death caps is a toxin called α-amanitin. By happenstance, researchers who were investigating α-amanitin's mode of action made a crucial discovery about transcription in eukaryotes. The researchers observed that mRNA synthesis stopped when eukaryotic cells were treated with low concentrations of α-amanitin. The production of other types of RNAs—specifically, transfer RNAs (tRNAs) and ribosomal RNAs (rRNAs)—continued, however. At very high concentrations of the toxin, tRNA production also stopped but transcription of rRNA genes continued. These observations eventually led to the discovery that eukaryotes have three RNA polymerases, each with a distinct role in the transcription of mRNA, tRNA, and rRNA. Low concentrations of α-amanitin inhibit RNA polymerase II, which synthesizes mRNAs. As a result, low concentrations of the toxin halt protein synthesis. High concentrations of this poison block RNA polymerase III, which synthesizes tRNAs. RNA polymerase I is not affected by α-amanitin at any concentration.

Toxins such as α-amanitin are useful in research on transcription and translation for the same reason that knock-out mutations, which completely disable genes, are useful in genetics. Toxins allow researchers to understand what happens when a molecule or a particular stage in a process does *not* work. This is similar, in concept, to learning how a car works by removing a part or a system and carefully recording the machine's response; or to learning how the brain works by studying the behavior of people whose brains have been damaged in specific, quantifiable ways by accidents. Toxins and knock-out mutations provide a molecular dissecting kit. They let biologists explore the intricacies of cell function by inactivating key components.

as do bacterial genes. Instead, the regions of eukaryotic genes that code for proteins come in pieces that are separated by hundreds or many thousands of intervening DNA bases. Although these intervening bases are part of the gene, they do not code for a product.

The Startling Discovery of Eukaryotic Genes in Pieces

When researchers realized that the protein-coding regions of eukaryotic genes are interrupted by stretches of noncoding DNA, it became clear that the nature of the gene and information processing are different in bacteria and eukaryotes. In eukaryotes, the conversion of information in DNA sequences to mRNA sequences does not occur directly. To make a functional mRNA, eukaryotic cells must dispose of certain sequences inside genes and then combine the separated coding sections into an integrated whole.

What sort of data would provoke such a startling claim? To answer this question, consider work that Phillip Sharp and colleagues carried out in the late 1970s to determine how DNA templates are transcribed. They began one of their experiments by heating DNA molecules enough to break the hydrogen bonds between complementary bases. This treatment separated the two strands. The single-stranded DNA was then incubated with the mRNA encoded by the sequence. The team's idea was to promote base pairing between the mRNA and the single-stranded DNA. Under these conditions, the research team expected the mRNA to form base pairs with the DNA sequences that act as the template for its synthesis. When the researchers examined the DNA-RNA hybrid molecules with the electron microscope, however, they observed the structure shown in **Figure 16.6a**. Instead of matching up exactly, parts of the DNA formed loops. As **Figure 16.6b** shows, Sharp and his coworkers interpreted these loops as stretches of nucleotides that are present in the DNA template strand but are *not* in the corresponding mRNA.

To make sense of these and other results, Sharp's group and a team headed by Richard Roberts proposed that there is not a one-to-one correspondence between the nucleotide sequence of a eukaryotic gene and its mRNA. By analogy, eukaryotic genes do not carry messages such as "Biology is my favorite course of all time." Instead, eukaryotic genes carry messages that read something like "BIOL*τηεπροτεινχοδινγ‴ ρεγιονσοφγενεσ*OGY IS MY FAVOR*αρειντερρυπτεδβψ‴ νονχοδινγδ*ITE COURSE OF*ανδηαωετοβεσπλιχεδτογετ‴ ηερ*ALL TIME." Here the sections of noncoding sequence are represented with Greek letters. They must be removed from the mRNA before it can carry an intelligible message to the translation machinery.

When it became clear that the genes-in-pieces hypothesis was correct, Walter Gilbert suggested that regions of eukaryotic genes that are part of the final mRNA be referred to as **exons** (because they are *ex*pressed) and the untranslated stretches as **introns** (because they are *int*ervening). Exons code for segments of functional proteins or RNAs; introns do not. Introns are sections of genes that are not represented in the final mRNA product. As a result, eukaryotic genes are much larger than their corresponding mature RNA transcripts (**Figure 16.6c**).

Exons, Introns, and RNA Splicing

The transcription of eukaryotic genes by RNA polymerase generates a **primary RNA transcript** that contains both the exon and intron regions (**Figure 16.7a**). As transcription proceeds, the introns are removed from the growing RNA strand by a process known as **splicing**. In this phase of information processing, pieces of the primary transcript are removed and the remaining segments are joined together. Splicing occurs while transcription is still under way and results in an RNA that contains an uninterrupted genetic message.

(a) Micrograph of DNA-RNA hybrid

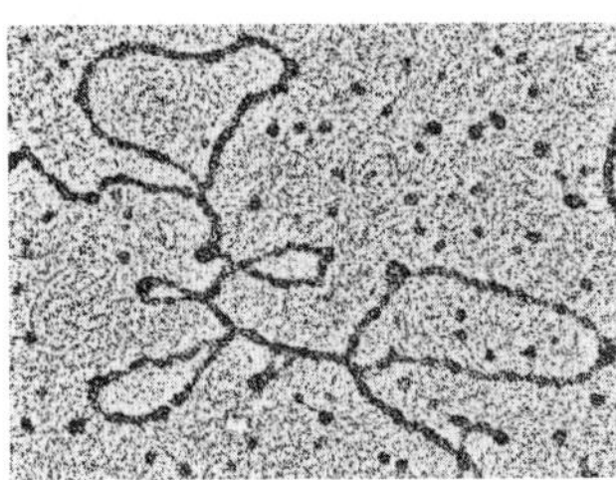

(b) Interpretation of micrograph

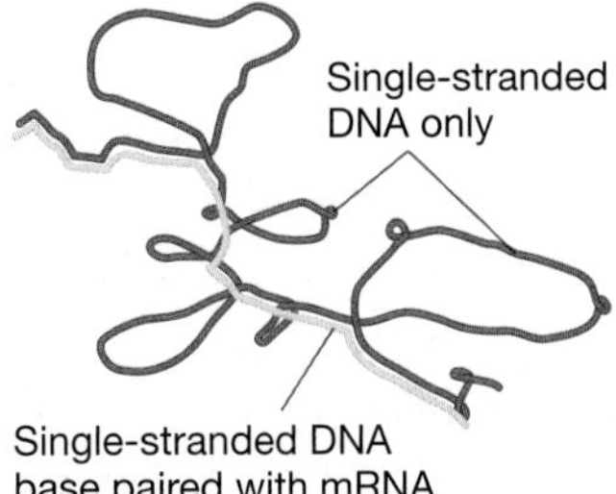

(c) Genes and RNA transcripts differ in length.

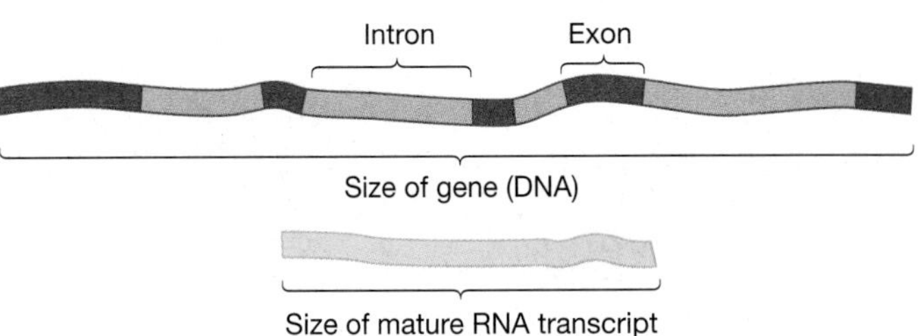

FIGURE 16.6 The Discovery of Noncoding Regions of DNA. (a) Electron micrograph of single-stranded DNA that is bonded via complementary base pairing to the mRNA it encodes. **(b)** Researchers' interpretation of the photograph in part (a). The loops represent regions of DNA that do not have an equivalent sequence in the mRNA. These intervening regions are "extra" DNA compared with the sequences in the mRNA. **(c)** Because eukaryotic genes contain both introns and exons, these genes are much larger than the corresponding mature mRNA.

● **QUESTION** If introns did not exist, what would the micrograph in part (a) look like?

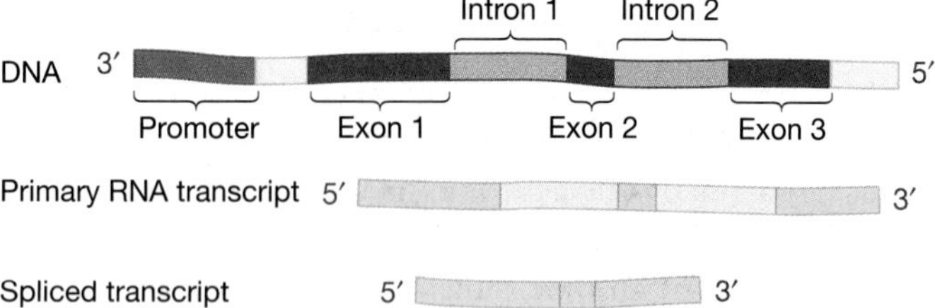

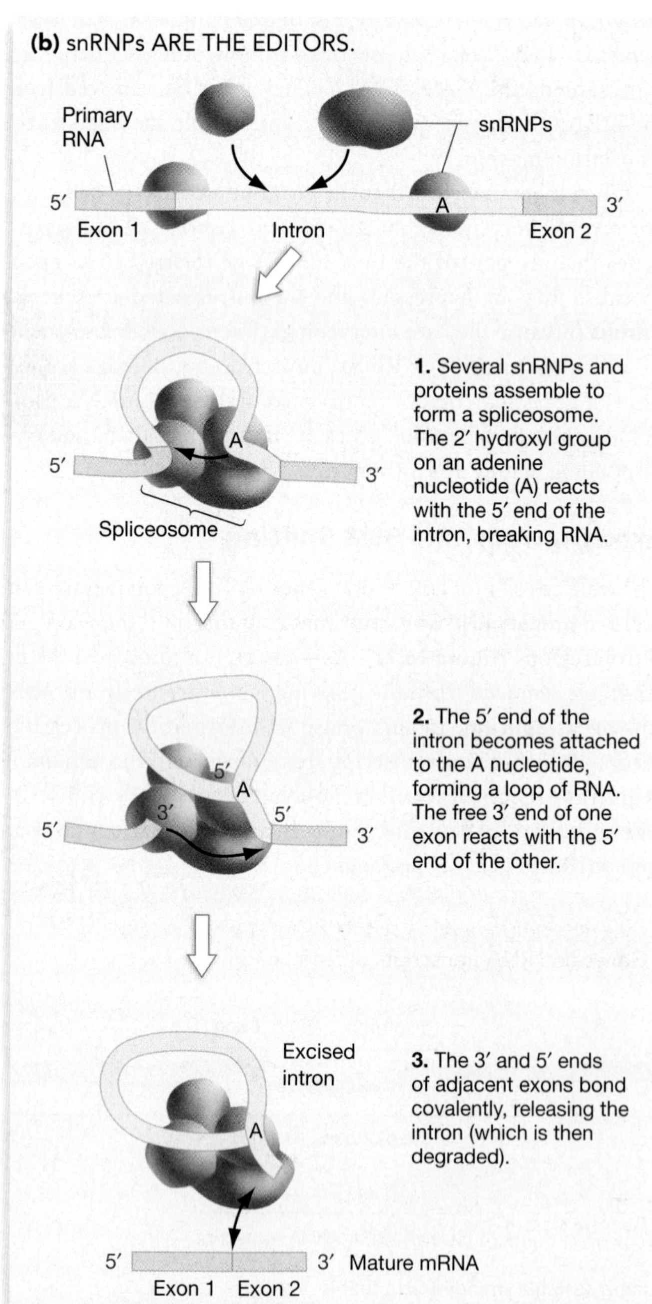

FIGURE 16.7 Introns Are Spliced Out of the Original mRNA. (a) RNA polymerase produces a primary RNA transcript that contains exons and introns. Subsequent processing generates a mature transcript. **(b)** snRNPs are responsible for the splicing reactions that take place in the nucleus of eukaryotes.

Figure 16.7b provides more detail about how introns are removed from genes. In the case illustrated here, splicing is catalyzed by a complex of proteins and small RNAs known as **small nuclear ribonucleoproteins,** or **snRNPs** (pronounced "snurps"). The process begins when a snRNP binds to the 5′ exon-intron boundary. Once the initial snRNPs are in place, other snRNPs arrive to form a multipart complex called a **spliceosome**. Recent data have shown that the spliceosomes found in human cells contain about 145 different proteins and RNAs, making them the most complex molecular machines known. Once the spliceosome forms, the intron forms a loop with an adenine ribonucleotide at its base. The adenine participates in a reaction that cuts the loop out. A phosphodiester bond then links the exons on either side, forming a contiguous coding sequence. Splicing is now complete. In most cases, the excised intron is degraded to ribonucleotide monophosphates.

Current data suggest that both the cutting and rejoining reactions that occur during splicing are catalyzed by RNA molecules in the spliceosome. Recall from Chapter 4 that RNA molecules with catalytic ability are called ribozymes. In eukaryotes, ribozymes play a key role in the production of proteins.

Adding Caps and Tails to Transcripts

Intron splicing is not the only way that primary RNA transcripts are processed in eukaryotes. As soon as the 5′ end of a eukaryotic RNA emerges from RNA polymerase, enzymes add a structure called the **5′ cap** (**Figure 16.8**). The cap consists of the molecule 7-methylguanylate and three phosphate groups. In addition, an enzyme cleaves the 3′ end of most RNAs once transcription is complete, and another enzyme adds a long tract of 100–250 adenine nucleotides. This sequence is known as the **poly(A) tail**. These poly(A) tails are not encoded on the template strand. ● With the addition of the cap and tail, processing of the primary RNA transcript is complete. The product is a mature mRNA. The molecule contains the coding sequence for a polypeptide flanked by sequences that are not destined to be translated. These 5′ and 3′ untranslated regions help stabilize the mature RNA and regulate its translation.

Not long after the caps and tails on eukaryotic mRNAs were described, evidence began to accumulate that they also protect mRNAs from degradation by ribonucleases and enhance the efficiency of translation. For example, experimental mRNAs that have a cap and a tail last longer when they are introduced into cells than do experimental mRNAs that lack a cap, a tail, or both a cap and a tail. The mRNAs with caps and tails also produce more proteins than do mRNAs without caps and tails.

Recent work has shown why the cap and tail are so critical to mRNAs. The 5′ cap serves as a recognition signal for the translation machinery, and the poly(A) tail extends the life span of an mRNA by protecting the message from degradation by ribonucleases in the cytosol.

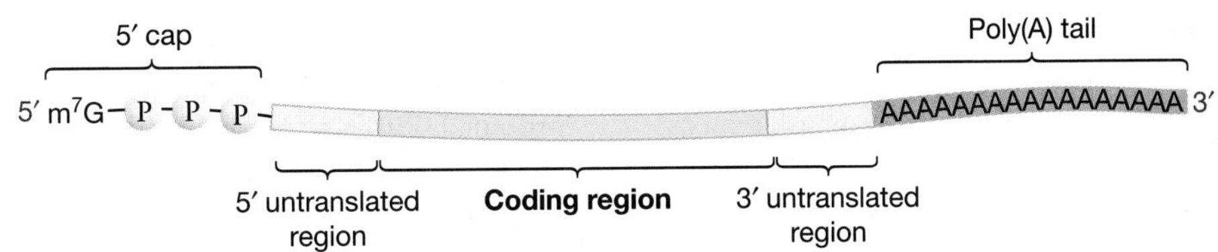

FIGURE 16.8 In Eukaryotes, mRNAs Are Given a Cap and a Tail. Eukaryotic mRNAs have a cap consisting of a molecule called 7-methylguanylate (symbolized as m^7G) bonded to three phosphate groups; the tail is made up of a long series of adenine residues.

Splicing and the addition of caps and tails are major differences in how mRNAs are produced in bacteria versus eukaryotes. It is not the only important contrast, however. **Table 16.2** summarizes some of the similarities and differences observed in transcription in these two lineages.

Check Your Understanding

If you understand that...

- Transcription occurs when RNA polymerase catalyzes the $5' \rightarrow 3'$ synthesis of an mRNA molecule.
- RNA synthesis is driven by the potential energy stored in nucleotide triphosphates.
- The sequence of an RNA transcript is complementary to the sequence in a template DNA strand.

You should be able to...

1) Compare and contrast how transcription occurs in bacteria and eukaryotes.
2) Diagram the capping, splicing, and tail-addition steps that occur as primary RNA transcripts are processed in eukaryotes.

16.3 An Introduction to Translation

To synthesize a protein, the sequence of bases in a messenger RNA molecule is translated into a sequence of amino acids in a polypeptide. The genetic code specifies the relationship between the bases of a triplet codon in mRNA and the amino acid it codes for (see Chapter 15). But how are the amino acids assembled into a polypeptide according to the information in messenger RNA?

Studying how translation occurred in cell-free systems proved to be an extremely productive approach to answering this question. Once in vitro translation systems had been developed from human cells, *E. coli*, and a variety of other organisms, it became clear that the basic mechanisms of translation are fundamentally the same throughout the tree of life. The sequence of events that occurs during protein synthesis is similar in bacteria, archaea, and eukaryotes.

Ribosomes Are the Site of Protein Synthesis

The first question that biologists answered about translation concerned where it occurs. The answer was inspired by a simple observation: There is a strong, positive correlation between the number of small structures known as **ribosomes** in a given type of cell and the rate at which that cell synthesizes proteins. For example, immature human red blood cells divide rapidly, synthesize millions of copies of the protein hemoglobin, and contain large numbers of ribosomes. In contrast, the same cells at maturity have low rates of protein synthesis and very few ribosomes. Based on this correlation, investigators proposed that ribosomes are the site of protein synthesis in the cell.

TABLE 16.2 **Comparing Transcription in Bacteria and Eukaryotes**

Aspect	Bacteria	Eukaryotes
RNA polymerase	One	Three; each produces a different class of RNA
Promoter structure	Typically contains a –35 box and a –10 box	Complex and variable; often includes a TATA box –30 from start of gene
Protein(s) involved in contacting promoter	Sigma; different versions of sigma bind to different promoters	Many basal transcription factors
RNA processing steps	None; translation occurs while transcription is still under way	Extensive; several processing steps occur in nucleus before RNA is exported to cytoplasm for translation: **1.** Enzyme-catalyzed addition of 5′ cap **2.** Splicing (intron removal) by spliceosome **3.** Enzyme-catalyzed addition of 3′ poly(A) tail

To test this hypothesis, Roy Britten and collaborators did a pulse-chase experiment similar in design to experiments introduced in Chapter 7. Recall that the goal of a pulse-chase experiment is to label a population of molecules as they are being produced. The location of the tagged molecules is then followed through time. The tagging is done by supplying a pulse of radioactive atoms that are incorporated into the molecule as it is being synthesized, followed by a chase of unlabeled atoms. Britten's group wanted to label amino acids with radioactive atoms and then track the molecules as translation occurred.

The biologists began by feeding a pulse of radioactive sulfate ($^{35}SO_4^{-2}$) to growing cultures of *E. coli.* They expected the cells to incorporate the radioactive sulfur into the amino acids methionine and cysteine—which contain sulfur (see Chapter 3)—and then into newly synthesized proteins. Fifteen seconds after adding the radioactive sulfate, the researchers "chased" the label by adding a large excess of nonradioactive sulfate to the culture medium. If the ribosome hypothesis was correct, then the radioactive signal should be associated with ribosomes for a short period of time (while they are making proteins). Later, all of the radioactivity should be found in proteins not in the ribosomes.

This is exactly what the researchers found. Soon after the pulse of labeled sulfate ended, the radioactive atoms were found in free amino acids or in ribosomes. Later, all of the radioactive atoms were found on completed proteins. Based on these data, biologists concluded that proteins are synthesized at ribosomes and then released.

About a decade after the ribosome hypothesis was confirmed, electron micrographs showed bacterial ribosomes in action (**Figure 16.9a**). The images confirmed that in bacteria, ribosomes attach to mRNAs and begin synthesizing proteins even before transcription is complete (**Figure 16.9b**). Transcription and translation can occur concurrently, or be coupled, in bacteria because there is no nuclear envelope to separate the two processes. Thus, transcription and translation are physically connected. But in eukaryotes, RNAs are processed in the nucleus and then mRNAs are exported to the cytoplasm (**Figure 16.10**). Once mRNAs are outside the nucleus, ribosomes attach to them and begin translation. In eukaryotes, transcription and translation are separated in time and space.

How Does an mRNA Triplet Specify an Amino Acid?

When an mRNA interacts with a ribosome, hereditary instructions encoded in nucleic acids are translated into a different chemical language—the amino acid sequences found in proteins. The discovery of the genetic code revealed that triplet codons in mRNA specify particular amino acids in a protein. How does this conversion happen?

(a) Bacterial ribosomes during translation

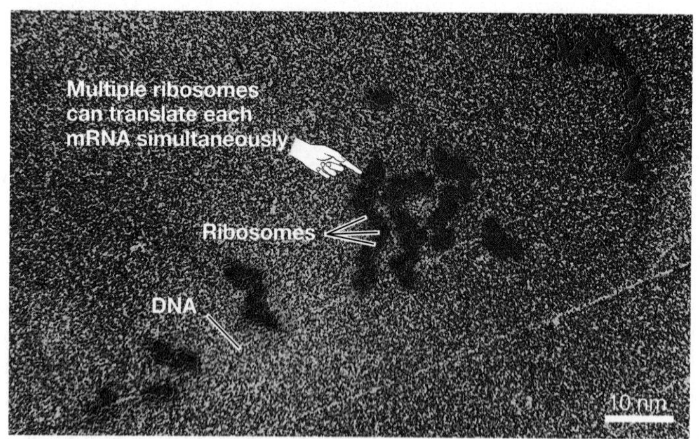

(b) In bacteria, transcription and translation are tightly coupled.

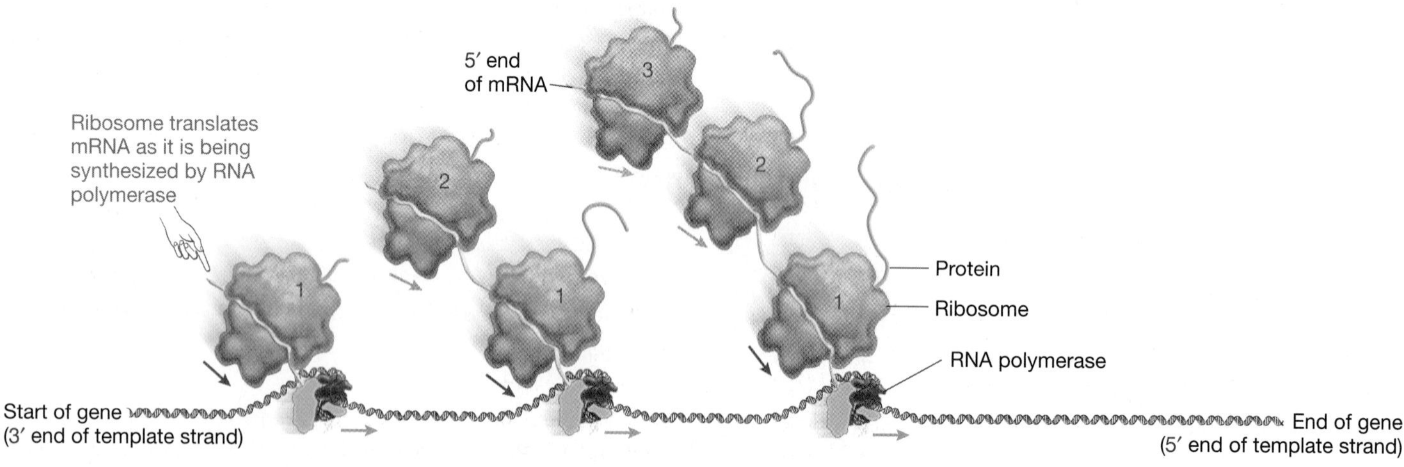

FIGURE 16.9 Transcription and Translation Occur Simultaneously in Bacteria. In bacteria, ribosomes attach to mRNA transcripts and begin translation while RNA polymerase is still transcribing the DNA template strand. Note that the colors in the micrograph in part (a) are artificial—they've been added to make the structures more visible.

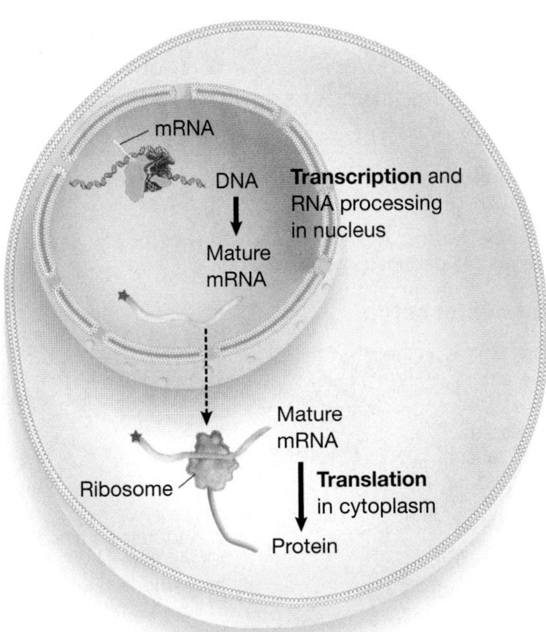

FIGURE 16.10 Transcription and Translation Are Separated in Space and Time in Eukaryotes. In eukaryotes, mRNA transcripts are processed inside the nucleus before they are exported to the cytoplasm. There, they are translated by ribosomes.

(a) Hypothesis 1: Amino acids interact directly with mRNA codons.

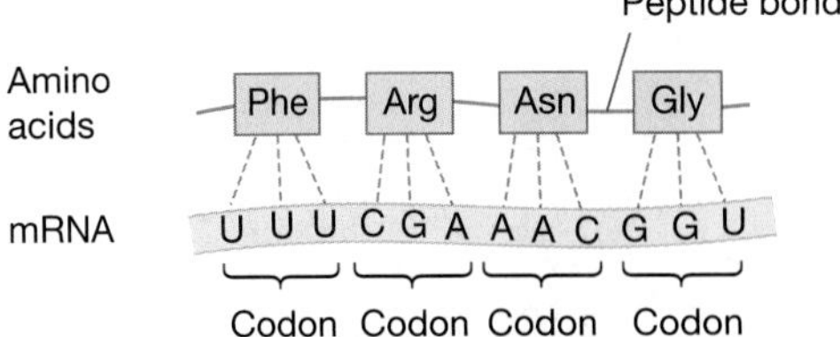

(b) Hypothesis 2: Adapter molecules hold amino acids and interact with mRNA codons.

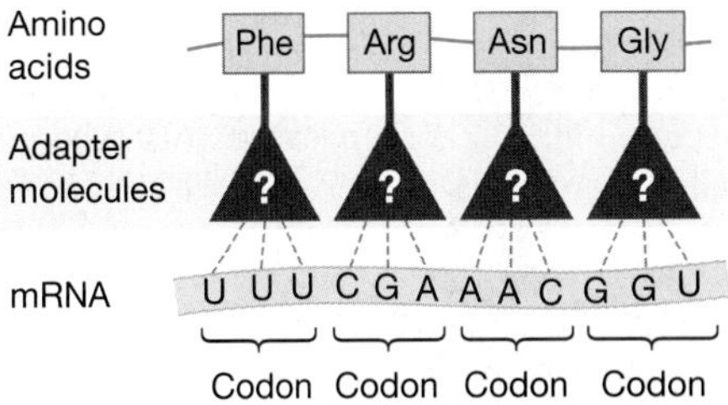

FIGURE 16.11 Two Hypotheses for How mRNA Codons Interact with Amino Acids.

One early hypothesis was that mRNA codons and amino acids interact directly. The proposal was that the bases in a particular codon were complementary in shape or charge to the side group of a particular amino acid (**Figure 16.11a**). But Francis Crick pointed out that the chemistry involved didn't make sense. For example, how could the nucleic acid bases interact with a hydrophobic amino acid side group, which does not form hydrogen bonds?

Crick proposed an alternative hypothesis. As **Figure 16.11b** shows, he suggested that some sort of adapter molecule holds amino acids in place while interacting directly and specifically with a codon in mRNA via hydrogen bonding. In essence, Crick predicted the existence of a chemical go-between that produced a physical connection between the two types of molecules. As it turns out, Crick was right.

16.4 The Role of Transfer RNA

Crick's adapter molecule was discovered by accident. Biologists were trying to work out a cell-free protein synthesis system derived from mammalian liver cells and had discovered that ribosomes, mRNA, amino acids, ATP, and a molecule called guanosine triphosphate, or GTP, had to be present for translation to occur. (GTP is similar to ATP but contains guanosine instead of adenosine.) These results were logical—ribosomes provide the catalytic machinery, mRNAs contribute the message to be translated, amino acids are the building blocks of proteins, and ATP and GTP supply potential energy to drive the endergonic polymerization reactions responsible for forming proteins. But in addition, a cellular fraction that contained a previously unknown type of RNA turned out to be indispensable. If this type of RNA is missing, protein synthesis does not occur. What is this mysterious RNA, and why is it essential to translation?

The novel class of RNAs eventually became known as **transfer RNA (tRNA)**. The role of tRNA in translation was a mystery until some researchers happened to add a radioactive amino acid—leucine—to an in vitro protein synthesis system. The treatment was actually done as a control for an unrelated experiment. To the researchers' amazement, some of the radioactive leucine attached to tRNA molecules. Follow-up experiments showed that the attachment of an amino acid to a tRNA requires an input of energy in the form of ATP. A tRNA molecule that becomes covalently linked to an amino acid is called an **aminoacyl tRNA**. More recent research has shown that enzymes called **aminoacyl tRNA synthetases** are responsible for catalyzing the addition of amino acids to tRNAs. For each of the 20 major amino acids, there is a different aminoacyl tRNA synthetase and one or more tRNAs. **Figure 16.12** summarizes how amino acids are transferred from aminoacyl tRNA synthetases to tRNAs.

What happens to the amino acid that is carried by an aminoacyl tRNA? Biologists who tracked the fate, over time, of radioactive leucine molecules that were attached to tRNAs discovered that the amino acids are transferred from aminoacyl tRNAs to proteins. The data supporting this conclusion are

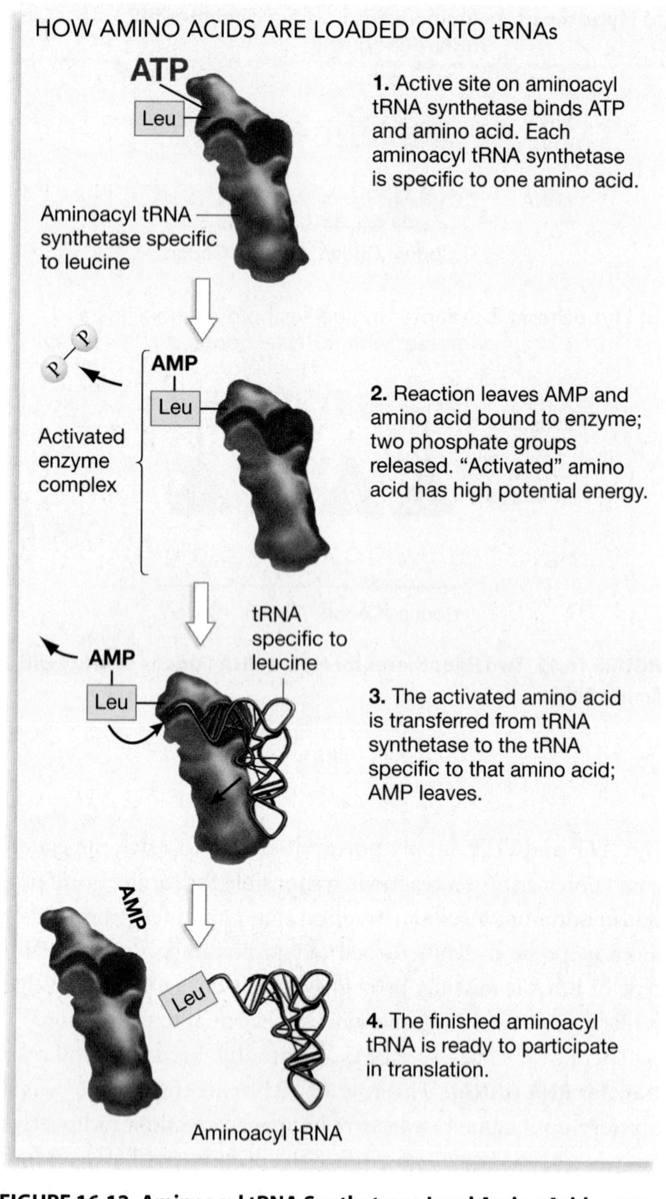

FIGURE 16.12 Aminoacyl tRNA Synthetases Load Amino Acids onto the Appropriate tRNA. An aminoacyl tRNA synthetase—in this case, for leucine—becomes activated in a reaction that requires ATP. Activated aminoacyl tRNA synthetase transfers an amino acid to a tRNA, producing an aminoacyl tRNA.

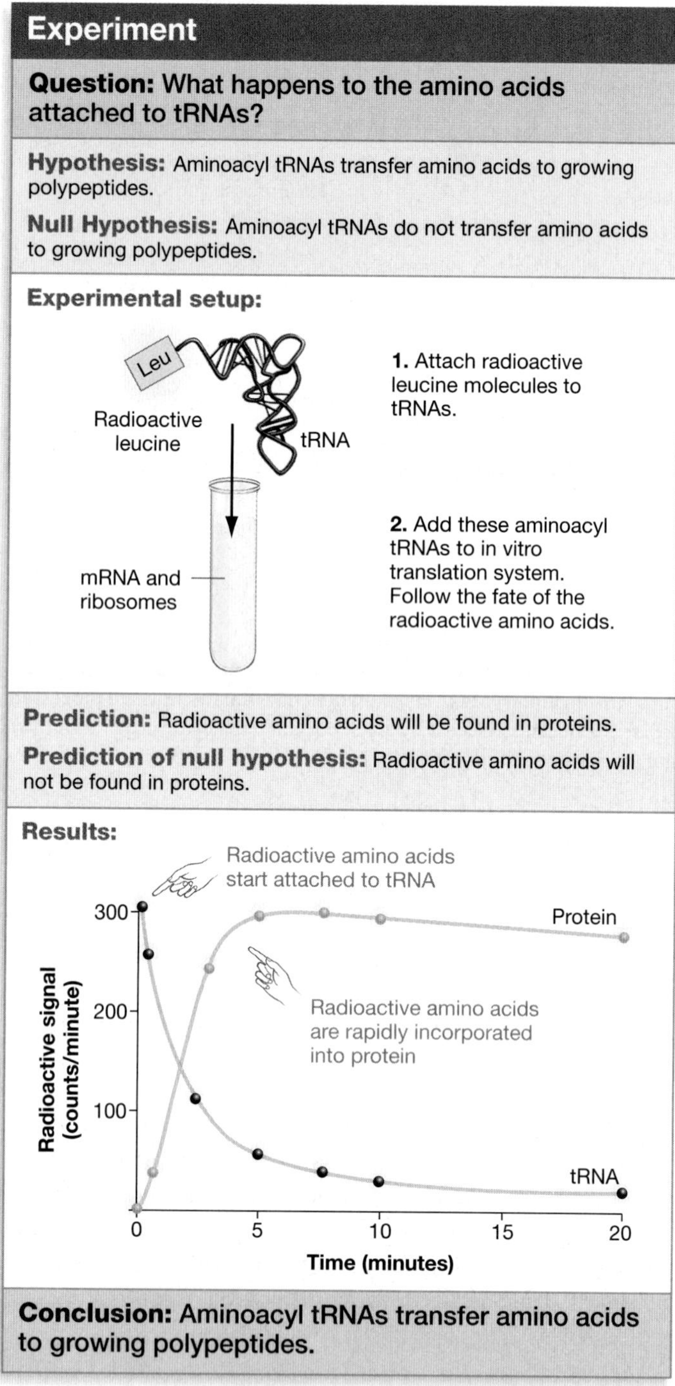

FIGURE 16.13 Experimental Evidence that Amino Acids Are Transferred from tRNAs to Proteins.

shown in **Figure 16.13**. The graph in the Results section of this figure shows that radioactive amino acids are lost from tRNAs and incorporated into polypeptides synthesized in ribosomes. These results inspired the use of transfer in tRNA's name. The experiment also confirmed that aminoacyl tRNAs act as the interpreter in the translation process. tRNAs are Crick's adapter molecules.

What Do tRNAs Look Like?

Transfer RNAs serve as chemical go-betweens that allow amino acids to interact with an mRNA template. But precisely how does the connection occur?

This question was answered by research on tRNA's molecular structure. The initial studies established the sequence of nucleotides in various tRNAs, or what is termed their primary structure. Transfer RNAs are relatively short, ranging from 75 to 85 nucleotides in length. When biologists studied the primary sequence closely, they noticed that certain parts of the molecules can form secondary structures. Specifically, some of the bases in the tRNA molecule can form hydrogen bonds with complementary bases in a different region of the same molecule. As a result, the entire tRNA molecule could assume the cloverleaf shape illustrated in **Figure 16.14a**. The stems in the cloverleaf are produced by complementary base pairing between different portions of the molecule running in antiparallel directions, while the loops consist of single-stranded RNA (see Chapter 4).

Two aspects of this secondary structure proved especially interesting. A CCA sequence at the 3′ end of each tRNA molecule offered a binding site for amino acids, while the triplet on a loop on the opposite end of the cloverleaf could serve as an anticodon. An **anticodon** is a set of three ribonucleotides that forms base pairs with the mRNA codon. **Figure 16.14b** depicts an early model of how the anticodon of a tRNA connects an mRNA codon with the appropriate amino acid.

This model had to be modified, however, when X-ray crystallography studies revealed the tertiary structure of tRNAs. Recall from Chapter 3 that the tertiary structure of a molecule is defined by the three-dimensional arrangement of its atoms and is usually a product of folding. According to the X-ray crystallography data, the cloverleaf structure folds over to produce an L-shaped molecule (**Figure 16.14c**). All of the tRNAs in a cell have the same L-shaped structure, but each has a distinct anticodon and attached amino acid. The tertiary structure of tRNAs is important because it results in a precise separation between the anticodon and the attached amino acid. As we'll see, this separation is key to the positioning of the amino acid and the anticodon in the ribosome.

● If you understand the structure of tRNAs, you should be able to make a rough sketch of a tRNA with and without an amino acid attached, and explain the relationship between the anticodon of a tRNA and a codon in an mRNA.

How Many tRNAs Are There?

When research succeeded in characterizing all of the different types of tRNAs available in cells, a paradox arose. According to the genetic code introduced in Chapter 15, the 20 most common amino acids found in proteins are specified by 61 different mRNA codons. Instead of containing 61 different tRNAs with 61 different anticodons, though, most cells contain only

(a) Secondary structure of tRNA

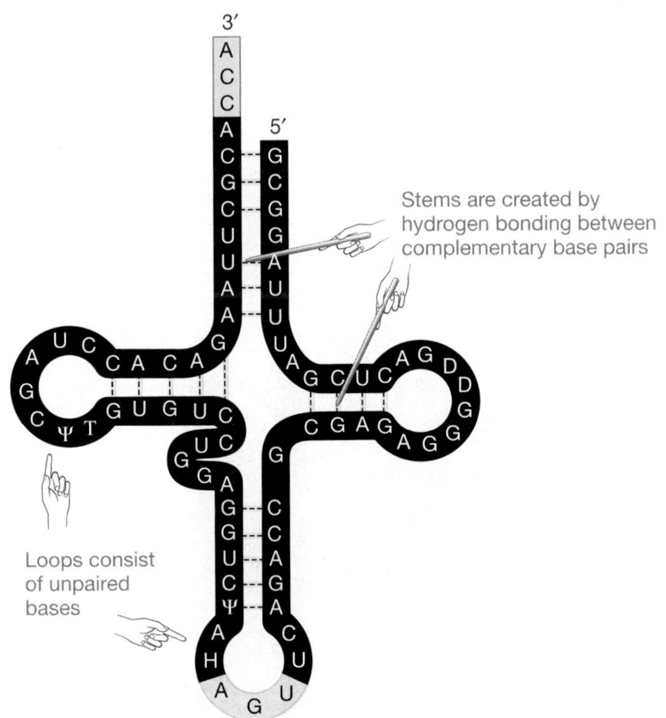

(b) Early model of aminoacyl tRNA function

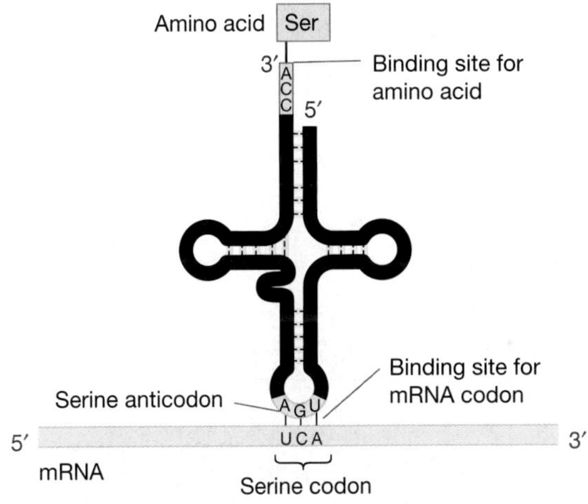

(c) Revised model incorporating tertiary structure of tRNA

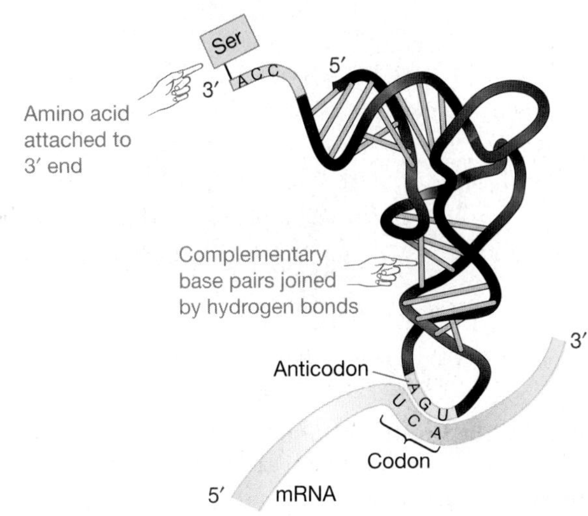

FIGURE 16.14 The Structure of Transfer RNA. (a) The secondary structure of tRNA resembles a cloverleaf. **(b)** If amino acids are attached to the 3′ end of a tRNA with an anticodon appropriate for that amino acid, then the tRNA anticodon will form a complementary base pair with the mRNA codon. **(c)** Recent data indicate that tRNAs have an L-shaped tertiary structure.

about 40. How can all 61 mRNA codons in the genome be translated with only two-thirds of the tRNAs required?

To resolve this paradox, Francis Crick proposed what is known as the **wobble hypothesis**. To understand his logic, recall from Chapter 15 that many amino acids are specified by more than one codon. Further, recollect that codons for the same amino acid tend to have the same nucleotides at the first and second positions but a different nucleotide at the third position. For example, both of the codons CAA and CAG code for the amino acid glutamine. (Codons are always written in the $5' \rightarrow 3'$ direction.) Surprisingly, experimental data have shown that a tRNA with an anticodon of GUU can base pair with both CAA and CAG in mRNA. (Anticodons are written in the $3' \rightarrow 5'$ direction.) The GUU anticodon matches the first two bases (C and A) in both codons, but the U in the third position forms a nonstandard base pair with a G in the CAG codon.

Crick proposed that inside the ribosome, tRNAs with nonstandard base pairing at the third position of the codon can still bind successfully to an anticodon. If so, it would allow a limited flexibility, or "wobbling," in the base pairing. According to the wobble hypothesis, a nonstandard base pair in the third position of a codon—such as G-U—is acceptable as long as it does not change the amino acid for which the codon codes. In this way, wobble in the third position of a codon allows just 40 or so tRNAs to bind to all 61 mRNA codons.

16.5 Ribosomes and the Mechanism of Translation

Protein synthesis occurs when the sequence of bases in an RNA message is translated into a sequence of amino acids in a polypeptide. The conversion begins when the anticodon of a tRNA that carries an amino acid binds to a codon in mRNA. The conversion is complete when a peptide bond forms between that amino acid and the growing polypeptide chain.

Both of these events occur inside a ribosome. Biologists have known since the 1930s that ribosomes contain a considerable amount of protein along with a great deal of **ribosomal RNA (rRNA)**. When high-speed centrifuges became available, researchers found that ribosomes could also be separated into two major substructures, called the large subunit and small subunit. Each subunit consists of a complex of RNA molecules and proteins. For example, the large subunit in *E. coli* contains two different rRNA molecules and 34 proteins. More recent work has shown that the small subunit holds the mRNA in place during translation and that peptide bond formation takes place in the large subunit.

The three-dimensional ribbon diagram of the ribosome in **Figure 16.15a** shows how all of the molecules required for translation fit together. Note that during protein synthesis, three distinct tRNAs are lined up inside the ribosome. All three are bound to their corresponding mRNA codon at the base of

(a) Ribbon model of ribosome during translation

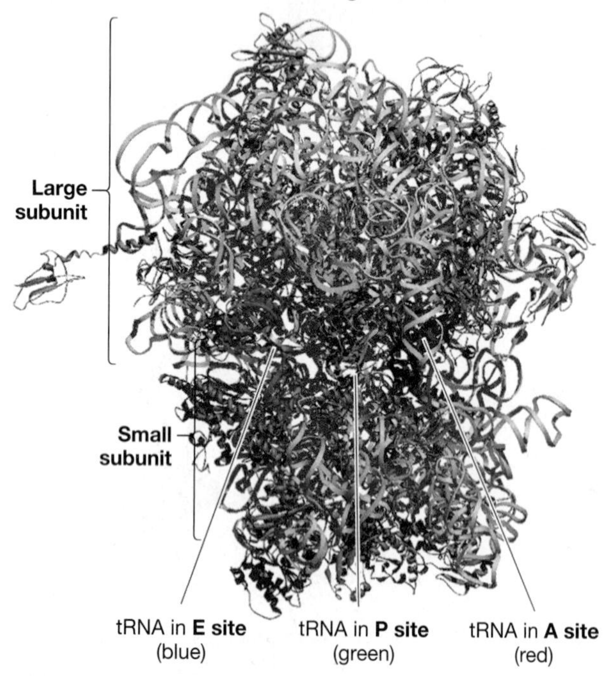

(b) Diagram of ribosome during translation

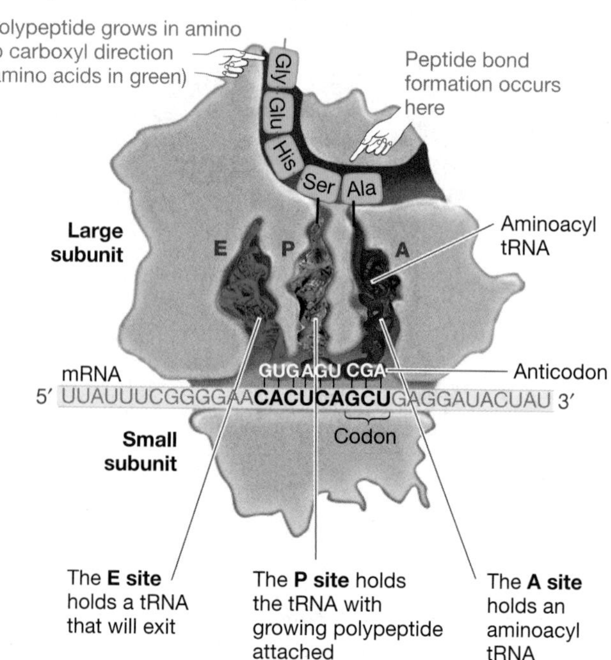

FIGURE 16.15 The Structure of the Ribosome. (a) During translation, three tRNAs line up side by side inside the ribosome. **(b)** Different events occur at each of the three sites inside the ribosome where tRNAs are found.

● **EXERCISE** Circle and label the active site in the ribosome in part (b).

the structure. The tRNA that is on the right in the figure, and colored red, carries an amino acid. This tRNA's position in the ribosome is called the A site—"A" for acceptor or aminoacyl. The tRNA that is in the middle (green) holds the growing polypeptide chain and occupies the P site, for peptidyl, inside the ribosome. (Think of "P" for peptide-bond formation.) The left-hand (blue) tRNA no longer has an amino acid attached and is about to leave the ribosome. It occupies the ribosome's E site—"E" for exit. **Figure 16.15b** summarizes what happens to each tRNA inside the ribosome. Because all tRNAs have similar secondary and tertiary structure, they all fit equally well in the A, P, and E sites.

The ribosome is a molecular machine that synthesizes proteins in a three-step sequence:

1. An aminoacyl tRNA diffuses into the A site; its anticodon binds to a codon in mRNA.

2. A peptide bond forms between the amino acid held by the aminoacyl tRNA in the A site and the growing polypeptide, which is held by a tRNA in the P site.

3. The ribosome moves ahead, and all three tRNAs move one position down the line. The tRNA in the E site exits; the tRNA in the P site moves to the E site; and the tRNA in the A site switches to the P site.

The protein that is being synthesized grows by one amino acid each time this three-step sequence repeats. The process occurs up to 20 times per second in bacterial ribosomes and about 2 times per second in eukaryotic ribosomes. Protein synthesis starts at the amino-end (N-terminus) of a polypeptide and proceeds to the carboxy end (C-terminus; see Chapter 3).

This introduction to how tRNAs, mRNAs, and ribosomes interact during protein synthesis leaves several key questions unanswered, however. How do mRNAs and ribosomes get together to start the process? Once protein synthesis is under way, how is peptide bond formation catalyzed inside the ribosome? And how does protein synthesis conclude when the ribosome reaches the end of the message? Let's consider each question in turn.

Initiation

To translate an mRNA, a ribosome must begin at a specific point in the message, translate the mRNA up to the message's termination codon, and then stop. Biologists call these three phases of protein synthesis **initiation**, **elongation**, and **termination**, respectively. One key to understanding initiation is to recall, from Chapter 15, that a start codon (usually AUG) is found near the 5′ end of all mRNAs and that it codes for the amino acid methionine. The presence of this start codon is an aspect of initiation that is common to both bacteria and eukaryotes.

Figure 16.16 shows how translation gets under way in bacteria. The process begins when a section of rRNA in the small subunit of a ribosome binds to a complementary sequence on an mRNA. The mRNA region is called the **ribosome binding site** or **Shine-Dalgarno sequence**, after the biologists who discovered it. The site is about six nucleotides upstream from the AUG start codon. It consists of all or part of the bases 5′-AGGAGGU-3′. The complementary sequence in the rRNA of the small subunit reads 3′-UCCUCCA-5′. The interaction between the small subunit and the message is mediated by proteins called **initiation factors** (Figure 16.16, step 1). In eukaryotes, initiation factors bind to the 5′ cap on mRNAs and guide it to the ribosome.

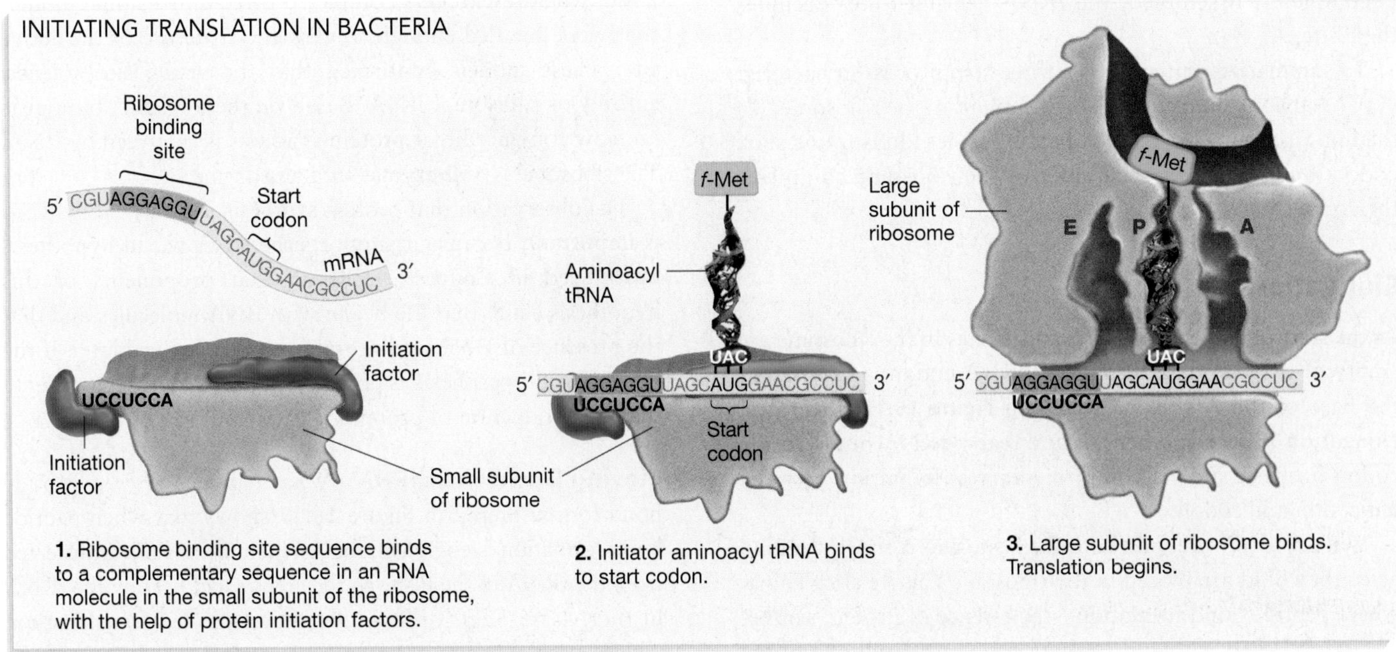

FIGURE 16.16 The Initiation Phase of Translation in Bacteria.

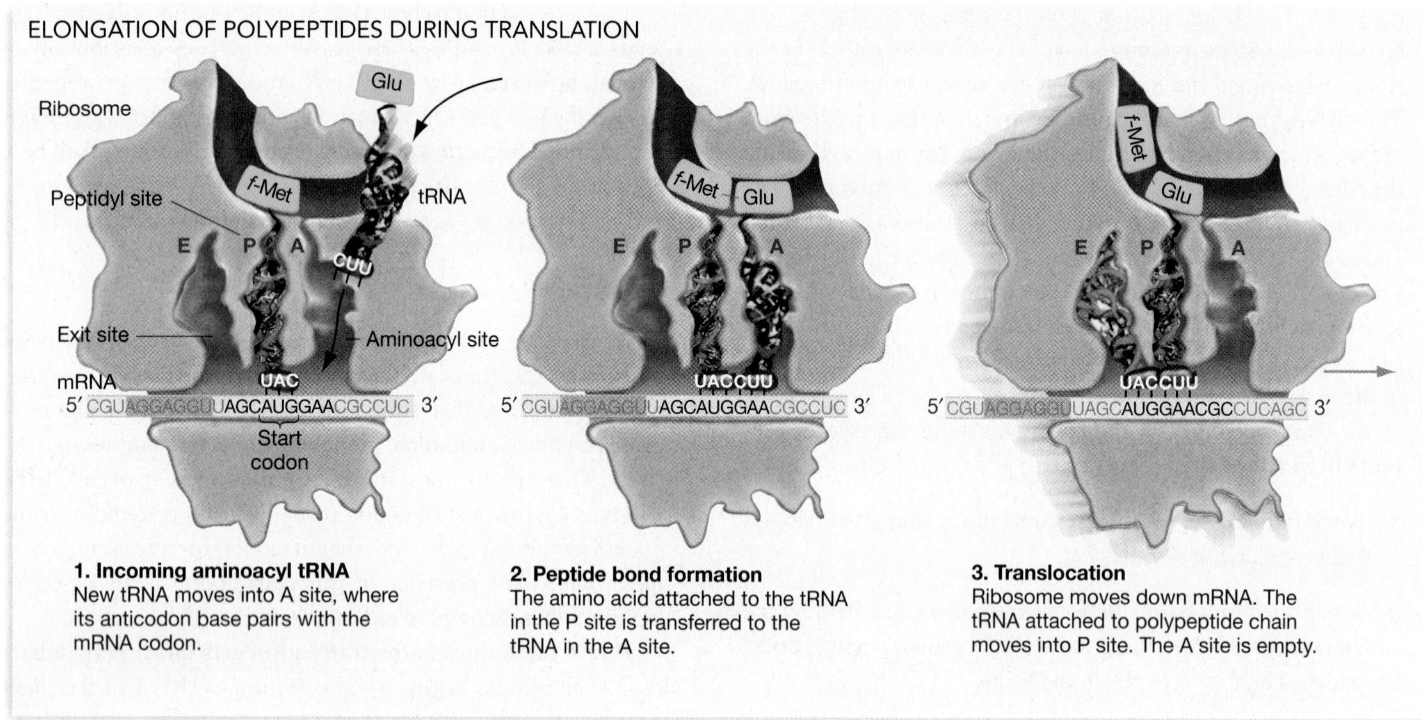

FIGURE 16.17 The Elongation Phase of Translation.

Once the Shine-Dalgarno sequence has attached to the small ribosomal subunit, an aminoacyl tRNA bearing a modified form of methionine called *N*-formylmethionine (abbreviated *f*-met) binds to the AUG start codon (Figure 16.16, step 2). In eukaryotes, this initial amino acid is normal methionine. Initiation is complete when the large subunit joins the complex (Figure 16.16, step 3). When the ribosome is completely assembled, the tRNA bearing *f*-met occupies the P site.

To summarize, initiation is a three-step process in bacteria: (1) The mRNA binds to the small subunit of the ribosome, (2) the initiator aminoacyl tRNA bearing f-Met binds to the start codon, and (3) the large subunit of the ribosome completes the complex.

Elongation

At the start of elongation, the E and A sites in the ribosome are empty of tRNAs. As a result, an mRNA codon is exposed at the base of the A site. As step 1 in **Figure 16.17** illustrates, elongation proceeds when an aminoacyl tRNA binds to the codon in the A site via complementary base pairing between anticodon and codon.

When a tRNA occupies both the P site and A site, the amino acids they hold are placed in the ribosome's active site. This is where peptide bond formation—the essence of protein synthesis—occurs. Peptide bond formation is considered one of the most important reactions that takes place in cells, because manufacturing proteins is among the most fundamental of all cell processes. The question is, how does it happen?

Is the Ribosome an Enzyme or a Ribozyme? Because ribosomes contain both proteins and RNA, researchers had argued for decades over whether the active site consisted of protein or RNA. The debate was not resolved until the year 2000, when researchers completed three-dimensional models that were detailed enough to view the structure of the active site. These models confirmed that the active site consists entirely of ribosomal RNA. Based on these results, biologists are now convinced that protein synthesis is catalyzed by RNA. The ribosome is a **ribozyme**—not an enzyme.

The observation that protein synthesis is catalyzed by RNA is important, because it supports the RNA world hypothesis introduced in Chapter 4. Recall that proponents of this hypothesis claim that life began with RNA molecules and that the presence of DNA and proteins in cells evolved later. If the RNA world hypothesis is correct, then it is logical to observe that the production of proteins is catalyzed by RNA.

Moving Down the mRNA What happens after a peptide bond forms? Step 2 in Figure 16.17 shows that when peptide bond formation is complete, the polypeptide chain is transferred from the tRNA in the P site to the amino acid held by the tRNA in the A site. Step 3 shows the process called **translocation**, which occurs when the ribosome moves down the mRNA in the $5' \rightarrow 3'$ direction. Translocation does several things: It moves the

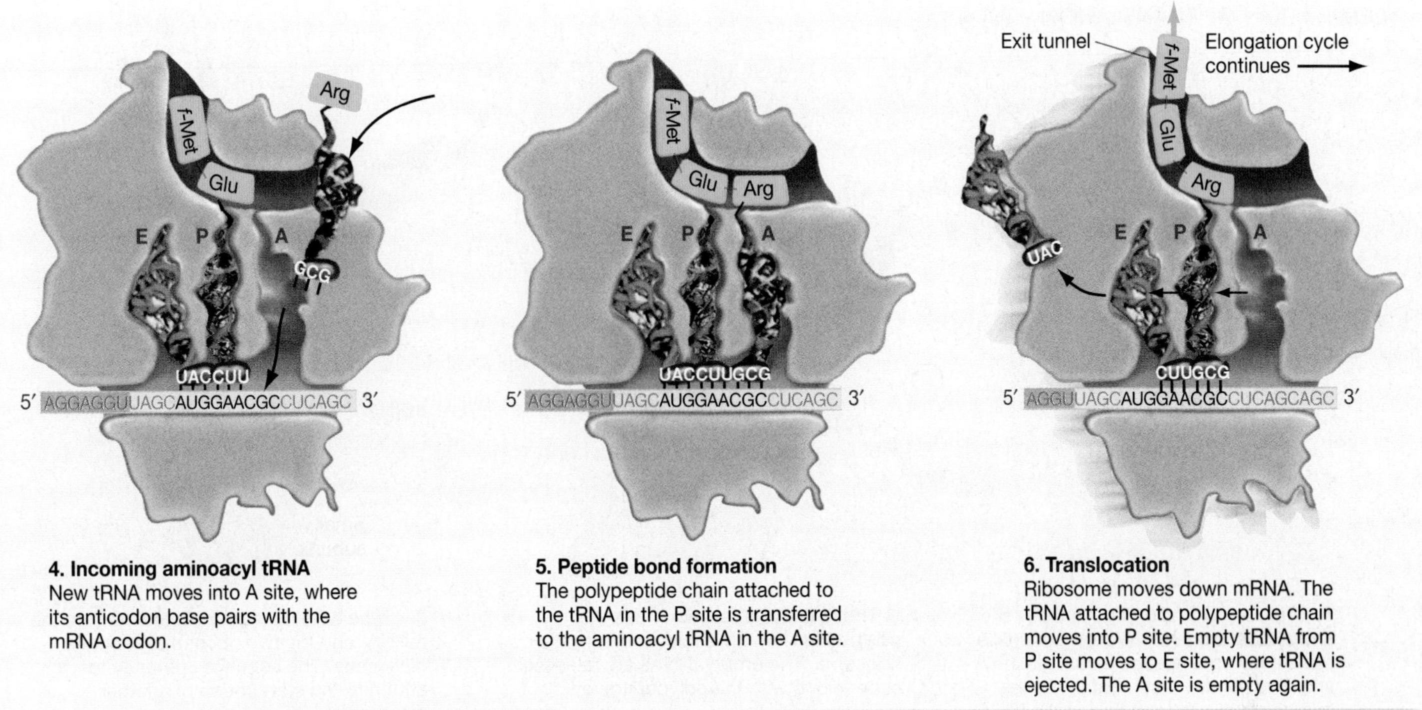

4. Incoming aminoacyl tRNA
New tRNA moves into A site, where its anticodon base pairs with the mRNA codon.

5. Peptide bond formation
The polypeptide chain attached to the tRNA in the P site is transferred to the aminoacyl tRNA in the A site.

6. Translocation
Ribosome moves down mRNA. The tRNA attached to polypeptide chain moves into P site. Empty tRNA from P site moves to E site, where tRNA is ejected. The A site is empty again.

empty tRNA into the E site; it moves the tRNA containing the growing polypeptide into the P site; and it opens the A site and exposes a new mRNA codon. If the E site is occupied when translocation occurs, the tRNA there is ejected into the cytosol.

The three steps in elongation—(1) arrival of aminoacyl tRNA, (2) peptide bond formation, and (3) translocation—repeat down the length of the mRNA. By adding specific molecules to in vitro translation systems, researchers have confirmed that each elongation cycle depends on an input of energy from several GTP molecules as well as assistance from proteins called **elongation factors**. Further, recent three-dimensional models of ribosomes in various stages of the translation sequence show that the machine as a whole is highly dynamic during the process. The ribosome constantly changes shape as tRNAs come and go and catalysis and translocation occur. Its structure changes in conjunction with its multipart function: coordinating interactions between tRNAs and an mRNA, and catalyzing peptide bond formation.

Electron microscopy has also confirmed that once elongation is under way, additional ribosomes bind to the start site in the same message and initiate translation. As this process continues, strings of ribosomes called **polyribosomes** assemble along an mRNA. Each ribosome synthesizes a copy of the protein (**Figure 16.18**). Even though it may take a minute or more for a single ribosome to produce a large polypeptide, the presence of polyribosomes can quicken the overall pace of protein production. Polyribosomes are observed routinely in both bacteria and eukaryotes.

Termination

How does protein synthesis end? To answer this question, recall from Chapter 15 that the genetic code includes three stop codons: UAA, UAG, and UGA. In most cells, no aminoacyl tRNA has an anticodon that binds to these sequences. When translocation opens the A site and exposes one of the stop

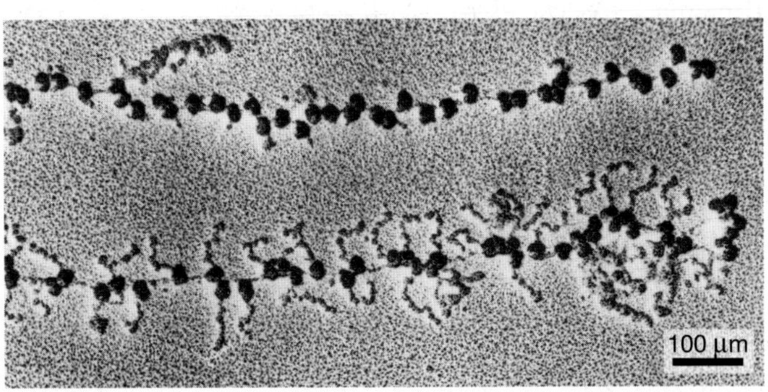

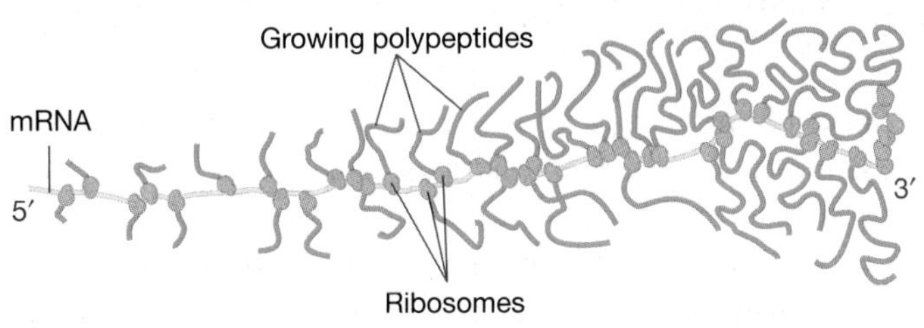

FIGURE 16.18 Polyribosomes. In both bacteria and eukaryotes, many ribosomes synthesize proteins from the same mRNA at the same time. The resulting structures are called polyribosomes.

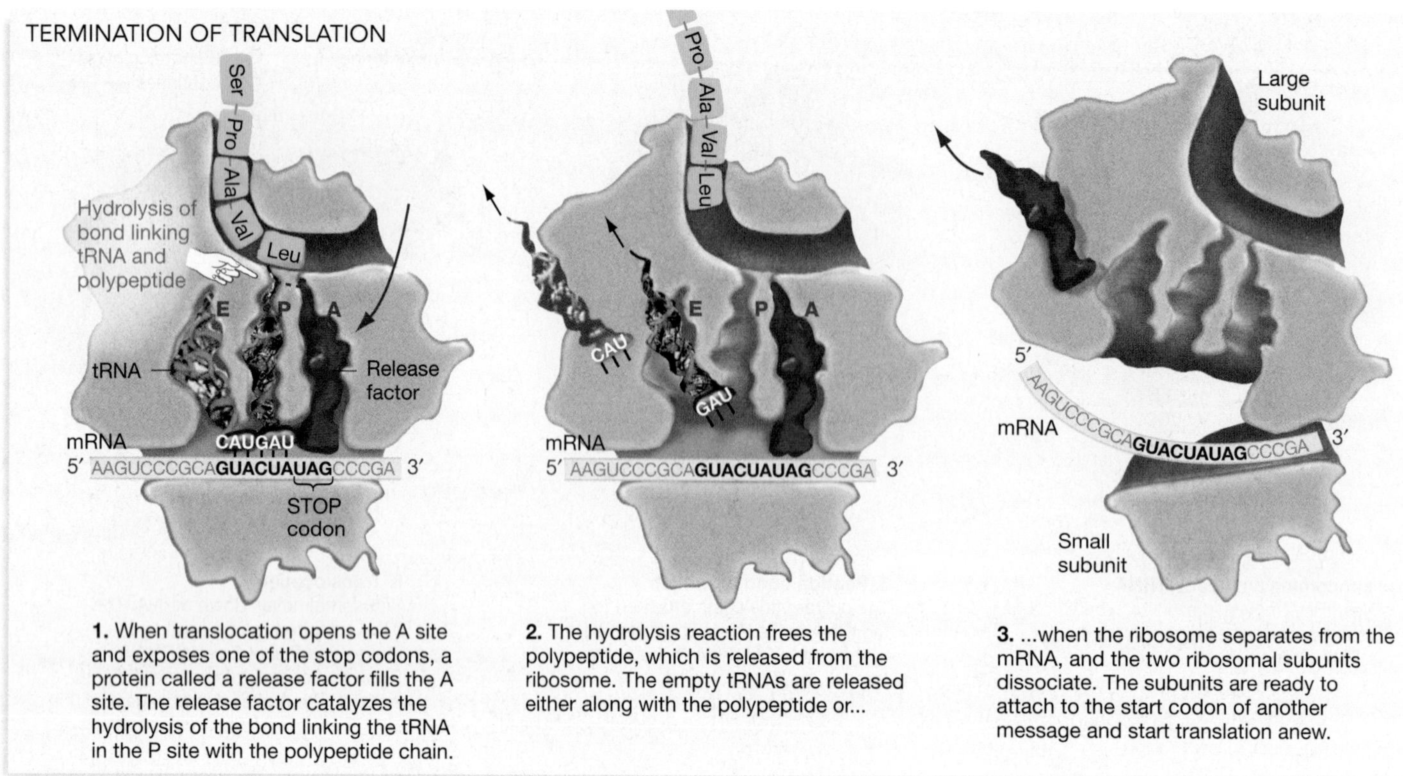

FIGURE 16.19 The Termination Phase of Translation.

codons, a protein called a **release factor** fills the A site (**Figure 16.19**). Release factors are proteins that fit tightly into the A site because their size, shape, and electrical charge are tRNA-like. Release factors do not carry an amino acid, however. When a release factor occupies the A site, the active site catalyzes the hydrolysis of the bond linking the tRNA in the P site with the polypeptide chain. This reaction frees the polypeptide. The newly synthesized polypeptide is released from the ribosome, the ribosome separates from the mRNA, and the two ribosomal subunits dissociate. The subunits are ready to attach to the start codon of another message and start translation anew.

Post-Translational Modifications Proteins are not fully formed and functional when termination occurs. From earlier chapters, it should be clear that most proteins go through an extensive series of processing steps, collectively called post-translational modification, before they are ready to go to work in a cell:

- Recall from Chapter 3 that a protein's function depends on its shape and that a protein's shape depends on how it folds. Folding actually begins during elongation, long before termination occurs and ribosomes disassemble. Although it occurs spontaneously, in the sense that no energy input is required, folding is frequently speeded up by proteins called **molecular chaperones**. Recent data have shown that in some bacteria, chaperone proteins actually bind to the ribosome near the "tunnel" where the growing polypeptide emerges from the ribosome. This result suggests that folding occurs as the polypeptide is emerging from the ribosome.
- Chapter 7 pointed out that many eukaryotic proteins are extensively modified after they are synthesized. For example, small chemical groups may be added to proteins in the organelles called the rough endoplasmic reticulum and the Golgi apparatus. Some proteins receive a sorting signal that serves as an address and ensures that the molecule will be carried to the correct location in the cell. Certain proteins are also augmented with sugar or lipid groups that are critical for normal functioning.
- Many proteins are altered by enzymes that add or remove a phosphate group. Phosphorylation (addition of phosphate) and dephosphorylation (removal of phosphate) of proteins were introduced in Chapter 9. Recall that because a phosphate group has two negative charges, adding or removing a phosphate group may cause major changes in the shape and chemical reactivity of proteins. These changes have a dramatic effect on the protein's activity—often switching it from an inactive state to an active state or vice versa.

The overall message is that the manufacture of a completely functional protein depends not only on translation in the ribosome but also on a wide array of other molecules and events that take place throughout the cell.

Check Your Understanding

If you understand that...

- Translation begins when (1) the ribosome binding site on an mRNA binds to an rRNA sequence in the small subunit of the ribosome, (2) the initiator aminoacyl tRNA binds to the start codon in the mRNA, and (3) the large subunit of the ribosome attaches to the small subunit to complete the ribosome.
- Translation elongation occurs when (1) an appropriate aminoacyl tRNA enters the A site, (2) a peptide bond forms between the amino acid held by that tRNA in the A site and the polypeptide held by the tRNA in the P site, and (3) the ribosome moves down the mRNA one codon.
- Translation ends when the ribosome reaches a stop codon.

You should be able to...

1) Diagram a ribosome during translation initiation, elongation, and termination.
2) Label each major structure in the diagrams (large and small ribosomal subunits, ribosome binding site and initiation factors, mRNA, E/P/A sites of ribosome, growing polypeptide, release factor).
3) Describe the function of each major structure.

Translation; Protein Synthesis

16.6 What Is the Molecular Basis of Mutation?

This chapter has been focusing on how the information in DNA is transcribed to RNA and then translated into a functioning protein. Now the question is, what happens if the information in DNA changes? What are the consequences for the cell?

A **mutation** is any permanent change in an organism's DNA. It is a modification in a cell's information archive—a change in its genotype. The molecular machinery introduced in this chapter faithfully transcribes and translates mutations in DNA sequences. The result is the production of novel types of proteins. In this way, mutations may lead to changes in proteins and RNAs that affect the organism's phenotype.

Point Mutation

Figure 16.20 shows how a common type of mutation occurs. If DNA polymerase mistakenly inserts the wrong base as it synthesizes a new strand of DNA, and if proofreading by DNA polymerase and the mismatch repair system fail to correct the mismatched base before another round of DNA replication occurs, a change in the sequence of bases in DNA results. A single base change such as this is called a **point mutation**.

What happens when point mutations are transcribed and translated? To answer this question, consider the first point mutation ever described. This mutation occurs in the human gene for the protein hemoglobin. Hemoglobin is abundant in red blood cells, and it carries oxygen to tissues. You might recall from Chapter 3 that each hemoglobin protein is made up of four polypeptides—two copies of two different polypeptides. In adult humans, the two distinct polypeptides are α-globin and β-globin.

Figure 16.21a shows a small section of DNA sequence from the normal gene for β-globin along with the same sequence from the mutant allele found in individuals who suffer from sickle-cell disease. Note that the mutant form of β-globin has a thymine in place of an adenine at the second position in the sixth codon specified by this gene. A glance back at the genetic code in Chapter 15 confirms why this point mutation is significant: During protein synthesis, the mutant codon specifies valine instead of glutamic acid in the amino acid chain of β-globin. Point mutations that cause changes in the amino acid sequence of proteins are called **missense mutations** or **replacement mutations**. In the case of hemoglobin, the single

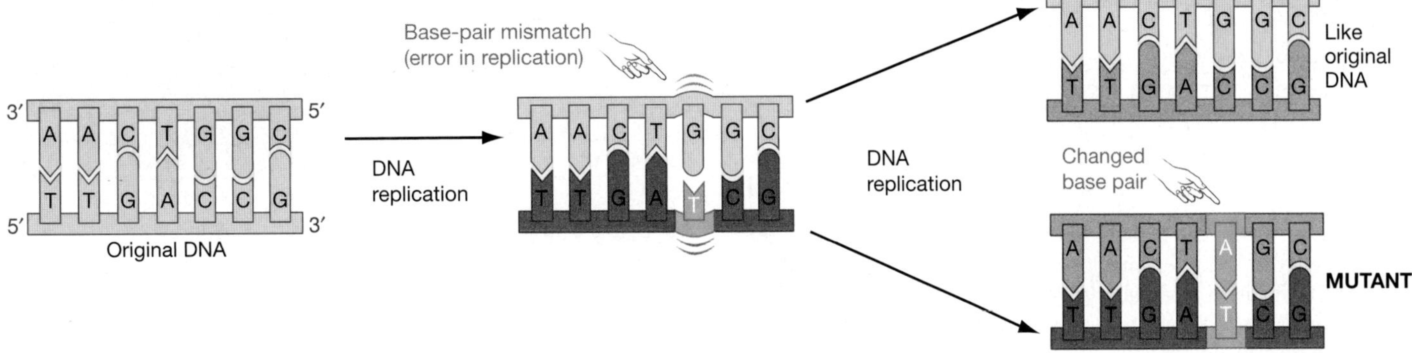

FIGURE 16.20 Unrepaired Mistakes in DNA Synthesis Lead to Point Mutations.

QUESTION Why is it logical that the type of mutation illustrated here is termed a point mutation?

change in primary structure of β-globin causes the protein to crystallize when oxygen levels in the blood are low. When hemoglobin crystallizes, it causes red blood cells to become sickle shaped (**Figure 16.21b**). The misshapen cells get stuck in blood vessels, so nearby cells are starved for oxygen. The result is intense pain. Sickled cells also have a much shorter life span in the body than normal red blood cells, reducing overall red blood cell counts and making people anemic.

Missense mutations are often **deleterious**, meaning they reduce an individual's fitness—the ability to survive and reproduce. Missense mutations may also be beneficial, however. For example, sickle cell disease develops only in people who are homozygous for the sickling allele. But people who are heterozygous for the sickling allele have a fitness advantage in regions of the world where the life-threatening disease malaria is common. The advantageous effect is possible because the parasite that causes malaria infects red blood cells. If an individual has one normal hemoglobin allele and one sickling allele, red blood cells that become infected with the parasite tend to sickle while others function normally. The deformed cells are quickly destroyed by the body, eliminating the parasitic cells inside the blood cells. As a result, individuals who have one copy of the mutant gene have lower parasite loads and tend to be much healthier than people who have two normal copies of the gene. Although having two copies of the sickling allele is disadvantageous in all environments because it leads to sickle cell disease, having one copy of the sickling allele is advantageous in malaria-prone areas of the world.

In contrast to having a negative or positive effect on the organism, many point mutations have virtually no effect at all. To understand why, suppose that at the next site in the hemoglobin gene of Figure 16.21 (the third position in the same sixth codon), a thymine had been substituted for the cytosine. This point mutation would have no consequence. Why? Both GAA and GAG specify glutamic acid. In this case, the amino acid sequence of the gene product does not change even though the DNA sequence is altered due to mutation. This type of alteration in the base sequence is called a **silent mutation**. Silent mutations are said to be **neutral** in their effect on an individual's fitness. It is possible for point mutations to be silent and neutral with respect to fitness because of the redundancy in the genetic code.

The take-home message from this analysis of hemoglobin, sickle cell disease, and malaria is simple: Point mutations may be deleterious, beneficial, or neutral. Recent studies have shown that the vast majority of point mutations are neutral or slightly deleterious with respect to fitness. **Table 16.3** summarizes the types of point mutations that have been documented and reviews their consequences for the amino acid sequences of proteins.

Chromosome-Level Mutations

Besides documenting various types of point mutations, biologists study mutations involving larger-scale changes in the com-

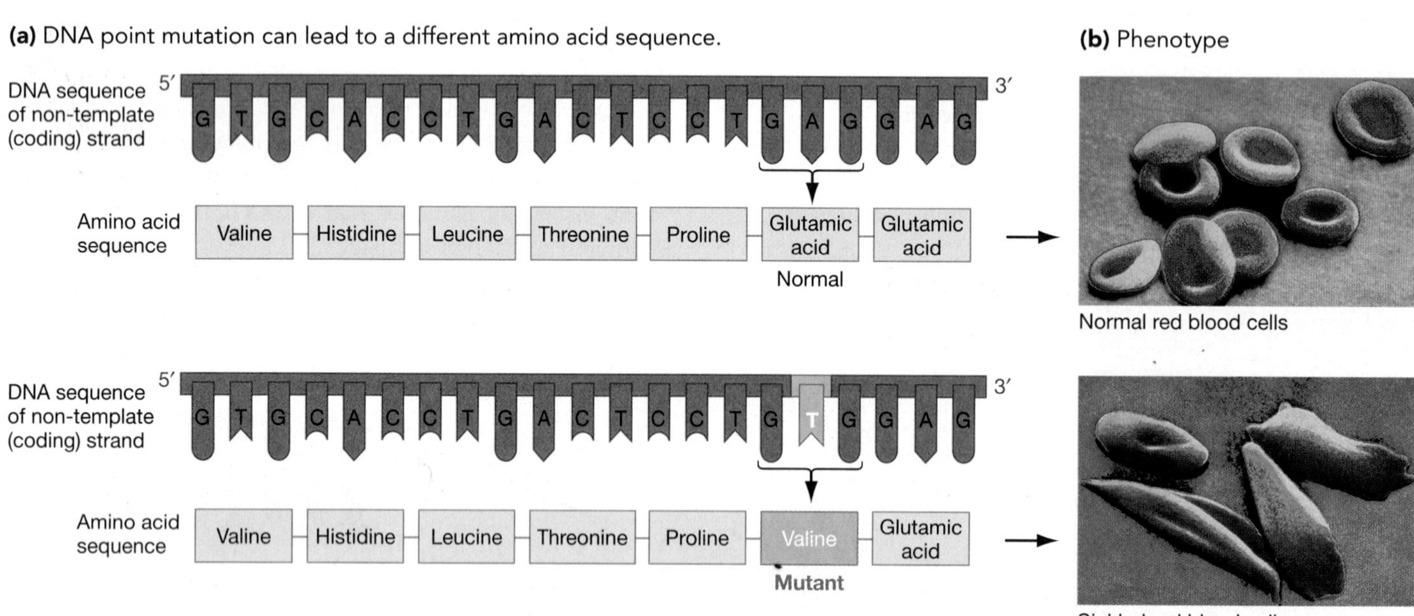

FIGURE 16.21 Sickle-Cell Disease Results from a Point Mutation in the Gene for Hemoglobin. (a) These normal and mutant genes that code for hemoglobin differ by a single nucleotide. The sequences shown are just a small portion of the gene. **(b)** The mutation in part (a) changes the primary sequence of the gene product. The change in amino acid sequence causes hemoglobin molecules to crystallize when oxygen levels in the blood are low. As a result, red blood cells sickle and get stuck in small blood vessels.

EXERCISE In part (a), add the base sequence of the template strand and label its 5′ and 3′ ends. Write out the base sequence of the mRNA transcribed from the template strand and label its 5′ and 3′ ends.

SUMMARY TABLE **16.3 Known Types of Point Mutations**

Name	Definition	Example	Consequence
	Original DNA sequence	TAT TGG CTA GTA CAT Tyr–Trp–Leu–Val–His — Original polypeptide	
Silent	Change in nucleotide that does not change amino acid specified by codon	TAC TGG CTA GTA CAT Tyr–Trp–Leu–Val–His	Change in genotype but no change in phenotype
Missense (Replacement)	Change in nucleotide that changes amino acid specified by codon	TAT TGT CTA GTA CAT Tyr–**Cys**–Leu–Val–His	Change in primary structure of protein
Nonsense	Change in nucleotide that results in early stop codon	TAT TGA CTA GTA CAT Tyr **STOP**	Premature termination—polypeptide is truncated
Frameshift	Addition or deletion of a nucleotide	TAT TCG GCT AGT ACA T Tyr–**Ser**–**Ala**–**Ser**–**Thr**	Reading frame is shifted (see Chapter 15)—massive missense

position of chromosomes. Chapter 12 introduced **polyploidy,** which is a change in the number of each type of chromosome present, and **aneuploidy,** which involves the addition or deletion of a chromosome. Polyploidy, aneuploidy, and other changes in chromosome number result from chance mistakes in the partitioning of chromosomes during meiosis or mitosis. In this sense they are similar to the random errors that DNA polymerase makes, resulting in point mutations.

But in addition to changes in overall chromosome number, the composition of individual chromosomes can change in important ways. For example, chromosome segments can become detached when accidental breaks in chromosomes occur. The segments may become flipped and rejoin—a phenomenon known as a chromosome **inversion**—or become attached to a different chromosome, causing a chromosome **translocation** (**Table 16.4**). These mutations affect the arrangement of genes on chromosomes. They are important because they influence the frequency of crossing over between chromosomes as well as how certain genes are turned on or off.

To summarize, point mutations and chromosome-level mutations are random changes in DNA that produce new genes, new alleles, and new traits. At the level of individuals, mutations can cause disease or death or lead to increases in fitness. At the level of populations, mutations furnish the heritable variation that Mendel and Morgan analyzed and that makes evolution possible.

SUMMARY TABLE **16.4 Some Types of Chromosome-Level Mutations**

Name	Description	Example	Consequence
Chromosome inversion	Chromosome segment flips and reattaches to same chromosome.	Genes A B C D → A B C D → A C B D	Changes gene order along chromosome. Frequency of crossing over is reduced (when homologs synapse, inverted segments can't cross-over with non-inverted segments).
Chromosome translocation	Chromosome segment detaches and attaches to different chromosome.	A B C D; E F G H I → E A B C D; F G H I	Leads to deletion or addition of chromosome segments. In many cases, expression of genes in translocated segments changes.

Chapter Review

SUMMARY OF KEY CONCEPTS

The instructions for making and operating a cell are archived in DNA, transcribed into messenger RNA, and then translated into protein.

After RNA polymerase binds DNA with the help of other proteins, it catalyzes the production of an RNA molecule whose sequence is complementary to the base sequence of the DNA template strand.

RNA polymerase begins transcription by binding to promoter sequences in DNA. In bacteria, this binding occurs in conjunction with a regulatory protein called sigma. Sigma recognizes particular sequences within promoters that are centered 10 bases and 35 bases upstream from the start of the actual genetic message. These binding sites ensure that the correct DNA strand is transcribed in the correct direction. Eukaryotic promoters are more complex and variable than bacterial promoters are, and they interact with many more transcription factors.

You should be able to predict the consequences of a mutation that changes the sequence of nucleotides in a promoter.

MB Web Animation at www.masteringbio.com
Transcription

Some sections of an RNA are encoded by gene regions called exons, while others are encoded by gene regions called introns. During RNA processing, introns are removed and the ends of the RNA receive a cap and tail.

Stretches of noncoding RNA called introns are spliced out by complex molecular machines called spliceosomes, a "cap" signal is added to the 5′ end, and a poly(A) tail is added to the 3′ end. The 5′ cap and the poly(A) tail serve as recognition signals for the translation machinery and protect the message from degradation by ribonucleases.

You should be able to draw the structure of a primary mRNA and a mature mRNA from a eukaryote, and contrast the two molecules with a bacterial mRNA.

Inside ribosomes, mRNAs are translated to proteins via intermediary molecules called transfer RNAs. Transfer RNAs carry an amino acid and have a three-base-pair anticodon, which binds to a three-base-long mRNA codon. The amino acid carried by the transfer RNA is then added to the growing protein via formation of a peptide bond.

Experiments with radioactively labeled amino acids confirmed that ribosomes are the site of protein synthesis, and that transfer RNAs (tRNAs) serve as the chemical bridge between the RNA message and the polypeptide product. tRNAs have an L-shaped tertiary structure. One leg of the L contains the anticodon, which forms a base pair with the mRNA codon, while the other leg holds the amino acid appropriate for that codon. Because imprecise pairing—or "wobbling"—is allowed in the third position of a codon-anticodon pair, only about 40 different tRNAs are required to translate the 61 codons that code for amino acids.

Inside the ribosome, protein synthesis involves three steps: (1) an incoming aminoacyl tRNA occupies the A site; (2) the growing polypeptide chain is transferred from a peptidyl tRNA in the ribosome's P site to the amino acid bound to the tRNA in the A site, and a peptide bond is formed; and (3) the ribosome is translocated to the next codon on the mRNA, accompanied by ejection of the empty tRNA from the E site. Peptide bond formation is catalyzed by a ribozyme (RNA), not an enzyme (protein).

While translation is in progress, proteins fold into their three-dimensional conformation (tertiary structure), sometimes with the aid of chaperone proteins. Some proteins are targeted to specific locations in the cell by the presence of signal sequences, while others remain inactive until modified by phosphorylation or the removal of certain amino acids.

You should be able to create a concept map (see **BioSkills 6**) describing the events in translation, starting with the addition of amino acids to tRNAs and including the major steps in initiation, elongation, and termination.

MB Web Animation *BioFlix* at www.masteringbio.com
Translation; Protein Synthesis

Mutations are random changes in DNA that may or may not produce changes in the phenotype.

Depending on the location and type of alteration in DNA and its impact on the resulting RNA or protein product, a mutation can be beneficial, deleterious, or neutral with respect to fitness. Mutations produce novel proteins and RNAs. They are the source of the heritable variation that makes evolution possible.

You should be able to explain why mutation is random. Your answer should be based on the molecular mechanisms responsible for point and chromosomal mutations.

QUESTIONS

Test Your Knowledge

1. How did the A site of the ribosome get its name?
   a. It is where amino acids are affixed to tRNAs, producing aminoacyl tRNAs.
   b. It is where the amino group on the growing polypeptide chain is available for peptide bond formation.
   c. It is the site occupied by incoming aminoacyl tRNAs.
   d. It is surrounded by α-helices of ribosomal proteins.

2. How did the P site of the ribosome get its name?
   a. It is where the promoter resides.
   b. It is made up of protein.
   c. It is where peptidyl tRNAs reside.
   d. It is where a growing polypeptide chain is phosphorylated.
3. What is a molecular chaperone?
   a. a protein that recognizes the promoter and guides the binding of RNA polymerase
   b. a protein that activates or deactivates another protein by adding or removing a phosphate group
   c. a protein that is a component of the large ribosomal subunit and that assists with peptide bond formation
   d. a protein that helps newly translated proteins fold into their proper three-dimensional configuration
4. The three types of RNA polymerase found in eukaryotic cells transcribe different types of genes. What does RNA polymerase II produce?
   a. rRNAs
   b. tRNAs
   c. mRNAs
   d. spliceosomes
5. What is an anticodon?
   a. the part of an mRNA that signals translation termination
   b. the part of an mRNA that signals the start of translation
   c. the part of a tRNA that binds to a codon in mRNA
   d. the part of a tRNA that accepts an amino acid, via a reaction catalyzed by tRNA synthetase
6. What happens during a chromosome inversion?
   a. Pieces of different chromosomes break and swap positions.
   b. Homologs stay together during cell division, so a daughter cell is a chromosome short.
   c. Chromosome number doubles.
   d. A segment of a chromosome breaks, flips in orientation, and reattaches.

Test Your Knowledge answers: 1. c; 2. c; 3. d; 4. c; 5. c; 6. d

Test Your Understanding

Answers are available at www.masteringbio.com

1. Explain the relationship among eukaryotic promoter sequences, basal transcription factors, and RNA polymerase. Explain the relationship among bacterial promoter sequences, sigma, and RNA polymerase.
2. According to the wobble rules, the correct amino acid can be added to a growing polypeptide chain even if the third base in an mRNA codon does not correctly match the corresponding base in a tRNA anticodon. How do the wobble rules relate to the redundancy of the genetic code?
3. Why does splicing occur in eukaryotic mRNAs? Where does it occur, and how are snRNPs involved?
4. Explain the sequence of events that occurs during translation as a protein elongates by one amino acid. At each step, specify what is happening in the ribosome's A site, P site, and E site.
5. What evidence supports the hypothesis that peptide bond formation is catalyzed by a ribozyme?
6. Explain why all point mutations change the genotype, but why only some point mutations change the phenotype.

Applying Concepts to New Situations

Answers are available at www.masteringbio.com

1. The 5′ cap and poly(A) tail on eukaryotic mRNAs protect the message from degradation by ribonucleases. But why do ribonucleases exist? What function would an enzyme that destroys messages serve? Answer this question using the example of an mRNA for a hormone that causes human heart rate to increase.
2. The nucleotide shown below is called cordycepin.

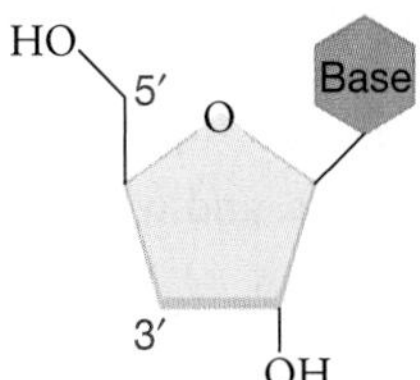

   If cordycepin triphosphate, which has three phosphate groups bonded to the 5′ hydroxyl group in the figure, is added to a cell-free transcription reaction, the nucleotide is added onto the growing RNA chain. This observation confirms that synthesis occurs in the 5′ → 3′ direction. Make a drawing, similar to Figure 16.1, showing why cordycepin triphosphate cannot be added to the 5′ end of an RNA.
3. Certain portions of the rRNAs in the large subunit of the ribosome are very similar in all organisms. To make sense of this result, Carl Woese suggests that the conserved sequences have an important functional role. His logic is that these conserved sequences are so important to cell function that any changes in the sequence cause death. Which specific portions of the ribosome would you expect to be identical or nearly identical in all organisms, and which would you expect to be more variable? Explain your logic.
4. Recent structural models show that α-amanitin inhibits translation by binding to a site inside RNA pol II, but not to the active site itself. Based on the model of bacterial RNA polymerase in Figure 16.2, predict where α-amanitin binds and why it inhibits transcription.

www.masteringbio.com is also your resource for • Answers to text, table, and figure caption questions and exercises • Answers to *Check Your Understanding* boxes • Online study guides and quizzes • Additional study tools including the *E-Book for Biological Science* 3rd ed., textbook art, animations, and videos.

17 Control of Gene Expression in Bacteria

KEY CONCEPTS

- Gene expression can be controlled at three levels: transcription, translation, or post-translation (protein activation).
- Changes in gene expression allow bacterial cells to respond to environmental changes.
- Transcriptional control can be negative or positive. Negative control occurs when a regulatory protein prevents transcription. Positive control occurs when a regulatory protein increases the transcription rate.
- Many regulatory proteins bind to specific sites in DNA. Because each type of regulatory protein has a distinct amino acid sequence, each type binds to different DNA sequences.

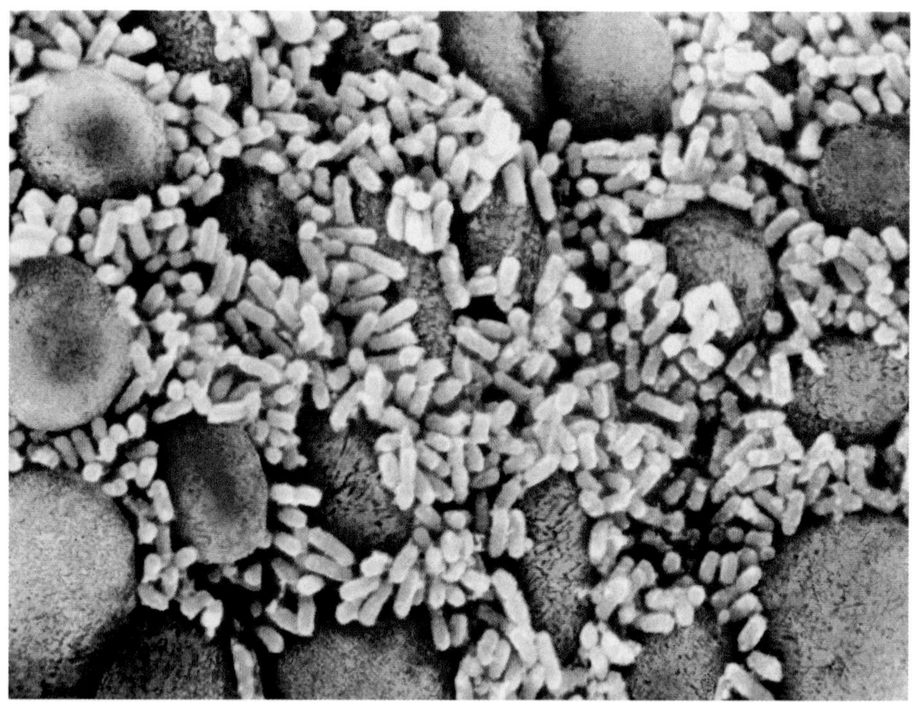

The red mounds in this micrograph are human intestinal cells; the yellow structures are bacteria. In the intestine, the nutrients available to bacteria constantly change. This chapter explores how changes in gene expression help bacteria respond to environmental changes.

Imagine waiting anxiously to hear the opening lines of a wonderfully melodic symphony, played by a renowned orchestra. The crowd applauds as the celebrated conductor comes onstage, then hushes as he takes the podium. He cocks the baton; the musicians raise their instruments. As the baton comes down, every instrument begins blaring a different tune at full volume. A tuba plays "Dixie"; a violinist renders "In-A-Gadda-Da-Vida"; and a cellist begins Mexico's national anthem. A snare drum lays down beats for Outkast's "Hey Ya" while the bass drum simulates the cannons in the "1812 Overture." Instead of music, there is pandemonium. The conductor staggers off stage, clutching his heart.

A cacophony like this would result if a bacterial cell "played" all its genes at full volume all the time. The *Escherichia coli* cells living in your gut right now have over 4300 genes. If all of those genes were expressed at the fastest possible rate at all times, the *E. coli* cells would stagger off the stage, too. But this does not happen. Cells are extremely selective about which genes are expressed, in what amounts, and when.

This chapter explores how bacterial cells control the activity of their genes. **Gene expression** is said to occur when a protein or other gene product is synthesized and is active in the cell.

Understanding how bacterial cells regulate gene expression is a fundamental issue in biological science, for two reasons. First, bacteria are the most abundant organisms on Earth, and they occupy virtually every habitat known. In many cases, bacterial cells are able to grow and reproduce successfully because they respond rapidly to dramatic changes in temperature, pH, light, competitors, and nutrients. Changes in gene expression give bacteria the ability to cope with these types of environ-

Key Concept   Important Information   Practice It

mental change. Second, questions about bacterial gene expression have enormous practical significance. For example, biologists are trying to understand which genes are expressed when disease-causing bacteria colonize a person and start an infection. The goal of this research is to develop drugs that alter the expression of key genes in the disease-causing cells. Bacterial cells are also used commercially to manufacture important products such as human insulin and growth hormone. Efficient use of these cells depends on a solid understanding of how transcription and translation are regulated.

This chapter's goal is to introduce key concepts in the regulation of gene expression. Let's begin by reviewing some of the environmental challenges that bacterial cells face and then explore how these organisms meet them.

17.1 Gene Regulation and Information Flow

Bacteria are found in virtually every habitat on Earth, from boiling hot springs to alpine snowfields and from the open ocean to crevices miles underground. The millions of bacterial species existing today have evolved a bewildering variety of ways to solve a central problem of living: obtaining the materials and energy required for growth and reproduction. Although some bacteria specialize by using just one type of food, the vast majority of species are able to switch among several different sources of carbon and energy, according to which nutrients are available in the environment at the time. The fundamental question addressed in this chapter is *how* this switching occurs. Each type of nutrient requires a different membrane transport protein to bring the nutrient molecule into the cell and a different suite of enzymes to process it. How does a bacterial cell start transcribing certain genes and end the transcription of others, so that it can take advantage of alternative food sources? More generally, how do bacteria regulate gene expression so that a cell makes only the products that it needs?

As a case study, this chapter focuses on strategies observed in *Escherichia coli*. These cells can use a wide array of carbohydrates to supply the carbon and energy they need. For example, *E. coli* are abundant in your intestinal tract and use carbohydrates found in your food. But as your diet changes from day to day, the availability of different sugars in your intestines varies. Precise control of gene expression gives *E. coli* the ability to respond to these changes in its environment and use the different sugars.

To appreciate why precise control over gene expression is so important to these cells, it is critical to realize that bacteria can be packed an inch thick along your intestinal walls. The organisms represent many different species that are competing for space and nutrients. For a cell to survive and reproduce in this environment, it must use resources efficiently—particularly resources that provide energy and carbon compounds. Using energy and carbon compounds to synthesize proteins that aren't needed would reduce a cell's ability to produce the proteins it *does* need, compete successfully for resources, and produce offspring.

Based on this reasoning, biologists predicted that the transcription and translation of individual genes in bacteria are triggered by specific signals from the environment, such as the presence of specific sugars. Did you drink milk at your last meal, or eat French fries and a candy bar? Each type of food contains different sugars. Each sugar should induce a different response by the *E. coli* cells in your intestine. Just as a conductor needs to regulate the orchestra's musicians, cells need to regulate which proteins they produce at any given time.

Mechanisms of Regulation—An Overview

Gene expression can be controlled at any step between the synthesis of RNA and the activation of the final gene product. Three steps occur as information flows from DNA to proteins, represented by arrows in the following expression:

$$\text{DNA} \rightarrow \text{mRNA} \rightarrow \text{protein} \rightarrow \text{activated protein}$$

The arrow from DNA to RNA represents transcription—the making of messenger RNA (mRNA). The arrow from RNA to protein represents translation, in which ribosomes read the information in mRNA and use that information to synthesize a protein. The arrow from protein to activated protein represents post-translational modifications—including folding, addition of carbohydrate or lipid groups, or perhaps phosphorylation.

How can a bacterial cell avoid producing proteins that are not needed at a particular time, and thus use resources efficiently? A look at the flow of information from DNA to protein suggests three possible mechanisms:

1. The cell could avoid making the mRNAs for particular enzymes. If there is no mRNA, then ribosomes cannot make the gene product. For example, various regulatory proteins affect the ability of RNA polymerase to bind to a promoter and initiate transcription. Genes that are controlled in this way are said to undergo **transcriptional control**:

$$\text{DNA} \nrightarrow \text{mRNA} \rightarrow \text{protein} \rightarrow \text{activated protein}$$

2. If the mRNA for an enzyme has been transcribed, the cell might have a way to prevent the mRNA from being translated into protein. Mechanisms that alter the length of time an mRNA survives before it is degraded by ribonucleases, that affect translation initiation, or that affect elongation factors and other proteins during the translation process are forms of **translational control**:

$$\text{DNA} \rightarrow \text{mRNA} \nrightarrow \text{protein} \rightarrow \text{activated protein}$$

3. Chapter 16 pointed out that some proteins are manufactured in an inactive form and have to be activated by chemical

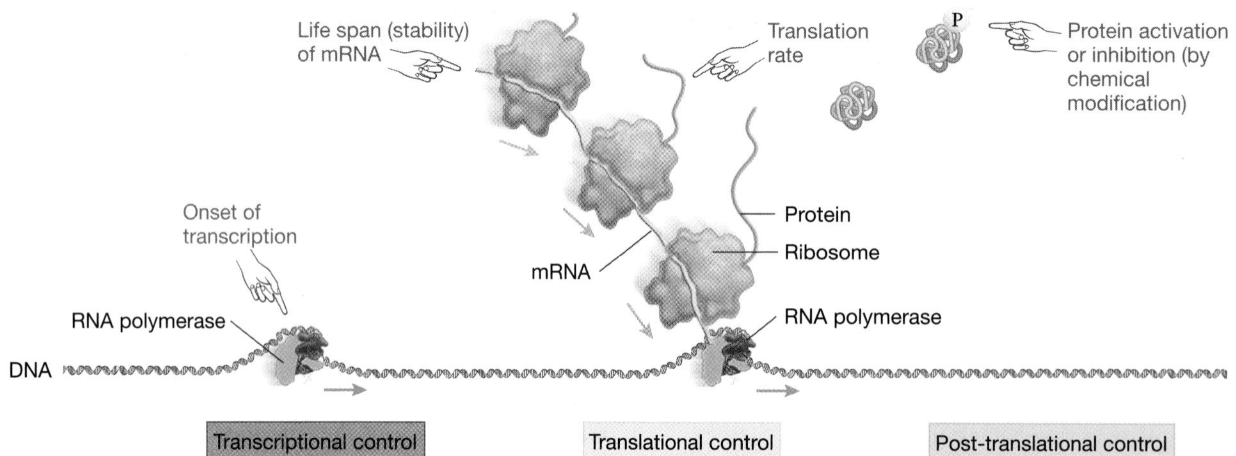

FIGURE 17.1 Gene Expression Can Be Regulated at Three Levels. Although this chapter focuses on how regulatory molecules affect the ability of RNA polymerase to initiate transcription, translational and post-translational controls also occur in bacteria.

● **EXERCISE** Label which mode of regulation is the slowest in response time, and which is fastest. Label which is the most efficient and least efficient in resource use.

modification, such as the addition of a phosphate group. This type of regulation is **post-translational control:**

DNA → mRNA → protein ⇸ activated protein

● Which of these three forms of control occur in bacteria? The short answer to this question is "all of the above." As **Figure 17.1** shows, many factors affect how much active protein is produced from a particular gene. Transcriptional control is particularly important due to its efficiency—it saves the most energy for the cell, because it stops the process at the earliest possible point. Translational control is advantageous because it allows a cell to make rapid changes in its array of proteins. Post-translational control is significant as well, and it provides the most rapid response of all three mechanisms. Among these mechanisms of gene regulation, there is a clear trade-off between the speed of response and the conservation of ATP, amino acids, and other resources. Transcriptional control is slow but efficient in resource use. Post-translational control is fast but energetically expensive.

Although this chapter focuses almost exclusively on mechanisms of transcriptional control in bacteria, it is important to keep in mind that both translational and post-translational controls occur in these organisms. ● It's also important to recognize that some genes—such as those that code for the enzymes required for glycolysis—are transcribed all the time, or **constitutively**. But the expression of other genes is regulated, meaning that they may be induced or repressed. Finally, it is critical to realize that gene expression is not an all-or-none proposition. Genes are not just "on" or "off"—instead, the level of expression is highly variable. Variation in gene expression allows cells to respond to changes in their environment. What factors determine how much a particular gene is expressed at a particular time?

Metabolizing Lactose—A Model System

As Chapters 13 through 16 have shown, many of the fundamental advances in genetics have been achieved through the analysis of various model systems. Studying the inheritance of seed shape in garden peas revealed the fundamental patterns of gene transmission. Exploring transcription in viruses and *E. coli* led to the discovery of RNA polymerase, transcription factors such as the sigma proteins, and promoters. In studies of gene regulation, the key model system has been the metabolism of the sugar lactose in *E. coli*.

Jacques Monod, François Jacob, and many colleagues introduced lactose metabolism in *E. coli* as a model system during the 1950s and 1960s. Although they worked with a single species of bacterium, their results had a profound effect on thinking about gene regulation in all organisms. Some details turned out to be specific to the *E. coli* genes responsible for lactose metabolism, but many of Monod and Jacob's results are universal.

Escherichia coli can use a wide variety of sugars for ATP production, via glycolysis and the electron transport chain or via fermentation. These sugars also serve as a raw material in the synthesis of amino acids, vitamins, and other complex compounds. Glucose is *E. coli*'s preferred carbon source, however, meaning that it is the source of energy and carbon atoms that the organism uses most efficiently. This observation is logical, because glycolysis begins with glucose and is the main pathway for the production of ATP. Lactose, the sugar found in milk, is also used by *E. coli*, but only when glucose supplies are depleted. Recall from Chapter 5 that lactose is a disaccharide made up of one molecule of glucose and one molecule of galactose.

To use lactose, *E. coli* must first transport the sugar into the cell. Once lactose is inside the cell, the enzyme β-galactosidase catalyzes a reaction that breaks the sugar down into glucose and galactose. The glucose released by this reaction is used directly via glycolysis; other enzymes convert the galactose to an intermediate in the glycolytic pathway.

In the early 1950s, biologists discovered that *E. coli* produces high levels of β-galactosidase only when lactose is present in the environment. If lactose is absent, little β-galactosidase is synthesized. Because lactose appears to induce the production of the enzyme that acts on lactose, researchers proposed that lactose itself regulates the gene for β-galactosidase. More formally, the hypothesis was that lactose acts as an inducer. An **inducer** is a molecule that stimulates the expression of a specific gene or genes.

In the late 1950s Jacques Monod investigated how the presence of glucose affects the regulation of the β-galactosidase gene. Would *E. coli* produce high levels of β-galactosidase when both glucose and lactose were present in the surrounding environment? As the experiment summarized in **Figure 17.2** shows, the answer was no. β-galactosidase is produced only when lactose is present and glucose is not present.

Monod teamed up with François Jacob to investigate exactly how lactose and glucose regulate the genes responsible for lactose metabolism. Research on how these genes are regulated opened a window on how genes in all organisms are controlled. Research on this system is still continuing, over 50 years later.

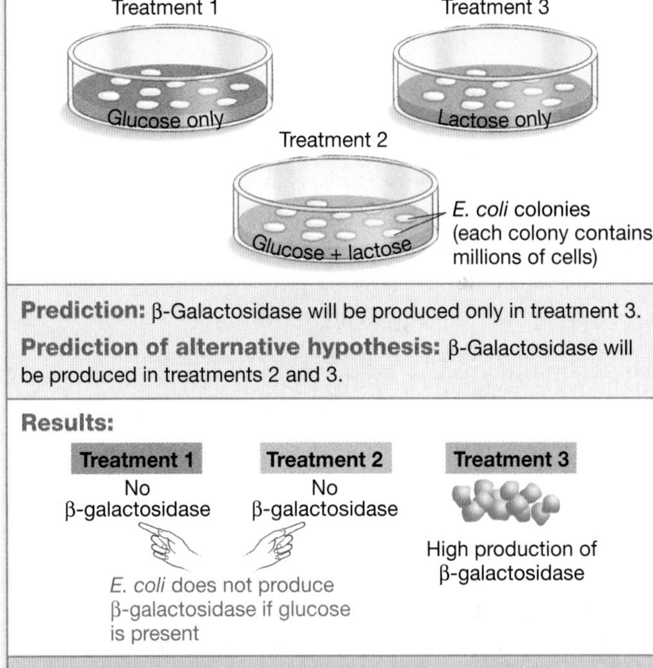

FIGURE 17.2 Glucose Affects the Regulation of the β-Galactosidase Gene.

17.2 Identifying the Genes Involved in Lactose Metabolism

To understand how *E. coli* controls production of β-galactosidase and the membrane transport protein that brings lactose into the cell, Monod and Jacob first had to find the genes that code for these proteins. To do this, they employed the same tactic used in the pioneering studies of DNA replication, transcription, and translation reviewed in earlier chapters: They isolated and analyzed mutant individuals. In this case, their goal was to find *E. coli* cells that were *not* capable of metabolizing lactose. Cells that can't use lactose must lack either β-galactosidase or the lactose-transporter protein.

Finding mutants with respect to a particular trait is a two-step process. A researcher's first step is to generate a large number of individuals with mutations at random locations in their genomes. To produce mutant cells, Monod and others exposed *E. coli* populations to X-rays, UV light, or chemicals that damage DNA and increase mutation rates. The second step is to screen the mutants to find individuals with defects in the process or biochemical pathway in question—in this case, defects in lactose metabolism. A technique that allows researchers to identify individuals with a particular type of mutation is called a **genetic screen.**

Monod and colleagues were looking for cells that cannot grow in an environment that contains only lactose as an energy source. Normal cells grow well in this environment. How could the researchers select cells on the basis of *lack* of growth?

How Were the Genes Found?

Replica plating and growth on indicator plates were key techniques in the search for mutants with defects in lactose metabolism. **Replica plating** begins with spreading mutagenized bacteria on a plate that is filled with gelatinous agar containing glucose (**Figure 17.3**). This plate is known as the master plate. It is important that the mutant cells you want to identify are able to grow on this master plate. The bacteria are then allowed to grow, so that each cell produces a single colony. (A bacterial colony consists of a large number of identical cells that are descended from a single cell.) Next, a block covered with a piece of sterilized velvet is pressed onto the master plate. Because of the contact, cells from each colony on the master

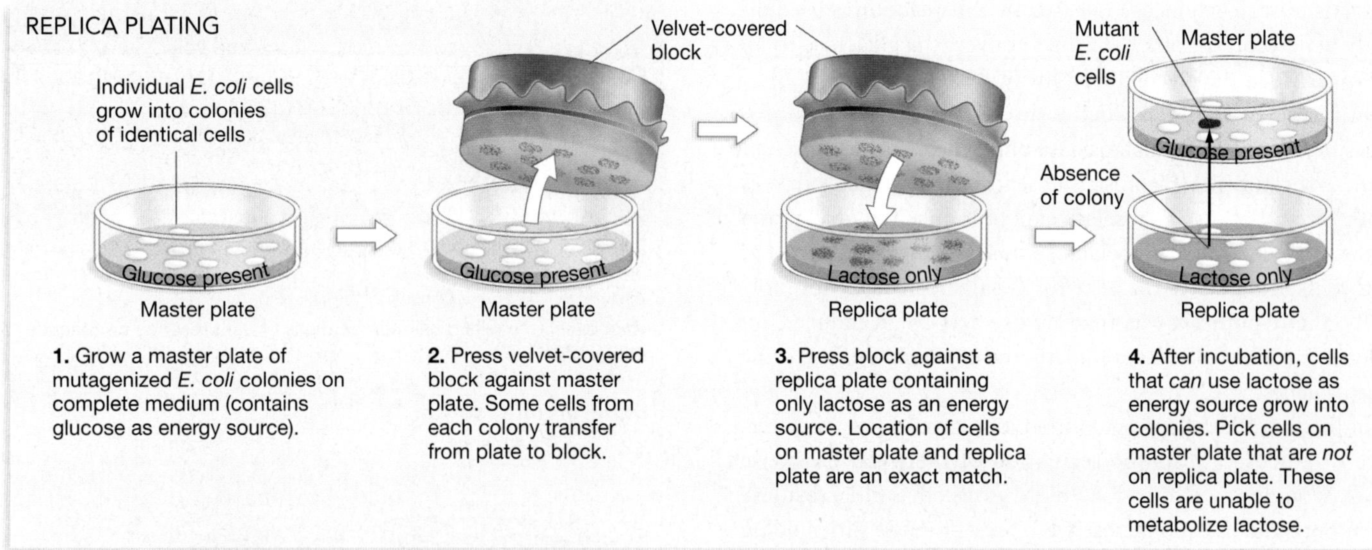

FIGURE 17.3 Replica Plating Is a Technique for Identifying Mutant Cells. Here, replica plating is used to isolate mutant *E. coli* cells with a deficiency in lactose metabolism. There are two keys: (1) The location of colonies on the master plate and replica plate must match exactly, and (2) the media in the two plates must differ by just one component.

QUESTION How would you alter this protocol to isolate mutant cells with a deficiency in the enzymes required to synthesize tryptophan?

plate are transferred to the velvet. Then the velvet is pressed onto a plate containing a medium that differs from that of the master plate by a single component. For instance, in Figure 17.3, step 3, the second medium has only lactose as a source of carbon and energy. Cells from the velvet stick to the plate's surface, producing an exact copy of the colonies on the master plate. This copy is called the replica plate. After these cells grow, an investigator can compare the colonies that thrive on the replica plate's medium with those on the master plate. In this case, colonies that grow on the master plate but are missing on the replica plate represent mutants deficient in lactose metabolism. By picking these particular colonies from the master plate, researchers build a collection of lactose mutants.

Monod also used an alternative strategy based on **indicator plates**, where mutants with metabolic deficiencies are observed directly. In this case, he added a compound that is acted on by β-galactosidase. The compound acts as an indicator molecule for the presence of functioning β-galactosidase because one of the molecules produced by the reaction is yellow. Colonies that stay white are unable to process the indicator molecule, meaning they have a defect in the β-galactosidase enzyme or its production.

Different Classes of Lactose Metabolism Mutants

The initial mutant screen yielded the three types of mutants summarized in **Table 17.1**. In one class, the mutant cells were unable to cleave the indicator molecule even if lactose was present inside the cells to induce production of the β-galactosidase protein. The investigators concluded that these mutants must lack a functioning version of the β-galactosidase protein—meaning the gene that encodes β-galactosidase is defective. This gene was designated *lacZ*, and the mutant allele *lacZ*$^-$.

In the second class of mutants, the cells failed to accumulate lactose inside the cell. In normal cells the concentration of lactose is about 100 times that of lactose in the surrounding

TABLE **17.1 Three Distinct Types of Mutants in Lactose Metabolism of *E. coli***

Mutant Phenotype	Interpretation	Inferred Genotype
Cells cannot cleave indicator molecule even if lactose is present as an inducer.	No β-galactosidase; gene for β-galactosidase is defective. Call this gene *lacZ*.	*lacZ*$^-$
Cells cannot accumulate lactose inside the cell.	No membrane protein (galactoside permease) required for import of lactose; gene for galactoside permease is defective. Call this gene *lacY*.	*lacY*$^-$
Indicator molecule is cleaved even if lactose is absent (no inducer).	Constitutive expression of *lacZ* and *lacY*; gene for regulatory protein that shuts down *lacZ* and *lacY* is defective—it does not need to be induced by lactose. Call this gene *lacI*.	*lacI*$^-$

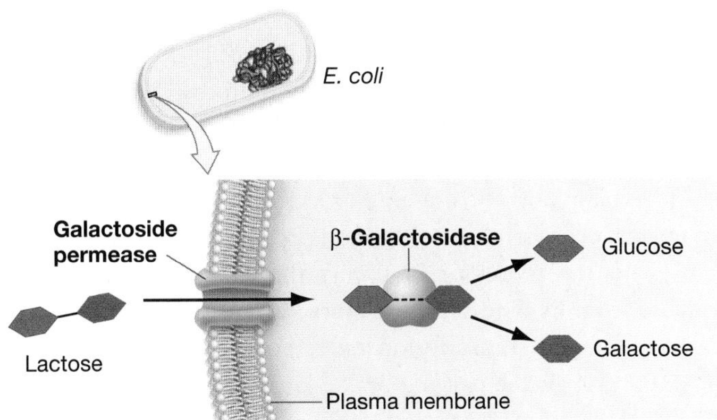

FIGURE 17.4 Two Proteins Are Critical for *E. coli* Cells to Use Lactose. For *E. coli* cells to use lactose, the membrane protein galactoside permease must be present to bring the sugar into the cell. Then the enzyme β-galactosidase must be present to break lactose into its glucose and galactose subunits.

environment, but in the mutant cells lactose concentrations were much lower. To interpret this result, Jacob and Monod hypothesized that the mutant cells had defective copies of the membrane protein responsible for transporting lactose into the cell. This protein was identified and named galactoside permease; the gene that encodes it was designated *lacY*. **Figure 17.4** summarizes the functions of β-galactosidase and galactoside permease.

The third and most surprising class of mutants did not regulate expression of β-galactosidase and galactoside permease normally. For example, when these mutant cells were grown on lactose alone and the indicator molecule was added, they turned yellow just as normal cells do. But if they were grown on a medium that contained glucose but no lactose, they still turned yellow after the indicator molecule was added. Normal cells, in contrast, remain white when grown on glucose and given the indicator molecule. Normal cells do not produce β-galactosidase when glucose is present.

Cells that are abnormal because they produce a product at all times are called **constitutive** mutants. Unlike normal cells, their product is always part of the cell's constitution. The gene that mutated to produce constitutive β-galactosidase expression was named *lacI*. The use of "I" was appropriate because these mutants did not need an inducer to express β-galactosidase or galactoside permease. Recall that in normal cells, the expression of these genes is induced by the presence of lactose. But in cells with a mutant form of *lacI* (*lacI*$^-$ mutants), gene expression occurred with or without lactose. This meant that *lacI*$^-$ mutants have a defect in gene regulation. In these mutants, the gene remains "on" when it should be turned off.

Based on these observations, it is logical to infer that the normal product of the *lacI* gene prevents the transcription of *lacZ* and *lacY* when lactose is absent. Because lactose acts as an inducer for the production of β-galactosidase, it is reasonable to expect that the *lacI* gene or gene product interacts with lactose in some way. (Later work showed that the inducer is actually a derivative of lactose called *allolactose*. For the sake of historical accuracy and simplicity, however, this discussion refers to lactose itself as the inducer.)

Several Genes Are Involved in Metabolizing Lactose

Jacob and Monod had succeeded in identifying three genes involved in lactose metabolism: *lacZ*, *lacY*, and *lacI*. They had concluded that *lacZ* and *lacY* code for proteins involved in the metabolism and import of lactose, while *lacI* is responsible for some sort of regulatory function. When lactose is absent, the *lacI* gene or gene product shuts down the expression of *lacZ* and *lacY*. But when lactose is present, the opposite occurred—transcription of *lacZ* and *lacY* was induced. ● If you understand the genes involved in lactose metabolism, you should be able to describe the specific function of *lacZ* and *lacY*. You should also be able to describe the effect of the *lacI* gene product when lactose is present versus absent, and explain why these effects are logical.

When Jacob and Monod followed up on these experiments by mapping the physical location of the three genes on *E. coli*'s circular chromosome, they found that the genes are close together (**Figure 17.5**). This was a crucial result, because it suggested that both *lacZ* and *lacY* might be controlled by *lacI*. Could one regulatory gene manage more than one protein-encoding gene? If so, how does *lacI* actually work? And why do lactose and glucose have opposite effects on it?

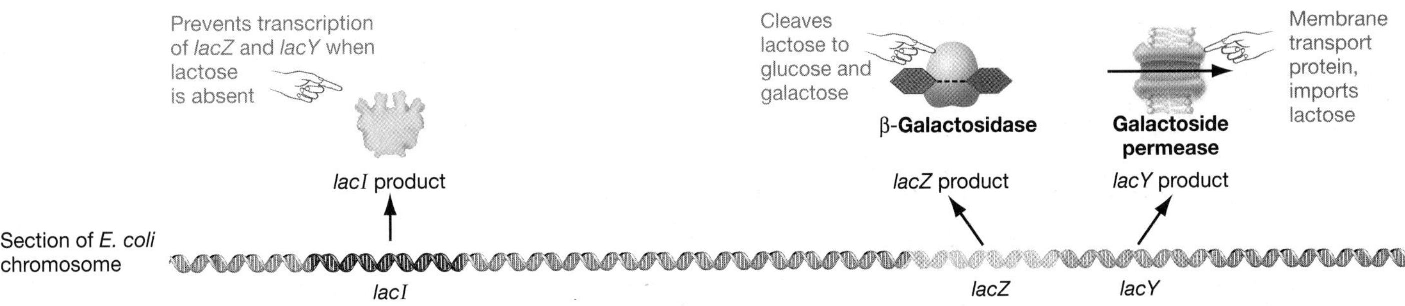

FIGURE 17.5 The *lac* Genes Are in Close Physical Proximity.

17.3 Mechanisms of Negative Control: Discovery of the Repressor

In principle, there are two ways that transcription can be regulated: via negative control or positive control. Negative control occurs when a regulatory protein binds to DNA and shuts down transcription (**Figure 17.6a**); positive control, also called **induction,** occurs when a regulatory protein binds to DNA and triggers transcription (**Figure 17.6b**). When a car is sitting at the curb, negative control occurs when the parking brake is set; positive control occurs when you step on the gas pedal. It turned out that the *lacZ* and *lacY* genes in *E. coli* are controlled by both a parking brake and a gas pedal. These genes are under both negative control and positive control.

The hypothesis that the *lacZ* and *lacY* genes might be under negative control originated with Leo Szilard in the late 1950s. Szilard suggested to Monod that the *lacI* gene codes for a product that represses transcription of the *lacZ* and *lacY* genes. Recall from Section 17.2 that in *lacI*$^-$ mutants, the production of β-galactosidase and galactoside permease is constitutive, even if lactose is not available. Szilard proposed that regulation fails because the lacI$^+$ protein—which normally prevents enzyme synthesis—is inactive. Stated another way, the *lacI* gene produces an inhibitor that exerts negative control over the *lacZ* and *lacY* genes. This transcription inhibitor was called a **repressor** and was thought to bind directly to DNA near or on the promoter for the *lacZ* and *lacY* genes (**Figure 17.7a**).

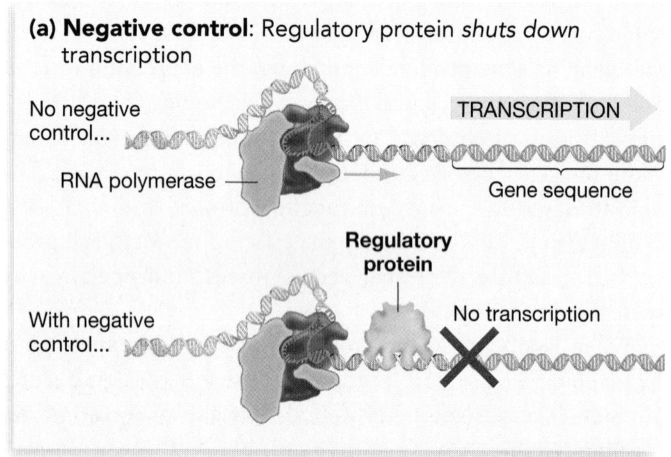

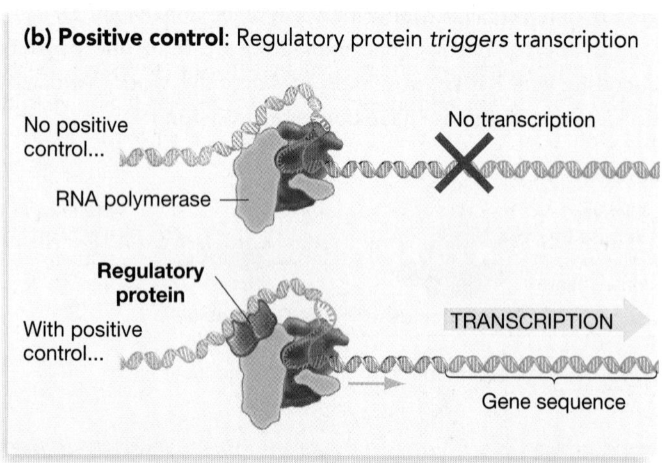

FIGURE 17.6 Genes Are Regulated by Negative Control, Positive Control, or Both.

To explain how the presence of lactose triggers transcription in normal cells, Szilard and Monod proposed that lactose interacts with the repressor in a way that makes the repressor release from its binding site (**Figure 17.7b**). The idea was that lactose induces transcription by removing negative control. The repressor is the parking brake; lactose releases the parking brake. As **Figure 17.7c** shows, it was logical to observe that constitutive transcription occurs in *lacI*$^-$ mutants because a functional repressor is absent—the parking brake is broken.

To test these hypotheses, Jacob and Monod and co-workers created *E. coli* cells that had functioning copies of the genes for β-galactosidase and galactoside permease but lacked a functional gene for the repressor. As predicted, these cells made β-galactosidase all the time. But when these cells received a functioning copy of the repressor gene, β-galactosidase production declined and then stopped. This result supported the hypothesis that the repressor codes for a protein that shuts down transcription—that it is indeed the "parking brake" on transcription. But if an inducer such as lactose was added to the experimental cells, β-galactosidase activity continued. This result supports the hypothesis that lactose removes the repressor—it releases the parking brake. The *lacI* gene codes for a repressor protein that exerts negative control on *lacZ* and *lacY*. Lactose acts as an inducer by removing the repressor and ending negative control.

The *lac* Operon

Jacob and Monod summarized the results of their experiments with a comprehensive model of negative control that was published in 1961. In essence, their experimental data had confirmed the hypotheses illustrated in Figure 17.7. One of their key conclusions was that the genes for β-galactosidase and galactoside permease are controlled together. To capture this point, they coined the term **operon** for a set of coordinately regulated bacterial genes that are transcribed together into one mRNA. Logically enough, the group of genes involved in lactose metabolism was termed the *lac* operon. Later, a gene called *lacA* was found to be tightly linked to *lacY* and *lacZ* and part of the same operon. The *lacA* gene codes for the enzyme transacetylase. The enzyme's function is protective in nature. It catalyzes reactions that allow certain types of sugars to be exported from the cell when they are too abundant.

Three hypotheses are central to the Jacob-Monod model of *lac* operon regulation:

1. The *lacZ*, *lacY*, and *lacA* genes are adjacent and are transcribed into one mRNA initiated from the single promoter of the *lac* operon. As a result, the expression of the three genes is coordinated.

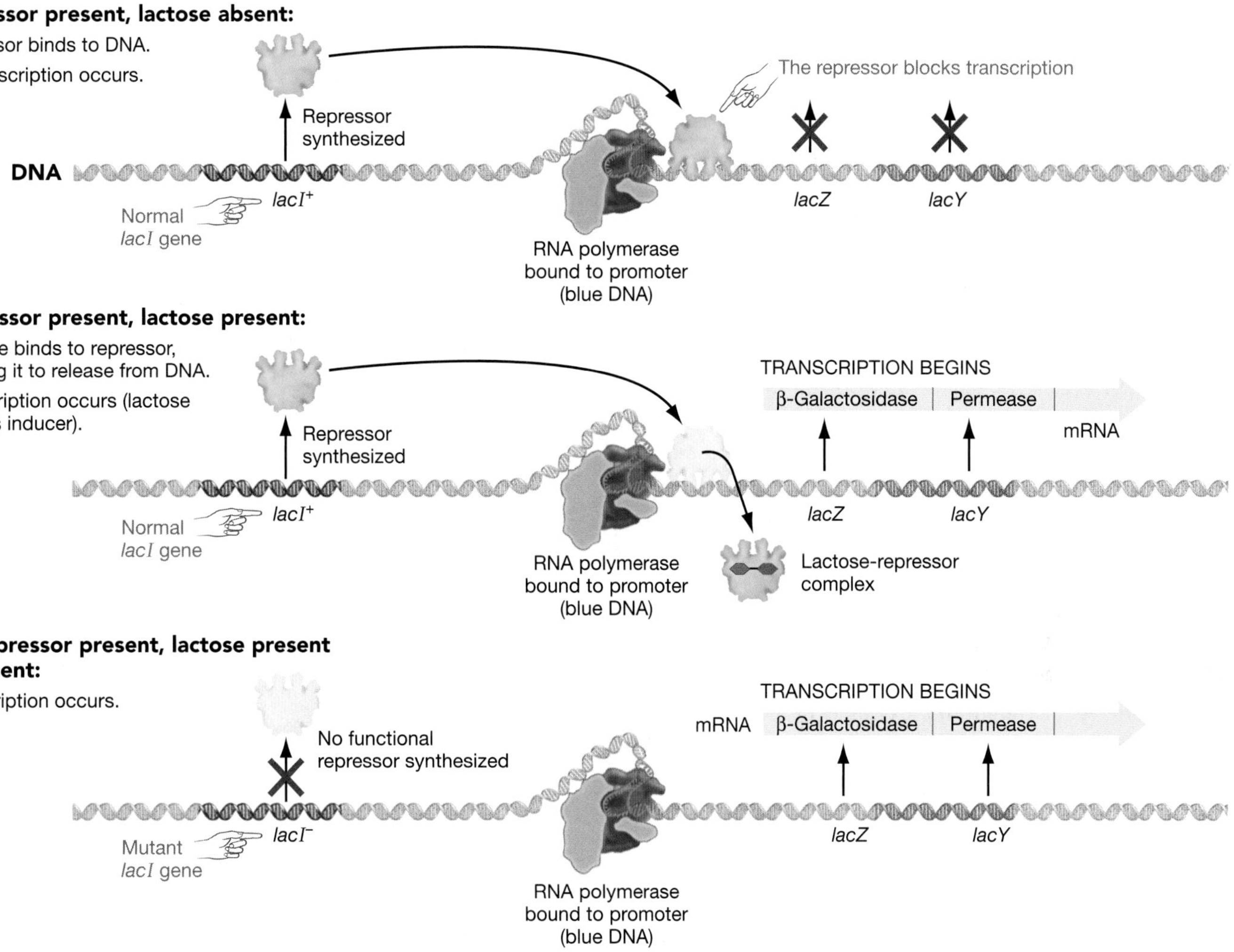

(a) Repressor present, lactose absent:

Repressor binds to DNA.

No transcription occurs.

The repressor blocks transcription

Repressor synthesized

DNA

Normal *lacI* gene

lacI^{+}

RNA polymerase bound to promoter (blue DNA)

lacZ

lacY

(b) Repressor present, lactose present:

Lactose binds to repressor, causing it to release from DNA.

Transcription occurs (lactose acts as inducer).

TRANSCRIPTION BEGINS

β-Galactosidase | Permease

mRNA

Repressor synthesized

Normal *lacI* gene

lacI^{+}

lacZ

lacY

RNA polymerase bound to promoter (blue DNA)

Lactose-repressor complex

(c) No repressor present, lactose present or absent:

Transcription occurs.

TRANSCRIPTION BEGINS

mRNA

β-Galactosidase | Permease

No functional repressor synthesized

Mutant *lacI* gene

lacI^{-}

lacZ

lacY

RNA polymerase bound to promoter (blue DNA)

FIGURE 17.7 The Hypothesis of Negative Control. The negative-control hypothesis maintains that **(a)** transcription of genes involved in lactose use is normally blocked by a repressor molecule that binds to DNA on or near the promoter for *lacZ* and *lacY*, **(b)** lactose induces transcription of *lacZ* and *lacY* by interacting with the repressor, and **(c)** when a functional repressor is absent, transcription proceeds.

2. The repressor is a protein encoded by *lacI* that binds to DNA and prevents transcription of *lacZ*, *lacY*, and *lacA*. Jacob and Monod proposed that *lacI* is expressed constitutively and that the repressor binds to a section of DNA in the *lac* operon called the **operator**.

3. The inducer (lactose) interacts directly with the repressor by binding to it. As a result, the repressor changes shape in a way that causes it to drop off the DNA strand. Recall from Chapter 3 that this form of control over protein function is called **allosteric regulation**. When allosteric regulation occurs, a small molecule binds directly to a protein and causes it to change its shape and activity. When the inducer binds to the repressor, negative control ends and transcription can proceed (**Figure 17.8**).

After their model of negative control was published, Jacob and Monod confirmed the existence of the operator by finding *E. coli* mutants that had normal forms of the repressor but that still expressed *lacZ* constitutively. In these new mutants, the repressor protein was unable to function because the nucleotide sequence of the operator was altered. In 1967 Walter Gilbert

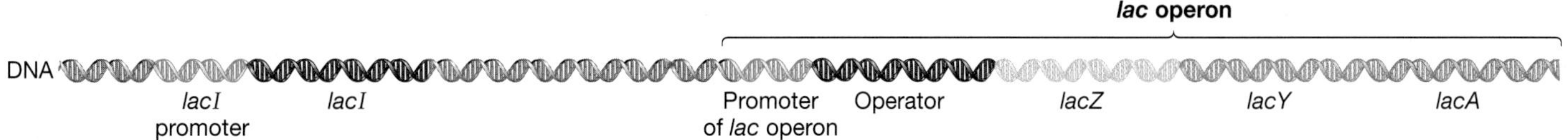

FIGURE 17.8 Components of Negative Control in the *lac* Operon.

EXERCISE Using small, colored bits of candy or paper, add RNA polymerase to the figure. Then add the repressor protein, and then add lactose. At each step, explain what happens after the molecule is added.

and Benno Müller-Hill were able to tag copies of the repressor protein with a radioactive atom. Their experiments showed that the repressor physically binds to the DNA sequences of the operator. This result confirmed that the operator is not a protein or an RNA product, but instead is part of the DNA sequence of the *lac* operon. **Box 17.1** explains how researchers eventually determined the operator's exact DNA sequence.

Why Has the *lac* Operon Model Been So Important?

Regulation of the *lac* operon provided an immensely important model system in genetics. Follow-up work has shown that numerous bacterial genes and operons are under negative control via repressor proteins. Even more important, the *lac* operon model introduced the idea that gene expression is regulated by physical contact between specific regions in regulatory proteins and specific regulatory sites in DNA. Publication of the *lac* operon model was a watershed event in the history of biological science. Its impact on thinking about gene regulation is hard to overestimate.

Besides confirming the existence of negative control, regulatory proteins, and regulatory sites in DNA, work on the *lac* operon offered an important example of post-translational control over gene expression. The key observation here was that the repressor protein is always present—because it is transcribed and translated constitutively at low levels. When a rapid change in *lac* operon activity is required, it does not occur via changes in the transcription or translation of the repressor. Instead, the activity of existing repressor proteins is altered. This turns out to be common. In virtually all cases, the activity of key regulatory proteins is controlled by post-translational modifications.

A New Twist on Negative Control: Comparing the *trp* and *lac* Operons

The success of the *lac* operon model inspired work on how a variety of other genes and operons are regulated in *E. coli*. Work on the *trp* operon (pronounced "trip") turned out to be particularly informative, because it extended and deepened the concept of negative control.

The *trp* operon, like the *lac* operon, produces a **polycistronic mRNA**—a message that contains more than one protein-encoding segment. In most cases, polycistronic mRNAs code for products devoted to a common end. The *lacZ* and *lacY* products are required for lactose metabolism, and the five genes in the *trp* operon in *E. coli* code for enzymes required for different steps in the synthesis of the amino acid tryptophan. Biologists interpret the polycistronic structure of many bacterial genes as an adaptation that increases the efficiency of gene expression. By binding to one promoter, regulatory proteins can control the expression of several genes that are required for the same process. Polycistronic mRNAs are common in bacteria but rare or nonexistent in most eukaryotes.

Like the *lac* operon, the *trp* operon is under negative control. There is a twist, however. Instead of coding for enzymes that perform catabolism—meaning that they break down a compound (see Chapter 9)—the *trp* operon codes for enzymes that perform anabolism. These enzymes synthesize a product. In this case, natural selection should favor cells where gene expression starts when the molecule is absent and is needed by the cell, and where transcription is reduced when the molecule is present. This is the opposite of what happens in the *lac* operon. How does negative control work in a case like this?

Early work on the *trp* operon established that it has an operator sequence that overlaps the promoter and a repressor protein that binds to the operator to prevent transcription—as in the *lac* operon. Experiments also showed that tryptophan binds to the repressor to regulate transcription, just as lactose binds to the *lac* operon's repressor to regulate transcription. But as **Figure 17.9a** shows, the *trp* repressor binds operator DNA and exerts negative control only when it is complexed with tryptophan. When tryptophan is absent, the repressor does not bind to the operator, and the operon is expressed constitutively (**Figure 17.9b**). This is exactly the opposite of how the presence of lactose affects the *lac* repressor (**Figure 17.10**). And it makes perfect sense for the *trp* operon: Bacterial cells will be more efficient, and have higher fitness, if the operon is transcribed when tryptophan levels are low but shut down when tryptophan is abundant.

After considering how the *lac* and *trp* operons are regulated in *E. coli*, it's easy to have new respect for "simple" forms of life. Bacteria may be small and unicellular, but they perform fantastically complex, finely controlled chemistry. Even the simplest of cells are marvelous machines.

(a) When tryptophan is present, transcription is blocked.

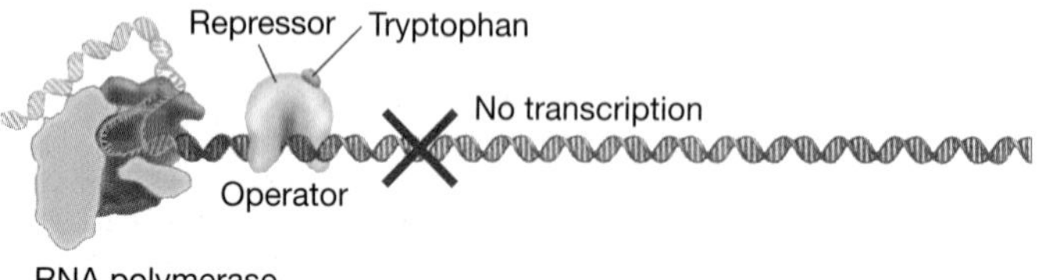

(b) When tryptophan is absent, transcription occurs.

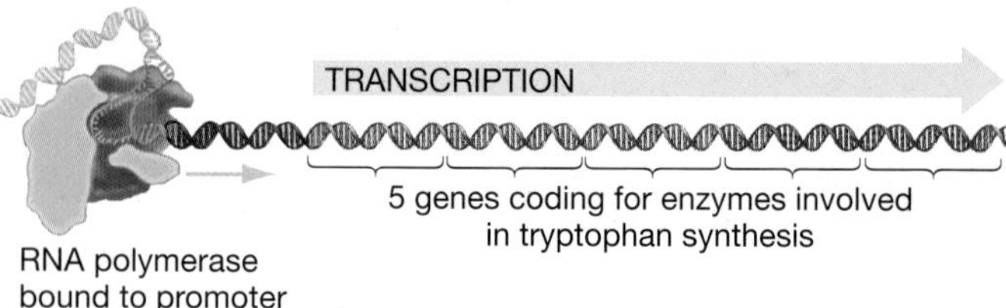

FIGURE 17.9 The *trp* Operon Is under Negative Control. (a) When tryptophan is present, it binds to a repressor protein. The tryptophan-repressor complex binds to the operator and shuts down transcription, which ends the synthesis of tryptophan. **(b)** The *trp* operon contains five coding genes.

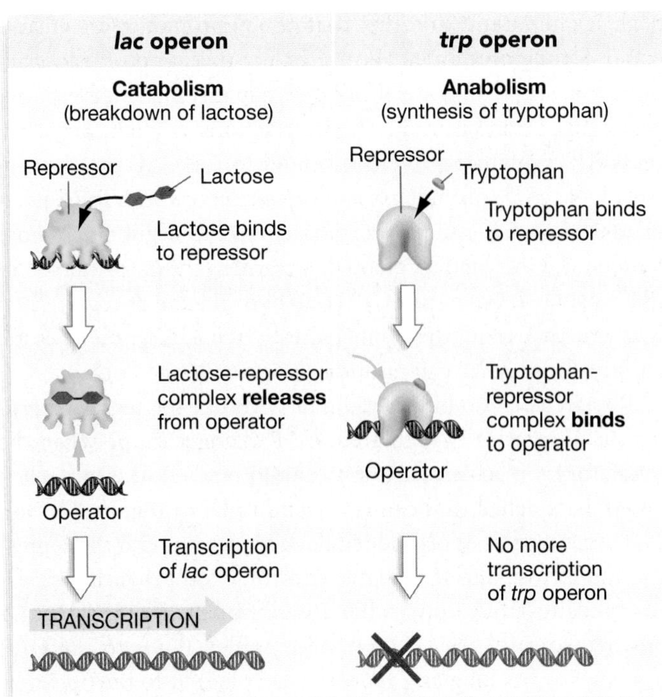

FIGURE 17.10 Comparing and Contrasting Negative Control of the *lac* and *trp* Operons. In contrast to the *lac* operon, negative control of the *trp* operon is released when tryptophan is absent.

Check Your Understanding

If you understand that...

- Negative control occurs when something must be taken away for transcription to occur.
- The *lac* operon repressor exerts negative control over three protein-coding genes by binding to the operator site in DNA, near the promoter.
- For transcription to occur in the *lac* operon, an inducer molecule (a derivative of lactose) must bind to the repressor, causing it to release from the operator.

You should be able to...

1) Explain why it is logical to observe that a molecule derived from lactose induces transcription of the *lac* operon.
2) Diagram the *lac* operon, showing the relative positions of the operator, the promoter, and the three protein-coding genes.
3) Show what is happening at the operon in the presence of lactose and in the absence of lactose.

BOX 17.1 DNA Footprinting

DNA footprinting is used to characterize DNA sequences that are bound by regulatory proteins **(Figure 17.11).** Researchers begin a footprinting study by obtaining DNA from a region of interest—the *lac* operon of *E. coli*, for example. Then they allow copies of the regulatory protein to bind to the DNA sequence being analyzed, and add a nuclease—an enzyme that cleaves DNA in random locations—to cut the DNA. Segments of DNA that are bound by the regulatory protein are protected from the enzyme and don't get cut. Segments of DNA that are not bound by the regulatory protein are not protected, meaning that they do get cut.

By analyzing the protected, uncut portion of the DNA segment, researchers can determine the exact nucleotide sequence of the regulatory site.

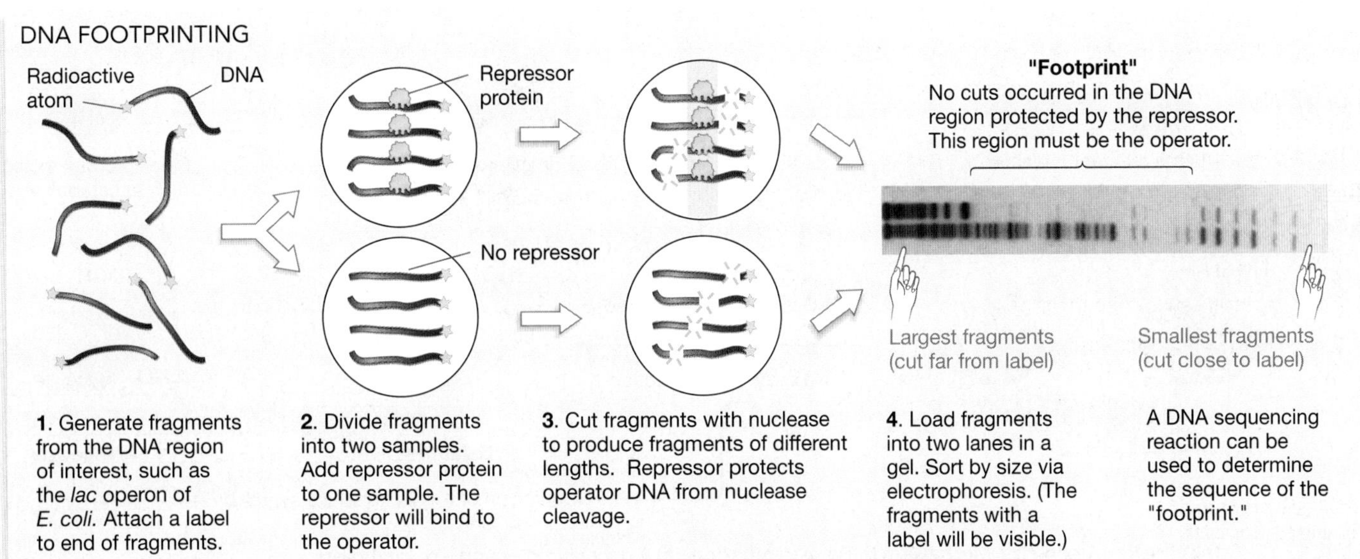

FIGURE 17.11 DNA Footprinting Allows Researchers to Identify Sites Where Proteins Bind.

17.4 Mechanisms of Positive Control: Catabolite Repression

The model of negative control over the *lac* operon, summarized in Figure 17.7, is elegant and successful in explaining experimental results. But it is not complete. After studying the model, you may think of an important question: Where does glucose fit in?

Transcription of the *lac* operon is drastically reduced when glucose is present in the environment—even when lactose is available to induce β-galactosidase expression. Given that glucose is *E. coli*'s preferred carbon source, this observation is logical. Glucose is produced when β-galactosidase cleaves lactose. When glucose is already present, it is not necessary for the cell to cleave lactose and produce still more glucose.

Biologists use the term **catabolite repression** to describe a situation like this. The term is appropriate because the hydrolysis of lactose into its glucose and galactose subunits is an example of catabolism (**Figure 17.12a**). In many cases, operons that encode catabolic enzymes are inhibited when the end product of the reaction, the catabolite, is abundant. Catabolite repression is a form of end-product inhibition. In the case of the *lac* operon, glucose is the catabolite. When glucose is abundant in the cell, transcription of the *lac* operon is decreased by catabolite repression (**Figure 17.12b**).

How does glucose prevent expression of the *lac* operon? An answer to this question began to emerge when researchers discovered a second major control element in the *lac* operon—an example of positive control. This regulatory switch consists of a DNA sequence known as the **CAP binding site**, which is located just upstream of the *lac* promoter, and a regulatory protein called **catabolite activator protein (CAP)**, which binds to this DNA sequence.

The CAP protein binds to the CAP binding site and triggers transcription of the *lac* operon. To understand how this happens, it's important to realize that not all promoters are created equal. Strong promoters allow efficient initiation of transcription by RNA polymerase; weak ones support much less efficient initiation of transcription. The *lac* promoter is weak. But when the CAP regulatory protein is bound to the CAP site just upstream of the *lac* promoter, the protein interacts with RNA polymerase in a way that allows transcription to begin much more frequently. CAP binding greatly strengthens the *lac* promoter. In this way, CAP exerts positive control of the *lac* operon. When CAP is active, transcription increases. If the inducer removes the parking brake, CAP can push the gas pedal to the floor.

Researchers also discovered that CAP, like the repressor protein, is allosterically regulated. CAP changes shape when the regulatory molecule **cyclic AMP (cAMP)** binds to it. Only when cAMP is attached to it can CAP bind to DNA (**Figure 17.13a**). The same type of allosteric regulation goes on with the repressor and lactose during negative control, except that the repressor binds to DNA only in the *absence* of the inducer. During positive control, the CAP-cAMP complex binds to the CAP binding site. As long as a repressor isn't bound to the operator, the complex increases the efficiency of transcription. If no cAMP is bound to CAP, then CAP has a conformation that does not allow binding to the CAP site (**Figure 17.13b**). cAMP is the green light that tells CAP to floor it.

How Does Glucose Influence Formation of the CAP-cAMP Complex?

Where does the elusive molecule glucose fit into this scheme? Glucose's role in positive control of the *lac* operon is indirect. More specifically, it is mediated by the influence that glucose has on the concentration of cAMP. Glucose levels outside the cell and cAMP levels inside the cell are inversely related. When extracellular glucose concentrations are high, intracellular cAMP concentrations are low; when extracellular glucose con-

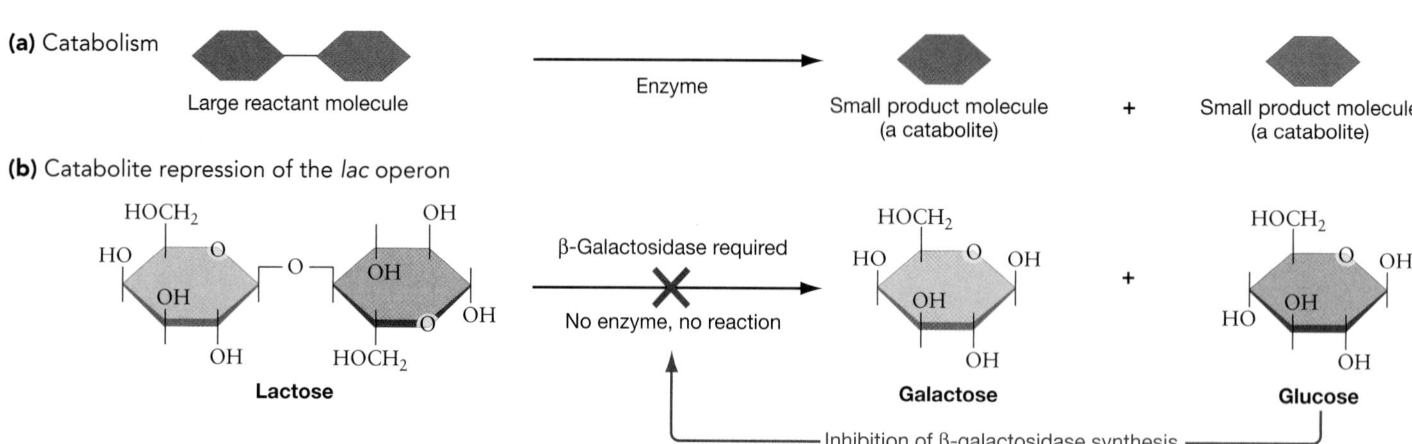

FIGURE 17.12 Catabolite Repression Is a Mechanism of Gene Regulation. (a) A generalized example of catabolism. **(b)** Catabolite repression occurs when one of the small product molecules represses the production of the enzyme responsible for the reaction. In the case of lactose metabolism, the production of β-galactosidase is suppressed when glucose is present.

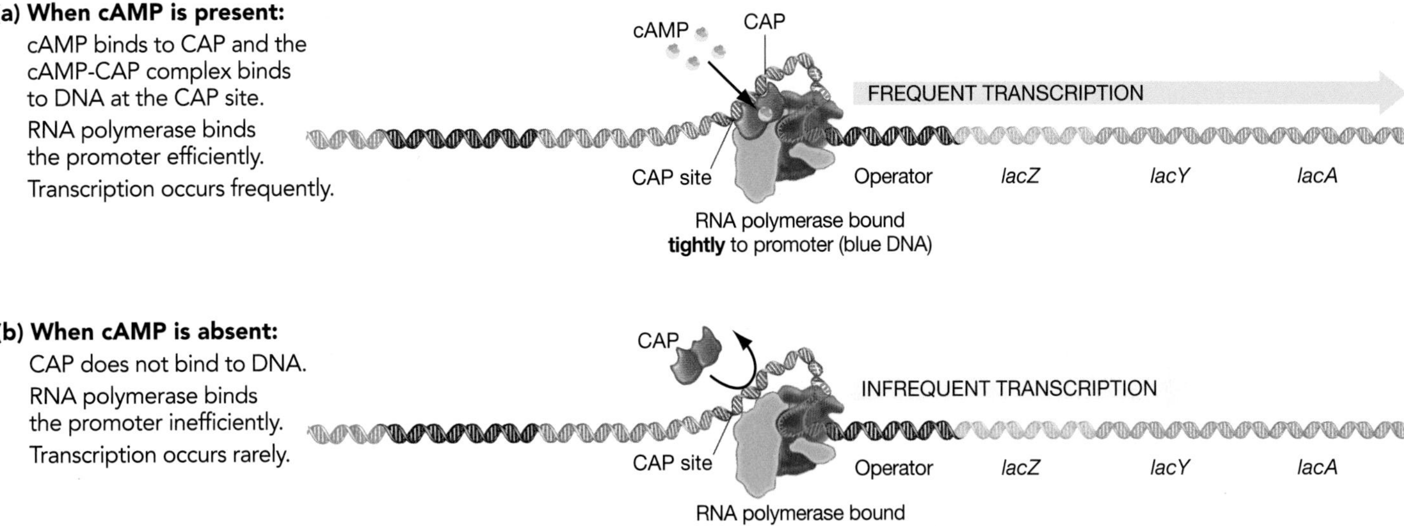

FIGURE 17.13 Positive Control of the *lac* Operon. **(a)** When glucose levels in an *E. coli* cell are low, cAMP is produced. cAMP then interacts with CAP to increase transcription of the *lac* operon. **(b)** When glucose is abundant, cAMP is rare in the cell and positive control does not occur.

centrations are low, intracellular cAMP concentrations are high. This seesaw is driven by the enzyme **adenylyl cyclase**, which produces cAMP from ATP (**Figure 17.14a**). Adenylyl cyclase's activity is inhibited by extracellular glucose. To see the consequences of this fact, imagine a situation in which glucose is abundant outside the cell (**Figure 17.14b**). In this state, adenylyl cyclase activity is low. Therefore, cAMP levels inside the cell are low. CAP is not in a CAP-cAMP complex, so it does not have the conformation that allows it to bind to the CAP site and stimulate *lac* operon transcription. Conversely, when the extracellular concentration of glucose is low, there is an increase in the intracellular level of cAMP. In this case, the CAP-cAMP complex forms, binds to the CAP site, and allows RNA polymerase to initiate transcription efficiently. As long as

(a) Glucose inhibits the activity of the enzyme adenylyl cyclase, which catalyzes production of cAMP from ATP.

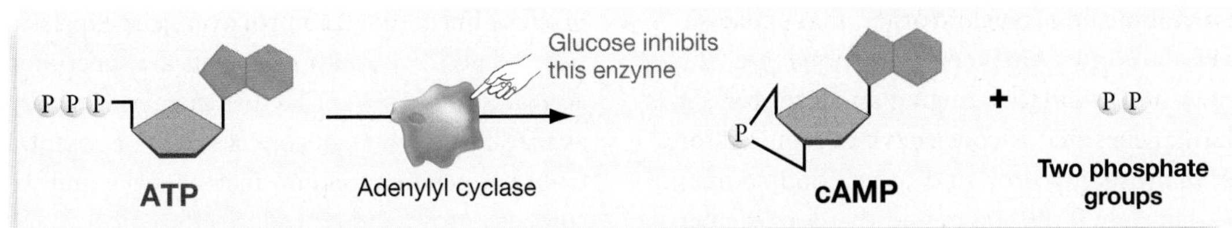

(b) The amount of cAMP and the rate of transcription of the *lac* operon are inversely related to the concentration of glucose.

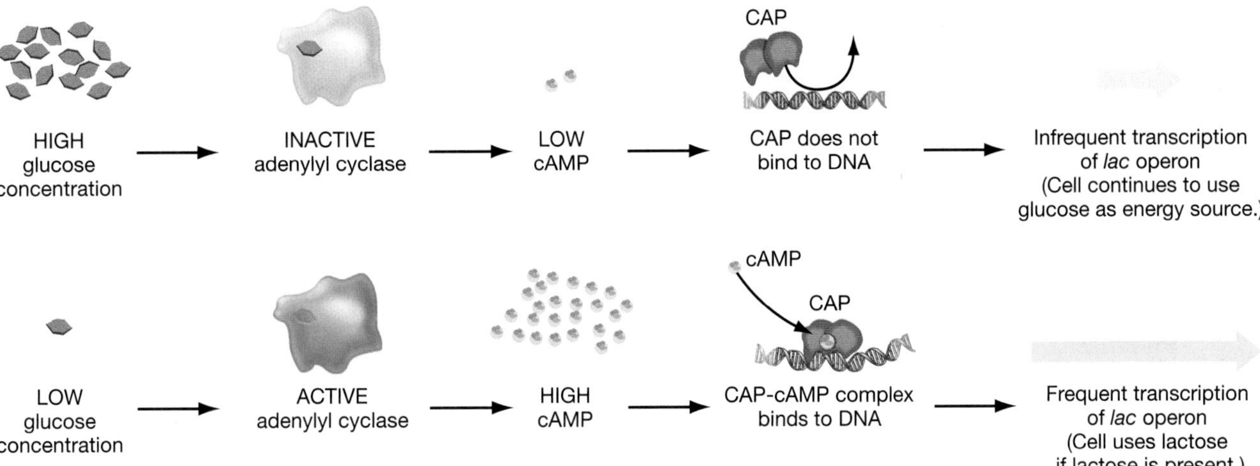

FIGURE 17.14 Cyclic AMP (cAMP) Is Synthesized When Glucose Levels Are Low.

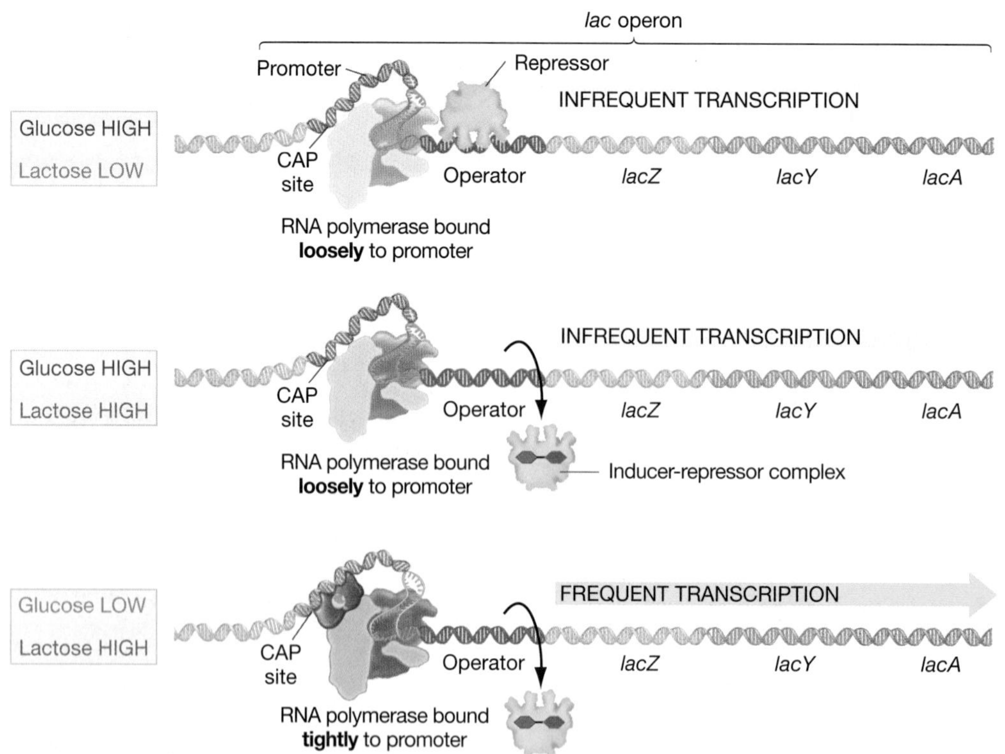

FIGURE 17.15 An Overview of Lactose Use.

the repressor is not bound to the operator, transcription of the *lac* operon is initiated frequently and lactose can be employed as an alternative energy source.

It is important to note that the CAP-cAMP system influences many genes in addition to the *lac* operon. CAP sites are found adjacent to the promoters for several operons that are required for the metabolism of sugars other than glucose. When glucose levels fall and cAMP concentrations rise, the effect on gene regulation is similar to ringing an alarm bell. In response to the alarm, genes that encode enzymes required for the use of lactose, maltose, glycerol, and other food sources may be turned on—but only if the compound that a particular operon acts on is available to the cell. Conversely, cAMP levels are low when glucose supplies are adequate. In this case, the genes for these enzymes are expressed only at low levels even when the appropriate metabolite is present.

Figure 17.15 summarizes how positive control and negative control combine to regulate the *lac* operon. **Figure 17.16** offers a similar summary in the form of a flowchart. The general message of these figures is that interactions among regulatory elements produce finely tuned control over gene expression. Because positive and negative control elements are superimposed, *E. coli* fully activates the genes for lactose metabolism only when lactose is available and when glucose is scarce or absent. In this way, control over gene expression increases the ability of these cells to compete, grow, and reproduce efficiently.

The *lac* Operon

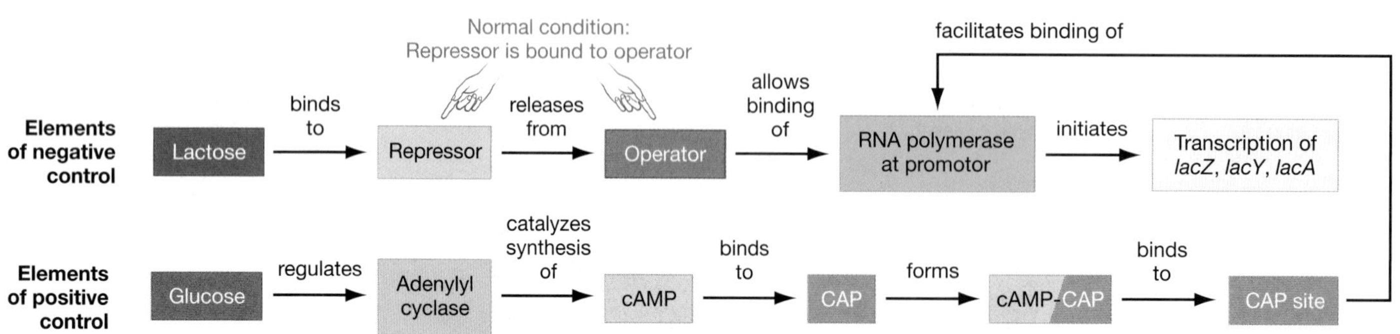

FIGURE 17.16 Negative and Positive Control in the *lac* Operon Interact: A Flowchart.

● **EXERCISE** Redraw this flowchart as a concept map (**BioSkills 6**), and add notes indicating what happens when concentrations of glucose and lactose are (1) glucose high/lactose low, (2) glucose low/lactose high, (3) glucose high/lactose high, and (4) glucose low/lactose low.

Check Your Understanding

If you understand that... 

- Positive control occurs when something must be added for transcription to occur.
- CAP exerts positive control over the *lac* operon by binding to the CAP site and increasing the transcription rate.
- CAP can bind only when it is complexed with cAMP—a signaling molecule indicating that glucose levels are low.

You should be able to...

1) Diagram what the operon looks like when cAMP levels are low and when cAMP levels are high.
2) Explain how positive control and negative control work together to control the *lac* operon in the presence or absence of glucose and lactose.

17.5 The Operator and the Repressor—An Introduction to DNA-Binding Proteins

The insights that biologists gained from studying the *lac* operon proved to be relevant to many other genes and species. For example, transcription of bacterial and eukaryotic genes is either constitutive or controlled. When regulation occurs, it is negative, positive, or based on a combination of negative and positive factors. Negative control is based on repressor proteins, such as the *lacI* product, whereas positive control depends on transcription activator proteins, such as CAP. In both bacteria and eukaryotes, as in the *lac* operon, it is particularly common for genes to be regulated by a combination of several control systems. Catabolite repression is also extremely common.

Currently, biologists who are interested in gene regulation are trying to understand how DNA-binding proteins such as CAP and the repressor can control transcription. Precisely how do these proteins interact with DNA? How do small regulatory molecules such as cAMP and lactose induce changes in the conformation of these large DNA-binding proteins? Once again, investigators are using the *lac* operon as a model system that offers insights applicable to all cells.

Before taking a detailed look at the repressor and operator, let's step back and think about what is required for a DNA-binding protein to control transcription. First, the protein must bind to a specific DNA sequence associated with a specific gene or operon. Often, an investigator's first task is to identify what that sequence is. Second, the protein's structure must allow it to bind to DNA, interact with a regulatory molecule such as cAMP or lactose, and affect RNA polymerase. What structural features make each of these functions possible? As it turns out, the *lac* operon's operator DNA sequence and repressor protein structure have features that are common to many regulatory elements in bacteria and eukaryotes.

How Was the Operator Found?

To understand how the *lac* repressor controls transcription, a logical first step is to identify the DNA sequences it targets: the operator. By mapping mutants with defects in the operator, researchers were able to confirm that the target site is just downstream from the promoter. But what is the actual DNA sequence?

Footprinting experiments with the *lac* repressor yielded two new discoveries. First, the *lac* operon actually contains three sites where the repressor protein may bind. As **Figure 17.17a** shows, these sites are called O_1, O_2, and O_3. When the repressor protein is bound to DNA, the protein simultaneously binds O_1 and one of the two other operators. This finding required a modification of the original Jacob–Monod model, which proposed the existence of a single operator. Second, all three operators have a similar DNA sequence that has an unusual characteristic: It contains an axis of symmetry. To understand what an axis of symmetry in DNA is, look at the O_1 sequence shown in **Figure 17.17b**. Note that if the sequence were rotated 180°, it would be little changed. This type of symmetry is known as **twofold rotational symmetry** or **dyad symmetry**. Follow-up work showed that the nucleotide sequences recognized by many DNA-binding proteins have twofold rotational symmetry.

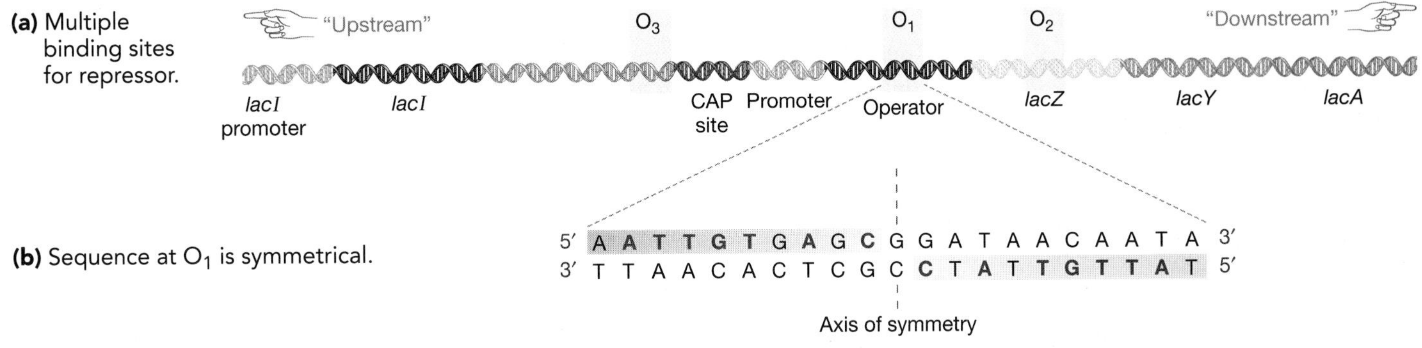

FIGURE 17.17 Operator Sequences. (a) The *lac* operon contains three operator sequences, called O_1, O_2, and O_3, where the repressor can bind. A repressor simultaneously binds O_1 and either O_2 or O_3. **(b)** Many sequences targeted by DNA-binding proteins are symmetrical. In this example, all but three bases are identical on the two sides of the plane of symmetry.

Now that we've examined the operator, we need to consider the repressor. How does this DNA-binding protein recognize specific sequences along the double helix? In particular, why is twofold rotational symmetry important?

How Does the Repressor Bind to the Operator?

Like other proteins, the *lac* operon's repressor resembles a hand tool in one important respect: Both a hand tool and a repressor protein have distinct regions or parts. Consider a screwdriver, which consists of a handle, a shaft, and a blade. Each of these parts has a distinctive three-dimensional structure and function. Proteins also have regions or parts, called **domains**, that have a distinctive three-dimensional structure and function. Just as screwdrivers of various shapes and sizes all have a handle, shaft, and blade, it is common to observe the same domain in proteins of various shapes and sizes. A domain that is observed in many different proteins is known as a **motif**. Each individual screwdriver has a handle domain, a shaft domain, and a blade domain. As a group, screwdrivers are characterized by a handle motif, a shaft motif, and a blade motif. The handle and shaft motifs are also found in hammers and socket wrenches, but the blade motif is unique to screwdrivers. Hundreds of different motifs have been characterized in proteins.

When investigators analyzed repressor proteins from different bacteria, they discovered that many of the bacteria contained the domain illustrated in **Figure 17.18a**. This structure is a **helix-turn-helix motif**. As its name indicates, it consists of two α-helices connected by a short stretch of amino acids that forms a turn. If you hold your thumb and index finger apart, they would resemble a helix-turn-helix motif. Physical models constructed during the 1980s suggested that the helix-turn-helix region is the section of the repressor that binds to DNA. This hypothesis was later supported by structural studies based on X-ray crystallography (see **BioSkills 8**).

To see how a helix-turn-helix motif works, recall from Chapter 4 that the exterior of the DNA double helix has a minor groove and a major groove. A protein that has a helix-turn-helix domain can bind to DNA because one of the helices interacts with the sugar-phosphate backbone of a DNA strand while the other helix binds to the base pairs in the major groove (**Figure 17.18b**). The "index finger" in the helix-turn-helix motif fits in the major groove; the "thumb" hooks around the sugar-phosphate backbone. Interactions between the amino acids in the α-helices and the bases inside the major groove allow the finger and thumb to "grab" DNA. ● If you understand this concept, you should be able to predict what would happen to the interaction illustrated in Figure 17.18b if either the amino acids in the "index finger" or the bases in the major groove were changed.

Notice that DNA does not have to become single stranded for the helix-turn-helix domain of the repressor to bind to the operator. This observation holds for nearly all proteins that recognize specific sequences of double-stranded DNA.

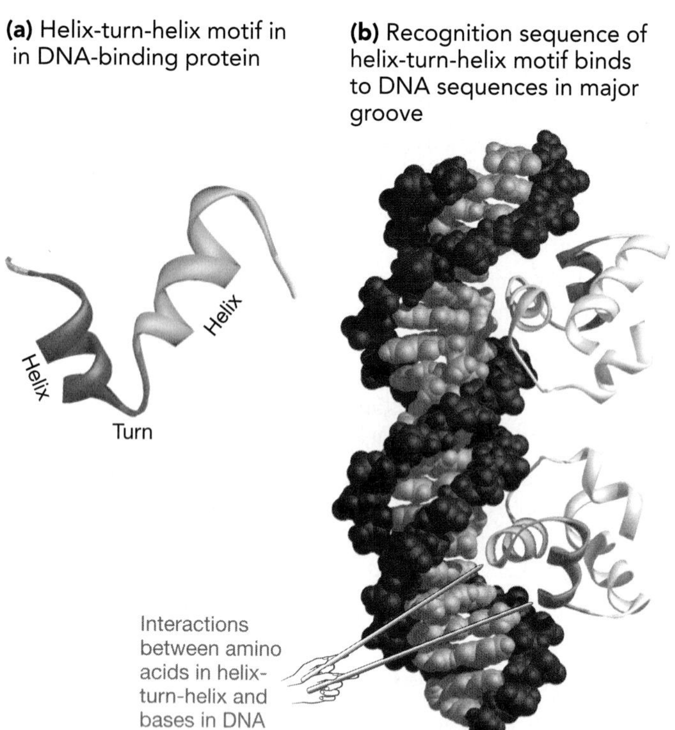

FIGURE 17.18 Helix-Turn-Helix Motifs in DNA-Binding Proteins.

● **QUESTION** The part of a helix-turn-helix motif that interacts directly with DNA is called the recognition sequence. Why?

The section of the helix-turn-helix motif that binds inside the major groove is called the recognition sequence. The amino acid sequence of this section is crucial because specific sequences of amino acids bind to specific sequences of DNA. Each type of regulatory protein with a helix-turn-helix motif has a unique sequence of amino acids in its recognition sequence. As a result, each of these regulatory proteins binds to a unique regulatory sequence in DNA. In many cases, precise control over gene expression is based on precise chemical interactions between the amino acid sequence in a recognition sequence and the DNA sequence in the major groove of a regulatory site.

Biologists used X-ray crystallography to document how the *lac* repressor binds to the DNA of the operator. Consider the structure of the repressor alone, shown in **Figure 17.19a**. Note that the protein consists of four identical polypeptide chains, and that four helix-turn-helix domains project from the main body of the molecule. When researchers analyzed the structure of the repressor while it was bound to the operator sequence, they confirmed that each of these helix-turn-helix motifs binds operator DNA inside the major groove. But when lactose molecules bind to the repressor, the helix-turn-helix motifs move in a way that disrupts their interaction with the operator **Figure 17.19b**. As a result, the repressor releases from the site. The parking brake is off, and transcription can begin.

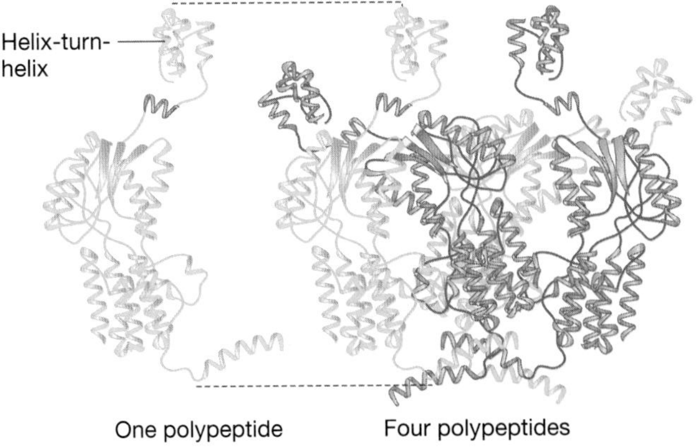

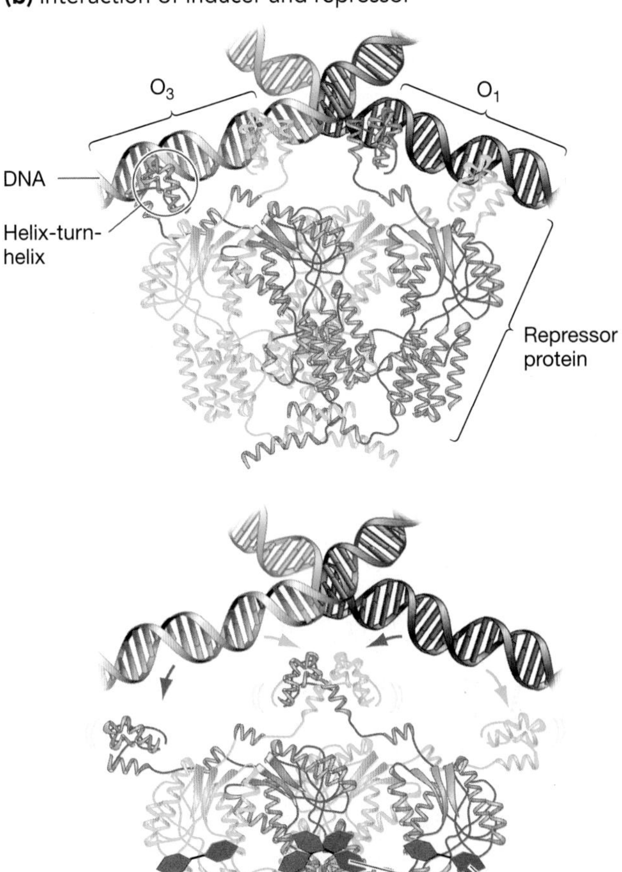

FIGURE 17.19 How Does the Repressor Interact with the Operator and the Inducer? (a) The repressor is a complex of four polypeptides, each a product of the *lacI* locus. **(b)** The repressor binds to the operator sequences only when the inducer is not present (top). When the inducer binds to the repressor, the protein changes shape and falls off the operator DNA (bottom).

What's Next?

Now that biologists understand the physical basis of negative control and induction, what does the future hold? One promising line of research is focused on how the activity of many genes and operons is coordinated. For example, suppose that an *E. coli* cell in your gut is faced with starvation. Does it adjust the activity of hundreds or even thousands of its genes in order to survive? If so, how are changes in gene expression coordinated throughout the genome? Because biologists have a fairly solid understanding of how individual genes and operons are regulated, the current challenge is to explore how multiple genes and operons are controlled in a way that increases the fitness of bacterial cells. Answering these questions may be important to our understanding of how certain bacteria are able to sustain infections that threaten human health, and how some bacterial species are able to detoxify pollutants by using them as food.

Chapter Review

SUMMARY OF KEY CONCEPTS

Gene expression can be controlled at three levels: transcription, translation, or post-translation (protein activation).

Among the three levels of gene regulation, there is a trade-off between the speed of response to changed conditions and the efficient use of resources. In post-translational control, for example, the change in gene expression is extremely fast but demands a considerable investment of resources to make the protein in question.

You should be able to predict whether the same three levels of control and the same trade-offs in speed versus cost occur in eukaryotes.

Changes in gene expression allow bacterial cells to respond to environmental changes.

Transcription in bacteria is constitutive or inducible. Constitutive expression occurs in genes whose products are required at all times, such as genes that encode glycolytic enzymes. Expression of most genes is induced by environmental signals, however, so gene products are produced on demand.

Precise regulation over gene expression is important to bacterial cells. For example, some species switch between different food sources to take advantage of the nutrients currently available in their environment. Efficient use of nutrients is beneficial because cells are usually competing for access to resources. In the intestinal bacterium *E. coli*, for example, changes in gene expression allow cells to use lactose as a food source when glucose is not available.

You should be able to discuss whether changes in the environment inside a multicellular eukaryote—such as a drop in oxygen or glucose availability—would trigger changes in gene expression in its cells.

Transcriptional control can be negative or positive. Negative control occurs when a regulatory protein prevents transcription. Positive control occurs when a regulatory protein increases the transcription rate.

Protein-coding genes involved in lactose metabolism are under both negative and positive control. Negative control occurs because a repressor protein is transcribed at low levels. The repressor binds to a DNA sequence—the operator—near the promoter of the protein-encoding genes. But when lactose is present, it binds to the repressor and induces a conformational change that causes the repressor to fall off the operator. Rapid transcription of the *lac* operon does not begin, however, unless glucose is also lacking. When glucose is scarce, cAMP levels rise in the cell, cAMP binds to the regulatory protein CAP, and the CAP-cAMP complex binds to a control sequence near the *lac* promoter. Binding by CAP-cAMP increases transcription and exerts positive control because it facilitates binding by RNA polymerase.

You should be able to extend this chapter's analogy for negative and positive control by identifying (1) how the operator, repressor, and inducer relate to a parking brake; and (2) how CAP, cAMP, and the CAP-binding site relate to a gas pedal.

MB Web Animation at www.masteringbio.com
The *lac* Operon

Many regulatory proteins bind to specific sites in DNA. Because each type of regulatory protein has a distinct amino acid sequence, each type binds to different DNA sequences.

The regulatory proteins involved in negative control and positive control are often allosterically regulated, meaning that they have the ability to switch between two different conformations in response to binding a small molecule—often a nutrient or product affected by the gene being regulated. The change in conformation affects the protein's ability to bind to DNA.

You should be able to make a drawing that predicts how CAP interacts with the CAP-binding site in the presence or absence of cAMP and explain why a mutation that changed the binding-site sequence might increase or decrease CAP's ability to bind.

QUESTIONS

Test Your Knowledge

1. *E. coli* expresses genes for glycolytic enzymes constitutively. Why?
   a. Their expression is controlled at three levels: transcriptional, translational, and post-translational.
   b. Cells always need fresh supplies of ATP, so glycolysis is occurring constantly.
   c. Transcription is activated only in response to signals indicating that ATP supplies are low.
   d. They do not need to be expressed when fermentation instead of cellular respiration is occurring.

2. Why are the genes involved in lactose metabolism considered to be an operon?
   a. They occupy adjacent locations on the *E. coli* chromosome.
   b. They have a similar function.
   c. They are all required for normal cell function.
   d. They are under the control of the same promoter.

3. In the *lac* operon, which regulatory molecule exerts negative control by inhibiting transcription?
   a. lactose (the inducer)
   b. the repressor protein
   c. the catabolite activator protein (CAP)
   d. cAMP

4. In the *lac* operon, which regulatory molecule is controlled at the post-translational level, by cAMP?
   a. lactose (the inducer)
   b. the repressor protein
   c. the catabolite activator protein (CAP)
   d. cAMP

5. What is catabolite repression?
   a. a mechanism that turns off the synthesis of enzymes responsible for catabolic reactions when the product is present
   b. a mechanism that turns off the synthesis of enzymes responsible for catabolic reactions when the product is absent
   c. repression that occurs because of allosteric changes in a regulatory protein
   d. repression that occurs because of allosteric changes in a DNA sequence

6. What is a helix-turn-helix motif?
   a. a protein domain involved in folding into the active conformation
   b. a protein domain involved in induction
   c. a protein domain involved in DNA binding
   d. a protein domain involved in catalysis

Test Your Knowledge answers: 1. b; 2. d; 3. b; 4. c; 5. a; 6. c

Test Your Understanding

Answers are available at www.masteringbio.com

1. Explain the difference between positive and negative control over transcription. Why is it advantageous for the *lac* operon in *E. coli* to be under both positive control and negative control? What would happen if only negative control occurred? What would happen if only positive control occurred?
2. In *E. coli*, rising levels of cAMP can be a considered a starvation signal. Explain.
3. Explain the role of post-translational control in the *lac* operon.
4. CAP is also known as the cAMP-receptor protein. Why?
5. Explain how the *lac* repressor binds to DNA. Then explain the relationship between the twofold rotational symmetry of the operator and the four-part structure of the repressor protein.
6. The galactose released when β-galactosidase cleaves lactose enters the glycolytic pathway in *E. coli* once a series of enzyme-catalyzed reactions has converted the galactose to glucose-6-phosphate. Why, then, is glucose, not lactose, the preferred sugar in *E. coli*?

Applying Concepts to New Situations

Answers are available at www.masteringbio.com

1. You are interested in using bacteria to metabolize wastes at an old chemical plant and convert them into harmless compounds. You find bacteria that are able to tolerate high levels of the toxic compounds toluene and benzene, and you suspect that it is due to the ability of the bacteria to break these compounds into less-toxic products. If that is true, these toluene- and benzene-resistant strains will be valuable for cleaning up toxic sites. How could you find out whether these bacteria have enzymes that allow them to metabolize toluene?
2. Assuming that the bacteria you examined in Applying Concepts to New Situations Problem 1 do have an enzymatic pathway to break down toluene, would you predict that the genes involved are constitutively expressed, under positive control, or under negative control? Why? What experiments could you conduct to test your hypothesis?
3. The *lacI* gene mutants produce β-galactosidase constitutively because no repressors are present to bind to the operator. Other repressor mutants have been isolated that are called *LacI*^{S} mutants. These repressor proteins continue to bind to the operator, even in the presence of the inducer. How would this mutation affect the function of the *lac* operon? Specifically, how well would *LacI*^{S} mutants do in an environment that has lactose as its sole sugar?
4. X-gal is a colorless, lactose-like molecule that can be split into two fragments by β-galactosidase. One of these product molecules is blue. The following photograph is a close-up of *E. coli* colonies growing in a medium that contains lactose.

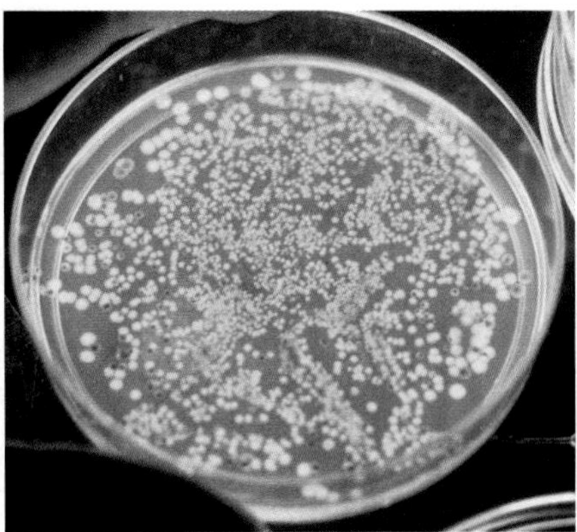

Draw a line to three colonies whose cells have functioning copies of β-galactosidase. Draw a line to three colonies whose cells have mutations in the *lacZ* locus or in one of the genes involved in regulation of the *lacZ*. Suppose you could analyze the sequence of the β-galactosidase gene from each of the mutant colonies. How would these data help you distinguish which cells are structural mutants and which are regulatory mutants?

www.masteringbio.com is also your resource for • Answers to text, table, and figure caption questions and exercises • Answers to *Check Your Understanding* boxes • Online study guides and quizzes • Additional study tools including the *E-Book for Biological Science* 3rd ed., textbook art, animations, and videos.

18 Control of Gene Expression in Eukaryotes

KEY CONCEPTS

- Changes in gene expression allow eukaryotic cells to respond to changes in the environment and cause distinct cell types to develop.
- In eukaryotes, DNA is packaged with proteins into complex structures that must be opened before transcription can occur.
- In eukaryotes, transcription can be initiated only when specific proteins bind to the promoter and to regulatory sequences that may be close to the promoter or far from it.
- Alternative splicing allows a single gene to code for several different products.
- Once an mRNA is in the cytoplasm, gene expression is controlled by molecules that regulate the (1) life span of mRNAs, (2) efficiency of translation, and (3) activation or inactivation of protein products.
- Cancer can develop when mutations disable genes that regulate cell-cycle control genes.

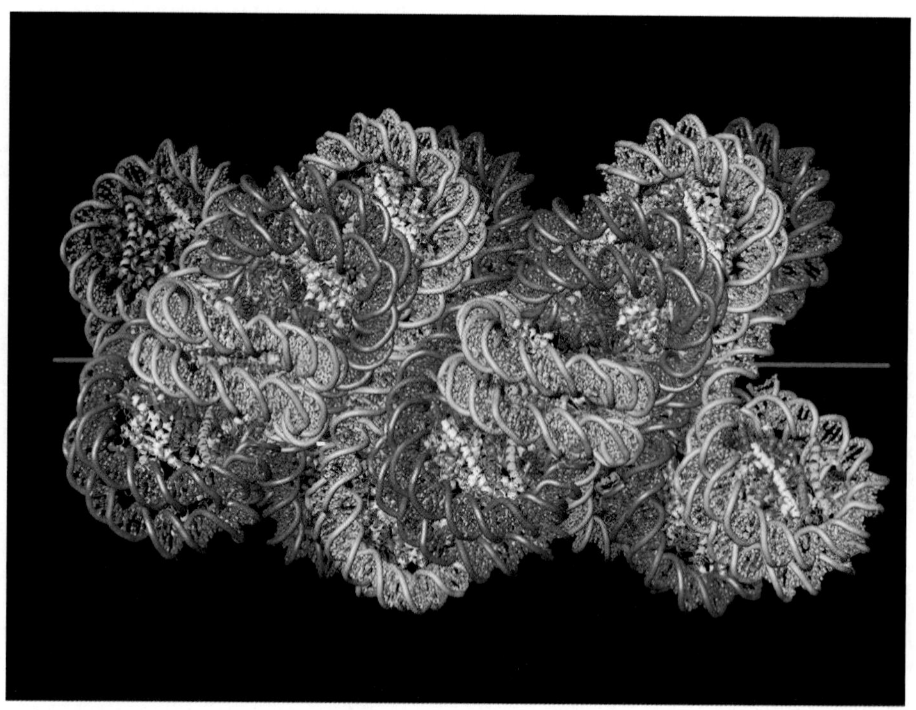

This model shows what eukaryotic DNA looks like in the condensed state. The DNA has to be uncoiled before transcription can take place.

In bacteria, precise regulation of genes is essential for cells to respond to changes in their environment, as Chapter 17 indicated. In *Escherichia coli*, the genes that are required to import and cleave lactose are rapidly expressed only when the cells have to rely on lactose as a source of energy—when glucose is absent and lactose is present.

Unicellular eukaryotes face similar challenges. Consider the yeast *Saccharomyces cerevisiae*, which is used extensively in the production of beer, wine, and bread. In nature, the cells live on the skins of grapes and other fruits. In the wild, there is often wide variation in temperature and humidity. In addition, the sugars that are available to the cells change dramatically in type and concentration as the fruit they are living on ripens, falls, and rots. For yeast cells to grow and reproduce efficiently, gene expression has to be modified in response to these environmental changes. Further, the offspring of a cell that lives on a grape may be blown to a nearby orchard and take up residence on an apple. Different genes are expressed in yeast cells that live on grapes versus those that live on apples.

The cells that make up multicellular eukaryotes face an even greater challenge. Consider that your body consisted of a single cell at one time—a fertilized egg. Now you have trillions of cells, each with a specialized structure and function. You have heart muscle cells, lung cells, nerve cells, skin cells, and so on. Even though these cells look and behave very differently, each

Key Concept Important Information Practice It

of them contains the same genes. Your bone cells are different from your blood cells not because they contain different genes, but because they express different genes. Your bone cells have blood cell genes—they just don't transcribe them.

How is it possible for the cells inside your body to express unique sets of genes? The answer is that your cells respond to changes in their environment, just as bacteria and unicellular eukaryotes do. But instead of responding to the presence of molecules such as glucose or lactose in the external environment, cells in a multicellular eukaryote respond to the presence of signals from other cells—from an internal environment. As a human being or an oak tree develops, cells that are located in different parts of the organism are exposed to different signals. As a result, they express different genes. **Differential gene expression** is responsible for creating different cell types, arranging them into tissues, and coordinating their activity to form the multicellular society we call an individual.

How does all this happen? Unit 4 explores the nature of the signals that trigger the formation of muscle, bone, leaf, and flower cells in multicellular organisms. Here we need to focus on what happens once a eukaryotic cell receives a signal. Let's begin by surveying the various ways that gene expression can be controlled and then plunge into a detailed look at key points in gene regulation. The chapter closes with a look at how defects in the control of gene expression can help trigger cancer.

18.1 Mechanisms of Gene Regulation—An Overview

Like bacteria, eukaryotes can control gene expression at the levels of transcription, translation, and post-translation (protein activation or inactivation). But as **Figure 18.1** shows, two additional levels of control occur in eukaryotes as genetic information flows from DNA to proteins. The first level involves the DNA-protein complex at the top of the figure. In eukaryotes, DNA is wrapped around proteins to create a protein-DNA complex called **chromatin**. Eukaryotic genes have promoters, just as bacterial genes do. Before transcription can begin in eukaryotes, DNA near the promoter must be released from tight interactions with proteins so that RNA polymerase can make contact with the promoter. To capture this point, biologists say that **chromatin remodeling** must occur prior to transcription.

The second level of regulation that is unique to eukaryotes involves **RNA processing**—the steps required to produce a mature, processed mRNA from a **primary RNA transcript**, the preliminary result of transcription. Recall from Chapter 16 that introns have to be spliced out of primary transcripts. In some cases, carefully orchestrated alterations in the splicing pattern occur. When the splicing events in a particular primary RNA transcript change, a different message emerges. The altered message, in turn, leads to the production of a different product.

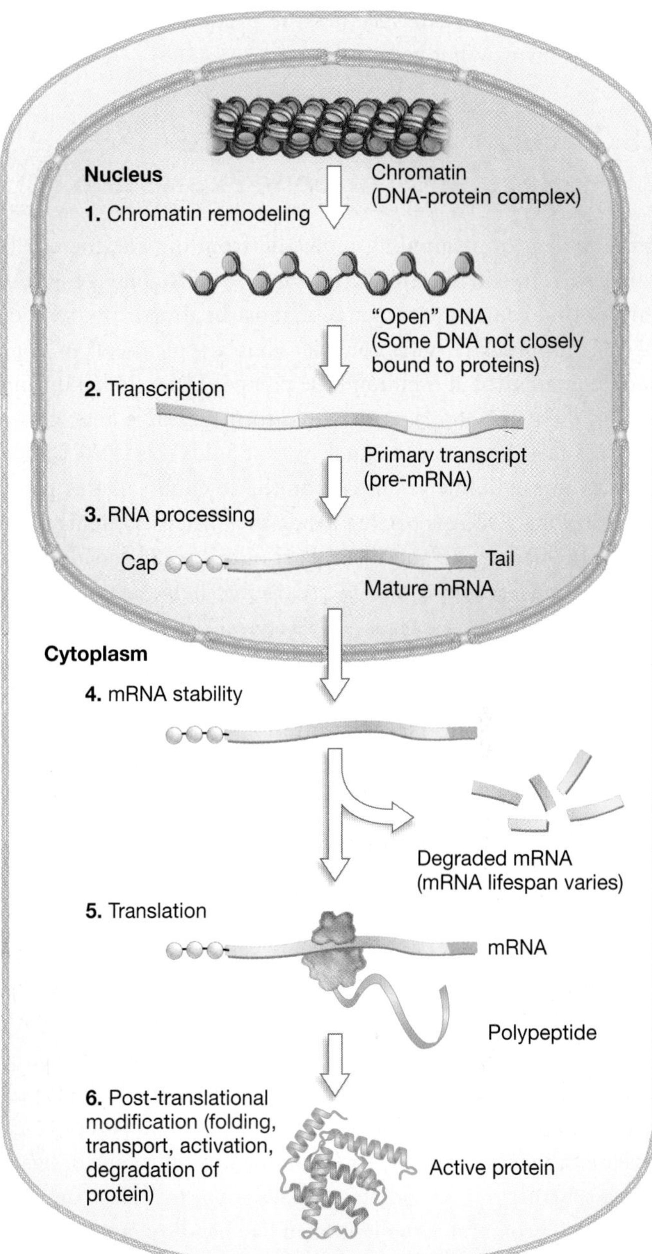

FIGURE 18.1 In Eukaryotes, Gene Expression Can Be Controlled at Many Different Steps. Numerous steps occur during protein synthesis in eukaryotes. Each step provides an opportunity for regulation of gene expression.

EXERCISE Place and label a horizontal line such that mechanisms of transcriptional control fall above the line and mechanisms of post-transcriptional control fall below it.

Each of the six potential control points shown in Figure 18.1 is employed at certain times in a typical eukaryotic cell. This chapter explores all six—chromatin remodeling, transcription initiation, RNA processing, mRNA stability, translation, and post-translational modification of proteins.

To appreciate the breadth and complexity of gene regulation in eukaryotes, let's follow the series of events that occurs as an embryonic cell responds to a developmental signal. Suppose a molecule arrives that specifies the production of a muscle-specific protein. What happens next?

18.2 Eukaryotic DNA and the Regulation of Gene Expression

If the arrival of a signaling molecule from outside the cell is going to result in the transcription of a particular gene, the chromatin around the target gene must be drastically remodeled. To appreciate why, consider that a typical cell in your body contains about 6 billion base pairs of DNA. Lined up end to end, these nucleotide pairs would form a double helix about 2 m (6.5 feet) long. But the nucleus that holds this DNA is only about 5 μm in diameter—less than the thickness of this page. In eukaryotes, DNA is packed inside the nucleus so tightly that RNA polymerase can't access it. Part of this packing is done by supercoiling—meaning the DNA double helix is twisted on itself many times, just as in bacterial chromosomes (see Chapter 7). The supercoiled DNA found in bacteria does not need to be altered extensively before it can be transcribed, however. Eukaryotic DNA is different—it has to undergo a series of significant physical changes before transcription can take place. The reason is that supercoiling is just part of the packing system found in eukaryotes.

How Is Chromatin Structured?

The first data on the physical nature of eukaryotic DNA were published in the early 1900s, when chemical analyses established that eukaryotic DNA is intimately associated with proteins. Later work documented that the most abundant DNA-associated proteins belong to a group called the **histones**. In the 1970s electron micrographs like the one in **Figure 18.2a** revealed that the protein-DNA complex, or chromatin, has a regular structure. In some preparations for electron microscopy, chromatin actually looked like beads on a string. The "beads" came to be called **nucleosomes**. More details emerged in 1984 when researchers determined the three-dimensional structure of eukaryotic DNA by using X-ray crystallography (a technique introduced in **BioSkills 8**). The X-ray crystallographic data indicated that each nucleosome consists of DNA wrapped almost twice around a core of eight histone proteins. As **Figure 18.2b** indicates, a histone called H1 "seals" DNA to each set of eight nucleosomal histones. Between each pair of nucleosomes there is a "linker" stretch of DNA.

The intimate association between DNA and histones occurs in part because DNA is negatively charged and histones are positively charged. DNA has a negative charge because of its phosphate groups; histones are positively charged because they contain many lysine and/or arginine residues (see Chapter 3).

More recent work has shown that there is another layer of complexity in eukaryotic DNA. As **Figure 18.2c** indicates, H1 histones interact with each other and with histones in other nucleosomes to produce a tightly packed structure. Based on its width, this structure is called the 30-nanometer fiber. (Recall that a nanometer is one-billionth of a meter and is abbreviated nm.) Often, 30-nm fibers are packed into still larger structures, though it is still not clear exactly what these higher-order structures look like.

A chromosome, then, is made up of chromatin—or DNA complexed with proteins—that has several layers of organization. DNA is wrapped around nucleosomes; nucleosomes are

(a) Nucleosomes in chromatin

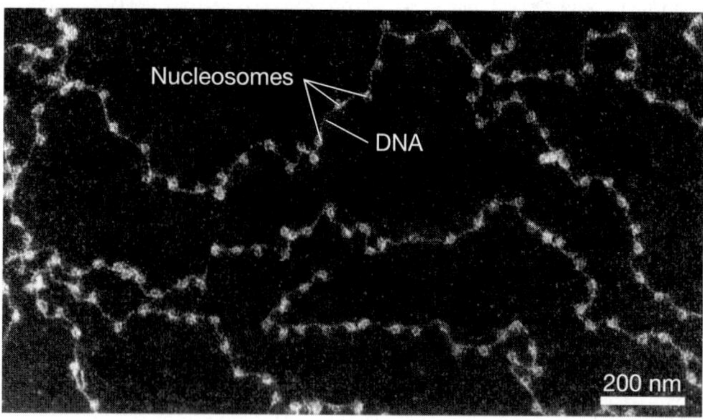

(b) Nucleosome structure

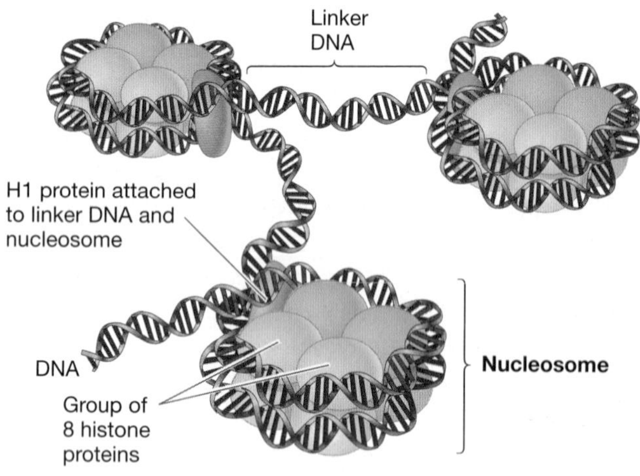

(c) In some cases, nucleosomes may be grouped into 30-nanometer fibers.

FIGURE 18.2 Chromatin Has Several Levels of Structure. (a) Electron micrograph of chromatin. **(b)** The nucleosome is the basic unit of chromosome structure. **(c)** Nucleosomes (shown as green spheres here) may become organized into 30-nm fibers, with H1 proteins forming a core in the center.

packed into 30-nm fibers, and 30-nm fibers are folded into structures that are still being studied.

The elaborate structure of chromatin does more than just package DNA into the nucleus. Chromatin structure has profound implications for the control of gene expression. To drive this point home, consider the 30-nm fiber illustrated in Figure 18.2c. If this tightly packed stretch of DNA contains a promoter, how can RNA polymerase bind to it and initiate transcription?

Evidence that Chromatin Structure Is Altered in Active Genes

Once the nucleosome-based structure of chromatin was established, biologists hypothesized that the close physical interaction between DNA and histones must be altered for RNA polymerase to make contact with DNA. More specifically, biologists hypothesized that a gene could not be transcribed until the chromatin near its promoter was remodeled. The idea was that the chromatin must be relaxed or decondensed for RNA polymerase to bind to the promoter. If so, then chromatin remodeling would represent the first step in the control of eukaryotic gene expression. Two types of studies have provided strong support for this hypothesis.

The first type of evidence comes from studies with the enzyme DNase, which cuts DNA at random locations. DNase cannot cut DNA efficiently if the molecule is tightly complexed with histones. As **Figure 18.3** shows, the enzyme works effectively only if DNA is in the "open" configuration. Harold Weintraub and Mark Groudine used this observation to test the hypothesis that the DNA of actively transcribed genes is in an open configuration. They performed the test by comparing chromatin structure in two genes in blood cells of chickens: the β-globin and ovalbumin genes. β-Globin is a protein that is part of the hemoglobin found in red blood cells; ovalbumin is a major protein of egg white. In blood cells, the β-globin gene is transcribed at high levels. The ovalbumin gene, in contrast, is not transcribed at all. After treating blood cells with DNase and then analyzing the state of the β-globin and ovalbumin genes, the researchers found that DNase cut up the β-globin gene much more readily than the ovalbumin gene. They interpreted this finding as evidence that chromatin in blood cells was in an open configuration at the β-globin gene but closed at the ovalbumin gene. Analogous studies using DNase on different genes and in different cell types yielded similar results.

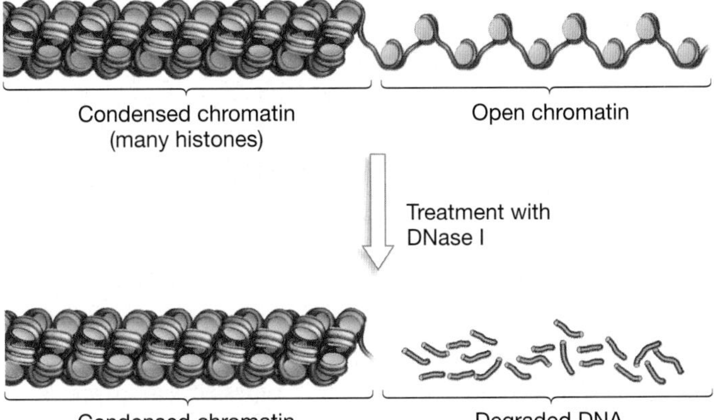

FIGURE 18.3 DNase Assay for Chromatin Structure. DNase is an enzyme that cuts DNA at random locations. It cannot cut intact condensed chromatin.

The second type of evidence in support of the chromatin-remodeling hypothesis comes from studies of mutant brewer's yeast cells that do not produce the usual complement of histones. Researchers found that many yeast genes that are normally never transcribed are instead transcribed at high levels at all times in these mutant cells. To interpret this finding, biologists hypothesized that the lack of histone proteins prevented the assembly of normal chromatin. If the absence of normal histone-DNA interactions promotes transcription, then the presence of normal histone-DNA interactions must prevent it.

Taken together, the data suggest that the normal or default state of eukaryotic genes is to be turned off. This is a new mechanism of negative control—different from the repressor proteins introduced in Chapter 17. When DNA is wrapped into a 30-nm fiber, the parking brake is on. If so, then gene expression depends on chromatin being opened up in the promoter region. Some sort of positive control must be at work.

How Is Chromatin Altered?

Although much remains to be learned about how chromatin is opened up prior to transcription, researchers have made important progress recently. One key finding is that two major types of protein are involved in modifying chromatin structure. One group of proteins creates structures called chromatin-remodeling complexes, which reshape chromatin through a series of reactions that are dependent on ATP. It is not known exactly how these ATP-dependent enzymes work, however. The second major type of chromatin-modifying protein works by adding small molecules such as acetyl (CH_3COOH) or methyl (CH_3) groups to histones. These processes are known as **acetylation** and **methylation**, respectively.

The **histone acetyl transferases (HATs)** are some of the best-studied enzymes that modify chromatin through acetylation or methylation. HATs acetylate the positively charged lysine residues in histones. When a HAT adds an acetyl group to selected histones, the number of positive charges on the histones is reduced. The result is less electrostatic attraction between histones and the negatively charged DNA. The association between nucleosomes and DNA is loosened and the chromatin decondenses.

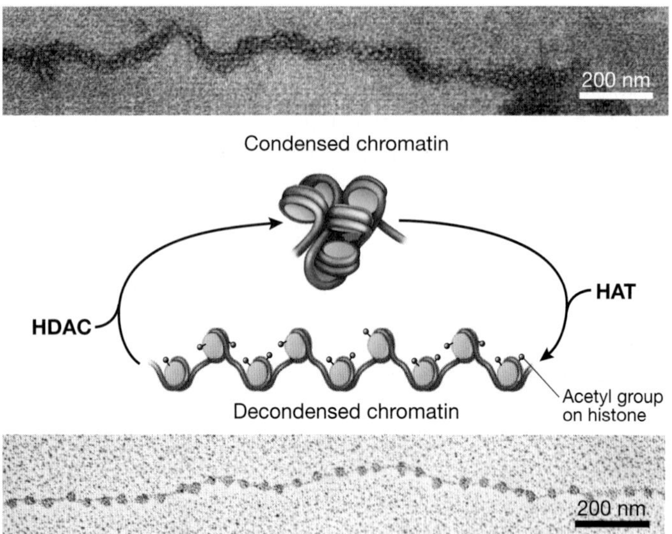

FIGURE 18.4 Chromatin Decondensation and Condensation Are Driven by Histone Acetyl Transferases and Histone Deacetylases. Histone acetyl transferases (HATs) cause chromatin to decondense, and histone deacetylases (HDACs) cause it to condense.

QUESTION Are HATs and HDACs elements in positive control or in negative control?

As **Figure 18.4** shows, chromatin is "recondensed" by a group of enzymes called **histone deacetylases (HDACs)**, which remove the acetyl groups added by HATs. Histone deacetylase activity reverses the effects of acetylation, returning chromatin to its default condensed state. If HATs are an on switch for transcription, HDACs are the off switch.

Chromatin Modifications Can Be Inherited

Work on chromatin remodeling has one fundamental message: The state of the histone proteins that are complexed with DNA is fundamental to determining whether transcription can occur. Acetylation of histones is usually associated with positive control—meaning, activation of genes. Methylation can be correlated with either activation or inactivation, depending on which histones are altered and where the methyl groups occur on the protein. Methylation can set the parking brake or release it, depending on conditions.

The pattern of chemical modifications occurring on histones varies from one cell type to another. For example, suppose you analyzed the same gene in a cell that was destined to give rise to muscle and in a cell that is a precursor to part of the brain. The gene would likely have a completely different pattern of acetylation and methylation in the two cell types. Just as important, some or most these chemical modifications are passed on to daughter cells when the muscle- and brain-associated cells divide. This is a key observation, because it means that daughter cells inherit patterns of gene expression from their parent cells. This is an example of **epigenetic inheritance**, or patterns of inheritance that are not due to differences in gene sequences. Muscle cells are different from brain cells in part because they inherited different types of modified histones—not different types of genes.

Now the question is, what happens once a section of DNA is opened up by chromatin remodeling and exposed to RNA polymerase? In bacteria, the protein sigma binds to RNA polymerase and allows it to contact the promoter. Then transcription begins if any required positive regulators, such as CAP, are in place and negative regulators, such as the repressor, are absent. Does anything similar happen in eukaryotes?

Check Your Understanding

If you understand that...

- Eukaryotic DNA is wrapped tightly around histones, forming nucleosomes, which are then coiled into dense structures called 30-nm fibers.
- Before transcription can begin, the chromatin (DNA-protein complex) must be relaxed by chromatin-remodeling complexes or HATs or methylation so that RNA polymerase can contact the promoter.

You should be able to...

1) Draw and label the structure of a nucleosome and a 30-nm fiber.
2) Compare and contrast the structure of DNA in bacteria versus that in eukaryotes.
3) Draw what happens to chromatin when histones are acetylated.

18.3 Regulatory Sequences and Regulatory Proteins

Chapter 16 introduced the **promoter**—the site in DNA where RNA polymerase binds to initiate transcription. Based on their position and function, eukaryotic promoters are similar to bacterial promoters. Most eukaryotic promoters are located just upstream of the point where RNA polymerase begins transcription, and all have a highly conserved element analogous to the –35 box and –10 box in bacterial promoters. For example, many genes that are transcribed by RNA polymerase II contain a specific base sequence, known as a **TATA box**, where a sigma-like protein binds and allows the enzyme to contact DNA. Recall from Chapter 16 that bacterial promoters may vary in sequence and bind different types of sigma proteins. Eukaryotic genes also have promoters that vary in sequence, but all eukaryotic promoters are bound by the same protein: the **TATA-binding protein (TBP)**.

If eukaryotic genes have promoters that interact with the same promoter-binding protein, how can transcription be controlled? The answer lies in interactions between regulatory sequences other than the promoter and regulatory proteins other than TBP. **Regulatory sequences** are sections of DNA that are involved in controlling the activity of genes, similar to the CAP site and operators introduced in Chapter 17. Eukaryotic regulatory proteins, which change gene activity when they bind to regulatory sites, are analogous to *E. coli*'s CAP and the repressor protein analyzed in Chapter 17.

As it turns out, some regulatory sequences found in eukaryotes are similar to those observed in bacteria. Others are radically different. Let's take a closer look.

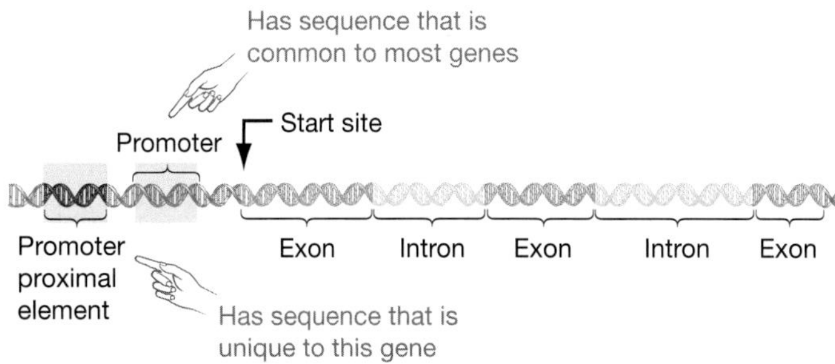

FIGURE 18.5 Promoter-Proximal Elements Regulate the Expression of Some Eukaryotic Genes. All eukaryotic genes have promoters, the sites where RNA polymerase initially is in contact with the gene. Exons and introns are not drawn to scale throughout this chapter. They are typically very large compared with promoters and promoter-proximal (regulatory) elements.

Some Regulatory Sequences Are Near the Promoter

The first regulatory sequences in eukaryotic DNA were discovered in the late 1970s, when Yasuji Oshima and co-workers developed an important model system to study control of gene expression in eukaryotes. Much as François Jacob and Jacques Monod had focused on lactose metabolism in *E. coli*, Oshima's team set out to understand how yeast cells control the metabolism of the sugar galactose. When galactose is absent, yeast cells produce tiny quantities of the enzymes required to metabolize it. But when galactose is present, transcription of the genes encoding these enzymes increases by a factor of 1000.

Oshima and co-workers focused their early work on mutants that were unable to use galactose. The team's first major result was the discovery of mutant cells that failed to produce any of the five enzymes required for galactose metabolism, even if galactose was present. To interpret this observation, they hypothesized that the cells had a knock-out (loss-of-function) mutation that completely disabled a regulatory protein. Like CAP, this hypothesized protein was thought to exert positive control over five genes—the five that coded for the enzymes involved in galactose metabolism. Although not on the same chromosome, the five genes appeared to be regulated together. The team called the hypothesized regulatory protein GAL4 and the gene that encodes it *GAL4*. (In *S. cerevisiae*, the names of genes are written in full caps and italicized. As in other species, gene names are italicized but protein names are not.)

This hypothesis was supported when other researchers isolated the regulatory protein and found that it has a DNA-binding domain, analogous to the helix-turn-helix motif introduced in Chapter 17. The domain binds to a short stretch of DNA located just upstream from the promoter for the five genes that GAL4 regulates. The location and structure of this regulatory sequence are comparable to those of the CAP binding site in the *lac* operon of *E. coli*. Similar regulatory sequences have now been found in a wide array of eukaryotic genes and species. Because such sequences are located close to the promoter and bind regulatory proteins, they are termed **promoter proximal elements**.

As **Figure 18.5** shows, promoter-proximal elements are just upstream from the promoter and the gene's start site; the exons and introns that make up the actual gene are farther downstream. Unlike the promoter itself, promoter-proximal elements have sequences that are unique to specific genes. In this way, they furnish a mechanism for eukaryotic cells to exert precise control over transcription.

The discovery of positive control and promoter-proximal elements provided a satisfying parallel between gene regulation in bacteria and in eukaryotes. This picture changed, however, when researchers discovered a new class of eukaryotic DNA regulatory sequences—sequences unlike anything in bacteria.

Some Regulatory Sequences Are Far from the Promoter

Susumu Tonegawa and colleagues made a discovery that may rank as the most startling in the history of research on gene expression. Tonegawa's group was exploring how human immune system cells regulate the genes involved in the production of antibodies. **Antibodies** are proteins that bind to specific sites on other molecules. In your immune system, antibodies bind to viruses and bacteria and mark them for destruction by other cells. The antibody gene that the researchers were working with is broken into many introns and exons. Introns are DNA sequences spliced out of the primary mRNA transcript; exons are regions of eukaryotic genes that are included in the mature RNA once splicing is complete. The biologists used techniques that will be introduced in Chapter 19 to place copies of an intron in new locations and had found that, when they placed the intron close to a gene, the gene's transcription rate increased. Based on this observation, Tonegawa and co-workers hypothesized that the intron contained some sort of regulatory sequence.

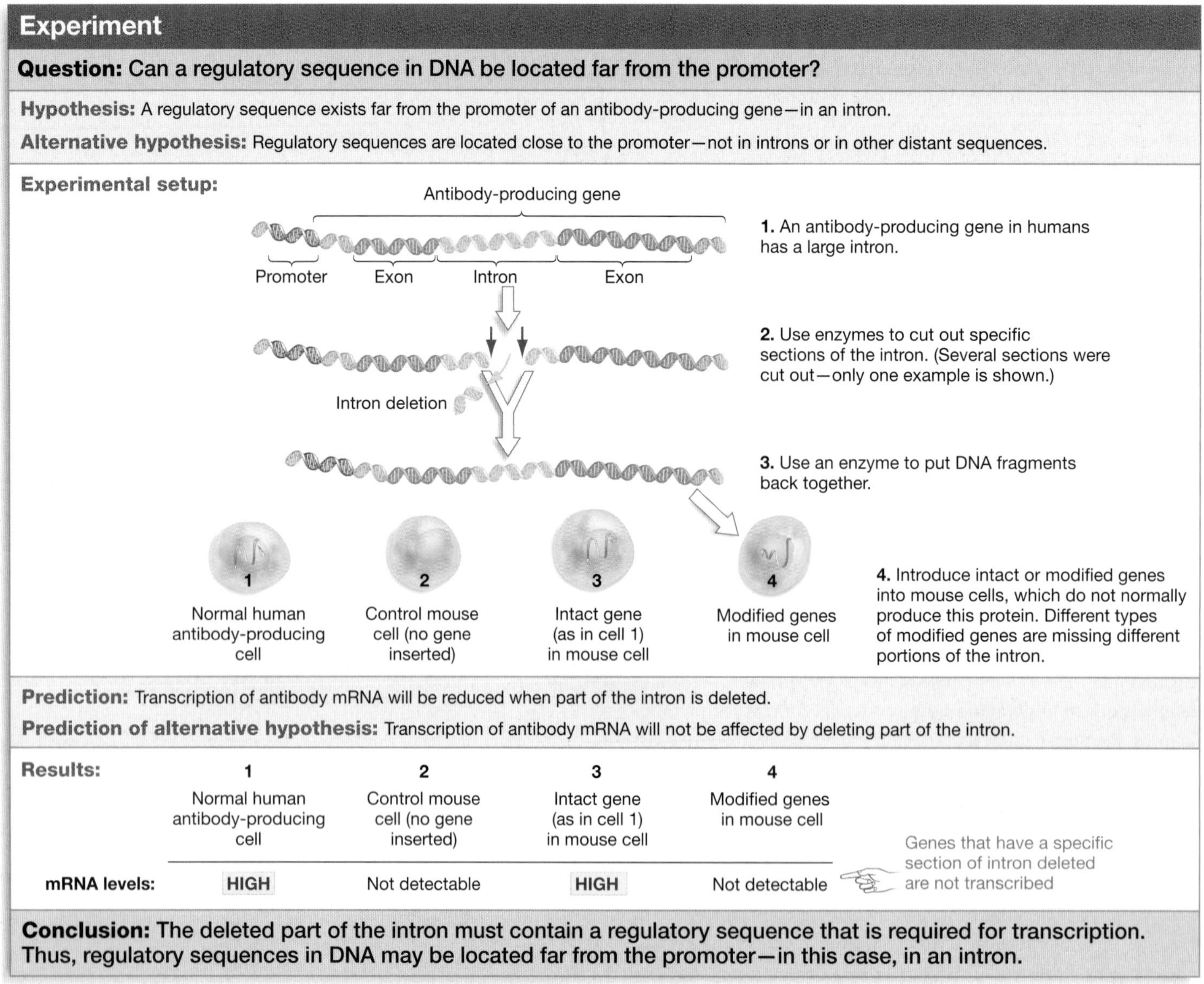

Experiment

Question: Can a regulatory sequence in DNA be located far from the promoter?

Hypothesis: A regulatory sequence exists far from the promoter of an antibody-producing gene—in an intron.

Alternative hypothesis: Regulatory sequences are located close to the promoter—not in introns or in other distant sequences.

Experimental setup:

Antibody-producing gene

Promoter Exon Intron Exon

1. An antibody-producing gene in humans has a large intron.

Intron deletion

2. Use enzymes to cut out specific sections of the intron. (Several sections were cut out—only one example is shown.)

3. Use an enzyme to put DNA fragments back together.

1 Normal human antibody-producing cell

2 Control mouse cell (no gene inserted)

3 Intact gene (as in cell 1) in mouse cell

4 Modified genes in mouse cell

4. Introduce intact or modified genes into mouse cells, which do not normally produce this protein. Different types of modified genes are missing different portions of the intron.

Prediction: Transcription of antibody mRNA will be reduced when part of the intron is deleted.

Prediction of alternative hypothesis: Transcription of antibody mRNA will not be affected by deleting part of the intron.

Results:

	1 Normal human antibody-producing cell	2 Control mouse cell (no gene inserted)	3 Intact gene (as in cell 1) in mouse cell	4 Modified genes in mouse cell
mRNA levels:	HIGH	Not detectable	HIGH	Not detectable

Genes that have a specific section of intron deleted are not transcribed

Conclusion: The deleted part of the intron must contain a regulatory sequence that is required for transcription. Thus, regulatory sequences in DNA may be located far from the promoter—in this case, in an intron.

FIGURE 18.6 Evidence that Enhancers Are Required for Transcription.

QUESTION Why did the researchers assess mRNA production in human cells that did *not* receive a modified or intact gene?

To test this hypothesis, the biologists performed what is now considered a classic experiment (**Figure 18.6**). The protocol was simple in concept: Starting with a human antibody-producing gene that included an intron flanked by two exons (step 1 of Figure 18.6), the team used enzymes to cut out several different specific pieces of the intron (step 2) and to ligate (link) the resulting sections of the gene back together (step 3). Each of the modified genes was missing a different section of the intron. Then the team inserted the various versions of the gene into mouse cells, which do not normally contain the antibody-producing gene. Some mouse cells received normal copies of the gene; others received copies of the gene missing different portions of the intron; still others received no gene (step 4). Finally, the researchers analyzed the mRNAs that were produced and compared them with mRNA from a normal human cell and from mouse cells that received the normal gene. If a regulatory sequence was located inside the intron as they predicted, then some of the modified genes should lack that sequence and fail to transcribe the antibody gene.

As Figure 18.6 shows, some of the modified copies of the gene were not transcribed at all. Based on these results, Tonegawa and co-workers proposed that the intron contains a regulatory sequence that is required for transcription to occur.

This result was remarkable for two reasons: (1) The regulatory sequence was thousands of bases away from the promoter, and (2) it was downstream of the promoter instead of upstream. Regulatory elements that are far from the promoter are termed **enhancers**. Follow-up work has shown that enhancers occur in all eukaryotes and that they have several key characteristics:

- Enhancers can be more than 100,000 bases away from the promoter. They can be located in introns or in untranscribed 5′ or 3′ sequences flanking the gene (see **Figure 18.7**). Researchers have yet to find enhancers located in exons.

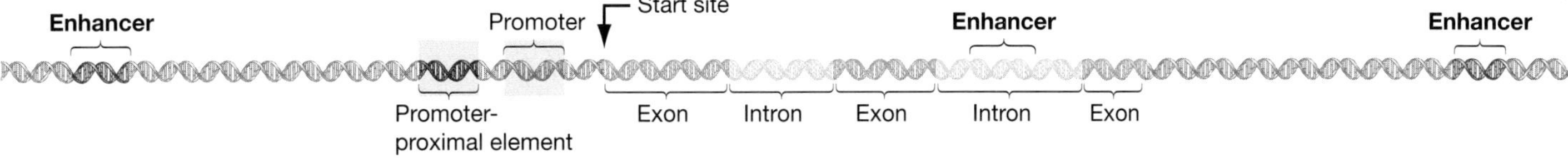

FIGURE 18.7 Enhancers Are Far from the Genes They Regulate. Eukaryotic genes often have more than one enhancer. Enhancers can be located either 5′ or 3′ to the gene they regulate or in introns and can be located tens of thousands of base pairs away from the promoter. All of the colored sections of the DNA strand shown here are part of the same gene.

QUESTION How does the structure of this typical eukaryotic gene compare with the structure of a bacterial operon?

- Like promoter-proximal elements, many types of enhancers exist. Different enhancers are associated with different genes.
- Enhancers can work even if their normal $5' \rightarrow 3'$ orientation is flipped.
- Enhancers can work even if they are moved to a new location in the vicinity of the gene, on the same chromosome.

Enhancers are regulatory sequences that are unique to eukaryotes. When regulatory proteins bind to enhancers, transcription begins. Enhancers are a gas pedal—an element in positive control. In addition, eukaryotic genomes contain regulatory sequences that are similar in structure to enhancers but opposite in function. These sequences are **silencers.** When regulatory proteins bind to silencers, transcription is shut down. Silencers are a brake—an element in negative control.

The discovery of enhancers and silencers expanded the catalog of regulatory sites known in organisms and inspired researchers to reconsider the nature of the gene. Biologists now define the **gene** as a section of DNA that codes for a functional polypeptide or RNA molecule along with the regulatory sequences required for expression.

Once enhancers and silencers had been characterized, it was clear that they represented a type of regulatory sequence that worked very differently than promoter-proximal elements did. For example, there was good evidence that GAL4 and other types of regulatory proteins bind to promoter-proximal elements and interact directly with TBP or RNA polymerase. These interactions stabilize binding and promote transcription. But how could a regulatory sequence that is distant from the promoter help initiate transcription?

What Role Do Regulatory Proteins Play?

An experiment that followed up on Tonegawa's work supported the hypothesis that enhancers are binding sites for proteins that regulate transcription. As **Figure 18.8** shows, researchers cut the enhancer sequence from an antibody-producing gene and spliced it into copies of the gene that codes for the protein β-globin. When this modified gene was inserted into antibody-producing cells, which normally do not express the gene for β-globin, the β-globin mRNA was produced. This result suggests that antibody-producing cells contain some factor that interacts specifically with the antibody-producing gene's enhancer to induce

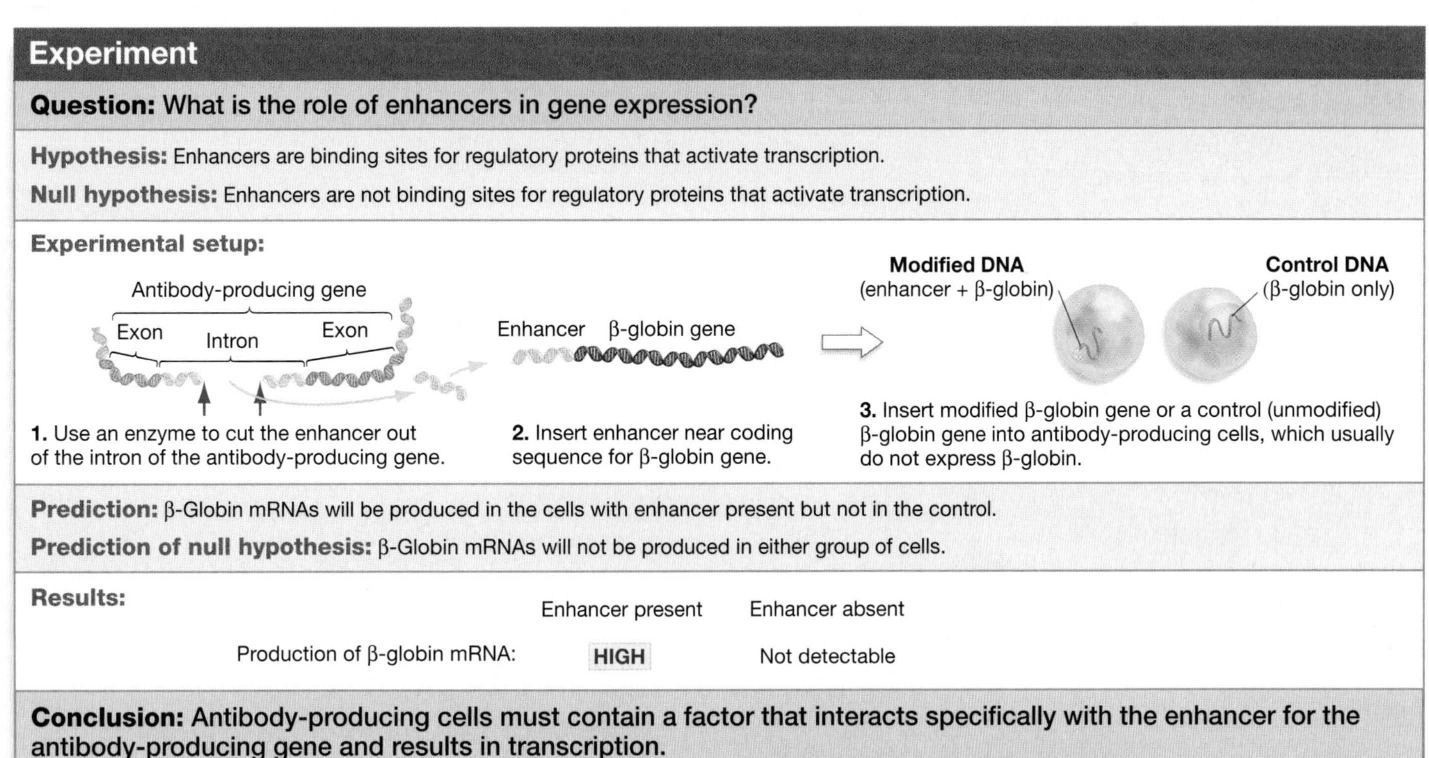

FIGURE 18.8 Evidence that Enhancers Are Involved in Tissue-Specific Expression of Genes

transcription. Follow-up work showed that the factor is a regulatory protein. Enhancers are regulatory sequences where regulatory proteins bind.

By analyzing mutant yeast, fruit flies, and roundworms that have defects in the expression of particular genes, biologists have identified a large number of regulatory proteins that bind to enhancers and silencers. This research program has supported one of the most general conclusions about gene regulation in eukaryotes: In multicellular species, different types of cells express different genes because they contain different regulatory proteins. The regulatory proteins, in turn, are produced in response to signals that arrive from other cells early in embryonic development. For example, a signaling molecule might arrive at a cell early in development and trigger the production of regulatory proteins that are specific to muscle cells. Because the regulatory proteins bind to specific enhancers and silencers and promoter-proximal elements, they trigger the production of muscle-specific proteins (**Figure 18.9**). If no "become a muscle cell" signal arrives, then no muscle-specific regulatory proteins are produced and no muscle-specific gene expression takes place.

To summarize, differential gene expression is based on the production of specific regulatory proteins. Eukaryotic genes are turned on when specific regulatory proteins bind to enhancers and promoter-proximal elements; the genes are turned off when regulatory proteins bind to silencers. Distinctive regulatory proteins are what make a muscle cell a muscle cell and a bone cell a bone cell.

How do the interactions between regulatory proteins and DNA sequences control transcription? As we'll see in the next section, the initiation of transcription is a marvelously complex process in eukaryotes. It requires the coordinated activity of many types of proteins interacting with each other and a variety of DNA sequences.

Check Your Understanding

If you understand that...

- Eukaryotic genes have regulatory sequences called promoter-proximal elements close to their promoters.
- Eukaryotic genes also have regulatory sequences called enhancers or silencers far from their promoters.

You should be able to...

1) Explain the logic of the experiment that led to the discovery of an enhancer in an intron.
2) Compare and contrast regulatory sequences in the DNA of bacteria to those in eukaryotes.
3) Describe how gene expression is dependent on the production of specific regulatory proteins in eukaryotes.

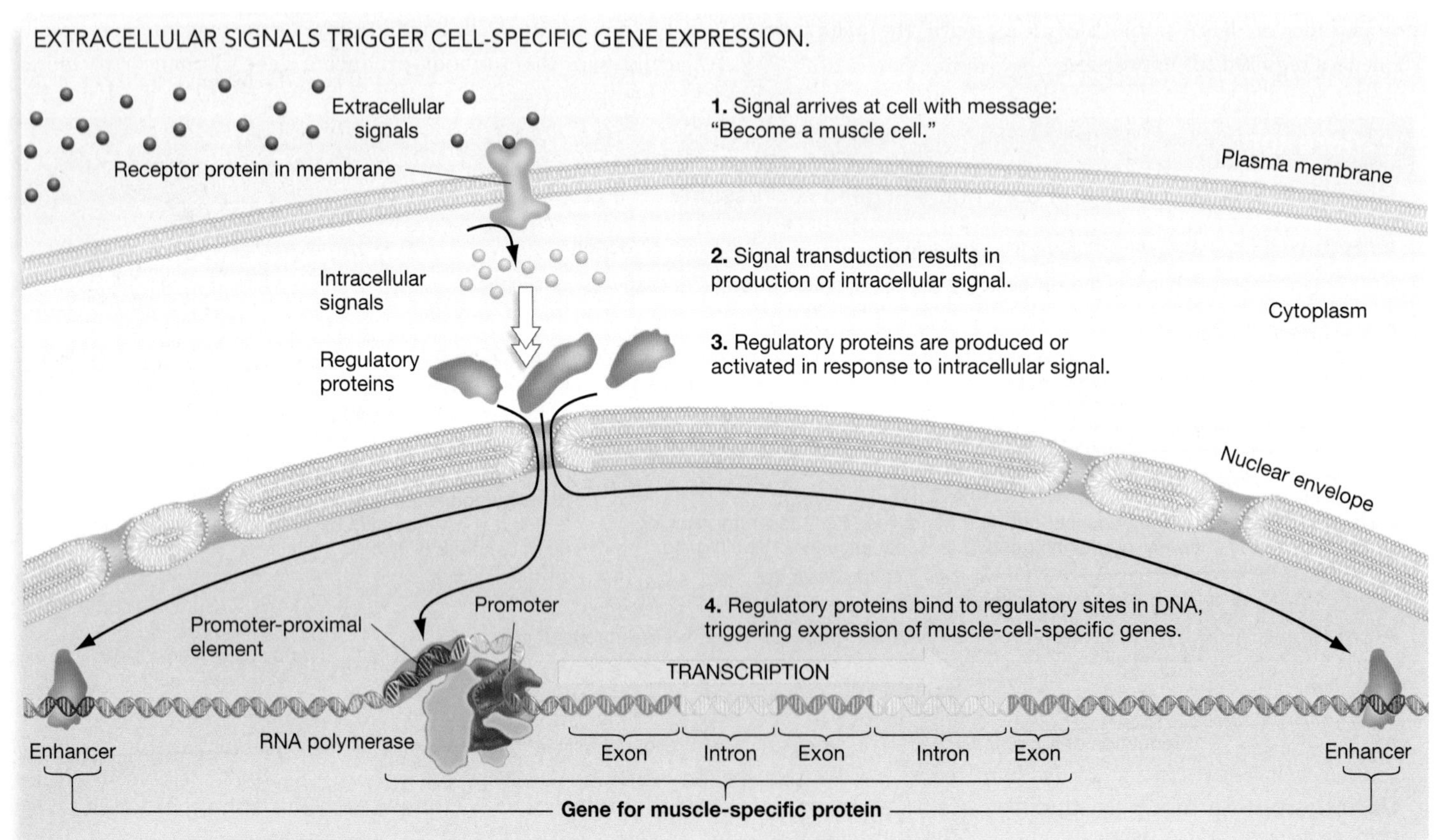

FIGURE 18.9 Why Are Certain Proteins Produced Only in Certain Types of Cells? Signaling molecules from outside the cell trigger the production or activation of cell-specific regulatory proteins, which then influence transcription by binding to enhancers, silencers, or promoter proximal elements in DNA. (See Chapter 8 to review how signal transduction occurs.)

18.4 Transcription Initiation

Many questions remain about how transcription is initiated in eukaryotes. It is clear, however, that two broad classes of regulatory proteins interact with regulatory sequences at the start of transcription: (1) regulatory transcription factors and (2) basal transcription factors. **Regulatory transcription factors** are proteins that bind to enhancers, silencers, or promoter-proximal elements. These transcription factors are responsible for the expression of particular genes in particular cell types and at particular stages of development. **Basal transcription factors,** in contrast, interact with the promoter and are not restricted to particular cell types. Basal transcription factors must be present for transcription to occur, but they do not provide much in the way of regulation. TBP, for example, is a basal transcription factor that is common to all genes. Other basal transcription factors are specific to promoters recognized by RNA polymerase I, II, or III. ● If you understand this concept, you should be able to compare and contrast the regulatory and basal transcription factors found in muscle cells versus nerve cells.

In addition, proteins called **coactivators** are involved in starting transcription. Coactivators do not bind to DNA. Instead, they link the proteins involved in initiating transcription—regulatory transcription factors and basal transcription factors. Their activity helps explain how regulatory sites can be so far from the transcription start site.

Figure 18.10 summarizes a current model for how transcription is initiated in eukaryotes. The first step is the binding of regulatory transcription factors to DNA; these factors recruit chromatin-remodeling complexes and histone acetyl transferases (HATs). The result is chromatin remodeling, a loosening of the chromatin structure. Once the chromatin-remodeling

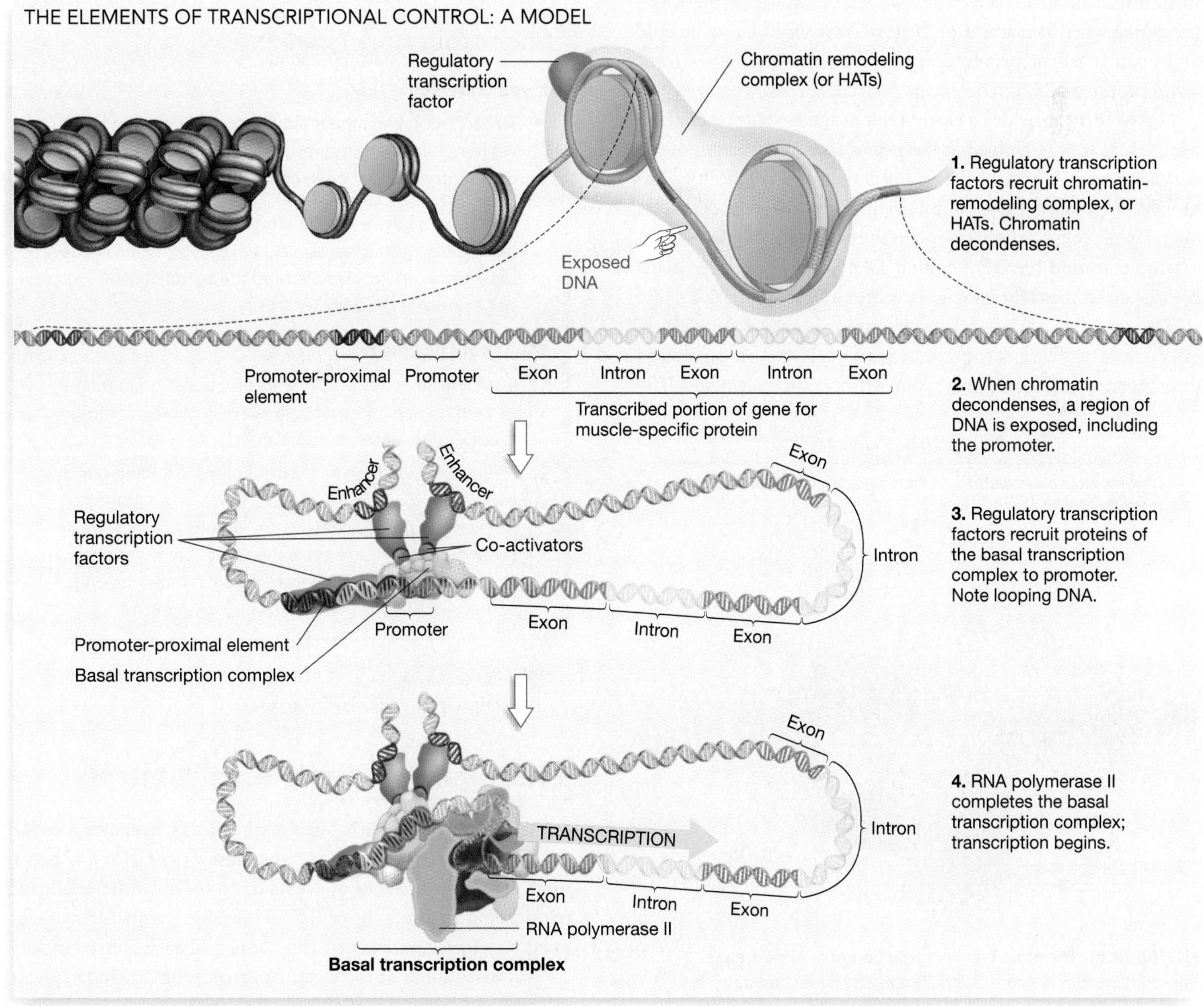

FIGURE 18.10 The Elements of Transcriptional Control: A Model. According to the currently accepted model, transcription is initiated through a series of steps.

complexes and HATs are in place, they open a broad swath of chromatin that includes the promoter region (step 2).

Once chromatin's grip is relaxed, other regulatory transcription factors can bind to enhancers or promoter-proximal elements. When these regulatory transcription factors are bound to DNA, they interact with basal transcription factors. Because enhancers are located far from the promoter, the regulatory transcription factors are normally far from the basal transcription factors. For the two sets of proteins to make contact, DNA has to loop out and away from the promoter, as shown in Figure 18.10 (step 3). When all of the basal transcription factors have assembled at the promoter in response to interactions with regulatory transcription factors and coactivators, they form a multi-protein machine called the **basal transcription complex**. The basal transcription complex then "recruits" RNA polymerase II so that transcription can begin (step 4).

If you understand transcriptional control in eukaryotes, you should be able to explain why DNA loops around the promoter when transcription begins. You should also be able to predict what the promoter looks like when regulatory transcription factors bind to silencers instead of enhancers.

Figure 18.11 provides a closer look at the basal transcription complex as it is assembled at the promoter. Construction of the complex begins when the TATA-binding protein (TBP) binds to the TATA box in the promoter. In a multistep process, as many as 60 other proteins—including RNA polymerase—then assemble around the DNA-bound TBP. Some of these proteins are associated directly with TBP; others are not.

The assembly of the basal transcription complex depends on interactions with regulatory transcription factors that are bound to enhancers, silencers, and promoter-proximal elements. The result is a large, multimolecular machine that is positioned at the start site and able to start transcription. Compared with the situation in bacteria, where just 3 to 5 proteins may interact at the promoter to initiate transcription, the state of affairs in eukaryotes is remarkably complicated.

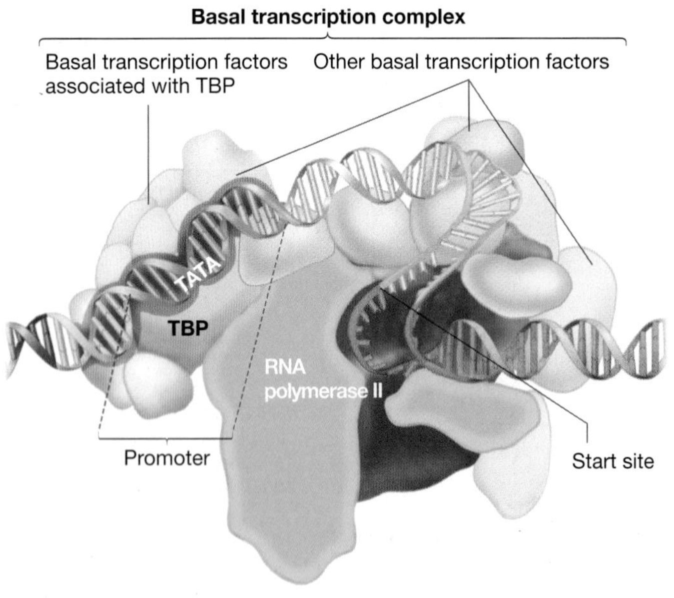

Currently, biologists are focused on understanding exactly how regulatory and basal transcription factors interact to control the formation of the basal transcription complex. Given the number of different proteins involved, progress may be slow. Progress is important, however, for transcription initiation lies at the heart of gene expression. Careful regulation of transcription is critical not only to the development of embryos but also to the daily life of eukaryotes. Right now, cells throughout your body are starting and stopping the transcription of specific genes in response to signals from nearby and distant cells. As the environment inside and outside your body continually changes, your cells continually change which genes are being transcribed.

FIGURE 18.11 The Basal Transcription Complex. The basal transcription complex consists of TBP and associated proteins, other basal transcription factors, and RNA polymerase II.

QUESTION How does this situation compare with the situation at bacterial promoters?

Check Your Understanding

If you understand that...

- Transcription initiation is a multistep process that begins when regulatory transcription factors bind to DNA and recruit proteins that open chromatin.
- Interactions between regulatory transcription factors and basal transcription factors result in the formation of the basal transcription complex and the arrival of RNA polymerase at the gene's start site.

You should be able to...

1) Compare and contrast the sigma proteins and regulatory proteins observed in bacteria with the basal and regulatory transcription factors found in eukaryotes.
2) Make a concept map with boxed terms including gene expression, bacteria, eukaryotes, sigma, regulatory sites, regulatory proteins, coactivators, promoter, promoter-proximal element, enhancers, silencers, basal transcription complex, TBP, RNA polymerase, chromatin remodeling complex/HATs, positive control, negative control.

MB Web Animation at www.masteringbio.com
Transcription Initiation in Eukaryotes

18.5 Post-Transcriptional Control

In the process of regulating gene expression, chromatin remodeling and transcription are just the start of the story. Once an mRNA is made, a series of events has to occur if the final product is going to affect the cell. Each of these events offers an opportunity to regulate gene expression, and each is used in some cells at least some of the time. Control points include splicing mRNAs in various ways, altering the rate at which translation is initiated, and modifying the life span of mRNAs and proteins after translation has occurred. Let's consider each in turn.

Alternative Splicing of mRNAs

Introns are spliced out of primary RNA transcripts while the message is still inside the nucleus. The mRNA that results from splicing consists of sequences that are encoded by exons and are protected by a cap on the 5′ end and a long poly(A) tail on the 3′ end. Recall from Chapter 16 that splicing is accomplished by the molecular machines called **spliceosomes.** What that chapter did not mention, however, is that splicing provides an opportunity for the regulation of gene expression.

During splicing, changes in gene expression are possible because selected exons, as well as introns, may be removed. As a result, the same primary RNA transcript can yield mature, processed mRNAs with several different combinations of transcribed exons. This is important. If the sequence of ribonucleotides in the mature mRNAs varies, then the polypeptides translated from those mature mRNAs will likewise vary. When the same primary RNA transcript is spliced in different ways to produce different mature mRNAs and thus different proteins, **alternative splicing** is said to occur.

To see how alternative splicing works, consider the muscle-cell protein tropomyosin. The tropomyosin gene is expressed in both skeletal muscle and smooth muscle, which are distinct muscle types. (Skeletal muscle is responsible for moving your bones; smooth muscle lines many parts of your gut and certain blood vessels. The two types of muscle are composed of distinct types of muscle cells.) As **Figure 18.12a** shows, the primary transcript from the tropomyosin gene contains 14 exons. In skeletal muscle and smooth muscle cells, different subsets of the 14 exons are spliced together to produce two different messages for translation (**Figure 18.12b**). Each mature mRNA contains information from a different combination of exons. As a result of alternative splicing, the tropomyosin proteins found in these two cell types are distinct. One of the reasons skeletal muscle and smooth muscle are different is that they contain different types of tropomyosin.

Alternative splicing is controlled by proteins that bind to mRNAs in the nucleus and interact with spliceosomes. When cells that are destined to become skeletal muscle or smooth muscle are developing, they receive signals leading to the production of specific proteins that are active in the regulation of splicing. Instead of transcribing different versions of the tropomyosin gene, the cells splice the same primary RNA transcript but in different ways.

Before the importance of alternative splicing was widely appreciated, a gene was considered to be a nucleotide sequence that encodes a single protein or RNA. Based on this view, estimates for the number of genes in the human genome were typically in the range of 60,000 to 100,000. Once the complete human genome sequence was available, however, researchers realized that we actually have fewer than 20,000 genes. Even though our genomes contain a relatively low number of genes, recent data indicate that at least 35 percent of human genes undergo alternative splicing. Thus, the number of different proteins that your cells can produce is believed to be between 100,000 and 1 million.

Thanks to results like these, genes now have to be thought of as DNA sequences that are capable of directing the production of one or more related polypeptides or RNAs. The current record holder for the number of mRNAs derived from one gene is the *Dscam* gene in the fruit fly *Drosophila melanogaster*. The products of this gene are involved in guiding growing nerve cells within the embryo. Because the primary transcript is spliced into about 38,000 distinct forms of mRNA, the *Dscam* gene can produce about 38,000 different products. Alternative splicing ranks as a major mechanism in the control of gene expression in multicellular eukaryotes. If you understand alternative splicing, you should be able to explain why it does not occur in bacteria, comment on its relationship to the evolution of the nuclear envelope, and indicate where it occurs in Figure 18.1.

mRNA Stability and RNA Interference

Once splicing is complete and processed mRNAs are exported to the cytoplasm, new regulatory mechanisms come into play. For example, it has long been known that the life span of an mRNA in the cell can vary. The mRNA for casein—the major protein in milk—is produced in the mammary gland tissue of female mammals. Normally, many of these mRNAs persist in the cell for just

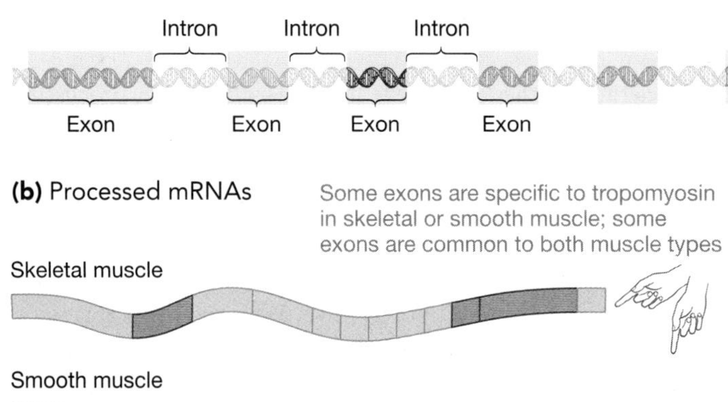

FIGURE 18.12 Alternative Splicing Produces More than One Mature mRNA from the Same Gene. (a) The tropomyosin gene has a large series of exons and introns. The purple exons are found in skeletal muscle tropomyosin, and the blue exons are found in smooth muscle tropomyosin. The orange exons are found in the tropomyosin of both types of muscle cells. **(b)** In skeletal and smooth muscle cells, alternative splicing leads to the production of distinct mature mRNAs for the tropomyosin protein.

an hour, and little casein protein is produced. But when a female mouse is lactating, regulatory molecules help the mRNAs persist almost 30 times longer—leading to a huge increase in the production of casein protein. In the case of casein, mRNA stability is associated with changes in the length of its poly (A) tail.

In many cases, the life span of an mRNA is controlled by tiny, single-stranded RNA molecules that bind to complementary sequences in the mRNA. Once part of an mRNA becomes double stranded in this way, specific proteins degrade the mRNA or prevent it from being translated into a polypeptide. This phenomenon is known as **RNA interference**. How does it work?

As **Figure 18.13** shows, RNA interference begins when RNA polymerase transcribes DNA sequences that code for an unusual product—a small RNA molecule that doubles back on itself to form a hairpin (step 1). Hairpin formation occurs because pairs of sequences within the RNA transcript are complementary. Some of the RNA is trimmed by enzymes in the nucleus, then the double-stranded segment that remains is exported to the cytoplasm (Figure 18.13, step 2). In the cytoplasm, the double-stranded RNA sequence is cut by another enzyme into molecules that are typically about 22 nucleotides long. One of these short RNA strands is taken up by a group of proteins called the RNA-induced silencing complex, or RISC (Figure 18.3, step 3). The single-stranded RNA held by the RISC is a called a **microRNA** (**miRNA**). Once it is part of a RISC, the miRNA binds to its complementary sequences in a target RNA (step 4). If the match between an miRNA and an mRNA is perfect, an enzyme in the RISC destroys the mRNA by cutting it in two (step 5). In effect, tight binding by an miRNA is a "kiss of death" for the mRNA. If the match isn't perfect, the mRNA is not destroyed. Instead, its translation is inhibited. In both cases, miRNAs are responsible for interfering with mRNAs—hence the phenomenon known as RNA interference.

The first papers on RNA interference were published in the mid-1990s, and the first miRNAs were characterized in 2001. In the short time since then, research on miRNAs and RNA interference has virtually exploded. Current data suggest that a typical animal or plant species has at least 500–1000 sequences that code for miRNAs. Because many miRNAs regulate more than one mRNA, it's estimated that some 20–30 percent of all animal and plant genes are regulated by these tiny molecules.

Although a great deal remains to be learned about what miRNAs do and how they are regulated, their discovery has had a tremendous impact on how biologists think about gene regulation in eukaryotes. Just when researchers thought they were starting to get a handle on how eukaryotic genes are controlled, RNA interference came along and added an entirely new layer of complexity. ● If you understand RNA interference, you should be able to label where it occurs in Figure 18.1.

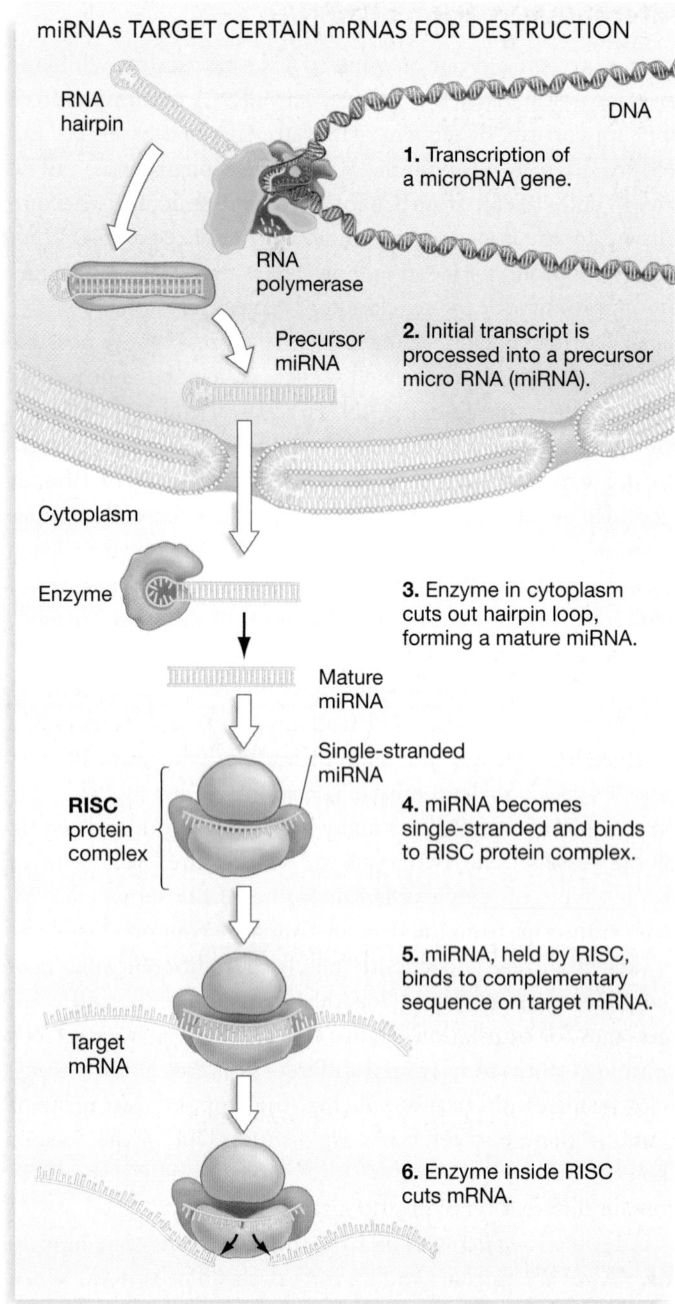

FIGURE 18.13 RNA Interference: Micro RNAs Target Certain mRNAs for Destruction.

How Is Translation Controlled?

Stopping or starting translation is a common mechanism for controlling gene expression. Although RNA interference usually operates at the level of the mRNA—after RNA processing is complete, but independently of the translation machinery—many of the small RNAs responsible for RNA interference disrupt translation directly. In other cases, mechanisms that do not involve miRNAs are responsible for controlling the timing and rate of translation. The best-studied examples depend on regulatory proteins that bind to mRNAs or the ribosome.

- In many species of animals, eggs are loaded with mRNAs that are not translated until fertilization occurs. The proteins produced from these messages are involved in directing the early development of the embryo, so they are not needed until fertilization is complete. Translation of the mRNAs in the egg is prevented when a regulatory protein binds to them, or when the mRNA's cap or tail is modified.

- The overall rate of translation in a cell may slow or stop in response to a sudden increase in temperature or infection by a virus. The slowdown occurs because regulatory proteins add a phosphate group to a protein that is part of the ribosome. You might recall from earlier chapters that phosphorylation frequently leads to changes in the shape and chemical reactivity of proteins. In the case of the phosphorylated ribosomal protein, the shape change slows or prevents translation. For the cell, this dramatic change in gene expression can mean the difference between life and death. If the danger is due to a sudden increase in ambient temperature that disrupts protein folding, shutting down translation prevents the production of improperly folded polypeptides; if the insult is a virus, the cell avoids manufacturing viral proteins.

Observations like these are a reminder that gene expression can be regulated at multiple points: at the level of chromatin structure, transcription initiation, RNA processing, mRNA life span, and translation. The sixth and last level occurs after translation is complete. In many cases, proteins are not activated as soon as they are produced.

Post-Translational Control

Control of gene expression may continue even after translation occurs and a protein product is complete. Recall from Chapter 17 that in bacteria, mechanisms of post-translational regulation are important because they allow the cell to respond to new conditions rapidly. The same is true for eukaryotes. Instead of waiting for transcription, RNA processing, and translation to occur, the cell can respond to altered conditions by quickly activating or inactivating existing proteins. Regulatory mechanisms occurring late in the flow of information from DNA to RNA to protein involve a trade-off between speed and resource use, however, because transcription and translation require energy and materials.

As an example of post-translational control, consider a group of regulatory transcription factors called **signal transducers and activators of transcription (STATs)**. STATs are common in the cytoplasm of white blood cells in mammals, where they normally reside as single polypeptide chains. In this form, the STATs are inactive. But a chain of events is triggered when a signaling molecule binds to a receptor on the cell surface (**Figure 18.14**). When the receptor is activated by the signal, it adds a phosphate to a STAT polypeptide, triggering the formation of a **dimer** (a unit made of two parts). The activated STAT dimer then moves into the nucleus, binds to an enhancer, and activates transcription of genes that trigger cell growth and division. In this way, STATs are key players in a signal

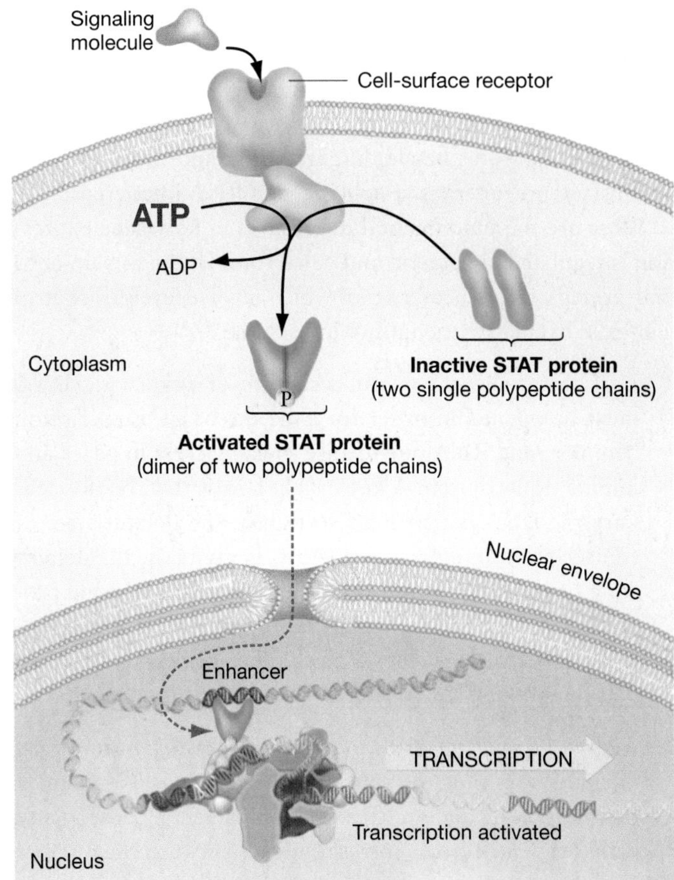

FIGURE 18.14 STATs Activate Transcription after They Are Phosphorylated. Signal transduction triggers a phosphorylation event that causes two STAT polypeptide chains to join and form an activated dimer. Activated STAT proteins move to the nucleus, bind to DNA, and activate the transcription of selected target genes.

transduction pathway required for your white blood cells to respond to invading bacteria, viruses, or other threats.

Phosphorylation is a common mechanism of post-translational control over gene expression, particularly in signal transduction pathways (see Chapter 8). It is only one of several post-translational control mechanisms that have been documented, however. The activity of a protein may also be modified by folding or by enzymes that cleave off a portion of the molecule or degrade the entire structure. But no matter what the mechanism, post-translational control is correlated with particularly rapid changes in gene expression.

A 50-Year Retrospective: How Does Gene Expression in Bacteria Compare with That in Eukaryotes?

For about five decades, biologists have been studying how gene expression is controlled. Almost as soon as they knew that information in DNA is transcribed into RNA and then translated into proteins, researchers began asking questions about how that flow of information is regulated. Early studies on bacteria established the existence of promoters, the structure and function of regulatory proteins and regulatory sequences, and the principles of positive control and negative control. More recent work on eukaryotes has highlighted the importance of chromatin structure, alternative splicing, and RNA interference.

What are the fundamental differences in how gene expression is regulated in bacteria and eukaryotes? Biologists point to four primary differences, two of which involve levels of control that exist in eukaryotes but not in bacteria:

1. *Packaging* The chromatin structure of eukaryotic DNA must be opened in order for TBP, the basal transcription complex, and RNA polymerase to gain access to genes and initiate transcription. A key insight here is that, because eukaryotic DNA is packaged so tightly, the default state of transcription in eukaryotes is off. In contrast, the default state of transcription in bacteria, which lack histone proteins and have freely accessible promoters, is on. Chromatin structure provides a mechanism of negative control that does not exist in bacteria.

2. *Alternative splicing* Prior to translation, primary transcripts in eukaryotes must be spliced—an event that is extremely rare in bacteria and archaea. The fundamental consequence is that the one-to-one correspondence between the number of genes and the number of gene products observed in bacteria and archaea is not seen in eukaryotes. Instead, each eukaryotic gene may code for one to thousands of distinct products.

3. *Complexity* Transcriptional control is much more complex in eukaryotes than in bacteria. The function of sigma proteins in bacteria is analogous to the role of the basal transcription complex in eukaryotes. Likewise, the function of CAP, the repressor, and other regulatory proteins is analogous to the role of regulatory transcription factors in eukaryotes. But the sheer number of eukaryotic proteins involved in regulating transcription—and the complexity of their interactions—dwarfs those in bacteria.

4. *Coordinated expression* In bacteria, genes that are involved in the same cellular response are organized into operons controlled by a single promoter. Because their polycistronic mRNAs are translated together, several proteins are produced in a coordinated fashion. In contrast, operons are rare in eukaryotes. In these organisms, genes that are physically scattered can be expressed at the same time because a single set of regulatory transcription factors can trigger the transcription of several genes. For example, muscle-specific genes found on several different chromosomes can be transcribed in response to the same muscle-specific regulatory transcription factor. In this way, eukaryotes coordinate the expression of functionally related genes.

To date, biologists do not have a good explanation for why gene expression is so much more complex in unicellular eukaryotes than it is in bacteria and archaea. All unicellular organisms have to respond to environmental changes in an appropriate way. It is easier to generate a logical hypothesis to explain why gene expression is complex in multicellular eukaryotes. In these organisms, cells have to differentiate as an individual develops. Changes in gene expression are re-sponsible for the differentiation of muscle cells, bone cells, leaf cells, and flower cells in response to signals from other cells. The need for each cell type to have a unique pattern of gene expression may explain why control of gene expression is so much more complex in multicellular eukaryotes than in bacteria.

One of the two great frontiers in research on gene expression is to understand how developmental signals produce cell-specific gene expression in multicellular organisms. The other major frontier is to understand how certain defects in gene regulation result in uncontrolled cell growth and the suite of diseases called cancer.

18.6 Linking Cancer with Defects in Gene Regulation

Normal regulation of gene expression results in the orderly development of an embryo and appropriate responses to environmental changes. Abnormal regulation of gene expression, in contrast, can lead to developmental abnormalities and diseases such as cancer. Hundreds of distinct cancers exist. These diseases are enormously variable regarding the tissues they affect, their rate of progression, and their outcome. Because the underlying defects, symptoms, and consequences are so diverse, cancer is not a single disease but a family of related diseases. Cancers are related because they all originate in uncontrolled cell growth. For a cancer to become dangerous, however, two other things have to happen: The rapidly growing cells must metastasize, meaning that some cells leave their point of origin and invade other tissues (see Chapter 11), and they must stimulate the growth of blood vessels that supply them with nutrients.

Here let's consider the first step in cancer formation, and the question of what causes uncontrolled cell growth. The short answer is that each type of cancer is caused by a different set of genetic defects that lead to uncontrolled cell growth. The longer and more accurate answer is based on two key observations. First, recall from Chapter 11 that cancer results from defects in the proteins that control the cell cycle. Second, cancer is associated with mutations that alter key genes. For example, Chapter 14 linked the cancer-causing disorder xeroderma pigmentosum with defects in DNA repair. This association supports the hypothesis that increased mutation rates are involved in triggering cancer. The hypothesis is also supported by the observation that people who have been exposed to **mutagens**—

that is, radiation or chemicals that induce mutation—have an increased risk of developing cancer. For example, cancer rates were extremely high in the tens of thousands of people who were exposed to massive doses of radiation from the atomic bombs dropped on Hiroshima and Nagasaki, Japan, at the end of World War II. The link between mutation and cancer is so strong that suspected **chemical carcinogens** (cancer-causing compounds) are first tested by their ability to cause mutations in bacteria.

The fundamental question of cancer biology comes into sharper focus once the role of cell-cycle regulators and mutation is acknowledged. Now we can ask: Which genes, when mutated, disrupt the cell cycle and trigger uncontrolled cell growth?

Intensive research over the past two decades has shown that many cancers are associated with mutations in regulatory transcription factors. These mutations lead to cancer when they affect one of two classes of genes: (1) genes that stop or slow the cell cycle, and (2) genes that trigger cell growth and division by initiating specific phases in the cell cycle.

Genes that stop or slow the cell cycle are called **tumor suppressor** genes. Their products prevent the cell cycle from progressing unless specific signals indicate that conditions are right for moving forward with mitosis and cell division. If a mutation disrupts normal function of a tumor suppressor gene, then a key "brake" on the cell cycle is eliminated.

Genes that encourage cell growth by triggering specific phases in the cell cycle are called **proto-oncogenes** (literally, "first-cancer-genes"). In normal cells, proto-oncogenes are required to initiate each phase in the cell cycle. They are active only when conditions are appropriate for growth, however. In cancerous cells, defects in the regulation of proto-oncogenes cause these genes to stimulate growth at all times. In cases such as this, a mutation has converted the proto-oncogene into an **oncogene**—an allele that promotes cancer development.

To gain a deeper understanding of how defects in gene expression can lead to cancer, consider research on the gene that is most often defective in human cancers. The gene is called ***p53*** because the protein it codes for has a molecular weight of approximately 53 kilodaltons (53,000 daltons, or 53,000 amu—see Chapter 2). DNA sequencing studies have revealed that mutant, nonfunctional forms of *p53* are found in over half of all human cancers. The *p53* gene codes for a regulatory transcription factor.

What is the link between a loss of p53 protein activity and cancer? A key observation suggested an answer to this question: When researchers exposed normal, noncancerous human cells to UV radiation, levels of p53 protein increased markedly. Recall from Chapter 14 that UV radiation damages DNA. Follow-up studies confirmed that there is a close correlation between DNA damage and the amount of p53 in a cell. In addition, analyses of the protein's structure showed that it contains a DNA-binding domain.

These observations inspired the hypothesis that p53 is a transcription factor that serves as the master brake on the cell cycle. In this model, p53 is activated after DNA damage occurs. The activated protein binds to the enhancers of genes that arrest the cell cycle. Once these genes are activated, the cell has time to repair its DNA before continuing to grow and divide.

Recent research has shown that this model of p53 function is correct in almost every detail. For example, consider the results of studies based on X-ray crystallography. The three-dimensional models generated by this technique confirmed that p53 binds directly to DNA (**Figure 18.15**). In addition, researchers who mapped the location of mutations that make p53 defective found that virtually all of the cancer-causing mutations were located in p53's DNA-binding site (see the amino acids highlighted in Figure 18.15). This observation supports the hypothesis that defective forms of the protein can't bind to enhancers. Further, investigators have documented that one of the genes induced by p53 encodes a protein that prevents cell-cycle regulatory proteins from triggering M (mitosis) phase.

Recent research has also shown that when a cell's DNA is extensively damaged and cannot be repaired, p53 activates the transcription of genes that cause the cell to take its own life via apoptosis (see Chapter 11). But if mutations in the *p53* gene make the protein product inactive, then damaged cells are not shut down or killed. They continue to move through the cell

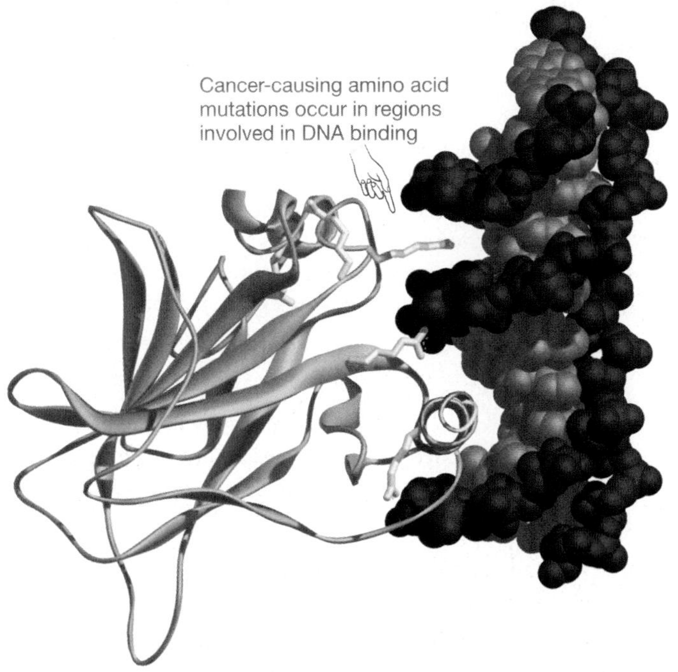

FIGURE 18.15 p53 Is a DNA-Binding Protein That Acts as a Tumor Suppressor. Ribbon model showing the DNA-binding domain of p53. In cancer patients, the amino acids highlighted in yellow often differ from the normal versions. The molecule shown in red and pink is DNA.

QUESTION Why would substituting one amino acid for another at one of the highlighted locations affect a protein's ability to bind to DNA? What happens to a cell when p53 cannot bind to DNA efficiently?

cycle, except now they are likely to contain many mutations because of the DNA damage they have sustained. If these mutations create oncogenes, the cells have taken a key step on the road to cancer.

To summarize, *p53* functions as a tumor suppressor gene. It prevents cancer onset by stopping the cell cycle when DNA is damaged. When *p53* is functioning normally, mutations that produce oncogenes are repaired or eliminated before uncontrolled cell growth can begin. The role of *p53* in preventing cancer is so fundamental that biologists call this gene "the guardian of the genome." Currently, research on the p53 protein is forging ahead on two fronts: Biologists are striving to identify more of the genes that are regulated by this protein and to find molecules that could act as anticancer drugs by mimicking p53's shape and activity.

Check Your Understanding

If you understand that...

- Cancer is associated with mutations and with loss of control over the cell cycle.
- Uncontrolled cell growth may occur when a protein that activates the cell cycle is activated constitutively because of a mutation in a regulatory gene.
- Uncontrolled cell growth may occur when tumor suppressor genes stop functioning, due to mutation.

You should be able to...

Explain why knock-out (loss-of-function) mutations in *p53* occur in so many cancers.

Chapter Review

SUMMARY OF KEY CONCEPTS

Changes in gene expression allow eukaryotic cells to respond to changes in the environment and cause distinct cell types to develop.

In a multicellular eukaryote, different types of cells are radically different in size, shape, and function—even though they have the same DNA. Cells are different not because they have different genes but because they express different genes.

In embryos, cells begin expressing specific genes in response to signals from other cells. Because different cells receive different signals, one cell might begin expressing muscle-specific genes while a nearby cell expresses bone-specific genes. Once cells mature, they continue to receive signals that are released by other cells and that carry information about changes in the environment. In response, changes in gene expression may occur.

In both embryonic cells and mature cells, gene expression is regulated at several levels. The transcription of specific genes may be initiated or repressed, mRNAs may be spliced in different ways to produce a different product, the life span of specific mRNAs may be extended or shortened, and the life span or activity of particular proteins may be altered.

You should be able to explain why muscle-specific genes are expressed in muscle cells while nearby nerve-specific genes are not.

In eukaryotes, DNA is packaged with proteins into complex structures that must be opened before transcription can occur.

Eukaryotic DNA is wrapped around histone proteins to form a bead-like nucleosome, which is then coiled into 30-nm fibers and higher-order chromatin structures. Transcription cannot be initiated until the interaction between DNA and histones in chromatin is relaxed. These changes depend on the acetylation or methylation of histones and the action of molecular machines called chromatin-remodeling complexes.

You should be able to draw what a gene looks like when chromatin is in a condensed versus uncondensed state.

In eukaryotes, transcription can be initiated only when specific proteins bind to the promoter and to regulatory sequences that may be close to the promoter or far from it.

Regulatory transcription factors are proteins that bind to regulatory sequences called enhancers and silencers, which are often located at a distance from the gene in question, or to promoter-proximal sequences near the start of the coding sequence. The first regulatory transcription factors that bind to DNA recruit proteins that loosen the histones' grip on the gene, making the promoter accessible to basal transcription factors. Interactions between regulatory and basal transcription factors lead to the formation of the basal transcription complex. Once this large, multi-protein complex is intact, RNA polymerase is recruited to the site and transcription begins.

You should be able to draw what a eukaryotic gene looks like when it is being transcribed. Label enhancers, promoter proximal elements, the promoter, regulatory transcription factors, basal transcription factors, and RNA polymerase.

(MB) Web Animation at www.masteringbio.com
Transcription Initiation in Eukaryotes

Alternative splicing allows a single gene to code for several different products.

Once a message has been transcribed, several other regulatory events come into play. Alternative splicing allows a single gene to produce more than one mRNA and more than one protein. It is regulated by proteins that interact with the spliceosome.

You should be able to explain why humans have so few genes.

Once an mRNA is in the cytoplasm, gene expression is controlled by molecules that regulate the (1) life span of mRNAs, (2) efficiency of translation, and (3) activation or inactivation of protein products.

RNA interference occurs when tiny strands of RNA, called microRNAs (miRNAs), bind to mRNAs in company with the protein complex called RISC and mark the mRNAs for degradation or when small RNAs inhibit translation. Once translation occurs, proteins may be activated or inactivated by the addition or removal of a phosphate group or other events.

You should be able to draw the interaction between an miRNA, RISC, and an mRNA.

Cancer can develop when mutations disable genes that regulate cell-cycle control genes.

The cell cycle is controlled by specific genes. But if mutations alter transcription factors that control the cell cycle, then uncontrolled cell growth and tumor formation may result. For example, the regulatory transcription factor *p53* is responsible for stopping the cell cycle when DNA is damaged. If *p53* is mutated in a way that prevents its protein product from binding to DNA, then the cell cycle is not arrested and the damaged DNA is not repaired—leading to mutations.

You should be able to explain why cancer is common in people who have been exposed to high levels of radiation, and why it is more common in older people than younger people.

QUESTIONS

Test Your Knowledge

1. What is chromatin?
   a. the protein core of the nucleosome, which consists of histones
   b. the 30-nm fiber
   c. the DNA-protein complex found in eukaryotes
   d. the histone *and* non-histone proteins in eukaryotic nuclei

2. What is a tumor suppressor?
   a. a gene associated with tumor formation when its product does not function
   b. a gene associated with tumor formation when its product functions normally
   c. a gene that accelerates the cell cycle and leads to uncontrolled cell growth
   d. a gene that codes for a transcription factor involved in tumor formation

3. Which of the following statements about enhancers is correct?
   a. They contain a unique base sequence called a TATA box.
   b. They are located only in 5′-flanking regions.
   c. They are located only in introns.
   d. They are found in a variety of locations and are functional in any orientation.

4. In eukaryotes, why are certain genes expressed only in certain types of cells?
   a. Different cell types contain different genes.
   b. Different cell types have the same genes but different promoters.
   c. Different cell types have the same genes but different enhancers.
   d. Different cell types have different regulatory transcription factors.

5. What is alternative splicing?
   a. the phosphorylation events that lead to different types of post-translational regulation
   b. mRNA processing events that lead to different combinations of exons being spliced together
   c. folding events that lead to proteins with alternative conformations
   d. action by regulatory proteins that leads to changes in the life span of an mRNA

6. What types of proteins bind to promoter-proximal elements?
   a. the basal transcription complex
   b. the basal transcription complex plus RNA polymerase
   c. basal transcription factors
   d. regulatory transcription factors

Test Your Knowledge answers: 1. c; 2. a; 3. d; 4. d; 5. b; 6. d

Test Your Understanding

Answers are available at www.masteringbio.com

1. Compare and contrast (a) enhancers and the CAP site; (b) promoter-proximal elements and the *lac* operon operator; (c) basal transcription factors and sigma; and (d) eukaryotic promoters and bacterial promoters.

2. Why does chromatin need to be "remodeled" for transcription to occur? Why was it logical to observe that DNA-protein interactions can be loosened by the addition of acetyl groups to histones?

3. Compare and contrast (a) enhancers and silencers; (b) promoter-proximal elements and enhancers; (c) transcription factors and coactivators; and (d) regulatory transcription factors and basal transcription factors.

4. Explain the relationship between complementary base pairing and RNA interference. Suggest a way that miRNAs that are complementary to viral RNAs could be used as drugs.

5. Now that the human genome has been sequenced, researchers estimate that it contains fewer than 20,000 genes. Previously, most investigators expected that the human genome held at least 100,000 genes. Explain how alternative splicing can explain the discrepancy between the observed and predicted numbers of genes.

6. Explain why mutations in regulatory proteins and regulatory sequences can lead to loss of control over the cell cycle and the development of cancer. Include a specific example.

Applying Concepts to New Situations

Answers are available at www.masteringbio.com

1. Histone proteins have been extremely highly conserved during evolution. The histones found in fruit flies and humans, for example, are nearly identical in amino acid sequence. Offer an explanation for this observation. (Hint: What are the consequences of a mutation in a histone?)
2. Cancers are most common in tissues where cell division is common, such as blood cells and cells in the lining of the lungs or gut. Why is this observation logical?
3. Levels of p53 protein in the cytoplasm increase after DNA damage. Design an experiment to determine whether this increase is due to increased transcription of the *p53* gene or to activation of preexisting p53 proteins via a post-translational mechanism such as phosphorylation.
4. Why are individuals who experience severe sunburn or deep tanning more likely to develop skin cancer?

www.masteringbio.com is also your resource for • Answers to text, table, and figure caption questions and exercises • Answers to *Check Your Understanding* boxes • Online study guides and quizzes • Additional study tools including the *E-Book for Biological Science* 3rd ed., textbook art, animations, and videos.

Analyzing and Engineering Genes

19

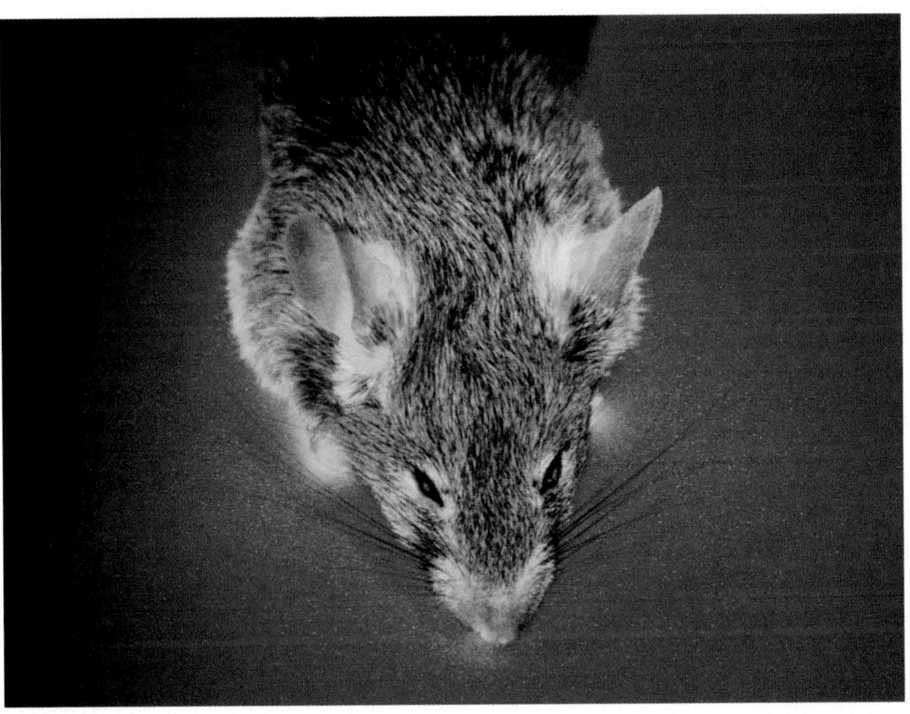

A mouse that has been genetically engineered to express a gene from jellyfish. The jellyfish gene codes for a protein that naturally luminesces, or emits light. The gene's product is called green fluorescent protein, or GFP.

KEY CONCEPTS

- The discovery of enzymes that cut DNA at specific locations, along with enzymes that piece DNA segments back together, gave biologists the ability to move genes from one location to another.
- To analyze a gene, biologists have to obtain many identical copies of it. This can be done by inserting the gene into a bacterial cell that copies the gene as it grows, or by performing a polymerase chain reaction.
- Once researchers have many identical copies of a gene, its sequence of bases can be determined by dideoxy sequencing.
- To find the genes associated with a particular trait, researchers start with a genetic map. If certain markers on the map are found only in individuals that have a distinctive phenotype, the gene responsible for that phenotype is likely to be near that marker.
- Researchers are attempting to insert genes into humans to cure genetic diseases. Efforts to insert genes into plants to provide them with novel traits, such as the ability to withstand insect attacks, have been much more successful.

The molecular revolution in biological science began when researchers confirmed that DNA is the hereditary material and understood the molecule's secondary structure. But when biologists discovered how to remove DNA sequences from an organism, manipulate them, and insert them into different individuals, the molecular revolution took off. Efforts to manipulate DNA sequences in organisms are often referred to as genetic engineering. The goal of this chapter is to introduce some of the key techniques and conceptual issues involved in analyzing and engineering DNA, using a case history approach.

Genetic engineering became possible with the discovery of bacterial enzymes that cut DNA at specific sites and that paste DNA sequences together. Although cutting and pasting DNA sequences sounds simple in concept, these new molecular tools were extremely powerful. Biologists no longer had to rely solely on controlled breeding experiments to change the genetic characteristics of individuals. Instead, they could quickly mix and match specific DNA sequences in the lab. Efforts to manipulate genes usually result in novel combinations of genes along chromosomes, or genetic recombination (see Chapter 12).

Appropriately enough, techniques used to engineer genes are often referred to as **recombinant DNA technology**.

The enzymes used to cut and paste DNA were initially discovered and used during the 1960s and 1970s. Over the subsequent two decades, researchers followed up by developing systems for transferring recombinant genes into various types of organisms and by working out techniques for controlling the expression of the introduced DNA. Since then, research has concentrated on applying these techniques to solve problems in medicine, industry, and agriculture. The goals of genetic engineering are to (1) improve our understanding of how genes work and (2) advance **biotechnology**—the manipulation of organisms to create products or cure disease.

To introduce the techniques that biologists are using to analyze and engineer genes, we're going to view them in action. More specifically, we'll consider a series of case histories that introduce the basic techniques in molecular biology in the context of solving problems. Our first example represents one of the initial efforts to use recombinant DNA technology to cure an inherited disease in humans. Section 19.1 focuses on how the search for a cure for pituitary dwarfism relied on techniques for identifying genes, copying them, moving them into a new host organism, and then controlling their expression. Later sections introduce techniques biologists use to produce many identical copies of a gene, determine a gene's sequence, and find the gene or genes associated with a particular disease or other trait. The chapter closes with a look at how genetic engineering has been applied to humans and crop plants in the hope of curing devastating genetic diseases and increasing the quality and quantity of food available to people in poverty-stricken areas of the globe.

In addition to exploring the techniques used to manipulate DNA, it is essential to consider the ethical, economic, ecological, and political issues involved. Manipulating the genetic makeup of organisms raises ethical concerns and has the potential to create harm. Gene therapy—the introduction of normal alleles to cure genetic defects—and the release of genetically modified plants are under intense scrutiny in the press. Lawmaking bodies around the world are debating legislation to regulate both efforts. What are the potential perils and benefits of introducing recombinant genes into human beings, food plants, and other organisms? This question represents one of the great ethical challenges of the twenty-first century, and is a recurrent theme in this chapter.

19.1 Using Recombinant DNA Techniques to Manufacture Proteins: The Effort to Cure Pituitary Dwarfism

To understand the basic techniques and tools that are used by genetic engineers, let's consider the effort to treat pituitary dwarfism in humans. The pituitary gland is a structure at the base of the mammalian brain that produces several important molecules, including a protein that stimulates growth. The molecule, which was found to be just 191 amino acids long, was named growth hormone. The gene that codes for this protein is called *GH1*.

Once growth hormone was discovered, researchers immediately suspected that at least some forms of inherited dwarfism might be due to a defect in the GH1 protein. This hypothesis was confirmed when it was established that people with certain types of dwarfism produce little growth hormone or none at all. These people have defective copies of *GH1* and exhibit pituitary dwarfism, type I.

By studying the pedigrees of families in which dwarfism was common, several teams of researchers established that pituitary dwarfism, type I, is an autosomal recessive trait (see Chapter 13). Stated another way, affected individuals have two copies of the defective allele. People who have only one defective allele are carriers of the trait, meaning that they can transmit the defective allele to their offspring but are not themselves affected. Affected individuals grow more slowly than average people, reach puberty from two to ten years later than average, and are short in stature as adults—typically no more than 120 cm (4 feet) tall.

Why Did Early Efforts to Treat the Disease Fail?

Once the molecular basis of pituitary dwarfism was understood, physicians began treating the disease with injections of growth hormone. This approach was inspired by the spectacular success that physicians had achieved in treating type I diabetes mellitus. Diabetes mellitus is due to a deficiency of the peptide hormone insulin, and clinicians had been able to alleviate the disease's symptoms by injecting patients with insulin from pigs. Early trials showed that people with pituitary dwarfism could be treated successfully with growth hormone therapy, but only if the protein came from humans. Growth hormones isolated from pigs, cows, or other animals were ineffective. Until the 1980s, however, the only source of human growth hormone was pituitary glands dissected from human cadavers. Up to 20,000 pituitary glands had to be collected from cadavers to supply enough growth hormone to treat the population of affected individuals. As a result, the drug was extremely scarce and expensive.

Meeting demand turned out to be the least of the problems with growth hormone therapy, however. To understand why, recall from Chapter 3 that infectious proteins called prions can cause degenerative brain disorders in mammals. Kuru and Creutzfeldt–Jakob disease are prion diseases that affect humans. Certain prion diseases, including some forms of Creutzfeldt–Jakob disease, are hereditary. In most cases, however, people and other animals become infected with prion diseases by directly ingesting prion proteins in food. When physicians found that some of the children treated with human growth hormone developed Creutzfeldt–Jakob disease in their teens and twenties, the physicians realized that the supply of growth hormone was contaminated with a prion protein from the brains of the cadavers supplying the hormone. In 1984, the use of growth hormone isolated from cadavers was banned.

Using Recombinant DNA Technology to Produce a Safe Supply of Growth Hormone

To replace natural sources of growth hormone, researchers turned to genetic engineering. The idea was to insert fully functional copies of human *GH1* into the bacterium *Escherichia coli*; the hope was that huge quantities of recombinant bacterial cells could be grown. If so, then the recombinant cells would produce uncontaminated growth hormone in sufficient quantities to meet demand at an affordable price.

To meet this goal, investigators had to find *GH1*, obtain many copies of the gene, and insert them into *E. coli* cells. Their ability to do this hinged on using three of the most basic tools in molecular biology. Let's consider each in turn.

Using Reverse Transcriptase to Produce cDNAs An enzyme called reverse transcriptase is responsible for a major exception to the central dogma of molecular biology: It allows information to flow from RNA to DNA. **Reverse transcriptase** can do this because it catalyzes the synthesis of DNA from an RNA template.

DNA that is produced from RNA is called **complementary DNA**, or **cDNA**. Although reverse transcriptase initially produces a single-stranded cDNA, it is also capable of synthesizing the complementary strand to yield a double-stranded DNA. In many cases, however, researchers add a primer to single-stranded cDNAs and use DNA polymerase to synthesize the second strand (**Figure 19.1**).

Reverse transcriptase played a key role in the search for the growth hormone gene because *GH1* is actively transcribed in cells from the pituitary gland. To get the gene hunt started, then, researchers isolated mRNAs from pituitary glands and used reverse transcriptase to reverse-transcribe the mRNAs to cDNAs. When this step was completed, researchers had a solution containing double-stranded cDNAs that corresponded to each gene that is actively expressed in pituitary cells. Their next move? Isolating each of these cDNAs and obtaining many identical copies of them. Efforts to produce many identical copies of a gene are referred to as **DNA cloning**. If a researcher says that she has cloned a gene, it means that she has isolated it and produced many identical copies.

How Are Plasmids Used in Cloning? In many cases, researchers can clone a gene by inserting it into a small, circular DNA molecule called a **plasmid**. You might recall from Chapter 7 that plasmids are common in bacterial cells but are physically distinct from the bacterial chromosome. They are not required by the cell for normal growth and reproduction, and most replicate independently of the chromosome. Some plasmids carry genes for antibiotic resistance or other traits that increase the cell's ability to grow in a particular environment. In natural populations of bacteria, plasmids are sometimes copied and transferred from one cell to another bacterial cell during conjugation (see Chapter 12). Plasmids have also been discovered in the nuclei of certain unicellular eukaryotes, including baker's yeast.

Researchers quickly realized that if they could splice a loose piece of DNA into a plasmid and then insert the modified plasmid into a bacterial cell, the engineered plasmid would be replicated and passed on to daughter cells as the bacterium grew and divided. If this recombinant bacterium were placed in a nutrient broth and allowed to grow overnight, billions of copies of the plasmid DNA would be produced in the resulting bacterial cells. When plasmids are used to carry DNA from another source, the plasmid is called a **cloning vector,** or simply a **vector.** Vector DNA can carry foreign DNA and make several copies of it. Plasmids are often used as vectors.

Biologists could harvest the recombinant genes by breaking the bacteria open, isolating all of the DNA, and then separating the plasmids from the main chromosomes. This is a promising strategy for obtaining many identical copies of a gene, but how can a gene be inserted into a plasmid in the first place?

Using Restriction Endonucleases and DNA Ligase to Cut and Paste DNA To cut a gene out for later insertion into a cloning vector, researchers use an enzyme called a restriction

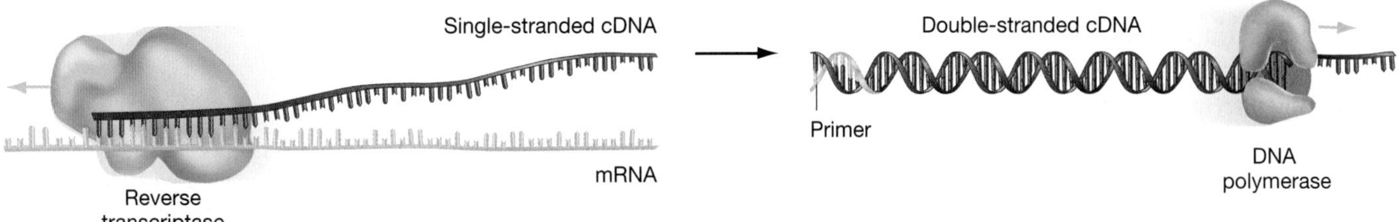

FIGURE 19.1 Reverse Transcriptase Catalyzes the Synthesis of DNA from RNA. The single-stranded DNA produced by reverse transcriptase is complementary to the RNA template. The cDNA can be made double stranded by reverse transcriptase or DNA polymerase. DNA polymerase requires a primer.

endonuclease. Restriction endonucleases are bacterial enzymes that cut DNA molecules at specific base sequences.

Most of the 400 known restriction endonucleases cut DNA only at sites that form palindromes. In English, a word or sentence is a palindrome if it reads the same way backward as it does forward. "Madam, I'm Adam" is an example. In biology, a stretch of double-stranded DNA forms a palindrome if the $5' \rightarrow 3'$ sequence of one strand is identical to the $5' \rightarrow 3'$ sequence on the antiparallel, complementary strand.

To insert the pituitary gland cDNAs into plasmids, researchers performed the series of steps outlined in **Figure 19.2**. They began with a plasmid containing a palindromic sequence that is cut by a specific restriction endonuclease. Notice in step 1 of the figure that the $5' \rightarrow 3'$ sequence on one strand of the endonuclease recognition site reads GAATTC and that the $5' \rightarrow 3'$ sequence on the complementary strand also reads GAATTC. Then they attached the same palindromic sequence to the ends of each cDNA in their sample (step 2).

The researchers' next step was to cut the recognition sites at the ends of each cDNA with a restriction endonuclease called EcoRI. (The name stands for *Escherichia coli* restriction I, because it was the first restriction endonuclease discovered in *E. coli*.) As the top part of step 3 in Figure 19.2 shows, however, EcoRI did not break its recognition sequence neatly in two. Like most restriction endonucleases, it made a staggered cut in the palindrome. The result was DNA fragments that had **sticky ends**. The ends of the fragments are said to be sticky because the single-stranded bases on one fragment are complementary to the single-stranded bases on the other fragment. As a result, the two ends will tend to pair up and hydrogen bond to one another.

The creation of sticky ends in DNA is significant. If restriction sites in different DNA sequences are cut with the same restriction endonuclease, the presence of sticky ends allows the resulting fragments to be spliced together via complementary base pairing. This is the essence of recombinant DNA technology—the ability to create novel combinations of DNA sequences by cutting specific sequences and pasting them into new locations.

Step 4 of Figure 19.2 demonstrates this aspect of genetic engineering. The base sequence of the sticky ends of the plasmid are shown in red, while the base sequence of the sticky ends of the cDNA are shown in black. As the figure shows, the complementary base pairs in the pairs of sticky ends bind to each other via complementary base pairing. Researchers then used **DNA ligase**—the enzyme introduced in Chapter 14 that connects Okazaki fragments during DNA replication—to seal the recombinant pieces of DNA together at the arrows marked in green (step 5).

At the end of this cutting and pasting step in the search for the growth hormone gene, researchers had a set of recombinant plasmids. Each contained a different cDNA made from a human pituitary gland mRNA.

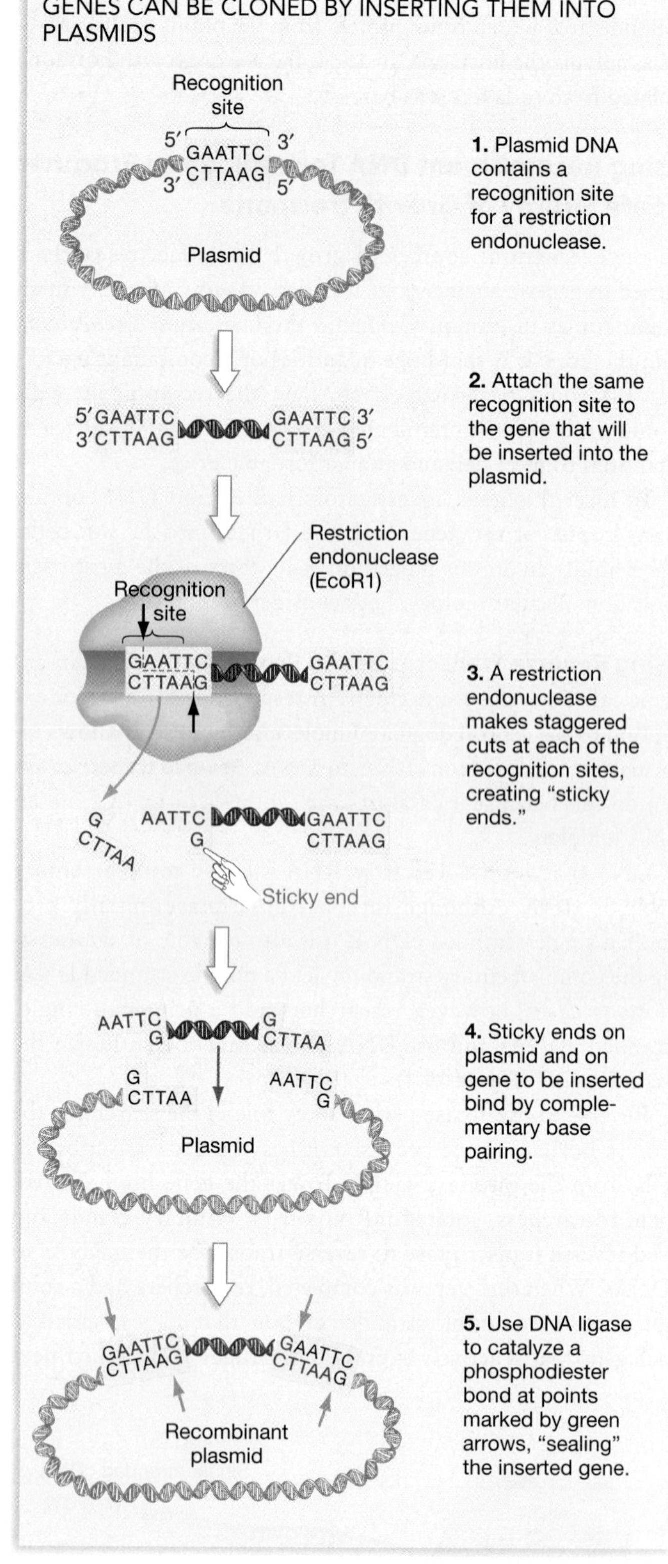

FIGURE 19.2 Genes Can Be Cloned into Plasmids. To clone a gene, researchers often start by inserting the sequence into a plasmid. Once the gene is in the plasmid, the plasmid can be introduced into bacterial cells and grown to produce many identical copies of the gene.

QUESTION Some restriction endonucleases make even cuts instead of staggered cuts. Are they useful for cloning?

Transformation: Introducing Recombinant Plasmids into Bacterial Cells Plasmids serve as a vector—meaning, a vehicle for transferring recombinant genes to a new host. If a recombinant plasmid can be inserted into a bacterial or yeast cell, the foreign DNA will be copied and transmitted to new cells as the host cell grows and divides. In this way, researchers can obtain millions or billions of copies of specific genes. Although viruses are used as vectors in some applications (see Section 19.5), plasmids are often the vector of choice in genetic engineering.

How can recombinant plasmids be introduced into cells? Cells that take up DNA from the environment and incorporate it into their genomes are said to undergo **transformation**. To transform bacterial cells with a plasmid, researchers increase the permeability of the cell's plasma membranes using a specific chemical treatment or an electrical shock. Typically, just a single plasmid enters the cell during this treatment. The resulting cells are spread out on plates in a layer thin enough to ensure that each cell is physically isolated. The cells are then allowed to grow into colonies containing millions of identical cells.

Producing a cDNA Library **Figure 19.3** summarizes the steps covered thus far in the hunt for the growth hormone gene. The result, shown in step 5, is a collection of transformed bacterial cells. Each of the cells contains a plasmid with one cDNA from a pituitary gland mRNA. A collection of genes like this, each of which is inserted into a vector, is called a **DNA library**. If the genes are cDNAs from a particular cell type or tissue, the library is called a **cDNA library**. If the genes are fragments of DNA that collectively represent the entire genome of an individual, the library is called a **genomic library**. In each case, the library is made up of cloned genes—each gene present can be produced in large quantity and isolated in pure form. If you understand this concept, you should be able to make a diagram showing the steps involved in making a genomic library, starting with DNA from your own cells and using the restriction endonuclease EcoRI to cut the genome into fragments.

DNA libraries are important because they give researchers a way to store information from a particular cell type or genome in a form that is accessible. But like a college or community library, a DNA library isn't very useful unless there is a way to retrieve specific pieces of information. At your college library, you use call numbers or computer searches to retrieve a particular book or article. How do you go about retrieving a particular gene from a DNA library? For example, how did researchers find the growth hormone gene in the cDNA library of the human pituitary gland?

How Is a DNA Library Screened? Molecular biologists are often faced with the task of finding one specific gene in a large collection of DNA fragments. To do this, you have to have a **probe**—a marked copy of the molecule you are looking for. A DNA probe, for example, is a single-stranded fragment of a labeled, known gene that binds to a single-stranded complementary sequence in the sample of DNA being analyzed. By binding to the target sequence, it marks it—distinguishing it from

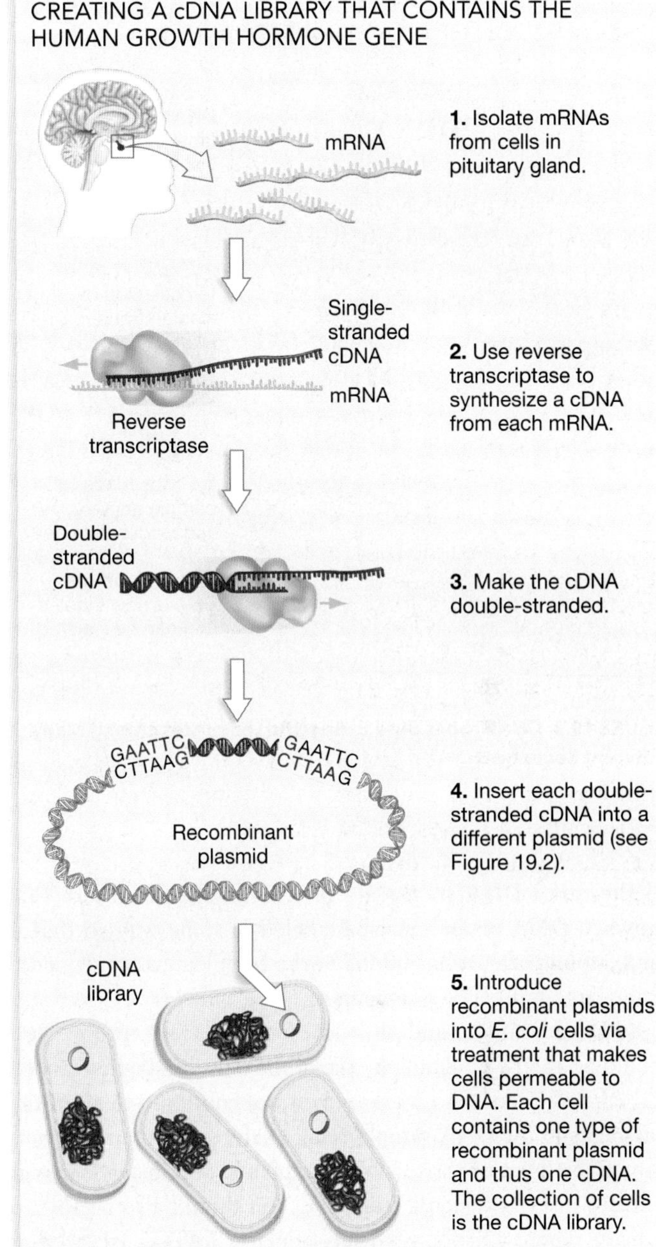

FIGURE 19.3 Complementary DNA (cDNA) Libraries Represent the mRNAs in a Cell. The hunt for growth hormone gene began with the creation of a cDNA library from pituitary-gland cells. Because growth hormone is produced in this tissue, the cDNA for growth hormone should exist somewhere in the library.

QUESTION Would each type of cDNA in the library be represented just once? Why or why not?

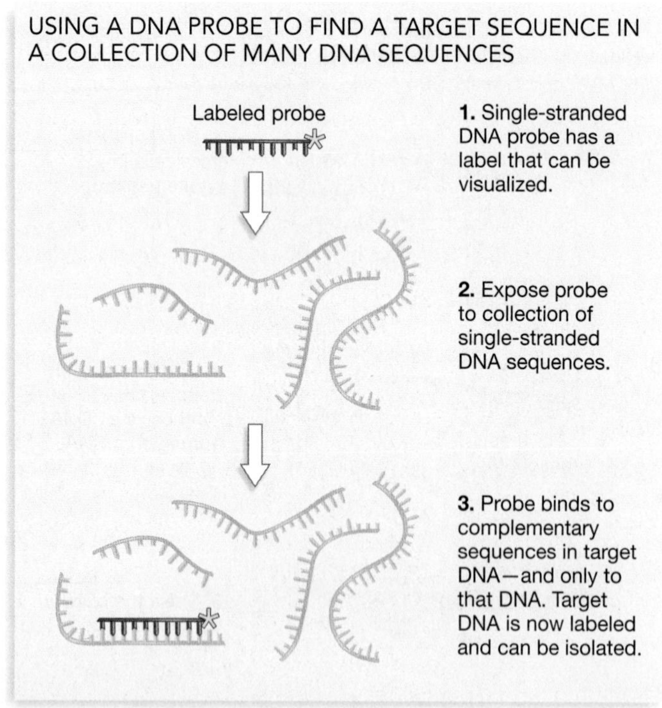

FIGURE 19.4 DNA Probes Bind to Specific Sequences among Many Different Sequences.

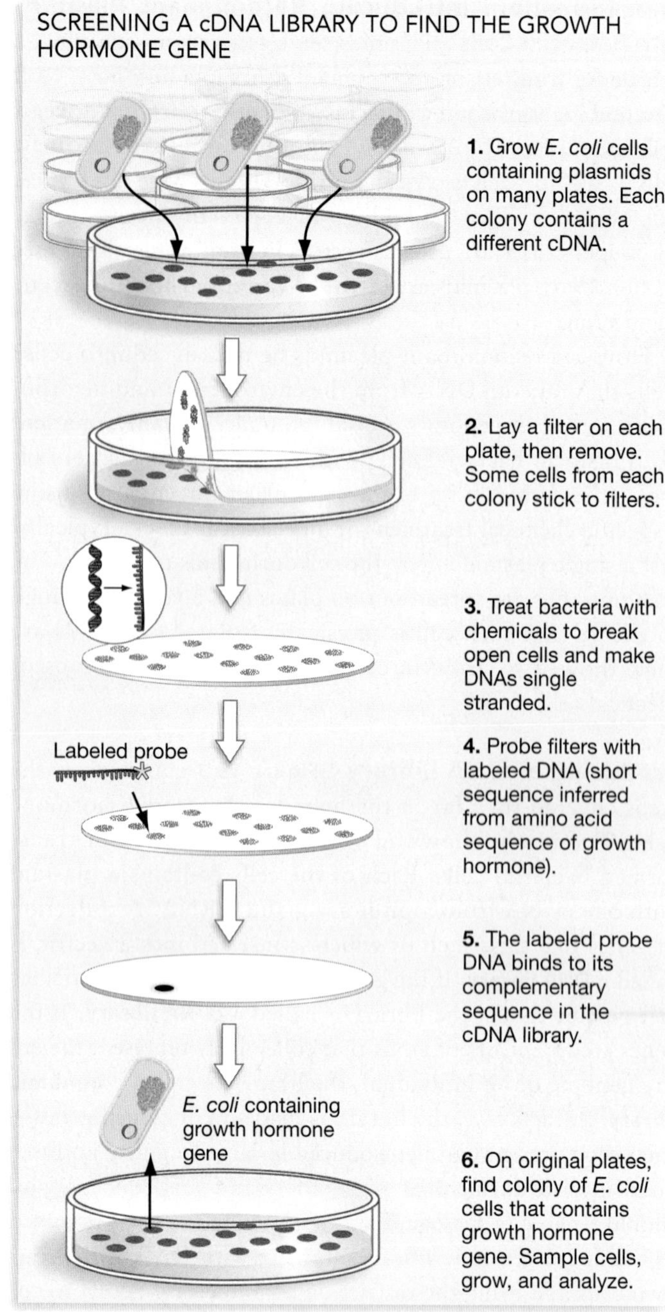

FIGURE 19.5 Finding Specific Genes by Probing a cDNA Library. Once bacteria with the growth hormone cDNA were isolated from the cDNA library, researchers could produce virtually unlimited quantities of the cDNA.

all the other DNA fragments in the sample. As **Figure 19.4** shows, a DNA probe must be labeled in some way so that it can be found after it has bound to the complementary sequence in the large sample of fragments.

● If you understand the concept of a DNA probe, you should be able to explain why the probe must be single stranded and labeled in order to work; why it does not bind to all of the fragments in the DNA sample being analyzed, or even to a random assortment of those DNA fragments; and instead, why it binds to just one specific fragment. You should also be able to indicate where a probe with the sequence AATGC (recall that sequences are always written 5′ to 3′) will bind to a target DNA with the sequence TTTTACCCATTTACGATTGGCCT (again written 5′ to 3′).

To find an appropriate probe for the human growth hormone gene, researchers began by using the genetic code to infer the approximate DNA sequence of *GH1*. This was possible because the sequence of amino acids in the polypeptide was known. Thus, the researchers could infer which mRNA codon and DNA sequences coded for each amino acid. You did similar exercises in Chapter 15. But recall from Chapter 15 that the genetic code is redundant, so it is possible for more than one codon to code for an amino acid. As a result, the researchers' inferred sequence for the growth hormone gene was actually a set of related sequences.

The next step was to synthesize many copies of a short, single-stranded stretch of DNA that was complementary to the inferred sequence. Because these molecules would bind to single-stranded fragments from the actual gene via complementary base pairing, they could act as a probe. In this case, researchers attached a radioactive atom to the probe.

Figure 19.5 shows how researchers used this probe to find the plasmid in the cDNA library that contained *GH1*. In a similar vein, **Box 19.1** (page 398) explains how DNA, RNA, or protein probes are used to find particular molecules in a collection of molecules separated on a gel by electrophoresis.

As predicted, the radioactively labeled growth hormone probe bound to its complementary sequence in the cDNA library. The probe had identified the recombinant cell that contained the human growth hormone.

Mass-Producing Growth Hormone To accomplish their goal of producing large quantities of the human growth hormone, the investigators used recombinant DNA techniques to transfer the growth hormone cDNA to a new plasmid. The plasmid in question was one containing a promoter sequence, like those introduced in Chapter 17, that is recognized by *E. coli*'s RNA polymerase holoenzyme. The recombinant plasmids were then introduced into *E. coli* cells.

The *E. coli* cells that resulted from these steps contained a gene for human growth hormone attached to an *E. coli* promoter. As a result, the transformed *E. coli* cells began to transcribe and translate the human growth hormone gene. (Note that because the cDNAs are synthesized from mature mRNAs, which do not contain introns, cDNAs do not contain introns either. Thus, bacterial cells can translate an mRNA transcribed from a cDNA directly—no splicing is required.)

The human growth hormone accumulated in the cells and could be isolated and purified. Bacterial cells containing the human growth hormone gene are now grown in huge quantities. These cells have proved to be a safe and reliable source of the human growth hormone protein. The effort to cure pituitary dwarfism using recombinant DNA technology was a spectacular success.

Ethical Concerns over Recombinant Growth Hormone

As supplies of growth hormone increased, physicians used it in treating not only people with pituitary dwarfism but also children who were short, although they had no actual growth hormone deficiency. Even though the treatment requires several injections per week until adult stature is reached, growth hormone therapy was popular because it often increased the height of these children by a few centimeters. In essence, the drug was being used as a cosmetic—a way to improve appearance in cultures where height is deemed attractive. But should a parent request growth hormone treatment for genetically normal children to change their appearance? If short people are discriminated against in a culture, is a medical treatment a better solution than education and changes in attitudes? And what if parents wanted a tall child to be even taller to enhance her potential success as, say, a basketball player? Currently, the U.S. Food and Drug Administration has approved the use of human growth hormone for only the shortest 1.2 percent of children, who are projected to reach adult heights of less than 160 cm (5′3″) in males and 150 cm (4′11″) in women.

Growth hormone has also been found to enhance the maintenance of bone density and muscle mass. As a result, it has become a popular performance-enhancing drug for athletes. Part of its popularity stems from the fact that it is virtually undetectable in the drug tests currently administered by governing bodies. Should athletes be able to enhance their physical skills by taking hormones or other types of drugs? And is the drug safe at the dosages being used by athletes? These types of questions are being debated by physicians and researchers, agencies that govern sports, and legislative bodies. Even though these questions are still unanswered, it is clear that, while solving one important problem, recombinant DNA technology created others. Throughout this chapter, an important theme is that genetic engineering has costs that must be carefully weighed against its benefits.

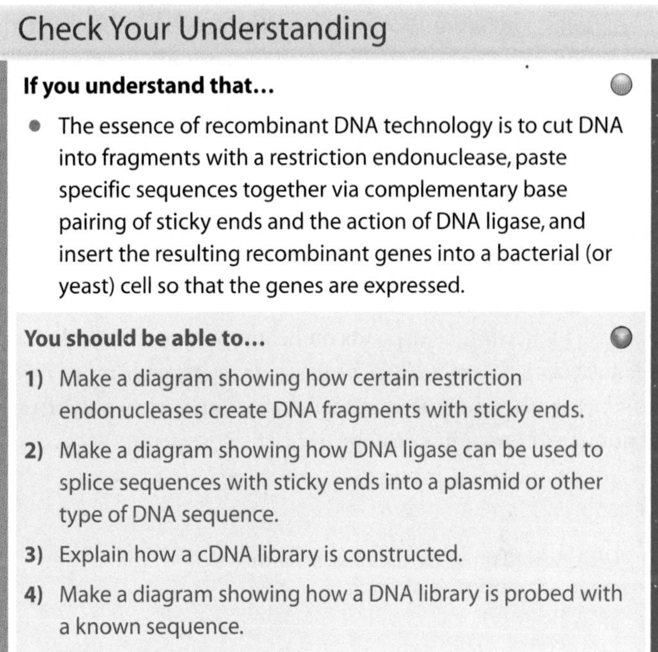

Check Your Understanding

If you understand that...

- The essence of recombinant DNA technology is to cut DNA into fragments with a restriction endonuclease, paste specific sequences together via complementary base pairing of sticky ends and the action of DNA ligase, and insert the resulting recombinant genes into a bacterial (or yeast) cell so that the genes are expressed.

You should be able to...

1) Make a diagram showing how certain restriction endonucleases create DNA fragments with sticky ends.
2) Make a diagram showing how DNA ligase can be used to splice sequences with sticky ends into a plasmid or other type of DNA sequence.
3) Explain how a cDNA library is constructed.
4) Make a diagram showing how a DNA library is probed with a known sequence.

Producing Human Growth Hormone

19.2 Another Approach to Cloning: The Polymerase Chain Reaction

Inserting a gene into a bacterial plasmid is one method for cloning DNA. The polymerase chain reaction is another. The **polymerase chain reaction (PCR)** is an in vitro DNA synthesis reaction in which a specific section of DNA is replicated over and over to amplify the number of copies of that sequence. It's a technique for generating many identical copies of a particular section of DNA.

Although PCR is much faster and technologically easier than cloning genes into a DNA library, there is a catch: PCR is possible only when a researcher already has some information about

DNA sequences near the gene in question. Sequence information is required because to do a polymerase chain reaction, you have to start with short lengths of single-stranded DNA that match sequences on either side of the gene of interest. These short segments act as primers for the synthesis reaction.

As **Figure 19.6a** shows, the primer sequences must be complementary to bases on either side of the gene in question. One primer is complementary to a sequence on one strand upstream of the target DNA—the sequence of interest; the other primer is complementary to a sequence on the other strand, downstream of the target DNA. If the target DNA molecule is made single stranded, then the primers will bond or anneal to their complementary sequences as shown in **Figure 19.6b**. You might recall that DNA polymerase cannot work unless a sequence is primed in this way. Once the primers are bound, DNA polymerase can extend each strand in the 5′ to 3′ direction.

As step 1 of **Figure 19.7** shows, a PCR procedure begins with a reaction mix containing an abundant supply of the four deoxynucleoside triphosphates (dNTPs; see Chapter 14), copies of the template DNA—that is, a DNA sample that includes the gene of interest—copies of the two primers, and an enzyme called *Taq* polymerase. *Taq* polymerase is a DNA polymerase found in the bacterium *Thermus aquaticus*, which was originally discovered in a hot spring inside Yellowstone National Park, Wyoming. *Taq* polymerase is the enzyme of choice in PCR because the technique depends on heating the reaction mix, and because *Taq* polymerase is heat stable. Although most DNA polymerases are destroyed at high temperature, *Taq* polymerase continues to function normally even when heated to 95°C.

(a) PCR primers must be located on either side of the target sequence, on opposite strands.

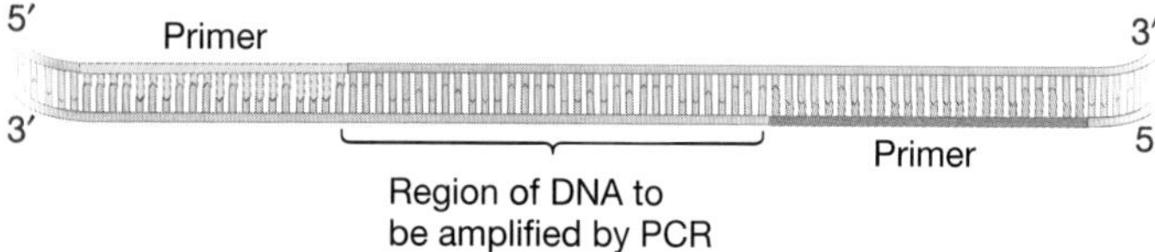

(b) When target DNA is single stranded, primers bind and allow DNA polymerase to work.

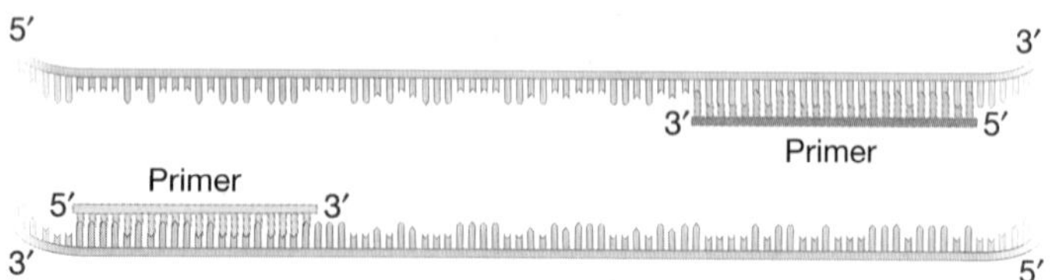

FIGURE 19.6 The Polymerase Chain Reaction (PCR) Depends on Having Specific Primers. (a) The orange and yellow sequences indicate a set of single-stranded primers, which bracket the region of DNA to be amplified. To do PCR, a researcher must have enough sequence information to be able to design the appropriate primers. **(b)** The primers bind to single-stranded target DNA as shown.

EXERCISE Indicate where DNA polymerase would begin to work on each strand and add arrows indicating the direction of DNA synthesis.

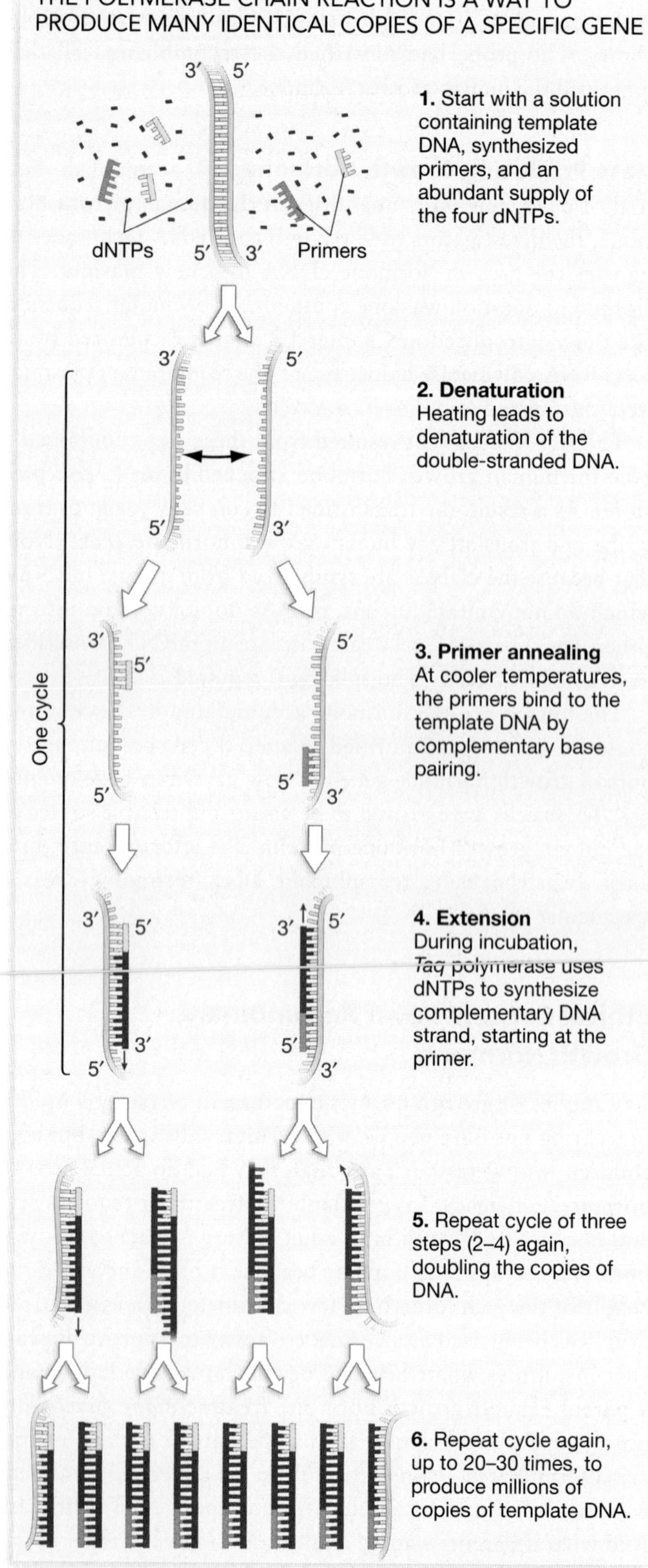

FIGURE 19.7 The Polymerase Chain Reaction (PCR) Is a Method for Producing Many Copies of a Specific Sequence. Each PCR cycle (denaturation, primer annealing, and extension) results in a doubling of the number of copies of the sequence between and including the primers.

EXERCISE Draw the events that produced the sequences in step 5.

As step 2 in Figure 19.7 shows, PCR gets under way when the reaction mix is heated to 94°C. At this temperature, the double-stranded template DNA denatures. This means that the two DNA strands separate, forming single-stranded templates. Then the mixture is allowed to cool to 50–60°C. In this temperature range, some of the denatured DNA strands re-form double helices. But some of the primers bond, or anneal, to complementary portions of the single-stranded template DNA (step 3). This step is called *primer annealing*. The reaction mix is then heated to 72°C. At this temperature, *Taq* polymerase synthesizes the complementary DNA strand from the dNTPs, starting at the primer. This step is called *extension* (step 4). The temperature changes required in each step are now automated by PCR machines.

The denaturation, primer annealing, and extension steps constitute a single PCR cycle. If one copy of the template sequence existed in the sample originally, then there are two copies at the end of the first cycle (see step 4 in Figure 19.7). The two copies present at the start of the second cycle then act as templates for another round of denaturation, primer annealing, and extension, and four copies of the target gene exist at the end of cycle number 2 (step 5). As the cycle repeats, the amount of template sequence doubles each time (step 6). Doubling occurs because each newly synthesized segment of DNA serves as a template in the subsequent cycle. Starting with a single copy, successive cycles result in the production of 2, 4, 8, 16, 32, 64, 128, 256 copies, and so on. A total of n cycles generates 2^n copies; so in just 20 cycles, one sequence can be amplified to over a million copies. By performing up to 30 cycles, researchers obtain enormous numbers of copies of the template sequence.

PCR in Action: Studying Fossil DNA

To understand why PCR is so valuable, consider a recent study by Svante Pääbo and colleagues. These biologists wanted to analyze DNA recovered from the 30,000-year-old bones of a fossilized human of the species *Homo neanderthalensis*. The individual involved was actually the first Neanderthal fossil ever discovered, from the Neander valley of Germany. Pääbo's goal was to determine the sequence of bases in the ancient DNA, compare it with DNA from modern humans (*Homo sapiens*), and analyze how similar the two species are. If some modern humans have sequences that are identical or almost identical to the base sequences found in Neanderthals, it would support the hypothesis that some of us inherited at least some of our DNA directly from a Neanderthal ancestor. That could happen only if *Homo sapiens* and *Homo neanderthalensis* interbred while they coexisted in Europe about 30,000 years ago.

The Neanderthal bone was so old, however, that most of the DNA in it had degraded into tiny fragments. The biologists could recover only a minute amount of DNA that was still in moderate-sized pieces. But two fortunate events occurred: (1) The Neanderthal DNA sample included a few fragments of the gene region that Pääbo's team wanted to study, and (2) the researchers were able to design primers that bracketed this gene region, based on the sequence of highly conserved sections of the same gene from *Homo sapiens*. Using PCR, the researchers were able to produce millions of copies of a specific DNA section from a Neanderthal who died over 30,000 years ago. Upon analyzing these sequences, the team found that they are highly distinct from the same gene segment found in modern humans. Subsequent work with DNA from seven other Neanderthal fossils, from locations throughout Europe, gave the same result. These data support the hypothesis that Neanderthals never interbred with modern humans—even though they lived in the same areas of Europe at the same time.

PCR has been used to study other fossil DNAs, with the current record holder for oldest DNA to be amplified via PCR coming from 17-million-year-old magnolia trees. But PCR is useful in a wide array of other situations as well. Forensic biologists, who use biological analyses to help solve crimes, clone DNA from tiny drops of blood or hair gathered at crime scenes. The copied DNA can then be analyzed to identify victims or implicate perpetrators. Genetic counselors, who advise pregnant couples on how likely their offspring are to suffer from inherited diseases, can use PCR to find out if an embryo being carried by a client has alleles associated with deadly illness. For example, it is possible to obtain a small sample of the embryo's cells, isolate the DNA, add primers on either side of the Huntington's disease gene, and clone enough copies to analyze the alleles and let the parents know if their offspring is at risk for the illness.

Because the complete genomes of a wide array of organisms have now been sequenced, researchers can find appropriate primer sequences to use in cloning almost any target gene, in a wide array of organisms, via PCR. The polymerase chain reaction is now one of the most basic and widely used techniques in molecular biology.

Check Your Understanding

If you understand that...

- PCR is a technique for amplifying a specific region of DNA into millions of copies, which can then be sequenced or used for other types of analyses.

You should be able to...

1) Explain the steps in a PCR cycle.
2) Write down the sequence of a DNA strand 50 base pairs long, then design 20-base-pair-long primers that would allow you to amplify the segment via PCR.
3) Explain why "chain reaction" is an appropriate part of PCR's name.

at www.masteringbio.com
The Polymerase Chain Reaction

BOX 19.1 Southern Blotting

Southern blotting was named after its inventor, Edwin Southern, and it is among the most basic techniques in molecular biology. It is a multistep procedure that allows researchers to identify and characterize specific genes within an organism's genome. If you have a DNA sample from an individual and wish to find a specific gene in that sample, Southern blotting is a way to do it.

The first step in Southern blotting is to obtain DNA from the cells of the organisms being studied and digest it with restriction endonucleases (**Figure 19.8**, step 1). The DNA fragments generated in this way can then be separated by gel electrophoresis (steps 2 and 3). Once the fragments are sorted by size, they are treated with an alkaline solution that breaks the hydrogen bonds between base pairs, resulting in the formation of single-stranded DNA (step 4). Next the fragments of single-stranded DNA are transferred from the gel to a piece of nitrocellulose or nylon filter, using the blotting technique illustrated in step 5.

The product of these steps is a series of single-stranded DNA fragments, sorted by size and permanently bound to the filter. In some cases, these fragments represent the entire genome of the organism being studied. To find a certain gene in this collection of fragments, a researcher must have a DNA sequence that is complementary to some region of that gene. Recall that such a sequence is called a probe. The

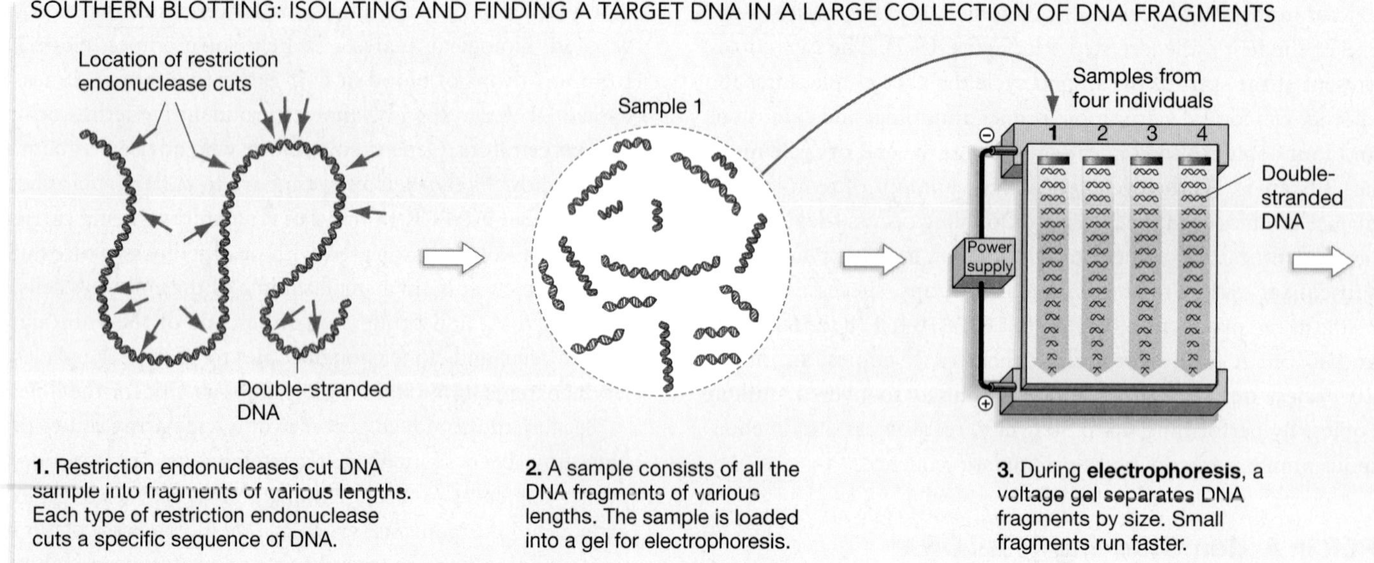

FIGURE 19.8 Southern Blotting Protocol. Southern blotting is a technique for locating a specific gene in a DNA sample containing many genes. In essence, the investigator cuts the DNA of interest into fragments, separates the fragments by size using gel electrophoresis, and then probes the fragments to find a specific gene.

19.3 Dideoxy DNA Sequencing

Once researchers have cloned a gene from a DNA library or via PCR, determining the gene's base sequence is usually one of the first things they want to do. Understanding a gene's sequence is valuable for a variety of reasons:

- Sequence differences among alleles can be analyzed to understand why some versions of the gene function better or differently than others do. The Huntington's disease allele, for example, has a different base sequence from alleles of the same gene that don't cause disease.
- Once a gene's sequence is known, it is straightforward to infer the amino acid sequence of its product from the genetic code. Understanding a protein's primary structure, in turn, is often helpful in inferring its function.
- A gene sequence can be compared with sequences of genes that have the same function in other species. These comparisons are often interesting. For example, bacteria, yeast, and humans are about as different as organisms can be; yet all three contain DNA polymerase genes with sections that are nearly identical in base sequence. Biologists explain this similarity by hypothesizing that the bacteria inhabiting your gut, the yeast in the bread you eat, and you are descended from a common ancestor that had DNA polymerase with the same sequence. Genes that are similar due to descent from a common ancestor are said to be **homologous**.

probe DNA is labeled—typically with a radioactive atom or fluorescent marker—and made single stranded by heating. The probe is then added to a solution bathing the filter (step 6). During incubation, the labeled probe binds to the fragment or fragments on the filter that have complementary base pairs. This step is called **nucleic acid hybridization.** In this way, the probe identifies the gene of interest.

To visualize which fragments hybridized with a radioactively labeled probe, a researcher lays X-ray film over the filter. As step 7 of Figure 19.8 shows, radioactive emissions from the probe DNA expose the film. The black band that results identifies the target gene. Fluorescent markers can be visualized and photographed under the appropriate wavelengths of light.

A variation on Southern blotting is based on separating RNAs via gel electrophoresis, transferring them to a filter paper, and probing them with a single-stranded and radioactively labeled DNA probe. This technique is used to identify the RNA fragments produced by a particular gene. It is called **Northern blotting,** in a lighthearted tribute to the protocol from which it was derived. The variation called **Western blotting** involves separating proteins via electrophoresis and then probing the resulting filter with an antibody probe that binds to the protein of interest. The use of antibodies in research is explored in detail in later chapters.

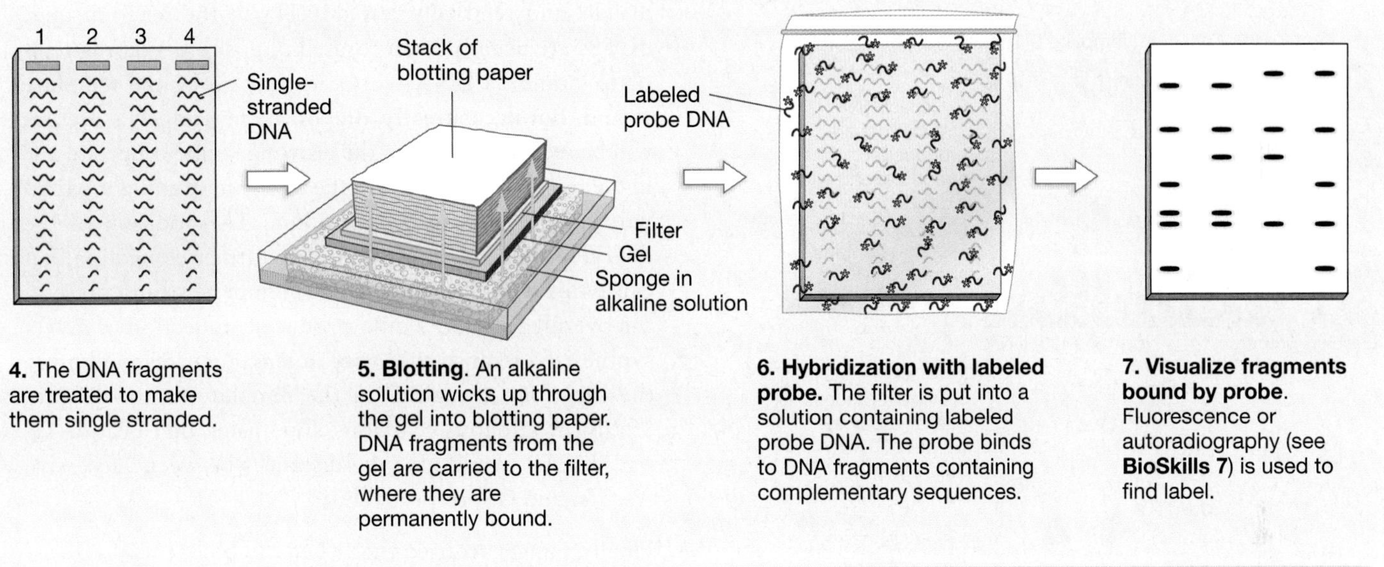

4. The DNA fragments are treated to make them single stranded.

5. Blotting. An alkaline solution wicks up through the gel into blotting paper. DNA fragments from the gel are carried to the filter, where they are permanently bound.

6. Hybridization with labeled probe. The filter is put into a solution containing labeled probe DNA. The probe binds to DNA fragments containing complementary sequences.

7. Visualize fragments bound by probe. Fluorescence or autoradiography (see **BioSkills 7**) is used to find label.

- Gene sequences can be compared to infer how closely related various species are. For example, analyzing DNA sequences isolated from fossil bones allowed researchers to evaluate whether our species, *Homo sapiens*, ever interbred with the species of human called *Homo neanderthalensis*, Neanderthals.

Given that sequence data are important, how can biologists obtain them? In 1975 Frederick Sanger developed **dideoxy sequencing** as a clever variation on the basic in vitro DNA synthesis reaction to determine exact base sequences. But saying "clever" is an understatement. Sanger had to link three important insights, summarized in **Figure 19.9**, to make his sequencing strategy work:

1. Sanger chose the name "dideoxy" because the method uses monomers called dideoxyribonucleoside triphosphates, or ddNTPs, along with deoxyribonucleoside triphosphates (dNTPs). These ddNTPs are identical to the dNTPs found in DNA, except that they lack a hydroxyl group at their 3′ carbon (Figure 19.9a). Sanger realized that if a ddNTP were added to a growing DNA strand, it would terminate synthesis. Why? A ddNTP has no hydroxyl group available on its 3′ carbon to link to the 5′ carbon on an incoming dNTP monomer. As a result, DNA polymerization stops once a ddNTP is added. Four types of ddNTPs are used in dideoxy sequencing, each named according to whether it contains adenine (ddATP), thymine (ddTTP), cytosine (ddCTP), or guanine (ddGTP).

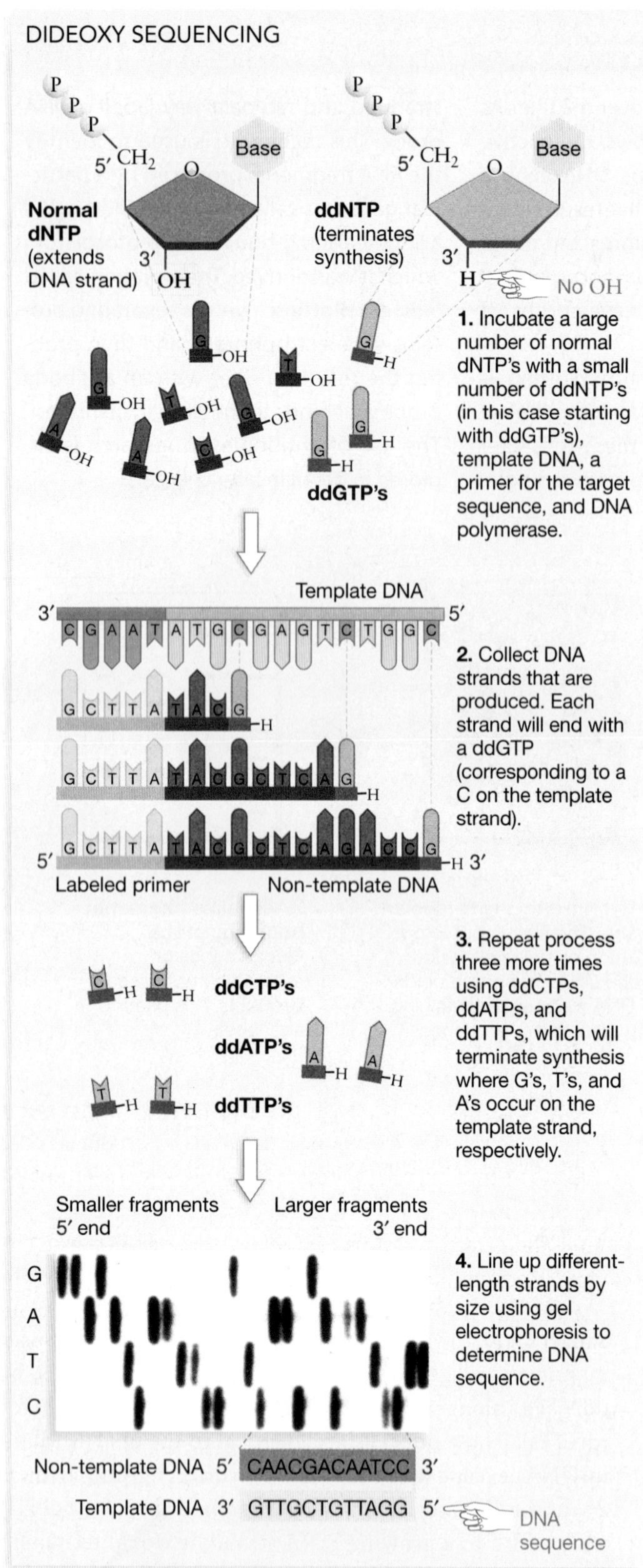

FIGURE 19.9 Dideoxy Sequencing Is a Method for Determining the Base Sequence of DNA.

● **EXERCISE** Starting at the far left end of the gel pictured in step 4 write down the entire sequence shown. Hint: The sequence starts with GG.

● **QUESTION** Why is it necessary to use a labeled DNA primer?

2. Sanger linked this property of ddNTPs to a second fundamental insight. Suppose, he reasoned, that a labeled primer could be attached to a template DNA, and that a large number of copies of this template DNA with a labeled primer attached could be added to a reaction mix. Suppose further that this population of labeled templates were incubated with DNA polymerase, the four dNTPs, and a small amount of ddGTP. Under these conditions, an array of different daughter strands would be synthesized. Each of these resulting daughter strands would be a different length, with each distinct DNA fragment ending with a ddGTP. To understand why, consider that DNA polymerase would synthesize a complementary strand from each labeled template in the reaction mix. The synthesis of each one of these complementary strands would start at the same point—the primer. Because there are many dGTPs and relatively few ddGTPs in the reaction mix, dGTPs are usually incorporated opposite C's on the template strand as DNA works its way along the template strand. But occasionally, one of the few ddGTPs present will be incorporated into the growing strand, opposite a C in the template. Which C in the template strand got paired with a ddGTP would be random. The addition of the ddGTP at this location would stop further elongation. Because this would happen in each template strand present, the overall reaction would produce a collection of newly synthesized strands that vary in length, corresponding to the location of each C in the template strand (Figure 19.9b). Analogous reactions done using ddTTP, ddATP, and ddCTP would give the distances between successive A's, T's, and G's, respectively.

3. Finally, Sanger realized that when the fragments produced by the four reactions are lined up by size, they reveal the sequence of bases in the template DNA. To line up fragments in order of size, he separated them using gel electrophoresis. As step 4 in Figure 19.9 shows, the sequence can then be read directly from the resulting gel.

● If you understand the essence of a DNA sequencing reaction, you should be able to consider a template strand with the sequence AGGCTACCATTCTGCTGAAG (remember, this is written 5′ to 3′) and write out, in the 5′ to 3′ direction, the sequence of each of the five fragments that would be produced in a sequencing reaction that includes ddGTP. (You should also be able to do the same exercise for reaction mixes containing ddCTP, ddATP, and ddTTP. There are five fragments from this template for each reaction mix, for a total of 20 from the four reactions.)

Several important innovations have enormously improved the speed and affordability of dideoxy sequencing. For example, researchers traditionally attached a radioactive isotope to the primer in the reaction mix to label the fragments of DNA produced by a sequencing reaction. This meant that each se-

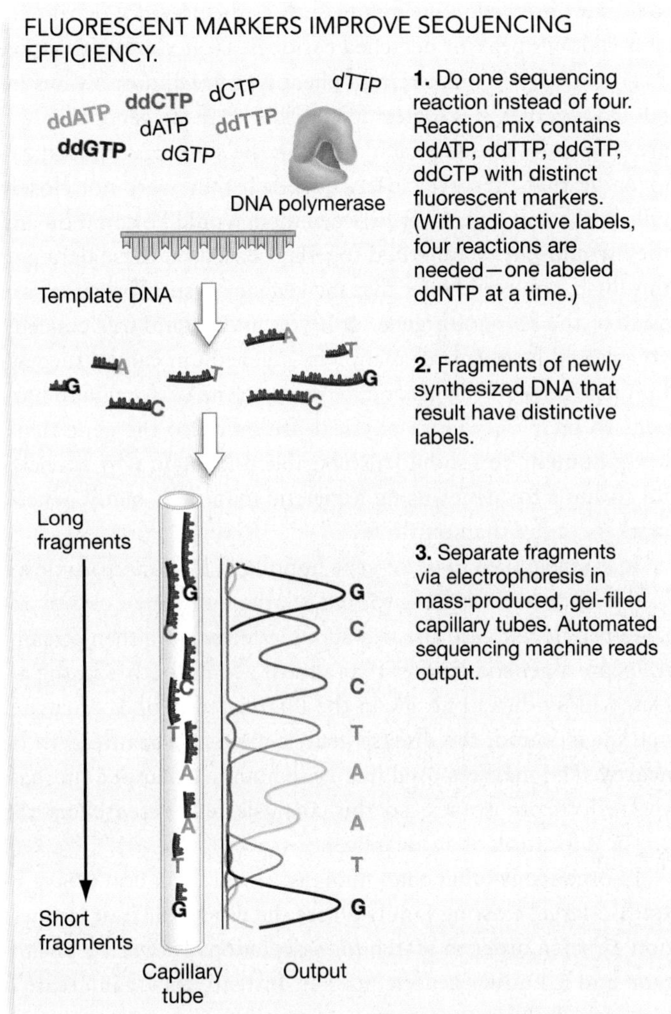

FIGURE 19.10 Use of Fluorescent Markers Simplifies Sequencing. Researchers made DNA sequencing more efficient by attaching distinctive fluorescent markers to the ddATP, ddGTP, ddCTP, and ddTTP used in sequencing reactions. Fluorescent markers allow researchers to use all four ddNTPs in a single reaction mix instead of having to perform four separate reactions. In addition, fluorescent markers are safer than radioactive labels and can be read by a machine.

EXERCISE State another advantage of this protocol compared with sequencing methods that rely on radioactive markers.

quencing experiment required four reactions—one with each of the four ddNTPs. Now fluorescent markers are bonded to the dideoxyribonucleoside triphosphates (ddNTPs) used in a sequencing reaction—with a different color attached to each ddNTP. As **Figure 19.10** shows, the switch to fluorescent markers was important for two reasons: (1) Each stretch of DNA could be sequenced with one dideoxy reaction instead of four separate ones, and (2) machines were developed to detect the fluorescence produced by each fragment and read the output of the sequencing reaction. In addition, instead of separating the fragments generated by sequencing reactions through electrophoresis in hand-prepared gels that are poured between glass plates, researchers now perform the electrophoresis step with mass-produced, gel-filled capillary tubes.

Dideoxy sequencing ranks among the greatest of all technological advances in the history of biological science. Its impact is comparable to the development of the light microscope, the electron microscope, and recombinant gene technology.

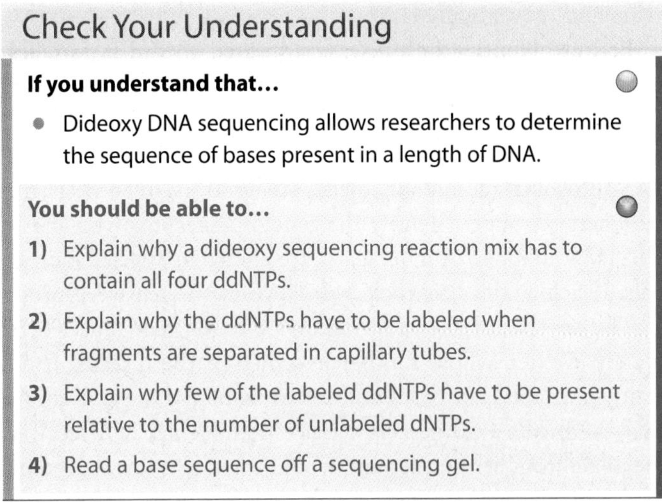
Check Your Understanding

If you understand that...

- Dideoxy DNA sequencing allows researchers to determine the sequence of bases present in a length of DNA.

You should be able to...

1) Explain why a dideoxy sequencing reaction mix has to contain all four ddNTPs.
2) Explain why the ddNTPs have to be labeled when fragments are separated in capillary tubes.
3) Explain why few of the labeled ddNTPs have to be present relative to the number of unlabeled dNTPs.
4) Read a base sequence off a sequencing gel.

19.4 Finding Genes by Mapping: The Huntington's Disease Story

For almost 100 years after Mendel's work launched the field of genetics, the nature of the gene was a black box. Then the confirmation of DNA as the genetic material and the discovery of the double helix changed everything. It launched an explosive era of work on the nature of DNA synthesis and the central dogma of molecular biology—including cracking the genetic code and investigating the mechanisms responsible for controlling transcription and translation.

Although work on the regulation of gene expression continues at a rapid clip, the field of genetics has also become focused on efforts to find and characterize individual genes—to understand the connection between genotype and phenotype as explicitly and directly as possible. Efforts to engineer genes, for example, depend on having copies of specific genes in hand. The question is, How do researchers find the genes associated with certain traits in the first place? How do you find the gene responsible for seed shape in peas, or white eyes in flies, or the mutS DNA repair enzyme in *E. coli*?

As an example of how this type of gene hunt is done, let's consider the first successful effort ever conducted in humans—the search for the gene associated with Huntington's disease.

How Was the Huntington's Disease Gene Found?

Huntington's disease is a rare but devastating illness. Typically, affected individuals first show symptoms between the ages of 35 and 45. At onset, an individual appears to be clumsier than normal and tends to develop small tics and abnormal movements. As the disease progresses, uncontrollable movements become more pronounced. Eventually the affected individual twists and writhes involuntarily. Personality and intelligence are also affected—to the extent that the early stage of this disease is sometimes misdiagnosed as the brain disorder schizophrenia. The illness may continue to progress for 10 to 20 years and is eventually fatal.

Because Huntington's disease appeared to run in families, physicians suspected that it was a genetic disease. Recall from Chapter 13 that an analysis of pedigrees from families affected by Huntington's disease suggested that the trait was due to a single, autosomal dominant allele. To test this hypothesis, researchers set out to identify the gene or genes involved and to document that one or more genes are altered in affected individuals. Reaching this goal took over 10 years of intensive effort.

The search for the Huntington's disease gene was led by Nancy Wexler, whose mother had died of the disease. If the trait was indeed due to an autosomal dominant allele, it meant that there was a 50 percent chance that Wexler had received the allele from her mother and would begin to show symptoms when she reached middle age.

Using Genetic Markers To locate the gene or genes associated with a particular phenotype, such as a disease, researchers start with a **genetic map,** also known as a **linkage map** or **meiotic map** (see Chapter 13). Recall that in fruit flies and other organisms, the relative positions of genes on the same chromosome can be determined by analyzing the frequency of recombination between pairs of genes. In contrast to a genetic map, a **physical map** of the genome records the absolute position—in numbers of base pairs—along a chromosome. Techniques for creating physical maps are introduced briefly in Chapter 20.

A genetic map is valuable in gene hunts because it contains **genetic markers**—genes or other genetic loci or sites that have known locations. Each genetic marker provides a landmark at a position along a chromosome that is known relative to other markers.

To understand how genetic markers can be used to locate the positions of unknown genes, suppose that you knew the position of a hair-color gene in humans relative to other genetic markers. Suppose further that various alleles of this gene contributed to the development of black hair, red hair, blond hair, and brown hair in the family that you were studying. This variation in phenotype associated with the marker is crucial. To be useful in a gene hunt, a genetic marker has to be **polymorphic**, meaning that the phenotype associated with the marker varies. In our hypothetical example, hair color is a polymorphic genetic marker.

Now suppose that the genetic disease called cystic fibrosis is common in the family you are studying. Then consider what it means if your data suggest that individuals who have cystic fibrosis almost always have black hair—even though they were just as likely as unaffected individuals to have any other inherited trait observed in the family, such as the presence or absence of a widow's peak or detached earlobes. If you observe that a certain marker and a certain phenotype are almost always inherited together, it is logical to conclude that the genes involved are physically close to each other on the same chromosome—meaning that they are closely linked. If they were not closely linked, then crossing over between them would be common and they would *not* be inherited together. Based on these data and this logic, you could infer that the gene for cystic fibrosis is very close to the hair-color gene. If you understand this concept, you should be able to explain why a gene hunt can be successful only if one of the genetic markers on the existing map happens to be in very close physical proximity to the gene that's being hunted. You should also be able to explain why it's helpful to hunt for genes using a genetic map with many genetic markers rather than only a few.

In essence, then, disease-gene hunting in humans boils down to this: Researchers have to find a large number of closely related people who are affected and unaffected, and then attempt to locate a genetic marker that almost always occurs in the affected individuals but not in the unaffected people. If such a marker is found, the disease gene is almost guaranteed to be nearby. The markers used in a gene hunt are mapped on their species' chromosomes, so this analysis tells researchers the physical location of a disease gene.

In organisms other than humans, researchers don't have to study a large, existing family where the disease or trait in question exists in order to search for associations between a phenotype and a known genetic marker. Instead, they can create a large family by arranging experimental crosses between individuals with different versions of the trait in question. Then they can look for associations between genetic markers in the offspring of the crosses and the presence of that trait in those offspring. The ability to make experimental crosses makes gene hunts in these organisms much more efficient than in humans.

It's also important to note that the types of genetic markers used in gene hunts have changed over time. Today, researchers often have a large catalog of genetic markers available, including the particularly abundant markers known as **single nucleotide polymorphisms** (**SNPs**, pronounced "snips"). A SNP is a site in DNA where some individuals in the population have different bases. For example, some might have an A at a certain site while others have a C. That site would qualify as a SNP. But in the late 1970s and early 1980s, when biologists were searching for the Huntington's gene, the best genetic markers available were restriction sites, the short stretches of DNA where restriction endonucleases cut the double helix. These sequences are also known as restriction endonuclease recognition sites. The sites that Wexler's team used were polymorphic—in the family they studied, some individuals had sites where cuts occurred; but in other individuals, the DNA sequences at the same site were different and no cuts occurred. Thus, each restriction site was either present in an individual or not present—just as an

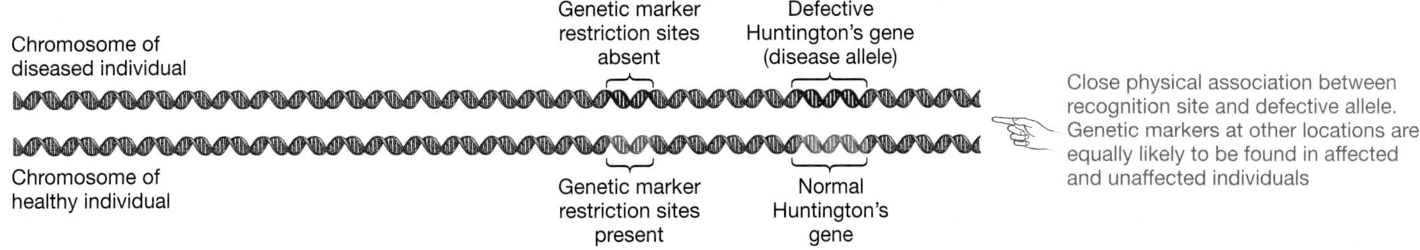

FIGURE 19.11 Genetic Markers and Disease Alleles Are Inherited Together if They Are Closely Linked. To find a gene, researchers look for an association between a phenotype and a genotype. The phenotype is associated with the gene of interest, and the genotype is a mapped genetic marker.

individual might have an A instead of a C at a certain SNP or a yellow versus black hair-color allele. Wexler's team was looking for restriction sites that are almost always present in diseased individuals but absent in healthy individuals, or absent in diseased individuals but present in healthy people (**Figure 19.11**).

Using a Pedigree Once a genetic map containing many genetic markers has been assembled, biologists need help from families affected by an inherited disease to find the gene in question. Recall that the fundamental goal is to find a genetic marker that is almost always inherited along with the disease-causing allele. Gene hunts are more likely to be successful if large families are involved. Large sample sizes minimize the probability that researchers will observe an association between one or more markers and the disease just by chance—not because they are closely linked. Wexler's Huntington's disease team was fortunate to find a large, extended family affected with the disease living along the shores of Lake Maracaibo, Venezuela.

From historical records, the researchers deduced that the Huntington's disease allele was introduced to this family by a European sailor or trader who visited the area in the early 1800s. When family members agreed to participate in the study, there were over 3000 of his descendants living in the area. One hundred of these people had been diagnosed with Huntington's disease. To help in the search for the gene, family members agreed to donate skin or blood samples for DNA analysis and to furnish information on who was related to whom.

This research resulted in the pedigree shown in **Figure 19.12**. The diagram includes information about the disease phenotype and the particular genetic markers observed in each family member. The combinations of markers called A, B, C, and D turned out to be particularly important. These

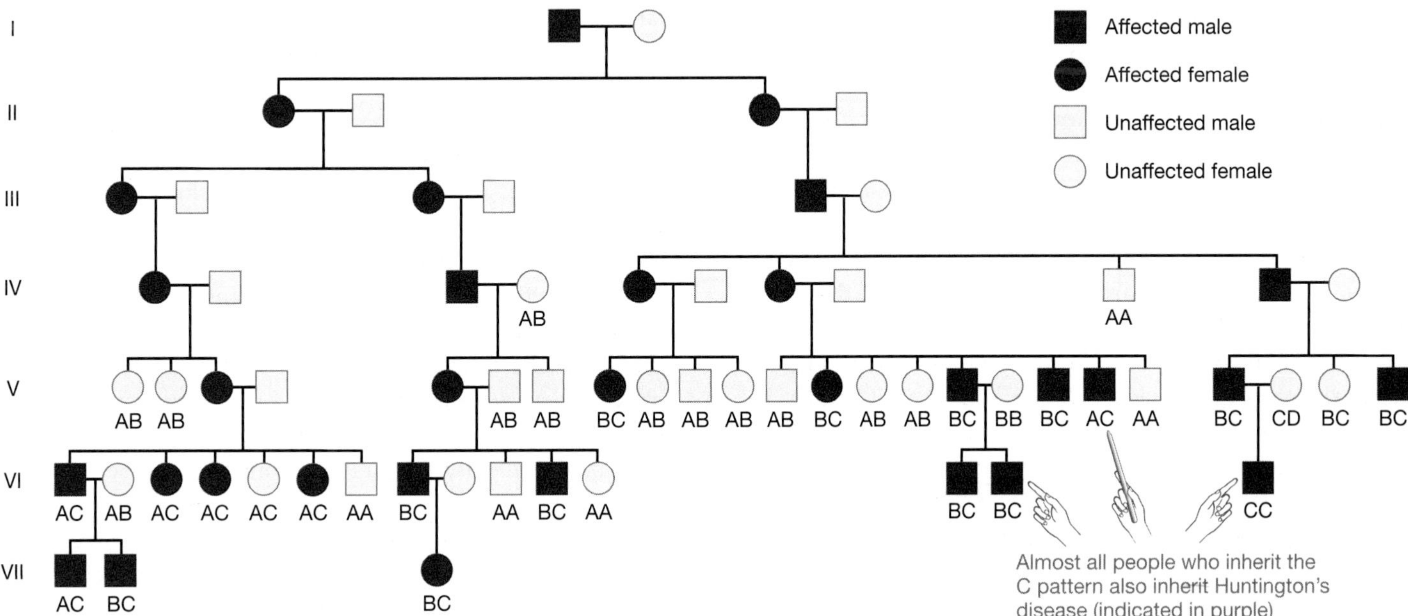

FIGURE 19.12 Certain Genetic Markers Are Inherited with the Huntington's Disease Gene. Part of a pedigree from seven generations (I–VII) of a large extended family. Individuals affected by Huntington's disease inherited the defective allele from a common ancestor. The letters A–D represent combinations of restriction sites found in each individual—in essence, each is a different allele of the same genetic marker. Two letters are indicated for each individual because humans are diploid.

QUESTION In generation VI, an individual with the genotype AC does not have Huntington's disease. How is this possible?

four combinations of restriction sites represent alleles. The key finding is that the C allele was almost always found in diseased individuals. Apparently, the English sailor who introduced the Huntington's disease allele also had the C allele—this particular combination of restriction sites—in his DNA. The C allele and the Huntington's disease gene are so close together that recombination between them—which would put an A or B or D allele next to the Huntington's disease allele in his descendants—has been extremely rare.

From the existing genetic map of humans, Wexler's team knew that the C restriction sites were on chromosome 4. Eventually the team succeeded in narrowing down the location of the Huntington's disease gene to a region about 500,000 base pairs long. Because the haploid human genome contains over 3 billion base pairs, this was a huge step in focusing the search for the gene.

Pinpointing the Defect Once the general location of the Huntington's disease gene was known, biologists looked in that region for exons that encode a functional mRNA. Then they sequenced exons from diseased and normal individuals, compared the data, and pinpointed specific bases that differed between the two groups of individuals. In this way, dideoxy sequencing played a key role in the gene hunt.

When this analysis was complete, the research team found that individuals with Huntington's disease have an unusual number of CAG codons near the 5′ end of a particular gene. CAG codes for glutamine. Healthy individuals have 11–25 copies of the CAG codon, while affected individuals have 42 or more copies. Although many genetic diseases are caused by single base changes that alter the amino acid sequence of a protein, diseases caused by the expansion of a particular codon repeat have also been observed.

When the Huntington's disease research team confirmed that the increase in the CAG codon was always observed in affected individuals, the team concluded that the long search for the Huntington's disease gene was over. They named the newly discovered gene *IT15* and its protein product huntingtin. In both affected and normal individuals, the huntingtin protein is involved in the early development of nerve cells. It is only later in life that mutant forms of the protein cause disease.

What Are the Benefits of Finding a Disease Gene?

Finding the genes associated with specific traits has helped biologists gain a better understanding of hundreds of issues, ranging from why some pea plants are dwarfed to how the repressor shuts down the *lac* operon. In general, how have efforts to find disease genes improved human health and welfare? More specifically, how has the effort to locate the Huntington's disease gene helped researchers and physicians understand and treat the illness? Biomedical researchers point to three major benefits of successful disease-gene hunts.

Improved Understanding of the Phenotype Once a disease gene is found and its sequence is known, researchers can usually figure out why its product causes disease. In the case of *IT15*, autopsies of Huntington's patients had shown that their brains actually decrease in size due to the death of neurons, and that the brain tissue contains clumps, or aggregates, of the huntingtin protein. These aggregates are thought to be a direct consequence of the changes in the number of CAG repeats in the *IT15* gene and thus the number of glutamine residues in the protein. Long stretches of polyglutamine (a polymer of the amino acid glutamine) are known to result in the formation of protein aggregates. The leading hypothesis to explain Huntington's disease proposes that a gradual buildup of the huntingtin protein aggregates triggers neurons to undergo apoptosis, or programmed cell death.

These results helped explain why Huntington's disease is pleiotropic. Patients suffer from abnormal movements *and* personality changes because neurons from throughout the brain are being killed. The results also help explain why the disease takes so long to appear, and why it is progressive: The defective huntingtin proteins take time to build up to deadly levels, but then continue to increase over time. Finally, understanding the molecular mechanism responsible for the illness explained why the disease allele is dominant. One copy of the defective gene is enough to produce fatal concentrations of polyglutamine aggregates.

Therapy Once the sequence of a gene is known, researchers can better understand the structure and function of its protein product. Information on a protein's mode of action is extremely helpful in developing drugs or other treatments.

The search for new therapies for Huntington's disease began with an effort to introduce the defective allele into mice, using the types of genetic engineering techniques discussed in Section 19.5. These mice are called **transgenic** (literally, "across-genes") because they have alleles that have been modified by genetic engineering. Transgenic mice that produce defective versions of the huntingtin protein develop tremors and abnormal movements, exhibit higher-than-normal levels of aggression toward litter and cage mates, and experience a loss of neurons in the brain. The transgenic mice get a version of Huntington's disease. Laboratory animals with disease symptoms that parallel those of a human disease are said to provide an **animal model** of the disease.

Animal models are valuable in disease research, because they can be used to test potential treatments before investigators try them on human patients. Now that an animal model for Huntington's disease is available, research groups have begun testing drugs that appear to prevent or reduce the aggregation of the huntingtin protein. Other biologists are attempting to design drugs that prevent the death of cells containing aggregated huntingtin proteins.

Genetic Testing When the Huntington's gene was found, biologists used the information to develop a test for the presence of the defective allele. The test involves obtaining a DNA sample from an individual and using the polymerase chain reaction to amplify the chromosome region that contains the CAG repeats responsible for the disease. If the number of CAG repeats is 35 or less, the individual is considered to be normal. Forty or more repeats results in a positive diagnosis for Huntington's.

As more becomes known about the human genome, more and more tests like this will be developed for diseases that have some genetic component. What types of genetic testing are done currently?

- *Carrier Testing* Before starting their own family, people from families affected by a genetic disease frequently want to know whether they carry the allele responsible. That is especially true for diseases, such as cystic fibrosis (CF), that are due to recessive alleles. If only one of the prospective parents has the allele, then none of the children they have together should develop CF. But if both prospective parents carry the allele, then each child they produce has a 25 percent chance of having CF. Carrier testing can determine whether an individual is a carrier for a genetic disease.

- *Prenatal Testing* Suppose that two parents, both carrying the CF allele, decide to have children but do not want to pass along that allele. Once the mother is pregnant, a physician can obtain fetal cells early in gestation. The cells can be cultured and DNA isolated. The CF allele can then be amplified by PCR and sequenced. Based on the test results, the couple may choose to continue or terminate a pregnancy.

- *Adult Testing* Because the symptoms of Huntington's disease do not appear until middle age, many affected people have married and had children by the time they develop symptoms. But people with a family history of Huntington's disease can now be tested as young adults. If they learn that they have the defective allele, they might make different decisions about starting their own families and passing on the illness. And Huntington's disease isn't the only trait that appears in adulthood. For example, 5–10 percent of the women who develop breast cancer have inherited a faulty gene. To date, more than five genes are known to create a predisposition to breast cancer. Adult women with a family history of the illness can be tested for the presence of these mutant genes. If they have a susceptibility allele, they can be tested for the onset of cancer more frequently than normal. They may also be able to make changes in their diet or activity that make them less likely to actually develop the illness.

Recent work on the Huntington's disease gene highlights just how far the field of molecular genetics has come in just a few short decades. Before the molecular revolution, the molecular nature of the gene was not well understood. Now, researchers can find and sequence the alleles associated with a wide array of traits in humans and other organisms and place them in other organisms. **Table 19.1** summarizes some of the most important tools and techniques that made these advances possible.

Along with any other type of change, however, the advances emerging from gene hunts create new problems that must be addressed. Like the misuse of human growth hormone produced with recombinant DNA technology, successful gene hunts can create serious moral and legal dilemmas.

Ethical Concerns over Genetic Testing

Genetic testing raises controversial ethical issues. For example, some individuals maintain that it is morally wrong to terminate any pregnancy, even if the fetus involved is guaranteed to be born with a debilitating genetic disease. Also, consider Nancy Wexler's position soon after the discovery of *IT15*. Would you choose to be tested for the defective allele and risk finding out that you were almost certain to develop Huntington's disease? Should physicians agree to test people for genetic diseases that have no cure? Should it be legal for health insurance companies to test clients for genetic diseases that will require expensive treatments so that the companies can refuse coverage?

These questions are still being debated by political and religious leaders, health-care workers, philosophers, and members of the general public. Difficult as they are, the issues raised by genetic testing pale in comparison with some of the questions raised by the effort to create transgenic humans.

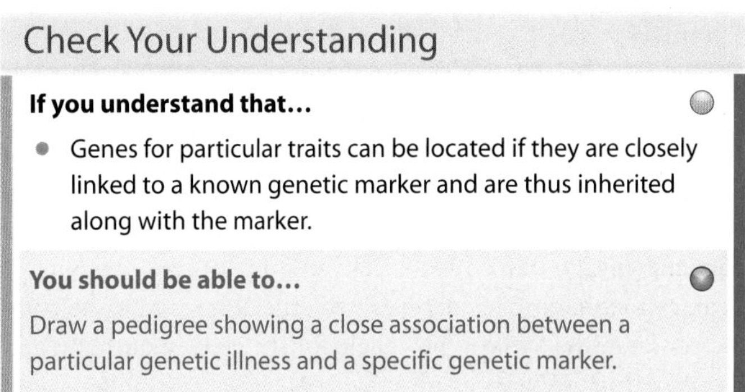

19.5 Can Gene Therapy Cure Inherited Diseases in Humans? Research on Severe Immune Disorders

For biomedical researchers interested in curing inherited diseases such as Huntington's, sickle-cell anemia, and cystic fibrosis, the

SUMMARY TABLE **19.1** **Some Common Tools and Techniques Used in Genetic Engineering**

Tools			
Name	**Description**	**How Used**	**Example or Illustration**
Reverse transcriptase	Enzyme that catalyzes synthesis of a complementary DNA (cDNA) from an RNA template	Making cDNAs used in constructing a genetic library (many other applications).	cDNA RNA
Restriction endonucleases	Enzyme that cuts DNA at a specific sequence—often a palindromic sequence that is 5 base pairs long.	Allows researchers to cut DNA at specific locations. Cuts in palindromic sites create "sticky ends."	G AATTC CTTAA G
DNA ligase	An enzyme that catalyzes the formation of a phosphodiester bond between nucleotides on the same DNA strand.	Ligates (joins) sequences that were cut with a restriction endonuclease. Gives researchers the ability to splice fragments of DNA together.	G AATTC C TTAAG
Plasmids	Small, extrachromosomal loops of DNA found in many bacteria and in some yeast.	After a target gene is inserted into a plasmid, the recombinant plasmid serves as a vector for transferring the gene into a bacterial or yeast cell, so the gene can be cloned.	GAATTC CTTAAG GAATTC CTTAAG
Taq polymerase	DNA polymerase from the bacterium *Thermus aquaticus*. Catalyzes synthesis of DNA from a primed DNA template; remains stable at 95°C.	Responsible for the "primer extension" step in the polymerase chain reaction. Heat stability allows enzyme to be active even after denaturation step of PCR cycle at 94°C.	
Single nucleotide polymorphisms (SNPs)	Sites in DNA where the identity of the base varies among individuals in a population.	One of many polymorphic types of DNA sequences that are useful in creating the genetic maps required for gene hunts.	Sequence of some individuals ... GATTCTTATTAACTACGC ... ... GATTCTTCTTAACTACGC ... Sequence of other individuals

ultimate goal is to replace or augment defective copies of the gene with normal alleles. This approach to treatment is called **gene therapy**.

For gene therapy to succeed, two crucial steps must be completed. First, the "wild-type" allele for the gene in question and its regulatory sequences must be sequenced and understood. This allele codes for a product associated with a healthy phenotype. Second, there must be a method for introducing a copy of the wild-type allele into affected individuals. The DNA has to be introduced in a way that ensures expression of the gene in the correct tissues, in the correct amount, and at the correct time. If the defective allele is dominant, then the introduction step may be even more complicated: In at least some cases, the introduced allele must physically replace or block the expression of the existing dominant allele.

Chapter 18 discussed aspects of gene regulation in humans and other eukaryotes; here, we focus on techniques for introducing DNA into humans. As a case study, let's consider the first successful application of gene therapy in humans.

How Can Novel Alleles Be Introduced into Human Cells?

Section 19.1 reviewed how recombinant DNA sequences are packaged into plasmids and taken up by *E. coli* cells. Humans and other mammals lack plasmids, however, and their cells do not take up foreign DNA in response to chemical or electric treatments as efficiently as bacterial cells do.

How can foreign genes be introduced efficiently into human cells? To date, researchers have focused on packaging

SUMMARY TABLE **19.1 Some Common Tools and Techniques Used in Genetic Engineering *(continued)***

Techniques			
Name	**Description**	**How Used**	**Example or Illustration**
Recombinant DNA technology ("genetic engineering")	Taking a copy of a gene from one individual and placing it in the genome of a different individual (often of a different species).	Many applications, including DNA cloning, gene therapy (see Section 19.5), and biotechnology (see Sections 19.1 and 19.6).	
Genetic library	A collection of all DNA sequences present in a particular source. The library consists of individual DNA fragments that are isolated and inserted into a plasmid or other vector, so they can be cloned.	cDNA libraries allow researchers to catalog the genes being expressed in a particular cell type. Genomic libraries allow researchers to archive all the DNA sequences present in a genome. Libraries can be screened to find a particular target gene.	
DNA Probing/Screening a genetic library	Use of a labeled, known DNA fragment to hybridize (via complementary base pairing) with a collection of unlabeled, unknown fragments.	Allows a researcher to find a particular DNA sequence in a large collection of sequences.	
Polymerase chain reaction	Requires known primer sequences on either side of a target gene. Reaction is based on many cycles of primer annealing, primer extension, and DNA denaturation	Produces many identical copies of a target sequence. A shortcut method for DNA cloning.	
Dideoxy sequencing	In vitro DNA synthesis reaction that includes dideoxyribonucleotide monomers (ddNTPs).	Determining the base sequence of a gene or other section of DNA.	
Genetic mapping	Creation of a map showing the relative positions of genes or specific DNA sequences on chromosomes. Done by analyzing the frequency of recombination between sequences (see Chapter 13).	Many applications, including use of mapped genetic markers to find unknown genes associated with diseases or other distinctive phenotypes.	

● **EXERCISE** Fill in the "Example or Illustration" cells in the Techniques column of the table.

foreign DNA into viruses for transport into human cells. A viral infection begins when a virus particle enters or attaches to a host cell and inserts its genome into that host cell (see Chapter 35). In some cases the viral DNA becomes integrated into a host-cell chromosome. As a result, viruses that infect human cells can be used as vectors to carry engineered alleles into the chromosomes of target cells. Potentially, the alleles delivered by the virus could be expressed and produce a product capable of curing a genetic disease.

Currently, the vector of choice in gene therapy is **retroviruses**. Retroviruses are viruses that have an RNA genome. The genomes include the gene for the enzyme reverse transcriptase, which catalyzes the production of a double-stranded DNA molecule from a single-stranded RNA template. (Recall from Section 19.1 that researchers use reverse transcriptase to produce cDNAs.) When a retrovirus infects a human cell, reverse transcriptase catalyzes the production of a DNA copy of the viral genome. Other viral enzymes catalyze the insertion of the viral DNA into a host-cell chromosome. If human genes can be packaged into a retrovirus, then the virus is capable of inserting the human alleles into a chromosome in a target cell (**Figure 19.13**).

Although viruses would appear to be ideal vectors for the delivery of normal human alleles, there are serious problems associated with their use. Viruses usually cause disease. The retroviruses include the human immunodeficiency virus (HIV), which causes AIDS, as well as an array of viruses that cause cancer in humans and other animals. Using these agents for

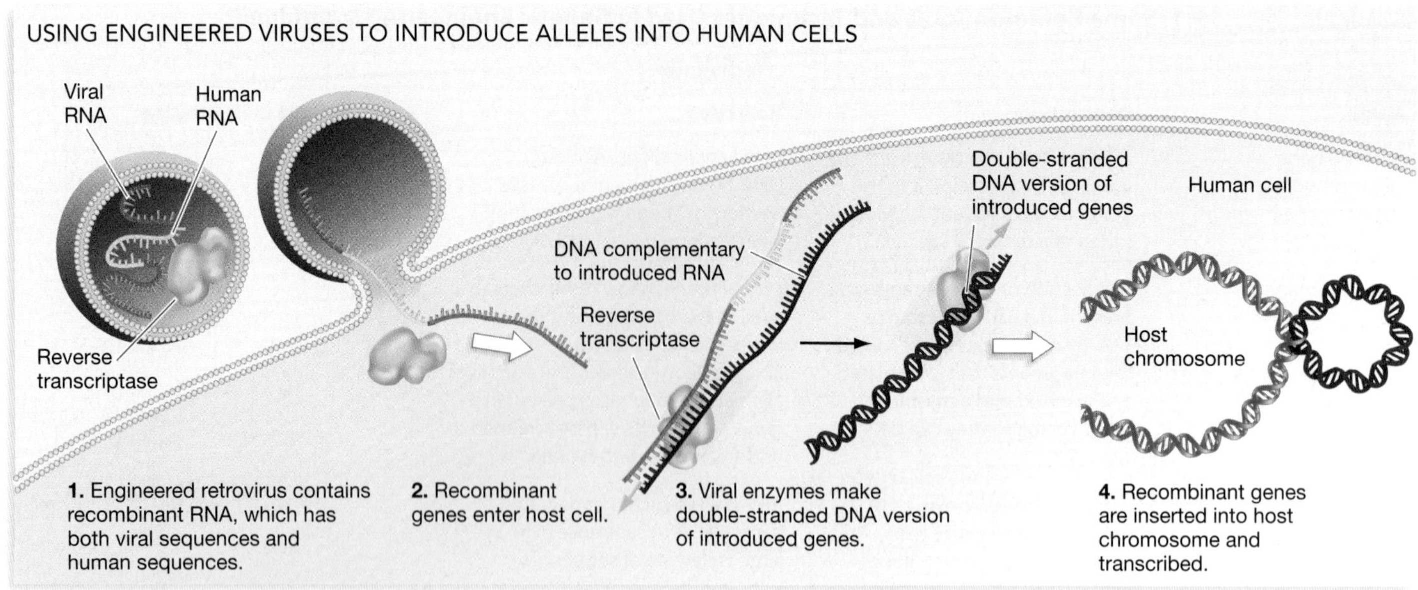

FIGURE 19.13 Retroviruses Insert Their Genes into Host-Cell Chromosomes. Retroviruses can be used as vectors to carry specific genes into cells and to insert these genes into the chromosomes of target cells.

QUESTION What happens if the recombinant DNA is inserted in the middle of a gene that is critical to the cell's normal function?

gene therapy requires that sequences responsible for causing disease be inactivated or removed from their genomes. Even if they have been inactivated, the altered particles may be able to recombine with viral DNA that already exists in the individual being treated and lead to the formation of a new infectious strain. In addition, the presence of viral proteins triggers a response by the immune system that can cause dangerous side effects during treatment. Finally, if viral genes happen to insert themselves in a position that disrupts the function of an important gene in the target cell, the consequences can be serious. Despite these risks, retroviruses are still the best vectors currently available for human gene therapy. If you understand this concept, you should be able to add a row called "Retroviruses" to the "Tools" section of Table 19.1 and fill in the "Description," "How Used," and "Illustration" columns.

Using Gene Therapy to Treat X-Linked Immune Deficiency

In 2000 a research team reported the successful treatment of a disease by means of gene therapy. The illness is called **severe combined immunodeficiency (SCID)**. Children who are born with SCID lack a normal immune system and are unable to fight off infections.

The type of SCID treated by the group is designated SCID-X1, because it is caused by mutations in a gene on the X chromosome. The gene codes for a receptor protein necessary for the development of immune system cells called T cells. T cells develop in bone marrow, from undifferentiated cells that divide continuously.

Traditionally, physicians have treated SCID-X1 by keeping the patient in a sterile environment, isolated from any direct human contact until the patient could receive bone-marrow tissue transplanted from a close relative (**Figure 19.14**). In most bone-marrow transplants, the T cells that the patient needs are produced by the transplanted bone-marrow cells and allow the individual to live normally. In some cases, though, no suitable donor is available. Could gene therapy cure this disease by furnishing functioning copies of the defective gene?

After extensive testing suggested that their treatment plan was safe and effective, the research team gained approval to

FIGURE 19.14 A "Bubble Child". Children who are born with SCID cannot fight off bacterial or viral infections. As a result, such children must live in a sterile environment.

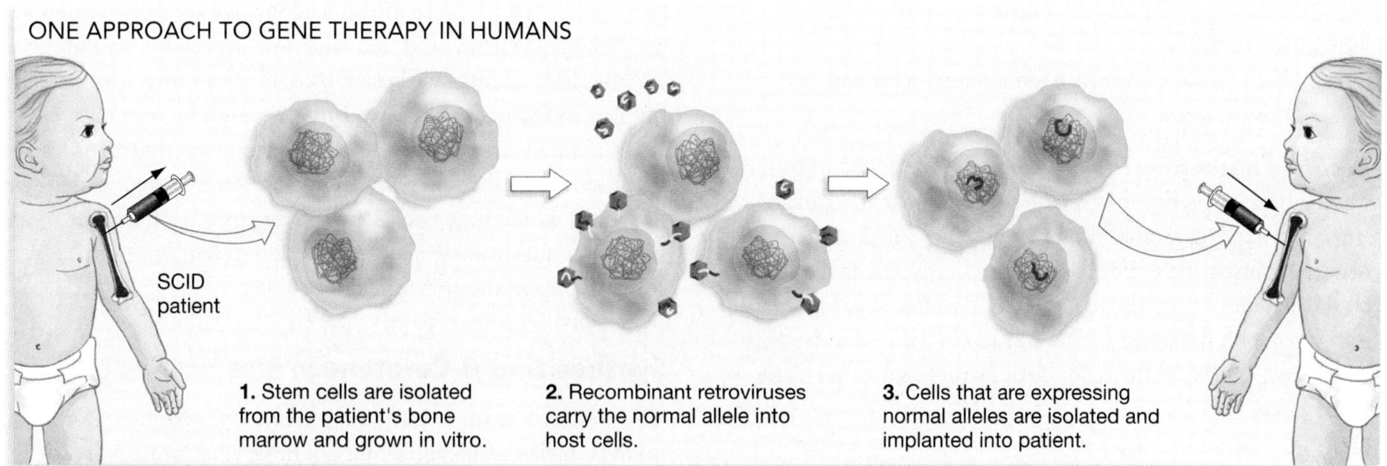

FIGURE 19.15 A Gene Therapy Protocol. For gene therapy to work, copies of a normal allele have to be introduced into a patient's cells and be expressed.

treat 10 boys, each less than 1 year old, who had SCID-X1 but no suitable bone-marrow donor. To implement gene therapy, the researchers removed bone marrow from each patient and collected the cells that produce mature T cells (**Figure 19.15**). The researchers then infected the cells with an engineered retrovirus that carried the normal receptor gene. Cells that were producing normal receptor protein were then isolated and transferred back into the patients.

Within four months after reinsertion of the transformed marrow cells, nine of the boys had normal levels of functioning T cells. These patients were removed from germ-free isolation rooms and began residing at home, where they grew and developed normally. During a routine check-up 30 months after gene therapy, however, the research team discovered that two of the boys had developed a type of cancer characterized by unchecked growth of T cells. Follow-up analyses of their bone-marrow cells showed that in both boys, a viral-borne receptor gene had been inserted near the gene for a regulatory transcription factor that triggers the growth of T cells. The viral sequences apparently acted as an enhancer and led to constitutive expression of the transcription factor. As this book goes to press, the two boys have responded to cancer chemotherapy and are healthy.

The tenth boy to receive gene therapy never succeeded in producing T cells at all. For unknown reasons, his recombinant stem cells failed to function normally when they were transplanted back into his bone marrow. Fortunately, physicians were later able to find a bone marrow donor whose cells were a close enough match to the boy's cells to make a successful transplant possible.

Ethical Concerns over Gene Therapy

Throughout the history of medicine, efforts to test new drugs, vaccines, and surgical protocols have always carried a risk for the patients involved. Gene therapy experiments are no different. The researchers who run gene therapy trials must explain the risks clearly and make every effort to minimize them. The initial report on the development of cancer in the boys who received gene therapy for SCID-X1 concluded with the following statement: "We have proposed ... a halt to our trial until further evaluation of the causes of this adverse event and a careful reassessment of the risks and benefits of continuing our study of gene therapy." Although recombinant gene therapy still holds promise for the treatment of a wide variety of devastating inherited diseases, fulfilling that promise is almost certain to require many years of additional research and testing, as well as the refinement of legal and ethical guidelines.

When recombinant DNA technology first became possible, many researchers thought they would live to see most or all of the serious inherited diseases in humans cured by gene therapy. Several decades later, that optimism is tempered. In humans, gene therapy is still highly experimental, extremely expensive, and intensely controversial.

19.6 Biotechnology in Agriculture: The Development of Golden Rice

Progress in human gene therapy has been slow, but progress in transforming crop plants with recombinant genes has been breathtakingly rapid. In the United States alone, genetically modified or transgenic strains of 15 major crop species are now approved for commercial use. Farmers growing maize (corn) in the United States currently have over 15 distinct transgenic varieties to choose from. In Mississippi, 97 percent of the land area devoted to cotton is now planted with recombinant strains. Globally, about 87.2 million hectares (192 million acres) of transgenic crops were grown in 2006. The use of genetically modified plants has become a major element of modern agriculture.

Recent efforts to develop transgenic plants have focused on three general objectives:

1. *Reducing losses to herbivore damage* The most popular strategy for reducing losses caused by plant-eating organisms is based on introducing the gene for a naturally occurring insecticide. For example, researchers have transferred a gene from the bacterium *Bacillis thuringiensis* into corn; the presence of the "Bt toxin" encoded by this gene protects the plant from corn borers and other caterpillar pests.
2. *Reducing competition with weeds* Competition with weeds is usually reduced by introducing a gene whose product makes the crop plant resistant to an herbicide (a chemical that destroys plants). For example, soybeans have been genetically engineered for resistance to the herbicide glyphosate. Soybean fields with the engineered strain can be sprayed with glyphosate to kill weeds without harming the soybeans.
3. *Improving the quality of the product consumed by people* Some crop plants, including soybeans and canola plants, have been engineered to produce a higher percentage of unsaturated fatty acids relative to saturated fatty acids (see Chapter 6). Saturated fatty acids can contribute to heart disease, so crops with less of them are healthier to eat.

To gain a better understanding of genetic engineering in plants, let's take a detailed look at efforts to produce transgenic rice with improved nutritional qualities.

Rice as a Target Crop

Although almost half the world's population depends on rice as its staple food, this grain is an extremely poor source of certain vitamins and essential nutrients. For example, rice contains no vitamin A. This is a serious issue. In addition to causing blindness in 250,000 Southeast Asian children each year, vitamin A deficiency can make children more susceptible to diarrhea, respiratory infections, and childhood diseases such as measles.

Humans and other mammals synthesize vitamin A from a precursor molecule known as β-carotene (beta-carotene). β-Carotene belongs to a family of plant pigments called the carotenoids, which were introduced in Chapter 10. Carotenoids are orange, yellow, and red and are especially abundant in carrots. Rice plants synthesize β-carotene in their chloroplasts but not in the part of the seed that is eaten by humans. Could genetic engineering produce a strain of rice that synthesized β-carotene in the carbohydrate-rich seed tissue called endosperm, which humans do eat? A research team set out to answer this question. If successful, their research could help to solve an important global health problem.

Synthesizing β-Carotene in Rice

The research team began their work by searching for compounds in rice endosperm that could serve as precursors for the synthesis of β-carotene. They found that maturing rice endosperm contains a molecule called geranyl geranyl diphosphate (GGPP), which is an intermediate in the synthetic pathway that leads to the production of carotenoids. As **Figure 19.16** shows, three enzymes are required to produce β-carotene from GGPP. If genes that encode these enzymes could be introduced into rice plants along with regulatory sequences that would trigger their synthesis in endosperm, the researchers could produce a transgenic strain of rice that would contain β-carotene.

Fortunately, genes that encode two of the required enzymes had already been isolated from daffodils, and the gene for the third enzyme had been purified from a bacterium. Because the sequences had been inserted into plasmids and grown in bacteria, many copies were available for manipulation. To each of the coding sequences in the plasmids, biologists added the promoter region from an endosperm-specific protein. This segment included a regulatory site that would promote transcription of the recombinant sequences in endosperm cells.

To develop transgenic rice strains that are capable of producing β-carotene, the three sets of sequences had to be inserted into rice plants. As you know from reading this chapter, introducing recombinant DNA is fairly straightforward into *E. coli* but difficult in humans. How are foreign genes introduced into plants?

The *Agrobacterium* Transformation System

Agrobacterium tumefaciens is a bacterium that infects plants. Plant tissues infected with this parasite form a tumorlike growth called a **gall**. Such plants are said to have crown gall disease (**Figure 19.17a**). When researchers looked into how

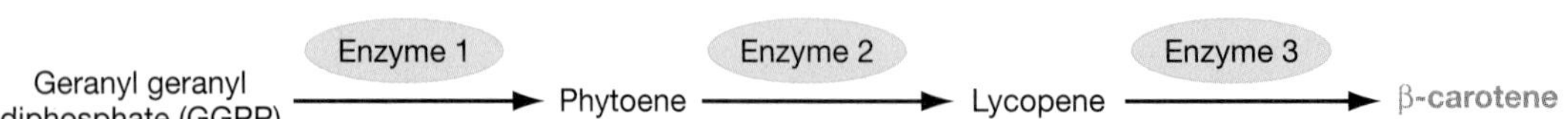

FIGURE 19.16 Synthetic Pathway for β-Carotene. GGPP is a molecule found in rice seeds. Three enzymes are required to produce β-carotene from GGPP.

(a) Plant with crown gall disease

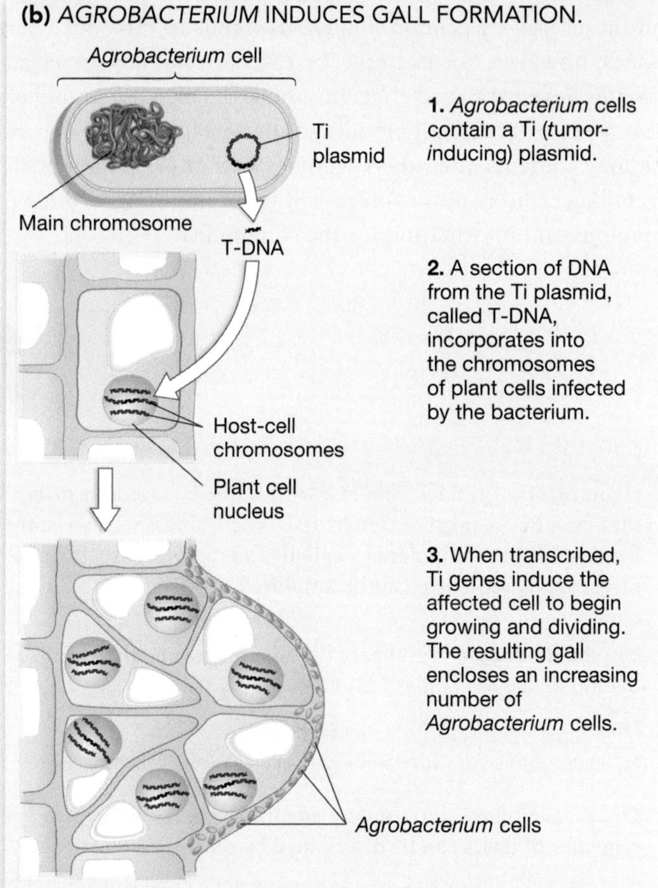

FIGURE 19.17 Agrobacterium Infections Introduce Foreign Genes into a Host-Cell Chromosome. (a) The gall, or tumorlike growth at the bottom of this plant's stem, is due to an infection by the bacterium *Agrobacterium*. **(b)** Ti plasmids of *Agrobacterium* cells induce gall formation.

these infections occur, they found that a plasmid carried by the *Agrobacterium* cells, called a **Ti** (tumor-inducing) **plasmid**, plays a key role (**Figure 19.17b**, step 1). Ti plasmids contain several functionally distinct sets of genes. One set encodes products that allow the bacterium to bind to the cell walls of a host. Another set, referred to as the virulence genes, encodes the proteins required to transfer part of the Ti DNA, called T-DNA (transferred DNA), into the interior of the plant cell. The T-DNA then travels to the plant cell's nucleus and integrates into its chromosomal DNA (step 2). When transcribed, T-DNA induces the infected cell to grow and divide. The result is the formation of a gall that encloses a growing population of *Agrobacterium* cells (step 3).

Researchers soon realized that the Ti plasmid offers an efficient way to introduce recombinant genes into plant cells. Follow-up experiments confirmed that recombinant genes could be added to the T-DNA that integrates into the host chromosome, that the gall-inducing genes could be removed, and that the resulting sequence is efficiently transferred and expressed in its new host plant.

Using the Ti Plasmid to Produce Golden Rice

To generate a strain of rice that produces all three enzymes needed to synthesize β-carotene in endosperm, the researchers exposed embryos to *Agrobacterium* cells containing genetically modified Ti plasmids (**Figure 19.18**). The transformed plants were then grown in the greenhouse. When the transgenic individuals had matured and set seed, the researchers found that some rice grains contained so much β-carotene that they appeared yellow. This "golden rice" is compared with unmodified

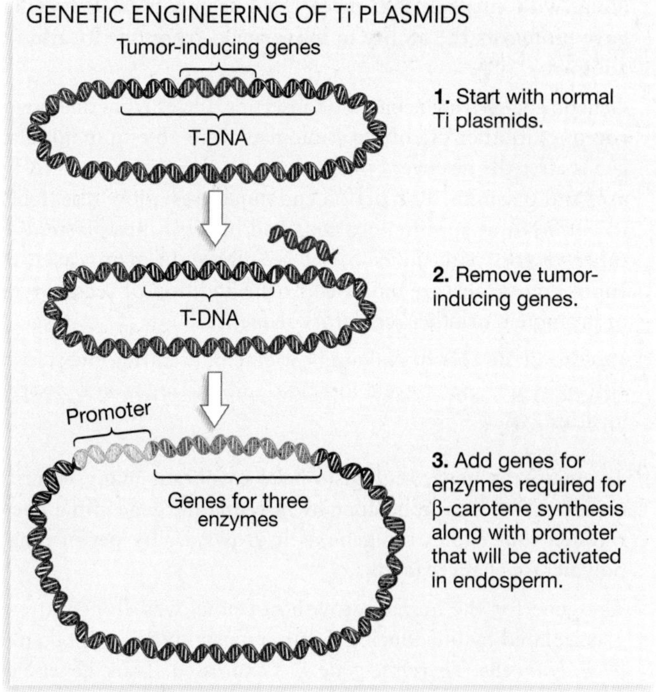

FIGURE 19.18 "Golden Rice" Is a Transgenic Strain. The construction of a Ti plasmid to produce a strain of rice capable of synthesizing β-carotene in the endosperm of their seeds.

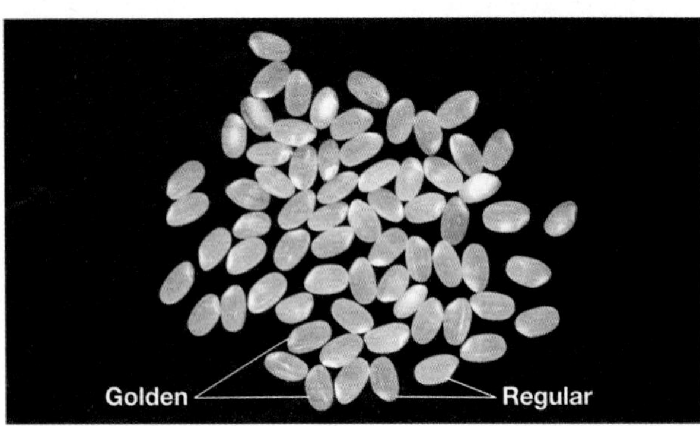

FIGURE 19.19 Comparing Regular Rice and Golden Rice Chemical analyses confirmed that the golden color in transformed seeds is due to large quantities of β-carotene.

rice in **Figure 19.19**. Testing is under way to determine whether the transgenic rice contains enough β-carotene to relieve vitamin A deficiency in children.

While the production of golden rice in the laboratory is promising, there is still much to do before the advance has an impact on public health. The transgenic rice must be grown successfully under normal field conditions. In addition, most or all of the rice strains grown in different parts of the world will need to be crossed with transgenic stock to acquire the appropriate genes. Finally, the transgenic strains must be made available to poor farmers at an affordable price.

In addition to the possibility of alleviating vitamin A deficiency, transgenic strains of rice may be able to ease other nutritional problems in rice-eating people—specifically, protein deficiency and iron deficiency. Early attempts to develop transgenic rice strains with augmented amounts of stored protein and available iron have been promising.

As the examples reviewed in this chapter show, recent advances in agricultural and medical biotechnology promise to increase the quality and quantity of food and to alleviate at least some inherited diseases that plague humans. Each solution offered by genetic engineering tends to introduce new issues, however. For example, researchers and consumer advocates have expressed concerns about the increased numbers and types of genetically modified foods. During this debate, biology students and others who are well informed about the techniques and issues involved will be an important source of opinions and information for the community.

Chapter Review

SUMMARY OF KEY CONCEPTS

- **The discovery of enzymes that cut DNA at specific locations, along with enzymes that piece DNA segments back together, gave biologists the ability to move genes from one location to another.**

Genetic engineering is based on inserting alleles from one organism into another organism. Biologists were able to manipulate genes after the discovery of restriction endonucleases, DNA ligase, and plasmids. Restriction endonucleases allow researchers to cut DNA at specific locations and insert it into plasmids or other vectors with the help of DNA ligase. In many cases, the engineered alleles are modified by the addition of certain types of promoters or other regulatory sequences.

You should be able to make a flowchart illustrating the steps in cutting a gene out of one location and pasting it into another location.

- **To analyze a gene, biologists have to obtain many identical copies of it. This can be done by inserting the gene into a bacterial cell that copies the gene as it grows, or by performing a polymerase chain reaction.**

The gene for the human growth hormone was cloned after it was isolated by introducing it into a plasmid that was taken up by *E. coli* cells. There the gene was expressed. Thus, genetic engineering and DNA cloning have resulted in large populations of recombinant *E. coli* cells that produce large, safe supplies of human growth hormone at an affordable price.

The most common alternative to cloning genes in plasmids is an in vitro DNA synthesis reaction called the polymerase chain reaction (PCR). The PCR technique is based on primers that bracket a target stretch of DNA. Using the heat-stable form of DNA polymerase called *Taq* polymerase, a single target DNA sequence can be amplified to millions of identical copies.

You should be able to compare the advantages and disadvantages of cloning genes via plasmids versus PCR.

MB Web Animation at www.masteringbio.com
Producing Human Growth Hormone; The Polymerase Chain Reaction

- **Once researchers have many identical copies of a gene, its sequence of bases can be determined by dideoxy sequencing.**

The most common way to analyze the gene copies produced by plasmid-based cloning or PCR is to sequence them by the dideoxy sequencing method. This is an in vitro synthesis reaction that employs dideoxyribonucleotides to stop DNA replication at each base in the sequence. By running the resulting DNA fragments out on a gel, the sequence of nucleotides in a gene can be determined.

You should be able to explain why the use of fluorescent markers and gel-filled capillary tubes increased the efficiency of sequencing studies.

- **To find the genes associated with a particular trait, researchers start with a genetic map. If certain markers on the map are found only in individuals that have a distinctive phenotype, the gene responsible for that phenotype is likely to be near that marker.**

To find the gene associated with Huntington's disease, investigators analyzed a large number of genetic markers and a large pedigree of an affected family. Their goal was to find a marker that was inherited along with the allele responsible for the disease. Once this strategy pinpointed the general area where the gene was located, biologists sequenced DNA from the region to determine exactly where the defect occurred.

You should be able to create a genetic map for a hypothetical organism and explain how you could use it to find a gene associated with body size.

Researchers are attempting to insert genes into humans to cure genetic diseases. Efforts to insert genes into plants to provide them with novel traits, such as the ability to produce important nutrients, have been much more successful.

Once genes are located and characterized, they can be introduced into other individuals or species in an effort to change their traits. Genetic transformation can occur in several ways, depending on the species involved. In humans, recombinant DNA must be introduced by viruses. Because introducing foreign genes into humans is difficult, and because complex ethical and safety issues are involved, progress in human gene therapy has been slow. Genetic engineering of crop plants is much easier technically because certain bacteria that infect plants have plasmids that integrate their genes into the host-plant genome. By adding recombinant alleles to these plasmids, researchers have been able to introduce alleles that improve the nutritional quality of crops, make them resistant to herbicides, or allow them to produce insecticides.

You should be able to explain what would have to happen for recombinant gene technology to become as feasible in humans as it is in plants.

QUESTIONS

Test Your Knowledge

1. What do restriction endonucleases do?
   a. They cleave bacterial cell walls and allow viruses to enter the cells.
   b. They join pieces of DNA by catalyzing the formation of phosphodiester bonds between them.
   c. They cut stretches of DNA at specific sites known as recognition sequences.
   d. They act as genetic markers in the chromosome maps used in gene hunts.
2. What is a plasmid?
   a. an organelle found in many bacteria and certain eukaryotes
   b. a circular DNA molecule that in some cases replicates independently of the main chromosome(s)
   c. a type of virus that has a DNA genome and that infects certain types of human cells, including lung and respiratory tract tissue
   d. a type of virus that has an RNA genome, codes for reverse transcriptase, and inserts a cDNA copy of its genome into host cells
3. When present in a DNA synthesis reaction, a ddNTP molecule is added to the growing chain. No further nucleotides can be added afterward. Why?
   a. There are not enough dNTPs available.
   b. A ddNTP can be inserted at various locations in the sequence, so fragments of different length form—each ending with a ddNTP.
   c. The 5′ carbon on the ddNTP lacks a hydroxyl group, so no phosphodiester bond can form.
   d. The 3′ carbon on the ddNTP lacks a hydroxyl group, so no phosphodiester bond can form.
4. Once the gene that causes Huntington's disease was found, researchers introduced the defective allele into mice to create an animal model of Huntington's disease. Why was this model valuable?
   a. It allowed them to test potential drug therapies without endangering human patients.
   b. It allowed them to study how the gene is regulated.
   c. It allowed them to make large quantities of the huntingtin protein.
   d. It allowed them to study how the gene was transmitted from parents to offspring.
5. To begin the hunt for the human growth hormone gene, researchers created a cDNA library from cells in the pituitary gland. What did this library contain?
   a. only the sequence encoding growth hormone
   b. DNA versions of all the mRNAs in the pituitary-gland cells
   c. all of the coding sequences in the human genome, but no introns
   d. all of the coding sequences in the human genome, including introns
6. What does it mean to say that a genetic marker and a disease gene are closely linked?
   a. The marker lies within the coding region for the disease gene.
   b. The sequence of the marker and the sequence of the disease gene are extremely similar.
   c. The marker and the disease gene are on different chromosomes.
   d. The marker and the disease gene are in close physical proximity and tend to be inherited together.

Test Your Knowledge answers: 1. c; 2. b; 3. d; 4. a; 5. b; 6. d

Test Your Understanding

Answers are available at www.masteringbio.com

1. Explain how restriction endonucleases and DNA ligase are used to insert foreign genes into plasmids and create recombinant DNA. Make a drawing that shows why sticky ends are sticky, and that identifies the exact location where DNA ligase catalyzes a key reaction.
2. Explain the function of a vector in genetic engineering. Then explain how recombinant plasmids are introduced into *E. coli* cells, how viruses are used to transport genes into human cells, and how the Ti plasmid in *Agrobacterium* cells is used to transport genes into plants. List the attributes of a "perfect" vector.
3. What is a cDNA library? How is one created? Would you expect the cDNA library from a human muscle cell to be different from the cDNA library from a human nerve cell in the same individual? Explain why or why not.

4. What are genetic markers, and how are they used to create a genetic map? Explain how researchers combine analyses of human pedigrees and genetic markers to narrow down the location of disease genes.
5. Researchers added the promoter sequence from an endosperm-specific gene to the Ti plasmids used in creating golden rice. Why was this step important? Comment on the roles of promoter and enhancer sequences in genetic engineering in eukaryotes.
6. List the molecules that are required for a typical PCR reaction, and explain their role in the primer-annealing, primer-extension, and denaturation steps of a PCR cycle. Then list the function of each molecule required for the replication of DNA in vivo. For each molecule on the in vivo list, explain why it is or is not required for the corresponding in vitro reaction.

Applying Concepts to New Situations

Answers are available at www.masteringbio.com

1. Suppose you had a large amount of sequence data, similar to the data that Nancy Wexler's team had in the region of the Huntington's disease gene, and that you knew that genes in the species being considered typically contain about 1500 bases. How would you use the genetic code (see Chapter 15) and information on the structure of promoters (see Chapters 17 and 18) to locate the precise location of one more genes in your sequence?
2. Should human gene therapy be restricted to somatic cells, or should individuals be able to alter their germ-line cells (meaning that they would alter the alleles they pass on to their offspring)? Should gene therapy be approved for disease-causing alleles only, or should parents also be able to pay to transform their children with alleles associated with height, intelligence, hair color, eye color, athletic performance, musical ability, or similar traits?
3. Describe similarities between how researchers screen a DNA library and how they perform a genetic screen—for example, for mutant *E. coli* cells that cannot metabolize lactose (see Chapter 17).
4. A friend of yours is doing a series of PCR reactions and comes to you for advice. She purchased three sets of primers, hoping that one set would amplify the template sequence shown here. (The dashed lines in the template sequence stand for a long sequence of bases.) None of the three primer pairs produced any product DNA, however.

	Primer a	Primer b
Primer Pair 1:	5′ GTCCAGC 3′ & 5′ CCTGAAC 3′	
Primer Pair 2:	5′ GGACTTG 3′ & 5′ GCTGCAC 3′	
Primer Pair 3:	5′ GTCCAGG 3′ & 5′ CAAGTCC 3′	

Template

5′ ATTCGGACTTG — — — GTCCAGCTAGAGG 3′
3′ TAAGCCTGAAC — — — CAGGTCGATCTCC 5′

a. Explain why each primer pair didn't work. Indicate whether both primers are at fault or just one primer is the problem.
b. Your friend doesn't want to buy new primers. She asks you whether she can salvage this experiment. What do you tell her to do?

www.masteringbio.com is also your resource for • Answers to text, table, and figure caption questions and exercises • Answers to *Check Your Understanding* boxes • Online study guides and quizzes • Additional study tools including the *E-Book for Biological Science* 3rd ed., textbook art, animations, and videos.

Genomics

20

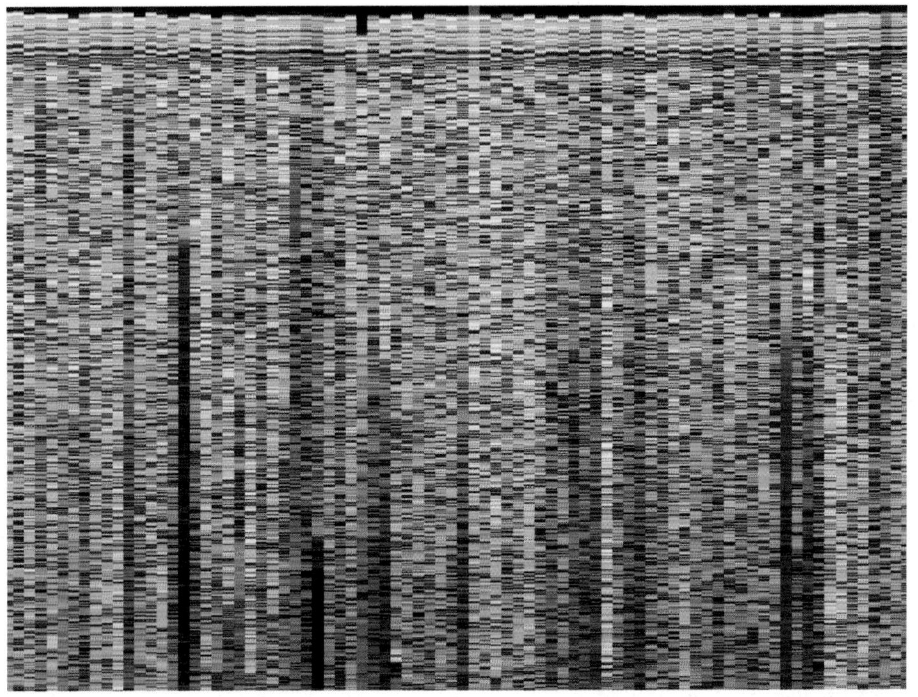

Output from an automated genome sequencing machine, representing about 48,000 bases from the human genome. Each vertical stripe represents the sequence of a stretch of DNA.

KEY CONCEPTS

- Once a genome has been completely sequenced, researchers use a variety of techniques to identify which sequences code for products and which act as regulatory sites.
- In bacteria and archaea, there is a positive correlation between the number of genes in a species and the species' metabolic capabilities. Gene transfer between species is also common.
- In eukaryotes, genomes are dominated by sequences that have little to no effect on the fitness of the organism.
- Data from genome sequencing projects are now being used in the development of new drugs and vaccines, and to search for alleles associated with inherited diseases.

The first data sets describing the complete DNA sequence, or **genome**, of humans were published in February 2001. These papers were immediately hailed as a landmark in the history of science. In just 50 years, biologists had gone from not understanding the molecular nature of the gene to knowing the molecular makeup of every gene present in our species.

Leading up to the 2001 papers and since then, the multinational effort called the **Human Genome Project** has produced a wide array of data on the locations and functions of genes and other types of DNA sequences found in humans. It's important to recognize, though, that research on *Homo sapiens* is part of a much larger and ongoing effort to sequence genomes from an array of other eukaryotes, hundreds of bacteria, and dozens of archaea. The effort to sequence, interpret, and compare whole genomes is referred to as **genomics**. The pace of research in this field is nothing short of explosive.

The arrival of genomics has also triggered the development of the field called functional genomics. Genomics supplies a list of the genes present in an organism; **functional genomics** asks when those genes are expressed and how their products interact. This effort is important. Besides providing new insights into how gene expression is coordinated as an embryo develops into a college student or an oak tree, functional genomics is helping researchers identify new drug targets in disease-causing species, design new vaccines, and speed efforts to locate the genes responsible for inherited diseases.

As an introductory biology student, you are part of the first generation trained in the genome era. Genomics is revolutionizing several fields within biological science and will almost certainly be an important part of your personal and professional life. Let's look at what genomics is, how and why it's being done, and what is being learned.

20.1 Whole-Genome Sequencing

Genomics has moved to the cutting edge of research in biology largely because a series of technological advances—including the fluorescent markers and gel-filled capillary tubes introduced in Chapter 19—have increased the speed of DNA sequencing and driven down the expense. Thanks largely to increased automation, the cost of sequencing DNA has declined by a factor of two every year and a half since the human genome project began in 1988. Researchers have now established factory-style DNA sequencing centers, each containing dozens of automated sequencing machines, at over 20 locations in the United States, the United Kingdom, Germany, France, Japan, and China. Some of these laboratories employ dozens of biologists and can conduct 100,000 sequencing reactions daily.

As data became less expensive and faster to obtain, the pace of whole-genome sequencing accelerated. The result is that an almost mind-boggling number of sequences are now being generated. As this book goes to press, the primary international repositories for DNA sequence data contain over 146 *billion* nucleotides. (By way of comparison, a haploid human genome contains about 3 billion bases.) The size of this database is increasing by about 20 percent every year (**Figure 20.1**).

How Are Complete Genomes Sequenced?

Genomes range in size from a few million base pairs to several billion. But a single sequencing reaction can analyze only about 1000 base pairs. How do investigators break a genome into sequencing-sized pieces, and then figure out how the thousands or millions of pieces go back together?

Most recent genome sequencing projects answer this question with a whole-genome shotgun approach, or simply **shotgun sequencing**. In shotgun sequencing, a genome is broken up into a set of overlapping fragments that are small enough to be sequenced. Using the regions of overlap, the sequenced fragments are then put back into the correct order.

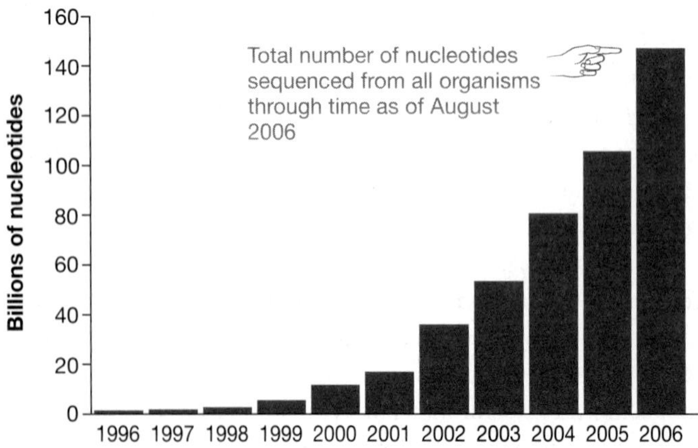

FIGURE 20.1 The Total Number of Bases Sequenced Is Growing Rapidly.

As step 1 of **Figure 20.2** shows, shotgun sequencing begins by using high-frequency sound waves, or sonication, to break up a genome into pieces about 160 kilobases (kb) long (1 kb = 1000 bases). Next, each 160-kb piece is inserted into a plasmid called a **bacterial artificial chromosome (BAC).** BACs are able to replicate large segments of DNA. Using transformation techniques introduced in Chapter 19, each BAC is then inserted into a different *Escherichia coli* cell, creating what researchers call a BAC library (step 2). A BAC library is a genomic library: a set of all the DNA sequences in a particular genome, split into small segments and inserted into cloning a vector (see Chapter 19). By separating the cells in a BAC library and then allowing each cell to grow into a large colony, researchers can isolate large numbers of each 160-kb fragment.

Once a research group has many copies of each 160-kb fragment, the DNA is again broken into fragments—but this time, segments that are about 1000 base pairs long (step 3). These small fragments are then inserted into plasmids and placed inside bacterial cells (step 4). In this way, a genome is broken down into two manageable levels: 160-kb fragments in BACs and 1-kb segments in plasmids. The plasmids are copied many times as the bacterial cells grow into a large population. Large numbers of each 1000-base-pair fragment are then available for sequencing reactions (step 5).

Once the 1000-base-pair fragments from each 160-kb BAC clone are sequenced, computer programs analyze regions where the ends of each 1000-base-pair segment overlap (step 6). Overlaps occur because there were many copies of each 160-kb segment, and each was fragmented randomly by sonication. The computer mixes and matches segments from a single BAC clone until an alignment consistent with all available data is obtained. Then the ends of each BAC are analyzed in a similar way (step 7). The goal is to arrange each 160-kb segment in its correct position along the chromosome, based on regions of overlap.

In essence, the shotgun strategy consists of breaking a genome into tiny fragments, sequencing the fragments, and then putting the sequence data back into the correct order. If you understand shotgun sequencing, you should be able to explain why it is essential for regions of overlap to exist between fragments that are adjacent to each other on a chromosome.

Once complete genome sequences became available, databases that could hold completed sequence information had to be created and managed in a way that made the raw data and a variety of annotations available to the international community of researchers. These sequence databases also had to be searchable, so that investigators could evaluate how similar newly discovered genes were to genes that had been studied previously.

Because the amount of data involved is so large, the computational challenges involved in genomics are formidable. Thus far, sophisticated algorithms and continually improving computer hardware have allowed researchers to keep pace with the rate of data acquisition. The vast quantity of data generated by

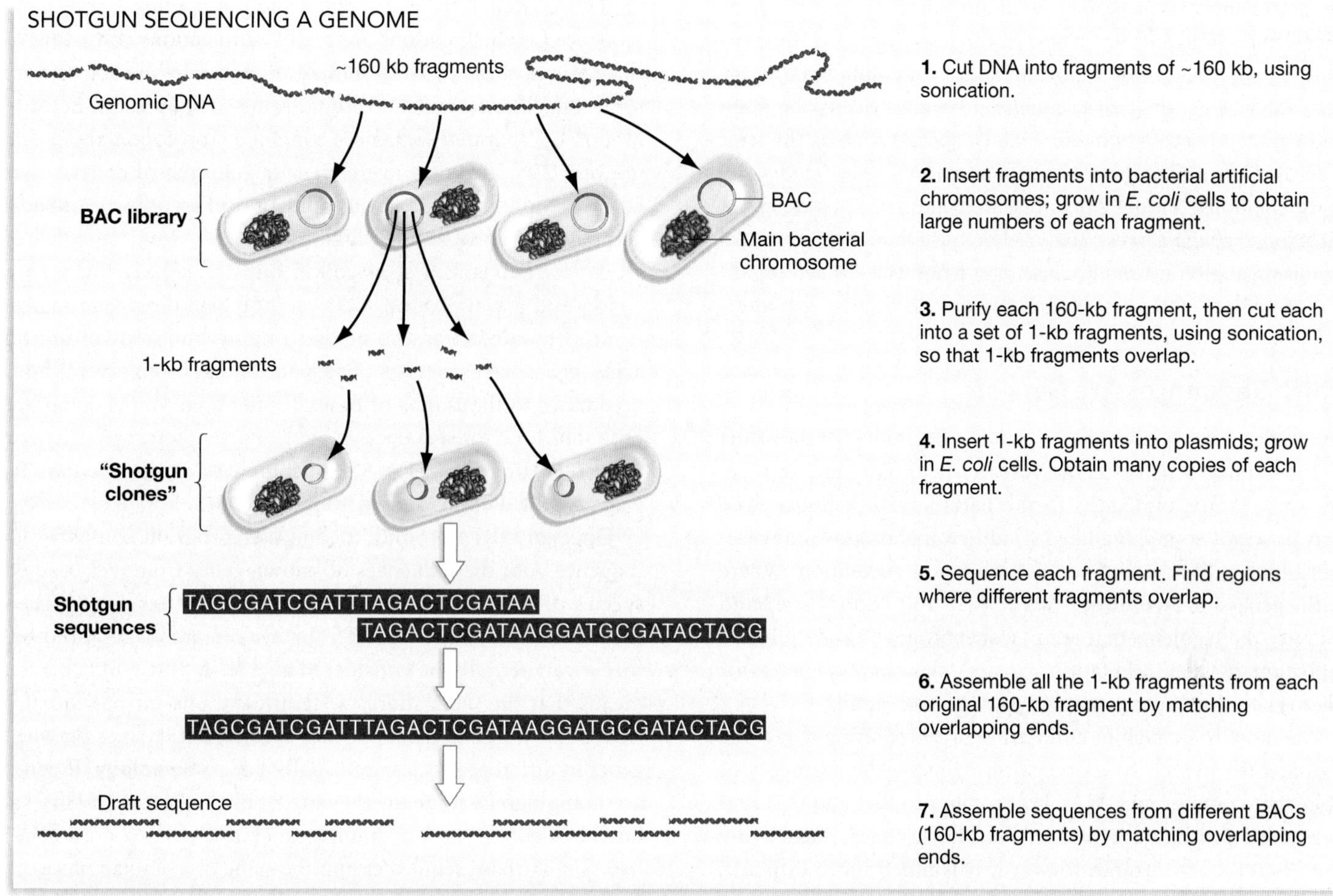

FIGURE 20.2 Shotgun Sequencing Breaks Large Genomes into Many Short Segments. Shotgun sequencing is a multistep process. A genome is initially fragmented into 160-kb sections via sonication and cloned into bacterial artificial chromosomes (BACs). Each 160-kb section is then cut into 1-kb fragments that can be cloned into plasmids, grown to a large number of copies, and sequenced.

QUESTION Why is "shotgun" an appropriate way to describe this sequencing strategy?

genome sequencing centers has made **bioinformatics**—the effort to manage, analyze, and interpret information in biology—a key to continued progress in the field.

Which Genomes Are Being Sequenced, and Why?

The first genome to be sequenced from an organism—not a virus—came from a bacterium that lives in the human upper respiratory tract. This bacterium, *Haemophilus influenzae*, has one circular chromosome and a total of 1,830,138 base pairs of DNA. This genome was small enough to sequence completely with a reasonable amount of time and money, given the technology available in the early 1990s. *H. influenzae* was an important research subject because it causes earaches and respiratory tract infections in children; what's more, one particular strain is capable of infecting the membranes surrounding the brain and spinal cord, causing meningitis.

Publication of the *H. influenzae* genome in 1995 was quickly followed by complete genomes sequenced from an assortment of bacteria and archaea. The first eukaryotic genome, from the yeast *Saccharomyces cerevisiae*, was finished in 1996. After that breakthrough, complete genome sequences were published from a variety of protists, plants, and animals. To date, complete genomes have been sequenced from over 510 bacterial species, 47 archaeal species, and 52 eukaryotic species. Incomplete genome sequence data are available from more than 200 other species.

Most of the organisms that have been selected for whole-genome sequencing cause disease or have other interesting biological properties. For example, genomes of bacteria and archaea that inhabit extremely hot environments have been sequenced in the hopes of discovering enzymes useful for high-temperature industrial applications and understanding how proteins can work under those conditions. Other bacteria and archaea were chosen for sequencing because they do interesting chemistry, such as producing methane (natural gas; CH_4) or other compounds. In some cases, researchers hope that these organisms might act as an important source of commercial products. The rice genome was sequenced because rice is the main food source for most humans. Finally, species such as the fruit fly *Drosophila melanogaster*, the roundworm *Caenorhabditis elegans*, the house mouse *Mus musculus*, and the mustard plant *Arabidopsis thaliana* were analyzed because they serve as model organisms in biology and because data from well-studied organisms promised to help researchers interpret the human genome.

Which Sequences Are Genes?

Obtaining raw sequence data is just the beginning of the effort to understand a genome. As researchers point out, raw sequence data are analogous to the parts list for a house. The parts list reads something like "windowwabeborogovestaircasedoorjubjub ...," however, because it has no punctuation. Where do the genes for "window," "staircase," and "door" start and end? Are the segments that read "wabeborogove" and "jubjub" important in gene regulation, or are they simply spacers or other types of sequences that have no function at all?

The most basic task in annotating or interpreting a genome is to identify which bases constitute genes—the segments of DNA that code for an RNA or a protein product and that regulate their production. In bacteria and archaea, identifying genes is relatively straightforward. It is much more difficult, however, in eukaryotes.

Identifying Genes in Bacterial and Archaeal Genomes

Biologists begin with computer programs that scan the sequence of a genome in both directions. These programs identify each reading frame that is possible on the two strands of the DNA. (Recall from Chapter 15 that a reading frame is the sequence in which codons are read.) With codons consisting of three bases, three reading frames are possible on each strand, for a total of six possible reading frames (**Figure 20.3**). Because randomly generated sequences contain a stop codon about one in every 20 codons on average, a long stretch of codons that lacks a stop codon is a good indication of a coding sequence. The computer program highlights any "gene-sized" stretches of sequence that lack a stop codon but are flanked by a stop codon and a start codon. Because polypeptides range in size from a few dozen amino acids to many hundreds of amino acids, gene-sized stretches of sequence range from several hundred bases to thousands of bases. ● In addition, the computer programs look for sequences typical of promoters, operators, or other regulatory sites. DNA segments that are identified in this way are called **open reading frames**, or **ORFs**.

Once an ORF is found, a computer program compares its sequence with the sequences of known genes from well-studied species. If the ORF appears to be a gene that has not yet been described in any other species, further research is required before it can actually be considered a gene. A "hit," in contrast, means that the ORF shares a significant amount of sequence with a known gene from another species. Similarities between genes in different species are usually due to **homology**. If genes are homologous, it means they are similar because they are related by descent from a common ancestor. Homologous genes have similar base sequences and the same or a similar function. For example, consider the genes introduced in Chapter 14 that code for enzymes involved in repairing mismatches in DNA. Recall that the mismatch repair genes in *E. coli*, yeast, and humans are extremely similar in structure, DNA sequence, and function. To explain this similarity, biologists hypothesize that the common ancestor of all cells living today had mismatch re-

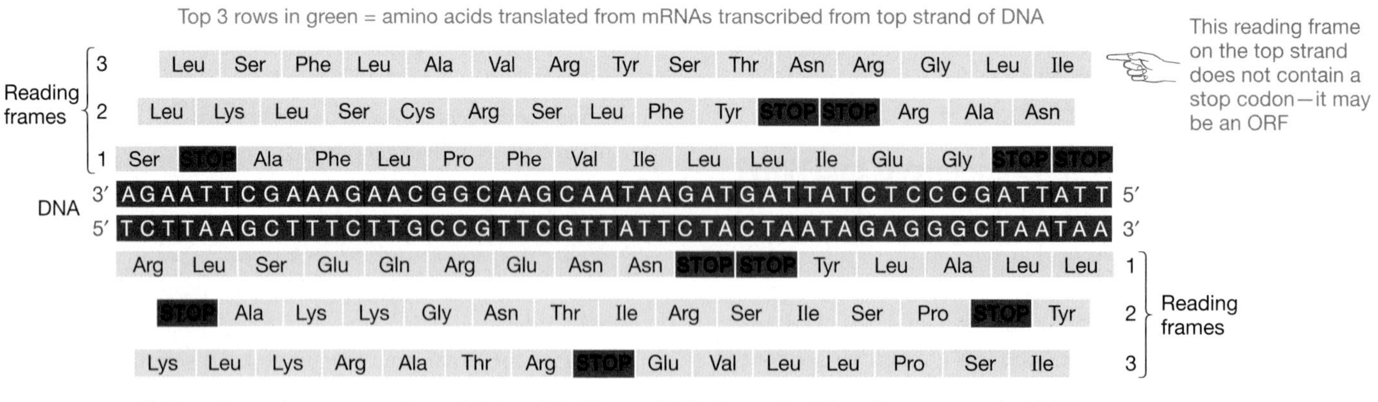

FIGURE 20.3 Open Reading Frames May Be the Locations of Genes. Computer programs scan the three possible reading frames on each strand of DNA and use the genetic code to translate each codon. A long stretch of codons that contains a start codon but lacks a stop codon may be an open reading frame (ORF)—a possible gene.

● **QUESTION** To predict the mRNA codons that would be produced by a particular reading frame, a computer analyzes the DNA in the 3′ to 5′ direction. Why?

pair genes—thus, the descendants of this ancestral species also have versions of these genes.

Based on this logic, researchers can confirm that an ORF is actually a gene by finding that it is homologous to a known gene. They can also analyze the product that would be produced by an ORF, and see if it conforms to a known gene. Unfortunately, finding and analyzing genes by identifying ORFs does not work well in eukaryotes.

Identifying Genes in Eukaryotic Genomes Mining eukaryotic sequence data for genes is complicated by two observations: Coding regions are broken up by introns, and the vast majority of eukaryotic DNA does not actually code for a product. In the human genome, for example, it is estimated that less than 2 percent of the DNA present actually codes for proteins, tRNAs, ribosomal RNAs, or other types of products. Finding a coding region in eukaryotic DNA is like finding a diamond in a huge pile of rock. To do so, researchers pursue several strategies:

- Computer programs can be written to search for sequences that are homologous to known genes. If a stretch of bases in the newly sequenced genome is similar to the sequence of a known gene, then researchers infer that it codes for a product whose function is similar to the function of the known gene.
- As Chapter 19 showed, investigators can isolate mRNAs from the organism being studied and then use enzymes to make the complementary DNAs (cDNAs). If the sequence of these cDNAs is determined, then a computer program can scan the genome sequence and pinpoint where each of the cDNAs is located. This approach allows researchers to identify genes that are expressed in certain cell types—the tissues where the original mRNA was found.
- To identify genes that have no known function, computers compare the genomes of closely related species and highlight sequences that are similar. Sequences that are shared by closely related species are hypothesized to be located in the protein-coding or regulatory regions of genes. The logic behind this claim runs as follows: Sequences that are part of a gene are expected to change much more slowly over time than sequences that are not actually part of a gene. Gene sequences change slowly over time because most gene products work less efficiently when they change randomly by mutation. Thus, it is logical to expect that natural selection eliminates most mutations in genes and that genes should change slowly over time. But changes in sequences that do not code for products or regulate gene expression do not affect the organism's phenotype. Mutations in these regions are much less likely to be eliminated by natural selection, so they change relatively rapidly over time.

Although each of these gene-finding strategies has been productive, it will probably be many years before biologists are convinced they have identified all of the coding regions in even a single eukaryotic genome. As that effort continues, though, researchers are analyzing the data and making some remarkable observations. Let's first consider what genome sequencing has revealed about the nature of bacterial and archaeal genomes and then move on to eukaryotes. Is the effort to sequence whole genomes paying off?

MB **Web Animation** at www.masteringbio.com
Human Genome Sequencing Strategies

20.2 Bacterial and Archaeal Genomes

By the time you read this paragraph, the genomes of over 600 bacterial and archaeal species will have been sequenced. In addition to this impressive array of different species, complete genome sequences are now available for several strains of the same bacterial species. For example, researchers have sequenced the genome of a laboratory population of *Escherichia coli*—derived from the harmless strain that lives in your gut—as well as the genome of a form that causes severe disease in humans. As a result, researchers can now compare the genomes of closely related cells that have different ways of life.

This section focuses on a simple question: Based on data published between 1995 and 2007, what general observations have biologists been able to make about the nature of bacterial and archaeal genomes?

The Natural History of Prokaryotic Genomes

In a sense, biologists who are working in genomics can be compared to the naturalists of the eighteenth and nineteenth centuries. These early biologists explored the globe, collecting the plants and animals they encountered. Their goal was to describe what existed. Similarly, the first task of a genome sequencer is to catalog what is in a genome—specifically, the number, type, and organization of genes. Several interesting conclusions can be drawn from relatively straightforward observations about the data obtained thus far:

- In bacteria, there is a general correlation between the size of a genome and the metabolic capabilities of the organism. For example, most parasites have much smaller genomes than nonparasitic organisms do. **Parasites** live off a host and thus reduce the host's fitness—its ability to produce offspring. Some of the smallest genomes are found in parasitic bacteria from the genus *Mycoplasma*. These bacteria live and multiply inside host cells. *Mycoplasma pneumoniae*, for example, parasitizes lung cells and causes pneumonia in humans. *Mycoplasma* lack the enzymes required to manufacture many essential compounds. Instead they acquire almost all of their nutrients from their hosts. In contrast, the genomes of nonparasitic strains of the bacteria *E. coli* and *Pseudomonas aeruginosa* are 8 to 10 times larger.

Their genes code for enzymes that synthesize virtually every molecule needed by the cell. Based on this observation, it is not surprising that *E. coli* is able to grow under a wide variety of environmental conditions. Using similar logic, researchers hypothesize that the large genome of *P. aeruginosa* explains why it is able to occupy a wide array of soil types, including marine and marshy habitats, as well as human tissues, where it can cause illness.

- Biologists still do not know the function of many of the genes that have been identified. Although *E. coli* probably qualifies as the most intensively studied of all organisms, the function of over 30 percent of its genes is unknown.
- There is tremendous genetic diversity among bacteria and archaea. About 15 percent of the genes in each species' genome appear to be unique. That is, about one in six genes in one of these species is found nowhere else.
- Redundancy among genes is common. For instance, the genome of E. coli has 86 pairs of genes whose DNA sequences are nearly identical—meaning that the proteins they produce are nearly alike in structure and presumably in function. Although the significance of this redundancy is unknown, biologists hypothesize that slightly different forms of the same protein are produced in response to slight changes in environmental conditions.
- Multiple chromosomes are more common than anticipated. Several species of bacteria and archaea have two different circular chromosomes instead of one. And at least some bacteria have linear chromosomes.
- Many species contain the small, extrachromosomal DNA molecules called plasmids. Recall from Chapter 19 that plasmids contain a small number of genes, though not genes that are absolutely essential for growth. In many cases, plasmids can be exchanged between cells of the same species or even of different species (see Chapter 12).

Perhaps the most surprising observation of all is that in many bacterial and archaeal species, a significant proportion of the genome appears to have been acquired from other, often distantly related, species. In some bacteria and archaea, 15–25 percent of the genetic material appears to be "foreign." This is a remarkable claim. What evidence backs up the assertion that prokaryotes acquire DNA from other species? How could this happen, and what are the consequences?

Evidence for Lateral Gene Transfer

Biologists use two general criteria to support the hypothesis that sequences in bacterial or archaeal genomes originated in another species: (1) when stretches of DNA are much more similar to genes in distantly related species than to those in closely related species and (2) when the proportion of G-C base pairs to A-T base pairs in a particular gene or series of genes is markedly different from the base composition of the rest of the genome. In many cases, the proportion of G-C bases in a genome is characteristic of a genus or species.

How can genes move from one species to another? In at least some cases, plasmids appear to be responsible. For example, most of the genes that are responsible for conferring resistance to antibiotics are found on plasmids. Researchers have documented the transfer of plasmid-borne antibiotic-resistance genes between very distantly related species of disease-causing bacteria. In some cases, genes from plasmids become integrated into the main chromosome of a bacterium, resulting in genetic recombination (see Chapter 12). The movement of DNA from one species to another species is called **lateral gene transfer** (**Figure 20.4**).

Some biologists hypothesize that lateral gene transfer also occurs via transformation—when bacteria and archaea take up raw pieces of DNA from the environment, perhaps in the course of acquiring other molecules. This may have occurred in the bacterium *Thermotoga maritima*, which occupies the high-temperature environments near deep-sea vents. Almost 25 percent of the genes in this species are extremely closely related to genes found in archaea that live in the same habitats. The archaea-like genes occur in distinctive clusters within the *T. maritima* genome, which supports the hypothesis that the sequences were transferred in large pieces from an archaean to the bacterium.

Similar types of direct gene transfer are hypothesized to have occurred in the bacterium *Chlamydia trachomatis*. This organism is a major cause of blindness in humans from Africa and Asia; it also causes chlamydia, the most common sexually transmitted bacterial disease in the United States. The *C. trachomatis* genome contains 35 genes that resemble eukaryotic genes in structure. Because *C. trachomatis* lives inside the cells that it parasitizes, the most logical explanation for this observation is that the bacterium occasionally takes up DNA directly from its host cell, resulting in a eukaryote-to-bacterium transfer.

In addition to being transferred between species by means of plasmids or DNA fragments, genes can be transported by viruses. For example, investigators who compared the sequences of laboratory and pathogenic (disease-causing) strains of *E. coli* found that the pathogenic cells have almost 1400 "extra" genes. Compared with the rest of the genome, most of these genes have a distinctive G-C to A-T ratio. Many are also extremely similar to sequences isolated from viruses that infect *E. coli*. Based on these observations, most researchers support the hypothesis that at least some of the disease-causing genes in *E. coli* were brought in by viruses.

To summarize, mutation and genetic recombination within species are not the only source of genetic variation in bacteria and archaea. Over the course of evolution, lateral gene transfer has been an important source of new genes and allelic diversity in these domains. This insight would not have been possible without data from whole-genome sequencing. Have efforts to sequence eukaryotic genomes led to similar types of insights?

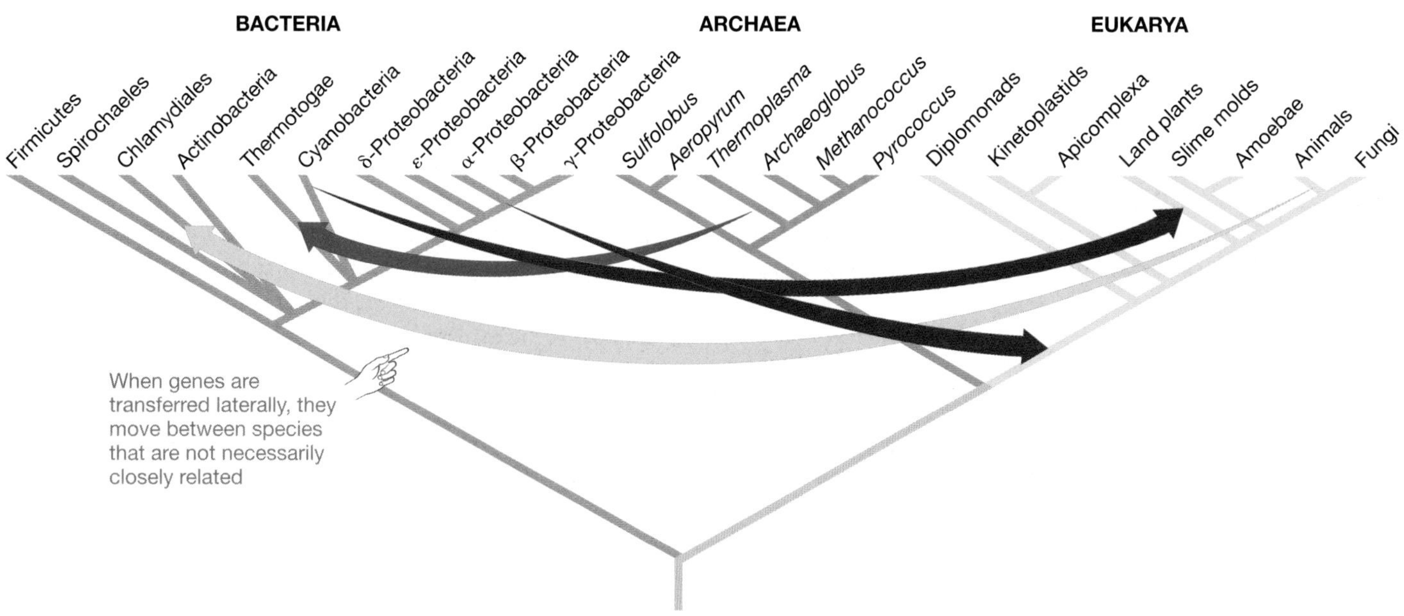

FIGURE 20.4 Lateral Gene Transfer Is Movement of DNA between Species. Lateral gene transfer can occur between very distantly related organisms. For tips on how to read an evolutionary tree like this, see **BioSkills 2**.

EXERCISE Beside this tree, list the mechanisms responsible for lateral gene transfer.

Check Your Understanding

If you understand that...

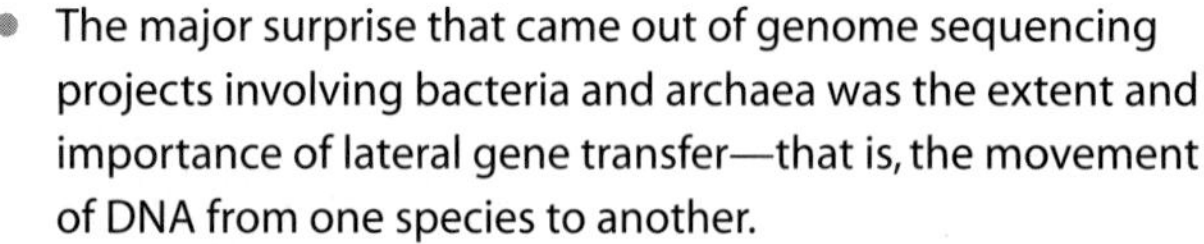

- The major surprise that came out of genome sequencing projects involving bacteria and archaea was the extent and importance of lateral gene transfer—that is, the movement of DNA from one species to another.

You should be able to...

1) Summarize how researchers infer that lateral gene transfer is responsible for the presence of a particular DNA sequence.
2) Summarize evidence that the size of a prokaryotic genome is correlated with the organism's metabolic capabilities.

20.3 Eukaryotic Genomes

Sequencing eukaryotic genomes presents two daunting challenges. The first is sheer size. Compared with the genomes of bacteria and archaea, which range from 580,070 base pairs in *Mycoplasma genitalium* to over 6.3 million base pairs in *Pseudomonas aeruginosa*, eukaryotic genomes are even larger. The haploid genome of *Saccharomyces cerevisiae* (baker's yeast), a unicellular eukaryote, contains over 12 million base pairs. The roundworm *Caenorhabditis elegans* has a genome of 97 million base pairs; the fruit-fly genome contains 180 million base pairs; the mustard plant *Arabidopsis thaliana*'s genome has 130 million base pairs; and humans, rats, mice, and cattle contain roughly 3 billion base pairs each.

The second great challenge in sequencing eukaryotic genes is coping with noncoding sequences that are repeated many times. Many eukaryotic genomes are dominated by repeated DNA sequences that occur between genes and do not code for products used by the organism. These repeated sequences pose serious problems in aligning and interpreting sequence data. What are they? If such sequences don't code for a product, why do they exist?

Natural History: Types of Sequences

In many eukaryotic genomes, the exons and regulatory sequences associated with genes make up a relatively small percentage of the genome. Recall from Section 20.1 that in humans, protein-coding sequences constitute less than 2 percent of the total genome while repeated sequences account for well over 50 percent. In contrast, over 90 percent of a bacterial or archaeal genome consists of genes—DNA sequences that code for a product needed by the cell and regulate its transcription.

When noncoding and repeated sequences were discovered, they were initially considered "junk DNA" that was nonfunctional and probably unimportant and uninteresting. But subsequent work has shown that many of the repeated sequences observed in eukaryotes are actually derived from sequences known as transposable elements. **Transposable elements** are segments of DNA that are capable of moving from one location to another, or transposing, in a genome. They are similar to viruses, except that viruses leave a host cell that they have infected and find a new cell to infect. In contrast, transposable elements never leave their host cell—they simply make copies of themselves and move to new locations in the genome. Transposable elements are passed from parents to offspring, generation after generation, because they are part of the genome.

Transposable elements are examples of what biologists call selfish genes. A selfish gene is a DNA sequence that survives and reproduces but does not increase the fitness of the host genome. Transposable elements and viruses are classified as parasitic because it takes time and resources to copy them along with the rest of the genome, and because they can disrupt gene function when they move and insert in a new location. As a result, they decrease their host's fitness. Transposable elements are genomic parasites.

How Do Transposable Elements Work? Transposable elements come in a wide variety of types and spread through genomes in a variety of ways. Different species—fruit flies, yeast, and humans, for example—contain distinct types of transposable elements.

As an example of how these selfish genes work, let's consider a well-studied type called a **long interspersed nuclear element (LINE)** that is found in humans and other eukaryotes. Because LINEs are so similar to the retroviruses, introduced in Chapter 19, biologists hypothesize that they are derived from them evolutionarily. Your genome contains tens of thousands of LINEs, each between 1000 and 5000 bases long.

An active LINE contains all the sequences required for it to make copies of itself and insert them into a new location in the genome (**Figure 20.5**, step 1): a gene that codes for the enzyme reverse transcriptase, a gene that codes for the enzyme integrase, and a single promoter that is recognized by RNA polymerase II (step 2). After a LINE is transcribed to an mRNA, reverse transcriptase and integrase are synthesized by ribosomes in the cytoplasm (steps 3 and 4). Reverse transcriptase makes a cDNA version of the LINE mRNA, and integrase inserts the newly synthesized LINE DNA into a new location in the genome (step 5). In this way, the parasitic sequence reproduces (step 6). If the transposition event occurs in reproductive cells that go on to form eggs or sperm, the copied LINE will be passed on to offspring. If the LINE happens to insert itself inside a gene or a regulatory sequence, it causes a mutation that is almost certain to reduce the host's fitness.

Most of the LINEs observed in the human genome do not actually function, however, because they don't contain a promoter or the genes for either reverse transcriptase or integrase. To make sense of this observation, researchers hypothesize that the insertion process illustrated in steps 6 and 7 of Figure 20.5 is usually disrupted in some way. Analyses of the human genome have revealed that only a handful of our LINEs appear to be complete and potentially active.

Virtually every prokaryotic and eukaryotic genome examined to date contains at least some transposable elements. They vary

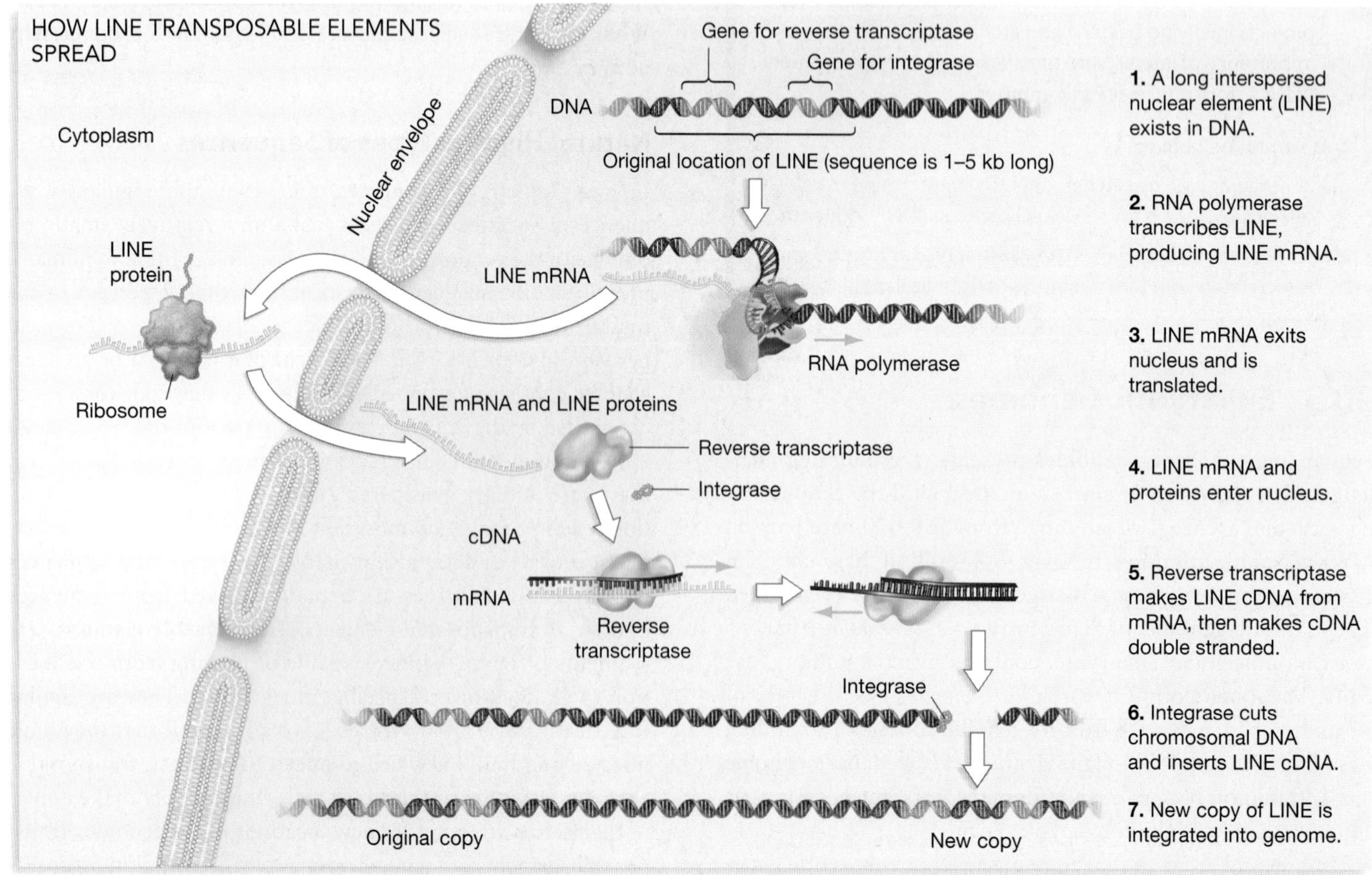

FIGURE 20.5 Transposable Elements Spread within a Genome.

widely in type and number, however, and bacterial and archaeal genomes have relatively few transposable elements compared to most eukaryotes studied thus far. This observation has inspired the hypothesis that bacteria and archaea either have efficient means of removing parasitic sequences or can somehow thwart insertion events. To date, however, this hypothesis has yet to be tested rigorously.

Research on transposable elements and lateral gene transfer has revolutionized how biologists view the genome. Many genomes are riddled with parasitic sequences, and others have undergone radical change in response to lateral gene transfer events. Genomes are much more dynamic and complex than previously thought. Their size and composition can change dramatically over time.

Repeated Sequences and DNA Fingerprinting In addition to containing repeated sequences from transposable elements, eukaryotic genomes have several thousand loci called simple tandem repeats (STRs). These are small sequences repeated one after another down the length of a chromosome. There are two major classes of STRs: Repeating units that are just 1 to 5 bases long are known as **microsatellites** or **simple sequence repeats**; repeating units that are 6 to 500 bases long are called **minisatellites** or **variable number tandem repeats** (**VNTRs**). Both types of repeated sequences make up 3 percent of the human genome. The most common type of microsatellite is a repeated stretch of the dinucleotide AC, giving the sequence ACACACAC.... Microsatellite sequences are thought to originate when DNA polymerase skips or mistakenly adds extra bases during replication; the origin of minisatellites is still unclear.

Soon after these sequences were first characterized, Alec Jeffreys and co-workers established that microsatellite and minisatellite loci are "hypervariable," meaning that they vary among individuals much more than any other type of sequence does. **Figure 20.6** illustrates one hypothesis for why microsatellites and minisatellites have so many different alleles: These highly repetitive stretches often align incorrectly when homologous chromosomes synapse and cross over during prophase of meiosis I. Instead of lining up in exactly the same location, the two chromosomes pair in a way that matches up bases in different repeated segments. Due to this misalignment, **unequal crossover** occurs. Chromosomes produced by unequal crossover contain different numbers of repeats. The key observation is that if a particular microsatellite or minisatellite locus has a unique number of repeats, it represents a unique allele. Each allele has a unique length. As with any allele, microsatellite and minisatellite alleles are transmitted from parents to offspring.

Misalignment or errors by DNA polymerase are so common at these loci that, in most eukaryotes, the genome of virtually every individual has at least one new allele. This variation in repeat number among individuals is the basis of DNA fingerprinting. **DNA fingerprinting** refers to any technique for identifying individuals based on the unique features of their genomes. Because microsatellite and minisatellite loci vary so much among individuals, they are now the loci of choice for DNA fingerprinting. To fingerprint an individual, researchers obtain a DNA sample and perform the polymerase chain reaction (PCR), using primers that flank a region containing an STR. Once many copies of the region are available, they can be analyzed to determine the number of repeats present (**Figure 20.7a**). Primers are now available

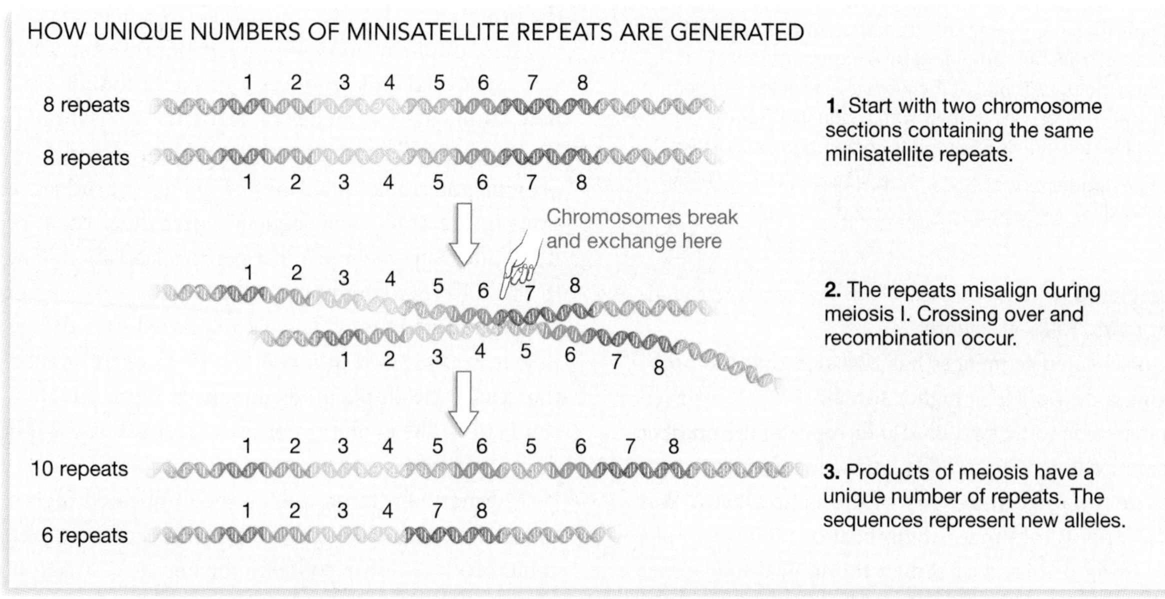

FIGURE 20.6 Unequal Crossover Changes the Numbers of Simple Sequence Repeats. The alignment of homologous chromosomes during prophase of meiosis I is driven by sequence similarity between homologs. Because simple sequence repeats are so similar, they are likely to misalign during synapsis.

(a) Using PCR to amplify minisatellite and microsatellite loci

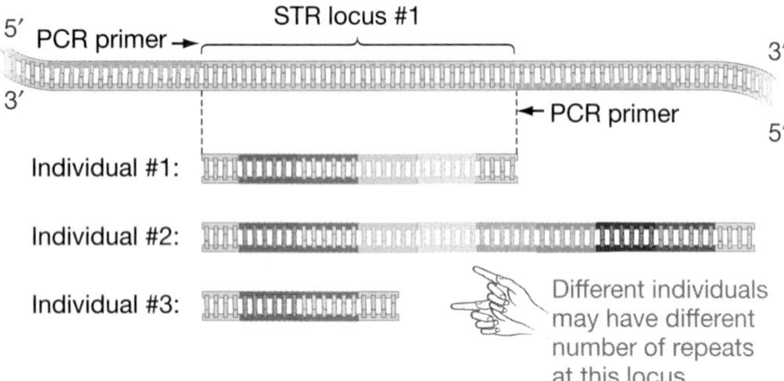

(b) A gel showing minisatellite sequences from unrelated and related individuals

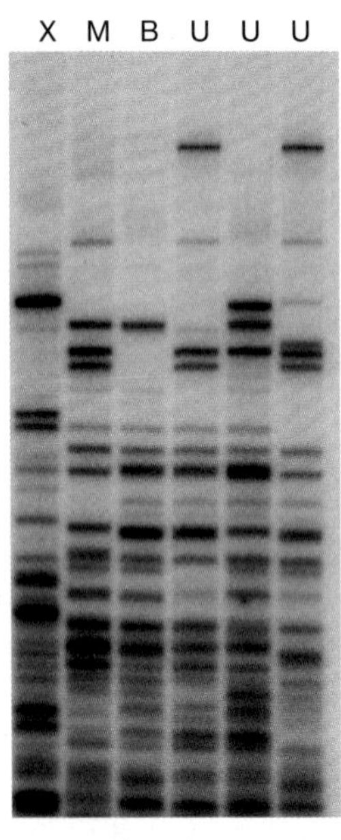

FIGURE 20.7 DNA Fingerprinting Can Be Used to Identify Fathers. **(a)** If a minisatellite locus contains different numbers of repeats in different individuals, then DNA fragments from those individuals will have different lengths. **(b)** A gel containing DNA fragments from a mother, a purported biological child of the woman, and several of her confirmed biological children was probed with sequences from a minisatellite locus. Related individuals tend to share fragment patterns.

EXERCISE Circle fragments in the M, B, and U lanes that support the hypothesis that the disputed boy actually is M's son.

for many different STR loci, so researchers can analyze the alleles present at many STRs efficiently.

Research on repeated sequences has revealed that the probability of getting a new allele is higher for shorter repeats than for longer repeats. For some two-base-pair repeats, the number of repeats present changes so quickly over time that only very close relatives are likely to share any of the same alleles. This observation has important practical implications. For example, DNA fingerprinting of blood or semen found at crime scenes has been used to show that people who were accused of crimes were actually innocent. DNA fingerprinting has also been used as evidence to convict criminals or assign paternity in birds, humans, and other species that have well-characterized microsatellite or minisatellite sequences (**Figure 20.7b**).

Now that we've reviewed the characteristics of some particularly prominent types of noncoding sequences in eukaryotes, it's time to consider the nature of the coding sequences in these genomes. Let's start with the most basic question of all: Where do eukaryotic genes come from?

Gene Duplication and the Origin of Gene Families

In eukaryotes, the major source of new genes is the duplication of existing genes. Biologists infer that genes have been duplicated recently when they find groups of similar genes clustered along the same chromosome. The genes are usually similar in structural aspects, such as the arrangement of exons and introns, and in their base sequence. The degree of sequence similarity among these clustered genes varies. In the genes that code for ribosomal RNAs (rRNAs) in vertebrates, the sequences are virtually identical—meaning that each individual has many exact copies of the same gene. In other cases, though, the proportion of bases that are identical is 50 percent or less.

Within a species, genes that are extremely similar to each other in structure and function are considered to be part of the same **gene family**. Genes that make up gene families are hypothesized to have arisen from a common ancestral sequence through gene duplication. When **gene duplication** occurs, an extra copy of a gene is added to the genome.

The most common type of gene duplication results from crossover during meiosis. As **Figure 20.8** shows, gene-sized segments of chromosomes can be deleted or duplicated if homologous chromosomes misalign during prophase of meiosis I and an unequal crossover occurs. The duplicated segments resulting from unequal crossover are arranged in tandem—one after the other.

Gene duplication is important because the original gene is still functional and produces a normal product. As a result, the new, duplicated stretches of sequence are redundant. In some cases the duplicated genes retain their original function and provide additional quantities of the same product. But if mutations in the duplicated sequence alter the protein product, and if the altered protein product performs a valuable new function in the cell, then an important new gene has been created. The duplicated gene may also be regulated in a different way, so that it is expressed in novel locations or at novel times. In either case, the duplicated sequences represent new genes and can lead to the evolution of novel traits. Gene duplication produces new genes and creates gene families.

Alternatively, mutations in the duplicated region may make expression of the new gene impossible. For example, a mutation could produce a stop codon in the middle of an exon. A member of a gene family that resembles a working gene but does not code for a functional product, due to early stop codons or other defects, is called a **pseudogene**. Pseudogenes have no function.

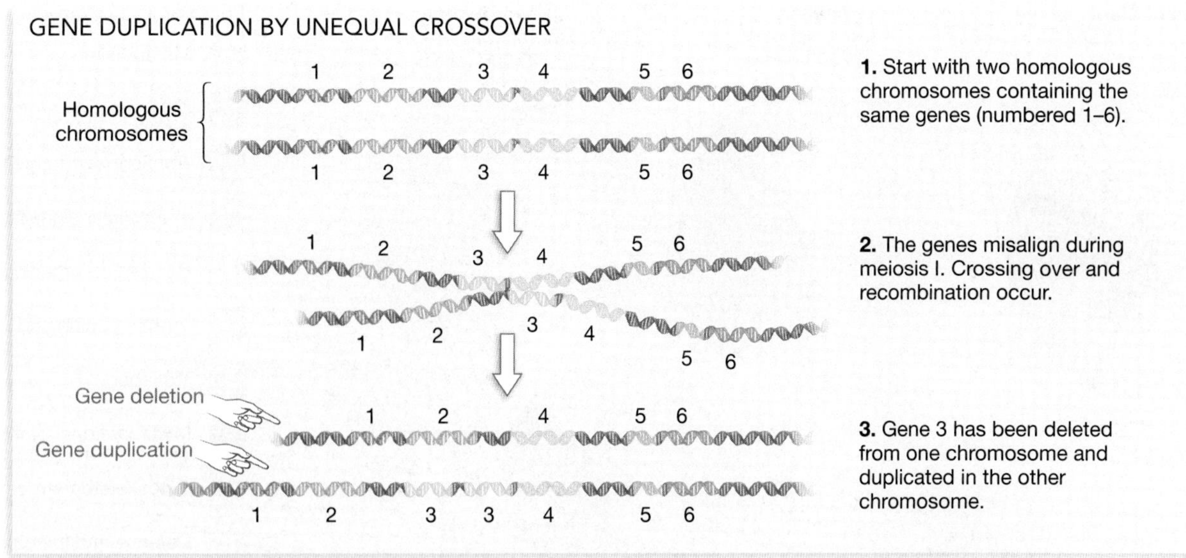

FIGURE 20.8 Unequal Crossover Changes the Numbers of Genes along a Chromosome. If unequal crossover occurs as diagrammed here, the resulting chromosomes contain one gene fewer than the original chromosome or an additional copy of a gene.

As an example of a gene family, consider the human globin genes diagrammed in **Figure 20.9**. These genes code for proteins that form part of hemoglobin—the oxygen-carrying molecule in your red blood cells. Analyzing the globin genes illustrates several important points about gene families. In humans, the globin gene family contains several pseudogenes, along with several genes that code for oxygen-transporting proteins. The various coding genes in the family serve slightly different functions. For example, some genes are active only in the fetus or the adult. Follow-up work showed that oxygen is much more likely to bind to the proteins encoded by the fetal genes compared to the proteins expressed in adults. As a result, oxygen is able to move from the mother's blood to the fetus's blood (see Chapter 44).

In addition to the gene duplication events resulting from unequal crossover, the entire complement of chromosomes may be duplicated due to a mistake in either mitosis or meiosis. In this case, the resulting organism contains double the normal complement of chromosomes. Recall from Chapter 12 that species with duplicated chromosome complements are said to be **polyploid**. When polyploidy occurs, every gene in the duplicated genome is redundant. As a result, each gene may experience mutations that lead to new functions or to the loss of function and creation of a pseudogene.

By comparing how many copies of gene families occur in eukaryotes that have completely sequenced genomes, researchers have concluded that a whole-genome duplication occurred early in the evolution of vertebrates. They further conclude that another genome duplication occurred early in the evolution of the ray-finned fish—a lineage of over 24,000 living species, including familiar groups such as the trout, tuna, and guppies. Genome duplication has also been a particularly important source of new genes in plants.

Insights from the Human Genome Project

The human genome is rapidly becoming the most intensively studied of all eukaryotic genomes. In many or even most cases, researchers are gaining insights into how the human genome works by comparing it to the genomes of other

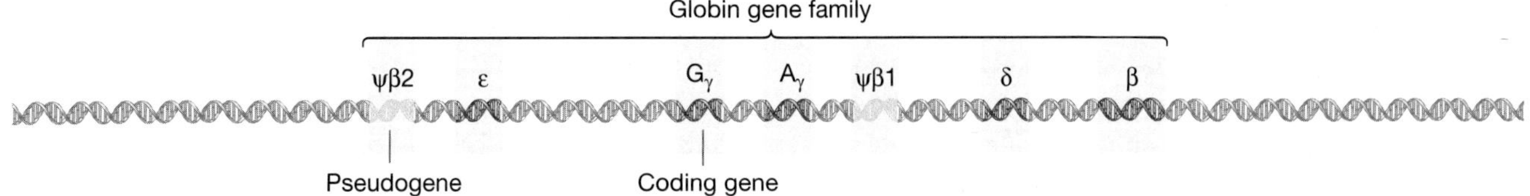

FIGURE 20.9 Gene Families Are Clusters of Closely Related Genes. Genes within the globin family. Red segments represent functioning genes, and yellow segments are pseudogenes. The members of a gene family are arranged one after the other, or in tandem. Most of these genes are expressed at different times during development.

EXERCISE Suppose that during prophase of meiosis I, the β locus on one chromosome aligned with the $\psi\beta2$ locus on another chromosome, and then crossing over occurred just to their left. Draw the chromosomes that would result.

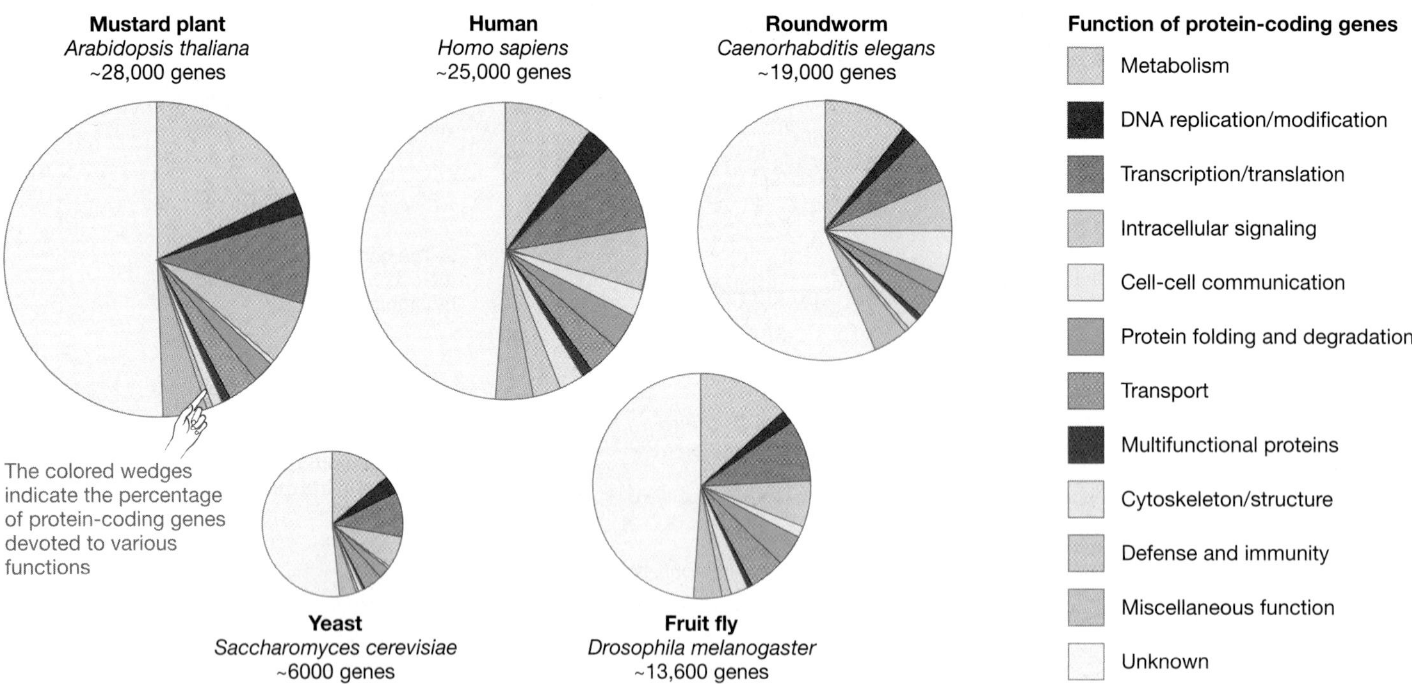

FIGURE 20.10 Comparing Gene Functions in Various Eukaryotes. [Adapted from Fig. 9.34, H. Lodish et al., *Molecular Cell Biology*, 5th ed., 2004. © W. H. Freeman and Company.]

● **QUESTION** In the organisms pictured here, about what percentage of the genes present have unknown function?

species. **Figure 20.10**, for example, shows the relative proportion of genes devoted to various functions in humans and four other species of eukaryotes. A careful look at this figure should convince you that humans have a particularly large proportion of their genome devoted to immunity (defense against bacteria, viruses, and other parasites), and that humans and the roundworm *Caenorhabditis elegans* have a larger percentage of their genome devoted to cell-cell signaling than do other eukaryotes studied thus far. But even a quick look at these diagrams carries an important message: No one knows the function of most genes found in humans and other eukaryotes.

Although it is clear that a great deal remains to be learned about the human genome, two important questions have emerged from the early studies. Let's consider each of them in turn.

Why Do Humans Have So Few Genes? Of all observations about the nature of eukaryotic genomes, perhaps the most striking is that organisms with complex morphology and behavior do not appear to have particularly large numbers of genes. **Table 20.1** indicates the estimated number of genes found in selected eukaryotes. Notice that the total number of genes in *Homo sapiens*, which is considered a particularly complex organism, is not that much higher than the total number of genes in fruit flies. Gene number in humans is about the same as in roundworms, mice, rats, puffer fish, and chickens and substantially less than in rice and the weedy mustard plant *Arabidopsis thaliana*. Before the human genome was sequenced, many biologists expected that humans would have at least 100,000 genes. We now know that we have only a fifth of that number—perhaps less.

How can this be? In prokaryotes there is a correlation between genome size, gene number, a cell's metabolic capabilities, and the cell's ability to live in a variety of habitats. Similarly, it is logical to observe that plants have exceptionally large numbers of genes because they synthesize so many different and complex molecules from just carbon dioxide, nitrate ions, phosphate ions, and other simple nutrients. The idea is that large numbers of genes enable plants to produce large numbers of enzymes. But why isn't there a stronger correlation between gene number and morphological and behavioral complexity in animals?

The leading hypothesis focuses on **alternative splicing**. Recall from Chapter 18 that the exons of a particular gene can be spliced in ways that produce distinct mature mRNAs. As a result, a single eukaryotic gene can code for multiple transcripts and thus multiple proteins. The alternative-splicing hypothesis claims that at least certain multicellular eukaryotes do not need enormous numbers of distinct genes. Instead, alternative splicing creates different proteins from the same gene. The alternative forms might be produced at different developmental stages or in response to different environmental conditions.

In support of the alternative-splicing hypothesis, researchers have analyzed the mRNAs produced by human genes and have estimated that each gene produces an average of slightly more

TABLE **20.1 Number of Genes in Selected Eukaryotes**

Species	Description	Genome Size (Millions of Base Pairs)	Estimated Number of Genes
Saccharomyces cerevisiae	Baker's and brewer's yeast; a unicellular fungus; an important model organism in biochemistry and genetics	12	6000
Plasmodium falciparum	Single-celled, parasitic eukaryote; causes malaria in humans	30	6500
Drosophila melanogaster	Fruit fly; an important model organism in genetics and developmental biology	180	13,600
Caenorhabditis elegans	A roundworm; an important model organism in developmental biology	97	19,000
Canis familiarus	Domestic dog	2410	19,300
Gallus gallus	Chicken	1050	20,000–23,000
Homo sapiens	Humans	3000	20,000
Rattus norvegicus	Norway rat; an important model organism in physiology and behavior	2750	21,000
Mus musculus	House mouse; an important model organism in genetics and developmental biology	2500	~30,000
Arabidopsis thaliana	A mustard plant; an important model organism in genetics and developmental biology	119	28,000
Oryza sativa	Rice	389	37,500

than three distinct transcripts. If this result is valid for the rest of the genome, the actual number of different proteins that can be produced is more than triple the gene number. Humans may have fewer than 20,000 genes according to current estimates, but these genes may have the ability to produce 100,000 different transcripts. Researchers are currently trying to assess whether alternative splicing is this frequent in other eukaryotes, as well.

How Can the Human and Chimp Genomes Be So Similar?

Comparing the numbers of genes found in humans and in mice created a paradox that may be resolved by the alternative-splicing hypothesis. Comparing the base sequences of genes in humans and in other species has created an analogous paradox.

Here is the issue: At the level of base sequences, human beings and chimpanzees are 98.8 percent identical on average. Of the homologous genes analyzed in humans and chimps, 29 percent are identical in amino acid sequence; the average difference between homologous proteins is just two amino acids. If humans and chimps are so similar genetically, why do they appear to be so different in their morphology and behavior?

The leading hypothesis to resolve this paradox focuses on the importance of regulatory genes and regulatory sequences. Recall from Chapter 18 that a **regulatory sequence** is a section of DNA involved in controlling the activity of other genes; it may be a promoter, a promoter-proximal element, an enhancer, or a silencer. The term **structural gene**, in contrast, refers to a sequence that codes for a tRNA, rRNA, protein, or other type of product. **Regulatory genes** code for regulatory transcription factors that alter the expression of specific genes.

To resolve the sequence-similarity paradox, biologists propose that even though many structural genes in closely related species, such as humans and chimps, are identical or nearly identical, regulatory sequences and regulatory genes might have important differences between the two species. Suppose that the structural gene for human growth hormone and chimp growth hormone are identical in base sequence. But if changes in transcription factors, enhancers, or promoters change the pattern of expression of that gene—perhaps turning it on later and longer in humans than in chimps—then height and other characteristics will change even though the structural gene is the same. Based on current analyses, biologists suggest that the human genome contains about 3000 different regulatory transcription factors. Subtle mutations in these proteins and the regulatory sites that they bind to could have a significant effect on gene expression and thus on the phenotype.

The regulatory hypothesis is certainly logical, and it is consistent with data suggesting that regulation of alternative splicing underlies phenotypic complexity in *Homo sapiens* and other large vertebrates. It may be true that most of the genetic changes responsible for the rapid evolution of humans over the past 5 million years have been due to changes in regulatory genes and sequences and alternative splicing rather than to changes in structural genes. To date, however, there are no specific examples of changes in the regulatory sequences responsible for the phenotypic differences observed between humans and chimps or other closely related species. The regulatory hypothesis still needs to be tested rigorously.

Check Your Understanding

If you understand that...

- Eukaryotic genomes are riddled with parasitic sequences that do not contribute to the fitness of the organism.
- Simple repeated sequences are also common in eukaryotic genomes.
- In eukaryotes, many of the coding sequences are organized into families of genes with related functions.

You should be able to...

1) Explain why transposable elements are considered selfish genes.
2) Explain why simple sequence repeats make DNA fingerprinting possible.
3) Explain how unequal crossover leads to duplicated sequences.
4) Explain why researchers hypothesize that human evolution has been dominated by alternative splicing and other types of regulatory changes.

20.4 Functional Genomics and Proteomics

To explain the impact of genomics on the future of biological science, Eric Lander has compared the sequencing of the human genome to the establishment of the periodic table of the elements in chemistry. Once the periodic table was established and validated, chemists focused on understanding how the elements combine to form molecules. Similarly, biologists now want to understand how the elements of the human genome combine to produce an individual.

In essence, a genome sequence is a parts list. Once that list is assembled, researchers delve deeper to understand how genes interact to produce an organism. Let's explore some of the ways in which researchers use whole-genome data to answer fundamental questions about how organisms work.

What Is Functional Genomics?

For decades, biologists have worked at understanding how and when individual genes are expressed. Research on the *lac* operon and *trp* operon, reviewed in Chapter 17, are typical of this effort. But now, with complete catalogs of the genes present in a variety of organisms whose genomes have been sequenced, researchers can ask how and when *all* of the genes in an organism are expressed. These types of large-scale analyses of gene expression are sometimes called functional genomics. The research is motivated by the realization that gene products do not exist in a vacuum. Instead, groups of RNAs and proteins act together to respond to environmental challenges such as extreme heat or drought. Similarly, distinct groups of genes are transcribed at different stages as a multicellular eukaryote grows and develops.

One of the most basic tools used in functional genomics is called a microarray. A **DNA microarray** consists of a large number of single-stranded DNAs that are permanently affixed to a glass slide. For example, the slide pictured in **Figure 20.11** contains thousands of spots, each of which contains single-stranded DNA from a unique exon found in the human genome.

To do an experiment with a DNA microarray, researchers follow the protocol outlined in **Figure 20.12**. The first step is to isolate the mRNAs being produced by two contrasting types of cells. In this example, the control cells are functioning at normal temperature. The other cells, in contrast, have been exposed to high temperatures. The goal of this experiment is to compare genes that are expressed during normal cell activity with those expressed under heat stress.

Once they've purified mRNAs from the two populations of cells, investigators use reverse transcriptase to make a single-stranded cDNA version of each RNA in the samples (see Chapter 19). In addition to the four standard dNTPs, one of the DNA building blocks used in this reaction carries a fluorescent label. The label used for normal cells glows green, while the label chosen for the heat-stressed cells glows red.

The labeled cDNAs can then be used to probe the microarray. As Chapter 19 noted, a probe allows an investigator to find a particular molecule in a sample containing many different molecules. In this case, the labeled cDNAs will bind to the single-stranded DNAs on the plate by complementary base pairing. Out of all the exons in the genome, then, only the exons that are being expressed will be labeled. In our example, genes that are expressed under normal conditions will be labeled green, while those expressed during heat stress will be labeled red. If one of

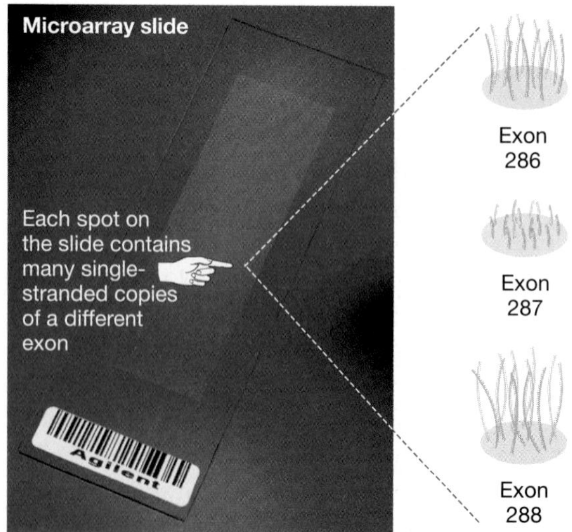

FIGURE 20.11 DNA Microarrays Represent Every Gene in a Genome. To create a DNA microarray, investigators spot thousands of short, single-stranded DNA sequences from coding sequences onto a glass plate. The DNAs typically represent every exon in the genome of a particular species.

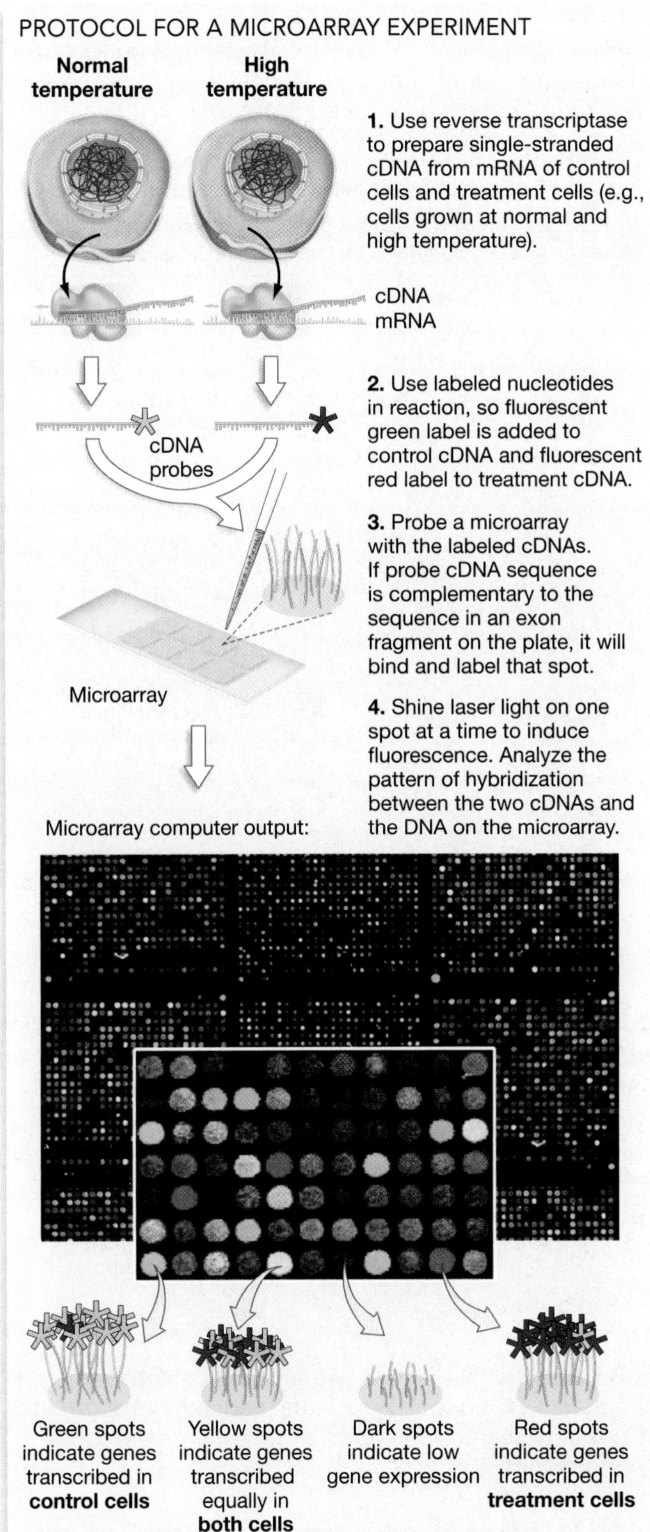

FIGURE 20.12 DNA Microarrays Are Used to Study Changes in Gene Expression. By probing a microarray with labeled cDNAs synthesized from mRNAs, researchers can identify which coding sequences are being transcribed. Here mRNAs from cells growing at normal temperature are green, while mRNAs from cells growing at high temperature are red.

QUESTION If every spot on a microarray represents a different exon, how could an experiment like this document the existence of alternative splicing?

the exons in the microarray is expressed under both sets of conditions, then both green- and red-labeled cDNAs will bind to that spot and make it appear yellow. In this way, a microarray lets researchers study the expression of thousands of genes at a time. As a result, they can identify which sets of genes are expressed in concert under specific sets of conditions.

Once a microarray has been used, the bound cDNA probes can be removed. The original DNAs remain in place, so the slide may then be reused to assess gene expression in a different type of cell, or in the same cell type under different conditions. Researchers can use microarrays to establish which genes are transcribed in different organs and tissues, during cancerous growth, or in response to changes in environmental conditions, such as starvation, the presence of a toxin, or a viral infection. If you understand the concept of how microarrays are used, you should be able to design an experiment that uses a DNA microarray to compare the genes expressed in brain cells versus liver cells of an adult human.

What Is Proteomics?

The Greek root *–ome*, meaning "all," inspired the term *genome*. Similarly, biologists use the term **transcriptome** in referring to the complete set of genes that are transcribed in a particular cell, and **proteome** in referring to the complete set of proteins that are produced. **Proteomics**, it follows, is the large-scale study of protein function. Proteomic studies begin by identifying the proteins present in a cell or organelle; then, researchers attempt to determine the locations and interactions of proteins and document how they change through time or compare with other cells.

Proteomics can be thought of as a branch of functional genomics. Instead of studying individual proteins or how two proteins might interact, biologists can study all of the proteins present at once. One approach to studying protein-protein interactions is similar to the use of DNA microarrays, except that large numbers of proteins, rather than DNA sequences, are affixed on a glass plate. This microarray of proteins is then treated with an assortment of proteins produced by the same organism. These proteins are labeled with a fluorescent or radioactive tag. If any labeled proteins bind to the proteins in the microarray, the two molecules may also interact in the cell. In this way, researchers hope to identify proteins that physically bind to one another—like the G proteins and associated enzymes introduced in Chapter 8, or the cyclin and Cdk molecules introduced in Chapter 11. Microarray technology is allowing biologists to study protein-protein interactions on a massive scale.

20.5 Can Genomics Help Improve Human Health and Welfare?

With the advent of microarray technology, the "periodic tables" provided by genome projects are having an important impact on research into gene expression and protein-protein interactions.

But the governments and corporations that fund genome projects have underwritten the expense primarily because of the potential benefits for improving human health and welfare. In this respect, is genomics living up to its promise?

Although large amounts of genome sequence data have been available for only a few years, early indications are that the investment may indeed pay off with substantial advances in biomedicine. Let's consider how whole-genome data are informing the development of new drugs and vaccines, and then look at a project focused on searching for the alleles associated with inherited diseases.

Identifying Potential Drug Targets

Currently, dozens of whole-genome sequencing projects are focused on species that cause disease in humans or the livestock and crops that we depend on. As each of these projects is completed, biologists begin comparing the genomes of pathogenic strains with closely related species that are harmless. The goal is to achieve a much more detailed understanding of the genetic basis of **virulence**—the tendency for a parasite to harm its host.

One of the first tasks biologists undertake is to identify genes that occur only in pathogens and that may be required for virulence. Typically, biologists are finding that "virulence genes" code for proteins that allow parasitic cells to adhere to host cells, produce enzymes that break down host-cell walls or membranes, or secrete toxins that poison host-cell enzymes. Identifying these genes is important because it gives investigators targets for the development of new drugs. If drugs that knock out the protein products of these genes can be formulated, they would inhibit disease-causing cells while leaving closely related but helpful species unharmed. For example, comparing the genomes of benign strains of *E. coli* with *E. coli* strains that cause food poisoning allowed researchers to find virulence genes in the pathogenic strain. These genes code for proteins that poison cells in the human intestine and cause severe diarrhea. If drugs that neutralize these proteins can be developed, they might provide an effective treatment for food poisoning.

Having the complete gene catalog from pathogenic species is giving biomedical researchers new targets for drug development and new possibilities for therapy. This work is particularly urgent, because disease-causing bacteria continue to evolve resistance to many of the antibiotics currently in use (see Chapter 24).

Designing Vaccines

Although efforts to exploit genome data in drug design are still in their infancy, genomics has already inspired important advances in vaccine development. In essence, researchers are testing proteins that are identified by genome sequencing to see if the molecules stimulate the immune system enough to function as vaccines.

To illustrate how this work is proceeding, consider recent research on the bacterium *Neisseria meningitidis*. This species is a major cause of meningitis and blood infections in children and was one of the first bacteria to have its genome sequenced. Although antibiotics can treat *N. meningitidis* infections effectively, the organism grows so quickly that it often injures or even kills the victim before a diagnosis can be made and drugs administered. As a result, biomedical researchers

Experiment

Question: Could vaccines be developed from the products of newly discovered genes?

Hypothesis: Some of the genes discovered through genome sequencing code for proteins that can be used in vaccines.

Null hypothesis: None of the genes discovered through genome sequencing code for proteins that can be used in vaccines.

Experimental setup:

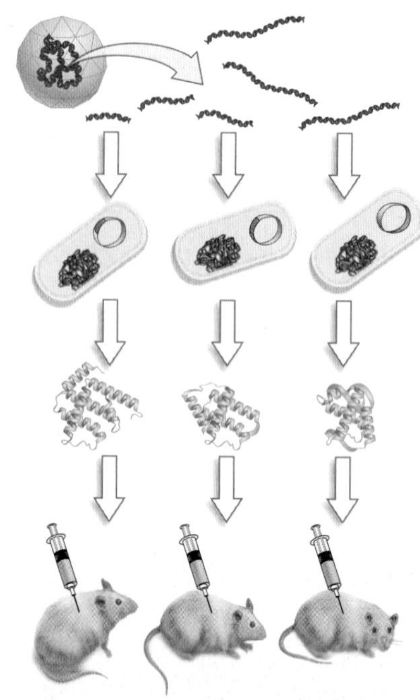

1. Isolate open reading frames (ORFs) from pathogen genome sequence.

2. Introduce ORFs into *E. coli* cells.

3. Isolate proteins that result from transcription and translation.

4. Inject proteins into mice. As a control, inject only the solution used to suspend the proteins.

Prediction: Some proteins will elicit an immune response similar to that elicited by vaccines. The control solution will not elicit an immune response.

Prediction of null hypothesis: No injections (control or treatments) will elicit an immune response similar to that elicited by vaccines.

Results:	Number of mice with immune response:	Number of mice with no immune response:
Strong immune responses	7	343

Conclusion: The seven proteins that elicited an immune response are potential vaccine components. Further research is needed to test their safety and effectiveness.

FIGURE 20.13 Newly Discovered Proteins Can Be Tested for Vaccine Development. Because all potential genes are identified after whole-genome sequencing, virtually all proteins produced by a pathogen can be tested for their ability to provoke an immune response and act as a vaccine.

have been interested in developing a vaccine that would prime the immune system and allow children to ward off infections.

Vaccine development has been difficult in this case, however. As Chapter 49 will explain in detail, the immune system usually responds to molecules on the outer surface of bacteria or viruses. When an immune system cell recognizes one of these surface molecules as foreign, the invading bacterium or viral particle is destroyed. This recognition step by the immune system takes time, however, and is speeded up by vaccination. Vaccines contain surface molecules from a bacterium or virus. Even though the vaccine is harmless, the immune system cells go through their normal recognition sequence. In this way, ingesting a vaccine alerts or primes the immune system. If an actual infection occurs later, the cells are ready to spring into action and destroy the pathogens before they can grow and cause disease.

Unfortunately, *N. meningitidis* is covered with a polysaccharide that is identical to a compound found on the surface of brain cells. Immune system cells normally do not attack compounds found on the body's own cells, so a vaccine composed of the *N. meningitidis* polysaccharide would elicit no response.

To circumvent this problem, biologists analyzed the genome sequence of *N. meningitidis* and tested 600 open reading frames for the ability to encode vaccine components. The researchers inserted the 600 DNA sequences into *E. coli* cells, following the steps shown in **Figure 20.13**. Later they succeeded in isolating 350 different *N. meningitidis* proteins from the transformed cells. The biologists injected these proteins into mice and then analyzed whether an immune response occurred. Their results show that seven of the proteins tested evoked a strong immune response and represent promising vaccine components. Follow-up work is now under way to determine whether one or more of these proteins could act as a safe and effective vaccine in humans.

Finding Genes Associated with Inherited Disease: The HapMap Project

Chapter 19 explained how gene hunters analyzed genetic markers to find the allele responsible for Huntington's disease. Data from human genome sequencing have made this approach to finding disease genes much more powerful.

Because DNA from many individuals was used as source material during the Human Genome Project, and because overlapping segments of genes were routinely sequenced, researchers were able to identify 1.42 million sites where single bases vary among individuals. Where you might have a "C" at a particular site, others may have a "T." Recall that these variable sites are called **single nucleotide polymorphisms,** or **SNPs,** and that they can serve as genetic markers—mapped sites in the genome that vary among individuals.

By sequencing DNA samples from dozens of people representing four ethnic groups from widely dispersed geographic areas, a recent research effort called the HapMap Project has extended this initial catalog of SNPs to a current total of 10 million sites that vary among humans. HapMap is short for haplotype mapping. A **haplotype** is the set of alleles found on a single chromosome. By mapping SNPs, researchers hope to be able to determine the haplotype of any individual at any of their chromosomes.

For disease gene hunters, this new catalog of SNPs is an enormously powerful resource. The fundamental idea is to compare the haplotypes of individuals who have an inherited disease with the haplotypes of unaffected people. If certain SNPs are extremely common in affected individuals but rare or absent in unaffected individuals, it is likely that those SNPs are near or even within a gene that contributes to the disease.

The possibility of analyzing the inheritance of millions of polymorphic sites all over the genome—instead of just a few hundred widely scattered ones like those used in the Huntington's disease gene hunt—makes it much more likely that researchers can locate genes associated with illnesses such as Alzheimer's disease, bipolar disorder, diabetes, rheumatoid arthritis, and cardiovascular disease. The database promises to be particularly important in understanding the genetic basis of diseases that involve many different genes, instead of a single allele as Huntington's disease does. As biologists continue their efforts to annotate the human genome, it's very likely that they will be able to track down the genes responsible for many or even most inherited diseases.

Chapter Review

SUMMARY OF KEY CONCEPTS

Once a genome has been completely sequenced, researchers use a variety of techniques to identify which sequences code for products and which act as regulatory sites.

Recent technical advances have allowed investigators to sequence DNA much more rapidly and cheaply than before, resulting in a flood of genome data. Researchers annotate genome sequences by finding genes and determining their function. To identify genes in bacteria and archaea, researchers use computers to scan the genome for start and stop codons that are in the same reading frame and that are separated by gene-sized stretches of sequence. Finding such open reading frames (ORFs) is difficult in eukaryotes, because exons are interrupted by introns and because most eukaryotic DNA does not code for a product. One approach to finding eukaryotic genes is to analyze the sequences of complementary DNAs (cDNAs) synthesized from mRNAs and then match these sequences to DNA found in the genome itself. Sequences that are highly conserved among species are also hypothesized to indicate the locations of genes.

You should be able to describe how a research group that discovered a gene for coat color in mice would determine whether a homologous gene existed in the human genome.

MB **Web Animation** at www.masteringbio.com
Human Genome Sequencing Strategies

In bacteria and archaea, there is a positive correlation between the number of genes in a species and the species' metabolic capabilities. Gene transfer between species is also common.

Species of bacteria and archaea are usually targeted for whole-genome sequencing because they cause disease or have interesting metabolic abilities. In these groups, the size of an organism's genome and its morphological complexity or biochemical capabilities are correlated. Parasites tend to have small genomes; organisms that live in a broad array of habitats or that use a wide variety of nutrients tend to have larger genomes. Many of the genes identified in bacteria and archaea still have no known function, however, and a significant percentage of them are extremely similar to other genes in the same genome. Another generalization about prokaryotic genomes is that genes are frequently transferred laterally, or between species. Lateral gene transfer appears to be common in genes responsible for causing disease.

You should be able to describe two mechanisms responsible for lateral gene transfer in bacteria and archaea.

In eukaryotes, genomes are dominated by sequences that have little to no effect on the fitness of the organism.

Compared with prokaryotic genomes, eukaryotic genomes are large and contain a high percentage of transposable elements, repeated sequences, and other noncoding sequences. There is no obvious correlation between morphological complexity and gene number in eukaryotes, although the number of distinct transcripts produced may be much larger than the actual gene number in certain species as a result of alternative splicing. Gene duplication and polyploidy have been the most important sources of new genes in eukaryotes.

You should be able to explain what biologists mean when they refer to "junk DNA," and whether these sequences lack function and are uninteresting, as originally proposed.

Data from genome sequencing projects are now being used in the development of new drugs and vaccines, and to search for alleles associated with inherited diseases.

The availability of whole-genome sequences is inspiring new research programs. Biologists are affixing exons or proteins to microarrays in order to study changes in gene expression or protein-protein interactions. In addition, the availability of whole-genome data has allowed investigators to find new drug targets, new proteins that may serve as vaccine candidates, and new genetic markers that should aid in the hunt for alleles associated with human disease.

You should be able to explain the difference between studies of gene expression in single genes versus microarrays, and expression of individual proteins versus microarrays.

QUESTIONS

Test Your Knowledge

1. What is an open reading frame?
   - **a.** a gene whose function is already known
   - **b.** a DNA section that is thought to code for a protein because it is similar to a complementary DNA (cDNA)
   - **c.** a DNA section that is thought to code for a protein because it has a start codon and a stop codon flanking hundreds of base pairs
   - **d.** any member of a gene family
2. What best describes the logic behind shotgun sequencing?
   - **a.** Break the genome into tiny pieces. Sequence each piece. Use overlapping ends to assemble the pieces in the correct order.
   - **b.** Start with one end of each chromosome. Sequence straight through to the other end of the chromosome.
   - **c.** Use a variety of techniques to identify genes and ORFs. Sequence these segments—not the noncoding and repeated sequences.
   - **d.** Break the genome into pieces. Map the location of each piece. Then sequence each piece.
3. What are minisatellites and microsatellites?
   - **a.** small, extrachromosomal loops of DNA that are similar to plasmids
   - **b.** parts of viruses that have become integrated into the genome of an organism
   - **c.** incomplete or "dead" remains of transposable elements in a host cell
   - **d.** short and simple repeated sequences in DNA
4. What is the leading hypothesis to explain the paradox that large, morphologically complex eukaryotes such as humans have relatively small numbers of genes?
   - **a.** lateral transfer of genes from other species
   - **b.** alternative splicing of mRNAs
   - **c.** polyploidy, or the doubling of the genome's entire chromosome complement
   - **d.** expansion of gene families through gene duplication
5. What evidence do biologists use to infer that a gene is part of a gene family?
   - **a.** Its sequence is exactly identical to that of another gene.
   - **b.** Its structure—meaning its pattern of exons and introns—is identical to that of a gene found in another species.
   - **c.** Its composition, in terms of percentage of A-T and G-C pairs, is unique.
   - **d.** Its sequence, structure, and composition are similar to those of another gene in the same genome.
6. What is a pseudogene?
   - **a.** a coding sequence that originated in a lateral gene transfer
   - **b.** a gene whose function has not yet been established
   - **c.** a polymorphic gene—meaning that more than one allele is present in a population
   - **d.** a gene whose sequence is similar to that of functioning genes but does not produce a functioning product

Test Your Knowledge answers: 1. c; 2. a; 3. d; 4. b; 5. d; 6. d

Test Your Understanding

Answers are available at www.masteringbio.com

1. Explain how open reading frames are identified in the genomes of bacteria and archaea. Why is it more difficult to find open reading frames in eukaryotes?
2. Why is the observation that parasitic organisms tend to have relatively small genomes logical?
3. Review how a LINE sequence transmits a copy of itself to a new location in the genome. Why are LINEs and other repeated sequences referred to as "genomic parasites"?
4. How does DNA fingerprinting work? Stated another way, how does variation in the size of microsatellite and minisatellite loci allow investigators to identify individuals?
5. Researchers can create microarrays of short, single-stranded DNAs that represent many or all of the exons in a genome. Explain how these microarrays are used to document changes in the transcription of genes over time or in response to environmental challenges.
6. Explain the concept of homology and how identifying homologous genes helps researchers identify the function of unknown genes. Are duplicated sequences that form gene families homologous? Explain.

Applying Concepts to New Situations

Answers are available at www.masteringbio.com

1. Parasites lack genes for many of the enzymes found in their hosts. Most parasites, however, have evolved from free-living ancestors that had large genomes. Based on these observations, W. Ford Doolittle claims that the loss of genes in parasites represents an evolutionary trend. He summarizes his hypothesis with the quip "use it or lose it." What does he mean?
2. According to eyewitness accounts, communist revolutionaries executed Nicholas II, the last czar of Russia, along with his wife and five children, the family physician, and several servants. Many decades after this event, a grave purporting to hold the remains of the royal family was identified. Biologists were asked to analyze DNA from each adult and juvenile skeleton and determine whether the bodies were indeed those of several young siblings, two parents, and several unrelated adults. If the grave was authentic, describe how similar the DNA fingerprints of each skeleton would be relative to the fingerprints of other individuals in the grave.
3. The human genome contains a gene that encodes a protein called syncytin. This gene is expressed in placental cells during pregnancy. The syncytin gene is nearly identical in DNA sequence to a gene in a virus that infects humans. In this virus, the syncytin-like gene codes for a protein found in the virus's outer envelope. State a hypothesis to explain the similarity between the two genes.
4. A recent study used microarrays to compare the patterns of expression of genes that are active in the brain, liver, and blood of chimpanzees and humans. Although the overall patterns of gene expression were similar in the liver and blood of the two species, expression patterns were strikingly different in the brain. How does this study relate to the hypothesis that most differences between humans and chimps involve changes in gene regulation?

www.masteringbio.com is also your resource for • Answers to text, table, and figure caption questions and exercises • Answers to *Check Your Understanding* boxes • Online study guides and quizzes • Additional study tools including the *E-Book for Biological Science* 3rd ed., textbook art, animations, and videos.

21 Principles of Development

KEY CONCEPTS

- As an embryo develops, cells have to proliferate, move or expand in a directed manner, become specialized, and interact with other cells.
- Cells become specialized because they express different genes, not because they contain different genes.
- Cells interact continuously during development via cell-cell signals.
- Early signals trigger a cascade of effects, including the production of transcription factors and other signals. Interactions among these signals, transcription factors, and regulatory DNA sequences cause cells to differentiate as development proceeds.
- The evolution of new body sizes, shapes, and structures is possible when mutations alter genes responsible for development.

A young fish, hatching from an egg. This chapter introduces the processes responsible for transforming a fertilized egg into an individual with specialized cells, tissues, and organs.

In 1859 Charles Darwin referred to the origin of species as "the mystery of mysteries." At that time the most urgent task confronting biologists was to explain how species come to be. Today, however, Darwin's theory of evolution by natural selection explains most of our fundamental questions about how new species arise and how they change through time. What question qualifies as the current mystery of mysteries in biological science? Although there are many candidates, one of the most compelling is the question addressed in this unit: How does a multicellular individual develop from a single cell—the fertilized egg?

It's important to pause for a moment and think about the magnitude of this problem. For example, at one time you consisted of a single cell. If you had been able to watch your own development, you would have seen that cell divide rapidly and form a ball of tiny, identical-looking cells. At that point, the fertilized egg had given rise to an **embryo**—a young, developing organism. After continued cell division, large groups of these cells would begin moving into the interior of the embryo, in a series of coordinated but mysterious-looking movements. Cell division would continue rapidly. After a week or two the embryo would elongate, and a recognizable head and tail portion would appear. Then tiny precursors of your vertebrae would be visible, along with your eyes. Eventually, buds would emerge and go on to form your limbs. As development continued, that embryo would eventually become you.

Biologists who have watched this process in humans or other organisms never cease to marvel at it. The question is, how does all the growth and body formation happen? In many species, the adult body contains trillions of cells and dozens of different types of highly organized tissues and organs. How does a fully formed individual that can move, eat, and reproduce develop from a tiny and formless-looking mass of cells?

People have pondered this question for centuries. Using light microscopes, early investigators observed the development of fertilized eggs and documented the events that occur as an individual grows from an embryo to an adult. Based on these studies, the physical pattern of development appeared to be highly variable among species. You've already thought for a moment about the events of early human development. But consider that a fertilized egg from a burr oak tree takes about four months to grow into an **embryo** that becomes encased in an acorn. After spending the winter in a dormant state, the embryo grows into a seedling. The organism will continue to grow and develop new leaves, roots, flowers, and seeds every year for the next 250–300 years. At another extreme, a fertilized egg of the fruit fly *Drosophila melanogaster* takes just a day to develop into a caterpillar-like larva that can move about and feed. About five days later, the larva stops feeding and forms a cocoon-like, stationary pupa. In another four days, an adult fly emerges from the pupal case. The adult flies about, feeds, courts, and mates—starting the cycle anew.

The variation in developmental patterns observed among species adds another layer of complexity to today's mystery of mysteries. Biologists have to explain the fundamental mystery of how highly integrated individuals, both large and small, can develop from a single cell. What's more, they have to make sense of all the variability that exists among species in the rate and pattern of development.

Fortunately, a century's worth of research has culminated in one of the great insights of contemporary biology: A few fundamental principles are common to all developmental sequences observed in multicellular organisms. Development is like jazz, where different members of a band play variations on the same theme. If you understand the theme and how chords and scales are put together, you can understand what each soloist is doing and how they are doing it. Although many critically important elements remain to be worked out, the discovery of the general principles underlying development has unified the vast variation observed in how embryos grow. And it has given biologists a framework for explaining how a single cell can give rise to a complex, multicellular individual.

The goal of this chapter is to explore the common principles that underlie development. As you study this material, you'll need to draw on what you've already learned about cell-cell interactions, the regulation of gene expression, and a host of other topics. Part of the excitement surrounding work on development is that it draws on insights from biochemistry, cell biology, genetics, and evolution. Developmental biology is among the most synthetic fields in all of biological science. Let's delve in.

21.1 Four Essential Developmental Processes

At the level of the cell, four general processes occur throughout development. Cells divide, move or expand in a directed way, begin to express certain genes rather than others, and signal to each other about where they are, what they are doing, and what type of cell they are becoming (**Table 21.1**). We'll consider each of these four processes in turn, and then go on to consider more specific issues about cell specialization and cell-cell interactions.

Cell Proliferation and Programmed Cell Death

For an organism to grow and develop, its cells have to proliferate—they have to divide and make more cells. This statement seems obvious. Perhaps less obvious is an equally important point: For an individual to develop from an undifferentiated mass of cells, the location, timing, and extent of cell division have to be tightly controlled.

Chapter 11 introduced the processes of mitosis and cytokinesis, which are responsible for cell proliferation in eukaryotes. That chapter also introduced the stages of the cell cycle and how they are controlled. You might recall that cells initiate mitosis in response to a regulatory protein complex called mitosis-promoting factor (MPF), that each stage of the cell cycle has checkpoints that are carefully regulated, and that cells continue to grow or stop growing in response to what biologists like to call "social controls"—meaning, signals from other cells. During development, all of these layers of regulation interact to determine when, where, and how much each embryonic cell proliferates.

What Chapter 11 did not mention is that death is also a highly regulated aspect of life for some cells. Programmed cell death in animals, or what biologists call **apoptosis** (literally, "falling away"), occurs as certain tissues and organs take shape. As the human hand and foot form, for example, cells that are initially present between the fingers and toes must die in order for separate digits to form. In plants, programmed cell death allows flower petals to fall after pollination has occurred, and for leaves or entire stems to be lost in autumn.

In these and many other cases, cell death is a normal part of development—it is carefully timed and regulated. **Box 21.1** introduces the genetic mechanisms responsible for apoptosis and highlights how defects in the normal patterns of programmed cell death can lead to abnormal development and disease.

To summarize, complex structures develop in multicellular organisms due in part to the growth of some cells and the regulated loss of others. Together, cell proliferation and programmed cell death are essential to the emergence of complex structures from the initial mass of embryonic cells.

SUMMARY TABLE 21.1 **Four Essential Developmental Processes**

Cell proliferation and programmed cell death	Cells divide by mitosis and cytokinesis. The timing, location, and amount of cell division and cell death is regulated.
Cell movement or differential expansion	Cells can move past one another within a block of animal cells, causing drastic shape changes within the embryo. Certain animal cells can break away from a block of cells and migrate to new locations. Plant cells can divide in certain planes and expand in specific directions, causing dramatic changes in shape.
Cell differentiation	Undifferentiated cells specialize at specific times and places in a stepwise fashion. Cells that do not undergo differentiation are called stem cells in animals. Many plant cells are capable of de-differentiating.
Cell-cell interactions	Embryonic cells grow, move, or differentiate in response to signals from other cells.

Cell Movement and Expansion

In addition to growing and dividing (and dying), many animal cells have to move for normal development to occur. Some of the most dramatic cell movements occur early in animal development, after rapid cell divisions have produced a mass of similar-looking cells. During a process called **gastrulation**, cells in different parts of the mass rearrange themselves into three distinctive types of embryonic tissues, which then give rise to the skin, gut, and other basic parts of the body (see Chapter 22). Throughout development, certain animal cells break away from their original sites and migrate to new locations in the embryo. There they give rise to many different types of cells, including germ cells (sperm or eggs), pigment-containing cells, precursors of blood cells, or certain nerve cells. If any of these cell movements is inhibited, or if migrating cells end up in the wrong place or move at the wrong time, the embryo is likely to be deformed or die before development is complete.

Plant cells, in contrast, are encased in stiff cell walls and do not move. Instead, plant cells control how the cleavage plane is oriented during cell division and the direction of cell growth that subsequently takes place. This directionality in cell division and growth results in the proper formation of straight and branched stems, leaf blades, and other structures.

Cell Differentiation

For development to occur, many cells have to proliferate, some cells have to die, and other cells have to move to new locations or grow in specific directions. In addition, it is essential for most cells to undergo **differentiation**—the process of becoming a specialized type of cell. As you developed, some of your embryonic cells differentiated to form muscle cells that contract and relax, while others became nerve cells that conduct electrical signals throughout the body. As an oak tree develops, some cells secrete thickened walls and transport water as part of a woody stem, while others become flattened and secrete the waxy coating found on the surface of leaves. In most multicellular eukaryotes, the fertilized egg gives rise to hundreds of distinctive cell types.

Differentiation may be the most basic process in the development of a multicellular organism, and it probably qualifies as the most intensively studied. In addition to the more detailed analysis upcoming in Sections 21.2 and 21.3, it's important to note several general points about differentiation.

- *Differentiation is a progressive, step-by-step process.* Cells initially are determined to become a certain cell type—meaning they are committed to a specific developmental pathway. Only later do they become differentiated—that is, they actually begin to look and behave like a specific cell type.
- *Some cells do not become specialized adult cells.* Undifferentiated cells are common in embryos, but even adult organisms have populations of cells that retain the ability to divide and give rise to an array of specialized cell types. Animal cells with this ability are called stem cells. The products of animal stem cell division can differentiate to help heal wounds, replace skin and blood cells that die, or perform other functions that maintain the body in good working order. In plants, populations of undifferentiated cells called **meristems** give rise to the flowers, roots, leaves, and other structures that develop throughout life.
- *Many or even most plant cells are capable of "de-differentiating"*—meaning they can change their structure and func-

BOX 21.1 Apoptosis: Programmed Cell Death

How does programmed cell death occur? An answer to this question emerged from studies of the roundworm *Caenorhabditis elegans*. This species is a popular model organism in developmental biology because it has a complex array of organs and tissues but only about a thousand cells. In addition, the cells are transparent (**Figure 21.1**). As a result, biologists are able to identify individual cells in both embryos and adults. Thanks to these features, biologists know the fate of each cell in the embryo at every stage of development.

As a *C. elegans* individual matures, 131 of its 1090 original somatic cells—cells that aren't involved in gamete formation—undergo apoptosis. To explore how this process occurs, Hillary Ellis and Robert Horvitz set out to identify mutants that do not undergo the normal pattern of cell deaths. You might recall from Unit 3 that this is a common research strategy in genetics. To find the genes responsible for a trait, researchers look for mutants that do not have the trait.

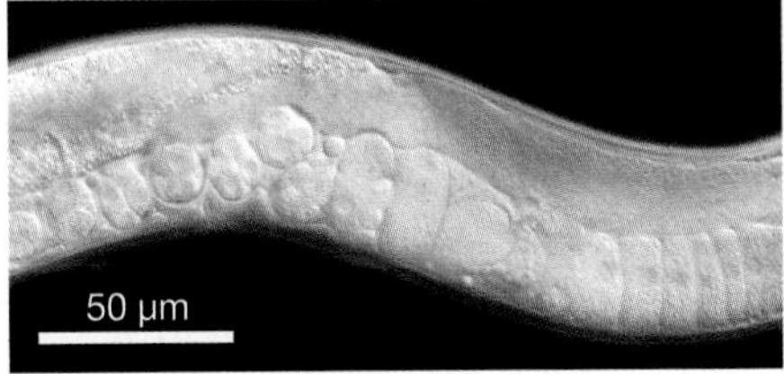

FIGURE 21.1 *Caenorhabditis elegans* Is an Important Model Organism. *C. elegans* has a defined number of transparent cells. Each cell can be identified and followed during development.

Ellis and Horvitz's initial work uncovered two genes that are essential for apoptosis. If a worm carries mutations that inactivate either gene, cells that would normally die survive. The researchers proposed that the genes are part of a genetic program that executes apoptosis—in short, that they were cell-suicide genes.

By searching databases of known DNA sequences, biologists found genes in mice that are similar to the worm genes. To test the hypothesis that these genes and their products are important in mouse development, a research team used genetic engineering techniques to produce mice in which both copies of one of their cell-suicide genes were disrupted. In contrast to the normal embryo in **Figure 21.2a**, the mutant embryos like the one in **Figure 21.2b** exhibited a severe malformation of the brain. The defect occurred because cells that would normally die early in development survived.

As an aside, it's interesting to note that the mutant mouse embryos had normal digit separation, even though they had abnormal brains. Based on this observation, the researchers suggested that other cell-suicide genes are involved in programmed cell death during the formation of mouse toes.

Are the same genes found in humans? The answer is yes. Genes that are homologous to the worm and mouse cell-suicide genes have been identified in the human genome. (Recall from earlier chapters that homologous genes are similar in structure and function because they are derived from the same gene in a common ancestor.) Normal apoptosis is important in the development of human embryos, and abnormal apoptosis has been implicated in certain diseases of adults. For example, inappropriate activation of programmed cell death is involved in some neurodegenerative diseases, including ALS (Lou Gehrig's disease). Work on worm genes opened a new frontier in developmental biology, inspiring new insights into some devastating human illnesses.

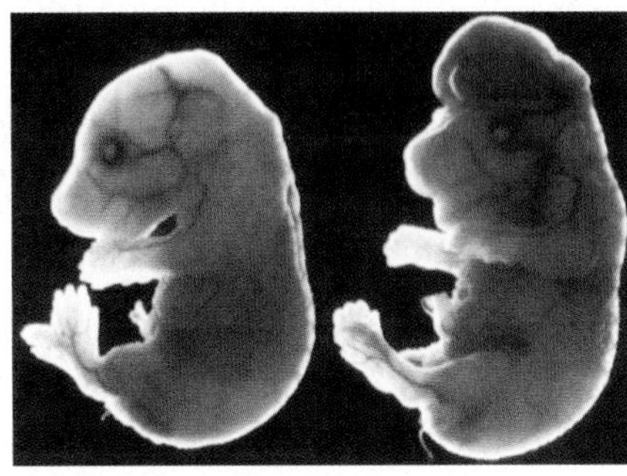

FIGURE 21.2 Defects Occur When Programmed Cell Death Fails. (a) A normal mouse embryo; **(b)** a mouse embryo that has two defective alleles of a cell suicide gene.

QUESTION If the defective version of the cell suicide gene is recessive, what would the embryo in part (b) look like if it were heterozygous instead of homozygous?

tion, even after they have specialized. For example, if a branch of a cedar tree droops down low enough to make contact with the soil, cells in the branch will de-differentiate and then re-differentiate to form root cells. A fully formed root will grow where a branch initially made contact with the ground.

Cell-Cell Interactions

Chapter 8 introduced the topic of cell-cell interactions by examining the extracellular matrix found between cells. That chapter also explored how adjacent cells are attached and exchange materials and how cells respond to signals from other cells. During development, the most important cell-cell interactions involve sending and receiving signals. You might recall from Section 8.3 that when a signal arrives at the surface of a cell, its message is transduced to a new molecular form and amplified inside the receiving cell. Chapter 8 concluded by noting that cells change their activity in response to the signals they receive. What the chapter did not explain is that cell-cell signals are essential for changing cell activity during development.

As Section 21.3 will show, embryonic cells grow, move, or differentiate in response to signals from other cells. In many cases, the signal transduction pathways introduced in Chapter 8 trigger the production of the transcription factors introduced in Chapter 18. As a result, the arrival of cell-cell signals changes patterns of gene expression and thus the embryonic cell's structure and behavior.

21.2 The Role of Differential Gene Expression in Development

Differential gene expression is the essence of cell differentiation during development. The muscle cells in your body are different from your nerve cells because they express different genes. The same is true for plants. The water-transport cells in an oak tree are different from its leaf-surface cells because they express different genes and produce different proteins.

If you think about these statements, you'll realize that they have to be true. The only way that cells can have different structures and functions is if they contain different molecules. What is less obvious is whether cells express different genes because they contain different genes, or whether all the cells in a body contain the same genes but express only a specialized subset.

As it turns out, researchers found this question relatively simple to answer for plants. You've already learned, for example, that cells from the stem of a cedar tree can de-differentiate to form roots. This differentiation would be possible only if the cells from a branch continue to contain the genes required by root cells. Gardeners and farmers have known for centuries that with many plant species, complete new individuals can be produced from a small section of a root or shoot. Based on observations like this, researchers strongly suspected that all plant cells contain the same genes—meaning that they are genetically equivalent. This result was confirmed in the 1950s when biologists were able to grow entire tobacco plants or carrots from a single, differentiated cell taken from an adult (see Chapter 39).

In contrast to plants, the issue of genetic equivalence was extremely difficult to resolve for animals. Although serious experimental work on the problem began in the 1950s, the question was not answered until the late 1990s.

Are Differentiated Animal Cells Genetically Equivalent?

Early experiments on genetic equivalence in animals focused on transferring nuclei from diploid cells in frog tadpoles or adults into unfertilized eggs whose nuclei had been removed. At least some of these transplanted nuclei were able to direct the development of tadpoles successfully. These results provided strong evidence that all mature cells are genetically equivalent.

In 1997 Ian Wilmut and colleagues reinforced this conclusion through nuclear transfer experiments in sheep. As shown in **Figure 21.3,** these researchers removed mammary-gland cells

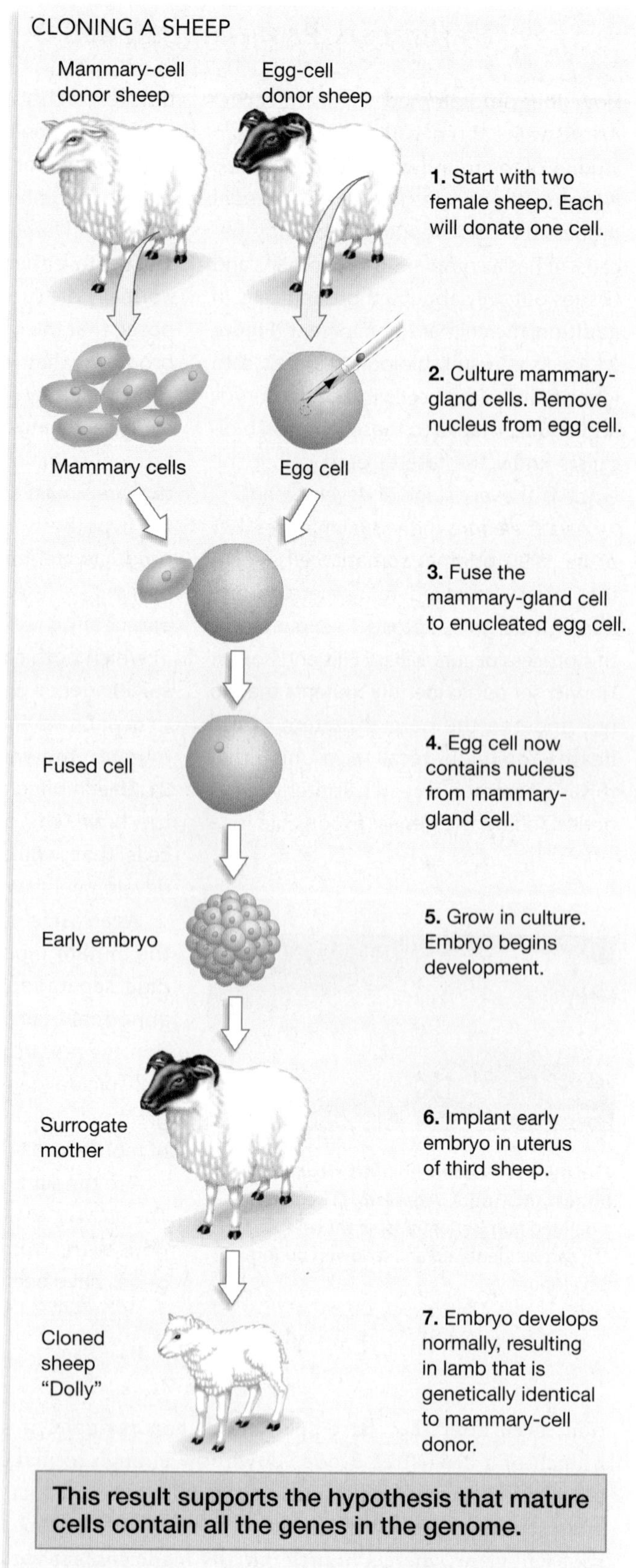

FIGURE 21.3 Mammals Can Be Cloned by Transplanting Nuclei from Mature Cells. The lamb that resulted from the nuclear transfer experiment shown here resembled the breed from which the donor *nucleus* came, not the breed of the egg donor or the surrogate mother. DNA testing also showed that the lamb was genetically identical to the individual that donated the nucleus.

from a 6-year-old pregnant female and grew them in culture (steps 1 and 2). Later they fused these cells with enucleated eggs—eggs whose nuclei had been removed. This procedure resulted in egg cells that contained nuclei from mammary-gland cells (steps 3 and 4). As the upper drawings show, the eggs came from a black-faced breed of sheep, while the donor nuclei came from a white-faced breed. After developing in culture, the resulting embryos were implanted in the uteri of surrogate mothers (steps 5 and 6). In one of several hundred such transfer attempts, a white-faced lamb named Dolly was born (step 7). Genetically, Dolly was a clone of the white-faced donor of the mammary-gland cell. (Recall from Chapter 12 that a **clone** is a genetically exact copy of a parent.) Dolly grew into a fertile adult and, by normal mating, produced her own lamb named Bonnie (**Figure 21.4**). In 1998 other research groups reported similar results in mice and cows; since then, cats, horses, a monkey, and individuals of several other species have been cloned (but see **Box 21.2**).

FIGURE 21.4 Dolly with an Offspring of Her Own.

Taken together, work on cloning plants and animals has shown that in most cases, the process of cellular differentiation does not involve changes in the genetic makeup of cells. Instead, it is based on differential gene expression. There are some important exceptions to this rule, however. For example, small stretches of DNA are rearranged in certain immune system cells in humans and other mammals, late in development. As a result, many immune cells are genetically unique. Chapter 49 explains how this happens.

What Is the Most Important Level of Control over Gene Expression?

As Chapter 18 emphasized, eukaryotic cells control gene expression at several different levels. A gene can be regulated by turning transcription on or off (or up or down), alternative splicing of mRNAs, selective destruction of mRNAs, speeding up or slowing down translation, and activating or deactivating proteins after they are translated. All of these processes occur during development. But which is most important?

BOX 21.2 Human Cloning?

Biologists can harvest nuclei from the somatic cells of mice, cows, and sheep and use the nuclei to produce embryos. These nuclear transfer experiments produce offspring that are genetically identical to the individual providing the donor nucleus. A group of genetically identical organisms are clones, so the process of creating individuals by transferring nuclei into enucleated eggs is called cloning. Can humans be cloned?

Before addressing the technical feasibility of cloning humans, let's consider a more practical issue. Why would anyone want to clone a human? Proponents of this technology contend that cloning by nuclear transfer would allow otherwise infertile couples to have offspring that are genetically related—in fact, identical—to one of the parents. Opponents of this technology are concerned that, in an attempt at immortality, dictators or wealthy eccentrics could finance clones of themselves in perpetuity.

Even benign uses of human cloning create ethical dilemmas that are yet to be resolved. What is the effect on a family of having children who are genetically identical to a parent? Would cloning be limited to individuals with "desirable" characteristics? Given the number and seriousness of the questions involved, several nations have proposed a moratorium on all research related to cloning humans.

Besides the legal and ethical issues involved, significant technical hurdles need to be overcome before human cloning would be feasible. Of the species that have been successfully cloned, monkeys and mice are most closely related to humans. Less than 3 percent of nuclear transfers result in viable mice, however. Most result in aborted pregnancies; some produce individuals that die soon after birth. Currently, the success rate is similar in cloning sheep and cows, and deformed newborns have been produced. Most people regard a failure rate this high as unacceptable for humans, especially given the high likelihood of producing grossly deformed offspring.

As cloning techniques improve, however, human cloning may become technologically feasible. Meanwhile, the nations of the world are debating the merits of the technology and deciding whether it will be regulated or banned outright.

The answer is transcriptional control. To understand why, ask yourself whether a muscle cell should produce mRNAs or proteins that are specifically required by nerve cells, and then remain a muscle cell by producing microRNAs that disable nerve-cell mRNAs or by producing regulatory molecules that keep nerve-cell proteins inactivated. It is much more logical to predict that muscle cells transcribe only genes required by muscle cells. This is exactly what researchers have found. Transcription is the fundamental level of control in differential gene expression during development. In eukaryotes, transcription is controlled primarily by the presence of **regulatory transcription factors** that influence chromatin remodeling and bind to promoter-proximal elements, enhancers, silencers, or other regulatory sites in DNA (see Chapter 18).

This simple insight is extremely important. To understand differentiation, researchers have to understand how and why regulatory transcription factors vary among cells.

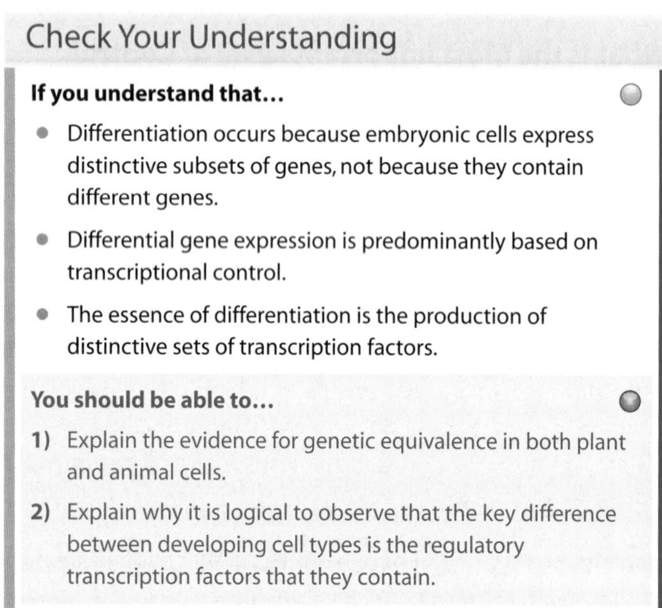

Check Your Understanding

If you understand that...

- Differentiation occurs because embryonic cells express distinctive subsets of genes, not because they contain different genes.
- Differential gene expression is predominantly based on transcriptional control.
- The essence of differentiation is the production of distinctive sets of transcription factors.

You should be able to...

1) Explain the evidence for genetic equivalence in both plant and animal cells.
2) Explain why it is logical to observe that the key difference between developing cell types is the regulatory transcription factors that they contain.

21.3 What Triggers Differential Gene Expression?

To understand development, you have to think like a cell. Suppose you were one of the hundreds or thousands of cells in a developing animal embryo. Your fate—whether you ended up as part of an arm or a kidney, and whether you differentiated into a nerve cell or a blood-vessel cell—would depend on your location in four dimensions: time (meaning the stage of development that is currently occurring) plus the three spatial dimensions of the body axis. The three body axes are illustrated in **Figure 21.5**.

1. One axis runs **anterior** (toward the head) to **posterior** (toward the tail).

(a) The three body axes observed in humans and other animals...

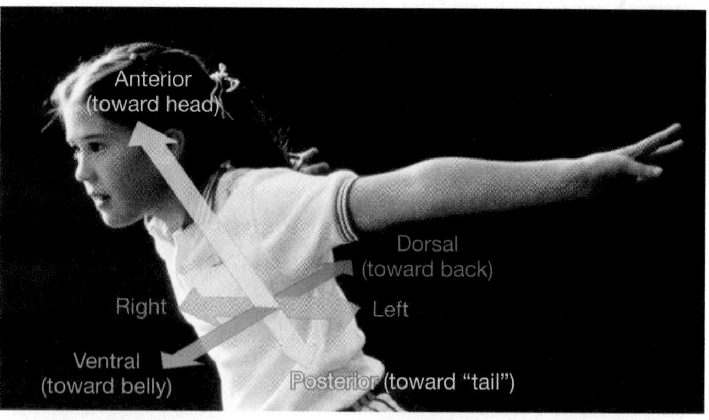

(b) ...are initially established in embryos (such as this mouse).

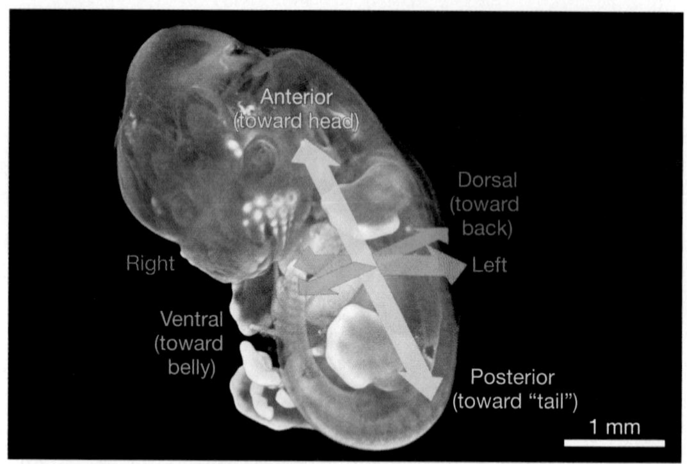

FIGURE 21.5 Most Animals Have Three Major Body Axes.

2. One axis runs **ventral** (toward the belly) to **dorsal** (toward the back).
3. One axis runs left to right.

Cells "know" where they are in time and space because they are constantly interacting via cell-cell signals. In effect, much of development is organized by signals that cells send and receive. These signals activate transcription factors that turn specific genes on or off. As development proceeds, the distinctive suite of genes that are activated at successive stages determines the fate of each cell. If you understand this concept, you should be able to explain (1) how a cell could "know" where it is in relation to the three body axes and (2) how the same cell could receive different information early versus late in development.

Let's consider how this process happens, beginning with one of the first developmental signals ever discovered. Although you'll be analyzing what happens as a fruit-fly embryo develops, keep an important point in mind: Principles that were discovered in fruit flies are relevant to virtually all multicellular organisms studied to date—from mustard plants to humans.

Master Regulators Set Up the Major Body Axes

Biologists refer to the events that determine the spatial organization of an embryo as **pattern formation**. If a molecule signals that a target cell is in the embryo's head, or tail, or dorsal side, or ventral side, that molecule is involved in pattern formation. As it turns out, certain early signals act as master regulators that set up the general anterior-posterior, dorsal-ventral, and left-right axes of an embryo. Then a network of genes activated by these master regulators sends signals with more specific information about the location of cells in space. The process occurs again and again: As development proceeds, a series of signals arrive and activate genes that specify finer and finer control over what a cell becomes.

The Discovery of *bicoid* The insights you just reviewed first began to emerge from work on the fruit fly *Drosophila melanogaster*—the model organism that has been so important in genetics (see Chapter 13). Because fruit flies produce large numbers of offspring rapidly, researchers could survey laboratory populations for rare mutant embryos in which the normal spatial relationships among cells are disrupted. A *Drosophila* larva that lacked a head, for example, was likely to have a mutation in a gene that helps pattern the head. Stated another way, biologists took a genetic approach to studying development. This approach has turned out to be extraordinarily productive. For example, studies of mutant fruit flies identified cell-cell signals that also direct pattern formation in mice, humans, and other animals.

The effort to dissect pattern formation in *Drosophila* began in the 1970s, when Christiane Nüsslein-Volhard and Eric Wieschaus exposed adult flies to treatments that cause mutations and then examined embryos or larvae descended from these individuals for body plan defects. After intensive effort, Nüsslein-Volhard and Wieschaus were able to identify over 100 genes that play fundamental roles in pattern formation.

One of the most dramatic mutations found by Nüsslein-Volhard and Wieschaus alters the normal arrangement of segments in the embryo. A **segment** is a distinct region of an animal body that is repeated along its length. In fly embryos, segments are grouped in three broad regions: head (anterior), thoracic (middle), and abdominal (posterior); these segments are pictured in **Figure 21.6a**. Embryos with the dramatic mutation were missing all of the segments and other structures normally found in the anterior end (**Figure 21.6b**). These had been replaced by duplicated posterior structures. The gene responsible for this phenotype was therefore called *bicoid*, meaning "two tailed." Based on its phenotype, Nüsslein-Volhard and Wieschaus suspected that the *bicoid* gene's product must provide positional information. The hypothesis was that the *bicoid* gene coded for a signal that tells cells where they are located along the anterior-posterior body axis.

When they mated flies that carried the mutant *bicoid* allele, Nüsslein-Volhard and Wieschaus found that the mutant allele was autosomal and recessive. More importantly, they found an interesting pattern in the inheritance and expression of this gene. Suppose that *b* stands for the mutant *bicoid* allele. The researchers found that some *bb* offspring are normal—even though they had two copies of the defective allele. In contrast, all mothers with the *bb* genotype looked normal but produced offspring that were all deformed. To interpret this finding, Nüsslein-Volhard and Wieschaus hypothesized that the *bicoid* gene is not expressed in embryos. Instead, it is expressed in mothers—as eggs are being formed. Mothers who were *bb* could not add any normal bicoid product to their eggs, resulting in offspring that couldn't develop normally. This hypothesis meant that *bicoid* mRNA or protein had to be present in the egg. Nüsslein-Volhard and Wieschaus thought they had found a master regulator that mothers loaded into their eggs.

Where Is the *bicoid* Product Found? To determine how the maternally endowed *bicoid* product works, Nüsslein-Volhard and colleagues cloned and sequenced the gene using

(a) A normal fruit-fly embryo

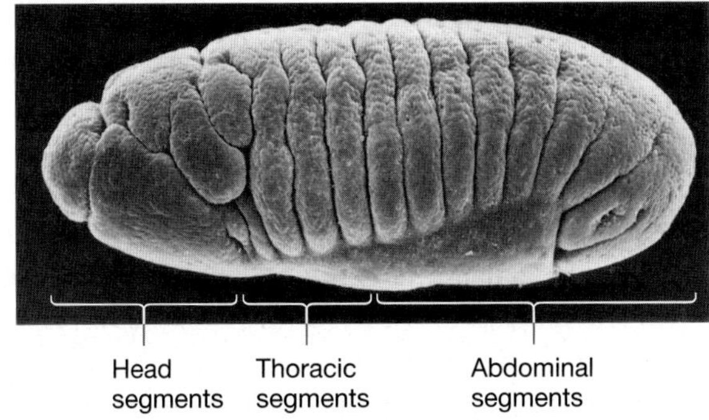

(b) A *bicoid* mutant

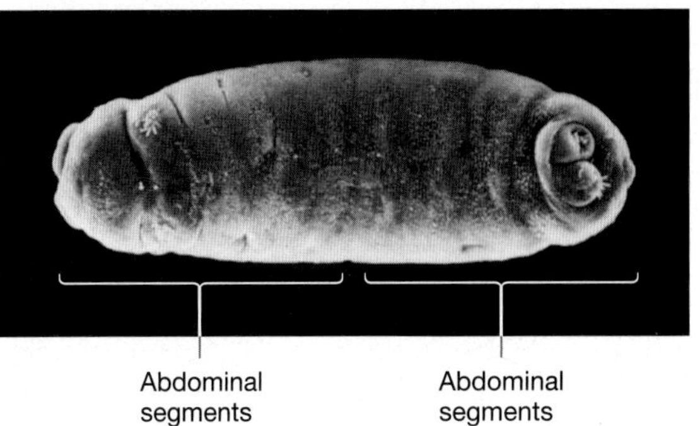

FIGURE 21.6 Pattern-Formation Mutants Have Misshapen Bodies. (a) Normal embryos have distinct head, thorax, and abdominal regions. **(b)** In *bicoid* mutants, embryos have no head or thoracic segments. Instead they have duplicated sets of the posteriormost regions, making them "two tailed."

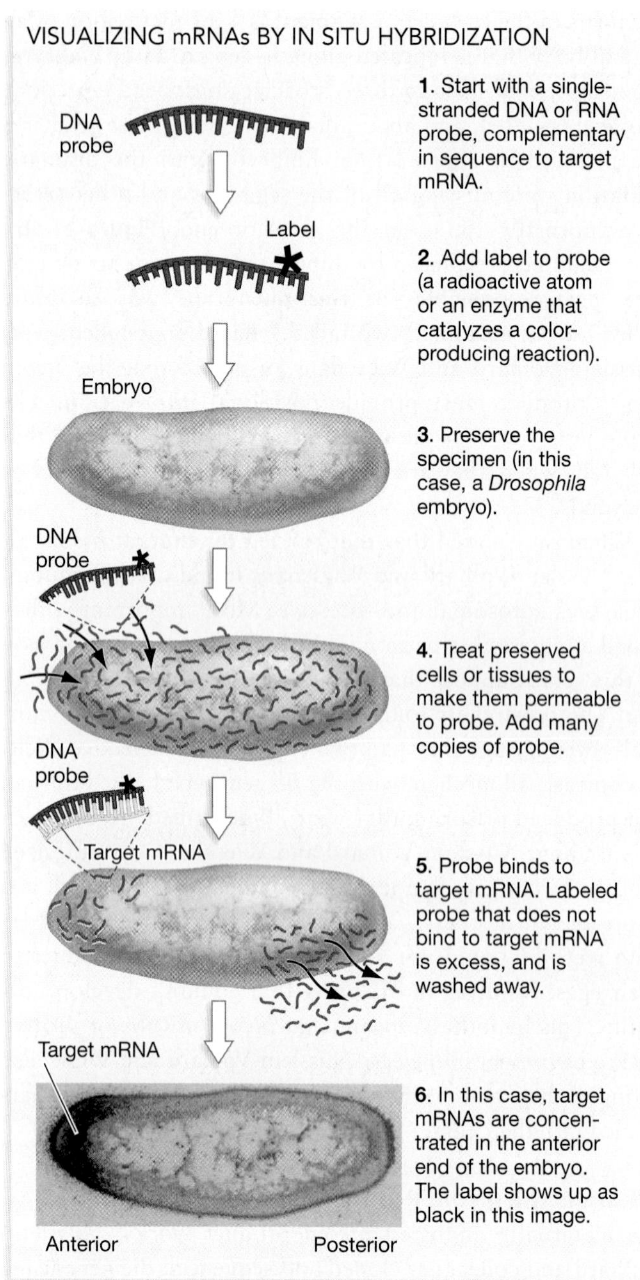

FIGURE 21.7 In Situ Hybridization Allows Researchers to Pinpoint the Location of Specific mRNAs. The micrograph in the last step shows the location of mRNAs from the *bicoid* gene in a fruit fly egg.

● **QUESTION** In situ hybridization is usually used to identify cells that are expressing a particular gene. Why is this method valid?

techniques introduced in Chapter 19. Then they used a technique called **in situ** ("in place") **hybridization** to find where *bicoid* mRNAs are located in embryos. As **Figure 21.7** shows, researchers perform an in situ hybridization by adding a label to single-stranded copies of DNA or RNA molecules that are complementary in sequence to the mRNA they want to find. In this case, the probes were designed to bind to *bicoid* mRNA inside the embryo—marking the location of the mRNAs.

In situ hybridization is enormously important in developmental biology because it allows researchers to pinpoint where and when specific genes are expressed. It is a way of visualizing how gene expression varies among cells in space. By doing in situ hybridizations with embryos of different ages, researchers can also track how gene expression changes through time. Differential gene expression is the essence of development, and in situ hybridization is an important way to document it.

When Nüsslein-Volhard's group treated eggs and early embryos with labeled copies of a probe that bound to *bicoid* RNA, they found that the mRNA is highly localized at the anterior end (see step 6 in Figure 21.7). Their next task was to determine where and when the Bicoid protein is produced. To do this, they made Bicoid protein from *bicoid* DNA (step 1 in **Figure 21.8**) and generated antibodies that bind specifically to the Bicoid protein (step 2). Recall that antibodies are proteins that bind to specific segments of a molecule. They attached a fluorescent compound to the Bicoid antibodies to mark the location of the Bicoid protein in embryos (step 3). These experiments indicated that Bicoid protein was produced from mRNAs localized at the anterior end. The protein then diffused away from the anterior end of the embryo. As the "Results" photograph in Figure 21.8 indicates, a steep concentration gradient is produced: The protein is abundant in the anterior end but declines to progressively lower concentrations in the posterior end.

To pull these observations together, Nüsslein-Volhard and co-workers hypothesized that high concentrations of Bicoid protein lead to the formation of anterior structures such as the head, with progressively lower concentrations giving rise to thoracic segments and the first abdominal segments. Absence of Bicoid, in contrast, contributes to formation of posterior structures. This hypothesis provided an explanation for the original *bicoid* mutants that the researchers had observed. Larvae that lack *bicoid* mRNA have posterior structures in their anterior end.

How Does Bicoid Work? DNA sequencing studies provided important insights into how the Bicoid protein acts as a master regulator. The *bicoid* gene contains sequences that are typical of regulatory transcription factors (see Chapter 18). Based on this observation, biologists hypothesized that the Bicoid protein must enter nuclei, bind to regulatory sites in DNA, and change the expression of specific genes (**Figure 21.9**). Follow-up work showed that Bicoid turns on genes responsible for forming anterior structures.

To summarize, Bicoid protein is a master regulator. Because it is present in a concentration gradient, it provides cells with information about their position along the anterior-posterior

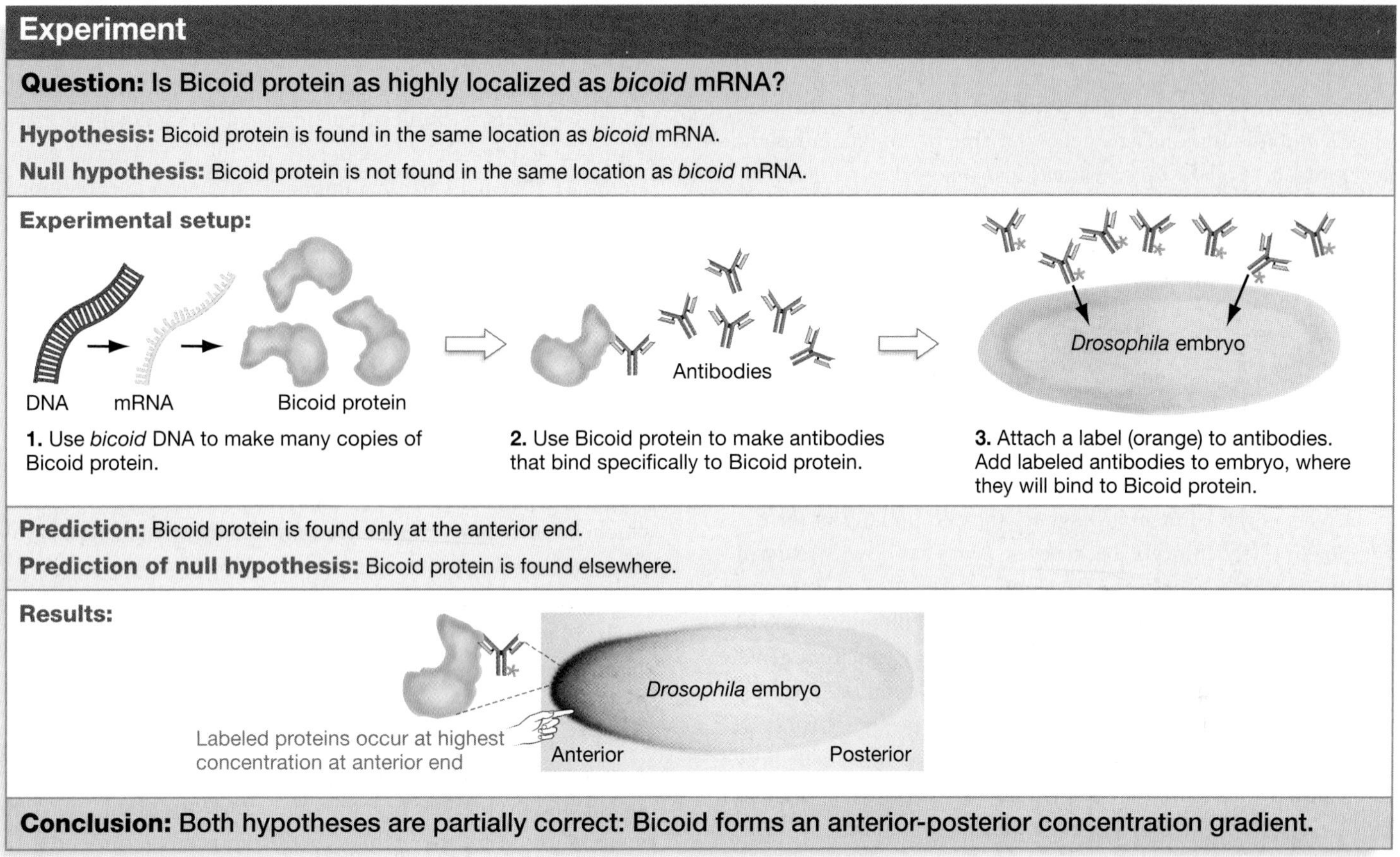

Experiment

Question: Is Bicoid protein as highly localized as *bicoid* mRNA?

Hypothesis: Bicoid protein is found in the same location as *bicoid* mRNA.

Null hypothesis: Bicoid protein is not found in the same location as *bicoid* mRNA.

Experimental setup:

DNA → mRNA → Bicoid protein

1. Use *bicoid* DNA to make many copies of Bicoid protein.

Antibodies

2. Use Bicoid protein to make antibodies that bind specifically to Bicoid protein.

Drosophila embryo

3. Attach a label (orange) to antibodies. Add labeled antibodies to embryo, where they will bind to Bicoid protein.

Prediction: Bicoid protein is found only at the anterior end.

Prediction of null hypothesis: Bicoid protein is found elsewhere.

Results:

Drosophila embryo

Labeled proteins occur at highest concentration at anterior end

Anterior

Posterior

Conclusion: Both hypotheses are partially correct: Bicoid forms an anterior-posterior concentration gradient.

FIGURE 21.8 The *bicoid* mRNA Is Localized, and Bicoid Protein Forms a Concentration Gradient. Bicoid protein forms an anterior-posterior concentration gradient.

body axis. Other initial signals are found in concentration gradients that tell cells where they are along the left-right and dorsal-ventral axes. Thanks to this information, cells have a general sense of where they are early in development. As cell division continues and the embryo grows and changes, how do the descendants of these cells get more specific information about where they are and what they should become?

(MB) **Web Animation** **at www.masteringbio.com**
Early Pattern Formation in *Drosophila*

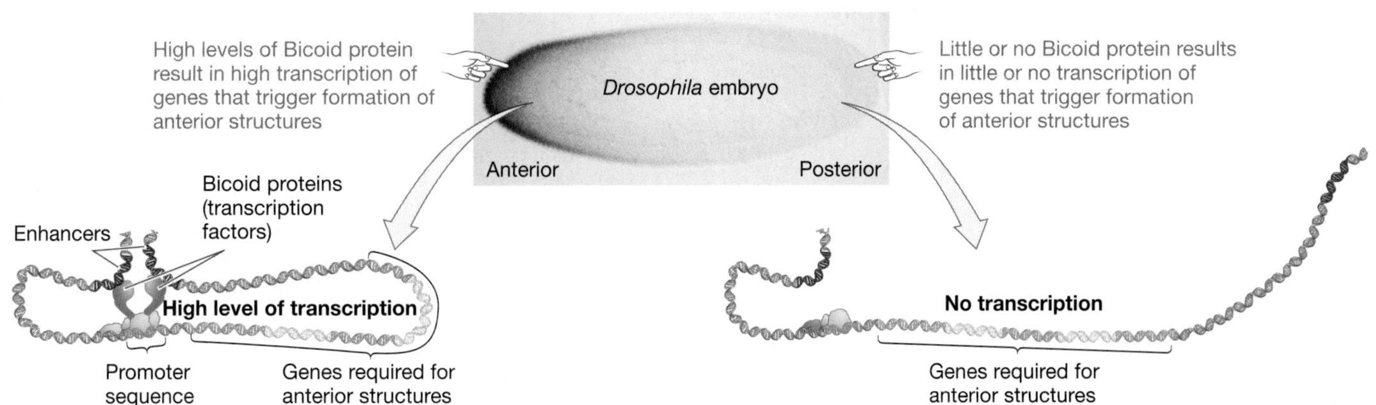

FIGURE 21.9 Differential Gene Expression Is Based on the Presence or Absence of Regulatory Transcription Factors. Bicoid protein is a regulatory transcription factor. It binds to DNA and triggers the expression of specific genes—those involved in the formation of anterior structures. Because Bicoid is present in a gradient, different genes are expressed in the anterior of the embryo versus the posterior.

Regulatory Genes Provide Increasingly Specific Positional Information

The initial work on *bicoid* illustrated the importance of cell-cell signals and interactions as a general theme in animal and plant development. Follow-up work was instrumental in focusing attention on a second fundamental developmental principle, common to both plants and animals: that differentiation is a progressive, step-by-step process.

Along with the "two-posteriored monsters" they named *bicoid* mutants, Nüsslein-Volhard and Wieschaus found an array of embryos that had normal anterior-posterior patterning but defects in how their body segments became organized later in development. These mutants had defective **segmentation genes**.

Defects in the segmentation genes showed up later in the development of fly embryos because the genes act downstream of *bicoid*—meaning later in the developmental sequence. Research on the relationships among these genes helped illuminate one of the great general principles of development: Regulatory genes act in a sequence that provides progressively detailed information on where cells are in time and space. One signal leads to another, and signals interact with other signals to tell cells how they should differentiate. This is another case where detailed work on fruit flies established principles that hold for other organisms as well. Let's take a closer look.

What Do Segmentation Genes Do? Researchers who first identified the segmentation genes noticed that they are expressed in sequence. Sequences called gap genes are expressed first. Experiments with antibodies to gap gene proteins showed that the genes are expressed in broad regions along the head-to-tail axis (**Figure 21.10a**), suggesting that they define the general position of segments in the anterior, middle, or posterior part of the body. Pair-rule genes are expressed next. In situ hybridization experiments showed that their mRNAs occur in alternating bands (**Figure 21. 10b**). This pattern of expression suggests that pair-rule genes demarcate the boundaries of individual segments within each general region. Segment polarity genes are expressed subsequently, in more restricted bands (**Figure 21.10c**). This pattern implies that they delineate boundaries within individual segments.

The general observation is that gap, pair-rule, and segment polarity genes are expressed in order and in increasingly restricted regions. In this way, the anterior to posterior axis set up by the Bicoid signal is progressively divided into what becomes a segmented body.

These data also suggested that the segmentation genes might interact directly. Specifically, investigators hypothesized that the gap genes may activate the transcription of pair-rule genes, which in turn may trigger the expression of other pair-rule genes and segment polarity genes.

Although research on interactions among segmentation gene products is continuing, an enormous amount of progress has already been made. One of the key early discoveries was that the Bicoid protein directly activates transcription of a gap gene. This gap gene is not expressed until the Bicoid protein binds to its regulatory DNA sequences. As a result, the gap gene's mRNA and protein are found in high concentration in the anterior part of the embryo, where the Bicoid protein is abundant.

Subsequent work showed that this and other gap genes encode transcription factors that regulate the expression of pair-rule genes. Each gap gene product regulates the production of a defined set of pair-rule proteins. In addition, several of the gap genes' products regulate the expression of other gap genes or even their own expression. The pair-rule genes activated by the gap genes also encode regulatory transcription factors, and some of the genes they regulate are segment polarity genes. In this way, cells along the anterior-posterior axis of the embryo have their fate determined in a step-by-step process.

(a) Early in development: Gap genes

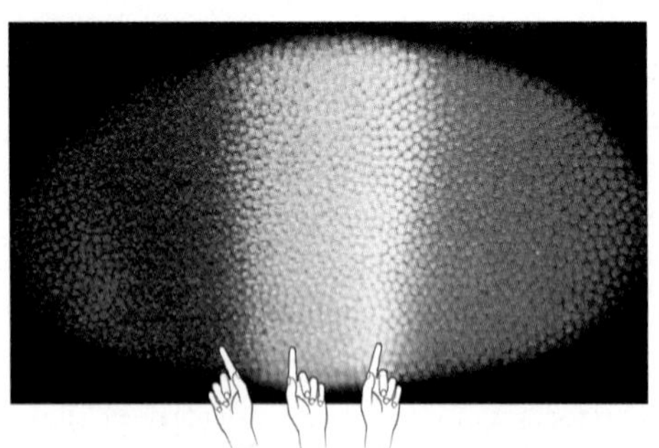

Each stripe indicates the position of a group of segments

(b) Later: Pair-rule genes

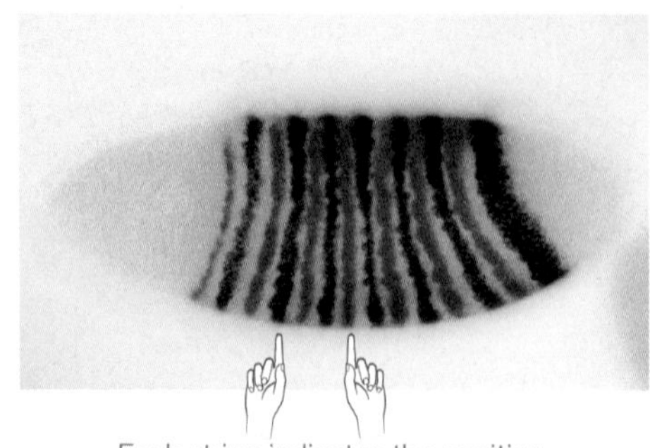

Each stripe indicates the position of a segment boundary

(c) Still later: Segment polarity genes

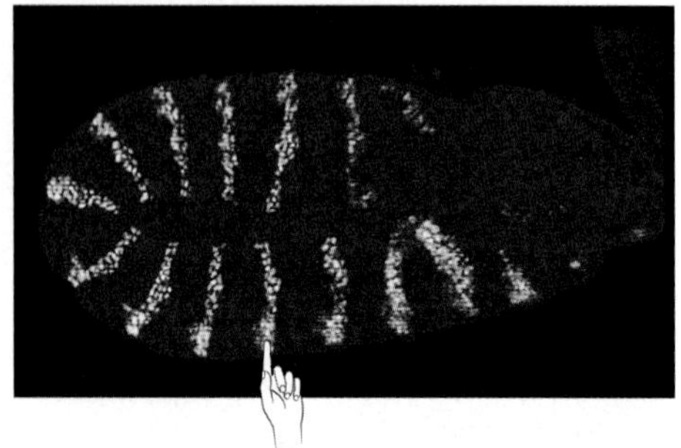

Each stripe indicates the position of regions within individual segments

FIGURE 21.10 A Series of Genes Demarcates Body Segments in Fruit Flies. Antibody-staining and in situ hybridizations show the location of segmentation gene products in fly embryos. The embryos in **(a)** and **(b)** were stained for two different gene products. The embryo in **(c)** was stained for one gene product.

QUESTION From (a) to (c), these photographs show embryos at progressively later stages of development. Why is the sequence logical?

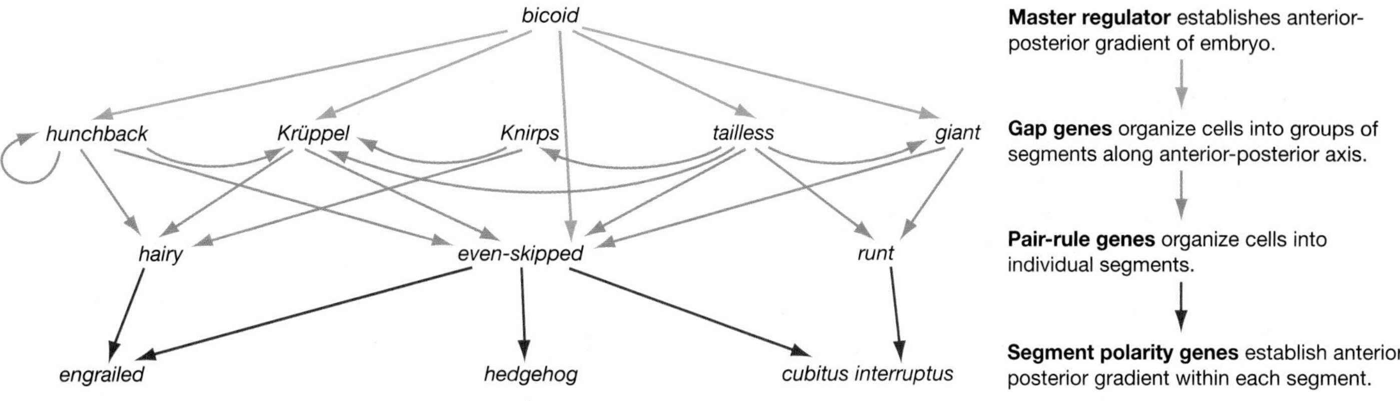

FIGURE 21.11 Segmentation Genes Regulate Each Other. The arrows indicate a small subset of the known interactions among the segmentation genes and their products. In each case, the outcome depends on the concentration of each protein involved. In reality, 9 gap genes, 8 pair-rule genes, and 9 segment polarity genes have been described thus far. Interactions among all their gene products are still being worked out. It's important to realize that analogous networks occur along the dorsal-ventral and right-left axes of an embryo.

Regulatory Genes Form Cascades, and Are Networked

As **Figure 21.11** shows, the interactions among *bicoid* and the segmentation genes can be described as a hierarchy called a **regulatory cascade**. The *bicoid* gene, gap genes, pair-rule genes, and segment polarity genes each define a level in the cascade. The cascade is triggered by the initial concentration of Bicoid protein and is later modified by other signals. For example, some of the segment polarity genes code for proteins that are secreted from the cell. These proteins bind to receptors on nearby cells, which modify their gene expression in response.

To illustrate how a regulatory cascade works, consider a cell that happens to be located at one end of a fly embryo. Very early in development, the Bicoid protein that binds to its DNA acts as a signal that means "You're in the anterior end." In response, the cell begins expressing genes that are expressed only in anterior cells. So do its descendants. Later, a gap gene signal communicates that "You're part of the anterior segments" and triggers other changes in transcription. Still later, pair-rule and segment polarity gene products send signals that specify, "Here's the boundary of the head segment, and here's the boundary of its anterior part." Step by step, the cell's progeny get increasingly specific information about where they are. In response, they become progressively committed to becoming brain cells or other components of the head segment. ● If you understand the relationship between regulatory cascades and segmentation, you should be able to describe the sequence of events that occur in a cell in the posterior of a fly embryo.

The full scope of interactions occurring in this and other regulatory cascades are marvelously complex and are still being worked out. But several general take-home messages are clear, and they are just as relevant to maize, mice, and other eukaryotes as they are to flies:

- Differential gene expression results from regulatory cascades that start early in development: Master regulators trigger the production of other regulatory signals and transcription factors, which trigger production of another set of signals and regulatory proteins, and so on.
- Because the identity and concentration of signals and transcription factors vary along the three major body axes, cells in different locations receive unique positional information.
- Each level in a regulatory cascade provides more specific information about where a cell is.
- As regulatory cascades proceed over time, a cell's fate becomes more and more finely determined.

Homeotic Genes: What Are They, and Where Do They Fit In? Once gap genes, pair-rule genes, and segment polarity genes have established the identity of each segment along the anterior-posterior axis of a fly embryo, the regulatory cascade continues with the activation of **homeotic genes**. Segmentation genes establish the boundaries of each segment; homeotic genes identify which segment is which. More specifically, homeotic genes trigger the development of structures that are appropriate to each type of segment, such as antennae, wings, or legs.

The homeotic genes were discovered when researchers found adult fruit flies with body parts in the wrong place. For example, a series of homeotic mutations can transform thoracic segment number 3 in flies into thoracic segment 2. Instead of bearing the pair of small stabilizer structures called halteres normally found on thoracic segment 3, the transformed segment bears a pair of wings. The mutant has four wings instead of two (**Figure 21.12**). This type of replacement of one structure by another is termed **homeosis** ("like-condition"). Homeosis occurs when cells get incorrect information about where they are in the body.

Studies conducted during the 1970s and 1980s initially identified eight genes in the *Drosophila* genome that lead to

Normal fruit fly

Antennae

Haltere

Homeotic mutant

Wings in place of halteres

Homeotic mutant

FIGURE 21.12 Homeotic Mutants in *Drosophila* Have Structures in the Wrong Locations. Among the most spectacular homeotic mutants in fruit flies are individuals with wings growing where small, stabilizing structures called halteres should be, or legs growing where antennae should be.

homeosis when they are defective. As **Figure 21.13a** shows, the eight genes are found in two clusters on the same chromosome. Together, these eight genes became known as the homeotic complex, or ***Hox* genes.** Once body segments are established in a *Drosophila* embryo, the *Hox* genes are expressed in a distinctive pattern along the anterior-posterior axis. These genes code for transcription factors that trigger the production of segment-specific structures. Subsequent work revealed that not all *Hox* genes cause homeosis when they are defective.

(a) The location of *Hox* genes on the fly chromosome correlates with their pattern of expression in fly embryos.

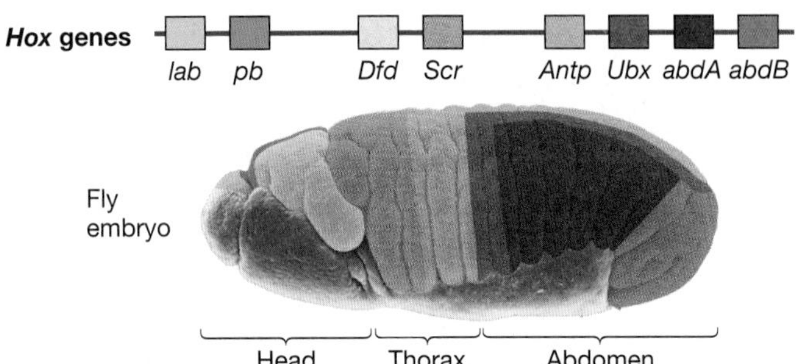

(b) The location of *Hox* genes on the mouse chromosome correlates with their pattern of expression in mouse embryos.

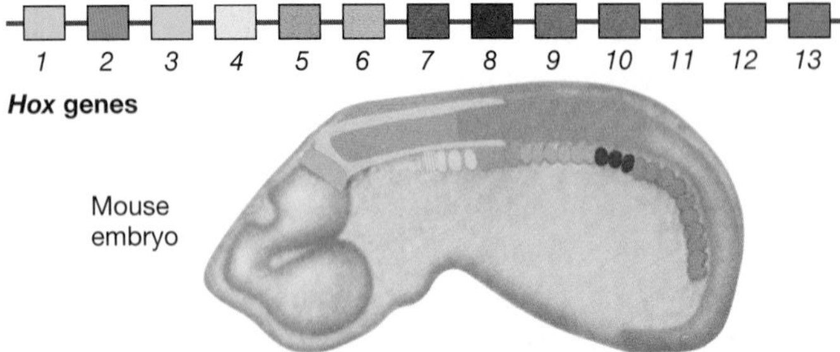

FIGURE 21.13 Organization and Expression of *Hox* Genes.

● **QUESTION** Suppose someone claimed that the anterior-posterior expression of *Hox* genes in a mouse embryo correlated with their positions along the length of the chromosome, as well as with the expression of the homologous genes in flies. Do these data support these assertions or contradict them?

Although it was exciting to establish that homeotic genes form the next link in the regulatory cascade responsible for building a fruit fly, that is not the reason they qualify as one of the most intensively studied groups of genes in developmental biology. Homeotic genes helped illuminate several other key principles of development.

Cell-Cell Signals and Regulatory Genes Are Evolutionarily Conserved

After homeotic genes were discovered and characterized in fruit flies, researchers began looking for similar genes in other animals. The results have been spectacular. Investigators have found that clusters of *Hox* genes occur in virtually every animal examined to date, including frogs, crustaceans (crabs and their relatives), birds, various types of worms, mice, and humans. Although the number of *Hox* genes varies widely among species, their chromosomal organization is similar to that of the fly homeotic genes shown in Figure 21.13a.

Recent studies of *Hox* genes have shown that they are expressed along the head-to-tail axis of the mouse embryo in the same sequence as in fruit flies (**Figure 21.13b**). In addition, experiments have shown that when mouse *Hox* genes are altered by mutation, defects in pattern formation result. Based on these data, biologists conclude that in flies, mice, humans, and most other animals, *Hox* genes play a key role in identifying the position of cells along the head-to-tail axis of the body. This conclusion was supported in dramatic fashion when researchers in William McGinnis's lab introduced the *Hoxb6* gene from mice into fruit-fly eggs. The *Hoxb6* gene in mice is similar in structure and sequence to the *Antp* gene of flies. Because it was introduced without its normal regulatory sequences, the *Hoxb6* gene was expressed throughout the treated fly embryos. The resulting larvae had defects identical to those observed in naturally occurring fly mutants, in which the *Antp* gene is mistakenly expressed throughout the embryo. This is a stunning result: A mouse allele not only affected the development of a fly but also mimicked the effect of a specific fly allele.

To interpret these observations, biologists hypothesize that the genes in *Hox* complexes of animals are homologous—meaning that they are similar because they are descended from genes in a common ancestor. If this hypothesis is correct, it means that the first *Hox* genes arose before the origin of animals. For the past billion years, *Hox* gene products have been helping to direct the development of animals.

Hox genes are by no means an isolated example of signals or regulatory proteins that direct development and are evolutionarily conserved. A well-studied gene called *Pax* has a similar structure and function in many animals—it is involved in the formation of eyes in species ranging from fruit flies to humans. Similarly, homologs of certain segment polarity genes have been found in a wide array of animals, where they perform similar cell-cell signaling functions during development.

The take-home message from these studies is that at least some of the molecular mechanisms of pattern formation have been highly conserved during animal evolution. The discovery of these shared mechanisms is one of the most significant results to have emerged from animal development studies to date. Although animal bodies are spectacularly diverse in size and shape, the underlying mechanisms responsible for their development are similar. What varies among species is less a matter of which genes are present and more a matter of when, where, and in what quantity similar genes are expressed.

Common Signaling Pathways Are Used in Many Contexts

The observation that key regulatory transcription factors and signals are conserved during evolution qualifies as a general principle of animal and plant development. But a related observation also qualifies as a general principle: During development, the same regulatory transcription factors and cell-cell signals are used in a variety of contexts. Regulatory proteins and signals are not only conserved, they are reused. For example, *Hox* genes are involved in the development of an array of animal structures, including appendages like your legs and arms, in addition to their role in defining the anterior-posterior axis of the animal body. But the most spectacular example of this phenomenon may be the *wingless* genes, abbreviated *Wnt* and pronounced "wint."

In humans and other mammals, the genome contains a family of 15 *Wnt* genes. Each of these genes codes for proteins responsible for cell-cell signaling during development. More specifically, they are involved in the regulatory cascade involved in setting up the anterior-posterior axis in the embryo—much like one of the roles of the *Wnt* genes of fruit flies. But in addition, signals from mammalian *Wnt* genes are involved in the formation of back muscles, the midbrain region, the limbs, the gonads (testes or ovaries), hair follicles, parts of the intestine, and structures inside the kidney, among other things. To capture this point, biologists like to say that multicellular organisms have a tool kit of common signals, signal transduction pathways, and regulatory proteins that are used over and over during development. The common tool kit can direct the development of dramatically different structures because its tools are deployed at different times and in different locations.

To create a human analogy for this developmental principle, consider the signal called a pinch. When you were little, a pinch on the cheek from your grandmother meant something very different from a pinch on the ear from your mother or a pinch on the arm from an older sibling. It also means something different from a pinch you might receive from a close friend, now. In human communication and in development, the context in which a signal is sent and received—its location, timing, and intensity—is crucial.

Check Your Understanding

If you understand that... 

- Cell-cell signals trigger the production of specific sets of transcription factors that change gene expression in receiving cells.
- Different cells express different genes because they receive different sets of signals, and because they produce different sets of transcription factors in response.

You should be able to... 

1) Explain why the gradient of Bicoid protein delivers information about where cells are along the anterior-posterior axis of a fly embryo.
2) Describe how the cascade of transcription factors triggered by Bicoid leads to the gradual differentiation of segments along the anterior-poster axis of a fly embryo.

21.4 Changes in Developmental Pathways Underlie Evolutionary Change

For an embryo to develop, cells have to proliferate, move, differentiate, and interact. Differentiation is caused by differential gene expression. It results from signals that tell cells where they are in time and space, and which trigger a complex cascade of interacting transcription factors.

If any of these processes are disrupted, the embryo is likely to die. But if one of these processes is modified in some slight way, the affected structure is likely to have a different size or shape or activity. As a result, the embryo will develop new features and the adult will have a novel phenotype.

Once biologists began working out the regulatory signals and cascades introduced earlier in the chapter, they realized that

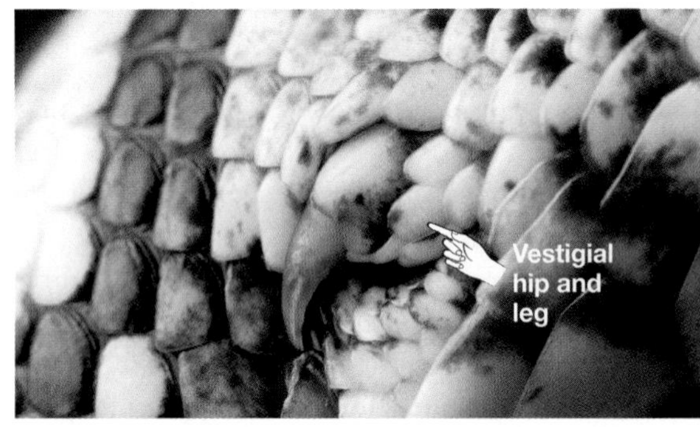

FIGURE 21.14 Some Snakes Have Rudimentary Hindlimbs. This claw, of a South African python, is attached to a rudimentary femur and hip.

the genetic changes altering these developmental processes must be the foundation of evolutionary change. For example, the increase in body size that has occurred during human evolution must have resulted from mutations that altered the signals, regulatory sequences in DNA, or transcription factors that are involved in the amount and timing of cell proliferation throughout the body.

A research field called evolution and development, or **evo-devo**, is focused on understanding how changes in developmentally important genes have led to the evolution of new phenotypes such as the limbs found in tetrapods (amphibians, reptiles, mammals) and arthropods (crabs, insects, millipedes). As an example of how this work is done, let's consider an instance of limb *loss*.

An array of analyses has shown that snakes are clearly tetrapods and that they are most closely related to lizards. Although some snakes do not develop any sort of forelimb or hindlimb at all, boas and pythons have tiny pelvic (hip) bones and a rudimentary femur (thigh bone) and claw, as **Figure 21.14** shows. The ancestors of all snakes had four functional legs along with feet and toes, however. How did snakes lose their limbs? What developmental changes were responsible for this dramatic evolutionary event?

Biologists were able to answer this question because the cell-cell signals and regulatory cascades involved in the development of the tetrapod limb are reasonably well understood. For example, researchers knew that two *Hox* genes, *Hoxc6* and *Hoxc8*, are expressed together in vertebrae that form ribs, but that *Hoxc6* is expressed without *Hoxc8* in the region that gives rise to the forelimbs (**Figure 21.15a**). In snakes, however, they found that *Hoxc6* and *Hoxc8* are expressed together throughout the part of the trunk where forelimbs should form (**Figure 21.15b**). These data suggest that a change in the regulation of *Hoxc8*—specifically, where the gene is expressed—led to the evolutionary loss of the forelimb. Hindlimb loss, in contrast, appears to be due to defects in a signaling molecule encoded by a gene called *Sonic hedgehog*. This gene product is homologous to that of the fruit fly's *hedgehog* gene, introduced in Section 21.3. In snakes, hindlimb loss is due to a genetic change in part of the signaling cascade that sets up the anterior-posterior body axis.

In the case of both forelimb and hindlimb loss, biologists can point to alterations in a specific transcription factor or cell-cell signal that explains why an important evolutionary change occurred. Changes in a regulatory cascade led to changes in the adult body of a lizard-like ancestor. Snakes have been legless ever since.

(a) Pattern of gene expression in tetrapods with forelimbs.

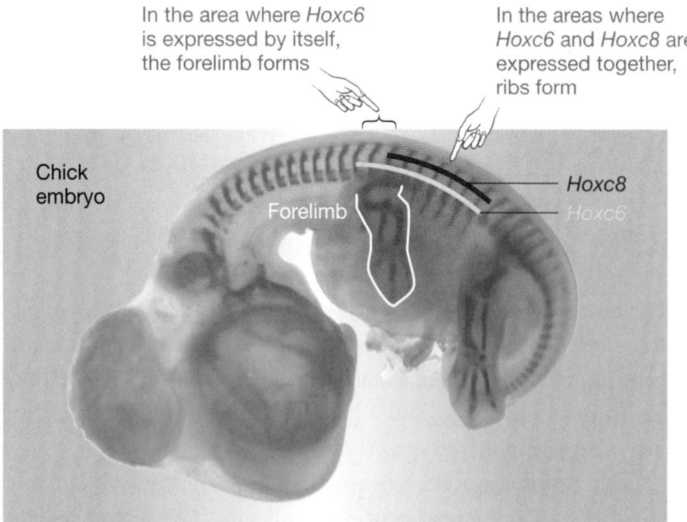

(b) Pattern of gene expression in snakes (no forelimbs).

Hoxc6 and *Hoxc8* are always expressed together, so no forelimbs form

Snake embryo

Hoxc8

Hoxc6

FIGURE 21.15 Changes in Homeotic Gene Expression Led to Limb Loss in Snakes. *Hoxc6* and *Hoxc8* code for transcription factors. If both transcription factors are present in a population of cells along the anterior-posterior axis of a tetrapod, genes that lead to the formation of vertebrae and ribs are activated. But if only *Hoxc6* is present, then genes that lead to the formation of forelimbs are activated. Contrasting patterns of gene expression are responsible for **(a)** the development of forelimbs in most tetrapods and **(b)** the lack of forelimbs in snakes.

Chapter Review

SUMMARY OF KEY CONCEPTS

As an embryo develops, cells have to proliferate, move or expand in a directed manner, become specialized, and interact with other cells.

Four essential processes underlie development: (1) Cells have to grow and divide in a regulated manner as the body develops, and some cells have to undergo programmed cell death. (2) In animals, masses of cells move in a coordinated way early in development; later, smaller populations of cells move to specific locations in the body. Plant cells do not move, but normal development depends on precise control over the plane of cell division and directed expansion of cells. (3) Cells undergo a step-by-step process that leads to differentiation, or the production of a specialized cell type. (4) Cell-cell signals provide a constant flow of information about where cells are in space and time.

You should be able to predict how the development of an oak seedling would be affected by a drug that (1) reduces the rate of cell division by half, or (2) prevents cells from expanding after mitosis, or (3) obstructs the first step in the sequence that leads to flower formation, or (4) blocks the signal that sets up the shoot-to-root axis.

Cells become specialized because they express different genes, not because they contain different genes.

In animals and plants, cells differentiate due to differential gene expression, not differential loss of genes. Cells of various cell types are specialized because they contain transcription factors that activate certain genes and not others. Differentiation occurs in a step-by-step manner because cell-cell signals trigger a cascade of effects that lead to certain transcription factors being present in the cell over time.

You should be able to explain why chromatin structure in a muscle cell is different from chromatin structure in a nerve cell.

Cells interact continuously during development via cell-cell signals.

Cells "know" where they are in the body and how far along development has progressed because they produce and receive a steady stream of signals. These signaling molecules operate through signal transduction pathways that alter gene expression—primarily transcription. Cells begin expressing a distinctive suite of proteins and differentiate because they receive a distinctive suite of signals.

You should be able to explain why development is based on transcriptional control rather than RNA processing, RNA stability, translation, or protein modification.

Early signals trigger a cascade of effects, including the production of transcription factors and other signals. Interactions among these signals, transcription factors, and regulatory DNA sequences cause cells to differentiate as development proceeds.

In fruit flies, a master regulator called Bicoid is responsible for setting up the overall anterior-posterior axis of the body. Bicoid triggers the production of transcription factors encoded by the gap genes, which activate the expression of pair-rule genes, which stimulate the transcription of segment polarity genes, which regulate the expression of homeotic genes. As this cascade continues, cell-cell signals from genes like *hedgehog* and *wingless* moderate the array of transcription factors produced in cells throughout the body. Complex interactions among these signals and transcription factors are responsible for differentiation.

You should be able to explain why *bicoid* mutants die as embryos, while most homeotic mutants can develop into adults.

Early Pattern Formation in *Drosophila*

The evolution of new body sizes, shapes, and structures is possible when mutations alter the genes responsible for development.

If the cell-cell signals, transcription factors, and regulatory DNA sequences that are active in development change, then the adult phenotype is likely to change as a result. Variation in phenotypes is the basis of evolutionary change. In several cases, researchers have been able to pinpoint changes in cell-cell signals, transcription factors, or regulatory DNA sequences that were responsible for major events in evolution, such as the loss of limbs in snakes.

You should be able to explain the significance of the four-winged homeotic mutant in Figure 21.12, in terms of efforts to explain the genetic mechanisms responsible for insect evolution. (The ancestors of fruit flies had two pairs of full-sized wings that functioned in flight.)

QUESTIONS

Test Your Knowledge

1. What is apoptosis?
   a. an experimental technique that biologists use to kill specific cells
   b. programmed cell death that is required for normal development
   c. a pathological condition observed only in damaged or diseased organisms
   d. a developmental mechanism unique to the roundworm *C. elegans*

2. What is the function of stem cells in adult mammals?
   a. Some of their daughter cells remain as stem cells and continue to divide throughout life.
   b. They give rise to hair, fingernails, and other structures that grow throughout life.
   c. They produce compounds that stem blood loss from wounds.
   d. They produce cells that differentiate to replace dead or damaged cells.

3. Why are in situ hybridizations such a valuable tool for studying development?
   a. They identify the location of specific mRNAs, and so provide a picture of differential gene expression.
   b. They allow researchers to understand how cell-cell signals and regulatory transcription factors interact.
   c. They provide data on homology—the presence of similar genes in different species.
   d. They can be done with RNA or DNA probes.
4. Why is it logical to observe that differentiation is based on transcriptional control, versus translational or post-translational control?
   a. There is a trade-off between speed and efficiency at the various levels of control over gene expression.
   b. Differentiation is based on differential gene expression.
   c. Differentiation is usually permanent, so cells should not produce proteins that will never be used.
   d. Transcriptional control is the only level of control that can be carefully regulated.
5. What is a homeotic mutant?
   a. an individual with a structure located in the wrong place
   b. an individual with an abnormal head-to-tail axis
   c. in flies, an individual that is missing segments; in *Arabidopsis*, an individual that is missing a hypocotyl or other embryonic structure
   d. an individual with double the normal number of structures or segments
6. Why are regulatory cascades in development also referred to as networks?
   a. The cascade is not linear—within and between levels in the cascade, various gene products interact to form a network.
   b. Signals as well as transcription factors are involved in the cascade.
   c. Each transcription factor and signal in the cascade has a limited life span.
   d. Each step in the cascade takes time, so the steps in the cascade are sequential.

Test Your Knowledge answers: 1. b; 2. d; 3. a; 4. c; 5. a; 6. a

Test Your Understanding

Answers are available at www.masteringbio.com

1. How did researchers go about looking for molecules that have important roles in establishing the anterior-posterior body axis and body segments in *Drosophila*?
2. What evidence suggests that at least some of the molecular mechanisms responsible for pattern formation have been highly conserved over the course of animal evolution?
3. Why is it significant that many of the genes involved in development encode regulatory transcription factors?
4. Explain the connection between the existence of regulatory gene and signaling cascades and the observation that differentiation is a step-by-step process.
5. Explain the logic behind using nuclear transplant experiments to test the hypothesis that all animal cells in the body are genetically equivalent.
6. Explain why cell proliferation and death, cell movement or growth, cell differentiation, and cell-cell interactions are considered the four fundamental processes in development.

Applying Concepts to New Situations

Answers are available at www.masteringbio.com

1. Recent research has shown that the products of two different *Drosophila* genes are required to keep *bicoid* mRNA concentrated at the anterior end of the egg. In individuals with mutant forms of these proteins, *bicoid* mRNA diffuses farther toward the posterior pole than it normally does. Predict what effect these mutations will have on segmentation of the larva.
2. In 1992 David Vaux and colleagues used genetic engineering technology to introduce an active human gene for a protein that inhibits apoptosis into embryos of the roundworm *C. elegans*. When the team examined the embryos, they found that cells that normally undergo programmed cell death survived. What is the significance of this observation?
3. Suppose that physicians implanted human stem cells into the brains of patients with Parkinson's disease, in the area of the brain where the patients had lost many nerve cells. What would have to happen for the stem cells to differentiate into functioning nerve cells?
4. According to the data presented in this chapter, segmentation genes set up the segmented body plan of a fruit fly while homeotic genes trigger the production of segment-specific structures such as antennae, legs, and wings. Sketch the body of a fruit fly next to the body of a spider, centipede, and earthworm. Generate hypotheses to explain the differences between the bodies of fruit flies and the other three species, based on changes in the segmentation genes and homeotic genes.

www.masteringbio.com is also your resource for • Answers to text, table, and figure caption questions and exercises • Answers to *Check Your Understanding* boxes • Online study guides and quizzes • Additional study tools including the *E-Book for Biological Science* 3rd ed., textbook art, animations, and videos.

An Introduction to Animal Development

22

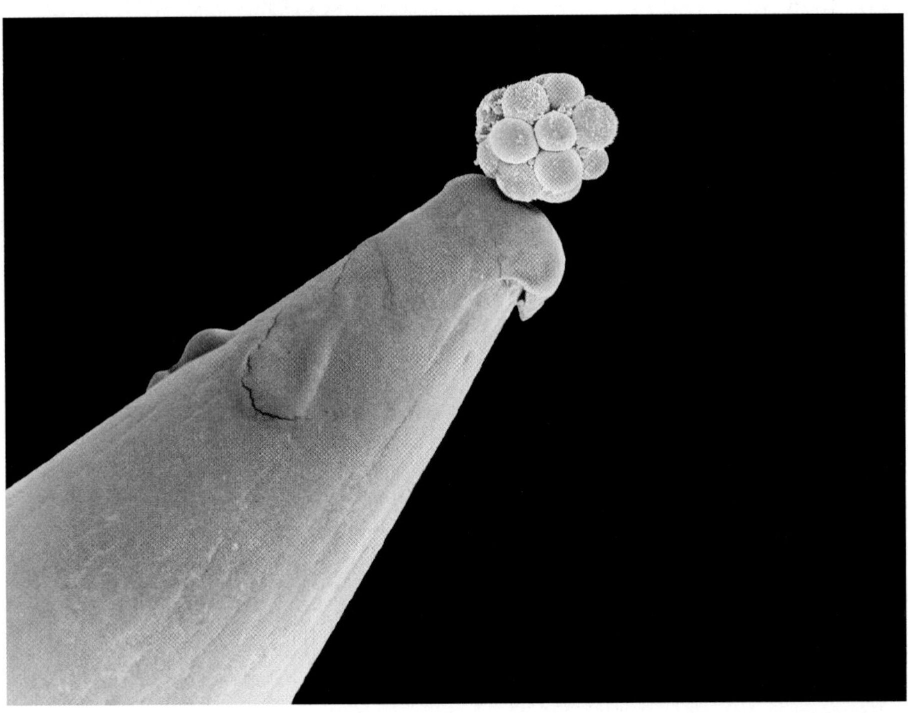

A human embryo, about 3 days old, on the head of a pin.

KEY CONCEPTS

- In most animals, fertilization depends on specific interactions between proteins on the plasma membranes of sperm and egg.
- The earliest cell divisions divide the fertilized egg into a mass of cells. The fate of these cells depends on the cytoplasmic determinants they contain, the signals they receive, or a combination of both.
- Early in development, cells undergo massive, coordinated movements to form distinct tissue layers. Each of these layers gives rise to distinct groups of tissues and organs in the adult.
- As development proceeds, the formation of specialized organs and other structures depends on cell-cell signals, cell proliferation, cell movements, and differentiation. Differentiation is complete when cells express tissue-specific proteins.

When the physician, researcher, and writer Lewis Thomas considered how a human being develops from a fertilized egg, he could only marvel: "You start out as a single cell derived from the coupling of a sperm and an egg, this divides into two, then four, then eight, and so on, and at a certain stage there emerges a single cell which will have as all its progeny the human brain. The mere existence of that cell should be one of the great astonishments of the earth. People ought to be walking around all day, all through their waking hours, calling to each other in endless wonderment, talking of nothing except that cell. It is an unbelievable thing, and yet there it is, popping neatly into its place amid the jumbled cells of every one of the several billion human embryos around the planet, just as if it were the easiest thing in the world to do."[1]

Many biologists share Thomas's feelings. The experience of watching a fertilized egg go through early development inspires awe and wonder. In animals, the fertilized egg undergoes a rapid series of cell divisions. The resulting cells begin a series of extensive, tightly coordinated movements, and the embryo is dramatically reorganized. Just a few hours or days later, a recognizable creature with internal organs, a head, eyes, back, belly, and other

[1]Lewis Thomas, *The Medusa and the Snail* (New York: Viking Press, 1979), 156. Thomas was exercising some poetic license here. The brain actually arises from a group of cells in the embryo rather than from a single cell.

body parts appears. In many cases, these structures continue to grow and develop for years after hatching or birth occurs.

Understanding how and why these events happen is among the greatest of all challenges facing biologists today. Chapter 21 introduced this challenge by reviewing the basic principles that unify animal and plant development. The most fundamental message of Chapter 21 was that cell-cell signals cause specific sets of transcription factors to be produced in various cells throughout the embryo, resulting in differential gene expression and differentiation.

This chapter's goal is to apply the general developmental principles to understand the sequence of events that occurs as an animal embryo develops. We begin by examining gametogenesis, the process responsible for producing gametes. Recall from Chapter 12 that **gametes** are haploid reproductive cells. In animals and many other organisms, male gametes are called **sperm** and female gametes are called **eggs**. As **Figure 22.1** shows, gametogenesis is followed by fertilization, or the union of a sperm and an egg. In animals, a fertilized egg undergoes a series of rapid cell divisions called cleavage, forming a mass of cells known as a blastula. The cells then move and rearrange themselves in the process of gastrulation. During this phase of development, the embryo is referred to as a gastrula. These early events in development are analyzed in Sections 22.1 through 22.3. They set the stage for the development of specialized tissues and organs during the process of organogenesis, introduced in Section 22.4.

The events summarized in Figure 22.1 represent the cycle of life in animals. As you study each step in this cycle, remember to keep an eye out for the cell-cell signals and regulatory proteins that allow developing cells to divide, move, interact, and differentiate—to do the things required for them to "pop into place" and contribute to the formation of a new individual.

22.1 Gametogenesis

The development of an offspring begins with **gametogenesis**, or the formation of gametes—sperm and egg—in the reproductive organs of adult organisms. The DNA and cytoplasm in these reproductive cells are the initial components of the new individual. Both sperm and egg contribute an equal number of chromosomes—usually a haploid genome containing one allele of each gene—to the offspring. But because egg cells are routinely hundreds or thousands of times larger than sperm cells, an egg contributes much more cytoplasm than does a sperm.

The sequence of mitotic and meiotic cell divisions leading up to the production of sperm and eggs are discussed in Chapter 40 (plant reproduction) and Chapter 48 (animal reproduction). Here, let's focus on the structure of the mature reproductive cells. Understanding how these specialized cells are put together will lay the groundwork for addressing questions about fertilization and early development.

Sperm Structure and Function

Sperm cells begin to develop after meiosis has resulted in the production of a haploid nucleus. As a mammalian sperm cell matures, the four main compartments shown in **Figure 22.2** develop: the head, neck, midpiece, and tail.

- The head region contains the nucleus and an enzyme-filled structure called the **acrosome**. The chromosomes inside the

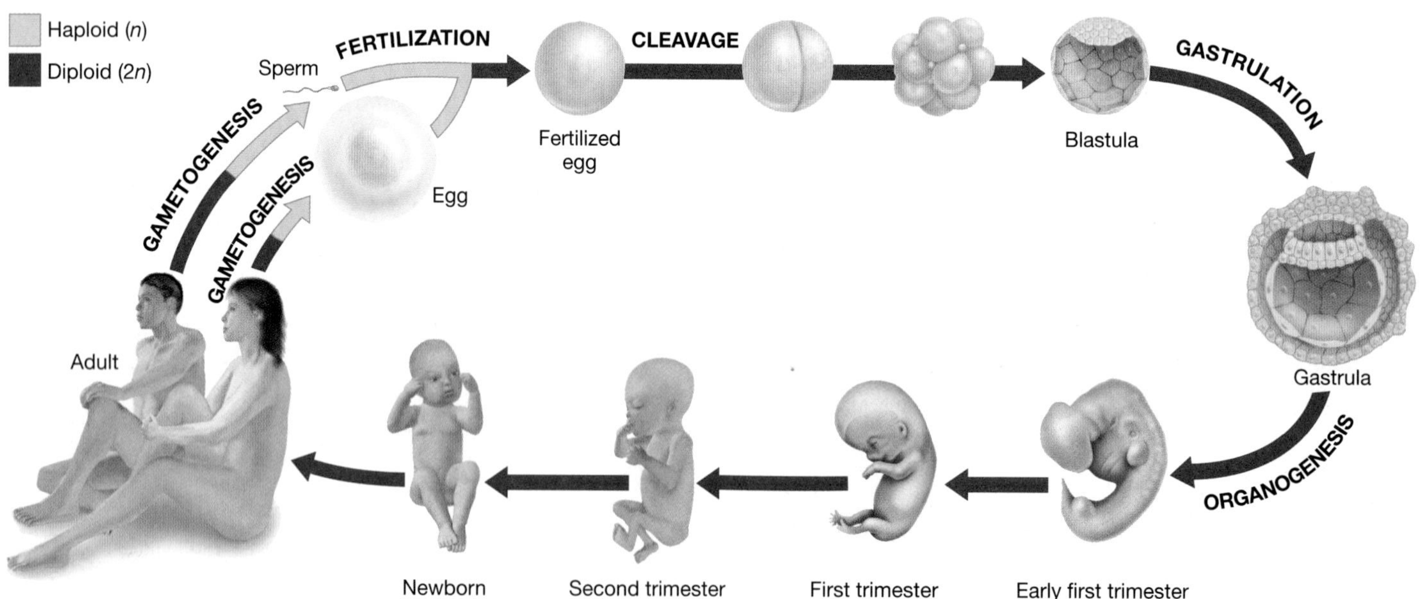

FIGURE 22.1 Development Proceeds in Ordered Phases. In animals, the development of a new individual starts with the formation of gametes and continues with fertilization and early cell divisions (cleavage), resulting in a blastula. Gastrulation then rearranges the blastula into a gastrula. Organogenesis follows and leads to the formation of adult tissues and structures.

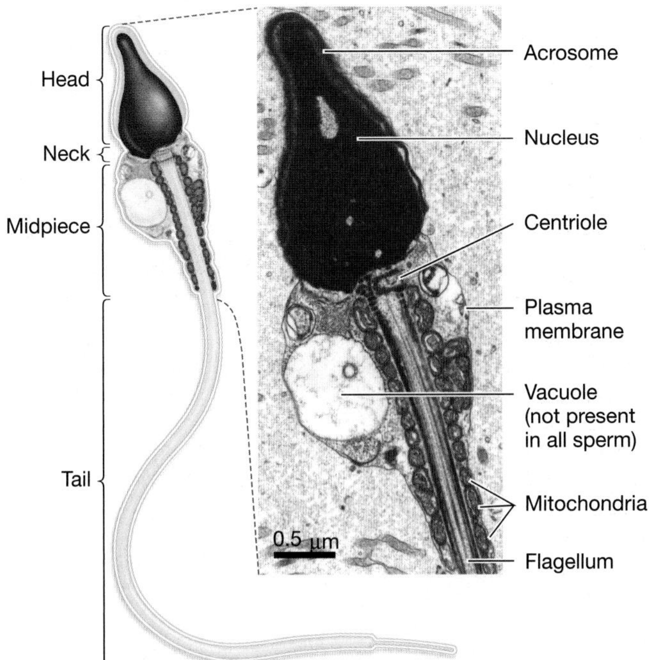

FIGURE 22.2 Animal Sperm Structure. The morphology of human sperm is typical of many animal species.

QUESTION If you were examining the gametes of an animal that you'd never seen before, how would you decide which cells are sperm and which are eggs?

nucleus are condensed into an even more densely packed arrangement than in other cells. The acrosome at the tip of the sperm is derived from vesicles produced by the Golgi apparatus (see Chapter 7). The enzymes stored in the acrosome allow the sperm to penetrate barriers that exist around the egg.

- The neck encloses a **centriole**. You might recall from Chapter 11 that animal centrioles play a key role in the formation of the spindle apparatus, which moves chromosomes to daughter cells during mitosis. The fertilized egg inherits one centriole from the sperm cell and one centriole from the egg.
- The midpiece is packed with **mitochondria** (see Chapter 9). Sperm swim long distances to reach an egg. This movement is powered by ATP molecules synthesized in the mitochondria of the midpiece.
- The tail region consists of a **flagellum**—the long structure composed of microtubules and surrounded by the plasma membrane that was introduced in Chapter 7. Sperm cells are simple enough to be characterized as "DNA with a propeller." The "propeller" is the flagellum, which arises from the centriole early in sperm-cell maturation. The flagellum powers movement by whipping back and forth.

Sperm are race cars—stripped down, streamlined, souped-up cells that are specialized for racing other sperm to the egg. Eggs, in comparison, are like semitrailers—hulking, locked, immobile storage containers that are packed with valuable merchandise.

Egg Structure and Function

Animal gametes are easy to tell apart: Sperm are small and motile; eggs are large and nonmotile. Eggs are large mainly because they contain the nutrients required for the embryo's early development. But in addition to nutrients, the eggs of many species contain key developmental regulatory molecules called **cytoplasmic determinants**—like the *bicoid* mRNA introduced in Chapter 21—that control the early events of development.

The quantity of nutrients present in the egg varies widely among species. In mammals, embryos start to obtain nutrition through a maternal organ called the placenta within a week or two after fertilization. Thus, the egg only has to supply nutrients for early development and is relatively small. But in species where females lay eggs directly into the environment, stores in the egg are the *only* source of nutrients until organs have formed and a larva or juvenile hatches and begins to feed. In these species, the nutrients required for early development are provided by **yolk**—a fat- and protein-rich cytoplasm that is loaded into egg cells as they mature. Yolk may be present as one large mass or as many small granules.

Yolk and the cytoplasmic determinants present in some species are not the only specialized material or structure found inside animal eggs. Many eggs also contain organelles called **cortical granules**, which are small vesicles filled with enzymes that are activated during fertilization. As the egg matures, cortical granules are synthesized in the Golgi apparatus, transported to the cell surface, and localized just under the inner surface of the plasma membrane.

Just outside the plasma membrane of eggs, a fibrous, mat-like sheet of glycoproteins called the **vitelline envelope** forms and surrounds the egg (**Figure 22.3**). In the eggs of humans and other mammals, this structure is unusually thick and is called the **zona pellucida.** In some species, a large gelatinous matrix known as a jelly layer surrounds the vitelline envelope to further enclose the egg.

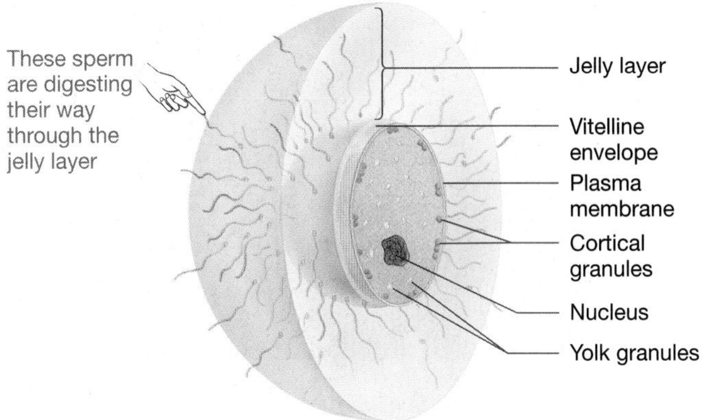

FIGURE 22.3 Sea Urchin Eggs Are Covered by a Jelly Layer. Animals that lay their eggs in water frequently surround their eggs with a jelly layer, a gelatinous coat.

The egg is loaded with rich stores of nutrients and guarded by one or more layers of protective material. How does a sperm cell penetrate these coatings to reach the egg's plasma membrane and fertilize the egg?

22.2 Fertilization

Fertilization seems like a simple process: A sperm cell fuses with an egg cell to form a diploid cell known as a **zygote** (a fertilized egg). Upon reflection, though, you begin to appreciate that the process is extraordinarily complex. For fertilization to take place, sperm and egg cells must be in the same place at the same time. Then they must recognize and bind to each other. Next they have to fuse—even though most of the other cells in the body do not fuse with cells they contact. In most species, fusion must also be limited to a single sperm so that the zygote does not receive extra chromosomes. Finally, the fusion of the two gametes has to trigger the onset of development. This complexity has made fertilization a fascinating research topic. The contact that takes place between sperm and egg at fertilization qualifies as the best studied of all cell-to-cell interactions.

Fertilization in Sea Urchins

Research on fertilization began in earnest early in the twentieth century, when biologists started to study the sperm-egg interaction in sea urchins. Sea urchins are marine organisms that are closely related to sea stars and sand dollars. Like most aquatic animals, they shed their gametes into the surrounding water and fertilization occurs externally (**Figure 22.4**). To maximize the probability that sperm and egg will meet, male and female sea urchins secrete huge quantities of gametes. As a result, this organism provides researchers with large numbers of gametes that can be studied in a test tube or culture dish filled with seawater. Sea urchins continue to be an intensively studied model system in research on fertilization and other aspects of development.

FIGURE 22.4 Many Aquatic Species Have External Fertilization. Sea urchins release their gametes into the ocean. Fertilization and early development take place in the open water.

EXERCISE Sea urchins cannot see. Suggest a hypothesis to explain how it is possible for many sea urchins to release gametes at the same time.

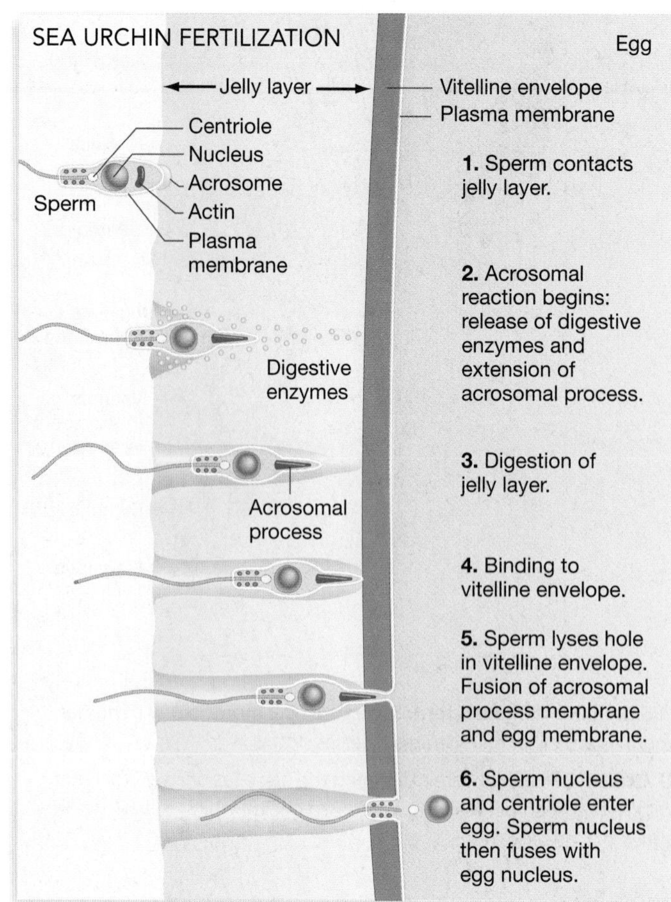

FIGURE 22.5 The Fertilization Sequence in Sea Urchins Is Based on Cell-Cell Interactions.

In sea urchins, the jelly layer surrounding the egg contains a molecule that attracts sperm. The attractant is a small peptide that diffuses away from the egg and into the surrounding seawater. Sperm respond to the attractant by swimming toward areas where its concentration is high. The same type of interaction between sperm and eggs has been observed in a wide variety of animals that lay their eggs in seawater.

When a sea urchin sperm and egg meet, the head of the sperm initially encounters the jelly layer of the egg cell (**Figure 22.5**). To reach the plasma membrane of the egg cell, the sperm head must digest its way through the egg's jelly layer and vitelline envelope. This process begins with the **acrosomal reaction**, which is triggered by contact between the sperm's head and the jelly layer. In the first part of this reaction, the contents of the acrosome—the structure at the tip of the sperm's head—are released. The enzymes that diffuse away digest a pathway through the egg's jelly layer so the sperm can reach the vitelline envelope of the egg. The second part of the acrosomal reaction involves polymerization of actin into microfilaments that form the acrosomal process, a protrusion that extends from the sperm's head until it makes contact with the vitelline envelope. Finally, the plasma membranes of the egg and sperm fuse. The sperm nucleus, mitochondria, and centriole enter the egg—though the sperm mitochondria later disintegrate—and the sperm and egg nuclei fuse to form the zygote nucleus. Fertilization is complete.

In 1910 Frank Lillie recognized that an important question was hidden in this sequence of events: What prevents cross-species fertilization and the production of dysfunctional hybrid offspring? After all, in many habitats sperm and eggs from a particular sea urchin species float in seawater along with eggs and sperm from other sea urchin species and many other organisms. The effort to answer this question blossomed into an extraordinarily productive line of research.

How Do Gametes from the Same Species Recognize Each Other? Lillie was the first researcher to identify a substance on the surface of egg cells that appeared to be involved in binding sperm. He called the compound *fertilizin* and showed that if the purified molecule were experimentally applied to a group of sperm, it caused the cells to clump together. This result supported the hypothesis that fertilizin is involved in sperm-egg coupling, because it implied that clumps of fertilizin molecules were binding to sperm, causing them to clump. Lillie also showed that clumping occurred only when fertilizin from the eggs of a particular sea urchin species was combined with sperm from the same species. Each species appeared to have its own version of fertilizin.

Based on these observations, Lillie proposed that fertilizin on the surface of an egg interacted with a substance on sperm in a lock-and-key fashion. Further, Lillie suggested that this interaction was a product of natural selection (see Chapter 1), because it increased the probability of fertilization occurring between sperm and eggs of the same species.

Decades passed before Lillie's hypothesis was extended and tested, however. Then, in the 1970s, Victor Vacquier and coworkers succeeded in identifying a protein on the head of sea urchin sperm that binds to the surface of sea urchin eggs in a species-specific manner. They called this protein *bindin*. Follow-up work showed that the bindin proteins from even very closely related species are distinct. As a result, bindin should ensure that a sperm binds only to eggs from the same species. The next question was, To what does bindin bind? If bindin acts as a key, what acts as the lock?

To address these questions, Kathleen Foltz and William Lennarz hypothesized that fertilizin is the "lock" protein, that it is found on the surface of sea urchin eggs, and that it acts as a receptor for bindin. To test these hypotheses, they set out to isolate the bindin receptor. More specifically, they set out to isolate the part of fertilizin that is exposed on the outside of the egg. They predicted that this region of the protein interacts with the bindin on sperm.

Figure 22.6 illustrates Foltz and Lennarz's experimental approach. They began by treating the surface of sea urchin eggs with a **protease**—an enzyme that cleaves peptide bonds. When the investigators isolated the protein fragments that were released from the egg surface, they found one that bound to sperm and to isolated bindin molecules. Further, this binding occurred in a species-specific manner. A protein fragment from the eggs of one species bound to sperm of its own species, but did not bind to sperm of different species. Based on these observations, the biologists claimed that they had found the outward-facing portion of the egg-cell receptor protein for sperm.

These experiments provided convincing evidence that an egg receptor for sperm exists on the surface of sea urchin eggs. Foltz and Lennarz succeeded in isolating the compound Lillie had

Experiment

Question: If bindin is the protein on the surface of sperm that acts like a key, what is the "lock"?

Hypothesis: The lock is a receptor protein on the surface of sea urchin eggs. Bindin binds to this receptor.

Null hypothesis: The lock is not a receptor protein on the surface of sea urchin eggs. Bindin does not bind to this receptor.

Experimental setup:

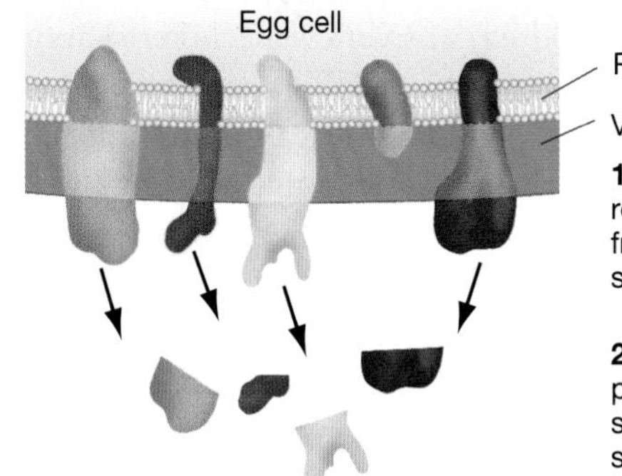

Prediction: One of the protein fragments isolated from egg surface will bind to sperm and to isolated bindin molecules.

Prediction of null hypothesis: The protein fragments isolated from egg surface will not bind to sperm or to isolated bindin molecules.

Results:

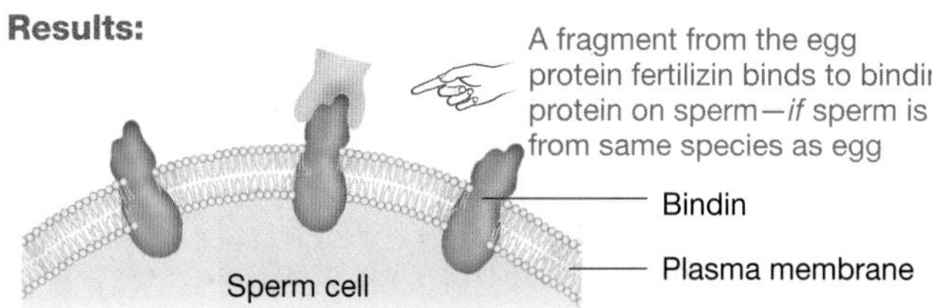

Conclusion: Fertilizin is a "lock" protein on the surface of eggs that binds to bindin (the "key") in a species-specific manner.

FIGURE 22.6 The Egg-Cell Receptor for Sperm Was Isolated and Characterized. By treating sea urchin egg cells with a protease, researchers isolated the fragment of the egg-cell receptor that binds to sperm.

named fertilizin. In doing so, they provided important support for Lillie's lock-and-key hypothesis. During sea urchin fertilization, species-specific bindin molecules on sperm interact with species-specific fertilizin receptors on the surface of the egg. This interaction is required for the plasma membranes of sperm and egg to fuse. As a result, cross-species fertilization is rare.

Why Does Only One Sperm Enter the Egg? Early in the history of studies on sea urchin fertilization, researchers noticed that only one sperm succeeded in fertilizing the egg, even when dozens or even hundreds of sperm were clustered around the vitelline envelope. From the standpoint of producing a viable offspring, this observation was logical. If multiple fertilization, or **polyspermy**, occurred, the resulting zygote would have more than two copies of each chromosome. Sea urchin embryos with more than two copies of each chromosome die. If many sperm are present, how is polyspermy avoided?

Research over the past 75 years has revealed a wide array of mechanisms that block polyspermy in various animal species. In sea urchins, for example, fertilization results in the erection of a physical barrier to sperm entry. As **Figure 22.7a** shows, the entry of a sperm causes calcium ions (Ca^{2+}) to be released from storage areas inside the egg. A wave of ions starts at the point of sperm entry and propagates throughout the egg. In response to this dramatic increase in Ca^{2+} concentration, a series of events occurs in the egg. For example, the cortical granules located just inside the egg cell's plasma membrane respond to the Ca^{2+} signal by fusing with the membrane and releasing their contents to the exterior. The contents include proteases that digest the exterior-facing fragment of the egg-cell receptor for sperm, which prevents any new sperm from binding to the egg surface. In addition, ions and other compounds released by the cortical granules accumulate between the egg cell's plasma membrane and the vitelline envelope. Because these solutes are at high concentration, they cause water to flow by osmosis into the space between the plasma membrane and vitelline envelope (see Chapter 6). The influx of water causes the envelope matrix to lift away from the cell and form a **fertilization envelope** (**Figure 22.7b**). The fertilization envelope, in turn, keeps additional sperm from contacting the egg's plasma membrane.

To summarize, a century of research on sea urchin fertilization has illuminated one of the most important cell-to-cell interactions in nature. The interaction between sperm and egg

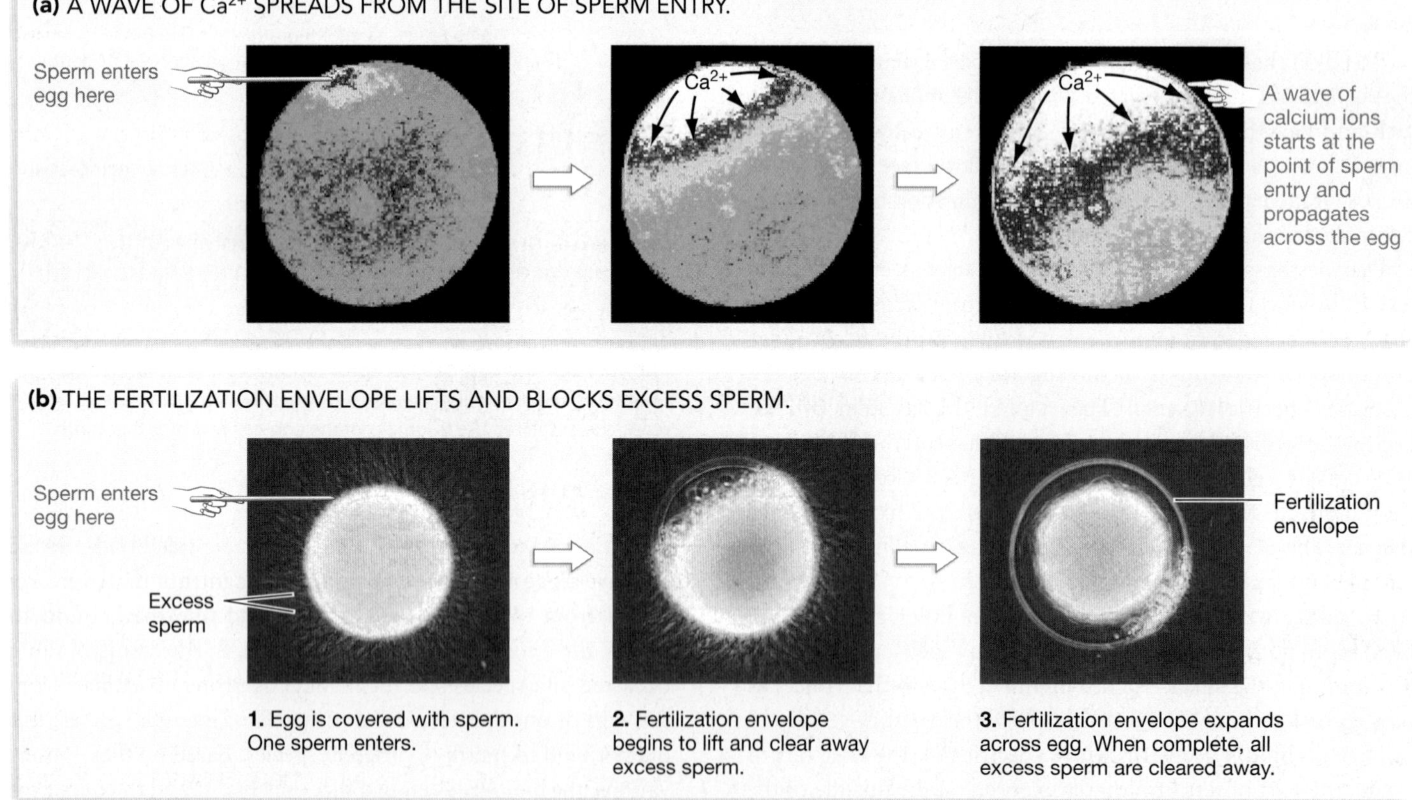

FIGURE 22.7 A Physical Barrier Erected after Fertilization Prevents Polyspermy. (a) During fertilization, a wave of Ca^{2+} begins at the point of sperm entry and spreads throughout the egg. The white dots are from a reagent that reacts with calcium ions. **(b)** In response to increased Ca^{2+} concentrations, a fertilization envelope rises and clears away excess sperm.

triggers a series of remarkable events, including the acrosomal reaction and mechanisms for blocking polyspermy.

Fertilization in Mammals

Although sea urchins have acted as a productive model system, an important question remains: How relevant are these findings to other species? In particular, how similar is sea urchin fertilization to events in humans and other mammals?

The most obvious difference between sea urchin and mammal fertilization is that the process occurs internally in mammals, not externally as in sea urchins. As a result, most female mammals actively choose a mate prior to the sperm-egg interaction. This simple difference means that a species-specific interaction between sperm and egg is much less important in mammals and other species with internal fertilization than it is in sea urchins and other species with external fertilization.

Internal fertilization also makes mammalian fertilization much more difficult to study than fertilization in sea urchins and similar species. With the advent of in vitro fertilization (IVF) methods, however, biologists have finally acquired the ability to study mammalian fertilization under laboratory conditions. What have they learned?

In humans and other mammals, the acrosomal reaction occurs after the sperm's head reaches the equivalent of the sea urchin vitelline envelope—the gelatinous extracellular matrix known as the **zona pellucida**. The enzymes released from the acrosome digest the glycoproteins that make up this protective layer. As a result, the sperm's head is able to reach the egg cell's plasma membrane and fuse with it.

Although as yet there is no convincing evidence for a bindin-like protein on the sperm head in mammals, molecules around and in the zona pellucida interact extensively with sperm. To search for these molecules in mouse eggs, researchers analyzed the three glycoproteins found in the zona pellucida. They found that one of the three glycoproteins binds to the heads of sperm. The interaction between the egg-cell glycoprotein and molecules on the surface of the sperm head causes sperm cells to stick to the zona pellucida. The interaction also triggers the acrosomal reaction.

Because male mammals typically introduce tens of millions of sperm during a single mating, sea urchin and mammal eggs share the problem of preventing polyspermy. Mammal eggs do not produce a fertilization envelope, but enzymes that are released from cortical granules perform a function similar to the sea urchin enzymes that destroy the egg-cell receptor for sperm. More specifically, the mammalian enzymes modify the egg cell glycoprotein in a way that releases sperm bound to the zona pellucida and prevents binding by additional sperm.

Once the sperm nucleus enters the egg, the two haploid nuclei fuse to form a diploid zygote. Development can begin.

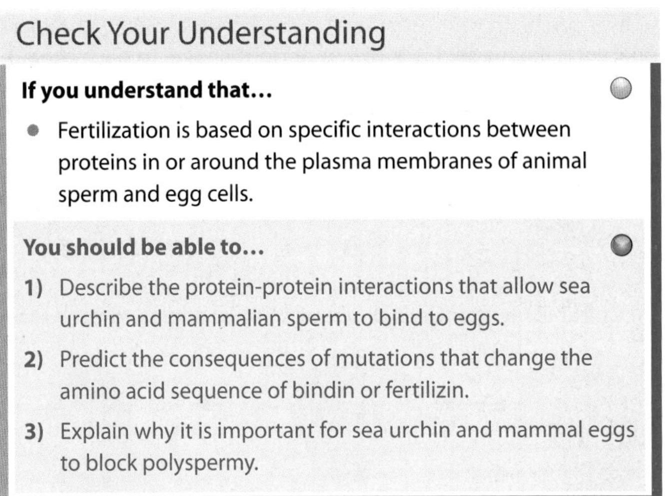
Check Your Understanding

If you understand that...

- Fertilization is based on specific interactions between proteins in or around the plasma membranes of animal sperm and egg cells.

You should be able to...

1) Describe the protein-protein interactions that allow sea urchin and mammalian sperm to bind to eggs.
2) Predict the consequences of mutations that change the amino acid sequence of bindin or fertilizin.
3) Explain why it is important for sea urchin and mammal eggs to block polyspermy.

22.3 Cleavage

Cleavage refers to the rapid cell divisions that take place in animals after fertilization. Cleavage is the first step in **embryogenesis**—the process by which a single-celled zygote becomes a multicellular embryo.

In most animals, cleavage partitions the egg cytoplasm without additional cell growth taking place. The zygote simply divides into two cells, then four cells, then eight, and so on, without concurrent growth overall. The process is fast—the fastest cell divisions recorded over an individual's lifetime. In frogs, cleavage produces 37,000 cells in under two days. When cleavage occurs in fruit flies, mitosis can occur every 10 minutes; about 50,000 cells are produced in half a day. The key point about cleavage is that the cytoplasm present in the egg is divided into a larger and larger number of smaller and smaller daughter cells.

The cells that are created by cleavage divisions are called **blastomeres** (literally, "bud-part"). When cleavage is complete, the embryo consists of a mass of blastomere cells called a **blastula** ("little-sprout, bud").

Variation in Cleavage Patterns

The principles of cleavage are the same in all animals: The egg cytoplasm is divided to form many cells that constitute a blastula. As **Figure 22.8** shows, however, the exact pattern of cleavage varies widely among species. In groups that undergo radial cleavage, cells divide at right angles to each other so that they form tiers, as shown in Figure 22.8a. In species that have spiral cleavage, cells divide at oblique angles to each other so that they pile up as shown in Figure 22.8b. (Notice that in these two parts of the figure, a part of the original egg cytoplasm is highlighted in color. This is done to illustrate a key point about cleavage: Certain parts of the egg cytoplasm end up in certain blastomeres.)

(a) Radial cleavage: Cells divide at right angles to each other.

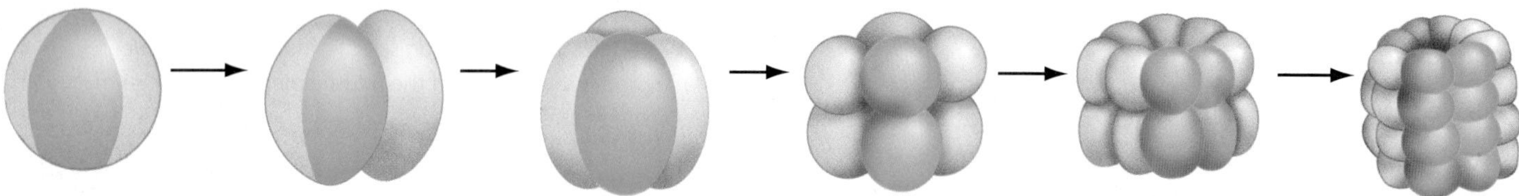

(b) Spiral cleavage: Cells divide at oblique angles to each other.

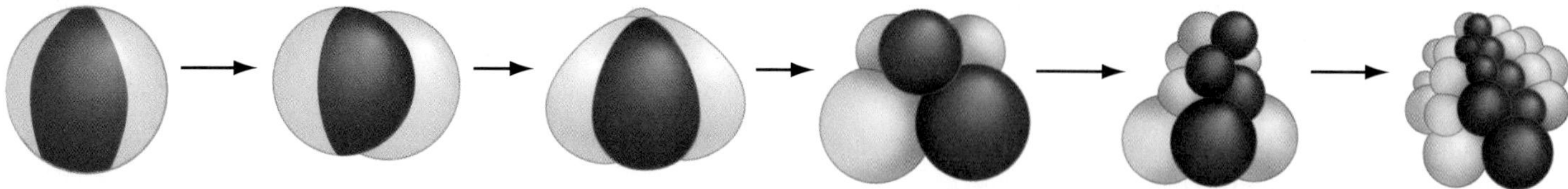

(c) Discoidal cleavage: A disk of small cells forms on top of a large cell containing the yolk.

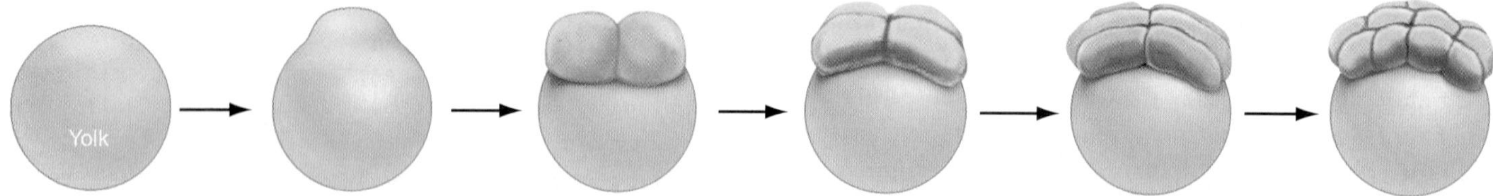

(d) Superficial cleavage: Mitosis is not immediately followed by cytokinesis, resulting in a multinucleate embryo.

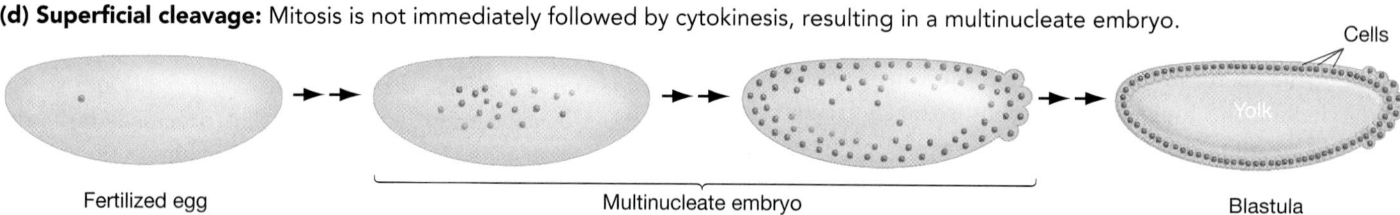

FIGURE 22.8 Cleavage Patterns Vary Widely among Species. During cleavage, cells divide but growth does not occur. Just some of the known cleavage patterns are shown here. **(a)** Radial cleavage is typical in echinoderms such as sea urchins. **(b)** Spiral cleavage is typical in segmented worms and mollusks. **(c)** Discoidal cleavage occurs in fishes, lizards, birds, and other species with extremely large amounts of yolk. **(d)** Superficial cleavage occurs in insects and produces multinucleate embryos.

The radial and spiral patterns are not the only variations on the theme of cleavage, however. For example, birds, fish, and other species have eggs with large, membrane-bound structures filled with yolk. In these animals, cleavage does not split the egg completely. Instead, it produces a mound of cells around the yolk or on top of it, as shown in Figure 22.8c. This pattern is called discoidal cleavage. Fruit flies and other insects have another distinctive cleavage pattern, called superficial cleavage. As Figure 22.8d shows, mitosis is not followed by cytokinesis early in cleavage. Instead, a large number of nuclei become scattered through the cytoplasm. Later, the nuclei migrate to the periphery of the embryo and plasma membranes form around them. **Box 22.1** analyzes the events that occur during cleavage in humans.

Although the cleavage patterns in Figure 22.8c are logical to observe in eggs with large, dense yolk deposits—which cannot be divided up efficiently during cytokinesis—the adaptive significance of the radial, spiral, and insect cleavage patterns is not well understood. Biologists still do not know if the variations in cleavage patterns affect fitness in any way or help individuals adapt to different environments—even though most of the patterns were described well over 100 years ago.

What Role Do Cytoplasmic Determinants Play?

The essence of cleavage is that the egg cytoplasm gets divided into many daughter cells. The key to understanding the process, then, is not to memorize the differences between spiral and radial and other cleavage patterns, but to analyze what is happening to the egg cytoplasm during all types of cleavage.

To drive this point home, think back to the Bicoid protein introduced in Chapter 21. Bicoid is a transcription factor

BOX 22.1 What Happens during Cleavage in Humans?

In humans, a special type of cleavage called rotational cleavage occurs when the fertilized egg is in a part of the female reproductive tract called the fallopian tube, or oviduct (**Figure 22.9**). The **fallopian tube, (oviduct)** is a structure that connects reproductive organs called the ovary and the uterus. The **ovary** is the organ in which the egg matures, and the **uterus** is the organ in which the embryo develops. Fertilization occurs near the ovary, and cleavage occurs as the embryo travels down the length of the fallopian tube toward the uterus. Cleavage results in a specialized type of blastula called a **blastocyst** ("sprout-bag"), which has two major populations of cells. The exterior of the blastocyst is a thin-walled, hollow structure called the **trophoblast** ("feeding-sprout"). The interior of the trophoblast contains a cluster of cells called the **inner cell mass** (**ICM**). After arriving at the uterus, the blastocyst implants into the uterine wall and a structure called the placenta begins to form. The **placenta** is an organ derived from a mixture of maternal cells and trophoblast cells. It allows nutrients and wastes to be exchanged between the mother's blood and the embryo's blood. The ICM, in contrast, contains the cells that undergo gastrulation and develop into the embryo. Later events in human pregnancy and birth are introduced in Chapter 48.

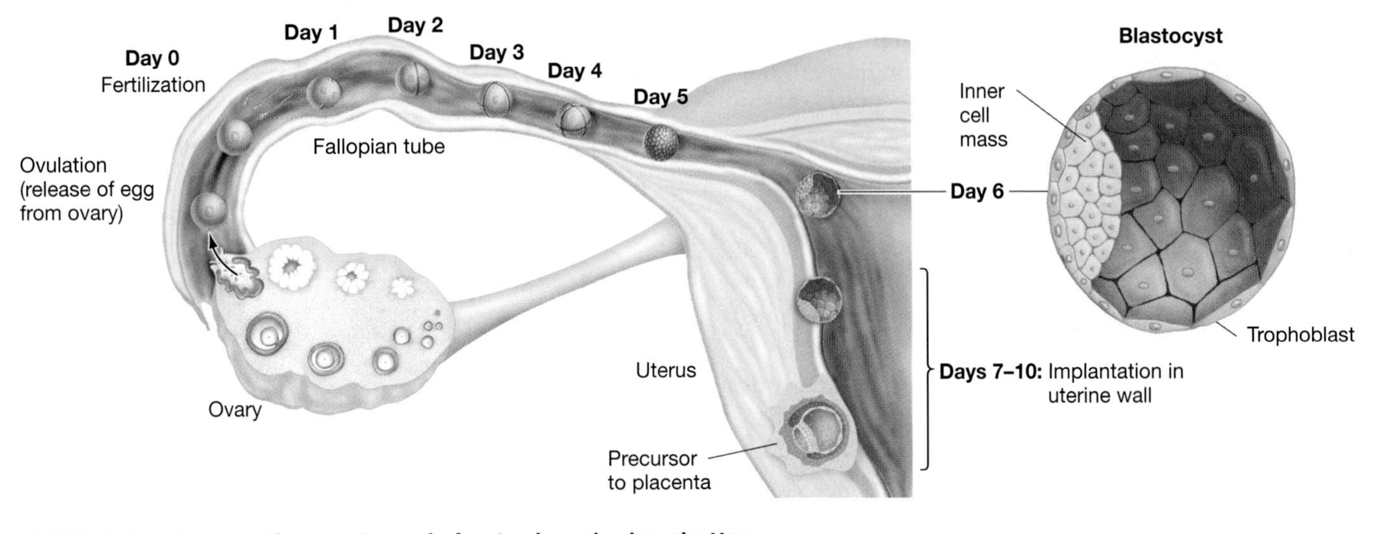

FIGURE 22.9 In Humans, Cleavage Occurs before Implantation into the Uterus.

found in an anterior-posterior concentration gradient. When the nuclei shown in Figure 22.8d form in the early stages of fruit-fly cleavage, only those in the anterior end of the egg are exposed to a high concentration of Bicoid. To extend this observation, consider what would happen if the cytoplasm that is highlighted in color in Figures 22.8a and b contained a key signal or transcription factor—a master regulator analogous to Bicoid. Only the colored cells in the figure would have this molecule.

A molecule that exists in eggs and helps direct early development is called a cytoplasmic determinant. *Bicoid* mRNA is a cytoplasmic determinant; the color highlighting in Figure 22.8 represents a cytoplasmic determinant. These examples underscore a key point: Cytoplasmic determinants are found in specific locations within the egg cytoplasm, so they end up in specific populations of blastomeres. As a result, only certain regulatory gene cascades are triggered in certain blastomeres. By dividing the egg cytoplasm to precisely distribute cytoplasmic determinants to certain cells, cleavage initiates the step-by-step process that, in combination with signals received from other cells, results in the differentiation of cells.

Check Your Understanding

If you understand that...

- Cleavage is a series of rapid cell divisions that result in formation of a blastula.
- During cleavage, the egg cytoplasm is divided.
- Because certain regulatory signals and transcription factors are found at specific locations in the egg cytoplasm of most species, different blastomeres end up with different sets of cytoplasmic determinants.

You should be able to...

Explain how the asymmetric distribution of cytoplasmic determinants in blastomeres affects the eventual differentiation of muscle cells, nerve cells, and other cell types.

22.4 Gastrulation

As cleavage nears completion, cell division slows. During the next phase of the developmental sequence, cell proliferation stops being the most important developmental process. Instead, cell movement becomes primary. During **gastrulation**, extensive and highly organized cell movements radically rearrange the embryonic cells into a structure called the gastrula.

Research on this phase of development started in the 1920s, with efforts to document the movement of individual cells during newt and frog gastrulation. These early experiments were based on soaking tiny blocks of agar (a gelatinous compound) with a nontoxic dye. The dyed blocks were then pressed against the surface of blastula-stage embryos, so that a small number of blastomeres became marked with dye. By allowing marked embryos to develop and then examining them at intervals during gastrulation, researchers were able to follow the movement of cells.

Subsequent research has shown that the pattern of gastrulation varies among species of animals almost as much as cleavage patterns do. But like cleavage, the variation in the details is underlain by a common outcome: Gastrulation results in the formation of embryonic tissue layers. A **tissue** is an integrated set of cells that function as a unit. Most adult animals have a wide array of tissues, including muscle tissue, connective tissue, and nerve tissue. But most early animal embryos have three primary tissue layers: (1) ectoderm ("outside skin"), (2) mesoderm ("middle skin"), and (3) endoderm ("inner skin"). These embryonic tissues are called **germ layers** because they give rise to the organs and tissues of the adult. As **Figure 22.10** shows, embryonic **ectoderm** forms the outer covering of the adult body and the nervous system; **mesoderm** gives rise to muscle, most internal organs, and connective tissues such as bone and cartilage; and **endoderm** produces the lining of the digestive tract or gut, along with some of the associated organs.

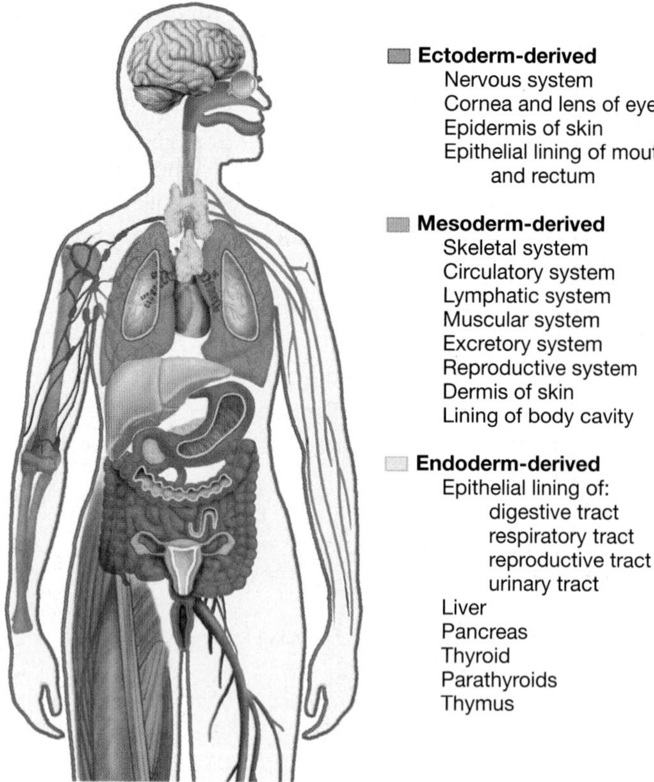

FIGURE 22.10 The Three Embryonic Tissues Give Rise to Different Adult Tissues and Organs.

● **QUESTION** In a gastrula, mesoderm is sandwiched between ectoderm on the outside of the embryo and endoderm on the inside. Relate this observation to the positions of adult tissues and organs derived from mesoderm, ectoderm, and endoderm.

Figure 22.11 shows how the cell movements of gastrulation result in the formation of the three embryonic tissue layers in a frog embryo. The drawings are based on experiments with dyed blastomeres, and they identify blastomeres according to the type of embryonic tissue they will become. Cells that will become ectoderm are shown in blue; cells destined to form mesoderm are pink, and cells that will form endoderm are colored yellow. These cell fates are determined by cell-cell signals, transcription factors, and cytoplasmic determinants in the blastomeres.

As you start to examine the steps of this figure, notice that the blastula contains a fluid-filled interior space called the **blastocoel** (step 1). In many animal species, the blastula is hollow or partially hollow like this. As gastrulation begins, cells begin moving into this space through an invagination—similar to the indentation that forms when you push a finger into a balloon (step 2). In frogs this invagination is slit-like and eventually forms the **blastopore**. Cells from the periphery continue to move to the interior of the embryo through the blastopore, forming the initial part of the gut or digestive tract (step 3). When gastrulation is complete, the ectoderm, mesoderm, and endoderm cells are arranged in three distinct layers (step 4).

The other major outcome of gastrulation is that the major body axes become visible. In frogs, for example, the blastopore becomes the anus, and the region just above the blastopore (as drawn in Figure 22.11) becomes the dorsal or back side of the embryo. In this way, the anterior-posterior and dorsal-ventral axes of the body become apparent. These axes were partially determined early in development, by Bicoid-like cytoplasmic determinants that were stored in the egg and partitioned into blastomeres during cleavage.

As a comparison with frog gastrulation, the series of micrographs in **Figure 22.12** shows the sequence of events during gastrulation in a sea urchin. ● If you understand the major concepts of gastrulation, you should be able to (1) describe the cell movements taking place in steps 2 and 3 of Figure 22.12, (2) label the three embryonic tissue layers and

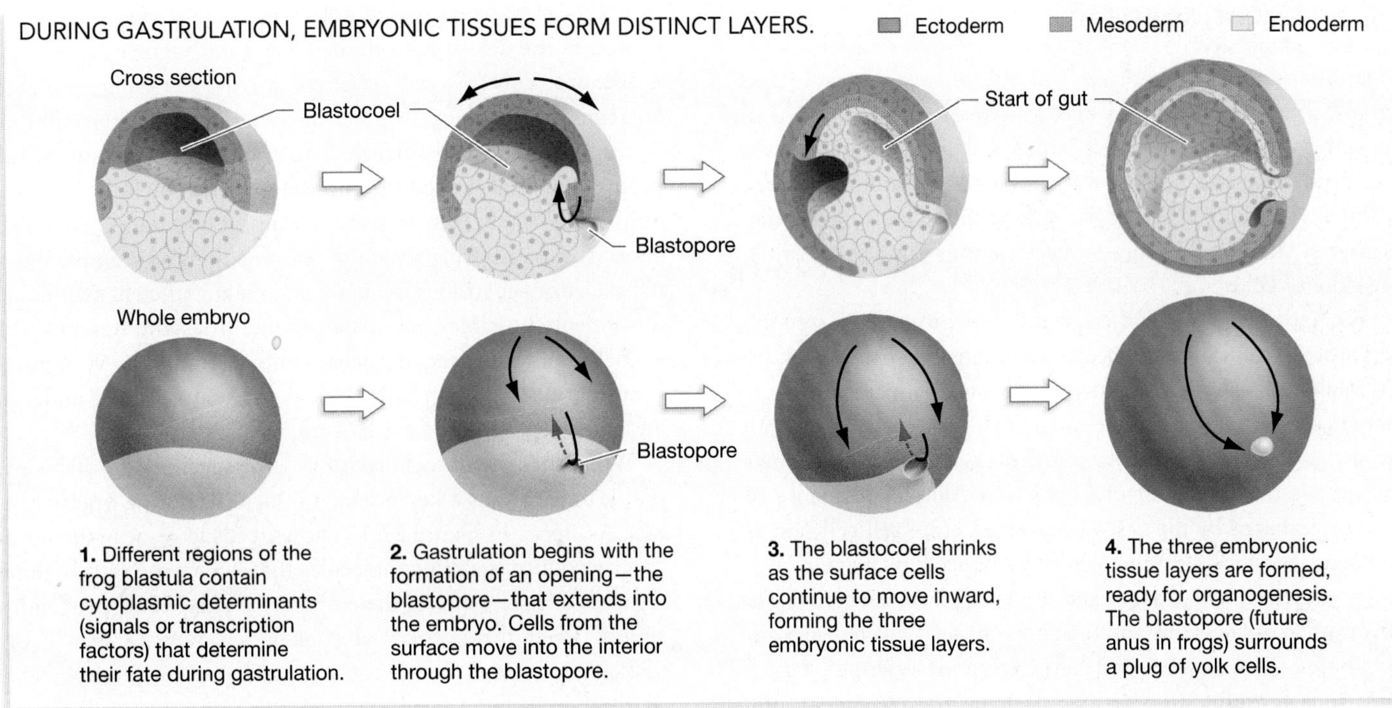

FIGURE 22.11 Precise Cell Movements Occur during Gastrulation.

EXERCISE In frogs, the initial opening of the blastopore is on what will become the dorsal side of the larva and adult. The blastopore becomes the anus of the larva and adult. On the last drawing, mark where the head, tail, back, and belly will develop.

the anterior-posterior axis visible at the end of gastrulation (step 6), and (3) label a type of cell movement that occurs in sea urchin gastrulation but not frog gastrulation—the migration of individual cells from the exterior to the interior of the embryo.

Current research on gastrulation is focused on understanding how cells move in such an organized way—on the mechanism responsible for the movement itself as well as how cells navigate. But like the embryo itself, we'll move on to the next phase in the developmental sequence—forming tissues, organs, and other basic structures as the body takes shape.

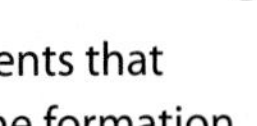

(MB) **Web Animation** at www.masteringbio.com
Early Stages of Animal Development

Check Your Understanding

If you understand that...

- Gastrulation consists of extensive cell movements that reorganize the embryonic cells and result in the formation of embryonic germ layers.
- Once gastrulation is complete, the major body axes are visible and organs and other structures can begin to form.

You should be able to...

1) Name the three embryonic germ layers.
2) Describe the tissues and organs that each of these layers gives rise to in the adult.

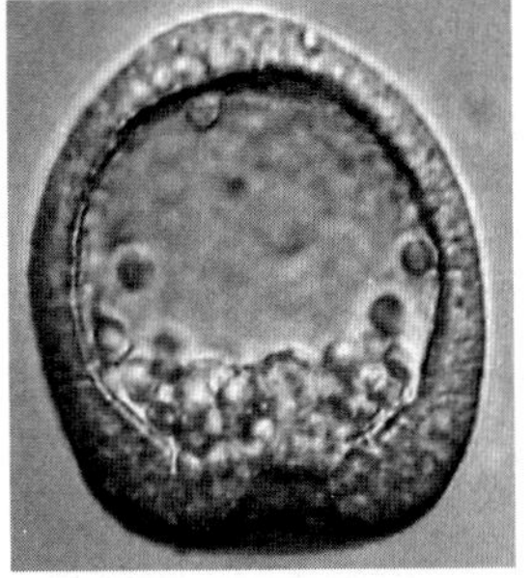
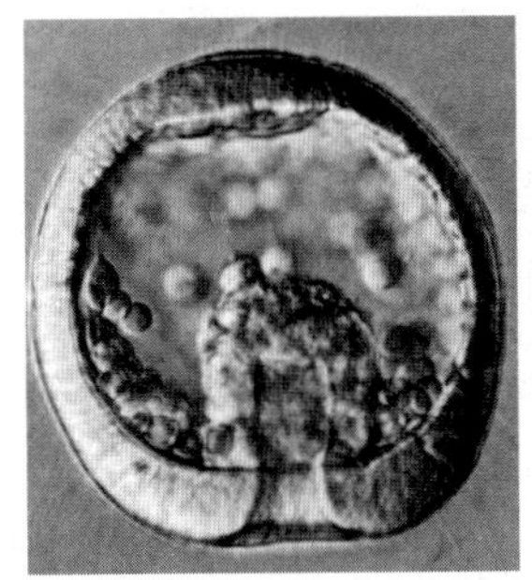

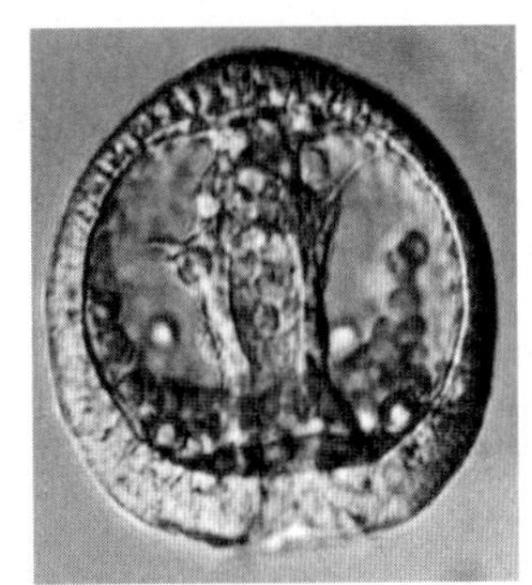
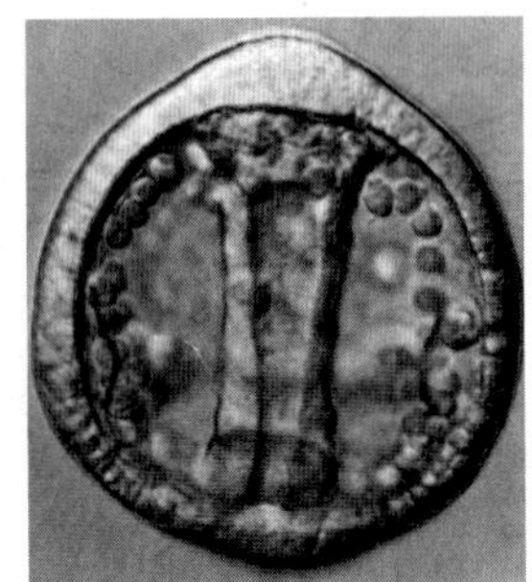

FIGURE 22.12 The Sequence of Events during Gastrulation Varies among Species. These micrographs show stages in sea urchin gastrulation. Notice that some cells break off from the embryo's outer cell layer and migrate into the interior individually.

22.5 Organogenesis

Organogenesis ("organ-origin") is the process of tissue and organ formation that begins once gastrulation is complete and the embryonic germ layers are in place. During organogenesis, cells proliferate and become differentiated—meaning that they become a specialized cell type. Differentiated cells have a distinctive structure and function because they express a distinctive suite of genes.

As Chapter 21 noted, differentiation is a progressive, step-by-step process. Cells become specified to a fate long before they are differentiated—meaning, before they begin producing products that are specific to a certain cell type. This observation is logical, given that an irreversible commitment to become a particular cell type is the end point of a long and complex sequence of events, mediated by the cascades of interacting cell-cell signals and regulatory transcription factors introduced in Chapter 21.

To explore how organogenesis takes place, let's consider how muscle tissue forms and then how muscle cells become differentiated. What causes undifferentiated mesodermal cells to form muscle tissue and begin producing the muscle-specific proteins that make breathing, walking, and turning the pages of textbooks possible?

Organizing Mesoderm into Somites: Precursors of Muscle, Skeleton, and Skin

Figure 22.13 illustrates some of the key events that occur as organogenesis begins in frogs, humans, and other vertebrates. In this cross section of a frog embryo, the dorsal (back) side of the individual is up and the ventral (belly) side is at the bottom. You should already be familiar with the ectoderm, mesoderm, and endoderm layers shown in step 1, along with the early gut cavity. But you should also note that a new structure has appeared in the dorsal mesoderm—a rod-like element called the **notochord**. This structure is shown in cross section in the figure and runs the length of the anterior-posterior axis of the embryo, just under the dorsal surface. The notochord is unique to the group of animals called the **chordates**, which includes humans and other vertebrates. In some species of chordates, the notochord is a long-lasting structure that functions as a simple internal skeleton—it stiffens the body and makes efficient swimming movements possible. But in species like frogs and humans, the notochord is transient. It appears only in embryos. As organogenesis proceeds, many of the cells in the notochord undergo apoptosis—programmed cell death.

What does the notochord do in these species? The answer is that it functions as a key organizing element during organogenesis. As step 2 in Figure 22.13 shows, cells in or near the notochord produce signaling molecules that induce the ectoderm on the dorsal (back) side of the embryo to fold. More specifically, signals from the notochord trigger massive changes in cytoskeletal elements inside each of the dorsal ectodermal cells. As the cytoskeleton is reorganized, the ectodermal cells extend in length and then constrict at their dorsal end and expand at their ventral end. The constriction above and expansion below makes the sheet of cells fold upward. As folding continues, a structure called the **neural tube** forms (step 3). The neural tube is the precursor of the brain and spinal cord. Besides producing signals that direct formation of the neural tube, the notochord furnishes physical support as the ectodermal cells fold into their final configuration.

Once the neural tube forms, organogenesis continues when nearby mesodermal cells become organized into blocks of tissue called **somites**. Somites form on both sides of the neural tube, down the length of the body (**Figure 22.14**). Somite formation is

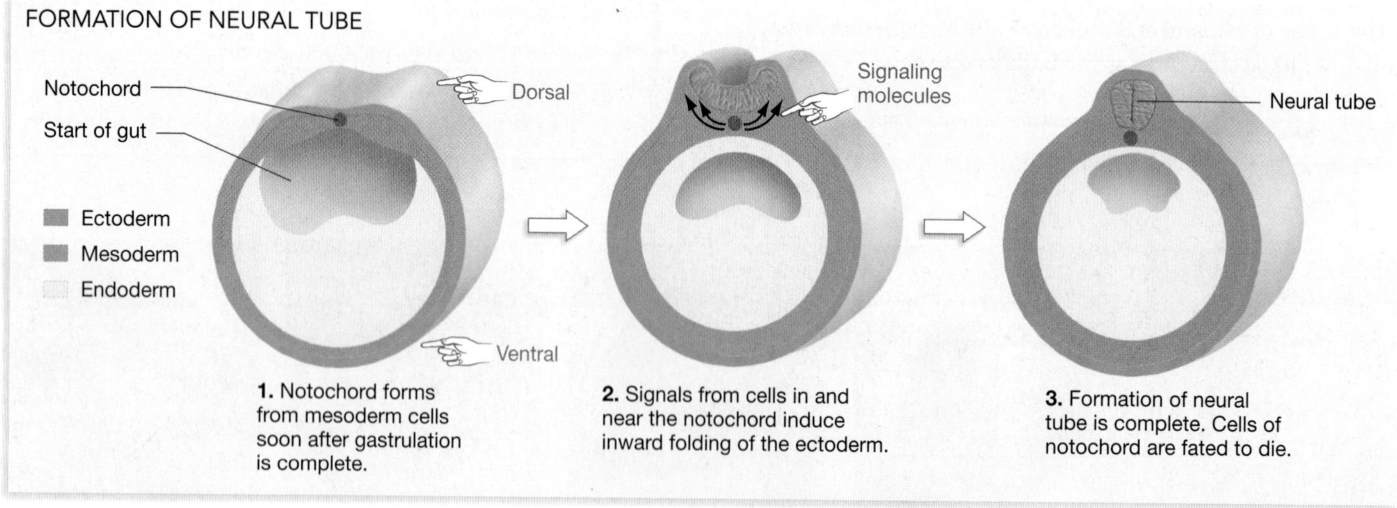

FIGURE 22.13 The Notochord and Neural Tube Form Early in Organogenesis. In vertebrates, the notochord forms from mesoderm cells soon after gastrulation is complete. Molecules produced in or near the notochord induce the formation of the neural tube and other structures along the dorsal (back) side of the embryo.

(a) Surface view of somites

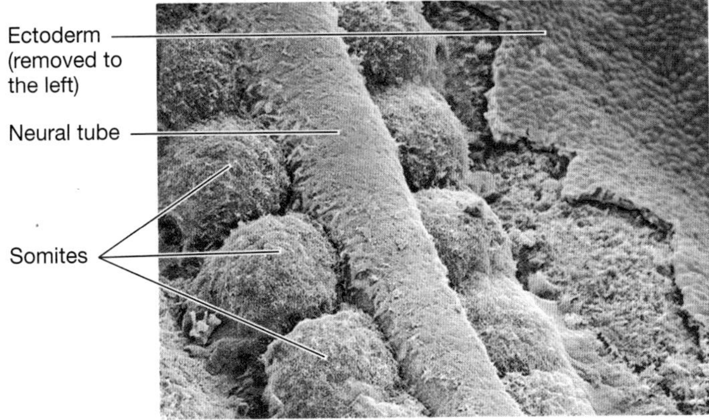

(b) Cross section of somites

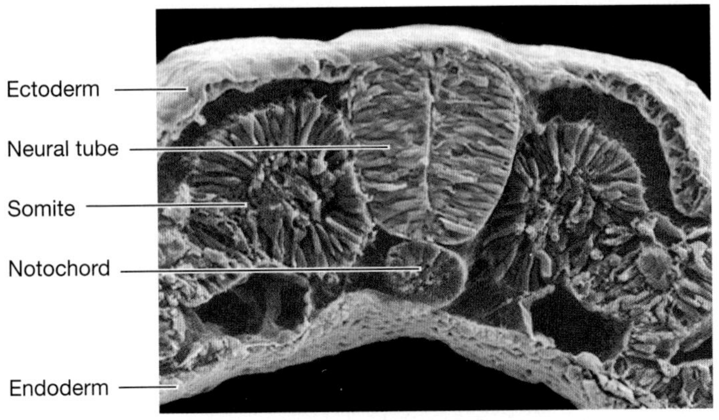

FIGURE 22.14 Somites Form on Both Sides of the Neural Tube. Somites are made of mesodermal cells.

a response to changes in the cell adhesion molecules that keep mesodermal cells attached to each other (see Chapter 8).

Somites are transient structures, just as the notochord is. But by marking cells and following them over time, researchers have found that the mesodermal cells in a somite are destined to form a variety of structures. As development proceeds, somite cells do not die as notochord cells do. Instead, they break away as distinct populations that migrate to their final location in the developing embryo and continue to proliferate. These cell movements are critical to organogenesis. Cells from somites build the vertebrae and ribs, the deeper layers of the skin that covers the back, and the muscles of the back, body wall, and limbs. As **Figure 22.15** shows, the destiny of a somite cell depends on its position within the somite at a particular time during organogenesis.

By transplanting cells from one location to another early after somite formation, researchers have found that initially any cell in a somite can become any of the somite-derived elements of the body. For example, biologists transplanted cells from various locations to the side of the somite farthest from the neural tube. All the transplants eventually became committed to form limb muscles (**Figure 22.16**). Cells transplanted later in development, however, failed to become the type of cell associated with their new position. Instead, the transplanted cells differentiated into the cell types they would normally form in their original position, even though they were now in a new and inappropriate location.

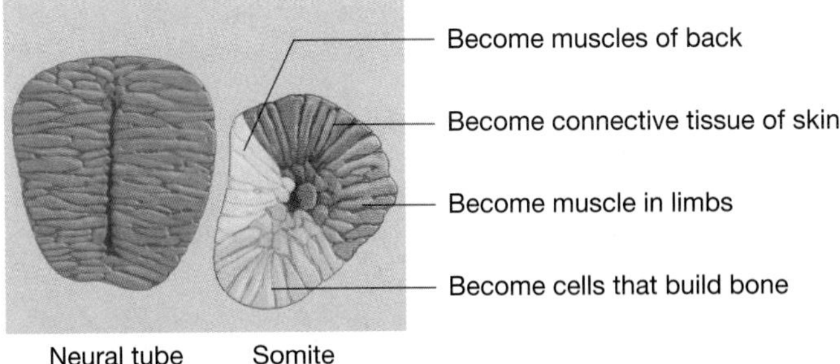

FIGURE 22.15 Cells in the Somite Give Rise to an Array of Tissues. Each somite eventually breaks up into distinct populations of cells, each with a different fate. Within a somite, the position of a cell determines its fate.

As the somite matures, then, somite cells become irreversibly determined—meaning that they will eventually differentiate into a specific cell type—based on their location within the somite. Experiments like this illustrate the general observation that a cell's fate becomes fixed in a step-by-step process.

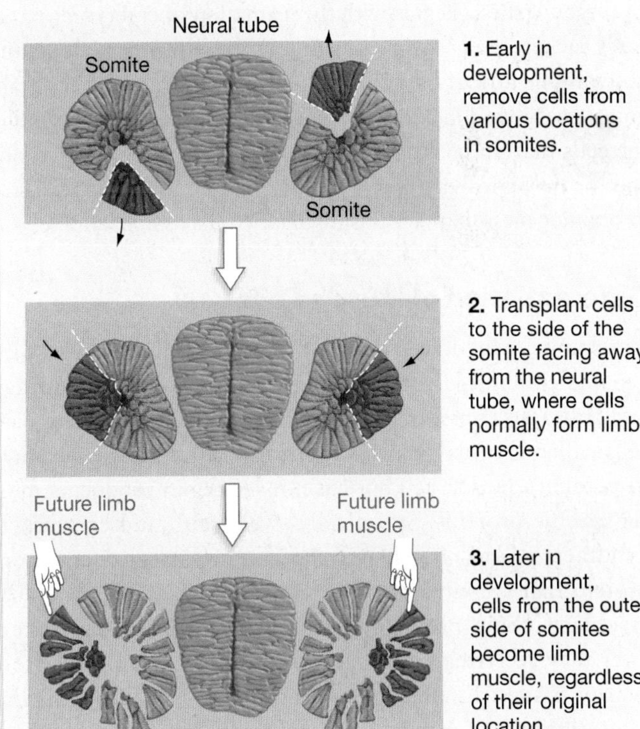

FIGURE 22.16 Experimental Evidence for Progressive Differentiation of Mesodermal Cells in Somites. By transplanting cells early in somite formation to new locations in this structure, researchers showed that the cells are not determined initially.

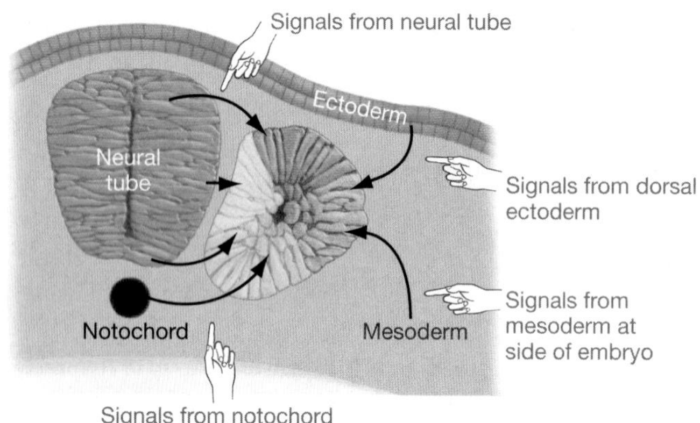

FIGURE 22.17 Somite Cells Differentiate in Response to Signals from Nearby Tissues. Somite cells become determined in response to signals from the neural tube, notochord, and nearby ectoderm and mesoderm.

Recent studies of this process, called **determination**, have shown that distinct populations of cells in a somite differentiate in response to distinct local combinations of cell-cell signals. **Figure 22.17** illustrates how these signals diffuse away from cells in the notochord, the neural tube, and nearby ectoderm and mesoderm to act on specific populations of target cells in the somite. These unique combinations of cell-cell signals direct the movement of somite cells when the structure breaks up and trigger the production of transcription factors required for the expression of muscle-, bone-, or skin-specific proteins.

Organogenesis begins with the formation of embryonic structures such as the notochord, neural tube, and somites. It continues with the formation of bones, muscles, skin, and other organs and tissues found in a larval or juvenile individual. At this point, the cells that make up these structures begin expressing tissue-specific proteins. Let's take a closer look at how this crucial step in organogenesis happens, using muscle cells as an example.

Differentiation of Muscle Cells

In a somite, why do cells located farthest from the neural tube become committed to producing muscle? Harold Weintraub and colleagues answered this question by experimenting with cells called myoblasts. A myoblast is a cell that is determined to become a muscle cell but has not yet begun producing muscle-specific proteins. Specifically, Weintraub and co-workers hypothesized that myoblasts contain at least one regulatory protein that commits them to their fate. Their idea was that myoblasts begin producing this muscle-determining protein after they receive an appropriate set of cell-cell signals from nearby tissues. In effect, they were looking for a regulatory transcription factor in myoblasts that dictates, "I will become a muscle cell."

Figure 22.18 outlines how the biologists went about searching for this hypothetical protein. They began by isolating mRNAs from myoblasts and using reverse transcriptase to con-

Experiment

Question: Why do cells on the outside of somites become committed to producing muscle in response to the signals they receive?

Hypothesis: Cell-cell signals trigger production of a regulatory protein (or proteins) that commits myoblasts to their fate.

Null hypothesis: Myoblasts do not contain a regulatory protein that commits them to their fate.

Experimental setup:

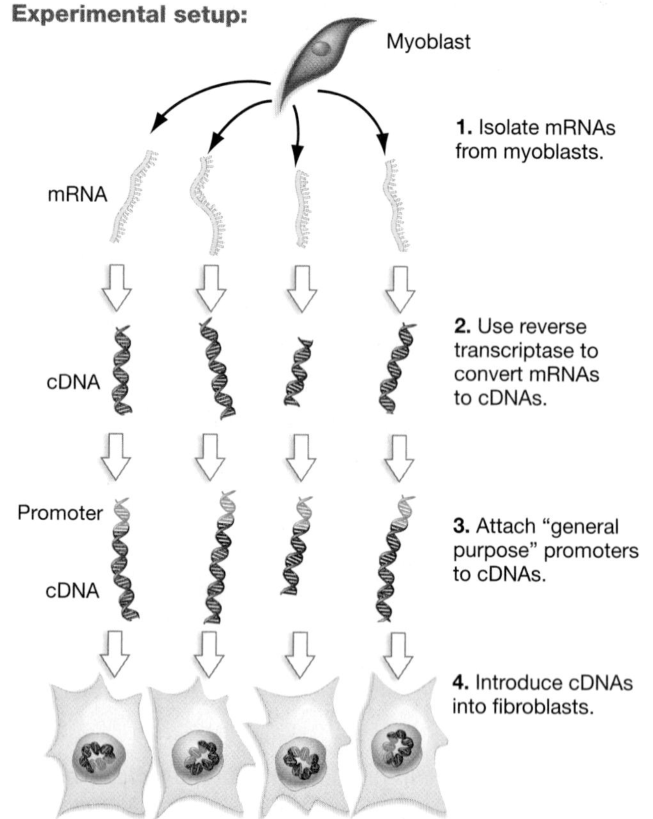

Prediction: One of the myoblast-derived cDNAs will convert fibroblasts into cells that produce muscle-specific proteins.

Prediction of null hypothesis: None of the myoblast-derived cDNAs will convert fibroblasts into cells that produce muscle-specific proteins.

Results:

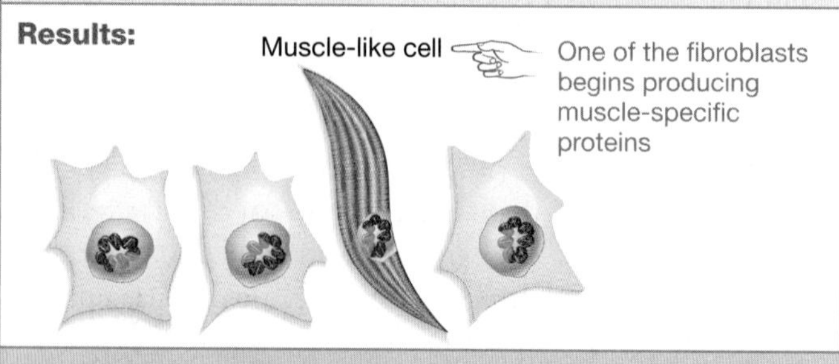

Conclusion: Cells on the outside of somites contain a regulatory protein (later called MyoD) that commits them to differentiate into muscle.

FIGURE 22.18 A Gene That Causes Muscle-Cell Differentiation.

QUESTION Why did the researchers have to attach a "general purpose" promoter to the cDNAs?

vert the mRNAs to cDNAs (steps 1 and 2). Because myoblasts transcribe genes required for a muscle cell to function, these cDNAs represented muscle-specific genes. Then the biologists attached a type of promoter to the cDNAs that would ensure the genes' expression in any type of cell (step 3). Finally, they introduced the recombinant genes into non-muscle cells called fibroblasts and monitored the development of the transformed cells (step 4). Just as predicted, one of the myoblast-derived cDNAs converted fibroblasts into muscle-like cells. Follow-up experiments showed that the same gene could convert pigment cells, nerve cells, fat cells, and liver cells into cells that produced muscle-specific proteins.

Weintraub's group called the protein product of this gene **MyoD**, for *myo*blast *d*etermination. Follow-up work showed that the *MyoD* gene encodes a regulatory transcription factor and that the MyoD protein binds to enhancer elements located upstream of muscle-specific genes. In addition, the MyoD protein activates expression of the *MyoD* gene. This was a key observation because it meant that once *MyoD* is turned on, it triggers its own expression—meaning that the gene continues to be transcribed. Other researchers have found that genes closely related to *MyoD* are also required for the differentiation of muscle cells.

To put this specific example into context, think back to the sequence of events occurring in early development and to the principles of development introduced in Chapter 21. To understand why the muscle cells in your bicep became muscle cells, you have to start with an egg inside your mother. After fertilization triggered the onset of cleavage, certain cells in the blastocyst began producing signals that triggered regulatory gene cascades—changes in gene expression that led specific cells to become mesoderm in your back. During gastrulation, these cells moved into position. Early in organogenesis, signals from the notochord and nearby cells induced the production of MyoD and other muscle-determining proteins in certain populations of cells from somites. In response, these target cells were committed to becoming muscle and moved into your upper arm as it formed. Later, the MyoD-containing cells began expressing muscle-specific proteins. All of these steps made it possible for you to move your arms after birth. The rest, as they say, is history.

Chapter Review

SUMMARY OF KEY CONCEPTS

In most animals, fertilization depends on specific interactions between proteins on the plasma membranes of sperm and egg.

Animal development begins with the formation of gametes in parents. Sperm cells contribute a haploid genome and a centriole to the embryo. Eggs, in contrast, contribute a large amount of cytoplasm to the embryo in addition to a haploid genome and a centriole. The egg cytoplasm usually contains cytoplasmic determinants and a large supply of nutrients.

Fertilization is the best known of all cell-cell interactions, primarily because of observational and experimental studies on sea urchin eggs and sperm. When a sea urchin sperm contacts the jelly layer surrounding the egg, enzymes released from the sperm digest the jelly layer and bindin proteins on the sperm head attach to sperm-binding receptors on the egg. Once the sperm nucleus enters the egg, a fertilization envelope lifts off the egg cell's plasma membrane and protects against multiple fertilizations.

You should be able to draw what the "lock and key" interaction between bindin and fertilizin might look like, and why interactions between specific amino acids in the two proteins would cause binding to occur.

The earliest cell divisions divide the fertilized egg into a mass of cells. The fate of these cells depends on the cytoplasmic determinants they contain, the signals they receive, or a combination of both.

Early embryonic development begins with cleavage—a series of cell divisions that divide the egg cytoplasm into a large number of cells. During cleavage, an array of signals and regulatory transcription factors are apportioned to different blastomeres. As a result, different blastomeres have different fates.

You should be able to describe how Bicoid mRNA is partitioned into cells during cleavage in a fruit-fly embryo.

Early in development, cells undergo massive, coordinated movements to form distinct tissue layers. Each of these layers gives rise to distinct groups of tissues and organs in the adult.

Once cleavage is complete, the embryo consists of a mass of cells. Studies with dyes and other tools have shown that embryonic cells undergo dramatic movements during gastrulation. These movements arrange the embryonic tissues into endoderm, mesoderm, and ectoderm layers and make the back-to-belly and head-to-tail axes of the body visible.

(MB) Web Animation at www.masteringbio.com
Early Stages of Animal Development

As development proceeds, the formation of specialized organs and other structures depends on cell-cell signals, cell proliferation, cell movements, and differentiation. Differentiation is complete when cells express tissue-specific proteins.

Tissues, organs, and other structures form during the developmental phase called organogenesis. Early in vertebrate organogenesis, cells in a transient structure called the notochord release cell-cell signals that induce the formation of a neural tube—precursor to the brain and spinal cord. Blocks of mesodermal cells, called somites, form next to the neural tube. In response to cell-cell signals from the notochord, neural tube, and other nearby tissues, cells in the somites move to new positions, proliferate, and begin expressing tissue-specific proteins.

You should be able to describe how the regulatory gene and signaling cascades described in Chapter 21 relate to the step-by-step process of differentiation described in this chapter.

QUESTIONS

Test Your Knowledge

1. How are the vitelline envelope of sea urchins and the zona pellucida of mammals similar?
   a. Both are gelatinous coats that protect the egg.
   b. Both hold cortical granules, which block polyspermy.
   c. Both hold stores of Ca^{2+}, which block polyspermy.
   d. Both are an extracellular matrix that sperm bind to and digest.
2. What happens during the acrosomal reaction?
   a. Bindin binds to the egg-cell receptor for sperm.
   b. The sperm and egg cells' plasma membranes fuse.
   c. Enzymes that digest the egg's jelly layer are released, and microfilaments in the tip of the sperm head polymerize to form a protrusion.
   d. Cortical granules fuse with the plasma membrane, triggering formation of the fertilization envelope.
3. What happens during cleavage?
   a. The neural tube—precursor of the spinal cord and brain—forms.
   b. Basal and apical cells—precursors of the suspensor and embryo, respectively—form.
   c. The fertilized egg divides without growth occurring, forming a mass of cells.
   d. Massive movements of cells make the primary body axes visible and organize the three embryonic tissues.
4. What happens during gastrulation?
   a. The neural tube—precursor of the spinal cord and brain—forms.
   b. Basal and apical cells—precursors of the suspensor and embryo, respectively—form.
   c. The fertilized egg divides without growth occurring, forming a ball of cells.
   d. Massive movements of cells make the primary body axes visible and organize the three embryonic tissues.
5. In animals, which adult tissues and organs are derived from ectoderm?
   a. lining of the digestive tract and associated organs
   b. blood, heart, kidney, bone, and muscle
   c. nerve cells and skin
   d. blastopore and blastocoel
6. During organogenesis in vertebrates, which of the following does *not* occur?
   a. establishment of the anterior-posterior axis
   b. differentiation of cells
   c. movement of cells into new positions
   d. extensive cell-cell signaling

Test Your Knowledge answers: 1.d; 2.c; 3.c; 4.d; 5.c; 6.a

Test Your Understanding

Answers are available at www.masteringbio.com

1. Draw a typical animal sperm and egg and label the major structures. List the contributions to offspring that are made by a sperm versus an egg.
2. Why is it logical to observe that sperm-egg interactions are extremely species specific in sea urchins but much less so in mammals?
3. Blastomeres look identical. Explain why they are not.
4. Describe three elements in the cascade of interacting cell-cell signals and regulatory transcription factors that lead to differentiation of muscle cells in human embryos.
5. Why are ectoderm, mesoderm, and endoderm called germ layers?
6. Give an example of how each of the four fundamental developmental processes—cell proliferation, cell movements, cell-cell signals, and differentiation—plays a role in animal development.

Applying Concepts to New Situations

Answers are available at www.masteringbio.com

1. At the level of the amino acid, explain why bindin on the sperm of the sea urchin *Strongylocentrotus purpuratus* attaches to the egg-receptor for sperm on eggs of *S. purpuratus*, but why bindin from other species of *Strongylocentrotus* cannot.
2. A molecule called α-amanitin inhibits translation. When researchers treat fruit-fly eggs with this compound, the early stages of cleavage proceed normally after fertilization. What can you conclude from this experiment?
3. In the marine organisms called sea squirts, eggs contain a yellow pigment. Blastomeres that contain this yellow pigment become muscle cells. State a hypothesis to explain this observation.
4. Would a molecule that bound to the glycoproteins in the human zona pellucida serve as an effective contraceptive? Explain why or why not.

wwww.masteringbio.com is also your resource for • Answers to text, table, and figure caption questions and exercises • Answers to *Check Your Understanding* boxes • Online study guides and quizzes • Additional study tools including the *E-Book for Biological* Science 3rd ed., textbook art, animations, and videos.

An Introduction to Plant Development

23

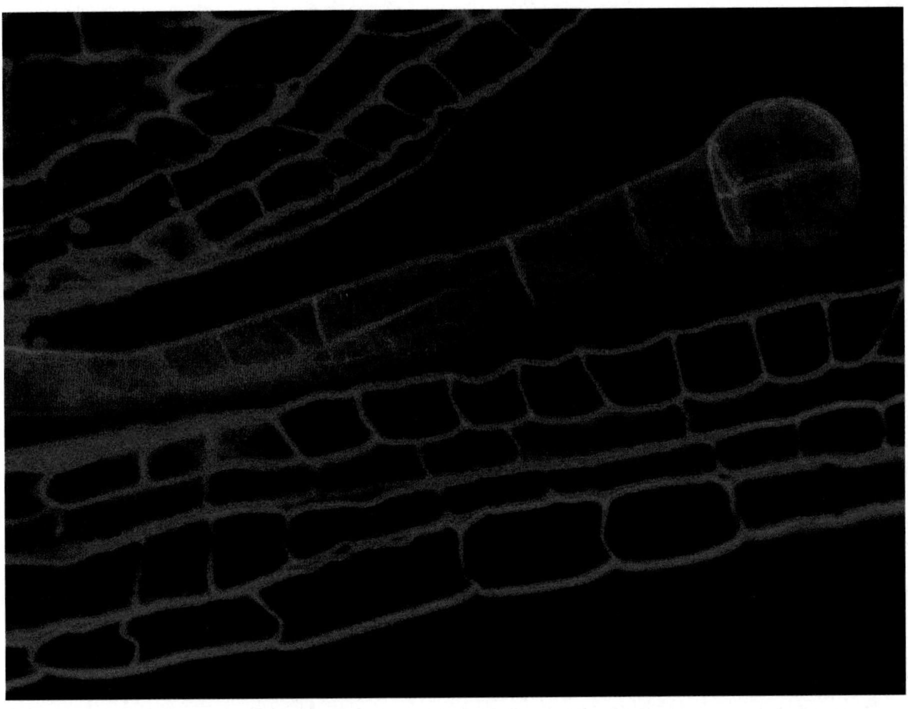

The red cells are an *Arabidopsis thaliana* embryo at the 8-cell stage; the purple cells are part of the seed coat. This chapter explores how plant embryos develop into adults that continue to grow throughout life and reproduce.

KEY CONCEPTS

- Several aspects of plant development contrast sharply with animals: Plants develop continuously, commit certain cells to gamete production late in development, and produce gametes by mitosis.
- In flowering plants, double fertilization results in the production of a zygote and a nutritive tissue that supports embryogenesis.
- Embryogenesis results in the formation of the major body axes and three types of embryonic tissue.
- Vegetative development is based on meristems where cell division occurs throughout life, producing cells that go on to differentiate.
- When a meristem is converted from vegetative to reproductive development, regulatory transcription factors are activated that control the position and identity of floral organs.

About the year 2700 B.C., perhaps less than a century before the first pyramids were constructed in ancient Egypt, a remarkable thing came to pass on a mountainside in southwestern North America. A sperm cell and an egg cell from a bristlecone pine tree fused to form a zygote. The offspring produced by that blessed event has been growing and developing ever since—for over 4700 years.

Unlike many animals, plants continue to grow and develop throughout their lives, whether that life lasts two weeks or thousands of years. As a dramatic example of this continuous development, the entire aboveground shoot system of many plant species dies each fall and grows back the next spring. One hallmark of plant development is that the essential developmental processes of cell proliferation and expansion, differentiation, and cell-cell interaction occur throughout life.

In addition, most plant cells retain the ability to de-differentiate—to stop producing a certain suite of tissue-specific proteins and begin producing proteins typical of a different type of tissue. Under the right circumstances, for example, a photosynthetic cell in a plant stem can de-differentiate and give rise to cells that function as storage cells in a root. But none of your muscle cells or nerve cells will ever give rise to liver cells. Certain aspects of animal development can be flexible, but animal development is rigid and inflexible in comparison with plants. This extraordinary flexibility is another hallmark of plant development.

The striking differences existing between plant and animal development should not be surprising, given their very different ways of life. Most animals move around to find food and mates while plants stand their ground, make their own food, and use

FIGURE 23.1 ***Arabidopsis thaliana*** **Is an Important Model Organism in Developmental Biology.**

● **QUESTION** What are the attributes of a model organism? In what sense is *Arabidopsis* a "model"?

water, wind, or insects to carry their gametes. In addition, multicellularity evolved independently in the two lineages. Chapter 8 reviewed evidence to back the claim that multicellularity has evolved many times during the history of life. That chapter also analyzed some consequences of multicellularity evolving independently in plants and animals, including striking differences in the chemical composition of plant versus animal extracellular matrices and the mechanisms of cell-cell contact and adhesion.

The goal of this chapter is to explore the sequence of events occurring as a plant develops from a fertilized egg into an adult that can reproduce, and that continues to grow and develop new tissues and organs throughout its life. The discussion in this chapter focuses on the flowering plants called eudicots (see Chapter 30), because they are the most species-rich and abundant group of plants. In many cases our model organism will be a small flowering plant called *Arabidopsis thaliana*, which qualifies as the best-studied of all plants (**Figure 23.1**). *Arabidopsis* has been a successful model organism in genetics and developmental biology because it is relatively easy to grow in the greenhouse, produces large numbers of offspring, and completes its entire life cycle in as little as six weeks. Even though their life span is short and their body size is small, individuals are complex enough to be interesting to study.

A plant's life story begins with gamete formation and fertilization (**Figure 23.2**). In *Arabidopsis* and other flowering plants, an egg is fertilized inside a protective, womb-like structure called the **ovule**. Development continues inside the ovule with **embryogenesis** (literally, "embryo-origin"). In many plant species embryogenesis ends with the maturation of the ovule into a **seed**—a structure that contains the embryo and a supply of nutrients surrounded by a protective coat. The embryo can remain in a nongrowing state inside the seed for months, years, or in some cases, centuries. When conditions are favorable, however, the seed undergoes **germination**—meaning that it resumes growth—to form a seedling. Organogenesis continues throughout life, forming the three vegetative organs: roots, leaves, and stems. **Vegetative organs** are the nonreproductive portions of the plant body. Later, cells in the stem will become converted to reproductive structures, initiating gamete production and the sexual phase of the plant life cycle.

A common theme should emerge as you study the steps in this sequence. Despite numerous differences between plant and animal development, both are underlain by the common prin-

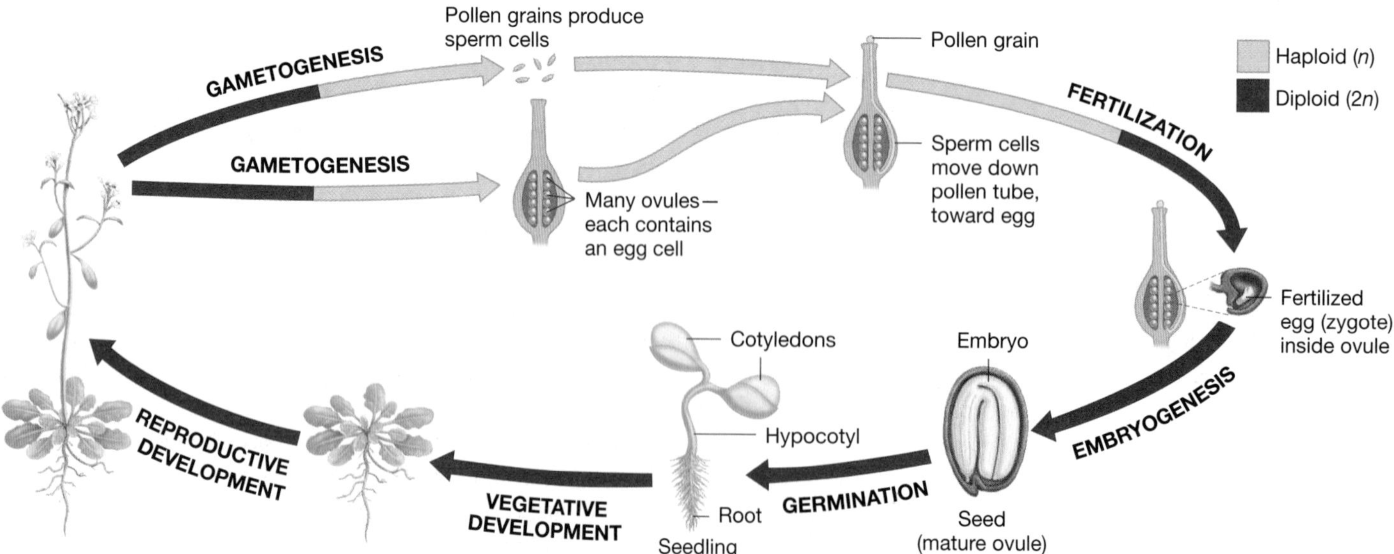

FIGURE 23.2 An Overview of Development in *Arabidopsis*. This chapter traces the events in the life of a plant, from fertilization and embryogenesis to vegetative and reproductive development.

ciples described in Chapter 21. As plant cells proliferate, they divide in precise orientations and expand in specific directions. In some cases the results are analogous to those achieved by the cell movements that occur during animal development. Developing plant cells also communicate constantly via cell-cell signals, and they differentiate due to specific combinations of cell-cell signals and regulatory transcription factors. This unity of underlying principles is one of the great results emerging from research on plant developmental biology.

23.1 Pollination and Fertilization

The story of pollination and fertilization starts with the production of a pollen grain. Inside a mature flower, diploid cells undergo meiosis to form haploid cells that give rise to pollen grains. More specifically, the haploid cells divide by mitosis to form a two-celled structure, which becomes encased in a tough, protective coat. As **Figure 23.3a** shows, the resulting **pollen grains** are tiny, multicellular individuals. One of the cells inside the pollen grain will give rise to a sperm cell that fertilizes the egg.

Pollen grains are carried to a mature flower by wind, water, insects, bats, birds, or some other agent. If the pollen grains land on a structure inside the flower called the stigma, one of the best-studied cell-cell interactions in all of plant biology takes place. For the pollination and fertilization process to continue successfully, proteins on the surface of the pollen grain have to interact with proteins on the surface of the stigma (**Figure 23.3b**). Like bindin attaching to the egg-cell receptor for sperm in sea urchins (see Chapter 22), these interactions prevent fertilization from occurring between members of different species. In many cases, the interactions also prevent self-fertilization—the union of a sperm and egg from the same individual (see Chapter 40 for details on these cell-cell interactions).

If the protein-protein interactions on the surface of the stigma are appropriate, the pollen begins to grow and a **pollen tube** extends toward the egg (**Figure 23.3c**). Recent research suggests that this growth is guided by cell-cell signals released from the egg at the base of the female reproductive structure, or **carpel**. Before the pollen tube is completed, one of the cells inside the pollen grain divides by mitosis to form two sperm cells.

If you think about that last sentence for a moment, you will realize that this aspect of gametogenesis is strikingly different from the pattern in animals. In plants, sperm and egg cells are produced from haploid cells via mitosis—not from diploid cells via meiosis, as in animals. This aspect of the plant life cycle is explored in detail in Chapter 30; the details of plant gametogenesis are analyzed in detail in Chapter 40.

Another dramatic contrast with animal development occurs when the pollen tube reaches the base of the carpel. The two sperm cells pass through the wall of the ovule and enter a structure that contains the egg cell and a maternal cell that in many species contains two haploid nuclei. One sperm fuses with the egg to form the diploid zygote, while the other sperm

(a) Pollen grains are microscopic but multicellular.

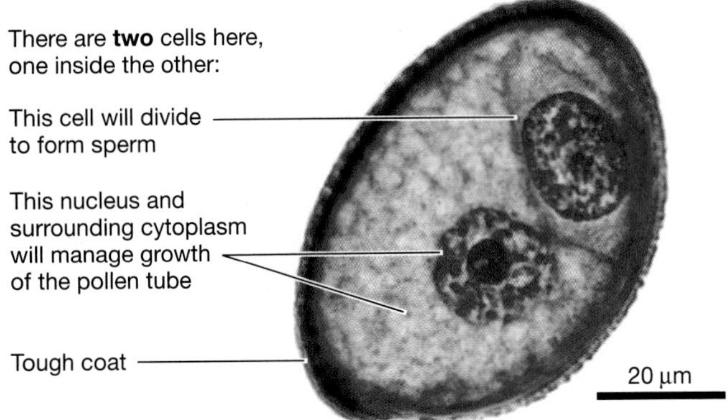

(b) Pollen grains interact with the stigma.

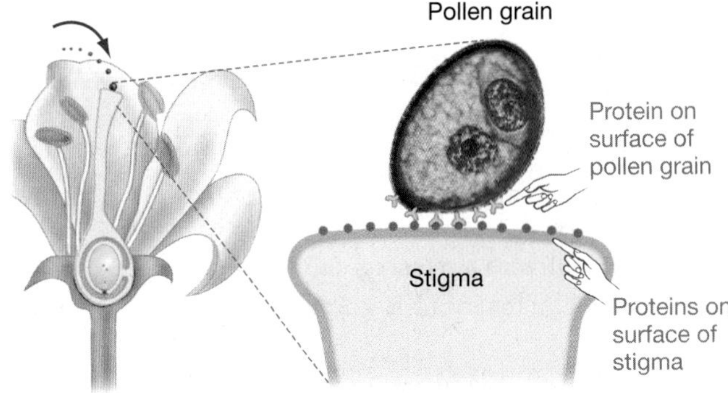

(c) Plant sperm move to the egg through a pollen tube.

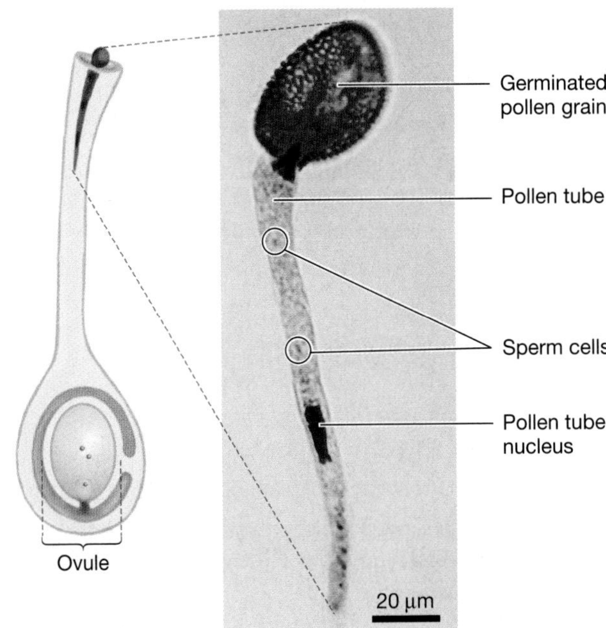

FIGURE 23.3 The Fertilization Process Begins When Pollen and Stigma Interact. In many species, sperm cells are produced by mitosis after the pollen grain germinates and the pollen tube begins growing toward the egg.

QUESTION How does the interaction between pollen and stigma in angiosperms compare to the interaction between bindin and fertilizin in sea urchins?

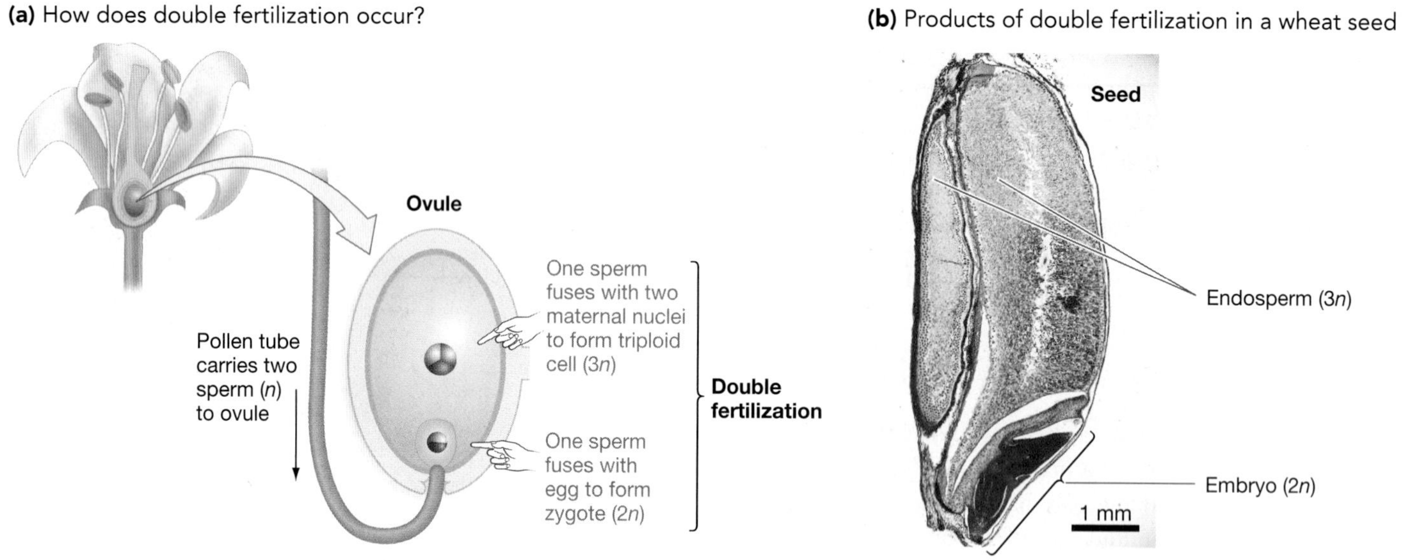

(a) How does double fertilization occur?

(b) Products of double fertilization in a wheat seed

FIGURE 23.4 Double Fertilization Leads to the Formation of Endosperm. (a) In flowering plants, a sperm (n) fuses with two or more haploid maternal nuclei near the egg to form a cell that gives rise to endosperm tissue. In many species, endosperm cells are $3n$. **(b)** Endosperm is a nutritive tissue packed with proteins and carbohydrates, and fats or oils.

fuses with the cell with two maternal haploid nuclei to form a triploid ($3n$) cell. This event is known as double fertilization (**Figure 23.4a**).

The triploid cell divides repeatedly by mitosis to form a triploid nutritive tissue called endosperm. **Endosperm** ("inside-seed") provides the proteins, carbohydrates, and fats or oils required for embryonic development, **seed germination**, and early seedling growth. In species with large seeds, the endosperm grows into a sizeable nutrient reservoir as the ovule matures (**Figure 23.4b**). When you eat wheat, rice, corn, or other grains, you are eating primarily endosperm. Functionally, endosperm is analogous to the yolk found in animal eggs.

Once double fertilization is complete, the stage is set for embryogenesis—the early development of a new individual.

Check Your Understanding

If you understand that...

- Fertilization is based on specific interactions between proteins on the surfaces of plant pollen grains and stigmas.
- In flowering plants, double fertilization results in the production of a zygote and a (usually) triploid cell that grows into a nutritive tissue.

You should be able to...

1) Explain why protein-protein interactions between pollen grains and stigmas are important to the individual's ability to reproduce. Your answer should consider what might happen if the protein-protein interactions did *not* occur.
2) Describe when meiosis and mitosis occur during the development of a sperm cell in *Arabidopsis*.

23.2 Embryogenesis

In flowering plants like *Arabidopsis*, embryogenesis takes place inside the ovule as the seed is maturing. In essence, embryogenesis produces a tiny, simplified precursor to a mature plant. The process is equivalent to the cleavage, gastrulation, and organogenesis phases of early animal development introduced in Chapter 22.

Let's first look at the sequence of events, which was worked out through careful observation of different-aged embryos. Then we can consider more research on the genetics of embryogenesis—studies that are focused on finding and characterizing the genes that make this elaborate process work.

What Happens during Embryogenesis?

Figure 23.5 illustrates the key events in embryogenesis, using *Arabidopsis* as a model organism. After fertilization, the zygote undergoes a highly asymmetric cell division. The cells resulting from this initial division are unlike in size, content, and fate. The bottom, or basal cell, is large and is dominated by an extensive vacuole. It gives rise to a column of cells called the suspensor, which anchors the embryo as it develops. The small cell above the basal cell, called the apical cell, is rich in cytoplasm and gives rise to the mature embryo.

The asymmetries in the basal and apical cell help establish one of the primary axes of the plant body: the **apical-basal axis. Apex** refers to the tip; **basal** refers to the base or foundation.

As Figure 23.5 shows, the basal cell divides perpendicular to the apical-basal axis, giving rise to the suspensor. Only one cell in the suspensor—the one closest to the apical cell—contributes

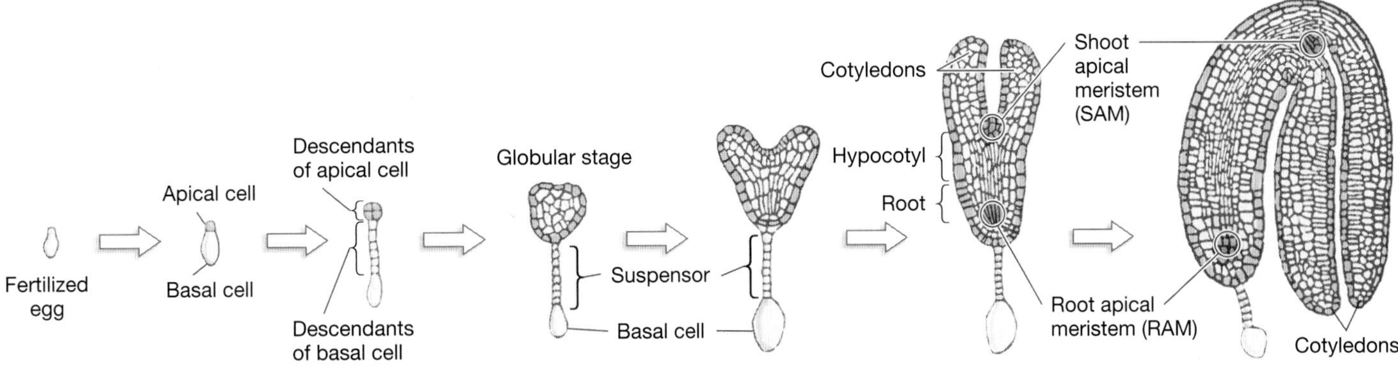

FIGURE 23.5 The Events of Embryogenesis. Embryogenesis takes place inside the developing seed.

EXERCISE Label the apical-basal and radial axes on the globular-stage embryo.

cells to the embryo and thus to the mature adult. The apical cell, in contrast, divides both perpendicularly to the apical-basal axis and parallel to it. These divisions give rise to a simple ball of cells at the tip of the suspensor. The embryo is said to be in the globular stage, and it includes distinctive-looking cells that cover the exterior. Thus, there is now a difference between cells in the interior of the embryo and cells on the surface. In this way, the second major body axis becomes visible during the globular stage. The radial axis extends from the interior of the body out to the exterior.

These initial events in embryogenesis illustrate two general, and critically important, points about plant development: (1) The shape of the body depends on how the planes of cell division are oriented and on the direction of subsequent cell expansion, and (2) the fate of a plant cell can be summed up in the old quip about the three keys to success in real estate—"location, location, location." Starting with the initial division that creates the apical and basal cells, plant cells differentiate based on where they are in the body.

As the ball of cells continues to grow and develop, the initial leaves, or **cotyledons**, begin to take shape. The cotyledons are connected to the developing root by a stem-like structure called the **hypocotyl** (**Figure 23.6**). Together, the cotyledons and hypocotyl make up the **shoot**, which will become the aboveground portion of the body. The shoot system functions in photosynthesis and reproduction. The **root**, in contrast, forms the belowground portion of the body. The root system anchors the individual and functions as a water- and nutrient-gathering structure.

Once the apical-basal and radial axes are established and as the cotyledons, hypocotyl, and root begin to take shape, groups of cells called the **shoot apical meristem** (**SAM**) and **root apical meristem** (**RAM**) form. A **meristem** consists of undifferentiated cells that divide repeatedly, with some of their daughter cells becoming specialized cells. Meristem cells are analogous to the stem cells found in animals, except that plant meristems are much more flexible. The root meristem can form all the underground portions of the plant, and the shoot meristem can form all of the aerial portions, including reproductive structures. Throughout a plant's life, meristematic tissues continue to produce embryonic cells that can differentiate into adult tissues and structures.

Note that all of this growth and development takes place without the aid of the cell migration seen in gastrulation in animals. Because plant cells have stiff cell walls, they do not move. For the cotyledons and other embryonic structures to take shape, then, cell divisions must occur in precise orientations, and the resulting cells must exhibit differential growth—meaning that some cells grow larger than others.

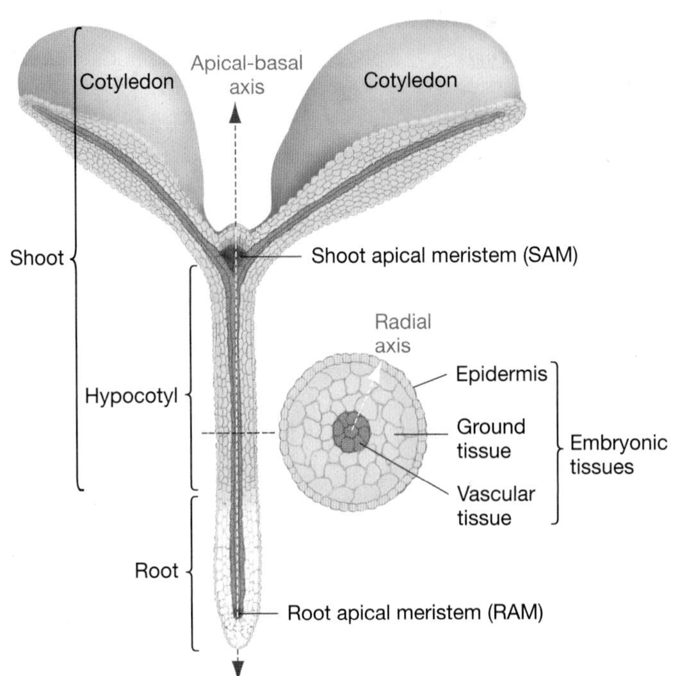

FIGURE 23.6 Embryogenesis Establishes the Two Body Axes and Three Embryonic Tissues. The apical-basal axis is the long axis of the individual; the radial axis extends from the center toward the exterior.

In addition to establishing the two body axes, early development in *Arabidopsis* produces three embryonic tissues. As Figure 23.6 shows, the embryo contains an **epidermis** ("over-skin"). This is an outer covering of specialized cells that protects the individual. Inside the epidermal layer of cells is **ground tissue**, a mass of cells that may later differentiate into cells that are specialized for photosynthesis, food storage, or other functions. The **vascular tissue** in the center of the plant will eventually differentiate into specialized cells that transport food and water between root and shoot. In the embryo, the three tissue systems are arranged in a radial pattern. These embryonic tissues make up the radial axis and are analogous to the ectoderm, mesoderm, and endoderm of developing animals (see Chapter 22).

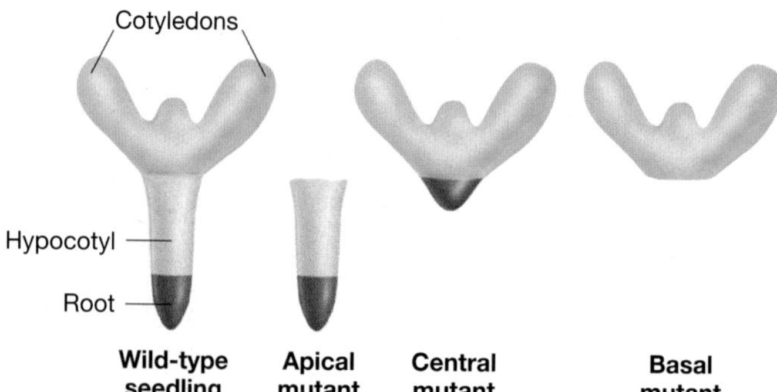

FIGURE 23.7 Mutant *Arabidopsis* Embryos with Misshapen Bodies. Researchers have identified *Arabidopsis* mutant individuals missing certain defined sections of the body along the apical-basal axis.

QUESTION How do these mutants compare and contrast with Bicoid mutants in fruit flies?

Which Genes and Proteins Set Up Body Axes?

The genetic approach to development that was pioneered with research on *Drosophila melanogaster* has also proven to be a powerful way of studying plant embryogenesis. Although the specific genes involved are different in plants than in animals, the basic logic of how genes direct the earliest events in development is similar in both.

Research on the genetics of early development in plants was pioneered by Gerd Jürgens and colleagues. This research group set out to identify genes that are transcribed in the zygote or embryo of *Arabidopsis* and that are responsible for establishing the apical-basal axis of the plant body. It's no surprise that this effort was similar to the work on anterior-posterior pattern formation in *Drosophila* introduced in Chapter 21. Jürgens had participated in the work with flies.

The biologists' initial goal was to identify individuals with developmental defects at the seedling stage. More specifically, they were looking for mutants that lacked particular regions along the apical-basal axis of the body. The team succeeded in finding several bizarre-looking mutants (**Figure 23.7**). Individuals called apical mutants lacked the first leaves, or cotyledons. Some individuals lacked the embryonic stem, or hypocotyl, and were named central mutants. Individuals dubbed basal mutants lacked both hypocotyls and roots.

To interpret these results, the researchers suggested that each type of *Arabidopsis* mutant had a defect in a different gene and that each gene played a role in specifying the position of cells along the apical-basal axis of the body. They hypothesized that these genes are analogous to the segmentation genes of fruit flies, which specify the identity of cells within well-defined regions along the anterior-posterior axis of insects.

What are these *Arabidopsis* genes, and what do they do? To answer these questions, consider the gene responsible for the mutants lacking hypocotyls and roots. This gene has been cloned and sequenced and named *MONOPTEROS*. Because its DNA sequence indicates that its protein product has a DNA-binding domain, *MONOPTEROS* is hypothesized to encode a transcription factor that regulates the activity of target genes. The MONOPTEROS protein, in turn, is manufactured in response to signals from **auxin**. Auxin is a cell-to-cell signal molecule that is produced in the shoot apical meristem and transported toward the basal parts of the individual. (For more detail on auxin's prominent role as a long-distance signal, or **hormone**, see Chapter 39). The concentration of auxin along the apical-basal axis of a plant forms a concentration gradient that provides positional information, not unlike the Bicoid concentration gradient found in fruit-fly embryos (see Chapter 22).

The take-home message of these results is that the auxin signal is part of a regulatory cascade that triggers the production of MONOPTEROS and other regulatory transcription factors specific to cells in the developing hypocotyl and roots. Although the genes and proteins involved in forming the cotyledons, hypocotyl, and root of *Arabidopsis* are not yet understood in as much detail as the segmentation genes of *Drosophila* are, several important similarities are clear. Both developmental pathways are based on cell-to-cell signals and regulatory cascades that result in the step-by-step specification of a cell's position and fate.

Many questions remain about embryogenesis in *Arabidopsis*. How is auxin production turned on as the shoot apical meristem first begins to form in embryos? Once production of the *MONOPTEROS* gene product begins, what target genes are affected? What genes other than *MONOPTEROS* are found in the regulatory cascade responsible for development along the apical-basal axis? Are any of them active in adults as well? These questions present a host of interesting challenges to current researchers.

Check Your Understanding

If you understand that...

- Early embryonic development results in the formation of the apical-basal and radial axes of the plant body.
- The early structures of the shoot and root systems form along the body's apical-basal axis.
- The genes responsible for setting up the body axes are currently the focus of intense research.

You should be able to...

1) Give an example of how an asymmetric cell division affects the fate of daughter cells.
2) Give an example of how variation in the plane of cell division affects the fate of daughter cells.
3) Predict the effect on the *monopteros* gene of adding auxin to embryonic root cells.
4) Predict the effect of adding a drug that blocks auxin transport away from the shoot apical meristem in embryos.

23.3 Vegetative Development

For a plant to thrive, it has to adjust to constantly changing conditions. Consider just one such condition—the availability of light. You might recall from Chapter 10 that plants use wavelengths in the blue and red portions of the spectrum to drive photosynthesis. Now suppose that you are an oak tree with a life expectancy of 300 years. The quality and quantity of light that your leaves receive depends on where you happened to germinate. Are you growing on flat ground? With a southern exposure in full sun? With a northern exposure rarely exposed to full sun? In addition, from the time you emerge from an acorn to the time of your death, the light you receive will depend on changes in climate and weather as well as on your size relative to the size and proximity of plants that compete with you for light and shade. Finally, the leaves in your bottommost branches experience a different light regime from the leaves at your apex.

How do plants cope with all this variation in their living conditions? Unlike most animals, they don't move around until they find a place that suits their requirements. Instead, a large part of the answer is the continuous growth and development of roots, stems, and leaves. If an oak tree is heavily shaded on one side, it stops growing in that direction and extends branches on the other side. If it is heavily shaded on all sides, its growth is directed upward. This constant adjustment to changing environmental conditions is possible because of the meristems that exist at the tips of shoots and roots.

Meristems Are Responsible for Continuous Growth and Development

When embryonic development is complete, the basic body axes are established and the initial structures in the root and shoot systems have formed. For the rest of the individual's life, further development is driven by the meristems. As **Figure 23.8a** shows, meristematic tissue exists at each tip in the shoot and root sys-

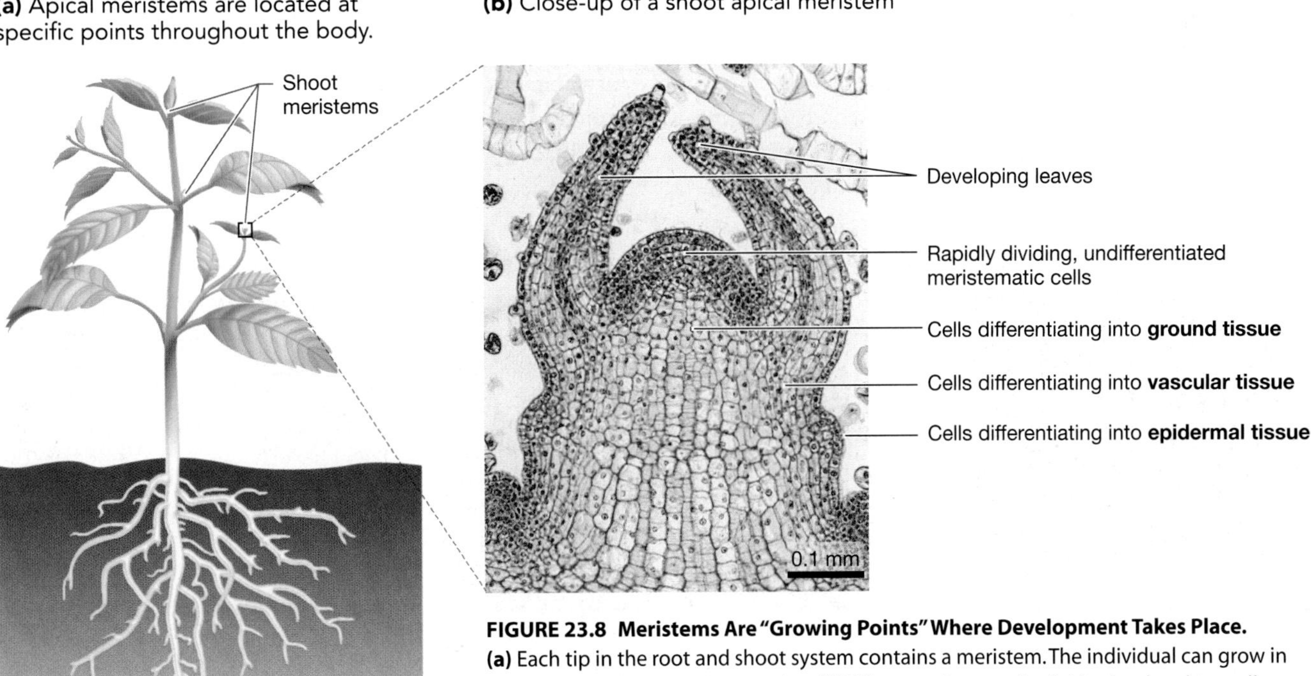

FIGURE 23.8 Meristems Are "Growing Points" Where Development Takes Place.
(a) Each tip in the root and shoot system contains a meristem. The individual can grow in any direction that a meristem exists. **(b)** When meristem cells divide, the daughter cells either remain undifferentiated and continue to function as meristem cells or differentiate into new epidermal, ground, or vascular cells.

tems. As a result, the individual is capable of growing in any direction aboveground or belowground, depending on conditions.

Figure 23.8b provides a close-up view of a shoot apical meristem, or SAM. The cells within the meristem are small and undifferentiated. Within each meristem, the rate and direction of cell growth are dictated by cell-cell signals that are produced in response to environmental cues like the arrival of spring, the presence of abundant water, and the amount of light striking the plant. Just below the meristem, daughter cells produced by mitosis and cytokinesis in the meristem initially differentiate into epidermal, ground, or vascular tissue. Eventually these cells will differentiate into specialized cell types.

Careful microscopy allowed biologists to tease out the sequence of events that occur as meristems grow, and intense research continues to explore how interactions between auxin and other cell-cell signals influence the fate of cells produced by meristems. But recently researchers have also turned to the question of which genes respond to these cell-cell signals. Specifically, which genes and gene products direct the formation of specific structures during vegetative development? Let's consider one particularly well-studied example—the genetic control of shape in developing leaves.

Which Genes and Proteins Determine Leaf Shape?

Leaves can be induced to grow by applying tiny amounts of auxin to a SAM. Based on this observation, it appears clear that the initiation of a leaf depends on the concentration of auxin in parts of a SAM, although other types of cell-cell signals are undoubtedly involved as well. Once a leaf begins to grow, the next key event in its development is the formation of three axes: the proximal-distal, lateral, and upper-lower axes shown in **Figure 23.9**. Proximal is toward the main body while distal is away from the main body; the lateral axis runs from the middle of a leaf toward its margin. The amount and direction of growth along these axes determines the shape of the leaf.

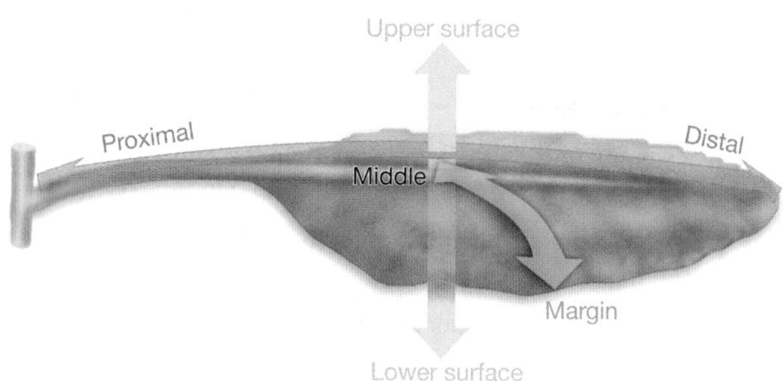

FIGURE 23.9 Leaves Have Three Axes. Overall, the plant body has just two axes: apical-basal and radial. Every leaf has the three axes shown here.

● **QUESTION** How do the body axes in plants compare and contrast with the body axes in animals?

Recent research has begun to identify the genes responsible for specifying these three leaf axes. For example, analyses of mutant individuals in *Arabidopsis* and other species has shown that a gene called *PHANTASTICA* (abbreviated *PHAN*) is critical in setting up the upper-lower axis of leaves. The protein product of *PHAN* has a DNA-binding domain and acts as a regulatory transcription factor. *PHAN* triggers the expression of genes that cause cells to form the upper surface of leaves and suppresses transcription of genes required for forming the lower leaf surface. It is part of a regulatory cascade that begins with auxin and other cell-cell signals and ends with the growth of a normal-shaped leaf.

Recent research on tomatoes also suggests that changes in *PHAN* expression may underlie at least some of the evolutionary changes observed in leaf shape. Leaf shape varies widely among species (see Chapter 36). Simple leaves consist of a single blade, but as **Figure 23.10a** shows, tomatoes have compound leaves—a single leaf blade divided into smaller units

(a) Normal leaves—compound

(b) Transgenic leaves (reduced *PHAN* expression)—simple or palmately compound

FIGURE 23.10 Changes in Genes That Establish Leaf Axes Can Change Leaf Shape. (a) Normal tomato leaves consist of a series of leaflets. **(b)** When the expression of *PHAN* is reduced experimentally, leaves become cup shaped (left) or have leaflets arising from a single point. [Reprinted by permission from *Nature* 424 (July 24, 2003), p. 439, by Kim, M. et al. ©2004 Macmillan Publishers Ltd. Photographs supplied by Dr. Neelima Sinah.]

● **QUESTION** Suppose that *PHAN* were overexpressed. What type of leaf do you predict would form?

called leaflets. Other species have palmately compound leaves, meaning that leaflets radiate from a single point.

To explore whether changes in *PHAN* might have a role in the evolution of various leaf shapes, a team of biologists used techniques introduced in Chapter 19 to create transgenic tomato plants. In these individuals, the *PHAN* gene product was blocked to a moderate or large extent. As **Figure 23.10b** shows, leaf shape in the transgenic individuals changed dramatically. Some of the individuals had cup-shaped (simple) leaves, while others were palmately compound. Although it is still uncertain why changing the specification of the upper-lower leaf axis affects overall shape, it appears clear that *PHAN* expression plays an important role.

These results have inspired the hypothesis that at least some evolutionary changes in leaf size and shape are due to mutations that created new alleles of genes that regulate *PHAN* expression. Alleles that resulted in lowered *PHAN* expression might lead to simple leaves with a single blade; alleles that increased the extent of *PHAN* expression might result in compound leaves like those of normal tomatoes. By experimentally altering the genes that regulate development, researchers are beginning to understand the genetic changes leading to novel types of leaves. If this research continues to be productive, biologists will have another example of how changes in regulatory pathways that direct development underlie evolutionary change (see Chapter 21).

Check Your Understanding

If you understand that...

- Vegetative development is continuous, and vegetative growth can occur in all directions.
- Vegetative growth and development occur in ways that increase an individual's ability to survive and produce offspring in the face of changing environmental conditions.
- Continuous development is possible in plants because meristematic tissue is present at the tips of each component of the root and shoot system.

You should be able to...

1) Make a sketch of a flowering plant and label the locations of SAMs and RAMs. Make another sketch of what the same plant would look like after extensive growth, given that a cow defecated near its left side.
2) Sketch the three axes of a leaf, and give an example of how changes in *PHAN* expression can result in changes in overall leaf shape.

23.4 Reproductive Development

Among the many startling contrasts observed in animal and plant development, one of the most important concerns the development of reproductive tissues and organs. In animals, the cells that give rise to sperm and egg cells are set aside or sequestered early in development. These **germ cells** migrate to the ovaries or testes once the reproductive organs have developed. As a result, the cells that give rise to animal gametes undergo relatively few rounds of mitosis—perhaps 20 to 50—prior to meiosis and gametogenesis.

In contrast, plants do not have germ cells that are set aside early in development. Instead, flowering and gametogenesis occur when a SAM converts from vegetative development to reproductive development. As a result, meristematic cells that have divided thousands of times can give rise to the reproductive organs of plants, and eventually to sperm and eggs. Because mutations occur during each cell cycle, plants generate much more genetic variation via mutation than animals do. Although biologists have only begun to explore the consequences of this fact, research on the mechanisms responsible for the formation of reproductive structures has been intense. To introduce this huge body of work, let's consider some highlights of research on *Arabidopsis*.

The Floral Meristem and the Flower

When specialized proteins in *Arabidopsis* sense that days are getting longer and the temperature is favorable, they trigger the production of signals that convert SAMs from vegetative to reproductive development. A **floral meristem** is a modified SAM that produces flowers containing the reproductive organs instead of vegetative structures. The genes that take part in the regulatory cascade responsible for the maturation of a floral meristem are now well characterized.

As an *Arabidopsis* flower develops, the floral meristem produces four organs arranged as shown in **Figure 23.11**: (1) sepals, (2) petals, (3) stamens, and (4) carpels. Each of these organs is a modified leaf. **Sepals** are located around the outside of the flower and help protect it. In some species, sepals are also colorful and function in attracting pollinators. Inside the sepals is a whorl or circular arrangement of **petals**, which enclose the male and female reproductive organs. If insects or other animals pollinate the species in question, the petals may be colored to help

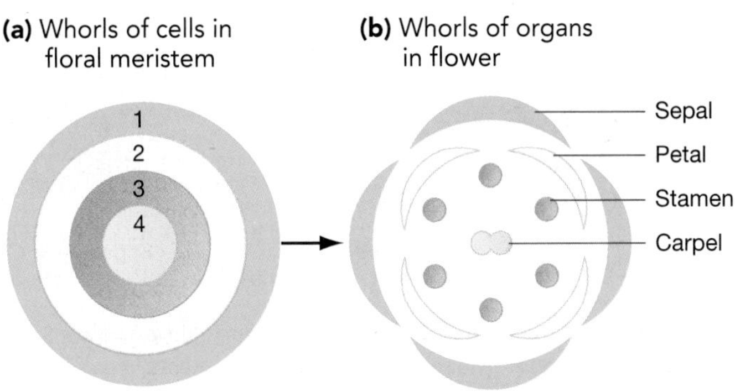

FIGURE 23.11 Flowers Are Composed of Four Organs. (a) In a floral meristem, whorls of cells produce **(b)** the four whorls of floral organs.

advertise the reproductive structures. The pollen-producing organs, or **stamens**, are located in a whorl inside the petals. In the center of the entire structure are egg-producing reproductive organs, or **carpels**. The question is, How does the floral meristem produce these four organs in the characteristic pattern of whorls within whorls?

The first hint of an answer came in the late 1800s, when researchers realized that several types of mutant flowering plants—including some that were popular garden plants—were homeotic. In the mutant individuals, one kind of floral organ was replaced by another, even though the total number and relative position of each organ did not change. For example, one homeotic mutant had flowers with sepals, petals, another ring of petals, and carpels instead of having sepals, petals, stamens, and carpels. The mutant phenotype was similar to the transformation of segments later observed in *Drosophila* homeotic mutants (see Chapter 21).

Just as an analysis of homeotic mutants in fruit flies triggered a breakthrough in understanding the genetic control of body axis formation and segmentation in animals, an analysis of homeotic floral mutants in *Arabidopsis* triggered a breakthrough in understanding the genetic control of flower structure in plants.

Which Genes Control the Development of Flowers?

Over 100 years after floral homeotic mutants were first described, Elliot Meyerowitz and colleagues assembled a large collection of *Arabidopsis* individuals with homeotic mutations in flower structure. The researchers' goal was to identify and characterize the genes responsible for specifying the four floral organs.

Meyerowitz's group found that the mutants could be sorted into the three general classes shown in **Figure 23.12**. The classes were distinguished by the type of homeotic transformation that occurred. Some mutants had only carpels and stamens; others had only sepals and carpels; still others had only petals and sepals. The key observation was that each type of mutant lacked the elements found in two of the four whorls.

What was going on? To begin searching for an answer, the biologists hypothesized that each of the three classes of homeotic mutants was due to a defect in a single gene. Meyerowitz then reasoned that if three genes are responsible for setting up the pattern of a flower, then the mutants suggested a hypothesis for how the three gene products interact. Because he referred to the three hypothetical genes as *A*, *B*, and *C*, his hypothesis is called the ABC model.

Three basic ideas underlie the ABC model. The first is that each of the three genes involved is expressed in two adjacent whorls. The second idea is that if each gene is expressed in two adjacent whorls, then a total of four different combinations of gene products can occur. The final idea is that each of these four combinations of gene products triggers the development of a different floral organ. Specifically, Meyerowitz proposed that (1) the A protein alone causes cells to form sepals, (2) a combination of A and B proteins sets up the formation of petals, (3) B and C combined specify stamens, and (4) the C protein alone designates cells as the precursors of carpels.

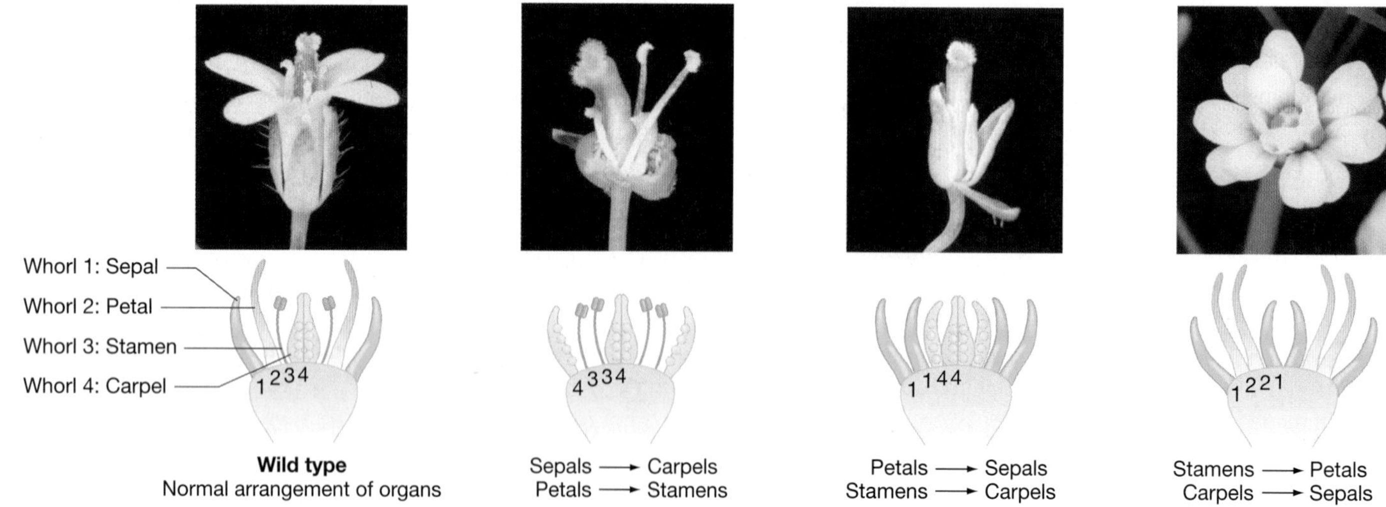

FIGURE 23.12 Homeotic Mutants in *Arabidopsis* Flowers Have Structures in the Wrong Locations. The photos show a normal (wild type) individual and three types of mutants. In each type of homeotic floral mutant, two adjacent whorls of organs are transformed into different organs. Sepals and petals, petals and stamens, or stamens and the carpel are transformed. The numbers indicate the wild-type whorls from outermost (1) to innermost (4).

(a) The ABC model

Idea 1: Three genes pattern the flower. Each gene is expressed in two adjacent whorls.

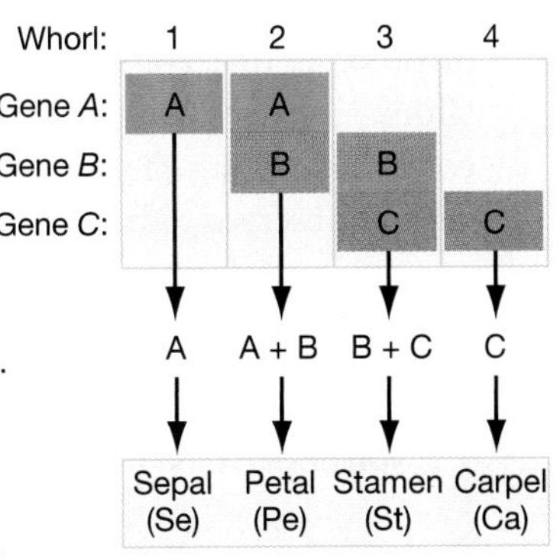

Idea 2: A total of four different combinations of gene products occurs.

Idea 3: Each of the four combinations of gene products triggers the development of a different floral organ.

Assumption: Presence of gene A product inhibits expression of gene C. Presence of gene C product inhibits expression of gene A.

No A + C

(b) Testing the ABC model

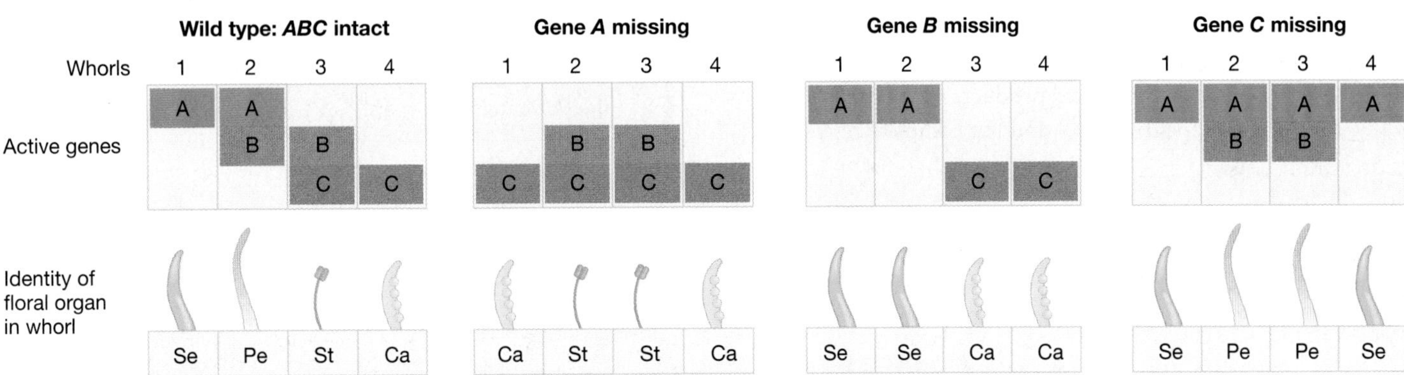

FIGURE 23.13 The ABC Model for Genetic Control of Flower Development. The ABC model is a hypothesis to explain why three types of homeotic mutants exist.

● **QUESTION** Why is it necessary to propose that the *A* gene product inhibits expression of the *C* gene, and that the *C* gene product inhibits expression of the *A* gene?

Does this model explain how the three classes of homeotic mutants occur? The answer is yes, if two additional elements are added to the model: The presence of the A protein inhibits the production of the C protein, and the presence of the C protein inhibits the production of the A protein (**Figure 23.13a**). Then the patterns of gene expression and mutant phenotypes in **Figure 23.13b** correspond. For example, if the *A* gene is disabled by mutation, then it no longer inhibits the expression of the *C* gene and all cells produce the C protein. As a result, cells in the outermost whorl express only C protein and develop into carpels, while cells in the whorl just to the inside produce B and C proteins and develop into stamens. ● If you understand the ABC model, you should be able to explain why biologists did not hypothesize that four genes are involved—one that specifies each of the four floral organs—and why they proposed that each gene involved is expressed in two adjacent whorls.

Although the model is plausible and appeared to explain the data, it needed to be tested directly. To accomplish this, Meyerowitz and co-workers mapped the genes responsible for the mutant phenotypes and identified the appropriate DNA sequences. Once they had isolated the genes, they were able to obtain and use single-stranded DNAs to perform in situ hybridizations (see Chapter 21). The goal was to document the pattern of expression of the *A*, *B*, and *C* genes and see if that pattern corresponded to the model's predictions. As anticipated, the mRNAs for each of the three genes showed up in the sets of whorls predicted by the model. The *A* gene is expressed in the outer two whorls, the *B* gene is expressed in the middle two whorls, and the *C* gene is expressed in the inner two whorls.

This result strongly supported the validity of the ABC model. Just as different combinations of *Hox* gene products specify the identity of fly segments, different combinations of floral identity genes specify the parts of a flower.

Similarities with principles of animal development did not end there. When Meyerowitz and others analyzed the DNA sequences of the floral organ identity genes, they discovered that all three genes contained a segment that coded for a long sequence of 58 amino acids, called the **MADS box**, that binds to DNA. Based on this observation, the researchers hypothesized that the floral genes encode regulatory transcription factors that bind to enhancers or other regulatory sequences and trigger the expression of genes required for sepal, petal, carpel, and stamen formation. The discovery highlighted remarkable similarities between the *MADS-box* genes and the homeotic genes found in *Drosophila* and other animals. The *Hox* genes of animals have a region called the homeobox that is functionally similar to the MADS-box: It encodes a DNA-binding domain in Hox proteins. Both *Hox* and *MADS-box* gene products are parts of regulatory cascades that lead to the specification of important structures. *Hox* genes regulate the expression of genes responsible for forming limbs or other structures in specific parts of the body; *MADS-box* genes regulate the expression of genes responsible for forming flowers. Both sets of genes tell cells where they are in the body and can produce homeotic mutants if they do not work properly. Although different genes are involved, the logic of how to put a multicellular body together is similar in plants and animals.

Biologists are currently working to identify the genes targeted by the ABC proteins, just as they are working to identify the genes controlled by the *Hox* products in animals. Eventually, researchers hope to understand the complete regulatory cascade from initiation of the floral meristem to the expression of proteins that are specific to petals or other floral organs. Ultimately, the goal is to explore how changes in this cascade could have led to the evolution of the spectacular diversity of flowers observed today.

Check Your Understanding

If you understand that... 

- Reproductive development begins when SAMs are converted to floral meristems.
- Development of the four floral organs depends on the expression of regulatory transcription factors encoded by the *A*, *B*, and *C* class genes.

You should be able to...

1) Describe a mechanism whereby an A protein could inhibit expression of a *C* gene.
2) Compare and contrast the *MADS-box* genes of plants with the *Hox* genes of animals.

Chapter Review

SUMMARY OF KEY CONCEPTS

Several aspects of plant development contrast sharply with that of animals: Plants develop continuously, commit certain cells to gamete production late in development, and produce gametes by mitosis.

The striking differences between plant and animal development are not surprising, given their different ways of life and the observation that multicellularity evolved independently in the two groups. Surprising similarities exist in plant and animal development, however: (1) Fertilization is mediated by protein-protein interactions between cells produced by each parent, (2) the body axes are established early in development, (3) three types of embryonic tissues give rise to the array of adult tissues and organs, (4) homeotic genes exist in both lineages, (5) meristem cells and stem cells are similar in function, and (6) development is based on cell-cell signals that trigger and modulate cascades of regulatory transcription factors.

You should be able to explain the similarities between auxin and Bicoid; and between the epidermal, vascular, and ground tissues of embryonic plants and the endoderm, mesoderm, and ectoderm of animals.

In flowering plants, double fertilization results in the production of a zygote and a nutritive tissue that supports embryogenesis.

The pollination and fertilization process begins with interactions between proteins on the surface of pollen grains and proteins on the surface of the stigma and continues with growth of the pollen tube and production of sperm cells. In flowering plants, one sperm fertilizes the egg to form a zygote while a second sperm fuses with a (usually) diploid cell to form a triploid cell. The triploid cell goes on to produce the nutritive tissue called endosperm.

You should be able to diagram the relationship between the zygote and endosperm nucleus right after double fertilization, and explain the relationship between an embryo and endosperm in a mature wheat seed.

Embryogenesis results in the formation of the major body axes and three types of embryonic tissue.

The earliest cell divisions in embryogenesis establish the apical-basal and radial axes of the individual. Later, the embryonic structures called the cotyledons, hypocotyl, and root develop and the embryonic epidermal, ground, and vascular tissues form.

You should be able to sketch a seedling and label the apical-basal and radial axes, cotyledons, hypocotyl, and root.

Vegetative development is based on meristems where cell division occurs throughout life, producing cells that go on to differentiate.

Plants can grow continuously because meristematic tissue exists at each tip in the root and shoot system. Continuous growth and development allows plants to respond to environmental conditions that change throughout their life.

You should be able to explain why plants don't have just one SAM and one RAM, at their apex and base, in terms of their ability to grow and reproduce.

When a meristem is converted from vegetative to reproductive development, regulatory transcription factors are activated that control the position and identity of floral organs. Meristems that undergo vegetative development can be converted to reproductive development as plants respond to changes in day length or other environmental cues. Once a floral meristem is established, combinations of regulatory transcription factors encoded by *A*, *B*, and *C* class genes interact to produce the flower's sepals, petals, stamens, and carpels.

You should be able to explain why individuals with defective alleles at the *B* gene are considered homeotic mutants.

QUESTIONS

Test Your Knowledge

1. What is the fate of the two cells found inside pollen grains prior to germination?
   a. One cell directs development of the pollen tube; the other gives rise to sperm cells via mitosis.
   b. One cell directs development of the pollen tube; the other gives rise to sperm cells via meiosis.
   c. One cell fertilizes the egg; the other fuses with a diploid cell to form triploid endosperm.
   d. One cell initiates germination by interacting with proteins on the surface of the stigma; the other gives rise to sperm cells.

2. Which of the following does *not* represent a contrast between plant and animal development?
   a. Under certain conditions, plant cells can "de-differentiate" readily.
   b. The fate of a cell is determined in part by its location in the embryo.
   c. Germ cells are set aside early in development.
   d. Plant cells do not move.

3. What evidence suggests that auxin concentrations help determine where leaves form near SAMs?
   a. Auxin is produced in SAMs and transported from there toward the root.
   b. Auxin is present in a concentration gradient, with higher concentrations apically and lower concentrations basally.
   c. Auxin concentrations are relatively constant along the radial axis of the body, and leaves form along the radial axis.
   d. Addition of small quantities of auxin to a SAM can induce leaf development.

4. Which of the following does *not* occur during embryogenesis?
   a. Formation of the radial axis.
   b. Production of the suspensor.
   c. Formation of the cotyledons and hypocotyl.
   d. Formation of the lateral and proximal-distal axes.

5. When does the apical-basal axis first become apparent?
   a. When the epidermal, ground, and vascular tissues form.
   b. When the cotyledons, hypocotyl, and root form.
   c. When the first cell division produces the apical cell and basal cell.
   d. During the globular stage, when the suspensor is complete.

6. What evidence suggests that changes in the way that the *PHAN* gene is expressed could be partly responsible for evolutionary changes in leaf shape?
   a. If *PHAN* gene expression is manipulated experimentally, individuals produce leaf types found in different species.
   b. Experiments have shown that *PHAN* plays a role in establishing the upper-lower surface axis in leaves.
   c. Sequencing studies and other data have shown that *PHAN* encodes a regulatory transcription factor.
   d. All plant species surveyed to date have a gene homologous to *PHAN*.

Test Your Knowledge answers: 1. a; 2. b; 3. d; 4. d; 5. c; 6. a

Test Your Understanding

Answers are available at www.masteringbio.com

1. Make a set of sketches showing the sequence of events occurring between the time a pollen grain lands on a stigma and fertilization occurs. Note where important protein-protein interactions occur, and indicate how cell-cell signals allow sperm and egg to communicate. Explain what is double about double fertilization.

2. Sketch the two major body axes found in plants and the three axes present in a leaf. Compare and contrast these axes with the three body axes of animals.

3. Compare and contrast the stem cells of animals with the meristems of plants. How are these cells similar? How are they different?

4. In what sense are the epidermal, ground, and vascular tissues produced in the SAMs and RAMs of a 300-year-old oak tree "embryonic"?

5. When in situ hybridization experiments documented where *A*, *B*, and *C* class genes were expressed in developing *Arabidopsis* flowers, it was considered strong support for the ABC model. Explain why. Also, explain the significance of the observation that the *A*, *B*, and *C* class genes all encode a MADS-box domain.

6. Give an example of how each of the four fundamental developmental processes—cell proliferation, cell expansion, cell-cell signals, and differentiation—plays a role in plant development.

Applying Concepts to New Situations

Answers are available at www.masteringbio.com

1. Transcription factors with MADS boxes also occur in fungi and animals. But the *MADS-box* genes of fungi and animals are not active in the formation of complex structures like flowers or segments, and they do not produce homeotic mutations when they fail to function properly. State a hypothesis to explain why the function of homologous DNA sequences like the *MADS-box* genes can be so different in plants versus these other groups.
2. When growth occurs in an unlimited or unrestricted way, it is said to be indeterminate. But when growth is of limited duration and then stops, it is said to be determinant. Which process is observed in vegetative development and which in reproductive development? Explain your logic.
3. Convergent evolution is said to occur when similar structures or processes are observed in different species, but when the similar structures and processes are not due to homology—meaning, not due to shared ancestry. Stated another way, convergence is due to the independent evolution of similar traits. Explain why the existence of homeotic mutants in both plants and animals could be considered a form of convergent evolution. In this case, what would cause convergence?
4. Make a sketch showing the locations of SAM and RAM in a young oak tree. Suppose that a concrete sidewalk gets poured on the left side of the tree, preventing water penetration below. A building goes up on the right side, cutting off light from that direction. But after a few years, an underground water pipe just next to the building begins to leak, providing abundant water year round. Draw the expected locations of SAM and RAM in the same tree 50 years after your initial sketch. Explain your logic.

www.masteringbio.com is also your resource for • Answers to text, table, and figure caption questions and exercises • Answers to *Check Your Understanding* boxes • Online study guides and quizzes • Additional study tools including the *E-Book for Biological Science* 3rd ed., textbook art, animations, and videos.

Evolution by Natural Selection

24

The color difference between these two peppered moths is largely due to the action of different alleles at a single gene. Biologists have documented changes in the frequencies of these alleles in several different populations around the world, due to natural selection.

KEY CONCEPTS

- Populations and species evolve, meaning that their heritable characteristics change through time. More precisely, evolution is defined as changes in allele frequencies over time.
- Natural selection occurs when individuals with certain alleles produce the most surviving offspring in a population. An adaptation is a genetically based trait that increases an individual's ability to produce offspring in a particular environment.
- Evolution by natural selection is not progressive, and it does not change the characteristics of the individuals that are selected—it changes only the characteristics of the population. Animals do not do things for the good of the species, and not all traits are adaptive. All adaptations are constrained by trade-offs and genetic and historical factors.

This chapter is about one of the great ideas in science. The theory of evolution by natural selection, formulated independently by Charles Darwin and Alfred Russel Wallace, explains how organisms have come to be adapted to environments ranging from arctic tundra to tropical wet forest. As an example of a revolutionary breakthrough in our understanding of the world, the theory of evolution by natural selection ranks alongside Copernicus's theory of the Sun as the center of our solar system, Newton's laws of motion and theory of gravitation, the germ theory of disease, the theory of plate tectonics, and Einstein's general theory of relativity. These ideas are the foundation stones of modern science; all are accepted on the basis of overwhelming evidence.

Evolution by natural selection has become one of the best-supported and most important theories in the history of scientific research. But like most scientific breakthroughs, this one did not come easily. When Darwin published his theory in 1859 in a book called *On the Origin of Species by Means of Natural Selection*, it unleashed a firestorm of protest throughout Europe. At that time, the leading explanation for the diversity of organisms was a theory called special creation. This theory held that all species were created independently, by God, perhaps as recently as 6000 years ago. The theory of special creation also maintained that species were immutable, or incapable of change, and thus had been unchanged since the moment of their creation. Darwin's ideas were radically different. He proposed that life on Earth was ancient and that species change through time.

To understand the contrast between the theory of special creation and the theory of evolution by natural selection more thoroughly, recall from Chapter 1 that scientific theories usually have two components: a pattern and a process. The first component is either a claim about a pattern that exists in nature or a statement that summarizes a series of observations about the natural world. In short, the pattern component is about facts—about how things *are* in nature. The second component of a scientific theory is a process that produces that pattern or set of observations. For example, the pattern component in the theory of special creation was that species were created independently of one another and that they do not change through time. The process that explained this pattern was the instantaneous and independent creation of living organisms by a supernatural being.

To help you understand the pattern and process components of evolution by natural selection, the chapter begins by examining the ideas that led to Darwin's breakthrough. Section 24.2 analyzes the pattern component of the theory—specifically, the evidence behind the claim that species are not independent but are instead related, and that they are not immutable but instead have changed through time. Darwin proposed that natural selection, introduced in Section 24.3, explains this pattern. The concluding sections review two recent studies of evolution by natural selection. These case studies illustrate how biologists test Darwin's theory by studying evolution in action and help clarify some common misunderstandings about how evolution by natural selection works.

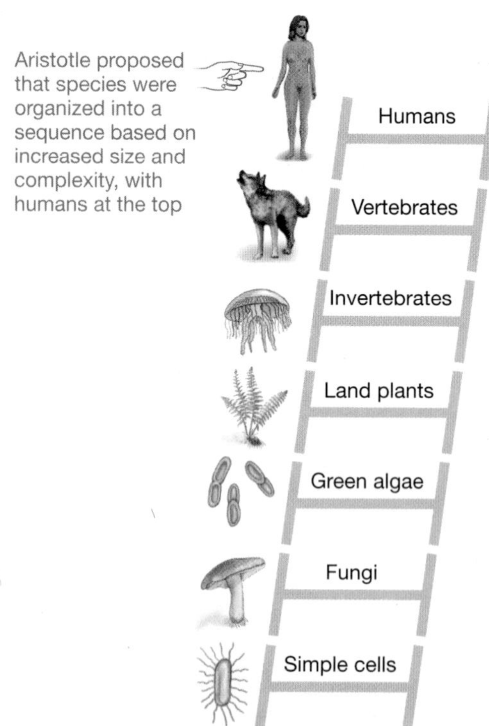

FIGURE 24.1 The Great Chain of Being Proposed by Aristotle.

24.1 The Evolution of Evolutionary Thought

People often use the word *revolutionary* to describe the theory of evolution by natural selection. Revolutions overturn things—they replace an existing entity with something new and often radically different. A political revolution removes the ruling class or group and replaces it with another. The industrial revolution replaced small shops for manufacturing goods by hand with huge, mechanized assembly lines. A scientific revolution, in contrast, overturns an existing idea about how nature works and replaces it with another, radically different, idea.

The idea that Darwin and Wallace overturned had dominated thinking about the nature of organisms for over 2000 years. The Greek philosopher Plato claimed that every organism was an example of a perfect essence, or type, created by God, and that these types were unchanging. Plato acknowledged that the individual organisms present on Earth might deviate slightly from the perfect type, but he said this was similar to seeing shadows created by the perfect type on a wall. The key to understanding life, in Plato's mind, was to ignore the shadows and focus on understanding each type of unchanging, perfect essence.

Not long after Plato developed his ideas, Aristotle ordered the types of organisms known at the time into a linear scheme called the great chain of being, also called the scale of nature or the ladder of life (**Figure 24.1**). Aristotle proposed that species were organized into a sequence based on increased size and complexity, with humans at the top.

Today, philosophers and biologists refer to ideas like these as typological thinking. Typological thinking is based on the idea that species are unchanging types and that variations within species are unimportant or even misleading. Typological thinking occurs in the Bible's book of Genesis, where God creates one of each type of organism, and inspired Linneaus's effort in the 1700s to classify each type of organism in the hierarchy of kingdoms, classes, orders, families, and genera introduced in Chapter 1.

The idea that types never change eventually began to break down, however. In 1809 the biologist Jean-Baptiste de Lamarck proposed that species are not static, but have changed through time. Lamarck was the first to propose a formal theory of **evolution**—the idea that species change through time.

The pattern component of Lamarck's theory was initially based on the great chain of being, however. When he started his work on evolution, Lamarck claimed that simple organisms originate at the base of the chain by spontaneous generation (see Chapter 1) and then evolve by moving up the chain over time. Thus, Lamarckian evolution is progressive in the sense of always producing larger and more complex, or "better," species. To capture this point, biologists like to say that Lamarck turned the ladder of life into an escalator.

Lamarck also contended that species change through time via the inheritance of acquired characters. The idea here is that individuals change as they develop in response to challenges posed by the environment, and they pass on these phenotypic changes to offspring. A classic scenario is that giraffes develop long necks as they stretch to reach leaves high in treetops, and they then produce offspring with elongated necks.

In contrast, Darwin and Wallace proposed that evolution does not follow this linear, progressive pattern. (**Box 24.1** discusses why the theory of evolution by natural selection became associated primarily with Darwin's name.) What is more important, they emphasized that the process responsible for change through time is based on variation among individuals in populations. This was a radical break from the typological thinking that had dominated scientific thought since Plato. Darwin claimed that instead of being unimportant or an illusion, variation among individuals in a population was the key to understanding the nature of species. Biologists refer to this view as population thinking. A **population** consists of individuals of the same species that are living in the same area at the same time. Darwin and Wallace proposed that evolution occurs because traits vary among the individuals in a population, and

BOX 24.1 Why Darwin Gets Most of the Credit

Although Charles Darwin and Alfred Russel Wallace formulated the same explanation for how species change over time, Darwin's name is much more prominently associated with the theory of evolution because he developed the idea more thoroughly and provided massive evidence for it in *On the Origin of Species*. But historians of science speculate about whether Darwin would have published his theory at all had Wallace not threatened to scoop him (**Figure 24.2**).

Darwin wrote a paper explaining evolution by natural selection in 1842—a full 17 years before the first edition of *On the Origin of Species* came out. He never submitted the work for publication, however. Why? Darwin claimed that he needed time to document all of the arguments for and against the theory and to examine its many implications. There is probably an element of truth in this—Darwin was a remarkably thorough thinker and writer. But many historians of science argue that he held off largely out of fear. Because his theory was inconsistent with the creation story in the Bible's Book of Genesis, Darwin knew that he would be exposed to scathing criticism from religious and scientific leaders. He was also an extremely private person, had a strong religious upbringing, and was frequently in poor health. He responded to stress or personal attacks by suffering long bouts of debilitating illness. The prospect of fighting for his ideas against the most powerful men in Europe was daunting. But Wallace forced Darwin's hand.

(a) Charles Darwin

(b) Alfred Russel Wallace

FIGURE 24.2 The Codiscoverers of Evolution by Natural Selection. (a) Charles Darwin in 1840, four years after he returned from the voyage of the *Beagle* and two years before he drafted his first paper explaining evolution by natural selection. **(b)** Alfred Russel Wallace, who in 1858 independently formulated the theory of natural selection.

[(b) Alfred Russel Wallace by unknown artist, after a photograph by Thomas Sims, fl. 1860s. Reg. No.: 1765. National Portrait Gallery, London.]

Wallace was also a native of England, but he had been making a living by collecting butterflies and other natural history specimens in Malaysia and selling them to private collectors. While recuperating from a bout of malaria there in 1858, he wrote a brief article outlining the logic of evolution by natural selection. He sent a copy to Darwin, who immediately recognized that they had formulated the same explanation for how populations change through time. The two had their papers read together before the Linnean Society of London, and Darwin then rushed *On the Origin of Species* into publication a year later. The first edition sold out in a day.

Fortunately for Darwin's health, a fellow biologist and friend named Thomas Huxley publicly defended the theory against criticism, which came from both scientific and religious quarters. Darwin continued to live quietly on his estate in Down, England, and actively continued a brilliant research career.

because individuals with certain traits leave more offspring than others do.

The theory of evolution by natural selection was revolutionary because it overturned the idea that species are static and unchanging, and because it replaced typological thinking with population thinking. It also proposed a mechanism that could account for change through time and be tested through observation and experimentation. Plato and his followers emphasized the existence of fixed types; evolution by natural selection is all about change and diversity.

Now the question is, what evidence backs the claim that species are not fixed types? What data convinced biologists that the theory of evolution by natural selection is correct?

24.2 The Pattern of Evolution: Have Species Changed through Time?

In *On the Origin of Species*, Darwin repeatedly used the phrase **descent with modification** to describe evolution. By this he meant that the species existing today have descended from other, preexisting species and that species are modified, or change, through time. This view was a radical departure from the pattern of independently created and immutable species as embodied in Plato's work and in the theory of special creation. In essence, the pattern component of the theory of evolution by natural selection makes two claims about the nature of species: (1) They change through time, and (2) they are related by common ancestry. Let's consider the evidence for each of these claims in turn.

Evidence for Change through Time

When Darwin began his work, biologists and geologists had just begun to assemble and interpret the fossil record. A **fossil** is any trace of an organism that lived in the past. These traces range from bones and branches to shells, tracks or impressions, and dung (**Figure 24.3**). The **fossil record** consists of all the fossils that have been found and described in the scientific literature.

Initially, fossils were organized according to their relative ages. This was possible because most fossils are found in **sedimentary rocks**, which form from sand or mud or other materials deposited at locations such as beaches or river mouths, and because sedimentary rocks are known to form in layers. Fossils from rocks underneath other rocks were judged to be older than the fossils found above them. In this way, researchers began putting fossils in an older-to-younger sequence. They also began naming different periods of geologic time, creating the sequence of eons, epochs, and periods called the **geologic time scale**. After the discovery of radioactivity in the late 1800s, researchers used radiometric dating techniques to assign absolute ages to the relative ages in the geologic time scale. According to data from radiometric dating, Earth is about 4.6 billion years old, and the earliest signs of life appear in rocks that formed 3.4–3.8 billion years ago. Instead of being 6000 years old as some proponents of the theory of special creation claimed, life on Earth is ancient.

The fossil record continues to expand in size and quality. Several observations about this data set convinced biologists that species have indeed changed through time.

Extinction In the early nineteenth century, researchers began discovering fossil bones, leaves, and shells that were unlike structures from any known animal or plant. At first, many scientists insisted that living examples of these species would be found in unexplored regions of the globe. But as research continued and the number and diversity of fossil collections grew, the argument became less and less plausible. After Baron Georges Cuvier published a detailed analysis of an **extinct**

(a) 110-million-year-old ammonite shell

(b) 50-million-year-old bird tracks

(c) 20,000-year-old sloth dung

FIGURE 24.3 A Fossil Is *Any* Trace of an Organism That Lived in the Past. In addition to **(a)** body parts such as shells or bones or branches, fossils may consist of **(b)** tracks or impressions, or even **(c)** pieces of dung.

species—that is, a species that no longer exists—called the Irish "elk" in 1812, most scientists accepted extinction as a reality. This gigantic deer was judged to be too large to have escaped discovery and too distinctive to be classified as a large-bodied population of an existing species.

Advocates of the theory of special creation argued that the fossil species were victims of the flood at the time of Noah. Darwin, in contrast, interpreted them as evidence that species are not static, immutable entities, unchanged since the moment of special creation. His reasoning was that if species have gone extinct, then the array of species living on Earth has changed through time. Recent analyses of the fossil record support the claim that more species have gone extinct than exist today. The data also indicate that species have gone extinct continuously throughout Earth's history—not just in one or even a few catastrophic events.

Transitional Forms Long before Darwin published his theory, researchers reported striking resemblances between the fossils found in the rocks underlying certain regions and the living species found in the same geographic areas. The pattern was so widespread that it became known as the "law of succession." The general observation was that extinct species in the fossil record were succeeded, in the same region, by similar species (**Figure 24.4a**). Early in the nineteenth century, the pattern was simply reported and not interpreted. But later, Darwin pointed out that it provided strong evidence in favor of the hypothesis that species had changed through time. His idea was that the extinct forms and living forms were related—that they represented ancestors and descendants.

As the fossil record improved, researchers discovered transitional forms that broadened the scope of the law of succession. A **transitional form** is a fossil species with traits that are intermediate between those of older and younger species. For example, intensive work over the past several decades has yielded fossils that document a gradual change over time from land-dwelling mammals that had limbs to ocean-dwelling mammals that had reduced limbs or no limbs (**Figure 24.4b**). All of the species in this sequence have distinctive types of ear bones that researchers use to identify whales. The oldest whale fossils found to date are from fox-sized animals that had eyes located at the tops of their heads. Based on this observation and the fact that the fossils were found in rocks that form only in ocean deposits, biologists suggest that the earliest whales were semiaquatic animals not unlike hippopotamuses. Over the subsequent 12 million years, the fossil record shows that the limbs of whale species became more reduced. These observations support the hypothesis that whales gradually became more strictly aquatic and more like today's whales in appearance and lifestyle. Whale species have clearly changed through time.

(a) Living species "succeed" fossil species in the same region.

Fossil sloth from South America

Present-day sloth from South America

(b) Transitional forms during the evolution of whales

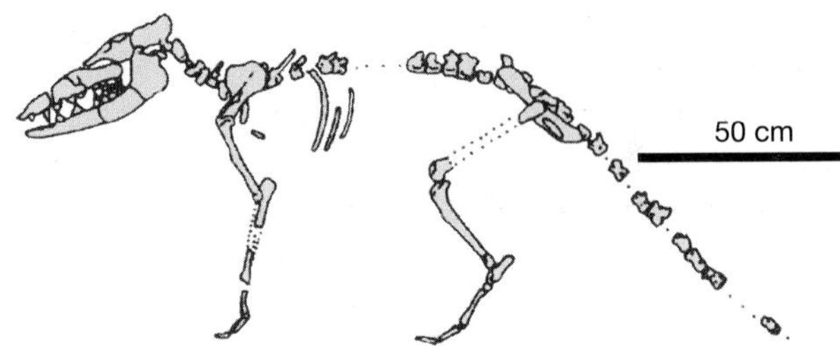

Pakicetus, about 50 million years old

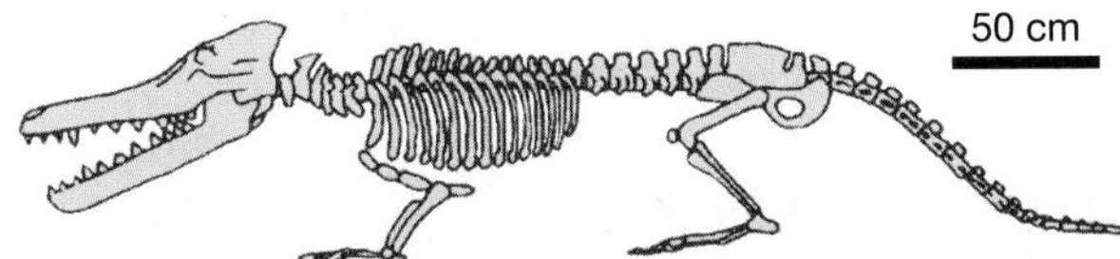

Ambulocetus, about 49 million years old

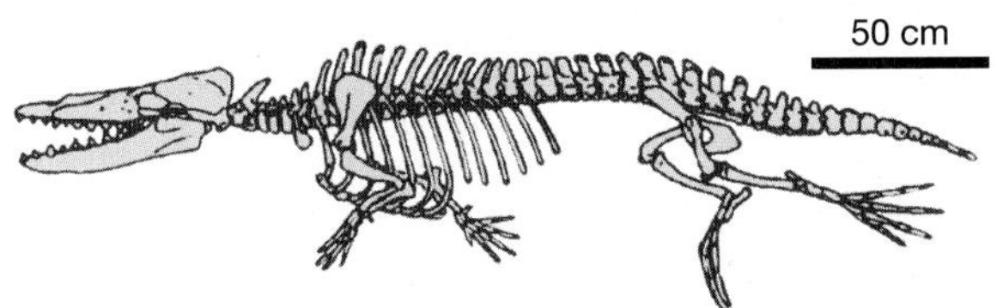

Rhodocetus, about 47 million years old

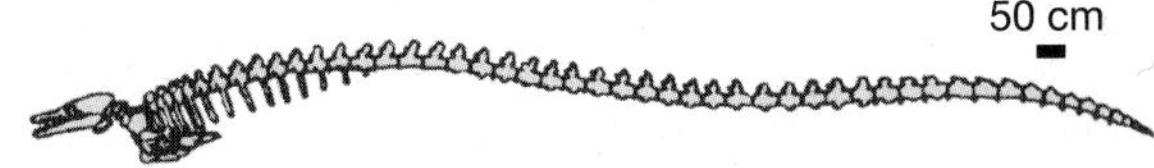

Basilosaurus, about 38 million years old

FIGURE 24.4 Evidence That Species Have Changed through Time. **(a)** Fossil and living sloths are found only in Central America and South America. Darwin argued that living species are descended from ancestors that lived in the same area. **(b)** Transitional forms document the changes that occurred as whales evolved from terrestrial mammals to the aquatic mammals of today.

● **QUESTION** How would these observations be explained under the theory of special creation?

Similar sequences of transitional forms document changes that led to the evolution of feathers and flight in birds, stomata and vascular tissue in plants, upright posture and large brains in humans, jaws in **vertebrates** (animals with backbones), limbs in amphibians and other vertebrates, the loss of limbs in snakes, and other traits. Each of these transitional forms provides strong evidence for change through time.

Vestigial Traits Darwin was the first to provide a widely accepted interpretation of vestigial traits. A **vestigial trait** is a reduced or incompletely developed structure that has no function or reduced function, but it is clearly similar to functioning organs or structures in closely related species.

Biologists have documented thousands of examples of vestigial traits. The genomes of humans and other organisms contain hundreds of pseudogenes—the functionless DNA sequences introduced in Chapter 20. Bowhead whales and rubber boas have tiny hip and leg bones that do not help them swim or slither; ostriches and kiwis have reduced wings and cannot fly; blind cave-dwelling fish still have eye sockets. The human appendix is a reduced version of the cecum—an organ found in other vertebrates that functions in digestion. Monkeys and many other primates have long tails; but our coccyx, illustrated in **Figure 24.5a**, is too tiny to help us maintain balance. Many mammals, including primates, are able to erect their hair when they are cold or excited. But our sparse fur does little to keep us warm, and goose bumps are largely ineffective in signaling our emotional state (**Figure 24.5b**).

The existence of vestigial traits is inconsistent with the theory of special creation, which maintains that species were perfectly designed by a supernatural being and that the characteristics of species are static. Instead, vestigial traits are evidence that the characteristics of species have changed over time.

Current Examples Biologists have documented hundreds of contemporary populations that are changing in response to changes in their environment. Bacteria have become resistant to drugs; insects have become resistant to pesticides; weedy plants have become resistant to herbicides. Section 24.4 provides a detailed analysis of research on two examples of evolution in action.

Biologists have also studied dozens of cases where new species are forming, right before our eyes. You'll be able to analyze data from one of these research projects in Chapter 26.

To summarize, change through time continues and can be measured directly. Evidence from the fossil record and living species indicates that life is ancient, that species have changed through the course of Earth's history, and that species continue to change. The take-home message is that species are dynamic—not static and unchanging, as claimed by the theory of special creation.

(a) The human tailbone is a vestigial trait.

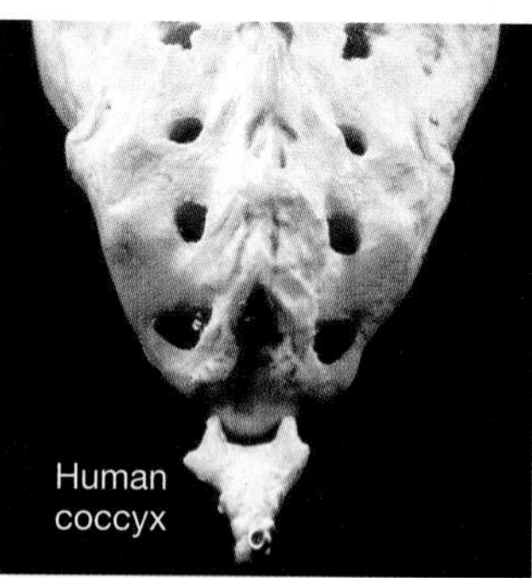

(b) Goose bumps are a vestigial trait.

Erect hair on chimp
(insulation, emotional display)

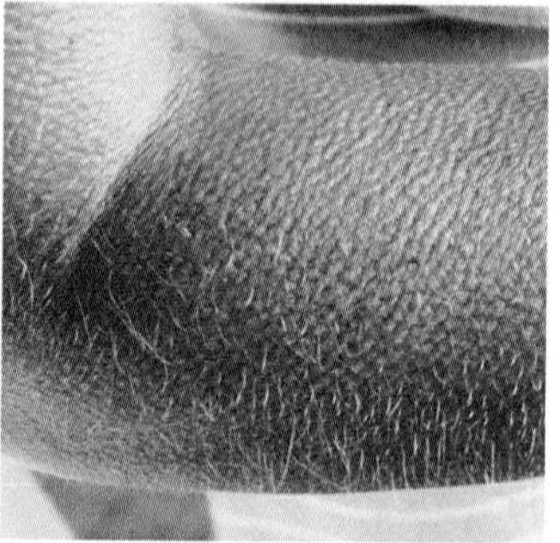

Human goose bumps

FIGURE 24.5 Vestigial Traits Are Reduced Versions of Traits in Other Species. (a) The tailbone and **(b)** goose bumps are human traits that have reduced function. They are similar to larger, fully functional structures in other species.

QUESTION How would these observations be explained if evolution occurred via inheritance of acquired characters?

Evidence That Species Are Related

Data from the fossil record and contemporary species refute the hypothesis that species are immutable. What about the claim that species were created independently—meaning that they are unrelated to each other?

Geographic Relationships Charles Darwin began to realize that species are related by common ancestry, just as individuals within a family are, during a five-year voyage he took aboard the English naval ship HMS *Beagle*. While fulfilling its mission to explore and map the coast of South America, the *Beagle* spent considerable time in the Galápagos Islands off the coast of present-day Ecuador. Darwin had taken over the role of ship's naturalist and gathered extensive collections of the plants and animals found in these islands. Among the birds he collected were what came to be known as the Galápagos mockingbirds, pictured in **Figure 24.6a**.

Several years after Darwin had returned to England, a naturalist friend in London pointed out that the mockingbirds Dar-

(a) Four mockingbird species on the Galápagos islands

Nesomimus parvulus

Nesomimus melanotis

Nesomimus trifasciatus

Nesomimus macdonaldi

(b) Darwin reasoned that they share a common ancestor.

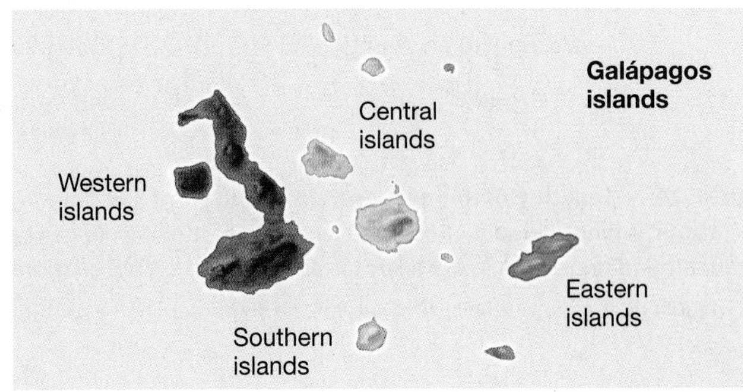

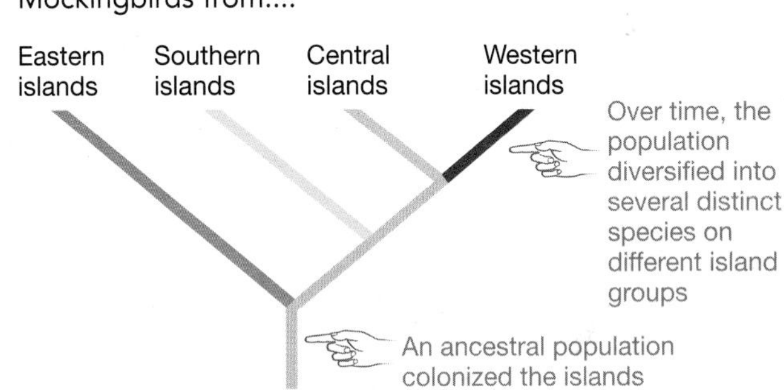

FIGURE 24.6 Close Relationships among Island Forms Argue for Shared Ancestry. (a) Darwin collected mockingbirds from several islands of the Galápagos. **(b)** Phylogeny illustrating Darwin's explanation for why mockingbirds from different islands are similar yet distinct.

QUESTION How would these observations be explained under the theory of special creation?

win had collected on different islands were distinct species, based on differences in coloration and beak size and shape. This struck Darwin as remarkable. Why would species that inhabit neighboring islands be so similar, yet clearly distinct? This turns out to be a very general pattern: In island groups across the globe, it is routine to find similar species on neighboring islands.

Darwin realized that this pattern—puzzling when examined as a product of special creation—made perfect sense when interpreted in the context of evolution, or descent with modification. He proposed that the mockingbirds were similar because they had descended from the same common ancestor. If so, then the mockingbird species are part of a phylogeny—a family tree of populations or species. Further, the mockingbirds can be placed on a phylogenetic tree, a branching diagram that describes the ancestor–descendant relationships among species, much as a genealogy describes the ancestor–descendant relationships among individual humans. **Figure 24.6b** illustrates this idea. Darwin's hypothesis was that instead of being created independently, mockingbird populations that colonized different islands had changed through time and formed new species. The presence of similar species in the same geographic area is still considered strong evidence that species are related by common ancestry—specifically, that their common ancestor lived in the same region.

Homology Translated literally, *homology* means "the study of likeness." When biologists first began to study the anatomy of humans and other vertebrates, they were struck by the remarkable similarity of their skeletons, muscles, and other structures. But because the biologists who did these early studies were advocates of the theory of special creation, they could not explain why striking similarities existed among certain organisms but not others. Today, biologists recognize that **homology** is a similarity that exists in species descended from a common ancestor. Human hair and dog fur are homologous. Humans have hair and dogs have hair because they share a common ancestor that was a mammal and also had hair.

Homology can be recognized and studied at three interacting levels. The most fundamental of these levels is **genetic homology**—a similarity in the DNA sequences of different species. As an example, consider the *eyeless* gene in fruit flies and the *Aniridia* gene in humans. Both genes act in determining where eyes will develop. The genes are so similar in DNA sequence that they code for proteins that are nearly identical in

Gene:	Amino acid sequence (single-letter abbreviations):
Aniridia (Human)	LQRNRTSFTQEQIEALEKEFERTHYPDVFARERLAAKIDLPEARIQVWFSNRRAKWRREE
eyeless (Fruit fly)	LQRNRTSFTNDQIDSLEKEFERTHYPDVFARERLAGKIGLPEARIQVWFSNRRAKWRREE

Only six of the 60 amino acids in these sequences are different. The two sequences are 90% identical.

FIGURE 24.7 Genetic Homology: Genes from Different Species May Be Similar in DNA Sequence or Other Attributes. Amino acid sequences from a portion of the *Aniridia* gene product found in humans and the *eyeless* gene product found in *Drosophila*. For a key to the single-letter abbreviations used for the amino acids, see Chapter 3.

QUESTION How would these observations be explained if evolution occurred via inheritance of acquired characters?

amino acid sequence (**Figure 24.7**). This observation is interesting because eye structure is so different in the two species—fruit flies have a compound eye with many lenses, and humans have a camera eye with a single lens. To explain this observation, biologists propose that fruit flies and humans descended from a common ancestor that had a gene similar to *eyeless* and *Aniridia*, and that this gene was involved in the formation of a simple light-gathering organ. The structure of the organ diverged as insects and mammals evolved, but the same gene remained responsible for where eyes are located.

Although Chapter 18 and Chapter 20 considered other examples of genetic homologies, the most remarkable of all genetic homologies is the genetic code introduced in Chapter 15. Except for one or two codons in a handful of species, the same 64 mRNA codons specify the same amino acids in all organisms that have been studied. To explain the existence of the universal genetic code, biologists hypothesize that today's code also existed in the common ancestor of all organisms alive today. Similarly, all organisms living today have a plasma membrane consisting of a phospholipid bilayer with interspersed proteins, transcribe the information coded in DNA to RNA via RNA polymerase, use ribosomes to synthesize proteins, employ ATP as an energy currency, and make copies of their genome via DNA polymerase. Like the genetic code, these traits undoubtedly existed in the cell that gave rise to all the species alive today.

The second level where biologists analyze homology was introduced in Chapters 21–23. A **developmental homology** is a similarity in embryonic traits. Developmental homologies are routinely observed in the overall **morphology,** or form, of embryos and in the fate of particular embryonic tissues.

Figure 24.8 illustrates the strong general resemblance among the embryos of vertebrates. Early in development, structures called gill pouches and tails that extend past the anus form in chicks, humans, and cats. Later in development, gill pouches are lost in all three species and tails are lost in humans. In fish, however, the gill pouches stay intact and give rise to functioning gills in adults. To explain this observation, biologists hypothesize that gill pouches and tails exist in chicks, humans, and cats because they existed in the fishlike species that was the common ancestor of today's fish, birds, and mammals. Embryonic gill pouches are a vestigial trait in chicks, humans, and cats; embryonic tails are a vestigial trait in humans.

Developmental homologies are also observed at the level of specific tissues. Even though the structure of the adult jaw is different in fish and mammals, the same group of embryonic cells develops into the jaw structure in both groups. This observation is logical if fish and mammals descended from the same common ancestor, and if this ancestor also had a jaw that developed from the same population of embryonic cells.

Developmental homologies are due to homologous genes and give rise to **structural homology**—similarities in adult morphology. A classic example is the common structural plan observed in the limbs of vertebrates (**Figure 24.9**). In Darwin's own words, "What could be more curious than that the hand of a man, formed for grasping, that of a mole for digging, the leg of the horse, the paddle of the porpoise, and the wing of the bat, should all be constructed on the same pattern, and should include the same bones, in the same relative positions?" Darwin raised the question because an engineer would

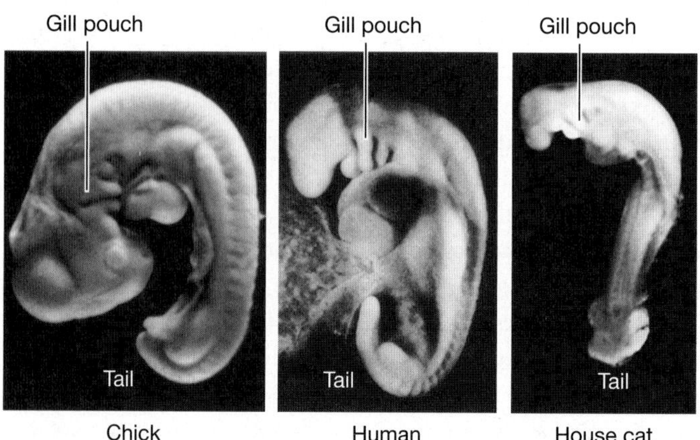

FIGURE 24.8 Developmental Homology: Structures That Appear Early in Development Are Similar. The early embryonic stages of a chick, a human, and a cat, showing a strong resemblance.

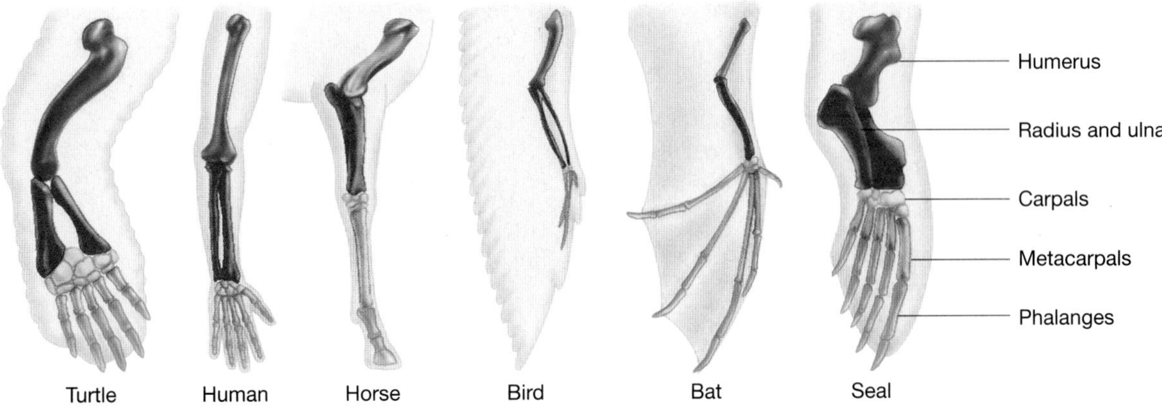

FIGURE 24.9 Structural Homology: Limbs with Different Functions Have the Same Underlying Structure. Even though their function varies, all vertebrate limbs are modifications of the same number and arrangement of bones. Darwin interpreted structural homologies like these as a product of descent with modification. (These limbs are not drawn to scale.)

QUESTION What is the relationship among genetic homologies, developmental homologies, and structural homologies?

never use the same underlying pattern to design the structure of a grasping tool, a digging implement, a walking device, a propeller, and a wing. But if all mammals descended from a common ancestor, and if that ancestor had a limb with the same basic plan shown in Figure 24.9, then it would be logical to observe that its descendants had modified forms of the same design.

The general point is that, in many cases, traits are similar in different species because the species in question are related to each other by common descent. If species were created independently of one another, these types of similarities would not occur.

Evolution Is Change through Time

Biologists draw upon data from several sources to challenge the hypothesis that species are immutable and were created independently. The data support the idea that species have descended, with modification, from a common ancestor. **Table 24.1** summarizes this evidence.

By the late 1800s, the vast majority of biologists were convinced that the pattern component of the theory of evolution was valid and that the theory of special creation was incorrect. This was long before genetic homologies had been described, before many of the most important transitional forms were discovered, and before contemporary examples of populations changing through time and new species being created had been documented. Among biologists, controversy over the fact of evolution ended more than 120 years ago.

As you evaluate the evidence supporting the pattern component of the theory of evolution, it's important to recognize that no single observation or experiment instantly "proved" the fact of evolution and swept aside belief in special creation. Rather, Darwin and others argued that the pattern called evolution was much more consistent with the data than was the pattern predicted by special creation. Descent with modification was a more successful and powerful scientific theory because it explained observations—such as vestigial traits and the close relationships among species on neighboring islands—that special creation could not.

What about the process component of the theory of evolution by natural selection? If the limbs of bats and humans were not created independently and recently, how did they come to be?

TABLE **24.1** **Evidence for Evolution**

Prediction 1: Species Are Not Static, but Change through Time
• Most species have gone extinct.
• Fossil (extinct) species frequently resemble living species found in the same area.
• Transitional forms document change in traits through time.
• Earth is ancient.
• Vestigial traits are common.
• Populations and species can be observed changing today.
Prediction 2: Species Are Related, Not Independent
• Closely related species often live in the same geographic area.
• Homologous traits are common and exist at three levels:
1. genetic (gene structure and the genetic code)
2. developmental (embryonic structures and processes)
3. structural (morphological traits)

24.3 The Process of Evolution: How Does Natural Selection Work?

Darwin's greatest contribution did not lie in recognizing the fact of evolution. Lamarck and other researchers had already proposed evolution as a pattern in nature long before Darwin began his work. Instead, Darwin's crucial insight lay in recognizing a process, called **natural selection**, that could explain the pattern of descent with modification.

In his original formulation, Darwin broke the process of evolution by natural selection into four simple postulates, or steps in a logical sequence:

1. The individual organisms that make up a population vary in the traits they possess, such as their size and shape.
2. Some of the trait differences are heritable, meaning that they are passed on to offspring genetically. For example, tall parents may tend to have tall offspring.
3. In each generation, many more offspring are produced than can possibly survive. Thus, only some individuals in the population survive long enough to produce offspring; and among the individuals that produce offspring, some will produce more than others.
4. The subset of individuals that survive best and produce the most offspring is not a random sample of the population. Instead, individuals with certain heritable traits are more likely to survive and reproduce. Natural selection occurs when individuals with certain characteristics produce more offspring than do individuals without those characteristics.

Because the selected traits are passed on to offspring, the frequency of the selected traits increases from one generation to the next. **Evolution**—a change in the genetic characteristics of a population over time—is simply an outcome of these four steps.

In studying these steps, you should realize that variation among individuals is essential if evolution is to occur. You should also recognize that Darwin had to introduce population thinking into biology because it is populations that change over time when evolution occurs. To come up with these postulates and understand their consequences, Darwin had to think in a revolutionary way.

Today, biologists usually condense Darwin's four postulates into two statements that communicate the essence of evolution by natural selection more forcefully: Evolution by natural selection occurs when (1) heritable variation leads to (2) differential success in survival and reproduction.

To illustrate this condensed version of Darwin's logic, consider a population of 10 moths where differences in wing coloration are due primarily to two alleles of a single gene (**Figure 24.10a**). These alleles are called A_1 and A_2. The moths are diploid so there are a total of 20 alleles present in the population, before selection starts. The A gene product is a protein

(a) If heritable variation...

Color varies among individuals primarily because of differences in their genotype

(b) ... leads to differential success...

Birds find and eat many more dark-winged moths than light-winged moths

(c) ... then evolution results.

	Frequency of A_1 allele	Frequency of A_2 allele
Start	0.5 (10 of 20 alleles present)	0.5 (10 of 20 alleles present)
End	0.625 (5 of 8 alleles present)	0.375 (3 of 8 alleles present)

Allele frequencies have changed in the surviving moths

FIGURE 24.10 Evolution by Natural Selection Occurs When Heritable Variation Leads to Differential Success. See text for explanation. Note that this example is not hypothetical. The allele frequency changes diagrammed here have occurred independently in populations of peppered moths (*Biston betularia*) in England and several locations in North America.

QUESTION Why is it important that similar changes in allele frequencies have been observed in different populations of this species, independently?

that is involved in the synthesis or deposition of a dark pigment called melanin. Individuals with the genotype A_1A_1 have a light gray coloration, while individuals with the A_1A_2 or A_2A_2 genotype are black. Now, it's important to recognize that some of the variation in wing color in this population is due to differences in the environment that the individuals experience, and some is due to differences in the alleles they carry. For example, not all individuals with the A_2A_2 genotype will be exactly the same shade of black, because each individual was exposed to different temperatures and had different amounts of nutrients available when its wings were developing and the *A* gene product was active. But because the A_1 and A_2 alleles have such different effects on wing color, there is heritable variation in the trait within the population.

These moths are active at night and spend the day resting on tree trunks and branches, where they are hunted by birds. In an environment where trees with light-colored bark are common, birds can find and eat dark-winged individuals much more readily than they can find and eat light-winged individuals. Because predation by birds causes natural selection on wing color, there is differential success: In this environment, light-winged individuals survive better than dark-winged individuals. In our example, only 4 individuals survived—meaning that only 8 alleles are now present in the population (**Figure 24.10b**). As a result, evolution occurs. Evolution is defined as a change in allele frequencies in a population over time. In this case, the frequency of the A_1 allele increases in this population over time (**Figure 24.10c**).

To explain the process of natural selection, Darwin referred to successful individuals as "more fit" than other individuals. He gave the word *fitness* a definition different from its everyday English usage. **Darwinian fitness** is the ability of an individual to produce offspring, relative to that ability in other individuals in the population. This is a measurable quantity. Researchers study populations in the lab or in the field and estimate the relative fitness of each individual by counting how many offspring it produces relative to other individuals. In environments where most trees have light gray bark, light-winged moths have higher fitness than dark-winged moths. But in environments dominated by trees with dark bark, black-winged moths have higher fitness than light-winged moths.

The concept of fitness, in turn, provides a compact way of formally defining adaptation. The biological meaning of adaptation, like the biological meaning of fitness, is quite different from its normal English usage. In biology, an **adaptation** is a heritable trait that increases the fitness of an individual in a particular environment relative to individuals lacking the trait. Adaptations increase fitness—the ability to produce offspring. Light-colored wings are an adaptation in environments where most trees have gray bark.

To summarize, evolution by natural selection occurs when heritable variation in traits leads to differential success in survival and reproduction. If you understand this concept, you should be able to make a figure analogous to Figure 24.10, illustrating what would happen to allele frequencies in the population if dark-barked trees were much more abundant than light-barked trees. You should also be able to explain what would happen if all of the variation in wing coloration were due to differences in nutrition or temperature that the individuals experienced—meaning that they all had the same genotype.

To help you get a thorough understanding of how evolution by natural selection works, Section 24.4 is devoted to data—specifically, to recent studies of how natural selection works in actual populations. Biologists accept Darwin's theory not only because of its explanatory power but also because evolution by natural selection has been observed directly.

24.4 Evolution in Action: Recent Research on Natural Selection

Darwin's theory of evolution by natural selection is testable. If the theory is correct, biologists should be able to test the validity of each of Darwin's postulates and actually observe evolution in natural populations.

This section summarizes two examples in which evolution by natural selection has been, or is being, observed in nature. Literally hundreds of other case studies are available, involving a wide variety of traits and organisms. To begin, let's explore the evolution of drug resistance—one of the great challenges facing today's biomedical researchers and physicians.

How Did *Mycobacterium tuberculosis* Become Resistant to Antibiotics?

Mycobacterium tuberculosis, the bacterium that causes tuberculosis, or TB, has long been a scourge of humankind. TB was responsible for almost 25 percent of all deaths in New York City in 1804; in nineteenth-century Paris, the figure was closer to 33 percent. To put these numbers in perspective, consider that all types of cancer, combined, currently account for about 30 percent of the deaths that occur in the United States. TB was once as great a public health issue as cancer is now.

Although tuberculosis still kills more adults than any other viral or bacterial disease in the world, TB attracted relatively little attention in the industrialized nations between about 1950 and 1990. During that time, TB was primarily a disease of developing nations.

The decline of tuberculosis in Western Europe, North America, Japan, Korea, and Australia is one of the great triumphs of modern medicine. In these countries, sanitation, nutrition, and general living conditions began to improve dramatically in the early twentieth century. When people are healthy and well nourished, their immune systems work well enough to stop most *M. tuberculosis* infections quickly—before the infection can harm the individual and before the bacteria can be transmitted to a new host. In addition, antibiotics such as rifampin started to become available in the industrialized countries in the early 1950s. These drugs allowed physicians to stop even advanced infections and saved millions of lives.

In the late 1980s, however, rates of *M. tuberculosis* infection surged in many countries, and in 1993 the World Health Organization (WHO) declared TB a global health emergency. Physicians were particularly alarmed because the strains of *M. tuberculosis* responsible for the increase were largely or completely resistant to rifampin and other antibiotics that were once extremely effective. How and why did the evolution of drug resistance occur? The case of a single patient—a young man who lived in Baltimore—will illustrate what is happening all over the world.

The story begins when the individual was admitted to the hospital with fever and coughing. Chest X-rays, followed by bacterial cultures of fluid ejected from the lungs, showed that he had an active TB infection. He was given several antibiotics for 6 weeks, followed by twice-weekly doses of rifampin and isoniazid for an additional 33 weeks. Ten months after therapy started, bacterial cultures from his chest fluid indicated no *M. tuberculosis* cells. His chest X-rays were also normal. The antibiotics seemed to have cleared the infection.

Just two months after the TB tests proved normal, however, the young man was readmitted to the hospital with a fever, severe cough, and labored breathing. Despite being treated with a variety of antibiotics, including rifampin, he died of respiratory failure 10 days later. Samples of material from his lungs showed that *M. tuberculosis* was again growing actively there. But this time the bacterial cells were completely resistant to rifampin.

Drug-resistant bacteria had killed this patient. Where did they come from? Is it possible that a strain that was resistant to antibiotic treatment evolved *within* him? To answer this question, a research team analyzed DNA from the drug-resistant strain and compared it with stored DNA from *M. tuberculosis* cells that had been isolated a year earlier from the same patient. After examining extensive stretches from each genome, the biologists were able to find only one difference: a point mutation in a gene called *rpoB*. This gene codes for a component of the enzyme RNA polymerase. Recall from Chapter 16 that RNA polymerase transcribes DNA to mRNA, and that a point mutation is a single base change in DNA. In this case, the mutation changed a cytosine to a thymine, altering the normal codon TCG to a mutant one, TTG. As a result, the RNA polymerase produced by the drug-resistant strain had leucine instead of serine at the 153rd amino acid in the polypeptide chain.

This result is meaningful. The drug that was being used to treat the patient works by binding to the RNA polymerase of *M. tuberculosis*. When the drug enters an *M. tuberculosis* cell and binds to RNA polymerase, it interferes with transcription. If sufficient quantities of rifampin are present for long enough and if the drug binds tightly, bacterial cells will not be able to make proteins efficiently and produce few offspring. But apparently the substitution of a leucine for a serine prevents rifampin from binding efficiently. Consequently, cells with the $C \rightarrow T$ mutation continue to produce offspring efficiently even in the presence of the drug.

These results suggest that a chain of events led to this patient's death (**Figure 24.11**). The researchers hypothesized that by chance, one or a few of the cells present early in the course of the infection happened to have an *rpoB* gene with the $C \rightarrow T$ mutation. Under normal conditions, mutant forms of RNA polymerase do not work as well as the more common form, so cells with the $C \rightarrow T$ mutation would not produce many offspring and would stay at low frequency—even while the overall population grew to the point of inducing symptoms that sent the young man to the hospital.

At that point, therapy with rifampin began. In response, cells in the population with normal RNA polymerase began to grow much more slowly or to die outright. As a result, the overall bacterial population declined in size so drastically that the patient appeared to be cured—his symptoms began to disappear. But cells with the $C \rightarrow T$ mutation had an advantage in the new environment. They began to grow more rapidly than the normal cells and continued to increase in number after therapy ended. Eventually the *M. tuberculosis* population regained its former abundance, and the patient's symptoms reappeared. However, drug-resistant cells now dominated the population. This is why the second round of rifampin therapy was futile. If health-care workers or the patient's family had contracted TB from him, rifampin therapy would have been useless on them, too, and the disease would have continued to spread.

Does this sequence of events mean that evolution by natural selection occurred? One way of answering this question is to review Darwin's four postulates and determine whether each was tested and verified:

1. *Did variation exist in the population?* The answer is yes. Due to mutation, both resistant and nonresistant strains of TB were present prior to administration of the drug. Most *M. tuberculosis* populations, in fact, exhibit variation for the trait; studies on cultured *M. tuberculosis* show that a mutation conferring resistance to rifampin is present in one out of every 10^7 to 10^8 cells.

2. *Was this variation heritable?* The answer is yes. The researchers showed that the variation in the phenotypes of the two strains—from drug susceptibility to drug resistance—was due to variation in their genotypes. Because the mutant *rpoB* gene is passed on to daughter cells when a *Mycobacterium* replicates, the allele and the phenotype it produces—drug resistance—are passed on to offspring.

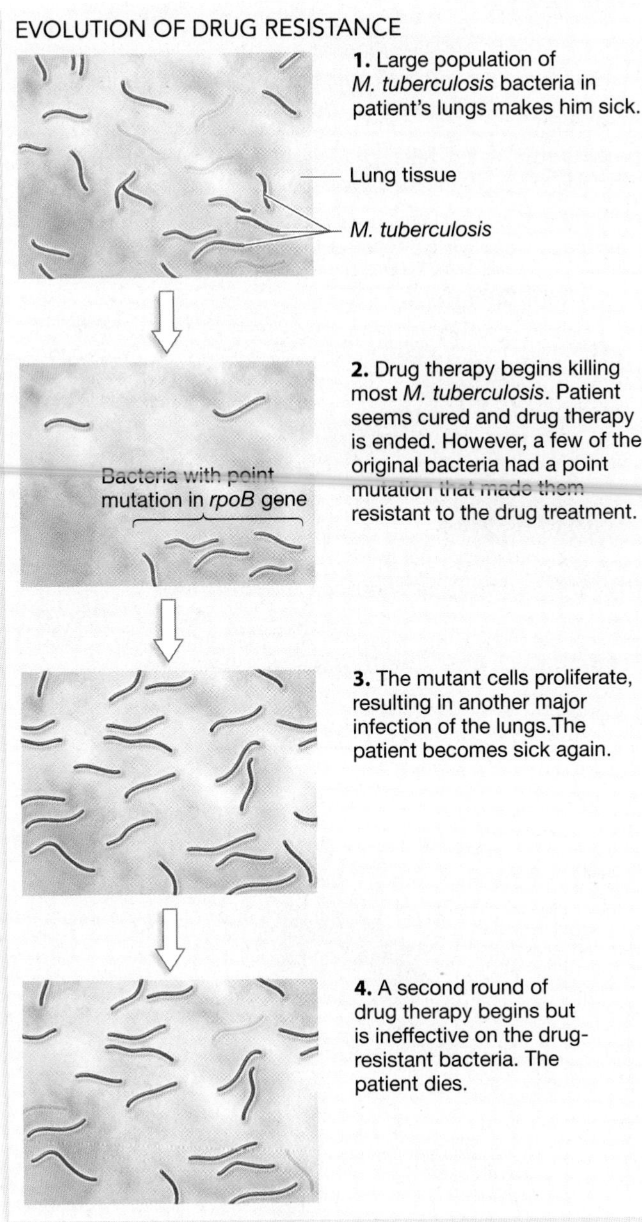

FIGURE 24.11 Alleles That Confer Drug Resistance Increase in Frequency When Drugs Are Used.

QUESTION In most individuals, the immune system is able to eliminate the few bacteria that remain at step 2. This individual had AIDS, however, so his immune system was compromised. Why did step 3 occur? If he had transmitted the infection to another person at step 3 or step 4, would they respond to drug therapy?

3. *Was there variation in reproductive success?* That is, did some *M. tuberculosis* individuals survive better and leave more offspring than other *M. tuberculosis* individuals? The answer is yes. Only a tiny fraction of M. tuberculosis cells in the patient survived the first round of antibiotics long enough to reproduce—so few that, after the initial therapy, his chest X-ray was normal and his fluid sample contained no *M. tuberculosis* cells. In contrast, a small number of drug-resistant bacterial cells were able to survive and keep reproducing after the onset of drug treatment.

4. *Did selection occur?* That is, did a nonrandom subset of the population produce the most offspring? The answer is yes. The *M. tuberculosis* population present early in the infection was different from the *M. tuberculosis* population present at the end. This could have occurred only if cells with the drug-resistant allele had higher reproductive success when rifampin was present than did cells with the normal allele. *M. tuberculosis* individuals with the mutant *rpoB* gene had higher fitness in an environment where rifampin was present. The mutant allele produces a protein that is an adaptation when the cell's environment contains the antibiotic.

This study verified all four postulates and confirmed that evolution by natural selection had occurred. The *M. tuberculosis* population evolved because the mutant *rpoB* allele increased in frequency. The individual cells themselves did not evolve, however. When natural selection occurred, the individual cells did not change through time; they simply survived or died, or produced more or fewer offspring. This is a fundamentally important point: Natural selection acts on individuals, because individuals experience differential success. But only populations evolve. Allele frequencies change in populations, not in individuals. Understanding evolution by natural selection requires population thinking—not typological thinking.

The events just reviewed have occurred many times in other patients. Recent surveys indicate that drug-resistant strains now account for about 10 percent of the *M. tuberculosis*–causing infections throughout the world. And the emergence of drug resistance in TB is far from unusual. Resistance to a wide variety of insecticides, fungicides, antibiotics, antiviral drugs, and herbicides has evolved in hundreds of insects, fungi, bacteria, viruses, and plants. In many cases, the specific mutations that lead to a fitness advantage and the spread of the resistance alleles are known.

(MB) **Web Animation** **at www.masteringbio.com**
Natural Selection for Antibiotic Resistance

Why Are Beak Size, Beak Shape, and Body Size Changing in Galápagos Finches?

The TB example is particularly satisfying, because the molecular basis of both heritable variation and differential success is understood. It is one of many examples of natural selection and rapid evolution induced by drugs, pesticides, herbicides, or other human-caused changes in the environment. But can biologists study evolution in response to natural environmental change—when humans are not involved? The answer is yes. As an example of how this work is done, let's review research led by Peter and Rosemary Grant. These biologists have been

FIGURE 24.12 The Medium Ground Finch Is a Seed-Eater.

investigating changes in beak size, beak shape, and body size that have occurred in finches native to the Galápagos Islands.

The medium ground finch, pictured in **Figure 24.12**, makes its living by eating seeds. Finches crack seeds with their beaks. For over three decades, the population of medium ground finches on Isle Daphne Major of the Galápagos has been studied intensively by the Grants' team. Because Daphne Major is small—about the size of 80 football fields—the researchers have been able to catch, weigh, and measure all individuals and mark each one with a unique combination of colored leg bands.

Early studies of the finch population established that beak size and shape and body size vary among individuals, and that beak morphology and body size are heritable. Stated another way, parents with particularly deep beaks tend to have offspring with deep beaks. Large parents also tend to have large offspring. Beak size and shape and body size are traits with heritable variation.

Not long after the team had established these results, a dramatic selection event occurred. In the annual wet season of 1977, Daphne Major received just 24 mm of rain instead of the 130 mm that normally falls. During the drought, few plants were able to produce seeds, and 84 percent (about 660 individuals) of the medium ground finch population disappeared.

Two observations support the hypothesis that most or all of these individuals died of starvation. The researchers found a total of 38 dead birds, and all were emaciated. Further, none of the missing individuals were spotted on nearby islands, and none reappeared once the drought had ended and food supplies returned to normal.

The research team realized that the die-off presented an opportunity to study natural selection. The change in the environment produced what biologists call a **natural experiment**. Instead of comparing groups created by direct manipulation under controlled conditions, natural experiments allow researchers to compare treatment groups created by an unplanned, natural change in conditions. In this case, the Grant's team could compare the population before and after the drought. Were the survivors different from nonsurvivors? When the biologists analyzed the characteristics of each group, they found that survivors tended to have much deeper beaks than did the birds that died (**Figure 24.13**). This was an impor-

Experiment

Question: Did natural selection on ground finches occur when the environment changed?

Hypothesis to be tested: Beak characteristics changed in response to changes in food availability.

Null hypothesis: No changes in beak characteristics occurred, even though food availability changed.

Experimental setup:

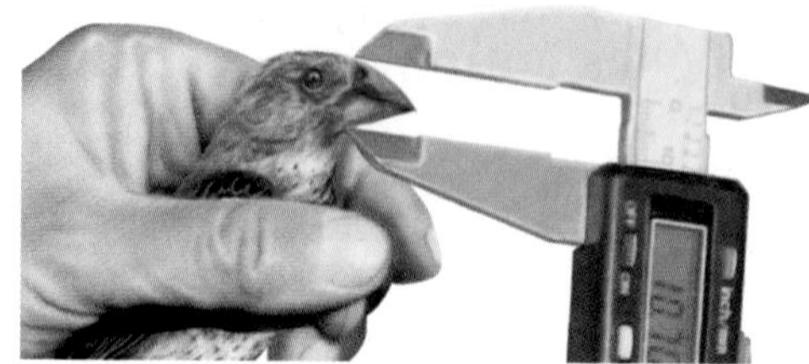

Weigh and measure all birds in the population before and after the drought.

Prediction of natural selection hypothesis:

Prediction of null hypothesis (no selection):

Results:

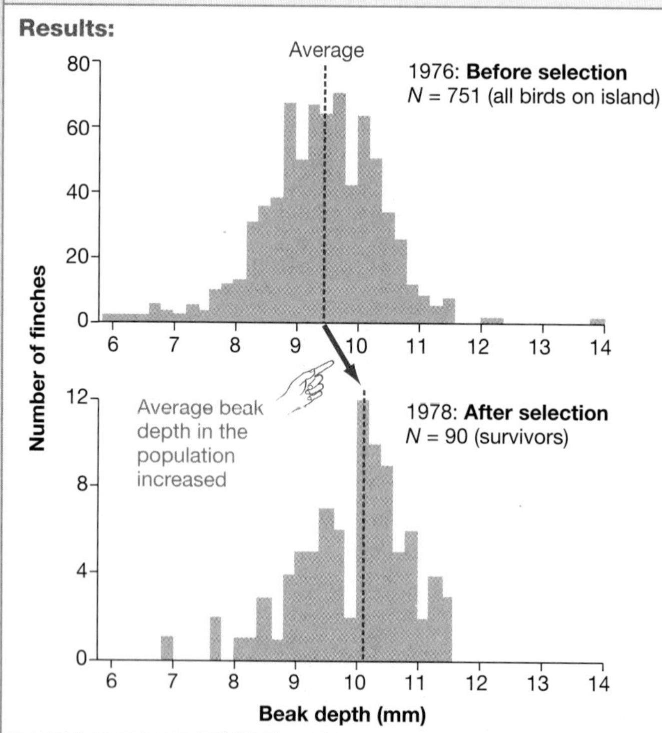

Conclusion: Natural selection occurred. The characteristics of the population have changed.

FIGURE 24.13 A Natural Experiment: Changes in a Medium Ground Finch Population in Response to a Change in the Environment (a Drought). The histograms in the Results section show the distribution of beak depth in medium ground finches on Daphne Major before and after the drought of 1977. *N* is the sample size.

● **QUESTION** Why was the sample size so much smaller in 1978?

● **EXERCISE** Fill in the predictions made by the two hypotheses.

tant finding, because the type of seeds available to the finches had changed dramatically as the drought continued. At the drought's peak, the tough fruits of a plant called *Tribulus cistoides* served as the finches' primary food source. These fruits are so difficult to crack that they are ignored in years when food supplies are normal. Grant's group hypothesized that individuals with particularly large and deep beaks were more likely to crack these fruits efficiently enough to survive.

At this point, the Grants had shown that natural selection led to an increase in average beak depth in the population. When breeding resumed in 1978, the offspring that were produced had beaks that were half a millimeter deeper, on average, than those in the population that existed before the drought. This result confirmed that evolution had occurred. In only one generation, natural selection had led to a measurable change in the characteristics of the population. Alleles that led to the development of deep beaks must have increased in frequency. Large, deep beaks were an adaptation for cracking large fruits and seeds. More recent work, reviewed in **Box 24.2** on page 499, has identified one of the genes in which allele frequencies may have changed during this event.

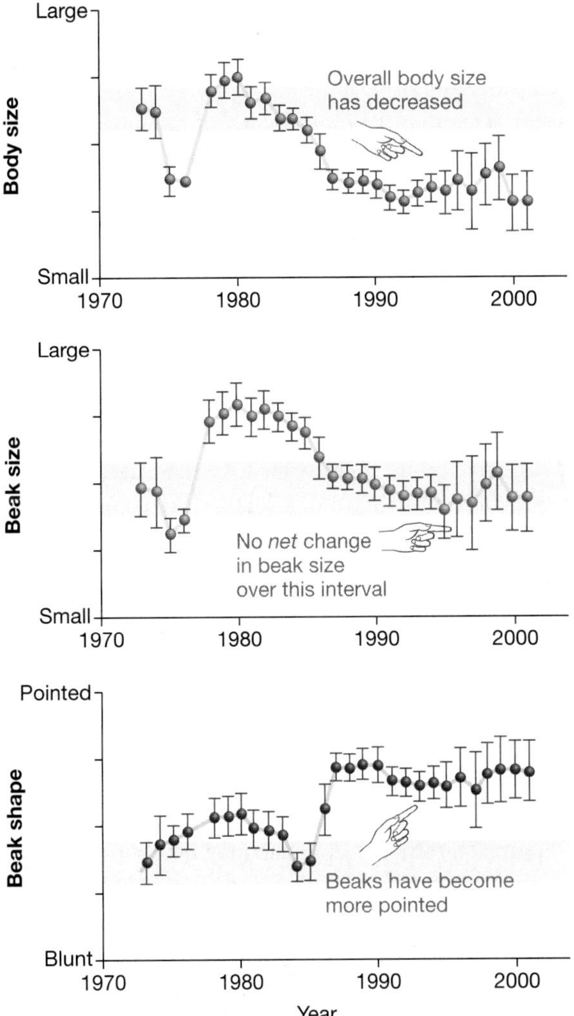

FIGURE 24.14 Body Size, Beak Size, and Beak Shape in Finches Changed over a 30-Year Interval.

EXERCISE Label the drought in 1977 and the wet year in 1983. Circle years when (1) average body size increased, (2) average beak size declined, and (3) beaks became pointier.

EXERCISE The Grants recently published more data on beak size (the middle graph). In 2005 and 2006, average beak size dropped to about the first hash mark above "Small" on the vertical axis. Place these data points on the graph. Do the new data change your interpretation?

In 1983, however, the environment changed again. Over a seven-month period, a total of 1359 mm of rain fell. Plant growth was luxuriant, and finches fed primarily on small, soft seeds that were being produced in abundance. During this interval, small individuals with small, pointed beaks had exceptionally high reproductive success. As a result, the characteristics of the population changed again. In fact, the Grants have documented continued evolution in response to continued changes in the environment. **Figure 24.14** documents changes that have occurred in average body size, beak size, and beak shape over the past 30 years. Although beak size showed no net change when the start and end of this interval are compared, beak shape changed dramatically. On average, finch beaks got much pointier over the course of the study. In addition, overall body size got smaller. Long-term studies such as this have been powerful, because they have succeeded in documenting natural selection in response to changes in the environment.

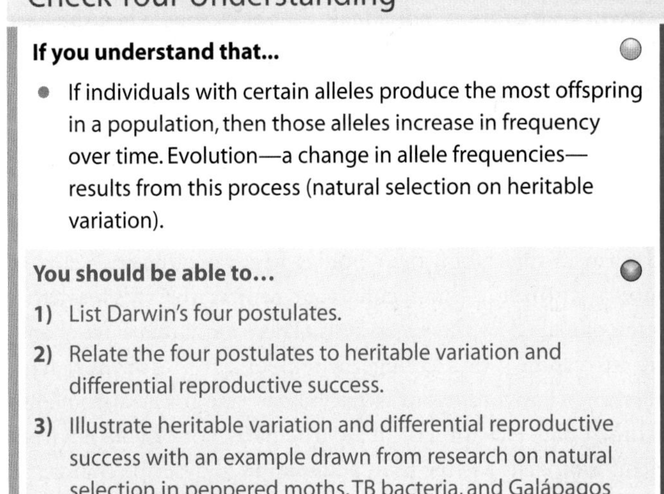

Check Your Understanding

If you understand that...

- If individuals with certain alleles produce the most offspring in a population, then those alleles increase in frequency over time. Evolution—a change in allele frequencies—results from this process (natural selection on heritable variation).

You should be able to...

1) List Darwin's four postulates.
2) Relate the four postulates to heritable variation and differential reproductive success.
3) Illustrate heritable variation and differential reproductive success with an example drawn from research on natural selection in peppered moths, TB bacteria, and Galápagos finch populations.

24.5 The Nature of Natural Selection and Adaptation

Natural selection appears to be a simple process, but appearances can be deceiving. Research has shown that it is often misunderstood. To help clarify how natural selection works, let's consider some of the more common misconceptions about natural selection in light of data on drug resistance in the TB bacterium and changes in finch populations.

Selection Acts on Individuals, but Evolutionary Change Occurs in Populations

Perhaps the most important point to clarify about natural selection is that during the process, individuals do not change—only the population does. During the drought, the beaks of individual finches did not become deeper. Rather, the average beak depth in the population increased over time, because deep-beaked individuals produced more offspring than shallow-beaked individuals did. Natural selection acted on individuals, but the evolutionary change occurred in the characteristics of the population.

In the same way, individual bacterial cells did not change when rifampin was introduced to their environment. Each *M. tuberculosis* cell had the same polymerase alleles all its life. But because the mutant allele increased in frequency over time, the characteristics of the bacterial population changed.

This point should make sense, given that evolution is defined as changes in allele frequencies. An individual's allele frequencies cannot change over time—it has the alleles it was born with all its life.

This point also highlights a sharp contrast between evolution by natural selection and evolution by the inheritance of acquired characters—the hypothesis promoted by Jean-Baptiste de Lamarck. Lamarck proposed that individuals change in response to challenges posed by the environment and that the changed traits are then passed on to offspring. In contrast, Darwin realized that individuals do not change when they are selected—they simply produce more offspring than other individuals do.

The issue is tricky because individuals often *do* change in response to changes in the environment. For example, wood frogs native to northern North America are exposed to extremely cold temperatures as they overwinter. When ice begins to form in their skin, their bodies begin producing a sort of natural antifreeze—molecules that protect their tissues from being damaged by the ice crystals. These individuals are changing in response to a change in temperature.[1] You may have observed changes in your own body as you got accustomed to living at high elevation or in a particularly hot or cold environment. Biologists use the term **acclimation** to describe changes in an individual's phenotype that occur in response to changes in environmental conditions. The key is to realize that phenotypic changes due to acclimation are not passed on to offspring. As a result, they cannot cause evolution. ● If you understand this concept, you should be able to explain the difference between the biological definition of adaptation and its use in everyday English, and then explain the difference between acclimation and adaptation.

[1]In some species of frogs, so much extracellular fluid freezes during cold snaps that individuals appear to be frozen solid. Their hearts also stop beating. When temperatures warm in the spring, their hearts start beating again, their tissues thaw, and they resume normal activities.

Evolution Is Not Goal-Directed or Progressive

It is tempting to think that evolution by natural selection is goal directed. For example, you might hear a fellow student say that *M. tuberculosis* cells "wanted" or "needed" the mutant, drug-resistant allele so that they could survive and continue to reproduce in an environment that included rifampin. This does not happen. The mutation that created the mutant allele occurred randomly, due to an error during DNA synthesis, and it just happened to be advantageous when the environment changed. Adaptations do not occur because organisms want or need them.

It is also tempting to think that evolution by natural selection is progressive—meaning organisms have gotten "better" over time. (In this context, better usually means more complex.) It is true that the groups appearing later in the fossil record are often more morphologically complex or "advanced" than closely related groups that appeared earlier. Flowering plants are considered more complex than mosses, and most biologists would agree that the morphology of mammals is more complex than that of amphioxus. But there is nothing predetermined or absolute about this tendency. In fact, complex traits are routinely lost or simplified over time as a result of evolution by natural selection. Populations that become parasitic are particularly prone to this trend. Tapeworms, for example, lack a mouth and digestive system. As parasites that live in the intestines of humans and other mammals, they simply absorb nutrients directly from their environment, across their plasma membranes. But tapeworms evolved from species with a sophisticated digestive tract. Tapeworms lost their digestive tract as a result of evolution by natural selection. In a similar vein, snakes evolved from lizard-like ancestors that had legs (see Chapter 21).

The nonprogressive nature of evolution by natural selection contrasts sharply with Lamarck's conception of the evolutionary process, in which organisms progress over time to higher and higher levels on a chain of being. Under Aristotle's and Lamarck's hypothesis, it is sensible to refer to "higher" and "lower" organisms (**Figure 24.15a**). But under evolution by natural selection, there is no such thing as a higher or lower organism (**Figure 24.15b**). Green algae may be a more ancient group than flowering plants, but neither group is higher or lower than the other. Green algae simply have a different suite of adaptations than do flowering plants, so they thrive in different types of environments.

The general message here is that all populations have evolved by natural selection based on their ability to gather resources and produce offspring. All organisms are adapted to their environment. Evolution by natural selection is not progressive, so no organism is any "higher" than any other organism.

To drive this point home, recall what happened when torrential rains fell on Daphne Major. Instead of continuing to

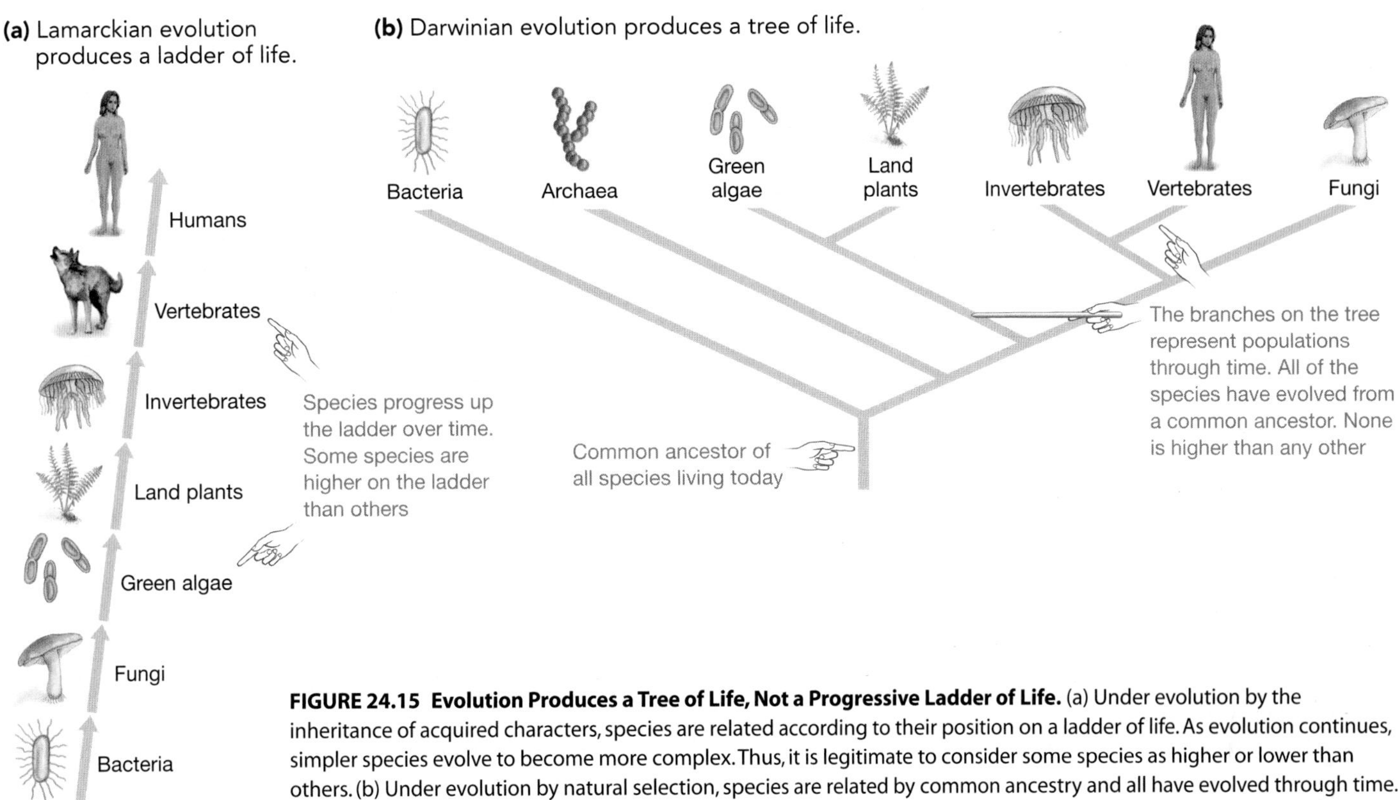

FIGURE 24.15 Evolution Produces a Tree of Life, Not a Progressive Ladder of Life. (a) Under evolution by the inheritance of acquired characters, species are related according to their position on a ladder of life. As evolution continues, simpler species evolve to become more complex. Thus, it is legitimate to consider some species as higher or lower than others. (b) Under evolution by natural selection, species are related by common ancestry and all have evolved through time. None is any higher or lower than any other. As evolution continues, species may become simpler or more complex, depending on what traits are favored by the environment. (For help in reading evolutionary trees, see **BioSkills 2**.)

increase in size, the average beak size of the finch population declined. Natural selection did not produce a progression of ever-larger beaks. Instead, the population simply responded to whatever change happened to occur in the environment. In many bacterial and viral populations, the frequency of drug-resistant individuals has declined dramatically when the drug in question was discontinued.

Natural selection is not goal directed or progressive. It simply favors individuals that happen to be better adapted to the environment existing at the time.

Animals Do Not Do Things for the Good of the Species

Consider the widely circulated story that rodents called lemmings sacrifice themselves for the good of their species. The story claims that when lemming populations are high, overgrazing is so extensive that the entire species is threatened with starvation and extinction. In response, some individuals throw themselves into the sea and drown. This lowers the overall population size and allows the vegetation to recover enough to save the species. Even though individuals suffer, the good-of-the-species hypothesis maintains that the behavior evolved because the group benefits.

The lemming suicide story is false. Although lemmings do disperse from areas of high population density in order to find habitats with higher food availability, they do not throw themselves into the sea. The individual wearing the inner tube in **Figure 24.16** represents the reason lemmings do not kill themselves for the good of the species. The cartoon assumes that certain alleles predispose lemmings to sacrifice themselves for others. But the inner tube represents what biologists call a "cheater," or "selfish," allele. Individuals with self-sacrificing alleles die and do not produce offspring. But individuals with selfish, cheater alleles survive and produce offspring. As a result, selfish alleles increase in frequency while self-sacrificing alleles decrease in frequency. Thus, it is not possible for individuals to sacrifice themselves for the good of the species. No instance of purely self-sacrificing behavior—where the individual received no fitness benefit in return—has ever been recorded in nature. Chapter 52 provides additional details on this point.

Not All Traits Are Adaptive

Although organisms are often exquisitely adapted to their environment, adaptation is far from perfect. Vestigial traits such as the human coccyx (tailbone), goose bumps, and appendix do not increase the fitness of individuals with those traits. The structures are not adaptive. They exist simply because they were present in the ancestral population.

THE FAR SIDE® By GARY LARSON

FIGURE 24.16 Self-Sacrificing Behavior Cannot Evolve if "Cheater" Alleles Exist. Most individuals in this population have alleles that lead to self-sacrificing behavior and result in death; the individual with the inner tube has alleles that prevent self-sacrificing behavior.

QUESTION Why is the individual with the inner tube smiling?

Vestigial traits are not the only types of structures with no function. Some adult traits exist as holdovers from structures that appear early in development. For example, human males have rudimentary mammary glands. The structures are not adaptive. They exist only because nipples form in the human embryo before sex hormones begin directing the development of male organs instead of female organs.

Perhaps the best example of nonadaptive traits involves evolutionary changes in DNA sequences. Recall from Chapter 16 that mutation may change a base in the third position of a codon without changing the amino acid sequence of the protein encoded by that gene. Changes such as these are said to be neutral, or silent. They occur because of the redundancy of the genetic code (see Chapter 16). Neutral changes in DNA sequences are extremely common, yet not adaptive.

The general point here is that not all traits are adaptive. Evolution by natural selection does not lead to "perfection." Besides carrying an array of traits that have no function, the adaptations that organisms have are constrained in a variety of important ways.

Genetic Constraints The Grants' team analyzed data on the characteristics of finches that survived the 1977 drought, and the team made an interesting observation: Although individuals with deep beaks survived better than individuals with shallow beaks, birds with particularly narrow beaks survived better than individuals with wider beaks. This observation made sense because finches crack *Tribulus* fruits by twisting them. Narrow beaks concentrate the twisting force more efficiently than wider beaks, so they are especially useful for cracking the fruits. But narrower beaks did not evolve in the population. To explain why, the biologists noted that parents with deep beaks tend to have offspring with beaks that are both deep and wide. This is a common pattern. Many alleles that affect body size have an effect on all aspects of size—not just one structure or dimension. As a result, selection for increased beak depth overrode selection for narrow beaks, even though a deep and narrow beak would have been more advantageous.

The general point here is that selection was not able to optimize all aspects of a trait. In the case of the finches, wider beaks were not the best possible beak shape for individuals living in an arid habitat. Wider beaks evolved anyway, due to a type of constraint called a **genetic correlation**. Genetic correlations occur because of pleiotropy (see Chapter 13). In this case, selection on alleles for one trait (increased beak depth) caused a correlated, though suboptimal, increase in another trait (beak width).

Genetic correlations are not the only genetic constraint on adaptation. Lack of genetic variation is also important. Consider that salamanders have the ability to regrow severed limbs. Some eels and sharks can sense electric fields. Birds can sense magnetic fields and see ultraviolet light. Even though it is possible that these traits would confer increased reproductive success in humans, they do not exist—because the requisite genes are lacking.

Fitness Trade-offs In everyday English, the term *trade-off* refers to a compromise between competing goals. It is difficult to design a car that is both large and fuel efficient, a bicycle that is both rugged and light, or a plane that is both fast and maneuverable.

In nature, selection occurs in the context of fitness trade-offs. A **fitness trade-off** is a compromise between traits, in terms of how those traits perform in the environment. During the drought in the Galápagos, for example, medium ground finches with large bodies had an advantage because they won fights over the few remaining sources of seeds. But individuals with large bodies also require large amounts of food to maintain their mass; they also tend to be slower and less nimble than smaller individuals. When food is short, large individuals are more prone to starvation. Even if large size is advantageous in an environment, there is always counteracting selection that prevents individuals from getting even bigger.

BOX 24.2 Which Allele Frequencies Changed When Finch Beaks Changed?

Researchers investigating the evolution of *M. tuberculosis* populations have strong evidence that alleles of the *rpoB* gene change in frequency when the antibiotic rifampin is introduced into the environment. In this and many other cases of evolution in response to drugs or herbicides or pesticides, biologists know exactly which base pairs in specific genes are favored by natural selection.

In cases like the evolution of finch beaks, however, it is much more difficult to understand the exact genes and alleles involved. Characteristics like body size and body shape are polygenic, meaning that many genes—each one exerting a relatively small effect—influence the trait (see Chapter 13). And because most work in molecular genetics has been done on organisms like *E. coli* and *Drosophila melanogaster*, we know relatively little about which genes might be causing evolutionary change in organisms like birds, fish, and mammals.

The situation is starting to change rapidly, however, thanks to advances in the field called evolution and development, or evo-devo (see Chapter 21). As an example, consider recent work done in Cliff Tabin's lab. These researchers normally work on topics like the genetic control of limb development in chickens and other vertebrates. They focus on understanding the cell-cell signals and regulatory transcription factors that regulate development. But developmental biologists are aware that different alleles exist for each of the cell-cell signals and regulatory transcription factors they identify. These alleles represent heritable variation that can influence the evolution of traits like limb size and shape.

Interesting things began to happen when Tabin's group started interacting with Peter and Rosemary Grant. The researchers began studying beak development in an array of Galápagos finch species. More specifically, they looked for variation in the pattern of expression of cell-cell signals that had already been identified as important in the development of chickens. They struck pay dirt when they did in situ hybridizations showing where a cell-cell signal called *Bmp4* is expressed. As **Figure 24.17a** shows, there is a strong correlation between the amount of *Bmp4* expression when beaks are developing in young Galápagos finches and the width and depth of adult beaks. And when the researchers experimentally increased *Bmp4* expression in young chickens, they found that beaks got wider and deeper (**Figure 24.17b**). Based on these data, the researchers suggest that alleles that increase *Bmp4* expression were selected for during the drought of 1977. *Bmp4* may have been one of several genes in which allele frequencies changed and caused evolution.

(a) Natural variation in *Bmp4* expression: Finches

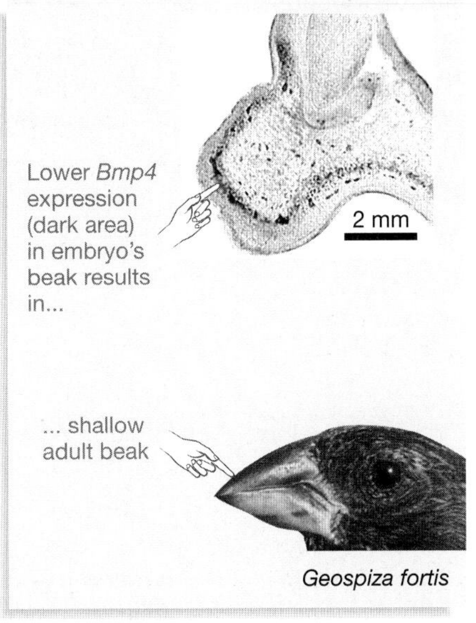

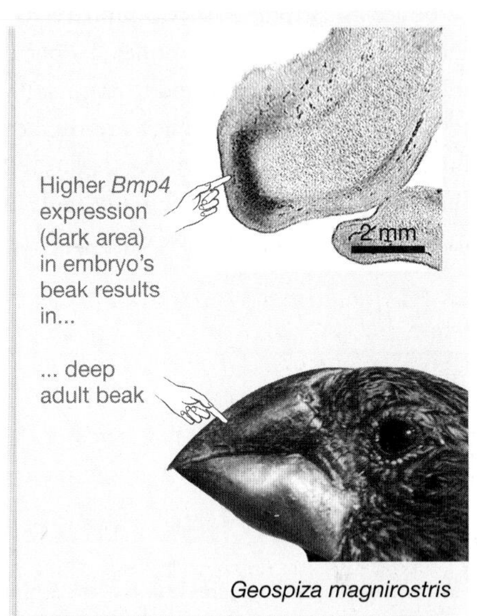

(b) Experimental variation in *Bmp4* expression: Chickens

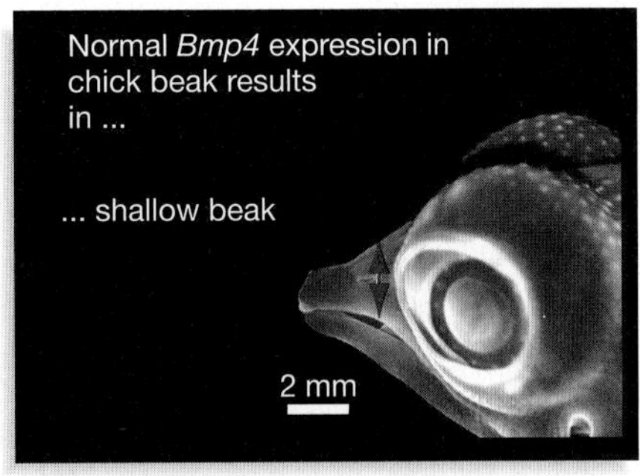

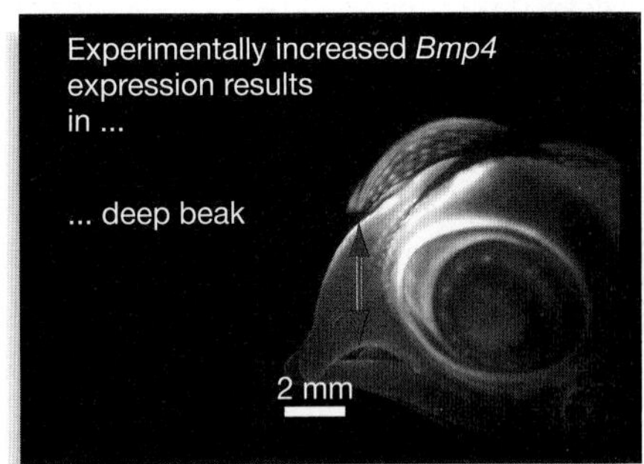

FIGURE 24.17 Changes in *Bmp4* Expression Change Beak Depth and Width. (a) The micrographs are in situ hybridizations (see Chapter 21) showing the location and extent of *Bmp4* expression in young *Geospiza fortis* and *G. magnirostris*. In these and four other species that were investigated, the amount of Bmp4 protein produced correlates with the depth and width of the adult beak. **(b)** When *Bmp4* expression in developing chickens is increased experimentally, beak depth and width increase.

Biologists have documented trade-offs between the size of eggs or seeds that an individual makes and the number of offspring it can produce, between rapid growth and long lifespan, and between bright coloration and tendency to attract predators. The message of this research is simple: Because selection acts on many traits at once, every adaptation is a compromise.

Historical Constraints In addition to being constrained by genetic correlations, lack of genetic variation, and fitness trade-offs, adaptations are constrained by history. The reason is simple: All traits have evolved from previously existing traits. For example, the tiny hammer, anvil, and stirrup bones found in your middle ear evolved from bones that were part of the jaw and braincase in the ancestors of mammals. These bones now function in the transmission and amplification of sound from your outer ear to your inner ear. Biologists routinely interpret these bones as adaptations that improve your ability to hear airborne sounds. But are the bones a "perfect" solution to the problem of transmitting sound from the outside of the ear to the inside? The answer is no. They are the best solution possible, given an important historical constraint. Natural selection was acting on structures that originally had a very different function. Other vertebrates have different structures involved in transmitting sound to the ear. In at least some cases, those structures may be more efficient than our hammer, anvil, and stirrup.

To summarize, not all traits are adaptive, and even adaptive traits are constrained by genetic and historical factors. In addition, natural selection is not the only process that causes evolutionary change. Chapter 25 introduces three other processes that change allele frequencies over time. Compared with natural selection, these processes have very different consequences.

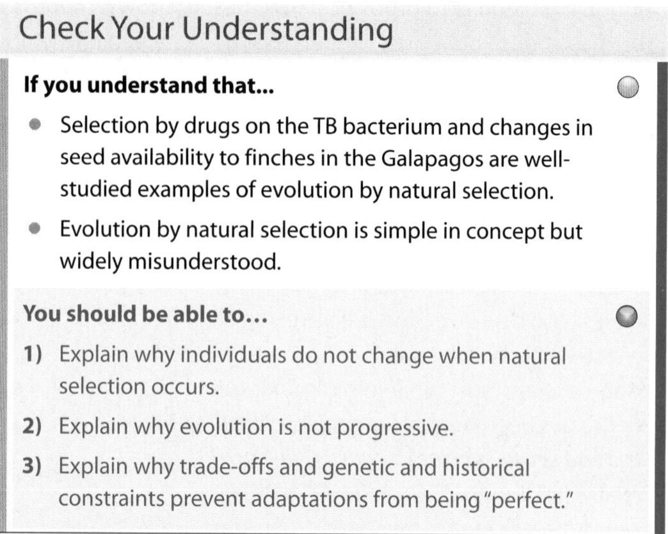

Check Your Understanding

If you understand that...

- Selection by drugs on the TB bacterium and changes in seed availability to finches in the Galapagos are well-studied examples of evolution by natural selection.
- Evolution by natural selection is simple in concept but widely misunderstood.

You should be able to...

1) Explain why individuals do not change when natural selection occurs.
2) Explain why evolution is not progressive.
3) Explain why trade-offs and genetic and historical constraints prevent adaptations from being "perfect."

Chapter Review

SUMMARY OF KEY CONCEPTS

Populations and species evolve, meaning that their heritable characteristics change through time. More precisely, evolution is defined as changes in allele frequencies over time.

Evidence for the fact of evolution—that species are related through shared ancestry and have changed over time—has accumulated over the past years. An array of observations is inconsistent with the alternative theory that species were formed instantaneously, recently, and independently, and have remained unchanged through time. These observations include the geographic proximity of closely related species such as the Galápagos mockingbirds; the existence of structural, developmental, and genetic homologies; the near universality of the genetic code; resemblances of modern to fossil forms; transitional fossils; the fact of extinction; and the presence of vestigial traits. These data support the hypothesis that populations and species change through time.

You should be able to predict how changes in finch populations and *Mycobacterium* populations would be explained under the theory of special creation and under evolution by inheritance of acquired characters.

Natural selection occurs when individuals with certain alleles produce the most surviving offspring in a population. An adaptation is a genetically based trait that increases an individual's ability to produce offspring in a particular environment.

Natural selection occurs whenever genetically based differences among individuals lead to differences in their ability to reproduce—when heritable variation leads to reproductive success in survival and reproduction. Alleles or traits that increase the reproductive success of an individual are said to increase the individual's fitness. A trait that leads to higher fitness, relative to individuals without the trait, is an adaptation. If a particular allele increases fitness and leads to adaptation, the allele will increase in frequency in the population.

Evolution is an outcome of natural selection. It is defined as changes in allele frequencies that occur in populations from one generation to the next. Evolution by natural selection has been confirmed by a wide variety of studies and has long been considered to be the central organizing principle of biology.

You should be able to explain the difference between the biological and everyday English definitions of the words *fitness* and *adaptation*.

MB Web Animation at www.masteringbio.com
Natural Selection for Antibiotic Resistance

Evolution by natural selection is not progressive, and it does not change the characteristics of the individuals that are selected—it changes only the characteristics of the population. Animals do not do things for the good of the species, and not all traits are adaptive. All adaptations are constrained by trade-offs and genetic and historical factors.

Individuals that are naturally selected are not changed by the process—they simply produce more offspring than other individuals do. Traits that increase in frequency under natural selection do so because they improve fitness, not because they are necessarily larger or more complex. Because natural selection acts only on existing traits, it does not lead to perfection.

You should be able to give an example of an adaptation—like the large brains of *Homo sapiens* or the ability of falcons to fly very fast—and discuss how it is constrained. **You should also be able to** explain why self-sacrificing alleles cannot increase in frequency and why individuals do not change when selection acts on them.

QUESTIONS

Test Your Knowledge

1. How can Darwinian fitness be estimated?
   a. Document how long different individuals in a population survive.
   b. Count the number of offspring produced by different individuals in a population.
   c. Determine which individuals are strongest.
   d. Determine which phenotype is the most common one in a given population.
2. Why are some traits considered vestigial?
   a. They improve the fitness of an individual who bears them, compared with the fitness of individuals without those traits.
   b. They change in response to environmental influences.
   c. They existed a long time in the past.
   d. They are reduced in size, complexity, and function compared with traits in related species.
3. What is an adaptation?
   a. a trait that improves the fitness of its bearer, compared with individuals without the trait
   b. a trait that changes in response to environmental influences within the individual's lifetime
   c. an ancestral trait—one that was modified to form the trait observed today
   d. the ability to produce offspring
4. Why does the presence of extinct and transitional forms in the fossil record support the pattern component of the theory of evolution by natural selection?
   a. It supports the hypothesis that individuals change over time.
   b. It supports the hypothesis that weaker species are eliminated by natural selection.
   c. It supports the hypothesis that species evolve to become more complex and better adapted over time.
   d. It supports the hypothesis that species have changed through time.
5. Why are homologous traits similar?
   a. They are derived from a common ancestor.
   b. They are derived from different ancestors.
   c. They result from convergent evolution.
   d. Their appearance, structure, or development is similar.
6. Which of the following statements is correct?
   a. When individuals change in response to challenges from the environment, their altered traits are passed on to offspring.
   b. Species are created independently of each other and do not change over time.
   c. Populations—not individuals—change when natural selection occurs.
   d. The Earth is young, and most of today's landforms were created during the floods at the time of Noah.

Test Your Knowledge answers: 1. b; 2. d; 3. a; 4. d; 5. a; 6. c

Test Your Understanding

Answers are available at www.masteringbio.com

1. Compare and contrast the theory of evolution by natural selection and the theory of special creation and evolution by inheritance of acquired characters. What are the central claims of each theory? What testable predictions does each make?
2. Some biologists encapsulate evolution by natural selection with the phrase "mutation proposes, selection disposes." Explain what they mean, using the formal terms introduced in this chapter.
3. Review the section on the evolution of drug resistance in *Mycobacterium tuberculosis*.
   - In *M. tuberculosis*, how does heritable variation arise for the trait of drug resistance?
   - What evidence do researchers have that a drug-resistant strain evolved in the patient analyzed in their study, instead of having been transmitted from another infected individual?
   - If the antibiotic rifampin were banned, would the mutant *rpoB* gene have lower or higher fitness in the new environment? Would strains carrying the mutation continue to increase in frequency in *M. tuberculosis* populations?
4. Compare and contrast typological thinking with population thinking. Why was Darwin's emphasis on the importance of variation among individuals so crucial to his theory, and why was it a revolutionary idea in Western science?
5. The evidence supporting the pattern component of the theory of evolution can be criticized on the grounds that it is indirect. For example, no one has directly observed the formation of a vestigial trait over time. Due to the indirect nature of the evidence, it could be argued that structural and genetic homologies are coincidental and do not result from common ancestry. Is indirect evidence for a scientific theory legitimate? Are you persuaded that descent with modification is the best explanation available for the data reviewed in Section 24.2? Why or why not?
6. Why isn't evolution by natural selection progressive? Why don't the strongest individuals in a population always produce the most offspring?

Applying Concepts to New Situations

Answers are available at www.masteringbio.com

1. The geneticist James Crow wrote that successful scientific theories have the following characteristics: (1) They explain otherwise puzzling observations; (2) they provide connections between otherwise disparate observations; (3) they make predictions that can be tested; and (4) they are heuristic, meaning that they open up new avenues of theory and experimentation. Crow added two other elements that he considered important on a personal, emotional level: (5) They should be elegant, in the sense of being simple and powerful; and (6) they should have an element of surprise.

   How well does the theory of evolution by natural selection fulfill these six criteria? Think of a theory you've been introduced to in another science course—for example, the atomic theory or the germ theory of disease—and evaluate it by using this list.

2. The average height of humans has increased steadily for the past 100 years in industrialized nations. This trait has clearly changed over time. Most physicians and human geneticists believe that the change is due to better nutrition and a reduced incidence of disease. Has human height evolved?

3. Genome sequencing projects may dramatically affect how biologists analyze evolutionary changes in quantitative traits. For example, suppose that the genomes of many living humans are sequenced and that genomes could be sequenced from many people who lived 100 years ago. (That might be possible with preserved tissue.) If 20 genes have been shown to influence height, how could you use the sequence data from these genes to test the hypothesis that human height has evolved in response to natural selection?

4. In some human populations, individuals tan in response to exposure to sunlight. Tanning is an acclimation response to a short-term change in the environment. It is adaptive because it prevents sunburn. The ability to tan varies among individuals in these populations, however. Is the ability to tan an acclimation or an adaptation? Explain your logic.

www.masteringbio.com is also your resource for • Answers to text, table, and figure caption questions and exercises • Answers to *Check Your Understanding* boxes • Online study guides and quizzes • Additional study tools including the *E-Book for Biological Science* 3rd ed., textbook art, animations, and videos.

Evolutionary Processes 25

A male raggiana bird of paradise, left, displays for a female, right. His long, colorful feathers and dramatic behavior result from sexual selection—a process introduced in this chapter.

KEY CONCEPTS

- The Hardy-Weinberg principle acts as a null hypothesis when researchers want to test whether evolution or nonrandom mating is occurring at a particular gene.
- Each of the four evolutionary mechanisms has different consequences. Only natural selection produces adaptation. Genetic drift causes random fluctuations in allele frequencies. Gene flow equalizes allele frequencies between populations. Mutation introduces new alleles.
- Inbreeding changes genotype frequencies but does not change allele frequencies.
- Sexual selection leads to the evolution of traits that help individuals attract mates. It is usually stronger on males than on females.

Chapter 24 defined evolution as a change in allele frequencies. One of the key concepts from that chapter was that even though natural selection acts on individuals, evolutionary change occurs in **populations**. A population is a group of individuals from the same species that live in the same area and regularly interbreed.

Natural selection is not the only process that causes evolution, however. There are actually four mechanisms that shift allele frequencies in populations:

1. *Natural selection* increases the frequency of certain alleles—the ones that contribute to success in survival and reproduction.
2. *Genetic drift* causes allele frequencies to change randomly. In some cases, drift may even cause alleles that decrease fitness to increase in frequency.
3. *Gene flow* occurs when individuals leave one population, join another, and breed. Allele frequencies may change when gene flow occurs, because arriving individuals introduce alleles to their new population and departing individuals remove alleles from their old population.
4. *Mutation* modifies allele frequencies by continually introducing new alleles. The alleles created by mutation may be beneficial or detrimental or have no effect on fitness.

This chapter has two fundamental messages: Natural selection is not the only agent responsible for evolution, and each of the four evolutionary processes has different consequences. Natural selection is the only mechanism that acting alone can result in adaptation. Mutation, gene flow, and genetic drift do not favor certain alleles over others. Mutation and drift introduce a nonadaptive component into evolution.

Let's take a closer look at the four evolutionary processes by examining a null hypothesis—what happens to allele frequencies when the evolutionary mechanisms are *not* operating.

25.1 Analyzing Change in Allele Frequencies: The Hardy-Weinberg Principle

To study how the four evolutionary processes affect populations, biologists take a three-pronged approach. First they create mathematical models that track the fate of alleles over time. Then they collect data to test predictions made by the models' equations. Finally, they apply the results to solve problems in human genetics, conservation of endangered species, or other fields.

This research strategy began in 1908, when both G. H. Hardy and Wilhelm Weinberg published a major result independently. At the time, it was commonly believed that changes in allele frequency occur simply as a result of sexual reproduction—meiosis followed by the random fusion of gametes (eggs and sperm) to form offspring. Some biologists claimed that dominant alleles inevitably increase in frequency. Others predicted that two alleles of the same gene inevitably reach a frequency of 0.5.

To test these hypotheses, Hardy and Weinberg analyzed what happens to the frequencies of alleles when many individuals in a population mate and produce offspring. Instead of thinking about the consequences of a mating between two parents with a specific pair of genotypes, as we did with Punnett squares in Chapter 13, Hardy and Weinberg wanted to know what happened in an entire population, when *all* of the individuals—and thus all possible genotypes—bred. Like Darwin, Hardy and Weinberg were engaged in population thinking.

To analyze the consequences of matings among all of the individuals in a population, Hardy and Weinberg invented a novel approach: They imagined that all of the gametes produced in each generation go into a single group called the **gene pool** and then combine at random to form offspring. Something very much like this happens in species like clams and sea stars and sea urchins, which release their gametes into the water, where they mix randomly with gametes from other individuals in the population and combine to form zygotes.

To determine which genotypes would be present in the next generation and in what frequency, Hardy and Weinberg simply had to calculate what happened when two gametes were plucked at random out of the gene pool, many times, and each of these gamete pairs was then combined to form offspring. These calculations would predict the genotypes of the offspring that would be produced, as well as the frequency of each genotype.

The researchers began by analyzing the simplest situation possible—that just two alleles of a particular gene exist in a population. Let's call these alleles A_1 and A_2. We'll use p to symbolize the frequency of A_1 alleles in the gene pool and q to symbolize the frequency of A_2 alleles in the same gene pool. Because there are only two alleles, the two frequencies must add up to 1; that is, $p + q = 1$. Although p and q can have any value between 0 and 1, let's suppose that the initial frequency of A_1 is 0.7 and that of A_2 is 0.3 (**Figure 25.1**, step 1). In this case, 70 percent of the gametes in the gene pool carry A_1 and 30 percent carry A_2 (Figure 25.1, step 2).

Because only two alleles are present, three genotypes are possible: A_1A_1, A_1A_2, and A_2A_2 (Figure 25.1, step 3). What will the frequency of these three genotypes be in the next generation? Figure 25.1, step 4, explains the logic of Hardy's and Weinberg's result:

- The frequency of the A_1A_1 genotype is p^2.
- The frequency of the A_1A_2 genotype is $2pq$.
- The frequency of the A_2A_2 genotype is q^2.

The genotype frequencies in the offspring generation must add up to 1, which means that $p^2 + 2pq + q^2 = 1$. In our numerical example, $0.49 + 0.42 + 0.09 = 1$. Figure 25.1, step 5, shows how the frequencies of alleles A_1 and A_2 are calculated from these genotype frequencies. In our example, the frequency of allele A_1 is still 0.7 and the frequency of allele A_2 is still 0.3. Thus, the frequency of allele A_1 in the next generation is still p and the frequency of allele A_2 is still q. No allele frequency change occurred. Even if A_1 is dominant to A_2, it does not increase in frequency (Figure 25.1, step 6). And there is no trend toward both alleles reaching a frequency of 0.5. **Figure 25.2** illustrates the same result a little differently. The figure uses a Punnett square in a novel way: to predict the outcome of random mating— meaning, random combinations of all gametes in a population. The outcome is the same as in Figure 25.1.

This result is called the **Hardy-Weinberg principle**. It makes two fundamental claims:

1. If the frequencies of alleles A_1 and A_2 in a population are given by p and q, then the frequencies of genotypes A_1A_1, A_1A_2, and A_2A_2 will be given by p^2, $2pq$, and q^2 for generation after generation.
2. When alleles are transmitted according to the rules of Mendelian inheritance, their frequencies do not change over time. For evolution to occur, some other factor or factors must come into play.

What are these other factors?

The Hardy-Weinberg Model Makes Important Assumptions

The Hardy-Weinberg model is based on important assumptions about how populations and alleles behave. Specifically, for a population to conform to the Hardy-Weinberg principle, none of the four mechanisms of evolution can be acting on the population. In addition, the model assumes that mating is random with respect

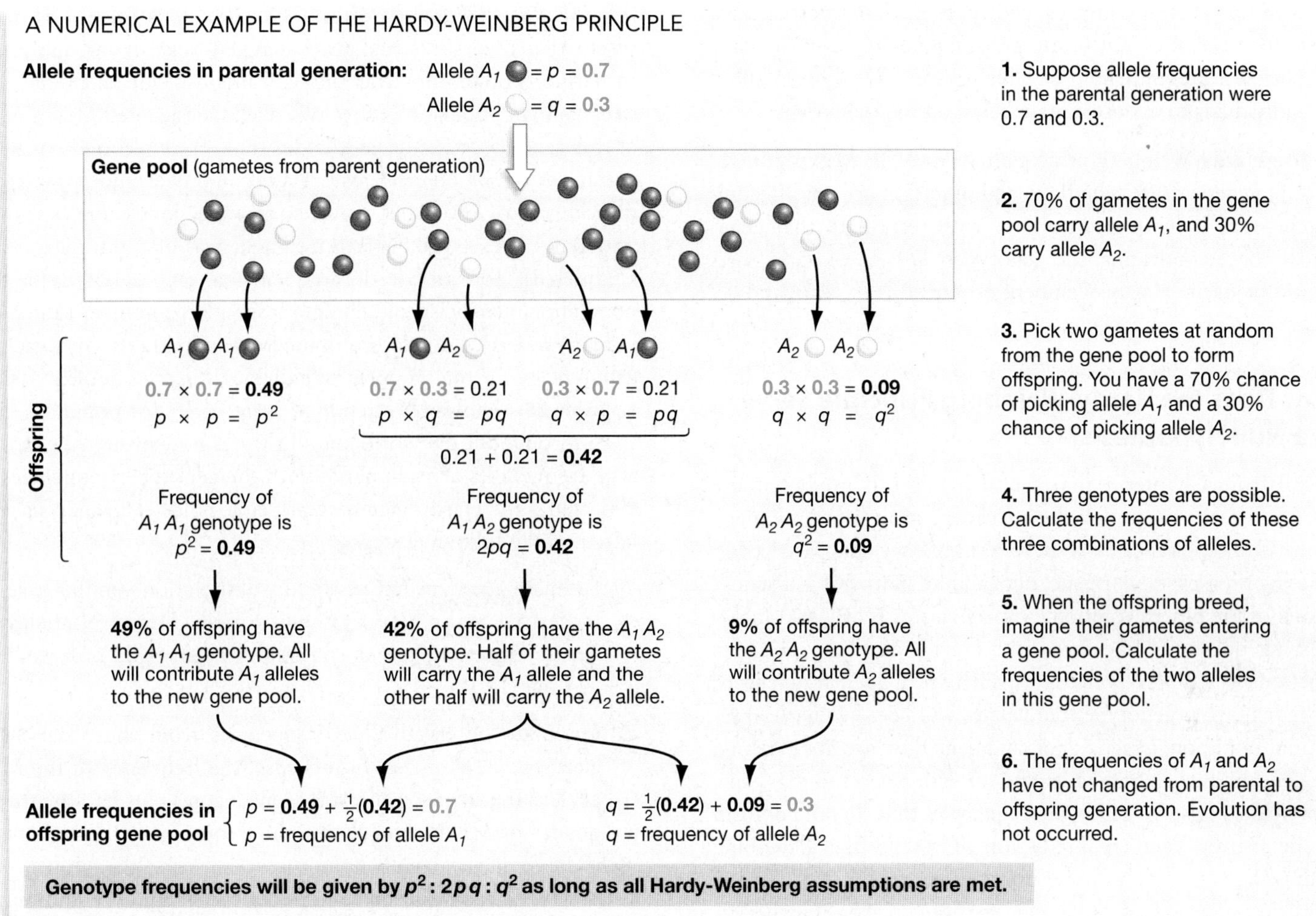

FIGURE 25.1 A Numerical Example of the Hardy-Weinberg Principle. To understand the logic behind calculating the frequency of A_1A_2 genotypes in step 4, see **BioSkills 9.**

to the gene in question. Thus, here are the five conditions that must be met:

1. *No natural selection* at the gene in question. In step 2 of Figure 25.1, the model assumed that all members of the parental generation survived and contributed equal numbers of gametes to the gene pool, no matter what their genotype.

2. *No genetic drift, or random allele frequency changes*, affecting the gene in question. We avoided this type of allele frequency change in step 4 of Figure 25.1 by assuming that we drew alleles in their exact frequencies p and q, and not at some different values caused by chance. For example, allele A_1 did not "get lucky" and get drawn more than 70 percent of the time. No random changes due to luck occurred.

3. *No gene flow*. No new alleles were added by immigration or lost through emigration anywhere in Figure 25.1. As a result, all of the alleles in the offspring population came from the original population's gene pool.

4. *No mutation*. We didn't consider that new A_1s or A_2s or other, new alleles might be introduced into the gene pool in step 2 or step 5 of Figure 25.1.

Allele frequencies in parental generation:

A_1 = p = 0.7 A_2 = q = 0.3

All eggs in gene pool

All sperm in gene pool	0.7 A_1	0.3 A_2
0.7 A_1	A_1A_1 0.49	A_1A_2 0.21
0.3 A_2	A_2A_1 0.21	A_2A_2 0.09

Genotype frequencies in offspring generation:

$A_1A_1 = p^2 = 0.49$

$A_1A_2 = 2pq = 0.42$

$A_2A_2 = q^2 = 0.09$

Allele frequencies in offspring generation:

$A_1 = p = 0.49 + \frac{1}{2}(0.42) = 0.70$

$A_2 = q = \frac{1}{2}(0.42) + 0.09 = 0.30$

Allele frequencies have not changed

FIGURE 25.2 A Punnett Square Illustrates the Hardy-Weinberg Principle.

5. *Random mating* with respect to the gene in question. We enforced this condition by picking gametes from the gene pool at random in step 3 of Figure 25.1. We did not allow individuals to choose a mate based on their genotype.

The Hardy-Weinberg principle tells us what to expect if selection, genetic drift, gene flow, and mutation are not affecting a gene, *and* if mating is random with respect to that gene. Under these conditions, the genotypes A_1A_1, A_1A_2, and A_2A_2 should be in the Hardy-Weinberg proportions p^2, $2pq$, and q^2, and no evolution will occur.

How Does the Hardy-Weinberg Principle Serve as a Null Hypothesis?

Recall from Chapter 1 that a null hypothesis predicts there are no differences among the treatment groups in an experiment. Biologists often want to test whether natural selection is acting on a particular gene, nonrandom mating is occurring, or one of the other evolutionary mechanisms is at work. In addressing questions like these, the Hardy-Weinberg principle functions as a null hypothesis. Given a set of allele frequencies, it predicts what genotype frequencies will be when natural selection, mutation, genetic drift, and gene flow are not affecting the gene; and when mating is random with respect to that gene. If biologists observe genotype frequencies that do not conform to the Hardy-Weinberg prediction, it means that something interesting is going on: Either nonrandom mating is occurring, or allele frequencies are changing for some reason. Further research is needed to determine which of the five Hardy-Weinberg conditions is being violated.

Let's consider two examples to illustrate how the Hardy-Weinberg principle is used as a null hypothesis: MN blood types and *HLA* genes, both in humans.

Are MN Blood Types in Humans in Hardy-Weinberg Equilibrium? One of the first genes that geneticists could analyze in natural populations was the MN blood group of humans. Most human populations have two alleles, designated *M* and *N*, at this gene. Because the gene codes for a protein found on the surface of red blood cells, researchers could determine whether individuals are *MM*, *MN*, or *NN* by treating blood samples with antibodies to each protein (this technique was first introduced in Chapter 8). To estimate the frequency of each genotype in a population, geneticists obtain data from a large number of individuals and then divide the number of individuals with each genotype by the total number of individuals in the sample.

Table 25.1 shows MN genotype frequencies for populations from throughout the world and illustrates how observed genotype frequencies are compared with the genotype frequencies expected if the Hardy-Weinberg principle holds. The analysis is based on the following steps:

1. Estimate genotype frequencies by observation—in this case, by testing many blood samples for the *M* and *N* alleles. These frequencies are given in the rows labeled "observed" in Table 25.1.

2. Calculate observed allele frequencies from the observed genotype frequencies. In this case, the frequency of the *M* allele is the frequency of *MM* homozygotes plus half the frequency of *MN* heterozygotes; the frequency of the *N* allele is the frequency of *NN* homozygotes plus half the frequency of *MN* heterozygotes. (You can review the logic behind this calculation in steps 5 and 6 of Figure 25.1.)

3. Use the observed allele frequencies to calculate the genotypes expected according to the Hardy-Weinberg principle. Under the null hypothesis of no evolution and random mating, the expected genotype frequencies are p^2: $2pq$: q^2.

TABLE **25.1** **The MN Blood Group of Humans: Observed and Expected Genotype Frequencies**

The expected genotype frequencies are calculated from the observed allele frequencies, using the Hardy-Weinberg principle.

		Genotype Frequencies			Allele Frequencies	
Population and Location		***MM***	***MN***	***NN***	***M***	***N***
Inuit (Greenland)	Observed	0.835	0.156	0.009	0.913	0.087
	Expected	0.834	0.159	0.008		
Native Americans (U.S.)	Observed	0.600	0.351	0.049	0.776	0.224
	Expected	0.602	0.348	0.050		
Caucasians (U.S.)	Observed	0.292	0.494	0.213	0.540	0.460
	Expected	0.290	0.497	0.212		
Aborigines (Australia)	Observed	0.025	0.304	0.672	0.178	0.825
	Expected	0.031	0.290	0.679		
Ainu (Japan)	Observed (Step 1 →)	0.179	0.502	0.319 (Step 2 →)		
	Expected			(← Step 3)		

EXERCISE Fill in the values for allele frequencies and expected genotype frequencies for the Ainu people of Japan.

4. Compare the observed and expected values. Researchers must use statistical tests to determine whether the differences between the observed and expected genotype frequencies are small enough to be due to chance or large enough to reject the null hypothesis of no evolution and random mating.

Although using statistical testing is beyond the scope of this text (see **BioSkills 3** for a brief introduction to the topic), you should be able to inspect the numbers and comment on them. In these populations, for example, the observed and expected *MN* genotype frequencies are almost identical. (A statistical test shows that the small differences observed are probably due to chance.) For every population surveyed, genotypes at the *MN* locus are in Hardy-Weinberg proportions. As a result, biologists conclude that the assumptions of the Hardy-Weinberg model are valid for this locus. The results imply that when these data were collected, the *M* and *N* alleles in these populations were not being affected by the four evolutionary mechanisms and that mating was random with respect to this gene—meaning that humans were not choosing mates on the basis of their *MN* genotype.

Before moving on, however, it is important to note that a study such as this does not mean that the *MN* gene has never been under selection or subject to nonrandom mating or genetic drift. Even if selection has been very strong for many generations, one generation of no evolutionary forces and of random mating will result in genotype frequencies that conform to Hardy-Weinberg expectations. The Hardy-Weinberg principle is used to test the hypothesis that currently no evolution is occurring at a particular gene and that in the previous generation, mating was random with respect to the gene in question.

Are *HLA* Genes in Humans in Hardy-Weinberg Equilibrium? A research team recently collected data on the genotypes of 125 individuals from the Havasupai tribe native to Arizona. These biologists were studying two genes that are important in the functioning of the human immune system. More specifically, the genes that they analyzed code for proteins that help immune system cells recognize and destroy invading bacteria and viruses. Previous work had shown that different alleles exist at both the *HLA-A* and *HLA-B* genes, and that the alleles at each gene code for proteins that recognize slightly different disease-causing organisms. Like the *M* and *N* alleles, *HLA* alleles are *codominant*—meaning that both are expressed and create the phenotype (see Chapter 13). As a result, the research group hypothesized that individuals who are heterozygous at one or both of these genes may have a strong fitness advantage. The logic is that heterozygous people have a wider variety of HLA proteins, so their immune systems can recognize and destroy more types of bacteria and viruses. They should be healthier and have more offspring than homozygous people do.

To test this hypothesis, the researchers used their data on observed genotype frequencies to determine the frequency of each allele present. When they used these allele frequencies to

TABLE **25.2 HLA Genes of Humans: Observed and Expected Genotypes**

	Observed Number	Expected Number under Hardy-Weinberg
HLA-A		
Homozygotes	38	48
Heterozygotes	84	74
HLA-B		
Homozygotes	21	30
Heterozygotes	101	92

SOURCE: T. Markow et al., *HLA* polymorphism in the Havasupai: Evidence for balancing selection. *American Journal of Human Genetics* 53 (1993): 943–952.

calculate the expected number of each genotype according to the Hardy-Weinberg principle, they found the observed and expected values reported in **Table 25.2**. When you inspect these data, notice that there are many more heterozygotes and many fewer homozygotes than expected under Hardy-Weinberg conditions. Statistical tests show it is extremely unlikely that the difference between the observed and expected numbers could occur purely by chance.

These results supported the team's prediction and indicated that one of the assumptions behind the Hardy-Weinberg principle was being violated. But which assumption? The researchers argued that mutation, migration, and drift are negligible in this case and offered two competing explanations for their data:

1. *Mating may not be random with respect to the* HLA *genotype*. Specifically, people may subconsciously prefer mates with *HLA* genotypes unlike their own and thus produce an excess of heterozygous offspring. This hypothesis is plausible. For example, experiments have shown that college students can distinguish each others' genotypes at genes related to *HLA* on the basis of body odor. Individuals in this study were more attracted to the smell of genotypes unlike their own. If this is true among the Havasupai, then nonrandom mating would lead to an excess of heterozygotes compared with the proportion expected under Hardy-Weinberg.
2. *Heterozygous individuals may have higher fitness*. This hypothesis is supported by data collected by a different research team, studying Hutterite people living in South Dakota. In this population, married women who have the same *HLA*-related alleles as their husbands have more trouble getting pregnant and experience higher rates of spontaneous abortion than do women with *HLA*-related alleles different from those of their husbands. The data suggest that homozygous fetuses have lower fitness than do fetuses heterozygous at these genes. If this were true among the Havasupai, selection would lead to an excess of heterozygotes relative to Hardy-Weinberg expectations.

Which explanation is correct? It is possible that both are. But the fact is, no one knows. Using the Hardy-Weinberg principle as a null hypothesis allowed biologists to detect an interesting pattern in a natural population. Research continues on the question of why the pattern exists.

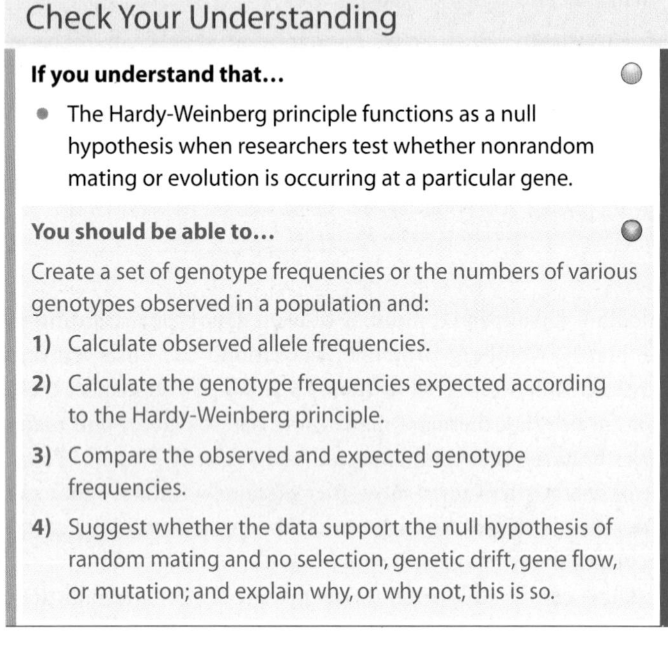

Check Your Understanding

If you understand that...

- The Hardy-Weinberg principle functions as a null hypothesis when researchers test whether nonrandom mating or evolution is occurring at a particular gene.

You should be able to...

Create a set of genotype frequencies or the numbers of various genotypes observed in a population and:

1) Calculate observed allele frequencies.
2) Calculate the genotype frequencies expected according to the Hardy-Weinberg principle.
3) Compare the observed and expected genotype frequencies.
4) Suggest whether the data support the null hypothesis of random mating and no selection, genetic drift, gene flow, or mutation; and explain why, or why not, this is so.

(MB) **Web Animation** at **www.masteringbio.com**
The Hardy-Weinberg Principle

25.2 Types of Natural Selection

Natural selection occurs when individuals with certain phenotypes produce more offspring than individuals with other phenotypes do. If certain alleles are associated with the favored phenotypes, they increase in frequency while other alleles decrease in frequency. The result is evolution. To use the language introduced in Chapter 24, evolution by natural selection occurs when heritable variation leads to differential success in survival and reproduction.

Although you should have a solid understanding of why evolution by natural selection occurs, it is important to recognize that natural selection occurs in a wide variety of patterns. Each of these patterns has different causes and consequences. For example, the second explanation for the data in Table 25.2 is a pattern of natural selection called **balancing selection** or **heterozygote advantage.** When balancing selection occurs, heterozygous individuals have higher fitness than homozygous individuals do. The consequence of this pattern is that genetic variation is maintained in populations. **Genetic variation** refers to the number and relative frequency of alleles that are present in a particular population.

When biologists analyze the consequences of different patterns of selection, they often focus on genetic variation. The reason is simple: Lack of genetic variation in a population is usually a bad thing. To understand why, recall from Chapter 24 that selection can occur only if heritable variation exists in a population. If genetic variation is low and the environment changes—perhaps due to the emergence of a new disease-causing virus, a rapid change in climate, or a reduction in the availability of a particular food source—it is unlikely that any alleles will be present that have high fitness under the new conditions. As a result, the average fitness of the population will decline. If the environmental change is severe enough, the population may even be faced with extinction.

Let's examine some of the different types of natural selection with this question in mind: How do they affect the level of genetic variation in the population?

Directional Selection

According to the data introduced in Chapter 24, natural selection has increased the frequency of drug-resistant strains of the tuberculosis bacterium and caused changes in beak shape and body size in medium ground finches. This type of natural selection is called **directional selection**, because the average phenotype of the populations changed in one direction.

Figure 25.3a illustrates how directional selection works when the trait in question has a bell-shaped, normal distribution in a population. Recall from Chapter 13 that when many different genes influence a trait, the distribution of phenotypes in the population tends to form a bell-shaped curve. In such cases, directional selection is acting on many different genes at once. In the case of selection on drug resistance in the TB bacterium, however, selection was acting on a single gene.

Most often, directional selection tends to reduce the genetic diversity of populations. If directional selection continues over time, the favored alleles will eventually reach a frequency of 1.0 while disadvantageous alleles will reach a frequency of 0.0. Alleles that reach a frequency of 1.0 are said to be fixed; those that reach a frequency of 0.0 are said to be lost. When disadvantageous alleles decline in frequency **purifying selection** is said to occur.

Fixation and loss may not always occur under directional selection, however. To appreciate why, consider recent data on the body size of cliff swallows native to the Great Plains of North America. In 1996 a population of cliff swallows endured a six-day period of exceptionally cold, rainy weather. Cliff swallows feed by catching mosquitoes and other insects in flight. Insects disappeared during this cold snap, however, and the biologists recovered the bodies of 1853 swallows that died of starvation. As soon as the weather improved, the researchers caught and measured the body size of 1027 survivors from the same population. As the histograms in **Figure 25.3b** show, survivors were much larger on average than the birds that died. Directional selection, favoring large body size, had occurred. To explain this observation, the investigators suggest that larger birds survived because they had larger fat stores and did not get as cold as the

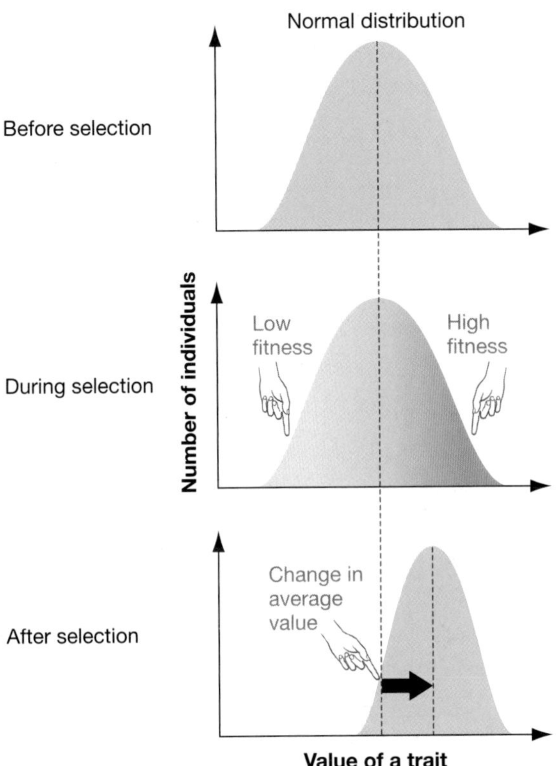

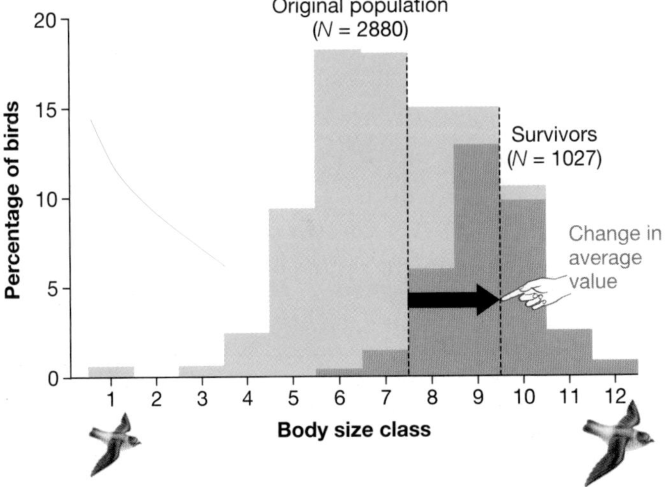

FIGURE 25.3 Directional Selection. (a) When directional selection acts on traits that have a normal distribution, individuals with one set of extreme values experience poor reproductive success. **(b)** The light green histogram shows the distribution of overall body size in cliff swallows prior to an extended cold snap that killed many individuals. (The various body-size classes were calculated from measurements of wing length, tail length, leg length, and beak size.) The dark green histogram shows the size distribution of individuals from the same population that survived the cold spell. Here *N* indicates the sample size.

smaller birds. As a result, the larger birds were less likely to die of exposure to cold and more likely to avoid starvation until the weather warmed up and insects were again available.

If the differences in body size among individuals were due in part to differences in their genotypes—meaning that heritable variation in body size existed—then the population evolved. If so, and if this type of directional selection continued, then alleles that contribute to small body size would quickly be eliminated from the cliff swallow population. It is not clear that this will be the case, however, because directional selection is rarely constant throughout a species' range and through time. By examining weather records, the researchers established that cold spells as severe as the one that occurred in 1996 are rare. Further, research on other swallow species suggests that smaller birds are more maneuverable in flight and thus more efficient when they feed. If so, then selection for feeding efficiency could counteract selection by cold weather. When this is the case, individuals with intermediate body size should be favored. Opposing patterns of directional selection will help maintain genetic variation for this trait.

When studies on a wide array of populations and species are considered, it is common to find that one cause of directional selection on a trait is counterbalanced by a different factor that causes selection in the opposite direction. This concept, known as a fitness trade-off, was introduced in Chapter 24. In such cases, the optimal phenotype is intermediate. The same pattern can result from an entirely different type of natural selection, called stabilizing selection.

Stabilizing Selection

When cliff swallows were exposed to cold weather, selection greatly reduced one extreme in the range of phenotypes and resulted in a directional change in the average characteristics of the population. But selection can also reduce both extremes in a population, as illustrated in **Figure 25.4a**. This pattern of selection is called **stabilizing selection**. It has two important consequences: There is no change in the average value of a trait over time, and genetic variation in the population is reduced.

Figure 25.4b shows a classical data set in humans illustrating stabilizing selection. Biologists who analyzed birth weights and mortality in 13,730 babies born in British hospitals in the 1950s found that babies of average size (slightly over 7 pounds) survived best. Mortality was high for very small babies and very large babies. This is persuasive evidence that birth weight was under strong stabilizing selection in this population. Alleles associated with high birthweight or low birthweight were subject to purifying selection.

Disruptive Selection

Disruptive selection has the opposite effect of stabilizing selection. Instead of favoring phenotypes near the average value and

(a) Stabilizing selection reduces the amount of variation in a trait.

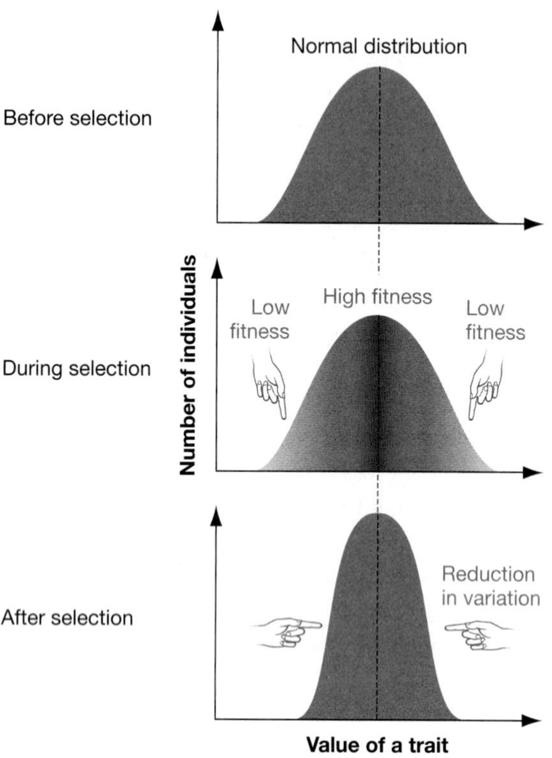

(b) For example, very small and very large babies are the most likely to die, leaving a narrower distribution of birth weights.

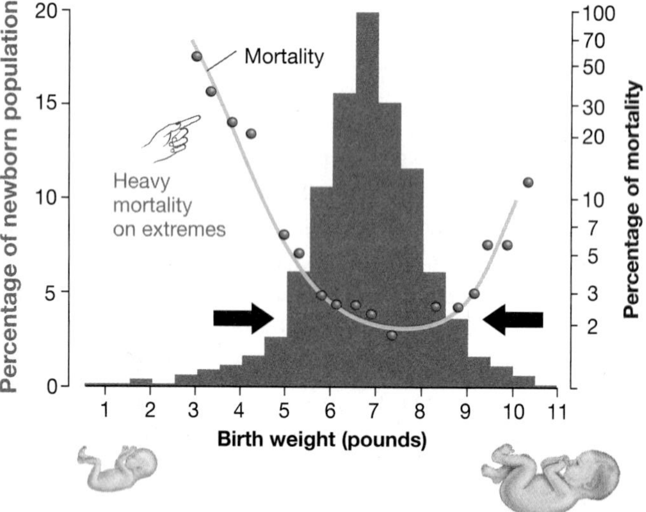

FIGURE 25.4 Stabilizing Selection. (a) When stabilizing selection acts on normally distributed traits, individuals with extreme phenotypes experience poor reproductive success. **(b)** Histogram showing the percentages of newborns with various birth weights on the left-hand axis. The purple dots indicate the percentage of newborns in each weight class that died, plotted on the logarithmic scale shown on the right.

(a) Disruptive selection increases the amount of variation in a trait.

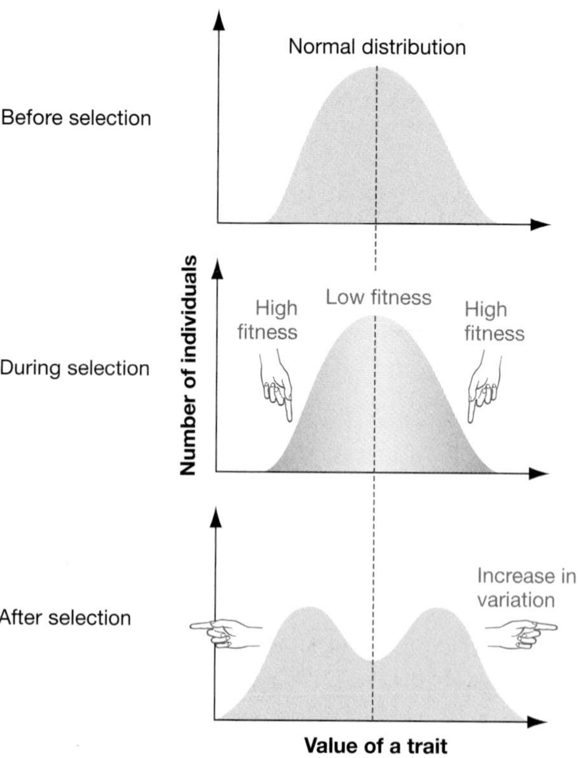

(b) For example, only juvenile black-bellied seedcrackers that had very long or very short beaks survived long enough to breed.

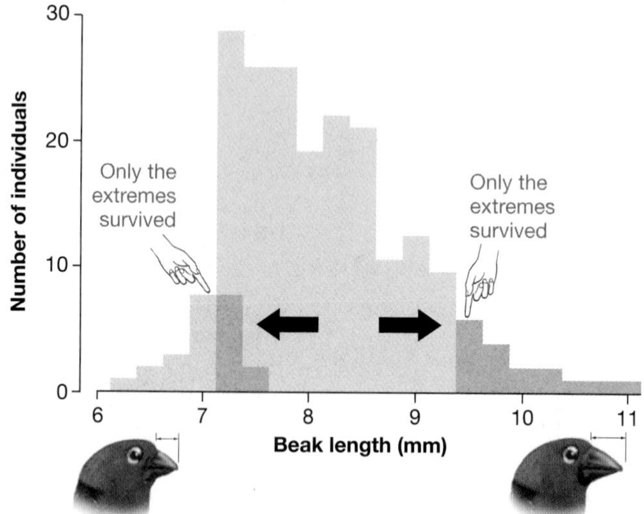

FIGURE 25.5 Disruptive Selection. (a) When disruptive selection occurs on traits with a normal distribution, individuals with extreme phenotypes experience high reproductive success. **(b)** Histogram showing the distribution of beak length in a population of black-bellied seedcrackers. The light orange bars represent all juveniles; the dark orange bars, juveniles that survived to adulthood.

eliminating extreme phenotypes, it eliminates phenotypes near the average value and favors extreme phenotypes (**Figure 25.5a**). When disruptive selection occurs, the overall amount of genetic variation in the population is maintained.

Recent research has shown that disruptive selection is responsible for the striking bills of black-bellied seedcrackers, pictured in **Figure 25.5b**. The data plotted in the figure show that individuals with either very short or very long beaks survive best and that birds with intermediate phenotypes are at a disadvantage. In this case, the agent that causes natural selection is food. At a study site in south-central Cameroon, West Africa, a researcher found that only two sizes of seed are available to the seedcrackers: large and small. Birds with small beaks crack and eat small seeds efficiently. Birds with large beaks handle large seeds efficiently. But birds with intermediate beaks have trouble with both, so alleles associated with medium-sized beaks are subject to purifying selection. Disruptive selection maintains high overall variation in this population.

Disruptive selection is important because it sometimes plays a part in **speciation**, or the formation of new species. If small-beaked seedcrackers began mating with other small-beaked individuals, their offspring would tend to be small beaked and would feed on small seeds. Similarly, if large-beaked individuals chose only other large-beaked individuals as mates, they would tend to produce large-beaked offspring that would feed on large seeds. In this way, selection would result in two distinct populations. Under some conditions, the populations may eventually form two new species. The process of species formation, based on disruptive selection and other mechanisms, is explored in detail in Chapter 26.

at www.masteringbio.com
Three Modes of Natural Selection

25.3 Genetic Drift

Natural selection is not random. It is directed by the environment and results in adaptation. **Genetic drift**, in contrast, is undirected. It is defined as any change in allele frequencies in a population that is due to chance. The process is aptly named, because it causes allele frequencies to drift up and down randomly over time. When drift occurs, allele frequencies change due to blind luck—what is formally known as **sampling error**.

To understand why genetic drift occurs, consider the group of people who founded the present population of Pitcairn Island in the South Pacific. The founding event occurred in 1789, when a small band of mutineers led by Fletcher Christian took over the British warship HMS *Bounty* and fled to Pitcairn. The six sailors were joined by two Tahitian men and six Tahitian women. Suppose that six couples formed and raised two children each. How will allele frequencies compare in the founding population that bred versus the next generation?

To answer this question, let's focus on a hypothetical gene A with alleles A_1, A_2, and A_3. Suppose that the six males and six females in the original couples have the genotypes given in **Table 25.3**. The gametes formed by each of these parents have an equal chance of carrying either allele. Further, eggs and sperm combine at random when fertilization occurs—meaning that each type of egg and sperm has an equal probability of combining, irrespective of its genotype. To simulate the fertilization process, you can flip a coin to decide which gametes combine. Let heads stand for the first allele

TABLE **25.3 A Thought Experiment on Genetic Drift**

The genotypes of the children in this table were generated by flipping a coin to simulate which alleles combined during fertilization. The frequencies of alleles A_1, A_2, and A_3 differ in the two generations, due to genetic drift.

	Father	**Mother**	**Child 1**	**Child 2**
Couple 1	A_3A_3	A_2A_3	A_3A_3	A_3A_3
Couple 2	A_2A_3	A_2A_3	A_2A_3	A_3A_2
Couple 3	A_1A_2	A_2A_3	A_2A_3	A_1A_2
Couple 4	A_1A_1	A_1A_2	A_1A_2	A_1A_1
Couple 5	A_1A_2	A_3A_3	A_1A_3	A_1A_3
Couple 6	A_1A_2	A_1A_3	A_2A_3	A_2A_1

listed in the parent's genotype, while tails represents the second allele listed. Doing these coin flips resulted in the offspring genotypes given in the "Child 1" and "Child 2" columns of Table 25.3.

Now let's count up alleles and calculate allele frequencies. Because there are 12 breeding individuals and 12 offspring, the total number of alleles present in each generation (ignoring the two founding individuals who did not breed) is 24. From the data in Table 25.3, the allele frequencies are as follows:

	A_1	A_2	A_3
Allele frequencies in the parents	$\frac{7}{24} = 29.2\%$	$\frac{8}{24} = 33.3\%$	$\frac{9}{24} = 37.5\%$
Allele frequencies in the offspring	$\frac{7}{24} = 29.2\%$	$\frac{7}{24} = 29.2\%$	$\frac{10}{24} = 41.6\%$

Although the frequency of allele A_1 did not change, the frequency of A_2 declined by over 4 percent and the frequency of A_3 increased by over 4 percent, purely by chance. Allele A_3 did not confer higher fitness. Instead, it just got lucky. Random chance caused evolution—a change in allele frequencies in a population. Instead of each allele being sampled in exactly its original frequency when offspring formed, as the Hardy-Weinberg principle assumes, a chance sampling error occurred.

This exercise helps to illustrate several important points:

- Genetic drift is random with respect to fitness. The allele frequency changes it produces are not adaptive.
- Genetic drift is most pronounced in small populations. If the original population had included only couples 1 and 2, then the frequency of allele A_2 would have declined by 12.5 percent and the frequency of A_3 would have increased by 12.5 percent. The smaller the sample, the larger the sampling error. Conversely, if the population had consisted of 500 couples who produced 1000 offspring (necessitat-

ing 2000 coin flips), it is extremely unlikely that drift would produce a change in allele frequency as large as 4 percent.

- Over time, genetic drift can lead to the random loss or fixation of alleles. If allele A_1 continued to be unlucky generation after generation, it would eventually drift to a frequency of 0.0 and be lost. If allele A_3 continued to be lucky for several more generations, it might drift to a frequency of 1.0. When random loss or fixation occurs, genetic variation in the population declines.

Figure 25.6 provides an example of these points. The graphs show the output from computer simulations that combine the alleles in a gene pool at random to create an offspring generation, calculate the allele frequencies in the offspring generation, and use those allele frequencies to create a new gene pool; this process was continued for 100 generations. The top graph shows eight replicates of this process with a population size of 4; the bottom graph shows eight replicates with a population of 400. Notice the striking differences between the effects of drift in the small versus large population and the consequences for genetic variation when alleles drift to fixation or loss.

Given enough time, drift can be an important factor even in large populations. To drive this point home, consider two types of alleles that were introduced in earlier chapters and that have no effect on fitness. Recall from Chapter 16 that alleles containing silent mutations, usually in the third position of a codon, do not change the gene product. As a result, they have no effect on the phenotype. Yet these alleles routinely drift to high frequency or even fixation over time. Similarly, recall from Chapter 20 that pseudogenes do not code for a product. Although their presence does not affect an individual's phenotype, dozens of pseudogenes in the human genome have reached fixation, due to drift.

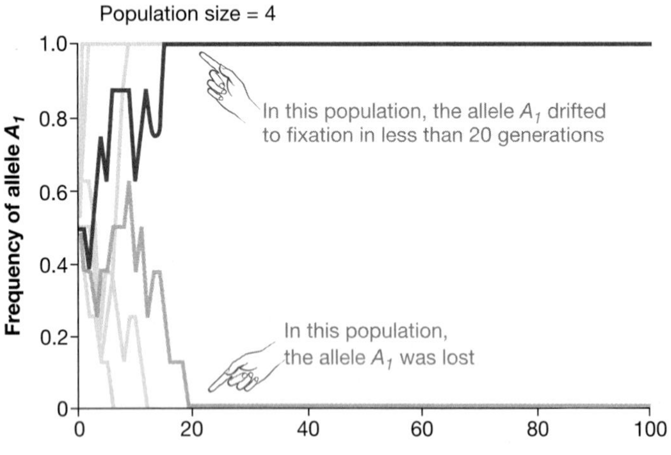

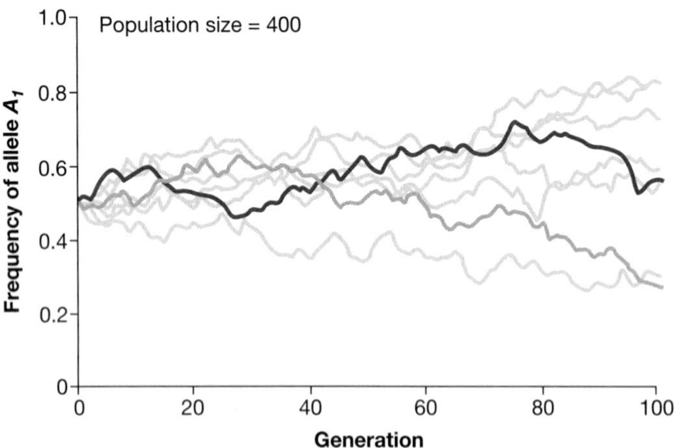

FIGURE 25.6 Genetic Drift Is More Pronounced in Small Populations than Large Populations.

EXERCISE Draw graphs predicting what these graphs would look like for a population size of 40 and a population size of 4000.

Experimental Studies of Genetic Drift

Research on genetic drift began with theoretical work in the 1930s and 1940s, which used mathematical models to predict the effect of genetic drift on allele frequencies and genetic variation. In the mid-1950s, Warwick Kerr and Sewall Wright did an experiment to show how drift works in practice. The biologists started with a large laboratory population of fruit flies that contained a **genetic marker**—a specific allele that causes a distinctive phenotype. In this case, the marker was the morphology of bristles. Fruit flies have bristles that can be either straight or bent. This difference in bristle phenotype depends on a single gene. Kerr and Wright's lab population contained just two alleles—normal (straight) and "forked" (bent).

To begin the experiment, the researchers set up 96 cages in their lab. Then they placed four adult females and four adult males of the fruit fly *Drosophila melanogaster* in each. They chose individual flies to begin these experimental populations so that the frequency of the normal and forked alleles in each of the 96 starting populations was 0.5. The two alleles do not affect the fitness of flies in the lab environment, so Kerr and Wright could be confident that if changes in the frequency of normal and forked phenotypes occurred, they would not be due to natural selection.

After these first-generation adults bred, Kerr and Wright reared their offspring. Then they randomly chose four males and four females—meaning that they simply grabbed individuals without regard to whether their bristles were normal or forked—from each of the 96 offspring populations and allowed them to breed and produce the next generation. The researchers repeated this procedure until all 96 populations had undergone a total of 16 generations. During the entire course of the experiment, no migration from one population to another occurred. Previous studies had shown that mutations from normal to forked are rare. Thus, the only evolutionary process operating during the experiment was genetic drift.

Their result? After 16 generations, the 96 populations fell into three groups (**Figure 25.7**). Forked bristles were found on all of the individuals in 29 of the experimental populations. Due to drift, the forked allele had been fixed in these 29 populations and the normal allele had been lost. In 41 other populations, however, the opposite was true: All individuals had normal bristles. In these populations, the forked allele had been lost due to chance. Both alleles were still present in 26 of the populations. The message of the study is startling: In 73 percent of the experimental populations (70 out of the 96), genetic drift had reduced allelic diversity at this gene to zero. As predicted, genetic drift decreased genetic variation within populations and increased genetic differences between populations. Is drift important in natural populations as well?

Genetic Drift in Natural Populations

The sampling process that occurs during fertilization and that caused changes in allele frequencies in Pitcairn islanders and fruit flies occurs in every population in every generation in every species that reproduces sexually. It is particularly important in small populations, where sampling error tends to be high. This is of enormous concern to conservation biologists, because populations all over the planet are being drastically reduced in size by habitat destruction and other human activities. Small populations that occupy nature reserves or zoos are particularly susceptible to genetic drift. If drift leads to a loss of genetic diversity, it could darken the already bleak outlook for some endangered species.

It is important to realize that because drift is caused by sampling error, it can occur by *any* process or event that involves sampling—not just the sampling of gametes that occurs during fertilization. Let's consider two such examples, called founder effects and bottlenecks.

How Do Founder Effects Cause Drift? When a group of individuals emigrates to a new geographic area and establishes a new population, a founder event is said to occur. If the group is small enough, the allele frequencies in the new population are almost guaranteed to be different from those in the source population, due to sampling error. A change in allele frequencies that occurs when a new population is established is called a **founder effect**. As soon as the mutineers and Tahitians set foot on Pitcairn Island and took up residence, evolution had occurred due to drift, via a founder effect. The allele frequencies in the new population were different from those in England or in Tahiti. (Once breeding took place, drift continued to act via sampling errors that occurred during fertilization.)

Fishermen on the island of Anguilla in the Caribbean recently witnessed a founder event involving green iguanas. A few weeks after Hurricane Luis and Hurricane Marilyn swept

(a) Bristle shape is a useful genetic marker in fruit flies

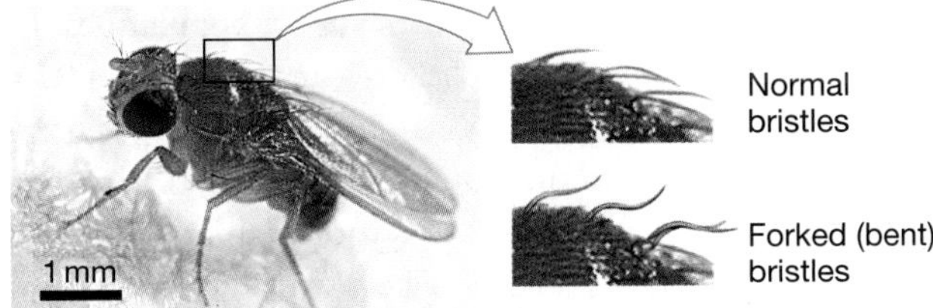

(b) In most experimental populations, the forked bristle allele drifted to fixation or loss

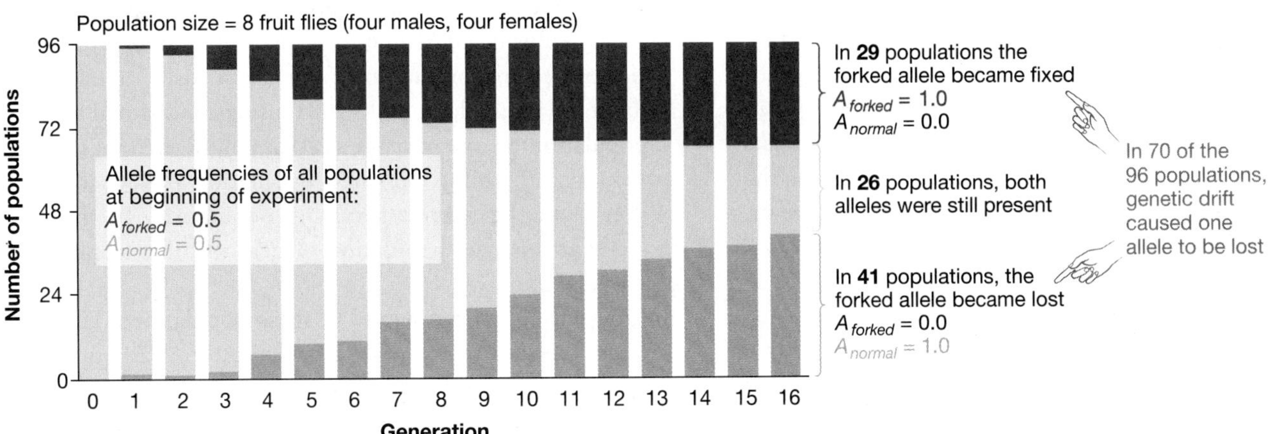

FIGURE 25.7 An Experiment on the Effects of Genetic Drift in Small Populations. In this experiment on fruit flies, population size in 96 separate populations was kept at eight individuals in each generation for 16 generations. The only evolutionary force acting in these populations was genetic drift. At the start of the experiment, the frequencies of the forked-bristle alleles and wild-type bristle alleles were each 50 percent.

● **QUESTION** Predict what the results of this experiment would be if the allele for forked-bristles was originally at a frequency of 90% instead of 50%.

through the region in September 1995, a large raft composed of downed logs tangled with other debris floated onto a beach on Anguilla. The fishermen noticed green iguanas on the raft and several on shore. Because green iguanas had not previously been found on Anguilla, the fishermen notified biologists. The researchers were able to document that at least 15 individuals had arrived; two years later they were able to confirm that at least some of the individuals were breeding. During this founder event, it is extremely unlikely that allele frequencies in the new Anguilla population of green iguanas exactly matched those of the source population, thought to be on the islands of Guadeloupe.

Founder events like these have been the major source of populations that occupy islands all over the world, as well as island-like habitats such as mountain meadows, caves, and ponds. Each time a founder event occurs, a founder effect is likely to accompany it, changing allele frequencies via genetic drift.

How Do Population Bottlenecks Cause Drift? If a large population experiences a sudden reduction in size, a population bottleneck is said to occur. The term comes from the metaphor of a few individuals passing through the neck of a bottle, by chance. Disease outbreaks, natural catastrophes such as floods or fires or storms, or other events can cause population bottlenecks.

Genetic bottlenecks follow population bottlenecks, just as founder effects follow founder events. A **genetic bottleneck** is a sudden reduction in the number of alleles in a population. Drift occurs during genetic bottlenecks and causes a change in allele frequencies. As an example of a genetic bottleneck, consider the human population of Pingelap Atoll in the South Pacific. On this island, only about 20 people out of a population of several thousand managed to survive the effects of a typhoon and a subsequent famine that occurred around 1775.

The survivors apparently included a person who carried a loss-of-function allele at a gene called *CNGB3*, which codes for a protein involved in color vision. The allele is recessive, and when it is homozygous it causes a serious vision deficit called achromatopsia. The condition is extremely rare in most populations, with the homozygous genotype and the affliction occurring in about 0.005 percent of the population. If genotypes at this locus are in Hardy-Weinberg proportions, then $q^2 = 0.00005$ and the frequency of the loss-of-function allele in most populations is about $\sqrt{0.00005}$, or 0.7 percent. In the population that survived the Pingelap Atoll disaster, however, the loss-of-function allele was at a frequency of about 1/40, or 2.5 percent. If the allele was at the typical frequency of 0.7 percent prior to the population bottleneck, then a huge frequency change occurred during the bottleneck, due to drift.

In today's population on Pingelap Atoll, about 1 in 20 people is afflicted with achromatopsia and the allele is at a frequency of over 20 percent. Because it is extremely unlikely that the loss-of-function allele is favored by directional selection or heterozygote advantage, researchers hypothesize that its frequency in this small population has continued to increase over the past 230 years due to drift.

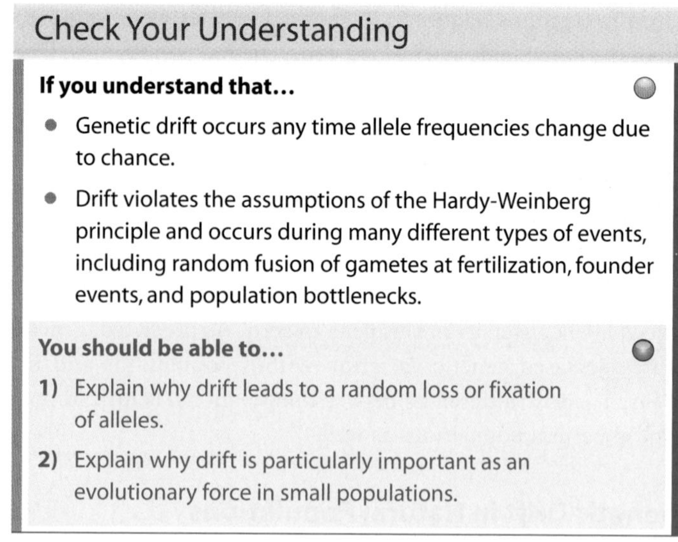

Check Your Understanding

If you understand that...

- Genetic drift occurs any time allele frequencies change due to chance.
- Drift violates the assumptions of the Hardy-Weinberg principle and occurs during many different types of events, including random fusion of gametes at fertilization, founder events, and population bottlenecks.

You should be able to...

1) Explain why drift leads to a random loss or fixation of alleles.
2) Explain why drift is particularly important as an evolutionary force in small populations.

25.4 Gene Flow

Gene flow is the movement of alleles from one population to another. It occurs when individuals leave one population, join another, and breed. As an evolutionary mechanism, gene flow usually has one outcome: It equalizes allele frequencies between the source population and the recipient population. When alleles move from one population to another, the populations tend to become more alike. To capture this point, biologists say that gene flow homogenizes allele frequencies among populations.

To see the consequences of gene flow in action, consider data from prairie lupine populations on Mount St. Helens in the state of Washington. Prairie lupine is a perennial, meaning that individuals grow from the same root system year after year. They are fairly short lived, however, with each individual rarely living for more than 5 years. Prairie lupines are most commonly found living at high altitude on volcanoes, in habitats like that shown in **Figure 25.8a**.

The explosion of Mount St. Helens in 1980 created thousands of hectares of mudflows and ash deposits that are ideal habitat for prairie lupines. In 1981 a single individual founded a population on an extensive ash plain in the area most highly affected by the blast. This new population was located over 4 km from the nearest surviving population. In the blast zone near this new population, dozens of other populations eventually appeared.

To study genetic diversity in these populations, biologists collected tissue samples from 532 individuals in 32 populations. The age of each of the study populations was known. The researchers made sure they analyzed tissues from the oldest individuals in populations that had been established just 1 to 3 years prior to the start of the study. These individuals were of

(a) Lupines colonize sites and form populations.

(b) Gene flow reduces genetic differences among populations.

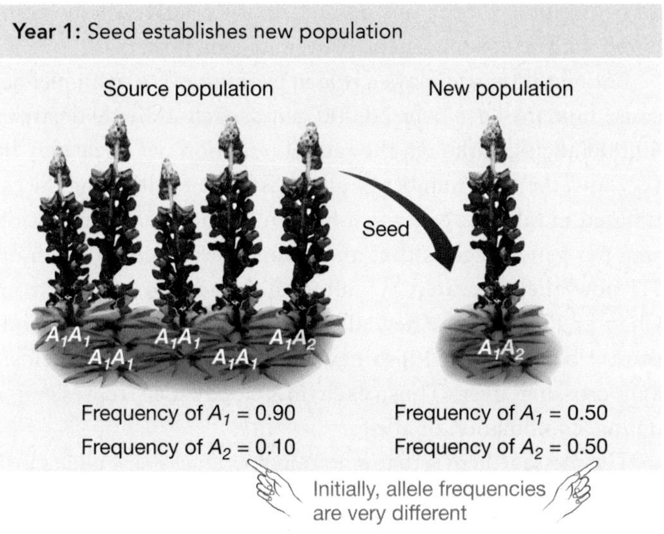

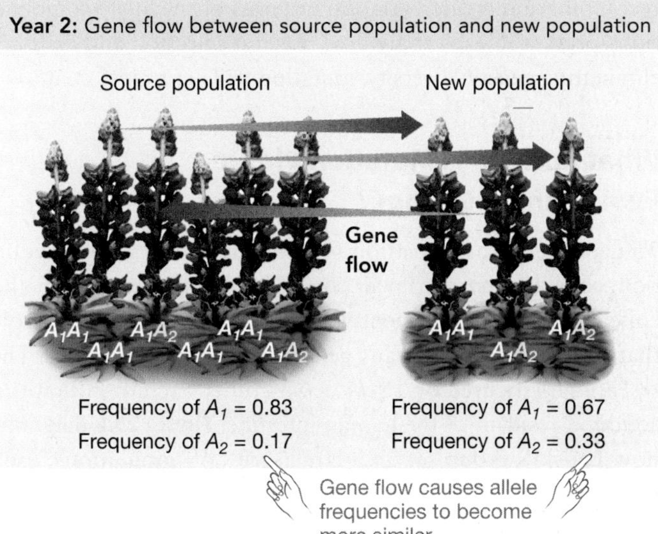

FIGURE 25.8 Gene Flow Equalizes Allele Frequencies between Populations. (a) Several species of lupine grow on the slopes of volcanoes. **(b)** The movement of alleles between populations tends to equalize allele frequencies over time.

special interest, because they were undoubtedly the founders of the new populations. Once tissues from all 532 individuals had been collected, the investigators used the polymerase chain reaction (see Chapter 19) to amplify the alleles of two genes in each individual. These data allowed them to estimate the number of alleles present in each population, as well as the frequency of each allele.

Their results? The data showed that the oldest individuals in newly established populations had allele frequencies that were very different from allele frequencies in the source populations. To interpret this observation, the biologists hypothesize that strong founder effects occurred each time a new lupine population became established. But their data also showed that as populations got older, allele frequencies became progressively more like those in the oldest populations—specifically, like the populations that had survived the eruption. Because selection due to differences in soil and moisture conditions, diseases, and predators present in each population would tend to make populations less similar over time, the researchers doubted that natural selection played a large role in equalizing allele frequencies. Instead, they propose that a continuous flow of alleles between populations, primarily via pollen carried by bees, was responsible for making the populations genetically more alike over time (**Figure 25.8b**).

What impact did gene flow have on the average fitness of individuals in these populations? The answer is not known, but theory suggests that it might vary. If a population had lost alleles due to genetic drift, then the arrival of new alleles via gene flow should increase genetic diversity. In this way, gene flow might increase the average fitness of individuals. But if natural selection had resulted in a population that was highly adapted to a specific habitat, then gene flow from other populations might introduce alleles that have low fitness in that particular environment. In this case, gene flow would have a negative effect on average fitness in the population.

To summarize, gene flow is random with respect to fitness—the arrival or departure of alleles can increase or decrease average fitness, depending on the situation. But a movement of alleles between populations tends to reduce genetic differences between them. This latter generalization is particularly important in our own species right now. Large numbers of people from Africa, the Middle East, Mexico, Central America, and Asia are emigrating to the countries of the European Union and the United States. Because individuals from different cultural and ethnic groups are intermarrying frequently and having offspring, allele frequencies in human populations are becoming more similar.

25.5 Mutation

To appreciate the role of mutation as an evolutionary force, let's return to one of the central questions that biologists ask about an evolutionary mechanism: How does it affect genetic variation in a population? Gene flow can increase genetic diversity in a recipient population if new alleles arrive with immigrating individuals. But gene flow can instead decrease genetic variation in the source population if alleles leave with emigrating individuals. The impact of genetic drift on genetic variation is much more clear cut than the impact of gene flow. Genetic drift tends to decrease genetic diversity over time, as alleles are randomly lost or fixed. Similarly, most forms of selection favor certain alleles and lead to a decrease in overall genetic variation.

If most of the evolutionary mechanisms lead to a loss of genetic diversity over time, what restores it? In particular, where do entirely new alleles come from? The answer to both of these questions is **mutation.**

As Chapter 16 noted, mutations occur when DNA polymerase makes an error as it copies a DNA molecule, resulting in a change in the sequence of deoxyribonucleotides. If a mutation occurs in a stretch of DNA that codes for a protein, the changed codon may result in a polypeptide with a novel amino acid sequence. Because errors are inevitable, mutation constantly introduces new alleles into populations in every generation. Mutation is an evolutionary mechanism that increases genetic diversity in populations.

The second important point to recognize about mutation is that it is random with respect to the fitness of the affected allele. Mutation just happens. Changes in DNA do not occur in a way that tends to increase fitness or decrease fitness. Because most organisms are well adapted to their current habitat, random changes in genes usually result in products that do not work as well as the alleles that currently exist. Stated another way, most mutations in sequences that code for a functional protein or RNA result in **deleterious** alleles—alleles that lower fitness. On rare occasions, however, mutation in these types of sequences produces a beneficial allele—an allele that allows individuals to produce more offspring. Beneficial alleles should increase in frequency in the population due to natural selection.

Because mutation produces new alleles, it can in principle change the frequencies of alleles through time. But does mutation occur often enough to make it an important factor in changing allele frequencies of a particular gene? The short answer is no.

Mutation as an Evolutionary Mechanism

To understand why mutation is not a significant mechanism of evolutionary change by itself, consider that the highest mutation rates that have been recorded at individual genes in humans are on the order of 1 mutation in every 10,000 gametes produced by an individual. This rate means that for every 10,000 alleles produced, on average one will have a mutation. When two gametes combine to form an offspring, then, at most about 1 in every 5000 offspring will carry a mutation at a particular gene. Will mutation affect allele frequencies in a species such as ours? To answer this question, suppose that 195,000 humans live in a population; that 5000 offspring are born one year; and that at the end of that year, the population numbers 200,000. Humans are diploid, so in a population this size, there is a total of 400,000 copies of each gene. Only one of them is a new allele created by mutation, however. Over the course of a year, the allele frequency change introduced by mutation is 1/400,000, or 0.0000025 (2.5×10^{-6}). At this rate, it would take 4000 years for mutation to produce a change in allele frequency of 1 percent.

These calculations support the conclusion that mutation does little to change allele frequencies on its own. Although mutation can be a significant evolutionary force in bacteria and archaea, which have extremely short generation times, mutation in eukaryotes rarely causes a change from the genotype frequencies expected under the Hardy-Weinberg principle. As an evolutionary mechanism, mutation is relatively slow compared with selection, genetic drift, and gene flow.

But mutation still plays a role in evolution. For example, because humans have over 20,000 genes, each individual carries 40,000 alleles. Although the rate of mutation per allele may be very low, the total number of alleles is high. Multiplying the estimated number of genes in a human by the average mutation rate per gene suggests that an average person contains about 1.1 new alleles created by mutation. In addition, recall from Chapter 12 that these new alleles and existing alleles are constantly being shuffled into new combinations during meiosis and crossing over. Thus, each individual also represents a unique combination of alleles.

The message here is that mutation introduces new alleles into every individual in every population in every generation. And in species that undergo sexual reproduction, meiosis and genetic recombination create variation in terms of the allele combinations present in each individual. Even if selection and drift are eliminating genetic diversity, mutation will act to restore it.

What Role Does Mutation Play in Evolutionary Change?

To get a better appreciation for how mutation affects evolution, consider an experiment conducted by Richard Lenski and colleagues. The experiment was designed to evaluate the role that mutation plays in many genes over many generations. The researchers focused on *Escherichia coli*, a bacterium that is a common resident of the human intestine. **Figure 25.9** diagrams how Lenski's group set up a large series of populations, each founded with a single cell. Half of the cells contained an allele that allowed them to metabolize the sugar arabinose. The populations founded by these cells were designated Ara^{+}. The other cells carried an allele that did not allow them to metabolize arabinose. The populations founded by these cells were designated Ara^{-}. When Ara^{+} cells are grown on special plates, they produce colonies that look red. When Ara^{-} colonies are

Experiment

Question: How does average fitness in a population change over time?

Hypothesis: Average fitness increases over time.

Null hypothesis: Average fitness does not increase over time.

Experimental setup:

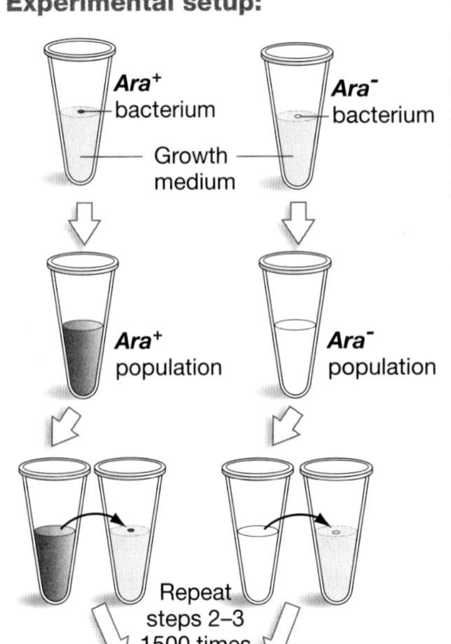

1. Place 10 mL of identical growth medium into many replicate tubes with one bacterium in each, some Ara^+ and some Ara^- so that they can be distinguished by color when grown on special plates.

2. Incubate overnight. Average population in each tube is now 5×10^8 cells. (Cells are red or white when grown on special plates.)

3. Remove 0.1 mL from each tube and move to 10 mL of fresh medium. Freeze remaining cells for later analysis.

4. Put an equal number of cells from generation 1 and a later generation in fresh growth medium (competition experiment).

5. Incubate overnight and count the cells. Which are more numerous?

Prediction: Descendant populations have higher average fitness. (There will be more individuals from descendant than ancestral populations on the plates.)

Prediction of null hypothesis: There will be no difference in fitness between descendant and ancestral populations.

Results:

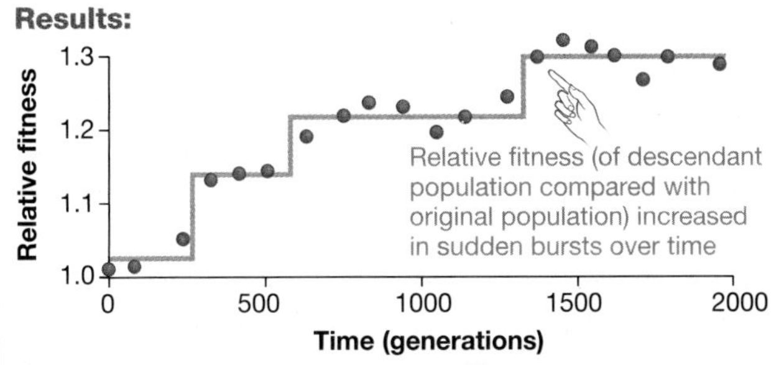

Conclusion: Descendant populations have higher fitness than do ancestral populations.

FIGURE 25.9 An Experiment to Test Changes in Fitness over Time.

grown on the same plates, they produce colonies that look white. Under the temperature and nutrient conditions used in the experiment, however, neither *Ara*$^-$ nor *Ara*$^+$ cells had a selective advantage. The presence of the *Ara* allele simply gave the researchers a way to identify particular populations by color. What happened to the descendants of these founding cells over time?

To answer this question, the researchers transferred a small number of cells from each of the populations into a new batch of the same growth medium, under the same light and temperature conditions, every day for over four years. In this way, each population grew continuously. Over the course of the experiment, the researchers estimated that each population underwent a total of 10,000 generations. This is the equivalent of over 200,000 years of human evolution. In addition, the biologists saved a sample of cells from each population at regular intervals and stored them in a freezer. The frozen cells served as a fossil record of cells that existed over the 10,000 generation time interval. But because frozen *E. coli* cells resume growth when they are thawed, the "fossil" individuals could be brought back to life.

The strain of *E. coli* used in the experiment is completely asexual and reproduces by cell division. Thus, mutation was the only source of genetic variation in these populations. Although no gene flow occurred, both selection and genetic drift were operating in each population.

Were cells from the older and newer generations of each population different? This question was addressed via competition experiments. *Ara*$^+$ cells from one generation and *Ara*$^-$ cells from a different generation were put together in the same flask. After allowing the cells to grow and compete, the researchers grew them on special plates and counted the number of red versus white colonies. At the end of these competition experiments, the population of *Ara*$^+$ or *Ara*$^-$ cells that was more numerous had grown the fastest, meaning that it was better adapted to the experimental environment. In this way the researchers could measure the fitness of descendant populations relative to ancestral populations. If relative fitness was greater than 1, it meant that recent-generation cells outnumbered older-generation cells when the competition was over.

The data from a series of competition experiments are graphed at the bottom of Figure 25.9. Notice that relative fitness increased dramatically—almost 30 percent—over time. But notice also that fitness increased in fits and starts. This pattern is emphasized by the solid line on the graph, which represents a mathematical function fitted to the data points.

What caused this stair-step pattern? Lenski's group hypothesizes that genetic drift was relatively unimportant in this experiment because population sizes were so large. Instead, they propose that each jump was caused by a novel mutation that conferred a fitness benefit. Their interpretation is that cells that happened to have the beneficial mutation grew rapidly and came to dominate the population. After a beneficial

SUMMARY TABLE **25.4** **Evolutionary Mechanisms**

Process	Definition and Notes	Effect on Genetic Variation	Effect on Average Fitness
Selection	Certain alleles are favored	Can lead to maintenance, increase, or reduction of genetic variation	Can produce adaptation
Genetic drift	Random changes in allele frequencies: most important in small populations	Tends to reduce genetic variation, via loss or fixation of alleles	Usually reduces average fitness
Gene flow	Movement of alleles between populations; reduces differences between populations	May increase genetic variation by introducing new alleles; may decrease it by removing alleles	May increase average fitness by introducing high-fitness alleles or decrease it by introducing low-fitness alleles
Mutation	Production of new alleles	Increases genetic variation by introducing new alleles	Random with respect to fitness; most mutations in coding sequences lower fitness

mutation occurred, the fitness of the population stabilized—sometimes for hundreds of generations—until another random but beneficial mutation occurred and produced another jump in fitness.

The experiment makes an important point: Mutation is the ultimate source of genetic variation. If mutation did not occur, evolution would eventually stop. Without mutation, there is no variation for natural selection to act on.

To summarize, mutation alone is usually inconsequential in changing allele frequencies at a particular gene. When considered across the genome and when combined with natural selection, however, it becomes an important evolutionary mechanism. Mutation is the ultimate source of the heritable variation that makes evolution possible.

Table 25.4 summarizes the four evolutionary forces and their consequences. If one or more of these processes affects a gene, then genotypes will not be in Hardy-Weinberg proportions. But we have yet to consider the effects of another assumption in the Hardy-Weinberg model—that mating takes place at random with respect to the gene in question. What happens when this assumption is violated?

25.6 Nonrandom Mating

In the Hardy-Weinberg model, gametes were picked from the gene pool at random and paired to create offspring genotypes. In nature, however, matings between individuals may not be random with respect to the gene in question. Even in species like clams that simply broadcast their gametes into the surrounding water, gametes from individuals who live close to each other are more likely to combine than gametes from individuals that live farther apart. In insects, vertebrates, and many other animals, females don't mate at random but actively choose certain males.

Two mechanisms that violate the Hardy-Weinberg assumption of random mating have been studied intensively. One is the phenomenon known as inbreeding; the other is the process called sexual selection.

Inbreeding

Mating between relatives is called **inbreeding**. By definition, relatives share a recent common ancestor. Individuals that inbreed are likely to share alleles they inherited from their common ancestor.

To understand how inbreeding affects populations, let's follow the fate of alleles and genotypes when it occurs. We'll again focus on a single locus with two alleles, A_1 and A_2. Suppose that these alleles initially have equal frequencies of 0.5. But now let's imagine that the individuals in the population don't produce gametes that go into a gene pool. Instead, they self-fertilize. Many flowering plants, for example, contain both male and female organs and routinely self-pollinate. Homozygous parents that self-fertilize produce all homozygous offspring. Heterozygous parents, in contrast, produce homozygous and heterozygous offspring in a 1:2:1 ratio (**Figure 25.10a**). **Figure 25.10b** shows the outcome for the population as a whole. In this figure, the width of the boxes represents the frequency of the three genotypes, which start out at the Hardy-Weinberg ratio of $p^2 : 2pq : q^2$. Notice that the homozygous proportion of the population increases each generation, while the heterozygous proportion is halved. At the end of the four generations illustrated, heterozygotes are rare.

This simple exercise demonstrates two fundamental points about inbreeding. First, inbreeding increases homozygosity. When inbreeding occurs, the frequency of homozygotes increases and the frequency of heterozygotes decreases. In essence, inbreeding takes alleles from heterozygotes and puts them into homozygotes. Second, inbreeding does not cause evolution, because allele frequencies do not change in the population as a whole. Inbreeding changes genotype frequencies—not allele frequencies. Self-fertilization, or selfing, is the most extreme form of inbreeding; but the same outcomes occur, more slowly, with less-extreme forms of inbreeding. If you understand this concept, you should be able to predict how observed genotype frequencies should differ from those expected under the Hardy-Weinberg principle, when inbreeding is occurring.

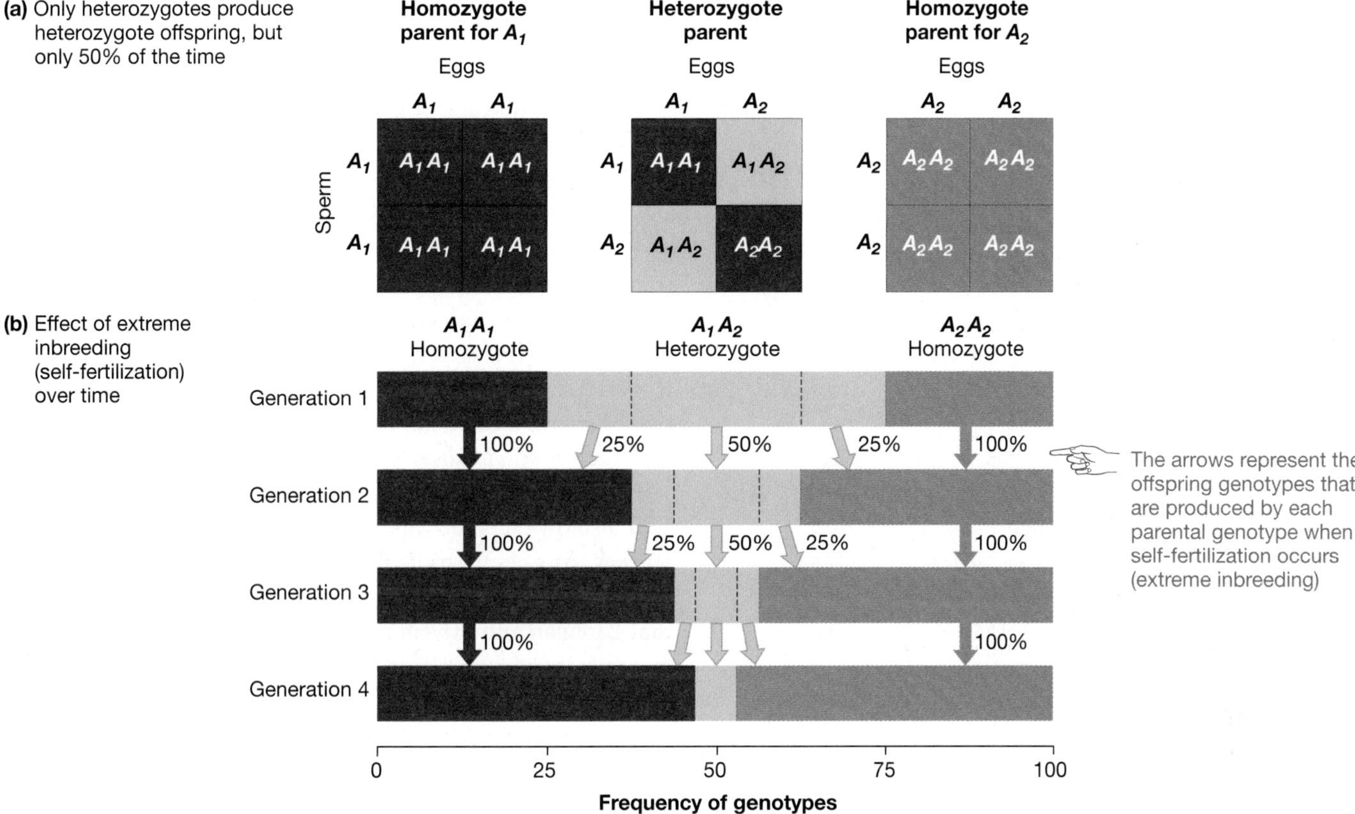

FIGURE 25.10 Inbreeding Increases Homozygosity and Decreases Heterozygosity. (a) Heterozygous parents produce homozygous and heterozygous offspring in a 1:2:1 ratio. **(b)** The width of the boxes corresponds to the frequency of each genotype.

QUESTION What proportion of this population is homozygous in generation 1? In generation 4?

Inbreeding has attracted a great deal of attention because of a phenomenon called inbreeding depression. Inbreeding depression is a decline in average fitness that takes place when homozygosity increases and heterozygosity decreases in a population. Inbreeding depression results from two processes:

1. Many recessive alleles represent loss-of-function mutations. Because these alleles are usually rare, there are normally very few homozygous recessive individuals in a population. Instead, most loss-of-function alleles exist in heterozygous individuals. The alleles have little or no effect when they occur in heterozygotes, because one normal allele usually produces enough functional protein to support a normal phenotype. But inbreeding increases the frequency of homozygous recessive individuals. Loss-of-function mutations are usually deleterious or even lethal when they are homozygous.
2. Many genes—especially those involved in fighting disease—are under intense selection for heterozygote advantage. If an individual is homozygous at these genes, then fitness declines.

The upshot here is that the offspring of inbred matings are expected to have lower fitness than the progeny of outcrossed matings. This prediction has been verified in a wide variety of species, often through laboratory or greenhouse studies that compare the fitnesses of offspring from controlled matings (**Figure 25.11**). As **Table 25.5** shows, inbreeding depression is also

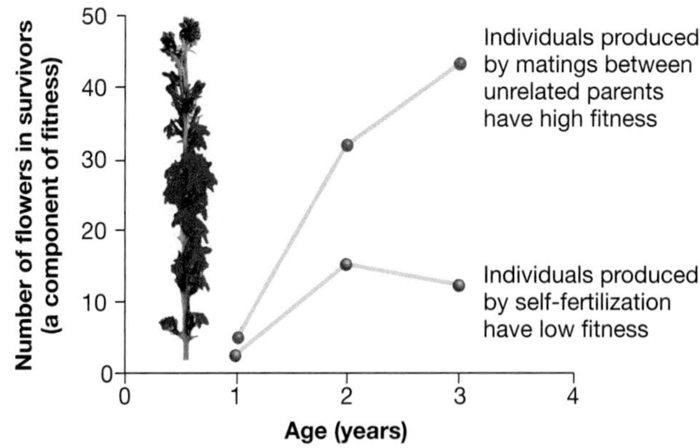

FIGURE 25.11 Inbreeding Depression Occurs in *Lobelia cardinalis*. Inbreeding depression is the fitness difference between non-inbred and inbred individuals.

EXERCISE Label the parts on this graph that indicate inbreeding depression. Does inbreeding depression increase with age in this species or remain constant throughout life?

TABLE **25.5 Inbreeding Reduces Fitness in Humans**

The percentages reported here give the mortality rate of children produced by first-cousin marriages versus marriages between nonrelatives. In every study, children of first-cousin marriages have a higher mortality rate.

		Deaths	
Age	**Period**	**Children of First Cousins (%)**	**Children of Nonrelatives (%)**
Children under 20 (U.S.)	18th–19th century	17.0	12.0
Children under 10 (U.S.)	1920–1956	8.1	2.4
At/before birth (France)	1919–1950	9.3	3.9
Children (France)	1919–1950	14.0	10.0
Children under 1 (Japan)	1948–1954	5.8	3.5
Children 1–8 (Japan)	1948–1954	4.6	1.5

SOURCE: C. Stern, *Principles of Human Genetics* (San Francisco: Freeman, 1973).

pronounced in humans. Because inbreeding has such deleterious consequences in humans, it is not surprising that many contemporary human societies have laws forbidding marriages between individuals who are related as first cousins or closer. If you understand inbreeding depression, you should be able to explain why it does not occur in species like garden peas, where self-fertilization has occurred routinely for many generations.

The trickiest point to grasp about inbreeding is that even though it does not cause evolution directly—because it does not change allele frequencies—it can speed the rate of evolutionary change. More specifically, it increases the rate at which purifying selection eliminates deleterious recessive alleles from a population. A moment's thought should convince you why this is so. Deleterious recessives are usually very rare in populations, because they lower fitness. Rare alleles are usually found in heterozygotes because it is more likely for individuals to have one copy of a rare allele than two. When no inbreeding is occurring, the deleterious recessives found in heterozygotes cannot be eliminated by natural selection. But when inbreeding occurs, most deleterious recessives are found in homozygotes and are quickly eliminated by selection. If you understand this concept, you should be able to explain why inbreeding helps "purge" deleterious recessive alleles.

Sexual Selection

If female peacocks choose males with the longest and most iridescent tails as mates, then nonrandom mating is occurring. Due to this type of nonrandom mating, the frequency of alleles that contribute to long, iridescent tails will increase in the population.

Charles Darwin was the first biologist to recognize that selection based on success in courtship is a mechanism of evolutionary change. The process is called **sexual selection**, and it can be considered a special case of natural selection. Sexual selection occurs when individuals within a population differ in their ability to attract mates. It favors individuals with heritable traits that enhance their ability to obtain mates.

In 1948 A. J. Bateman contributed a fundamental insight about how sexual selection works. His idea was elaborated by Robert Trivers in 1972. The Bateman-Trivers theory contains two elements: a claim about a pattern in the natural world and a mechanism that causes the pattern.

The pattern component of their theory is that sexual selection usually acts on males much more strongly than on females. As a result, traits that attract members of the opposite sex are much more highly elaborated in males. The mechanism that Bateman and Trivers proposed to explain this pattern can be summarized with a quip: "Eggs are expensive, but sperm are cheap." That is, the energetic cost of creating a large egg is enormous, whereas a sperm contains few energetic resources. Thus, in most species, females invest much more in their offspring than do males. This phenomenon is called the fundamental asymmetry of sex. It is characteristic of almost all sexual species and has two important consequences:

1. Because eggs are large and energetically expensive, females produce relatively few young over the course of a lifetime. A female's fitness is limited primarily by her ability to gain the resources needed to produce more eggs and healthier young—not by the ability to find a mate.

2. Sperm are so simple to produce that a male can father an almost limitless number of offspring. For males, fitness is limited not by the ability to acquire the resources needed to produce sperm, but by the number of females they can mate with.

The theory of sexual selection makes strong predictions. If females invest a great deal in each offspring, then they should protect that investment by being choosy about their mates. Conversely, if males invest little in each offspring, then they should be willing to mate with almost any female. If there are an equal number of males and females in the population, and if males are trying to mate with any female possible, then males will have to compete with each other for mates. As a result, sexual selection should act more strongly on males than on females. Traits that evolve due to sexual selection—meaning traits that are useful only in courtship or in competition for mates—should be found primarily in males.

Do data from experimental or observational studies agree with these predictions? Let's consider each of them in turn.

Sexual Selection via Female Choice If females are choosy about which males they mate with, what criteria do females use to make their choice? Recent experiments have shown that in several bird species, females prefer to mate with males that are well fed and in good health. These experiments were motivated by three key observations: (1) In many bird species, the existence of colorful feathers or a colorful beak is due to the presence of the red and yellow pigments called carotenoids. (2) Carotenoids protect tissues and stimulate the immune system to fight disease more effectively. (3) Animals cannot synthesize their own carotenoids, but plants can. Animals have to obtain carotenoids by eating carotenoid-rich plant tissues.

These observations suggest that the healthiest and best-nourished birds in a population have the most colorful beaks and feathers. Sick birds have dull coloration because they are using all of their carotenoids to stimulate their immune system. Poorly fed birds have dull coloration because they have few carotenoids available. By choosing a colorful male as the father of her offspring, a female is likely to have offspring with alleles that will help the offspring fight disease effectively and feed efficiently.

To test the hypothesis that females prefer to mate with colorful males, a team of researchers experimented with zebra finches (**Figure 25.12a**). They fed one group of male zebra finches a diet that was heavily supplemented with carotenoids, and they fed a second group of male zebra finches (the control group) a diet that was similar in every way except for the additional carot-enoids. To control for other differences between the groups, the researchers used individuals that had been raised and maintained in environments that were as similar as possible, identified pairs of brothers, and randomly assigned one brother to the control group and one brother to the treatment group.

As predicted, the males eating the carotenoid-supplemented diet developed more colorful beaks than did the males fed the carotenoid-poor diet (**Figure 25.12b**). When given a choice of mating with either of the two brothers, 9 out of 10 females tested preferred the more-colorful male. These results are strong evidence that females of this species are choosy about their mates and that they prefer to mate with healthy, well-fed males.

Enough experiments have been done on other bird species to support a general conclusion: Colorful beaks and feathers, along with songs and dances and other types of courtship displays, carry the message "I'm healthy and well fed because I have good alleles. Mate with me."

Choosing "good alleles" is not the entire story in sexual selection via female choice, however. In many species, females prefer to mate with males that care for young or that provide the resources required to produce eggs. Brown kiwi females make an enormous initial investment in their offspring—their eggs routinely represent over 15 percent of the mother's total body weight (**Figure 25.13**)—but choose to mate with males that take over all of the incubation and other care of the offspring. It is common to find that female fish prefer to mate

(a) Male zebra finches have orange beaks.

(b) Males fed carotenoids get more colorful beaks.

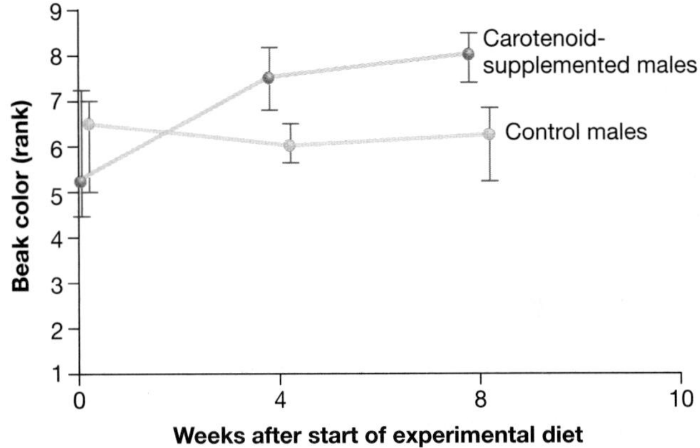

FIGURE 25.12 If Male Zebra Finches Are Fed Carotenoids, Their Beaks Get More Colorful.

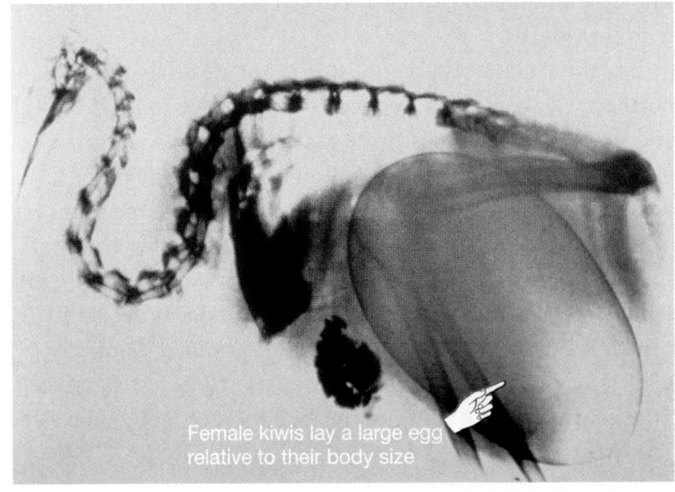

FIGURE 25.13 In Many Species, Females Make a Large Investment in Each Offspring. X-ray of a female kiwi, ready to lay an egg.

● **QUESTION** Suppose you are examining the gametes of a newly discovered species. How do you know which are sperm and which are eggs?

with males that protect a nest site and care for the eggs until they hatch. In humans and many species of birds, males provide food, protection, and other resources required for rearing young.

To summarize, females may choose mates on the basis of (1) physical characteristics that signal male genetic quality, (2) resources or parental care provided by males, or (3) both. In some species, however, females do not have the luxury of choosing a male. Instead, competition among males is the primary cause of sexual selection.

Sexual Selection via Male-Male Competition As an example of research on how males compete for mates, consider data from a long-term study of a northern elephant seal population breeding on Año Nuevo Island, off the coast of California. Elephant seals feed on marine fish and spend most of the year in the water. But when females are ready to mate and give birth, they haul themselves out of the water onto land. Females prefer to give birth on islands, where newborn pups are protected from terrestrial and marine predators. Because elephant seals have flippers that are ill suited for walking, females can haul themselves out of the water only on the few beaches that have gentle slopes. As a result, large numbers of females congregate in tiny areas to breed.

Male elephant seals establish territories on breeding beaches by fighting (**Figure 25.14a**). A **territory** is an area that is actively defended and that provides exclusive use by the owner. Males that win battles with other males monopolize matings with the females residing in their territories. Females can't choose among males—they simply have to mate with the winning male. Males that lose battles are relegated to territories with few females or are excluded from the beach. Fights are essentially slugging contests and are usually won by the larger male. The males stand face to face, bite each other, and land blows with their heads.

Based on these observations, it is not surprising that male northern elephant seals frequently weigh three tons (2700 kg) and are over four times more massive, on average, than females. The logic here runs as follows: Males that own beaches with large congregations of females will father large numbers of offspring. Males that lose fights will father few or no offspring. As a result, the alleles of territory-owning males will rapidly increase in frequency in the population. If the ability to win fights and produce offspring is determined primarily by body size, then alleles for large body size will have a significant fitness advantage, leading to the evolution of large male size. The fitness advantage is due to sexual selection.

Figure 25.14b provides evidence for intense sexual selection in males. Biologists have marked many of the individuals in the seal population on Año Nuevo to track the lifetime reproductive success of a large number of individuals. As the data show, in this population a few males father a large number of offspring, while most males father few or none. Among females, variation in reproductive success is also high; but it is much lower than in males (**Figure 25.14c**). In this species, most sexual selection is driven by male-male competition rather than female choice.

(a) Males compete for the opportunity to mate with females.

(b) Variation in reproductive success is high in males.

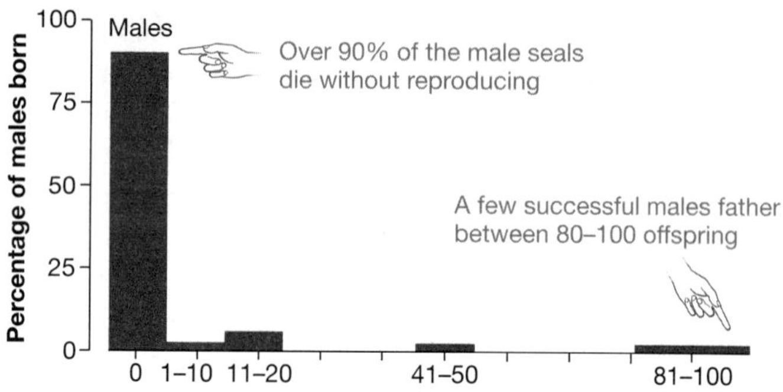

(c) Variation in reproductive success is relatively low in females.

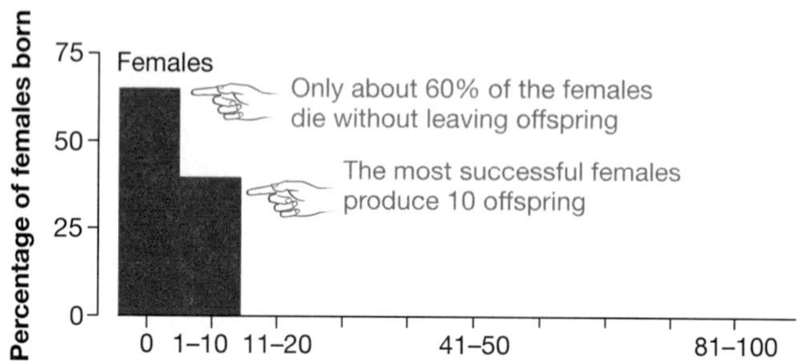

FIGURE 25.14 Intense Sexual Selection on Male Elephant Seals. **(a)** Two bull northern elephant seals compete for the opportunity to mate with females that will breed on this beach. The histograms show that variation in lifetime reproductive success is even higher **(b)** in male northern elephant seals than it is **(c)** in females.

QUESTION Consider an allele that increases reproductive success in males versus an allele that increases reproductive success in females. Which allele will increase in frequency faster, and why?

FIGURE 25.15 Sexually Selected Traits Are Used to Compete for Mates. Males often have exaggerated traits that they use in fighting or courtship. In many species, females lack these traits.

[(b) top ©B. Schorre/VIREO; (b) bottom ©R. & A. Simpson/VIREO]

QUESTION How would you set up an experiment to test the hypothesis that the exaggerated traits shown here help males of these species attract mates?

What Are the Consequences of Sexual Selection? In elephant seals and most other animals studied, most females that survive to adulthood get a mate. In contrast, many males do not. Because sexual selection tends to be much more intense in males than females, males tend to have many more traits that function only in courtship or male-male competition. Stated another way, sexually selected traits often differ sharply between the sexes.

Sexual dimorphism (literally "two-forms") refers to any trait that differs between males and females. **Figure 25.15** illustrates sexually dimorphic traits. They range from weapons that males use to fight over females, such as antlers and horns, to the elaborate ornamentation and behavior used in courtship displays. Humans are sexually dimorphic in size, distribution of body hair, and many other traits.

Like inbreeding and other forms of nonrandom mating, sexual selection violates the assumptions of the Hardy-Weinberg principle. Unlike inbreeding, however, it causes certain alleles to increase or decrease in frequency and results in evolution.

Check Your Understanding

If you understand that...

- Inbreeding and sexual selection are forms of nonrandom mating.
- Inbreeding is mating between relatives. It is not an evolutionary process, because it changes genotype frequencies—not allele frequencies. Inbreeding increases homozygosity and may lead to inbreeding depression.
- Sexual selection is based on differential success in obtaining mates. It causes evolution by increasing the frequency of alleles associated with successful courtship.

You should be able to...

1) Define the fundamental asymmetry of sex.
2) Explain why males are usually the sex with exaggerated traits used in courtship.
3) Explain how male-male competition for mates differs from female choice of mates.

Chapter Review

SUMMARY OF KEY CONCEPTS

The Hardy-Weinberg principle acts as a null hypothesis when researchers want to test whether evolution or nonrandom mating is occurring at a particular gene.

Biologists study the consequences of the different evolutionary mechanisms through a combination of mathematical modeling and experimental or observational research. The Hardy-Weinberg principle can serve as a null hypothesis in evolutionary studies because it predicts what genotype and allele frequencies are expected to be if mating is random with respect to the gene in question and none of the four evolutionary processes is operating on that gene.

You should be able to predict how genotype frequencies differ from Hardy-Weinberg proportions under selection for heterozygote advantage, and directional selection.

(MB) Web Animation at www.masteringbio.com
The Hardy-Weinberg Principle

Each of the four evolutionary mechanisms has different consequences. Only natural selection produces adaptation. Genetic drift causes random fluctuations in allele frequencies. Gene flow equalizes allele frequencies between populations. Mutation introduces new alleles.

Natural selection occurs in a wide variety of patterns and a wide variety of intensities. Directional selection may lead to certain alleles becoming fixed—and thus reduces allelic diversity in populations. Stabilizing selection eliminates phenotypes with extreme characteristics and decreases allelic diversity in populations. The rate of evolution under natural selection depends on both the intensity of selection and the amount of genetic variation available.

Genetic drift results from sampling error and is an important evolutionary force in small populations. Drift leads to the random fixation of alleles and tends to reduce overall allelic diversity.

Gene flow is the movement of alleles between populations. Gene flow tends to homogenize allele frequencies and decrease differentiation among populations, but it can also serve as an important source of new variation in populations.

Mutation is too infrequent to be a major cause of allele frequency change. But because mutation continually introduces new alleles at all genes, it is essential to evolution. Without mutation, natural selection and genetic drift would eventually eliminate genetic variation, and evolution would cease.

You should be able to explain how the evolutionary forces affect the management of endangered species. What effect will drift have? How will gene flow from captive populations affect wild populations?

(MB) Web Animation at www.masteringbio.com
Three Modes of Natural Selection

Inbreeding changes genotype frequencies but does not change allele frequencies.

Inbreeding, or mating among relatives, is a form of nonrandom mating. Inbreeding does not change allele frequencies, so it is not an evolutionary mechanism. It does, however, change genotype frequencies by leading to an increase in homozygosity and a decrease in heterozygosity. These patterns can accelerate natural selection and can cause inbreeding depression.

You should be able to predict how extensive inbreeding during the 1700s and 1800s affected the royal families of Europe.

Sexual selection leads to the evolution of traits that help individuals attract mates. It is usually stronger on males than on females.

Sexual selection is a form of nonrandom mating. It occurs when certain traits help males succeed in contests over mates or when certain traits are attractive to females. It is responsible for the evolution of phenotypic differences between males and females.

You should be able to predict the pattern of sexual dimorphism observed in animal species where males make a much larger investment in offspring than females do.

QUESTIONS

Test Your Knowledge

1. Why isn't inbreeding considered an evolutionary mechanism?
   a. It does not change genotype frequencies.
   b. It does not change allele frequencies.
   c. It does not occur often enough to be important in evolution.
   d. It does not violate the assumptions of the Hardy-Weinberg principle.
2. Why is genetic drift aptly named?
   a. It causes allele frequencies to drift up or down randomly.
   b. It is the ultimate source of genetic variability.
   c. It is an especially important mechanism in small populations.
   d. It occurs when populations drift into new habitats.
3. How do sexual selection and inbreeding differ?
   a. Unlike inbreeding, sexual selection changes allele frequencies and affects only genes involved in attracting mates.
   b. Unlike sexual selection, inbreeding changes allele frequencies and involves any mating between relatives—not just self-fertilization.
   c. Unlike inbreeding, sexual selection results from the random fusion of gametes during fertilization. It is particularly important in small populations, where few mates are available.
   d. Inbreeding occurs only in small populations, while sexual selection can occur in any size population.

4. What does it mean when an allele reaches "fixation"?
   a. It is eliminated from the population.
   b. It has a frequency of 1.0.
   c. It is dominant to all other alleles.
   d. It is adaptively advantageous.

5. In what sense is the Hardy-Weinberg principle a null hypothesis, similar to the control treatment in an experiment?
   a. It defines what genotype frequencies should be if nonrandom mating is occurring.
   b. Expected genotype frequencies can be calculated from observed allele frequencies and then compared with the observed genotype frequencies.
   c. It defines what genotype frequencies should be if natural selection, genetic drift, gene flow, or mutation is occurring *and* if mating is random.
   d. It defines what genotype frequencies should be if evolutionary mechanisms are *not* occurring.

6. Mutation is the ultimate source of genetic variability. Why is this statement correct?
   a. DNA polymerase (the enzyme that copies DNA) is remarkably accurate.
   b. "Mutation proposes and selection disposes."
   c. Mutation is the only source of new alleles.
   d. Mutation occurs in response to natural selection. It generates the alleles that are required for a population to adapt to a particular habitat.

Test Your Knowledge answers: 1.b; 2.a; 3.a; 4.b; 5.d; 6.c

Test Your Understanding

Answers are available at www.masteringbio.com

1. Create concept map summarizing the effects of selection, drift, gene flow, mutation, and inbreeding on genetic variation and fitness. In addition to indicating whether each of the five processes increases or decreases genetic variation and average fitness, your diagram should indicate why.
2. Directional selection can lead to the fixation of favored alleles. When this occurs, genetic variation is zero and evolution stops. Explain why this rarely occurs.
3. Why does sexual selection often lead to sexual dimorphism? Why are males usually the sex that exhibits exaggerated characteristics?
4. Is it possible for one gene in a population *not* to have genotype frequencies that are in Hardy-Weinberg proportions, even though all other genes in the population are in Hardy-Weinberg proportions? Explain why or why not.
5. Explain why small populations become inbred.
6. Explain why genetic drift is much more important in small populations than large populations.

Applying Concepts to New Situations

Answers are available at www.masteringbio.com

1. In humans, albinism is caused by loss-of-function mutations in genes involved in the synthesis of melanin, the dark pigment in skin. Only people homozygous for a loss-of-function allele have the relevant phenotype. In Americans of northern European ancestry, albino individuals are present at a frequency of about 1 in 10,000 (or 0.0001). Knowing this genotype frequency, we can calculate the frequency of the loss-of-function alleles. If we let p_2 stand for this frequency, we know that $p_2^2 = 0.0001$; therefore $p_2 = \sqrt{0.0001} = 0.01$. By subtraction, the frequency of normal alleles is 0.99. If the genes responsible for albinism conform to the conditions required by the Hardy-Weinberg principle, what is the frequency of "carriers"—or people who are heterozygous for this condition? Your answer indicates the percentage of Caucasians in the United States who carry an allele for albinism.
2. Conservation managers frequently use gene flow, in the form of transporting individuals or releasing captive-bred young, to counteract the effects of drift on small, endangered populations. Explain how gene flow can also mitigate the effects of inbreeding.
3. Suppose you were studying several species of human. In one, males never lifted a finger to help females raise children. In another, males did just as much parental care as females except for actually carrying the baby during pregnancy. How does the fundamental asymmetry of sex compare in the two species? How would you expect sexual dimorphism to compare between the two species?
4. You are a conservation biologist charged with creating a recovery plan for an endangered species of turtle. The turtle's habitat has been fragmented into small, isolated but protected areas by suburbanization and highway construction. Some evidence indicates that certain turtle populations are adapted to marshes that are normal, whereas others are adapted to acidic wetlands or salty habitats. Further, some turtle populations number less than 25 breeding adults, making genetic drift and inbreeding a major concern. In creating a recovery plan, the tools at your disposal are captive breeding, the capture and transfer of adults to create gene flow, or the creation of habitat corridors between wetlands to make migration possible. Write a two-paragraph essay outlining the major features of your proposal.

wwww.masteringbio.com is also your resource for • Answers to text, table, and figure caption questions and exercises • Answers to *Check Your Understanding* boxes • Online study guides and quizzes • Additional study tools including the *E-Book for Biological Science* 3rd ed., textbook art, animations, and videos.

26 Speciation

KEY CONCEPTS

- Speciation occurs when populations of the same species become genetically isolated by lack of gene flow and then diverge from each other due to selection, genetic drift, or mutation.
- Populations can be recognized as distinct species if they are reproductively isolated from each other, if they have distinct morphological characteristics, or if they form independent branches on a phylogenetic tree.
- Populations can become genetically isolated from each other if they occupy different geographic areas, if they use different habitats or resources within the same area, or if one population is polyploid and cannot breed with the other.
- When populations that have diverged come back into contact, they may fuse, continue to diverge, stay partially differentiated, or have offspring that form a new species.

The Bullock's oriole was once considered a member of the same species as the Baltimore oriole, because the two populations interbreed in certain regions. Recent analyses, based on the phylogenetic species concept introduced in this chapter, have shown that Bullock's orioles are an independent species.

Although Darwin called his masterwork *On the Origin of Species by Means of Natural Selection*, he actually had little to say about how new species arise. Instead, his data and analyses focused on the process of natural selection and the changes that occur within populations over time. He spent much less time considering changes that occur *between* populations.

Since Darwin, however, biologists have realized that populations of the same species may diverge from each other when they are isolated in terms of gene flow. Recall from Chapter 25 that gene flow equalizes allele frequencies between populations. When gene flow ends, allele frequencies in the isolated populations are free to diverge—meaning that the populations begin to evolve independently of each other. If mutation, selection, and genetic drift cause isolated populations to diverge sufficiently, distinct types, or species, form—that is, the process of **speciation** takes place. Speciation is a splitting event that creates two or more distinct species from a single ancestral group (**Figure 26.1**). When speciation is complete, a new branch has been added to the tree of life.

In essence, then, speciation results from genetic isolation and genetic divergence. Isolation results from lack of gene flow, and divergence occurs because selection, genetic drift, and mutation proceed independently in the isolated populations. How does genetic isolation come about? And how do selection, drift, and mutation cause divergence?

This chapter is devoted to exploring these questions. Our first task is to examine how species are defined and identified. Subsequent sections focus on how speciation occurs in two contrasting situations: when populations are separated into

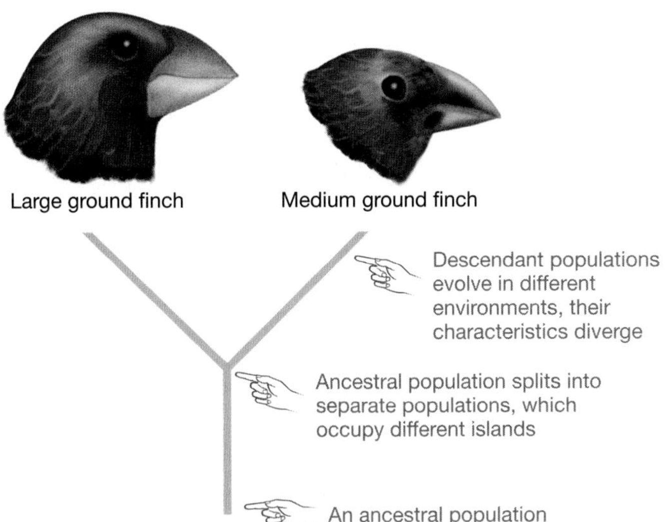

FIGURE 26.1 Speciation Creates Evolutionarily Independent Populations. The large ground finch and medium ground finch are derived from the same ancestral population. This ancestral population split into two populations isolated by lack of gene flow. Because the populations began evolving independently, they acquired the distinctive characteristics observed today.

distinct regions and when they occupy the same geographic area. The chapter concludes with a look at a classical question in speciation research: What happens when populations that have been isolated from one another come back into contact? Do they interbreed and merge back into the same species, or do they remain independent and form new species?

26.1 How Are Species Defined and Identified?

Species are distinct types of organisms and represent evolutionarily independent groups. Like the Galápagos finches in Figure 26.1, species are distinct from one another in appearance, behavior, habitat use, or other traits. These characteristics differ among species because their genetic characteristics differ. Genetic distinctions occur because mutation, selection, and drift act on each species independently of what is happening in other populations.

What makes one species "evolutionarily independent" of other species? The answer begins with *lack* of gene flow. As Chapter 25 explained, gene flow eliminates genetic differences among populations. Allele frequencies in populations, and thus the populations' characteristics, become more alike when gene flow occurs between them. If gene flow between populations is extensive and continues over time, it eventually causes even highly distinct populations to coalesce into the unit known as a species. Conversely, if gene flow between populations stops, then mutation, selection, and drift begin to act on the populations independently. If a new mutation creates an allele that changes the phenotype of individuals in one population, there is no longer any way for that allele to appear in the other population. As a result, allele frequencies and other characteristics in the populations diverge. When allele frequencies change sufficiently over time, populations become distinct species.

Formally, then, a **species** is defined as an evolutionarily independent population or group of populations. Even though this definition sounds straightforward, it can be exceedingly difficult to put into practice. How can evolutionarily independent populations be identified in the field and in the fossil record? There is no single, universal answer. Even though biologists agree on the definition of a species, they frequently have to use different sets of criteria to identify them. Three criteria for identifying species are in common use: (1) the biological species concept, (2) the morphospecies concept, and (3) the phylogenetic species concept.

The Biological Species Concept

According to the **biological species concept**, the critical criterion for identifying species is reproductive isolation. This is a logical yardstick because no gene flow occurs between populations that are reproductively isolated from each other. Specifically, if two different populations do not interbreed in nature, or if they fail to produce viable and fertile offspring when matings take place, then they are considered distinct species. Groups that naturally or potentially interbreed, and that are reproductively isolated from other groups, belong to the same species. Biologists can be confident that reproductively isolated populations are evolutionarily independent.

Reproductive isolation can result from a wide variety of events and processes. To organize the various mechanisms that stop gene flow between populations, biologists distinguish (1) **prezygotic** (literally, "before-zygote") **isolation**, which prevents individuals of different species from mating, and (2) **postzygotic** (literally, "after-zygote") **isolation**, in which the offspring of matings between members of different species do not survive or reproduce. In prezygotic isolation, reproductive isolation occurs before mating can occur. In postzygotic isolation, interspecies mating does occur, but any hybrid offspring produced have low fitness. **Table 26.1** summarizes some of the more important mechanisms of prezygotic and postzygotic isolation.

Although the biological species concept has a strong theoretical foundation, it has disadvantages. The criterion of reproductive isolation cannot be evaluated in fossils or in species that reproduce asexually. In addition, it is difficult to apply when closely related populations do not happen to overlap with each other geographically. In this case, biologists are left to guess whether interbreeding and gene flow would occur if the populations happened to come into contact.

The Morphospecies Concept

How do biologists identify species when the criterion of reproductive isolation cannot be applied? Under the **morphospecies** ("form-species") **concept**, researchers identify evolutionarily

TABLE **26.1** **Mechanisms of Reproductive Isolation**

Type	Description	Example
Prezygotic Isolation		
Temporal	Populations are isolated because they breed at different times.	Bishop pines and Monterey pines release their pollen at different times of the year.
Habitat	Populations are isolated because they breed in different habitats.	Parasites that begin to exploit new host species are isolated from their original population.
Behavioral	Populations do not interbreed because their courtship displays differ.	To attract male fireflies, female fireflies give a species-specific sequence of flashes.
Gametic barrier	Matings fail because eggs and sperm are incompatible.	In sea urchins, a protein called bindin allows sperm to penetrate eggs. Differences in the amino acid sequence of bindin cause matings to fail between closely related populations.
Mechanical	Matings fail because male and female genitalia are incompatible.	In many insects, the male copulatory organ and female reproductive canal fit like a "lock and key." Changes in either organ initiate reproductive isolation.
Postzygotic Isolation		
Hybrid viability	Hybrid offspring do not develop normally and die as embryos.	When ring-necked doves mate with rock doves, less than 6% of eggs hatch.
Hybrid sterility	Hybrid offspring mature but are sterile as adults.	Eastern meadowlarks and western meadowlarks are almost identical morphologically, but hybrid offspring are largely infertile.

independent lineages by differences in size, shape, or other morphological features. The logic behind the morphospecies concept is that distinguishing features are most likely to arise if populations are independent and isolated from gene flow.

The morphospecies concept is compelling simply because it is so widely applicable. It is a useful criterion when biologists have no data on the extent of gene flow, and it is equally applicable to sexual, asexual, or fossil species. Its disadvantage is that the features used to distinguish species are subjective. In the worst case, different researchers working on the same populations disagree on the characters that distinguish species. For example, some researchers who work on the fossil record of humans argue that the specimens currently named *Homo habilis* and *Homo rudolfensis* (**Figure 26.2**) actually belong to the same species. Disagreements like these often end in a stalemate, because no independent criteria exist for resolving the conflict.

The Phylogenetic Species Concept

The **phylogenetic species concept** is a recent addition to the tools available for identifying evolutionarily independent lineages and is based on reconstructing the evolutionary history of populations. Proponents of this approach argue that it is widely applicable and precise.

The reasoning behind the phylogenetic species concept begins with Darwin's claim that all species are related by common ancestry. Chapter 1 and Chapter 25 introduced this claim and the phylogenetic trees that are used to represent the

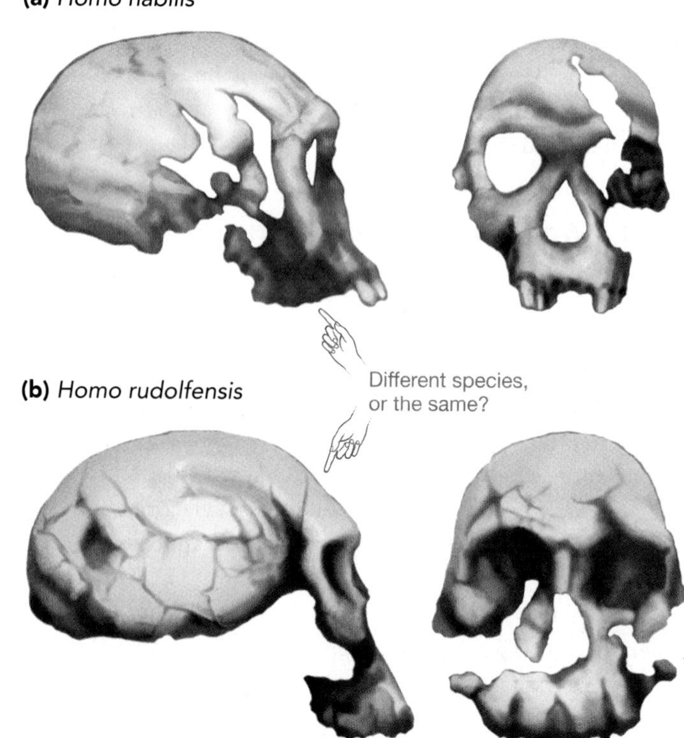

FIGURE 26.2 Morphospecies May Be Difficult to Distinguish from Each Other. Fossils from **(a)** *Homo habilis* and **(b)** *Homo rudolfensis* have been recovered from the same region in Africa, in rocks of the same age. Biologists argue over whether the two populations were distinct enough morphologically to be considered separate species or whether they should be considered the same species.

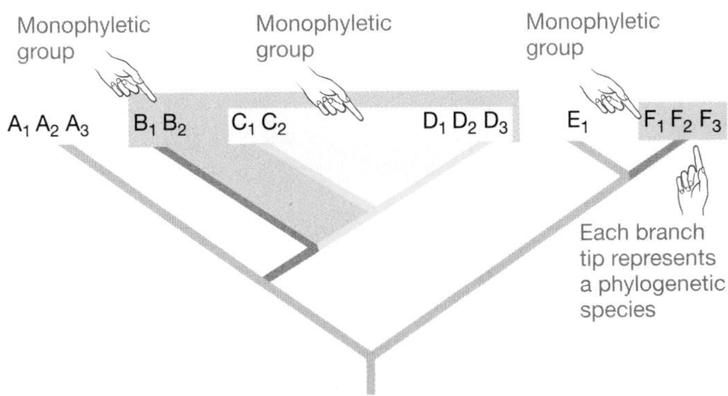

FIGURE 26.3 Monophyletic Groups. The color-coded lineages on this phylogenetic tree are monophyletic because they contain a common ancestor and all its descendants.

EXERCISE Circle the six monophyletic groups that represent phylogenetic species.

genealogical relationships among populations. (For help with interpreting phylogenetic trees, see **BioSkills 2**.) **Figure 26.3** shows such a tree. On this and all other phylogenetic trees, branches represent populations that are changing through time. Notice that each branch ends in a tip. On this tree, there are six tips.

Each of the tips on the tree in Figure 26.3 represents a phylogenetic species. To understand why, it is essential to understand the concept of a monophyletic ("one-tribe") group. A **monophyletic group**, also called a **clade** or **lineage**, consists of an ancestral population, all of its descendants, and *only* those descendants. Three monophyletic groups are color-coded on the tree in Figure 26.3, including one (containing species B, C, and D) that overlaps another monophyletic group (the one containing just species C and D). On any given evolutionary tree, there are many monophyletic groups. Under the phylogenetic species concept, a species is defined as the smallest monophyletic group in a phylogenetic tree that compares populations—as opposed to larger groups such as "pine trees" or "mammals." On a tree of populations, each tip is a phylogenetic species.

The tree in Figure 26.3, for example, is based on data from an array of populations; but many of the populations are not distinctive enough to represent separate tips. The phylogenetic species on the tree are labeled A, B, C, D, E, and F. These are the smallest monophyletic groups on the tree. The clusters at some of the tips (A_1, A_2, A_3, etc.) represent populations within those species. These may be separated geographically, but their characteristics are so similar that they do not form independent tips on the tree. They are simply part of the same monophyletic group containing other populations.

If you understand the phylogenetic species concept, you should be able to draw a tree showing that the relationships among gorillas, common chimps, and humans are like species B, C, and D in Figure 26.3. On this tree, add labels to represent populations of gorillas from Rwanda and Congo, chimps from East Africa versus West Africa, and humans from Siberia, Australia, and North America.

The phylogenetic species concept has two distinct advantages: (1) It can be applied to any population (fossil, asexual, or sexual), and (2) it is logical because populations are distinct enough to be monophyletic only if they are isolated from gene flow and have evolved independently. The approach has a distinct disadvantage, however: Carefully estimated phylogenies are available only for a tiny (though growing) subset of populations on the tree of life. Critics of this approach also point out that it would probably lead to recognition of many more species than either the morphospecies or biological species concept. Proponents counter that, far from being a disadvantage, the recognition of increased numbers of species might better reflect the extent of life's diversity.

In actual practice, researchers use all three species concepts summarized here **(Table 26.2)**. Conflicts have occurred, however, when different species concepts are applied to the real world. To appreciate this point, let's consider the case of the dusky seaside sparrow.

Species Definitions in Action: The Case of the Dusky Seaside Sparrow

Seaside sparrows live in salt marshes along the Atlantic and Gulf Coasts of the United States. Recall from Chapter 1 that

SUMMARY TABLE **26.2 Species Concepts**

Species Concept	Criterion for Recognizing Species	Advantages	Disadvantages
Biological	Reproductive isolation between populations (they don't breed and don't produce viable offspring)	Reproductive isolation = evolutionary independence	Not applicable to asexual or fossil species; difficult to assess if populations do not overlap geographically
Morphospecies	Morphologically distinct populations	Widely applicable	Subjective (researchers often disagree about how much or what kinds of morphological distinction = speciation)
Phylogenetic	Smallest monophyletic group on phylogenetic tree	Widely applicable; based on testable criteria	Relatively few well-estimated phylogenies are currently available

(a) Each subspecies of seaside sparrow has a restricted range.

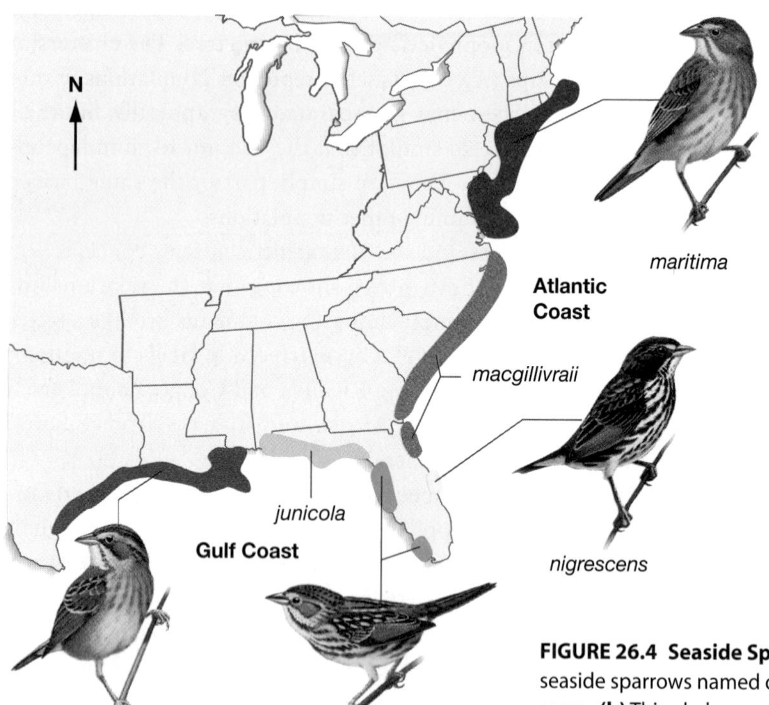

(b) The six subspecies form two monophyletic groups when DNA sequences are compared.

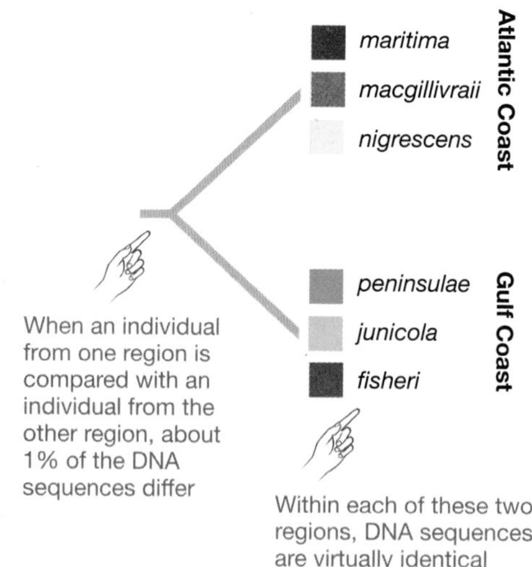

FIGURE 26.4 Seaside Sparrows Form Two Monophyletic Groups. (a) The "subspecies" of seaside sparrows named on this map are distinguished by their distinctive coloration and songs. **(b)** This phylogenetic tree was constructed by comparing DNA sequences. The tree shows that seaside sparrows represent two distinct monophyletic groups, one native to the Atlantic Coast and the other native to the Gulf Coast.

● **QUESTION** If you were a conservation biologist and could save only two subspecies of seaside sparrows from extinction, would you choose two subspecies from the Atlantic Coast, two from the Gulf Coast, or one from the Atlantic and one from the Gulf? Explain why.

scientific names consist of a genus name followed by a species name; the scientific name of this species is *Ammodramus maritimus*. Researchers had traditionally named a variety of seaside sparrow "subspecies" under the morphospecies concept. **Subspecies** are populations that live in discrete geographic areas and have distinguishing features, such as coloration or calls, but are not considered distinct enough to be called separate species.

Because salt marshes are often destroyed for agriculture or oceanfront housing, by the late 1960s biologists began to be concerned about the future of some seaside sparrow populations. A subspecies called the dusky seaside sparrow (*Ammodramus maritimus nigrescens*) was in particular trouble; by 1980 only six individuals from this population remained. All were males.

At this point government and private conservation agencies sprang into action under the auspices of the Endangered Species Act, a law whose goal is to prevent the extinction of species. The law uses the biological species concept to identify species and calls for the rescue of endangered species through active management. Because current populations of seaside sparrows are physically isolated from one another, and because young seaside sparrows tend to breed near where they hatched, researchers believed that little to no gene flow occurred among populations. Under the biological species concept and morphospecies concept, there may be as many as six species of seaside sparrow (**Figure 26.4a**). The dusky seaside sparrow subspecies became a priority for conservation efforts because it was reproductively isolated.

To launch the rescue program, the remaining male dusky seaside sparrows were taken into captivity and bred with females from a nearby subspecies: *A. maritimus peninsulae*. Officials planned to use these hybrid offspring as breeding stock for a reintroduction program. The goal was to preserve as much genetic diversity as possible by reestablishing a healthy population of dusky-like birds. The plan was thrown into turmoil, however, when a different group of biologists estimated the phylogeny of the seaside sparrows by comparing gene sequences. This tree, shown in **Figure 26.4b**, shows that seaside sparrows represent just two distinct monophyletic groups: one native to the Atlantic Coast and the other native to the Gulf Coast. Under the phylogenetic species concept, only two species of seaside sparrow exist. Far from being an important, reproductively isolated population, the phylogeny showed that the dusky sparrow is part of the same monophyletic group that includes the other Atlantic Coast sparrows. Further, officials had unwittingly crossed the dusky males with females from the Gulf Coast lineage. Because the goal of the conservation effort was to preserve existing genetic diversity, this was the wrong population to use.

The researchers who did the phylogenetic analysis maintained that the biological and morphospecies concepts had misled a well-intentioned conservation program. Under the phylogenetic species concept, they claimed that officials should have allowed the dusky sparrow to go extinct and then concentrated their efforts on simply preserving one or more populations from each coast. In this way, the two monophyletic groups of sparrows—and the most genetic diversity—would be preserved. Under the morphospecies concept, however, officials did the right thing by preserving distinct types. They argue that dusky seaside sparrows had distinctive, heritable traits like coloration and songs that are now lost forever.

When conservation funding is scarce, life-and-death decisions like these are crucial. Now our task is to consider an even more fundamental question: How do isolation and divergence produce the event called speciation?

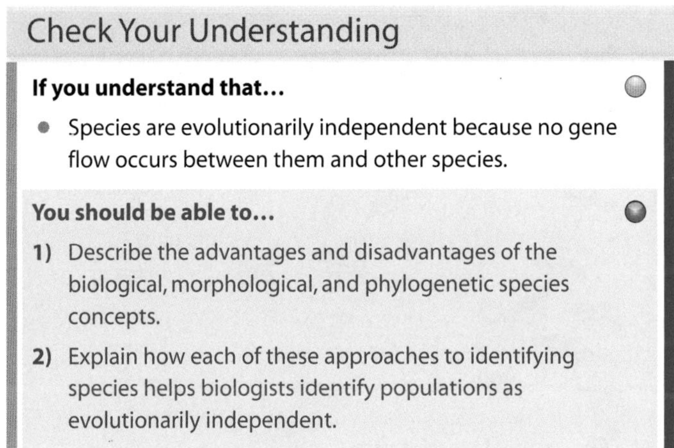

Check Your Understanding

If you understand that...

- Species are evolutionarily independent because no gene flow occurs between them and other species.

You should be able to...

1) Describe the advantages and disadvantages of the biological, morphological, and phylogenetic species concepts.
2) Explain how each of these approaches to identifying species helps biologists identify populations as evolutionarily independent.

26.2 Isolation and Divergence in Allopatry

Speciation begins when gene flow between populations is reduced or eliminated. Genetic isolation happens routinely when populations become physically separated. Physical isolation, in turn, occurs in one of two ways: dispersal or vicariance. As **Figure 26.5a** illustrates, a population can disperse to a new habitat, colonize it, and found a new population. Alternatively, a new physical barrier can split a widespread population into two or more subgroups that are physically isolated from each other (**Figure 26.5b**). A physical splitting of habitat is called **vicariance**. Speciation that begins with physical isolation via either dispersal or vicariance is known as **allopatric** ("different-homeland") **speciation**. Populations that live in different areas are said to be in **allopatry**.

The case studies that follow address two questions: How do colonization and range-splitting events occur? Answering this question takes us into the field of **biogeography**—the study of how species and populations are distributed geographically. Once populations are physically isolated, how do genetic drift and selection produce divergence?

Dispersal and Colonization Isolate Populations

Peter Grant and Rosemary Grant witnessed a colonization event while working in the Galápagos Islands off the coast of South America. Recall from Chapter 24 that they had been studying medium ground finches on the island of Daphne Major since 1971. In 1982 five members of a new species, called the large ground finch, arrived and began nesting. These colonists had apparently dispersed from a population that lived on a nearby island in the Galápagos. Because finches normally stay on the same island year-round, the colonists represented a new population, allopatric with their source population.

The large ground finches' arrival gave the researchers a chance to test a long-standing hypothesis about dispersal and colonization. Decades ago Ernst Mayr suggested that colonization events are likely to trigger speciation, for two reasons: (1) The physical separation between populations reduces or eliminates gene flow, and (2) genetic drift will cause the old and new populations to diverge rapidly. Drift occurs during the colonization event itself via the founder effect introduced in Chapter 25. And if the number of individuals in the new population remains small for several generations, genetic drift will continue to alter allele frequencies. In addition, natural selection may cause divergence if the newly colonized environment is different from the original habitat.

To evaluate whether genetic drift occurred when large ground finches colonized Daphne Major, Grant and Grant caught, weighed, and measured most of the parents and offspring produced on Daphne Major over the succeeding 12 years. When they compared these data with measurements of large ground finches in other populations, they discovered that the average bill size in the new population was much larger. As predicted, the colonists represented a nonrandom sample of the original population. Genetic drift produced a colonizing population with characteristics significantly different from those of the source population.

The novel environment experienced by the colonizers will also expose them to new forms of natural selection. In many cases, the agent of selection is changes in available food. In finches, the size and shape of the beak are closely correlated with the types of seeds or insects that individuals eat. Based on this observation, it is logical to predict that if large seeds are particularly common on Daphne Major, then large ground finches with large beaks will survive and reproduce well, and a new large-beaked species will evolve.

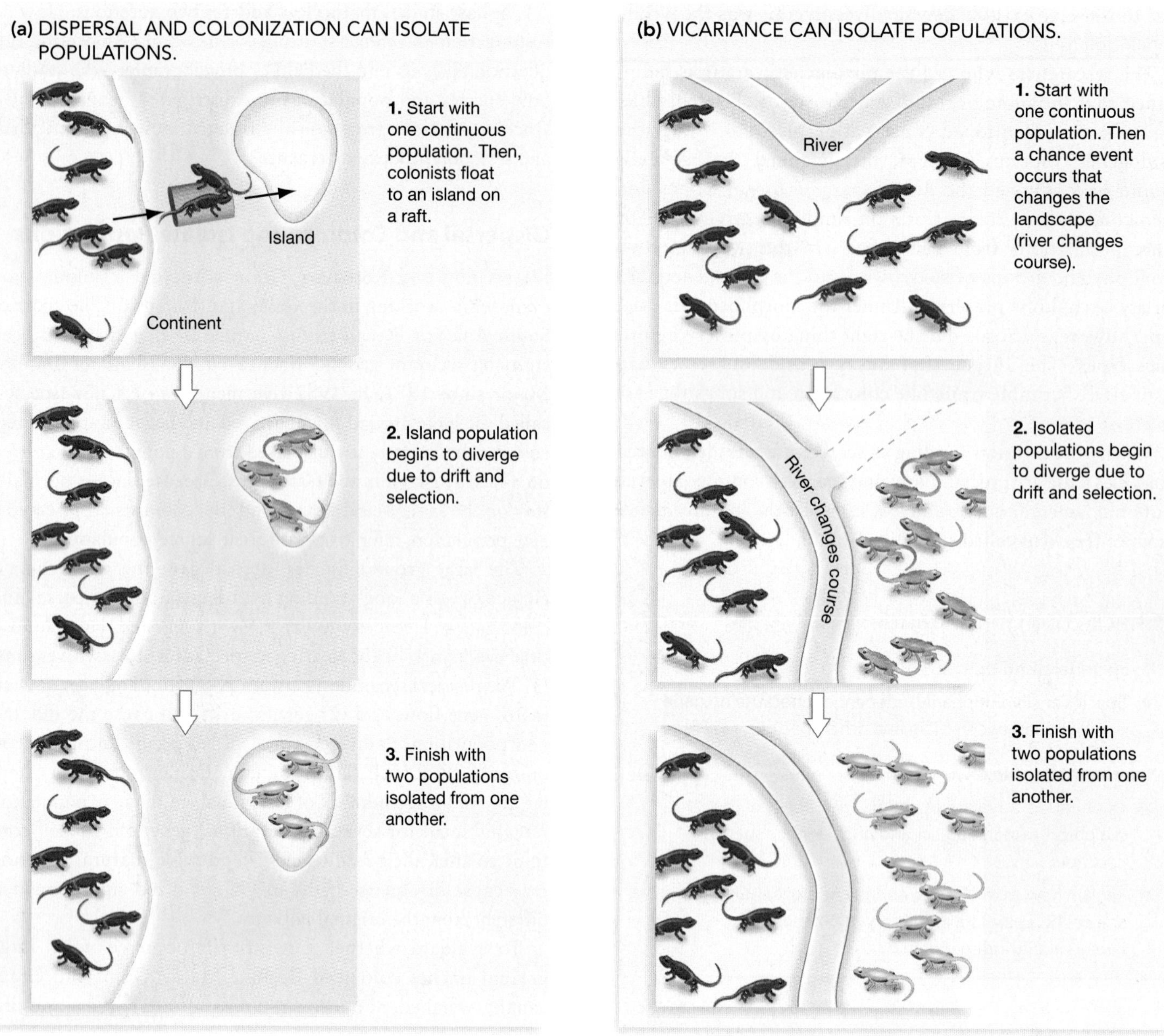

FIGURE 26.5 Allopatric Speciation Begins via Dispersal or Vicariance. (a) When dispersal occurs, colonists establish a new population in a novel location. **(b)** In vicariance, a widespread population becomes fragmented into isolated subgroups.

The general message here is that the characteristics of a colonizing population are likely to be different from the characteristics of the source population due to founder effects. Subsequent natural selection may extend the rapid divergence that begins with genetic drift. Colonization, followed by genetic drift and natural selection, is thought to be responsible for speciation in Galápagos finches and many other island groups.

Vicariance Isolates Populations

If a new physical barrier such as a mountain range or river splits the geographic range of a species, vicariance has taken place. Vicariance events during the most recent ice age are thought to be responsible for the origin of many of the species observed today. Over the past several million years, glaciers that covered large regions of the northern continents advanced and retreated repeatedly. During most of the glacial advances, the growing ice fields fragmented existing forest and grassland habitats into smaller regions that were isolated from each other by expanses of ice. If populations of the same species had occupied these isolated regions, then an inability of individuals to migrate over the ice fields would have left the populations genetically isolated. The advancing ice would have split turtle, flowering plant, insect, and fish species into geographically separated populations that then might have undergone speciation.

Another example of speciation by vicariance involves the group of large, flightless birds called the ratites. You probably are most familiar with the ostrich; kiwis, emus, rheas, and cassowaries are also ratites. Today, ratites are found in South America, Africa, Australia, and New Zealand. Unfortunately, habitat destruction by humans and hunting recently extinguished ratites called elephant birds that lived on the island of Madagascar, off the southeast coast of Africa, as well as 11 species of moas that were native to New Zealand. The elephant bird is the largest bird species ever recorded, with a maximum height of 3.5 m (11 ft), a maximum mass of 454 kg (1000 lbs), and eggs with a volume of up to 7 L (2 gallons). Moas may have been the tallest birds that ever lived, with some species possibly reaching 4 m (13 ft) when they held their necks erect.

The earliest ratites in the fossil record lived about 150 million years ago, on a landmass called Gondwana. As **Figure 26.6a** shows, Gondwana was made up of a number of physically distinct landmasses. The theory of plate tectonics holds that Earth's entire crust—the layer at its surface—is made up of moving blocks of rock called plates. Landmasses, including Gondwana, make up continental plates. The term continental drift is used to describe the ongoing motion of continental plates through time.

Although the "supercontinent" Gondwana existed for tens of millions of years, the continents in it began to drift apart about 140 million years ago (**Figure 26.6b**). Geologic data indicate that first, a landmass consisting of today's South America and Africa split off from a slightly smaller landmass composed of Antarctica, Madagascar, India, and Australia. Later, each of these large landmasses split up to form today's configuration of islands and continents. Ratites are flightless, so each vicariance event isolated distinct populations. Initially, each ratite population would have diverged via genetic drift. These differences would have been extended by natural selection as environments changed on each plate, leading to the evolution of the species observed today and their current distributions (**Figure 26.6c**). In this way, a series of vicariance events triggered a series of speciation events.

To summarize, physical isolation of populations via dispersal or vicariance produces genetic isolation—the first requirement of speciation. When genetic isolation is accompanied by genetic divergence due to mutation, selection, and genetic drift, speciation results.

(MB) **Web Animation** at www.masteringbio.com
Allopatric Speciation

(a) Gondwana was the original home of ratites.

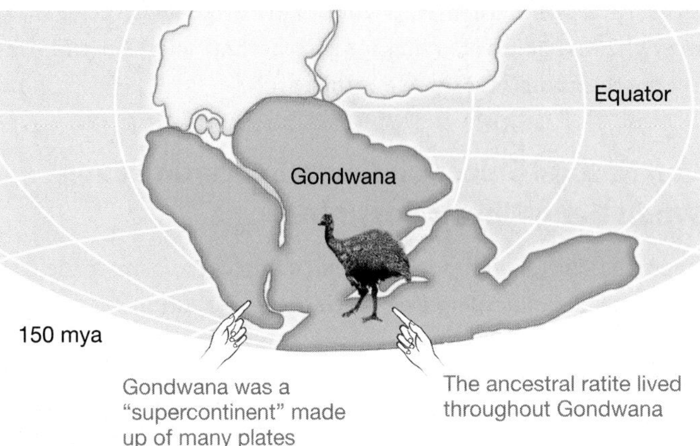

(b) Gondwana began to break up into separate continents.

(c) Ratites speciated as the continents moved apart.

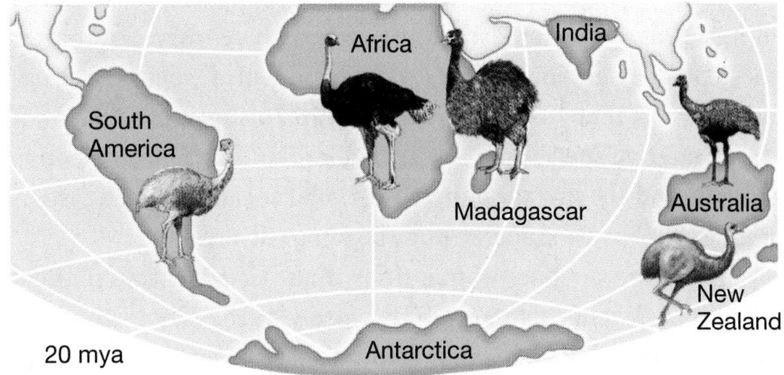

FIGURE 26.6 Continental Drift Caused Vicariance in Ratite Birds. **(a)** The ancestors of today's ratites lived on the supercontinent of Gondwana. **(b)** Continental drift led to the breakup of Gondwana starting about 140 million years ago. This was a vicariance event that isolated ratite populations. **(c)** Continued continental drift brought the continents and islands into their present positions about 20 million years ago (though they continue to move). Ratites in each area diverged in response to mutation, selection, and drift.

QUESTION Did speciation in ratites take 120 million years? Explain your answer.

26.3 Isolation and Divergence in Sympatry

When populations or species live in the same geographic area, or at least close enough to one another to make interbreeding possible, biologists say that they live in **sympatry** ("together-homeland"). Traditionally, researchers have predicted that speciation could not occur among sympatric populations, because gene flow is possible. The prediction was that gene flow would easily overwhelm any differences among populations created by genetic drift and natural selection. As Chapter 25 showed, gene flow can homogenize gene frequencies even between populations that are allopatric, such as an island population close to a population on a continent. In general, gene flow overwhelms the diversifying force of natural selection and prevents speciation. Is this always the case?

Can Natural Selection Cause Speciation Even When Gene Flow Is Possible?

Recently, several well-documented examples have upset the traditional view that **sympatric speciation**—speciation that occurs without physical isolation—is rare or nonexistent. These studies are fueling a growing awareness that under certain circumstances, natural selection that causes populations to diverge can overcome gene flow and cause speciation. The key realization is that even though sympatric populations are not physically isolated, they may be isolated by preferences for different habitats. As an example, let's consider research on speciation in soapberry bugs.

The soapberry bug is a species of insect, illustrated in **Figure 26.7a**, native to the south-central and southeastern United States. The bugs feed on plants in a family called Sapindaceae, including the soapberry tree, serjania vine, and balloon vine. As the figure shows, the bugs feed by piercing fruits with their beaks, reaching in to penetrate the coats of the seeds located deep inside the fruit, and then sucking up the contents of the seeds through their beaks. The bugs also mate on their host plants.

The soapberry bug's story began to get interesting when horticulturists brought three new species of sapindaceous plants to North America from Asia in the twentieth century. Soon after these plants were introduced to the New World, soapberry bugs began using them as food. As **Figure 26.7b** shows, the fruits of the nonnative species are much smaller than the fruits of native species. Did the arrival of new host plants lead to genetic isolation? If so, have soapberry bugs begun to diverge?

In soapberry bug populations that feed on native host plants, beak length corresponds closely to the size of the host fruit. For example, bugs that feed on species with big fruit tend to have long beaks. The correlation between fruit size and beak length is logical, because it should allow individuals to reach the seeds inside the fruit efficiently. It also prompted a biologist to ask a simple question: In populations of soapberry bugs that exploit the introduced plant species, have beak lengths evolved to match the sizes of the new fruits? If so, it would imply that populations have become genetically isolated on different host plants and that natural selection is currently causing soapberry bug populations to diverge.

To answer this question, researchers measured large samples of bugs found on both native and nonnative hosts. Some of the resulting data are shown in **Figure 26.7c**. The histogram shows that bugs collected on native plants growing in south Florida have much longer beaks than those collected on nonnative plants growing in central Florida. The data support the argument that soapberry bug populations exploiting exotic species have indeed changed—presumably in response to natural selection for efficient use of host fruits. More specifically, the data support the hypothesis that disruptive selection has occurred on beak length ever since some soapberry bug populations switched to new host plant species.

The researchers who are following the story expect that soapberry bugs adapted to nonnative species with small fruit will continue to diverge from populations that exploit native species with large fruits, and that they will eventually form distinct species. Because soapberry bugs mate on or near their host plants, switching to a new host species should reduce gene flow among populations at the same time that it sets up disruptive selection. As a result, natural selection may be able to overwhelm gene flow and cause speciation even when populations are sympatric—meaning that they are close enough geographically to make mating physically possible.

(a) Soapberry bugs use their beaks to reach seeds inside fruits.

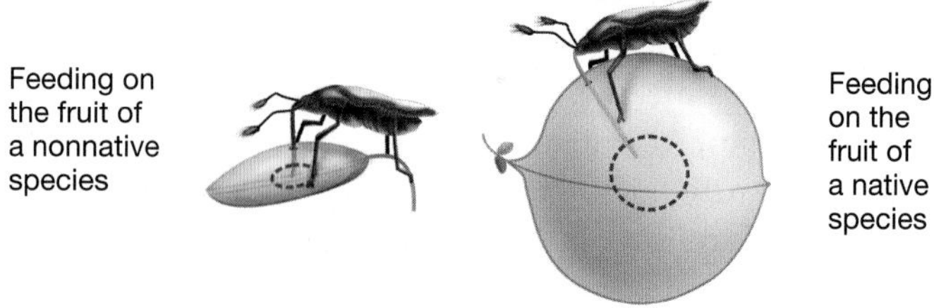

(b) Nonnative fruits are much smaller than native fruits.

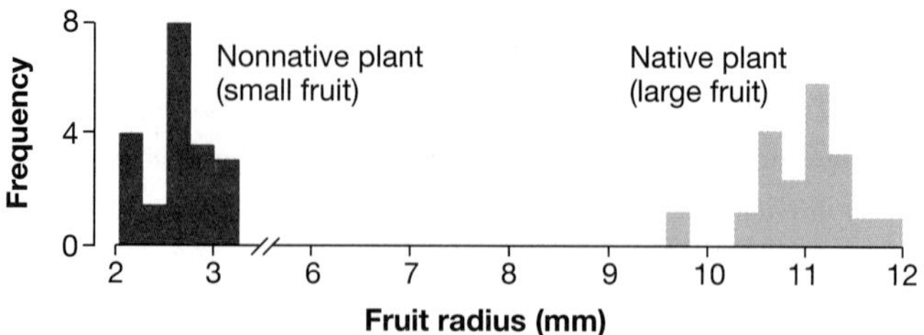

(c) Evidence for disruptive selection on beak length

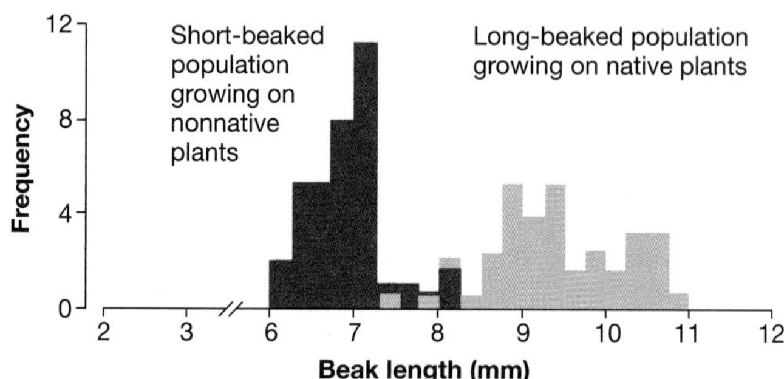

FIGURE 26.7 Disruptive Selection on Beak Length in Soapberry Bugs. (a) A soapberry bug feeding on a nonnative plant called the flat-podded golden rain tree (left) and one feeding on a native plant called the balloon vine (right). **(b)** Nonnative species of soapberry, with small fruits, were introduced to areas where soapberry bugs live in the 1940s and 1950s. **(c)** Soapberry bug populations that feed on these two host plants have very different beak lengths, which correspond to differences in the size of the host fruit.

QUESTION To promote divergence, why is it important for soapberry bugs to mate on their food plants?

Although the soapberry bug's story might seem localized and specific, the events may be common. Biologists currently estimate that over 3 million insect species exist. Most of these species are associated with specific host plants. Based on these observations, it is reasonable to hypothesize that switching host plants, as soapberry bugs have done, has been a major trigger for speciation throughout the course of insect evolution.

How Can Polyploidy Lead to Speciation?

Based on the theory and data reviewed thus far, it is clear that gene flow, genetic drift, and natural selection play important roles in speciation. Can the fourth evolutionary process—mutation—influence speciation as well? The answer might appear to be no. Chapter 25 emphasized that even though mutation is the ultimate source of genetic variation in populations, it is an inefficient mechanism of evolutionary change. If populations become isolated, it is unlikely that mutation, on its own, could cause them to diverge appreciably.

There is a particular type of mutation, though, that turns out to be extremely important in speciation—particularly in plants. The key is that the mutation reduces gene flow between mutant and normal, or wild-type, individuals. It does so because mutant individuals have more than two sets of chromosomes. This condition is known as **polyploidy**.

Polyploidy occurs when an error in meiosis or mitosis results in a doubling of the chromosome number. For example, chromosomes in a diploid ($2n$) species may fail to pull apart during anaphase of mitosis, resulting in a tetraploid cell ($4n$) instead of a diploid cell.

To understand why polyploid individuals are genetically isolated from wild-type individuals, consider what happens when that tetraploid cell undergoes meiosis to form gametes and mates with a diploid individual. By meiosis, the normal individuals produce haploid gametes, while the mutant individuals produce diploid gametes. These gametes unite to form a triploid ($3n$) zygote. Even if this offspring develops normally and reaches sexual maturity, it is rare that it will be able to form functional gametes. The sketch at the bottom of **Figure 26.8** illustrates why: When meiosis occurs in a triploid individual, homologous chromosomes cannot synapse and separate correctly. Thus, they are not distributed to daughter cells evenly, and virtually all of the gametes produced by the triploid individual end up with an uneven number of chromosomes. Because its gametes contain a dysfunctional set of chromosomes, the triploid individual is virtually sterile. As a result, the tetraploid and diploid individuals rarely produce fertile offspring when they mate. Tetraploid and diploid populations are reproductively isolated.

How do the polyploid individuals involved in speciation form? There are two general mechanisms:

1. **Autopolyploid** ("same-many-form") individuals are produced when a mutation results in a doubling of chromosome number and the chromosomes all come from the same species.

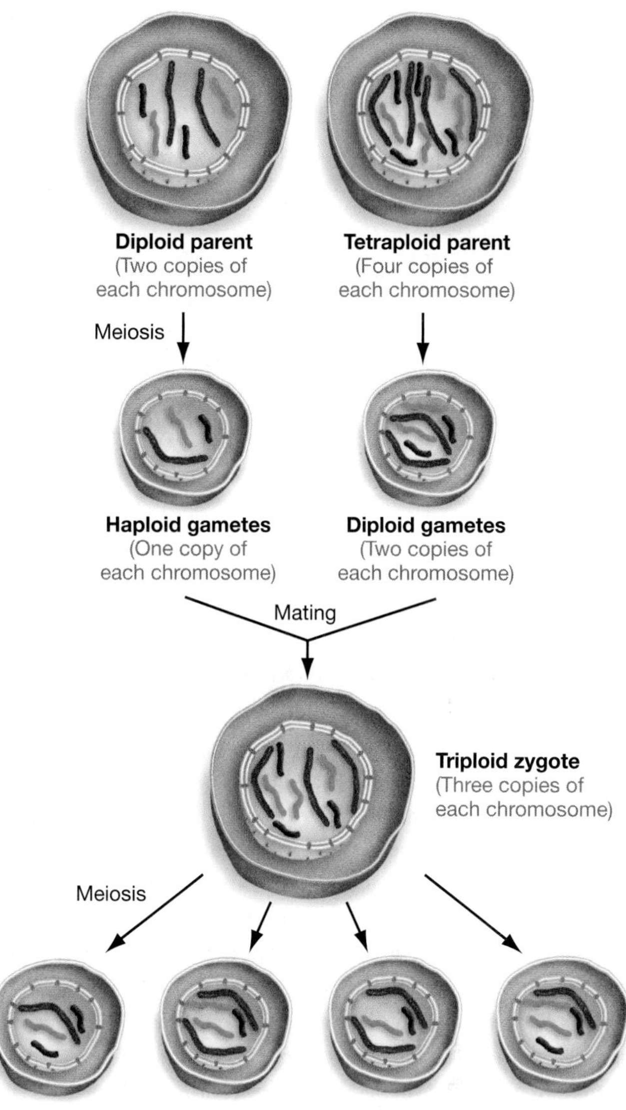

FIGURE 26.8 Polyploidy Can Lead to Reproductive Isolation. The mating diagrammed here illustrates why tetraploid individuals are reproductively isolated from diploid individuals.

EXERCISE Identify gamete combinations that *would* result in viable offspring. Would the likelihood of triploid individuals producing viable gametes increase or decrease if the number of chromosomes was 10 instead of 4?

2. **Allopolyploid** ("different-many-form") individuals are created when parents that belong to different species mate and produce an offspring where chromosome number doubles. Allopolyploid individuals have chromosome sets from different species.

Let's consider specific examples to illustrate how speciation by polyploidy occurs.

Autopolyploidy Although autopolyploidy is thought to be much less common than allopolyploidy, biologists recently

FIGURE 26.9 Maidenhair Ferns. This is the diploid form of the maidenhair fern; a recently discovered population of tetraploid individuals is indistinguishable morphologically.

documented autopolyploidy in the maidenhair fern. This plant inhabits woodlands across North America (**Figure 26.9**). During the normal life cycle of a fern, individuals alternate between a haploid (n) stage and a diploid ($2n$) stage. Biologists initially set out to do a routine survey of allelic diversity in a population of these ferns. They happened to be examining individuals in the haploid stage and found several individuals that had *two* versions of each gene instead of just one. These individuals were diploid even though they had the "haploid" growth form. The biologists followed these individuals through their life cycle and confirmed that when the ferns mated, they produced offspring that were tetraploid ($4n$). The researchers had stumbled upon polyploid mutants within a normal population.

To follow up on the observation, they located the parent of the mutant individuals. The parent turned out to have a defect in meiosis. Instead of producing normal, haploid cells as a result of meiosis, the mutant individual produced diploid cells. These diploid cells eventually led to the production of diploid gametes. Because maidenhair ferns can self-fertilize, the diploid gametes could combine to form tetraploid offspring. The tetraploid offspring could then self-fertilize or mate with their tetraploid parent. If the process continued, a polyploid population of maidenhair ferns would be established. The polyploid individuals would be genetically isolated from the original population and thus evolutionarily independent, because tetraploid individuals can breed with other tetraploids but not with diploids. If genetic drift and selection then caused the two populations to diverge, speciation would be under way.

● If you understand how autopolyploidy works, you should be able to create a scenario explaining how the process gave rise to a tetraploid grape with extra-large fruit, from a diploid population with smaller fruit. (You've probably seen both types of fruit in the supermarket.)

This autopolyploidy study documented the critical first step in speciation—the establishment of genetic isolation. In this population of maidenhair ferns, as in soapberry bugs, speciation is under way right before our eyes.

Allopolyploidy New tetraploid species may be created when two diploid species hybridize. **Figure 26.10** illustrates the sequence of events involved. If a diploid offspring that forms from a mating between two different species has chromosomes that do not pair normally during meiosis, the offspring is sterile. But if a mutation occurs that doubles the chromosome number, then homologs synapse, meiosis can proceed, and diploid gametes are produced. When diploid gametes fuse during self-fertilization, a tetraploid individual results.

Exactly this chain of events occurred repeatedly after three European species of weedy plants in the genus *Tragopogon* were introduced to western North America in the early 1900s. In 1950 a biologist described the first of two tetraploid species

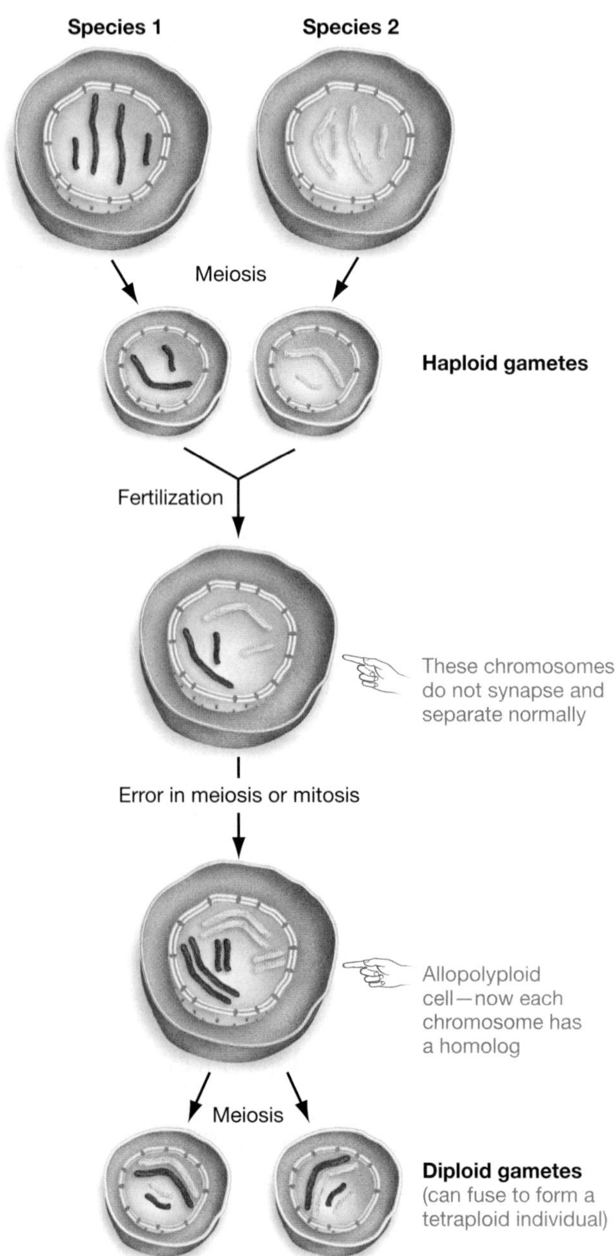

FIGURE 26.10 Allopolyploidy May Occur after Two Species Hybridize. Allopolyploid individuals contain chromosomes from two different species.

that have been discovered. Both were clearly the descendants of the introduced diploids. Further, the nature and amount of genetic variation in the tetraploid species suggests that the allopolyploid sequence occurred repeatedly—meaning that each tetraploid species of *Tragopogon* originated independently multiple times. Follow-up work has shown that at least one of the new tetraploid species is expanding its geographic range.

● If you understand how alloploidy works, you should be able to create a scenario explaining how a cross between a tetraploid population called Emmer wheat and a wild, diploid wheat gave rise to the hexaploid bread wheat grown throughout the world today.

The claim that speciation by polyploidization has been particularly important in plants is backed by the observation that many diploid species have close relatives that are polyploid. Three properties of plants have been noteworthy in making this mode of speciation possible: (1) Reproductive cells and somatic cells are not separated early in development, as they are in animals (see Chapter 23). Instead, plant somatic cells that have undergone many rounds of mitosis can undergo meiosis and produce gametes. If sister chromatids separate during anaphase of one of these mitotic divisions but do not migrate to opposite poles, the result can be a tetraploid daughter cell that later undergoes meiosis to form diploid gametes. (2) The ability of some plant species to self-fertilize makes it possible for diploid gametes to fuse and create genetically isolated tetraploid populations. (3) Hybridization between species is common, creating opportunities for speciation via formation of allopolyploids.

To summarize, speciation by polyploidization is driven by chromosome-level mutations and occurs in sympatry (**Table 26.3**). Compared to the gradual process of speciation by geographic isolation or by disruptive selection in sympatry, speciation by polyploidy is virtually instantaneous. It is fast, sympatric, and common.

Check Your Understanding

If you understand that...

- Speciation occurs when populations become isolated genetically and then diverge due to selection, genetic drift, or mutation.

You should be able to...

1) Give an example of at least three different types of events that lead to the genetic isolation of populations.
2) Explain why selection and drift cause the populations in each example to diverge.

 **Web Animation** at www.masteringbio.com

Speciation by Changes in Ploidy

26.4 What Happens When Isolated Populations Come into Contact?

Suppose two populations that have been isolated come into contact again. If divergence has taken place and if divergence has affected when, where, or how individuals in the populations mate, then it is unlikely that interbreeding will take place. In cases such as this, prezygotic isolation exists. When it does, mating between the populations is rare, gene flow is minimal, and the populations continue to diverge.

But what if prezygotic isolation does not exist, and the populations begin interbreeding? The simplest outcome is that the populations fuse over time, as gene flow erases any distinctions between them. Several other possibilities exist, however. Let's explore three of them: reinforcement, hybrid zones, and speciation by hybridization.

SUMMARY TABLE **26.3 Mechanisms of Sympatric Speciation**

	Process	Notes and/or Example
Disruptive selection	Natural selection for different habitats or resources causes divergence.	Must be accompanied by some mechanism of genetic isolation (e.g., soapberry bugs selected to feed on different-sized fruits mate on those fruits, so little to no interbreeding occurs between small- and large-beaked populations).
Polyploidization	Genetic isolation is created by formation of polyploid individuals that can breed only with each other.	Particularly common in plants, because frequent hybridization occurs between species, and many mitotic divisions occur prior to meiosis.
Autopolyploidy	Polyploids have duplicate chromosome sets from same species (e.g., a chromosome doubling to produce a tetraploid).	Maidenhair fern individual became tetraploid due to an error in meiosis.
Allopolyploidy	Polyploids have chromosome sets from different species (they originate with a hybridization event, followed by chromosome doubling).	*Tragopogon* species introduced to North America hybridized and formed offspring that became tetraploid and formed new species.

Reinforcement

If two populations have diverged extensively and are distinct genetically, it is reasonable to expect that their hybrid offspring will have lower fitness than their parents. The logic here is that if organisms have evolved distinctive developmental sequences or reproductive systems, then a hybrid offspring will not be able to develop or reproduce normally. Recall from Table 26.1 that hybrid offspring may die early in development or survive to sexual maturity but be infertile. In such cases, postzygotic isolation exists. When it occurs, there should be strong natural selection against interbreeding. The hypothesis is that hybrid offspring represent a wasted effort on the part of parents. Individuals that do not interbreed, due to a different courtship ritual or pollination system or other form of prezygotic isolation, should be favored because they produce more viable offspring.

Natural selection for traits that isolate populations in this way is called **reinforcement**. The name is descriptive because the selected traits reinforce differences that developed while the populations were isolated from one another.

Some of the best data on reinforcement come from laboratory studies of closely related fruit fly species in the genus *Drosophila*. Researchers analyzed a large series of experiments that tested whether members of closely related fly species are willing to mate with one another. The biologists found an interesting pattern. If closely related species are sympatric—meaning that they live in the same area—individuals from the two species are seldom willing to mate with one another. But if the species are allopatric—meaning that they live in different areas—then individuals are often willing to mate with one another. This is exactly the pattern that is expected if reinforcement is occurring. The pattern is logical because natural selection can act to reduce mating between species only if their ranges overlap. Thus, it is reasonable to find that sympatric species exhibit prezygotic isolation but that allopatric species do not. There is a long-standing debate, however, over just how important reinforcement is in groups other than the genus *Drosophila*.

Hybrid Zones

Hybrid offspring are not always dysfunctional. In some cases they are capable of mating and producing offspring and have features that are intermediate between those of the two parental populations. When this is the case, hybrid zones can form. A **hybrid zone** is a geographic area where interbreeding occurs and hybrid offspring are common. Depending on the fitness of hybrid offspring and the extent of breeding between parental species, hybrid zones can be narrow or wide, and long or short lived. As an example of how researchers analyze the dynamics of hybrid zones, let's consider recent work on two bird species.

Townsend's warblers and hermit warblers live in the coniferous forests of North America's Pacific Northwest. In southern Washington State, where their ranges overlap, the two species hybridize extensively. As **Figure 26.11a** shows, hybrid offspring have characteristics that are intermediate relative to the two parental species. To explore the dynamics of this hybrid zone, a team of biologists examined gene sequences in the mitochondrial DNA (mtDNA) of a large number of Townsend's, hermit, and hybrid warblers collected from forests throughout the region. The team found that each of the parental species has certain species-specific mtDNA sequences. This result allowed the researchers to infer how hybridization was occurring. To grasp the reasoning here, it is critical to realize that mtDNA is maternally inherited in most animals and plants. If a hybrid individual has Townsend's mtDNA, its mother had to be a Townsend's warbler while its father had to be a hermit warbler. In this way, identifying mtDNA types allowed the research team to infer whether Townsend's females were mating with hermit males, or vice versa, or both.

Their data presented a clear pattern: Most hybrids form when Townsend's males mate with hermit warbler females. One of the investigators followed up on this result with experiments showing that Townsend's males are extremely aggressive in establishing territories and that they readily attack hermit warbler males. Hermit males, in contrast, do not challenge Townsend's males. The hypothesis, then, is that Townsend's males invade hermit territories, drive off the hermit males, and mate with hermit females.

The team also found something completely unexpected. When they analyzed the distribution of mtDNA types along the Pacific Coast and in the northern Rocky Mountains, they found that many Townsend's warblers actually had hermit mtDNA. **Figure 26.11b** shows that in some regions—such as the larger islands off the coast of British Columbia—*all* of the warblers had hermit mtDNA, even though they looked like full-blooded Townsend's warblers. To explain this result, the team hypothesized that hermit warblers were once found as far north as Alaska and that Townsend's warblers have gradually taken over their range. Their logic is that repeated mating with Townsend's warblers over time made the hybrid offspring look more and more like Townsend's, even while maternally inherited mtDNA kept the genetic record of the original hybridization event intact.

If this hypothesis is correct, then the hybrid zone should continue moving south. If it does so, hermit warblers may eventually become extinct. In many cases, however, hybridization does not lead to extinction but rather leads to the opposite—the creation of new species.

New Species through Hybridization

A team of researchers recently examined the relationships of three sunflower species native to the American West: *Helianthus annuus*, *H. petiolaris*, and *H. anomalus*. The first two of these species are known to hybridize in regions where their ranges overlap. The third species, *H. anomalus*, resembles these hybrids. In fact, because some gene sequences in *H. anomalus* are remarkably similar to those found in *H. annuus*, while other gene sequences are almost identical to those found in *H. petiolaris*, biologists have suggested that *H. anomalus* originated in hybridization between *H. annuus* and *H. petiolaris* (**Figure 26.12**). All three species have the same number of

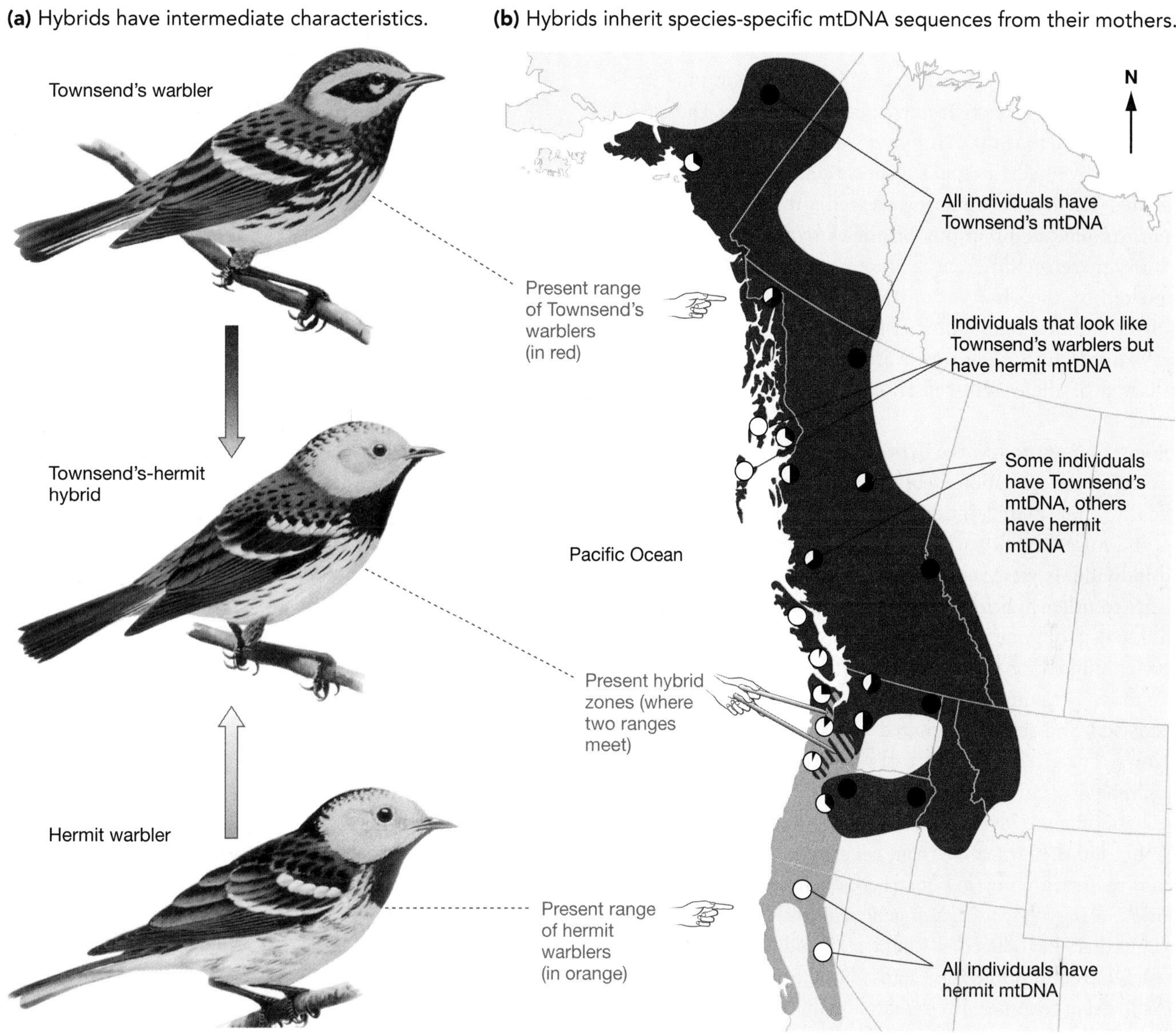

FIGURE 26.11 Analyzing a Hybrid Zone. (a) When Townsend's warblers and hermit warblers hybridize, the offspring have intermediate characteristics. **(b)** Map showing the current range of Townsend's and hermit warblers. The small pie charts show the percentage of individuals with Townsend's warbler mtDNA (in black) and hermit warbler mtDNA (in white).

Hybridization hypothesis for the origin of a new species:

Helianthus annuus x *H. petiolaris* ? → *H. anomalus*

FIGURE 26.12 Hybridization Occurs between Sunflower Species. The sunflower *Helianthus anomalus* may have originated in hybridization events between *H. annuus* and *H. petiolarus*.

chromosomes, so neither allopolyploidy nor autopolyploidy was involved.

If this interpretation is correct, then hybridization must be added to the list of ways that new species can form. The specific hypothesis here is that *H. annuus* and *H. petiolaris* were isolated and diverged as separate species, and later began interbreeding. The hybrid offspring created a third, new species that had unique combinations of alleles from each parental species and therefore different characteristics. This hypothesis is supported by the observation that *H. anomalus* grows in much drier habitats than either of the parental species and is distinct in appearance. This observation suggests that a unique combination of alleles allowed *H. anomalus* to thrive in certain habitats.

Biologists set out to test the hybridization hypothesis by trying to re-create the speciation event experimentally (**Figure 26.13**). Specifically, they mated individuals from the two parental species and raised the offspring in the greenhouse. When these hybrid individuals were mature, the researchers either mated the plants to other hybrid individuals or "backcrossed" them to individuals from one of the parental species. This breeding program continued for four more generations before the experiment ended. Ultimately, the experimental lines were backcrossed twice, and they were mated to other hybrid offspring three times. The goal of these crosses was to simulate matings that might have occurred naturally.

The experimental hybrids looked like the natural hybrid species, but did they resemble them genetically? To answer this question, the research team constructed genetic maps of each population, using a large series of genetic markers similar to the types of markers introduced in Chapter 19 and Chapter 20. Because each parental population had a large number of unique markers in their genomes, the research team hoped to identify which genes found in the experimental hybrids came from which parental species.

Their results are diagrammed in the "Results" section of Figure 26.13. The bottom bar in the illustration represents a region called S in the genome of the naturally occurring species *Helianthus anomalus*. As the legend indicates, this region contains three sections of sequences that are also found in *H. petiolaris* (indicated with the color orange) and two that are also found in *H. annuus* (indicated with the color red). The top bar shows the composition of this same region in the genome of the experimental hybrid lines. The key observation here is that the genetic composition of the synthesized hybrids matches that of the naturally occurring hybrid species.

In effect, the researchers had succeeded in re-creating a speciation event. Their results provide strong support for the hybridization hypothesis for the origin of *H. anomalus*. The data also highlight the dynamic range of possible outcomes as a result of secondary contact of two populations: fusion of the populations, reinforcement of divergence, founding of stable hybrid zones, extinction of one population, or the creation of new species (**Table 26.4**).

Experiment

Question: Can new species arise by hybridization between existing species?

Hypothesis: *Helianthus anomalus* originated by hybridization between *H. annuus* and *H. petiolaris.*

Null hypothesis: *Helianthus anomalus* did not originate by hybridization between *H. annuus* and *H. petiolaris.*

Experimental setup:

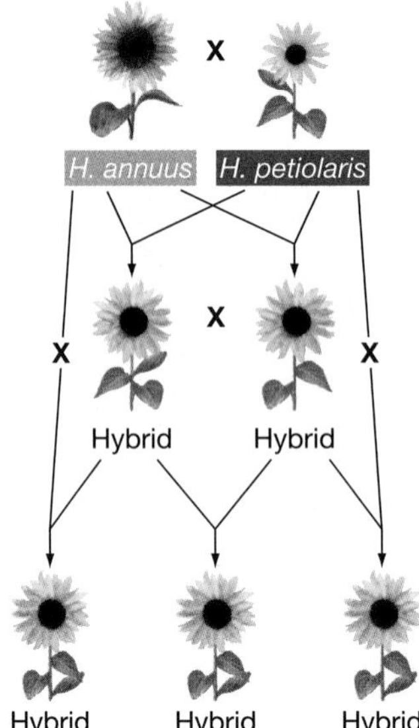

1. Mate *H. annuus* and *H. petiolaris* and raise offspring.

2. Mate F_1 hybrids or backcross F_1s to parental species; raise offspring.

3. Repeat for four more generations.

Prediction: Experimental hybrids will have the same mix of *H. annuus* and *H. petiolaris* genes as natural *H. anomalus.*

Prediction of null hypothesis: Experimental hybrids will not have the same mix of *H. annuus* and *H. petiolaris* genes as natural *H. anomalus.*

Results:

DNA comparison of a chromosomal region called S:

(Only colored portions of S region were analyzed)

Experimental hybrid (cross of *H. annuus* and *H. petiolaris*)

H. anomalus (naturally occurring species)

Gene region similar to *H. annuus*

Gene region similar to *H. petiolaris*

Genetic composition of experimental hybrids matches that of *H. anomalus*

Conclusion: New species may arise via hybridization between existing species.

FIGURE 26.13 Experimental Evidence That New Species Can Originate in Hybridization Events.

SUMMARY TABLE 26.4 **Outcomes of Secondary Contact**

Possible Outcome	Description	Example
Fusion of the populations	The two populations freely interbreed.	Occurs whenever populations of the same species come into contact.
Reinforcement	If hybrid offspring have low fitness, natural selection favors the evolution of traits that prevent interbreeding between the populations.	Appears to be common in fruit fly species that occupy the same geographic areas.
Hybrid zone formation	There is a well-defined geographic area where hybridization occurs. This area may move over time or be stable.	Many stable hybrid zones have been described; the hybrid zone between hermit and Townsend's warblers appears to have moved over time.
Extinction of one population	If one population or species is a better competitor for shared resources, then the poorer competitor may be driven to extinction.	Townsend's warblers may be driving hermit warblers to extinction.
Creation of new species	If the combination of genes in hybrid offspring allows them to occupy distinct habitats or use novel resources, they may form a new species.	Hybridization between sunflowers gave rise to a new species with unique characteristics.

Chapter Review

SUMMARY OF KEY CONCEPTS

Speciation occurs when populations of the same species become genetically isolated by lack of gene flow and then diverge from each other due to selection, genetic drift, or mutation.

When gene flow does not occur between populations, natural selection, genetic drift, and mutation act on the populations independently. As a result, the genetic and physical characteristics of the populations change over time. Eventually, the populations become so different that they are recognized as distinct species—evolutionarily independent populations. Speciation is a splitting event in which one lineage gives rise to two or more independent descendant lineages.

You should be able to design an experiment that would, given enough time, result in the production of two species from a single ancestral population.

Populations can be recognized as distinct species if they are reproductively isolated from each other, if they have distinct morphological characteristics, or if they form independent branches on a phylogenetic tree.

Researchers use several criteria to test whether populations represent distinct species. The biological species concept focuses on the degree of hybridization between species to determine whether gene flow is occurring. The morphospecies concept infers that speciation has occurred if populations have distinctive morphological traits. The phylogenetic species concept defines *species* as the smallest monophyletic groups on evolutionary trees containing populations.

You should be able to explain whether human populations would be considered separate species under the biological, morphological, and phylogenetic species concepts.

Populations can become genetically isolated from each other if they occupy different geographic areas, if they use different habitats or resources within the same area, or if one population is polyploid and cannot breed with the other.

Speciation often begins when small groups of individuals colonize a new habitat or when a large, continuous population becomes fragmented into isolated habitats. Colonization is thought to be a major mode of speciation on islands, while range splitting is thought to be a major mode of speciation in glaciated areas and in species that occupied the supercontinent Gondwana during its breakup.

Contrary to traditional expectations, biologists have found that speciation occurs frequently due to disruptive selection when populations are sympatric. Similarly, mutations that produce polyploidy can trigger rapid speciation in sympatry because they lead to reproductive isolation between diploid and tetraploid populations.

You should be able to evaluate whether your experiment on speciation (see above) represents a case of vicariance, dispersal, different habitat use, or polyploidy.

(MB) **Web Animation** at www.masteringbio.com
Allopatric Speciation; Speciation by Changes in Ploidy

When populations that have diverged come back into contact, they may fuse, continue to diverge, stay partially differentiated, or have offspring that form a new species.

If prezygotic isolation exists, populations that come back into contact will probably continue to diverge. Alternatively, interbreeding can cause diverged populations to fuse into the same species. Researchers have also documented that secondary contact can lead to reinforcement and complete reproductive isolation, the formation of hybrid zones, or the creation of a new hybrid species.

You should be able to predict the outcome of hybridization when hybrid offspring have higher, lower, or equal fitness to the parental populations.

QUESTIONS

Test Your Knowledge

1. What distinguishes a morphospecies?
   a. It has distinctive characteristics, such as size, shape, or coloration.
   b. It represents a distinct twig in a phylogeny of populations.
   c. It is reproductively isolated from other species.
   d. It is a fossil from a distinct time in Earth history.
2. When does vicariance occur?
   a. small populations coalesce into one large population
   b. a population is fragmented into isolated subpopulations
   c. individuals colonize a novel habitat
   d. individuals disperse and found a new population
3. Why is "reinforcement" an appropriate name for the concept that natural selection should favor divergence and genetic isolation if populations experience postzygotic isolation?
   a. Selection should reinforce high fitness for hybrid offspring.
   b. Selection should reinforce the fact that they are "good species" under the morphological species concept.
   c. Selection acts because hybrid offspring do not develop at all or are sterile when mature.
   d. It reinforces selection for divergence that began when the species were geographically isolated.
4. The biological species concept can be applied only to which of the following groups?
   a. bird species living today
   b. dinosaurs
   c. bacteria
   d. archaea
5. Why are genetic isolation and genetic divergence occurring in soapberry bugs, even though populations occupy the same geographic area?
   a. Different populations feed and mate on different types of fruit.
   b. One population is tetraploid; others are diploid.
   c. Nonnative host plants caused a vicariance event.
   d. Beak length has changed due to disruptive selection.
6. When the ranges of different species meet, a stable "hybrid zone" occupied by hybrid individuals may form. How is this possible?
   a. Hybrid individuals may have intermediate characteristics that are advantageous in a given region.
   b. Hybrid individuals are always allopolyploid and are thus unable to mate with either of the original species.
   c. Hybrid individuals may have reduced fitness and thus be strongly selected against.
   d. One species has a selective advantage, so as hybridization continues, the other species will go extinct.

Test Your Knowledge answers: 1. a; 2. b; 3. d; 4. a; 5. a; 6. a

Test Your Understanding

Answers are available at www.masteringbio.com

1. Make an outline listing the sections and subsections in this chapter. Fill it in with notes on the experimental and analytical approaches used in the case studies provided for each topic. Which studies represent direct observation of speciation, and which are indirect studies of historical events?
2. In the case of the seaside sparrow, how did the species identified by the biological species concept, the morphospecies concept, and the phylogenetic species concept conflict?
3. Explain why genetic drift occurs during colonization events. Explain why natural selection occurs after colonization events.
4. Explain how isolation and divergence are occurring in soapberry bugs. Of the four evolutionary processes (mutation, gene flow, drift, and selection), which two are most important in causing this event?
5. Unlike animal gametes, plant reproductive cells do not differentiate until late in life. Explain why plants are much more likely to produce diploid gametes and produce polyploid offspring than are animals.
6. Summarize the possible outcomes when populations that have been separated for some time come back into contact and hybridize. Explain why the fitness of hybrid offspring determines each outcome.

Applying Concepts to New Situations

Answers are available at www.masteringbio.com

1. A large amount of gene flow is now occurring among human populations due to intermarriage among people from different ethnic groups and regions of the world. Is this phenomenon increasing or decreasing racial differences in our species? Explain.
2. Humans have introduced thousands of species to new locations around the globe. Few, if any, of these colonization events have resulted in speciation. Use these data to evaluate the hypothesis that founder events trigger speciation.
3. Recall from Chapter 25 that 15 green iguanas recently colonized the island of Anguilla. Iguanas were not found on Anguilla previously. It is likely that the 15 iguanas originally lived on the island of Guadalupe.
   - Outline a short-term study designed to test the hypothesis that genetic drift produced allele frequency differences in the two populations (the "old" green iguana population of Guadalupe and the new population on Anguilla).
   - Outline a long-term study designed to test the hypothesis that natural selection will produce changes in the characteristics of the two populations over time.
4. All over the world, natural habitats are being fragmented into tiny islands as suburbs, ranches, and farms expand. Predict how allele frequencies will change in these fragmented populations, and then predict the long-term outcome of the global fragmentation of habitats. In each case, explain your logic.

www.masteringbio.com is also your resource for • Answers to text, table, and figure caption questions and exercises • Answers to *Check Your Understanding* boxes • Online study guides and quizzes • Additional study tools including the *E-Book for Biological Science* 3rd ed., textbook art, animations, and videos.

Phylogenies and the History of Life

27

A fossilized shell from an ammonite. Ammonites were large, ocean-dwelling animals related to today's squid. The last ammonites disappeared during a mass extinction event analyzed in section 27.5.

KEY CONCEPTS

- Phylogenies and the fossil record are the major tools that biologists use to study the history of life.
- The Cambrian explosion was the rapid morphological and ecological diversification of animals that occurred during the Cambrian period.
- Adaptive radiations are a major pattern in the history of life. They are instances of rapid diversification associated with new ecological opportunities and new morphological innovations.
- Mass extinctions have occurred repeatedly throughout the history of life. They rapidly eliminate most of the species alive in a more or less random manner.

This chapter is about time and change. More specifically, it's about vast amounts of time and profound change in organisms. Both of these topics can be difficult for humans to grasp. Our lifetimes are measured in decades, and our knowledge of history is usually measured in centuries or millennia. But this chapter analyzes events that occurred over millions and even billions of years. A million years is completely beyond our experience and almost beyond our imagination.

It takes practice to get comfortable analyzing the profound changes that occur in organisms over deep time. To help you get started, the chapter begins by introducing the two major analytical tools that biologists use to reconstruct the history of life: phylogenetic trees and the fossil record. In the remaining three sections of the chapter, we'll explore three of the great events in the history of life: the initial diversification of animals, the phenomenon called adaptive radiation, and the phenomenon known as mass extinction.

27.1 Tools for Studying History: Phylogenetic Trees

The evolutionary history of a group of organisms is called its **phylogeny**. Phylogenies are usually summarized and depicted in the form of a phylogenetic tree. A **phylogenetic tree** shows the ancestor-descendant relationships among populations or species, and clarifies who is related to whom. In a phylogenetic tree, a **branch** represents a population through time; the point where two branches diverge, called a **node** (or fork) represents the point in time when an ancestral group split into two or more descendant groups; and a **tip** (or terminal node), the endpoint of a branch, represents a group (a species or larger taxon) living today or one that ended in extinction.

Evolutionary trees have been introduced at various points in earlier chapters—often with just enough information to help you understand a new concept. Now let's focus on the trees themselves. How do you read a finished tree, and how are trees put together in the first place? We'll consider each question in turn.

BioSkills 2 introduces the parts of a phylogenetic tree and how to read one. (It would be a *very* good idea to review BioSkills 2 right now!) Here let's focus on how biologists go about building them.

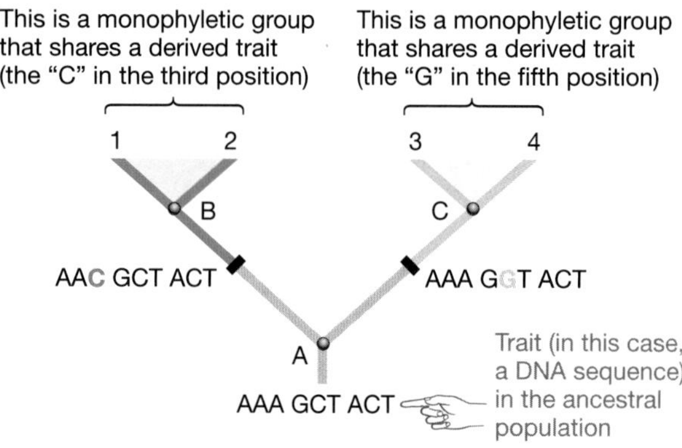

FIGURE 27.1 Synapomorphies Identify Monophyletic Groups.

How Do Researchers Estimate Phylogenies?

Phylogenetic trees are an extremely effective way of summarizing data on the evolutionary history of a group of organisms. But like any other pattern or measurement in nature, from the average height of a person in a particular human population to the speed of a passing airplane, the genealogical relationships among species cannot be known with absolute certainty. Instead, the relationships depicted in an evolutionary tree are estimated from data.

To infer the historical relationships among species, researchers analyze the species' morphological or genetic characteristics, or both. For example, to reconstruct relationships among fossil species of humans, scientists analyze aspects of tooth, jaw, and skull structure. To reconstruct relationships among contemporary human populations, investigators usually compare the sequences of bases in a particular gene.

The fundamental idea in phylogeny inference is that closely related species should share many of their characteristics, while distantly related species should share fewer characteristics. But there are two general strategies for using data to estimate trees: the phenetic approach and the cladistic approach.

The **phenetic approach** to estimating trees is based on computing a statistic that summarizes the overall similarity among populations, based on the data. For example, researchers might use gene sequences to compute an overall "genetic distance" between two populations. A genetic distance summarizes the average percentage of bases in a DNA sequence that differ between two populations. A computer program then builds a tree that clusters the most similar populations and places more-divergent populations on more-distant branches.

The **cladistic approach** to inferring trees is based on the realization that relationships among species can be reconstructed by identifying shared derived characters, or **synapomorphies** (literally, "union-forms"), in the species being studied. A synapomorphy is a trait that certain groups of organisms have that exists in no others. Synapomorphies allow biologists to recognize **monophyletic groups**—also called **clades** or **lineages**. (See Chapter 26 and BioSkills 2.) For example, fur and lactation are synapomorphies that identify mammals as a monophyletic group. Synapomorphies are characteristics that are shared because they are derived from traits that existed in their common ancestor.

Figure 27.1 illustrates the logic behind a cladistic analysis. When the ancestral population at the bottom of the figure splits into two descendant lineages at node A, each descendant group begins evolving independently and acquires unique traits. These traits are derived from their common ancestor via mutation, selection, and genetic drift. As an example, the traits might be a particular region of DNA that changes as follows:

AAA GCT ACT ancestral population

AAC GCT ACT a descendant population

AAA GGT ACT another descendant population

When the two lineages themselves split at nodes B and C, the species that result share the derived characteristics. In this way, the ancestor at node B and species 1 and 2 can be recognized as a monophyletic group. Similarly, the ancestor at node C and species 3 and 4 can be recognized as a different monophyletic group. When many such traits have been measured, a computer program can be used to identify which traits are unique to each monophyletic group and then place the groups in a tree in the correct relationship to each other.

How Can Biologists Distinguish Homology from Homoplasy? Although the logic behind phenetic and cladistic analyses is elegant, problems arise. The issue is that traits can be similar in two species not because those traits were present in a common ancestor, but because similar traits evolved independently in two distantly related groups. In the example given in Figure 27.1, it is possible that species 2 is not at all closely related to species 1. Its ancestors may have had the sequence TAT GGT AGT, which happened to change to AAC GCT ACT due to mutation, selection, and drift that took place independently of the changes that took place in the ancestors of species 1.

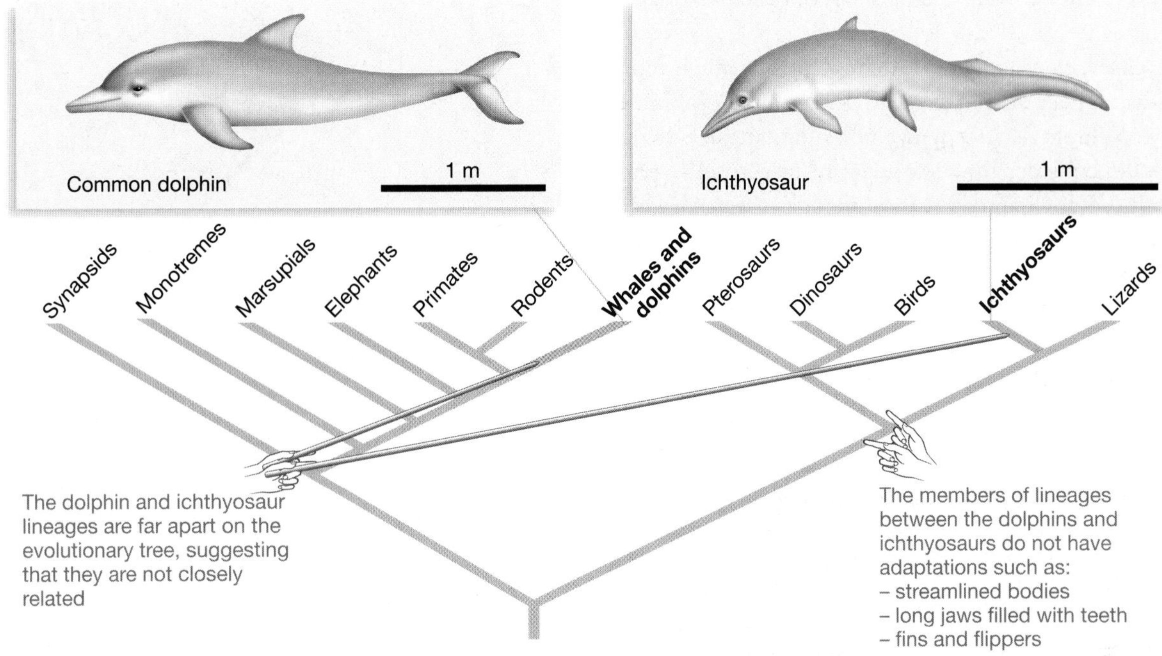

(b) Homology: Similarities are inherited from a common ancestor.

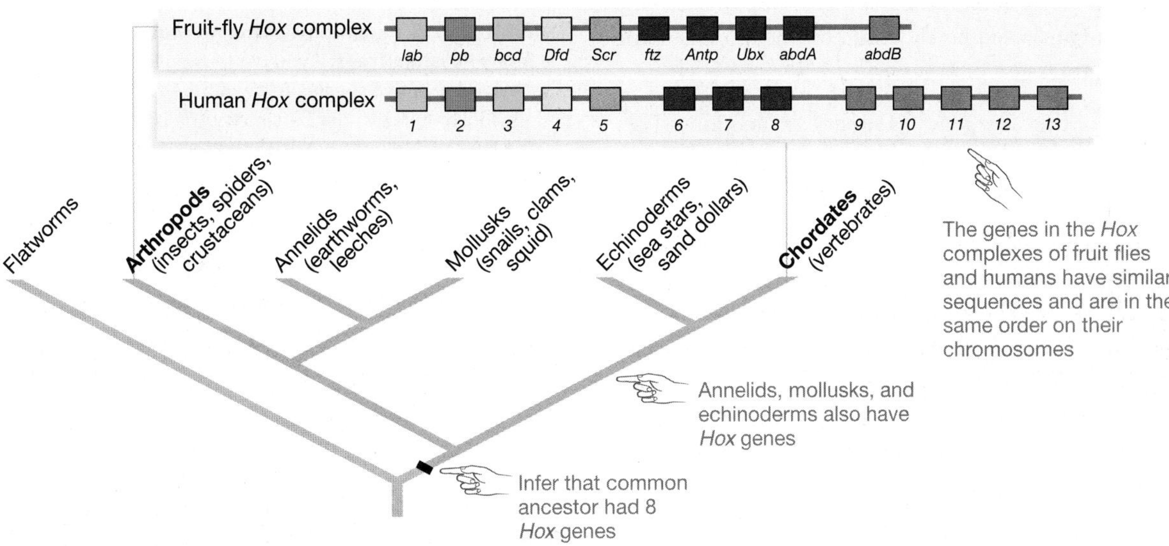

FIGURE 27.2 Homoplasy and Homology: Different Causes of Similarity in Traits. (a) Dolphins and ichthyosaurs look similar but are not closely related—dolphins are mammals; ichthyosaurs are reptiles. Because their similar traits did not exist in the common ancestor of mammals and reptiles, biologists infer that traits such as streamlined bodies, sharp teeth, and flippers evolved independently in these two groups. **(b)** All of the animal groups illustrated on this phylogeny have *Hox* complexes that are similar to those illustrated for fruit flies and humans.

Homology ("same-source") occurs when traits are similar due to shared ancestry (see Chapter 24); **homoplasy** ("same-form") occurs when traits are similar for reasons other than common ancestry. For example, the aquatic reptiles called ichthyosaurs were strikingly similar to modern dolphins (**Figure 27.2a**). Both are large marine animals with streamlined bodies and large dorsal fins. Both chase down fish and capture them between elongated jaws filled with dagger-like teeth. But no one would argue that ichthyosaurs and dolphins are similar because the traits they share existed in a common ancestor. As the phylogeny in Figure 27.2a shows, analyses of other traits show that ichthyosaurs are reptiles whereas dolphins are mammals.

Based on these data, it is logical to argue that the similarities between ichthyosaurs and dolphins result from convergent evolution. **Convergent evolution** occurs when natural selection favors similar solutions to the problems posed by a similar way of making a living. But convergent traits do not occur in the common ancestor of the similar species. Streamlined bodies and elongated jaws filled with sharp teeth help *any* species—whether it is a reptile or a mammal—chase down fish in open water. Convergent evolution is a common cause of homoplasy; it results in what biologists once called analogous traits.

In many cases, homology and homoplasy are much more difficult to distinguish than in the ichthyosaur and dolphin example. How do biologists recognize homology in such cases? As an example, consider the *Hox* genes of insects and vertebrates introduced in Chapter 21. Even though insects and vertebrates last shared a common ancestor some 600–700 million years ago, biologists argue that their *Hox* genes are derived from the same ancestral sequences. There are several lines of evidence to support this hypothesis:

- The genes are organized in a similar way. **Figure 27.2b** shows that these genes in both insects and vertebrates are found in gene complexes, with similar genes found adjacent to one another on the chromosome. Recall from Chapter 20 that genes with these characteristics are called gene families. The organization of the gene families is nearly identical among insect and vertebrate species.
- All of the *Hox* genes share a 180-base-pair sequence called the *homeobox*, introduced in Chapter 21. The polypeptide encoded by the homeobox is almost identical in insects and vertebrates and has a similar function: It binds to DNA and regulates the expression of other genes.
- The products of the *Hox* genes have similar functions: They identify the locations of cells in embryos. They are also expressed in similar patterns in time and space.

In addition, many other animals, on lineages that branched off between insects and mammals, have similar genes. This is a crucial observation: If similar traits found in distantly related lineages are indeed similar due to common ancestry, then similar traits should be found in many intervening lineages on the tree of life. (Examine the phylogeny in Figure 27.2a and ask yourself whether the analogous traits found in ichthyosaurs and dolphins fulfill this criterion.)

Now suppose that a researcher set out to infer a phylogenetic tree from a set of morphological traits or DNA sequences. Without already having a tree in hand, it can be difficult or impossible to tell which traits are homologous and qualify as synapomorphies. Just as any data set contains unavoidable errors, or "noise," the data sets used to infer phylogenies inevitably include homoplasy. To reduce the chance that homoplasy will lead to erroneous conclusions about which species are most closely related, biologists who are using cladistic approaches invoke the logical principle of **parsimony**. Under parsimony, the most likely explanation or pattern is the one that implies the least amount of change. For example, a biologist might compare all of the branching patterns that are theoretically possible and count the number of changes in DNA sequences that would be required to produce each pattern. For example, the tree in **Figure 27.3a** requires two changes in base sequence; the tree in **Figure 27.3b** requires four. Convergent evolution and other causes of homoplasy should be rare compared with similarity due to shared descent, so the tree that implies the fewest overall evolutionary changes should be the one that most accurately reflects what really happened during evolution. If the branching pattern in Figure 27.3a is the most parsimonious of all possible trees, then biologists conclude that it is the most likely representation of actual phylogeny based on the data in hand.

(a) Two changes

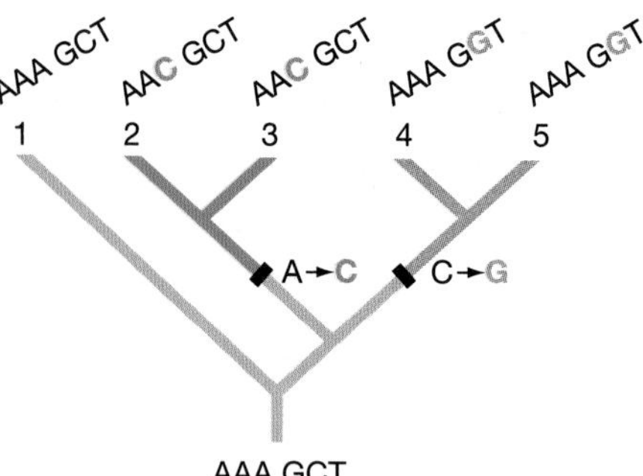

(b) Four changes

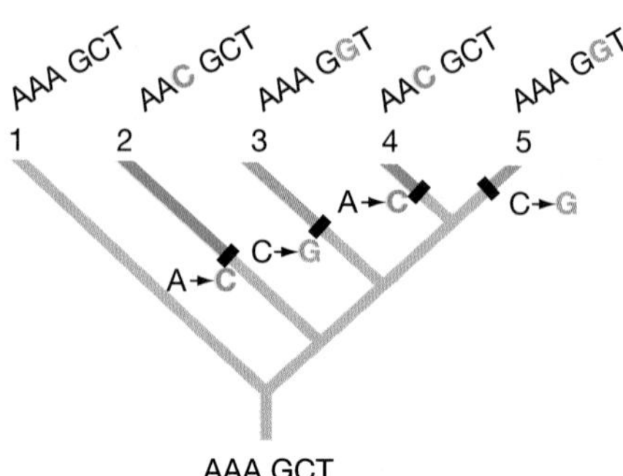

FIGURE 27.3 Parsimony Is One Method for Choosing among the Many Possible Trees. If you were using DNA sequence data to infer the phylogeny of five species, many different trees are possible. (Only two are shown here). Under parsimony, the best tree is the one that requires the fewest changes to explain the sequences observed in the five species.

Whale Evolution: A Case History As an example of how a cladistic approach works, consider the evolutionary relationships of the whales and the lineage of mammals called the Artiodactyla. Cows, deer, and hippos are artiodactyls. Members of this group have hooves and an even number of toes. They also share another feature: the unusual pulley shape of an ankle bone called the astragalus. Along with having feet with hooves and an even number of toes, the shape of the astragalus is a shared, derived character that identifies the artiodactyls as a monophyletic group. These data support the tree shown in **Figure 27.4a**. Whales do not have an astragalus and are shown as an outgroup on this tree—that is, a species or group that is closely related to the monophyletic group but not part of it.

(a) The astragalus is a synapomorphy that identifies artiodactyls as a monophyletic group.

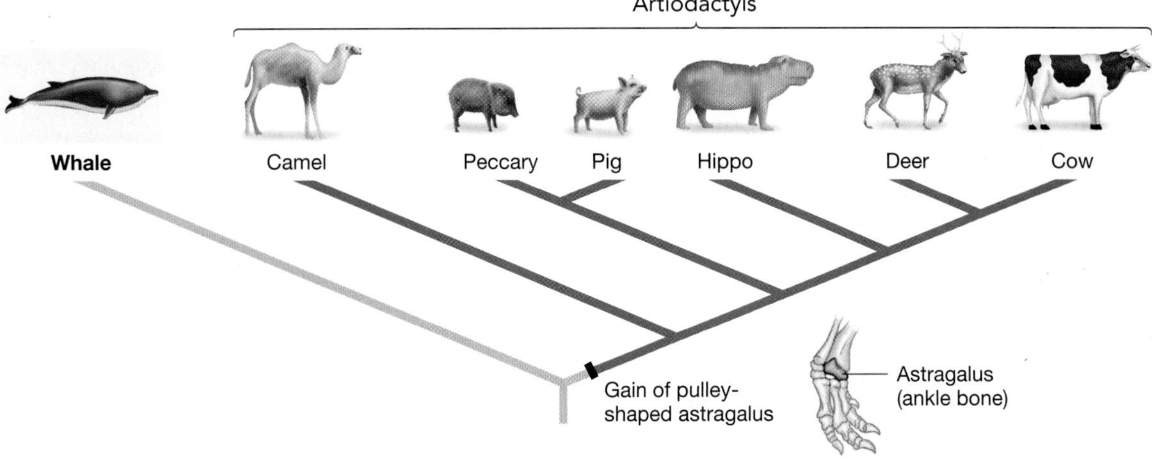

(b) If whales are related to hippos, then two changes occurred in the astragalus.

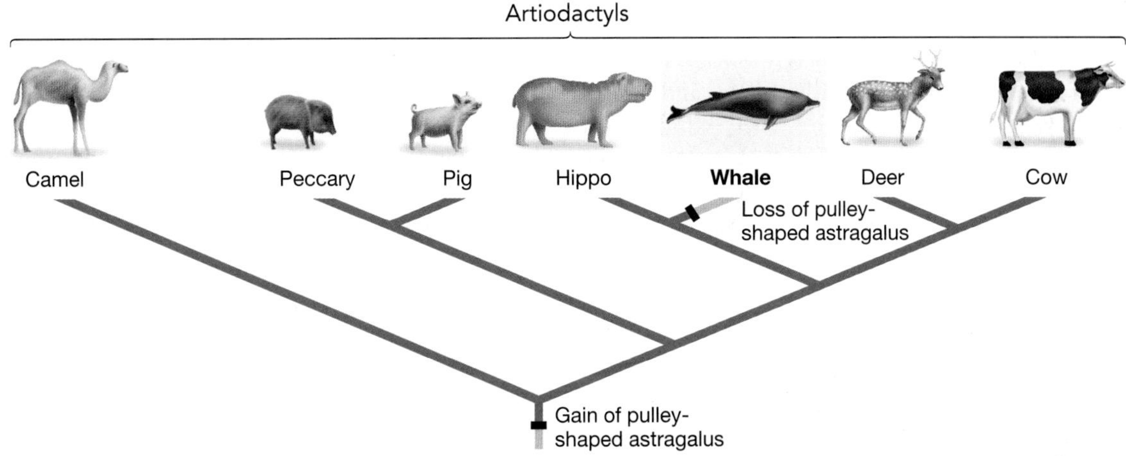

(c) Data on the presence and absence of SINE genes support the close relationship between whales and hippos.

Locus	1	2	3	4	5	6	7	8	9	10	11	12	13	14	15	16	17	18	19	20
Cow	0	0	0	0	0	0	0	1	1	1	1	1	1	1	1	1	1	1	0	0
Deer	0	0	0	0	0	0	0	1	?	1	1	1	1	1	1	?	1	1	0	0
Whale	1	1	1	1	1	1	1	0	?	1	0	1	1	0	0	0	?	1	0	0
Hippo	0	?	0	1	1	1	1	0	1	1	0	1	1	0	0	0	?	1	0	0
Pig	0	0	0	?	0	0	0	0	?	0	0	0	?	?	0	0	0	1	1	1
Peccary	?	?	?	?	?	?	?	?	?	?	?	?	?	?	?	?	?	?	1	1
Camel	0	0	0	0	0	0	0	0	0	0	0	0	0	0	0	0	0	0	0	0

1 = gene present
0 = gene absent
? = still undetermined

Whales and hippos share four *unique* SINE genes (4, 5, 6, and 7)

FIGURE 27.4 Evidence That Whales and Hippos Form a Monophyletic Group. Based on a parsimony analysis of the pulley-shaped astragalus, biologists favored **(a)** a tree excluding whales from the Artiodactyla over **(b)** the hypothesis that whales are artiodactyls and are closely related to hippos. **(c)** Data on the presence and absence of SINE genes support the whales + hippo hypothesis.

EXERCISE The presence of SINE genes 4, 5, 6, 7 identifies hippos and whales as part of a monophyletic group. The presence of SINE genes 8, 11, 14, 15, 17 identifies deer and cows as part of a monophyletic group. SINE genes 10, 12, 13 identify hippos, whales, deer, and cows as part of a monophyletic group. Map the origin of these SINE genes on the tree in part (b). Where did SINE genes 19 and 20 first insert themselves into the genomes of artiodactyls?

When researchers began comparing DNA sequences of artiodactyls and other species of mammals, however, the data showed that whales share many similarities with hippos. These results supported the tree shown in **Figure 27.4b**. The tree supported by the DNA data conflicts with the tree supported by morphological data because it implies that the pulley-shaped astragalus evolved in artiodactyls and then was lost during whale evolution. The tree in Figure 27.4b implies two changes in the astragalus (a gain *and* a loss of the astragalus), while the tree in Figure 27.4a implies just one (a gain only). In terms of the evolution of the astragalus, the "whale + hippo" tree is less parsimonious than the "whales-are-not-artiodactyls" tree.

The conflict between the data sets was resolved when researchers analyzed the distribution of the parasitic gene sequences called **SINEs (short interspersed nuclear elements)**, which occasionally insert themselves into the genomes of mammals. SINEs are transposable elements, similar to the LINEs introduced in Chapter 20. As the data in **Figure 27.4c** show, whales and hippos share several types of SINES that are not found in other groups. Specifically, whales and hippos share the SINEs numbered 4, 5, 6, and 7. Other SINE genes are present in some artiodactyls but not in others; camels have no SINE genes at all. To explain these data, biologists hypothesize that no SINEs were present in the population that is ancestral to all of the species in the study. After the branching event that led to the split between the camels and all the other artiodactyls however, different SINEs became inserted into the genomes of descendant populations. As a result, the presence of a particular SINE represents a derived character. Because whales and hippos share four of these derived characters, it is logical to conclude that these animals are closely related. If you understand this concept, you should be able to explain why SINES numbered 4–7 are synapomorphies that identify whales and hippos as a monophyletic group, and why the similarity in these SINES is unlikely to represent homoplasy.

Based on these data, most biologists accepted the phylogeny shown in Figure 27.4b as the most accurate estimate of evolutionary history. According to this phylogeny, whales are artiodactyls and share a relatively recent common ancestor with hippos. This observation inspired the hypothesis that both whales and dolphins are descended from a population of artiodactyls that spent most of their time feeding in shallow water, much as hippos do today.

Recently this hypothesis was supported in spectacular fashion. In 2001 two teams of researchers announced the independent discoveries of fossil artiodactyls that were clearly related to whales and yet had a pulley-shaped astragalus. One of these species is illustrated at the top of Figure 24.4b. The combination of phylogenetic data and data from the fossil record has clarified how a particularly interesting group of mammals evolved. What else do fossils have to say?

Check Your Understanding

If you understand that...

- Phylogenies can be estimated by finding synapomorphies that identify monophyletic groups.

You should be able to...

1) Explain whether the following traits represent homoplasy or homology: hair in humans and whales; extensive hair loss in humans and whales; limbs in humans and whales; social behavior in certain whales (e.g., dolphins) and humans.
2) Explain why the presence and absence of certain SINE genes serve as synapomorphies supporting the hypothesis that species of cows and deer form a monophyletic group.

27.2 Tools for Studying History: The Fossil Record

Phylogenetic analyses are powerful ways to infer the order in which events occurred during evolution and to understand how particular groups of species are related. But only the fossil record provides direct evidence about what organisms that lived in the past looked like, where they lived, and when they existed. A **fossil** is a piece of physical evidence from an organism that lived in the past. The **fossil record** is the total collection of fossils that have been found throughout the world. The fossil record is housed in thousands of private and public collections.

Let's review how fossils form, analyze the strengths and weaknesses of the fossil record, and then summarize major events that have taken place in life's approximately 3.5-billion-year history. (Although most evidence suggests that life originated 3.4–3.6 billion years ago, research continues.)

How Do Fossils Form? Most of the processes that form fossils begin when part or all of an organism is buried in ash, sand, mud, or some other type of sediment. Consider a series of events that begins when a tree falls into a swamp. **Figure 27.5** illustrates the leaves of a tree falling onto a patch of mud, where they are buried by soil and debris before they decay. Pollen and seeds settle into the muck at the bottom of the swamp, where decomposition is slow. The stagnant water is too acidic and too oxygen poor to support large populations of bacteria and fungi, so much of this material is buried intact before it decomposes. The trunk and branches that sit above the water line rot fairly quickly, but as pieces break off they, too, sink to the bottom and are buried.

Once burial occurs, several things can happen. If decomposition does not occur, the organic remains can be preserved intact—like the fossil pollen in **Figure 27.6a**. Alternatively, if

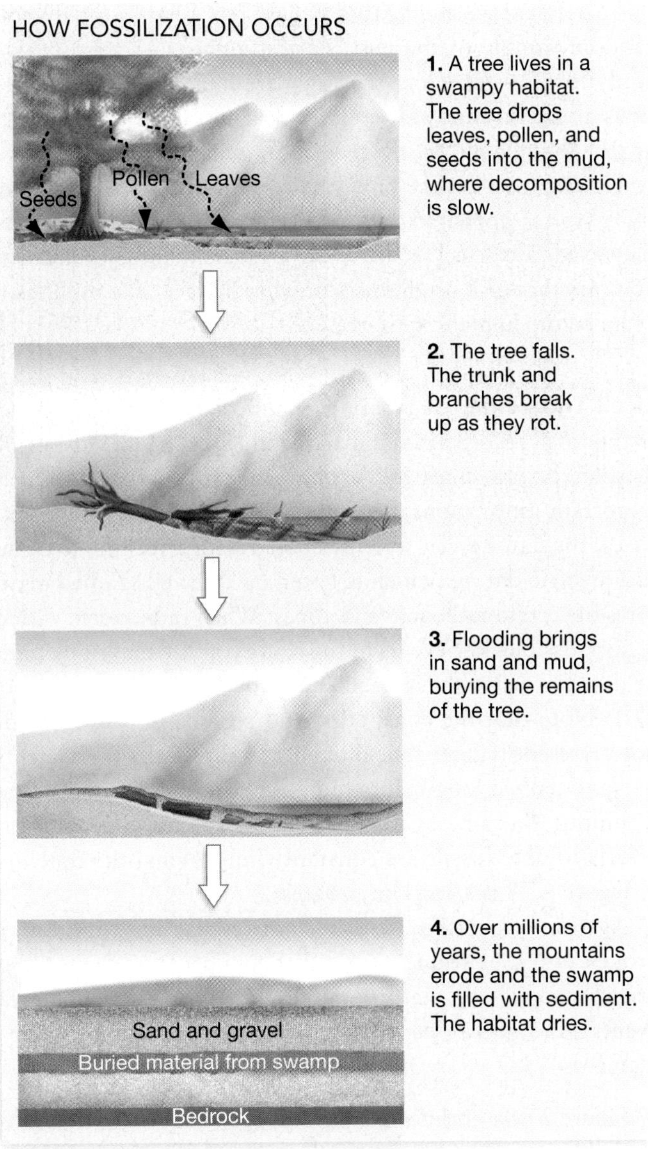

FIGURE 27.5 Fossilization Preserves Traces of Organisms That Lived in the Past. Fossilization occurs most readily when the remains of an organism are buried in sediments, where decay is slow.

sediments accumulate on top of the material and become cemented into rocks such as mudstone or shale, the sediments' weight can compress the organic material below into a thin, carbonaceous film. This happened to the leaf in **Figure 27.6b**. If the remains decompose *after* they are buried—as did the branch in **Figure 27.6c**—the hole that remains can fill with dissolved minerals and faithfully create a **cast** of the remains. If the remains rot extremely slowly, dissolved minerals can gradually infiltrate the interior of the cells and harden into stone, forming a permineralized fossil, such as petrified wood (**Figure 27.6d**).

After many centuries have passed, fossils can be exposed at the surface by many mechanisms, including erosion, a road cut, or quarrying. If researchers find a fossil, they can prepare it for study by painstakingly clearing away the surrounding rock. If the species represented is new, researchers describe its morphology in a scientific publication, name the species, estimate the fossil's age based on dates assigned to nearby rock layers, and add the specimen to a collection so that it is available for study by other researchers. It is now part of the fossil record. This is the information database that supports much of the research reviewed in this chapter.

The scenario just presented is based on conditions that are ideal for fossilization: The tree fell into an environment where decomposition was slow and burial was rapid. In most habitats the opposite situation occurs—decomposition is rapid and burial is slow. In reality, then, fossilization is an extremely rare event. To appreciate this point, consider that there are 10 specimens of the first bird to appear in the fossil record, *Archaeopteryx*. All were found at the same site in Germany where limestone is quarried for printmaking (the bird's specific name is *lithographica*). If you accept an estimate that crow-sized birds native to wetland habitats in northern Europe would have a population size of around 10,000 and a life span of 10 years, and if you accept the current estimate that the species existed for about 2 million years, then you can calculate that about 2 billion *Archaeopteryx*

(a) Intact fossil

The pollen was preserved intact because no decomposition occurred.

(b) Compression fossil

Sediments accumulated on top of the leaf and compressed it into a thin carbon-rich film.

(c) Cast fossil

The branch decomposed after it was buried. This left a hole that filled with dissolved minerals, faithfully creating a cast of the original.

(d) Permineralized fossil

The wood decayed very slowly, allowing dissolved minerals to infiltrate the cells gradually and then harden into stone.

FIGURE 27.6 Fossils Are Formed in Several Ways. Different preservation processes give rise to different types of fossils.

lived. But as far as researchers currently know, only 1 out of every 200,000,000 individuals fossilized. For this species, the odds of becoming a fossil were almost 40 times worse than your odds are of winning the grand prize in a state lottery.

Limitations of the Fossil Record Before looking at how the fossil record is used to answer questions about the history of life, it is essential to review the nature of this archive and recognize several features:

- *Habitat bias* Because burial in sediments is so crucial to fossilization, there is a strong habitat bias in the database. Organisms that live in areas where sediments are actively being deposited—including beaches, mudflats, and swamps—are much more likely to form fossils than are organisms that live in other habitats. Within these habitats, burrowing organisms such as clams are already underground—pre-buried—at death and are therefore much more likely to fossilize. Organisms that live aboveground in dry forests, grasslands, and deserts are much less likely to fossilize.

- *Taxonomic bias* Slow decay is almost always essential to fossilization, so organisms with hard parts such as bones or shells are most likely to leave fossil evidence. This requirement introduces a strong taxonomic bias into the record. Clams, snails, and other organisms with hard parts have a much higher tendency to be preserved than do worms. A similar bias exists for tissues within organisms. For instance, pollen grains are encased in a tough outer coat that resists decay, so they fossilize much more readily than do flowers. Teeth are the most common mammalian fossil, simply because they are so hard and decay resistant. Shark teeth are abundant in the fossil record; but shark bones, which are made of cartilage, are almost nonexistent.

- *Temporal bias* Recent fossils are much more common than ancient fossils. To understand why, consider that when two of Earth's tectonic plates converge, the edge of one plate usually sinks beneath the other plate. The rocks composing the edge of the descending plate are either melted or radically altered by the increased heat and pressure they encounter as they move downward into Earth's interior. These alterations obliterate any fossils in the rock. In addition, fossil-bearing rocks on land are constantly being broken apart and destroyed by wind and water erosion. The older a fossil is, the more likely it is to be demolished.

- *Abundance bias* Because fossilization is so improbable, the fossil record is weighted toward common species. Organisms that are abundant, widespread, and present on Earth for long periods of time leave evidence much more often than do species that are rare, local, or ephemeral.

To summarize, the fossil database represents a highly nonrandom sample of the past. **Paleontologists**—scientists who study fossils—recognize that they are limited to asking questions about tiny and scattered segments on the tree of life. Yet, as this chapter shows, the record is a scientific treasure trove. Analyzing fossils is the only way scientists have of examining the physical appearance of extinct forms and inferring how they lived. The fossil record is like an ancient library, filled with volumes that give us glimpses of what life was like millions of years before humans appeared.

Life's Time Line A few of the more significant data points in the fossil record are summarized in **Figure 27.8**. Most of the data on the time lines are "evolutionary firsts" that document important innovations during the history of life. In almost all cases, the dates given were estimated with a technique called radiometric dating, which is based on the well-studied decay rates of certain radioactive isotopes. When radiometric dating is not possible, key events in the history of life may sometimes be dated by using the molecular approaches introduced in **Box 27.1**. Note that the eons, eras, and periods in the figure do not represent regular time intervals, because they were identified and named long before radiometric and molecular dating techniques became available. Because the fossil record and efforts to date fossils are constantly improving, the time line in Figure 27.8 is a work in progress.

As you inspect Figure 27.8, notice that the time line is broken into four segments and that comments on the types of organisms present, nature of the climate, major geological events, and positions of the continents accompany most segments:

- Figure 27.8a outlines the interval between the formation of Earth about 4.6 billion years ago and the appearance of most animal groups about 542 million years ago (abbreviated mya). The entire interval is called the **Precambrian**; it is divided into the Hadean, Archaean, and Proterozoic eons. The important things to note about this time are that (1) life was exclusively unicellular for most of Earth's history and (2) oxygen was virtually absent from the oceans and atmosphere for almost 2 billion years after the origin of life.

- The interval between 542 mya and the present is called the Phanerozoic eon and is divided into three eras. Each of these eras is further divided into intervals called periods.

- Figure 27.8b summarizes the **Paleozoic** ("ancient life") **era**, which begins with the appearance of many animal lineages and ends with the obliteration of almost all multicellular life-forms at the end of the Permian period. The Paleozoic saw the origin and initial diversification of the animals, land plants, and fungi, as well as the appearance of land animals.

BOX 27.1 The Molecular Clock

Several researchers have proposed that the fossil record can be supplemented with information from molecules in living species. This hypothesis, called the **molecular clock**, is based on analyzing silent mutations (Chapter 16) and other changes in DNA that do not affect gene function. These changes are not affected by natural selection, but can increase to fixation by genetic drift. Research has shown that if mutation rates stay roughly constant over time, drift fixes mutations at a steady rate. For example, consider data on the protein hemoglobin. A biologist compared the amino acid sequence of this molecule in vertebrates ranging from sharks to humans. When he plotted the number of amino acid differences between species against the date that the species diverged, according to the fossil record, the graph shown in **Figure 27.7** resulted. The evolution of hemoglobin seems to proceed at a constant rate over time.

To see how researchers use molecular clocks to supplement the fossil record, let's consider some recent analyses of human evolution. Seven fossil species of *Homo* have been identified to date. The first fossils of our species, *H. sapiens*, appear about 100,000 years ago. But *sapiens* had to be around before that. The question is, when did *Homo sapiens* actually arise?

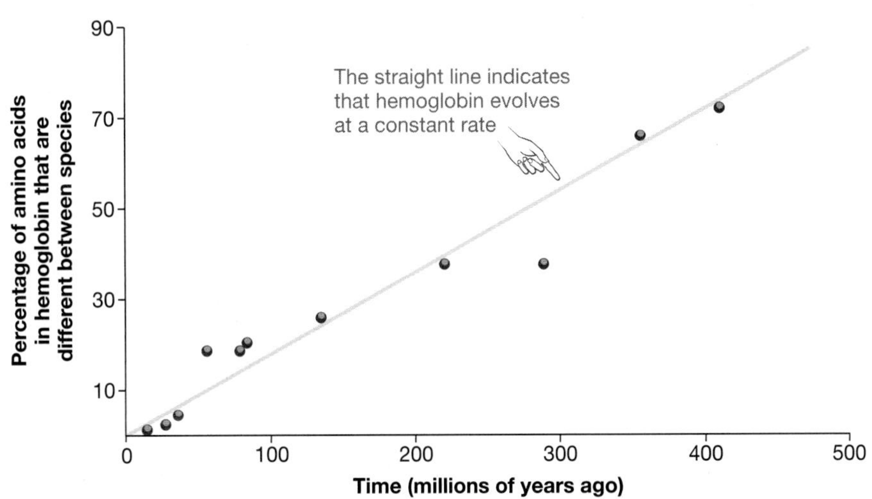

FIGURE 27.7 A Data Set That Supports a Molecular Clock. To interpret this graph, consider that the point at the upper right represents a comparison of hemoglobin proteins in sharks and mammals. According to the fossil record, sharks branched off from other vertebrates about 450 million years ago. The other data points represent comparisons between other species of vertebrates.

● **QUESTION** How fast does this clock tick? That is, what percentage of amino acids in hemoglobin change per 100 million years, on average?

A team of investigators used a molecular clock to answer this question. The group compared the complete mitochondrial genomes sequenced from African, European, and Japanese individuals. These data allowed the team to estimate the number of bases in mitochondrial DNA (mtDNA) that had changed as human populations diverged. To estimate how long it took these genetic differences to develop, the researchers also sequenced the mtDNA of our close relatives, the orangutan and African apes. According to the fossil record, orangutans and African apes diverged 13 million years ago. By dividing the number of bases that differ in the mtDNA of orangutans and African apes by 13 million years, the researchers arrived at an estimate that mtDNA changes at the rate of 7×10^{-8} per base per year.

The total amount of sequence divergence observed between two species should be equal to the rate of sequence evolution × the time since divergence occurred × 2. When the researchers used this equation with their estimate of the number of bases in mtDNA that had changed as human populations diverged, they concluded that the most recent common ancestor of all living humans lived between 125,000 and 161,000 years ago. Large species of mammals typically last in the fossil record for about 1.5 million years prior to extinction. *H. sapiens*, then, is a relatively young species.

- The **Mesozoic** ("middle life") **era** is outlined in Figure 27.8c. This interval is nicknamed the Age of Reptiles. It begins with the end-Permian extinction events and ends with the extinction of the dinosaurs and other groups at the boundary between the Cretaceous period and Paleogene period. In terrestrial environments of the Mesozoic, gymnosperms were the dominant plants and dinosaurs were the dominant vertebrates.
- Figure 27.8d highlights the **Cenozoic** ("recent life") **era**, which is divided into the Paleogene period and the Neogene period. The Cenozoic is sometimes nicknamed the Age of Mammals, because mammals diversified after the disappearance of the dinosaurs. On land, angiosperms were the dominant plants and mammals were the dominant vertebrates. Events that occur today are considered to be part of the Cenozoic era.

(a) The **Precambrian** (Hadean, Archaean, and Proterozoic Eons) included the origin of life, photosynthesis, and the oxygen atmosphere.

Formation of solar system
Moon forms
Earth formation complete
Liquid water on Earth
First oceans; heavy bombardment from space ends
Origin of life
First evidence of photosynthetic cells
First evidence of oxygenic photosynthesis
First rocks containing oxygen (in atmosphere and ocean)
First eukaryotic fossils
First photosynthetic eukaryotes
First red algae; first evidence of sexual structures
First lichen-like organism
First sponges; first bilaterally symmetric animals; ocean completey oxygenated

Hadean Eon
Archaean Eon
Proterozoic Eon

4570
3800
2500
542

Millions of years ago (mya)
All life is unicellular
Multicellular organisms begin to diversify slowly

Position of the continents unknown

Most of Earth is covered in ocean and ice.

(c) Phanerozoic Eon: The **Mesozoic Era** is sometimes called the Age of Reptiles.

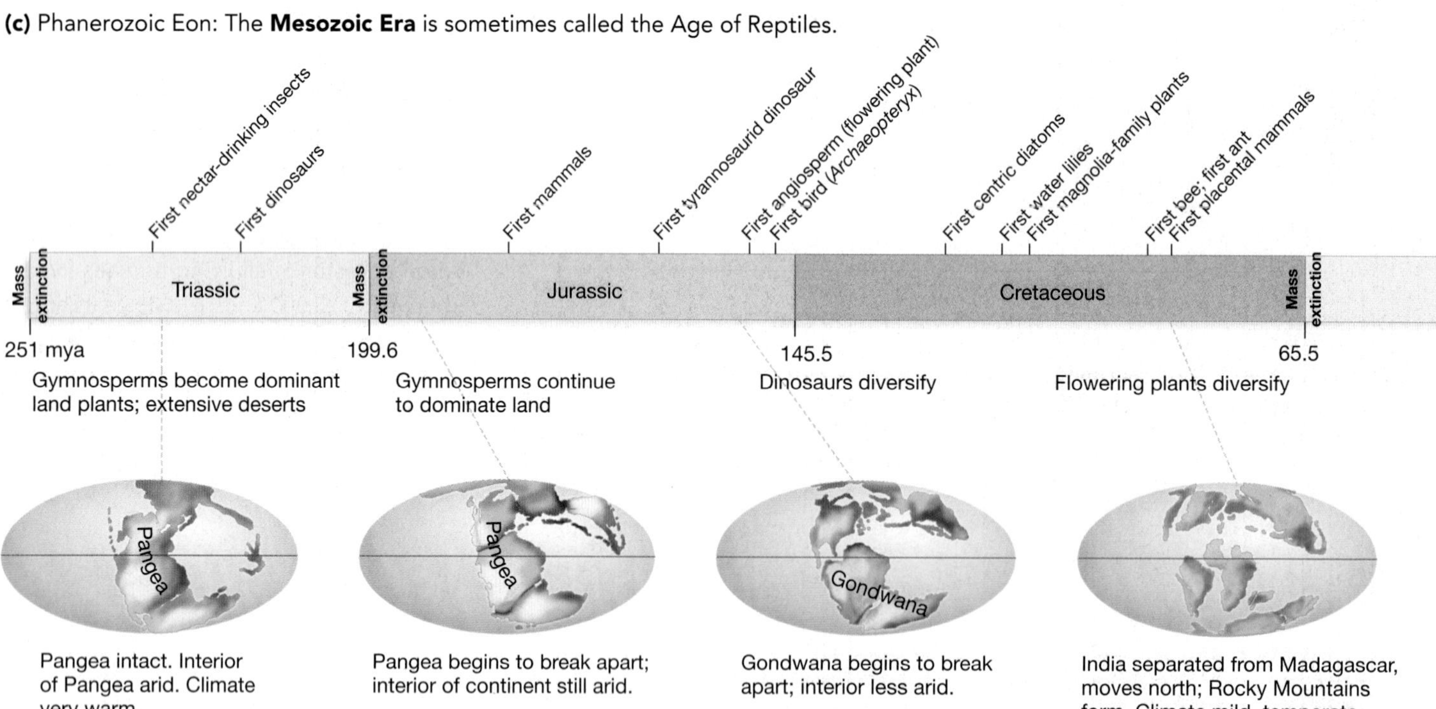

FIGURE 27.8 Life's Time Line. Significant events in Earth's history and the history of life are plotted for the **(a)** Precambrian, **(b)** Paleozoic era, **(c)** Mesozoic era, and **(d)** Cenozoic era (*mya* stands for "million years ago"). The Hadean, Archaean, and Proterozoic are vast stretches of time called eons. The Paleozoic, Mesozoic, and Cenozoic eras make up the Phanerozoic eon.

(b) Phanerozoic Eon: The **Paleozoic Era** included the origin and early diversification of animals, land plants, and fungi.

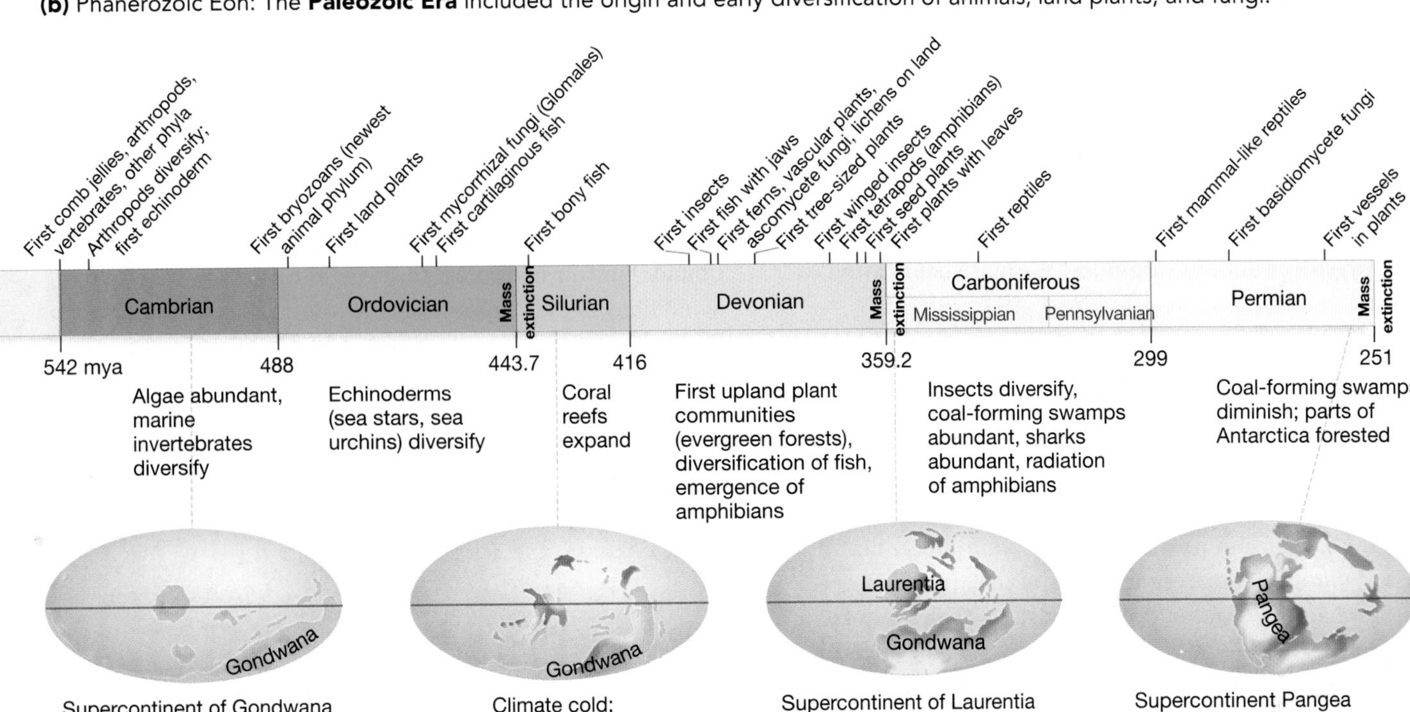

(d) Phanerozoic Eon: The **Cenozoic Era** is nicknamed the Age of Mammals.

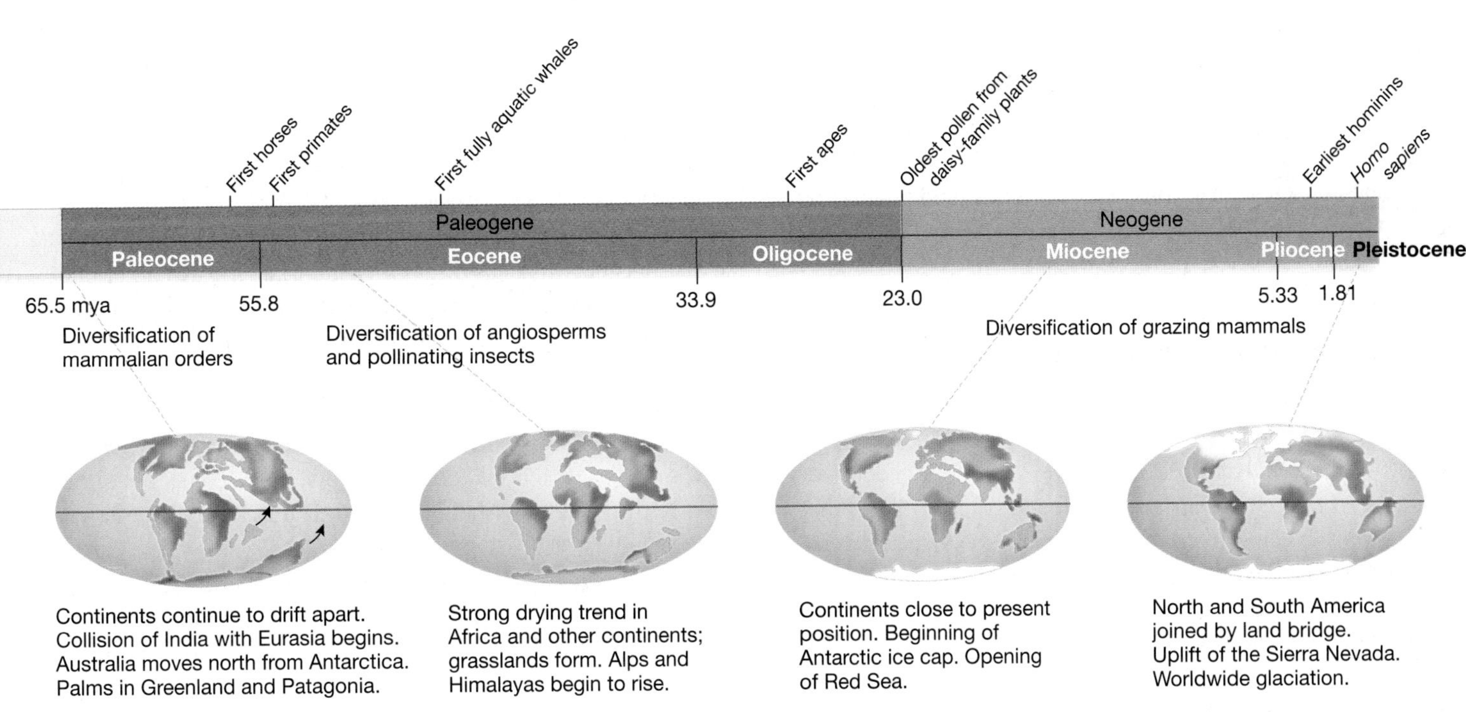

The data in Figure 27.8 show that the nature of life has changed radically since the first cell appeared on Earth. Although the changes that have taken place over the past 3.4 billion years are well documented in the fossil record, the sweep of time involved is difficult for the human mind to comprehend. A semester can seem long to a college student; but in relation to Earth's history, 100,000 years or even a million years is the blink of an eye.

27.3 The Cambrian Explosion

Phylogenetic trees and the fossil record give researchers powerful tools for understanding what life was like during the millions of years before biologists existed. They tell thousands of stories about the sweeping changes that have occurred throughout Earth's history. But here there is space to consider just a few of the most dramatic events. Let's begin with research on the origin and early diversification of animals.

The first animals appear in the fossil record about 570 million years ago. A burst of diversification occurred soon after that, at the start of the Cambrian period. This diversification happened so rapidly that it earned a nickname: the **Cambrian explosion**.

It is impossible to appreciate the drama of this event without recalling that almost all life-forms were unicellular for almost 3 *billion* years after the origin of life. The exceptions were several lineages of multicellular algae, which show up in the fossil record about 1 billion years ago. But the ancestors of animals were still unicellular, marine creatures with cilia or flagella—probably not too different from some of the aquatic eukaryotes living today. Then, about 565 million years ago, the first animals—sponges, jellyfish, and perhaps simple worms—appear in the fossil record. Just 50 million years later, virtually every major group of animals had appeared. In a relatively short time, creatures with shells, exoskeletons, internal skeletons, legs, heads, tails, eyes, antennae, jaw-like mandibles, segmented bodies, muscles, and brains had evolved. It was arguably the most spectacular period of evolutionary change in the history of life.

By combining evidence from the fossil record and phylogeny reconstruction, researchers have begun to clarify the timing and sequence of key events.

Cambrian Fossils: An Overview

The Cambrian explosion is documented by three major fossil assemblages that record the state of animal life at 570 mya, at 565 to 542 mya, and at 525 to 515 mya. The species collected from each of these intervals are referred to respectively as the Doushantuo fossils (from the Doushantuo formation in China), Ediacaran fossils (from Ediacara Hills, Australia), and Burgess Shale fossils (from British Columbia, Canada), as **Figure 27.9a** shows. Fossils from the Ediacaran interval and Burgess Shale interval have now been found at localities throughout the world.

Fortunately for biologists, the glimpses of life provided by the three groups of organisms are extraordinarily clear. These three assemblages all break one of the cardinal rules of fossil preservation. Soft-bodied animals, which usually do not fossilize efficiently, are well represented in all three groups. By sheer luck, during each of these time intervals there happened to be a few habitats in which burial occurred so rapidly and decomposition occurred so slowly that organisms without shells were able to fossilize.

To appreciate why this extraordinarily efficient preservation is so important, consider the Burgess Shale. In several localities around the world dated to 525–515 million years ago, fossilization processes were typical—only shelled organisms are found. But soft-bodied organisms also fossilized in the atypical conditions of the Burgess Shale. At this locality, over *five times* as many species are represented. The database is 500 percent better than usual.

The presence of these exceptionally rich deposits before, during, and after the Cambrian explosion makes the fossil record for this event extraordinarily complete. Let's take a brief look at the **faunas**—the animal types—found in the Doushantuo, Ediacaran, and Burgess Shale fossils.

The Doushantuo Microfossils

Two papers, published within days of each other in 1998, introduced the world to the fossils of the Doushantuo rock formation. Researchers were able to identify several dozen tiny sponges, ranging from 150 μm to 750 μm across, in samples dated at approximately 580 million years ago. In deposits dated to about 570 million years ago, a different team of biologists found clusters of cells that they interpreted as animal embryos. This conclusion was based on a simple observation: Their samples contained one-celled, two-celled, four-celled, and eight-celled fossils, along with individuals containing larger cell numbers whose overall size was the same (**Figure 27.9b**). Recall from Chapter 22 that this is exactly the pattern that occurs during cleavage in today's animals. In other words, cell number increases but total volume remains constant. What type of animal did such embryos develop into? The answer is still unknown. In addition, a different team of researchers has presented evidence that a large fraction of the Doushantuo "embryos" are actually fossils from large, sulfur-oxidizing bacteria that today live on the seafloor. As this book goes to press, the debate over the nature of the fossils in Figure 27.9b is unresolved. Research continues.

The sponges and possible embryos are scattered among abundant cyanobacterial cells as well as among multicellular algae that are the ancestors of today's seaweeds. These bacteria and algae were undoubtedly photosynthetic. The composite picture, then, is of a shallow-water marine habitat dominated by photosynthetic organisms. Scattered among other organisms in the Doushantuo formation were sponges and possibly other tiny creatures that probably made their living by filtering organic debris from the water. These were the first animals on Earth.

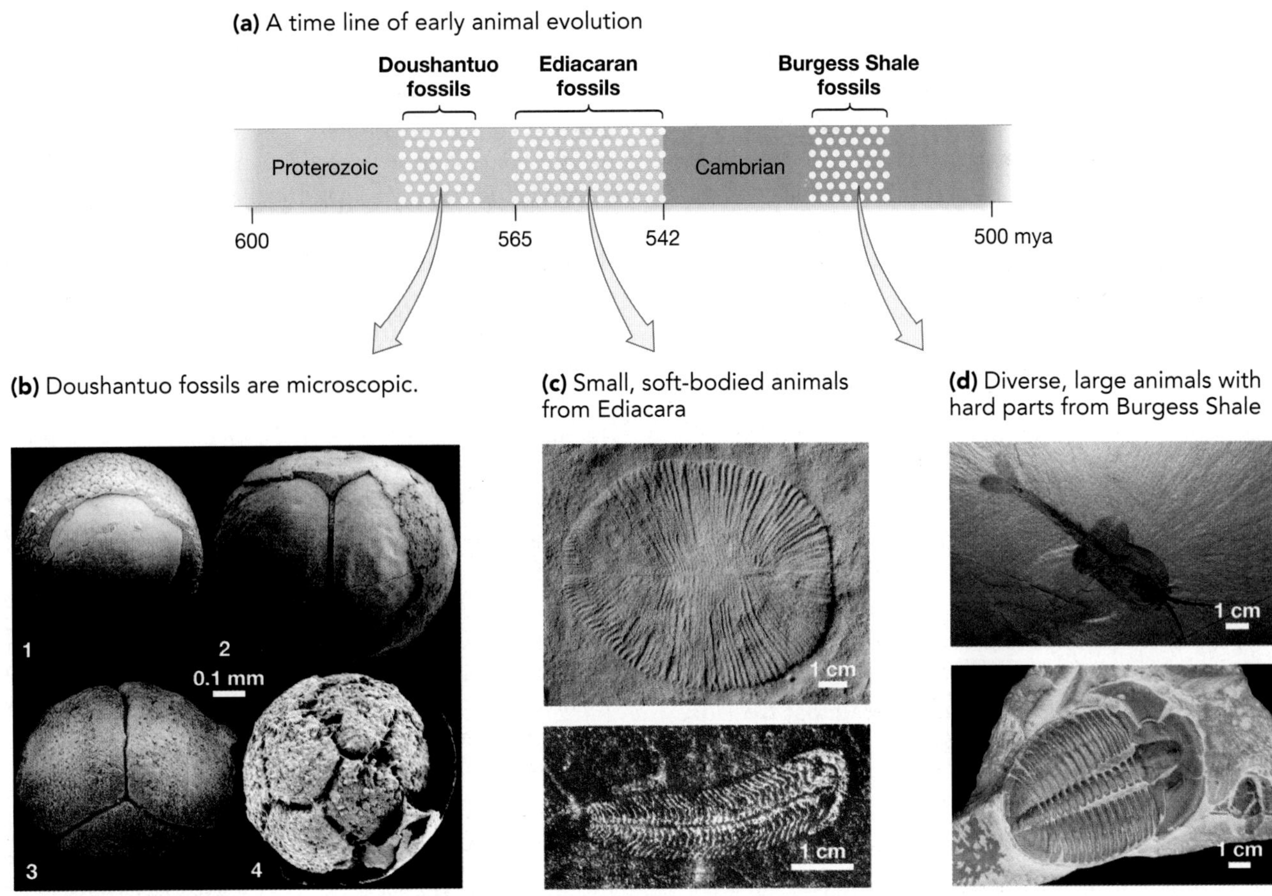

FIGURE 27.9 Fossils Document the Cambrian Explosion. **(a)** The origin of animals and their diversification during the Cambrian explosion is documented by three major fossil assemblages. **(b)** These fossilized cells from the Doushantuo deposits are arranged according to the hypothesis that they represent (1) a fertilized egg and then embryos at (2) the 2-cell, (3) the 4-cell, and (4) the 16-cell stages. **(c)** These 560-million-year-old fossils of small, soft-bodied animals were found in the Ediacara Hills of Australia. **(d)** Similar to this shrimplike arthropod (top) and trilobite (bottom), many of the animals fossilized in the Burgess Shale had heads, tails, shells (exoskeletons), and appendages.

The Ediacaran Faunas

Paleontologists discovered animal fossils in the Ediacara Hills of southern Australia in the 1940s. The specimens included the compressed bodies of sponges, jellyfish, and comb jellies as well as many burrows, tracks, and other traces from unidentified animal species (**Figure 27.9c**). In the decades since the initial discovery, similar faunas that are dated between 565 and 544 million years ago have been found at sites around the world. None of the organisms that fossilized during this period have shells or limbs, however, and none have heads or mouths or feeding appendages. As the scale bars in Figure 27.9c show, these organisms were also little in comparison to today's animals. These observations suggest that Ediacaran animals were small individuals that burrowed in sediments, sat immobile on the sea floor, or floated in the water. There is no evidence that they had structures associated with actively hunting and capturing food. Instead, it is likely that Ediacaran animals simply filtered or absorbed organic material from their surroundings.

The Burgess Shale Faunas

The discovery of fossils in the Burgess Shale formation of British Columbia, Canada, early in the twentieth century ranks among the most sensational additions ever made to the fossil record. Combined with the later unearthing of an extraordinary fossil assemblage in the Chengjiang deposits of China, the Burgess Shale gives researchers a compelling picture of life in the oceans 525–515 million years ago. Sponges, jellyfish, and comb jellies are abundant in these rocks; but entirely new lineages are present as well. Principal among these are the arthropods and mollusks. Today, the arthropods include the spiders, insects, and crustaceans (crabs, shrimp, and lobsters); mollusks include the clams, mussels, squid, and octopi. But echinoderms (sea stars and sea urchins), several types of worm and wormlike creatures, and even a chordate—the group that includes today's vertebrates—are found in these fossil faunas. In short, virtually every major animal lineage is documented there.

This tremendous increase in the size and morphological complexity of animals, illustrated in **Figure 27.9d**, was accompanied by diversification in how they made a living. The Cambrian seas were filled with animals that had eyes, mouths, limbs, and shells. They swam, burrowed, walked, ran, slithered, clung, or floated; there were predators, scavengers, filter feeders, and grazers. The diversification created and filled many of the ecological niches found in marine habitats today.

The Cambrian explosion still echoes. Animals that fill today's teeming tide pools, beaches, and mudflats trace their ancestry to species preserved in the Burgess Shale.

Did Gene Duplications Trigger the Cambrian Explosion?

The Doushantuo, Ediacaran, and Burgess Shale faunas document what happened during Cambrian explosion. Now the question is: *How* did all this morphological change come about? In particular, what genetic changes were responsible for the increased size and complexity of animals?

Chapter 21 introduced the homeotic genes that play a key role in laying out the three-dimensional pattern of multicellular organisms as they develop. You might recall that the *Hox* genes found in animals help specify the location and shape of limbs, antenna, and many other key structures. As soon as developmental biologists discovered how important these genes are in specifying morphological traits, they began asking about their role in evolution. Was the increase in morphological complexity during the Cambrian possible because of an increase in the number and complexity of homeotic genes?

To answer this question, biologists are working to determine the number and identity of homeotic genes found in different animal lineages. The idea is to look for correlations between the evolutionary history of animal lineages, their genetic makeup, and their morphology. Many researchers predicted there would be a strong association between the order in which animal lineages appeared during evolutionary history, the number of *Hox* genes present in each lineage, and each lineage's morphological complexity and body size. Some biologists even suggested that each major animal lineage would have unique homeotic genes associated with its unique body plans and appendages.

The logic behind this "new genes, new bodies" hypothesis was that gene duplication events could have occurred before and during the Cambrian explosion and produced new copies of existing homeotic genes. (Look back at Chapter 20 to review how gene duplication events occur.) These new genes would make possible the new body plans and appendages recorded in the Burgess Shale fauna. Again, the idea was that the number of homeotic genes present would correlate directly with morphological complexity.

To see if the predictions made by this hypothesis hold up, study the data in **Figure 27.10**. The phylogenetic tree represents the best current estimate of the evolutionary relationships among major animal lineages, and the colored boxes represent the *Hox* genes documented in each lineage. Note that *Hox* genes are found in clusters, with genes lined up on the chromosome one after the other. Note, too, that each gene in the *Hox* cluster has a distinct function as an embryo develops. Specifically, each *Hox* gene is involved in a different aspect of pattern formation—the events that organize cells in the space inside an embryo. To reflect these observations, the colors of the boxes indicate homology with other genes. If genes have the same color, it means their DNA sequences are so similar that researchers are confident the genes are related by common descent.

The following conclusions can be made from the data in the figure:

- As predicted by the new genes, new bodies hypothesis, the number and identity of *Hox* genes varies widely among animals. Groups that branched off early according to the fossil record and that have relatively small, simple bodies, such as jellyfish, have fewer *Hox* genes than do groups that branched off later—such as vertebrates. The number of genes in the *Hox* cluster appears to have expanded during the course of evolution (from top to bottom in Figure 27.10).

- It is sensible to argue that the new Hox genes were indeed created by gene duplication events, because genes within the cluster are similar in their structure and base sequence. For example, consider the genes that are colored green in Figure 27.10. These genes don't appear in jellyfish and probably originated when a mutation resulted in a duplication of the blue-colored gene, which is present in jellyfish. The two important points here are that (1) biologists can use the phylogeny to understand the order in which different genes within the cluster appeared, and (2) gene duplication provides a mechanism for the new genes, new bodies hypothesis.

- The entire *Hox* cluster was duplicated, and then duplicated again, in the lineage leading to vertebrates. All of the animals shown in Figure 27.10 have a single *Hox* cluster except for mice and zebra fish. Mice have 4 distinct clusters; zebra fish have 7. Because all vertebrates examined to date have more than one *Hox* cluster, the mutations that led to the duplication of the cluster probably occurred close to the origin of vertebrates. In addition, zebrafish and other ray-finned fishes have enough distinct *Hox* clusters to suggest that the entire cluster was duplicated yet again in their ancestors. Because vertebrates include some of the largest and most complex animals, and because ray-finned fishes are the most species rich and among the most morphologically diverse vertebrates, these observations also support the new genes, new bodies hypothesis. But it is important to note that in some cases, copies of genes within the duplicated clusters have been lost. Mice, for ex-

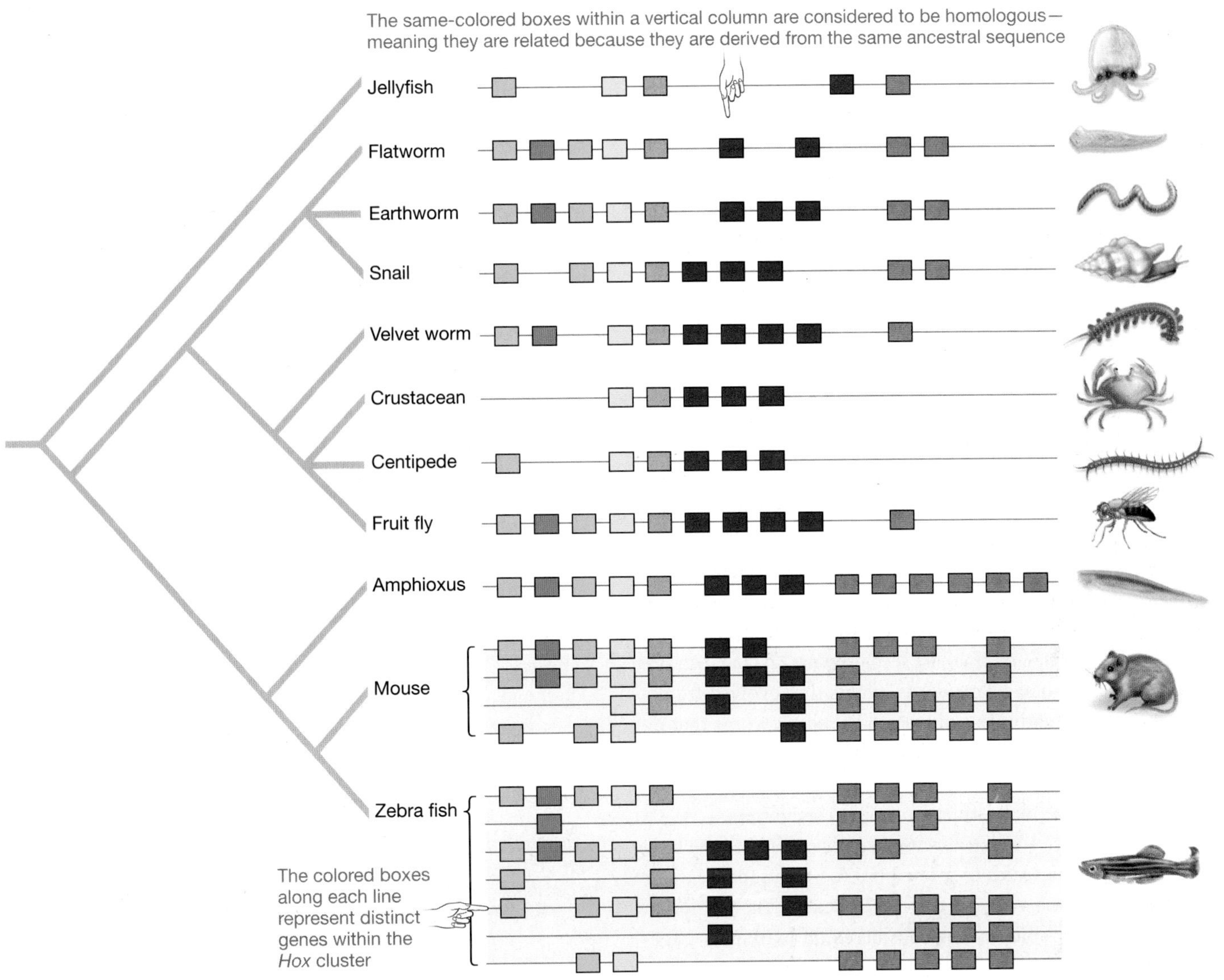

FIGURE 27.10 ***Hox*** **Genes in Animals.**

EXERCISE Next to the illustration of each animal on this evolutionary tree, write the number of Hox genes it has.

ample, have no blue-colored gene in their third set of *Hox* genes.

The data would certainly appear to support a simple-minded version of the new genes, new bodies hypothesis, except for one crucial observation: In arthropods (represented by the fruit fly, centipede, and crustacean), there is no correspondence at all between the number of *Hox* genes and the complexity of the resulting organisms. Crabs have just 5 *Hox* genes, while centipedes have 6 and the closely related but morphologically simpler velvet worms have 9.

Clearly, the situation is more involved than initially predicted. Duplication of *Hox* genes has undoubtedly been important in making the elaboration of animal body plans possible. But new genes are not the whole story: Changes in the expression and function of existing genes have been equally or even more important (see Chapter 22).

Check Your Understanding

If you understand that...

- The Cambrian explosion saw the rise of virtually every major animal lineage, and the evolution of a wide array of morphological innovations and food-getting strategies, in the relatively short time frame of 50 million years.

You should be able to...

1) Compare and contrast the animals documented in the Doushantuo, Ediacara, and Burgess Shale faunas.
2) Identify trends in early animal evolution—for example, whether animals tended to get larger or smaller, or less or more complex in their overall morphology.
3) Evaluate the hypothesis that the morphological evolution recorded in the Cambrian explosion was made possible by the duplication of *Hox* genes.

27.4 Adaptive Radiations

Suppose that biologists succeed in their goal of estimating the evolutionary relationships of all major lineages of organisms. The resulting phylogenetic tree would represent the complete tree of life. Now imagine that you could look at this tree from afar. One of the patterns that would jump out is that dense, bushy outgrowths are scattered among the branches. As **Figure 27.11a** shows, this shape results when many species branch off from a lineage in a short amount of time. Biologists sometimes call this pattern a star phylogeny, because of its starburst shape. Why does this pattern exist? One of the leading causes is a phenomenon known as an adaptive radiation.

An **adaptive radiation** occurs when a single lineage produces many descendant species that live in a wide diversity of habitats and find food in a variety of ways. **Figure 27.11b** shows a few of the Hawaiian honeycreepers—a diverse lineage of songbirds that evolved after a finchlike ancestor happened to colonize the Hawaiian islands. Considering how varied honeycreeper beaks are in size and shape, you should not be surprised that different types of honeycreepers obtain food in different ways—by eating insects, sucking nectar, or cracking seeds. **Figure 27.11c** illustrates a few of the Hawaiian silverswords, plants that evolved from a species of tarweed native to California. The silverswords that resulted from the adaptive radiation in Hawaii live in habitats ranging from lush rain forests to austere lava flows, and they range from mosslike mat-formers to vines to small trees.

Although adaptive radiations are usually studied at the level of species and time scales of a few hundred thousand years or less, they can also be analyzed at broader scales. The Cambrian explosion, for example, can be considered an extremely large-scale adaptive radiation. Another classical example is the diversification of mammals that took place between 65 and 50 million years ago. During this relatively short interval, the primates (monkeys and apes), bats, carnivores, deer, whales, horses, and rodents originated. The organisms resulting from this rapid divergence represent a remarkable array of adaptive forms. They swim, fly, glide, burrow, swing through trees, walk on four legs, or walk on two legs. They occupy habitats from the open ocean to mountaintops and from rain forests to deserts. They eat fruit, nuts, leaves, twigs, bark, insects, crustaceans, mollusks, fish, and other mammals.

The hallmark of an adaptive radiation is rapid speciation and ecological diversification within a single lineage. What makes adaptive radiations occur?

(a) Adaptive radiations produce star phylogenies.

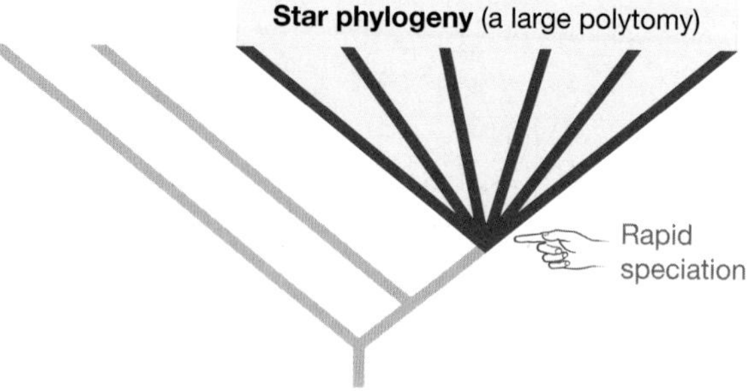

(b) Hawaiian honeycreepers underwent adaptive radiation.

(c) Hawaiian silverswords underwent adaptive radiation.

FIGURE 27.11 Adaptive Radiations Often Produce "Star Phylogenies." (a) A phylogeny has a starburst shape when speciation has been so rapid that the order of branching can't be resolved. If speciation is followed by divergence into many different adaptive forms, then an adaptive radiation has taken place. On the Hawaiian islands, adaptive radiations followed the arrival of both a finch from North America and tarweed seeds from California. These radiations produced the Hawaiian **(b)** finches and **(c)** silverswords.

EXERCISE On each tree, add a North American finch or a Californian tarweed as an outgroup. (To review what an outgroup is, see BioSkills 2.)

Ecological Opportunity as a Trigger

Whether adaptive radiations are analyzed at a large or small scale, biologists find that one of the most consistent themes in these events is ecological opportunity—meaning the availability of new or novel types of resources. The radiation of mammals, for example, occurred immediately after the extinction of dinosaurs. As mammals diversified, they took over the ecological roles formerly filled by dinosaurs and swimming reptiles. Adaptive radiations often occur when habitats are unoccupied by competitors.

Recently a group of biologists documented this process in detail. They did not study a radiation that followed an extinction event, however. Instead they analyzed radiations triggered by colonization events on islands that had distinct habitats and were free of competitors.

The study focused on the *Anolis* lizards of the Caribbean. Biologists have interpreted the history of this group of lizards as an adaptive radiation for two reasons: The lineage includes 150 species, and there is a strong correspondence between the size and shape of each species and the habitat it occupies. Most *Anolis* species that are twig-dwellers, for example, have relatively short legs and tails, while those that spend most of their time clinging to broad tree trunks or running along the ground tend to have long legs and tails (**Figure 27.12a**). These data suggest that lizard species have diversified in a way that allows them to occupy many different habitats.

Exactly how did the diversification occur? As the first step in answering this question, the biologists estimated the phylogeny of *Anolis* from DNA sequence data. Then they compared the habitats occupied by each species with their relationships on the phylogenetic tree. The results shown in **Figure 27.12b**, for species found on two different islands, are typical. Notice that the original colonist on each island belonged to a different ecological type. The initial species on Hispaniola lived on the trunks and crowns of trees, while the original colonist on Jamaica occupied twigs. From different evolutionary starting points, then, an adaptive radiation occurred on both islands. The key point is that on both islands, the same four ecological types eventually evolved. New species arose on each island independently, but because both islands had similar varieties of habitats, each island ended up with a complement of species that was similar in lifestyle and appearance. The researchers found the same pattern on many other islands, meaning that the same type of convergent evolution occurred repeatedly. In other words, a series of "miniature adaptive radiations" occurred—one on each island—within the overall *Anolis* radiation. The small-scale and large-scale radiations were triggered by two conditions: opportunity in the form of both available habitat when new islands were colonized, and lack of competitors.

Morphological Innovation as a Trigger

The other major trigger for adaptive radiations, in addition to ecological opportunity, is morphological innovation. Important new traits such as multicellularity, shells, exoskeletons, and limbs were a driving force behind the adaptive radiation called the Cambrian explosion. The evolution of the limb triggered

(a) Species of *Anolis* vary in leg length and tail length. Some species are ground dwelling; others live in distinct regions of shrubs or trees.

(b) The same adaptive radiation of *Anolis* has occurred on different islands, starting from different types of colonists.

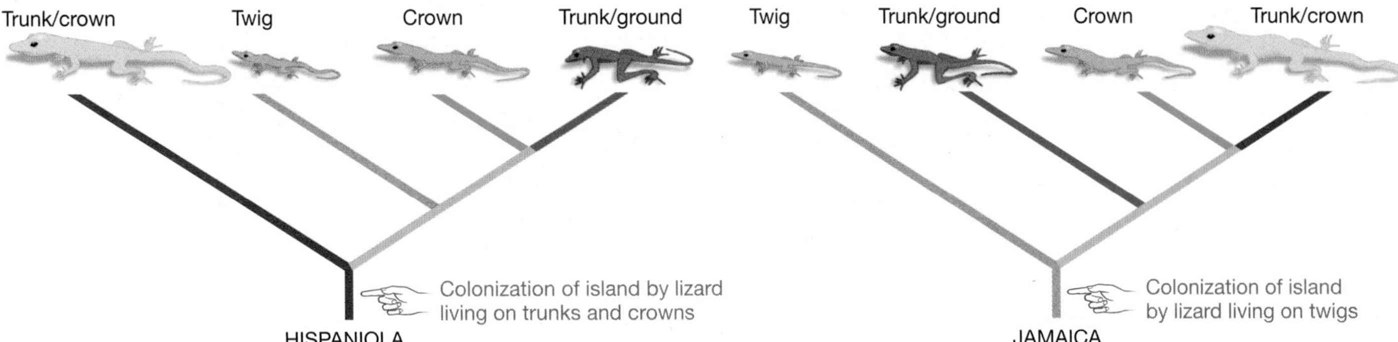

FIGURE 27.12 Adaptive Radiations of *Anolis* Lizards. (a) Short-legged lizard species (left, right) spend most of their time on the twigs of trees and bushes; long-legged species (center) live on tree trunks and the ground. **(b)** The evolutionary relationships among lizard species on the islands of Hispaniola and Jamaica. The initial colonist species was different on these islands; but in terms of how they look and where they live, a similar suite of four species evolved. According to these data, similar adaptive radiations took place independently on the two islands.

the diversification of land-dwelling vertebrates called the tetrapods. Similarly, many of the other important diversification events in the history of life started off with the evolution of a key morphological trait that allowed descendants to live in new areas, exploit new sources of food, or move in new ways:

- The evolution of wings, three pairs of legs, and a protective external skeleton helped make the insects the most diverse lineage on Earth, with perhaps well over 3 million species in existence today (**Figure 27.13a**).
- Flowers are a unique reproductive structure that helped trigger the diversification of angiosperms (flowering plants). Because flowers are particularly efficient at attracting pollinators, the evolution of the flower made angiosperms more efficient in reproduction. Today angiosperms are far and away the most species-rich lineage of land plants, with over 250,000 species known (**Figure 27.13b**).
- Cichlids are a lineage of fish that evolved a unique set of jaws in their throat. These second jaws make food processing extremely efficient. Different species have throat jaws specialized for crushing snail shells, shredding tissue from other fish, or mashing bits of algae. Over 300 species of cichlid live in Africa's Lake Victoria alone (**Figure 27.13c**).
- Feathers and wings gave some dinosaurs the ability to fly (**Figure 27.13d**). Today the lineage called birds contains about 10,000 species, with representatives that live in virtually every habitat on the planet.

In sum, adaptive radiation is a key pattern in the history of life and is usually associated with a new ecological opportunity or a morphological innovation. During an adaptive radiation, rapid speciation and morphological divergence are tightly linked.

MB **Web Animation** at www.masteringbio.com
Adaptive Radiation

27.5 Mass Extinctions

Mass extinction events are evolutionary hurricanes. They buffet the tree of life, snapping twigs and breaking branches. They are catastrophic episodes that wipe out huge numbers of species and lineages in a short time, giving the tree of life a drastic pruning. One mass extinction event, about 251 million years ago, nearly uprooted the tree entirely. The end-Permian extinction came close to ending multicellular life on Earth (**Box 27.2**).

Mass extinction events need to be distinguished from background extinctions. A **mass extinction** refers to the rapid extinction of a large number of lineages scattered throughout the tree of life. More specifically, a mass extinction occurs when at least 60 percent of the species present are wiped out within 1 million years. **Background extinction** refers to the lower, average rate of extinction observed when a mass extinction is not occurring. Although there is no hard-and-fast rule for distinguishing between the two extinction rates, paleontologists traditionally recognize and study five mass extinction events. **Figure 27.14**, for example,

(a) Insects have a distinctive body plan.

(b) Flowering plants have a unique reproductive structure (flower).

(c) Cichlids have "throat jaws" that can bite and process food.

(d) Feathers evolved in dinosaurs.

FIGURE 27.13 Some Adaptive Radiations Are Associated with Morphological Innovations. (a) The insect body includes several important innovations, including wings, compound eyes, three pairs of legs, and a segmented body organized into three general regions. **(b)** The flower was a morphological innovation that made pollen transfer and reproduction more efficient. **(c)** Cichlid species have distinctive mouthparts, including a pair of specialized jaws located in their throat. [©Don P. Northup, www.africancichlidphotos.com] **(d)** The evolution of feathers and flight in dinosaurs triggered the adaptive radiation of birds. This dinosaur, *Caudipteryx,* was covered with short downy body feathers and had longer feathers on its forelimbs and tail. Feathers originally evolved for display or insulation; later they were used in gliding and in powered flight.

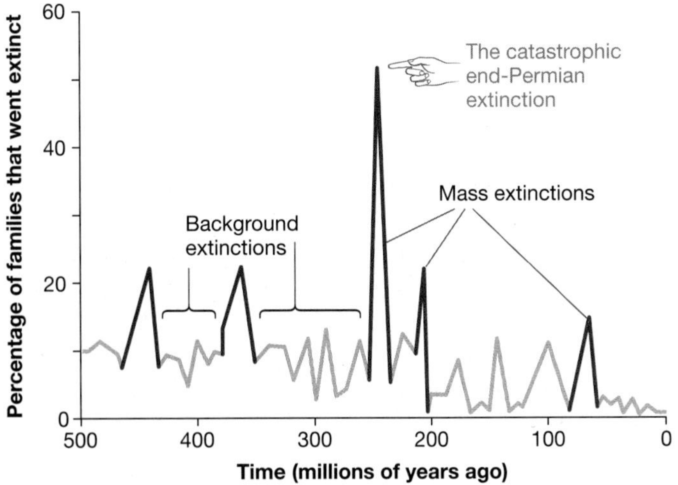

FIGURE 27.14 The Big Five Mass Extinction Events. This graph shows the percentage of families that went extinct over each interval in the fossil record since the Cambrian explosion. The five mass extinction events are drawn in dark red.

EXERCISE Circle the end-Cretaceous extinction event that occurred 65 million years ago, ending the era of the dinosaurs.

plots the percentage of plant and animal families that died out during each stage in the geologic time scale since the Cambrian explosion. (A plant or animal family is a grouping of similar genera in Linnean taxonomy—see Chapter 1.) Five spikes in the graph—denoting a large number of extinctions within a short time—are drawn in red. These are referred to as "The Big Five."

How Do Background and Mass Extinctions Differ?

Biologists are interested in distinguishing between background and mass extinctions because these events have contrasting causes and effects. Background extinctions are thought to occur when normal environmental change, emerging diseases, or competition with other species reduces certain populations to zero. Mass extinctions, in contrast, are thought to result from extraordinary, sudden, and temporary changes in the environment. During a mass extinction, species do not die out because individuals are poorly adapted to normal or gradually changing environmental conditions. Rather, species die out from exposure to exceptionally harsh, short-term conditions—such as huge volcanic

BOX 27.2 The End-Permian Extinction

The end-Permian has been called the Mother of Mass Extinctions. To appreciate the scale of what happened, imagine that you took a walk along a seashore and identified 100 different species of algae and animals living on the beach and tide pools and the shallow water offshore. Now imagine that you snapped your fingers and 90 of those species disappeared forever. Only 10 species are left. An area that was teeming with diverse forms of life would look barren. This is what happened, all over the world, during the end-Permian extinction. The event was a catastrophe of almost unimaginable proportions. On a personal level, it would be like nine of your ten best friends dying.

Although biologists have long appreciated the scale of the end-Permian extinction, research on its causes is ongoing. Consider the following:

- Flood basalts are outpourings of molten rock that flow across the Earth's surface. The largest flood basalts on Earth, called the Siberian traps, occurred during the end-Permian. They added enormous quantities of heat, CO_2, and sulfur dioxide to the atmosphere (**Figure 27.15**). The CO_2 led to intense global warming, and sulfur dioxide reacted with water to form sulfuric acid, which is toxic to most organisms.
- There is convincing evidence that sea level dropped dramatically during the extinction event, reducing the amount of habitat available for marine organisms, and that the oceans became completely or largely anoxic—meaning that they lacked oxygen.
- Terrestrial animals may have been restricted to low-elevation habitats, due to low oxygen concentrations and high CO_2 levels in the atmosphere.

In short, it is clear that both marine and terrestrial environments deteriorated dramatically for organisms that depend on oxygen to live. What biologists don't understand is *why* the environment changed so radically, and so quickly. Research continues.

FIGURE 27.15 The Siberian Traps Formed at the End-Permian. The largest flood basalt on Earth, the Siberian traps probably resembled this fissure eruption in Iceland, but on a much larger scale. The Siberian traps formed a layer of rock between 400-3,000 m thick that covered up to 7,000,000 km^2 of the end-Permian landscape, in what is now northeast Asia.

eruptions or catastrophic sea-level changes. In a general sense, background extinctions are thought to result primarily from natural selection. Mass extinctions, in contrast, function like genetic drift. The extinctions they cause are largely random with respect to the fitness of individuals under normal conditions.

To drive these points home, and to see what happens after a mass extinction has occurred, let's examine one of The Big Five in detail. The event we'll analyze, the mass extinction at the end of the Cretaceous period, was not the largest in history—the mass extinction at the end of the Permian period wiped out 90 percent of the multicellular organisms alive at the time. However, the end-Cretaceous extinction is the best understood and is among the most dramatic: It extinguished the dinosaurs and ushered in the diversification of mammals.

What Killed the Dinosaurs?

The end-Cretaceous extinction of 65 million years ago is as satisfying a murder mystery as you could hope for, but the butler didn't do it. The impact hypothesis for the extinction of the dinosaurs, first put forth in the early 1970s, proposed that an asteroid struck Earth and snuffed out an estimated 60–80 percent of the multicellular species alive.

The impact hypothesis was intensely controversial at first. As researchers set out to test its predictions, however, support began to grow:

- Worldwide, sedimentary rocks that formed at the Cretaceous-Paleogene (K–P)[1] boundary were found to contain extremely high quantities of the element iridium. Iridium is vanishingly rare in Earth rocks, but it is an abundant component of asteroids and meteorites (**Figure 27.16a**).
- Shocked quartz and microtektites are minerals that are found only at documented meteorite impact sites (**Figure 27.16b**). Shocked quartz forms when shock waves from an asteroid impact alter the structure of sand grains. Microtektites form when minerals are melted at an impact site and then cool and resolidify. In Haiti and an array of other locations, both shocked quartz and microtektites have been discovered in abundance in rock layers dated to 65 million years ago.
- A crater the size of Sicily was found just off the northwest coast of Mexico's Yucatán peninsula (**Figure 27.16c**). Microtektites are abundant in sediments from the crater's walls, and the crater dates to the K-P boundary.

[1]Geologists use *K* to abbreviate Cretaceous, because *C* refers to the Cambrian period.

Taken together, these data provided conclusive evidence in favor of the impact hypothesis. Researchers now agree that the mystery is solved.

Based on currently available data, astronomers and paleontologists estimate that the asteroid that struck Earth 65 million years ago was about 10 km across. To get a sense of this scale, consider that Mt. Everest is about 10 km above sea level and that planes cruise at an altitude of about 10 km. Imagine Earth being hit by a rock the size of Mt. Everest, or a rock that would fill the space between you and a jet in the sky.

The distribution of shocked quartz and microtektites dated to 65 mya indicates that the asteroid hit Earth at an angle and splashed material over much of southeastern North America. To understand the impact's consequences, consider the results of the Tunguska event. On June 30, 1908, a piece of a comet about 30 m across and about 2 megatons in mass exploded at least 5 km above Earth's surface near the Tunguska River in Siberia. The explosion released about 1000 times as much energy as the atomic bomb that destroyed Hiroshima in World War II. The event incinerated vegetation over hundreds of square kilometers, leveled trees over thousands of square kilometers, and significantly increased dust levels across the Northern Hemisphere. A deafening blast was heard 500 km (300 mi) away; people standing 60 km away were thrown to the ground or knocked unconscious. These events pale in comparison to what happened 65 million years ago. The energy the K-P asteroid unleashed was 37 *million* times greater than the Tunguska event.

According to both computer models and geologic data, the consequences of the K-P asteroid strike were nothing short of devastating. A tremendous fireball of hot gas would have spread from the impact site; large soot and ash deposits in sediments dated to 65 million years ago testify to catastrophic wildfires, worldwide. The largest tsunami in the last 3.5 billion years would have disrupted ocean sediments and circulation patterns. The impact site itself is underlain by a sulfate-containing rock called anhydrite. The SO_4^{2-} released by the impact would have reacted with water in the atmosphere to form sulfuric acid (H_2SO_4), triggering extensive acid rain. Massive quantities of dust, ash, and soot would have blocked the Sun for long periods, leading to rapid global cooling and a crash in plant productivity.

Selectivity The asteroid impact did not kill indiscriminately. Perhaps by chance, certain lineages escaped virtually unscathed while others vanished. Among vertebrates, for example, the dinosaurs, pterosaurs (flying reptiles), and all of the large-bodied marine reptiles (mosasaurs, ichthyosaurs, and plesiosaurs) expired; mammals, crocodilians, amphibians, and turtles survived.

Why? Answering this question has sparked intense debate. For years the leading hypothesis was that the K-P extinction event was size selective. The logic here was that the extended

(a) Iridium is present at high concentration in rocks formed 65 million years ago.

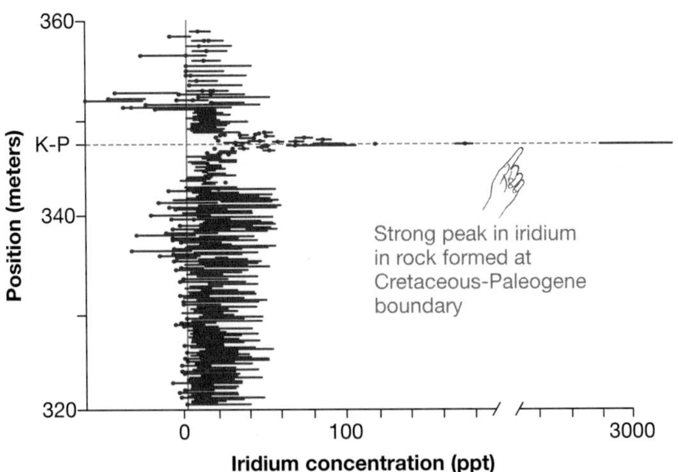

(b) Minerals that form during asteroid impacts

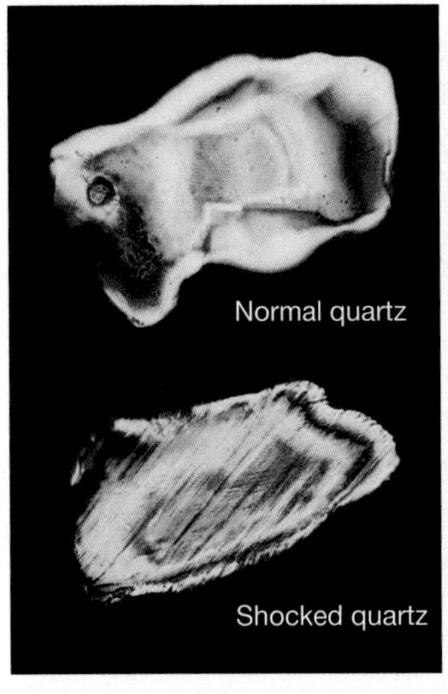

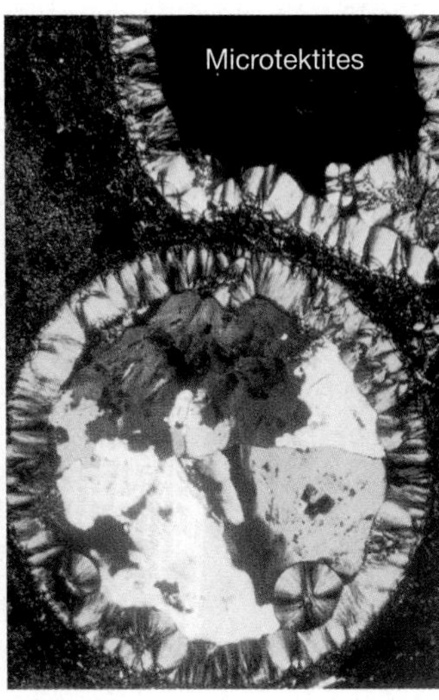

(c) The asteroid left a crater 180 km (112 miles) wide.

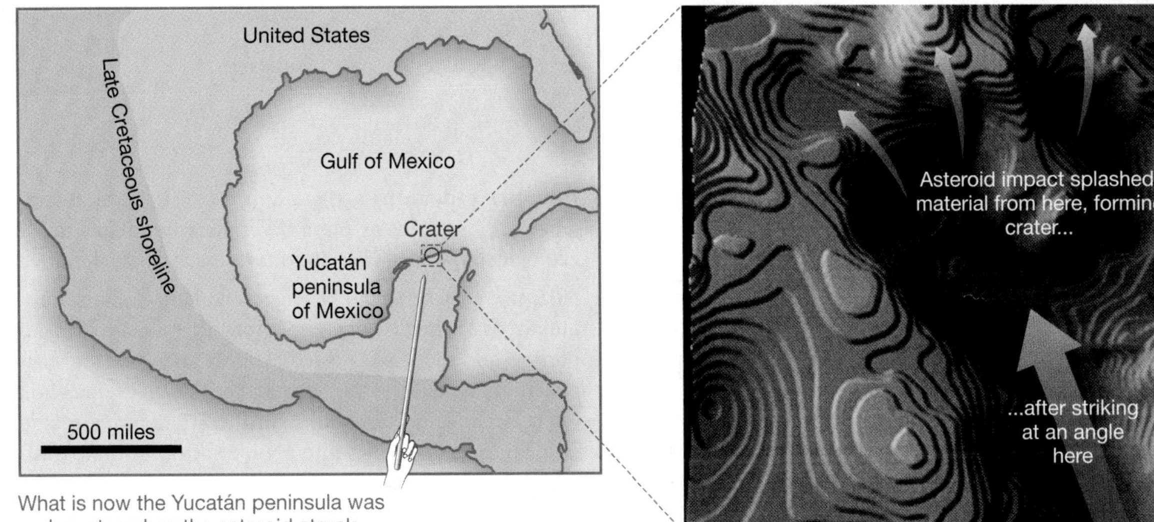

FIGURE 27.16 Evidence of an Asteroid Impact 65 Million Years Ago. (a) The concentration of iridium in rocks that formed on either side of the Cretaceous-Paleogene (K-P) boundary, in parts per trillion (ppt). **(b)** Normal quartz grains are markedly different from shocked quartz grains. The striations in the shocked quartz are caused by a sudden increase in pressure. Microtektites (right) are tiny glass particles formed when minerals melt at an impact site and then recrystallize. **(c)** Geologists have identified the walls of a crater off the northwest coast of the Yucatán Peninsula, dated to 65 million years ago.

● **EXERCISE** On the map in part (c), label the "splash zone" where microtektites and debris from the impact would have landed. Label where the tsunami generated by the impact would have struck land.

darkness and cold would affect large organisms disproportionately, because they require more food than do small organisms. But extensive data on the survival and extinction of marine clams and snails have shown no hint of size selectivity, and small-bodied and juvenile dinosaurs perished along with large-bodied and adult forms. One hypothesis currently being tested is that organisms that were capable of inactivity for long periods—by hibernating or resting as long-lived seeds or spores—were able to survive the catastrophe. But this aspect of the mystery is still unsolved.

Recovery After the K-P extinction, fern fronds and fern spores dominate the plant fossil record from North America and Australia. These data suggest that extensive stands of ferns replaced diverse assemblages of cone-bearing and flowering plants after the impact. The fundamental message here is that terrestrial ecosystems around the world were radically simplified. In marine environments, some invertebrate groups do not exhibit normal levels of species diversity in the fossil record until 4–8 million years past the K-P boundary. Recovery was slow.

The organisms present in the Paleogene were markedly different from those of the preceding period. The lineage called Mammalia, which had consisted largely of rat-sized predators and scavengers in the heyday of the dinosaurs, exploded after the impact and took the place of the dinosaurs. Within 10–15 million years, all of the major mammalian orders observed today had appeared—from pigs to primates. Why? A major branch on the tree of life had disappeared. With competitors removed, mammals flourished.

The Mesozoic is sometimes called the Age of Reptiles, and the Cenozoic the Age of Mammals. The change was not due to a competitive superiority conferred by adaptations such as fur and lactation. Rather, it was due to a chance event: a once-in-a-billion-years collision with a massive rock from outer space.

Check Your Understanding

If you understand that...

- Mass extinctions have occurred repeatedly throughout the history of life.

You should be able to...

1) Evaluate the strength of the evidence supporting the impact hypothesis for the extinction of the dinosaurs, and identify which evidence is strongest and which is weakest. (In science, evidence is considered strong if it cannot be explained by alternative hypotheses.)
2) Explain why mass extinctions wipe out species more or less randomly—much the way genetic drift affects changes in allele frequencies.

Chapter Review

SUMMARY OF KEY CONCEPTS

Phylogenies and the fossil record are the major tools that biologists use to study the history of life.

Phylogenetic trees can be estimated by grouping species based on overall similarity in traits or by analyzing shared, derived characters (synapomorphies) that identify monophyletic groups. Phylogenetic trees document the evolutionary relationships among species and identify the order in which events occurred. To minimize the impact of homoplasy when inferring phylogenies, researchers use parsimony or other approaches to decide which of the many trees that are possible is most likely to reflect actual evolutionary history. The fossil record is used in conjunction with phylogenetic analyses because it is the only direct source of data about what extinct organisms looked like and where they lived.

You should be able to describe how biologists use phylogenetic trees and the fossil record to infer when traits—like upright posture in humans—evolved.

The Cambrian explosion was the rapid morphological and ecological diversification of animals that occurred during the Cambrian period.

The diversification of animals over a 5-million-year period, starting about 565 million years ago, is perhaps the best-studied event in the history of life. The animals that lived in marine environments just before and after the Cambrian explosion are documented in the Doushantuo, Ediacaran, and Burgess Shale faunas. During this period the first heads, tails, appendages, shells, exoskeletons, and segmented bodies evolved. Data on the number and identity of homeotic genes in different animal lineages suggest that at least some of the animal radiation was possible because gene duplication events created new copies of *Hox* genes.

You should be able to explain why the Cambrian explosion can be considered an example of an adaptive radiation.

Adaptive radiations are a major pattern in the history of life. They are instances of rapid diversification associated with new ecological opportunities and new morphological innovations.

Adaptive radiations can be triggered by the colonization of a new habitat or the demise of competitors after a mass extinction. Morphological innovations, such as limbs, flowers, and feathers can also initiate adaptive radiations. Speciation events and morphological change occur rapidly during an adaptive radiation, as a single lineage diversifies into a wide variety of ecological roles.

You should be able to to generate and defend a hypothesis for an ecological opportunity or morphological innovation that triggered the Cambrian explosion.

(MB) Web Animation at www.masteringbio.com
Adaptive Radiation

Mass extinctions have occurred repeatedly throughout the history of life. They rapidly eliminate most of the species alive in a more or less random manner.

Mass extinctions have altered the course of evolutionary history at least five times. They prune the tree of life more or less randomly and have marked the end of several prominent lineages and the rise of new branches. The end-Cretaceous extinction killed 60 to 80 percent of existing species and was caused by an asteroid impact. After the devastation of a mass extinction, it can take 10–15 million years for ecosystems to recover their former levels of diversity. Data introduced in Chapter 55 suggest that a sixth mass extinction event—caused by humans—is currently under way.

You should be able to evaluate whether environmental changes caused by humans are eliminating species in a random manner, as opposed to those species being poorly adapted to the environment.

QUESTIONS

Test Your Knowledge

1. Choose the best definition of a fossil.
   a. any trace of an organism that has been converted into rock
   b. a bone, tooth, shell, or other hard part of an organism that has been preserved
   c. any trace of an organism that lived in the past
   d. the process that leads to preservation of any body part from an organism that lived in the past

2. Why are the Doushantuo, Ediacaran, and Burgess Shale fossil deposits unusual?
   a. Soft-bodied animals are preserved in them.
   b. They are easily accessible to researchers.
   c. They are the only fossil-bearing rock deposits from their time period.
   d. They include terrestrial, instead of just marine, species.

3. Which of the following best characterizes an adaptive radiation?
   a. Speciation occurs extremely rapidly, and descendant populations occupy a large geographic area.
   b. A single lineage diversifies rapidly, and descendant populations occupy many habitats and ecological roles.
   c. Natural selection is particularly intense, because disruptive selection occurs.
   d. Species recover after a mass extinction.

4. Which of the following is most accurate?
   a. Mass extinctions are due to asteroid impacts; background extinctions may have a wide variety of causes.
   b. Mass extinctions focus on particularly prominent groups, such as dinosaurs; background extinctions affect species from throughout the tree of life.
   c. Only five mass extinctions have occurred, but hundreds of background extinctions have occurred.
   d. Mass extinctions extinguish groups rapidly and randomly; background extinctions are slower and often result from natural selection.

5. Why do molecular clocks exist?
   a. Natural selection is not important at the molecular level.
   b. Homologous genes have the same structure.
   c. They can be calibrated.
   d. Some DNA sequences, in some lineages, change at a steady rate over time.

6. Why is burial a key step in fossilization?
   a. It slows the process of decay by bacteria and fungi.
   b. It allows tissues to be preserved as casts or molds.
   c. It protects tissues from wind, rain, and other corrosive elements.
   d. All of the above.

Test Your Knowledge answers: 1. c; 2. a; 3. b; 4. d; 5. d; 6. d

Test Your Understanding

Answers are available at www.masteringbio.com

1. The text claims that the fossil record is biased in several ways. What are these biases? If the database is biased, is it still an effective tool to use in studying the diversification of life? Explain.

2. The initial diversification of animals took place over some 50 million years, at the start of the Cambrian period. Why is the diversification called an "explosion"?

3. What is the "new genes, new bodies" hypothesis? Explain whether the data summarized in this chapter support or contradict the hypothesis.

4. Give an example of an adaptive radiation that occurred after a colonization event, after a mass extinction, and after a morphological innovation. In each case, provide a hypothesis to explain why the adaptive radiation occurred.

5. Summarize the evidence that supports the impact hypothesis for the K-P extinction.

6. Why are monophyletic groups identified by shared, derived traits?

Applying Concepts to New Situations

Answers are available at www.masteringbio.com

1. Suppose that the dying wish of a famous eccentric was that his remains be fossilized. His family has come to you for expert advice. What steps would you recommend to maximize the chances that his wish will be fulfilled?

2. Using data from the molecular phylogenies presented in this section and data on the fossil record of whales presented in Chapter 24, summarize how whales evolved from the common ancestor they share with today's hippos.

3. Some researchers contend that the end-Permian extinction event was also caused by an impact with a large extraterrestrial object. List the evidence, ordered from least convincing to most convincing, that you would like to see before you accept this hypothesis. Explain your rankings.

4. One of the "triggers" proposed for the Cambrian explosion is a dramatic rise in oxygen concentrations that occurred in the oceans about 800 mya. Review material in Chapter 9 on how oxygen compares with other atoms or compounds as an electron acceptor during cellular respiration (look near the end of Section 9.6, where aerobic and anaerobic respiration are compared). Then explain the logic behind the "oxygen-trigger" hypothesis for the Cambrian explosion.

www.masteringbio.com is also your resource for • Answers to text, table, and figure caption questions and exercises • Answers to *Check Your Understanding* boxes • Online study guides and quizzes • Additional study tools including the *E-Book for Biological Science* 3rd ed., textbook art, animations, and videos.

28 Bacteria and Archaea

KEY CONCEPTS

- Bacteria and archaea have a profound impact on humans and global ecosystems. A few bacteria cause important infectious diseases; some bacterial and archaeal species are effective at cleaning up pollution; photosynthetic bacteria were responsible for the evolution of the oxygen atmosphere; bacteria and archaea cycle nutrients through every terrestrial and aquatic environment.

- Bacteria and archaea have been evolving for billions of years and are extremely sophisticated organisms. Although they are small and relatively simple morphologically, they live in virtually every habitat known and use remarkably diverse types of compounds in cellular respiration and fermentation.

Although this hot spring looks devoid of life, it is actually teeming with billions of bacterial and archaeal cells from a wide variety of species. As this chapter will show, bacteria and archaea occupy virtually every environment on Earth.

Bacteria and Archaea form two of the three largest branches on the tree of life (**Figure 28.1**). The third major branch or domain consists of eukaryotes and is called the Eukarya. Virtually all members of the Bacteria and Archaea domains are unicellular, and all are prokaryotic—meaning that they lack a membrane-bound nucleus.

Although their relatively simple morphology makes bacteria and archaea appear similar to the untrained eye, they are strikingly different at the molecular level (**Table 28.1**). Organisms in the Bacteria and Archaea domains are distinguished by the types of molecules that make up their plasma membranes and cell walls, and by the machinery they use to transcribe DNA and translate messenger RNA into proteins. Most notably, **Bacteria** have a unique compound called peptidoglycan in their cell walls, and **Archaea** have unique phospholipids—compounds containing hydrocarbons called isoprenes in their tails—in their plasma membranes (see Chapters 5 and 6). In addition, the structures of the RNA polymerases and ribosomes found in Archaea and Eukarya are distinct from those found in Bacteria and similar to each other. This is why antibiotics that poison bacterial ribosomes do not affect the ribosomes of archaea or eukaryotes. If you were unicellular, bacteria and archaea would look and act as different as mammals and fish do to you now.

The lineages in the domains Bacteria and Archaea are ancient, diverse, ubiquitous, and abundant. The oldest fossils of any type found to date are 3.4-billion-year-old carbon-rich deposits derived from bacteria. Because eukaryotes do not appear

Key Concept Important Information Practice It

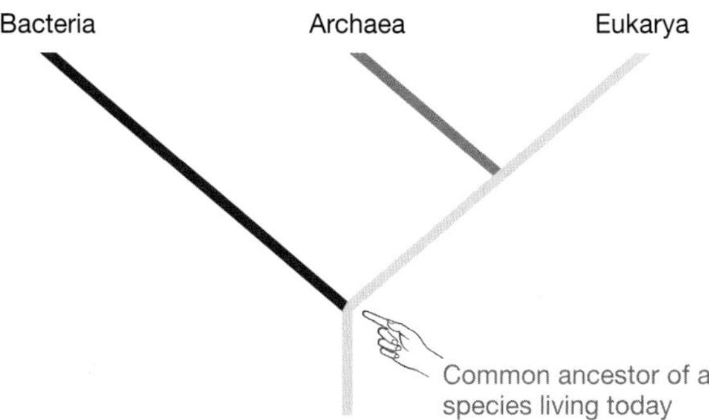

FIGURE 28.1 Bacteria, Archaea, and Eukarya Are the Three Domains of Life. Archaea are more closely related to eukaryotes than they are to bacteria.

● **EXERCISE** Circle the prokaryotic lineages on the tree of life.

● **QUESTION** Was the common ancestor of all species living today prokaryotic or eukaryotic? Explain your reasoning.

in the fossil record until 1.75 billion years ago, biologists infer that prokaryotes were the only form of life on Earth for almost 1.7 billion years. The Bacteria and Archaea have diversified into species that now blanket the planet.

Just how many bacteria and archaea are alive today? Although a mere 5000 species have been formally named and described to date—most by the morphological species concept introduced in Chapter 26—it is virtually certain that millions exist. Consider that over 400 species of prokaryotes are living in your gastrointestinal tract right now. Another 500 species live in your mouth, of which just 300 have been described and named. Norman Pace points out that there may be tens of millions of different insect species but notes, "If we squeeze out any one of these insects and examine its contents under the microscope, we find hundreds or thousands of distinct microbial species." Most of these **microbes** (microscopic organisms) are bacteria or archaea, and virtually all are unnamed and undescribed. If you want to discover and name new species, then study bacteria or archaea.

In addition to recognizing how diverse bacteria and archaea are in terms of numbers of species, it's critical to appreciate their abundance.

- The approximately 10^{13} (10 trillion) cells in your body are vastly outnumbered by the bacterial and archaeal cells living on and in you. An estimated 10^{12} bacterial cells live on your skin, and an additional 10^{14} bacterial and archaeal

SUMMARY TABLE **28.1 Characteristics of Bacteria, Archaea, and Eukarya**

	Bacteria	Archaea	Eukarya
Nuclear envelope?	No	No	Yes
Circular chromosome?	Yes (but linear in some species)	Yes	No
DNA associated with histone proteins? (see Chapter 18)	No	Yes	Yes
Organelles present?	Some in limited number of species	None described to date	Extensive in number and diversity
Flagella present?	Yes—spin like propeller	Yes; spin like bacterial flagella, but distinctive in molecular composition	Yes, but undulate back and forth and have completely different molecular composition compared with bacteria and archaea
Unicellular or multicellular?	Almost all unicellular	All unicellular	Many multicellular
Sexual reproduction?	No*	Not known	Common
Structure of lipids in plasma membrane	Glycerol bonded to straight-chain fatty acids via ester linkage	Glycerol bonded to branched fatty acids (synthesized from isoprene subunits) via ether linkage	Glycerol bonded to straight-chain fatty acids via ester linkage
Cell-wall material	Almost all include peptidoglycan, which contains muramic acid	Varies widely among species, but no peptidoglycan and no muramic acid	When present, usually made of cellulose or chitin
Transcription and translation machinery	One relatively simply RNA polymerase; translation begins with formylmethionine; translation poisoned by several antibiotics that do not affect archaea or eukaryotes	One relatively complex RNA polymerase; translation begins with methionine	Several relatively complex RNA polymerases; translation begins with methionine

*Sexual reproduction begins with meiosis and often involves the exchange of haploid genomes between individuals of the same species. In bacteria, meiosis does not occur. Small numbers of genes can be transferred from one bacterial cell to another, however, and genetic recombination may occur. For more detail, see Chapter 12.

● **EXERCISE** Using the data in this table, add labeled marks to Figure 28.1 indicating where the following traits evolved: peptidoglycan in cell wall, archaeal-type plasma membrane, archaeal and eukaryote-type ribosomes, nuclear envelope.

cells occupy your stomach and intestines. You, along with most other mammals, would not be able to digest food properly without these gut bacteria. You are a walking, talking habitat—one that is teeming with bacteria and archaea.

- A mere teaspoon of good-quality soil contains *billions* of microbial cells, most of which are bacteria and archaea.
- In sheer numbers, species in a lineage called the Group I marine archaea may be the most successful organisms on Earth. Biologists routinely find these cells at concentrations of 10,000 to 100,000 individuals per milliliter of seawater, at depths from 200 to 4000 or more meters below the surface, in most of the world's oceans. At these concentrations, a drop of seawater contains a population equivalent to that of a large human city. Yet this lineage was first described in the early 1990s.
- Biologists estimate the total number of individual bacteria and archaea alive today at 5×10^{30}. If they were lined up end to end, they would make a chain longer than the Milky Way galaxy. These cells contain 50 percent of all the carbon and 90 percent of all the nitrogen and phosphorus found in organisms. In terms of the total volume of living material on our planet, bacteria and archaea are dominant life-forms.

Bacteria and archaea are also found almost everywhere. They live in environments as unusual as oxygen-free mud, hot springs, and salt flats. They have been discovered living in bedrock to a depth of 1500 meters below Earth's surface. In the ocean they are found from the surface to depths of 10,000 m and at temperatures ranging from near 0°C in Antarctic sea ice to over 121°C near submarine volcanoes.

Although there are far more prokaryotes than eukaryotes, much more is known about eukaryotic diversity than about prokaryotic diversity. Researchers who study prokaryotic diversity are exploring one of the most wide-open frontiers in all of science. So little is known about the extent of these domains that recent collecting expeditions have turned up two entirely new **kingdoms** and numerous **phyla** (singular: **phylum**). These are names given to major lineages within each domain. To a biologist, this achievement is equivalent to the sudden discovery of a new group of eukaryotes as distinctive as the flowering plants or animals with backbones.

The physical world has been explored and mapped, and many of the larger plants and animals are named. But in **microbiology**—the study of organisms that can be seen only with the aid of a microscope—this is an age of exploration and discovery.

28.1 Why Do Biologists Study Bacteria and Archaea?

Biologists study bacteria and archaea for the same reasons they study any organisms. First, these organisms are intrinsically fascinating. Discoveries such as finding bacterial cells living a kilometer underground or in 95°C hot springs keep biologists awake at night, staring at the ceiling. They can't wait to get into the lab in the morning and start trying to figure out how those cells are staying alive. Second, researchers know how important understanding diversity is to the rest of biology. In fields from cell biology and genomics to ecology, the ability to compare characteristics in a diverse array of organisms is fundamental to increased understanding. Third, almost all biologists are aware that a mass extinction is currently under way (see Chapter 55). To preserve biodiversity, we have to understand it.

Those reasons aren't necessarily why governments, businesses, and foundations fund research on biological diversity, however. These institutions usually have more practical goals in mind. In many cases they want to know how studying organismal diversity improves human health and welfare. To answer this question, each chapter in this unit starts with a section on how the organisms in question affect humans and other species. The second section of each chapter focuses on methods—on *how* biologists study the group in question. The third section in each chapter is the heart of the matter: It delves into how the organisms make a living and analyzes the unique evolutionary innovations that were responsible for the group's origin and diversification. It also explains what's cool about the group, in a biological sense. The final section in each of this unit's chapters summarizes the key characteristics of prominent lineages within the group. It's written in a succinct, note-like style and is intended to give you an overview of who's who in the group.

And with that, let's plunge in. The ubiquity and abundance of bacteria and archaea make them exceptionally important in both human and natural economies. Let's consider their role in biomedicine, pollution control, industry, and global environmental change.

Some Bacteria Cause Disease

No archaea are known to cause disease in humans. But of the hundreds or thousands of bacterial species living in and on your body, a tiny fraction can disrupt normal body functions enough to cause illness. Bacteria that cause disease are said to be **pathogenic** (literally, "disease-producing"). Pathogenic bacteria have been responsible for some of the most devastating epidemics in human history.

Robert Koch was the first biologist to establish a link between a particular species of bacterium and a specific disease. When Koch began his work on the nature of disease in the late 1800s, microscopists had confirmed the existence of the particle-like organisms we now call bacteria, and Louis Pasteur had shown that bacteria and other microorganisms are responsible for spoiling milk, wine, broth, and other foods. Koch hypothesized that bacteria might also be responsible for causing infectious diseases, which spread by being passed from an infected individual to an uninfected individual.

Koch set out to test this hypothesis by identifying the organism that causes anthrax. Anthrax is a disease of cattle and other grazing mammals that can result in fatal blood poisoning. The disease also occurs infrequently in humans and mice.

To establish a causative link between a specific microbe and a specific disease, Koch proposed that four criteria had to be met:

1. *The microbe must be present in individuals suffering from the disease and absent from healthy individuals.* By careful microscopy, Koch was able to show that the bacterium *Bacillus anthracis* was always present in the blood of cattle suffering from anthrax, but absent from asymptomatic individuals.

2. *The organism must be isolated and grown in a pure culture away from the host organism.* Koch was able to grow pure colonies of *B. anthracis* in glass dishes on a nutrient medium, using gelatin as a substrate.

3. *If organisms from the pure culture are injected into a healthy experimental animal, the disease symptoms should appear.* Koch demonstrated this crucial causative link in mice injected with *B. anthracis*. The symptoms of anthrax infection appeared, and then the infected mice died.

4. *The organism should be isolated from the diseased experimental animal, again grown in pure culture, and demonstrated by its size, shape, and color to be the same as the original organism.* Koch did this by purifying *B. anthracis* from the blood of diseased experimental mice.

These criteria, now called **Koch's postulates**, are still used to confirm a causative link between new diseases and a suspected infectious agent. Koch's experimental results also became the basis for the **germ theory of disease**. This theory, which laid the foundation for modern medicine, holds that infectious diseases are caused by bacteria and viruses. (Viruses are acellular particles that parasitize cells and are analyzed in detail in Chapter 35.) Some of the bacteria that cause illness in human beings are listed in **Table 28.2**. The important things to note about the list are that pathogenic forms come from many lineages in the domain Bacteria and that pathogenic bacteria tend to affect tissues at the entry points to the body, such as wounds or pores in the skin, the respiratory and gastrointestinal tracts, and the urogenital canal.

In the industrialized countries, improvements in sanitation and nutrition have caused dramatic reductions in mortality rates due to infectious diseases (**Figure 28.2**), most of which are due to bacterial and viral infections. In addition, the discovery of antibiotics in 1928, their development over subsequent decades, and widespread use starting in the late 1940s gave physicians effective tools to combat most bacterial infections. **Antibiotics** are molecules that kill bacteria. Extensive use of antibiotics in the late twentieth century led to the evolution of drug-resistant strains of bacteria, however (see Chapter 24). Most of the bacterial species listed in Table 28.2 now include strains that are resistant to one or more of the commonly prescribed antibiotics. Coping with antibiotic resistance in pathogenic bacteria has become a great challenge of modern medicine.

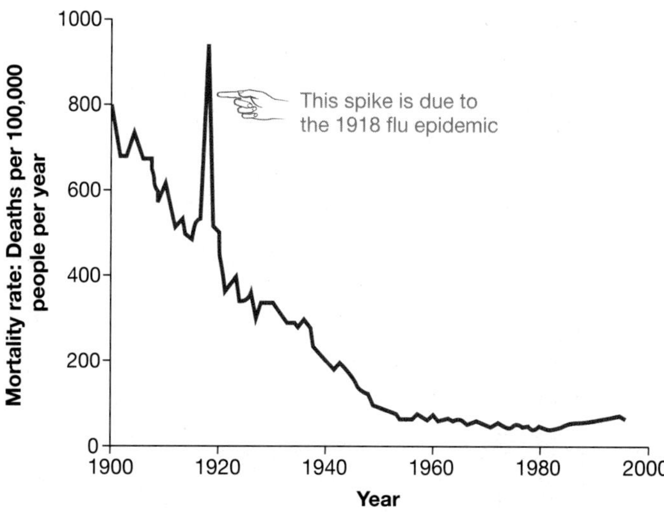

FIGURE 28.2 Deaths due to Bacterial Infections Have Declined Dramatically in Some Countries. This graph shows the death rate due to infectious diseases—meaning, bacterial or viral infections—in a country that industrialized in the late 1800s (the United States).

EXERCISE The first widely prescribed antibiotic in this country was available in the late 1940s. Label this event on the graph.

QUESTION Why was there a slight rise in the graph starting in the mid-1980s?

Although Pasteur, Koch, and other biologists began studying bacteria because of the role these prokaryotes play in disease, the scope of microbiology research has broadened enormously. Let's consider some topics that are currently inspiring studies of bacteria and archaea.

Bacteria Can Clean Up Pollution

Throughout the industrialized world, some of the most serious pollutants in soils, rivers, and ponds consist of organic compounds that were originally used as solvents or fuels but leaked or were spilled into the environment. Most of these compounds are highly hydrophobic. Because they do not dissolve in water, they tend to accumulate in sediments. If the compounds are subsequently ingested by burrowing worms or clams or other organisms, they can be passed along to fish, insects, humans, birds, and other species. Most of these compounds are toxic to eukaryotes in moderate to high concentrations. Petroleum from oil spills and compounds that contain ring structures and chlorine atoms, such as the family of compounds called dioxins, are particularly notorious because of their toxicity to humans.

Biologists who are responsible for cleaning up sites polluted with organic solvents and fuels are faced with a challenge, because at least some of the toxic compounds present are highly resistant to decomposition. Instead of being broken down into

TABLE **28.2 Some Diseases Caused by Bacteria**

Lineage	Species	Tissues Affected	Disease
Firmicutes	*Clostridium botulinum*	Gastrointestinal tract, nervous system	Food poisoning (botulism)
	Clostridium tetani	Wounds, nervous system	Tetanus
	Staphylococcus aureus	Skin, urogenital canal	Acne, boils, impetigo, toxic shock syndrome
	Streptococcus pneumoniae	Respiratory tract	Pneumonia
	Streptococcus pyogenes	Respiratory tract	Strep throat, scarlet fever
Spirochaetes	*Borrelia burgdorferi*	Skin and nerves	Lyme disease
	Treponema pallidum	Urogenital canal	Syphilis
Actinomycetes	*Mycobacterium leprae*	Skin and nerves	Leprosy
	Mycobacterium tuberculosis	Respiratory tract	Tuberculosis
	Propionibacterium acnes	Skin	Acne
Chlamydiales	*Chlamydia trachomatis*	Urogenital canal	Genital tract infection
Proteobacteria (ε group)	*Helicobacter pylori*	Stomach	Ulcer
Proteobacteria (β group)	*Neisseria gonorrhoeae*	Urogenital canal	Gonorrhea
Proteobacteria (γ group)	*Haemophilus influenzae*	Ear canal, nervous system	Ear infections, meningitis
	Pseudomonas aeruginosa	Urogenital canal, eyes, ear canal	Infections of eye, ear, urinary tract
	Salmonella enteritidis	Gastrointestinal tract	Food poisoning
	Vibrio parahaemolyticus	Gastrointestinal tract	Food poisoning
	Yersinia pestis	Lymph and blood	Plague

harmless compounds, certain toxic molecules tend to just "sit there." These compounds pose a long-term threat to nearby fish, birds, people, and other organisms.

To clean up sites like these, researchers have begun to explore more extensive use of **bioremediation**, the use of bacteria and archaea to degrade pollutants. Bioremediation is often based on complementary strategies:

- *Fertilizing contaminated sites* to encourage the growth of existing bacteria and archaea that degrade toxic compounds. After several recent oil spills, researchers added nitrogen to affected sites as a fertilizer, but left nearby beaches untreated as controls. Dramatic increases occurred in the growth of bacteria and archaea that use hydrocarbons in cellular respiration, probably because the added nitrogen was used to synthesize enzymes and other key compounds. In at least some cases, the fertilized sediments cleaned up much faster than the unfertilized sites (**Figure 28.3**).
- *"Seeding," or adding, specific species of bacteria and archaea* to contaminated sites shows promise of alleviating pollution in some situations. For example, researchers have recently discovered bacteria that are able to render certain chlorinated, ring-containing compounds harmless. Instead of being poisoned by the pollutants, these bacteria use ring-containing, chlorinated compounds as electron acceptors during cellular respiration. In at least some cases, the by-product is dechlorinated and nontoxic to humans and other eukaryotes. To follow up on these discoveries, researchers are now growing the bacteria in quantity and testing them in the field, to test the hypothesis that seeding can speed the rate of decomposition in contaminated sediments. Initial reports suggest that seeding may help clean up at least some polluted sites.

Extremophiles

Bacteria or archaea that live in high-salt, high-temperature, low-temperature, or high-pressure habitats are called **extremophiles**

FIGURE 28.3 Bacteria and Archaea Can Play a Role in Cleaning Up Pollution. On the left is a rocky coast that was polluted by an oil spill but fertilized to promote the growth of oil-eating bacteria. The portion of the beach on the right was untreated.

("extreme-lovers"). Studying them has been extraordinarily fruitful for understanding the tree of life, developing industrial applications, and exploring the structure and function of enzymes.

As an example of these habitats, consider hot springs at the bottom of the ocean, where water as hot as 300°C emerges and mixes with 4°C seawater. At locations like these, archaea are abundant forms of life. Researchers recently discovered an archaean that grows so close to these hot springs that its surroundings are at 121°C—a record for life at high temperature. This organism can live and grow in water that is heated past its boiling point (100°C) and at pressures that would instantly destroy a human being.

Extremophiles have become a hot area of research. The genomes of a wide array of extremophiles have been sequenced, and expeditions regularly seek to characterize new species. Why? Based on models of conditions that prevailed early in Earth's history, it appears likely that the first forms of life were extremophiles. Thus, understanding extremophiles may help explain how life on Earth began. In a similar vein, astrobiologists ("space-biologists") use extremophiles as model organisms in the search for extraterrestrial life. The idea is that if bacteria and archaea can thrive in extreme habitats on Earth, it is likely that cells might be found in similar environments on other planets or moons of planets. And because enzymes that function at extreme temperatures and pressures are useful in many industrial processes, extremophiles are of commercial interest as well. Chapter 19 introduced *Taq* polymerase, which is a DNA polymerase that is stable up to 95°C. Recall that *Taq* polymerase is used to run the polymerase chain reaction (PCR) in research and commercial settings. This enzyme was isolated from a bacterium called *Thermus aquaticus* ("hot water"), which was discovered in hot springs in Yellowstone National Park.

How Do Small Cells Affect Global Change?

Certain bacteria and archaea can live in extreme environments and use toxic compounds as food because they produce extremely sophisticated enzymes. The complex chemistry that they carry out, combined with their numerical abundance, has made them potent forces for global change throughout Earth's history. Bacteria and archaea have altered the composition of the oceans, atmosphere, and terrestrial environments for billions of years. They continue to do so today.

The Oxygen Revolution Today, oxygen represents almost 21 percent of the molecules in Earth's atmosphere. But researchers who study the composition of the atmosphere are virtually certain that no free molecular oxygen (O_2) existed for the first 2.3 billion years of Earth's existence. This conclusion is based on two observations: (1) There was no plausible source of oxygen at the time the planet formed; and (2) the oldest Earth rocks indicate that, for many years afterward, any oxygen that formed reacted immediately with iron atoms to produce iron oxides, such as hematite (Fe_2O_3) and magnetite (Fe_3O_4) Early in Earth's history, the atmosphere was dominated by nitrogen and carbon dioxide. Where did the oxygen we breathe come from? The answer is cyanobacteria.

Cyanobacteria are a lineage of photosynthetic bacteria (**Figure 28.4**). According to the fossil record, species of cyanobacteria first became numerous in the oceans about 2.7–2.55 billion years ago. Their appearance was momentous, because cyanobacteria were the first organisms to perform oxygenic ("oxygen-producing") photosynthesis. Oxygenic photosynthesis depends on the proteins and pigments in photosystem II. You might recall from Chapter 10 that photosystem II includes enzymes capable of stripping electrons from water molecules. The reaction that "splits" water results in the production of oxygen as well as electrons. The electrons are required for photosynthesis to continue; the oxygen molecules are simply released as a waste product.

The fossil record and geological record indicate that oxygen concentrations in the oceans and atmosphere began to increase 2.3–2.1 billion years ago. Once oxygen was common in the oceans, cells could begin to use it as the final electron acceptor during cellular respiration. **Aerobic** respiration was now a possibility. Prior to this, organisms had to use compounds other than oxygen as a final electron acceptor—only **anaerobic** respiration was possible.

The evolution of aerobic respiration was a crucial event in the history of life. Because oxygen is extremely electronegative, it is an efficient electron acceptor. Much more energy is released as electrons move through electron transport chains with oxygen as the ultimate acceptor than is released with other substances as the electron acceptor (**Figure 28.5**). Once oxygen was available, much more ATP could be produced for each electron donated by NADH or $FADH_2$. As a result, the rate of energy production could rise dramatically.

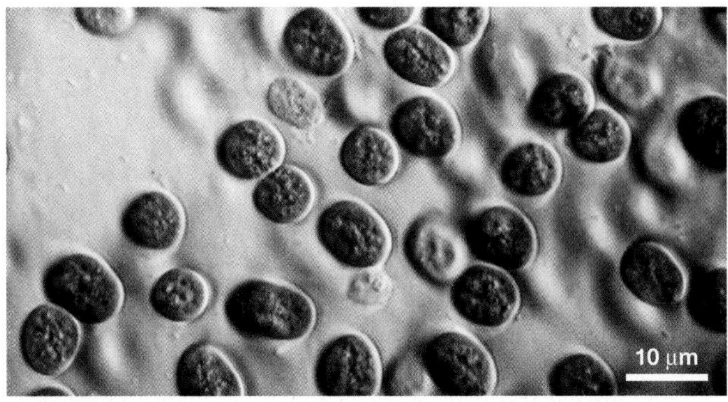

FIGURE 28.4 Cyanobacteria Were the First Organisms to Perform Oxygenic Photosynthesis. Life as we know it today would not have evolved if cyanobacteria had not begun producing oxygen as a by-product of photosynthesis about 2.7 billion years ago.

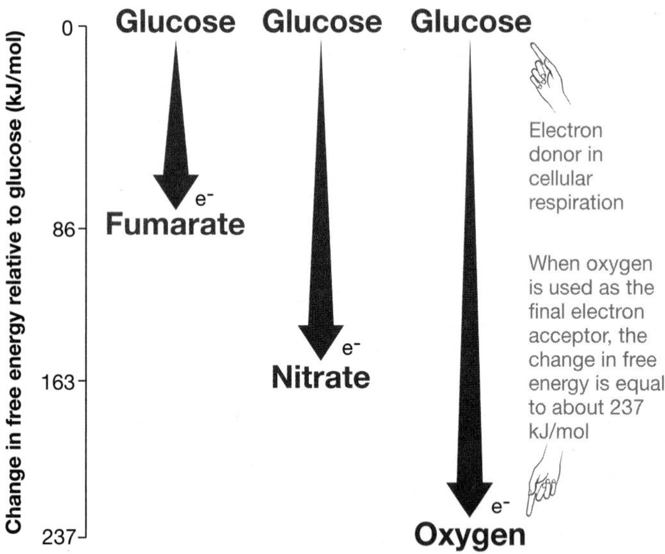

FIGURE 28.5 Cellular Respiration Can Produce More Energy When Oxygen Is the Final Electron Acceptor. Because oxygen has such high electronegativity, the potential energy of the electrons used in cellular respiration is much lower when oxygen is the final acceptor compared to other molecules or ions. As a result, a larger amount of free energy is released during cellular respiration with oxygen as the final electron acceptor.

QUESTION Which organisms grow faster—those using aerobic respiration or those using anaerobic respiration? Explain your reasoning.

Once oxygen levels had built up to significant levels in the oceans and atmosphere, the first eukaryotes—which were macroscopic algae—appear in the fossil record. Biologists hypothesize that a causal link was involved with the availability of oxygen. The claim is that large cell size, then multicellularity, and eventually large body size were made possible by the high metabolic rates and rapid growth fueled by aerobic respiration.

To summarize, data indicate that cyanobacteria were responsible for a fundamental change in Earth's atmosphere—to one with a high concentration of oxygen. Never before, or since, have organisms done so much to alter the nature of our planet.

The Nitrogen Cycle In many environments, fertilizing forests or grasslands with nitrogen results in increased growth. Researchers infer from these results that plant growth is limited by the availability of nitrogen.

Organisms must have nitrogen to synthesize proteins and nucleic acids. Although molecular nitrogen (N_2) is extremely abundant in the atmosphere, most organisms cannot use it. To incorporate nitrogen atoms into amino acids and nucleotides, all eukaryotes and many bacteria and archaea have to obtain N in a form such as ammonia (NH_3) or nitrate (NO_3^-).

The only organisms that are capable of converting molecular nitrogen to ammonia are bacteria. The steps in the process, called **nitrogen fixation**, are complex and highly endergonic reduction-oxidation (redox) reactions (see Chapter 9). The enzymes required to accomplish nitrogen fixation are found only in selected bacterial lineages. Certain species of cyanobacteria that live in surface waters of the ocean or in association with water plants are capable of fixing nitrogen. In terrestrial environments, nitrogen-fixing bacteria that are not cyanobacteria live in close association with plants—often taking up residence in special root structures called nodules (**Figure 28.6**).

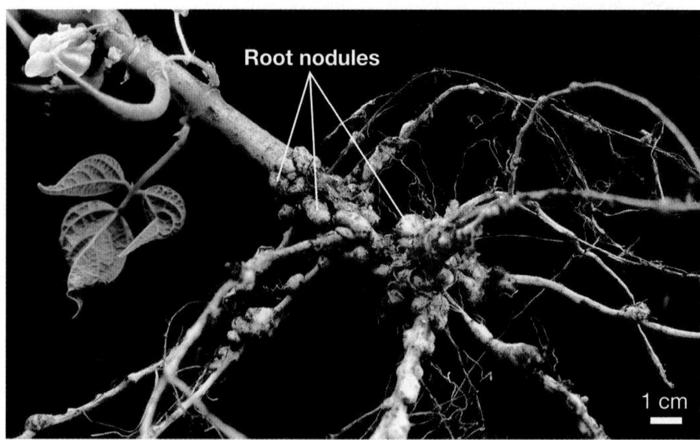

FIGURE 28.6 Some Nitrogen-Fixing Bacteria Live in Association with Plants. Root nodules form a protective structure for bacteria that fix nitrogen.

If bacteria could not fix nitrogen, it is virtually certain that only a tiny fraction of life on Earth would exist today. Large or multicellular organisms would probably be rare to nonexistent, because too little nitrogen would be available to make large quantities of proteins and build a large body.

Nitrate Pollution Corn, rice, wheat, and many other crop plants do not live in association with nitrogen-fixing bacteria. To increase yields of these crops, farmers use fertilizers that are high in nitrogen. In parts of the world, massive additions of nitrogen in the form of ammonia are causing serious pollution problems. **Figure 28.7** shows why. When ammonia is added to a cornfield—in midwestern North America, for example—much of it never reaches the growing corn plants. Instead, a significant fraction of the ammonia molecules is used as food by bacteria in the soil. Bacteria that use ammonia as an electron donor to fuel cellular respiration release nitrite (NO_2^-) as a waste product. Other bacteria use nitrite as an electron donor and release nitrate (NO_3^-). Nitrate molecules are extremely soluble in water and tend to be washed out of soils into groundwater or streams. From there they eventually reach the ocean, where they can cause pollution.

To understand why nitrates can pollute the oceans, consider the Gulf of Mexico. Nitrates carried by the Mississippi River are used as a nutrient by cyanobacteria and algae that live in the Gulf. These cells explode in numbers in response. When they die and sink to the bottom of the Gulf, bacteria and archaea and other decomposers use them as food. The decomposers use so much oxygen as an electron acceptor in cellular respiration that oxygen levels in the sediments and even in Gulf waters decline. Nitrate pollution has been so

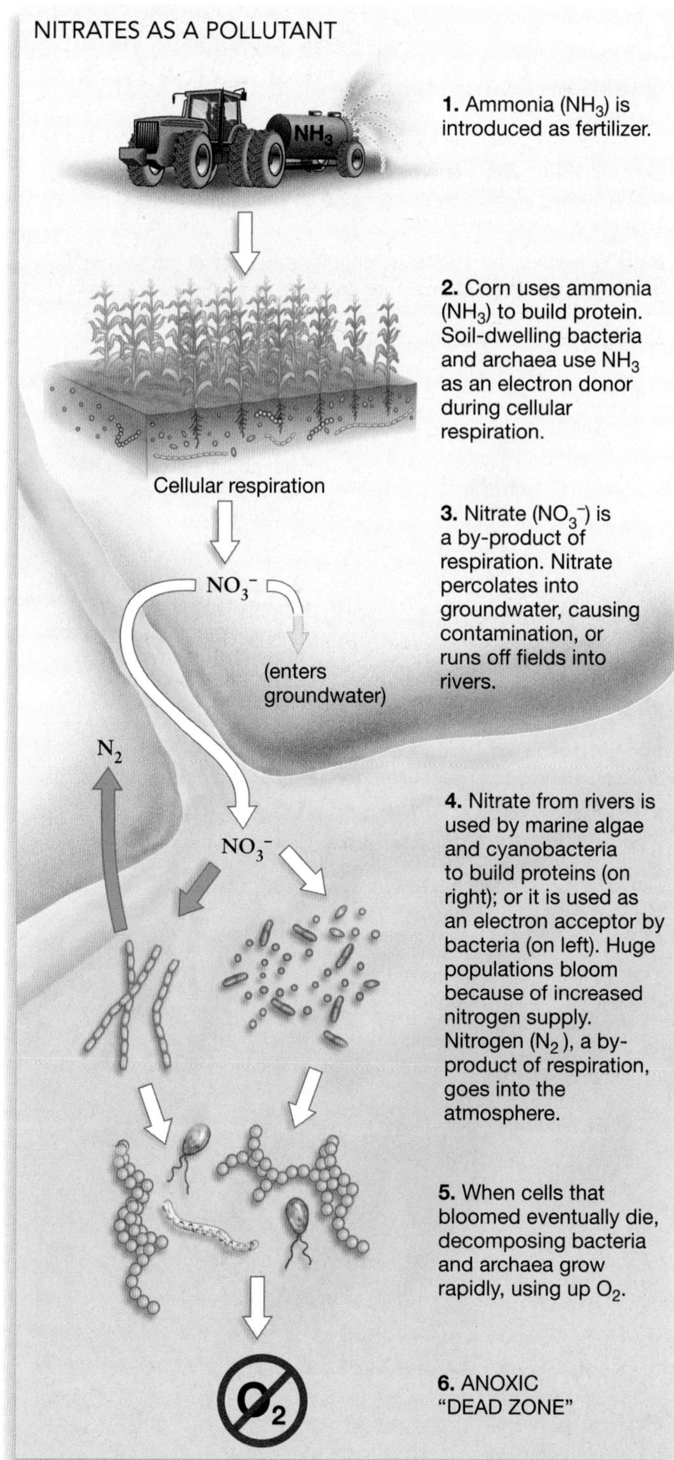

FIGURE 28.7 Nitrates Act as a Pollutant in Aquatic Ecosystems.

severe that large areas in the Gulf of Mexico are anoxic (lacking in oxygen). The oxygen-free "dead zone" in the Gulf of Mexico is devoid of fish, shrimp, and other organisms that require oxygen. Lately, the dead zone has encompassed about 18,000 km^2—roughly the size of New Jersey. Similar problems are cropping up in other parts of the world. Virtually every link in the chain of events leading to nitrate pollution involves bacteria and archaea.

The general message of this section is simple: Bacteria and Archaea may be small in size, but because of their abundance, ubiquity, and ability to do sophisticated chemistry, they have an enormous influence on the global environment.

28.2 How Do Biologists Study Bacteria and Archaea?

Biologists who study bacteria and archaea claim that they are investigating some of the most interesting and important organisms on Earth. Bacteria have been important model organisms in biological science for many decades (see **Box 28.1** on page 576). It is legitimate to state, however, that our understanding of the domains Bacteria and Archaea is advancing more rapidly right now than at any time during the past 100 years—and perhaps faster than our understanding of any other lineages on the tree of life.

As an introduction to the domains Bacteria and Archaea, let's examine a few of the techniques that biologists use to answer questions about them. Some of these research strategies have been used since bacteria were first discovered; some were invented less than 10 years ago.

Using Enrichment Cultures

Which species of bacteria and archaea are present at a particular location, and what do they use as food? To answer questions like these, biologists rely heavily on their ability to culture organisms in the lab. Of the 5000 species of bacteria and archaea that have been described to date, almost all were discovered when they were isolated from natural habitats and grown under controlled conditions in the laboratory.

One classical technique for isolating new types of bacteria and archaea is called **enrichment culture**. Enrichment cultures are based on establishing a specified set of growing conditions—temperature, lighting, substrate, types of available food, and so on. The idea is to sample cells from the environment and grow them under extremely specific conditions. Cells that thrive under the specified conditions will increase in numbers enough to be isolated and studied in detail.

To appreciate how this strategy works in practice, consider research on bacteria that live deep below Earth's surface. One recent study began with samples of rock and fluid from drilling operations in Virginia and Colorado. The samples came from sedimentary rocks at depths ranging from 860 to 2800 meters below the surface, where temperatures are between 42°C and 85°C. The questions posed in the study were simple: Is anything alive down there? If so, what do the organisms use to fuel cellular respiration?

The research team hypothesized that if organisms were living deep below the surface of the Earth, the cells might use hydrogen molecules (H_2) as an electron donor and the ferric ion (Fe^{3+}) as an electron acceptor (**Figure 28.8**). (Recall from Chapter 9 that most eukaryotes use sugars as electron donors and use oxygen as an electron acceptor during cellular respiration.) Fe^{3+} is the oxidized form of iron, and it is abundant in the rocks the biologists collected from great depths. It exists at great depths below the surface in the form of ferric oxyhydroxide. The researchers predicted that if an organism in the samples reduced the ferric ions during cellular respiration, a black, oxidized, and magnetic mineral called magnetite (Fe_3O_4) would start appearing in the cultures as a by-product of cellular respiration.

What did their enrichment cultures produce? In some culture tubes, a black compound began to appear within a week. Using a variety of tests, the biologists confirmed that the black substance was indeed magnetite. As the Results section of Figure 28.8 shows, microscopy revealed the organisms themselves—previously undiscovered bacteria. Because they grow only when incubated at between 45°C and 75°C, these organisms are considered **thermophiles** ("heat-lovers"). The discovery was spectacular, because it hinted that Earth's crust may be teeming with organisms to depths of over a mile below the surface. Enrichment culture continues to be a productive way to isolate and characterize new species of bacteria and archaea.

Using Direct Sequencing

Researchers estimate that of all the bacteria and archaea living today, less than 1 percent have been grown in culture. To augment research based on enrichment cultures, researchers are employing a technique called direct sequencing. **Direct sequencing** is a strategy for documenting the presence of bacteria and archaea that cannot be grown in culture. It is based on identifying phylogenetic species—populations that have enough distinctive characteristics to represent an independent twig on an evolutionary tree (see Chapter 26).

Direct sequencing allows biologists to identify and characterize organisms that have never been seen. The technique has revealed huge new branches on the tree of life and produced revolutionary data on the habitats where archaea are found.

Figure 28.9 outlines the steps performed in a direct sequencing study. To begin, researchers collect a sample from a habitat—often seawater or soil. Next, the cells in the sample are broken open (lysed), and their DNA is purified. Using the polymerase chain reaction introduced in Chapter 19, researchers can isolate specific genes from the species in the sample. After sequencing these genes, biologists compare the data with sequences in existing databases. If the sequences are markedly different, the sample probably contains previously undiscovered organisms.

Direct sequencing studies have produced new and sometimes startling results. For two decades after the discovery of the

Experiment

Question: Can bacteria live a mile below Earth's surface?

Hypothesis: Bacteria are capable of cellular respiration deep below Earth's surface by using H_2 as an electron donor and Fe^{3+} as an electron acceptor.

Null hypothesis: Bacteria from this environment are not capable of using H_2 as an electron donor and Fe^{3+} as an electron acceptor.

Experimental setup:

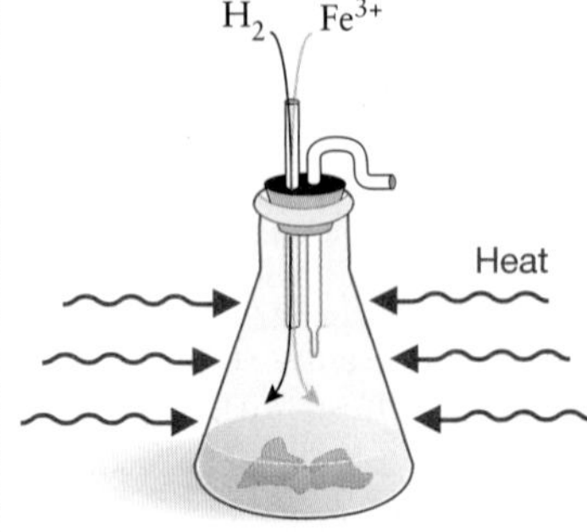

Enrichment culture method:

1. Create culture conditions with abundant H_2 and Fe^{3+}; raise temperatures above 45°C.

2. Add rock and fluid samples extracted from drilling operations at depths of about 1000 m below Earth's surface.

Prediction: Black, magnetic grains of magnetite (Fe_3O_4) will accumulate because Fe^{3+} is reduced by growing cells and shed as waste product. Cells will be visible.

Prediction of null hypothesis: No magnetite will appear. No cells will grow.

Results:

Conclusion: At least one bacterial species that can live deep below Earth's surface grew in this enrichment culture. Different culture conditions might result in the enrichment of different species present in the same sample.

FIGURE 28.8 Enrichment Cultures Isolate Large Populations of Cells That Grow under Specific Conditions. Bacteria that grow underground can withstand high temperatures. The species enriched in this experiment needs hydrogen (H_2) and ferric ions (Fe^{3+}) to live.

QUESTION How could the researchers determine to which bacterial lineage this newly discovered species belongs?

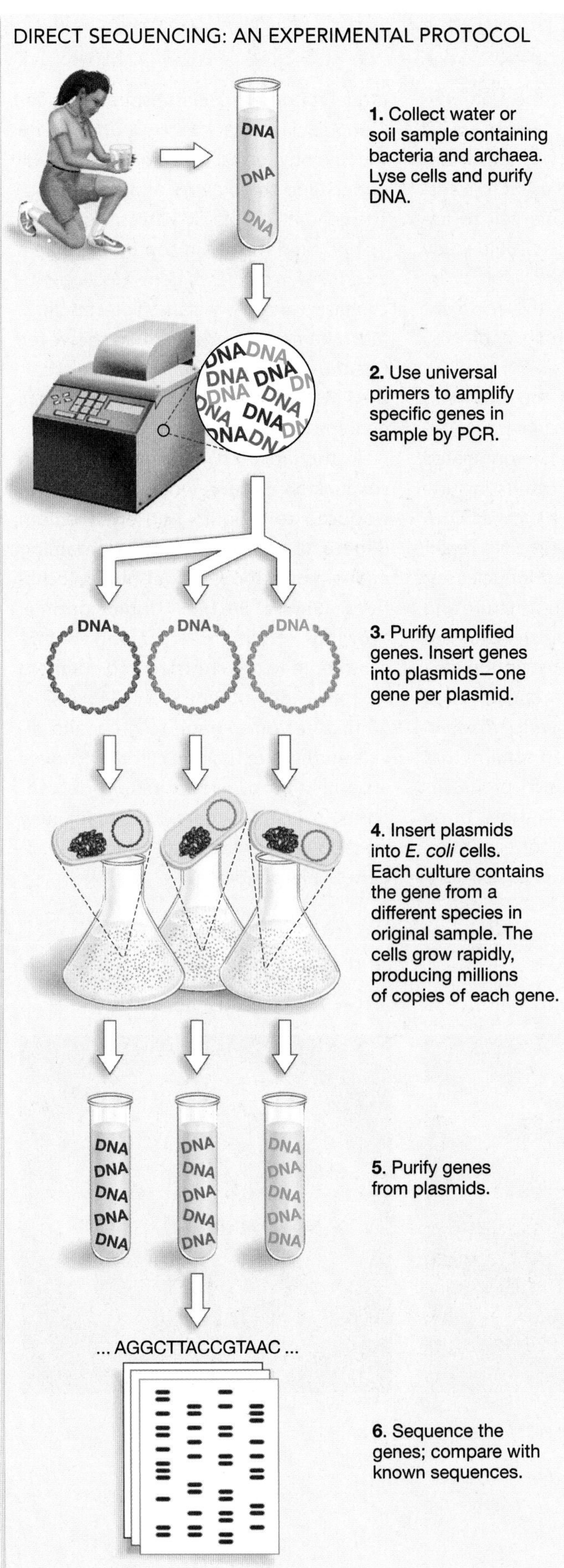

FIGURE 28.9 **Direct Sequencing Allows Researchers to Identify Species That Have Never Been Seen.** Direct sequencing allows biologists to isolate specific genes from the organisms present in a sample. The polymerase chain reaction is used to generate enough copies of these genes so that the DNA from different species can be sequenced.

Archaea, for example, researchers thought that these organisms could be conveniently grouped into just four categories: extreme halophiles, sulfate-reducers, methanogens, and extreme thermophiles. Extreme **halophiles** ("salt-lovers") live in salt lakes, salt ponds, and salty soils. **Sulfate reducers** are species that produce hydrogen sulfide (H_2S) as a by-product of cellular respiration. (H_2S may be familiar because it smells like rotten eggs.) **Methanogens** produce methane (natural gas; CH_4) as a by-product of cellular respiration. Extreme **thermophiles** grow best at temperatures above 80°C. Based on these early data, researchers thought that archaea were restricted to extreme environments and that the four phenotypes corresponded to separate lineages within the domain Archaea. As a result of direct sequencing studies, however, these generalizations have been discarded. Beginning in the mid-1990s, direct sequencing revealed archaea in habitats as diverse as rice paddies and the Arctic Ocean. Some of these newly discovered organisms appear to belong to entirely new lineages, tentatively called the **Korarchaeota** and **Nanoarchaeota**. These species' DNA sequences are so distinctive that they might represent "kingdoms" analogous to plants or animals. Yet both groups were identified and named before any of their members had actually been observed. Follow-up research is focused on confirming these initial results. Thanks to studies like these, direct sequencing is revolutionizing our understanding of bacterial and archaeal diversity.

Evaluating Molecular Phylogenies

To put data from enrichment culture and direct sequencing studies into context, biologists depend on the accurate placement of species on phylogenetic trees. Recall from Chapter 1, Chapter 27, and **BioSkills 2** that phylogenetic trees illustrate the evolutionary relationships among species and lineages. They are a pictorial summary of which species are more closely or distantly related to others.

Some of the most useful phylogenetic trees for the Bacteria and the Archaea have been based on studies of the RNA molecule found in the small subunit of ribosomes, or what biologists call SSU RNA. (See Chapter 16 for more information on the structure and function of ribosomes.) In the late 1960s Carl Woese and colleagues began a massive effort to determine and compare the base sequences of SSU RNA molecules from a wide array of species. The result of their analysis was the **universal tree**, or the **tree of life**, illustrated in Figure 28.1.

BOX 28.1 A Model Organism: *Escherichia coli*

Of all model organisms in biology, perhaps none has been more important than the bacterium *Escherichia coli*—a common inhabitant of the human gut. The strain that is most commonly worked on today, called K-12 (**Figure 28.10a**), was originally isolated from a hospital patient in 1922.

Escherichia coli K-12 did not begin to rise to its present prominence as a model organism until 1945, however. In that year, biologists observed that some K-12 cells were capable of a process called conjugation. Like all bacteria and archaea, K-12 and other strains of *E. coli* reproduce by **fission**—the splitting of a cell into two daughter cells. DNA replication precedes fission, so the daughter cells are genetically identical. But as Chapter 12 noted, some bacterial cells are also capable of transferring copies of the extracellular loops of DNA called **plasmids**. When **conjugation** occurs, a copy of a plasmid from one bacterial cell—sometimes joined by one or more genes from the main bacterial chromosome—is transferred to a recipient cell through a structure called a conjugation tube. Conjugation is sometimes referred to as bacterial sex. Its discovery in K-12 was important because researchers could use it to transfer specific alleles from one cell to another. For example, biologists could study the function of specific genes in K-12 by documenting changes in the recipient cell's phenotype once the transfer process was complete.

During the last half of the twentieth century, key results in molecular biology—introduced in Chapters 14–19—originated in studies of *E. coli*. These results include the discovery of enzymes such as DNA polymerase, RNA polymerase, DNA repair enzymes, and restriction endonucleases; the elucidation of ribosome structure and function; and the initial characterization of promoters, regulatory transcription factors, regulatory sites in DNA, and operons. In many cases, initial discoveries made in *E. coli* allowed researchers to confirm that homologous enzymes and processes existed in an array of organisms, often ranging from other bacteria to yeast, mice, and humans. The success of *E. coli* as a model for other species inspired Jacques Monod's claim that "Once we understand the biology of *Escherichia coli*, we will understand the biology of an elephant." The genome of *E. coli* K-12 was sequenced in 1997, and the strain continues to be a workhorse in studies of gene function, biochemistry, and, particularly, biotechnology. Much remains to be learned, however. Despite over 60 years of intensive study, the function of about a third of the *E. coli* genome is still unknown.

In the lab, *E. coli* is usually grown in **suspension culture**, where cells are introduced to a liquid nutrient medium (**Figure 28.10b**), or on plates containing **agar**—a gelatinous mix of polysacccharides (**Figure 28.10c**). Under optimal growing conditions—meaning before cells begin to get crowded and compete for space and nutrients—a cell takes just 30 minutes on average to grow and divide. At this rate, a single cell can produce a population of over a million descendants in just 10 hours. Except for new mutations, all of the descendant cells are genetically identical.

(a) *Escherichia coli*, strain K-12

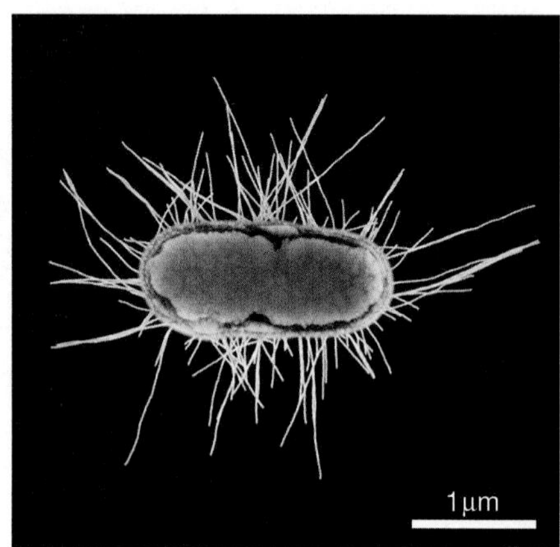

(b) Growth in liquid medium

(c) Growth on solid medium

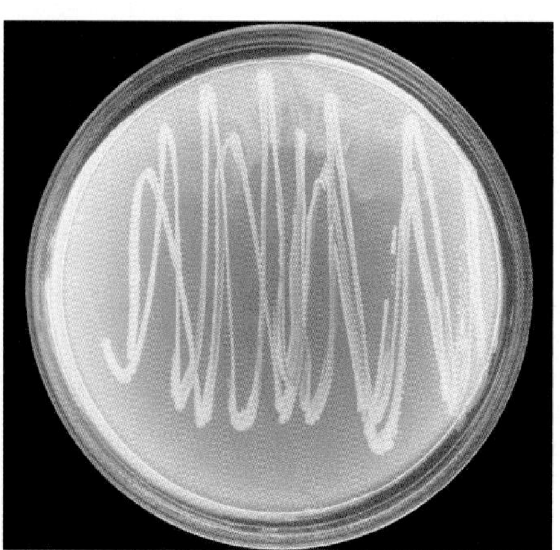

FIGURE 28.10 *E. coli* Is Readily Cultured in the Laboratory.

QUESTION *Escherichia coli* is grown at a temperature of 37°C. Why?

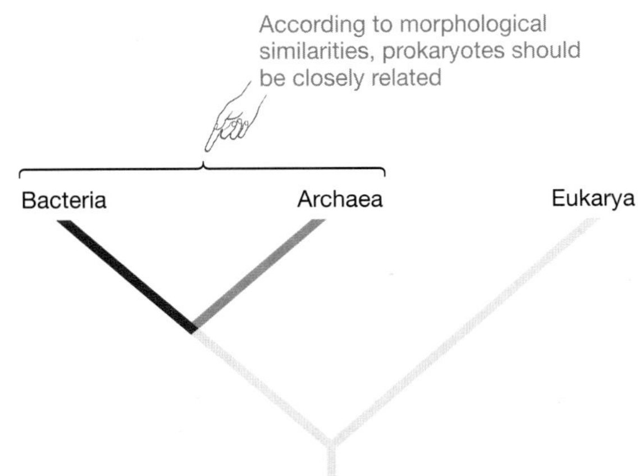

FIGURE 28.11 The Tree of Life Based on Morphology Was Incorrect. Until recently, the major division among organisms was thought to be between those without a membrane-bound nucleus (the prokaryotes) and those with a membrane-bound nucleus (the eukaryotes). Comparisons of RNA sequences have shown that this tree is not correct.

● **EXERCISE:** Draw the correct tree next to this one. Check your sketch against Figure 28.1. On the correct and incorrect trees, indicate where the nuclear envelope evolved.

Woese's tree is now considered a classic result. Prior to its publication, biologists thought that the major division among organisms was between prokaryotes and eukaryotes—between cells that lacked a membrane-bound nucleus and cells that possessed a membrane-bound nucleus (**Figure 28.11**). But based on data from the ribosomal RNA molecule, the major divisions of life-forms are the three groups that Woese named the Bacteria, Archaea, and Eukarya. Follow-up work documented that Bacteria were the first of the three lineages to diverge from the common ancestor of all living organisms. This result means that the Archaea and Eukarya are more closely related to each other than they are to the Bacteria.

Although virtually all biologists accept the three-domains system, understanding the relationships of major lineages within the Bacteria and Archaea has proven difficult. Analyses of both morphological and molecular characteristics have succeeded in identifying a large series of monophyletic groups within the domains. Recall from Chapter 27 that a **monophyletic group** consists of an ancestral population and all of its descendants. Monophyletic groups can also be called clades, or lineages. Rapid progress is being made on the question of how the major clades within Bacteria and Archaea are related to each other, and work continues. The phylogenetic tree in **Figure 28.12** summarizes recent results but is still considered highly provisional. As more data become available, it is almost certain that at least some of the branches on this tree will change position. In addition, recall that direct sequencing studies have recently led to the discovery of major new groups. If the existence of these groups is confirmed, entirely new branches will be added to this tree. Work on molecular phylogenies continues at a brisk pace.

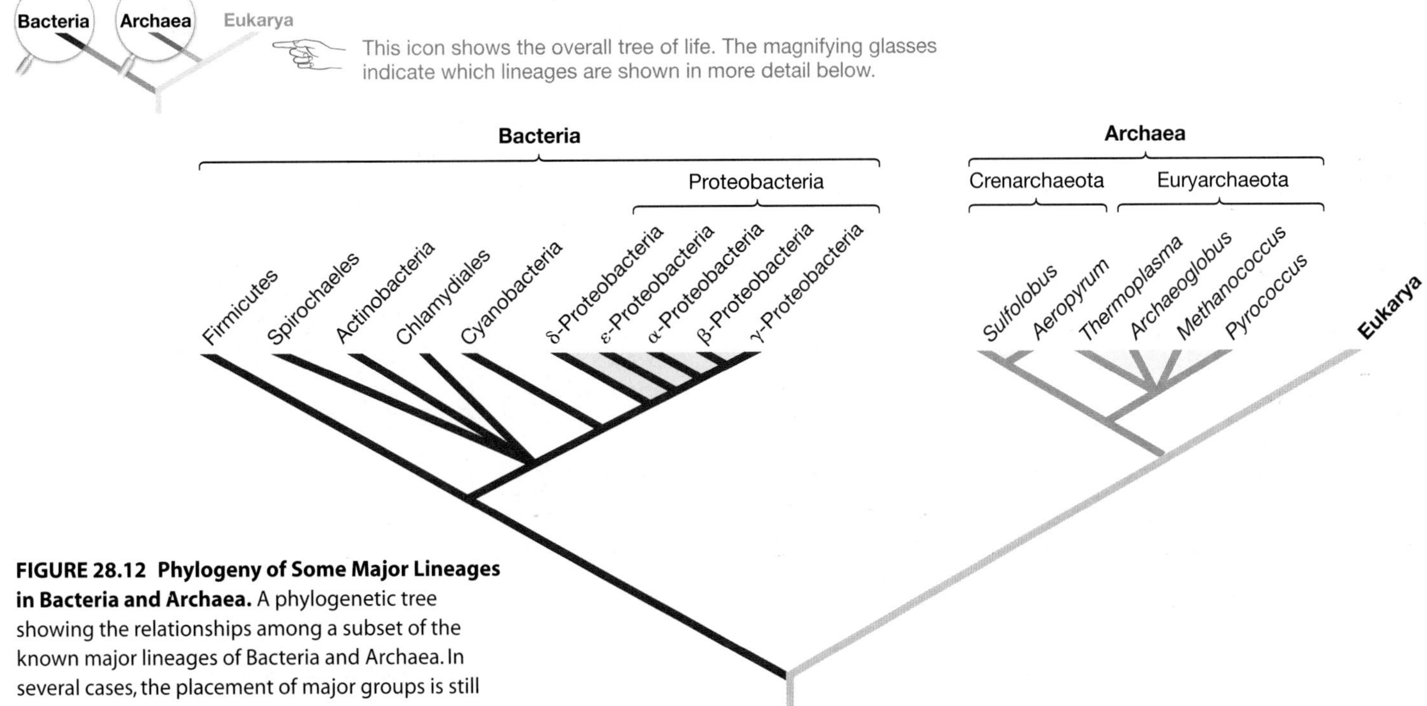

FIGURE 28.12 Phylogeny of Some Major Lineages in Bacteria and Archaea. A phylogenetic tree showing the relationships among a subset of the known major lineages of Bacteria and Archaea. In several cases, the placement of major groups is still highly uncertain.

Check Your Understanding

If you understand that...

- Enrichment cultures are based on setting up specified conditions in the laboratory and isolating the cells that grow rapidly in response. They create an abundant, pure sample of bacteria that thrive under particular conditions and can be studied further.
- Direct sequencing is based on isolating DNA from samples taken directly from the environment, purifying and sequencing specific genes, and then analyzing where those DNA sequences are found on the phylogenetic tree of Bacteria and Archaea.

You should be able to...

1) Design an enrichment culture that would isolate species that could be used to clean up oil spills.
2) Outline a study designed to identify the bacterial and archaeal species present in a soil sample near the biology building on your campus.

at www.masteringbio.com
The Tree of Life

28.3 What Themes Occur in the Diversification of Bacteria and Archaea?

Initially, the diversity of bacteria and archaea can seem almost overwhelming. To make sense of the variation among lineages and species, biologists focus on two themes: diversification in morphology and metabolism. Regarding metabolism, the key question is which molecules are used as food. Bacteria and archaea are capable of living in a wide array of environments because they vary in cell structure and in how they make a living.

Morphological Diversity

Because we humans are so large, it is hard for us to appreciate the morphological diversity that exists among bacteria and archaea. To us, they all look small and similar. But at the scale of a bacterium or archaean, different species are wildly diverse in morphology. For example, bacteria and archaea range in size from the smallest of all free-living cells—bacteria called mycoplasmas with volumes as small as 0.03 μm^3—to the largest bacterium known, *Thiomargarita namibiensis*, with volumes as large as 200×10^6 μm^3. Over a billion *Mycoplasma* cells could fit inside an individual *Thiomargarita* (**Figure 28.13a**). Bacteria

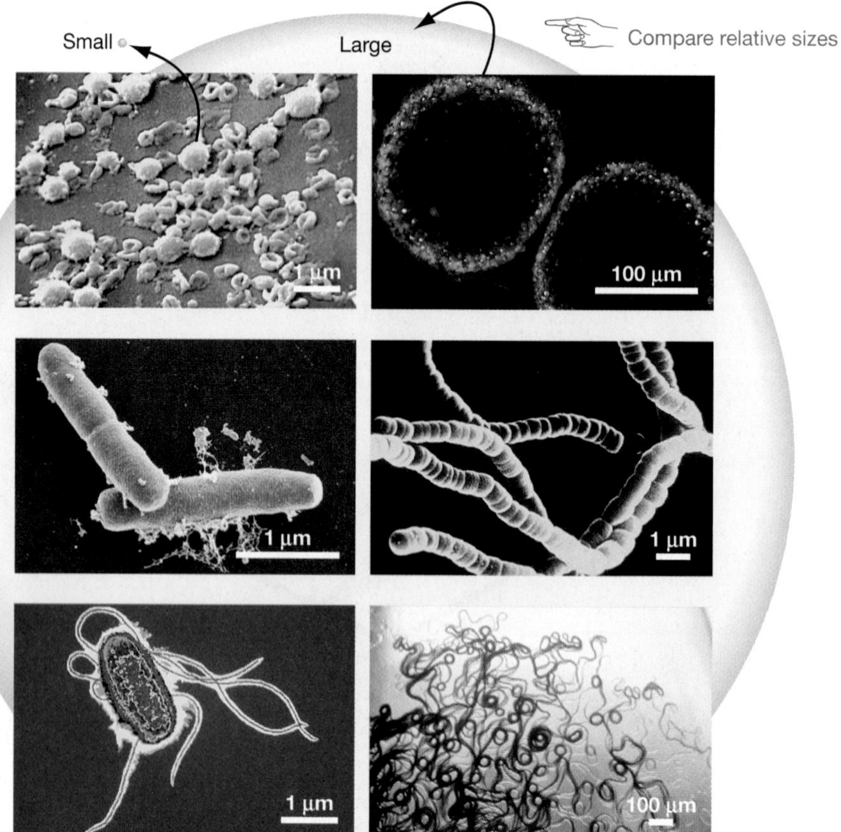

(a) Size varies
The sizes of bacteria and archaea vary. *Mycoplasma* cells (left) are about 0.5 μm in diameter, while *Thiomargarita namibiensis* cells (right) are about 150 μm in diameter.

(b) Shape varies
The shapes of bacteria and archaea vary from rods such as *Bacillus anthracis* (left) and spheres to filaments or spirals such as *Rhodospirillum*. In some species, such as *Streptococcus faecalis* (right), cells attach to one another and form chains.

(c) Mobility varies
A wide variety of bacteria and archaea use flagella (left) to power swimming movements. These cyanobacterial cells (right) move by gliding across a substrate.

FIGURE 28.13 Morphological Diversity among Bacteria and Archaea Is Extensive.

and archaea exhibit a variety of shapes as well, including filaments, spheres, rods, chains, and spirals (**Figure 28.13b**). Many cells are motile, with swimming movements powered by flagella (**Figure 28.13c**). Some cells can swim 10 more body lengths per second—much faster than any human can sprint. Although gliding movement occurs in several groups, the molecular mechanism responsible for this form of motility is still unknown.

At this scale, the composition of the plasma membrane and cell wall are particularly important. The introduction to this chapter highlighted the dramatic differences between the plasma membranes and cells walls of bacteria versus archaea. And within bacteria, two general types of cell wall exist. These types can be readily distinguished because they react differently when treated with a dye called the **Gram stain**. As **Figure 28.14a** shows, Gram-positive cells look purple but Gram-negative cells look pink. At the molecular level, cells that are **Gram-positive** have a plasma membrane surrounded by a cell wall with extensive peptidoglycan (**Figure 28.14b**). You might recall from Chapter 5 that peptidoglycan is a complex substance composed of cross-linked carbohydrate strands. Cells that are **Gram-negative**, in contrast, have a plasma membrane surrounded by a cell wall that has two components—a thin gelatinous layer containing peptidoglycan, and an outer phospholipid bilayer (**Figure 28.14c**).

To summarize, members of the Bacteria and the Archaea are remarkably diverse in their overall size, shape, and motility, as well as in the composition of their cell walls and plasma membranes. But when asked to name the innovations that were most responsible for the diversification of these two domains, biologists do not point to their morphological diversity. Instead, they point to metabolic diversity—variation in the chemical reactions that go on inside these cells. The most important thing to remember about bacteria and archaea is how diverse they are in the types of compounds they can use as food.

Metabolic Diversity

Bacteria and archaea are the masters of metabolism. Taken together, they can use almost anything as food—from hydrogen molecules to crude oil. Bacteria and archaea look small and relatively simple to us in their morphology, but their biochemical capabilities are dazzling.

Just how varied are bacteria and archaea when it comes to making a living? To appreciate the answer, recall from Chapters 9 and 10 that organisms have two fundamental nutritional needs—acquiring chemical energy in the form of adenosine triphosphate (ATP) and obtaining molecules with carbon-carbon bonds that can be used as building blocks for the synthesis of fatty acids, proteins, DNA, RNA, and other large, complex compounds required by the cell.

(a) Gram-positive cells retain Gram stain more than Gram-negative cells do.

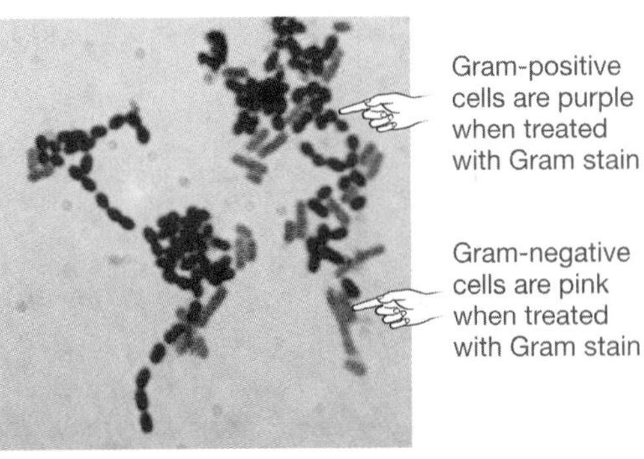

(b) Cell walls in Gram-positive bacteria have extensive peptidoglycan.

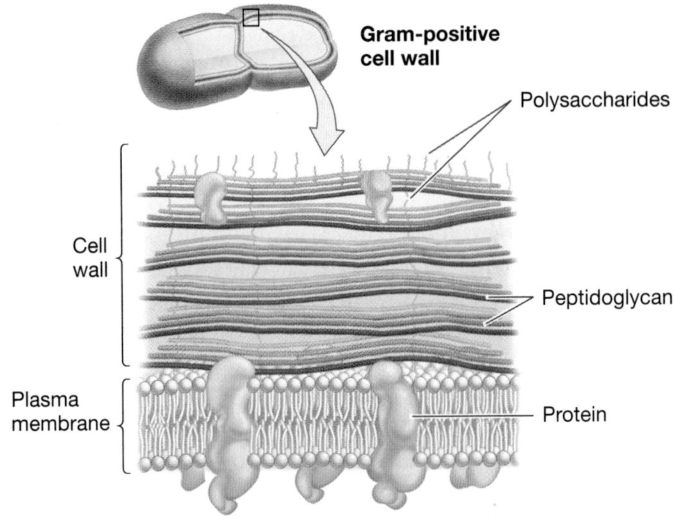

(c) Cell walls in Gram-negative bacteria have some peptidoglycan and an outer membrane.

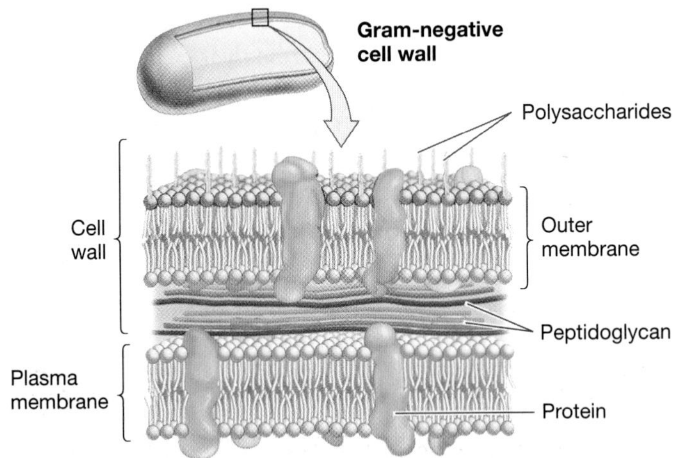

FIGURE 28.14 Gram Staining Distinguishes Two Types of Cell Walls in Bacteria. (a) When treated with the Gram stain, some cells retain a large amount of stain and look purple while other cells retain little stain and look pink. The amount of Gram stain retained is high in cells with walls that contain a large amount of peptidoglycan **(b)** and small in cells that contain little peptidoglycan but have an outer membrane layer **(c)**.

SUMMARY TABLE **28.3** **Six General Methods for Obtaining Energy and Carbon**

	Source of Building-Block Compounds with Carbon-Carbon Bonds (for synthesis of complex organic compounds)	
Source of Energy (for ATP production)	***Autotrophs*** (Synthesize their own building-block compounds from CO_2, CH_4, or other simple molecules.)	***Heterotrophs*** (Use organic compounds with high potential energy produced by other organisms.)
Light (phototrophs)	Photoautotrophs	Photoheterotrophs
Organic Molecules with High Potential Energy (organotrophs)	***Chemoorganoautotrophs***	***Chemoorganoheterotrophs***
Inorganic Molecules with High Potential Energy **(lithotrophs)**	***Chemolithotrophs***	***Chemolithotrophic heterotrophs***

Bacteria and archaea produce ATP in three ways:

1. **Phototrophs** ("light-feeders") use light energy to promote electrons to the top of electron transport chains. ATP is produced by photophosphorylation (see Chapter 10).
2. **Chemoorganotrophs** oxidize organic molecules with high potential energy, such as sugars. ATP may be produced by cellular respiration—with sugars serving as electron donors—or via fermentation pathways (see Chapter 9).
3. **Chemolithotrophs** ("rock-feeders") oxidize inorganic molecules with high potential energy, such as ammonia (NH_3) or methane (CH_4). ATP is produced by cellular respiration, with inorganic compounds serving as the electron donor.

Bacteria and archaea fulfill their second nutritional need—obtaining building block compounds with carbon-carbon bonds—in two ways: (1) by synthesizing their own from simple starting materials such CO_2 and CH_4, or (2) by absorbing ready-to-use organic compounds from their environment. Organisms that manufacture their own building-block compounds are termed **autotrophs** ("self-feeders"). Organisms that acquire building-block compounds from other organisms are called **heterotrophs** ("other-feeders").

Because there are three distinct ways of producing ATP and two general mechanisms for obtaining carbon, there are a total of six methods for producing ATP and obtaining carbon. The names that biologists use for organisms that use these six "feeding strategies" are given in **Table 28.3**. Of the six possible ways of producing ATP and obtaining carbon, just two are observed in eukaryotes. But bacteria and archaea do them all. In their metabolism, eukaryotes are simple compared with bacteria and archaea. ● If you understand the essence of metabolic diversity in bacteria and archaea, you should be able to match the six example species described in **Table 28.4** to the appropriate category in Table 28.3.

What makes this remarkable diversity possible? Bacteria and archaea have evolved dozens of variations on the basic processes you learned about in Chapters 9 and 10. They use compounds with high potential energy to produce ATP via cellular respiration (electron transport chains) or fermentation, they use light to produce high-energy electrons, and they reduce carbon from CO_2 or other sources to produce sugars or

TABLE **28.4** **Examples of Metabolic Diversity in Bacteria and Archaea**

Bacteria/Archaea	**How ATP Is Produced**	**How Building-Block Molecules Are Synthesized**
Cyanobacteria	via photosynthesis	from CO_2 via the Calvin cycle
Clostridium aceticum	fermentation of glucose	from CO_2 via reactions called the acetyl-CoA pathway
Nitrifying bacteria (e.g., *Nitrosomonas* sp.)	via cellular respiration, using ammonia (NH_3) as an electron donor	from CO_2 via the Calvin cycle
Heliobacteria	via photosynthesis	absorb carbon-containing building-block molecules from the environment
Escherichia coli	fermentation of organic compounds or cellular respiration, using organic compounds as electron donors	absorb carbon-containing building-block molecules from the environment
Beggiatoa	via cellular respiration, using hydrogen sulfide (H_2S) as an electron donor	absorb carbon-containing building-block molecules from the environment

other building-block molecules with carbon-carbon bonds. The story of bacteria and archaea can be boiled down to two sentences: The basic chemistry required for photosynthesis, cellular respiration, and fermentation originated in these lineages. Then the evolution of variations on each of these processes allowed prokaryotes to diversify into millions of species that occupy diverse habitats. Let's take a closer look.

Producing ATP via Photosynthesis: Variation in Electron Sources and Pigments Instead of using molecules as a source of high-energy electrons, phototrophs pursue a radically different strategy: **photosynthesis**. Phototrophs use the energy in light to raise electrons to high-energy states. As these electrons are stepped down to lower energy states by electron transport chains, the energy released is used to generate ATP.

Chapter 10 introduced an important feature of photosynthesis: The process requires a source of electrons. Recall that in cyanobacteria and plants, the required electrons come from water. When these organisms "split" water molecules apart to obtain electrons, they generate oxygen as a by-product. Species that use water as a source of electrons for photosynthesis are said to complete **oxygenic** photosynthesis. In contrast, many phototrophic bacteria use a molecule other than water as the source of electrons. In many cases, the electron donor is hydrogen sulfide (H_2S); a few species can use the ion known as ferrous iron (Fe^{2+}). Instead of producing oxygen as a by-product of photosynthesis, these cells produce elemental sulfur (S) or the ferric ion (Fe^{3+}). They are said to complete **anoxygenic** photosynthesis and live in habitats where oxygen is rare.

Chapter 10 also introduced the photosynthetic pigments found in plants and explored the light-absorbing properties of chlorophylls *a* and *b*. Cyanobacteria have these two pigments. But researchers have isolated seven additional chlorophylls from bacterial phototrophs. Each major group of photosynthetic bacteria has one or more of these distinctive chlorophylls, and each type of chlorophyll absorbs light best at a different wavelength.

Why are bacterial chlorophylls so diverse? The leading hypothesis is that photosynthetic species with different absorption spectra are able to live together without competing for light. If so, then the diversity of photosynthetic pigments observed in bacteria has been an important mechanism for generating species diversity among phototrophs.

Producing ATP via Cellular Respiration: Variation in Electron Donors and Acceptors Millions of bacterial, archaeal, and eukaryotic species—including animals and plants—are organotrophs. These organisms obtain the energy required to make ATP by oxidizing organic compounds such as sugars, starch, or fatty acids. As Chapter 9 showed, cellular enzymes can strip electrons from organic molecules that have high potential energy and then transfer these high-energy electrons to the electron carriers NADH and $FADH_2$. These compounds feed electrons to an electron transport chain (ETC), where electrons are stepped down from a high-energy state to a low-energy state (**Figure 28.15a**). The energy that is released allows components of the ETC to generate a proton gradient across the plasma membrane (**Figure 28.15b**). The resulting flow of protons through the enzyme ATP synthase results in the production of ATP, via the process called chemiosmosis.

(a) Electrons pass from an electron donor through an electron transport chain, to a final electron acceptor.

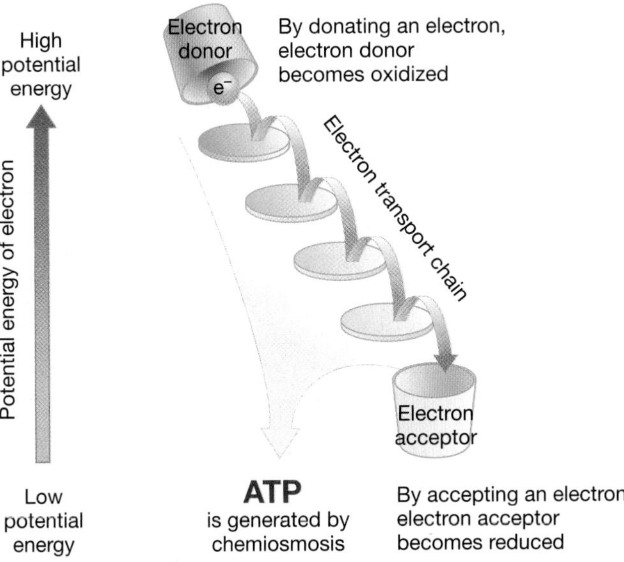

(b) The electron transport chain is organized in a series of multi-molecular complexes in the plasma membrane.

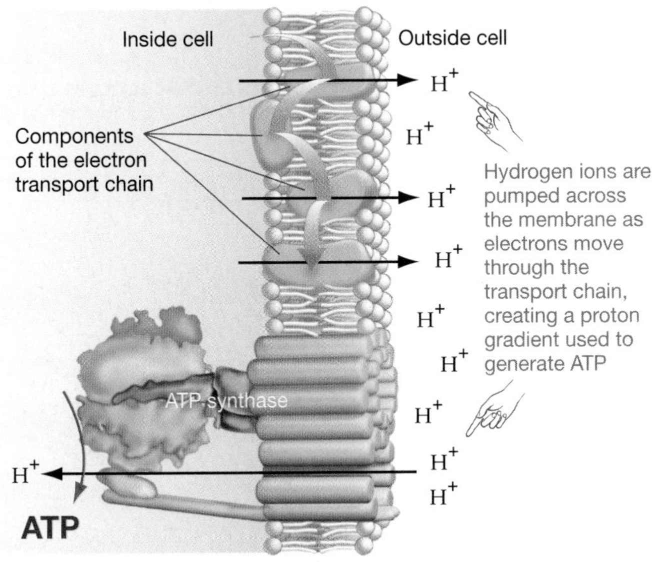

FIGURE 28.15 Cellular Respiration Is Based on Electron Transport Chains.

EXERCISE In part (a), add the chemical formula for a specific electron donor, electron acceptor, and reduced by-product for a species of bacteria or archaea. Then write in the electron donor, electron acceptor, and reduced by-product observed in humans.

The essence of this process, called **cellular respiration**, is that a molecule with high potential energy serves as an original electron donor and is oxidized, while a molecule with low potential energy serves as a final electron acceptor and becomes reduced. Much of the potential energy difference between the electron donor and electron acceptor is transformed into chemical energy in the form of ATP.

In introducing respiration, Chapter 9 focused on the role of organic compounds with high potential energy—such as glucose—as the original electron donor and oxygen as the final electron acceptor (see Figure 9.22). Many bacteria and archaea, as well as all eukaryotes, rely on these molecules. When cellular respiration is complete, glucose is completely oxidized to CO_2, which is given off as a by-product. When oxygen acts as the final electron acceptor, water is also produced as a by-product.

However, many other bacteria and archaea employ an electron donor other than sugars and an electron acceptor other than oxygen during cellular respiration, and they produce by-products other than carbon dioxide and water. As **Table 28.5** shows, many bacteria and archaea are lithotrophs—they use inorganic ions or molecules as electron donors. The substances used as electron donors range from hydrogen molecules (H_2) and hydrogen sulfide (H_2S) to ammonia (NH_3) and methane (CH_4). And instead of using oxygen, some organisms use compounds with low potential energy—such as sulfate (SO_4^{2-}), nitrate (NO_3^-), carbon dioxide (CO_2), or ferric ions (Fe^{3+})—as electron acceptors. Instead of producing CO_2 and H_2O as by-products, these organisms might produce CH_4 or NO_3^-. It is only a slight exaggeration to claim that researchers have found bacterial and archaeal species that can use almost any compound with relatively high potential energy as an electron donor and almost any compound with relatively low potential energy as an electron acceptor.

Because the electron donors and electron acceptors used by bacteria and archaea are so diverse, one of the first questions biologists ask about a species is how it accomplishes cellular respiration. The best way to answer this question is through the enrichment culture technique introduced in Section 28.2. Recall that in an enrichment culture, researchers supply specific electron donors and electron acceptors in the medium and try to isolate cells that can use those compounds to support growth.

The remarkable metabolic diversity of bacteria and archaea is important. First, it explains their ecological diversity. Bacteria and archaea are found almost everywhere because they exploit an almost endless variety of molecules as electron donors and electron acceptors. Second, it explains why they play such a key role in cleaning up some types of pollution. Species that use organic solvents or petroleum-based fuels as electron donors or electron acceptors may be effective agents in bioremediation efforts if the waste products are less toxic than the solvents or fuels. Finally, metabolic diversity is what makes bacteria and archaea major players in global change. Nitrogen, phosphorus, sulfur, carbon, and other crucial nutrients cycle from one organism to another because bacteria and archaea can use them in almost any molecular form. The nitrite (NO_2^-) that some bacteria produce as a by-product of respiration does not build up in the environment, because it is used as an electron acceptor by other species and converted to molecular nitrate (NO_3^-). This molecule, in turn, is converted to molecular nitrogen (N_2) by yet another suite of bacterial and archaeal species. Bacteria can convert N_2 in the atmosphere to NH_3 in proteins (by bacterial or archaeal fixation and decomposition). Similar types of interactions occur with molecules that contain phosphorus, sulfur, and carbon. In this way, bacteria and archaea play a key role in the cycling of nutrients (**Figure 28.16**).

TABLE **28.5** **Some Electron Donors and Acceptors Used by Bacteria and Archaea**

		By-Products		
Electron Donor	**Electron Acceptor**	**From Electron Donor**	**From Electron Acceptor**	**Category***
Sugars	O_2	CO_2	H_2O	Organotrophs
H_2 or organic compounds	SO_4^{2-}	H_2O or CO	H_2S	Sulfate reducers
H_2	CO_2	H_2O	CH_4	Methanogens
CH_4	O_2	CO_2	H_2O	Methanotrophs
S^{2-} or H_2S	O_2	SO_4^{2-}	H_2O	Sulfur bacteria
Organic compounds	Fe^{3+}	CO_2	Fe^{2+}	Iron reducers
NH_3	O_2	NO_2^-	H_2O	Nitrifiers
Organic compounds	NO_3^-	CO_2	N_2O, NO, or N_2	Denitrifiers (or nitrate reducers)
NO_2^-	O_2	NO_3^-	H_2O	Nitrosifiers

*The name biologists use to identify species that use a particular metabolic strategy.

QUESTION Explain why the terms *organotrophs*, *sulfate reducers*, and *methanogens* are appropriate. (The word root *–gen* means source or origin; *–troph* refers to feeding.)

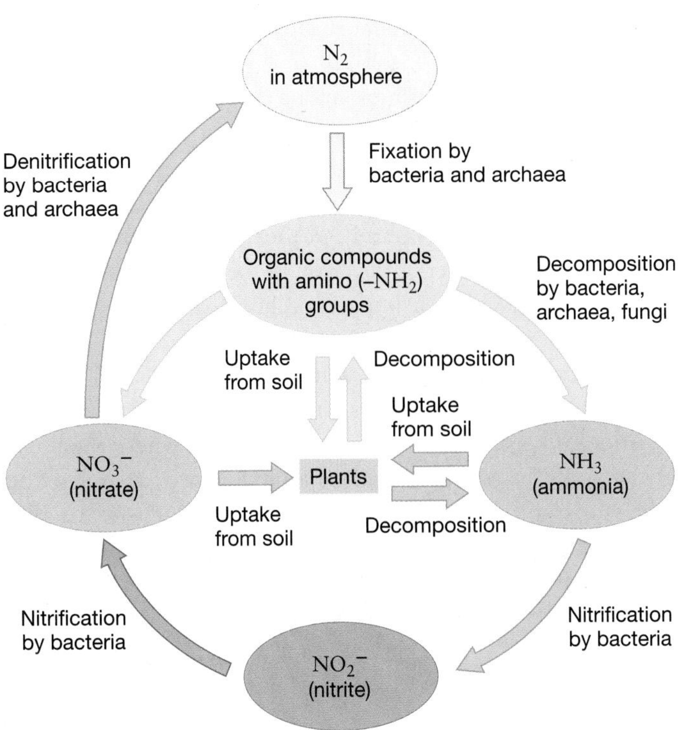

FIGURE 28.16 Nitrogen Atoms Cycle through Environments in Different Molecular Forms

● **EXERCISE** Suppose that bacteria and archaea were no longer capable of fixing nitrogen. Draw an X through the part(s) of the cycle that would be most directly affected.

● **EXERCISE** Add arrows and labels to indicate that animals ingest amino groups from plants or other animals and release amino groups or ammonia via excretion and decomposition.

Producing ATP via Fermentation: Variation in Substrates Chapter 9 introduced **fermentation** as a strategy for making ATP that does not involve electron transport chains. In fermentation, no outside electron acceptor is used, and the redox reactions required to produce ATP are internally balanced. Because fermentation is a much less efficient way to make ATP compared with cellular respiration, in many species it occurs as an alternative metabolic strategy when no electron acceptors are available to make cellular respiration possible. In other species, fermentation does not occur at all. In still other species, fermentation is the only way for cells to make ATP.

Although the presentation in Chapter 9 focused on how glucose is fermented to ethanol or lactic acid, some bacteria and archaea are capable of using other organic compounds as the starting point for fermentation. Bacteria and archaea that produce ATP via fermentation are still classified as organotrophs, but they are much more diverse in the substrates used. For example, the bacterium *Clostridium aceticum* can ferment ethanol, acetate, and fatty acids as well as glucose. Other species of *Clostridium* ferment complex carbohydrates (including cellulose or starch), proteins, amino acids, or even purines. Species that ferment amino acids produce by-products with names such as cadaverine and putrescine. These molecules are responsible for the odor of rotting flesh. Other bacteria can ferment lactose, a prominent component of milk. This fermentation has two end products: propionic acid and CO_2. Propionic acid is responsible for the taste of Swiss cheese; the CO_2 produced during fermentation creates the holes in cheese.

The diversity of enzymatic pathways observed in bacterial and archaeal fermentations extends the metabolic repertoire of these organisms and supports the claim that as a group, bacteria and archaea can use virtually any molecule with relatively high potential energy as a source of high-energy electrons for producing ATP. Given this diversity, it is no surprise that bacteria and archaea are found in such widely varying habitats. Different environments offer different energy-rich molecules. Various species of bacteria and archaea have evolved the biochemical machinery required to exploit most or all of these food sources.

Obtaining Building-Block Compounds: Variation in Pathways for Fixing Carbon In addition to acquiring energy, organisms must obtain building-block molecules that contain carbon-carbon bonds. Chapters 9 and 10 introduced the two mechanisms that organisms use to procure usable carbon—either making their own or getting it from other organisms. Autotrophs make their own building-block compounds; heterotrophs don't.

In many autotrophs, including cyanobacteria and plants, the enzymes of the Calvin cycle transform carbon dioxide (CO_2) to organic molecules that can be used in synthesizing cell material. The carbon atom in CO_2 is reduced during the process and is said to be "fixed." Animals and fungi, in contrast, obtain carbon from living plants or animals or by absorbing the organic compounds released as dead tissues decay.

Bacteria and archaea pursue these same two strategies. Some interesting twists occur among bacterial and archaeal autotrophs, however. Not all of them use the Calvin cycle to make building-block molecules, and not all start with CO_2 as a source of carbon atoms. For example,

- Several groups of bacteria fix CO_2 using pathways other than the Calvin cycle. Three of these distinctive pathways have been discovered to date.
- Some proteobacteria are called **methanotrophs** ("methane-eaters") because they use methane (CH_4) as their carbon source. (They also use CH_4 as an electron donor in cellular respiration.) Methanotrophs process CH_4 into more complex organic compounds via one of two enzymatic pathways, depending on the species.
- Some bacteria can use carbon monoxide (CO) or methanol (CH_3OH) as a starting material.

These observations drive home an important message from this chapter: Compared with eukaryotes, the metabolic capabilities of bacteria and archaea are remarkably sophisticated and complex.

Check Your Understanding

If you understand that...

- As a group, Bacteria and Archaea can use a wide array of raw materials for cellular respiration and fermentation, perform non-oxygenic as well as oxygenic photosynthesis, and fix carbon from several different sources via a variety of pathways.

You should be able to ...

1) Describe the six ways of synthesizing ATP and obtaining carbon-containing building blocks.
2) Give an example of the following:
   - An electron donor other than a sugar and an electron acceptor other than oxygen used by a bacterium or archaean.
   - A fermentation that does not start with pyruvate and yields a by-product other than lactic acid or ethanol.
   - A photosynthetic pathway that relies on an electron source other than water.
   - A starting point for carbon fixation other than CO_2.

28.4 Key Lineages of Bacteria and Archaea

In the decades since the phylogenetic tree identifying the three domains of life was first published, dozens of studies have confirmed the result. It is now well established that all organisms alive today belong to one of the three domains, and that archaea and eukaryotes are more closely related to each other than either group is to bacteria.

Although the relationships among the major lineages within Bacteria and Archaea are still uncertain in some cases (see Figure 28.12), many of the lineages themselves are well studied. Let's survey the attributes of species from selected major lineages within the Bacteria and Archaea, with an emphasis on themes explored earlier in the chapter: their morphological and metabolic diversity, their impacts on humans, and their importance to other species and to the environment.

Each chapter in this unit will feature a similar "key lineages" section containing notes on what species in selected lineages look like, how they make a living, and where they live. Often you'll be asked to analyze a phylogenetic tree and add labels to indicate the origin of synapomorphies—traits that identify groups as monophyletic (see Chapter 27).

The overall goal of these "key lineages" sections is to give you a concise summary of who's who in major groups on the tree of life, and to highlight the innovations or features that allowed particular groups to arise and diversify.

Bacteria

The name *bacteria* comes from the Greek root *bacter*, meaning "rod" or "staff." The name was inspired by the first bacteria to be seen under a microscope, which were rod shaped. But as the following descriptions indicate, bacterial cells come in a wide variety of shapes. If the group were to be named now, biologists might use roots from the Latin words *diversus* or *abundantia*.

Biologists who study bacterial diversity currently recognize at least 16 major lineages, or phyla, within the domain. Some of these lineages were recognized by distinctive morphological characteristics; others by phylogenetic analyses of gene sequence data. The lineages reviewed here are just a sampling of bacterial diversity (**Figure 28.17**).

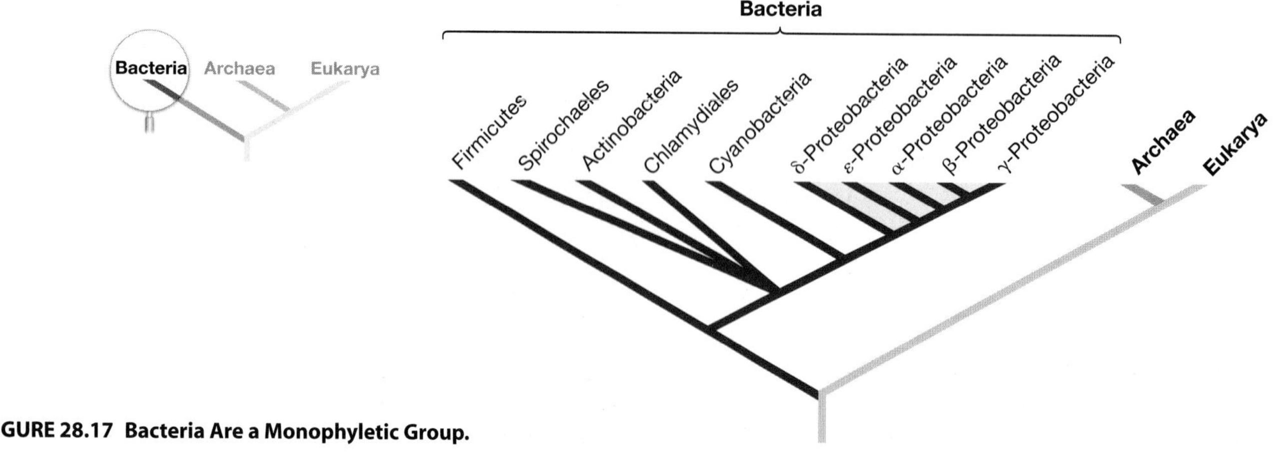

FIGURE 28.17 Bacteria Are a Monophyletic Group.

Bacteria > Firmicutes

The Firmicutes have also been called "low-GC Gram positives" because their cell walls react positively to the Gram stain—meaning that they lack a membrane outside their cell wall—and because they have a relatively low percentage of guanine and cytosine (G and C) in their DNA. In some species, G and C represent less than 25 percent of the bases present. There are over 1100 species. ● You should be able to mark the origin of the Gram-positive cell wall and low-GC genome on Figure 28.17 (only Firmicutes have a low-GC genome; Actinobacteria are the only other Gram-positive lineage).

Morphological diversity Most are rod shaped or spherical. Some of the spherical species form chains or tetrads (groups of four cells). A few form a durable resting stage called a **spore**. One subgroup lacks cell walls entirely; another synthesizes a cell wall made of cellulose.

Metabolic diversity Some species can fix nitrogen; some perform non-oxygenic photosynthesis. Others make all of their ATP via various fermentation pathways; still others perform cellular respiration, using hydrogen gas (H_2) as an electron donor.

Human and ecological impacts Species in this group cause a variety of diseases, including anthrax, botulism, tetanus, walking pneumonia, boils, gangrene, and strep throat. *Bacillus thuringiensis* produces a toxin that is one of the most important insecticides currently used in farming. Species in the genus *Lactobacillus* are used to ferment milk products into yogurt or cheese (**Figure 28.18**). Species in this group are important components of soil, where they speed the decomposition of dead plants, animals, and fungi.

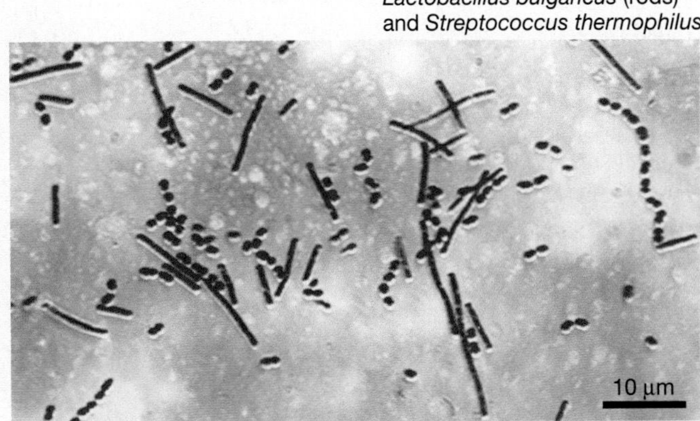

FIGURE 28.18 Firmicutes in Yogurt.

Bacteria > Spirochaetes (Spirochetes)

The spirochetes are one of the smaller bacterial phyla in terms of numbers of species: only 13 genera and a total of 62 species have been described to date. Recent analyses place spirochetes near the base of the bacterial phylogenetic tree (**Figure 28.19**).

Morphological diversity Spirochetes are distinguished by their unique corkscrew shape and unusual flagella. Instead of extending into the water surrounding the cell, spirochete flagella are contained within a structure called the outer sheath, which surrounds the cell. When the flagella beat, the cell lashes back and forth and swims forward. ● You should be able to mark the origin of the spirochete flagellum on Figure 28.17.

Metabolic diversity Most spirochetes manufacture ATP via fermentation. The substrate used in fermentation varies among species and may consist of sugars, amino acids, starch, or the pectin found in plant cell walls. A spirochete that lives only in the hindgut of termites can fix nitrogen.

Human and ecological impacts The sexually transmitted disease syphilis is caused by a spirochete. So is Lyme disease, which is transmitted to humans by deer ticks. Spirochetes are extremely common in freshwater and marine habitats; many live only under anaerobic conditions.

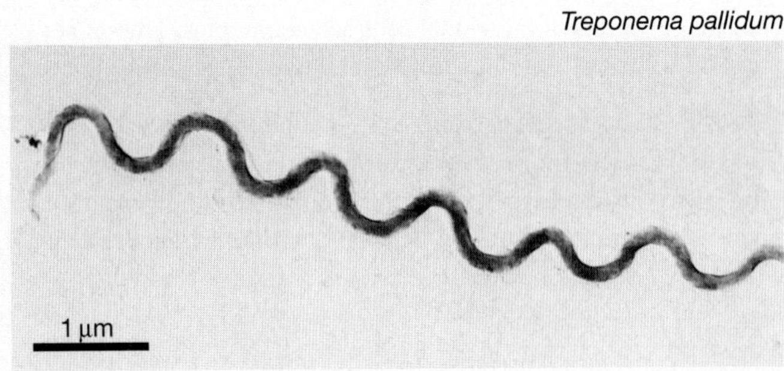

FIGURE 28.19 Spirochetes Are Corkscrew-Shaped Cells Inside an Outer Sheath.

Bacteria > Actinobacteria

Actinobacteria are sometimes called the "high-GC Gram positives" because (1) their cell-wall material appears purple when treated with the Gram stain—meaning that they have a peptidoglycan-rich cell wall and lack an outer membrane—and (2) their DNA contains a relatively high percentage of guanine and cytosine. In some species, G and C represent over 75 percent of the bases present. Over 1100 species have been described to date (**Figure 28.20**). ● You should be able to mark the origin of the high-GC genome in Actinobacteria on Figure 28.17.

Morphological diversity Cell shape varies from rods to filaments. Many of the soil-dwelling species are found as chains of cells that form extensive branching filaments called **mycelia**.

Metabolic diversity Many are heterotrophs that use an array of organic compounds as electron donors and oxygen as an electron acceptor. There are a handful of parasitic species. Like other parasites, they get most of their nutrition from host organisms.

Human and ecological impacts Over 500 distinct antibiotics have been isolated from species in the genus *Streptomyces*; 60 of these—including streptomycin, neomycin, tetracycline, and erythromycin—are now actively prescribed to treat diseases in humans or domestic livestock. Tuberculosis and leprosy are caused by members of this group. One species is critical to the manufacture of Swiss cheese. Species in the genus *Streptomyces* and *Arthrobacter* are abundant in soil and are vital as decomposers of dead plant and animal material. Some species in these genera live in association with plant roots and fix nitrogen; others can break down toxins such as herbicides, nicotine, and caffeine.

Streptomyces griseus

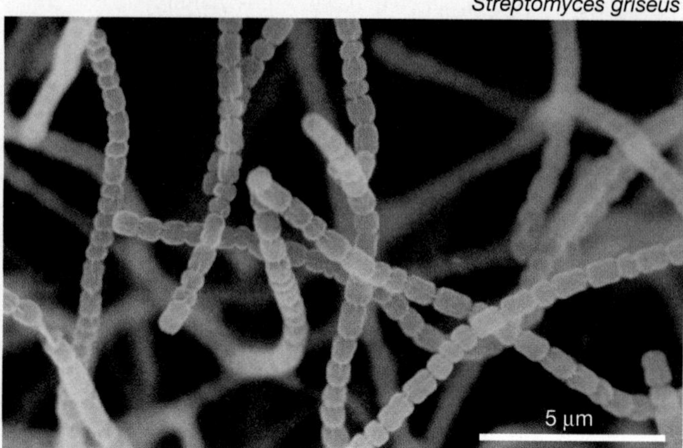

FIGURE 28.20 A *Streptomyces* Species That Produces the Antibiotic Streptomycin.

Bacteria > Cyanobacteria

The cyanobacteria were formerly known as the "blue-green algae"—even though algae are eukaryotes. Only about 80 species of cyanobacteria have been described to date, but they are among the most abundant organisms on Earth. In terms of total mass, cyanobacteria dominate the surface waters in many marine and freshwater environments.

Morphological diversity Cyanobacteria may be found as independent cells, in chains that form filaments (**Figure 28.21**), or in the loose aggregations of individual cells called colonies. The shape of colonies varies from flat sheets to ball-like clusters of cells.

Metabolic diversity All perform oxygenic photosynthesis; many can also fix nitrogen. Because cyanobacteria can synthesize virtually every molecule they need, they can be grown in culture media that contain only CO_2, N_2, H_2O, and a few mineral nutrients. ● You should be able to mark the origin of oxygenic photosynthesis on Figure 28.17.

Human and ecological impacts If cyanobacteria are present in high numbers, their waste products can make drinking water smell bad. Some species release molecules called microcystins that are toxic to plants and animals. Cyanobacteria were responsible for the origin of the oxygen atmosphere on Earth. Today they still produce much of the oxygen and nitrogen and many of the organic compounds that feed other organisms in freshwater and marine environments. A few species live in association with fungi, forming lichens.

Nostoc species

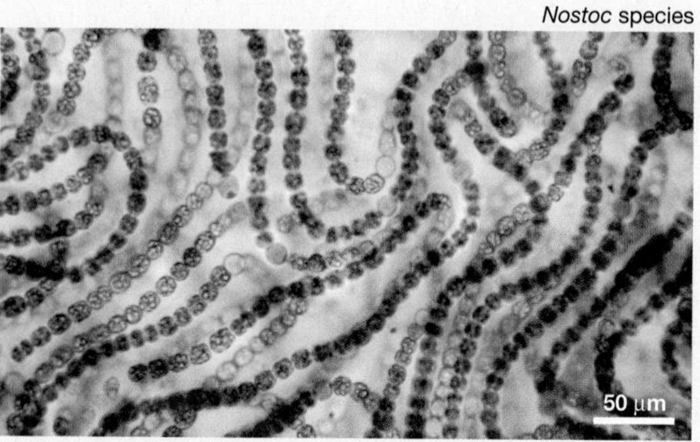

FIGURE 28.21 Cyanobacteria Contain Chlorophyll and Are Green.

Bacteria > Chlamydiales

In terms of numbers of species living today, Chlamydiales may be the smallest of all major bacterial lineages. Although chlamydiae are highly distinct phylogenetically, only four species in one genus (*Chlamydia*) are known.

Morphological diversity Chlamydiae are spherical. They are tiny, even by bacterial standards.

Metabolic diversity All known species live as parasites *inside* host cells and are termed **endosymbionts** ("inside-together-living"). Chlamydiae contain few enzymes of their own and get almost all of their nutrition from their hosts. In **Figure 28.22,** the chlamydiae are the pink-stained cells, which are living inside blue-stained animal cells. ● You should be able to mark the origin of the endosymbiotic lifestyle in this lineage on Figure 28.17. (The endosymbiotic lifestyle has also arisen in other bacterial lineages, independently of Chlamydiales.)

Human and ecological impacts *Chlamydia trachomatis* infections are the most common cause of blindness in humans. When the same organism is transmitted from person to person via sexual intercourse, it can cause serious urogenital tract infections. One species causes epidemics of a pneumonia-like disease in birds.

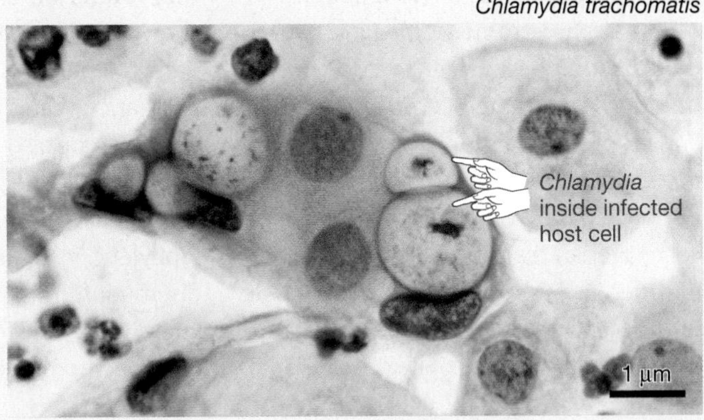

FIGURE 28.22 Chlamydiae Live Only inside Animal Cells.

Bacteria > Proteobacteria

The approximately 1200 species of proteobacteria form five major subgroups, designated by the Greek letters α (alpha), β (beta), γ (gamma), δ (delta), and ε (epsilon). The lineage is named after the Greek god Proteus, who could assume many shapes, because they are so diverse in their morphology and metabolism.

Morphological diversity Proteobacterial cells can be rods, spheres, or spirals. Some form stalks (**Figure 28.23a**). Some are motile. In one group, cells may move together to form colonies, which then transform into the specialized cell aggregate shown in **Figure 28.23b**. This structure is known as a **fruiting body**. Cells that are surrounded by a durable coating are produced at the tips of fruiting bodies. These spores sit until conditions improve, and then they resume growth.

Metabolic diversity Proteobacteria make a living in virtually every way known to bacteria—except that none perform oxygenic photosynthesis. Various species may perform cellular respiration by using organic compounds, nitrite, methane, hydrogen gas, sulfur, or ammonia as electron donors and oxygen, sulfate, or sulfur as an electron acceptor. Some perform non-oxygenic photosynthesis (**Box 28.2**).

Human and ecological impacts Pathogenic proteobacteria cause Legionnaire's disease, cholera, food poisoning, dysentery, gonorrhea, Rocky Mountain spotted fever, typhus, ulcers, and diarrhea. *Wolbachia* infections are common in insects and are often transmitted from mothers to offspring via eggs. Biologists use *Agrobacterium* cells to transfer new genes into crop plants. Certain acid-loving species of proteobacteria are used in the production of vinegars. Species in the genus *Rhizobium* (α-proteobacteria) live in association with plant roots and fix nitrogen. The bdellovibrios are a group in the δ-proteobacteria that are predators—they drill into bacterial cells and digest them. Proteobacteria are critical players in the cycling of nitrogen atoms through terrestrial and aquatic ecosystems.

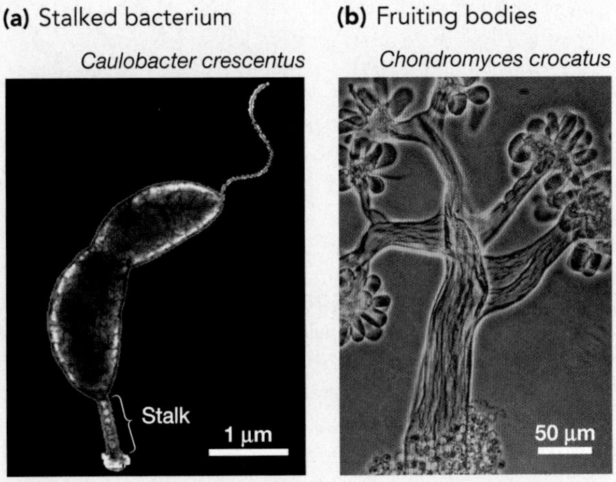

(a) Stalked bacterium

Caulobacter crescentus

Stalk

1 µm

(b) Fruiting bodies

Chondromyces crocatus

50 µm

FIGURE 28.23 Some Proteobacteria Grow on Stalks or Form Fruiting Bodies.

BOX 28.2 Lateral Gene Transfer and Metabolic Diversity in Bacteria

If you read the notes on metabolic diversity in Section 28.4 carefully, you will realize that species capable of performing various types of photosynthesis are scattered among many bacterial lineages. The same is true for species that can fix nitrogen.

The "scattered among lineages" pattern is interesting because it is reasonable to predict that extremely complex structures and processes such as photosystem I, photosystem II, and nitrogen fixation evolved just once. If they did, then it would be logical to predict that photosynthesizers and nitrogen-fixers would form monophyletic groups, each group consisting of an ancestral species and all of its descendants. In contrast, a group that consists of a common ancestor and some *but not all* of its descendants is said to be **paraphyletic** ("beside-group"). If species that perform photosynthesis were monophyletic, then they would be arranged in the pattern shown in **Figure 28.24a**. But instead, the data in **Figure 28.24b** shows that photosynthetic species are paraphyletic. Nitrogen fixers are also paraphyletic.

How could the photosystems and nitrogen fixation evolve just once and yet be paraphyletic in their present phylogenetic distribution? This answer is a process called **lateral gene transfer**—the physical transfer of genes from species in one lineage to species in another lineage. Chapter 20 described several of the mechanisms responsible for lateral gene transfer. According to the lateral gene transfer hypothesis for the distribution of photosynthesis and nitrogen fixation between lineages, it is correct to claim that photosystem I, photosystem II, and the nitrogen-fixing enzymes evolved just once. But over the past 3.4 billion years, the genes responsible for these processes have been picked up by organisms from a wide array of bacterial lineages and incorporated into their genomes. Lateral gene transfer is thought to be an important mechanism for generating metabolic diversity in the Bacteria.

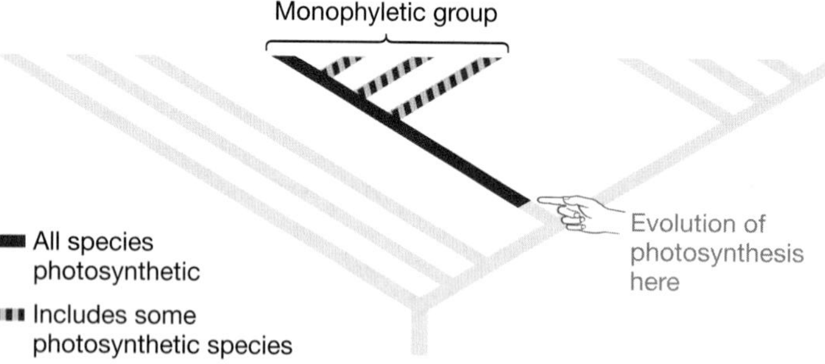

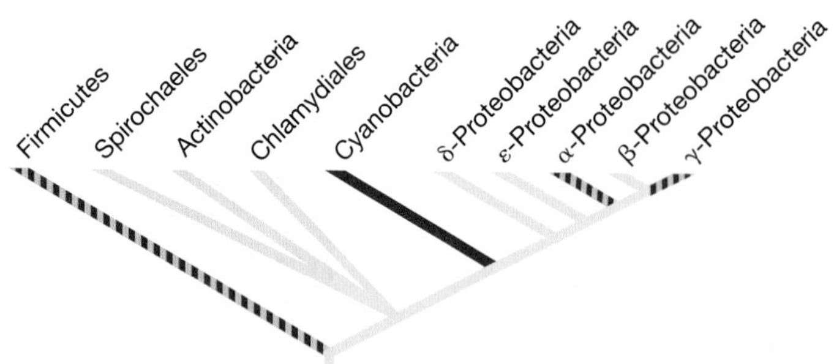

FIGURE 28.24 Photosynthetic Bacteria Are Paraphyletic. (a) If photosynthetic bacteria were monophyletic, their distribution on the phylogenetic tree of Bacteria would look something like this. **(b)** Species that can perform photosynthesis are scattered throughout the phylogenetic tree. Two bacterial lineages not shown—green sulfur bacteria and green nonsulfur bacteria—also perform a type of photosynthesis.

Archaea

The name *archaea* comes from the Greek root *archae*, for "ancient." The name was inspired by the hypothesis that this is a particularly ancient group, which turned out to be incorrect. Also incorrect was the hypothesis that archaeans are restricted to hot springs, salt ponds, and other extreme habitats. If the group were to be named now, biologists might start with the Latin root *ubiquit*, for "everywhere." Archaea live in virtually every habitat known. As far as biologists currently know, however, there are no parasitic archaea.

Phylogenies based on DNA sequence data have consistently shown that the domain is composed of at least two major phyla, called the Crenarchaeota and Euryarchaeota (**Figure 28.25**). Although it is clear that these two groups are highly differentiated at the DNA sequence level, biologists are still searching for shared, derived morphological traits that help define each lineage as a monophyletic group. In addition, the domain Archaea was discovered so recently that major groups are still being discovered and described. As mentioned earlier in this chapter, preliminary data indicate that two additional phyla may exist.

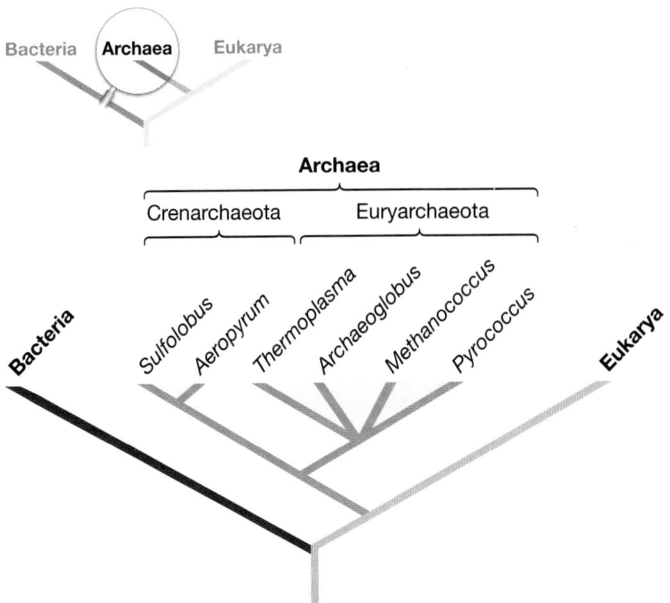

FIGURE 28.25 Archaea Are Monophyletic.

Archaea > Crenarchaeota

The Crenarchaeota got their name because they are considered similar to the oldest archaeans; the word root *cren-* refers to a source or fount. Although only 37 species have been named to date, it is virtually certain that thousands are yet to be discovered.

Morphological diversity Crenarchaeota cells can be shaped like filaments, rods, discs, or spheres. One species that lives in extremely hot habitats has a tough cell wall consisting solely of glycoprotein.

Metabolic diversity Depending on the species, cellular respiration can involve organic compounds, sulfur, hydrogen gas, ammonia, or Fe^{2+} ions as electron donors and oxygen, nitrate, sulfate, sulfur, carbon dioxide, or Fe^{3+} ions as electron acceptors. Some species make ATP exclusively through fermentation pathways.

Human and ecological impacts Crenarchaeota have yet to be used in the manufacture of commercial products. In certain extremely hot, high-pressure, cold, or acidic environments, crenarchaeota may be the only life-form present (**Figure 28.26**). Acid-loving species thrive in habitats with pH 1–5; some species are found in ocean sediments at depths ranging from 2500 to 4000 m below the surface.

FIGURE 28.26 Some Crenarchaeota Live in Sulfur-Rich Hot Springs.

Archaea > Euryarchaeota

The Euryarchaeota are aptly named, because the word root *eury-* means "broad." Members of this phylum live in every conceivable habitat. Some species are adapted to high-salt habitats with pH 11.5—almost as basic as household ammonia (**Figure 28.27**). Other species are adapted to acidic conditions with a pH as low as 0. Species in the genus *Methanopyrus* live near hot springs called black smokers that are 2000 m (over 1 mile) below sea level. About 170 species have been identified thus far, and more are being discovered each year.

Morphological diversity Euryarchaeota cells can be spherical, filamentous, rod shaped, disc shaped, or spiral. Rod-shaped cells may be short or long or arranged in chains. Spherical cells can be found in ball-like aggregations. Some species have several flagella. Some species lack a cell wall; others have a cell wall composed entirely of glycoproteins.

Metabolic diversity The group includes a variety of methane-producing species. These methanogens can use up to 11 different organic compounds as electron acceptors during cellular respiration; all produce CH_4 as a by-product of respiration. In other species of Euryarchaeota, cellular respiration is based on hydrogen gas or Fe^{2+} ions as electron donors and nitrate or sulfate as electron acceptors. Species that live in high-salt environments can use the molecule retinal—which is responsible for light reception in your eyes—to capture light energy and perform photosynthesis.

Human and ecological impacts Species in the genus *Ferroplasma* (literally, "hot iron") live in piles of waste rock near abandoned mines. As a by-product of metabolism, they produce acids that drain into streams and pollute them. Methanogens live in the soils of swamps and the guts of mammals (including yours). They are responsible for adding about 2 billion tons of methane to the atmosphere each year.

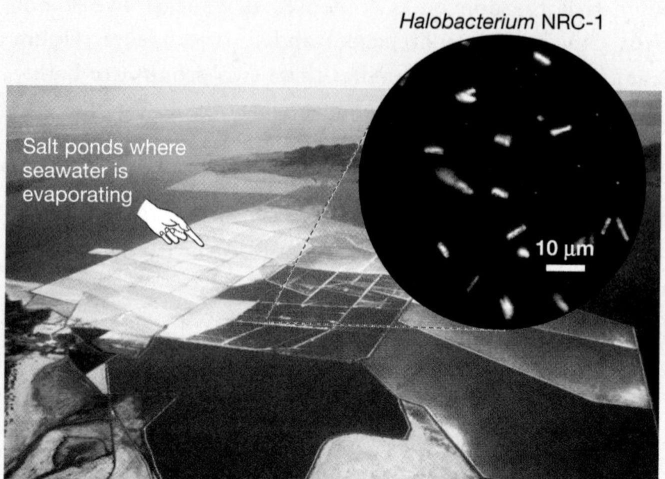

FIGURE 28.27 Some Euryarchaeota Live in High-Salt Habitats.

Chapter Review

SUMMARY OF KEY CONCEPTS

A few bacteria cause important infectious diseases; and collectively, bacteria and archaea play a key role in global ecosystems. Some species are effective at cleaning up pollution; photosynthetic bacteria were responsible for the evolution of the oxygen atmosphere; and bacteria and archaea cycle nutrients through terrestrial and aquatic environments.

Bacteria and archaea may be tiny, but they have a huge impact on global ecosystems and human health. Cyanobacteria produce much of the oxygen in the oceans and atmosphere, and nitrogen-fixing bacteria and archaea keep the global nitrogen cycle running. Bacteria also cause some of the most dangerous human diseases, including plague, syphilis, botulism, cholera, and tuberculosis. Disease results when bacteria kill host cells or produce toxins that disrupt normal cell functions.

Enrichment cultures are used to grow large numbers of bacterial or archaeal cells that thrive under specified conditions, such as in the presence of certain electron donors and electron acceptors. To study bacteria and archaea that cannot be cultured, biologists frequently take advantage of direct sequencing. In this research strategy, DNA sequences are extracted directly from organisms in the environment—without first culturing the organisms in the lab. By analyzing where these sequences are placed on the tree of life, biologists can determine whether the branches represent organisms that are new to science. If so, then information on where the original sample was collected can expand knowledge about the types of habitats used by bacteria and archaea.

You should be able to explain what the composition of the atmosphere and what the nitrogen cycle would be like if bacteria and archaea did not exist.

Many bacterial and archaeal species are restricted in distribution and have a limited diet, but as a group they live in virtually every habitat known and use remarkably diverse types of compounds in cellular respiration and fermentation. Although their

overall morphologies are small and relatively simple, the chemistry they can do is extremely sophisticated.

Metabolic diversity and complexity are the hallmarks of the bacteria and archaea, just as morphological diversity and complexity are the hallmarks of the eukaryotes. Like eukaryotes, many bacteria and archaea can extract energy from carbon-containing compounds with high potential energy, such as sugars. These molecules are processed through fermentation pathways or by transferring high-energy electrons to electron transport chains with oxygen as the final electron acceptor. But among the bacteria and archaea, many other inorganic or organic compounds with high potential energy serve as electron donors, and a wide variety of inorganic or organic molecules with low potential energy serve as electron acceptors. Dozens of distinct organic compounds are fermented, including proteins, purines, alcohols, and an assortment of carbohydrates.

Photosynthesis is also widespread among bacteria. In cyanobacteria, water is used as a source of electrons during photosynthesis, and oxygen gas is generated as a by-product. But in other species, the electron excited by photon capture comes from a reduced substance such as ferrous iron (Fe^{2+}) or hydrogen sulfide (H_2S) instead of water (H_2O); the oxidized by-product is the ferric ion (Fe^{3+}) or elemental sulfur (S) instead of oxygen (O_2). These organisms also contain chlorophylls not found in plants or cyanobacteria.

To acquire building-block molecules containing carbon-carbon bonds, some species use the enzymes of the Calvin cycle to reduce CO_2. But biologists have also discovered three additional biochemical pathways in bacteria and archaea that transform carbon dioxide (CO_2), methane (CH_4), or other sources of inorganic carbon into organic compounds such as sugars or carbohydrates.

You should be able to describe a habitat on Earth where bacteria and archaea would *not* be found, and explain why.

at www.masteringbio.com
The Tree of Life

QUESTIONS

Test Your Knowledge

1. How do molecules that function as electron donors and those that function as electron acceptors differ?
   a. Electron donors are almost always organic molecules; electron acceptors are always inorganic.
   b. Electron donors are almost always inorganic molecules; electron acceptors are always organic.
   c. Electron donors have relatively high potential energy; electron acceptors have relatively low potential energy.
   d. Electron donors have relatively low potential energy; electron acceptors have relatively high potential energy.
2. What do some photosynthetic bacteria use as a source of electrons instead of water?
   a. oxygen (O_2)
   b. hydrogen sulfide (H_2S)
   c. organic compounds (e.g., CH_3COO^-)
   d. nitrate (NO_3^-)
3. What is distinctive about the chlorophylls found in different photosynthetic bacteria?
   a. their membranes
   b. their role in acquiring energy
   c. their role in carbon fixation
   d. their absorption spectra
4. What are organisms called that use inorganic compounds as electron donors in cellular respiration?
   a. phototrophs
   b. heterotrophs
   c. organotrophs
   d. lithotrophs
5. What has direct sequencing allowed researchers to do for the first time?
   a. identify important morphological differences among species
   b. study organisms that cannot be cultured (grown in the lab)
   c. It is based on sampling organisms from an environment and sequencing DNA from a particular gene.
   d. It is based on sampling organisms from an environment and sequencing the entire genomes present.
6. Koch's postulates outline the requirements for which of the following?
   a. showing that an organism is autotrophic
   b. showing that a bacterium's cell wall lacks an outer membrane and consists primarily of peptidoglycan
   c. showing that an organism causes a particular disease
   d. showing that an organism can use a particular electron donor and electron acceptor

Test Your Knowledge answers: 1. c; 2. b; 3. d; 4. d; 5. b; 6. c

Test Your Understanding

Answers are available at www.masteringbio.com

1. Biologists often use the term *energy source* as a synonym for "electron donor." Why?
2. The text claims that the tremendous ecological diversity of bacteria and archaea is possible because of their impressive metabolic diversity. Do you agree with this statement? Why or why not?
3. Suppose that universal PCR primers were available for genes involved in electron transport chains or for some of the different types of chlorophyll found in bacteria. Why would it be interesting to use these genes in a direct sequencing study?
4. The text claims that the evolution of an oxygen atmosphere paved the way for increasingly efficient cellular respiration and higher growth rates in organisms. Explain.
5. Look back at Table 28.5 and note that the by-products of respiration in some organisms are used as electron donors or acceptors by other organisms. In the table, draw lines between the dual-use molecules listed in the "Electron Donor," "Electron Acceptor," and "By-products" columns.
6. Explain the statement, "Prokaryotes are a paraphyletic group."

Applying Concepts to New Situations

Answers are available at www.masteringbio.com

1. The researchers who observed that magnetite was produced by bacterial cultures from the deep subsurface carried out a follow-up experiment. These biologists treated some of the cultures with a drug that poisons the enzymes involved in electron transport chains. In cultures where the drug was present, no more magnetite was produced. Does this result support or undermine their hypothesis that the bacteria in the cultures perform anaerobic respiration with Fe^{3+} serving as the electron acceptor? Explain your reasoning.
2. *Streptococcus mutans* obtains energy by oxidizing sucrose. This bacterium is abundant in the mouths of Western European and North American children and is a prominent cause of cavities. The organism is virtually absent in children from East Africa, where tooth decay is rare. Propose a hypothesis to explain this observation. Outline the design of a study that would test your hypothesis.
3. Suppose that you've been hired by a firm interested in using bacteria to clean up organic solvents found in toxic waste dumps. Your new employer is particularly interested in finding cells that are capable of breaking a molecule called benzene into less toxic compounds. Where would you go to look for bacteria that can metabolize benzene as an energy or carbon source? How would you design an enrichment culture capable of isolating benzene-metabolizing species?
4. Would you predict that disease-causing bacteria, such those listed in Table 28.2, obtain energy from light, organic molecules, or inorganic molecules? When they perform cellular respiration, which substance would you predict that they use as an electron acceptor? Explain your answer.

www.masteringbio.com is also your resource for • Answers to text, table, and figure caption questions and exercises • Answers to *Check Your Understanding* boxes • Online study guides and quizzes • Additional study tools including the *E-Book for Biological Science* 3rd ed., textbook art, animations, and videos

Protists

29

KEY CONCEPTS

- Protists are a paraphyletic grouping that includes all eukaryotes except the green plants, fungi, and animals. Biologists study protists to understand how eukaryotes evolved, because they are important in freshwater and marine ecosystems and global warming, and because some species cause debilitating diseases in plants, humans, and other organisms.
- Protists are diverse morphologically. They vary in the types of organelles they contain; they may be unicellular or multicellular, and they may have a cell wall or other external covering, or no such covering.
- Protists vary widely in terms of how they find food. Many species are photosynthetic, while others obtain carbon compounds by ingesting food or parasitizing other organisms.
- Protists vary widely in terms of how they reproduce. Sexual reproductive evolved in protists, and many protist species can reproduce both sexually and asexually.

Diatoms are single-celled protists that live inside a glassy case. They may be the most abundant of all eukaryotes found in aquatic environments.

This chapter introduces the third domain on the tree of life: the **Eukarya**. Eukaryotes range from single-celled organisms that are the size of bacteria to sequoia trees and blue whales. The largest and most morphologically complex organisms on the tree of life—algae, plants, fungi, and animals—are eukaryotes.

Although species in the Eukarya are astonishingly diverse, they share fundamental features that distinguish them from bacteria and archaea. Eukaryotes are defined by the presence of the shared, derived character called the nuclear envelope. Chapter 7 pointed out other key distinctions between eukaryotes and prokaryotes. You might recall that most eukaryotic cells are much larger than bacterial or archaeal cells. They also have many more organelles and have a much more extensive system of structural proteins called the cytoskeleton. Multicellularity is rare in bacteria and unknown in archaea, but has evolved several times in eukaryotes (see Chapter 8). Unlike bacteria and archaea, which reproduce by fission, eukaryotes undergo cell division via mitosis; many eukaryotes can also undergo meiosis. Like archaea, most eukaryotes have chromosomes where DNA is complexed with proteins called histones (see Chapter 18).

One of this chapter's fundamental goals is to explore how these morphological innovations—features like the nuclear envelope and organelles—evolved. Another goal is to analyze how morphological innovations allowed eukaryotes to pursue novel ways of performing basic life tasks such as feeding, moving, and reproducing.

Key Concept Important Information Practice It

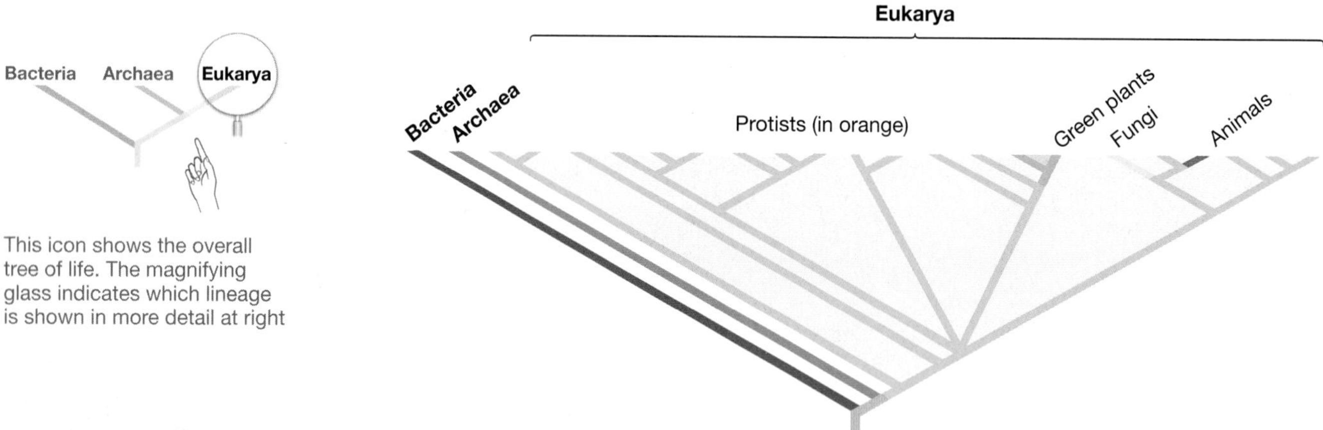

FIGURE 29.1 Protists Are Paraphyletic. The group called protists includes some, but not all, descendants of a single common ancestor.

EXERCISE On the tree, indicate the common ancestor of all eukaryotes.

In introducing the Eukarya, this chapter focuses on an informal grab bag of lineages known as the protists. The term **protist** refers to all eukaryotes that are not green plants, fungi, or animals. Protist lineages are colored yellow-orange in **Figure 29.1**. As you study this tree, note that protists do not make up a monophyletic group. Instead, protists constitute a **paraphyletic group**—meaning that they represent some, but not all, of the descendants of a single common ancestor (see **Box 29.1**). To use the vocabulary introduced in Chapter 27, no synapomorphies define the protists. There is no trait that is found in protists but no other organisms.

By definition, then, the protists are a diverse lot. The story of the eukaryotes as a whole revolves around the evolution of complex new cell structures and methods for reproducing; the story of protists is how these new features diversified over time. The organelles and cytoskeletons found in protists vary widely in structure. Where and when meiosis occurs is highly variable among species. Many protists are microscopic single cells; others are multicellular organisms up to 60 meters long. Their lifestyles are just as diverse. Some are parasitic while others are predatory or photosynthetic. They may be stationary all their lives or in virtually constant motion. Their cells can change shape almost continuously or be encased in rigid, glassy shells. The common feature among protists is that they tend to live in environments where they are surrounded by water (**Figure 29.2**). Most plants, fungi, and animals are terrestrial, but protists are found in wet soils, aquatic habitats, or the bodies of other organisms.

BOX 29.1 How Should We Name the Major Branches on the Tree of Life?

Taxonomy is the branch of biology devoted to describing and naming new species and classifying groups of species. Carolus Linnaeus founded this field and published the first work in it in 1735. Linnaeus invented the system of Latin binomials, still being used today, in which each organism is given a unique name consisting of its genus and species. He also invented a hierarchy of more general taxonomic categories that included kingdoms, classes, orders, and families (see Chapter 1). Species were placed into different families, orders, classes, and kingdoms based on their morphological similarity.

Linnaeus worked long before Darwin had discovered the principle of evolution by natural selection, however. As a result, Linnaeus viewed the groups he described as entities that had been created separately and independently from one another. When biologists came to realize that all organisms are related by common descent, they had to reinterpret the categories that Linnaeus had established. Instead of representing neat "bins" holding species with similar characteristics, biologists recognized that any groupings that are named should actually be designated as twigs, branches, and stems on a tree of life.

According to the rules currently being adopted, most biologists assign names only to monophyletic groups—that is, to branches on the tree of life that include all the descendants of a common ancestor. Because it violates this rule, the name *protist* will probably be abandoned before long.

According to a Chinese proverb, "The first step in wisdom is to call things by their right name." In taxonomy, a massive effort is under way to do just that. The initial task is to produce accurate estimates of where each branch occurs on the tree of life. Then monophyletic groups can be named, with confidence that each represents a distinct stem, branch, or twig on the tree.

(a) Open ocean:

Surface waters teem with microscopic protists, such as these diatoms.

(b) Shallow coastal waters:

Gigantic protists, such as these kelp, form underwater forests.

(c) Intertidal habitats:

Protists such as these red algae are particularly abundant in tidal habitats.

FIGURE 29.2 Protists Are Particularly Abundant in Aquatic Environments.

To begin, let's ask why biologists are spending so much time and energy studying protists, and which techniques are proving most useful. Section 29.3 delves into the heart of the protist story by analyzing the innovations and subsequent diversification that took place in their morphology and their methods of feeding, moving, and reproducing. The chapter concludes with a survey of characteristics in key lineages—a who's who of protists that you are almost sure to encounter later on in your career.

29.1 Why Do Biologists Study Protists?

Biologists study protists in part because protists are intrinsically interesting, in part because they are so important medically and ecologically, and in part because they are critical to understanding the evolution of plants, fungi, and animals. The remainder of the chapter will focus on why protists are interesting in their own right and how they evolved; here let's consider their impact on human health and the environment.

Impacts on Human Health and Welfare

The most spectacular crop failure in history, the Irish potato famine, was caused by a protist. In 1845 most of the 3 million acres that had been planted to grow potatoes in Ireland became infested with *Phytophthora infestans*—a parasite that belongs to a lineage of protists called Oomycota. Potato tubers that were infected with *P. infestans* rotted in the fields or in storage.

As a result of crop failures in Ireland in two consecutive years, an estimated 1 million people out of a population of less than 9 million died of starvation or starvation-related illnesses. Several million others emigrated. Many people of Irish heritage living in North America, New Zealand, and Australia trace their ancestry to relatives who left Ireland to evade the famine. As devastating as the potato famine was, however, it does not begin to approach the misery caused by the protist *Plasmodium*.

Malaria Physicians and public health officials point to three major infectious diseases that are currently afflicting large numbers of people worldwide: tuberculosis, HIV, and malaria. Tuberculosis is caused by a bacterium and was introduced in Chapter 24; HIV is caused by a virus and is analyzed in Chapter 35. Here we consider malaria—a disease caused by several species in the eukaryotic lineage called Apicomplexa.

Malaria ranks as the world's most chronic public health problem. In India alone, over 30 million people each year suffer from debilitating fevers caused by malaria. At least 300 million people worldwide are sickened by it each year, and over 1 million die from the disease annually. The toll is equivalent to eight 747s, loaded with passengers, crashing every day. Most of the dead are children of preschool age.

Four species of the protist *Plasmodium* are capable of parasitizing humans and causing malaria. Infections start when *Plasmodium* cells enter a person's bloodstream during a mosquito bite. As **Figure 29.3** shows, *Plasmodium* initially infects liver cells; later, some cells change into a distinctive cell type that infects red blood cells. The *Plasmodium* cells multiply inside the host cells and kill them as they exit to infect additional liver cells or red blood cells. If cells released from infected red blood cells are transferred to a mosquito during a bite, they differentiate to form gametes. Inside the mosquito, gametes fuse to form a diploid cell called an oocyst, which undergoes meiosis. The haploid cells that result from meiosis can infect a human when the mosquito bites again.

Because each *Plasmodium* species spends part of its life cycle inside mosquitoes, most antimalaria campaigns have focused on controlling these insects. This strategy has become less and less effective over time, however, because natural selection has favored mosquito strains that are resistant to the insecticides that have been sprayed in their breeding habitats. Further, *Plasmodium* itself has evolved resistance to most of the drugs used to control its growth in infected people. Efforts to develop a vaccine against *Plasmodium* have also been fruitless to date, in part because the parasite evolves so quickly. (Chapter 35 explains why it has not been possible to develop effective vaccines against rapidly evolving viruses and organisms such as *Plasmodium*, cold and influenza viruses, and HIV.) Although *Plasmodium* is arguably the best studied of all protists, researchers have still not been able to devise effective and sustainable measures to control it.

Unfortunately, malaria is not the only important human disease caused by protists. **Table 29.1** lists protists that have been the cause of human suffering and economic losses. Parasitic protists affect hundreds of millions of people every year and are a major concern for physicians worldwide. Harmful protists also worry biologists who manage the fisheries that many people depend on for their food or livelihood.

Harmful Algal Blooms When a unicellular species experiences rapid population growth and reaches high densities in an aquatic environment, it is said to "bloom." Unfortunately, a handful of the many protist species involved in blooms can be harmful. Harmful algal blooms are usually due to photosynthetic protists called dinoflagellates. Certain dinoflagellates are harmful at high population density because they synthesize toxins to protect themselves from predation by small animals called copepods. Because toxin-producing dinoflagellates have high concentrations of red accessory pigments called xanthophylls, their blooms are known informally as red tides (**Figure 29.4**)

Algal blooms can be harmful to people because clams and other shellfish filter photosynthetic protists out of the water as food. During a bloom, high levels of toxins can build up in the flesh of these shellfish. Typically, the shellfish themselves are not harmed. But if a person eats contaminated shellfish, several

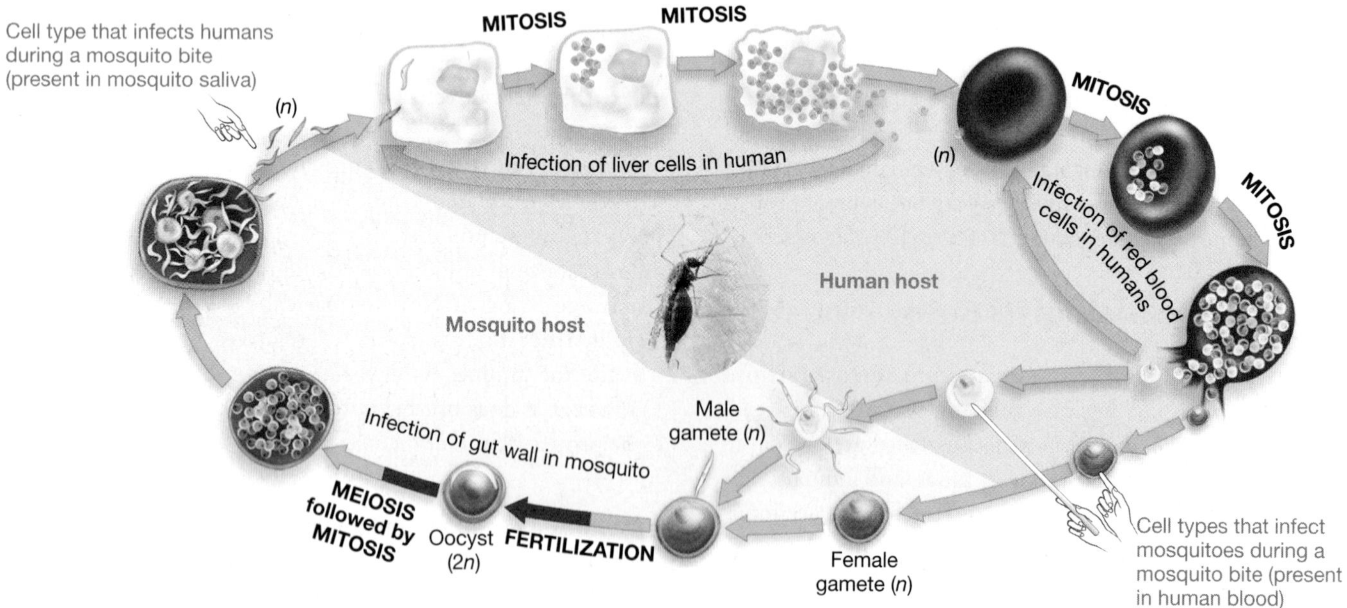

FIGURE 29.3 *Plasmodium* Lives in Mosquitoes and in Humans, where It Causes Malaria. Over the course of its life cycle, *Plasmodium falciparum* develops into a series of distinct cell types. Each type is specialized for infecting a different host cell in mosquitoes or humans. In mosquitoes, the protist lives in the gut and salivary glands. In humans, it infects and kills liver cells and red blood cells, contributing to anemia and high fever.

TABLE 29.1 **Human Health Problems Caused by Protists**

Species	Disease
Four species of *Plasmodium*, primarily *P. falciparum* and *P. vivax*	Malaria has the potential to affect 40 percent of the world's total population.
Toxoplasma	Toxoplasmosis may cause eye and brain damage in infants and in AIDS patients.
Many species of dinoflagellates	Toxins released during "red tides" accumulate in clams and mussels and poison people if eaten.
Giardia	Diarrhea due to giardiasis (beaver fever) can last for several weeks.
Trichomonas	Trichomoniasis is a reproductive tract infection and one of the most common sexually transmitted diseases. About 2 million young women are infected in the United States each year; some of them become infertile.
Leishmania	Leishmaniasis can cause skin sores or affect internal organs—particularly the spleen and liver.
Trypanosoma gambiense and *T. rhodesiense*	Trypanosomiasis ("sleeping sickness") is a potentially fatal disease transmitted through bites from tsetse flies. Occurs in Africa.
Trypanosoma cruzi	Chagas disease affects 16–18 million people and causes 50,000 deaths annually, primarily in South and Central America.
Entamoeba histolytica	Amoebic dysentery results from severe infections.
Phytophthora infestans	An outbreak of this protist wiped out potato crops in Ireland in 1845–1847, causing famine.

types of poisoning can result. For example, paralytic shellfish poisoning occurs when people eat shellfish that have fed heavily on protists that synthesize poisons called saxitoxins. Saxitoxins block ion channels that have to open for electrical signals to travel through nerve cells (see Chapter 45). In humans, high dosages of saxitoxins cause unpleasant symptoms such as prickling sensations in the mouth or even life-threatening symptoms such as muscle weakness and paralysis.

No antidote exists to the poisons secreted by protists during harmful blooms. As a result, biologists prevent poisonings by carefully monitoring protist populations in regions where shellfish are harvested for food. If harmful protist species begin to bloom, the shellfish beds are immediately closed to harvest until toxins are at lower levels.

FIGURE 29.4 Harmful Algal Blooms Are Sometimes Called Red Tides. Dinoflagellates with red coloration have "bloomed" along this beach.

Ecological Importance of Protists

As a whole, the protists represent just 10 percent of the total number of named eukaryote species. Although the species diversity of protists may be relatively low, their abundance is extraordinarily high. The number of individual protists found in some habitats is astonishing. One milliliter of pond water can contain well over 500 single-celled protists that swim with the aid of flagella. Under certain conditions, dinoflagellates can reach concentrations of 60 million cells per liter of seawater.

The great abundance of protists is important to an array of ecological events. For example, photosynthetic organisms take in carbon dioxide from the atmosphere and reduce, or "fix," it to form sugars or other organic compounds with high potential energy (Chapter 10). Photosynthesis transforms some of the energy in sunlight into chemical energy that organisms can use to grow and produce offspring. Species that produce chemical energy in this way are called **primary producers**. Diatoms, for example, are photosynthetic protists that rank among the leading primary producers in the oceans, simply because they are so abundant. Production of organic molecules in the world's oceans, in turn, represents almost half of the total carbon dioxide that is fixed on Earth. Why is this important?

Protists Play a Key Role in Aquatic Food Chains Small organisms that live near the surface of oceans or lakes and that drift along or swim only short distances are called **plankton**. The sugars and other organic compounds produced by **phytoplankton**—that is, photosynthetic species of plankton—are the basis of food chains in freshwater and marine environments. A **food chain** describes nutritional relationships

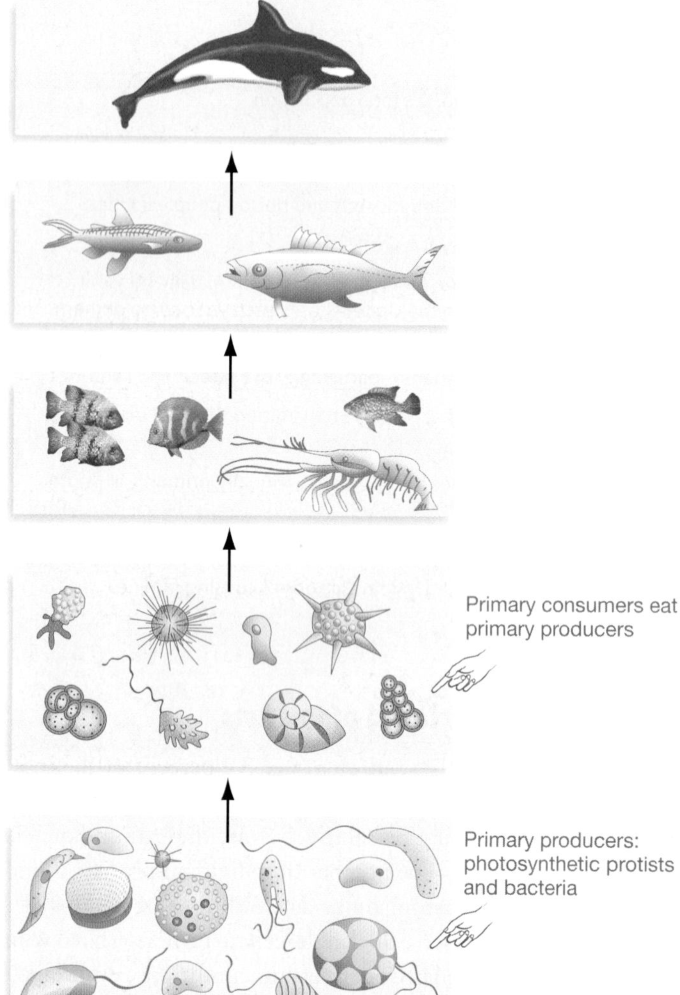

FIGURE 29.5 Protists Are Key Primary Producers in Marine and Freshwater Habitats. In aquatic habitats, photosynthetic protists form the base of the food chain along with photosynthetic bacteria.

EXERCISE Indicate the levels at which humans feed

among organisms. Thus, it also describes how chemical energy flows within ecosystems.

Figure 29.5 shows a food chain for a marine environment. At the bottom of the chain are photosynthetic protists, photosynthetic bacteria, and other primary producers. These species are eaten by primary consumers, many of which are protists. Primary consumers are eaten by secondary consumers, which in turn are eaten by tertiary consumers, and so on. The fish and shellfish at the middle levels of the food chain are important sources of protein not only for whales, squid, and very large fish (such as tuna) at the uppermost levels but also for people.

Many of the species at the base of food chains in aquatic environments are protists. It is not an exaggeration to say that without protists, most food chains in freshwater and marine habitats would collapse.

Could Protists Help Reduce Global Warming? As Chapter 54 will explain in detail, carbon dioxide levels in the atmosphere are increasing rapidly due to human activities such as the burning of fossil fuels and forests. Because carbon dioxide traps heat that is radiating from Earth back out to space, high CO_2 levels in the atmosphere contribute to global warming—an issue that many observers consider today's most pressing environmental problem.

To understand global warming, it is critical to analyze how carbon atoms move among organisms and molecules in both terrestrial and aquatic habitats. The movement of carbon atoms from carbon dioxide molecules in the atmosphere to organisms in the soil or the ocean, and then back to the atmosphere is called the **global carbon cycle**. To reduce global warming, researchers are trying to figure out ways to decrease carbon dioxide concentrations in the atmosphere and increase the amount of carbon stored in terrestrial and marine environments.

Recall from Chapter 10 that in terrestrial environments, trees take CO_2 from the atmosphere and use some of the carbon atoms to make wood. Wood is a storage area, or "sink," for carbon, because carbon atoms tend to remain in wood for decades or centuries—until trees burn or decay. As a result, the growth of trees reduces the amount of carbon in the atmosphere and increases the amount of carbon stored in forests. Efforts to plant trees in deforested areas might play a role in efforts to reduce atmospheric CO_2 concentrations and slow the rate of global warming.

In the world's oceans, attempts to fertilize photosynthetic protists and bacteria might have a similar effect. To understand why, first consider the carbon cycle diagrammed in **Figure 29.6**. The cycle starts when CO_2 from the atmosphere dissolves in water and is taken up by phytoplankton. The phytoplankton are eaten by primary consumers, die and are consumed by decomposers or scavengers, or die and sink to the bottom of the ocean. There they may enter one of two long-lived repositories—sedimentary rocks or petroleum:

1. Several lineages of protists have shells made of calcium carbonate ($CaCO_3$) When these shells rain down from the ocean surface and settle in layers at the bottom, the deposits that result are compacted by the weight of the water and by sediments accumulating above them. Eventually they turn into rock. The limestone used to build the pyramids of Egypt consists of protist shells. Limestone and other carbon-containing rocks that form on the ocean floor can potentially lock up carbon atoms for tens of millions of years.

2. Although the process of petroleum (oil) formation is not well understood, it begins with accumulations of dead bacteria, archaea, and protists at the bottom of the ocean. CO_2 that passes from the atmosphere to the body of a photosynthetic protist and then to petroleum is removed from the atmosphere for millions of years—unless humans pump the petroleum out of the ground and burn it.

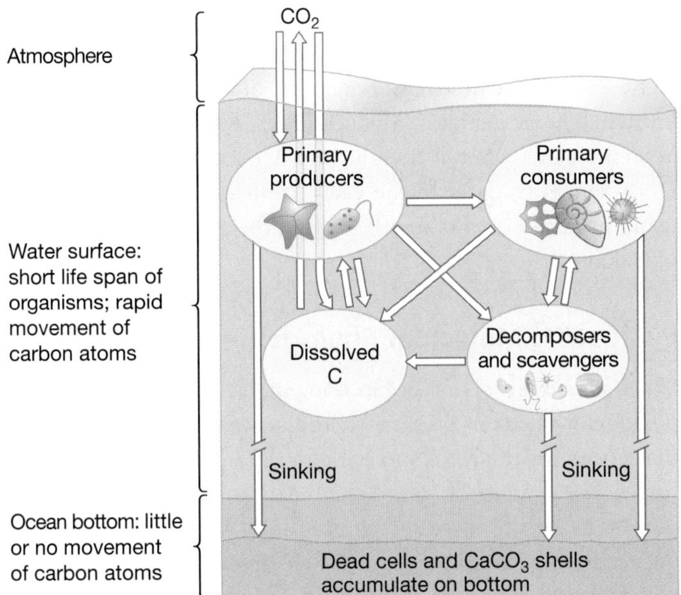

FIGURE 29.6 Protists Play a Key Role in the Marine Carbon Cycle. Arrows indicate the movement of carbon atoms. At the surface, carbon atoms tend to shuttle quickly among organisms. But if carbon atoms sink to the bottom of the ocean in the form of shells or dead cells, they may be locked up for long periods in carbon sinks. (Arrows "Sinking" are broken to indicate that the bottom may be miles below the surface.)

QUESTION Why are there arrows leading from primary producers and primary consumers to dissolved C?

When carbon atoms are removed from CO_2 in the atmosphere and tied up in limestone or petroleum, they have entered a particularly long-lived carbon sink.

How do these observations relate to global warming? Recent experiments have shown that dramatic things happen when habitats in the middle of the ocean are fertilized with iron. Iron is a critical component of the electron transport chains responsible for photosynthesis and respiration, but it is in particularly short supply in the open ocean. After iron is added to ocean waters, it is not uncommon to see populations of protists and other primary producers increase by a factor of 10. Some researchers hypothesize that when these blooms occur, the amount of carbon that rains down into carbon sinks in the form of shells and dead cells may increase. If so, then fertilizing the ocean to promote blooms might be an effective way to reduce CO_2 concentrations in the atmosphere.

The effectiveness of iron fertilization is hotly debated, however. As some researchers point out, fertilizing the ocean with iron might lead to large accumulations of dead organic matter and the formation of anaerobic dead zones like those described in Chapter 28. Consequently, many biologists are not convinced that adding iron and inducing massive blooms of protists, bacteria, and archaea would be a good idea. But if further research shows that iron fertilization is safe and effective, it could be added to the list of possible approaches for reducing carbon dioxide levels in the atmosphere.

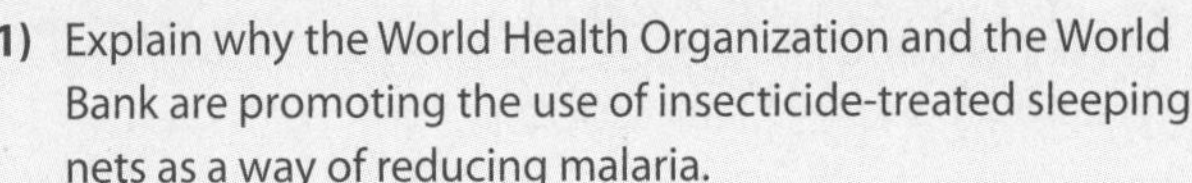

If you understand that...

- Biologists study protists because they cause disease and harmful algal blooms, and because they are key primary producers in aquatic environments.

You should be able to...

1) Explain why the World Health Organization and the World Bank are promoting the use of insecticide-treated sleeping nets as a way of reducing malaria.
2) Make a diagram of the carbon cycle that predicts the consequences of adding massive amounts of iron to marine plankton over many years.

29.2 How Do Biologists Study Protists?

Although biologists have made great strides in understanding pathogenic protists and the role that protists play in the global carbon cycle, it has been extremely difficult to gain any sort of solid insight into how the group as a whole diversified over time. The problem is that protists are so diverse that it has been difficult to find any overall patterns in the evolution and diversification of the group.

Recently, researchers have made dramatic progress in understanding protist diversity by combining data on the morphology of key groups and phylogenetic analyses of DNA sequence data. For the first time, a clear picture of how the Eukarya diversified may be within sight. Let's analyze how this work is being done, beginning with classical results on the morphological traits that distinguish major eukaryote groups.

Microscopy: Studying Cell Structure

Using light microscopy, biologists were able to identify and name many of the protist species known today. When transmission electron microscopes became available, a major breakthrough in understanding protist diversity occurred: Detailed studies of cell structure revealed that protists could be grouped according to characteristic overall form, according to organelles with distinctive features, or both. For example, both light and electron microscopy confirmed that the species that caused the Irish potato famine has reproductive cells with an unusual type of **flagellum**. Flagella are organelles that project from the cell and whip back and forth to produce swimming movements (Chapter 7). In reproductive cells of *Phytophthora infestans*, one of the two flagella present has tiny, hollow, hairlike projections. Biologists noted that kelp and other forms of brown algae also have cells with this type of flagellum.

To make sense of these results, researchers interpreted these types of distinctive morphological features as **synapomorphies**—shared, derived traits that distinguish major monophyletic

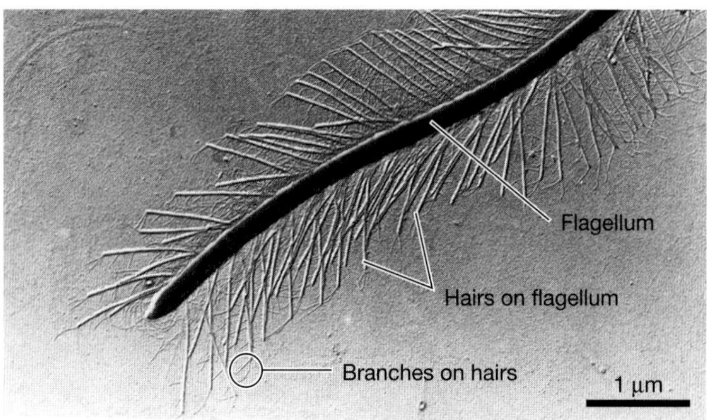

FIGURE 29.7 Species in the Lineage Called Stramenopiles Have a Distinctive Flagellum. The unusual, hollow "hairs" that decorate the flagella of stramenopiles often have three branches at the tip.

groups (see Chapter 27). Species that have a flagellum with hollow, hairlike projections became known as stramenopiles ("straw-hairs"); the hairs typically have three branches at the tip (**Figure 29.7**). In recognizing this group, investigators hypothesized that because an ancestor had evolved a distinctive flagellum, all or most of its descendants also had this trait. The qualifier *most* is important, because it is not unusual for certain subgroups to lose particular traits over the course of evolution, much as humans are gradually losing fur and tailbones.

Eventually, eight major groups of eukaryotes came to be identified on the basis of diagnostic morphological characteristics. These groups and the synapomorphies that identify them are listed in **Table 29.2**. Note that in almost every case, the synapomorphies listed in the table represent changes in structures that protect or support the cell or that influence the organism's ability to move or feed. Note also that the plants, fungi, and animals analyzed in Chapters 30 through 34 represent subgroups within two of the eight major eukaryotic lineages.

Although individual groups of protists were well characterized on the basis of morphology, the relationships among the eight eukaryotic lineages remained almost completely unknown. The next major advance in understanding eukaryotic diversity came when it became possible to obtain and analyze DNA sequence data. In the early 1990s, investigators began using molecular traits to reconstruct the evolutionary history of Eukarya.

Evaluating Molecular Phylogenies

When investigators first began using molecular data to infer the phylogeny of eukaryotes, most studies were based on the gene that codes for the RNA molecule in the small subunit of ribosomes. Recall from Chapter 1 that analyses of this gene revealed the nature of the three domains of life. When researchers sequenced this gene from an array of eukaryotes and used the data to infer a phylogeny of the domain Eukarya, the analysis suggested that the eight groups identified on the basis of distinctive morphological characteristics were indeed monophyletic groups. This was important support for the hypothesis that the distinctive morphological features were shared, derived characters that existed in a common ancestor of each lineage.

To understand the relationships among the eight lineages and their subgroups, biologists began analyzing sequence data from other genes and combining these results with data from morphological traits. Although estimating the phylogeny of Eukarya is still a work in progress, the phylogenetic tree in **Figure 29.8** is the current best estimate of the group's evolutionary history.

One of the emerging results is that the Amoebozoa and the Opisthokonta—which include fungi and animals—appear to form a monophyletic group recently given the provisional name Unikonta. Similarly, the Alveolata and Stramenopila appear to form a monophyletic group that biologists are call-

SUMMARY TABLE **29.2 Major Lineages of Eukaryotes**

Lineage	Distinguishing Morphological Features (synapomorphies)
Excavata	Cells have a pronounced "feeding groove" where prey or organic debris is ingested. No functioning mitochondria are present, although genes derived from mitochondria are found in the nucleus.
Discicristata	Cells have mitochondria with distinctive disc-shaped cristae.
Alveolate	Cells have sac-like structures called alveoli that form a continuous layer just under the plasma membrane. Alveoli are thought to provide support.
Stramenopila	If flagella are present, cells usually have two—one of which is covered with hairlike projections.
Rhizaria	Cells lack cell walls, although some produce an elaborate shell-like covering. When portions of the cell extend outward to move the cell, they are slender in shape.
Plantae	Cells have chloroplasts with a double membrane.
Opisthokonta	Reproductive cells have a single flagellum at their base. The cristae inside mitochondria are flat, not tube shaped as in other eukaryotes. (This lineage includes protists as well as the fungi and the animals. Fungi and animals are discussed in detail in Chapters 31 through 34.)
Amoebozoa	Cells lack cell walls. When portions of the cell extend outward to move the cell, they form large lobes.

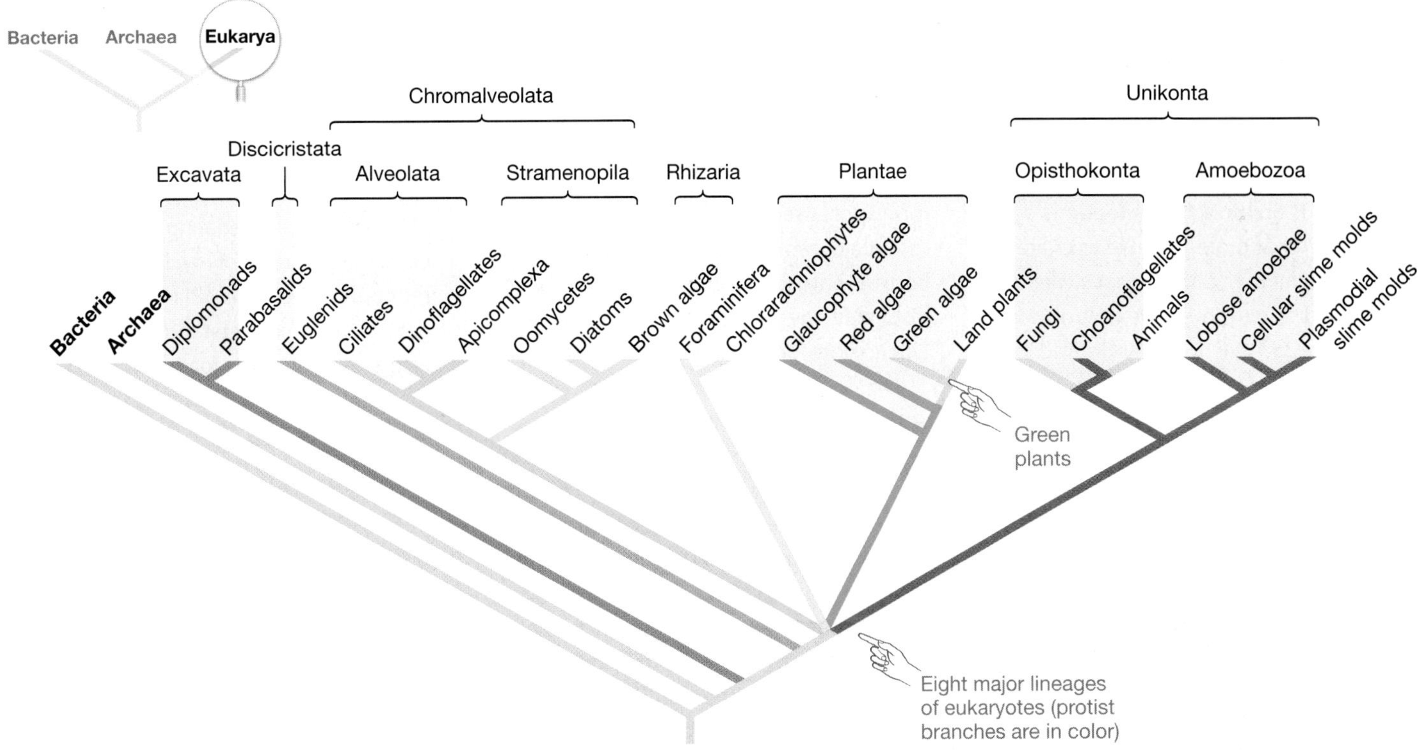

FIGURE 29.8 Phylogenetic Analyses Have Identified Eight Major Lineages of Eukaryotes. This tree shows selected subgroups from the eight major lineages discussed in this chapter. Many other lineages of protists have been identified.

EXERCISE Put a bar and label across the branch where the unusual flagellum of stramenopiles evolved. Put another bar and label to indicate where the distinctive mitochondria of the Discicristata evolved.

ing the Chromalveolata. Understanding where the root or base of the tree lies has been more problematic, however. The latest data suggest that a group of single-celled organisms called Excavates, which have a distinctive "feeding groove," forms the most basal lineage of the Eukarya. Groups with chloroplasts, specialized types of flagella, or unusual mitochondria evolved later. As more data become available, our understanding of eukaryote phylogeny will continue to improve.

Discovering New Lineages via Direct Sequencing

The effort to refine the phylogeny of the Eukarya is ongoing, and protists continue to be used as important model organisms in biology (see **Box 29.2**). But of all the research frontiers in eukaryotic diversity, the most exciting may be the one based on the technique called direct sequencing.

As Chapter 28 explained, **direct sequencing** is based on collecting organisms from a habitat and analyzing the DNA sequence of specific genes without growing larger populations of individuals in laboratory culture. The approach is based on using the polymerase chain reaction (PCR; see Chapter 19) to amplify certain genes in the organisms that have been collected. The resulting DNA sequence data are then used to place the organisms on a phylogenetic tree.

Recall that direct sequencing led to the discovery of previously unknown but major lineages of Archaea. To the amazement of biologists all over the world, the same thing happened when researchers used direct sequencing to survey eukaryotes.

The first direct sequencing studies that focused on eukaryotes were published in 2001 and were motivated by the hypothesis that direct sequencing could detect the presence of previously unknown species. To test this hypothesis, researchers sampled cells at various depths and locations in the oceans, used PCR primers to amplify genes that encode the RNA component of the ribosome's small subunit, sequenced the genes, and compared the data with the sequences of previously studied eukaryotes. One study sampled organisms at depths from 250 to 3000 m below the surface in waters off Antarctica; another focused on cells at depths of 75 m in the Pacific Ocean, near the equator. Both studies found a wide array of distinctive ribosomal RNA sequences—new species under the phylogenetic species concept introduced in Chapter 26.

Investigators who followed up on these results by examining the samples under the microscope were astonished to find that many of the newly discovered eukaryotes were tiny—from $0.2\mu m$ to $5\mu m$ in diameter. Other direct sequencing studies have now confirmed the existence of protists that are less than $0.2\mu m$ in diameter. These eukaryotic cells overlap in size with bacteria, which typically range from $0.5\mu m$ to $2\mu m$ in

BOX 29.2 A Model Organism: *Dictyostelium discoideum*

Several species of protists have served as model organisms in biology. Among the most important has been the cellular slime mold *Dictyostelium discoideum*. *Dictyostelium* is not always slimy, and it is not a mold—meaning a type of fungus. Instead, it is an amoeba. **Amoeba** is a general term that biologists use to characterize a unicellular protist that lacks a cell wall and is extremely flexible in shape. *Dictyostelium* has long fascinated biologists because it is a social organism. Independent cells sometimes aggregate to form a multicellular structure.

Figure 29.9 shows the *Dictyostelium discoideum* **life cycle**—the sequence of events that occurs over an individual's life span. Note that under most conditions, *D. discoideum* cells are haploid (*n*) and move about in decaying vegetation on forest floors or other habitats. They feed on bacteria by engulfing them whole. When these cells reproduce, they do so asexually, by mitosis. If food begins to run out, however, the cells begin to aggregate. In many cases, tens of thousands of cells cohere to form a 2-mm-long mass called a **slug**. (This is not the slug that is related to snails.) After migrating to a sunlit location, the slug stops and individual cells differentiate according to their position in the slug. Some form a stalk; others form a mass of spores at the tip of the stalk. A spore is a single cell that develops into an adult organism, but it is not formed from gamete fusion like a zygote. The entire structure, stalk plus mass of spores, is called a **fruiting body**. Cells that form spores secrete a tough coat and represent a durable resting stage. The fruiting body eventually dries out, and the wind disperses the spores to new locations, where more food might be available.

On occasion, *Dictyostelium* may also undergo sexual reproduction. When an aggregation is forming, two cells inside it may fuse to form a diploid (2*n*) zygote called a *giant cell*. The giant cell grows by feeding on the haploid amoebae in the aggregation. It then secretes a tough, protective coat. Later the zygote undergoes meiosis to form haploid offspring, which undergo mitosis. Eventually the haploid cells break out of the coating and begin to move about in search of food.

Dictyostelium discoideum has been an important model organism for investigating questions about eukaryotes:

- Cells in a slug are initially identical in morphology but then differentiate into distinctive stalk cells and spores. Studying this process helped biologists better understand how cells in plant and animal embryos differentiate into distinct cell types.
- The process of slug formation has helped biologists study how animal cells move and how they aggregate as they form specific types of tissues. In addition, the discovery that aggregating *D. discoideum* cells follow trails of cyclic adenosine monophosphate (cAMP) helped investigators better understand the mechanisms responsible for **chemotaxis**—movement in response to a chemical—in other organisms.
- When *D. discoideum* cells aggregate to form a slug, they stick to each other. The discovery of membrane proteins responsible for cell-cell adhesion helped biologists understand some of the general principles of multicellular life highlighted in Chapter 8.
- Cells that aggregate to form a slug are not genetically identical. About 20 percent of the cells help other individuals produce offspring by forming the stalk, but die without producing offspring themselves. Why? Researchers are using *D. discoideum* to answer questions about why some individuals appear to sacrifice themselves to help others.

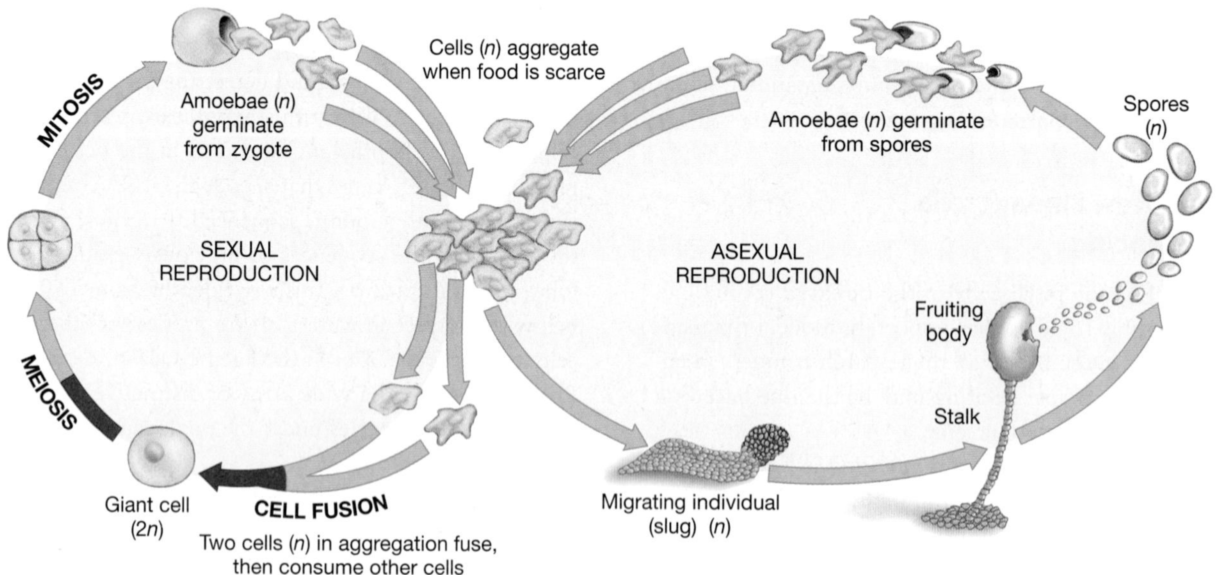

FIGURE 29.9 The Life Cycle of *Dictyostelium discoideum*.

QUESTION In slime molds, why isn't spore production considered a form of sexual reproduction?

QUESTION Most biologists consider the slime mold slug to be a simple form of multicellularity. Why?

diameter. The take-home message is that eukaryotic cells are much more variable in size than previously imagined. A whole new world of tiny protists has just been discovered.

The initial analyses suggest that at least some of the new species may represent important new lineages of dinoflagellates, stramenopiles, or alveolates. As additional direct sequencing studies are done, it is virtually certain that our understanding of eukaryotic diversity will change—perhaps radically—and improve.

Check Your Understanding

If you understand that...

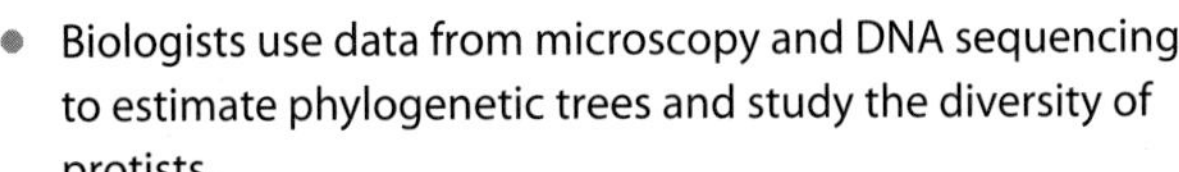

- Biologists use data from microscopy and DNA sequencing to estimate phylogenetic trees and study the diversity of protists.
- According to the most recent analyses, the domain Eukarya comprises eight major lineages. Members of each lineage have distinctive aspects of cell structure.
- Direct sequencing has allowed investigators to recognize large numbers of previously undescribed eukaryotes, some of which are extremely small.

You should be able to...

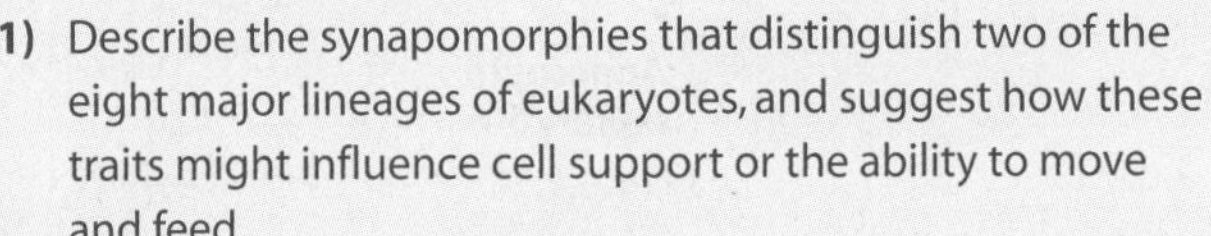

1) Describe the synapomorphies that distinguish two of the eight major lineages of eukaryotes, and suggest how these traits might influence cell support or the ability to move and feed.
2) Outline the steps in a direct sequencing study that would allow you to characterize the protists present in the waters surrounding a coral reef.

29.3 What Themes Occur in the Diversification of Protists?

The protists range in size from bacteria-sized single cells to giant kelp. They live in habitats from the open oceans to dank forest floors. They are almost bewildering in their morphological and ecological diversity. Because they are a paraphyletic group, they do not share derived characteristics that set them apart from all other lineages on the tree of life.

Fortunately, one general theme helps tie protists together. Once an important new innovation arose in protists, it triggered the evolution of species that live in a wide array of habitats and make a living in diverse ways. Let's first consider novel morphological traits that arose in protists and then analyze new methods of feeding, moving, and reproducing that evolved in these lineages. Throughout, you should focus on understanding how each innovation helped make the fantastic diversity of protists possible.

What Morphological Innovations Evolved in Protists?

What did the earliest eukaryotes look like? Because the most ancient eukaryotic groups are unicellular, and because virtually all bacteria and all archaea are also unicellular, biologists infer that the first eukaryote was a single-celled organism. Further, all eukaryotes alive today have a nucleus and endomembrane system, mitochondria or genes that are normally found in mitochondria, and a cytoskeleton. Based on these observations, biologists conclude that their common ancestor also had these structures. And because the most ancient eukaryotic lineages lack cell walls, biologists suggest that the common ancestor of all eukaryotes living today also lacked this feature.

In sum, the earliest eukaryotes were probably single-celled organisms with a nucleus and endomembrane system, mitochondria, and a cytoskeleton, but no cell wall. It is also likely that these cells swam using a novel type of flagellum. Eukaryotic flagella are completely different structures from bacterial flagella and evolved independently. The flagella found in protists and other eukaryotes are made up of microtubules, and dynein is the major motor protein. An undulating motion occurs as dynein molecules walk down microtubules. The flagella of bacteria and archaea, in contrast, are composed primarily of a protein called flagellin (see Chapter 7). Instead of undulating, these flagella rotate to produce movement.

If you understand the synapomorphies that identify the eukaryotes as a monophyletic group, you should be able to map the origin of the nuclear envelope and the eukaryotic flagellum on Figure 29.8. Once you've done that, let's consider how several of these key new morphological features arose and influenced the subsequent diversification of protists, beginning with the trait that defines the Eukarya—the nuclear envelope.

The Nuclear Envelope Although researchers are still gathering evidence on how the nuclear envelope originated, the leading hypothesis is that it is derived from infoldings of the plasma membrane. As the drawings in **Figure 29.10** show, these infoldings are thought to have given rise to the nuclear envelope and the endoplasmic reticulum (ER) together. Two lines of evidence support this hypothesis: Infoldings of the plasma membrane occur in some bacteria living today, and the nuclear envelope and ER of today's eukaryotes are continuous (see Chapter 7). If you understand the infolding hypothesis, you should be able to explain why these observations support it.

According to current thinking, the evolution of the nuclear envelope was advantageous because it separated transcription and translation. RNA transcripts are processed inside the nucleus but translated outside the nucleus. In bacteria and archaea, transcription and translation occur together. Once a simple nuclear envelope was in place, alternative splicing and other forms of RNA processing could occur—giving the early

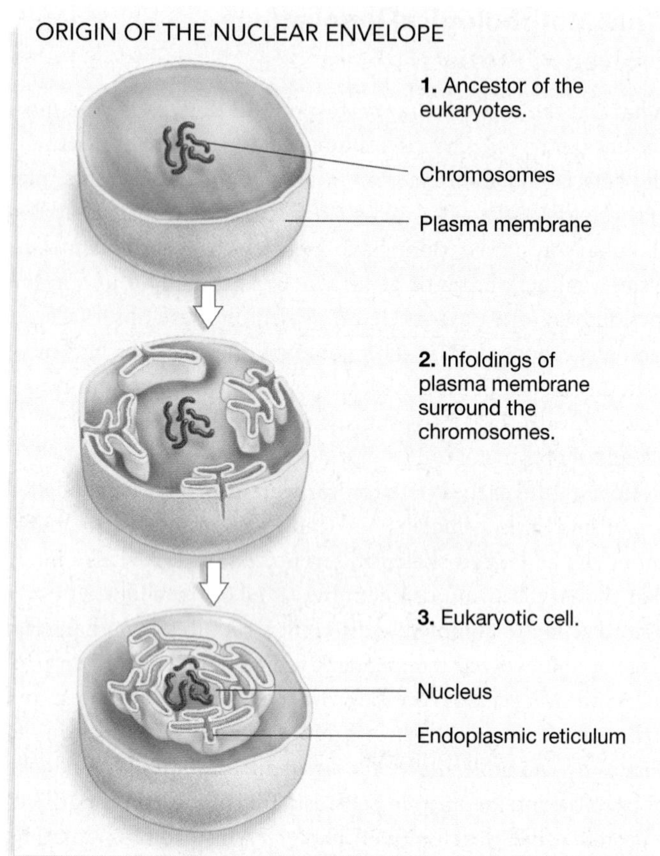

FIGURE 29.10 A Hypothesis for the Origin of the Nuclear Envelope. Infoldings of the plasma membrane, analogous to those shown here, have been observed in bacteria living today.

eukaryotes a novel way to control gene expression (see Chapter 18). The take-home message here is that an important morphological innovation gave the early eukaryotes a new way to manage and process genetic information.

Once a nucleus had evolved, it underwent diversification. In some cases, unique types of nuclei are associated with the founding of important lineages of protists. Ciliates, for example, have a diploid micronucleus that is involved only in reproduction and a polyploid macronucleus where transcription occurs. Diplomonads have two nuclei that look identical; it is not known how they interact. In foraminifera, red algae, and plasmodial slime molds, certain cells may contain many nuclei. Dinoflagellates have chromosomes that lack histones and attach to the nuclear envelope. In each of these groups the distinctive structure of the nucleus—and presumably, differences in how genetic information is processed—represent synapomorphies present in a common ancestor.

The Mitochondrion Mitochondria are organelles that generate ATP using pyruvate as an electron donor and oxygen as the ultimate electron acceptor (see Chapter 9). In 1981 Lynn Margulis expanded on a radical hypothesis—first proposed in the nineteenth century—to explain the origin of mitochondria. The **endosymbiosis theory** proposes that mitochondria originated when a bacterial cell took up residence inside a eukaryote about 2 billion years ago. Its name is inspired by the Greek word roots *endo*, *sym*, and *bio* (literally, "inside-together-living"). **Symbiosis** is said to occur when individuals of two different species live in physical contact; **endosymbiosis** occurs when an organism of one species lives inside an organism of another species.

In its current form, the endosymbiosis theory proposes that mitochondria evolved through a series of steps. As **Figure 29.11** shows, the process began when eukaryotic cells started to use their cytoskeletal elements to surround and engulf smaller prey.

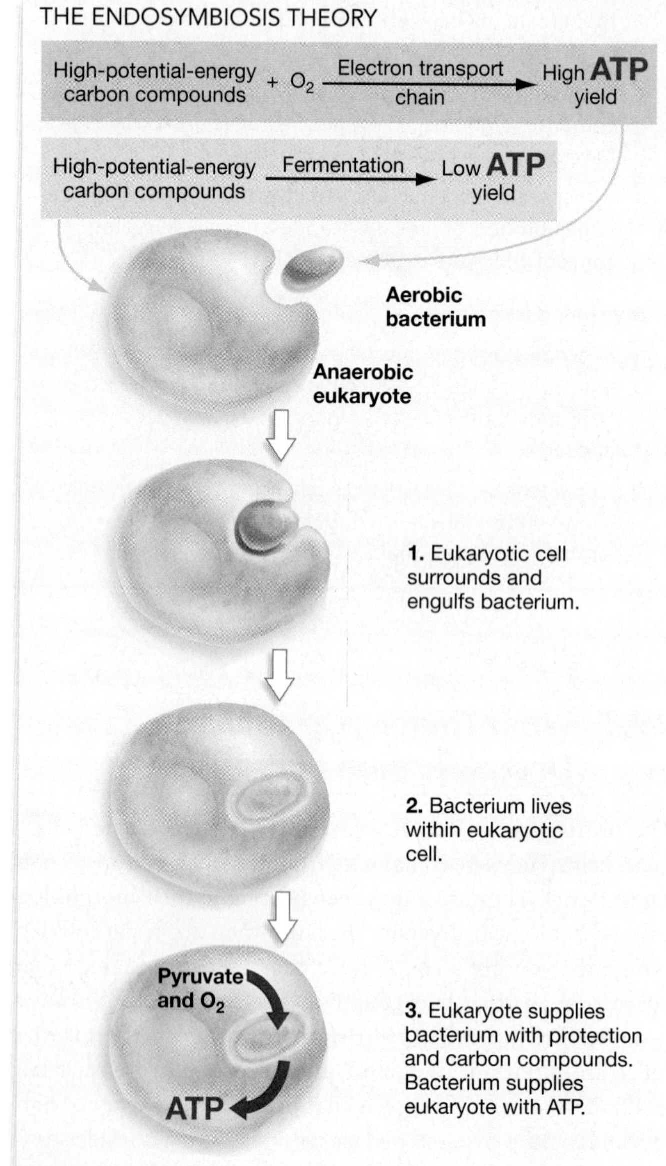

FIGURE 29.11 Proposed Initial Steps in the Evolution of the Mitochondrion.

QUESTION According to this hypothesis, how many membranes should surround a mitochondrion? Explain your logic.

The theory proposes that instead of being digested, an engulfed bacterium began to live inside its eukaryotic host. Specifically, the theory maintains that the engulfed cell survived by absorbing carbon molecules with high potential energy from its host and oxidizing them, using oxygen as a final electron acceptor. The host cell, in contrast, is proposed to be a predator capable only of anaerobic fermentation—meaning it could not use oxygen as an electron acceptor in cellular respiration. The relationship between the host and the engulfed cell was presumed to be stable because a mutual advantage existed between them: The host supplied the bacterium with protection and carbon compounds from its other prey, while the bacterium produced much more ATP than the host cell could synthesize on its own. Cells that can use oxygen during cellular respiration are able to produce much more ATP than are cells that cannot (see Chapter 28).

When Margulis first began promoting the theory, it met with a storm of criticism—largely because it seems slightly preposterous. But gradually biologists began to examine it rigorously. For example, endosymbiotic relationships between protists and bacteria exist today. Among the α-probacteria alone, three major groups are found *only* inside eukaryotic cells. Several observations about the structure of mitochondria are also consistent with the endosymbiosis theory:

- Mitochondria are about the size of an average bacterium and replicate by fission, as do bacterial cells. The duplication of mitochondria takes place independently of division by the host cell. When eukaryotic cells divide, each daughter cell receives some of the many mitochondria present.
- Mitochondria have their own ribosomes and manufacture their own proteins. Mitochondrial ribosomes closely resemble bacterial ribosomes in size and composition and are poisoned by antibiotics such as streptomycin that inhibit bacterial, but not eukaryotic, ribosomes.
- Mitochondria have double membranes, consistent with the engulfing mechanism of origin illustrated in Figure 29.11.
- Mitochondria have their own genomes, which are organized as circular molecules—much like a bacterial chromosome. Mitochondrial genes code for the enzymes needed to replicate and transcribe the mitochondrial genome.

Although these data are impressive, they are only consistent with the endosymbiosis theory. Stated another way, they do not exclude other explanations. This is a general principle in science: Evidence is considered strong when it cannot be explained by reasonable alternative hypotheses. In this case, the key was to find data that tested predictions made by Margulis's idea against predictions made by an alternative theory: that mitochondria evolved within eukaryotic cells, separately from bacteria.

A breakthrough occurred when researchers realized that according to the "within-eukaryotes" theory, the genes found in mitochondria had to have been derived from some of the nuclear genes of ancestral eukaryotes. Margulis's theory, in contrast, proposed that the genes found in mitochondria were bacterial in origin.

These predictions were tested by studies on the phylogenetic relationships of mitochondrial genes (**Figure 29.12**). For example, researchers compared gene sequences isolated from the

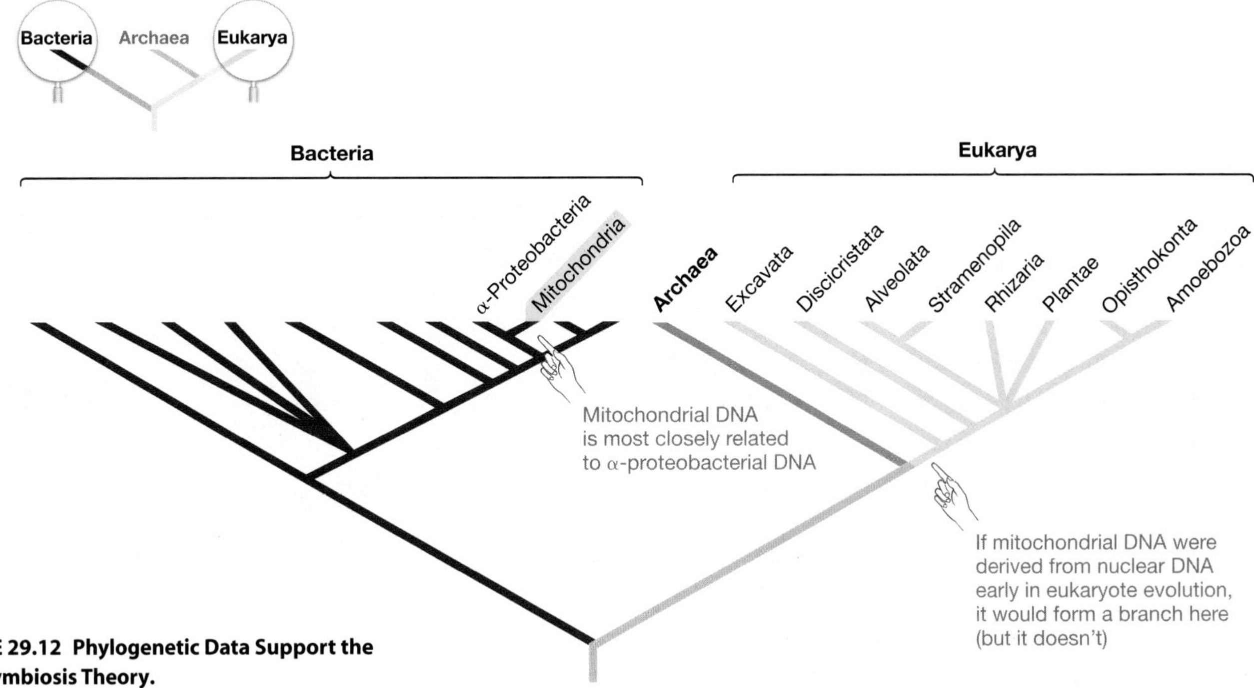

FIGURE 29.12 Phylogenetic Data Support the Endosymbiosis Theory.

nuclear DNA of eukaryotes, mitochondrial DNA from eukaryotes, and DNA from several species of bacteria. Exactly as the endosymbiosis theory predicted, the mitochondrial gene sequences turned out to be much more closely related to the sequences from the α-proteobacteria than to sequences from the nuclear DNA of eukaryotes. The result was considered overwhelming evidence that the mitochondrial genome came from an α-proteobacterium rather than from a eukaryote. The endosymbiosis theory was the only reasonable explanation for the data. The results were a stunning vindication of a theory that had once been intensely controversial. Mitochondria evolved via endosymbiosis. ● If you understand the endosymbiosis theory and the evidence for it, you should be able to describe how the chloroplast—the organelle where photosynthesis takes place in eukaryotes—could arise via endosymbiosis. You should also be able to list the types of evidence that would support an endosymbiotic origin for the chloroplast.

Once the mitochondrion was present in the ancestor of today's protists, it underwent diversification. In Diplomonads and Parabasalids, the organelle was lost entirely or is now a vestigial trait. In the Discicristata and Opisthokonta, the sac-like cristae inside the mitochondrion—where the electron transport chain and ATP synthase are located—have distinctive shapes (see Table 29.2). Presumably, variation in the presence or structure of mitochondria reflects variation in how ATP is produced by protists. In these cases, the mitochondrion's absence or structure qualifies as a synapomorphy that identifies a monophyletic group. The mitochondrion was a morphological innovation that subsequently diversified.

Structures for Support and Protection Bacterial and archaeal cells contain protein filaments that provide a rudimentary cytoskeleton. In contrast, the microfilaments, intermediate filaments, and microtubules found in eukaryotes form an extensive and dynamic internal skeleton (see Chapter 7). Understanding how these cytoskeletal elements evolved is a subject of current research. And as far as is known, the basic structure of the cytoskeleton does not vary much among protists. What does vary significantly is the presence and nature of other structures that provide support and protection for the cell. Many protists have cell walls outside their plasma membrane; others have hard external structures called a **test** or a **shell**; others have rigid structures inside the plasma membrane. In many cases, these novel structures represent synapomorphies that identify monophyletic groups among protists. For example:

- Diatoms are surrounded by a glass-like, silicon-oxide shell (**Figure 29.13a**). The shell is made up of two pieces that fit together in a box-and-lid arrangement, like the petri plates you may have seen in lab.
- Dinoflagelletes have a cell wall made up of cellulose plates (**Figure 29.13b**).
- Within Foraminifera, some lineages secrete an intricate, chambered test of calcium carbonate ($CaCO_3$; **Figure 29.13c**).
- Members of other Foraminifera lineages, and some amoebae, cover themselves with tiny pebbles.
- The parabasalids have a distinctive internal support rod, consisting of cross-linked microtubules that run the length of the cell.
- The euglenids have a collection of protein strips located just under the plasma membrane. The strips are supported by microtubules and stiffen the cell.
- The alveolates have distinctive sac-like structures called alveoli, located just under the plasma membrane, that help stiffen the cell.

In many cases, the diversification of protists has been associated with the evolution of innovative structures for support and protection.

(a) Diatom

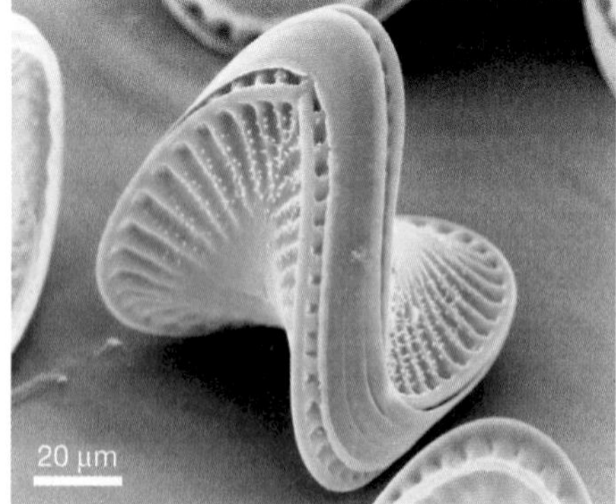

Test made of silicon oxides

(b) Dinoflagellate

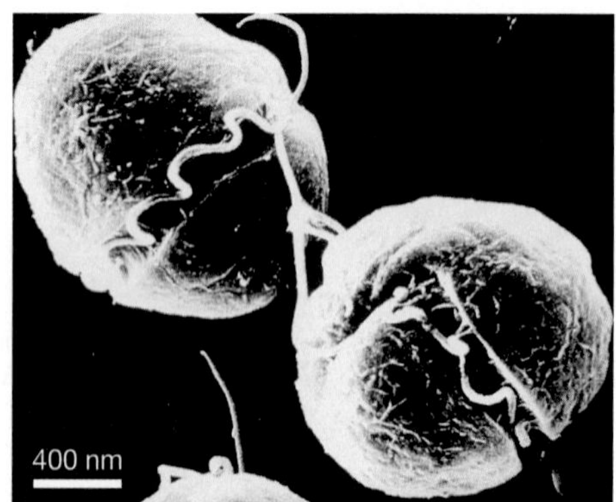

Plates made of cellulose

(c) Foraminiferan

Calcium carbonate test, with chambers

FIGURE 29.13 Hard Outer Coverings in Protists Vary in Composition.

Multicellularity The story of many protist lineages starts with the evolution of novel structures associated with managing genetic information, producing ATP, and supporting or protecting the cell. In other lineages of protists, though, the key morphological innovation was **multicellularity**. Multicellular individuals contain more than one cell. At least some of the cells are attached to each other, and some are specialized for different functions. In the simplest multicellular species, certain cells are specialized for producing or obtaining food while other cells are specialized for reproduction. The key point about multicellularity is that not all cells express the same genes.

A few species of bacteria are capable of aggregating and forming fruiting bodies similar to those observed in the protist *Dictyostelium discoideum* (see Box 29.2). Because cells in the fruiting bodies of these bacteria differentiate into specialized stalk cells and spore-forming cells, they are considered multicellular. But the vast majority of multicellular species are members of the Eukarya. Based on data presented in Figure 8.8, biologists are convinced that multicellularity evolved several times as protists diversified. The condition also evolved independently in the green plants, fungi, and animals. Multicellularity is a synapomorphy shared by all of the brown algae and all of the plasmodial and cellular slime molds. It also arose in some lineages of red algae.

To summarize, an array of novel morphological traits played a key role as protists diversified: the nucleus and endomembrane system, the mitochondrion, structures for protection and support, and multicellularity. Protists invented a variety of new ways to build and manage an individual. Once a new type of eukaryotic cell or multicellular individual existed, subsequent diversification was often triggered by novel ways of finding food, moving, or reproducing. Let's consider each of these life processes in turn.

How Do Protists Find Food?

According to Chapter 28, bacteria and archaea can use a wide array of molecules as electron donors and electron acceptors during cellular respiration. They get these molecules by absorbing them directly from the environment. Other bacteria don't absorb their nutrition—instead, they make their own food via photosynthesis.

Many groups of protists are similar to bacteria in the way they find food: They perform photosynthesis or absorb their food directly from the environment. But one of the most important stories in the diversification of protists was the evolution of a novel method for finding food. Many protists ingest their food—they eat bacteria, archaea, or even other protists whole. When ingestive feeding occurs, an individual takes in packets of food much larger than individual molecules. Thus, protists feed by (1) ingesting packets of food, (2) absorbing organic molecules directly from the environment, or (3) performing photosynthesis. Some ingest food as well as performing photosynthesis—meaning that they use a combination of feeding strategies.

Ingestive Feeding Ingestive lifestyles are based on eating live or dead organisms or on scavenging loose bits of organic debris. Protists such as the cellular slime mold *Dictyostelium discoideum* are large enough to engulf bacteria and archaea; many protists are large enough to surround and ingest other protists or microscopic animals. The engulfing process is possible in protists that lack a cell wall. A flexible membrane and dynamic cytoskeleton give these species the ability to surround and "swallow" prey using long, fingerlike projections called **pseudopodia** ("false-feet"). The engulfing process is illustrated in **Figure 29.14a**; the evolution of large cell size in eukaryotes is analyzed in **Box 29.3**.

(a) Pseudopodia engulf food

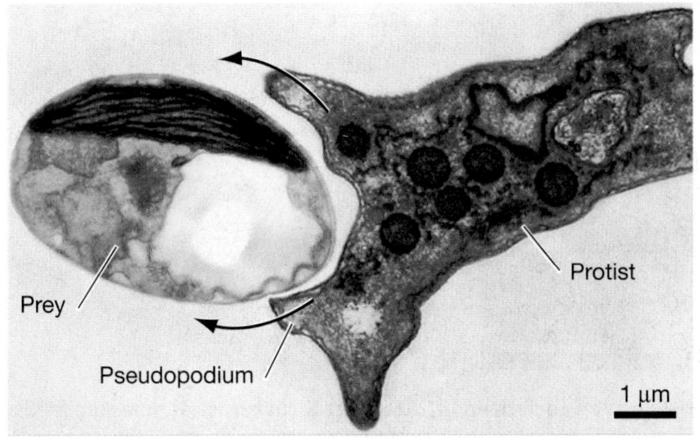

(b) Ciliary currents sweep food into gullet

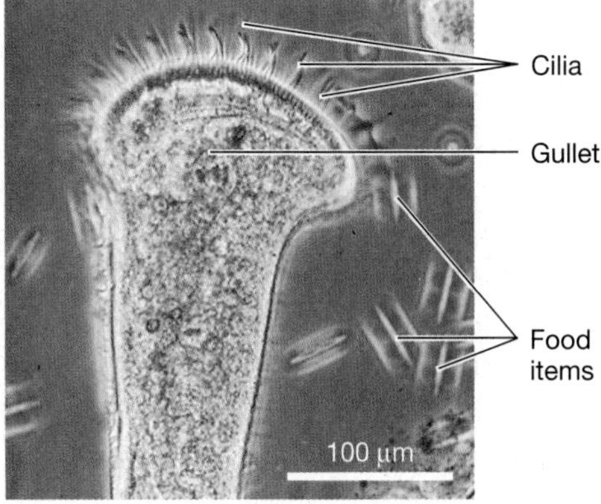

FIGURE 29.14 Ingestive Feeding. Methods of prey capture vary among ingestive protists. **(a)** Some predators engulf prey with pseudopodia; **(b)** other predators sweep them into their gullets with water currents set up by the beating of cilia.

Losing a cell wall and gaining the ability to move their plasma membrane around prey items or food particles was an important innovation during the evolution of protists. Recall from Chapter 28 that bacteria and archaea are abundant in wet soils and aquatic habitats, where protists also live. Instead of competing with bacteria and archaea for sunlight or food molecules, protists could eat them.

Although many ingestive feeders actively hunt down prey and engulf them, others do not. Instead of taking themselves to food, these species attach themselves to a surface. Protists that feed in this way have cilia that surround the mouth and beat in a coordinated way. The motion creates water currents that sweep food particles into the cell (**Figure 29.14b**).

Absorptive Feeding When nutrients are taken up directly from the environment, across the plasma membrane, absorptive feeding occurs. Absorptive feeding is common among protists. It's important to recognize, though, that protists are not nearly

BOX 29.3 Why Can Eukaryotes Have Such Large Cells?

Although eukaryotic cells vary widely in size, the domain Eukarya is distinguished by the evolution of the largest cells known. An average-sized eukaryotic cell is 10 times larger in diameter than an average-sized bacterial cell.

How could eukaryotic cells get so big? The answer to this question isn't obvious, because large cell size presents an important physical challenge: As cells become larger, their volume increases much more rapidly than does their surface area. This is because a cell's volume increases as the cube of a sphere's diameter (volume $\propto$ diameter^3) while its surface area increases as the square of a sphere's diameter (surface area $\propto$ diameter^2). As cells get larger, then, the proportion of surface area available gets smaller and smaller relative to the volume present (see Section 42.3). This creates a problem because food, gases, and waste molecules must diffuse across the cell's surface, while the volume is filled with biosynthetic machinery that requires raw materials and generates waste. As cell size increases, metabolism in the cell's interior can outstrip the transport and exchange processes at the surface—where raw materials enter and waste products leave.

Eukaryotic cells manage this dilemma in part because they are divided into membrane-bound compartments, each of which has a relatively high surface area and relatively low volume. As an example of how the compartmentalization of the eukaryotic cell works, consider the ciliate called *Paramecium* (**Figure 29.15**). This organism makes a living by eating bacteria, which it sweeps into an indentation known as the gullet and then into the cell mouth. After ingesting a bacterium, a *Paramecium* surrounds it with an internal membrane, forming a compartment called a **food vacuole**. Food vacuoles merge with membrane-bound structures called lysosomes, which hold digestive enzymes, and circulate around the cell. When the food has been digested and nutrients have diffused out of the food vacuole, the vacuole merges with the plasma membrane at a special organelle—the anal pore—and expels waste molecules. The food vacuoles provide a large area of internal membrane, allowing nutrients to be delivered efficiently throughout the volume of the cell and waste products to be expelled rapidly.

Although not all eukaryotic cells contain food vacuoles and anal pores, they all have many other internal compartments. Each of the internal structures introduced in Chapter 7—including the nucleus, lysosomes, peroxisomes, mitochondria, chloroplasts, central vacuole, Golgi apparatus, and rough and smooth endoplasmic reticulum (ER)—has a distinct function. Many are membrane bound and are devoted to the synthesis, transport, and distribution of molecules. These organelles were morphological innovations that made the evolution of large size possible. Large cell size was important because it made ingestive feeding possible. Without ingestive feeding, endosymbiosis and the evolution of mitochondria and chloroplasts never would have occurred.

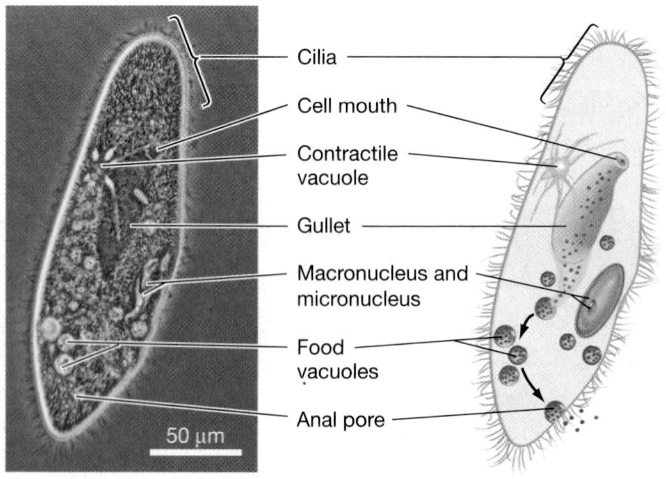

FIGURE 29.15 Protists Have Specialized Intracellular Structures. *Paramecium* feeds by sweeping bacteria into a subcellular structure called the gullet.

EXERCISE Next to the drawing of *Paramecium,* list other organelles and intracellular structures that are found in protist cells.

as diverse as bacteria and archaea in terms of the electron donors and electron acceptors they absorb to use in cellular respiration. Most protists use glucose or other sugars as electron donors and oxygen as an electron acceptor. Unlike bacteria and archaea, which can ferment many different types of molecules, most protists can only ferment sugars.

Some protists that live by absorptive feeding are decomposers. A **decomposer** is an organism that feeds on dead organic matter, or **detritus**. But many of the protists that absorb their nutrition directly from the environment live inside other organisms. If they damage their host, the absorptive species is called a **parasite**. Most of the protists responsible for the diseases listed in Table 29.1 are parasites.

Photosynthesis You might recall from Chapter 10 and Chapter 28 that photosystems I and II evolved in bacteria, and that both photosystems occur in cyanobacteria. None of the basic machinery required for photosynthesis evolved in eukaryotes. Instead, they "stole" it via endosymbiosis.

The endosymbiosis theory contends that the organelle where photosynthesis takes place in eukaryotes originated when a protist engulfed a cyanobacterium. Once inside the protist, the photosynthetic bacterium provided its eukaryotic host with oxygen and glucose in exchange for protection and access to light. If the endosymbiosis theory is correct, today's chloroplasts trace their ancestry to cyanobacteria.

All of the photosynthetic eukaryotes have both a chloroplast and a mitochondrion, and the evidence for an endosymbiotic origin for the chloroplast is even more persuasive than for mitochondria. Chloroplasts have the same list of bacteria-like characteristics presented earlier for mitochondria. There are many examples of endosymbiotic cyanobacteria living inside protists or animals today, and the DNA sequences inside chloroplasts are extremely similar to cyanobacterial genes. In addition:

- The photosynthetic organelle of one group of protists, called the glaucophyte algae, has an outer layer containing the same constituent (peptidoglycan) found in the cell walls of cyanobacteria.
- Like chloroplasts, the cyanobacterium *Prochloron* contains chlorophylls *a* and *b* and has a system of internal membranes where the photosynthetic pigments and enzymes are located.

● If you understand the evidence for the endosymbiotic origin of the chloroplast, you should be able to explain why these latter two observations support the theory. You should also be able to add a branch indicating the location of chloroplast genes on the phylogenetic tree in Figure 29.12.

The acquisition of the chloroplast, like the acquisition of the mitochondrion, is called primary endosymbiosis. Because all of the species in the Plantae have chloroplasts with two membranes and similar molecular composition, biologists infer that primary endosymbiosis occurred in their common ancestor. Stated another way, the chloroplast is the synapomorphy that identifies the glaucophyte algae, red algae, and green plants (green algae and land plants) as part of the same monophyletic group.

The plot thickens, however, because chloroplasts also occur in three of the other major lineages of protists—the Discicristata, Chromalveolates, and Rhizaria. But in these species, the chloroplast is surrounded by more than two membranes—usually four. To explain this observation, researchers hypothesize that the ancestors of these groups acquired their chloroplasts by ingesting photosynthetic protists that already had chloroplasts. This process, called secondary endosymbiosis, occurs when an organism engulfs a photosynthetic eukaryotic cell and retains its chloroplasts as intracellular symbionts (**Figure 29.16**).

Figure 29.17 shows where primary and secondary endosymbiosis occurred on the phylogenetic tree of eukaryotes. Once protists obtained the chloroplast, it was "swapped around" to new lineages via secondary endosymbiosis. Many lineages of protists acquired the ability to photosynthesize independently of each other. In each case, the acquisition of a chloroplast triggered a radiation of photosynthetic species. ● If you under-

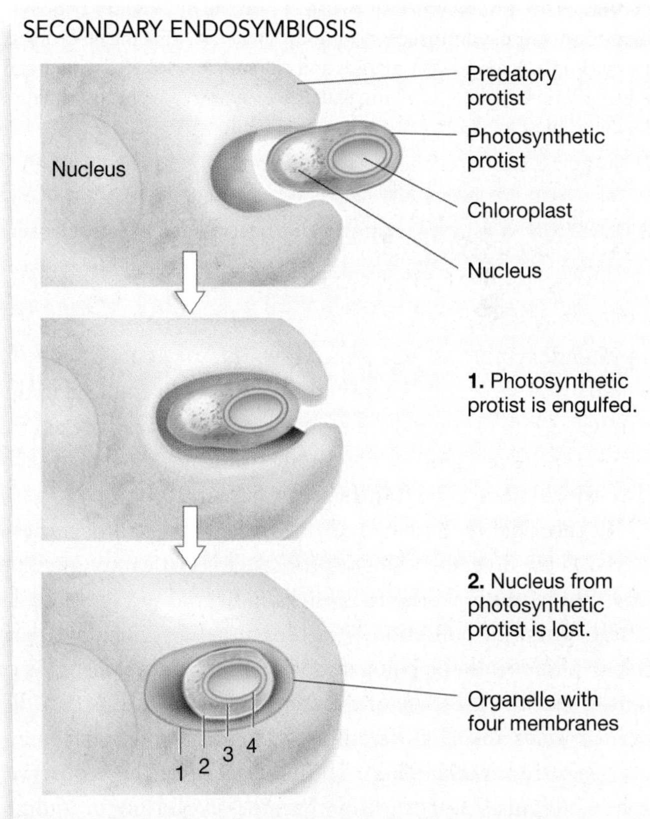

FIGURE 29.16 Secondary Endosymbiosis Leads to Organelles with Four Membranes. The chloroplasts found in some protists have four membranes and are hypothesized to be derived by secondary endosymbiosis. In species where chloroplasts have three membranes, biologists hypothesize that secondary endosymbiosis was followed by the loss of one membrane.

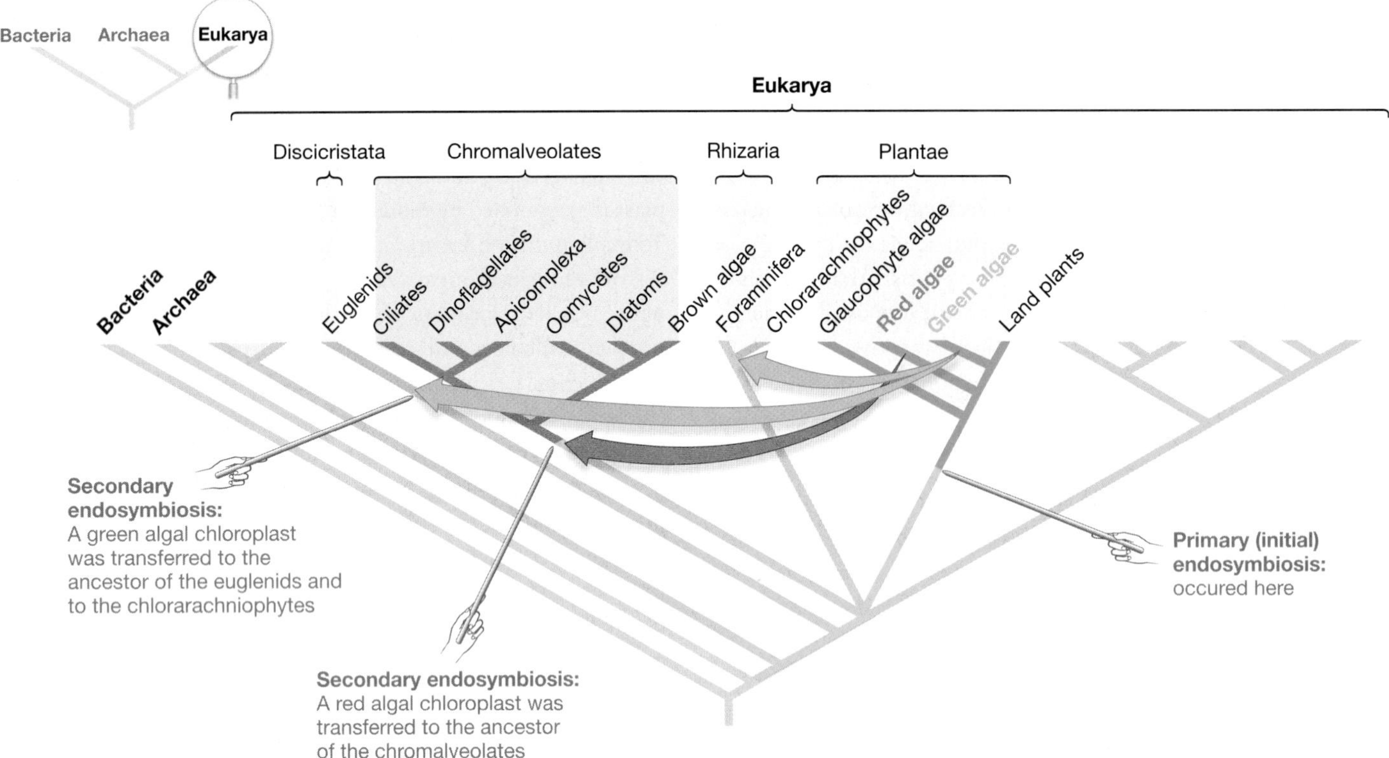

FIGURE 29.17 Photosynthesis Arose in Protists by Primary Endosymbiosis, Then Spread among Lineages via Secondary Endosymbiosis. Biochemical similarities link the chloroplasts found in chromalveolates with red algae and the chloroplasts found in euglenids and chlorarachniophytes with green algae. Note that among the chromalveolates, only the dinoflagellates, diatoms, and brown algae have photosynthetic species. In ciliates, apicomplexa, and oomycetes, the chloroplast has been lost or changed function.

EXERCISE Mark and label the primary endosymbiosis that gave rise to the mitochondrion, based on the observation that all eukaryotes living today have a mitochondrion or a vestige of one.

QUESTION Why haven't mitochondria been gained by secondary endosymbiosis?

stand endosymbiosis, you should be able to explain why the primary and secondary endosymbiosis events introduced in this chapter represent the most massive lateral gene transfers in the history of life, in terms of the number of genes moved at once. (To review lateral gene transfer, see Chapter 20.)

To thoroughly understand the role that photosynthesis played in the diversification of protists, it is critical to note that most of the photosynthetic lineages indicated in Figure 29.17 contain a unique collection of photosynthetic pigments. A particular photosynthetic pigment absorbs specific wavelengths of light. Thus, the presence of different pigments means that different species absorb different wavelengths in the electromagnetic spectrum (**Table 29.3**). To make sense of these observations, biologists hypothesize that the evolution of unique combinations of pigments gave photosynthetic eukaryotes the ability to harvest unique wavelengths of light and avoid competition. Chapter 28 pointed out that a similar phenomenon appears to have occurred among photosynthetic bacteria. Like the nuclear envelope, mitochondrion, multicellularity, ingestive feeding, and other traits that originated with protists, the acquisition of the chloroplast was an innovation that triggered subsequent diversification.

It is hard to overstate the importance of photosynthesis in the diversification of protists, and it is hard to overstate the importance of photosynthetic protists to other species. For example, many photosynthetic protists live symbiotically with animals or other protists—either inside or outside host cells. Although there are hundreds of examples, perhaps the most important involve the animals called corals. Virtually all reef-building corals harbor single-celled, photosynthetic protists in their skin or gut tissue. The concentrations of endosymbiotic cells can be impressive—up to a million protist cells per square centimeter of coral surface (**Figure 29.18**). The sheer abundance of these and other photosynthetic protists is significant. Plants may be the dominant producers of sugars and other high-energy organic compounds on land, but photosynthetic protists and bacteria are the dominant producers of carbon compounds in oceans and lakes.

Diversity in Lifestyles Before leaving the topic of obtaining food in protists, it's important to recognize that all three

TABLE 29.3 **Pigments in Photosynthetic Protists**

Lineage	Photosynthetic pigments	Accessory pigments
Glaucophytes	Chlorophyll *a*	Phycocyanin, allophycocyanin, β-carotene
Red algae	Chlorophyll *a*	Primarily phycoerythrin
Brown algae, diatoms, dinoflagellates	Chlorophyll *a* and *c*	Fucoxanthin
Green algae	Chlorophyll *a* and *b*	Fucoxanthin
Chlorarachniophytes	Chlorophyll *a* and *b*	Xanthophylls
Euglenids	Chlorophyll *a* and *b*	Xanthophylls, β-carotene

QUESTION Are the pigments listed in this table consistent with the primary endosymbiosis and secondary endosymbiosis hypotheses described in Figure 29.17? Explain your answer.

lifestyles—ingestive, absorptive, and photosynthetic—occur in many different eukaryote lineages. All three types of food getting can occur within a single clade.

To drive this point home, consider the monophyletic group called the alveolates. There are three major subgroups of this lineage, called dinoflagellates, apicomplexa, and ciliates. Within each subgroup, species vary in how they make a living. About half of the dinoflagellates are photosynthetic, while many others are parasitic. Apicomplexa are parasitic. Ciliates include many species that ingest prey, but some ciliates live in the guts of cattle or the gills of fish and absorb nutrients from their hosts. Other ciliate species make a living by holding algae or other types of photosynthetic symbionts inside their cells.

FIGURE 29.18 Many Photosynthetic Protists Live Symbiotically. The coral pictured here contains thousands of symbiotic dinoflagellates.

Within each of the eight major lineages of eukaryotes, different methods for feeding helped trigger diversification.

How Do Protists Move?

Many protists actively move to find food. Predators such as *Dictyostelium discoideum* crawl over a substrate in search of prey. Most of the unicellular, photosynthetic species are capable of swimming to sunny locations. How are these crawling and swimming movements possible?

Amoeboid motion is a sliding movement observed in some protists. In the classic mode illustrated in **Figure 29.19a**, pseudopodia stream forward over a substrate, with the rest of the cytoplasm, organelles, and plasma membrane following. The motion requires ATP and involves interactions between proteins called actin and myosin inside the cytoplasm. The mechanism is related to muscle movement in animals, which is detailed in Chapter 47. But at the level of the whole cell, the precise sequence of events during amoeboid movement is still somewhat uncertain.

The other major mode of locomotion in protists involves flagella or cilia (**Figure 29.19b and c**). Both have identical structures. Recall from Chapter 7 that both consist of nine sets of doublet (paired) microtubules arranged around two central, single microtubules. Flagella, however, are long and are usually found alone or in pairs; cilia are short and usually occur in large numbers on any one cell. Flagella and cilia can also be distinguished by the types of structures associated with the basal bodies where they originate.

Even closely related protists can use radically different forms of locomotion. For example, consider again the lineage of protists called the alveolates and the subgroups known as ciliates, dinoflagellates, and apicomplexa. The ciliates swim

(a) Amoeboid motion via pseudopodia

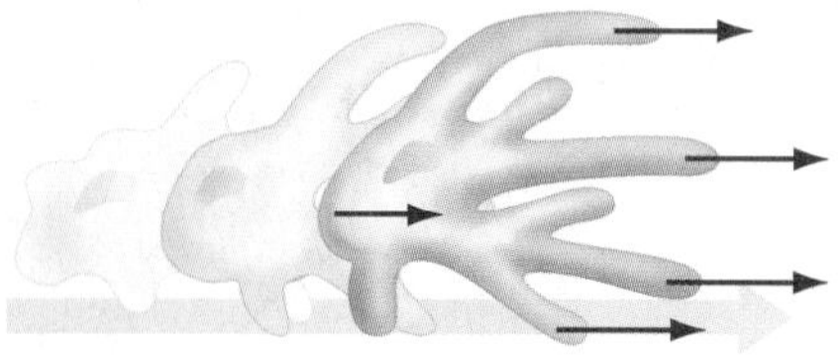

(b) Swimming via flagella

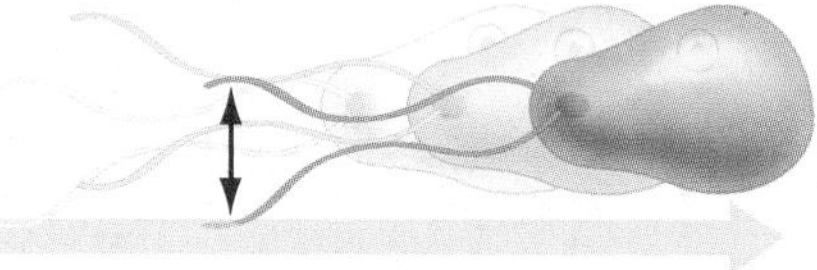

(c) Swimming via cilia

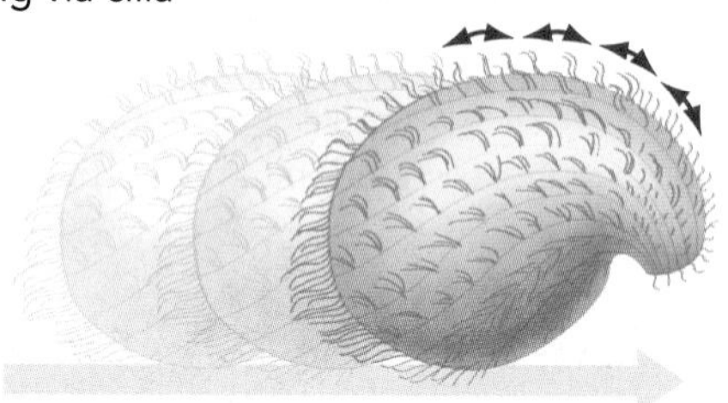

FIGURE 29.19 Modes of Locomotion in Protists Vary. (a) In amoeboid motion, long pseudopodia stream out from the cell. The rest of the cytoplasm, organelles, and external membrane follow. **(b)** Flagella are long and few in number, and they power swimming movements. **(c)** Cilia are short and numerous. In many cases they are used in swimming.

by beating their cilia. Dinoflagellates swim by whipping their flagella. Mature cells in apicomplexa move by amoeboid motion, but their gametes swim via flagella. Within the other major lineages of protists, it is common to observe variation in how cells move and common to find species that do not exhibit active movement but instead float passively in water currents. Movement is yet another example of the extensive diversification that occurred within each of the eight major eukaryote lineages.

How Do Protists Reproduce?

Sexual reproduction evolved in protists. As Chapter 12 pointed out, sexual reproduction can best be understood in contrast to asexual reproduction. Asexual reproduction is based on mitosis in eukaryotic organisms and on fission in bacteria and archaea, and it results in offspring that are genetically identical to the parent. Most protists undergo asexual reproduction routinely. When sexual reproduction occurs, in contrast, offspring are genetically different from their parents and from each other. Sexual reproduction is based on meiosis and fusion of gametes and occurs only intermittently in many protists—often at one particular time of year, or when individuals are crowded or food is scarce—and is based on meiosis. The evolution of sexual reproduction ranks among the most significant evolutionary innovations observed in eukaryotes.

Sexual versus Asexual Reproduction When reproduction occurs via meiosis, the resulting offspring are genetically variable—meaning their genotypes are different from those of other offspring and from those of their parents. The leading hypothesis to explain why meiosis evolved states that genetically variable offspring may be able to thrive if the environment changes. For example, offspring with genotypes different from those of their parents may be better able to withstand attacks by parasites that successfully attacked their parents (see Chapter 12). This is a key point because many types of parasites, including bacteria and viruses, have short generation times and evolve very quickly. Because the genotypes and phenotypes of parasites are constantly changing, natural selection is constantly favoring host individuals with new genotypes. The idea is that new offspring genotypes generated by meiosis may contain combinations of alleles that allow hosts to withstand attack by new strains of parasites. In short, many biologists view sexual reproduction as an adaptation to fight disease.

Variation in Life Cycles A life cycle describes the sequence of events that occur as individuals grow, mature, and reproduce. The evolution of meiosis introduced a new event in the life cycle of protist species; what's more, it created a distinction between haploid and diploid phases in the life of an individual. Recall from Chapter 12 that diploid individuals have two of each type of chromosome inside each cell, while haploid individuals have just one of each type of chromosome inside each cell. When meiosis occurs in diploid cells, it results in the production of haploid cells.

The life cycle of most bacteria and archaea is extremely simple: A cell divides, feeds, grows, and divides again. Bacteria and archaea do not undergo sexual reproduction, and they are always haploid. In contrast, among protists virtually every aspect of a life cycle is variable—whether meiosis occurs, whether asexual reproduction occurs, and whether the haploid or the diploid phase of the life cycle is the longer and more prominent phase.

Figure 29.20 illustrates some of the wide variation that occurs in life cycles of unicellular protists. Figure 29.20a depicts the haploid-dominated life cycle observed in many unicellular protists. The specific example given here is the dinoflagellate *Gymnodinium fuscum*. To analyze a life cycle, start with **fertilization**—the fusion of two gametes to form a diploid zygote. Then trace what happens to the zygote. In this case, the diploid zygote undergoes meiosis. The haploid products

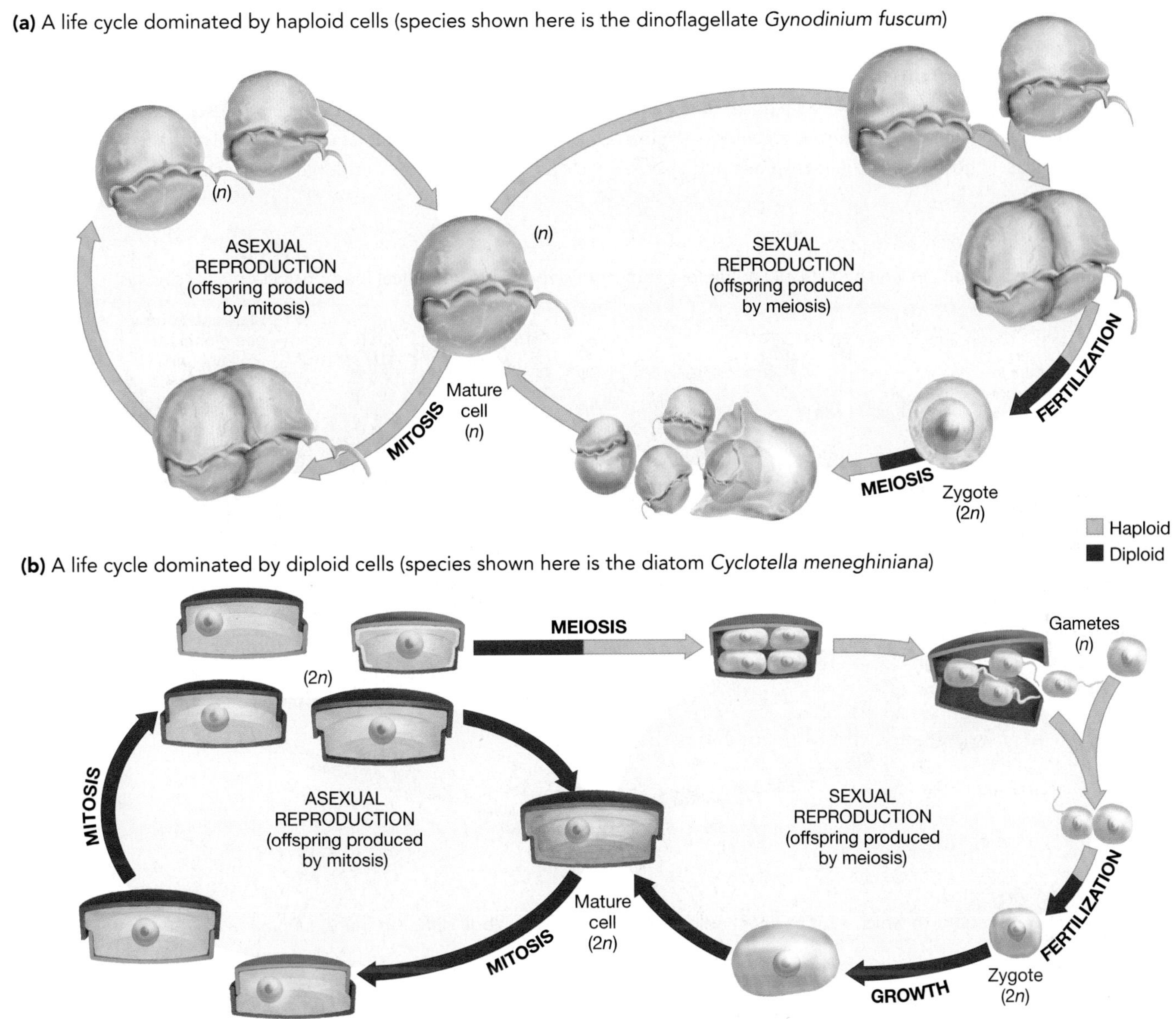

FIGURE 29.20 Life Cycles Vary Widely among Protists. Many protists can reproduce by both asexual reproduction and sexual reproduction. In unicellular species, the cell may be **(a)** haploid for most of its life or **(b)** diploid for most of its life.

of meiosis then grow into mature cells that eventually undergo asexual reproduction or produce gametes by mitosis. Contrast that cycle with the diploid-dominated life cycle of Figure 29.20b. The specific organism shown here is the diatom *Cyclotella meneghiniana*. To analyze this life cycle, start again with fertilization. In this case, the diploid zygote develops into a sexually mature, diploid adult cell. Meiosis occurs in the adult and results in the formation of haploid gametes, which then fuse to form a diploid zygote. The important contrasts are that meiosis occurs in the adult cell rather than in the zygote and that gametes are the only haploid cells in the life cycle.

In contrast to the relatively simple life cycles of Figure 29.20, many multicellular protists have one phase in their life cycle that is based on a multicellular haploid form and another phase that is based on a multicellular diploid form. This alternation of multicellular haploid and diploid forms is known as **alternation of generations**. The multicellular haploid form is called a **gametophyte**, because specialized cells in this individual produce gametes by mitosis. The multicellular diploid form is called a **sporophyte**, because it has specialized cells that undergo meiosis to produce haploid cells called spores. A **spore** is a single cell that develops into an adult organism but is not a product of fusion by gametes. When alternation of generation occurs, a spore divides by mitosis to form a haploid, multicellular gametophyte. The haploid gametes produced by the gametophyte then fuse to form a diploid zygote, which grows into the diploid, multicellular sporophyte. Among the protists, alternation of generations evolved independently in brown algae, red algae, and other groups.

Gametophytes and sporophytes may be identical in appearance, as in the brown alga called *Ectocarpus siliculosus* (**Figure 29.21a**). In many cases, however, the gametophyte and sporophyte look different, as in the brown alga called *Laminaria solidungula* (**Figure 29.21b**).

To get a better understanding of alternation of generations, compare and contrast the events described in Figure 29.20a and Figure 29.21a, starting with fertilization. Specifically, notice that zygotes undergoing alternation of generations do

(a) Alternation of generations in which multicellular haploid and diploid forms look identical (here, *Ectocarpus siliculosus*)

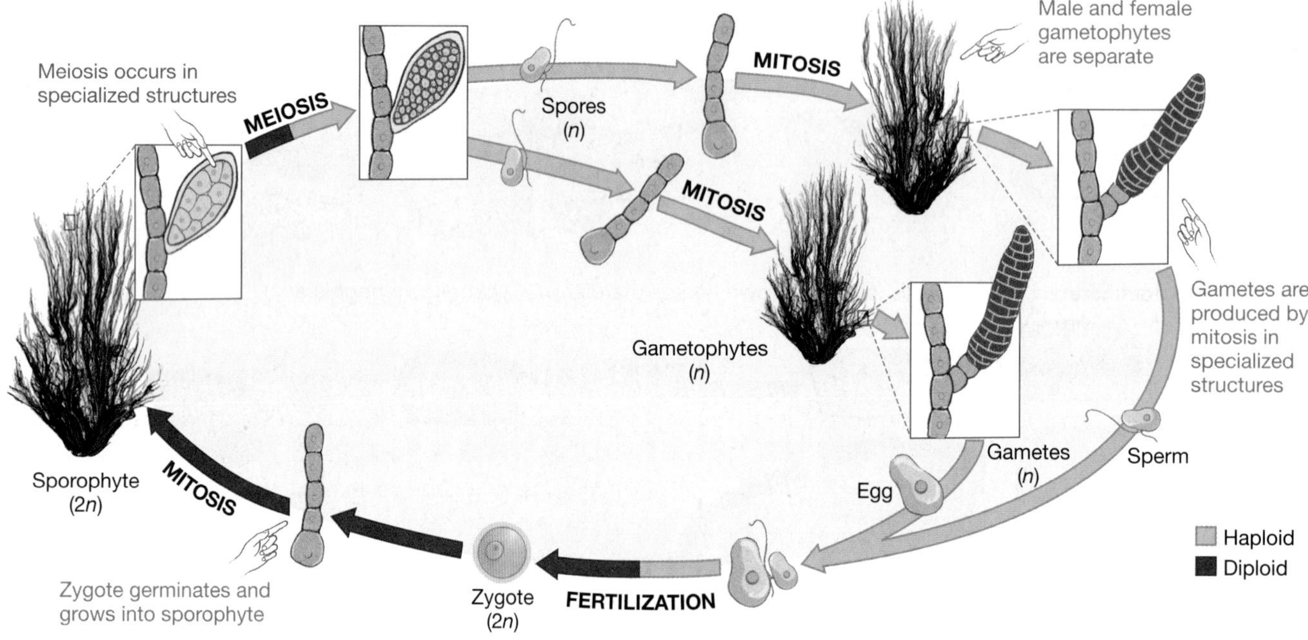

(b) Alternation of generations in which multicellular haploid and diploid forms look different (here, *Laminaria solidungula*)

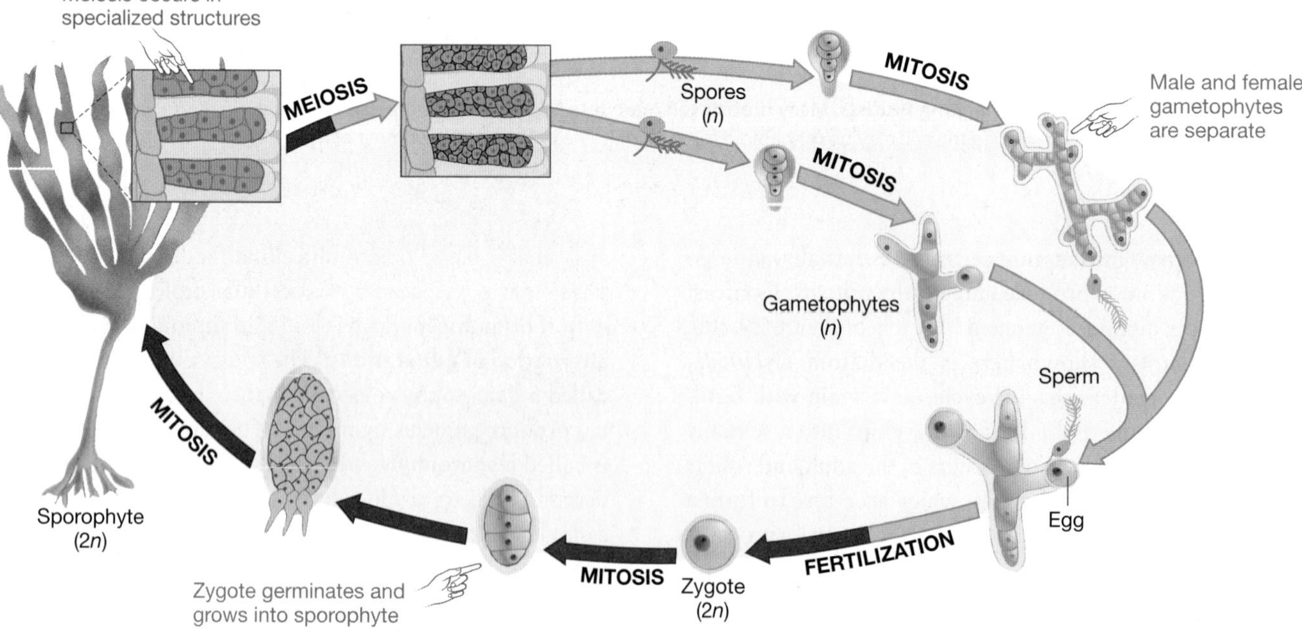

FIGURE 29.21 Alternation of Generations Occurs in an Array of Multicellular Protist Lineages.

not undergo meiosis immediately. Instead they germinate into diploid individuals that function as sporophytes. Cells in the sporophyte then undergo meiosis to form haploid spores, which germinate and grow into haploid gametophytes.

● If you understand how life cycles vary among multicellular protists, you should be able to define the terms *alternation of generations*, *gametophyte*, *sporophyte*, and *spore*. You should also be able to diagram a life cycle where alternation of generation occurs, without looking at Figure 29.21.

Why does so much variation occur in the types of life cycles observed among protists? The answer is not known. Variation in life cycles is a major theme in the diversification of protists. Explaining why that variation exists remains a topic for future research.

Check Your Understanding

If you understand that...

- Protist diversity can be understood by focusing on key innovations that were instrumental in the founding of major lineages or in their subsequent diversification.
- Important morphological innovations include the nucleus and endomembrane system, the mitochondrion, novel structures that provide protection or support, and multicellularity.
- Protists obtain chemical energy and building-block molecules via ingestive feeding, photosynthesis, or absorptive feeding (often, as parasites).
- Protists move via amoeboid movement, flagella, or cilia.
- Meiosis and sexual reproduction were crucial evolutionary innovations that originated in protists.

You should be able to...

1) Pick two of the innovations listed above and describe how each was involved in the founding of a major lineage or the subsequent diversification of a major lineage.
2) Explain why organelles made the evolution of large cell size possible.
3) Provide evidence that traits such as multicellularity and photosynthesis arose repeatedly over the course of eukaryote evolution.

(MB) **Web Animation** at www.masteringbio.com
Alteration of Generations in a Protist

29.4 Key Lineages of Protists

An important generalization jumps out of the Eukarya phylogenetic tree and data on variation among protists in morphology, feeding method, locomotion, and mode of reproduction: Each of the eight major Eukarya lineages has at least one distinctive morphological characteristic. But once an ancestor evolved a set of distinctive characteristics, its descendants diversified into a wide array of lifestyles. For example, parasitic species evolved independently in all eight major lineages. Photosynthetic species exist in most of the eight, and multicellularity evolved independently in four. Similar statements could be made about the evolution of life cycles and modes of locomotion.

In effect, each of the eight lineages represents a similar radiation of species into a wide array of lifestyles. In each case, the radiation began with a morphological innovation—often a change in the structure or function of one or more organelles. Let's take a more detailed look at some representative taxa from seven of the eight major groups. The eighth group—the opisthokonts—includes fungi, animals, and several protist lineages. The fungi and animals are discussed in Chapters 31 through 34. The lineage called green plants is analyzed in Chapter 30.

Excavata

The unicellular species that form the Excavata are named for the morphological feature that distinguishes them—an "excavated" feeding groove found on one side of the cell. Because all excavates lack mitochondria, they were once thought to trace their ancestry to eukaryotes that existed prior to the origin of mitochondria. But researchers have found that excavates either have nuclear genomes containing genes that are normally found in mitochondria or unusual organelles that appear to be vestigial mitochondria. These observations support the hypothesis that the ancestors of excavates had mitochondria, but that these organelles were lost or reduced over time. There are several major lineages of Excavata, two of which are detailed here (**Figure 29.22**). ● You should be able to mark the loss of the mitochondrion on Figure 29.22.

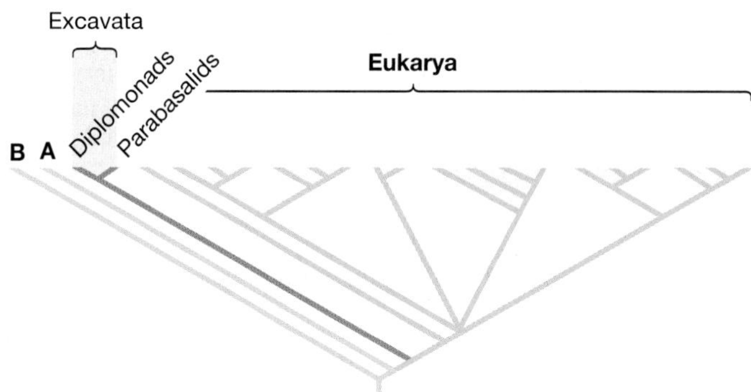

FIGURE 29.22 Excavates Are a Monophyletic Group.

Excavata > Diplomonadida

The first species of diplomonad known was described by the early microscopist Anton van Leeuwenhoek, who found *Giardia intestinalis* while examining samples of his own feces. About 100 species have been named to date. Many live in the guts of animal species without causing harm to their host; other species live in stagnant water habitats. In both types of environment, oxygen availability tends to be very low.

Morphology Each cell has two nuclei, which resemble eyes when viewed under the microscope (**Figure 29.23**). Each nucleus is associated with four flagella, for a total of eight flagella per cell. Some species lack the organelles called peroxisomes and lysosomes in addition to lacking functional mitochondria. All diplomonads lack a cell wall. ● You should be able to mark the origin of the double nucleus on Figure 29.22.

Feeding and locomotion Some are parasitic, but most ingest bacteria whole. They swim using their flagella.

Reproduction Only asexual reproduction occurs; meiosis has yet to be observed in members of this lineage.

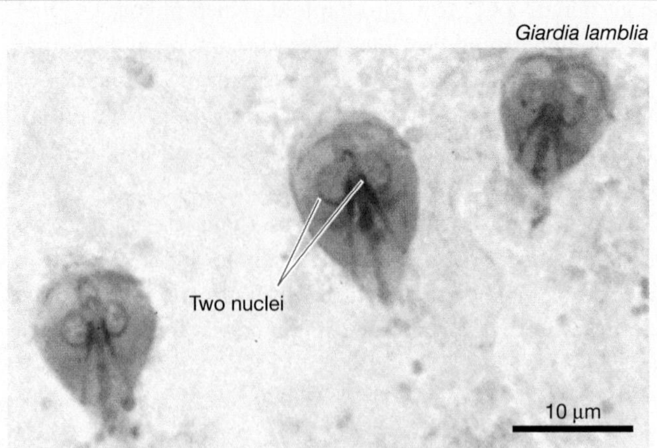

FIGURE 29.23 ***Giardia*** **Causes Intestinal Infections in Humans.**

Human and ecological impacts Both *Giardia intestinalis* and *G. lamblia* are common intestinal parasites in humans and cause giardiasis, or beaver fever. *Hexamida* is an agricultural pest that causes heavy losses in turkey farms each year.

Excavata > Parabasalida

No free-living parabasalids are known; all of the species described to date live inside animals. Of the 300 species of parabasalid, several live only in the guts of termites. Termites eat wood but cannot digest it themselves. Instead, the parabasalids produce enzymes that digest the cellulose in wood and release compounds that can be used by the termite host. The relationship between termites and parabasalids is considered mutualistic, because both parties benefit from the symbiosis.

Morphology Parabasalid cells lack a cell wall and mitochondria, and they have a single nucleus. A distinctive rod of cross-linked microtubules runs the length of the cell. The rod is attached to the basal bodies where a cluster of flagella arise (**Figure 29.24**). Although the number of flagella present varies widely, four or five is typical. ● You should be able to mark the origin of the parabasalid structure for supporting flagella on Figure 29.22.

Feeding and locomotion Parabasalids feed by engulfing bacteria, archaea, and organic matter. They swim using their flagella.

Reproduction All parabasalids reproduce asexually. Sexual reproduction also has been observed in a few species.

Human and ecological impacts *Trichomonas* infections can sometimes cause reproductive tract problems in humans, although members of this genus may also live in the gut or mouth of humans without causing harm. *Histomonas meleagridis* is an agricultural pest that causes disease outbreaks on chicken and turkey farms.

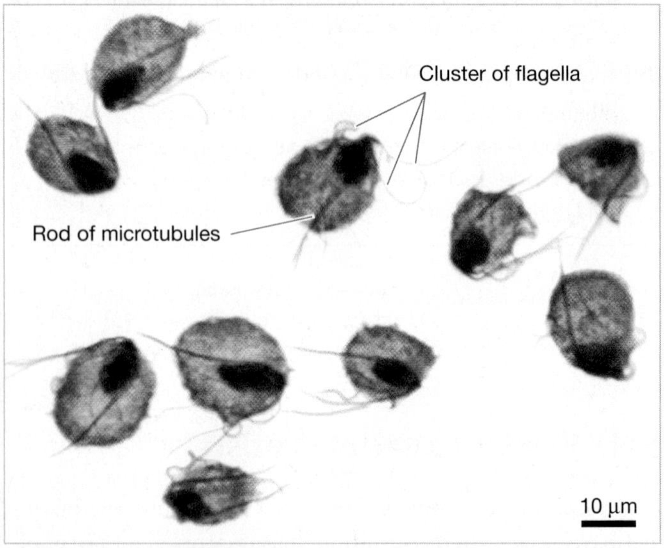

Trichomonas vaginalis

Cluster of flagella

Rod of microtubules

10 μm

FIGURE 29.24 ***Trichomonas*** **Causes the Sexually Transmitted Disease Trichomoniasis.**

Discicristata

The discicristates are all unicellular and were named for the distinctive disc shape of the cristae within their mitochondria. Several major subgroups have been identified, such as amoeboid forms and lineages that include the species responsible for the diseases leishmaniasis and trypanosomiasis in humans (see Table 29.1). Because they are extremely common in ponds and lakes, the euglenids are the best-studied discicristates. ● You should be able to mark the origin of disc-shaped cristae in mitochondria on **Figure 29.25**.

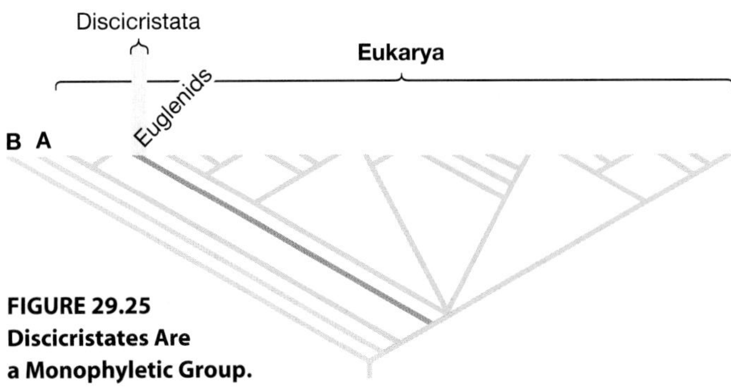

FIGURE 29.25 Discicristates Are a Monophyletic Group.

Discicristata > Euglenida

There are about 1000 known species of euglenid. Although most live in freshwater, a few are found in marine habitats. Fossil euglenids have been found in rocks over 410 million years old.

Morphology Most euglenids lack an external wall but have a unique system of interlocking protein molecules lying under the plasma membrane, which stiffen and support the cell.

Feeding and locomotion About one-third of the species have chloroplasts and perform photosynthesis, but most ingest bacteria and other small cells or particles. ● You should be able to explain how the observation of ingestive feeding in euglenids relates to the hypothesis that this lineage gained chloroplasts via secondary endosymbiosis. Instead of storing starch, the photosynthetic species synthesize a unique storage carbohydrate called paramylon. Some cells have a light-sensitive "eyespot" and use flagella to swim toward light (**Figure 29.26**). Other euglenids can inch along substrates using a unique form of movement based on extension of microtubules.

Reproduction Only asexual reproduction is known to occur among euglenids.

Human and ecological impacts Euglenids are important components of freshwater plankton and food chains.

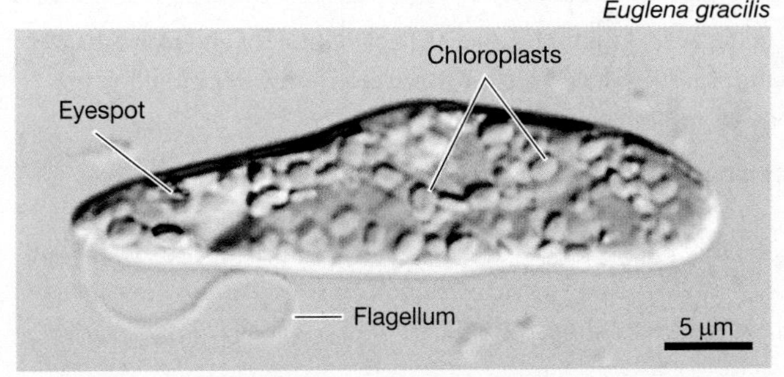

FIGURE 29.26 Euglenids Are Common in Ponds and Lakes.

Alveolata

Alveolates are distinguished by small sacs, called alveoli, that are located just under their plasma membranes. Although all members of this lineage are unicellular, the groups highlighted here—the Ciliata, the Dinoflagellata, and the Apicomplexa—are remarkably diverse in morphology and lifestyle. ● You should be able to mark the origin of alveoli on **Figure 29.27**.

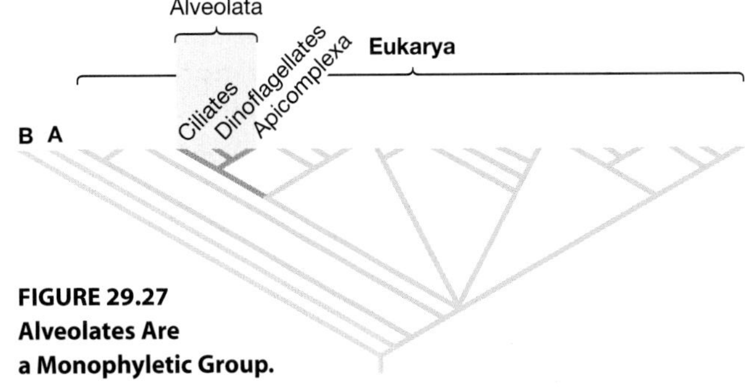

FIGURE 29.27 Alveolates Are a Monophyletic Group.

Alveolata > Apicomplexa

All of the 5000 known species of Apicomplexa are parasitic. Species in the genus *Plasmodium* are well studied because they cause malaria in humans and other vertebrates.

Morphology Apicomplexa cells have a system of organelles at one end, called the apical complex, that is unique to the group. The apical complex allows the apicomplexan to penetrate the plasma membrane of its host. ● You should be able to mark the origin of the apical complex on Figure 29.27. The cells have a chloroplast-derived, nonphotosynthetic organelle with four membranes, indicating that apicomplexans descended from an ancestor that gained a chloroplast via secondary endosymbiosis.

(continued on next page)

Alveolata > Apicomplexa *continued*

Feeding and locomotion All absorb nutrition directly from their host. They lack cilia or flagella, but some species can move by amoeboid motion.

Reproduction Apicomplexans can reproduce sexually or asexually. In some species, the life cycle involves two distinct hosts, and cells must be transmitted from one host to the next.

Human and ecological impacts Various species in the genus *Plasmodium* infect birds, reptiles, or mammals and cause malaria. *Toxoplasma* is an important pathogen in people infected with HIV (**Figure 29.28**), and chicken farmers estimate that they lose $600 million per year controlling the apicomplexan *Eimeria*. Apicomplexans that parasitize insects have been used to control pests.

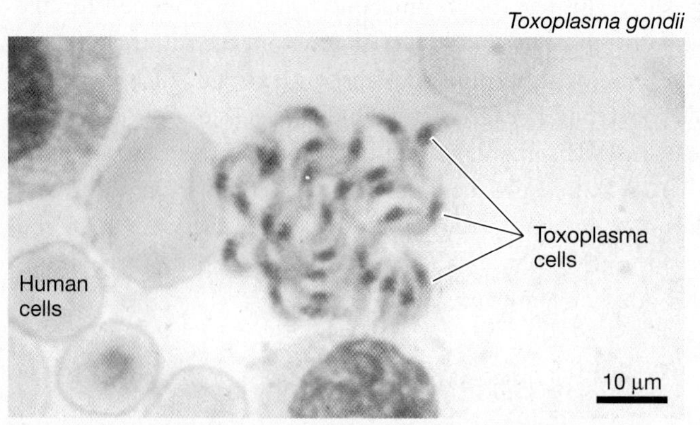

FIGURE 29.28 ***Toxoplasma*** **Causes Infections in AIDS patients.**

Alveolata > Ciliata

Ciliates were named for the cilia that cover them (**Figure 29.29**). Some 12,000 species are known from freshwater habitats, marine environments, and wet soils.

Morphology Ciliata cells have two distinctive nuclei: a large macronucleus and a small micronucleus. The macronucleus is polyploid and is actively transcribed. The micronucleus is diploid and is involved only in reproduction. A few ciliates secrete an external skeleton. ● You should be able to mark the origin of the macronucleus/micronucleus structure on Figure 29.27.

Feeding and locomotion Depending on the species, ciliates may be filter feeders, predators, or parasites. They use cilia to swim and have a mouth area where food is ingested.

Reproduction Ciliates divide to produce daughter cells asexually. They also undergo an unusual type of sexual reproduction called conjugation. During conjugation in ciliates, two cells line up side by side and physically connect. Micronuclei exchange between cells and fuse. The resulting nucleus eventually forms a new macronucleus and micronucleus.

Human and ecological impacts Ciliates are abundant in marine plankton and are important consumers. They are common in the digestive tracts of goats, sheep, cattle, and other grazers, where they feed on plant matter and help the host animal digest it.

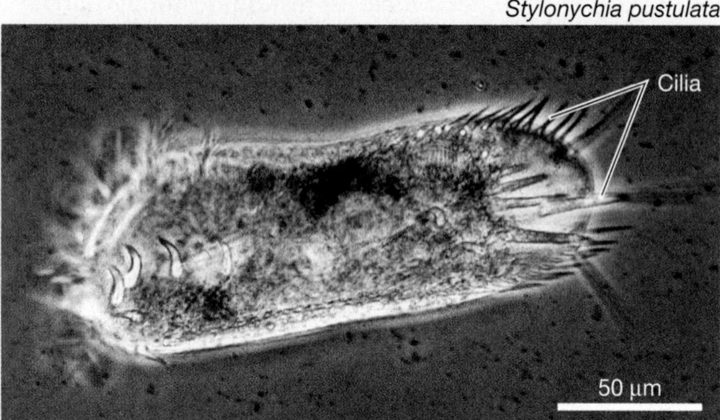

FIGURE 29.29 Ciliates Are Abundant in Freshwater Plankton.

Alveolata > Dinoflagellata

Most of the 4000 known species of dinoflagellates are ocean-dwelling plankton, although they are abundant in freshwater as well. Some species are capable of **bioluminescence**, meaning they emit light via an enzyme-catalyzed reaction (**Figure 29.30**).

Morphology Most dinoflagellates are unicellular, although some live in the aggregations of individual cells called colonies. Each species has a distinct shape maintained by plates of cellulose inside the cell. Unlike chromosomes of other eukaryotes, chromosomes in dinoflagellates are attached to the nuclear envelope at all times and do not contain histones.

Feeding and locomotion About half of the dinoflagellates are photosynthetic. Some of the photosynthetic species live in association with corals or sea anemones, but most are planktonic. The other species are predatory or parasitic. Dinoflagellates are distinguished by the arrangement of their two flagella: One flagel-

(continued on next page)

Alveolata > Dinoflagellata *continued*

lum projects out from the cell while the other runs across the cell in a groove (see Figure 29.20a). The two flagella are perpendicular to each other. Cells swim in a spinning motion using the two flagella. ● You should be able to mark the origin of perpendicularly oriented flagella on Figure 29.27.

Reproduction Both asexual and sexual reproduction occur. Cells that result from sexual reproduction may form tough cysts. A cyst is a resistant structure that can remain dormant until environmental conditions improve.

Human and ecological impacts Photosynthetic dinoflagellates are important primary producers in marine ecosystems—second only to diatoms in the amount of carbon they fix per year. A few species are responsible for harmful algal blooms, or red tides (see Figure 29.4).

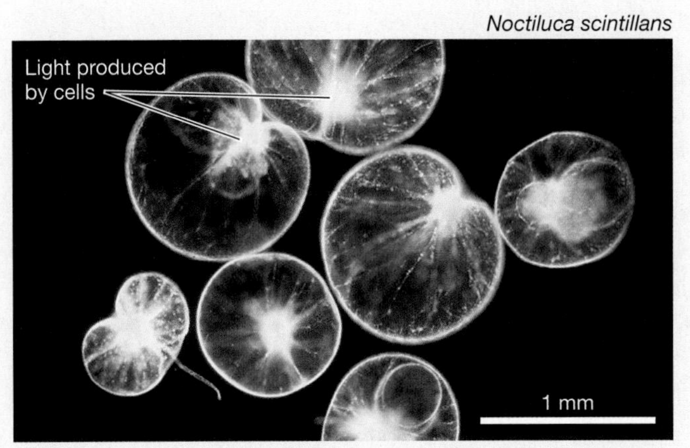

FIGURE 29.30 Some Dinoflagellates Species Are Bioluminescent.

Stramenopila (Heterokonta)

Stramenopiles are sometimes called *heterokonts*, which translates as "different hairs." At some stage of their life cycle, all stramenopiles have flagella that are covered with distinctive hollow "hairs." The structure of these flagella is unique to the stramenopiles (**Figure 29.31**). ● You should be able to mark the origin of the hairy flagellum on Figure 29.31. The lineage includes a large number of unicellular forms, although the brown algae are multicellular and include the world's tallest marine organisms, the kelp.

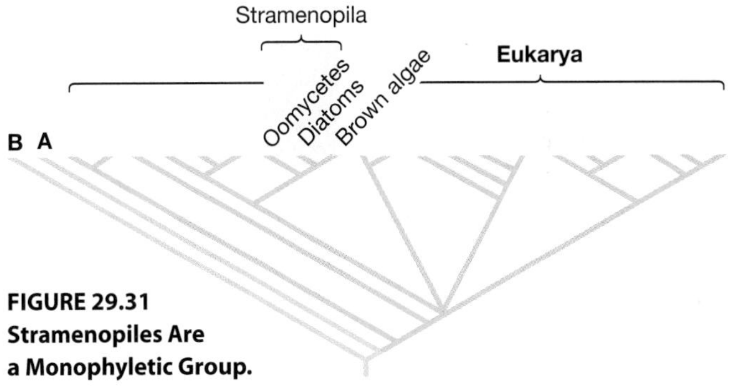

FIGURE 29.31 Stramenopiles Are a Monophyletic Group.

Stramenopila > Diatoms

Diatoms are unicellular or form chains of cells. About 10,000 species have been described to date, but some researchers claim that millions of species exist and have yet to be discovered.

Morphology Diatom cells are supported by external, silicon-rich, glassy shells that form a box-and-lid arrangement (**Figure 29.32** and Figure 29.20b). ● You should be able to mark the origin of the glassy test on Figure 29.31.

Feeding and locomotion Diatoms are photosynthetic. Only sperm cells have flagella and are capable of powered movement. In many species, the adult cells float in the water. But species living on a surface can glide via microtubules that project from their shells and that move in response to motor proteins.

Reproduction Diatoms divide by mitosis to reproduce asexually or by meiosis to form gametes that fuse to form a new individual. Many species can produce spores that are dormant during unfavorable growing conditions.

Human and ecological impacts In abundance, diatoms dominate the plankton of cold, nutrient-rich waters. They are found in virtually all aquatic habitats and are considered the most important producer of carbon compounds in fresh and salt water. Their shells settle into massive accumulations that are mined and sold commercially as diatomaceous earth, which is used in filtering applications and as an ingredient that adds bulk to polishes, paint, cosmetics, and other products.

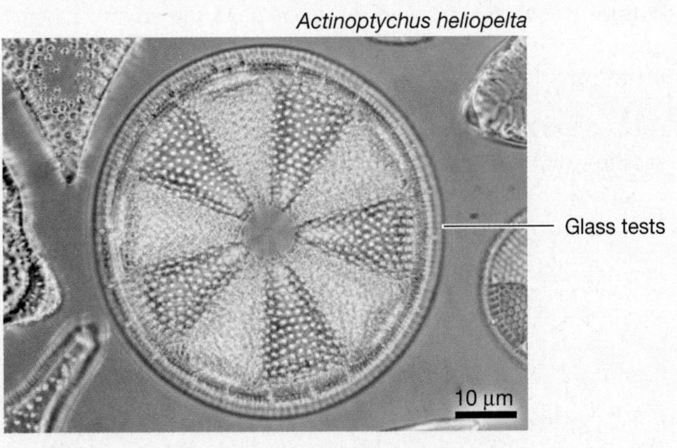

FIGURE 29.32 Diatoms Have Glass-like Shells.

Stramenopila > Oomycota (Water Molds)

Based on morphology, oomycetes were thought to be fungi and were given fungus-like names, such as downy mildew and water molds. They are readily distinguished from fungi at the DNA level, are primarily aquatic, have cellulose instead of chitin as the primary carbohydrate in the cell walls, and have gametes with stramenopile-like flagella. The morphological similarities between oomycetes and fungi result from convergent evolution, because both groups make their living absorbing nutrition from living or dead hosts. ● You should be able to mark the origin of the absorptive lifestyle in oomycetes on Figure 29.31.

Morphology Some oomycetes are unicellular and some form long, branching filaments called **hyphae**. Species that form hyphae often have multinucleate cells. Cells have walls containing cellulose.

Feeding and locomotion Most species feed on decaying organic material in freshwater environments; a few are parasitic. Mature individuals are sessile.

Reproduction Most species are diploid throughout the majority of their life cycle. In aquatic species, spores that are produced via asexual or sexual reproduction have flagella and swim to find new food sources; in terrestrial species, spores swim in rainwater or are dispersed by the wind in dry conditions. Spores form in special structures, like those shown in **Figure 29.33**.

Human and ecological impacts Oomycetes are extremely important decomposers in aquatic ecosystems. Along with certain bacteria and archaea, they are responsible for breaking down dead organisms and releasing nutrients for use by other species. The parasitic species can be harmful to humans, however. The organism that caused the Irish potato famine was an oomycete. An oomycete parasite almost wiped out the French wine industry in the 1870s, and other species are responsible for epidemic diseases of trees, including the diebacks currently occurring in oaks found in Europe and the western United States, and in eucalyptus in Australia.

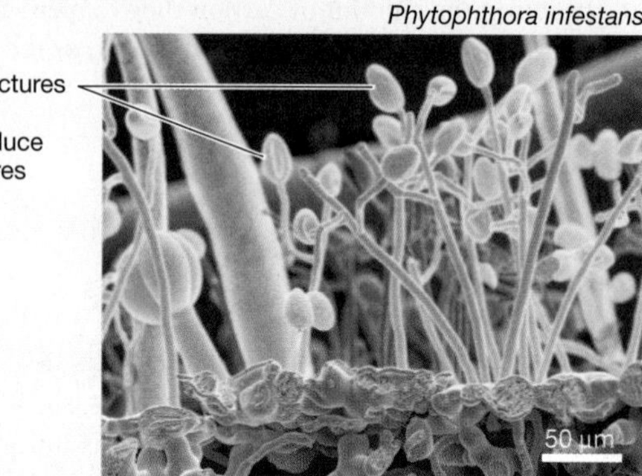

FIGURE 29.33 *Phytophthora infestans* Infects Potatoes.

Stramenopila > Phaeophyta (Brown Algae)

The color of brown algae is due to their unique suite of photosynthetic pigments. Over 1500 species have been described, most of them living in marine habitats.

Morphology The walls of Phaeophyta cells contain cellulose in addition to other complex polymers. Brown algae are unique among protists, because all species are multicellular. The body typically consists of leaflike blades, a stalk known as a stipe, and a rootlike holdfast, which attaches the individual to a substrate (**Figure 29.34**). ● You should be able to mark the origin of multicellularity and brown algal photosynthetic pigments on Figure 29.31.

Feeding and locomotion Brown algae are photosynthetic and **sessile**—that is, permanently fixed to a substrate—though reproductive cells may have flagella and be motile—capable of locomotion.

Reproduction Sexual reproduction occurs via the production of swimming gametes, which fuse to form a zygote. Most species exhibit alternation of generations. Depending on the species, the gametophyte and sporophyte stages may look similar or different.

Human and ecological impacts In many coastal areas, brown algae form forests or meadows that are important habitats for a wide variety of animals. In the Sargasso Sea, off Bermuda, floating brown algae form extensive rafts that harbor an abundance of animal species. The compound *algin* is purified from kelp and used in the manufacture of cosmetics and paint.

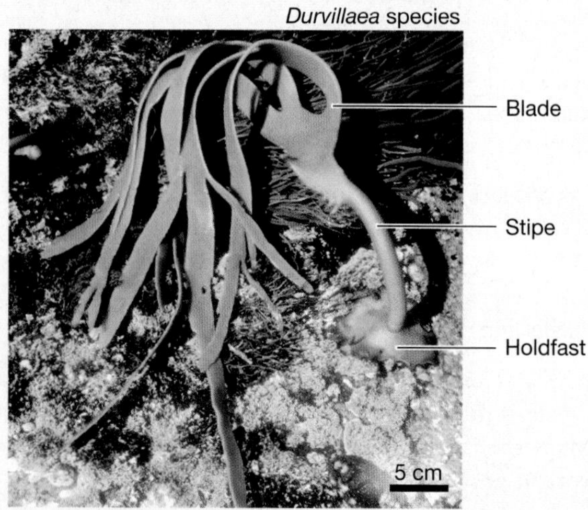

FIGURE 29.34 Many Brown Algae Have a Holdfast, Stalk, and Leaflike Blades.

Rhizaria

The rhizarians are single-celled amoebae that lack cell walls, though some species produce elaborate shell-like coverings. They move by amoeboid motion and produce long, slender pseudopodia. Over 11 major subgroups have been identified and named, including the planktonic organisms called actinopods, which synthesize glassy, silicon-rich skeletons, and the chlorarachniophytes, which obtained a chloroplast via secondary endosymbiosis and are photosynthetic. The best-studied and most abundant group is the Foraminifera (**Figure 29.35**). ● You should be able to mark the origin of the amoeboid form on Figure 29.35.

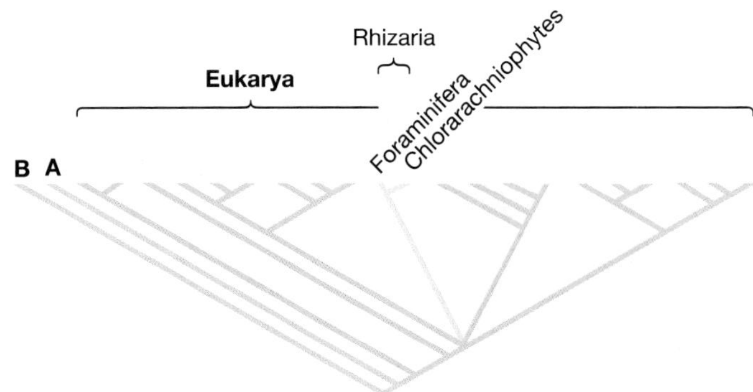

FIGURE 29.35 Rhizaria Are a Monophyletic Group.

Rhizaria > Foraminifera

Foraminifera, or forams, got their name from the Latin *foramen*, meaning "hole." Forams produce tests or shells that have holes through which pseudopodia emerge (**Figure 29.36**; see Figure 29.13c for a dead test on which holes are visible). ● You should be able to mark the origin of hard shells on Figure 29.35. Fossil tests of foraminifera are abundant in marine sediments—there is a continuous record of fossilized forams that dates back 530 million years. Although they are abundant in marine plankton as well as bottom habitats, their biology is relatively poorly known.

Morphology Foraminifera cells generally have multiple nuclei. The tests of forams are usually made of organic material stiffened with calcium carbonate ($CaCO_3$), and most species have several chambers. One species known from fossils was 12 cm long, but most species are much smaller. The size and shape of the test are traits that distinguish foram species from each other.

Feeding and locomotion Like other rhizarians, forams feed by extending their pseudopodia and using them to capture and engulf bacterial and archaeal cells or bits of organic debris, which are digested in food vacuoles. Some species have symbiotic algae that perform photosynthesis and contribute sugars to their host. Forams simply float in the water.

Reproduction Asexual reproduction occurs by mitosis. When meiosis occurs, the resulting gametes are released into the open water, where pairs fuse to form a new individual.

Human and ecological impacts The tests of dead forams commonly form extensive sediment deposits when they settle out of the water, producing layers that eventually solidify into chalk, limestone, or marble. Geologists use the presence of certain foram species to date rocks—particularly during petroleum exploration.

FIGURE 29.36 Forams Are Shelled Amoebae.

Plantae

Biologists are beginning to use the name **Plantae** to refer to the monophyletic group that includes red algae, green algae, land plants, and glaucophyte algae (**Figure 29.37**). All of these lineages are descended from a common ancestor that engulfed a cyanobacterium, beginning the endosymbiosis that led to the evolution of the chloroplast—their distinguishing morphological feature. This initial endosymbiosis probably occurred in an ancestor of today's glaucophyte algae. To support this hypothesis, biologists point to several important similarities between cyanobacterial cells and the chloroplasts that are found in the glaucophytes. Both have cell walls that contain peptidoglycan and that can be disrupted by molecules called lysozyme and penicillin. The glaucophyte chloroplast also has a membrane outside its wall that is similar to the membrane found in Gram-negative bacteria (see Chapter 28). Consistent with these observations, phylogenetic analyses of DNA sequence data place the glaucophytes as the most basal group of the Plantae. ● You should be able to mark the origin of the plant chloroplast on Figure 29.37.

The glaucophyte algae are unicellular or colonial. They live in plankton or attached to substrates in freshwater environments—particularly in bogs or swamps. Some glaucophyte species have flagella or produce flagellated spores, but sexual reproduction has never been observed in the group. The chloroplasts of glaucophytes have a distinct bright blue-green color. Chapter 30 introduces the green algae and land plants; here we consider just the red algae in detail.

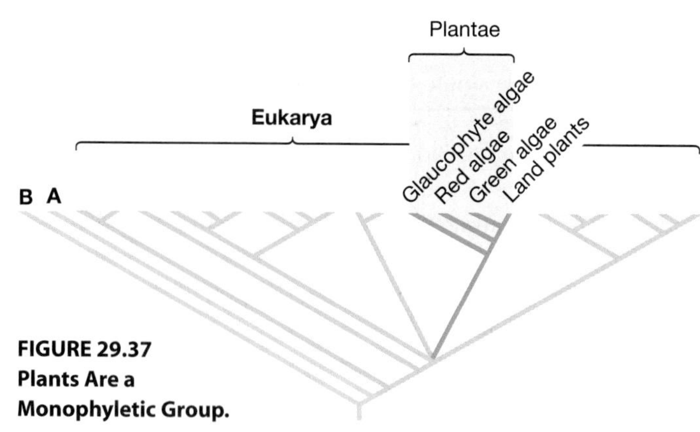

FIGURE 29.37 Plants Are a Monophyletic Group.

Plantae > Rhodophyta (Red Algae)

The 6000 species of red algae live primarily in marine habitats. One species lives over 200 m below the surface; another is the only eukaryote capable of living in acidic hot springs. Although their color varies, many species are red because their chloroplasts contain the accessory pigment phycoerythrin, which absorbs strongly in the blue and green portions of the visible spectrum. Because blue light penetrates water better than other wavelengths, red algae are able to live at considerable depth in the oceans. ● You should be able to mark the origin of red algal photosynthetic pigments on Figure 29.37.

Morphology Red algae cells have walls that are composed of cellulose and other polymers. A few species are unicellular, but most are multicellular. Many of the multicellular species are filamentous, but others grow as thin, hard crusts on rocks or coral (**Figure 29.38**). Some species have erect, leaf-like structures called thalli. Some species have cells with many nuclei.

Feeding and locomotion The vast majority of red algae are photosynthetic, though a few parasitic species have been identified. Red algae are the only type of algae that lack flagella.

Reproduction Asexual reproduction occurs through production of spores by mitosis. Alternation of generations is common, but the types of life cycles observed in red algae are extremely variable.

Human and ecological impacts On coral reefs, some red algae become encrusted with calcium carbonate. These species contribute to reef building and help stabilize the entire reef structure. Cultivation of *Porphyra* (or nori) for sushi and other foods is a billion-dollar-per-year industry in East Asia.

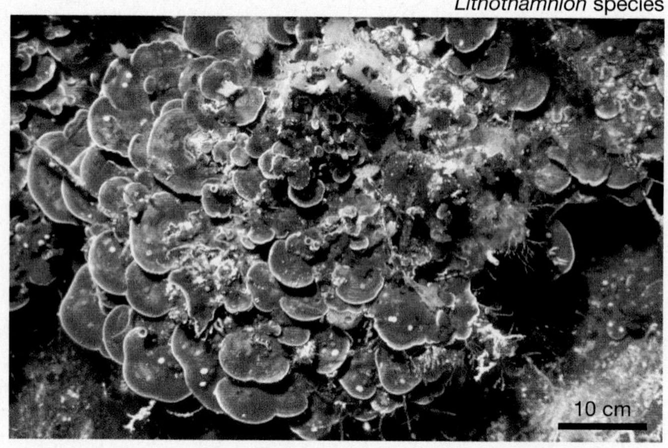

FIGURE 29.38 Red Algae Adopt an Array of Growth Forms.

Amoebozoa

Species in the Amoebozoa lack cell walls and take in food by engulfing it. They move via amoeboid motion and produce large, lobe-like pseudopodia. Major subgroups in the lineage are lobose amoebae, cellular slime molds, and plasmodial slime molds (**Figure 29.39**). ● You should be able to mark the origin of the amoeboid form on Figure 29.39 and explain whether it evolved independently of the amoeboid form in Rhizaria. The cellular slime mold *Dictyostelium discoideum* was described in detail in Box 29.2. Amoebae are abundant in freshwater habitats and in wet soils; some are parasites of humans and other animals. More details on plasmodial slime molds follow.

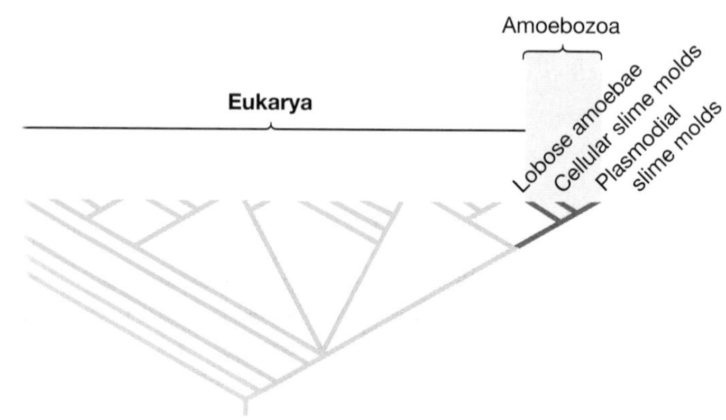

FIGURE 29.39 Amoebozoids Are a Monophyletic Group.

Amoebozoa > Myxogastrida (Plasmodial Slime Molds)

The plasmodial slime molds got their name because individuals form a large, weblike structure that consists of a single cell containing many diploid nuclei (**Figure 29.40**). Like oomycetes, the plasmodial slime molds were once considered fungi on the basis of their general morphological similarity (see Chapter 31).

Morphology The huge "supercell" form, with many nuclei in a single cell, occurs in few protists other than plasmodial slime molds. You should be able to mark the origin of the supercell on Figure 29.39.

Feeding and locomotion Myxogastrida cells feed on decaying vegetation and move by amoeboid motion.

Reproduction When food becomes scarce, part of the amoeba forms a stalk topped by a ball-like structure in which nuclei undergo meiosis and form spores. The spores are then dispersed to new habitats by the wind or small animals. After spores germinate to form amoebae, two amoebae fuse to form a diploid cell that begins to feed and eventually grows into a supercell.

Human and ecological impacts Like cellular slime molds, plasmodial slime molds are important decomposers in forests. They help break down leaves, branches, and other dead plant material, releasing nutrients that they and other organisms can use.

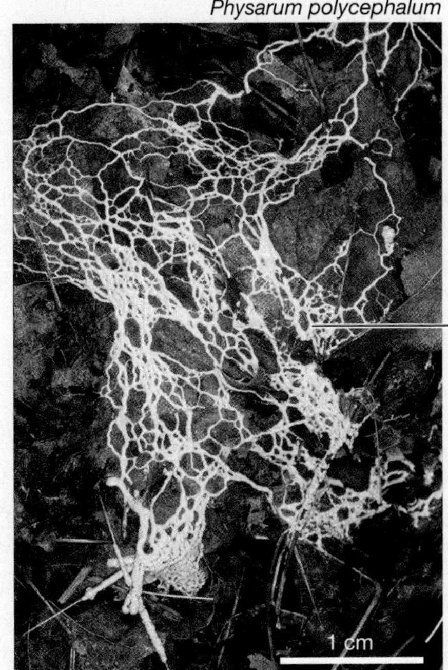

FIGURE 29.40 Plasmodial Slime Molds Are Important Decomposers in Forests.

Chapter Review

SUMMARY OF KEY CONCEPTS

Protists are a paraphyletic grouping that includes all eukaryotes except the green plants, fungi, and animals. Biologists study protists to understand how eukaryotes evolved, because they are important in freshwater and marine ecosystems and global warming, and because some species cause debilitating diseases in humans and other organisms.

Protists are often tremendously abundant in marine and freshwater plankton and other habitats. As a result, protists provide food for many organisms in aquatic ecosystems and fix so much carbon that they have a large impact on the global carbon budget. Toxin-producing protists that grow to high densities result in a harmful algal bloom. Parasitic protists cause several important diseases in humans, including malaria.

You should be able to explain what terms biologists would substitute for *protists* if only monophyletic groups were named.

Protists are diverse morphologically. They vary in the types of organelles they contain; they may be unicellular or multicellular, and they may have a cell wall or other external covering or no such covering.

Several morphological features are common to all or almost all eukaryotes, including the nucleus, endomembrane system, cytoskeleton, and flagellum. In addition to these common features, many aspects of morphology and lifestyle are extremely variable among the protist lineages. The protists include many unicellular organisms as well as multicellular slime molds, red algae, and brown algae. Multicellularity and structures that provide support or protection have evolved in many different protist groups independently.

Eukaryotes also contain mitochondria or have genes indicating that their ancestors once contained mitochondria. Several types of data support the hypothesis that mitochondria originated as endosymbiotic bacteria. The symbiosis is thought to have been successful because the endosymbiotic bacterium provided its host with larger amounts of ATP than it could produce on its own, while the host provided the bacterium with carbon compounds and protection. Similarly, the chloroplast's size, DNA structure, ribosomes, double membrane, and evolutionary relationships are consistent with the hypothesis that this organelle originated as an endosymbiotic cyanobacterium. After primary endosymbiosis occurred, chloroplasts were "passed around" to new lineages of protists via secondary endosymbiosis.

You should be able to make a drawing that shows how primary symbiosis would result in an organelle with two membranes containing genes that are closely related to bacteria, and how secondary endosymbiosis would result in an organelle with four membranes containing genes that are closely related to bacteria.

Protists vary widely in the way they find food. Many species are photosynthetic, while others obtain carbon compounds by ingesting food or parasitizing other organisms.

Protists exhibit predatory, parasitic, or photosynthetic lifestyles, which evolved in many groups independently. The evolution of ingestive feeding was important for two reasons: (1) It allowed eukaryotes to obtain resources in a new way—by eating bacteria, archaea, and other eukaryotes; and (2) it made endosymbiosis and the evolution of mitochondria and chloroplasts possible.

You should be able to propose a hypothesis to explain the wide diversity of photosynthetic pigments observed among protists.

Protists vary widely in the way they reproduce. Sexual reproduction evolved in protists, and many protist species can reproduce both sexually and asexually.

Protists undergo cell division based on mitosis and reproduce asexually. Many protists also undergo meiosis and sexual reproduction at some phase in their life cycle. Alternation of generations is common in multicellular species—meaning there are separate haploid and diploid forms of the same species. When alternation of generations occurs, haploid gametophytes produce gametes by mitosis; diploid sporophytes produce spores by meiosis.

You should be able to explain how you can tell a gametophyte from a sporophyte, and how you can distinguish a spore from a zygote.

Web Animation at www.masteringbio.com
Alternation of Generations in a Protist

QUESTIONS

Test Your Knowledge

1. Why are protists considered paraphyletic?
   a. They include many extinct forms, including lineages that no longer have any living representatives.
   b. They include some but not all, descendants of their most recent common ancestor.
   c. They represent all of the descendants of a single common ancestor.
   d. Not all protists have all of the synapomorphies that define the Eukarya, such as a nucleus.
2. What material is *not* used by protists to manufacture hard outer coverings?
   a. cellulose
   b. lignin
   c. glass-like compounds that contain silicon
   d. mineral-like compounds such as calcium carbonate ($CaCO_3$)
3. What does amoeboid motion result from?
   a. interactions among actin, myosin, and ATP
   b. coordinated beats of cilia
   c. the whiplike action of flagella
   d. action by the mitotic spindle, similar to what happens during mitosis and meiosis
4. According to the endosymbiosis theory, what type of organism is the original ancestor of the chloroplast?
   a. a photosynthetic archaean
   b. a cyanobacterium
   c. an algal-like, primitive photosynthetic eukaryote
   d. a modified mitochondrion
5. Multicellularity is defined in part by the presence of distinctive cell types. At the cellular level, what does this criterion imply?
   a. Individual cells must be extremely large.
   b. The organism must be able to reproduce sexually.
   c. Cells must be able to move.
   d. Different cell types express different genes.
6. Why are protists an important part of the global carbon cycle and marine food chains?
   a. They have high species diversity.
   b. They are numerically abundant.
   c. They have the ability to parasitize humans.
   d. They have the ability to undergo meiosis.

Test Your Knowledge answers: 1. b; 2. b; 3. a; 4. b; 5. d; 6. b

Test Your Understanding

Answers are available at www.masteringbio.com

1. What is the connection between the evolution of large cell size in protists and the evolution of ingestive feeding? Why is an advanced cytoskeleton and the lack of a cell wall required for ingestive modes of feeding? How does it relate to the acquisition of mitochondria and chloroplasts by endosymbiosis?
2. Is the evolution of novel structures for support and protection a synapomorphy that identifies the eukaryotes as a monophyletic group? Explain why or why not.
3. What is the relationship between meiosis and the alternation of generations? Why doesn't alternation of generations occur in unicellular species?
4. Outline the steps in the endosymbiosis theory for the origin of the mitochondrion. What did each partner provide the other, and what did each receive in return? Answer the same questions for the chloroplast.
5. Why was finding a close relationship between mitochondrial DNA and bacterial DNA considered particularly strong evidence in favor of the endosymbiosis theory? What evidence suggests that some protists acquired chloroplasts via secondary endosymbiosis?
6. The text claims that the evolutionary history of protists can be understood as a series of innovations that founded new lineages and/or triggered the diversification of existing lineages. Give an example that supports this claim.

Applying Concepts to New Situations

Answers are available at www.masteringbio.com

1. Consider the following:
   - All living eukaryotes have mitochondria or have evidence in their genomes that they once had these organelles. Thus it appears that eukaryotes acquired mitochondria very early in their history.
   - The first eukaryotic cells in the fossil record correlate with the first appearance of rocks formed in an oxygen-rich ocean and atmosphere.

   How are these observations connected? (HINT: Before answering, glance at Chapter 9 and remind yourself what happens in a mitochondrion.)

2. Consider the following:
   - *Plasmodium* has an unusual organelle called an apicoplast. Recent research has shown that apicoplasts are derived from chloroplasts via secondary endosymbiosis and have a large number of genes encoded by chloroplast DNA.
   - Glyphosate is one of the most widely used herbicides. It works by poisoning an enzyme encoded by a gene in chloroplast DNA.
   - Biologists are testing the hypothesis that glyphosate could be used as an antimalarial drug in humans.

   How are these observations connected?

3. Suppose a friend says that we don't need to worry about global warming. Her claim is that increased temperatures will make planktonic algae grow faster and that carbon dioxide (CO_2) will be removed from the atmosphere faster. According to her, this carbon will be buried at the bottom of the ocean in calcium carbonate tests. As a result, the amount of carbon dioxide in the atmosphere will decrease and global warming will decline. Comment.

4. Biologists are beginning to draw a distinction between "species trees" and "gene trees." A *species tree* is a phylogeny that describes the actual evolutionary history of a lineage. A *gene tree*, in contrast, describes the evolutionary history of one particular gene, such as a gene required for the synthesis chlorophyll *a*. In some cases, species trees and gene trees don't agree with each other. For example, the species tree for green algae indicates that their closest relatives are protists and plants. But the gene tree based on chlorophyll *a* from green algae suggests that this gene's closest relative is a bacterium, not a protist. What's going on? Why do these types of conflicts exist?

www.masteringbio.com is also your resource for • Answers to text, table, and figure caption questions and exercises • Answers to *Check Your Understanding* boxes • Online study guides and quizzes • Additional study tools including the *E-Book for Biological Science* 3rd ed., textbook art, animations, and videos.

30 Green Plants

KEY CONCEPTS

- The green plants include both the green algae and the land plants. Green algae are an important source of oxygen and provide food for aquatic organisms; land plants hold soil and water in place, build soil, moderate extreme temperatures and winds, and provide food for other organisms.
- Land plants were the first multicellular organisms that could live with most of their tissues exposed to the air. A series of key adaptations allowed them to survive on land. In terms of total mass, plants dominate today's terrestrial environments.
- Once plants were able to grow on land, a sequence of important evolutionary changes made it possible for them to reproduce efficiently—even in extremely dry environments.

Mosses are common in moist habitats and share many similarities with the earliest land plants. According to data reviewed in this chapter, the earliest land plants evolved from green algae that inhabited ponds, streams, and other freshwater habitats.

In terms of their total mass and their importance to other organisms, the **green plants** dominate terrestrial and freshwater habitats. When you walk through a forest or meadow, you are surrounded by green plants. If you look at pond or lake water under a microscope, green plants are everywhere.

The green plants comprise two major types of organisms: the green algae and the land plants. Green algae are important photosynthetic organisms in aquatic habitats—particularly lakes, ponds, and other freshwater settings—while land plants are the key photosynthesizers in terrestrial environments.

Although green algae have traditionally been considered protists, it is logical to study them along with land plants for two reasons: (1) They are the closest living relative to land plants and form a monophyletic group with them, and (2) the transition from aquatic to terrestrial life occurred when land plants evolved from green algae.

Land plants were the first organisms that could thrive with their tissues completely exposed to the air instead of being partially or completely submerged. When they evolved, multicellular organisms began to occupy dry terrestrial habitats that had been largely barren of visible life for over 3 billion years. Land plants made the Earth green. According to the fossil record, plants colonized the land in conjunction with fungi that grew in a mutually beneficial association. The fungi grew belowground and helped provide land plants with nutrients from the soil; in return, the plants provided the fungi with sugars and other products of photosynthesis. Not long after fungi and land plants evolved and began to diversify, animals also accomplished the feat of moving from aquatic to terrestrial habitats. But they could do so only because plants were there first and provided them something to eat. Close associations between fungi and land plants continue to this day and are detailed in Chapter 31; the importance of plants for humans and other animals is a major theme of this chapter.

Before land plants evolved, it is likely that the only life on the continents consisted of bacteria, archaea, and single-celled protists that thrive in wet soils. By colonizing the continents, plants transformed the nature of life on Earth. In the words of Karl Niklas, the movement of green plants from water to land ranks as "one of the greatest adaptive events in the history of life." They were the first multicellular organisms on land. Let's look at why so many biologists have devoted their lives to studying plants, then analyze how these organisms made the momentous transition to life on land.

30.1 Why Do Biologists Study the Green Plants?

Biologists study plants because people could not live without them and because they are fascinating. Along with most other animals and fungi, humans are almost completely dependent on plants for food. People rely on plants for other necessities of life as well—oxygen, fuel, building materials, and the fibers used in making clothing, paper, ropes, and baskets. But we also rely on land plants for important intangible values, such as the aesthetic appeal of landscaping and bouquets. To drive this point home, consider that the sale of cut flowers generates over $1 billion each year in the United States alone.

Based on these observations, it is not surprising that agriculture, forestry, and horticulture are among the most important endeavors supported by biological science. Tens of thousands of biologists are employed in research designed to increase the productivity of plants and to create new ways of using them in ways that benefit people. Research programs also focus on two types of land plants that cause problems for people: weeds that decrease the productivity of crop plants and newly introduced species that invade and then degrade natural areas.

Plants Provide Ecosystem Services

An **ecosystem** consists of all the organisms in a particular area, along with physical components of the environment such as the atmosphere, precipitation, surface water, sunlight, soil, and nutrients. Plants are said to provide **ecosystem services** because they add to the quality of the atmosphere, surface water, soil, and other physical components of an ecosystem. Stated another way, plants alter the landscape in ways that benefit other organisms:

- *Plants produce oxygen.* Recall from Chapter 10 that plants perform oxygenic (literally, "oxygen-producing") photosynthesis. In this process, electrons that are removed from water molecules are used to reduce carbon dioxide and produce sugars. In the process of stripping electrons from water, plants release oxygen molecules (O_2) as a by-product. As Chapter 28 noted, oxygenic photosynthesis evolved in cyanobacteria and was responsible for the origin of an oxygen-rich atmosphere. The evolutionary success of plants continued this trend because plants add huge amounts of oxygen to the atmosphere. Without the green plants, we and other terrestrial animals would be in danger of suffocating for lack of oxygen.
- *Plants build soil.* Leaves and roots and stems that are not eaten when they are alive fall to the ground and provide food for worms, fungi, bacteria, archaea, protists, and other decomposers in the soil. These organisms add organic matter to the soil, which improves soil structure and the ability of soils to hold nutrients and water.
- *Plants hold soil.* The extensive network of fine roots produced by trees, grasses, and other land plants helps hold soil particles in place. And by taking up nutrients in the soil, plants prevent the nutrients from being blown or washed away. When areas are devegetated by grazing, farming, logging, or suburbanization, large quantities of soil and nutrients are lost to erosion by wind and water (**Figure 30.1**).

(a) Wind erosion

(b) Water erosion

FIGURE 30.1 Plants Hold Soil in Place.

- *Plants hold water.* Plant tissues take up and retain water. Intact forests, prairies, and wetlands also prevent rain from quickly running off a landscape, for several reasons: Plant leaves soften the physical impact of rainfall on soil; plant roots hold soil particles in place during rainstorms; and plant organic matter builds the soil's water-holding capacity. When areas are devegetated, streams are more prone to flooding and groundwater is not replenished efficiently. It is common to observe streams alternately flooding and then drying up completely when the surrounding area is deforested.
- *Plants moderate the local climate.* By providing shade, plants reduce temperatures beneath them and increase relative humidity. They also reduce the impact of winds that dry out landscapes or make them colder. When plants are removed from landscapes to make way for farms or suburbs, habitats become much dryer and are subject to more extreme temperature swings.

Perhaps the most important ecosystem service provided by plants, however, involves food. Land plants are the dominant primary producers in terrestrial ecosystems. (As Chapter 29 indicated, primary producers convert energy in sunlight into chemical energy.) The sugars and oils that land plants produce by photosynthesis provide the base of the food chain in the vast majority of terrestrial habitats. As **Figure 30.2** shows, plants are

Tertiary Consumers: Secondary carnivores eat carnivores.

Secondary Consumers: Carnivores eat animals.

Primary Consumers: Herbivores eat plants.

Producers: Plants form the base of the terrestrial food chain.

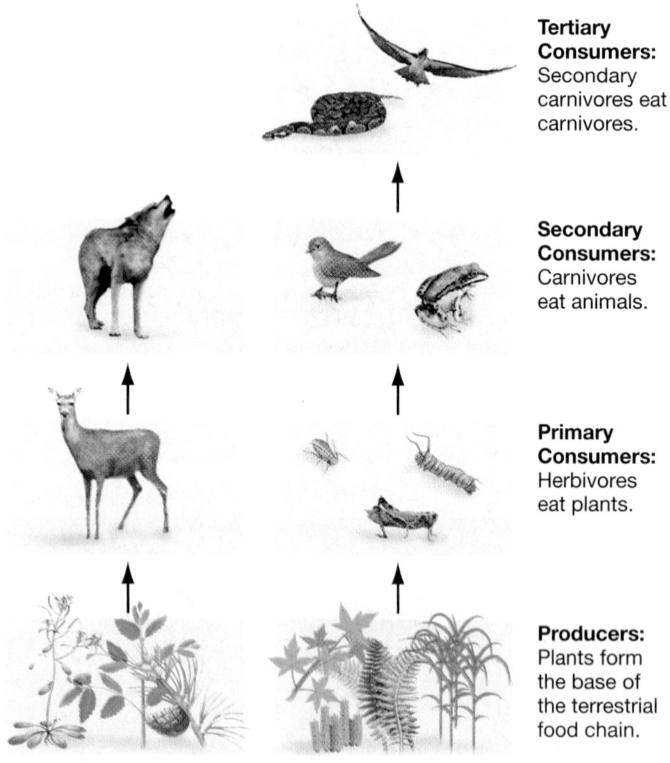

FIGURE 30.2 Plants Are the Basis of Food Chains in Terrestrial Environments. Virtually every organism that lives on land depends on plants for food, either directly or indirectly.

EXERCISE Label the levels in this food chain where humans feed.

eaten by **herbivores** ("plant-eaters"), which range in size from insects to elephants. These consumers are eaten by **carnivores** ("meat-eaters"), ranging in size from the tiniest spiders to polar bears. Humans are an example of **omnivores** ("all-eaters"), organisms that eat both plants and animals. Omnivores feed at several different levels in the terrestrial food chain. For example, people consume plants, herbivores such as chicken and cattle, and carnivores such as salmon and tuna.

Finally, just as photosynthetic protists and bacteria are the key to the carbon cycle in the oceans, green plants are the key to the carbon cycle on the continents. Plants take CO_2 from the atmosphere and reduce it to make sugars. Although both green algae and land plants also produce a great deal of CO_2 as a result of cellular respiration, they fix much more CO_2 than they release. The loss of plant-rich prairies and forests, due to fires or logging or suburbanization, has contributed to increased concentrations of CO_2 in the atmosphere. Higher carbon dioxide levels, in turn, are responsible for the rapid warming that is occurring worldwide (see Chapter 54).

Plants Provide Humans with Food, Fuel, Fiber, Building Materials, and Medicines

It is difficult to overstate the importance of plant research to the well-being of human societies. Plants provide our food supply as well as a significant percentage of the fuel, fibers, building materials, and medicines that we use.

- Agricultural research began with the initial domestication of crop plants, which occurred independently at several locations around the world between 10,000 and 2000 years ago. By actively selecting individuals with the largest and most nutritious seeds or leaves or stalks year after year, our ancestors gradually changed the characteristics of several wild species **(Figure 30.3a)**. This process is called **artificial selection**, and it continues today (see Chapter 1). Over the past 100 years, for example, artificial selection has been responsible for dramatic increases in the oil content of corn kernels (**Figure 30.3b**). (Corn oil serves as a cooking oil and is used in other products.) Chapter 20 highlighted a current focus in agricultural research—the improvement of crop varieties through genetic engineering.
- For perhaps 100,000 years, wood burning was the primary source of energy used by all humans. As **Figure 30.4a** shows, however, wood has been replaced in the industrialized countries by other sources of energy. The first fuel to replace wood in many cases has been coal, which forms from partially decayed plant material that is compacted over time by overlying sediments and hardened into rock (**Figure 30.4b**). Starting in the mid-1800s, people in England, Germany, and the United States began to mine coal deposits that originally formed during the Carboniferous period some 350–275 million years ago. The coal fueled blast furnaces that smelted

(a) Plants were domesticated at an array of locations.

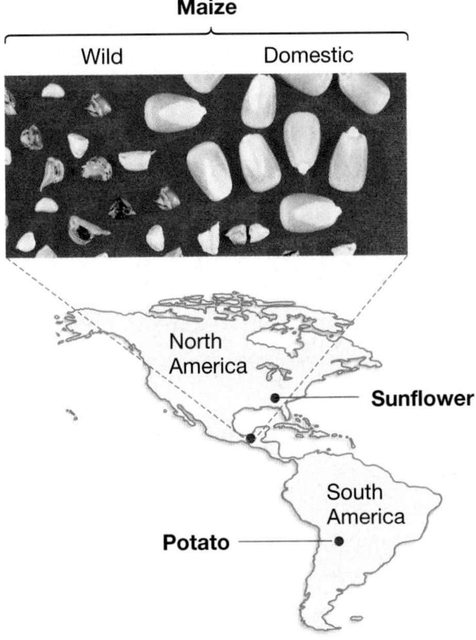

(b) ARTIFICIAL SELECTION CHANGES THE TRAITS OF DOMESTICATED SPECIES.

Less oil rich

Oil rich

1. Observe variation in kernel oil content.

2. Plant oil-rich seeds and grow to maturity.

3. Harvest kernels from mature plants. Repeat steps 1–3.

4. After many generations, kernel oil content increases.

Oil content (%)

0 5 10 15 20 25

0 20 40 60 80 100

Generations selected

FIGURE 30.3 Crop Plants Are Derived from Wild Species via Artificial Selection. (a) Crop plants have originated on virtually every continent, including millet from Africa, wheat and barley from the Middle East, and rice and soybean from east Asia. **(b)** Artificial selection can lead to dramatic changes in plant characteristics.

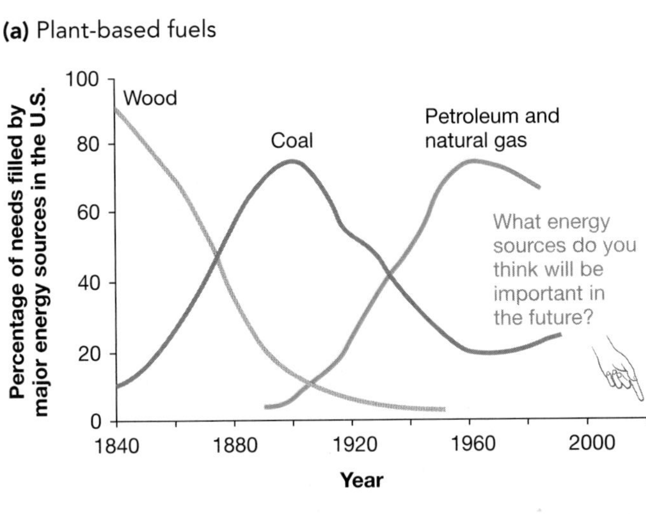

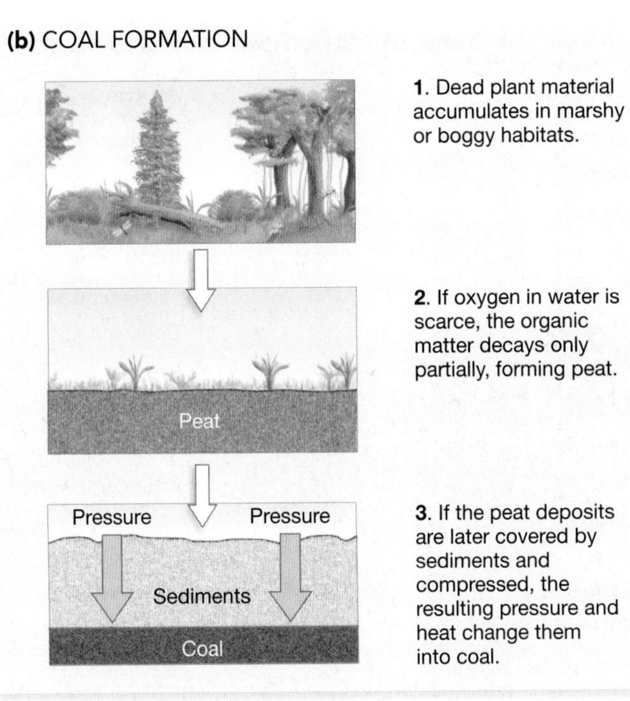

FIGURE 30.4 Humans Have Relied on Plant-Based Fuels. (a) Although wood has declined in importance in the industrialized countries, it is still the primary cooking and heating fuel in many areas of the world. **(b)** Coal is formed by terrestrial deposits of partially decayed organisms.

QUESTION Is coal a renewable resource? Explain why or why not.

vast quantities of iron ore into steel and powered the steam engines that sent trains streaking across Europe and North America. It is no exaggeration to claim that the sugars synthesized by plants more than 300 million years ago laid the groundwork for the Industrial Revolution. Coal still supplies about 20 percent of the energy used in Japan and Western Europe and 80 percent of the energy used in China. Current research is focused on finding cleaner and more efficient ways of mining and burning coal.

- Although nylon and polyester derived from petroleum are increasingly important in manufacturing, cotton and other types of plant fibers are still important sources of raw material for clothing, rope, and household articles like towels and bedding.
- Woody plants provide lumber for houses and furniture and most of the fibers used in papermaking. Relative to its density, wood is a stiffer and stronger building material than concrete, cast iron, aluminum alloys, or steel. The cellulose fibers refined from trees or bamboo and used in paper manufacturing are stronger under tension (pulling) than nylon, silk, chitin, collagen, tendon, or bone—even though cellulose is 25 percent less dense. Current research in forestry focuses on maintaining the productivity and increasing the diversity of forests that are managed for wood and pulp production.
- In both traditional and modern medicine, plants are a key source of drugs. **Table 30.1** lists some of the more familiar medicines derived from land plants; overall, it has been estimated that about 25 percent of the prescriptions written in the United States each year include at least one molecule derived from plants. In most cases, plants synthesize these compounds in order to repel insects, deer, or other types of herbivores. For example, experiments have confirmed that morphine, cocaine, nicotine, caffeine, and other toxic compounds found in plants are effective deterrents to insect or mammalian consumers. Researchers continue to isolate and test new plant compounds for medicinal use in humans and domesticated animals.

Given the importance of plants to the planet in general and humans in particular, it is not surprising that understanding plant diversity is an important component of biological science. Let's first consider how biologists go about analyzing the diversity of green plants; then go on to explore the evolutionary innovations that made the diversification possible.

30.2 How Do Biologists Study Green Plants?

To understand the genetics and developmental biology of plants, researchers use thale cress, a weedy mustard, as a model organism (**Box 30.1**). To understand how green plants originated and diversified, biologists use three tools: They (1) compare the fundamental morphological features of various green algae and green plants; (2) analyze the fossil record of the lineage; and (3) assess similarities and differences in DNA sequences from homologous genes to estimate phylogenetic trees. The three approaches are complementary and have produced a remarkably clear picture of how land plants evolved from green algae and then diversified. Let's consider each of these research strategies.

Analyzing Morphological Traits

By comparing and contrasting the morphological traits of green plants, biologists have identified several distinct groups

TABLE 30.1 Some Drugs Derived from Land Plants

Compound	Source	Use
Atropine	Belladonna plant	Dilating pupils during eye exams
Codeine	Opium poppy	Pain relief, cough suppressant
Digitalin	Foxglove	Heart medication
Ipecac	Ipecac	Treating amoebic dysentery, poison control
Menthol	Peppermint tree	Cough suppressant, relief of stuffy nose
Morphine	Opium poppy	Pain relief
Papain	Papaya	Reduce inflammation, treat wounds
Quinine	Quinine tree	Malaria prevention
Quinidine	Quinine tree	Heart medication
Salicin	Aspen, willow trees	Pain relief (aspirin)
Steroids	Wild yams	Precursor compounds for manufacture of birth control pills and cortisone (to treat inflammation)
Taxol	Pacific yew	Ovarian cancer
Tubocurarine	Curare vine	Muscle relaxant used in surgery
Vinblastine, vincristine	Rosy periwinkle	Leukemia (cancer of blood)

BOX 30.1 A Model Organism: *Arabidopsis thaliana*

In the early days of biology, the best-studied plants were agricultural varieties such as maize (corn), rice, and garden peas. When biologists began to unravel the mechanisms responsible for oxygenic photosynthesis in the early to mid-1900s, they relied on green algae that were relatively easy to grow and manipulate in the lab—often the unicellular species *Chlamydomonas reinhardii*—as an experimental subject. Although crop plants and green algae continue to be the subject of considerable research, a new model organism emerged in the 1980s and now serves as the preeminent experimental subject in plant biology. That organism is *Arabidopsis thaliana*, commonly known as thale cress or wall cress (**Figure 30.5**).

FIGURE 30.5 *Arabidopsis thaliana* Is the Most Important Model Organism in Plant Biology. Like fruit flies and *Escherichia coli*, *A. thaliana* is small and short lived enough to grow easily in the laboratory, yet it is complex enough to be an interesting organism for study.

Arabidopsis is a member of the mustard family, or Brassicaceae, so it is closely related to radishes and broccoli. In nature it is a **weed**—meaning a species that is adapted to thrive in habitats where soils have been disturbed. In Europe, for example, *Arabidopsis* is common along roadsides and the edges of agricultural fields. It is also an **annual** plant, which means that individuals do not live from year to year but overwinter as seeds. (Plants that survive from one year to the next are said to be **perennial**.) Indeed, one of the most attractive aspects of working with *A. thaliana* is that individuals can grow from a seed into a mature, seed-producing plant in just four to six weeks. Several other attributes make it an effective subject for study: It has just five chromosomes, has a relatively small genome with limited numbers of repetitive sequences, can self-fertilize as well as undergo cross-fertilization, can be grown in a relatively small amount of space and with a minimum of care in the greenhouse, and produces up to 10,000 seeds per individual per generation.

Arabidopsis has been instrumental in a variety of studies in plant molecular genetics and development, and it is increasingly popular in ecological and evolutionary studies. In addition, the entire genome of the species has now been sequenced, and studies have benefited from the development of an international "*Arabidopsis* community"—a combination of informal and formal associations of investigators who work on *Arabidopsis* and use regular meetings, e-mail, and the Internet to share data, techniques, and seed stocks.

of green algae and a series of major lineages, or **phyla** (singular: **phylum**), of land plants.

The green algae have long been hypothesized to be closely related to land plants on the basis of several key morphological traits. Both green algae and land plants have chloroplasts that contain the photosynthetic pigments chlorophyll *a* and *b* and the accessory pigment β-carotene as well as similar arrangements of the internal, membrane-bound sacs called thylakoids (see Chapter 10). The cell walls of green algae and land plants are similar in composition, both groups synthesize starch as a storage product in their chloroplasts, and their sperm and peroxisomes are similar in structure and composition. (Recall from Chapter 7 that peroxisomes are organelles in which specialized oxidation reactions take place.)

The green algae include species that are unicellular, colonial, or multicellular and that live in marine, freshwater, or moist terrestrial habitats. Of all the green algal groups, the two most similar to land plants are the Coleochaetophyceae (coleochaetes) and Charaphyceae (stoneworts; **Figure 30.6**). Because the species that make up these groups are multicellular and live in ponds and other types of freshwater environments, biologists hypothesize that land plants evolved from multicellular green algae that lived in freshwater habitats.

FIGURE 30.6 Most Green Algae Are Aquatic. Examples of species from the green algal lineages most closely related to the land plants.

Although some land plants live in ponds or lakes or rivers, the vast majority of species in this lineage live on land. Based on morphology, the most important phyla are traditionally

(a) Nonvascular plants do not have vascular tissue to conduct water and provide support.

Hepaticophyta (liverworts)

Anthocerophyta (hornworts)

Bryophyta (mosses)

(b) Seedless vascular plants have vascular tissue but do not make seeds.

Lycophyta (lycophytes or club mosses)

Psilotophyta (whisk ferns)

Sphenophyta (horsetails)

Pteridophyta (ferns)

FIGURE 30.7 Morphological Diversity in Land Plants.

clustered into three broad categories:

1. **Nonvascular plants**, which include the groups called Hepaticophyta (liverworts), Anthocerophyta (hornworts), and Bryophyta (mosses). Liverworts and hornworts lack **vascular tissue**—meaning specialized groups of cells that conduct water or dissolved nutrients from one part of the plant body to another. Some moss species have specialized tissues that conduct water and food, but the cells that make up these tissues do not have the reinforced cell walls that define true vascular tissue. Nonvascular plants are extremely abundant in certain habitats but are usually small and grow close to the ground (**Figure 30.7a**).

2. **Seedless vascular plants**, which have well-developed vascular tissue but do not make seeds. A **seed** consists of an embryo and a store of nutritive tissue, surrounded by a tough protective layer. Lycophyta (lycophytes or club mosses), Psilophyta (whisk ferns), Sphenophyta (horsetails), and Pteridophyta (ferns) are among the major groups of seedless vascular plants (**Figure 30.7b**). Although most of the living representatives of these lineages are relatively small in stature, some fossil and some other living representatives of the seedless vascular plants are tree sized.

3. **Seed plants**, which have vascular tissue and make seeds. Although the members of this group vary a great deal in size and shape, seed plants encompass some of the world's largest organisms. Biologists recognize six major lineages in the group: Cycadophyta (cycads), Ginkgophyta (ginkgos), Gnetophyta (gnetophytes), Pinophyta (pines, spruces, firs), other conifers (redwoods, junipers, yews—species which bear seeds in cones, like the pinophytes, but represent a distinct lineage), and Anthophyta (angiosperms or flowering plants), as pictured in **Figure 30.7c.** (The group denoted "other conifers" was discovered so recently that it has not yet received a formal name.) The cycads, ginkgos, gnetophytes, pines, and other conifers are collectively known as **gymnosperms** ("naked-seeds"), because their seeds do not develop in an enclosed structure. In the flowering plants, or **angiosperms** ("encased-seeds"), seeds develop inside a protective structure called a carpel. Today, angiosperms are far and away the most important lineage of land plants in terms of species diversity. Almost 90 percent of the land plant species alive today, and virtually all of the domesticated forms, are angiosperms.

This quick overview of morphological diversity in land plants raises a number of questions:

(c) Seed plants have vascular tissue and make seeds.

Cycadophyta (cycads)

Ginkgophyta (ginkgo)

Other conifers (redwoods, junipers, yews)

Gnetophyta (gnetophytes)

Pinophyta (pines, spruces, firs)

Anthophyta (angiosperms or flowering plants)

FIGURE 30.7 *(continued)*

- Does other evidence support the hypothesis that green algae are ancestral to land plants, and that coleochaetes and stoneworts are the closest living relatives to land plants?
- Did the green algae evolve first, followed by nonvascular plants, then seedless vascular plants, and finally seed plants?
- Are both green algae and land plants monophyletic—meaning each is a distinct lineage tracing back to a single common ancestor?
- Are the nonvascular plants, seedless vascular plants, and seed plants monophyletic?

Answers are emerging from data in the fossil record and DNA sequences used to infer the green plant phylogeny.

Using the Fossil Record

The first green plants that appear in the fossil record are green algae in rocks that formed 700–725 million years ago. The first land plant fossils are much younger—they are found in rocks that are about 475 million years old. Because green algae appear long before land plants, the fossil record supports the hypothesis that land plants are derived from green algae.

The appearance and early diversification of green algae about 700 million years ago is also significant because at roughly the same time, the oceans and atmosphere were starting to become oxygen-rich—as never before in Earth's history. Based on this time correlation, it is reasonable to hypothesize that the evolution of green algae contributed to the rise of oxygen levels on Earth. The origin of the oxygen atmosphere occurred not long before the appearance of animals in the fossil record and may have played a role in their origin and early diversification.

The fossil record of the land plants themselves is massive. In an attempt to organize and synthesize the database, **Figure 30.8** breaks it into five time intervals—each encompassing a major event in the diversification of land plants.

The oldest interval begins 475 million years ago (mya), spans 30 million years, and documents the origin of the group. Most of the fossils dating from this period are microscopic. They consist of the reproductive cells called **spores** and sheets of a waxy coating called **cuticle**. Several observations support the hypothesis that these fossils came from green plants that were growing on land. First, cuticle is a watertight barrier that coats today's land plants and helps them resist drying. Second, the fossilized spores are surrounded by a sheetlike coating. Under the electron microscope, the coating material appears

FIGURE 30.8 The Fossil Record of Land Plants Can Be Broken into Five Major Intervals.

EXERCISE Add the following to the time line: (1) first terrestrial vertebrate animals (370 Ma); (2) first mammals (195 Ma).

almost identical in structure to a watertight material called **sporopollenin**, which encases spores and pollen from modern land plants and helps them resist drying. Third, fossilized spores that are 475 million years old have recently been found in association with spore-producing structures called **sporangia** (singular: **sporangium**). The fossilized sporangia are similar in appearance to the sporangia observed in some of today's liverworts.

The second major interval in the fossil record of land plants is called the "Silurian-Devonian explosion." In rocks dated 445–359 mya, biologists find fossils from most of the major plant lineages. Virtually all of the adaptations that allow plants to occupy dry, terrestrial habitats are present, including water-conducting tissue and roots.

The third interval in the fossil history of plants spans the aptly named Carboniferous period. In sediments dated from about 359 to 299 mya, biologists find extensive deposits of coal. Coal is a carbon-rich rock packed with fossil spores, branches, leaves, and tree trunks. Most of these fossils are derived from lycophytes, horsetails, and ferns. Although the only living lycophytes and horsetails are small, during the Carboniferous these groups were species rich and included a wide array of tree-sized forms. Because coal formation is thought to start only in the presence of water, the Carboniferous fossils indicate the presence of extensive forested swamps.

The fourth interval in land plant history is characterized by gymnosperms. Recall that gymnosperms include the cycads, ginkgos, gnetophytes, pines, and other cone-bearing trees. Because gymnosperms grow readily in dry habitats, biologists infer that both wet and dry environments on the continents became blanketed with green plants for the first time during this interval. Gymnosperms are particularly prominent in the fossil record from from 251 to 145 mya.

The fifth interval in the history of land plants is still under way. This is the age of flowering plants—the angiosperms. The first flowering plants in the fossil record appear about 150 mya. The plants that produced the first flowers are the ancestors of today's grasses, orchids, daisies, oaks, maples, and roses.

According to the fossil record, then, the green algae appear first, followed by the nonvascular plants, seedless vascular plants, and seed plants. Organisms that appear late in the fossil record are often much less dependent on moist habitats than are groups that appear earlier. For example, the sperm cells of mosses and ferns swim to accomplish fertilization, while gymnosperms and angiosperms produce pollen grains that are transported via wind or insects and that then produce sperm. These observations support the hypotheses that green plants evolved from green algae and that in terms of habitat use, the evolution of green plants occurred in a wet-to-dry trend.

To test the validity of these observations, biologists analyze data sets that are independent of the fossil record. Foremost among these are the DNA sequences and morphological data used to infer phylogenetic trees. Does the phylogeny of land plants confirm or contradict the patterns in the fossil record?

Evaluating Molecular Phylogenies

Understanding the phylogeny of green plants is a monumental challenge. The challenge is being met by the biologists who use both morphological traits and DNA sequence data to infer the evolutionary relationships among green plants.

The phylogenetic tree in **Figure 30.9** is a recent version of results coming out of laboratories from around the world. The black bars across some branches on this tree show when key innovations occurred, based on the fossil record and their oc-

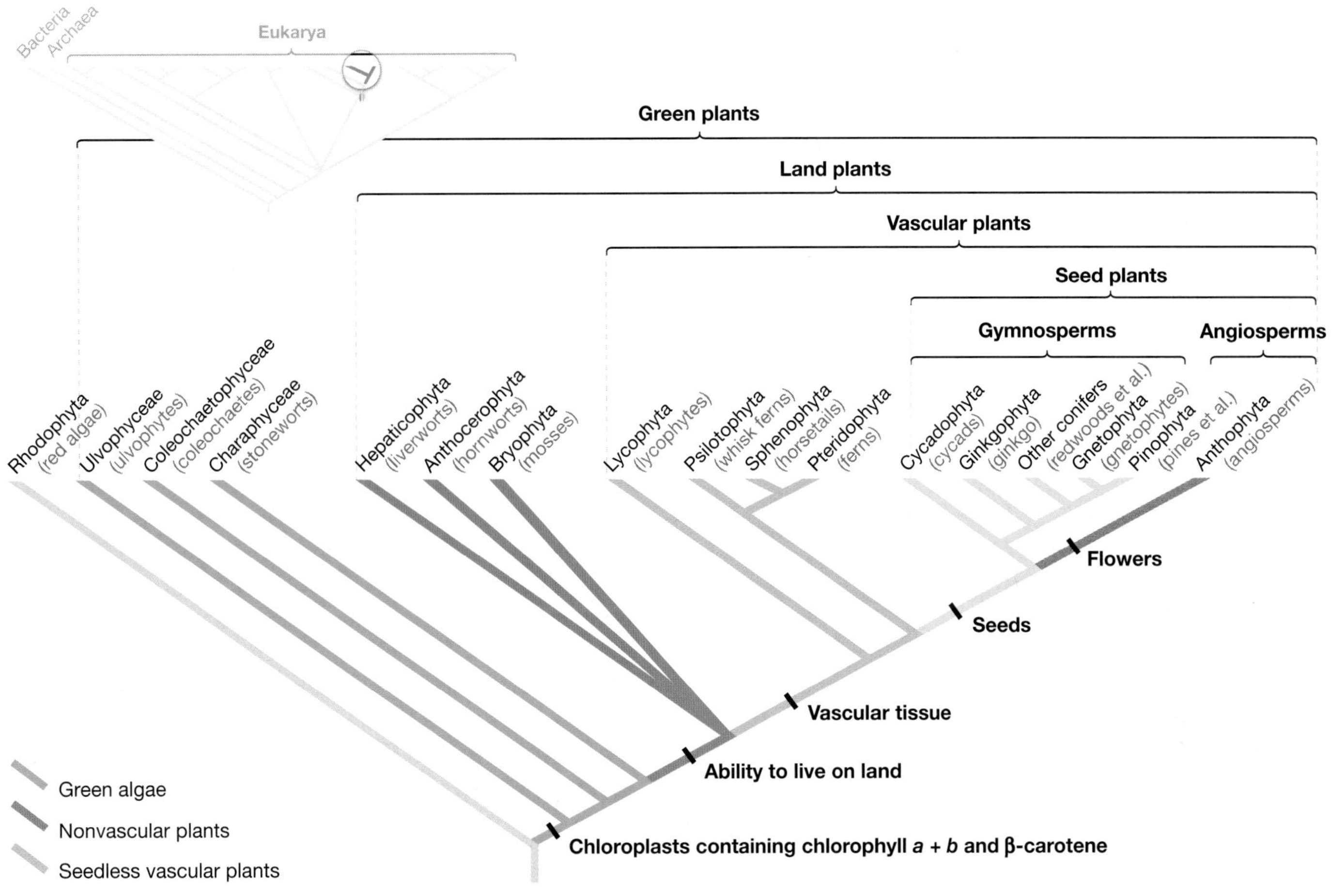

FIGURE 30.9 The Phylogeny of Green Plants. The blue branches on this figure indicate three of many major lineages collectively called green algae.

EXERCISE On the figure, identify why the groups called green algae and seedless vascular plants are paraphyletic.

currence in species living today. There are several important points to note about the relationships implied by this tree:

- The most basal branches on the tree, near the root, lead to coleochaetes, stoneworts, and other groups of green algae. This result supports the hypothesis that land plants evolved from green algae.
- The green algal group called Charaphyceae is the **sister group** to land plants—meaning that Charaphyceae are their closest living relative. Because living members of the Charaphyceae are multicellular and dwell in freshwater, the data support the hypothesis that land plants evolved from a multicellular ancestor that lived in ponds or lakes or other freshwater habitats.
- The green plants are monophyletic, meaning that a single common ancestor gave rise to all of the green algae and land plants. In contrast, the group called green algae is paraphyletic. Stated another way, the green algae include some but not all of the descendants of a single common ancestor.
- The land plants are monophyletic. This result supports the hypothesis that the transition from freshwater environments to land occurred just once. According to the phylogenetic tree, the population that made the transition then diversified into the array of land plants observed today.
- The nonvascular plants—that is, liverworts, hornworts, and mosses—are the earliest-branching, or most basal, groups among land plants. Although recent data confirm that each of these three lineages is monophyletic, the relationships among the three are still uncertain. For example, it is still not clear whether the nonvascular plants are monophyletic or paraphyletic.
- Some mosses have simple water-conducting tissues, and liverworts and hornworts have none at all. This finding suggests that the earliest land plants lacked water-conducting cells and vascular tissue, and that these traits evolved later. The phylogenetic tree supports the hypothesis that water-conducting cells and tissues evolved in a gradual fashion, with simpler structures preceding the evolution of more complex structures.
- The lycophytes are the sister group to all other seedless vascular plants. The whisk ferns, horsetails, and ferns form a monophyletic group; but as a whole the seedless vascular

plants form a **grade**—meaning a sequence of lineages that are not monophyletic. In contrast, the vascular plants are monophyletic—meaning that vascular tissue evolved once during the diversification of land plants.

- Because they lack roots and leaves, the whisk ferns were traditionally thought to be a basal group in the land plant radiation. Molecular phylogenies challenge this hypothesis and support an alternative hypothesis: The morphological simplicity of whisk ferns is a derived trait—meaning that complex structures have been lost in this lineage.
- The seed plants consist of the gymnosperms and the angiosperms and are a monophyletic group, meaning that seeds evolved once.
- The gymnosperms are monophyletic. Within the gymnosperms, the firs, spruces, and other species closely related to pines form an independent lineage to another group of cone-bearing plants (redwoods, junipers, yews, cypresses).
- The fossil record and the phylogenetic tree agree on the order in which groups appeared. Land plant evolution began with nonvascular plants, proceeded to seedless vascular plants, and continued with the evolution of seed plants.

The tree in Figure 30.9 will undoubtedly change and improve as additional data accumulate. Perhaps the most urgent need now is to clarify the relationships among the nonvascular plants—the liverworts, hornworts, and mosses. Because they are the most basal groups on the tree, understanding their relationships may help clarify how the earliest events in land plant evolution occurred.

Check Your Understanding

If you understand that...

- Biologists use the fossil record and phylogenetic analyses to study how green plants diversified.
- The data analyzed to date support the hypotheses that green plants are monophyletic and that land plants evolved from multicellular green algae that inhabited freshwater.
- The fossil record and molecular analyses agree that the nonvascular plants evolved first, followed by the seedless vascular plants and the seed plants.

You should be able to...

1) Take a phylogenetic tree of green plants and label the groups called green plants, green algae, land plants, nonvascular plants, vascular plants, seed plants, and gymnosperms.
2) Explain which of these named groups are monophyletic and which are paraphyletic.
3) For monophyletic groups, name a shared, derived trait that distinguishes the lineage.

30.3 What Themes Occur in the Diversification of Green Plants?

Land plants have evolved from algae that grew on the muddy shores of ponds 475 million years ago to organisms that enrich the soil, produce much of the oxygen you breathe and most of the food you eat, and serve as symbols of health, love, and beauty. How did this happen?

Answering this question begins with recognizing the most striking trend in the phylogeny and fossil record of green plants: The most ancient groups in the lineage are dependent on wet habitats, while more recently evolved groups are tolerant of dry—or even desert—conditions. The story of land plants is the story of adaptations that allowed photosynthetic organisms to move from aquatic to terrestrial environments. Let's first consider adaptations that allowed plants to grow in dry conditions without drying out and dying, and then analyze the evolution of traits that allowed plants to reproduce efficiently on land. This section closes with a brief look at the radiation of flowering plants, which are the most important plants in many of today's terrestrial environments.

The Transition to Land, I: How Did Plants Adapt to Dry Conditions?

For aquatic green algae, terrestrial environments are deadly. Compared with a habitat in which the entire organism is bathed in fluid, in terrestrial environments only a portion, if any, of the plant's tissues are wet. Tissues that are exposed to air tend to dry out and die.

Once green plants made the transition to survive out of water, though, growth on land offered a bonanza of resources. Take light, for example. The water in ponds, lakes, and oceans absorbs and reflects light. As a result, the amount of light available to drive photosynthesis is drastically reduced even a meter or two below the water surface. In addition, the most important molecule required by photosynthetic organisms, carbon dioxide, is much more readily available in air than it is in water. Not only is it more abundant in the atmosphere, but it diffuses more readily there than it does in water.

Natural selection favored early land plants with adaptations that solved the water problem. These adaptations arose in two steps: (1) prevention of water loss from cells, which kept the cells from drying out and dying; and (2) transportation of water from tissues with direct access to water to tissues without access. Let's examine both of these steps in turn.

Preventing Water Loss: Cuticle and Stomata Section 30.2 pointed out that sheets of the waxy substance called cuticle are present early in the fossil record of land plants, along with encased spores. This observation is significant because the presence of cuticle in fossils is a diagnostic indicator of land plants. Cuticle is a waxy, watertight sealant that covers the aboveground parts of plants and gives them the ability to survive in dry environments (**Figure 30.10a**). If biologists had to point to one innovation that made the transition to land possible, it would be the production of cuticle.

Covering surfaces with wax creates a problem, however, regarding the exchange of gases across those surfaces. Plants need to take in carbon dioxide (CO_2) from the atmosphere in order to perform photosynthesis. But cuticle is almost as impervious to CO_2 as it is to water. Most modern plants solve this problem with a structure called a **stoma** ("mouth"; plural: **stomata**), consisting of an opening surrounded by specialized **guard cells** (**Figure 30.10b**). The opening, called a pore, opens or closes as the guard cells change shape. When guard cells become soft, they close the stomata. Pores are closed in this way to limit water loss from the plant. When guard cells become taut, in contrast, they open the pore, not only allowing CO_2 to diffuse into the interior of leaves and stems where cells are actively photosynthesizing but also allowing excess O_2 to diffuse out. (The mechanism behind guard-cell movement is explored in Chapter 39.)

Stomata are present in all land plants except the liverworts, which have pores but no guard cells. These data suggest that the earliest land plants evolved pores that allowed gas exchange to occur at breaks in the cuticle-covered surface. Later, the evolution of guard cells gave land plants the ability to regulate gas exchange—and control water loss—by opening and closing their pores.

Transporting Water: Vascular Tissue and Upright Growth Once cuticle and stomata had evolved, plants could keep from drying out and thus keep photosynthesizing while exposed to air. Cuticle and stomata allowed plants to grow on the saturated soils of lake or pond edges. The next challenge? Defying gravity.

Multicellular green algae can grow erect because they float. They float because the density of their cells is similar to water's density. But outside of water, the body of a multicellular green alga collapses. The water that fills its cells is 1000 times denser than air. Although the cell walls of green algae are strengthened by the presence of cellulose, their bodies lack the structural support to withstand the force of gravity and to keep an individual erect in air.

Based on these observations, biologists hypothesize that the first land plants were small or had a low, sprawling growth habit. Besides lacking rigidity, the early land plants would have had to obtain water through pores or through a few cells that lacked cuticle—meaning they would have had to grow in a way that kept many or most of their tissues in direct contact with moist soil.

The sprawling-growth hypothesis is supported by the observation that the most basal groups of land plants living today (liverworts, hornworts, and mosses) are all low-growing forms. If this hypothesis is correct, then competition for space and light would have become intense soon after the first plants began growing on land. To escape competition, plants would have to grow upright. Plants with adaptations that allowed some tissues to remain in contact with wet soil while other tissues grew erect would have much better access to sunlight compared with individuals that were incapable of growing erect.

For plants to adopt erect growth habits on land, though, two problems had to be overcome. The first is transporting

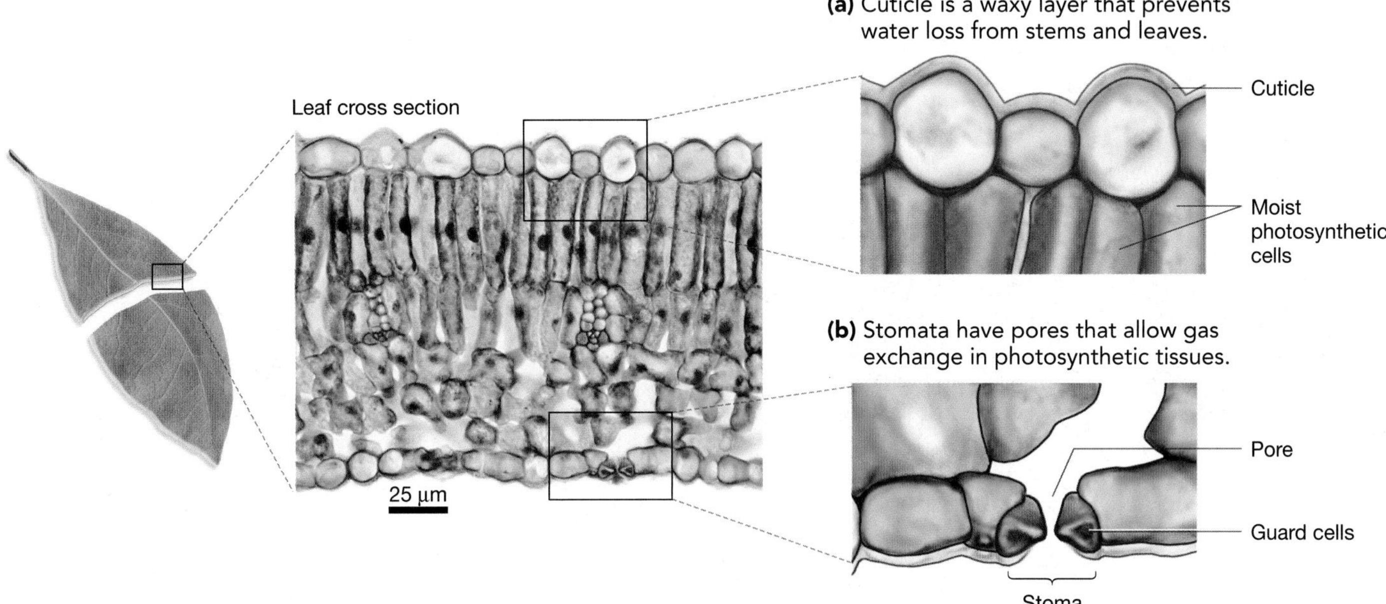

FIGURE 30.10 Cuticle and Stomata Are the Most Fundamental Plant Adaptations to Life on Land. In these micrographs, leaf cells have been stained blue to make their structure more visible. **(a)** The interior of plant leaves and stems is extremely moist; cuticle prevents water from evaporating away. **(b)** Stomata create pores to allow CO_2 to diffuse into the interior of leaves and stems where cells are actively photosynthesizing, and to allow excess O_2 to diffuse out.

QUESTION Why was the evolution of guard cells important?

water from tissues that are in contact with wet soil to tissues that are in contact with dry air, against the force of gravity. The second is becoming rigid enough to avoid falling over in response to gravity and wind. As it turns out, vascular tissue helped to solve both problems.

Paul Kenrick and Peter Crane explored the origin of water-conducting cells and erect growth in plants by examining the extraordinary fossils found in a rock formation in Scotland called the Rhynie Chert. These rocks formed about 400 million years ago and contain some of the first large plant specimens in the fossil record—as opposed to the microscopic spores and cuticle found in older rocks. The Rhynie Chert also contains numerous plants that fossilized in an upright position. This indicates that many or most of the Rhynie plants grew erect. How did they stay vertical?

By examining fossils with the electron microscope, Kenrick and Crane established that species from the Rhynie Chert contained elongated cells that were organized into tissues along the length of the plant. Based on these data, the biologists hypothesized that the elongated cells were part of water-conducting tissue and that water could move from the base of the plants upward to erect portions through these specialized water-conducting cells. Some of the water-conducting cells had simple, cellulose-containing cell walls like the water-conducting cells found in today's mosses (**Figure 30.11a**). But in addition, some of the water-conducting cells present in the early fossils had cell walls with thickened rings containing a molecule called lignin (**Figure 30.11b**). **Lignin** is a complex polymer built from six-carbon rings. It is extraordinarily strong for its weight and is particularly effective in resisting compressing forces such as gravity.

These observations inspired the following hypothesis: The evolution of lignin rings gave stem tissues the strength to remain erect in the face of wind and gravity. Today, the presence of lignin in the cell walls of water-conducting cells is considered the defining feature of vascular tissue. The evolution of vascular tissue allowed early plants to support erect stems and transport water from roots to aboveground tissues.

Once simple water-conducting tissues evolved, evolution by natural selection elaborated them. In rocks that are about 380 million years old, biologists find the advanced water-conducting cells called tracheids. **Tracheids** are long, thin, tapering cells that have (1) a thickened, lignin-containing **secondary cell wall** in addition to a cellulose-based **primary cell wall**; and (2) gaps in the secondary cell wall, in the sides and ends of the cell, where water can flow efficiently from one tracheid to the next (**Figure 30.11c**). The secondary cell wall gave tracheids the ability to provide better structural support, but water could still move through the cells easily because of the gaps. Today, all vascular plants contain tracheids.

In fossils dated to 250–270 million years ago, biologists have documented the most advanced type of water-conducting cells observed in plants; these cells are known as vessel elements. **Vessel elements** are shorter and wider than tracheids. More importantly, the upper and lower ends of vessel elements have gaps in both the primary and secondary cell wall, which

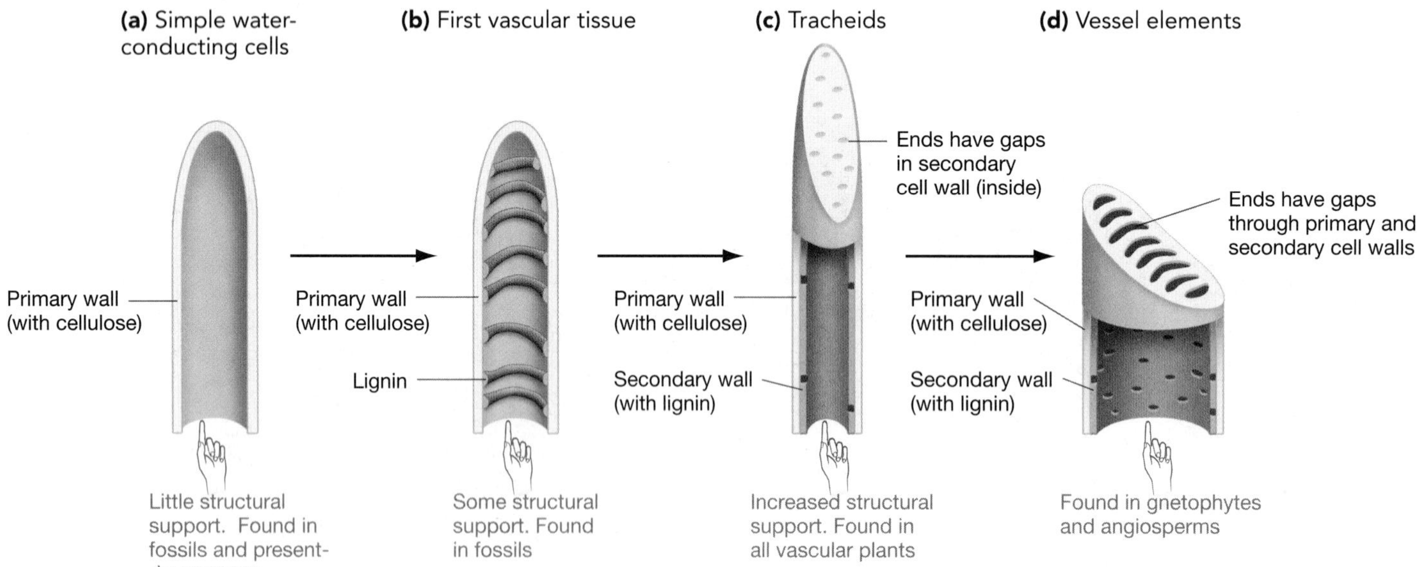

FIGURE 30.11 Evolutionary Sequence Observed in Water-Conducting Cells. According to the fossil record and the phylogeny of green plants, water-conducting cells became stronger over time due to the evolution of lignin and secondary cell walls. Efficient water transport was maintained through gaps in the secondary and/or primary cell wall.

QUESTION Biologists claim that vessels are more efficient than tracheids at transporting water, in part because vessels are shorter and wider than tracheids. Why does this claim make sense?

makes water movement extremely efficient (**Figure 30.11d**). In vascular tissue, vessel elements are lined up end to end to form a continuous pipelike structure.

In the stems and branches of some vascular plant species, tracheids or a combination of tracheids and vessels can form the extremely strong support material called **wood**. The anatomy of wood is explained in detail in Chapter 36.

All of the cell types shown in Figure 30.11 are dead when they mature, which means they lack cytoplasm. This feature allows water to move through the cells more efficiently. Taken together, the data summarized in the figure indicate that vascular tissue evolved in a series of gradual steps that provided increased structural support and increased efficiency in water transport.

Mapping Evolutionary Changes on the Phylogenetic Tree

Figure 30.12 summarizes how land plants adapted to dry conditions by mapping where major innovations occurred as the group diversified. Fundamentally important adaptations to dry conditions—such as the cuticle, pores, stomata, vascular tissue, and tracheids—evolved just once. In contrast, current analyses suggest that vessels evolved independently in gnetophytes and angiosperms.

The evolution of cuticle, stomata, and vascular tissue made it possible for plants to avoid drying out and to grow upright, while moving water from the base of the plant to its apex. Plants gained adaptations that allowed them not only to survive on land, but thrive. Now the question is, how did they reproduce?

at www.masteringbio.com
Plant Evolution and the Phylogenetic Tree

The Transition to Land, II: How Do Plants Reproduce in Dry Conditions?

Section 30.2 introduced one of the key adaptations for reproducing on land: spores that resist drying because they are encased in a tough coat of sporopollenin. Sporopollenin-like compounds are found in the walls of some green algal zygotes; thick-walled, sporopollenin-rich spores appear early in the fossil record of land plants and occur in all land plants living today. Based on these observations, biologists infer that sporopollenin-encased spores were one of the innovations that made the initial colonization of land possible. Two other innovations also occurred early in land plant evolution and were instrumental for efficient reproduction in a dry environment: (1) Gametes were produced in complex, multicellular structures; and (2) the embryo was retained on the parent (mother) plant and was nourished by it.

Retaining and Nourishing Offspring: Land Plants as Embryophytes The fossilized gametophytes of early land plants contain specialized reproductive organs called **gametangia** (singular: **gametangium**). Although members of the Charales also develop gametangia, the gametangia found in land plants are larger and more complex. The evolution of an elaborate gametangium was important because it protected gametes from drying and from mechanical damage. Gametangia are present in all land plants living today except angiosperms, where structures inside the flower perform the same functions.

In both the Charales and the land plants, individuals produce distinctive male and female gametangia. The sperm-producing structure is called an **antheridium** (plural:

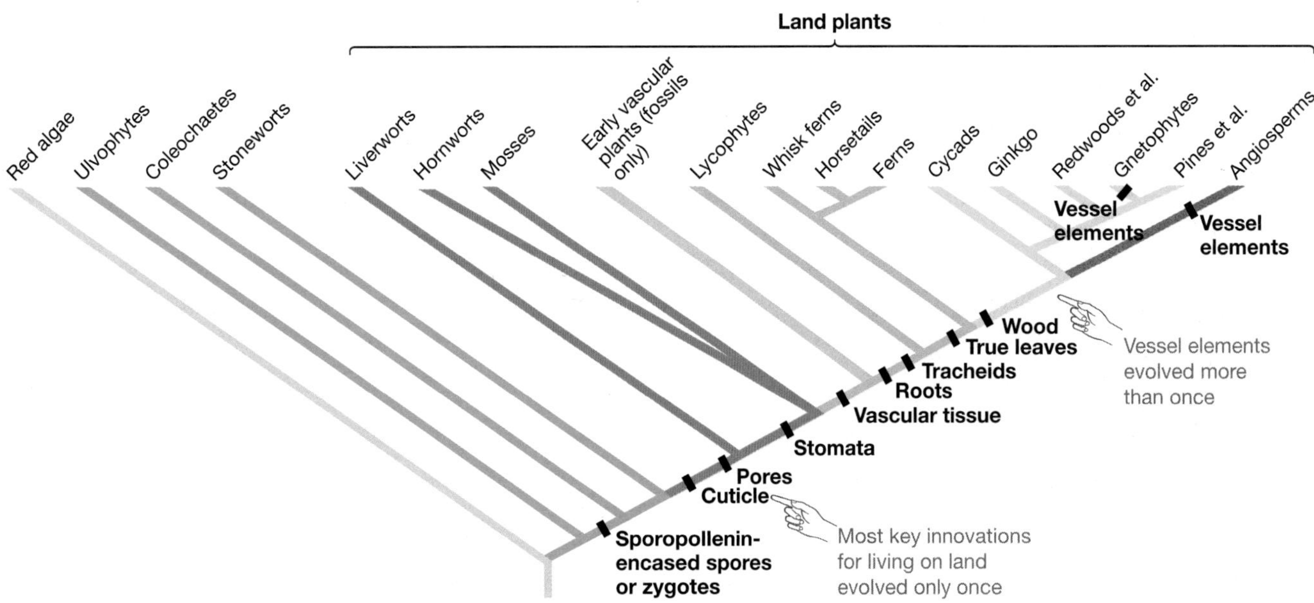

FIGURE 30.12 A Series of Evolutionary Innovations Allowed Plants to Adapt to Life on Land.

QUESTION Explain the logic that biologists used to map the location of each innovation indicated here.

(a) Sperm form in antheridia.

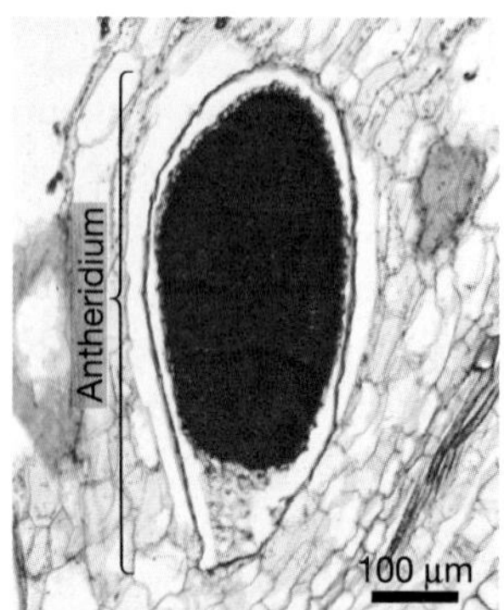

(b) Eggs form in archegonia.

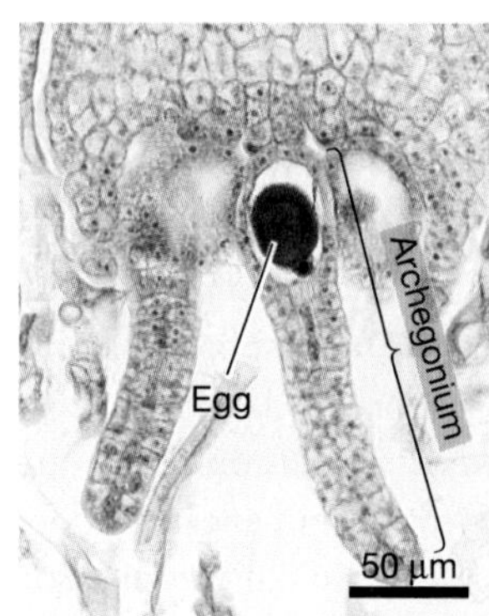

FIGURE 30.13 In All Land Plant Groups but Angiosperms, Gametes Are Produced in Gametangia. Gametangia are complex, multicellular structures that protect developing gametes from drying and mechanical damage. Note that there is one egg per archegonium.

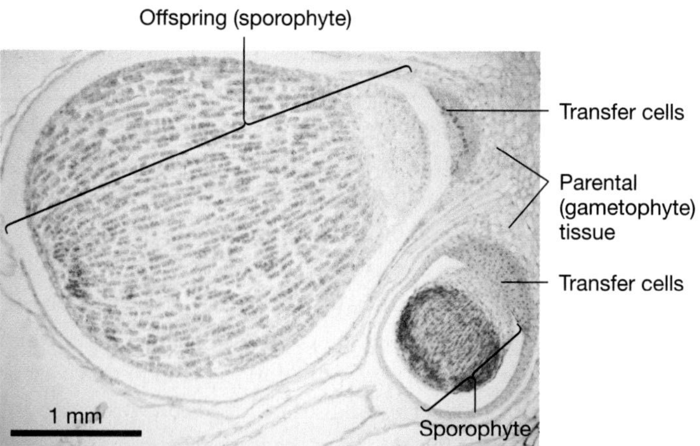

FIGURE 30.14 Land Plants Are Also Known as Embryophytes Because Parents Nourish Their Young. In land plants, fertilization and early development take place on the parent plant. As embryos develop, transfer cells carry nutrients from the mother to the offspring.

antheridia; Figure 30.13a, and the egg-producing structure is called an **archegonium** (plural: **archegonia; Figure 30.13b**). In terms of their function, the antheridium and archegonium are analogous to the testes and ovaries of animals.

The second innovation that occurred early in land plant evolution involved the eggs that formed inside archegonia. Instead of shedding their eggs into the water or soil, land plants retain them. Eggs are also retained in the green algal lineages that are most closely related to land plants: In Charales and other closely related groups, sperm swim to the egg, fertilization occurs, and the resulting zygote stays attached to the parent. Either before or after fertilization, the egg or zygote receives nutrients from the mother plant. But because these algae live in northern latitudes, the parent plant dies each autumn as the temperature drops. The zygote remains on the dead tissue from the parent, settles to the bottom of the lake or pond, and overwinters. In spring, meiosis occurs, and the resulting spores develop into haploid adult plants.

In land plants, the zygote is also retained on the parent plant after fertilization. But in contrast to the zygotes of most green algae, the zygotes of all land plants begin to develop on the parent plant, forming a multicellular embryo that remains attached to the parent and can be nourished by it. This is important because land plant embryos do not have to manufacture their own food early in life. Instead, they receive most or all of their nutrients from the parent plant.

The retention of the embryo was such a key event in land plant evolution that the formal name of the group is Embryophyta—literally, the "embryo-plants." The retention in these plants, commonly called **embryophytes**, is analogous to pregnancy in mammals, where offspring are retained by the mother and nourished through the initial stages of growth. Land plant embryos even have specialized **transfer cells**, which make physical contact with parental cells and facilitate the transfer of nutrients (**Figure 30.14)** much like the placenta that develops in a pregnant mammal.

Thick-walled spores, elaborate gametangia, and the embryophyte condition weren't the only key innovations associated with reproducing on land, though. In addition, all land plants undergo the phenomenon known as **alternation of generations**, introduced in Chapter 29.

Alternation of Generations When alternation of generations occurs, individuals have a multicellular haploid phase and a multicellular diploid phase. The multicellular haploid stage is called the **gametophyte**; the multicellular diploid stage is called the **sporophyte**. The two phases of the life cycle are connected by distinct types of reproductive cells—gametes and spores.

Although alternation of generations is observed in a wide array of protist lineages and in some groups of green algae, it does not occur in the algal groups most closely related to land plants. In the coleochaetes and stoneworts, the multicellular form is haploid. Only the zygote is diploid. As **Figure 30.15** shows, the zygote undergoes meiosis to form haploid spores. After dispersing with the aid of flagella, the spores begin dividing by mitosis and eventually grow into an adult, haploid individual. You might recall that this haploid-dominant life cycle is common in protists and was diagrammed in Figure 29.20a.

These data suggest that alternation of generations originated in land plants independently of its evolution in other groups of eukaryotes, and that it originated early in their history—soon after they evolved from green algae. The adaptive significance of alternation of generations and its role in the successful colonization of land is still being debated, however. Keep this in mind as you study alternation of generations in land plants: You are analyzing one of the great unsolved problems in contemporary biology.

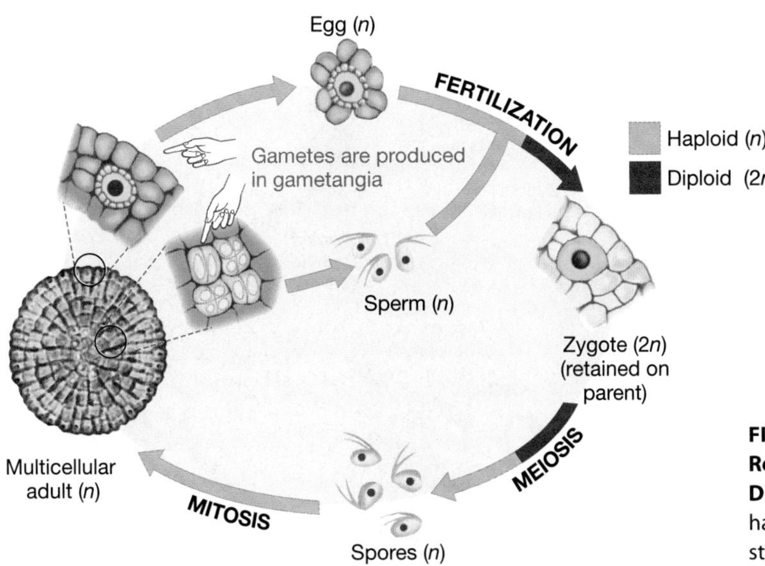

FIGURE 30.15 In Green Algae That Are Closely Related to Land Plants, Only the Zygote Is Diploid. The coleochaetes and stoneworts do not have alternation of generations. The multicellular stage is haploid.

Alternation of generations always involves the same basic sequence of events, illustrated in **Figure 30.16**. To review how this type of life cycle works, put your finger on the gametophyte in the figure and trace the cycle clockwise to find the following five key events:

1. Gametophytes produce gametes by mitosis. Both the gametophyte and the gametes are haploid.
2. Two gametes unite during fertilization to form a diploid zygote.
3. The zygote divides by mitosis and develops into a multicellular, diploid sporophyte.
4. The sporophyte produces spores by meiosis. Spores are haploid.
5. Spores divide by mitosis and develop into a haploid gametophyte.

Once you've traced the cycle successfully, take a moment to review the differences between a zygote and a spore. Both are single cells that divide by mitosis to form a multicellular individual. Zygotes, however, result from the fusion of two cells, such as a sperm and an egg. Spores are not formed by the fusion of two cells. In addition, zygotes produce sporophytes while spores produce gametophytes. Spores are produced inside structures called sporangia while gametes are produced inside gametangia.

Alternation of generations can be a difficult topic to master, for two reasons: (1) It is unfamiliar because it does not occur in humans or other animals, and (2) gamete formation begins with mitosis—not meiosis, as it does in animals (see Chapter 12). If you understand the basic principles of alternation of generations, you should be able to draw a life cycle starting with a gametophyte that produces gametes, which then fuse to form a zygote, which germinates to form a sporophyte, which produces spores via meiosis, which germinate to form a gametophyte. You should also be able to label the gametophyte, gametes, zygote, sporophyte, and spores, and indicate where meiosis takes place.

The Gametophyte-Dominant to Sporophyte-Dominant Trend in Life Cycles The five steps illustrated in Figure 30.16 occur in all species with alternation of generations. But in land plants, the relationship between the gametophyte and sporophyte is highly variable.

In nonvascular plants such as mosses, the sporophyte is small and short lived and is largely dependent on the gametophyte for nutrition (**Figure 30.17a**). When you see leafy-looking mosses growing on a tree trunk or on rocks like those in the photo at the start of this chapter, you are looking at gametophytes. Because

FIGURE 30.16 All Land Plants Undergo Alternation of Generations. Alternation of generations always involves the same sequence of five events: (1) Gametophytes produce haploid gametes by mitosis; (2) gametes fuse to form a diploid zygote; (3) zygotes develop into a diploid sporophyte; (4) sporophytes produce haploid spores by meiosis; and (5) spores develop into haploid gametophytes.

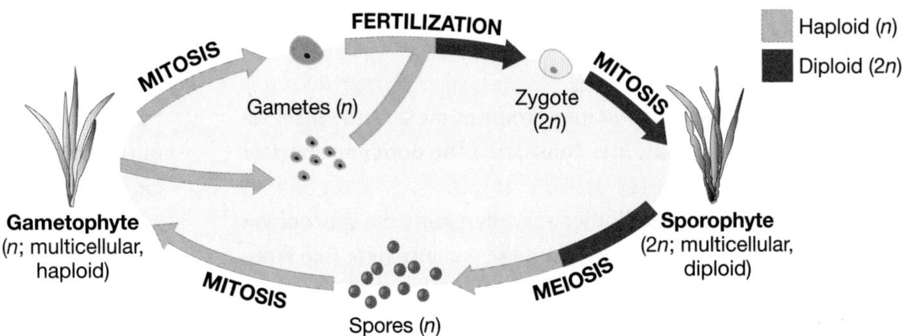

(a) Mosses: Gametophyte is large and long lived; sporophyte depends on gametophyte for nutrition.

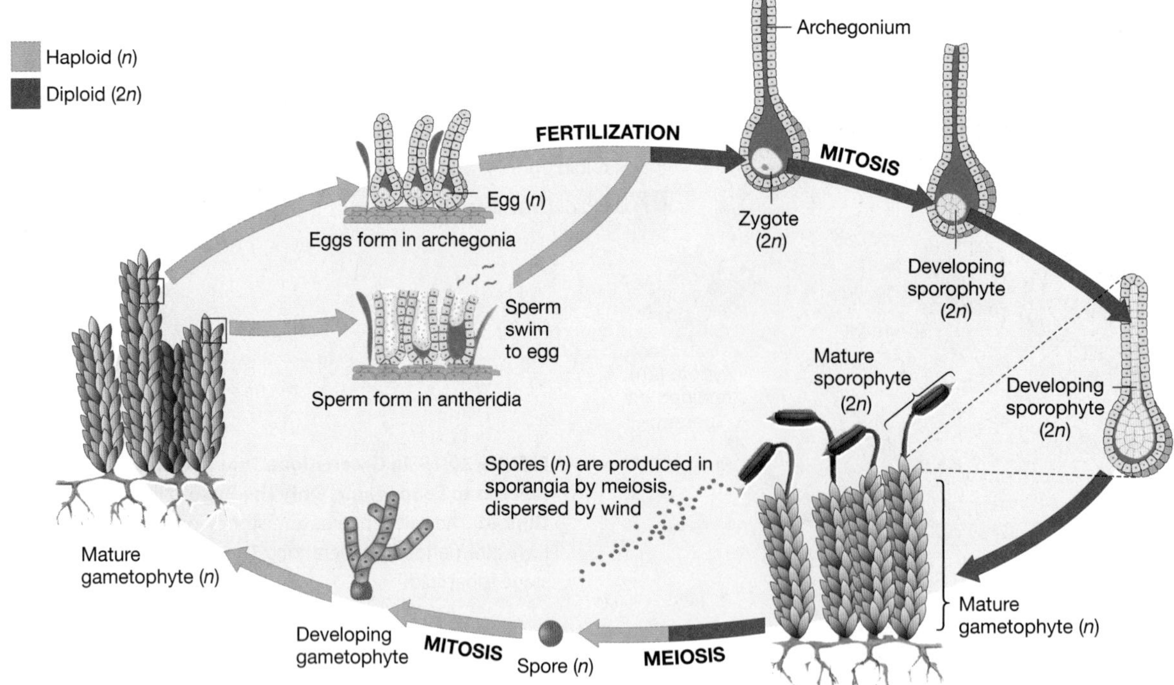

(b) Ferns: Sporophyte is large and long lived but, when young, depends on gametophyte for nutrition.

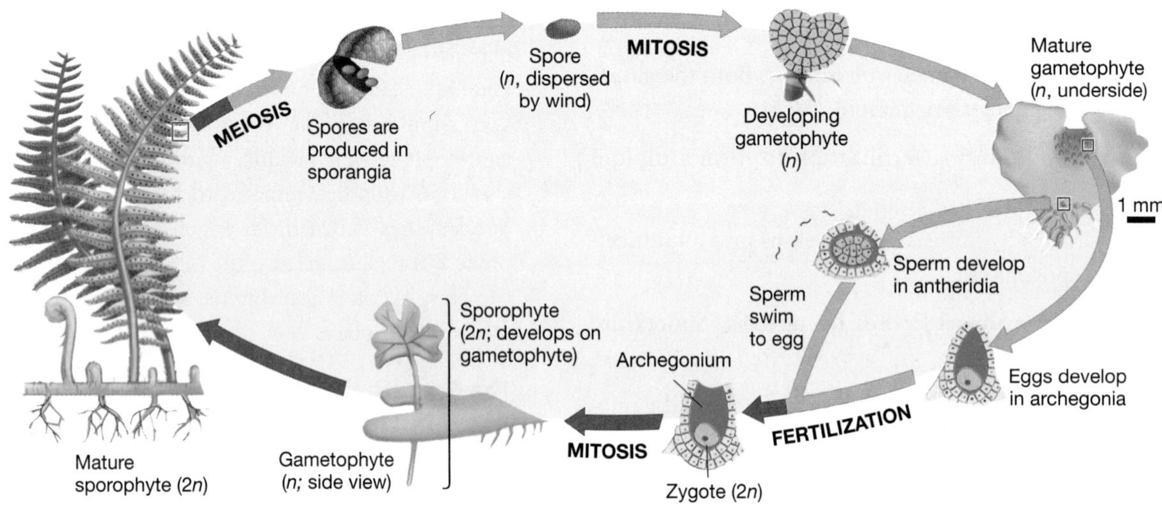

FIGURE 30.17 Gametophyte-Dominated Life Cycles Evolved Early; Sporophyte-Dominated Life Cycles Evolved Later. The most basal lineages of land plants, like mosses, have gametophytes that are much larger and longer lived than the sporophyte. In more derived lineages, such as ferns, the sporophyte is much larger and longer lived than the gametophyte.

● **QUESTION** How can you tell that alternation of generations occurs in these species?

● **QUESTION** Spores and zygotes are single cells that grow into an adult via mitosis. How do they differ?

the gametophyte is long lived and produces most of the food required by the individual, it is considered the dominant part of the life cycle.

In contrast, in ferns and other vascular plants the sporophyte is much larger and longer lived than the gametophyte (see **Figure 30.17b**). The ferns you see growing in gardens or forests are sporophytes. You'd have to hunt on your hands and knees to find their gametophytes, which are typically just a few millimeters in diameter. As you'll learn later in the chapter, the gametophytes of gymnosperms and angiosperms are even smaller—they are microscopic. Ferns and other vascular plants are said to have a sporophyte-dominant life cycle. ● If you understand the difference between gametophyte-dominant and sporophyte-dominant life cycles, you should be able to examine

(a) Hornwort gametophytes and sporophytes

(b) Horsetail gametophytes and sporophytes

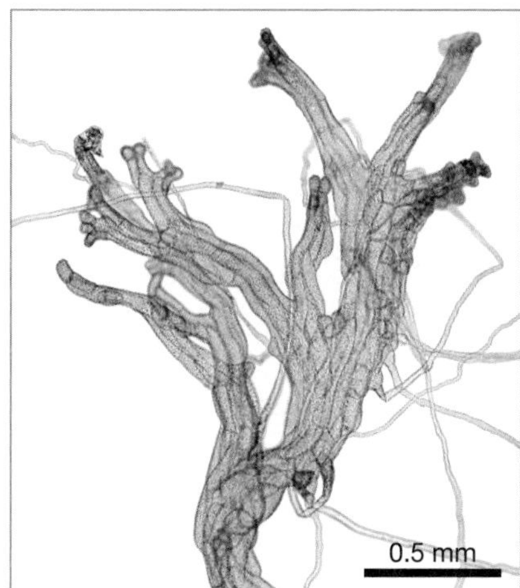

FIGURE 30.18 The Reduction of the Gametophyte Is One of the Strongest Trends in Land Plant Evolution. (a) A hornwort showing the sporophyte emerging from the gametophyte. **(b)** The same horsetail species as a tiny gametophyte and macroscopic sporophyte.

● **EXERCISE** In parts (a) and (b), identify which is the gametophyte and which is the sporophyte.

the photos of hornworts (a nonvascular plant) and horsetails (a vascular plant) in **Figure 30.18**, and identify which is the gametophyte and which is the sporophyte.

The transition from gametophyte-dominated life cycles to sporophyte-dominated life cycles is one of the most striking of all trends in land plant evolution. To explain why it occurred, biologists hypothesize that sporophyte-dominated life cycles were advantageous because diploid cells can respond to varying environmental conditions more efficiently than haploid cells can—particularly if the individual is heterozygous at many genes. This idea has yet to be tested rigorously, however, and the trend is still considered something of a mystery. If you could come up with an explanation for why alternation of generations evolved in land plants and why the gametophyte-dominant to sporophyte-dominant trend occurred, and if your ideas stood up to intensive testing, the result would be celebrated worldwide.

Heterospory In addition to sporophyte-dominated life cycles, another important innovation found in seed plants is **heterospory**—the production of two distinct types of spore-producing structures and thus two distinct types of spores. Some lycophytes and a few ferns are also heterosporous, but all of the nonvascular plants and most of the seedless vascular plants are **homosporous**—meaning that they produce a single type of spore. Homosporous species produce spores that develop into bisexual gametophytes. Bisexual gametophytes produce both eggs and sperm (**Figure 30.19a**).

The two types of spore-producing structures found in heterosporous species are often found on the same individual. Microsporangia are spore-producing structures that produce microspores. **Microspores** develop into male gametophytes, which produce the small gametes called sperm. Megasporangia are spore-producing structures that produce megaspores. **Megaspores** develop into female gametophytes, which produce the large gametes called eggs (**Figure 30.19b**). Thus, the gametophytes of seed plants are either male or female, but never both.

The evolution of heterospory was a key event in land plant evolution because it made possible one of the most important adaptations for life in dry environments—pollen.

Pollen The nonvascular plants and the most ancient vascular plant lineages, such as ferns, have male gametes that swim to the egg to perform fertilization. For a sperm cell to swim to the egg and fertilize it, there has to be a continuous sheet of water between the male and female gametophyte, or a raindrop has to splash sperm onto a female gametophyte. In species that live in dry environments, these conditions are rare. The land plants made their final break with their aquatic origins and were able to reproduce efficiently in dry habitats when a structure evolved that could move their gametes without the aid of water.

(a) Non-vascular plants and most seedless vascular plants are homosporous.

Sporangium → Spores → Bisexual gametophyte → Sperm / Eggs

(b) Seed plants are heterosporous.

FIGURE 30.19 Heterosporous Plants Produce Male and Female Spores That Are Morphologically Distinct.

In heterosporous seed plants, the microspore germinates to form a tiny male gametophyte that is surrounded by a tough, desiccation-resistant coat of sporopollenin. The resulting structure is called a **pollen grain**. Pollen grains can be exposed to the air for long periods of time without dying from dehydration. They are also tiny enough to be carried to female gametophytes by wind or animals. Upon landing near the egg, the male gametophyte releases the sperm cells that accomplish fertilization.

When pollen evolved, then, heterosporous plants lost their dependence on water to accomplish fertilization. Instead of swimming to the egg as a naked sperm cell, their tiny gametophytes took to the skies.

Seeds The evolution of large gametangia protected the eggs and sperm of land plants from drying. Embryo retention allowed offspring to be nourished directly by their parent, and pollen enabled fertilization to occur in the absence of water. Although retaining embryos ensures that off-spring are nourished, it has a downside: In ferns and horsetails, sporophytes have to live in the same place as their parent gametophyte. Seed plants overcome this limitation, however, because their embryos are portable and can disperse to new locations.

A **seed** is a structure that includes an embryo and a food supply surrounded by a tough coat (**Figure 30.20a**). Seeds allow embryos to be dispersed to a new habitat, away from the parent plant. Like a bird's egg, the seed provides a protective case for the embryo and a store of nutrients provided by the mother. In addition, seeds are often attached to structures that aid in dispersal by wind, water, or animals (**Figure 30.20b**).

The evolution of heterospory, pollen, and seeds triggered a dramatic radiation of seed plants starting about 290 million years ago. To make sure that you understand these key processes and structures, study the life cycle of the pine tree pictured in **Figure 30.21**. Starting with the sporophyte on the left, note that this and many other gymnosperm species have separate structures, called cones, where microsporangia and megasporangia develop. In this case, the two types of spores associated with heterospory develop in separate cones. The microsporangia contain a cell that divides by meiosis to form microspores, which then divide by mitosis to form pollen grains—tiny male gametophytes. Megasporangia are found inside protective structures called ovules, and they contain a mother cell that divides by meiosis to form a megaspore. The megaspore undergoes mitosis to form the female gametophyte, which contains egg cells. Notice that the female gametophyte stays attached to the sporophyte as pollen grains arrive and produce sperm that fertilize the eggs. Seeds mature as the embryo develops and cells derived from the female gametophyte become packed with nutrients provided by the sporophyte. When the seed disperses and germinates, the cycle of life begins anew.

Flowers Flowering plants, or angiosperms, are the most diverse land plants living today. About 250,000 species have been described, and more are discovered each year. Their success in terms of geographical distribution, number of individuals, and number of species revolves around a reproductive organ—the **flower**.

(a) Seeds package an embryo with a food supply.

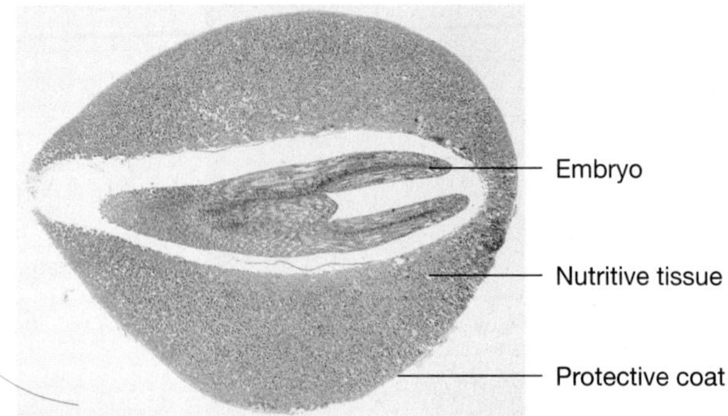

(b) Seeds are dispersed by wind, water, and animals.

FIGURE 30.20 Seeds Contain an Embryo and a Food Supply and Can Be Dispersed. (a) Seeds consist of an embryo, a store of carbohydrate or oil-rich nutritive tissue, and a protective coat. **(b)** Seeds are often found in structures that facilitate dispersal by wind, water, or animals.

Flowers contain two key reproductive structures: the stamens and carpels illustrated on the left-hand side of **Figure 30.22**. Stamens and carpels are responsible for heterospory. A **stamen** includes a structure called an anther, where microsporangia develop. Meiosis occurs inside the microsporangia, forming microspores. Microspores then divide by mitosis to form pollen grains. A flower's **carpel**, in contrast, contains a protective structure called an **ovary** where the ovules are found. The presence of enclosed ovules inspired the name *angiosperm* ("encased-seed") as opposed to *gymnosperm* ("naked-seed"). As in gymnosperms, ovules contain the megasporangia. A cell inside the megasporangium divides by meiosis to form the megaspore, which then divides by mitosis to form the female gametophyte. When a pollen grain lands on a carpel and produces sperm, fertilization takes place, as shown on the right-hand side of Figure 30.22. In angiosperms, though, fertilization involves two sperm cells. One sperm fuses with the egg to form the zygote, while a second

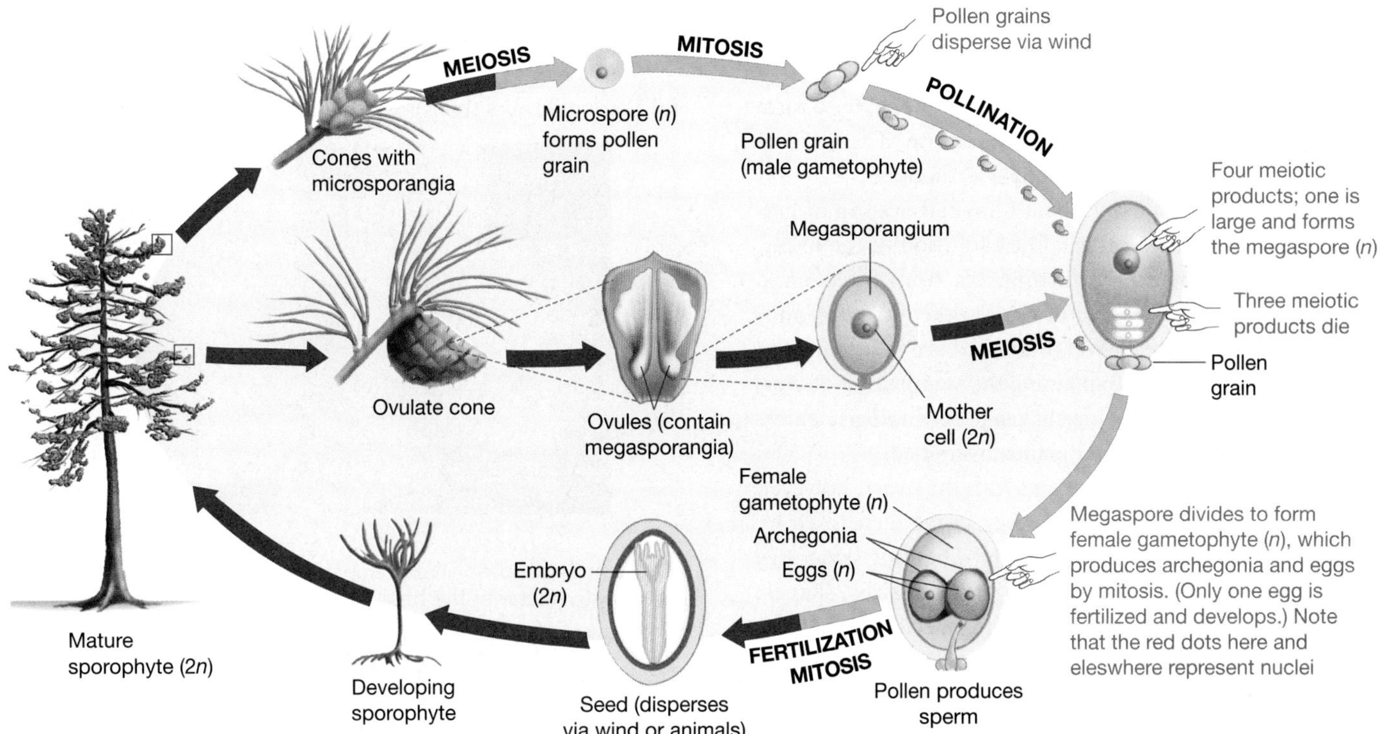

FIGURE 30.21 Heterospory in Gymnosperms: Microspores Produce Pollen Grains; Megaspores Produce Female Gametophytes.

● **QUESTION** Compare this life cycle with that of the fern pictured in Figure 30.17b. Is the gymnosperm gametophyte larger than, smaller than, or about the same size as a fern gametophyte? Compared with ferns, is the gymnosperm gametophyte more or less dependent on the sporophyte for nutrition?

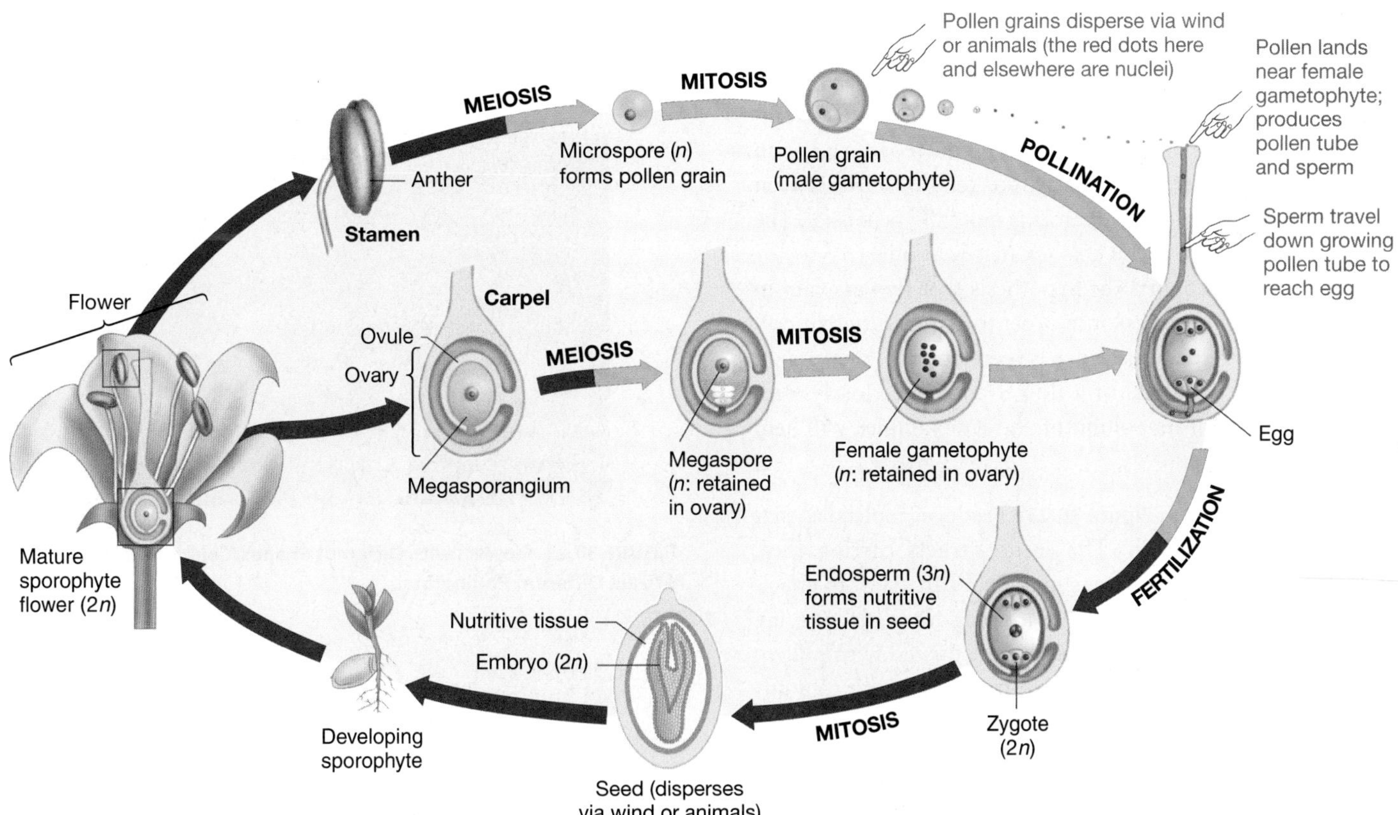

FIGURE 30.22 Heterospory in Angiosperms: Flowers Contain Microspores and Megaspores.

● **QUESTION** Gymnosperm pollen grains typically contain from 4 to 40 cells; mature angiosperm pollen grains contain three cells. Gymnosperm female gametophytes typically contain hundreds of cells; angiosperm female gametophytes typically contain seven. Do these observations conflict with the trend of reduced gametophytes during land plant evolution, or are they consistent with it? Explain your logic.

sperm fuses with two nuclei in the female gametophyte to form a triploid ($3n$) nutritive tissue called **endosperm**. The involvement of two sperm nuclei is called **double fertilization**.

The evolution of the flower, then, is an elaboration of heterospory. The key innovation was the evolution of the ovary, which helps protect female gametophytes from insects and other predators. Double fertilization is another striking innovation associated with the flower, but its adaptive significance is still not well understood. Explaining the significance of double fertilization is another major challenge for biologists interested in understanding how land plants diversified.

The story of the flower doesn't end with the ovary, however. Once stamens and carpels evolved, they became enclosed by modified leaves called **sepals** and **petals**. The four structures then diversified to produce a fantastic array of sizes, shapes, and colors—from red roses to blue violets. Specialized cells inside flowers also began producing a wide range of scents.

To explain these observations, biologists hypothesize that flowers are adaptations to increase the probability that an animal will perform **pollination**—the transfer of pollen from one individual's stamen to another individual's carpel. Instead of leaving pollination to an undirected agent such as wind, the hypothesis is that natural selection favored structures that reward an animal—usually an insect—for carrying pollen directly from one flower to another. Under the directed-pollination hypothesis, natural selection has favored flower colors and shapes and scents that are successful in attracting particular types of pollinators. A pollinator is an animal that disperses pollen. Pollinators are attracted to flowers because flowers provide the animals with food in the form of protein-rich pollen or a sugar-rich fluid known as **nectar.** In this way, the relationship between flowering plants and their pollinators is mutually beneficial. The pollinator gets food; the plant gets sex (fertilization).

What evidence supports the hypothesis that flowers vary in size, structure, scent, and color in order to attract different pollinators? The first type of evidence is correlational in nature. In general, the characteristics of a flower correlate closely with the characteristics of its pollinator. A few examples will help drive this point home:

- The carrion flower in **Figure 30.23a** produces molecules that smell like rotting flesh. The scent attracts carrion flies, which normally lay their eggs in animal carcasses. In effect, the plant tricks the flies. While looking for a place to lay their eggs on a flower, the flies become dusted with pollen. If the flies are already carrying pollen from a visit to a different carrion flower, they are likely to deposit pollen grains near the plant's female gametophyte. In this way, the carrion flies pollinate the plant.
- Flowers that are pollinated by hummingbirds typically have petals that form a long, tubelike structure corresponding to the size and shape of a hummingbird's beak (**Figure 30.23b**). Nectar-producing cells are located at the base of the tube. When hummingbirds visit the flower, they insert their beaks and harvest the nectar. In the process, th ey transfer pollen grains attached to their throats or faces.
- Hummingbird-pollinated flowers also tend to have red petals (Figure 30.23b), while bee-pollinated flowers tend to be purple or yellow (**Figure 30.23c**). Hummingbirds are attracted to red; bees have excellent vision in the purple and ultraviolet end of the spectrum.

(a) Carrion flowers smell like rotting flesh and attract carrion flies.

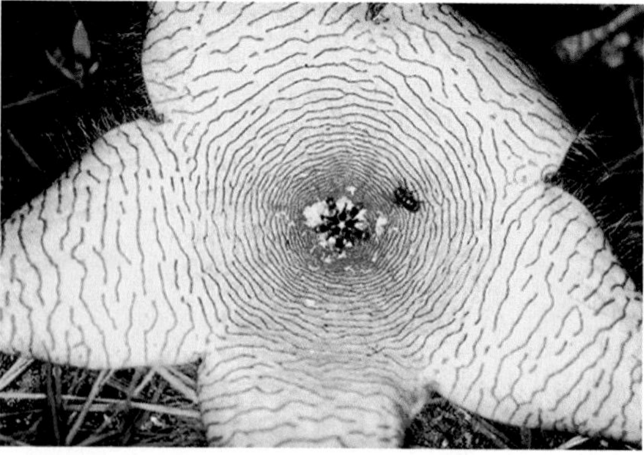

(b) Hummingbird-pollinated flowers are red and have long tubes with nectar at the base.

(c) Bumble-bee-pollinated flowers are often bright purple.

FIGURE 30.23 Flowers with Different Shapes, Colors, and Fragrances Attract Different Pollinators.

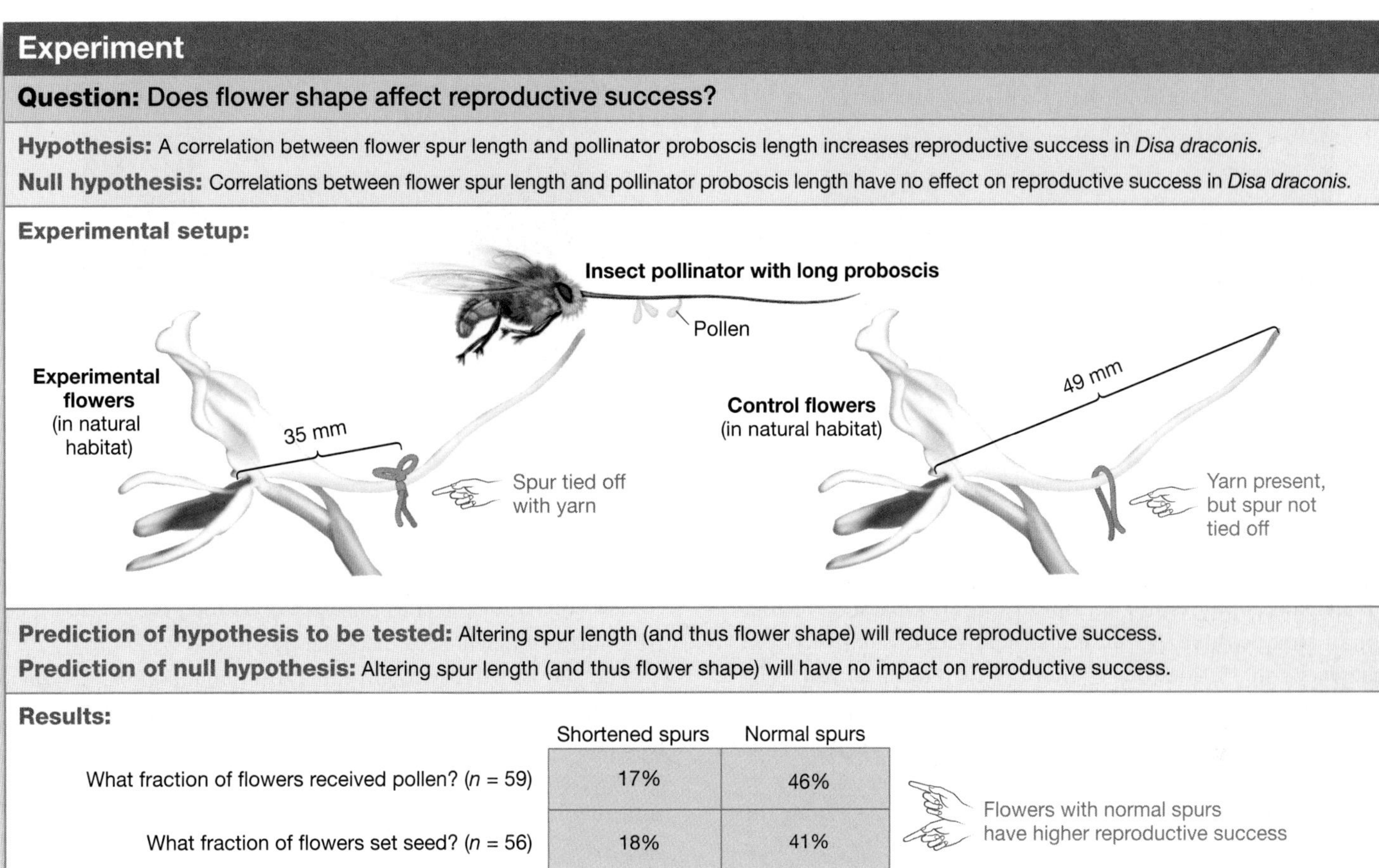

	Shortened spurs	Normal spurs
What fraction of flowers received pollen? (n = 59)	17%	46%
What fraction of flowers set seed? (n = 56)	18%	41%

Conclusion: In this species, flower shape—specifically spur length—affects reproductive success.

FIGURE 30.24 The Adaptive Significance of Flower Shape: An Experimental Test. The spurs of experimental *D. draconis* flowers were tied off with yarn 35 mm from the opening; the spurs of control flowers had a piece of yarn loosely tied around them, too, but had normal spurs averaging 49 mm.

● **QUESTION** If spur length has no effect on reproductive success, what would the data in this table look like?

● **QUESTION** Why did the researchers bother to put yarn around the control flowers?

The directed-pollination hypothesis also has strong experimental support. Consider, for example, recent work on a South African orchid called *Disa draconis*. The length of the long tube, or spur, located at the back of this flower varies among populations of this species. As predicted by the directed-pollination hypothesis, each population is pollinated by a different insect. Short-spurred plants that grow in mountain habitats are pollinated by insect species that have a relatively short proboscis. (A *proboscis* is a specialized mouthpart found in some insects. When extended, it functions like a straw in sucking nectar or other fluids.) Long-spurred orchids that grow on low-lying sandplain habitats are pollinated by insects that have a particularly long proboscis.

To test the hypothesis that spur length affects pollination success, researchers artificially shortened the spurs of individuals in a long-spurred population (**Figure 30.24**). The biologists did this by tying off the spurs of randomly selected flowers with a piece of yarn, so that flies could not reach the end of the spur. The idea was that flies would not make contact with the flower's reproductive organs when they inserted their proboscis into the shortened tube. The biologists left nearby flowers alone but tied a piece of yarn near the spur. As the data in the "Results" box of Figure 30.24 indicate, flowers with short spurs received much less pollen and set much less seed than did flowers with normal-length spurs. These data strongly support the hypothesis that spur length is an adaptation that increases the frequency of pollination by particular insects. Based on results like this, biologists contend that the spectacular diversity of angiosperms resulted, at least in part, from natural selection exerted by the equally spectacular diversity of insect, mammal, and bird pollinators.

Fruits The evolution of the ovary was an important event in land plant diversification, but not only because it protected the

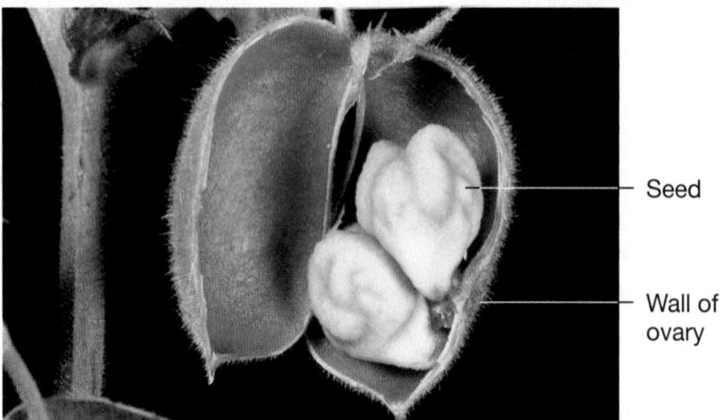

FIGURE 30.25 Fruits Are Derived from Ovaries Found in Angiosperms. (a) A pea pod is one of the simplest types of fruit. **(b)** The ovary wall often becomes thick, fleshy, and nutritious enough to attract animals that disperse the seeds inside.

female gametophytes of angiosperms. It also made the evolution of fruit possible. A **fruit** is a structure that is derived from the **ovary** and encloses one or more seeds (**Figure 30.25a**). Tissues derived from the ovary are often nutritious and brightly colored. Animals eat these types of fruits, digest the nutritious tissue around the seeds, and disperse seeds in their feces (**Figure 30.25b**). In other cases, the tissues derived from the ovary help fruits disperse via wind or water. The evolution of flowers made efficient pollination possible; the evolution of fruits made efficient seed dispersal possible.

The list of adaptations that allow land plants to reproduce in dry environments is impressive; **Figure 30.26** summarizes them. Once land plants had vascular tissue and could grow efficiently in dry habitats, the story of their diversification revolved around traits that allowed sperm cells to reach eggs efficiently and helped seeds disperse to new locations.

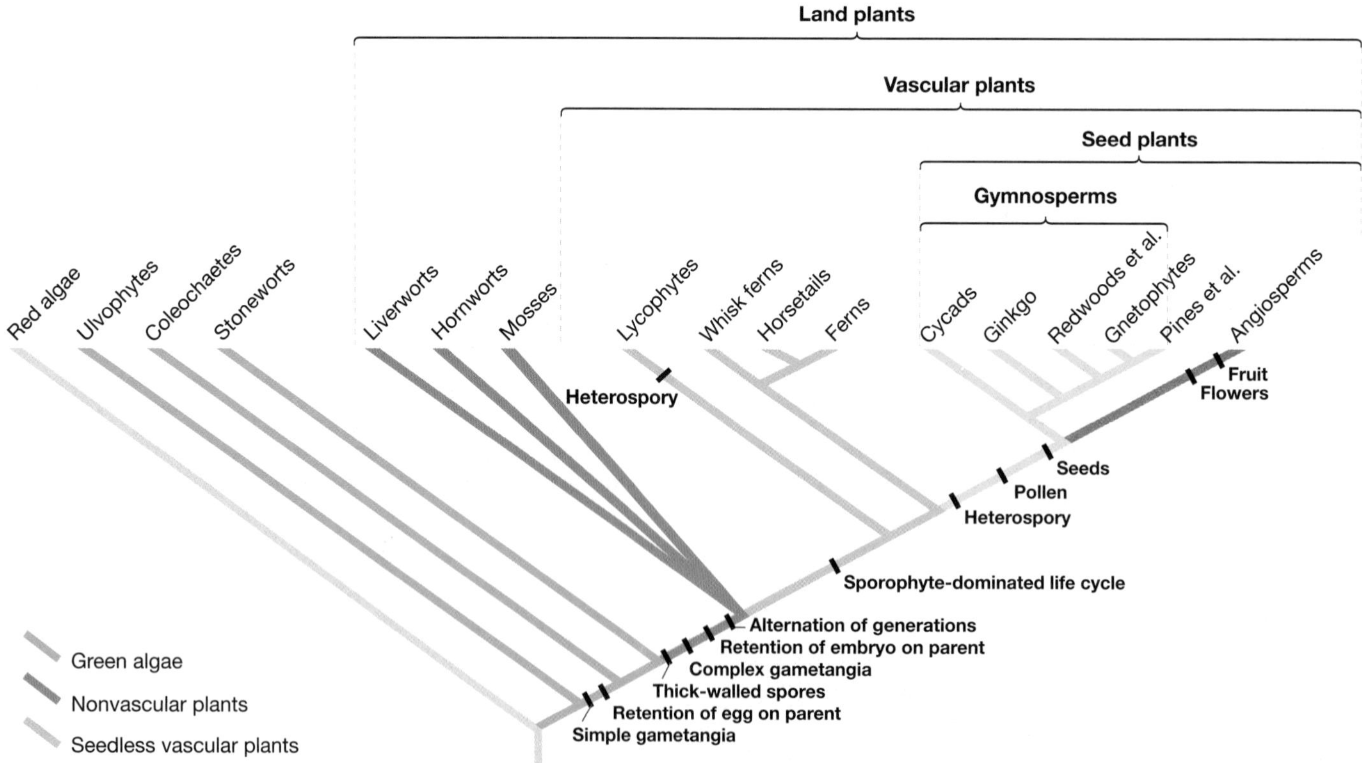

FIGURE 30.26 Evolutionary Innovations Allowed Plants to Reproduce Efficiently on Land.

EXERCISE Conifers, pines, gnetophytes, and angiosperms are the only land plants that do not have flagellated sperm that swim to the egg (at least a short distance). Mark the loss of flagellated sperm on the tree.

FIGURE 30.27 Four Morphological Differences between Monocots and Dicots.

The Angiosperm Radiation

For the past 125 million years, land plant diversification has really been about angiosperms. Angiosperms represent one of the great adaptive radiations in the history of life. As Chapter 27 noted, an **adaptive radiation** occurs when a single lineage produces a large number of descendant species that are adapted to a wide variety of habitats. The diversification of angiosperms is associated with three key adaptations: (1) vessels, (2) flowers, and (3) fruits. In combination, these traits allow angiosperms to transport water, pollen, and seeds efficiently. Based on these observations, it is not surprising that most land plants living today are angiosperms.

On the basis of morphological traits, the 250,000 species of angiosperms identified to date have traditionally been classified into two major groups: the monocotyledons, or **monocots**, and the dicotyledons, or **dicots**. Some familiar monocots include the grasses, orchids, palms, and lilies; familiar dicots include the roses, buttercups, daisies, oaks, and maples. The names of the two groups were inspired by differences in a structure called the cotyledon. A **cotyledon** is the first leaf that is formed in an embryonic plant. As **Figure 30.27** shows, monocots have a single cotyledon (visible inside the seed) while dicots have two cotyledons (visible in newly germinated plants). The figure also highlights other major morphological differences observed in monocots and dicots, concerning the arrangement of vascular tissue and leaf veins and the characteristics of flowers.

It would be misleading, however, to think that all species of flowering plants fall into one of these two groups—either monocots or dicots. Recent work has shown that dicots do not form a natural group consisting of a common ancestor and all of its descendants. To drive this point home, consider the phylogeny illustrated in **Figure 30.28**. These relationships were estimated by comparing the sequences of several genes that are shared by all angiosperms. Notice that species with dicot-like

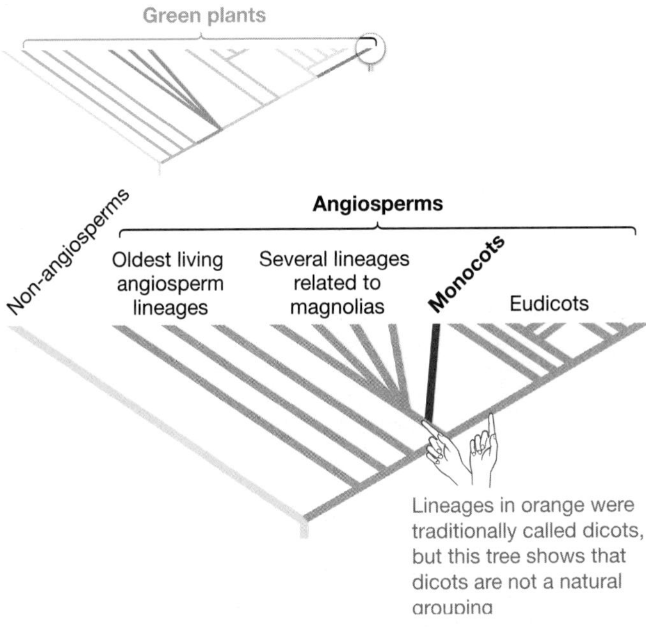

FIGURE 30.28 Monocots Are Monophyletic, but Dicots Are Paraphyletic.

characters are scattered around the angiosperm phylogenetic tree. Based on this analysis, biologists have concluded that although monocots are monophyletic, dicots are not. Instead, they are paraphyletic.

Biologists have adjusted the names assigned to angiosperm lineages to reflect this new knowledge of phylogeny. The most important of these changes was identifying the **eudicots** ("true dicots") as a lineage that includes roses, daisies, and maples. Plant systematists continue to work toward understanding relationships throughout the angiosperm phylogenetic tree, and there will undoubtedly be more name changes as knowledge grows.

Check Your Understanding

If you understand that...

- Land plants were able to make the transition to growing in terrestrial environments, where sunlight and carbon dioxide are abundant, based on a series of evolutionary innovations.
- Adaptations for growing on land included cuticle, stomata, and vascular tissue.
- Adaptations for effective reproduction on land included gametangia, the retention of embryos on the parent, pollen, seeds, flowers, and fruits.

You should be able to...

1) Explain why each of these adaptations was important in survival or reproduction.
2) Map where each of these evolutionary innovations occurred on the phylogenetic tree of green plants.

30.4 Key Lineages of Green Plants

The evolution of cuticle, pores, stomata, and water-conducting tissues allowed green plants to grow on land, where resources for photosynthesis are abundant. Once the green plants were on land, the evolution of gametangia, retained embryos, pollen, seeds, and flowers enabled them to reproduce efficiently even in very dry environments. The adaptations reviewed in Section 30.3 allowed the land plants to make the most important water-to-land transition in the history of life.

To explore green plant diversity in more detail, let's take a closer look at some major groups of green algae. Then we'll discuss the major phyla of land plants in the context of their broad morphological groupings: nonvascular plants, seedless vascular plants, and seed plants.

Green Algae

The **green algae** are a paraphyletic group that totals about 7000 species. Their bright green chloroplasts are similar to those found in land plants, with a double membrane and chlorophylls *a* and *b* but relatively few accessory pigments. And like land plants, green algae synthesize starch in the chloroplast as a storage product of photosynthesis and have a cell wall composed primarily of cellulose. Green algae are important primary producers in nearshore ocean environments and in all types of freshwater habitats. They are also found in several types of more exotic environments, including snowfields at high elevations, pack ice, and ice floes. These habitats are often splashed with bright colors due to large concentrations of unicellular green algae (**Figure 30.29a**). Although these cells live at near-freezing temperatures, they make all their own food via photosynthesis.

In addition, green algae live in close association with an array of other organisms:

- **Lichens** are stable associations between green algae and fungi or between cyanobacteria and fungi, and are often found in terrestrial environments that lack soil, such as tree bark or bare rock (**Figure 30.29b**). The algae or cyanobacteria are

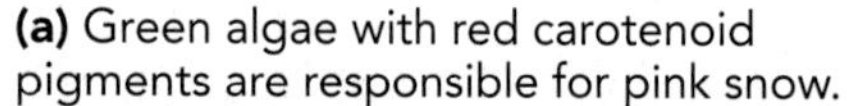

(a) Green algae with red carotenoid pigments are responsible for pink snow.

(b) Most lichens are an association between fungi and green algae.

(c) Many unicellular protists harbor green algae.

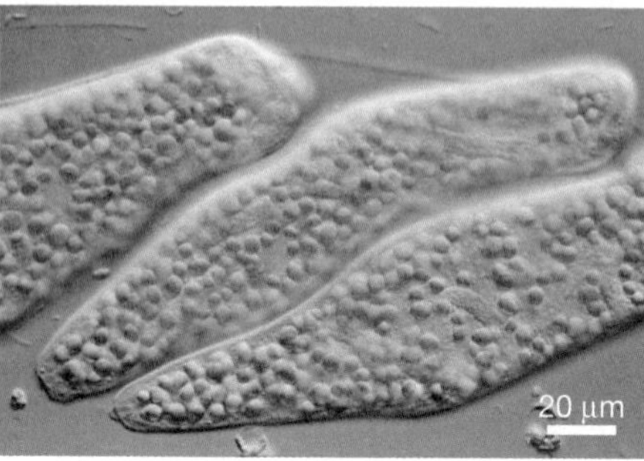

FIGURE 30.29 Some Green Algae Live in Unusual Environments.

protected from drying by the fungus; in return they provide sugars produced by photosynthesis. Of the 17,000 species of lichens described to date, about 85 percent involve green algae. The green algae that are involved are unicellular or grow in long filaments. Lichens are explored in more detail in Chapter 31.

- Unicellular green algae are common endosymbionts in planktonic protists that live in lakes and ponds (**Figure 30.29c**). The association is considered mutually beneficial: The algae supply the protists with food; the protists provide protection to the algae.

Green algae are a large and fascinating group of organisms. Let's take a closer look at just three of the many lineages (**Figure 30.30**).

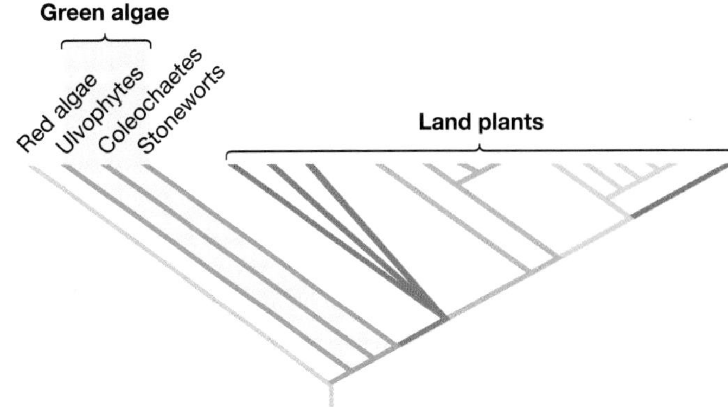

FIGURE 30.30 Green Algae Are Paraphyletic.

Green Algae > Coleochaetophyceae (Coleochaetes)

There are 19 species in this group. Most coleochaetes are barely visible to the unaided eye and grow as flat sheets of cells (**Figure 30.31**). They are considered multicellular because they have specialized photosynthetic and reproductive cells and because they contain **plasmodesmata**—structures introduced in Chapter 8 that connect adjacent cells.

The coleochaetes are strictly freshwater algae. They grow attached to aquatic plants such as water lilies and cattails or over submerged rocks in lakes and ponds. When they grow near beaches, they are often exposed to air when water levels drop in late summer.

Reproduction Asexual reproduction is common in coleochaetes and involves production of flagellated spores. During sexual reproduction, eggs are retained on the parent and are nourished after fertilization with the aid of transfer cells—a situation very similar to that observed in land plants. In some species certain individuals are male and produce only sperm, while other individuals are female and produce only eggs.

Life cycle Alternation of generations does not occur. Multicellular individuals are haploid; the only diploid stage in the life cycle is the zygote.

Human and ecological impacts Because they are very closely related to land plants, coleochaetes are studied intensively by researchers interested in how land plants made the water-to-land transition.

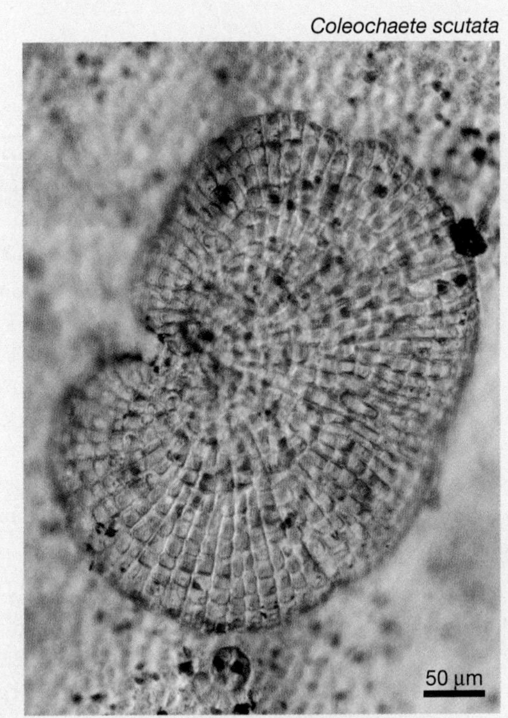

FIGURE 30.31 Coleochaetes Are Thin Sheets of Cells.

Green Algae > Ulvophyceae (Ulvophytes)

The Ulvophyceae are a monophyletic group composed of several diverse and important subgroups, with a total of about 4000 species. Members of this lineage range from unicellular to multicellular.

Many of the large green algae in habitats along ocean coastlines are members of the Ulvophyceae. *Ulva*, the sea lettuce (**Figure 30.32**), is a representative marine species. But there are also large numbers of unicellular or small multicellular species that inhabit the plankton of freshwater lakes and streams.

Reproduction Most ulvophytes reproduce both asexually and sexually. Asexual reproduction often involves production of spores that swim with the aid of flagella. Sexual reproduction usually results in production of a resting stage—a cell that is dormant in winter. In many species the gametes are not called eggs and sperm, because they are the same size and shape. In most species gametes are shed into the water, so fertilization takes place away from the parent plants.

Life cycle Many unicellular forms are diploid only as zygotes. Alternation of generations occurs in multicellular species. When alternation of generation occurs, gametophytes and sporophytes may look identical or different.

Human and ecological impacts Ulvophyceae are important primary producers in freshwater environments and in coastal areas of the oceans.

FIGURE 30.32 Green Algae Are Important Primary Producers in Aquatic Environments.

Green Algae > Charaphyceae (Stoneworts)

There are several hundred species in this group. They are collectively known as stoneworts, because they commonly accumulate crusts of calcium carbonate ($CaCO_3$) over their surfaces. Like the coleochaetes, they have plasmodesmata and are multicellular. ● You should be able to mark the origin of plasmodesmata on Figure 30.30. (They do not occur in ulvophytes.) Some species of stonewort can be a meter or more in length.

The stoneworts are freshwater algae. Certain species are specialized for growing in relatively deep waters, though most live in shallow water near lake beaches or pond edges.

Reproduction Sexual reproduction is common and involves production of prominent, multicellular gametangia similar to those observed in early land plants. As in coleochaetes, in stoneworts the eggs are retained on the parent plant, which supplies eggs with nutrients prior to fertilization. ● You should be able to mark the origin of egg retention on Figure 30.30. (It does not occur in ulvophytes.)

Life cycle Alternation of generations does not occur. Multicellular individuals are haploid; the only diploid stage in the life cycle is the zygote.

Human and ecological impacts Some species form extensive beds in lake bottoms or ponds and provide food for ducks and geese as well as food and shelter for fish (**Figure 30.33**). They are a good indicator that water is not polluted.

FIGURE 30.33 Stoneworts Can Form Beds on Lake Bottoms.

Nonvascular Plants ("Bryophytes")

The most basal lineages of land plants are collectively known as the **bryophytes**. The evolutionary relationships among the three lineages with living species—liverworts, hornworts, and mosses—are still unclear (**Figure 30.34**).

All of the nonvascular plant species present today have a low, sprawling growth habit. In fact, it is unusual to find bryophytes more than 5 to 10 centimeters tall. Individuals are anchored to soil, rocks, or tree bark by structures called **rhizoids**. No bryophytes have vascular tissue with lignin-reinforced cell walls. In the lineages present today, simple water-conducting cells and tissues are found only in some mosses. All bryophytes have flagellated sperm that swim to eggs through raindrops or small puddles on the plant surface. Spores are dispersed by wind.

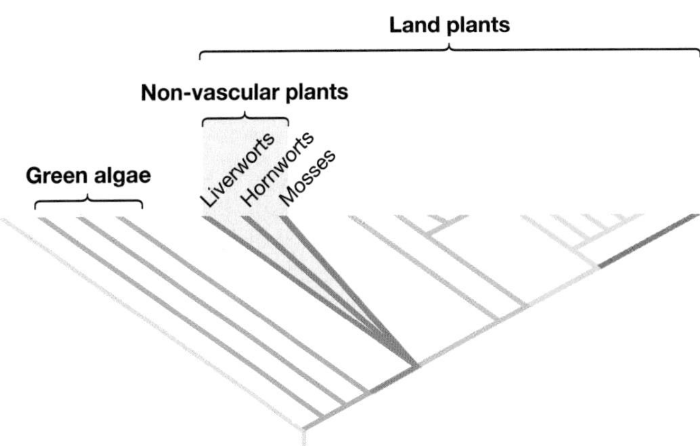

FIGURE 30.34 It Is Unclear Whether Bryophytes Are Monophyletic or Paraphyletic.

Nonvascular Plants > Bryophyta (Mosses)

Over 12,000 species of mosses have been named and described to date, and more are being discovered every year—particularly in the tropics. Mosses are informally grouped with other "bryophytes" (liverworts and hornworts) but are formally classified in their own monophyletic group: the phylum Bryophyta.

Although mosses are common in moist forests, they can also be abundant in more extreme environments, such as deserts and windy, treeless habitats in the Arctic, Antarctic, or mountaintops. In these severe conditions, mosses are able to thrive because their bodies can become extremely dry without dying. When the weather makes photosynthesis difficult, individuals dry out and become dormant or inactive (**Figure 30.35a**). Then when rains arrive or temperatures warm, the plants rehydrate and begin photosynthesis and reproduction (**Figure 30.35b**).

Adaptations to land One subgroup of mosses contains simple conducting tissues consisting of cells that are specialized for the transport of water or food. But because these cells do not have walls that are reinforced by lignin, they are not considered true vascular tissue. ● You should be able to mark the origin of the simple water-conducting cells and tissues in this moss sub-group on Figure 30.34. Because they lack true vascular tissue, most mosses are not able to grow much taller than a few centimeters.

Reproduction Asexual reproduction often occurs by fragmentation, meaning that pieces of gametophytes that are broken off by wind or a passing animal can begin growing independently. In many species, sexual reproduction cannot involve self-fertilization because the sexes are separate—meaning that an individual plant produces only eggs in archegonia or only sperm in antheridia. A typical sporophyte produces up to 50 million tiny spores. Spores are usually distributed by wind.

Life cycle The moss life cycle is similar to that of liverworts and hornworts: The sporophyte is retained on the much larger and longer-lived gametophyte and gets most of its nutrition from the gametophyte.

Human and ecological impacts Species in the genus *Sphagnum* are often the most abundant plant in wet habitats of northern environments. Because *Sphagnum*-rich environments account for 1 percent of Earth's total land area, equivalent to half the area of the United States, *Sphagnum* species are among the most abundant plants in the world. *Sphagnum*-rich habitats have an exceptionally short growing season, however, so the decomposition of dead mosses and other plants is slow. As a result, large deposits of semi-decayed organic matter, known as **peat**, accumulate. Researchers estimate that the world's peatlands store about 400 billion metric tons of carbon. If peatlands begin to burn or decay rapidly due to global warming, the CO_2 released will exacerbate the warming trend (see Chapter 54).

Peat is harvested as a traditional heating and cooking fuel in some countries. It is also widely used as a soil additive in gardening, because *Sphagnum* can absorb up to 20 times its dry weight in water. This high water-holding capacity is due to the presence of large numbers of dead cells in the leaves of these mosses, which readily fill with water via pores in their walls.

(a) Moss in dry weather

(b) Moss in wet weather

FIGURE 30.35 Many Moss Species Can Become Dormant When Conditions Are Dry.

Nonvascular Plants > Hepaticophyta (Liverworts)

Liverworts got their name because some species native to Europe have liver-shaped leaves. According to the medieval *Doctrine of Signatures*, God indicated how certain plants should be used by giving them a distinctive appearance. Thus, liverwort teas were hypothesized to be beneficial for liver ailments. (They are not.) About 6500 species are known. They are commonly found growing on damp forest floors or riverbanks, often in dense mats (**Figure 30.36**), or on the trunks or branches of tropical trees.

Adaptations to Land Liverworts are covered with cuticle. Some species have pores that allow gas exchange; in species that lack pores, the cuticle is very thin.

Reproduction Asexual reproduction occurs when fragments of a plant are broken off and begin growing independently. Some species also produce small structures called **gemmae** asexually, during the gametophyte phase. Mature gemmae are knocked off the parent plant by rain and grow into independent gametophytes. During sexual reproduction, sperm and eggs are produced in gametangia.

Life cycle The gametophyte is the largest and longest-lived phase in the life cycle. Sporophytes are small, grow directly from the gametophyte, and depend on the gametophyte for nutrition. Spores are shed from the sporophyte and are carried away by wind or rain.

Human and ecological impacts When liverworts grow on bare rock or tree bark, their dead and decaying body parts contribute to the initial stages of soil formation.

Marchantia bryophyta

FIGURE 30.36 Liverworts Thrive in Moist Habitats.

Nonvascular Plants > Anthocerophyta (Hornworts)

Hornworts got their name because their sporophytes have a horn-like appearance (**Figure 30.37**) and because *wort* is the Anglo-Saxon word for *plant*. About 100 species have been described to date.

Adaptations to land Hornwort sporophytes have stomata. Research is under way to determine if they can open their pores to allow gas exchange or close their pores to avoid water loss during dry intervals.

Reproduction Depending on the species, gametophytes may contain only egg-producing archegonia or only sperm-producing antheridia, or both. Stated another way, individuals of some species are either female or male, while in other species each individual has both types of reproductive organs.

Life cycle The gametophyte is the longest-lived phase in the life cycle. Although sporophytes grow directly from the gametophyte, they are green because their cells contain chloroplasts. Sporophytes manufacture some of their own food but also get nutrition from the gametophyte. Spores disperse from the parent plant via wind or rain.

Human and ecological impacts Some species harbor symbiotic cyanobacteria that fix nitrogen.

Phaeocerus leavis

FIGURE 30.37 Hornworts Have Horn-Shaped Sporophytes.

Seedless Vascular Plants

The seedless vascular plants are a paraphyletic group that forms a grade between the nonvascular plants and the seed plants (**Figure 30.38**). All species of seedless vascular plants have conducting tissues with cells that are reinforced with lignin, forming vascular tissue. Tree-sized lycophytes and horsetails are abundant in the fossil record, and tree ferns are still common inhabitants of certain habitats, such as mountain slopes in the tropics.

The sporophyte is the larger and longer-lived phase of the life cycle in all of the seedless vascular plants. The gametophyte is physically independent of the sporophyte, however. Eggs are retained on the gametophyte, and sperm swim to the egg with the aid of flagella. Thus, seedless vascular plants depend on the presence of water for reproduction—they need enough water to form a continuous layer that "connects" gametophytes and allows sperm to swim to eggs. Sporophytes develop on the gametophyte and are nourished by the gametophyte when they are small.

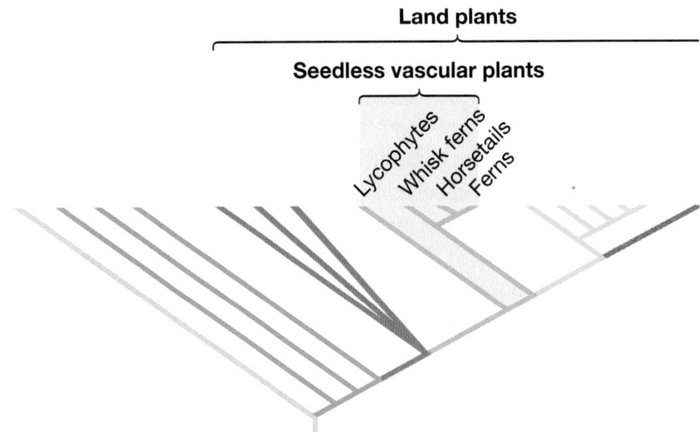

FIGURE 30.38 The Seedless Vascular Plants Are Paraphyletic.

Vascular Plants > Lycophyta (Lycophytes, or Club Mosses)

Although the fossil record documents lycophytes that were 2 m wide and 40 m tall, the 1000 species of lycophytes living today are all small in stature (**Figure 30.39**). Most live on the forest floor or on the branches or trunks of tropical trees. Because of their appearance, they are often called ground pines or **club mosses**—even though they are neither pines nor mosses.

Adaptations to land Lycophytes are the most ancient land plant lineage with **roots**—a belowground system of tissues and organs that anchors the plant and is responsible for absorbing water and mineral nutrients. Roots differ from the rhizoids observed in bryophytes, because roots contain vascular tissue and thus are capable of conducting water and nutrients from belowground to the upper reaches of the plant. Lycophytes are distinguished by small leaflike structures called microphylls that extend from the stems. ● You should be able to mark the origin of microphylls on Figure 30.38.

Reproduction Asexual reproduction can occur by fragmentation or gemmae. During sexual reproduction, spores of some species give rise to bisexual gametophytes—meaning that each gametophyte produces both eggs and sperm. In the genera called *Selaginella* and *Isoetes*, however, heterospory occurs and gametophytes are male or female. Self-fertilization is extremely rare in most of these species, however. Some club mosses have separate male and female gametophytes.

FIGURE 30.39 Lycophytes Living Today Are Small in Stature.

Life cycle The gametophytes of some species live entirely underground and get their nutrition from symbiotic fungi. In certain species, gametophytes live 6 to 15 years and give rise to a large number of sporophytes over time.

Human and ecological impacts Tree-sized lycophytes were abundant in the coal-forming forests of the Carboniferous period. In coal-fired power plants today, electricity is being generated by burning fossilized lycophyte trunks and leaves.

Vascular Plants > Psilotophyta (Whisk Ferns)

Only two genera of whisk ferns are living today, and there are perhaps six distinct species. Whisk ferns are restricted to tropical regions and have no fossil record. They are extremely simple morphologically, with aboveground parts consisting of branching stems that have tiny, scale-like outgrowths instead of leaves (**Figure 30.40**).

Adaptations to land Whisk ferns lack roots. Some species gain most of their nutrition from fungi that grow in association with the whisk ferns' extensive underground stems called **rhizomes**. Other species grow in rock crevices or are **epiphytes** ("upon-plants"), meaning that they grow on the trunks or branches of other plants—in this case, in the branches of tree ferns. ● After reviewing where leaves and roots originated during land plant evolution (see Figure 30.12), you should be able to mark the loss of leaves and roots in whisk ferns on Figure 30.38.

FIGURE 30.40 Psilotophytes Are Extremely Simple Morphologically.

Reproduction Asexual reproduction is common in sporophytes via the extension of rhizomes and the production of new aboveground stems. When spores mature, they are dispersed by wind and germinate into gametophytes that contain both archegonia and antheridia.

Life cycle Sporophytes may be up to 30 cm tall, but gametophytes are less than 2 mm long and live under the soil surface. Gametophytes absorb nutrients directly from the surrounding soil and from symbiotic fungi. Fertilization takes place inside the archegonium, and the sporophyte develops directly on the gametophyte.

Human and ecological impacts Some whisk fern species are popular landscaping plants, particularly in Japan. The same species can be a serious pest in greenhouses.

Vascular Plants > Sphenophyta (or Equisetophyta) (Horsetails)

Although horsetails are prominent in the fossil record of land plants, just 15 species are known today. All 15 are in the genus *Equisetum*; the phylum as a whole is sometimes called Equisetophyta. Translated literally, *Equisetum* means "horse-bristle." Both the scientific name and the common name, horsetail, come from the brushy appearance of the stems and branches in some species (**Figure 30.41**). Horsetails may be locally abundant in wet habitats such as stream banks or marsh edges.

Adaptations to land Horsetails have an interesting adaptation that allows them to flourish in waterlogged, oxygen-poor soils. Horsetail stems are hollow, so oxygen readily diffuses down the stem to reach roots that cannot obtain oxygen from the surrounding soil. Horsetails are also distinguished by having whorled leaves and branches. ● You should be able to mark the origin of hollow stems, whorled branches, and whorled leaves on Figure 30.38.

Reproduction Asexual reproduction is common in sporophytes and occurs via fragmentation or the extension of rhizomes. From these rhizomes, two types of erect, specialized stems may grow—stems that contain tiny leaves and chloroplast-rich branches and that are specialized for photosynthesis, or stems that bear clusters of sporangia and produce huge numbers of spores by meiosis (Figure 30.41).

Life cycle Gametophytes perform photosynthesis but are small and short lived. They normally produce both antheridia and archegonia, but in most cases the sperm-producing structure matures first. This pattern is thought to be an adaptation that minimizes self-fertilization and maximizes cross-fertilization.

Human and ecological impacts Horsetail stems are rich in silica granules. The glass-like deposits not only strengthen the stem but also make these plants useful for scouring pots and pans—hence these plants are often called "scouring rushes."

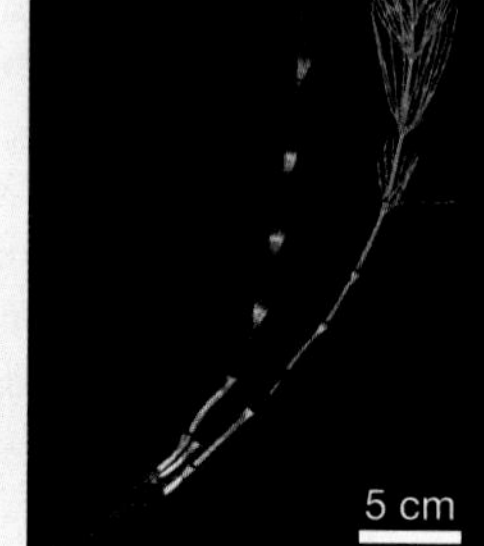

FIGURE 30.41 Horsetails Have Separate Reproductive and Vegetative Stalks.

Vascular Plants > Pteridophyta (Ferns)

With 12,000 species, ferns are by far the most species-rich group of seedless vascular plants. They are particularly abundant in the tropics. About a third of the tropical species are epiphytes, usually growing on the trunks or branches of trees. Species that can grow epiphytically live high above the forest floor, where competition for light is reduced, without making wood and growing tall themselves. The growth habits of ferns are highly variable among species, however, and ferns range in size from rosettes the size of your smallest fingernail to 20-meter-tall trees (**Figure 30.42a**).

Adaptations to land Ferns are the only seedless vascular plants that have large, well-developed leaves—commonly called **fronds**. Leaves give the plant a large surface area, allowing it to capture sunlight for photosynthesis efficiently.

Reproduction In a few species, gametophytes reproduce asexually via production of gemmae. Typically, species that can reproduce via gemmae never produce gametes or sporophytes. In most species, however, sexual reproduction is the norm.

Life cycle Although fern gametophytes contain chloroplasts and are photosynthetic, the sporophyte is typically the larger and longer-lived phase of the life cycle. In mature sporophytes, sporangia are usually found in clusters called **sori** on the undersides of leaves (**Figure 30.42b**). The structure of the sporangia is a distinctive feature of ferns: It arises from a single cell and has a wall composed of a single cell layer. ● You should be able to mark the evolution of the distinctive fern sporangium on Figure 30.38.

Human and ecological impacts In many parts of the world, people gather the young, unfolding fronds, or "fiddleheads," of ferns in spring as food. Ferns are also widely used as ornamental plants in landscaping.

(a) Ferns range from small to tree sized.

Gonocormus minutus

Dicksonia antarctica

(b) Fern sporangia are often located on the underside of fronds.

Polypodium vulgare

FIGURE 30.42 Ferns Are the Most Species-Rich Group of Seedless Vascular Plants.

Seed Plants

The seed plants are a monophyletic group consisting of the gymnosperms—cycads, ginkgo, conifers, gnetophytes, and pines—and the angiosperms (**Figure 30.43**). Seed plants are defined by the production of seeds and the production of pollen grains. Recall from Section 30.3 that seeds are a specialized structure for dispersing embryonic sporophytes to new locations. Seeds are the mature form of a fertilized ovule, the female reproductive structure that encloses the female gametophyte and egg cell. Also recall from Section 30.3 that pollen grains are tiny, sperm-producing gametophytes that are easily dispersed through air as opposed to water. The structure and function of seed plants is the focus of Chapters 36 through 40.

Seed plants are found in virtually every type of habitat, and they adopt every growth habit known in land plants. Their forms range from mosslike mats to shrubs and vines to 100-meter-tall trees. Seed plants can be annual or perennial, with life spans ranging from a few weeks to almost five thousand years.

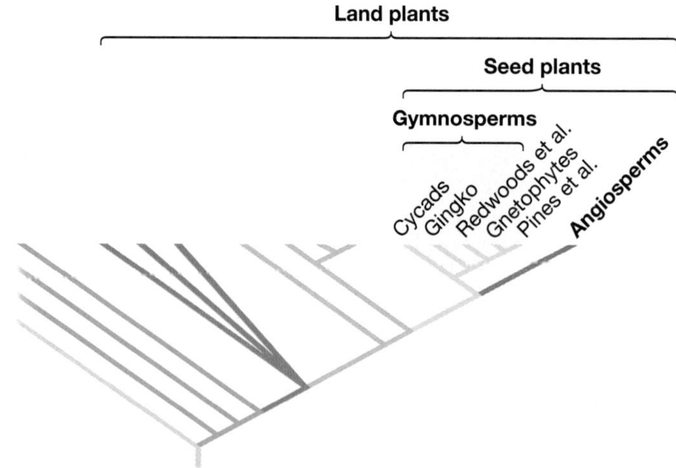

FIGURE 30.43 The Seed Plants Are a Monophyletic Group.

Seed Plants > Gymnosperms > Cycadophyta (Cycads)

The cycads are so similar in overall appearance to palm trees, which are angiosperms, that cycads are sometimes called "sago palms." Although cycads were extremely abundant when dinosaurs were present on Earth 150–65 million years ago, only about 140 species are living today. Most are found in the tropics (**Figure 30.44**).

Cycas revoluta

FIGURE 30.44 Cycads Resemble Palms but Are Not Closely Related to Them.

Adaptations to land Cycads have woody stems; many are tree sized. They are unique among gymnosperms in having compound leaves—meaning that each leaf is divided into many smaller leaflets. ● You should be able to mark the origin of the distinctive cycad leaf on Figure 30.43.

Reproduction and life cycle Like other seed plants, cycads are heterosporous. Each sporophyte individual bears either microsporangia or megasporangia, but not both. Pollen is carried by insects (usually beetles or weevils) or, in some species, wind. Cycad seeds are large and often brightly colored. The colors attract birds and mammals, both of which eat and disperse the seeds.

Human and ecological impacts Cycads harbor large numbers of symbiotic cyanobacteria in specialized, aboveground root structures. The cyanobacteria are photosynthetic and fix nitrogen. The nitrogen acts as an important nutrient for nearby plants as well as the cycads themselves. Cycads are popular landscaping plants in some parts of the world.

Seed Plants > Gymnosperms > Ginkgophyta (Ginkgos)

Although ginkgos have an extensive fossil record, just one species is alive today. Leaves from the ginkgo, or maidenhair, tree are virtually identical in size and shape to those observed in fossil ginkgos that are 150 million years old (**Figure 30.45**).

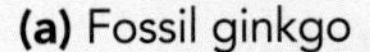

FIGURE 30.45 The Ginkgo Tree Is a "Living Fossil."

Adaptations to land Unlike most gymnosperms, the ginkgo is **deciduous**—meaning that it loses its leaves each autumn. This adaptation allows plants to be dormant during the winter, when photosynthesis and growth are difficult.

Reproduction and life cycle Sexes are separate—individuals are either male or female. Pollen is transported by wind. Sperm have flagella, however. Once pollen grains land near the female gametophyte and mature, the sperm cells leave the pollen grain and swim to the egg cells.

Human and ecological impacts Although today's ginkgo trees are native to southeast China, they are planted widely as an ornamental all over the world. They are especially popular in urban areas, as they are tolerant of air pollution. In some countries, the inside of the seed is eaten as a delicacy.

Seed Plants > Gymnosperms > Gnetophyta (Gnetophytes)

The gnetophytes comprise about 70 species in three genera. One genus consists of vines and trees from the tropics. A second is made up of desert-dwelling shrubs, including what may be the most familiar gnetophyte—the shrub called Mormon tea, which is common in the deserts of southwestern North America (see Figure 30.7c). The third genus contains a single species that probably qualifies as the world's most bizarre plant (**Figure 30.46**)—*Welwitschia mirabilis*, which is native to the deserts of southwest Africa. Although it has large belowground structures, the aboveground part consists of just two strap-like leaves, which grow continuously from the base and die at the tips. The leaves also split lengthwise as they grow and age.

FIGURE 30.46 *Welwitschia* Is an Unusual Plant.

Adaptations to land Gnetophytes have vessel elements in addition to tracheids. All of the living species make wood as a support structure.

Reproduction and life cycle The microsporangia and megasporangia are arranged in clusters at the end of stalks, similar to the way flowers are clustered in some angiosperms. Pollen is transferred by the wind or by insects. Double fertilization occurs in two of the three genera. As in other gymnosperms, seeds do not form inside an encapsulated structure.

Human and ecological impacts The drug ephedrine was originally isolated from species in the gnetophyte genus *Ephredra*, which is native to northern China and Mongolia. Ephedrine is used in the treatment of hay fever, colds, and asthma.

Seed Plants > Gymnosperms > Pinophyta (Pines, Spruces, Firs)

The gymnosperms include two major lineages of cone-bearing species: the pines and allies discussed here, and a group that includes redwoods, junipers, yews, and cypresses, featured below. Species in both lineages have a reproductive structure called the cone, in which microsporangia and megasporangia are produced (**Figure 30.47**).

The Pinophyta includes the familiar pines, spruces, firs, Douglas fir, tamaracks, and true cedars. These are among the largest and most abundant trees on the planet, as well as some of the most long-lived. One of the bristlecone pines native to southwestern North America is at least 4900 years old.

Adaptations to land Pinophyta have needle-like leaves, with a small surface area that allows them to thrive in habitats where water is scarce. Pines are common on sandy soils that have poor water-holding capacity, and spruces and firs are common in cold environments where water is often frozen. All of the living species make wood as a support structure.

Reproduction and life cycle Pollen is transferred by the wind.

Human and ecological impacts Pines, spruces, firs, and other species in this group dominate forests that grow at high latitudes and high elevations. Their seeds are key food sources for a variety of birds, squirrels, and mice, and their wood is the basis of the building products and paper industries in many parts of the world. The paper in this book was made from species in this group.

(a) Cones that produce microsporangia and pollen
Picea abies
Pollen
1 cm

(b) Cones that produce macrosporangia and eggs
Picea abies

FIGURE 30.47 Pollen-Bearing Cones Produce Microsporangla; Ovulate Cones Produce Megasporangla.

Seed Plants > Gymnosperms > Other Conifers (Redwoods, Junipers, Yews)

The species in this lineage vary in growth form from sprawling juniper shrubs to the world's largest plants. Redwood trees growing along the Pacific Coast of North America can reach heights of up to 115 m (375 ft) and trunk diameters of over 11 m (36 ft).

Adaptations to land All of the species in this lineage are trees or large shrubs. Most have narrow leaves, which in many cases are arranged in overlapping scales (**Figure 30.48**). Narrow leaves have a small amount of surface area, which is not optimal for capturing sunlight and performing photosynthesis. But because the small surface area reduces water loss from leaves, many species in this lineage thrive in dry habitats or in cold environments where water is often frozen.

Reproduction and life cycle The species in this group are wind pollinated. As in all seed plants, the female gametophyte is retained on the parent. Thus, fertilization and seed development take place in the female cone. Like other gymnosperms, seeds do not form inside an encapsulated structure. Depending on the species, the seeds are dispersed by wind or by seed-eating birds or mammals.

Human and ecological impacts Redwoods, redcedar, whitecedar, and yellowcedar have wood that is highly rot-resistant and thus prized for making furniture, decks, house siding, or other applications where wood is exposed to the weather. Yew wood is often preferred for making traditional archery bows, and juniper berries are used to flavor gin.

Thuja plicata

FIGURE 30.48 Some Species in This Group Have Scale-like Leaves.

Seed Plants > Anthophyta (Angiosperms)

The flowering plants, or angiosperms, are far and away the most species-rich lineage of land plants. Over 250,000 species have already been described. They range in size from *Lemna gibba*—a floating, aquatic species that is less than half a millimeter wide—to massive oak trees. Angiosperms thrive in desert to freshwater to rain forest environments and are found in virtually every habitat except the deep oceans. They are the most common and abundant plants in terrestrial environments other than northern and high-elevation coniferous forests.

The defining adaptation of angiosperms is the flower. Flowers are reproductive structures that hold either pollen-producing microsporangia or the megasporangia that produce megaspores and eggs, or both. Nectar-producing cells are often present at the base of the flower, and the color of petals helps to attract insects, birds, or bats that carry pollen from one flower to another (**Figure 30.49a**). Some angiosperms are pollinated by wind, however. Wind-pollinated flowers lack both colorful petals and nectar-producing cells (**Figure 30.49b**).

Adaptations to land In addition to flowers, angiosperms evolved vessels, the conducting cells that make water transport particularly efficient. Most angiosperms contain both tracheids and vessels.

Reproduction and life cycle Unlike gymnosperms, angiosperms have a carpel, a structure within the flower that contains an ovary. The ovary encloses the ovule, which in turn encloses the female gametophyte. In most cases, male gametophytes are carried to female gametophytes by pollinators that are inadvertently dusted with pollen as they visit flowers to find food. Depending on the angiosperm species, self-fertilization may be common or absent. When the egg produced by the female gametophyte is fertilized, the ovule develops into a seed. When the ovary matures it forms a fruit, which contains the seed or seeds.

Human and ecological impacts It is almost impossible to overstate the importance of angiosperms to humans and other organisms. In most terrestrial habitats today, angiosperms supply the food that supports virtually every other species. For example, many insects eat flowering plants. Historically, the diversification of angiosperms correlated closely with the diversification of insects, which are by far the most species-rich lineage on the tree of life. It is not unusual for a single tropical tree to support dozens or even hundreds of insect species. Angiosperm seeds and fruits have also supplied the staple foods of virtually every human culture that has ever existed.

(a) Animal-pollinated flower (this species produces both pollen and eggs in the same flower)

Ornithogalum dubium

(b) Wind-pollinated flower (this species has separate male and female flowers)

Acer negundo

Acer negundo

FIGURE 30.49 Wind-Pollinated Flowers Lack Colorful Petals and Nectar.

Chapter Review

SUMMARY OF KEY CONCEPTS

The green plants include both the green algae and the land plants. Green algae are an important source of oxygen and provide food for aquatic organisms; land plants hold soil and water in place, build soil, moderate extreme temperatures and winds, and provide food for other organisms.

Plants improve the quality of the environment, and humans depend on them for food, fiber, and fuel. This dependence became extreme after the domestication of plant species beginning about 10,000 years ago. Artificial selection techniques have produced huge increases in the yields and dramatic changes in the characteristics of domesticated plants.

You should be able to predict how the current and massive loss of plant species and plant communities will affect the environment.

Land plants were the first multicellular organisms that could live with most of their tissues exposed to the air. A series of key adaptations allowed them to survive on land. In terms of total mass, plants dominate today's terrestrial environments.

The land plants evolved from green algae and colonized terrestrial environments in conjunction with fungi. The evolution of cuticle allowed plant tissues to be exposed to air without dying. The evolution of pores provided breaks in the cuticle and facilitated gas exchange, with CO_2 diffusing into leaves and O_2 diffusing out. Later, the evolution of guard cells allowed plants to control the opening and closing of pores in a way that maximizes gas exchange and minimizes water loss.

Vascular tissue evolved in a series of steps, beginning with simple water-conducting cells and tissues such as those observed in today's mosses. True vascular tissue has cells that are dead at maturity and that have secondary cell walls reinforced with lignin. As a result, vascular tissue conducts water and provides structural support that makes erect growth possible. Erect growth is important because it reduces competition for light. Tracheids are water-conducting cells found in all vascular plants; in addition, angiosperms and gnetophytes have water-conducting cells called vessels.

You should be able to describe plant adaptations that solved the following problems posed by terrestrial life: keeping tissues moist, holding the body erect, transporting water, and exchanging gases. Based on your answer, explain why it is logical that algae made the transition to terrestrial life just once.

Web Animation at www.masteringbio.com

Plant Evolution and the Phylogenetic Tree

Once plants were able to grow on land, a sequence of important evolutionary changes made it possible for them to reproduce efficiently—even in extremely dry environments.

All land plants are embryophytes, meaning that eggs and embryos are retained on the parent plant. Consequently, the developing embryo can be nourished by its mother. Seed plant embryos are then dispersed from the parent plant to a new location, encased in a protective housing, and supplied with a store of nutrients.

All land plants have alternation of generations. Over the course of land plant evolution, the gametophyte phase became reduced in size and life span and the sporophyte phase became more prominent. In seed plants, male gametophytes are reduced to pollen grains and female gametophytes are reduced to tiny structures that produce an egg. The evolution of pollen was an important breakthrough in the history of life, because sperm no longer needed to swim to the egg—tiny male gametophytes could be transported through the air via wind or insects.

You should be able to describe plant adaptations that solved the following problems posed by terrestrial life: transporting sperm, nourishing embryos, and dispersing embryos. Explain the disadvantage of having spores serve as the dispersal stage in a plant life cycle.

QUESTIONS

Test Your Knowledge

1. Which of the following groups is definitely monophyletic?
   a. nonvascular plants
   b. green algae
   c. green plants
   d. seedless vascular plants

2. What is a difference between tracheids and vessels?
   a. Tracheids are dead at maturity; vessels are alive and are filled with cytoplasm.
   b. Vessels have gaps in the primary and secondary cell wall; tracheids have gaps only in the secondary cell wall.
   c. Only tracheids have a thick secondary cell wall containing lignin.
   d. Only vessels have a thick secondary cell wall containing lignin.

3. Which of the following statements is *not* true?
   a. Green algae in the lineage called Charales are the closest living relatives of land plants.
   b. "Bryophytes" is a name given to the land plant lineages that do not have vascular tissue.
   c. The horsetails and the ferns form a distinct clade, or lineage. They have vascular tissue but reproduce via spores, not seeds.
   d. According to the fossil record and phylogenetic analyses, angiosperms evolved before the gymnosperms. Angiosperms are the only land plants with vessels.

4. The appearance of cuticle and stomata correlated with what event in the evolution of land plants?
   a. the first erect growth forms
   b. the first woody tissues
   c. growth on land
   d. the evolution of the first water-conducting tissues
5. What do seeds contain?
   a. male gametophyte and nutritive tissue
   b. female gametophyte and nutritive tissue
   c. embryo and nutritive tissue
   d. mature sporophyte and nutritive tissue
6. What is a pollen grain?
   a. male gametophyte
   b. female gametophyte
   c. male sporophyte
   d. sperm

Test Your Knowledge answers: 1. c; 2. b; 3. d; 4. c; 5. c; 6. a

Test Your Understanding

Answers are available at www.masteringbio.com

1. Soils, water, and the atmosphere are major components of the abiotic (nonliving) environment. Describe how green plants affect the abiotic environment in ways that are advantageous to humans.
2. The evolution of cuticle presented land plants with a challenge that threatened their ability to live on land. Describe this challenge and explain why stomata represent a solution. Compare and contrast stomata with the pores found in liverworts. Explain why it is logical to observe that liverworts that lack pores have extremely thin cuticle.
3. Diagram four steps in the evolution of vascular tissue. Why was the evolution of lignin-reinforced cell walls significant? In dry habitats, why are vascular plants more common than nonvascular plants, and why are most of them taller?
4. Land plants may have reproductive structures that (1) protect gametes as they develop; (2) nourish developing embryos, (3) allow sperm to be transported in the absence of water, (4) provide stored nutrients and a protective coat so that offspring can be dispersed away from the parent plant, and (5) provide nutritious tissue around seeds that facilitates dispersal by animals. Name each of these five structures, and state which land plant group or groups have each structure.
5. What does it mean to say that a life cycle is gametophyte-dominant versus sporophyte-dominant? Give an example of each type of life cycle.
6. Explain the difference between homosporous and heterosporous plants. Where are the microsporangium and a megasporangium found in a tulip? What happens to the spores that are produced by these structures?

Applying Concepts to New Situations

Answers are available at www.masteringbio.com

1. What is the significance of the observation that some members of the Coleochaetales and Charales synthesize sporopollenin and/or lignin?
2. Vessel elements transport water much more efficiently than tracheids, but are much more susceptible than tracheids to being blocked by air bubbles. Suggest a hypothesis to explain why the vascular tissue of angiosperms consists of a combination of vessel elements and tracheids.
3. Angiosperms such as grasses, oaks, and maples are wind pollinated. The ancestors of these subgroups were probably pollinated by insects, however. As an adaptive advantage, why might a species "revert" to wind pollination? (Hint: Think about the costs and benefits of being pollinated by insects versus wind.) Why is it logical to observe that wind-pollinated species usually grow in dense stands containing many individuals of the same species? Why is it logical to observe that in wind-pollinated deciduous trees, flowers form very early in spring—before leaves form?
4. You have been hired as a field assistant for a researcher interested in the evolution of flower characteristics in orchids. Design an experiment to determine whether color, size, shape, scent, or amount of nectar is the most important factor in attracting pollinators to a particular species. Assume that you can change any flower's color with a dye and that you can remove petals or nectar stores, add particular scents, add nectar by injection, or switch parts among species by cutting and gluing.

www.masteringbio.com is also your resource for • Answers to text, table, and figure caption questions and exercises • Answers to *Check Your Understanding* boxes • Online study guides and quizzes • Additional study tools, including the *E-Book for Biological Science* 3rd ed., textbook art, animations, and videos.

31 Fungi

KEY CONCEPTS

- Fungi are important in part because many species live in close association with land plants. They supply plants with key nutrients and decompose dead wood. They are the master recyclers of nutrients in terrestrial environments.
- All fungi make their living by absorbing nutrients from living or dead organisms. Fungi secrete enzymes so that digestion takes place outside their cells. Their morphology provides a large amount of surface area for efficient absorption.
- Many fungi have unusual life cycles. It is common for species to have a long-lived heterokaryotic stage, in which cells contain haploid nuclei from two different individuals. Although most species reproduce sexually, very few species produce gametes.

Amanita muscaria lives in association with living tree roots. The mushrooms you see are only reproductive structures—the vast majority of this individual exists underground. *Amanita muscaria* mushrooms contain molecules that can induce hallucinations in mammals. This species has been used in religious ceremonies by some native cultures but is considered highly toxic.

Fungi are eukaryotes that grow as single cells or as large, branching networks of multicellular filaments. Familiar fungi include the mushrooms you've encountered in woods or lawns, the molds and mildews in your home, the organism that causes athlete's foot, and the yeasts used in baking and brewing.

Along with the land plants and animals, the **fungi** are one of three major lineages of large, multicellular eukaryotes that occupy terrestrial environments. When it comes to making a living, the species in these three groups use radically different strategies. Land plants make their own food through photosynthesis. Animals eat plants, protists, fungi, or each other. Fungi absorb their nutrition from other organisms—dead or alive.

Fungi that absorb nutrients from dead organisms are the world's most important decomposers. Although a few types of organisms are capable of digesting the cellulose in plant cell walls, fungi and a handful of bacterial species are the only organisms capable of completely digesting both the lignin and cellulose that make up wood. Without fungi, Earth's surface would be piled so high with dead tree trunks and branches that there would be almost no room for animals to move or plants to grow.

Other fungi specialize in absorbing nutrients from living organisms. When fungi absorb these nutrients without providing any benefit in return, they lower the fitness of their host organism and act as parasites. If you've ever had athlete's foot or a vaginal yeast infection, you've hosted a parasitic fungus.

Key Concept Important Information Practice It

The vast majority of fungi that live in association with other organisms benefit their hosts, however. In these cases, fungi are not parasites but **mutualists**. The roots of virtually every land plant in the world are colonized by an array of mutualistic fungi. In exchange for sugars that are synthesized by the host plant, the fungi provide the plant with key nutrients such as nitrogen and phosphorus. Without these nutrients, the host plants grow much more slowly or even starve. It is not possible to overstate the importance of these relationships between living land plants and the fungi that live in their roots. In the soils beneath every prairie, forest, and desert, an underground economy is flourishing. Plants are trading the sugar they manufacture for nitrogen or phosphorus atoms that are available from fungi. These plant-fungal associations are the world's most extensive bartering system. The soil around you is alive with an enormous network of fungi that are fertilizing the plants you see above ground.

In short, fungi are the master traders and recyclers in terrestrial ecosystems. Some fungi release nutrients from dead plants and animals; others transfer nutrients they obtain to living plants. Because they recycle key elements such as carbon, nitrogen, and phosphorus and because they transfer key nutrients to plants, fungi have a profound influence on productivity and biodiversity. In terms of nutrient cycling on the continents, fungi make the world go around.

31.1 Why Do Biologists Study Fungi?

Given their importance to life on land and their intricate relationships to other organisms, it's no surprise that fungi are fascinating to biologists. But there are important practical reasons for humans to study fungi as well. They nourish the plants that nourish us. They affect global warming, because they are critical to the carbon cycle on land. Unfortunately, a handful of species can cause debilitating diseases in humans and crop plants. Let's take a closer look at some of the ways that fungi affect human health and welfare.

Fungi Provide Nutrients for Land Plants

Fungi that live in close association with plant roots are said to be **mycorrhizal** (literally, "fungal-root"). When biologists first discovered how extensive these fungal-plant associations are, they asked an obvious question: Does plant growth suffer if mycorrhizal fungi are absent? **Figure 31.1** shows a result typical of many experiments and provides a convincing answer. In this case, potato seedlings were grown in the presence and absence of the mycorrhizal fungi normally found on their roots. The photographs document that this species grows three to four times faster in the presence of its normal fungal associates than it does without them. For farmers, foresters, and ranchers, the presence of normal mycorrhizal fungi can mean the difference between profit and loss. Fungi are critical to the productivity of forests, croplands, and rangelands.

Fungi Speed the Carbon Cycle on Land

The introduction to this chapter claims that fungi are master decomposers and recyclers. To back up this assertion, consider two particularly dramatic events documented in the fossil record. One episode was based on a lack of fungi, the other on an abundance of fungi:

1. Researchers who examine the fossils present in coal from the Carboniferous period find remarkably few fungi that are capable of degrading dead plant material. Fungi that

FIGURE 31.1 Plants Grow Better in the Presence of Mycorrhizal Fungi. (Left) Typical experimental results when plants are grown with and without their normal mycorrhizal fungi (fungi are not visible in the photo). (Right) Root system of a larch tree seedling, with the mycelium from a mycorrhizal fungus visible.

make their living by digesting dead plant material are called **saprophytes** ("rotten-plants"). Noting the "dip" in the fossil record of saprophytic fungi, researchers hypothesized that their absence was responsible for the enormous buildup of dead plant material that occurred during the Carboniferous period. Recall from Chapter 30 that deposits of compressed, partially decayed plant material are called peat and that coal formed when the peat produced during the Carboniferous period was buried under other sediments and subjected to heat and pressure. Because the peatlands existing today are water-logged and highly acidic, biologists hypothesize that fungi did not grow in the coal-forming swamps of the Carboniferous period because the pH was too low. The message is that coal exists today because conditions were too acidic for fungi to do their job.

2. At the end of the Permian period, 250 million years ago, the greatest mass extinction in the history of life occurred. It is estimated that in substantially less than a million years, over 90 percent of multicellular species were wiped out. When the end-Permian extinction was first recognized, biologists thought it had affected primarily marine organisms. But then other researchers documented a huge "fungal spike" in rocks that formed in terrestrial environments. As **Figure 31.2** shows, the spike is a dramatic but short-lived increase in the number of fungal fossils that coincided with the mass extinction event. The explosion of fungal fossils was the first indication that land plants were also devastated during this interval. The hypothesis is that a massive die-off in trees and shrubs produced gigantic quantities of rotting wood and led to the explosion of fungal abundance documented in the spike.

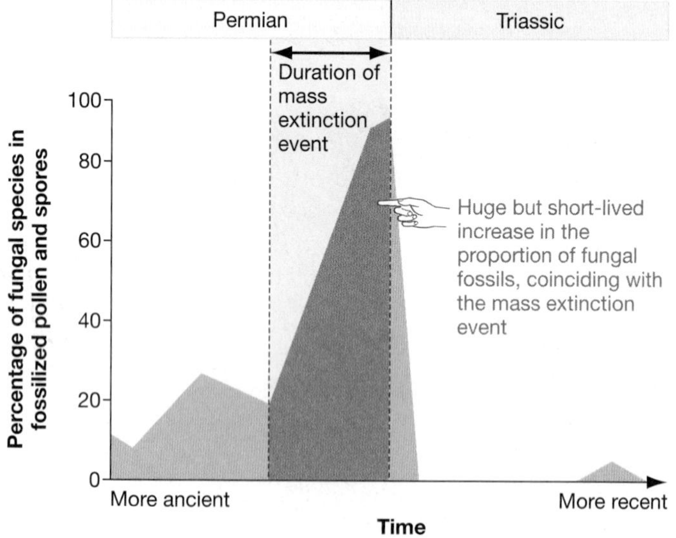

FIGURE 31.2 Fungal Fossils Spike during the End-Permian Extinction. The time interval documents changes in the abundance of fungi before, during, and after the mass extinction event that occurred at the boundary of the Permian and Triassic periods about 250 million years ago.

Saprophytic fungi play a key role in today's terrestrial environments as well. To understand why, recall from Chapter 30 that cells in the vascular tissues of land plants have secondary cell walls containing both lignin and cellulose. Wood forms when stems grow in girth by adding layers of lignin-rich vascular tissue. When trees die, fungi are the organisms that break down wood into sugars and other small organic compounds that they and other organisms can use as food.

Figure 31.3 illustrates the consequences of these facts by highlighting the role that fungi play as carbon atoms cycle through today's terrestrial environments. Note that there are two basic components of the **carbon cycle** on land: (1) the fixation of carbon by land plants—meaning that carbon in atmospheric CO_2 is converted to cellulose, lignin, and other complex organic compounds in the bodies of plants; and (2) the release of CO_2 from plants, animals, and fungi as the result of cellular respiration—meaning the oxidation of glucose and production of the ATP that sustains life. The fundamental point is that, for most carbon atoms, fungi connect the two components. If fungi had not evolved the ability to digest lignin and cellulose soon after land plants evolved the ability to make these compounds, carbon atoms would have been

FIGURE 31.3 Fungi Speed Up the Cycling of Carbon Atoms on Land. Carbon atoms cycle through terrestrial ecosystems. If fungi could not degrade lignin to CO_2 and cellulose to glucose, most carbon would eventually be tied up in indigestible woody tissues. As a result, the cycle would slow dramatically.

● **EXERCISE** Draw an X through the arrow(s) that would not exist if fungi could not digest lignin and cellulose.

sequestered in wood for millennia instead of being rapidly recycled into glucose molecules and CO_2. Terrestrial environments would be radically different than they are today and probably much less productive. On land, fungi make the carbon cycle turn much more rapidly than it would without fungi. The nutrients that fungi release feed a host of other organisms.

Fungi Have Important Economic Impacts

In humans, parasitic fungi cause athlete's foot, vaginitis, diaper rash, ringworm, pneumonia, and thrush, among other miseries. But even though these maladies can be serious, in reality only about 31 species of fungi—out of the hundreds of thousands of existing species—regularly cause illness in humans. Compared with the frequency of diseases caused by bacteria, viruses, and protists, the incidence of fungal infections in humans is low. In addition, soil-dwelling fungi have been the source of many of the most important antibiotics currently being prescribed against bacterial infections. On balance, fungi have been much more helpful than harmful in human and veterinary medicine.

The major destructive impact that fungi have on people is through the food supply. Fungi known as rusts, smuts, mildews, wilts, and blights cause annual crop losses computed in the billions of dollars. These fungi are particularly troublesome in wheat, corn, barley, and other grain crops (**Figure 31.4a**). Saprophytic fungi are also responsible for enormous losses due to spoilage—particularly for fruit and vegetable growers (**Figure 31.4b**).

(a) Parasitic fungi infect corn and other crop plants.

(b) Saprophytic fungi rot fruits and vegetables.

FIGURE 31.4 Fungi Cause Problems with Crop Production and Storage. (a) A wide variety of grain crops are parasitized by fungi. Corn smut is a serious disease in sweet corn, although in Mexico the smut fungus is eaten as a delicacy. **(b)** Fungi decompose fruits and vegetables as well as leaves and tree trunks.

Fungi also have important positive impacts on the human food supply. Mushrooms are consumed in many cultures; in the industrialized nations they are commonly used in sauces, salads, and pizza. The yeast *Saccharomyces cerevisiae* was domesticated thousands of years ago; today it and other fungi are essential to the manufacture of bread, soy sauce, tofu, cheese, beer, wine, whiskey, and other products. In most cases, domesticated fungi are used by food and beverage producers in conditions where the cells grow via fermentation, creating by-products like the CO_2 that causes bread to rise and beer and champagne to fizz. In addition, enzymes derived from fungi are used to improve the characteristics of foods ranging from fruit juice and candy to meat.

In nature, recent epidemics caused by fungi have killed 4 billion chestnut trees and tens of millions of American elm trees in North America. The fungal species responsible for these epidemics were accidentally imported on species of chestnut and elm native to other regions of the world. When the fungi arrived in North America and began growing in chestnuts and elms native to North America, the results were catastrophic. The local chestnut and elm populations had virtually no genetic resistance to the pathogens and quickly succumbed. The epidemics radically altered the composition of upland and floodplain forests in the eastern United States. Before these fungal epidemics occurred, chestnuts and elms dominated these habitats.

Fungi Are Key Model Organisms in Eukaryotic Genetics

When biologists want to answer basic questions about how eukaryotic cells work, they usually turn to fungi. The filamentous fungus called *Neurospora crassa* was introduced in Chapter 15 because it was the study organism in classic experiments that supported the one-gene, one-enzyme hypothesis. The yeast *Saccharomyces cerevisiae* has been even more important in basic research on cell biology and molecular genetics.

S. cerevisiae is unicellular and relatively easy to culture and manipulate in the lab. In good conditions, yeast cells grow and divide almost as rapidly as bacteria. As a result, the species has become the organism of choice for experiments on control of the cell cycle and regulation of gene expression in eukaryotes. The morphology of this organism is so simple that it provides an example of a "pure" eukaryotic cell type—one that is suitable for experiments on how cell division occurs and how particular genes are turned on and off. Just as *Escherichia coli* serves as the model bacterial cell, *S. cerevisiae* serves as the model eukaryotic cell. For example, research has confirmed that several of the genes controlling cell division and DNA repair in yeast have homologs in humans; when mutated, these genes contribute to cancer. Strains of yeast that carry these mutations are now being used to test drugs that might be effective against cancer.

S. cerevisiae has become even more important as an increasing number of eukaryotic genomes are being sequenced (see Chapter 20). To begin interpreting the genomes of organisms like rice, mice, zebrafish, and humans, researchers turn to yeast. It is much easier to investigate the function of particular genes in *S. cerevisiae* by creating mutants or transferring specific alleles among individuals than it is to do the same experiments in mice or zebrafish. Once the function of a gene has been established in yeast, biologists can look for the homologous gene in other eukaryotes. If such a gene exists, they can usually infer that it has a function similar to its role in *S. cerevisiae*. In this way, yeast is serving as a key resource in the field called proteomics and functional genomics. *S. cerevisiae* was also the first eukaryote with a completely sequenced genome.

To summarize, biologists study fungi because they provide a window for understanding eukaryotic cells and because they affect a wide range of species in nature, including humans. How do biologists go about studying them? More specifically, what tools are helping researchers understand the diversity of fungi?

31.2 How Do Biologists Study Fungi?

About 80,000 species of fungi have been described and named to date, and about 1000 more are discovered each year. But the fungi are so poorly studied that the known species are widely regarded as a tiny fraction of the actual total. To predict the authentic number of species alive today, David Hawksworth looked at the ratio of vascular plant species to fungal species in the British Isles—the area where the two groups are the most thoroughly studied. According to Hawksworth's analysis, there is an average of six species of fungus for every species of vascular plant on these islands. If this ratio holds worldwide, then the estimated total of 275,000 vascular plant species implies that there are 1.65 million species of fungi.

Although this estimate sounds large, recent data on fungal diversity suggest that it may be an underestimate. Consider what researchers found recently when they analyzed fungi growing on Barro Colorado Island, Panama: Living on the healthy leaves of just two tropical tree species were a total of 418 distinct morphospecies of fungi. (Recall from Chapter 26 that morphospecies are distinguished from each other by some aspect of morphology.) Because over 310 species of trees and shrubs grow on Barro Colorado, the data suggest that tens of thousands of fungi may be native to this island alone. If further work on fungal diversity in the tropics supports these conclusions, there may turn out to be many millions of fungal species.

This viewpoint of fungal diversity was reinforced by a recent analysis of the fungi living in conjunction with the roots of a single species of grass native to Eurasia. In this study, researchers used the direct sequencing approach, introduced in Chapter 28, to analyze the gene that codes for the RNA molecule in the small subunit of fungal ribosomes. The data showed that a total of 49 phylogenetic species were living in conjunction with the grass roots. (The phylogenetic species concept was introduced in Chapter 26.) Most of the species had never before been described, and several represented completely new lineages of fungi. Biologists are only beginning to realize the extent of species diversity in fungi.

Let's consider how biologists are working to make sense of all this diversity, beginning with an overview of fungal morphology.

Analyzing Morphological Traits

Compared with animals and land plants, fungi have very simple bodies. Only two growth forms occur among them:

1. Single-celled forms called **yeasts** (**Figure 31.5a**), and
2. Multicellular, filamentous structures called **mycelia** (singular: **mycelium; Figure 31.5b**).

Many species of fungus grow only as a yeast or as a mycelium, but some regularly adopt both growth forms.

(a) Single-celled fungi are called yeasts.

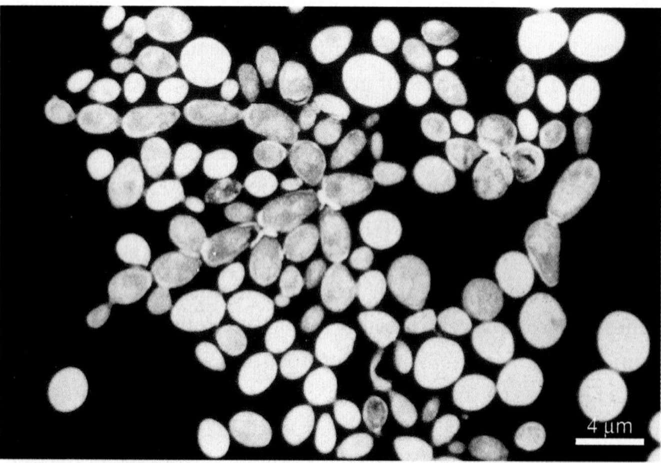

(b) Multicellular fungi have weblike bodies called mycelia.

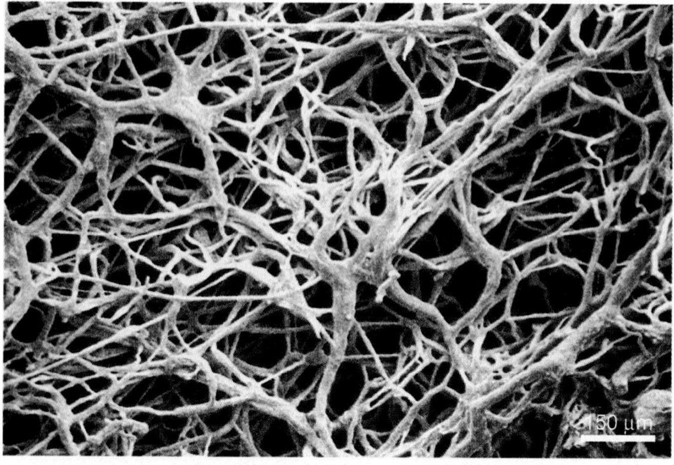

FIGURE 31.5 Fungi Have Just Two Growth Forms. Fungi grow **(a)** as single-celled yeasts and/or **(b)** as multicellular mycelia made up of long, thin, highly branched filaments.

Because most fungi form mycelia and because this body type is so fundamental to the absorptive mode of life, most studies of fungal morphology have focused on them. Let's take a closer look at the structure and function of a fungal mycelium.

The Nature of the Fungal Mycelium If food sources are plentiful, mycelia can be long lived and grow to be extremely large. Researchers recently discovered a mycelium growing across 1310 acres (6.5 km^2) in Oregon. This is an area substantially larger than most college campuses. The biologists estimated the individual's weight at hundreds of tons and its age at thousands of years, making it one of the largest and oldest organisms known.

Although most mycelia are much smaller and shorter lived than the individual in Oregon, all mycelia are dynamic. Mycelia constantly grow in the direction of food sources and die back in areas where food is running out. The body shape of a fungus can change almost continuously throughout its life.

The individual filaments that make up a mycelium are called **hyphae** (singular: **hypha**). Most hyphae are haploid though some are **heterokaryotic** ("different-kernel"), meaning that each cell contains two haploid nuclei—one from each parent. As **Figure 31.6a** shows, hyphae are long, narrow filaments that branch frequently. In most fungi, each filament is broken into cell-like compartments by cross-walls called **septa** (singular: **septum; Figure 31.6b**). Septa do not close off segments of hyphae completely. Instead, gaps called pores enable a wide variety of materials, even organelles and nuclei, to flow from one compartment to the next. Septa may have single large openings or a series of small gaps that give the septum a sieve-like appearance. Because nutrients, mitochondria, and even genes can flow though the entire mycelium—at least to a degree—the fungal mycelium is intermediate between a multicellular land plant or animal and an enormous single-celled organism. Some fungal species are even **coenocytic** ("common-celled"; pronounced *see-no-SIT-ick*)—meaning that they lack septa entirely. Coenocytic fungi have many nuclei scattered throughout the mycelium. In effect, they are a single, gigantic cell.

It's also important to appreciate just how thin hyphae are. Plant root tips range from 100 to 500 μm across, but fungal hyphae are typically less than 10 μm in diameter. This is equivalent to comparing the width of a piece of spaghetti to a railroad boxcar. Fungal mycelia can penetrate tiny fissures in soil and absorb nutrients that are inaccessible to plant roots.

Perhaps the most important aspect of mycelia and hyphae, however, is their shape. Because mycelia are composed of complex, branching networks of extremely thin hyphae, the body of a fungus has the highest surface-area-to-volume ratio possible in a multicellular organism. To drive this point home, consider that the hyphae found in any fist-sized ball of rich soil typically have a surface area equivalent to half a page of this book. This surface area is important because it makes absorption extremely efficient, and because fungi make their living via absorption.

The extraordinarily high surface area in a mycelium has a downside, however: Fungi are prone to drying out, because the amount of water that evaporates from an organism is a function of its surface area. Due to the high surface area of mycelia, fungi are most abundant in moist habitats. Fungi often endure dry conditions in the form of tough, watertight spores.

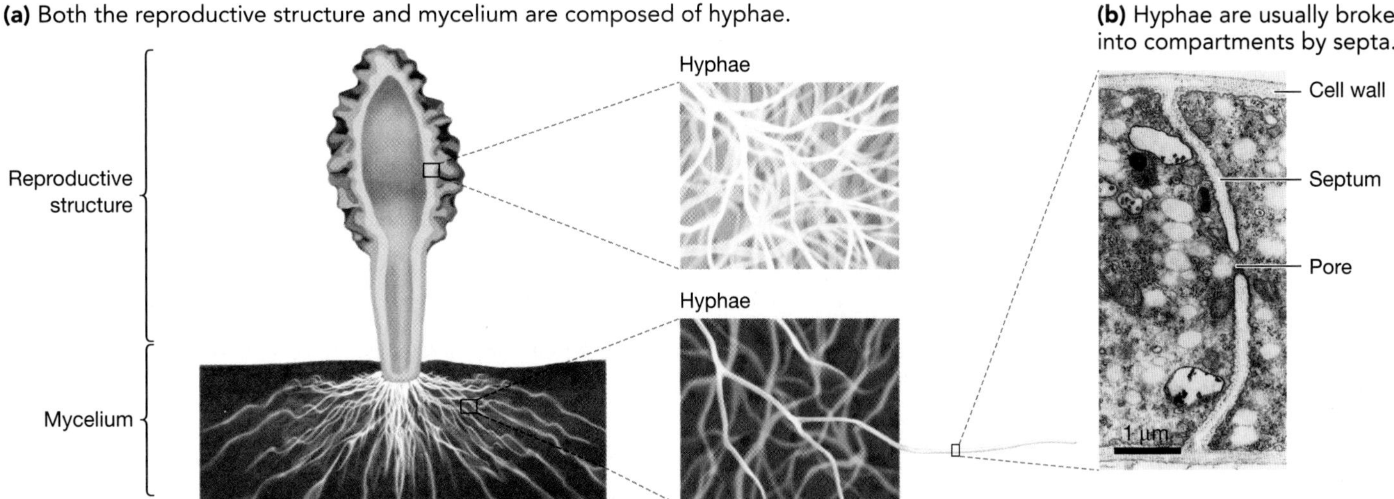

FIGURE 31.6 Multicellular Fungi Have Unusual Bodies. (a) The feeding structure of a fungus is a mycelium, which is made up of hyphae. In some species, hyphae come together to form multicellular structures such as mushrooms, brackets, or morels that emerge from the ground. **(b)** Hyphae are often divided into cell-like compartments by partitions called septa, which are broken by pores. As a result, the cytoplasm of different compartments is continuous.

EXERCISE The reproductive structure in part (a) is actually a tiny proportion of the entire fungal body. To give an idea of the correct scale, draw hyphae extending from below the reproduction structure across both pages of this text.

Mycelia are an adaptation that supports the absorptive lifestyle of fungi. The only thick, fleshy structures that fungi produce are reproductive organs—not feeding structures. Mushrooms, puffballs, and other dense, multicellular structures that arise from mycelia do not absorb food. Instead, they function in reproduction. Typically they are the only part of a fungus that is exposed to the atmosphere, where drying is a problem. The mass of filaments on the inside of mushrooms is protected from drying by the densely packed hyphae forming the surface. However, few species of fungi make the reproductive structures called mushrooms. Instead, each fungal species produces one of four types of distinctive reproductive structures—only one of which is found inside mushrooms.

Reproductive Structures When biologists study diversity in a lineage, they normally begin by comparing the morphologies of various species. Because fungal mycelia are so simple and are so similar among species, researchers have focused on the distinctive morphological structures that fungi produce during sexual reproduction. On the basis of these reproductive structures, fungi fall into four major groups:

1. Members of the Chytridiomycota, or **chytrids**, live primarily in water or wet soils, and they are the only fungi that have motile cells. The spores that chytrids produce during asexual reproduction have flagella, as do the gametes produced by members of this group during sexual reproduction. These swimming cells are reproductive structures that distinguish chytrids from other fungi (**Figure 31.7a**). Structurally, chytrid flagella are similar to the flagella in the sperm cells of animals.

2. The hyphae belonging to the Zygomycota, are haploid and come in several mating types. Instead of having morphologically distinct males and females that produce sperm and eggs, hyphae of different mating types look identical but will not combine unless the individuals have different alleles of one or more genes involved in mating. If chemical messengers released by two hyphae indicate that they are of different mating types, the individuals may become yoked together as shown in **Figure 31.7b**. (The Greek root *zygos* means to be yoked together like oxen. Translated literally, *Zygomycota* means "yoked-together fungi.") Cells from the yoked hyphae fuse to form a spore-producing structure called a **zygosporangium**. Yoked hyphae that form a zygosporangium are the reproductive structure unique to this group.

3. Mushrooms, bracket fungi, and puffballs are among the complex reproductive structures produced by members of the Basidiomycota, or **club fungi**. Inside these structures, specialized cells called **basidia** ("little-pedestals") form at the ends of hyphae and produce spores (**Figure 31.7c**). Only members of the Basidiomycota produce basidia. It is common for multiple mating types to occur in these species; the mushroom-forming *Schizophyllum commune* is estimated to have 28,000 mating types.

4. Members of the Ascomycota, also called **sac fungi**, produce complex reproductive structures—the largest of which are often cup shaped. The tips of hyphae inside these structures produce distinctive sac-like cells called **asci** (singular: **ascus**; **Figure 31.7d**). An ascus is a spore-producing structure found only in Ascomycota.

In sum, morphological studies allowed biologists to describe and interpret the growth habit of mycelia as an adaptation that makes absorption extremely efficient. Careful analyses of morphological features also allowed researchers to identify four major lineages of fungi, based on the nature of their reproductive structures. Do the Chytridiomycota, Zygomycota, Basidiomycota, and Ascomycota each represent a monophyletic group? If so, how are these four major groups related? And how are fungi related to other eukaryotes?

Evaluating Molecular Phylogenies

Researchers have sequenced and analyzed an array of genes to establish where fungi fit on the tree of life. The top part of **Figure 31.8** shows where fungi occur among the many eukaryote lineages; the close-up in the bottom part of the figure emphasizes that fungi are much more closely related to animals than they are to land plants. In addition to the DNA sequence

(a) Chytridiomycota make swimming gametes and spores.

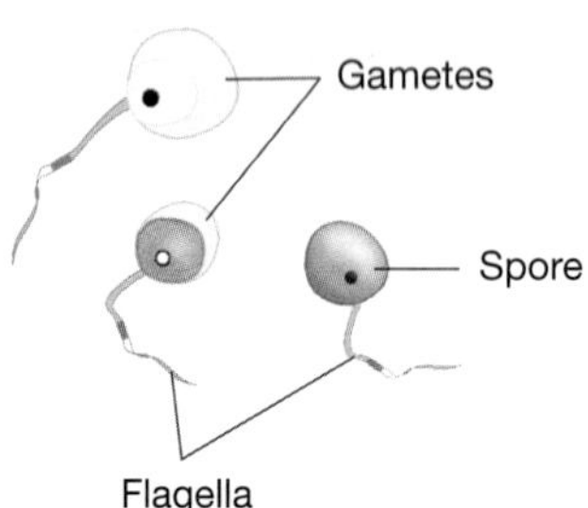

(b) Zygomycota hyphae yoke together and form a zygosporangium.

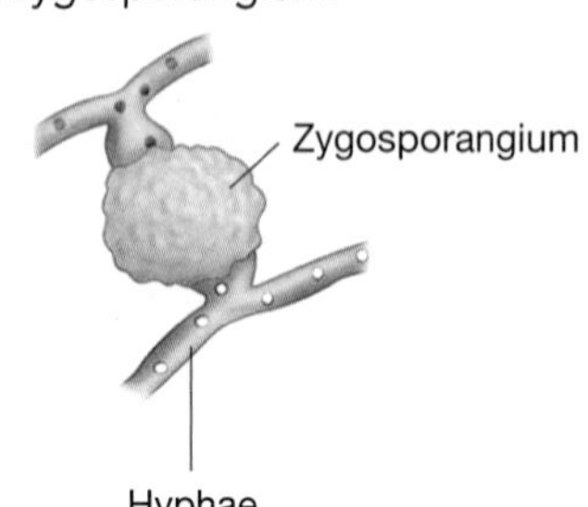

(c) Basidiomycota form spores on basidia (little pedestals).

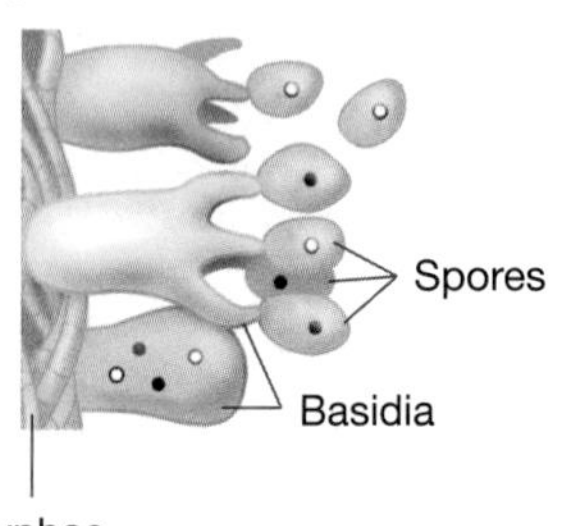

(d) Ascomycota form spores in asci (sacs).

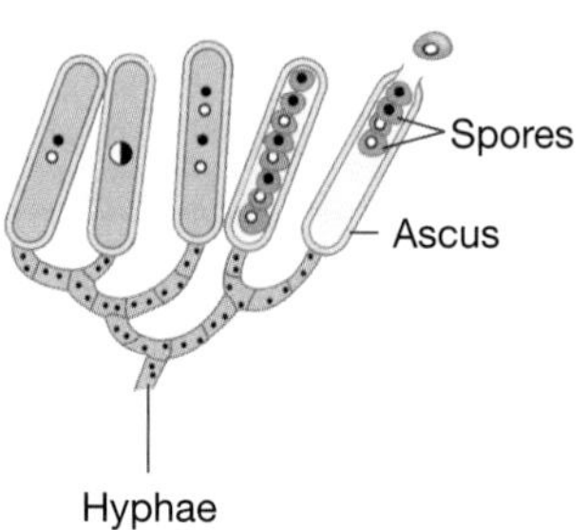

FIGURE 31.7 Four Distinct Reproductive Structures Are Observed in Fungi.

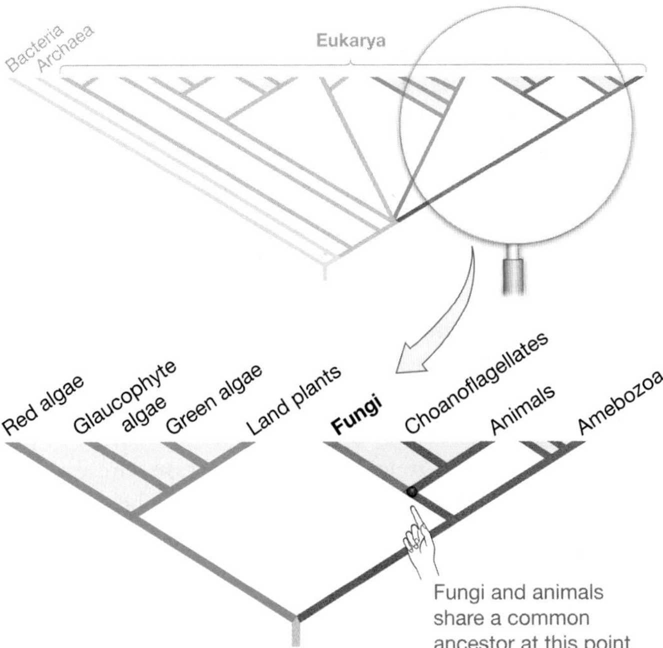

FIGURE 31.8 Fungi Are More Closely Related to Animals than to Land Plants. Phylogenetic tree showing the evolutionary relationships among the green plants, animals, fungi, and some groups of protists. (Choanoflagellates are solitary or colonial protists found in freshwater; they are introduced in Chapter 32.)

data, several morphological traits link animals and fungi. For example, most animals and fungi synthesize the tough structural material called chitin (see Chapter 5). **Chitin** is a prominent component of the cell walls of fungi. Also, the flagella that develop in chytrid spores and in chytrid gametes are very similar to those observed in animals: The flagella are single, are located at the back of the body, and move in a whiplash manner. Further, both animals and fungi store food by synthesizing the polysaccharide glycogen. Green plants, in contrast, synthesize starch as their storage product.

To understand the relationships among chytrids, zygomycetes, basidiomycetes, and ascomycetes, biologists have sequenced a series of genes from an array of fungal species and used the data to estimate the phylogeny of the group. The results, shown in **Figure 31.9**, support several important conclusions:

- Chytrids include the most basal groups of fungi. This result is consistent with the hypothesis that fungi evolved from aquatic ancestors. It also suggests that fungi made the transition to land early in their evolution.
- The Chytridiomycota and Zygomycota are paraphyletic. Neither grouping represents a single common ancestor and all of its descendants. This result means that either swimming gametes or yoked hyphae evolved more than once, or both were present in a common ancestor but then were lost in certain lineages.
- Although early work on molecular phylogenies suggested that organisms called microsporidians are the closest living relatives of fungi, subsequent analyses indicate that they are actually *within* the monophyletic group called fungi. This

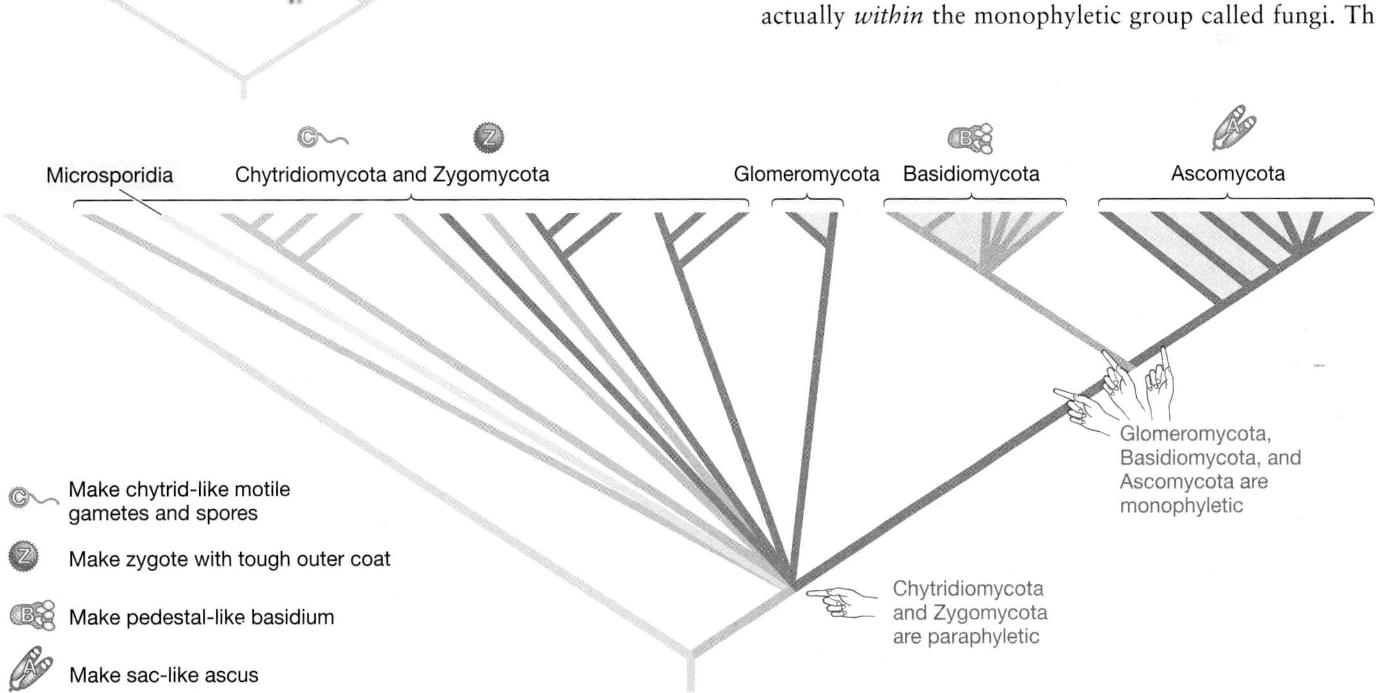

FIGURE 31.9 Phylogeny of the Fungi. A recent phylogenetic tree based on analyses of DNA sequence data. The key indicates the types of sexual reproductive structures observed in each major lineage.

was an important result because microsporidians cause serious disease in bee colonies, silkworm colonies, and people with AIDS. Based on the molecular phylogeny, researchers are testing the hypothesis that **fungicides**—meaning molecules that are lethal to fungi—may prove to be effective in combating microsporidian infections.

- An important group called the Glomeromycota is monophyletic. Sexual reproduction has yet to be observed in this lineage, so it is not known if they ever make spores via meiosis inside a zygosporangium.
- The Basidiomycota and Ascomycota are monophyletic and are the most highly derived groups of fungi.

Although progress on understanding the evolutionary history of fungi has been rapid, the phylogenetic tree in Figure 31.9 is still a work in progress. For example, it is not yet clear where microsporidians are placed relative to several lineages of chytrids and zygomycetes. (Microsporidians lack both swimming gametes and yoked hyphae.) In addition, efforts to understand the relationships of subgroups within the Basidiomycota and Ascomycota are in their infancy. Future work should clarify exactly how the masters of the absorptive lifestyle diversified and how the diversification of fungi relates to the diversification of land plants.

Experimental Studies of Mutualism

It is estimated that 90 percent of land plants live in close physical association with fungi. Stated another way, fungi and land plants often have a **symbiotic** ("together-living") relationship. Although some species of fungi live in association with an array of different land plant species, many documented fungal-plant associations are specific. It is not unusual for fungi to live in only a particular type of tissue, in one plant species. Based on this observation, biologists hypothesize that the evolution of symbiotic associations has played a large role in the diversification of fungi. Scientists categorize these symbiotic relationships as **mutualistic**, meaning they benefit both species; **parasitic**, meaning one species benefits at the expense of the other; or **commensal**, meaning one species benefits while the other is unaffected.

To understand the nature of the association, biologists turn to experimental approaches. Recall from Section 31.1 that in early experiments, researchers grew potatoes, trees, or other types of land plants with or without their normal mycorrhizal fungi. In these experiments, treatments lacking fungal symbionts are created by sterilizing soils with heat or by treating soils and seeds with fungicides. Presence-absence experiments have generally shown that plants grow much larger with their normal symbiotic fungi than they do without. Similarly, fungi that are typically symbiotic are usually unable to grow and reproduce if their regular host plant is absent.

To explore the nature of fungi-plant symbioses in more detail, researchers have used isotopes as tracers for specific elements (**Figure 31.10**). For example, to test the hypothesis that fungi obtain food in the form of carbon-containing compounds from their plant associates, biologists have introduced radioactively labeled carbon dioxide into the air surrounding plants that do or do not contain symbiotic fungi. The experimental plants are usually grown in pots inside laboratory growth chambers. The labeled CO_2 molecules are incorporated into the sugars produced during photosynthesis, and the location of the radioactive atoms can then be followed over time by means of a device that detects radioactivity. If plants feed their fungal symbionts, then labeled carbon compounds should be transferred from the plant to the fungi.

To test the hypothesis that plants are receiving nutrients from their symbiotic fungi in return for sugars, researchers have added radioactive phosphorus atoms or the heavy isotope of nitrogen (^{15}N) to potted plants that do or do not contain symbiotic fungi. If fungi facilitate the transfer of nutrients from soil to plants, then plants grown in the presence of their symbiotic fungi should receive much more of the radioactive phosphorus or heavy nitrogen than do plants grown in the absence of fungi.

As the "Results" section of Figure 31.10 shows, experiments with isotopes used as tracers have shown that sugars and other carbon-containing compounds produced by plants via photosynthesis are transferred to their fungal symbionts. In some cases, as much as 20 percent of the sugars produced by a plant end up in their symbiotic fungi. In exchange, the symbiotic fungi facilitate the transfer of phosphorus or nitrogen or both from soil to the plant. Because phosphorus and nitrogen are in extremely short supply in most environments, the nutrients supplied by symbiotic fungi are critical to the success of the plant. In this way, studies with isotopes have supported the hypothesis that most relationships between fungi and land plants are mutually beneficial.

Check Your Understanding

If you understand that...

- The bodies of fungi are either single-celled yeasts or multicellular mycelia.
- During sexual reproduction, different groups of fungi produce distinct reproductive structures.

You should be able to...

1) Describe the nature of a mycelium and explain why mycelia are interpreted as an adaptation to an absorptive lifestyle.
2) Identify the four types of reproductive structures observed in fungi.
3) Explain whether the Chytridiomycota, Zygomycota, Basidiomycota, and Ascomycota are paraphyletic or monophyletic.

Experiment

Question: Are mycorrhizal fungi mutualistic?

Hypothesis: Host plants provide mycorrhizal fungi with sugars and other photosynthetic products. Mycorrhizal fungi provide host plants with phosphorus and/or nitrogen from the soil.

Null hypothesis: No exchange of food or nutrients occurs between plants and mycorrhizal fungi. The relationship is not mutualistic.

Experimental setup:

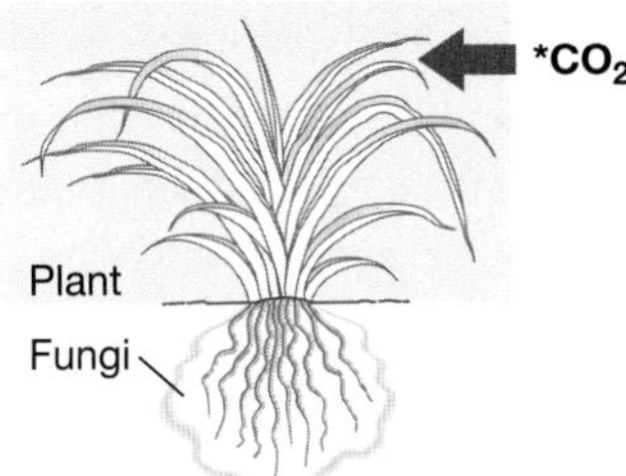

Labeled carbon treatment: Plant leaves are exposed to radioactive CO_2. Mycorrhizal fungi present.

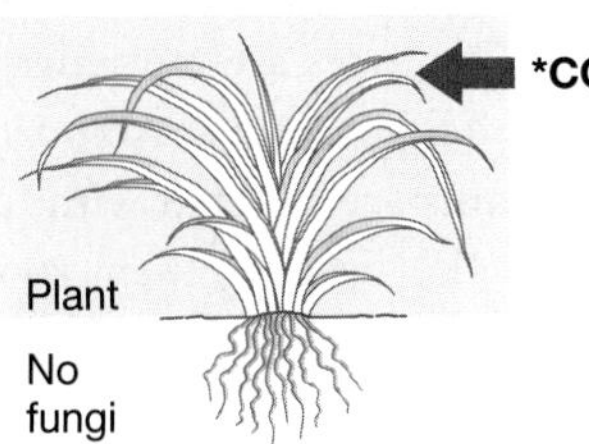

Labeled carbon control: Plant leaves are exposed to radioactive CO_2. Mycorrhizal fungi absent.

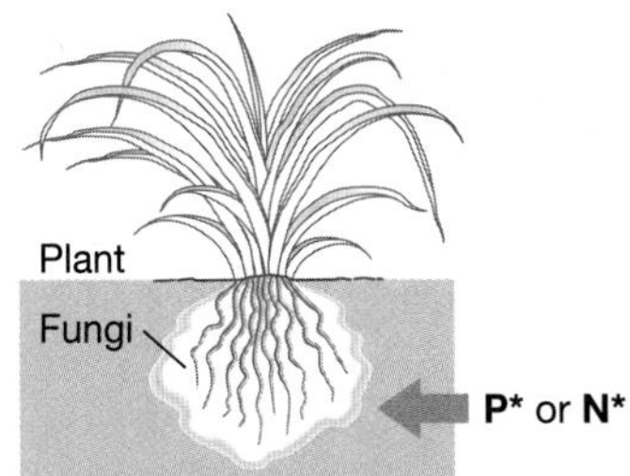

Labeled P or N treatment: Plant roots are exposed to radioactive P or heavy isotope of N. Mycorrhizal fungi present.

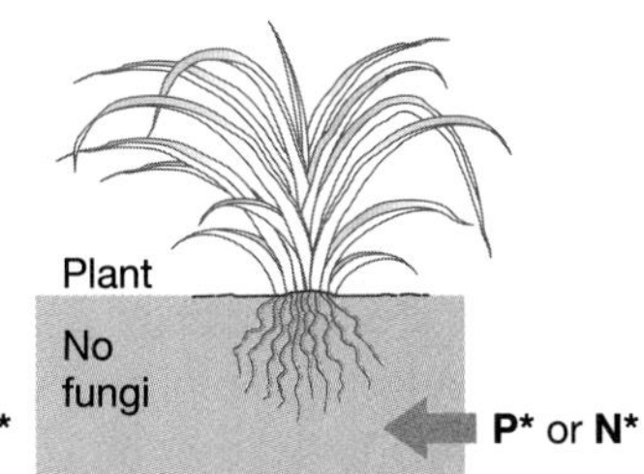

Labeled P or N control: Plant roots are exposed to radioactive P or heavy isotope of N. Mycorrhizal fungi absent.

Prediction for labeled carbon: A large percentage of the labeled carbon taken up by the plant will be transferred to mycorrhizal fungi. In the control, little labeled carbon will be present in the soil surrounding the roots.

Prediction of null hypothesis, labeled carbon: There will be no difference between amounts of labeled carbon in mycorrhizal fungi versus in soil when fungi are absent.

Prediction for labeled P or N: A large percentage of the labeled P or N taken up by the fungi will be transferred to the plant. In the control, little or no labeled P or N will be taken up by the plant.

Prediction of null hypothesis, labeled P or N: There will be no difference between amounts of labeled P or N found in plant in presence or absence of fungi.

Results:

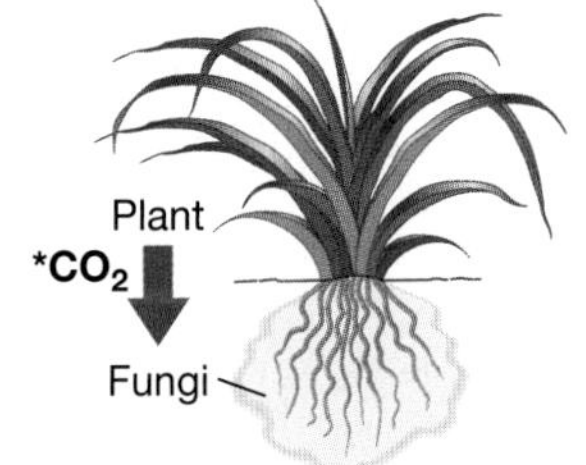

Labeled carbon treatment: Up to 20% of labeled carbon taken up by plant is transferred to mycorrhizal fungus.

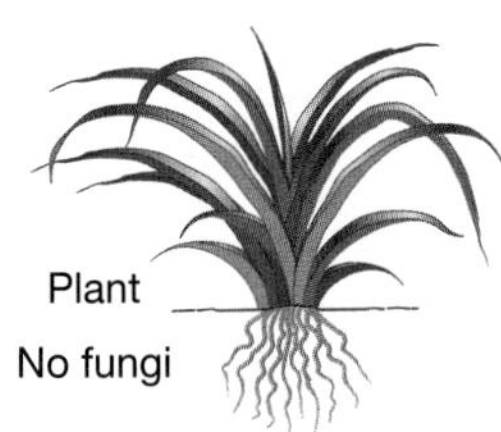

Labeled carbon control: Little to no labeled carbon is found in soil surrounding plant roots.

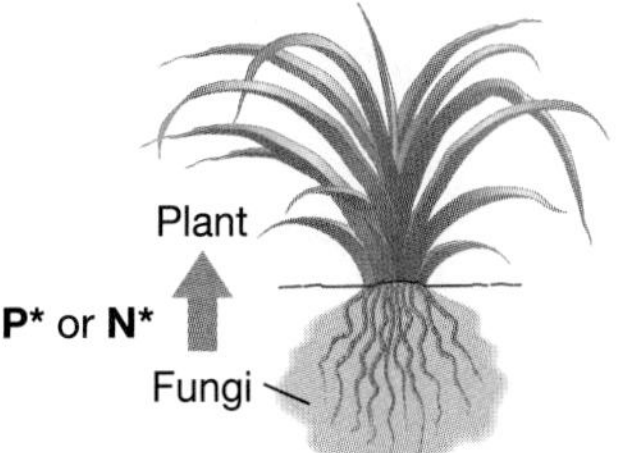

Labeled P or N treatment: Large amount of labeled P or N is found in host plant.

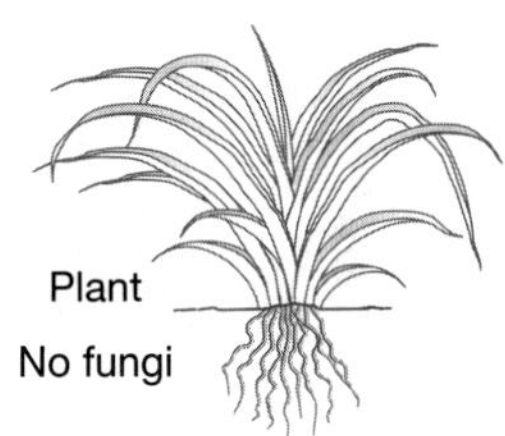

Labeled P or N control: Little labeled P or N is found in host plant.

Conclusion: The relationship between plants and mycorrhizal fungi is mutualistic. Plants provide mycorrhizal fungi with carbohydrates. Mycorrhizal fungi supply host plants with nutrients.

FIGURE 31.10 Experimental Evidence That Mycorrhizal Fungi and Plants Are Mutualistic. Nutrient transfer experiments indicate that sugars flow from plants to mycorrhizal fungi and that key nutrients flow from mycorrhizal fungi to plants.

31.3 What Themes Occur in the Diversification of Fungi?

Why are there so many different species of fungi? This question is particularly puzzling given that fungi share a common attribute: They all make their living by absorbing food directly from their surroundings. In contrast to the diversity of food-getting strategies observed in bacteria, archaea, and protists, all fungi make their living in the same basic way. In this respect, fungi are like plants—virtually all of which make their own food via photosynthesis.)

Chapter 30 showed that the diversification of land plants was driven not by novel ways of obtaining food, but by adaptations that allowed plants to grow and reproduce in a diverse array of terrestrial habitats. What drove the diversification of fungi? The answer is the evolution of novel methods for absorbing nutrients from a diverse array of food sources.

This section introduces a few of the ways that fungi go about absorbing nutrients from different food sources, as well as how they produce offspring. Let's explore the diversity of ways that fungi do what they do.

Fungi Participate in Several Types of Mutualisms

Not long after associations between fungi and the roots of land plants were discovered and shown to be mutualistic, researchers found that two types of mycorrhizal interactions are particularly common. The two major types of mycorrhizae have distinctive morphologies, geographic distributions, and functions. One type involves species from the Glomeromycota; the other usually involves species from the Basidiomycota, though some ascomycetes participate. But mycorrhizae are not the only type of symbiotic fungi found in plants. Researchers have also become interested in fungi that live in close association with the aboveground tissues of land plants—their leaves and stems. Fungi that live in the aboveground parts of plants are said to be **endophytic** ("inside-plants"). Recent research has shown that endophytic fungi are much more common and diverse than previously suspected. Further, data indicate that at least some species of endophytes are mutualistic.

These results support the general realization that most plants are covered with fungi—from the tips of their branches to the base of their roots. Many or even most plants are involved in several distinct types of mutualistic relationships with fungi.

Ectomycorrhizal Fungi The type of mycorrhizal fungus illustrated in **Figure 31.11a** is found on many of the tree species growing in the temperate latitudes of both hemispheres, as well

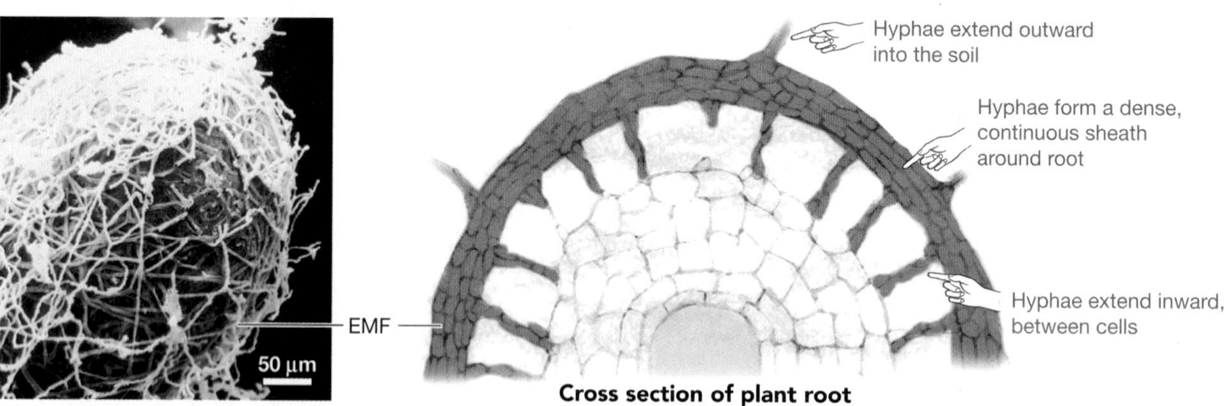

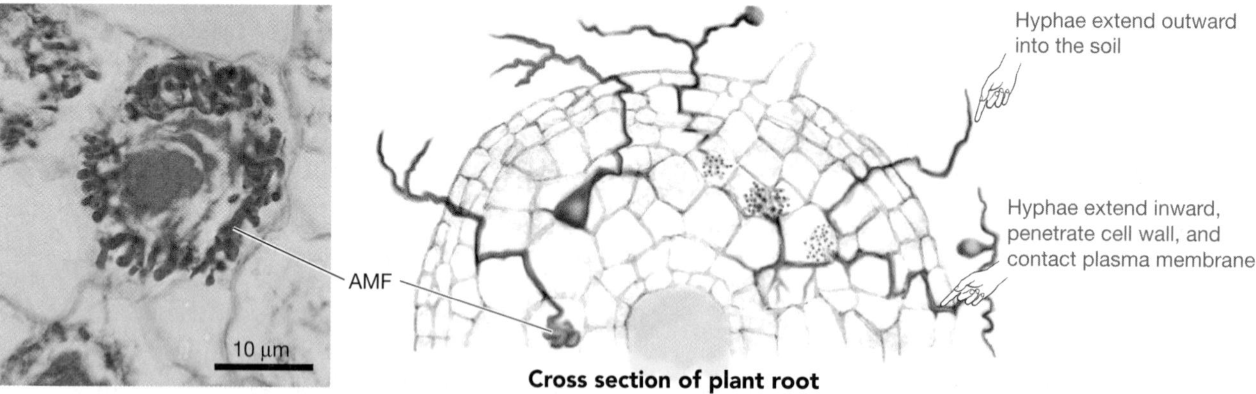

FIGURE 31.11 Mutualistic Fungi Interact with the Roots of Plants in Two Distinct Ways. (a) Ectomycorrhizal fungi (EMF) form a dense network around the roots of plants. Their hyphae penetrate the intercellular spaces of the root but do not enter the root cells. **(b)** The hyphae of arbuscular mycorrhizal fungi (AMF) penetrate the walls of root cells, where they branch into bushy structures or balloon-like vesicles that contact the root cell's plasma membrane.

● **EXERCISE** In the cross sections, add arrows and labels showing the direction of movement of N (nitrogen-containing compounds), P (phosphorus-containing compounds), and C (carbon-containing compounds).

as on tree species in northern coniferous forests. In this type of association, hyphae form a dense network that covers a plant's root tips. As the cross section in Figure 31.11a shows, individual hyphae penetrate between cells in the outer layer of the root, but hyphae do not enter the root cells. Fungi with this growth form are called **ectomycorrhizal fungi** (**EMF**). The Greek root *ecto*, which refers to "outer," is appropriate because the fungi form an outer sheath that is often 0.1 mm thick. Hyphae also extend out from the sheath-like portion of the mycelium into the soil. Most EMF are basidiomycetes; a few are ascomycetes.

How and why do these trees and fungi interact? In the cold, northern habitats where EMF are abundant, the growing season is so short that the decomposition of needles, leaves, twigs, and trunks can be extremely sluggish. As a result, nitrogen atoms tend to remain tied up in amino acids and nucleic acids inside dead tissues instead of being available in the soil. Fortunately, the hyphae of EMF penetrate decaying material and release enzymes called peptidases that cleave the peptide bonds between amino acids in the dead tissues. The nitrogen released by this reaction is absorbed by the hyphae and then transported to the spaces between the root cells of trees, where it can be absorbed by the plant. EMF are also able to acquire phosphate ions that are bound to soil particles and transfer the ions to host plants. In return, the fungi receive sugars and other complex carbon compounds from the tree.

Researchers have found that when northern species such as birch tree seedlings are grown with and without their normal EMF in pots filled with forest soil, only the seedlings with EMF are able to acquire significant quantities of nitrogen and phosphorus. Inspired by such data, a biologist has referred to EMF as the "dominant nutrient-gathering organs in most temperate forest ecosystems." The hyphae of EMF are like an army of miners that discover, excavate, and deliver precious nuggets of nitrogen to the trees of northern forests.

Arbuscular Mycorrhizal Fungi (AMF) In contrast to the hyphae of EMF, the hyphae of **arbuscular mycorrhizal fungi** (**AMF**) grow *into* the cells of root tissue. The name *arbuscular* ("little-tree") was inspired by the bushy, highly branched hyphae, shown in **Figure 31.11b**, that form inside root cells. AMF are also called endomycorrhizal fungi, because they penetrate the interior of root cell walls, or vesicular-arbuscular mycorrhizae (VAM), because the hyphae of some species form large, balloon-like vesicles inside root cells.

The key point is that the hyphae of AMF penetrate the cell wall and contact the plasma membrane of root cells directly. The highly branched hyphae inside the plant cell wall are thought to be an adaptation that increases the surface area available for exchange of molecules between the fungus and its host. However, AMF do not form a tight sheath around roots, as do EMF. Instead, they form a pipeline extending from inside plant cells in the root to the soil well beyond the root.

Most AMF species belong to the lineage called Glomeromycota (see Figure 31.9). AMF are found in a whopping 80 percent of all land plant species, and are particularly common in grasslands and in the forests of warm or tropical habitats. Just as EMF are the dominant type of mycorrhizal association in cool, high-latitude environments, AMF are the dominant type in grasslands and in the tropics.

Besides being extremely common, AMF are ancient. Researchers have found AMF in fossilized root cells that are 400 million years old. This discovery confirms that mycorrhizal associations existed in the most ancient of all land plants. It also supports the hypothesis that plants and fungi colonized terrestrial environments together—meaning that fungi have been nourishing land plants since land plants first evolved.

What do AMF do? Plant tissues decompose quickly in the grasslands and tropical forests where AMF flourish because the growing season is long and warm. As a result, nitrogen is often readily available to plants. Phosphorus is often in short supply, though, because it is present as negatively charged phosphate ions that cling tightly to mineral particles. Based on these observations, biologists hypothesized that AMF transfer phosphorus atoms from the soil to the host plant. Experiments with radioactive atoms confirmed the phosphate-transfer hypothesis by showing that while AMF supply host plants with phosphorus, host plants supply fungi with reduced carbon. EMF mine nitrogen and some phosphorus in the north; AMF mine phosphorus in the south.

Are Endophytes Mutualists? Although endophytic fungi are relatively new to science, they are turning out to be both extremely common and highly diverse. Biologists in Brazil who examine tree leaves for the presence of fungi routinely find several previously undiscovered species of endophytes. Recall from Section 31.2 that a study on fungi in Panama found hundreds of fungal morphospecies living in the leaves of just two tree species. These newly discovered species are endophytes.

Recent research has shown that the endophytes found in some grasses produce compounds that deter or even kill herbivores in exchange for absorbing sugars from the plant. Based on these results, biologists have concluded that the relationship between endophytes and grasses is mutualistic. In other types of plants, however, researchers have not been able to document benefits for the plant host. The current consensus is that at least some endophytic fungi may be commensals—meaning the fungi and the plants simply coexist with no observable effect, either deleterious or beneficial, on the host plant. Although research on endophytes continues, it is already clear that most land plants are colonized by a wide array of fungi and that fungi have evolved an array of relationships with their plant symbionts.

What Adaptations Make Fungi Such Effective Decomposers?

The saprophytic fungi are master recyclers. Although bacteria and archaea are also important decomposers in terrestrial environments, fungi and a few bacterial species are the only organisms that can digest wood completely. Given enough time, fungi can turn even the hardest, most massive trees into soft soils that nourish an array of plants (**Figure 31.12**). You've already analyzed how extremely thin hyphae and the large surface area of a mycelium make absorption exceptionally efficient. Saprophytic fungi are also capable of growing toward the dead tissues that supply their food. What other adaptations help fungi decompose plant tissues?

Extracellular Digestion Large molecules such as starch, lignin, cellulose, proteins, and RNA cannot diffuse across the plasma membranes of hyphae. Only sugars, amino acids, nucleic acids, and other small molecules can enter the cytoplasm. As a result, fungi have to digest their food before they can absorb it.

Instead of digesting food inside a stomach or food vacuole, as most animals and some protists do, respectively, fungi synthesize digestive enzymes and then secrete them outside their hyphae, into their food. Fungi perform **extracellular digestion**—digestion that takes place outside the organism. The simple compounds that result from enzymatic action are then absorbed by the hyphae.

FIGURE 31.12 Fungi Recycle Nutrients. The fungi that are decomposing this section of tree trunk are breaking up its proteins, nucleic acids, lignin, and cellulose. In doing so, the fungi release nitrogen, phosphorus, and other nutrients that can be used by other organisms.

EXERCISE At least two species of basidiomycetes inhabit the area in this photo. Circle their reproductive structures.

As a case study in how this process occurs, consider the enzymes responsible for digesting lignin and cellulose. These are the two most abundant organic molecules on Earth. The term **lignin** refers to a family of extremely strong, complex polymers built from monomers that are six-carbon rings. Recall from Chapter 30 that lignin is found in the secondary cell walls of plant vascular tissues, where it furnishes structural support. **Cellulose** is a polymer of glucose and is found in the primary and secondary cell walls of plant cells. Basidiomycetes can degrade lignin completely—to CO_2—as well as digest cellulose. Let's take a closer look at how they do it.

Lignin Degradation Biologists have been keenly interested in understanding how basidiomycetes digest lignin. Paper manufacturers are also interested in this process because they need safe, efficient ways to degrade lignin in order to make soft, absorbent paper products.

To find out how lignin-digesting fungi do it, biologists began analyzing the proteins that these species secrete into extracellular space. After purifying these molecules, the investigators tested each protein for the ability to degrade lignin. Using this approach, investigators from two labs independently discovered an enzyme called lignin peroxidase.

The researchers who followed up on the discovery of lignin peroxidase found that the enzyme catalyzes the removal of a single electron from an atom in the ring structures of lignin. This oxidation step creates a free radical—an atom with an unpaired electron (see Chapter 2). This extremely unstable electron configuration leads to a series of uncontrolled and unpredictable reactions that end up splitting the polymer into smaller units.

Biologists have referred to this mechanism of lignin degradation as enzymatic combustion. The phrase is apt because the uncontrolled oxidation reactions triggered by lignin peroxidase are analogous to the uncontrolled oxidation reactions that occur when gasoline burns in a car engine. The nonspecific nature of the reaction is remarkable because virtually all of the other reactions catalyzed by enzymes are extremely specific. The lack of specificity in the lignin degradation reaction actually makes sense, however, given the nature of the compound. Unlike proteins, nucleic acids, and most other polymers with a regular and predictable structure, lignin is extremely heterogeneous. Over 10 types of covalent linkages are routinely found between the monomers that make up lignin. But once lignin peroxidase has created a free radical in the aromatic ring, any of these linkages can be broken.

The uncontrolled nature of the reactions has an important consequence. Instead of being stepped down an orderly electron transport chain as described in Chapter 9, the electrons involved in these reactions lose their potential energy in large, unpredictable jumps. As a result, the oxidation of lignin cannot be harnessed to drive the production of ATP and fuel the growth of hyphae. This conclusion is supported by an experimental observation: Fungi cannot grow with lignin as their sole source of food.

If wood-rotting fungi don't use lignin as food, why do they produce enzymes to digest it? The answer is simple. In wood, lignin forms a dense matrix around long strands of cellulose. Degrading the lignin matrix gives hyphae access to huge supplies of energy-rich cellulose. Saprophytic fungi are like miners. But instead of seeking out rare, gem-like nitrogen or phosphorus atoms as do EMF and AMF, the saprophytes use lignin peroxidase to blast away enormous lignin molecules, exposing rich veins of cellulose that can fuel growth and reproduction.

Cellulose Digestion Once lignin peroxidase has softened wood by stripping away its lignin matrix, the long strands of cellulose that remain can be attacked by enzymes called cellulases. Like lignin peroxidase, cellulases are secreted into the extracellular environment by fungi. But unlike lignin peroxidase, cellulases are extremely specific in their action. For example, biologists have purified seven different cellulases from the fungus *Trichoderma reesei*. Two of these enzymes catalyze a critical early step in digestion—they cleave long strands of cellulose into a disaccharide called cellobiose. The other cellulases are equally specific and also catalyze hydrolysis reactions. In combination, the suite of seven enzymes in *T. reesei* transforms long strands of cellulose into a simple monomer—glucose—that the fungus uses as a source of food.

Now that the enzymes responsible for lignin and cellulose degradation have been characterized, researchers would like to pursue a practical question: Can lignin peroxidase and cellulases be used to make paper production more efficient or perhaps speed efforts to clean up waste from old sawmill and paper mill operations? If the answer turns out to be yes, fungi could be used along with bacteria and archaea in the bioremediation efforts introduced in Chapter 29.

Variation in Life Cycles

Fungi evolved an array of ways to absorb nutrition as they diversified. But in addition, fungi evolved an array of ways to reproduce. Recall from Section 31.2 that fungi may produce swimming gametes and spores, yoked hyphae in which nuclei from different individuals fuse to form a zygote inside a protective structure, or specialized spore-producing cells called basidia and asci.

The **spore** is, in fact, the most fundamental reproductive cell in fungi. Spores are the dispersal stage in the fungal life cycle and are produced during both asexual and sexual reproduction. (Recall from Chapter 12 that asexual reproduction is based on mitosis, while sexual reproduction is based on meiosis.) Fungi produce spores in such prodigious quantities that it is not unusual for them to outnumber pollen grains in air samples. If a spore falls on a food source and is able to germinate, a mycelium begins to form. As the fungus expands, hyphae grow in the direction in which food is most abundant. If food begins to run out, mycelia respond by making spores, which are dispersed by wind or animals. Why would mycelia reproduce when food is low? To answer this question, biologists hypothesize that spore production is favored by natural selection when individuals are under nutritional stress. The logic is that spore production allows starving mycelia to disperse offspring to new habitats where more food might be available.

Unique Aspects of Fungal Life Cycles Compared with the life cycles of green plants, protists, and animals, the life cycles of fungi have important unique features. For example, only members of the Chytridiomycota produce gametes, and no species of fungi produce gametes that are so different in size that they are called sperm and egg. In addition, fertilization occurs in two distinct steps in many fungi: (1) fusion of cells and (2) fusion of nuclei from the fused cells. These two steps can be separated by long time spans and even long distances.

In many fungi, the process of sexual reproduction begins when hyphae from two individuals fuse to form a hybrid hypha. When the cytoplasm of two individuals fuses in this way, **plasmogamy** is said to occur (**Figure 31.13a**). But in some cases, the nuclei from the two or more individuals stay independent and grow into a **heterokaryotic** mycelium. The distinct nuclei that are present in heterokaryotic hyphae function independently, even though gene expression must be coordinated in order for growth and development to occur. For example, the two nuclei divide as the hyphae expand, so each compartment that is divided by a septum contains one of each of the two nuclei. How the activities of the two nuclei are coordinated is almost a complete mystery. Eventually, however, one or more pairs of unlike nuclei may fuse to form a diploid zygote. The fusion of nuclei is called **karyogamy**. The nuclei that are produced by karyogamy then divide by meiosis to form haploid spores.

As Figure 29.20 showed, most eukaryotes have life cycles dominated by haploid cells or diploid forms. Many fungi, in contrast, have a life cycle dominated by heterokaryotic cells. If you understand the relationship between plasmogamy, heterokaryosis, and karyogamy, you should be able to compare and contrast these events with the life cycles of other eukaryotes. For example, which human cells undergo plasmogamy and karyogamy? Does heterokaryosis occur?

Most fungal species can reproduce asexually as well as sexually. There are, in fact, large numbers of ascomycetes that have never been observed to reproduce sexually. During asexual reproduction, spore-forming structures are produced by a haploid mycelium, and spores are generated by mitosis. As a result, offspring are genetically identical to their parent.

Among the sexually reproducing species of fungi, the presence of a heterokaryotic stage and the morphology of the spore-producing structure vary. Let's take a closer look at each of the four major types of life cycles that have been observed in fungi:

1. The Chytridiomycota are the only fungal group with any species that exhibit alternation of generations. **Figure 31.13b** shows how alternation of generations occurs in the well-studied chytrid *Allomyces*. The key points are that swimming gametes are produced in haploid adults by mitosis and that gametes from the same individual or different individuals then fuse to form a diploid zygote, which grows into a sporophyte. The life cycle is completed when meiosis occurs in the sporophyte, inside a structure called a **sporangium**. The haploid spores produced by meiosis disperse by swimming.

2. In Zygomycota, sexual reproduction starts when hyphae from different individuals fuse, as shown in **Figure 31.13c**. Plasmogamy forms a zygosporangium that develops a tough, resistant coat and is followed by karyogamy.

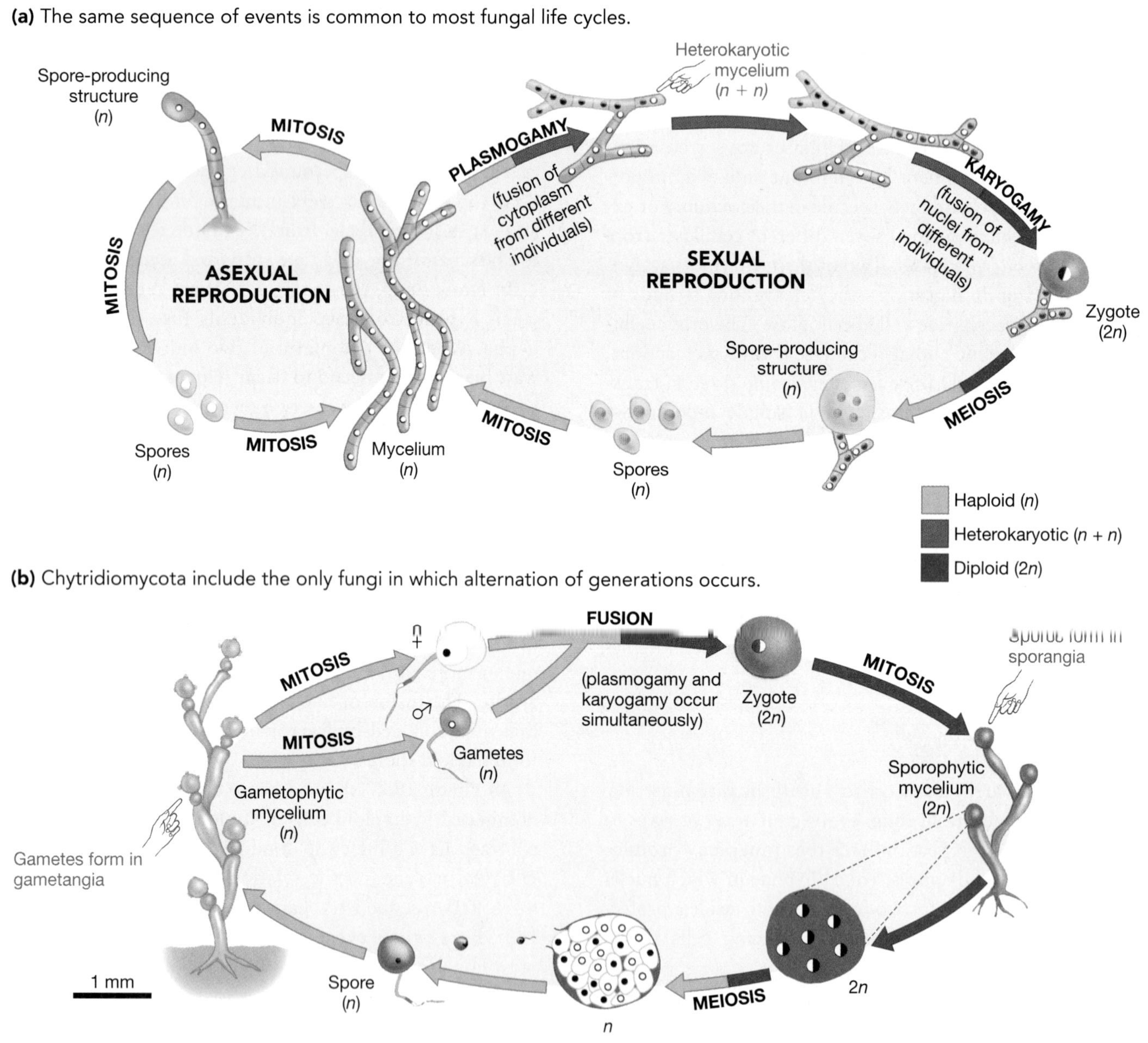

FIGURE 31.13 Fungi Have Unusual Life Cycles. (a) A generalized fungal life cycle, showing both asexual and sexual reproduction. **(b–e)** The sexual part of the life cycle in the four major groups of fungi: **(b)** Chytridiomycota, **(c)** Zygomycota, **(d)** Basidiomycota, and **(e)** Ascomycota. Asexual reproduction is particularly common in Zygomycota and Ascomycota, and is the only form of reproduction currently known in Glomeromycota.

● **EXERCISE** Fungi spend most of their lives feeding. On each drawing, indicate the longest-lived component of the life cycle. In parts (c) and (e), add a loop labeled "Asexual reproduction" similar to that in part (a). This loop should start at the mycelium and return to it, via the growth of asexually produced spores.

(c) Zygomycota form yoked hyphae that produce a spore-forming structure (zygosporangium).

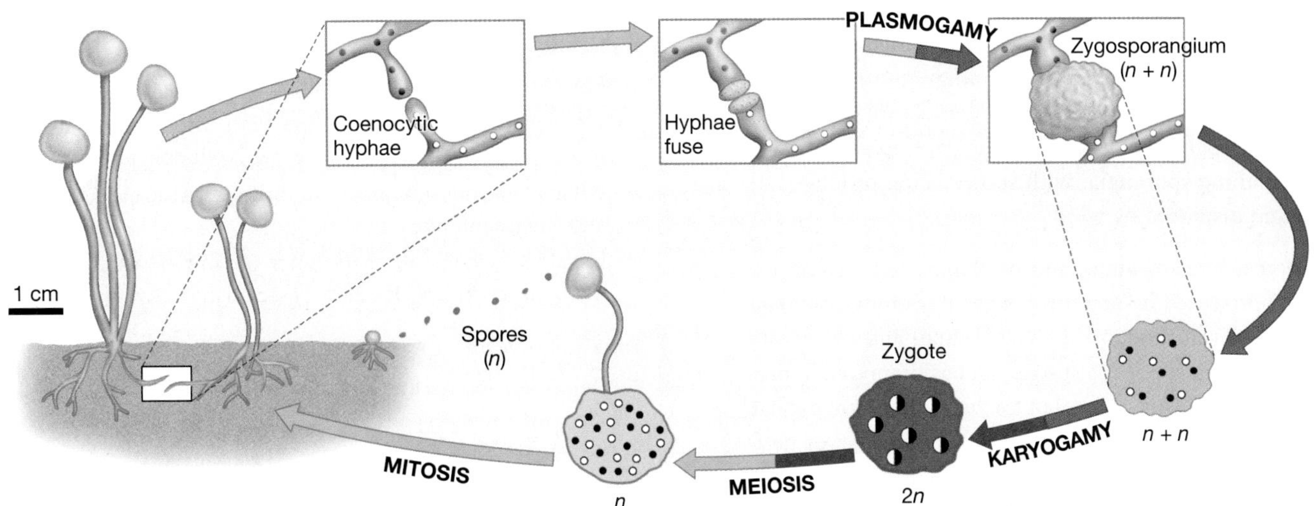

(d) Basidiomycota have reproductive structures with many spore-producing basidia.

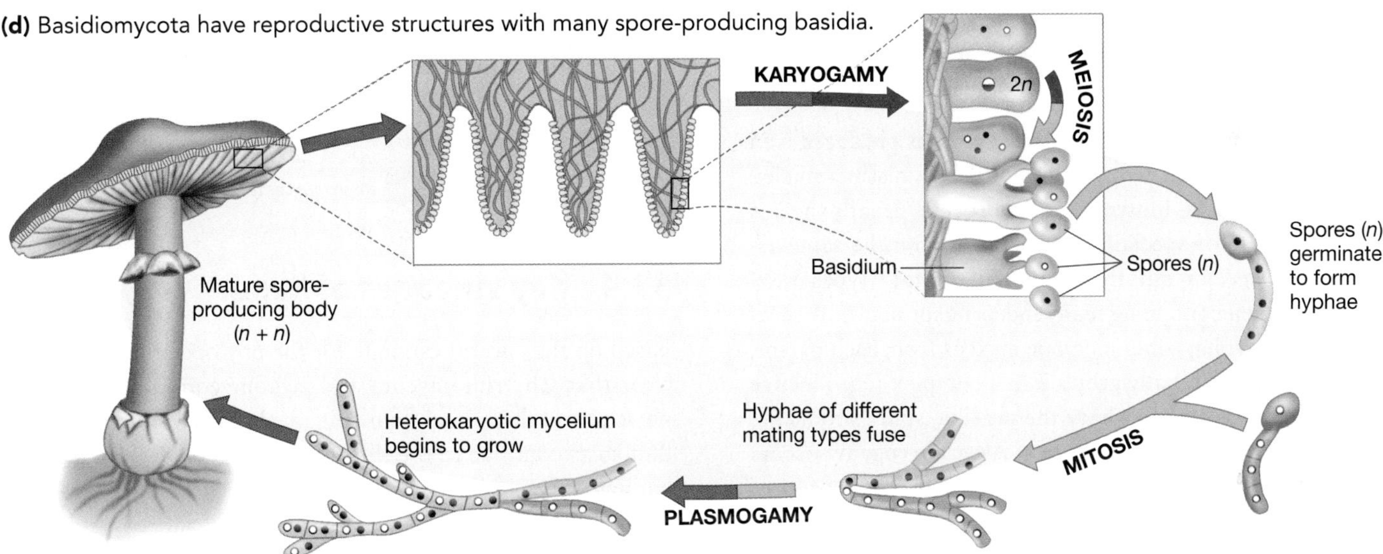

(e) Ascomycota have reproductive structures with many spore-producing asci.

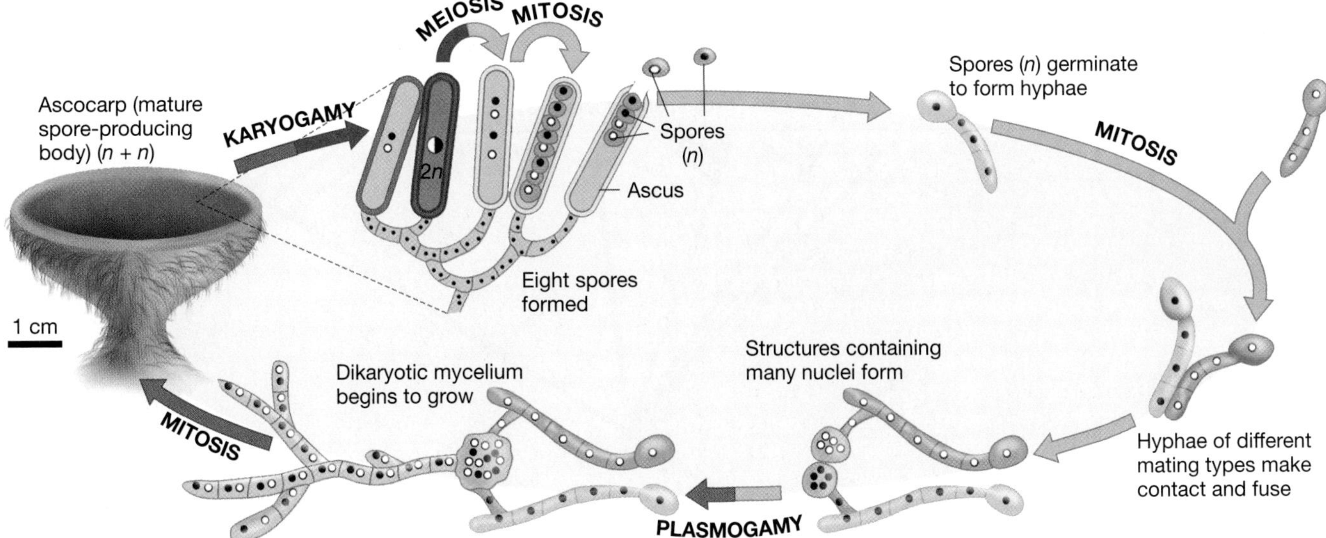

FIGURE 31.13 ***(continued)***

The zygosporangium can persist if conditions become too cold or dry to support growth. When temperature and moisture conditions are again favorable, meiosis occurs. The meiotic products within the sporangium produce haploid spores. When spores are released and germinate, they grow into new mycelia. These mycelia can reproduce asexually by making sporangia, with spores being produced by mitosis and dispersed by wind.

3. Mushrooms, bracket fungi, and puffballs are reproductive structures produced by members of the Basidiomycota (**Figure 31.13d**). Even though their size, shape, and color vary enormously from species to species, all basidiomycete reproductive structures originate from the heterokaryotic hyphae of mated individuals. Inside a mushroom or bracket or puffball, the pedestal-like, spore-producing cells called basidia form at the ends of hyphae. Karyogamy occurs within the basidia. The diploid nucleus that results undergoes meiosis, and haploid spores mature. Spores are ejected from the end of the basidia and are dispersed by the wind. It is not unusual for a single puffball or mushroom to produce a billion spores.

4. The reproductive structure in Ascomycota is produced by a dikaryotic mycelium—one containing two distinct nuclei. As **Figure 31.13e** illustrates, the process usually begins when hyphae or specialized structures from the same ascomycete species but from different mating types fuse, forming a cell containing many independent nuclei. A short dikaryotic hypha, containing one nucleus from each parent, emerges and eventually grows into a complex reproductive structure whose hyphae have the sac-like, spore-producing structures called asci at their tips. After karyogamy occurs inside each ascus, meiosis takes place and haploid spores are produced. When the ascus matures, the spores inside are forcibly ejected. The spores are often picked up by the wind and dispersed.

Check Your Understanding

If you understand that... 

- Most fungi live in close association with land plants.
- When plants are alive, mycorrhizal fungi associate with their roots and many fungal species grow as endophytes.
- When plants die, saprophytic fungi degrade their tissues and release nutrients.
- Instead of being based on the fusion of gametes, sexual reproduction in fungi is usually based on the fusion of hyphae.

You should be able to... 

1) Describe evidence that mutualism occurs in EMF, AMF, and the endophytes of grasses.
2) Explain how extracellular digestion works, using the decomposition of lignin and cellulose as an example.
3) Draw a generalized fungal life cycle.
4) Explain the difference between plasmogamy and karyogamy, and relate them to the process of fertilization in plants and animals.
5) Explain what a heterokaryotic mycelium is.

at www.masteringbio.com
Life Cycle of a Mushroom

31.4 Key Lineages of Fungi

Based on the current estimate for the phylogeny of Fungi, it is clear that Chytridiomycota and Zygomycota are not monophyletic and thus should not be named as phyla (**Figure 31.14**). Biologists continue to use the old names, though, because new naming schemes for these groups are still being developed. Although the following summaries treat the chytrids and zygomycetes as coherent groups, keep in mind that they are not and that they will be renamed soon.

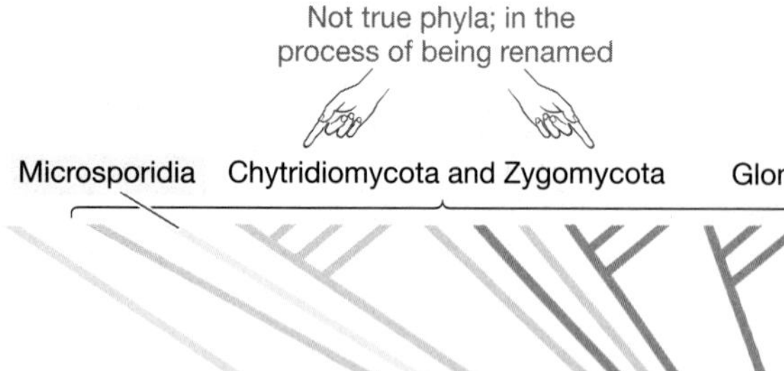

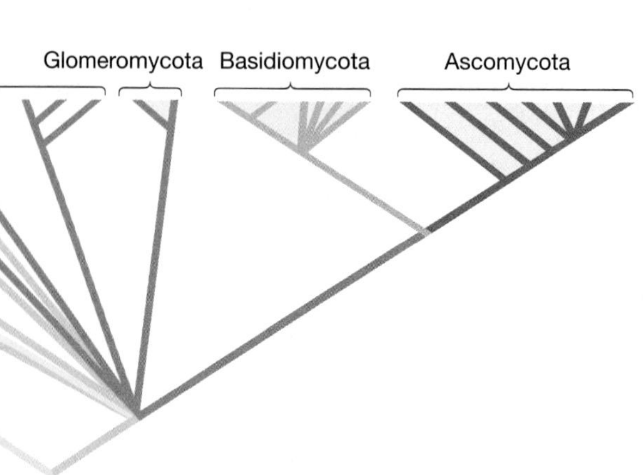

FIGURE 31.14 Major Lineages of Fungi.

EXERCISE Mark the origin of the mycelia growth habit, ascus, and EMF association.

Chytridiomycota (Chytrids)

Chytrids are largely aquatic and are particularly common in freshwater environments. Some species live in wet soils, however, and a few have been found in desert soils that are wet only during a short rainy season. Species that are found in dry soils have resting spores that allow them to endure harsh conditions. Resting spores from chytrids have been shown to germinate normally after a resting period of 31 years. Members of this group are the only fungi that produce motile cells. In each case, the motile cells use flagella to swim.

Absorptive lifestyle Many species of chytrids have enzymes that allow them to digest cellulose. As a result, these species are important decomposers of plant material in wet soils, ponds, and lakes. Many of the freshwater species are parasitic, and on occasion parasitic chytrids cause disease epidemics in algae or aquatic insects (including mosquitoes). Other species parasitize mosses, ferns, or flowering plants. Mutualistic chytrids are among the most important of the many organisms living in the guts of deer, cows, elk, and other mammalian herbivores, because the chytrids help these mammals digest their food (**Figure 31.15**). Chytrids produce cellulases that degrade the cell walls of grasses and other plants ingested by the animal, releasing sugars that are used by both the chytrids and their hosts.

Life cycle During asexual reproduction, most chytrid species produce spores that swim to new habitats via a flagellum. A few species reproduce sexually as well as asexually and exhibit alternation of generations. (Recall from Section 31.3 that these chytrids are the only fungal species to do so.) In species that reproduce sexually, plasmogamy and karyogamy may occur in a variety of ways: through fusion of hyphae, gamete-forming structures, or gametes.

Human and ecological impacts Chytrids are important decomposers in aquatic habitats and key mutualists in the guts of large, plant-eating mammals. Parasitic chytrids are also common. Biologists are investigating the possibility of using a chytrid that parasitizes mosquito larvae as a biological control agent; potato tubers are sometimes invaded and spoiled by a parasitic chytrid that causes black wart disease; and chytrids may be responsible for catastrophic declines that have been occurring recently in frog populations all over the world.

Neocallimastix species

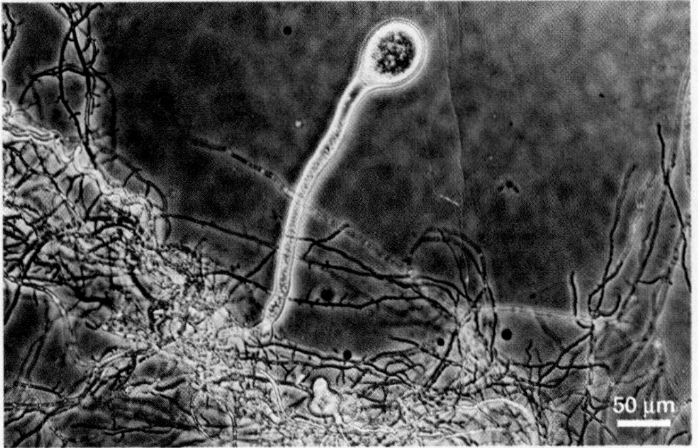

FIGURE 31.15 A Chytrid That Lives in the Gut of Cows and Digests the Cellulose in Grass Leaves.

Microsporidia

All of the estimated 1200 species of microsporidians are single celled and parasitic. They are distinguished by a unique structure called the polar tube (**Figure 31.16**), which allows them to enter the interior of the cells they parasitize. The tube shoots out from the microsporidian, penetrates the membrane of the host cell, and acts as a conduit for the contents of the microsporidian cell to enter the host cell. Once inside, the microsporidian replicates and produces a generation of daughter cells, which go on to infect other host cells. ● You should be able to mark the origin of the polar tube on Figure 31.14.

Microsporidians have a dramatically reduced genome—the smallest known among eukaryotes—and lack functioning mitochondria. Like other fungi, the polysaccharide chitin is a prominent component of their cell wall.

Absorptive lifestyle Most microsporidians parasitize insects or fish. Because they enter the interior of host cells, they are called intracellular parasites.

Nosema tractabile

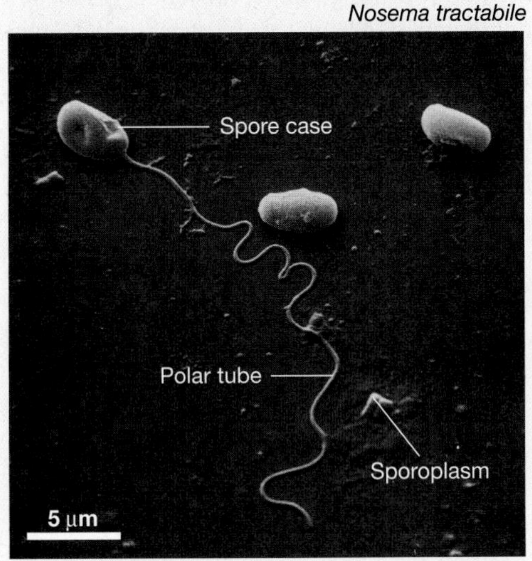

FIGURE 31.16 Microsporidia Infect Other Cells via a Polar Tube.

(Continued on next page)

Microsporidia *continued*

Life cycle Variation in life cycles is extensive. Some species appear to reproduce only asexually. Others produce several different types of sexual or asexual spores, and some must successfully infect several different host species to complete their life cycle.

Human and ecological impacts Species from eight different genera can infect humans. In most cases, however, microsporidians cause serious infections only in AIDS patients and other individuals whose immune systems are not functioning well. Some microsporidians are serious pests in honeybee and silkworm colonies, while others cause severe infections in grasshoppers and are marketed as a biological control agent.

Zygomycota

The Zygomycota ("yoked-fungi") are primarily soil-dwellers. Their hyphae yoke together and fuse during sexual reproduction and then form the durable, thick-walled zygosporangium that is characteristic of the group.

Absorptive lifestyle Many members of zygomycete lineages are saprophytes and live in plant debris. Some parasitize other fungi, however, or are important parasites of insects and spiders. About 65 species in one subgroup are predatory—primarily on amoebae and other unicellular protists, though some are large enough to capture the microscopic animals called roundworms. The predatory zygomycetes trap their prey by means of sticky substances on their cell walls or catch prey in snares consisting of looped hyphae (**Figure 31.17**). Once a prey individual is captured, hyphae invade its body, digest it, and absorb the nutrients that are released.

Life cycle Asexual reproduction is extremely common. Ball-like sporangia are produced at the tips of hyphae that form stalks, and mitosis results in the production of spores, which are dispersed by wind. During sexual reproduction, fusion of hyphae occurs only between individuals of different mating types. Fusion of hyphae from the same individual mycelium does not occur.

Human and ecological impacts The black bread mold *Rhizopus stolonifer* is a common household pest—probably the zygomycete that is most familiar to you. Saprophytic and parasitic members of these lineages are responsible for rotting strawberries and other fruits and vegetables and causing large losses in the fruit and vegetable industries. Some species of *Mucor* are used in the production of steroids for medical use. Species of *Rhizopus* and *Mucor* are used in the commercial production of organic acids, pigments, alcohols, and fermented foods.

FIGURE 31.17 Some Zygomycetes Are Predatory.

Glomeromycota

Recent phylogenetic analyses have shown that the arbuscular mycorrhizal fungi (AMF) form a distinct phylum, indicating that they are a major monophyletic group. Before phylogenies based on molecular characters were available, the lineage now called Glomeromycota was called Glomales and considered part of the Zygomycota. ● You should be able to mark the origin of the AMF association on Figure 31.14.

Absorptive lifestyle Recall from Section 31.3 that AMF absorb phosphorus-containing ions or molecules in the soil and transfer them into the roots of trees, grasses, and shrubs in grassland and tropical habitats (**Figure 31.18**). In exchange, the host plant provides the symbiotic fungi with sugars and other organic compounds.

Life cycle Most species form spores underground. Glomeromycetes are difficult to grow in the laboratory, so their life cycle is not well known. No one has yet discovered a sexual phase in these species.

Human and ecological impacts Because grasslands and tropical forests are among the most productive habitats on Earth, the AMF are enormously important to both human and natural economies.

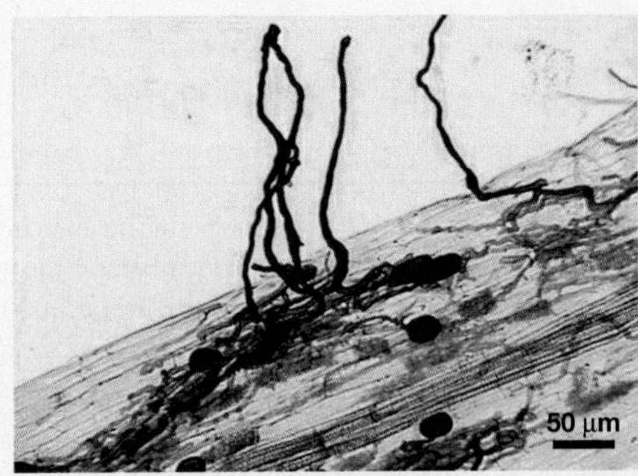

FIGURE 31.18 AMF Penetrate the Walls of Plant Root Cells.

Basidiomycota (Club Fungi)

Although most basidiomycetes form mycelia and produce multicellular reproductive structures, some species have the unicellular growth form. The group is named for basidia, the club-like or pedestal-like cells where meiosis and spore formation occur. ● You should be able to mark the origin of the basidium on Figure 31.14. About 31,000 species have already been described, and more are being discovered each year.

Absorptive lifestyle Basidiomycetes are important saprophytes. Along with a few soil-dwelling bacteria, they are the only organisms capable of synthesizing lignin peroxidase and completely digesting wood. Some basidiomycetes are ectomycorrhizal fungi (EMF) that associate with trees in temperate and northern forests. One subgroup consists entirely of parasitic forms called rusts, including species that cause serious infections in wheat and rye fields. The plant parasites called smut fungi are also basidiomycetes. Smuts specialize in infecting grasses; a few infect other fungi. Thus, the entire array of absorptive lifestyles found in fungi—saprophytic, mutualistic, and parasitic—is found within Basidiomycota.

Life cycle Asexual reproduction through production of spores is common in the Basidiomycota, although not as prevalent as in other groups of fungi. Asexual reproduction also occurs through growth and fragmentation of mycelia in the soil or in rotting wood, resulting in genetically identical individuals that are physically independent. During sexual reproduction, all basidiomycetes—even unicellular ones—produce basidia. In the largest subgroup in this lineage, basidia form in large, aboveground reproductive structures called mushrooms, brackets, earthstars, or puffballs (**Figure 31.19**).

Human and ecological impacts Because temperate and northern forests provide most of the hardwoods and softwoods used in furniture-making, building construction, and papermaking, the EMF are enormously important in forest management. Throughout the world, mushrooms are cultivated or collected from the wild as a source of food. The white button, crimini, and portabella mushrooms you may have seen in grocery stores are all varieties of the same species, *Agaricus bisporus*. Some of the toxins found in poisonous mushrooms are used in biological research; others have hallucinogenic effects on people and are used and traded illegally.

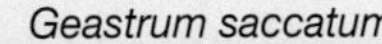

FIGURE 31.19 Earthstars Can Produce Billions of Spores.

Ascomycota (Sac Fungi)

Over half of all known fungi belong to the phylum Ascomycota. This group is named for the sac-like cells, or asci, where meiosis and spore formation take place.

The Ascomycota is extremely large and diverse, and the phylogenetic relationships among major subgroups are still being worked out. As a result, the discussion of "key lineages" that follows is based on two lifestyles observed in ascomycetes and not on distinct monophyletic groups. Work on the phylogenetic relationships within Ascomycota continues.

Ascomycota > Lichen-Formers

About half of the ascomycetes grow in symbiotic association with cyanobacteria and/or single-celled members of the green algae, forming the structures called **lichens**. Over 15,000 different lichens have been described to date; in most, the fungus involved is an ascomycete (although a few basidiomycetes participate as well). To name a lichen, biologists usually use the genus and species name assigned to the fungus that participates in the association.

Absorptive lifestyle It is not yet clear whether all of the relationships observed in lichens are mutualistic. In habitats where lichens are common, neither partner can exist as a free-living organism. The fungus in lichens appears to protect the photosynthetic bacterial or algal cells. The fungal hyphae form a dense protective layer that shields the photosynthetic species and reduces water loss. In return, the cyanobacterium or alga provides carbohydrates that the fungus uses as a source of carbon and energy. However, the hyphae of some lichen-forming fungi have been observed to invade algal cells and kill them. This observation suggests a partially parasitic relationship in at least some lichens. The nature of lichen-forming associations is the subject of ongoing research.

Life cycle As **Figure 31.20a** shows, many lichens reproduce asexually via the production of small "mini-lichen" structures called **soredia** that contain both symbionts. Soredia disperse to a new location via wind or water and then develop into a new, mature individual via the growth of the algal or bacterial and fungal symbionts. In addition, the fungal partner may form asci. Spores that are shed from asci germinate to form a small mycelium. If the growing hyphae encounter enough algal cells, a new lichen can form.

Human and ecological impacts In terms of their abundance and diversity, lichens dominate the Arctic and Antarctic tundras and are extremely common in boreal forests (see Chapter 50). They are the major food of caribou, as well as the most prevalent colonizers of bare rock surfaces throughout the world (**Figure 31.20b**). Rock-dwelling lichens are significant, because they break off mineral particles from the rock surface as they grow—launching the first step in soil formation. About 10,000 tons of lichens are processed annually and used in perfume production—either as a source of fragrant molecules or as a source of molecules that keep the fragrant components of perfumes from evaporating too rapidly.

(a) Cross section of a lichen, showing three layers

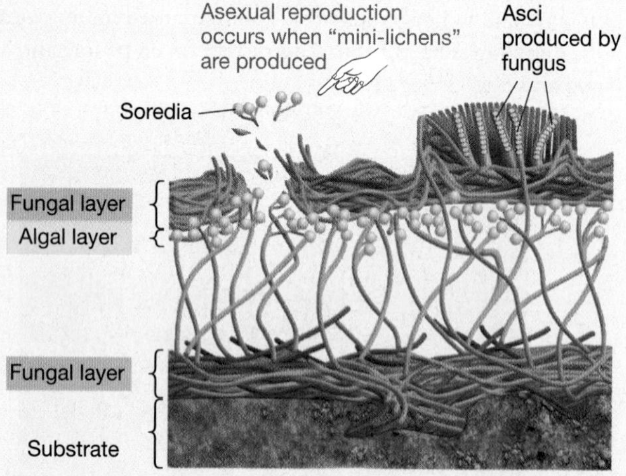

(b) Top view of lichens on a rock

FIGURE 31.20 Lichens Are Associations between a Fungus and a Cyanobacterium or Green Alga. (a) In a lichen, cyanobacteria or green algae are enmeshed in a dense network of fungal hyphae. **(b)** Lichens often colonize surfaces, such as tree bark or bare rock, where other organisms are rare.

Ascomycota > Non-Lichen-Formers

The ascomycetes that do not form lichens are found in virtually every terrestrial habitat, as well as some freshwater and marine environments. Although most ascomycetes form mycelia, many are single-celled yeasts.

Absorptive lifestyle A few members of the Ascomycota form mutualistic ectomycorrhizal fungi (EMF) associations with tree roots. Ascomycetes are also the most common endophytic fungi on aboveground tissues. Large numbers are saprophytic and are abundant in forest floors and in grassland soils. Parasitic forms are common as well. The entire array of absorptive lifestyles found in fungi has evolved within Ascomycota as well as within the Basidiomycota.

Life cycle The aboveground, ascus-bearing reproductive structures of these fungi may be a cup or saucer shape called an **ascocarp** (**Figure 31.21**). For that reason, these ascomycetes are known as **cup fungi**. It is also routine for spores called conidia to be produced asexually, at the ends of specialized hyphae called conidiophores.

Human and ecological impacts Some saprophytic ascomycetes can grow on jet fuel or paint and are used to help clean up contaminated sites. *Penicillium* is an important source of antibiotics, and *Aspergillus* produces citric acid used to flavor soft drinks and candy. Truffles and morels are so highly prized that they can fetch $1000 per pound, and the multibillion-dollar brewing, baking, and wine-making industries would collapse without the yeast *Saccharomyces cerevisiae*. A few parasitic ascomycetes cause infections in humans and other animals. In land plants, parasites from this group cause diseases including Dutch elm disease and chestnut blight.

Sarcoscypha coccinea

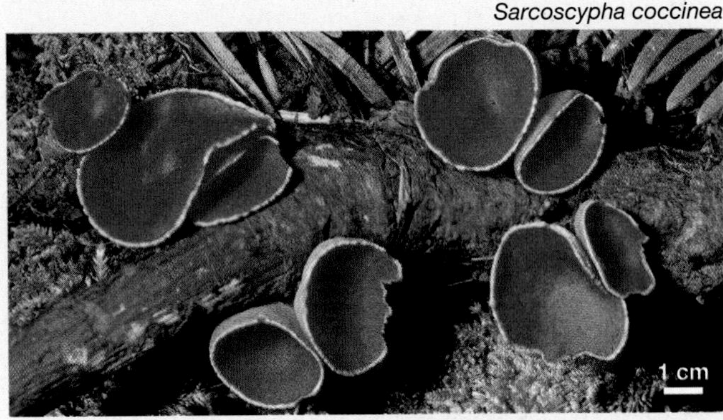

FIGURE 31.21 Ascomycetes Are Sometimes Called Cup Fungi.

Chapter Review

SUMMARY OF KEY CONCEPTS

Fungi are important in part because many species live in close association with land plants. They supply plants with key nutrients and decompose dead wood. They are the master recyclers of nutrients in terrestrial environments.

Living plants are colonized by fungi. The roots of grasses and trees that grow in warm or tropical habitats are infiltrated by symbiotic glomeromycetes that supply the plant with phosphorus. Similar mycorrhizal associations, based on an exchange of nitrogen for sugars and other products of photosynthesis, are found between basidiomycetes and the roots of trees in cold, northern habitats. Many fungi also live in leaves and stems; in some grasses, these endophytic fungi secrete toxins that discourage herbivores.

Although mutualistic associations between fungi and living plants are common, so are parasitic relationships. Parasitic fungi are responsible for devastating blights in crops and other plants.

Once plants die, saprophytic fungi degrade the lignin and cellulose in wood and release nutrients from decaying plant material. Because they free up carbon atoms that would otherwise be locked up in wood, fungi speed up the carbon cycle in terrestrial habitats.

You should be able to predict the results of using fungicides to experimentally exclude fungi from 10m × 10m plots in a northern forest, a tropical forest, and a grassland.

All fungi make their living by absorbing nutrients from living or dead organisms. Fungi secrete enzymes so that digestion takes place outside their cells. Their morphology provides a large amount of surface area for efficient absorption.

Several adaptations make fungi exceptionally effective at absorbing nutrients from the environment. Thin hyphae can penetrate tiny openings in soil, and the two growth habits found among Fungi—single-celled "yeasts" or mycelia composed of long, filamentous hyphae—give fungal cells extremely high surface-area-to-volume ratios. Extracellular digestion, in which enzymes are secreted into food sources, is another important adaptation in fungi. Perhaps the most important reactions catalyzed by these enzymes are the degradations of lignin and cellulose found in wood. Lignin decomposes through a series of uncontrolled oxidation reactions triggered by the enzyme lignin peroxidase. This "enzymatic combustion" contrasts sharply with the breakdown of cellulose, which occurs in a carefully regulated series of steps, each catalyzed by a specific cellulase.

You should be able to draw a mycelium penetrating soil or rotting wood and label the site of extracellular digestion and nutrient absorption.

Many fungi have unusual life cycles. It is common for species to have a long-lived heterokaryotic stage, in which cells contain haploid nuclei from two different individuals. Although

most species reproduce sexually, very few species produce gametes.

Many species of fungi have never been observed to reproduce sexually, though most species can produce haploid spores either sexually or asexually. Sexual reproduction usually starts when hyphae from different individuals fuse—an event called plasmogamy. If the fusion of nuclei, or karyogamy, does not occur immediately, a heterokaryotic mycelium forms. Later, heterokaryotic cells produce spore-forming structures where karyogamy and meiosis take place.

The four traditional groupings of fungi are based on distinct reproductive structures: (1) the Chytridiomycota, which are aquatic fungi with motile gametes; (2) the Zygomycota, which are soil-dwelling fungi with tough sporangia; (3) the Basidiomycota, which have club-like, spore-forming structures; and (4) the Ascomycota, which have sac-like, spore-forming structures. Recent analyses of DNA sequence data have revealed that only the Basidiomycota and Ascomycota are monophyletic, however. Researchers are still trying to understand how the various groups of chytrids and zygomycetes are related to each other and to the microsporidians.

You should be able to explain why most fungi don't need to have gametes to accomplish sexual reproduction.

at www.masteringbio.com
Life Cycle of a Mushroom

QUESTIONS

Test Your Knowledge

1. The mycelial growth habit leads to a body with a high surface-area-to-volume ratio. Why is this important?
   a. Mycelia have a large surface area for absorption.
   b. The hyphae that make up mycelia are long, thin tubes.
   c. Most hyphae are broken up into compartments by walls called septa, although some exist as single, gigantic cells.
   d. Hyphae can infiltrate living or dead tissues.

2. What do researchers hypothesize to be the cause of the "fungal spike" at the end of the Permian period?
   a. unusual soil conditions, which made it more likely that fungal spores and hyphae could fossilize
   b. wet, acidic environments, which slowed the rate of decomposition and allowed peat to form
   c. a switch to saprophytic lifestyles
   d. a mass extinction event that created a huge quantity of dead plant material

3. The Greek root *ecto* means "outer." Why are ectomycorrhizal fungi, or EMF, aptly named?
   a. Their hyphae form tree-like branching structures inside plant cell walls.
   b. They are mutualistic.
   c. Their hyphae form dense mats that envelop roots but do not penetrate the walls of cells inside the root.
   d. They transfer nitrogen from outside their plant hosts to the interior.

4. The hyphae of AMF form bushy or balloon-like structures after making contact with the plasma membrane of a root cell. Why?
   a. They anchor the fungus inside the root, so the association is more permanent.
   b. They increase the surface area available for the transfer of nutrients.
   c. They produce toxins that protect the plant cells against herbivores.
   d. They break down cellulose and lignin in the plant cell wall.

5. What does it mean to say that a hypha is dikaryotic or heterokaryotic?
   a. Two nuclei fuse during sexual reproduction to form a zygote.
   b. Two or more independent nuclei, derived from different individuals, are present.
   c. The nucleus is diploid or polyploid—not haploid.
   d. It is extremely highly branched, which increases its surface area and thus absorptive capacity.

6. Very few organisms besides fungi have a heterokaryotic stage in their life cycle. Which of the following is another unusual aspect of the fungal life cycle?
   a. Some fungi exhibit alternation of generations—meaning that there is a multicellular diploid stage and a multicellular haploid stage.
   b. They produce eggs and sperm in approximately equal numbers, instead of many sperm and a few eggs.
   c. Spores have to fuse with each other before they develop into a new mycelium.
   d. Most varieties undergo sexual reproduction without producing eggs or sperm.

Test Your Knowledge answers: 1. a; 2. d; 3. c; 4. b; 5. b; 6. d

Test Your Understanding

Answers are available at www.masteringbio.com

1. Explain why fungi that degrade dead plant materials are important to the global carbon cycle. Do you accept the text's statement that, without these fungi, "Terrestrial environments would be radically different than they are today and probably much less productive"? Why or why not?

2. Lignin and cellulose provide rigidity to the cell walls of plants. But in most fungi, chitin performs this role. Why is it logical that most fungi don't have lignin or cellulose in their cell walls?

3. Biologists claim that EMF and AMF species are better than plants at acquiring nutrients because they have a higher surface area and because they are more effective at acquiring phosphorus (P) and/or nitrogen (N). Compare and contrast the surface area of mutualistic fungi and plant roots. Explain why fungi are particularly efficient at acquiring P and N, compared to plants.

4. Using information from Chapters 3 through 6, list three key macromolecules found in plants that contain phosphorus. Explain why plant growth might be limited by access to P.

5. Compare and contrast the way that fungi degrade lignin with the way that they digest cellulose.

6. In most eukaryotes that reproduce sexually, only two sexes exist. But in some fungal species, thousands of different mating types exist. Explain why this contrast can occur.

Applying Concepts to New Situations

Answers are available at www.masteringbio.com

1. The box on chytrids in section 31.4 mentions that they may be responsible for massive die-offs currently occurring in frogs. Review Koch's postulates in Chapter 28, then design a study showing how you would use Koch's postulates to test the hypothesis that chytrid infections are responsible for the frog deaths.
2. Some biologists contend that the ratio of plant species to fungus species worldwide is on the order of 1:6. Explain why you agree or disagree with this claim. In doing so, consider the analyses of endophytic, parasitic, lichen-forming, mycorrhizal, and saprophytic strategies presented in this chapter. Also, consider the diversity of tissues and organs available in plants.
3. Experiments indicate that cellulase genes are transcribed and translated together. If cells are selected to be extremely efficient at digesting cellulose, is this result logical? Would you predict that the gene that codes for lignin peroxidase is transcribed along with the cellulase genes? How would you test your prediction?
4. Many mushrooms are extremely colorful. Fungi do not see, so it is unlikely that colorful mushrooms are communicating with one another. One hypothesis is that the colors serve as a warning to animals that eat mushrooms, much like the bright yellow and black stripes on wasps. Design an experiment capable of testing this hypothesis.

32 An Introduction to Animals

KEY CONCEPTS

- Animals are a particularly species-rich and morphologically diverse lineage of multicellular organisms.
- Major groups of animals are recognized by their basic body plan, which differs in the number of tissues observed in embryos, symmetry, the presence or absence of a body cavity, and the way in which early events in embryonic development proceed.
- Recent phylogenetic analyses of animals have shown that there were three fundamental splits as animals diversified, resulting in two protostome groups (Lophotrochozoa and Ecdysozoa) and the deuterostomes.
- Within major groups of animals, evolutionary diversification was based on innovative ways of feeding and moving. Most animals get nutrients by eating other organisms, and most animals move under their own power at some point in their life cycle.
- Methods of sexual reproduction vary widely among animal groups, and many species can reproduce asexually. It is common for individuals to undergo metamorphosis during their life cycle.

Jellyfish are among the most ancient of all animals—they appear in the fossil record over 560 million years ago. Compared with most animals living today, they have relatively simple bodies. But like most other animals, they make their living by eating other organisms and are able to move under their own power.

The **animals** are a monophyletic group of eukaryotes that can be recognized by three traits: They are multicellular, they ingest their food, and they move under their own power at some point in their life cycle. Many unicellular protists also ingest other organisms or dead organic material (detritus) but are small, so they are limited to eating microscopic prey. Animals, in contrast, are the largest and most abundant predators, herbivores, and detritivores in virtually every ecosystem—from the deep ocean to alpine ice fields and from tropical forests to arctic tundras. Animals find food by tunneling, swimming, filtering, crawling, creeping, slithering, walking, running, or flying. They eat nearly every organism on the tree of life.

Over 1.2 million species of animals have been described and given scientific names to date, and biologists predict that tens of millions more have yet to be discovered. To analyze the almost overwhelming number and diversity of animals, this chapter presents a broad overview of how they diversified. It also provides information on the characteristics of the first groups of animals that evolved. The next two chapters provide a more detailed exploration of two major phylogenetic groups in animals: protostomes and deuterostomes. Chapter 33 explores the protostomes, which include familiar organisms such as the insects, crustaceans (crabs and shrimp), and mollusks (clams and snails). Chapter 34 features the deuterostomes, which range from the sea stars to the vertebrates, including humans.

32.1 Why Do Biologists Study Animals?

If you ask biologists why they study animals, the first answer they'll give is, "Because they're fascinating." It's hard to argue with this statement. Consider ants. Ants live in colonies that routinely number millions of individuals. But colony-mates cooperate so closely in tasks such as food-getting, colony defense, and rearing young that each ant seems like a cell in a multicellular organism instead of an individual. Other species of ant parasitize this cooperative behavior, however. Parasitic ants look and smell like their host species but enslave them, forcing the hosts to rear the young of the parasitic species instead of their own. Ant colonies also vary widely in habitat. Some species live in trees and protect their host plants by attacking giraffes and other grazing animals a million times their size. Other species are ranchers or farmers. Rancher ants protect the plant-sucking insects called aphids and eat the sugar-rich honeydew that aphids secrete from their anus (**Figure 32.1a**). Farmer ants eat fungi that they plant, fertilize, and harvest in underground gardens (**Figure 32.1b**). New ant species are discovered every year.

Based on observations like these, most people would agree that ants—and by extension, other animals—are indeed fascinating. But beyond pure intellectual interest, there are other compelling reasons that biologists study animals:

(a) "Rancher ants" tend aphids and eat their sugary secretions.

(b) "Farmer ants" cultivate fungi in gardens.

FIGURE 32.1 Examples of Sophisticated Behavior in Ants.

- Animals are **heterotrophs**—meaning they obtain the carbon compounds they need from other organisms. Recall from earlier chapters that photosynthetic protists and bacteria are **primary producers**, which form the base of the food chain in most marine environments. Land plants play the same role in most terrestrial habitats. Heterotrophs eat producers and other organisms and are called **consumers**. Animals are consumers that occupy the upper levels of food chains in both marine and terrestrial regions. As a result, it is not possible to understand or preserve ecosystems without understanding and preserving animals.
- Animals are a particularly species rich and morphologically diverse lineage of multicellular organisms on the tree of life. Current estimates suggest that there are between 10 million and 50 million species of animals, although only a fraction of these have been formally described and named. Animals range in size and complexity from tiny sponges, which attach to a substrate and contain just a few cell types and simple tissues, to blue whales, which migrate tens of thousands of kilometers each year in search of food and contain trillions of cells, dozens of distinct tissues, an elaborate skeleton, and highly sophisticated sensory and nervous systems. A great deal of diversification has occurred in this lineage. To understand the history of life, it is important to understand how animals came to be so diverse.
- Humans in every country depend on wild and domesticated animals for food. Horses, donkeys, oxen, and other domesticated animals also provide most of the transportation and power used in preindustrial societies.
- Efforts to understand human biology depend on advances in animal biology. For example, most drug testing and other types of biomedical research are done on mice, rats, or primates.

Given that studying animals is interesting and valuable, let's get started. What makes an animal an animal, and how do biologists go about studying them?

32.2 How Do Biologists Study Animals?

Animals are distinguished by several traits other than multicellularity, eating, and moving. The cells of animals lack walls but have an extensive extracellular matrix, which includes proteins specialized for cell-cell adhesion and communication (see Chapter 8). Animals are the only lineage on the tree of life with species that have muscle tissue and nervous tissue. Although many animals reproduce both sexually and asexually, no animals undergo alternation of generations. During an animal's life cycle, adults of most species are diploid; the only haploid cells are gametes produced during sexual reproduction.

Beyond these shared characteristics, animals are almost overwhelmingly diverse—particularly in morphology. Biologists currently recognize about 34 **phyla**, or major lineages, of animals—including those listed in **Table 32.1**. Each animal phylum has distinct morphological features.

Analyzing Comparative Morphology

In essence, animals are moving and eating machines. A quick glance at the diversity of ways that animals find and capture food, like those illustrated in **Figure 32.2**, should convince you that evolution by natural selection has indeed produced a wide array of ways to move and eat. This diversity is possible because of extensive variation in appendages used in movement and in mouthparts or other organs used to capture and process food. Limbs and mouths are specialized structures that make particular ways of moving and eating possible.

In contrast to the spectacular diversity observed among animal limbs and mouthparts, the basic architecture of the animal body has been highly conserved throughout evolution. Just as there are only a few basic ways to frame a house—with posts and beams, stud walls, or cement blocks, for example—evolution has produced just a handful of ways to organize an animal body. Once these basic body plans evolved, an extraordinary radiation of species ensued—based on elaborations of limbs and mouthparts or other structures for moving and capturing food.

The most fundamental groups of animals are defined by variation in the basic body plan; the phyla and sub-phyla within these groups are usually defined by variation in appendages and mouthparts. A **body plan** is an animal's architecture—the major features of its structural and functional design. Four features define the basic elements of an animal's body plan: (1) the number of tissue layers found in embryos, (2) the type of body

TABLE **32.1** **An Overview of Major Animal Phyla**

Group and Phylum	Common Name or Example Taxa	Estimated Number of Species
Porifera	Sponges	7,000
Cnidaria	Jellyfish, corals, anemones, hydroids, sea fans	10,000
Ctenophora	Comb jellies	100
Acoelomorpha	Acoelomate worms	10
Protostomes (basal group)		
Chaetognatha	Arrow worms	100
Protostomes: Lophotrochozoa		
Rotifera	Rotifers	1,800
Platyhelminthes	Flatworms	20,000
Nemertea	Ribbon worms	900
Gastrotricha	Gastrotrichs	450
Acanthocephala	Acanthocephalans	1,100
Entoprocta	Entroprocts	150
Gnathostomulida	Gnathostomulids	80
Sipuncula	Peanut worms	320
Echiura	Spoon worms	135
Annelida	Segmented worms	16,500
Mollusca	Molluscs (clams, snails, octopuses)	94,000
Phoronida	Horseshoe worms	20
Ectoprocta	Ectoprocts	4,500
Brachiopoda	Brachiopods; lamp shells	335
Protostomes: Ecdysozoa		
Nematoda	Roundworms	25,000
Kinorhyncha	Kinorhynchs	150
Nematomorpha	Hair worms	320
Priapula	Priapulans	16
Onychophora	Velvet worms	110
Tardigrada	Water bears	800
Arthropoda	Arthropods (spiders, insects, crustaceans)	1,100,000
Deuterostomes		
Echinodermata	Echinoderms (sea stars,sea urchins, sea cucumbers)	7000
Hemichordata	Acorn worms	85
Chordata	Chordates (tunicates, lancelets, sharks, bony fish, amphibians, reptiles, mammals)	50,000

(a) Caterpillar mandibles harvest leaves.

(b) Feather worm tentacles filter debris.

(c) Shark jaws and teeth capture prey.

FIGURE 32.2 Animals Move and Eat in Diverse Ways. Variation in limbs and mouthparts allows animals to move and harvest food in a wide variety of ways.

symmetry and degree of cephalization (informally, this means the formation of a head region), (3) the presence or absence of a fluid-filled body cavity, and (4) the way in which the earliest events in the development of an embryo proceed. The origin and early evolution of animals was based on the origin and elaboration of these four features. Let's consider each in detail.

The Evolution of Tissues All animals have groups of similar cells that are organized into the tightly integrated structural and functional units called **tissues**. Although sponges lack many types of tissues found in other animals, they do have **epithelium**—a layer of tightly joined cells that covers the surface. Other animals have an array of other tissue types as well as epithelium.

In animals other than sponges, the number of tissue layers that exist in an embryo is a key trait. Animals whose embryos have two types of tissue layers are called **diploblasts** (literally, "two-buds"); animals whose embryos have three types are called **triploblasts** ("three-buds"). You might recall from Chapter 22 that these embryonic tissues are organized in layers, called **germ layers**. In diploblasts these germ layers are called **ectoderm** and **endoderm**. The Greek roots *ecto* and *endo* refer to *outer* and *inner*, respectively; the root *derm* means "skin." In most cases the outer and inner "skins" of diploblast embryos are connected by a gelatinous material that may contain some cells. In triploblasts, however, there is a tissue layer called **mesoderm** between the ectoderm and endoderm. (The Greek root *meso* refers to *middle*.)

The embryonic tissues found in animals develop into distinct adult tissues, organs, and organ systems (see Figure 22.10). In triploblasts, for example, ectoderm gives rise to skin and the nervous system. Endoderm gives rise to the lining of the digestive tract. The circulatory system, muscle, and internal structures such as bone and most organs are derived from mesoderm. In general, then, ectoderm produces the covering of the animal and endoderm generates the digestive tract. Mesoderm gives rise to the tissues in between. The same pattern holds in diploblasts, except that (1) muscle is simpler in organization and is derived from ectoderm, and (2) reproductive tissues are derived from endoderm.

Traditionally, two groups of animals have been recognized as diploblasts: the cnidarians and the ctenophorans (**Figure 32.3a**). The Cnidaria (pronounced *ni-DARE-ee-uh*) include jellyfish, corals, anemones, sea pens (**Figure 32.3b**), and hydra—which

(a) Cnidarians and ctenophores are diploblastic.

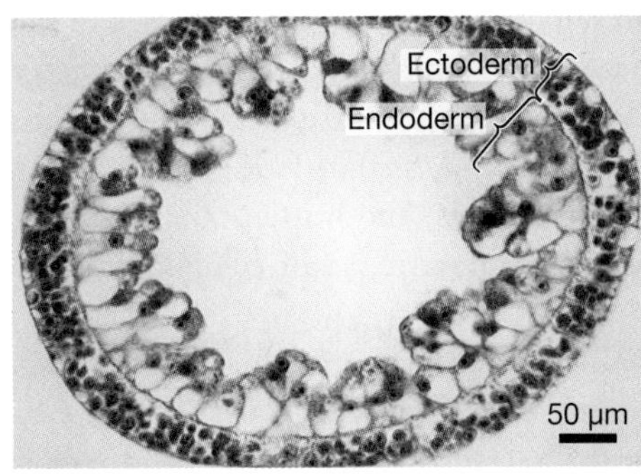

(b) Cnidaria include hydra, jellyfish, corals, and sea pens (shown).

(c) Ctenophora are the comb jellies.

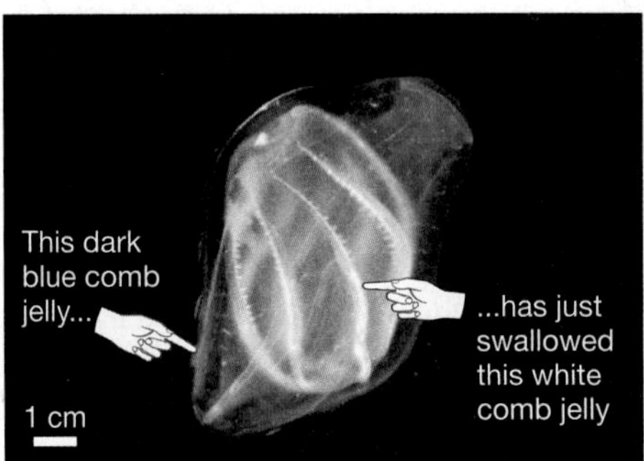

FIGURE 32.3 Diploblastic Animals Have Bodies Built from Ectoderm and Endoderm. (a) Diploblasts have just two embryonic tissue types. **(b)** Like most members of the Cnidaria, this sea pen lives in marine environments. **(c)** Comb jellies, belonging to the Ctenophora, are a major component of planktonic communities in the open ocean.

BOX 32.1 A Model Organism: *Hydra*

Cnidarians are of particular interest to biologists because they are the most ancient living lineage of animals with complex tissues. The ectoderm and endoderm present in their embryos give rise to a number of cell types and tissues in the adults. Cnidarian tissues may be composed of sensory cells that initiate electrical signals in response to environmental stimuli, nerve cells that process those electrical signals and conduct them throughout the body, or muscle cells that contract or relax in response to electrical signals. Cnidarians also have epithelium which consists of a tightly joined layer of cells that is attached to an extensive extracellular matrix. In animals, epithelium covers the outside of the body and lines the surfaces of internal organs.

To understand why the presence of these tissues is interesting, consider the freshwater cnidarian called hydra (**Figure 32.4**). Most species in the genus *Hydra* are about half a centimeter long, live attached to rocks or other firm substrates, and make their living by catching small prey or pieces of organic debris with a cluster of long tentacles. An adult hydra has three major body regions: (1) a basal disk, which attaches the individual to a rock; (2) a tubular section that makes up the bulk of the body; and (3) a "head" that contains the mouth and tentacles.

For over 100 years, biologists have been doing experiments based on cutting hydra bodies apart in various ways and studying how missing tissues and body regions **regenerate**—that is, re-form. This work has led to a deeper understanding of how nerve cells, muscles, epithelia, and other specialized cells and tissues arise from unspecialized cells called stem cells (see Chapter 21). In particular, hydra experiments provided fundamental insights into how cell-to-cell signals organize cells into tissues and, later in development, trigger the differentiation of cells into specialized cell types.

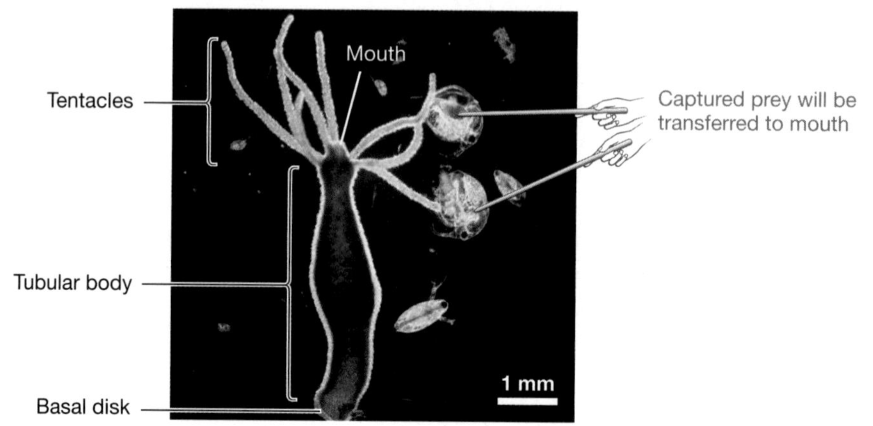

FIGURE 32.4 Hydra Is a Model Organism in Biology. Hydra grow quickly and are relatively easy to maintain in the lab. If an adult is cut into pieces, missing body parts can regenerate in some fragments to form complete adults.

have long been an important source of model organisms in developmental biology (**Box 32.1**). The Ctenophora (pronounced *ten-AH-for-ah*) are the comb jellies (**Figure 32.3c**). All other animals, from leeches to humans, are triploblastic.

Symmetry and Cephalization A basic feature of a multicellular body is the presence or absence of a plane of symmetry. An animal's body is symmetrical if it can be divided by a plane such that the resulting pieces are nearly identical. Animal bodies can have 0, 1, 2, or more planes of symmetry. Most sponges, including the one illustrated in **Figure 32.5a**, are asymmetrical—that is, having no planes of symmetry. They cannot be sectioned in a way that produces similar sides.

All other animals exhibit radial ("spoke") symmetry or bilateral ("two-sides") symmetry. Organisms with **radial symmetry** have at least two planes of symmetry (**Figure 32.5b**). Most of the radially symmetric animals living today either float in water or live attached to a substrate.

Organisms with **bilateral symmetry**, in contrast, have one plane of symmetry and tend to have a long, narrow body. The evolution of bilateral symmetry was a critical step in animal evolution because it triggered a series of associated changes that are collectively known as **cephalization**: the evolution of a head, or anterior region, where structures for feeding, sensing the environment, and processing information are concentrated (**Figure 32.5c**). Feeding and sensory structures on the head face the environment, while posterior regions, at the opposite end of the organism, are specialized for locomotion. All triploblastic animals have bilateral symmetry except for species in the phylum Echinodermata—a group where radial symmetry evolved independently of the diploblastic groups. The echinoderms (pronounced *ee-KINE-oh-derms*) include species such as sea stars, sea urchins, feather stars, and brittle stars. Although their larvae are bilaterally symmetric, adult echinoderms are radially symmetric.

To explain the pervasiveness of bilateral symmetry, biologists point out that locating and capturing food is particularly efficient when movement is directed by a distinctive head region and powered by a long posterior region. In combination with the origin of mesoderm, which made the evolution of extensive musculature possible, a bilaterally symmetric body plan enabled rapid, directed movement and hunting. Lineages with a

(a) Asymmetry

(b) Radial symmetry

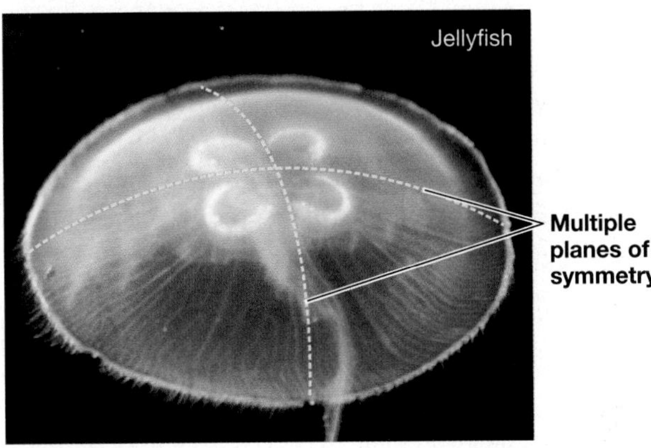

(c) Bilateral symmetry

FIGURE 32.5 There Are Three Types of Body Symmetry in Animals.

QUESTION Are humans, sea stars, the comb jelly in Figure 32.3c, and the hydra in Figure 32.4 asymmetric, radially symmetric, or bilaterally symmetric?

(a) Acoelomates have no enclosed body cavity.

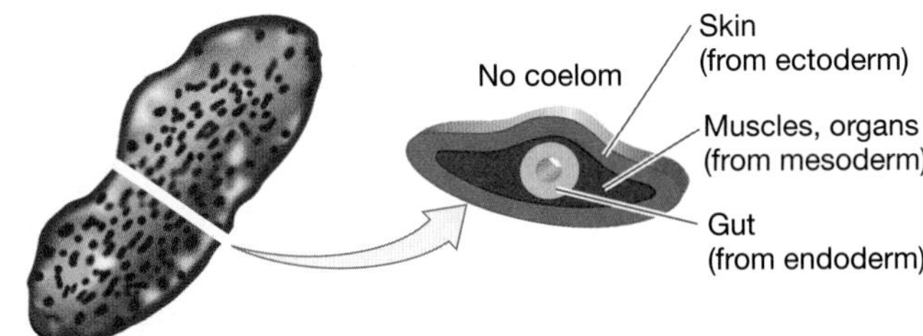

(b) Pseudocoelomates have an enclosed body cavity partially lined with mesoderm.

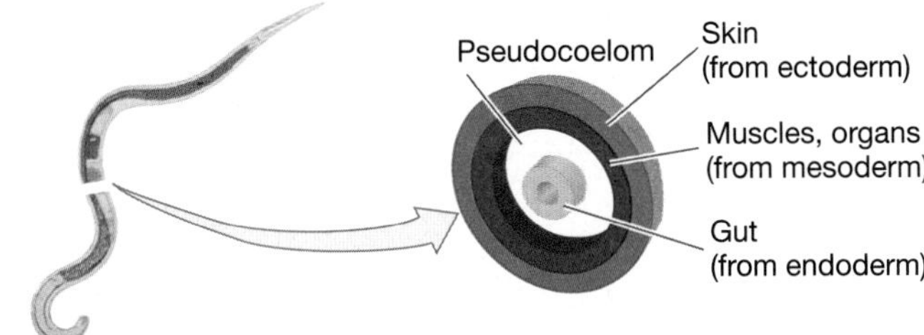

(c) Coelomates have an enclosed body cavity completely lined with mesoderm.

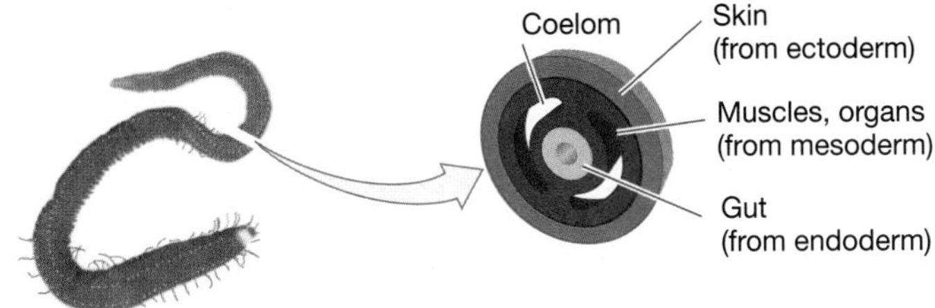

FIGURE 32.6 Animals May or May Not Have a Body Cavity.

triploblastic, bilaterally symmetric body had the potential to diversify into an array of formidable eating and moving machines.

Why Was the Evolution of a Body Cavity Important? A third architectural element that distinguishes animal phyla is the presence of an enclosed, fluid-filled cavity called a **coelom** (pronounced *SEE-lum*). Although diploblasts have a central canal that functions in digestion and circulation, they do not have a coelom.

Biologists were able to determine the nature of the body cavity in various triploblastic phyla through careful observation and dissection of developing embryos and adults. Triploblasts that do not have a coelom are called **acoelomates** ("no-cavity-form"; see **Figure 32.6a**); those that possess a coelom are known as **coelomates**. In a few coelomate groups, such as the roundworms and rotifers, the enclosed cavity forms between the endoderm and mesoderm layers in the embryo. This design is called a **pseudocoelom**, meaning "false-hollow" (**Figure 32.6b**). The term is unfortunate, because there is nothing false about the fluid-filled cavity—it exists. It is simply structured in a different way than a "true" coelom, which forms from within the mesoderm itself and thus is lined on both sides with cells from the mesoderm (**Figure 32.6c**). A pseudocoelom has mesoderm on the outer side of the fluid-filled cavity; a coelom has mesoderm on both the inner and outer sides. As a result, muscle and blood vessels can form on either side of the coelomate's body cavity. In this respect, the coelom represents a more sophisticated design than the pseudocoelom.

The coelom creates a container for the circulation of oxygen and nutrients, along with space where internal organs can move independently of each other. But the coelom is considered a critically important innovation during animal evolution because an enclosed, fluid-filled chamber can act as an efficient **hydrostatic skeleton**. Soft-bodied animals with hydrostatic skeletons can move even if they do not have fins or limbs. Movement is possible because the fluid inside the coelom stretches the body wall—much like a water balloon—meaning that it is under tension (**Figure 32.7a**). This force exerts pressure on the fluid inside the coelom. When muscles in the body wall contract against the pressurized fluid, the fluid moves because the water cannot be compressed.

Using a nematode (roundworm) as an example, **Figure 32.7b** shows how coordinated muscle contractions and relaxations produce changes in the shape of a hydrostatic skeleton that make writhing or swimming movements possible. ● If you understand how a hydrostatic skeleton works, you should be able to demonstrate its function with a long, tube-shaped water balloon, using your fingers to pinch each side and simulate muscle contractions. You should also be able to explain whether your balloon example models a coelom or a pseudocoelom. By providing a hydrostatic skeleton, the evolution of the coelom gave bilaterally symmetric organisms the ability to move efficiently in search of food.

(a) Hydrostatic skeleton of a nematode

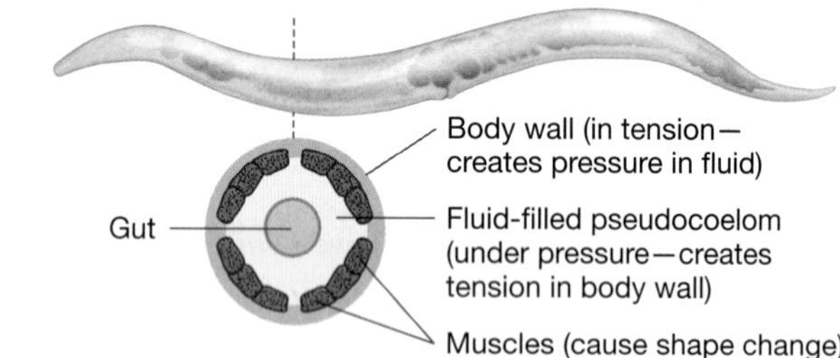

(b) Coordinated muscle contractions result in locomotion.

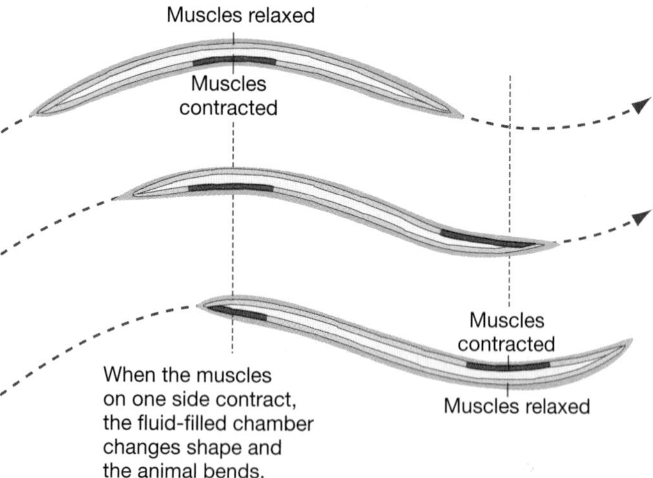

FIGURE 32.7 Hydrostatic Skeletons Allow Limbless Animals to Move. The nematode (roundworm) moves with the aid of its hydrostatic skeleton.

● **QUESTION** Suppose muscles on both sides of this nematode contracted at the same time. What would happen?

What Are the Protostome and Deuterostome Patterns of Development? Except for adult echinoderms, all of the coelomates—including juvenile forms of echinoderms—are bilaterally symmetric and have three embryonic tissue layers. This huge group of organisms is formally called the **Bilateria**, because they are bilaterally symmetric at some point in their life cycle. The bilaterians, in turn, can be split into two subgroups based on distinctive events that occur early in the development of the embryo. The two groups are the protostomes and the deuterostomes. The vast majority of animal species, including the arthropods, molluscs, and annelids, are protostomes; chordates and echinoderms are deuterostomes. (The phylum Arthropoda includes insects, spiders, and crustaceans; Annelida are the segmented worms; Chordata includes bony fishes, amphibians, and mammals.)

To understand the contrasts in how protostome and deuterostome embryos develop, recall from Chapter 22 that the development of an animal embryo begins with cleavage. **Cleavage** is a rapid series of mitotic divisions that occurs in the absence of growth. Cleavage divides the egg cytoplasm and results in a mass of cells. In many protostomes, these cell divisions take place in a pattern known as **spiral cleavage**. When spiral cleavage occurs, the mitotic spindles of dividing cells orient at an angle to the main axis of the cells and result in a helical arrangement of cells. In many deuterostomes, the mitotic spindles of dividing cells orient parallel or perpendicular to the main axes of the cells, resulting in cells that stack directly on top of each other in a pattern called **radial cleavage** (**Figure 32.8a**).

After cleavage has created a ball of cells, the process called gastrulation occurs. As you might recall from Chapter 22, **gastrulation** is a series of cell movements that forms the three embryonic tissue layers. In both protostomes and deuterostomes, gastrulation begins when cells move into the center of the embryo. The migration of the cells creates a pore that opens to the outside (**Figure 32.8b**). In **protostomes**, this pore becomes the mouth. The other end of the gut, the **anus**, forms later. In **deuterostomes**, however, this initial pore becomes the anus; the mouth forms later. Translated literally, *protostome* means "first-mouth" and *deuterostome* means "second-mouth."

The final developmental difference between the groups arises as gastrulation proceeds and the coelom begins to form. Protostome embryos have two blocks of mesoderm beside the gut. As the left side of **Figure 32.8c** indicates, their coelom begins to form when cavities open within each of the two blocks of mesoderm. In contrast, deuterostome embryos have layers of mesodermal cells located on either side of the gut. As the right side of Figure 32.8c shows, their coelom begins to form when these layers bulge out and pinch off to form fluid-filled pockets lined with mesoderm.

To summarize, the protostome and deuterostome patterns of development result from differences in three processes: cleavage, gastrulation, and coelom formation. In essence, the

PROTOSTOMES DEUTEROSTOMES

(a) Cleavage
(zygote undergoes rapid divisions, eventually forming a mass of cells)

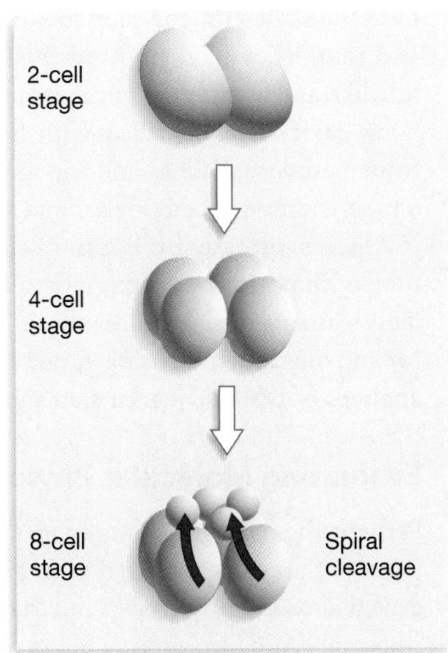

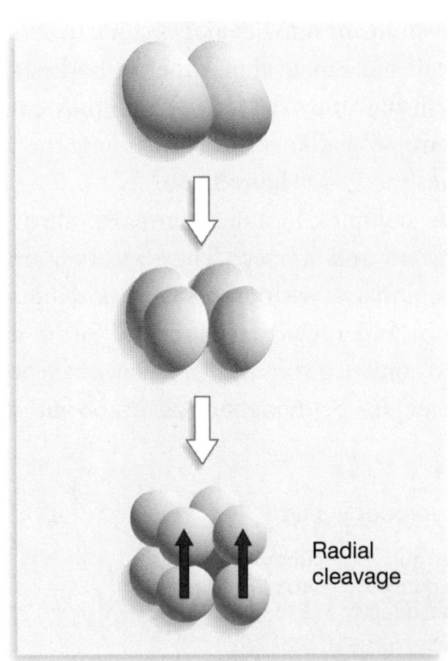

(b) Gastrulation
(mass of cells formed by cleavage is rearranged to form gut and embryonic tissue layers)

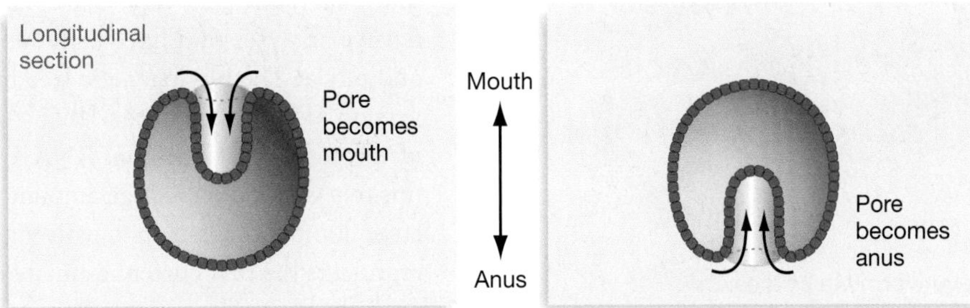

(c) Coelom formation
(body cavity lined with mesoderm develops)

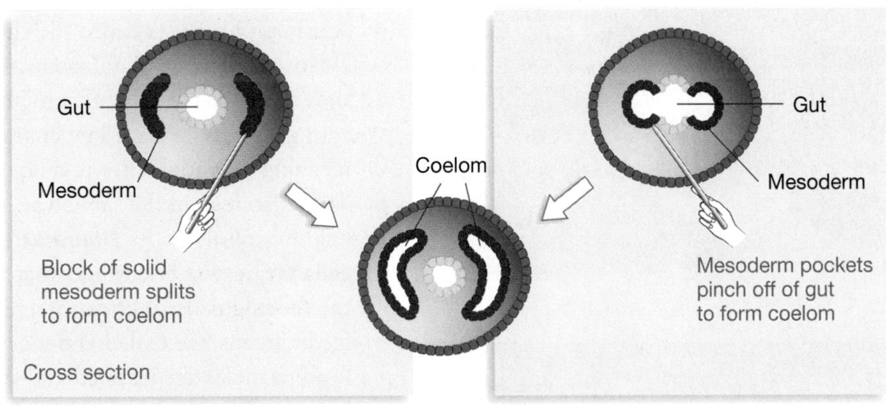

FIGURE 32.8 In Protostomes and Deuterostomes, Three Events in Early Development Differ. The differences between protostomes and deuterostomes show that there is more than one way to build a bilaterally symmetric, coelomate body plan.

protostome and deuterostome patterns of development represent two distinct ways of achieving the same end—the construction of a bilaterally symmetric body that contains a cavity lined with mesoderm.

The Tube-within-a-Tube Design Over 99 percent of the animal species alive today are bilaterally symmetric triploblasts that have coeloms and follow either the protostome or deuterostome pattern of development. This combination of features turned out to be a spectacularly successful way to organize a moving and eating machine.

Although it sounds complex to call a certain animal a "bilaterally symmetric, coelomic triploblast with protostome [or deuterostome] development," the bodies of most animals are actually extremely simple in form. The basic animal body is a tube within a tube. The inner tube is the individual's gut, and the outer tube forms the body wall, as illustrated in **Figure 32.9a**. The mesoderm in between forms muscles and organs. In several animal phyla, individuals have long, thin, tubelike bodies that lack limbs. Animals with this body shape are commonly called **worms**. There are many wormlike phyla, including the nemerteans and sipunculids shown in **Figure 32.9b**.

What about more complex-looking animals, such as grasshoppers and lobsters and horses? They are bilaterally symmetric, coelomic triploblasts with protostome or deuterostome development, too, and they aren't worms. But a moment's reflection should convince you that, in essence, the body plan of these animals can also be thought of as a tube within a tube, except that the tube is mounted on legs. Consider that most animals with complex-looking bodies are relatively long and thin. They have an outer body wall that is more or less tubelike and an internal gut that runs from mouth to anus. The body cavity itself is filled with muscles and organs derived from mesoderm. Wings and legs are just efficient ways to move a tube-within-a-tube body around the environment.

Once evolution by natural selection produced the basic tube-within-a-tube design, most of the diversification of animals was triggered by the evolution of novel types of structures for moving and capturing food. Do data from phylogenetic analyses of DNA sequence data support this view?

(a) The tube-within-a-tube body plan

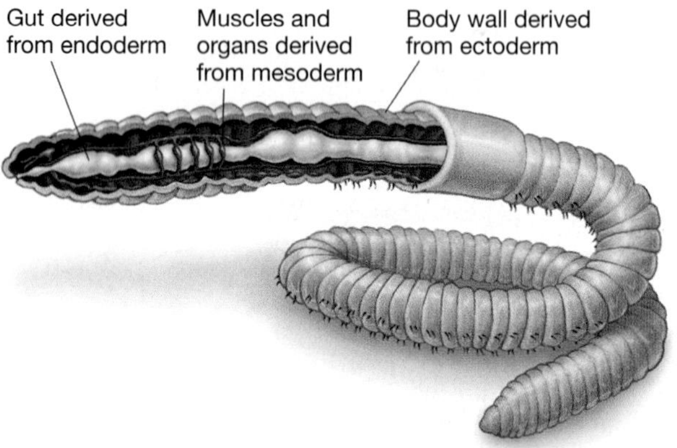

(b) Many animal phyla have wormlike bodies.

Nemertean (ribbon worm)

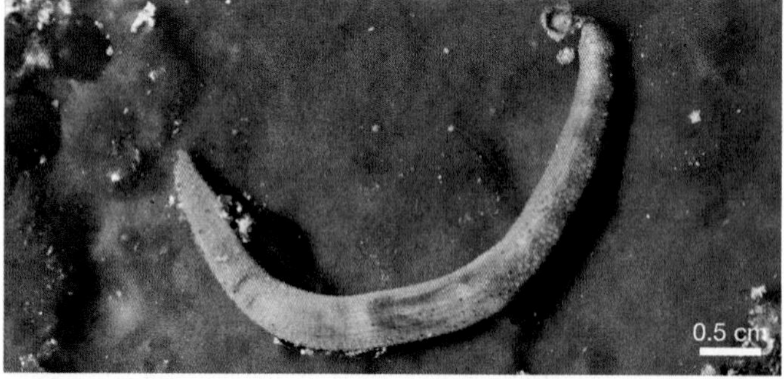

Sipunculid (peanut worm)

FIGURE 32.9 The Tube-within-a-Tube Body Plan Is Common in Animals.

QUESTION How is the term *worm* similar to the term *yeast* (see Chapter 31)?

Evaluating Molecular Phylogenies

Perhaps the most influential paper ever published on the phylogeny of animals appeared in 1997. Using sequences from the gene that codes for the RNA molecule in the small subunit of the ribosome, Anna Marie Aguinaldo and colleagues estimated the phylogeny of species from 14 animal phyla. The results were revolutionary, and they have been verified by more recent and extensive analyses that have included data from additional genes and phyla. The phylogenetic tree in **Figure 32.10** is an updated version of the result of the 1997 study, based on further studies of the genes for ribosomal RNA and several proteins. Because this tree is based on a large amount of sequence data, and thus a large number of traits that evolve independently of each other, it represents the best current estimate of animal phylogeny.

To begin analyzing this tree, start from the root and work your way up. You should note several key points:

- A group of protists called the choanoflagellates are the closest living relatives of animals, and the Porifera (sponges) are the most basal animal phylum. Choanoflagellates and sponges share several key characteristics. Both are **sessile**, meaning that adults live permanently attached to a substrate. They also feed in the same way, using cells with nearly identical morphology. As **Figure 32.11** shows, the beating of flagella creates water currents that bring organic debris toward the feeding cells of choanoflagellates and sponges. Sponge feeding cells are called **choanocytes**. In these feeding cells, food particles are trapped and ingested. The key distinction between the two groups is that sponges are multicellular.
- Sponges are paraphyletic. Stated another way, not all sponges are derived from the same common ancestor. Instead, sponges represent several independent lineages.
- The radially symmetric, diploblastic phyla are placed just up from sponges on the tree, meaning they split off from other animals slightly later than sponges. This conclusion is bolstered by the fossil record of animals, because sponges appear prior to cnidarians and ctenophores. The placement of diploblasts at the base of the tree implies that endoderm and ectoderm were the first tissue types to evolve, and that radial symmetry evolved before bilateral symmetry.

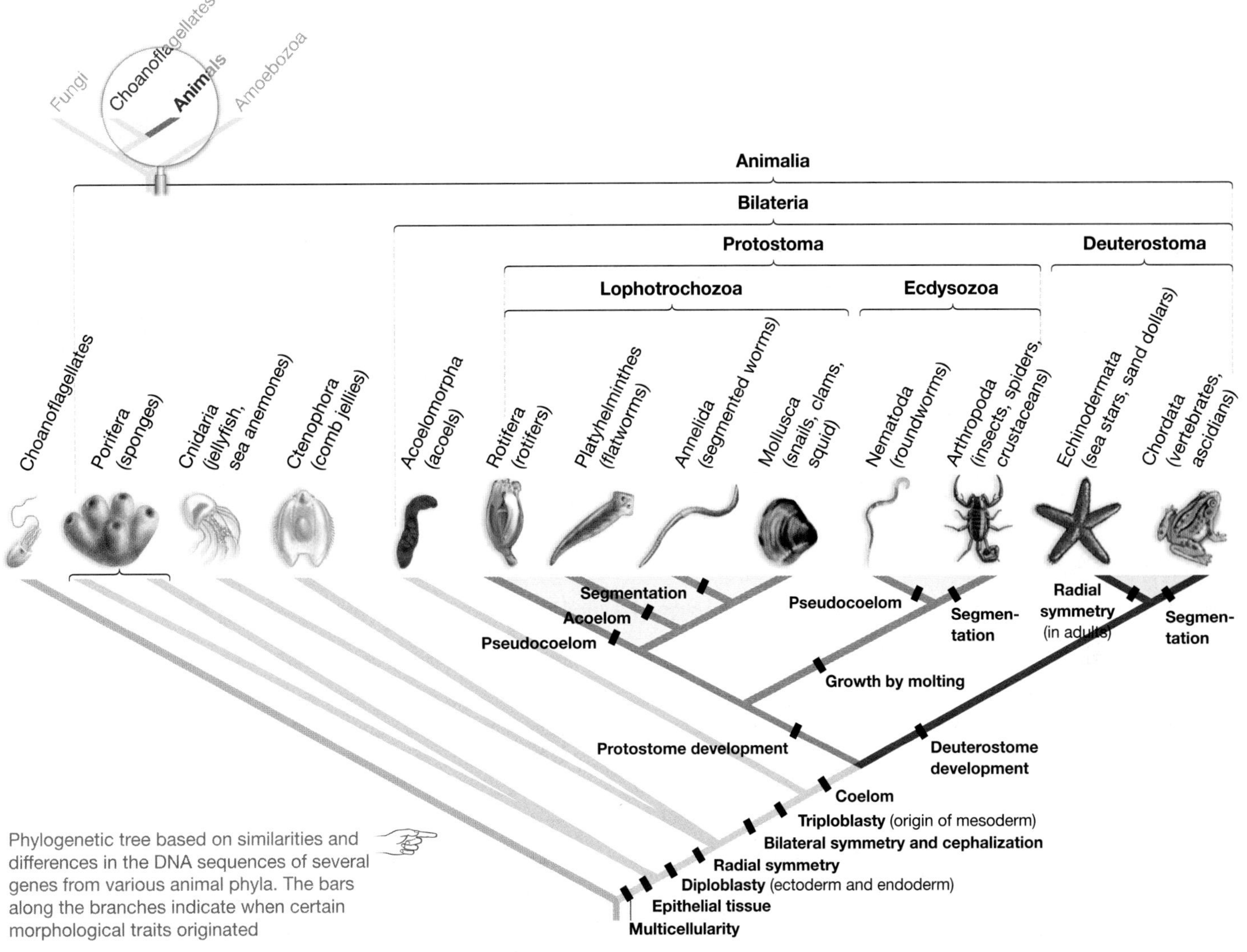

FIGURE 32.10 A Phylogeny of Animal Phyla Based on DNA Sequence Data. Phylogenetic tree based on similarities and differences in the DNA sequences of several genes from various animal phyla. The bars along the branches indicate when certain morphological traits originated.

EXERCISE Recent work suggests that segmentation may also have arisen independently within molluscs. Add a "Segmentation?" label to the tree to indicate this possibility.

- The most ancient groups of bilaterally symmetric triploblasts, the Acoelomorpha, lack a coelom. This result supports an important hypothesis—that animal bodies usually evolved from simpler to more complex forms.
- Based on morphology, biologists had thought that the major event in the evolution of the Bilateria was the split between the protostomes and the deuterostomes. The molecular data concur but show that an additional, equally fundamental split occurred within protostomes, forming two major subgroups with protostome development: (1) The Ecdysozoa (pronounced *eck-die-so-ZOH-ah*) includes the arthropods and the nematodes; (2) the Lophotrochozoa (pronounced *low-foe-tro-ko-ZOH-ah*) includes the molluscs and the annelids. **Ecdysozoans** grow by shedding their external skeletons and expanding their bodies, while **lophotrochozoans** grow by extending the size of their skeletons.
- Species in the phylum Platyhelminthes (flatworms) do not have a coelom but are lophotrochozoans. To interpret this result, biologists point out that platyhelminths had to have evolved from an ancestor that had a coelom. Stated another way, the acoelomate condition in these species is a derived condition. It represents the *loss* of a complex trait, indicated by the "Acoelom" label on the tree. In this case, animals did not evolve from simpler to more complex forms.
- Twice during the course of evolution, bodies with pseudocoeloms arose from ancestors that had "true" coeloms. A change from coelom to pseudocoelom occurred in the ancestors of today's (1) nematodes (roundworms) and (2) rotifers.
- When a body is divided into a series of repeated structures, such as an earthworm's segments or a fish's vertebral column and ribs, it is said to be segmented. Segmentation evolved

(a) Choanoflagellates are sessile protists; some are colonial.

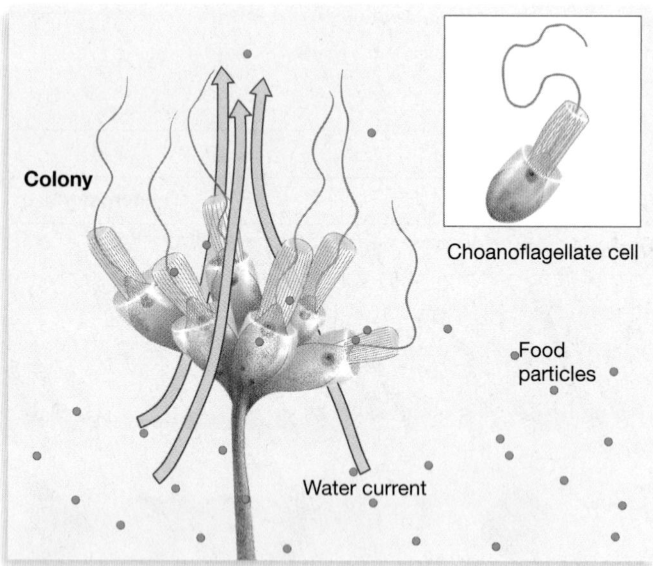

(b) Sponges are multicellular, sessile animals.

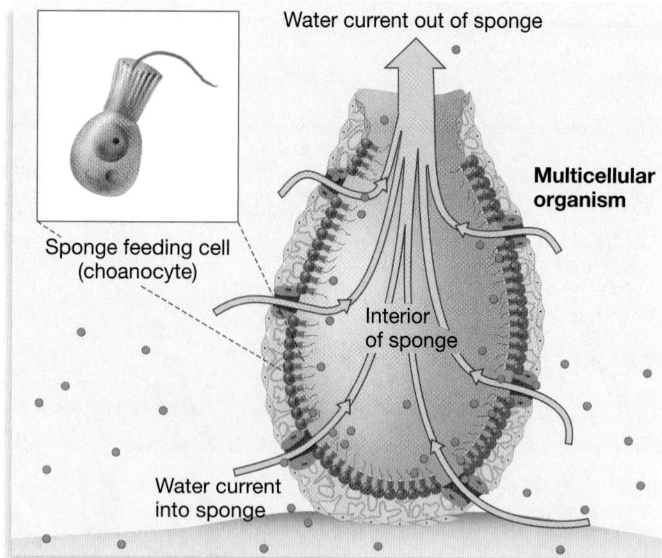

FIGURE 32.11 Choanoflagellates and Sponge Feeding Cells Are Almost Identical in Structure and Function.
(a) Choanoflagellates are suspension feeders. **(b)** A cross section of a simple sponge as it suspension feeds. The beating of flagella produces a water current that brings food into the body of the sponge, where it can be ingested by feeding cells.

independently in annelids (earthworms and other segmented worms) and arthropods (insect, spiders, and crustaceans), as well as in vertebrates. **Vertebrates** are a monophyletic lineage within the Chordata that is defined by the presence of a skull; many vertebrate species also have a backbone (see Chapter 34). The group called the **invertebrates**, which is defined as all animals that are not vertebrates, is paraphyletic—meaning that they include some, but not all, of the descendants of a common ancestor. Recent evidence suggests that segmentation also evolved independently in molluscs (snails, clams, squid).

Although biologists are increasingly confident that most or all of these conclusions are correct, the phylogeny of animals is still very much a work in progress. As data sets expand, it is likely that new analyses will not only confirm or challenge these results but also contribute other important insights into how the most species-rich lineage on the tree of life originated and diversified. Stay tuned.

MB **Web Animation** at www.masteringbio.com
The Architecture of Animals

Check Your Understanding

If you understand that . . .

The origin and early diversification of animals was marked by changes in four fundamental features: body symmetry, the number of embryonic tissues present, the evolution of a body cavity, and protostome versus deuterostome patterns of development.

You should be able to . . .

1) Explain why bilateral symmetry in combination with triploblasty and a coelom is responsible for the "tube-within-a-tube" design observed in most animals living today.
2) Explain why cephalization was important.
3) Make a rough sketch of the phylogeny of animals, showing choanoflagellates as an out-group, sponges and jellyfish as basal groups, Acoelomorpha as the most ancient members of the Bilateria, and then three branches showing the relationships among the Lophotrochozoa, Ecdysozoa, and deuterostomes.
4) On the tree you sketched, mark the origin of multicellularity, triploblasty, bilateral symmetry, protostome development, and deuterostome development.

32.3 What Themes Occur in the Diversification of Animals?

Within each animal phylum, the basic features of the body plan do not vary from species to species. For example, molluscs are triploblastic, bilaterally symmetric protostomes with a reduced coelom; their body plan features a muscular foot, a cavity called the visceral mass, and a structure called a mantle. But there are over 100,000 species of mollusc. If the major animal lineages are defined by a particular body plan, what triggered the diversification of species within each lineage?

In most cases, the answer to this question is the evolution of innovative methods for feeding and moving. Recall that most animals get their food by ingesting other organisms. Animals are diverse because there are thousands of ways to find and eat the millions of different organisms that exist.

(a) Krill filter feed using their legs.

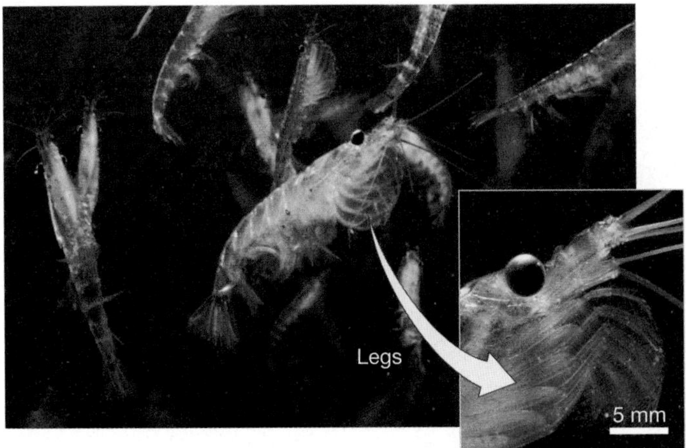

(b) Baleen whales filter feed using their baleen.

FIGURE 32.12 Suspension Feeders Harvest Suspended Food. (a) Krill and **(b)** baleen whales filter food particles from water, using the trapping structures shown in the close-ups.

Feeding

To organize the diversity of ways that animals find food, biologists distinguish *how* individuals eat from *what* they eat. In terms of how they eat, animals obtain food in four general ways: suspension feeding, deposit feeding, fluid feeding, and feeding on food masses. In terms of what they eat, animals have three general sources of food: plants or algae, other animals, and detritus—dead organic material. In many cases, species that pursue different food-getting strategies and food sources are found within a single lineage. Let's review the diversity of ways that animals eat, then consider the diversity of ways that animals move around to find food.

How Animals Feed: Four General Tactics Two simple ideas are key to understanding how animals from the same lineage can have the same basic body plan but feed in radically different ways: (1) Their mouthparts vary, and (2) the structure of an animal's mouthparts correlates closely with its method of feeding. Keep these concepts in mind as you review the general tactics that animals use to obtain food.

- **Suspension feeders**, also known as **filter feeders**, capture food by filtering out particles suspended in water or air. **Figure 32.12a** shows the small marine animals called krill, which suspension feed as they swim. As individuals move forward, their legs wave in and out. Projections on their legs trap food particles that flow past. The food particles are then moved up the body to the mouth, where they are ingested. Krill, in turn, are eaten by suspension feeders called the baleen whales **(Figure 32.12b)**. These whales have a series of long plates, made from a horny material called baleen, hanging from their jaws. Baleen whales feed by gulping water that contains krill, squeezing the water out between their baleen plates, and trapping the krill inside their mouths. Suspension feeders employ a wide array of structures to trap suspended particles—usually small algae or animals or bits of detritus—and bring them to their mouths.
- **Deposit feeders** eat their way through a substrate. Earthworms, for example, are annelids that swallow soil as they tunnel through it **(Figure 32.13a)**. Many deposit feeders digest organic matter in the soil; their food consists of soil-dwelling

(a) Earthworms (Annelida) eat their way through soil.

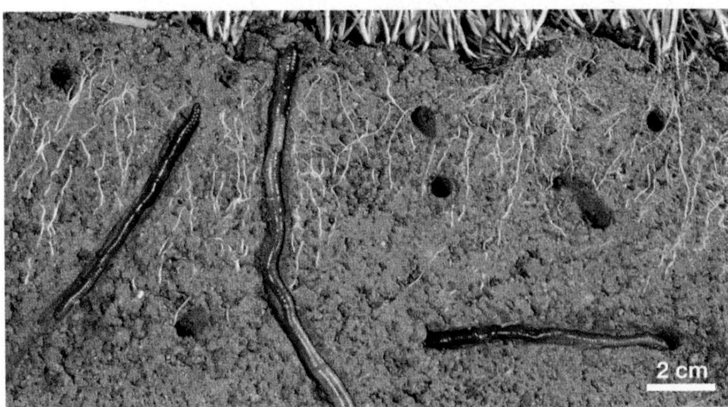

(b) Sea cucumbers eat their way through organic deposits.

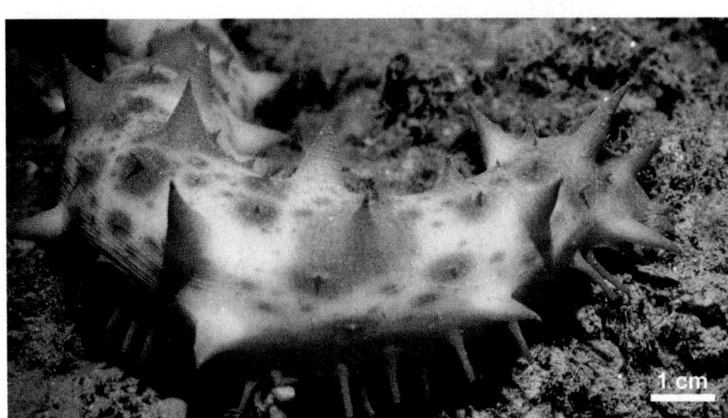

FIGURE 32.13 Deposit Feeders Eat Their Way through a Substrate.

QUESTION Why is it logical that deposit feeders tend to have tubelike bodies?

(a) Butterflies have an extensible, hollow proboscis.

(b) Blowflies have an extensible, sponge-like mouthpart.

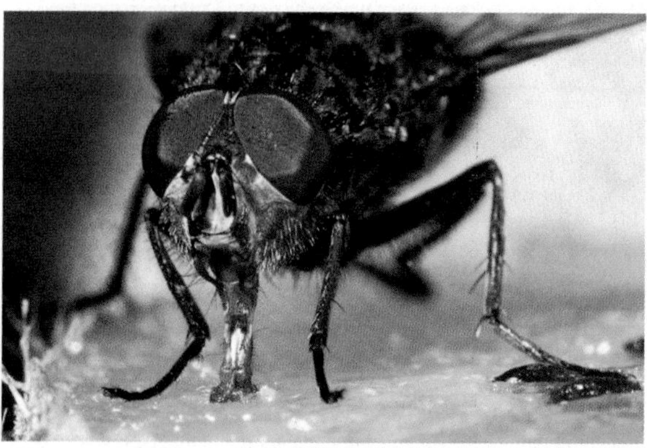

FIGURE 32.14 Fluid Feeders Drink Liquids.

QUESTION Hummingbirds, ticks, and mosquitoes are also fluid feeders. How does the structure of their mouthparts correlate with function in feeding?

bacteria, archaea, protists, and fungi, along with detritus. Insects that burrow through plant leaves and stems, bore through piles of feces, or mine the carcasses of dead animals or plants can also be considered deposit feeders because they eat through a substrate. Perhaps the most common and important deposit feeders, though, are the echinoderms called sea cucumbers **(Figure 32.13b)**. The floor of the sea is rich in organic matter that rains down from the surface and collects in food-rich deposits, which sea cucumbers exploit.

Unlike suspension feeders, which are diverse in size and shape and use various trapping or filtering systems, deposit feeders are similar in appearance. They usually have simple mouthparts if they eat soft substrates, and their body shape is wormlike. Like suspension feeders, however, deposit feeders occur in a wide variety of lineages including roundworms (Nematoda), molluscs (Mollusca), peanut worms (Sipunculida), and chordates such as hagfish.

- **Fluid feeders** suck or mop up liquids like nectar, plant sap, blood, or fruit juice. They range from butterflies that feed on nectar with a straw-like proboscis (**Figure 32.14a**) to blowflies that feed on rotting fruit using a sponge-like mouthpart (**Figure 32.14b**). They are found in a wide array of lineages and often have mouthparts that allow them to pierce seeds, stems, skin, or other structures in order to withdraw the fluids inside.
- **Mass feeders** take chunks of food into their mouths. In these species, the structure of the mouthparts correlates with the type of food pieces that are harvested and ingested. Horses, for example, have sharp teeth in the front of the jaw for biting off grass stems and broad, flat molar teeth in the back of the jaw for mashing the coarse stems into a soft wad that can be swallowed (**Figure 32.15a**). In many snails, a feeding structure called a **radula** functions like a rasp or a file. The sharp plates on the radula move

(a) Horses feed on grass stems.

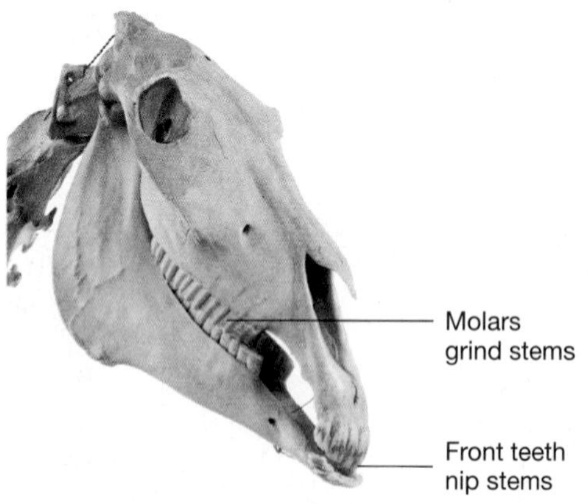

(b) Terrestrial snails feed on leaves.

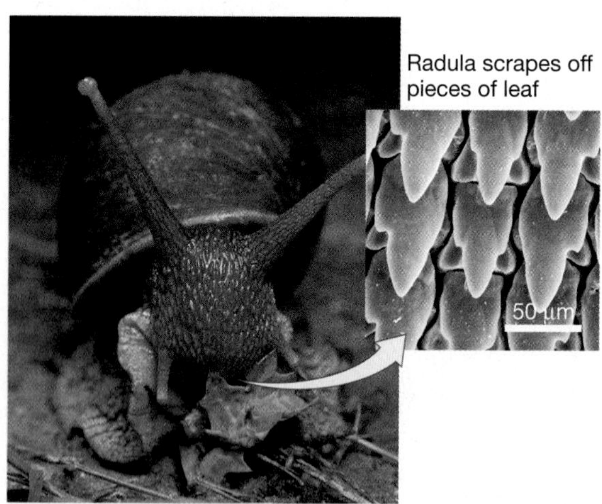

FIGURE 32.15 Food-Mass Feeders Ingest Bites or Lumps of Food.

QUESTION Humans, dogs, and eagles are also food-mass feeders. How does the structure of their mouthparts correlate with function in feeding?

(a) Many frogs (Chordata) sit and wait for prey.

(b) Wolves (Chordata) chase prey.

FIGURE 32.16 Predators Kill and Eat Organisms.

back and forth to scrape material away from a plant or alga so that it can be ingested (Figure 32.15b).

What Animals Eat: Three General Sources Whether they feed by filtering, eating through deposits, or taking in fluids or masses, animals can be classified as **herbivores** that feed on plants or algae, **carnivores** that feed on animals, or **detritivores** that feed on dead organic matter. In addition, herbivores and carnivores can be sub-classified as predators or parasites; animals that eat both plants and animals are called **omnivores**.

Predators kill other organisms for food, using an array of mouthparts and hunting strategies. Many types of frogs, for example, are sit-and-wait predators. They sit still and wait for an insect or worm to move close, then capture it with a lightning-quick extension of their long, sticky tongue (**Figure 32.16a**). In contrast, wolves hunt by locating a prey organism and then running it down during an extended, long-distance chase (**Figure 32.16b**). Mountain lions stalk their prey slowly, then pounce on it or run it down in a short sprint. Although most predators kill other animals to live, species that eat seeds are commonly referred to as seed predators.

Unlike predators, **parasites** are usually much smaller than their victims and often harvest nutrients without causing death. **Endoparasites** live inside their hosts and usually have simple, wormlike bodies. Tapeworms, for example, are platyhelminths with no digestive system. Instead of a mouth, they have hooks or other structures on their head, called a scolex, that attach to their host's intestinal wall (**Figure 32.17a**). Instead of digesting food themselves, they absorb nutrients directly from their surroundings. **Ectoparasites,** in contrast, live outside their hosts. They usually have limbs or mouthparts that allow them to grasp the host and mouthparts that allow them to pierce their host's skin and suck the nutrient-rich fluids inside. The louse in **Figure 32.17b** is an example of an insect (Arthropoda) ectoparasite that afflicts humans.

(a) Tapeworms (Platyhelminthes) are endoparasites.

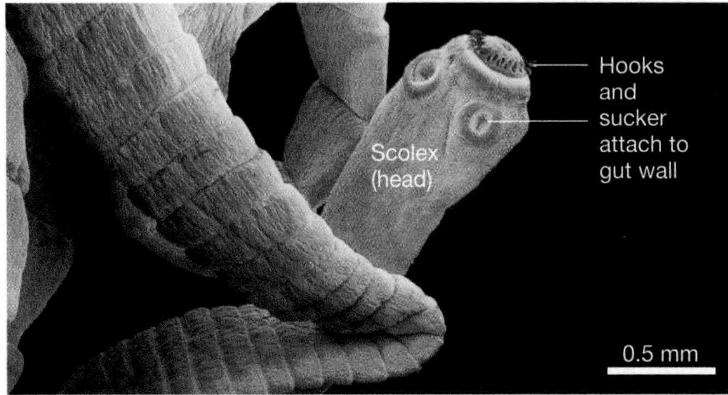

(b) Lice (Arthropoda) are ectoparasites.

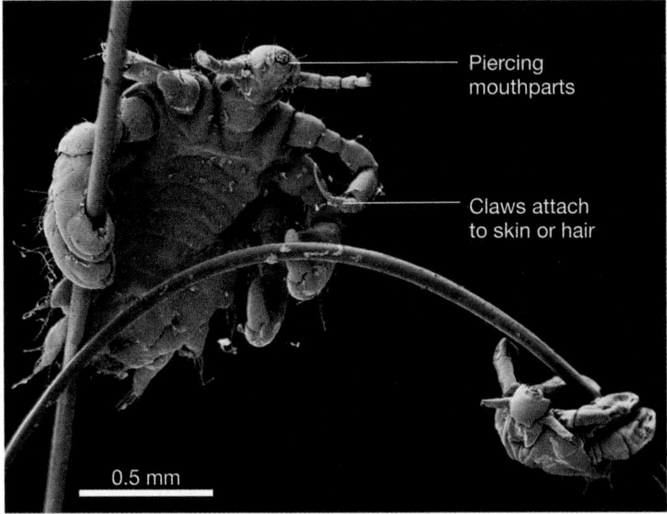

FIGURE 32.17 Parasites Take Nutrients from Living Animals.
(a) Tapeworms are common intestinal parasites of humans and other vertebrates. They attach to the wall of the digestive tract, using the barbed hooks and suckers on their anterior end, and absorb nutrition directly across their body wall. **(b)** Lice are insects that parasitize mammals and birds. This louse, *Phthirus pubis,* uses its clawlike legs to attach to the pubic region of humans. The animal pierces the host's skin with its mouthparts and feeds by sucking body fluids.

(a) Motile larval anemone

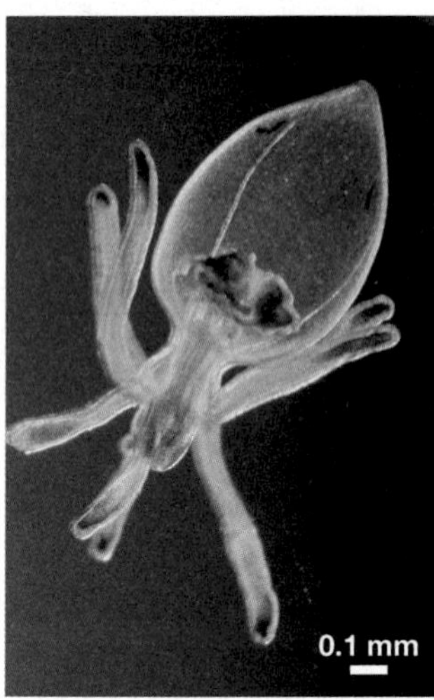

(b) Sessile adult anemone

FIGURE 32.18 If Adults Are Sessile, Then Larvae Disperse. (a) Larval anemones swim under their own power and can disperse to new habitats. **(b)** Anemones are sessile most of their lives.

Movement

Many animals are sit-and-wait predators, and some are sessile throughout their adult lives. But the vast majority of animals move under their own power either as a juvenile or as an adult. For example, the eggs of sea anemones hatch into larvae that swim (**Figure 32.18a**). Adult sea anemones, however, spend most of their lives attached to a rock or another substrate and make their living by capturing and eating fish or other organisms that pass by (**Figure 32.18b**). In species such as these, larvae function as a dispersal stage. They are a little like the seeds of land plants—a life stage that allows individuals to move to new habitats, where they will not compete with their parents for space and other resources.

In animals that move as adults, locomotion has three functions: (1) finding food, (2) finding mates, and (3) escaping from predators. The ways that animals move in search of food and sex are highly variable; they burrow, slither, swim, fly, crawl, walk, or run. The structures that power movement are equally variable—they include cilia, flagella, and muscles that attach to a hard skeleton or compress a hydrostatic skeleton, enabling wriggling movements. The hydrostatic skeleton is an evolutionary innovation unique to animals and is responsible for locomotion in the many animal phyla with wormlike bodies. Another major innovation occurred in animals, however, that made highly controlled, rapid movement possible: the limb.

Types of Limbs: Unjointed and Jointed Limbs are a prominent feature of species in many phyla and are particularly important in two major lineages: the ecdysozoans and the vertebrates. Some members of the Ecdysozoa, such as onychophorans (velvet worms), have unjointed, sac-like limbs (**Figure 32.19a**); others, such as crabs and other arthropods, have more complex, jointed limbs (**Figure 32.19b**). Jointed limbs make fast, precise movements possible and are a prominent type of limb in vertebrates and arthropods.

(a) Onychophorans are ecdysozoans with sac-like limbs.

(b) Crabs (Arthropoda) have jointed limbs.

(c) Polychaetes (Annelida) have parapodia.

(d) Sea urchins (Echinodermata) have tube feet.

FIGURE 32.19 Various Animal Appendages Function in Locomotion. Lineages within the Ecdysozoa have **(a)** sac-like legs or **(b)** jointed limbs. **(c)** Some species in the Lophotrochozoa have small projections called parapodia. **(d)** Echinoderms have unusual structures called tube feet.

Are All Animal Appendages Homologous? Chapter 24 introduced the concept of homology, which is defined as similarity in traits due to inheritance from a common ancestor. Traditionally, biologists have hypothesized that the major types of jointed and unjointed animal limbs were not homologous. To appreciate the logic behind this hypothesis, it's important to recognize just how diverse animal appendages are. Animals in a wide array of phyla have structures that stick out from the main body wall and function in locomotion. These appendages range from the human arm to bristle-like structures called parapodia in segmented worms (**Figure 32.19c**) and the soft, extensible tube feet found in echinoderms (**Figure 32.19d**). Because the structure of animal appendages is so diverse, it was logical to maintain that at least some appendages evolved independently of each other. As a result, biologists predicted that completely different genes are responsible for each major type of appendage.

Recent results have challenged this view, however. The experiments in question involve a gene called *Distal-less,* which was originally discovered in fruit flies. (*Distal* means "away from the body.") *Distal-less,* or *Dll,* is aptly named. In fruit flies that lack this gene's normal protein product, only the most rudimentary limb buds form. The mutant limbs are "distal-less." Based on the morphology of *Dll* mutants, the protein seems to deliver a simple message as a fruit-fly embryo develops: "Grow appendage out this way."

A group of biologists working in Sean Carroll's lab set out to test the hypothesis that *Dll* might be involved in limb or appendage formation in other animals. As **Figure 32.20** shows, they used a fluorescent marker that sticks to the *Dll* gene product to locate tissues where the gene is expressed. When they introduced the fluorescent marker into embryos from annelids, arthropods, echinoderms, chordates, and other phyla, they found that it bound to *Dll* in all of them. More important, the highest concentrations of *Dll* gene products were found in cells that form appendages—even in phyla with wormlike bodies that have extremely simple appendages. Other experiments have shown that *Dll* is also involved in limb formation in vertebrates.

Based on these findings, biologists are concluding that at least a few of the same genes are involved in the development of *all* appendages observed in animals. To use the vocabulary introduced in Chapter 24, the hypothesis is that all animal appendages have some degree of genetic homology and that they are all derived from appendages that were present in a common ancestor. The idea is that a simple appendage evolved early in the history of the Bilateria and that, subsequently, evolution by natural selection produced the diversity of limbs, antennae, and wings observed today. This hypothesis is controversial, however, and research continues at a brisk pace.

Reproduction and Life Cycles

An animal may be efficient at moving and eating, but if it does not reproduce, the alleles responsible for its effective locomotion

Experiment

Question: Is the gene *Dll* involved in limb formation in species other than insects?

Hypothesis: In all animals, *Dll* signals "grow appendage out here."

Null hypothesis: *Dll* is not involved in the development of appendages in species other than insects.

Experimental setup:

Add a stain to developing embryos that will attach to *Dll* gene products (proteins), revealing their location. The stain can be fluorescent green or dark brown.

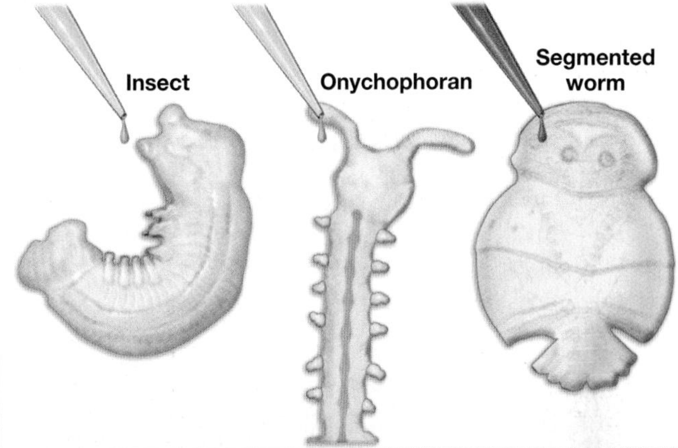

Prediction: In embryos from a wide array of species, stained *Dll* gene products will be localized to areas where appendages are forming.

Prediction of null hypothesis: Stained *Dll* gene products will be localized to areas where appendages are forming only in insects.

Results:

Insect

Developing legs

Onychophoran

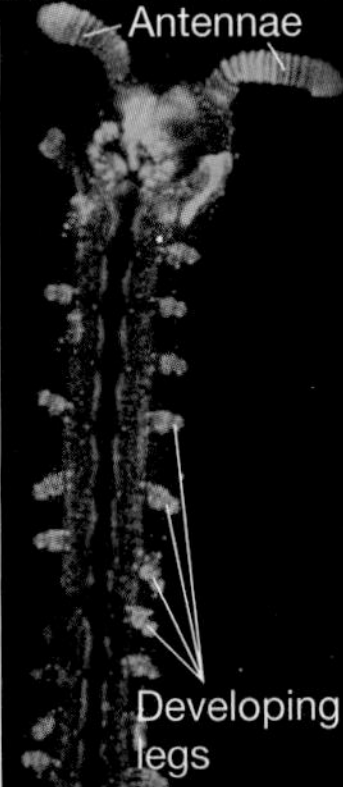

Segmented worm

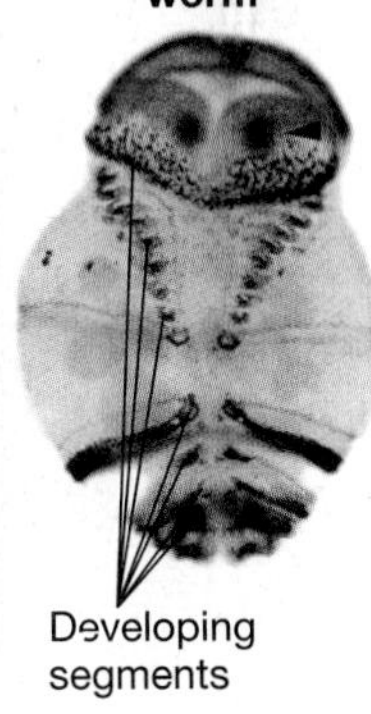

In species representing both Ecdysozoa and Lophotrochozoa, *Dll* is localized in areas of the embryo where appendages are forming.

Conclusion: The gene *Dll* is involved in limb formation in diverse species. The results suggest that all animal appendages may be homologous.

FIGURE 32.20 Experimental Evidence That All Animal Appendages Are Homologous.

● **QUESTION** What results would have supported the null hypothesis?

and feeding will not increase in frequency in the population. As Chapter 24 emphasized, natural selection occurs when individuals with certain alleles produce more offspring than other individuals do. Organisms live to reproduce.

Given the array of habitats and lifestyles pursued by animals, it is not surprising that they exhibit a high degree of variation in how they reproduce. Although animal reproduction will be explored in detail in Chapter 48, a few examples will help drive home just how variable animal reproduction is:

- At least some species in most animal phyla can reproduce asexually, via mitosis, as well as sexually—via meiosis and fusion of gametes. In the lophotrochozoan phylum Rotifera, an entire lineage called the bdelloids (pronounced *DELL-oyds*) reproduces only asexually. Even certain fish, lizard, and snail species have never been observed to undergo sexual reproduction.

- When sexual reproduction does occur, fertilization may be internal or external. When internal fertilization takes place, males typically insert a sperm-transfer organ into the body of a female (**Figure 32.21a**). In some cases, males produce sperm in packets, which females then pick up and insert into their own bodies. But in seahorses, females insert eggs into the male's body, where they are fertilized. (The male is pregnant for a time and then gives birth to live young.) External fertilization is extremely common in aquatic species. Females lay eggs onto a substrate or into open water. Males then shed sperm on or near the eggs (**Figure 32.21b**).

- Eggs or embryos may be retained in the female's body during development, or eggs may be laid outside to develop independently of the mother. Mammals and other species that nourish embryos inside the body and give birth to live young are said to be **viviparous** ("live-bearing"; **Figure 32.22a**); species that deposit fertilized eggs are **oviparous** ("egg-bearing"; **Figure 32.22b**); and some species are **ovoviviparous** ("egg-live-

(a) Internal fertilization

(b) External fertilization

FIGURE 32.21 Fertilization Can Be Internal or External in Animals. **(a)** Internal fertilization in dragonflies. **(b)** A giant clam emitting sperm that will fertilize a clutch of eggs released by a different individual.

(a) Viviparity ("live-bearing"): the birth of a shark

(b) Oviparity ("egg-bearing")

FIGURE 32.22 Some Animal Species Give Birth to Live Young, but Most Lay Eggs. **(a)** A lemon shark is viviparous, meaning its embryos develop for a period of time inside the female's body. **(b)** A corn snake is oviparous, meaning it lays fertilized eggs.

bearing"). In ovoviviparous species, the females retain eggs inside their body during early development; but the growing embryos are nourished by yolk inside the egg and not by nutrients transferred directly from the mother, as in viviparous species. Ovoviviparous females then give birth to well-developed young. Mammals and a few species of sea stars, onychophorans, fish, and lizards are viviparous; some snails, insects, reptiles, fishes, and sharks are ovoviviparous. But the vast majority of animals are oviparous.

Besides reproducing in a variety of ways, animal life cycles vary widely. Perhaps the most spectacular innovation in animal life cycles involves the phenomenon known as **metamorphosis** ("change-form")—a change from a juvenile to an adult body type. Juveniles are sexually immature, meaning their reproductive organs are undeveloped and the individual cannot breed. Adults are the reproductive stage in the life cycle.

Metamorphosis occurs in two basic ways. In **holometabolous** ("whole change") **metamorphosis**, the juvenile form is called a **larva** (plural: **larvae**) and looks substantially different from the adult form. Mosquito larvae, for example, live in quiet bodies of freshwater, where they suspension feed on bacteria, algae, and detritus (**Figure 32.23a,** top). When a larva has grown sufficiently, the individual stops feeding and moving and secretes a protective case. The individual is now known as a **pupa** (plural: **pupae**; Figure 32.23a, middle). During **pupation**, the pupa's body is completely remodeled into a new, adult form. In mosquitoes, the adult individual flies and gets its nutrition as a parasite—taking blood meals from mammals and sucking fluids from plants (Figure 32.23a, bottom). Holometabolous metamorphosis is also called complete metamorphosis. The transformation of butterfly and moth caterpillars to flying adults is another spectacular example of holometabolous metamorphosis.

In **hemimetabolous** ("half-change") **metamorphosis**, the juvenile form is called a **nymph** and looks like a miniature version of the adult (**Figure 32.23b**, top). For example, a grasshopper nymph sheds its external skeleton several times and grows—gradually changing from a wingless, sexually

(a) Mosquito: **Holo**metabolous metamorphosis

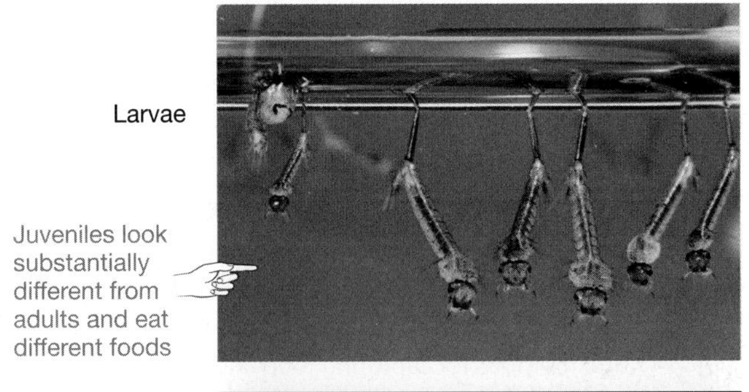

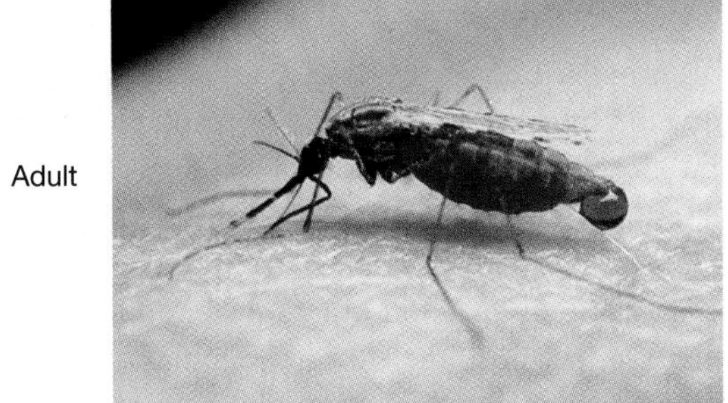

(b) Grasshopper: **Hemi**metabolous metamorphosis

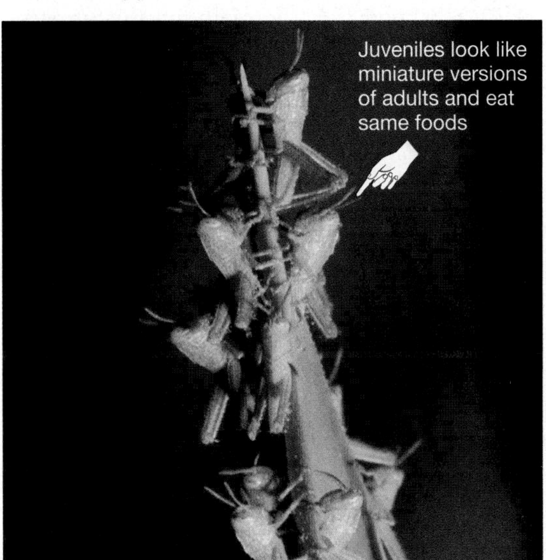

FIGURE 32.23 During Metamorphosis, Individuals May or May Not Change Form Completely. (a) When holometabolous metamorphosis occurs, juvenile forms are called larvae. During pupation, the juvenile body is remodeled into the adult form. Adult forms are sexually mature. (b) When hemimetabolous metamorphosis occurs, juvenile forms are called nymphs. Adult forms are sexually mature.

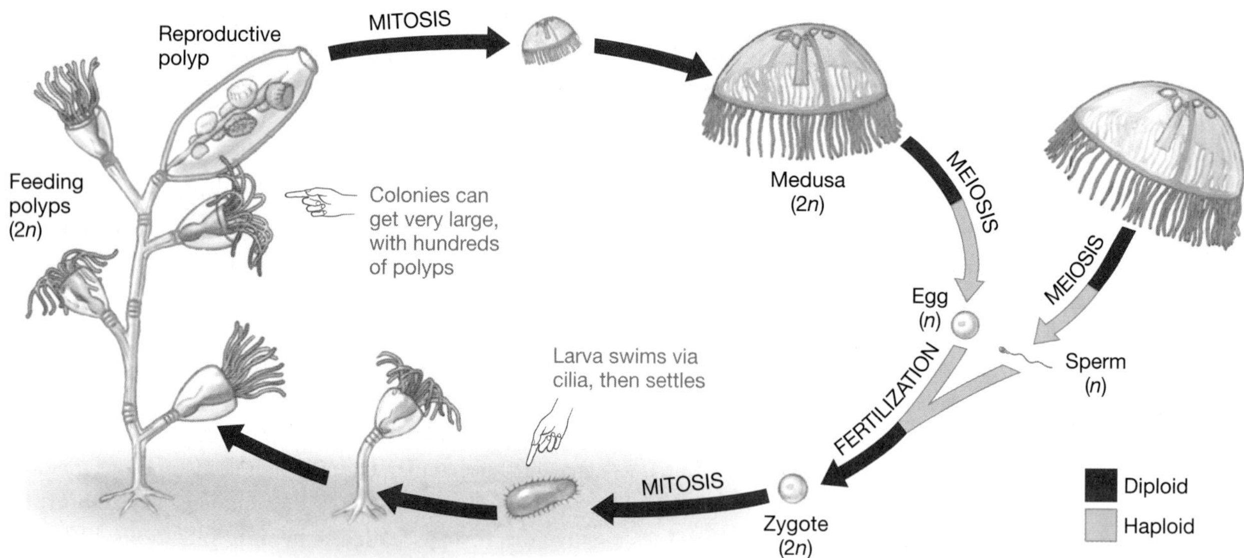

FIGURE 32.24 Cnidarian Life Cycles May Include a Polyp and Medusa Form. This is an example of a common life cycle in jellyfish.

QUESTION Does alternation of generations occur in this species? Explain why or why not.

immature nymph to a sexually mature adult that is capable of flight (Figure 32.23b, bottom). Throughout this process, grasshoppers feed on the same food source in the same way: They chew leaves. Hemimetabolous metamorphosis is also called incomplete metamorphosis.

In insects, holometabolous metamorphosis is 10 times more common than hemimetabolous metamorphosis. One hypothesis to explain this observation is based on efficiency in feeding. Because juveniles and adults from holometabolous species feed on different materials in different ways and sometimes even in different habitats, they do not compete with each other. When complete metamorphosis is part of the life cycle, individuals of the same species but different ages often show dramatic variations in their mode of feeding. An alternative hypothesis to explain the evolutionary success of complete metamorphosis is based on specialization in feeding and mating. In many moths and butterflies, for example, larvae are specialized for feeding, whereas adults are specialized for mating. Larvae are largely sessile, whereas adults are highly mobile. If specialization leads to higher fitness, then complete metamorphosis would be advantageous. These hypotheses are not mutually exclusive, however—meaning that both could be correct—and they are still being tested.

Complete metamorphosis is also extremely common in marine animals. For example, some cnidarians have two distinct body types during their life cycle: (1) A largely sessile form called a **polyp** alternates with (2) a free-floating stage called a **medusa** (plural: **medusae**; **Figure 32.24**). Polyps usually live attached to a substrate, suspension feed on detritus, and frequently form large clusters of individuals called colonies. A **colony** is a group of identical individuals that are physically attached. Medusae, in contrast, float freely in the plankton and feed on crustaceans. Because polyps and medusae live in different habitats, the two stages of the life cycle exploit different food sources.

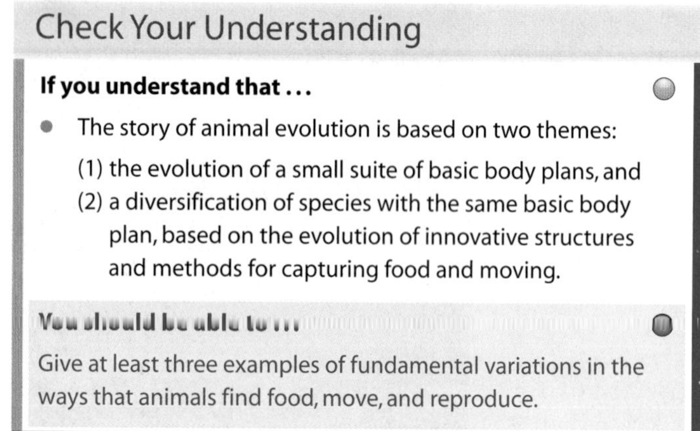

Check Your Understanding

If you understand that ...

- The story of animal evolution is based on two themes:
  (1) the evolution of a small suite of basic body plans, and
  (2) a diversification of species with the same basic body plan, based on the evolution of innovative structures and methods for capturing food and moving.

You should be able to ...

Give at least three examples of fundamental variations in the ways that animals find food, move, and reproduce.

32.4 Key Lineages of Animals: Basal Groups

The goal of this chapter is to provide a broad overview of how animals diversified into such a morphologically diverse lineage. According to recent phylogenetic analyses, the closest living relatives to the animals are the choanoflagellates, a group of protists. Phylogenetic analyses and the fossil record also indicate that the phyla Porifera (sponges), Cnidaria (jellyfish and others), Ctenophora, and Acoelomorpha are the most ancient of all animal groups (**Figure 32.25**). Let's explore the origins of animals by taking a more detailed look at each of the most basal lineages. The protostomes and deuterostomes are explored in Chapters 33 and 34, respectively.

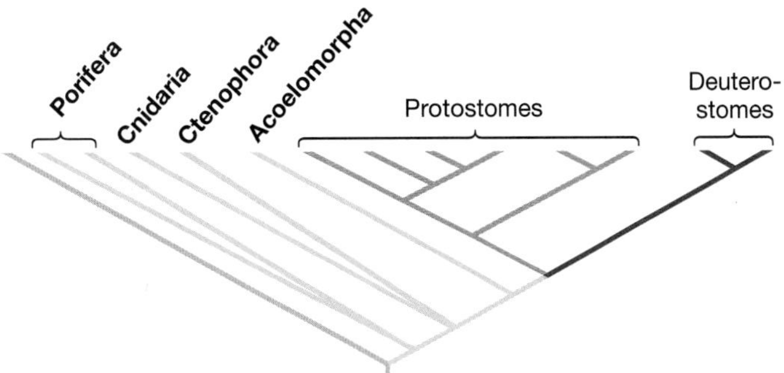

FIGURE 32.25 Animals Are a Monophyletic Group. Use this tree to map the origin of new features discussed in the following key lineage boxes.

EXERCISE Write in the name of the out-group to animals.

Porifera (Sponges)

About 7000 species of sponges have been described to date. Although a few freshwater species are known, most are marine. All sponges are **benthic**, meaning that they live at the bottom of aquatic environments. Sponges are particularly common in rocky, shallow-water habitats of the world's oceans and in coastal multicellularity and epithelial tissue areas of Antarctica.

The architecture of sponge bodies is built around a system of tubes and pores that create channels for water currents. Body symmetry varies among sponge species; most are asymmetrical, but some species are radially symmetric (**Figure 32.26**). Sponges have specialized cell types and well-organized epithelial tissue layers lining the inside and outside of the body, sealing a middle layer in between. You should be able to indicate the origin of multicellularity and epithelial tissue on Figure 32.25. They lack the other tissue types found in other animals, however. The "looseness" of sponge tissues is thought to be an adaptation that facilitates regeneration after wounding, the ability to remodel the body if water currents change, and the ability to move into and colonize new substrates as they become available. In many species, either flexible collagen fibers or **spicules**—stiff spikes of silica or calcium carbonate ($CaCO_3$)—provide structural support for the body. One sponge species native to the Caribbean can grow to heights of 2 m.

Sponges have commercial and medical value to humans. The dried bodies of certain sponge species are able to hold large amounts of water and thus are prized for use in bathing and washing. In addition, researchers are increasingly interested in the array of toxins that sponges produce to defend themselves against predators and bacterial parasites. Some of these compounds have been shown to have antibacterial properties or to promote wound healing in humans.

Feeding Most sponges are suspension feeders. Their cells beat in a coordinated way to produce a water current that flows through small pores in the outer body wall, into chambers inside the body, and out through a single larger opening. As water passes by feeding cells, organic debris and bacteria, archaea, and small protists are filtered out of the current and then digested.

Movement Most adult sponges are sessile, though a few species are reported to move at rates of up to 4 mm per day. Most species produce larvae that swim with the aid of cilia.

Reproduction Asexual reproduction occurs in a variety of ways, depending on the species. Sponge cells are totipotent, meaning that an isolated adult cell has the capacity to develop into a complete adult organism. Thus, any fragment that breaks off an adult sponge has the potential to grow into a new individual. Although individuals of most species produce both eggs and sperm, self-fertilization is rare because individuals release their male and female gametes at different times. Fertilization usually takes place in the water, but some ovoviviparous species retain their eggs and then release mature, swimming larvae after fertilization and early development have occurred.

Pseudoceratina crassa

FIGURE 32.26 Some Sponges Form Radially Symmetric Tubes.

Cnidaria (Jellyfish, Corals, Anemones, Hydroids)

Although a few species of Cnidaria inhabit freshwater, the vast majority of the 11,000 species are marine. They are found in all of the world's oceans, occupying habitats from the surface to the substrate, and are important predators. The phylum comprises four main lineages: Hydrozoa (hydroids—see Figure 32.4), Cubozoa (box jellyfish), Scyphozoa (jellyfish—see the chapter opening photo and Figure 32.24), and Anthozoa (anemones, corals, and sea pens—see Figure 32.3b).

Cnidarians are radially symmetric diploblasts consisting of ectoderm and endoderm layers that sandwich gelatinous material known as **mesoglea**, which contains a few scattered ectodermal cells. They have a gastrovascular cavity instead of a flow-through gut—meaning there is only one opening to the environment for both ingestion and elimination of wastes.

Most cnidarians have a life cycle that includes both a sessile polyp form (**Figure 32.27a**) and a free-floating medusa (**Figure 32.27b**). Anemones, hydroids, and coral, however, exist only as polyps—never as medusae. Reef-building corals secrete outer skeletons of calcium carbonate that create the physical structure of a coral reef—one of the world's most productive habitats (see Chapter 54).

Feeding The morphological innovation that triggered the diversification of the cnidarians is a specialized cell, called a **cnidocyte**, which is used in prey capture. When cnidocytes brush up against a fish or other type of prey, the cells forcibly eject a barbed, spear-like structure called a nematocyst, which is coated with toxins. The barbs hold the prey, and the toxins subdue it until it can be brought to the mouth and ingested. Cnidocytes are commonly located near the mouths of cnidarians or on elongated structures called tentacles. Cnidarian toxins can be deadly to humans as well as to prey organisms; in Australia, twice as many people die each year from stings by box jellyfish as from shark attacks. ● You should be able to indicate the origin of cnidocytes on Figure 32.25. Besides capturing prey actively, most species of coral and many anemones host photosynthetic dinoflagellates. The relationship is mutually beneficial, because the protists supply the cnidarian host with food in exchange for protection.

Movement Both polyps and medusae have simple, muscle-like tissue derived from ectoderm or endoderm. In polyps, the gut cavity acts as a hydrostatic skeleton that works in conjunction with the muscle-like cells to contract or extend the body. Many polyps can also creep along a substrate, using muscle cells at their base. In medusae, the bottom of the bell structure is ringed with muscle-like cells. When these cells contract rhythmically, the bell pulses and the medusa moves by jet propulsion—meaning a forcible flow of water in the opposite direction of movement. Cnidarian larvae swim by means of cilia.

Reproduction Polyps may produce new individuals asexually by (1) budding, in which a new organism grows out from the body wall of an existing individual; (2) fission, in which an existing adult splits lengthwise to form two individuals; or (3) fragmentation, in which parts of an adult regenerate missing pieces to form a complete individual. During sexual reproduction, gametes are usually released from the mouth of a polyp or medusa and fertilization takes place in the open water. Eggs hatch into larvae that become part of the plankton before settling and developing into a polyp.

(a) Polyps attach to substrates.

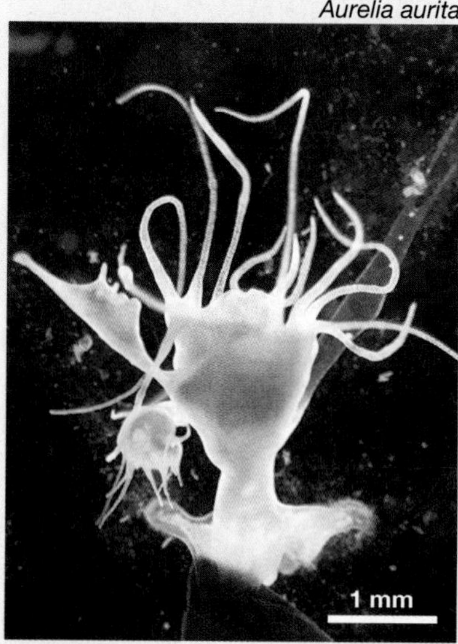

(b) Medusae float near the water surface.

FIGURE 32.27 Most Cnidarians Have Polyp and Medusa Stages.

Ctenophora (Comb Jellies)

Ctenophores are transparent, ciliated, gelatinous diploblasts that live in marine habitats (**Figure 32.28**). Although a few species live on the ocean floor, most are planktonic—meaning that they live near the surface. Only about 100 species have been described to date, but some are abundant enough to represent a significant fraction of the total planktonic biomass.

Feeding Ctenophores are predators. Feeding occurs in several ways, depending on the species. Some comb jellies have long tentacles covered with cells that release an adhesive when they contact prey. These tentacles are periodically wiped across the mouth so that captured prey can be ingested. In other species, prey can stick to mucus on the body and be swept toward the mouth by cilia. Still other species ingest large prey whole. Figure 32.3c shows a comb jelly that has just swallowed a second comb jelly almost as large as itself.

Movement Adults move via the beating of cilia, which occur in comblike rows running the length of the body. Ctenophores are the largest animals known to use cilia for locomotion. ● You should be able to indicate the origin of cilia-powered swimming on Figure 32.25.

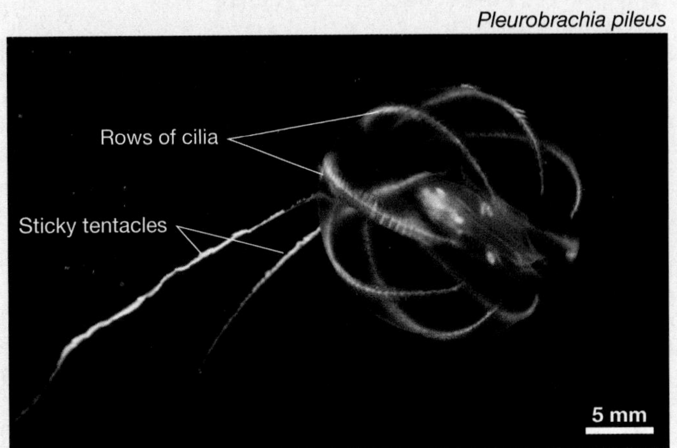

FIGURE 32.28 Ctenophores Are Planktonic Predators.

Reproduction Most species have both male and female organs and routinely self-fertilize, though fertilization is external. Larvae are free swimming. The few species that live on the ocean floor undergo internal fertilization and brood their embryos until they hatch into larvae.

Acoelomorpha

As their name implies, the acoelomorphs lack a coelom. They are bilaterally symmetric worms that have distinct anterior and posterior ends and are triploblastic. ● You should be able to indicate the origin of bilateral symmetry and mesoderm on Figure 32.25. They have simple digestive tracts or no digestive system at all. In some species with a digestive tract, the mouth is the only opening for the ingestion of food and excretion of waste products. Species that lack a digestive tract absorb nutrients across their body wall. Most acoelomorphs are only a couple of millimeters long and live in mud or sand in marine environments (**Figure 32.29**).

Feeding Acoelomorphs feed on detritus or prey on small animals or protists that live in mud or sand.

Movement Acoelomorphs swim, glide along the surface, or burrow through substrates with the aid of cilia that cover either the entire body or other than epithelium the ventral surface.

Reproduction Adults can reproduce asexually by fission (splitting in two) or by direct growth (budding) of a new individual from the parent's body. Individuals produce both sperm and eggs. Fertilization is internal, and fertilized eggs are laid outside the body.

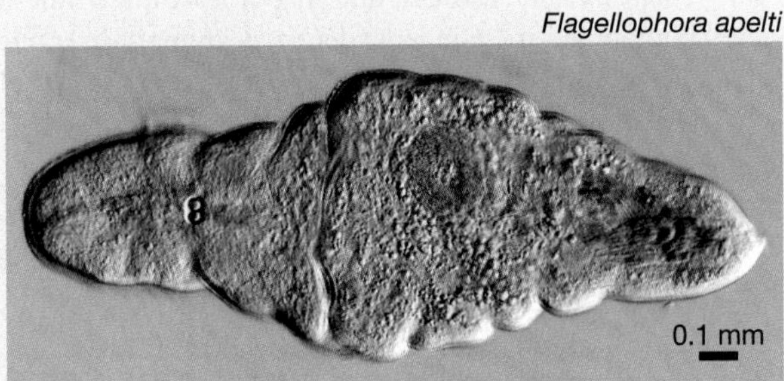

FIGURE 32.29 Acoelomorphs Are Small Worms That Live in Mud or Sand.

Chapter Review

SUMMARY OF KEY CONCEPTS

Animals are a particularly species rich and morphologically diverse lineage of multicellular organisms.

The animals consist of about 34 phyla and may number 10 million or more species. Biologists study animals because they are key consumers and because humans depend on them for transportation, power, or food.

You should be able to predict how a series of 10m × 10m plots in a forest would change if all animals were excluded.

Major groups of animals are recognized by nature of their basic body plan, which differs in the number of tissues observed in embryos, symmetry, the presence or absence of a body cavity, and the way in which early events in embryonic development proceed.

Sponges lack highly organized, complex tissues other than epithelium and may be asymmetric. The cnidarians and ctenophores have radial symmetry and two embryonic tissue layers. A handful of species have bilateral symmetry and three embryonic tissues but lack a body cavity, or coelom. The vast majority of animal species have bilateral symmetry, three embryonic tissues, and a coelom—features that gave rise to a "tube-within-a-tube" body plan. Depending on the species involved, the tube-within-a-tube design is built in one of two fundamental ways—via the protostome or deuterostome patterns of development.

You should be able to draw an asymmetric, radially symmetric, and bilaterally symmetric animal, and make a labeled drawing that shows how animals with tube-within-a-tube body plans move with limbs versus without limbs.

Recent phylogenetic analyses of animals have shown that there were three fundamental splits as animals diversified, resulting in two protostome groups (Lophotrochozoa and Ecdysozoa) and the deuterostomes.

Phylogenetic data support the hypothesis that triploblasty, bilateral symmetry, coeloms, and protostome and deuterostome development all evolved just once. However, phylogenetic data also suggest that coeloms were lost in the phylum Platyhelminthes; that pseudocoeloms evolved independently in the phyla Nematoda and Rotifera; and that at least three times, segmented body plans arose independently as animals diversified.

You should be able to describe how protostomes and deuterostomes differ, and how lophotrochozoans and ecdysozoans differ.

at www.masteringbio.com
The Architecture of Animals

Within major groups of animals, evolutionary diversification was based on innovative ways of feeding and moving. Most animals get nutrients by eating other organisms, and most animals move under their own power at some point in their life cycle.

Among animals, there is a wide variety of feeding strategies. Individuals capture food in one of four ways: Suspension feeders filter organic material or small organisms from water; deposit feeders swallow soils or other materials and digest the food particles they contain; fluid feeders lap or suck up liquids; food-mass feeders harvest packets of material. In addition, the sources of food that animals exploit vary from plant and algal material to other animals to detritus.

The evolution of the coelom made efficient movement possible because an enclosed, fluid-filled body cavity could act as a hydrostatic skeleton. Limbs and other types of support structures (skeletons) evolved later. Recent research suggests that, even though the types of appendages used in animal locomotion range from simple sac-like limbs to complex lobster legs, all appendages may be homologous to some degree.

You should be able to (1) give an example—other than one provided in the text—of a suspension, fluid, deposit, and food-mass feeder; (2) explain how the animal's mouthparts or appendages make this mode of feeding efficient.

Methods of sexual reproduction vary widely among animal groups, and many species can reproduce asexually. It is common for individuals to undergo metamorphosis during their life cycle.

Sexual reproduction may involve laying eggs or giving birth to well-developed young. In most cases, metamorphosis involves a dramatic change between the juvenile and adult body form, as well as stark contrasts between juveniles and adults in the habitat used and feeding strategy.

You should be able to (1) give an example—other than one provided in the text—of an animal that undergoes holometabolous metamorphosis; (2) explain how the food and habitats exploited by juveniles versus adults differ.

QUESTIONS

Test Your Knowledge

1. What synapomorphy distinguishes animals as a monophyletic group, distinct from choanoflagellates?
   a. Multicellularity and tissues.
   b. Movement via a hydrostatic skeleton.
   c. Growth by molting.
   d. Ingestive feeding.

2. Which of the following patterns in animal evolution is correct?
   a. All triploblasts have a coelom.
   b. All triploblasts evolved from an ancestor that had a coelom.
   c. Sponges have epithelial tissues that line an enclosed, fluid-filled body cavity.
   d. All bilaterally symmetric animals are triploblastic.

3. Which of the following patterns in animal evolution is correct?
   a. Segmentation evolved once.
   b. Pseudocoeloms evolved several times.
   c. Sponges lack tissues and are asymmetrical.
   d. All triploblasts are bilaterally symmetric as adults.
4. Why do some researchers maintain that the limbs of all animals are homologous?
   a. Homologous genes, such as *Dll*, are involved in their development.
   b. Their structure—particularly the number and arrangement of elements inside the limb—is the same.
   c. They all function in the same way—in locomotion.
   d. Animal appendages are too complex to have evolved more than once.
5. In a "tube-within-a-tube" body plan, what is the interior tube?
   a. ectoderm
   b. mesoderm
   c. either the coelom or the pseudocoelom
   d. the gut
6. What is the key difference between choanoflagellates and sponges?
   a. Sponges are multicellular.
   b. Sponges are asymmetrical and do not have tissues.
   c. Choanoflagellates have distinctive flagellated cells that function in suspension feeding.
   d. Choanoflagellates are strictly aquatic.

Test Your Knowledge answers: 1. a; 2. d; 3. b; 4. a; 5. d; 6. a

Test Your Understanding

Answers are available at www.masteringbio.com

1. Explain the difference between a diploblast and a triploblast. Why was the evolution of a third embryonic tissue layer important?
2. Explain how a hydrostatic skeleton works. Compare the types of movement that are possible with a hydrostatic skeleton to the types of movement observed in animals that lack a hydrostatic skeleton.
3. Draw a hypothetical phylogenetic tree of an animal lineage that radiated into a diverse array of species. Indicate where the group's basic body plan evolved. Indicate where adaptations for suspension, deposit, fluid, and food mass feeding may have evolved.
4. Compare and contrast the types of mouthparts and body shapes you would expect to find in herbivorous insect species that suck plant fluids from stems, bore through stems, bite leaves, and suck nectar from flowers.
5. Explain the differences in the protostome and deuterostome patterns of development. Do you agree with the text's claim that these are just different variations on the tube-within-a-tube body plan? Explain why or why not.
6. Why was the evolution of cephalization correlated with the evolution of bilateral symmetry? Why were both of these features significant?

Applying Concepts to New Situations

Answers are available at www.masteringbio.com

1. Would you expect internal fertilization to be more common in aquatic or terrestrial environments? Explain your answer. What types of differences would you expect to find in the structure of eggs that are laid in aquatic versus terrestrial environments?
2. Ticks are arachnids (along with spiders and mites); mosquitoes are insects. Both ticks and mosquitoes are ectoparasites that make their living by extracting blood meals from mammals. Ticks undergo incomplete metamorphosis, while mosquitoes undergo complete metamorphosis. Based on these observations, would you predict ticks or mosquitoes to be the more successful group in terms of number of species, number of individuals, and geographic distribution? Explain why. How could you test your prediction?
3. Suppose you are walking along an ocean beach at low tide and find an animal that is unlike any you have ever seen before. How would you go about determining how the animal feeds and how it moves? How would you go about determining the major features of its body plan?
4. Suspension feeding is extremely common in aquatic organisms but rare in terrestrial organisms. Generate a hypothesis to explain this observation.

www.masteringbio.com is also your resource for • Answers to text, table, and figure caption questions and exercises • Answers to *Check Your Understanding* boxes • Online study guides and quizzes • Additional study tools including the *E-Book for Biological Science* 3rd ed., textbook art, animations, and videos.

33 Protostome Animals

KEY CONCEPTS

- Molecular phylogenies support the hypothesis that protostomes are a monophyletic group divided into two major subgroups: the Lophotrochozoa and the Ecdysozoa.
- Although the members of many protostome phyla have limbless, wormlike bodies and live in marine sediments, the most diverse and species-rich lineages—Mollusca and Arthropoda—have body plans with a series of distinctive, complex features. Molluscs and arthropods inhabit a wide range of environments.
- Key events triggered the diversification of protostomes, including several lineages making the water-to-land transition, a diversification in appendages and mouthparts, and the evolution of metamorphosis in both marine and terrestrial forms.

A small sample of the insects collected from a single tree in the Amazonian wet forest. In numbers of individuals and species richness, protostomes are the most abundant and diverse of all animals.

Protostomes include some of the most familiar and abundant organisms on Earth. The phylum Arthropoda alone includes the insects, chelicerates (spiders and mites), crustaceans (shrimp, lobster, crabs, barnacles), and myriapods (millipedes, centipedes). Biologists estimate that there are over 10 million arthropod species—most of them unnamed and undescribed. Among insects alone, about 925,000 species have been formally identified to date. Well over a third of these insect species are beetles. In addition to being a species-rich clade, beetles are abundant. A typical acre of pasture in England is home to almost 18 *million* individual beetles. In tropical rain forests, beetles and other insects make up 40 percent of the total mass of organisms present. And beetles are not the only super-abundant type of arthropod. There may be over 6.5 billion humans on Earth, but the world population of ants is estimated to be 1 million billion individuals. The phylum Mollusca is almost as impressive. This lineage comprises over 93,000 snail, clam, chiton, and cephalopod (octopuses and squid) species.

It is important to recognize, though, that not all protostome lineages have been as spectacularly successful as the Arthropoda and Mollusca. The 22 phyla of protostomes include some of the most obscure lineages on the tree of life as well as some of the most prominent. There are just 135 species in the phylum Echiura (spoon worms), for example, and 150 species in the Kinorhyncha (mud dragons). The 80 species in the Gnathostomulida (pronounced *nath-oh-stoh-MEW-lida*) and the 450 species in the Gastrotricha (*GAS-troh-trika*) average less than 2 mm long.

BOX 33.1 Model Organisms: *Caenorhabditis elegans* and *Drosophila melanogaster*

Protostomes include two of the most important model organisms in the life sciences: the fruit fly *Drosophila melanogaster* and the roundworm *Caenorhabditis elegans*. (*Caenorhabditis* is pronounced *see-no-rab-DIE-tiss*.) If you walk into a biology building on any university campus around the world, you are almost certain to find at least one lab where fruit flies or roundworms are being studied.

Since the early 1900s, *Drosophila melanogaster* has been a key experimental subject in genetics. It was initially chosen as a focus for study by T. H. Morgan because it can be reared in the laboratory easily and inexpensively (**Figure 33.1a**), matings can be arranged, the life cycle is completed in less than two weeks, and females lay a large number of eggs. These traits made fruit flies valuable subjects for breeding experiments designed to test hypotheses about how traits are transmitted from parents to offspring (see Chapter 13). More recently, *D. melanogaster* has also become a key model organism in the field of developmental biology. The use of flies in developmental studies was inspired in large part by the work of Christianne Nüsslein-Volhard and Eric Wieschaus, who in the 1980s isolated flies with genetic defects in early embryonic development. By investigating the nature of these defects, researchers have gained valuable insights into how various gene products influence the development of eukaryotes (see Chapter 21).

Caenorhabditis elegans emerged as a model organism in developmental biology in the 1970s, due largely to work by Sydney Brenner and colleagues. This roundworm was chosen for three reasons: (1) Its cuticle (soft outer layer) is transparent, making individual cells relatively easy to observe (**Figure 33.1b**); (2) adults have exactly 959 nonreproductive cells; and, most important, (3) the fate of each cell in an embryo can be predicted because cell fates are invariant among individuals. For example, when researchers examine a 33-cell *C. elegans* embryo, they know exactly which of the 959 cells in the adult will be derived from each of those 33-embryonic cells. In addition, *C. elegans* are small (less than 1 mm long), are able to self-fertilize or cross-fertilize, and undergo early development in just 16 hours. The entire of genome of *C. elegans*, and that of *D. melanogaster*, has now been sequenced (see Chapter 20).

(a) Fruit flies can be reared in bottles.

(b) *Caenorhabditis elegans* is transparent and can be reared in petri dishes.

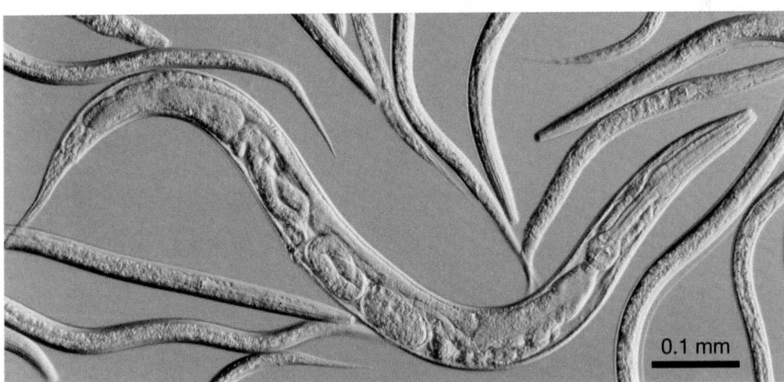

FIGURE 33.1 Model Organisms Can Be Readily Reared and Studied.

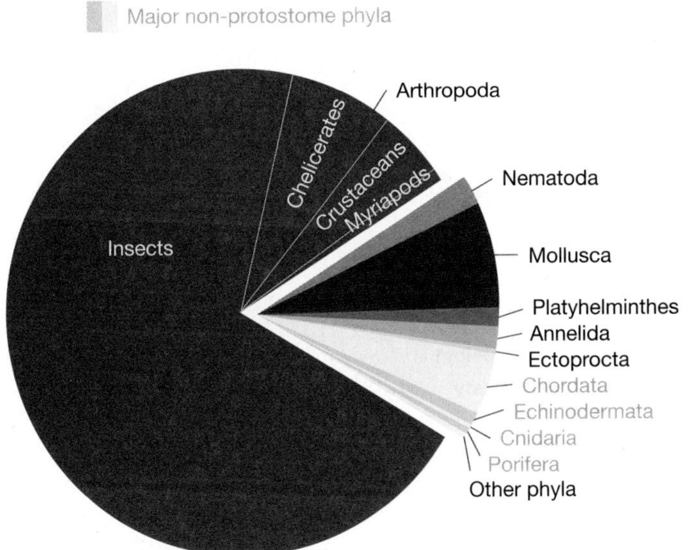

Figure 33.2 is a pie chart showing the relative numbers of species in various animal phyla. Except for the small wedges labeled Chordata, Cnidaria, Echinodermata, Porifera, and some of the "Other invertebrates," all animal species are protostomes.

As a group, protostomes live in just about every habitat that you might explore and include some of the most important model organisms in all of biological science (**Box 33.1**). If one of biology's most fundamental goals is to understand the diversity of life on Earth, then protostomes—particularly the arthropods and molluscs—demand our attention.

FIGURE 33.2 The Relative Abundances of Animal Lineages. Most animals are protostomes. About 70 percent of all known species of animals on Earth are insects, most of them beetles. (Humans and other vertebrates are deuterostomes, in the phylum Chordata.) This chapter focuses on the most species-rich lineages of protostomes.

● **QUESTION** About what percentage of known animal phyla are protostomes?

33.1 An Overview of Protostome Evolution

There are two major monophyletic groups of bilaterally symmetric, triploblastic, coelomate animals: the protostomes and deuterostomes. You might recall from Chapter 32 that protostomes and deuterostome embryos develop in dramatically different ways. The protostome and deuterostome patterns of early development represent distinctive pathways for building a bilaterally symmetric body with a coelom. In protostomes, the pattern of cell divisions called spiral cleavage occurs after fertilization. During gastrulation, the initial pore that forms in the embryo becomes the mouth. If a **coelom** (a body cavity) forms later in development, it forms from openings that arise within blocks of mesodermal tissue.

Phylogenetic studies have long supported the hypothesis that protostomes are a monophyletic group. This result means that the protostome developmental sequence arose just once. But when recent analyses of DNA sequence data indicated that two major subgroups existed within the protostomes, it was considered a key insight into animal evolution. The two monophyletic groups of protostomes are called the Lophotrochozoa and Ecdysozoa, and they were initially identified by analyzing DNA sequences from the gene for the small subunit of ribosomal RNA. (The lineages are pronounced *low-foe-troe-koe-ZOE-an* and *eck-dye-so-ZOE-an*.) Newer work based on sequence data from several additional genes has supported this result. Although research continues, an emerging consensus holds that the divergence of protostomes into two major subgroups was an important event as protostomes diversified (**Figure 33.3**).

What Is a Lophotrochozoan?

The 14 phyla of lophotrochozoans include the molluscs, annelids, and flatworms (Platyhelminthes; pronounced *plah-tee-hell-MIN-theez*). Although **lophotrochozoans** can be recognized by several distinct features, including growth by incremental additions to their body, the group's name was inspired by morphological traits that are found in some, but not all, of the phyla in the lineage: (1) a feeding structure called a lophophore, which is found in three phyla, and (2) a type of larva called a trochophore, which is common to many of the phyla in the lineage.

As **Figure 33.4a** shows, a **lophophore** (literally, "tuft-bearer") is a specialized structure that rings the mouth and functions in suspension feeding. Lophophores consist of tentacles that have ciliated cells along their surface. The beating of the cilia generates a water current that sweeps protists and other algae, detritus, and animal larvae into the region above the mouth, where they are trapped by the tentacles, swept into the mouth opening, and ingested. Lophophores are found in bryozoans ("moss animals"), brachiopods ("lamp shells"), and phoronids ("horseshoe worms").

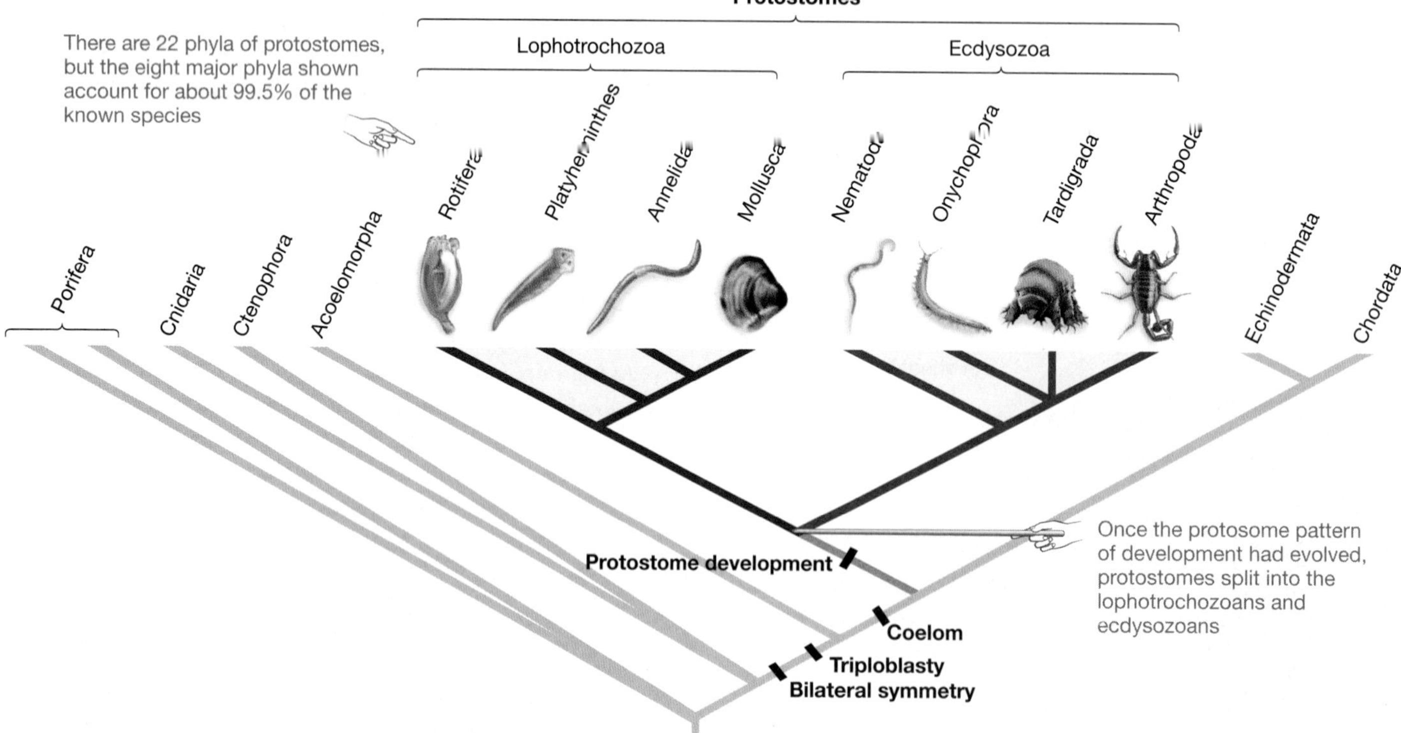

FIGURE 33.3 Protostomes Are a Monophyletic Group Comprising Two Major Lineages.

(a) Lophophores function in suspension feeding in adults.

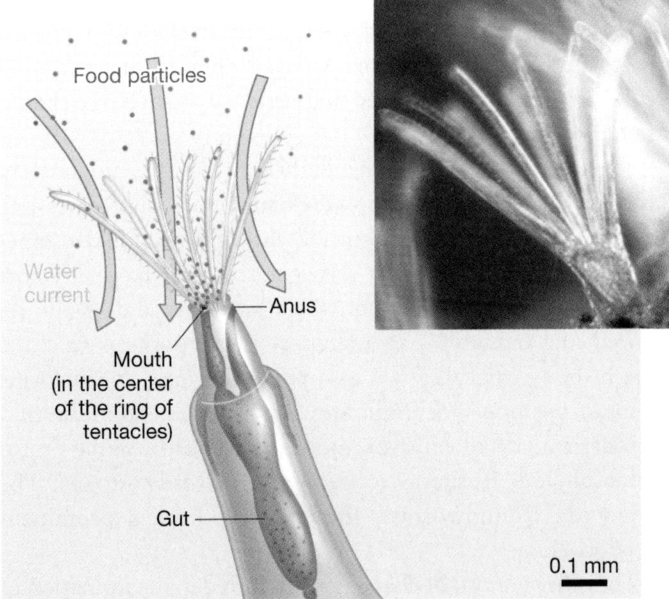

(b) Trochophore larvae swim and feed.

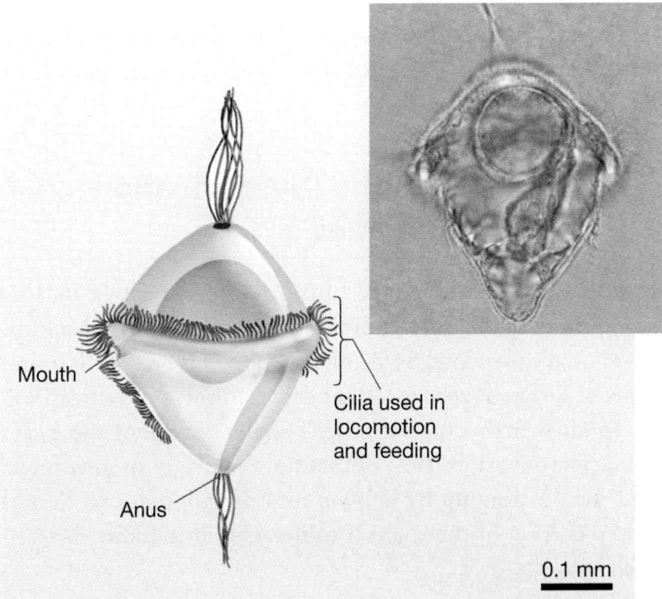

FIGURE 33.4 Lophotrochozoans Have Distinctive Traits. (a) Three phyla of lophotrochozoans have the feeding structure called a lophophore. **(b)** Many phyla of lophotrochozoans have the type of larva called a trochophore.

EXERCISE *Lopho–* means "tuft"; *trocho–* means *wheel*. Label the tuft and wheel in this figure.

Trochophores are a type of larvae common to marine molluscs, annelids that live in the ocean, and several other phyla in the Lophotrochozoa. As **Figure 33.4b** shows, a **trochophore** ("wheel-bearer") larva has a ring of cilia around its middle. These cilia aid in swimming. The beating of the cilia may also produce water currents that sweep food particles into the mouth, though many trochophores feed primarily on yolk provided by the mother. Thus, trochophores may represent a feeding stage as well as a dispersal stage in the life cycle. Recent analyses suggest that the trochophore may have originated early in the evolution of Lophotrochozoans, though different types of larva evolved later in some groups.

What Is an Ecdysozoan?

The primary contrast between lophotrochozoans and ecdysozoans involves the methods of growth used by organisms in each group. Instead of growing by steady and incremental additions to the body, **ecdysozoans** grow by **molting**—that is, by shedding an exoskeleton or external covering. The Greek root *ecdysis*, which means "to slip out or escape," is appropriate because during a molt, an individual sheds its outer layer—called a **cuticle** if it is soft or an **exoskeleton** if it is hard—and slips out of it (**Figure 33.5**). Once the old covering is gone, the body expands and a larger cuticle or exoskeleton then forms. As ecdysozoans grow and mature, they undergo a succession of molts. The most prominent of the seven ecdysozoan phyla are the roundworms (Nematoda) and arthropods (Arthropoda).

Growth by molting was required in ecdysozoans once they had evolved tough outer cuticles and thick exoskeletons. The leading hypothesis to explain the evolution of these stiff body coverings is based on protection from predators. To support this claim, biologists point to the secretive behavior of ecdysozoans while individuals are molting. When crabs and other crustaceans have shed an old exoskeleton, their new exoskeleton

FIGURE 33.5 Ecdysozoans Grow by Molting. Once an ecdysozoan has left its old exoskeleton, hours or days pass before the new exoskeleton is hardened. During this time, the individual is highly susceptible to predation or injury.

takes several hours to harden. During this interval, individuals hide and do not feed or move about. Experiments have shown that it is much easier for predators to attack and subdue individuals that are not protected by an exoskeleton during the intermolt period.

33.2 Themes in the Diversification of Protostomes

Protostomes have diverged into 22 different phyla that are recognized by distinctive body plans or specialized mouthparts used in feeding. Some of these phyla then split into millions of different species. What drove all this diversification?

To answer this question, let's analyze some of the evolutionary innovations that resulted in the origin of new phyla, and then follow up by delving into the adaptations that allowed certain of these phyla to diversify into many different species.

How Do Body Plans Vary among Phyla?

All protostomes are triploblastic and bilaterally symmetric, and all protostomes undergo embryonic development in a similar way. In contrast, radical changes occurred in coelom formation as protostomes diversified. **Figure 33.6** summarizes recent results on the relationships among protostome phyla and indicates where major changes occurred in the evolution of protostome body plans. Only a few of the 22 protostome phyla are included in this phylogenetic tree, however, because the relationships of most groups within the Lophotrochozoa and Ecdysozoa are still poorly resolved. Stated another way, it is not yet clear how most of the phyla within these two groups are related. As investigators analyze data from additional genes, a more highly resolved and better-supported tree should gradually emerge.

The most dramatic change that occurred after the coelom evolved was a reversion to an **acoelomate** body plan—meaning the lack of a body cavity—in Platyhelminthes. Flukes, tapeworms, and other types of flatworms do not have any sort of body cavity, even though their ancestors did. Data reviewed in Chapter 32 also support the hypothesis that the type of body cavity known as a **pseudocoelom**, which forms from an opening that originates between the endoderm and mesoderm layers of embryos, arose independently in the protostome phylum Rotifera (rotifers) and the ecdysozoans. The Nematoda (roundworms), for example, have a prominent pseudocoelom.

The other important change that occurred in the formation of body cavities in certain phyla was a drastic reduction of the coelom. To put this change in context, you might recall from Chapter 32 that most protostome phyla have wormlike bodies with a basic tube-within-a-tube design. The outside tube is the skin, which is derived from ectoderm; the inside tube is the gut, which is derived from endoderm. Muscles and organs derived from mesoderm are located between the two tubes. In the wormlike phyla, the coelom is well developed and functions as a hydrostatic skeleton that is the basis of movement. But in the most species-rich and morphologically complex protostome phyla—

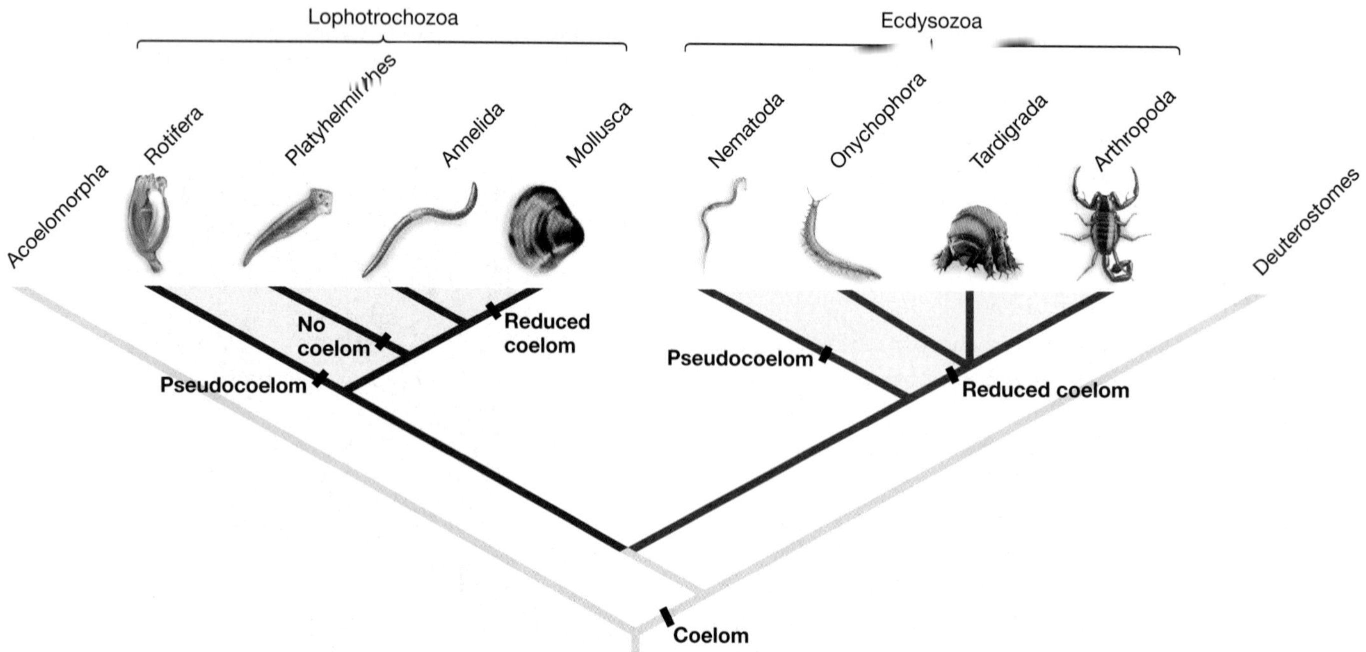

FIGURE 33.6 The Nature of the Coelom Varies Among Protostomes. This phylogenetic tree is based on analyses of DNA sequences from several genes. Most of the 14 protostome phyla not included in this tree have complete coeloms. The position of these phyla on the tree is still too uncertain to include them here, however.

the Arthropoda and the Mollusca (snails, clams, squid)—the coelom is drastically reduced. To use a term introduced in Chapter 24, the arthropod and mollusc coeloms are vestigial traits.

A fully functioning coelom has two roles: providing space for fluids to circulate among organs, and providing a hydrostatic skeleton for movement. In arthropods and molluscs, other structures fulfill these functions. Species in these phyla don't have a fully functioning coelom for a simple reason: They don't need one. Let's take a closer look.

The Arthropod Body Plan As **Figure 33.7a** shows, arthropods have segmented bodies that are organized into prominent regions called tagma. In the grasshopper shown, the tagma are called the head, thorax, and abdomen. In addition, arthropods are distinguished by their jointed limbs and an exoskeleton made primarily of the polysaccharide chitin (see Chapter 5). In crustaceans, the exoskeleton is strengthened by the addition of calcium carbonate ($CaCO_3$).

Instead of being based on muscle contraction against a hydroskeleton, arthropod locomotion is based on muscles that apply force against the exoskeleton to move legs or wings. Species in the closely related phyla called Onychophora and Tardigrada (see Figure 33.3) also have limbs and reduced coeloms. Based on these observations, biologists suggest that the evolution of limbs made a hydrostatic skeleton unnecessary.

Besides their limbs and the other features labeled in Figure 33.7a, arthropods have a spacious body cavity called the **hemocoel** ("blood-hollow"; pronounced *HEE-mah-seal*) that provides space for internal organs and circulation of fluids. In caterpillars and other types of arthropod larva, the hemocoel also functions as a hydrostatic skeleton.

The Molluscan Body Plan Dramatic changes also occurred during the origin of molluscs, although the nature of the body plan that evolved is very different from that of arthropods. As **Figure 33.7b** shows, the mollusc body plan is based on three major components: (1) the **foot**, a large muscle located at the base of the animal and usually used in movement; (2) the **visceral mass**, the region containing most of the main internal organs and external gill; and (3) the **mantle**, a tissue layer that covers the visceral mass and that in many species secretes a shell made of calcium carbonate. Some mollusc species have a single shell; others have two, eight, or none at all.

In molluscs, the coelom's functions are replaced by the visceral mass and by the muscular foot. The visceral mass provides space for organs and the circulation of fluids. In some species the fluid enclosed by the visceral mass also functions as a hydrostatic skeleton. In other molluscs the muscular foot functions as a hydrostatic skeleton as it moves. Molluscs do not have limbs, but they did evolve ways of moving around that don't require a coelom.

Variation Among Body Plans of the Wormlike Phyla New types of body plans distinguish most of the phyla illustrated in Figure 33.6. In contrast, the general body plan is extremely similar in most of the protostome phyla that have wormlike bodies. In many cases these lineages are distinguished by specialized mouthparts or feeding structures. To drive this point home, consider the three phyla illustrated in **Figure 33.8**.

- Echiurans (spoon worms) burrow into marine mud and suspension feed using an extended structure called a **proboscis**, which forms a gutter leading to the mouth (Figure 33.8a). Cells in the gutter secrete mucus, which is sticky enough to capture pieces of detritus. The combination of mucus and detritus is then swept into the mouth by cilia on cells in the gutter.

(a) Arthropod body plan (external view)

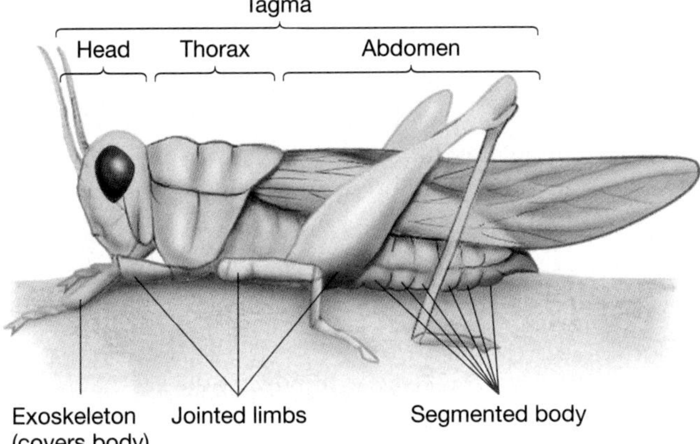

(b) Mollusc body plan (internal view)

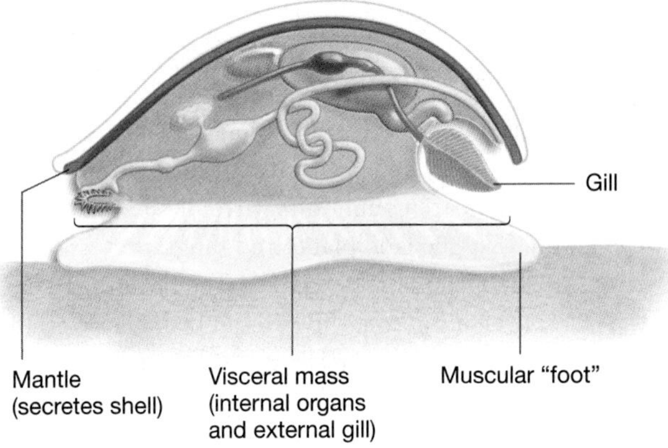

FIGURE 33.7 Arthropods and Molluscs Have Specialized Body Plans. (a) Arthropods have segmented bodies and jointed limbs, which enable these animals to move despite their hard outer covering, the exoskeleton. **(b)** The mollusc body plan is based on a foot, a visceral mass, and a mantle.

(a) Echiurans ("spoon worms")

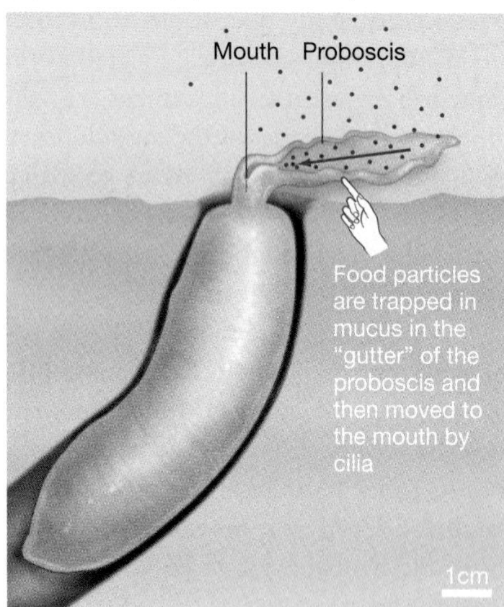

(b) Priapulids ("penis worms")

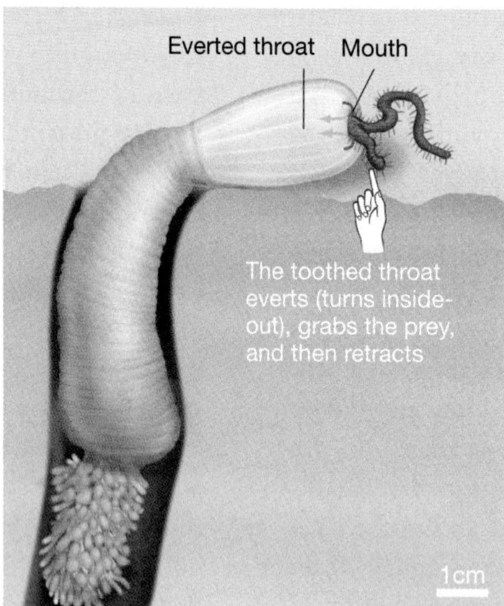

(c) Nemerteans ("ribbon worms")

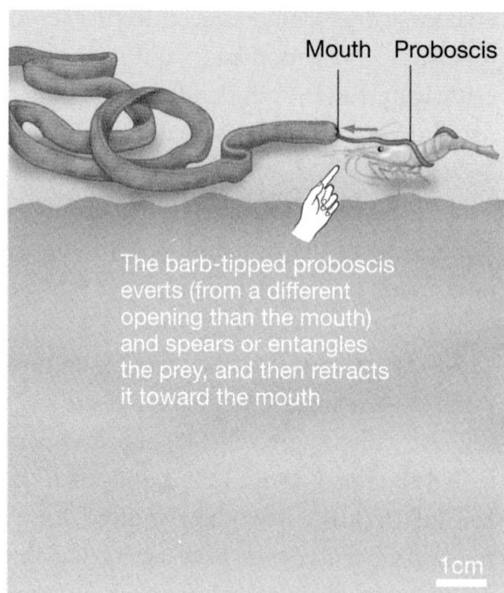

FIGURE 33.8 Some Protostome Phyla with Wormlike Bodies Have Specialized Mouthparts.

QUESTION It is common to find species from an array of wormlike phyla in the same habitat. Suggest a hypothesis to explain why this is possible.

- Priapulids (penis worms) also burrow into the substrate, but act as sit-and-wait predators. When a polychaete (Annelida) or other prey item approaches, the priapulid everts its toothed, cuticle-lined throat—meaning that it turns its throat inside out—grabs the prey item, and retracts the structure to take in the food (Figure 33.8b).
- The nemerteans (ribbon worms) are active predators that move around the ocean floor in search of food. Instead of everting their throat to capture prey, they have a proboscis that can extend or retract (Figure 33.8c). Nemerteans spear small animals with their proboscis or wrap the extended proboscis around prey. The food is then pulled into the mouth.

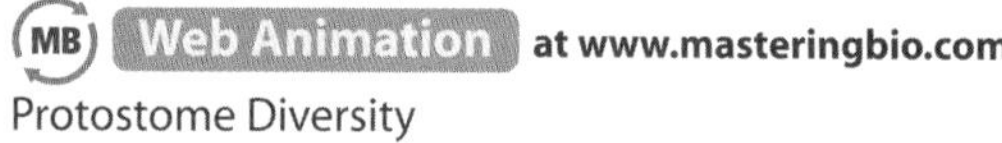

at www.masteringbio.com
Protostome Diversity

The Water-to-Land Transition

Protostomes are the most abundant animals in the surface waters and substrates of many marine and freshwater environments. But protostomes are common in virtually every terrestrial setting as well. Like the land plants and fungi, protostomes made the transition from aquatic to terrestrial environments. To help put this achievement in perspective, recall from Chapter 30 that green plants made the move from freshwater to land just once. From Chapter 31, recall that it is not yet clear whether fungi moved from aquatic habitats to land once or several times. Chapter 34 will show that only one lineage among deuterostomes moved onto land. But even given the current uncertainty in estimating the protostome phylogeny, it is clear that a water-to-land transition occurred multiple times as protostomes diversified. The ability to live in terrestrial environments evolved independently in arthropods (at least twice), molluscs, roundworms, and annelids.

Repeated water-to-land transitions are a prominent theme in the evolution of protostomes. In each case, the evidence for this claim is based on phylogenetic analyses, which support the hypothesis that the ancestors of the terrestrial lineages in each major subgroup were aquatic.

Why were so many different protostome groups able to move from water to land independently? The short answer is that it was easier for protostomes to accomplish this than for plants to do so. For example, land plants had to evolve roots and vascular tissue to transport water and support their stems. Similarly, the first terrestrial animals had to be able to support their body weight on land and move about. But the protostome groups that made the water-to-land transition already had hydrostatic skeletons, exoskeletons, appendages, or other adaptations for support and locomotion that happened to work on land as well as water.

To make the transition to land, new adaptations allowed protostomes to (1) exchange gases and (2) avoid drying out. As detailed in Chapter 44, land animals exchange gases with the atmosphere readily as long as a large surface area is exposed to the air. The bigger challenge is to prevent the gas-exchange surface and other parts of the body from drying out. Roundworms and earthworms solve this problem by living in wet soils or other moist environments. Arthropods and many molluscs have a watertight exoskeleton or a shell that minimizes water loss, and their respiratory structures are located inside the body. In insects, the openings to respiratory passages can be closed to minimize water loss. And unlike land plants and fungi, all land animals can move to moister habitats if the area they are in gets too dry.

Water-to-land transitions are important because they open up entirely new habitats and new types of resources to exploit. For example, plants moved onto land about 450 million years ago. But the first land-dwelling vertebrates don't appear in the fossil record until about 365 million years ago. For almost 90 million years, then, any protostome species that moved into terrestrial environments had an abundance of food available to eat, with a minimal number of competitors. Natural selection would then favor mutations that changed morphology in ways that allowed different species of protostomes to exploit different types of plant resources. Based on this reasoning, biologists claim that the ability to live in terrestrial environments was a key event in the diversification of several protostome phyla.

How Do Protostomes Feed, Move, and Reproduce?

Some of the protostome phyla have undergone dramatic amounts of speciation, while others have diversified hardly at all. Based on species counts, biologists can claim that certain body plans have proven much more successful than others at making efficient eating and moving machines.

● Once the wormlike, arthropod, and mollusc body plans had evolved, subsequent diversification was largely driven by adaptations that allowed protostomes to feed, move, or reproduce in novel ways. Recall that an **adaptation** is a trait that increases the fitness of individuals relative to individuals without the trait.

Adaptations for Feeding Protostomes include suspension, deposit, liquid, and food-mass feeders. Besides exploiting detritus, they prey on or parasitize plants, algae, or other animals. Exploiting a diversity of foods is possible because protostomes have such a wide variety of mouthparts for capturing and processing food. Figure 33.8 highlighted mouthparts that distinguish some of the wormlike phyla. But within phyla, arthropods take the prize for mouthpart diversity. The mouthparts observed in this phylum vary in structure from tubes to pincers and allow the various species to pierce, suck, grind, bite, mop, chew, engulf, cut, or mash (**Figure 33.9**). All arthropods have the same basic body plan, but their mouthparts and food sources are highly diverse.

In many species of arthropods, the jointed limbs also play a key role in getting food. The crustaceans called krill use their legs to sweep food toward their mouths as they suspension feed; certain insects, crustaceans, spiders, and molluscs use their appendages to capture prey or hold food as it is being chewed or bitten by the mouthparts.

In most cases, juvenile and adult forms of the same species exploit different food sources. Metamorphosis is the predominant mode of development in protostomes and usually results in larvae and adults that live in different habitats and have different overall morphology and mouthparts.

Adaptations for Moving In protostomes, variation in movement depends on variation in two features: (1) the

(a) Leaf-cutter ants cut leaves.

(b) Deer ticks pierce mammalian skin.

(c) Nut weevils bore into hard fruit.

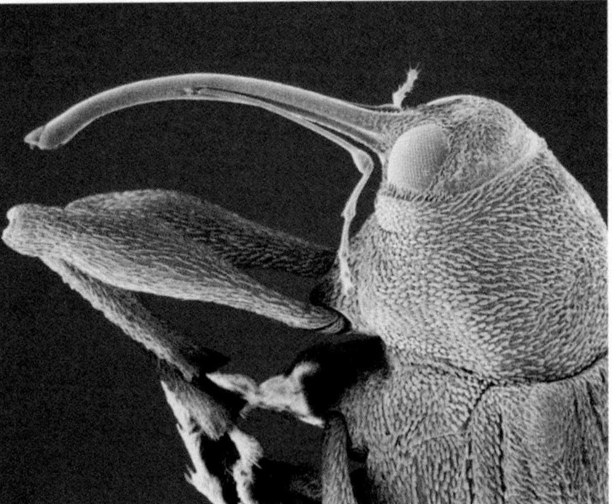

FIGURE 33.9 Arthropod Feeding Structures Are Diverse. Even though their mouthparts and food sources are diverse, all of these arthropods have segmented bodies that are organized into tagma, with an exoskeleton and jointed appendages.

● **QUESTION** How do the phyla with similar bodies but diverse mouthparts illustrated here compare with the wormlike phyla shown in Figure 33.8?

(a) Walking, running, and jumping

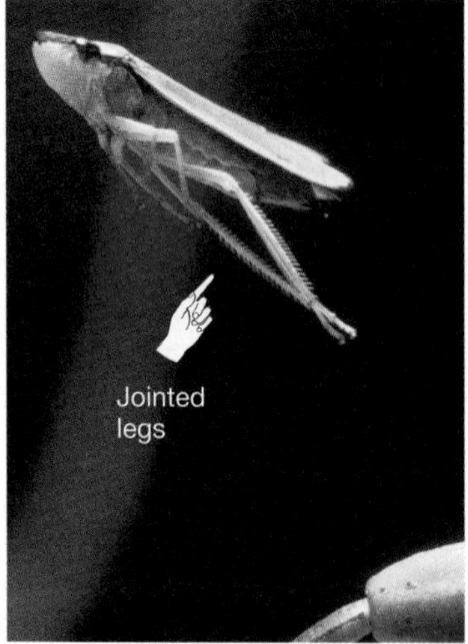

(b) Flying

(c) Gliding and crawling

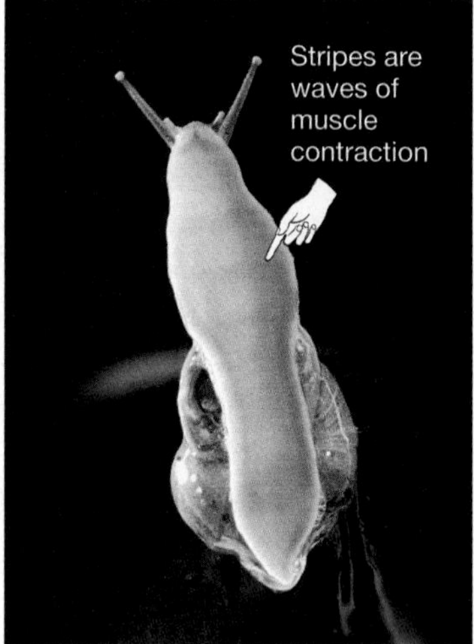

(d) Jet propulsion

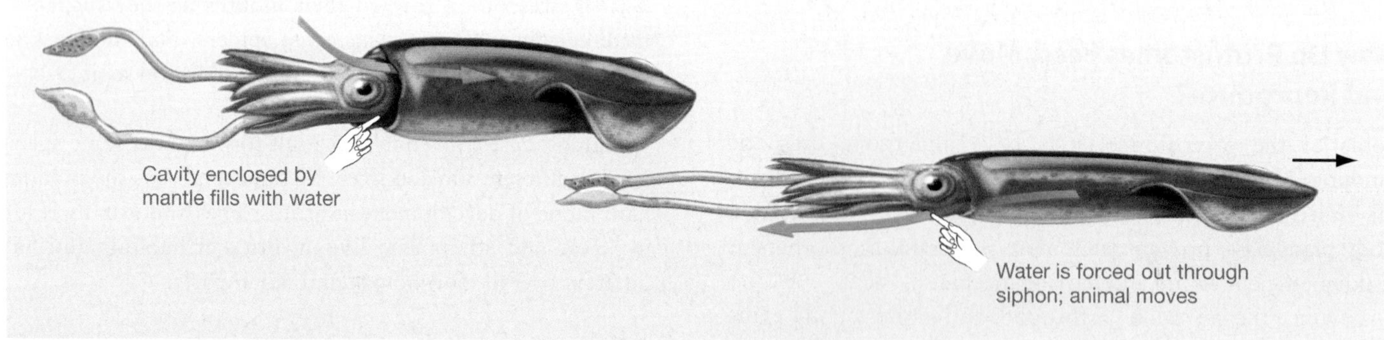

FIGURE 33.10 Protostome Locomotion Is Diverse. The evolution of **(a)** jointed limbs and **(b)** wings were key innovations in arthropod movement. **(c)** A wave of muscle contractions, down the length of the foot, allows molluscs to glide along a substrate. **(d)** In jet propulsion, muscular contractions force the water out through a muscular siphon. The siphon can aim outgoing water in a specific direction.

presence or absence of limbs and (2) the type of skeleton that is present. Wormlike protostomes that lack limbs move with the aid of a hydrostatic skeleton. This type of movement is also observed in caterpillars, grubs, maggots, and other types of insect larvae—even though adults of the same species have a hard exoskeleton and move with the aid of jointed limbs (**Figure 33.10a**). Although insect larvae have a highly reduced coelom, they do have an enclosed, fluid-filled body cavity that functions as a hydrostatic skeleton.

The evolution of jointed limbs made rapid, precise swimming and running movements possible and was a key innovation in the evolution of protostomes. The jointed limb is a major reason that arthropods have been so spectacularly successful in terms of species diversity, abundance of individuals, geographic range, and duration in the fossil record.

But the jointed limb is just one of several evolutionary innovations that allowed protostomes to move in unique ways:

- The insect wing is one of the most important adaptations in the history of life. About two-thirds of the multicellular species living today are winged insects. According to data in the fossil record, insects were the first organisms that had wings and could fly. Like most insects today, the earliest insects had two pairs of wings (**Figure 33.10b**). In most four-winged insects living today, however, the four wings function as two. Beetles fly with only their hindwings; butterflies, moths, bees, and wasps have hooked structures that make their two pairs of wings move together. Flies appear later in the fossil record. They have a single pair of large wings that power flight along with a pair of small, winglike structures called halteres that provide stability during flight.

- In molluscs, waves of muscle contractions sweep down the length of the large, muscular foot, allowing individuals to glide along a surface (**Figure 33.10c**). Cells in the foot secrete a layer of mucus, which reduces friction with the substrate and increases the efficiency of gliding.

- Squid are molluscs that have a mantle lined with muscle. When the cavity surrounded by the mantle fills with water and the mantle muscles contract, a stream of water is forced out of a tube called a **siphon**. The force of the water propels the squid forward (**Figure 33.10d**). This mechanism, jet propulsion, evolved in squid long before human engineers thought of using the same principle to power aircraft.

Adaptations in Reproduction When it comes to variation in reproduction and life cycles, protostomes do it all. Asexual reproduction by splitting the body lengthwise or by fragmenting the body is common in many of the wormlike phyla. Many crustacean and insect species reproduce asexually via **parthenogenesis** ("virgin-origin")—the production of unfertilized eggs that develop into offspring. Sexual reproduction is the predominant mode of producing offspring in most protostome groups, however. Sexual reproduction usually begins with external fertilization in sessile forms, such as clams, bryozoans, and brachiopods. It begins with internal fertilization in groups that are capable of movement—such as crustaceans, snails, and insects—because males and females can meet. Females of ovoviviparous insects and snails bear live young, although they do so by retaining fully formed eggs inside their bodies and then nourishing them via the nutrient-rich yolk that is inside the egg (see Chapter 32).

Two unique reproductive innovations occurred during protostome diversification: (1) the evolution of metamorphosis and (2) an egg that would not dry out on land. Metamorphosis is common in marine protostomes—especially in lineages where adults move little, if at all. In these species, metamorphosis is hypothesized to be an adaptation that allows juveniles to disperse to new habitats by floating or swimming in the plankton. Metamorphosis is also common in terrestrial insects, where it is hypothesized to be an adaptation that reduces competition for food between juveniles and adults. Regarding the ability to colonize terrestrial environments, though, the most critical adaptation was an egg that resists drying. Insect eggs have a thick membrane that keeps moisture in, and the eggs of slugs and snails have a thin calcium carbonate shell that helps retain water. Desiccation-resistant eggs evolved repeatedly in populations that made the transition to life on land. Recall from Chapter 30 that an analogous adaptation arose in land plants, when the evolution of the gametangia helped protect eggs and embryos from drying. And as Chapter 34 will show, a membrane-bound amniotic egg, analogous to the membrane-bound eggs of insects, evolved in the terrestrial vertebrates.

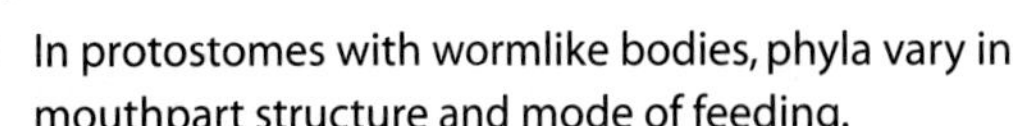

If you understand that...

- In protostomes with wormlike bodies, phyla vary in mouthpart structure and mode of feeding.
- In protostomes that do not have wormlike bodies, phyla vary in the nature of the coelom—whether the coelom is complete, a pseudocoelom, vestigial, or absent.
- Arthropods and molluscs have vestigial coeloms but specialized body plans that perform the functions of a coelom.
- A water-to-land transition occurred in several lineages independently.

You should be able to...

1) Describe the major features of the arthropod and mollusc body plans, and explain which structures perform the functions of a coelom.
2) Give examples of variation in mouthpart structure and feeding, structures used in locomotion, and metamorphosis in protostomes.
3) Give two examples of adaptations that facilitated the water-to-land transition in protostomes.

33.3 Key Lineages: Lophotrochozoans

Although analyses of extensive DNA sequences and the presence of unique, derived features such as lophophores and trochophore larvae helped biologists identify the Lophotrochozoa, the lineages within this group (**Figure 33.11**) are highly diverse in morphology. To drive this point home, let's take a closer look at four of the key phyla in the group: (1) Rotifera, (2) Platyhelminthes, (3) Annelida, and (4) Mollusca.

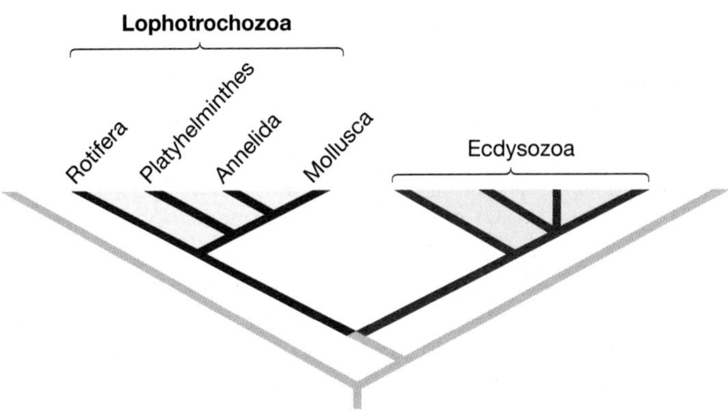

FIGURE 33.11 Lophotrochozoans Are a Monophyletic Group.

Lophotrochozoans > Rotifera (Rotifers)

The 1800 rotifer species that have been identified thus far live in damp soils as well as marine and freshwater environments. They are important components of the plankton in freshwater and in brackish waters, where rivers flow into the ocean and water is slightly salty. **Rotifers** have pseudocoeloms, and most species are less than 1 mm long. Although rotifers do not have a lophophore or a trochophore larval stage, extensive similarities in DNA sequence identify them as a member of the lophotrochozoan lineage.

Feeding Rotifers have a cluster of cilia at their anterior end called a **corona** (**Figure 33.12**). In many species, the beating of the cilia in the corona makes suspension feeding possible by creating a current that sweeps microscopic food particles into the mouth. The corona is the signature morphological feature of this group. ● You should be able to indicate the origin of this structure on Figure 33.11.

Movement Although a few species of rotifers are sessile, most swim via the beating of cilia in the corona.

Reproduction Females produce unfertilized eggs by mitosis; the eggs then hatch into new, asexually produced individuals. Recall that the production of offspring via unfertilized eggs is termed parthenogenesis. An entire group of rotifers—called the bdelloids—reproduces only asexually via parthenogenesis. Both sexual reproduction and asexual reproduction are observed in most species, however. Development is direct, meaning that fertilized eggs hatch and grow into adults without going through metamorphosis.

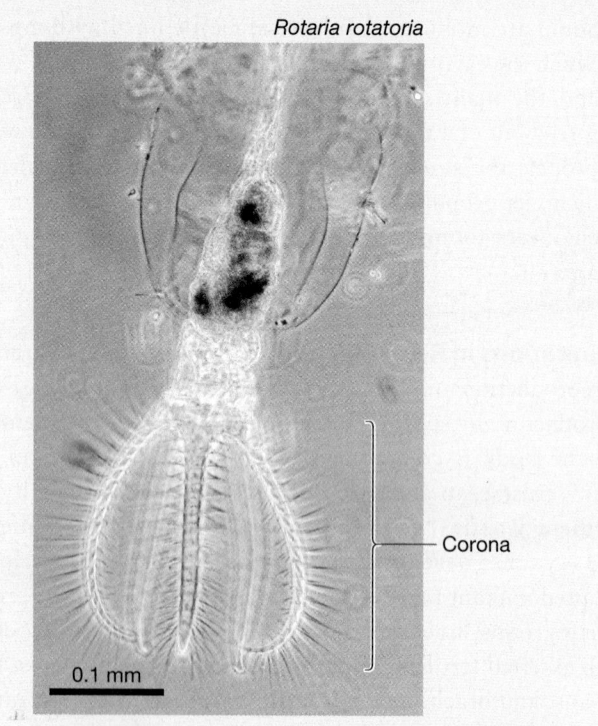

FIGURE 33.12 Rotifers Are Tiny, Aquatic Suspension Feeders.

Lophotrochozoans > Platyhelminthes (Flatworms)

The flatworms are a large and diverse phylum. More than 20,000 species have been described in three major subgroups. In traditional classification schemes like those described in Chapter 1, each of these lineages was referred to as a class. They are (1) the free-living species called Turbellaria (**Figure 33.13a**), (2) the endoparasitic tapeworms called Cestoda (**Figure 33.13b**), and (3) the endoparasitic or ectoparasitic flukes, called Trematoda (**Figure 33.13c**). Although a few turbellarian species are terrestrial, most live on the substrates of freshwater or marine environments. Tapeworms and other cestodes parasitize fish, mammals, or other vertebrates. Flukes parasitize vertebrates or molluscs.

Flatworms are named for the broad, flattened shape of their bodies. (The Greek roots *platy* and *helminth* mean "flatworm.") Species in the Platyhelminthes are unsegmented and lack a coelom. They also lack structures that are specialized for gas exchange—taking in oxygen and ridding the body of carbon dioxide. Further, they do not have blood vessels or any other type of system for circulating oxygen and nutrients to their cells. Based on these observations, biologists interpret the flattened bodies of these animals as an adaptation that gives flatworms an extremely high surface-area-to-volume ratio. Because they have so much surface area, a large amount of gas exchange can occur

(Continued on next page)

Lophotrochozoans > Platyhelminthes (Flatworms) *continued*

directly across their body wall, with oxygen diffusing into the body from the surrounding water and carbon dioxide diffusing out. Because the volume of the body is so small relative to the surface area available, nutrients and gases can diffuse efficiently to all of the cells inside the animal. This body plan has a downside, however: Because the body surface has to be moist for gas exchange to take place, flatworms are restricted to environments where they are surrounded by fluid. ● You should be able to indicate the origin of the flattened, acoelomate body plan on Figure 33.11.

Feeding Platyhelminthes lack a lophophore and have a digestive tract that is "blind"—meaning it has only one opening for ingestion of food and elimination of wastes. Most turbellarians are hunters that prey on protists or small animals; others scavenge dead animals. Tapeworms and flukes, in contrast, are strictly parasitic and feed on nutrients provided by hosts. Flukes gulp host tissues and fluids through a mouth and have a blind digestive tract. Tapeworms do not have a mouth, and they do not have a digestive tract. They obtain nutrients solely by diffusion through their body wall.

Movement Some turbellarians can swim a little by undulating their bodies, and most can creep along substrates with the aid of cilia on their ventral surface. Tapeworms and flukes move much less; adult cestodes have hooked attachment structures at their anterior end that permanently attach them to the interior of their host.

Reproduction Turbellarians can reproduce asexually by splitting themselves in half. If they are fragmented as a result of a predator attack, the body parts can regenerate into new individuals. Most turbellarians contain both male and female organs and reproduce sexually by aligning with another individual and engaging in mutual and simultaneous fertilization. Flukes (trematodes) and tapeworms (cestodes) also reproduce sexually and either cross-fertilize or self-fertilize. The reproductive systems and life cycles of flukes and tapeworms are extremely complex, in many cases involving two or even three distinct host species, with sexual reproduction occurring in the **definitive host** and asexual reproduction occurring in one or more **intermediate hosts**. For example, humans are the definitive host of the blood fluke *Schistosoma mansoni*. Fertilized eggs are shed in a human host's feces and enter aquatic habitats if sanitation systems are poor. The eggs develop into larvae that infect snails. Inside the snail, asexual reproduction results in the production of a different type of larva, which emerges from the snail and burrows into the skin of humans who wade in infested water. Once inside a human, the parasite lives in blood and develops into a sexually mature adult.

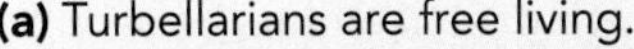

(a) Turbellarians are free living.

Pseudoceros ferrugineus

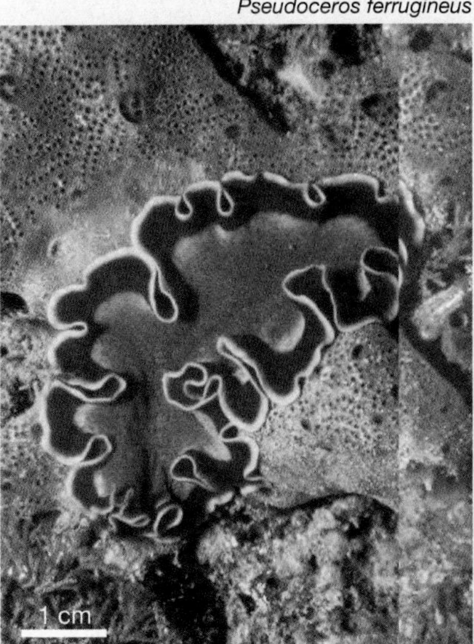

(b) Cestodes are endoparasitic.

Taenia species

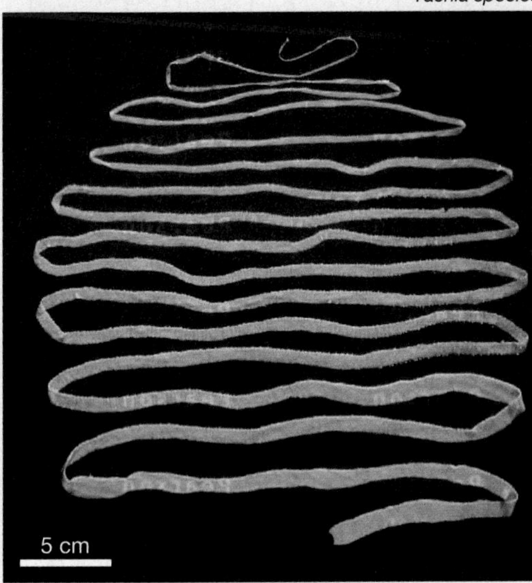

(c) Trematodes are endoparasitic.

Dicrocoelium dendriticum

1 mm

FIGURE 33.13 Flatworms Have Simple, Flattened Bodies.

Lophotrochozoans > Annelida (Segmented Worms)

All **annelids** have a segmented body and a coelom that functions as a hydrostatic skeleton. ● You should be able to indicate the origin of annelid segmentation on Figure 33.11. The 16,500 species that have been described thus far can be divided into two major lineages, traditionally called Polychaeta and Clitellata:

1. The Polychaeta (*pol-ee-KEE-ta*), or polychaetes ("many-bristles"), are named for their numerous, bristle-like extensions called **chaetae** (**Figure 33.14a**). The chaetae extend from appendages called **parapodia**. Polychaetes are mostly marine, and they range in size from species that are less than 1 mm long to species that grow to lengths of 3.5 m. They include a large number of burrowers in mud or sand, sedentary forms that secrete chitinous tubes, and active, mobile species.
2. The Clitellata is composed of the oligochaetes (*oh-LIG-oh-keetes*) and leeches, or Oligochaeta and Hirudinea, respectively. Oligochaetes ("few-bristles") include the earthworms, which burrow in moist soils (**Figure 33.14b**); an array of freshwater species; and a few marine forms. Oligochaetes lack parapodia and, as their name implies, have many fewer chaetae than do polychaetes. The coelom of leeches is much reduced in size compared with that of other annelids and consists of a series of connected chambers. Leeches live in freshwater as well as marine habitats (**Figure 33.14c**).

Feeding Polychaetes have a wide variety of methods for feeding: The burrowing forms deposit feed; the sedentary forms suspension feed with the aid of a dense crown of tentacles; and the active forms hunt small animals, which they capture by everting their throats. Virtually all oligochaetes, in contrast, make their living by deposit feeding in soils. About half of the leeches are ectoparasites that attach themselves to fish or other hosts and suck blood and other body fluids. Hosts are usually unaware of the attack, because leech saliva typically contains an anesthetic. The host's blood remains liquid as the leech feeds, because the parasite's saliva also contains an anticoagulant. Parasitic leeches are still used by physicians to remove blood from particularly large bruises. The nonparasitic leech species are predators or scavengers.

Movement Polychaetes and oligochaetes crawl or burrow with the aid of their hydrostatic skeletons; the parapodia of polychaetes also act as paddles or tiny feet that aid in movement. Leeches can swim by using their hydrostatic skeletons to make undulating motions of the body.

Reproduction Asexual reproduction occurs in polychaetes and oligochaetes by transverse fission or fragmentation—meaning that body parts can regenerate a complete individual. Sexual reproduction in polychaetes may begin with internal or external fertilization, depending on the species. Polychaetes have separate sexes and usually release their eggs directly into the water. Some species produce eggs that hatch into trochophore larvae; others undergo direct development. In oligochaetes and leeches, individuals produce both sperm and eggs and engage in mutual, internal cross-fertilization. Eggs are enclosed in a mucus-rich, cocoon-like structure; after fertilization, offspring develop directly into miniature versions of their parents.

(a) Most polychaetes are marine.

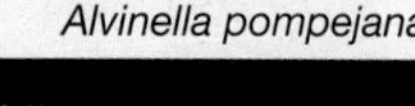

Alvinella pompejana

(b) Most oligochaetes are terrestrial.

Paranais litoralis

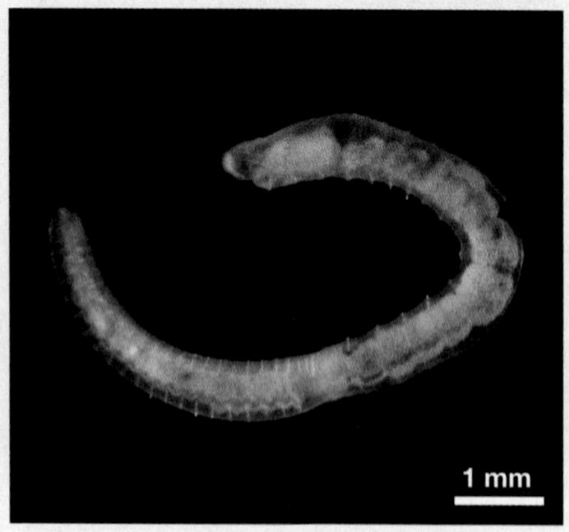

(c) Most leeches live in freshwater.

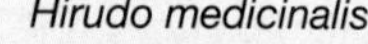

Hirudo medicinalis

FIGURE 33.14 Annelids Are Segmented Worms.

Mollusca (Molluscs)

The **molluscs** are far and away the most species-rich and morphologically diverse group in the Lophotrochozoa. They have a specialized body plan based on a muscular foot, a visceral mass, and a mantle that may or may not secrete a calcium carbonate shell. The coelom is much reduced or absent. ● You should be able to indicate the origin of the molluscan body plan on Figure 33.11. Over 93,000 species have been described thus far. Although most molluscs live in marine environments, there are some terrestrial and freshwater forms.

Because molluscs are so diverse, let's consider each of the major subgroups or classes in the phylum separately and analyze the traits responsible for their diversification. The four most important lineages of molluscs are (1) **bivalves** (clams and mussels), (2) **gastropods** (slugs and snails), (3), **chitons,** and (4) **cephalopods** (squid and octopuses). The bivalves are suspension feeders; the other three groups of molluscs are herbivores or predators.

Mollusca > Bivalvia (Clams, Mussels, Scallops, Oysters)

The bivalves are so named because they have two separate shells made of calcium carbonate secreted by the mantle. The shells are hinged, and they open and close with the aid of muscles attached to them (**Figure 33.15a**). When the shell is closed, it protects the mantle, visceral mass, and foot. The bivalve shell is an adaptation that reduces predation.

Most bivalves live in the ocean, though there are many freshwater forms. Clams burrow into mud, sand, or other soft substrates and are sedentary as adults. Oysters and mussels are largely sessile as adults, but most of them live above the substrate, attached to rocks or other solid surfaces. Scallops are mobile and live on the surface of soft substrates. The smallest bivalves are freshwater clams that are less than 2 mm long; the largest bivalves are giant marine clams that may weigh more than 400 kg. All bivalves can sense gravity, touch, and certain chemicals, and scallops even have eyes.

Because most bivalves live on or under the ocean floor and because they are covered by a hard shell, their bodies are often buried in sediment after death. Bivalves thus fossilize readily, and the Bivalvia lineage has the most extensive fossil record of any animal, plant, or fungal group. This large database has allowed biologists to conduct thorough studies of the evolutionary history of bivalves.

Bivalves are important commercially. Clams, mussels, scallops, and oysters are farmed or harvested from the wild in many parts of the world and used as food by humans. Pearl oysters that are cultivated or collected from the wild are the most common source of natural pearls used in jewelry.

Feeding Bivalves are suspension feeders that take in any type of small animal or protist or detritus. Suspension feeding is based on a flow of water across gas-exchange structures called **gills.** The gills are part of the visceral mass and consist of a series of thin membranes where particles are trapped. A water current exists because siphons, which are tubes formed by edges of the mantle, extend from the shell and form a plumbing system. The siphons conduct water into the gills and then back out of the body, powered by the beating of cilia on the gills (**Figure 33.15b**). Bivalves are the only major group of molluscs that lack the feeding structure called a radula (see Chapter 32).

(a) Scallops live on the surface of the substrate and suspension feed.

Lima scabra

(b) Most clams burrow into soft subtrates and suspension feed

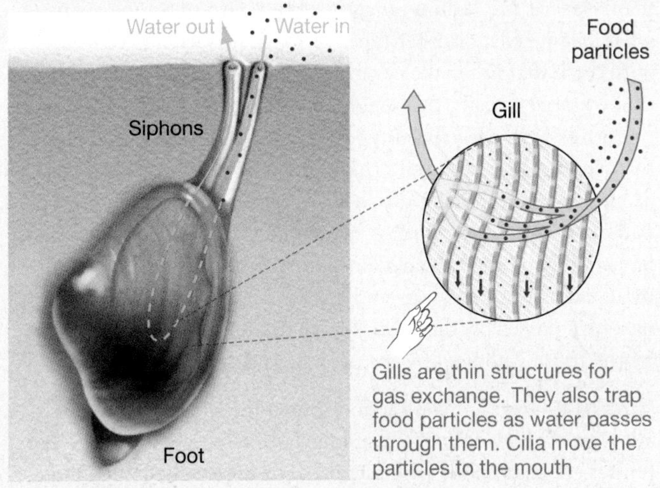

FIGURE 33.15 Bivalves Have Two Shells.

Movement Clams burrow with the aid of their muscular foot, which functions as a hydrostatic skeleton. But they are otherwise sedentary. Scallops are able to swim by clapping their shells together and forcing water to jet out, pushing them back-

(Continued on next page)

Mollusca > Bivalvia (Clams, Mussels, Scallops, Oysters) *continued*

ward. Scallop locomotion is similar to the swimming of cnidarian medusae, which move when muscular contractions force water out of their bell (see Chapter 32). Bivalves produce a swimming trochophore larva that is responsible for dispersing individuals to new locations.

Reproduction Only sexual reproduction occurs in bivalves. Eggs and sperm are shed into the water, and fertilized eggs develop into trochophore larvae. Trochophores then metamorphose into a distinct type of larva called a **veliger**, which continues to feed and swim before settling to the substrate and metamorphosing into an adult form that secretes a shell.

Mollusca > Gastropoda (Snails, Slugs, Nudibranchs)

The gastropods ("belly-feet") are named for the large, muscular foot on their ventral side. Their most startling feature, though, is a developmental process called torsion. During gastropod development, the visceral mass rotates. The rotation twists the mantle cavity and digestive tract and results in the anus being located above the head.

Most snails can retract their foot and body into a shell when they are attacked or when their tissues begin to dry out (**Figure 33.16a**). Land slugs and nudibranchs (pronounced *NEW-da-branks*) lack shells but often contain toxins or foul-tasting chemicals to protect them from being eaten. The bright colors of nudibranchs, or sea slugs, are thought to act as a warning to potential predators (**Figure 33.16b**). Gastropods are used as food in some cultures and are important medically because they serve as a host for flukes that also infect humans before completing their life cycle. About 70,000 species of gastropods are known.

Feeding Gastropods and other molluscs have a unique structure in their mouths called a radula. Recall from Chapter 32 that in many species the **radula** functions like a rasp to scrape away algae, plant cells, or other types of food. It is usually covered with teeth that are made of chitin and that vary in size and shape among species. Although most gastropods are herbivores or detritivores, specialized types of teeth allow some gastropods to act as predators. Species called drills, for example, use their radula to bore a hole in the shells of oysters or other molluscs and expose the visceral mass, which they then eat. Cone snails have highly modified, harpoon-like "teeth" mounted at the tip of an extensible proboscis and armed with poison. When a fish or worm passes by, the proboscis shoots out. The prey is speared by the tooth, subdued by the poison, and consumed.

Movement Waves of contractions down the length of the foot allow gastropods to move by gliding (see Figure 33.10c). Sea butterflies are gastropods with a reduced or absent shell but a large, winglike foot that flaps and powers swimming movements.

Reproduction Females of some gastropod species can reproduce asexually by producing eggs parthenogenetically, but most reproduction is sexual. Sexual reproduction in gastropods begins with internal fertilization. Some marine gastropods produce a trochophore larva that may disperse up to several hundred kilometers from the parent. But in most marine species and all terrestrial forms, larvae are not free living. Instead, offspring remain in an egg case while passing through several larval stages and then hatch as nymphs—miniature versions of the adults.

(a) Snails have a single shell, which they use for protection.
Maxacteon flammea

(b) Land slugs and sea slugs (nudibranchs) lack shells.
Chromodoris geminus

FIGURE 33.16 Gastropods Have a Single Shell or Lack Shells.

Mollusca > Polyplacophora (Chitons)

The Greek word roots that inspired the name *Polyplacophora* mean "many-plate-bearing." The name is apt because chitons (pronounced *KITE-uns*) have eight calcium carbonate plates along their dorsal side (**Figure 33.17**). The plates form a protective shell. The approximately 1000 species of chitons are marine. They are usually found on rocky surfaces in the intertidal zone, where rocks are periodically exposed to air at low tides.

Feeding Chitons have a radula and use it to scrape algae and other organic matter off rocks.

Movement Chitons move by gliding on their broad, muscular foot—as gastropods do.

Reproduction Chiton sexes are separate, and fertilization is external. In some species, however, sperm that are shed into the water enter the female's mantle cavity and fertilize eggs inside the body. Depending on the species involved, eggs may be enclosed in a membrane and released or retained until hatching and early development are complete. Most species have trochophore and veliger larvae.

Tonicella lineata

FIGURE 33.17 Chitons Have Eight Shell Plates.

Mollusca > Cephalopoda (Nautilus, Cuttlefish, Squid, Octopuses)

The cephalopods ("head-feet") have a well-developed head and a foot that is modified to form arms and/or tentacles. **Tentacles** are long, thin, muscular extensions that aid in movement and prey capture (**Figure 33.18**). Except for the nautilus, cephalopods have either highly reduced shells or none at all. They also have large brains and image-forming eyes with sophisticated lenses.

Octopus dofleini

FIGURE 33.18 Cephalopods Have Highly Modified Bodies.

Feeding Cephalopods are highly intelligent predators that hunt by sight and use their arms or tentacles to capture prey—usually fish or crustaceans. They have a radula as well as a structure called a **beak**, which can exert powerful biting forces. Some squid and octopuses also inject poisons into their prey to subdue them.

Movement Cephalopods can swim by moving their fins to "fly" through the water, or by jet propulsion using the mantle cavity and siphon. They draw water into their mantle cavity and then force it out through a siphon (see Figure 33.10d). Squid are built for speed and hunt small fish by chasing them down. Octopuses, in contrast, crawl along the substrate by using their long, armlike tentacles. They chase down crabs or other crustaceans, or they pry mussels or clams from the substrate and then use their beaks to crush the exoskeletons of their prey.

Reproduction Cephalopods have separate sexes, and some species have elaborate courtship rituals that involve color changes and interaction of tentacles. When a male is accepted by a female, he deposits sperm that are encased in a structure called a **spermatophore**. The spermatophore is taken up by the female, and fertilization is internal. Females lay eggs. When they hatch, juveniles develop directly into adults.

33.4 Key Lineages: Ecdysozoans

The ecdysozoans were first recognized as monophyletic when investigators began using DNA sequence data to estimate the phylogeny of protostomes (**Figure 33.19**). Seven phyla are currently recognized in the lineage (see Table 32.1), including the Onychophora (*on-ee-KOFF-er-uh*) and the Tardigrada (**Figure 33.20**). They are not species-rich phyla, but both are closely related to arthropods. Onychophorans and tardigrades are similar to arthropods in having a segmented body and limbs. Unlike arthropods, their limbs not jointed and they do not have an exoskeleton. The onychophorans, or velvet worms, are small, caterpillar-like organisms that live in moist leaf litter and prey on small invertebrates. Onychophorans have sac-like appendages and segmented bodies with a hemocoel.

The tardigrades, or water bears, are microscopic animals that live in benthic (bottom) habitats of marine or freshwater environments. Large numbers of water bears can also be found in the film of water that covers moss or other land plants in moist habitats. Tardigrades have a reduced coelom but a prominent hemocoel, and they walk on their clawed, sac-like legs. Most feed by sucking fluids from plants or animals; others are **detritivores**.

Let's take a closer look at the two most diverse and abundant phyla of the Ecdysozoa: (1) Nematoda and (2) Arthropoda.

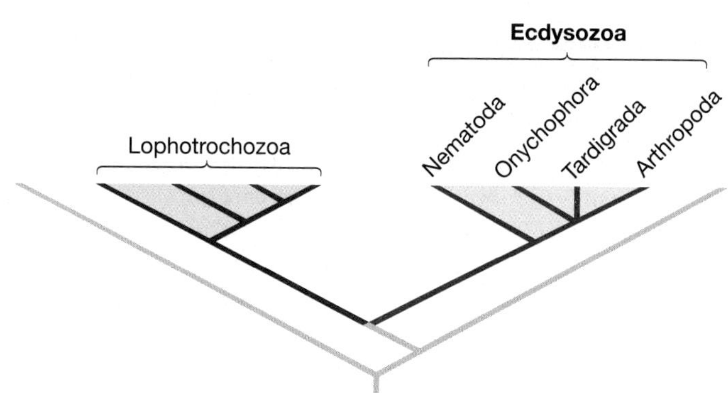

FIGURE 33.19 Ecdysozoans Are a Monophyletic Group.

(a) Onychophorans have lobe-like limbs.

Peripatus species

(b) Tardigrades have lobe-like limbs with claws.

Echiniscus species

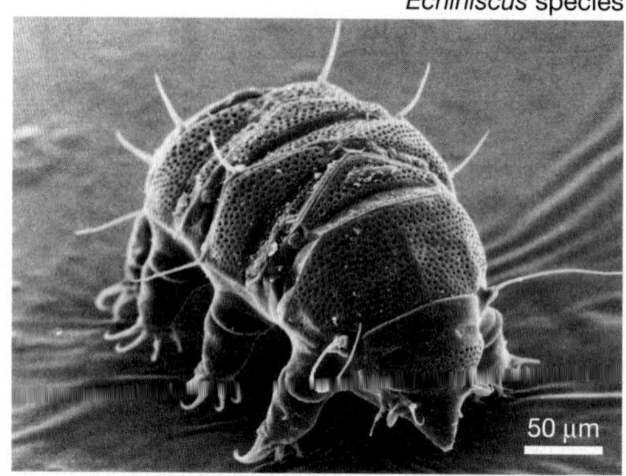

FIGURE 33.20 Onychophorans and Tardigrades Have Limbs and Are Closely Related to Arthropods. (a) The Onychophora are called velvet worms because they have wormlike bodies and a velvety appearance. **(b)** The Tardigrada are known as water bears because they live in moist habitats and, when viewed under the microscope, resemble bears.

Ecdysozoans > Nematoda (Roundworms)

Species in the phylum Nematoda are commonly called **nematodes** or **roundworms** (**Figure 33.21**). Roundworms are unsegmented worms with a pseudocoelom, a tube-within-a-tube body plan, and no appendages. They have no muscles that can change the diameter of the body—their body musculature consists solely of longitudinal muscles that shorten or lengthen the body upon contracting or relaxing, respectively. ● You should be able to indicate the evolution of a pseudocoelom and loss of circular muscles on Figure 33.19. Although some nematodes can grow to lengths of several meters, the vast majority of species are tiny—most are much less than 1 mm long. They lack specialized systems for exchanging gases and for circulating nutrients and wastes. Instead, gas exchange occurs across the body wall, and nutrients and wastes move by simple diffusion.

The nematode *Caenorhabditis elegans* is one of the most thoroughly studied model organisms in biology (see Box 33.1). Species that parasitize humans have also been studied intensively. Pin-

(Continued on next page)

Ecdysozoans > Nematoda (Roundworms) *continued*

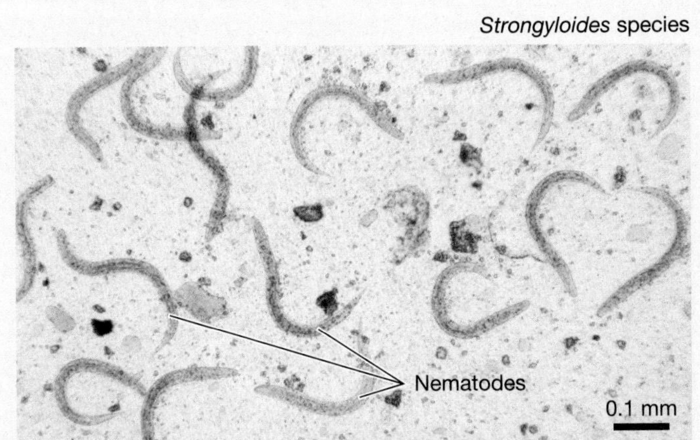

FIGURE 33.21 Most Nematodes Are Free Living, but Some Are Parasites.

worms, for example, infect about 40 million people in the United States alone, and *Onchocerca volvulus* causes an eye disease that infects around 20 million people in Africa and Latin America. Advanced infections by the roundworm *Wuchereria bancrofti* result in the blockage of lymphatic vessels, causing fluid accumulation and massive swelling—the condition known as elephantiasis.

Parasites actually represent a tiny fraction of the 25,000 nematode species that have been described to date, however. The vast majority of nematodes are free living. They are important ecologically because they are found in virtually every habitat known.

Nematodes are also fabulously abundant. Biologists have found 90,000 roundworms in a single rotting apple and have estimated that rich farm soils contain up to 9 billion roundworms per acre. Although they are not the most species-rich animal group, they are the most abundant. The simple nematode body plan has been extraordinarily successful.

Feeding Roundworms feed on a wide variety of materials, including bacteria, fungi, plant roots, small protists or animals, or detritus. In most cases, the structure of their mouthparts is specialized in a way that increases the efficiency of feeding on a particular type of organism or material.

Movement Roundworms move with the aid of their hydrostatic skeleton. Most roundworms live in soil or inside a host, so when contractions of their longitudinal muscles cause them to wriggle, the movements are resisted by a stiff substrate. As a result, the worm pushes off the substrate, and the body moves.

Reproduction Sexes are separate in most nematode species, and asexual reproduction is rare or unknown. Sexual reproduction begins with internal fertilization and culminates in egg laying and the direct development of offspring. Individuals go through a series of four molts over the course of their lifetime.

Arthropoda (Arthropods)

In terms of duration in the fossil record, species diversity, and abundance of individuals, arthropods are easily the most successful lineage of eukaryotes. They appear in the fossil record over 520 million years ago and have long dominated the animals observed in both marine and terrestrial environments. Well over a million living species have been described, and biologists estimate that millions or perhaps even tens of millions of arthropod species have yet to be discovered.

Morphologically, **arthropods** are distinguished by segmented bodies, a sophisticated, jointed exoskeleton. They have a highly reduced coelom but possess an extensive body cavity called a hemocoel, enclosed by the exoskeleton, and paired, jointed appendages. The body is organized into distinct head and trunk regions in all arthropods; in many species there is an additional grouping of segments into two distinct trunk regions, usually called an abdomen and a thorax. ● You should be able to indicate the origin of the elements of the arthropod body plan on Figure 33.19.

Metamorphosis is common in arthropods. Their larvae have segmented bodies but may lack a hardened exoskeleton. As in other ecdysozoans, larval and adult forms grow by molting.

At least some segments along the arthropod body produce paired, jointed appendages. Arthropod appendages have an array of functions: sensing aspects of the environment, exchanging gases, feeding, or locomotion via swimming, walking, running, jumping, or flying. Arthropod appendages provide the ability to sense environmental stimuli and make sophisticated movements in response. Most species also have sophisticated, image-forming compound eyes. A **compound eye** contains many lenses, each associated with a light-sensing, columnar structure. (Human and cephalopod eyes are **simple eyes**, meaning they have just one lens.) Most arthropods also have a pair of antennae on the head. **Antennae** are long, tentacle-like appendages that contain specialized receptor cells used to touch or smell.

Although the phylogeny of arthropods is still being worked out, most data sets agree that the phylum as a whole is monophyletic; that the myriapods (millipedes and centipedes), chelicerates (spiders), insects, and crustaceans represent four major subphyla within the phylum; and that crustaceans and insects are closely related. The phylum Arthropoda is so large and diverse that a detailed treatment would fill a book this size; space permits just a few notes about the four major lineages.

Arthropoda > Myriapods (Millipedes, Centipedes)

The **myriapods** have relatively simple bodies, with a head region and a long trunk featuring a series of short segments, each bearing one or two pairs of legs (**Figure 33.22**). If eyes are present, they consist of a few to many simple structures clustered on the sides of the head. The 11,600 species that have been described to date inhabit terrestrial environments all over the world.

Feeding Millipedes and centipedes have mouthparts that can bite and chew. These organisms live in downed, rotting logs and other types of dead plant material that litters the ground in forests and grasslands. Millipedes are detritivores. Centipedes, in contrast, use a pair of poison-containing fangs just behind the mouth to hunt an array of insects. Large centipedes can inject enough poison to debilitate a human.

Movement Myriapods walk or run on their many legs; a few species burrow. Some millipedes have over 190 trunk segments, each with two pairs of legs, for a total of over 750 legs. Centipedes usually have fewer than 30 segments, with one pair of legs per segment.

Reproduction Myriapod sexes are separate, and fertilization is internal. Males deposit sperm in packets that are picked up by the female or transferred to her by the male. After females lay eggs in the environment, the eggs hatch into juveniles that develop into adults via a series of molts.

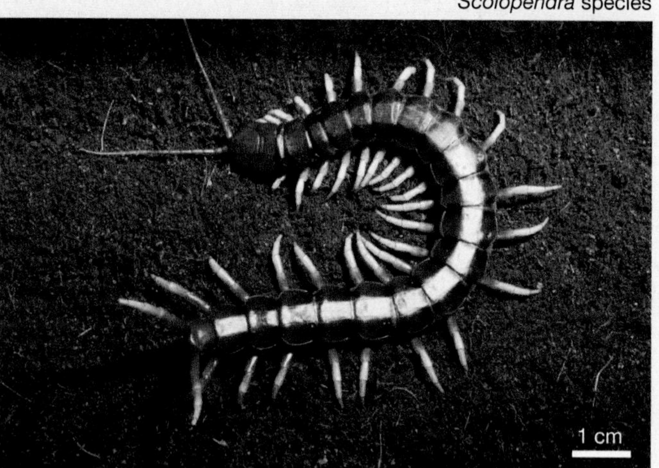

FIGURE 33.22 Myriapods Have a Pair of Legs on Each Body Segment.

Arthropoda > Chelicerata (Spiders, Ticks, Mites, Horseshoe Crabs, Daddy Longlegs, Scorpions)

Most of the 70,000 species of chelicerates are terrestrial, although the horseshoe crabs and sea spiders are marine. The lineage is considered a class in most of the traditional classification schemes and has several subgroups or subclasses. The most prominent of these chelicerate lineages is the arachnids (spiders, scorpions, mites, and ticks).

The chelicerate body consists of anterior and posterior regions (**Figure 33.23a**). The anterior region lacks antennae for sensing touch or odor but usually contains eyes. The group is named for appendages called **chelicerae,** found near the mouth. Depending on the species, the chelicerae are used in feeding, defense, copulation, movement, or sensory reception.

Feeding Spiders, scorpions, and daddy longlegs capture and sting insects or other prey. Although some of these species are active hunters, most spiders are sit-and-wait predators. They create sticky webs to capture prey, which fly or walk into the web and are subsequently trapped. The spider senses the vibrations of the struggling prey, pounces on it, and administers a toxic bite. Mites and ticks are ectoparasitic and use their piercing mouthparts to feed on host animals (**Figure 33.23b**). Horseshoe crabs eat a variety of protostomes as well as detritus. Most scorpions feed on insects; the largest scorpion species occasionally eat snakes and lizards.

Movement Like other arthropods, chelicerates move with the aid of muscles attached to an exoskeleton. They walk or crawl on their four pairs of jointed walking legs; some species are also capable of jumping. Horseshoe crabs and some other marine forms can swim slowly. Newly hatched spiders spin long, silken threads that serve as balloons and that can carry them on the wind more than 400 kilometers from the point of hatching.

Reproduction Sexual reproduction is the rule in chelicerates, and fertilization is internal in most groups. Courtship displays are extensive in many arthropod groups and may include both visual displays and the release of chemical odorants. In spiders, males use organs that are located on their legs to transfer sperm to females. These organs fit into the female reproductive tract in a "lock-and-key" fashion. Differences in the size and shape of male genitalia are often the only way to identify closely related species of spider.

(Continued on next page)

Arthropoda > Chelicerata (Spiders, Ticks, Mites, Horseshoe Crabs, Daddy Longlegs, Scorpions) *continued*

(a) Spider, showing general chelicerate features

Dolomedes fimbriatus

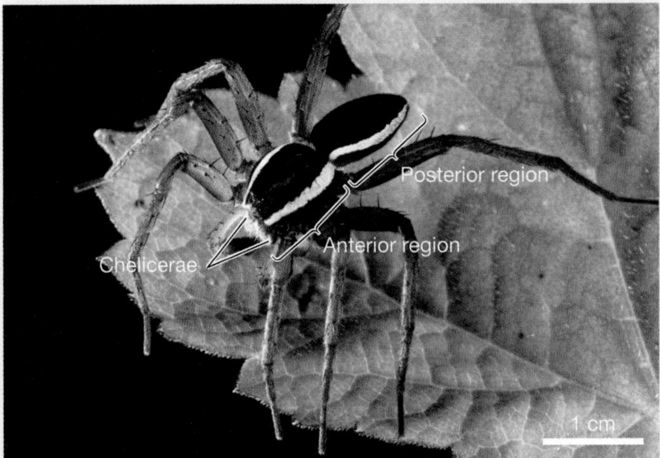

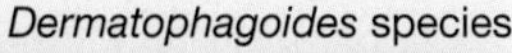

(b) Mites are ectoparasitic.

Dermatophagoides species

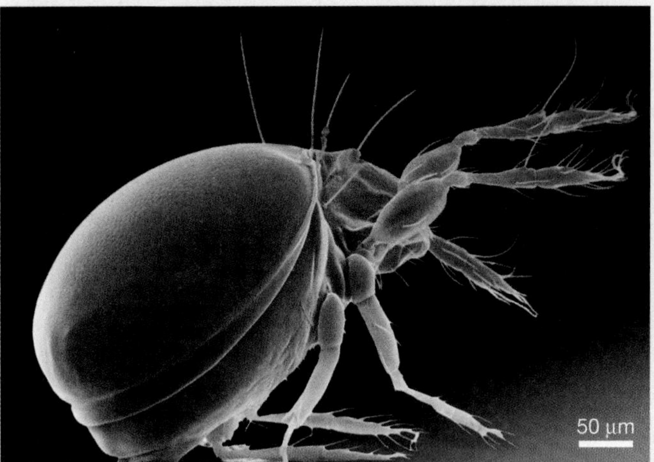

FIGURE 33.23 Chelicerates Have Two Body Regions.

EXERCISE Label the two major body regions and the chelicerae in part (b).

Development is direct, meaning that metamorphosis does not occur. In spiders, males may present a dead insect as a gift that the female eats as they mate; in some species, the male himself is eaten as sperm is being transferred. In scorpions, females retain fertilized eggs. After the eggs hatch, the young climb on the mother's back. They remain there until they are old enough to hunt for themselves.

Arthropoda > Insecta (Insects)

About 925,000 species of insects have been named thus far, but it is certain that many more exist. In terms of species diversity and numbers of individuals, insects dominate terrestrial environments. In addition, the larvae of some species are common in freshwater streams, ponds, and lakes. **Table 33.1** provides detailed notes about eight of the most prominent lineages, called orders, of insects.

Insects are distinguished by having three body regions—(1) the **head**, (2) **thorax**, and (3) **abdomen** (see Figure 33.7). Three pairs of walking legs are located on the ventral surface of the thorax. Most species have one or two pairs of wings, mounted on the dorsal (back) side of the thorax. Typically the head contains three sets of mouthparts that are derived from jointed appendages, a fourth pair of mouthparts, a pair of antennae that are used to touch and smell, and a pair of compound eyes.

Feeding Because most insect species have four sets of mouthparts (labrum, mandible, maxilla, and labium) that vary greatly in structure among species, insects are able to feed in every conceivable manner and on almost every type of food source available on land. In species with holometabolous metamorphosis, larvae have wormlike bodies; most are deposit feeders, though some are leaf eaters. Adults are predators or parasites on plant or animal tissues. Because so many insects make their living by feeding on plant tissues or fluids, the diversification of insects was closely correlated with the diversification of land plants. Insect predators usually eat other insects; insect parasites usually victimize other arthropods or mammals.

Movement Insects use their legs to walk, run, or swim, or they use their wings to fly. When insects walk or run, the sequence of movements usually results in three of their six legs maintaining contact with the ground at all times.

Reproduction Insect sexes are separate. Mating usually takes place through direct copulation, with the male inserting a sperm-transfer organ into the female. Most females lay eggs, but in a few species eggs are retained until hatching. Many species are also capable of reproducing asexually, through the production of unfertilized eggs via mitosis. In the vast majority of species, either incomplete or complete metamorphosis occurs, with complete metamorphosis being most common.

TABLE 33.1 **Prominent Orders of Insects**

Order	Common Name	Number of Known Species	Description
Coleoptera ("sheath-winged")	Beetles	350,000	*Key traits:* Hardened forewings, called elytra, protect the membranous hindwings that power flight. During flight, the elytra are held to the side and act as stabilizers. *Feeding:* Adults are important predators and scavengers. Larvae are called grubs and often have chewing mouth-parts. *Reproduction:* All have complete metamorphosis. *Notes:* The most species-rich lineage on the tree of life. Range in length from 0.25 mm to 10 cm. Scarab beetles were worshipped in ancient Egypt.
Lepidoptera ("scale-winged") © Gustav W. Verderber	Butterflies, moths	180,000	*Key traits:* Wings are covered with tiny, often colorful scales. The forewings and hindwings hook together and move as a unit. *Feeding:* Larvae usually have chewing mouthparts and are herbivorous deposit feeders or food-mass feeders; adults often feed on nectar. *Reproduction:* All have complete metamorphosis. *Notes:* Some species migrate long distances.
Diptera ("two-winged")	Flies (including mosquitoes, gnats, midges)	120,000	*Key traits:* Reduced hindwings, called halteres, act as stabilizers during flight. *Feeding:* Adults are usually liquid feeders; larvae are often parasitic. *Reproduction:* All have complete metamorphosis. *Notes:* First flies in fossil record appear 225 million years ago.
Hymenoptera ("membrane-winged")	Ants, bees, wasps	115,000	*Key traits:* Membranous forewings and hindwings lock together via tiny hooks, thus acting as a single wing during flight. They have club-like antennae. *Feeding:* Most ants feed on plant material; most bees feed on nectar; wasps are predatory and often have parasitic larvae. *Reproduction:* All have complete metamorphosis. Males are haploid (they hatch from unfertilized eggs) and females are diploid (they hatch from fertilized eggs). Females deposit eggs via an ovipositor, which in some species is modified into a stinger used in defense. Larvae are often fed and protected by adults. *Notes:* Most species live in colonies, and many are "eusocial"—meaning that some individuals in the colony help raise the queen's offspring but never reproduce themselves.

(Continued on next page)

TABLE **33.1 Prominent Orders of Insects** *(continued)*

Order	Common Name	Number of Known Species	Description
Hemiptera ("different-winged") 	Bugs (including leaf hoppers, aphids, cicadas, scale insects)	85,000	*Key traits:* Have a thickened forewing with a membranous tip; also mouthparts that are modified for piercing and sucking. *Feeding:* Most suck plant juices, but some are predatory. *Reproduction:* All have complete metamorphosis. *Notes:* Range in length from 1 mm to 11 cm.
Orthoptera ("straight-winged") 	Grasshoppers, crickets	20,000	*Key traits:* Large, muscular hind legs power movement by jumping. *Feeding:* Most have chewing mouthparts and are leaf eaters. *Reproduction:* All have incomplete metamorphosis. In many species, males give distinctive songs to attract mates. *Notes:* Range in length from 5 mm to 11.5 cm.
Trichoptera ("hairy-winged") 	Caddisflies	12,000	*Key traits:* Wings are covered with minute hairs. *Feeding:* Adults lack mouthparts and do not feed; larvae are all aquatic and feed on detritus. Most larvae build a protective case of tiny stones and sticks. *Reproduction:* All have complete metamorphosis. *Notes:* Larvae and adults are important food sources for fish.
Odonata ("toothed") 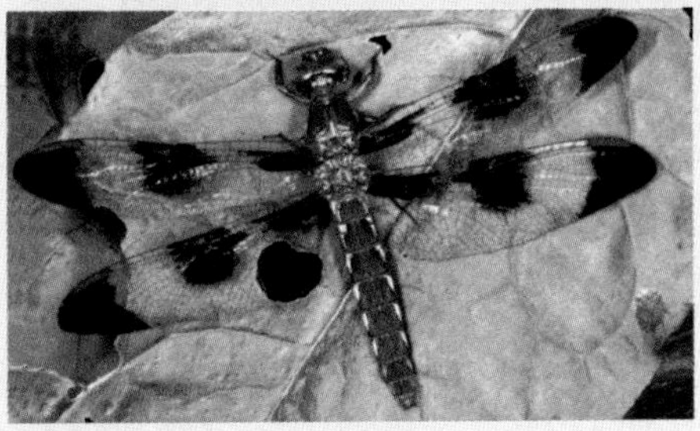	Dragonflies, damselflies	6,500	*Key traits:* Four membranous wings and long, slender abdomens. *Feeding:* Larvae and adults are predatory, often on flies. The "odon" in their name refers to strong, toothlike structures on their mandibles. *Reproduction:* All have complete metamorphosis. *Notes:* They hunt by sight and have reduced antennae (used for touch and smell) but enormous eyes with up to 28,000 lenses.

Arthropoda > Crustaceans (Shrimp, Lobster, Crabs, Barnacles, Isopods, Copepods)

The 67,000 species of **crustaceans** that have been identified to date live primarily in marine and freshwater environments. A few species of crab and some isopods are terrestrial, however. (Terrestrial isopods are known as sowbugs, pillbugs, or roly-polies). Crustaceans are common in surface waters, where they are important consumers. They are also important grazers and predators in shallow-water benthic environments.

The segmented body of most crustaceans is divided into two distinct regions: (1) the cephalothorax, which combines the head and thorax, and (2) the abdomen. Many crustaceans have a **carapace**—a platelike section of their exoskeleton that covers and protects the cephalothorax (**Figure 33.24a**). They are the only type of arthropod with two pairs of antennae, and they have sophisticated, compound eyes—usually mounted on stalks (**Figure 33.24b**). Crustaceans and trilobites were the first arthropods to appear in the fossil record.

Feeding Most crustaceans have five pairs of mouthparts that are derived from jointed appendages, and as a group they use every type of feeding strategy known. Barnacles (**Figure 33.24c**) and many shrimp are suspension feeders that use feathery structures located on head or body appendages to capture passing prey. Crabs and lobsters are active hunters, herbivores, and scavengers. Typically they have a pair of mouthparts called **mandibles** that bite or chew. Individuals capture and hold their food source with claws or other types of feeding appendages near their mouth and then use their mandibles to shred the food into small bits that can be ingested. As herbivores or detritivores, many species of crustaceans depend on algae for food.

Movement Crustacean limbs are highly diverse, and it is common for species to have many pairs of limbs and more than one type of limb. Limb structures in crustaceans include paddle-shaped forms used in swimming, feathery structures used in capturing food that is suspended in water, and jointed appendages that make sophisticated walking or running movements possible. Barnacles are one of the few types of sessile crustaceans. Adult barnacles cement their heads to a rock or other hard substrate, secrete a protective shell of calcium carbonate, and use their legs to create water currents that bring food to the mouth.

Reproduction An individual crustacean is either male or female, and sexual reproduction is the norm. Fertilization is usually internal, and eggs are usually retained by the female until they hatch. Most crustaceans pass through several distinct larval stages; many species include a larval stage called a **nauplius**, which is usually planktonic. A nauplius has a single eye and appendages that develop into the two pairs of antennae and the mouthparts of the adult.

(a) Deep-sea lobster *Enoplometopus occidentalis*

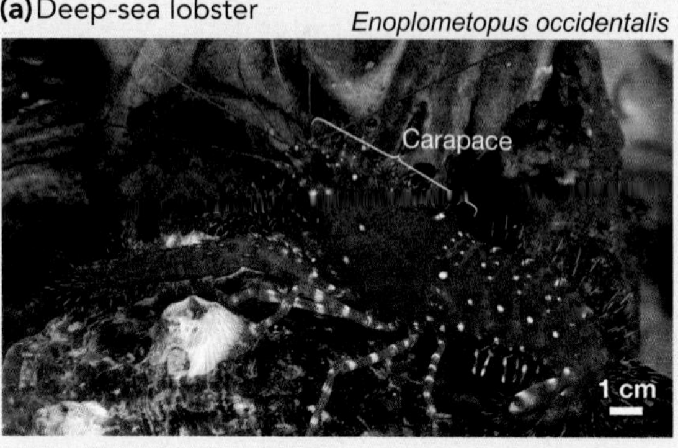

(b) Red barnacle *Tetraclita species*

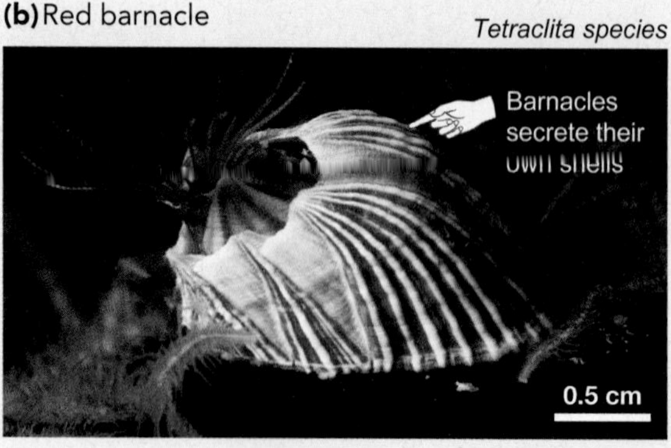

(c) Fiddler crab *Uca vocans*

FIGURE 33.24 Most Crustaceans Are Aquatic.

EXERCISE Circle the lobster's cephalothorax (combined head and thorax).

Chapter Review

SUMMARY OF KEY CONCEPTS

Molecular phylogenies support the hypothesis that protostomes are a monophyletic group divided into two major subgroups: the Lophotrochozoa and the Ecdysozoa.

The protostomes comprise about 20 phyla and were originally identified because their embryos undergo early development in the same way. Only recently did biologists recognize the existence of the lophotrochozoans and ecdysozoans, however. Several phyla in the Lophotrochozoa have characteristic feeding structures called lophophores, and many have a distinctive type of larvae called trochophores. Species in the Ecdysozoa grow by molting—meaning that they shed their old external covering and grow a new, larger one.

You should be able to draw the phylogenetic relationships among the three major lineages of bilaterians, and label the origin of growth by molting, the trochophore larva, and the protostome and deuterostome patterns of development.

at www.masteringbio.com

Protostome Diversity

Although the members of many protostome phyla have limbless, wormlike bodies and live in marine sediments, the most diverse and species-rich lineages—Mollusca and Arthropoda—have body plans with a series of distinctive, complex features. Molluscs and arthropods inhabit a wide range of environments.

All protostomes are bilaterally symmetric and triploblastic, but the nature of the body cavity is variable. The wormlike phyla of protostomes have well-developed coeloms that provide a hydrostatic skeleton used in locomotion. Flatworms lack a coelom, however, and both nematodes and rotifers evolved a pseudocoelom independently. The coelom is also much reduced or absent in the Mollusca and the Arthropoda lineages, although a body cavity is present. The mollusc's body is made up of a muscular foot, a visceral mass of organs, and a protective mantle. The arthropod body is segmented, divided into specialized regions such as the head, thorax, and abdomen of insects, and protected by an exoskeleton made of chitin.

You should be able to explain why flatworms lack a coelom, and why arthropods have a drastically reduced coelom.

Key events triggered the diversification of protostomes, including several lineages making the water-to-land transition, a diversification in appendages and mouthparts, and the evolution of metamorphosis in both marine and terrestrial forms.

The transition to living on land occurred several times independently during the evolution of protostomes. It was facilitated by watertight shells and exoskeletons and the ability to move to moist locations. Diversification in appendages and mouthparts gave protostomes the ability to move and find food in innovative ways. The evolution of metamorphosis was significant because it allowed juveniles in sessile species to disperse to new habitats and because it enabled juveniles and adults to find food in different ways.

You should be able to suggest hypotheses to explain why metamorphosis can be advantageous in both marine and terrestrial environments.

QUESTIONS

Test Your Knowledge

1. Why is it logical to observe that molluscs have a drastically reduced coelom or no coelom at all?
   a. They evolved from flatworms, which also lack a coelom.
   b. They are the most advanced of all the protostomes.
   c. Their bodies are encased by a mantle, which may or may not secrete a shell.
   d. They have a muscular foot and visceral mass that fulfill the functions of a coelom.

2. What major feature or features distinguish most of the wormlike phyla of protostomes from each other?
   a. Their mouthparts and feeding methods vary.
   b. Their modes of locomotion differ—they burrow, swim, crawl, or walk.
   c. Metamorphosis may be lacking, incomplete, or complete.
   d. They have a well-developed coelom and a tube-within-a-tube body plan.

3. What is the function of the arthropod exoskeleton?
   a. Because hard parts fossilize more readily than do soft tissues, the presence of an exoskeleton has given arthropods a good fossil record.
   b. It has no well-established function. (Trilobites had an exoskeleton, and they went extinct.)
   c. It provides protection and functions in locomotion.
   d. It makes growth by molting possible.

4. Why is it logical to observe that Platyhelminthes have flattened bodies?
   a. They have simple bodies and evolved early in the diversification of protostomes.
   b. They lack a coelom, so their body cannot form a rounded tube-within-a-tube design.
   c. A flat body provides a surface area for gas exchange, which compensates for their lack of gas-exchange organs.
   d. They are sit-and-wait predators that hide from passing prey by flattening themselves against the substrate.

5. In number of species, number of individuals, and duration in the fossil record, which of the following phyla of wormlike animals has been the most successful? What feature is hypothesized to be responsible for this success?
   a. Annelida—a segmented body plan
   b. Platyhelminthes—bilateral symmetry but acoelomate condition
   c. Phoronida—a lophophore used as a feeding apparatus
   d. Echiura—a grooved proboscis used in suspension feeding

6. Which protostome phylum is distinguished by having body segments organized into distinct regions?
   a. Mollusca
   b. Arthropoda
   c. Annelida
   d. Nematoda

Test Your Knowledge answers: 1. d; 2. a; 3. c; 4. c; 5. a; 6. b

Test Your Understanding

Answers are available at www.masteringbio.com

1. Describe the traits that distinguish the Lophotrochozoa and Ecdysozoa. Describe the traits that are common to both groups.
2. Define a segmented body plan. Did segmentation evolve once or multiple times during the evolution of protostomes? Provide evidence to justify your answer.
3. The nature of the coelom changed drastically during the evolution of protostomes. Describe the structure and function of the coelom in nematodes, rotifers, and the wormlike phyla of protostomes other than flatworms.
4. Compare and contrast the role of metamorphosis in the life cycle of a sedentary, marine protostome such as a clam and a terrestrial, mobile protostome such as a butterfly.
5. Explain why the evolution of the exoskeleton and the jointed limb was such an important event in the evolution of arthropods.
6. The text claims that the diversification of protostomes was largely driven by diversification in mouthpart structure and feeding strategies. Considering only the wormlike phyla, give an example of this pattern.

Applying Concepts to New Situations

Answers are available at www.masteringbio.com

1. Recall that the phylum Platyhelminthes includes three major groups: the free-living turbellarians, the parasitic cestodes, and the parasitic trematodes. Because the parasitic forms are so simple morphologically, researchers suggest that they are derived from more complex, free-living forms. Draw a phylogeny of Platyhelminthes that would support the hypotheses that ancestral flatworms were morphologically complex, free-living organisms and that parasitism is a derived condition.
2. Brachiopoda is a phylum within the Lophotrochozoa. Even though they are not closely related to molluscs, brachiopods look and act like bivalve molluscs (clams or mussels). Specifically, brachiopods suspension feed, live inside two calcium carbonate shells that hinge together in some species, and attach to rocks or other hard surfaces on the ocean floor. How is it possible for brachiopods and bivalves to be so similar if they did not share a recent common ancestor?
3. The Mollusca includes a group of about 370 wormlike species called the Aplacophora. Although aplacophorans do not have a shell, their cuticle secretes calcium carbonate scales or spines. They lack a well-developed foot, and some species have a simple radula.
   - Predict where Aplacophora are found in the phylogenetic tree of Mollusca relative to bivalves, gastropods, chitons, and cephalopods. Explain your reasoning.
   - Predict where Aplacophora live and how they move. Explain your reasoning.
4. Consider the following: (a) The first arthropods to appear in the fossil record are marine crustaceans and trilobites, (b) terrestrial insects appeared relatively late in arthropod evolution, (c) the crustacean group called isopods includes both aquatic and terrestrial forms, (d) myriapods are strictly terrestrial, and (e) analyses of DNA sequence data suggest that crustaceans and insects are closely related. Based on these observations, do you agree or disagree with the hypothesis that arthropods made the water-to-land transition more than once? Explain your reasoning.

www.masteringbio.com is also your resource for • Answers to text, table, and figure caption questions and exercises • Answers to *Check Your Understanding* boxes • Online study guides and quizzes • Additional study tools including the *E-Book for Biological Science* 3rd ed., textbook art, animations, and videos.

Deuterostome Animals

34

In most habitats the "top predators"—meaning animals that prey on other animals and aren't preyed upon themselves—are deuterostomes.

KEY CONCEPTS

- The most species-rich deuterostome lineages are the echinoderms and the vertebrate groups called ray-finned fishes and tetrapods.
- Echinoderms and vertebrates have unique body plans. Echinoderms are radially symmetric as adults and have a water vascular system. All vertebrates have a skull and an extensive endoskeleton made of cartilage or bone.
- The diversification of echinoderms was triggered by the evolution of appendages called podia; the diversification of vertebrates was driven by the evolution of the jaw and limbs.
- Humans are a tiny twig on the tree of life. Chimpanzees and humans diverged from a common ancestor that lived in Africa 6–7 million years ago. Since then, at least 14 humanlike species have existed.

The **deuterostomes** include the largest-bodied and some of the most morphologically complex of all animals. They range from the sea stars that cling to dock pilings, to the fish that dart in and out of coral reefs, to the wildebeests that migrate across the Serengeti Plains of East Africa.

Biologists are drawn to deuterostomes in part because of their importance in the natural and human economies. Deuterostomes may not be as numerous as insects and other protostomes, but they act as key predators and herbivores in most marine and terrestrial habitats. If you diagram a food chain that traces the flow of energy from algae or plants up through several levels of consumers (see Chapter 30), deuterostomes are almost always at the top of the chain. In addition, humans rely on deuterostomes—particularly **vertebrates**, or animals with backbones—for food and power. Fish and domesticated livestock are key sources of protein in most cultures, and in the developing world agriculture is still based on the power generated by oxen, horses, water buffalo, or mules. In industrialized countries, millions of people bird-watch, plan vacations around seeing large mammals in national parks, or keep vertebrates as pets. Humans are deuterostomes—we have a special interest in our closest relatives, and a special fondness for them.

To introduce this key group, let's begin with an overview of the morphological traits that distinguish major lineages of deuterostomes and then delve into the phylogenetic relationships among the major vertebrate groups and the evolutionary innovations that led to their diversification. After summarizing the key characteristics of deuterostome lineages, the chapter closes with an introduction to human evolution.

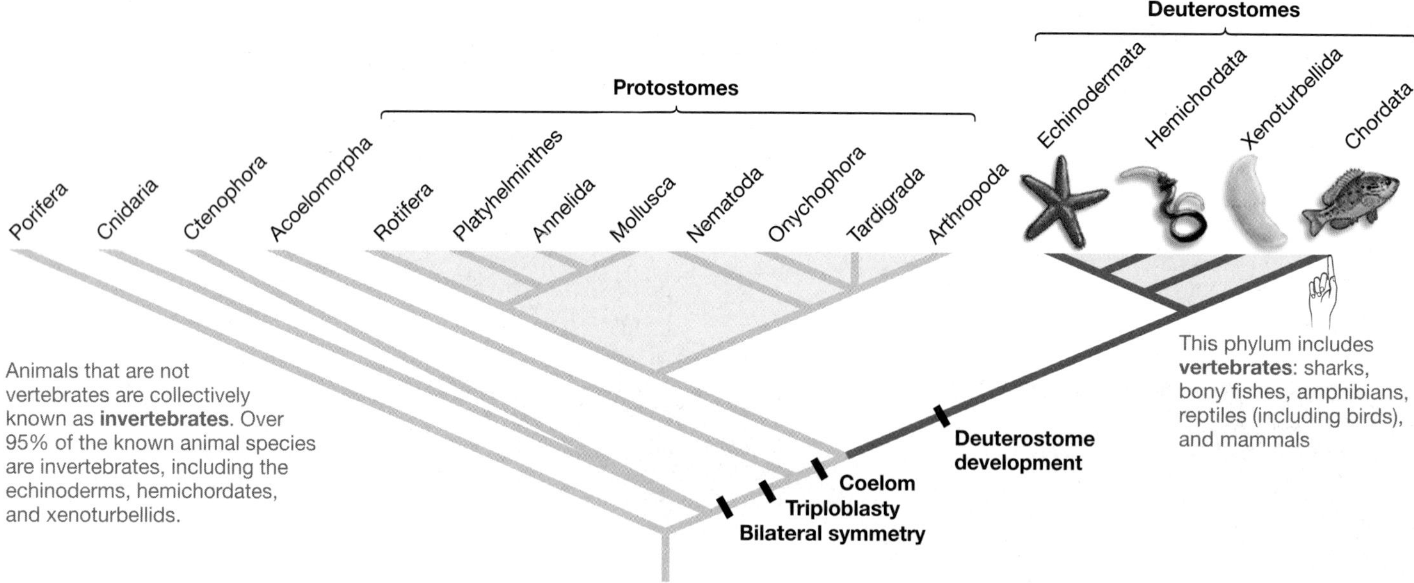

FIGURE 34.1 There Are Four Phyla of Deuterostomes. The deuterostomes comprise the phyla Echinodermata, Hemichordata, Xenoturbellida, and Chordata. The vertebrates are a subphylum of chordates.

QUESTION Are invertebrates a monophyletic or a paraphyletic group?

34.1 An Overview of Deuterostome Evolution

Most biologists recognize just four phyla of deuterostomes: the Echinodermata, the Hemichordata, the Xenoturbellida, and the Chordata (**Figure 34.1**). The echinoderms include the sea stars and sea urchins. The hemichordates, or "acorn worms," are probably unfamiliar to you—they burrow in marine sands or muds and make their living by deposit feeding or suspension feeding. A lone genus with two wormlike species, called *Xenoturbella,* was recognized as a distinct phylum in 2006. The chordates include the vertebrates. The vertebrates, in turn, comprise the sharks, bony fishes, amphibians, reptiles (including birds), and mammals. Animals that are not vertebrates are collectively known as **invertebrates**. Echinoderms, hemichordates, and *Xenoturbella* are considered invertebrates, even though they are deuterostomes.

The deuterostomes were initially grouped together because they all undergo early embryonic development in a similar way. When a humpback whale, sea urchin, or human is just beginning to grow, cleavage is radial, the gut starts developing from posterior to anterior—with the anus forming first and the mouth second—and a coelom (if present) develops from outpocketings of mesoderm (see Chapter 32).

Although deuterostomes share important features of embryonic development, their adult body plans and their feeding methods, modes of locomotion, and means for reproduction are highly diverse. Let's explore who the deuterostome animals are and how they diversified, starting with the morphological traits that distinguish the echinoderms, chordates, and vertebrates.

What Is an Echinoderm?

All deuterostomes are considered bilaterians because they evolved from an ancestor that was bilaterally symmetric (see Chapter 32). But a remarkable event occurred during the evolution of echinoderms: There was a reversion to a type of radial symmetry. Adult **echinoderms** have bodies with five-sided radial symmetry, called pentaradial symmetry (**Figure 34.2a**), even though both their larvae and their ancestors are bilaterally symmetric (**Figure 34.2b**)

Recall from Chapter 32 that radially symmetric animals do not have well-developed head and posterior regions. As a result,

(a) Adult echinoderms are radially symmetric.

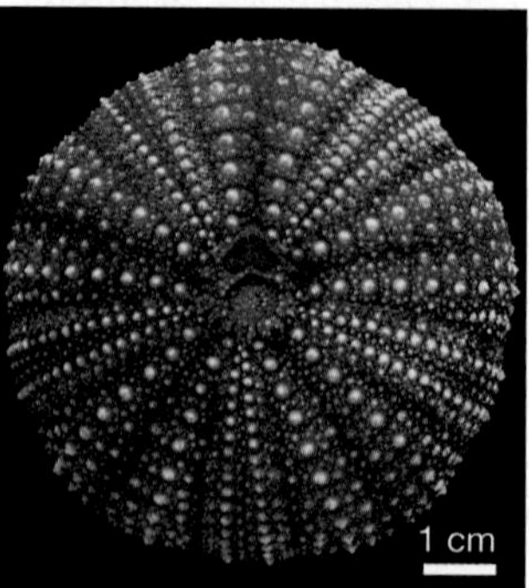

(b) Echinoderm larvae are bilaterally symmetric.

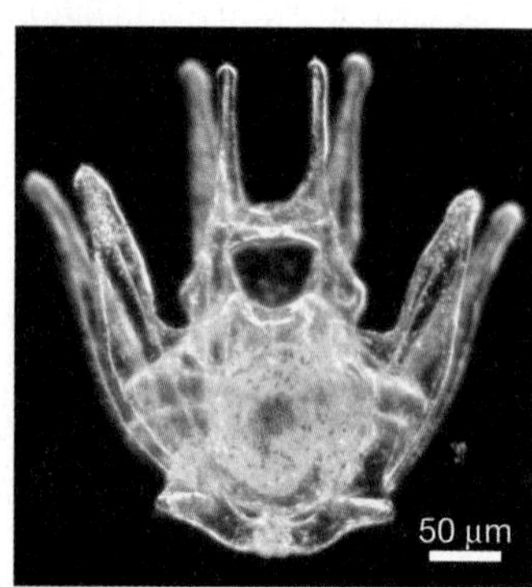

FIGURE 34.2 Body Symmetry Differs in Adult and Larval Echinoderms. (a) The skeleton of an adult and **(b)** the larva of sea urchin *Strongylocentrotus franciscanus*. Bilaterally symmetric larvae undergo metamorphosis and emerge as radially symmetric adults.

QUESTION Why are echinoderms considered members of the lineage called Bilateria (bilaterally symmetric animals)?

(a) Echinoderms have a water vascular system.

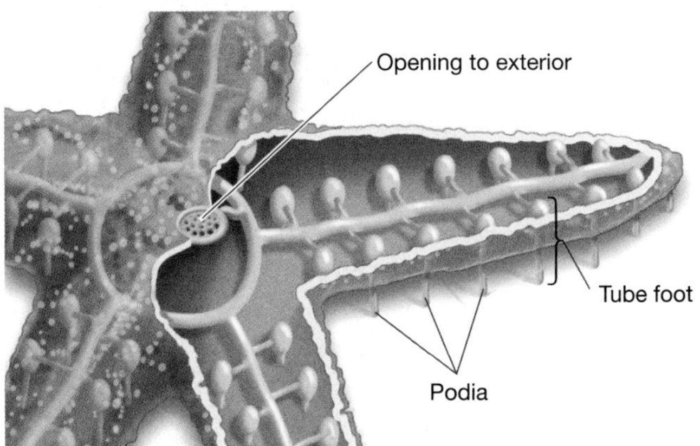

(b) Podia project from the underside of the body.

FIGURE 34.3 Echinoderms Have a Water Vascular System. (a) The water vascular system is a series of tubes and reservoirs that radiates throughout the body, forming a sophisticated hydrostatic skeleton. **(b)** Podia aid in movement because they extend from the body and can grab and release the substrate

EXERCISE Echinoderms have an endoskeleton just under their skin, which functions in protection. Label it in part (a).

they tend to interact with the environment in all directions at once instead of facing the environment in one direction. If adult echinoderms are capable of movement, they tend to move equally well in all directions instead of only headfirst.

The other remarkable event in echinoderm evolution was the origin of a unique morphological feature: a series of branching, fluid-filled tubes and chambers called the **water vascular system** (**Figure 34.3a**). One of the tubes is open to the exterior where it meets the body wall, so seawater can flow into and out of the system. Inside, fluids move via the beating of cilia that line the interior of the tubes and chambers. In effect, the water vascular system forms a sophisticated hydrostatic skeleton.

Figure 34.3a highlights a particularly important part of the system called tube feet. **Tube feet** are elongated, fluid-filled structures. **Podia** (literally, "feet") are sections of the tube feet that project outside the body (**Figure 34.3b**) and make contact with the substrate. As podia extend and contract in a coordinated way along the base of an echinoderm, they alternately grab and release from the substrate. As a result, the individual moves.

The other noteworthy feature of the echinoderm body is its **endoskeleton**, which is a hard, supportive structure located just inside a thin layer of epidermal tissue. (The structure in Figure 34.2a is an endoskeleton). As an individual is developing, cells secrete plates of calcium carbonate inside the skin. Depending on the species involved, the plates may remain independent and result in a flexible structure or fuse into a rigid case. Along with radial symmetry and the water vascular system, this type of endoskeleton is a synapomorphy—a trait that identifies echinoderms as a monophyletic group. If you understand this concept, you should be able to indicate the origin of pentaradial symmetry in adults, the water vascular system, and the echinoderm endoskeleton on Figure 34.1.

Two other phyla form a monophyletic group with echinoderms: Xenoturbellida and Hemichordata. The xenoturbellids ("strange-flatworms") have extremely simple, wormlike body plans. They don't have a gut, coelom, or brain, and they make their living absorbing nutrients in aquatic sediments. The hemichordates got their "half-chordates" name because they have an unusual feature found in chordates—openings into the throat called **pharyngeal gill slits**. The hemichordates are suspension feeders that live buried in muddy habitats on the ocean floor. As **Figure 34.4** shows, their pharyngeal gill slits function in feeding and gas exchange. Water enters the mouth, flows through structures where oxygen and food particles are extracted, and exits through the pharyngeal gill slits. They are not members of the phylum Chordata, however, because they lack several defining features shared by chordates. If you understand this point, you should be able to indicate two events on Figure 34.1: the origin of the pharyngeal gill slits observed in hemichordates and chordates, and the loss of pharyngeal gill slits in echinoderms and *Xenoturbella*.

What Is a Chordate?

The **chordates** are defined by the presence of four morphological features: (1) pharyngeal gill slits; (2) a stiff and supportive but flexible rod called a **notochord**, which runs the length of the body; (3) a bundle of nerve cells that runs the length of the body

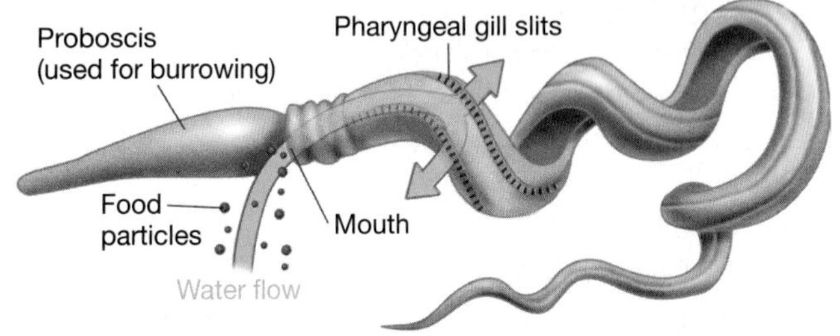

FIGURE 34.4 Hemichordates Suspension Feed Via Pharyngeal Gill Slits. The hemichordates, also known as acorn worms, burrow in marine muds and suspension feed as shown.

and forms a dorsal hollow **nerve cord**; and (4) a muscular, post-anal tail—meaning a tail that contains muscle and extends past the anus.

The phylum Chordata is made up of three major lineages, traditionally called subphyla: the (1) urochordates, (2) cephalochordates, and (3) vertebrates. All four defining characteristics of chordates—pharyngeal gill slits, notochords, dorsal hollow nerve cords, and tails—are found in these species.

- **Urochordates** are also called tunicates. As **Figure 34.5a** shows, pharyngeal gill slits are present in both larvae and adults and function in both feeding and gas exchange, much as they do in hemichordates. The notochord, dorsal hollow nerve cord, and tail are present only in the larvae, however.

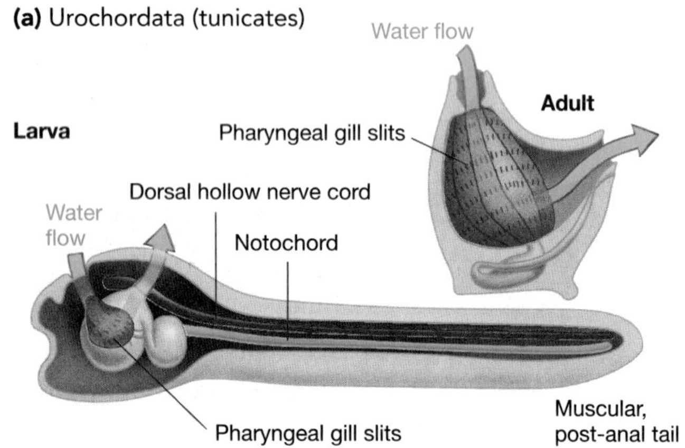

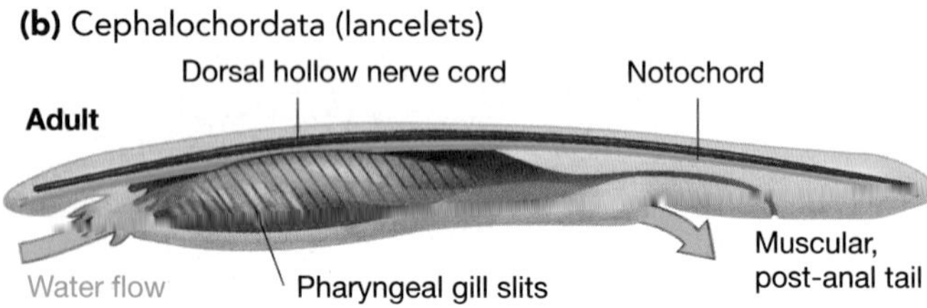

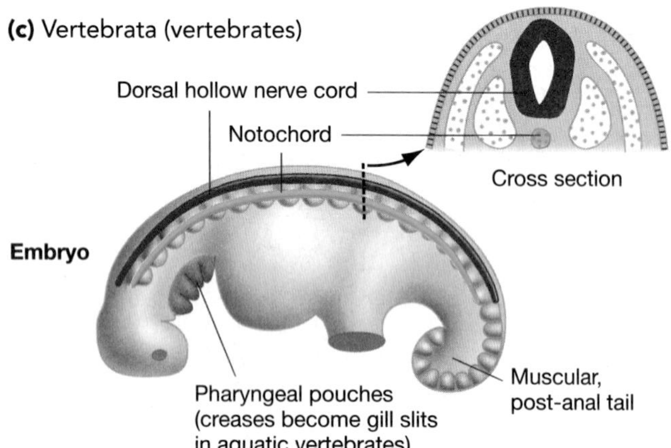

FIGURE 34.5 Four Traits Distinguish the Chordates. In chordates, either larvae or adults or both have notochords and muscular tails in addition to pharyngeal gill slits and dorsal hollow nerve cords. The three major chordate lineages are (a) tunicates, (b) lancelets, and (c) vertebrates.

QUESTION In tunicate larvae and in lancelets, what is the function of the notochord and tail?

Because the notochord stiffens the tail, muscular contractions on either side of a larva's tail wag it back and forth and result in swimming movements. As larvae swim or float in the upper water layers of the ocean, they drift to new habitats where food might be more abundant.

Urochordates underline one of the most fundamental conclusions about chordate evolution: The four features that distinguish the group enabled new types of feeding and movement. Pharyngeal gill slits function in suspension feeding. The notochord functions as a simple endoskeleton, while nerves organized into the dorsal cord stimulate muscles in the tail that make efficient swimming movements possible.

- **Cephalochordates** are also called lancelets or amphioxus; they are small, mobile suspension feeders that look a little like fish (**Figure 34.5b**). Adult cephalochordates live in ocean-bottom habitats, where they burrow in sand and suspension feed with the aid of their pharyngeal gill slits, much like urochordate larvae do. The cephalochordates also have a notochord that stiffens their bodies, so that muscle contractions on either side result in fishlike movement when they swim during dispersal or mating.

- Vertebrates (**Figure 34.5c**) include the sharks, several lineages of fishes (**Box 34.1**, page 744), amphibians, reptiles (including birds), and mammals. In vertebrates, the dorsal hollow nerve cord is elaborated into the familiar spinal cord—a bundle of nerve cells that runs from the brain to the posterior of the body. Structures called pharyngeal pouches are present in all vertebrate embryos. In aquatic species, the creases between pouches open into gill slits and develop into part of the main gas-exchange organ—the **gills**. In terrestrial species, however, gill slits do not develop after the pharyngeal pouches form. A notochord also appears in all vertebrate embryos. Instead of functioning in body support and movement, though, it helps organize the body plan. You might recall from Chapter 22 that cells in the notochord secrete proteins that help induce the formation of **somites**, which are segmented blocks of tissue that form along the length of the body. Although the notochord itself disappears, cells in the somites later differentiate into the vertebrae, ribs, and skeletal muscles of the back, body wall, and limbs. In this way, the notochord is instrumental in the development of the trait that gave vertebrates their name.

What Is a Vertebrate?

The vertebrates are a monophyletic group distinguished by two traits: a column of cartilaginous or bony structures, called **vertebrae**, that form along the dorsal sides of most species, and a **cranium**, or skull—a bony, cartilaginous, or fibrous case that encloses the brain. The vertebral column is important because it protects the spinal cord. The cranium is important because it protects the brain along with sensory organs such as eyes.

The vertebrate brain develops as an outgrowth of the most anterior end of the dorsal hollow nerve cord and is important to the vertebrate lifestyle. Vertebrates are active predators and herbivores that can make rapid, directed movements with the aid of their endoskeleton. Coordinated movements are possible in part because vertebrates have large brains divided into three distinct regions: (1) a **forebrain**, housing the sense of smell; (2) a **midbrain**, associated with vision; and (3) a **hindbrain**, responsible for balance and hearing. In vertebrate groups that evolved more recently, the forebrain is a large structure called the **cerebrum**, and the hindbrain consists of enlarged regions called the **cerebellum** and **medulla oblongata**. (The structure and functions of the vertebrate brain are analyzed in detail in Chapter 45.) The evolution of a large brain, protected by a hard cranium, was a key innovation in vertebrate evolution.

Check Your Understanding

If you understand that...

- The deuterostomes include four phyla: the echinoderms, hemichordates, xenoturbellids, and chordates.
- Within the chordates, the vertebrates are the most species-rich and morphologically diverse subphylum.

You should be able to...

1) Draw a simple phylogenetic tree showing the relationships among the echinoderm, hemichordate, chordate, and vertebrate lineages.
2) Identify the key innovations that distinguish each of these four groups.
3) Explain why each of these innovations is important in feeding or movement.

34.2 An Overview of Vertebrate Evolution

Vertebrates have been the focus of intense research for well over 300 years, in part because they are large and conspicuous and in part because they include the humans. All this effort has paid off in an increasingly thorough understanding of how the vertebrates diversified. Let's consider what data from the fossil record have to say about key events in vertebrate evolution, then examine the current best estimate of the phylogenetic relationships among vertebrates.

The Vertebrate Fossil Record

Both echinoderms and vertebrates are present in the Burgess Shale deposits that formed during the Cambrian explosion 542–488 million years ago (see also Chapter 27). The vertebrate fossils in these rocks show that the earliest members of this lineage lived in the ocean about 540 million years ago, had streamlined, fishlike bodies, and appear to have had a skull made of cartilage. **Cartilage** is a stiff tissue that consists of scattered cells in a gel-like matrix of polysaccharides and protein fibers. Several groups of early vertebrates had endoskeletons made of cartilage, as do the sharks and rays living today.

Following the appearance of vertebrates, the fossil record documents a series of key innovations that occurred as this lineage diversified (**Figure 34.6**):

- Fossil vertebrates from the early part of the Silurian period, about 440 million years ago, are the first fossils to have bone. **Bone** is a tissue consisting of cells and blood vessels encased in a matrix made primarily of a calcium- and phosphate-containing compound called hydroxyapatite, along with a small amount of protein fibers. When bone first evolved, it did not occur in the endoskeleton.

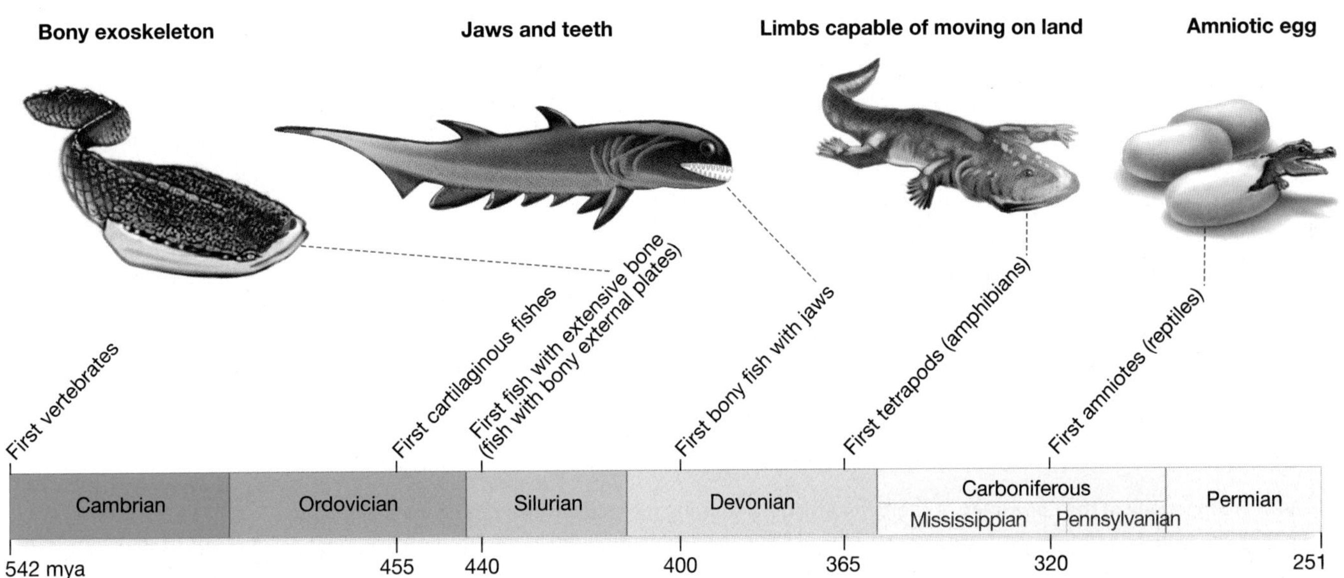

FIGURE 34.6 Timeline of the Vertebrate Fossil Record.

Instead, bone was deposited in scalelike plates that formed an **exoskeleton**—a hard, hollow structure that envelops the body. Based on the fossils' overall morphology, biologists infer that these animals swam with the aid of a notochord and that they breathed and fed by gulping water and filtering it through their pharyngeal gill slits. Presumably, the bony plates helped provide protection from predators.

- The first bony fish with jaws show up in the fossil record about 400 million years ago. The evolution of jaws was significant because it gave vertebrates the ability to bite, meaning that they were no longer limited to suspension feeding or deposit feeding. Instead, they could make a living as herbivores or predators. Soon after, jawbones with teeth appear in the fossil record. With jaws and teeth, vertebrates became armed and dangerous. The fossil record shows that a spectacular radiation of jawed fishes followed, filling marine and freshwater habitats.

- The next great event in the evolution of vertebrates was the transition to living on land. The first animals that had limbs and were capable of moving on land are dated to about 365 million years ago. These were the first of the **tetrapods** ("four-footed")—animals with four limbs (Figure 34.6).

- About 20 million years after the appearance of tetrapods in the fossil record, the first amniotes are present. The Amniota is a lineage of vertebrates that includes all tetrapods other than amphibians. The **amniotes** are named for a signature adaptation: the amniotic egg. An **amniotic egg** is an egg that

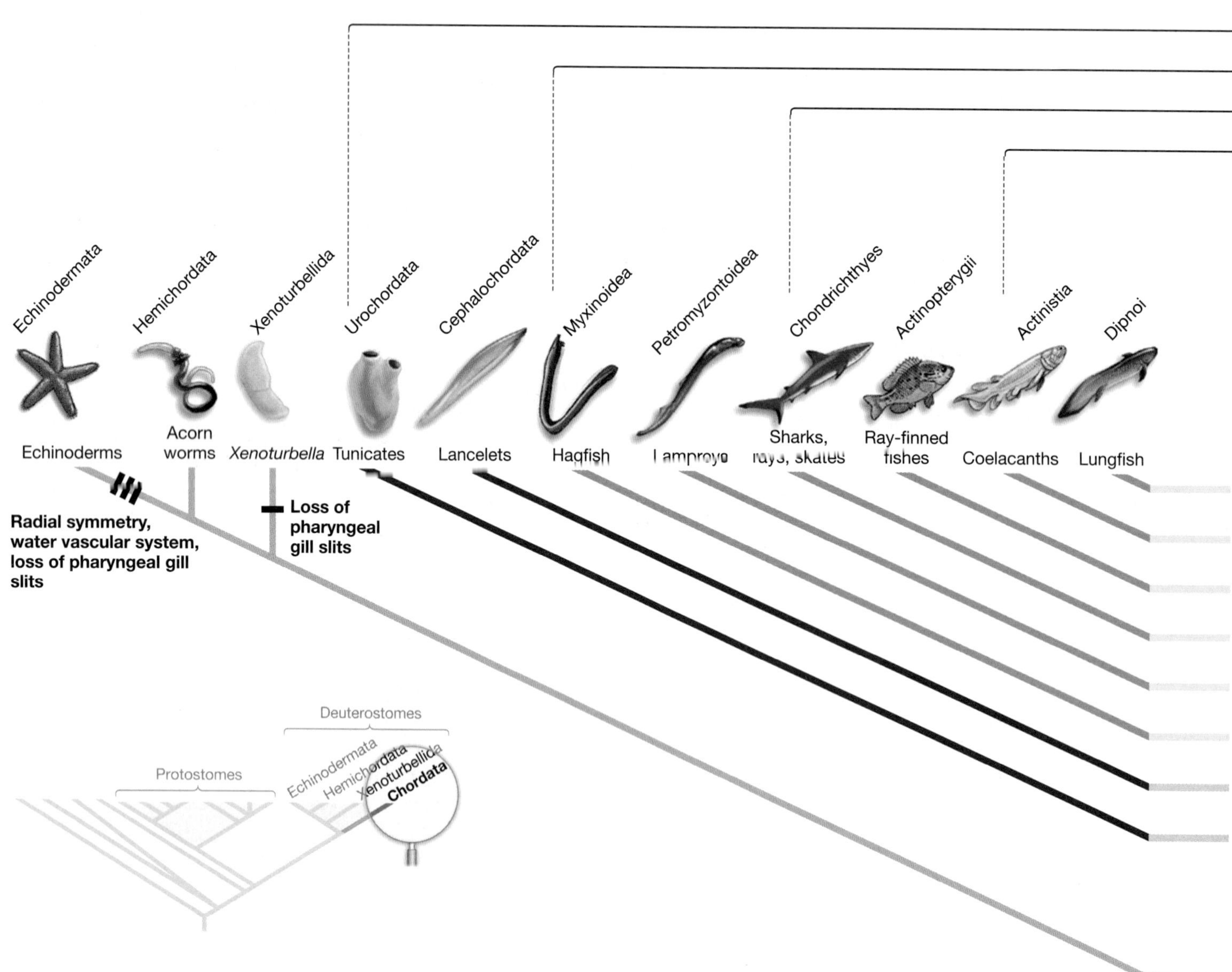

FIGURE 34.7 A Phylogeny of the Chordates. A chordate phylogeny showing the relationships among the major groups within the subphylum vertebrata.The short bars indicate where major innovations occurred as the lineages diversified.

QUESTION Which group is more closely related to the amphibians: mammals or birds?

has a watertight shell or case enclosing a membrane-bound food supply, water supply, and waste repository. The evolution of the amniotic egg was significant because it gave vertebrates the ability to reproduce far from water. Amniotic eggs resist drying out, so vertebrates that produce amniotic eggs do not have to return to aquatic habitats to lay their eggs.

To summarize, the fossil record indicates that vertebrates evolved through a series of major steps, beginning about 540 million years ago with vertebrates whose endoskeleton consisted of a notochord. The earliest vertebrates gave rise to cartilaginous fishes (sharks and rays) and fish with bony skeletons and jaws. After the tetrapods emerged and amphibians resembling salamanders began to live on land, the evolution of the amniotic egg paved the way for the evolution of the first truly terrestrial vertebrates. The fossil record indicates that a radiation of reptiles followed, along with the animals that gave rise to mammals. Do phylogenetic trees estimated from analyses of DNA sequence data agree or conflict with these conclusions?

Evaluating Molecular Phylogenies

Figure 34.7 provides a phylogenetic tree that summarizes the relationships among vertebrates, based on DNA sequence data. The labeled bars on the tree indicate where major innovations occurred. Although the phylogeny of deuterostomes continues to be a topic of intense research, researchers are increasingly confident that the relationships described in Figure 34.7 are accurate. Box 34.1 discusses some of the implications of this tree, in terms of how lineages on the tree of life are named.

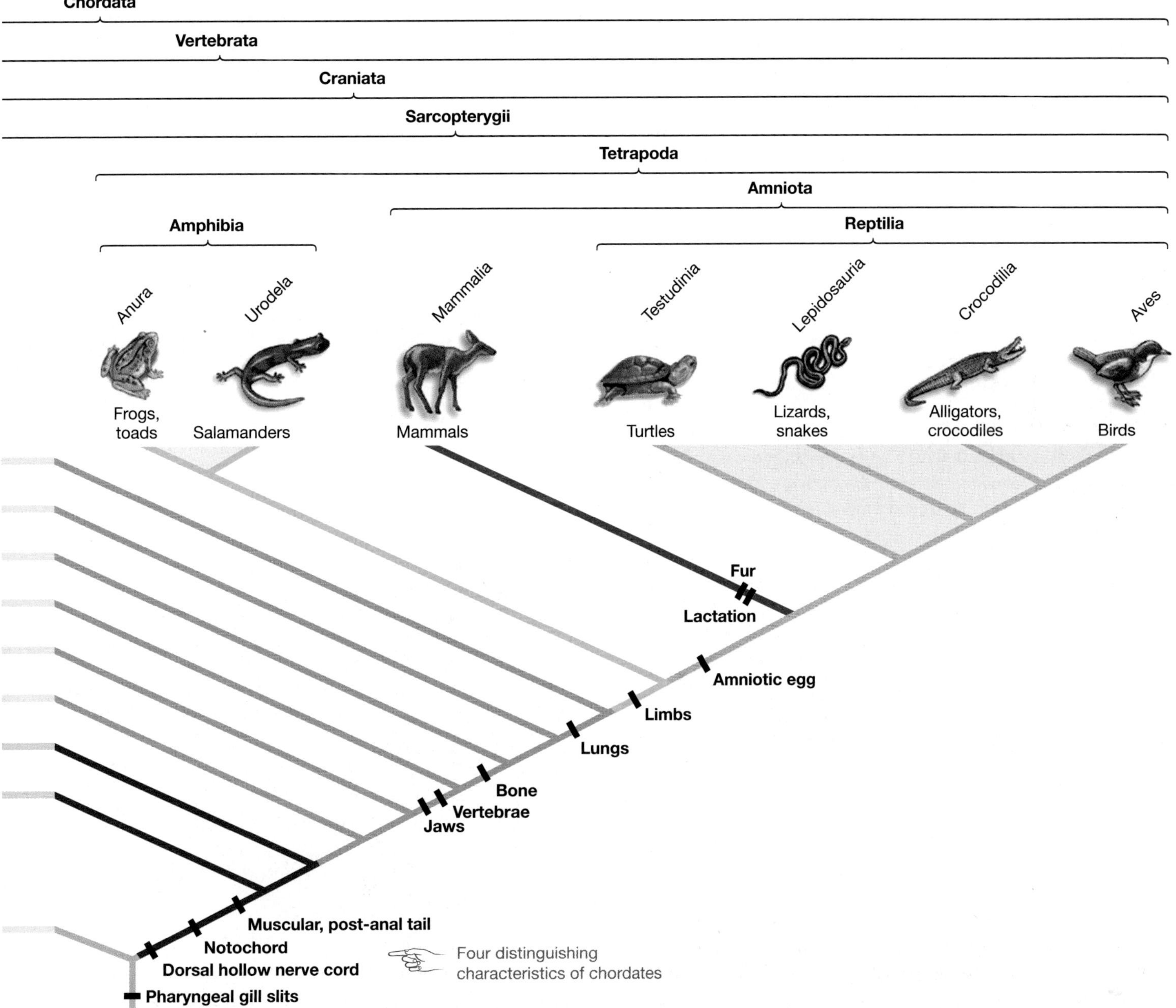

BOX 34.1 What Is a Fish?

In everyday English, people use the word *fish* to mean just about any vertebrate that lives in water and has a streamlined body and fins. But to a biologist, saying that an organism is a fish prompts a question: Which fish?

As the blue branches on Figure 34.7 show, there is no monophyletic group that includes all of the fishlike lineages. Instead, "fishy" organisms are a series of independent monophyletic groups. Taken together, they form what biologists call a **grade**—a sequence of lineages that are paraphyletic. You might recall from Chapter 30 that the seedless vascular plants also form a grade and are paraphyletic.

In naming schemes that recognize only monophyletic groups, there is no such thing as a fish. Instead, there are jawless fishes, cartilaginous fishes (sharks and rays), lobe-finned fishes, lungfishes, and ray-finned fishes. Each of the monophyletic fish groups will be explored in more detail in Section 34.5. ● If you understand the difference between a paraphyletic and monophyletic group, you should be able to draw what Figure 34.7 would look like if there actually were a monophyletic group called "The Fish."

The overall conclusion from this tree is that the branching sequence inferred from morphological and molecular data correlates with the sequence of groups in the fossil record. Reading up from the base of the tree, it is clear that the closest living relatives of the vertebrates are the cephalochordates. The most basal groups of chordates lack the skull and vertebral column that define the vertebrates, and the most ancient lineages of vertebrates lack jaws and bony skeletons. ● You should be able to indicate the origin of the cranium on Figure 34.7.

To understand what happened during the subsequent diversification of echinoderms and vertebrates, let's explore some of the major innovations involved in feeding, movement, and reproduction in more detail.

Check Your Understanding

If you understand that...

- The fossil record documents a series of innovations that occurred as vertebrates diversified, including a cartilaginous skeleton, bone, jaws, limbs, and a membrane-bound egg.

You should be able to...

1) Explain how each of these innovations allowed vertebrates to move, find food, or reproduce efficiently.
2) Indicate the origin of each of these key traits on a phylogenetic tree of vertebrates.

(MB) **Web Animation** at www.masteringbio.com
Deuterostome Diversity

34.3 What Themes Occur in the Diversification of Deuterostomes?

In terms of numbers of species and range of habitats occupied, the most successful lineages of deuterostomes are the echinoderms, with 7000 named species, and the vertebrates, with 54,000 known species. Echinoderms are widespread and abundant in marine habitats. In some deepwater environments, echinoderms represent 95 percent of the total mass of organisms. Among vertebrates, the most species-rich and ecologically diverse lineages are the ray-finned fishes and the tetrapods (**Figure 34.8**). Ray-finned fishes occupy habitats ranging from deepwater environments, which are perpetually dark, to shallow ponds that dry up each year. Tetrapods include the large herbivores and predators in terrestrial environments all over the world.

Today there are about 7000 species of echinoderms, about 27,000 species of ray-finned fish, and about 27,000 species of tetrapods. To understand why these lineages have been so successful, it's important to recognize that they have unique body plans. Recall that echinoderms are radially symmetric and have a water vascular system, and that ray-finned fishes and tetrapods have a bony endoskeleton. In this light, the situation in deuterostomes appears to be similar to that in protostomes.

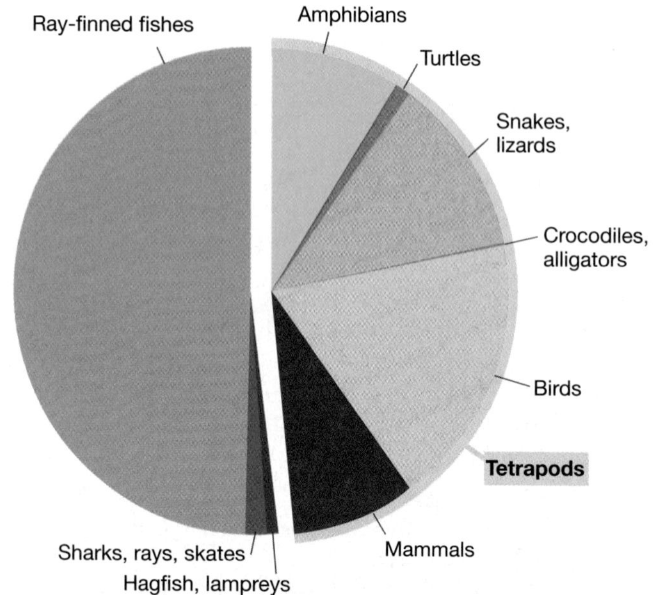

FIGURE 34.8 Relative Species Abundance Among Vertebrates.

Recall from Chapter 33 that the most evolutionarily successful protostome lineages are the arthropods and molluscs. A major concept in that chapter was that arthropods and molluscs have body plans that are unique among protostomes. Once their distinctive body plans evolved, evolution by natural selection led to extensive diversification based on novel methods for eating, moving, and reproducing.

Feeding

Animals eat to live, and it is logical to predict that each of the tens of thousands of echinoderm and vertebrate species eat different things in different ways. It is also logical to predict that echinoderms and vertebrates have traits that make diverse ways of feeding possible.

Both predictions are correct. Echinoderms have feeding strategies that are unique among marine animals. Many are based on the use of their water vascular system and podia. Ray-finned fishes and tetrapods, in contrast, depend on their jaws.

How Do Echinoderms Feed? Depending on the lineage and species in question, echinoderms make their living by suspension feeding, deposit feeding, or harvesting algae or other animals. In most cases, an echinoderm's podia play a key role in obtaining food. Many sea stars, for example, prey on bivalves. Clams and mussels respond to sea star attacks by contracting muscles that close their shells tight. But by clamping onto each shell with their podia and pulling, sea stars are often able to pry the shells apart a few millimeters (**Figure 34.9a**). Once a gap exists, the sea star extrudes its stomach from its body and forces the stomach through the opening between the bivalve's shells. Upon contacting the visceral mass of the bivalve, the stomach of the sea star secretes digestive enzymes. It then begins to absorb the small molecules released by enzyme action. Eventually, only the unhinged shells of the prey remain.

Podia are also involved in echinoderms that suspension feed (**Figure 34.9b**). In most cases, podia are extended out into the water. When food particles contact them, the podia flick the food down to cilia, which sweep the particles toward the animal's mouth. In deposit feeders, podia secrete mucus that is used to sop up food material on the substrate. The podia then roll the food-laden mucus into a ball and move it to the mouth.

The Vertebrate Jaw The most ancient groups of vertebrates have relatively simple mouthparts. For example, hagfish and lampreys lack jaws and cannot bite algae, plants, or animals. They have to make their living as ectoparasites or as deposit feeders that scavenge dead animals.

Vertebrates were not able to harvest food by biting until jaws evolved. The leading hypothesis for the origin of the jaw proposes that natural selection acted on mutations that affected the morphology of **gill arches**, which are curved regions of tissue between the gills. The jawless vertebrates have bars of cartilage that stiffen these gill arches. The gill-arch hypothesis proposes that mutation and natural selection increased the size of an arch and modified its orientation slightly, producing the first working jaw (**Figure 34.10**). Three lines of evidence, drawn from comparative anatomy and embryology, support the gill-arch hypothesis:

1. Both gill arches and jaws consist of flattened bars of bony or cartilaginous tissue that hinges and bends forward.
2. During development, the same population of cells gives rise to the muscles that move jaws and the muscles that move gill arches.
3. Unlike most other parts of the vertebrate skeleton, both jaws and gill arches are derived from specialized cells in the embryo called neural crest cells.

(a) Podia adhere to bivalve shells and pull them apart.

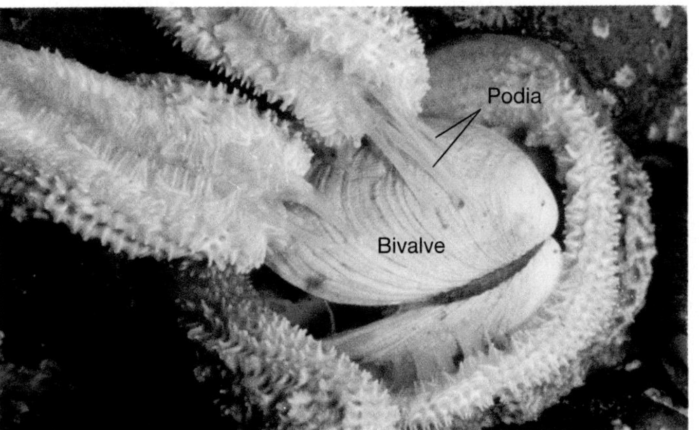

(b) Podia trap particles during suspension feeding.

FIGURE 34.9 Echinoderms Use Their Podia in Feeding. (a) A sea star pries open a bivalve slightly and then everts its stomach into the bivalve to digest it. **(b)** Feather stars extend their podia when they suspension feed.

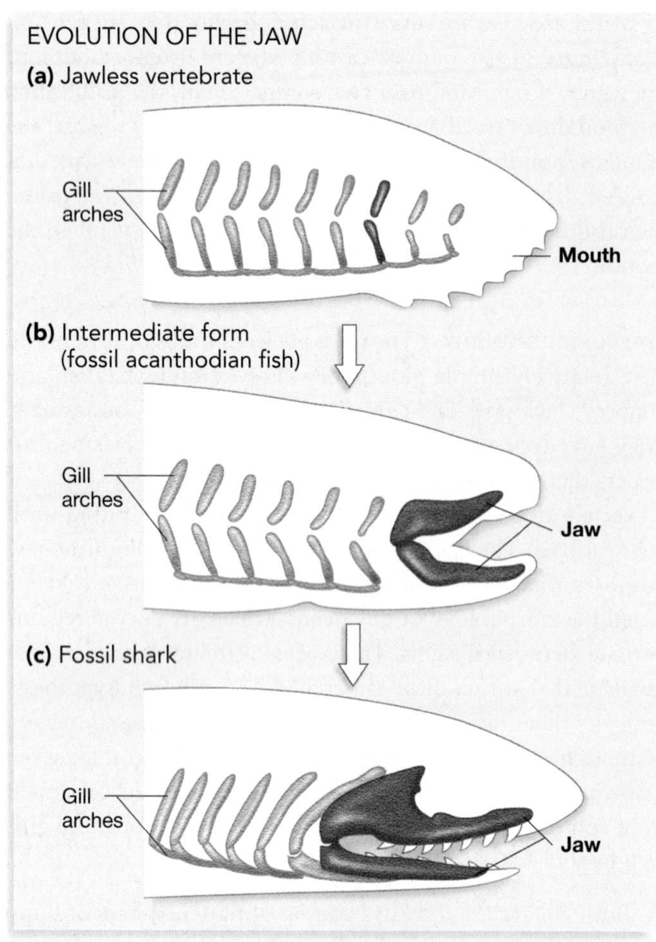

FIGURE 34.10 A Hypothesis for the Evolution of the Jaw. (a) Gill arches support the gills in jawless vertebrates. **(b)** In the fossil record, jawbones appear first in fossil fish called acanthodians. **(c)** Fossil sharks that appeared later had more elaborate jaws.

QUESTION The transition from gill arches to the jaws of acanthodian fishes is complex. Intermediate forms have yet to be found in the fossil record. Would intermediate stages in the evolution of the jaw have any function?

Taken together, these data support the hypothesis that jaws evolved from gill arches.

To explain why ray-finned fishes are so diverse in their feeding methods, biologists point to important modifications of the jaw. In most ray-finned fishes, for example, the jaw is protrusible—meaning it can be extended to nip or bite out at food. In addition, several particularly species-rich lineages of ray-finned fishes have a set of pharyngeal jaws. The **pharyngeal** ("throat") **jaw** consists of modified gill arches that function as a second set of jaws, located in the back of the mouth. Pharyngeal jaws are important because they make food processing particularly efficient. (For more on the structure and function of pharyngeal jaws, see Chapter 43.)

To summarize, the radiation of ray-finned fishes was triggered in large part by the evolution of the jaw, by modifications that made it possible to protrude the jaw, and by the origin of the pharyngeal jaw. The story of tetrapods is different, however. Although jaw structure varies somewhat among tetrapod groups, the adaptation that triggered their initial diversification involved the ability to get to food, not to bite it and process it.

Movement

The signature adaptations of echinoderms and tetrapods involve locomotion. We've already explored the water vascular system and tube feet of echinoderms; here let's focus on the tetrapod limb.

Most tetrapods live on land and use their limbs to move. But for vertebrates to succeed on land, they had to be able not only to move out of water but also to breathe air and avoid drying out. To understand how this was accomplished, consider the morphology and behavior of their closest living relatives, the lungfish (**Figure 34.11**). Most living species of lungfish inhabit shallow, oxygen-poor water. To supplement the oxygen taken in by their gills, they have lungs and breathe air. Some also have fleshy fins supported by bones and are capable of walking short distances along watery mudflats or the bottoms of ponds. In addition, some species can survive extended droughts by burrowing in mud.

Fossils provide strong links between lungfish and the earliest land-dwelling vertebrates. **Figure 34.12** shows three of the species involved. The first, an aquatic animal related to today's lungfishes, is from the Devonian period—about 382 million years ago. The second is one of the oldest tetrapods, or limbed vertebrates, found to date. This animal appears in the fossil record about 365 million years ago. The third is a more recent tetrapod fossil, dated to about 350 million years ago. The figure highlights the numbers and arrangement of bones in the fossil fish fin and the numbers and arrangement of bones in the limbs of early tetrapods. The color coding emphasizes that each fin or limb has a single bony element that is proximal (closest to the body) and then two bones that are

FIGURE 34.11 Lungfish Have Limb-like Fins. Some species of lungfish can walk or crawl short distances on their fleshy fins.

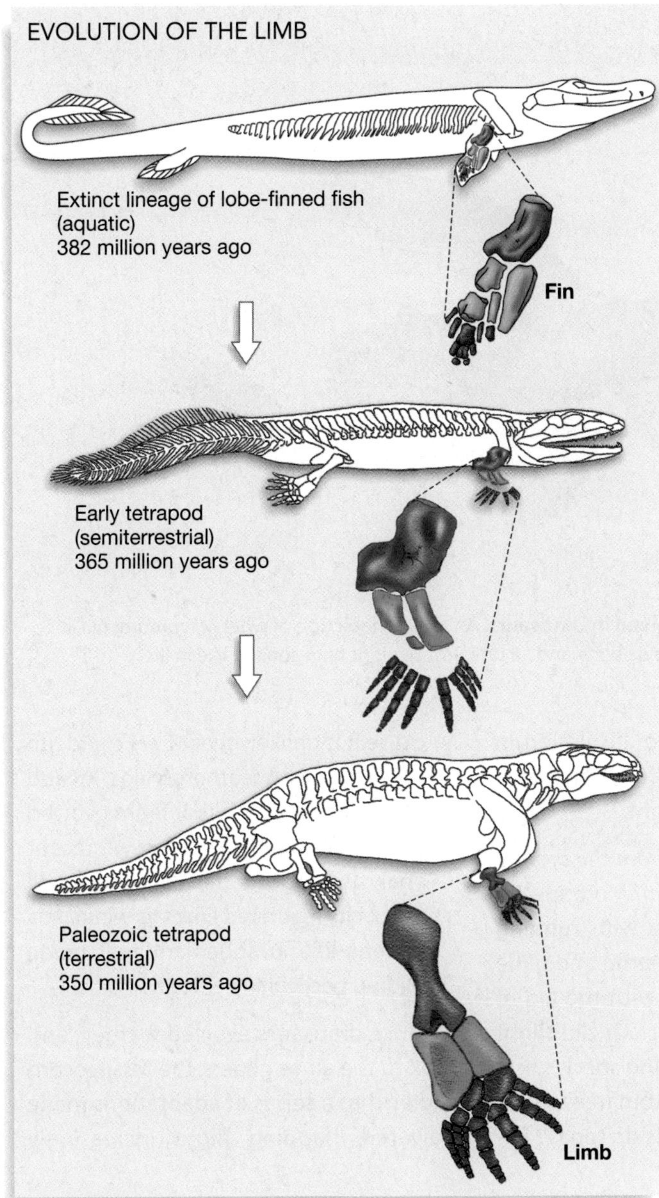

FIGURE 34.12 Fossil Evidence for a Fin-to-Limb Transition. The number and arrangement of bones in the fins and limbs of these fossil organisms agree with the general form of the modern tetrapod limb. The color coding indicates homologous elements.

distal (farther from the body) and arranged side by side, followed by a series of distal elements. Because the structures are similar, and because no other animal groups have limb bones in this arrangement, the evidence for homology is strong. Based on the lifestyle of living lungfish, biologists suggest that mutation and natural selection gradually transformed fins into limbs as the first tetrapods became more and more dependent on terrestrial habitats.

The hypothesis that tetrapod limbs evolved from fish fins has also been supported by molecular genetic evidence. Recent work has shown that several regulatory proteins involved in the development of zebrafish fins and the upper parts of mouse limbs are homologous. Specifically, the proteins produced by several different *Hox* genes are found at the same times and in the same locations in fins and limbs. These data suggest that these appendages are patterned by the same genes. As a result, the data support the hypothesis that tetrapod limbs evolved from fins.

Once the tetrapod limb evolved, natural selection elaborated it into structures that are used for running, gliding, crawling, burrowing, or swimming. In addition, wings and the ability to fly evolved independently in three lineages of tetrapods: the extinct flying reptiles called pterosaurs (pronounced *TARE-oh-sors*), birds, and bats. Tetrapods and insects are the only animals that have taken to the skies. **Box 34.2** explores how flight evolved in birds.

To summarize, the evolution of the jaw gave tetrapods the potential to capture and process a wide array of foods. With limbs, they could move efficiently on land in search of food. What about the other major challenge of terrestrial life? How did tetrapods produce offspring that could survive out of water?

Reproduction

Among the various lineages of fish, a few species undergo only asexual reproduction. When sexual reproduction occurs, it may be based on internal or external fertilization and oviparity, viviparity, or ovoviviparity. In addition, parental care is extensive in some fish species, and it often involves guarding eggs from predators and fanning them to keep oxygen levels high. All fish lay their eggs or give birth in water, however. Tetrapods were the first vertebrates that were able to breed on land.

Three major evolutionary innovations gave tetrapods the ability to produce offspring successfully in terrestrial environments: (1) the amniotic egg, (2) the placenta, and (3) elaboration of parental care. Let's explore each of these innovations in turn.

What Is an Amniotic Egg? Amniotic eggs have shells that minimize water loss as the embryo develops inside. The first tetrapods, like today's amphibians (frogs, toads, and salamanders), lacked amniotic eggs. Although their eggs were encased by a membrane, the first tetrapods laid eggs that would dry out and die unless they were laid in water. This fact limited the range of habitats these animals could exploit. Like today's amphibians, the early tetrapods were largely restricted to living in or near marshes, lakes, or ponds.

In contrast, reptiles (including birds) and the egg-laying mammals produce amniotic eggs. The outer membrane of the amniotic egg is leathery in turtles, snakes, and lizards but hard—due to deposition of the calcium carbonate—in crocodiles, birds, and

BOX 34.2 The Evolution of Flight in Birds

FIGURE 34.13 Feathers Evolved in Dinosaurs. An artist's depiction of what *Microraptor gui,* a dinosaur that had feathers on its body and all four limbs, might have looked like in life.

In 2003 Xing Xu and colleagues announced the discovery of a spectacular fossilized dinosaur called *Microraptor gui.* As **Figure 34.13** shows, *M. gui* had feathers covering its wings, body, and legs. But as impressive as it is, *M. gui* was just one in an exciting series of feathered dinosaur species unearthed by Xu's group. Taken together, the newly discovered species answer several key questions about the evolution of birds, feathers, and flight:

- *Did birds evolve from dinosaurs?* On the basis of skeletal characteristics, all of these recently discovered fossil species clearly belong to a lineage of dinosaurs called the dromaeosaurs. The fossils definitively link the dromaeosaurs and the earliest known fossil birds.
- *How did feathers evolve?* The fossils support Xu's model that feathers evolved in a series of steps, beginning with simple projections from the skin and culminating with the complex structures observed in today's birds (**Figure 34.14**). *M. gui,* for example, had two distinct types of feathers but lacked the complex feathers found in contemporary birds. It is controversial, though, whether the original function of feathers was for courtship, other types of display, or insulation. In today's birds, feathers function in display, insulation, and flight.
- *Did birds begin flying from the ground up or from the trees down?* More specifically, did flight evolve with running species that began to jump and glide or make short flights, with the aid of feathers to provide lift? Or did flight evolve from tree-dwelling species that used feathers to glide from tree to tree, much as flying squirrels do today? Because it is unlikely that *M. gui* could run efficiently with feathered legs, Xu and colleagues propose that flight evolved from tree-dwelling gliders. A recent analysis suggests that *M. gui* used all four of its feathered limbs as wings, in a biplane-like arrangement that made gliding particularly efficient.

Once dinosaurs evolved feathers and took to the air as gliders, the fossil record shows that a series of adaptations made powered, flapping flight increasingly

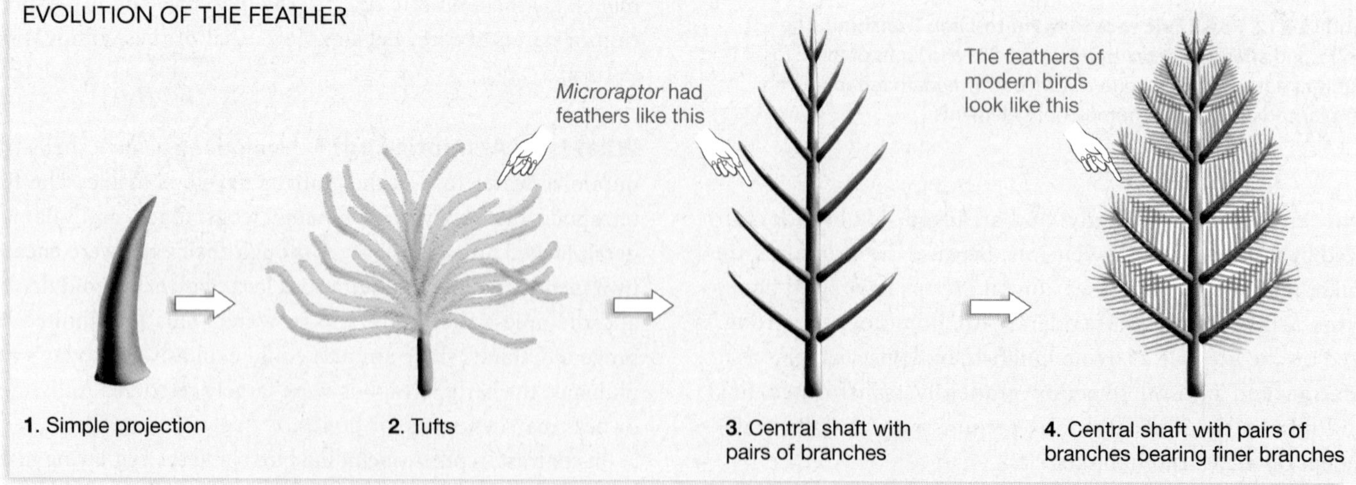

FIGURE 34.14 Feathers Evolved through Intermediate Stages. This model for the evolution of feathers is supported by the fossil record.

QUESTION Suggest a function for simple projections and tufts in the ancestors of birds, prior to the evolution of more complex feathers that made gliding or flapping flight possible.

efficient. Although most dinosaurs have a flat sternum (breastbone), the same structure in birds has an elongated projection called the keel, which provides a large surface area to which flight muscles attach (**Figure 34.15**; note that on a chicken or turkey, the flight muscles are called "breast meat"). Birds are also extraordinarily light for their size, primarily because they have a drastically reduced number of bones and because their larger bones are thin-walled and hollow—though strengthened by bony "struts" (see Figure 34.15). Birds are also capable of long periods of sustained activity year-round, because they are **endotherm**—meaning that they maintain a high body temperature by producing heat in their tissues. From dinosaurs that jumped and glided from tree to tree, birds have evolved into extraordinary flying machines.

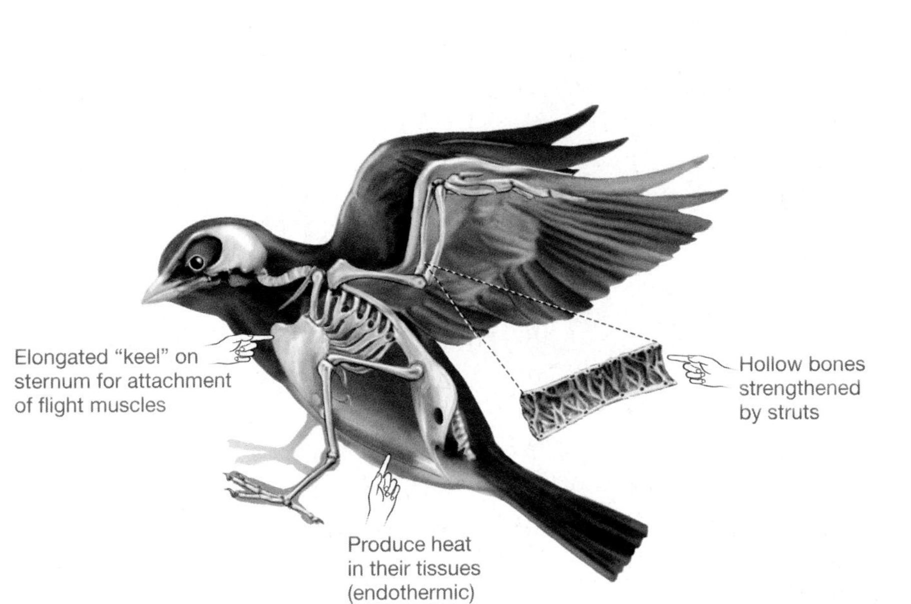

FIGURE 34.15 In Addition to Feathers, Birds Have Several Adaptations That Allow for Flight.

the egg-laying mammals. Besides having an outer shell or membrane that is largely watertight, an amniotic egg contains a membrane-bound supply of water in a protein-rich solution called **albumen** (**Figure 34.16**). The embryo itself is enveloped in a protective inner membrane known as the **amnion**. Inside an amniotic egg, the embryo is bathed in fluid. The egg itself is highly resistant to drying.

The evolution of the amniotic egg was a key event in the diversification of tetrapods because it allowed turtles, snakes, lizards, crocodiles, birds, and the egg-laying mammals to reproduce in any terrestrial environment—even habitats as dry as deserts. Members of the lineage called Amniota now occupy all types of terrestrial environments. But during the evolution of mammals, a second major innovation in reproduction occurred that eliminated the need for any type of egg laying: the placenta.

The Placenta Recall from Chapter 32 that egg-laying animals are said to be **oviparous**, while species that give birth are termed **viviparous**. In many viviparous animals, females produce an egg that contains a nutrient-rich yolk. Instead of laying the egg, however, the mother retains it inside her body. In these **ovoviviparous** species, the developing offspring depends on the resources in the egg yolk. In most mammals, however, the eggs that females produce lack yolk. After fertilization occurs and the egg is retained, the mother produces a placenta within her uterus. The **placenta** is an organ that is rich in blood vessels and that facilitates a flow of oxygen and

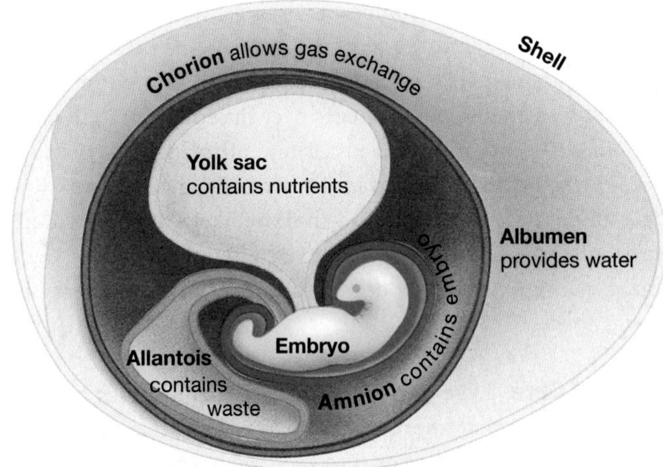

FIGURE 34.16 An Amniotic Egg. Amniotic eggs have membrane-bound sacs that hold nutrients, water, and waste and that allow gas exchange.

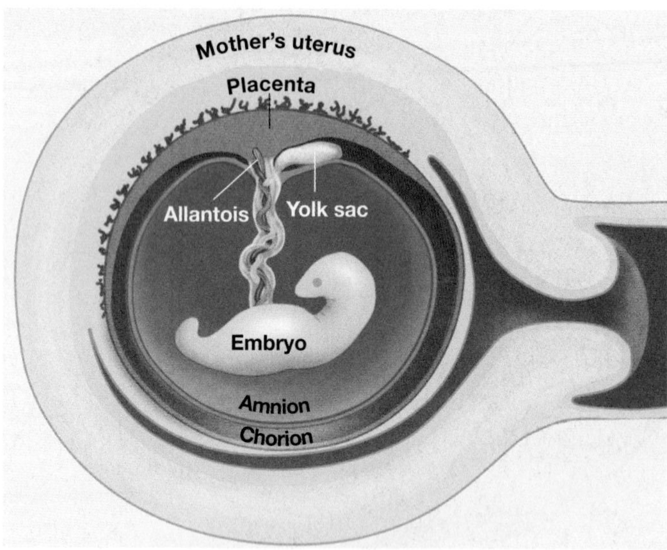

FIGURE 34.17 The Placenta Allows the Mother to Nourish the Fetus Internally.

QUESTION Compare the relative positions of the chorion and amnion here with those in the amniotic egg (Figure 34.16). Are they the same or different?

nutrients from the mother to the developing offspring (**Figure 34.17**). After a development period called **gestation**, the embryo emerges from the mother's body.

Why did viviparity and the placenta evolve? Biologists have formulated an answer to this question by pointing out that females have a finite amount of time and energy available to invest in reproduction. As a result, a female can produce a large number of small offspring or a small number of large offspring but not a large number of large offspring. Stated more formally, every female faces a **trade-off**—an inescapable compromise—between the quantity of offspring she can produce and their size. In some lineages, natural selection has favored traits that allow females to produce a small number of large, well-developed offspring. Viviparity and the placenta are two such traits. Compared with female insects or echinoderms, which routinely lay thousands or even millions of eggs over the course of a lifetime, a female mammal produces just a few offspring. But because those offspring are protected inside her body and fed until they are well developed, they are much more likely to survive than sea star or insect embryos are. Even after the birth of their young, many mammals continue to invest time and energy in rearing them.

Parental Care The term **parental care** encompasses any action by a parent that improves the ability of its offspring to survive, including incubating eggs to keep them warm during early development, keeping young warm and dry, supplying young with food, and protecting them from danger. In some insect and frog species, mothers carry around eggs or newly hatched young; in fishes, parents commonly guard eggs during development and fan them with oxygen-rich water.

The most extensive parental care observed among animals is provided by mammals and birds. In both groups, the mother and often the father continue to feed and care for individuals after birth or hatching—sometimes for many years (**Figure 34.18**). Female mammals also **lactate**—meaning that they produce a nutrient-rich fluid called milk and use it to feed their offspring after birth. With the combination of the placenta and lactation, placental mammals make the most extensive investment of time and energy in offspring known. Among large animals, the evolution of extensive parental care is hypothesized to be a major reason for the evolutionary success of mammals and birds.

(a) Mammal mothers feed and protect newborn young.

(b) Many bird species have extensive parental care.

FIGURE 34.18 Parental Care in Mammals and Birds.
(a) Female mammals feed and protect embryos inside their bodies until the young are well developed. Once the offspring is born, the mother feeds it milk until it is able to eat on its own. In some species, parents continue to feed and protect young for many years. **(b)** In birds, one or both parents may incubate the eggs, protect the nest, and feed the young after hatching occurs.

Check Your Understanding

If you understand that...

- Echinoderms and vertebrates have distinctive body plans. Subsequent diversification in each lineage was based on innovations that made it possible for species to feed, move, and reproduce in novel ways. Most echinoderms use their podia to move, but they feed in a wide variety of ways—including using their podia to pry open bivalves, suspension feed, or deposit feed.
- An array of key innovations occurred during the evolution of vertebrates: Jaws made it possible to bite and process food, limbs allowed tetrapods to move on land, and amniotic eggs could be laid on land.

You should be able to...

1) Summarize the leading hypotheses to explain how the jaw and limb evolved.
2) Diagram the structure of an amniotic egg.
3) Explain the role of increased parental care in the evolution of birds and mammals.

34.4 Key Lineages: Echinodermata

The echinoderms ("spiny-skins") were named for the spines or spikes observed in many species. They are bilaterally symmetric as larvae but undergo metamorphosis and develop into radially symmetric adults. As adults they all have a water vascular system and produce calcium carbonate plates under their skin to form an endoskeleton.

The echinoderms living today make up five major lineages, traditionally recognized as classes: (1) feather stars and sea lilies, (2) sea stars, (3) brittle stars and basket stars, (4) sea urchins and sand dollars, and (5) sea cucumbers (**Figure 34.19**). Most feather stars and sea lilies are sessile suspension feeders. Brittle stars and basket stars have five long arms that radiate out from a small central disk. They use these arms to suspension feed, deposit feed by sopping up material with mucus, or capture small prey animals. Sea cucumbers are sausage-shaped animals that suspension feed or deposit feed with the aid of tentacles arranged in a whorl around their mouths. Sea stars, sea urchins, and sand dollars are described in detail in the text that follows.

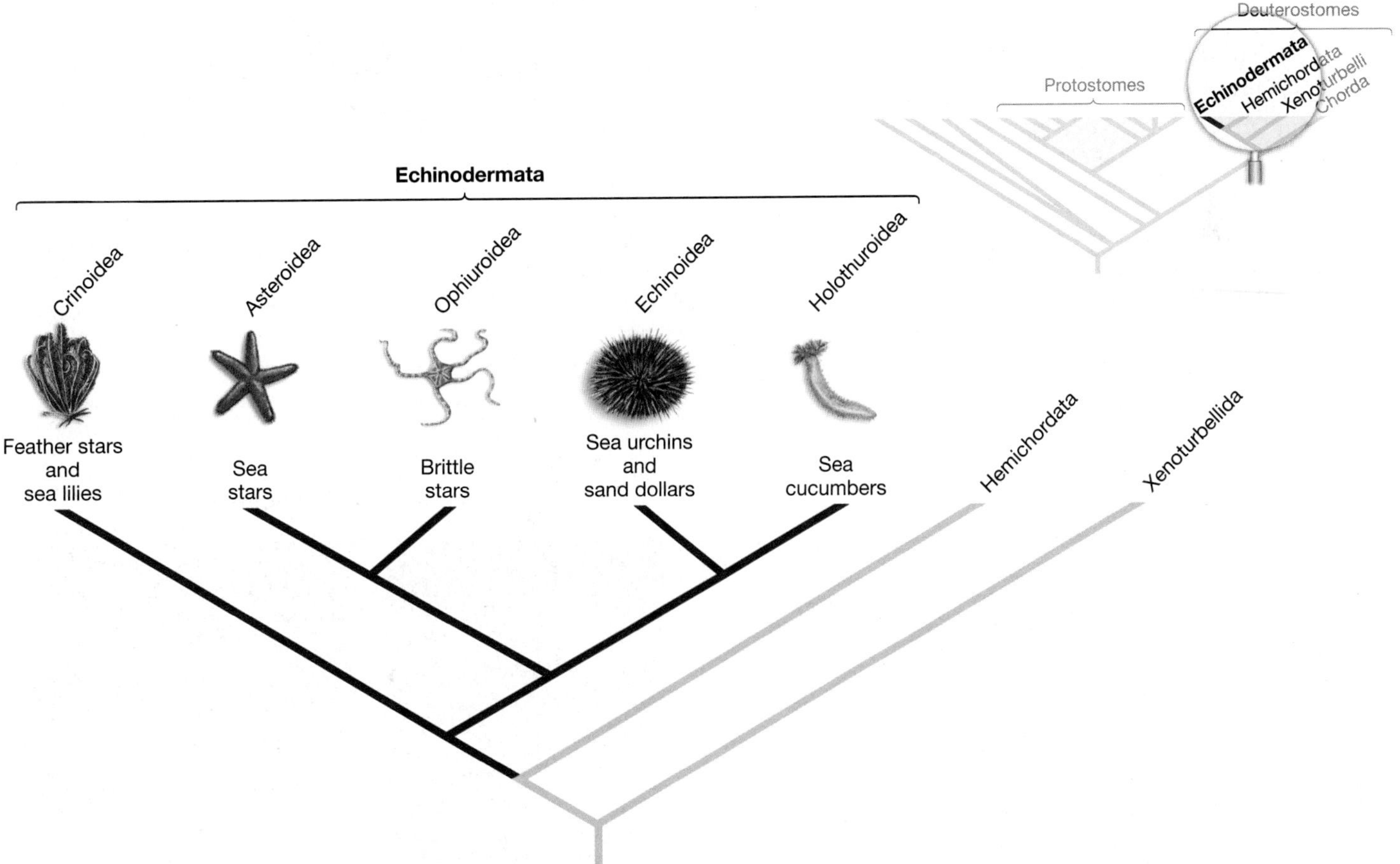

FIGURE 34.19 There Are Five Major Lineages of Echinoderms.

Echinodermata > Asteroidea (Sea Stars)

The 1700 known species of sea stars have bodies with five or more long arms—in some species up to 300—radiating from a central region that contains the mouth, stomach, and anus (**Figure 34.20**). Unlike brittle stars, though, the sea star's arms are not set off from the central region by clear, joint-like articulations. ● You should be able to indicate the origin of five arms, continuous with the central region, on Figure 34.19.

When fully grown, sea stars can range in size from less than 1 cm in diameter to 1 m across. They live on hard or soft substrates along the coasts of all the world's oceans. Although the spines that are characteristic of echinoderms are reduced to knobs on the surface of most sea stars, the crown-of-thorns star and a few other species have prominent, upright, movable spines.

Feeding Sea stars are predators or scavengers. Some species pull bivalves apart with their tube feet and evert their stomach into the prey's visceral mass. Sponges, barnacles, and snails are also common prey. The crown-of-thorns sea star specializes in feeding on corals and is native to the Indian Ocean and western Pacific Ocean. Its population has skyrocketed recently—possibly because people are harvesting their major predator, a large snail called the triton, for its pretty shell. Large crown-of-thorns star populations have led to the destruction of large areas of coral reef.

Movement Sea stars crawl with the aid of their tube feet. Any one of the five or more arms may be in "front" as the animal moves.

Reproduction Sexual reproduction predominates in sea stars, and sexes are separate. At least one sea star arm is filled with reproductive organs that produce massive amounts of gametes—millions of eggs per female, in some species. Species that are native to habitats in the far north, where conditions are particularly harsh, care for their offspring by holding fertilized eggs on their body until the eggs hatch. Most sea stars are capable of regenerating arms that are lost in predator attacks or storms. Some species can reproduce asexually by dividing the body in two, with each of the two individuals then regenerating the missing half.

Pycnopodia helianthoides

FIGURE 34.20 Sea Stars May Have Many Arms.

Echinodermata > Echinoidea (Sea Urchins and Sand Dollars)

There are about 800 species of echinoids living today; most are sea urchins or sand dollars. Sea urchins have globe-shaped bodies and long spines and crawl along substrates (**Figure 34.21a**). Sand dollars are flattened and disk-shaped, have short spines, and burrow in soft sediments (**Figure 34.21b**). ● You should be able to indicate the origin of globular or disc-shaped bodies on Figure 34.19.

Feeding Sand dollars use their mucus-covered podia to collect food particles in sand or in other soft substrates. Most types of sea urchins are herbivores. In some areas of the world, urchins are extremely important grazers on kelp and other types of algae. In fact, when urchin populations are high, their grazing can prevent the formation of kelp forests. Most echinoids have a unique, jaw-like feeding structure in their mouths that is made up of five calcium carbonate teeth attached to muscles. In many species, this apparatus can extend and retract as the animal feeds.

Movement Using their podia, sea urchins crawl and sand dollars burrow. Sea urchins can also move their spines to aid in crawling along a substrate.

Reproduction Sexual reproduction predominates in sea urchins and sand dollars. Fertilization is external, and sexes are separate.

(a) Sea urchin *Echinus tylodes*

Teeth at center of underside

1 cm

(b) Sand dollar *Dendraster excentricus*

1 cm

FIGURE 34.21 Sea Urchins and Sand Dollars Are Closely Related.

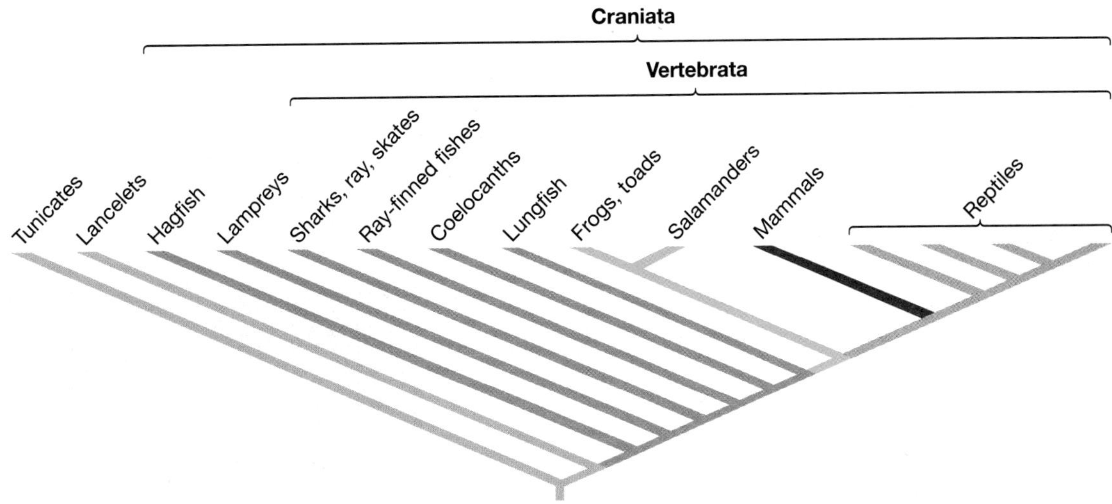

FIGURE 34.22 Craniates and Vertebrates Are Monophyletic Groups. All vertebrates are craniates, but not all craniates are vertebrates.

EXERCISE Draw a brace at the top of the tree to indicate which groups belong to the Chordata.

34.5 Key Lineages: Chordata

The chordates comprise three major subgroups or subphyla: (1) the urochordates (also called tunicates, or sea squirts and salps), (2) the cephalochordates (lancelets), and (3) the craniates and vertebrates (**Figure 34.22**). There are about 1600 species of tunicates, 24 species of lancelets, and over 50,000 vertebrates. At some stage in their life cycle, all species in the phylum Chordata have a dorsal hollow nerve cord, a notochord, pharyngeal gill slits, and a muscular tail that extends past the anus.

Chordata > Urochordata (Tunicates)

The urochordates have several common names. In addition to being called tunicates, they are also known as the sea squirts (or ascidians) and salps. All of the approximately 300 species described to date live in the ocean. Sea squirts live on the ocean floor (**Figure 34.23a**), while salps live in open water (**Figure 34.23b**).

The distinguishing characters of urochordates include an exoskeleton-like coat of polysaccharide, called a tunic, that covers and supports the body; a U-shaped gut; and two openings, called siphons, where water enters and leaves an individual during feeding. You should be able to indicate the origin of the tunic, siphons, and the U-shaped gut on Figure 34.22.

Feeding Adult urochordates use their pharyngeal gill slits to suspension feed. The slits trap particles present in the water that enters one siphon and leaves through the other siphon.

Movement Larvae swim with the aid of the notochord, which stiffens the body and functions as a simple endoskeleton. Larvae are a dispersal stage and do not feed. Adults are sessile or float in currents.

Reproduction In most species, individuals produce both sperm and eggs. In some species, both sperm and eggs are shed into the water and fertilization is external; in other species, sperm are released into the water but eggs are retained, so that fertilization and early development are internal. Asexual reproduction by budding is also common in some groups.

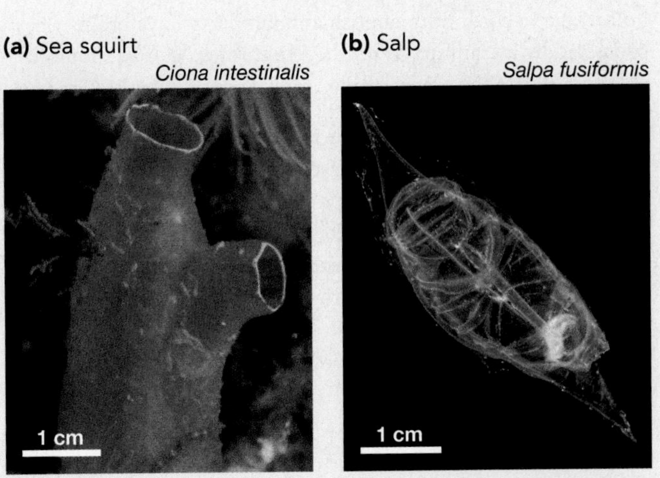

FIGURE 34.23 Sea Squirts and Salps Live in Different Habitats.

Chordata > Cephalochordata (Lancelets)

About two dozen species of cephalochordate have been described to date, all of them found in marine sands. Lancelets—also called amphioxus—have several characteristics that are intermediate between invertebrates and vertebrates. Chief among these is a notochord that is retained in adults, where it functions as an endoskeleton. ● You should be able to indicate the origin of the notochord that is retained in adults on Figure 34.22.

Feeding Adult cephalochordates feed by burrowing in sediment until only their head is sticking out into the water. They take water in through their mouth and trap food particles with the aid of their pharyngeal gill slits.

Movement Adults have large blocks of muscle arranged in a series along the length of the notochord. Lancelets are efficient swimmers because the flexible, rod-like notochord stiffens the body, making it wriggle when the blocks of muscle contract (**Figure 34.24**).

Reproduction Asexual reproduction is unknown, and individuals are either male or female. Gametes are released into the environment and fertilization is external.

Branchiostoma lanceolatum

FIGURE 34.24 Lancelets Look like Fish but Are Not Vertebrates.

Chordata > Craniata > Myxinoidea (Hagfish) and Petromyzontoidea (Lampreys)

Although recent phylogenetic analyses indicate that hagfish and lampreys may belong to two independent lineages, some data suggest that they are a single group called the Agnatha ("not-jawed"). Because these animals are the only vertebrates that lack jaws, the 110 species in the two groups are still referred to as the jawless fishes. The hagfish and lampreys are the only surviving members of the earliest branches at the base of the vertebrates.

Hagfish and lampreys have long, slender bodies and are aquatic. Most species are less than a meter long when fully grown. Hagfish lack any sort of vertebral column, but lampreys have small pieces of cartilage along the length of their dorsal hollow nerve cord. Both hagfish and lampreys have brains protected by a cranium, as do the vertebrates. ● You should be able to indicate the origin of the cranium on Figure 34.22.

Feeding Hagfish are scavengers and predators (**Figure 34.25a**). They deposit feed on the carcasses of dead fish and whales, and some are thought to burrow through ooze at the bottom of the ocean, feeding on polychaetes and other buried prey. Lampreys, in contrast, are ectoparasites. They attach to the sides of fish or other hosts by suction, then use spines in their mouth and tongue to rasp a hole in the side of their victim (**Figure 34.25b**). Once the wound is open, they suck blood and other body fluids.

FIGURE 34.25 Hagfish and Lampreys Are Jawless Vertebrates.

(Continued on next page)

Chordata > Craniata > Myxinoidea (Hagfish) and Petromyzontoidea (Lampreys) *continued*

Movement Hagfish and lampreys have a well-developed notochord and swim by making undulating movements. Lampreys can also move themselves upstream, against the flow of water, by attaching their suckers to rocks and looping the rest of their body forward, like an inchworm. Although lampreys have fins that aid in locomotion, they do not have the paired lateral appendages—meaning fins or limbs that emerge from each side—found in vertebrates.

Reproduction Virtually nothing is known about hagfish mating or embryonic development. Some lampreys live in freshwater; others are **anadromous**—meaning they spend their adult life in the ocean, but swim up streams to breed. Fertilization is external, and adults die after breeding once. Lamprey eggs hatch into larvae that look and act like lancelets. The larvae burrow into sediments and suspension feed for several years before metamorphosing into free-swimming adults.

Chordata > Vertebrata > Chondrichthyes (Sharks, Rays, Skates)

The 970 species in this lineage are distinguished by their cartilaginous skeleton (*chondrus* is the Greek word for cartilage), the presence of jaws, and the existence of paired fins. Paired fins were an important evolutionary innovation because they stabilize the body during rapid swimming—keeping it from pitching up or down, yawing to one side or the other, or rolling. ● You should be able to indicate the origin of paired fins—which are also found in other groups of fishlike organisms—on Figure 34.22.

Most sharks, rays, and skates are marine, though a few species live in freshwater. Sharks have streamlined, torpedo-shaped bodies and an asymmetrical tail—the dorsal portion is longer than the ventral portion (**Figure 34.26a**) In contrast, the dorsal-ventral plane of the body in rays and skates is strongly flattened (**Figure 34.26b**).

Feeding A few species of ray and shark suspension feed on plankton, but most species in this lineage are predators. Skates and rays lie on the ocean floor and ambush passing animals; electric rays capture their prey by stunning them with electric discharges of up to 200 volts. Most sharks, in contrast, are active hunters that chase down prey in open water and bite them. The larger species of shark feed on large fish or marine mammals. Sharks are referred to as the "top predator" in many marine ecosystems, because they are at the top of the food chain—nothing eats them. Yet the largest of all sharks, the whale shark, is a suspension feeder. Whale sharks filter plankton out of water as it passes over their gills.

Movement Rays and skates swim by flapping their greatly enlarged pectoral fins. (Pectoral fins are located on the sides of an organism; dorsal fins are located on the dorsal surface.) Sharks swim by undulating their bodies from side to side and beating their large tails.

Reproduction Sharks use internal fertilization, and fertilized eggs may be shed into the water or retained until the young are hatched and well developed. In some viviparous species, embryos are attached to the mother by specialized tissues in a mammal-like placenta, where the exchange of gases, nutrients, and wastes takes place. Skates are oviparous, but rays are viviparous.

(a) Sharks are torpedo shaped. *Prionace glauca*

(b) Skates and rays are flat. *Taeniura melanospila*

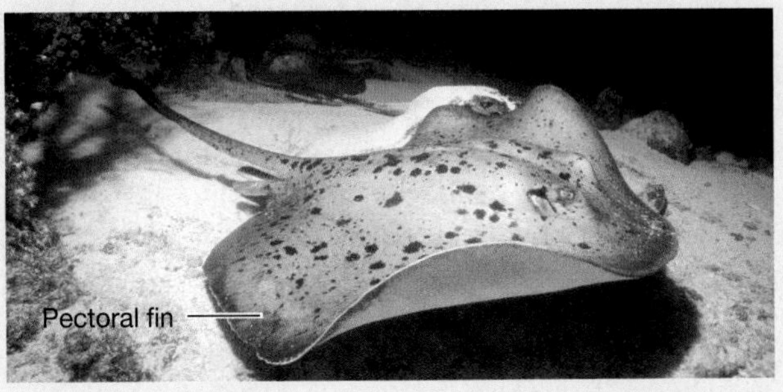

FIGURE 34.26 Sharks and Rays Have Cartilaginous Skeletons.

Chordata > Vertebrata > Actinopterygii (Ray-Finned Fishes)

Actinopterygii (pronounced *ack-tin-op-teh-RIJ-ee-i*) means "ray-finned." Logically enough, these fish have fins that are supported by long, bony rods arranged in a ray pattern. They are the most ancient living group of vertebrates that have a skeleton made of bone. Their bodies are covered with interlocking scales that provide a stiff but flexible covering, and they have a gas-filled **swim bladder**. The evolution of the swim bladder was an important innovation because it allowed ray-finned fishes to avoid sinking. Tissues are heavier than water, so the bodies of aquatic organisms tend to sink. Sharks and rays, for example, have to swim to avoid sinking. But ray-finned fishes have a bladder that changes in volume, depending on the individual's position. Gas is added to the bladder when a ray-finned fish swims down; gas is removed when the fish swims up. In this way, ray-finned fishes maintain neutral buoyancy in water of various depths and thus various pressures. ● You should be able to indicate the origin of rayed fins and the swim bladder on Figure 34.22.

The actinopterygians are the most successful vertebrate lineage based on number of species, duration in the fossil record, and extent of habitats occupied. Almost 27,000 species of ray-finned fishes have been identified thus far. In traditional classifications, Actinopterygii is considered a class.

The most important major lineage of ray-finned fishes is the Teleostei. About 96 percent of all living fish species, including familiar groups like the tuna, trout, cod, and goldfish, are teleosts (**Figure 34.27**).

Feeding Teleosts can suck food toward their mouths, grasp it with their protrusible jaws, and then process it with teeth on their jaws and with pharyngeal jaws in their throat. The size and shape of the mouth, the jaw teeth, and the pharyngeal jaw teeth all correlate with the type of food consumed. For example, most predatory teleosts have long, spear-shaped jaws armed with spiky teeth, as well as bladelike teeth on their pharyngeal jaws. Besides being major predators, ray-finned fishes are the most important large herbivores in both marine and freshwater environments.

Movement Ray-finned fishes swim by alternately contracting muscles on the left and right sides of their bodies from head to tail, resulting in rapid, side-to-side undulations. Their bodies are streamlined to reduce drag in water. Teleosts have a flexible, symmetrical tail, which reduces the need to use their pectoral (side) fins as steering and stabilizing devices during rapid swimming.

Reproduction Most ray-finned fish species rely on external fertilization and are oviparous; some species have internal fertilization with external development; still others have internal fertilization and are viviparous. Although it is common for fish eggs to be released in the water and left to develop on their own, parental care occurs in some species. Parents may carry fertilized eggs on their fins, in their mouth, or in specialized pouches to guard them until the eggs hatch. In freshwater teleosts, offspring develop directly; but marine species have larva that are very different from adult forms. As they develop, marine fish larvae undergo a metamorphosis to the juvenile form, which then grows into an adult.

FIGURE 34.27 Teleosts Are Ray-Finned Fishes That Have a Flexible Tail.

Chordata > Vertebrata > Sarcopterygii > Actinistia (Coelacanths) and Dipnoi (Lungfish)

Although coelacanths (pronounced *SEEL-uh-kanths*) and lungfish represent independent lineages, they are sometimes grouped together and called **lobe-finned fishes**. Lobe-finned fishes are common and diverse in the fossil record in the Devonian period, about 400 million years ago, but only eight species are living today. They are important, however, because they represent a crucial evolutionary link between the ray-finned fishes and the tetrapods. Instead of having fins supported by rays of bone, their fins are fleshy lobes supported by a linear—not radial—array of bones and muscles, similar to those observed in the limbs of tetrapods (**Figure 34.28**). ● You should be able to indicate the origin of a linear array of fin bones on Figure 34.22.

Coelacanths are marine and occupy habitats 150–700 m below the surface. In contrast, lungfish live in shallow, freshwater ponds and rivers (see Figure 34.11). As their name implies, lungfish have lungs and breathe air when oxygen levels in their habitats drop.

(Continued on next page)

Chordata > Vertebrata > Sarcopterygii > Actinistia (Coelacanths) and Dipnoi (Lungfish) *continued*

FIGURE 34.28 Coelacanths Are Lobe-Finned Fishes.

Some species burrow in mud and enter a quiescent, sleeplike state when their habitat dries up during each year's dry season.

Feeding Coelacanths prey on fish. Lungfish are **omnivorous** ("all-eating"), meaning that they eat algae and plant material as well as animals.

Movement Coelacanths swim by waving their pectoral and pelvic ("hip") fins in the same sequence that tetrapods use in walking with their limbs. Lungfish swim by waving their bodies, and they can use their fins to walk along pond bottoms.

Reproduction Sexual reproduction is the rule, with fertilization internal in coelacanths and external in lungfish. Coelacanths are ovoviviparous; lungfish lay eggs. Lungfish eggs hatch into larvae that resemble juvenile salamanders.

Chordata > Vertebrata > Amphibia (Frogs, Salamanders, Caecilians)

The 5500 species of **amphibians** living today form three distinct clades traditionally termed orders: (1) frogs and toads, (2) salamanders, and (3) caecilians (pronounced *suh-SILL-ee-uns*). Amphibians are found throughout the world and occupy ponds, lakes, or moist terrestrial environments (**Figure 34.29a**). Translated literally, their name means "both-sides-living." The name is appropriate because adults of most species of amphibian feed on land but lay their eggs in water. In many species of amphibians, gas exchange occurs exclusively or in part across their moist, mucus-covered skin. ● You should be able to indicate the origin of "skin-breathing" on Figure 34.22.

Feeding Adult amphibians are carnivores. Most frogs are sit-and-wait predators that use their long, extendible tongues to capture passing prey. Salamanders also have an extensible tongue, which some species use in feeding. Terrestrial caecilians prey on earthworms and other soil-dwelling animals; aquatic forms eat vertebrates and small fish.

Movement Most amphibians have four well-developed limbs. In water, frogs and toads move by kicking their hind legs to swim; on land they kick their hind legs out to jump or hop. Salamanders walk on land; in water they undulate their bodies to swim. Caecilians lack limbs and eyes; terrestrial forms burrow in moist soils (**Figure 34.29b**).

Reproduction Frogs are oviparous and have external fertilization, but salamanders and caecilians have internal fertilization. Most salamanders are oviparous, but many caecilians are viviparous. In some species of frogs, parents may guard or even carry eggs. In many frogs, young develop in the water and suspension feed on plant or algal material. Salamander larvae are carnivorous. Later the larvae undergo a dramatic metamorphosis into land-dwelling adults. For example, the fishlike tadpoles of frogs and toads develop limbs, and their gills are replaced with lungs.

(a) Frogs and other amphibians lay their eggs in water.

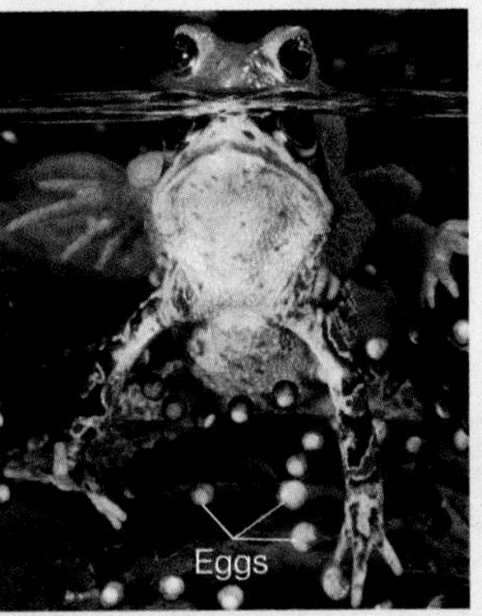

(b) Caecilians are legless amphibians.

FIGURE 34.29 Amphibians Are the Most Ancient Tetrapods.

Mammalia (Mammals)

Mammals are easily recognized by the presence of hair or fur, which serves to insulate the body. Like birds, mammals are endotherms that maintain high body temperatures by oxidizing large amounts of food and generating large amounts of heat. Instead of insulating themselves with feathers, though, mammals retain heat because the body surface is covered with layers of hair or fur. Endothermy evolved independently in birds and mammals. In both groups, endothermy is thought to be an adaptation that enables individuals to maintain high levels of activity— particularly at night or during cold weather.

In addition to being endothermic and having fur, mammals have **mammary glands**—a unique structure that makes lactation possible. The evolution of mammary glands gave mammals the ability to provide their young with particularly extensive parental care. Mammals are also the only vertebrates with facial muscles and lips and the only vertebrates that have a lower jaw formed from a single bone. In traditional classifications, Mammalia is designated as a class (**Figure 34.30**).

Mammals evolved when dinosaurs and other reptiles were the dominant large herbivores and predators in terrestrial and aquatic environments. The earliest mammals in the fossil record appear about 195 million years ago; most were small animals that were probably active only at night. Many of the 4800 species of mammal living today have good nocturnal vision and a strong sense of smell, as their ancestors presumably did. The adaptive radiation that gave rise to today's diversity of mammals did not take place until after the dinosaurs went extinct about 65 million years ago. After the dinosaurs were gone, the mammals diversified into lineages of small and large herbivores, small and large predators, or marine hunters—ecological roles that had once been filled by dinosaurs and the ocean-dwelling, extinct reptiles called mosasaurs.

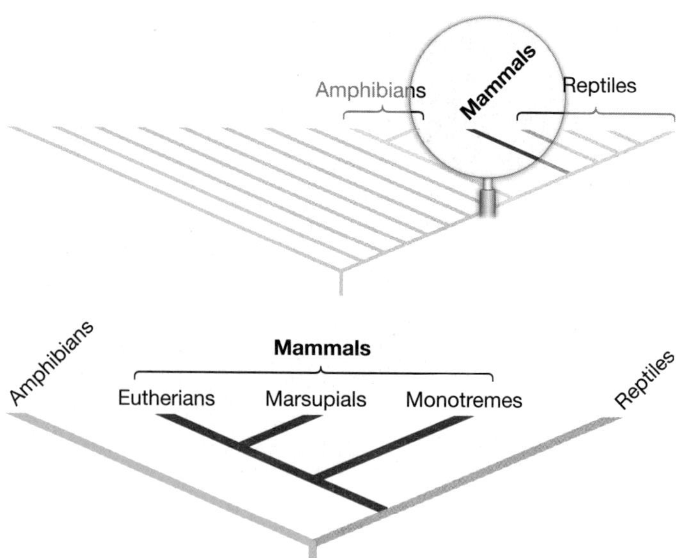

FIGURE 34.30 Mammals Are a Monophyletic Group.

Mammalia > Monotremata (Platypuses, Echidnas)

The **monotremes** are the most ancient lineage of mammals living today, and they are found only in Australia and New Zealand. They lay eggs and have metabolic rates—meaning a rate of using oxygen and oxidizing sugars for energy—that are lower than other mammals. Three species exist: one species of platypus and two species of echidna. ● You should be able to indicate the origin of fur and lactation on Figure 34.30.

Feeding Monotremes have a leathery beak or bill. The platypus feeds on insect larvae, molluscs, and other small animals in streams (**Figure 34.31a**). Echidnas feed on ants, termites, and earthworms (**Figure 34.31b**).

Movement Platypuses swim with the aid of their webbed feet. Echidnas walk on their four legs.

Reproduction Platypuses lay their eggs in a burrow, while echidnas keep their eggs in a pouch on their belly. The young hatch quickly, and the mother must continue keeping them warm and dry for another four months.

(a) Platypus *Ornithorhynchus anatinus*

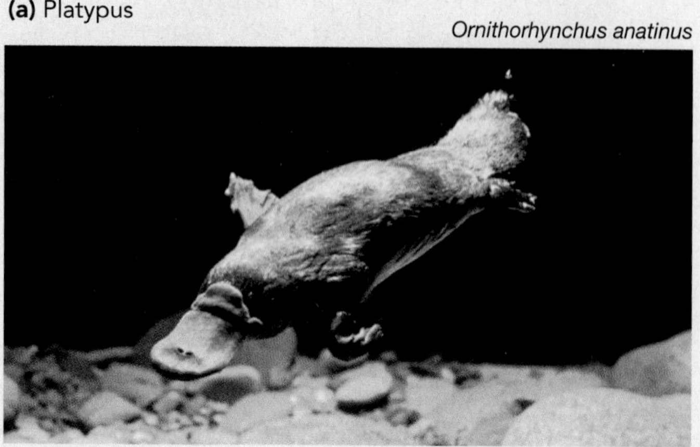

(b) Echidna *Tachyglossus aculeatus*

FIGURE 34.31 Platypuses and Echidnas Are Egg-Laying Mammals.

Mammalia > Marsupiala (Marsupials)

The 275 known species of **marsupials** live in the Australian region and the Americas (**Figure 34.32**) and include the familiar opossums, kangaroos, wallabies, and koala. Although females have a placenta that nourishes embryos during development, the young are born after a short embryonic period and are poorly developed. They crawl from the opening of the female's reproductive tract to the female's nipples, where they suck milk. They stay attached to their mother until they grow large enough to move independently. ● You should be able to indicate the origin of the placenta and viviparity—traits that are also found in Eutherian mammals—on Figure 34.30.

Didelphis virginiana

FIGURE 34.32 Marsupials Give Birth after a Short Embryonic Period. Opossums are the only marsupials in North America.

Feeding Marsupials are herbivores, omnivores, or carnivores. In many cases, convergent evolution has resulted in marsupials that are extremely similar to placental species in overall morphology and way of life. For example, a recently extinct marsupial called the Tasmanian wolf was a long-legged, social hunter similar to the timber wolves of North America and northern Eurasia. A species of marsupial native to Australia specializes in eating ants and looks and acts much like the South American anteater, which is not a marsupial.

Movement Marsupials move by crawling, gliding, walking, running, or hopping.

Reproduction Marsupial young spend more time developing while attached to their mother's nipple than they do inside her body being fed via the placenta.

Mammalia > Eutheria (Placental Mammals)

The approximately 4300 species of **placental mammals, eutherians**, are distributed worldwide. They are far and away the most species-rich and morphologically diverse group of mammals.

Biologists group placental mammals into 18 lineages called orders. The six most species-rich orders are the rodents (rats, mice, squirrels; 1814 species), bats (986 species), insectivores (hedgehogs, moles, shrews; 390 species), artiodactyls (pigs, hippos, whales, deer, sheep, cattle; 293 species), carnivores (dogs, bears, cats, weasels, seals; 274 species), and primates (lemurs, monkeys, apes, humans; 235 species).

Hylobates lar

FIGURE 34.33 Eutherians Are the Most Species-Rich and Diverse Group of Mammals.

Feeding The size and structure of the teeth correlate closely with the diet of placental mammals. Herbivores have large, flat teeth for crushing leaves and other coarse plant material; predators have sharp teeth that are efficient at biting and tearing flesh. Omnivores, such as humans, usually have several distinct types of teeth. The structure of the digestive tract also correlates with the placental mammals' diet. In some plant-eaters, for example, the stomach hosts unicellular organisms that digest cellulose and other complex polysaccharides.

Movement In placental mammals, the structure of the limb correlates closely with the type of movement performed. Eutherians fly, glide, run, walk, swim, burrow, or swing from trees (**Figure 34.33**). Limbs are reduced or lost in aquatic groups such as whales and dolphins, which swim by undulating their bodies.

Reproduction Eutherians have internal fertilization and are viviparous. An extensive placenta develops from a combination of maternal and fetal tissues; and at birth, young are much better developed than in marsupials—some are able to walk or run minutes after emerging from the mother. ● You should be able to indicate the origin of delayed birth (extended development prior to birth) on Figure 34.30. All eutherians feed their offspring milk until the young have grown large enough to process solid food. A prolonged period of parental care, extending beyond the nursing stage, is common as offspring learn how to escape predators and find food on their own.

Reptilia (Turtles, Snakes and Lizards, Crocodiles, Birds)

The **reptiles** are a monophyletic group and represent one of the two major living lineages of amniotes—the other lineage consists of the extinct mammal-like reptiles and today's mammals. The major feature distinguishing the reptilian and mammalian lineages is the number and placement of openings in the side of the skull. Jaw muscles that make possible sophisticated biting and chewing movements pass through these openings and attach to bones on the upper part of the skull.

Several features adapt reptiles for life on land. Their skin is made watertight by a layer of scales made of the protein keratin, which is also a major component of mammalian hair. Reptiles breathe air through well-developed lungs and lay shelled, amniotic eggs that resist drying out. In turtles, the egg has a leathery shell; in other reptiles, the shell is made of stiff calcium carbonate. Fertilization is internal because the sperm and egg have to meet and form the zygote before the amniotic membrane and shell form. If fertilization were external, sperm would have to pass through the shell and amnion to reach the egg cell.

The reptiles include the dinosaurs, pterosaurs (flying reptiles), mosasaurs (marine reptiles), and other extinct lineages that flourished from about 250 million years ago until the mass extinction at the end of the Cretaceous Period, 65 million years ago. Today the Reptilia are represented by four major lineages, traditionally recognized as subclasses: (1) turtles, (2) snakes and lizards, (3) crocodiles and alligators, and (4) birds (**Figure 34.34**). Except for birds, all of these groups are **ectothermic** ("outside-heated")—meaning that individuals do not use internally generated heat to regulate their body temperature. It would be a mistake, however, to conclude that reptiles other than birds do not regulate their body temperature closely. Reptiles bask in sunlight, seek shade, and perform other behaviors to keep their body temperature at a preferred level.

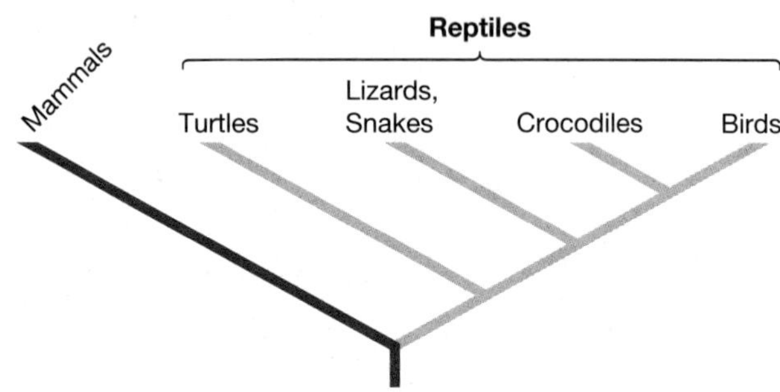

FIGURE 34.34 Reptiles Are a Monophyletic Group.

Reptilia > Testudinia (Turtles)

The 300 known species of turtles inhabit freshwater, marine, and terrestrial environments throughout the world. The testudines are distinguished by a shell composed of bony plates, covered with a material similar in composition to human fingernails, that fuse to the vertebrae and ribs (**Figure 34.35**). ● You should be able to indicate the origin of the turtle shell on Figure 34.34. The turtles' skulls are highly modified versions of the skulls of other reptiles. Turtles lack teeth, but their jawbone and lower skull form a bony beak.

Feeding Turtles are either carnivorous—feeding on whatever animals they can capture and swallow—or herbivorous. They may also scavenge dead material. Most marine turtles are carnivorous. Leatherback turtles, for example, feed primarily on jellyfish, and they are only mildly affected by the jellyfish's stinging cnidocytes (see Chapter 32). In contrast, species in the lineage of terrestrial turtles called the tortoises are plant-eaters.

Movement Turtles swim, walk, or burrow. Aquatic species usually have feet that are modified to function as flippers.

Reproduction All turtles are oviparous. Other than digging a nest prior to depositing eggs, parental care is lacking. The sex of a baby turtle is often not determined by sex chromosomes. Instead, in many species gender is determined by the temperature at which the egg develops. High temperatures produce mostly males, while low temperatures produce mostly females.

FIGURE 34.35 Turtles Have a Shell Consisting of Bony Plates.

Reptilia > Lepidosauria (Lizards, Snakes)

Most lizards and snakes are small reptiles with elongated bodies and scaly skin. Most lizards have well-developed jointed legs, but snakes are limbless (**Figure 34.36**). The hypothesis that snakes evolved from limbed ancestors is partially supported by the presence of vestigial hip and leg bones in boas and pythons. There are about 7000 species of lizards and snakes alive now. ● You should be able to indicate the origin of scaly skin on Figure 34.34.

Feeding Small lizards prey on insects. Although most of the larger lizard species are herbivores, the 3-meter-long monitor lizard from the island of Komodo is a predator that kills and eats deer. Snakes are carnivores; some subdue their prey by injecting poison through modified teeth called fangs. Snakes prey primarily on small mammals, amphibians, and invertebrates, which they swallow whole—usually headfirst.

Movement Lizards crawl or run on their four limbs. Snakes and limbless lizards burrow through soil, crawl over the ground, or climb trees by undulating their bodies.

Reproduction Although most lizards and snakes lay eggs, many are oviviviparous. Most species reproduce sexually, but asexual reproduction, via the production of eggs by mitosis, is known to occur in six groups of lizards and one snake lineage.

Morelia viridis

FIGURE 34.36 Snakes Are Limbless Predators.

Reptilia > Crocodilia (Crocodiles, Alligators)

Only 24 species of crocodile and alligator are known. Most are tropical and live in freshwater or marine environments. They have eyes located on the top of their heads and nostrils located at the top of their long snouts—adaptations that allow them to sit semi-submerged in water for long periods of time (**Figure 34.37**).

Feeding Crocodilians are predators. Their jaws are filled with conical teeth that are continually replaced as they fall out during feeding. Their usual method of killing small prey is by biting through the body wall. Large prey are usually subdued by drowning. Crocodilians eat amphibians, turtles, fish, birds, and mammals.

Movement Crocodiles and alligators walk or gallop on land. In water they swim with the aid of their large, muscular tails.

Reproduction Although crocodilians are oviparous, parental care is extensive. Eggs are laid in earth-covered nests that are guarded by the parents. When young inside the eggs begin to vocalize, parents dig them up and carry the newly hatched young inside their mouths to nearby water. Crocodilian young can hunt when newly hatched but stay near their mother for up to three years. ● You should be able to indicate the origin of extensive parental care—which also occurs in birds (and dinosaurs)—on Figure 34.34.

Alligator mississippiensis

FIGURE 34.37 Alligators Are Adapted for Aquatic Life.

Reptilia > Aves (Birds)

The fossil record provides conclusive evidence that birds descended from a lineage of dinosaurs that had a unique trait: **feathers.** In dinosaurs, feathers are hypothesized to have functioned as insulation and in courtship or aggressive displays. In birds, feathers insulate and are used for display but also furnish the lift, power, and steering required for flight. Birds have many other adaptations that make flight possible, including large breast muscles used to flap the wings. Bird bodies are lightweight because they have a reduced number of bones and organs and because their hollow bones are filled with air sacs linked to the lungs. Instead of teeth, birds have a horny beak. They are endotherms ("within-heating"), meaning that they have a high metabolic rate and use the heat produced, along with the insulation provided by feathers, to maintain a constant body temperature. The 9100 bird species alive today occupy virtually every habitat, including the open ocean (**Figure 34.38**). ● You should be able to indicate the origin of feathers, endothermy, and flight on Figure 34.34.

Feeding Plant-eating birds usually feed on nectar or seeds. Most birds are omnivores, although many are predators that capture insects, small mammals, fish, other birds, lizards, molluscs, or crustaceans. The size and shape of a bird's beak correlate closely with its diet. For example, predatory species such as falcons have sharp, hook-shaped beaks; finches and other seedeaters have short, stocky bills that can crack seeds and nuts; fish-eating species such as the great blue heron have spear-shaped beaks.

Movement Although flightlessness has evolved repeatedly during the evolution of birds, almost all species can fly. The size and shape of birds' wings correlate closely with the type of flying they do. Birds that glide or hover have long, thin wings; species that specialize in explosive takeoffs and short flights have short, stocky wings. Many seabirds are efficient swimmers, using their webbed feet to paddle or flapping their wings to "fly" under water. Ground-dwelling birds such as ostrich and pheasants can run long distances at high speed.

Reproduction Birds are oviparous but provide extensive parental care. In most species, one or both parents build a nest and incubate the eggs. After the eggs hatch, parents feed offspring until they are large enough to fly and find food on their own.

Diomedea melanophris

FIGURE 34.38 Birds Are Feathered Descendants of Dinosaurs.

34.6 Key Lineages: The Hominin Radiation

Although humans occupy a tiny twig on the tree of life, there has been a tremendous amount of research on human origins. This section introduces the lineage of mammals called the Primates, the fossil record of human ancestors, and data on the relationships among human populations living today.

The Primates

The lineage called Primates consists of two main groups: prosimians and anthropoids. The **prosimians** ("before-monkeys") consist of the lemurs, found in Madagascar, and the tarsiers, pottos, and lorises of Africa and south Asia. Most prosimians live in trees and are active at night (**Figure 34.39a**). The Anthropoidea or **anthropoids** ("human-like") include the New World monkeys found in Central and South America, the

(a) Prosimians are small, tree-dwelling primates.

Loris tardigradus

(b) New World monkeys are anthropoids.

Ateles geoffroyi

FIGURE 34.39 There Are Two Main Lineages of Primates.
(a) Prosimians live in Africa, Madagascar, and south Asia. **(b)** Anthropoids include Old World monkeys, New World monkeys, and great apes.

Old World monkeys that live in Africa and tropical regions of Asia, the gibbons of the Asian tropics, and the Hominidae, or **great apes**—orangutans, gorillas, chimpanzees, and humans (**Figure 34.39b**). The phylogenetic tree in **Figure 34.40** shows the evolutionary relationships among these groups.

Primates are distinguished by having eyes located on the front of the face. Eyes that look forward provide better depth perception than do eyes on the sides of the face. Primates also tend to have hands and feet that are efficient at grasping, flattened nails instead of claws on the fingers and toes, brains that are large relative to overall body size, complex social behavior, and extensive parental care of offspring.

The lineage in Figure 34.40 that is composed of the great apes, including humans, is known as the Hominidae or **hominids.** From extensive comparisons of DNA sequence data, it is now clear that humans are most closely related to the chimpanzees and that our next nearest living relatives are the gorillas.

Compared with most types of primate, the great apes are relatively large bodied and have long arms, short legs, and no tail. Although all of the great ape species except for the orangutans live primarily on the ground, they have distinct ways of walking. When orangutans do come to the ground, they occasionally walk with their knuckles pressed to the ground. More commonly, though, they fist-walk—that is, they walk with the backs of their hands pressed to the ground. Gorillas and chimps, in contrast, only knuckle-walk. They also occasionally rise up on two legs—usually in the context of displaying aggression. Humans are the only great ape that is fully **bipedal** ("two-footed")—meaning they walk upright on two legs. Bipedalism is, in fact, the shared derived character that defines the group called hominins. The Homininae, or **hominins**, are a monophyletic group comprising *Homo sapiens* and more than a dozen extinct, bipedal relatives.

Fossil Humans

According to the fossil record, the common ancestor of chimps and humans lived in Africa about 7 million years ago. As a

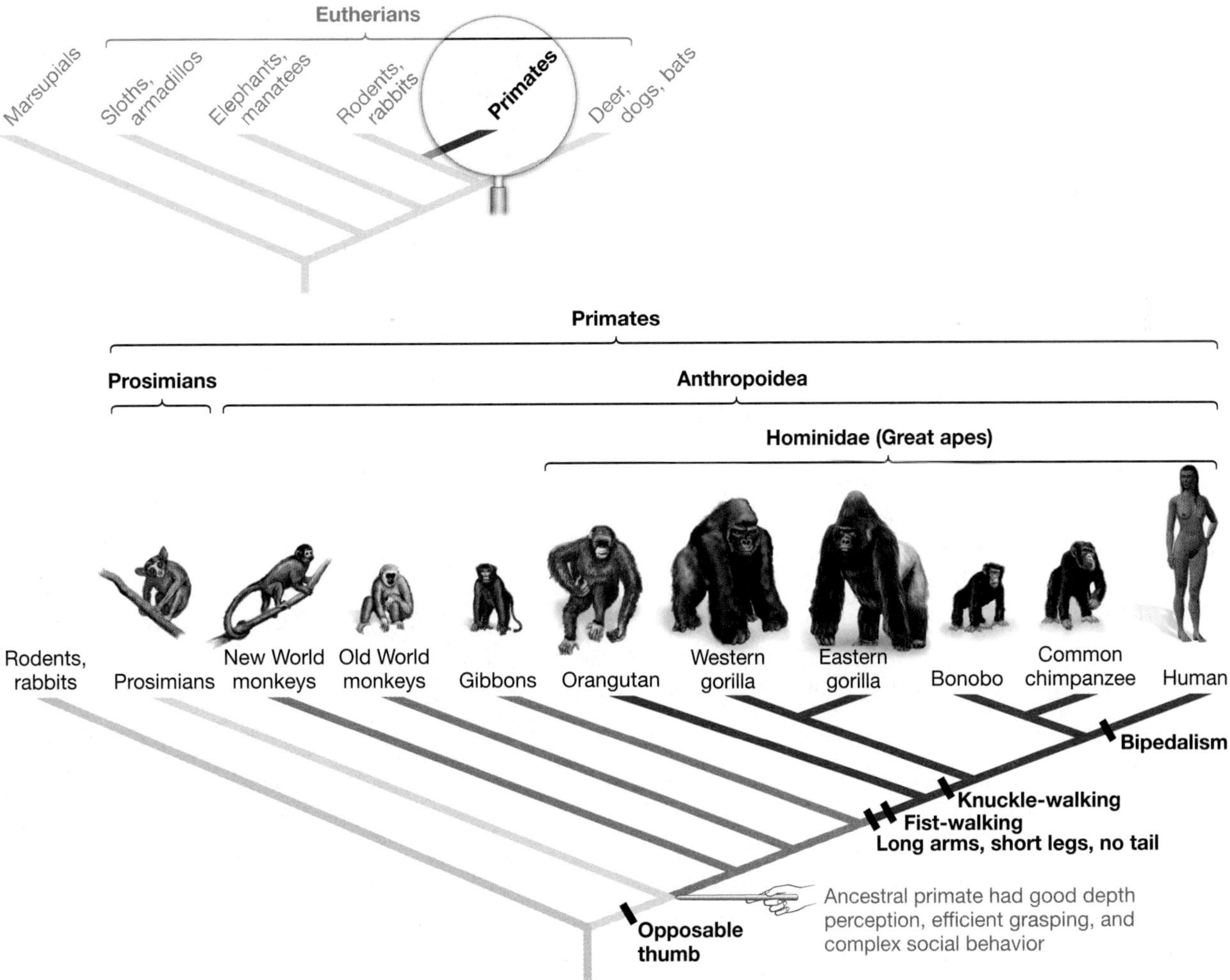

FIGURE 34.40 A Phylogeny of the Monkeys and Great Apes. Phylogenetic tree estimated from extensive DNA sequence data. According to the fossil record, humans and chimps shared a common ancestor 6 to 7 million years ago.

group, all the species on the branch leading to contemporary humans are considered hominins. The fossil record of hominins, though not nearly as complete as investigators would like, is rapidly improving. About 14 species have been found to date, and new fossils that inform the debate over the ancestry of humans are discovered every year. Although naming the hominin species and interpreting their characteristics remain intensely controversial, most researchers agree that they can be organized into four groups:

1. ***Australopithecus*** Four species of small apes called gracile australopithecines have been identified thus far (**Figure 34.41a**). The adjective *gracile*, or "slender," is appropriate because these organisms were slightly built. Adult males were about 1.5 meters tall and weighed about 36 kg. The genus name *Australopithecus* ("southern ape") was inspired by the earliest specimens, which came from South Africa. Several lines of evidence support the hypothesis that the gracile australopithecines were bipedal. For example, the hole in the back of their skulls where the spinal cord connects to the brain is oriented downward, just as it is in our species, *Homo sapiens*. In chimps, gorillas, and other vertebrates that walk on four feet, this hole is oriented backward.

2. ***Paranthropus*** Three species are grouped in the genus *Paranthropus* ("beside-human"). Like the gracile australopithecines, these robust australopithecines were bipedal. They were much stockier than the gracile forms, however—about the same height but an estimated 8–10 kilograms (20 pounds) heavier on average. In addition, their skulls were much broader and more robust (**Figure 34.41b**). All three species had massive cheek teeth and jaws, very large cheekbones, and a sagittal crest—a flange of bone at the top of the skull. Because muscles that work the jaw attach to the sagittal crest and cheekbones, researchers conclude that these organisms had tremendous biting power and made their living by crushing large seeds or coarse plant materials. One species is nicknamed "nutcracker man." The name *Paranthropus* was inspired by the hypothesis that the three known species are a monophyletic group that was a side branch during human evolution—an independent lineage that went extinct.

3. Early ***Homo*** Species in the genus *Homo* are called **humans**. As **Figure 34.41c** shows, species in this genus have flatter and narrower faces, smaller jaws and teeth, and larger braincases than the earlier hominins do. (The **braincase** is the portion of the skull that encloses the brain.) The appearance of early members of the genus *Homo* in the fossil record coincides closely with the appearance of tools made of worked stone—most of which are interpreted as handheld choppers or knives. Although the fossil record does not exclude the possibility that *Paranthropus* made tools, many researchers favor the hypothesis that extensive toolmaking was a diagnostic trait of early *Homo*.

(a) Gracile australopithecines (*Australopithecus africanus*)

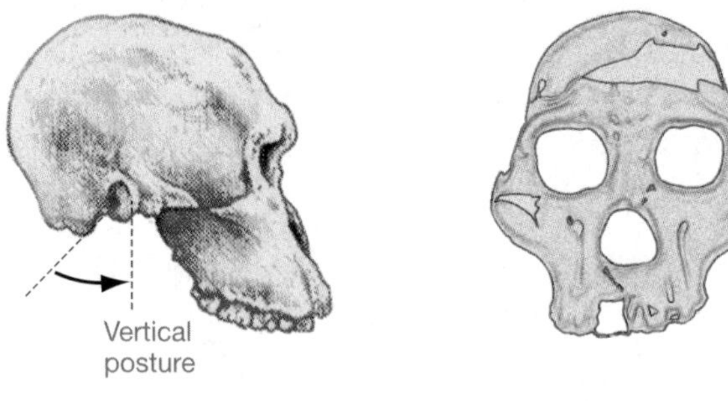

(b) Robust australopithecines (*Paranthropus robustus*)

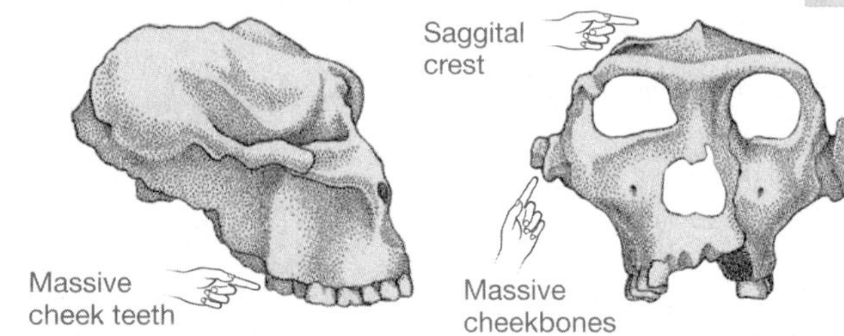

(c) Early *Homo* (*Homo habilis*)

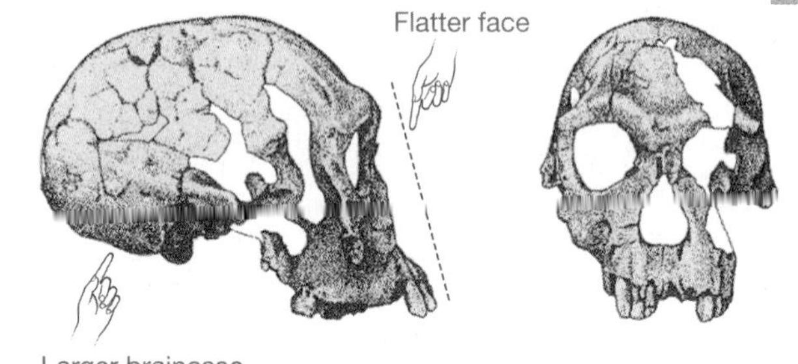

(d) Recent *Homo* (*Homo sapiens*, Cro-Magnon)

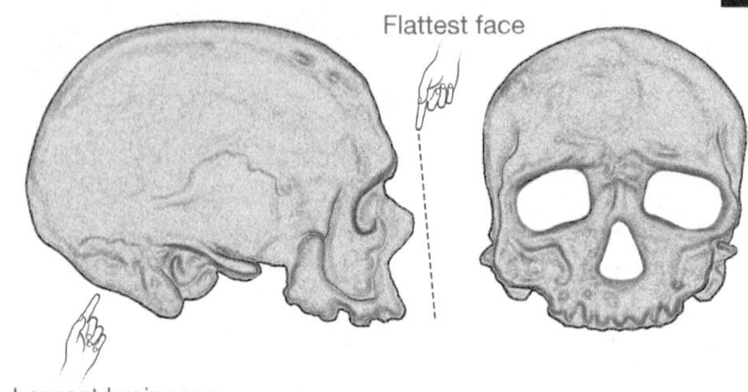

FIGURE 34.41 African Hominins Comprise Four Major Groups.

QUESTION The skulls are arranged as they appear in the fossil record, from most ancient to most recent—(a) to (d). How did the forehead and brow ridge of hominins change through time?

TABLE **34.1 Characteristics of Selected Hominins**

Species	Location of Fossils	Estimated Average Braincase Volume (cm^3)	Estimated Average Body Size (kg)	Associated with Stone Tools?
Australopithecus afarensis	Africa	450	36	no
A. africanus	Africa	450	36	no
Paranthropus boisei	Africa	510	44	no?
Homo habilis	Africa	550	34	yes
H. ergaster	Africa	850	58	yes
H. erectus	Africa, Asia	1000	57	yes
H. heidelbergensis	Africa, Europe	1200	62	yes
H. neanderthalensis	Middle East, Europe, Asia	1500	76	yes
H. floresiensis	Flores (Indonesia)	380	28	yes
H. sapiens	Middle East, Europe, Asia	1350	53	yes

4. **Recent *Homo*** The recent species of *Homo* date from 1.2 million years ago to the present. As **Figure 34.41d** shows, these species have even flatter faces, smaller teeth, and larger braincases than the early *Homo* species do. The 30,000-year-old fossil in the figure, for example, is from a population of *Homo sapiens* (our species) called the **Cro-Magnons**. The Cro-Magnons were accomplished painters and sculptors who buried their dead in carefully prepared graves. There is also evidence that another species, the **Neanderthal** people (*Homo neanderthalensis*) made art and buried their dead in a ceremonial fashion. Perhaps the most striking recent *Homo*, though, is *H. floresiensis*. This species has been found only on the island of Flores in Indonesia, which was also home to a species of dwarfed elephants. *H. floresiensis* consisted of individuals that had braincases smaller than those of gracile australopithecines and were about a meter tall. Fossil finds suggest that the species inhabited the island from about 100,000 to 12,000 years before present and that dwarfed elephants were a major source of food.

Table 34.1 summarizes data on the geographic range, braincase volume, and body size of selected species within these four groups. **Figure 34.42** provides the time range of each species in the fossil record. Although researchers do not have a solid understanding of the phylogenetic relationships among the hominin species, several points are clear from the available data. First, the shared, derived character that defines the hominins is bipedalism. Second, several species from the lineage were present simultaneously during most of hominin evolution. For example, about 1.8 million years ago there may have been as many as five hominin species living in eastern and southern Africa. Because fossils from more than one species have been found in the same geographic location in rock strata of the same age, it is almost certain that different hominin species lived in physical contact. Finally, compared with the gracile and robust australopithecines and the great apes, species in the genus *Homo* have extremely large brains relative to their overall body size.

Why did humans evolve such gigantic brains? The leading hypothesis on this question is that early *Homo* began using

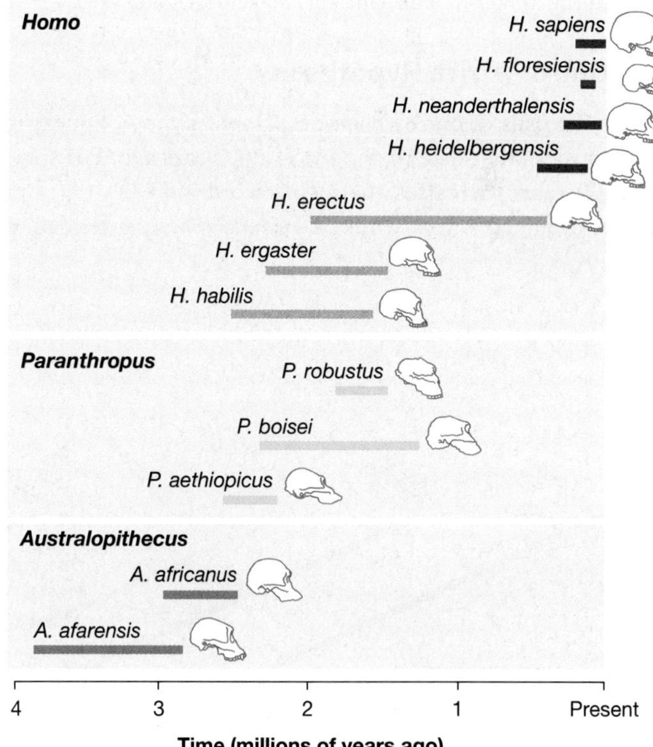

FIGURE 34.42 A Timeline of Human Evolution. Plot of the ages of selected fossil hominins.

● **QUESTION** How many species of hominin existed 2.2 million years ago, 1.8 million years ago, and 100,000 years ago?

symbolic spoken language along with initiating extensive tool use. The logic here is that increased toolmaking and language use triggered natural selection for the capacity to reason and communicate, which required a larger brain. To support this hypothesis, researchers point out that, relative to the brain areas of other hominins, the brain areas responsible for language were enlarged in the earliest *Homo* species. There is even stronger fossil evidence for extensive use of speech in *Homo neanderthalensis* and early *Homo sapiens*:

- The hyoid bone is a slender bone in the voice box of modern humans that holds muscles used in speech. In Neanderthals and early *Homo sapiens*, the hyoid is vastly different in size and shape from a chimpanzee's hyoid bone. Researchers recently found an intact hyoid bone associated with a 60,000-year-old Neanderthal individual and showed that it is virtually identical to the hyoid of modern humans.
- *Homo sapiens* colonized Australia by boat between 60,000 and 40,000 years ago. Researchers suggest that an expedition of that type could not be planned and carried out in the absence of symbolic speech.

To summarize, *Homo sapiens* is the sole survivor of an adaptive radiation that took place over the past 7 million years. From a common ancestor shared with chimpanzees, hominins evolved the ability to walk upright, make tools, and talk.

The Out-of-Africa Hypothesis

The first fossils of our own species, *Homo sapiens*, appear in African rocks that date to about 195,000 years ago. For some 130,000 years thereafter, the fossil record indicates that our species occupied Africa while *H. neanderthalensis* resided in Europe and the Middle East. Some evidence suggests that *H. erectus* may still have been present in Asia at that time. Then, in rocks dated between 60,000 and 30,000 years ago, *H. sapiens* fossils are found throughout Europe, Asia, Africa, and Australia. *H. erectus* had disappeared by this time, and *H. neanderthalensis* went extinct after coexisting with *H. sapiens* in Europe for perhaps a thousand years.

Phylogenies of *H. sapiens* estimated with DNA sequence data agree with the pattern in the fossil record. In phylogenetic trees that show the relationships among human populations living today, the first lineages to branch off lead to descendant populations that live in Africa today (**Figure 34.43**). Based on this observation, it is logical to infer that the ancestral population of modern humans also lived in Africa. The tree shows that lineages subsequently branched off to form three monophyletic groups. Because the populations within each of these clades live in a distinct area, the three lineages are thought to descend from three major waves of migration that occurred as a *Homo sapiens* population dispersed from east Africa to (1) north Africa, Europe, and central Asia, (2) northeast Asia and the Americas, and (3) southeast Asia and the South Pacific (**Figure 34.44**). To summarize, the data suggest that (1) modern humans originated in Africa; and (2) a population that left Africa split into three broad groups, which then spread throughout the world.

This scenario for the evolution of *H. sapiens* is called the **out-of-Africa hypothesis**. It contends that *H. sapiens* evolved independently of the earlier European and Asian species of *Homo*—meaning there was no interbreeding between *H. sapiens* and Neanderthals, *H. erectus*, or *H. floresiensis*. Stated another way, the out-of-Africa hypothesis proposes that *H. sapiens* evolved its distinctive traits in Africa and then dispersed throughout the world.

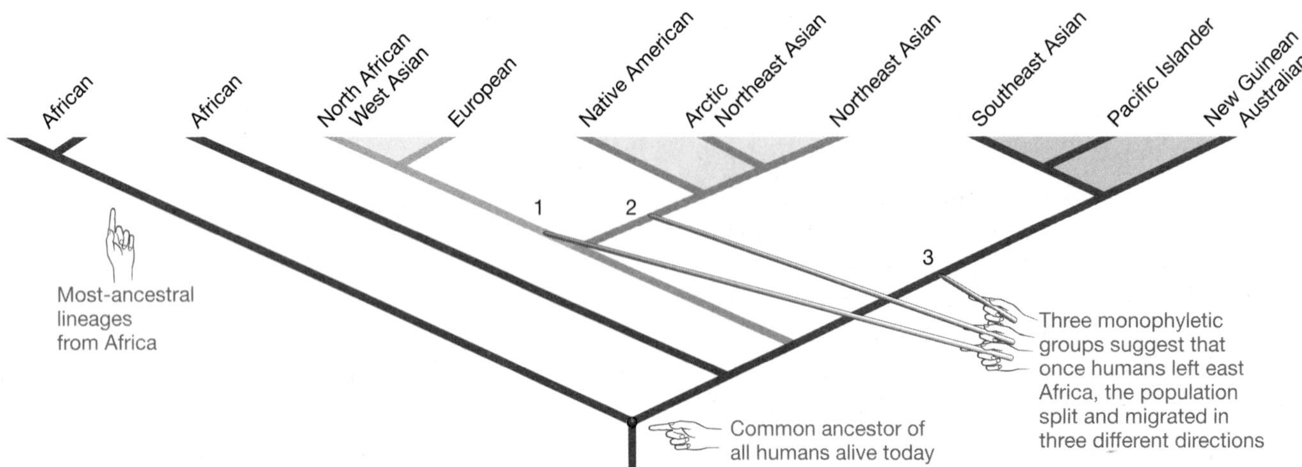

FIGURE 34.43 Phylogeny of Human Populations Living Today. Phylogeny of modern human populations, as estimated from DNA sequence data.

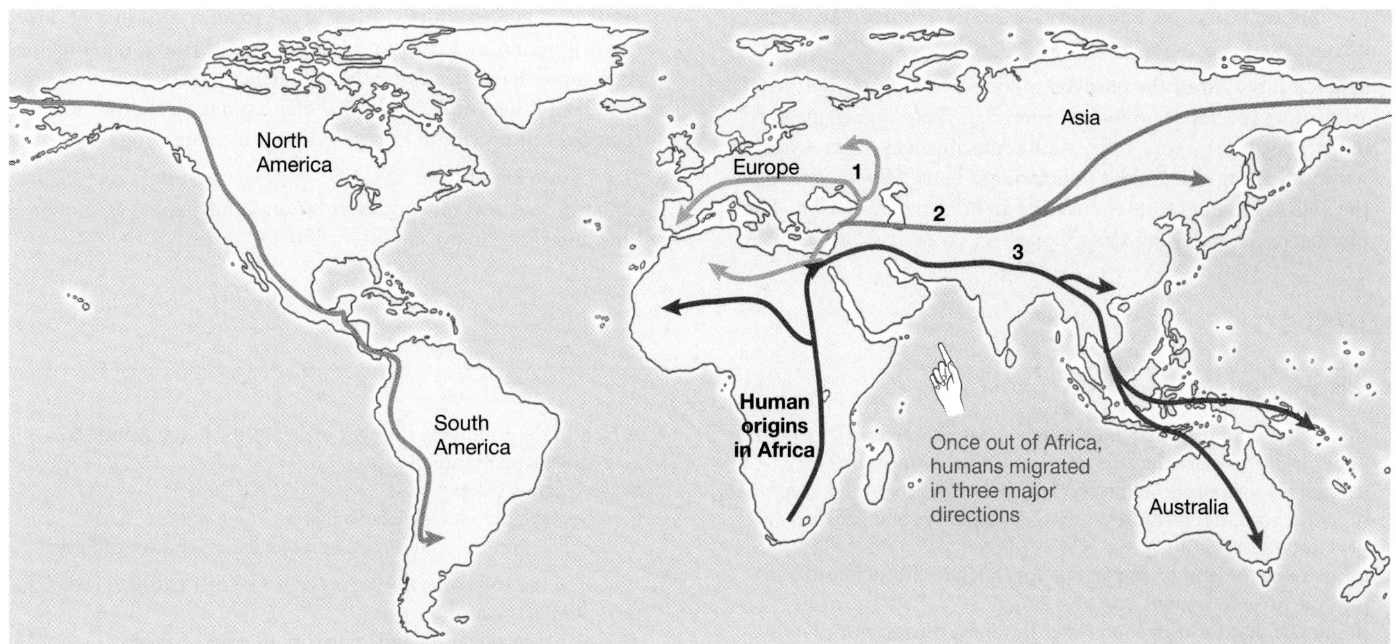

FIGURE 34.44 *Homo sapiens* Originated in Africa and Spread throughout the World. The phylogeny in Figure 34.43 supports the hypothesis that humans originated in Africa and spread out in three major migrations: to southeast Asia and the Pacific Islands, to Europe, and to northeast Asia and the New World.

Chapter Review

SUMMARY OF KEY CONCEPTS

The most species-rich deuterostome lineages are the echinoderms and the vertebrate groups called ray-finned fishes and tetrapods.

Echinoderms, ray-finned fishes, and tetrapods are also the most important large-bodied predators and herbivores in marine and terrestrial environments.

You should be able to describe what food chains in aquatic and terrestrial environments would look like if echinoderms, ray-finned fishes, and tetrapods did not exist.

Echinoderms and vertebrates have unique body plans. Echinoderms are radially symmetric as adults and have a water vascular system. All vertebrates have a skull and an extensive endoskeleton made of cartilage or bone.

Echinoderm larvae are bilaterally symmetric but undergo a metamorphosis into radially symmetric adults. Their water vascular system is composed of fluid-filled tubes and chambers and extends from the body wall in projections called podia. Podia can extend and retract in response to muscle contractions that move fluid inside the water vascular system.

Chordates are distinguished by the presence of a notochord, a dorsal hollow nerve cord, pharyngeal gill slits, and a muscular tail that extends past the anus. Vertebrates are distinguished by the presence of a cranium; most species also have vertebrae. In more derived groups of vertebrates, the body plan features an extensive endoskeleton composed of bone.

You should be able to sketch a sea star and a cephalochordate and label aspects of the body plan that qualify as synapomorphies.

(MB) Web Animation at www.masteringbio.com
Deuterostome Diversity

The diversification of echinoderms was triggered by the evolution of appendages called podia; the diversification of vertebrates was driven by the evolution of the jaw and limbs.

Most echinoderms move via their podia, and many species suspension feed, deposit feed, or act as predators with the aid of their podia.

Ray-finned fishes and tetrapods use their jaws to bite food and process it with teeth. Species in both groups move when muscles attached to their endoskeletons contract or relax. Tetrapods can move on land because their limbs enable walking, running, or flying. The evolution of the amniotic egg allowed tetrapods to lay eggs on land. Parental care was an important adaptation in some groups of ray-finned fishes and tetrapods—particularly mammals.

You should be able to explain why adaptations for efficient movement and feeding led to evolutionary success in terms of numbers and diversity of species in echinoderms and vertebrates.

Humans are a tiny twig on the tree of life. Chimpanzees and humans diverged from a common ancestor that lived in Africa

6–7 million years ago. Since then, at least 14 humanlike species have existed.

The fossil record of the past 3.5 million years contains at least 14 distinct species of hominins. Several of these organisms lived in Africa at the same time, and some lineages went extinct without leaving descendant populations. Thus, *Homo sapiens* is the sole surviving representative of an adaptive radiation. The phylogeny of living humans, based on comparisons of DNA sequences, agrees with evidence in the fossil record that *H. sapiens* originated in Africa and later spread throughout Europe, Asia, and the Americas. DNA sequences recovered from the fossilized bones of *H. neanderthalensis* suggest that *H. sapiens* replaced this species in Europe without interbreeding.

You should be able to describe evidence supporting the hypotheses that several species of hominin have lived at the same time and that *Homo sapiens* originated in Africa.

QUESTIONS

Test Your Knowledge

1. If you found an organism on a beach, what characteristics would allow you to declare that the organism is an echinoderm?
   a. radially symmetric adults, spines, and presence of tube feet
   b. notochord, dorsal hollow nerve cord, pharyngeal gill slits, and muscular tail
   c. exoskeleton and three pairs of appendages; distinct head and body (trunk) regions
   d. mouth that forms second (after the anus) during gastrulation
2. What is the diagnostic trait of vertebrates?
   a. skull
   b. jaws
   c. endoskeleton constructed of bone
   d. endoskeleton constructed of cartilage
3. Why are the pharyngeal jaws found in many ray-finned fishes important?
   a. They allow the main jaw to be protrusible (extendible).
   b. They make it possible for individuals to suck food toward their mouths.
   c. They give rise to teeth that are found on the main jawbones.
   d. They help process food.
4. Which of the following lineages make up the living Amniota?
   a. reptiles and mammals
   b. viviparous fishes
   c. frogs, salamanders, and caecilians
   d. hagfish, lampreys, and cartilaginous fishes (sharks and rays)
5. Which of the following does *not* occur in either cartilaginous fishes or ray-finned fishes?
   a. internal fertilization and viviparity or ovoviviparity
   b. external fertilization and oviparity
   c. formation of a placenta
   d. feeding of young
6. Most species of hominins are known only from Africa. Which species have been found in other parts of the world as well?
   a. early *Homo—H. habilis* and *H. ergaster*
   b. *H. erectus*, *H. neanderthalensis*, and *H. floresiensis*
   c. gracile australopithecines
   d. robust australopithecines

Test Your Knowledge answers: 1. a; 2. a; 3. d; 4. a; 5. d; 6. b

Test Your Understanding

Answers are available at www.masteringbio.com

1. Explain how the water vascular system of echinoderms functions as a type of hydrostatic skeleton.
2. List the four morphological traits that distinguish chordates. How are these traits involved in locomotion and feeding in larvae or adults?
3. Describe evidence that supports the hypothesis that jaws evolved from gill arches in fish.
4. Describe evidence that supports the hypothesis that the tetrapod limb evolved from the fins of lobe-finned fishes.
5. The text claims that "*Homo sapiens* is the sole survivor of an adaptive radiation that took place over the past 7 million years." Do you agree with this statement? Why or why not? Identify three major trends in the evolution of hominins, and suggest a hypothesis to explain each.
6. Explain how the evolution of the placenta and lactation in mammals improved the probability that their offspring would survive. Over the course of a lifetime, why are female mammals expected to produce fewer eggs than do female fish?

Applying Concepts to New Situations

Answers are available at www.masteringbio.com

1. Describe the conditions under which it might be possible for a new species of *Homo* to evolve from current populations of *H. sapiens*.
2. Compare and contrast adaptations that triggered the diversification of the three most species-rich animal lineages: molluscs, arthropods, and vertebrates.
3. Aquatic habitats occupy 73 percent of Earth's surface area. How does this fact relate to the success of ray-finned fishes? How does it relate to the success of coelacanths and other lobe-finned fishes?
4. Mammals and birds are endothermic. Did they inherit this trait from a common ancestor, or did endothermy evolve independently in these two lineages? Provide evidence to support your answer.

Viruses

35

Photomicrograph created by treating seawater with a fluorescing compound that binds to nucleic acids. The smallest, most abundant dots are viruses. The larger, numerous spots are bacteria and archaea. The largest splotches are protists.

KEY CONCEPTS

- Viruses are tiny, noncellular parasites that infect virtually every type of cell known. They cannot perform metabolism on their own—meaning outside a parasitized cell—and are not considered to be alive. Different types of viruses infect particular species and types of cells.
- Although viruses are diverse morphologically, they can be classified as two general types: enveloped and nonenveloped.
- The viral infection cycle can be broken down into six steps: (1) entry into a host cell, (2) production of viral proteins, (3) replication of the viral genome, (4) assembly of a new generation of virus particles, (5) exit from the infected cell, and (6) transmission to a new host.
- In terms of diversity, the key feature of viruses is the nature of their genetic material. The genomes of viruses may consist of double-stranded DNA, single-stranded DNA, double-stranded RNA, or one of several types of single-stranded RNA.

A **virus** is an obligate, intracellular parasite. The adjective *obligate* is appropriate because viruses are completely dependent on host cells. The adjective *intracellular* is appropriate because viruses cannot replicate unless they enter a cell. The noun *parasite* is appropriate because viruses reproduce at the expense of their host cells.

Viruses can also be defined by what they are not. They are not cells and are not made up of cells, so they are not considered organisms. They cannot manufacture their own ATP or amino acids or nucleic acids, and they cannot produce proteins on their own. Viruses enter a **host cell**, take over its biosynthetic machinery, and use that machinery to manufacture a new generation of viruses. Outside of host cells, viruses cannot do anything—they simply exist.

Because they are not organisms, viruses are referred to as particles or agents and are not given scientific (genus + species) names. Most biologists would argue that viruses are not even alive. Yet viruses have a genome, they are superbly adapted to exploit the metabolic capabilities of their host cells, and they evolve. **Table 35.1** summarizes some characteristics of viruses.

The diversity and abundance of viruses almost defy description. Each type of virus infects a specific unicellular species or cell type in a multicellular species, and nearly all organisms examined thus far are parasitized by at least one kind of virus. The bacterium *Escherichia coli*, which resides in the human intestine, is afflicted by several dozen types of **bacteriophage** (literally, "bacteria-eater"). A bacteriophage is a virus that

Key Concept Important Information Practice It

SUMMARY TABLE **35.1 Characteristics of Viruses versus Characteristics of Organisms**

	Viruses	Organisms
Hereditary material	DNA or RNA; can be single stranded or double stranded	DNA; always double stranded
Plasma membrane present?	No	Yes
Can carry out transcription independently?	No—even if a viral polymerase is present, transcription of viral genome requires use of ATP and nucleotides provided by host cell	Yes
Can carry out translation independently?	No	Yes
Metabolic capabilities	Virtually none	Extensive—synthesis of ATP, reduced carbon compounds, vitamins, lipids, nucleic acids, etc.

infects bacteria. The surface waters of the world's oceans teem with bacteria and archaea, yet viruses outnumber them in this habitat by a factor of 10 to 1. A wine bottle filled with seawater taken from the ocean's surface contains about 10 billion virus particles—over one and a half times the world's population of humans.

35.1 Why Do Biologists Study Viruses?

Any study of life's diversity would be incomplete unless it included a look at the acellular parasites that exploit that diversity. But viruses are also important from a practical standpoint. To health-care workers, agronomists, and foresters, these parasites are a persistent—and sometimes catastrophic—source of misery and economic loss. The nature of viruses has been understood only since the 1940s, but they have been the focus of intense research ever since. Biologists study viruses because they cause illness and death. In the human body, virtually every system, tissue, and cell can be infected by one or more kinds of virus (**Figure 35.1**). Research on viruses is motivated by the desire to minimize the damage they can cause.

In addition, biologists study viruses with the goal of exploiting their ability to enter cells. Recall from Chapter 19 that viruses are being tested as possible therapeutic agents in the treatment of genetic diseases. If viruses can be engineered to carry normal copies of human genes into the cells of patients, it is possible that the gene products could cure symptoms.

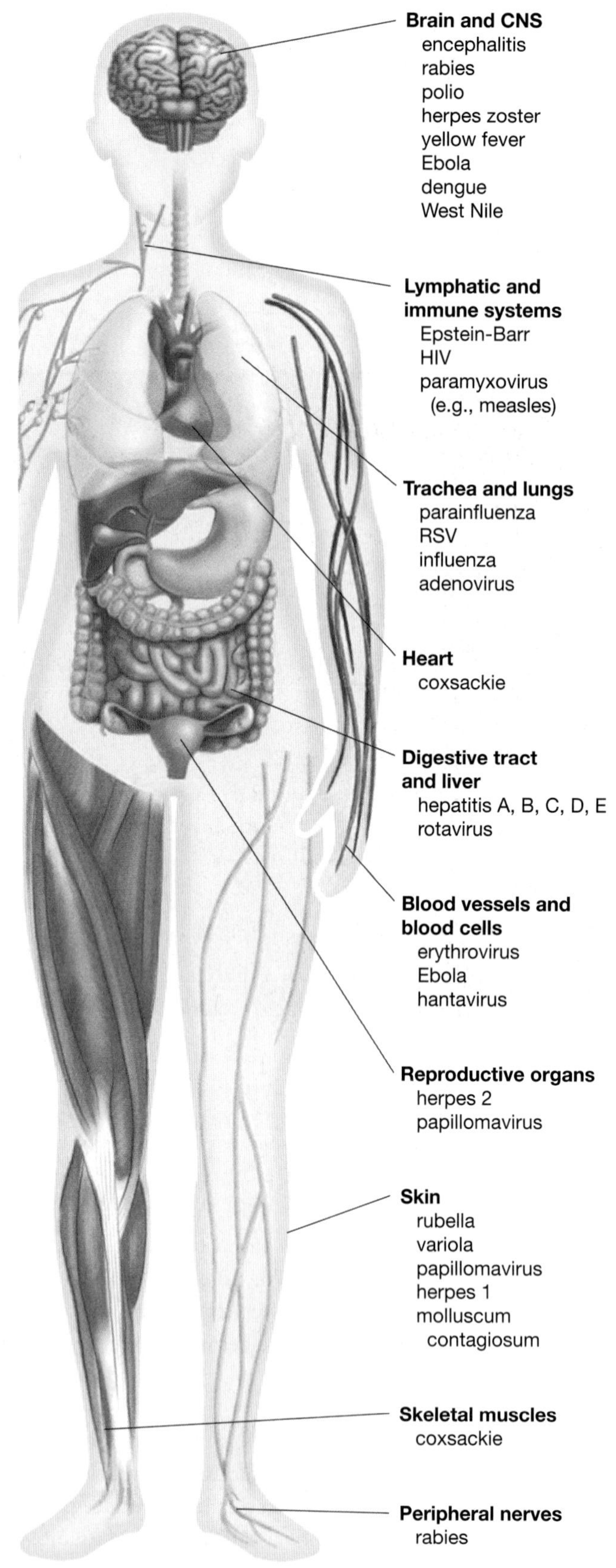

FIGURE 35.1 Human Organs and Systems That Are Parasitized by Viruses.

● **EXERCISE** Choose two viruses that you are familiar with. Next to each, write the symptoms caused by an infection with this virus.

Recent Viral Epidemics in Humans

Physicians and researchers use the term **epidemic** (literally, "upon-people") to describe a disease that rapidly affects a large number of individuals over a widening area. Viruses have caused the most devastating epidemics in recent human history. During the eighteenth and nineteenth centuries, it was not unusual for Native American tribes to lose 90 percent of their members over the course of a few years to measles, smallpox, and other viral diseases spread by contact with European settlers. To appreciate the impact of these epidemics, think of 10 close friends and relatives—then remove nine.

An epidemic that is worldwide in scope is called a **pandemic**. The influenza outbreak of 1918–1919, called "Spanish flu," qualifies as the most devastating pandemic recorded to date. Influenza is a virus that infects the upper respiratory tract. The strain of influenza virus that emerged in 1918 infected people worldwide and was particularly **virulent**—meaning it tended to cause severe disease. Within hours of showing symptoms, the lungs of previously healthy people often became so heavily infected that affected individuals suffocated to death. Most victims were between the ages of 20 and 40. The viral outbreak occurred just as World War I was drawing to a close and killed far more people than did the conflict itself. For example, ten times as many Americans died of influenza than were killed in combat in the war. Worldwide, the Spanish flu is thought to have killed 20–50 million people.

Current Viral Epidemics in Humans: HIV

In terms of the total number of people affected, the measles and smallpox epidemics among native peoples of the Americas and the 1918 influenza outbreak are almost certain to be surpassed by the incidence of AIDS. **Acquired immune deficiency syndrome** (**AIDS**) is an affliction caused by the **human immunodeficiency virus** (**HIV**; **Figure 35.2**). HIV is now the most intensively studied of all viruses. Since the early 1980s, governments and private corporations from around the world have spent hundreds of millions of dollars on HIV research. More biologists are working on HIV than on any other type of virus. Given this virus's current and projected impact on human populations around the globe, the investment is justified.

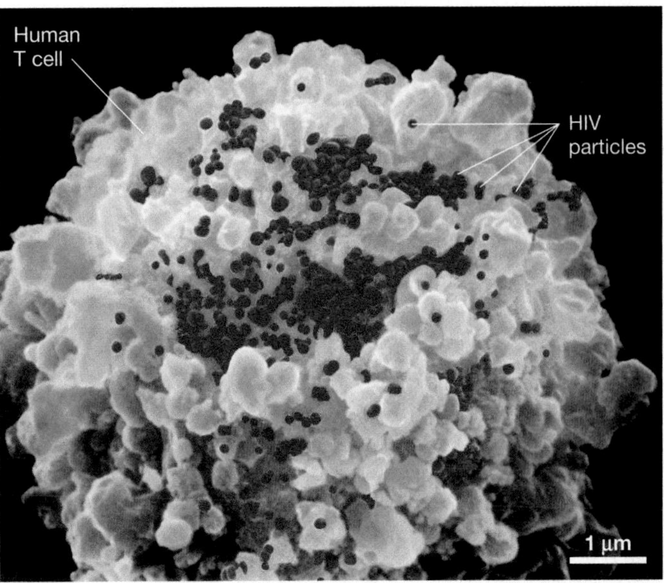

FIGURE 35.2 The Human Immunodeficiency Virus (HIV). Colorized scanning electron micrograph showing HIV particles emerging from an infected human T cell (a type of immune system cell).

How Does HIV Cause Disease? Like other viruses, HIV parasitizes specific types of cells. The cells most affected by HIV are called helper T cells and macrophages. These cells are components of the **immune system**, which is the body's defense system against disease. Chapter 49 explains just how crucial helper T cells and macrophages are to the immune system's response to invading bacteria and viruses. If an HIV particle succeeds in infecting one of these cells and reproducing inside, however, the cell dies as hundreds of new virus particles break out and infect more cells. Although the body continually replaces helper T cells and macrophages, the number produced does not keep pace with the number being destroyed by HIV. As a result, the total number of helper T cells in the bloodstream gradually declines as an HIV infection proceeds (**Figure 35.3**). When the T-cell count drops, the immune system's responses to invading bacteria and viruses become less and less effective. Eventually, too few helper T cells are left to fight off pathogens efficiently, and pathogenic bacteria and viruses

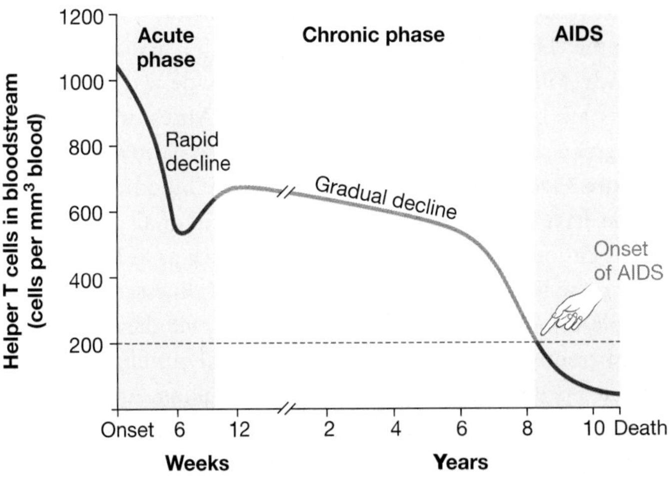

FIGURE 35.3 T-Cell Counts Decline during an HIV Infection. Graph of changes in the number of T cells that are present in the bloodstream over time, based on data from a typical patient infected with HIV. The acute phase of infection occurs immediately after infection and is sometimes associated with symptoms such as fever. Infected people usually show no disease symptoms in the chronic phase, even though their T-cell counts are in slow, steady decline. AIDS typically occurs when T-cell counts dip below 200/mm^3 of blood.

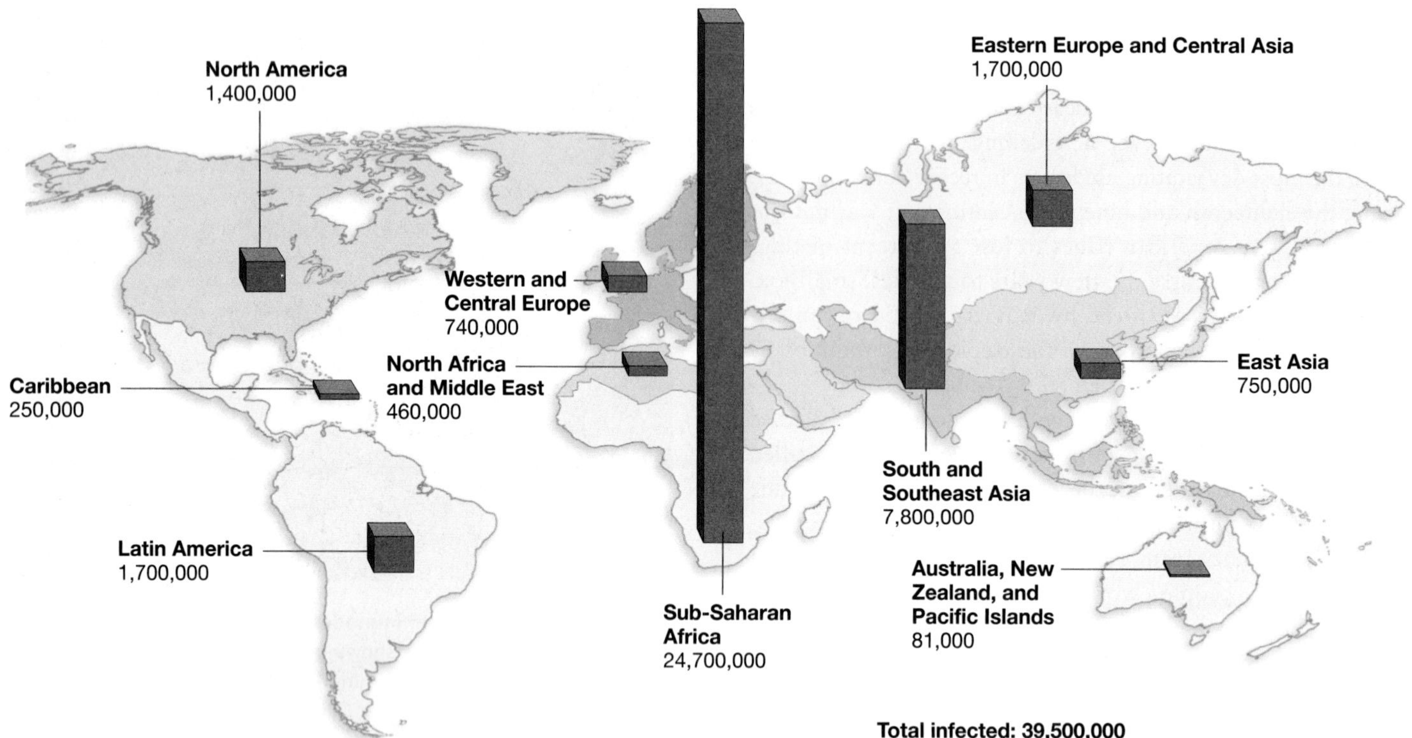

FIGURE 35.4 Geographic Distribution of HIV Infections. Data, compiled by the United Nations AIDS program, showing the numbers of people living with HIV in December 2006, by geographic area.

begin to multiply unchecked. In almost all cases, one or more of these infections proves fatal. HIV kills people indirectly—by making them susceptible to pneumonia, fungal infections, and unusual types of cancer.

What Is the Scope of the AIDS Epidemic? Researchers with the United Nations AIDS program estimate that AIDS has already killed 28 million people worldwide. HIV infection rates have been highest in east and central Africa, where one of the greatest public health crises in history is now occurring (**Figure 35.4**). In Botswana and elsewhere, blood-testing programs have confirmed that more than a third of all adults carry HIV. Although there may be a lag of as much as 8–12 years between the initial infection and the onset of illness, virtually all people who become infected with the virus will die of AIDS.

Currently, the UN estimates the total number of HIV-infected people worldwide at about 39.5 million. An additional 4.4 million people are infected each year, and the pandemic is growing. Researchers are particularly alarmed because the focus of the epidemic is shifting from its historical center of incidence—central and southern Africa—to south and east Asia. Infection rates are growing rapidly in some of the world's most populous countries—particularly India and China.

Most viral and bacterial diseases afflict the very young and the very old. But because HIV is a sexually transmitted disease, young adults are most likely to contract the virus and die. People who become infected in their late teens or twenties die of AIDS in their twenties or thirties. Tens of millions of people are being lost in the prime of their lives. Physicians, politicians, educators, and aid workers all use the same word to describe the epidemic's impact: staggering.

35.2 How Do Biologists Study Viruses?

Researchers who study viruses focus on two goals: (1) developing vaccines that help hosts fight off disease if they become infected and (2) developing antiviral drugs that prevent a virus from replicating efficiently inside the host. Both types of research begin with attempts to isolate the virus in question.

Isolating viruses takes researchers into the realm of "nanobiology," in which structures are measured in billionths of a meter. (One nanometer, abbreviated nm, is 10^{-9} meter.) Viruses range from about 20 to 300 nm in diameter. They are dwarfed by eukaryotic cells and even by bacterial cells (**Figure 35.5**). Millions of viruses can fit on the head of a pin.

If virus-infected cells can be grown in culture or harvested from a host individual, researchers can usually isolate the virus by passing the cells through a filter. The filters used to study viruses have pores that are large enough for viruses to pass through but are too small to admit cells. To test the hypothesis that the solution passing through the filter contains viruses that cause a specific disease, researchers expose uninfected host cells to this filtrate. If exposing host cells to the filtrate results in infection, then the virus-causation hypothesis is correct. In this

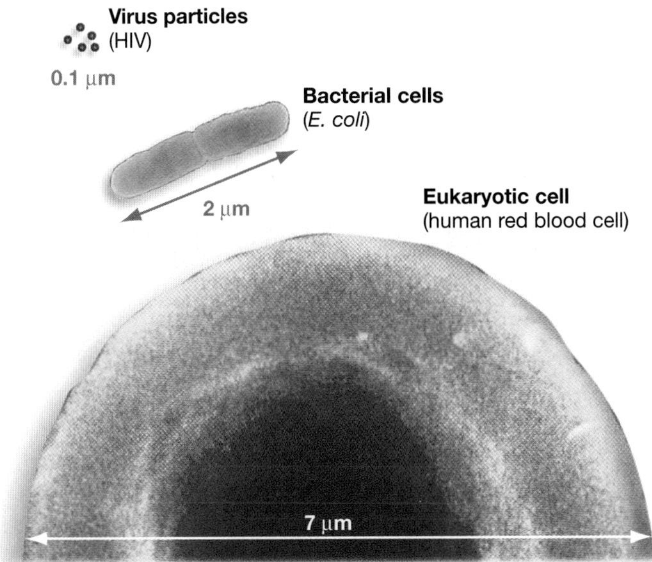

FIGURE 35.5 Viruses Are Tiny. Viruses are much smaller than bacterial cells or eukaryotic cells.

QUESTION What is the diameter of these virus particles in nanometers?

way, researchers can isolate a virus and confirm that it is the causative agent of infection. Recall from Chapter 28 that these steps are inspired by Koch's postulates, which established the criteria for linking a specific infectious agent with a specific disease.

Once biologists have isolated a virus, how do they study and characterize it? Let's begin with morphological traits, then consider how viral replication cycles vary.

Analyzing Morphological Traits

To see a virus, researchers usually rely on transmission electron microscopy (see **BioSkills 8**). Only the very largest viruses, such as the smallpox virus, are visible with a light microscope. Electron microscopy has revealed that viruses come in a wide variety of shapes, and many viruses can be identified by shape alone (**Figure 35.6**). In overall structure, however, viruses fall into just two general categories. Viruses can either be (1) enclosed by just a shell of protein called a **capsid** or (2) enclosed by both a capsid and a membrane-like **envelope**. Regarding their morphology, then, the important distinction among viruses is whether they are nonenveloped or enveloped.

Nonenveloped viruses consist of genetic material and possibly one or more enzymes inside a capsid—a protein coat. The nonenveloped virus illustrated in **Figure 35.7a** is an adenovirus. You undoubtedly have adenoviruses on your tonsils or in other parts of your upper respiratory passages right now. As the micrograph in Figure 35.6d shows, the morphology of nonenveloped viruses may be complex.

Enveloped viruses also have genetic material inside a capsid, but the capsid is surrounded by an envelope. The envelope consists of a phospholipid bilayer with a mixture of viral proteins and proteins derived from a membrane found in a host cell—specifically, the host cell in which the virus particle was manufactured (**Figure 35.7b**). Later sections in the chapter will detail how most of these viruses obtain their envelope from an infected host cell.

Once a virus has been isolated and its overall morphology characterized, researchers usually focus on understanding the nature of the virus's replication cycle and on attempts to develop a vaccine. A **vaccine** is a preparation that primes a host's immune system to respond to a specific type of virus. **Box 35.1** explains how vaccines work and why it has not been possible to develop a vaccine for certain viruses—particularly the flu viruses or HIV. Here let's focus on variation in virus replication cycles.

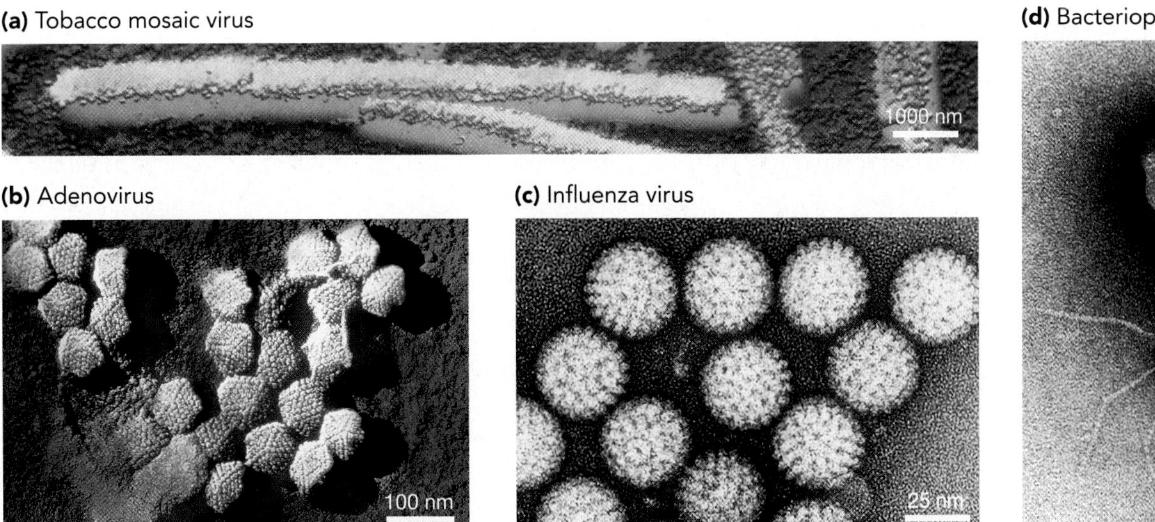

FIGURE 35.6 Viruses Vary in Size and Shape. Virus shapes include **(a)** rods, **(b)** polyhedrons, **(c)** spheres, and **(d)** complex shapes with "heads" and "tails."

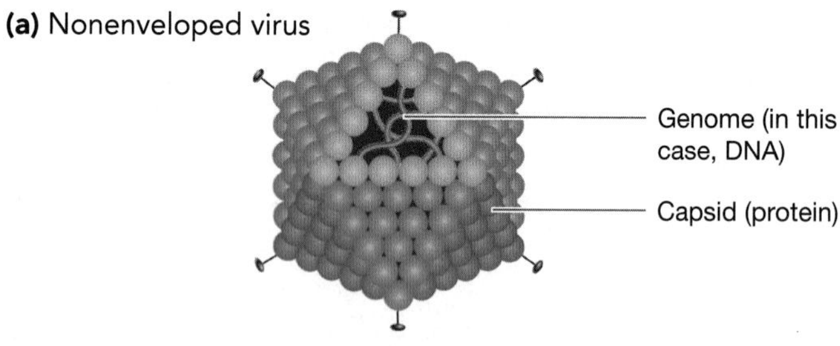

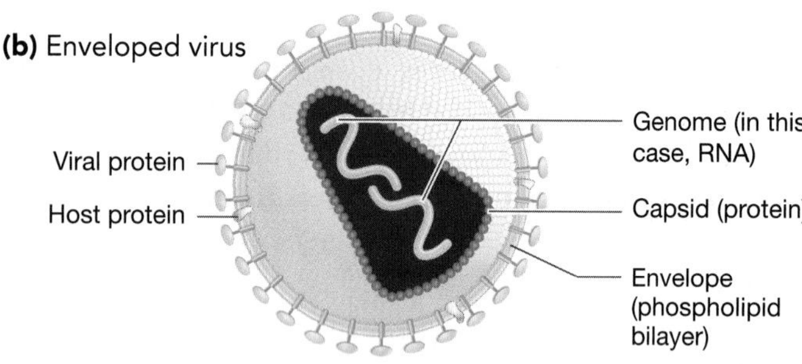

FIGURE 35.7 Viruses Are Nonenveloped or Enveloped. (a) A protein coat forms the exterior of nonenveloped viruses. **(b)** In enveloped viruses, the exterior is composed of a membranous sphere. Inside this envelope, the hereditary material is enclosed by a protein coat.

Analyzing Variation in Growth Cycles: Replicative and Latent Growth

Although it is likely that millions of types of virus exist, they all infect their host cells in one of two general ways: via replicative growth or in a dormant form referred to as latency in animal viruses or lysogeny in bacteriophages. All viruses undergo replicative growth; some can halt the replication cycle and enter a dormant state as well. Both types of viral infection begin when part or all of a virus particle enters the interior of a host cell.

Figure 35.8a shows a **lytic cycle**, or replicative growth. As illustrated, the viral genome enters the host cell and viral or host enzymes make copies of it, using nucleotides and ATP provided by the host. The host cell also manufactures viral proteins. When synthesis of the viral genome and viral proteins is complete, a new generation of virus particles assembles inside the host cell. A mature virus particle is called a **virion**. The replicative cycle is complete when the virions exit the cell—usually killing the host cell in the process.

Figure 35.8b diagrams a **lysogenic cycle**, or lysogeny, in a bacteriophage. Only certain types of viruses are capable of a latent infection. During lysogenic or latent growth, viral DNA

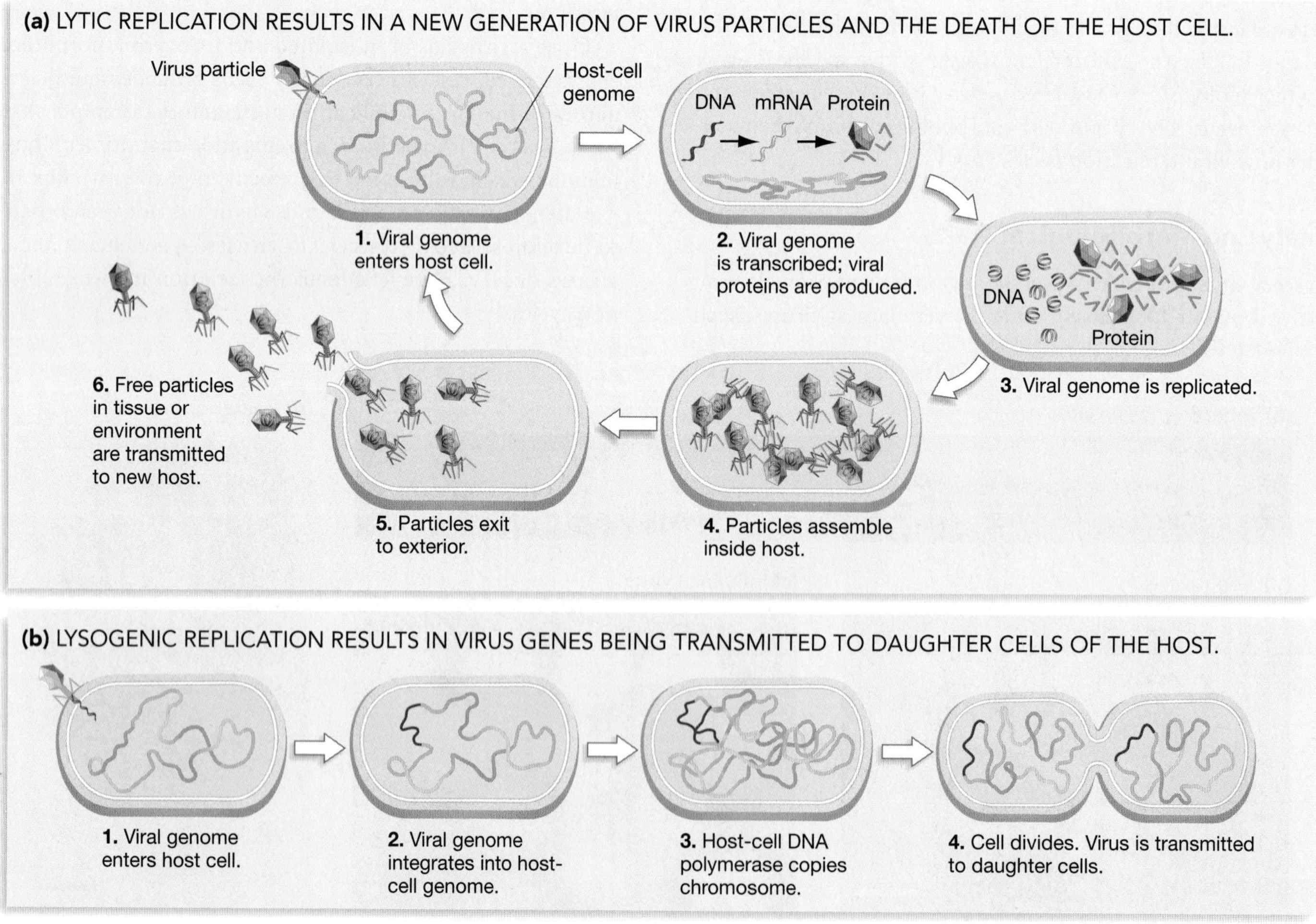

(a) LYTIC REPLICATION RESULTS IN A NEW GENERATION OF VIRUS PARTICLES AND THE DEATH OF THE HOST CELL.

Virus particle
Host-cell genome
1. Viral genome enters host cell.
DNA mRNA Protein
2. Viral genome is transcribed; viral proteins are produced.
DNA
Protein
3. Viral genome is replicated.
4. Particles assemble inside host.
5. Particles exit to exterior.
6. Free particles in tissue or environment are transmitted to new host.

(b) LYSOGENIC REPLICATION RESULTS IN VIRUS GENES BEING TRANSMITTED TO DAUGHTER CELLS OF THE HOST.

1. Viral genome enters host cell.
2. Viral genome integrates into host-cell genome.
3. Host-cell DNA polymerase copies chromosome.
4. Cell divides. Virus is transmitted to daughter cells.

FIGURE 35.8 Viruses Replicate via Lytic or Lysogenic Cycles, or Both. (a) All viruses follow the same general lytic replication cycle. **(b)** Some viruses are also capable of lysogeny, meaning that their genome can become integrated into the host-cell chromosome.

EXERCISE Compare and contrast a lysogenic virus to the transposable elements introduced in Chapter 20.

becomes incorporated into the host's chromosome. Often this integration occurs without serious damage to the host cell. Once the viral genome is in place, it is replicated by the host's DNA polymerase each time the cell divides. Copies of the viral genome are passed on to daughter cells just like one of the host's own genes. In the lysogenic or latent state, a virus is quiescent. This means that no new particles are being produced and no unrelated cells are being infected. The virus is transmitted from one generation to the next along with the host's genes. HIV and other retroviruses are also capable of being transmitted to daughter cells when they are latent. ● If you understand this concept, you should be able to explain the contrast in how viral genes are transmitted to offspring via lytic versus lysogenous growth.

It is not possible to treat a latent infection with drugs called **antivirals**, because the viruses are quiescent—they just sit there. But even lytic infections are notoriously difficult to treat because viruses use so many of the host cell's enzymes during the lytic replication cycle. Drugs that disrupt these enzymes usually harm the host much more than they harm the virus. To understand viral diversity and how antiviral drugs are developed, let's consider each phase of the lytic cycle in more detail.

Analyzing the Phases of the Replicative Cycle

● Six phases are common to replicative growth in virtually all viruses: (1) entry into a host cell, (2) transcription of the viral genome and production and processing of viral proteins, (3) replication of the viral genome, (4) assembly of a new generation of virions, (5) exit from the infected cell, and (6) transmission to a new host; corresponding steps are depicted in Figure 35.8a. Each virus has a particular way of entering a host cell and completing the subsequent phases of the cycle. Let's take a closer look.

BOX 35.1 How Are Vaccines Developed?

As Chapter 49 will show, one of the primary ways that an individual's immune system responds to parasite attack is by producing antibodies. An **antibody** ("against-body") is a protein that binds with high specificity to a particular site on another molecule. Any molecule that elicits the production of antibodies—and that antibodies stick to—is called an **antigen** ("against-produce").

Vaccines contain antigens. The antigens are usually components of a virus's exterior—the capsid from a nonenveloped virus or the envelope proteins from an enveloped virus. Exterior proteins are effective antigens because they are the part of the virus that can be attacked by antibodies, outside the cell.

To vaccinate a person or a domestic animal, viral antigens are injected or swallowed. Once inside the individual's blood, the antigens stimulate immune system cells called B cells (see Chapter 49), which produce antibodies to the antigen (**Figure 35.9**). Once antigens are coated with antibodies, the antigens are destroyed by other cells and components of the immune system.

(Continued on next page)

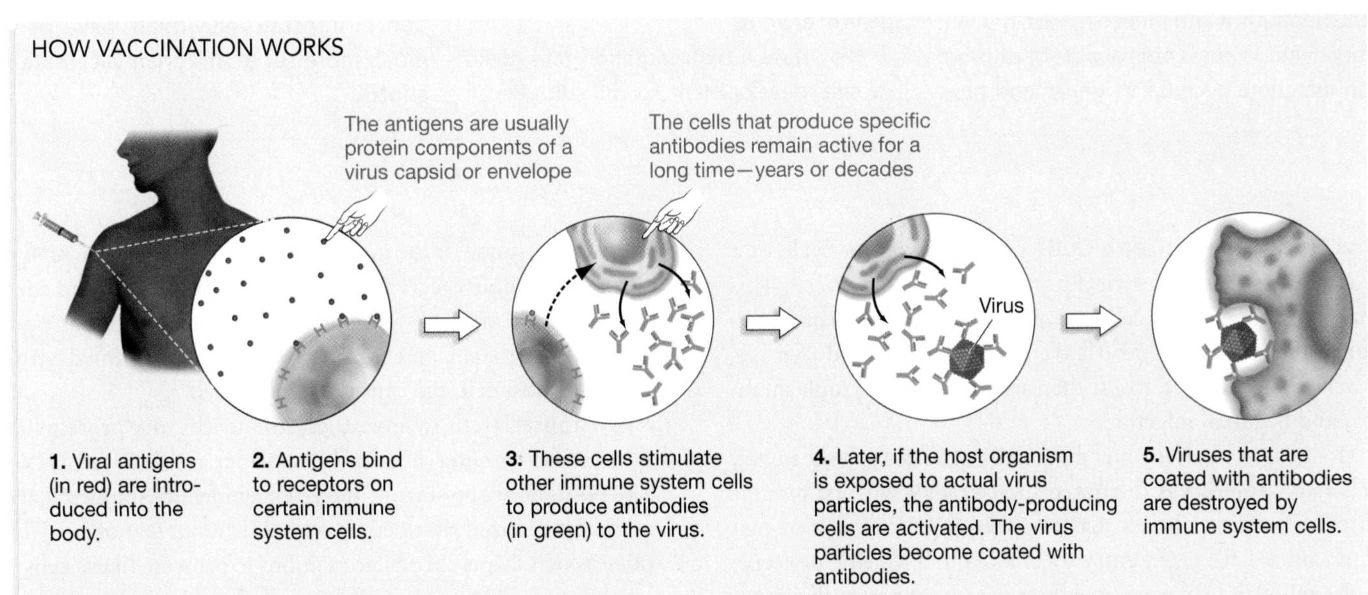

FIGURE 35.9 Vaccination Induces a Response from the Immune System. The immune system cells that respond to a vaccination are "immortalized," meaning that they stay active for a long time. Thus, the body can respond quickly to any future infections by the same virus.

Vaccinations cause fake or abortive infections. They function like fire drills or earthquake preparedness drills—they prepare the immune system to respond to a specific type of threat. Immune system cells that are stimulated to produce anti-bodies by a vaccination remain active in the vaccinated person for a long time—years or decades. If a vaccinated person is later exposed to the antigen in the form of active virus particles, the immune system can respond quickly and effectively enough to stop the infection before it threatens the individual's health.

In many cases, vaccines have been spectacularly successful at curbing or even eliminating viral diseases in humans and domestic animals. In humans alone, effective vaccines have been developed against smallpox, yellow fever, polio, measles, and some forms of hepatitis. These diseases used to kill or sicken hundreds of thousands of people each year.

There Are Three General Types of Vaccines

Successful vaccines consist of isolated viral proteins, inactivated viruses, or attenuated viruses. Isolated viral proteins are called subunit vaccines; familiar examples include the hepatitis B and influenza vaccines. An inactivated virus is not capable of causing an infection, because its genes and proteins have been damaged by chemical treatments—often exposure to formaldehyde—or irradiation with ultraviolet light. If you have been vaccinated for hepatitis A, you may have received an inactivated virus.

Attenuated viruses are also called "live" virus vaccines, because they consist of complete virus particles. The adjective attenuated means that the viruses lack virulence. Researchers can make a virus harmless to a host by culturing the virus on cells from species other than that host. In adapting to growth on the atypical cells, viral strains usually lose the ability to grow rapidly in their normal host cells. Although attenuated viruses still provoke a strong immune response, they are not capable of sustaining an infection in a vaccinated individual. The smallpox, polio, and measles vaccines consist of attenuated viruses.

Why Isn't There an Effective Vaccine for HIV and the Flu Viruses?

Researchers have not succeeded in developing vaccines against HIV and have been only moderately successful in developing a vaccine against flu viruses. The reason is that flu viruses and HIV have exceptionally high mutation rates. HIV actually has the highest mutation rate observed in any organism or virus.

Why does a high mutation rate make vaccine development so difficult? Recall from Chapter 16 that a mutation is defined as a change in the genetic material—DNA in organisms; DNA or RNA in viruses. Because the enzymes that copy the genes found in flu viruses and HIV are exceptionally error prone, many of the mature virus particles that are produced contain mutations in the genes for the virus's envelope proteins. When these genes are transcribed and translated, the resulting envelope proteins are likely to have an altered structure. Due to the high mutation rates of flu viruses and HIV, the antigens presented by these viruses constantly change through time. Stated another way, new strains of these viruses—with novel antigens—are constantly evolving. The antibodies produced against the envelope proteins of earlier strains do not work against strains that appear later, because the antigens presented by the strains are different. A vaccination that protected an individual against certain strains of flu virus or HIV would not help against other strains.

To date, it has not been possible to design antigens from the flu viruses or HIV that protect vaccinated individuals effectively. In the fight against HIV, condom use and other methods to prevent infection, as well as drugs that inhibit viral enzymes and thus stop or slow viral replication in infected individuals, have been much more successful than vaccination efforts.

How Do Viruses Enter a Cell? The replication cycle of a virus begins when a free viral particle enters a target cell. This is no simple task. All cells are protected by a plasma membrane, and many cells also have a cell wall. How do viruses breach these defenses, insert themselves into the cytoplasm inside, and begin an infection?

Most plant viruses enter host cells after a sucking insect, such as an aphid, has disrupted the cell wall with its mouthparts. In contrast, viruses that parasitize bacterial cells or that attack animal cells gain entry by binding to a specific molecule on the cell wall or plasma membrane. In response to this binding event, viruses that attack bacteria release an enzyme called lysozyme that degrades the bacterial cell wall. (Lysozyme is also found in human tears, where it acts as an antibiotic in the eye.) The genome of the nonenveloped virus enters the host cell through the hole created by lysozyme, a portion of the capsid seals the hole, and the remainder of the capsid remains on the cell wall or plasma membrane. But when enveloped viruses bind to a host cell, the capsid enters the cell.

To appreciate how investigators identify the proteins that viruses use to enter host cells, consider research on HIV. In 1981—right at the start of the AIDS epidemic—biomedical researchers realized that people with AIDS had few or no T cells possessing **CD4**, a particular membrane protein. These cells are symbolized $CD4^+$. This discovery led to the hypothesis that CD4 functions as the "doorknob" that HIV uses to enter host cells. The doorknob hypothesis predicts that if CD4 is blocked, then HIV will not be able to enter host cells.

Experiment

Question: Does the CD4 protein function as the "doorknob" that HIV uses to enter host cells?

Hypothesis: CD4 is the membrane protein that HIV uses to enter cells.

Null hypothesis: CD4 is not the membrane protein that HIV uses to enter cells.

Experimental setup:

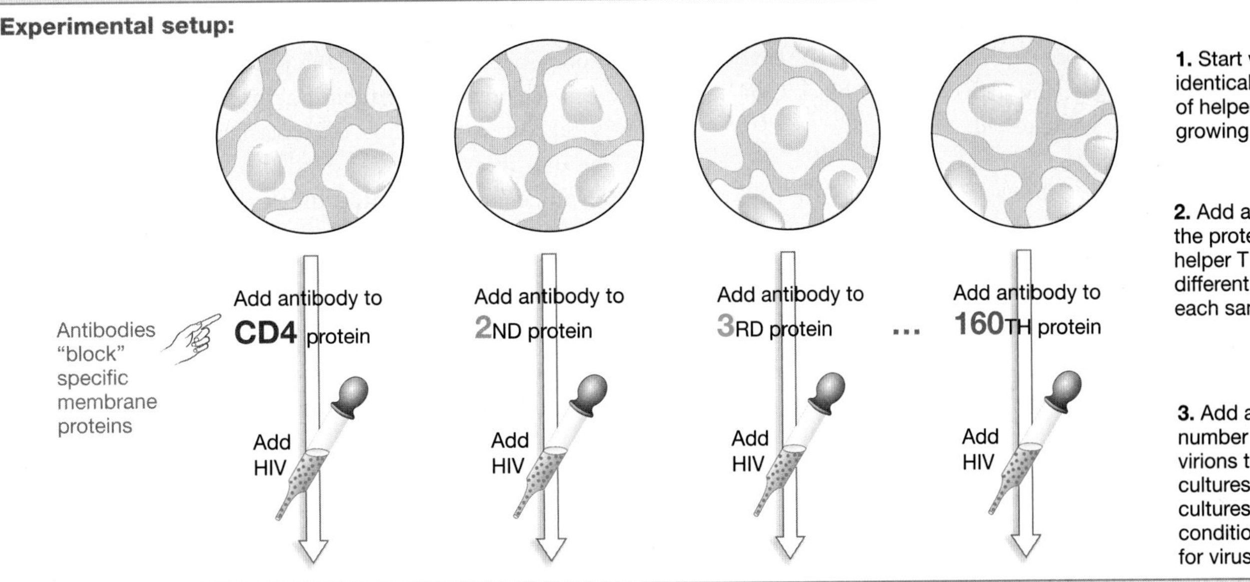

Prediction: HIV will not infect cells with antibody to CD4 but will infect other cells.

Prediction of null hypothesis: HIV will infect cells with antibody to CD4.

Results:

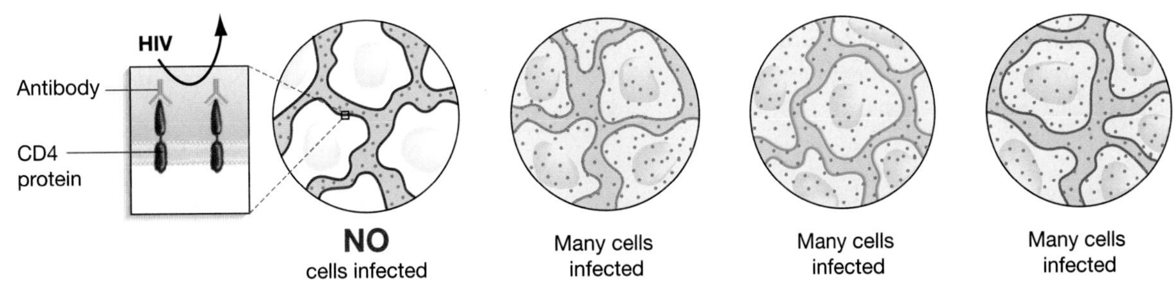

Conclusion: HIV uses CD4 proteins as the "doorknob" to enter helper T cells. Thus, only cells with CD4 on their surface can be infected by HIV.

FIGURE 35.10 Experiments Confirmed that CD4 Is the Receptor Used by HIV to Enter Host Cells. In this experiment, the antibodies added to each culture bound to a specific protein found on the surface of helper T cells. Antibody binding blocked the membrane protein, so the protein could not be used by HIV to gain entry to the cells.

Two teams used the same experimental strategy to test this hypothesis (**Figure 35.10**). They began by growing large populations of helper T cells in culture. Then they took a sample of the cells and added an antibody to one of the cell-surface proteins found on helper T cells, along with HIV particles. They repeated this experiment 160 times but used a different sample of cells and added a different antibody each time. The key point here is that each of the 160 antibodies bound to and effectively blocked a different cell-surface protein. If one of the antibodies used in the experiment happened to bind to the receptor used by HIV, that antibody would cover up the receptor. In this way, the antibody would prevent HIV from entering the cell and protect that cell from infection. This approach led both research teams to reach exactly the same result: Only antibodies to CD4 protected the cells from viral entry.

Later work confirmed that HIV particles can enter cells only if the virions bind to a second membrane protein, called a **co-receptor**, in addition to CD4. In most individuals, proteins called CXCR4 and CCR5 function as co-receptors. If the proteins in a virion's envelope bind to both CD4 and a co-receptor, the lipid bilayers of the particle's envelope and the plasma membrane of the helper T cell fuse (**Figure 35.11**). When fusion occurs, HIV has breached the cell boundary. The viral capsid enters the cytoplasm and infection proceeds.

The discovery of the proteins required for viral entry inspired a search for compounds that would block them and prevent HIV from entering cells. Drugs that act in this way are called fusion inhibitors. A molecule that blocks the HIV envelope protein is currently in use, and molecules that block CD4, CXCR4, or CCR5 are now being tested.

Producing Viral Proteins Viruses cannot make the ribosomes and tRNAs necessary for translating their own mRNAs into proteins. For a virus to make the proteins required to produce a new generation of virus particles, it must exploit the host cell's biosynthetic machinery. Production of viral proteins begins soon after a virus enters a cell and continues after the viral genome is replicated. Viral mRNAs and proteins are produced and processed in one of two ways, depending on whether the proteins end up in the outer envelope of a particle or in the capsid.

RNAs that code for a virus's envelope proteins follow a route through the cell identical to that of the RNAs of the cell's own membrane proteins (see Chapter 7). As **Figure 35.12a** shows, these viral mRNAs are translated by ribosomes attached to the endoplasmic reticulum (ER). Afterward the resulting proteins are transported to the Golgi apparatus, where carbohydrate groups are attached, producing glycoproteins. The finished glycoproteins are then inserted into the plasma membrane, where they are ready to be assembled into new virions.

In contrast, a different route is taken by RNAs that code for proteins that make up the capsid or inner core of a virus particle (see **Figure 35.12b**). These RNAs are translated by ribosomes in the cytoplasm, just as non-membrane-bound cellular mRNAs are. In some viruses, long polypeptide sequences are later cut into functional proteins by a viral enzyme called **protease**. This enzyme cleaves viral polypeptides at specific locations—a critical step in the production of finished viral proteins. The resulting protein fragments are assembled into a new viral core near the host cell's plasma membrane.

The discovery that HIV produces a protease triggered a search for drugs that would block the enzyme. This search got a huge boost when researchers succeeded in visualizing the three-dimensional structure of HIV's protease, using the X-ray crystallographic techniques introduced in **BioSkills 8**. The molecule has an opening in its interior adjacent to the active site (**Figure 35.13a**). Polypeptides fit into the opening and are cleaved at the active site. Based on these data, researchers immediately began searching for molecules that could fit into the opening and prevent protease from functioning by blocking the active site (**Figure 35.13b**). Several HIV protease inhibitors are now on the market.

How Do Viruses Copy Their Genomes? Viruses must copy their genes to make a new generation of particles and continue an infection. Most viruses contain a polymerase that copies the viral genome inside the infected host cell, using nucleotides and ATP provided by the host cell. Many DNA viruses, for example, contain a DNA polymerase that makes a copy of the viral genome.

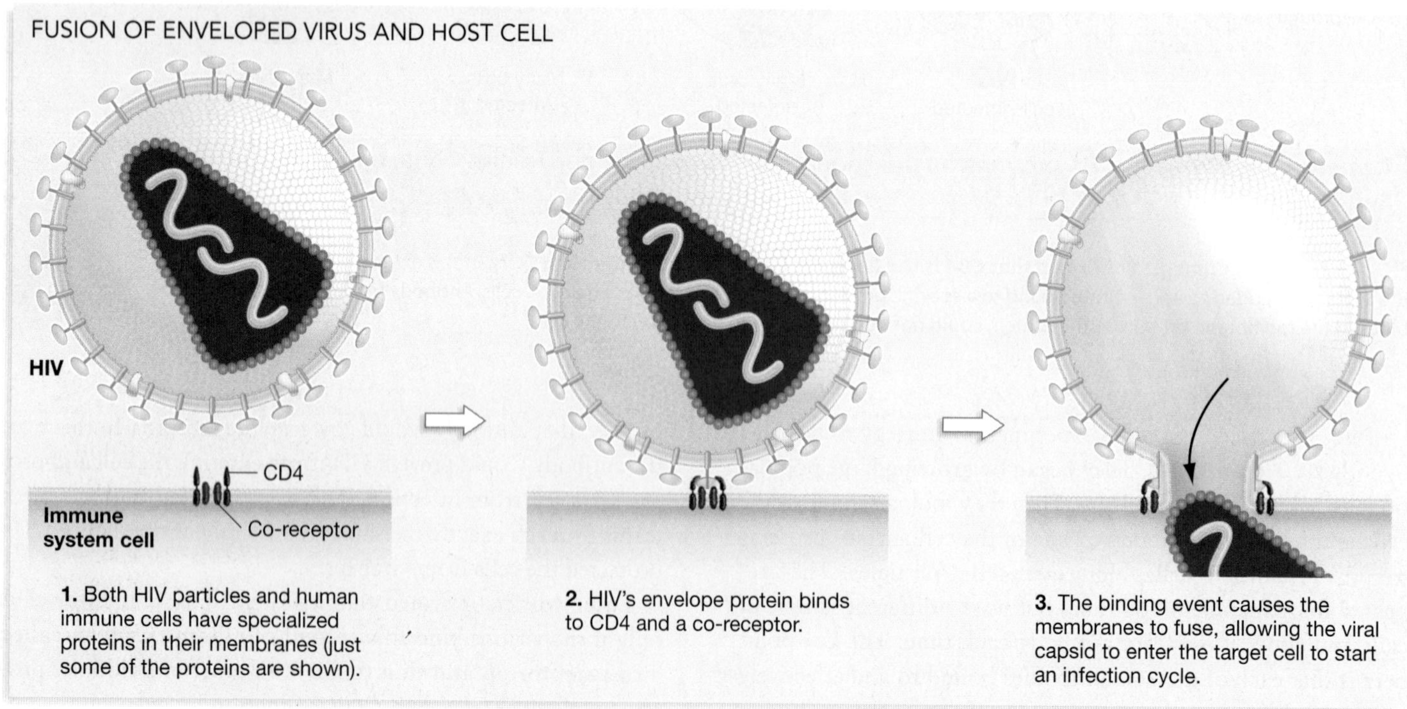

FIGURE 35.11 Enveloped Viruses Bind to Membrane Proteins of Target Cells.

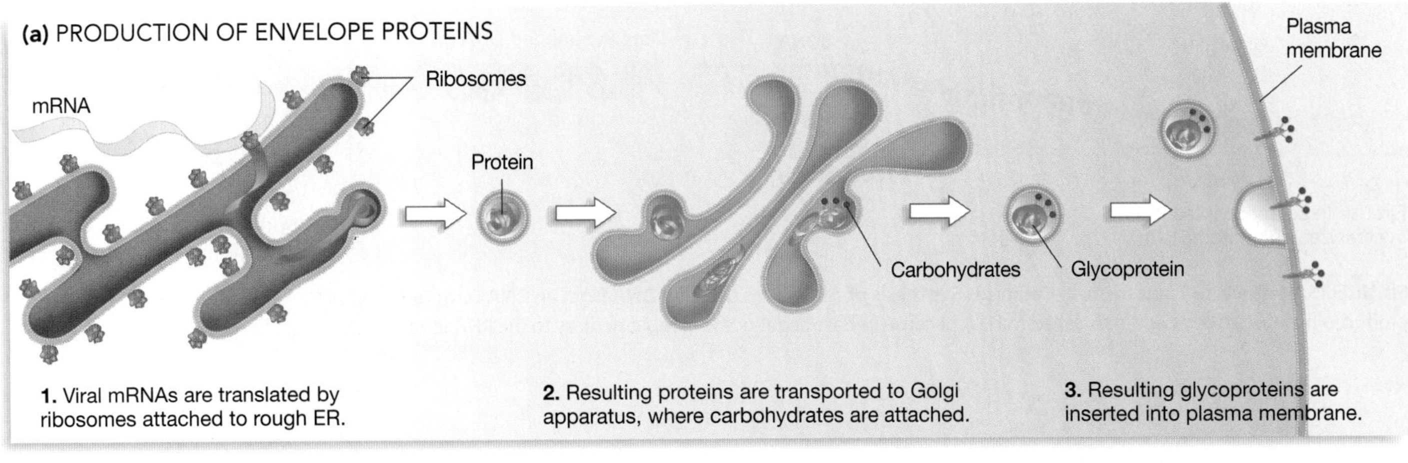

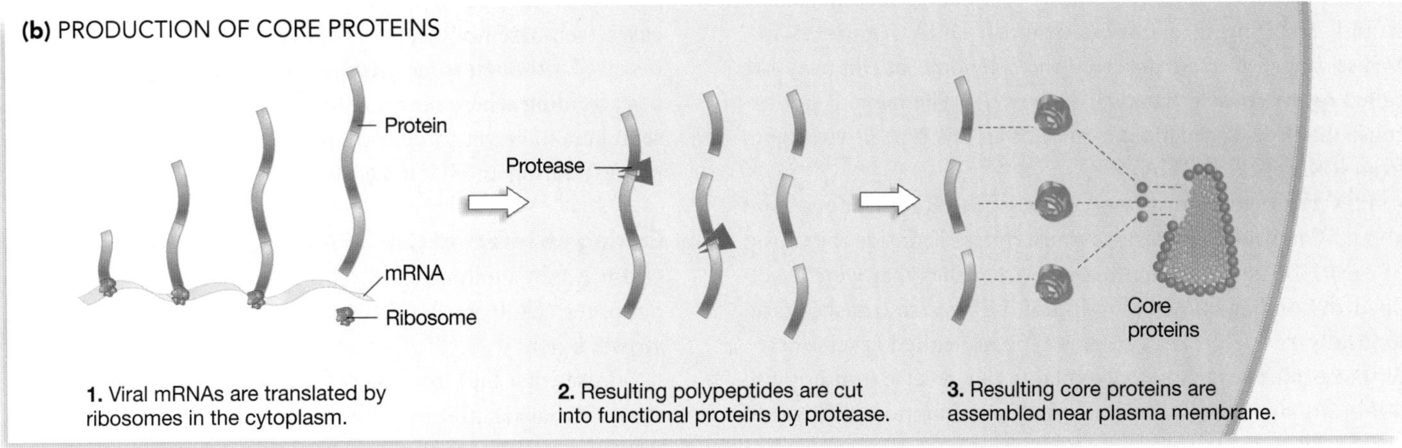

FIGURE 35.12 Production of Viral Proteins. (a) After being synthesized on the rough ER, envelope proteins are inserted into the plasma membrane. **(b)** After being synthesized by ribosomes in the cytoplasm and processed, core proteins assemble near the host cell's plasma membrane.

In some viruses, however, the genome consists of RNA. In most viruses that have an RNA genome, copies of the genome are synthesized by the viral enzyme **RNA replicase**, which is an RNA polymerase. RNA replicase synthesizes RNA from an RNA template, using ribonucleotides provided by the host cell.

In other RNA viruses, however, the genome is transcribed from RNA to DNA by a viral enzyme called **reverse transcriptase**. This enzyme is a DNA polymerase that makes a single-stranded **complementary DNA**, or **cDNA**, from a single-stranded RNA template. Reverse transcriptase then removes the RNA

(a) HIV's protease enzyme

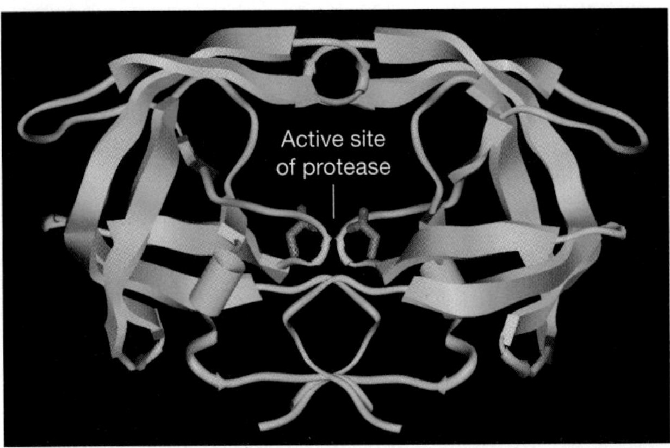

(b) Could a drug block the active site?

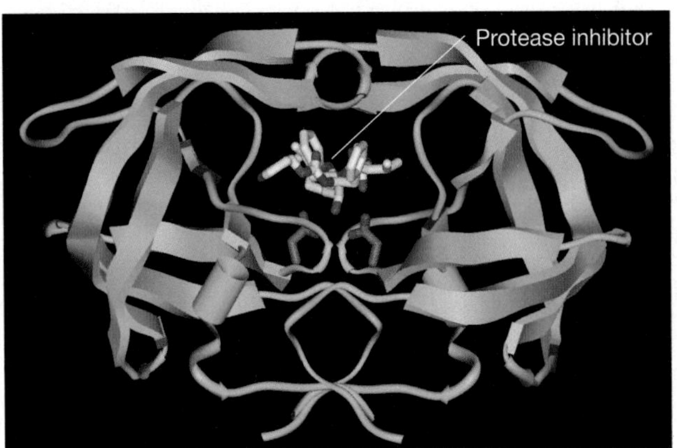

FIGURE 35.13 The Three-Dimensional Structure of Protease. (a) Ribbon diagram depicting the three-dimensional shape of HIV's protease enzyme. **(b)** Once protease's structure was solved, researchers began looking for compounds that would fit into the active site and prevent the enzyme from working.

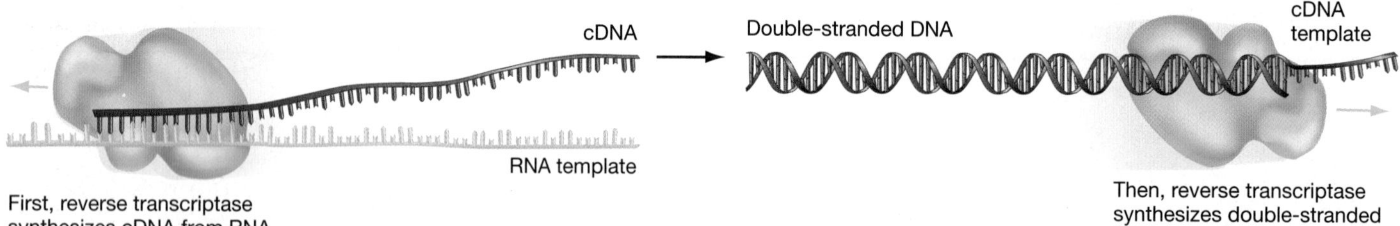

FIGURE 35.14 Reverse Transcriptase Catalyzes Synthesis of a Double-Stranded DNA from an RNA Template. The DNA produced by reverse transcriptase is called a cDNA because its base sequence is complementary to the RNA template.

strand and catalyzes the synthesis of a complementary DNA strand, resulting in a double-stranded DNA (**Figure 35.14**). Viruses that reverse-transcribe their genome in this way are called **retroviruses** ("backward viruses"). The name is apt because the flow of genetic information in this type of virus goes from RNA back to DNA.

HIV is a retrovirus. Two copies of the RNA genome and about 50 molecules of reverse transcriptase lie inside the capsid of each HIV particle. The first antiviral drugs that were developed to combat HIV act by inhibiting reverse transcriptase. Logically enough, drugs of this type are called reverse transcriptase inhibitors. **Box 35.2** explains why reverse transcriptase inhibitors are usually prescribed in combination with other antiviral drugs.

After reverse transcriptase makes a cDNA copy of the viral genome, another viral enzyme called integrase catalyzes the insertion of the double-stranded cDNA into a host-cell chromosome. Once it is integrated into the host's genome in this way, HIV may stay in a latent state for a period—its genes may "just sit there." More commonly, though, the viral genes are actively transcribed to mRNA, using the cell's RNA polymerase, and then translated into proteins by the host cell's ribosomes. In addition, some of the RNA transcripts serve as genomes for the next generation of HIV virions.

Assembly of New Virions Once the viral genome has been replicated and viral proteins are produced, they are transported to locations where a new generation of virions assembles inside the infected host cell. During assembly, the capsid forms around the viral genome. In many cases, copies of non-capsid proteins like DNA or RNA polymerases or reverse transcriptases are also packaged inside the capsid. And in enveloped viruses, envelope proteins become inserted into the host cell's plasma membrane.

In many cases, the details of the assembly process are not yet well understood. Bacteriophages and other nonenveloped viruses with complex morphologies appear to assemble in a step-by-step process resembling an assembly line. HIV and other enveloped viruses appear to assemble while attached to the inside surface of the host's plasma membrane. In most cases, self-assembly occurs—meaning that no enzymes are involved—though some viruses produce proteins that provide a scaffolding where new virions are put together. To date, researchers have yet to develop drugs that inhibit the assembly process during an HIV infection.

Exiting an Infected Cell Viruses leave a host cell in one of two ways: by budding from cellular membranes or by bursting out of the cell. In general, enveloped viruses bud; nonenveloped viruses burst.

Viruses that bud from one of the host cell's membranes take some of that membrane with them. As a result, they incorporate host-cell phospholipids into their envelope, along with envelope proteins encoded by the viral genome (**Figure 35.15a**). Most budding viruses infect host cells that lack a cell wall. In contrast, viruses that burst are able to infect host cells that have a cell wall. Bacteriophages produce lysozyme to create a hole in the cell wall of bacterial cells and then escape through the hole (**Figure 35.15b**).

How Are Viruses Transmitted to New Hosts? Once particles are released, the infection cycle is successful. Hundreds of newly assembled virions are now in extracellular space. What happens next?

If the host cell is part of a multicellular organism, the new generation of particles begins traveling through the body—often via the bloodstream or lymph system. There, they may be bound by antibodies produced by the immune system. In vertebrates, this binding marks the particles for destruction. But if a particle contacts an appropriate host cell before it encounters antibodies, then the particle will infect that cell. This starts the replication cycle anew.

What if the virus has infected a unicellular organism, or if the virus leaves a multicellular host entirely? For example, when people cough, sneeze, spit, or wipe a runny nose, they help rid their body of viruses and bacteria. But they also project the pathogens into the environment, sometimes directly onto an uninfected host. From the virus's point of view, this new host represents an unexploited habitat brimming with resources in

(a) Budding of enveloped viruses

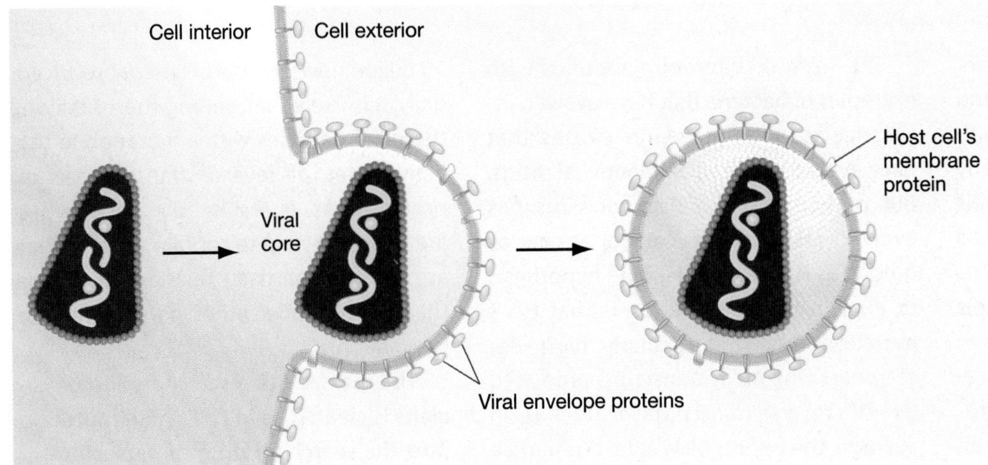

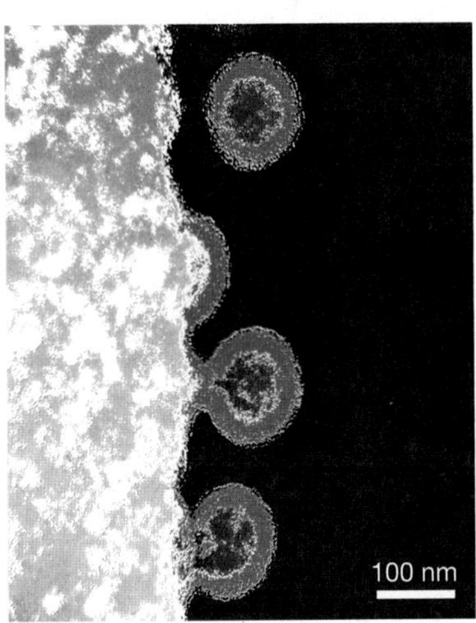

(b) Bursting of nonenveloped viruses

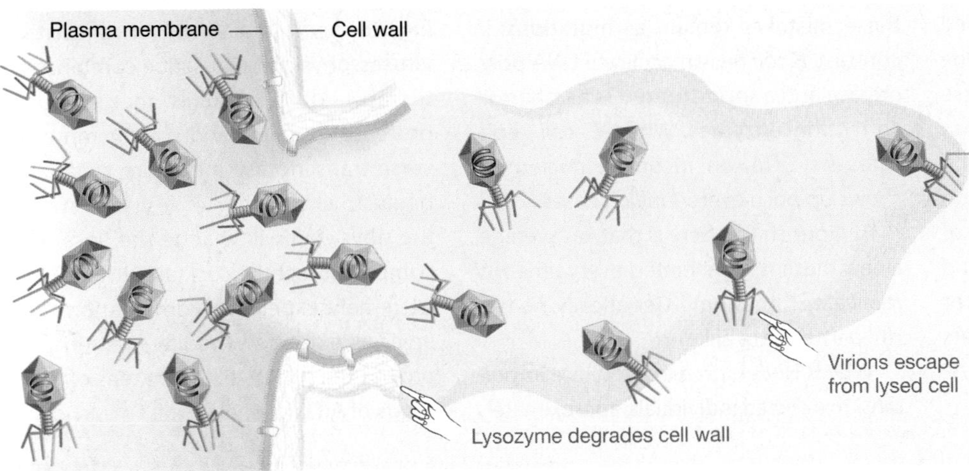

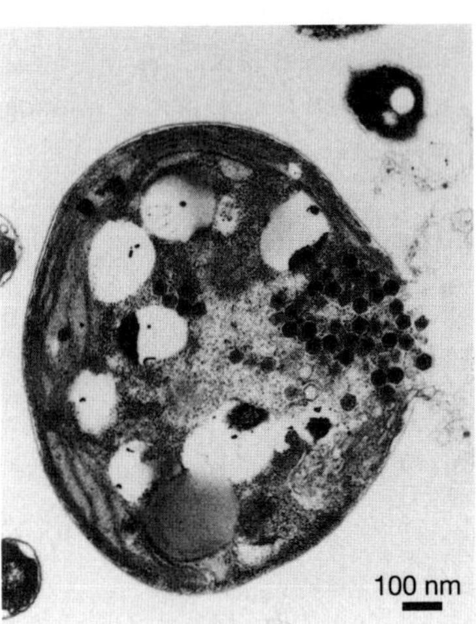

FIGURE 35.15 Viruses Leave Infected Cells by Budding or Bursting.

QUESTION Both budding and bursting kill the host cell. Propose a hypothesis to explain why infection with a budding virus is fatal.

the form of target cells. The situation is analogous to that of a multicellular animal dispersing to a new habitat and colonizing it. Viruses that successfully colonize a new host replicate and increase in number. The alleles carried by these successful colonists increase in frequency in the total population. In this way, natural selection favors alleles that allow viruses to do two things: (1) replicate within a host and (2) be transmitted to new hosts.

For agricultural scientists, physicians, and public health officials, reducing the likelihood of **transmission**—the spread of pathogens from one individual to another— is often an effective way to combat a viral disease. For example, HIV particles are transmitted from person to person via body fluids such as blood, semen, or vaginal secretions. Faced with decades of disappointing results in drug and vaccine development, public health officials are aggressively promoting preventive medicine. Condom use reduces sexual transmission of HIV. Aggressive treatment of venereal diseases may also help; the lesions caused by chlamydia, genital warts, and gonorrhea encourage the transmission of HIV-contaminated blood during sexual intercourse. The most effective forms of preventive medicine are sexual abstinence or monogamy. In areas where blood supplies are routinely screened, HIV is rarely contracted by means other than unprotected sex with an infected person or needle sharing among drug abusers.

The effectiveness of preventive medicine underscores one of this chapter's fundamental messages: Viruses are a fact of life. Every organism is victimized by viruses; every organism has defenses against them. But the tree of life will never be free of these parasites. Mutation and natural selection guarantee that viral genomes will continually adapt to the de-

BOX 35.2 HIV Drug Cocktails and the Evolution of Drug Resistance

HIV reverse transcriptase inhibitors were dispensed widely in North America and Europe beginning in the mid-1990s, with spectacular results. After therapy, many patients no longer had detectable levels of HIV in their blood. The drugs knocked HIV populations down.

Within two years, however, HIV levels in many of the patients taking reverse transcriptase inhibitors began to rebound. To investigate why this was happening, researchers sequenced the HIV reverse transcriptase gene in these patients. The researchers found that a series of mutations had occurred in the reverse transcriptase gene. These mutations led to changed amino acid sequences in the enzyme's active site. Because reverse transcriptase inhibitor molecules did not fit as well into the altered version of the active site, the enzyme could function reasonably well even in the presence of the inhibitor molecules. Almost as soon as a new reverse transcriptase inhibitor went into widespread use, researchers had documented the evolution of resistance to the drug.

The scientific literature abounds with examples of bacteria that have evolved resistance to antibiotics and viruses that have evolved resistance to antiviral drugs. But perhaps no organism or virus has evolved resistance to control agents as quickly as HIV has. The leading hypothesis to explain this observation is that HIV's mutation rate is particularly high. Researchers who assay transcripts produced by HIV's reverse transcriptase find that, on average, the wrong base is inserted once every 8000 nucleotides. HIV's genome does not code for any of the error-correcting enzymes introduced in Chapter 14, so these mistakes remain as mutations. In contrast, *E. coli* has an accurate DNA polymerase and a sophisticated suite of error-correcting enzymes. When *E. coli* replicates its DNA, an incorrect nucleotide shows up once every 1 *billion* bases.

The punch line here is that, on average, a new mutant is generated every time HIV replicates its genome. Genetically, no two HIV particles are alike.

Why is HIV's high mutation rate important? In infected individuals, approximately 10 billion new viral particles are produced daily. It is likely that, among the 10 billion, there are particles with a mutation in the active sites of reverse transcriptase or protease. As a result, HIV populations are almost certain to contain variants that are at least partially resistant to drugs that cripple most other particles in the population.

The message to researchers and physicians is clear: Due to HIV's high mutation rate, the search for drug therapies promises to be an "arms race"—a constant battle between novel drugs and novel, resistant strains of the virus. In attempting to keep ahead of drug-resistant viruses, physicians prescribe combination therapy—drug "cocktails" that include a protease inhibitor and one or more reverse transcriptase inhibitors. If patients begin to show signs of resistant strains, the physician will change the dosage or composition of the cocktail. Although it is extremely expensive, combination therapy has extended the life span and improved the quality of life for tens of thousands of AIDS patients.

fenses offered by their hosts, regardless of whether those defenses are devised by an immune system or by biomedical researchers. Viruses are a constant threat for every organism alive.

Check Your Understanding

If you understand that...

- All organisms and cell types are parasitized by some type of virus.
- After infecting a cell, viruses may replicate or become latent.

You should be able to...

1) Diagram a lysogenic replication cycle.
2) Diagram the six phases of a lytic replication cycle: entry, protein synthesis, genome replication, viral assembly, exit, and transmission to a new host.
3) Explain how HIV performs each step in its replicative cycle.

(MB) Web Animation at www.masteringbio.com
The HIV Replicative Cycle

35.3 What Themes Occur in the Diversification of Viruses?

If viruses can infect virtually every type of organism and cell known, how can biologists possibly identify themes that help organize viral diversity? The answer is that, in addition to being identified as enveloped or nonenveloped, viruses can be categorized by the nature of their hereditary material—in essence, the type of molecule their genes are made of. The single most important aspect of viral diversity is the variation that exists in their genetic material.

Two other points are critical to recognize about viral diversity: (1) Biologists do not have a solid understanding of how viruses originate, but (2) it is certain that viruses will continue to diversify. Let's analyze the diversity of molecules that make up viral genes, then consider hypotheses to explain where viruses come from and recent data on new or "emerging" viruses.

The Nature of the Viral Genetic Material

DNA is the hereditary material in all cells. As cells synthesize the molecules they need to function, information flows from

SUMMARY TABLE 35.2 **The Diversity of Viral Genomes**

Key: ss = single stranded; ds = double stranded; (+) = positive sense (genome sequence is the same as viral mRNA); (−) = negative sense (genome sequence is complementary to viral mRNA)

Genome		Example(s)	Host	Result of Infection	Notes
(+)ssRNA	+	TMV	Tobacco plants	Tobacco mosaic disease (leaf wilting)	TMV was the first RNA virus to be discovered.
(−)ssRNA	−	Influenza	Many mammal and bird species	Influenza	The negative-sense ssRNA viruses transcribe their genomes to mRNA via RNA replicase.
dsRNA	+ −	Phytovirus	Rice, corn, and other crop species	Dwarfing	Double-stranded RNA viruses are transmitted from plant to plant by insects. Many can also replicate in their insect hosts.
ssRNA that requires reverse transcription for replication	+	Rous sarcoma virus	Chickens	Sarcoma (cancer of connective tissue)	Rous sarcoma virus was identified as a cancer-causing agent in 1911, decades before any virus was seen.
ssDNA—can be (+), (−), or (+) and (−)	+ or −	$\varphi \times 174$	Bacteria	Death of host cell	The genome for $\varphi \times 174$ is circular and was the first complete genome ever sequenced.
dsDNA	+ −	Baculovirus Smallpox Bacteriophage	Insects Humans Bacteria	Death Smallpox Death	These are the largest viruses in terms of genome size and overall size.

DNA to mRNA to proteins (Chapter 15). Although all cells follow this pattern, which is called the central dogma of molecular biology, some viruses break it. This conclusion traces back to work done in the 1950s, when biologists were able to separate the protein and nucleic acid components of a particle known as the tobacco mosaic virus, or TMV. Surprisingly, the nucleic acid portion of this virus consisted of RNA, not DNA. Later experiments demonstrated that the RNA of TMV, by itself, could infect plant tissues and cause disease. This was a confusing result because it showed that, in this virus at least, RNA—not DNA—functions as the genetic material.

Subsequent research revealed an amazing diversity of viral genome types. In some groups of viruses, such as the agents that cause measles and flu, the genome consists of RNA. In others, such as the particles that cause herpes and smallpox, the genome is composed of DNA. Further, the RNA and DNA genomes of viruses can be either single stranded or double stranded. The single-stranded genomes can also be classified as "positive sense" or "negative sense" or "ambisense." In a **positive-sense virus**, the genome contains the same sequences as the mRNA required to produce viral proteins. In a **negative-sense virus**, the base sequences in the genome are complementary to those in viral mRNAs. In an **ambisense virus**, some sections of the genome are positive sense while others are negative sense. **Table 35.2** summarizes the diversity of viral genome types. If you understand the concept of diversity in viral genomes, you should be able to name two types of viral genomes that are not based on double-stranded DNA. For each of these genomes, you should be able to diagram the steps in information flow from gene to protein.

Finally, the number of genes found in viruses varies widely. The tymoviruses that infect plants contain as few as three genes, but the genome of smallpox can code for up to 353 proteins.

Where Did Viruses Come From?

No one knows where viruses originated, but many biologists suggest that they may be derived from the plasmids and transposable elements introduced in Chapter 19 and Chapter 20. Viruses, plasmids, and transposable elements are all acellular, mobile genetic elements that replicate with the aid of a host cell. Simple viruses are actually indistinguishable from plasmids except for one feature: They encode proteins that form a capsid and allow the genes to exist outside of a cell.

Some biologists hypothesize that simple viruses, plasmids, and transposable elements represent "escaped gene sets." This hypothesis states that mobile genetic elements are descended from clusters of genes that physically escaped from bacterial or eukaryotic chromosomes long ago. According to this hypothesis, the escaped gene sets took on a mobile, parasitic existence because they happened to encode the information needed to replicate themselves at the expense of the genomes that once held them. In the case of viruses, the hypothesis is that the escaped genes included the instructions for making a protein capsid and possibly envelope proteins. According to the escaped-gene hypothesis, it is likely that each of the distinct types of RNA viruses and some of the DNA viruses represent distinct "escape events."

The same researchers contend that DNA viruses with large genomes may have originated in a very different manner, however. Here the hypothesis is that the large DNA viruses trace their ancestry back to free-living bacteria that once took up residence inside eukaryotic cells. The idea is that these organisms degenerated into viruses by gradually losing the genes required to synthesize ATP, nucleic acids, amino acids, and other compounds. Although this idea sounds speculative, it cannot be dismissed lightly. Chapter 28 introduced species in the genus *Chlamydia*, which are bacteria that live as parasites inside animal cells. And Chapter 29 provided evidence that the organelles called mitochondria and chloroplasts, which reside inside eukaryotic cells, originated as intracellular symbionts. Investigators contend that, instead of evolving into intracellular symbionts that aid their host cell, DNA viruses became parasites capable of destroying the host.

To date, neither of these hypotheses has been tested rigorously. To support the escaped-genes hypothesis, researchers would probably have to discover a brand-new virus that originated in this way, or viruses that had so recently derived from intact bacterial or eukaryotic genes that the viral DNA sequence still strongly resembled the DNA sequence of those genes. To support the degeneration hypothesis, researchers point to mimivirus, which still contains some genes involved in protein synthesis. A third alternative is also being discussed—the possibility that viruses coevolved with the first, RNA-based, forms of life (see Chapter 4). Currently, there is no widely accepted view of where viruses came from. Because viruses are so diverse, it is logical to predict that all three hypotheses may be valid.

Emerging Viruses, Emerging Diseases

Although it is not known how the various types of virus originated, it is certain that viruses will continue to diversify. With alarming regularity, the front pages of newspapers carry accounts of deadly viruses that are infecting humans for the first time. In 1993 a hantavirus that normally infects mice suddenly afflicted dozens of people in the southwestern United States. Nearly half of the people who developed hantavirus pulmonary syndrome died. Still higher fatality rates were recorded in 1995 when the Ebola virus, a variant of a monkey virus, caused a wave of infections in the Democratic Republic of Congo. By the time the outbreak subsided, over 200 cases had been reported; 80 percent were fatal. During a 1976 outbreak of Ebola, 90 percent of cases were fatal. Numerous reports of avian flu infecting and killing humans caused worldwide alarm in 2005–2006.

Hantavirus pulmonary syndrome, Ebola, and avian flu are examples of **emerging diseases:** new illnesses that suddenly affect significant numbers of individuals in a host population. In these cases, the causative agents were considered emerging viruses because they had switched from their traditional host species to a new host—humans. Many investigators consider HIV to be an emerging virus because it originated in chimps and was first transmitted to humans in the early to mid-twentieth century (see **Box 35.3**).

Physicians become alarmed when they see a large number of patients with identical and unusual disease symptoms in the same geographic area over a short period of time. The physicians report these cases to public health officials, who take on two urgent tasks: (1) identifying the agent that is causing the new illness and (2) determining how the disease is being transmitted.

Several strategies can be used to identify a pathogen. In the case of the outbreak of hantavirus pulmonary syndrome, officials recognized strong similarities between symptoms in the U.S. cases and symptoms caused by a hantavirus called Hantaan virus native to northeast Asia. The Hantaan virus rarely causes disease in humans; its normal host is rodents. To determine whether a Hantaan-like virus was responsible for the U.S. outbreak, researchers began capturing mice in the homes and workplaces of afflicted people. About a third of the captured rodents tested positive for the presence of a Hantaan-like virus. (The test that was done is explained in Chapter 49.) DNA sequencing studies confirmed that the virus was a previously undescribed type of hantavirus. Further, the sequences found in the mice matched those found in infected patients. Based on these results, officials were confident that a rodent-borne hantavirus was causing the wave of infections.

The next step in the research program, identifying how the agent is being transmitted, is equally critical. If a virus that normally parasitizes a different species suddenly begins infecting humans, if it can be transmitted efficiently from person to person, and if it causes serious illness, then the outbreak has the potential to become an epidemic. But if transmission takes place only between the normal host and humans, as is the case with rabies, then the number of cases will probably remain low. To date, for example, avian flu has not been transmitted efficiently from person to person—only inefficiently from birds to people.

Determining how a virus is transmitted takes old-fashioned detective work. By interviewing patients about their activities, researchers called epidemiologists decide whether each patient could have acquired the virus independently. Were the individuals infected with hantavirus in contact with

BOX 35.3 Where Did HIV Originate?

The key to discovering where HIV comes from lies in understanding the evolutionary history, or phylogeny, of the virus. Researchers have reconstructed this history by comparing the composition of HIV genes with sequences from closely related viruses that parasitize chimps, monkeys, and other mammals (**Figure 35.16**). The relationships among these gene sequences support several conclusions:

- HIV belongs to a group of viruses called the lentiviruses, which infect a wide range of mammals, including house cats, horses, goats, and primates. (*Lenti* is a Latin root that means "slow"; here it refers to the long period observed between the start of an infection by these viruses and the onset of the diseases they cause.)
- Many of HIV's closest relatives also have *immunodeficiency* in their names. Like HIV, these agents parasitize cells that are part of the immune system. Several of them cause diseases with symptoms reminiscent of AIDS. For unknown reasons, though, HIV's closest relatives don't appear to cause disease in their hosts. These viruses infect monkeys and chimpanzees and are called simian immunodeficiency viruses (SIVs).
- There are two distinct types of human immunodeficiency viruses, called HIV-1 and HIV-2. Although both can cause AIDS, HIV-1 is far more virulent and is the better studied of the two. It serves as the focus of this chapter.
- HIV-1's closest relatives are immunodeficiency viruses isolated from chimpanzees that live in central Africa. In contrast, HIV-2's closest relatives are immunodeficiency viruses that parasitize monkeys called sooty mangabeys. In central Africa, where HIV-1 infection rates first reached epidemic proportions, contact between chimpanzees and humans is extensive. Chimps are hunted for food and kept as pets. Similarly, sooty mangabeys are hunted and kept as pets in west Africa, where HIV-2 infection rates are highest. To make sense of these observations, biologists suggest that HIV-1 is a descendant of viruses that infect chimps, and HIV-2 is a descendant of viruses that infect sooty mangabeys. The "jumps" between host species probably occurred when a human cut up a monkey or chimpanzee for use as food.
- Several strains of HIV-1 exist. A virus **strain** consists of populations that have similar characteristics. The most important HIV-1 strains are called O for *out-group* (meaning, the most basal group relative to other strains), N for *new*, and M for *main*. Because few mechanisms exist for the virus to jump from humans back to chimps or monkeys, it is likely that each of these strains represents an independent origin of HIV from a chimp SIV strain. This is a key point. The existence of these distinct strains suggests that HIV-1 has jumped from chimps to humans several times and may do so again in the future.

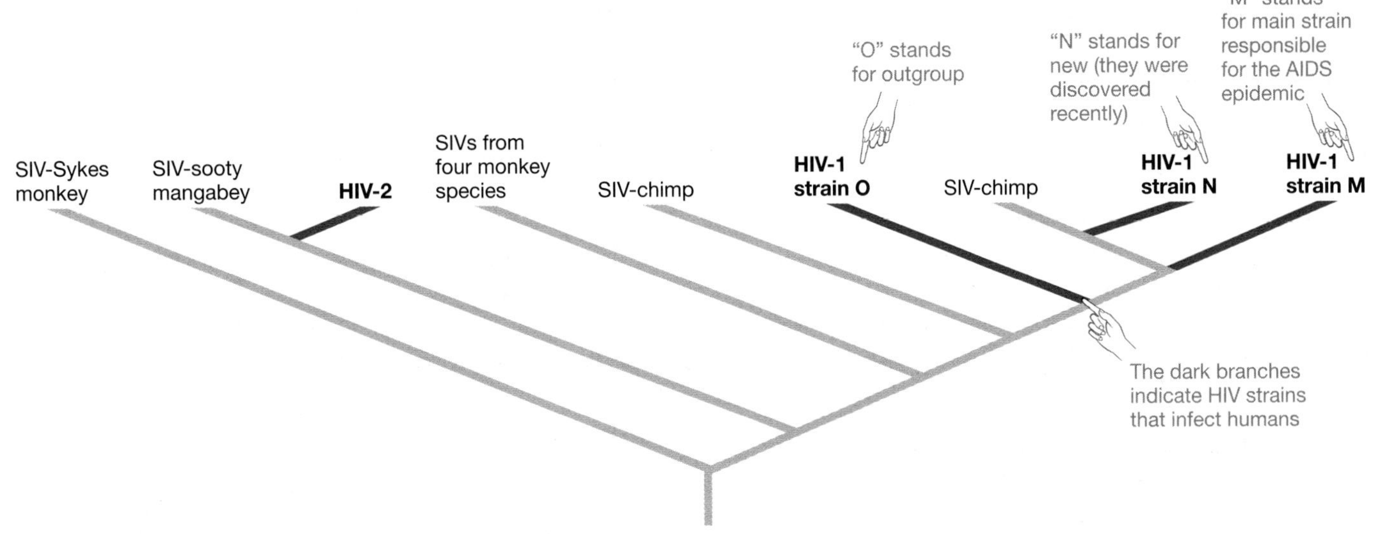

FIGURE 35.16 Phylogeny of HIV Strains and Types. Phylogenetic tree showing the evolutionary relationships among some of the immunodeficiency viruses that infect primates—including chimpanzees, humans, and several species of monkeys.

EXERCISE On the appropriate branches, indicate where an SIV jumped to humans (draw and label bars across the appropriate branches).

mice? If so, were they bitten? Did they handle rodent feces or urine during routine cleaning chores or come into contact with contaminated food? Was the illness showing up in health-care workers, implying that it was being transmitted from human to human?

In the hantavirus outbreak, public health officials concluded that no human-to-human transmission was taking place. Health-care workers did not become ill; because patients had not had extensive contact with one another, it was likely that each had acquired the virus independently. The outbreak also coincided with a short-term, weather-related explosion in the local mouse population. The most likely scenario was that people had acquired the pathogen by inhaling dust or handling food that contained remnants of mouse feces or urine. In short, hantavirus did not have the potential to cause an epidemic. The best medicine was preventive: Homeowners were advised to trap mice in living areas, wear dust masks while cleaning any area where mice might have lived, and store food in covered jars.

The Ebola virus, in contrast, was clearly transmitted from person to person. Many doctors and nurses were stricken after tending to patients with the virus. Infections that originate in hospitals usually spread when carried from patient to patient on the hands of caregivers. But by carefully observing the procedures that were being followed by hospital staff, researchers concluded that the Ebola virus was being transmitted only through direct contact with body fluids (blood, urine, feces, or sputum). The outbreak was brought under control when hospital workers raised their standards of hygiene and insisted on the immediate disposal or disinfection of all contaminated bedding, utensils, and equipment. The situation with the avian flu outbreak of 2005–2006 was similar—person-to-person transmission was extremely rare and based only on extensive, direct contact with body fluids. Had the Ebola virus or avian flu virus been transmitted by casual contact, such as touch or inhalation—as is the common cold virus—a massive epidemic could have ensued.

Check Your Understanding

If you understand that...

Among viruses, several different types of molecules serve as the genetic material.

You should be able to...

1) Explain how the negative sense, single-stranded RNA genome of influenza viruses, which are complementary in sequence to viral mRNA, is used to produce a new generation of flu virions.
2) State whether a mutation that allowed avian flu virus to spread via airborne particles coughed out by infected individuals would make the virus more or less dangerous to humans? Explain your logic.

35.4 Key Lineages of Viruses

Because scientists are almost certain that viruses originated multiple times throughout the history of life, there is no such thing as the phylogeny of all viruses. Stated another way, there is no single phylogenetic tree that represents the evolutionary history of viruses as there is for the organisms discussed in previous chapters. Instead, researchers focus on comparing base sequences in the genetic material of small, closely related groups of viruses and using these data to reconstruct the phylogenies of particular lineages. Phylogenetic trees for viruses are usually inferred from comparisons of nucleic acid sequence data, using techniques introduced in Chapter 27.

The phylogenetic tree of the simian immunodeficiency viruses and human immunodeficiency viruses in Figure 35.16 is a good example of how researchers construct and interpret the phylogenies of particular groups of viruses. These types of phylogenies have been important in understanding viral diversity and the sources of emerging viruses. For example, phylogenetic analyses allowed biologists to recognize that there are two distinct types of HIV. The phylogenetic data diagrammed in Figure 35.16 indicate that HIV-1 and HIV-2 are distinct types of virus that originated from different host species (see Box 35.3). These data correlate with the observation that HIV-1 and HIV-2 differ in important ways. For instance, although both types of HIV infect immune system cells and are usually transmitted through sexual contact, HIV-2 is much less virulent and much less easily transmitted from person to person than is HIV-1.

To organize the diversity of viruses on a larger scale, researchers group them into seven general categories based on the nature of their genetic material. Within these broad groupings, biologists also identify a total of about 70 virus families that are distinguished by (1) the structure of the virion (often whether it is enveloped or nonenveloped), and (2) the nature of the host species.

Although they do not have formal scientific names, viruses within families are grouped into distinct genera for convenience. Within genera, biologists identify and name types of virus, such as HIV, the measles virus, and smallpox. Within each of these viral types, populations with distinct characteristics may be identified and named as strains. The strain is the lowest, or most specific, level of taxonomy for viruses. The O, N, and M strains of HIV-1, which are highlighted in Figure 35.16, are examples of separate virus strains. In the case of HIV-1, the named strains resulted from independent "jumps" of a simian immunodeficiency virus to a new host: humans. Each of these strains has distinguishing characteristics. The M strain, for example, was named *main* because it is responsible for most of the HIV infections known to date.

To get a sense of viral diversity, let's survey a few of the major groups that can be identified by the nature of their genetic material.

Double-Stranded DNA (dsDNA) Viruses

The double-stranded DNA viruses are a large group, composed of some 21 families and 65 genera. Smallpox (**Figure 35.17**) is perhaps the most familiar of these viruses. Although smallpox had been responsible for millions of deaths throughout human history, it was eradicated by vaccination programs. Smallpox is currently extinct in the wild; the only remaining samples of the virus are stored in research labs.

Genetic material As their name implies, the genes of these viruses consist of a single molecule of double-stranded DNA. The molecule may be linear or circular.

Host species These viruses parasitize hosts from throughout the tree of life, with the notable exception of land plants. They include virus families called the T-even and λ bacteriophages, some of which infect *E. coli*. In addition, the pox viruses, herpesviruses, and adenoviruses—some types of which parasitize humans—have genomes consisting of double-stranded DNA.

Replication cycle In most double-stranded DNA viruses that infect eukaryotes, viral genes have to enter the nucleus to be replicated. The viral genes are replicated only during S phase, when the host cell's chromosomes are being replicated. As a result, these types of viruses can sustain an infection only in cells that are actively dividing, such as the cells lining the respiratory tract or urogenital canal. Some of these viruses are capable of inducing replication in infected cells.

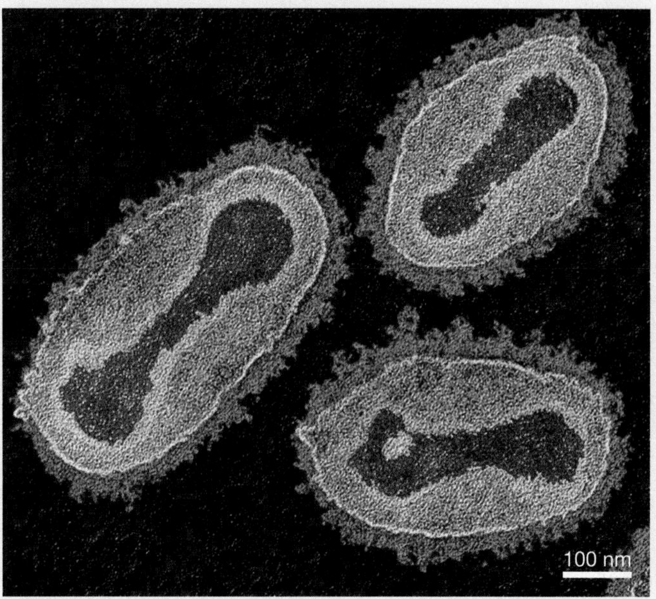

FIGURE 35.17 Smallpox Is a Double-Stranded DNA Virus.

RNA Reverse-Transcribing Viruses (Retroviruses)

The genomes of the RNA reverse-transcribing viruses are composed of single-stranded RNA. There is only one family, called the retroviruses.

Genetic material Virus particles have two copies of their single-stranded RNA genome, so they are diploid.

Host species Species in this group are known to parasitize only vertebrates—specifically birds, fish, or mammals. HIV is the most familiar virus in this group. The Rous sarcoma virus, the mouse mammary tumor virus, and the murine (mouse) leukemia virus are other retroviruses that have also been studied intensively. Rous sarcoma virus was the first virus shown to be associated with the development of cancer (in chickens); the mouse viruses were the first viruses known to increase the risk of cancer in mammals (**Figure 35.18**). In most cases, viruses that are associated with cancer development carry genes that contribute to uncontrolled growth of the cells they infect.

Replication cycle Retroviruses contain the enzyme reverse transcriptase inside their capsid. (A typical HIV particle contains about 50 reverse transcriptase molecules.) When the virus's RNA genome and reverse transcriptase enter a host cell's cytoplasm, the enzyme catalyzes the synthesis of a single-stranded cDNA from the original RNA. Reverse transcriptase then makes this cDNA double stranded. The double-stranded DNA version of the genome enters the nucleus with a viral protein called integrase. Integrase catalyzes the integration of the viral genes into a host chromosome. The virus may remain quiescent for a period, but eventually the genes are transcribed to RNA to begin lytic growth and the production of a new generation of virus particles.

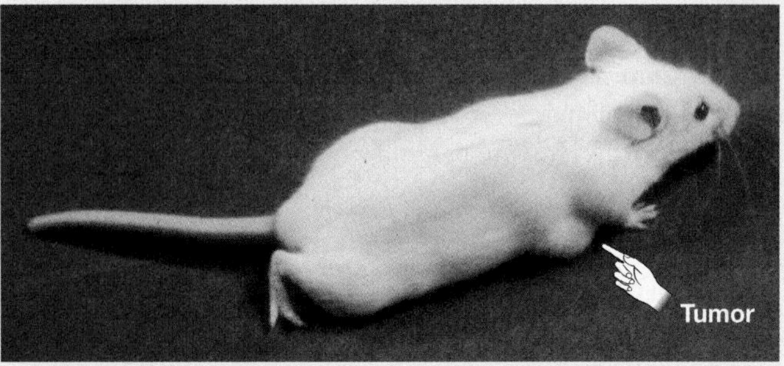

FIGURE 35.18 Some Retroviruses, Such as the Mouse Mammary Tumor Virus, Are Associated with Cancer.

Double-Stranded RNA (dsRNA) Viruses

There are 7 families of double-stranded RNA viruses and a total of 22 genera. Most of the viruses in this group are nonenveloped.

Genetic material In some families, virus particles typically have a genome consisting of 10–12 double-stranded RNA molecules; in other families, the genome is composed of just 1–3 RNA molecules.

Host species A wide variety of hosts, including fungi, land plants, insects, vertebrates, and bacteria, are victimized by viruses with double-stranded RNA genomes. Particularly prominent are viruses that cause disease in rice, corn, sugar cane, and other crops. **Figure 35.19** shows rice plants that are being attacked by a double-stranded RNA virus. Infections are also common in *Penicillium*, a filamentous fungus that produces the antibiotic penicillin. Reovirus and rotavirus infections, the leading cause of infant diarrhea in humans, are responsible for over 110 million cases and 440,000 deaths each year.

Replication cycle Once inside the cytoplasm of a host cell, the double-stranded genome of these viruses serves as a template for the synthesis of single-stranded RNAs, which are then translated into viral proteins. The proteins form the capsids for a new generation of virus particles. Copies of the genome are created when a viral enzyme makes the original single-stranded RNA double stranded.

FIGURE 35.19 Double-Stranded RNA Viruses, Such as the Ragged Stunt Virus, Parasitize a Wide Array of Organisms—Here, Rice.

Negative-Sense Single-Stranded RNA ([−]ssRNA) Viruses

There are 7 families and 30 genera in this group. Most members of this group are enveloped, but some negative-sense single-stranded RNA viruses lack an envelope.

Genetic material The sequence of bases in a negative-sense RNA virus is opposite in polarity to the sequence in a viral mRNA. Stated another way, the single-stranded virus genome is complementary to the viral mRNA. Depending on the family, the genome may consist of a single RNA molecule or up to eight separate RNA molecules.

Host species A wide variety of plants and animals are parasitized by viruses that have negative-sense single-stranded RNA genomes. If you have ever suffered from the flu, the mumps, or the measles, then you are painfully familiar with these viruses (**Figure 35.20**). The Ebola, Hantaan, and rabies viruses also belong to this group.

Replication cycle When the genome of a negative-sense single-stranded RNA virus enters a host cell, a viral RNA polymerase uses that genome as a template to make new viral mRNA. These viral mRNAs are then translated to form viral proteins. The viral mRNAs also serve as a template for the synthesis of new copies of the negative-sense single-stranded RNA genome.

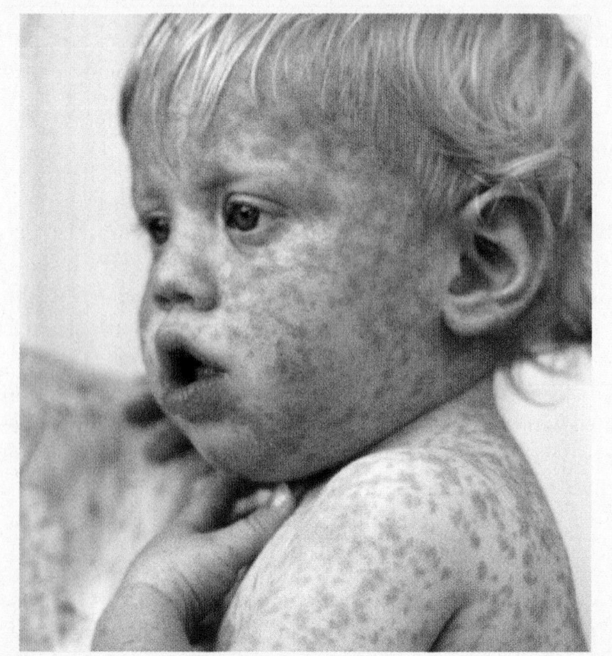

FIGURE 35.20 Negative-Sense Single-Stranded RNA Viruses, Such as the Measles Virus, Cause Some Common Childhood Diseases.

Positive-Sense Single-Stranded RNA ([+]ssRNA) Viruses

This is the largest group of viruses known, with 81 genera organized into 21 families.

Genetic material The sequence of bases in a positive-sense RNA virus is the same as that of a viral mRNA. Stated another way, the genome does not need to be transcribed in order for proteins to be produced. Depending on the species, the genome consists of one to three RNA molecules.

Host species Most of the commercially important plant viruses belong to this group. Because they kill groups of cells in the host plant and turn patches of leaf or stem white, they are often named mottle viruses, spotted viruses, chlorotic (meaning, lacking chlorophyll) viruses, necrotic (meaning, killed cells surrounded by intact tissue) viruses, or mosaic viruses. **Figure 35.21** shows a healthy cowpea leaf and a cowpea leaf that has been attacked by a positive-sense single-stranded RNA virus. Some species in this group of viruses specialize in parasitizing bacteria, fungi, or animals, however. A variety of human maladies, including the common cold, polio, and hepatitis A, C, and E, are caused by positive-sense RNA viruses.

Replication cycle When the genome of these viruses enters a host cell, the single-stranded RNA is immediately translated into viral proteins. These proteins include enzymes that make copies of the genome. The new generation of virus particles is assembled in a complex structure that is associated with host cell membranes.

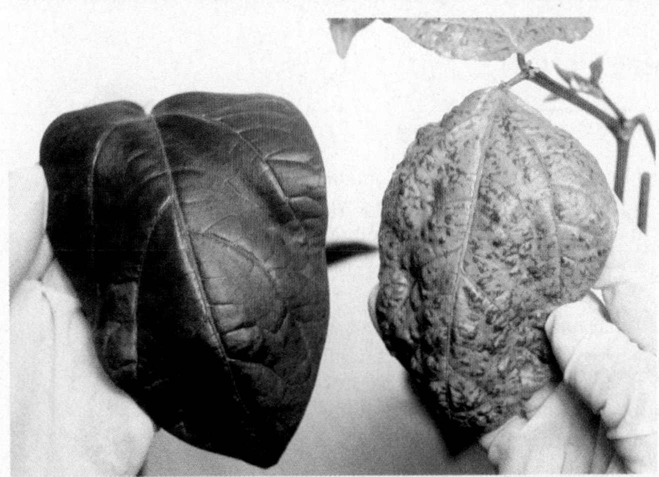

FIGURE 35.21 Positive-Sense Single-Stranded RNA Viruses, Such as the Cowpea Mosaic Virus, Cause Important Plant Diseases.

Chapter Review

SUMMARY OF KEY CONCEPTS

- **Viruses are tiny, noncellular parasites that infect virtually every type of cell known. They cannot perform metabolism on their own—meaning outside a parasitized cell—and are not considered to be alive. Different types of viruses infect particular species and types of cells.**

Viruses cause illness and death in plants, fungi, bacteria, and archaea, as well as in humans and other animals. Vaccination is frequently effective in preventing virus epidemics. Unfortunately, it is difficult or impossible to design a vaccine that can prepare the immune system for viruses that mutate very rapidly, such as HIV and the flu viruses.

You should be able to explain why, despite the diversity and abundance of viruses, most cells are healthy.

- **Although viruses are diverse morphologically, they can be classified as two general types: enveloped and nonenveloped.**

Both nonenveloped and enveloped viruses have a capsid made of protein. The capsid usually encloses viral enzymes as well as the viral genome. Nonenveloped viruses consist of a naked capsid, but in enveloped viruses, the capsid is surrounded by a membranous envelope. Nonenveloped viruses exit a cell by lysis; enveloped viruses exit a cell by budding.

You should be able to make sketches comparing the general morphologies of a nonenveloped and enveloped virus.

- **The viral infection cycle can be broken down into six steps: (1) entry into a host cell, (2) production of viral proteins, (3) replication of the viral genome, (4) assembly of a new generation of virus particles, (5) exit from the infected cell, and (6) transmission to a new host.**

A viral infection begins when the contents of a virus particle enter a host cell. For example, when HIV binds to a transmembrane protein called CD4 and a co-receptor, the virus's membrane-like envelope fuses with the host cell's plasma membrane, and the viral genome and proteins enter the cytoplasm. In the second phase of the replication cycle, viral proteins are produced. The viral genome is replicated in the third phase, and the viral proteins and genome then assemble into complete particles (the fourth phase of the infection cycle). The new generation of complete particles then buds or bursts from the cell, usually killing the host cell in the process. Once they are released, particles can infect new cells in the same multicellular organism or be transmitted to a new host.

You should be able to explain how at least three of the six phases of a viral replicative cycle can be stopped, using HIV as an example.

The HIV Replicative Cycle

In terms of diversity, the key feature of viruses is the nature of their genetic material. The genomes of viruses may consist of double-stranded DNA, single-stranded DNA, double-stranded RNA, or one of several types of single-stranded RNA.

Viral genomes are small and can consist of either RNA or DNA, but not both. Viral genomes do not code for ribosomes, and most viral genomes do not code for the enzymes needed to translate their own proteins or to perform other types of biosynthesis. When a virus infects a host cell, it uses that cell's enzymes, nutrients, ATP, and ribosomes to manufacture a new generation of virus particles. This observation explains why viral diseases are difficult to treat with drugs: Molecules that incapacitate enzymes needed by the virus are likely to damage host cells as well.

You should be able to explain how a viral genome that consists of single-stranded DNA is copied by a viral DNA polymerase.

QUESTIONS

Test Your Knowledge

1. How do viruses that infect animals enter an animal's cells?
   a. The viruses pass through a wound.
   b. The viruses bind to a membrane protein.
   c. The viruses puncture the cell wall.
   d. The viruses lyse the cell.
2. What does reverse transcriptase do?
   a. It synthesizes proteins from mRNA.
   b. It synthesizes tRNAs from DNA.
   c. It synthesizes DNA from RNA.
   d. It synthesizes RNA from DNA.
3. What do host cells provide for viruses?
   a. nucleotides and amino acids
   b. ribosomes
   c. ATP
   d. all of the above
4. When do most enveloped virus particles acquire their envelope?
   a. during entry into the host cell
   b. during budding from the host cell
   c. as they burst from the host cell
   d. as they integrate into the host cell's chromosome
5. What reaction does protease catalyze?
   a. polymerization of amino acids into peptides
   b. cutting of long peptide chains into functional proteins
   c. folding of long peptide chains into functional proteins
   d. assembly of viral particles
6. Why is it difficult to design a vaccine for viruses with high mutation rates, such as HIV and flu viruses?
   a. The vaccines tend to be unstable and deteriorate over time.
   b. So many protein fragments are presented by these viruses that the immune system overreacts.
   c. They have no protein fragments that can be recognized by a host cell.
   d. New mutations constantly change viral proteins.

Test Your Knowledge answers: 1. b; 2. c; 3. d; 4. b; 5. b; 6. d

Test Your Understanding

Answers are available at www.masteringbio.com

1. The outer surface of a virus consists of either a membrane-like envelope or a protein capsid. Which type of outer surface does HIV have? Which type does adenovirus have? How does the outer surface of a virus correlate with its mode of exiting a host cell, and why?
2. Compare the morphological complexity of HIV with that of bacteriophage T4. Which virus would you predict has the larger genome? Explain the logic behind your hypothesis.
3. Compare and contrast lytic growth with growth during the latent state. Is it possible for viral populations to increase if virions remain in the latent state?
4. Explain why viral diseases are more difficult to treat than diseases caused by bacteria.
5. Draw the lytic cycle of a nonenveloped virus with a positive-sense single-stranded RNA genome that infects cells in the roots of rice plants. Describe the modes of action of two drugs that could be developed to treat plants infected with this virus.
6. What type of data convinced researchers that HIV originated when a simian immunodeficiency virus "jumped" to humans? Do you agree with this conclusion? Why or why not?

Applying Concepts to New Situations

Answers are available at www.masteringbio.com

1. Suppose you could isolate a virus that parasitizes the pathogen *Staphylococcus aureus*—a bacterium that causes acne, boils, and a variety of other afflictions in humans. How could you test whether this virus might serve as a safe and effective antibiotic?
2. If you were in charge of the government's budget devoted to stemming the AIDS epidemic, would you devote most of the resources to drug development, vaccine development, or preventive medicine? Defend your answer.
3. Bacteria fight viral infections with restriction endonucleases (bacterial enzymes; introduced in Chapter 19). Restriction endonucleases cut up, or break, viral DNA at specific sequences. The enzymes do not cut a bacterium's own DNA, because the bases in the bacterial genome are protected from the enzyme by methylation (the addition of a CH_3 group). Generate a hypothesis to explain why members of the Eukarya do not have restriction endonucleases.
4. Consider these two contrasting definitions of life:
   a. An entity is alive if it is capable of replicating itself via the directed chemical transformation of its environment.
   b. An entity is alive if it is an integrated system for the storage, maintenance, replication, and use of genetic information.

   According to these definitions, are viruses alive? Explain.

Plant Form and Function 36

Cross section from the trunk of a redwood tree that lived 1710 years. This chapter explores features of plant anatomy, including how wood forms and why growth rings appear in tree trunks.

KEY CONCEPTS

- The vascular plant body consists of (1) a root system that anchors the individual and absorbs water and key ions, and (2) a shoot system that absorbs carbon dioxide and sunlight. Both systems are dynamic—they grow and change throughout life.
- Because body size and shape varies so much among species and individuals, different plants are able to harvest light and other resources in unique ways.
- Primary growth occurs when cells located at the tips of each root and shoot divide. Primary growth extends the body and gives rise to three primary tissue systems. These tissue systems are specialized for protection, food production and storage, and transport.
- In some species, secondary growth makes roots and shoots widen throughout life. Secondary growth occurs when cells near the perimeter of a root or shoot divide. Secondary growth adds transport tissue and provides additional structural support.

Photosynthetic plants do the most remarkable chemistry of any terrestrial organism. Using the energy in sunlight and the simplest of starting materials—carbon dioxide, water, and ions containing nitrogen, phosphorus, potassium, or other key atoms—plants synthesize thousands of different carbohydrates, proteins, nucleic acids, and lipids. They use these compounds to build bodies that may live for thousands of years. The feat is even more impressive when you consider that the simple starting materials that plants need to grow are tiny and diffuse—carbon dioxide molecules, ammonium ions, and other resources are found at low concentrations over a large area. To gather the simple raw materials required for their sophisticated biosynthetic machinery, a plant's roots and shoots grow outward, extending the individual into the soil and atmosphere.

In essence, a plant's body harvests diffuse resources and concentrates them in cells and tissues. The structure of the body is dynamic, though, because most plants grow throughout their lives. A 5000-year-old bristlecone pine has roots and shoots that are still growing. To describe this phenomenon, biologists say that plants can exhibit **indeterminate growth**. In response to favorable conditions, a plant sends roots and shoots in the most promising directions, seeking out more and more light and the simple compounds it requires.

The contrast between the plant and animal way of life is striking. Most animals move around and eat concentrated sources of food. But plants stay in one place, extend their roots and shoots to harvest diffuse resources, and make their own food.

This chapter focuses on three fundamental questions: (1) How is the plant body organized? (2) Why are plants so diverse in size and shape? and (3) How do plants grow throughout their lives? Instead of surveying the entire catalog of land plants, though, the focus here is on the angiosperms. Recall from Chapter 30 that angiosperms are the flowering plants. They are the most recent major group of plants to appear in the fossil record and the most abundant and geographically widespread on Earth today. They are also the most species-rich plant lineage and are of huge economic and medical importance to humans. Most of the food we eat and many of the drugs we use are derived from angiosperms.

If you look outside or down the produce aisle of a grocery store, you're almost certain to see a wide diversity of plants or plant products. By the time you finish this chapter, you'll understand how these plant bodies are put together and how they grow. Exploring questions about the anatomy of flowering plants is vital to understanding the world at large as well as the other chapters in this unit.

36.1 Plant Form: Themes with Many Variations

Chapter 10 detailed how plants—along with algae, cyanobacteria, and a variety of protists—obtain the energy and carbon they need to grow and reproduce. Plants use light energy (photons) to synthesize carbohydrates using carbon dioxide from the air and water from the soil. For photosynthesis to occur, plants need large amounts of light and carbon dioxide and a small amount of water as an electron source (**Figure 36.1**). Plants also need large amounts of water to fill their cells and maintain them at normal volume and pressure. To synthesize nucleic acids, enzymes, phospholipids, and the other macromolecules needed to build and run cells, plants must obtain nitrogen (N), phosphorus (P), potassium (K), magnesium (Mg), and a host of other nutrients. Most of these key elements exist in nature as ions that dissolve in water found in soil.

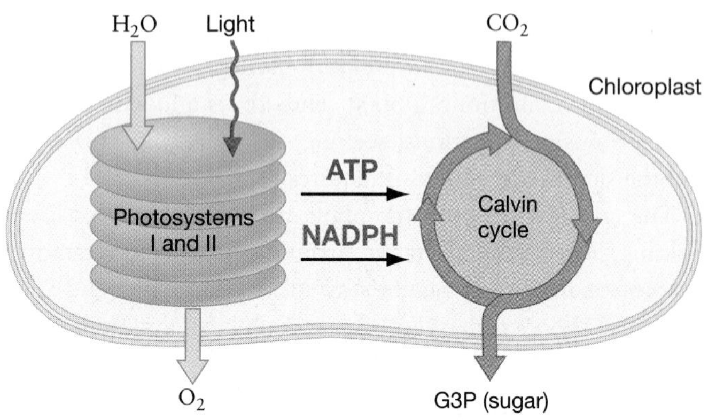

FIGURE 36.1 Plants Need Resources to Perform Photosynthesis.

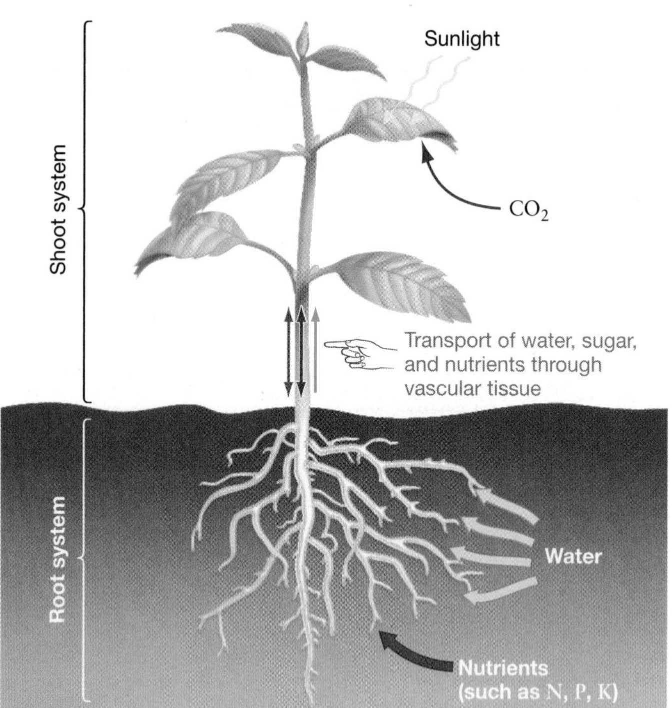

FIGURE 36.2 Root and Shoot Systems Acquire and Transport Resources. Shoot systems are specialized for harvesting light and CO_2; root systems absorb water and key nutrients such as nitrogen (N), phosphorus (P), and potassium (K).

EXERCISE Suppose that there was a new deposit of nutrient-rich soil to the left and that much more sunlight was suddenly available to the right. Draw what you would expect this individual to look like in 1 month.

To acquire the resources they need for photosynthesis, plants use the two basic systems shown in **Figure 36.2.** A belowground portion called the **root system** anchors the plant and takes in water and nutrients from the soil, while an aboveground portion called the **shoot system** harvests light and carbon dioxide from the atmosphere. These systems grow throughout the life of the individual, allowing it to increase in size, overtop competitors, and forage for the resources it needs.

In most plants, the root and shoot systems are connected by vascular tissue. Water is transported from roots to shoots through vascular tissue; sugars and other nutrients are transported in both directions.

Figure 36.3 provides a more detailed look at the anatomy of roots and shoots. The "generic" plant shown in Figures 36.2 and 36.3 is an **herbaceous plant** or herb, meaning a seed plant that lacks wood and has a relatively short-lived stem. Most land plants conform to the basic themes illustrated in the figures—meaning they have bodies organized into root and shoot systems joined by vascular tissue. There are many variations on this theme, however. The organization of the vascular tissue is not identical in all plants, and the morphology of root and shoot systems and leaves varies widely.

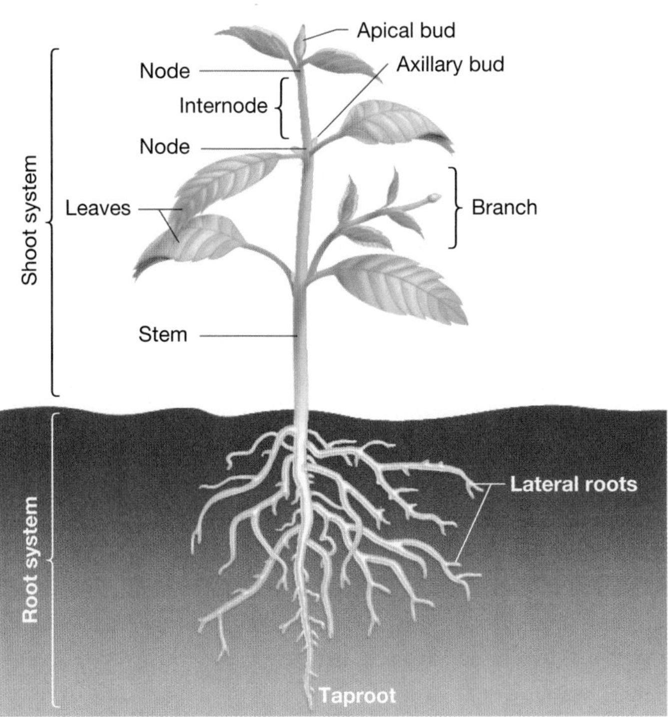

FIGURE 36.3 Most Root and Shoot Systems Have the Same General Structures.

The Root System

Many root systems have a vertical section called a **taproot**, as well as numerous **lateral roots** that run more or less horizontally. The root system functions to anchor the plant in the soil, absorb water and ions from the soil, conduct water and selected ions to the shoot, and store material produced in the shoot for later use.

Root systems can be impressive in extent. For example, a researcher grew a winter rye plant in a container full of soil for four months, then unearthed the plant and meticulously measured the extent of its roots. He found that the root system of this single individual contained more than 13 million identifiable structures with a combined length of over 11,000 km—almost one-third of Earth's circumference. Other studies have shown that the roots of trees routinely extend wider than their aboveground canopy and that it is not unusual for a plant's root system to represent over 80 percent of its total mass. Many plants devote a great deal of energy and resources to the growth of their root systems.

The root systems observed in different species are diverse, as well. This diversity can be analyzed on three levels: morphological diversity among species, changes in the structure of an individual's root system over time, and modified types of roots that are specialized for unusual functions. Let's consider each level in turn.

Morphological Diversity Most people don't appreciate just how diverse root systems are in size and shape, simply because they have never seen them. As an example of morphological diversity in root systems, consider the prairie plants of North America. Prairies are grassland ecosystems found in areas of the world such as the Serengeti Plain of East Africa, the Pampas region of Argentina, and the steppes of central Asia. Rain is abundant enough in these areas to support a lush growth of grasses and herbaceous plants but scarce enough to exclude trees and most shrubs (**Figure 36.4a**). The growth of woody species is also discouraged by the fires that regularly sweep through these

(a) Prairies are dominated by grasses and herbaceous plants.

(b) Root systems in prairie plants are diverse.

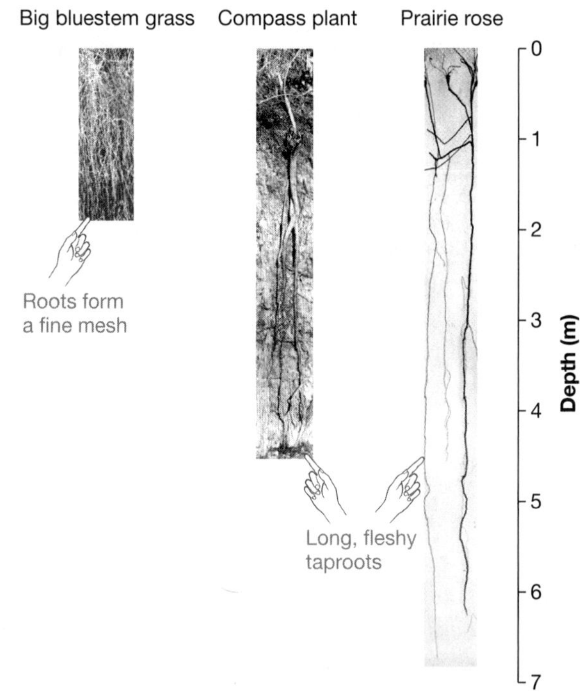

FIGURE 36.4 Prairie Plants Have Diverse Root Systems.
(a) Prairies, such as this one in Kansas, are grassland ecosystems.
(b) The roots of prairie plants that live side by side can be very different. For example, the roots of these three plants reach depths of 6 ft, 14 ft, and 21 ft, respectively.

EXERCISE Roots function as storage organs as well as resource-harvesting systems and anchoring devices. In part (b), draw arrows to thickened areas of roots that store starch. How might you confirm that they have high starch concentrations?

ecosystems. Although the aboveground portions of prairie plants are burned during fires and die back during the winter or dry season, their root systems are **perennial**, meaning that they live for many years. The root system sends up a new shoot system after a fire and each spring.

To examine the root systems of prairie plants, researchers dig deep trenches and examine the exposed root systems, or excavate around a particular plant to reveal its roots. Some of the photographs that have resulted from efforts such as these are shown in **Figure 36.4b**. The figure shows that the root systems of prairie plants that live side by side can be very different from one another. Notice that the dense, fibrous root system in the grass called big bluestem does not have a taproot. The prairie rose, in contrast, has a taproot that can be up to 6.4 m (21 ft) long. Other species, such as the compass plant, have thick roots that contain particularly large reservoirs of food stored in the form of starch.

Even though these root systems have similar functions—they anchor the plant, store starch, and absorb water and nutrients—they are highly diverse in structure. For example, the tips of roots are where most water absorption and nutrient absorption take place, but the absorptive portions of roots of different species are found at different depths in the soil. To explain this observation, biologists suggest that natural selection has favored a diverse array of root systems among species that grow in the same habitat, to minimize competition for water and nutrients.

The diversity of root systems observed in prairie plants has two important consequences: (1) An array of prairie plants can coexist in the same area with less competition for soil resources, and (2) most individuals can survive intense water stress during drought years. Each type of root system is specialized for acquiring resources from a different region of the soil. Due in part to the diversity of root systems illustrated in Figure 36.4b, a small patch of prairie is able to sustain a relatively large number of species.

Phenotypic Plasticity Morphological diversity in roots occurs within species as well as among species. Some of the within-species variation is due to genetic diversity among individuals, but some is due to how roots respond to the environment. Roots show a great deal of **phenotypic plasticity**—meaning that they are plastic or changeable, depending on environmental conditions. For example, spruce trees that grow in water-logged soils tend to have flattened or "pancaked" root systems (**Figure 36.5**). Their roots are shallow because the wet soil lacks oxygen, and because root cells suffocate in the anoxic conditions less than a meter below the surface. The same tree growing in drier soils would develop a root system that extends several meters deep. The key point here is that even genetically identical individuals will have very different-looking root systems if they are growing in different types of environments.

Phenotypic plasticity is particularly important in plants because they grow throughout their lives. Thanks to this mechanism, plants can respond when environmental conditions change over the course of their lifetime. Root systems that grow into nutrient-rich septic fields or leaky sewer pipes are a prime example. Roots actively grow into areas of soil where resources are abundant; roots do not grow—or die back—in areas where resources are used up or lacking.

Modified Roots The taproots and fibrous roots illustrated earlier do not begin to exhaust the types of roots found among plants. **Figure 36.6a** shows the prop roots of corn, which help brace the plant in windy weather. Prop roots are an example of an **adventitious root**: a root that develops from the shoot system instead of the root system. The roots that grow from the nodes of ivy plants and help them cling to brick walls are another example of an adventitious root.

Roots can also be modified for functions other than anchoring the individual and absorbing water and ions from the soil. The "snorkel roots" or pneumatophores of a mangrove, for example, are specialized lateral roots that function in gas exchange (**Figure 36.6b**). Mangroves grow in swamps where water often becomes stagnant and deoxygenated. Their root cells do not suffocate, however, because oxygen from the atmosphere can diffuse into the root system via the pneumatophores. Unlike most roots, mangrove pneumatophores grow upward—not downward—in response to gravity.

To summarize, the structure and function of root systems varies widely among species and individuals. Taproots and fibrous roots are common structures, and most roots function to anchor the individual, harvest water and nutrients from the soil, conduct water and nutrients to the shoot, and store

FIGURE 36.5 Phenotypic Plasticity in Root Systems. This fallen spruce tree has shallow, "pancaked" roots because it was growing in an oxygen-poor, waterlogged soil. In a drier habitat, the root system of the same tree would extend several meters deep in the soil.

(a) Prop roots function in support.

(b) Snorkel roots function in gas exchange.

FIGURE 36.6 Modified Roots Have Unusual Structures or Functions. **(a)** The prop roots of corn plants that help stabilize the stem are adventitious—they develop from the shoot system. **(b)** The pneumatophores of mangrove trees allow gas exchange to occur between root tissues and the atmosphere.

● **QUESTION** Why do root cells need oxygen?

materials made in the shoot. But roots can arise from shoot systems, and the roots of certain species are specialized for oxygen exchange or other tasks. Among individuals, the overall size and shape of the root system is variable and dynamic.

The Shoot System

The shoot system consists of one or more **stems**, which are vertical aboveground structures. Each stem consists of **nodes**, where leaves are attached, and **internodes**, or segments between nodes. A **leaf** is an appendage that projects from a stem laterally. Leaves usually function as photosynthetic organs. The nodes where they attach to the stem are also the site of **axillary buds.** If conditions are appropriate, an axillary bud may develop into a **branch**—a lateral extension of the shoot system. At the tip of each stem and branch is an **apical bud**, where growth occurs that extends the length of the stem or branch. If conditions are appropriate, apical or axillary buds may develop into flowers or other reproductive structures.

In essence, the shoot system is a repeating series of nodes, internodes, leaves, and apical and axillary buds. As plants grow, the number of nodes, internodes, and leaves increases. Once a leaf, node, or internode forms, however, it does not increase much in size over time. Thus, the shoot system of a plant grows by adding more parts rather than by increasing the size of each part.

As with root systems, diversity in shoots can be analyzed on three levels: morphological diversity among species, phenotypic plasticity within individuals, and modified shoots with specialized functions.

Morphological Diversity The shoot systems of land plants range in size from species like Scotch moss that are just a few millimeters tall to redwood trees that reach heights of over 100 m (300 ft) and giant sequoia trunks that are estimated to weigh as much as 2.6 million kg (over 5.7 million lbs)—about the same as 10 diesel locomotives. It is not unusual for plants to invest much of their energy and resources in the growth of their light-gathering apparatus.

The shape of the shoot system also varies a great deal among species. For example, the manner in which new branches are added as the shoot system grows affects the shape of the individual and its ability to compete for light. As **Figure 36.7** shows, a plant growing with wide branching angles and short internodes has a very different shape than that of a plant with narrow branching angles and long internodes.

● Variation in the size and shape of the shoot system is important: it allows plants of different species to harvest light at different locations and thus minimize competition. It also allows them to thrive in a wide array of habitats.

As an example of how the shape of a shoot system varies among species in different environments, consider the silversword plants native to Hawaii. Phylogenetic analyses like those introduced in Chapter 27 have shown that all of the silverswords are descended from the same ancestor—a species of tarweed that arrived in Hawaii from California. Today, silverswords can be found from sea level to mountaintops and from rain forests to the desertlike conditions of exposed lava flows.

Besides occupying a wide variety of habitats, silverswords are wonderfully diverse in size, shape, and growth habit. Some have thick, woody stems and grow as shrubs or trees

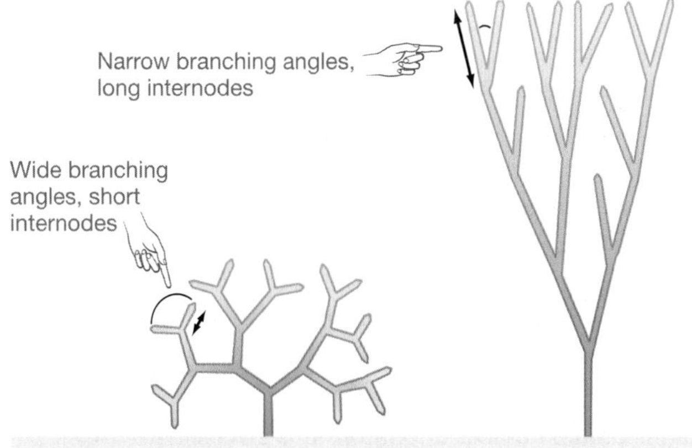

FIGURE 36.7 Plant Form Can Vary as Function of Branch Angle and Internode Length.

● **EXERCISE** Add leaves to the stems in both plants (the length of leaves is independent of the length of the internodes). Draw the shape of a plant where internode length is almost zero.

(a) Tree-sized silversword

(b) Mat-forming silversword

(c) Rosette-forming silversword

FIGURE 36.8 Silverswords Have Diverse Shoot Systems.

(**Figure 36.8a**). Others colonize lava flows and form sprawling cushions or mats (**Figure 36.8b**). Several species grow as compact forms called **rosettes** (**Figure 36.8c**). Rosettes form in species that have extremely reduced internodes between leaves. Dandelions and cabbages are other examples of rosette-forming species.

Biologists interpret this diversity of shoot systems as a suite of adaptations for harvesting light and carbon dioxide in different environments. In lush habitats, where competition for light is intense, woody individuals grow tall and are favored by natural selection. But in dry, windblown habitats, individuals with short stems or the rosette growth form thrive because they require less water than taller individuals do and don't blow over. The adaptive radiation of silverswords has been based in part on diversification in shoot systems.

Phenotypic Plasticity Just as roots respond to variation in water and nutrient availability, shoots respond to variation in light availability. The sugar maple trees shown in **Figure 36.9** are a dramatic example. The tree on the left grew in an open, grassy habitat where light was abundant, and its shoot system is relatively short and broad. The individuals on the right, in contrast, grew in dense forest where competition for light is severe. Their shoot systems are extremely tall and narrow. In each case, the size and shape of the shoot system allows an individual to harvest a maximum number of photons in a given environment.

Because an individual's shoot system continues to grow over the course of its lifetime, it can respond to changes in environmental conditions just as the root system can. Experiments highlighted in Chapter 39, for example, established that shoot systems can bend toward light if an individual is shaded on one side. Plants also undergo differential growth, with more branches and leaves being produced in regions of the body that are exposed to the highest light levels. In effect, a plant's shoot system moves around to maximize its chances of capturing light.

Modified Shoots Even though they are a single lineage, silverswords illustrate many of the types of shoot systems found among land plants. But still other variations occur. Not all stems grow vertically, and not all stems are involved in acquiring carbon dioxide and photons:

- Many desert cacti have highly modified stems. Instead of functioning primarily to support leaves, cactus stems are often enlarged into water-storage organs (**Figure 36.10a**). Water accounts for up to 98 percent of the weight of a cactus stem. A cactus stem also contains the plant's photosynthetic tissue. Instead of being the main food-producing organ, its leaves are modified into the protective structures called **spines**.
- **Stolons** are modified stems that run over the soil surface, producing roots and leaves at each node (**Figure 36.10b**). Because new plants form at these nodes, stolons function in asexual reproduction (see Chapter 12).
- **Rhizomes** are similar to stolons—they are stems that grow horizontally instead of vertically. But while stolons grow aboveground, rhizomes spread belowground (**Figure 36.10c**). Rhizomes are also similar to stolons in function: They produce new plants at nodes and thus aid asexual reproduction. Rhizomes also store starch.
- **Tubers** are compact, modified stems that function as carbohydrate-storage organs (**Figure 36.10d**). The eyes of a potato—a typical tuber—are nodes in the stem where new branches may arise.
- **Thorns** are modified stems that help protect the plant from attacks by large **herbivores**, or plant-eaters, such as deer, giraffe, or cattle (**Figure 36.10e**).

FIGURE 36.9 Phenotypic Plasticity in Shoot Systems. These maple trees are similar genetically but have different shapes. The one on the left is growing in an open, sunny habitat while the individuals on the right are growing in a forest where competition for light is intense.

(a) Cactus stems (shown here in cross section) store water.

(b) Stolons produce new individuals at nodes aboveground.

(c) Rhizomes produce new individuals at nodes belowground.

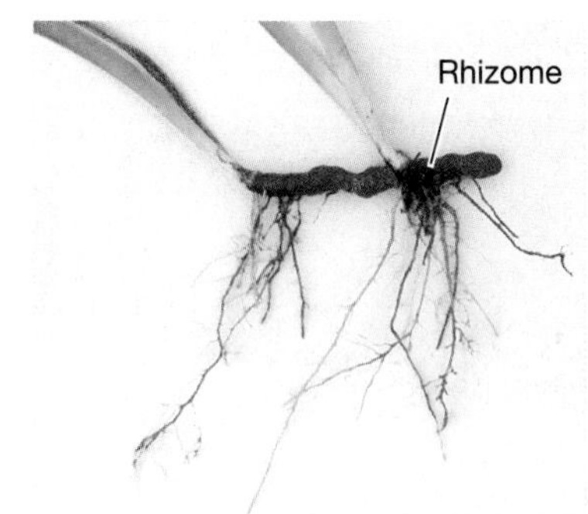

(d) Tubers store carbohydrates.

(e) Thorns provide protection.

FIGURE 36.10 Modified Stems Have Unusual Structures or Functions. In some species, stems are modified to perform functions other than producing leaves that absorb carbon dioxide and sunlight.

EXERCISE In part (a), label the modified leaves called spines.

The Leaf

Leaves are particularly important parts of the shoot system, for a simple reason: In most plant species, the vast majority of photosynthesis occurs in leaves. The total area of leaf produced by a single plant can be enormous—a single tree can have hundreds of thousands of leaves with a total leaf surface area equivalent to that of a football field. All of this area is available for absorbing photons and supporting photosynthesis.

Figure 36.11a illustrates the external anatomy of a typical leaf. A simple leaf like that in Figure 36.11a is composed of just two major structures: an expanded portion called the **blade** and a stalk called the **petiole**. But like the root and shoot systems, leaves exhibit an enormous number of variations on the central theme of a flattened structure specialized for performing photosynthesis.

Morphological Diversity If you glance outside or stroll through a garden, you'll be able to find many types of leaves that have an easily recognizable blade and petiole. (Grass leaves will stump you, though, because they lack petioles entirely.) You should also be able to find compound leaves that have blades divided into a series of leaflets (**Figure 36.11b**). You may even encounter doubly compound leaves, which have leaflets that are again divided (**Figure 36.11c**).

(a) Simple leaves have a petiole and a single blade.

(b) Compound leaves have blades divided into leaflets.

(c) Doubly compound leaves are large yet rarely damaged by wind or rain.

(d) Species from very cold or hot climates have needle-like leaves.

FIGURE 36.11 Leaves Vary in Size and Shape. The structures in parts **(a)** through **(c)** represent single leaves; part **(d)** shows two leaves.

QUESTION For capturing photons, what is the advantage of having a leaf with a large surface area? In terms of wind damage and water loss, what is the disadvantage of having a leaf with a large surface area?

FIGURE 36.12 The Arrangement of Leaves on Stems Varies. The position of leaves on stems affects their ability to absorb light efficiently.

Not all leaf blades are thin and have a large surface area that is advantageous for absorbing large numbers of photons, however. For example, plants that thrive in deserts and in northern habitats tend to have leaves that are needle-shaped (**Figure 36.11d**). The leading hypothesis to explain this pattern is based on two observations: (1) Water is often in short supply in these environments because it is absent in deserts or frozen and thus unavailable in cold habitats, and (2) leaves with large surface areas lose large amounts of water through an evaporative process called **transpiration** (discussed in Chapter 37). Thus, needlelike leaves are interpreted as adaptations that minimize transpiration in water-short habitats. Small, narrow leaves are also much less susceptible to wind damage than are large, broad leaves.

The arrangement of leaves on a stem can vary as much as their overall shape. For example, leaves can be paired opposite each other on the stem (**Figure 36.12a**), be arranged in a whorl (**Figure 36.12b**), alternate on either side of the stem (**Figure 36.12c**), or grow in a compact basal arrangement where internodes are extremely short—leading to the rosette growth form (**Figure 36.12d**). Different plant species have dramatically different ways to arrange their leaves in space, presumably to maximize the efficiency of light capture.

Phenotypic Plasticity Even though leaves do not grow continuously, they exhibit phenotypic plasticity just as root and shoot systems do. Leaves from the same individual that grow in shade versus sun, like the oak tree leaves shown in **Figure 36.13**, are a prominent example. Shade leaves are relatively large and broad, providing a high surface area that maximizes absorption of rare photons. Sun leaves, in contrast, have a smaller surface area to reduce water loss in areas of the body where light is abundant. Water loss is not as much of a problem for shade leaves, because humidity is higher than in bright sun.

Modified Leaves It's important to recognize that not all leaves function primarily in photosynthesis. Recall that cactus spines are modified leaves that protect the stem. Onion bulbs consist of thickened leaves, separated by highly condensed internodes, that store nutrients (**Figure 36.14a**). The thick leaves of species called succulents store water (**Figure 36.14b**), the tendrils that garden peas and other vines use to climb are modified leaflets or leaves (**Figure 36.14c**), and the bright red leaves of poinsettias attract pollinators to the flower nearby (**Figure 36.14d**). The leaves of some carnivorous plants help trap prey as well as performing photosynthesis. The pitcher plant shown in **Figure 36.14e**, for example, has leaves that curl into a tube. When insects enter the tube, they are discouraged from flying up and out by the dark "hood" that covers the opening. As they feed on nectar provided by the plant, they appear to become dizzy. Eventually they fall into the bottom of the tube,

FIGURE 36.13 Phenotypic Plasticity in Leaves. These leaves came from the same tree.

(a) Onion leaves store food.

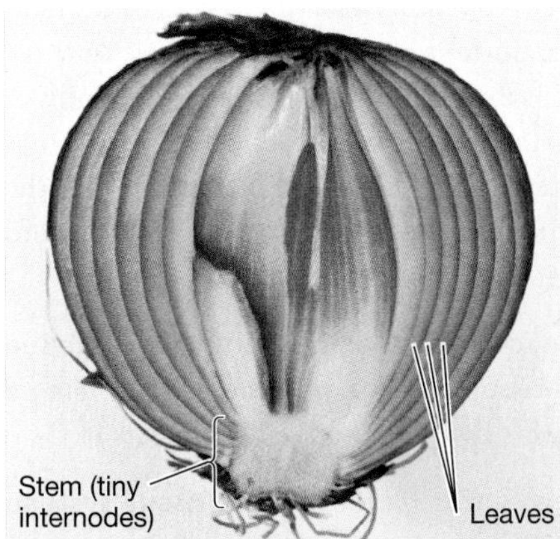

(b) Aloe vera leaves store water.

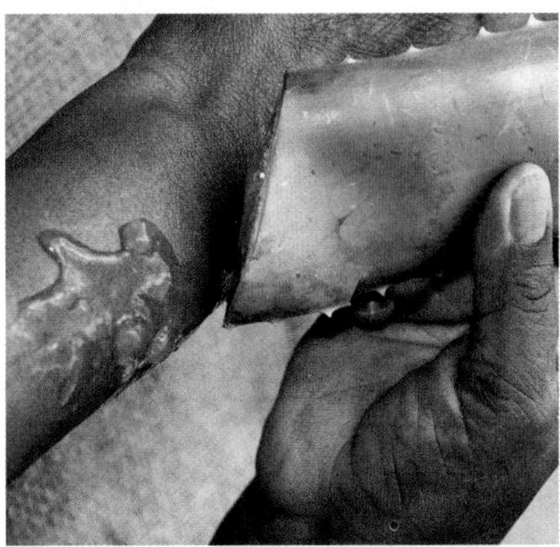

(c) Pea tendrils aid in climbing.

(d) Poinsettia leaves attract pollinators.

(e) Pitcher plant leaves trap insects.

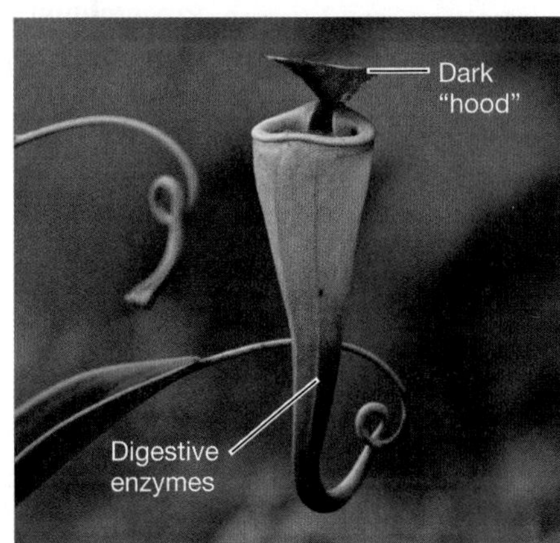

(f) Flowerpot plant leaves collect soil.

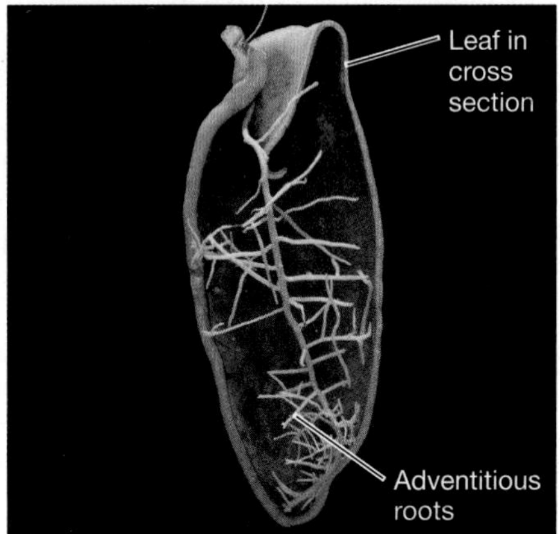

FIGURE 36.14 Modified Leaves Have Unusual Structures or Functions.

QUESTION How does the structure of each of these leaves correlate with its function?

where they drown in water that has accumulated. After they are digested by enzymes secreted by the plant, the nutrients are taken up by epidermal cells. Some leaves even move. For example, the leaves of the Venus flytrap snap shut on insects that are subsequently digested.

The most unusual of all leaf types known, however, may be the balloon-like leaves of the flowerpot plant (**Figure 36.14f**). These leaves are used as homes by ant colonies and develop a network of adventitious roots. These observations inspired Bert Hölldobler and Edward O. Wilson to propose that the adventitious roots help feed the plant by absorbing nutrients from ant feces and other material that accumulates inside the hollow leaf. This "flowerpot leaf hypothesis" was supported when other researchers added radioactive isotopes of carbon and nitrogen to the debris inside leaves, and found that the nutrients were readily taken up by the adventitious roots and incorporated into molecules used in other locations in the plant.

Thanks to the variability of their root systems, shoot systems, and leaves, plants are exquisitely adapted to the environments they occupy. Diversity, plasticity, and dynamism are recurring themes in the study of plant anatomy.

Check Your Understanding

If you understand that... 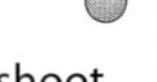

- The plant body is organized into a root system and a shoot system.
- Roots and shoots explore the environment via continuous growth and efficiently absorb diffuse resources like water, ions, carbon dioxide, and sunlight.
- Roots and shoots may also function to anchor the plant, store water, produce offspring asexually, provide protection, or store carbohydrates.
- Leaves vary among species and within individuals, and may be modified to function in food or water storage, insect capture, or pollinator attraction.

You should be able to... 

1) Diagram a generalized version of the angiosperm body, labeling each major part.
2) Provide two examples each of root systems, shoot systems, and leaves that differ from the generalized body in structure and/or function.

36.2 Primary Growth

Plants grow continuously because they have **meristems**—populations of undifferentiated cells that retain the ability to undergo mitosis and produce new cells. When a meristematic cell divides, one of the daughter cells remains in the meristem, allowing the meristem to persist. The other cell, though, undergoes differentiation. You might recall from Chapter 21 that differentiation is a developmental process that produces a specialized cell—one that expresses only certain genes and has a distinctive structure and function.

Apical meristems are located at the tip of each root and shoot. As cell number increases in response to cell division and as cell size increases during differentiation, the meristems at each root and shoot tip extend the plant body outward, allowing it to explore new space. Biologists refer to this process as **primary growth**. Primary growth mostly increases the length of roots and shoots. Cells that are derived from apical meristems form the primary plant body.

To understand how primary growth occurs, let's look at the overall organization of the primary plant body and then delve into a detailed look at the tissues and cells present.

How Do Apical Meristems Produce the Primary Plant Body?

Whether they are located in the root or the shoot, apical meristems give rise to three distinct populations of cells. These cells are partially differentiated but retain the character of meristematic cells because they keep dividing. The three types of primary meristematic cells are important because they give rise to three major tissue systems that extend throughout the plant body. A **tissue** is a group of cells that functions as a unit.

Figure 36.15 indicates where the apical meristems and the three primary meristematic cell types are found in shoots and roots. Note that:

- **Protoderm** gives rise to the **dermal** (literally, "skin") **tissue system**. The dermal tissue system, or **epidermis**, is a single layer of cells that covers the plant body and protects it.
- **Ground meristem** gives rise to the **ground tissue system**, which makes up the bulk of the plant body and is responsible for photosynthesis and storage.
- **Procambium** gives rise to the **vascular tissue system**, which provides support and transports water, nutrients, and photosynthetic products between the root system and shoot system. Vascular tissue runs through ground tissue, so the cells that make up ground tissue are adjacent to cells that conduct the water and nutrients they need.

Figure 36.16 shows how the dermal, ground, and vascular tissues are distributed in the plant body. In contrast to these tissues, meristematic cells are highly localized at the tips of shoots and roots. The key point to remember is that the dermal, ground, and vascular tissue systems are originally derived from cells in apical meristems. Thus, they represent the primary plant body.

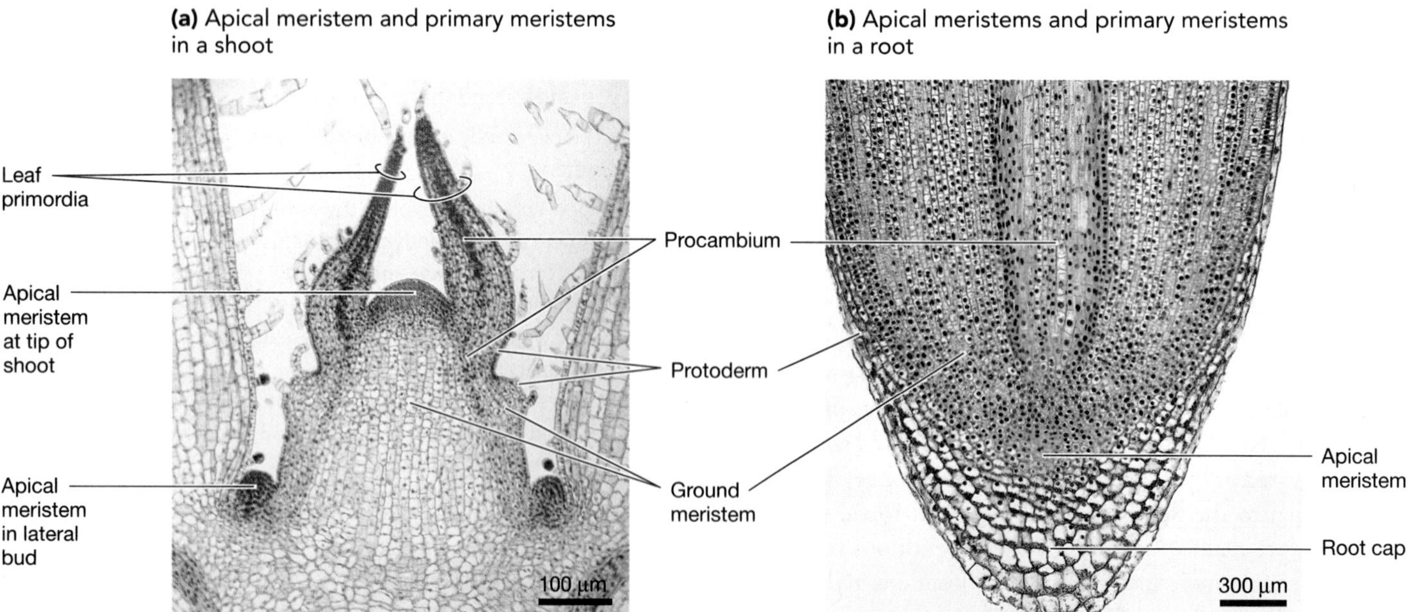

FIGURE 36.15 The Structure of Apical Meristems in the Shoot and Root. Apical meristems consist of small, similar-looking cells that divide rapidly when water and nutrients are plentiful. Three types of cells—procambium, protoderm, and ground meristem—are derived from the apical meristem and consist of partially differentiated cells that can still divide.

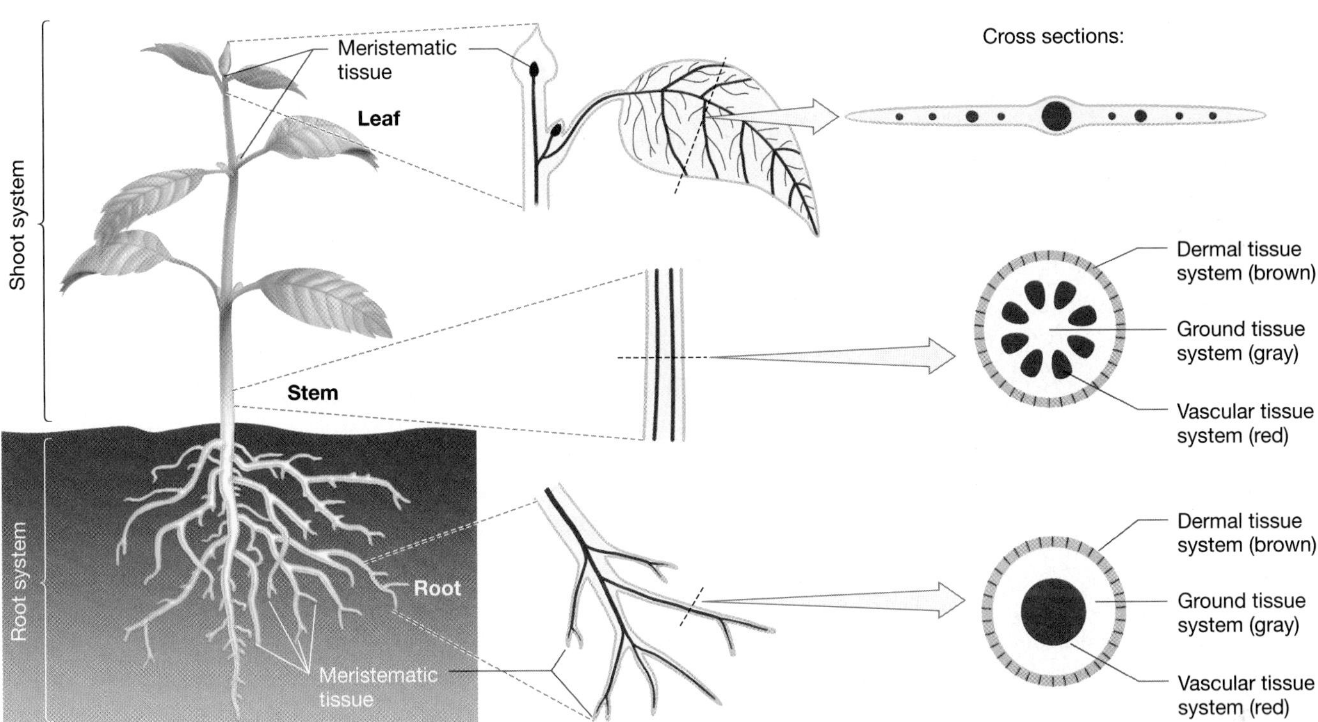

FIGURE 36.16 The Primary Plant Body Comprises the Dermal, Ground, and Vascular Tissue Systems. The dermal, ground, and vascular tissue systems arise from the protoderm, ground meristem, and procambium illustrated in Figure 36.15.

EXERCISE Label the dermal, ground, and vascular tissue systems in the cross section of the leaf.

How Is the Primary Root System Organized? Roots have several features that allow them to grow into new regions of the soil, so they can furnish cells throughout the body with water and key nutrients. As **Figure 36.17** shows, the root apical meristem is protected by a group of cells called the **root cap**. Cells produced by the meristem constantly replenish the cap, which regularly loses cells through abrasion by soil. In addition to protecting the root tip, root cap cells are important in sensing gravity and determining the direction of growth. They also synthesize and secrete a slimy substance called **mucigel**, which helps lubricate the root tip, reducing friction and making movement more efficient.

Three distinct populations of cells exist behind the root cap:

1. The **zone of cellular division** (0.5–1.5 mm behind the root tip) contains the apical meristem, where cells are actively dividing, along with the protoderm, ground meristem, and procambium, where additional cell division occurs;
2. The **zone of cellular elongation** (4–10 mm behind the root tip) is made up of cells that are recently derived from the primary meristematic tissues and actively increasing in length; and
3. The **zone of cellular maturation** (1–5 cm from the root tip) is where older cells complete their differentiation into dermal, vascular, and ground tissues.

The zone of cellular elongation is the region most responsible for the movement of roots through the soil. The cells in this

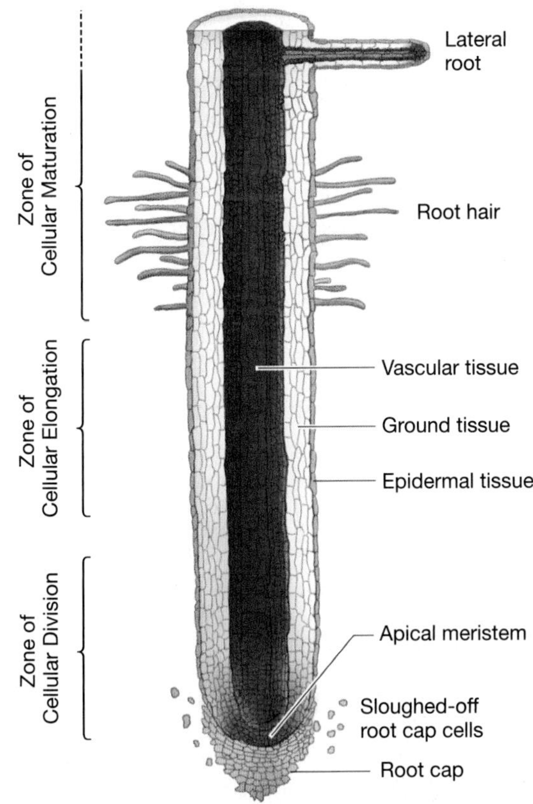

FIGURE 36.17 Roots Extend into the Soil via Growth of Apical Meristems and Cell Elongation. This is a longitudinal (lengthwise) section. The zone of cellular maturation is actually much larger than can be shown here. Most absorption of water and nutrients occurs at root hairs.

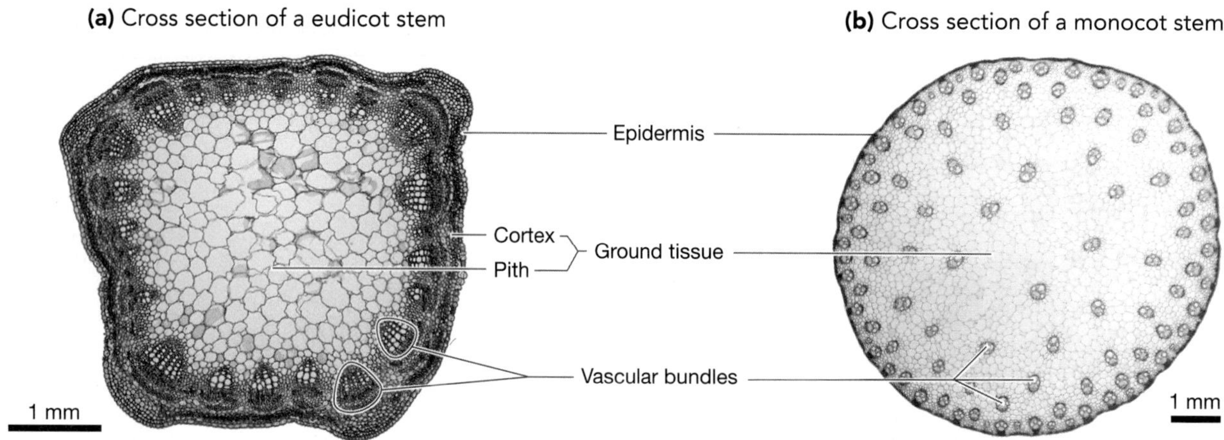

FIGURE 36.18 Stems Contain a Variety of Cell and Tissue Types. As these cross sections show, vascular bundles are **(a)** arranged in a ring near the perimeter of eudicot stems but **(b)** are scattered throughout the pith in monocots.

region increase in length by taking up water. Their expansion provides the force that pushes the root cap and apical meristem through the soil. When conditions are good, roots can extend by as much as 4 centimeters per day.

The zone of cellular maturation is the most important root segment in terms of water and nutrient absorption. In this region, epidermal cells produce outgrowths called **root hairs**, which greatly increase the surface area of the dermal tissue. Root hairs furnish the actual sites of water and nutrient absorption; the rest of the root system provides structural support for the root hairs, conducts water and ions to the shoot, stores the products of photosynthesis, and anchors the plant in the soil. Uptake of water and nutrients in root hairs is vital to plants; portions of Chapters 37 and 38 focus on how these processes occur.

The zone of cellular maturation is also where lateral roots begin to grow. In contrast to lateral branches in the shoot, which arise from apical meristems in axillary buds (see Figure 36.3), lateral roots arise from a ring of cells in the center of the root and erupt through the surrounding ground tissue.

How Is the Primary Shoot System Organized? If you visit a garden regularly over the course of a growing season, you can only imagine the movement of root tips as they penetrate the soil and expand to form complex networks deep underground. Even a casual observer can watch the growth of shoot systems directly, though, as the tips of stems extend and branch over time and as new leaves form and expand.

Just behind each shoot apical meristem, the primary meristematic cells are giving rise to dermal, ground, and vascular tissues. **Figure 36.18a** shows how these tissues are arranged in the stem of an alfalfa plant—a species widely planted for hay—when they mature. Note that the vascular tissues are found in groups called **vascular bundles**, which form strands running the length of the stem. In alfalfa and other eudicots, the vascular bundles are arranged in a ring near the stem's perimeter. The ground tissue that the vascular tissue runs through is divided into two major regions: ground tissue inside the vascular bundles is called **pith**; ground tissue outside the vascular bundles is called the **cortex**. The arrangement of the vascular bundles and ground tissue is dramatically different in the stems of monocots, however. As **Figure 36.18b** shows, vascular bundles tend to be scattered throughout the ground tissue of monocot stems.

Now let's drill down a bit deeper, and look at the composition of the dermal, ground, and vascular tissue systems. Each of these tissue systems is made up of an array of distinct cell and tissue types. What are they?

A Closer Look at the Cells and Tissues of the Primary Plant Body

Chapter 7 introduced a generalized version of the plant cell, similar to the cell in **Figure 36.19a.** Recall that plant cells differ from animal cells in several important ways:

1. All plant cells are surrounded by a stiff, cellulose-rich **cell wall** that supports the cell and defines its shape.

2. Plant cells often contain several types of organelles that are not found in animals—specifically chloroplasts and a large, membrane-bound organelle called a vacuole, which fills most of the cell's volume. **Chloroplasts** are the site of photosynthesis; non-photosynthetic cells may have organelles that are related to chloroplasts but are specialized for storing pigments, starch, oils, or protein. **Vacuoles**, which contain an aqueous solution called **cell sap**, store wastes and in some cases also digest wastes, as do animal lysosomes. In addition, plant vacuoles store water and nutrients. They may also hold pigments that provide color or poisons that deter plant-eating animals.

3. You might recall from Chapter 8 that the cytoplasm of adjacent plant cells is connected via **plasmodesmata** (singular: **plasmodesma**). Plasmodesmata consist of cytoplasm and segments of smooth endoplasmic reticulum (ER) that run through tiny, membrane-lined gaps in the cell wall (**Figure 36.19b**).

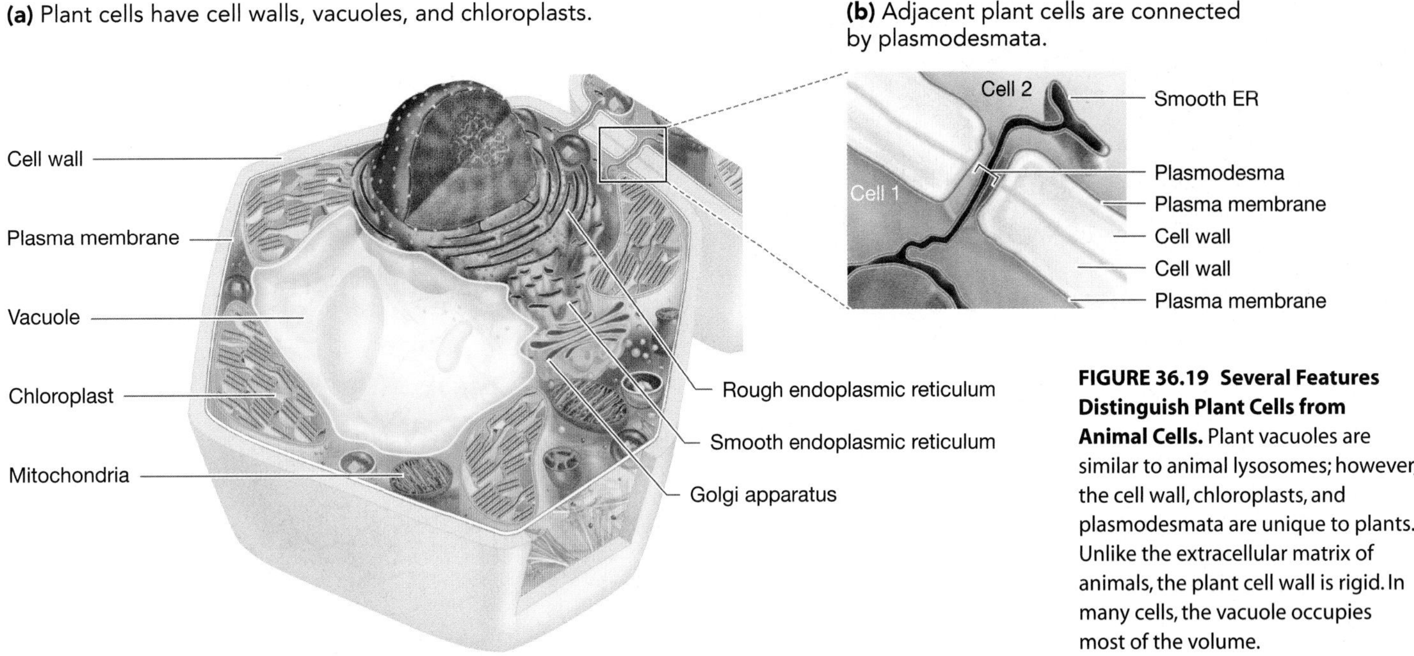

FIGURE 36.19 Several Features Distinguish Plant Cells from Animal Cells. Plant vacuoles are similar to animal lysosomes; however, the cell wall, chloroplasts, and plasmodesmata are unique to plants. Unlike the extracellular matrix of animals, the plant cell wall is rigid. In many cells, the vacuole occupies most of the volume.

Another important distinction between plant cells and animal cells is that no plant cells change position in the plant once they form. Some animal cells change positions either early in the development of an individual or as mature cells. But all plant cells, like most animal cells, are stationary. Once they are created by mitosis in a meristem, they stay in the same place during their entire life span. This is a key point, because a plant cell's location in the body determines which tissue system it contributes to, and what type of cell it becomes. Let's consider the cells and tissues of the three primary tissue systems in turn.

(MB) ***BioFlix*** **at www.masteringbio.com**
Tour of a Plant Cell

The Dermal Tissue System **Dermal tissue** is the interface between the individual and the external environment. Its primary function is to protect the plant body—from water loss, disease-causing agents, and herbivores.

In the dermal tissue system, most of the cells are epidermal cells, which are flattened and lack chloroplasts. Epidermal cells secrete the **cuticle**: a waxy layer that overlays the shoot system (see Chapter 30). Waxes are lipids and thus highly hydrophobic. As a result, the presence of cuticle on stems and leaves drastically reduces the amount of water that is lost by evaporation from the interior of the plant. From a human perspective, the water-repellent properties of cuticle also make it a valuable ingredient in polishes and lipsticks. The carnauba wax used in car and floor polishes, for example, is secreted by epidermal cells in the leaves of carnauba palms native to Brazil.

Besides minimizing water loss, cuticle forms a barrier to protect the plant from virus particles, bacterial cells, and the spores or growing filaments of parasitic fungi. In this way, the plant epidermis forms the first line of defense against disease-causing agents, or **pathogens**. For pathogens to enter the plant body and initiate an infection, they must either secrete enzymes that digest the cuticle or enter via a wound where the cuticle has been torn away.

The waxes found in cuticle can also be detrimental to the plant, however, by reducing gas exchange. This can be a serious problem because photosynthesis depends on the free flow of carbon dioxide to the photosynthetic cells. To allow carbon dioxide to enter photosynthetically active tissues, most land plants have structures called **stomata** (singular: **stoma**). Stomata consist of two specialized cells called **guard cells**, which change shape to open or close an opening in the epidermis known as a pore (**Figure 36.20**). If stomata are open, then CO_2, O_2, water vapor, and other gases can move between the atmosphere and the interior of the plant by diffusion. Stomata open when conditions for gas exchange are favorable. They close in dry conditions when large amounts of water are lost by transpiration.

In addition to minimizing water loss and regulating gas exchange, cells in dermal tissue may be involved in protecting

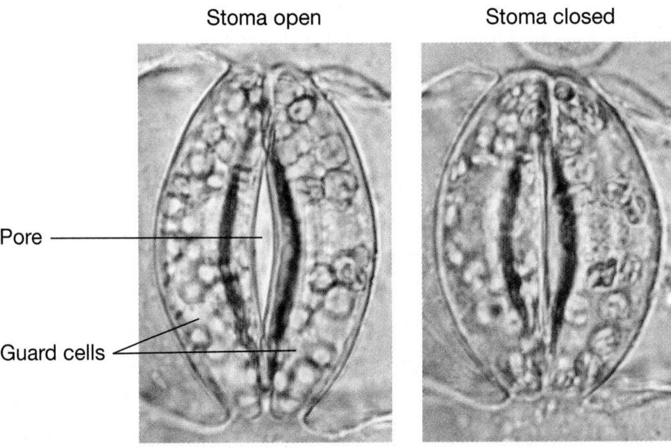

FIGURE 36.20 In Stomata, Guard Cells Regulate the Opening of Pores.

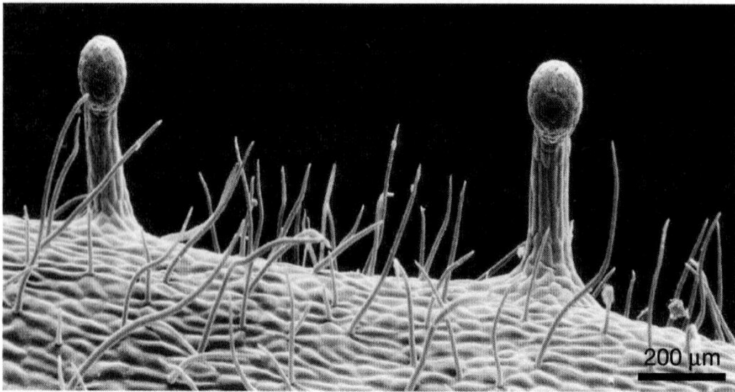

FIGURE 36.21 Epidermal Cells Produce Trichomes That Provide Protection. Some trichomes on this leaf are hairlike extensions of a single cell, while other types of trichomes are multicellular structures.

EXERCISE Label spearlike trichomes that discourage insects or other small herbivores from eating leaf tissue. Label trichomes that are filled with toxic compounds, creating a "minefield" on the leaf surface.

the individual from the damaging effects of intense sunlight and from attacks by herbivores. These functions are accomplished by **trichomes**, appendages made up of specialized epidermal cells. Trichomes come in a wide variety of shapes, sizes, and abundance. Depending on the species, trichomes may (1) keep the leaf surface cool by reflecting sunlight, (2) reduce water loss by forming a dense mat that limits transpiration, (3) provide barbs or store toxic compounds that thwart herbivores, or even (4) trap and digest insects (**Figure 36.21**).

The Ground Tissue System Most photosynthesis, as well as most carbohydrate storage, takes place in ground tissue. Cells in ground tissue are also responsible for most of the synthesis and storage of specialized products such as the chemical signals called hormones, colorful pigments, and toxins required for defense. If the business of the dermal tissue system is protection, the ground tissue is all about producing and storing valuable molecules.

Ground tissue is made up of three distinct cell types: parenchyma (pronounced *pa-REN-ki-ma*), collenchyma (*ko-LEN-ki-ma*), and sclerenchyma (*skle-REN-ki-ma*).

1. **Parenchyma cells** have relatively thin primary cell walls and are the most abundant and versatile plant cells. For example, the parenchyma tissue found in leaves is filled with chloroplasts and is the primary site of photosynthesis (**Figure 36.22a**). But in other organs the function of parenchyma cells is to store starch deposits (**Figure 36.22b**). When you eat a salad, a potato, or an apple, you are ingesting primarily parenchyma cells in ground tissue.

   Many parenchyma cells are **totipotent**, meaning they retain the capacity to divide and develop into a complete, mature plant. The totipotency of parenchyma cells is important in wound healing and in asexual reproduction via stolons or rhizomes. In each case, parenchyma cells may begin to divide, grow, and differentiate to form new roots and shoots. This property of parenchyma cells allows gardeners to clone plants by making cuttings. For example, if you cut a piece of coleus stem and place it in water, parenchyma cells will divide to produce a mass of undifferentiated cells called a **callus**. Roots develop from the callus (**Figure 36.23**), and the new individual can be planted in

(a) In leaves, parenchyma cells function in photosynthesis and gas exchange.

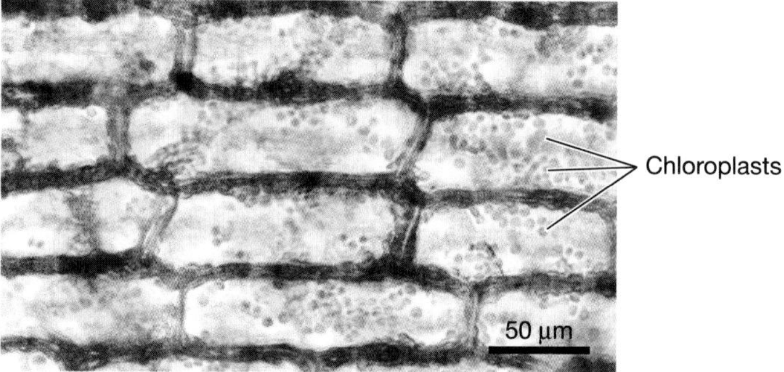

(b) In roots, parenchyma cells function in carbohydrate storage.

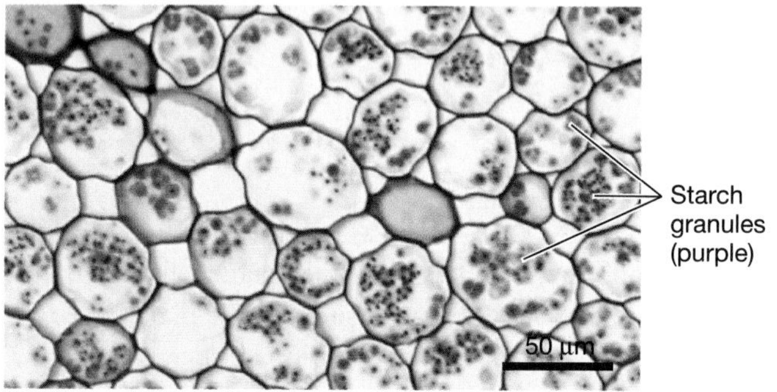

FIGURE 36.22 Parenchyma Cells Perform a Wide Array of Tasks. Parenchyma are "workhorse cells."

EXERCISE Give an example of a gene that is likely to be expressed in these leaf cells, but not the root cells, and a gene that is likely to be expressed in the root cells but not the leaf cells.

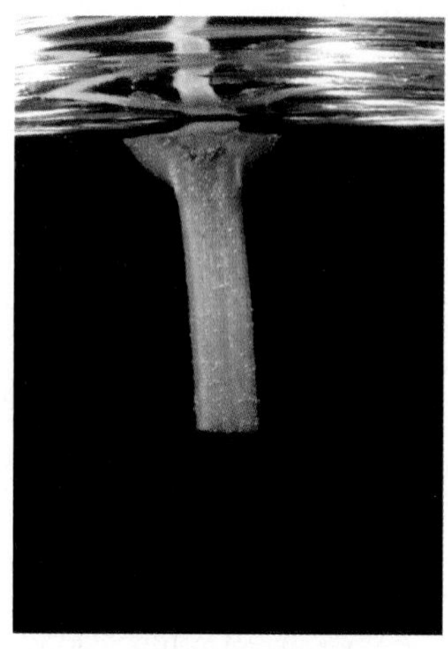

FIGURE 36.23 Parenchyma Cells in Cut Stems Can Form Adventitious Roots. Parenchyma cells in a cut coleus stem (left) divide to form a mass of undifferentiated cells called a callus, which then sprouts roots (right).

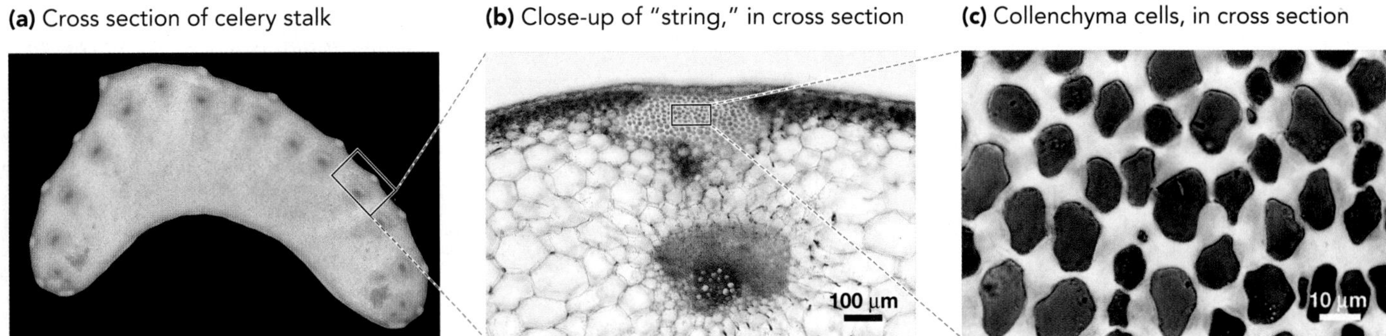

FIGURE 36.24 Collenchyma Cells Support Growing Tissues. A celery stalk is actually a petiole; the strands you can peel from it are columns of collenchyma cells.

soil. Bananas and several other commercially important species are propagated entirely by cuttings.

2. **Collenchyma cells** function mainly in support. They have primary cell walls that are thicker in some areas than others, and their overall shape is longer and thinner than that of parenchyma cells. Even when collenchyma cells are mature, their cell walls retain the ability to stretch and elongate. As a result, collenchyma cells can continue to lengthen as they provide structural support to the growing regions of shoots. Thus, it should not be surprising to learn that collenchyma cells are particularly abundant in elongating stems and in the stalk portions of leaves. The "strings" you may have peeled from a stalk of celery or rhubarb include many strands of collenchyma cells (**Figure 36.24**; note that a celery or rhubarb stalk is actually a petiole).

3. **Sclerenchyma cells** are specialized support cells that produce a thick **secondary cell wall** in addition to the relatively thin primary cell wall found in all cells. Unlike a primary cell wall, the secondary cell wall contains the tough, rigid compound **lignin** in addition to cellulose (see Chapter 30). Whereas collenchyma cells can support actively growing parts of the plant because they have an expandable primary cell wall, the nonexpendable secondary cell wall of sclerenchyma cells specializes them for supporting stems and other structures after active growth has ceased. Another key difference between collenchyma and sclerenchyma is that sclerenchyma cells are usually dead at maturity—meaning that they contain no cytoplasm.

Ground tissue typically contains two types of sclerenchyma cells. **Fibers** are extremely elongated (**Figure 36.25a**). Depending on the plant species they are isolated from, fibers may be used in the manufacture of paper, hemp or jute ropes, or linen and other fabrics. The fiber cells from ramie plants, also known as China grass, can be over half a meter long. **Sclereids**, in contrast, are relatively short, have variable shapes, and often function in protection (**Figure 36.25b**). The tough coats of seeds and the thick shells of nuts are composed of sclereids. The gritty texture of pears is caused by clumps of sclereids.

In many, or even most, cases, the parenchyma, collenchyma, and sclerenchyma cells in ground tissue are found in masses of a single cell type that function as a unit. As a result, biologists recognize the existence of parenchyma tissue, collenchyma tissue, and sclerenchyma tissue. Plant tissues that consist of a single cell type like this are called simple tissues. Tissues that contain several types of cells are termed complex tissues. The three tissue systems—dermal, ground, and vascular—are made up of simple tissues, complex tissues, or a combination of the two.

The Vascular Tissue System The vascular tissue system functions in support and in the long-distance transport of water and dissolved nutrients. It moves the products made and stored in ground tissue.

The vascular tissue system is made up of two complex tissues called xylem and phloem. **Xylem** (pronounced *ZYE-lem*) conducts water and dissolved ions from the root system to the shoot system. **Phloem** (*FLO-em*) is made up primarily of parenchyma cells and conducts sugar, amino acids, chemical signals, and other substances throughout the plant body.

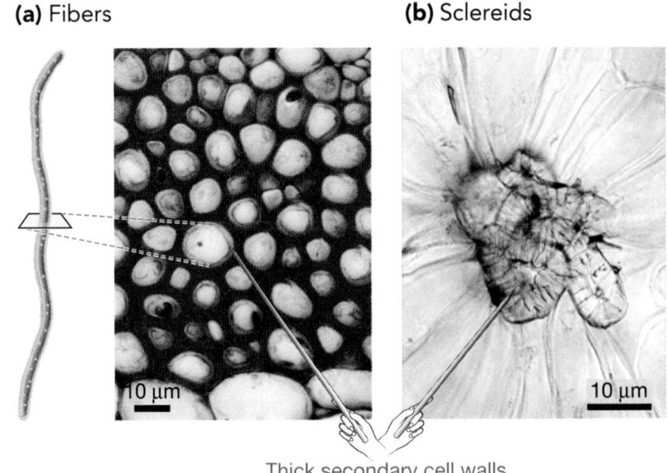

FIGURE 36.25 Sclerenchyma Cells Support Mature Tissues. (a) Fibers and **(b)** sclereids have thickened secondary cell walls. These cells provide support for tissues that are no longer growing.

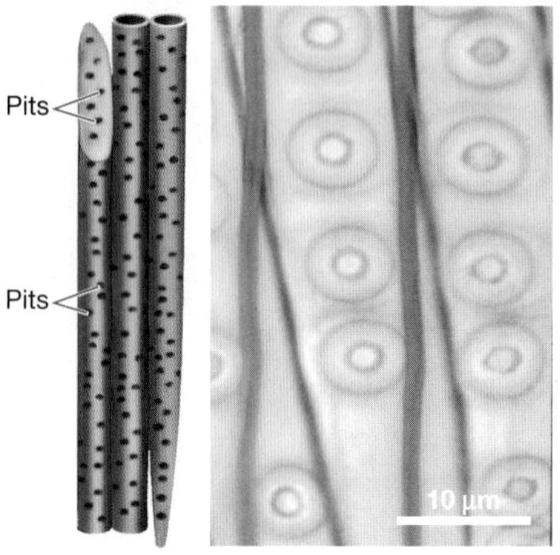

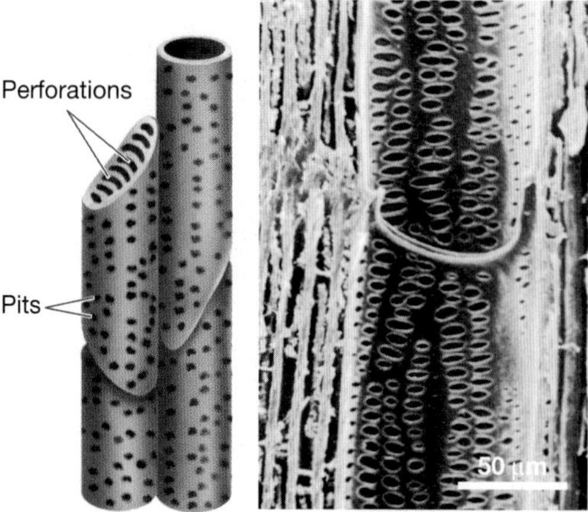

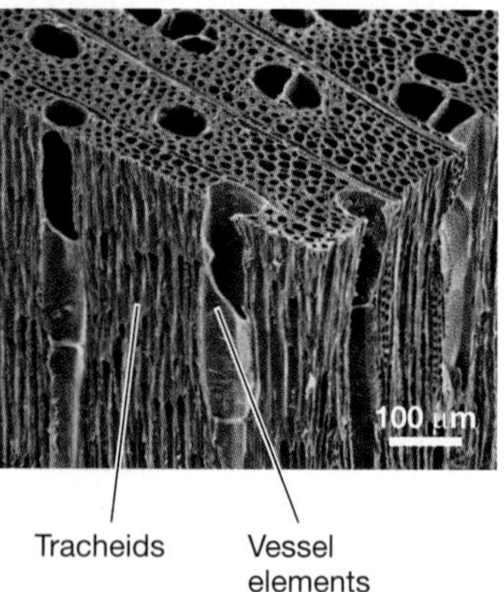

FIGURE 36.26 Xylem May Contain Two Types of Water-Conducting Cells. (a) Tracheids are long and thin compared to **(b)** vessel elements, which are much shorter and wider. **(c)** Both types of water-conducting cells are found in the vascular tissue of angiosperms.

In all vascular plants, xylem contains water-conducting cells called **tracheids** (*TRAY-kee-ids*). In angiosperms and species in the group Gnetophyta, xylem also contains conducting cells called **vessel elements**. Tracheids and vessel elements have thick, lignin-containing secondary cell walls that are often deposited in ringlike or spiral patterns. Both tracheids and vessel elements are dead at maturity. As a result, they are filled with the fluids that they conduct instead of with cytoplasm.

As **Figure 36.26a** shows, tracheids are long, slender cells with tapered ends. The sides and ends of tracheids have structures called **pits**, which are gaps in the secondary cell wall where only the primary cell wall is present. When water is moving up a plant through tracheids, it moves from cell to cell both vertically and laterally through pits, because that is where resistance to flow is lowest. Vessel elements, in contrast, are shorter and wider than tracheids (**Figure 36.26b**). In addition to having pits, vessel elements have **perforations**—openings that lack both primary and secondary cells walls. In some species, the ends of vessel elements lack any cell wall at all, and stacked cells form open pipes called vessels. Vessel elements conduct water much more efficiently than do tracheids because their width and perforations offer less resistance to flow. In angiosperms, tracheids and vessel elements are found adjacent to each other (**Figure 36.26c**). Xylem also contains some parenchyma cells that transport materials laterally in the stem—not vertically.

Phloem is made up primarily of two specialized types of parenchyma cells: sieve-tube members and companion cells. Both are alive at maturity, lack lignified secondary cell walls, and arise from division of a common precursor cell. **Sieve-tube members** (also called sieve-tube elements) are long, thin cells that have perforated ends called **sieve plates** (**Figure 36.27**).

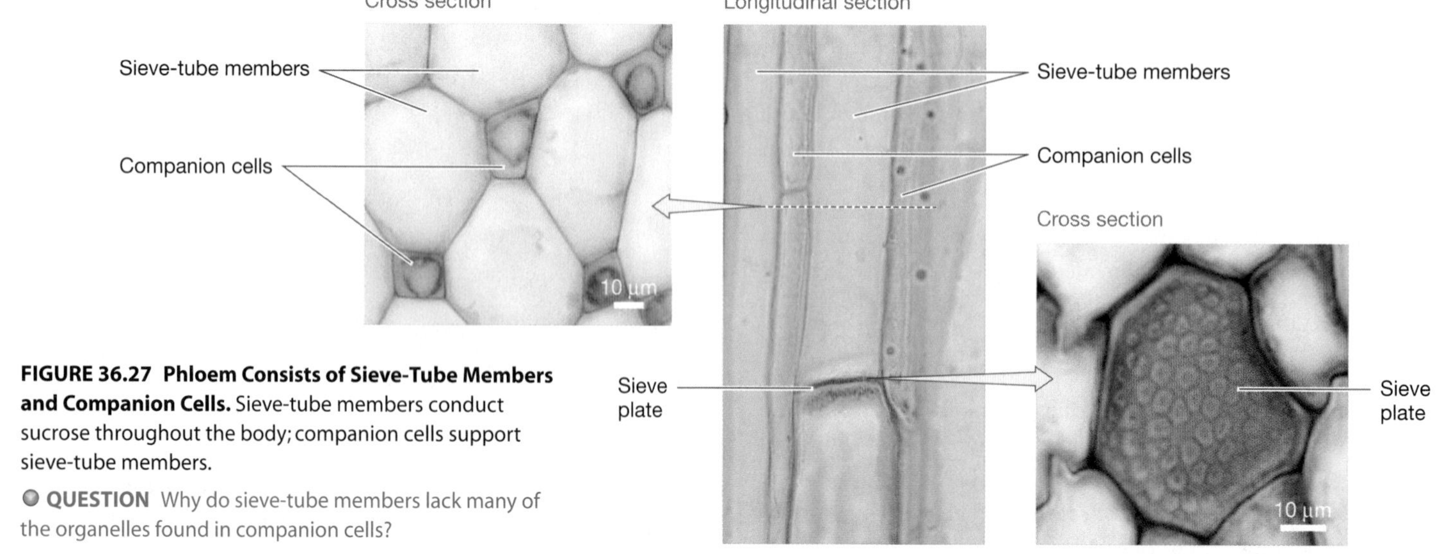

FIGURE 36.27 Phloem Consists of Sieve-Tube Members and Companion Cells. Sieve-tube members conduct sucrose throughout the body; companion cells support sieve-tube members.

QUESTION Why do sieve-tube members lack many of the organelles found in companion cells?

SUMMARY TABLE **36.1 Components of the Primary Plant Body**

Tissues Present	Description of Tissue	Function
Dermal Tissue System (arises from protoderm)		
Epidermal	A complex tissue consisting of epidermal cells, guard cells, and trichome cells	Protection, gas exchange
Ground Tissue System (arises from ground meristem)		
Parenchyma	A simple tissue consisting of parenchyma cells	Synthesis and storage
Collenchyma	A simple tissue consisting of collenchyma cells	Support (expandable in size)
Sclerenchyma	Simple tissues consisting of sclereids or fibers	Support (fixed in size)
Vascular Tissue System (arises from procambium)		
Xylem	A complex tissue consisting of tracheids, vessels, and parenchyma cells and sclerenchyma cells (fibers)	Transport of water and ions; support
Phloem	A complex tissue consisting of parenchyma cells (sieve-tube members, companion cells) and sclerenchyma cells (fibers, sclereids)	Transport of sugars, amino acids, hormones, etc.; support

EXERCISE Create a summary table similar to this one that summarizes the general structure and function of meristematic cells, parenchyma cells, collenchyma cells, and sclerenchyma cells.

Sieve-tube members are responsible for transporting sugars and other nutrients and lack nuclei, chloroplasts, and most other major organelles. However, they are directly connected by means of numerous plasmodesmata to adjacent **companion cells**, which contain all of the organelles normally found in a plant cell. Companion cells are not conducting cells, but instead provide materials to maintain the cytoplasm and plasma membrane of sieve-tube members. As Chapter 37 will show, companion cells are also involved in the loading and unloading of carbohydrates and other nutrients from the solution inside sieve-tube members. Phloem also contains fibers and sclereids, which provide support.

Table 36.1 summarizes the major tissue types found in the three primary tissue systems and the cell types found in these tissues. Once you've mastered the structure and function of the primary plant body, you're ready to consider the next level of complexity: secondary growth.

Check Your Understanding

If you understand that...

- Primary growth results from cell division in apical meristems. Its function is to extend the shoot system into the air and the root system into the soil.
- Apical meristems contain three types of primary meristematic cells: protoderm, ground meristem, and procambium. The dermal, ground, and vascular tissue systems that arise from these meristematic cells extend throughout the individual and make up the primary plant body.
- The dermal system protects the individual; the ground system makes and stores the molecules that make life possible; the vascular system moves those molecules from place to place and holds the plant up. Each of these systems consists of an array of distinctive cell and tissue types.

You should be able to...

1) Diagram the relationship between an apical meristem and the three primary meristematic tissues.
2) Diagram the distribution of the dermal, ground, and vascular tissue systems in the primary plant body.
3) Describe the cells and tissues that allow the dermal system to protect the plant, the ground system to synthesize and store materials, and the vascular system to transport molecules and support the body.

36.3 Secondary Growth

Whereas primary growth increases the length of roots and shoots, **secondary growth** increases their width. The function of primary growth is to extend the reach of the root and shoot system and thus increase a plant's ability to absorb photons and acquire carbon dioxide, water, and ions. The function of secondary growth, in contrast, is to increase the amount of conducting tissue available and provide the structural support required for extensive primary growth. Without the support provided by secondary growth, roots would not be massive enough to anchor extremely large shoot systems, and long stems would fall over or break.

Secondary growth of xylem produces **wood** and occurs in species that have lateral meristems in addition to apical meristems. A **lateral meristem** is also called a secondary meristem or **cambium**. Lateral meristems differ from apical meristems in two ways: (1) Lateral meristems form cylinders that run the length of a root or stem and are made up of thin layers of

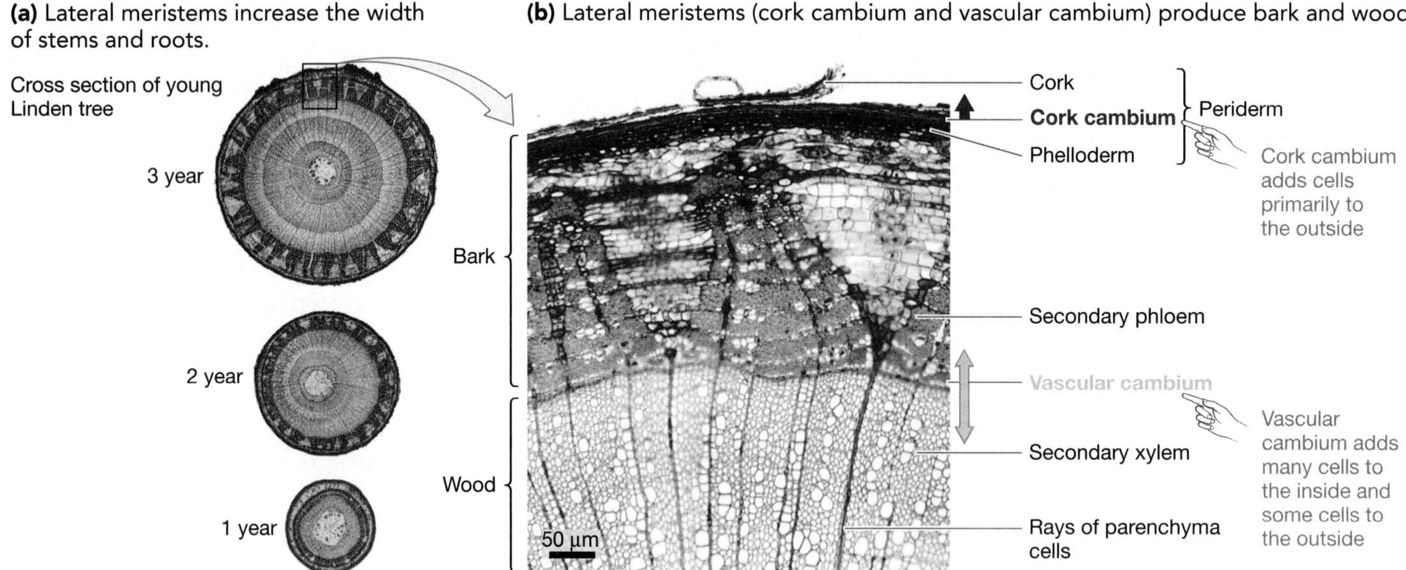

FIGURE 36.28 Lateral Meristems Increase the Width of Stems and Roots and Produce Wood and Bark. (a) Tree trunks contain two types of lateral meristems: cork cambium and vascular cambium. **(b)** Bark consists of the secondary phloem produced by vascular cambium and the cork cells produced by the cork cambium. Wood consists of the secondary xylem and the parenchyma cells produced by vascular cambium.

meristematic cells. In contrast, apical meristems are localized at root tips and shoot tips and are dome shaped. (2) In lateral meristems, cells divide in a way that increases the width of roots and shoots (**Figure 36.28a**). Cells in an apical meristem divide in a way that extends the root and shoot tips.

As **Figure 36.28b** shows, there are two distinct types of lateral meristem in plants that undergo secondary growth. A ring of meristematic cells called the **vascular cambium** forms between the secondary xylem and phloem, inside the stem. A second ring of meristematic cells called the **cork cambium** is located near the perimeter of the stem. One other observation is critical to understanding how lateral meristems work: The cork cambium produces new cells primarily to the outside. Vascular cambium, in contrast, generates new layers of cells both to the inside and outside. The new cells formed to the inside push all of the other cells toward the outside, causing an increase in girth. ● If you understand this concept, you should be able to draw a cross section of a stem, draw rings representing the vascular cambium and the cork cambium, and add arrows showing the direction of growth in each meristem. Now let's consider how each of these meristems works.

What Does Vascular Cambium Produce?

Vascular cambium produces both phloem and xylem (Figure 36.28b). New cells that are produced to the outside of the meristem differentiate into phloem, while new cells produced to the inside differentiate into xylem. More specifically, cells produced by vascular cambium develop into secondary phloem and secondary xylem—as opposed to the primary phloem and primary xylem produced by the procambium at each apical meristem. These cells are not always produced simultaneously. In most cases, the vascular cambium produces many more secondary xylem cells than secondary phloem cells. ● If you understand this concept, you should be able to add labels reading "Secondary phloem" and "Secondary xylem" to the arrows on your first diagram, and make one arrow fatter than the other to reflect the relative amount of cell division that occurs.

Primary phloem and xylem are found throughout the roots and shoots of all vascular plants, but secondary phloem and xylem are found only in gymnosperms and certain angiosperms.

Functionally, primary and secondary phloem are similar; primary and secondary xylem are also similar. Secondary phloem functions in sugar transport. In combination with the tissues associated with the cork cambium, it forms the outer layers called bark. Secondary xylem functions in water transport and structural support, forming the structural material called wood.

Primary and secondary phloem and primary and secondary xylem are complex tissues, made up of more than one cell type. Besides producing conducting cells such as sieve-tube members, tracheids, and vessel elements, the vascular cambium produces fibers for additional strength, along with parenchyma cells. The parenchyma cells radiate laterally across the xylem and form structures called **rays** (see Figure 36.28b). The parenchyma cells in rays are important because they transport water and nutrients laterally across the stem.

Secondary growth in roots is similar to secondary growth in stems. In both portions of the plant, width increases as cells produced on the inside of the vascular cambium form second-

SUMMARY TABLE 36.2 **Components of Secondary Growth**

Tissues Present	Description of Tissue	Function
Periderm* (arises from and includes cork cambium)		
Cork	A simple tissue consisting of cork cells	Protection
Cork cambium	A simple tissue consisting of meristematic cells	Production of cork and phelloderm
Phelloderm	A simple tissue consisting of parenchyma cells	Synthesis and storage
Secondary Phloem* (arises from vascular cambium)		
Phloem	A complex tissue consisting of parenchyma cells (sieve-tube members, companion cells) and sclerenchyma cells (fibers, sclereids)	Transport of sugars, amino acids, hormones, etc.; support
Secondary Xylem† (arises from vascular cambium)		
Xylem	A complex tissue consisting of tracheids, vessels, parenchyma cells (arranged in rays) and sclerenchyma cells (fibers)	Transport of water and ions; support

*Bark comprises the periderm and secondary phloem.
†Secondary xylem is also called wood.

ary xylem, and cells produced on the outside form secondary phloem. It's important to realize, though, that the results of cell division in lateral meristems are highly asymmetrical. As the vascular cambium grows, all of the secondary xylem is retained and accumulates, but the inner region of secondary phloem is eventually crushed and replaced. In addition, the outermost secondary phloem and cork layers are sloughed off as the stem increases in diameter. As a result, mature woody roots and stems are dominated by secondary xylem, or wood. **Table 36.2** summarizes the major tissue types and cell types involved in secondary growth.

What Does Cork Cambium Produce?

The cork cambium produces **cork cells** to the outside and a smaller layer of cells called the **phelloderm** ("cork-skin") to the inside (Figure 36.28b). Taken together, the cork cambium, cork cells, and phelloderm make up the **periderm** ("around-skin"). ● If you understand the structure of the periderm, you should be able to add labels to your diagram that read "Cork cells," "Phelloderm," and "Periderm." You should also be able to make one arrow from the cork cambium fatter than the other to reflect the relative amount of cell division that occurs.

Periderm is important because it allows continued protection of the woody stem when it increases in girth. As a woody stem or root matures, the epidermal tissue produced by the apical meristem during primary growth is replaced by the periderm, which takes over the role of preventing water loss and protecting the stem and root from pathogens and herbivores. In some species, exceptionally thick periderm can also protect the shoot system from fire damage.

The periderm provides a particularly tough barrier in species where cork cells secrete a strong secondary cell wall containing lignin. Periderm also helps prevent water loss because cork cells produce a layer of wax and other molecules inside their cell walls, making them impermeable to water and gases. Gas exchange can still occur between the atmosphere and living tissues inside the stem, though—through small, spongy segments of the periderm called lenticels.

Cork cells die when they mature. As a stem continues to increase in width, the cork layer often cracks and flakes.

Cells produced by the cork cambium are an important component of the protective material called bark. But the term **bark** actually refers to all cells outside the vascular cambium. As a result, it includes some of the cells produced by the vascular cambium as well as the periderm. ● If you understand this concept, you should be able to add a label to your diagram that reads "Bark." You should also be able to label the areas of the stem that function in structural support, transport of water and nutrients, and protection.

The Structure of a Tree Trunk

Trees are perennial plants, meaning that they live for many years. As a tree matures and grows in width, the innermost xylem layers stop functioning in water transport—only the xylem from the most recent years actually transports fluid. Xylem that no longer transports begins accumulating protective compounds secreted by other tissues. These compounds form resins, gums, and other complex mixtures. The deposition of these molecules causes the oldest portions of secondary xylem to become darker than the younger portions. The inner xylem region is called **heartwood** while the outer xylem is called **sapwood** (**Figure 36.29a**). If you look closely at a piece of wood furniture or flooring, you may be able to find some sapwood and note how it differs in color from heartwood.

(a) Heartwood and sapwood have different functions.

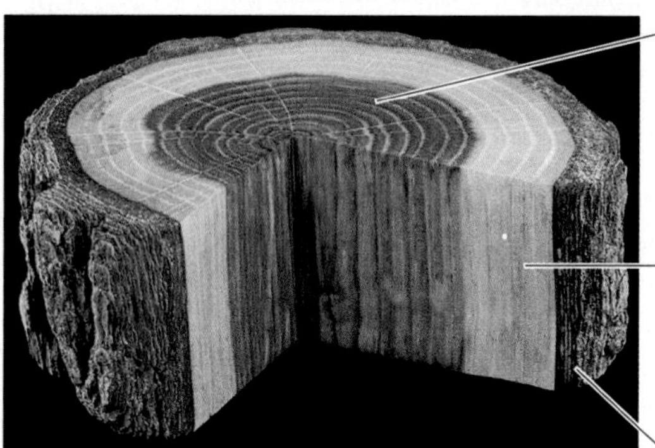

(b) Growth rings result from variation in cell size.

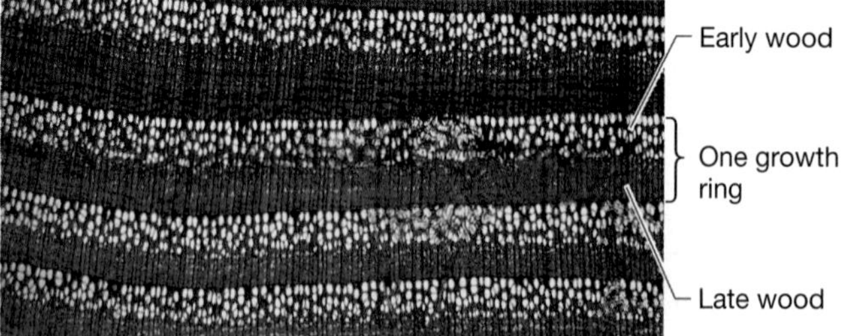

(c) Patterns in growth rings can tell a tree's history.

FIGURE 36.29 Anatomy of a Tree Trunk. (a) Unstained section of wood, showing the marked color difference between heartwood and sapwood. **(b)** Section of wood, stained to show individual cells. Growth rings appear because the size of secondary xylem cells that are produced changes during the course of a growing season. Cells produced at the start of the season are large; those produced near the end of the growing season are small. **(c)** Unstained section through a fir tree from Germany's Black Forest, showing the effects of acid rain on growth.

● **QUESTION** About how old is the tree in part (a)?

● **EXERCISE** In part (b), label a tree ring produced during a dry year and a ring produced during a wet year.

Another important phenomenon occurs in environments where the vascular cambium ceases growth for a portion of each year. This period of no growth, or **dormancy**, occurs during the winter in cold climates and during the dry season in tropical habitats. When the vascular cambium resumes growth in the spring or at the start of the rainy season, it produces large, relatively thin-walled cells. As the growing season nears its end, conditions tend to dry out or become cooler; the secondary xylem cells that are produced at this time tend to be smaller, thicker walled, and darker in appearance. Thus when growth is seasonal, regions of large, thin-walled cells alternate with layers of small, thick-walled cells. The result is annual growth rings that can be observed in cross sections of wood from seasonal habitats (**Figure 36.29b**).

Analyzing patterns in tree growth rings is an important field of study in biology. Because trees grow faster when moisture and nutrients are plentiful, wide tree rings are reliable indicators of wet years. In contrast, narrow rings signal drought years—or in the case of the fir tree shown in **Figure 36.29c**, years when abundant acid rain, due to air pollution, reduced growth. By studying the growth rings found in fossil trees and extremely old living trees, biologists can often assemble a continuous record that dates back thousands of years. In doing so, they gain a better understanding of climate changes that occurred in the past. With continued research, researchers also hope to predict how forests might respond to the global warming that is currently under way.

Check Your Understanding

If you understand that... 

- Secondary growth occurs in species with lateral meristems and results in a broadening of the shoot and root systems.
- Secondary growth results from cell division in the vascular cambium and the cork cambium. Vascular cambium gives rise to secondary vascular tissues. Cork cambium gives rise to the protective tissue called periderm.

You should be able to... 

1) Explain how secondary growth relates to primary growth.
2) Redo the diagram you made while reading the text, showing what the same stem looks like 10 years later.

at www.masteringbio.com

Primary and Secondary Growth

Chapter Review

SUMMARY OF KEY CONCEPTS

The vascular plant body consists of (1) a root system that anchors the individual and absorbs water and key ions, and (2) a shoot system that absorbs carbon dioxide and sunlight. Both systems are dynamic—they grow and change throughout life.

The root and shoot systems of plants are specialized for harvesting the light, water, and nutrients required for performing photosynthesis. Roots extract water and nutrients such as nitrogen, phosphorus, and potassium from the soil; shoots capture light and carbon dioxide from the atmosphere. Leaves, the major organ for performing photosynthesis, usually consist of a flattened blade that extends from a petiole. Roots, stems, and leaves may be modified to perform a variety of other functions, however, including nutrient storage, water storage, protection, and asexual reproduction. Because roots and shoots grow throughout life, a plant is able to respond appropriately to changes in environmental conditions.

You should be able to explain whether phenotypic plasticity in roots, shoots, and leaves is expected to be more important (1) in environments where conditions are variable versus stable, and (2) in short-lived versus long-lived species.

Because body size and shape varies so much among species and individuals, different plants are able to harvest light and other resources in unique ways.

The overall morphology of root and shoot systems varies widely among plant species. In the prairie plants of North America, for example, root systems range from long, linear taproots to shallow, dense mats. Among the silverswords of Hawaii, shoot systems vary from low mats or rosettes to woody, highly branched tree trunks. Among species, variation in plant size and shape allows individuals to reduce competition for resources and thrive in a particular habitat.

You should be able to describe a habitat where (1) plants would be expected to have relatively large root systems and small shoot systems, and (2) plants would be expected to have relatively small root systems and large shoot systems.

Primary growth occurs when cells located at the tips of each root and shoot divide. Primary growth extends the body and gives rise to three primary tissue systems. These tissue systems are specialized for protection, food production and storage, and transport.

Each apical meristem gives rise to three primary meristematic tissues: the protoderm, ground meristem, and procambium. These meristematic tissues then give rise to the dermal, ground, and vascular tissue systems, which extend throughout the plant body.

The dermal tissue system is usually one cell layer thick and plays a role in protection and water conservation. The ground tissue system performs photosynthesis and stores carbohydrates and other compounds. The vascular tissue system transports materials throughout the plant. Within the vascular system, xylem tissue transports water and dissolved ions while phloem tissue transports sugars.

Each tissue system is made up of simple tissues that contain a single cell type, complex tissues that contain two or more cell types, or a combination of simple and complex tissues. Some of the more prominent cell types are parenchyma cells, which function in materials synthesis and storage, collenchyma cells, which provide structural support for growing regions, and sclerenchyma cells, which strengthen regions of the body that have stopped growing.

You should be able to predict the results of an experiment where a drug was used to (1) poison an apical meristem in the shoot and an apical meristem in the root, and (2) selectively poison protoderm cells in a shoot apical meristem.

BioFlix at www.masteringbio.com
Tour of a Plant Cell

In some species, secondary growth occurs when a ring of cells located near the perimeter of each root and shoot divide. Secondary growth increases width and provides additional structural support.

In some plant species, shoots and roots are widened by lateral meristems that produce secondary xylem, secondary phloem, and periderm. Wood consists of secondary xylem, while bark consists of all tissue outside of the vascular cambium.

You should be able to predict the results of an experiment where a drug was used to (1) slow the growth of the vascular cambium but not the cork cambium, and (2) slow the growth of both the vascular and cork cambia on one side of a tree trunk.

Web Animation at www.masteringbio.com
Primary and Secondary Growth

QUESTIONS

Test Your Knowledge

1. Which of the following functions is not performed by parenchyma cells?
   a. transporting sucrose
   b. transporting water
   c. performing photosynthesis
   d. storing nutrients

2. How do tracheids differ from vessel elements, in addition to their overall shape?
   a. Tracheids are stacked end to end to form continuous, open columns.
   b. In tracheids, water flows from cell to cell primarily through gaps in the secondary cell wall called pits.
   c. Tracheids are dead at maturity.
   d. Tracheids have secondary cell walls reinforced with lignin.

3. What is a sieve-tube member?
   a. the sugar-conducting cell found in phloem
   b. the widened, perforation-containing, water-conducting cell found only in angiosperms
   c. the nutrient- and water-absorbing cell found in root hairs
   d. the nucleated and organelle-rich support cell found in phloem

4. What is an adventitious root?
   a. one that performs a function other than anchoring the plant or absorbing water and ions
   b. one that erupts through the root cortex and spreads laterally
   c. a long, filamentous extension that increases the surface area of the root system, for efficient absorption
   d. one that arises from the shoot system

5. Which statement best characterizes primary growth?
   a. It does not occur in roots, only in shoots.
   b. It leads to the development of cork.
   c. It produces the dermal, ground, and vascular tissues.
   d. It produces rings of xylem and phloem tissue as well as rings of cork tissue.

6. Which statement best characterizes secondary growth?
   a. It results from divisions of the vascular cambium cells.
   b. It increases the length of the plant stem.
   c. It results from divisions in the apical meristem cells.
   d. It often produces phloem cells to the inside and xylem cells to the outside of the vascular cambium.

Test Your Knowledge answers: 1. b; 2. b; 3. a; 4. d; 5. c; 6. a

Test Your Understanding

Answers are available at www.masteringbio.com

1. Describe the general function of the shoot system and of the root system. What tissues are continuous throughout these two systems? Suggest a hypothesis to explain why the shoot and root systems of various species are so variable in size and shape.
2. Explain why continuous growth enhances the phenomenon known as phenotypic plasticity. Give an example of phenotypic plasticity in the root system and in the shoot system.
3. Give examples of (a) adventitious roots; (b) stems that are modified for nutrient storage, water storage, or asexual reproduction; and (c) leaves that are modified for nutrient storage, defense, climbing, water storage, or pollinator attraction.
4. What does cuticle do? What do stomata do? Predict how the thickness of cuticle and the number of stomata differ in plants from wet habitats versus dry habitats. How would you test your predictions?
5. Compare and contrast the roles of parenchyma cells in the ground tissue system versus the vascular tissue system. Do the same for sclerenchyma cells.
6. Describe how the vascular cambium produces secondary xylem and secondary phloem. Why are growth rings apparent when tree trunks are viewed in cross section? Why do trees that grow in tropical wet forests lack annual growth rings?

Applying Concepts to New Situations

Answers are available at www.masteringbio.com

1. Chapter 1 explains how the shoot systems of domesticated broccoli family plants (broccoli, cauliflower, cabbage, kohlrabi, and others) changed in response to artificial selection. Make a sketch of three varieties; then predict which cells, tissues, and systems changed as ancestral populations underwent artificial selection. For example, would you predict that leafy populations such as kale and savoy have more or fewer sclerenchyma cells in their stems and leaves than the wild population has? How would you test your predictions?
2. Look out the nearest window, and pick out a plant to observe carefully. Sketch it and label the prominent features of its external anatomy. Make a cutaway view that predicts the major features of its internal anatomy. If possible, obtain a stem or section of root. Examine and label its parts.
3. What is the physical and functional relationship between the primary vascular system and secondary vascular system of an oak tree?
4. Trees can be killed by girdling—meaning the removal of bark and vascular cambium in a narrow ring all the way around the tree. Explain why.

www.masteringbio.com is also your resource for • Answers to text, table, and figure caption questions and exercises • Answers to *Check Your Understanding* boxes • Online study guides and quizzes • Additional study tools including the *E-Book for Biological Science* 3rd ed., textbook art, animations, and videos.

Water and Sugar Transport in Plants

37

A giant sequoia that is 82.9 m (272 ft) tall. This chapter explores how plants move water from their roots to their leaves and how they transport sugars to all of their tissues—sometimes over great distances.

KEY CONCEPTS

- Water moves from areas of high water potential to areas of low water potential. Water's potential energy in plants is a combination of (1) its tendency to move in response to differences in solute concentration and (2) the pressure exerted on it.
- Plants do not expend energy to replace water that is lost to transpiration when stomata are open and photosynthesis is occurring. Instead, water moves from soil and roots to leaves along a water potential gradient. The gradient exists because water at the air-water surface in leaves is under negative pressure (tension).
- In phloem, sugars are transported from "sources"—tissues that release sugars for use elsewhere—to "sinks"—tissues in which sugars are being used or stored. Movement occurs because cells expend energy to load sucrose actively into phloem cells near source tissues. Water follows by osmosis, creating a pressure gradient that favors the movement of water and sucrose to sinks.

On a hot summer day, a single, large deciduous tree can lose enough water to fill three 55-gallon drums. To understand why, recall from Chapter 36 that the surfaces of leaves are dotted with structures called stomata. Stomata are open during the day so that gas exchange can occur between the atmosphere and the cells inside the leaf. This exchange is crucial. For photosynthesis and food production to continue, leaf cells must acquire carbon dioxide (CO_2). There's a catch, however. While stomata are open so that gas exchange can take place, the moist interior of the leaf is exposed to the dry atmosphere. As a result, large quantities of water evaporate from the leaf surface.

In essence, then, water loss is an inevitable consequence of a plant's need to obtain carbon and release oxygen. Water loss is a side effect of photosynthesis. The evaporation can actually be beneficial under some conditions because it cools the plant, just as sweating cools your body. Heavy rates of evaporation can lower leaf temperatures by as much as 10–15°C, which is one reason that forests often feel cool on a hot day.

The question remains, however: How do plants make up for lost water? If water that is lost to evaporation is not replaced, cells will dry out and die. In the case of a redwood tree, the leaves that are losing water may be located 100 m from the root hairs that are absorbing water. How do plants transport water against the force of gravity—in some cases, the length of a football field? Similarly, how do plants move the sugar they produce from active photosynthetic sites to storage sites in roots? These questions are the heart and soul of this chapter. Answering them is a fundamental part of understanding how plants work.

37.1 Water Potential and Water Movement

Loss of water via evaporation from the aerial parts of a plant is called **transpiration**. Transpiration occurs whenever two conditions are met: (1) Stomata are open, and (2) the air surrounding leaves is drier than the air inside leaves. The first condition is usually met during the day, when photosynthesis is occurring. The second condition is met whenever atmospheric humidity is less than 100 percent.

Plants replace water that is lost from leaves with water that is absorbed by roots. One of the most astonishing observations in biology is that water moves from roots to leaves passively—that is, with no expenditure of ATP. Plants do not need a heart muscle to pump water from roots to shoots. Even in 100-m-tall redwood trees, water flows passively from the root system to the shoot system. The movement occurs because of differences in the potential energy of water. Recall from Chapter 2 that potential energy is stored energy. Changes in potential energy are associated with changes in position, such as the position of a molecule or an electron.

Biologists use the term **water potential** to indicate the potential energy that water has in a particular environment compared with the potential energy of pure water at room temperature and atmospheric pressure. Under these conditions, pure water has a water potential of 0.

Water potential is symbolized by the Greek letter ψ (psi, pronounced *sigh*). Differences in water potential determine the direction that water moves. Water always flows from areas of high water potential to areas of lower water potential.

Water potential is key to understanding the movement of water throughout the plant. Because water potential is high in soil and roots but low in leaves and the atmosphere, there is a **water potential gradient** between roots and shoots. To move *up* a plant, water moves *down* this gradient. When it does so, it replaces the water lost to transpiration.

What Factors Affect Water Potential?

To understand how water moves in a plant, consider the cell in the beaker on the left of **Figure 37.1a**. Notice that it is sitting in a **solution**—a homogenous, liquid mixture containing several substances. In this case, the solution consists of water and dissolved substances, or **solutes**. In the beaker on the left, the solute concentrations in the cell and in the surrounding solution are the same. Recall from Chapter 6 that such a solution is said to be **isotonic** to the cell.

Now consider what happens when the cell is transferred to the beaker on the right of Figure 37.1a. This beaker contains pure water, which has no solutes. As a result, the solution surrounding the cell is strongly **hypotonic** relative to the cell. The concentration of the solution is important, because when solutions are separated by a selectively permeable membrane, such as the plasma membrane of a cell, water passes through the membrane from regions of low solute concentration to regions of high solute concentration. This movement of water across membranes, in response to differences in water potential, is called **osmosis** (see Chapter 6). The tendency for water to move in response to differences in solute concentrations, like those in Figure 37.1a, is called the **solute potential** (ψ_S) or **osmotic potential**. The solute potential of a solution is defined by its solute concentration relative to pure water. If water contains a high concentration of solutes, then it has a low solute potential compared with pure water.

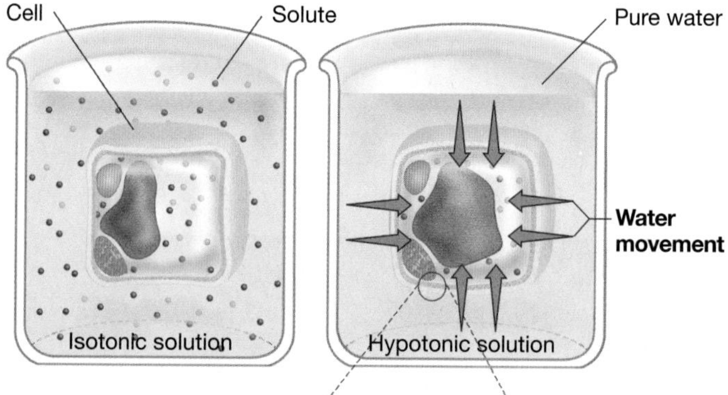

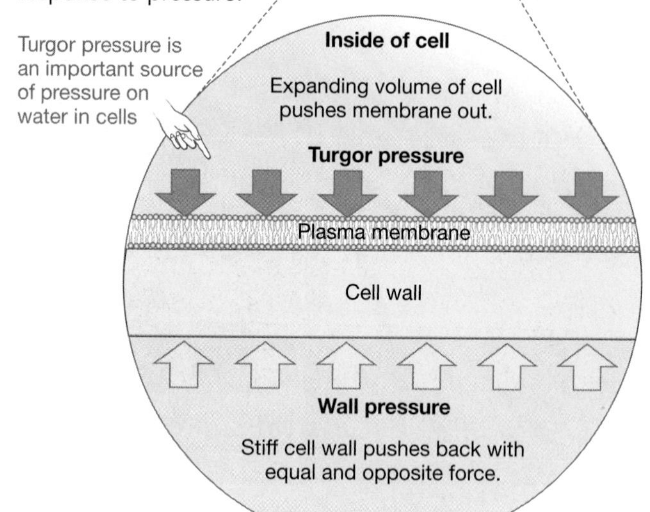

FIGURE 37.1 Water Potential Has Two Major Components: Solute Potential and Pressure Potential.

BOX 37.1 How Do Plants Adapt to Salty Soils?

Plants that are able to thrive in salty environments are called **halophytes** (literally, "salt-plants"). Salty environments are found throughout the world, primarily along ocean coastlines and in inland deserts. In such habitats, the water potential of the soil is often −4 MPa or less, which is much lower than the water potential typically found inside plant roots. If water moved along the water potential gradient, it would leave plant roots and enter the soil.

To cope with soils that have extremely low water potentials and to avoid an outflow of water, many halophytes lower the water potential of their root cells by lowering their solute potential. More specifically, these plants have enzymes and transport proteins that increase the concentration of ions in their vacuoles and increase the concentration of sugars and other organic molecules in the cytoplasm. As a result, they can keep the water potential of their tissues even lower than the water potential of salty soil.

The other major problem that halophytes face is acquiring toxic levels of certain ions present in salty soils—particularly sodium (Na^+) and chloride (Cl^-)—which tend to enter root cells along their electrochemical gradients. Some halophytes have mechanisms for preventing the entry of ions, but others transport them to specialized structures located at the surface of their leaves, where Na^+ and Cl^- are crystallized as salt (NaCl). Once the ions are out of solution, they are harmless to the plant. Halophiles that crystallize salt often appear grayish due to the salt crystals deposited on their stems and leaves (**Figure 37.2**).

FIGURE 37.2 Some Halophytes Secrete the Salt They Take In. Black mangroves secrete excess sodium and chloride ions from special structures in the leaves. The ions form salt crystals.

When an animal cell is placed in a hypotonic solution and water enters the cell via osmosis, the volume of the cell increases until the cell bursts. This does not happen to plant cells, however. If a plant cell swells in response to incoming water, its plasma membrane pushes against the relatively rigid cell wall. The cell wall resists expansion of the cell volume by pushing back, much as the walls of a basketball push back when the ball is inflated. The force exerted by the wall is called **wall pressure** (**Figure 37.1b**). As water moves into the cell, the pressure inside the cell, known as **turgor pressure**, increases until wall pressure is induced. Cells that are firm and that experience wall pressure are said to be **turgid**.

Turgor pressure is important, because it counteracts the movement of water due to osmosis. In the example on the right side in Figure 37.1a, the solute potential favors water moving into the cell. However, the rigid cell walls limit the amount of water that can enter the cell.

The collective term for any kind of physical pressure on water is termed its **pressure potential** (ψ_P). Inside a cell, the pressure potential consists of turgor pressure.

To summarize, water moves in response to (1) differences in the concentration of solutes and (2) pressure. Water potential is primarily made up of a solute potential and a pressure potential. Now, how do these potentials interact? How do biologists determine the direction of water movement when osmosis and pressure affect a cell at the same time?

Calculating Water Potential

Water potential (ψ) is a form of potential energy, and can be thought of as water's stored energy or its tendency to move to a new position. Thus, a water potential summarizes the stored energy that will tend to make water move—in response to the combined effects of a pressure potential and a solute potential. The pressure potential and solute potentials that act on water are measured at water's location relative to other locations.

When selectively permeable membranes are present, water tends to move by osmosis from areas of high solute potential to areas of low solute potential. When no membranes are present to stop it, water moves from areas of high pressure potential to areas of low pressure potential.

If we ignore the effects of gravity, water potential is defined algebraically as

$$\psi = \psi_P + \psi_S$$

In words, the potential energy that water has in a particular location is the sum of the pressure potential and the solute potential that it experiences. Water potential is measured in units called **megapascals** (**MPa**, 10^6 Pa). A **pascal** (**Pa**) is a unit of measurement commonly applied to pressures—force per unit area. A car tire is inflated to about 0.2 MPa, and the water pressure in home plumbing is usually 0.2 to 0.3 MPa.

By convention, solute potentials (ψ_S) are always negative. They are measured relative to the solute potential of pure water. Because pure water contains no dissolved substances, the solute potential of pure water is defined as 0. And because there are always some solutes inside a cell, the water inside always has a potential energy lower than that of pure water. Therefore pure water will tend to move *into* the cell. Increasing the concentration of solutes in a cell lowers its water potential even more.

In contrast to the solute potentials in cells, the pressure potential (ψ_P) from turgor pressure is positive inside cells. It increases the potential energy of the water inside by exerting pressure on the water, making it more likely to move *out of* the cell. But pressure potentials can also be negative.

To see how solute potential and pressure potential interact, consider the U-shaped tube diagrammed in **Figure 37.3a**. Notice that the two sides of the tube are separated by a selectively permeable membrane and that pure water, which has a solute potential of 0, is on the left side. The system is open to the atmosphere and thus is not under additional pressure, meaning ψ_P = 0 MPa. The ψ_S for the solution on the right side of the membrane is −1.0 MPa. Because water potential is higher on the left side of the tube than on the right side, water moves from left to right. ● If you understand the concept of solute potential, you should be able to explain (1) how you would change the solute potential of the pure water on the left of Figure 37.3a to make it lower than the solute potential in the solution to the right, and (2) what the consequences for water movement would be.

In **Figure 37.3b**, the concentrations are the same as in Figure 37.3a, but the solution on the right-hand side experiences pressure exerted by a plunger. If the force on the plunger produces a pressure potential of 1.0 MPa on the right side and the ψ_S for the solution on the right side of the membrane is still −1.0 MPa, then the water potential of the right side is −1.0 MPa + 1.0 MPa = 0. In this case, the water potential on both sides of the membrane is equal and there will be no net movement of water. If the force on the plunger were greater than 1.0 MPa, then the solution on the right side would have a *higher* water potential and water would flow from right to left. ● If you understand the concept of pressure potential, you should be able to explain how you would use the plunger in Figure 37.3b to create a negative pressure instead of a positive pressure.

Figure 37.3c introduces an analogous situation in a cell. In this case a cell that is **flaccid**—meaning it has no turgor pressure and thus a pressure potential of 0—has been placed into a solution of pure water. Because the cell has low solute potential (−1.0 MPa) and the pure water has a higher water potential than the cell, water enters the cell via osmosis. But, as **Figure 37.3d** shows, the incoming water creates turgor pressure. When the positive turgor pressure (+1.0 MPa) plus the cell's negative solute potential (−1.0 MPa) equals 0 MPa—the water potential of pure water—the system reaches equilibrium and there is no additional net movement of water. In this way, turgor pressure acts like the plunger in Figure 37.3b.

(a) Solute potentials differ.

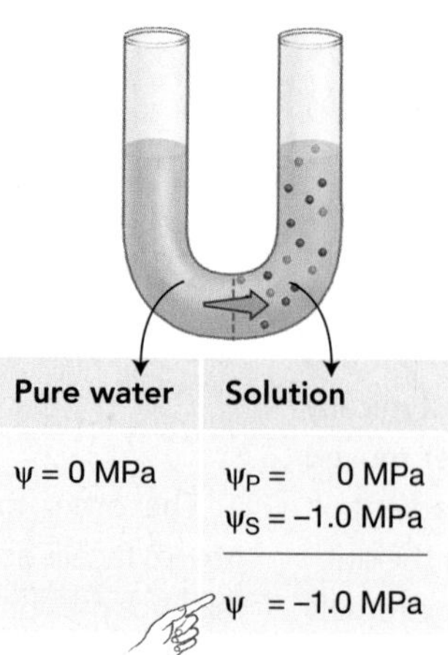

Pure water	Solution
$\psi = 0$ MPa	$\psi_P =$ 0 MPa $\psi_S = -1.0$ MPa
	$\psi = -1.0$ MPa

Water moves left to right—from area with high water potential to area with low water potential

(b) Solute and pressure potentials differ.

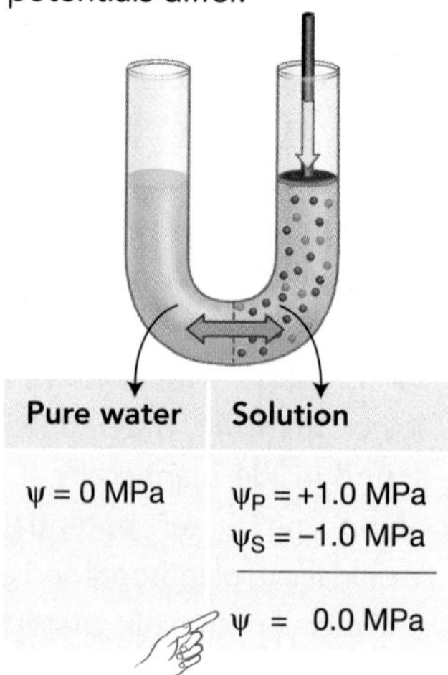

Pure water	Solution
$\psi = 0$ MPa	$\psi_P = +1.0$ MPa $\psi_S = -1.0$ MPa
	$\psi = 0.0$ MPa

Water potentials are equal—no net movement

(c) Solute potentials differ.

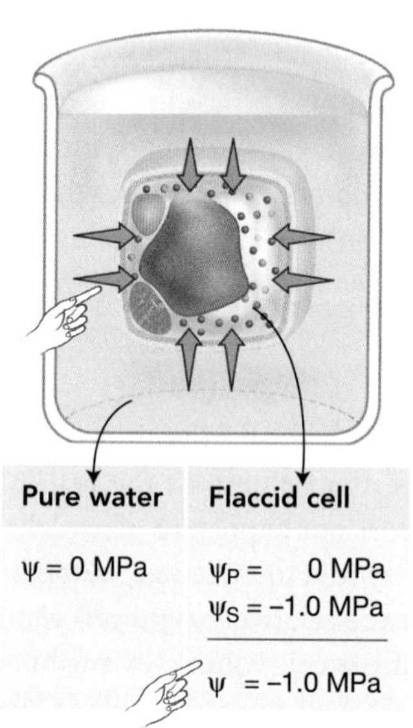

Pure water	Flaccid cell
$\psi = 0$ MPa	$\psi_P =$ 0 MPa $\psi_S = -1.0$ MPa
	$\psi = -1.0$ MPa

Water moves into cell—from area with high water potential to area with low water potential

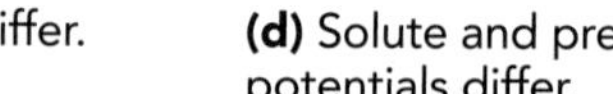

(d) Solute and pressure potentials differ.

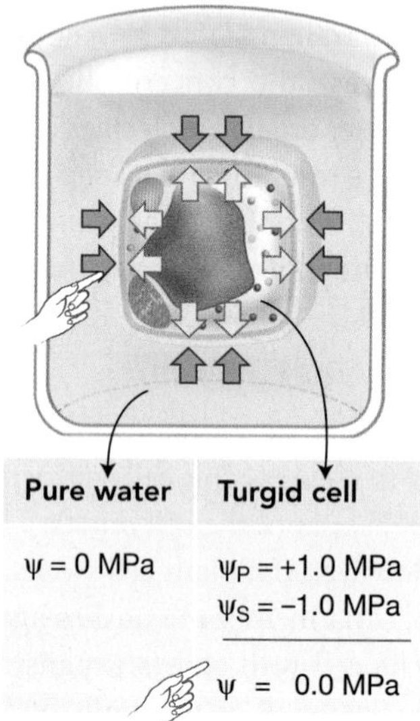

Pure water	Turgid cell
$\psi = 0$ MPa	$\psi_P = +1.0$ MPa $\psi_S = -1.0$ MPa
	$\psi = 0.0$ MPa

Water potentials are equal—no net movement

FIGURE 37.3 Solute Potential and Pressure Potential Interact.

● **EXERCISE** In each of these diagrams, there is a semipermeable membrane that the solutes (red dots) cannot cross. Label them.

Water Potentials in Soils, Plants, and the Atmosphere

Although Figures 37.1 and 37.2 focus on the water potential of individual cells, it's important to realize that ψ can also be measured for tissues, root and shoot systems, and entire plants. The water contained within a leaf, root system, or plant has a pressure potential and a solute potential, just as the water inside a cell does. Likewise, the soil surrounding the root system has a water potential. So does the air around the shoot system. In the atmosphere, water exists as a vapor with a solute potential of 0. But the concentration of water vapor in the atmosphere can be low or high.

In soil, the water that fills crevices between soil particles usually contains relatively few solutes and normally is not under much pressure. As a result, its water potential tends to be high relative to the water found in a plant's roots, which is high in solutes and under considerable turgor pressure. Similarly, the water potential of air is usually extremely low relative to the moist interior of a leaf. These contrasts set up a series of water potential differences between the soil, a plant, and the atmosphere. Because water moves along this water potential gradient, plants tend to gain water from the soil and lose it to the atmosphere (**Figure 37.4**).

It's important to note, though, that the water potentials of plant tissues, air, and soil are dynamic. They routinely change from day to day and even hour to hour in response to changes in the weather. Water potentials may also vary by location or season. To drive these points home, consider the following examples:

- If a fog rolls in, the concentration of water vapor in the atmosphere increases and water is less likely to leave plants via transpiration.
- The water in salty soils contains large amounts of solutes. In these types of habitats, soil has a very low solute potential and water potential. **Box 37.1** on page 815 explores how plants adapt to life in salty soils.
- When soils dry, water no longer fills the spaces between soil particles. Instead of some water floating freely in these spaces, all of the water that is left adheres tightly to soil particles. When water adheres to soil in this way, it creates a negative pressure that lowers the pressure potential of soil water. When the water potential in soil is low, water is less likely to move from soil into plants. **Box 37.2** analyzes how some species of plants can continue gaining water from drying soils by lowering the water potential of their tissues.

Let's return to typical conditions, however. What happens when a water potential gradient exists between the soil, a plant, and the dry atmosphere? When the cells in a leaf or stem lose water to the atmosphere faster than it can be replaced, their plasma membranes contract and the cells shrivel like grapes drying into raisins. If the cells do not regain turgor soon, they will be at risk of dehydration and death. When an entire tissue loses turgor, it is said to **wilt** (**Figure 37.5**). Turgor pressure is required for growth to occur; without it, cells cannot expand once cell division is complete. In addition, turgor pressure provides structural support—which is why wilted plants droop. Unless corrected, wilting leads to the death of the tissue and, eventually, the plant.

Low water potential

Atmosphere ψ: –95.2 MPa
(Changes with humidity; usually very low)

Leaf ψ: –0.8 MPa
(Depends on transpiration rate; low when stomata are open)

Root ψ: –0.6 MPa
(Medium–high)

Soil ψ: –0.3 MPa
(High if moist; low if extremely dry)

High water potential

FIGURE 37.4 A Water Potential Gradient Exists between the Soil, Plants, and the Atmosphere. Water moves from regions of high water potential to regions of low water potential.

Loss of turgor pressure in cells ... leads to wilting.

FIGURE 37.5 Wilting Occurs when Water Loss Leads to Loss of Turgor Pressure. Plants wilt when turgor pressure falls and cells become flaccid. Wilting is a life-threatening condition in plants, analogous to severe dehydration in humans.

BOX 37.2 How Do Plants Adapt to Dry Soils?

Some plant species can thrive in habitats that are dry enough to kill other species. To understand how this is possible, consider research on some of the trees and shrubs growing in the Rocky Mountains of western North America. Water availability varies enormously there. Mountainsides that face the south are much drier than northern exposures, and steep slopes are much drier than gentle slopes. To explore how species that occupy dry sites acquire water, biologists measured the water potential and solute potential of leaves from several species over the course of the growing season.

Some of the data that resulted from this research are shown in **Figure 37.6a**. The graph documents how the solute potential of leaf tissue in ninebark shrubs varied during one growing season. This species is found primarily on dry sites. Notice that the solute potential of the leaf tissue is relatively high in June, at the start of the growing season. In the Rockies, June tends to be rainy and cool. July and August are progressively hotter and drier, however. As the summer progressed, ninebark shrubs were able to keep growing because they experienced a huge drop in solute potential.

Figure 37.6b plots changes in the water potential of ninebarks and Sitka alder shrubs, which grow on wet north-facing slopes, over the course of a single day during August. Notice that both plants have high water potentials at night, when stomata are closed, and lower water potentials during the day. By midday, however, the water potential of ninebark is almost 1.5 MPa lower than that of Sitka alder.

To understand why this difference is important, study **Figure 37.6c** and ask yourself, "Which direction would soil water tend to move if a Sitka alder and ninebark were growing side by side?" Water always moves toward regions of lower water potential, so the ninebark would be able to compete for soil water much more efficiently than the Sitka alder. Ninebark plants are adapted to growing in dry soils because they can withstand extremely low water potentials—their cells are able to tolerate low solute potentials—without dying.

How do the enzymes, membrane proteins, and phospholipids of dry-adapted plants allow them to tolerate such low solute potentials? Is there a cost to ninebark and similar species, in terms of how fast they can grow when water is readily available? These questions remain unanswered. Like most good research, the data in Figure 37.6 generate as many questions as they answer.

(a) Changes in solute potential during a growing season

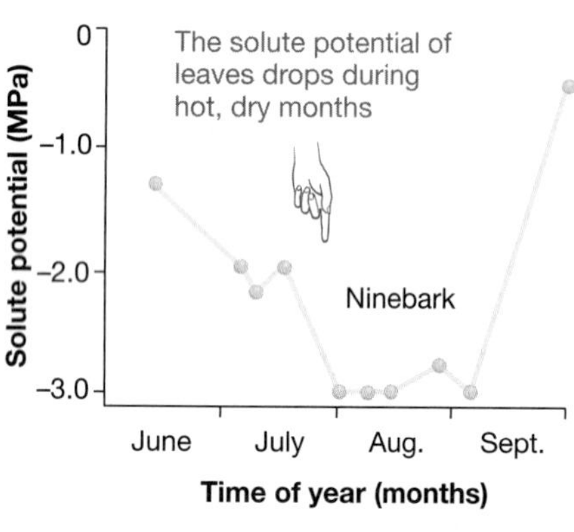

(b) Changes in water potential during a day

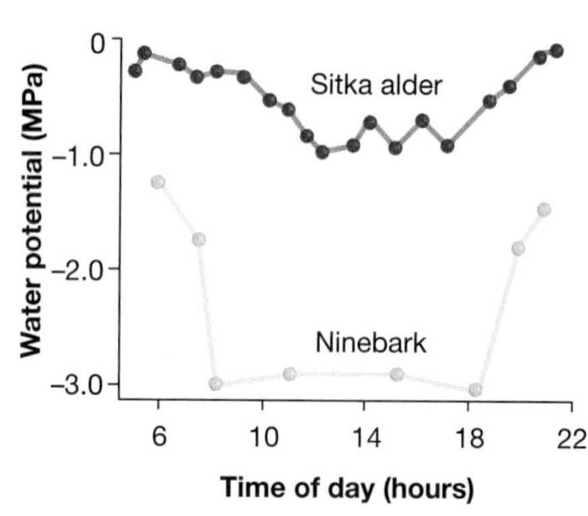

(c) Sitka alder and ninebark compete.

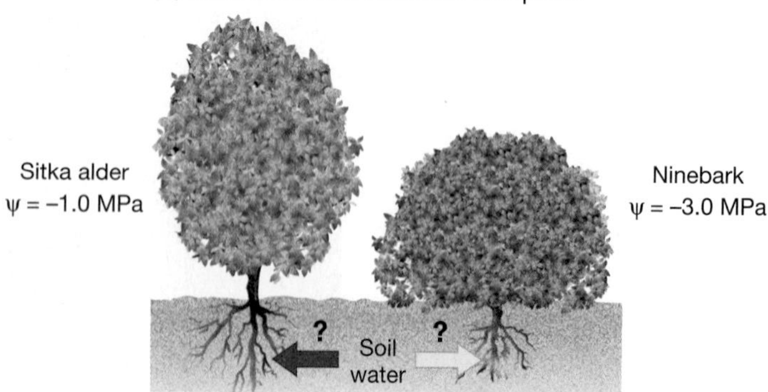

FIGURE 37.6 Plants with Low Solute Potentials Can Grow in Dry Soils.

QUESTION If Sitka alder and ninebark were competing for water, which plant would gain water more effectively?

Check Your Understanding

If you understand that...

- Water moves from areas where its potential energy is high to areas where its potential energy is low.
- In areas separated by a selectively permeable membrane, part of water's potential energy is made up of its solute potential—its tendency to move via osmosis.
- Water also has a pressure potential. In plant cells, for example, the cell wall can exert pressure on water and affect its pressure potential.

You should be able to...

Describe the solute potential, pressure potential, and overall water potential of each of the following:

1) Wet soils and dry soils.
2) The roots, stems, and leaves of a plant.
3) The atmosphere.

37.2 How Does Water Move from Roots to Shoots?

Suppose that you are in charge of caring for the wilted plant shown on the right in Figure 37.5. If you add water to the soil, the water potential of the soil increases and water should begin to move into the plant along a water potential gradient. Biologists have tested three major hypotheses for how the water could be transported: (1) root pressure—a pressure potential that develops in roots—could drive water up against the force of gravity; (2) capillary action could draw water up the cells of xylem; and (3) cohesion-tension, a force generated in leaves, could pull water up from roots. Let's consider each in turn.

Root Pressure and Short-Distance Transport

To understand how water enters a root, consider the cross section through a young buttercup root shown in **Figure 37.7**. Starting at the outside of the root and working inward, notice that several distinct tissues are present:

- The **epidermis** ("outside skin") is a single layer of cells. In addition to protecting the root, some epidermal cells produce root hairs, which greatly increase the total amount of surface area of the root (see Chapter 36).
- The **cortex** consists of ground tissue—usually parenchyma cells—and functions in the storage of carbohydrates.
- The **endodermis** ("inside skin") is a cylindrical layer of cells that separates the cortex from the vascular tissue. The function of the endodermis is to control ion uptake.
- The **pericycle** ("around-circle") is a layer of cells that can become meristematic and produce lateral roots.
- **Vascular tissues** function in transport and are located in the center of roots in buttercups and other eudicots. Notice that, in these species, phloem is situated between each of four arms formed by xylem, which is arranged in a cross-shaped pattern.

When water enters a root along a water potential gradient, it does so through root hairs. As water is absorbed, it moves through the root cortex toward the vascular tissue—where it can be transported up to the shoot system—through two distinct pathways (**Figure 37.8a**). The first pathway is called the **apoplast** ("away-formed"). The apoplastic pathway consists of cell walls, which are porous, and the spaces that exist between cells. The second pathway is called the **symplast** ("together-formed") and consists of the continuous connection through

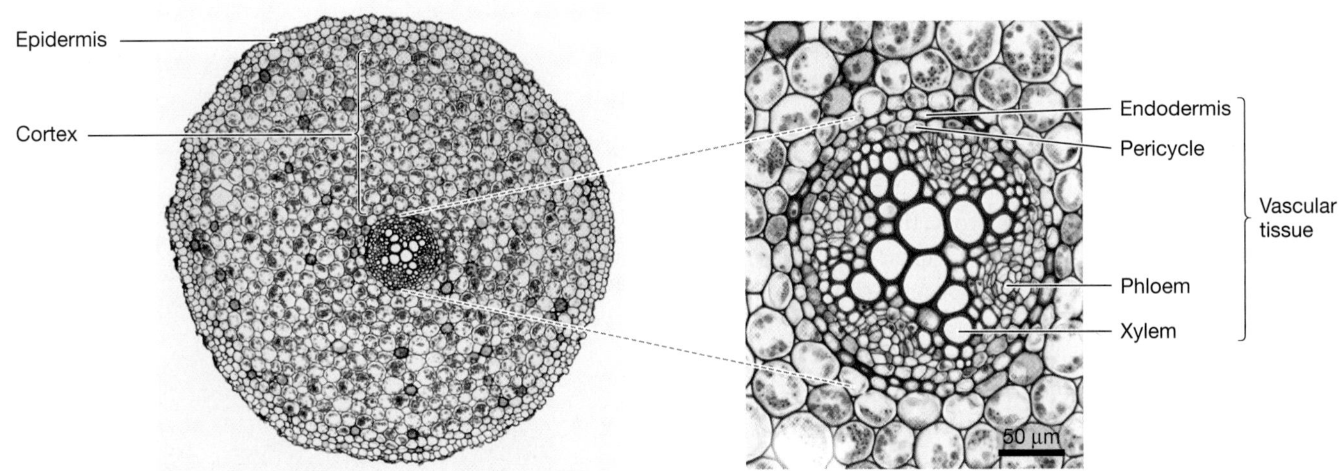

FIGURE 37.7 In Roots, Water Has to Travel through Several Tissue Layers to Reach Vascular Tissue. Cross section through a buttercup root, showing the anatomy that is typical of roots in eudicots. The purple dots inside cells are starch granules.

QUESTION What type of cell makes up the root cortex?

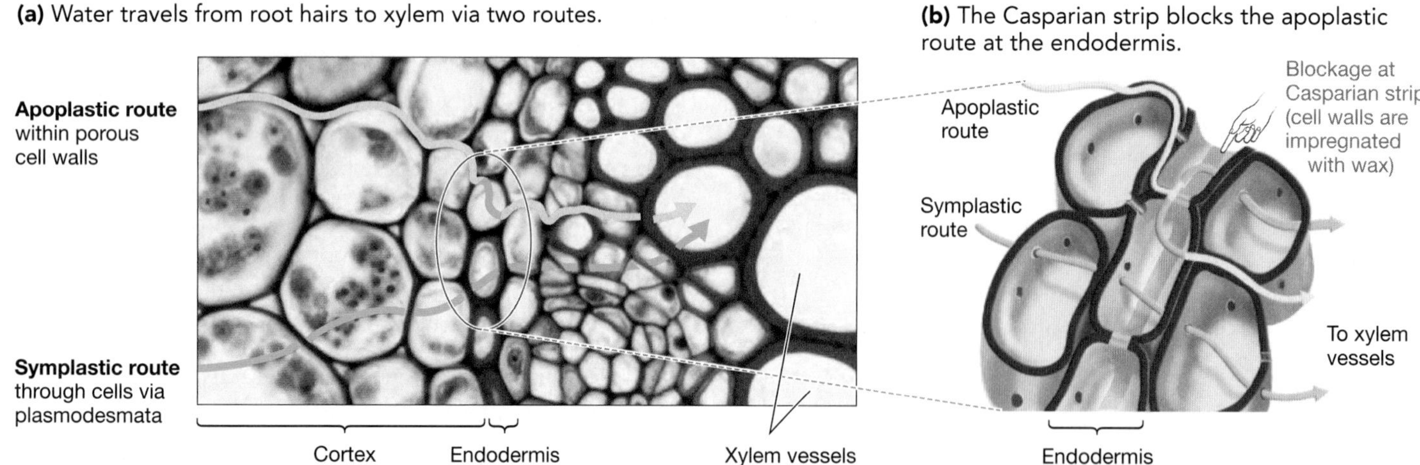

FIGURE 37.8 In Roots, Water Moves to Xylem via Cell Walls or Cytoplasm. (a) The apoplastic route takes water through cell walls; the symplastic route takes water from cell to cell through cytoplasm via plasmodesmata. **(b)** At the endodermis, the apoplastic route is blocked by the Casparian strip—a ring of waxy suberin in the cell walls of endodermal cells. Water in the apoplast *must* pass through the plasma membrane of endodermal cells before entering xylem.

EXERCISE Label the plasmodesmata in part (b). Also add another arrow showing water moving via the apoplastic pathway, and add another arrow showing water moving through the symplast.

cells that exists via plasmodesmata (see Chapter 8). In essence, water can flow around cells by the apoplastic route, through cells by the symplastic route, or both.

The situation changes when water reaches the endodermis, however. Endodermal cells are tightly packed and secrete a narrow band of wax called the **Casparian strip**. This layer is composed primarily of a compound called **suberin**, which forms a water-repellent cylinder at the endodermis. In this way, the Casparian strip temporarily blocks the apoplastic pathway. More specifically, it prevents water from creeping around the walls of endodermal cells and proceeding to the vascular tissue (**Figure 37.8b**). Water that has been moving through the symplastic pathway is not affected by the waxy strip in the endodermal cell walls.

The Casparian strip is important because it means that for water and solutes to reach the vascular tissue, they have to cross the membrane of an endodermal cell. Endodermal cells, in turn, act as filters. They allow ions such as potassium (K^+) that are needed by the plant to pass through to the vascular tissue but can prevent the passage of ions such as sodium (Na^+) that are not needed or may be harmful. In addition, if conditions are dry and the water potential of the cortex drops, the Casparian strip prevents water from flowing out of the vascular tissue and into the cortex. If you understand this concept, you should be able to predict what would happen if a plant had a mutation that prevented synthesis of suberin and formation of the Casparian strip. You should also be able to diagram what would happen to water movement in a plant that secreted suberin *throughout* its endodermal cell walls, instead of in a strip.

Movement of ions and water is also responsible for the phenomenon known as **root pressure**. Stomata close during the night, which minimizes water loss and slows the movement of water into roots. But roots often continue to accumulate ions that their epidermal cells acquire from the soil as nutrients. These nutrients move into xylem. The influx of ions lowers the water potential of xylem below the water potential in the surrounding cells. As water flows into xylem from other root cells in response, a positive pressure is generated that forces fluid up the xylem. When this happens, more water moves up xylem and into leaves than is being transpired from the leaves. In low-growing plants, enough water can move to force water droplets out of the leaves, a phenomenon known as **guttation**. If you are ever up and around early in the morning, you may have observed water drops on leaf tips formed by guttation (**Figure 37.9**).

At one time, positive root pressure was a leading hypothesis to explain how water moves from roots to leaves in trees.

FIGURE 37.9 Root Pressure Causes Guttation. When ions accumulate in the xylem of roots at night, enough water may enter xylem via osmosis to force water up and out of low-growing leaves.

However, research showed that root pressure cannot push water all the way up a tall tree. Over long distances, the force of root pressure is not enough to overcome the force of gravity on the water inside xylem. In addition, researchers were able to demonstrate that cut stems, which have no contact with the root system, are still able to transport water to leaves. Biologists concluded that there must be some other mechanism responsible for the long-distance transport of water. What is it?

Water Movement via Capillary Action

The root-pressure hypothesis was not the only proposed explanation for the long-distance transport of water in plants. Researchers have also evaluated a hypothesis based on the phenomenon of **capillarity**, or movement of water up a narrow tube. When a thin glass tube is placed upright in a pan of water, water creeps up the tube (**Figure 37.10a**). The movement occurs in response to three forces: (1) surface tension, (2) adhesion, and (3) cohesion. Let's consider each force in turn.

Surface tension is a downward pull that exists on water molecules at an air-water interface. As **Figure 37.10b** shows, a concave boundary layer called a **meniscus** (plural: **menisci**) forms at most air-water interfaces. Menisci form because of a property of water that was introduced in Chapter 2: hydrogen bonding. In the body of a solution, all the water molecules present are surrounded by other water molecules and form hydrogen bonds in all directions. The water molecules at a surface, however, can form hydrogen bonds in one direction only—with the water molecules below them. As a result, the topmost layer of water molecules is pulled downward by the bonds and the surface forms a meniscus. Because pulling forces create tension, the pulling force on the water molecules at the air-water interface is called surface tension.

Cohesion is a mutual attraction among like molecules, such as the hydrogen bonding that occurs among molecules in water. Because water molecules cohere, the tension at the surface of a thin tube is transmitted downward through the water column. Water molecules are pulled up because the tension at the surface is much greater than hydrogen bonding that pulls molecules to the side or down. **Adhesion**, in contrast, is the attraction of unlike molecules. In this case, water interacts with a solid substrate—such as the glass walls of a capillary tube or the cell walls of tracheids or vessel elements—through hydrogen bonding. As water molecules cohere to each other and adhere to the side of the tube, they are pulled upward. The result is capillarity.

Like root pressure, however, the amount of water that can be transported by capillarity is limited. Capillary action is responsible for moving water along the surfaces of mosses and other low-growing, nonvascular plants; but capillarity alone is capable of raising the water in the xylem of a vertical stem only about 1 m. Thus, root pressure and capillarity cannot begin to explain how water moves from soil to the top of a redwood tree—an organism that can grow 5 to 6 stories higher than the Statue of Liberty. How does it happen?

The Cohesion-Tension Theory

The leading hypothesis to explain water movement in vascular plants is the **cohesion-tension theory**. This theory claims that water is pulled to the tops of trees along a water potential gradient, via forces generated by transpiration at leaf surfaces.

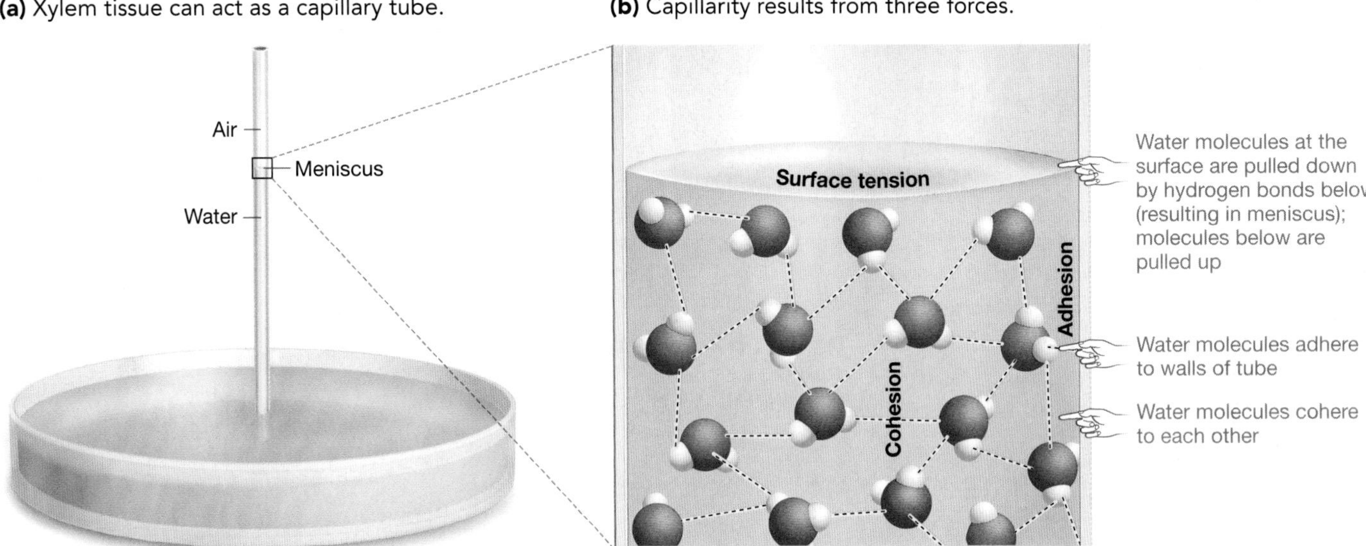

FIGURE 37.10 Water Can Rise in Xylem via Capillarity. (a) Capillarity causes water to rise in thin tubes like xylem. **(b)** Capillarity occurs through a combination of three forces, all of which are generated by hydrogen bonding: the surface tension generated at the meniscus, the adhesion of water molecules to the sides of the tube, and the cohesion of water molecules (which transmits the pulling force of surface tension).

To understand how cohesion-tension works, follow the sequence of events diagrammed in **Figure 37.11**. Start with step 1, in the upper right-hand corner, and notice that spaces in the middle of the leaf are filled with moist air as a result of evaporation from the surfaces of the surrounding cells. When a stoma opens, this humid air is exposed to the atmosphere. In most cases, the atmosphere is much drier than the space inside the leaf, creating a steep concentration gradient between the leaf interior and its surroundings. Consequently, water vapor diffuses out through the stomata.

Step 2 shows that as water is lost from the leaf and enters the atmosphere, the humidity of the gas-filled space inside the leaf drops. In response, more water begins to evaporate from the walls of the parenchyma cells, where menisci exist at the air-water interface. If water molecules leave these menisci rapidly due to a steep water potential gradient with the atmosphere, then fewer water molecules are available at the surface than before. The inward pull on the remaining water molecules at the surface becomes stronger, which increases the curve of each meniscus. As menisci deepen, the surface tension increases

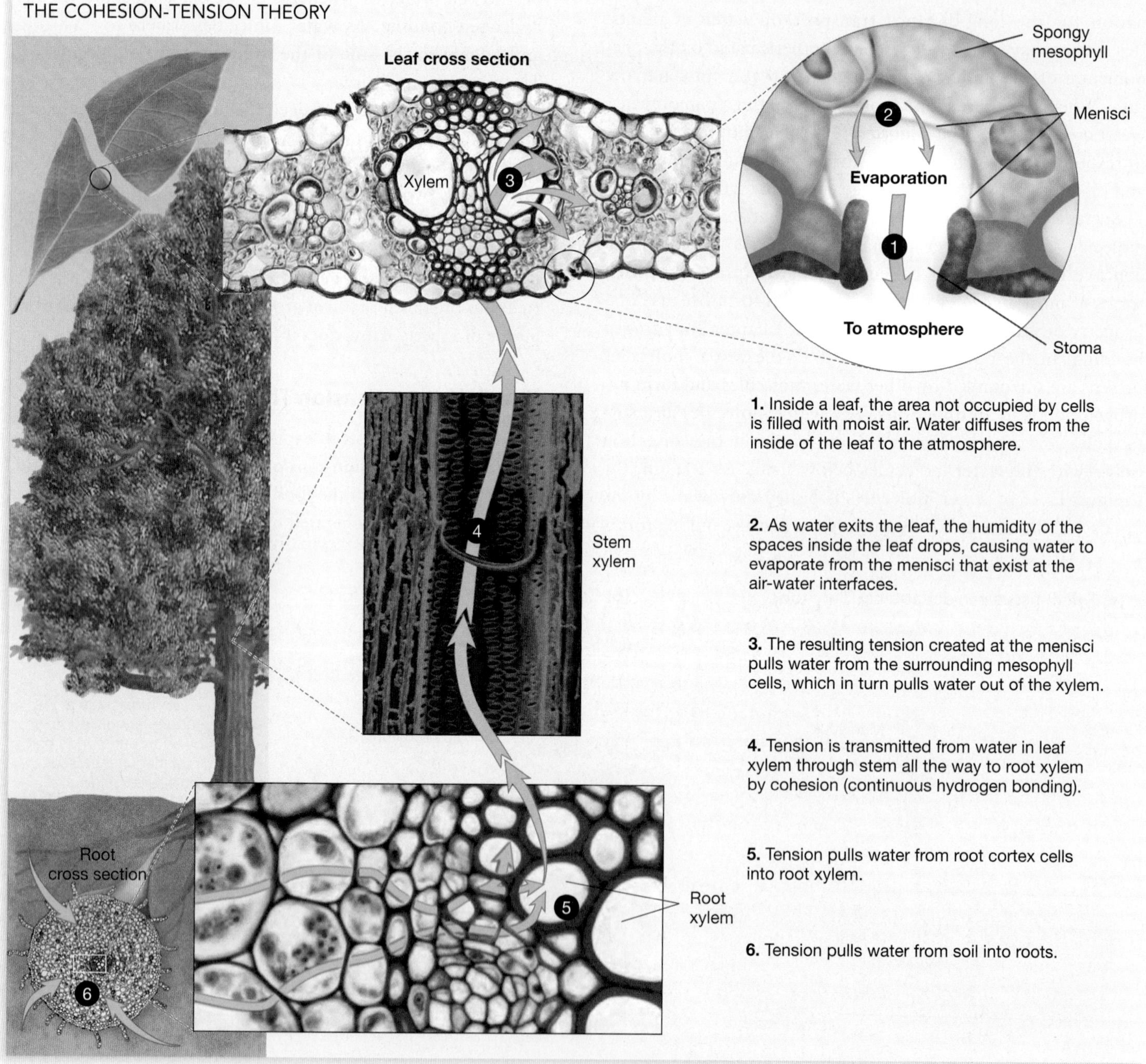

FIGURE 37.11 Transpiration Creates Tension That Is Transmitted from Leaves to Roots.

EXERCISE Using the data in Figure 37.4, label typical values for the water potential present in soil, roots, leaves, and the atmosphere.

dramatically—meaning the pull on water molecules below the surface increases. In 1894 Henry Dixon and John Joly hypothesized that the formation of these steep menisci produces a force capable of pulling water up from the roots, dozens of meters into the air.

At first glance, Dixon and Joly's idea seems fantastic. Is it really possible?

The Role of Surface Tension in Water Transport The key concept in the cohesion-tension theory is that the negative force or pull (tension) generated at the air-water interface is transmitted through the water present in leaf cells (step 3 in Figure 37.11), to the water present in xylem (step 4), on to the water in the vascular tissue of roots (step 5), and finally on to the water in the soil (step 6). This continuous transmission of pulling force is possible because water is present throughout the plant and because all of the water molecules present hydrogen bond to one another in a continuous fashion (cohesion). Note that the plant does not expend energy to create the pulling force—the force is generated by energy from the Sun, which drives evaporation from the leaf surface.

In effect, the cohesion-tension theory of water movement claims that, because of the hydrogen bonding between water molecules, water is pulled up through xylem in continuous columns. You can also think about the cohesion-tension theory in terms of water potentials. The pulling force generated at menisci lowers the pressure potential of water in leaves enough to create a strong water potential gradient—one that is steep enough to overcome the force of gravity and pull water up long distances. Even though the tension created at each meniscus is relatively small, there are millions or billions of menisci in the leaves of the entire plant. The tension created by summing many small pulling forces is remarkable.

To appreciate just how great the forces involved are, think of the vessel elements or tracheids in xylem as groups of straws. When you use a straw, the negative pressure or tension that you exert with your mouth causes liquid to rise. Sucking on a straw can create a pressure difference of about −0.1 MPa, which can draw water up a maximum of about 10 m. In contrast, the negative pressure exerted by the menisci in leaves can be as high as −2.0 MPa—enough to draw water up as much as 100 m.

How can vascular tissue withstand such large negative pressures? If you suck on a straw too hard, atmospheric pressure makes the straw collapse. The same problem would occur in tracheids and vessel elements, were it not for a key adaptation: The secondary thickenings characteristic of cell walls in tracheids and vessel elements (see Chapter 36)—reinforced with tough lignin molecules—keep xylem from collapsing.

Unlike the forces involved in root pressure and capillarity, the forces generated by the cohesion-tension mechanism appear to be great enough to transport water long distances. Calculations based on the negative pressures measured at menisci in leaves make the theory appear plausible, but it is important to evaluate it rigorously. To evaluate a hypothesis, researchers set up experiments that test predictions made by that hypothesis and by one or more alternative hypotheses.

What Evidence Do Biologists Have for the Cohesion-Tension Theory? If the cohesion-tension theory is correct, then the water present in xylem should experience a strong pulling force. This may be the most fundamental of all predictions made by the theory. Let's consider some of the ways that biologists have tested it.

A simple experiment supports the idea that the fluid in xylem is under tension, as predicted by the cohesion-tension theory. If you find a leaf that is actively transpiring and cut its petiole, the watery fluid in the xylem, or **xylem sap**, withdraws from the edge toward the inside of the leaf (**Figure 37.12**). According to the cohesion-tension theory, this observation is due to a transpirational pull at the air-water interface in spongy mesophyll cells of leaves.

The apparatus in **Figure 37.13** was used in one of the more unusual tests of the cohesion-tension theory. Notice that the instrument used, called a dendrograph, is mounted on a tree. It has a pin that is free to move and that is in contact with the exterior of the tree. This pin is attached to a recording device. Large spikes anchor the instrument into the tree's heartwood, which does not transport water. As a result, the pin on the outside of the trunk can move in response to changes in the thickness of the tree's sapwood. If the cohesion-tension theory is correct, then xylem vessels in sapwood should be under a large amount of tension on hot days—when transpiration rates are particularly high—and actually shrink. The "Results" section

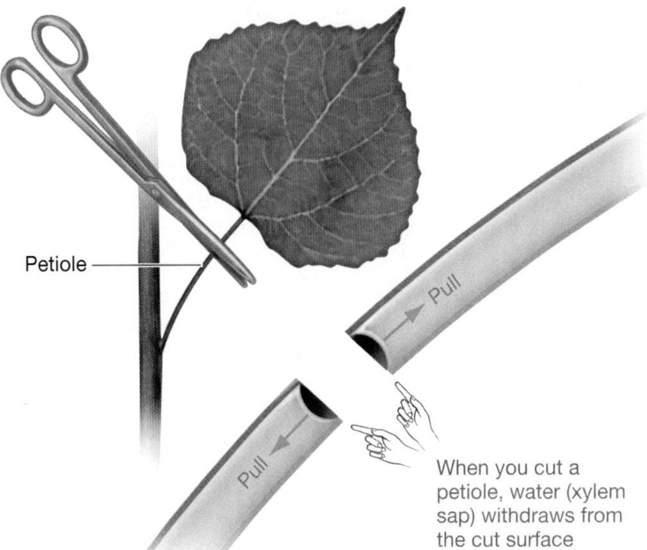

FIGURE 37.12 Xylem Sap in Cut Stems "Snaps Back."

QUESTION Why does this happen? How much of this "snap-back" would occur in identical leaves on a foggy day versus a sunny, dry day? Explain your logic.

Experiment

Question: Do daily changes in the diameter of tree trunks support the cohesion-tension theory?

Hypothesis: When transpiration is occurring, xylem is under enough tension to make tree trunks shrink slightly.

Null hypothesis: When transpiration is occurring, xylem is not under enough tension to make tree trunks shrink slightly.

Experimental setup:

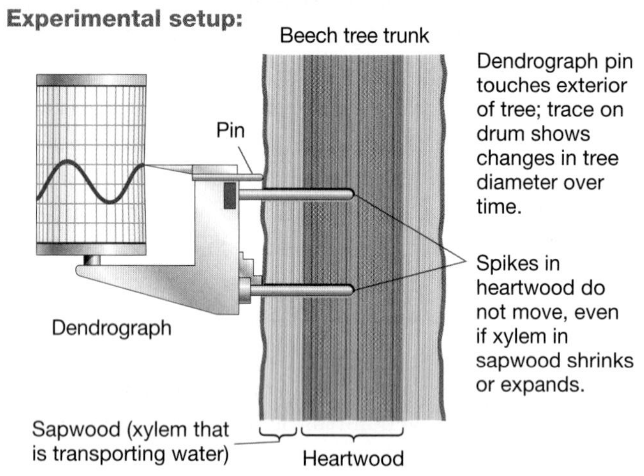

Prediction: Tree trunks should shrink during the day, when transpiration is occurring, and should expand at night, when transpiration stops.

Prediction of null hypothesis: Tree trunks expand only as they grow, not due to changes in transpiration.

Results:

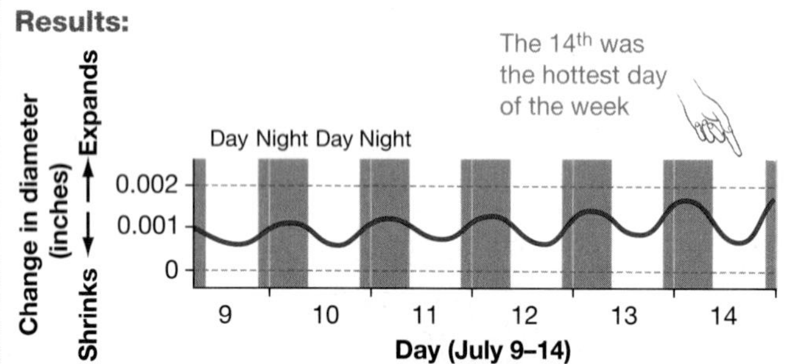

Conclusion: On hot summer days, xylem is under enough tension to shrink tree trunks slightly. This observation supports the cohesion-tension theory.

FIGURE 37.13 Trees Shrink when They Transpire. A dendrograph records changes in the diameter of trees. The data shown represent a trace from a dendrograph monitoring a beech tree in midsummer.

QUESTION The pin on the dendrograph touches the exterior of the tree, while the bolts that hold the instrument are anchored to the interior of the trunk. Why don't the exterior and interior of the tree trunk move together?

QUESTION What would the data from the dendrograph look like if xylem sap were pushed rather than pulled up the tree?

in Figure 37.13 shows data from one tree collected over the course of a week. As the cohesion-tension theory predicts, the tree expanded in diameter at night and shrank during the day. On the hottest day of that week, July 14, transpiration rates were particularly high, and the total change in trunk diameter was over a thousandth of an inch. The correlation between changes in trunk diameter and changes in transpiration suggests that the tension created in xylem is great enough to make tree trunks shrink in diameter. If you understand how this experiment tests the cohesion-tension theory, you should be able to make a graph showing the results expected if the null hypothesis of no tension in xylem is correct.

Additional evidence supporting the cohesion-tension theory has relied on an instrument called a *pressure bomb*, which measures the water potential of plant tissues. Researchers take a leaf or branch that has just been cut from a transpiring plant and place it in an airtight container. Depending on the tension in the stem at the time it was cut, the xylem sap will withdraw from the stem a great deal or a small amount. Researchers then apply a steadily increasing external pressure and record the positive pressure required to push the xylem sap back to the cut surface. This pressure is equal to the negative pressure (tension) exerted on the xylem sap. To understand why, it is critical to realize that the solute potential of the tissues involved does not change during the experiment—only the pressure potential does. If ψ_S is constant, then the positive ψ_P recorded in the experiment is equal to ψ in the leaf or branch. In effect, the pressure that researchers apply to a pressure bomb acts like the plunger in Figure 37.3b.

To see how pressure bombs are used to study water transport, consider data that researchers collected on the water potential of Sitka spruce leaves. They collected branches from the tops of 10-m-tall spruce trees several times a day for several days. Environmental conditions varied throughout this interval, so the tension on the xylem sap was predicted to vary as well. When the group used a pressure bomb to measure the water potential of the tissues, they produced the data shown in the "Results" section of **Figure 37.14**. The fundamental observation is that the water potential of these tissues dropped during the day and rose at night. This result is logical, given that stomata are open during the day, when light is available and photosynthesis is under way. At night, stomata close and transpiration virtually ceases.

Although the observation that xylem sap is under tension was important, the cohesion-tension theory remained controversial. For example, several studies failed to document rapid changes in xylem pressure when temperature and humidity conditions were changed experimentally. Changes in light availability, temperature, or humidity should change either the degree to which stomata open or the transpiration rate or both and thus should lead to rapid changes in xylem pressure. Failure to observe these changes undermined biologists' confidence in the cohesion-tension theory, because the model predicts that the tension on xylem sap should change almost immediately when transpiration slows down or speeds up. But according to advocates of the cohesion-tension theory, the negative results were due to instruments that could not document small and rapid changes in pressure potential accurately. Who was right?

Chunfang Wei, Melvin Tyree, and Ernst Steudle answered this question in the late 1990s with instruments called a *root*

Experiment

Question: Do daily changes in the water potential of leaf tissue support the cohesion-tension theory?

Hypothesis: Increased transpiration during daylight hours decreases the water potential of shoot tissues.

Null hypothesis: Transpiration rate does not affect the water potential of shoot tissues.

Experimental setup:

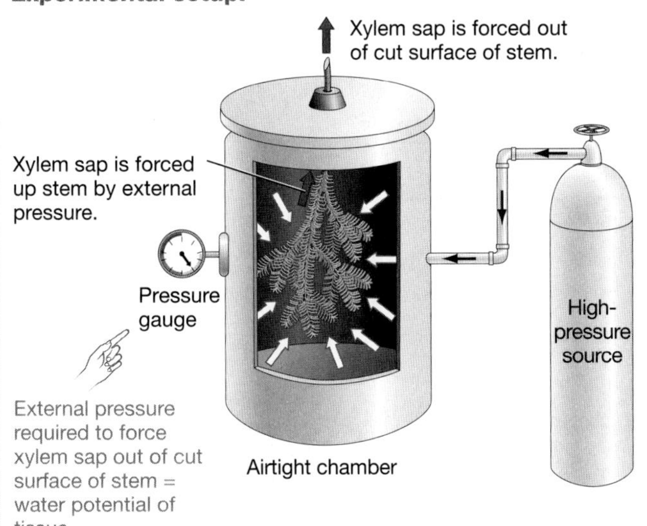

Prediction: Water potential of shoots will decrease during the day as transpiration increases.

Prediction of null hypothesis: Water potential will be unaffected by transpiration rate.

Results:

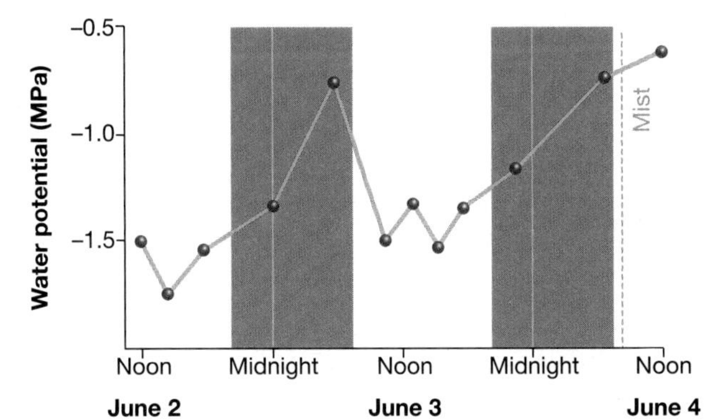

Conclusion: Increased transpiration rates during the day lead to lower water potential in shoot tissues. This observation supports the cohesion-tension theory.

FIGURE 37.14 The Water Potential of Spruce Tree Leaves Varies over the Course of a Day. The figure gives data for branches from the tops of spruce trees, over a two-day period in early summer.

QUESTION In the graph at the bottom of the figure, "Mist" indicates that a light rain fell on the morning of June 4. Why didn't the water potential of the tree leaves decrease that morning, as it had the previous two days when it didn't rain?

bomb and a *xylem pressure probe*. Their apparatus is diagrammed in the "Experimental setup" section of **Figure 37.15**. The xylem pressure probe includes an oil-filled glass tube that Wei and co-workers inserted into the xylem of a corn leaf. The oil transmitted changes in pressure to a gauge, allowing them to record any changes in xylem pressure instantly and directly.

To test the cohesion-tension theory, the researchers altered xylem pressure in two ways. First, they used the root bomb illustrated in Figure 37.15 to add pressure to the root systems of corn. As predicted by the cohesion-tension theory, pulses of pressure applied through the root bomb produced sharp increases in xylem pressure. Second, the investigators released pressure on the root system and began to alter light levels in order to alter transpiration rates in the leaves. By weighing the entire plant (including soil and bomb) and documenting that it became lighter—due to water loss—after experiencing high light intensities, the experimenters supported their claim that transpiration rates actually did increase. What did the xylem pressure probe record as the researchers increased light intensities? The "Results" section of Figure 37.15 shows the data. While reading this graph, remember that tension is a pull, or a negative pressure. Thus, the xylem pressure probe showed that the tension acting on xylem sap increased each time that light intensity was increased—just as the cohesion-tension theory predicts.

These results were considered convincing evidence that increased transpiration leads to increased tension on xylem sap. Increased tension, in turn, lowers the water potential of leaves enough to pull water up from the roots and soil, where the water potential is high. On the basis of these and other results, most biologists now accept the cohesion-tension theory.

Check Your Understanding

If you understand that...

- Water can move a short distance in xylem via root pressure or capillarity.
- Long-distance transport of water depends on movement along a steep water potential gradient. This gradient is created primarily by the negative pressure potential of water in leaves, due to surface tension that develops in response to transpiration.

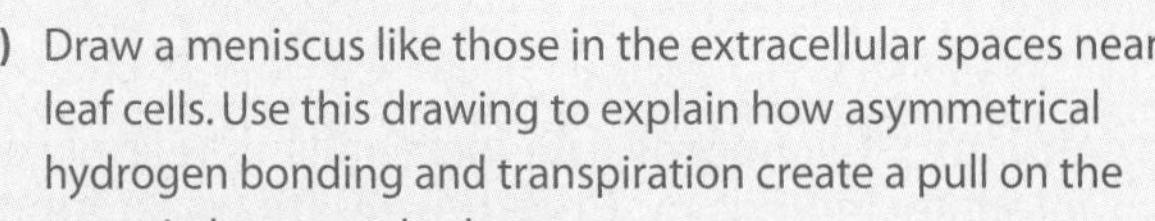

You should be able to...

1) Draw a meniscus like those in the extracellular spaces near leaf cells. Use this drawing to explain how asymmetrical hydrogen bonding and transpiration create a pull on the water in leaves and xylem.
2) Predict what happens to the meniscus when each of the following events occurs: a nearby stoma closes, a nearby stoma opens, a rain shower starts, and weather changes and dry air blows in.

BioFlix at www.masteringbio.com
Water Transport in Plants

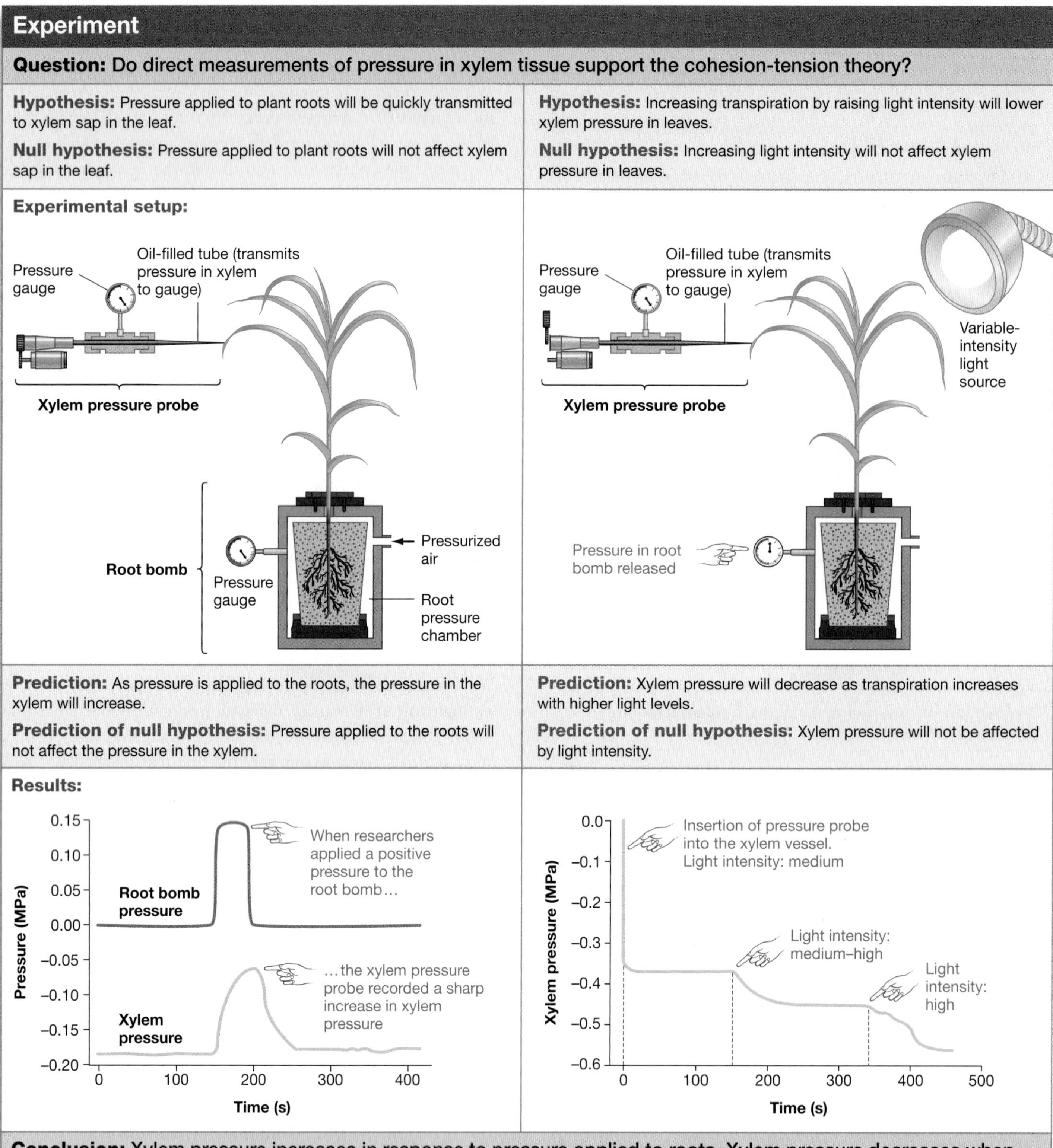

FIGURE 37.15 Direct Measurements of Pressure in Xylem Support the Cohesion-Tension Theory.

37.3 Water Absorption and Water Loss

One of the most important features of cohesion-tension is that it does not require plants to expend energy. Instead, the Sun furnishes the energy required to pull water from roots to shoots. Energy from the Sun heats water molecules at the air-water interface inside leaves enough to break the hydrogen bonds between them and cause transpiration. Rapid transpiration creates deep menisci in the walls of leaf cells, causing tension that lowers the water potential of leaves. Hydrogen bonding between water molecules transmits this tension down to water molecules in the root and soil. Xylem is a passive conduit—a set of pipes that allows water to move from a region of high water potential (the soil) to a region of low water potential (the leaves). Water flows from roots to shoots as long as the water potential gradient—from soil to root to leaf to atmosphere—is intact.

When soils begin to dry, however, it becomes difficult for plants to replace water being lost via transpiration. If water is not being replaced fast enough, the solute potentials of leaves begin to drop and leaves and branches may begin to wilt. In response, stomata may close down partially or completely to reduce transpiration rates and conserve water. Closing stomata also affects the ability of plants to carry on photosynthesis, however, because CO_2 acquisition slows or stops. The balance between conserving water and maximizing photosynthesis is termed the photosynthesis-transpiration compromise. This compromise is particularly delicate for species that grow on dry sites. How do they cope?

(a) Oleander

(b) Cross section of oleander leaf

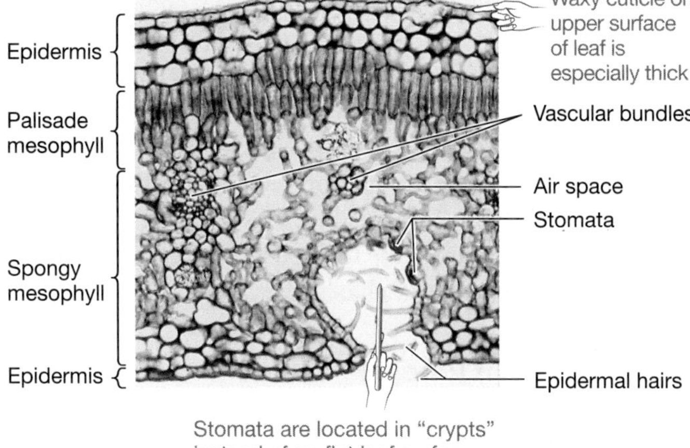

FIGURE 37.16 Species That Are Adapted to Dry Habitats Have Modified Leaf Structures. Oleander grows in desertlike conditions such as those in south-central Asia.

Limiting Water Loss

Plants that thrive in dry sites have several adaptations that help them slow transpiration and limit water loss. As an example, consider the oleander plant illustrated in **Figure 37.16a**. This organism is native to the dry shrub-grassland habitats of southern Eurasia. The micrograph in **Figure 37.16b** shows a cross section through an oleander leaf, which has the following special features:

- The upper surface of oleander leaves is covered by a particularly thick cuticle. This waxy layer helps minimize water loss from cells that are directly exposed to sunlight. In general, species that are adapted to dry soils have much thicker cuticles than do species adapted to wet soils.
- The epidermis of most plants is a single cell layer thick, but the epidermis in oleanders is several cell layers deep. The thick epidermis of species found in dry environments is thought to reduce water loss from parenchyma cells in the leaf, where most photosynthesis takes place.
- The stomata of oleanders are located on the undersides of their leaves, inside deep pits in the epidermis. These pits are shielded from the atmosphere by hairlike extensions of epidermal cells called trichomes (see Section 36.2). The leading hypothesis to explain these traits is that they slow the loss of water vapor from stomata to the dry air surrounding the leaf.

Other species have other adaptations for limiting water loss. For example, recall from Chapter 30 that many species adapted to water-short habitats—either cold environments where water is often frozen or deserts where rainfall is rare—have needlelike leaves. Long, thin leaf shapes minimize the surface area exposed to sunlight and thus help minimize transpiration. Plants adapted to wetter habitats, in contrast, tend to have broad leaves with a large surface area.

Obtaining Carbon Dioxide under Water Stress

Many of the species that thrive in deserts and other hot, dry habitats can continue photosynthesizing even when soil moisture content is low. Recall from Chapter 10 that two novel biochemical pathways are found in these species: **crassulacean acid metabolism** (**CAM**) and **C_4 photosynthesis**. Although some of the enzymes involved are different, both CAM and C_4 are mechanisms that allow plants to conserve water.

CAM plants open their stomata at night and store the CO_2 that diffuses into their tissues by adding the carbon dioxide molecules to organic acids. When sunlight is available during the day and photosynthesis begins, the CO_2 molecules are released from the organic acids and transferred to **rubisco**—the enzyme that initiates the Calvin cycle. In this way, CAM plants can photosynthesize and grow even with their stomata closed.

C_4 plants minimize the width their stomata are open because they use CO_2 so efficiently. More specifically, mesophyll cells in C_4 plants take up CO_2 and add it to organic acids. The CO_2 is then transferred to specialized cells called **bundle-sheath cells** (see Chapter 10), where rubisco is abundant. In effect, the C_4 pathway is a mechanism for concentrating carbon dioxide in cells deep inside the leaf, so stomata do not have to be wide open continuously. CAM and C_4 photosynthesis are similar to the cuticle and stomata-containing pits of oleanders: They are examples of adaptations that help plants conserve water by limiting transpiration.

To summarize, water loss via transpiration is an inevitable side effect of the need to open stomata and obtain CO_2 for photosynthesis. The water that is lost can be replaced by water in the soil without an energy expenditure by the plant, however, because transpiration creates a strong enough water potential gradient to pull water up against the force of gravity. Because plants that live in chronically dry soils must limit water losses or risk wilting, they have evolved adaptations to reduce transpiration.

37.4 Translocation

Translocation is the movement of sugars throughout a plant—specifically, from sources to sinks. In vascular plants, a **source** is a tissue where sugar enters the phloem; a **sink** is a tissue where sugar exits the phloem. Sources contain a high concentration of sugar; sinks have a low concentration of sugar.

Where do sources and sinks occur in a plant? The answer often depends on the time of year. During the growing season, mature leaves and stems that are actively photosynthesizing and producing sugar in excess of their own needs act as sources. Sugar moves from these tissues to a variety of sinks, where sugar use is high and production is low: apical meristems, lateral meristems, developing leaves, flowers, developing seeds and fruits, and storage cells in roots (**Figure 37.17**). However, early in the growing season, the situation is different. The storage cells in roots act as sources, and developing leaves act as sinks.

To explore the relationship between sources and sinks in more detail, consider research on sugar-beet plants that were exposed to carbon dioxide molecules containing the radioactive isotope ^{14}C. The goal of these experiments was to track where carbon atoms moved after they were incorporated into sugars via photosynthesis. The location of ^{14}C atoms inside a plant can be documented in one of two ways: (1) by measuring the number of radioactive emissions emanating from different tissues or (2) by laying plant parts on X-ray film and allowing the radioactivity to expose and blacken the film.

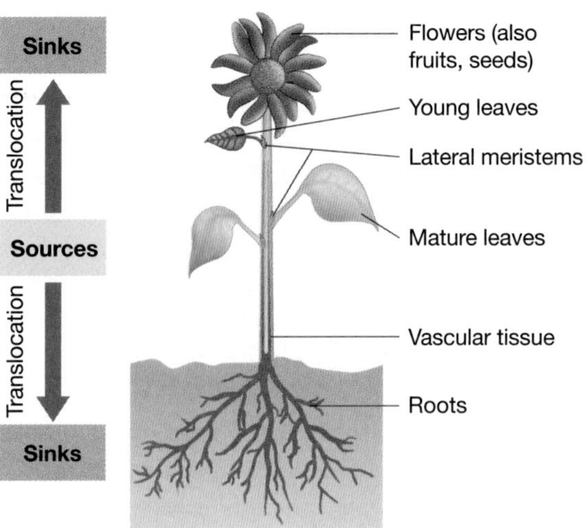

FIGURE 37.17 Sugars Move from Sources to Sinks.

EXERCISE The relationships shown apply during the middle of the growing season. Redraw the figure to indicate the relationship of sources to sinks early in spring, when leaves are first developing and before the plant flowers.

Do carbon atoms move from sources to sinks, as expected? When a researcher enclosed individual leaves of intact plants in a bag and then introduced a fixed amount of radioactive CO_2 for a fixed amount of time, he found that mature leaves retained just over 9 percent of the labeled carbon. In contrast, growing leaves retained 67 percent. These data are consistent with the prediction that fully expanded leaves act as sources of sugar, while actively growing leaves and roots act as sinks. When the researcher exposed all the leaves on a growing plant to labeled carbon, over 16 percent of the total carbon was translocated to root tissue within 3 hours. This result is consistent with the prediction that, during the growing season, roots also act as sinks.

Similar experiments have made two important generalizations possible. First, sugars can be translocated rapidly—typically 50–100 cm/hr. Second, there is a strong correspondence between the physical locations of certain sources and sinks. When mature leaves act as sources, for example, they send sugar to tissues on the same side of the plant (**Figure 37.18a**). In addition, experiments with tall herbaceous plants show that leaves on the upper part of the stem send sugar to apical meristems, but that leaves on the lower part of the plant send sugar to the roots (**Figure 37.18b**). Why would leaves send sugar to tissues on a certain side or part of the body? The answer hinges on understanding the anatomy of phloem.

The Anatomy of Phloem

Chapter 36 introduced the two specialized parenchyma cell types that make up phloem: **sieve-tube members** and **companion cells**. Unlike the tracheids and vessel elements that make up most of the xylem, both sieve-tube members and companion

(a) Source leaves send sugar to the same side of the plant.

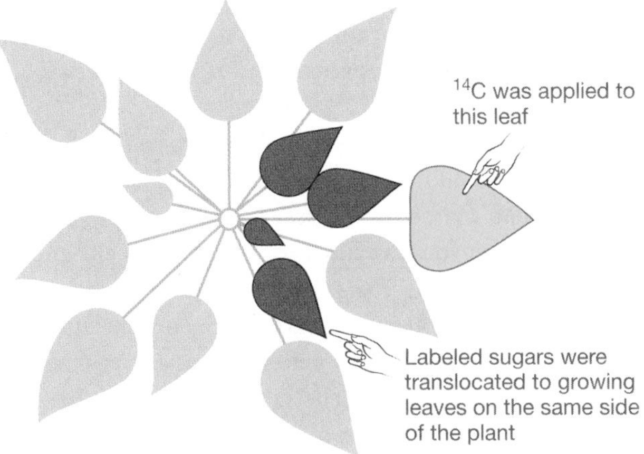

(b) Source leaves send sugar to tissues on the same end of the plant.

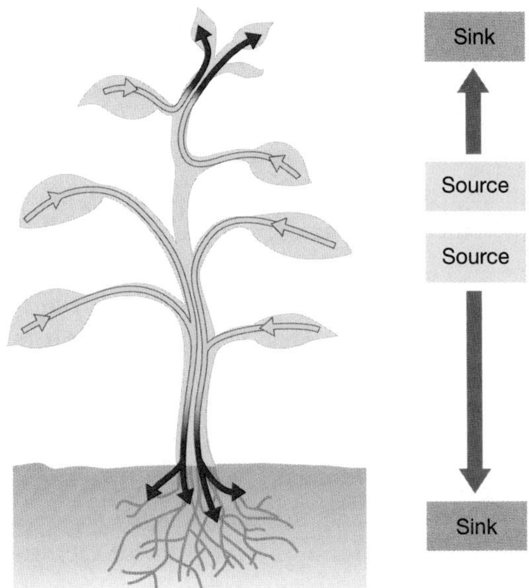

FIGURE 37.18 Sources Supply Sinks on the Same Side and Same End of the Body.

cells are alive at maturity. Recall that, in most plants, sieve-tube members lack nuclei and many major organelles. They are connected to one another, end to end, by perforated structures called **sieve plates** (**Figure 37.19**). The pores create a direct connection between the cytoplasms of adjacent cells. Companion cells, in contrast, have nuclei and a rich assortment of ribosomes, mitochondria, and other organelles.

Sieve-tube members were discovered by T. Hartig in 1837. Hartig was also the first to establish that a sugary solution called phloem sap moves through phloem. He arrived at this conclusion by removing a ring of bark from a tree trunk during the summer and noting that sugars accumulated above this so-called girdle. It had long been known that cutting away a ring of tree bark would kill a tree; Hartig's result showed why. Recall from Chapter 36 that phloem is part of a tree's bark. When phloem tissues are interrupted all the way around a tree, sugars cannot be transported between roots and shoots, and the roots starve.

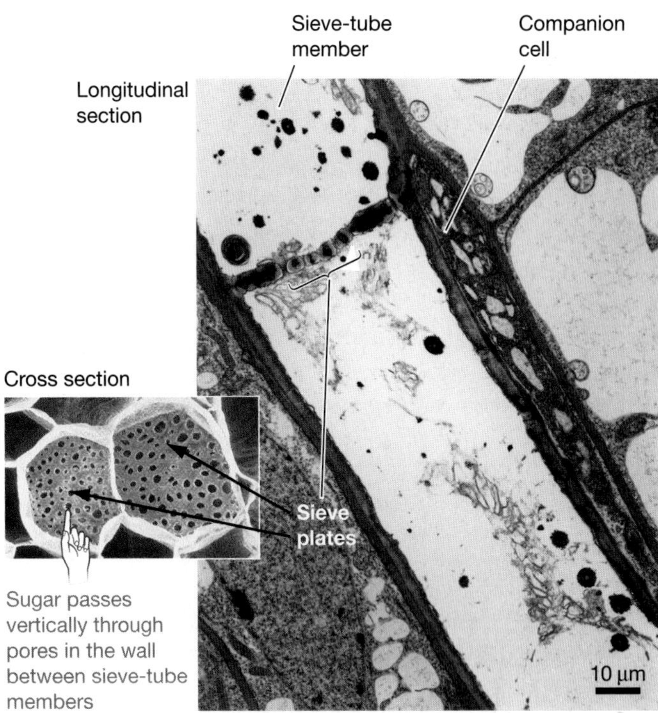

FIGURE 37.19 Sieve-Tube Members Are Connected by Pores. The micrographs show the connections between sieve-tube members in cross section and in longitudinal section.

Subsequent work documented that the phloem present in bark has a high concentration of sugar and that direct anatomical connections link the phloem in trunks, stems, branches, and roots. As a result, the sieve-tube members in bark phloem represent a continuous system for transporting sugar throughout the plant body. Recall from Chapter 36, however, that in tissues that don't form wood, phloem is restricted to discrete vascular bundles. Each vascular bundle runs the length of stems and roots, and certain bundles branch off into specific branches, leaves, and lateral roots. But phloem sap does not move from one vascular bundle to another; each bundle is independent.

Based on these results, the physical relationships observed between sources and sinks in herbaceous plants are logical. For example, the phloem in the leaves on one side of an herbaceous plant connects directly with the phloem of branches, stems, and roots on the same side of the individual, through a specific set of vascular bundles.

By 1930 most major physical components of the sugar transport system in angiosperms had been described in detail, and chemical analyses had shown that phloem sap is often dominated by the disaccharide sucrose—table sugar. Phloem sap can contain small amounts of minerals, amino acids, hormones, and other compounds as well. Now the question is, how does this solution move? What mechanism is responsible for translocating sugars from sources to sinks?

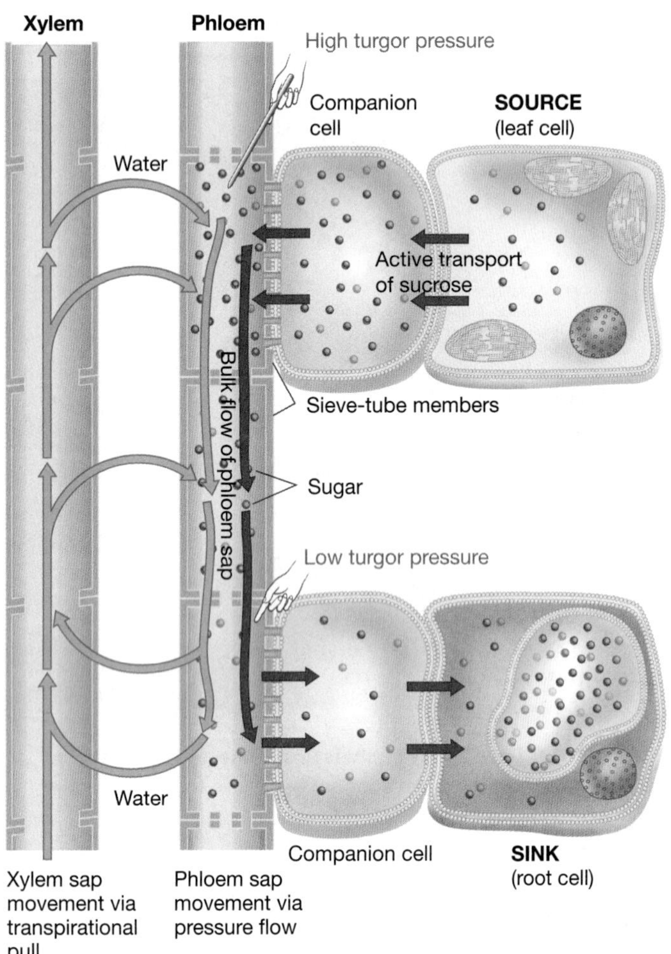

FIGURE 37.20 The Pressure-Flow Hypothesis: High Turgor Pressure Near Sources Causes Phloem Sap to Flow to Sinks. The pressure-flow hypothesis predicts that water cycles between xylem and phloem and that water movement in phloem is a response to a gradient in pressure potential.

EXERCISE This diagram shows how water moves between xylem and phloem in the middle of the growing season, when leaves are sources and roots are sinks. Add new arrows, in new colors, to indicate the direction of water and phloem sap flow in spring, when roots act as sources and leaves act as sinks.

The Pressure-Flow Hypothesis

Ernst Münch proposed a mechanism for translocation in 1926. His idea, called the **pressure-flow hypothesis**, is diagrammed in **Figure 37.20**. The fundamental idea is that events at source tissues and at sink tissues create a steep pressure potential gradient in phloem. The water in phloem sap moves down this gradient, and sugar molecules are carried along by **bulk flow**—a mass movement of molecules along a pressure gradient. Like the cohesion-tension theory for water transport, the pressure-flow hypothesis is based on movement along a water potential gradient—specifically, a gradient created by changes in pressure potential. Unlike the cohesion-tension theory for water transport, however, the driving force for movement in phloem sap is not provided by transpiration. Instead, the force responsible for movement is generated by large differences between the turgor pressure in the phloem in source tissues and the turgor pressure in the phloem in sink tissues. Creating these differences in turgor pressure requires an expenditure of ATP.

To understand how Münch's model works, take a closer look at Figure 37.20, starting with the source cell at the upper right. The small red arrows reflect Münch's proposal that sucrose moves from source cells into companion cells and from there into sieve-tube members. Because of this phloem loading, the phloem sap near the source has a high concentration of sucrose, resulting in a very low water potential compared with that of the adjacent xylem cells in a vascular bundle. As the blue arrows in the upper part of the diagram show, water then moves passively from xylem across the selectively permeable plasma membrane of sieve-tube members along a water potential gradient. In response, pressure begins to build in the sieve-tube members nearest the source region.

What is happening at the sink? Münch proposed that cells in the sink (bottom right in Figure 37.20) remove sucrose from the phloem sap by active transport. As a result of this phloem unloading, the water potential in sieve-tube members increases until it is higher than the water potential in adjacent xylem cells. As the blue arrows at the bottom of the figure show, water flows across the selectively permeable membranes of sieve-tube members into xylem along a water potential gradient. Turgor pressure in the sieve-tube members near the sink drops in response.

The net result of these events is high turgor pressure near the source and low turgor pressure near the sink, created by the loading and unloading of sugars. This difference in pressure potential drives phloem sap from source to sink via bulk flow and is responsible for a one-way flow of sucrose molecules. All these events produce a continuous loop of water flow, with water being supplied from the xylem. The cycle illustrated in Figure 37.20 is driven by water potential gradients between xylem and phloem and pressure potential gradients within phloem. In spring, when roots act as sources and growing leaves act as sinks, the same cycle occurs in the opposite direction.

The pressure-flow hypothesis is logical, given the anatomy of vascular tissue and the principles that govern water movement. But has any experimental work supported the theory? Some of the best tests have relied on aphids—small insects that make their living ingesting phloem sap. Aphids insert a syringe-like mouthpart, called a stylet, into sieve-tube members. The pressure on the fluid in these cells forces it through the stylet, into the aphid's digestive tract, and out its anus as droplets of "honeydew" (**Figure 37.21**). If the aphids are then severed from their stylets, sap continues to flow out through the stylets. This phenomenon allows researchers to collect phloem sap efficiently for analysis. It also confirms that the aphids aren't actively sucking the fluid. Instead, phloem is indeed under pressure.

FIGURE 37.21 Aphids Feed on Phloem Sap. The tip of this aphid's mouthpart (the stylet) is in a sieve-tube member within the plant stem. The droplet emerging from the aphid's anus is honeydew, which consists of sugary phloem sap.

Experiments with aphids confirmed an important prediction of the pressure-flow hypothesis. But what about the mechanism of phloem loading and unloading? How does sucrose enter and leave phloem in a way that sets up the water potential gradient?

Phloem Loading

In contrast to the cohesion-tension model of water movement in plants, pressure flow requires that plants expend energy to set up a water potential gradient in phloem. To create a high pressure potential in sieve-tube members near source cells, large amounts of sugar have to be transported actively into the phloem sap—enough to make the concentration of sugar higher in the sieve-tube members than it is in the source cells (see Figure 37.20, top right). As a result, loading requires an expenditure of ATP and some sort of membrane transport system. Conversely, sugar must sometimes be unloaded against its concentration gradient at sinks. Unloading requires another expenditure of ATP and a second membrane transport mechanism. How do phloem loading and unloading occur? What specific membrane proteins are involved?

To answer these questions, let's look first at events at source tissues, where the active transport of sucrose into sieve-tube members results in a high pressure potential and high water potential. The chapter then concludes with a look at how the same molecules are offloaded to maintain low water potentials at sinks.

How Are Sugars Concentrated in Sieve-Tube Members at Sources? Molecules and ions can be transported across membranes in a variety of ways. Recall from Chapter 6 that transport is said to be passive if ions or molecules move along their concentration gradients and active if substances move against their concentration gradients. **Box 37.3** summarizes the mechanisms responsible for passive transport and active transport.

Because measurements showed that sucrose is more highly concentrated in companion cells than in photosynthetic cells where it is produced, researchers hypothesized that sucrose transport from source cells must be active. Another key observation—that strong pH differences exist between the interior and exterior of phloem cells—inspired a hypothesis for the way active transport occurs in companion cells. These data suggested that sucrose might enter companion cells along with protons. **Figure 37.22** explains the logic behind this proposal: A membrane protein in companion cells hydrolyzes ATP and uses the energy released to transport protons (H^+) across the membrane to the exterior of the cell. Proteins like these are called **proton pumps**, or more formally, **H^+-ATPases.** Proton pumps establish a large difference in charge and in hydrogen ion concentration on the two sides of the membrane. The resulting electrochemical gradient favors the entry of protons into the cell. A second protein,

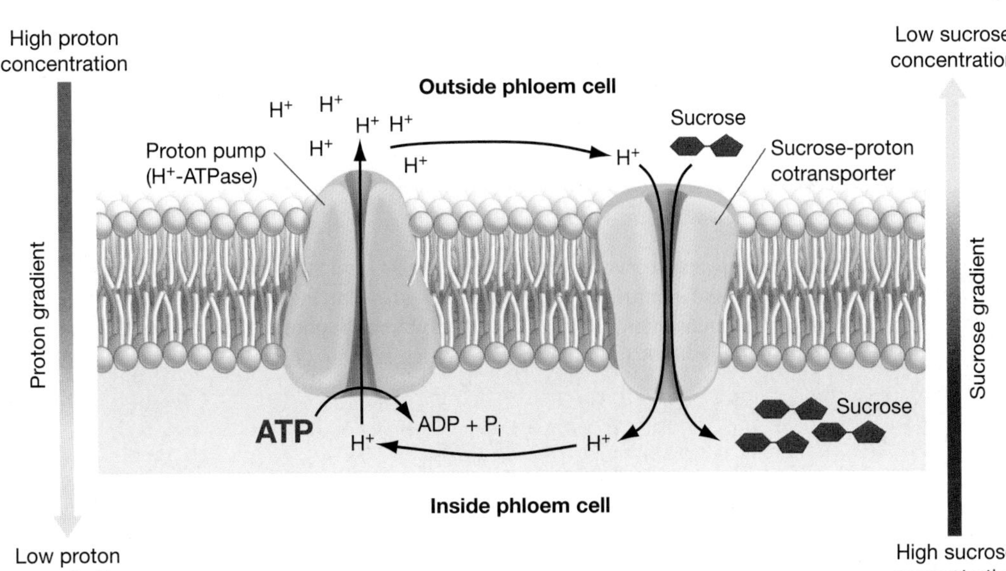

FIGURE 37.22 A Model for Cotransport of Protons and Sucrose. According to the model of cotransport, a proton pump hydrolyzes ATP to move hydrogen ions to the exterior of the cell. The resulting high concentration of H^+ establishes an electrochemical gradient that allows the transport of sucrose into the cell against its concentration gradient.

BOX 37.3 How Are Solutes Transported across Membranes?

Passive transport (**Figure 37.23a**) occurs when ions or molecules move across a plasma membrane by diffusion—that is, with their electrochemical gradient. The adjective *passive* is appropriate because no expenditure of energy is required for the movement to occur.

Recall from Chapter 6 that small, nonpolar molecules diffuse across phospholipid bilayers rapidly (Figure 37.23a, left). But ions and many large molecules diffuse across phospholipid bilayers very slowly, even when their movement is favored by a strong electrochemical gradient. To diffuse rapidly, they must avoid direct contact with the phospholipid bilayer by passing through a membrane protein. As Chapter 3 explained, two general types of membrane protein permit the passive diffusion of specific ions or molecules. **Channels** form pores that selectively admit certain ions (Figure 37.23a, middle). **Carriers** are also membrane proteins; but instead of forming a pore, they work like enzymes, undergoing a conformational change that transports a bound substrate across the lipid bilayer (Figure 37.23a, right). Carriers are responsible for **facilitated diffusion**.

Active transport (**Figure 37.23b**) occurs when ions or molecules move across a plasma membrane against their electrochemical gradient. The adjective *active* is appropriate because cells must expend energy in the form of ATP to move solutes in an energetically unfavorable direction.

Active transport always involves membrane proteins. **Pumps** are proteins that change shape when they bind ATP or a phosphate group from ATP. As they move, pumps transport ions or molecules against an electrochemical gradient. In this way, pumps can establish an electrochemical gradient. For example, the pump on the far left side of Figure 37.23b has established an electrochemical gradient for the ions or molecules symbolized by the light gray balls.

The electrochemical gradients established by pumps may be used to transport other molecules or ions by two types of membrane proteins called **cotransporters**: symporters and antiporters. **Symporters** transport solutes *against* a concentration gradient (like the red molecules in Figure 37.23b, middle), using the energy released when a different solute moves in the same direction *down* its electrochemical gradient. Antiporters work in a similar way, except that the solute being transported against its concentration gradient (like the yellow ions in Figure 37.23b, right) moves in the direction *opposite* that of the solute moving down its concentration gradient.

(a) Passive transport is based on diffusion.

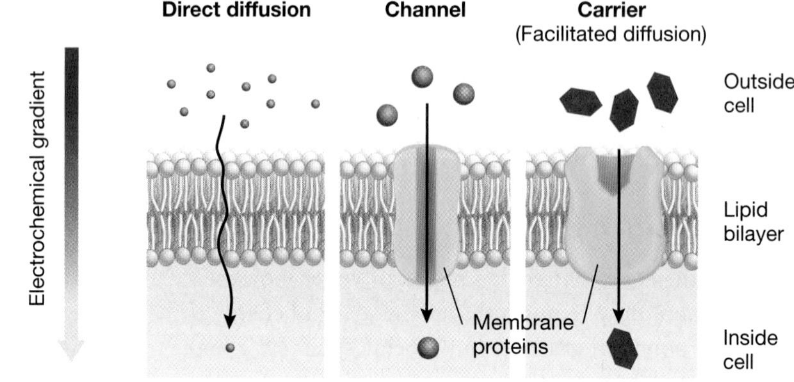

(b) Active transport moves ions or molecules against an electrochemical gradient.

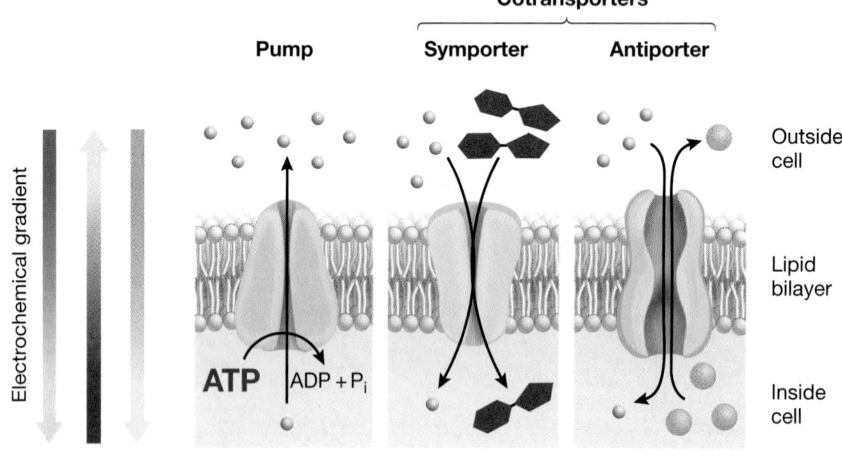

FIGURE 37.23 Passive and Active Transport Occur via Several Mechanisms. (a) In passive transport, ions or molecules diffuse across membranes—meaning they follow their electrochemical gradient. The movement can occur directly through the phospholipid bilayer or via a channel or carrier protein. **(b)** Active transport requires an expenditure of ATP.

called a cotransporter, acts as a conduit for protons and sucrose to enter the cell together—protons along their electrochemical gradient, and sucrose against its concentration gradient.

If phloem loading depends on the activity of a proton pump, then researchers should be able to find and characterize the pump proteins.

Where Are H^+-ATPases Located? Proton pumps are found in the plasma membranes of a wide variety of organisms, including bacteria, fungi, and animals. To analyze the proton pumps that are found in plants, researchers from several laboratories began focusing on a small member of the mustard family called *Arabidopsis thaliana.*

Researchers determined the amino acid sequence of a proton pump purified from the plasma membranes of *Arabidopsis* and used these data to infer the DNA sequence of the corresponding gene. The biologists found that the *Arabidopsis* genome actually codes for 10 different pump proteins. One of these genes, called *AHA3*, appeared to be expressed primarily in vascular tissues.

These observations led Natalie DeWitt and Michael Sussman to hypothesize that *AHA3* encodes the proton pump responsible for phloem loading. To test this hypothesis, they produced antibodies to the *AHA3* protein. You might recall from earlier chapters that an antibody is a protein that binds to a specific location on a molecule—typically, a specific protein. To track the antibody, the investigators attached gold particles to it. When viewed with the electron microscope, a gold particle looks like a black dot. DeWitt and Sussman's goal was to treat *Arabidopsis* leaves with the *AHA3* antibody, examine treated leaves under the electron microscope, and determine exactly where the pump proteins are located.

Figure 37.24 shows the resulting data. The key observation is that the proton pumps responsible for phloem loading are found almost exclusively in the plasma membranes of companion cells. This result supported the following model for phloem loading: (1) Proton pumps in the membranes of companion cells create a strong gradient that favors a flow of protons into companion cells; (2) a cotransporter protein in the membranes of companion cells uses the proton gradient to bring sucrose into companion cells from the surrounding cell walls and intercellular spaces around source cells; and (3) once inside companion cells, sucrose travels into sieve-tube members via a direct cytoplasmic connection. ● If you understand this model, you should be able to add proton pumps and H^+-sucrose cotransporters to the appropriate locations in Figure 37.20. You should also be able to label the location where pH is particularly low. Finally, label where you would expect to find direct cytoplasmic connections responsible for the transit of sucrose from companion cells to sieve-tube members.

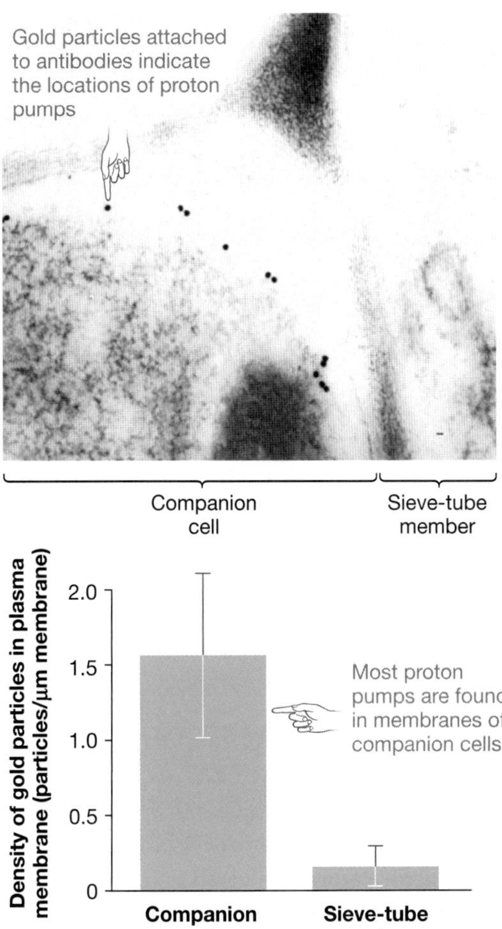

FIGURE 37.24 The Membranes of Companion Cells Contain an H^+-ATPase. The micrograph shows that the proton pumps responsible for phloem loading are located in the plasma membranes of companion cells. The histogram compares the average density of proton pumps in the membranes of companion cells versus those in the membranes of sieve-tube members. These pumps create a proton gradient that allows sucrose to be transported into companion cells against a concentration gradient.

Although work on the mechanism of phloem loading in *Arabidopsis* and other species continues, most researchers are convinced that proton pumps and proton-sucrose cotransporters play a key role. Now, once sucrose has been loaded into sieve-tube members near sources and follows a water potential gradient to sinks, how is it unloaded?

Phloem Unloading

Research on the mechanism of phloem unloading at sinks can be summarized with one general statement: The membrane proteins that are involved in transporting sucrose molecules and the mechanism of movement vary among different types of sinks within the same plant, as well as among different species.

(a) Phloem unloading into growing leaves of sugar beets

Sieve-tube member
Companion cell
Leaf cell
Sucrose
Movement along concentration gradient
Sucrose + other substrate + ATP
Proteins and nucleic acids
Passive transport across membrane, then use of ATP within cell indirectly

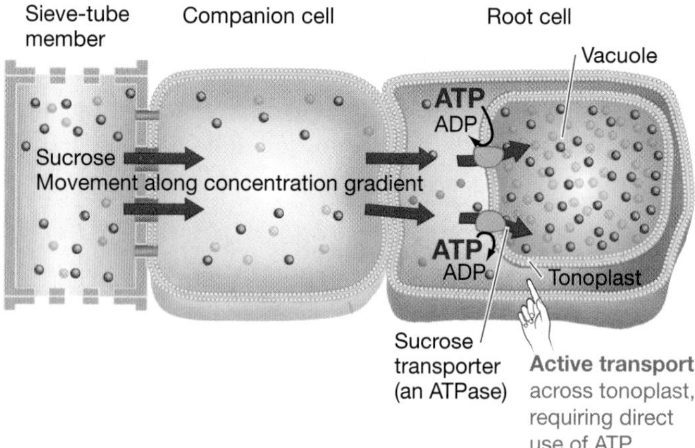

FIGURE 37.25 Phloem Unloading Occurs when Sucrose Is Taken Up by Cells. The mechanism of phloem unloading can vary from sink to sink, such as **(a)** young leaves and **(b)** roots of the same plant.

EXERCISE In each part, add arrows indicating the concentration gradient that favors the movement of sucrose.

To appreciate this diversity, consider how sucrose is unloaded in the tissues of sugar beets—a crop grown for the storage tissues in its root, which are a major source of the granulated and powdered sucrose sold in grocery stores. In sugar beets, sucrose crosses the membranes of cells in young, growing leaves along a concentration gradient. This passive transport occurs because sucrose is rapidly used up inside the cells to provide carbon for the synthesis of proteins, nucleic acids, and phospholipids needed by the growing cells (**Figure 37.25a**). In the roots of the same plant, however, an entirely different mechanism is responsible for offloading sucrose. Root cells in this species have a large vacuole that stores sucrose. The membrane surrounding this organelle is called the **tonoplast**. It contains a protein that hydrolyzes ATP and uses the energy released to transport sucrose into the vacuole, against its concentration gradient (**Figure 37.25b**). The movement of sucrose into the vacuole keeps the water potential of the phloem sap near the sink low, as required by the Münch pressure-flow model.

In short, more than seven decades of research on plants has provided convincing evidence that the pressure-flow hypothesis is fundamentally correct. The cohesion-tension theory for water movement and the pressure-flow model for phloem sap movement represent major advances in our understanding of how plants work.

Check Your Understanding

If you understand that... 

- Phloem sap moves from areas of high water potential to areas of low water potential.
- In phloem, high water potential is due to the high turgor pressure observed in the sieve-tube members near source cells. This pressure is created by pumps that actively load sucrose into companion cells against a concentration gradient. Water follows by osmosis, creating high turgor pressure inside sieve tubes.
- At sinks, turgor pressure is much lower than it is at sources because sucrose is removed from phloem sap by growing tissues or by storage cells. The decreased concentration of solutes in phloem causes water to leave phloem and enter xylem.

You should be able to... 

Predict the water potential of phloem sap near leaves and near roots under each of the following conditions:

1) At the start of growth in spring

2) At midday when growing conditions are ideal in summer

3) Just after the first frost in autumn

4) In midwinter

at www.masteringbio.com
Solute Transport in Plants

Chapter Review

SUMMARY OF KEY CONCEPTS

Water moves from areas of high water potential to areas of low water potential. Water's potential energy in plants is a combination of (1) its tendency to move in response to differences in solute concentration and (2) the pressure exerted on it.

Plants lose water as an inevitable side effect of exchanging gases with the atmosphere. The flow of water from soil to air via plant tissues follows a water potential gradient. Water potential (ψ) is a measure of the tendency of water to move down its potential energy gradient. In plants, water potential has two components: (1) a solute, or osmotic, potential, created by the concentration of solutes in a cell or tissue; and (2) a pressure potential, provided by the cell wall and other factors. The water potential of a cell, tissue, or plant is the sum of its solute potential and pressure potential. When selectively permeable membranes are present, water moves by osmosis from areas of high potential to areas of low potential. When no membranes are present, water moves by bulk flow from areas of high pressure to areas of low pressure, independently of differences in solute potential.

You should be able to describe conditions that would create a water potential gradient capable of moving water out of a plant and into the soil.

Plants do not expend energy to replace water that is lost to transpiration when stomata are open and photosynthesis is occurring. Instead, water moves from soil and roots to leaves along a water potential gradient. The gradient exists because water at the air-water surface in leaves is under negative pressure (tension).

According to the cohesion-tension theory, water lost from stomata is replaced without any expenditure of energy by the plant. Water is pulled in one continuous column from the soil to roots to shoots, against the force of gravity, by the surface tension caused by transpiration from leaves. The surface tension occurs at menisci that form as water evaporates from the walls of leaf cells and is transmitted downward via hydrogen bonding between water molecules. In this way, the energy in sunlight is responsible for the movement of water from roots to shoots.

Several types of data support predictions made by the cohesion-tension theory. For example, on hot, sunny days when transpiration is rapid, xylem sap is under enough tension to make tree trunks shrink slightly in diameter. The measured water potential of leaf tissue drops during the day, when stomata are open; it rises at night, when transpiration is low. Finally, direct measurements show that the tension on xylem sap changes rapidly in response to changes in light intensity and thus in transpiration rate. All these results support the cohesion-tension theory because they are consistent with the prediction that xylem sap is under tension.

Plants that occupy dry habitats have several ways of limiting the amount of water they lose to transpiration. In some species, stomata are located in pits on the undersides of their leaves, helping protect the pits from the dry atmosphere. There is also evidence that the tissues of dry-habitat plants can tolerate higher solute potentials than can tissues of wet-habitat plants. As a result, dry-habitat plants have lower water potentials and can compete more successfully for scarce water in the soil. The CAM and C_4 photosynthetic pathways are also interpreted as adaptations that limit water loss in dry habitats.

You should be able to explain why xylem cells are dead at maturity, and why they have thickened secondary cell walls.

(MB) ***BioFlix*** at www.masteringbio.com
Water Transport in Plants

In phloem, sugars are transported from "sources"—tissues that release sugars for use elsewhere—to "sinks"—tissues in which sugars are being used or stored. Movement occurs because cells expend energy to load sucrose actively into phloem cells near source tissues. Water follows by osmosis, creating a pressure gradient that favors the movement of water and sucrose to sinks.

Translocation, the movement of sucrose and other products through the plant, has been studied by labeling sugars with ^{14}C According to the Münch pressure-flow model, sugars move from sources to sinks via bulk flow along a pressure gradient that develops in the sieve tubes. The gradient is generated by the transport of sugars into sieve-tube members in source tissues, coupled with the transport of sucrose out of sieve-tube members at sink tissues. Water moves from xylem into sieve-tube members near sources and cycles back to xylem near sinks. Research on *Arabidopsis* indicates that the membrane proteins responsible for phloem loading likely include a proton pump and a proton-sucrose cotransporter.

You should be able to explain why phloem cells must be alive at maturity to function, and why they lack secondary cell walls.

(MB) **Web Animation** at www.masteringbio.com
Solute Transport in Plants

QUESTIONS

Test Your Knowledge

1. Under what conditions does the rate of transpiration increase?
   a. in species in which stomata are located in pits on the bottom of leaves
   b. when stomata close at night
   c. during rainstorms, when atmospheric pressure is low
   d. when the weather changes and air becomes drier

2. Which of the following does *not* affect the pressure potential in water?
   a. High sucrose concentrations in companion cells create turgor pressure.
   b. Menisci around spongy mesophyll cells create tension when transpiration rates are high.
   c. Few ions or other solutes are present in soil water.
   d. Water tends to adhere tightly to soil particles when soils dry.
3. The cells of a certain plant species can tolerate extremely high solute potentials. Which of the following statements is correct?
   a. Transpiration rates will tend to be extremely low.
   b. It can compete for water effectively and live in dry soils.
   c. It will grow most effectively in soils that are saturated with water year round.
   d. Leaves will wilt easily.
4. What forces are responsible for capillarity?
   a. adhesion of water molecules to the sides of xylem cells, cohesion of water molecules to each other, and surface tension
   b. surface tension created by transpiration and cohesion of water molecules in a continuous flow from leaf to root
   c. high solute potentials created by the entry of ions during the night, when transpiration rates are low, followed by an influx of water
   d. gravity and wall pressure (from the sides of xylem cells)
5. What is a proton pump?
   a. a membrane protein that transports sucrose against a concentration gradient
   b. a membrane protein that transports protons against an electrochemical gradient
   c. a membrane protein that transports protons *with* an electrochemical gradient and sucrose *against* a concentration gradient
   d. any membrane protein that acts as a channel—meaning it does not consume ATP
6. Why is the transport of phloem sap considered an active process?
   a. The manufacture of sucrose via photosynthesis is driven by the energy in sunlight.
   b. Transpiration is driven by the energy in sunlight.
   c. ATP is used to transport sucrose into companion cells near sources against a concentration gradient.
   d. In spring, phloem sap moves against the force of gravity.

Test Your Knowledge answers: 1. d; 2. c; 3. b; 4. a; 5. b; 6. c

Test Your Understanding

Answers are available at www.masteringbio.com

1. Draw a plant cell in pure water. Add dots to indicate solutes inside the cell. Now add dots to indicate an increase in solute potential inside the cell. Add an arrow showing the direction of water movement in response. Add arrows showing the direction of wall pressure in response to water movement. Repeat the same exercise, but this time add solutes to the solution outside the cell. In general, why does water move from regions of high water potential to regions of low water potential?
2. Compare and contrast the forces involved in transporting water in xylem via capillarity, root pressure, and transpiration. Which of these mechanisms are passive? Can they occur at the same time, or are they mutually exclusive?
3. Why are "cohesion-tension" and "pressure-flow" sensible names for the hypotheses analyzed in this chapter?
4. Suppose Aphid A and Aphid B are sitting on the same plant. A inserts her stylet into the phloem of a large, mature leaf, while B prefers to probe the phloem of the young growing tissue near the shoot's apical meristem. Which aphid is attacking a source, and which is preying on a sink? Which aphid is getting a higher concentration of sugar, A or B? Where would you expect aphids to be found on a plant growing in the wild?
5. How does cotransport result in phloem loading? Where does this process occur? What data support your answers to these questions?
6. How does a pressure bomb work, and what does it measure?

Applying Concepts to New Situations

Answers are available at www.masteringbio.com

1. The text claims that water loss is an inevitable side effect of gas exchange. What data or observations support or challenge this claim? Would the same statement be true in bacteria? In animals?
2. Suppose that plants over 1 meter tall had to expend energy to transport water from their roots to their leaves. What would be the consequences in terms of growth rates and overall height?
3. When young trees are transplanted to a new site, it takes them several weeks or months for their root system to grow and establish a high capacity to take up water. If a heat wave occurs during this period, the trees are likely to die—but not of starvation. What kills them?
4. A seed is a sink when it is forming inside the parent plant. When is it a source?

www.masteringbio.com is also your resource for • Answers to text, table, and figure caption questions and exercises • Answers to *Check Your Understanding* boxes • Online study guides and quizzes • Additional study tools including the *E-Book for Biological Science* 3rd ed., textbook art, animations, and videos.

Plant Nutrition

38

In most plants, roots obtain the water and key elements required for individuals to survive and thrive.

KEY CONCEPTS

- In addition to needing carbon dioxide and water, plants require an array of essential nutrients to support growth. These nutrients are available as ions dissolved in soil water and are taken up by roots.
- Nutrient absorption occurs via specialized proteins in the plasma membranes of root cells. Most plants also obtain nitrogen or phosphorus from fungi associated with their roots. Toxins that enter roots are excluded from xylem by endodermal cells or are actively transported into cell vacuoles and stored.
- Some species of plants have specialized methods of obtaining nutrients, including associations with nitrogen-fixing bacteria, parasitism, and carnivory.

The most urgent tasks facing any organism are to acquire (1) carbon-containing molecules that will be used as cellular building blocks and (2) the chemical energy required to make ATP. Plants acquire both by producing sugar through the process of photosynthesis. Yet plants cannot live on sugar alone. Besides making the carbohydrates they need, plants synthesize all of their own nucleic acids, amino acids, enzymes, chlorophylls, enzyme cofactors, and other macromolecules. Plants can do some of the world's most impressive synthetic organic chemistry.

A plant's ability to perform sophisticated reactions depends on its capacity to harvest a wide variety of simple ions and elements as raw materials. In addition to needing carbon dioxide and water, plants have to obtain nitrogen, phosphorus, potassium, sulfur, magnesium, and other elements. Soil provides most of these nutrients, the majority existing as ions that are dissolved in soil water at low concentrations—sometimes extremely low concentrations.

The plant body is an efficient machine for harvesting these types of diffuse resources and concentrating them in cells and tissues. Chapter 36 introduced how the organization and growth of the plant root and shoot systems make resource acquisition possible; Chapter 37 focused on how water and nutrients are transported throughout the plant body. This chapter concentrates on how plants take up simple ions and elements from the soil, so that these nutrients can be transported to the cells that need them. Questions about nutrition are fundamental to understanding how plants work and how people who grow plants—from farmers, horticulturists, and foresters to backyard gardeners—manage their land. Let's begin by analyzing the basic nutritional needs of plants—the equivalent of their minimum daily requirements.

38.1 Nutritional Requirements

What do plants need to live? In the early 1600s, Jean-Baptiste van Helmont performed a classic experiment that started a long research tradition focused on answering this question. Van Helmont wanted to know where the mass of a growing plant comes from, and he used a willow tree as a study organism. He began the study by weighing out 200 pounds of soil and placing it in a pot with a 5-pound willow sapling (**Figure 38.1**). He allowed the plant to grow for five years, adding only water. At the end of the experiment, he weighed the willow and the soil. The willow weighed 169 lb, 3 ounces; the soil weighed 199 lb, 14 ounces.

Where had the additional 164 pounds, 3 ounces, of tree come from? Because he was not aware that gases have mass, van Helmont concluded that the new plant material came from water. He also ignored the loss of 2 ounces in the soil, chalking it up to measurement error. As it turned out, his measurements were not the problem—his conclusions were. Follow-up work by other researchers established that most of the mass of the tree came from carbon dioxide in the atmosphere. The 2 ounces removed from the soil contained vital ions and elements—the nutrients that are the focus of this chapter. What are they?

Experiment

Question: Where does the mass of a growing plant come from?

Hypothesis: The mass of a growing plant comes from soil.

Alternate hypothesis: The mass of a growing plant comes from water.

Experimental setup:

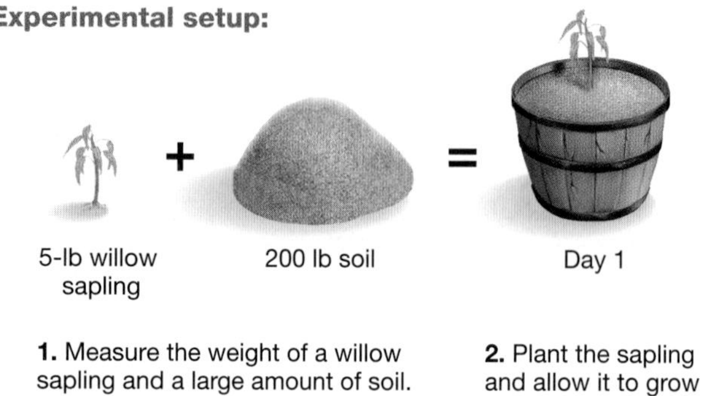

1. Measure the weight of a willow sapling and a large amount of soil.

2. Plant the sapling and allow it to grow for 5 years.

Prediction: After 5 years, the soil mass will decrease by the same amount that the plant mass increased.

Prediction of alternate hypothesis: The soil mass will not decrease.

Results:

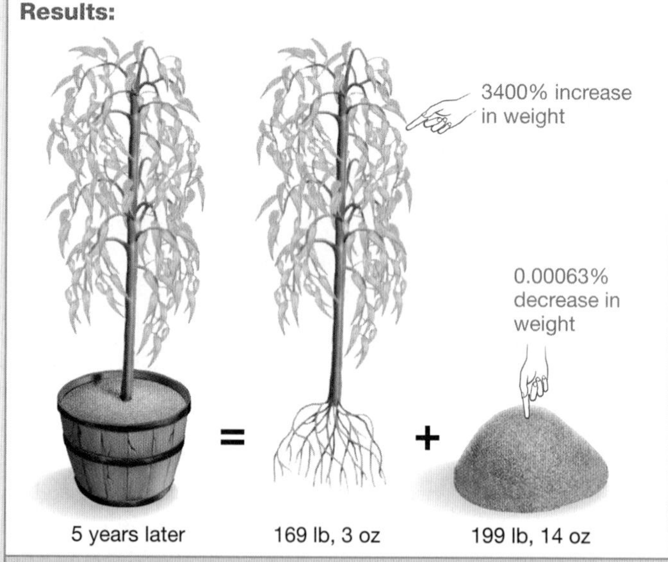

Conclusion: The mass of a growing plant comes from water. [Note: This conclusion was later found to be incorrect.]

FIGURE 38.1 An Early Experiment on the Role of Soil in Plant Nutrition.

Which Nutrients Are Essential?

About half the elements in the periodic table—more than 60—can be found in the tissues of one or more plant species. The question that farmers, gardeners, and biologists ask is, Which of these elements are essential for growth and reproduction in most species, and in what relative quantities? To answer this question, biologists define an **essential nutrient** as an element or compound that fulfills both of the following criteria:

1. It is required for normal growth and reproduction—meaning that the plant cannot complete its life cycle without this nutrient; and
2. It is required for a specific structure or metabolic function.

By denying specific nutrients to plants and documenting the effects or lack of effect, biologists have determined that 17 elements are essential for most vascular plants. Just three of these—carbon, hydrogen, and oxygen—typically make up about 96 percent of the dry weight of a plant.

Although different classification schemes for the essential elements have been proposed, perhaps the most common is based on distinguishing nutrients that are available in water or carbon dioxide from nutrients that are available in the soil, and then dividing soil nutrients into macronutrients and micronutrients (**Table 38.1**).

Macronutrients—the building blocks of nucleic acids, proteins, carbohydrates, phospholipids, and other key molecules—are required in relatively large quantities. Among the macronutrients, nitrogen (N), phosphorus (P), and potassium (K) are particularly important because they often act as **limiting nutrients**, meaning that plant growth is limited by their availability. If N, P, and/or K are added to soil as fertilizer, plant growth usually increases. This observation explains why the leading ingredients in virtually every commercial fertilizer are N, P, and K—usually listed in that order on the bag or box.

In contrast to macronutrients, **micronutrients** are required in very small quantities. When plant tissues are dried and then analyzed, micronutrients are typically present in 1–100 parts

TABLE **38.1** **Essential Nutrients**

Element	Form Available to Plants	Functions	Average % Dry Weight*	Deficiency Symptoms
Obtained from CO_2 or H_2O				
Oxygen	O_2, H_2O	Electron acceptor in cellular respiration; major component of organic compounds	45	Usually affects roots—cells suffocate, leading to root rot and wilting
Carbon	CO_2	Substrate for photosynthesis; major component of organic compounds	45	Slow growth (starvation)
Hydrogen	H_2O	Major component of organic compounds; electrical balance and establishment of electrochemical gradients	6	Slow growth due to cell death (desiccation)
Obtained from soil				
(a) Macronutrients				
Nitrogen	NO_3^- (nitrate) NH_4^+ (ammonium ion)	Component of nucleic acids, proteins, hormones, and coenzymes	1.5	Failure to thrive; chlorosis (yellowing of leaves)
Potassium	K^+	Cofactor for many enzymes; necessary for osmotic adjustment in cells; required for synthesis of organic molecules	1.0	Chlorosis at margins of leaves or in mottled pattern; weak stems; short internodes
Calcium	Ca^{2+}	Regulatory functions; role in cell wall structure; stabilizes membranes; second messenger in signal transduction	0.5	Necrosis (small spots of dead cells) in meristems; deformation of young leaves; stunted, highly branched root system
Magnesium	Mg^{2+}	Chlorophyll component; activates many enzymes	0.2	Chlorosis between leaf veins; premature leaf drop
Phosphorus	$H_2PO_4^-$ (dihydrogen phosphate ion) HPO_4^{2-} (hydrogen phosphate ion)	Component of ATP nucleic acids, phospholipids, and several coenzymes	0.2	Stunted growth in young plants; dark green leaves with necrosis
Sulfur	SO_4^{2-} (sulfate ion)	Component of protein and coenzymes	0.1	Stunted growth; chlorosis
(b) Micronutrients				
Chlorine	Cl^-	Needed for water-splitting step of photosynthesis; functions in water balance and electrical balance	0.01	Wilting at leaf tips; general chlorosis and necrosis of leaves or development of bronze color
Iron	Fe^{3+} (ferric ion) Fe^{2+} (ferrous ion)	Necessary for chlorophyll synthesis; component of cytochromes and ferredoxin; enzyme cofactor	0.01	Chlorosis between veins of young leaves
Manganese	Mn^{2+}	Involved in photosynthetic O_2 evolution; enzyme activator; important in electron transfer	0.005	Chlorosis between leaf veins and small necrotic spots
Zinc	Zn^{2+}	Involved in synthesis of the plant hormone auxin; maintenance of ribosome structure; enzyme activation	0.002	Small internodes; stunted and and distorted ("puckered") leaves
Boron	$H_2BO_3^-$ (borate ion)	Cofactor in chlorophyll synthesis; possible role in sugar transport; aids in regulation of enzyme function	0.002	Black necrosis in young leaves and buds
Copper	Cu^+ (cuprous ion) Cu^{2+} (cupric ion)	Cofactor of some enzymes; present in lignin of xylem	0.0006	Dark-green leaves with necrotic spots; twisted and malformed leaves
Nickel	Ni^{2+}	Cofactor for enzyme functioning in nitrogen metabolism	[no data]	Necrosis at leaf tips
Molybdenum	MoO_4^{2-} (molybdate ion)	Cofactor in nitrogen reduction; essential for nitrogen fixation	0.00001	Chlorosis between veins; necrosis of older leaves

*These percentages were obtained by drying vascular plants and then documenting what proportion of the waterless mass consists of various elements.

per trillion. The difference in abundance springs from another contrast with macronutrients: Instead of acting as components of macromolecules, micronutrients usually function as cofactors for specific enzymes—substances that are required for normal enzyme function (Chapter 3). Micronutrients may be required in tiny amounts, but their importance should not be underestimated. For example, a typical plant contains just one molybdenum atom for every 60 million hydrogen atoms in its body, not including water. Yet plants die in the absence of molybdenum, because it functions as a cofactor for several enzymes involved in nitrogen processing. What happens to plants when other essential nutrients are missing?

What Happens when Key Nutrients Are in Short Supply?

In at least some cases, biologists can examine a plant that is growing poorly and diagnose a nutrient deficiency problem (**Figure 38.2**). For example, nutrients like N, P, K, and magnesium are mobile—meaning they are readily transferred from older leaves to newer leaves when they are in short supply. As a result, deficiencies in mobile elements initially affect older leaves. Immobile nutrients like iron or calcium, in contrast, stay tied up in older leaves. When they are in short supply, newer leaves are the first to show symptoms.

In large part, the ability to diagnose specific nutrient deficiencies is based on the results of studies that have used hydroponic growth systems to examine the effects of specific nutrient deficiencies. **Hydroponic growth** takes place in liquid cultures, without soil. Because experimental plants are grown in the absence of soil, the availability of nutrients can be precisely controlled.

As an example of this work, consider the experiment on copper deficiency in tomatoes outlined in **Figure 38.3**. Researchers grew seedlings in two types of treatments. One treatment consisted of flasks containing water and all the essential nutrients listed in Table 38.1 in the relative concentrations that are optimal for tomato growth. The second treatment was identical, except that the nutrient solution lacked copper.

As the "Results" section of Figure 38.3 shows, copper-deprived individuals have stunted shoots, unnaturally light foliage, curled leaves, and no flowers. Given copper's role as a cofactor or component of several enzymes involved in redox reactions required for ATP production, it is understandable that all tissues in the plant were severely affected. And because copper is a micronutrient, it is reasonable to expect that a relatively small amount would cure the deficiency. In line with this

(a) Normal barley **(b)** N deficiency **(c)** P deficiency

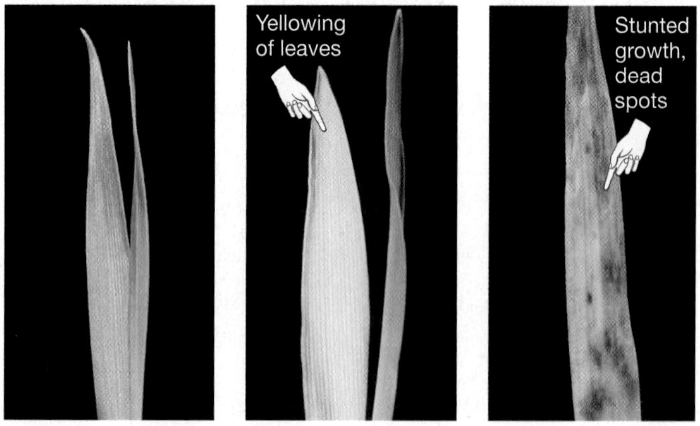

Yellowing of leaves

Stunted growth, dead spots

FIGURE 38.2 Nutrient Deficiencies Can Have Distinctive Symptoms. These photos show leaves of normal, (N) nitrogen-deficient, and (P) phosphorus-deficient barley plants.

● **QUESTION** When nitrogen is in short supply, which molecules or processes are affected?

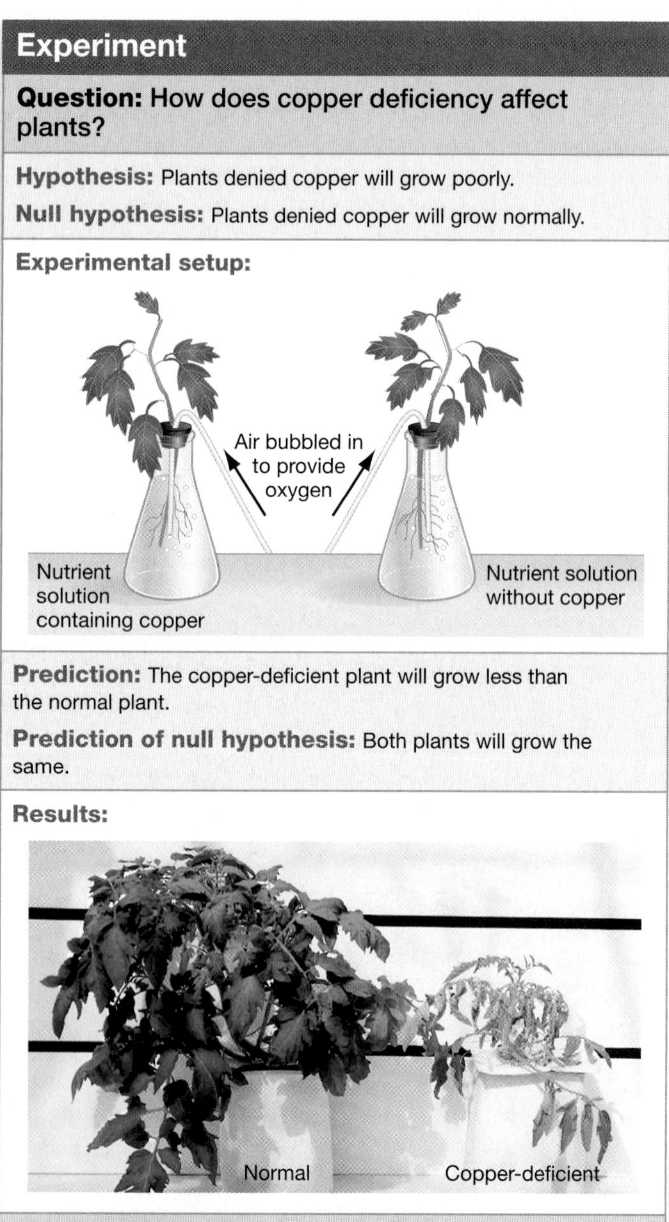

FIGURE 38.3 Hydroponics Is Used to Study Nutrient Deficiencies. Culturing plants in a solution instead of in soil allows researchers to control precisely which nutrients are available and in what concentrations.

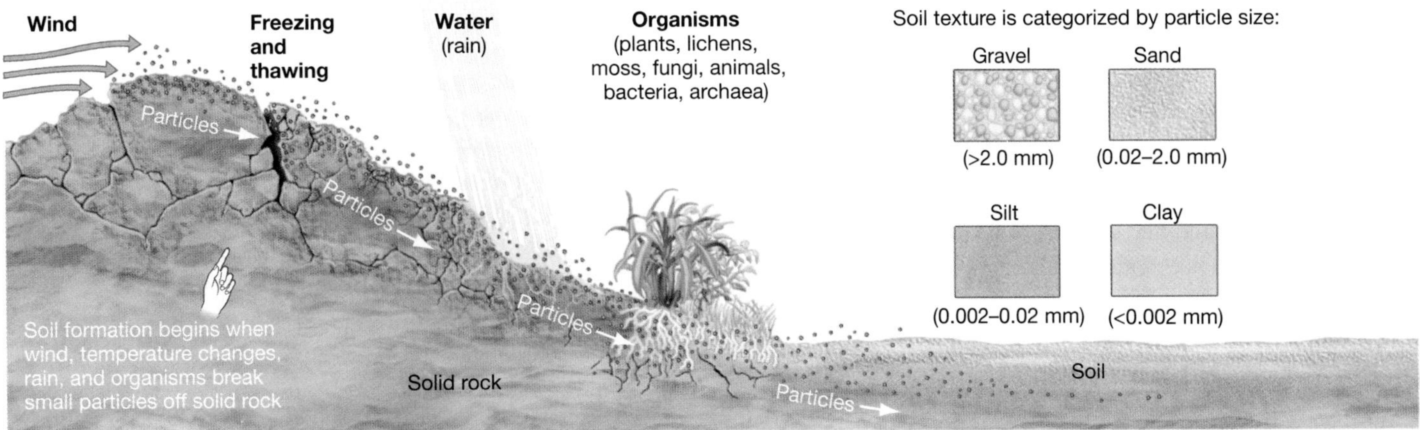

FIGURE 38.4 Soil Formation Begins with Erosion of Rock. Soil formation begins when wind, rain, and organisms break small particles off solid rock.

QUESTION Soil erosion due to wind and rain is much more severe on devegetated versus vegetated sites. Why?

prediction, the researchers found that the symptoms were prevented if the plants were cultured in a solution containing just 0.002 mg/L of this element. Analogous studies have now been done on the other essential nutrients.

For farmers, gardeners, and plant ecologists, understanding which nutrients are essential, and why, is basic to understanding why certain plants thrive and others fail. Now, where do these nutrients come from? The answer—soil—is simple. But soil itself is astonishingly complex.

38.2 Soil

The process of soil building begins with solid rock. As **Figure 38.4** shows, **weathering**—the forces applied by rain, running water, and wind—continually breaks tiny pieces off large rocks. The process is accelerated if small cracks develop in the rock. If plant roots grow into the crack, they expand as they grow, widen the crack, and break off small flakes or pebbles. A similar effect occurs in high latitudes or at high elevations when water enters the cracks, freezes in winter, and expands.

Depending on their size and composition, the particles resulting from these processes are called gravel, sand, silt, or clay. These rock fragments are the first ingredient in soil. As organisms occupy the substrate, they add dead cells and tissues and feces. This decaying organic matter is called **humus** (pronounced *HEW-muss*). With time, soil eventually becomes complex and dynamic mixture of inorganic particles, organic particles, and living organisms. It is commonplace to find thousands of species living in the top few centimeters of a single square meter of soil. In addition to plants, soil-dwelling organisms include a variety of fungi and animals, along with vast numbers of bacteria, archaea, and microscopic protists (**Figure 38.5**).

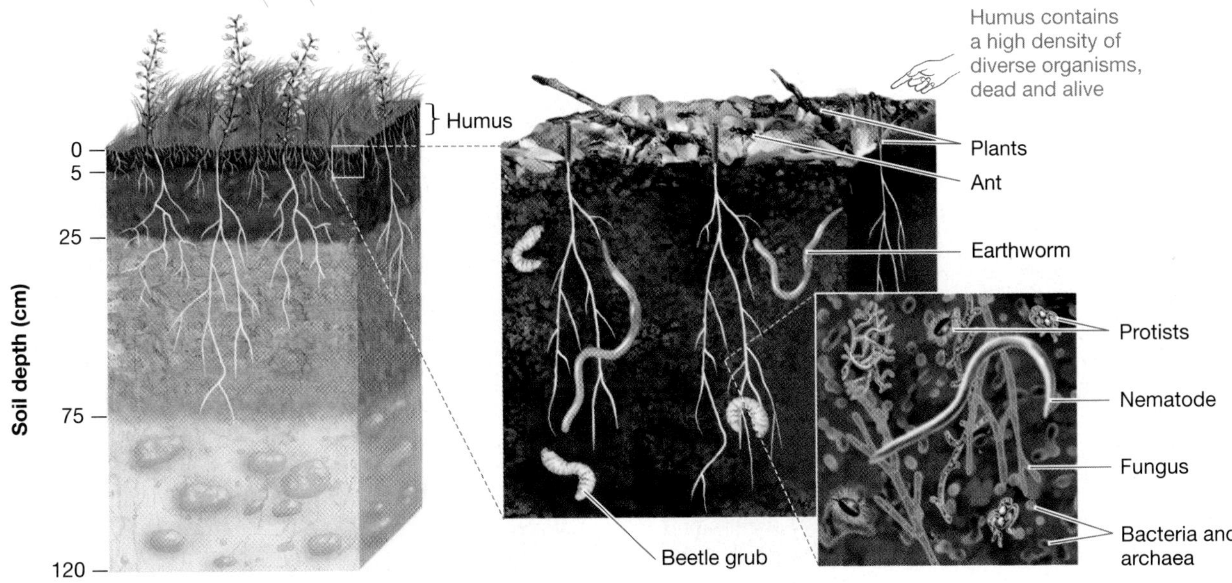

FIGURE 38.5 Mature Soils Are a Complex Mixture of Organic and Inorganic Components. In addition to mineral particles, soil contains humus—organic material derived from dead organisms—and a wide array of living organisms.

Both the parent rock that contributes inorganic soil components and the organisms and organic matter that occupy soils vary from one site to another. Texture and other qualities of soil vary as well. **Texture** refers to the proportions of gravel, sand, silt, and clay present in soil; it is important for several reasons:

- Texture affects the ability of roots to penetrate more deeply to obtain water and nutrients, as well as to anchor and support the body. For example, soil that is dominated by clay-sized particles tends to compact and resist root penetration.
- Texture affects a soil's ability to hold water and make it available to plants. Water tends to adhere to clay and silt particles but runs through sands and gravels.
- A soil's texture and its water content dictate the availability of oxygen. Like other eukaryotes, plants have to take in oxygen to use as an electron acceptor during cellular respiration. The oxygen used by plant root cells is found in air pockets among soil particles. This explains why overwatering a plant is just as detrimental as underwatering it: Overwatering drowns a plant's roots.

Topsoils that have good texture and large amounts of organic matter can take thousands of years to develop through the weathering of rocks and the continual addition of humus. Unfortunately, it can take just a few years of abuse by humans for them to blow or wash away.

Soil Conservation

Soil erosion occurs when soil is carried away from a site by wind or water. Soil erosion occurs naturally when rivers cut away at their banks and carry material downstream, as well as in other situations. In most natural environments, though, the rate of soil formation exceeds the rate of soil erosion, so soils build up over time. Unfortunately, the situation can change dramatically when areas are exploited by humans.

When plant cover is removed for forestry, farming, or suburbanization, soil particles are exposed to wind and rain without plant roots to hold them in place. The results can be devastating. At some locations in the United States, 8–10 cm of topsoil were blown away during the Dust Bowl of the 1930s, when drought and poor farming practices left thousands of acres of soil unprotected (**Figure 38.6a**). Although U.S. soil erosion rates have declined dramatically since then, in 2003 researchers estimated that almost 30 percent of all croplands in the country were eroding too fast to maintain their long-term productivity. Elsewhere, deforestation has exposed soils in the tropics and contributed to disasters such as the mudslides and flooding that occurred in the Dominican Republic and Haiti during May and September of 2004, killing close to 5000 people (**Figure 38.6b**). Worldwide, it is estimated that 36 billion tons of soil are being lost to erosion every year.

Erosion rates are increasing as human populations expand and as land that was previously thought to be unsuitable for agriculture or intensive forestry is brought into production. In addition to outright loss of soils, soil quality can diminish over time in areas subjected to continual farming or tree cutting. In these cases, farmers and foresters remove organic matter that otherwise would have been returned to the soil to form humus.

Fortunately, soil is a renewable resource if managed carefully. Farmers can reduce soil loss dramatically by planting rows of trees as windbreaks and using techniques that minimize the amount of plowing and tilling that needs to be done to control weeds. They can maintain soil quality by adding organic material in the form of manures and planting cover crops that are plowed in and allowed to decompose. Ranchers and foresters can employ similar management approaches. Techniques that maintain long-term soil quality and produc-

(a) Wind erosion in the United States, 1930s

(b) Mudslides and flooding in Haiti, 2004

FIGURE 38.6 Soil Erosion Can Have Devastating Consequences.

tivity are the basis of **sustainable agriculture** and sustainable forestry.

Productivity is influenced by soil texture and organic matter but is primarily a function of nutrient availability. Let's take a closer look.

What Factors Affect Nutrient Availability?

The elements required for plant growth are not found in the soil as atoms. Instead, they exist as **ions**. The second column of Table 38.1 lists some of the ions that contain essential nutrients and are found in soil. Notice that some of these nutrients are available as elemental ions, such as K^+ or Cl^-, while others exist as molecular ions, such as HPO_4^{2-} or NO_3^-.

The ions present in soil tend to behave in one of two ways, depending on their charge (**Figure 38.7**). Anions—ions with negative charges—usually dissolve in soil water, because they interact with water molecules via hydrogen bonding. (Phosphate is an exception to this rule; phosphate ions tend to form insoluble complexes with iron, aluminum, calcium, or other positively charged cations.) Because they exist as solutes, negatively charged anions are readily available to plants for absorption, but they are also easily washed out of the soil by rain. The loss of nutrients via the movement of water through soil is called **leaching**.

Cations—ions with positive charges—are not as immediately available as anions. This pattern occurs because cations interact with the negative charges found on two types of soil particles: (1) organic matter that is rich in negatively charged organic acids; and (2) the surfaces of the tiny, sheetlike particles called clay, which are rich in mineral anions (see Figure 38.7). Organic soils that contain clay tend to retain nutrients, because few positively charged cations leach away and because these soils hold water (and thus anions) better than sandy soils do. The presence of clay makes cations more difficult for plants to extract and use, however, because they are tightly bound.

Several other factors can influence the availability of essential elements present in soil. Perhaps the most important of these is soil pH. Recall from Chapter 2 that pH indicates the relative concentration of hydrogen ions in a solution and that the pH scale ranges from 0 to 14. Soils with low pH have a relatively high concentration of hydrogen ions and are considered acidic. Soils with a high pH contain relatively few hydrogen ions and are termed basic or alkaline. Acidic soils tend to be found in regions such as conifer forests, where the decomposition of organic matter results in the production of carbonic acid, phosphoric acid, or nitric acid. Alkaline soils are common in regions where limestone ($CaCO_3$) is abundant. When limestone reacts with water, the calcium ions that are released take the place of protons that cling to soil particles. The protons then react with CO_3^{2-} to form bicarbonate ions (HCO_3^-) lowering the hydrogen ion concentration of the soil and raising the pH.

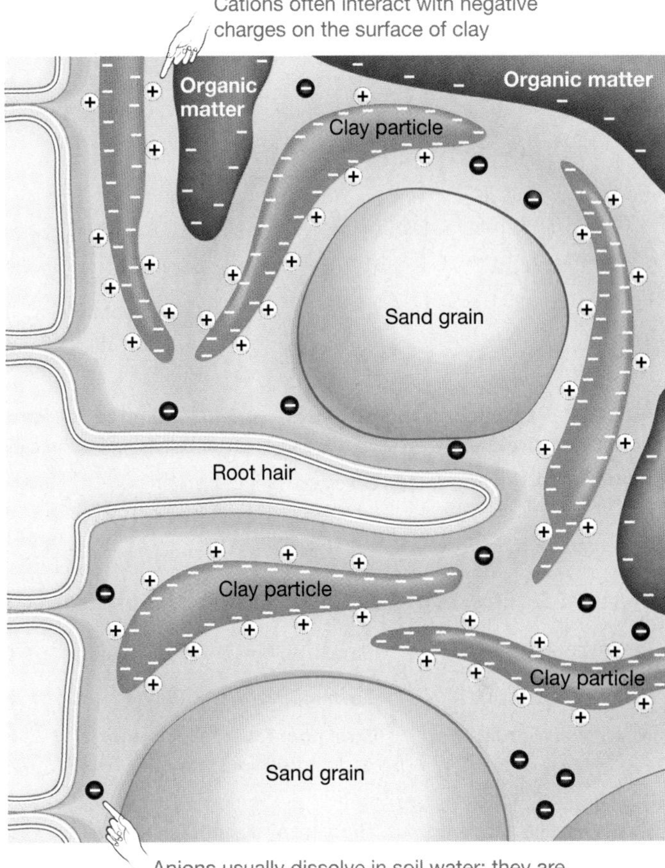

FIGURE 38.7 Cations Tend to Bind to Soil Particles; Anions Stay in Solution. Cations bind to organic matter in soil as well as to clay particles. Anions, in contrast, tend to go into solution.

EXERCISE Add symbols indicating that positive ions bind to organic matter.

Most plants prefer to grow in soils with a relatively neutral pH—between 6 and 7. But species such as rhododendron and blueberries grow best in acidic soils (pH around 3 to 5), while some plants require an alkaline soil (pH above 8). Certain species have enzymes that are adapted to work well in low-pH or high-pH environments.

Soil pH affects the availability of plant nutrients in a number of ways. Acids trigger acid-base reactions that increase the rate at which elemental ions are released from rocks, and many molecular ions are more soluble in solutions with low pH. The presence of protons in soil water can also cause the release of cations that are bound to soil particles, via a process called **cation exchange.** Cation exchange occurs when protons or other cations bind to negative charges on soil particles and cause bound cations such as magnesium or calcium to be released (**Figure 38.8a**), making those nutrients available to nearby plant roots (**Figure 38.8b**). Plants influence cation exchange by pumping protons out of root hairs. But if soil is too acidic, then rain can wash cations away before the nutrients can be taken up by roots (**Figure 38.8c**).

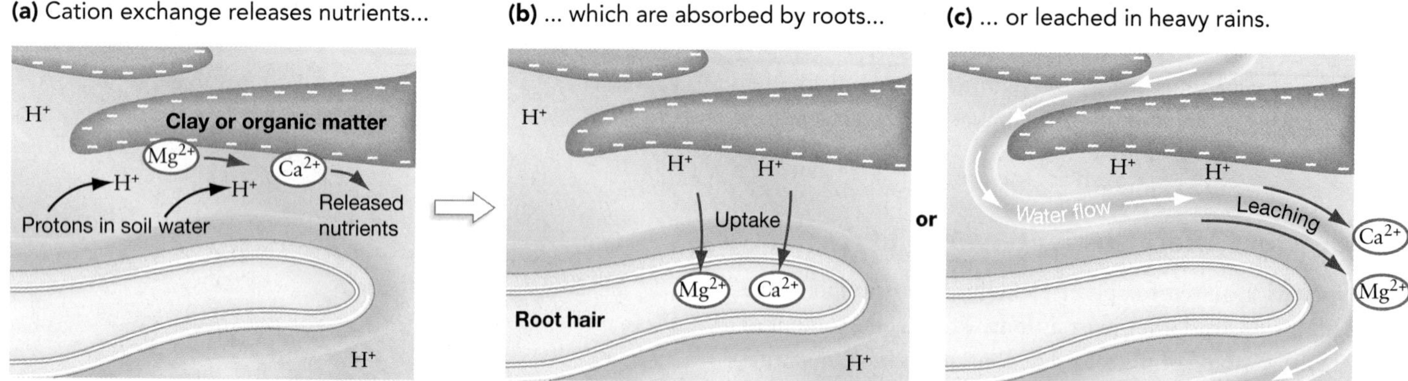

FIGURE 38.8 Cation Exchange Releases Nutrients Bound to Soil Particles. When cation exchange occurs, a proton binds to negative charges on clay or organic matter, releasing bound cations.

QUESTION Why is the term *cation exchange* appropriate?

TABLE **38.2 Effects of Soil Composition on Soil Properties**

Soil Property	Sand	Clay	Organic Matter
Water availability	**Low:** water drains through	**High:** water clings to charged surface	**High:** water clings to charged surface
Nutrient availability	**Low:** poor capacity for cation exchange; anions leave in solution	**High:** large capacity for cation exchange; anions remain in solution	**High:** source of nutrients; large capacity for cation exchange; anions remain in solution
Oxygen availability	**High:** many air-containing spaces	**Low:** few air-containing spaces	**High:** many air-containing spaces
Root penetration ability	**High:** does not pack tight	**Low:** packs tight	**High:** does not pack tight

To summarize, negative ions stay in solution in soil water, and they are readily available to plants but may wash away easily. Positive ions tend to bind to soil particles but can be released by cation exchange. **Table 38.2** details how the presence of sand, clay, and organic matter affects ion availability and other soil properties.

Now the question is, Once ions are in solution and available for absorption by plants, how do they actually enter root cells? Nutrients are found at extremely low concentrations in soil but at high concentrations in cells. How are plants able to bring N, P, K, and other key elements into their bodies against a concentration gradient?

38.3 Nutrient Uptake

In most species of plants, the root system is the site of nutrient uptake. The general features of root anatomy were introduced in Chapter 36, and the role of roots in water uptake was analyzed in Chapter 37. Now we need to explore the fine anatomy of roots and analyze events that occur in the plasma membranes of their epidermal cells.

Nutrient uptake occurs just above the growing root tip, in the region called the **zone of maturation**. Recall from Chapter 36 that epidermal cells in this part of the root have extensions called **root hairs** (**Figure 38.9**). Root hairs dramatically increase the surface area available for nutrient and water absorption. When a biologist examined the root system of a single annual rye plant, he estimated the total surface area to be the size of a basketball court. Most of this area—60 percent—was found in an estimated 10^{10} root hairs. Root

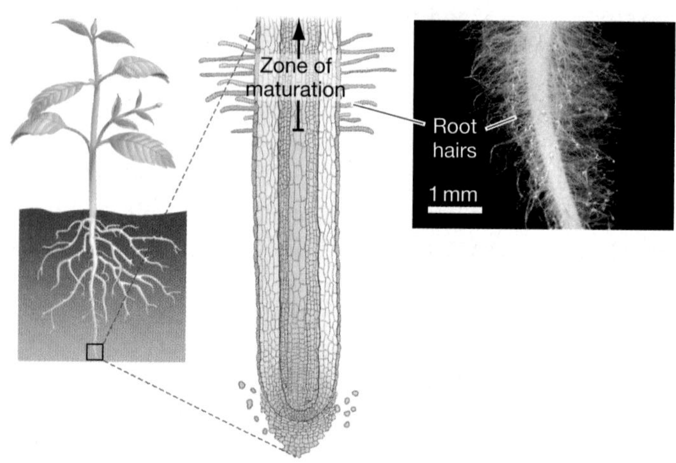

FIGURE 38.9 Root Hairs Increase the Surface Area Available for Nutrient Absorption.

hairs are so numerous and so efficient at absorbing nutrients from soil that, over time, they create a "zone of nutrient depletion" in soil immediately surrounding them. The creation of this mined-out region is why continued root growth is vital to a plant's health. Because roots continue to grow throughout a plant's life, the zone of maturation is continually entering new and potentially nutrient-rich areas of soil.

If a root hair happens to encounter soil areas where nutrients are available, what happens? How do the ions cross the plasma membrane and enter the cell?

Mechanisms of Nutrient Uptake

Ions, small molecules, and even large molecules can pass through the plant cell wall freely. The plasma membrane, in contrast, is highly selective. Recall from Chapter 6 that the plasma membrane is a fluid, sheetlike structure consisting of a phospholipid bilayer studded with proteins. Because the interior of the phospholipid bilayer is nonpolar and hydrophobic, it resists the passage of ions—even ions that are required by the plant. Some membrane proteins span the bilayer, however, and allow specific ions to cross the membrane.

Because root hairs have such a large surface area, they have large numbers of membrane proteins that bring nutrients into the cortex of the root. These membrane proteins cannot import the ions that the plant needs completely on their own, however.

Establishing and Using a Proton Gradient Plants harvest diffuse nutrients and concentrate them in their tissues. In a root-hair cell, ions such as potassium (K^+), hydrogen phosphate (HPO_4^{2-}), and nitrate (NO_3^-) are many times more concentrated than they are in the soil water outside the cell. For these and other ions to enter the cell, they have to cross the plasma membrane against a strong concentration gradient. How is this possible?

As **Figure 38.10a** shows, the answer hinges on **proton pumps**, or H^+-ATPases, in the plasma membranes of root-hair cells. These H^+-ATPases are similar to the pumps that make it possible for companion cells to load sucrose into phloem against a strong concentration gradient. Recall from Chapter 37 that when a phosphate group from ATP binds to the pumps, they change conformation in a way that allows them to transport protons to the exterior of the cell (see Figure 37.23b). The activity of these pumps leads to a strong excess of protons on the exterior of the plasma membrane relative to the interior. In the case of root hairs, this differential results in a strong concentration gradient favoring the movement of protons into the epidermal cell. In addition, the outside of the root-hair membrane is positively charged relative to the inside. Stated another way, there is a separation of charge—a **voltage**—across the membrane.

Proton pumps are found in all types of plant cells, and all plant cell membranes carry a voltage. Because voltage is a form of potential energy, the charges that are separated by the membrane create a **membrane potential**, or difference in electrical charge across a cell membrane. By convention, membrane potentials are expressed as inside-cell relative to outside-cell. Typically, the membrane potential of an active plant cell is about -200 mV (millivolts).

In addition to concentration gradients that affect nutrients, then, the membranes of root hairs create a strong electrical gradient. Recall from Chapter 6 that the combination of concentration and electrical gradients is called an **electrochemical gradient**. In the membranes of root hairs, the electrical gradient

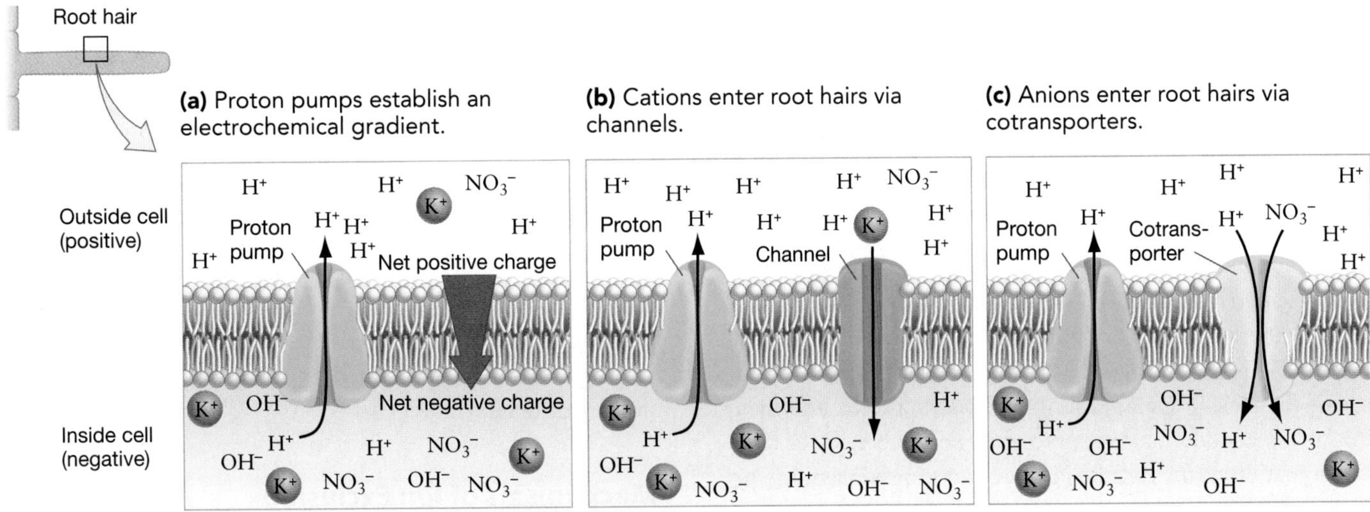

FIGURE 38.10 Ions Enter Roots along Electrochemical Gradients Created by Proton Pumps.

EXERCISE In part (b), add an arrow labeled "Proton gradient." In part (c), add an arrow labeled "Electrical gradient" and indicate its positive-negative polarity.

established by proton pumps, which favors the entry of positive ions, is strong enough to overcome the concentration gradient, which opposes the entry of these cations. In essence, plant cells are batteries that are charged up to attract nutritionally necessary cations.

To drive this point home, consider recent research on how root-hair membranes import potassium cations. Potassium is an essential nutrient in plants because it is required as a cofactor by over 40 enzymes. In addition, it is found at relatively high concentrations inside plant cells and plays a key role in bringing water into cells via osmosis and maintaining normal turgor pressure. Researchers who added radioactive potassium ions to the solution outside of a root cell and followed their movement found that K^+ does indeed flow into cells along an electrochemical gradient (**Figure 38.10b**). In follow-up experiments, biologists were able to isolate the membrane protein responsible for the uptake of K^+ and to sequence the gene that encodes the protein. ● If you understand this concept, you should be able to make a diagram showing how cations are imported into a cell via ion channels; and add other diagrams showing (1) what happens if H^+-ATPases fail, and (2) what happens if a molecule blocks a cation channel.

Cations like K^+ enter root hairs along an electrochemical gradient, but how is it possible for anions to enter? The negatively charged interior of the cell should repel these ions. As **Figure 38.10c** shows, though, anions such as NO_3^- are able to enter root hairs for two reasons: (1) A strong electrochemical gradient favors the entry of protons; and (2) membrane proteins called **cotransporters** use the energy released by the entry of a proton along its electrochemical gradient to transport nitrate, phosphate ions, or other anions *against* their electrochemical gradients. An array of cations also enters cells via cotransporters. Ions that are missed by root hairs and that move into the root through cell walls can be taken up in a similar fashion by the membranes of cortical cells or of endodermal cells. ● If you understand this concept, you should be able to make a diagram showing how anions are imported into a cell via ion channels; and add other diagrams showing (1) what happens if H^+-ATPases fail, and (2) what happens if a molecule blocks an anion cotransporter.

● To summarize, the electrochemical gradient set up by proton pumps makes it possible for plant roots to absorb key cations and anions via ion channels and cotransporters. Once ions have entered the root, they are transported across the root cortex via the symplastic or apoplastic pathways introduced in Chapter 37. To enter xylem, however, nutrients have to pass through the plasma membranes of endodermal cells and then diffuse from endodermal cells into the xylem tubes. Once they become part of xylem sap, the ions are transported passively to tissues throughout the plant.

(MB) **Web Animation** at www.masteringbio.com
Soil Formation and Nutrient Uptake

Nutrient Transfer via Mycorrhizal Fungi To synthesize proteins and nucleic acids, plants need to extract large quantities of nitrogen and phosphorus from the soil. Although the vast majority of plants take up nutrients by establishing a protein gradient across the membranes of their root hairs, most can only absorb enough N and P to satisfy their nutritional needs with help from fungi that live in close association with their roots. You might recall from Chapter 31 that these types of fungi are called **mycorrhizae** (literally, "fungus-root"). Mycorrhizae and plants are said to be **symbiotic** ("living together"), meaning that they live in physical contact with each other.

Trees and shrubs that grow in northern forests receive large quantities of nitrogen from mycorrhizal fungi. The fungal symbionts are particularly efficient at digesting amino acids in decaying plant material and absorbing ammonium ions (NH_4^+) and other forms of usable nitrogen. Because these fungi have hyphae—filaments—that wrap around the epidermal cells of roots and radiate out into the surrounding soil, they are known as **ectomycorrhizal fungi**, or **EMF**; see Figure 31.11a. (The Greek root *ecto* means "outside.") In contrast, plants that live in grasslands and in tropical forests receive much of the phosphorus they need from species of fungi whose hyphae can actually penetrate the walls of plant root cells. Because some of these fungal hyphae form branched clusters that look like little trees when viewed under a microscope, these symbionts are called **arbuscular mycorrhizal fungi**, or **AMF**; see Figure 31.11b. (The Latin root *arbor* means "tree.") Biologists estimate that more than 80 percent of all vascular plant species associate with mycorrhizae.

Chapter 31 presented experimental evidence that EMF and AMF transfer nitrogen and phosphorus from soil to plant roots. In exchange, the plants that they associate with transfer sugars and other photosynthetic products to the fungi. Because there is a reciprocal exchange of nutrients, the symbiotic relationship is considered **mutualistic**, or mutually beneficial (**Figure 38.11**). Chapter 31 also emphasized two reasons that fungi are particularly efficient at acquiring the nutrients required by plants: First, they form networks of filamentous hyphae that increase the surface area available for absorbing nutrients by up to 700 percent. Second, they synthesize and secrete a wide array of digestive enzymes. These enzymes break down dead plant material and other sources of nutrients, releasing ions that can be absorbed and used by the fungus itself or by its plant partner. Most plant species grow slowly and are overwhelmed by competitors if denied their mycorrhizal associates.

Mechanisms of Ion Exclusion

Plants have sophisticated systems for absorbing nutrients via the proton gradients they generate and mutualistic relationships with fungi. Not all ion uptake is beneficial, however. For

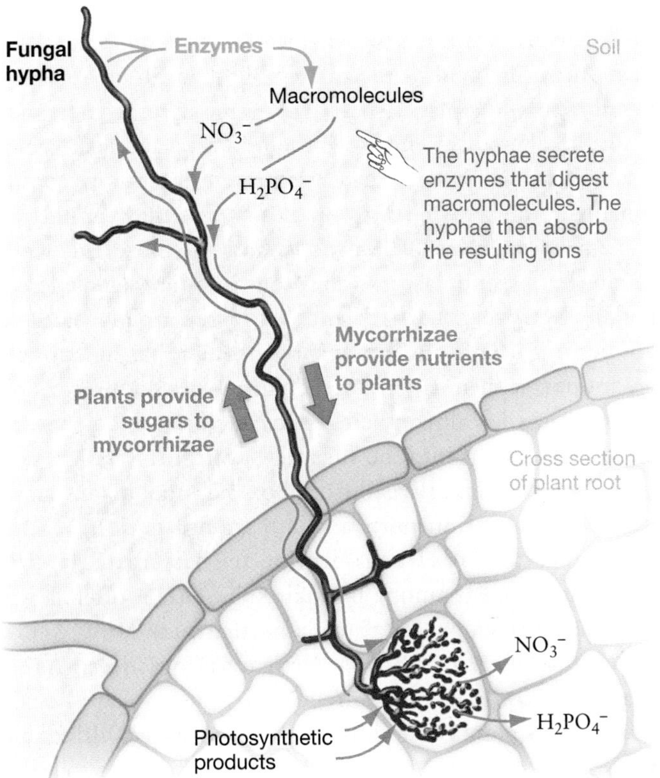

FIGURE 38.11 Most Mycorrhizae and Plants Are Mutualists. Although some mycorrhizal fungi are parasitic, most provide nutrients to their host plants in exchange for photosynthetic products.

example, certain types of natural soils—as well as soils that have been contaminated by waste products from mining or smelting operations—contain enough cadmium, zinc, nickel, lead, or other metals to poison most plants. Sodium is also detrimental at high concentrations; an excess of Na^+ inside cells can disrupt enzyme functioning, while a surplus of sodium in extracellular spaces can create a solute potential that pulls enough water out of cells to result in a loss of turgor. How do plants exclude ions that are detrimental?

Passive Exclusion Ions that enter roots via the apoplastic pathway may never make it to xylem. To understand why, recall from Chapter 37 that the Casparian strip forces all solutes that are traveling through the root apoplast to cross the plasma membrane of endodermal cells and enter their cytoplasm. Ions that enter endodermal cells in this way may remain there—excluded from entering the rest of the plant. This occurs because the membrane of an endodermal cell contains only certain types of transporters and channels, meaning that it acts as a selective filter. In this way, the presence of the Casparian strip allows the endodermal cells to prevent some ions from entering the symplast and reaching the xylem. The excluded ions are not actively pumped out of the root cortex, so the Casparian strip qualifies as a mechanism for the passive exclusion of metals, sodium, or other types of ions (**Figure 38.12**).

Passive exclusion also occurs in root hairs. This is a second location where solutes enter the symplast on their way to xylem. If root hairs lack the membrane protein required for a certain ion to enter the cell, the ion won't enter. As an example, consider recent research on variation in salt tolerance. Some crop species are much more tolerant of salty soils than others are. For example, rice is notoriously sensitive to salt buildup, while barley is relatively tolerant. Even within the same species, the degree of salt tolerance can vary widely. Researchers grew seedlings from salt-tolerant and salt-intolerant varieties of corn (*Zea mays*) in hydroponic cultures containing a high concentration of NaCl. They found that salt-intolerant individuals took up almost twice

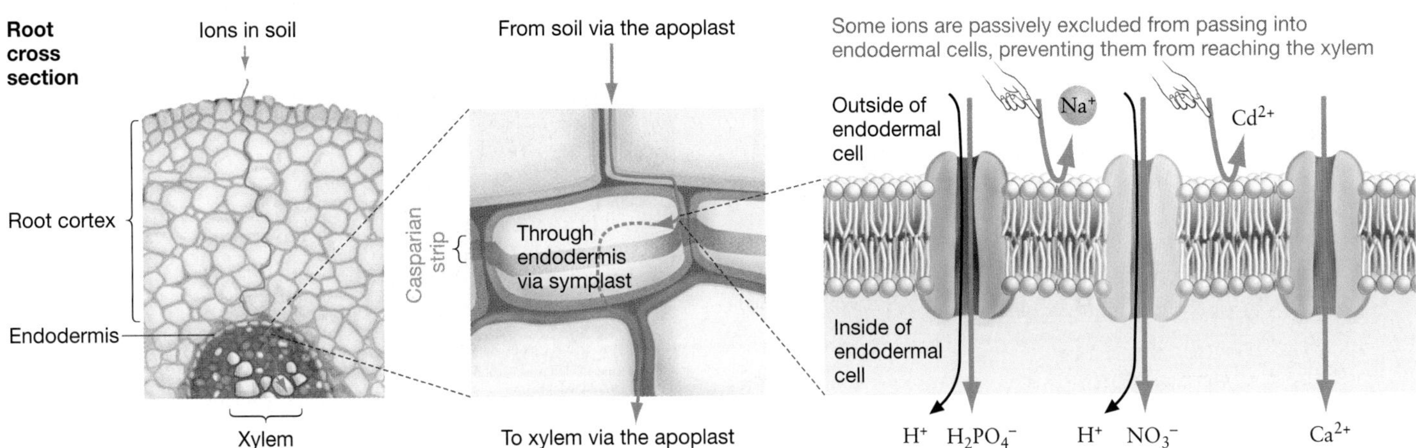

FIGURE 38.12 Passive Exclusion Occurs in Endodermal Cells. Certain ions are excluded from endodermal cells because it is difficult for them to cross the plasma membrane.

EXERCISE Label the phosphate ion cotransporter, the nitrate cotransporter, and the calcium ion channel illustrated here.

as much Na^+ as did salt-tolerant individuals. To interpret this result, the researchers hypothesized that individuals from salt-tolerant populations have fewer sodium channels in their roots than do salt-intolerant individuals. If this hypothesis is confirmed, it implies that salt tolerance is at least partly based on variation in the abundance of membrane proteins in hair roots.

Active Exclusion Plants also have mechanisms for coping with toxins once they are inside their cells. This is important, because all nutrients are toxic if they are present at high enough concentrations. Consider copper, for example. Copper channels admit the small amounts required for normal cell function. But plants that grow on soils near copper-mining operations experience large concentration gradients that favor an influx of this nutrient, so a surplus is eventually likely to build up inside the plant body. How do plants neutralize excess nutrients before they begin to poison key enzymes?

One mechanism for coping with toxic concentrations of metals involves small proteins called metallothioneins. **Metallothioneins** bind to metal ions and prevent them from acting as a poison. Genes for these proteins have been found in a wide variety of organisms—bacteria, fungi, and animals as well as plants. Recent research on the mustard-family species *Arabidopsis thaliana* has shown that individuals from populations with a high tolerance for copper produce many more metallothionein proteins than do individuals from populations with a low tolerance for copper. Producing metallothionein proteins requires an expenditure of energy and represents a form of active exclusion.

A second mechanism for actively neutralizing specific toxins involves transport proteins located in the **tonoplast**—the membrane surrounding the vacuole. Proteins in the tonoplast membrane allow plants to actively remove toxic substances from the cytoplasm and store them in the vacuole. Perhaps the best-studied example involves proteins that move sodium ions from the cytosol into vacuoles, where they cannot poison enzymes. **Figure 38.13a** shows how these proteins work. Pumps in the tonoplast move protons into the vacuole, creating an electrochemical gradient that favors the movement of H^+ out of the vacuole. A transport protein that functions as an **antiporter** then uses this gradient to conduct protons out of the vacuole, along their electrochemical gradient, and bring sodium ions into the vacuole—against the sodium concentration gradient. (Recall from Box 37.3 that an antiporter is a protein that transports a solute against its electrochemical gradient.) This H^+/Na^+ antiporter has created a great deal of excitement among biologists, because it offers a way to genetically engineer crop plants that may grow well in salty soils created by poor irrigation practices. As an example, consider an experiment in which normal *Arabidopsis* plants and individuals transformed with copies of the gene for the H^+/Na^+ antiporter were exposed to different levels of NaCl. As **Figure 38.13b** shows, the genetically transformed individuals were able to grow efficiently even when exposed to high concentrations of salt. In a similar vein, **Box 38.1** on page 852 explains how plants with the ability to sequester toxins in their vacuoles might be used to clean up contaminated soils.

(a) In the tonoplast, antiporters send H^+ out and Na^+ in.

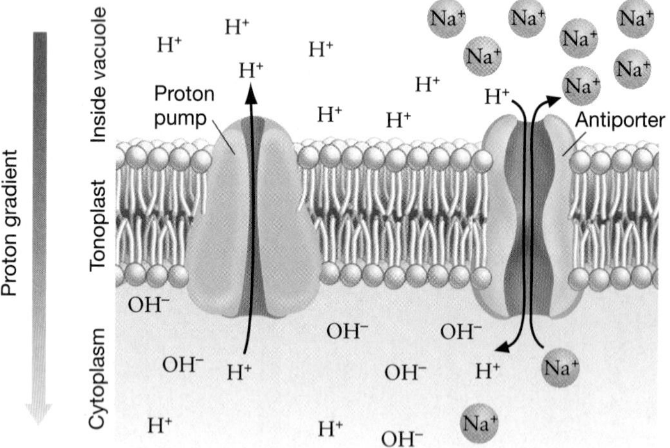

(b) Plants with H^+/Na^+ antiporters tolerate salt.

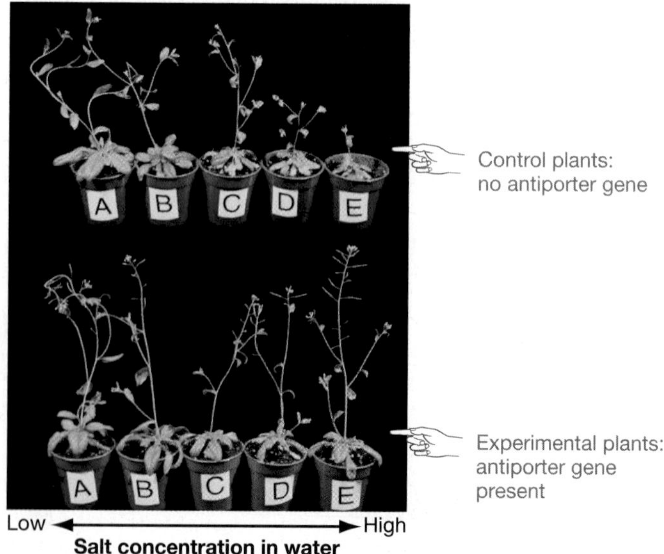

FIGURE 38.13 In Salt-Tolerant Plants, an Antiporter Concentrates Sodium in Vacuoles. (a) In the tonoplast of salt-tolerant species, a proton gradient is established by proton pumps. This gradient allows antiporters to concentrate sodium ions inside the vacuole, where they do not poison enzymes in the cytoplasm. **(b)** *Arabidopsis* plants that were transformed with the gene for the antiporter protein grow well even when watered with salty water.

Check Your Understanding

If you understand that...

- Plants absorb most of the nutrients they need via membrane proteins in root hairs.
- Proteins in the plasma membranes of root-hair cells pump protons out of the cells, creating an electrochemical gradient that favors the entry of selected ions via carriers or cotransporters.
- In many species, mycorrhizal fungi are important for bringing ions that contain nitrogen or phosphorus atoms into the root.

You should be able to...

1) Explain how plants generate and use electrical power to import key ions.
2) Make a diagram that traces a cadmium ion from the soil into the plant, showing how it is excluded at different locations by a total of two passive and two active mechanisms of ion exclusion.

38.4 Nitrogen Fixation

Nitrogen gas (N_2) makes up 80 percent of the atmosphere. Unfortunately, plants and other eukaryotes cannot use nitrogen in this form. Nitrogen gas is unreactive because the triple bond between the two nitrogen atoms is extremely strong and hard to break. To synthesize amino acids, nucleic acids, and other nitrogen-containing compounds, plants have to absorb nitrogen in more accessible forms, such as ammonium or nitrate ions. But these ions are in short supply in many soils, meaning that plant growth is often restricted by the availability of usable nitrogen. Most plants grow much faster when they receive a nitrogen-containing fertilizer.

Using nitrogen-based fertilizers has drawbacks, however. Fertilizer production is extremely energy-intensive; and in many parts of the world, ammonia and other nitrogen-based fertilizers are too expensive for farmers. In more affluent regions, these fertilizers are used so extensively that they are causing serious pollution problems (see Chapter 28). For these and other reasons, there is intense interest in understanding the molecular basis of a biological phenomenon called nitrogen fixation.

Among all the organisms on the tree of life, only a few species of bacteria and archaea are able to absorb N_2 from the atmosphere and convert it to ammonia, nitrites, or nitrates. This process is called **nitrogen fixation.** Nitrogen fixation requires a series of specialized enzymes and cofactors, including a large multi-enzyme complex called nitrogenase. The process is extremely energy demanding. An expenditure of up to 24 ATP molecules is required for nitrogenase to reduce one molecule of N_2 to two molecules of NH_3.

In many cases, bacterial cells that are capable of nitrogen fixation take up residence *inside* plant root cells. Although several different bacteria and plant hosts can be involved, the best-studied nitrogen-fixing bacteria are members of the genus *Rhizobium* that associate with plants in the pea family. Members of the genus *Rhizobium* and closely related species are often called **rhizobia**; pea family plants are often called **legumes**.

As **Figure 38.14** shows, the root cells of legumes form distinctive structures called **nodules**, where nitrogen-fixing rhizobia are found. Like mycorrhizal fungi and their host plants, legumes and rhizobia have a mutualistic relationship. The nitrogen-fixing bacteria provide the plant with ammonia, while the legume provides the bacteria with carbohydrates and protection. The association is costly for the host plant, however. For the bacteria to synthesize the enormous amounts of ATP required to manufacture ammonia, plants have to supply them with large quantities of sugar.

Research groups from all over the world have been trying to understand which genes and proteins are involved as legumes and rhizobia interact. The research efforts are motivated in part by the hope that the required genes could be introduced into species such as wheat or rice, using techniques described in Chapter 19. The roots of wheat and rice could then be occupied by nitrogen-fixing bacteria and the use of nitrogen-based fertilizers reduced.

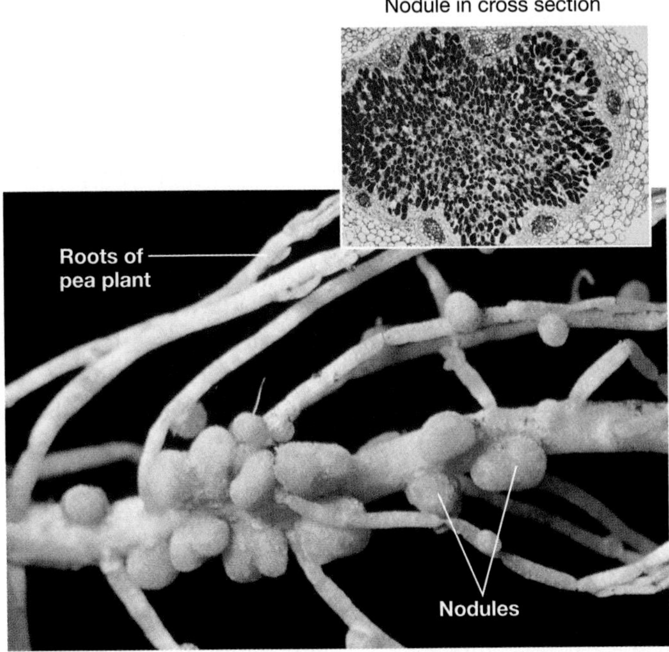

FIGURE 38.14 In Some Plants, Roots Form Nodules where Nitrogen-Fixing Bacteria Live.

How Do Nitrogen-Fixing Bacteria Colonize Plant Roots?

When a pea seed germinates, its roots do not contain a population of rhizobia. Instead, the root must make contact with bacterial cells existing in the soil, the rhizobia must colonize root cells, and the root cells must grow into a nodule. Colonization is a complex process involving a series of specific interactions between the rhizobia and the legume.

The first event in colonization is a recognition step that occurs between a pea family plant and its symbiotic bacterium. Recognition begins with compounds called flavonoids that are released by root hairs. When rhizobia contact the flavonoids, the bacteria respond by producing sugar-containing molecules called **Nod factors** (for *nod*ule-formation). Nod factors, in turn, bind to proteins on the membrane surface of root hairs.

The recognition step in the interaction is species specific. Each legume species produces a different flavonoid that acts as a recognition signal, and each rhizobium species responds with one or more unique Nod factors. When investigators have performed experiments based on switching recognition signals or Nod factors between species, the recognition step fails.

How Do Host Plants Respond to Contact from a Symbiotic Bacterium?

When Nod factors bind to the root-hair surface, they set off a chain of events that leads to dramatic morphological changes in the host legume. As **Figure 38.15** shows, rhizobia multiply and then enter the epidermal cell through an invagination of the root-hair membrane called an **infection thread**. After invading the root cortex, the infection thread bursts and releases rhizobia into the interior of cortex cells. The infected cortex cells begin to divide rapidly, forming root nodules.

Genetic changes occur in concert with these morphological changes. For example, rhizobia need to obtain oxygen to carry out aerobic respiration and produce the large quantities of ATP required for nitrogen fixation. But the key enzyme complex involved, nitrogenase, is poisoned by the presence of oxygen. The problem is solved by an iron-containing molecule called leghemoglobin (short for legume hemoglobin). **Leghemoglobin** is produced by host cells in response to infection with rhizobia, and consists of a globin protein complexed with an oxygen-binding heme group (see Chapter 3). The molecule is similar to the hemoglobin in your red blood cells and causes the pinkish color of root nodules (see Figure 38.14). The iron atom in the heme group binds to oxygen molecules, just the way the heme group in hemoglobin binds to oxygen in your lungs. Oxygen molecules inside rhizobia bind to leghemoglobin instead of binding to nitrogenase. After binding to leghemoglobin, oxygen is delivered directly to enzymes in the electron transport chain in rhizobia, where it acts as an electron acceptor during cellular respiration (**Figure 38.16**).

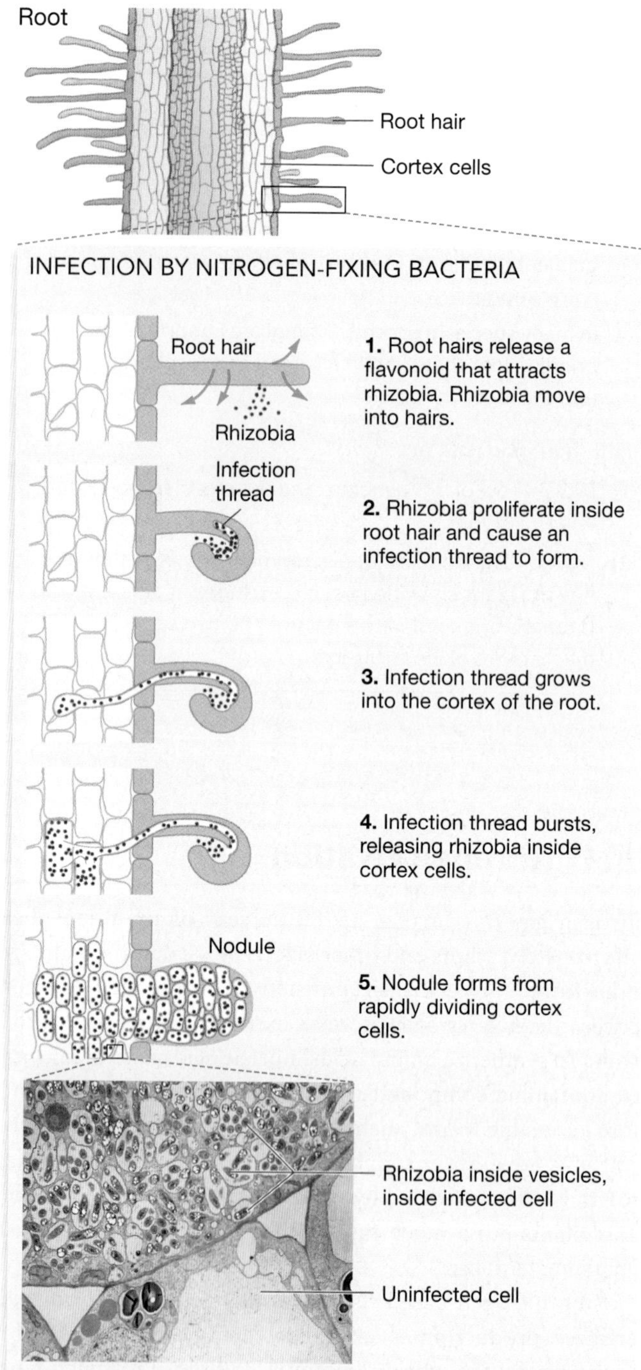

FIGURE 38.15 Infection by Nitrogen-Fixing Bacteria Is a Multistep Process. After rhizobia bind to a root hair, they enter the cytoplasm and travel down an infection thread into the root cortex. If the bacteria enter cortex cells, root cells respond by dividing and forming a nodule.

To summarize, a complex sequence of steps makes it possible for rhizobia to colonize legume root hairs, form nodules, and fix nitrogen. Each phase of the process involves genes in both the host and the symbiont. Identifying all the critical structural genes and regulatory sequences in DNA and introducing them into wheat, rice, or corn will be no easy task.

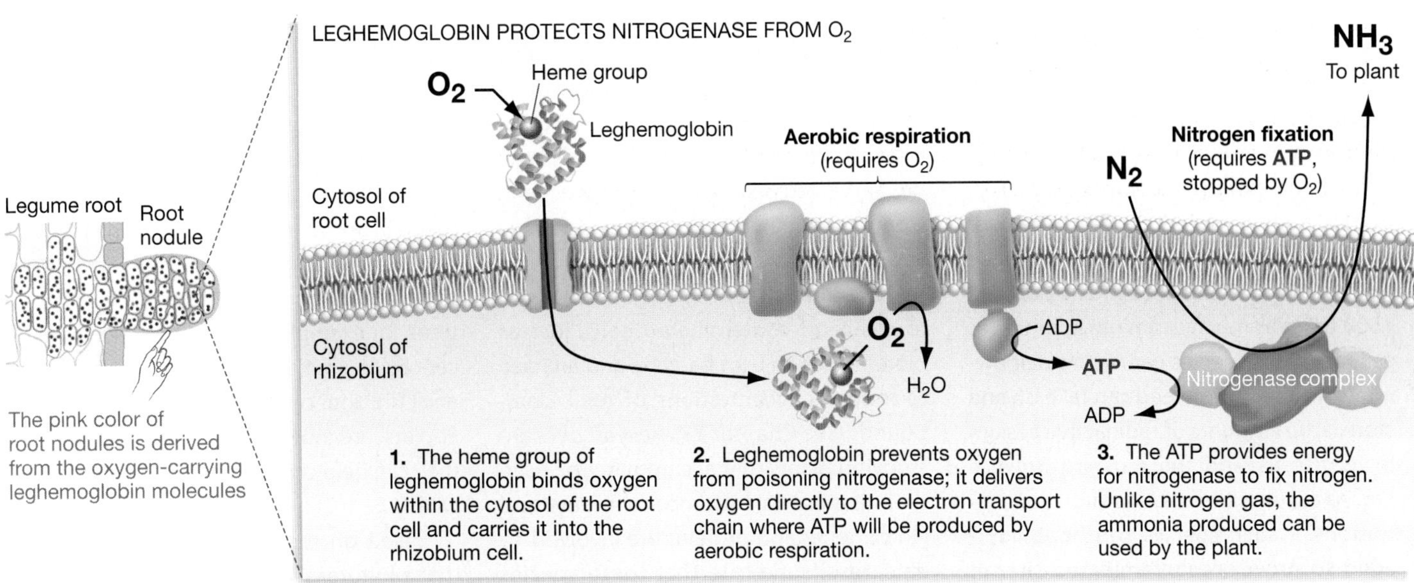

FIGURE 38.16 Leghemoglobin Protects Nitrogenase from O_2.

Although developing a strain of nitrogen-fixing rice or corn would be considered a crowning achievement in biological science, investigators have realized that it will be extremely difficult to replicate the recognition and response steps in nonlegume species.

Check Your Understanding

If you understand that...

- Certain species of bacteria infect plant root cells and fix nitrogen, which is then made available to the plant.

You should be able to...

1) Make a flowchart summarizing the steps in host-symbiont recognition and nodule formation.
2) Generate hypotheses addressing two additional questions about nitrogen fixation: Why don't all plants have symbiotic bacteria that fix nitrogen? Why is the interaction that establishes the infection by nitrogen-fixing bacteria so complex?

38.5 Nutritional Adaptations of Plants

Based on the data available so far, it is reasonable to claim that over 95 percent of vascular plants use proton pumps as a mechanism for importing nutrients from soil. Over 80 percent supplement their "diet" with nutrients acquired from mycorrhizal fungi, and a small but significant fraction associate with nitrogen-fixing bacteria. Perhaps 99 percent of all living plants species make their own sugar through the process of photosynthesis.

What about the small number of plant species that don't follow these rules? Some appear to live on air, some parasitize other plants, and others catch insects and digest them.

Epiphytic Plants

Species from a diverse array of plant lineages do not absorb nutrients from soil. In fact, these species never even make contact with soil. As **Figure 38.17a** shows, they often grow in the leaves or branches of trees. For this reason they are called **epiphytes** ("upon-plants"). The Spanish moss that hangs from oak trees in the southern United States is a familiar example—although this species is actually not a moss but a bromeliad, a relative of the pineapple. In northern forests, it is common for mosses and ferns to grow epiphytically on tree trunks. In the tropics and subtropics, one-third of all ferns grow as epiphytes; and there are thousands of species of epiphytic orchids, bromeliads, and lycophytes.

Epiphytes absorb most of the water and nutrients they need from rainwater, dust, and particles that collect in their tissues or in the crevices of bark. As **Figure 38.17b** shows, some epiphytes have leaves that grow in rosettes and form "tanks" that accumulate and hold water and organic debris. In such cases, nutrients are actually absorbed through the leaves themselves.

Parasitic Plants

Organisms that live in close physical contact with individuals from another species and that lower the fitness of those individuals, usually by obtaining water or nutrients from them, are called **parasites**. Based on the data currently available, biologists estimate that about 3000 species of angiosperms are

BOX 38.1 Can Phytoremediation Help Clean Up Contaminated Soils?

Some species of plants are able to grow on sites contaminated with toxic cleaning agents, the explosive compound TNT, or metals such as cadmium, zinc, nickel, or lead. Alpine pennycress, for example, can accumulate 300 times as much zinc and 1500 times as much cadmium in its leaves as most plant species can, without showing toxic effects. Pigweed can take up and store large amounts of radioactive cesium in its tissues, and the mustard *Brassica juncea* accumulates lead at high concentrations. If such species can be encouraged to grow on contaminated sites, it might be possible to reduce contamination in the soil gradually by harvesting the aboveground plant parts and disposing of them. Some researchers even suggest that if plants accumulate metal ions and can be harvested like hay, the leaves and stems could be burned and the ash used as a source of metal ore.

Phytoremediation is the use of plants to clean up contaminated soils, just as bioremediation uses bacteria and archaea to reduce concentrations of toxic compounds (see Chapter 28). Researchers are studying plants that accumulate toxins in their tissues in an attempt to understand which genes and proteins are involved. To date, results indicate that these species have specialized transport proteins that bring certain ions or molecules into the xylem and subsequently concentrate the toxins in the vacuoles of cells found in leaves and stems.

In addition to species that accumulate toxins in aboveground tissues, some plants have enzymes that act on toxic compounds and render them harmless. Poplar trees, for example, are able to break a cancer-causing molecule called trichloroethylene (TCE) into carbon dioxide and chlorine. Poplars are now being tested to see if they can help clean up sites contaminated with TCE.

Based on recent research, it appears that phytoremediation and bioremediation may turn out to be useful tools in the effort to reduce contamination in at least some types of soils.

parasitic. This number represents less than 1 percent of the total number of plant species that have been studied and named to date.

Some plant parasites are non-photosynthetic and obtain all of their nutrition by tapping into the vascular tissue of the host individual. But most parasitic plants make their own sugars through photosynthesis and tap the xylem of other species for water and essential nutrients. The mistletoe shown in **Figure 38.19** is an example. This parasite is green and photosynthetic but has structures called haustoria that penetrate a host's xylem and extract water and ions that the mistletoe uses as nutrients. In the forests of western North American, mistletoe infection is estimated to cause more economic loss than any herbivore or pathogen (disease-causing agent).

Carnivorous Plants

Carnivorous plants trap insects and other animals, kill them, and absorb the prey's nutrients. Carnivorous species make their own carbohydrates via photosynthesis but use carnivory as a way to supplement the nitrogen available in the environment. Most are found in habitats where nitrogen is scarce or unavailable, such as the bog environments featured in **Box 38.2**.

(a) Epiphytes grow on trees.

(b) Water-holding "tanks" formed by leaves of an epiphytic bromeliad

FIGURE 38.17 Epiphytes Are Adapted to Grow in the Absence of Soil. The environments where epiphytes grow—tree trunks and branches—are dry and nutrient poor.

BOX 38.2 Why Are Bogs So Nutrient Poor?

Bogs are wetlands that have poor drainage and a pH between 3 and 4. They are most common in northern environments, where winters are long and summers are cool. Because bog water is largely stagnant, the rate of oxygen diffusion into the water is often outpaced by the rate of oxygen consumption by aerobic organisms that live on or in the water. Thus, bog water tends to be oxygen poor. The short growing season and cold, oxygen-poor water discourage the growth of fungi and aerobic bacteria that efficiently decompose dead organic matter. As a result, decomposition rates are slow. Bogs gradually fill with the partially decayed organic matter called **peat**.

Bogs are nutrient poor for two reasons. First, few nutrients are released by decomposition. Second, the scarcity of oxygen in the water makes it difficult for root cells to make the large amounts of ATP required to power proton pumps and bring nutrients into the plant against a concentration gradient. Relatively few plant species thrive in bogs. Besides carnivorous plants, bogs are home to sphagnum moss, certain shrubs such as cranberries, and trees such as tamarack and black spruce. Although the remains of all these species contribute to peat, the major component is sphagnum moss.

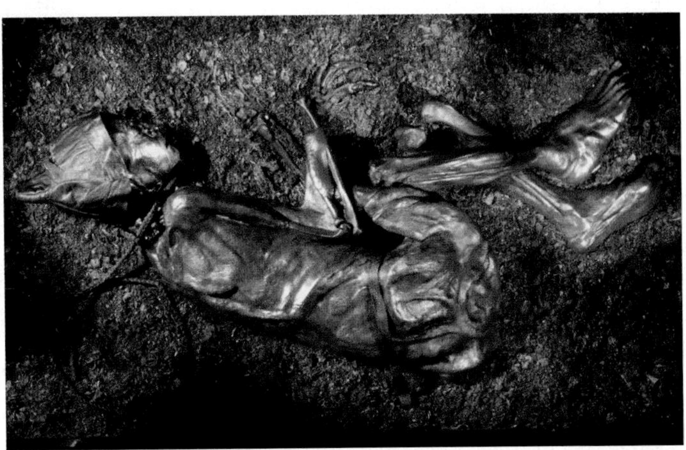

FIGURE 38.18 The Acidic, Low-Oxygen Conditions of Bogs Preserve Materials Well. Hundreds of corpses have been found preserved in bogs. This one, called "Tollund man," was found in a peat bog near Tollund, Denmark.

Dried peat can be burned and has been used for cooking and heating in northwestern Europe for centuries. Occasionally peat-cutters discover objects that have been well preserved by the acidic, low-oxygen conditions. Among these finds are the corpses of hundreds of humans, dating from between 8000 BCE and the early medieval period. The body that researchers call Tollund man is a good example of these "bog people" (**Figure 38.18**). Tollund man is so well preserved that when peat-cutters discovered him in 1950, they thought he'd been murdered recently. He is estimated to be around 2000 years old, however, meaning that he lived during the Early Iron Age. He was naked except for a leather cap and belt. Plant material recovered from his intestines indicated that his most recent meal was a soup made of barley and other plants. A leather rope around his neck suggests that he died by strangulation and was thrown in the bog after being murdered or executed. Many of the bog people, in fact, show clear evidence of a violent death.

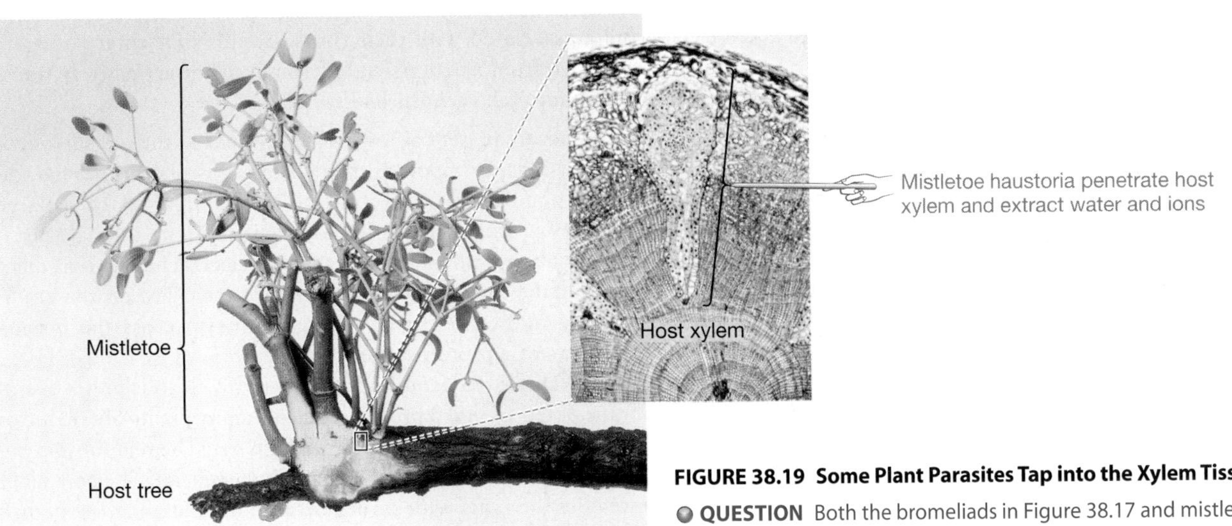

FIGURE 38.19 Some Plant Parasites Tap into the Xylem Tissue of Their Hosts.

● **QUESTION** Both the bromeliads in Figure 38.17 and mistletoe are epiphytes. How are they different in their relationship to the trees on which they grow?

Carnivorous plants use modified leaves or roots to trap insects. Chapter 36 highlighted the leaves of pitcher plants, which form tubes (see Figure 36.14e). Prey are enticed to the tube by an attractive odor, then have a difficult time climbing back out. Eventually they fall into the pool of water below, where they drown. Resident bacteria and fungi then digest the prey, and the plant absorbs the nutrients obtained from their dead bodies.

In sundews, modified leaves develop hairs that exude a sticky substance and function like flypaper (**Figure 38.20**). After insects that become trapped on the surface die, glands near the sundew's trap release enzymes that slowly digest the prey. The leaf then absorbs the nutrients that are released.

FIGURE 38.20 Sundews Have Modified Leaves with a Sticky Surface That Catches Insects.

In the Venus flytrap, modified leaves trap insects mechanically. Each leaf has several sensory hairs protruding from the epidermis. When an insect lands on the trap and bumps two or three of these hairs, the hair cells produce an electrical signal and the leaf responds by snapping shut. If the prey continues to stimulate hairs, the leaf begins to secrete digestive enzymes that slowly digest the prey.

Initially, research on carnivorous plants focused on documenting that meat-eating was adaptive. Early experiments in natural environments supported this hypothesis by showing that, compared to individuals of the same species that are not fed fruit flies, carnivorous plants that are fed fruit flies grow faster and flower more often. Follow-up work focused on exploring the costs as well as the benefits of carnivory. For example, several teams of researchers showed that when carnivorous plants are provided with nitrogen-based fertilizers, they produce fewer of their specialized insect-trapping leaves and a higher proportion of leaves that function primarily in photosynthesis. These results suggest carnivory is a trait that shows phenotypic plasticity (see Chapter 36): Plants increase investment in prey-capture devices when nitrogen is rare, but they decrease investment in prey-capturing structures when nitrogen is more readily available. Currently, some research groups are doing "evo-devo" research (see Chapter 23) on carnivory. They are working to identify the novel alleles that made the evolution of carnivory possible in certain groups. Meat-eating may be relatively rare in plants, but it has inspired fruitful research.

Chapter Review

SUMMARY OF KEY CONCEPTS

In addition to needing carbon dioxide and water, plants require an array of essential nutrients to support growth. These nutrients are available as ions dissolved in soil water and are taken up by roots.

Typically about 96 percent of the dry weight of a plant consists of carbon, hydrogen, and oxygen. Plants obtain these elements by absorbing carbon dioxide from the atmosphere and water from the soil. The other 4 percent of the plant body consists of a complex suite of elements. The most important of these are the essential nutrients, which are required for normal growth and reproduction. Essential elements must be absorbed from the soil, generally in the form of ions like nitrate and phosphate.

Soil is a complex and dynamic mixture of inorganic particles such as clay and sand, organic particles, and living organisms. Soil provides plants with oxygen, water, and nutrients as well as a physical substrate for anchoring and supporting the body.

You should be able to design a field experiment that would determine which of the atoms listed in Table 38.1 are the top three limiting nutrients in a habitat near your campus.

Nutrient absorption occurs via specialized proteins in root membranes. Most plants also obtain nitrogen or phosphorus from fungi associated with their roots. Toxins that enter roots are excluded from xylem by endodermal cells or are actively transported into cell vacuoles and stored.

Nutrients are present at low concentrations in the soil surrounding roots. Plants import nutrients against a concentration gradient by pumping protons into the extracellular space. A large excess of protons outside the root hair creates a strongly negative membrane potential and thus an electrochemical gradient favoring the entry of positively charged ions. The proton gradient also allows negatively charged ions to cross the plasma membrane via proton cotransporters.

In addition to acquiring nutrients by establishing a membrane potential and a proton gradient, many plants obtain nitrogen or phosphorus from mycorrhizal fungi. The relationship between plants and the fungi is mutually beneficial. The host plant provides the fungi with carbohydrates; the fungi harvest N or P from the surrounding soil and transport it to the root system.

Passive and active systems also exist for excluding certain ions. This is important because all nutrients are toxic if they are present at high concentration. Passive exclusion occurs when ions cannot pass through the Casparian strip. Active exclusion often results from the production of specialized proteins that bind a particular ion, or the active transport of toxic ions into vacuoles.

You should be able to predict the consequences of mutations that (1) allowed root cells to establish a membrane voltage that was much more negative than normal, or (2) prevented root cells from establishing a highly negative membrane voltage.

MB Web Animation at www.masteringbio.com
Soil Formation and Nutrient Uptake

Some species of plants have specialized methods of obtaining nutrients, including associations with nitrogen-fixing bacteria, parasitism, and carnivory.

Certain plants, including those in the pea family, are capable of symbiosis with nitrogen-fixing bacteria. This association begins when flavonoids produced by the host plant are detected by bacteria in the soil. In response, bacterial cells produce sugar-containing molecules called Nod factors, which bind to a receptor protein on a root hair. A suite of host-plant genes is transcribed in response to this binding event, and the proteins that are produced result in the formation of a root nodule. The bacteria eventually take up residence inside the root cells that form the nodule, where they receive protection and sugar in exchange for producing ammonia from nitrogen gas.

Not all plants take up nutrients through membrane proteins or bacterial or mycorrhizal associations. Instead, some species are parasitic, and some are carnivorous. Most parasitic plants produce their own carbohydrates through photosynthesis but steal water and nutrients by infecting the xylem of host plants. Carnivorous plants have evolved mechanisms for trapping and digesting insects. The primary benefit of carnivory is to obtain nitrogen in low-nitrogen environments.

You should be able to design an experiment, using radioactive carbon and the heavy isotope of nitrogen (^{15}N), that would test whether the rhizobia-pea plant interaction is mutualistic. Your experiment should document the nature of the association when nitrate and ammonium ions are (1) scarce or (2) abundant in the soil.

QUESTIONS

Test Your Knowledge

1. Which of the following characteristics defines an element as essential for a particular species?
   a. It has to be added as fertilizer to achieve maximum seed production.
   b. If it is missing, a plant cannot grow or reproduce normally.
   c. If it is present in high concentration, plant growth increases.
   d. If it is absent, other nutrients may be substituted for it.
2. Why is the presence of clay particles important in soil?
   a. They provide macronutrients—particularly nitrogen, phosphorus, and potassium.
   b. They bind metal ions, which would be toxic if absorbed by plants.
   c. They allow water to percolate through the soil, making oxygen-rich air pockets available.
   d. The negative charges on clay bind to positively charged ions and prevent them from leaching.
3. Where does most nutrient uptake occur in roots?
   a. at the root cap, where root tissue first encounters soil away from the zone of nutrient depletion
   b. at the Casparian strip, where ions must enter the symplast prior to entering xylem cells
   c. in the symplastic and apoplastic pathways
   d. in root hairs, in the zone of maturation
4. Why is the activity of proton pumps in root hair membranes important?
   a. They create a strongly positive membrane potential (voltage).
   b. They allow toxins to be concentrated in vacuoles, so the toxins do not poison enzymes in the cytoplasm.
   c. They set up an electrochemical gradient that makes it possible to absorb cations and anions.
   d. They set up the membrane voltage required for action potentials to occur.
5. Phytoremediation is based on which of the following properties?
   a. Fungal hyphae have an extremely large surface area, and they secrete enzymes that digest dead plant material efficiently.
   b. Some plants can take up toxins from the soil and concentrate them in their tissues.
   c. Nitrogen-fixing bacteria live symbiotically *inside* root cortex cells, which divide to form protective nodules.
   d. The cold, oxygen-poor conditions in bogs slow decomposition rates, allowing peat to form.
6. How do epiphytes obtain nutrients?
   a. by trapping and digesting insects
   b. from their host plant
   c. via symbiotic bacteria called rhizobia
   d. from dust or other airborne particles

Test Your Knowledge answers: 1. b; 2. d; 3. d; 4. c; 5. b; 6. d

Test Your Understanding

Answers are available at www.masteringbio.com

1. A farmer is concerned that her corn crop is suffering from iron deficiency. Design a greenhouse experiment that uses hydroponic cultures to (1) identify the symptoms of iron deficiency in corn and (2) identify the minimum amount of iron required to support normal growth rates.
2. Explain why carnivorous and parasitic plants are most common in nutrient-poor habitats.

3. Sandy soils contain very little clay. They are often nutrient poor because of extensive leaching and tend to be very dry. Why? Would you expect plant productivity to be higher on sandy soils or on soils containing both sand and clay? Why?
4. Sketch the plasma membrane of a root hair. Include proton pumps, ion channels, and cotransporters. Now add ATP to the system, then protons in their correct relative concentration, and ions that move across the membrane via channels or cotransporters. Explain why plant cells have a negative membrane potential.
5. Why is it important for plants to exclude certain ions? Summarize the difference between active and passive exclusion mechanisms.
6. Fill in the table below.

Nutrient	How Used in Cell?	What Form Available to Plants?	How Obtained by Plants?
Nitrogen			
Phosphorus			
Carbon			
Oxygen			
Hydrogen			
Iron			
Magnesium			

Applying Concepts to New Situations

Answers are available at www.masteringbio.com

1. There is a conflict between van Helmont's data on willow tree growth and the data on essential nutrients listed in Table 38.1. According to the table, nutrients other than C, H, and O should make up about 4 percent of a willow tree's weight. Most or all of these nutrients should come from soil. But van Helmont claimed that the soil in his experiment lost just 2 ounces, while the tree gained 2627 ounces. If so, then soil contributed just 0.08 percent of the added weight instead of 4 percent. List three hypotheses to explain the conflict. How would you test these hypotheses?
2. Acid rain occurs when sulfur oxides and nitrous oxides released by cars or factories react with water vapor, forming sulfuric and nitric acids, respectively. In areas affected by acid rain, cations quickly leach out of the soil. Explain why.
3. Design an experiment using the drug vanadate, which poisons proton pumps, to test the hypothesis that phosphate ions enter cells via an H^+-HPO_4^{2-} cotransporter.
4. Spruce trees normally associate with mycorrhizal fungi. Predict the consequences of fertilizing a stand of spruce with nitrogen and phosphorus in terms of the extent of the mycorrhizal association. Explain your logic.

www.masteringbio.com is also your resource for • Answers to text, table, and figure caption questions and exercises • Answers to *Check Your Understanding* boxes • Online study guides and quizzes • Additional study tools including the *E-Book for Biological Science* 3rd ed., textbook art, animations, and videos.

Plant Sensory Systems, Signals, and Responses

39

Plants have sophisticated information processing systems. These radish seedlings sense the presence of light and are bending toward it.

KEY CONCEPTS

- Plants are selective about the information they process. They perceive a wide variety of environmental stimuli that affect their ability to grow and reproduce.
- When sensory cells receive a stimulus, they transduce the signal and respond by producing hormones that carry information to target cells elsewhere in the body.
- Target cells respond to hormonal stimulation in ways that increase the ability of the plant to survive and reproduce.
- Hormones are also responsible for regulating how plants grow throughout their lives—especially in response to changes in environmental conditions. Each type of plant growth regulator plays a general role in the life of a plant, and growth responses are usually affected by interactions among several different hormones.

It is difficult for a human being to understand the sensory world of a plant. But imagine standing in place for several hundred years, like an oak tree. Each spring you produce flowers. All summer you absorb light from the Sun, carbon dioxide from the atmosphere, and water and nutrients from the soil. Water, ions, and sugars flow up and down your vascular tissue. For six months or more your body grows upward, outward, and downward. In fall, thousands of your offspring drop to the ground as acorns. Then as winter approaches, your metabolism slows and you stop growing. Like a hibernating animal, you spend the long, hard months of winter in a state of suspended animation, before awakening the following spring.

To stay alive, an oak needs to gather information about its environment. It has to be able to sense the season of the year, the time of day, the pull of gravity, the force of wind, and attacks by viruses, bacteria, or insects. It needs to sense not only when its leaves are being shaded but also which leaves are receiving the wavelengths of light that are required to support photosynthesis. Plants may not have eyes or ears, but they can sense light, gravity, pressure, and wounds. They have the equivalent of a sense of smell, because they can perceive certain airborne molecules. It could even be argued that they have a sense of taste, because their roots sense the presence of nutrients in the soil.

In addition to gathering information about the conditions around them, plants have to respond in an appropriate way. They do not jump or swim or run, but their shoot systems grow toward light or become shorter and stockier in response to wind. After measuring the length of nights, plants use the information to determine whether they will flower. In response to gravity, shoots grow up and roots grow down. In response to touch, the modified leaves of a Venus flytrap shut fast enough to catch flying insects. If a plant is being attacked or if it senses that a neighboring individual is under attack, it may lace its tissues with toxic compounds or mobilize other defenses.

The fundamental message of this chapter is simple: Plants have sophisticated systems for collecting information about their environment and for responding in ways that maximize their chances of surviving, thriving, and producing offspring.

39.1 An Overview of Information Processing

Every environment is full of information. But, like other organisms, plants monitor only aspects of the environment that matter to them—that affect their ability to stay alive and reproduce.

Figure 39.1 provides an overview of how plants gather, process, and respond to the information they monitor. The sequence of events can be analyzed in three steps: (1) a receptor cell receives an external signal and transduces it to an intracellular signal; (2) the receptor cell sends a signal to cells in another part of the body that can respond to the information; (3) responder cells receive this long-distance signal, transduce it to an intracellular signal, and change their activity in a way that produces an appropriate response. Let's briefly consider how each of these steps works, then delve into the details of how plants sense and respond to light, gravity, and other types of information.

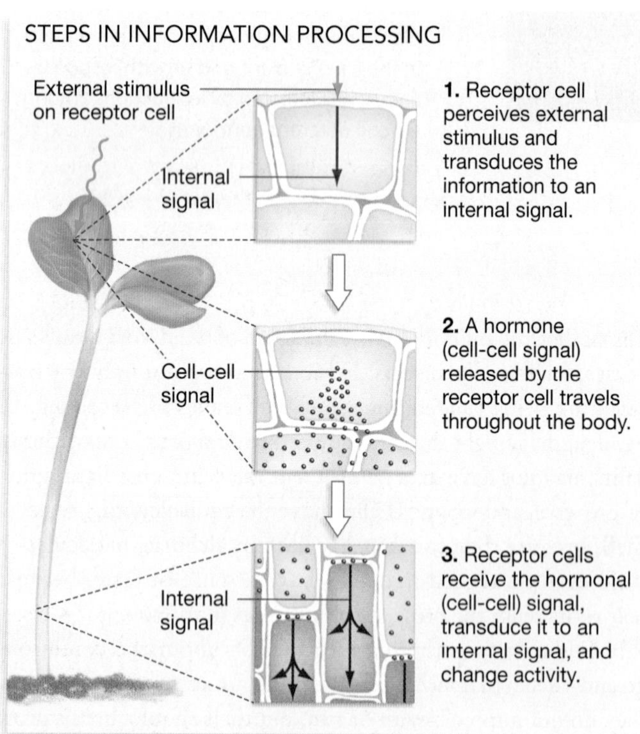

FIGURE 39.1 Steps in Information Processing. Receptor cells perceive environmental stimuli and, via cell-cell signals called hormones, communicate the information to cells that can respond in an appropriate way.

How Do Cells Receive and Transduce an External Signal?

When you text-message a friend, the signal travels from you to the receiver via airwaves. When the message arrives, the receiving cell phone changes the information in the airwaves into electrical signals and then into words that your friend can understand. Plants work in much the same way. When light strikes a receptor cell, for example, the information that it carries has to be changed into a form that is meaningful to the receiving cell. Signals from the environment are usually received by a protein specialized for that function. Receptor proteins change shape in response to being struck by a particular wavelength of light, having pressure applied, or binding to a particular type of airborne molecule.

When a receptor changes shape in response to a stimulus, the information changes form—from an external signal to an intracellular signal. This process is called **signal transduction** (**Figure 39.2**). The name was inspired by the verb transduce, which means "to convert energy from one form to another." Once the information has been transduced to an intracellular form, it begins to travel down what biologists call a signal transduction pathway.

You might recall from Chapter 8 that there are two basic types of signal transduction pathways: phosphorylation cascades and second messengers. Both begin with a receptor protein in the plasma membrane.

- **Phosphorylation cascades** are triggered when the change in the receptor protein's shape leads to the addition of a phosphate group (PO_4^{3-}) from ATP to the receptor or a nearby protein. To understand why this is significant, recall from Chapters 8 and 9 that many proteins switch from inactive to active states, or vice versa, when a phosphate group is added to them or removed—when phosphorylation or dephosphorylation occurs. As the left side of Figure 39.2 shows, phosphorylation events activate proteins involved in signal transduction cascades and cause them to phosphorylate and activate a different set of proteins, which in turn catalyze the phosphorylation and activation of still other proteins, and so on.
- **Second messengers** are produced when hormone binding results in the release of an intracellular signal from storage areas. As the right side of Figure 39.2 indicates, calcium ions (Ca^{2+}) stored in the vacuole usually function as the second messenger in plants.

Signal transduction primes the receptor cell for action. In most cases, though, the cells that receive information from the environment are located in a part of the body that is

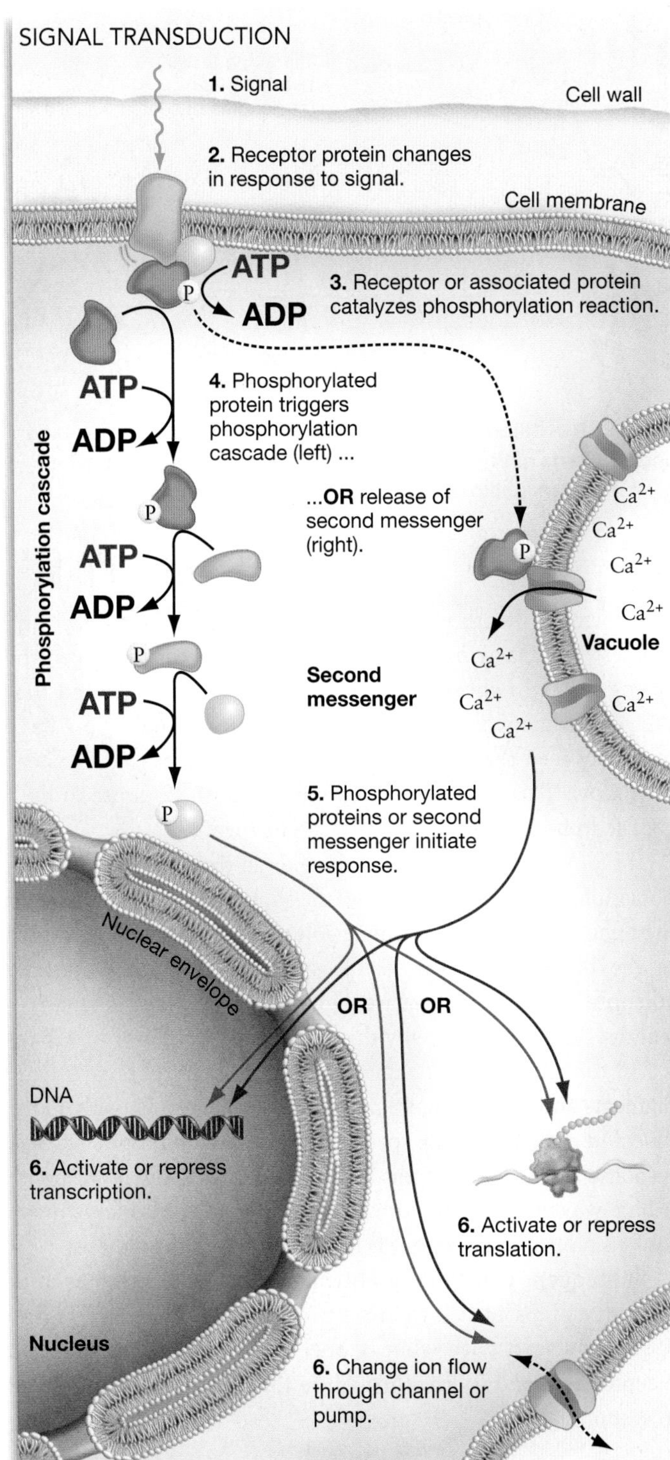

FIGURE 39.2 Signal Transduction Changes an External Signal to an Internal Signal. When a signal is transduced, it changes form—from an environmental or cell-cell signal to an intracellular signal. The intracellular signal then triggers a change in the cell's activity.

distant from the cells that need to respond to the information. How does information from an activated receptor cell get to responder cells?

In most cases, the answer is a **hormone**—an organic compound that is produced in small amounts in one part of a plant and transported to target cells in another region of the individual, where it causes a physiological response. Signal transduction in a receptor cell often results in the release of a hormone that carries information to responder cells.

How Are Cell-Cell Signals Transmitted?

Because plants perceive such a wide array of stimuli, they have a wide array of hormones coursing through their bodies. In most cases these signals act on a variety of target tissues throughout the body. Target cells routinely receive information from several different hormones at the same time, so it is common for different types of hormones to interact with each other and modulate the cell's response.

The molecules that function as hormones are wildly diverse in structure. They range from ethylene (C_2H_4)—an extraordinarily simple molecule that is a gas at room temperature—to the lipids called steroids (see Chapter 6) to small peptides. Plant hormones have several important properties in common, however: (1) They can elicit a response only if a cell has an appropriate receptor, and (2) they are active at extremely small concentrations. To understand why, let's look at what happens when a hormone arrives at a cell.

How Do Cells Respond to Cell-Cell Signals?

Cells are exposed to a constant stream of hormones, many of which have little to no effect on what is happening inside. But if a receptor on or in the cell—usually a protein located in the plasma membrane—binds to a hormone and changes shape in response, the effect is like a knock on the door in the middle of the night. The signal at the cell's periphery is rapidly transduced to increased activity inside the cell, via a signal transduction pathway. In this case, however, the phosphorylated proteins or second messengers that result from the signal transduction cascade don't result in the production of a hormone that is then released from the cell. Instead, the cell itself changes activity.

To understand why a few of these binding events can have dramatic effects on cell activity, look again at Figure 39.2. Activation of a signal transduction cascade results in the production of many phosphorylated proteins or the release of many second messengers. In this way, the original signal is amplified many times. The result is that tiny concentrations of plant hormones can have a large impact on target cells. Hormones are tiny molecules in tiny concentrations, but they produce big results.

As Figure 39.2 shows, the response to hormone binding can include activation of membrane transport proteins (Chapter 3), the translation machinery (Chapter 16), or transcription activators or repressors (Chapter 18). The end result may be a change in ion flows into or out of the cell—and thus a change in its electrical potential or pH—or the production of new proteins or RNAs. Hormones can activate existing proteins or stimulate the production of new proteins. In this way, a signal received at a distant target cell triggers a response. When cells respond to a hormone, the change in their activity helps the plant cope with the environmental change sensed by the receptor cell.

Responses are the end-point of information processing for plants, but they are the starting point for research: To show that an organism can perceive a particular aspect of its environment, researchers have to document a response to it. Once it is clear that plants can respond to gravity or water stress or other environmental conditions, researchers can start trying to figure out how the information is received and processed. Let's delve in, beginning with how plants respond to blue light.

Check Your Understanding

If you understand that...

- Plants can respond to changes in their environment because sensory cells receive information from the environment, sensory cells produce hormones or other cell-cell signals, and target cells respond to hormones.
- Information processing in sensory cells and the cells that respond to cell-cell signals involves three steps:
  (1) A receptor molecule changes in response to the stimulus.
  (2) A signal transduction pathway transforms the stimulus into an intracellular signal.
  (3) The intracellular signal triggers a response—the transcription of target genes, the activation of specific enzymes, or other changes in cell activity.

You should be able to...

1) Diagram the reception, cell-cell signal, and target cell response components of a system that allows plants to sense and respond to a specific wavelength of light. Label two locations where signal transduction occurs.
2) Add a "higher magnification" drawing to your diagram that describes a generalized signal transduction pathway beginning with a light receptor protein in the plasma membrane.
3) Explain why only certain cells in the body—not all—respond to an environmental signal, and why only certain cells respond to a hormone.

39.2 Blue Light: The Phototropic Response

Most of the general principles of plant communication emerged from studies of how plants respond to light. For example, consider the claim that plants are highly selective about the information they process. Light is made up of a wide array of wavelengths (**Figure 39.3**), but plants sense and respond to only a few.

This conclusion traces back to experiments that Charles Darwin and his son Francis did in 1881 with coleoptiles of a plant called reed canary grass. A **coleoptile** is a modified leaf

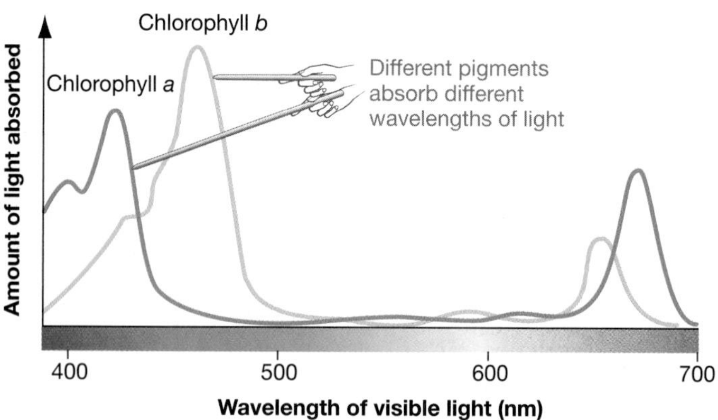

FIGURE 39.3 Chlorophyll *a* and *b* Absorb Most Strongly in the Blue and Red Parts of the Visible Spectrum. Chlorophyll *a* and *b* are the most important photosynthetic pigments in land plants.

that forms a sheath protecting the emerging shoots of young grasses. The Darwins germinated seeds in the dark, placed the young, straight shoots next to a light source, and noted that the shoots bent toward the light (**Figure 39.4a**). You have probably seen the same response in houseplants that are near a window. This type of directed movement in response to light is called **phototropism** (literally, "light-turn").

If they exposed shoots to light filtered through a solution of potassium dichromate, however, the shoots did not bend toward the light. Potassium dichromate solutions filter out wavelengths in the blue part of the visible spectrum—meaning that the coleoptiles bent only toward light that contained blue wavelengths. To understand why this is important, recall from Chapter 10 that chlorophylls *a* and *b* are the primary photosynthetic pigments and note in Figure 39.3 that they absorb strongly in the blue and red parts of the spectrum. Plants exhibit a phototropic response if blue wavelengths are available, but show no response if blue wavelengths are not present (**Figure 39.4b**). Plants move toward blue light because it is important in photosynthesis.

Subsequent work on phototropism has illuminated each of the three steps in information processing outlined in Figure 39.1. Let's start with what happens when blue light strikes a receptor cell, and follow through to the cells that actually bend in response.

Phototropins as Blue-Light Receptors

Although biologists knew that the blue-light receptor must be a **pigment**—a molecule that absorbs certain wavelengths of light—it took decades to find it. A key breakthrough finally came in the early 1990s, when researchers found a membrane protein that is abundant in the tips of emerging shoots and that gains a phosphate group in response to blue light. Researchers hypothesized that the membrane protein becomes activated when it is phosphorylated in response to blue light and that the activated protein then triggers the phototropic response.

(a) Shoots bend toward full-spectrum light.

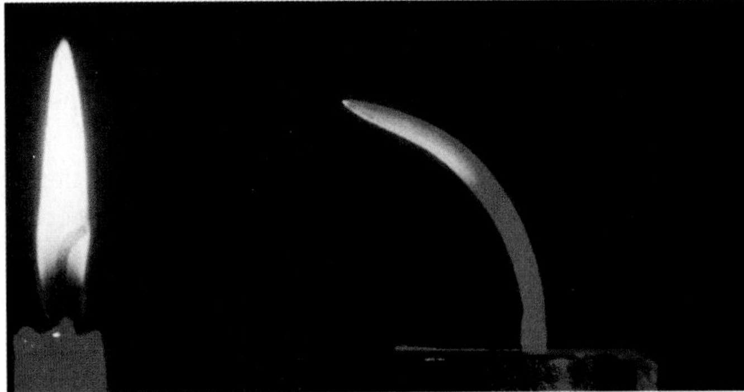

(b) Shoots bend specifically toward blue light.

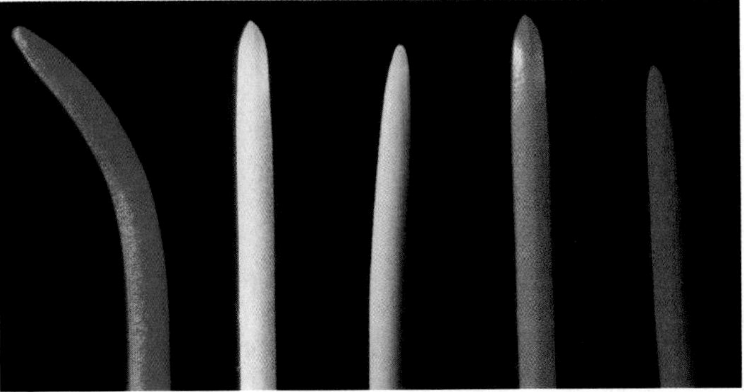

FIGURE 39.4 Experimental Evidence That Plants Sense Specific Wavelengths of Light. Though a candle is shown here, the Darwins used paraffin lamps or other sources of artificial light in their original experiments.

Subsequent work succeeded in isolating the gene that codes for the membrane protein. The gene was named *PHOT1*; it was found by analyzing mutant *Arabidopsis thaliana* individuals that do *not* show a phototropic response to blue light. **Figure 39.5** summarizes the experiment that convinced most biologists that *PHOT1* codes for a blue-light receptor. When researchers inserted copies of the *PHOT1* gene into insect cells that were growing in culture, they found that the PHOT1 protein product became phosphorylated in response to blue light. Because no other plant proteins were present in the experimental insect cells, the results suggest that PHOT1 phosphorylates itself in response to blue light. This result has been supported by follow-up experiments. The current consensus is that *PHOT1* encodes a blue-light detector in plants, and that a phototropic response is initiated when this receptor protein is phosphorylated (**Figure 39.6**). Researchers are currently investigating how the receptor is arranged in the plasma membrane.

Even more recent research indicates that there are multiple blue-light receptors related to PHOT1. Collectively, photoreceptors that detect blue light and initiate phototropic responses are known as **phototropins**. It's important to note, though, that not all blue-light receptors function in photo-

Experiment

Question: Does *PHOT1* encode a blue-light receptor?

Hypothesis: The *PHOT1* gene codes for a blue-light receptor.

Null hypothesis: The *PHOT1* gene does not code for a blue-light receptor.

Experimental setup:

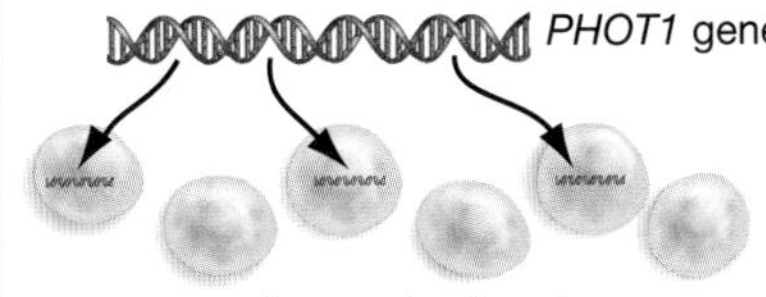

1. Insert *PHOT1* genes into half of insect cells in culture. These cells should produce PHOT1 protein. The other half serve as the control.

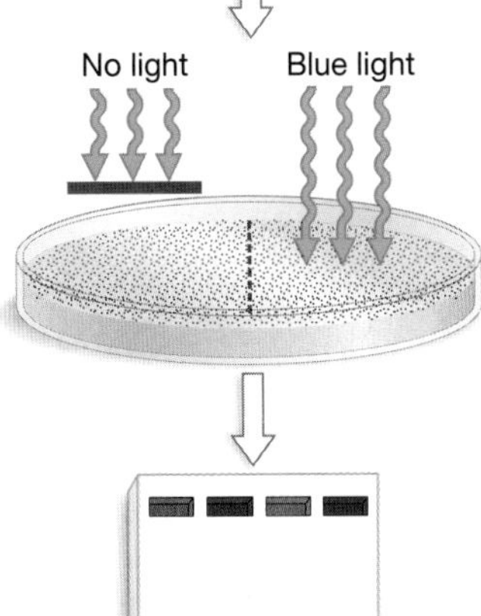

2. Grow cells in a medium containing radioactive phosphorus, and subject cells to darkness or to blue light.

3. Isolate proteins and separate via electrophoresis. See BioSkills 7 for an introduction to this technique.

Prediction: Insect cells containing the *PHOT1* gene will have a radiolabeled band when treated with blue light.

Prediction of null hypothesis: Insect cells containing the *PHOT1* gene, like control cells, will not have a radiolabeled band.

Results:

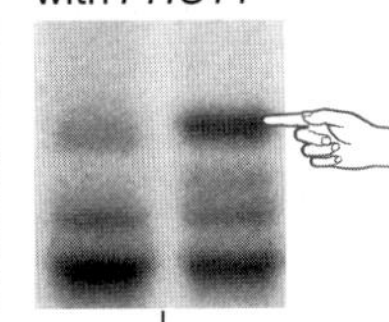

Conclusion: PHOT1 is phosphorylated in response to blue light. Because no other plant proteins were present in the experimental cells, PHOT1 must phosphorylate itself.

FIGURE 39.5 Experimental Evidence That PHOT1 Is a Blue-Light Receptor That Phosphorylates Itself. The results show that the transformed insect cells produced the PHOT1 protein and that the proteins became phosphorylated if the cells were exposed to blue light.

● **QUESTION** Why did the researchers bother to analyze cells that were not transformed with the *PHOT1* gene and that were not exposed to blue light?

BOX 39.1 Plants Have an Array of Receptors and Responses to Blue Light

Because of its importance to photosynthesis, it's not surprising that blue light triggers a variety of responses in plants in addition to bending. For example, when phototropins are activated by blue light, they trigger a signal transduction cascade that results in chloroplast movements inside leaf cells. These movements put chloroplasts in positions that make optimal light absorption possible.

What's more, phototropins are not the only types of blue-light receptors. The carotenoid pigment zeaxanthin, for example, initiates the opening of stomata in response to blue light (phototropins are also involved in this response). As a result, carbon dioxide can diffuse into cells as blue light triggers photosynthesis. A group of photoreceptors called cryptochromes ("hidden-colors") are blue-light photoreceptors involved in stem growth under shady conditions, in flower induction, and in monitoring day length.

The central messages here are that plants have an array of blue-light receptors, and that each receptor triggers a response that increases a plant's ability to harvest light, make food, and produce offspring.

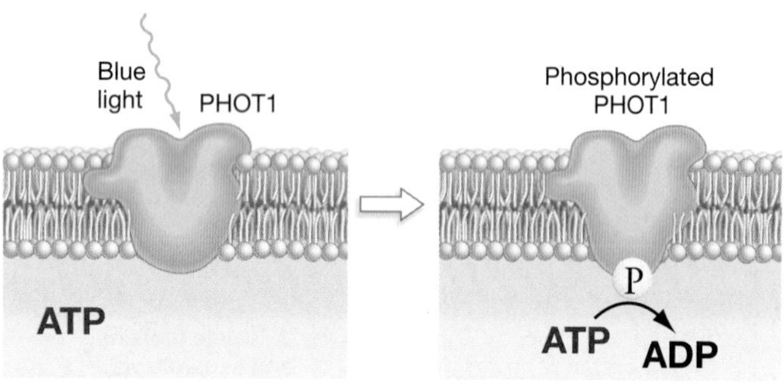

FIGURE 39.6 Phototropins Autophosphorylate when They Absorb Blue Light. When phototropins absorb blue light, they change shape in a way that catalyzes the addition of a phosphate group. The process is called autophosphorylation because it occurs without involvement from an associated protein.

● **QUESTION** Where does this event fit into the signal transduction cascade outlined in Figure 39.2?

tropism. As **Box 39.1** indicates, phototropins and other blue-light receptors trigger an array of responses in plants.

Auxin as the Phototropic Hormone

Long before PHOT1 and the other phototropins were identified, biologists knew that receptor cells responded to blue light by releasing a hormone. The Darwins established this result when they followed up on their initial experiments. When they removed the tips of coleoptiles, they found that the decapitated seedlings stopped bending toward the light (**Figure 39.7**). To extend this observation, they covered the tips of coleoptiles with opaque covers or put opaque collars below the tips, in the area where bending occurs. The coleoptiles with caps did not bend, but the coleoptiles with collars did. These data provided convincing evidence that the blue-light sensors were located in the tips.

How did the sensory cells in the tip communicate with the cells that actually elongate? The Darwins proposed that phototropism depends on "some matter in the upper part which is acted on by light, and which transmits its effects to the lower part." Their hypothesis was that a substance produced at the tip of the coleoptile acts as a signal and is transported to the

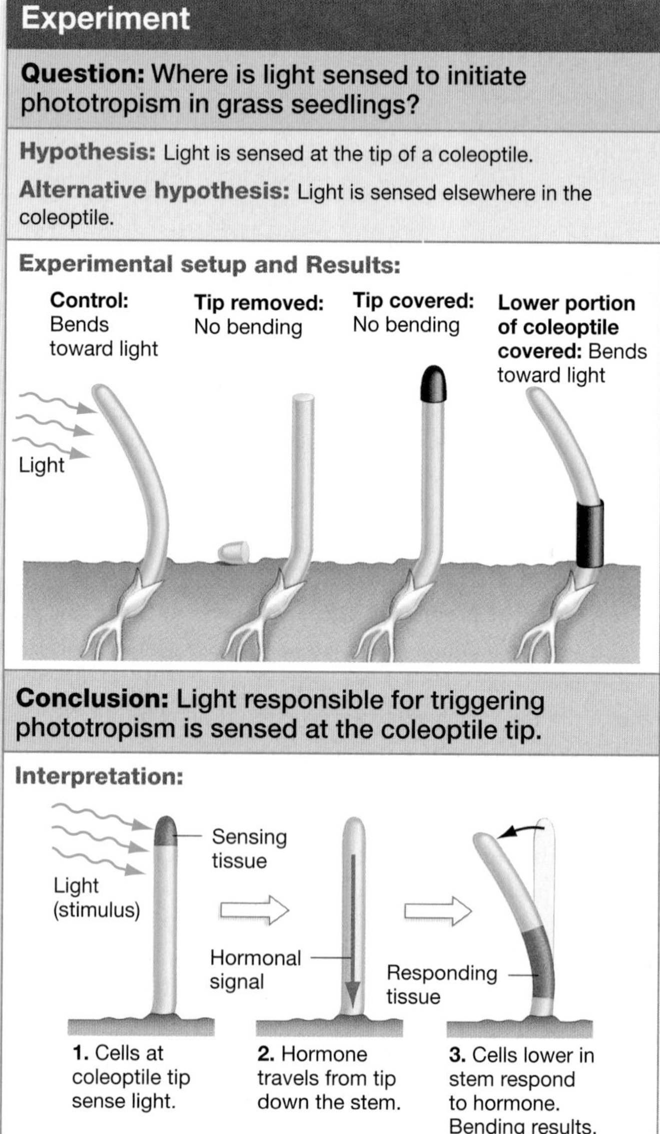

FIGURE 39.7 The Sensory and Response Cells Involved in Phototropism Are Not the Same.

● **QUESTION** A critic could argue that this experiment lacked appropriate controls for the treatments labeled "Tip removed" and "Tip covered." Suggest better controls for these treatments than the unmanipulated individual at the far left.

area of bending. This was the first explicit hypothesis stating that hormones must exist.

The hormone hypothesis was not tested rigorously until 1913, when Peter Boysen-Jensen published the experiments summarized in **Figure 39.8a**. Boysen-Jensen cut the tips off young oat shoots and placed either a porous block of the gelatinous compound called agar or a nonporous substance between the tip and the shoot. Because only the stems treated with the porous agar block showed normal phototropism, Boysen-Jensen concluded that the phototropic signal was indeed a chemical and that it could diffuse. Further, because the agar that he used was a water-based gelatin, he determined that the molecule was water soluble.

Twelve years later, Frits Went extended these results. He placed the decapitated tips of oat coleoptiles on agar blocks, with the goal of collecting the hypothesized hormone for phototropism (**Figure 39.8b**). Then he placed agar blocks that had or had not been exposed to oat tips off-center on the decapitated coleoptiles of other individuals. If the agar blocks had been exposed to oat tips, then the stems responded by bending—even though the stems were kept in the dark during the entire experiment. In this way, Went succeeded in producing the phototropic response without the stimulus of light. Because it promotes cell elongation in the shoot, Went named the hormone **auxin** (from the Greek *auxein*, "to increase"). Auxin was the first plant hormone ever discovered.

(a) The phototropic signal is a chemical.

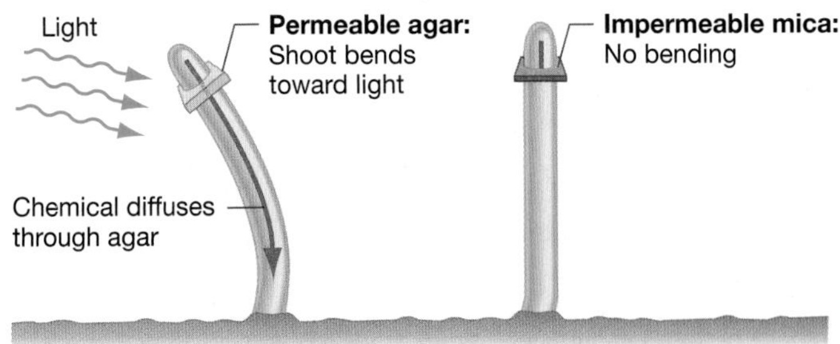

(b) The hormone can cause bending in darkness.

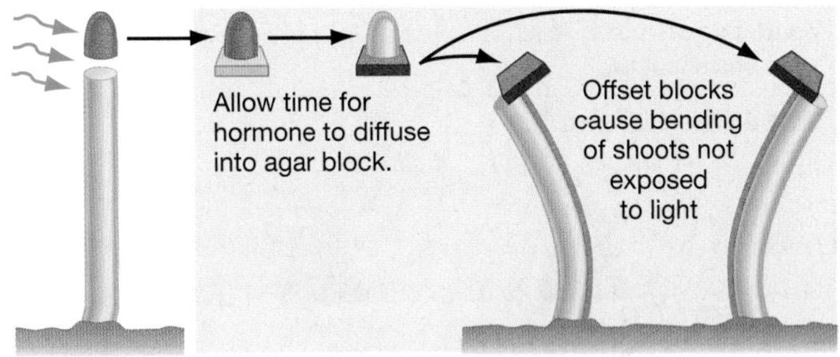

(c) The hormone causes bending by elongating cells.

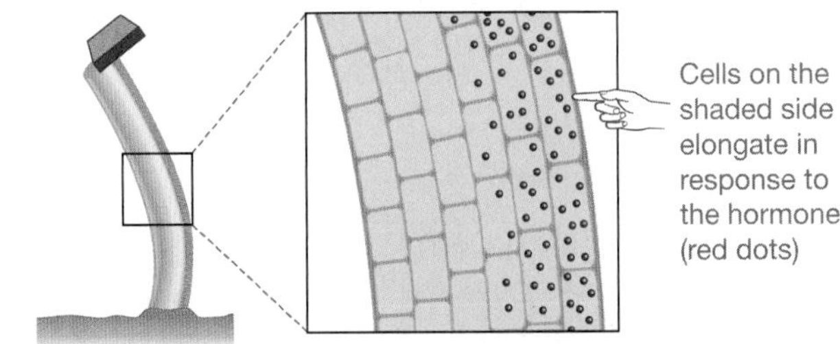

FIGURE 39.8 Experimental Evidence Supports the Hormone Hypothesis for Phototropism. (a) Coleoptiles bend in response to light if substances from the tip are allowed to move downward. **(b)** If bending can take place in darkness, then light is not directly required for the response. Only the hormone is required. **(c)** During the phototropic response, bending occurs because cells on the shaded side of the shoot elongate.

Isolating and Characterizing Auxin

Went's work gave researchers a place to look for the phototropic hormone—it was somewhere in the agar blocks used in his experiments—along with a way to identify it. If a molecule could be found that produced the bending response in decapitated shoots, it was likely to be the auxin hormone.

After years of effort, researchers in two laboratories independently succeeded in isolating and characterizing auxin. The hormone turned out to be indole acetic acid, or IAA. In a typical plant, about 50 ng of IAA are present for every 50 g of fresh tissue. This is a concentration of about 280 nM (nanomolar). (The prefix *nano*– refers to billionths.)

Why are these observations about auxin's structure and concentration interesting? Like the other plant hormones introduced in this chapter, auxin is a small molecule with a relatively simple structure. It is present in quantities so small that its concentration is difficult to measure. Yet its impact is huge. Auxin can bend stems, in some cases leading to tree trunks that are permanently bowed.

The Cholodny-Went Hypothesis

Went's experiments were a breakthrough in research on information processing—they confirmed the hormone hypothesis and led to the discovery and characterization of IAA. But Went's experiments also inspired an important hypothesis for *how* the hormone produces the bending response. Working independently, both N. O. Cholodny and Went proposed that phototropism results from an asymmetric distribution of auxin. The Cholodny-Went hypothesis contends that auxin produced in the tips of coleoptiles is shunted from one side of the tip to the other in response to light. The auxin is then transported straight down one side of the shoot. Because auxin concentration is higher on one side than the other, cells on one side of the coleoptile elongate more than cells on the other side. Bending results, as shown in **Figure 39.8c**.

Other researchers proposed a simpler hypothesis, however. They suggested that auxin is broken down or otherwise inactivated by blue light. If so, then an asymmetric distribution would result not from a transport process but from the destruction of auxin on one side of the tip.

To test these alternatives, Winslow Briggs grew corn seedlings in the dark, cut off their tips, and placed the tips on agar blocks. As **Figure 39.9** shows, he either kept the tips and agar blocks in the dark or exposed them to light from one side. Then he put the agar blocks from each treatment on one side

Experiment

Question: How does an asymmetric distribution of auxin in shoot tips, which causes bending, develop?

Hypothesis: Auxin is destroyed on the sunny side of a shoot tip, resulting in an asymmetric distribution.

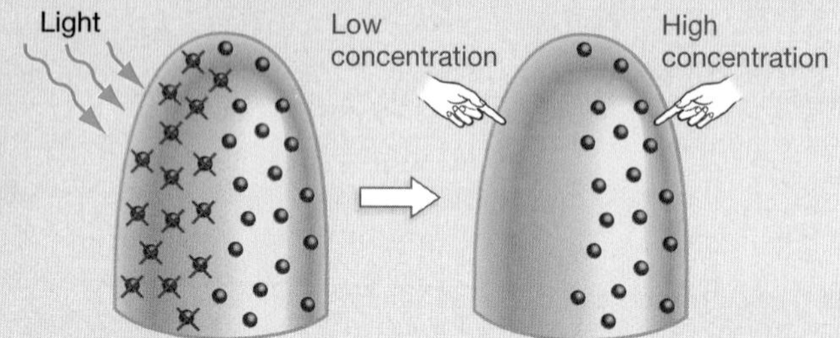

Null hypothesis: Auxin is not destroyed by light on the sunny side of a shoot tip.

Experimental setup:

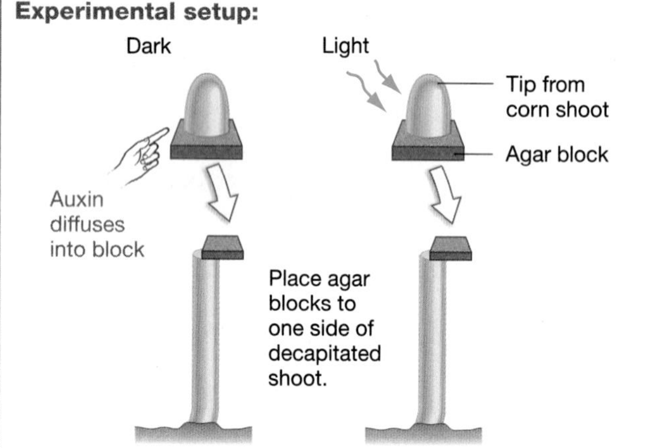

Prediction: The agar block beneath the light-treated shoot tip will elicit a smaller bending response.

Prediction of null hypothesis: Agar blocks from dark-treated and light-treated shoot tips will lead to similar bending responses.

Results:

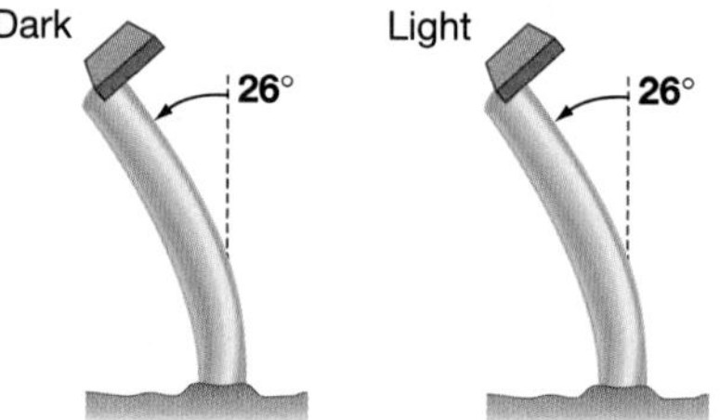

Bending response is similar in each treatment.

Conclusion: Auxin is not destroyed by light on the sunny side of a shoot tip. Null hypothesis is supported.

FIGURE 39.9 Experimental Test of the Auxin Destruction Hypothesis.

Experiment

Question: How does an asymmetric distribution of auxin in shoot tips, which causes bending, develop?

Hypothesis: Auxin moves from the sunny side to the shady side of shoot tip, resulting in an asymmetric distribution.

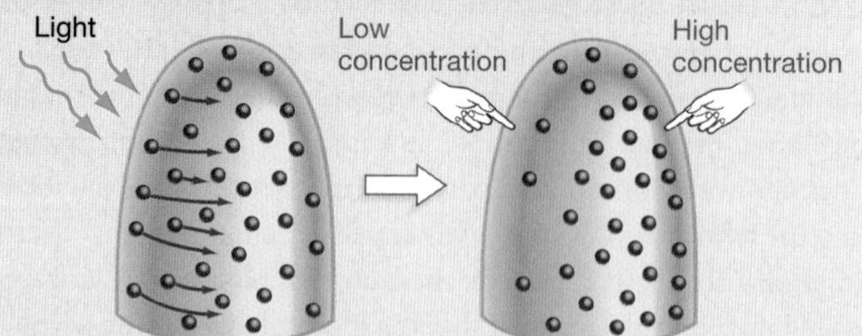

Null hypothesis: Auxin does not move from the sunny side to the shady side of the shoot tip.

Experimental setup:

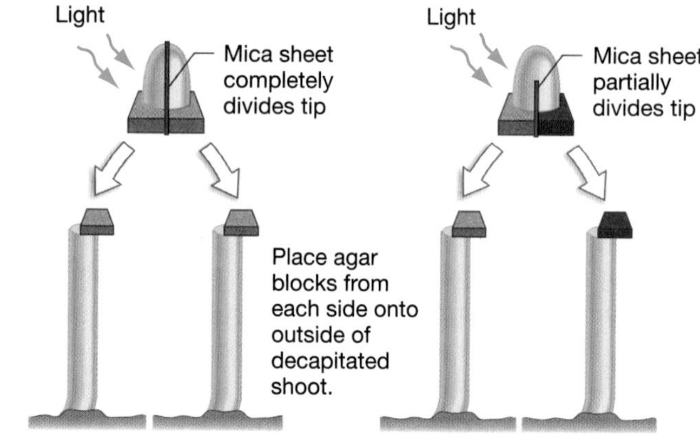

Prediction: In completely divided tip, both agar blocks will elicit the same degree of bending. In partially divided tip, block receiving the most light will elicit less bending.

Prediction of null hypothesis: The same degree of bending will occur in all treatments.

Results:

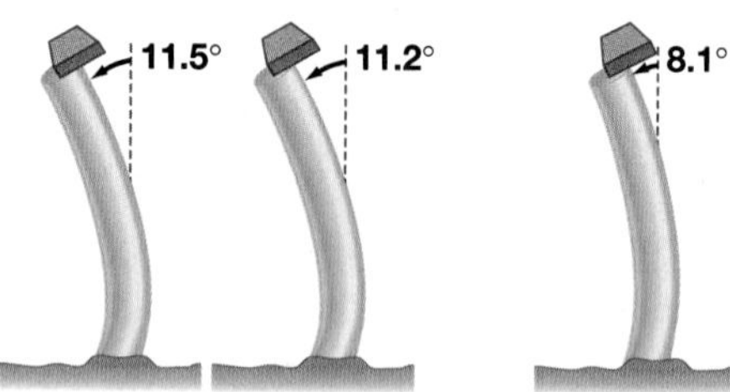

Interpretation: Mica prevented flow of auxin to shaded side, and shoots bent the same amount.

Interpretation: Auxin was redistributed to shaded side, which bent more than sunny side.

Conclusion: Asymmetric distribution of auxin results from lateral redistribution in tip.

FIGURE 39.10 Testing the Auxin Redistribution Hypothesis.

QUESTION Consider the hypothesis that cutting coleoptile tips and inserting mica sheets disrupts normal cell activity. What prediction does this hypothesis make?

of decapitated coleoptiles. In response, the coleoptiles from each treatment bent the same amount—about 26° from vertical. This result is inconsistent with the auxin destruction hypothesis. If light destroys auxin, then the block exposed to light should be much less effective in inducing bending.

To test the Cholodny-Went model, Briggs divided coleoptile tips completely or partially in half with a thin piece of mica, which is impermeable to dissolved molecules (**Figure 39.10**). The idea here was that the movement of auxin would be stopped in the completely divided tips and agar blocks, but it would not be stopped in the partially divided tips and agar blocks. To test this prediction, Briggs placed the resulting blocks on one side of decapitated shoots and recorded the bending response. The result from the completely divided tip was consistent with the earlier result. There was no difference in bending induced by the sunlit or shaded side of the tip. But as the Cholodny-Went hypothesis predicted, the bending responses differed in the partially divided tip. The side away from light induced much more bending, indicating that auxin had been transported from one side of the tip to the other.

Figures 39.9 and 39.10 summarize classic work: Two plausible competing hypotheses made clear but contrasting predictions about auxin's behavior. Briggs's experimental designs elegantly tested the predictions. The asymmetric distribution model has been successful in explaining how auxin leads to asymmetric cell elongation, and thus the bending response called phototropism.

The Auxin Receptor Decades of experimentation convinced biologists that cells in the lower parts of coleoptiles must contain a receptor for auxin. Researchers pursued a straightforward strategy to find the receptor molecule: They attached a radioactive label to the hormone, treated cells with the labeled molecule, and purified the protein to which it bound.

Although purifying the bound molecule turned out to be extraordinarily difficult, in 1985 biologists succeeded in isolating auxin-binding protein 1, or ABP1, from corn plants. Since then, experiments have confirmed that ABP1 is an auxin receptor that is located in the plasma membrane of cells in the stem and leaves. The gene that encodes ABP1 has been sequenced, and researchers are hunting for other proteins that may act as auxin receptors.

When it binds auxin, ABP1 triggers cell elongation and leads to phototropism. The question is, how?

The Cell-Elongation Response To explain how auxin causes cells to elongate, researchers proposed that once ABP1 has bound to auxin, the signal transduction cascade that follows results in an increase in the number of membrane H^+-ATPases, or proton pumps. Recall from Chapter 38 that **proton pumps** use the energy in ATP to drive protons out of the cell against an electrochemical gradient. Because the pH of the cell wall goes down when H^+-ATPases are active, the idea that these pumps are responsible for cell elongation became known as the **acid-growth hypothesis**.

To understand the rationale behind the acid-growth hypothesis, it's important to realize that two things have to happen for a plant cell to get larger:

1. Water has to enter the cell to increase its volume and generate turgor pressure on the cell wall (**Figure 39.11a**). Proton pumps could lead to water entry, because potassium (K^+) or other positively charged ions often enter a cell after protons are pumped out, and because sugar enters via proton cotransporters. As the concentrations of these ions inside the cell increase, water follows via osmosis.
2. The cell wall has to expand to create a larger volume (**Figure 39.11b**). A connection between acidification and cell-wall expansion is now well established, because researchers have discovered two classes of cell-wall proteins that actively increase cell length when the pH of the wall falls below 4.5. These proteins, called **expansins**, have been found in a wide variety of species and tissues, and the genes that code for them have been identified. What expansins actually do has yet to be determined, however. One possibility is that they make cell walls extensible by interfering with the ionic bonds between cellulose microfibrils or other wall components. Cell elongation occurs when the existing cell wall is loosened and new cellulose microfibrils are produced to extend the cell's length.

To test the acid-growth hypothesis, researchers have used antibodies that bind specifically to H^+-ATPase. When the antibodies are tagged with a fluorescent molecule, they can be used to quantify the number of proton pumps in a plasma membrane. Biologists have found that, when corn shoots are treated with additional auxin along with fluorescently tagged antibodies to H^+-ATPase, the number of proton pumps increases by 80 percent relative to untreated controls. Other researchers have shown that treating cells with auxin can lower the pH of the cell wall by as much as a full unit—from a normal value of 5.5 to 4.5.

Based on these results, most biologists agree that the acid-growth hypothesis explains how plant cells elongate. Auxin's role in the phototropic response may qualify as the best-understood example of information processing in plants.

Check Your Understanding

If you understand that...

The chain of events involved in phototropism can be summarized as follows:

- When phototropins in shoot-tip cells absorb blue light, auxin is redistributed to the shaded side of the tip.
- Auxin is transported down the shoot and binds to ABP1 receptors in the plasma membrane of target cells.
- These cells elongate when activated ABP1 leads to the installation of additional proton pumps and the activation of expansin proteins.

You should be able to...

1) Make a diagram, analogous to the one you made at the end of Section 39.1, that describes each of these steps.
2) Annotate the diagram with notes summarizing the experimental evidence behind each of these steps in the information processing pathway.
3) Based on the information in this section, ask a question about the pathway that has yet to be answered.

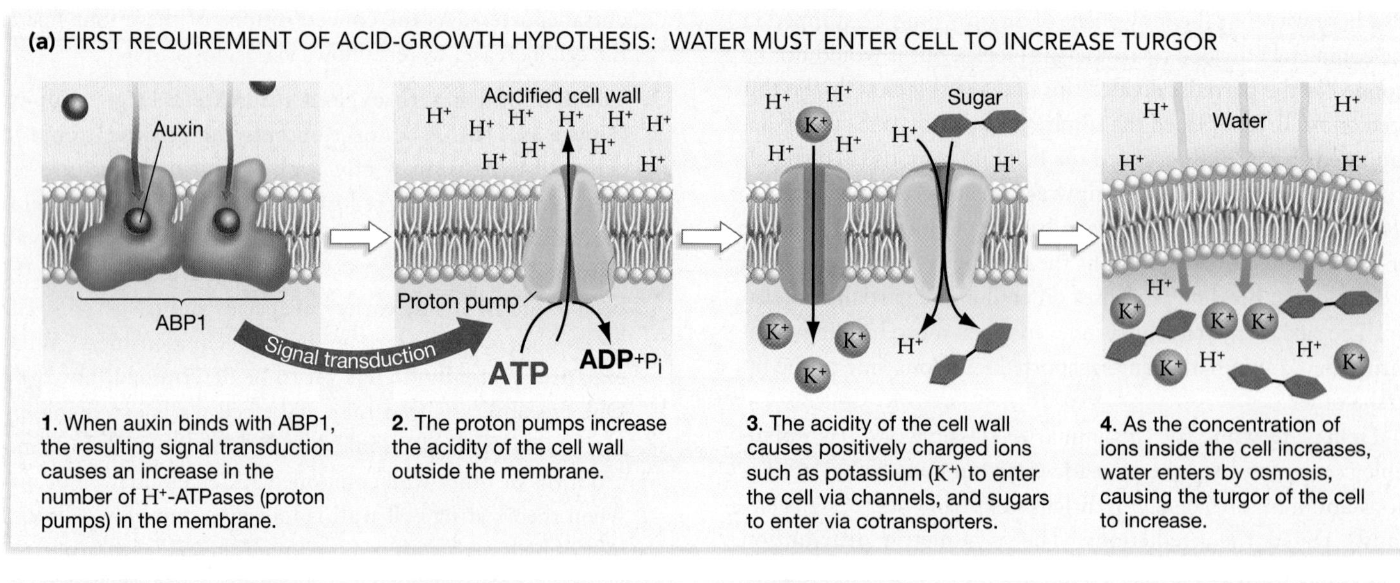

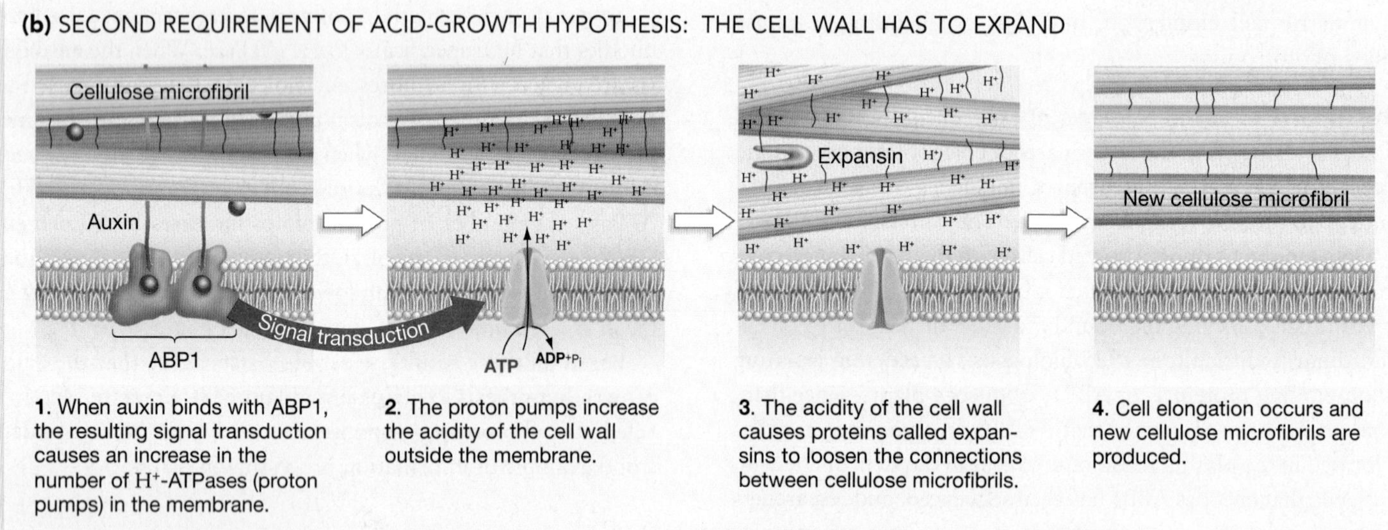

FIGURE 39.11 The Acid-Growth Hypothesis for Cell Elongation. When activated proton pumps lower the pH outside the cell membrane in response to auxin, a series of events leads to **(a)** intake of water and **(b)** expansion of the cell wall. The result is elongation of the cell.

QUESTION What would happen if turgor pressure increased but the cell wall didn't expand? What would happen if the cell wall expanded but turgor pressure did not increase?

39.3 Red and Far-Red Light: Germination, Flowering, and Stem Elongation

Studying phototropism is satisfying because all three components of information processing are relatively well understood. In many cases, though, only part of an information processing pathway is well characterized, and researchers are actively trying to deepen our understanding of signal transduction, or hormone interactions, or response mechanisms. Even so, the general principles of information processing ring true: Plants are selective about the information they receive, and they process it in a stepwise fashion.

For an illustration of this point, consider work showing that in addition to blue wavelengths, plants are sensitive to wavelengths in the red and far-red portions of the visible spectrum. This sensitivity is interesting, because red wavelengths and far-red wavelengths signal very different things to a plant. Red light (wavelengths from about 660 to 700 nm) drives photosynthesis, just as blue light does (see Figure 39.3). But far-red wavelengths (over 710 nm) are not absorbed strongly by photosynthetic pigments. Instead, they act as an important signal about the local environment. Because far-red wavelengths tend to pass through leaves instead of being absorbed, they are prominent in light that is filtered through tree leaves before it reaches the forest floor. Far-red light indicates shade.

TABLE 39.1 **How Do Red Light and Far-Red Light Affect the Germination of Lettuce Seeds?**

Biologists exposed lettuce seeds to flashes of light containing one of two wavelengths: red or far red (FR). After exposure to light, the seeds were moistened and held in the dark for several days.

Light Exposure	Germination (%)
None	9
Red	98
Red + FR	54
Red + FR + Red	100
Red + FR + Red + FR	43
Red + FR + Red + FR + Red	99
Red + FR + Red + FR + Red + FR	54
Red + FR + Red + FR + Red + FR + Red	98

QUESTION According to the data above, what is the average germination rate of lettuce seeds that were last exposed to red light? To far-red light? How do these values compare with the germination rate of seeds that are buried underground and receive no light at all?

(MB) **Web Animation** at www.masteringbio.com
Sensing Light

The first hint that plants monitor red and far-red light emerged from studies on how lettuce seed germinates. In lettuce, seed sprouting is affected by light. By exposing seeds to various specific wavelengths and plotting their frequency of germination, researchers discovered that germination rates peak when seeds receive red light (about 660 nm). To interpret this observation, biologists noted that lettuce thrives best when it grows in bright sunlight. But the stimulatory effect of red light disappeared when seeds were later exposed to far-red light, which indicates that the seeds are shaded. Germination was inhibited most effectively by wavelengths near 735 nm.

As the data in **Table 39.1** show, red and far-red light act like an on-off switch for lettuce seed germination. In this species, red light promotes germination, while far red inhibits it. The last wavelength sensed by the seed determines whether germination occurs at a high rate. This was a critically important observation. It implied that plants sense red and far-red light *together*. How could this happen?

Phytochromes as Red/Far-Red Receptors

To interpret the red/far-red switch in seed germination, biologists hypothesized that the same pigment absorbs both types of wavelength. Further, they suggested that the pigment exists in two shapes, or conformations: one shape absorbs red light, and one shape absorbs far-red light. The idea was that switching behavior, or what the researchers called **photoreversibility**, occurs because light absorption makes the photoreceptor pigment change shape, like a light switch moving up or down in response to touch. Each conformation would be responsible for a different response.

The biologists who suggested these hypotheses called the proposed pigment **phytochrome** ("plant-color"). Phytochrome was thought to be a specialized light receptor, different from any of the pigments involved in absorbing light during photosynthesis. This hypothesis was logical, because photosynthetic pigments aren't photoreversible. They don't undergo a dramatic shape change when they absorb light—instead, one of their electrons is promoted to a high energy state (see Chapter 10). Further, light-dependent responses are observed in seeds that have no chlorophyll as well as in dark-grown plants, which have so little chlorophyll that they are extremely pale.

Figure 39.12 illustrates the photoreversibility hypothesis. The idea was that a conformation of phytochrome called P_r (phytochrome red) absorbs red light. Another conformation of the same molecule, called P_{fr} (phytochrome far-red), absorbs far-red light. According to the photoreversibility hypothesis, each conformation switches to the other when it absorbs its preferred wavelength.

As **Box 39.2** shows, another response to red and far-red light—flowering in response to changes in day length—also showed photoreversibility. And researchers wondered whether yet another response to far-red light—stem elongation—might

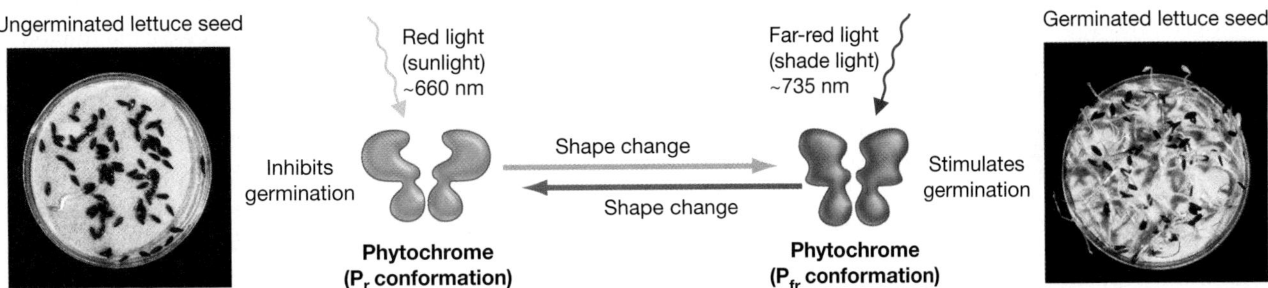

FIGURE 39.12 The Photoreversibility Hypothesis for Phytochrome Behavior. According to the photoreversibility hypothesis, (1) the phytochrome protein has two distinct conformations; (2) the P_r conformation absorbs red light, while the P_{fr} conformation absorbs far-red light; (3) when either form absorbs light, it converts to the other form; and (4) the P_{fr} form stimulates germination, while the P_r form inhibits it.

be triggered by phytochrome. If you have grown bean seeds indoors and watched the shoots get long and spindly, you have seen this response to far-red light firsthand. Beans normally grow in open sunlight. If they are grown indoors, the plants react as though they are being shaded and are attempting to grow high enough to reach open sunlight.

Based on these observations, biologists proposed that phytochrome might be the start of information processing pathways that can lead to germination, stem elongation, or flowering. Were any of these ideas correct?

How Were Phytochromes Isolated?

Young corn plants, like bean seeds, lengthen their stems in response to light deprivation or exposure to an excess of far-red light. This behavior suggested that young corn plants have a receptor protein for far-red light. To follow up on this observation, researchers purified proteins from corn coleoptiles and succeeded in isolating one that was photoreversible. Specifically, when the protein was placed in solution and exposed to alternating red and far-red light, the color of the solution switched from blue to blue-green and back. The color

BOX 39.2 Photoperiodism, Flowering, and Photoreversibility

Photoperiodism is any response by an organism that is based on photoperiod—that is, the relative lengths of day and night. In plants, the ability to measure photoperiod is important because it allows individuals to respond to seasonal changes in climate—for example, to flower when pollinators are available and when resources for producing seed are abundant. Experiments on photoperiodism in plants have shown that, in regard to flowering, plants fall into three main categories (**Figure 39.13a**):

1. **Long-day plants** bloom in midsummer, when days are longest and nights shortest. Radishes, lettuce, spinach, corn, irises, and other long-day plants flower only when days are longer than a certain length—usually between 10 and 16 hours, depending on the species.
2. **Short-day plants** bloom in spring, late summer, or fall. Asters, chrysanthemums, poinsettias, and other short-day plants flower only if days are shorter than a certain species-specific length.
3. **Day-neutral plants** flower without regard to photoperiod. Day length has no effect on flowering in roses, snapdragons, dandelions, tomatoes, cucumbers, and many weeds (plants that are adapted to grow in soils that have been disturbed enough to remove or damage existing vegetation).

Even though researchers still refer to "long-day" and "short-day" species, plants do not actually sense day length. Instead, they monitor the length of the night. This hypothesis was tested in classic experiments on flowering in a short-day plant called cocklebur, which requires less than 16 hours of light to flower. To explore what plants are sensing when they sense photoperiod, researchers grew cocklebur in growth chambers where all aspects of the environment could be controlled and where photoperiod could be manipulated. The researchers found that interrupting day periods with dark made no difference to flowering. In contrast, interrupting the night period with light did matter. To interpret these results, the researchers hypothesized that when nights were interrupted with a short period of light, plants responded to the environment as a short night. Follow-up experiments showed that long-day plants also respond to interruptions in night length, but that day-neutral species do not (**Figure 39.13b**). Based on these observations, it could be argued that plants should be classified as short-night instead of long-day, long-night instead of short-day, and night-neutral instead of day-neutral.

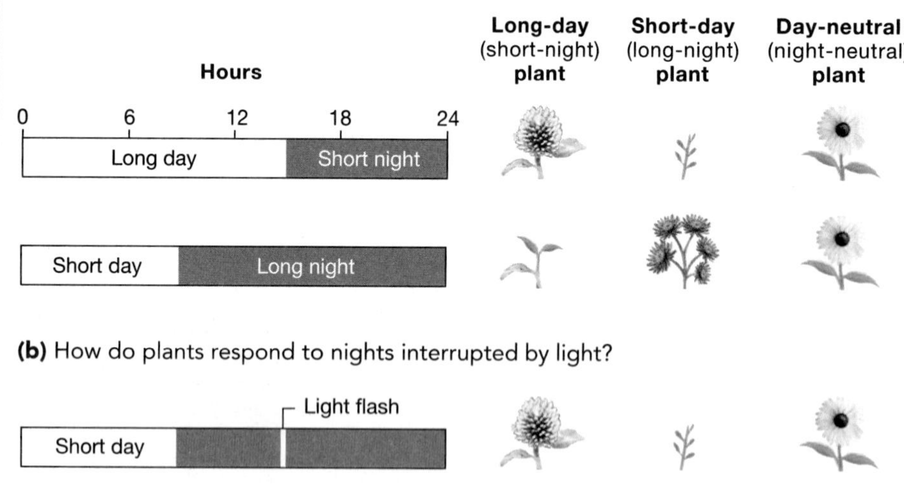

FIGURE 39.13 Different Species Respond to Photoperiod in Different Ways. (a) Flowering responses to photoperiod fall into three categories. **(b)** Flashes of light during the night alter the photoperiod response in short-day (long-night) species.

EXERCISE Add a "part (c)" to this figure, showing an experiment where a long day is interrupted by a flash of dark. Hint: A flash of dark during the day does not change the plant's perception of night length.

When researchers explored the interrupted-night phenomenon in more detail, they made an interesting discovery. In short-day (long-night) plants, flowering was inhibited most effectively by interrupting the dark period with red light—wavelengths around 660 nm. But a subsequent flash of far-red light—wavelengths around 735 nm—erased the effect (**Figure 39.14**). Analogous reversals occurred in long-day (short-night) plants. This was the first hint that flowering might be triggered by a photoreversible pigment like phytochrome.

Subsequent work confirmed the phytochrome hypothesis and explained how this pigment allows plants to measure night length: In the dark, P_{fr} reverts to P_r. The amount of P_r that accumulates due to reversion depends on the length of the night.

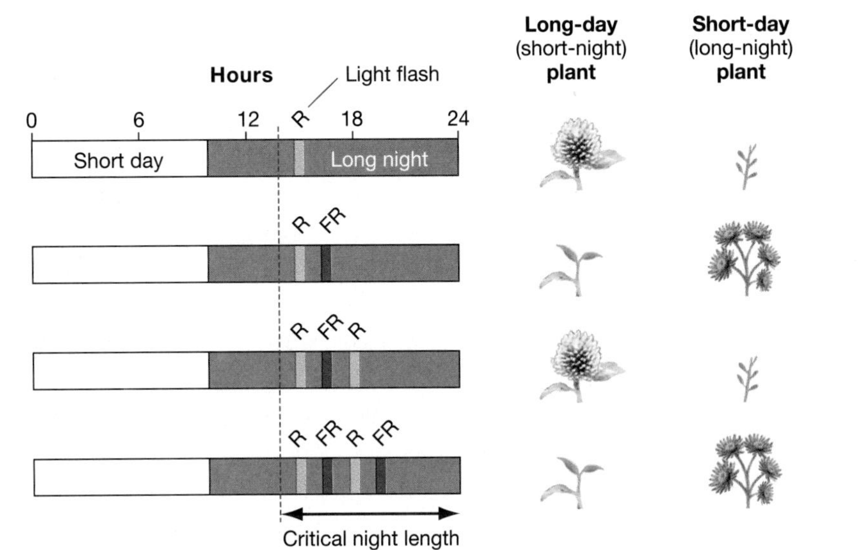

FIGURE 39.14 Flashes of Red Light and Far-Red Light Switch the Photoperiod Response On and Off. If flashes of red light and far-red light are alternated during the night, the plant's flowering response correlates with the last light it experienced.

switches supported the hypothesis that the protein absorbs red as well as far-red light. Based on these data, the biologists claimed that they had succeeded in isolating phytochrome. The year was 1959.

More recently, the gene that codes for phytochrome has been isolated and sequenced. In fact, the genome of *Arabidopsis thaliana* has five distinct genes that encode phytochrome proteins. The genes are referred to as phytochromes A through E and are symbolized *PHYA, PHYB*, and so on. The proteins encoded by these genes hold a small pigment molecule that absorbs light in the red and far-red parts of the spectrum. In addition, all five proteins appear to be photoreversible. As **Figure 39.15** shows, a portion of the phytochrome molecule changes shape in response to red and far-red light. The shape changes cause phytochrome to take on or lose a phosphate group, just as the shape changes triggered by blue light results in phosphorylation of the phototropins.

Finally, recent data suggest that different phytochromes within the same individual may be responsible for the various types of responses to red/far-red light. For example, the *PHYB* gene product is involved in the regulation of germination in *Arabidopsis*. But *Arabidopsis* individuals with mutant forms of *PHYB* also have abnormal stem elongation responses, indicating that this particular phytochrome plays a role in stem elongation as well as in germination. Stem elongation is also affected by the *PHYA, PHYD*, and *PHYE* gene products. *PHYD* and *PHYE* are responsible for monitoring night length and for triggering an appropriate flowering response.

How is the information present in each of the P_r-P_{fr} switches translated into action that affects germination, stem elongation, flowering, and other responses to red and far-red light? This question brings us to the forefront of research on phytochromes.

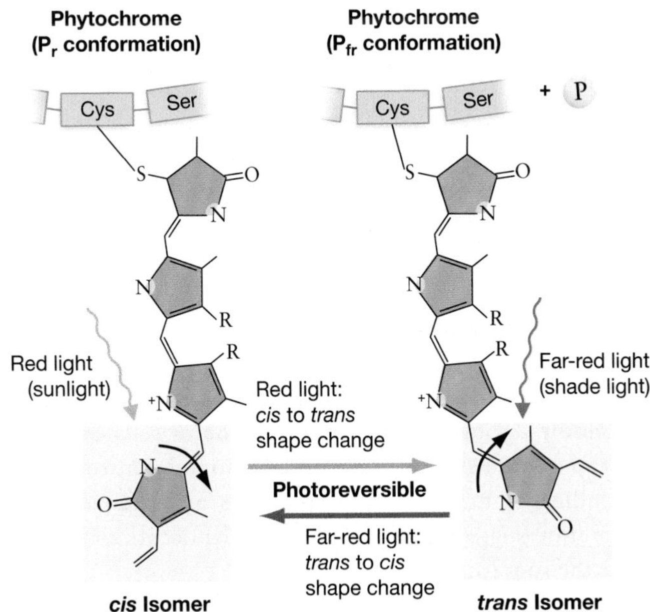

FIGURE 39.15 Phytochrome Changes Shape after Absorbing Red and Far-Red Light. This diagram shows how a subunit of phytochrome changes shape in response to absorbing red or far-red light. The shape changes correspond to phytochrome's P_r and P_{fr} conformations.

QUESTION How does this shape change relate to signal transduction?

The answer appears to be complex; one study suggests that almost one-third of all *Arabidopsis* genes are involved in the elongated growth characteristic of shade avoidance. Work on linking the phytochrome switch to the elongation and germination responses continues.

Check Your Understanding

If you understand that...

- Plants respond to red and far-red light via phytochromes, which change shape and activity when they absorb red or far-red light.
- Red light acts as a sunlight or day indicator, while far-red light acts as a shade or night indicator.

You should be able to...

1) Explain why it is adaptive for red light to trigger germination in lettuce seeds while far-red light inhibits it.
2) Explain why it is adaptive for plants that normally grow in sunny habitats to elongate their stems in response to far-red light.
3) Explain why it is adaptive for changes in night length to affect the tendency to flower in some species.

39.4 Gravity: The Gravitropic Response

The wavelengths, quantity, and direction of light that a plant receives change with the season, the weather, the time of day, and shading by other plants. Gravity could not be more different. Light is variable and can come from almost any direction, but gravity is constant and unidirectional. Light means food; gravity provides information about how the plant should orient itself in space. Shoots usually respond to gravity by growing in an upward direction; roots usually respond by growing downward (**Figure 39.16**). How do plants sense gravity, so that it can be used as a signal to orient the body?

In 1881 Charles and Francis Darwin published one of the first experimental results on **gravitropism** ("gravity-turn")—the ability to move in response to gravity. To understand the result, recall from Chapter 36 that the ends of root tips are covered by a tough, protective collection of cells called the **root cap** (**Figure 39.17a**). The Darwins found that roots stop responding to gravity if the root caps are removed. This observation suggested that gravity sensing occurs somewhere inside the root cap.

Recently biologists demonstrated precisely which cells are involved in gravity sensing in *Arabidopsis* roots. By killing tiny blocks of cells with laser beams, researchers showed that the cells illustrated in **Figure 39.17b**, directly at the center of the root cap, are the most important for regulating the gravitropic response. Based on these results, biologists are confident that they know where gravity sensing occurs in roots. Cells in the tips of stems respond to blue light and initiate phototropism; cells in the tips of roots respond to gravity and initiate gravitropism.

(a) Roots grow down.

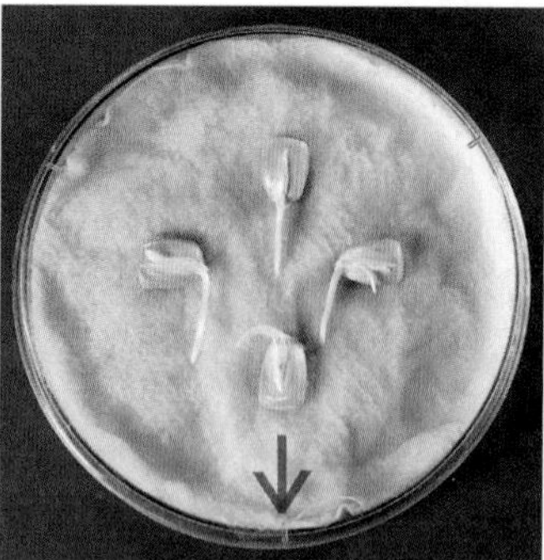

(b) Shoots grow up (or out, in some species).

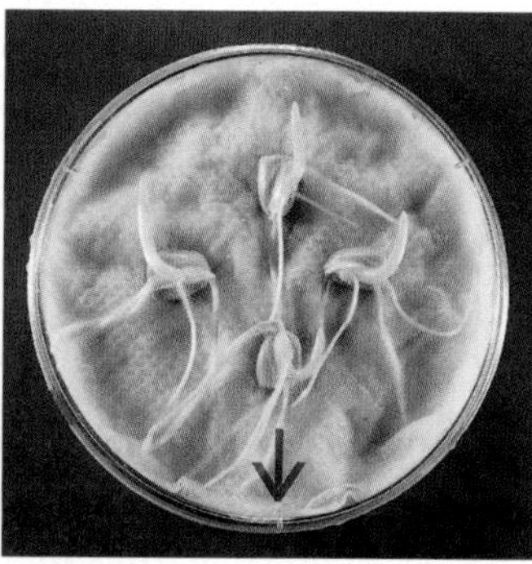

FIGURE 39.16 Gravitropism Occurs in Both Shoots and Roots. The photos show germinating corn seeds. The blue arrows indicate the direction of gravity. Roots always grow in the direction of gravity. In species with upright growth habits, shoots respond to gravity by growing away from it. In species that form mats or spread laterally, shoots respond to gravity by growing horizontally.

The Statolith Hypothesis

To explain how plants sense gravity, biologists proposed that a molecule located in root cap cells acts as a pressure receptor or a stretch receptor. More specifically, the **statolith hypothesis** contends that the dense, starch-storing organelles called **amyloplasts** function as gravity receptors (**Figure 39.18**). The idea is that gravity pulls the heavy amyloplasts to the bottom of cells,

(a) Root tips have a protective cap.

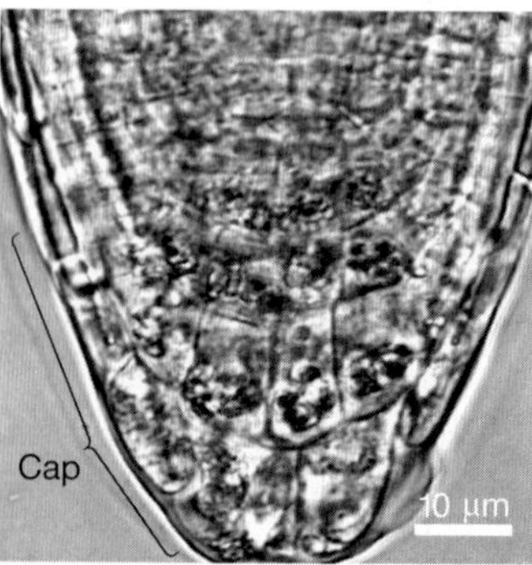

(b) Gravity-sensing cells are in the center of the cap.

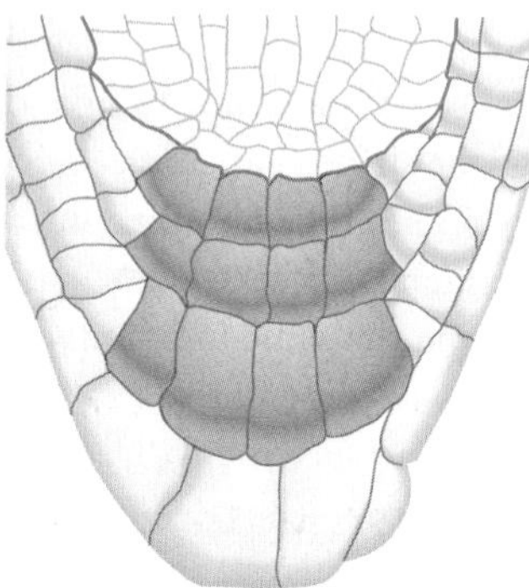

FIGURE 39.17 Experimental Evidence that Gravity-Sensing Cells Are Located in the Root Cap. (a) The root cap is a protective structure. **(b)** When the cells marked in orange are killed experimentally, the gravitational response in roots is dramatically reduced.

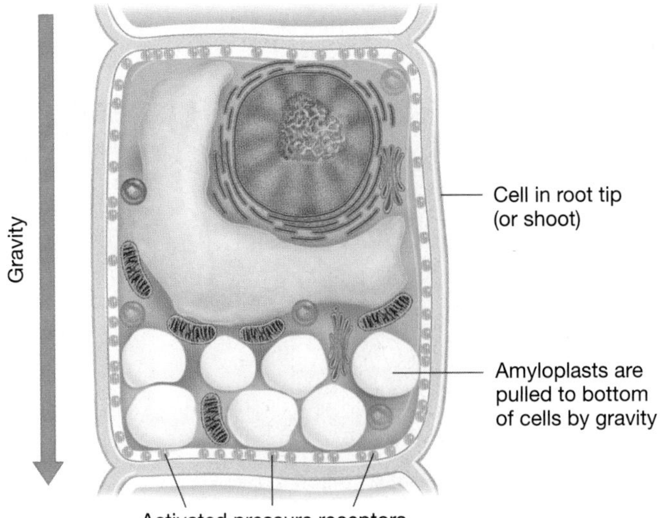

FIGURE 39.18 The Statolith Hypothesis States that Amyloplasts Stimulate Sensory Cells. Amyloplasts are filled with starch. They are dense, so they sink in response to gravity. The statolith hypothesis predicts that pressure receptors in the plasma membrane become activated as a result.

● **EXERCISE** Assuming that the receptor is a protein, suggest a hypothesis to explain how signal transduction occurs in response to pressure from a statolith.

where somehow the weight of the amyloplast activates a sensory protein that initiates the gravitropic response.

The statolith hypothesis was inspired by the way that many animals sense gravity. Lobsters, for example, take up grains of sand that become positioned in specialized gravity-sensing organs in their antennae. The grains of sand are called **statoliths** ("place-stones"). When the animal tilts or flips over, the statolith moves in response to gravity. Inside the organ, the sand grain ends up pushing against a sensory cell. When this cell is activated, it indicates that the animal is no longer upright. The same thing would happen in root cap cells if a plant were tipped over by the wind. The amyloplasts would settle onto the new "lower" cell walls and signal that the root was no longer facing in the correct direction.

Although recent experiments strongly support the statolith hypothesis, the search for the gravity receptor itself continues. Some researchers contend that the weight of the statoliths on receptors in the cell's plasma membrane elicits a gravitropic response. Other researchers propose that the key event occurs when amyloplasts pull on cytoskeletal elements, which in turn activate receptors in the plasma membrane. The nature of the gravity receptor remains one of the great unsolved problems in research on plant sensory systems.

Auxin as the Gravitropic Signal

Research continues on the first step in information processing in response to gravity. The second and third steps, however, are much better understood. As it turns out, they closely parallel the phototropic response in stems.

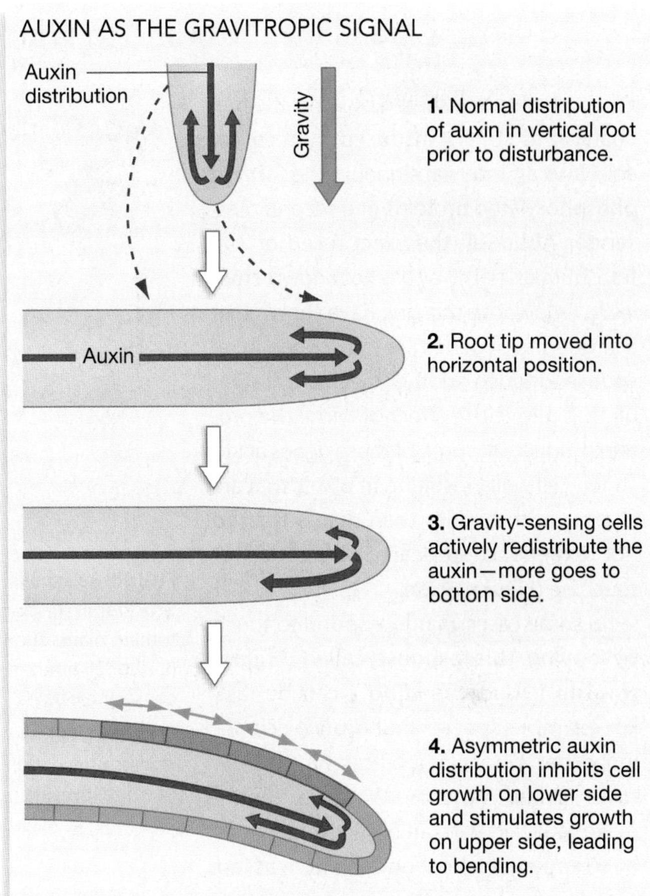

FIGURE 39.19 The Auxin Redistribution Hypothesis for Gravitropism. This sequence of events might begin when a growing root tip hits a rock and is displaced horizontally or when a plant is tipped in a windstorm and is partially uprooted.

Root cap cells that sense changes in the direction of gravitational pull respond by changing the distribution of auxin in the root tip. Under normal conditions, auxin flows down the middle of the root and then toward the perimeter and away from the root cap (**Figure 39.19**, step 1). But if the root is tipped (step 2), auxin is redistributed in much the same way as it is in response to blue light. The lower portion of the root receives increased concentrations of auxin, while the upper portion receives lower concentrations (step 3). In response to the differences in auxin concentration, cells in the lower portion of the root grow more slowly and cells in the upper portion grow more quickly. The result is bending (step 4).

Light and gravity are not the only physical forces that plants respond to. As **Box 39.3** shows, plants respond to mechanical stresses such as wind and touch. Plants also sense stimuli from the biotic world—specifically, the presence of danger in the form of herbivores or pathogenic bacteria and viruses.

BOX 39.3 How Do Plants Respond to Wind and Touch?

When plants are buffeted by wind, the mechanical force is transduced by receptor cells into an internal signal in the form of phosphorylated proteins or a second messenger. Although the exact receptor and transduction pathway are not known, studies in *Arabidopsis thaliana* have shown that a large suite of genes are transcribed in response to touch or other mechanical stimuli that mimic the effect of wind. The protein products of some of these genes act to stiffen cell walls, resulting in plants that are shorter and stockier than plants that do not experience repeated vibrations or touching (**Figure 39.20**).

In some cases, plants respond to touch by moving. This response, called **thigmotropism** ("touch-bending"), can be fast. For example, species that grow by climbing other plants or objects have structures called tendrils. When a tendril makes contact with an object, it responds by wrapping itself around the item as fast as one or more times per hour (**Figure 39.21**). Movement is even faster in "touch-sensitive" plants. A Venus flytrap, for example, closes fast enough to catch insects. Extremely rapid movements like this are possible when a touch-receptor cell transduces the mechanical signal to an electrical signal.

To understand how plants use electrical signaling, recall from Chapter 38 that proton pumps in the plasma membrane of most plant cells give them a negative charge relative to the exterior environment. The separation of charges creates a membrane voltage, or membrane potential. If a receptor protein responds to touch by allowing ions to flow across the membrane—which changes the amount of charge on either side—then the membrane potential changes. In this way, the mechanical signal (touch) can be transduced to an electrical signal.

To get from a sensory cell to a response cell, electrical signals are propagated in a characteristic form called an **action potential**. In a Venus flytrap, action potentials race across the leaf at a rate of about 10 cm/sec. When the action potentials reach cells on the outer surface of the trap, the cells swell and push the trap shut (**Figure 39.22**).

Recent experiments have clarified the mechanism responsible for trap closing. When the electrical signal arrives, H^+ is pumped out of the epidermal cells in response. The subsequent change in extracellular pH causes the cell wall to loosen. It also sets up a strong electrochemical gradient that leads to a rapid influx of potassium ions. When K^+ enters the cell, water follows by osmosis and causes the cell to swell. The response is rapid enough to resemble the way an animal's muscle responds to an action potential.

FIGURE 39.20 Plants Get Shorter and Stockier in Response to Wind or Touch. The tomato plants shown here were touched lightly 0, 10, or 20 times per day each day for 10 consecutive days.

QUESTION What is the adaptive significance of this response? Give an example of when it would occur in nature.

FIGURE 39.21 Thigmotropism Is Movement in Response to Touch. Portions of pea plants called tendrils wind around support structures after contacting them. The attachment provided by the tendrils allows peas to climb.

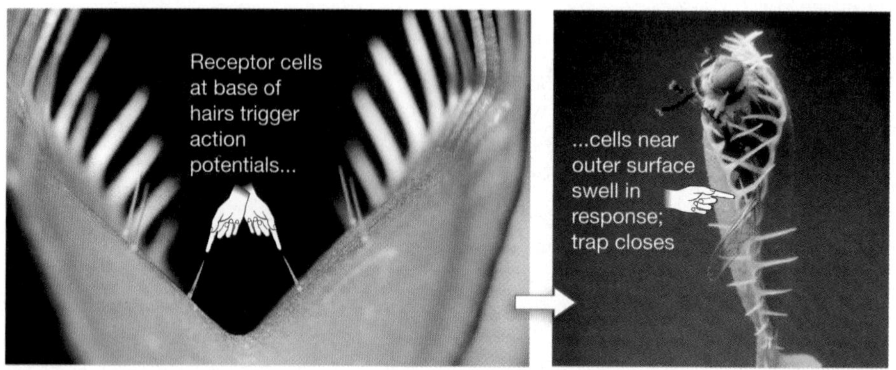

FIGURE 39.22 Venus Flytraps Close in Response to Action Potentials Generated by Sensory Hairs.

Check Your Understanding

If you understand that...

- Cells in root caps sense gravity via pressure that amyloplasts exert on membrane proteins.
- Changes in gravity sensing result in a redistribution of auxin and changes in the growth rate of root tips.

You should be able to...

1) Explain the parallel that exists between auxin redistribution in shoots and roots and the responses called phototropism and gravitropism.
2) Explain why auxin could be considered a gravitropic hormone.

at www.masteringbio.com
Plant Hormones

(a) Apical meristem intact

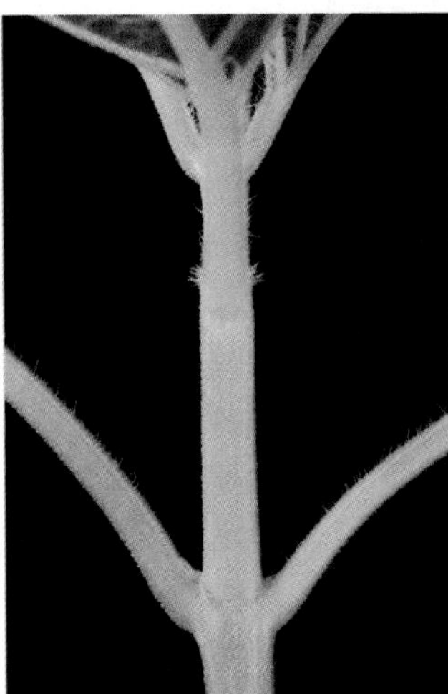

(b) Apical meristem cut off

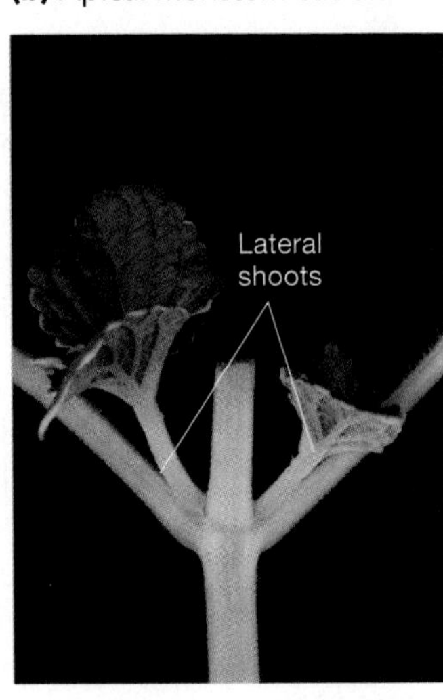

FIGURE 39.23 When Apical Dominance Occurs, Growth of Lateral Buds Is Suppressed. (a) The stem of a coleus plant is shown still intact. **(b)** The same plant, several weeks after the stem was cut. The lateral shoots will orient themselves vertically.

39.5 Youth, Maturity, and Aging: The Growth Responses

Plants grow throughout their lives, from the time they germinate until the time they die. As the plant body grows, it matures into an efficient machine for absorbing sunlight, water, nutrients, and other diffuse resources. But growth is not constant. It speeds up when water, light, and nutrients are abundant and slows or stops when conditions are poor or when it is time for leaves to drop or fruits to ripen. Controlling growth in response to changes in age or environmental conditions is one of the most basic aspects of information processing in plants. Hormones play a key role in regulating growth.

To explore how plants grow in response to changing conditions, let's consider five of the best-studied hormones involved in growth responses. We'll start with auxin—the molecule responsible for phototropism and gravitropism.

Auxin and Apical Dominance

One of the most basic aspects of plant growth is called apical dominance. When **apical dominance** occurs, most of a stem's growth occurs at the apical meristem of the shoot. Further, the presence of the topmost meristem inhibits the growth of apical meristems that are present lower down on the plant, at the plant's nodes. The photographs in **Figure 39.23** show this phenomenon in action. Intact coleus plants have one main shoot and an apical meristem, extending vertically. But if the main shoot is cut (Figure 39.23a), lateral branches begin to grow vigorously and vertically from **lateral buds**, which form at the angles between leaves and the stem.

What caused the change? Because auxin is produced in shoot tips, researchers suspected that it might have a role in apical dominance as well as phototropism. This hypothesis was confirmed when it was shown that apical dominance could be sustained by adding auxin to a shoot's cut surface after its tip had been removed. This result suggests that tip cells send a constant stream of information down to other organs and tissues, in the form of auxin. If the signal stops, it means that apical growth has been interrupted. In response, lateral branches sprout and begin to take over for the main shoot. Now the question is, How does this signal move?

Polar Transport of Auxin Auxin transport is **polar**, or unidirectional. If radioactively labeled auxin is added to the top of a cut stem, the hormone is transported toward the base. But if labeled auxin is added to the base of a cut stem, it is not transported toward the apex. Auxin is the only plant hormone known to be transported in one direction only.

Studies with labeled auxin have also shown that the hormone is transported all the way down the stem to the root, via parenchyma cells in the ground tissue and vascular tissue. Labeled auxin moves at about 10 cm/hr—approximately 10 times more slowly than substances traveling in the phloem or xylem. Recent research has shown that auxin enters the apical end of cells via a cotransport protein, which imports the hormone along with two protons that enter the cell along the electrochemical gradient—set up by H^+-ATPase—that favors proton entry. Once inside the cytoplasm, auxin diffuses to the basal portion of the cell, where it is transported out by carrier proteins located in the basal portion of the plasma membrane. In this way, auxin flows down stems in one direction only: into

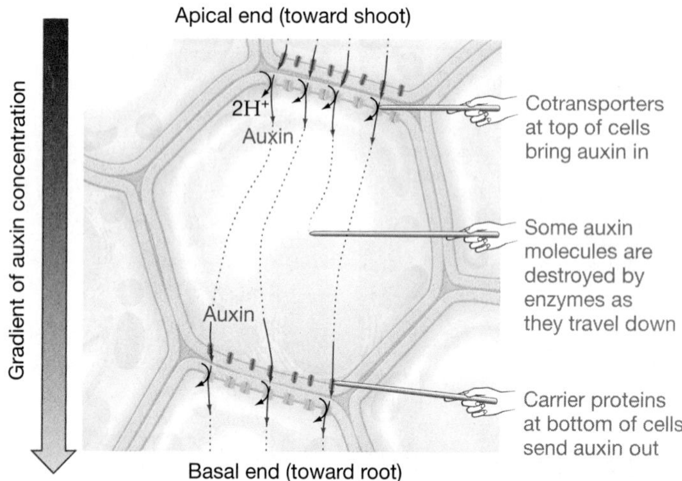

FIGURE 39.24 Auxin Transport is Polar: From Shoot to Root. Auxin transport is unidirectional because different types of auxin carriers are located at the tops and bottoms of cells.

EXERCISE Add cotransporters and carriers to the membranes of the partial cells shown at the right and left of this figure.

cells via a cotransporter in the apical end and out via a carrier protein in the basal end (**Figure 39.24**). Because some auxin molecules are destroyed by enzymes as they travel down the long axis of the plant, polar transport sets up a strong gradient in auxin concentration. Auxin concentrations are much higher in shoots than they are in roots.

What Is Auxin's Overall Role? Auxin clearly plays a key role in controlling growth via apical dominance, phototropism, and gravitropism. But it is critical to recognize that this chemical messenger has other important effects as well:

- Fruit development is influenced by auxin produced by seeds within the fruit.
- Falling auxin concentrations are involved in the **abscission**, or the shedding of leaves and fruits, associated with the genetically programmed aging process called senescence.
- The presence of auxin in growing roots and shoots is essential not only for the proper differentiation of xylem and phloem cells in vascular tissue but also for the development of vascular cambium.
- Auxin stimulates the development of adventitious roots in tissue cultures and cuttings.

Auxin has so many different effects on plants that it has been difficult for biologists to understand its overall role. Recently, several investigators have proposed that auxin's overall function is to signal where cells are in space. More specifically, the idea is that auxin concentration identifies where a cell is located relative to the long axis of the plant body. In this way, auxin plays a role in determining the overall shape of a plant. If conditions relating to the long axis change—because of changes in light availability, because a windstorm tips the plant and shifts the position of the roots, or because a deer eats the shoot apex—changes in auxin concentration effectively signal how the individual's tissues should respond. Phototropism, gravitropism, apical dominance, and the production of adventitious roots are all ways of coping with changes in the long axis of the plant.

It is important to realize that auxin does not act alone. Instead, its effects are modified by the action of other plant hormones. Researchers are particularly interested in how auxin interacts with the group of hormones called cytokinins, because there is evidence that plant growth is regulated by an antagonistic relationship between auxin and cytokinins. Auxin travels down from the shoot to the root, suppressing the growth of apical meristems in lateral buds along the way. In contrast, cytokinins are produced in the roots and travel upward, stimulating growth of axillary buds. From shoot to root, the two hormones are present in gradients of differing concentrations. Based on this observation, biologists hypothesize that interactions between auxin and cytokinins may help plants maintain an optimal proportion of root growth to shoot growth. Let's take a closer look at what cytokinins are and what they do.

Cytokinins and Cell Division

Cytokinins are a group of plant hormones that promote cell division. (*Cyto* is the Greek root for *cell*; *kinin* refers to *kinesis*, meaning "movement" or "division.") The initial hint that certain molecules can promote cell division in plants was the finding that coconut milk, which stores nutrients used by growing coconut embryos, promotes the growth of plant cells and embryos in culture. Later experiments showed that molecules derived from the nitrogenous base adenine also stimulate the growth of cells in culture. Eventually, naturally occurring adenine derivatives that stimulate growth were discovered in corn and apples. These hormones, which biologists found to be similar in structure to the growth-promoting ingredient in coconut milk, were named cytokinins. Cytokinins are synthesized in root tips, young fruits, seeds, growing buds, and other developing organs. **Box 39.4** introduces the use of cytokinins to promote the growth and differentiation of cells in culture.

The cytokinin that has been found in the most species is derived from adenine and is called zeatin. The other cytokinins that have been isolated are very similar to zeatin in chemical structure. Most of the zeatin and other cytokinins that are active in plants are synthesized in the apical meristems of roots and transported up into the shoot system via the xylem. A group of closely related proteins that act as cytokinin receptors has now been isolated and characterized. When cytokinin binds to these receptors in the plasma membranes of target cells, the receptors trigger the activation of genes that regulate growth.

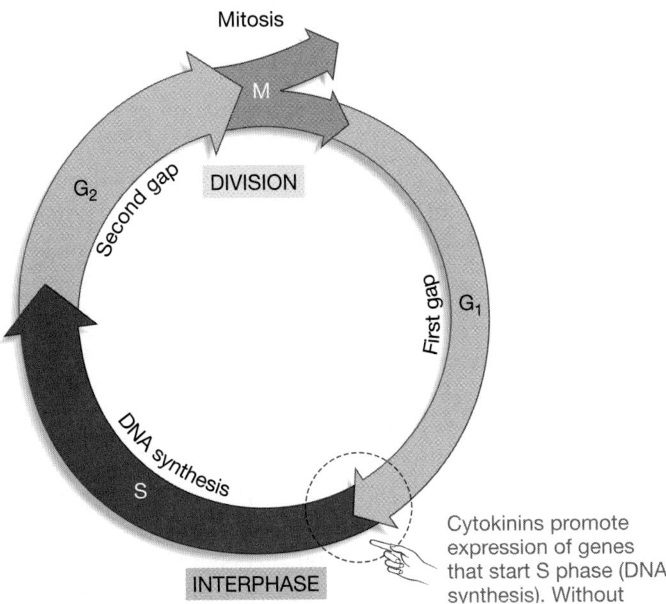

FIGURE 39.25 Cytokinins Affect the Cell Cycle.

Recent research on cytokinins has focused on understanding exactly how they function at the molecular level. Specifically, biologists have explored whether cytokinins might affect molecules that regulate the cell cycle. Some of these cell-cycle regulators, including the cyclins and the cyclin-dependent kinases (Cdk's), were introduced in Chapter 11. Recall that these proteins must be activated for cells to progress through checkpoints in the cell cycle. In this way, activated cyclins and Cdk's allow cell division to continue.

To assess whether cytokinins affect cell-cycle genes, researchers grew *Arabidopsis* cells in culture so that the nutrients and other molecules available could be carefully controlled. The biologists starved the cells of cytokinins for a day, then added the hormones again to half of the cells. When they assessed the level of mRNA from a cyclin gene called CycD3, they documented significant increases in the cells that were exposed to cytokinins again compared with the level in cells that were not re-exposed to cytokinins. This is strong evidence that cytokinins regulate growth by activating genes that keep the cell cycle going. In the absence of cytokinins, cells arrest at the G_1 checkpoint in the cell cycle and cease growth (**Figure 39.25**).

Gibberellins and ABA: Growth and Dormancy

In northern latitudes and at high elevation, most seeds and mature plants initiate growth in spring. Conditions for growth are good at that time of year, because temperatures are warming and soil moisture levels are usually at their peak. Seedlings and mature plants continue to grow throughout the summer and early fall if moisture and nutrients are still available. But during drought conditions, growth stops. A similar type of growth stoppage occurs in the embryos inside seeds. Embryos begin to develop as seeds mature, but cease this initial growth and remain dormant throughout the cold winter months. **Dormancy** is a temporary state of reduced metabolic activity or no metabolic activity.

What signals are responsible for initiating growth in plants in response to changes in environmental conditions, and what signals are responsible for stopping it? Two hormones provide the answer. The first is the family of closely related compounds called the **gibberellins**. The second is **abscisic acid**, commonly abbreviated **ABA**. In general, gibberellins stimulate growth. ABA inhibits it. In at least some cases, the two hormones interact like start and stop signals.

Gibberellins Stimulate Shoot Elongation Over 100 years ago, Japanese farmers noticed that some of their rice seedlings grew exceptionally quickly but fell over before they could be harvested. Biologists who investigated the situation found that the diseased plants were infected with the fungus *Gibberella fujikuroi*. A causal connection between infection with *Gibberella* and rapid stem elongation was confirmed when researchers treated rice seedlings with an extract from the fungus. As predicted, the treated seedlings produced abnormally long shoots.

The active component in the extract was eventually isolated and named gibberellic acid (GA), which is a gibberellin. Follow-up research showed that rice plants produce their own gibberellin but respond to applications of additional hormone by elongating their stems. In effect, the infected rice seedlings were suffering from a gibberellin overdose.

Gibberellins make up a large family of structurally related compounds and are found in a wide array of fungi and plants. Most plant species produce several different gibberellins that are active as hormones. Even though these molecules have dramatic effects on growth, they are present in vanishingly small concentrations. In stems and leaves, active forms of gibberellin may be present at just one part per billion—in 1 billion grams of tissue, there would be just 1 gram of gibberellin.

To find the genes that are responsible for producing gibberellins and that encode the proteins involved in the response to GA, biologists focused on analyzing mutant plants with abnormal stem length (**Figure 39.27**). Recall from Chapter 13 that Gregor Mendel analyzed the transmission of two alleles at a single gene that affected stem height in garden peas. One allele was associated with the production of tall stems; the other was associated with dwarfed growth. The tall allele was dominant to the dwarf allele.

The gene responsible for the stem-length differences came to be known as *Le* (for *le*ngth). Early work on *Le* dwarf mutants showed that they attain normal height if they are treated with the gibberellin called GA_1. This observation suggested that dwarf peas can respond to gibberellins normally but are unable to manufacture normal quantities of their own GA_1. Follow-up

BOX 39.4 How Are Hormones Used to Grow Plant Cells in Culture?

Recall from Chapter 36 that the parenchyma cells found in plant ground tissue are totipotent—meaning they retain the ability to divide and differentiate into a complete, mature plant, including new types of tissue. This property of parenchyma cells is important in wound healing and asexual reproduction. But it also allows researchers to grow complete adult plants from a small number of parenchyma cells.

Biologists who grow plants in tissue culture begin by placing parenchyma cells in a liquid or solid medium containing all the nutrients required for cell maintenance and growth. In the early days of plant tissue culture, investigators found not only that specific hormones were required for successful growth and differentiation but also that the relative abundance of hormones present was critical to success. This was one of the first hints that plant hormones interact with each other to produce a particular outcome.

The earliest experiments on hormone interactions in tissue cultures were done with tobacco cells in the 1950s by Folke Skoog and co-workers. These researchers found that when auxin alone was added to the culture, the cells enlarged but did not divide. But if the team added roughly equal amounts of auxin and cytokinin to the cells, the cells began to divide and eventually formed a **callus**, or an undifferentiated mass of parenchyma cells (**Figure 39.26**). By varying the proportion of auxin to cytokinins in different parts of the callus and through time, the team could stimulate the growth and differentiation of root and shoot systems and produce whole new plants. A high ratio of auxin to cytokinin led to the differentiation of a root system, while a high ratio of cytokinin to auxin led to the development of a shoot system. Eventually Skoog's team was able to produce a complete plant from just one parenchyma cell.

The ability to grow whole new plants in tissue culture from just one cell has been instrumental in the development of genetic engineering (see Chapter 19). Researchers insert recombinant genes into target cells, test the cells to identify those that successfully express the recombinant genes, and then use tissue culture techniques to grow those cells into adult individuals with a novel genotype and phenotype.

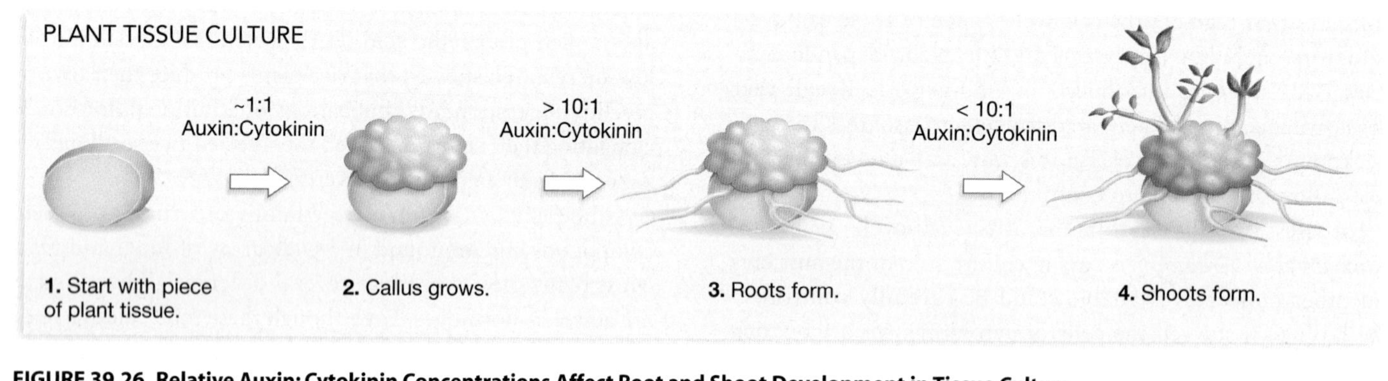

FIGURE 39.26 Relative Auxin: Cytokinin Concentrations Affect Root and Shoot Development in Tissue Culture.

experiments focused on treating dwarf peas with a radioactively labeled molecule used in the synthesis of GA_1. These plants did not succeed in producing radioactively labeled GA_1, but plants with the normal allele did. Based on these results, researchers became convinced that the *Le* locus encodes an enzyme involved in GA synthesis.

Investigators recently provided support for this hypothesis by finding a gene in peas that encodes an enzyme called 3β-hydrolase. This enzyme adds a hydroxyl group (–OH) to a gibberellin called GA_{20}, producing the biologically active molecule GA_1. Researchers who compared the DNA sequences of this gene from normal and dwarf pea plants found an important difference: In a part of the enzyme near the active site, the mutant DNA sequence codes for the amino acid threonine instead of alanine. Follow-up tests showed that the mutant enzymes are unable to convert GA_{20} to GA_1. This is strong evidence that the *Le* gene encodes 3β-hydrolase and that a single amino acid change renders the enzyme largely ineffective and causes dwarfing.

Some 130 years after Mendel's experiments, biologists finally understand the dwarfing trait at the molecular level. Now researchers would like to know how the gibberellins involved in shoot elongation interact with other hormones. In stems, gibberellins appear to promote both cell elongation and rates of cell division. But it is well established that auxin also promotes cell elongation and that cytokinins also promote cell division. Research continues on how GAs, cytokinins, and auxin interact on the molecular level to control plant growth and development.

Gibberellins and ABA Interact during Seed Dormancy and Germination Many plants produce seeds that have to undergo a period of drying or a period of cold, wet conditions before they are able to germinate in response to warm, wet

FIGURE 39.27 Dwarfed Individuals May Have Mutations That Affect Gibberellins. Dwarfed plants have much shorter stems than do normal individuals of the same age. By analyzing dwarfed individuals, biologists were able to identify genes involved in gibberellin synthesis.

QUESTION In many cases, dwarfed individuals produce more flowers and seeds than "normal" individuals do. Why?

conditions. A requirement for drying ensures that mature seeds will not sprout on the parent plant; an obligatory cold period protects seeds from germinating just before the onset of winter. Recall that some seeds also have to receive a dose of red light, which indicates that they are in a sunny location. In essence, then, seeds have an "off" setting that discourages germination and an "on" setting that initiates growth. The appropriate state for the on/off switch is determined by environmental cues such as temperature, moisture, and light.

By applying hormones to seeds, researchers learned that in many plants, ABA is the signal that inhibits seed germination and gibberellins are the signal that triggers embryonic development. To understand how these messengers work and how they interact, researchers have concentrated on studying a specific event: the production of an enzyme called α-amylase (alpha-amylase) in germinating oat or barley seeds.

α-Amylase acts as a digestive enzyme that breaks the bonds between the sugar subunits of starch. (Your saliva contains an amylase that acts on the starch in food. In humans and other mammals, this enzyme initiates carbohydrate digestion in the mouth.) **Figure 39.28** shows that, during the germination of a barley seedling, α-amylase is released from a tissue called the **aleurone layer**. The enzyme diffuses into the carbohydrate-rich storage tissue in the seed and releases sugars that can be transported and used by the growing embryo. Adding GA to the aleurone layer increases the production and release of α-amylase; adding ABA to that layer decreases α-amylase levels.

Box 39.5 details how GA stimulates α-amylase production and how ABA prevents it. Extensive work on the molecular mechanism of this hormone interaction carries several important messages:

1. A cell's response to a hormone often occurs because specific genes are turned on or off.
2. Hormones rarely act on genes directly. Instead, a receptor on the surface of a cell or in the cytosol receives the message and responds by initiating a signal transduction cascade, which activates specific gene regulatory proteins—the transcription activators and repressors introduced in Chapter 18.
3. Different hormones interact at the molecular level because they induce different gene regulatory proteins, which increase or decrease expression of key genes. The relative amounts of these regulatory factors reflect the relative amount of hormone signal received and determine which proteins are produced by the cell and in what quantities.

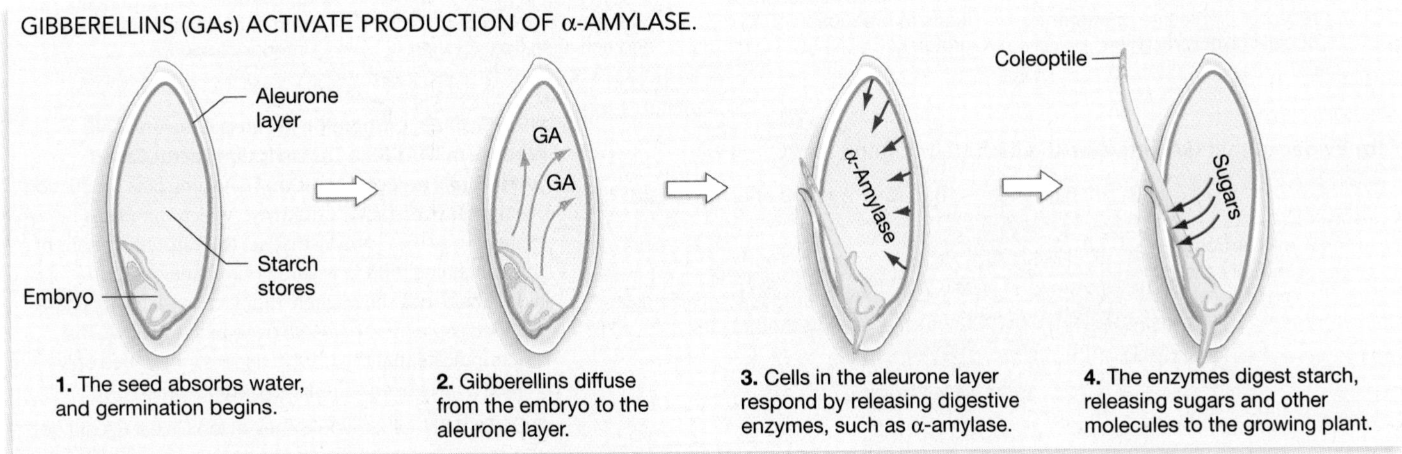

FIGURE 39.28 The Molecular Mechanism of Gibberellin Action.

BOX 39.5 The Molecular Mechanism of Hormone Action: A Case Study

To understand how α-amylase production is regulated at the molecular level, researchers followed up on a key observation: The α-amylase gene's promoter resembles the DNA sequences targeted by transcription factors called Myb proteins. Mybs (pronounced *mibs*) are DNA-binding proteins that turn gene transcription on or off. To use the vocabulary introduced in Chapter 18, Mybs function as transcription activators or repressors.

Based on these observations, the team proposed that a Myb might turn α-amylase on in response to a signal from GA or off in response to a signal from ABA. It had already been established that, as germination begins, GA produced by the embryo reaches cells in the aleurone layer. The researchers hypothesized that a GA receptor on the membrane receives this signal and that the receptor responds by activating the production of a Myb. The Myb, in turn, travels to the nucleus, binds to the α-amylase promoter, and triggers transcription **(Figure 39.29a)**.

To test this idea, the researchers began a search for a Myb that is activated by GA. Their first step was to isolate all the mRNAs produced in the aleurone layer of a germinating barley seed. Then they used the enzyme reverse transcriptase to make a DNA copy of each of these mRNAs. At that point, they had a copy of all the genes being transcribed in activated aleurone tissue. Were any of these sequences related to the Myb family? To answer this question, the team tested each DNA sequence to see if it would bind to the section of Myb DNA that encodes the DNA-binding region. One did. This DNA sequence had to be from a gene that encodes a Myb. The result confirmed that a Myb protein exists in activated aleurone tissue.

Is this transcription factor produced specifically in response to GA? To answer this question, the team isolated mRNAs from barley aleurone layers that were either exposed or not exposed to GA. They ran these mRNAs out on a gel and then probed the sequences with single-stranded DNA from the *Myb* gene they had discovered in aleurone tissue. As the gels in **Figure 39.29b** show, the mRNA for the Myb protein appeared only in cells that were exposed to GA. Follow-up experiments confirmed that this protein, which the team named GAMyb, binds to the α-amylase promoter. Just as their hypothesis predicted, GAMyb acts as a transcription activator that stimulates α-amylase production.

These data tell only part of the story, though, because ABA also induces the production of Myb proteins. But preliminary data suggest that in this case, ABA-dependent transcription factors bind to the α-amylase promoter and shut down transcription.

If these experiments are confirmed, the results suggest a molecular basis for how hormones with opposite effects interact. The hypothesis is that ABA enforces dormancy by activating transcription repressors that shut down the production of key enzymes, while GA breaks dormancy and supports germination through transcription activators that trigger the production of the same enzymes. If ABA is present in high concentration, repression dominates and dormancy occurs. If GA is present in high concentration, activators prevail and germination proceeds.

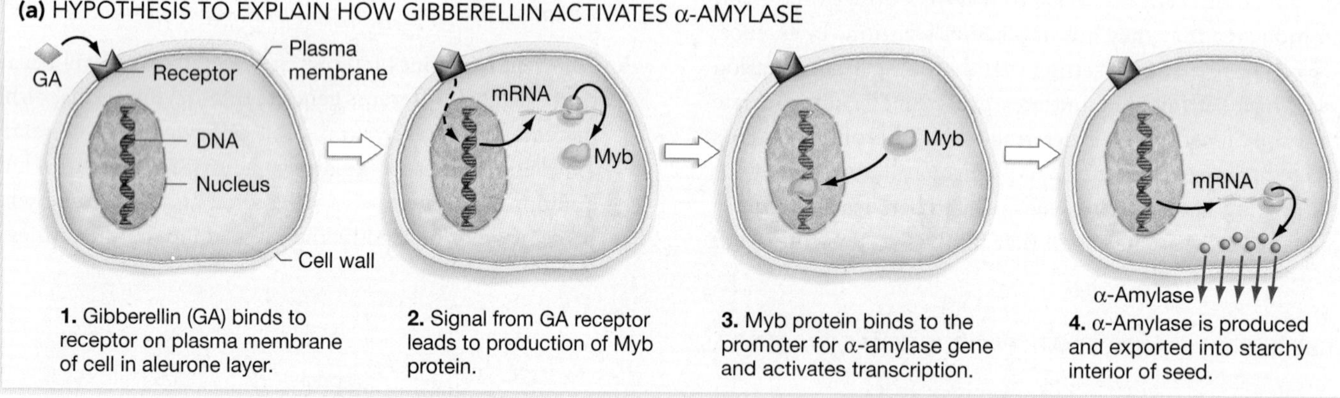

FIGURE 39.29 Gibberellin-Treated Aleurone Cells Produce mRNAs for a Transcription Factor Called GAMyb. **(a)** The logic here is that GA stimulates production of a transcription factor called Myb, which stimulates expression of the α-amylase gene. **(b)** Autoradiographs of gels containing mRNAs isolated from barley aleurone layers. The bands show where radioactively labeled, single-stranded copies of the *GAMyb* gene bind to mRNA. The results indicate that the *GAMyb* gene is transcribed only after cells have been stimulated by gibberellin (GA).

QUESTION What would the gel look like if GA did not trigger production of the Myb protein?

In addition to inspiring research on hormone interactions, GA's role in seed germination has important commercial applications. Brewers, for example, routinely use gibberellins in the malting process—the conversion of starches stored in seeds to sugars. The sugars that are released in response to GA treatment support fermentation by yeast and provide flavor in the finished beer.

ABA Closes Guard Cells in Stomata Chapter 37 introduced one of the major problems faced by land plants: replacing water that is lost to the atmosphere when the **stomata** are open. Because stomata open in response to blue light, they allow gas exchange to occur while the plant is receiving the wavelengths of light used in photosynthesis. However, if plant roots are unable to obtain enough water to replace the fluid being lost at the leaves, stomata close. Closing stomata is an adaptive response when roots cannot find adequate water, because continued transpiration would lead to wilting and potential tissue damage.

Early work on the mechanism of stomatal closing suggested that ABA is involved. For example, applying ABA to the exterior of stomata causes them to close. To explore the hypothesis that a hormone regulates stomatal closing, researchers performed a series of experiments based on the system diagrammed in **Figure 39.30**. The fundamental idea was to grow plants whose roots had been divided. Only one side of the experimental plants was watered, while both sides of the control plants were watered. During this treatment, investigators documented that the water potential of the leaves remained the same in both control and experimental plants. Yet the stomata of experimental plants began to close. This result suggested that roots from the dry side of the pot were signaling drought stress, even though the leaves were not actually experiencing a water shortage.

Follow-up experiments have supported two important predictions: ABA concentrations in roots on the dry side of the pot are extraordinarily high relative to the watered side, and ABA concentrations in the leaves of experimental plants are much higher than in the leaves of control plants. These results suggested that ABA from roots is transported to leaves and that it serves as an early warning of drought stress. In doing so, it overrides the signal from the blue-light photoreceptors introduced earlier in this chapter.

To understand the molecular mechanism involved in stomata opening and closing, recall from Chapter 36 that stomata consist of two **guard cells**. When the vacuoles of guard cells are filled with water, the cells are turgid. The shape of turgid guard cells results in an open pore, which allows gas exchange between the atmosphere and the interior of the leaf. But when vacuoles lose water and guard cells become flaccid and lacking in turgor, cell shape changes in a way that closes the pore and stops both gas exchange and loss of water via transpiration.

Based on these observations, the question of how stomata open and close is the same as asking how guard cells become turgid or flaccid, or how water flows into or out of vacuoles. Activation of PHOT by blue light leads to water entry and stomatal opening; activation of ABA receptors leads to water exit and stomatal closing.

Experiment

Question: Can roots communicate with shoots?

Hypothesis: Roots that are dry can signal shoots to close stomata.

Null hypothesis: Roots cannot communicate with the shoot.

Experimental setup:

1. Divide roots of many plants into two sides.

2. In experimental group, water one side.

3. In control group, water both sides.

4. In both groups, measure water potential of leaves and observe stomata.

Prediction: Stomata in experimental plants will close; stomata in control plants will stay open.

Prediction of null hypothesis: Stomata in both experimental and control plants will stay open.

Results: No difference between experimental and control plants in water potential of leaves.

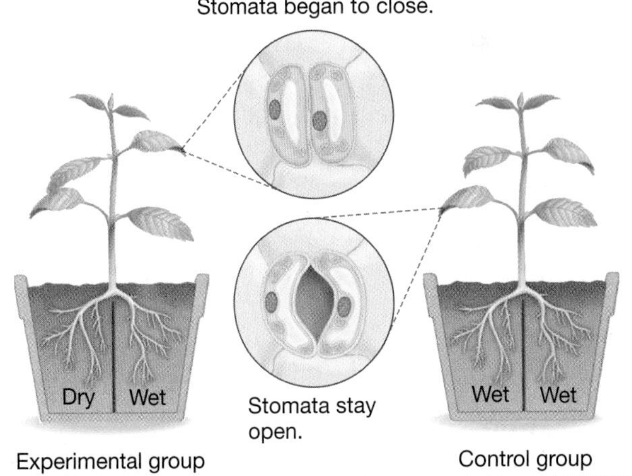

Conclusion: Roots can communicate with shoots. Dry roots signal the shoot and cause stomata to close, even though leaves are receiving sufficient water (from roots on the wet side of the plant).

FIGURE 39.30 Can Roots Communicate with Shoots?

Guard cell opening and closing is based on changes in the activity of H^+-ATPases in the plasma membrane. Activation of PHOT receptors by blue light leads to increased pumping of protons out of each guard cell. The pump's activity creates a strong electrochemical gradient that brings potassium and chloride ions into the interior of the guard cells (**Figure 39.31a**). Water follows the incoming ions via osmosis, causing the cells to swell. But as **Figure 39.31b** shows, guard cells respond to ABA in a very different way. When ABA causes calcium ions to be released from storage in the vacuole, two things happen: (1) Channels that allow chloride and other anions to leave along their electrochemical gradients are opened, and (2) H^+-ATPases and inward-directed potassium channels are inhibited. When the anions leave guard cells, the change in membrane potential causes outward-directed potassium channels to open. Large amounts of K^+ leave the cells, with water following by osmosis. The result is a loss of turgor and closing of the pore.

Whether it acts on guard cells or seeds, ABA fulfills a general role in plants as a dormancy or "no-growth" signal. In many cases, its action depends on input from other hormones and photoreceptors. To survive and reproduce successfully, plants have to integrate information from a variety of sources.

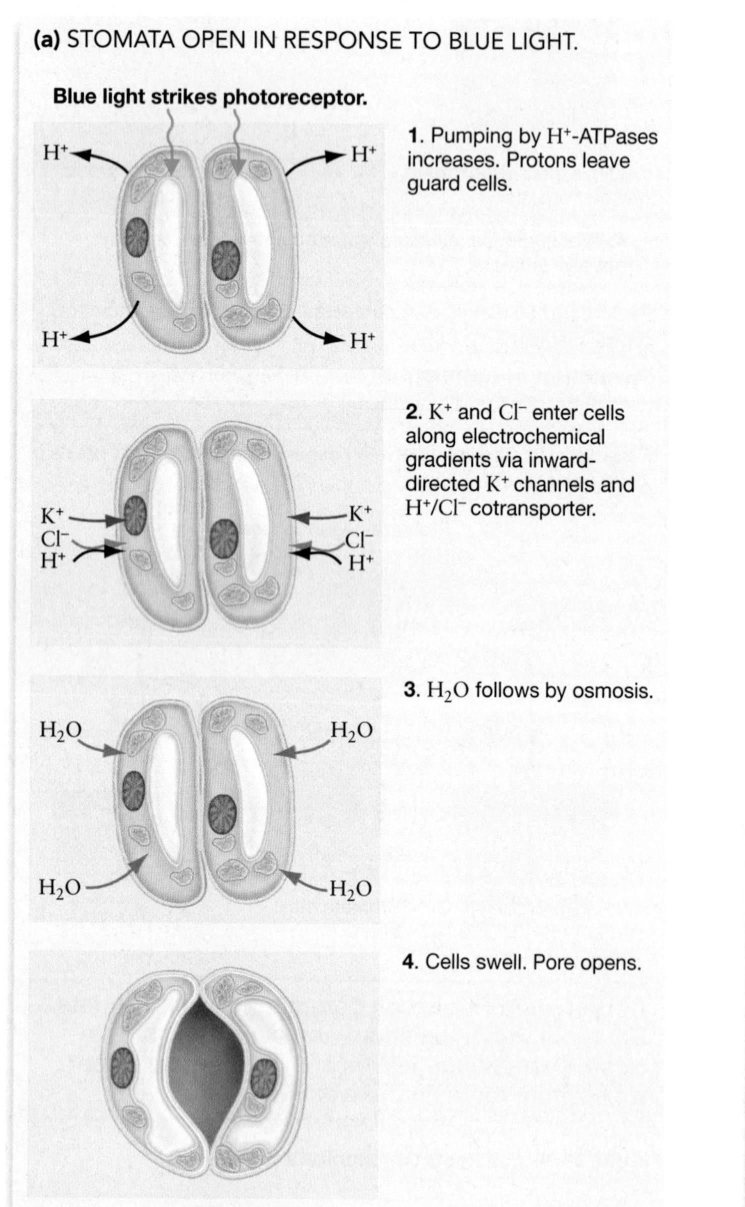

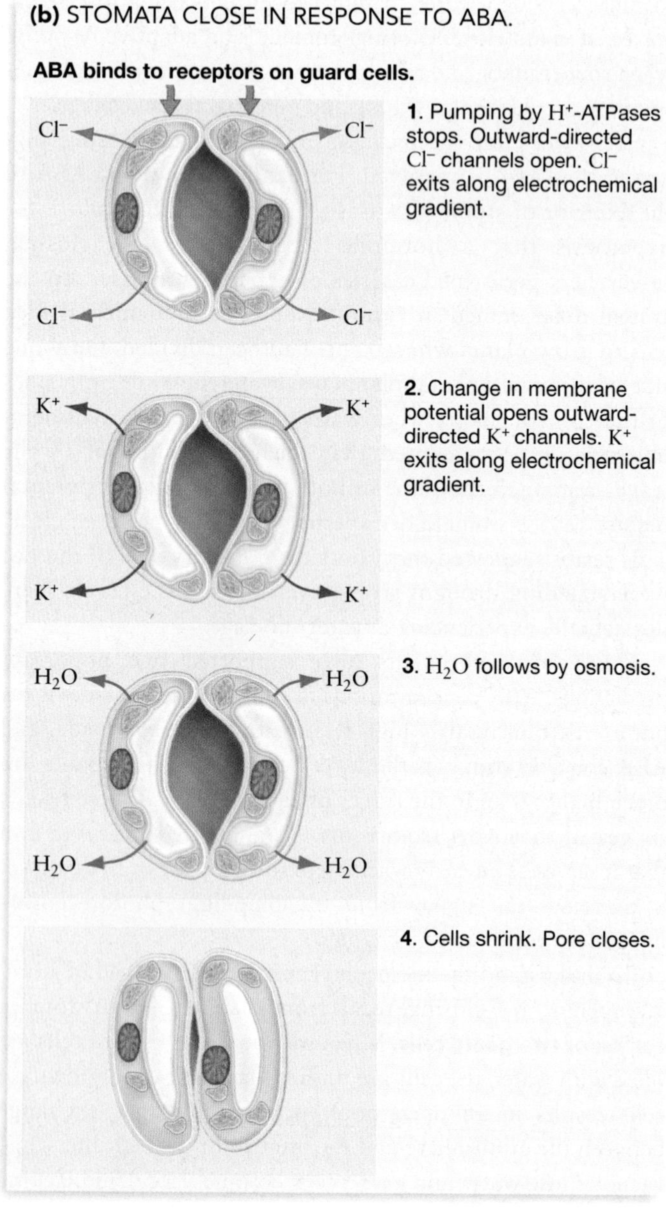

FIGURE 39.31 Changes in Ion Flows Are Responsible for the Opening and Closing of Stomata. (a) Activated blue-light receptors trigger ion flows into guard cells. The cells swell when water follows by osmosis. **(b)** Activated ABA receptors trigger ion flows out of guard cells. The cells shrink when water follows by osmosis.

Ethylene and Senescence

Senescence is the regulated process of aging, decline, and eventual death of an entire organism or particular organs, such as plant fruits and leaves. Like most other aspects of plant growth and development, senescence is triggered and regulated by complex interactions between several different hormones in response to changes in temperature, light, and other factors.

The hormone most strongly associated with senescence is **ethylene**. Like other plant hormones, ethylene is simple in structure and is active at small concentrations. Unlike other plant hormones, however, ethylene is a gas at normal physiological temperatures. Ethylene is synthesized from the amino acid methionine and is strongly involved in three aspects of senescence in plants: (1) fruit ripening, which eventually leads to the aging and rotting of fruit; (2) the fading of flowers; and (3) abscission, or the detachment of leaves. In addition, ethylene influences plant growth and is a stress hormone induced by drought and other conditions. The range of physiological responses regulated by ethylene is surprisingly large.

Ethylene was initially "discovered" in ancient China, when fruit growers noticed that burning incense in closed rooms hastened the ripening of pears. Westerners made a similar observation in the late 1800s, when gas street lamps came into wide use in cities and premature leaf drop subsequently occurred in the plants growing near leaky gas lines. Researchers followed up on the latter observation by showing that ethylene in lamp gas was the molecule responsible for the leaf loss; ethylene is also present in incense smoke. In the 1930s ethylene was found in the gases that are released by ripening apples.

Subsequently, biologists documented sharp spikes in ethylene production during fruit ripening in tomatoes, bananas, apples, and certain other species. Follow-up research on these species showed that ethylene induces the production of some of the enzymes required for the ripening processes, as well as an increase in cellular respiration, which furnishes ATP. During ripening, stored starch is converted to sugar, enhancing sweetness; protective toxins are removed or destroyed; cellulose is broken down, degrading cell walls and softening the fruit; chlorophyll is broken down; and pigments and aromas that signal ripeness are produced. Biologists interpret fruit ripening as an adaptation that enhances the attractiveness of fruits to birds, mammals, and other animals that disperse seeds to new locations.

Today, fruit growers manipulate ethylene levels to control fruit ripening. For example, they treat green bananas with ethylene after the bananas have been shipped to encourage ripening (**Figure 39.32**). Conversely, apples are stored in warehouses with high concentrations of CO_2 and low concentrations of O_2, which inhibits ethylene production in the fruit. Apples stored under these conditions can be sold long after their original harvest date, when untreated fruits have rotted.

FIGURE 39.32 Ethylene Speeds Ripening and Other Aspects of Senescence. These bananas are identical, except that the bunch on the right was exposed to the plant growth regulator ethylene.

Ethylene's effects on leaf senescence and leaf abscission involve complex interactions with auxin and cytokinins. Healthy leaves produce auxin, which is transported from the leaf to the stem through the petiole. In response to age or to changes in ambient temperature or day length, the amount of auxin that is produced by the leaf decreases. As a result, cells in a region of the leaf petiole called the **abscission zone** become more sensitive to ethylene in the tissue (**Figure 39.33**). This increase in ethylene sensitivity activates enzymes that weaken the cell walls of cells near the base of the petiole. At the same time, chlorophyll in the leaf begins to degrade, and nutrients are withdrawn and stored in parenchyma cells in the stem. Eventually the cell walls at the base of the petiole are degraded enough that the leaf falls. Applications of cytokinin, in contrast, reverse these effects and dramatically extend the life span of leaves. As a result, ethylene and cytokinins are thought to have opposite effects on at least some of the processes involved in senescence.

An Overview of Plant Growth Regulators

The quest to understand how ethylene, cytokinins, and other hormones interact in senescence brings us to the frontier of research on how plant growth regulators work. Current studies are focused on several questions. Primary among these is how cells perceive changes in temperature, moisture, age, or other conditions, and how the hormonal signals that result are received by target cells and transduced into changes in gene expression or protein activation. To date, only a handful of genes that are directly regulated by auxin, cytokinins, ABA, GAs, and ethylene have been identified. In short, a great deal remains to be learned about the third step in information processing during growth responses. The effort to understand chemical signaling in plants is an exceptionally dynamic field of inquiry.

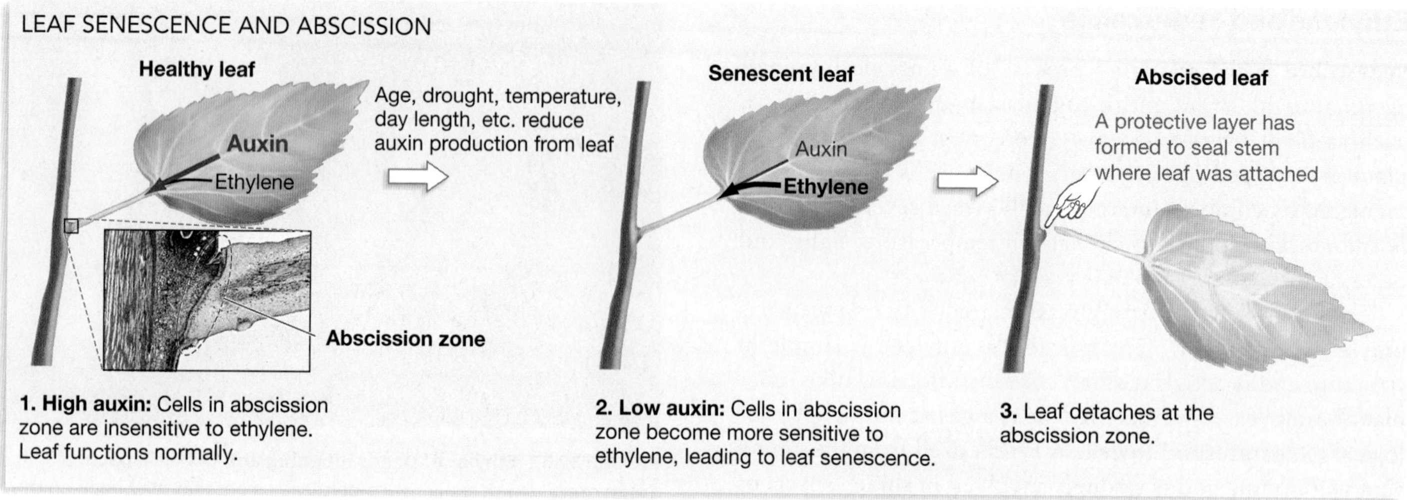

FIGURE 39.33 Leaves Drop in Response to Signals from Auxin and Ethylene. Young leaves produce much larger amounts of auxin than old leaves do. The combination of low auxin and high ethylene concentrations triggers leaf senescence and abscission.

QUESTION When ethylene levels in leaves are high relative to auxin, nutrients are transported from leaves to the stem and chlorophyll synthesis stops (this is why leaves change color in the fall). What is the adaptive significance of these two events?

To help you review some of the major plant hormones, **Table 39.2** provides notes on the structure and function of hormones discussed in this section, along with a recently discovered class of plant hormones called the brassinosteroids. As you study this table, two key observations should emerge:

1. It is common for a single hormone to affect many different target tissues. This means that there can be an array of responses to the same cell-cell signal. To interpret this pattern, biologists point out that most hormones carry a common message to a variety of tissues and organs. Auxin defines the long axis of the body; gibberellins trigger growth; cytokinins promote cell division; ABA slows or prevents growth; ethylene signals senescence.
2. In most cases, several hormones affect the same response. Stated another way, hormones do not work independently—they interact with each other. For example, auxin, the gibberellins, and the brassinosteroids are all involved in cell elongation—a basic physiological response required for growth and movement in stems, roots, and leaves. To interpret this pattern, biologists point out that individual hormones tend to be produced by an environmental cue at a certain location, such as water availability at root tips. Many environmental cues may be changing at the same time, however. For plants to respond appropriately, they need to synthesize information from various environmental cues perceived at various locations in the body.

The complex interactions among hormones involved in the growth response have a purpose: allowing individuals to survive and thrive long enough to reproduce. The same can be said for the hormones involved in protecting plants from danger.

Check Your Understanding

If you understand that...

- Auxin's primary role is to signal the position of cells along the long axis of the plant body. This is possible because auxin is produced in shoot apical meristems and undergoes polar transport to the roots, forming a concentration gradient.
- GAs are general signals to initiate or continue growth.
- ABA is a general signal to stop growth or remain dormant.
- Ethylene is a signal that controls senescence.

You should be able to...

1) Explain how GAs and ABA interact at the molecular level to control seed germination and dormancy.
2) Explain how signals from blue light and ABA interact to control the opening and closing of stomata.

39.6 Pathogens and Herbivores: The Defense Responses

Plants cannot run away from danger. Instead, they have to stand and fight.

Like humans and other animals, plants are constantly threatened by an array of viruses, disease-causing bacteria, and parasitic fungi. In addition, the roots of plants are susceptible to attacks by nematodes—the soil-dwelling roundworms introduced in Chapter 33. The ability to cause disease is called **virulence**; disease-causing agents are termed **pathogens**. If plants

SUMMARY TABLE **39.2 Plant Growth Regulators**

Hormone	Function(s)	Notes	Chemical Structure
Auxin	• Helps define long axis of body (phototropism and gravitropism responses) • Involved in cell elongation and apical dominance • Promotes cell division and leaf expansion • Induces ethylene production • Development of adventitious roots and secondary growth • Differentiation of xylem and phloem.	• First plant hormone ever characterized and isolated. • Produced in apical meristems and young leaves.	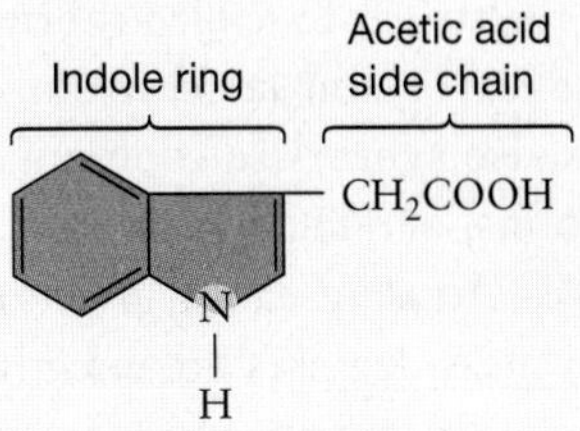 Indole ring; Acetic acid side chain; CH_2COOH; N; H
Gibberellins (GA)	• Promote stem growth via both cell elongation and division • Encourage seed germination • Involved in flowering.	• Fungi that produce gibberellins infect rice plants and induce hyper-elongated stems. • Analysis of these fungi led to discovery of gibberellins. • Produced in apical meristems, immature seeds, anthers (pollen-producing organs).	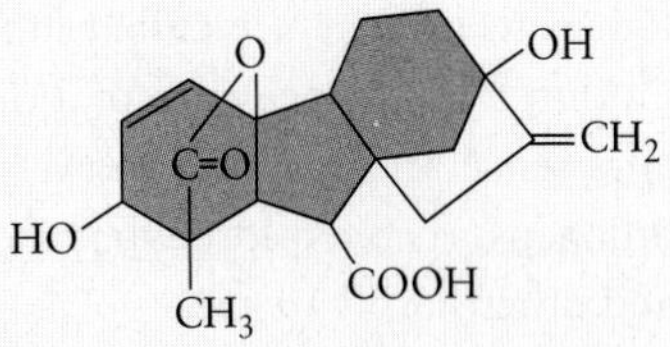 O; OH; C=O; $=CH_2$; HO; COOH; CH_3
Cytokinins	• Promote cell division in the presence of auxin • Promote chloroplast development and development of lateral buds • Delay senescence (aging).	• New data indicate they may act on cell cycle regulators. • Produced in root apical meristems, many other tissues.	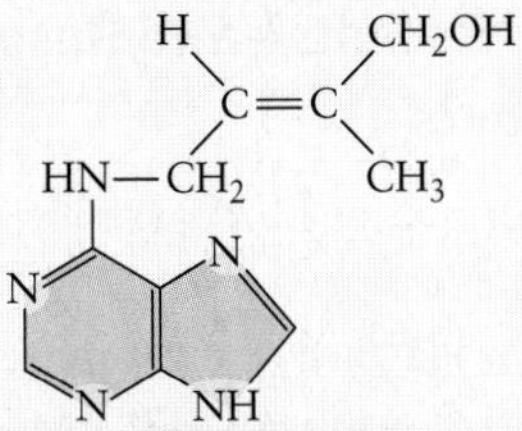 H; CH_2OH; C=C; HN—CH_2; CH_3; N; N; N; NH
Abscisic acid (ABA)	• Inhibits bud growth and seed germination • Induces closure of stomata in response to water stress • Involved in leaf abscission (drop).	• Acts as a stress hormone analogous to cortisol in humans • Produced in almost all cells	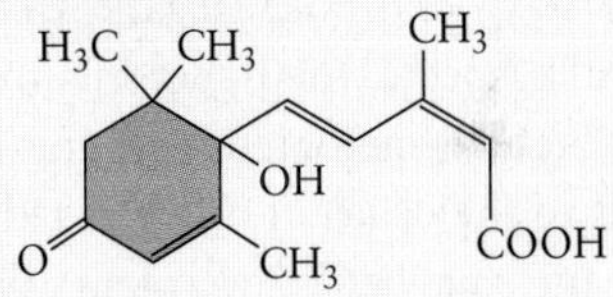 H_3C; CH_3; CH_3; OH; O; CH_3; COOH
Ethylene	• Involved in fruit ripening • Induces senescence of fruits, flowers, and leaves. • Produced when plants are under stress	• A gas; first identified through unusual morphology of plants growing near gas lines for illuminating streets. • Produced in all organs but highest in aging tissues and fruits	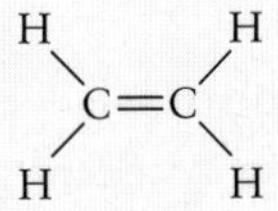 H, H; C=C; H, H
Brassinosteroids	• Promote cell elongation in stems (mutants that lack these hormones or their receptors are dwarfed).	• First steroid hormones discovered in plants • Structurally related to steroid hormones in animals • Produced in almost all tissues.	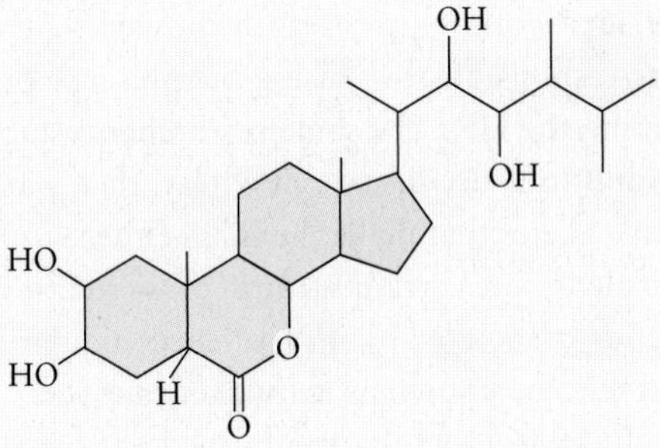 OH; OH; HO; HO; O; H; O

were not able to sense attacks by pathogens and respond to them quickly and effectively, the landscape would be littered with dead and dying vegetation.

The waxy cuticle that covers epidermal cells is an effective barrier to viruses, bacteria, fungi, and other disease-causing agents, and the structures called thorns, spines, and trichomes help protect leaves and stems from damage by herbivores. Most plants also synthesize compounds like opium, cocaine, tannins, nicotine, or THC that protect plant tissues by making the tissues toxic to plant-eating insects and vertebrates.

Although these defenses are effective, they are also expensive in terms of the energy and materials required to produce them. It is not surprising that plants may produce defenses or increase their existing defenses only in direct response to attacks by pathogens or herbivores.

Responses to attacks are called induced defenses, because they are induced by the presence of a threat. Let's first consider how plants sense and respond to the presence of viruses and other pathogens, then explore what they do when attacked by insects and other herbivores.

How Do Plants Sense and Respond to Pathogens?

If a virus, bacterium, or fungus is able to get inside a plant, the cells at the infection site respond by committing suicide. The rapid and localized death of one or a few infected cells is called the **hypersensitive response (HR)**. If the HR is successful, the pathogen is starved as host cells die around it.

In several respects, the hypersensitive response in plants is similar to the cell-mediated immune response in mammals, which leads to the death of infected cells (see Chapter 49). The HR is also extremely effective; plants that mount a hypersensitive response rarely succumb to disease. How do cells sense the presence of pathogens, so that the HR can kill them?

The Gene-for-Gene Hypothesis In the early decades of the twentieth century, crop breeders established that plants have disease-resistance genes that are inherited according to Mendel's rules. These genes came to be known as the **resistance (*R*) loci**. Follow-up research showed that many of the *R* genes are responsible for sensing the presence of pathogens and triggering the HR. The same researchers established that the fungi found to cause disease in wheat, flax, barley, and other crops have alleles that make the fungi either virulent or avirulent (not virulent) on certain strains or varieties of these crops. The genes associated with virulence or avirulence in pathogens came to be known as **avirulence (*avr*) loci**.

In 1956 H. H. Flor published data demonstrating a one-to-one correspondence between the resistance alleles found in host plants and the avirulence alleles found in pathogens. Because each *R* allele in the host corresponded to an *avr* allele in the pathogen, it made sense to suggest that each *R* gene product acted as a receptor for a particular pathogen. The idea was that an interaction between *R* and *avr* gene products triggers a hypersensitive response.

The idea that *R* and *avr* gene products interact in a specific way came to be called the **gene-for-gene hypothesis**. What molecular mechanism could be responsible for this pattern? Researchers who followed up on Flor's work suggested that the HR begins when proteins produced by host plants bind to proteins or other molecules produced by the pathogen. The hypothesis was that *R* genes produce receptors and that *avr* genes produce **ligands**—molecules that bind to receptors. **Figure 39.34** provides a general overview of the gene-for-gene model.

The first breakthrough in testing the gene-for-gene hypothesis occurred when researchers from a variety of laboratories around the world were able to clone and sequence a series of *R* genes from crop plants and *avr* genes from bacterial and fungal pathogens. These results confirmed that *R* genes and *avr* genes exist and that they code for products that could interact at the

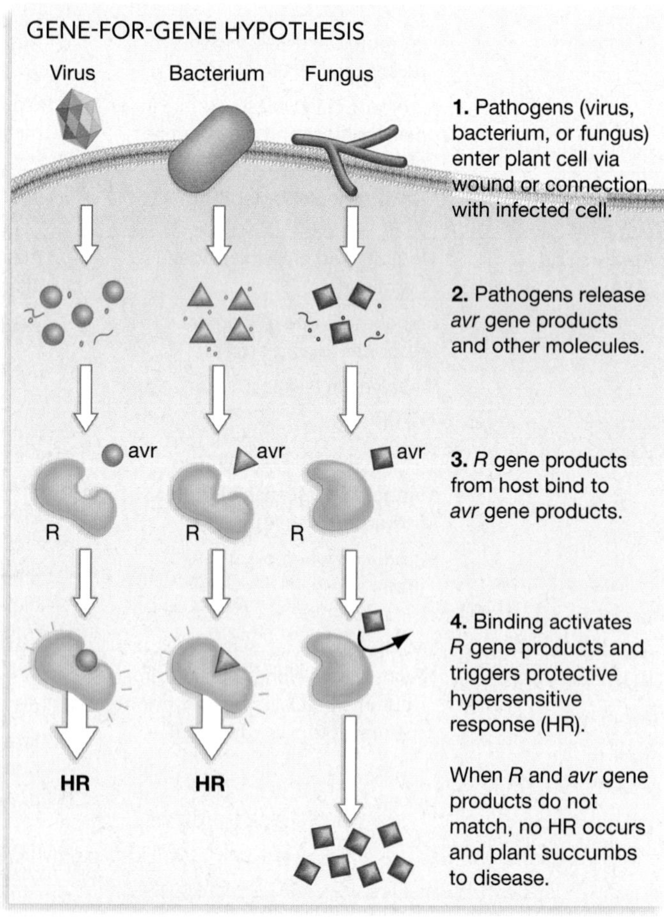

FIGURE 39.34 The Hypersensitive Response Begins when *R* Gene Products Bind to *avr* Gene Products. The gene-for-gene hypothesis predicts that *R* gene products act as receptors for *avr* gene products and that this interaction initiates the HR, which protects plants from pathogens.

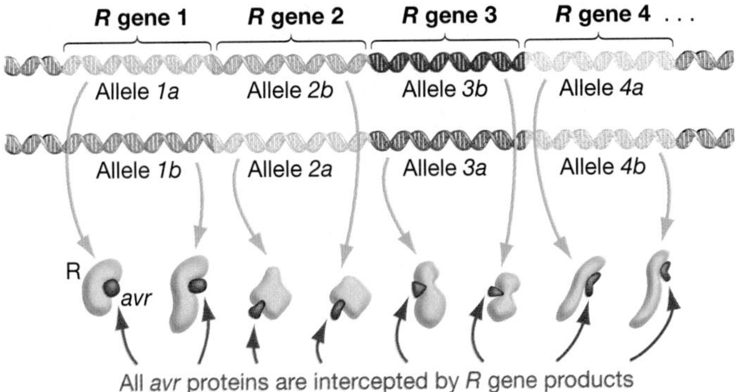

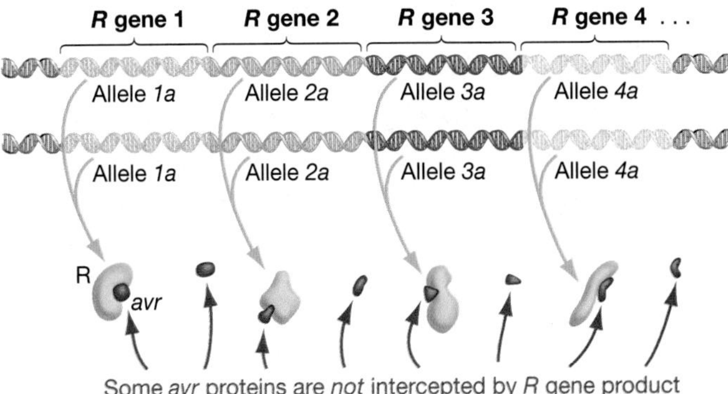

FIGURE 39.35 Polymorphic R Genes Provide Better Protection from Pathogens. Individuals tend to be heterozygous at *R* genes because so many different alleles exist in most populations. When diverse *R* gene products are present, more types of *avr* products can be recognized. Thus, more pathogens can be destroyed.

start of an infection. A second major advance, published in 1996, supported the gene-for-gene hypothesis by showing that *R* products and *avr* products actually do bind to one another. *R* gene products function as "disease receptors." ● If you understand the gene-for-gene hypothesis, you should be able to explain what happens when none of the proteins released by a pathogen bind to an *R* gene product.

Why Is the Existence of So Many Resistance Genes and Alleles Significant? A large number of *R* genes have now been identified in *Arabidopsis*, tomato, flax, tobacco, and other plants. Two general patterns are emerging as data on these genes accumulate: First, *R* genes that are similar in sequence and structure tend to cluster together on the same chromosome. Second, within a population of plants, there are usually many different alleles at each *R* locus. To use a term introduced earlier in the text, *R* loci are highly polymorphic.

These observations are important because they provide hints about the history and function of these genes. For example, clusters of similar genes, or what biologists call **gene families**, are thought to originate through errors in recombination. You might recall from Chapter 20 that **gene duplication** events occur when chromosomes misalign during crossing over. The result of this mutation is a chromosome with an extra copy of a gene. Because the organism with this chromosome already has a functioning copy of the original gene, mutations occurring in the extra copy of the gene do not damage the individual. Instead, the new copy may acquire mutations that make a new function possible. In the case of *R* genes, the hypothesis is that mutations in the new copies gave individuals the ability to recognize and respond to novel *avr* products and thus to new types of pathogens.

Why is it significant that many different alleles exist at each *R* locus? The hypothesis here is that different alleles allow plants to recognize different proteins from the same pathogen. Plants are diploid or polyploid, so they have at least two copies of each *R* gene. If many different alleles exist in a population, then each individual is likely to have at least two different alleles of each gene. The different alleles allow the host to recognize different *avr* products. This is important because new *avr* products constantly arise in pathogen populations via mutation. Plants with different alleles for each of many *R* genes should be able to sense the presence of a wide variety of disease-causing agents and respond by triggering the HR (**Figure 39.35**). ● If you understand this concept, you should be able to explain why bananas—which are propagated asexually and are thus genetically identical—are considered particularly vulnerable to disease epidemics.

The HR Response Once an *R* gene product is activated by binding to a ligand from a pathogen, what happens? Recent work on the HR has established that the response consists of a number of different events (**Figure 39.36**). For example, the binding of an *R* gene product to an avr protein triggers the production of nitric oxide (NO) and reactive oxygen intermediates. **Reactive oxygen intermediates** (**ROIs**) such as hydrogen peroxide H_2O_2 and superoxide ions (O_2^-) trigger reactions that help reinforce cell walls. In addition, nitric oxide and ROIs disrupt enzymes in cells at the point of infection and thus kill those cells. In much the same way, immune system cells in your body frequently use a lethal combination of NO and ROIs to kill cells that have become infected. But plants don't stop there. The R-avr binding event in plant cells also leads to the production of antibacterial and antifungal compounds that are collectively known as **phytoalexins**. As a result, the HR leads to the walling-off of the infected area, the direct killing of pathogens by phytoalexins, and the starvation of pathogens as nearby cells commit suicide. Although the signal transduction events

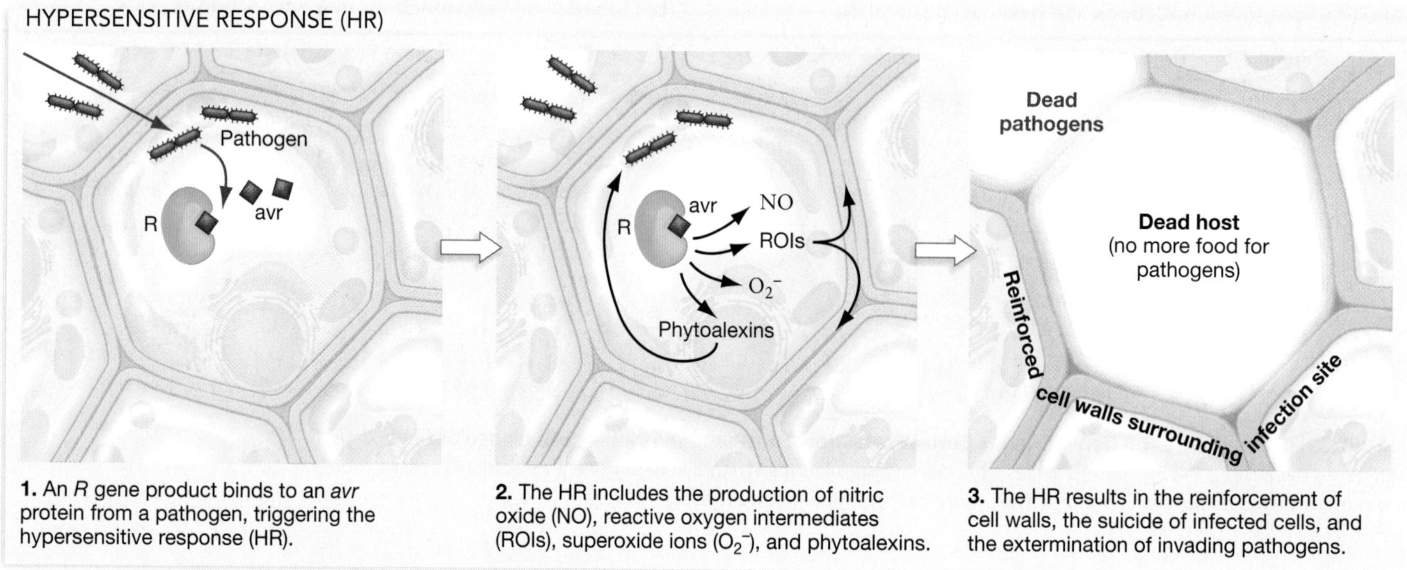

FIGURE 39.36 The Hypersensitive Response Protects Plants from Pathogens.

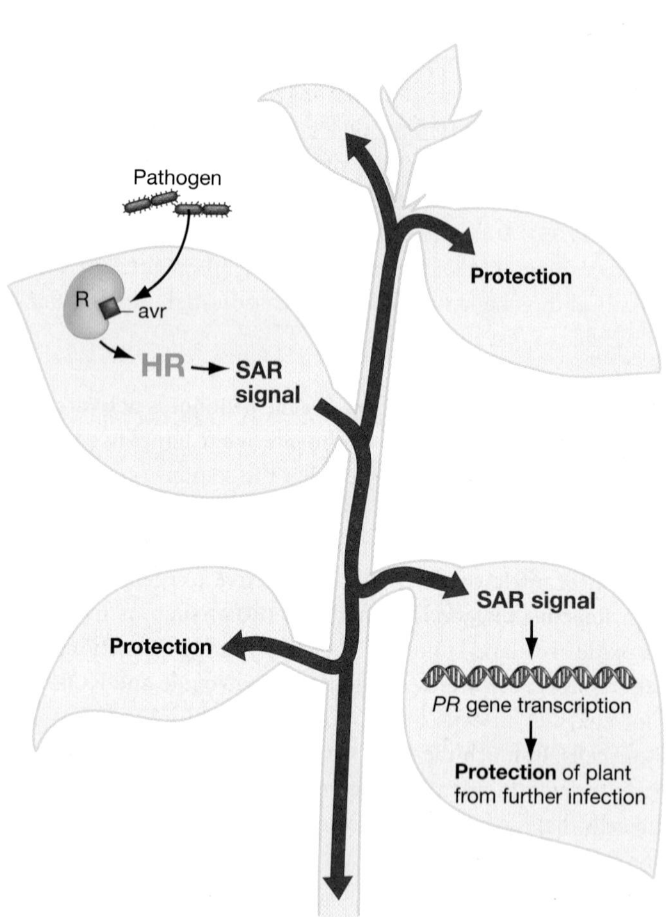

FIGURE 39.37 The Hypersensitive Response Produces a Signal That Induces Systemic Acquired Resistance. This diagram summarizes the current consensus on how the HR and SAR interact.

responsible for these events are not well understood, they are currently the focus of intense research.

Salicylic Acid Extends the HR Once the HR is under way in a localized area of infection, a hormone produced at the infection site travels throughout the body and triggers a slower and more widespread set of events called **systemic acquired resistance (SAR).** Over the course of several days, SAR primes cells throughout the root or shoot system for resistance to assault by a pathogen—even cells that have not been directly exposed to the disease-causing agent.

Figure 39.37 illustrates how the HR and SAR are thought to work together. In addition to triggering the HR, an interaction between the *R* and *avr* gene products leads to the production of a signal that initiates SAR. This hormone acts globally as well as locally—that is, at the point of infection—and results in the expression of a large suite of genes called the *PR* (pathogenesis-related) genes.

When biologists set out to locate the hormone responsible for SAR, their attention quickly turned to salicylic acid. **Salicylic acid (SA)** concentration was found to increase dramatically after tissues are infected with a pathogen. Several lines of evidence suggest that salicylic acid is a defense hormone:

- Direct application of salicylic acid to tissues triggers SAR in a variety of plant species.
- Biologists have isolated a gene that codes for the enzyme salicylate hydroxylase. This enzyme is not normally found in plants, but its product leads to the breakdown of SA. In tobacco plants that received the gene experimentally, both SA accumulation and SAR were abolished. As a result, the

transformed plants were susceptible to infection by a wide variety of pathogens.

- Biologists from several research groups have labeled SA with a radioactive atom and confirmed that it is transported from infected to uninfected tissues. This observation is consistent with the hypothesis that SA is a signal that travels throughout the plant.

It is still controversial, however, whether salicylic acid acts as the SAR hormone or is only a local signal that triggers the expression of genes involved in SAR. Research on the mechanisms responsible for systemic acquired resistance continues.

How Do Plants Sense and Respond to Herbivore Attack?

Over a million species of insects have been discovered and named so far. Most of them make their living by eating leaves, stems, phloem sap, seeds, roots, or pollen. Plants have effective systems for responding to pathogens like viruses, bacteria, and fungi. But how do they withstand the onslaught of insect herbivores?

When researchers started studying why some plant tissues are more palatable and digestible than others, biochemists discovered that many seeds and some storage organs, such as potato tubers, contain proteins called proteinase inhibitors. Proteinase inhibitors block the enzymes—found in the mouths and stomachs of animals—that are responsible for digesting proteins. When an insect or a mammalian herbivore ingests a large dose of a proteinase inhibitor, the herbivore gets sick. As a result, herbivores learn to detect proteinase inhibitors by taste and avoid plant tissues containing high concentrations of these molecules.

Although proteinase inhibitors are found in low concentrations in an array of plant tissues, biologists wanted to test the hypothesis that these proteins might also be part of an induced defense by the plant. To evaluate this idea, researchers allowed herbivorous beetles to attack one leaf on each of several potato plants. In liquid extracted from the other leaves on the attacked plants, proteinase inhibitor concentrations averaged 336μg per mL. In leaves of control plants, where no insect damage had occurred, proteinase inhibitor levels averaged just 103μg per mL of leaf juice. This result supports the idea that a hormone produced by wounded cells travels to undamaged tissues and induces the production of proteinase inhibitors.

What is this wound-response hormone? After years of effort, biologists succeeded in isolating the molecule by purifying the compounds found in tomato leaves and testing them for the ability to induce proteinase inhibitor production. The hormone turned out to be a polypeptide, just 18 amino acids long, called **systemin**. It was the first peptide hormone ever described in plants. Researchers who labeled copies of systemin with a radioactive carbon atom, injected the hormone into plants, and then monitored its location confirmed that systemin moves from damaged tissues to undamaged tissues.

Currently, work on the production of systemin and proteinase inhibitors is focused on determining each step in the signal transduction pathway that alerts undamaged cells to danger. As **Figure 39.38** shows, the data indicate that systemin binds to a receptor on the membrane of an undamaged cell. The activated receptor triggers a long series of chemical reactions that eventually results in the synthesis of a molecule called jasmonic acid. Jasmonic acid, in turn, activates the production of at least 15 new gene products, including proteinase inhibitors. In this way, plants build potent concentrations of insecticides in tissues that are in imminent danger of an attack.

Although proteinase inhibitors are potent herbicides, they are not the only line of plant defense. In some cases of insect attack, plants also call for help from other organisms.

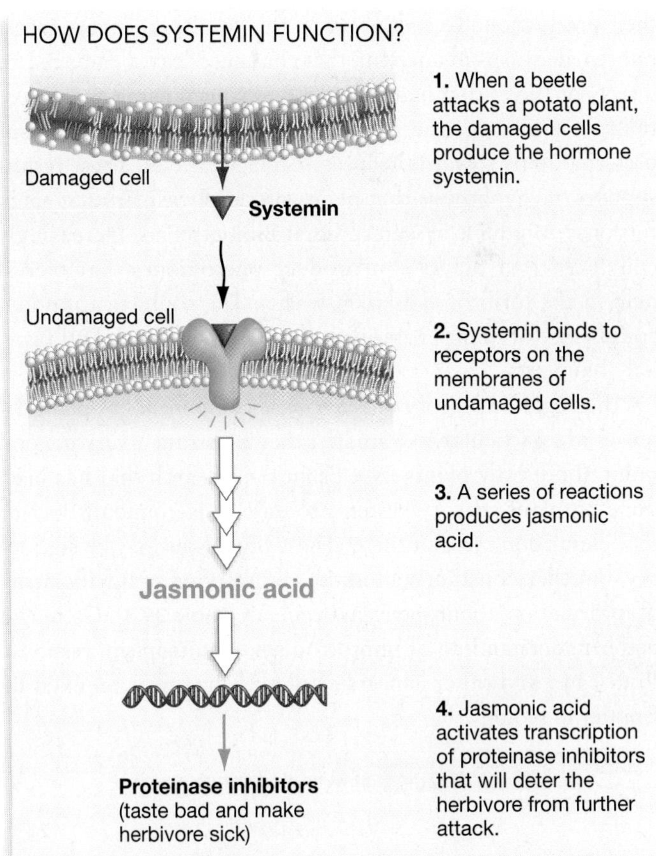

FIGURE 39.38 Signals from Insect-Damaged Cells Prepare Other Cells for Attack. Systemin is a hormone produced by herbivore-damaged cells that initiates a protective response in undamaged cells.

QUESTION Why is "systemin" an appropriate name for this hormone?

Pheromones Released from Plant Wounds Recruit Help from Wasps Caterpillars and other herbivorous insects have enemies of their own—often wasps that lay their eggs in the insects' bodies. When a wasp egg hatches inside a caterpillar, a wasp larva begins eating its host from inside (**Figure 39.39**). An organism that is free living as an adult but parasitic as a larva, such as these wasps, is called a **parasitoid.**

Biologists observed that parasitoids are particularly common when insect outbreaks occurred in croplands. They wondered whether wounded plants might release compounds that actively recruit parasitoids. More specifically, they hypothesized that plants might produce **pheromones**—chemical messengers that are synthesized by an individual and released into the environment, and that then elicit a response from a different individual. Hormones act on cells inside an individual; pheromones act on another individual.

To explore this idea, researchers began collecting compounds that are released from corn seedlings during attacks by caterpillars. Biologists who analyzed the compounds chemically found that insect-damaged leaves produced 11 molecules that were not produced by undamaged leaves. These compounds were not produced by leaves that had been cut with scissors or crushed with a tool; only insect damage triggered their production. To follow up on this result, the investigators put female wasps in an arena that contained leaves damaged by insects or leaves that had suffered mechanical damage. In more than two-thirds of the tests that were performed, the wasps preferred to fly toward the insect-damaged leaves. These results support the hypothesis that plants produce wasp attractants in response to attack by caterpillars. Biologists are increasingly convinced that plants can produce pheromones that recruit help in the form of egg-laden wasps. The sophistication and complexity of plant hormones continue to astonish and fascinate biologists.

Although the sensory and response components of plant defenses are particularly dramatic, they illustrate a very general point about how plants live: From the research that has been done on plant sensory systems to date, it is abundantly clear that plants don't just sit there. These organisms may be stationary, but they constantly monitor and respond to a wide array of information about their environment (**Table 39.3**). Gaining a better understanding of phototropism, gravitropism, response to disease, and other aspects of plant behavior is an exciting frontier in biology.

at www.masteringbio.com
Plant Defenses

FIGURE 39.39 Parasitoids Kill Herbivores. When this plant began to be attacked by this caterpillar, it produced pheromones that attracted a female wasp. The wasp laid her eggs in the caterpillar. As the wasp larvae grew, they devoured the caterpillar.

Check Your Understanding

If you understand that...

- The presence of *R* gene products allows plants to recognize attacks by parasites such as viruses, bacteria, fungi, and nematodes.
- When the protein product of an *R* allele binds to a specific *avr* gene product from a pathogen, a signal transduction sequence results in the hypersensitive response.
- When plants are attacked by insect herbivores, damaged tissues release signals that trigger the production of toxins in undamaged leaves and that recruit enemies of the insects.

You should be able to...

1) Explain the fitness advantage of having a wide array of *R* alleles in a single individual.
2) Explain how you would improve the quality of a popular, high-producing strain of wheat that is susceptible to infection by a particular strain of fungus. Assume that another strain of wheat is available that is not as productive but shows strong resistance to the fungal strain.
3) Explain why plants do not maintain high proteinase inhibitor concentrations in their tissues at all times.

SUMMARY TABLE **39.3 Selected Sensory Systems in Plants**

Stimulus	Receptor	Signal Transduction	Response	Adaptive Significance
Blue light	PHOT1 in stem and leaves	PHOT1 self-phosphorylates; remainder of signal transduction systems unknown.	Phototropism occurs. Also involved in stomatal opening.	Stems grow toward light with wavelengths needed for photosynthesis.
Blue light	Cryptochromes, zeanthin in stems and leaves	Details are unknown.	Stem elongation, stomatal opening, chloroplast movement	Increases efficiency of photosynthesis
Red light	Phytochrome in seeds and elsewhere	Phytochrome changes to P_{fr} form and activates responses.	Seed germinates.	Sunlight triggers sprouting in species that require full sunlight.
Far-red light	Phytochrome in stem	Phytochrome changes to P_r form and induces responses; mechanism is unknown.	Stems lengthen.	Species that require full sunlight attempt to escape shade.
Gravity	Proteins located in plasma membrane or in amyloplast membrane?	Details unknown.	Cells on opposite side of root or shoot elongate; tissue curves.	Roots grow down; shoots grow up.
Touch or wind	Stretch receptors; location unknown	Details unknown, but result is transcription activation in target genes.	Stems grow shorter and thicker.	Individual is more resistant to damage.
Touch	Receptor hair cell in Venus flytrap	Electrical changes in receptor cell's plasma membrane triggers action potential.	Target cells swell; trap shuts.	Plant can capture prey.
Pathogens	*R* gene products	Details unknown.	Hypersensitive response (HR); death of infected cells.	Pathogens starve, so infection is slowed or stopped.
Herbivores	Unknown; activated in response to molecule from herbivore	Details unknown.	Insecticide production; signals to parasitoids.	Herbivores are sickened or killed.

Chapter Review

SUMMARY OF KEY CONCEPTS

Plants are selective about the information they process. They perceive a wide variety of environmental stimuli that affect their ability to grow and reproduce.

In most cases, information processing starts when a receptor protein changes shape in response to blue light, red light, far-red light, gravity, or some other aspect of the environment that affects a plant's ability to thrive. For example, a portion of the phytochrome protein changes shape when it absorbs red light. The same protein changes to an alternate shape when it absorbs far-red light.

You should be able to state a hypothesis to explain why animals do not have phytochromes (a red/far-red light switch) and why plants do not have statoliths that allow them to sense balance.

(MB) **Web Animation** at www.masteringbio.com
Sensing Light

When sensory cells receive a stimulus, they transduce the signal and respond by producing hormones that carry information to target cells elsewhere in the body.

When signal transduction occurs, an external signal is changed into an internal signal. The process starts with a shape change in a receptor protein and continues with the activation of selected proteins via phosphorylation or the production of a second messenger such as Ca^{2+} ions. In receptor cells, signal transduction culminates in the production of hormones that are transported throughout the plant body.

You should be able to explain the analogy between how a plant sensory cell works and the following events: A person sees a barn on fire and calls the rural fire department; the dispatcher rings a siren; in response, members of the volunteer fire department race to the station to get the pumper truck.

Target cells respond to hormonal stimulation in ways that increase the ability of the plant to survive and reproduce.

If a hormone binds to a receptor on a target cell, signal transduction occurs and culminates in changes in gene expression, altered translation rates, or changes in the activity of specific membrane pumps, channels, or ion carriers. For example, cells near the tips of shoots sense changes in blue light and respond by altering the distribution of the hormone auxin. Cells on one side of the shoot accumulate high concentrations of auxin, which causes an increase in the number of proton pumps in those cells' plasma membranes. The resulting acidification of their cell walls activates expansin proteins and allows these cells to elongate much more than do cells on the other side of the stem. In this way, plants bend toward sunlight.

You should be able to predict whether the phototropic response differs in plants that require high-light conditions versus plants that thrive best in low-light conditions.

Hormones are also responsible for regulating how plants grow throughout their lives—especially in response to changes in environmental conditions. Each type of plant growth regulator plays a general role in the life of a plant, and growth responses are usually affected by interactions among several different hormones.

Auxin was the first plant growth regulator to be isolated and is among the best studied. The general function of auxin is to establish and maintain the long axis of the plant body. Besides playing a key role in phototropism and gravitropism, auxin is involved in apical dominance. All of these responses rely on the polar transport of auxin, which establishes a gradient in auxin from the plant's apex to its roots.

Gibberellins signal that conditions for growth are good and promote the initiation or continuation of growth and development. Abscisic acid (ABA), in contrast, signals that environmental conditions are bad by suppressing growth and enforcing dormancy. Regulation of dormancy and growth by ABA and by gibberellins (GAs) are examples of how hormones interact—allowing plants to integrate information from several different stimuli and respond appropriately.

You should be able to suggest a hypothesis explaining why hormones other than auxin are not transported in a single direction.

MB Web Animation at www.masteringbio.com
Plant Hormones; Plant Defenses

QUESTIONS

Test Your Knowledge

1. Which of the following statements about phytochrome is *not* correct?
   a. It is photoreversible.
   b. Its function was understood long before the protein itself was isolated.
   c. The P_{fr} form activates the responses to light.
   d. It phosphorylates itself in response to specific wavelengths of light.

2. Why was it logical to predict that amyloplasts function as statoliths?
   a. They are dense and sediment to the bottom of gravity-sensing cells.
   b. They are present in all plants, and in both shoot and root tissues where gravity sensing occurs.
   c. They make a direct physical connection with membrane proteins called integrins, which have been shown to be the gravity receptor molecule.
   d. Their density changes in response to gravity.

3. If a plant is touched repeatedly over the course of many days or if it experiences long-term exposure to wind, what happens?
   a. Growth of the root system is accelerated, making the plant more stable.
   b. Large-scale changes in gene expression occur, resulting in shoots that are short and stout.
   c. Electrical signals cause leaves to fold up, avoiding damage.
   d. Continued mechanical stimulation indicates that the individual is threatened with destruction, so it initiates flowering in an attempt to reproduce before it dies.

4. Which of the following statements about hormones is *not* correct?
   a. They tend to be small molecules.
   b. They exert their effects only on the same cells that produce them.
   c. They can exert strong effects even when they are present in extremely low concentrations.
   d. They trigger a response by binding to receptors in target cells.

5. In order for auxin to stimulate cell elongation, what two things have to happen?
   a. Transport proteins on the apical and basal ends of the cell must be activated.
   b. Myb proteins must be activated and α-amylase transcription increased.
   c. Water must flow into the cell, and the cell wall must become more extensible.
   d. Water must flow out of the cell, and the cell wall must become more extensible.

6. What evidence suggests that ABA from roots can signal guard cells to close?
   a. If roots are given sufficient water, guard cells close anyway.
   b. If roots are dry, guard cells begin to close—even though leaves are not experiencing water stress.
   c. Applying ABA on guard cells directly causes them to close.
   d. If roots are dry, ABA concentrations in leaf cells drop dramatically.

Test Your Knowledge answers: 1. d; 2. a; 3. b; 4. b; 5. c; 6. b

Test Your Understanding

Answers are available at www.masteringbio.com

1. Phytochromes can be considered "shade detectors," while phototropins such as *PHOT1* can be considered "sunlight detectors." Explain why these characterizations are valid.
2. In the experiment that confirmed the self-phosphorylation hypothesis for the blue-light receptor, researchers inserted the *PHOT1* gene into insect cells. Why was it important for the gene to be expressed in an organism other than a plant?
3. What does *transduce* mean? Give an example of a signal transduction event in plant sensory systems.
4. A plant's response to a given hormone depends on the cells or tissue that receive the signal, the plant's developmental stage or age, the concentration of the hormone, and the concentration of other plant hormones that are present. Provide examples that support each of these claims.
5. Relate the following two observations: (a) Different cytokinin-to-auxin ratios are required to induce the formation of shoot systems and root systems in tissue culture calluses; and (b) auxin is produced in shoot apical meristems and transported to roots, while cytokinins are produced in roots and transported via xylem.
6. Discuss the general role that cytokinins or ethylene serve in plants. Provide evidence that supports your claim.

Applying Concepts to New Situations

Answers are available at www.masteringbio.com

1. In general, small seeds that have few food reserves must be exposed to red light before they will germinate. (Lettuce is an example.) In contrast, large seeds that have substantial food reserves typically do not depend on red light as a stimulus to trigger germination. State a hypothesis to explain these observations.
2. To explore how hormones function, researchers have begun to transform plants with particular genes. In one experiment, a gene involved in cytokinin synthesis was introduced into tobacco plants. When the recombinant individuals matured, they produced more than the usual number of lateral branches. How does this experiment inform our understanding of how cytokinins function? Plants can also be transformed with antisense genes, which code for products that disable specific mRNAs by binding to them. Suggest an experiment that uses an antisense gene to explore cytokinin function. Predict the result.
3. In many species native to tropical wet forests, seeds do not undergo a period of dormancy. Instead, they germinate immediately. Make a prediction about the role of ABA in these seeds. How would you test your predictions?
4. Researchers have shown that stomata do not close if ABA is injected into guard cells. Stomata do close, however, if ABA is applied to the surface of guard cells. Based on these results, researchers claim that the ABA receptor must be on the surface of guard cells, not in the interior. Do you agree? Why or why not?

www.masteringbio.com is also your resource for • Answers to text, table, and figure caption questions and exercises • Answers to *Check Your Understanding* boxes • Online study guides and quizzes • Additional study tools including the *E-Book for Biological Science* 3rd ed., textbook art, animations, and videos.

40 Plant Reproduction

KEY CONCEPTS

- Plants undergo alternation of generations, in which a diploid sporophyte phase alternates with a haploid gametophyte phase. Sporophytes produce spores by meiosis. Gametophytes produce gametes by mitosis.
- In angiosperms, male and female gametophytes are microscopic and are produced inside flowers. Male gametophytes (pollen grains) are portable. Female gametophytes are encased in an ovary and are retained in the flower. When pollen grains land on a flower, they deliver sperm cells that fertilize the egg produced by the female gametophyte.
- Seeds contain an embryo and a food supply surrounded by a coat. In angiosperms, the walls of the ovary develop into a fruit that encloses the seed or seeds.

This chapter focuses on the structure and function of plant reproductive structures, such as those of this chicory flower.

It would be difficult to overemphasize the importance of the reproductive organs and processes that are analyzed in this chapter—for plants, for biologists, and for you.

- For plants, every structure in the body and every physiological process—from the acquisition and transport of water and nutrients to photosynthesis to systems for sensing environmental stimuli and responding—exist for one reason: to maximize the chances that the individual will produce offspring. Reproduction is the unconscious goal of everything that an organism does.
- For biologists, plant reproduction is not only fundamental to understanding how plants work but also the basis for major industries. Agriculture, horticulture, forestry, biotechnology, and ecological restoration draw extensively on what biologists know about plant reproduction.
- For you, plant reproduction means food. Human diets are based on consuming plant reproductive structures—primarily the seeds and fruits derived from flowers. A **flower** is a reproductive structure that produces gametes, attracts gametes from other individuals, nourishes embryos, and develops seeds and fruits. **Seeds** consist of an embryo and nutrient stores surrounded by a protective coat. **Fruits** develop from the flower's seed-producing organ and contain seeds.

This chapter's analysis of plant reproduction begins with a look at sexuality in plants, a review of the life cycle called

Key Concept Important Information Practice It

alternation of generations, and a discussion of asexual reproduction. Then we'll trace how the angiosperms or flowering plants reproduce, beginning with pollination and fertilization and ending with seed dispersal and germination. Angiosperms are the focus of this chapter for three reasons: (1) They represent over 260,000 of the approximately 300,000 land plants described to date, (2) aspects of reproduction in other land plant lineages were introduced in Chapter 30, and (3) virtually every important domesticated plant is an angiosperm. By the end of this chapter, you'll be able to appreciate the practical aspects of flowers as well as their beauty.

40.1 An Introduction to Plant Reproduction

The first thing to recognize about plant reproductive structures and processes is that they are highly variable among species. Consider just one aspect of reproductive organs: their size. As **Figure 40.1** shows, flowers vary in size from microscopic to the size of a small child; seeds and fruits range in size from dustlike particles to coconuts.

Fortunately for students of plant biology, several basic principles unify this diversity of reproductive systems. Let's begin by defining sex.

Sexual Reproduction

Most plants reproduce sexually. As Chapter 12 pointed out, **sexual reproduction** is based on meiosis and fertilization. To review briefly, **meiosis** is a type of nuclear division that cuts the chromosome number in half. **Fertilization** is the fusion of haploid cells termed **gametes**. The result of fertilization is the production of a single, diploid cell called the **zygote**, which will develop into a multicellular individual. Male gametes, or **sperm**, are small cells that contribute genetic information in the form of DNA but few or no nutrients to the offspring. Female gametes, or **eggs**, contain not just DNA but also nutrient stores. The important point is that although both sperm and eggs contribute DNA, they contribute vastly different amounts of other resources to the offspring.

One of the main aspects of sexual reproduction is that meiosis and fertilization result in offspring that are genetically unlike the parents. This is true even if individuals **self-fertilize**, or

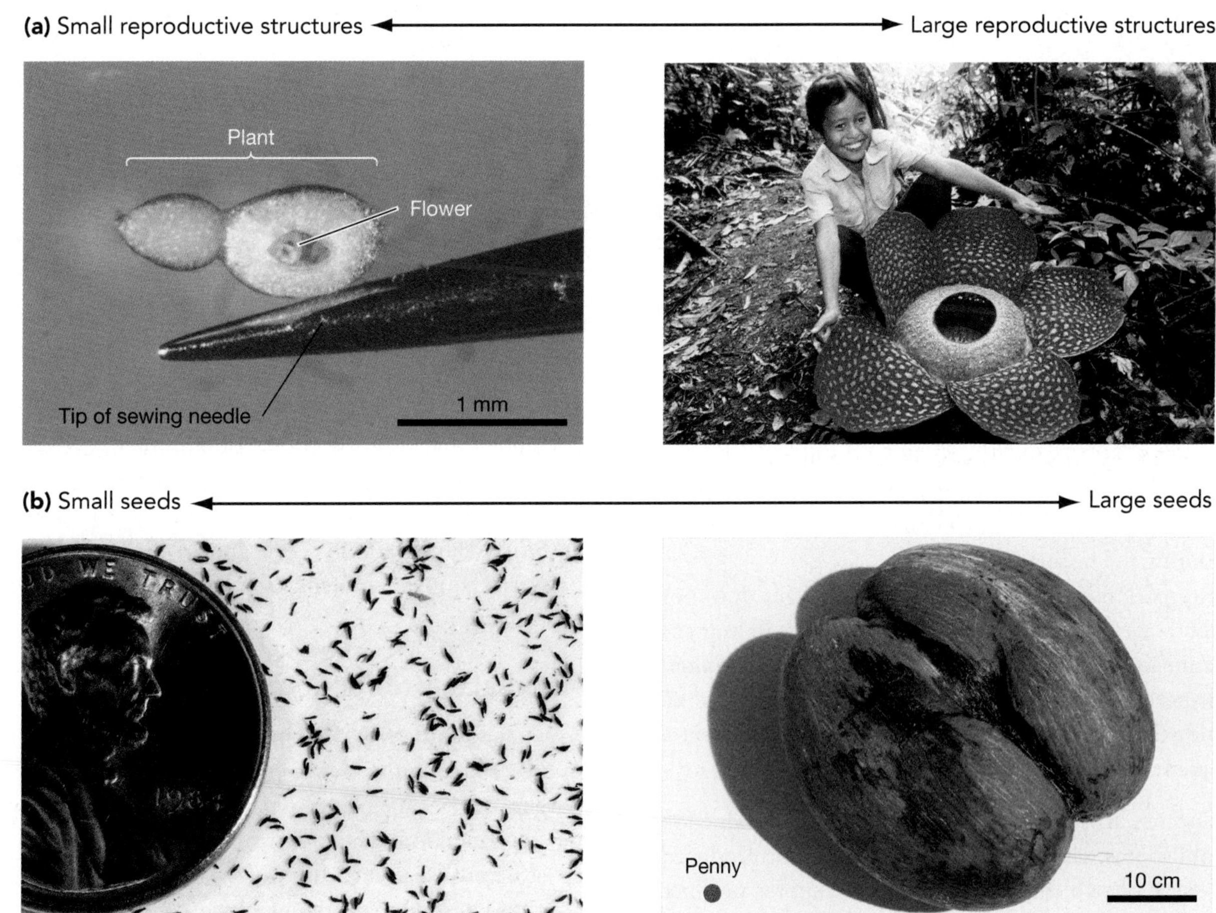

FIGURE 40.1 The Size of Plant Reproductive Structures Is Highly Variable. (a) Reproductive structures in plants range in size from (left) the tiny flowers of *Wolffia* to (right) the gigantic flowers of *Rafflesia arnoldii.* **(b)** All seeds contain an embryo and a food supply, but they vary in size from (left) orchid seeds to (right) the two large seeds inside this double coconut fruit.

self. Selfing occurs when a sperm and an egg from the same individual combine to produce an offspring. In most cases, though, plants **outcross**—meaning that sperm and eggs from different individuals combine to form a viable zygote.

Selfing and outcrossing each have advantages and disadvantages. The primary advantage of selfing is that successful **pollination**, or transfer of sperm-producing pollen grains to the egg, is virtually assured—it doesn't depend on agents other than the plant itself. Biologists have documented this benefit by hand-pollinating plants that normally outcross. In most cases, the hand-pollinated individuals produce far more seed than do individuals that are pollinated naturally. The disadvantage of selfing is that offspring are usually much less diverse genetically than outcrossed offspring are. Selfed offspring can also suffer from inbreeding depression (see Chapter 25). Though outcrossing is riskier in terms of the chances that pollination will occur, it results in genetically diverse offspring that may be much more successful at warding off attacks from viruses, bacteria, and other pathogens (see Chapter 12).

The advantages of sexual reproduction are common to all eukaryotes that undergo meiosis. When and where meiosis occurs, however, is highly variable among species. Let's take a closer look.

The Land Plant Life Cycle

In most animals, meiosis leads directly to the formation of gametes. In plants, the situation is much different. Land plants are characterized by a life cycle with two distinct multicellular forms—one diploid and one haploid. An individual in the diploid phase of the life cycle is called a **sporophyte**, while an individual in the haploid phase of the life cycle is called a **gametophyte**. This type of life cycle is called **alternation of generations** and was introduced in Chapters 29 and 30. Figures 29.21, 30.16, 30.17, 30.21, and 30.22 detail how alternation of generations works in various protists and land plant groups. It is a life cycle that has evolved several times independently during the history of life.

When alternation of generations occurs, meiosis does not lead directly to the formation of gametes as it does in humans and other animals. Instead it leads to the production of haploid cells called spores. A **spore** is a cell that grows directly into an adult individual. Several structures and processes are common to all land plant life cycles:

- Meiosis occurs in sporophytes and results in the production of haploid spores. Unlike zygotes, spores are not produced by the fusion of two cells. Unlike gametes, spores produce an adult without fusing with another cell. Meiosis and spore production occur inside structures called **sporangia**.
- Spores divide by mitosis to form multicellular, haploid gametophytes. Gametophytes produce gametes by mitosis.
- Fertilization occurs when two gametes fuse to form a diploid zygote. The zygote then grows by mitosis to form the sporophyte.

A good way to keep these terms straight is to remember that *sporophyte* means "spore-plant," while *gametophyte* means "gamete-plant." Sporophytes produce spores by meiosis. Gametophytes produce gametes by mitosis.

You might also recall from Chapter 30 that land plant species vary a great deal in how large and long lived the gametophyte and sporophyte are relative to each other, and in how the two phases relate to each other in terms of nutrition. To drive these points home, consider the life-cycle diagrams in **Figure 40.2**. Figure 40.2a shows a liverwort, which is a nonvascular plant. Notice that the largest stage in the liverwort life cycle is the gametophyte. The sporophyte is dependent on the gametophyte for nutrition. Now consider the angiosperm life cycle in Figure 40.2b. A close examination of this figure should convince you that male and female gametophytes are microscopic, physically separate, and completely dependent on the sporophyte for nutrition.

● If you understand the angiosperm life cycle, you should be able to (1) circle the following structures in Figure 40.2b: sporophyte, male spore, female spore, male gametophyte, female gametophyte, sperm, egg, zygote; (2) compare and contrast the same structures in Figure 40.2a; (3) provide evidence to support the statement that in angiosperms, female gametophytes never leave their parent plant; and (4) explain the relationship between the female gametophyte, the seed, and the adult sporophyte.

Liverworts are the most ancient, or basal, lineage on the evolutionary tree of land plants living today. Angiosperms, in contrast, are the most recent lineage of land plants to appear in the fossil record and the most derived branch on the evolutionary tree. Based on these observations, biologists have concluded that over the course of land plant evolution, gametophytes became reduced while sporophytes became more conspicuous. Why this trend occurred is still unknown. The adaptive significance of the gametophyte and sporophyte phases remains a challenge for biologists interested in plant reproduction.

Asexual Reproduction

Asexual reproduction does not involve fertilization and results in the production of **clones**—genetically identical copies of the parent plant. Some plants extend their life indefinitely by asexual reproduction. The oldest of all known plants is a ring of creosote bushes in the Mojave Desert of California. The bushes comprise a clone that originated from a parent plant estimated to have germinated 12,000 years ago.

Although all asexual reproduction is based on mitosis, a wide array of mechanisms are involved. For example, underground horizontal stems called **rhizomes** connect many of the individual

(a) Liverworts: Gametophytes are large and long lived; the sporophyte is small and short lived.

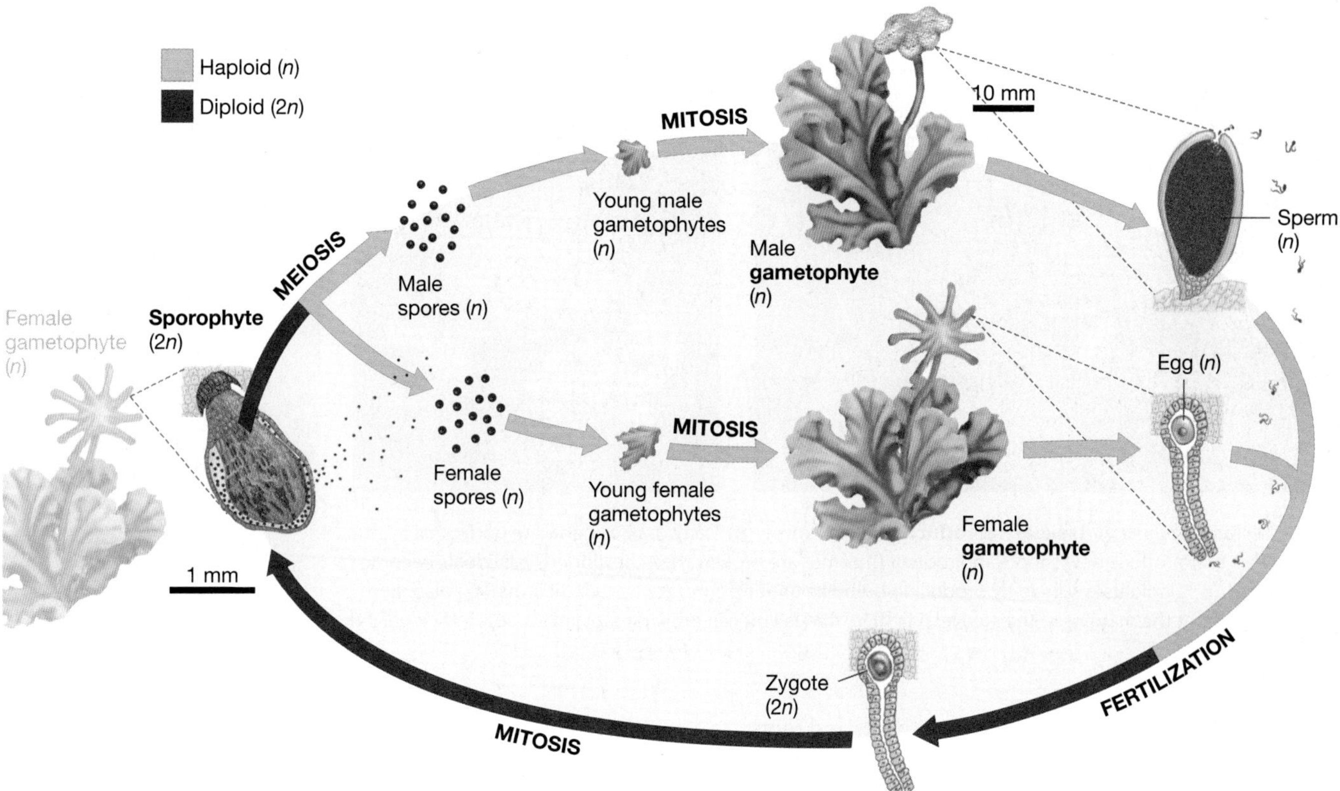

(b) Angiosperms: Sporophyte is large and long lived; gametophytes are small (microscopic) and short lived.

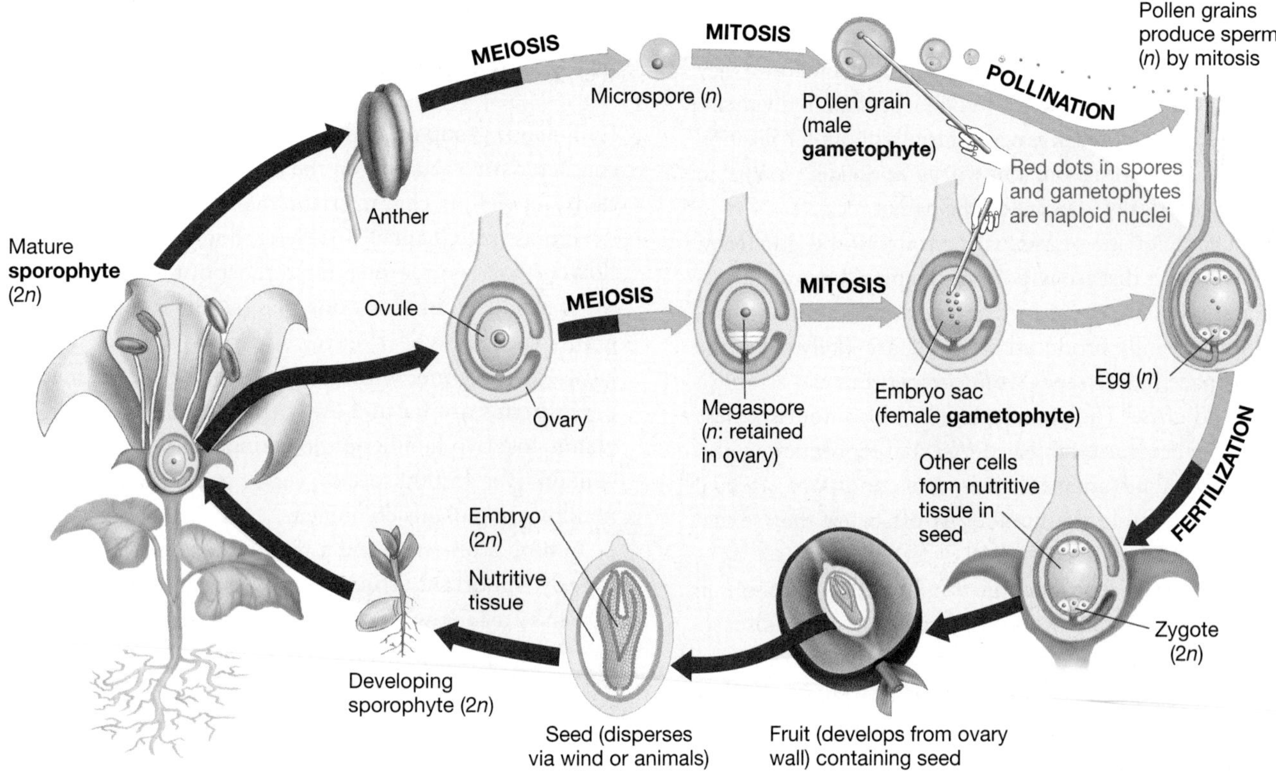

FIGURE 40.2 All Plants Undergo Alternation of Generations. (a) In liverworts and other basal groups of land plants, the sporophyte depends on the gametophyte for nutrition. **(b)** In angiosperms and other more recent groups of land plants, the gametophytes are dependent on the sporophyte for nutrition.

(a) Rhizome

(b) Corm

(c) Plantlets

FIGURE 40.3 The Mechanisms of Asexual Reproduction Are Diverse. (a) Many grasses generate horizontal underground stems that produce new shoots and roots. If rhizomes are broken, then the cloned individuals become physically separated. **(b)** Gladioluses spread by producing bulb-like underground stems called corms. **(c)** Kalanchoe produce small plantlets at the margins of their leaves. Plantlets that fall off can grow directly into a new individual if the location is favorable.

plants in a grass sod. Notice in **Figure 40.3a** that shoots and roots have sprouted from nodes on rhizomes; each of the individuals emerging from the nodes represents an asexually produced offspring. Similarly, the gladiolus plant in **Figure 40.3b** has propagated itself via modified stems called **corms**, which grow under the surface of the soil. The kalanchoe pictured in **Figure 40.3c** produces **plantlets**, which form from meristematic tissue located along the margins of its leaves. When the plantlets mature, they drop off the parent plant and grow into independent individuals. In some species, mature seeds can also form without fertilization occurring. This phenomenon, known as **apomixis,** results in seeds that are genetically identical to the parent.

The key characteristic of asexual reproduction is efficiency. If an herbivore or a disease wipes out the plants growing surrounding a grass plant, the grass can quickly send out horizontal stems. Its asexually produced offspring are likely to fill the unoccupied space before seeds from competitors can establish themselves and grow. The parent plant can also nourish these progeny as they become established. Asexual reproduction also carries a numerical advantage: Under ideal conditions, asexual species produce double the number of offspring that sexual species do (see Chapter 12).

Although asexual reproduction is extremely common in plants, it does have a downside. A fungus or other disease-causing agent that infects an individual plant will probably succeed in infecting the plant's cloned offspring as well, even if they are not still physically connected. This hypothesis is based on the observation that plants fight disease with a wide variety of molecules (see Chapter 39). Because sexually produced offspring are genetically unlike their parents, these offspring have unique combinations of the disease-fighting molecules. Consequently, the offspring may be able to resist infections that devastate their parents. This is an important point in agriculture and horticulture because asexually propagated apples, bananas, and other crops are more susceptible to epidemics caused by diseases or pests than are sexually propagated species.

Given that sexual reproduction is an important way to produce offspring, how do plants accomplish it?

40.2 Reproductive Structures

Each major group of plants, from mosses to angiosperms, has a characteristic variation on the theme of alternation of generations, as well as characteristic male and female reproductive structures (see Chapter 30). Here, though, we will focus on the flower—for two reasons. First, most of the food that humans eat is produced by flowering plants. The staples of our diet—grains such as wheat, corn, and rice—are the fruits of angiosperms. Second, with over 250,000 species named to date, angiosperms are far and away the most species-rich group of plants. Just two families of flowering plants, orchids and asters, contain over 45,000 species, versus about 28,500 species of all nonflowering plants combined.

In both a practical and a biological sense, flowers are enormously important structures. Let's begin our analysis of the flower by asking when plants produce them.

When Does Flowering Occur?

Anatomically, a flower is a modified shoot that develops from a compressed stem and highly modified leaves. In essence, flower formation begins when an apical meristem stops making energy-harvesting stems and leaves and begins to produce the modified stems and leaves that make up flowers. Instead of making more food through photosynthesis, a sporophyte commits to investing energy in sexual reproduction—by producing gametophytes that will produce gametes. When does this happen?

Early experiments on the environmental signals that promote flowering focused on the number of hours of light and dark during a day. Recall from Chapter 39 that changes in night length trigger flowering in many species and that phytochrome is the photoreceptor responsible for detecting night length.

For the intensively studied mustard plant, *Arabidopsis thaliana*, which occupies higher latitudes, night length is an external cue containing important information about the time of year. In contrast, for plants growing at the equator, night length provides no information at all, because it does not vary over the course of the year. In tropical habitats, it is reasonable to expect that external cues other than night length—such as the onset of seasonal rains—might trigger flowering. Based on observations like these, it is not surprising that different species respond to different external cues to induce flowering.

The situation is complex, because some species initiate flowering in response to several independent cues. In *Arabidopsis*, flowering can be triggered by events other than the short night lengths typical of summer. Even if individuals are exposed to long nights, flowering can begin in response to surges of the hormone gibberellin. Recall from Chapter 39 that gibberellins encourage growth. Gibberellins are an internal cue—a signal about the individual's condition.

The advantage of flowering in response to an external cue such as the short nights of summer seems clear. But what sorts of internal conditions might trigger a commitment to flowering? One possibility is the individual's nutritional status. In the case of *Arabidopsis*, flowering could be advantageous when light, water, and nutrient conditions are favorable. An individual that is growing under optimal conditions might be able to flower long before night length signals the arrival of midsummer. If nutritional status or other internal conditions affect the likelihood of flowering, then gibberellins or other hormones could serve as messengers that trigger the production of reproductive organs (see **Box 40.1**). In this way, nutritional status could explain why flowering occurs in the absence of normal external cues.

To summarize, flowering can be stimulated by either internal or external cues, or both. Now the question is, What does the resulting flower look like? How are flowers put together?

The General Structure of the Flower

Structurally, all flowers are variations on a theme. They are made up of four basic organs that are essentially modified leaves: (1) sepals, (2) petals, (3) stamens, and (4) one or more carpels (**Figure 40.4a**). These organs are attached to a compressed portion of stem called the receptacle. Not all four organs are necessarily present in all flowers, however. As **Figure 40.4b** shows, the coloration, size, and shape of these four components are fabulously diverse. Let's consider each part, in turn.

Sepals are leaflike structures that make up the outermost parts of a flower. Sepals are usually green and photosynthetic, and they are relatively thick compared with other parts of the flower. Because they attach to the receptacle in a circle or whorled arrangement, sepals enclose the flower bud as it develops and grows—protecting young buds from damage by insects or disease-causing agents. The entire group of sepals in the flower is called the **calyx**.

Petals are also arranged around the receptacle in a whorl. They are often brightly colored and scented, and they function to advertise the flower to bees, wasps, hummingbirds, and other pollinators. In some cases, the color of the petals correlates with the visual abilities of particular animals. Bees, for

(a) Basic parts of a flower

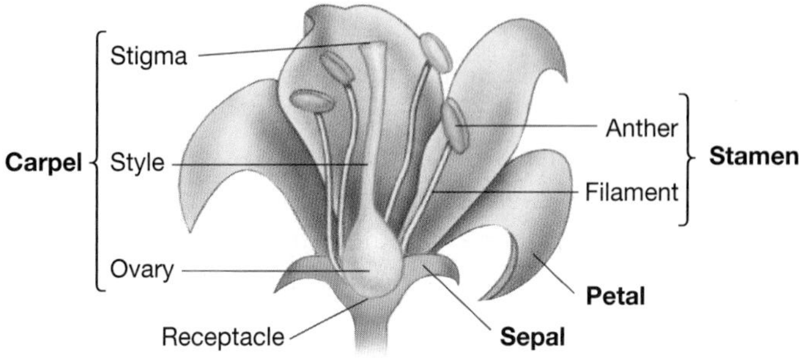

(b) Examples of flower diversity

FIGURE 40.4 The Basic Structures in Flowers Are Highly Variable. **(a)** The basic parts of a flower are the sepals, petals, stamens, and carpels. Not all four parts are present in all flowers. **(b)** The characteristics of the four parts of a flower vary enormously from species to species.

EXERCISE In the photographs, label at least 1 sepal, 1 petal, 1 stamen, and 1 carpel.

BOX 40.1 Is There a Flowering Hormone?

Since the 1930s, biologists have known that exposing even one leaf on a plant to the conditions necessary to induce flowering may result in the whole plant flowering (**Figure 40.5a**). This experimental result suggested that the signal to flower comes from leaves and travels to the apical meristem. Grafting experiments, in which an organ from one individual is physically attached to a different individual, confirmed this result. If you provide one plant with the appropriate flower-triggering night length and then cut off a leaf or stem and graft it onto a second, experimental plant that has never been exposed to the correct conditions for flowering, the experimental plant will flower (**Figure 40.5b**). Like the experiments on phototropism with agar blocks reviewed in Chapter 39, this result supported the hypothesis that some substance or hormone produced in the foreign, transplanted leaf must travel up the plant and trip a developmental switch in the apical meristem, causing it to switch from vegetative growth to flowering mode. Biologists were so convinced that flowering must be induced by a hormone that they gave it the name **florigen,** even though the actual hormone had not been discovered.

Almost 80 years later, researchers have found the florigen molecule. This work began by focusing on the *FLOWERING LOCUS* gene in *Arabidopsis thaliana*, which is known to promote flowering when activated. When researchers exposed leaves to long nights, they found that the gene was activated in the leaf vascular tissue. Biologists have now been able to show that the protein product of the *FLOWERING LOCUS* gene is transported from leaves to the shoot apical meristem, and that the protein's presence triggers the activation of genes required for converting the stem to a flower.

Researchers are increasingly impressed with the complexity of this flowering response. In *Arabidopsis thaliana*, for example, more than 40 genes are now known to be involved in the initiation of flowering. Mutations in some of these genes produce individuals that flower too soon or too late, or that alter the plant's response to night length. Working out the signals, receptors, and response system involved in flowering remains an exciting frontier in biological science.

Experiment

(a) Question: Can signals from single leaves induce flowering?

Hypothesis: Signals from single leaves can induce flowering.

Null hypothesis: Signals from single leaves cannot induce flowering.

Experimental setup:

Prediction: Plant will flower.

Prediction of null hypothesis: Plant will not flower.

Results:

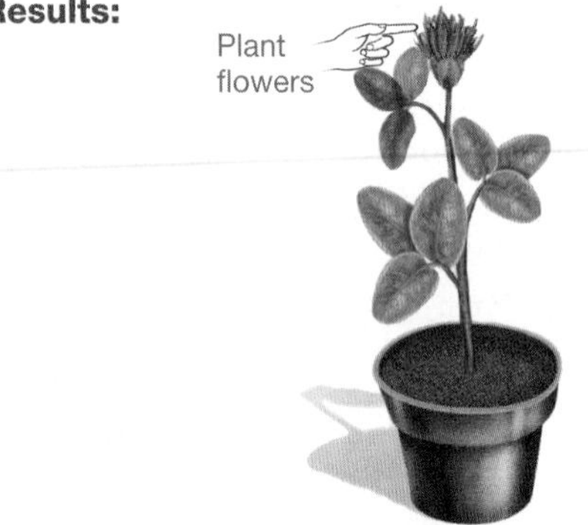

Conclusion: A signal from a single leaf can induce flowering.

Experiment

(b) Question: Can signals from different plants induce flowering?

Hypothesis: Signals from grafted leaves can induce flowering.

Null hypothesis: Signals from grafted leaves cannot induce flowering.

Experimental setup:

Prediction: Plant will flower.

Prediction of null hypothesis: Plant will not flower.

Results:

Conclusion: A signal from a grafted leaf can induce flowering.

FIGURE 40.5 Experimental Support for the Hypothesis that a Hormone for Flowering Exists.

QUESTION For these experiments to support the claim that a flowering hormone exists, control treatments need to be done in each experiment. What are these controls? If the hypothesis is correct, what is the predicted outcome of each control treatment?

(a) What you and a bee see

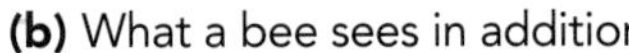

(b) What a bee sees in addition

FIGURE 40.6 Insects See in the Ultraviolet Range. (a) The inflorescence (flower cluster) of a black-eyed Susan, seen by the unaided human eye. **(b)** The same structure, photographed with a camera that records ultraviolet wavelengths that are visible to bees but invisible to humans.

example, respond strongly to wavelengths in the blue and purple regions of the light spectrum. Flowers that attract bees, in turn, often have blue or purple petals with ultraviolet patches. The ultraviolet sections of these petals, which are invisible to humans but visible to bees, frequently highlight the center of the flower **(Figure 40.6)**. Why? In these flowers, the base of the petals contains a gland called a **nectary**. A nectary produces the sugar-rich fluid called **nectar**, which is harvested by many of the animals that visit flowers. Wind-pollinated angiosperms such as oaks, birches, pecans, and grasses have flowers that lack nectaries and have small petals or no petals at all.

The entire group of petals in a flower is called the **corolla**. In some species, the petals within the corolla vary in size, shape, and function. For example, flattened petals may provide a landing pad for flying insects, while elongated, tubelike petals frequently have a nectary at their base that can be reached only by animals with long beaks or tongue-like probosci. Other petals protect the reproductive organs located inside the corolla. Specialized cells in the petals may also synthesize and release molecules that provide a scent attractive to certain species of pollinating insects.

Stamens are reproductive structures that produce male gametophytes—also known as pollen grains. These male gametophytes, in turn, produce sperm. Each stamen consists of two components: (1) a slender stalk termed the **filament** and (2) the pollen-producing organs called **anthers** (see Figure 40.4a).

The fourth reproductive structure is the **carpel**, which produces female gametophytes. A carpel consists of three regions: (1) the **stigma**, which is a moist tip that receives pollen; (2) a slender stalk called the **style**; and (3) the **ovary** at the base of the carpel (see Figure 40.4a). Inside the ovary, female gametophytes are produced in structures called **ovules**. The female gametophytes, in turn, produce eggs. An ovary may contain more than one ovule.

In most angiosperm species, stamens and carpels are produced on the same individual. Flowers that contain both

(a) Monoecy

(b) Dioecy

FIGURE 40.7 Male and Female Flowers Can Occur on the Same Individual or on Different Individuals. (a) The tassels of corn are male flowers; ears are female flowers. **(b)** In *Cannabis sativa*, male and female flowers are found on different individuals.

stamens and carpels are referred to as **perfect**. Flowers can also be **imperfect**, meaning they contain either stamens *or* carpels, but not both. In some cases, separate stamen- or carpel-producing flowers occur on the same individual. This condition is termed monoecy. The corn plant illustrated in **Figure 40.7a** is **monoecious** (literally, "one-house"). In corn, the tassel is a collection of stamen-producing "male" flowers, and the ear is a group of carpel-producing "female" flowers.[1] In contrast, some species with imperfect flowers are **dioecious** ("two-houses")—meaning each individual plant produces either stamen-bearing flowers only, and could be considered "male," or carpel-bearing flowers only and is "female." The *Cannabis sativa* illustrated in **Figure 40.7b** are examples of a dioecious plant.

Sepals protect the other organs and petals attract pollinators, but stamens and carpels are the business end of the enterprise—they are the structures where meiosis and gametophyte formation take place. Let's look at them in more detail.

[1]Technically, flowers are not referred to as male and female. Instead, they are staminate or carpellate. Staminate flowers produce stamens, which produce pollen grains, which produce male gametes (sperm). Carpellate flowers produce carpels, which contain ovaries. Female gametophytes develop inside ovaries and produce female gametes (eggs). For convenience, though, the text will sometimes refer to male and female flowers and reproductive structures.

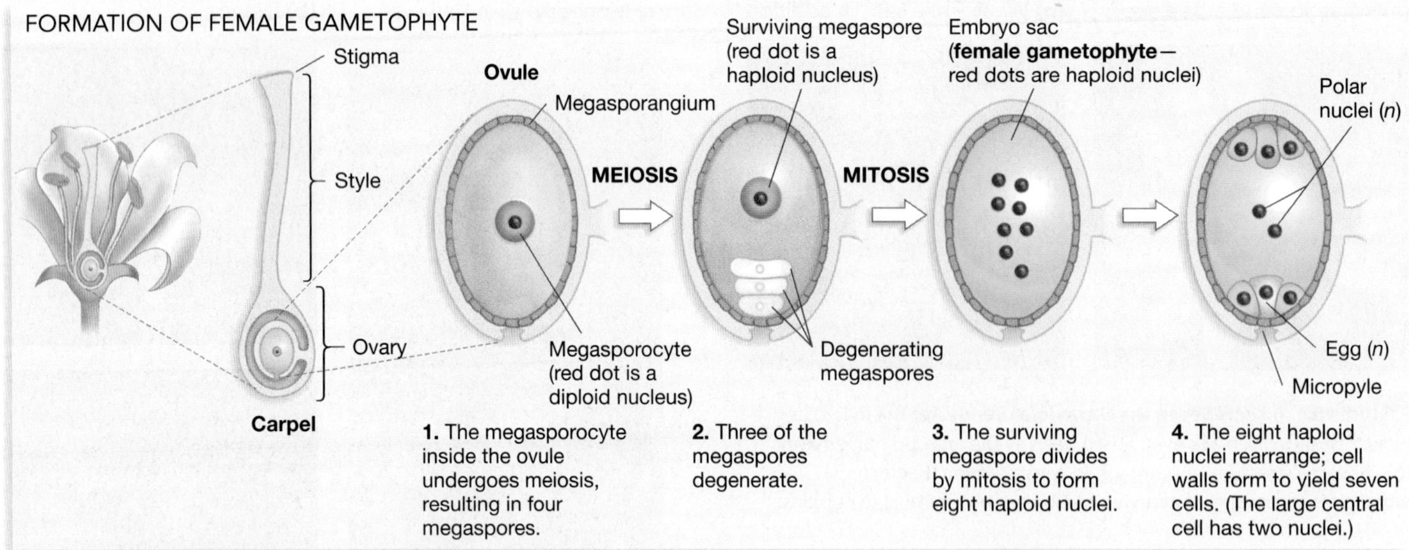

FIGURE 40.8 In Angiosperms, Female Gametophytes Are Produced by Megaspores. In a carpel, the ovary encloses one or more ovules. A megasporocyte inside the ovule divides by meiosis, but only one of the four products, called megaspores, survives. The surviving megaspore divides by mitosis to form the egg and a variety of other cells within the female gametophyte.

QUESTION Give the definition of a spore. Why do megaspores conform to this definition? Give the definition of a gametophyte. Why does the embryo sac conform to this definition?

How Are Female Gametophytes Produced?

What purposes do the three parts of the carpel serve? The function of the stigma and style will become clear in Section 40.3; for now let's concentrate on what happens inside the ovary. **Figure 40.8** provides a cross-sectional view of the inside of a typical angiosperm ovary. Notice that it contains one or more structures called ovules. Each ovule contains a cell called the megasporocyte. The megasporocyte is found inside a structure called the megasporangium. (The use of "mega" is appropriate, because these structures are much larger than their counterparts in the stamen.) The megasporangium is comparable to spore-producing organs found in other plants, such as the sporangia found on the back of fern leaves.

Figure 40.8 illustrates the steps involved in production of the female gametophyte. There are three important things to note:

1. The megasporocyte divides by meiosis.
2. Four haploid cells called **megaspores** result from meiosis, but three degenerate. No one is sure how or why this happens.
3. The surviving megaspore divides by mitosis to produce a multicellular, haploid structure. This is the female gametophyte—usually known as the **embryo sac**. In many angiosperms the embryo sac contains eight haploid nuclei. Notice in Figure 40.8, step 4, that these nuclei segregate to different positions. Cell walls form around them, resulting in seven cells. In the species shown here two of the nuclei, which are called **polar nuclei**, stay together within one central cell. When the female gametophyte is mature, a gamete (egg cell) has formed and is ready to be fertilized.

To summarize, a diploid cell in the sporophyte divides by meiosis to form a spore, which then divides by mitosis to form the female gametophyte. The most important components of the female gametophyte are the polar nuclei and the egg cell. There are two polar nuclei in many angiosperms, but the number varies among species. The polar nuclei are found in the large central cell—the largest cell in the ovule. The egg is located at one end of the female gametophyte, near an opening in the ovule called the **micropyle** ("little-gate").

How Are Male Gametophytes Produced?

Figure 40.9 provides a detailed look at the stamen and the steps that occur in the production of male gametophytes. Notice that a stamen consists of two major parts: an anther and a filament. Inside the anther, structures known as microsporangia contain diploid cells called microsporocytes, which undergo meiosis. Each haploid cell that results is called a **microspore**. Each of the four resulting microspores divides mitotically to form a haploid, immature male gametophyte also known as the **pollen grain**.

At this immature stage—before it has produced sperm—the male gametophyte consists of two cells: a small generative cell enclosed within a larger cell called the tube cell. The male gametophyte is considered mature when the haploid generative cell produces two sperm cells via mitosis. In some species, this maturation step occurs while pollen is still in the anther. In other species, maturation and sperm production don't occur until after the pollen grain lands on a stigma and begins to grow.

The wall of a pollen grain develops a tough outer coat that includes the watertight compound called sporopollenin, intro-

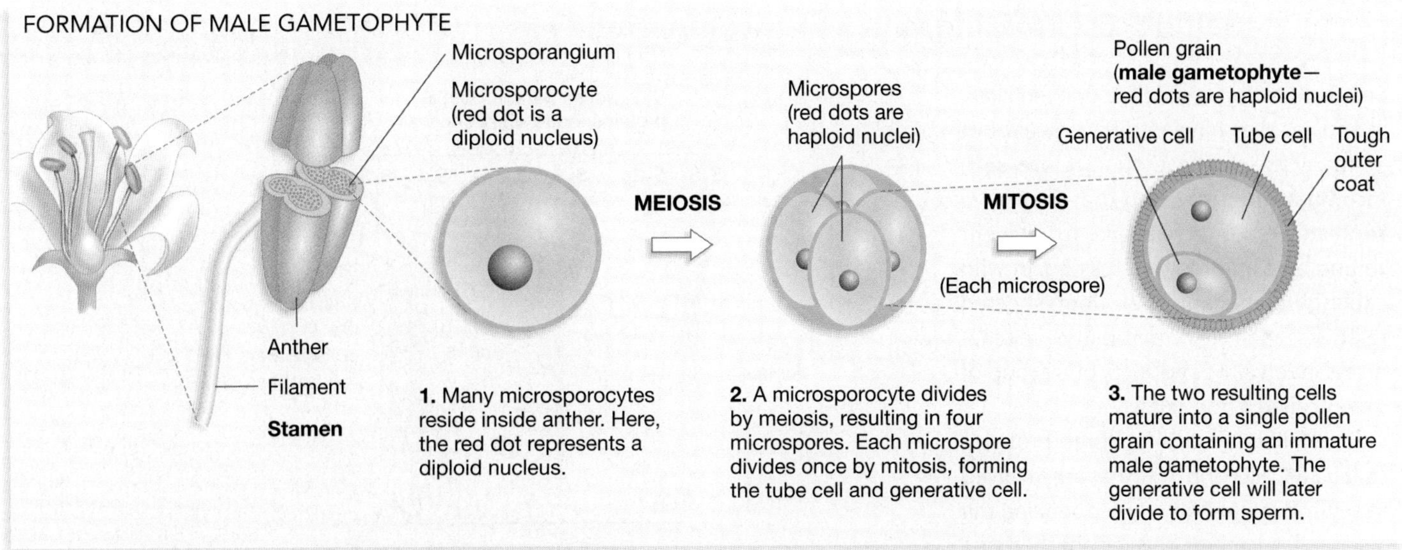

FIGURE 40.9 In Angiosperms, Male Gametophytes Are Produced by Microspores. Microsporocytes inside the anther divide by meiosis. The resulting cells, called microspores, each divide once by mitosis to form a male gametophyte.

QUESTION Give the definition of a spore. Why do microspores conform to this definition? Give the definition of a gametophyte. Why do pollen grains conform to this definition?

duced in Chapter 30. This coat protects the male gametophyte when the pollen is released from the parent plant into the environment. Depending on the species, pollen grains may be dispersed by an animal, the wind, or water currents.

Check Your Understanding

If you understand that...

In angiosperm sporophytes, flowers produce spores that develop into female and male gametophytes.

- The female reproductive structures called carpels contain ovaries. Ovaries enclose structures called ovules. Female gametophytes are produced inside ovules.
- Formation of a female gametophyte begins when a diploid megasporocyte inside an ovule undergoes meiosis. The product of meiosis is a haploid megaspore. The megaspore divides by mitosis to form the female gametophyte—including the egg and polar nuclei.
- Male gametophytes are produced inside reproductive structures called anthers.
- Formation of the male gametophyte begins when a diploid microsporocyte undergoes meiosis to form haploid microspores. Microspores divide by mitosis to form the male gametophyte—including a generative cell that will divide by mitosis to form sperm.

You should be able to...

1) Make diagrams showing the major steps in the formation of eggs and sperm in angiosperms.
2) Add labels your diagrams that indicate where meiosis occurs and that identify each of the structures mentioned in the bulleted points above.

And now we're finally ready to explore the moment of truth: What happens when a pollen grain arrives at a mature carpel of the same species, where an egg cell is waiting?

40.3 Pollination and Fertilization

Pollination is the transfer of pollen grains from an anther to a stigma; fertilization occurs when a sperm and an egg actually unite to form a diploid zygote. The two events are separated in space and time.

Pollination is not restricted to angiosperms. The gymnosperms introduced in Chapter 30 also package their male gametophytes into pollen grains. This section will focus on pollination and fertilization in flowering plants, however, because managing pollination and fertilization in angiosperms is a critical challenge for fruit growers and plant breeders. In addition, pollination and fertilization systems are thought to be key to the angiosperm's evolutionary success.

What is so remarkable about pollination and fertilization in flowering plants? Why have these innovations allowed angiosperms to become so successful in terms of their numbers of species?

Pollination

Cross-pollination is said to occur when pollen is carried from the anther of one individual to the stigma of a different individual. Self-pollination, in contrast, occurs when pollen falls from the anther of one individual onto the stigma of the same individual. Self-pollination leads to self-fertilization. **Box 40.2** discusses some of the molecular mechanisms that prevent self-fertilization in certain species.

BOX 40.2 Selfing, Outcrossing, and Self-Incompatibility

Self-fertilization is possible or even likely in species that have perfect flowers. But in many cases these species are **self-incompatible**—meaning that they have sophisticated mechanisms to prevent selfing and enforce outcrossing. In wild cabbage plants and their domesticated relatives, self-incompatibility mechanisms involve the products of a group of genes called *S* (for "self").

The *S* locus codes for three proteins, called SRK, SLG, and SCR, that are involved in self-incompatibility. By sequencing the *S* genes, researchers have established that up to 50 different *S* alleles may be present in a single population. This observation is remarkable—*S* is among the most highly polymorphic loci known in plants. (Recall from Chapter 13 that polymorphism is the occurrence of more than one allele at a particular gene.) Is there some connection between polymorphism and self-incompatibility?

To answer this question, researchers set out to determine what the *S* gene products do. Analyses showed that the SRK protein is located in the membranes of cells covering the surface of the stigma and that the SLG protein is in the walls of stigma cells. *SCR*, in contrast, is expressed by diploid cells in the anther. SCR proteins are found on the surfaces of pollen grains.

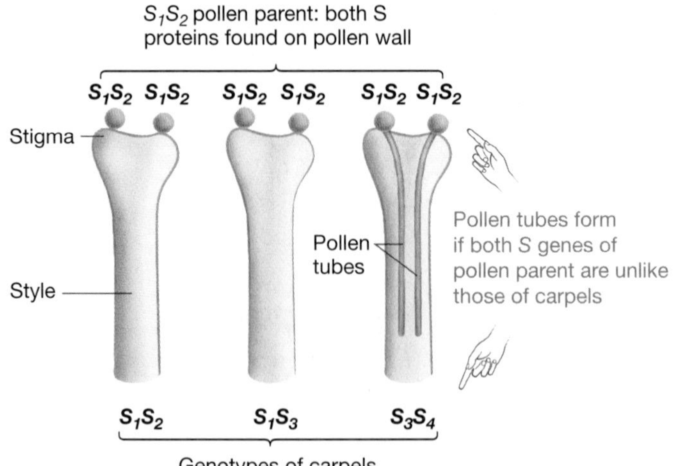

FIGURE 40.10 The Products of the S Gene Help Cabbage-Family Plants Avoid Inbreeding. If a pollen grain carries S proteins that are encoded by alleles *different* from those for the S proteins found on a stigma, then the pollen is allowed to germinate. If the pollen grain carries S proteins that match the *S* alleles found in the carpel, then a reaction takes place that prevents the pollen from germinating.

These observations were consistent with the model for the self-incompatibility system shown in **Figure 40.10**. The fundamental hypothesis is that, for a cross to be compatible, both alleles of the male parent must be different from both alleles of the female parent. For example, suppose the SRK and SLG proteins on a stigma come from S_1 and S_2 alleles (out of the 50 *S* allele types in the population). If the SCR proteins from a pollen grain are from S_3 and S_4 alleles, they must be from a different individual. In this case, pollen tubes—projections that grow out of the pollen grain—would be allowed to grow into the stigma. Sperm would travel down the tube and fertilization could occur. But if the SCR proteins were from S_1 and S_2 alleles, it would likely represent pollen from the same flower. In this case, the model maintains that the SRK and SLG proteins prevent the pollen tube from growing. In this way, self-fertilization is avoided. Investigators are now focused on understanding exactly how the S proteins on stigma and pollen interact.

Cross-pollination can be accomplished in a number of ways: Pollen can be carried from flower to flower by abiotic agents such as wind or water or by biotic intermediaries such as insects, birds, or bats. Animals visit flowers to eat pollen grains, harvest nectar, or both. As an animal feeds from a flower, pollen grains adhere to its body incidentally. When the same individual visits another flower of the same species to feed, some of these grains are deposited on a stigma of the second flower. In this way, cross-pollination is accomplished.

In most cases, animal pollination is an example of **mutualism**: a mutually beneficial relationship between two species. Pollinators usually benefit by receiving food; flowering plants gain by having their male gametophytes transferred to a different individual so that outcrossing takes place.

What Is the Adaptive Significance of Pollination? In mosses, ferns, and other groups that do not form pollen, sperm have flagella and swim to the egg through droplets of water. In conifers and most other gymnosperms, pollen is transmitted from male to female by wind. In some of the other groups that produce pollen, such as the cycads, gnetophytes, and angiosperms, many species are pollinated by animals—particularly by insects.

When these observations are placed in an evolutionary context, several important patterns emerge. Recall from Chapter 30 that researchers have compared DNA sequences from several genes found in all plants and have then used the observed similarities and differences to reconstruct which plant groups are most closely related and which are most distantly related.

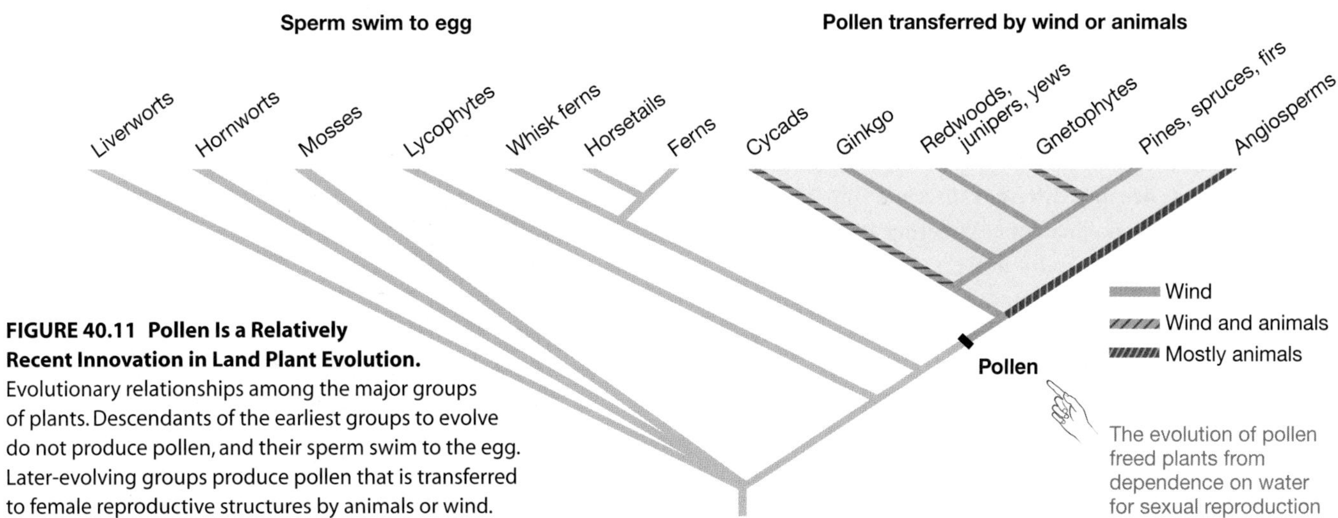

FIGURE 40.11 Pollen Is a Relatively Recent Innovation in Land Plant Evolution. Evolutionary relationships among the major groups of plants. Descendants of the earliest groups to evolve do not produce pollen, and their sperm swim to the egg. Later-evolving groups produce pollen that is transferred to female reproductive structures by animals or wind.

The most recent phylogenetic tree estimated from this type of analysis is shown in **Figure 40.11**. Notice that mosses and other groups that do not form pollen are the most basal groups on the tree. They also appear first in the fossil record. Conifers and other groups that are strictly or primarily wind pollinated evolved later but before angiosperms.

The first important pattern here is that the more recently evolved plant groups do not need water in order for sexual reproduction to occur. As a result, the evolution of pollen has allowed these species to be much less dependent on wet habitats. Along with the evolution of the seed—highlighted in Section 40.4—pollen paved the way for the colonization of drier environments.

The second important pattern is that pollination became a much more precise process when plants began to recruit animals to act as pollinators. Wind-borne pollen grains have a low probability of landing successfully on a flower stigma. Insect-borne pollen, in contrast, is much more likely to be successfully transferred to flowers of the same species. In effect, plants "pay" nectar- and pollen-eating insects to work for them. Wind is free, but insects are more precise. Insect pollination is an important adaptation because it makes sexual reproduction much more efficient.

Does Pollination by Animals Encourage Speciation?

Biologists have recently begun to explore another major consequence of pollination by insects and by other animals: the evolution of new species. In addition to affecting the fitness of individual plants, does pollination by animals make the formation of new species more likely?

As an example of how insect pollination could lead to the evolution of new species, consider the situation shown in **Figure 40.12**. A biologist has documented that two populations of a mountain-dwelling species called the alpine skypilot have

FIGURE 40.12 Interactions between Angiosperms and Animal Pollinators May Lead to Speciation. Flowers in alpine skypilot populations from timberline and tundra habitats have very different characteristics. Because the flowers attract different pollinators in the two habitats, interbreeding is rare and the two populations may eventually become different enough to be considered distinct species.

flowers with different characteristics. Alpine skypilots that grow in forested habitats at or below timberline have small flowers with short stalks and an aroma described as "skunky." Individuals that grow in the tundra habitats above timberline have large flowers with long stalks and smell sweet. These differences are interesting, because different insects pollinate the two populations. Small flies pollinate the timberline individuals; large bumblebees pollinate the tundra flowers. Experiments on these populations have shown that bumblebees prefer to pollinate large flowers. Because flies and bumblebees prefer to visit different types of flowers, the two skypilot populations do not breed together routinely and are evolving distinct characteristics. They are on their way to becoming different species.

The message here is that evolutionary changes in the size or food-finding habits of a pollinator affect the angiosperm populations they pollinate. In return, changes in flower size and shape affect the insects pollinating that population. Because mutation continuously introduces variations in these traits, insect and angiosperm populations frequently change, diverge, and form new species. It is no surprise that insects and angiosperms are exceptionally species-rich groups.

It is clear that pollination was a crucial innovation during plant evolution. Now let's get down to mechanics. What happens once a pollen grain is deposited on a stigma?

Fertilization

After landing on the stigma of a mature flower from the same species, a pollen grain absorbs water and germinates. **Germination** is a resumption of growth. The male gametophyte produces a long filament called a **pollen tube**, which grows down the length of the style (**Figure 40.13**). This growth is directed by chemical attractants released by **synergids**—cells in the female gametophyte that lie close to the egg. In the species illustrated here, the generative cell travels down the length of the tube and divides to form two sperm. When the pollen tube reaches the micropyle of the ovule, it grows through it and enters the interior of the female gametophyte. This event sets the stage for fertilization.

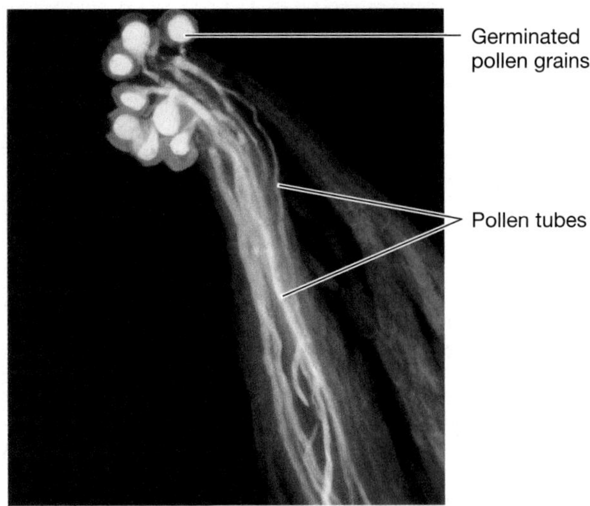

FIGURE 40.13 The Pollen Tube Grows toward the Egg. The male gametophyte germinates on the stigma and grows toward the egg. The sperm formed when the generative pollen nucleus divides travel down the tube toward the female gametophyte.

Recall that fertilization is the fusion of sperm and egg to form a diploid zygote. In most other plant groups, fertilization is straightforward—sperm and egg simply combine, and a diploid nucleus is formed. But in angiosperms, an unusual event called **double fertilization** takes place (**Figure 40.14**). One sperm nucleus unites with the egg nucleus to form the zygote. The

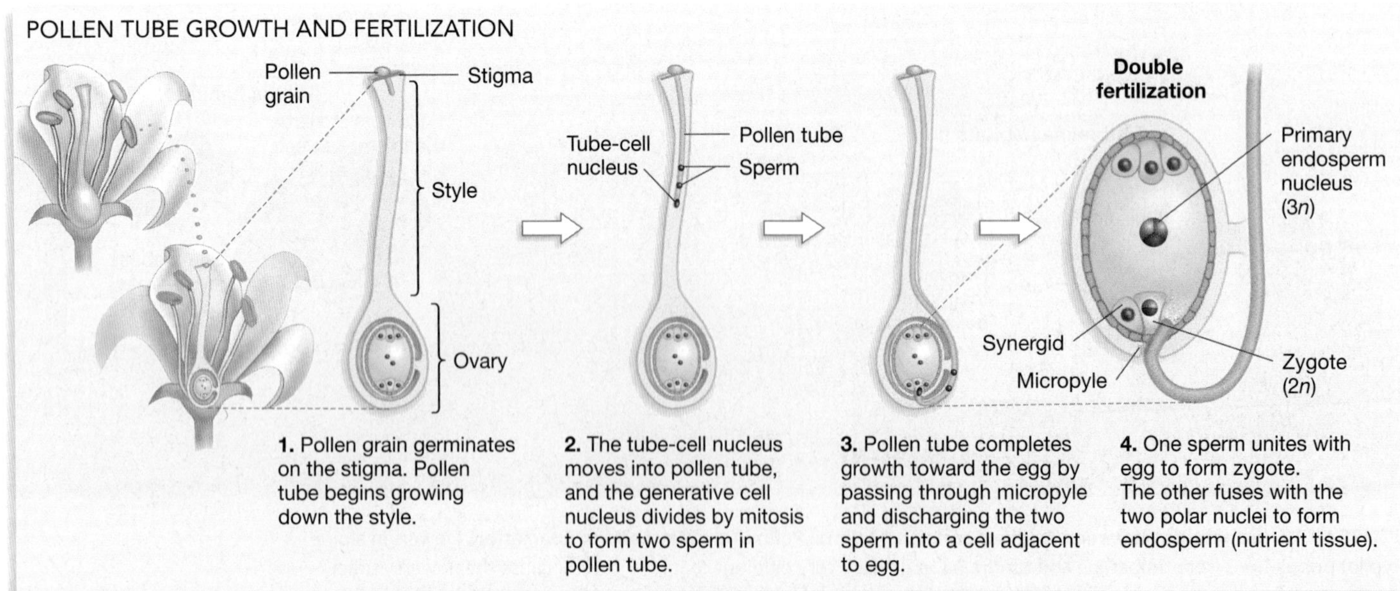

FIGURE 40.14 Double Fertilization Produces a Zygote and an Endosperm Nucleus. When the pollen tube reaches the female gametophyte, one sperm nucleus fertilizes the egg while the second fuses with the polar nuclei.

other sperm nucleus moves through the female gametophyte and fuses with the polar nuclei in the central cell. In most cases, two polar nuclei are present and a large triploid ($3n$) cell forms. If you understand double fertilization, you should be able to draw a female gametophyte immediately after fertilization and identify cells that are haploid, diploid, and triploid.

The triploid cell resulting from this second fertilization is called the primary endosperm nucleus. It undergoes a series of mitotic divisions that produce a tissue called **endosperm** ("inside-seed"). In most species, endosperm is triploid and functions to store nutrients. Endosperm cells are loaded with starch or oils (lipids) plus proteins and other nutrients that will be needed by the embryo after it germinates. But before germination can occur, seeds must develop and be dispersed.

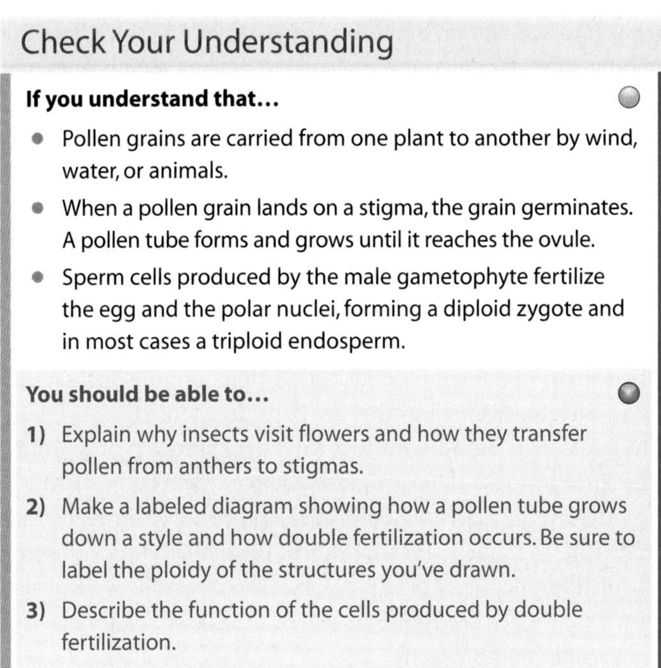

Check Your Understanding

If you understand that...

- Pollen grains are carried from one plant to another by wind, water, or animals.
- When a pollen grain lands on a stigma, the grain germinates. A pollen tube forms and grows until it reaches the ovule.
- Sperm cells produced by the male gametophyte fertilize the egg and the polar nuclei, forming a diploid zygote and in most cases a triploid endosperm.

You should be able to...

1) Explain why insects visit flowers and how they transfer pollen from anthers to stigmas.
2) Make a labeled diagram showing how a pollen tube grows down a style and how double fertilization occurs. Be sure to label the ploidy of the structures you've drawn.
3) Describe the function of the cells produced by double fertilization.

40.4 The Seed

Fertilization triggers the development of a young sporophyte. In angiosperms, the first stage in the sporophyte's life is the maturation of the seed. As a seed matures, the embryo and endosperm develop inside the ovule and become surrounded by a covering called a **seed coat**. At the same time, the ovary around the ovule develops into a fruit, which encloses and helps protect the seed (or seeds, if many ovules are contained within a single ovary). In addition to providing protection, fruits often aid in dispersing seeds away from the parent plant.

The mature seed consists of an embryo, a food supply—often in the form of endosperm—and a seed coat. Along with pollen, the evolution of this structure was a crucial innovation as land plants diversified. Because seeds contain stored nutrients, they allow offspring to be much more successful in colonizing dry habitats than are offspring produced from spores, which are single cells. As a young plant emerges from the seed, it can subsist on stored nutrients until it is well enough established to absorb water from the soil and feed itself via photosynthesis. Let's analyze how seeds work, beginning with a closer look at how the embryo develops.

Embryogenesis

Recall from Chapter 23 that **embryogenesis** is the process by which a single-celled zygote becomes a multicellular embryo. When an angiosperm zygote divides, it produces the two daughter cells shown in step 1 of **Figure 40.15**. The lower daughter cell, called the basal cell, divides to form a cell that forms part of the root tip and a cell that produces a row of cells. This row of cells is called the suspensor and provides a route for nutrient transfer from the parent plant to the developing embryo. The upper daughter cell, or terminal cell, is the parent of almost all the cells in the embryo.

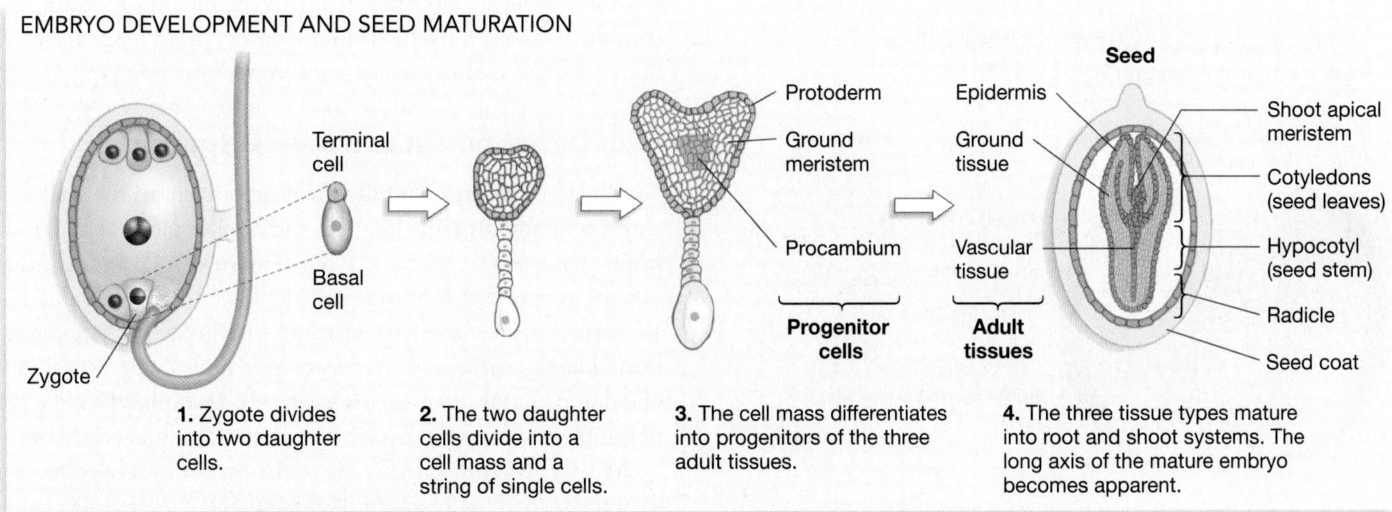

FIGURE 40.15 Embryonic Tissues and Structures Develop inside Seeds. The embryo inside a seed has the beginnings of root and shoot systems and the plant's first leaves, or cotyledons. The embryonic epidermis and the ground and vascular tissues are organized in distinct layers.

As the terminal cell and its progeny divide, the mass of cells in step 2 of Figure 40.15 forms. As step 3 in the figure shows, these cells then differentiate into three groups, each conforming to one of the three adult tissue types introduced in Chapter 36. The exterior layer of embryonic cells, called the **protoderm**, is the progenitor of the adult dermal tissue, or epidermis. The **ground meristem**—the cells just within the protoderm—gives rise to the ground tissue. And the **procambium** is a group of cells in the core of the embryo that becomes the vascular tissue. ● If you understand this critical step in embryonic development, you should be able to explain the relationship between the populations of embryonic cells labeled in step 3 of Figure 40.15 and the tissues labeled in step 4.

As the embryo continues to develop, the long axis of the plant begins to emerge and several important structures take shape (step 4, Figure 40.15). Particularly important are the **cotyledons**, or seed leaves. Recall from Chapter 30 that one prominent group of angiosperms—called the monocotyledons, or monocots—has just one seed leaf, whereas eudicotyledons, or eudicots, have two. In some eudicots, the cotyledons take up the nutrients in the endosperm and store them. In these species, there is no endosperm left by the time the seed matures—instead the cotyledons function as the nutrient storage organ. Another important structure within the seed is the **hypocotyl** ("under-cotyledon"), or seed stem, which is the embryonic stem. The hypocotyl terminates in the **radicle**, or embryonic root. Some embryos also have an **epicotyl** ("above-cotyledon"), which is a portion of the embryonic stem that extends above the cotyledons. **Figure 40.16** compares the seed structure in beans and corn—a representative eudicot and monocot, respectively.

By the time a seed matures, then, the three major tissue types have differentiated in the embryo. The root and shoot systems, along with the first leaves, have formed. Once these events are accomplished, the seed tissues dry and the embryo becomes quiescent—meaning it stops growing.

(MB) **Web Animation** at www.masteringbio.com
Reproduction in Flowering Plants

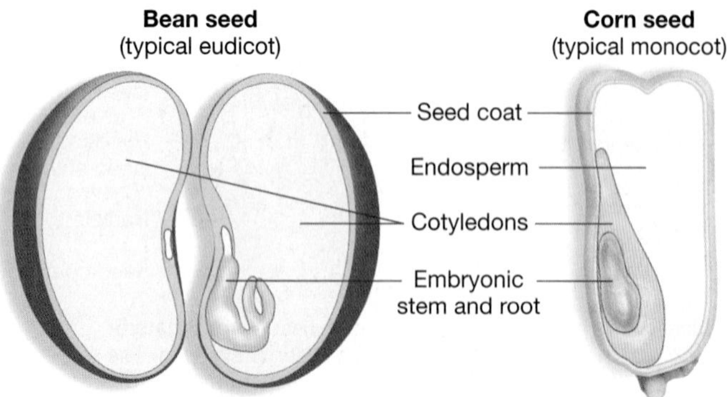

FIGURE 40.16 Seeds Contain an Embryo and a Food Supply Surrounded by a Tough Coat. In beans (left), the nutrients in the endosperm are absorbed by the cotyledons and stored. In corn (right), the endosperm is intact.

The Role of Drying in Seed Maturation

The seeds of many species dry out as they mature. Water makes up 90 percent of normal plant cells, but dried seeds contain just 5–20 percent water. Loss of water is interpreted as an adaptation that prevents seeds from germinating on the parent plant, where they would compete with the parent for light, water, and nutrients. In addition, the dry condition of seeds ensures that once they have dispersed from the parent plant, they will not germinate until water is available in the environment. This is logical, because water availability is crucial to the survival of germinated seedlings. Dry seeds are also less susceptible than wet seeds to damage from freezing, and they are lighter and more easily transported.

The dry state of seeds raises an important question, however: How do the plasma membranes and proteins in the embryo and endosperm survive? When researchers reduce the amount of water surrounding isolated plasma membranes and proteins to the levels observed in extremely dry seeds, some of the membranes and proteins disintegrate. Clearly, something is happening at the molecular level in seeds to keep these cell components intact.

Researchers have recently established that this "something" involves sugars. As water leaves the seed during drying, sugars act as a replacement for water, maintaining the integrity of plasma membranes and proteins. If drying is extreme, the sugars form an extremely viscous liquid that contains little if any water. Substances such as this are considered vitrified, or glass-like (glass is a liquid solution with the viscosity of a solid). Biologists propose that this glassy, sugary state helps maintain the integrity of plasma membranes and proteins in seeds that experience extremely dry conditions. (**Box 40.3** points out some practical implications of this work.) The researchers have also shown that when seeds imbibe water, the glassy sugars dissolve and germination proceeds.

Drying is only one part of the seed maturation process, however. Equally important is the development of tissues surrounding the seed itself. In many cases, these tissues are required for the seed to be dispersed from the parent plant.

Fruit Development and Seed Dispersal

Fertilization initiates not only the development of the seed and embryo in angiosperms but also the development of the fruit. Fruits come in several basic types (**Figure 40.17**). **Simple fruits** like the apricot develop from a single flower that contains a single carpel or several carpels that are fused together. **Aggregate fruits** like the raspberry also develop from a single flower, but one that contains many separate carpels. **Multiple fruits** like the pineapple develop from many flowers and thus many carpels.

As a fruit matures, the walls of the ovary thicken to form a structure called the pericarp (**Figure 40.18**). The **pericarp** is the part of the fruit that surrounds and protects the seed or seeds. Fruits can be dry when they are mature, as in nuts, or fleshy, as in cherries and tomatoes.

Flower | Developing fruit | Ripe fruit

Simple fruit (Apricot)

Develops from a single flower with one carpel or fused carpels

Aggregate fruit (Raspberry)

Develops from a single flower with many separate carpels

Multiple fruit (Pineapple)

Develops from many flowers with many carpels

FIGURE 40.17 Three Major Types of Fruits. Fruits come in many types. The structure of a fruit depends on the number of ovules found in each carpel and whether ovaries fuse during fruit maturation.

QUESTION When fruits ripen, their color changes in a way that makes them more conspicuous to fruit eaters. State a hypothesis to explain why this color change might increase the fitness of an individual. Design an experiment to test your hypothesis.

Fruits have two functions: They protect seeds from physical damage and seed predators, and they frequently aid in seed dispersal (see Figure 30.20). Dispersal is important to the fitness of the young sporophyte. This is especially true in long-lived species, in which the offspring may compete with the parent plant for light, water, and nutrients if there is no dispersal. Although fruits sometimes split open and release seeds to be dispersed directly, in many cases seeds are dispersed to new locations while they are still enclosed in the fruit.

Dry fruits are dispersed by wind, animals, or mechanical action—or the fruits simply fall to the ground. Fruits that are dispersed by wind often have external structures to catch

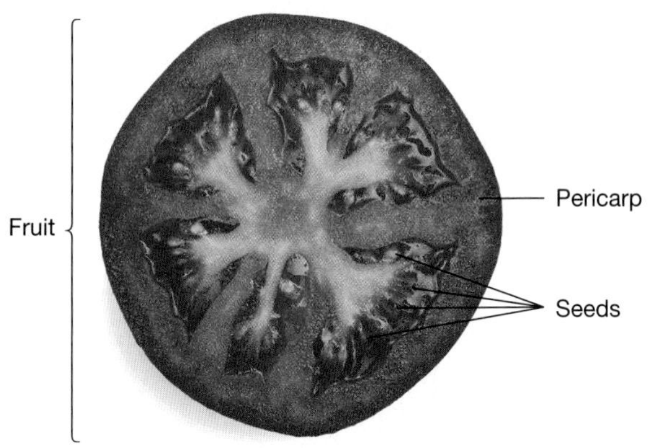

FIGURE 40.18 As Fruits Mature, the Ovary Wall Develops into a Pericarp That Surrounds the Seed or Seeds. A fruit consists of a pericarp and the enclosed seeds.

BOX 40.3 Pure and Applied Science

People sometimes draw a sharp distinction between pure research and applied research. One characterization is that pure science is inspired solely by inquisitiveness about how the natural world works, without regard to commercial payback. Applied science is described as research that is focused on solving a specific problem in business, technology, agriculture, or medicine.

The discovery that sugars can enter a glassy state as seeds dry shows just how blurry the line between pure and applied research can be, however. The initial financial support for research on the state of dry seeds was inspired by a practical argument: If biologists could discover why certain seeds have such a long life span, the knowledge could have practical benefits for commercial growers who want to preserve certain seeds for long periods. Yet the actual experiments that were done on dry seeds were "pure" in nature. Investigators simply wanted to know how molecules that normally exist in solution maintain their structural integrity when water is removed. Independently of the plant researchers, other biologists discovered that the formation of glassy sugars also helps certain insects tolerate extremely cold temperatures.

The twist to the story is that the discovery of glass-like sugars has potentially important practical applications that are completely different from those originally envisioned. Because proteins can be preserved in an extremely dry state by encasing them in sugars, it is possible to preserve certain drugs, vaccines, and dietary supplements without refrigeration. One of the first applications of this discovery was to sugar-coat the insulin manufactured for use by people with diabetes. Instead of injecting the polypeptide in solution into their bloodstream, at least some diabetics are now able to take their medicine as an easy-to-use inhalant. Once sugar-coated insulin molecules hit the wet surface of lung cells, the sugars would dissolve and the insulin would be taken up and transported to the bloodstream.

What are the messages of this story? For one thing, few research programs turn out to be completely pure or completely applied. What's more, experience has shown that it is next to impossible to predict the practical implications of research.

the breeze and extend the distance they travel. The fruits of dandelions and maple trees are familiar examples of wind-dispersed fruits. Other dry fruits have hooks or barbs that adhere to passing animals, while nuts are dispersed by seed predators. Still other types of dry fruits are actually dispersed by the plant via propulsion. The sandbox tree, for example, produces a seed pod that shrinks as it dries. Eventually the pod splits apart violently, spraying seeds in all directions. The pod bursts with so much force that the plant is sometimes called the dynamite tree. The sound of a seed pod bursting resembles a pistol shot, and seeds can be scattered as much as 40 m away from the parent plant. Similarly, the dwarf mistletoe fruit fills with sugars as it matures. Enough water follows via osmosis to make the fruit explode and shoot seeds as far as 5 m.

Animals are the most common dispersal agent for fleshy fruits. In animal-dispersed fruits, the seed coat has to be tough enough to resist the mechanical forces and chemical conditions found in the animal's mouth and digestive tract, so that seeds can emerge in the feces unscathed. In cases like this, seed dispersal is an example of mutualism. The plant provides a fruit rich in sugars and other nutrients; in return, the animal carries the fruit to a new location and excretes the seeds along with a supply of fertilizer.

Fruit Structure and Development

Seed Dormancy

Once they have dispersed from the parent plant, seeds may not germinate for a period of time. This condition is known as **dormancy**. Dormancy is usually a feature of seeds from species that inhabit seasonal environments, where conditions may be too cold or dry for seedlings to thrive for extended periods of time. Based on this observation, dormancy is interpreted as an adaptation that allows seeds to remain viable until conditions improve. In contrast, dormancy is rare or nonexistent in seeds produced by plants that inhabit tropical wet forests or other areas where conditions are suitable for germination year-round.

Let's consider two questions about dormancy: What molecular mechanisms are responsible for the condition? How does dormancy cease so that germination can begin?

What Role Does ABA Play in Dormancy? Chapter 39 introduced the hormone abscisic acid (ABA) and described its role in preventing germination. In some species, seeds that enter dormancy have a high concentration of this hormone. The seeds of desert plants, for example, have high concentrations of ABA in their seed coats. When these seeds are exposed to large amounts of water during rare or seasonal rains, the hormone literally washes out of the seed's outer tissues. Once the ABA concentration is reduced in this way, germination proceeds.

There is not a strict correlation between ABA concentration and degree of seed dormancy, however. In peas and many other species, seeds routinely contain high levels of ABA and yet are not dormant. In *Arabidopsis*, ABA concentrations rise as seeds mature and appear to impose dormancy. ABA levels eventually fall, however, so dormant, mature seeds contain only trace amounts of the hormone.

Faced with observations such as these, researchers have concluded that there is no single, universal mechanism for initiating and maintaining seed dormancy. In some cases, changing ABA levels or the ratio of ABA to gibberellin present controls the initiation or maintenance of the dormant state. In other cases, changes in sensitivity to ABA, rather than the sheer amount of the hormone present, appear to be important. It is also likely that novel mechanisms for initiating and maintaining dormancy are still to be discovered.

How Is Dormancy Broken? The coats of some seeds are thick enough to prevent water and oxygen from physically reaching the embryo. For germination to occur, these seed coats must be disrupted, or **scarified**. Crop seeds that require scarification are placed in large, revolving drums along with pieces of sandpaper. The abrasion from the sandpaper scarifies the seeds. In nature, seed coats can be disrupted by a fire, by the passage of the seed through an animal's digestive tract, or by abrasion against wind- or water-driven soil particles. The basic principle is that the seed coat has to be broken for water to enter the seed.

Other types of seeds must experience particular environmental conditions in addition to exposure to water. Species native to northern or alpine habitats often produce seeds that must undergo a time of cool, wet conditions before they will germinate. No one knows how these seeds perceive cold or what molecular mechanisms are involved in breaking dormancy when the cool, wet period ends.

Because small seeds have few nutrient reserves in their cotyledons or endosperm, many small-seeded species need to germinate near the soil surface, where individuals are exposed to light and can feed themselves via photosynthesis. As Chapter 39 indicated, lettuce seeds and other small seeds must be exposed to red light before they will break dormancy and germinate. Red light is an important environmental cue, because wavelengths in the red portion of the light spectrum support photosynthesis. Red light and blue light indicate that sunlight is abundant.

Finally, many of the seeds produced by species native to habitats where wildfires are frequent, such as the California chaparral and South African fynbos, have an unusual chemical requirement to break dormancy. These seeds must be exposed to fire or smoke before they will germinate. In fact, the commercial food product "liquid smoke" induces germination in these seeds as well as actual smoke does. Again, it is logical for seeds in this habitat to germinate after fire has cleared away old vegetation and has thinned existing trees and shrubs.

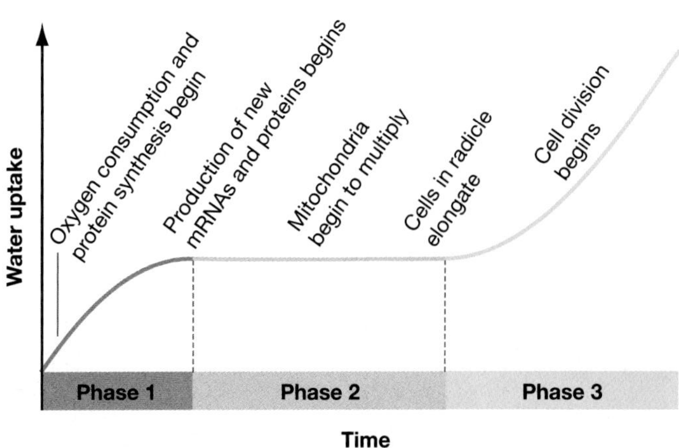

FIGURE 40.19 Germination Begins with Three Distinct Phases. Graph plotting the rate of water uptake as a typical seed germinates. (The graph is conceptual, meaning it represents a general pattern observed in data from many species, so the axes have no units.)

The message here is that dormancy can be broken in response to a wide variety of environmental cues. In general, the cue that triggers germination is a reliable signal that conditions for seedling growth are favorable for a particular species in a particular environment.

Seed Germination

Even if specific environmental signals are required to break dormancy, seeds do not germinate without water. Water uptake is the first event in germination. Once the seed coat allows water penetration, water enters by a steep water-potential gradient, because the seed is so dry.

Figure 40.19 graphs the uptake of water in a typical angiosperm seed. The important point is that the curve describing water uptake has three distinct phases. Germination begins with a rapid influx of water. This influx is followed by an extended period in which no further water uptake occurs. Phase 3 is marked by water intake starting up again and rising steadily.

During the first phase of water uptake, oxygen consumption and protein synthesis in the seed increase dramatically, but no new messenger RNAs are transcribed. Based on these observations, biologists have concluded that the earliest events in germination are driven by mRNAs that are stored in the seed prior to maturation. During the second phase, when water uptake stops, newly transcribed mRNAs appear and are translated into protein products. Mitochondria also begin to multiply. In effect, seeds take up enough water to hydrate their existing proteins and membranes, and then begin to manufacture the proteins and mitochondria needed to support growth. Water uptake resumes as growth begins. The second phase of water uptake enables cells to develop enough turgor pressure to enlarge. Eventually, the embryo bursts from the seed coat.

(a) Beans are eudicots with cotyledons that emerge aboveground.

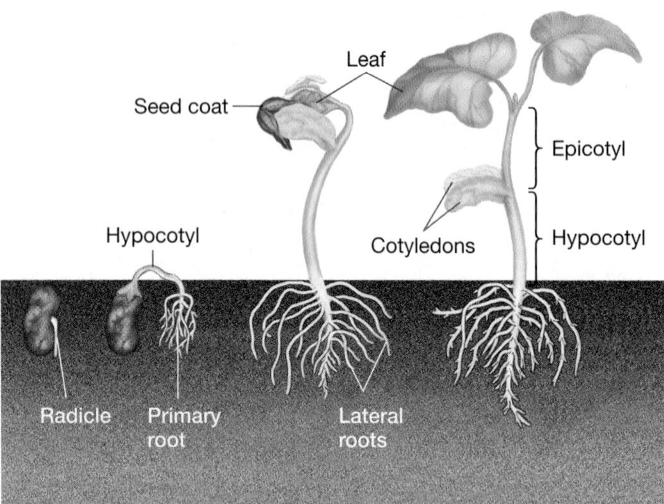

(b) Peas are eudicots with cotyledons that remain belowground.

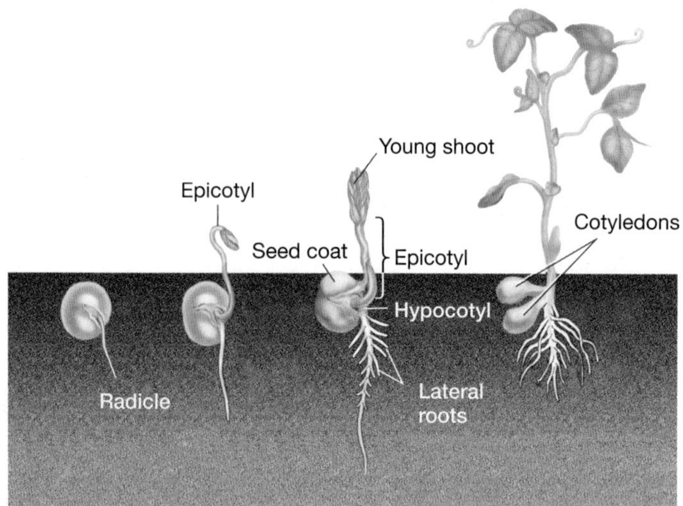

(c) Corn is a monocot with a cotyledon that remains belowground.

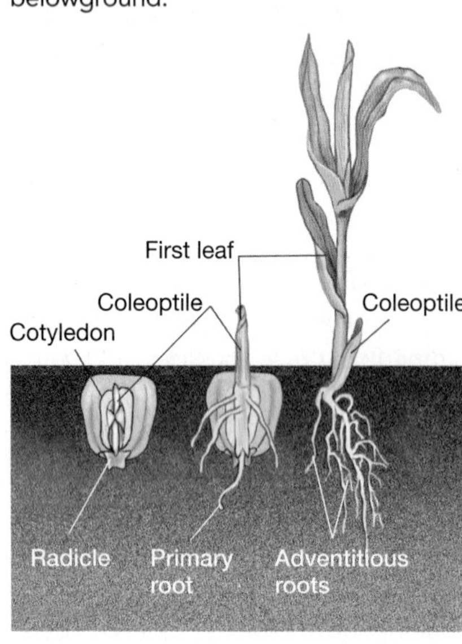

(d) Onion is a monocot with a cotyledon that emerges aboveground.

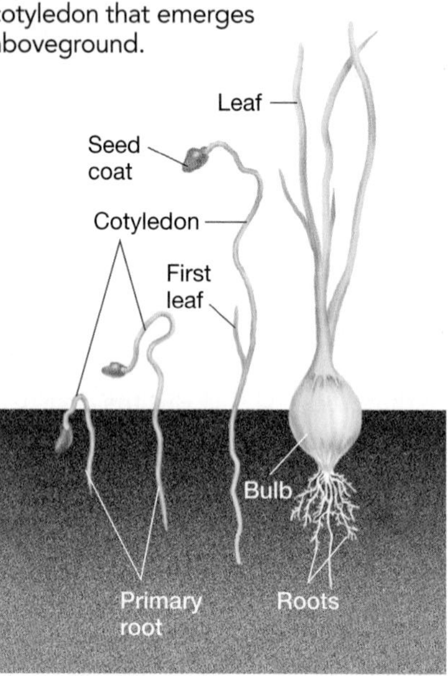

FIGURE 40.20 The Germination Sequence Varies among Species.

● **QUESTION** In which of these four species are cotyledons photosynthetic?

Figure 40.20 shows what happens as eudicot and monocot embryos emerge from the seed. The process begins with the emergence of the radicle, which subsequently develops into the mature root system. In eudicots, the shoot system with its cotyledons usually emerges shortly after the radicle appears. In corn, the radicle and the coleoptile, which covers the young shoot, emerge at the same time. Note that in eudicots, the emerging stem has a hook shape. Like the coleoptile of monocots, this trait is thought to protect the apical meristem from being damaged as the shoot works its way upward through rough soil particles.

The next major event in the seedling's life occurs when either the cotyledons or the earliest leaves produced by the growing seedling commence photosynthesis. The seedling is said to be established when the young plant no longer relies on food reserves in its endosperm or cotyledons; instead, it receives all of its nourishment from its own photosynthetic products. With this, a new generation is under way.

Chapter Review

SUMMARY OF KEY CONCEPTS

Most plants reproduce asexually as well as sexually. Asexual reproduction, or cloning, is based on mitosis and results in offspring whose genetic makeup is identical to that of the parent. Sexual reproduction is based on meiosis and results in offspring that are genetically unlike their parent.

Plants undergo alternation of generations, in which a diploid sporophyte phase alternates with a haploid gametophyte phase. Sporophytes produce spores by meiosis. Gametophytes produce gametes by mitosis.

The relative size and life span of the gametophyte and sporophyte phases vary a great deal among plant groups. In the most basal groups of land plants, gametophytes are larger and longer-lived than sporophytes, and sporophytes are dependent on gametophytes for nutrition. The opposite is true in angiosperms, or flowering plants, in which sporophytes are the large and long-lived phase where photosynthesis takes place.

You should be able to draw a generalized diagram of a plant life cycle, labeling the sporophyte, gametophyte, spore, gametes, and zygote and indicating the ploidy of each.

In angiosperms, male and female gametophytes are microscopic and are produced inside flowers. Male gametophytes (pollen grains) are portable. Female gametophytes are encased in an ovary and are retained in the flower. When pollen grains land on a flower, they deliver sperm cells that fertilize the egg produced by the female gametophyte.

Angiosperms initiate flowering and sexual reproduction in response to external cues from the environment as well as internal cues based on the individual's condition. Frequently, these cues allow individuals to flower when environmental conditions are favorable.

Flowers are made up of sepals, petals, stamens, and one or more carpels. The lower part of the carpel is called the ovary and contains one to many ovules. Within the ovule, a megasporocyte undergoes meiosis, producing a megaspore that develops into the female gametophyte. Similarly, microsporocytes within the anthers of the stamens undergo meiosis. The resulting microspores develop into male gametophytes, which are enclosed in pollen grains. Pollination occurs when pollen grains are transported to the stigma of the carpel.

If allowed to germinate on the stigma, a pollen grain sends a long pollen tube down the style. Two sperm nuclei are produced by mitosis, and double fertilization takes place: One sperm fuses with the egg to form a zygote, while the other fuses with polar nuclei within the female gametophyte. The fusion of sperm and polar nuclei produces endosperm—nutritive tissue that in most species is triploid.

You should be able to diagram the steps in the formation of a male gametophyte and female gametophyte in angiosperms and then modify the drawing to show what would happen if 3 of the 4 meiotic products in megasporangia did *not* degenerate.

Seeds contain an embryo and a food supply surrounded by a coat. In angiosperms, the walls of the ovary develop into a fruit that encloses the seed or seeds.

The development of an angiosperm embryo begins with the formation of dermal tissue (epidermis), ground tissue, and vascular tissue layers. This event is followed by the development of the radicle, hypocotyl, and cotyledons. As the embryo develops, endosperm cells divide to form a nutrient-rich tissue. In addition, cells along the outside of the ovules form a protective seed coat, and the ovary develops into a fruit. In many cases, the mature fruit contains structures that aid in dispersal of the mature seed via wind, water, propulsion, or animals.

Many seeds do not germinate immediately but instead experience a period of dormancy. In some species, dormancy is initiated or maintained by the hormone ABA. A wide variety of conditions, ranging from scarification to exposure to red light, may break seed dormancy. In many cases, the event that triggers germination ensures that the seed germinates when environmental conditions are favorable. Germination begins when the seed takes up water and mRNAs already present in the seed are translated. It ends when the radicle breaks the seed coat and begins to penetrate the soil.

You should be able to explain (1) the relationship between the products of double fertilization and the structures inside a mature seed, and (2) the relationships among carpels, ovaries, ovules, fruits, and seeds.

MB **Web Animation** at www.masteringbio.com
Reproduction in Flowering Plants; Fruit Structure and Development

QUESTIONS

Test Your Knowledge

1. What is the major evolutionary trend in land plant life cycles?
   a. Instead of being approximately the same size and shape, gametophytes and sporophytes began to look different.
   b. Sporophytes became larger and long lived while gametophytes became drastically reduced.
   c. In lineages that evolved more recently, such as angiosperms, spores are no longer produced.
   d. Sporophytes began to rely on gametophytes for all of their nutritional needs.

2. What happens when double fertilization occurs?
   a. Two zygotes are formed, but only one survives.
   b. Two sperm fertilize the egg, forming a triploid zygote.
   c. One sperm fertilizes the egg, while another sperm fuses with the polar nuclei.
   d. One sperm fertilizes the egg, while two other sperm fuse with a polar nucleus.
3. What is a fruit?
   a. a structure formed from the ovary wall that contains a seed or seeds
   b. a structure consisting of an embryo and a food supply surrounded by a tough coat
   c. a female gametophyte
   d. a male gametophyte
4. Which of the following is a key event during embryogenesis?
   a. The seed coat takes on water so that germination can begin.
   b. Starches are hydrolyzed, providing sugars that fuel the early stages of germination.
   c. The megasporocyte divides by mitosis, forming the cells that will become the embryonic female gametophyte.
   d. Distinct groups of cells form that will become dermal, ground, and vascular tissues.
5. Why is the interaction between angiosperms and pollinators considered mutualistic?
   a. New species can form if mutant flowers attract new types of pollinators.
   b. Flowers may have an array of traits, including corolla shape, color, scent, and the presence of nectar, to attract a specific type of pollinator.
   c. Wind pollination is much "cheaper," but animal pollination is much more precise.
   d. Angiosperms get their pollen dispersed, while pollinators get food.
6. What happens when outcrossing occurs?
   a. Inbred offspring are produced.
   b. The proteins produced by *S* loci block pollen grains from interacting with the stigma.
   c. Gametes from different individuals fuse to form a zygote.
   d. Gametes from the same individual fuse.

Test Your Knowledge answers: 1. b; 2. c; 3. a; 4. d; 5. d; 6. c

Test Your Understanding

Answers are available at www.masteringbio.com

1. List three mechanisms of asexual reproduction in plants. In terms of maximizing reproductive success, what is the advantage of asexual reproduction? What is the disadvantage?
2. In the angiosperm life cycle, which cells undergo meiosis? Which cells are spores? Which structures are gametophytes?
3. Draw a generalized flower. Indicate the function of each part during reproduction. Provide detailed diagrams of the reproductive structures that produce male gametophytes and female gametophytes. Then diagram flowers that are pollinated by wind, by hummingbirds, and by bumblebees. How is each flower part modified from the generalized flower you first drew, and why?
4. What are the advantages and disadvantages of self-fertilization versus those of outcrossing?
5. Explain the difference between a carpel, an ovary, and an ovule. Be sure to clarify the physical relationship among them (which is enclosed by which), and identify which is responsible for producing the female gametophyte versus the pericarp of a fruit. Are these structures part of the sporophyte, the gametophyte, or a combination of the two?
6. Make a labeled diagram of a bean seed and a corn seed. What is the relationship between the endosperm of corn and the cotyledons of beans? Suggest a hypothesis to explain how the seed coats of these species are broken so that germination can begin.

Applying Concepts to New Situations

Answers are available at www.masteringbio.com

1. Suppose you discovered an angiosperm that was new to science. The population grows on an island near the equator. The island is dry 10 months of the year but experiences a 2-month period of frequent rains. Predict what cues trigger flowering and seed germination in the newly discovered species. Design experiments to test your predictions.
2. Some flowering plants "cheat" their pollinators because they offer no food reward. Likewise, certain pollinators cheat plants by removing nectar from flowers without picking up pollen. (In some cases, they do so by chewing through the petals that hold the store of nectar.) Speculate on the types of mutations that might modify insect behavior and/or plant structure in a way that limits cheating and enforces mutualism.
3. Pollinators frequently deposit pollen from more than one individual on a stigma. When they do, pollen grains from different males compete to fertilize the egg. Design an experiment to test the hypothesis that (a) pollen grains produce a toxin that slows the growth of pollen tubes from different individuals and (b) pollen tubes grow faster when pollen grains germinate in the presence of pollen from a different individual.
4. Consider the following fruits: an acorn, a cherry, a burr, and a milkweed seed. Based on the structure of each of these fruits, predict how the seed is dispersed. How would you test this prediction? Design a study that would estimate the average distance that each type of seed is dispersed from the parent plant.

www.masteringbio.com is also your resource for • Answers to text, table, and figure caption questions and exercises • Answers to *Check Your Understanding* boxes • Online study guides and quizzes • Additional study tools including the *E-Book for Biological Science* 3rd ed., textbook art, animations, and videos.

Animal Form and Function

41

Oryx are adapted to desert life. They have an exceptional ability to withstand heat, and they acquire all of their water from the food they eat.

KEY CONCEPTS

- Biologists analyze the structure and function of animals at a variety of levels: molecules, cells, tissues, organs, and systems.
- Body size has a strong influence on how animals work. For example, because small animals have a large amount of surface area relative to their volume, they lose heat faster than do large animals. The relationship between surface area and volume also plays a significant role in diffusion-mediated processes, such as gas exchange, nutrient absorption, and waste excretion.
- Animals maintain a relatively constant environment inside their bodies because they have systems that sense changes in internal conditions and trigger responses that return conditions to normal.

The Sahara and Arabian deserts that cover much of northern Africa and the Arabian peninsula are extreme environments. In some parts of the Sahara, it is not unusual for several years to pass between rainfalls. In a single day, temperatures can fluctuate between –0.5°C and 37.5°C. In summer, midday temperatures can exceed 50°C. Yet few places in the Sahara or Arabian deserts are completely devoid of life. Even large animals, such as the oryx, thrive in parts of both deserts.

How do the organisms native to these areas cope? Plants may endure as seeds for months or years, then germinate, flower, and produce a new generation of seeds immediately after a rain. Small animals avoid the midday heat by retreating to a cool burrow deep underground or to the shade of a small shrub, where they wait until the temperature moderates in the evening. Large mammals have a harder time hiding from the Sun but possess traits that allow them to keep cool and conserve water. The Arabian oryx, for example, never drinks. It gets 86 percent of the water it needs from the vegetation it eats and the other 14 percent from water synthesized as a by-product of cellular respiration—what biologists call **metabolic water**. To conserve what water they have, Arabian oryx produce extremely concentrated urine and exceptionally dry fecal pellets. In addition, oryx do not sweat to cool off in hot weather. Instead they allow their body temperature to rise from a normal 37°C to just over 40°C as temperatures increase during the day. The excess body heat is then given off as temperatures fall during the cool desert nights, and their body temperature returns to normal.

In contrast to oryx, humans die of dehydration when denied water for just a week. And for a human, a body temperature of 40°C is a life-threatening situation—equivalent to running a fever of 104°F. How do oryx do it? More specifically, what aspects of their anatomy and physiology allow them to thrive in such an extreme environment? **Anatomy** is the study of an organism's physical structure. **Physiology** is the study of how the physical structures in an organism function. The anatomy and physiology of an oryx are clearly different from those of a human—or those of a shark, frog, tuna, fruit fly, or crab, for that matter. This chapter is an introduction to how biologists approach the study of anatomy and physiology in animals.

Biologists who study animal anatomy and physiology are studying **adaptations**—heritable traits that allow individuals to survive and reproduce in a certain environment better than individuals that lack those traits (see Chapter 24). The chapters in this unit explore traits that allow animals to digest food, breathe, move, hear, see, and fight infections. This chapter introduces some basic elements of animal anatomy and identifies fundamental themes that unify the diversity of anatomical and physiological traits observed in animals. Its goal is to help you begin to see animals in a new light—as efficient systems for gathering resources and producing offspring. After millions of years of evolution by natural selection, animals are finely tuned to thrive in a diversity of habitats.

41.1 Form, Function, and Adaptation

Recall from Chapter 24 that adaptation results from evolution by natural selection. Natural selection, in turn, occurs whenever individuals with certain alleles leave more offspring than do individuals with different alleles. Because of this difference in reproductive success, the frequency of the selected alleles increases from one generation to the next. Oryx with alleles that allow them to extract more water from their feces survive better and produce more offspring than do oryx with alleles that allow water to be lost in feces. The ability to produce extremely dry fecal pellets is an adaptation that helps oryx thrive in water-short environments.

But as Chapter 25 explained, natural selection is not the only process that leads to changes in allele frequencies over time. Evolution also occurs through random changes in allele frequencies called genetic drift, through gene flow—meaning the movement of alleles into and out of a population by migration—and through the constant introduction of new alleles by mutation. Of these processes however, natural selection is the only mechanism that increases the ability of organisms to survive and reproduce in a particular environment.

The Role of Fitness Trade-Offs

No adaptation is "perfect." Instead, adaptations are limited by which alleles are present in a population and by the nature of the traits that already exist—because all adaptations are derived from preexisting traits. For example, the human spine is a highly modified form of the vertebral column observed in human ancestors that walked on all fours (see Chapter 34). The modifications in the human spine can be considered adaptations to support our upright posture, but they are far from perfect—85 percent of U.S. adults under the age of 50 report that they have experienced back pain. The human spine is not an optimal trait or perfect adaptation. Its evolution has been constrained by the nature of the preexisting trait in our ancestors and by a lack of alleles that would improve its structure and function.

Perhaps the most important constraint on adaptation, though, involves the phenomenon known as a trade-off. A **trade-off** is an inescapable compromise between traits. For example, every female animal has a finite amount of time and energy available for producing offspring. In species that do not care for their young, a female's entire investment in offspring consists of the eggs she lays. Given that the total amount of energy available for egg production is limited, there should be a trade-off between the number of eggs a female produces and the quality of those eggs. Egg quality is determined by egg size—specifically by the amount of yolk, or nutrient-rich cytoplasm, in the egg.

How have biologists been able to study trade-offs in animal anatomy and physiology? To answer this question, let's consider experimental work on egg size and egg number in the side-blotched lizard, a reptile that lives in the deserts of western North America (**Figure 41.1**). Like many animals, side-blotched lizards lay eggs in groups called clutches. Theory predicts that there has to be a trade-off between the number and size of eggs in a clutch. Can this trade-off actually be documented in a population in nature?

To begin this work, biologists set out to create a large amount of variation in egg size and egg number. For example, they induced the production of small eggs by catching females, surgically removing yolk from their eggs early in development, replacing the eggs, and releasing the mothers back into the wild. To create clutches with small numbers of eggs, biologists caught females and removed all but two or three of their eggs early in development. As an experimental control, the researchers did "sham operations." They caught a large number of females and performed surgery to expose their eggs, but they left the eggs alone. These manipulations gave them a study population with a large variation in egg size and egg number.

The graph labeled "Results 1" of Figure 41.1 plots the average mass of eggs that were laid against clutch size, meaning the number of eggs laid by each female. The data show that as egg size increases, clutch size decreases. The pattern confirms that there is a trade-off between egg size and egg number in this species. It is not possible for a female lizard to produce large numbers of large eggs.

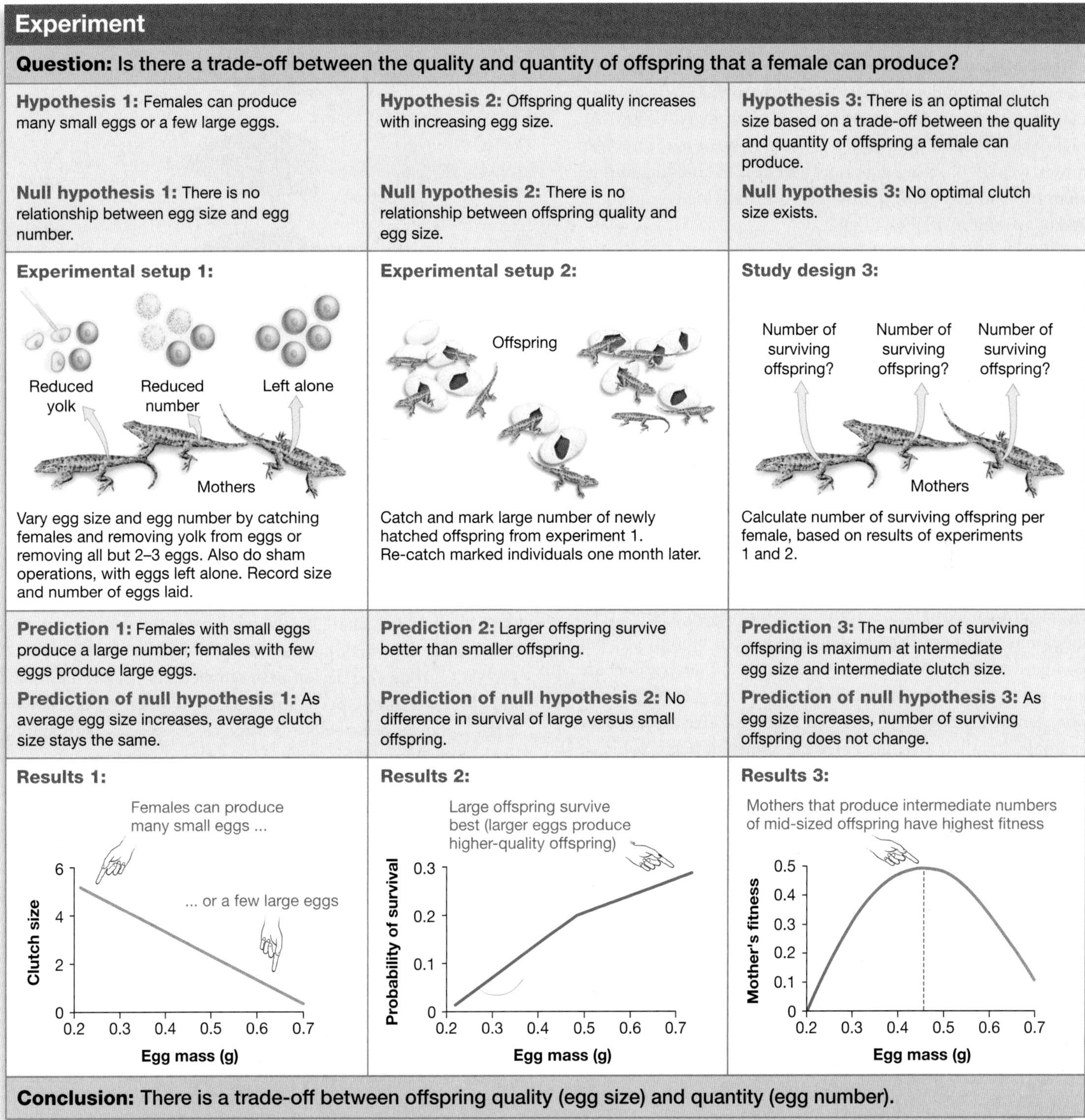

FIGURE 41.1 Trade-Offs Are Pervasive in Animal Anatomy and Physiology.

What about the prediction that large offspring are higher in quality, meaning that they survive better? To test this idea, the researchers marked 1668 newly hatched lizards in the experimental population, released them, and caught many of them again a month later. The graph labeled "Results 2" of Figure 41.1 confirms the prediction that larger offspring survive much better than do smaller offspring.

Given this trade-off between offspring quantity and offspring quality, does natural selection favor mothers that produce large numbers of small offspring or those that produce small numbers of large offspring? The graph labeled "Results 3" of Figure 41.1 shows the combined effect of egg number and offspring survival. In this population, mothers that produced an intermediate number of offspring of intermediate size generated the highest total number of surviving offspring.

Trade-offs such as the inescapable compromise between egg size and egg number are pervasive in nature. Desert animals that sweat to cool off are threatened with dehydration. An eagle's beak is superbly adapted for tearing meat but is not useful for weaving nesting materials together. In studying animal anatomy and physiology, biologists are studying compromise and constraint as well as adaptation.

Adaptation and Acclimatization

Although adaptation is a key concept in understanding animal form and function, the term is often used incorrectly. The problem is that in everyday English, the word *adaptation* is used to describe short-term, reversible responses to environmental fluctuations. In biology, physiological and biochemical changes like these are referred to as **acclimatization**, or **acclimation**. Acclimatization is a phenotypic change in an individual in response to short-term changes in the environment. Adaptation refers only to a genetic change in a population in response to natural selection exerted by the environment.

If you moved to Tibet, your body would acclimatize to high elevation by making more of the oxygen-carrying pigment hemoglobin and more hemoglobin-carrying red blood cells. But populations that have lived in Tibet for many generations are adapted to this environment. To increase fitness at high elevation, genetic changes have occurred. Among native Tibetans, for example, an allele that increases the ability of hemoglobin to hold oxygen has increased to high frequency. In populations that do not live at high elevation, this allele is rare or nonexistent.

In addition to understanding the distinction between adaptation and acclimatization, it's important to realize that the ability to acclimatize is an adaptation. Light-skinned humans, for example, vary in the ability to tan in response to sunlight. Some individuals in these populations tan easily—they have alleles that allow them to acclimate efficiently to environments with intense sun. Others do not. In this and many other cases, the ability to acclimatize is a genetically variable trait that can respond to natural selection.

41.2 Tissues, Organs, and Systems: How Does Structure Correlate with Function?

If a structure found in an animal is adaptive—meaning that it helps the individual survive and produce offspring—it is common to observe that the structure's size, shape, or composition correlates closely with its function. For example, you might recall from Chapter 24 that biologists have documented extensive changes in beak size and shape in medium ground finches from the Galápagos Islands. Such changes are due to natural selection. Individuals with deep beaks are better able to crack the large fruits that predominate during drought years, while individuals with small beaks are better able to harvest the small seeds that predominate during wet years. As **Figure 41.2** shows, a strong correlation between diet and beak structure is also found *among* species of Galápagos finch. Species with large, cone-shaped beaks eat large seeds; species with small, cone-shaped beaks eat small seeds. Species with long, tweezer-like beaks pick insects off tree trunks or other surfaces.

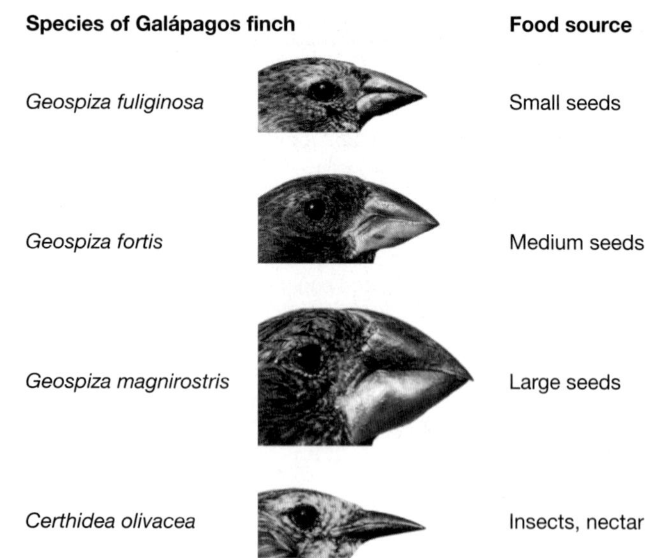

FIGURE 41.2 In Animal Anatomy and Physiology, Form Often Correlates with Function.

The mechanism responsible for these types of structure-function correlations is straightforward: If a mutant allele alters the size or shape of a structure in a way that makes it function more efficiently, then individuals who have that allele will produce more offspring than other individuals will. As a result, the allele will increase in frequency in the population over time.

Correlating structure and function is a pervasive theme in research on animal anatomy and physiology, and it is one of the most basic ways that biologists have to study adaptation. Correlations between form and function start at the molecular level. For example, earlier chapters emphasized that the shape of proteins correlates with their role as enzymes, structural components of the cell, or transporters. You might recall from Chapter 6 that membrane proteins called channels form pores that allow specific ions or molecules to pass in or out of cells (see Figure 6.25). The ends and interior of a channel are hydrophilic, which allows the protein to interact with the surrounding solution or the interior of the cell, while the outside of its middle portion is hydrophobic—allowing it to interact with the lipid bilayer. The protein's structure fits its function.

Similar correlations between structure and function occur within cells. Cells that manufacture and secrete hormones or digestive enzymes are packed with rough ER and Golgi; cells that store energy are dominated by large fat droplets; cells that hunt down, ingest, and destroy invading bacteria have numerous lysosomes. The overall shape of a cell can also correlate with its function. For example, cells that are responsible for transporting materials into or out of the body often have extremely large areas of plasma membrane. As a result, they have room to accommodate the thousands of membrane channels, transporters, and pumps required for extensive transport.

Structure-function correlations are not limited to the molecular and cellular level, however. They also exist at the levels of tissues, organs, and systems.

Tissues

Animals are **multicellular**, meaning that their bodies contain distinct types of cells that are specialized for different functions. Sponges are considered to be the simplest of all animals in overall morphology, but even they contain cells that are distinct structurally and that perform specialized functions. In a sponge, some types of sponge cells are responsible for covering the body; others are responsible for secreting material to support the body, for feeding, or for reproduction.

Frequently, animal cells that are similar in structure and function are physically attached to each other and form a tissue. A **tissue** is a group of similar cells that function as a unit. Sponges have just one type of tissue: an epithelial tissue that forms the exterior and interior surface of the individual.

The embryonic tissues called ectoderm, mesoderm, and endoderm are found in most animals and were introduced in Chapter 22 and Chapter 32. As an individual develops, the embryonic tissues give rise to four adult tissue types: (1) connective tissue, (2) nervous tissue, (3) muscle tissue, and (4) epithelial tissue. In each case, the structure of the tissue correlates closely with its function. Let's consider each in turn.

Connective Tissue

Connective tissue consists of cells that are loosely arranged in a liquid, jellylike, or solid extracellular matrix, which is secreted by the connective tissue cells themselves. Each type of connective tissue secretes a distinct type of extracellular matrix. As a result, the structure and function of various connective tissues vary widely. For example, **loose connective tissue** contains an array of fibrous proteins in a soft matrix and serves as a packing material between organs or padding under the skin (**Figure 41.3a**). Other connective tissues, such as **cartilage** and **bone**, have a firmer extracellular matrix (**Figure 41.3b**). Cartilage and bone provide structural support for the vertebrate body or protective enclosures for the brain and other components of the nervous system. **Blood** is a connective tissue that has cells surrounded by a liquid extracellular matrix called plasma; blood functions in the transport of materials throughout the body (**Figure 41.3c**).

(a) Loose connective tissue has a soft extracellular matrix; it provides padding.

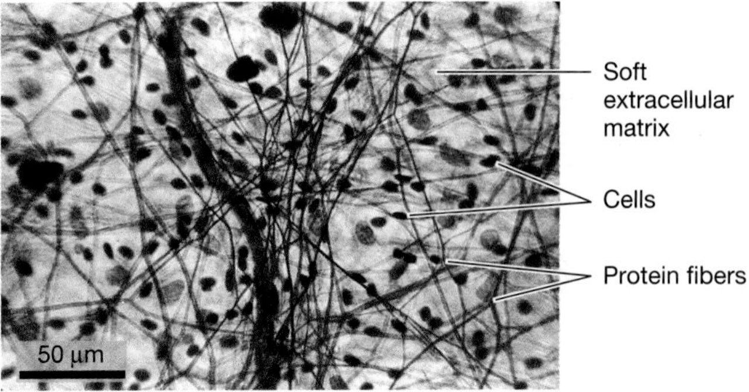

(b) Bone and cartilage have a hard (bone) or stiff (cartilage) extracellular matrix; they support the body.

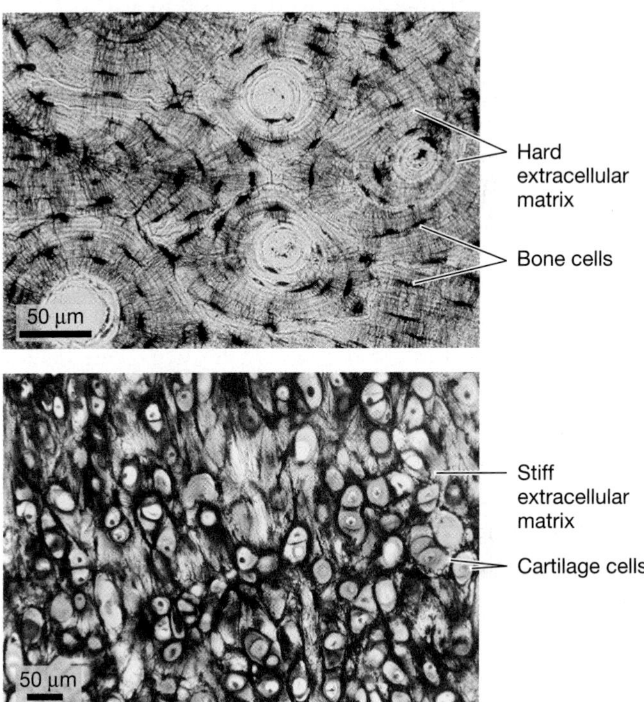

(c) Blood has a liquid extracellular matrix; it functions in transport.

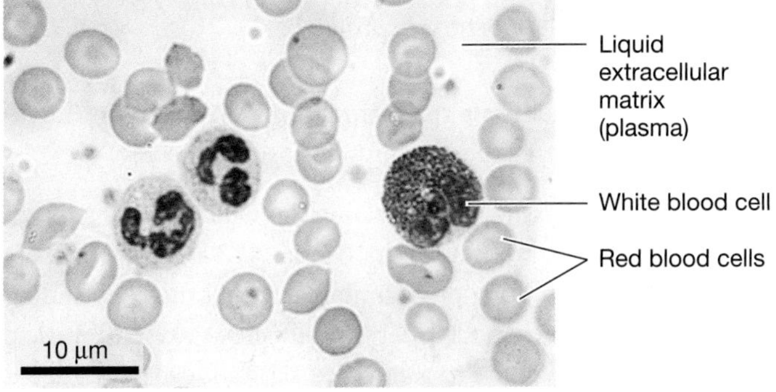

FIGURE 41.3 Connective Tissue Consists of Cells Loosely Arranged in an Extracellular Matrix.

QUESTION How does the stiffness of the extracellular matrix found in a connective tissue correlate with that tissue's function?

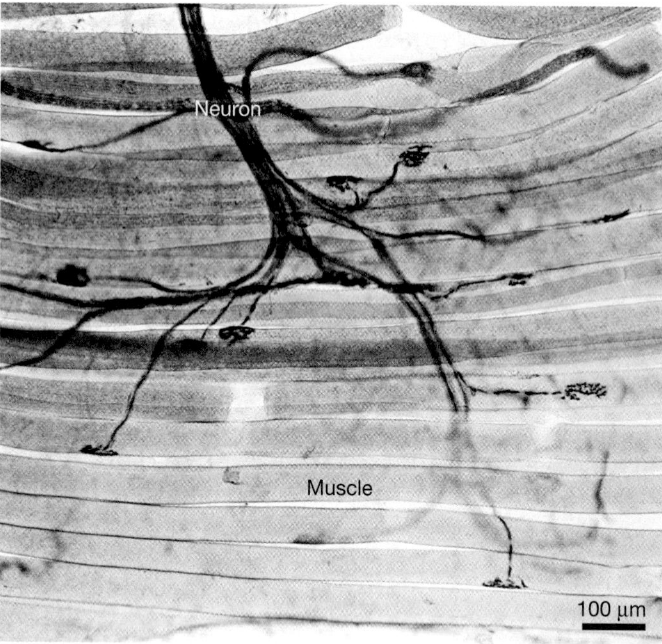

FIGURE 41.4 Nervous Tissue Contains Cells with Long Projections. Projections from nerve cells transmit electrical signals—in this case, to a muscle. Although some nerve cell projections are extremely short, some are over a meter long.

QUESTION How does the presence of projections on nerve cells support their function in electrical signaling?

(a) Skeletal muscle

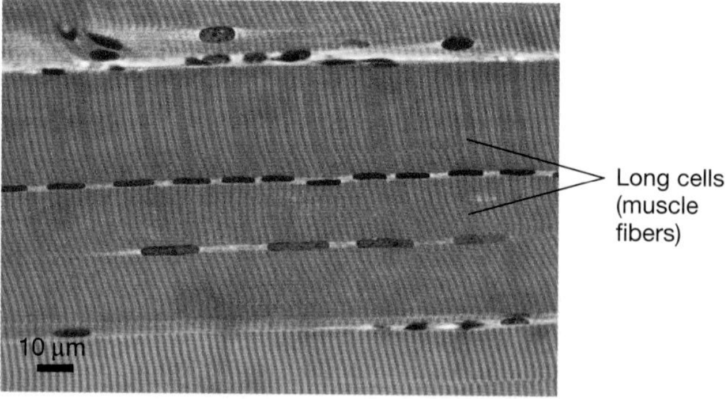

(b) Cardiac muscle

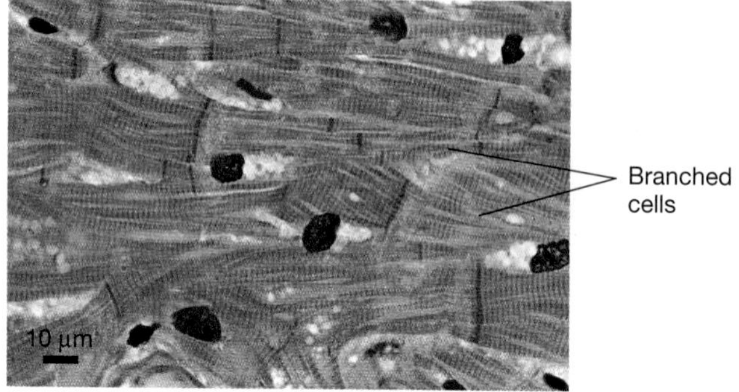

(c) Smooth muscle

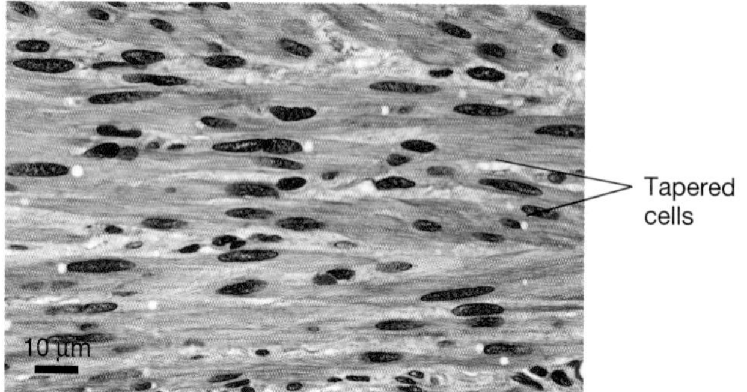

FIGURE 41.5 Muscle Tissues Comprise Cells with Fibers That Contract. The three types of muscle tissue have distinctive structures and functions.

Nervous Tissue **Nervous tissue** consists of nerve cells, which are also called **neurons**, and several types of supporting cells. Although they vary widely in shape, all neurons have projections that contact other cells (**Figure 41.4**). These projections deliver electrochemical signals, which are produced by changes in the permeability of the neuron's plasma membrane to ions. Although support cells outnumber neurons in the nervous system, they do not conduct electrical signals as neurons do. Instead, support cells regulate ion concentrations in the space surrounding neurons, supply neurons with nutrients, or serve as scaffolding or support for neurons.

Muscle Tissue **Muscle tissue** functions primarily in movement. There are three types of muscle tissue:

1. Most muscle tissue is **skeletal muscle**, which consists of the long cells called **muscle fibers** (**Figure 41.5a**). Muscle fibers are packed with long protein filaments that move by sliding past each other. As Chapter 45 will show, skeletal muscle contracts in response to a complex series of events triggered by electrical signals arriving from nerve cells.
2. **Cardiac muscle** makes up the walls of the heart. Although it is similar to skeletal muscle in some respects, each cardiac muscle cell branches and makes direct, end-to-end physical and electrical contact with other cardiac muscle cells (**Figure 41.5b**). These connections help transmit signals from one cardiac muscle cell to another during a heartbeat.
3. **Smooth muscle** cells are tapered at each end and form a muscle tissue that lines the walls of the digestive tract and the walls of the blood vessels (**Figure 41.5c**). Different types of neurons control the contraction of smooth muscle cells versus striated muscle cells. Smooth muscle is responsible for involuntary movements such as the passage of food down the digestive tract or the dilation (opening) of arteries near the skin in hot weather.

(a) Epithelium forms a surface layer.

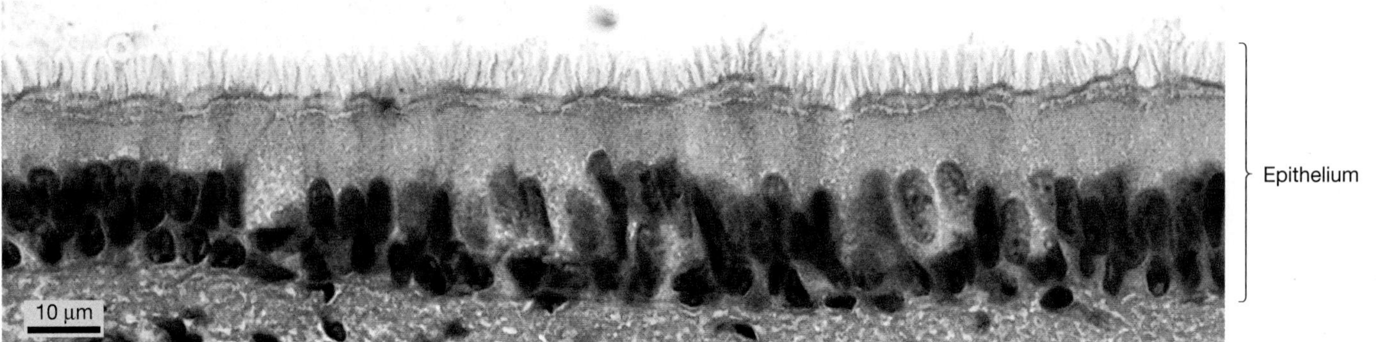

(b) Cells in epithelial tissues are joined tightly and have polarity.

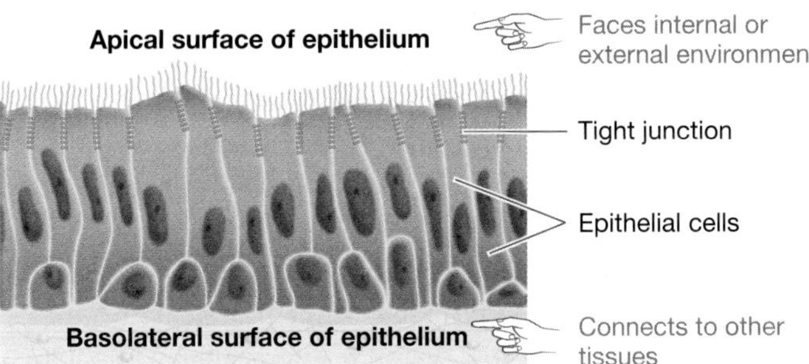

FIGURE 41.6 Epithelial Cells Provide Protection and Regulate Which Materials Pass across Body Surfaces.

QUESTION The inner surfaces of the lungs, stomach, and small intestine are lined with epithelium. What is the nature of the environment faced by the apical surface of these epithelia?

QUESTION Would you expect that the phosopholipids, glycolipids, glyoproteins, and proteins found in the basolateral and apical membranes of epithelial cells are similar or different? Why or why not?

Epithelial Tissues **Epithelial tissues** are also called **epithelia** (singular: **epithelium**). Epithelium covers the outside of the body, lines the surfaces of organs, and forms glands. An **organ** is a structure that serves a specialized function and consists of several tissues; a **gland** is a group of cells that secrete specific molecules or solutions. Epithelia form the interface between the inside of an organ or body and the outside. In addition to providing protection, they are gatekeepers. Epithelia regulate the transfer of heat between the interior and exterior of structures, as well as the transfer of water, nutrients, and other substances.

Because the primary function of epithelium is to act as a barrier and protective layer, it's not surprising to observe that epithelial cells typically form layers of closely packed cells (**Figure 41.6a**). In many cases, adjacent epithelial cells are joined by structures that hold them tightly together, such as tight junctions and desmosomes (introduced in Chapter 8).

Figure 41.6b shows the location of tight junctions in epithelial cells that line the surface of the trachea and illustrates a second key point about epithelium: No matter where it is found in the body, epithelial tissue has a polarity, or sidedness. More specifically, an epithelium has an apical side and a basolateral side. The **apical** side faces away from other tissues and toward the environment, while the **basolateral** side faces the interior of the animal and is connected to other tissues. The apical and basolateral sides of an epithelium have distinct structures and functions. A few examples will help drive this point home:

- The interior space of hollow organs such as your small intestine is called the **lumen**. The epithelial cells that line the lumen of your small intestine have an apical surface that is folded into fingerlike projections. These projections are studded with membrane proteins that take up glucose and other nutrients from the food you have digested. The basolateral membrane of these epithelial cells contains different proteins from those found on the apical side. On the basolateral surface, glucose and other nutrients are then transported across the plasma membrane to nearby blood vessels, which distribute them throughout the body.

- In your trachea (windpipe), the apical side of epithelial cells secretes mucus and is covered with cilia that help sweep away dust, bacteria, and viruses. The basolateral side does not secrete mucus and does not contain cilia.

- Glandular epithelial cells produce secretions that are released onto the apical surface of the epithelium or across the apical surface into the bloodstream. Secretion is unidirectional and does not occur across the basolateral surface.

Another major characteristic of epithelium is that epithelial cells tend to have short life spans. The cells that line your esophagus—the tube connecting your mouth and stomach—live for 2 to 3 days, while the cells that line your large intestine live for a maximum of 6 days. Epithelial cells have these short life spans because they are exposed to a harsh environment, where they are likely to be killed or scraped away. The tissue as a whole does not wear away, however, because it includes cells that actively undergo mitosis and cytokinesis—producing new cells to replace those lost on the side that faces the environment.

Organs and Systems

Cells with similar functions are organized into tissues, and tissues are organized into specialized structures called organs. Recall that an organ is a structure that serves a specialized function and consists of several types of tissues. The heart, for example, consists of cardiac muscle tissue, nervous tissue, connective tissue, and epithelial tissue. All four of the basic tissue types are also found in the small intestine, as illustrated in **Figure 41.7a**.

Organs such as the heart or small intestine are also part of larger functional units called systems. A **system** consists of groups of tissues and organs that work together to perform one or more functions. Using the digestive system as an example, **Figure 41.7b** illustrates how the structure of organs correlates with their function and how the components of a system work together in an integrated fashion.

Because an animal's body contains molecules, cells, tissues, organs, and systems, biologists who study animal anatomy and physiology must work at various levels of organization to gain a thorough understanding of how that body operates. **Figure 41.8** illustrates these levels of organization, using human neurons and the nervous system as an example. The key point is that, because the structure and function of each component in the body are integrated with other components, and because each level of organization is integrated with other levels of organization, the organism as a whole is greater than the sum of its parts. In other words, an organism is more than just a collection of individual systems, and each system is more than just a collection of individual cells or tissues or even organs.

In effect, each subsequent chapter in this unit focuses on a different system found in animals, beginning with the excretory system and ending with the immune system. The presentations introduce the structures that are found in each system and explore how biologists investigate their functions. Each of these systems can be interpreted as a suite of adaptations and trade-offs. Each system accomplishes a specific task required for survival and reproduction, and each works in conjunction with other systems.

Before delving into the various systems, however, it's essential to examine general phenomena that affect all systems in animals. Let's start by looking at how body size affects animal physiology.

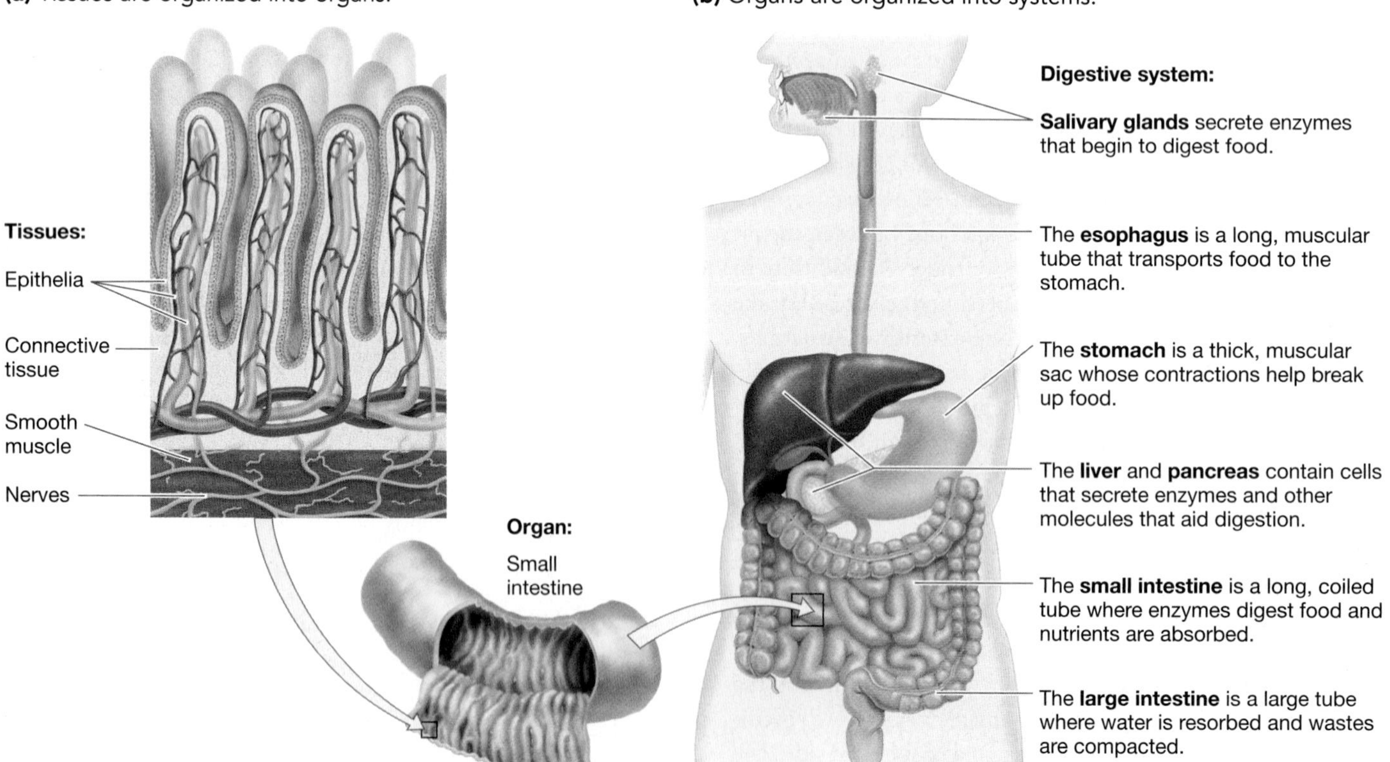

FIGURE 41.7 Organs Are Composed of Tissues; Systems Are Made Up of Organs. (a) The human small intestine is an organ composed of all four major tissue types: epithelium, connective tissue, smooth muscle, and nerves. **(b)** The human digestive system is essentially one long tube divided into chambers where food is processed and nutrients are absorbed. The salivary glands, liver, and pancreas are organs that secrete specific enzymes or compounds into the tube.

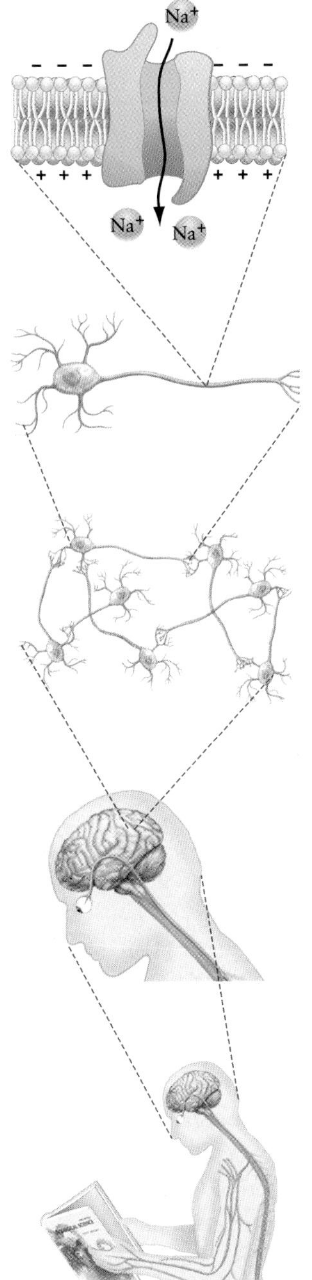

FIGURE 41.8 Biologists Study Anatomy and Physiology at Many Levels. The levels of organization within an organism are not independent of each other. Instead, they are tightly integrated.

Check Your Understanding

If you understand that ...

- Biologists study animal anatomy and physiology at the molecular, cellular, tissue, organ, and system levels.
- Events at each level of organization in an individual interact to form an integrated whole that responds to the environment in appropriate ways.

You should be able to ...

1) Give an example of a molecule that is part of a cell, name the tissue where that cell is found, name an organ that contains that tissue type, and identify the system that includes that organ.
2) Compare and contrast the structure and function of the four major types of animal tissues.
3) Explain why the polarity of epithelial cells is important.

41.3 How Does Body Size Affect Animal Physiology?

The laws of physics affect the anatomy and physiology of organisms in fundamental ways. The force of gravity, for example, limits how large an animal can be and still move efficiently. Or consider the forces exerted by the medium in which animals live. Because water is much denser than air, it is harder for animals to move through water efficiently. Based on this observation, it is not surprising to find that fish and aquatic mammals have much more streamlined bodies than terrestrial animals do.

Physical laws clearly affect body size. Just as clearly, body size has pervasive effects on how animals function. Large animals need more food than small animals do. Large animals also produce more waste, take longer to mature, reproduce more slowly, and live longer. Conversely, small animals are more susceptible to damage from cold and dehydration than large animals are, because they lose heat and water faster. Juveniles and adults of the same species face different challenges simply because their body sizes are different.

Why is body size such an important factor in how animals work? How do biologists study the consequences of size? Let's consider each question in turn.

Surface Area/Volume Relationships

From microscopic roundworms to gigantic blue whales, animals span an incredible range of body masses—a total of twelve orders of magnitude. Many of the challenges posed by increasing size are based on the asymmetrical relationship between surface area and volume.

To understand why the relationship between surface area and volume is important, recall from Chapter 6 that diffusion

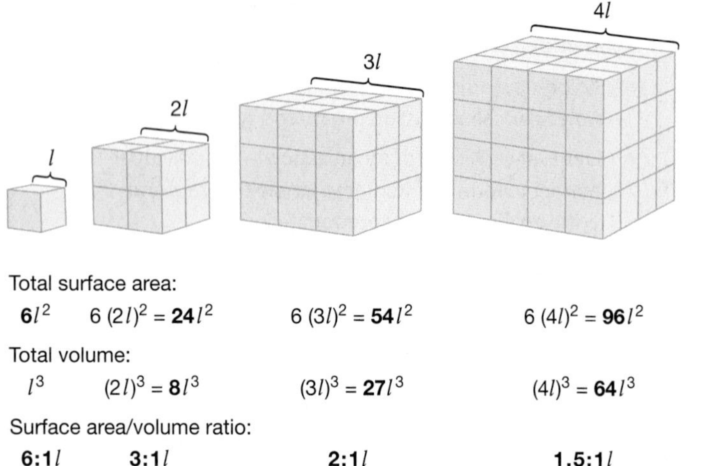

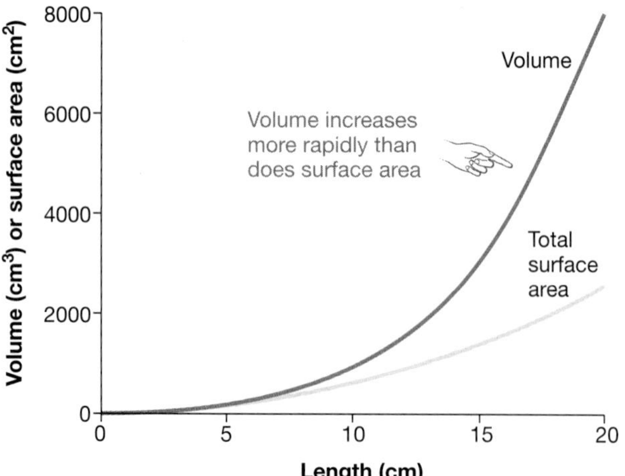

FIGURE 41.9 Surface Area and Volume Change as a Function of Overall Size. **(a)** The total surface area of an object increases as the square of the length (*l*). The volume increases as the cube of that linear dimension. **(b)** Volume increases much more rapidly than does total surface area as linear dimensions increase.

QUESTION Which animals have a higher surface area/volume ratio: large or small ones?

takes place across the surface of the plasma membrane. Oxygen and nutrients such as glucose must diffuse into the cell, and waste products such as urea and carbon dioxide must diffuse out. The rate at which these and other molecules and ions diffuse depends in part on the amount of surface area available for diffusion. In contrast, the rate at which nutrients are used and waste products are produced depends on the volume of the cell.

The contrast between processes that depend on surface area and those that depend on volume is important for a simple reason. As a cell gets larger, its volume increases much faster than its surface area does. Reviewing a little basic geometry will convince you why this is so. As **Figure 41.9a** shows, the surface area of a cube increases as a function of its linear dimension *squared*. Because cubes have six sides, the surface area of a cube of length l is $6l^2$ (six times the area of any one side). The volume of the same structure, however, increases as a function of its linear dimension *cubed*. Hence, the volume of a cube of length l is l^3. Area has two dimensions; volume has three. In general:

$$\text{Surface area} \propto (\text{length})^2$$

$$\text{Volume (or mass)} \propto (\text{length})^3$$

$$\text{Surface area} \propto (\text{volume})^{2/3}$$

Figure 41.9b graphs the consequences of these relationships. As a cube gets bigger, its surface area increases more slowly than does its volume (or mass); surface area increases at only the 2/3 power of its volume. The same general relationship holds for cells, tissues, organs, and systems. The general point is that quantities that are based on volume, such as body mass, increase disproportionately fast with increases in linear dimensions.

If you understand the relationship between surface area and volume, you should be able to predict which of the following has the higher surface area/volume ratio: a newborn versus an adult human; a flatworm versus a roundworm; a roundworm versus an anemone (see Chapter 32); the simple (unfolded) lungs of a frog versus the alveolar (highly folded) lungs of a mouse.

The relationship between surface area and volume is fundamental. How do changes in this relationship affect the form and function of animals?

Comparing Mice and Elephants As an example of how surface area/volume relationships affect an animal's physiology, consider the metabolic rate of mammals. **Metabolic rate** is the overall rate of energy consumption by an individual. Because consumption and production of energy in mammals depend largely on aerobic respiration, metabolic rate is often measured in terms of oxygen consumption, and is typically reported in units of milliliters of O_2 consumed per hour.

Because it is so much larger, an elephant consumes a great deal more oxygen per hour than a mouse does. But what is going on at the levels of cells and tissues? To furnish a more useful way to compare metabolic rates in different species, biologists divide metabolic rate by overall mass and report a mass-specific metabolic rate in units of mL O_2/gram/hour. This mass-specific metabolic rate gives the rate of oxygen consumption per gram of tissue. (Because an individual's metabolic rate varies dramatically with its activity, the accepted convention is to report the **basal metabolic rate** [**BMR**]—the rate at which an animal consumes oxygen while at rest, with an empty stomach, under normal temperature and moisture conditions.)

On a per-gram basis, small animals have much higher BMRs than do large animals (**Figure 41.10**). An elephant has much more mass than a mouse, but a gram of elephant tissue consumes much less energy than a gram of mouse tissue does. The leading hypothesis to explain this pattern is based on surface area/volume ratios. Many aspects of metabolism—including oxygen

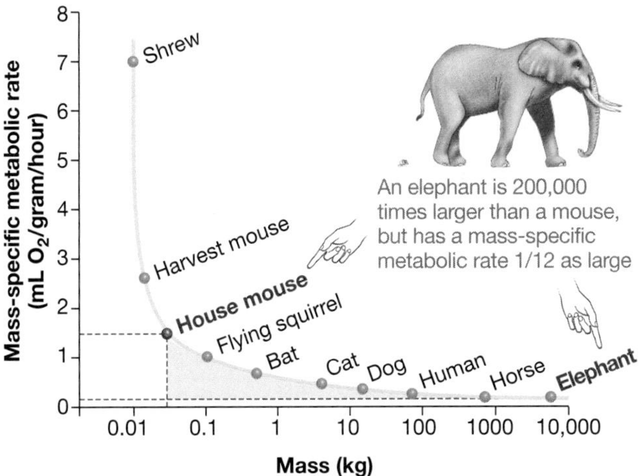

FIGURE 41.10 Small Animals Have Higher Metabolic Rates than Large Animals Do. Overall body mass, plotted on a logarithmic scale, versus metabolic rate per gram of tissue.

QUESTION Which mammal has to eat more to support each gram of its tissue: a dog or a human?

consumption, the digestion of food, the delivery of nutrients to tissues, and the removal of wastes and excess heat—depend on exchange across surfaces. As an organism's size increases, its mass-specific metabolic rate must decrease, or the surface area available for exchange of materials would fail to keep up with the metabolic demands generated by the enzymes in the organism. Small animals can "live fast" because they have enough surface area to support rapid metabolism; large animals don't have enough surface area to keep up and have to "live slow."

The same issues influence the development of an individual over the course-4p11.721 of its lifetime. A king salmon, for example, weighs a few milligrams or less at hatching but grows into an adult weighing 50 kg or more. This represents a million-fold increase in body mass. To explore the consequences of this

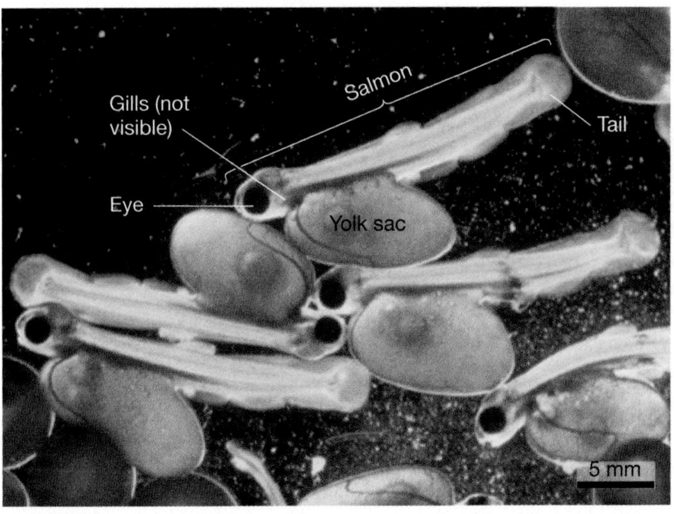

FIGURE 41.11 Newly Hatched Salmon Can Breathe through Their Skin and Their Gills. The gills of newly hatched salmon are so rudimentary that they are not visible here.

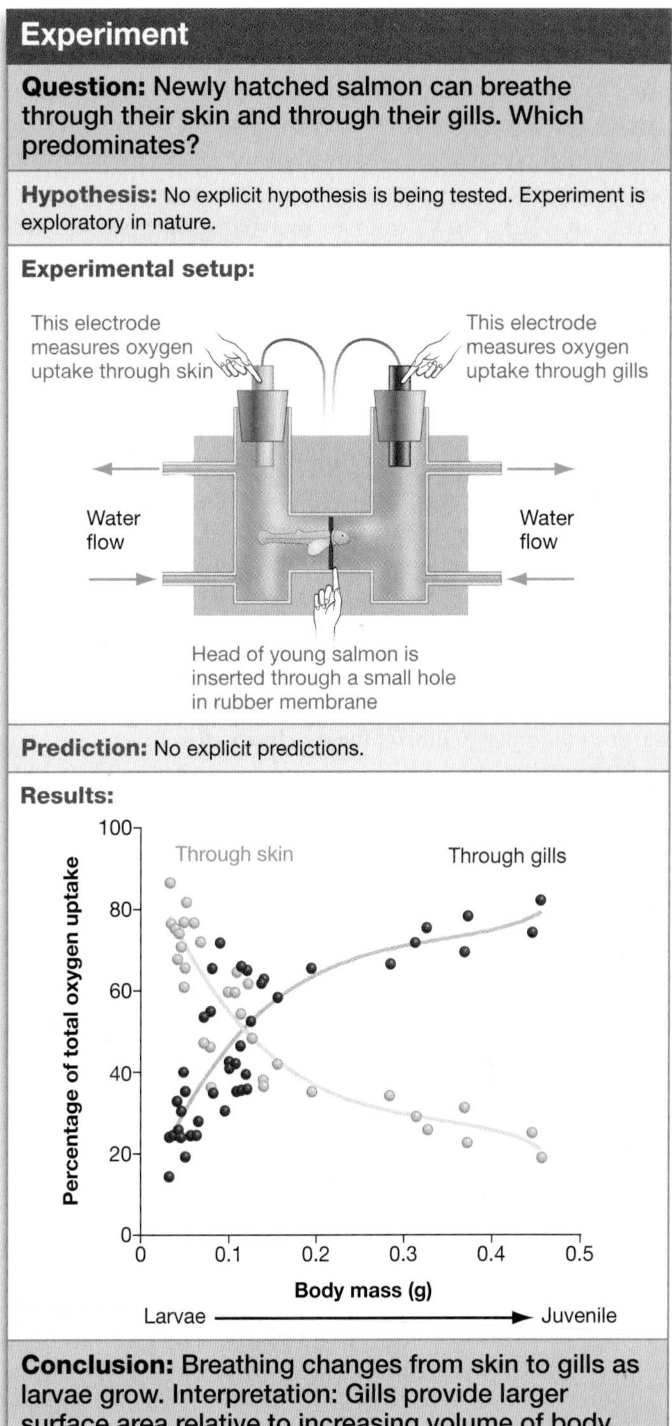

FIGURE 41.12 Do Young Salmon Breathe Mainly through Their Gills or Their Skin?

change, biologists have studied how gas exchange—uptake of oxygen and removal of carbon dioxide—occurs in newly hatched Atlantic salmon. Like most fish species, young salmon have rudimentary gills but also exchange gases across their skin and yolk sac (**Figure 41.11**). To document the amount of gas exchange that occurs in the gills versus the general body surface, researchers inserted the heads of individual salmon through a pinhole in a soft rubber membrane (**Figure 41.12**). After waiting for the fish to adjust to the apparatus, the biologists recorded the rate of oxygen uptake on either side of the membrane.

The graph in Figure 41.12 shows the results of such experiments. The data indicate that newly hatched larvae take up most of the oxygen they need by diffusion across the body surface. As an individual grows, however, its skin surface area decreases in relation to its volume. To avoid suffocation, the fish must develop gills to take over the bulk of the gas exchange activity. The individual breathes through its gills for the rest of its life. This outcome raises an interesting question: What makes gills so effective as the site of oxygen uptake?

Adaptations That Increase Surface Area In aquatic animals, gills are organs that allow the exchange of gases and dissolved substances between the animals' blood and the surrounding water. **Figure 41.13a** shows a close-up of a gill. The important points to notice are that a gill consists of sheetlike structures called **lamellae**. This is significant, because the sheets of epithelial cells provide this organ with an extremely high surface area relative to its volume. Because of these traits, gases are able to diffuse across the gills rapidly enough to keep up with the growth in the volume of the developing fish.

In general, if the function of a cell or tissue depends on diffusion, its structure has a shape that increases its surface area relative to its volume. In addition to flattening, folding and branching are effective ways for structures to have a high surface area/volume ratio:

- **Figure 41.13b** illustrates the extensive folding observed in portions of the digestive tract where nutrients diffuse into the body. These folds and the narrow, tubelike projections called **villi** make the surface area available for diffusion extensive. Folded surfaces are common in diffusion-dependent organs.
- The highly branched network shown in **Figure 41.13c** is a system of small, thin-walled blood vessels called **capillaries**. Capillaries have a high surface area available for gases, nutrients, and waste products to diffuse into and out of blood. In general, highly branched structures increase the surface area available for diffusion.

In almost every case, structures that are involved in diffusion are also thin. As Chapter 44 will detail, thin surfaces make diffusion more efficient because substances have a shorter distance to travel.

Surface area/volume relationships have a pervasive influence on the structure and function of animals. They will be an issue in almost every chapter in this unit.

Check Your Understanding

If you understand that... 

- An animal's overall size is important in part because body mass is affected by an array of physical forces.
- In addition, the amount of heat and waste that an animal produces and the amount of food and oxygen that it requires are proportional to its mass.
- The amount of surface area available relative to that mass is critical, because heat exchange and other important processes take place across surfaces.

You should be able to... 

1) Explain why large animals have a relatively small surface area/volume ratio.
2) Explain the relationship between body mass and basal metabolic rate.
3) Diagram three structures that increase surface area.

 Web Animation at www.masteringbio.com
Surface Area/Volume Relationships

(a) Flattened structures

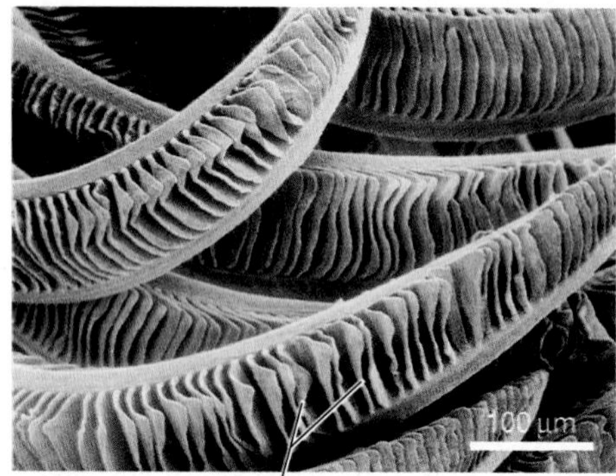

Lamellae of fish gills

(b) Folded surfaces with projections

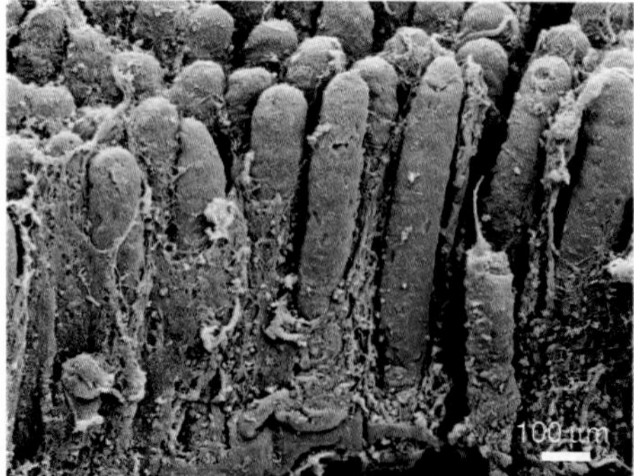

Lining of small intestine, showing villi

(c) Highly branched structures

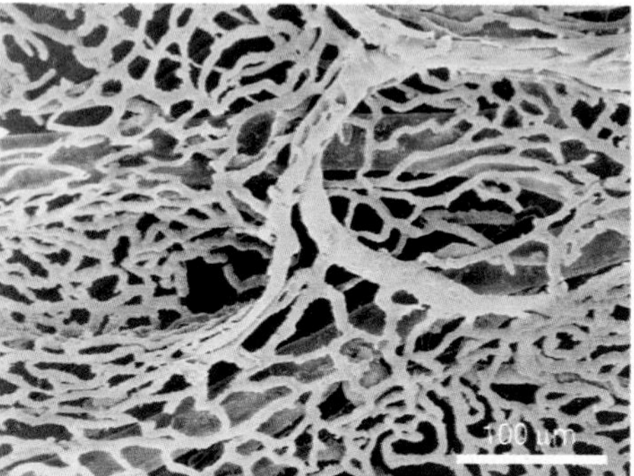

Capillaries

FIGURE 41.13 Certain Structures Have High Surface Area/Volume Ratios. (a) Flattened, **(b)** folded, and **(c)** branched structures have extremely high surface areas relative to their volume.

EXERCISE Draw and label arrows that indicate the diffusion-based process that takes place in each of the structures pictured.

41.4 Homeostasis

In analyzing the relationship between animal form and function, biologists focus on understanding adaptations as a product of natural selection, exploring the constraints on adaptation, and interpreting the influence of general principles such as surface area/volume ratios. Another general principle that researchers use in analyzing animal physiology is called homeostasis.

Homeostasis (literally, "alike-standing") is the array of relatively stable chemical and physical conditions in an animal's cells, tissues, and organs. Homeostasis is all about constancy. Although internal conditions may vary as an animal's environment changes, chemical and physical states are kept within a tolerable range.

Many of the structures and processes observed in animals can be interpreted as mechanisms for achieving homeostasis with respect to some quantity, such as pH or calcium ion concentration. It is important to understand, though, that constancy of physiological state can be achieved by two processes: (1) conformation or (2) regulation. The body temperature of Antarctic rock cod is an example of conformational homeostasis. The body temperature of these fish closely matches that of the surrounding seawater, which is typically −1.9°C. The rock cod is not regulating its body temperature to match that of seawater. Instead, its body temperature remains constant simply because it conforms to the temperature of its surroundings. Regulatory homeostasis, in contrast, is based on mechanisms that adjust the internal state. These adjustments keep the internal state within limits that can be tolerated, no matter what the external conditions. For example, a dog maintains a body temperature of about 38°C whether it's cold or hot outside. If its body temperature begins to rise, it might pant to cool off and maintain homeostasis. If its body temperature begins to drop, it might shiver to bring its temperature back up to the target value.

Because epithelium is the interface between the internal and external environments, it plays a key role in achieving homeostasis. Epithelium is responsible for creating an internal environment that is dramatically different from the external environment, and for maintaining physical and chemical conditions inside an animal that are relatively constant. One of its most basic functions is to control the exchange of materials across its surface in a way that is consistent with homeostasis. As subsequent chapters will show, many epithelial cells are studded with membrane proteins that regulate the transport of ions, water, nutrients, and wastes. No molecule can enter or leave the body without crossing an epithelium. Homeostasis is possible because epithelia control this exchange.

Why is homeostasis important? Much of the answer is based on enzyme function. Recall from Chapter 3 that enzymes are proteins that catalyze chemical reactions within cells. You might also recall that temperature, pH, and other physical and chemical conditions have a dramatic effect on the structure and function of enzymes. More specifically, most enzymes function best under a fairly narrow range of conditions. Other processes depend on homeostasis, too. Temperature changes affect membrane permeability and how quickly solutes diffuse. To take an extreme case, the expansion of water as it freezes can rip cells apart if tissues are allowed to drop much below 0°C. Conversely, extremely high temperatures can cause proteins to denature—meaning that they lose their tertiary structure and cease to function. When homeostasis occurs, conditions inside the body allow molecules, cells, tissues, organs, and systems to function at an optimal level.

The Role of Regulation and Feedback

To achieve homeostasis, animals have regulatory systems that constantly monitor internal conditions such as temperature, blood pressure, blood pH, and blood glucose. If one of these variables changes, a homeostatic system acts quickly to modify it. Like the thermostat in a home heating system, each of these systems has a **set point**—a normal or target value for the controlled variable. Animals have a set point for blood pH, blood oxygen concentration, nutrient availability, and other parameters. In most mammals, the set point for body temperature is somewhere between 35 and 39°C. How does an individual maintain its tissues at the set point in the face of changes in activity and the environment?

The key to answering this question is to recognize that homeostatic systems are based on the three general components shown in **Figure 41.14**. A **sensor** is a structure that senses

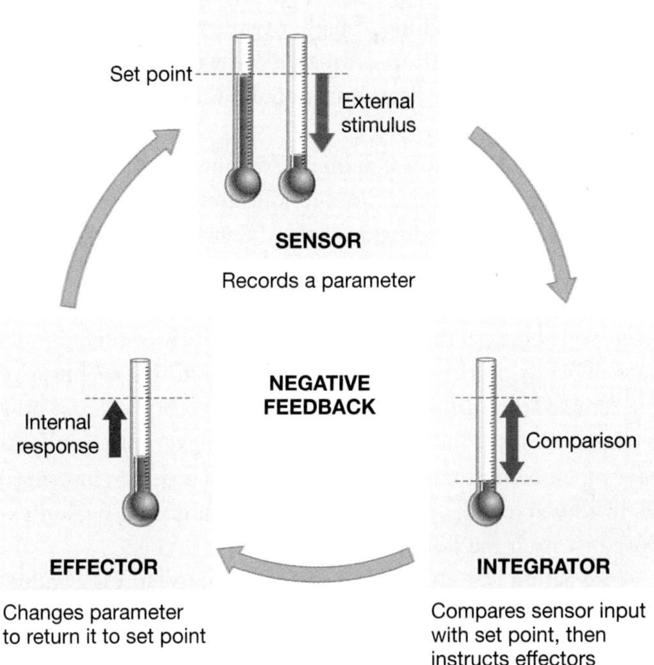

FIGURE 41.14 Animals Achieve Homeostasis through Negative Feedback. The interactions that occur among the three components of a generalized system for attaining homeostasis. Many animals use homeostatic systems similar to this one to maintain a preferred range of hydration (water concentration), blood pH, blood pressure, calcium ion concentration, body temperature, and so on.

some aspect of the external or internal environment. An **integrator** is a component that evaluates the incoming sensory information and "decides" whether a response is necessary to achieve homeostasis. (The word *decides* is in quotation marks because the decision is not a conscious one.) An **effector** is any structure that helps restore the desired internal condition. Without these three elements, homeostatic control systems are unable to maintain a desired set point and homeostasis is impossible.

Homeostatic systems are based on negative feedback. When **negative feedback** occurs, effectors reduce or oppose the change in internal conditions. For example, a rise in blood pH triggers effectors that act to reduce that rise. Blood pH returns to the set point in response to this negative feedback.

Subsequent chapters in this unit explore how animals use sensor-integrator-effector systems to achieve homeostasis with respect to the solute concentrations of their cells and tissues, their oxygen supply, and nutrient availability. To explore the concept further, let's focus on how different animals achieve homeostasis with respect to temperature.

MB Web Animation at www.masteringbio.com
Homeostasis

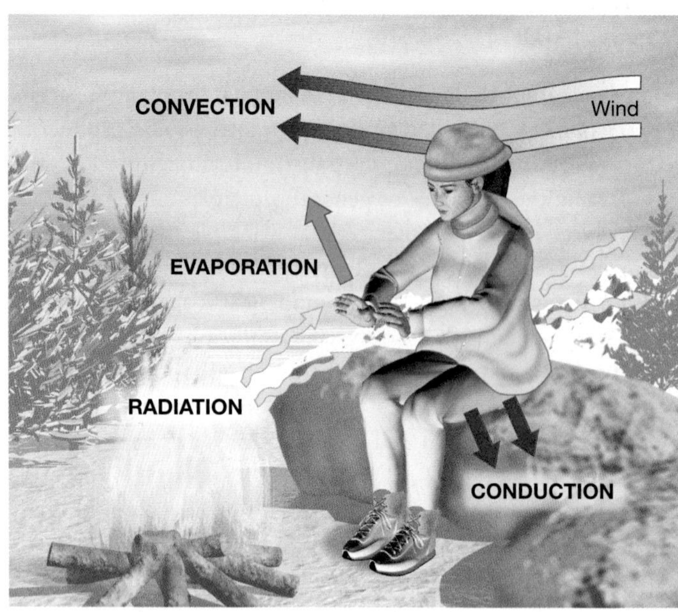

FIGURE 41.15 Four Methods of Heat Exchange. The arrows indicate the direction of heat exchange from the warmer body to the cooler body.
EXERCISE Sketch a similar diagram that features a butterfly in summer.

41.5 How Do Animals Regulate Body Temperature?

All animals exchange heat with their environment. Heat flows "downhill," from regions of higher temperature to regions of lower temperature. If an individual is warmer than its surroundings, it will lose heat; if it is cooler than its environment, it will gain heat.

As **Figure 41.15** shows, animals exchange heat with the environment in four ways: conduction, convection, radiation, and evaporation. **Conduction** is the direct transfer of heat between two physical bodies that are in contact with each other. For instance, when you sit on a cold rock, heat is transferred from your body to the rock. The rate at which conduction occurs depends on the surface area of transfer, the steepness of the temperature difference between the two bodies, and how well each body conducts heat. Water, for example, conducts heat much better than air does. As a result, a person immersed in 15°C water loses heat much faster than does a person exposed to air at the same temperature.

Convection is a special case of conduction. During conduction, heat is transferred between two solids; but during convection, heat is exchanged between a solid and a liquid or gas. For example, the heat loss that occurs when wind blows on your skin is due to convection. As the speed of the air or water flow increases, so does the rate of heat transfer.

Radiation is the transfer of heat between two bodies that are not in direct physical contact. All objects, including animals, radiate energy as a function of their temperature. A campfire radiates heat; so does your body, although to a much lesser degree. The major source of radiant energy is the Sun.

Evaporation is the phase change that occurs when liquid water becomes a gas. Conduction, convection, and radiation can cause heat gain or loss, but evaporation leads only to heat loss. You lose heat when water evaporates off your skin. Because of the extensive hydrogen bonding in liquid water, a large amount of energy is needed to heat water and produce evaporation. If you get overheated on a summer day, splashing water on your skin and sweating will absorb a large amount of heat and cool you off. Conversely, getting wet can be deadly on a cold day, because the water on your skin absorbs so much heat from your body.

Heat exchange is critical in animal physiology because individuals that get too hot or too cold may die. Overheating can cause enzymes and other proteins to denature and cease functioning, and it may lead to excessive water loss and dehydration. A sharp drop in body temperature, in contrast, can slow enzyme function and energy production. In humans, heat stroke and hypothermia ("under-heating") are life-threatening conditions.

Although most organisms cannot regulate their body temperature, some animals can. Let's take a closer look.

Body Temperature and Homeostasis

The control of body temperature is called **thermoregulation** and is an important aspect of homeostasis in some animals. **Figure 41.16** illustrates how the sensor-integrator-effector components of mammals regulate body temperature. Information about body temperature is constantly monitored by temperature

FIGURE 41.16 Mammals Regulate Temperature through Negative Feedback. In mammals, a set point for temperature is maintained by a complex negative feedback system that includes integrators in the anterior and posterior hypothalamus and sensors located throughout the body. The set point varies among species, from 30°C in monotremes to over 39°C in rabbits.

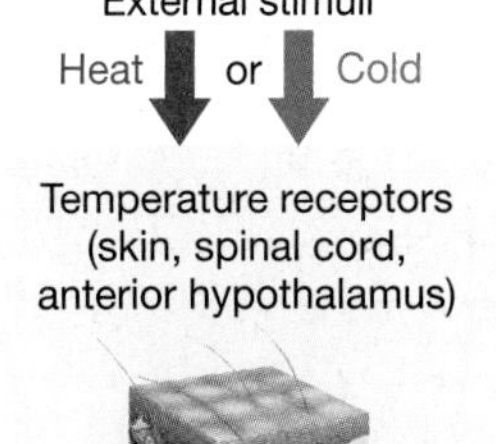

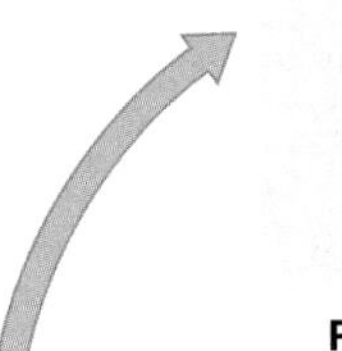

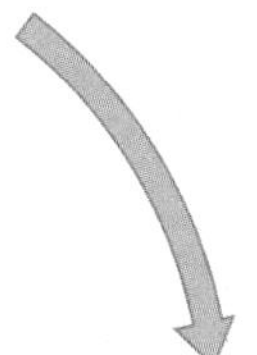

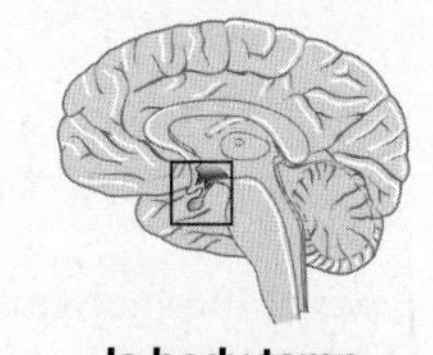

receptors located throughout the body. For example, temperature receptors in the skin respond to cooling or to heating. In response, these cool-sensors and heat-sensors alter the pattern of electrical signals that they send to adjacent neurons. Receptors in the brain region called the anterior **hypothalamus** respond in a similar fashion to changes in blood temperature.

The electrical signals that originate with temperature receptors in the skin, spinal cord, and anterior hypothalamus are transmitted to an integrator located in the brain. Current evidence indicates that separate centers in the hypothalamus of the brain integrate and respond to increases and decreases in body temperature. If a mammal becomes cold, cells in the posterior hypothalamus region of the brain send signals to effectors that return body temperature to the set point. For example, signals from the posterior hypothalamus might induce shivering to generate warmth, and fluffing of fur or feathers to improve insulation and retain heat. Signals from the same or nearby cells can also result in the release of blood-borne chemical signals that increase the rate at which cellular respiration is taking place throughout the body—thus generating more body heat. But if the individual becomes too hot, an integrator in the anterior hypothalamus sends signals that initiate sweating or panting—responses that cool the body. Signals from the same or nearby cells can also induce behavioral changes that slow heat production, such as seeking shade or a cool burrow and then resting.

In response to either cooling or heating, behavioral and physiological responses move the body temperature back toward the set point via negative feedback. The homeostatic system diagrammed in Figure 41.16 makes several general points about the effectors that maintain homeostasis:

- It is common to observe redundancy in feedback systems—there are usually several ways to increase or decrease a parameter;
- Feedback systems work in "antagonistic pairs": one set of responses increases a parameter while another set decreases it;
- Input from sensors and integrators is constant, so feedback systems are constantly making fine adjustments relative to the set point.

Gaining and Losing Heat

To understand temperature regulation in animals, it is critical to recognize that there is wide variation in how these organisms obtain heat. For example, an **endotherm** ("inner-heat") produces adequate heat to warm its own tissues, while an **ectotherm** ("outer-heat") relies principally on heat gained from the environment.

Endotherms and ectotherms represent two opposite extremes along a continuum of heat sources. Many animals are

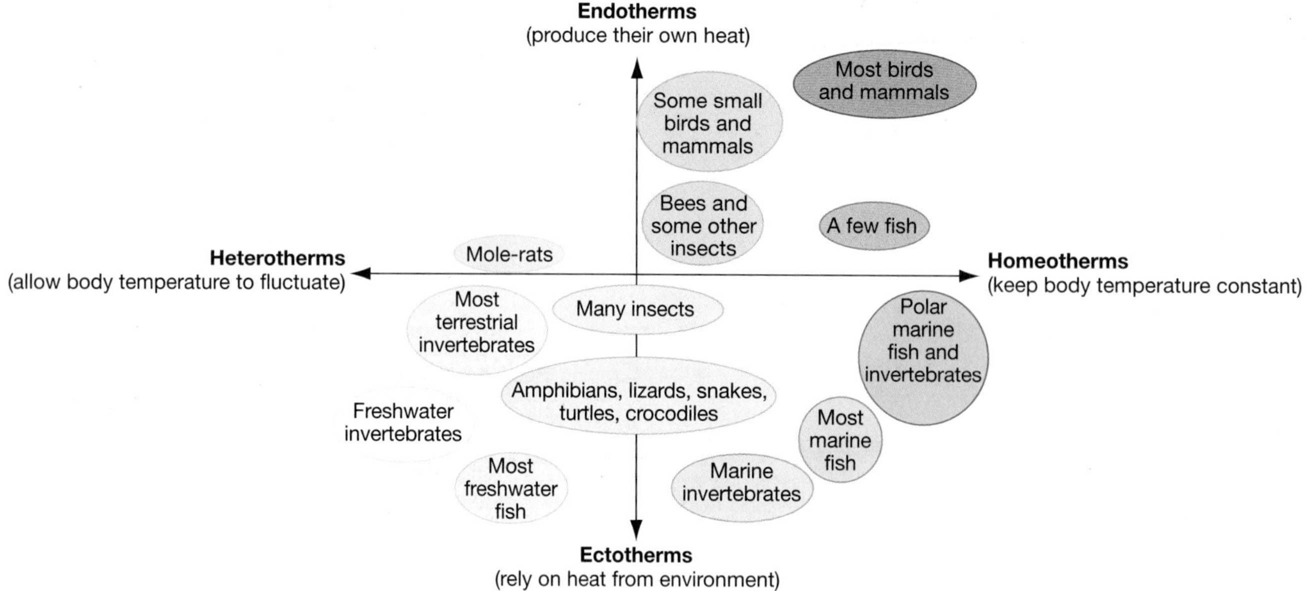

FIGURE 41.17 Types of Thermoregulation. Many types of thermoregulation occur in animals. Most mammals are endotherms and homeotherms. Most invertebrates are ectotherms and heterotherms.

● **EXERCISE** Add labels to the graph indicating the location of animals featured in this chapter, including oryx, side-blotched lizard, bumblebee, Galápagos finches.

partially endothermic and partially ectothermic. There are also two extremes on a continuum describing whether animals hold their body temperature constant: **Homeotherms** ("alike-heat") keep their body temperature constant, while **heterotherms** ("different-heat") allow their body temperature to rise or fall depending on environmental conditions.

The sources of heat that an animal uses and the degree to which its body temperature varies can be treated as axes on a graph, as shown in **Figure 41.17**. As strictly endothermic homeotherms, humans and other mammals, along with most birds, occupy the upper right-hand corner of the graph. These species produce their own heat metabolically and maintain a constant body temperature. As ectothermic heterotherms, mosquito larvae and other freshwater invertebrates occupy the lower left-hand corner of the graph. But many animal species lie somewhere between these extremes:

- Some desert-adapted mammals, such as the oryx featured in the introduction to this chapter, allow their body temperature to rise during the hotter part of the day—meaning they are somewhat heterothermic.
- Small mammals that inhabit cold climates lose heat rapidly because their surface area is large relative to their volume. To survive when temperatures are cold, species such as ground squirrels reduce their metabolic rate and allow their body temperature to drop. This condition is called **torpor**. If torpor persists for weeks or months, it is called **hibernation** (**Figure 41.18a**; see also **Box 41.1**).

(a) Ground squirrels like this chipmunk use torpor to save energy when food is scarce.

(b) Naked mole-rats' body temperature matches that of their underground burrows.

(c) Bumblebees shiver to warm up on cold mornings.

FIGURE 41.18 Different Strategies for Thermoregulation.

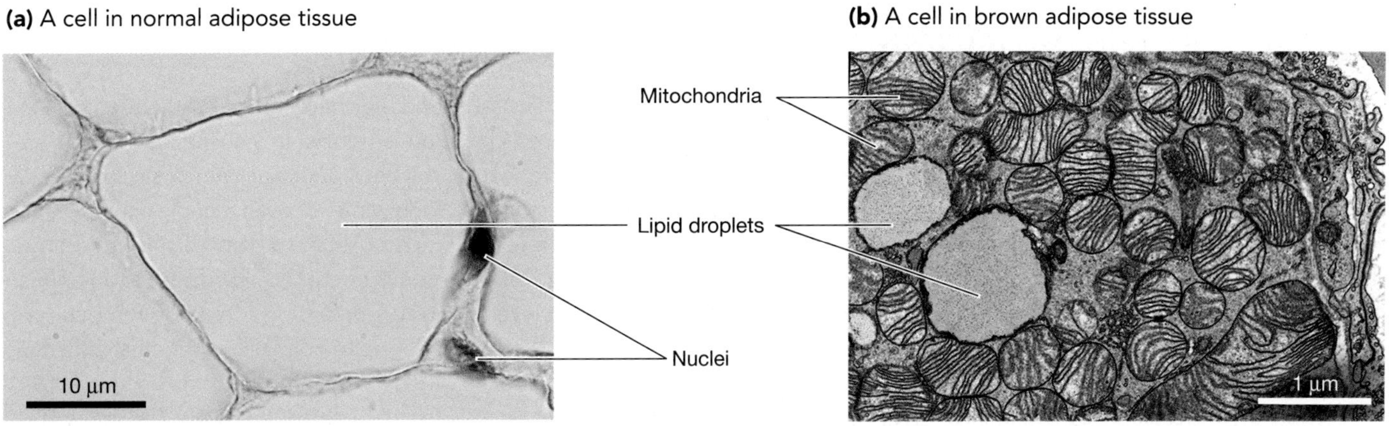

(a) A cell in normal adipose tissue

(b) A cell in brown adipose tissue

FIGURE 41.19 The Structure of Cells in Brown Adipose Tissue Correlates with Their Function. (a) Cells in normal adipose tissue store fat. **(b)** Cells in brown adipose tissue store fat but are also packed with mitochondria. Instead of producing ATP, these mitochondria produce primarily heat.

- Naked mole rats are mammals but lack insulation, because they have no fur (**Figure 41.18b**). They live in underground tunnels and allow their body temperature to rise and fall with burrow temperatures. They are heterothermic and intermediate between ectotherms and endotherms.
- On cold mornings, bumblebees "shiver" by contracting their flight muscles together. (During flight, these same muscles contract in an alternating pattern to beat the wings.) Shivering generates enough heat to raise the individual's body temperature in preparation for flight (**Figure 41.18c**).

Even in a homeothermic endotherm such as a mammal or bird, body temperature can vary widely in different body regions. When a Canada goose is standing on ice, its feet may be at a temperature of just 9°C, even though its body core is at 35°C. Similar types of variation exist in tuna and mackerel, which are ectotherms but generate heat to warm certain sections of their bodies, such as their eyes or the muscles that are responsible for rapid swimming.

Where Does Body Heat Come From? In addition to gaining heat directly by basking in sunlight or lying on warm rocks or soil, ectotherms generate some heat as a by-product of metabolism. The amount of heat they generate is small compared with the amount generated by endotherms, however, because ectotherms have low metabolic rates. Endotherms, in contrast, can warm themselves because their basal metabolic rates are extremely high. They can also maintain high body temperatures because they have elaborate insulating structures such as feathers or fur.

The contrast in metabolic rates and heat-producing abilities of endotherms and ectotherms is reflected in the structure of their tissues. For example, researchers find that mitochondrial density and mitochondrial enzyme activity are three to four times higher in endotherms' cells than in cells of ectotherms of similar size. This comparison is meaningful because most reactions that generate ATP and heat occur in mitochondria.

In addition, some endotherms have special heat-generating tissues. The best-studied example is the brown adipose tissue found in some mammals. **Adipose tissue** is a connective tissue whose cells store fats (**Figure 41.19a**). **Brown adipose tissue** is a specialized type of adipose tissue that features a high density of mitochondria as well as stored fats (**Figure 41.19b**). When the fat molecules are oxidized by the mitochondria in brown adipose tissue, no ATP is produced. Instead, all of the chemical energy stored in the fats is released as heat. As a result, brown adipose tissue produces almost 10 times as much heat as do other tissues.

Brown adipose tissue is particularly common in small animals and in the infants of large species, including humans. Heat production helps small animals come out of torpor (see Box 41.1). In humans, patches of brown adipose tissue are found in the neck and chest regions of newborns. Because infants are small, they have a high surface area/volume ratio and lose heat quickly. Based on such observations, brown adipose tissue is interpreted as an adaptation for small endotherms to achieve homeostasis with respect to temperature.

How Do Animals Conserve Heat? In terrestrial environments, heat is readily lost through evaporation, convection, and radiation. Air conducts heat poorly, however, so it is a good insulator. Thus, it is no coincidence that endothermic animals have elaborate external structures that trap air, slow the rate of heat exchange through conduction, and conserve body heat. Birds have down feathers that perform this function; mammals have underfur that forms a fine mesh, traps air, and prevents heat lost by conduction, convection, and radiation.

In aquatic environments, no heat is lost due to evaporation. Little heat is lost or gained due to radiation, because water transmits radiation poorly. But water is such an effective

BOX 41.1 A Closer Look at Torpor and Hibernation

As temperatures drop and days shorten in the fall, animals that live at high latitudes and high elevations face serious physiological challenges. For an endothermic individual to maintain normal body temperatures when the ambient temperature drops, it has to burn (oxidize) large amounts of food. But for species that eat seasonal foods such as insects, seeds, or nectar, food supplies are declining just as food demands are increasing. The problem is particularly acute for species that lose heat rapidly, because their body size is small. Yet squirrels, voles, and mice are abundant in northern and mountainous habitats. How do these animals cope?

In some cases, small mammals retreat to underground burrows that are much warmer than the ground surface and that have a supply of grasses and seeds stored during the summer. But other species endure by becoming inactive and allowing their body temperature to drop to levels just above ambient temperatures. Hibernating pocket mice and chipmunks expend only enough energy to maintain a body temperature of about 2°C, just above lethal limits, for weeks or months. On cold nights, hummingbirds may allow their body temperature to drop over 25°C, to less than 10°C.

Animals in torpor dramatically reduce their heart rate and oxygen consumption and may have long periods when breathing is temporarily suspended. Torpor results in significant energy savings over a long winter or even a single night.

Arousal from daily torpor or hibernation is energetically costly. For an animal in torpor to bring its body temperature back up to normal levels, it must either activate heat production from specialized tissue or shiver intensely. The amount of energy an animal uses during arousal from torpor can deplete energy stores to dangerously low levels. Yet it is clear that torpor confers a fitness advantage in some mammal and bird species, allowing them to thrive in otherwise inhospitable habitats.

conductor of heat that metabolic heat produced by aquatic organisms is lost rapidly—particularly through structures such as gills, which have a large surface area and a high rate of blood flow. As a result, almost all aquatic invertebrates and fish are ectotherms. Their body temperature is the same as that of the water they inhabit.

How do endotherms that live in aquatic habitats cope with heat loss due to conduction? If you've ever gone swimming in cold water, you can appreciate the problem faced by seals, otters, and whales. To conserve heat, otters have dense, water-repellent fur that maintains a layer of trapped air next to the skin. Seals and whales are insulated by thick layers of fatty blubber. Some whales have a feature that minimizes heat loss from their tongue, which is exposed to cold water during feeding. As **Figure 41.20a** shows, the tongue of the gray whale contains a series of arteries and veins that are juxtaposed. Each central artery carries warm, oxygenated blood from the body core to the tongue and is encircled by several smaller veins,

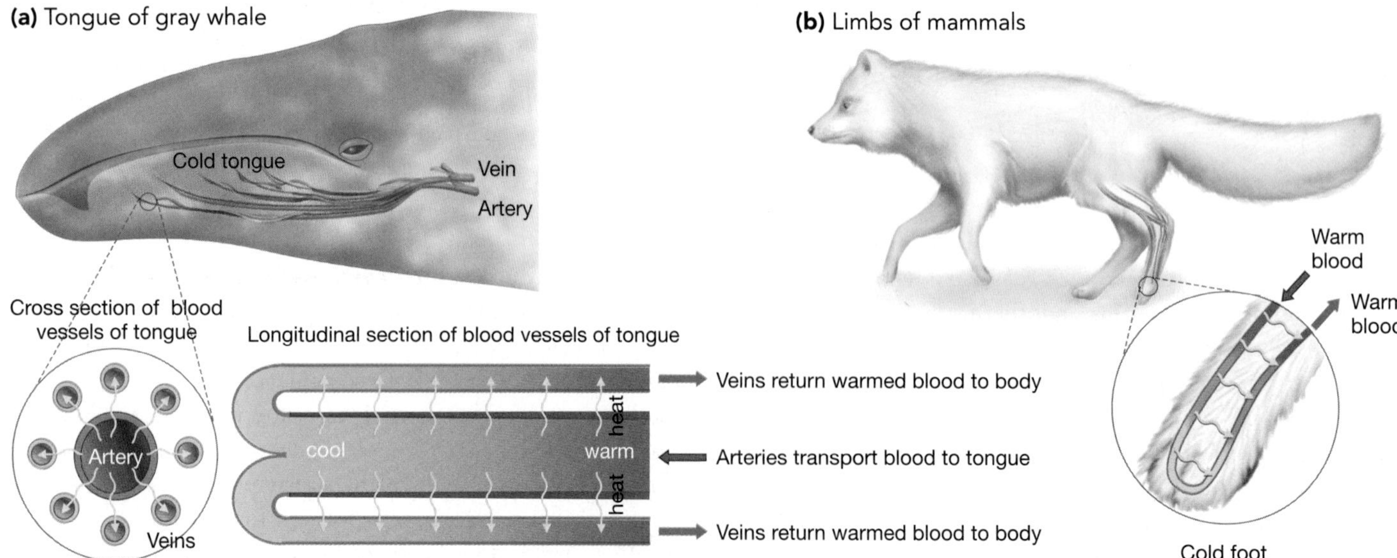

FIGURE 41.20 Countercurrent Exchangers Conserve Heat. (a) In the tongue of a gray whale, blood runs in opposite directions through juxtaposed veins and arteries. Heat flows from arteries to veins. That heat is returned to the body instead of being lost to the surrounding water. **(b)** Countercurrent arrangements are common in the limbs of mammals and birds that live in cold habitats.

which transport cool blood from the tongue surface back toward the body core. Because the two types of blood vessel are arranged in an antiparallel fashion, so that they transport blood in opposite directions, heat is transferred efficiently between them. This arrangement is called a countercurrent heat exchanger. Similar heat-conserving arrangements of arteries and veins are found in the flippers of whales and dolphins and in the legs of arctic foxes (**Figure 41.20b**).

Countercurrent exchangers are common in animals. The basic design principle is that two adjacent fluids flow through adjacent pipes in opposite directions. As the longitudinal section in Figure 41.20a shows, the fluid that enters the countercurrent heat exchanger is initially warm but steadily transfers heat to adjacent, cooler fluid flowing in the opposite direction. In a countercurrent heat exchanger, there is a warmer-to-cooler gradient between the two currents at every point along the length of the exchanger. The countercurrent arrangement makes it possible for a gradient to be maintained between the two pipes. If the two solutions were to run the same direction, the gradient between the two solutions would disappear as the source current cooled and the recipient current heated. Instead, a gradient exists along the entire length of a countercurrent exchanger.

A second key point about countercurrent heat exchangers is that, even though the temperature differential between the solutions at any point along the pipes is relatively small, there is a large temperature differential from one end of each pipe to the other. In effect, small differences in heat along the length of the exchanger sum up to create a large overall temperature gradient. The longer the system, the greater the overall differential will be. To highlight this property, the systems are sometimes called countercurrent multipliers.

● If you understand how countercurrent exchangers work, you should be able to diagram one that exchanges sodium ions (Na^+) via diffusion. Your diagram should include labels indicating areas of high versus low Na^+ concentration and arrows that indicate (1) the direction that solutions are moving in the exchanger and (2) the direction of Na^+ diffusion. You should also be able to diagram what would happen in this countercurrent exchanger if the two solutions involved ran in the same direction, instead of the opposite direction.

Ectothermy versus Endothermy

Analyzing animal thermoregulation provides a convenient way to review some of this chapter's major themes. Biologists interpret homeostatic systems as adaptations that have arisen via natural selection. The structure of cells and tissues involved in homeostasis, such as mitochondria-rich brown adipose tissue and countercurrent heat exchangers, correlates with their function. Because heat is gained and lost across surfaces, surface area/volume relationships are important. Small organisms lose heat much faster than large objects do, so it is difficult for small endotherms to stay warm without expending prohibitive amounts of energy. Surface area/volume relationships explain why there are no endotherms smaller than a shrew or a hummingbird, even though there are millions of centimeter- or millimeter-sized ectotherms.

Contrasting endothermy with ectothermy also provides a superb example of trade-offs in adaptation. Because endotherms maintain enzymes at optimal temperatures at all times, mammals and birds are able to remain active in winter and at night. Due to their high metabolic rates and insulation, endotherms are also able to sustain very high levels of aerobic activities, such as running or flying. These abilities come at a cost, however: To fuel their high metabolic rates, endotherms have to obtain large quantities of energy-rich food. The energy used to produce heat is then unavailable for other energy-demanding processes, such as reproduction and growth.

Ectotherms, in contrast, are able to thrive with much lower intakes of food. They can also use a greater proportion of their total energy intake to support reproduction. But because chemical reactions are temperature dependent, muscle activity and digestion slow down dramatically as the body temperature of an ectotherm drops. As a result, ectotherms are more vulnerable to predation in cold weather and in general are not as successful as endotherms at inhabiting cold environments or maintaining activity in cool nighttime temperatures.

In short, each suite of adaptations has advantages and disadvantages. Like all adaptations, endothermy and ectothermy involve trade-offs. Analyzing these trade-offs has also inspired a spirited debate on why endothermy evolved in the first place. Birds evolved from ectothermic dinosaurs, and mammals evolved from a lineage of ectothermic reptiles that is now extinct. Why did natural selection favor the evolution of endothermy in each case? Was it to make sustained activity such as flying possible? Or did endothermy allow birds and mammals to colonize cold habitats or remain active at night? These questions are unresolved. Research on animal form and function and the nature of adaptation continues.

Check Your Understanding

If you understand that...

- Two major aspects of temperature regulation vary among animal species: the amount of heat generated by the animal's own tissues and the degree to which body temperature varies over time.

You should be able to...

1) Describe three ways that animals gain heat and four ways that animals lose heat.
2) Diagram how the arteries and veins in a whale's tongue would be arranged for them *not* to function as a countercurrent heat exchanger.
3) Discuss the advantages and disadvantages of endothermy and ectothermy.

Chapter Review

SUMMARY OF KEY CONCEPTS

An adaptation is a heritable trait that allows individuals to produce more offspring in a particular environment than can individuals without the trait. Adaptations are genetic changes that occur in populations over time and that result from natural selection exerted by the environment. If the phenotype of an individual changes during its lifetime in response to environmental change, the change in the individual is due to acclimatization—not adaptation. In adaptive traits, there is often a strong correlation between a structure and its function.

Biologists analyze the structure and function of animals at a variety of levels: molecules, cells, tissues, organs, and systems.

Cells with a similar structure and common function are grouped together into four general types of tissue: epithelial tissue, connective tissue, muscle tissue, and nervous tissue. Epithelium is a particularly important type of tissue because it defines the interface between the animal's external and internal environments. Epithelial cells and tissues have a distinct polarity.

Organs are structures that are composed of several tissues and that perform specific tasks. Organs are organized into systems that work together in an integrated fashion to perform a function.

You should be able to (1) draw cells in the surface layer of an epithelium, (2) explain how their structure correlates with their function as an interface between the external and internal environments, and (3) predict the effect of a drug that made tight junctions between these cells loosen.

Body size has a strong influence on how animals work. For example, because small animals have a large amount of surface area relative to their volume, they lose heat faster than do large animals. The relationship between surface area and volume also plays a significant role in diffusion-mediated processes, such as gas exchange, nutrient absorption, and waste excretion.

Many of the effects of body size on the way animals work result from the relationship between the surface area of a structure and its volume. For example, large animals have low metabolic rates, because they have a relatively small surface area for exchanging the oxygen and nutrients required to support metabolism. The relatively high surface area of small animals, in contrast, means that they lose heat extremely rapidly. As a result, there are no tiny endothermic animals.

You should be able to explain why it would be impossible for a gorilla the size of King Kong to have fur.

(MB) Web Animation at www.masteringbio.com
Surface Area/Volume Relationships

Animals can maintain a relatively constant environment inside their bodies because they have systems that sense changes in internal conditions and trigger responses that return conditions to normal.

Homeostasis is the maintenance of relatively constant physical and chemical conditions inside the body. Animals have a set point, or target value, for blood pH, tissue oxygen concentration, nutrient availability, and other parameters. For example, most mammals have a set point for body temperature of about 37°C. If an individual starts to overheat, it will pant or sweat and seek a cool environment in response. If its body temperature begins to drop, it will respond by shivering, basking in sunlight, or fluffing its fur. Animals have adapted to life in "extreme" environments by evolving mechanisms of temperature regulation to cope with wide fluctuations in environmental temperatures.

You should be able to explain how, due to acclimatization as well as adaptation, the homeostatic system for body temperature in a species of mammal might change as global temperatures rise.

(MB) Web Animation at www.masteringbio.com
Homeostasis

QUESTIONS

Test Your Knowledge

1. How do biologists measure an animal's metabolic rate?
   a. by taking its temperature
   b. by measuring how rapidly it uses oxygen
   c. by measuring how rapidly it uses glucose
   d. by measuring how rapidly it produces wastes

2. How is the structure of a connective tissue most closely correlated with its function?
   a. The density of cells in the tissue correlates with the tissue's function.
   b. The surface area of the tissue correlates with the tissue's function.
   c. The origin of the tissue (from endoderm, mesoderm, or ectoderm) correlates with the tissue's function.
   d. The nature of the extracellular matrix correlates with the tissue's function.

3. As an animal gets larger, which of the following occurs?
   a. Its surface area grows more rapidly than its volume.
   b. Its volume grows more rapidly than its surface area.
   c. Its volume and surface area increase in perfect proportion to each other.
   d. Its volume increases, but its total surface area decreases.

4. Which of the following best describes the set point in a homeostatic system?
   a. the cells that collect and transmit information about the state of the system
   b. the cells that receive information about the state of the system and that direct changes to the system
   c. the various components that produce appropriate changes in the system
   d. the target or "normal" value of the parameter in question

5. What does it mean to say that an animal is a heterothermic endotherm?
   a. Its body temperature can vary, but it produces heat from its own tissues.
   b. Its body temperature varies because it gains most of its heat from sources outside its body.
   c. Its body temperature does not vary, because it produces heat from its own tissues.
   d. Its body temperature does not vary, even though it gains most of its heat from sources outside its body.

6. Which of the following is an advantage that ectotherms have over endotherms of the same size?
   a. They require much less food.
   b. They can save energy in cold weather by hibernating (entering torpor for long periods).
   c. They can remain active in cold weather or at nighttime—when temperatures cool.
   d. They have higher metabolic rates and grow much more quickly.

Test Your Knowledge answers: 1. b; 2. d; 3. b; 4. d; 5. a; 6. a

Test Your Understanding

Answers are available at www.masteringbio.com

1. Why is epithelium a particularly important tissue in achieving homeostasis?

2. The metabolic rate of a frog in summer (at 35°C) is about eight times higher than that in winter (at 5°C). Compare and contrast the individual's ability to move, exchange gases, and digest food at the two temperatures. During which season will the frog require more food energy, and why?

3. Consider the following cells, tissues, and organs:
   Nerve cells
   Brown adipose tissue
   Lung tissue in humans
   Absorptive sections of digestive tract
   Capillaries
   Beaks of Galápagos finches
   Fish gills

   In each case, how does the structure relate to the function?

4. Draw a large sphere and a small sphere. Explain why the surface area/volume ratio is different in the two structures. If materials diffuse into and out of each sphere, in which case will diffusion occur more efficiently relative to the volume? Explain why.

5. Consider a day in which daytime temperatures reach 30°C and nighttime temperatures drop to 18°C. Analyze how an ant might gain and lose heat by conduction, convection, radiation, and evaporation to avoid overheating during the day and escape cold-induced lethargy in the early morning and evening.

6. Explain how a countercurrent heat exchanger works. (Include a diagram.)

Applying Concepts to New Situations

Answers are available at www.masteringbio.com

1. When food is scarce, bigger medium ground finches win contests over seeds. Biologists have documented that there is strong natural selection in favor of large body size under these conditions, and that the finch population evolves in response. If so, why aren't medium ground finches a lot bigger than they are?

2. What data would you need to collect in order to document that acclimatization and/or adaptation is occurring in a population of lizards, as global warming intensifies?

3. An engineer has to design a system for dissipating heat from a new type of car engine that runs particularly hot. Recall that heat is gained and lost as a function of surface area. Suggest ideas to consider that are inspired by biological structures with exceptionally high surface area/volume ratios.

4. Suppose a friend of yours is trying to decide whether to buy a pet turtle or a pet mouse. In making the decision, all he cares about is how much it will cost to feed the animal. Your friend says he can't decide because the two animals weigh the same and their food costs the same per pound. As a biologist, what's your advice?

www.masteringbio.com is also your resource for • Answers to text, table, and figure caption questions and exercises • Answers to *Check Your Understanding* boxes • Online study guides and quizzes • Additional study tools including the *E-Book for Biological Science* 3rd ed., textbook art, animations, and videos.

42 Water and Electrolyte Balance in Animals

KEY CONCEPTS

- Freshwater, marine, and terrestrial habitats pose different challenges to animals with regard to maintaining water and electrolyte balance.
- In marine animals, specialized epithelial cells have membrane proteins that remove excess salt (NaCl) from the body so that it can be excreted. The same types of cells are found in the kidneys of mammals.
- In terrestrial insects, the hindgut and Malpighian tubules are responsible for excreting water-soluble waste products and achieving homeostasis with respect to water and electrolyte concentrations.
- In terrestrial vertebrates, the kidney is responsible for excreting water-soluble waste products and achieving homeostasis with respect to water and electrolyte concentrations.

Terrestrial animals lose water every time they breathe and urinate. For many animals, drinking is an important way to gain water and achieve homeostasis. This chapter explores how terrestrial and aquatic animals maintain water balance.

The chemical reactions that make life possible occur in an aqueous solution. If the balance of water and solutes in the aqueous solution is disturbed, those chemical reactions—and life itself—may stop. For example, humans can stay alive for weeks without eating but survive at most 8 days without drinking water. Hurricanes occasionally introduce so much freshwater to the ocean shore that normal salt concentrations are disrupted and marine animals die. Maintaining water balance is a matter of life or death.

An animal achieves water balance when its intake of water equals its loss of water. Maintaining water balance is an important element in homeostasis—the ability to keep cells and tissues in constant and favorable conditions. Water balance is also intimately associated with sustaining a balanced concentration of solutes throughout the body. In many animals, the most abundant solutes are ions: sodium (Na^+), chloride (Cl^-), potassium (K^+), and calcium (Ca^{2+}). As a group, these substances are called electrolytes. An **electrolyte** is a compound that dissociates into ions when dissolved in water. Because cells require precise concentrations of Na^+, Cl^-, K^+, and Ca^{2+} to function normally, maintaining electrolyte balance is crucial to survival. In humans, severe electrolyte imbalances can lead to muscle spasms, confusion, irregular heart rhythms, fatigue, paralysis, or even death.

Water balance, ion balance, and excretion of waste products are tightly integrated processes. If water enters an organism or a cell, ion concentrations drop. When water is lost during excretion of soluble wastes, ion concentrations rise and a water deficit may be created.

How do animals maintain water and electrolyte balance? Animals gain water in four ways: (1) by absorbing it via osmosis, (2) by drinking, (3) by eating, and (4) as a by-product of cellular respiration. They lose water via urine and feces, evaporation, or osmotic loss. To maintain water balance, gains and losses must be equal. Ion budgets have to be balanced in a similar fashion.

Our investigation into water and electrolyte balance begins with a look at how water moves into and out of cells. We'll consider why saltwater, freshwater, and terrestrial environments affect animals differently, and then we'll delve into research on the cellular and molecular mechanisms responsible for maintaining water and electrolyte balance in each of these environments. Although aquatic and terrestrial animals face very different challenges, in several cases these challenges are met by common molecular mechanisms.

42.1 Osmoregulation and Osmotic Stress

Chapter 6 introduced diffusion and osmosis. **Diffusion** is the movement of substances from regions of higher concentration to regions of lower concentration. Ions and molecules diffuse along their **concentration gradient**. **Figure 42.1a** illustrates a particularly important example of diffusion. In this case, two dissolved substances, or **solutes**, move down their concentration gradients via diffusion across a selectively permeable membrane. A **selectively permeable membrane**, such as a phospholipid bilayer, is a membrane that some solutes can cross more easily than other solutes can. When the solutes are randomly distributed throughout the solutions on both sides of the membrane, an equilibrium is established. At equilibrium, molecules continue to move back and forth across the membrane, but at equal rates.

Water can also move down its concentration gradient. To understand how this happens, notice that the concentration of

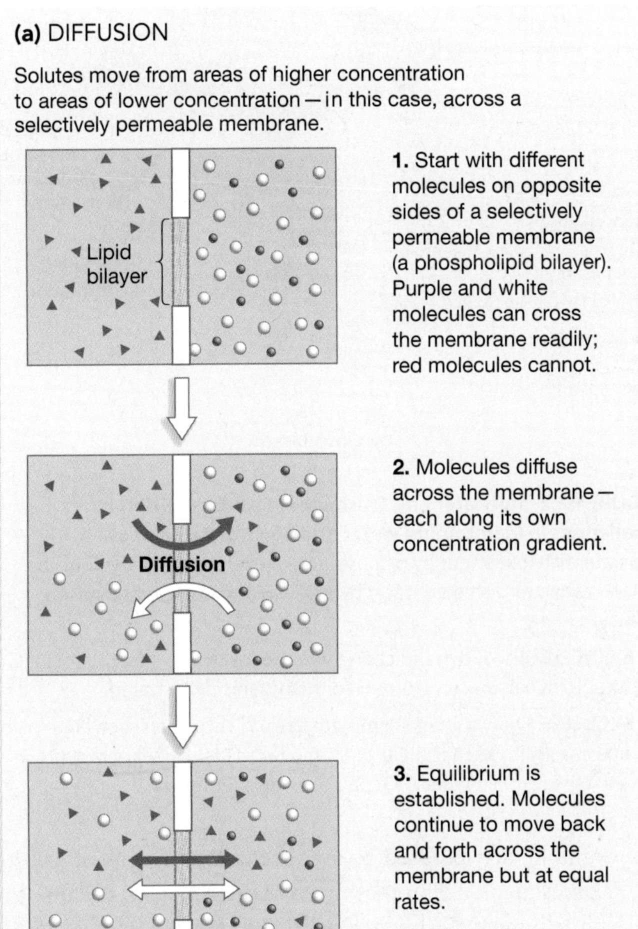

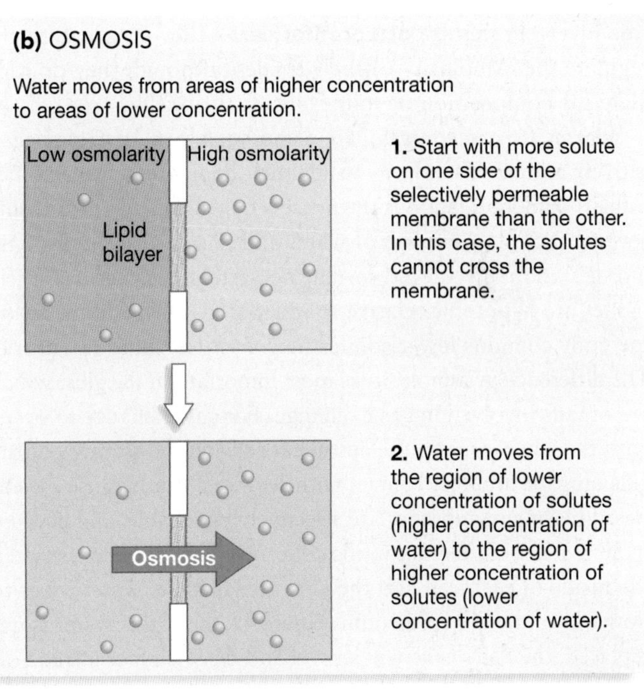

FIGURE 42.1 Solutes Move Down a Concentration Gradient via Diffusion; Water Moves Down a Concentration Gradient via Osmosis. **(a)** Diffusion occurs any time a solute is at higher concentration in one location than another. **(b)** Osmosis is a special case of diffusion involving the movement of water across a selectively permeable membrane.

QUESTION Why doesn't the presence of red molecules on the right side of part (a) affect the movement of the purple and white molecules?

EXERCISE In the upper drawing of part (b), label regions of high and low water concentration.

dissolved substances in a solution, measured in moles per liter, is referred to as the solution's **osmolarity**. When dissolved substances are separated by a selectively permeable membrane and the solutes cannot cross that membrane, water moves from areas where it is more concentrated—that is, solute concentrations are lower—to areas where it is less concentrated—that is, solute concentrations are higher. Stated another way, water moves across a selectively permeable membrane from regions of lower osmolarity to regions of higher osmolarity. The movement of water from areas of higher water concentration to areas of lower water concentration is called **osmosis** (**Figure 42.1b**).

Because diffusion and osmosis affect animals differently in freshwater, marine, and terrestrial habitats, these environments pose different challenges to animals in maintaining water and electrolyte balance. For example, achieving homeostasis with respect to water and electrolyte balance is straightforward in marine invertebrates such as sponges, jellyfish, and flatworms. Seawater is a fairly constant ionic and osmotic environment and nearly matches the electrolyte concentrations found within these animals. Relative to seawater, their tissues are **isotonic**: Solute concentrations inside and outside these animals are equal (see Chapter 6). As a result, diffusion and osmosis are relatively unimportant. Most marine invertebrates are **osmoconformers**—they do not actively regulate the osmolarity of their tissues, although they do regulate the composition of solutes inside their cells.

Marine fish, in contrast, are **osmoregulators**—meaning they actively regulate osmolarity inside their bodies to achieve homeostasis. **Osmoregulation** is the process by which living organisms control the concentration of water and salt in their bodies. Osmoregulation is required in marine vertebrates, because their tissues are **hypotonic** relative to salt water—the solution inside the body contains fewer solutes than does the solution outside. The difference in osmolarity is most important in the gills, which are organs involved in gas exchange. For gas exchange to occur with the environment, the epithelial cells on the surfaces of the gills must be in direct contact with seawater. But because the exposed plasma membranes are selectively permeable, and because there is a large difference in the concentration of solutes between the inside of each cell and the seawater outside, water tends to flow out of the gill epithelium (**Figure 42.2a**). If this water is not replaced, the fish's cells will shrivel and die. Fish face a trade-off between gas exchange and water and electrolyte balance.

Marine fish replace the lost water by drinking large quantities of seawater. Drinking brings in excess electrolytes, however. Ions and other solutes also diffuse into the fish across the gills, following a concentration gradient from seawater to tissues. To rid themselves of excess electrolytes, marine fish excrete urine that has a much higher concentration of electrolytes than seawater does.

Freshwater animals osmoregulate in a dramatically different environment. Marine fish gain salt and lose water, but freshwater animals lose salt and gain water. The gills of freshwater fish,

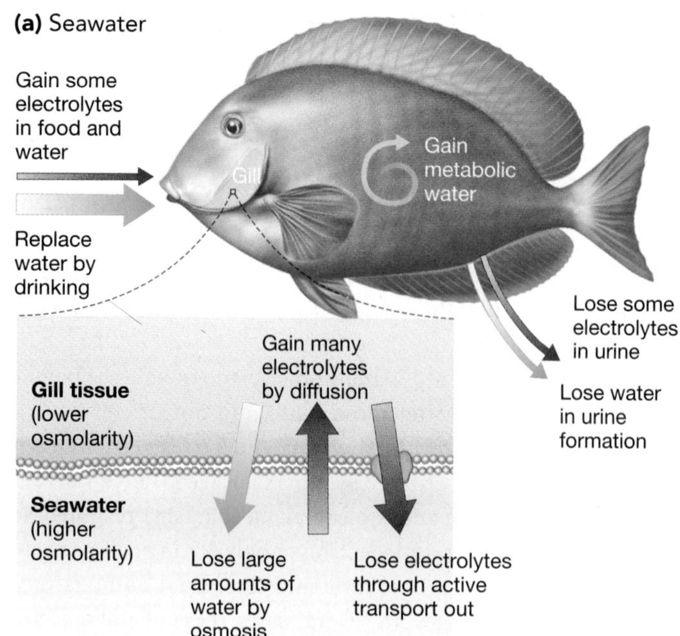

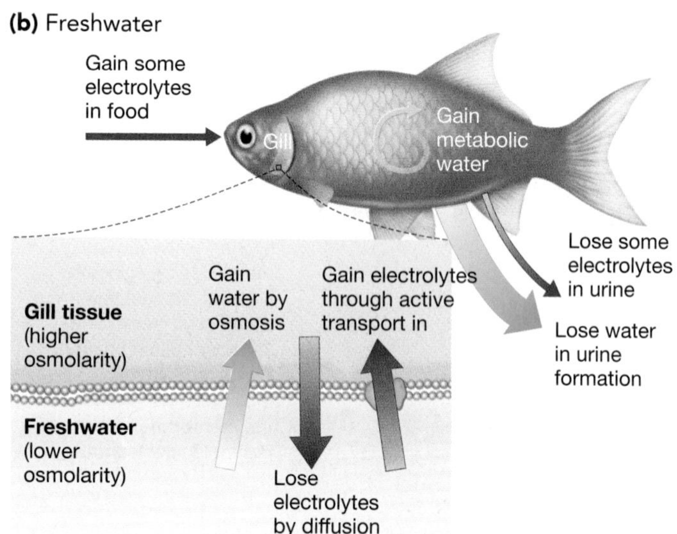

FIGURE 42.2 Saltwater and Freshwater Fish Face Different Challenges in Maintaining Water and Electrolyte Balance. (a) In seawater, fish lose water by osmosis and gain electrolytes by diffusion. **(b)** In freshwater, fish gain water by osmosis and lose electrolytes by diffusion.

EXERCISE Add labels indicating whether each water environment is hypotonic, hypertonic, or isotonic to the organism illustrated.

EXERCISE Draw a line to complete the path between each large pair of blue arrows for each fish. Do the same for the large pair of red arrows for each fish. Explain what each pair of arrows represents.

for example, are exposed to a concentration gradient 100 to 1000 times greater than that faced by marine fish. Because the gill epithelium of freshwater fish is **hypertonic** relative to the surrounding water—meaning the solution inside the cells contains more solutes than does the solution outside—the epithelial cells gain water via osmosis (**Figure 42.2b**). This water then moves from the epithelium into adjacent tissues, putting them under osmotic stress. Osmotic stress occurs when the concentration of dissolved substances in a cell or tissue is abnormal. If a freshwater fish does not get rid of incoming water, its cells will eventually burst and the individual will die. To achieve homeostasis and survive, freshwater fish excrete large amounts of water in their urine and do not drink.

In addition to gaining water, freshwater fish undergo osmotic stress because they lose electrolytes to the environment. Fresh water is hypotonic to the gill epithelium, so ions and other solutes tend to diffuse out of gill cells. Freshwater animals must replace electrolytes that are lost by obtaining them in food or by actively transporting them from the surrounding water—usually across the gills.

What about land animals? In terms of water balance, terrestrial environments are similar to the ocean. Land animals constantly lose water to the environment (**Figure 42.3**), just as many marine animals do. In this case, however, the process involved is not osmosis but evaporation (see Chapter 41). The epithelial cells that line a turtle's lung and a fruit fly's gas exchange structures have a moist surface, in order to protect the integrity of their plasma membranes. Because the atmosphere is almost always drier than the wet gas exchange surface, terrestrial animals lose water to evaporation. There is a trade-off between breathing and water and electrolyte balance.

In this respect, water balance in land animals mirrors the situation in land plants. As Chapter 37 pointed out, land plants lose huge quantities of water in the course of obtaining CO_2 through their stomata—as an inevitable by-product of gas exchange. Water balance is further complicated in terrestrial animals because all species lose water in the form of urine and because some species lose water when they sweat or pant to lower their body temperature. The lost water has to be replaced by drinking, ingesting water contained in food, or gaining metabolic water.

FIGURE 42.3 Maintaining Water and Electrolyte Balance in Terrestrial Environments Is a Challenge. In terrestrial environments, animals lose water by evaporation from the body surface and as water vapor during breathing. Electrolytes are lost primarily in the urine and feces.

What molecular mechanisms allow animals to cope with these diverse challenges? Let's start answering this question by focusing on how fish from marine or freshwater environments either excrete or take up electrolytes through their gills.

42.2 Water and Electrolyte Balance in Aquatic Environments

Whether they live in seawater or freshwater, virtually all of the 26,000 species of fish living today experience osmotic stress. Early research on how these species maintain water and electrolyte balance focused on sharks as a model organism.

In contrast to most marine fish, sharks and other cartilaginous fishes have tissues that are isotonic to seawater. The situation is complex, though, because the concentrations of certain electrolytes in their cells and extracellular fluids—specifically sodium ions, potassium ions, and chloride ions—are much less than their concentrations in seawater. Shark blood is still isotonic to seawater because it contains large quantities of soluble compounds such as urea and trimethylamine oxide (TMAO). In combination, the concentration of TMAO, urea, and ions is high enough to match the osmotic concentration of seawater and prevent the loss of water by osmosis. Sharks still need to secrete salt (NaCl), however, because sodium and chloride ions diffuse into their gill cells from seawater down the concentration gradients for the ions (**Figure 42.4**).

How do sharks excrete salt? Answering this question turned out to have two wide-ranging consequences.

1. The proteins and processes that are responsible for transporting salt in sharks also occur in a wide range of other marine animals, including oceangoing birds. Remarkably enough, the same mechanisms turned out to be crucial to the function of the kidney in humans and other terrestrial animals.

2. Salt excretion in sharks uncovered a general concept in animal and plant physiology: In many cases, cells do not transport ions against their electrochemical gradient directly. Instead, cells move key ions indirectly. They do this by using active transport to set up a strong electrochemical gradient for a different ion—typically Na^+ in animals and H^+ in plants. The sodium or proton gradient is then used to transport an array of other substances without a further expenditure of energy.

This is fundamental stuff. Let's dig in.

How Do Sharks Osmoregulate?

In studying shark anatomy, biologists discovered an organ called the **rectal gland**, which secretes a concentrated salt solution. To determine how this gland works, researchers studied it in vitro—meaning outside the shark's body. Specifically, they

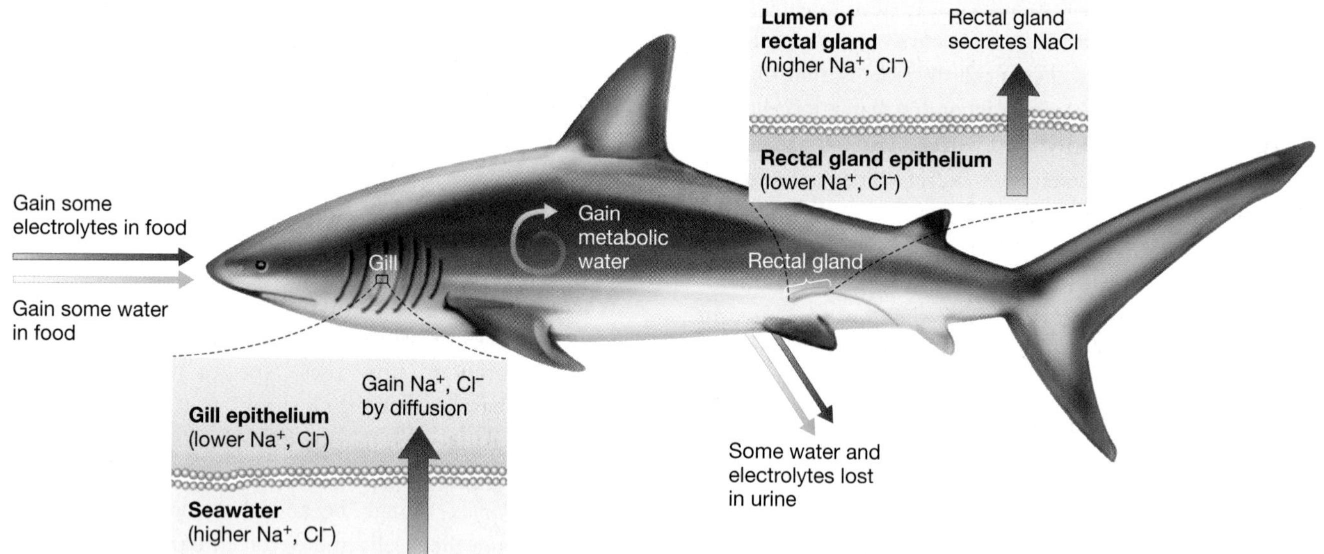

FIGURE 42.4 Sharks Face an Unusual Challenge in Maintaining Water and Electrolyte Balance. Even though shark tissues are isotonic to seawater, sodium ions and chloride ions diffuse into the gills along a concentration gradient.

EXERCISE Add a label indicating whether seawater is hypotonic, hypertonic, or isotonic to shark tissues.

dissected rectal glands and immersed them in a solution with a composition and osmolarity similar to those of normal shark tissue. Then the biologists connected a tube to the blood vessel that carries blood into the rectal gland. They also attached a collecting tube to the end of the gland that empties into the environment. This experimental setup allowed them to introduce solutions of defined composition and evaluate the fluid that the rectal gland produced in response.

To establish that the rectal gland functioned normally in this preparation, the biologists introduced a solution containing Na^+, K^+, and Cl^- in concentrations that might be observed in an intact animal: 280 mM (millimolar), 5 mM, and 270 mM, respectively. The fluid that left the gland contained the same ions in concentrations of 449 mM, 12 mM, and 446 mM instead. Because the product solution was more concentrated than the initial solution, the result gave the researchers confidence that the organ was functioning normally in vitro. And because they got this result only when ATP molecules were added to the preparation, the data supported the hypothesis that salt excretion is an energy-demanding activity. Ions can be concentrated in this way only if they are actively transported against a concentration gradient. (To review how molecules are transported across membranes, see **Box 42.1**). The question was, How does concentration occur?

The Role of Na^+/K^+-ATPase An energy-demanding mechanism for salt excretion implies that a protein in the plasma membrane of epithelial cells is involved in pumping Na^+, Cl^-, or both ions out of the extracellular fluid toward the lumen of the gland, where they are excreted. (Recall that the cavity inside any tube or sac-shaped organ is called the *lumen*.) The best-characterized membrane protein pump involving these ions is the **sodium-potassium pump**, also known as Na^+/K^+-ATPase.

To test the hypothesis that the sodium-potassium pump is involved in salt excretion by sharks, biologists attached a radioactive atom to a plant defense compound called **ouabain** (pronounced *WAA-bane*), which poisons the sodium-potassium pump. Ouabain is toxic to animals because it binds to Na^+/K^+-ATPase and prevents it from functioning. The biologists treated cells from the shark rectal gland with this radioactively labeled poison and then examined the distribution of the radioactive particles. These studies showed that the ouabain bound to Na^+/K^+-ATPase localized along the basolateral (bottom-side) membrane of the cells, which face the inside of the body and thus the blood supply. No pumps appeared in the apical (top) membrane of the cells, which face the gland's lumen—where salt is actually secreted.

The pumps' location was paradoxical, because Na^+/K^+-ATPase pumps Na^+ out of the cell and K^+ into it. (Three Na^+ go out for every two K^+ that come in.) The paradox arose because the rectal gland cells pump Na^+ on the opposite side of the cell and in the direction opposite the direction where sodium ions are actually secreted (**Figure 42.5**). The researchers also confirmed that Na^+/K^+-ATPase is essential for salt excretion by treating the gland with ouabain in vivo. Rectal glands treated with this drug stopped producing a concentrated salt solution. What was happening?

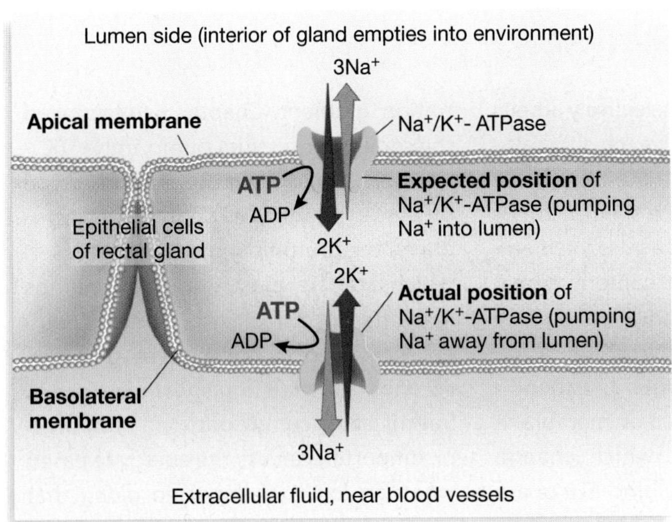

FIGURE 42.5 In Cells That Excrete Salt, the Na^+/K^+-ATPase Has an Unusual Location.

A Molecular Model for Salt Excretion The paradox of Na^+ being pumped in the direction opposite of excretion was solved by the model shown in **Figure 42.6,** which hinges on the coordinated action of four different membrane proteins:

1. Na^+/K^+-ATPase pumps sodium ions out of epithelial cells across the basolateral surface and into the surrounding extracellular fluid. The pump creates electrical and concentration (electrochemical) gradients favoring the diffusion of Na^+ into the cell and K^+ out of the cell. These gradients are used to transport other types of ions and molecules.
2. A $Na^+/Cl^-/K^+$ cotransporter, powered by the electrochemical gradient favoring Na^+ diffusion into the cell, brings these three ions from the extracellular fluid into epithelial cells across their basolateral membranes. **Cotransporters** are membrane proteins that transport more than one type of ion or molecule at a time. The cotransporter in the shark rectal gland is a **symporter** that uses the inward diffusion of Na^+ to power inward transport of $2Cl^-$ and K^+ against their gradients.
3. Although Na^+/K^+-ATPase pumps sodium ions back out, K^+ and Cl^- concentrations build up inside the cell as a result of the cotransport process. A potassium channel located on the basolateral membrane allows K^+ to diffuse back across the basolateral surface after it has been pumped in. A chloride channel located in the apical membrane of the epithelial cells allows Cl^- to diffuse down its concentration gradient into the lumen of the gland.
4. Sodium ions also diffuse into the lumen of the gland, following their charge and concentration gradient. But instead of passing through the epithelial cells as Cl^- does, Na^+ diffuses through spaces between the cells.

A Common Molecular Mechanism Underlies Many Instances of Salt Excretion Once the mechanism of salt excretion illustrated in Figure 42.6 had been worked out,

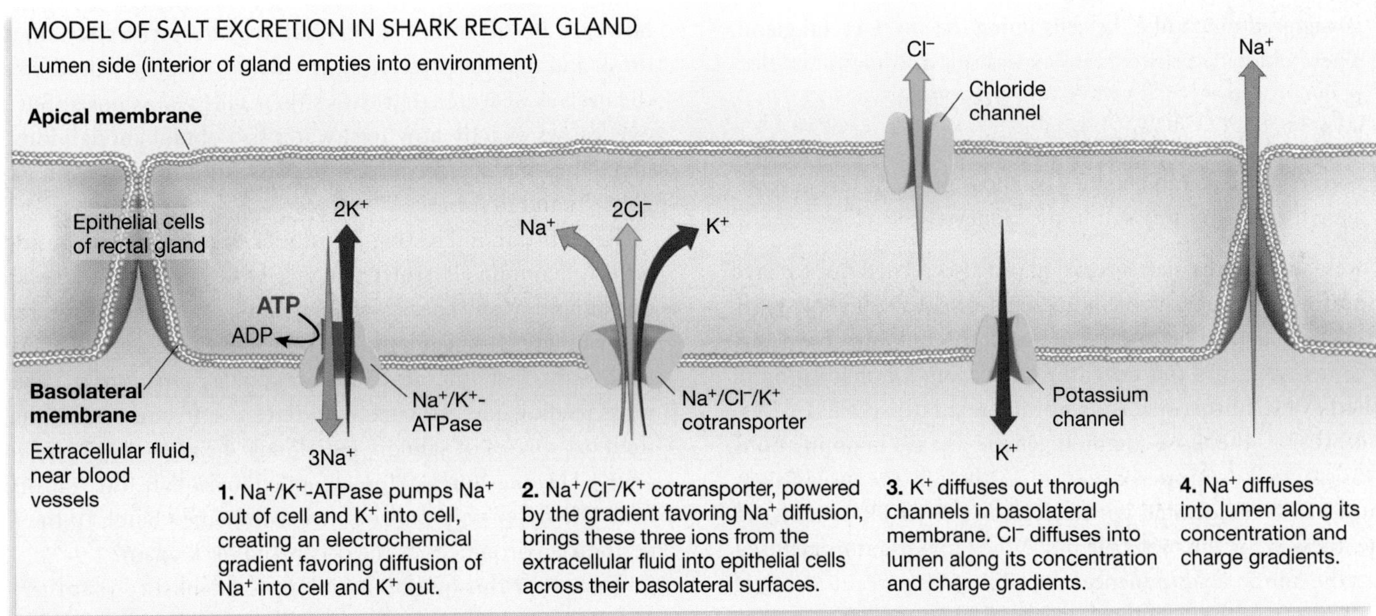

FIGURE 42.6 The Shark Rectal Gland Rids the Body of Excess Salt.

EXERCISE Label membrane proteins that are involved in (1) active transport, (2) secondary active transport, and (3) passive transport.

BOX 42.1 Passive and Active Transport: A Quick Review

Solutes move across membranes by passive or active transport. You might recall from earlier chapters that passive transport is driven by diffusion along an electrochemical gradient and does not require an expenditure of energy in the form of ATP. Active transport, in contrast, occurs when ATP powers the movement of a solute against its electrochemical gradient.

Because ions and large molecules such as glucose do not cross phospholipid bilayers readily, both passive and active transport take place via membrane proteins. In many cases, passive transport occurs through **channels**—proteins that form a pore, or an opening, that selectively admits a specific ion or ions. Passive transport also occurs via **carriers**, which are transmembrane proteins that bind a specific ion or molecule and transport it across the membrane by undergoing a conformational change. When solutes move via carriers, **facilitated diffusion** is said to occur.

Active transport requires an expenditure of energy and is based on membrane proteins called pumps, which change conformation when they bind ATP or are phosphorylated. This energy-demanding change in shape allows them to transport ions or molecules against their concentration gradient. Chapter 6 introduced the sodium-potassium pump, or Na^+/K^+-ATPase, which is the most important type of pump in animals; Chapter 38 introduced the proton pumps, or H^+-ATPases, which are the most important pumps in plants. Once an electrochemical gradient has been established by a pump, **secondary active transport** can occur. Specifically, a cotransporter (a symporter or antiporter) can use the energy released when an ion is transported *along* that electrochemical gradient to transport a different solute *against* its electrochemical gradient.

subsequent research showed that the same combination of membrane proteins is found in epithelial cells that transport sodium and chloride ions in many animals:

- Marine birds and reptiles drink salt water and must excrete NaCl. They have salt-excreting glands in their nostrils that function much like the shark rectal gland.
- Because marine fish with bony skeletons are hypotonic to seawater, salt constantly enters through their gills. Their gills contain specialized cells called chloride cells, which are configured much like the cells lining the shark rectal gland. These chloride cells excrete excess salt and maintain electrolyte balance.
- In mammals, cells with the same configuration of pumps, cotransporters, and channels as those in sharks are responsible for transporting salt in the kidneys.

Research on the shark rectal gland also turned out to have an unforeseen benefit for biomedical research. Several years after the shark chloride channel was isolated and characterized, investigators succeeded in identifying a human protein called cystic fibrosis transmembrane regulator (CFTR). Cystic fibrosis is the most common genetic disease in populations of northern European extraction. Although the disease was known to be associated with defects in the CFTR protein, no one knew what the molecule did. When investigators realized that the amino acid sequence of CFTR is 80 percent identical to that of the shark chloride channel, it was their first hint that CFTR is involved in Cl^- transport. Subsequent studies supported the hypothesis that cystic fibrosis results from a defect in a chloride channel. In this way, studies on water and electrolyte balance in sharks shed light on an important human disease.

How Do Salmon Osmoregulate?

Research on the shark rectal gland and the salt-secreting cells found in the gills of marine fish has been spectacularly successful. As a result, the molecular mechanisms of salt balance in marine animals are now well known. How freshwater fish achieve homeostasis with respect to electrolytes is still unclear, however. Recall from Figure 42.2b that freshwater fish do not drink and that they secrete large volumes of watery urine to rid themselves of water that enters their gills via osmosis. But no one knows exactly how freshwater fish obtain enough ions to make up for the electrolytes they lose to diffusion from their gill cells and in urine.

Recently, an interesting hypothesis to explain how freshwater fish maintain electrolyte balance has emerged from research on salmon. Many species of salmon are **anadromous** (literally, "up-running")—meaning that young develop from eggs laid in freshwater, then migrate to the ocean where they spend several years feeding and growing, then return to freshwater to breed. (In most species of salmon, individuals die after they breed just once.) The question is, How do individuals make the transition from the hypotonic environment where they hatch to the hypertonic environment of the ocean and back again?

To answer this question, biologists established that there is a significant increase in Na^+/K^+-ATPase activity in the gills of young salmon preparing to migrate to salt water, as well as a dramatic increase in the number of chloride cells in the gills. This observation is logical, because these cells are responsible for excreting salt when the fish are living in the ocean. But researchers have also found that the gills of young salmon living in freshwater versus salt water have two distinct populations of cells containing Na^+/K^+-ATPase. As **Figure 42.7** shows, individuals that live in freshwater and that have to import electrolytes have cells with Na^+/K^+-ATPase in the sheetlike lamellae that extend from the base of structures called gill filaments. In contrast, individuals that live in salt water and excrete electrolytes have a large number of sodium-potassium pumps in cells at the base of their gill filaments. Adult salmon from freshwater versus salt water show the same pattern, and similar changes have been observed in other fish species that switch between freshwater and saltwater habitats.

These observations inspired the hypothesis that salmon have two distinct types of cells involved in maintaining electrolyte balance. Researchers have proposed that chloride cells in the gill lamellae import electrolytes, while chloride cells at the base of gills secrete salt. Both cells contain sodium-potassium pumps, but they operate in opposite directions. This chloride-cell-switch hypothesis implies that salmon sense changes in the osmolarity of their environment and respond by producing or destroying the appropriate populations of chloride cells. The molecular mechanisms involved in the sensory reception and switching response are a mystery, however, and it is still unknown exactly how cells located in the gill lamellae take up electrolytes from freshwater. Identifying the mechanisms of electrolyte uptake in freshwater fish is an important challenge for researchers who want to know how aquatic organisms maintain water and electrolyte balance.

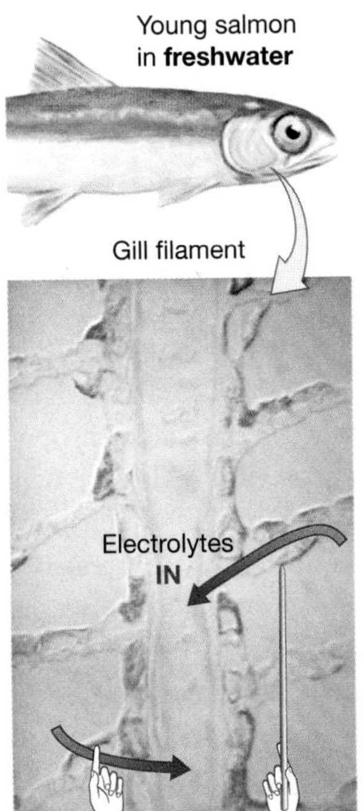

Chloride cells located on lamellae of gill filaments import electrolytes

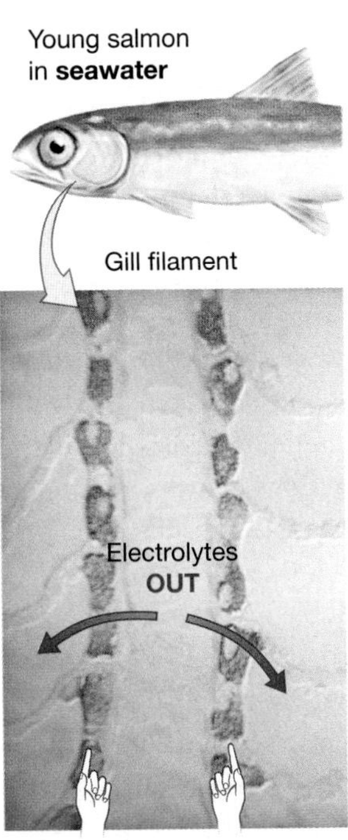

Chloride cells located at base of gill filaments secrete electrolytes

FIGURE 42.7 In Salmon Gills, Certain Cells Import Ions in Freshwater; Other Cells Secrete Ions in Salt Water. These gills have been treated with a molecule that binds to Na^+/K^+-ATPase and leaves a dark stain, labeling the chloride cells.

Check Your Understanding

If you understand that…

- Marine fish lose water by osmosis. To replace it, they have to drink salt water.
- Marine fish have to rid themselves of salt. They gain salt when they drink salt water or when sodium and chloride ions diffuse into their cells along a concentration gradient.
- Freshwater fish gain water by osmosis. They have to rid themselves of excess water by urinating.
- Freshwater fish lose electrolytes to the surrounding water by diffusion. They have to gain electrolytes in their food or by active transport from the surrounding water.

You should be able to…

1) Explain why Na^+/K^+-ATPase is important in the transport of salt in both saltwater and freshwater fish.
2) Predict what happens when cells in the gills of fish or the rectal glands of sharks are treated with ouabain.

42.3 Water and Electrolyte Balance in Terrestrial Insects

To understand how terrestrial invertebrates maintain water and electrolyte balance, researchers have focused on species that inhabit particularly dry environments. The success of this work has validated a general research strategy in biological science: By studying extreme situations or unusual organisms, biologists can often gain insight into how organisms cope with more moderate environments.

In studies on the molecular mechanisms of water and electrolyte balance in terrestrial insects, the most valuable model organisms have been the desert locust and the common household pest called the flour beetle. (You may have seen the larvae of flour beetles, called mealworms, in bags of flour that were not shut tightly enough to keep adults from entering and breeding.) Desert locusts and flour beetles rarely, if ever, drink—simply because little or no water is available in the habitats they occupy.

These insects live in environments where osmotic stress is severe. How do they maintain water and electrolyte balance? The answer has two parts: They minimize water loss from their body surface, and they carefully regulate the amount of water and electrolytes that they excrete in their urine and feces. Let's look at each issue in turn.

How Do Insects Minimize Water Loss from the Body Surface?

As Chapter 44 will show, terrestrial animals breathe by exposing an extremely thin layer of epithelium to the atmosphere. Oxygen diffuses into this epithelium, and carbon dioxide diffuses out. But water constantly leaks across the thin respiratory surface and is lost to the atmosphere via evaporation. Evaporation from the body surface itself is another threat—a particular challenge to insects, because they are small. As Chapter 41 emphasized, small organisms have a high surface area/volume ratio. Insects have a relatively large surface area from which to lose water but a small volume in which to retain it.

How do desert locusts, flour beetles, and other insects minimize water loss during gas exchange? In these species, gas exchange occurs across the membranes of epithelial cells that line an extensive system of tubes called **tracheae** (pronounced *TRAY-kee-ee*). The insect tracheal system connects with the atmosphere at openings called **spiracles** (**Figure 42.8a**). Muscles just inside each spiracle open or close the pore, much as guard cells open or close the pores in plant leaves and stems. When investigators manipulated bugs called *Rhodnius* so that their spiracles stayed open and then placed the insects in a dry environment, the bugs died within three days. These data support the hypothesis that the ability to close spiracles is an important adaptation for minimizing water loss. If an insect is under osmotic stress, it may be able to close its spiracles and wait until conditions improve before resuming activity.

Figure 42.8b shows how insects minimize evaporation from their body surfaces. This diagram is a cross-sectional view of the exoskeleton of an insect, which consists of a tough, nitrogen-containing polysaccharide called chitin and layers of protein. This combination of chitin and protein is known as **cuticle**. As the figure shows, the surface of the exoskeleton is covered with a layer of wax. Recall from Chapter 6 that waxes are a type of lipid. They are highly hydrophobic and thus highly impermeable to water. Researchers who removed the wax from insect exoskeletons have confirmed that the rate of water loss from the body surface increases sharply. Based on this observation, the wax layer is interpreted as an adaptation that minimizes evaporative water loss.

(a) Spiracles can be closed to minimize water loss from tracheae.

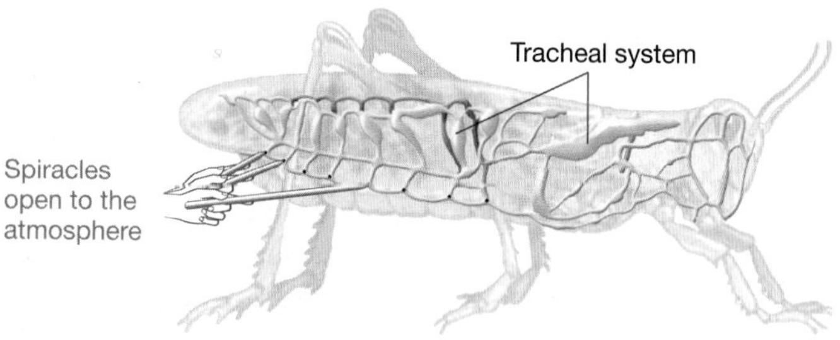

(b) Except at spiracles, the insect body is covered with wax.

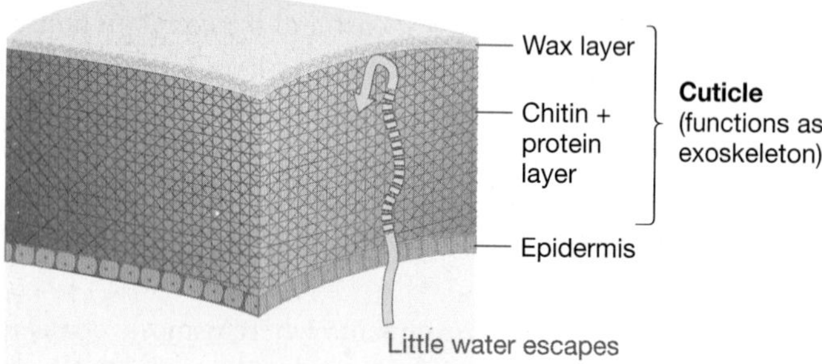

FIGURE 42.8 In Desert Locusts, Adaptations Limit Water Loss during Respiration and from the Body Surface.

QUESTION In desert grasshoppers, spiracles are located in a row along the underside of the abdomen. Why would this location help minimize water loss?

Types of Nitrogenous Wastes: Impact on Water Balance

Animal cells contain amino acids and nucleic acids that are used to synthesize proteins, RNA, and DNA. Both amino acids and nucleic acids are nitrogenous (nitrogen-containing) monomers. But if the monomers are present in excess of a cell's needs, they are broken down in catabolic reactions that result in the production of **ammonia** (NH_3). Ammonia is a strong base; it readily gains a proton to form an ammonium ion (NH_4^+). Ammonia is toxic to cells, because at high concentrations it raises the pH of intracellular and extracellular fluids enough to poison enzymes.

How do animals get rid of this highly toxic molecule safely and efficiently? Different species solve the problem in different ways. In freshwater fish, ammonia is diluted to low concentration and excreted in a watery urine. In addition, both freshwater and saltwater fish rid themselves of ammonia as it diffuses across their gills into the surrounding water along a concentration gradient. In humans, enzyme-catalyzed reactions convert ammonia to a much less toxic compound called **urea**, which is excreted in urine. In birds, reptiles, and terrestrial arthropods, reactions convert ammonia to **uric acid**. Uric acid is the white, paste-like substance that you have probably seen in bird feces. Compared with urea and ammonia, uric acid is extremely insoluble in water. Its low water solubility explains why washing bird droppings off a car is so difficult, and it provides a mechanism for birds, snakes, lizards, and terrestrial arthropods to rid themselves of excess nitrogen while losing a minimum amount of water. Many birds and some insects, in fact, do not produce any urine at all.

Why do different types of animals have different ways of detoxifying ammonia and excreting nitrogenous wastes? As **Table 42.1** shows, the type of nitrogenous waste produced by an animal correlates with its lineage—its evolutionary history. For example, mammals excrete urea while reptiles (including birds) and insects excrete uric acid. But waste production also correlates with the amount of osmotic stress a species endures.

SUMMARY TABLE 42.1 **Attributes of Nitrogenous Wastes Produced by Animals**

Attribute	Ammonia	Urea	Uric Acid
Solubility in water (moles/liter)	high	medium	very low
Water loss (amount required for excretion of waste)	high	medium	very low
Energy cost (amount of ATP required)	low	high	high
Toxicity	high	medium	low
Groups where it is the primary waste	fish, aquatic invertebrates	mammals,* sharks	birds† and other reptiles, most terrestrial insects and spiders
Method of synthesis	breakdown of amino acids and nucleic acids	synthesized in liver, starting with amino groups from amino acids	synthesis starts with amino acids and nucleic acids
Method of excretion	in urine and diffuses across gills	in urine (mammals); diffuses across gills (sharks)	in feces (in birds, uric acid is derived from the urine but excreted with the feces)

*Mammals also excrete a small amount of uric acid, synthesized from excess nucleic acids.
†Birds also excrete a small amount of ammonia.

Terrestrial birds conserve water by excreting about 90 percent of their nitrogenous waste as uric acid and only 3–4 percent as NH_3; but ducks and other birds with ready access to water excrete just 50 percent of their excess nitrogen as uric acid and 30 percent as NH_3. Tadpoles are aquatic and excrete ammonia, but adult frogs are terrestrial and excrete urea. Production of urea and uric acid is often interpreted as an adaptation that allows animals to cope with dry habitats. There is a fitness trade-off between the energetic cost of excreting urea or uric acid and the benefit of conserving water.

Maintaining Homeostasis: The Excretory System

For insects, minimizing water loss is only half the battle in avoiding osmotic stress. To maintain homeostasis, insects must also carefully regulate the composition of a blood-like fluid called **hemolymph**. In vertebrates, blood flows through vessels in a completely closed circulatory system, but this is not the case in insects. Instead, the heart pumps hemolymph through vessels and into the body cavity (see Chapter 44). In this way, hemolymph bathes tissues directly. Nutrients pass from the hemolymph into cells; waste products such as ammonia diffuse out of cells and into the hemolymph. Hemolymph also contains a wide variety of electrolytes.

How do insects regulate the composition of the hemolymph? This is an important question, because the nitrogenous wastes present in hemolymph must be removed before they build up to toxic concentrations. Similarly, if excess electrolytes enter the hemolymph after a meal, the ions must be excreted before they lead to osmotic stress. Water balance also must be regulated constantly.

To maintain water and electrolyte balance, insects rely on **Malpighian tubules**, which are an excretory organ, and on their hindgut—the posterior portion of their digestive tract (**Figure 42.9a**). Let's consider how each portion of the insect excretory system works in turn.

Pre-Urine Formation in the Malpighian Tubules As the enlarged section of Figure 42.9a shows, Malpighian tubules have a large surface area, are in direct contact with the hemolymph, and empty into the hindgut. Research has shown that the Malpighian tubules are responsible for forming a "pre-urine." This initial urine then passes into the hindgut, where it is processed and modified prior to excretion.

To explore how the Malpighian tubules work, biologists collected fluid from the lumen of Malpighian tubules in mealworms and compared the composition of the pre-urine with that of hemolymph from the same individuals. The analysis determined that the two solutions were roughly isotonic. They were not identical, however. Instead, the concentration of sodium ions inside the tubules was low relative to that of the hemolymph, while K^+ concentrations were extraordinarily high inside the tubules compared with concentrations in the hemolymph. This observation suggested that the epithelial cells in the Malpighian tubules are relatively impermeable to sodium ions but contain a pump that actively transports potassium ions into the tubules. To test this hypothesis, researchers dissected the tubules, rinsed them, and bathed their interior and exterior in a solution containing a known concentration of potassium ions. When the biologists measured the concentration of electrolytes on either side of the membrane over time, they found that K^+ accumulated on the inner face of the tubules against its concentration gradient.

This result supported the hypothesis that cells in the membranes of Malpighian tubules contain a pump that transports potassium ions into the lumen of the organ. Subsequent work

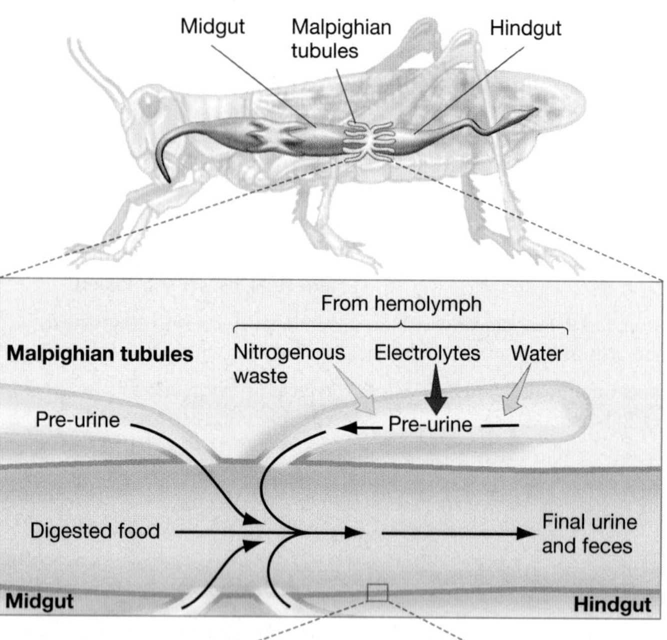

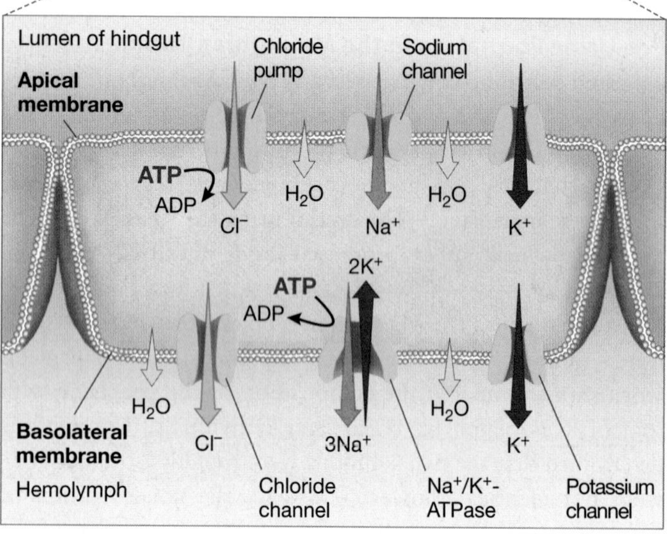

FIGURE 42.9 In Insects, Urine Formation Takes Place in the Malpighian Tubules and Hindgut. (a) The isotonic pre-urine that forms in the Malpighian tubules empties into the hindgut. **(b)** In the hindgut, the primary driving forces for reabsorption of electrolytes and water are chloride pumps in the apical membrane and Na^+/K^+-ATPases in the basolateral membrane.

EXERCISE In part (a), add a protein to the Malpighian tubule membrane that pumps K^+ into the structure from the hemolymph. How does this pump drive the movement of water, wastes, and other electrolytes?

EXERCISE In part (b), add a protein to the apical membrane that cotransports Na^+ and amino acids from the pre-urine. Why would it be advantageous for hindgut cells to recover amino acids from pre-urine?

has shown that a high concentration of potassium ions brings water into the tubules by osmosis. Other electrolytes and nitrogenous wastes then diffuse into the tubules along their concentration gradients.

The Hindgut: Selective Reabsorption of Electrolytes and Water The pre-urine that accumulates inside the Malpighian tubules flows into the hindgut, where it joins material emerging from the digestive tract. If an insect is osmotically stressed due to a shortage of electrolytes and water, electrolytes and water from the pre-urine are reabsorbed in the hindgut and returned to the hemolymph. Reabsorption results in the formation of a hypertonic final urine, the conservation of water, and the efficient elimination of nitrogenous wastes. In desert locusts, flour beetles, and other species that live in extremely dry environments, 80 to 95 percent of the water in the pre-urine is recovered and kept inside the body. The ability to recover this water allows these insects to live in dry habitats such as deserts and flour bins. How does reabsorption happen?

Experiments on desert locusts established that the mechanism involves a series of specific membrane pumps and channels, not unlike the system found in the chloride cells of fish. The nature of these pumps and channels was revealed by mounting the rectal epithelium from a desert locust as a sheet dividing two solutions. This preparation allowed biologists to manipulate electrolyte concentrations on either side of the rectal wall and then measure changes in the solutions over time. For example, when investigators removed K^+ and Na^+ from the solution on the lumen side of the organ, they observed that water reabsorption stopped. These data established that the hindgut's ability to recover water from urine is dependent on ion movement. More specifically, the epithelial cells in the hindgut must transport ions out of the pre-urine and into the hemolymph. Water would then follow by osmosis, leading to the formation of a concentrated urine.

How do the ions move? By poisoning the experimental membranes with ouabain, biologists confirmed that Na^+/K^+-ATPase is involved in moving ions out of the lumen and into the hemolymph. Ouabain-treated membranes continued to transport Cl^-, however. Later experiments confirmed that the rectal epithelium transports Cl^- against electrical and concentration gradients. These results support the hypothesis that the hindguts of insects have two active pumps: a chloride pump and Na^+/K^+-ATPase.

Years of experiments on the locust hindgut resulted in the model that is diagrammed in **Figure 42.9b**. Notice that Cl^- is pumped into cells from the hindgut lumen, with K^+ following through potassium channels along an electrochemical gradient and water following via osmosis. In the basolateral membrane, Na^+/K^+-ATPase sets up electrochemical and osmotic gradients that favor movement of Cl^-, K^+, and H_2O into the hemolymph. Although the diagram looks complex, it is actually a simplified version of the cells that are involved. There is, for example, strong experimental evidence for the existence of several other channels, pumps, and cotransporters in the epithelium of the insect hindgut. These membrane proteins transport protons, ammonia, amino acids, and other molecules across the tissue and help insects maintain water and electrolyte balance.

Regulating Water and Electrolyte Balance: An Overview

Much remains to be learned about the molecular mechanisms involved in the transport of ions and solutes in the hindgut of insects. Fortunately, the overall pattern in the regulation of water and electrolyte balance in insects is much simpler than the detailed mechanisms involved. Several general principles that have emerged from studies of insect excretion turn out to be relevant to vertebrate systems as well:

- There are no mechanisms for actively pumping water. Instead, water moves between cells or body compartments via osmotic gradients. These gradients are set up by the active transport of ions.
- The formation of the pre-urine is not particularly selective. Most of the molecules present in the hemolymph are also present in the Malpighian tubules.
- In contrast to pre-urine formation, reabsorption is highly selective. The protein pumps and channels involved in reabsorption are highly specific for certain ions and molecules. Waste products do not pass through the rectal membrane. Instead, they remain in the urine and feces and are eliminated from the body. Only valuable ions and molecules are reabsorbed.
- In contrast to pre-urine formation, reabsorption is tightly regulated. The membrane pumps and channels involved in reabsorption are activated and deactivated in response to osmotic stress. If an insect is dehydrated, then virtually all of the water in the pre-urine is reabsorbed. But if the same individual has plenty to drink, reabsorption does not occur and the urine is watery and hypotonic to the individual's hemolymph. The system is dynamic and allows precise control over water and electrolyte balance.
- Reabsorption of water is based on osmotic gradients created by ion pumps. Insects can form a concentrated urine and conserve water because cells expend energy in the form of ATP to move water out of the pre-urine, making it hypertonic relative to tissues.

Given the success of insects in terms of the numbers of species and individuals that exist and the array of habitats they occupy, it is clear that their systems for maintaining water and electrolyte balance are remarkably effective.

Check Your Understanding

If you understand that...

- Terrestrial insects are prone to dehydration, primarily via evaporation from their respiratory surfaces.
- Terrestrial insects have a cuticle, respiratory system, and excretory system that are designed to conserve water.

You should be able to...

Explain how the following traits are involved in water retention:

1) wax covering the body surface;
2) tracheae and spiracles;
3) the excretion of ammonia in the form of uric acid;
4) selective reabsorption of electrolytes and water in the hindgut.

42.4 Water and Electrolyte Balance in Terrestrial Vertebrates

With respect to water loss, terrestrial vertebrates face the same hazards as terrestrial insects do. Crocodiles, turtles, lizards, frogs, birds, and mammals lose water from their body surfaces and from the surface of their lungs every time they breathe. To replace water that is lost, most terrestrial vertebrates drink. Vertebrates that live on land also ingest electrolytes in food and excrete them in their wastes. Because sweat formation is based on the production of a salty solution, animals that sweat to cool off lose electrolytes when they do so.

Like other terrestrial organisms, land-dwelling vertebrates must carefully regulate the osmolarity of their tissues. Depending on conditions, terrestrial animals may need either to conserve or to excrete water and electrolytes to achieve homeostasis. How do land-dwelling vertebrates maintain water and electrolyte balance?

Due to intensive research on the mammalian kidney, a detailed answer to this question began to emerge in the mid-1970s. The **kidney** is the organ responsible for regulating water and electrolyte balance in terrestrial vertebrates. In this respect, the kidney's function is similar to the shark rectal gland. In addition, the kidney is where nitrogenous wastes—usually in the form of urea—are removed from the blood and eliminated in urine. As the organ responsible for both water and electrolyte balance and the excretion of water-soluble waste products, the kidney is equivalent to the Malpighian tubules and hindgut of insects.

The Structure of the Kidney

The general anatomy of the mammalian kidney was described in the late nineteenth century. As **Figure 42.10a** shows, kidneys

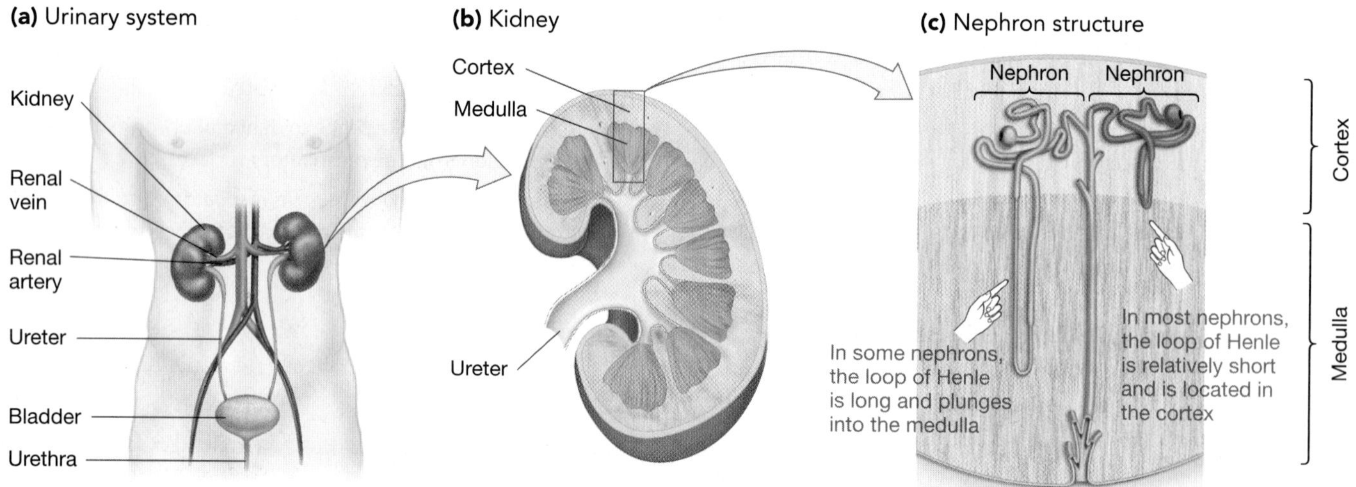

FIGURE 42.10 Anatomy of the Human Urinary System, Kidney, and Nephron. (a) In mammals, kidneys are paired and are located near the spinal column. **(b)** The kidney has an outer region called the cortex and an inner area called the medulla. **(c)** Most nephrons are located within the cortex, but some have a long loop that extends into the medulla.

occur in pairs and tend to be bean shaped. A large blood vessel called the *renal artery* brings blood that contains nitrogenous wastes into the organ; a large blood vessel called the *renal vein* carries blood away. The urine that is formed in the kidney is transported via a long tube called the **ureter** to a storage organ called the **bladder**. From the bladder, urine is transported to the body surface via a tube called the **urethra**, then excreted. In most vertebrates, the kidneys are located near the dorsal (back) side of the body.

Most of the kidney's mass is made up of small structures called nephrons. The **nephron** is the basic functional unit of the kidney. Nephrons perform the work involved in maintaining water and electrolyte balance. Most of the approximately 1 million nephrons in a human kidney are located in the outer region of the organ, or **cortex** (**Figure 42.10b**). As **Figure 42.10c** shows, however, some nephrons extend from the cortex into the kidney's inner region, or **medulla**.

To understand how the kidney maintains water and electrolyte balance, it is essential to understand how the nephron functions. **Figure 42.11a** provides a detailed view of the nephron, with its four major regions and the associated structure called the collecting duct labeled, and **Figure 42.11b** shows how each nephron is served by blood vessels. Let's follow the flow of material through each of the four regions of a nephron and the collecting duct, starting with the blood that arrives at the structure and ending with the urine that exits it. The changes occurring along the way are like those observed as water and solutes from insect hemolymph pass into the Malpighian tubules and from there to the hindgut. The important point to remember is that urine formation results from three processes: (1) formation of a pre-urine via filtration; (2) reabsorption of water, nutrients, and some solutes; and (3) removal of water from the final urine via osmosis.

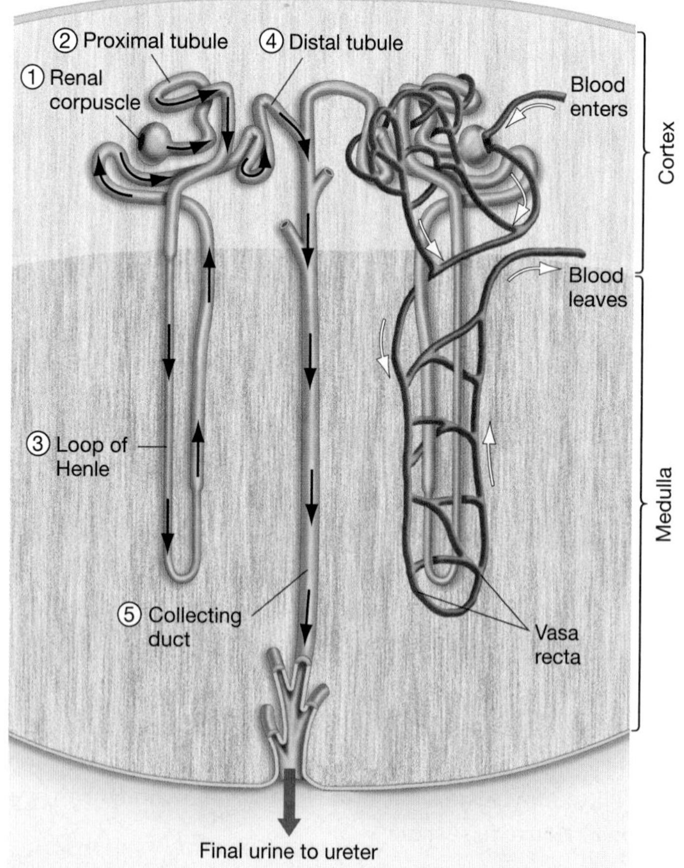

FIGURE 42.11 Nephrons Have Four Major Parts, Empty Into a Collecting Duct, and Are Served by Blood Vessels. Urine formation begins in the renal corpuscle and ends in the collecting duct.

EXERCISE Label the parts of the nephron in part (b).

Filtration: The Renal Corpuscle

In vertebrates, urine formation begins in a structure called the **renal corpuscle** ("kidney-little-body"). To understand how this part of the nephron works, examine Figure 42.11a again and notice that the nephron is a tube that is closed at one end and open at the other. The closed end is the beginning of the nephron; the open end is the terminus. As **Figure 42.12a** shows, the closed end forms a capsule that encloses a cluster of tiny blood vessels, or capillaries, that bring blood to the nephron from the renal artery. This cluster of capillaries is called the **glomerulus** ("ball of yarn"), and the region of the nephron that surrounds the glomerulus is named **Bowman's capsule**. The renal corpuscle, then, is composed of the glomerulus and Bowman's capsule.

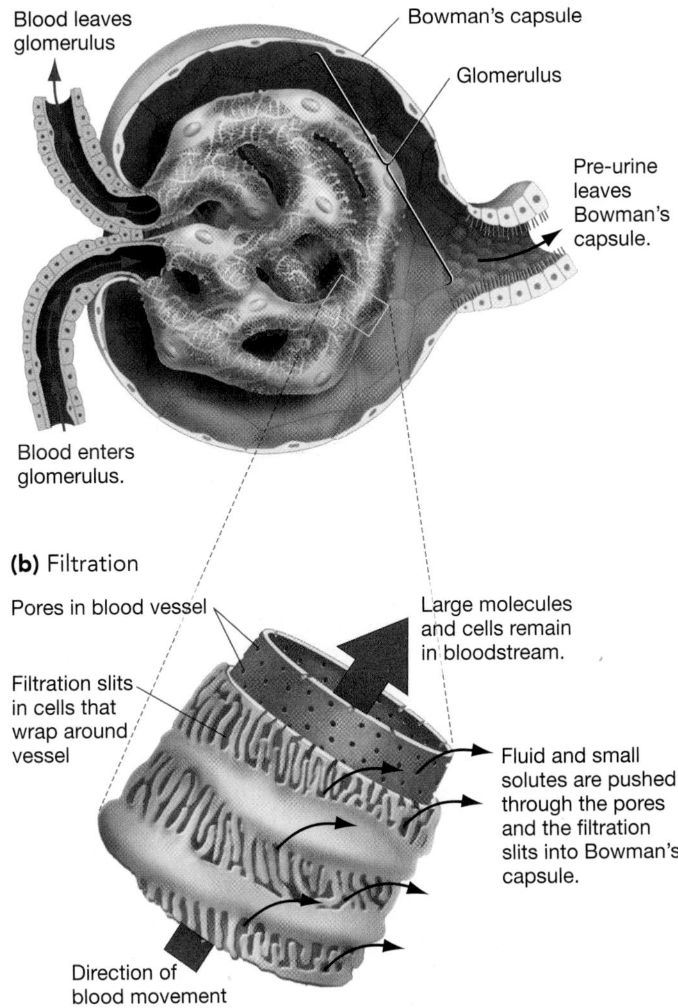

FIGURE 42.12 Urine Formation Begins when Blood Is Filtered in the Renal Corpuscle. (a) The renal corpuscle consists of Bowman's capsule and the glomerulus. **(b)** The capillaries in the glomerulus have pores and are surrounded by cells that have filtration slits. Blood pressure forces water and small molecules out of the capillaries, through the slits, and into Bowman's capsule.

By carefully sectioning and examining the renal corpuscle, biologists established that the glomerular capillaries have large pores, or openings. These blood vessels are also surrounded by unusual cells whose membranes fold into a series of slits and ridges (**Figure 42.12b**).

Based on these observations, researchers working in the late 1800s proposed that the renal corpuscle serves as a **filtration** device, allowing water and small solutes from the blood to pass into the nephron, and that filtration was based on the size of molecules and ions. The hypothesis was that water and small solutes are pushed out of the capillary's pores, through the slits in the surrounding cells, and into the fluid-filled space inside Bowman's capsule. Because proteins, cells, and other large components of blood would not fit through the pores, they could not enter the nephron. They would stay in the blood instead. Stated another way, urine formation was thought to start with a size-selective filtration step. According to this hypothesis, the force required to perform filtration is supplied by blood pressure. In vertebrates, blood is under higher pressure than the surrounding tissues because it is pumped by the heart through a closed system of vessels. This pressure is enough to force water and small solutes through the pores in the glomerulus, so the renal corpuscle strains large volumes of fluid without expending energy in the form of ATP. The result is the formation of a **filtrate**—in this case, a **pre-urine**.

In mammals, the filtration hypothesis was confirmed almost 100 years after it was initially proposed. In 1971 biologists were able to make direct measurements of the pressure inside the glomerular capillaries and in the tubule leading away from Bowman's capsule in rats. The investigators found that pressure was much higher inside the capillaries than it was in the surrounding capsule. As the filtration hypothesis predicts, the pressure differential forces water and solutes out of the blood and into the capsule space.

To summarize, the renal glomerulus filters the blood to create a pre-urine comprising water, electrolytes, and other small substances. During the formation of this filtrate, up to 25 percent of the water and solutes present in the blood is removed. ● If you understand this concept, you should be able to diagram a cross section of the glomerular filter, listing the contents of blood on one side and the contents of the pre-urine on the other, and labeling one side "higher pressure" and the other side "lower pressure."

It is critical to note two facts about the filtration step in urine formation. First, the renal corpuscles of a human kidney are capable of producing about 180 liters of pre-urine per day. This is an impressive volume—think of 180 one-liter bottles of soft drink arranged on a supermarket shelf. Second, almost all of the pre-urine is recycled, with only a tiny fraction of the original volume actually excreted. Approximately 99 percent of the volume is reabsorbed and returned to the bloodstream. Filtering large volumes from the blood allows wastes to be removed effectively; pairing this process with reabsorption allows waste excretion to occur with a minimum of water and nutrient loss.

Reabsorption: The Proximal Tubule

Where does pre-urine reabsorption occur? Fluid leaves Bowman's capsule and enters a convoluted structure called the **proximal tubule**. Researchers who analyzed samples of the fluid inside this tubule found that it contained the components predicted by the filtration hypothesis: water and small solutes such as urea, glucose, amino acids, vitamins, and electrolytes. It is important to remember that some of these molecules are waste products while others are valuable nutrients.

The first clue to the proximal tubule's function came from anatomical studies. As **Figure 42.13a** shows, the epithelial cells of this tubule have a prominent series of small projections, called **microvilli** ("little shaggy hairs"), facing the lumen. The presence of microvilli greatly expands the surface area of an epithelium. Because greater surface area provides more space for membrane proteins that act as pumps, channels, and cotransporters, it is logical to find that cells with microvilli are often associated with transport processes. The observation that epithelial cells in the proximal tubule are also packed with mitochondria suggests that ATP-demanding active transport is occurring. Based on these data, anatomists hypothesized that the proximal tubule functions in the active transport of selected molecules out of the pre-urine.

The selective active-transport hypothesis was supported by experiments on proximal tubules that were dissected from the kidney of rabbits and rats and isolated in vitro. By injecting solutions of known composition into proximal tubules in the presence or absence of ATP, researchers were able to confirm that selected electrolytes and nutrients are actively reabsorbed from the pre-urine that enters the tubules. When these solutes leave the proximal tubule and enter epithelial cells, water follows along the osmotic gradient. In this way, valuable solutes and water are reabsorbed and returned to the body.

Figure 42.13b summarizes the current model of the molecular mechanisms involved in selective reabsorption. Notice that the basolateral membrane of epithelial cells in the proximal tubule contains Na^+/K^+-ATPase and that the apical membrane contains a variety of cotransporters. The key to the model is

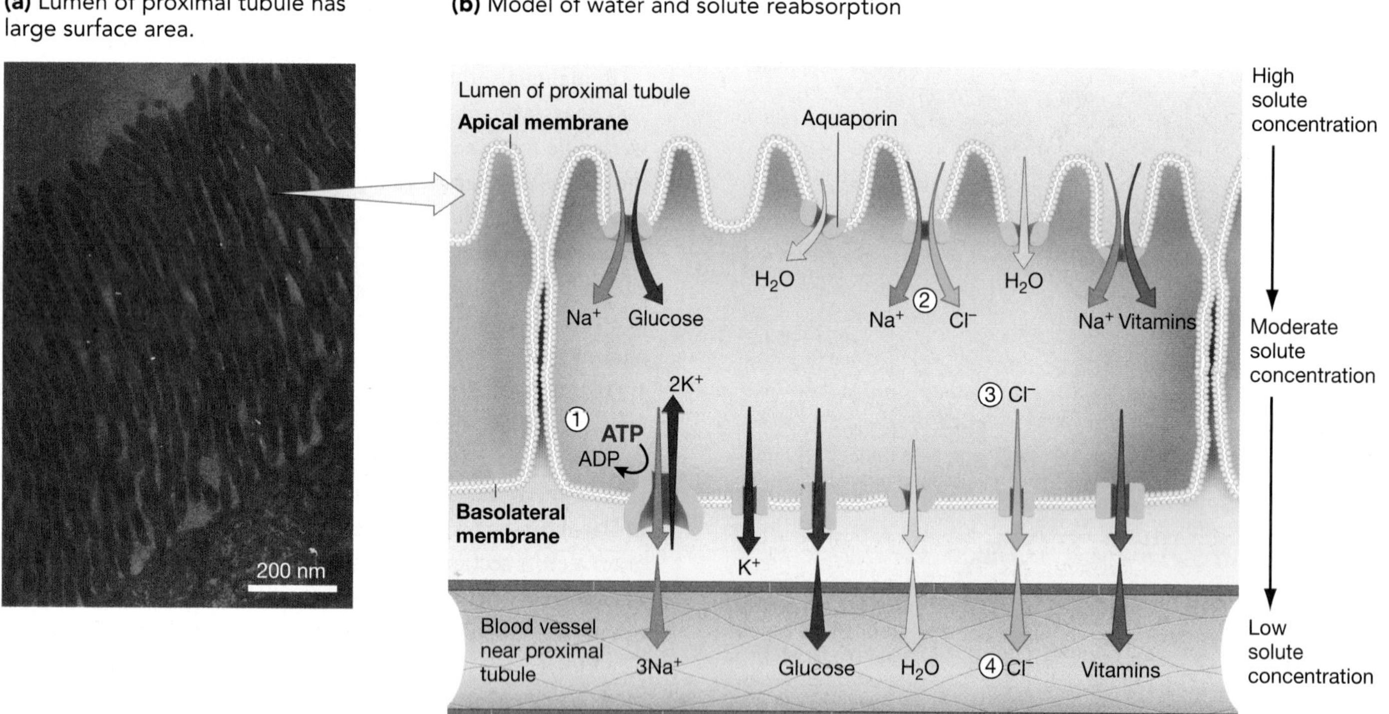

FIGURE 42.13 Water and Electrolytes Are Reabsorbed in the Proximal Tubule. (a) Microvilli from epithelial cells extend into the lumen of the proximal tubule. **(b)** Reabsorption of water and solutes occurs because (1) Na^+/K^+-ATPase in the basolateral membrane of epithelial cells sets up a strong concentration gradient to bring Na^+ into the cell across the apical membrane; (2) Na^+ cotransporters in the apical membrane result in reabsorption of nutrients and electrolytes, with water following via osmosis; (3) solutes also exit the cell through channels, with water following; and (4) solutes and water enter nearby blood vessels.

EXERCISE Aquaporins are responsible for water movement in the hindgut of insects as well. Add aquaporins to Figure 42.9b.

EXERCISE In part (b), label which membrane proteins are pumps, cotransporters, and channels.

that the Na^+/K^+-ATPase in the basolateral membranes removes Na^+ from the interior of the cell. The active transport of sodium ions out of the cell creates a gradient favoring the entry of Na^+ from the lumen. In the apical membrane adjacent to the lumen, Na^+-dependent cotransporters use this gradient to remove valuable ions and nutrients selectively from the pre-urine. Specifically, the cotransporters simultaneously bind Na^+ and another solute, such as glucose, an amino acid, or Cl^-. As with other examples of secondary active transport, the movement of Na^+ into the cell, *with* its concentration gradient, provides the means for moving other solutes *against* a concentration gradient. ● If you understand this concept, you should be able to compare and contrast the proximal tubule cell in Figure 42.13b with the shark rectal gland cell in Figure 42.6 and the Malpighian tubule cell in Figure 42.9b. You should also be able to explain why caffeine increases urine output, based on recent data indicating that caffeine may inhibit sodium ion reabsorption in the proximal tubule.

Another key point is that water follows the movement of these solutes from the pre-urine back to blood, by osmosis. Recent work has shown that water moves from the lumen of the proximal tubule into epithelial cells through specialized membrane proteins called **aquaporins** ("water-pores"; see Figure 42.13b). When an aquaporin is open, 3 billion water molecules pass through it per second. Aquaporins are water channels. Aquaporins located in the basolateral membranes of epithelial cells facilitate the movement of water back into the blood.

Careful measurements have confirmed that two-thirds of the NaCl and water that is originally filtered by the renal corpuscle is reabsorbed in the proximal tubule. The osmolarity of the tubular fluid is unchanged despite this huge change in volume, however, because water reabsorption is proportional to solute reabsorption. The electrolytes, nutrients, and water that are reabsorbed eventually enter blood vessels on the basolateral side of the tubule and are returned to the body.

To summarize, the cells that line the proximal tubule act as a recycling center. The filtration step in the renal corpuscle is based on size; the reabsorption step in the proximal tubule selectively retrieves small substances that are valuable. The pumps and cotransporters in the proximal tubule recover water, nutrients, and electrolytes but leave wastes. Compared with the fluid that enters the proximal tubule, the pre-urine that leaves is isotonic but greatly reduced in volume. As it flows into the next part of the nephron, it has a relatively high concentration of waste molecules and a relatively low concentration of nutrients.

Creating an Osmotic Gradient: The Loop of Henle

In mammals, the fluid that emerges from the proximal tubule enters a loop. In most nephrons, the loop is short and does not leave the cortex. But in about 20 percent of the nephrons present in a human kidney, the loop is long and plunges from the cortex of the kidney deep into the medulla (look back at Figure 42.10c). Jacob Henle described this part of the mammalian nephron in the early 1860s, and it was named the **loop of Henle**. It was not until 1942, however, that Werner Kuhn offered a hypothesis to explain what the structure does.

Kuhn's proposal was inspired by the use of countercurrent heat exchangers in chemistry and physics. Recall from Chapter 41 that a countercurrent heat exchanger is a system in which two adjacent fluids flow through pipes in opposite directions. Kuhn proposed that the loop of Henle functions as a **countercurrent exchanger** and multiplier—but not one that sets up a heat gradient like those diagrammed in Figure 41.20. Instead, Kuhn hypothesized that the loop of Henle sets up an osmotic gradient. Although he did not know the molecular mechanisms involved in setting up this gradient, he proposed that the osmolarity of the fluid inside the loop of Henle is low in the cortex and high in the medulla. Further, Kuhn maintained that the osmolarity in tissues surrounding the loop mirrors the gradient inside the loop. This is a key point. Kuhn's hypothesis proposed that an exchange of water and solutes occurs between the cells in each segment of the loop and the cells surrounding the loop of Henle—not between the solutions in the descending and ascending portions of the loop. ● If you understand this concept, you should be able to sketch the loop of Henle and label regions of low osmolarity and high osmolarity inside and outside the loop.

If Henle was correct, the situation in the loop contrasts with a conventional countercurrent system, in which exchange occurs between the two segments themselves. According to Kuhn, the exchange results in a steep osmotic gradient existing both inside and outside the loop of Henle and running from the cortex deep into the medulla. Was he correct?

Testing Kuhn's Hypothesis A series of papers published during the 1950s supplied important experimental support for the countercurrent exchange model. Two particularly important data sets, obtained by analyzing the osmolarity of tissue slices cut into sections perpendicular to the loop of Henle, are reproduced in **Figure 42.14**. The left-hand graph of the figure shows data on the osmolarity of fluid inside the loop of Henle. As predicted by Kuhn's model, a strong gradient in osmolarity exists from the cortex to the medulla. The data in the right-hand graph of Figure 42.14 show that outside the loop of Henle, the concentrations of Na^+, Cl^-, and urea also increase sharply from the cortex to the medulla. This was an important observation. It not only confirmed the prediction that an osmotic gradient exists in the tissue surrounding the loop of Henle but also suggested that the solutes responsible for this gradient are Na^+, Cl^-, and urea. As we'll see, the change in concentration of urea turned out to be particularly important.

Before considering how the steep osmotic gradient created by the loop of Henle affects urine formation, let's take a look

Experiment

Question: What is the function of the loop of Henle?

Hypothesis: The loop of Henle sets up an osmotic gradient for countercurrent exchange of water and solutes.

Null hypothesis: The loop of Henle does not set up an osmotic gradient for countercurrent exchange.

Experimental setup:

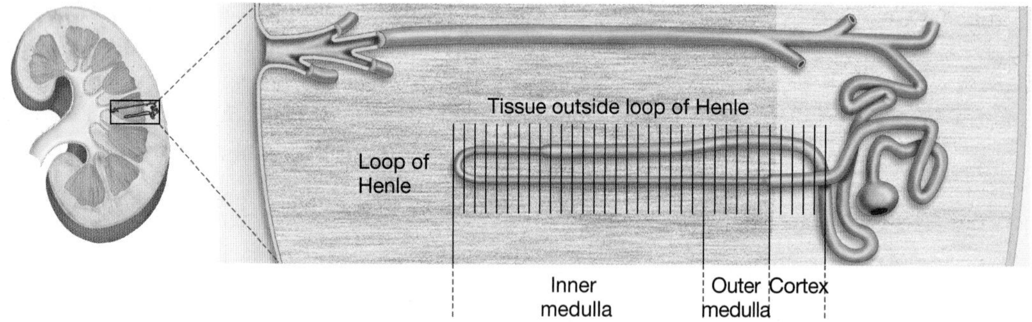

1. Slice kidney tissue in section perpendicular to the loop of Henle.

2. Measure the osmolarity inside each section of loop of Henle. (This experiment was performed on a rat nephron.)

3. Measure the concentration of urea, sodium ions, and chloride ions in each section of tissue outside the loop. (This experiment was performed on a dog nephron.)

Prediction: A strong gradient in osmolarity exists from the medulla to the cortex, both within the loop of Henle and in the tissue surrounding the loop of Henle.

Prediction of null hypothesis: A strong gradient in osmolarity does not exist from the medulla to the cortex, both within and outside the loop of Henle.

Results:

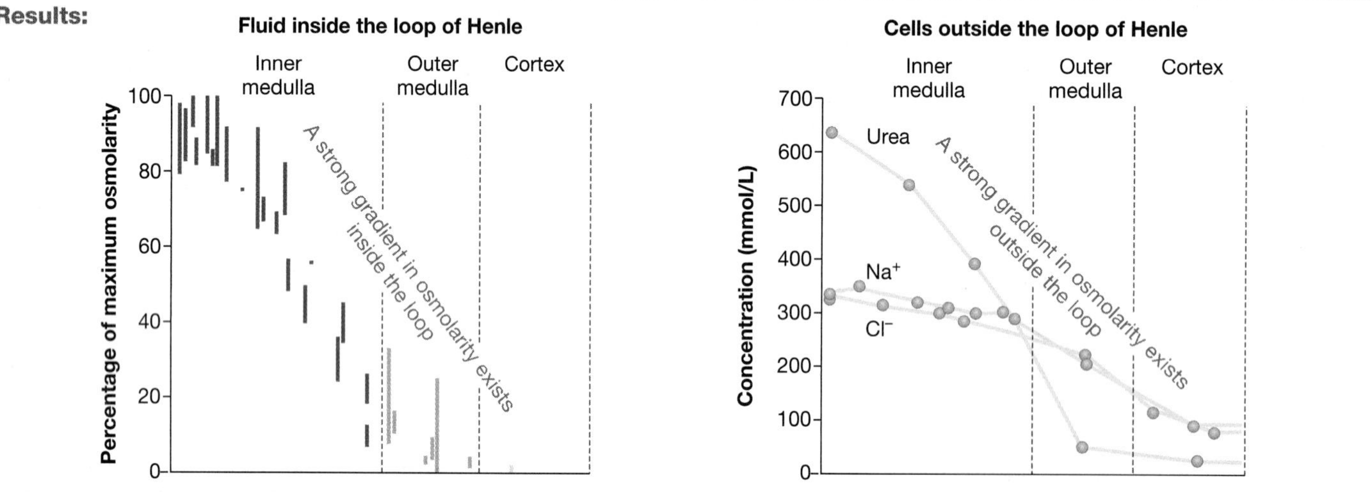

Conclusion: The loop of Henle sets up a countercurrent exchange system for the exchanges of solutes—specifically, urea, sodium, and chloride. Its function is to maintain a strong osmolarity gradient from the cortex to the inner medulla.

FIGURE 42.14 Data Confirm the Existence of a Strong Osmotic Gradient Inside and Outside the Loop of Henle. The graphs show the relative osmolarity, or solute concentrations, of the fluid inside (left) or outside (right) the nephron as a function of its location in the kidney.

at research on the molecular mechanisms involved in setting up the countercurrent exchange. How do water and solutes move into and out of the loop in a way that establishes the dramatic osmotic gradient that researchers observed?

How Is the Osmotic Gradient Established? **Figure 42.15a** provides a detailed look at the morphology of the loop of Henle. Notice that there are three distinct regions in the loop: the descending limb, the thin ascending limb, and the thick ascending limb. The thin and thick ascending limbs differ in the thickness of their walls. Do the three regions also differ in their permeability to water and solutes?

It took over 15 years of experiments performed in laboratories around the world to formulate a definitive answer to that question. One important result began to emerge after researchers were able to puncture Henle's loop with a micropipette, analyze the composition of the fluid inside, and compare it with the nephron's final product—urine. In the ascending limb of the loop of Henle, Na^+ and Cl^- constituted at least 60 percent of the solutes; urea constituted about 10

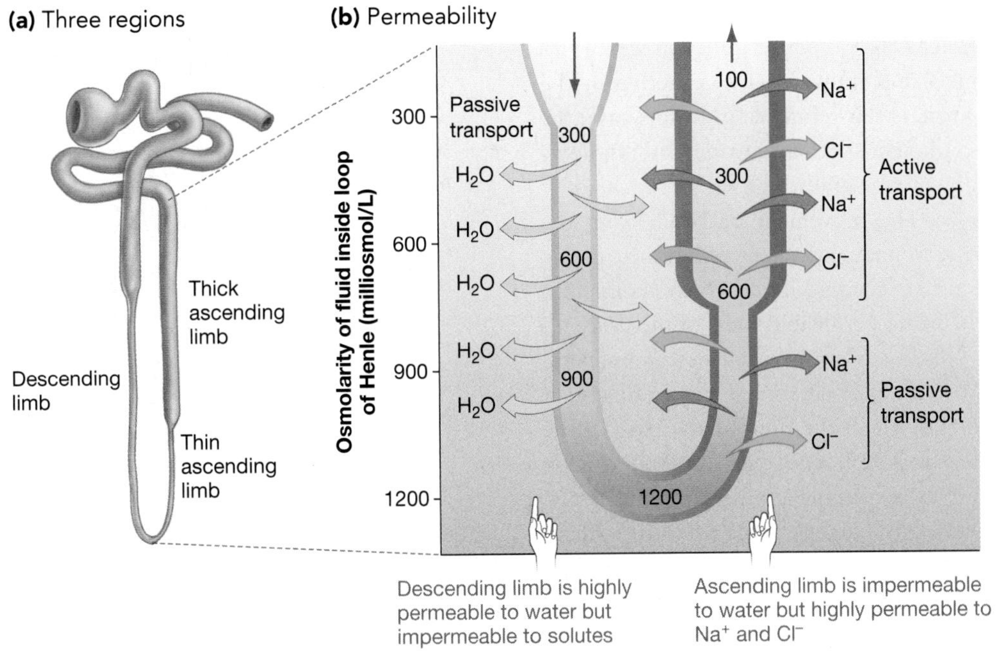

FIGURE 42.15 The Loop of Henle Maintains an Osmotic Gradient because Water Leaves the Descending Limb and Salt Leaves the Ascending Limb. The numbers inside the nephron in part (b) represent the osmolarity of the pre-urine.

percent. But in the final portion of the nephron, both ions were less common—urea was the major solute. These data were important because they suggested that valuable Na^+ and Cl^- ions were being retained while waste products in the form of urea were being discarded.

What mechanism is responsible for the removal of Na^+ and Cl^-? Because Na^+ and Cl^- were also present at high concentration in the tissue surrounding the thick ascending limb of the loop of Henle, researchers hypothesized that sodium might be actively pumped out of this portion of the nephron. The hypothesis was that the active transport of Na^+ out of the thick ascending limb would create an electrical gradient that would also favor the loss of Cl^- (**Figure 42.15b**).

Follow-up experiments using ouabain and other poisons supported the hypothesis that sodium ions are actively transported out of the solution inside the thick ascending limb. The epithelial cells responsible for sodium excretion in this portion of the nephron are configured almost exactly like the epithelium of the shark rectal gland (see Figure 42.6). A combination of Na^+/K^+-ATPase, Cl^- channels, and $Na^+/2Cl^-/K^+$ cotransporters in these cells results in the active transport of salt out of the loop and into the surrounding tissue. Once these electrolytes are removed, urea remains as the primary solute in the pre-urine.

What is happening in the descending limb and the thin ascending limb of the loop of Henle? An important series of experiments established how water and solutes move across these two remaining segments. By injecting solutions of known concentration into the nephrons of rabbits, biologists documented that the descending limb is highly permeable to water but almost completely impermeable to solutes. The thin ascending limb of the loop, in contrast, is highly permeable to Na^+ and Cl^-, moderately permeable to urea, and almost completely impermeable to water (Figure 42.15b).

A Comprehensive View of the Loop of Henle All of the observations just summarized came together in 1972 when two papers, published simultaneously and independently, proposed the same comprehensive model for how the loop of Henle works. To understand this model, follow the events that occur as fluid leaves the proximal tubule and moves through the loop, using Figure 42.15b:

1. As it flows down the descending limb, the fluid inside the loop loses water to the tissue surrounding the nephron. This movement of water is passive, meaning that it does not require an expenditure of ATP. The water moves because it follows an osmotic gradient. This osmotic gradient is created by the ascending limb.

2. At the bottom of the loop—in the inner medulla—the fluids inside and outside the nephron have high osmolarity. The pre-urine does not continue to lose water, though, because the membrane in the ascending limb is nearly impermeable to water. Instead, the fluid inside the nephron begins to lose Na^+ and Cl^- as it ascends and encounters surrounding tissues with progressively lower osmolarity. The ions move out of the loop passively along their concentration gradients.

3. In the thick ascending limb, where the osmolarity of the surrounding solution is low, additional Na^+ and Cl^- ions are actively transported out of the nephron.

The key point in this model is that the countercurrent flow of fluid, combined with changes in permeability to water and in the types of channels and pumps that are active in the epithelium of the nephron, is self-reinforcing. In this case, the presence of an osmotic gradient stimulates water and ion flows that in turn maintain an osmotic gradient.

Here's how it works: The movement of NaCl from the ascending limb into the surrounding tissue increases the osmotic concentration outside the descending limb, which results in a flow of water out of the water-permeable walls of the descending limb, via osmosis. This loss of water in the descending limb increases the osmolarity in the fluid entering the ascending limb. The high concentration of salt in the fluid at the base of the ascending limb triggers a passive flow of ions out—reinforcing the osmotic gradient. ● If you understand this concept, you should be able to predict what happens to the osmotic gradient when the drug furosemide inhibits membrane proteins that pump sodium and chloride ions out of the thick ascending limb. How does this drug affect urine formation?

The loop of Henle maintains an osmotic gradient, but a key feature of the nephron's final portion—the **collecting duct**—helps establish it. If you look back at the data in Figure 42.14, you'll note that urea is responsible for the steep osmotic gradient in the space surrounding the nephron. Urea is at high concentration in the inner medulla and low concentration in the outer medulla. This gradient exists because the innermost section of the collecting duct is permeable to urea. As **Figure 42.16** shows, the urea that diffuses out of the collecting duct is responsible for the high osmolarity of the inner medulla.

Although this system is extremely complex, its outcome is extremely simple: the creation and maintenance of a strong osmotic gradient. Due to the countercurrent flow of material through the nephron, the osmotic gradient is maintained with the minimum possible expenditure of energy.

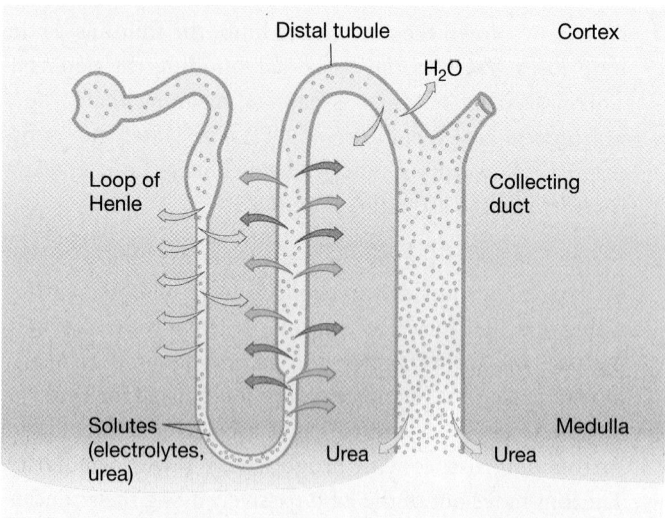

FIGURE 42.16 Urea from the Collecting Duct Is Responsible for the High Osmolarity of the Inner Medulla. Most of the nephron is impermeable to urea, except for the innermost portion of the collecting duct.

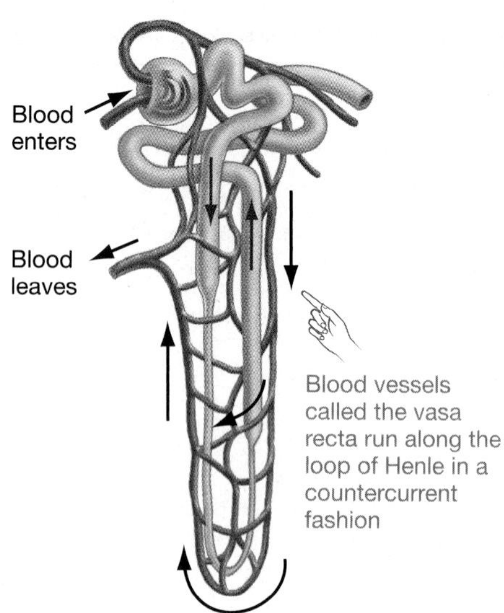

FIGURE 42.17 Blood Supply to the Loop of Henle. Water and solutes from the loop of Henle move into blood vessels (vasa recta).

What happens to the water and salt that move out of the loop? They quickly diffuse into the **vasa recta**, a network of blood vessels that runs along the loop. As a result, the water and electrolytes are returned to the body (**Figure 42.17**). The removal of water that leaves the descending limb is particularly important. If it were not drawn off into the bloodstream, it would dilute the concentrated fluid outside the loop of Henle and quickly destroy the osmotic gradient.

Regulating Water and Electrolyte Balance: The Distal Tubule and the Collecting Duct

The first three steps in urine formation—filtration, reabsorption, and establishment of an osmotic gradient—result in a fluid that is slightly hypotonic to blood. Once the pre-urine has passed through the loop of Henle, the major solutes that it contains are urea and other wastes.

Although the events that take place in the renal corpuscle, proximal tubule, and loop of Henle produce a urine that is distinctly different from the blood that entered the kidney, the events that take place are largely unregulated—meaning they do not vary with changes in the body's condition. There is one exception: The volume of blood filtered at the renal corpuscle can change dramatically, depending on circumstances. But the proximal tubule and loop of Henle function in about the same way at all times. As a result, the fluid entering the **distal tubule**, the portion of a nephron immediately following the loop of Henle, is relatively constant in composition over time. In contrast, the urine that leaves the collecting duct is highly variable in osmolarity and in Na^+ and Cl^- concentration.

How is it possible for the characteristics of urine to vary over time? The answer is based on two observations about the distal tubule and the collecting duct: Their activity is highly regulated, and their activity changes in response to the osmotic stress that an individual experiences. Although electrolytes and water are always reabsorbed in the distal tubule, the amount of Na^+, Cl^-, and water that is reabsorbed in the distal tubule and in the collecting duct varies with the animal's condition.

The activity of the distal tubule and collecting duct is controlled by hormones—signaling molecules that were introduced in Chapter 8 and are explored further in Chapter 47. Specifically:

- If Na^+ levels in the blood are low, the hormone **aldosterone** is released from the adrenal glands and leads to activation of sodium pumps and the reabsorption of Na^+ in the distal tubule. Aldosterone retains sodium.
- If an individual is dehydrated, a molecule called **antidiuretic hormone (ADH)** is released from the brain. (The term *diuresis* refers to increased urine production, so *antidiuresis* means inhibition of urine production. ADH is also referred to as vasopressin or arginine vasopressin.) ADH retains water.

ADH's role in urine formation is particularly well studied. When ADH arrives at epithelial cells lining the collecting duct, the hormone triggers the insertion of aquaporins into the plasma membrane. Because cells become much more permeable to water as a result, large amounts of water are reabsorbed. ADH also increases the collecting duct's permeability to urea, which increases the osmolarity of the surrounding fluid and thus water loss from the pre-urine. As **Figure 42.18a** shows, water leaves the collecting duct passively—following the concentration gradient maintained by the loop of Henle. The result is conservation of water and the formation of urine that is strongly hypertonic in relation to blood. When ADH is absent, however, few aquaporins are found in the epithelium of the collecting duct, and the structure is relatively impermeable to water (**Figure 42.18b**). In this case, a hypotonic urine is produced.

People with defective forms of ADH or aquaporins produce copious amounts of urine—up to 30 liters per day. They suffer from the condition **diabetes insipidus**. (*Diabetes* means "to run through"; *insipidus* means "tasteless." The disease diabetes mellitus is due to problems with regulating glucose concentrations in blood, rather than water retention in the kidneys, and is discussed in Chapter 43.) ● If you understand ADH's effect on the collecting duct, you should be able to predict how urine formation is affected by ingestion of ethanol, which inhibits ADH release, versus nicotine, which stimulates ADH release.

Table 42.2 reviews the functions of the four major regions of the nephron and the collecting duct. To summarize, the first three segments of the structure—the renal corpuscle, the proximal tubule, and the loop of Henle—accomplish two objectives: (1) They effectively concentrate nitrogenous wastes, and (2) they create the possibility for Na^+, Cl^-, and water to be

(a) ADH present: Collecting duct is highly permeable to water.

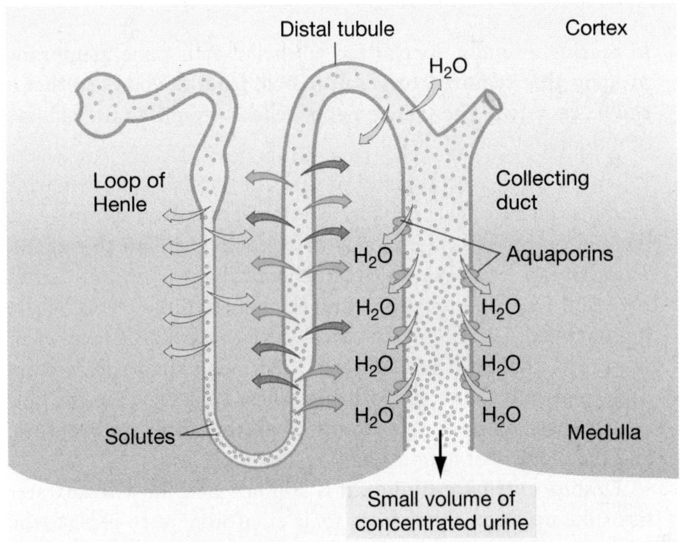

(b) No ADH present: Collecting duct is not permeable to water.

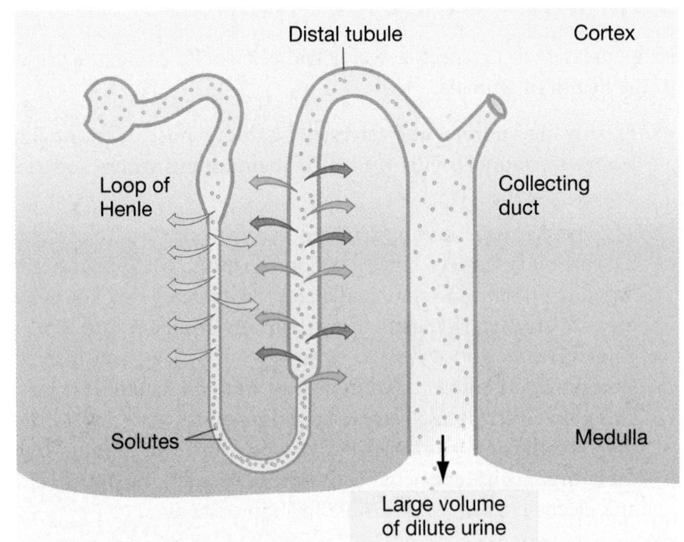

FIGURE 42.18 Water Reabsorption by the Collecting Duct Is Regulated by ADH. (a) Antidiuretic hormone (ADH) makes the collecting ducts permeable to water and leads to the production of small volumes of concentrated urine. The water that leaves the collecting ducts passes into the vasa recta (see Figure 42.17). **(b)** In the absence of ADH, urine increases in volume and contains a low concentration of solutes.

SUMMARY TABLE **42.2 Structure and Function of the Nephron and Collecting Duct**

Structure	Function
Renal corpuscle (Bowman's capsule and glomerulus)	Size-selective filtration: formation of pre-urine from blood (water and other small substances enter nephron)
Proximal tubule	Reabsorption of electrolytes, nutrients, water (active transport)
Loop of Henle	Maintenance of osmotic gradient from outer to inner medulla
• Descending limb	• Permeable to water (passive transport out of pre-urine)
• Thin ascending limb	• Permeable to Na^+, Cl^- (passive transport out of pre-urine)
• Thick ascending limb	• Active transport of Na^+, Cl^- out of pre-urine
Distal tubule	With aldosterone: reabsorption of Na^+. Without aldosterone: no reabsorption of Na^+.
Collecting duct	Regulation of water retention and loss
• Main portion	• With ADH: water leaves pre-urine; production of urine that is hypertonic to blood (reduced urine volume). Without ADH: water stays in pre-urine; production of urine that is hypotonic to blood (increased urine volume).
• Innermost portion	• Urea leaks out to establish/maintain high osmolarity of inner medulla.

either excreted or reabsorbed by the nephron's final two segments: the distal tubule and collecting duct. Of the 180 liters of primary urine that enters the renal corpuscle, all but 1.2 liters is reabsorbed and returned to the body each day. The nephron is a remarkably effective mechanism for regulating water and electrolyte balance and achieving homeostasis.

Check Your Understanding

If you understand that...

- The vertebrate kidney is specialized for the production of hypertonic urine.
- The loop of Henle is a countercurrent system that creates a strong osmotic gradient in the medulla.

You should be able to...

1) Describe the structure and function of the following parts of a nephron: renal corpuscle, proximal tubule, loop of Henle, distal tubule, and collecting duct.
2) Explain how the characteristics of urine are regulated by hormones that act in the distal tubule and collecting duct.
3) Predict how the following events would affect urine production: drinking massive amounts of water, eating large amounts of salt, and refraining from drinking for 48 hours.

MB Web Animation at www.masteringbio.com
The Mammalian Kidney

Chapter Review

SUMMARY OF KEY CONCEPTS

Homeostasis with respect to water and electrolyte balance is critical to the health of animals.

Freshwater, marine, and terrestrial habitats pose different challenges to animals with regard to maintaining water and electrolyte balance.

The mechanisms involved in regulating water and electrolyte balance vary widely among animal groups, because different habitats present different challenges. Freshwater fish, for example, are strongly hypertonic to their environment and tend to gain water and lose electrolytes. Marine fish, in contrast, are strongly hypotonic in relation to seawater and tend to lose water and gain electrolytes. Terrestrial animals lose water every time they breathe and from their body surfaces by evaporation. In all these organisms, epithelial cells that selectively transport water and electrolytes are responsible for homeostasis.

You should be able to describe the water and electrolyte balance challenges faced by river otters that live near the ocean and move from freshwater to saltwater habitats and back on a daily or weekly basis.

In marine animals, specialized epithelial cells have membrane proteins that remove excess salt (NaCl) from the body so that it can be excreted. The same types of cells are found in the kidneys of mammals.

Research on the rectal gland of sharks established how marine animals excrete excess salt. Sodium-potassium pumps are located on the basolateral side of epithelial cells in this gland. These pumps establish an electrochemical gradient that carries Na^+ and Cl^- from the blood into the lumen of the gland, where it is excreted. Later research revealed that a similar arrangement of pumps, channels, and cotransporters occurs in a wide variety of animal cells that transport salt. These cells occur in the gills of saltwater fish, the salt glands of marine birds and reptiles, and the kidneys of mammals.

Despite extensive studies, it is still not clear how freshwater fish take up electrolytes from their environment to replace the ions that they lose from their gills. Research on salmon has shown that the gills of freshwater fish contain cells with sodium-potassium pumps. Some researchers suggest that these cells may be involved in the active transport of ions into the gills.

You should be able to explain how an electrochemical gradient for sodium ions makes it possible for water and ions to move across a membrane passively.

In terrestrial insects, the hindgut and Malpighian tubules are responsible for excreting water-soluble waste products and achieving homeostasis with respect to water and electrolyte concentrations.

Terrestrial insects limit water loss in several ways. Their exoskeleton is covered with wax to limit evaporation, and the openings to their respiratory organs close when osmotic stress is severe. Insects are also capable of forming a hypertonic urine that minimizes water loss during the excretion of nitrogenous wastes. A pre-urine that is isotonic with the hemolymph forms in the Malpighian tubules of insects. If pumps in the epithelium of the hindgut are activated, then electrolytes and water are reabsorbed from the pre-urine and returned to the hemolymph. A hypertonic final urine results.

You should be able to predict whether desert locusts or water bugs expend more energy during urine formation, and explain your reasoning.

In terrestrial vertebrates, the kidney is responsible for excreting water-soluble waste products and achieving homeostasis with respect to water and electrolyte concentrations.

Nephrons in the mammalian kidney form a filtrate and then reabsorb valuable nutrients, electrolytes, and water. The resulting solution contains urea and electrolytes and flows through the loop of Henle, where changes in the permeability of epithelial cells to water and salt—along with active transport of salt—create a steep osmotic gradient. If ADH triggers changes in the water permeability of the collecting duct of the nephron, water is reabsorbed along the osmotic gradient and a hypertonic urine is produced.

In animals, maintaining water and electrolyte balance is an active, energy-demanding process based on the action of membrane proteins. Water and electrolyte excretion and reabsorption are carefully controlled to achieve homeostasis. As a result, animals are able to occupy habitats that present a diversity of osmotic challenges.

You should be able to explain why desert-dwelling kangaroo rats have extremely long loops of Henle, relative to their body size.

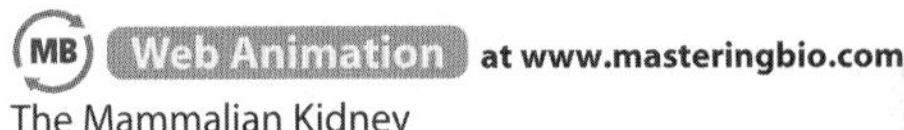

at www.masteringbio.com

The Mammalian Kidney

QUESTIONS

Test Your Knowledge

1. Which one of the following statements is true of fish that live in freshwater?
   a. Their tissues are isotonic with respect to their environment. As a result, they do not require a specialized organ to maintain water and electrolyte balance.
   b. They lose water to their environment primarily through the gills. They replace this water by drinking.
   c. Water enters epithelial cells in their gills via osmosis. Electrolytes leave the same cells via diffusion.
   d. They have specialized cells that actively pump Na^+ and Cl^- from blood into epithelial cells, so the ions can be excreted.

2. Na^+/K^+-ATPase is found in epithelial cells that produce urine or other excretions that are hypertonic relative to tissues. Why?
   a. To form a hypertonic solution, ions or molecules must be actively transported across their concentration gradient.
   b. Cells similar to those found in the shark rectal gland occur in many other species that need to excrete salt (NaCl).
   c. The sodium-potassium pump is directly responsible for moving sodium ions from blood or body tissues into the urine or salt excretion, against a concentration gradient.
   d. They are usually located in the basolateral membrane of those epithelial cells, along with other channels and cotransporters.

3. Regarding urine formation, the function of the insect hindgut is closest to which of the following structures in the vertebrate kidney?
   a. renal corpuscle
   b. proximal tubule
   c. loop of Henle
   d. distal tubules and collecting duct

4. Which of the following gives the correct sequence of the regions of the nephron with respect to fluid flow?
   a. distal tubule → ascending limb of Henle → descending limb of Henle → proximal tubule → collecting duct
   b. ascending limb of Henle → descending limb of Henle → proximal tubule → distal tubule → collecting duct
   c. proximal tubule → ascending limb of Henle → descending limb of Henle → distal tubule → collecting duct
   d. proximal tubule → descending limb of Henle → ascending limb of Henle → distal tubule → collecting duct

5. Which of the following statements about kidney function is *not* correct?
   a. The loop of Henle acts as a countercurrent exchanger, so no energy is required to maintain the osmotic gradient.
   b. The descending limb of the loop of Henle is highly permeable to water.
   c. The thin ascending limb of the loop of Henle is highly permeable to salt.
   d. Reabsorption of water and solutes takes place primarily in the proximal tubule.

6. What effect does antidiuretic hormone (ADH) have on the nephron?
   a. It increases water permeability of the descending limb of the loop of Henle.
   b. It decreases water permeability of the descending limb of the loop of Henle.
   c. It increases water permeability of the collecting duct.
   d. It decreases water permeability of the collecting duct.

Test Your Knowledge answers: 1. c; 2. a; 3. d; 4. d; 5. a; 6. c

Test Your Understanding

Answers are available at www.masteringbio.com

1. Explain the changes that need to occur in water and electrolyte balance as a salmon moves from freshwater to salt water and back. Specifically, state when the animal should drink or not drink, when cells in the gill epithelium should excrete or import electrolytes, and why each change occurs.
2. The chloride cells of fish gills are sometimes called mitochondria-rich cells due to their high density of mitochondria. How does high mitochondrial density relate to the functional role of chloride cells? Would you expect other epithelial cells involved in ion transport to contain large numbers of mitochondria? Explain.
3. Why is it significant that cells involved in transport processes often have microvilli?
4. This chapter introduced a number of features that help terrestrial animals reduce water loss. These traits include the layer of wax found on insect exoskeletons, the ability of insects to close the openings to their respiratory passages, the excretion of nitrogenous wastes as insoluble uric acid, and long loops of Henle in the mammalian kidney. Predict how each of these traits differs in animals that live in very humid versus very dry habitats. How would you test your predictions?
5. In insects, active transport of electrolytes into the Malpighian tubules leads to the formation of a pre-urine that is isotonic with respect to the hemolymph. In mammals, urine formation begins with blood pressure in the renal corpuscle that leads to the formation of a filtered pre-urine that is isotonic with respect to the blood. In insects, the pre-urine is processed in the hindgut; in mammals, the pre-urine is processed in the remainder of the nephron. How are the processing steps in insects and mammals similar? How are they different?
6. Compare and contrast the types of nitrogenous wastes observed in animals. Identify which compound can be excreted with a minimum of water, which is most toxic, and which waste product is found in fish, mammals, and insects. Which type of nitrogenous waste would you expect to find produced by embryos inside eggs laid on land?

Applying Concepts to New Situations

Answers are available at www.masteringbio.com

1. Examine Figure 42.17 again, and notice that the network of blood vessels that runs along the mammalian nephron is arranged so that blood flows in a countercurrent arrangement relative to the flow of fluid in the loop of Henle. The water and electrolytes that leave the loop of Henle and collecting duct diffuse into these blood vessels and are returned to the body. Why is the countercurrent arrangement for blood flow important?
2. You have isolated a segment of a rat nephron and introduced a solution of known composition. When you compare the fluid collected at the end of the segment with the test solution introduced at the beginning, you find that the volume has decreased by 30 percent and that the Na^+ concentration has decreased by 30 percent, but the urea concentration has increased by 50 percent. What processes in this segment might account for these changes?
3. In some areas of the world, highway maintenance crews make extensive use of salt to keep roads free of ice in winter. Biologists are concerned about the effect of this salt on terrestrial and freshwater organisms. Why?
4. Ethanol is the active ingredient in alcoholic beverages. One of its many effects is to inhibit the release of antidiuretic hormone (ADH) by the brain. How does the ingestion of ethanol affect urine formation? Why do people who ingest large amounts of ethanol suffer from headaches, body aches, and other symptoms of dehydration?

Animal Nutrition

43

A young crocodile has just caught a reed frog. Most animals obtain nutrients by ingesting them.

KEY CONCEPTS

- Animals require an array of nutrients to stay healthy, including specific amino acids, vitamins, and elements, as well as organic compounds that have high potential energy.
- Animals feed on a wide variety of items. In animals that ingest their food, there is usually a close correspondence between the structure of the animal's mouthparts and the function of those mouthparts in capturing and processing food.
- Digestion occurs in the digestive tract, which is compartmentalized into organs that have specialized functions in the digestion of food, absorption of nutrients and water, or excretion of wastes.
- Lack of homeostasis with respect to nutrients such as glucose and cholesterol can cause disease.

Animals get the two basic requirements for life—the chemical energy required to synthesize ATP and the carbon-containing compounds required for synthesizing complex macromolecules—from other organisms. They are **heterotrophs**. They eat to live.

The types of food that are available to different animals vary widely, however, and food is often in dangerously short supply. Based on these observations, it is logical to predict that animals have a variety of means for obtaining food and that animals are under intense natural selection to make efficient use of the food they have. How do animals get their food, and how do they process it? Which substances in food are used as nutrients, and how do humans and other animals maintain appropriate levels of key nutrients in their bodies?

If you're like most people, you probably give a fair amount of thought to the types of food that you eat but little thought to what happens to that food once it's inside you. As a meal moves through the digestive system, its chemical composition and physical characteristics change dramatically. Structurally and chemically, complex raw materials are broken down into simple components that can be used to build and maintain cells. Large packets of food that enter the mouth must be reduced to monomers that can be absorbed by cells and enter the bloodstream. The digestive system is analogous to an oil refinery or a lumber mill, where bulky, complex raw materials are processed into smaller and more usable products.

Research on feeding and digestion is fundamental to understanding many of the most basic aspects of animal anatomy and physiology. But research on animal nutrition has important practical applications as well. For example, this chapter addresses questions about whether certain nutrients can be used to boost athletic performance, and why type II diabetes and other nutrition-related diseases are on the rise in many human populations. Just as plant nutrition is fundamental to the productivity of agricultural and natural ecosystems, nutrition is a basic component of animal health and welfare. Let's begin with a look at what animals have to eat to live.

43.1 Nutritional Requirements

Animals get the chemical energy and carbon-containing building blocks they need from carbohydrates and fats. The structures of carbohydrates and fats were analyzed in Chapters 5 and 6; Chapter 9 detailed how these compounds are used in the synthesis of ATP and key macromolecules. But in addition to obtaining carbohydrates and fats, animals require an array of other nutrients. A **nutrient** is a substance that an organism needs to remain alive. Food is any material that contains nutrients. Animals eat a wide array of foods, and not all animals have the same nutritional requirements.

When an animal takes in enough nutrients to perform day-to-day tasks and stay healthy, it is said to be in **nutritional balance**. Disease results when an individual is out of nutritional balance. Among humans who live in the developing world, nutrition-related diseases are caused by deficits in total intake or specific nutrients. In many parts of the developed world, though, nutrition-related diseases are caused by an excess of nutrients. In the United States, for example, the obesity epidemic discussed in Section 43.4 may qualify as the most serious of all public health problems.

Understanding which nutrients an individual needs, and in what amounts, are basic issues in research on animal nutrition. Determining nutritional requirements is seldom simple, however. To illustrate how biologists go about the task, let's first consider what humans need to maintain good health and then what they need in the special case of enhancing athletic performance.

Meeting Basic Needs

The United States government instituted a program to establish basic nutritional requirements as part of its war effort during World War II. In 1943 the Food and Nutrition Board of the National Academy of Sciences* published the first Recommended Dietary Allowances (RDAs). The goal of the RDAs was to specify the amount of each essential nutrient that an individual must ingest to meet the needs of most healthy people.

In addition to obtaining chemical energy from carbohydrates and other compounds, humans require several other types of **essential nutrients**, that is, nutrients that cannot be synthesized and must be obtained in the diet:

- Proteins provide amino acids that are used to synthesize the body's own polypeptides and are sometimes oxidized to provide energy. Of the 21 amino acids required to manufacture proteins, humans can synthesize 12. The others must be obtained from food and are called **essential amino acids**.
- **Vitamins** are organic compounds that are vital for health but are required in only minute amounts. They have a variety of roles; several function as coenzymes in critical reactions. **Table 43.1** lists a few of the vitamins for which RDAs have

*The National Academy of Sciences is a group of scientists and engineers that advises the U.S. Congress on scientific and technical matters. Its Food and Nutrition Board is made up of biologists who specialize in animal nutrition.

TABLE **43.1** **Essential Vitamins (Required by Humans)**

Name	Source in Diet	Function	Symptoms if Deficient
Vitamin B_1 (thiamine)	legumes, whole grains, potatoes, peanuts	formation of coenzyme in Krebs cycle	beriberi (fatigue, nerve disorders, anemia)
Vitamin B_{12}	red meat, eggs, dairy products; also synthesized by bacteria in intestine	coenzyme in synthesis of proteins and nucleic acids; formation of red blood cells	anemia (fatigue and weakness due to low hemoglobin content in blood)
Niacin	meat, whole grains	component of coenzymes NAD^+ and $NADP^+$	pellagra (digestive problems, skin lesions, nerve disorders)
Folate	green vegetables, oranges, nuts, legumes, whole grains; also synthesized by bacteria in intestine	coenzyme in nucleic acid and amino acid metabolism	anemia
Vitamin C (ascorbic acid)	citrus fruits, tomatoes, broccoli, cabbage, green peppers	used in collagen synthesis, prevents oxidation of cell components, improves absorption of iron	scurvy (degeneration of teeth and gums)
Vitamin D	fortified milk, egg yolk; also synthesized in skin exposed to sunlight	aids absorption of calcium and phosphorus in small intestine	rickets (bone deformities) in children; bone softening in adults

TABLE 43.2 **Essential Elements (Required by Humans)**

Name	Source in Diet	Function	Symptoms if Deficient
Calcium (Ca)	dairy products, green vegetables, legumes	bone and tooth formation, nerve signaling, muscle response	loss of bone mass, slow growth
Fluorine (F)	fluoridated water, seafood	maintenance of tooth structure	higher frequency of tooth decay
Iodine (I)	iodized salt, algae, seafood	component of the thyroid hormones thyroxin and T_3	goiter (enlargement of thyroid gland)
Iron (Fe)	meat, eggs, whole grains, green leafy vegetables, legumes	enzyme cofactor; synthesis of hemoglobin and electron carriers	anemia, weakness
Magnesium (Mg)	whole grains, green leafy vegetables	enzyme cofactor	nerve disorders
Phosphorus (P)	dairy products, meat, grains	bone and tooth formation, synthesis of nucleotides and ATP	weakness, loss of bone
Sulfur (S)	any source of protein	amino acid synthesis	swollen tissues, degeneration of liver, mental retardation

been established, notes their functions, and indicates the problems that develop if they are missing in the diet.

- **Electrolytes** are inorganic ions that influence osmotic balance and are required for normal membrane function. Sodium (Na^+), potassium (K^+), and chloride (Cl^-) are the major ions in the body.
- Essential nutrients also include inorganic substances that fulfill a wide variety of functions not performed by electrolytes—often because they are important components of enzyme cofactors or structural materials (see **Table 43.2**). Some, such as calcium and phosphorus, are needed in relatively large quantities. Others, such as iron and magnesium, are required in small or trace amounts.

In the United States, packaged food must display how a serving of the contents meets the RDA values (**Figure 43.1**). In many cases, labels must also furnish information on the amount of chemical energy available in the food. In nutrition, the **Calorie** is a unit of energy. If a single helping of food contains 150 Calories, then 150 **kilocalories (kcal)** of energy are released when the food is completely oxidized by cells. (It is important to realize, though, that organisms rarely extract all the calories potentially available, because food is not completely digested.)

Most energy in food is in the form of carbohydrates and fats, which are carbon compounds with high potential energy (see Figure 5.9). Because fats and other lipids contain more C–H bonds than do carbohydrates such as starch and sugar, lipids provide over twice the energy per gram. Lipids yield 9 kcal per gram when completely oxidized, compared with only 4 kcal/gm for carbohydrates and proteins.

RDAs frequently change as research advances. In many cases, members of the Food and Nutrition Board are cautious about making recommendations, due to a lack of reliable data. Studying nutrient requirements in humans is difficult. The types of experiments that would be most enlightening would assign different, carefully controlled diets to several groups of subjects. Some of these diets might lead to illness, however, so they are unethical. As a result, researchers rely on observational studies that document the health of people whose intake levels for specific nutrients are known.

The scientists who publish the RDAs constantly remind the public that a balanced diet fulfills all nutrient requirements. In the vast majority of cases, nutritional supplements are not

Nutrition Facts
Serving Size 1 slice (69g)
Servings Per Container about 9

Amount per Serving	
Calories 150	Calories from Fat 25
	% Daily Value*
Total Fat 3g	5%
Saturated Fat 1.5	8%
Trans Fat 0g	
Cholesterol 5mg	2%
Sodium 270mg	11%
Total Carbohydrate 21g	7%
Dietary Fiber 2g	6%
Sugars 3g	
Protein 7g	
Vitamin A 4% •	Vitamin C 0%
Calcium 10% •	Iron 4%

* Percent Daily Values are based on a 2,000 calorie diet. Your daily values may be higher or lower depending on your calorie needs.

	Calories:	2,000	2,500
Total Fat	Less than	65g	80g
Sat Fat	Less than	20g	25g
Cholesterol	Less than	300mg	300mg
Sodium	Less than	2,400mg	2,400mg
Total Carbohydrate		300g	347g
Dietary Fiber		25g	30g

Calories per gram:
Fat 9 • Carbohydrates 4 • Protein 4

FIGURE 43.1 Nutrition Labeling on Food Packaging. A label from a frozen pizza.

needed. And if certain types of nutritional supplements are abused, health problems can result. An important exception to this rule exists: Some women may need to take iron supplements as recommended by their physician. This is because hemoglobin—the protein that carries oxygen in the blood—contains iron and significant amounts of iron are lost during menstruation.

Can Nutrition Enhance Athletic Performance?

RDAs identify the nutrients that most people need to stay healthy and the amounts that are required to maintain nutrient balance. These recommendations target people with normal needs and activities. How do researchers determine nutrient requirements for extraordinary circumstances? And under what circumstances can biologists take an experimental approach to evaluating nutritional requirements?

To answer these questions, let's consider research on athletes. These individuals do extraordinary amounts of physical work. What nutrients do they require to perform at an optimal level?

For a researcher interested in the nutrient requirements of athletes, at least some types of experiments are possible. As long as experimental subjects are provided with a diet that fulfills RDAs, and as long as safe dosages of experimental compounds are used, studies should be able to meet accepted ethical guidelines.

In 1967 Jonas Bergström and colleagues published a classic experiment in nutrition research. They were interested in the performance of endurance athletes, such as cross-country skiers, bicyclists, and distance runners. At the time, most researchers supported the hypothesis that fatty acids—lipids consisting of a hydrocarbon chain bonded to a carboxyl (COOH—) group—provide the fuel required for extended physical labor or athletic activity. This hypothesis was logical, because fatty acids provide more than twice as much energy as do carbohydrates on a per-gram basis.

Glycogen is an important energy storage molecule in animals. Recall from Chapter 5 that glycogen is a polysaccharide made up of glucose molecules and from Chapter 9 that glucose is the preferred starting compound for the production of ATP through cellular respiration. Most of the body's glycogen is stored in skeletal muscles and the liver. During exercise, glycogen is catabolized (broken down) into glucose. Muscle cells use the glucose that is released to manufacture ATP; glucose from the liver enters the bloodstream and is delivered to all cells that are doing work.

Based on these observations, Bergström and associates hypothesized that an individual's ability to run, ski, or bicycle fast and for a long time might depend primarily on the amount of glycogen stored in the body rather than on the amount of fatty acids. The team also predicted that a diet rich in carbohydrates would maximize the amount of stored glycogen.

To test these hypotheses, the biologists set up the experiment outlined in **Figure 43.2**. Nine student volunteers began by eating a mixed diet of carbohydrate, fat, and protein for a day. After the researchers removed a tiny sample of leg muscle from each volunteer to measure the amount of stored glycogen, the volunteers rode exercise bikes at the same relative work rate until they were exhausted. During the exercise period, the biologists removed blood samples from each student at regular intervals to check the volunteers' concentration of glucose.

Six students then switched to a diet high in fat and protein for three days. The remaining three students switched to a high-carbohydrate diet. After the researchers checked the volunteers' glycogen stores, all nine volunteers again rode a stationary bike to exhaustion while their blood-glucose levels were recorded. Finally, the volunteers switched diets for a three-day period: The six individuals who had been eating a diet high in fat and protein switched to a high-carbohydrate diet; the three students who had been eating a high-carbohydrate diet switched to a diet high in fat and protein. To conclude the experiment, all nine volunteers had their glycogen reserves documented and then rode the stationary bike to exhaustion as their blood-glucose concentrations were monitored.

The experimental design effectively controlled for two potentially complicating factors: (1) Each volunteer was subjected to all three types of diet, so the outcome of the experiment was not due to a biased sample; and (2) both groups of volunteers experienced the two experimental diets in a different order, to test the hypothesis that the sequence of food types ingested might affect the results.

The left-hand graph of the "Results" section of Figure 43.2 plots the amount of glycogen present in leg muscle versus "work time"—the amount of time each volunteer was able to ride after consuming each type of diet. (Each line represents one individual, and the color of each data point indicates the diet the individual had eaten prior to the test.) Notice that in every case, both the amount of stored glycogen and the work time were much higher after individuals had eaten a high-carbohydrate diet. Now examine the right-hand graph in the "Results" section of the figure, which plots the changes in blood-glucose levels that occurred in one of the subjects during each of the three rides. These data also support the hypothesis that a diet high in carbohydrate supports optimal performance.

Subsequent experiments, performed in a variety of labs around the world, confirmed that performance in endurance sports is enhanced by a high-carbohydrate diet. Does ingesting glucose or other carbohydrates just before a race also improve performance? A thorough set of studies has failed to support this hypothesis. The current consensus among nutrition researchers is that ingesting sugars right before a contest actually reduces performance.

The experiments that were conducted with student volunteers revolutionized how athletes all over the world train for

Experiment

Question: What types of diet optimize performance in endurance sports?

Hypothesis: Carbohydrate-rich diets should lead to better performance than normal or high-protein–high-fat diets, by maximizing glycogen stores.

Null hypothesis: There is no difference in performance on the basis of diet.

Experimental setup:

1. Start with nine volunteers, split into two groups.
2. Arrange for the volunteers to follow three types of diets:
   - **Mixed normal diet:** Mixed carbohydrate, protein, fat (2800 kcal total)
   - **Protein-fat diet:** 1500 kcal protein, 1300 kcal fat
   - **High-carbohydrate diet:** 2300 kcal carbohydrate, 500 kcal protein
3. Each group eats each type of diet for 1 to 3 days, but in different orders.
4. After eating each diet, volunteers undergo a performance test: Ride bike to exhaustion.

Prediction: Glycogen stores and performance will be best after rider has eaten high-carbohydrate diet before test ride.

Prediction of null hypothesis: No difference in glycogen stores or performance on the basis of diet preceding test.

Results:

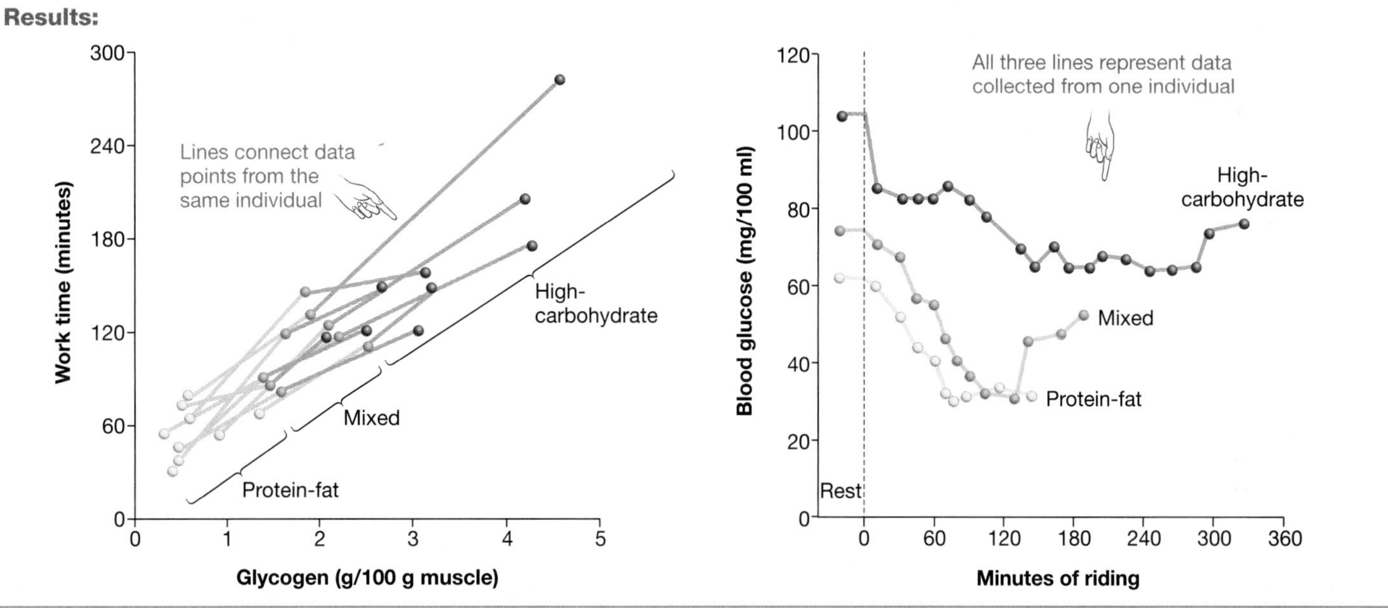

Conclusion: Individuals perform better after eating a diet high in carbohydrates than after eating a normal diet or a diet high in protein and fats.

FIGURE 43.2 Experimental Evidence That Carbohydrate Loading Increases Performance in Endurance Sports.

QUESTION Why were the researchers careful to provide the same amount of energy, in kcal, in each diet?

endurance events. The dietary regime known as carbohydrate loading, based on eating large amounts of carbohydrate 1–3 days before an event, is now a routine part of race preparations for endurance athletes such as marathoners, triathletes, distance swimmers, and cross-country skiers.

43.2 Obtaining Food: The Structure and Function of Mouthparts

Animals are heterotrophs. Instead of making their own food as plants do, animals obtain the energy and nutrients they need from other organisms. The question is, How?

As noted in Chapters 32 through 34, some animals obtain the food they need via suspension feeding or by absorbing it directly from the surrounding environment. **Suspension feeders**, such as sponges and tube worms, filter small organisms, or bits of organic debris from water by means of cilia or other structures. Most species that absorb nutrients directly are **endoparasites**, meaning that they live inside the body of a host individual. But by far the greatest number of animals are **mass feeders**: They seize and manipulate chunks of food by using mouthparts such as jaws and teeth, beaks, or special toxin-injecting organs.

The types of food that animals harvest range from soupy solutions in decaying carcasses to hard nuts. Given the diversity

FIGURE 43.3 Hundreds of Cichlid Species Live in the Rift Lakes of East Africa. Rift Lake cichlids are famous for their species diversity, dramatic coloration, and the diversity of food sources they exploit.

of food sources that animals exploit, it is not surprising that animals have a wide variety of mouthpart structures involved in capturing and processing food. Natural selection has tightly matched the structure of animal mouthparts to their function in obtaining food. Wolves have large canine teeth that function in tearing meat; horses have large, flat molars that mash coarse plant material efficiently (see Chapter 32).

Let's pursue the correlation between the size and shape of mouthparts and the size and shape of food sources further, by analyzing the structure and function of jaws and teeth in what may be the most diverse lineage among vertebrates: the cichlid fishes of Africa (**Figure 43.3**).

The cichlids that inhabit the Rift Lakes of East Africa are a spectacular example of adaptive radiation—the diversification of a single ancestral lineage into many species, each of which lives in a different habitat or employs a distinct feeding method (see Chapter 27). Lake Victoria, for example, is home to 300 **endemic** cichlids—meaning species that live nowhere else. Most feed on a specific item, but as a group they exploit almost every food source in the lake: planktonic organisms, crust-forming algae, leaflike algae, eggs, fish scales, fish fins, whole fish, plants, insects, and snails.

How can a group of very closely related species exploit so many different food sources? To answer this question, biologists point to a structure called the pharyngeal jaw. Many fish species have pharyngeal (throat) jaws located well behind the normal oral (mouth) jaws. Non-cichlids use their pharyngeal jaws to move food down their throats, but cichlids can also use theirs to bite. This is possible because the upper pharyngeal jaw attaches to the skull in cichlids, and because the muscles of their lower pharyngeal jaw allow it to move against the upper jaw (**Figure 43.4a**).

(a) Pharyngeal (throat) jaws are located behind the oral jaws.

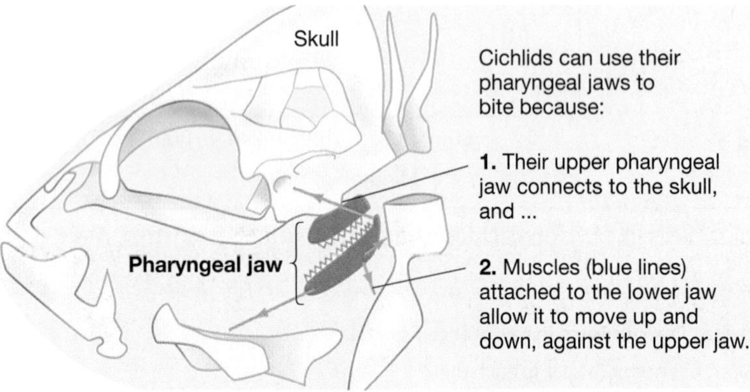

(b) Teeth on pharyngeal jaws vary among species of cichlids and correlate with the type of food eaten.

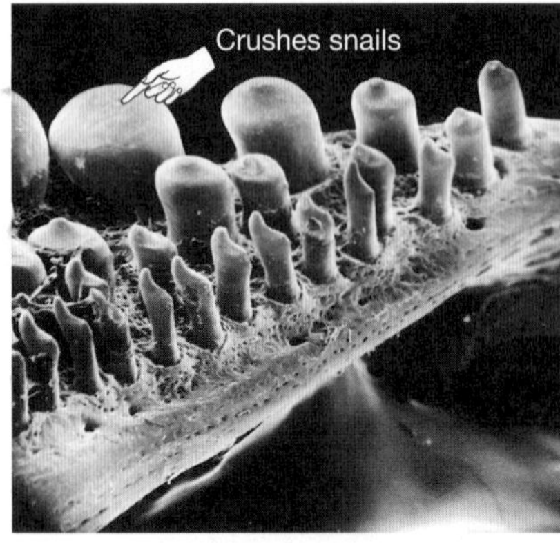

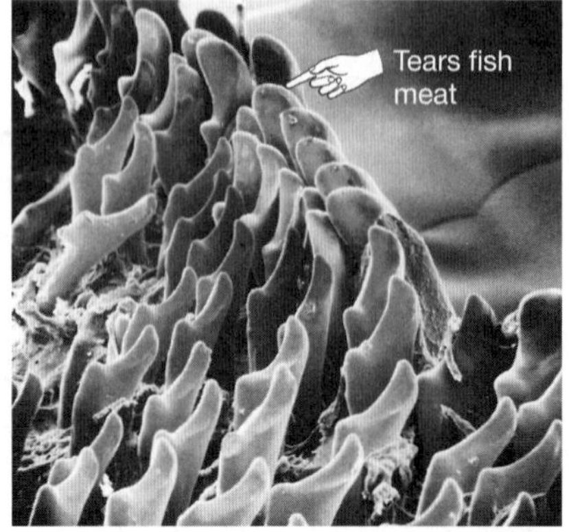

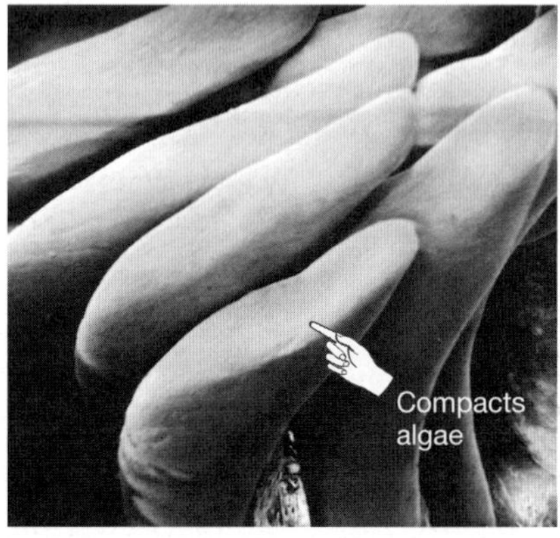

FIGURE 43.4 Rift Lake Cichlids Have Pharyngeal Jaws That Make Feeding Particularly Efficient.

EXERCISE In part (a), indicate where the teeth illustrated in part (b) are located. Label the oral jaw in the skull in part (a).

In addition to furnishing a second set of biting jaws that make food processing more efficient, cichlid pharyngeal jaws provide an extra set of toothlike structures. Protuberances on the pharyngeal jaws of cichlids vary in size and shape. The structure of these "teeth" correlates with their function in crushing snail shells, tearing meat, or compacting algae (**Figure 43.4b**). These observations add to a large body of evidence supporting a general pattern in animal evolution: Mouthparts have diversified in response to natural selection for exploiting a diversity of food sources. The structure of jaws, teeth, and other types of mouthparts correlates with their function in harvesting and processing food.

43.3 How Are Nutrients Digested and Absorbed?

Animals ingest just about every type of food conceivable. Ingestion is only the first of four processes needed to obtain energy, however. For an animal to stay alive, ingestion must be followed by digestion, absorption, and elimination of wastes.

Digestion is the breakdown of food into small enough pieces to allow for **absorption**—the uptake of specific ions and molecules that act as nutrients. Digestion is a key process in animals that ingest their food, because, unlike plants, unicellular organisms, and certain parasites, animals do not acquire nutrients as individual molecules. Instead, they take in packets of food that must be broken down into small pieces. Nutrients must be extracted from the small pieces and waste materials must be eliminated. How and where does this processing occur?

Digestion takes place in a tube called the **digestive tract**—also known as the alimentary (literally, "nourishment") canal or gastrointestinal (GI) tract. As **Figure 43.5** shows, the digestive tract has two openings: It starts at the mouth and ends at the anus. The interior of this tube communicates directly with the external environment via these openings. In embryos, it derives from the hollow tube that forms as cells invaginate during gastrulation (see Chapter 22).

A tubelike digestive tract has three main advantages. First, the digestion of material inside the tract allows animals to feed on large pieces of food. Second, different chemical and physical processes can be separated within the digestive tract, so that

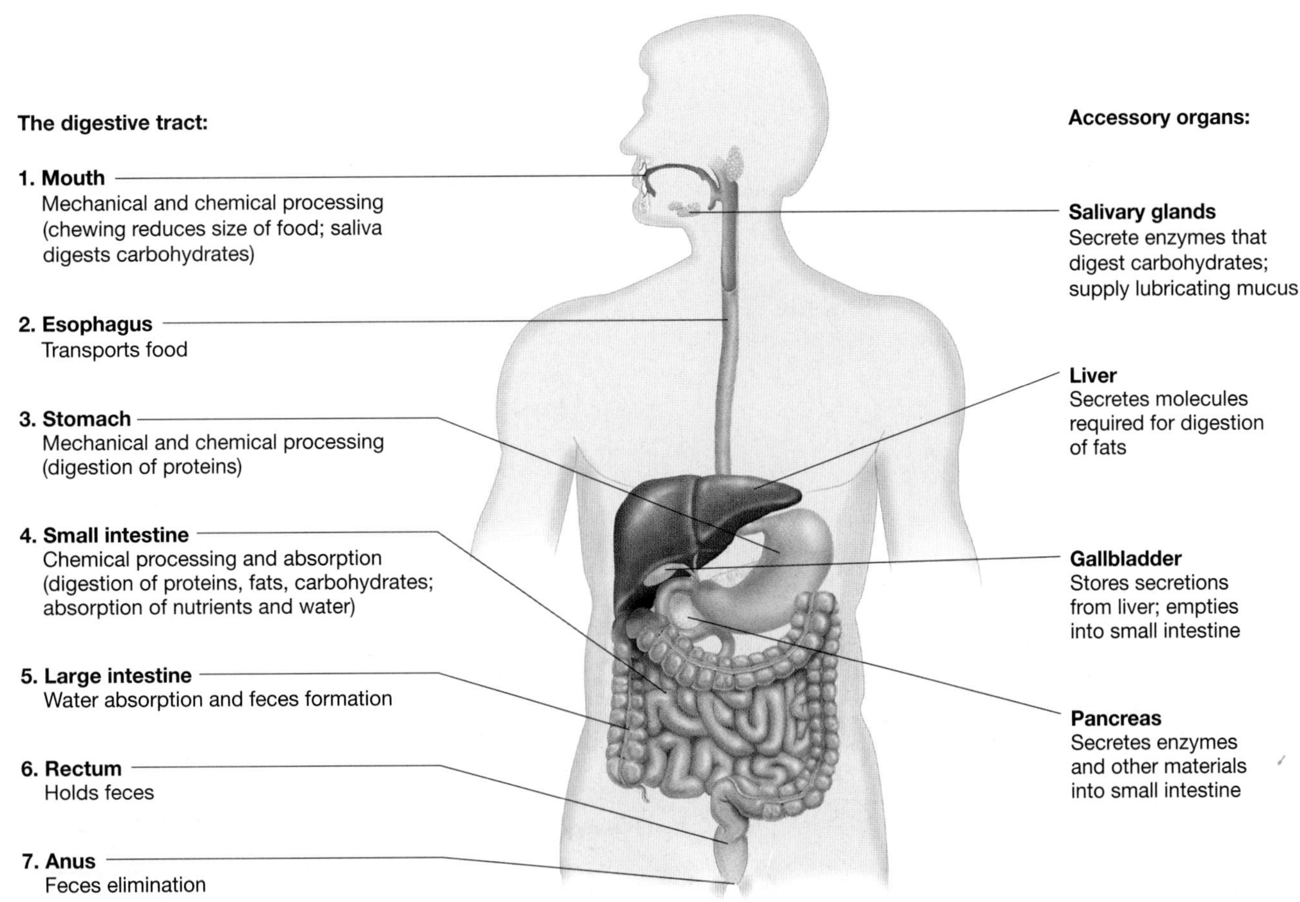

FIGURE 43.5 The Digestive System (Alimentary Canal) and Accessory Organs. The digestive tract is a tube that runs from the mouth to the anus. The salivary glands, liver, gallbladder, and pancreas are not part of the tract itself. Instead, they secrete material into the tract at specific points.

they occur independently of each other and in a prescribed sequence. The stomach, for example, provides an acidic environment for digestion. Material is transferred from there to the small intestine, where enzymes are specialized to function in an alkaline environment. Third, a tubelike digestive tract provides a one-way flow of food and allows wastes to exit the body at the anus.

The digestive tract is only part of the entire digestive system, however. Several vital organs and glands are connected to the digestive tract. These structures contribute digestive enzymes and other products to specific portions of the tract. They include the salivary glands, liver, gallbladder, and pancreas (see Figure 43.5).

Before analyzing the function of each component of the digestive system in detail, let's consider the general changes that happen to food on its way through the digestive tract. In this brief overview and in the detailed discussion that follows, humans will serve as a model species—simply because so much is known about human digestion.

Digestion begins with the tearing and crushing activity of teeth. Chewing reduces the size of food particles and softens them. In humans, the mechanical breakdown of food is augmented by the use of knives and cooking. In fact, the invention of cutting tools and cooking, which make food easier to chew, is the leading hypothesis to explain why average tooth size has declined steadily over the past several million years of human evolution.

Figure 43.6 summarizes the chemical changes that occur as food moves through the digestive tract. The chemical breakdown of carbohydrates begins in the mouth, through enzymes in saliva. Chemical processing continues in the stomach and is completed in the small intestine. Each of the three

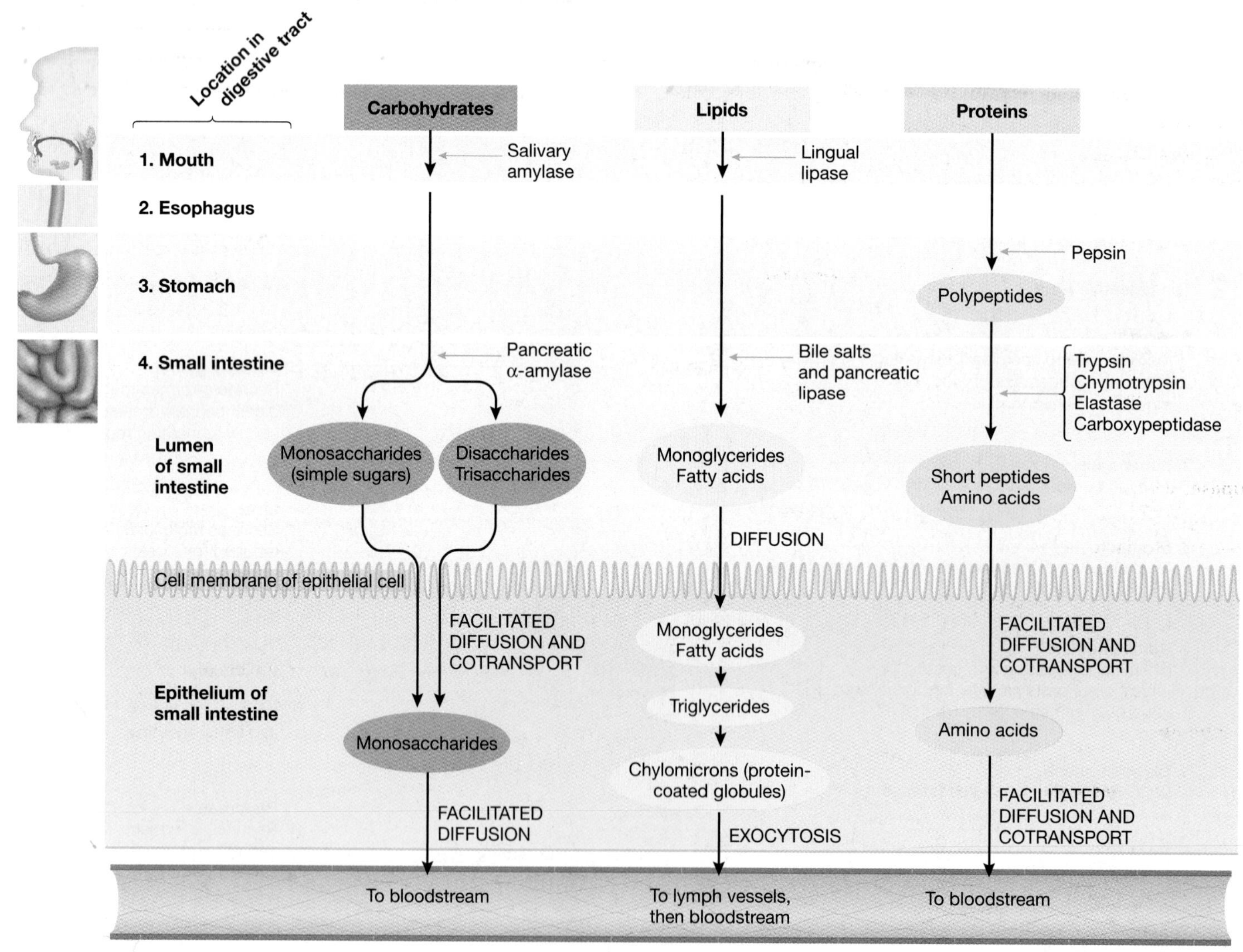

FIGURE 43.6 Carbohydrates, Lipids, and Proteins Are Processed in a Series of Steps. Three key types of macromolecules enter the digestive system (top of diagram). As they proceed through the digestive tract, they are broken apart by a variety of enzymes. Simple sugars, fatty acids, and amino acids then enter epithelial cells in the small intestine and are transported to the bloodstream.

major types of macromolecules—carbohydrates, lipids, and proteins—must be broken down during digestion. The small molecules that result from the digestion of these macromolecules are absorbed in the small intestine, along with water, vitamins, and ions. In the large intestine, or colon, more water is absorbed. The result is **feces**, which are eventually excreted.

For a human to stay healthy, each step in this process must be completed correctly. Problems in the digestive tract can lead to heartburn, ulcers, nausea, constipation, and other maladies. By far the most serious of these is diarrhea, which kills more than 4 million children each year. Because digestion is so important to understanding how animals work, let's analyze each step in more detail.

The Mouth and Esophagus

If you hold a cracker in your mouth long enough, it will start to taste sweet. The sensation occurs because starch molecules in the cracker are being hydrolyzed to maltose by an enzyme in saliva. Maltose is a disaccharide that is split in the small intestine to form two glucose monomers.

Starch breakdown was actually the first enzyme-catalyzed reaction ever discovered. In the early 1800s several researchers found that a component of certain plant extracts digested starch; in 1831 the same activity was discovered in human saliva. The enzyme responsible is called **amylase**. Amylase ranks as one of the best-studied enzymes; in animals it is the most important catalyst in the breakdown of carbohydrates. Amylase cleaves bonds that release maltose dimers from starch, glycogen, and other glucose polymers.

Amylase is not the only important ingredient in saliva, however. Cells in the tongue synthesize and secrete an enzyme called **lipase**, which begins the digestion of lipids. More specifically, the lingual lipase produced in the mouth begins breaking triglycerides into fatty acids and monoglycerides (see Chapter 6). **Salivary glands** in the mouth also release water and glycoproteins called mucins. When mucins contact water, they form the slimy substance called **mucus**. The combination of water and mucus makes food soft and slippery enough to be swallowed.

Once food is swallowed, it enters a muscular tube called the **esophagus**, which connects the mouth and stomach. Food is propelled down the esophagus by a wave of muscular contractions called **peristalsis**. About six seconds after being swallowed, food reaches the bottom of the esophagus. Because material is actively moved along the esophagus, you can swallow even when your mouth is lower than your stomach, such as when you bend over to drink from a stream. How does peristalsis occur?

An answer to this question began to emerge when anatomical studies established that the upper third of the esophagus consists of skeletal muscle, the lower third is composed of smooth muscle, and the middle third contains a mix of both muscle types. By cutting specific nerves in experimental animals, researchers established that peristalsis in the upper third of the esophagus occurs when a series of nerve cells that originate at the base of the brain sends electrical signals in a precise sequence. Each of these nerves terminates at skeletal muscle at a different location along the esophagus. Some of these muscles wrap around the esophagus; others run along its length. In response to nerve signals, the muscle contracts or relaxes in a coordinated fashion (**Figure 43.7**). In this way, a wave of muscle contractions propagates down the tube, propelling the food mass ahead of it. The action of these nerves is not the result of a conscious choice. The system is a **reflex**—an automatic reaction to a stimulus—that is stimulated by the act of swallowing.

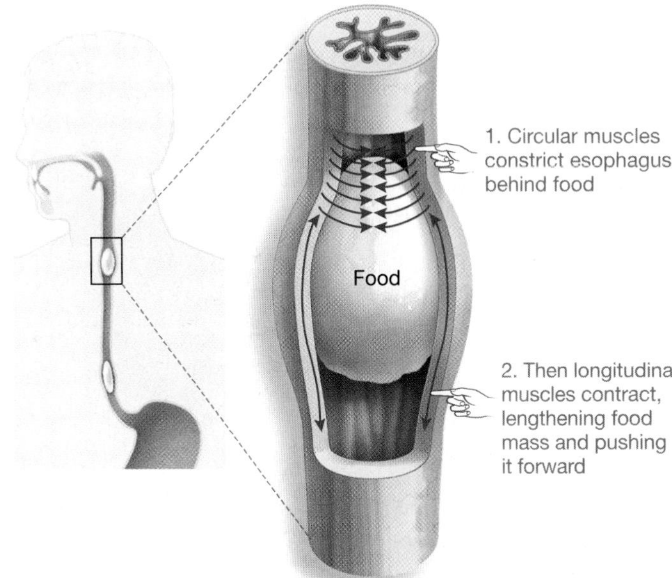

FIGURE 43.7 In the Esophagus, Peristalsis Transports Food to the Stomach. Alternating contractions of the circular and longitudinal muscles in the esophageal wall constrict the lumen of the esophagus behind food and expand the lumen in front of food. The wave of contraction begins at the oral end of the esophagus and propels the food toward the stomach.

Even after decades of experiments, though, the mechanism of peristalsis in the lower third of the esophagus is still not well understood. Researchers can induce normal peristalsis in this section of the esophagus in the absence of stimulation by nerves from the brain or spinal cord. Based on these results, most researchers agree that peristalsis in the lower region of the esophagus is coordinated at least in part by cells in the wall of the tube itself. A clear understanding of the mechanism has yet to emerge, however.

The Stomach

Although little if any digestion occurs in the esophagus, the situation changes dramatically when food reaches the stomach. Early anatomical studies showed that the **stomach** is a tough,

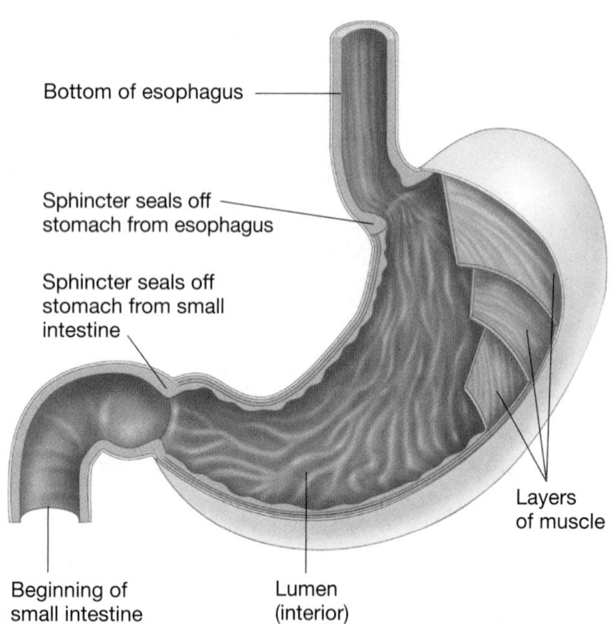

FIGURE 43.8 The Stomach Is an Acidic Environment Where Food Is Mixed and Proteins Are Digested. The stomach is a muscular outpocketing of the digestive tract. Muscle contractions mix food and break it into smaller pieces.

muscular pouch in the digestive tract, bracketed on both ends by valves called **sphincters** (**Figure 43.8**). When a meal fills the stomach, muscular contractions result in churning that mixes the contents and reduces food to a uniform consistency and solute concentration, or osmolarity. A certain amount of mechanical breakdown of food also occurs in response to this churning. The other main function of the stomach is the partial digestion of proteins.

Compared with the environment of the mouth or esophagus (or virtually any other tissue), the lumen of the stomach is highly acidic. Researchers in the late 1700s and early 1800s documented this fact by analyzing vomit or the contents of sponges that were tied to strings, swallowed, and pulled back up. In the early 1800s chemists were able to document that the predominant acid in the stomach is hydrochloric acid (HCl).

Not long after, a physician named William Beaumont established that digestion takes place in the stomach. He reached this conclusion through an extraordinary series of experiments on a young man named Alexis St. Martin. In 1822, when St. Martin was 19 years old, a shotgun accidentally discharged into his abdomen and created a series of wounds. Despite repeated attempts, Beaumont was unable to close a hole in St. Martin's stomach. Eventually Beaumont inserted a small tube through the opening; the tube remained in St. Martin's body for the rest of his life. (Today, biologists insert tubes into various parts of the digestive tract of cows or sheep to study how these animals digest different types of feed.) With the tube in place, Beaumont was able to tie a string onto small pieces of meat or vegetables, insert the food directly into St. Martin's stomach, and draw it out after various intervals. Beaumont was also able to remove liquid from inside the stomach and observe how this gastric (stomach) juice acted on food in vitro. His experiments showed that gastric juice digests food—particularly meat.

Theodor Schwann later purified the enzyme that is responsible for digesting proteins in the stomach and named it **pepsin**. Because it destroys proteins, biologists hypothesized that pepsin must be synthesized and stored in cells while it is in an inactive form—otherwise it would kill the cells where it is synthesized. In 1870 a biologist established through careful microscopy that granules occur in specialized stomach cells called chief cells. These granules were hypothesized to be a pepsin precursor. Follow-up work showed that this hypothesis was correct. The precursor compound found in chief cells, which came to be called **pepsinogen**, is converted to active pepsin by contact with the acidic environment of the stomach. Secretion of a protein-digesting enzyme in inactive form is important, because it prevents destruction of proteins in the cells where the enzyme is synthesized.

Which Cells Produce Stomach Acid? Researchers who were studying the anatomy of the stomach wall noticed clusters of distinctive **parietal cells** located along canals that communicate with the lumen of the stomach (**Figure 43.9a**). An investigator also documented that the shape and activity of these cells appeared to vary as the digestion of a meal proceeded. On the basis of these observations, he inferred that parietal cells are the source of the HCl in gastric juice, which may have a pH as low as 1.5. Earlier microscopists had shown that another type of cell, called a **goblet cell**, secretes the mucus that is found in gastric juice. Mucus lines the gastric epithelium and protects the stomach from damage by HCl. To summarize, these anatomical studies showed that the epithelium of the stomach contains several types of secretory cells, each of which is specialized for a particular function.

To test whether parietal cells secrete HCl directly or in an inactive form, a biologist injected a dye into the stomach of dogs. The dye that was used in this experiment turns blue in the presence of acid. When the biologist examined the stained stomach tissue with the microscope, she found blue strands running through the canals in the gastric epithelium and connecting to parietal cells. On the basis of these results, she claimed that parietal cells secrete protons and chloride ions directly. But because other researchers found inconclusive results with dye experiments, the hypothesis of direct secretion remained controversial for decades. It was resolved only when researchers began to work out the molecular mechanism of HCl production.

How Do Parietal Cells Secrete HCl? The first clues about how parietal cells manufacture hydrochloric acid emerged in the late 1930s, when a researcher found a high concentration of an enzyme called carbonic anhydrase in parietal cells. This

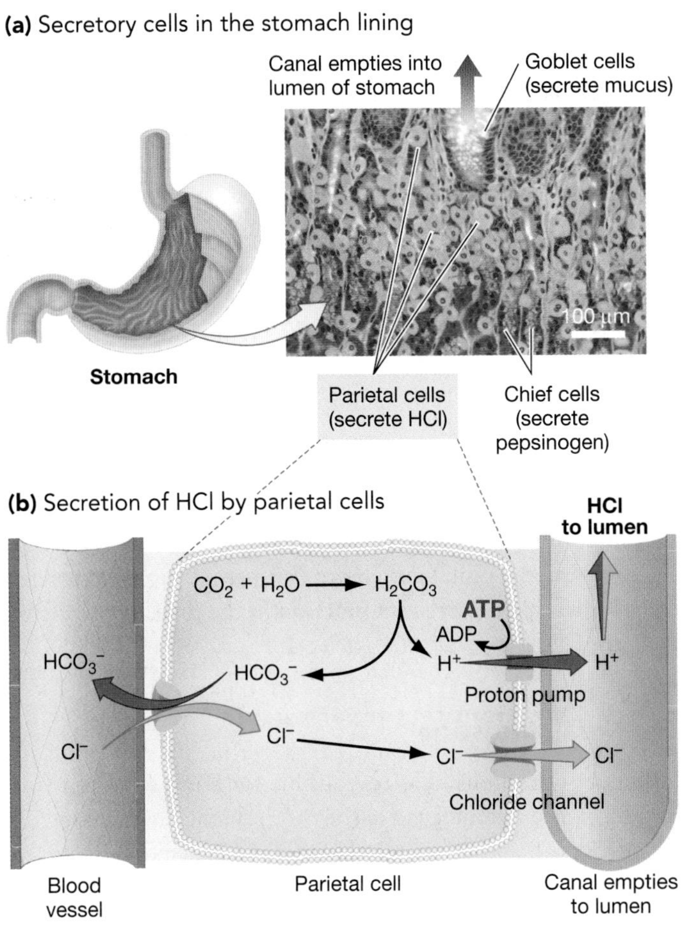

FIGURE 43.9 Parietal Cells in the Stomach Lining Secrete Hydrochloric Acid.

EXERCISE In part (b), label the reaction catalyzed by carbonic anhydrase.

result was interesting because **carbonic anhydrase** catalyzes the formation of carbonic acid (H_2CO_3) from carbon dioxide and water. In solution, the carbonic acid that is formed immediately dissociates to form a proton and the bicarbonate ion (HCO_3^-):

$$CO_2 + H_2O \rightleftharpoons H_2CO_3 \rightleftharpoons H^+ + HCO_3^-$$

A second clue to the formation of HCl came in the 1950s, when the development of the transmission electron microscope allowed researchers to analyze parietal cells at high magnification (see **BioSkills 8**). The biologists observed that parietal cells are packed with mitochondria. Based on this observation, researchers suspected that HCl secretion involves active transport. Later work confirmed this hypothesis by showing that the protons formed by the dissociation of carbonic acid are actively pumped into the lumen of the stomach. Subsequent studies showed that chloride ions from the blood enter parietal cells in exchange for bicarbonate ions, via an antiporter protein, and then move into the lumen through a chloride channel. The current model for HCl production is diagrammed in **Figure 43.9b**.

Although the acidic environment of the human stomach denatures (unfolds) proteins and makes their digestion highly efficient, it has a downside: When cells of the stomach wall are damaged by the acidity, holes called ulcers can develop (see **Box 43.1**). Fortunately, the stomach lining is usually well protected by a thick layer of mucus, by the rapid replacement of epithelial cells, and by secretion of pepsinogen instead of active pepsin.

The Ruminant Stomach Although the human stomach is a complex and fascinating organ, it is relatively simple compared with the stomachs of cattle, sheep, goats, deer, antelope, giraffe, and pronghorn. These species are called **ruminants.**

BOX 43.1 What Causes Gastric and Duodenal Ulcers?

When holes form in the lining of the stomach or in the duodenum—the initial section of the small intestine—individuals can suffer intense abdominal pain. For decades, the development of gastric and duodenal ulcers was thought to result from the production of excess acid in the stomach. As a result, physicians treated ulcers by prescribing compounds that neutralized hydrochloric acid in the stomach.

In 1983, however, Robin Warren and Barry Marshall published data indicating that ulcers were associated with infections of a bacterium called *Helicobacter pylori*. Instead of being caused by environmental influences like acid-rich diets and psychological factors like anxiety, Warren and Marshall hypothesized that ulcers were an infectious disease.

Physicians reacted to this hypothesis with intense skepticism, for two reasons: (1) the low acidity of the stomach was thought to sterilize the environment, and (2) treating ulcers as a bacterial rather than environmental disease required a radical change in thinking. In science, radical changes in thinking require exceptionally high standards of evidence.

To help provide this evidence, Marshall performed an experiment on himself. After a colleague succeeded in culturing *H. pylori* in his lab, Marshall drank fluid from a petri plate where the bacteria were growing. As predicted, he developed gastritis—an inflammation of the stomach that precedes development of ulcers. This experiment fulfilled several requirements of Koch's postulates for linking an organism to a specific disease (see Chapter 28). Subsequent work, in labs throughout the world, helped cement the link between *H. pylori* infection and ulcers.

Over a decade after the original paper was published, standards for treating ulcers finally began to change. Physicians now routinely prescribe antibiotics to relieve the symptoms of ulcers. In 2005, Warren and Marshall were awarded the Nobel Prize in Medicine.

They are distinct because they have stomachs that are specialized for digesting cellulose—not proteins. Recall from Chapter 5 that cellulose is a polymer of glucose and that it is the main structural component of plant cell walls. Digesting cellulose is not an easy task for ruminants, because mammals lack the enzymes called cellulases required to digest cellulose. Yet cellulose is the main carbohydrate in the leaves, stems, and twigs that ruminants ingest.

Ruminants are able to harvest energy from cellulose thanks to a combination of specialized anatomical structures, unique enzymes, and symbiotic relationships with bacteria and unicellular protists. When **symbiosis** occurs, members of two different species live in close physical contact with each other.

As **Figure 43.10** shows, ruminant mammals have four-chambered stomachs that are folded in complex ways. The largest chamber is called the rumen and serves as a fermentation vat. The rumen and the adjacent chamber called the reticulum are packed with symbiotic bacteria and protists. These organisms have enzymes capable of breaking apart the chemical bonds in cellulose, yielding glucose. The rumen is an oxygen-free environment, and the symbiotic organisms produce ATP from this glucose via fermentation, releasing fatty acids as a by-product (see Chapter 9). After the plant material has been partially digested in the rumen and in the reticulum, the ruminant regurgitates portions of that material into its mouth, forming a cud. The animal chews that regurgitated material further to enhance mechanical breakdown, then re-swallows it. As it passes into the stomach for the second time, the food bypasses the rumen and reticulum and enters the third chamber, or **omasum**, where water is removed. The final chamber is the **abomasum**, which contains the ruminant's own digestive enzymes and corresponds to a true stomach. A key point here is that most of a ruminant's food consists of (1) fatty acids and other compounds that are released as waste products of fermentation reactions in symbiotic organisms, and (2) the symbiotic cells themselves.

Although ruminants are the only animals that house symbionts in a modified stomach, they are not the only animals that use fermentation by symbionts to break down the cellulose in plant material. In rabbits and some species of rodents, for example, fermentation by symbionts occurs in an organ called the cecum. The **cecum** is a dead-ended or "blind" sac between the small intestine and the colon. It is greatly enlarged in species that use it as a fermentation chamber for processing cellulose. Specialized digestive systems such as the ruminant stomach and the rabbit cecum are another example of correlations between structure and function in the digestive tracts of animals.

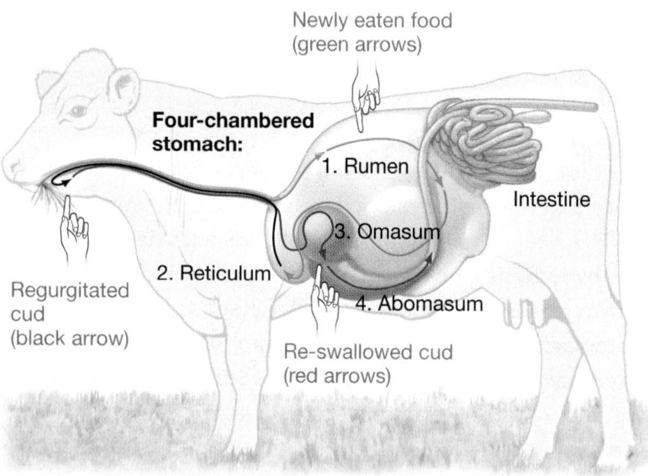

FIGURE 43.10 Ruminant Stomachs Facilitate the Digestion of Cellulose by Symbiotic Organisms. Ruminants obtain many of their nutrients from symbiotic bacteria and protists that live in the rumen and reticulum chambers of the stomach.

QUESTION The ability to regurgitate and re-chew food is thought to be an adaptation that reduces the amount of time that grazing animals have to spend harvesting food (because it can be thoroughly chewed later). Why would this be advantageous?

The Small Intestine

In humans, the stomach is responsible for mixing the contents of a meal into a homogenous slurry, mechanically breaking up food material, and providing the acid and enzymes required to partially digest proteins. Peristalsis in the stomach wall then moves small amounts of material through the valve created by a sphincter muscle at the base of the stomach and into the small intestine.

The **small intestine** is a long tube that is folded into a compact space between the stomach and the last major section of the gastrointestinal tract—the large intestine. In the small intestine, partially digested food mixes with secretions from the pancreas and the liver and begins a journey of about 6 m (20 ft). When passage through this structure is complete, digestion is finished and most nutrients—along with large quantities of water—have been absorbed.

In the small intestine, the surface area available for nutrient and water absorption is nothing short of remarkable. As **Figure 43.11** shows, the organ's epithelial tissue is folded and covered with fingerlike projections called **villi** (singular: **villus**). In turn, the cells that line the surface of villi have projections on their apical surfaces called **microvilli** (singular: **microvillus**). (Recall from Chapter 41 that an apical surface faces away from other tissues and toward the lumen of an organ or an external environment. A basolateral surface in an epithelium faces other tissues.) If the small intestine lacked folds, villi, and microvilli, it would have a surface area of about 3300 cm^2 (3.6 ft^2). Instead, the epithelium covers about 2 million cm^2 (over 2200 ft^2)—an area about the size of a tennis court. This enormous area increases the efficiency of nutrient absorption. And because each villus contains blood vessels and a lymphatic vessel called

(a) The lining of the small intestine has extensive folds.

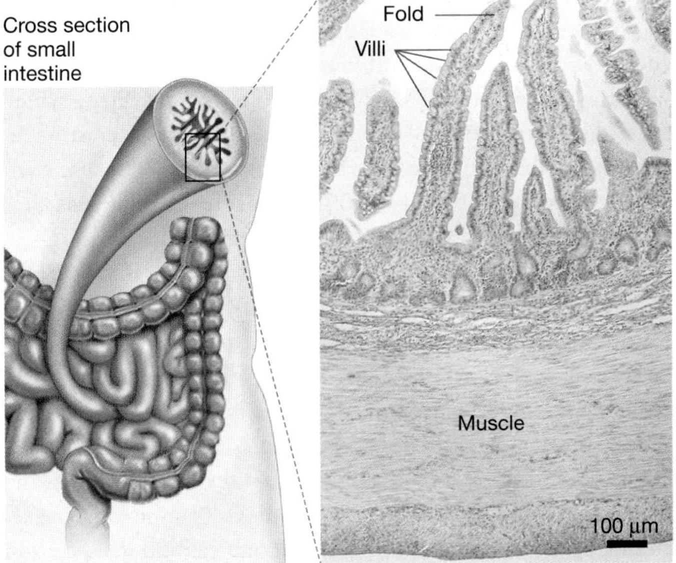

(b) Three-dimensional view of fold

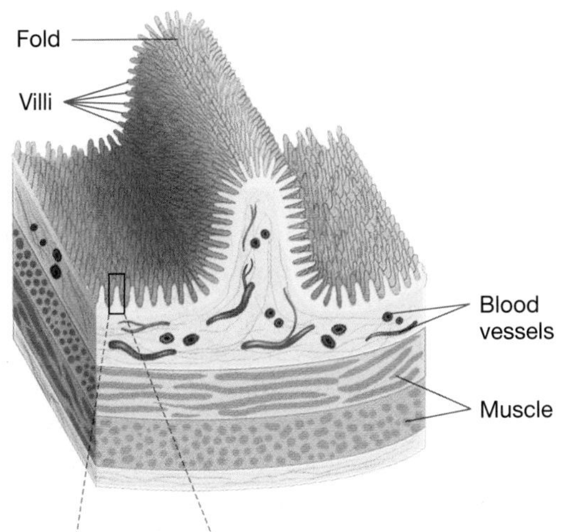

(c) Microvilli are extensions of epithelial cells in villi.

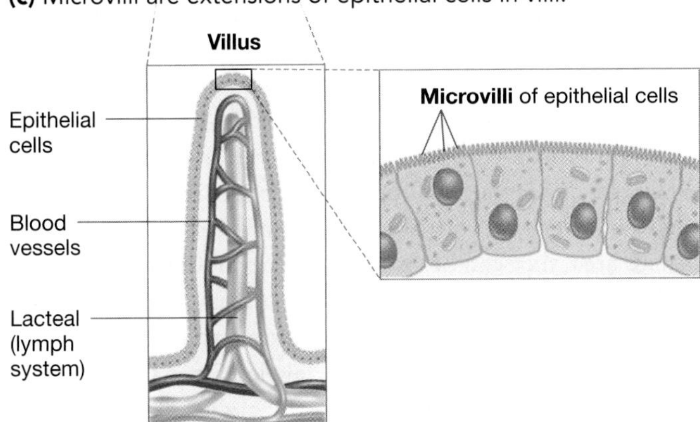

FIGURE 43.11 The Small Intestine Has an Extremely Large Surface Area.

EXERCISE In part (c), add transport proteins to the microvilli in the inset. Then draw arrows showing the path of nutrients from the lumen of the small intestine into the lacteals and blood vessels of a villus.

a **lacteal**, nutrients pass quickly from epithelial cells into the body's transport systems. (The circulatory system and lymphatic system are analyzed in detail in Chapters 44 and 49, respectively.) If you understand the importance of surface area in the small intestine, you should be able to explain (1) why surface area is so much higher there than in the stomach or esophagus, and (2) what the impact of surgical shortening of the small intestine would be.

To understand how digestion is completed and absorption occurs, let's explore what happens to proteins, lipids, and carbohydrates as they move through this section of the digestive tract.

Protein Processing by Pancreatic Enzymes The acidic environment of the stomach denatures proteins, destroying their secondary and tertiary structures. In addition, pepsin cleaves the peptide bonds next to certain amino acids so that long polypeptides are reduced to relatively small chains of amino acids. In the small intestine, protein digestion is completed so that individual amino acids can enter the bloodstream and be transported to cells throughout the body.

How does protein digestion occur? By the end of the nineteenth century, it had been established that enzymes in the small intestine digest polypeptides to monomers. Later work showed that each of these protein-digesting enzymes, or **proteases**, is specific to certain types or configurations of amino acids in a polypeptide chain. Thus, a suite of proteases is required to completely digest polypeptides to amino acid monomers.

In addition, by 1900 biologists had determined that proteases are synthesized in an inactive form in the **pancreas**, which is connected to the small intestine by the pancreatic duct. Like the production of inactive pepsinogen by chief cells in the stomach, the production of digestive enzymes in an inactive conformation prevents pancreatic cells from digesting themselves.

It took decades of work to understand how pancreatic enzymes are activated in the small intestine. In 1900 a researcher showed that pancreatic enzymes are activated by contact with juice from the upper part of the small intestine. Because activation did not occur when he heated the intestinal juice, and because heat denatures proteins, he hypothesized that the agent responsible for activating the pancreatic enzymes was also an enzyme. He called the unknown enzyme enterokinase. Decades later, a researcher succeeded in purifying a pancreatic enzyme called **trypsinogen** and demonstrated that enterokinase activates it in vitro. Follow-up work showed that enterokinase activates trypsinogen by phosphorylating it. The active enzyme that results from this reaction is called **trypsin**. Trypsin then triggers the activation of other protein-digesting enzymes, such as chymotrypsin, elastase, and carboxypeptidase. These enzymes are also synthesized by the pancreas and secreted in an inactive form.

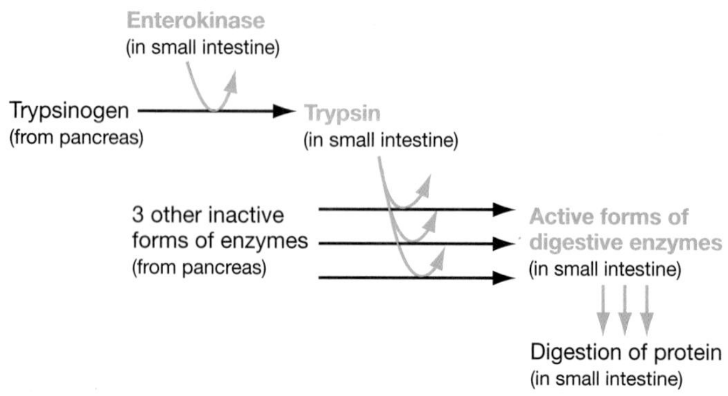

FIGURE 43.12 Enterokinase Triggers an Enzyme-Activation Cascade in the Small Intestine.

Figure 43.12 summarizes how the sequential activation of digestive enzymes proceeds in the small intestine. Enterokinase triggers the activation of trypsinogen to trypsin, which in turn triggers the activation of the three other protein-digesting enzymes. Once these enzymes are activated in the upper reaches of the small intestine, each begins cleaving specific peptide bonds. Eventually polypeptides are broken up into amino acid monomers.

What Regulates the Release of Pancreatic Enzymes?

Digestive enzymes are needed only when food reaches the small intestine. Based on this simple observation, it was logical for researchers to predict that their release would be carefully controlled. The leading hypothesis was that pancreatic secretion is controlled by the nervous system, possibly in response to the sight or smell of food. This hypothesis was inspired by an extensive series of experiments based on cutting specific nerves in experimental animals. These experiments had established that the nervous system is involved in regulating an array of digestive processes, including the release of saliva, swallowing, and the secretion of gastric juice in the stomach.

The hypothesis of nervous control in the small intestine was refuted, however, in a classic experiment that William Bayliss and Ernest Starling published in 1902. To study the mechanism responsible for stimulating secretion from the pancreas, Bayliss and Starling cut the nerves that connect to the pancreas and small intestine of a dog. Nervous control of the two organs was now impossible. But when the researchers introduced a weak HCl solution into the upper reaches of the animal's small intestine, to simulate the arrival of material from the stomach, the pancreas secreted in response. This observation was startling. The small intestine had successfully signaled the pancreas that food had arrived, even though the nerves connecting the two organs had been cut.

To explain this observation, Starling hypothesized that a chemical messenger must be involved, and that the chemical messenger must originate in the small intestine and travel to the pancreas—perhaps via the blood. He tested this idea by cutting off a small piece of the small intestine, grinding it up, and injecting the resulting solution into a vein in the animal's neck. Minutes later, the pancreas sharply increased secretion.

Bayliss and Starling had discovered the first **hormone**—a chemical messenger that influences physiological processes at very low concentrations. The molecule they detected, which they called **secretin**, is produced by the small intestine in response to the arrival of food from the stomach. The discovery of secretin was important because it confirmed that digestion is under both neural control and hormonal control.

Follow-up work showed that secretin's primary function is to induce a flow of bicarbonate ions (HCO_3^-) from the pancreas to the small intestine. The arrival of bicarbonate is important because it neutralizes the acid arriving from the stomach. Researchers also discovered a second hormone produced in the small intestine, called cholecystokinin (pronounced *ko-la-sis-toe-KIN-in*), that induces secretion from the liver as well as the pancreas. **Cholecystokinin** ("bile-bag-mover") stimulates the secretion of digestive enzymes from the pancreas and the secretion of molecules from the liver that are involved in the digestion of lipids. Hormones are involved in stomach function as well. For example, after being stimulated by nerves or the arrival of food, certain stomach cells produce the hormone **gastrin**. In response, parietal cells begin secreting HCl.

Digesting Lipids: Bile and Transport The pancreatic secretions include digestive enzymes that act on fats and carbohydrates as well as proteins. Like the lingual lipase added to saliva in the mouth, the enzyme **pancreatic lipase** breaks certain bonds present in complex fats and results in the release of fatty acids and other small lipids.

Recall from Chapter 6 that fats are insoluble in water. As a result, they tend to form large globules as they are churned in the stomach. Before pancreatic lipase can act, large fat globules that emerge from the stomach must be broken up, or emulsified. In the small intestine, **emulsification** results from the action of small lipids called bile salts. As **Figure 43.13** shows, bile salts function like the detergents that researchers use to break up the lipids in plasma membranes (see Chapter 6). Bile salts are synthesized in the liver and secreted in a complex solution called **bile**, which is stored in the **gallbladder**. When bile enters the small intestine, it raises the pH and emulsifies fats. Once fats are broken into small globules with high surface area, they can be attacked by enzymes and digested.

What happens to the monoglycerides and fatty acids released by lipase activity? An answer began to emerge when researchers injected radioactively labeled fatty acids into the small intestines of laboratory rats. The biologists analyzed the epithelial cells that line the small intestine of the experimental animals and found that most of the radioactive label had entered the cells and attached to a protein they named **fatty-acid binding protein**. Later, other researchers established that a fatty-acid binding protein also occurs in the membranes of these cells. Based on these and other results, fatty-acid binding

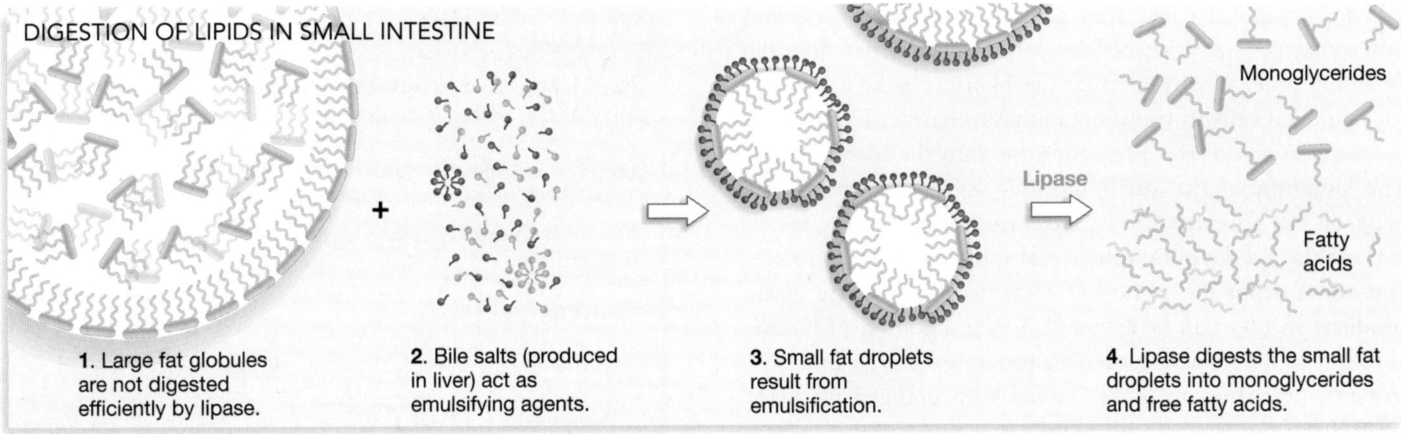

FIGURE 43.13 Emulsifying Agents and Lipases Digest Lipids in the Small Intestine. Once bile salts break up large fat globules, lipase can digest fats efficiently.

proteins are thought to bring lipids into the cell. There they are processed into protein-coated globules called **chylomicrons**. Chylomicrons diffuse into lacteals near the epithelial cells (see Figure 43.11c). The lacteals merge with larger lymph vessels, which merge with larger blood vessels. In this way, fats enter the bloodstream without clogging small blood vessels. Eventually, the products of fat digestion end up in fat-storage cells and other tissues.

How Are Carbohydrates Digested and Transported? In addition to manufacturing lipase and protein-digesting enzymes, the pancreas produces **nucleases** and an amylase that is similar to the salivary enzyme introduced earlier. **Nucleases** digest the RNA and DNA in food, while pancreatic amylase continues the digestion of carbohydrates that began in the mouth. **Table 43.3** summarizes the suite of digestive enzymes observed in the mammalian digestive tract.

SUMMARY TABLE **43.3 Digestive Enzymes**

Digestion is accomplished by the enzymes listed here, by HCl produced in the stomach in response to the hormone gastrin, and by bile salts from the liver. Bile salts are stored in the gallbladder. They are released in response to the hormone cholecystokinin and emulsify fats in the small intestine.

Name	Where synthesized	Regulation	Function
Carboxypeptidase	Pancreas	Released in response to cholecystokinin from small intestine; activated by trypsin	In small intestine, breaks peptide bonds in polypeptides—releasing amino acids
Chymotrypsin	Pancreas	Released in response to cholecystokinin from small intestine; activated by trypsin	In small intestine, breaks peptide bonds in polypeptides—releasing amino acids
Elastase	Pancreas	Released in response to cholecystokinin from small intestine; activated by trypsin	In small intestine, breaks peptide bonds in polypeptides—releasing amino acids
Lingual lipase	Salivary glands	Released in response to taste and smell stimuli	In mouth, breaks bonds in fats—releasing fatty acids and monoglycerides
Nucleases	Pancreas	Released in response to cholecystokinin from small intestine	In small intestine, break apart nucleic acids—releasing nucleotides
Pancreatic α-amylase	Pancreas	Released in response to cholecystokinin from small intestine	In small intestine, breaks apart carbohydrates—releasing sugars
Pancreatic lipase	Pancreas	Released in response to cholecystokinin from small intestine	In small intestine, breaks bonds in fats—releasing fatty acids and monoglycerides
Pepsin	Stomach	Released in inactive form (pepsinogen); activated by low pH in stomach lumen	In stomach, breaks peptide bonds between certain amino acids in proteins—releasing polypeptides
Salivary amylase	Salivary glands	Released in response to taste and smell stimuli	In mouth, breaks apart carbohydrates—releasing sugars
Trypsin	Pancreas	Released in inactive form (trypsinogen) in response to cholecystokinin from small intestine; activated by hormone enterokinase from small intestine	In small intestine, breaks specific peptide bonds in polypeptides—releasing amino acids

The monosaccharides that are released by amylase and other enzymes are absorbed by epithelial cells that line the small intestine. What molecular mechanisms make it possible for epithelial cells to transport monosaccharides like glucose from the lumen of the small intestine into the bloodstream? The situation in the small intestine is reminiscent of the nephron (see Chapter 42), where epithelial cells in the proximal tubule and the distal tubule reabsorb valuable electrolytes and water from the urine but leave urea and other waste products so they can be excreted. When digestion of food is complete in the small intestine, the resulting slurry is a treasure trove of nutrients and water mixed with undigestible plant fibers from food and bacterial cells that live symbiotically in the gut. It is logical to predict two general principles about nutrient absorption: (1) It is highly selective, meaning proteins in the plasma membranes of microvilli are responsible for bringing specific nutrients into the cell; and (2) it is active, meaning it requires an expenditure of ATP to bring nutrients into the epithelium across a concentration gradient.

Work over the past several decades has shown that both predictions are correct. One of the key results grew out of a series of experiments during the 1980s, which established that glucose absorption depends on the presence of an electrochemical gradient favoring an influx of sodium ions into the epithelium. Based on this observation, biologists hypothesized that the apical membranes of these cells must contain a series of cotransporters—proteins that would bring a nutrient molecule into the cell along with sodium ions. Work focused on glucose, because it is such a fundamentally important nutrient. To confirm that a sodium-glucose cotransporter exists, investigators set out to find the gene that codes for the hypothesized membrane protein.

The strategy that the researchers used is outlined in **Figure 43.14**. They began by purifying mRNAs from rabbit intestinal cells, which presumably were transcribing the cotransporter genes. Then the team separated the mRNAs by size via gel electrophoresis (see **BioSkills 7**) and injected one of each type of mRNA into a series of frog eggs—cells that do not normally transport glucose. The frog cells translated the rabbit mRNAs into proteins. When tested, one of the experimental eggs was able to import Na^+ and glucose in tandem. Based on these data, the biologists inferred that this egg had received the mRNA for the rabbit Na^+-glucose cotransporter. Using techniques introduced in Chapter 19, the researchers were able to make a DNA copy of the mRNA, determine the sequence of the gene, and infer the amino acid sequence of the membrane protein.

The discovery of the Na^+-glucose cotransporter inspired the model for glucose absorption that is summarized in **Figure 43.15**. Notice that Na^+/K^+-ATPase (sodium-potassium pumps) in the basolateral membrane of the epithelial cells creates an electrochemical gradient that favors the entry of Na^+. Glucose from digested food enters the cell along with sodium via the cotransporter, then diffuses into nearby blood vessels through a glucose carrier in the basolateral membrane.

Experiment

Question: How is glucose transported into epithelial cells of the small intestine?

Hypothesis: Glucose enters epithelial cells along with sodium ions via a Na^+-glucose cotransporter protein.

Null hypothesis: Glucose transport does not depend on Na^+ transport.

Experimental setup:

mRNA mRNA mRNA mRNA

1. Purify mRNA from intestinal cells.

2. Separate mRNAs by size via gel electrophoresis.

3. Inject individual mRNAs into frog eggs. Test each egg—can it absorb Na^+ and glucose?

Prediction: An egg will be able to absorb Na^+ and glucose, because it received the mRNA that codes for the Na^+-glucose cotransporter.

Prediction of null hypothesis: None of the eggs will be able to absorb Na^+ and glucose.

Results:

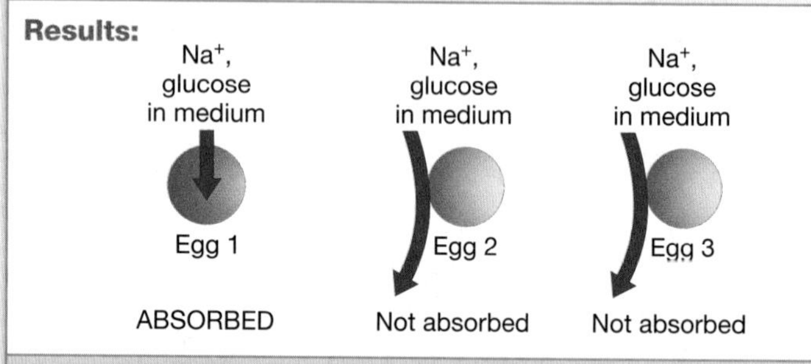

Conclusion: The egg that absorbs Na^+ and glucose received the mRNA from the Na^+-glucose cotransporter gene.

FIGURE 43.14 The Experimental Protocol for Locating the Na^+/Glucose Cotransporter Gene.

If the configuration of pumps, cotransporters, and channels looks familiar, there is a good reason: The same combination of membrane proteins occurs in the proximal tubule of the kidney, where the proteins are responsible for the reabsorption of sodium and glucose from urine (see Chapter 42).

Follow-up work showed that in the small intestine—just as in the proximal tubule—other cotransporters are responsible for the absorption of other nutrients, with specific channels in the basolateral membrane responsible for their transport to the blood.

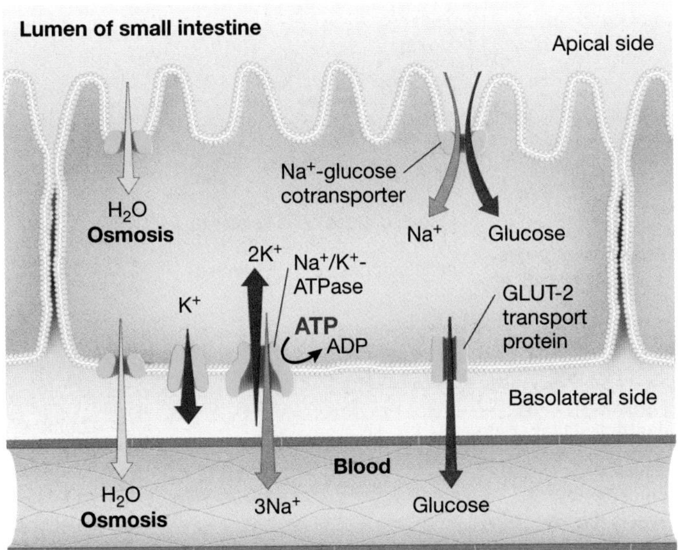

FIGURE 43.15 Glucose Is Actively Transported into Epithelial Cells. A summary of the current understanding of how glucose is absorbed in the small intestine. The GLUT-2 carrier protein facilitates diffusion of glucose across the basolateral membrane, into the bloodstream.

When solutes are brought into the epithelium via active transport, water follows passively by osmosis. This is an important mechanism to absorb water that has been ingested as well as reclaim liquid secreted into the digestive tract in the form of saliva, mucus, and pancreatic fluid. The mechanism of water absorption inspired an important medical strategy called oral rehydration therapy. If a patient has diarrhea, clinicians frequently prescribe dilute solutions of glucose and electrolytes to be taken orally. When the glucose in the drink is absorbed in the small intestine through the sodium-glucose cotransporter, enough water and sodium follow to prevent the life-threatening effects of dehydration. This simple medication saves thousands of lives every year. ● If you understand this concept, you should be able to predict two effects of a molecule that selectively blocks the sodium-glucose cotransporter.

MB Web Animation at www.masteringbio.com
The Digestion and Absorption of Food

The Large Intestine

By the time digested material reaches the large intestine, a large amount of water (approximately 5 liters per day in total) and virtually all of the available nutrients have been absorbed. The primary function of the **large intestine** is to compact the wastes that remain and absorb enough water to form feces. These processes occur in the colon—the main section of the structure. Feces are held in the **rectum**, which is the final part of the large intestine, until they can be excreted. Although the kidneys are responsible for maintaining water balance in the body, water absorption in the large intestine is important for individuals to remain well hydrated.

In an attempt to identify the mechanism of water absorption in the large intestine, researchers have recently focused on the membrane proteins called aquaporins, which were introduced in Chapter 42. Recall that **aquaporins** are water channels in plasma membranes that provide a mechanism for increasing the rate of water movement via osmosis. The best studied of these proteins, called AQP1 for aquaporin 1, is common in the nephrons of the kidney. It is one of the proteins responsible for water reabsorption from urine along the osmotic gradient described in Chapter 42.

To date, four distinct aquaporins have been found in the large intestines of rats, mice, and humans. The aquaporins called AQP3 and AQP4, for example, are located in the basolateral membrane of cells in the epithelium of the large intestine in mice. Researchers are working to unravel exactly how the activity of these aquaporins is regulated, however. To date, the molecular mechanisms responsible for water absorption in the large intestine are not as well understood as are those in the collecting duct of nephrons.

Check Your Understanding

If you understand that... 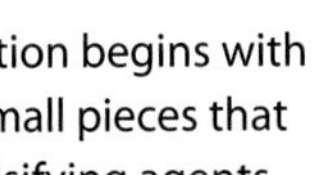

- In humans and many other animals, digestion begins with the mechanical breakdown of food into small pieces that are then acted on by acids, enzymes, emulsifying agents, and other chemical treatments.
- Nutrients are actively and selectively absorbed through specific membrane proteins in epithelial cells of the digestive tract.

You should be able to... 

1) Diagram the human digestive tract.
2) Explain how each compartment in the digestive system aids the digestion of food, absorption of nutrients, and excretion of wastes.
3) Predict the consequences of treating a person with drugs that inhibit the release of bile salts, that inactivate trypsin, or that block the action of Na^+/K^+-ATPase in the epithelial cells of the small intestine.

43.4 Nutritional Homeostasis—Glucose as a Case Study

When digestion is complete, amino acids, fatty acids, ions, and sugars enter the bloodstream and are delivered to the cells that need them. Too much of a nutrient, or too little, can be problematic or even fatal, however. A classic example is the illness called **diabetes mellitus**. People with diabetes experience abnormally high levels of glucose in their blood. Over the course of a lifetime, the chronic glucose imbalance associated with diabetes mellitus can lead to blindness, heart failure, and a reduction of circulation in the legs.

Why does diabetes occur? A key hint came to light in 1879, when researchers removed the pancreas from a dog and observed that diabetes developed. This experiment suggested that the pancreas secretes a compound involved in the removal of glucose from the blood. When other investigators cut up pancreatic tissues and injected extracts into diabetic dogs, however, it did not cure the disease. In fact, the investigators observed no response. Frustrated, researchers realized that digestive enzymes in the pancreas were probably destroying the active agent during the extraction process.

In 1921 Frederick Banting and Charles Best conducted a breakthrough experiment. They began by tying off a dog's pancreatic duct—the tube where digestive enzymes collect and flow to the small intestine. The logic was that blocking the secretion of digestive enzymes might kill the cells that synthesize them. The investigators waited several weeks for the cells near the duct to die and then removed the gland. They froze the pancreas tissue, ground it up, and injected an extract into a diabetic dog. To their delight, the dog's blood-sugar levels stabilized and the dog became more active and healthy looking. After Banting and Best repeated the experiment and observed the same result, they grew increasingly confident that they had located the source of the molecule responsible for controlling diabetes. The molecule came to be called insulin.

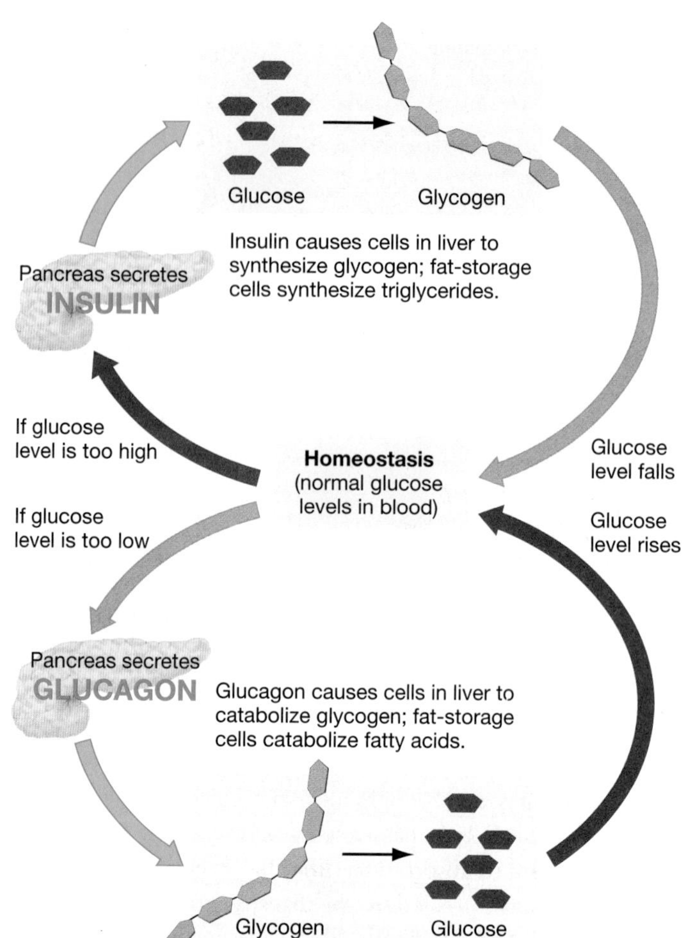

FIGURE 43.16 Insulin and Glucagon Provide Negative Feedback in a Homeostatic System. Both insulin and glucagon are secreted by cells in the pancreas but have opposite effects on blood-glucose concentrations.

EXERCISE Draw a large X through the arrow that is disrupted in individuals with diabetes mellitus. Next to the X, jot down notes indicating the nature of the defect in individuals with type I diabetes mellitus versus type II diabetes mellitus.

Insulin's Role in Homeostasis

Within a few years after Banting and Best's discovery, insulin had been isolated in pure form. Drug companies began purifying it from the pancreas of calves and other domestic animals in large enough quantities to administer to people suffering from diabetes mellitus. Later, insulin's mode of action was revealed. **Insulin** is a hormone that is produced in the pancreas when blood-glucose levels are high. It travels through the bloodstream and binds to receptors on cells throughout the body. In response, the cells increase their rate of glucose uptake and processing. For example, cells in the liver and skeletal muscle respond to insulin by synthesizing more glycogen from glucose monomers; cells that store lipids respond by synthesizing more storage forms of fat, using glucose as a precursor. The result is that glucose levels in the blood decline.

If blood-glucose levels fall, as they do after hard exercise or when food is lacking, cells in the pancreas secrete a hormone called **glucagon**. In response to glucagon, cells in the liver catabolize glycogen and produce glucose via **gluconeogenesis** (the synthesis of glucose from non-carbohydrate compounds). The result is that glucose levels in the blood rise. **Figure 43.16** summarizes how insulin and glucagon interact to form a negative feedback system capable of achieving homeostasis with respect to glucose concentrations in the blood.

Diabetes mellitus develops in people who do not synthesize insulin or who have defective versions of the receptor for insulin. The first condition is called type I diabetes mellitus, or insulin-dependent diabetes. The second condition is called type II diabetes mellitus, or non-insulin-dependent diabetes. Type I diabetes mellitus results from a lack of the hormone itself; type II diabetes mellitus results from defects in the hormone's receptor. In both cases, effector cells do not receive the signal that would result in a drop in blood-glucose levels.

Both types of diabetes mellitus lead to the production of large volumes of urine. Normally, signals from insulin keep blood-glucose levels low enough that all of the glucose can be reabsorbed from the pre-urine formed in the kidney. But when blood glucose levels are high, so much glucose enters the pre-urine that it cannot all be reabsorbed. High glucose concentrations in the pre-urine prevent water from being reabsorbed by osmosis, leading to high urine volume in diabetics.

Production of copious urine actually inspired the illness's name. Recall from Chapter 42 that diabetes means "to run through," and that the illness diabetes insipidus is also characterized by production of high urine volume. As that chapter pointed out, people with diabetes insipidus produce large amounts of urine because the collecting ducts of their kidneys fail to reabsorb water. *Insipidus* means "tasteless," while *mellitus* means "honeyed (sweet)." Before chemical methods of analyzing urine were available, physicians would distinguish between diabetes insipidus and diabetes mellitus by tasting the urine of their patients. This test was effective because a common symptom of diabetes mellitus is the excretion of glucose in urine just after a meal.

Currently, type I diabetes mellitus is treated with insulin injections and careful attention to diet; type II diabetes is managed primarily through prescribed diets and the monitoring of blood-glucose levels, as well as by the use of drugs that increase cellular responsiveness to insulin. For physicians and patients, the challenge is to achieve homeostasis with respect to blood glucose in the absence of the body's normal regulatory mechanisms. ● If you understand the difference between type I and type II diabetes mellitus, you should be able to draw the surface of a liver cell from patients with each disease and without disease, showing how insulin receptors in the plasma membrane interact with insulin and how they respond to high glucose concentrations in the blood.

The Type II Diabetes Mellitus Epidemic

The discovery of insulin qualifies as one of the great medical advances of the twentieth century, because it led to dramatic improvements in the quality of life for people who suffer from type I diabetes mellitus. Unfortunately, an epidemic of type II diabetes mellitus is currently under way in certain human populations. In the United States, about 6.6 percent of people aged 20 to 74 have type II diabetes. The frequency of this illness is much higher in African American, Latino, and Native American people, however, and the frequency of type II diabetes mellitus in teenagers of all ethnic backgrounds is increasing rapidly.

Because there is a strong association between the prevalence of type II diabetes mellitus in parents and children, researchers have concluded that some individuals have a genetic predisposition for developing the disease. But there is also strong evidence that environmental conditions have an important impact.

As an example of how environmental differences affect the incidence of type II diabetes mellitus, consider data from the Pima Indians of North America. The Pima consist of two main populations—one in southwest Arizona in the United States and one in a remote area of the Sierra Madre mountains of Mexico. Although researchers have been unable to find any significant genetic differences between the two groups, there are dramatic differences in the incidence of type II diabetes mellitus (**Figure 43.17a**). As in other populations around the world, the incidence of type II diabetes is correlated with obesity. Pima who live in the United States are much more likely than Pima who live in Mexico to have type

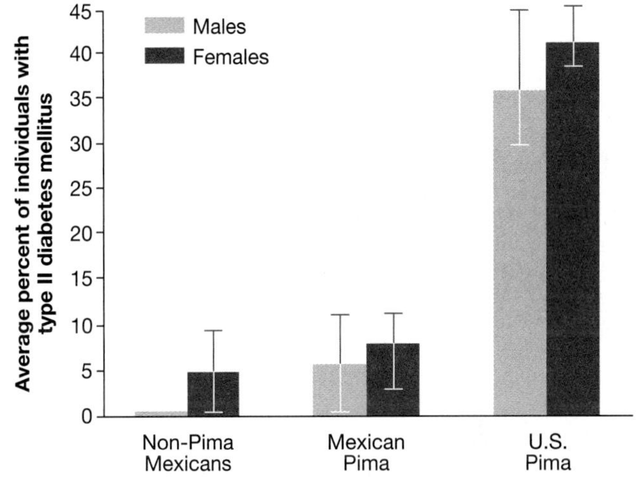

(b) Variation in average body mass index (kg/m^2)

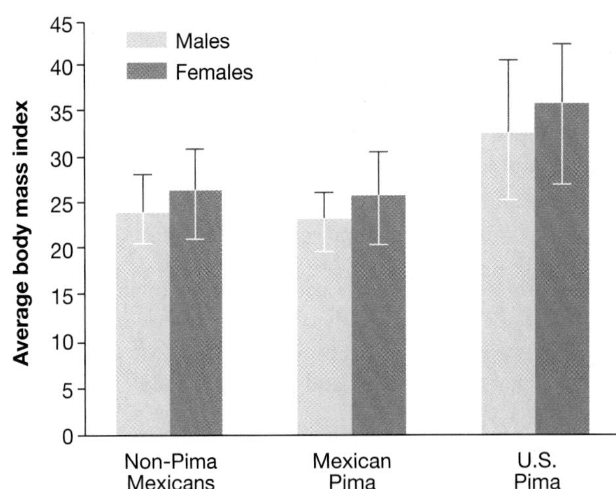

FIGURE 43.17 The Incidence of Type II Diabetes Mellitus Is Correlated with Obesity.

● **QUESTION** In many human populations, incidence of type II diabetes mellitus has been increasing rapidly—particularly among young people. Why?

II diabetes mellitus and to be obese. A person who has a body mass index greater than or equal to 30 is considered obese. The **body mass index** is calculated as weight (kg) divided by height (m) squared. Among the Pima, the differences in body mass index documented in **Figure 43.17b** are explained by differences in physical activity: Pima men in Mexico are physically active an average of 33 hours a week, while their counterparts in the United States are active just 12.1 hours per week. Pima women in Mexico average 22 hours of physical activity a week; Pima women in the United States average just 3.1.

Although Pima people in the United States have the highest rates of type II diabetes mellitus recorded to date, the same trends are occurring in populations all over the world. Nutrition-related diseases have become a major public health concern—especially in young people.

Check Your Understanding

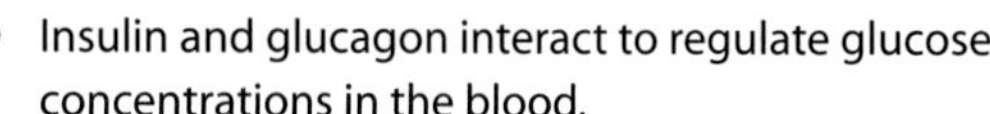

If you understand that...

- Insulin and glucagon interact to regulate glucose concentrations in the blood.
- Diabetes mellitus results from a lack of homeostasis in the concentration of the nutrient glucose in the blood.

You should be able to...

1) Explain the difference between diabetes insipidus, diabetes mellitus I, and diabetes mellitus II.
2) Predict what a person with type I diabetes mellitus should do when the blood-glucose level is too high and when the blood-glucose level is too low.

 Web Animation at www.masteringbio.com
Understanding Diabetes Mellitus

Chapter Review

SUMMARY OF KEY CONCEPTS

Animals require an array of nutrients to stay healthy, including specific amino acids, vitamins, and elements, as well as organic compounds that have high potential energy.

The diets of animals include fats and carbohydrates that provide energy, proteins that furnish amino acids, vitamins that serve as coenzymes and perform other functions, ions required for water balance and for nerve and muscle function, and elements that are incorporated into molecules synthesized by cells.

To determine the levels of nutrients that are needed for a particular animal species to sustain normal activities, researchers monitor the relationship between nutrient intake, the levels of nutrients maintained in the body, and health. To determine how nutrition can affect athletic performance in humans, researchers alter the intake of specific nutrients experimentally and assess how athletes respond. For example, investigators have shown that athletes who ingest a high-carbohydrate diet for three days outperform individuals who ingest a normal diet or a high-fat, high-protein diet.

You should be able to design an experiment to document the effects of magnesium deprivation in laboratory mice.

Animals feed on a wide variety of items. In animals that ingest their food, there is usually a close correspondence between the structure of the animal's mouthparts and the function of those mouthparts in capturing and processing food.

All animals are heterotrophs. Some species are suspension feeders, meaning they filter small food items from water using cilia or other structures. Others absorb their nutrition directly across their body wall. But most animals get food by seizing and manipulating it with mouthparts such as teeth, jaws, or beaks or with special toxin-injecting organs. Natural selection has modified mouthparts in different species to act as efficient tools for obtaining particular types of food.

You should be able to predict the types of mouthparts and nature of the digestive tract found in an insect and a mammal that eat only nectar from flowers.

Digestion occurs in the digestive tract, which is compartmentalized into organs that have specialized functions in the digestion of food, absorption of nutrients and water, or excretion of wastes.

Once food is ingested, it is processed in the digestive tract, which begins at the mouth and ends at the anus. Food processing frequently begins with tearing or chewing by a beak, teeth, or other mouthparts. In many species, chemical digestion of food also begins in the mouth. For example, the salivary enzyme α-amylase begins to hydrolyze the bonds linking glucose monomers in starch, glycogen, and other carbohydrates. Once food is swallowed, it passes down the esophagus via peristalsis.

Digestion continues in the stomach. In humans, the stomach is a highly acidic environment that denatures proteins and in which the enzyme pepsin begins the cleavage of peptide bonds that link amino acids.

Food passes from the stomach into the small intestine, where it is mixed with secretions from the pancreas and liver. Carbohydrate digestion is continued by pancreatic amylase; fats are emulsified by bile salts and digested by lipase; protein digestion is completed by a suite of pancreatic enzymes that are activated by enterokinase. Secretions from the liver and pancreas are triggered by the hormones cholecystokinin and secretin, which are produced in the small intestine.

Cells that line the small intestine absorb the nutrients released by digestion through transporters or cotransporters in their membranes. In many cases, uptake is driven by an electrochemical gradient established by Na^+/K^+-ATPase that favors a flow of Na^+ into the cell. As solutes leave the lumen of the intestine and enter cells, water follows by osmosis. Water reabsorption is completed in the large intestine, where feces form.

You should be able to explain why gastric bypass surgery, which makes the stomach smaller and allows food to bypass part of the small intestine, often leads to weight loss.

 Web Animation at www.masteringbio.com
The Digestion and Absorption of Food

Lack of homeostasis with respect to nutrients such as glucose and cholesterol can cause disease.

Maintaining homeostasis with respect to nutrients is critical to health. Diabetes mellitus develops when concentrations of glucose in the blood are chronically too high. This illness is caused by a defect in the production of insulin—a hormone secreted by the pancreas that promotes the uptake of glucose from the blood—or by a defect in the insulin receptor on the surface of cells. The development of type II diabetes is correlated with obesity and is reaching epidemic proportions in some populations.

You should be able to explain what happens to blood glucose concentration when a person with and without diabetes mellitus eats a candy bar.

Web Animation at www.masteringbio.com
Understanding Diabetes Mellitus

QUESTIONS

Test Your Knowledge

1. What does secretin stimulate?
   a. secretion of HCO_3^- from the pancreas
   b. secretion of HCl from the stomach epithelium
   c. secretion of digestive enzymes from the pancreas
   d. uptake of glucose from the bloodstream by cells throughout the body
2. Why does carbohydrate loading increase the performance of distance athletes?
   a. It increases glycogen stores in the muscle and liver.
   b. It increases fat storage.
   c. It increases the blood's capacity to deliver oxygen to tissues.
   d. It increases fatty-acid levels in the blood.
3. Why are the pharyngeal jaws of cichlid fish unique?
   a. No other group of fish has pharyngeal jaws.
   b. They are made of bone instead of cartilage, so they are harder and more durable.
   c. They function only in transporting food from the front of the mouth to the throat.
   d. They articulate with the braincase and have extra muscles attached, so they are capable of biting.
4. How are carbohydrates digested?
   a. by lipases in the small intestine
   b. by pepsin and HCl in the stomach
   c. by aquaporins in the large intestine
   d. by amylases in the mouth and small intestine
5. What role do bile salts play in the digestion of complex fats?
   a. They catalyze the cleavage of bonds leading to the release of fatty acids and other small lipids.
   b. They emulsify lipids, meaning that large masses of fat molecules are broken into smaller masses.
   c. They include fatty-acid binding proteins, which are involved in fat absorption.
   d. They activate the enzymes that are responsible for digesting fats.
6. How is water absorbed in the small intestine?
   a. through aquaporins
   b. The exact mechanism is not known.
   c. by sodium cotransporters
   d. by osmosis (following solutes)

Test Your Knowledge answers: 1. a; 2. a; 3. d; 4. d; 5. b; 6. d

Test Your Understanding

Answers are available at www.masteringbio.com

1. Describe a human disease that results from a nutritional deficiency. How can researchers establish nutritional requirements for humans, if it is unethical to do experiments that would reduce the levels of specific nutrients?
2. Why is it logical that digestive enzymes are produced in an inactive form and then activated in the lumen of the digestive tract?
3. Do you accept the conclusion that cichlid pharyngeal jaws are adaptations that increase feeding efficiency? Why or why not?
4. How was it established that gastric juice is acidic and that the stomach's primary function is the digestion of proteins?
5. What features are responsible for the large surface area of the small intestine? Why is it important that this organ have such a large surface area?
6. Diabetes mellitus results from the disruption of mechanisms for maintaining homeostasis in blood-glucose concentrations. Explain this statement. Make a diagram explaining how insulin and glucagon provide negative feedback in a homeostatic system.

Applying Concepts to New Situations

Answers are available at www.masteringbio.com

1. Predict how the nutritional requirements of female mammals change during pregnancy and breastfeeding. Design studies that would test your predictions, using humans as a study subject and using laboratory mice as a study subject.
2. Predict the physical symptoms that would result from defects in each of the following molecules: pancreatic amylase, pepsin, Na^+-glucose cotransporter, fatty-acid binding protein, and aquaporin.
3. Scientists who backed the hypothesis that secretion from the pancreas is under nervous control strenuously objected to the experiment that led to the discovery of hormonal control. They claimed that the experiment was inconclusive because it was very likely that not all of the nerves innervating the small intestine had been cut. The biologists who did the experiment replied that even if not all the nerves had been cut, the result was still valid. In your opinion, who is correct? Why?
4. Among vertebrates, the large intestine occurs only in lineages that are primarily terrestrial (amphibians, reptiles, and mammals). Propose a hypothesis to explain this observation.

www.masteringbio.com is also your resource for • Answers to text, table, and figure caption questions and exercises • Answers to *Check Your Understanding* boxes • Online study guides and quizzes • Additional study tools including the *E-Book for Biological Science* 3rd ed., textbook art, animations, and videos.

44 Gas Exchange and Circulation

KEY CONCEPTS

- Animals have to take in oxygen and expel carbon dioxide to sustain cellular respiration and stay alive. Terrestrial animals and aquatic animals face different challenges in performing gas exchange.
- The organs responsible for gas exchange maximize the rate of oxygen and carbon dioxide diffusion by presenting a large, thin surface area to the environment. These organs also function in a way that maintains a steep partial-pressure gradient that favors entry of oxygen and elimination of carbon dioxide.
- Blood is a specialized tissue that transports gases, along with nutrients and wastes, in some animals. Hemoglobin is an oxygen-carrying protein with characteristics that make it extremely efficient at taking up oxygen in the lungs and delivering it to tissues.
- Circulatory systems use the pressure generated by one or more hearts to transport blood and other substances throughout the body.

During intense exercise, animal circulatory systems must deliver large amounts of oxygen to tissues and remove large amounts of carbon dioxide. This chapter explores how gas exchange occurs in animals that live in aquatic or terrestrial environments.

Animal cells are like factories that run 24 hours a day. Inside the plasma membrane, the chemical reactions that sustain life produce a steady stream of wastes. Those reactions also require a steady input of raw materials. Chapter 42 analyzed how waste materials are removed from the body and excreted; Chapter 43 examined how nutrients enter the body. This chapter focuses on two major questions: (1) How are two of the most important molecules in the economy of the cell—the oxygen (O_2) required for cellular respiration and the carbon dioxide (CO_2) produced by cellular respiration—exchanged with the environment? and (2) How are these gases—along with wastes, nutrients, and other types of molecules—transported throughout the body?

Understanding gas exchange and circulation is fundamental to understanding how animals work. If either process fails, the consequences are dire. In human populations, for example, anemia, lung cancer, tuberculosis, malaria, heart disease, stroke, and other diseases caused by faulty gas exchange and circulation are leading causes of illness and death. Let's begin with an overview of animal respiratory and circulatory systems, then plunge into the details of how gases are exchanged and transported.

Key Concept Important Information Practice It

44.1 The Respiratory and Circulatory Systems

When the mitochondria inside animal cells are producing ATP via cellular respiration, they consume oxygen and produce carbon dioxide. To support continued ATP production, cells have to obtain oxygen and expel excess carbon dioxide continuously.

How does gas exchange occur between an animal's environment and its mitochondria? In most cases, gas exchange involves the four steps illustrated in **Figure 44.1**: (1) Ventilation occurs when air or water moves through a specialized gas-exchange organ, such as lungs or gills; (2) gas exchange takes place as O_2 and CO_2 diffuse between air or water and the blood at the ventilatory surface; (3) through circulation, the dissolved O_2 and CO_2 are transported throughout the body—along with nutrients, wastes, and other types of molecules—via the circulatory system; and (4) in tissues, where cellular respiration has led to low O_2 levels and high CO_2 levels, O_2 and CO_2 diffuse between blood and cells.

Steps 1 and 2 are accomplished by the **respiratory system**: the collection of cells, tissues, and organs responsible for gas exchange between the individual and its environment. In essence, a respiratory system consists of structures for conducting air or water to a surface where gas exchange takes place. In some animals the gas exchange surface is the skin, but in most species it is located in a specialized organ like the lungs of tetrapods, the tracheae of insects, or the gills found in molluscs, arthropods, and many types of fish. **Figure 44.2** provides an example: the respiratory system of humans. The structure and function of gills, trachaea, and lungs are analyzed in more detail in section 44.3.

An animal's **circulatory system** is responsible for moving O_2, CO_2, and other materials around the body. Not all animals have a circulatory system, however. Some animals are so small or have flattened bodies with so much surface area that materials can be exchanged directly between the interior and

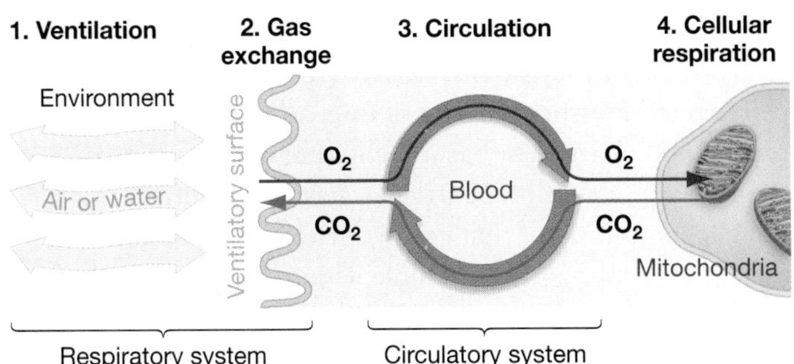

FIGURE 44.1 Gas Exchange Involves Ventilation, Circulation, and Respiration. In most animals, oxygen and carbon dioxide are exchanged across the surface of a lung, a gill, the skin, or some other gas exchange organ. These gases are transported to and from cells—where gas exchange again takes place—in blood.

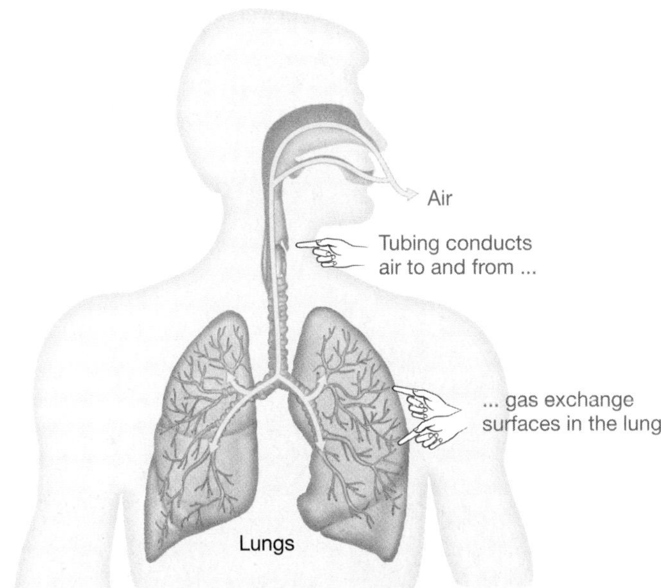

FIGURE 44.2 The Human Respiratory System. A respiratory system consists of structures that conduct air or water to a gas exchange surface.

the environment. In most cases, however, a specialized, liquid transport tissue is propelled throughout the body by a muscular heart, through a system of vessels.

Circulatory systems are classified as open or closed. In a **closed circulatory system**, the system of vessels is continuous (**Figure 44.3a**). In an **open circulatory system**, however, at least some vessels open into portions of the body cavity (**Figure 44.3b**). The transport tissue inside continuous systems is called **blood**; the transport tissue in open systems is called **hemolymph**. The distinction is necessary because animals with closed circulation usually have two subsystems: a system of blood vessels that carries blood, and a system of lymph vessels filled with lymph. **Lymph** consists of fluid that flows into lymph vessels from the spaces surrounding cells; it also contains specialized disease-fighting cells. The two circulatory subsystems are connected because lymph vessels eventually empty into blood vessels. Blood and lymph are not separated in open circulatory systems, so hemolymph is aptly named (the Greek root *heme* refers to blood).

Given this broad overview of gas exchange and circulatory systems, let's plunge into the details of how they work—starting with the question of how oxygen and carbon dioxide move between an animal's body and its environment.

44.2 Air and Water as Respiratory Media

Gas exchange between the environment and cells is based on diffusion. Under normal conditions, oxygen concentrations are relatively high in the environment and low in tissues, while carbon dioxide levels are relatively high in tissues and low in the

(a) Closed system: Blood never leaves vessels.

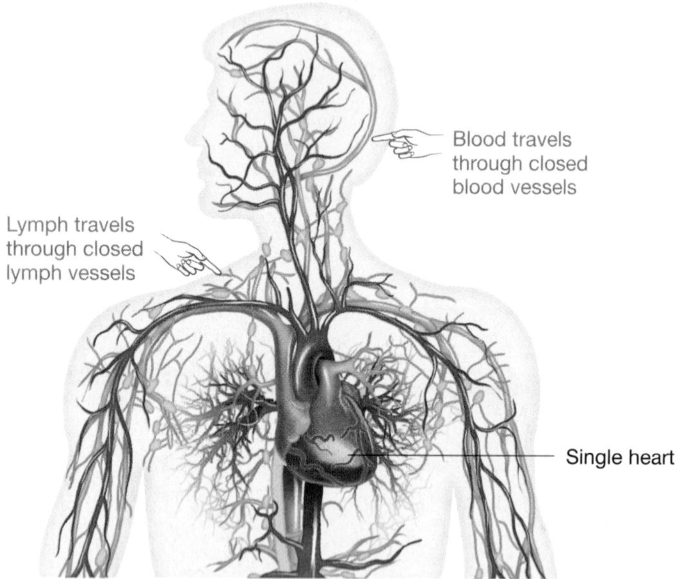

(b) Open system: Hemolymph leaves vessels and comes into direct contact with tissues.

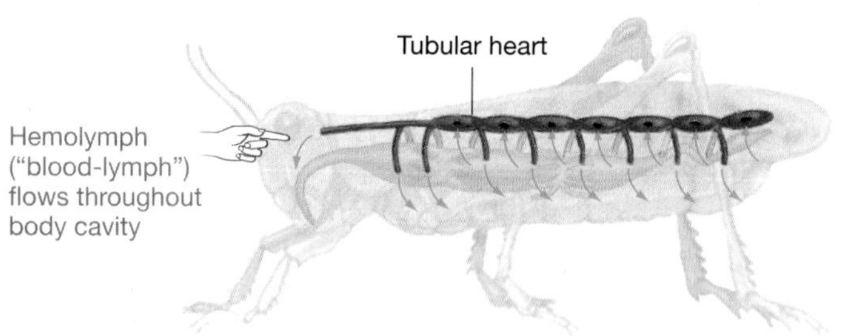

FIGURE 44.3 Closed and Open Circulatory Systems. The arrows in these figures indicate the direction of blood flow.

(a) What gases make up the atmosphere?

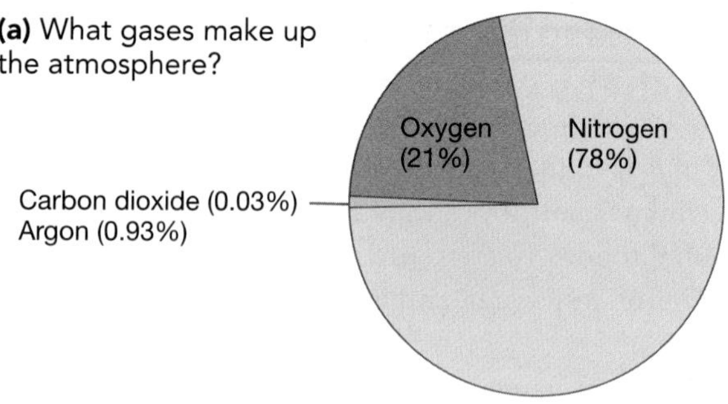

(b) The partial pressure of oxygen falls with increasing elevation.

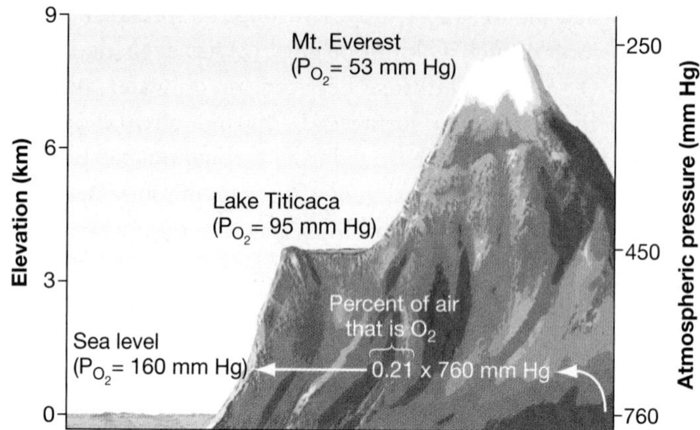

FIGURE 44.4 Oxygen Makes Up 21 Percent of the Atmosphere, but Its Partial Pressure Varies Widely. (a) Earth's atmosphere is dominated by nitrogen and oxygen. Oxygen makes up 21 percent of the atmosphere at all elevations. **(b)** Atmospheric pressure, and thus oxygen partial pressure (P_{O_2}), falls with increasing elevation.

● **EXERCISE** Show how the partial pressure of oxygen at the peak of Mt. Everest was calculated.

environment. Thus, oxygen tends to move from the environment into tissues, and carbon dioxide tends to move from tissues into the environment.

To get a more precise understanding of how gas exchange occurs by means of diffusion, it's important to consider some fundamental questions about how diffusion occurs in both terrestrial and aquatic environments. How much oxygen and carbon dioxide are present in the atmosphere versus in the ocean? What factors influence how quickly these gases move by diffusion?

How Do Oxygen and Carbon Dioxide Behave in Air?

As **Figure 44.4a** shows, the atmosphere is composed primarily of nitrogen (N_2) and oxygen, with trace amounts of argon and CO_2. Nitrogen and argon are not important to animals living at sea level and are usually ignored in analyses of gas exchange.

The data in Figure 44.4a are actually slightly misleading. To understand why, consider that the percentage of O_2 in the atmosphere does not vary with elevation. The atmosphere at the top of Mt. Everest is composed of 21 percent O_2, just as it is at sea level. The key difference is that far fewer molecules of oxygen and other atmospheric gases are present at high elevations than at sea level. Air at the top of Mt. Everest is much less dense than air at sea level, so much less oxygen is present.

To understand how gases move by diffusion, it is important to express their presence in terms of partial pressures instead of percentages. Pressure is a type of force. A **partial pressure** is the pressure of a particular gas in a mixture of gases. To calculate the partial pressure of a particular gas, you multiply the fractional composition of that gas by the total pressure exerted by the entire mixture.* For example, **Figure 44.4b** shows that the total atmospheric pressure at sea level is 760 mm Hg (millimeters of mercury). If you multiply this value by 0.21, which is

*This calculation is valid because the total pressure in a mixture of gases is the sum of the partial pressures of all the individual gases. This relationship is called Dalton's law.

the fraction of air that is O_2, you obtain a partial pressure of oxygen, abbreviated P_{O_2}, at sea level of 160 mm Hg. Because the atmospheric pressure is only about 250 mm Hg at the top of Mt. Everest, the P_{O_2} there is only $0.21 \times 250 = 53$ mm Hg.

Oxygen and carbon dioxide diffuse between the environment and cells along their respective partial-pressure gradients, just as solutes diffuse along their concentration gradients. In both air and water, O_2 and CO_2 move from regions of high partial pressure to regions of low partial pressure. It is hard to breathe at the top of Mt. Everest because the diffusion gradient between the atmosphere and your lung tissues is not large there.

How Do Oxygen and Carbon Dioxide Behave in Water?

To obtain oxygen, water breathers face a much more challenging environment than air breathers do. Aquatic animals live in an environment that contains much less oxygen than the environments inhabited by terrestrial animals. At 15°C, a liter of air can contain up to 209 ml of O_2, while a liter of water may contain up to a maximum of only 7 ml of O_2. To extract a given amount of oxygen, an aquatic animal has to process 30 times more water than the amount of air a terrestrial animal breathes. In addition, water is about a thousand times denser than air and flows much less easily. As a result, water breathers have to expend much more energy to ventilate their respiratory surfaces than do air breathers.

The partial pressure of oxygen varies in different types of aquatic habitats, just as it varies with altitude on land. Oxygen and carbon dioxide diffuse into water from the atmosphere, but the amount of gas that dissolves depends on several factors:

- *The solubility of the gas in water* Oxygen has very low solubility in water. Only 0.003 ml of oxygen dissolves in 100 ml of water for each 1-mm-Hg increase in oxygen partial pressure. Because of this low solubility, blood contains a molecule that binds to oxygen and delivers it to tissues. Without this carrier molecule, the rate of blood flow to tissues would have to increase dramatically in order to meet oxygen demand.

- *The temperature of the water* As the temperature of water increases, the amount of gas that dissolves in it decreases. Other things being equal, warm-water habitats have much less oxygen available than cold-water habitats do. For a fish, breathing in warm water is comparable to a land-dwelling animal breathing at high elevation.

- *The presence of other solutes* Because seawater has a much higher concentration of solutes than does freshwater, seawater can hold less dissolved gas. At 10°C, up to 8.02 ml of O_2 can be present per liter of freshwater versus only 6.35 ml of O_2 per liter of seawater. As a result, freshwater habitats tend to be more oxygen rich than marine environments are.

- *The partial pressure of the gas in contact with the water* Increasing the external partial pressure increases the rate at which a gas diffuses into a liquid. But if the partial pressure in the liquid phase exceeds that in the gas phase, the gas will "boil" out of the liquid. This is what happens when the cap is removed from a bottle of carbonated beverage. The drop in partial pressure in the gas phase that occurs when the bottle is opened allows carbon dioxide to escape the liquid phase.

Other important aspects of oxygen availability in water include the surface area of the water, the production of oxygen by photosynthesis, and the depletion of oxygen by organisms using it in cellular respiration. Surface area is important because it affects the ability of oxygen to diffuse into water. For example, shallow ponds and streams tend to be much better oxygenated than are deep bodies of water, because shallower bodies have a higher ratio of surface area to volume. Rapids, waterfalls, and other types of whitewater tend to be the most highly oxygenated of all aquatic environments, because a large surface area is exposed to the atmosphere as water splashes over rocks and logs and because air bubbles are incorporated into the water. In contrast, oxygen content tends to be extremely low in bogs and other stagnant-water habitats. The small amounts of oxygen that diffuse into stagnant water habitats are quickly used up by decomposers that use oxygen as an electron acceptor in cellular respiration.

Now let's consider how ventilatory organs are designed. How do the gills of fish, the tracheae of insects, and the lungs of mammals cope with the differences between air and water?

Check Your Understanding

If you understand that...

- O_2 and CO_2 move from regions of high partial pressure to regions of low partial pressure.
- Water breathing is much more difficult than air breathing, in part because the partial pressure of oxygen in water is much lower than its partial pressure in air.

You should be able to ...

1) Describe aquatic habitats where oxygen partial pressures are relatively high and relatively low.

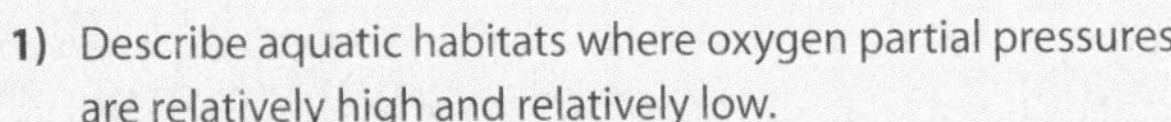

2) Explain whether a large, medium, or small amount of air should be bubbled into aquaria containing warm-water species, cold-water species, vigorous algal growth, highly active animals, or sedentary animals.

44.3 Organs of Gas Exchange

Many small animals do not have specialized gas exchange organs, such as gills or lungs. Instead, they obtain O_2 and eliminate CO_2 by diffusion directly across the body surface. For these animals, diffusion across the body surface is rapid enough to fulfill their requirements for taking in O_2 and expelling CO_2. Most of these animals are restricted to living in wet environments, however. Their skin needs to be extremely thin for diffusion to take place efficiently, but thin skin is prone to water loss. Living in wet environments allows these animals to exchange gases across their outer surface while avoiding dehydration.

In contrast, animals that are large or that live in dry habitats need some sort of specialized respiratory organ. To understand why the presence of gills or lungs is associated with large size, recall from Chapter 41 that the volume of an animal's body increases much faster than does its surface area. Respiratory organs provide a greater surface area for gas exchange—enough to meet the demands of a large body filled with cells.

Biologists have long marveled at the efficiency of gills and lungs. To appreciate why, let's first examine the physical factors that control diffusion rates and then look at the structure and function of these respiratory organs.

Design Parameters: The Law of Diffusion

In 1855 Adolf Fick derived an equation regarding diffusion, based on the results of experiments he had performed on the rate at which gases diffuse. As **Figure 44.5** shows, **Fick's law of diffusion** states that the rate of diffusion of a gas depends on five parameters: the solubility of the gas in the aqueous film lining the gas-exchange surface, the temperature, the surface area available for diffusion, the difference in partial pressures of the gas across the gas-exchange surface, and the thickness of the barrier to diffusion.

Fick's law provides a compact way of identifying traits that allow animals to maximize the rate at which oxygen and carbon dioxide diffuse across the surfaces of their gills or lungs.

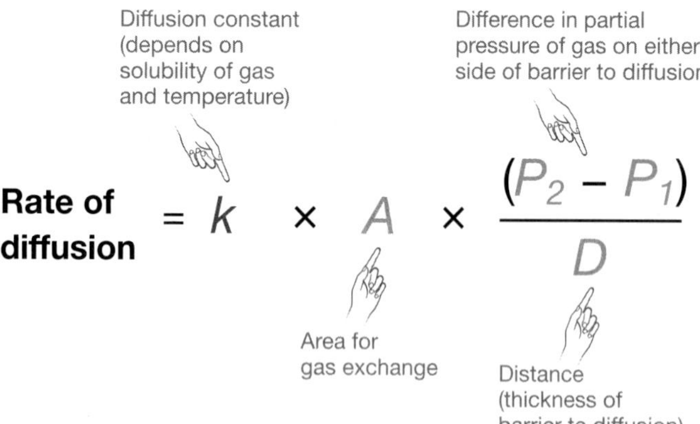

FIGURE 44.5 Fick's Law of Diffusion.

QUESTION Suppose you were going to design a system for the rapid diffusion of oxygen across a phospholipid bilayer. According to Fick's law, which aspects of the system should be maximized in order to maximize the diffusion rate? Which should be minimized?

Specifically, the law states that all gases, including O_2 and CO_2, diffuse in the largest amounts when three conditions are met:

1. A is large, meaning a large surface area is available for gas exchange. Based on Fick's law, it is not surprising that the respiratory surface in the human lungs would cover about 100 m^2—or almost a quarter of a basketball court—if the epithelium were spread flat.
2. D is small, meaning the respiratory surface is extremely thin. To appreciate just how thin the gas-exchange surface is, consider that researchers who asked bicyclists to ride up a steep hill for seven minutes observed a marked increase in blood on the respiratory surface of the athletes' lungs. To explain this observation, the researchers proposed that the cyclists' high heart rates (up to 177 beats/minute during their sprint up the hill) increased blood pressure to the point at which thin-walled vessels in the lungs ruptured and leaked blood into the structure.
3. $P_2 - P_1$ is large, meaning the partial-pressure gradient of the gas across the surface is large. High partial-pressure gradients are maintained in part by having an efficient circulatory system in close contact with the gas-exchange surface. When blood flows close to the respiratory surface, oxygen is rapidly taken away from the area where inward diffusion is occurring, and carbon dioxide is rapidly brought into the area where outward diffusion is occurring. As a result, $P_2 - P_1$ stays high.

What other aspects of gill and lung design conform to the dictates of Fick's law? What other factors affect the structure of these respiratory organs? To answer these questions, let's delve into the anatomy of gills and lungs.

How Do Gills Work?

Gills are outgrowths of the body surface or throat that are used for gas exchange in aquatic animals. To appreciate the structure and function of these organs, it is important to recognize just how challenging water is as a respiratory medium compared with air. Recall that water contains much less oxygen than air does and that water is much more difficult to move across respiratory surfaces because of its density and viscosity.

Gills are efficient solutions to the problems posed by water breathing, primarily because they present an extremely large surface area for oxygen to diffuse across an extremely thin epithelium. In some species of invertebrates, such as the nudibranch mollusc pictured in **Figure 44.6a**, gills project from the body surface and contact the surrounding water directly. In other invertebrate species, such as the crayfish in **Figure 44.6b**, gills are located inside the exoskeleton or body wall. If gills are internal, water must be driven over them by cilia, the limbs, or other specialized structures.

(a) External gills are in direct contact with water.

(b) Internal gills must have water brought to them.

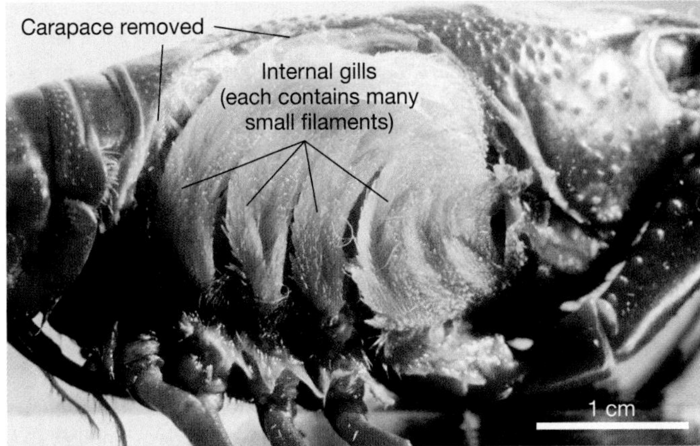

FIGURE 44.6 Gills Can Be External or Internal. (a) Nudibranchs are marine snails with gills that are outside the main body wall. **(b)** Crayfish gills are located inside the main body wall. A portion of the crayfish exoskeleton has been removed to expose the gills.

QUESTION Regarding fitness, what are the advantages and disadvantages of external versus internal gills? Why is it logical to expect that the orange-colored gills in part (a) are protected by toxins?

In contrast to the diversity of gills found in invertebrates, the gills of bony fishes are all very similar in structure. Fish gills are located on both sides of the head and in teleosts (see Chapter 34) consist of four arches, as **Figure 44.7** shows. To move water through these structures, most fish open and close their mouths and a stiff flap of tissue, called the **operculum**, that covers the gills. The pumping action of the mouth and operculum creates a pressure gradient that moves water over the gills. But tuna and other fish that are particularly fast swimmers force water through their gills by swimming with their mouths open. This process is called *ram ventilation*.

Regardless of how fish gills are ventilated, the flow of water over gills is unidirectional. To see why, note that long, thin structures called **gill filaments** extend from each gill arch. Each gill filament is composed of hundreds or thousands of **gill lamellae**. These are sheetlike structures, shown in detail at the bottom of Figure 44.7, through which a bed of small blood vessels called capillaries runs. The key point is that the flow of blood through the capillary bed is in the opposite direction to the flow of water over the gill surface. As a result, a **countercurrent exchanger** exists. Recall from Chapter 41 that countercurrent exchangers are based on two adjacent fluids flowing in opposite directions. The arrangement is important because in a co-current system, in which two adjacent fluids flowing through pipes in the same direction, oxygen transfer would stop. That is because the partial-pressure gradient driving diffusion would fall to zero partway along the length of the capillary **(Figure 44.8)**.

Countercurrent flow makes fish gills extremely efficient at extracting oxygen from water, because it ensures that the difference in the partial pressure of oxygen and carbon dioxide in water versus blood is large over the entire gas-exchange surface. If you understand this concept, you should be able to calculate how much more oxygen is extracted from water via countercurrent versus co-current exchange, using the data in Figure 44.8. Stated another way, the effect of countercurrent exchange is to maximize the $P_2 - P_1$ term in Fick's law of diffu-

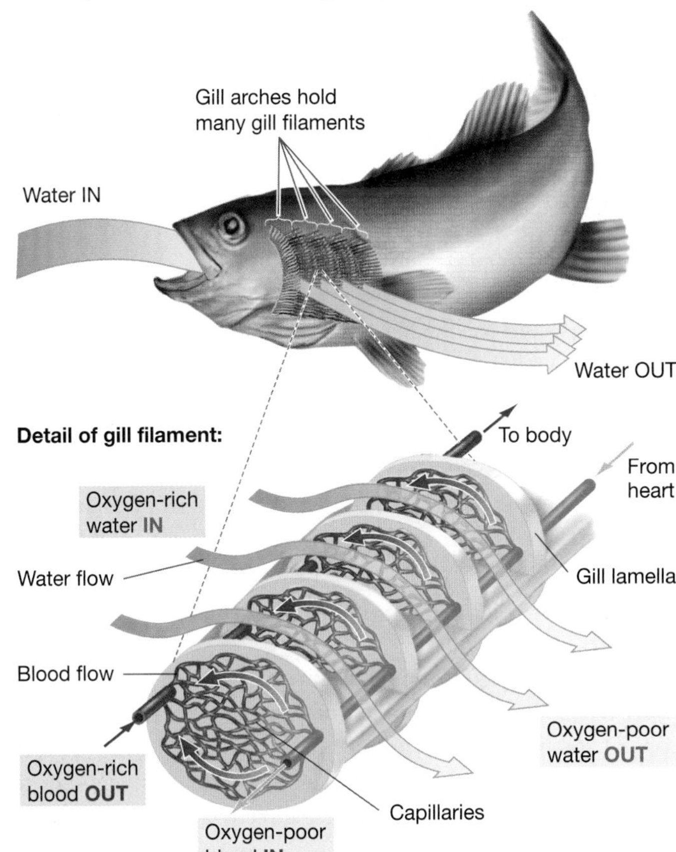

FIGURE 44.7 Fish Gills Are a Countercurrent Exchange System. In fish gills, water and blood flow in opposite directions. (Note that the operculum has been removed in the top drawing to reveal the gills.)

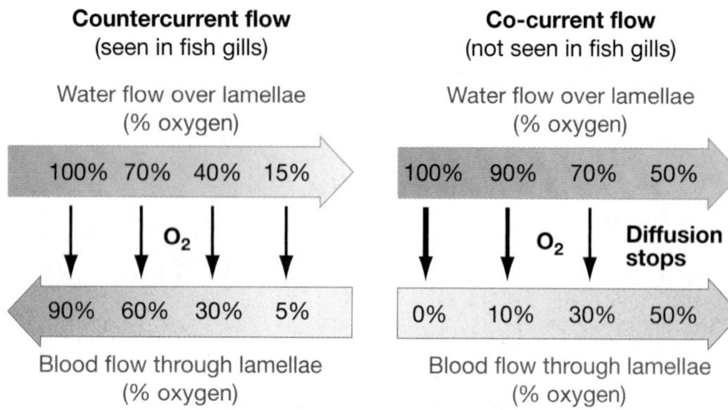

FIGURE 44.8 Countercurrent Exchange Is Much More Efficient than Co-current Exchange. In the countercurrent system of fish gills, oxygen is transferred along the entire length of the capillaries. If co-current flow occurred, oxygen transfer would stop, because the partial-pressure gradient driving diffusion would fall to zero partway along the length of the capillary.

sion over the entire gill surface. Based on this observation, biologists cite countercurrent exchange as another example of how gills are optimized for efficient gas exchange.

How Do Tracheae Work?

Air and water are dramatically different ventilatory media, because they have different densities, viscosities, and abilities to hold oxygen and carbon dioxide. In addition, the consequences of exposing the gas-exchange surface to air versus water differ. In aquatic habitats, ventilation tends to disrupt water and electrolyte balance, and homeostasis must be maintained by an active osmoregulatory system. As explained in Chapter 42, osmosis causes marine animals to lose water across their gas exchange surface and freshwater animals to gain water. Diffusion tends to cause marine animals to gain sodium, chloride, and other ions, and freshwater animals to lose them. In contrast, breathing leads to a loss of water by evaporation in terrestrial environments.

How do terrestrial animals minimize water loss while maximizing the efficiency of gas exchange? To answer this question, consider the tracheal system of insects. Recall from Chapter 42 that **tracheae** are an extensive system of tubes located well within the body. They connect to the exterior through openings called **spiracles**, which can be closed to minimize the loss of water by evaporation.

A key point about the insect tracheal system is that it consists of a series of tubes extending throughout the body. The tips of tracheae are tiny, fluid filled, and highly branched. In this way, the tracheal system transports air close enough to cells for gas exchange to take place directly across their plasma membranes. Consequently, insects do not use their circulatory system to transport gases to and from tissues, even though the insect circulatory system transports nutrients and wastes.

Recent research on the tracheal system has focused on how air moves into and out of the tubes through spiracles. Is simple diffusion efficient enough to ventilate the system, or is some type of breathing mechanism involved?

The short answer to this question is that breathing movements may play a role in at least some insects. To understand why this claim is valid, consider recent work on the sweet potato hawkmoth. A researcher set out to investigate how the amount of O_2 delivered to muscles changes during flight. To document the partial pressure of oxygen in flight muscles, he inserted a needlelike electrode into the wing muscles of a hawkmoth. The electrode was attached to an instrument that measured the partial pressure of oxygen, and the hawkmoth was tethered to a stand (**Figure 44.9**).

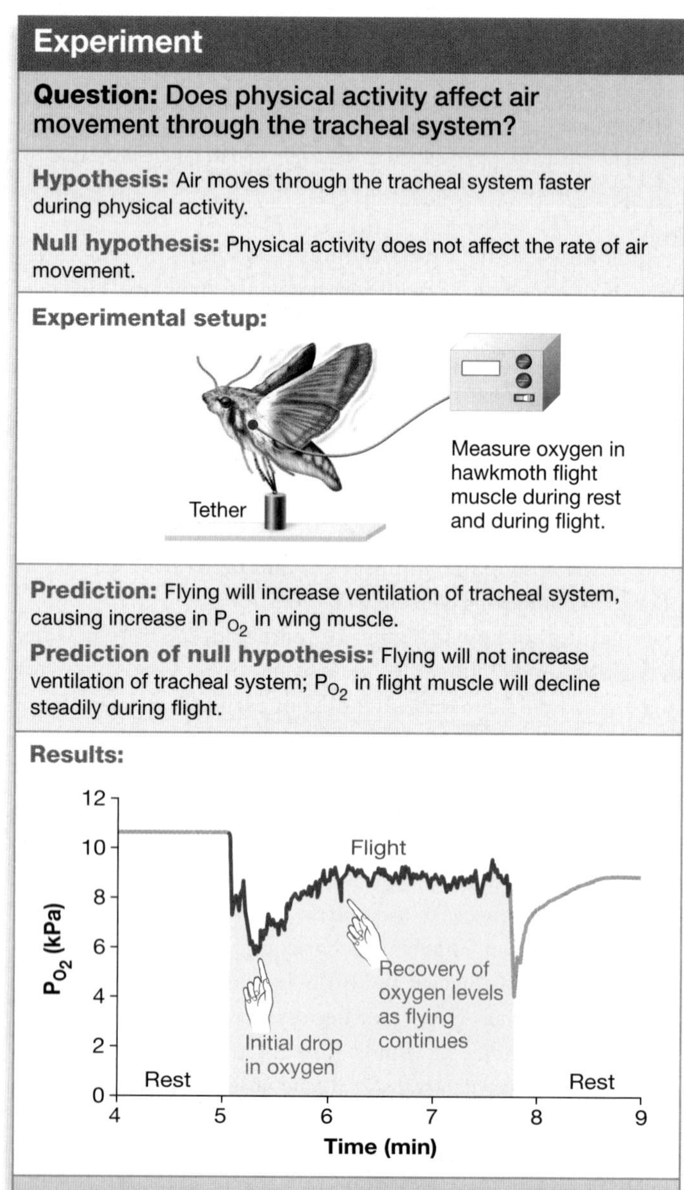

FIGURE 44.9 Evidence That the Trachael System Is Ventilated During Movement. On the graph's y-axis, kPa stands for a unit of pressure—or force per unit area—called the kilopascal.

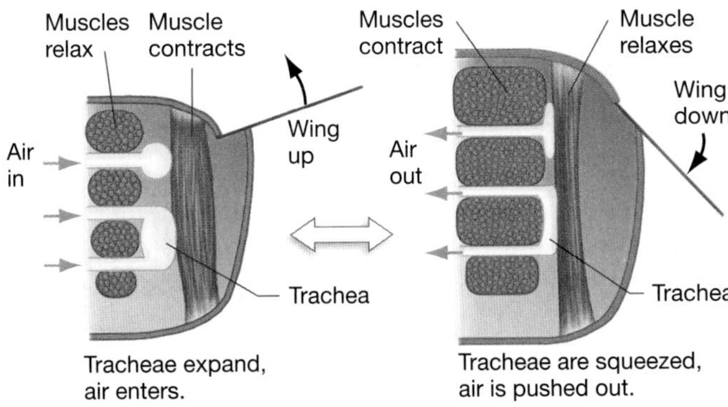

FIGURE 44.10 Muscle Activity May Ventilate Tracheae. To explain the result in Figure 44.9, biologists propose that tracheae change shape as wings beat.

When the hawkmoth recovered from the procedure and began to beat its wings normally, the biologist began recording P_{O_2} as the insect rested. Then the researcher stimulated the moth to fly by exposing it to wind. The "Results" section of Figure 44.9 shows how P_{O_2} changed during one such experiment. (The procedure was repeated on several individuals, and the same pattern was observed.) Because flight is an energetically demanding activity, it is not surprising that P_{O_2} levels in the flight muscles dropped initially. As flying continued, however, P_{O_2} levels recovered until they were nearly as high as when they were at rest.

These results can be explained by the breathing mechanism proposed in **Figure 44.10**. The key observation is that the tracheae of flying insects are sandwiched between muscles. When these muscles contract and relax, tracheae should constrict and expand in response. As a result, wing beats could force air into and out of the tracheae and lead to efficient ventilation of the respiratory surface.

How Do Lungs Work?

In most mammals, air enters the body through both the nose and mouth. A tube known as the **trachea** (not to be confused with the tracheae of insects) carries the inhaled air to narrower tubes called **bronchi** (singular: **bronchus**). The bronchi branch off into yet narrower tubes, the **bronchioles.** The organs of ventilation, the lungs, enclose the bronchioles and part of the bronchi (**Figure 44.11a**). **Lungs** are infoldings of the throat that are used for gas exchange. They occur in amphibians, reptiles (including birds), mammals, and certain fish and invertebrates. The amount of surface area available for gas exchange varies a great deal among species with lungs, however. In frogs and other amphibians, the lung is a simple sac lined with blood vessels. The lungs of mammals, in contrast, are finely divided into tiny sacs called **alveoli** (singular: **alveolus; Figure 44.11b**). Each human lung contains approximately 150 million of these sacs. Due to the presence of alveoli, mammalian lungs have about 40 times more surface area for gas exchange than does an equivalent volume of frog lung tissue.

As **Figure 44.11c** shows, an alveolus provides an interface between air and blood that consists of a thin aqueous film, a layer of epithelial cells, some extracellular matrix (ECM) material, and the wall of a capillary. In the human lung, this barrier to diffusion is only 0.2 μm thick—about 1/200th of the thickness of this page.

In addition to total surface area, the other major feature of lungs that varies among species is mode of ventilation. In the lungs of snails and spiders, air movement takes place by diffusion only. Vertebrates, in contrast, actively ventilate their lungs by pumping air via muscular contractions. One mechanism for pumping air is **positive pressure ventilation**, used by frogs and related animals. A frog lowers the floor of its throat to draw in air from the atmosphere, through the nasal

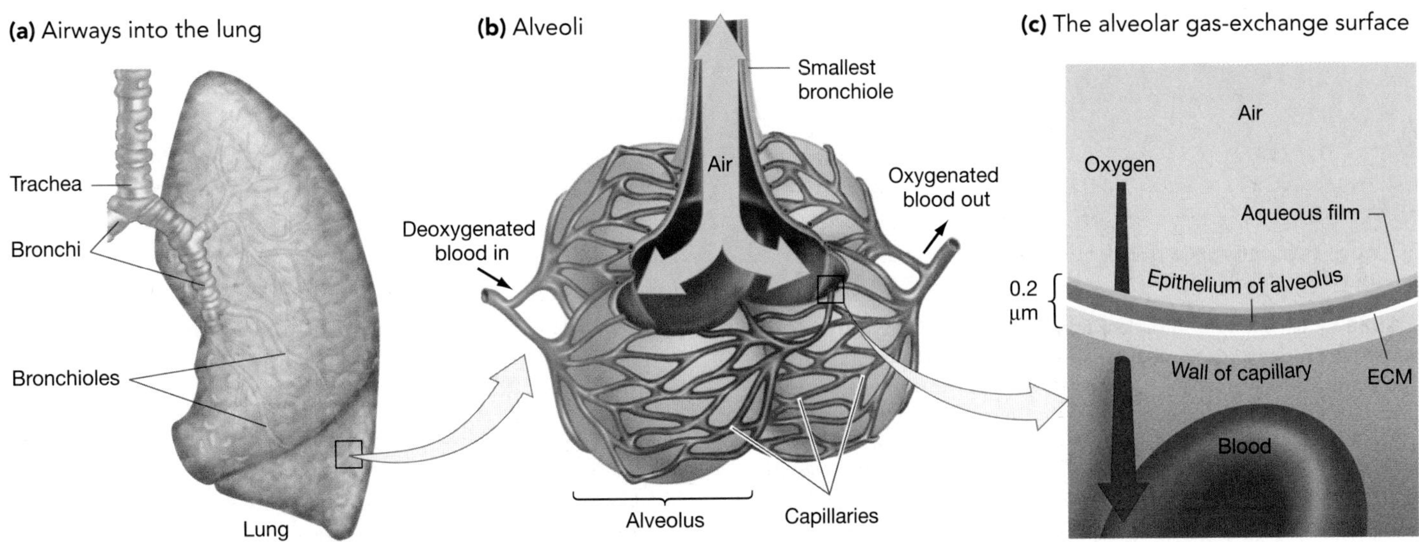

FIGURE 44.11 Lungs Offer a Large, Thin Surface Area for Gas Exchange between Air and Blood. (a) The human respiratory tract branches repeatedly from the largest airway, the trachea, to the smallest, called bronchioles. Gas exchange does not occur across these airways. The system of airways ends in clusters of tiny sacs called alveoli. **(b)** Alveoli are covered with capillary networks and **(c)** are the site of gas exchange.

passages, and into the oral cavity. The animal then closes the nasal passages and contracts its throat muscles. These actions increase the pressure on air in the oral cavity and force it into the lungs.

In effect, frogs push air into their lungs. In contrast, humans and other mammals pull air into their lungs. How does this **negative pressure ventilation** work?

(a) Lungs expand and contract in response to changes in pressure inside the chest cavity.

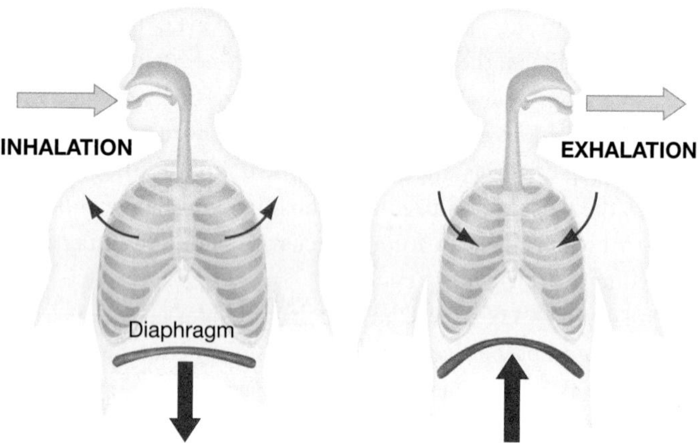

(b) Ventilatory forces can be modeled by a balloon in a jar.

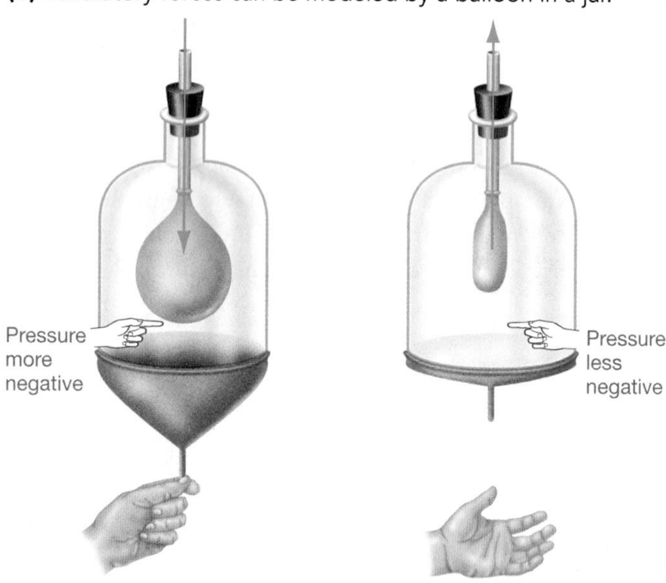

FIGURE 44.12 Changes in the Volume of the Chest Cavity Drive Negative Pressure Ventilation. (a) Inhalation: When the diaphragm and rib muscles contract, the volume of the lung cavity increases, causing internal pressure to become more negative. In response, the lungs expand and air flows in. Exhalation: When the diaphragm and rib muscles relax, the volume of the lung cavity decreases, causing internal pressure to become less negative. In response, lung volume decreases—due to elasticity of the lungs—and air flows out. **(b)** A model of negative pressure ventilation.

● **QUESTION** What happens when a person sighs or takes a deep breath?

Ventilation of the Human Lung The pressure inside the human chest cavity is about 5 mm Hg less than atmospheric pressure. Even though the lung is **elastic**—it returns to its original, collapsed shape if it is not stretched or compressed—the negative pressure surrounding the lung is just enough to keep it expanded. If a wound results in a hole in the wall of the chest and the pressure differential between the chest cavity and the atmosphere disappears, the lung on the side of the injury will collapse like a deflated balloon.

Humans ventilate their lungs by changing the pressure within the chest cavity between about −5 mm Hg and −8 mm Hg relative to the atmosphere. As **Figure 44.12a** shows, inhalation is based on lowering the pressure in the chest cavity. The change is caused by a downward motion of the thin muscular sheet called the **diaphragm** and an outward motion of the ribs. These movements increase the volume of the chest cavity and thus lower its internal pressure. As the pressure surrounding the lungs drops, air flows into the airways along a pressure gradient. Exhalation, in contrast, is a passive process driven by the elastic recoil of the lungs and chest wall as the diaphragm and rib muscles relax. (This is true at rest—exhalation is an energy-demanding, active process during exercise.) The changes in pressure that occur during negative pressure ventilation are analogous to changing the volume of a jar, as shown in **Figure 44.12b.**

About 450 ml of air moves into and out of the lungs in an average breath. Only about two-thirds of this volume actually participates in gas exchange, however, because 150 ml of the air occupies **dead space**—portions of the air passages that do not have a respiratory surface. The trachea and bronchi shown in Figure 44.11a, for example, represent dead space. Breathing is much more efficient during exercise, when the chest cavity undergoes larger changes in volume. When a person is breathing hard, over 2500 ml of air can move with each inhalation-exhalation cycle, but the 150 ml in dead space stays the same.

The volume and pressure changes that occur during a ventilatory cycle are fairly straightforward. But what accounts for the elasticity of the human lung? Stated another way, what force is responsible for returning lung tissue to its former shape? An early hypothesis was inspired by the thin aqueous film that exists on the surface of alveoli. The idea was that hydrogen bonding between water molecules in the film would create a high degree of surface tension and thus create a pull or tension on the alveolar wall (**Figure 44.13**). Once an alveolus finished expanding during inhalation, the contractile force created by surface tension would return the structure to its original size, imparting elasticity to the entire lung.

To test this hypothesis, a biologist realized that he could quantify the elasticity of a lung by measuring the pressure required to force a known volume of air into the structure. If the surface tension hypothesis is correct, then much less pressure should be required to force the same volume of saline solution, rather than air, into the same lung. This prediction is logical, because the presence of the solution would eliminate the air-water

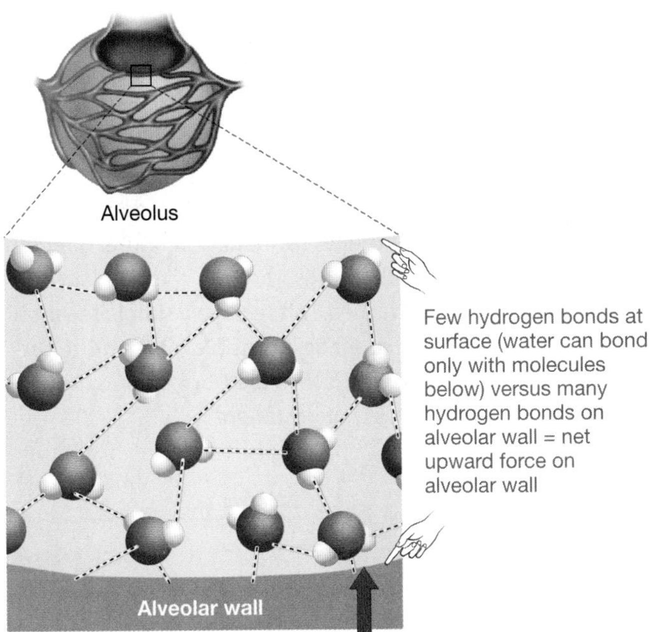

FIGURE 44.13 Surface Tension Creates a Pull on the Surface of the Lung.

interface inside the alveolus and thus eliminate surface tension. The biologist tested this prediction on lungs from a cadaver and showed that it was correct. Much less pressure was required to inflate the lungs with saline solution than with air. When other researchers repeated the experiment and got similar results, biologists accepted the hypothesis that most of the lung's elasticity is due to surface tension. Follow-up work has shown that surface tension within alveoli is so strong that it must actually be diminished somewhat for breathing to occur at all (see **Box 44.1**).

Ventilation of the Bird Lung Flight is one of the most energy-demanding activities performed by animals. Nevertheless, birds may fly tens of thousands of miles during annual migrations. Even more impressive is the observation that geese regularly fly over the top of Mount Everest. At this elevation, the partial pressure of oxygen is so low that most humans would immediately black out if they were flown there and dropped off. How are birds able to extract enough oxygen from the atmosphere to support long flights and flights at high elevations?

Biologists began working on this question by analyzing the anatomy of the avian respiratory system. As **Figure 44.14a** shows, a bird's trachea bypasses the lungs almost entirely and enters two large air sacs posterior to the lungs. These sacs are connected to the posterior portion of the lung by a series of tubes. The tubes branch into tiny airways, called **parabronchi**, within the lung. The parabronchi pass straight through the lung and terminate in another system of air sacs. These sacs link up with the trachea near the anterior portion of the lung. Gas exchange occurs in millions of spongelike air tubes, which extend laterally along the length of the parabronchi.

(a) Anatomy of the avian respiratory system

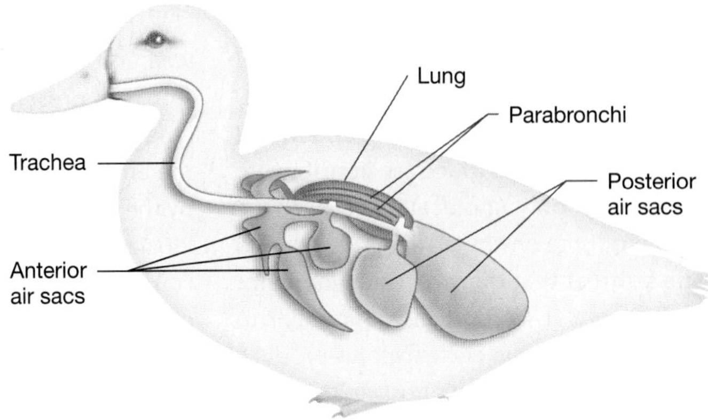

(b) One-way airflow through the avian lung

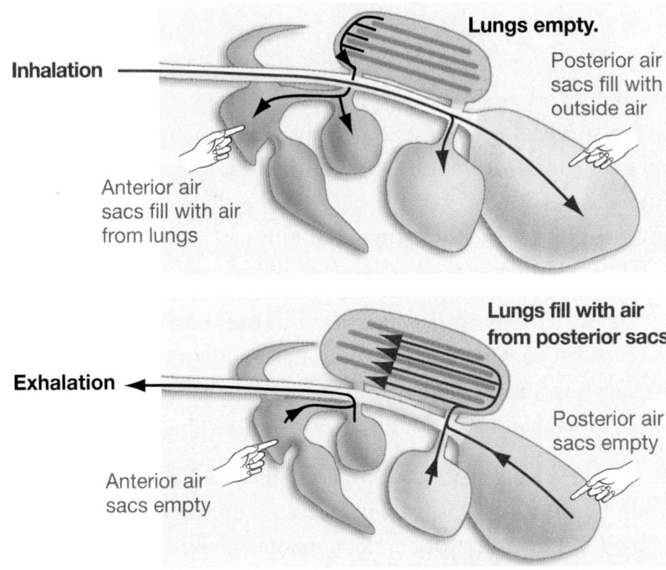

FIGURE 44.14 Air Flows in One Direction through the Bird Lung.

To explore how this respiratory system functions, Knut Schmidt-Nielsen and co-workers used a variation on the pulse-chase technique introduced in Chapter 7. In this case, the pulse was a single breath of pure oxygen, followed by a "chase" that consisted of a breath of normal air. To the researchers' surprise, the oxygen-enriched air was not detected at the bird's mouth until the *second* exhalation following the pulse. Follow-up experiments showed that air moves through a bird's respiratory system as diagrammed in **Figure 44.14b**. The key conclusion was that airflow through the avian lung is unidirectional.

To drive home how birds breathe, biologists make an analogy with how the Scottish bagpipe is played. Air is introduced intermittently into a large reservoir, such as the bag of a bagpipe or the posterior air sacs of a bird. Then the air is expelled continuously through air passages, such as the drone pipes of the instrument or the parabronchi of the bird lung. In birds, the air expelled from the parabronchi is collected in the anterior

BOX 44.1 Respiratory Distress Syndrome in Premature Infants

Because alveoli have a large amount of surface area in contact with gas, the watery film that covers them generates a large amount of surface tension. If water were the only molecule present, the pulling force on the alveolar wall would be so great that the structure would not inflate readily during inhalation. Fortunately, cells within each alveolus produce a mixture of soap-like phospholipids and proteins that act as a **surfactant**—that is, they disrupt the aqueous film lining the wall and greatly reduce its surface tension. The presence of the surfactant decreases lung stiffness enough to reduce the amount of work that has to be done by the diaphragm and rib muscles during inhalation.

Human infants acquire the ability to produce adequate amounts of surfactant at between 32 and 34 weeks of gestation, and birth normally occurs at 36 weeks. If born prior to 32 weeks of gestation, an infant may have such stiff lungs that it quickly fatigues from the effort of breathing and suffocates. This condition is called **respiratory distress syndrome**. Fortunately, it can be treated by administering a synthetic surfactant mixture directly into the lungs, through a tube placed in the trachea. This relatively simple technology has made it possible for infants to survive even if they are born after as little as 24 weeks of gestation.

air sacs and ultimately exhaled into the trachea through an opening that prevents the mixing of inhaled and exhaled air. One-way airflow is possible because the chest and air sacs do not compress at the same time, but alternately.

Three features make the avian respiratory system efficient:

1. Dead space in birds is restricted to the short stretch of trachea between the mouth and the opening of the anterior air sacs. As a result, inhaled air is used much more efficiently in birds than in mammals.
2. Birds exchange gases during both inhalation and exhalation. In contrast, no gas exchange occurs during the exhalation half of the respiratory cycle in mammals. Bird ventilation resembles the continuous ventilation of fish gills in this respect.
3. Blood circulates through the bird lung in capillaries that cross the parabronchi perpendicularly. This crosscurrent pattern is not as efficient as the countercurrent circulation of fish gills, but it is far more efficient than the weblike arrangement of capillaries that surrounds mammalian alveoli.

Homeostatic Control of Ventilation

An animal is in trouble if homeostasis with respect to blood oxygenation or the elimination of carbon dioxide fails. To ensure adequate ATP production, the partial pressures of oxygen and carbon dioxide must be maintained within a narrow range, both at rest and during vigorous exercise. How is ventilation controlled to achieve this critical homeostasis?

When mammals are resting, the rate of breathing is established by the medullary respiratory center, an area at the base of the brain, just above the spinal cord. This center stimulates the rib and diaphragm muscles to contract about 12–14 times per minute in humans. But during exercise, things change. Active muscle tissue takes up more oxygen from the blood. As a result, the partial pressure of oxygen (P_{O_2}) in blood drops. Those same muscles release quantities of carbon dioxide to the blood, tending to raise its partial pressure (P_{CO_2}) in blood. When it reaches the brain, carbon dioxide rapidly diffuses from the blood into the cerebrospinal fluid that bathes the brain. In both blood and cerebrospinal fluid, CO_2 reacts with water to form carbonic acid (H_2CO_3), which then dissociates to release a hydrogen ion (H^+) and a bicarbonate ion (HCO_3^-):

$$CO_2 + H_2O \rightleftharpoons H_2CO_3 \rightleftharpoons H^+ + HCO_3^-$$

The result is a slight drop in the pH of blood and cerebrospinal fluid. The change is sensed by specialized nerves located near the large arteries that travel from the heart into the neck and to the base of the brain itself. Signals from these nerves or from pH detectors in the medullary respiratory center cause the breathing rate to increase. The resulting rise in ventilation rate increases the rate of oxygen delivery to the tissues and the rate at which carbon dioxide is eliminated from the body, restoring P_{O_2} and P_{CO_2} to their resting levels.

This control system is so effective that it can maintain stable levels of oxygen and carbon dioxide in the blood even in the face of intense exercise. In addition, it allows the body to regulate blood pH carefully. To see why this is so, it's important to recognize that carbon dioxide is transported in the blood in the form of bicarbonate ions. When CO_2-rich blood reaches the lungs, the bicarbonate ions have to react with hydrogen ions to regenerate carbon dioxide, which then diffuses across the walls of alveoli and is exhaled. When breathing rate increases, then, excess hydrogen ions that are produced during rapid metabolism are eliminated along with the excess CO_2 that is generated, and blood pH is stabilized.

Now let's look in greater depth at blood itself. What is this substance, and how does it transport oxygen and carbon dioxide between the gas-exchange surface and an animal's tissues?

Check Your Understanding

If you understand that...

Most large-bodied animals exchange gases via gills, tracheae, or lungs.

You should be able to ...

1) Identify at least three features that are common to all three of these types of organs as well as one trait that is unique to each.
2) Explain the contrast between airflow in the mammalian lung and in the bird lung.
3) Make a diagram summarizing the sensor, integrator, and effector(s) involved in homeostatic control of the ventilation rate.

44.4 How Are Oxygen and Carbon Dioxide Transported in Blood?

Blood is a tissue—a collection of cells that functions as an integrated unit. More specifically, **blood** is a connective tissue that consists of cells in a watery extracellular matrix. In an average human, 50–60 percent of the blood volume is composed of an extracellular matrix called **plasma**. The remainder of the volume comprises a variety of cells and cell fragments that are collectively called formed elements.

Given the wide variety of functions that blood serves, it is not surprising that this tissue is complex. Besides carrying oxygen and carbon dioxide between cells and the lungs, blood transports nutrients from the digestive tract to other tissues in the body, moves waste products to the kidney and liver for processing, conveys hormones from glands to target tissues, delivers immune system cells to sites of infection, and distributes heat from deeper organs to the surface.

The formed elements in blood include platelets, several types of white blood cells, and red blood cells. **Platelets** are cell fragments that minimize blood loss from ruptured blood vessels by releasing material that assists in the formation of blockages known as clots. **White blood cells** are part of the immune system. As Chapter 49 will explain in detail, white blood cells are responsible for fighting infections. Although both platelets and white blood cells play an essential role in the body, they are vastly outnumbered by red blood cells. In humans, red blood cells make up 99.9 percent of the formed elements. **Red blood cells** transport oxygen from the lungs to tissues throughout the body. They also play a critical role in transporting carbon dioxide from tissues to the lungs.

In humans, new red blood cells are synthesized at the rate of 2.5 million per second to replace old red blood cells, which are dying at the same rate. These red blood cells live about 120 days. They develop along with white blood cells and platelets from stem cells located in the tissue inside bone (bone marrow). Vertebrates other than mammals transport oxygen in red blood cells that retain their nuclei. But in most mammals, red blood cells lose their nuclei as they mature, along with their mitochondria and most other organelles. Mammalian red blood cells are essentially bags filled with some 280 million copies of the oxygen-carrying molecule **hemoglobin**. (**Box 44.2** on page 993 analyzes some of the membrane proteins found on human red blood cells and their importance in managing blood transfusions.)

Structure and Function of Hemoglobin

Even though oxygen is not very soluble in water, it is often found in high concentrations in blood. Blood has a high oxygen-carrying capacity because O_2 readily binds to the hemoglobin molecules found in red blood cells. The evolution of hemoglobin was a key event in the diversification of animals. By increasing the oxygen-carrying capacity of blood, hemoglobin made it possible for cellular respiration rates to increase. High rates of ATP production, in turn, support high rates of growth, movement, digestion, and other activities.

Hemoglobin is a tetramer, meaning that it consists of four polypeptide chains (**Figure 44.15**). Each of the four polypeptide

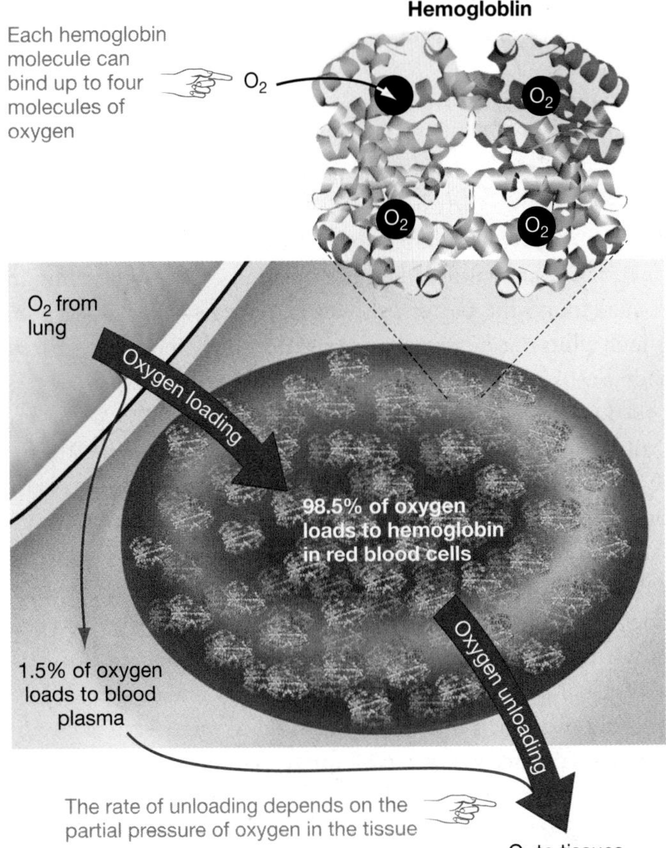

FIGURE 44.15 Hemoglobin Transports Oxygen to Tissues.

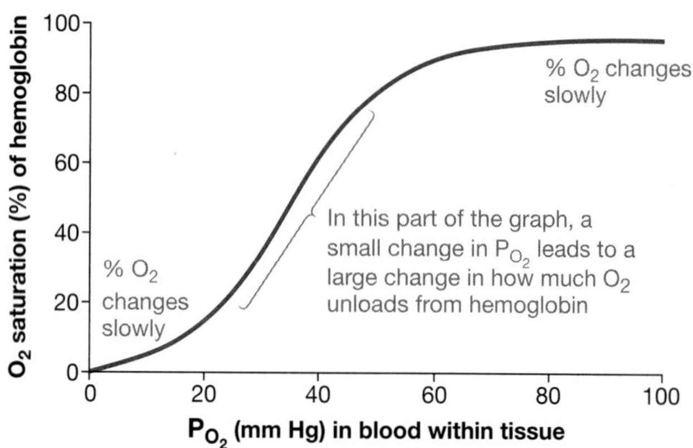

FIGURE 44.16 The Oxygen-Hemoglobin Equilibrium Curve. Partial pressure of oxygen in blood and tissues versus the percentage of binding sites on hemoglobin that hold oxygen molecules. The sigmoidal curve has three distinct regions.

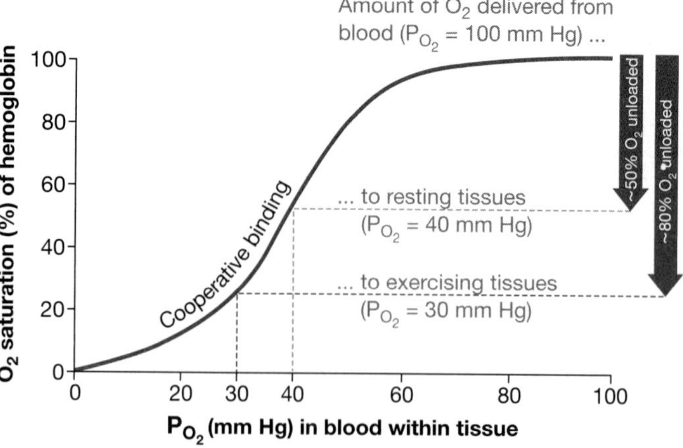

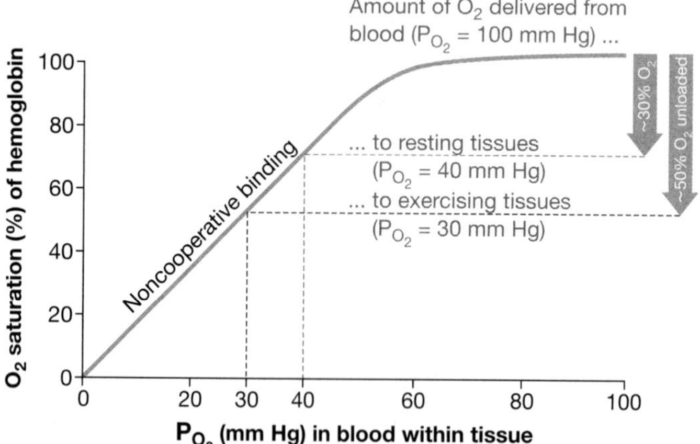

FIGURE 44.17 Cooperative Binding of O_2 by Hemoglobin Results in Greater O_2 Delivery than Noncooperative Binding. (a) Effect of cooperative versus **(b)** noncooperative oxgyen binding. As blood moves from a P_{O_2} of 100 mm Hg in the lungs to 30 mm Hg in exercising tissue, cooperative binding results in much more oxygen being transferred to tissues.

chains binds to a nonprotein group called **heme**. Each heme molecule, in turn, contains an iron ion (Fe^{2+}) that can bind to an oxygen molecule. As a result, each hemoglobin molecule can bind up to four oxygen molecules. In blood, 98.5 percent of the oxygen is bound to hemoglobin; only 1.5 percent is dissolved in plasma.

What happens when red blood cells carrying oxygen-rich hemoglobin reach cells that contain little oxygen?

Cooperative Binding Blood leaving the human lungs has a P_{O_2} of about 100 mm Hg, while muscles and other tissues have P_{O_2} levels of about 40 mm Hg at rest. This partial-pressure difference creates a diffusion gradient that results in the unloading of O_2 from hemoglobin to the tissues.

Researchers who studied the dynamics of O_2 unloading in tissues found the pattern shown in **Figure 44.16**. This graph, which plots the percentage saturation of hemoglobin in red blood cells versus the P_{O_2} levels in the blood within tissues, is called the **oxygen-hemoglobin equilibrium curve**. Each 25 percent change in saturation corresponds to an average of one additional oxygen molecule bound per hemoglobin molecule or one oxygen molecule delivered to tissues.

The most remarkable feature of the oxygen-hemoglobin equilibrium curve is that it is sigmoidal, or S-shaped. The pattern occurs because the binding of each successive oxygen molecule to a subunit of the hemoglobin molecule causes a conformational change in the protein that makes the remaining subunits much more likely to bind oxygen. This phenomenon is called **cooperative binding**. Conversely, the loss of a bound oxygen molecule changes hemoglobin's conformation in a way that makes the loss of additional oxygen molecules more likely.

Cooperative binding is important because it makes hemoglobin exquisitely sensitive to changes in the P_{O_2} of tissues. Consider what happens to hemoglobin saturation when tissue P_{O_2} changes from its resting level of about 40 mm Hg to 30 mm Hg, as might be seen during exercise (**Figure 44.17a**). In response to this small change in tissue P_{O_2}, there is an enormous change in the percentage saturation of hemoglobin—from about 52 percent to about 22 percent. This means that when blood encounters tissues at rest, hemoglobin gives up about 50 percent of the oxygen it is carrying (from almost 100 percent saturation to about 50 percent). But when blood encounters tissues during exercise, it gives up about 80 percent of its oxygen (from almost 100 percent saturation to about 20 percent). Because of cooperative binding, the loss of one oxygen molecule to the surrounding tissues means that others are more likely to follow. Stated another way, the sigmoidal shape of the oxygen equilibrium curve means that a small change in tissue P_{O_2} produces a large change in how much oxygen stays attached to hemoglobin. ● If you understand this concept, you should be able to explain the consequences of an oxygen-hemoglobin equilibrium curve with a middle section that is even steeper than the one shown in Figure 44.17a.

If cooperative binding did not occur, all four subunits of hemoglobin would load or unload oxygen independently of each other. They would lose or gain oxygen in direct proportion to the partial pressure of oxygen in the blood, until the molecule was either completely saturated or unsaturated with oxygen. The relationship shown in **Figure 44.17b** would result. As the dotted lines in that figure indicate, much less oxygen would be delivered when tissue P_{O_2} changes from its resting level of about 40 mm Hg to 30 mm Hg—specifically, the percentage of O_2 saturation would drop from about 68 percent to about 52 percent. The total amount of oxygen delivered would only increase from about 30 percent to about 50 percent, instead of the 50 percent to 80 percent change observed with cooperative binding.

How Do Temperature and pH Affect Hemoglobin?

Cooperative binding is only part of the story behind oxygen delivery. Hemoglobin—like other proteins—is sensitive to changes in pH and temperature. During exercise, the temperature and partial pressure of CO_2 in active muscle tissue rise. The CO_2 produced by exercising muscle reacts with the water in blood to form carbonic acid, which dissociates and releases a hydrogen ion. As a result, the pH of the blood in exercising muscle drops.

Decreases in pH and increases in temperature alter hemoglobin's conformation. As the graph in **Figure 44.18** shows, these shape changes make hemoglobin more likely to release O_2 at any given value of tissue P_{O_2}. This phenomenon is known as the **Bohr shift**. The name honors Christian Bohr, who published on the phenomenon in 1904, and highlights the rightward shift of the oxygen-hemoglobin equilibrium curve with decreasing pH. The Bohr shift is important because it makes hemoglobin more likely to release oxygen during exercise or other conditions in which P_{CO_2} is high, pH is low, and tissues are under oxygen stress. ● If you understand the Bohr shift and cooperative binding, you should be able to add notes to Figure 44.15 indicating (1) what happens to the other three oxygen molecules that are bound to heme groups once one O_2 molecule leaves, and (2) what happens to bound O_2 molecules in tissues that are acidic.

To appreciate the effect of cooperative binding and the Bohr shift in action, consider a recent experiment on how the oxygen transport system in rainbow trout responds to sustained exercise. To explore changes in blood O_2 content during exercise, biologists had fish swim continuously against a current in a water tunnel. As the researchers increased the speed of the current and thus the swimming speed of the fish, they periodically sampled the O_2 content of both the arterial and venous blood of experimental individuals. Arterial blood is freshly oxygenated and moving away from the gills; venous blood is returning to the gills from the rest of the body. Not surprisingly, the biologists found that arterial O_2 levels remained fairly constant as swimming speed increased—meaning the gills continued to work efficiently enough to saturate hemoglobin with oxygen. In contrast, the O_2 content of venous blood, which had undergone gas exchange with the tissues, dropped steadily as swimming speed increased. When the fish had reached their maximum sustainable speed, virtually all the oxygen available in hemoglobin in the venous blood had been extracted. The data show that in hard-working tissues, the combination of increased temperature, lower pH, and lower P_{O_2} caused hemoglobin to become almost completely deoxygenated. Oxygen-delivery systems based on hemoglobin are extremely efficient.

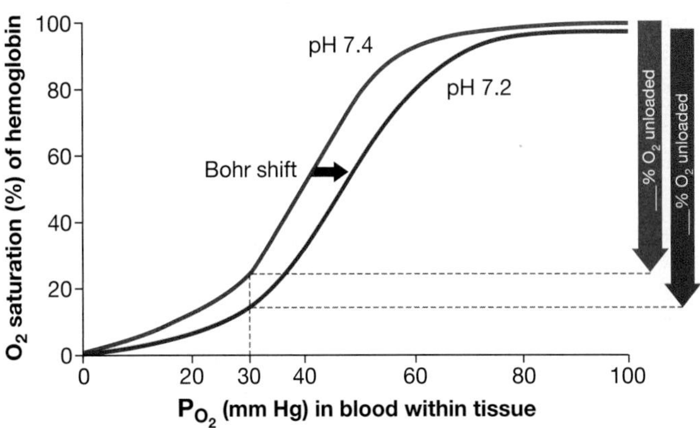

FIGURE 44.18 The Bohr Shift Makes Hemoglobin More Likely to Release Oxygen to Tissues with Low pH. As pH drops, oxygen becomes less likely to stay bound to hemoglobin at all values of tissue P_{O_2}. Exercising tissues have lower pH than resting tissues have and thus receive more oxygen from hemoglobin.

● **EXERCISE** Estimate how much more oxygen is released from hemoglobin at pH 7.2 than at pH 7.4 when the tissue has a P_{O_2} of 30 mm Hg.

Comparing Hemoglobins In addition to recognizing that individual hemoglobin molecules can change in structure and function depending on conditions in the body, it's important to appreciate that hemoglobin molecules from different individuals or species may vary in ways that affect fitness—the ability to survive and produce offspring. For example, consider the oxygen-hemoglobin equilibrium curves in **Figure**

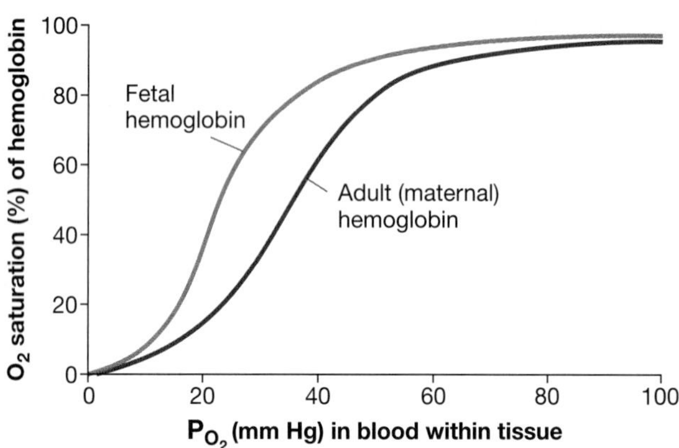

FIGURE 44.19 Some Types of Hemoglobin Bind Oxygen More Tightly than Others.

QUESTION Suppose you compare oxygen-hemoglobin equilibrium curves for hemoglobins found in closely related animal species that live at high elevation versus low elevation. Would the two curves differ? If so, how and why? If not, why not?

44.19. The curve in dark red is from a pregnant human; the curve in light red is from a fetus she is carrying. The hemoglobin found in fetuses is encoded by different genes than adult hemoglobin and has a distinctive structure and function. Specifically, the oxygen-hemoglobin equilibrium curve for fetal hemoglobin is shifted to the left with respect to the curve for adult hemoglobin. This means that at every P_{O_2}, fetal hemoglobin has a higher affinity for oxygen than adult hemoglobin. As a result, oxygen is transferred from the mother's blood to the fetus's blood. This difference in hemoglobin structure and function between mother and fetus ensures an adequate supply of oxygen as the fetus develops.

Now let's consider the other side of gas exchange: What happens as CO_2 is released from tissues to the blood? How does the body cope with the decline in pH that results? After all, if hemoglobin's conformation and activity are affected so dramatically by pH changes, it is likely that many other proteins will be altered as well. If these proteins function less well at low pH, then carbon dioxide buildup should result in illness.

CO_2 Transport and the Buffering of Blood pH

The carbon dioxide that is produced by cellular respiration in the tissues enters the blood, where it reacts with water to form carbonic acid. Recall that the resulting drop in blood pH stimulates an increase in breathing rate. Rapid exhalation of CO_2 then counteracts the drop in blood pH.

Changes in blood pH are also diminished by an elegant series of reactions that takes place inside red blood cells. Biologists were able to work out what was happening when they discovered large amounts of the enzyme **carbonic anhydrase** in red blood cells. Recall from Chapter 43 that carbonic anhydrase catalyzes the formation of carbonic acid from carbon dioxide in water. Consequently, CO_2 that diffuses into red blood cells is quickly converted to bicarbonate ions and protons. This is the mechanism responsible for the observation, made in Section 44.3, that most CO_2 is transported in blood (specifically in plasma) in the form of HCO_3^-—the bicarbonate ion. The same reaction occurs in the plasma surrounding red blood cells, but much more slowly in the absence of the enzyme.

Why is the carbonic anhydrase activity in red blood cells so important? The answer has two parts. First, the protons produced by the enzyme-catalyzed reaction induce the Bohr shift, which makes hemoglobin more likely to release oxygen. Second, the partial pressure of CO_2 in blood drops when carbon dioxide is converted to soluble bicarbonate ion, maintaining a strong partial-pressure gradient favoring the entry of CO_2 into red blood cells. Thus, carbonic anhydrase activity makes CO_2 uptake from tissues more efficient (**Figure 44.20**).

Once bicarbonate ions form in the red blood cell, they diffuse into the blood plasma along a concentration gradient. Bicarbonate ions move via an antiporter that exchanges bicarbonate with chloride ions. (Antiporters were introduced in Chapter 37.) In contrast, the protons produced by the reaction stay inside red blood cells. Relatively small amounts of H^+ build up in the plasma.

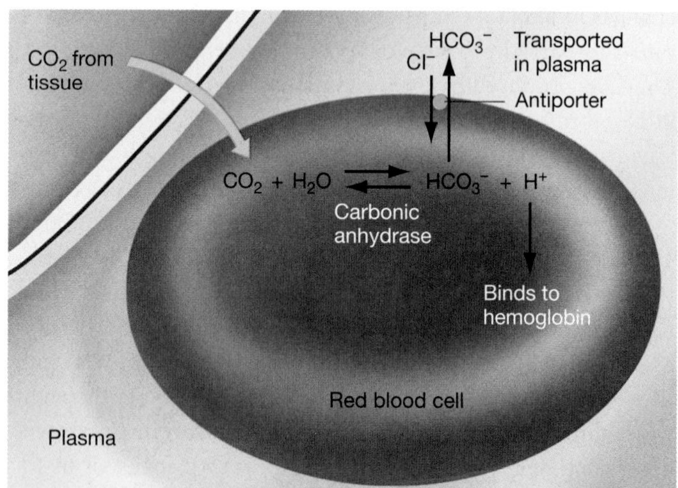

FIGURE 44.20 How Is Carbon Dioxide Transported in Blood? When CO_2 diffuses into red blood cells, carbonic anhydrase quickly converts it to carbonic acid, which dissociates into bicarbonate ion (HCO_3^-) and a proton (H^+). This reaction maintains the partial-pressure gradient favoring the entry of CO_2 into red blood cells. The protons produced by the reaction bind to deoxygenated hemoglobin. Most CO_2 in blood is transported to the lungs in the form of HCO_3^-.

QUESTION This diagram shows the sequence of events in tissues. Explain what happens when the red blood cell in the diagram reaches the lungs.

What happens to the protons inside red blood cells? The answer to this question emerged when researchers studied the composition of deoxygenated hemoglobin. These analyses showed that when hemoglobin is not carrying oxygen molecules, it has a high affinity for protons. As a result, much of the H^+ that is produced by the dissociation of carbonic acid is taken up by hemoglobin. In this way, hemoglobin acts as a **buffer**—a compound that minimizes changes in pH. In addition, analyses of hemoglobin in deoxygenated blood revealed that the CO_2 that is not transported as HCO_3^- binds to amino

BOX 44.2 ABO and Rh Blood Types

As red blood cells are maturing in bone marrow, their plasma membranes gain glycolipids that function like the security badges worn by hospital staff—they help identify the cells as legitimate components of the body. Their presence prevents red blood cells from being attacked by immune system cells, which are responsible for detecting and eliminating viruses, bacteria, and other types of dangerous invaders.

Some of the key glycolipids on human red blood cells are encoded by a single gene. Three alleles of this gene, symbolized i, I^A, and I^B, are common in most human populations and are responsible for the **ABO blood types.** The allele i is a loss-of-function allele that leads to a lack of glycoprotein on the cell surface; I^A encodes the A glycolipid; and I^B encodes the B glycolipid. Individuals with the genotype ii have blood type O; I^AI^A and I^Ai individuals have blood type A; I^BI^B and I^Bi individuals have blood type B; and I^AI^B individuals, who produce both types of glycolipid, have blood type AB.

The ABO blood types are well studied because they are critical to the success of the most common type of tissue transplant: blood transfusions. If blood from an individual with blood type A is introduced to a person with blood type O or B, the recipient's immune system produces proteins called antibodies that bind to the A glycolipids and cause the red blood cells to clump together. The clumping reaction inhibits blood flow and can endanger the recipient's life.

A similar issue occurs in response to a protein on the surface of red blood cells called the rhesus or Rh factor—so called because it was discovered in the rhesus macaque. Individuals with the **Rh factor** are said to be Rh positive (Rh^+); those without the Rh factor are typed Rh negative (Rh^-). Rh^- individuals produce antibodies to the Rh factor. If Rh^+ blood is introduced to an Rh^- individual, clumping may occur.

To prevent these problems, agencies that collect and distribute blood for transfusions do simple tests to determine the blood type of the donor, so it can be matched to the recipient. **Table 44.1** summarizes which ABO and Rh blood types can be successfully transferred from donor to recipient.

TABLE **44.1** **Blood Transfusions and the ABO/Rh Blood Types**

An "OK" indicates a successful transfusion;
an "X" indicates a transfusion that causes red blood cells to clump.

Recipient Blood Type (antibodies present)	**Donor Blood Type**							
	A⁺	**A⁻**	**B⁺**	**B⁻**	**AB⁺**	**AB⁻**	**O⁺**	**O⁻**
A^+ (anti-B)	OK	OK	X	X	X	X	OK	OK
A^- (anti-B, anti-Rh)	X	OK	X	X	X	X	X	OK
B^+ (anti-A)	X	X	OK	OK	X	X	OK	OK
B^- (anti-A, anti-Rh)	X	X	X	OK	X	X	X	OK
AB^+ (none)	OK	OK	OK	OK	OK	OK	OK	OK
AB^- (anti-Rh)	X	OK	X	OK	X	OK	X	OK
O^+ (anti-A, anti-B)	X	X	X	X	X	X	OK	OK
O^- (anti-A, anti-B, anti-Rh)	X	X	X	X	X	X	X	OK

QUESTION Why is type O^- called the "universal donor" and type AB^+ considered the universal recipient?

groups on the protein portion of the molecule. The overall effect of these events is efficient transport of CO_2, with a minimal effect on blood pH.

When deoxygenated blood reaches the alveoli, however, the environment changes dramatically. In the alveoli, a partial-pressure gradient favors the diffusion of CO_2 from plasma and red blood cells to the atmosphere. As CO_2 diffuses from the blood into the alveoli, P_{CO_2} in blood declines. The drop promotes the reversal of the chemical reactions that occurred in tissues: Hydrogen ions leave their binding sites on hemoglobin and combine with bicarbonate to form CO_2, which then diffuses into the alveoli and is exhaled from the lungs.

In the meantime, hemoglobin has picked up O_2. Oxygen transfer from air to blood is efficient. Hemoglobin's affinity for oxygen is high in alveoli because blood pH rises as P_{CO_2} declines.

When blood leaves the lungs, it has unloaded carbon dioxide and hemoglobin is saturated with oxygen. The cycle begins anew.

Check Your Understanding

If you understand that...

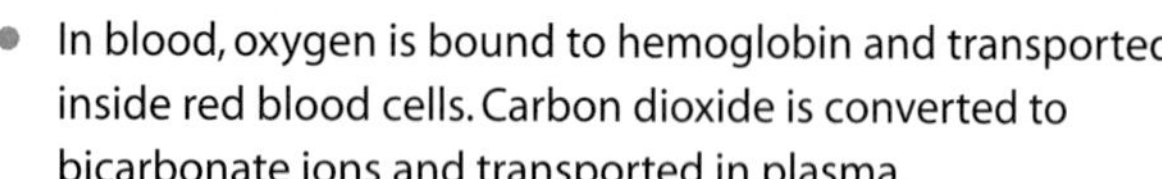

- In blood, oxygen is bound to hemoglobin and transported inside red blood cells. Carbon dioxide is converted to bicarbonate ions and transported in plasma.
- Hemoglobin has several properties that make it an effective transport protein, including cooperative binding of oxygen, the Bohr shift response to low pH, and the ability to bind the protons that are generated when carbon dioxide is converted to bicarbonate ions.

You should be able to...

Draw an oxygen-hemoglobin equilibrium curve and explain how it increases the efficiency of oxygen delivery to tissues, relative to an equilibrium curve in which cooperative binding does not occur.

MB Web Animation at www.masteringbio.com
Gas Exchange in the Lungs and Tissues

44.5 The Circulatory System

Recall from Chapter 41 that large animals have a much smaller amount of surface area relative to their volume than do small animals. In most animals that are more than a few millimeters in size, the available surface area is too small for gases, nutrients, and wastes to diffuse efficiently between the animal's surface and its interior. As a result, diffusion across the individual's exterior surface is not rapid enough to support metabolism. In large animals, this problem is solved by a circulatory system.

The function of a circulatory system is to carry blood or hemolymph into close contact with every cell in the body. In this case, "close contact" is a distance of about 0.1 mm between the smallest blood vessels and cells within tissues. Diffusion is efficient at this scale. But an enormous amount of tubing is required to distribute blood this close to cells throughout the body. If all the blood vessels in a human body were laid end to end, they would stretch about 100,000 km (over 60,000 miles).

The most sophisticated circulatory systems consist of one or more pumps called **hearts**; tough, thick-walled **arteries**, which are vessels that take blood away from the heart under high pressure; vessels called **capillaries**, whose walls are just one cell thick, allowing gases and other molecules to exchange with tissues in networks called **capillary beds**; and **veins**, which are vessels that return blood to the heart under low pressure. It's important to recognize that animals may lack some or all of these components or may have no circulatory system at all. In jellyfish and corals, for example, a large, internal gastrovascular cavity offers enough surface area that adequate exchange of molecules with the environment occurs via diffusion. Circulatory systems are also lacking in flatworms and tapeworms, whose bodies are flattened enough that their surface area/volume ratio is high. In these animals, molecules are exchanged with the environment directly across the outer body surface. Roundworms lack hearts but facilitate exchange between their fluid-filled body cavity and the environment through muscular contractions in their body wall. In effect, roundworms circulate fluids by sloshing them back and forth.

To explore how more sophisticated circulatory systems work, let's start by distinguishing the two most general types of systems—open and closed.

What Is an Open Circulatory System?

In an open circulatory system, hemolymph is actively pumped throughout the body but is not confined exclusively to blood vessels. The key observation about hemolymph is that it comes into direct contact with the tissues—meaning the molecules being exchanged between hemolymph and tissues do not have to diffuse across the wall of a blood vessel. **Hemolymph** transports wastes and nutrients and may also contain oxygen-carrying pigments, some cells, and clotting agents.

Figure 44.21 illustrates the open circulatory system in a clam. Note that the heart pumps hemolymph into blood vessels that empty into an open, fluid-filled space. Hemolymph is returned to the heart when the heart relaxes and its internal pressure drops below that in the body cavity. General body movements also move hemolymph to and from the heart(s).

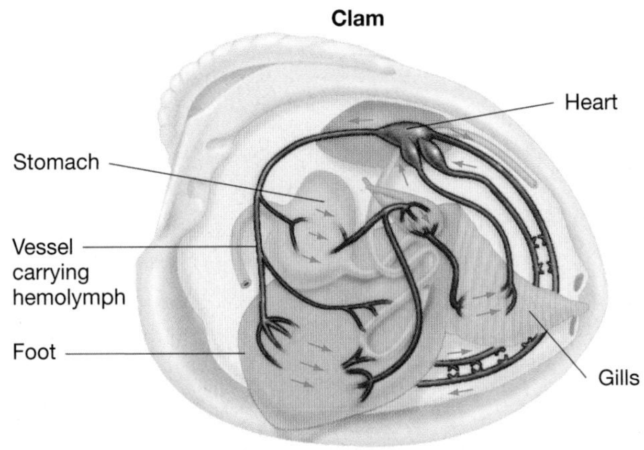

FIGURE 44.21 Open Circulatory Systems in Clams.

EXERCISE Circle the locations where hemolymph is in direct contact with tissues.

QUESTION What is "open" about an open circulatory system?

One salient characteristic of an open circulatory system is that hemolymph is under relatively low pressure. As a result, hemolymph flow rates may also be low. This feature makes open systems most suitable for relatively sedentary organisms, which do not have high oxygen demands. Insects, with their rapid movements and more active lifestyles, are an obvious exception to this rule. Insects overcome the limitations imposed by low hemolymph pressure via their tracheal respiratory system, which delivers oxygen directly to the tissues.

Another characteristic of open circulatory systems is that without discrete, continuous vessels, the flow of hemolymph cannot be directed toward tissues that have a high oxygen demand and CO_2 buildup. An open circulatory system moves hemolymph throughout an animal's body in much the same way that a ceiling fan moves air throughout a room in a house.

What Is a Closed Circulatory System?

In a closed circulatory system, blood flows in a continuous circuit through the body under pressure generated by the heart. Because blood is confined to vessels, a closed system can generate enough pressure to maintain a high flow rate. Blood flow can also be directed in a precise way. Regulatory systems can direct blood to specific vessels and thus to specific locations, such as muscles during exercise, the intestines after a meal, or regions of the brain engaged in particular mental tasks.

Closed circulatory systems are found in vertebrates and a few other lineages where individuals tend to be active. Earthworms and other annelids, for example, have closed circulatory systems. Annelids exchange gases with the environment across their thin, moist skin, which has a dense supply of capillaries. With this ventilatory surface and closed circulatory system, annelids are able to obtain and circulate enough oxygen to support intense muscular activity. Based on these observations, it is logical to find that most annelids live as active burrowers and hunters. A similar situation occurs in squid, octopuses, and other cephalopods that hunt down prey. The closed circulatory system of these animals generates high rates of blood flow, which oxygenates their muscles well enough to support rapid movements and a predatory lifestyle. And even though the circulatory system of crustaceans is classified as open, crustaceans have a fine network of vessels that can preferentially send blood to tissues with the highest oxygen demands.

Closed circulatory systems contain an array of blood vessel types, each of which has a distinct structure and function. Let's review some of these major types and then consider how the vessels of a closed circulatory system interact with the lymphatic system.

Types of Blood Vessels In a closed circulatory system, the heart ejects blood into a large artery, usually called the **aorta**. Although all arteries have both muscle fibers and elastic fibers in their walls, the walls of the aorta are dominated by elastic fibers. As a result, the aorta can expand when blood enters it under high pressure from the heart. When contraction of the heart ends, the diameter of the aorta returns to its resting state. This elastic response propels blood away from the heart and augments the force generated by heart contraction. Similar types of secondary pumping action occur to some extent in other large and medium-sized arteries as well and help maintain forward blood flow in the period between cardiac contractions.

In the walls of smaller blood vessels called **arterioles**, a continuous sheet of smooth muscle fibers is broken up by distinct bands of fibers that wrap around the circumference of the vessel. These muscle fibers are called **sphincters** and allow the diameter of the vessel to be carefully regulated in response to signals from the nervous system. When the muscle fibers are relaxed, the arteriole diameter increases, resistance to flow in the arteriole is reduced, and blood flow increases in the tissues served by the vessel. But when these arteriolar sphincters constrict, they increase resistance to flow, slow blood passage, and thus divert blood flow for use by other tissues.

(a) Capillaries are small and extremely thin walled.

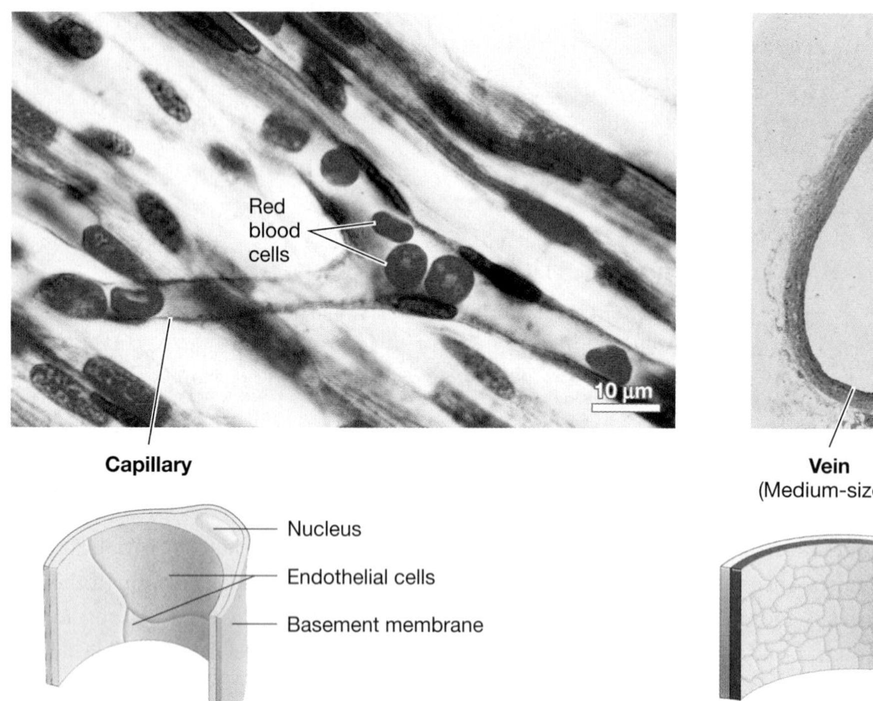

(b) Veins and arteries differ in structure.

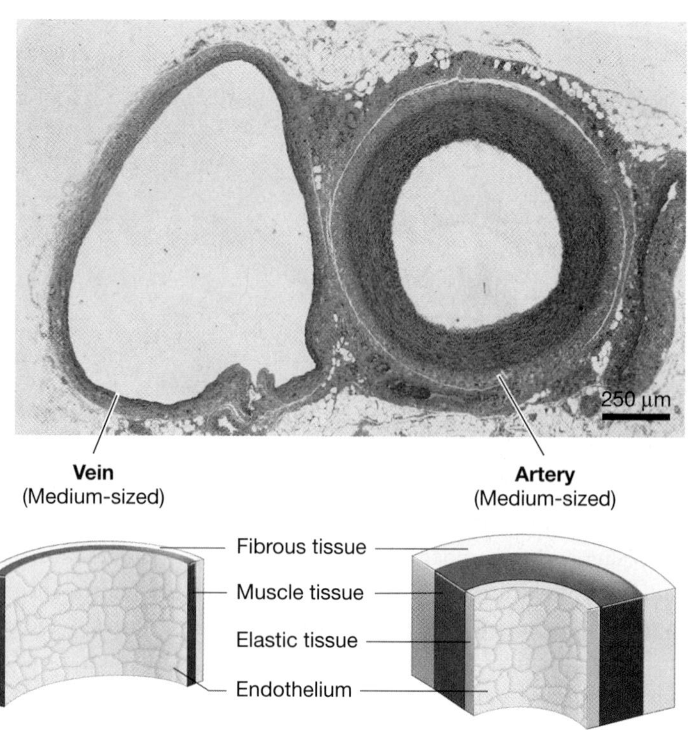

FIGURE 44.22 Capillaries, Veins, and Arteries. Micrographs of **(a)** a capillary and **(b)** a cross section through a vein and an artery. Notice the difference in relative wall thickness and overall size of the vein and artery.

Capillaries are the smallest blood vessels. They have walls that are only one cell layer thick, and they are just wide enough to let red blood cells through one at a time (**Figure 44.22a**). Because capillaries are extremely thin and because they form an extremely dense network throughout the body, they are the structure where gases, nutrients, and wastes are exchanged between the blood and other tissues. In some organs, such as the liver, the walls of capillaries contain numerous small openings, which further diminish the barrier to diffusion between blood and the tissues. Despite their thinness, it is rare for capillaries to rupture, because blood pressure drops dramatically as blood passes through arterioles on its way to capillary beds.

Veins carry blood back to the heart after the blood passes through capillaries. Because blood is under relatively low pressure by the time it exits the tissues, veins have thinner walls and larger interior diameters than arteries do (**Figure 44.22b**). Large veins are compressed by muscle activity in the extremities (say, the hands and feet), which helps speed the return of blood to the heart. Larger veins also contain one-way **valves**, which are thin flaps of tissue that prevent any backflow of blood. All veins contain some muscle fibers, which contract in response to signals from the nervous system, decreasing the diameter and overall volume of the vessels. Blood pressure in a closed circulatory system is regulated, in part, by active adjustment of the volume of blood contained within the veins.

Blood, Interstitial Fluid, and the Lymphatic System The relatively high operating pressure of closed circulatory systems, combined with the thinness of capillaries, produces a small but steady leakage of fluid from the blood vessels into the surrounding space. The area between cells is called interstitial space, and the fluid that fills it is called **interstitial fluid**. Because blood cells are retained within capillaries, the interstitial fluid resembles plasma in its electrolyte composition.

In 1896 Ernest Starling proposed that two forces affect fluid flow across capillaries (**Figure 44.23**). The first is the outward-directed hydrostatic force created by the pressure on blood generated by the heart. This force is analogous to the pressure that drives water through the wall of a leaky garden hose. The second is an inward-directed osmotic force, created by the higher concentration of solutes in the blood plasma than in the interstitial space.

Starling reasoned that at the end of the capillary nearest to an arteriole, the hydrostatic force would exceed the osmotic force and fluid would move out of the capillary into the interstitial space. But because blood pressure drops as fluid passes through a long, thin tube, he proposed that the osmotic force would exceed the hydrostatic force at the venous end of the capillary. Thus, the fluid lost at the arteriolar end of the capillary would be largely reclaimed at the venous end.

Starling also proposed that any interstitial fluid that was not reclaimed in the capillary would be collected in the **lym-**

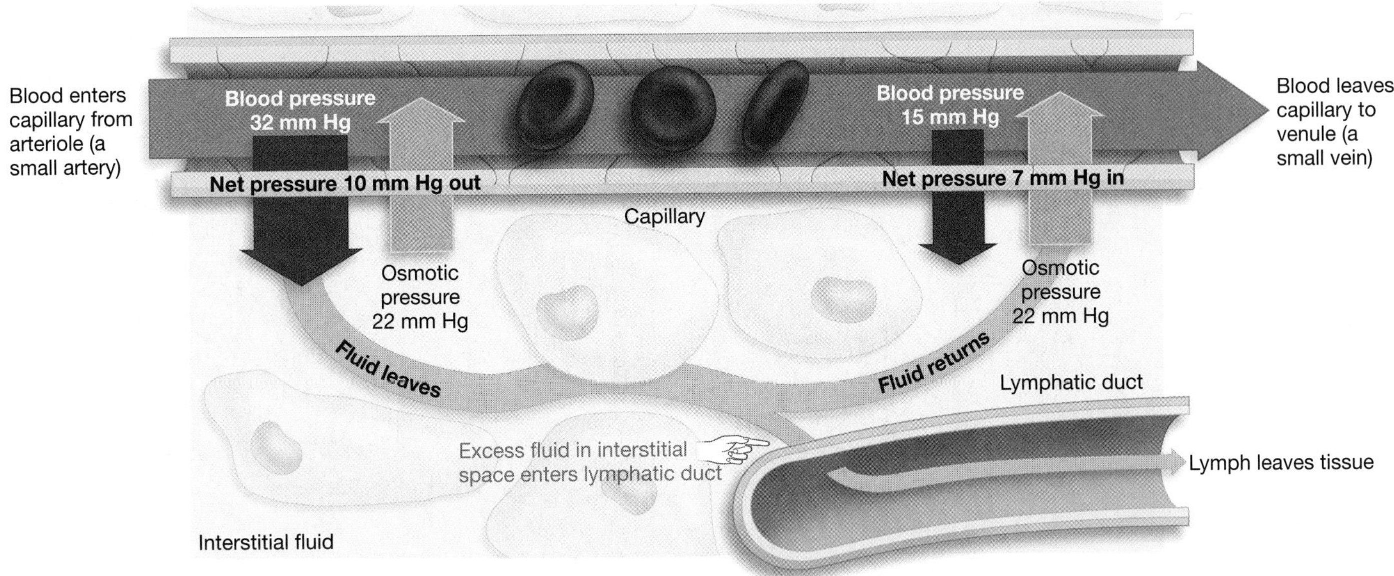

FIGURE 44.23 Pressure Differences in Capillaries Create Interstitial Fluid and Lymph. The balance of blood pressure and osmotic forces favors fluid loss from the beginning (inflow end) of capillaries and fluid recovery at the other (outflow) end. Fluid that is not recovered by the capillaries is transported out of the tissue as lymph.

phatic system, consisting of thin-walled, branching tubules called lymphatic ducts or vessels. It had been known for many years that these tiny ducts permeate all tissues and that they eventually join with one another like the tributaries of a river, to form larger vessels. The largest lymphatic vessels return excess interstitial fluid, in the form of **lymph**, to the major veins entering the heart. Starling's model was a comprehensive proposal for how blood, lymph, and the lymphatic ducts interact.

This model rested on a key assumption, however: For Starling's hypothesis to work, the total solute concentrations in plasma and interstitial fluid must be different enough to bring fluid into capillaries via osmosis. For this to be true, capillaries would have to act as filters. Specifically, large protein molecules or other solutes would have to be retained inside capillaries as fluid leaked out. To test this assumption, a group of biologists placed small tubes in the lymphatic ducts of animals so that they could collect lymph. Then the team introduced radioactively labeled proteins of various sizes into the blood inside veins. They found that even though small proteins crossed readily from the blood to the lymph, larger proteins did not. Enough large proteins were retained in the capillaries to set up the osmotic gradient required by Starling's model.

Follow-up work has shown that the most important of these large proteins is **albumin**. Albumin's large size and net negative charge effectively excludes it from interstitial fluid. The presence of albumin in blood keeps solute concentrations high, maintaining a strong osmotic gradient that brings fluid into capillaries. As a result, the total daily formation of lymph is limited to just 2–5 percent of plasma volume. This amount is easily taken up by lymph vessels and returned to the blood. To summarize, the lymphatic system consists of thin-walled vessels that take up excess fluid from throughout the body and return it to the veins of the circulatory system.

How Does the Heart Work?

In animals with closed circulatory systems, the heart contains at least two chambers. There is at least one thin-walled **atrium** (plural: **atria**), which is a chamber that receives blood, and at least one thick-walled chamber called a **ventricle**, which generates the force required to propel blood out of the heart and through the system. Atria are separated from ventricles by atrioventricular valves.

When studies of vertebrate hearts and circulatory systems had advanced enough to make comparisons among groups possible, biologists noticed several important patterns. To understand these, study **Figure 44.24**. This phylogenetic tree shows the evolutionary relationships among some major vertebrate groups. The sketches for each group offer a simplified version of the heart and circulatory system found in each lineage. Two points are particularly important to note:

1. The number of distinct chambers in the heart increased as vertebrates diversified. Fish hearts have two chambers; amphibians have three; turtles and lizards have five (if the partially divided ventricle is counted as three chambers); crocodiles, birds, and mammals have four.

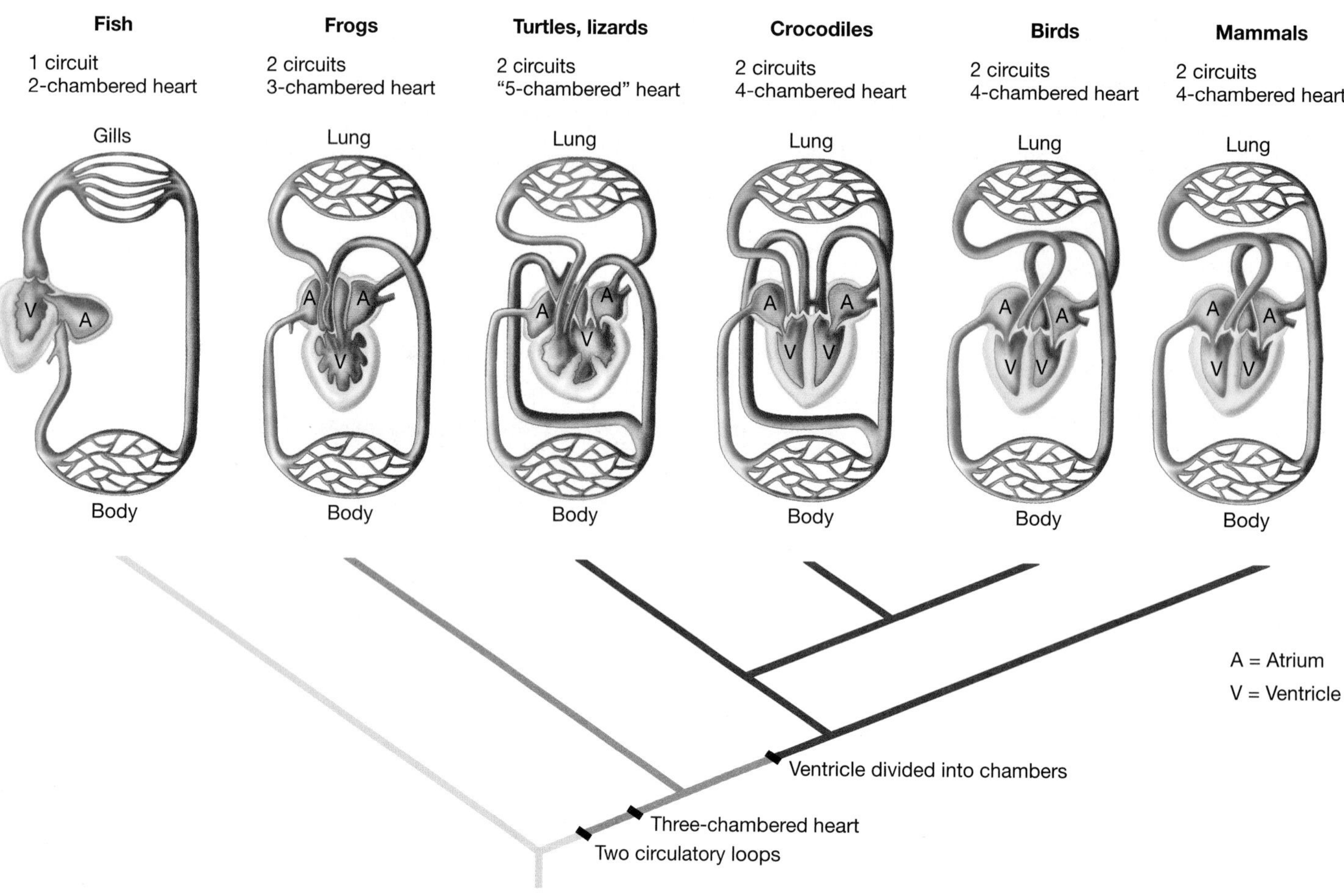

FIGURE 44.24 The Evolution of the Vertebrate Circulatory System. "A" denotes the atria—chambers that receive blood coming into the heart from the body and the lungs; "V" denotes the ventricles—chambers that pump blood out to the lungs and the body.

● **EXERCISE** Turtles (and relatives) and crocodiles have a blood vessel that shunts blood between their two circuits, as needed. Label these shunts on the diagram.

● **EXERCISE** Add arrows to major blood vessels, showing the direction of blood flow.

2. In fish, the circulatory system forms a single circuit—one loop services the gills and the body. In other groups, there are separate circuits to the lungs and to the body.

To understand these evolutionary patterns, let's first consider the relatively simple heart and circulatory system found in fish. A two-chambered heart and single circuit are adequate in fish. Even though blood pressure drops as blood passes through the gills—due to the mechanical resistance to flow that occurs in the gills' capillary beds—blood pressure stays high enough to move blood throughout the body. This is largely because fish live in the neutrally buoyant environment of water, where gravity does not have a large impact on blood flow.

In contrast, circulation in land-dwelling vertebrates is much more affected by the force of gravity. Gravity is particularly important in opposing blood flow to elevated portions of the body. To overcome gravity in terrestrial environments, blood must be pumped at high pressure. But high pressure blood flow presents a problem, because the capillaries and alveoli of the lungs are too thin to withstand high pressures.

The successful solution used by land-dwelling vertebrates was the evolution of two separate pumping circuits. The **pulmonary circulation** is a lower-pressure circuit to and from the lung, while the **systemic circulation** is a higher-pressure circuit to and from the rest of the body. The pulmonary and systemic circulations are completely separated in the four-chambered hearts of birds and mammals, which are essentially two fish hearts side by side.

The pulmonary and systemic circulations are only partially separated in the three-chambered hearts of amphibians and the "five-chambered" hearts of turtles and lizards, however. In addition, some of these animals have a bypass vessel running from the right side of the ventricle directly into the systemic circulation. This trait is also observed in the unusual four-chambered hearts of crocodiles. The bypass vessel has an important function: It allows blood to be shunted from the pulmonary to the systemic circulation when the animal is underwater. As a result,

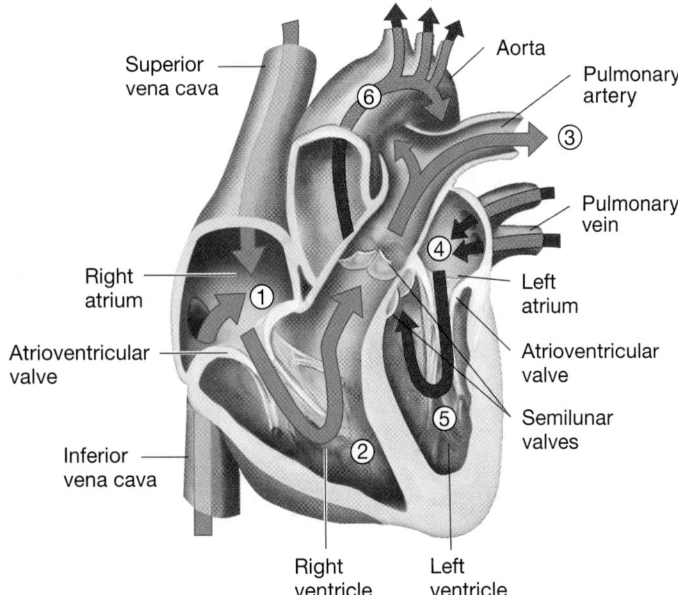

FIGURE 44.25 The Human Heart. Blood flows through a four-chambered heart in the sequence shown.

EXERCISE Circle the valves that separate the atria and ventricles, and the valves that separate the ventricles from the pulmonary artery (to lungs) and aorta (to body).

the individual greatly reduces blood flow to the lungs when it is not breathing.

The Human Heart Your heart is located in the chest cavity between your lungs and is roughly the size of your fist. As **Figure 44.25** shows, the circulatory system returns blood from the body to the right atrium of the human heart through two large veins called the inferior (lower) and superior (upper) **venae cavae** (singular: **vena cava**). This blood is low in oxygen. When the muscles that line the right atrium contract, this deoxygenated blood is sent to the right ventricle. The right ventricle, in turn, contracts and sends blood out to the lungs, via the **pulmonary artery**. In this way, the right ventricle powers the movement of blood through the pulmonary circulation. Blood that returns to the heart from the body is sent to the lungs.

Blood flows from atrium to ventricle to artery only one way, because one-way valves separate the heart's chambers from each other and from the adjacent blood vessels. As Figure 44.25 indicates, the valves are flaps, oriented in a way that ensures a one-way flow of blood with little or no backflow. If heart valves are damaged or defective, the resulting backflow can be heard through a stethoscope. The backflow reduces the organ's efficiency and is called a **heart murmur**.

After blood has circulated through the capillary beds in the lung's alveoli and becomes oxygenated, it returns to the heart through the **pulmonary veins**. The oxygenated blood enters the left atrium. When this chamber contracts, blood is pushed into the left ventricle. The walls of the left ventricle are so thick with muscle cells that their contraction sends oxygenated blood at high pressure out the aorta and into the arteries and capillaries that form the systemic circulation.

Figure 44.26 summarizes the flow pattern through the human circulatory system and the blood gas concentrations at various points in the pulmonary and systemic circulations. Notice that blood vessels are called arteries or veins according to the direction of blood flow relative to the heart, not because of the oxygen content of the blood in them. Thus, the pulmonary artery is called an artery because it takes blood away from the heart, even though this blood is low in oxygen content.

The data on gas partial pressures in Figure 44.26 make another important point: In healthy individuals, gas exchange in the lungs and tissues is rapid relative to the rate at which blood flows through capillaries. This situation ensures that the maximum amount of oxygen is taken up and the maximum amount of carbon dioxide is released. Thus, the partial pressures of oxygen and carbon dioxide in the alveoli and the pulmonary veins are equal. Similarly, the maximum amount of oxygen is released from blood to tissues and the maximum amount of carbon dioxide is taken up—resulting in partial pressures for oxygen and carbon dioxide in the tissues and the systemic veins that are equal.

The Cardiac Cycle Although the preceding discussion suggests that each chamber in the heart acts independently, this is not the case. The contraction phase of the atria and the ventricles, called **systole**, is closely coordinated with their relaxation phases, or **diastole**. For example, both atria contract simultaneously about 0.1 second ahead of ventricular con-

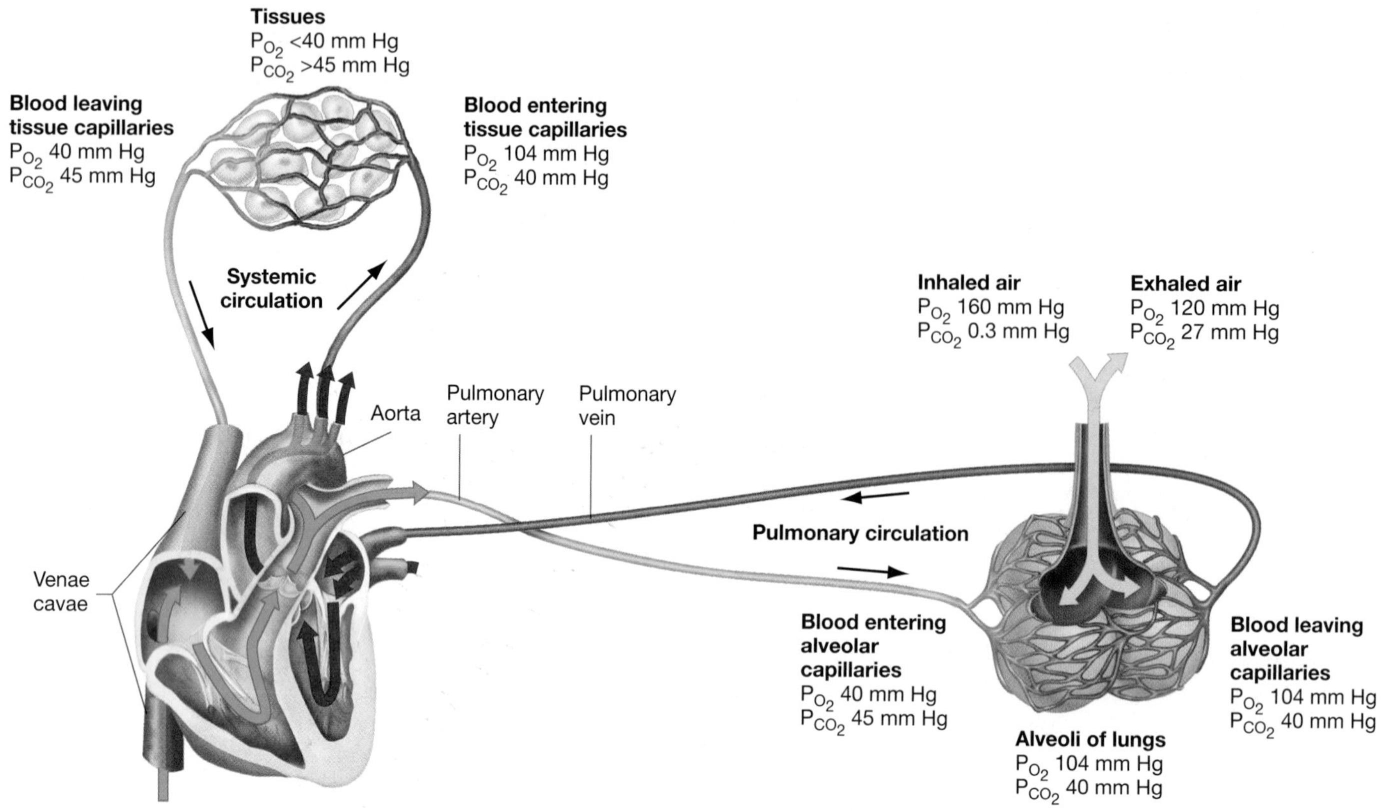

FIGURE 44.26 Partial Pressures of Gases Vary throughout the Human Circulatory System. Partial pressures of oxygen and carbon dioxide at various points in the human circulatory system.

QUESTION Why are the partial pressures of oxygen and carbon dioxide in exhaled air intermediate in magnitude between the partial pressures in inhaled and alveolar air?

traction. Because both ventricles are still in diastole and relaxed when the atria contract, the ventricles fill with blood from the atria just before they enter systole and contract. This sequence of contraction and relaxation is called the **cardiac cycle**. It consists of one diastole and one systole for both atria and ventricles.

Ventricular contraction (ventricular systole) leads to a rapid increase in pressure within both ventricles (**Figure 44.27**). Blood is ejected into the pulmonary artery and the aorta when ventricular pressure equals the pressure within each respective artery. Blood pressure measured in the systemic arterial circulation at the peak of ventricular ejection into the aorta is called the **systolic blood pressure.** Blood pressure measured just prior to ventricular ejection is called the **diastolic blood pressure**. Clinicians report blood pressure measurements in fractional notation, with systolic pressure as the numerator and diastolic pressure as the denominator. For the example shown in Figure 44.27, blood pressure is 120/80 mm Hg. People with blood pressures in excess of 150/90 mm Hg are considered to have high blood pressure, or **hypertension**. The measurement of blood pressure and some of the health consequences of hypertension are discussed further in **Box 44.3**.

at www.masteringbio.com
The Human Heart

Electrical Activation of the Heart How is the activity of each chamber of the heart coordinated? Cardiac muscle cells, like other muscle cells, contract in response to electrical signals. In invertebrates, the electrical impulses that trigger contraction come directly from the nervous system. But in vertebrates, a group of cells in the heart itself is responsible for generating the initial signal. The cells that initiate contraction, known as **pacemaker cells**, are located in a region of the right atrium called the **sinoatrial (SA) node** (**Figure 44.28**). The SA node and the muscle cells of the heart receive input from the nervous system and from chemical messengers carried in the blood. These inputs are important for regulating both the heart rate and the strength of ventricular contraction. However, a vertebrate heart will beat even if all nerves supplying it are severed.

The electrical impulse generated by the SA node is rapidly conducted throughout the right and left atria. The signal spreads quickly from cell to cell because of a striking property of cardiac muscle cells: They form physical and electrical connections with each other. More specifically, all cardiac muscle

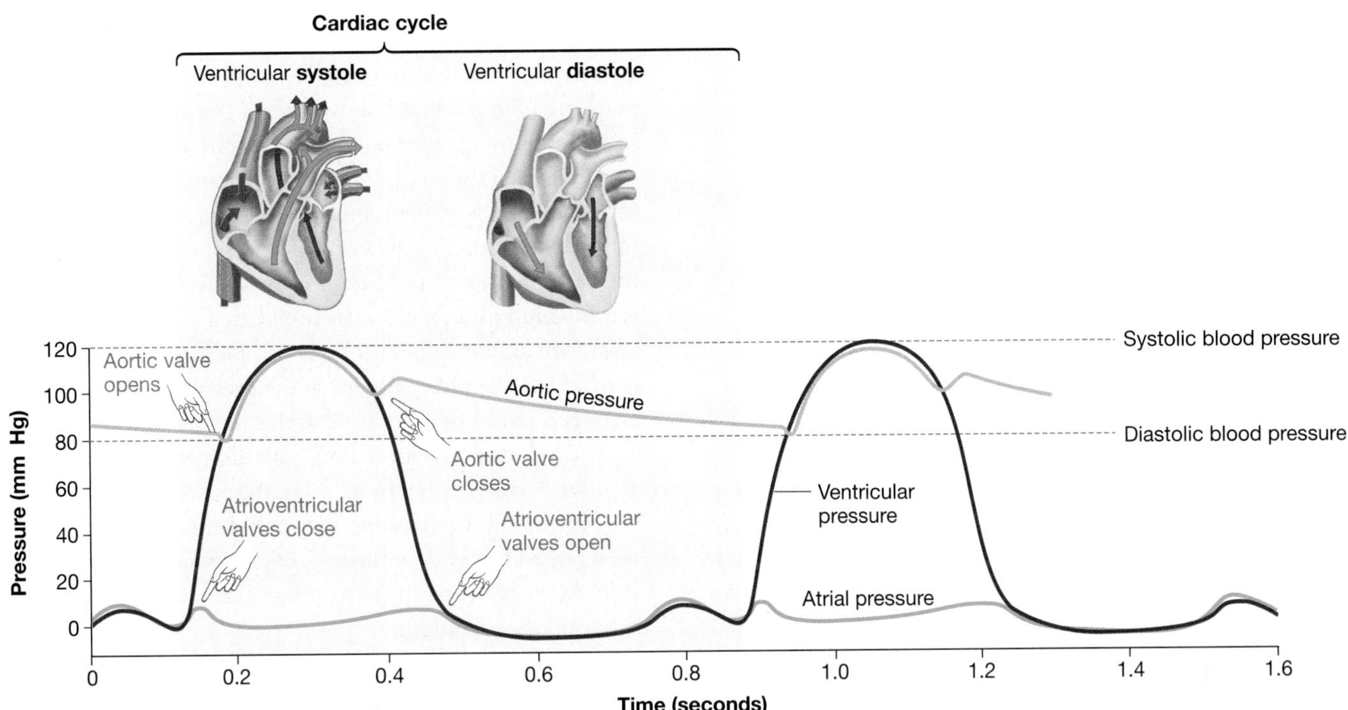

FIGURE 44.27 Blood Pressure Changes during the Cardiac Cycle. Pressures created by the left ventricle and aorta in the cardiac cycle. In this case, the blood pressure measured in the upper arm would be 120/80 mm Hg. The right ventricle and pulmonary artery pressures follow a similar pattern, but the blood pressure measured in the pulmonary artery would be closer to 25/8 mm Hg.

QUESTION Why does the left aortic valve open shortly after the beginning of ventricular systole and close slightly before the end of systole?

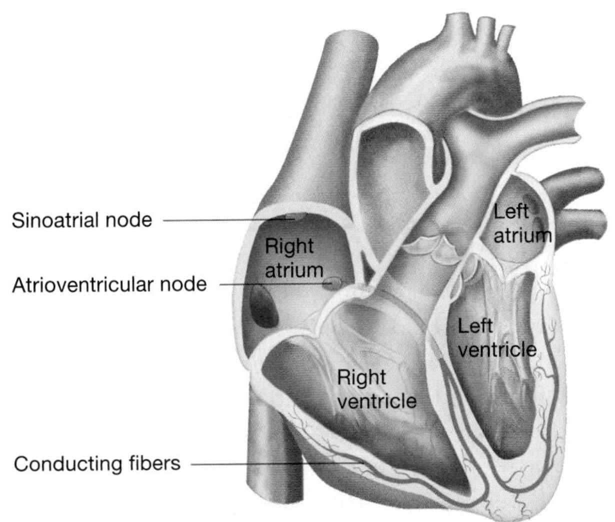

FIGURE 44.28 Sequential Electrical Activation Leads to Coordinated Contraction of the Heart. Electrical activation of the heart begins with the pacemaker cells in the sinoatrial node, spreads through the electrically connected muscle cells of the atria, and converges on the atrioventricular node, where it is delayed long enough to allow the atria to eject blood into the ventricles. The ventricles are then activated rapidly by the spread of electricity through specialized conducting fibers in the ventricular walls.

EXERCISE In this diagram, the heart is sectioned through the pulmonary artery. Circle the conducting fibers just dorsal to this artery.

cells branch to contact several other cells, are joined end to end with these neighboring cells, and are connected by specialized structures called **intercalated discs**. Because these discs contain numerous gap junctions—cell-to-cell connections described in Chapter 8—electrical signals pass directly from one cardiac muscle cell to the next. (In Chapter 45, we'll explore how electrical signals in animals are initiated and propagated.) As a result, the atria contract simultaneously and fill the ventricles.

Once the electrical signal has swept over the atria, however, it is conducted to an area of the heart called the **atrioventricular (AV) node**, which delays the signal slightly before passing it to the ventricles. This delay allows the ventricles to fill completely before they contract. After the delay, the electrical impulse is rapidly transmitted through specialized conducting fibers lying within the muscular wall that separates the ventricles. The impulse spreads through both ventricles, causing them to contract as the atria relax. The ventricles empty efficiently because the signal and the resulting muscular contraction move up to the top of each ventricle and then move forward to the arteries that allow blood to exit.

The electrical events responsible for the synchronized contraction of the atria and ventricles are summarized in

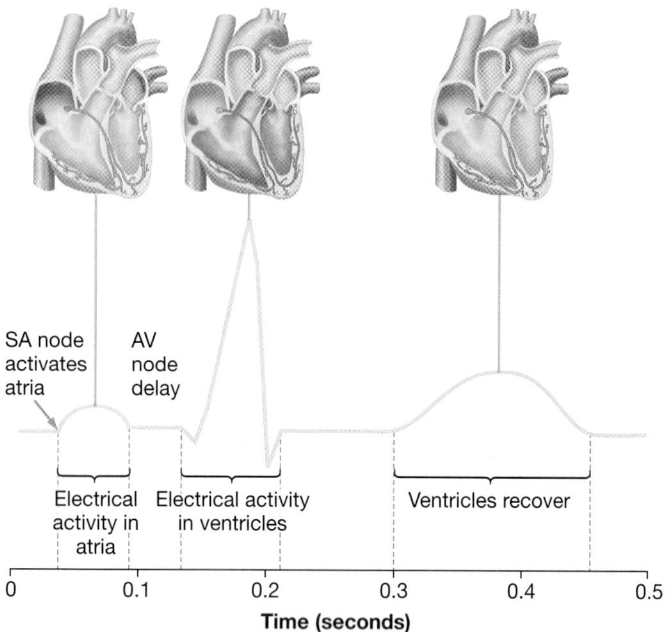

FIGURE 44.29 EKGs Record Electrical Events Associated with Cardiac Muscle Contraction.

QUESTION The electrical activity generated by ventricular contraction is much larger and more compressed than the electrical activity generated by atrial contraction. Why?

Figure 44.29. This figure also shows the recording known as an **electrocardiogram**, or **EKG**, which corresponds to the electrical activity associated with cardiac muscle contraction. The EKG recording is generated by amplifying the overall electrical signal conducted from the heart to the chest wall through the tissues of the body. By inspecting the waves generated by the EKG, physicians can diagnose disturbances of heart rhythm and detect damage to the heart muscle.

Patterns in Blood Pressure and Blood Flow

As blood moves through capillaries, blood pressure drops dramatically. **Figure 44.30** shows why. The key point is that as arteries branch, rebranch, and eventually form networks of capillaries, the total cross-sectional area of blood vessels in the circulatory system increases—even though the size of individual blood vessels decreases as blood moves away from the heart. Because the amount of mechanical resistance to flow is a function of total cross-sectional area of the vessels, friction losses are high in capillary beds and blood pressure drops dramatically. The pulsing nature of the pressure noted in the larger arteries diminishes as a result of this pressure drop.

The velocity of blood flow also decreases significantly in capillary beds relative to arteries and veins, because the same amount of fluid is passing through a much larger area. The slow rate of blood flow through capillaries is important: It provides sufficient time for gases, nutrients, and wastes to diffuse between tissues and blood.

These general patterns of blood pressure and blood flow don't tell the entire story, however, because blood movement throughout the body is also carefully regulated. Recall that arterioles have muscular sphincters in their walls, and that changes in the size of these sphincters can constrict or allow blood flow. This means that the nervous system, along with certain chemical messengers in the circulation, can accurately control blood flow to various tissues by contracting or relaxing the arteriolar sphincters. For example, arterioles in the skin may dilate during exercise. When they do, they divert blood flow to the skin for the purpose of eliminating excess heat. This accounts for the flushed facial appearance often induced by vigorous exercise.

As another example of how blood pressure and blood flow are regulated, consider what happens if you sit long enough for blood to pool in your lower extremities, under the influence of

BOX 44.3 Hypertension and the Measurement of Blood Pressure

Hypertension is a serious concern to physicians because it can lead to a variety of defects in the heart and circulatory system. Abnormally high blood pressure puts mechanical stress on arteries. If the walls of an artery fail, the individual may experience heart attack, stroke, kidney failure, and burst or dilated blood vessels.

Physicians often rely on blood pressure measurements to diagnose hypertension. A blood pressure reading—especially when combined with a measurement of resting pulse rate and cholesterol counts, gives clinicians relatively quick, accurate, and inexpensive data on the health risk to a patient's heart and circulatory system.

To measure a person's blood pressure, a clinician wraps an inflatable cuff attached to a pressure gauge around the upper arm. When the cuff is fully inflated, the pressure it exerts is transmitted evenly throughout the arm and closes down the main artery in this part of the body. The clinician then uses a stethoscope to listen for the sounds of blood flow downstream from the cuff.

At first, the clinician hears no pulse sounds through the stethoscope, because no blood is flowing downstream of the cuff. As the clinician slowly releases air from the cuff, however, the pressure in the cuff begins to fall until it is just below the systolic pressure in the artery. At this point, blood begins to spurt through the artery with each systolic peak. The flow is turbulent, so the clinician can detect a thumping pulse sound in the stethoscope. The cuff pressure when sound is first detected is equal to the systolic pressure. As more air is released from the cuff, its pressure falls below the diastolic pressure in the artery. At this point, blood flow in the artery is smooth and the sound of turbulent flow fades away. The cuff pressure when the sound becomes muffled or disappears represents the diastolic pressure.

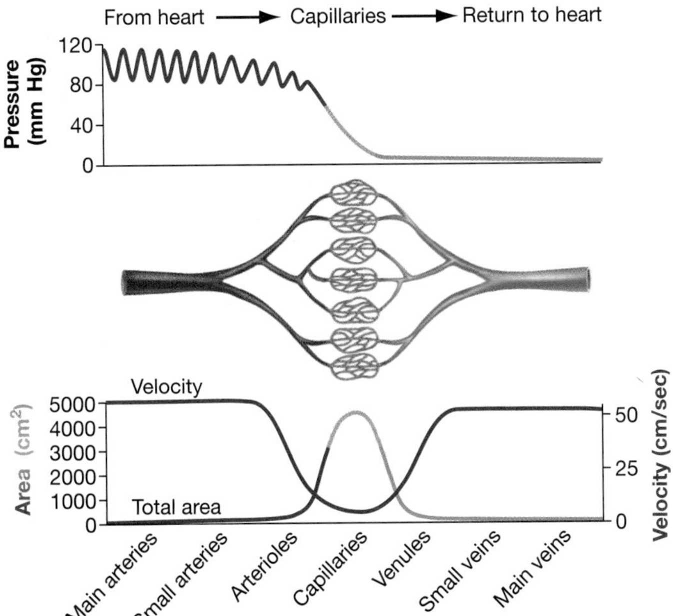

FIGURE 44.30 Blood Pressure Drops Dramatically in the Circulatory System. The top graph shows how blood pressure changes as blood leaves the heart and travels through arteries, capillaries, and veins, as in the branching pattern of the middle diagram. In arteries near the heart, each heartbeat causes fluctuations in blood pressure. These pressure pulses disappear in the capillaries, so blood flows there at a steady speed. The bottom graph plots the total area of blood vessels shown in the diagram, as well as the velocity of blood flow through the vessels.

EXERCISE Label the systolic and diastolic blood pressures in the top graph.

gravity. If you stand up rapidly, your blood pressure can drop enough to limit blood flow to the brain and cause dizziness, lightheadedness, or even a blackout. More serious drops in blood pressure, due to severe dehydration or blood loss following an injury, can lead to death. Fortunately, decreases in blood pressure elicit a powerful homeostatic response.

Recall from Chapter 41 that all homeostatic responses involve (1) sensors that detect the change in condition, (2) an integrator that processes information about the change, and (3) effectors that diminish the impact of the change. In the case of the circulatory system, falling blood pressure is detected by specialized pressure-sensing nerve cells called **baroreceptors**. Baroreceptors are found in the walls of the heart and the major arteries, both as they leave the heart and as they enter the neck. This distribution is logical, because the head tends to be the most elevated part of land-dwelling animals and because the preservation of blood flow to the brain is the highest priority.

When baroreceptors transmit signals to the brain indicating a serious fall in blood pressure, a rapid, three-component effector response ensues:

1. An increase in cardiac output, meaning the volume of blood leaving the left ventricle, is brought about by both an increase in heart rate and an increase in stroke volume—the amount of blood ejected from the ventricles during each cardiac cycle. (Cardiac output = heart rate × stroke volume.)

2. The constriction of arterioles serving the capillaries of tissues such as the skin and intestines, which can endure short-term restrictions in their blood supply. This arteriolar constriction diverts blood to more critical organs.

3. A generalized constriction of the veins, resulting in a decrease in their overall volume. Because more than half of the blood in the circulatory system is contained within the veins, constriction of these vessels shifts blood volume toward the heart and arteries to maintain blood pressure and flow to vital organs.

This coordinated response is mediated both by a portion of the nervous system called the sympathetic nervous system (Chapter 45) and by hormones (Chapter 47) produced by the adrenal glands. Sympathetic nerves and the hormones involved in blood pressure regulation deliver their messages directly to the SA node to increase heart rate, the ventricles of the heart to increase stroke volume, the sphincters of the arterioles to modify their resistance, and the muscular walls of the veins to modify their total volume.

The homeostatic response to falling blood pressure resembles the **fight-or-flight response**, which is triggered by a threatening stimulus (see Chapter 47). During the fight-or-flight response, an organism prepares to defend itself or rapidly escape from danger. Part of this response involves directing blood to the brain and muscles in preparation for quick action. The side effects include the "cold sweat" and nausea often induced by a fearful situation. These feelings occur because blood flow is directed away from the skin and intestines.

Check Your Understanding

If you understand that...

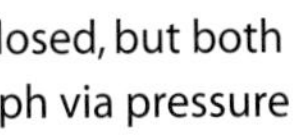

- Animal circulatory systems may be open or closed, but both types of systems circulate blood or hemolymph via pressure generated by one or more hearts.
- In closed systems, regulated changes in the diameter of blood vessels can direct blood to specific regions, and overall blood pressure is carefully regulated through changes in cardiac output.

You should be able to...

1) Explain how the mammalian lymphatic system and circulatory system interact.
2) Make a labeled diagram showing how blood circulates through the mammalian heart.
3) Explain why blood pressure drops dramatically as blood moves from arterioles through capillary beds.

Chapter Review

SUMMARY OF KEY CONCEPTS

Animals have to take in oxygen and expel carbon dioxide to sustain cellular respiration and stay alive. Terrestrial animals and aquatic animals face different challenges in performing gas exchange.

As media for exchanging oxygen and carbon dioxide, air and water are dramatically different. Compared with water, air contains much more oxygen and is much less dense and viscous. As a result, terrestrial animals have to process a much smaller volume of air to extract the same amount of O_2 and don't have to work as hard to do so.

Both aquatic and terrestrial animals pay a price for exchanging gases. Land-dwellers lose water to evaporation during ventilation. Freshwater animals lose ions through diffusion and gain excess water by osmosis. Marine animals gain sodium and chloride through diffusion and lose water by osmosis.

You should be able to explain why no water breathers have the extremely high rates of metabolism required for endothermy.

The organs responsible for gas exchange maximize the rate of oxygen and carbon dioxide diffusion by presenting a large, thin surface area to the environment. These organs also function in a way that maintains a steep partial-pressure gradient that favors entry of oxygen and elimination of carbon dioxide.

Natural selection has resulted in the evolution of gills, tracheae, lungs, and other gas-exchange organs that minimize the cost of ventilation while maximizing the rate at which O_2 and CO_2 diffuse. Consistent with predictions made by Fick's law of diffusion, respiratory epithelia tend to be extremely thin and folded to increase surface area. In fish gills, countercurrent exchange ensures that the partial-pressure differences between O_2 and CO_2 in water and blood are high over the entire length of the ventilatory surface. In bird lungs, structural adaptations lead to a much greater ratio of useful ventilatory space to dead space than is observed in the lungs of mammals. The rate of breathing is regulated by an area in the brain to keep the carbon dioxide content of the blood stable during both rest and exercise.

You should be able to explain why no large animals have gas exchange occurring only across the skin.

Blood is a specialized tissue that transports gases, along with nutrients and wastes, in some animals. Hemoglobin is an oxygen-carrying protein with characteristics that make it extremely efficient at taking up oxygen in the lungs and delivering it to tissues.

About half the blood volume in a typical human consists of red blood cells that are specialized for gas exchange. Red blood cells lack most organelles but contain hundreds of millions of hemoglobin molecules. Each hemoglobin molecule can carry up to four oxygen molecules. The tendency of hemoglobin to give up these oxygen molecules varies as a function of the P_{O_2} in surrounding tissue in a sigmoidal fashion. As a result, a relatively small change in tissue P_{O_2} causes a large change in the amount of oxygen released from hemoglobin. In addition, oxygen binds less tightly to hemoglobin when pH is low. Because CO_2 tends to react with water to form carbonic acid, the existence of high CO_2 partial pressures in exercising muscle tissues lowers their pH and makes oxygen less likely to stay bound to hemoglobin and more likely to be released into tissues.

The CO_2 that diffuses into red blood cells from tissues is rapidly converted to carbonic acid by the enzyme carbonic anhydrase. The protons that are released as carbonic acid dissociates bind to deoxygenated hemoglobin. In this way, hemoglobin acts as a buffer that takes protons out of solution and prevents large fluctuations in blood pH.

You should be able to explain how oxygen and carbon dioxide are transported in blood in the icefish native to the Antarctic, since these fish lack hemoglobin.

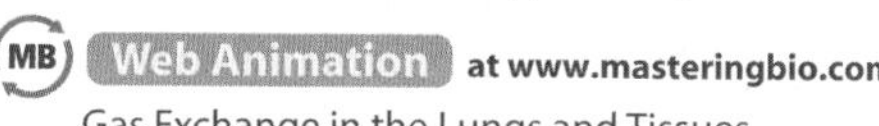

at www.masteringbio.com

Gas Exchange in the Lungs and Tissues

Circulatory systems use the pressure generated by one or more hearts to transport blood and other substances throughout the body.

In many animals, blood or hemolymph moves through the body via a circulatory system consisting of a pump (heart) and vessels. In open circulatory systems, overall pressure is low and tissues are bathed directly in hemolymph. In closed circulatory systems, blood is contained in vessels that form a continuous circuit. Containment of blood allows higher pressures and flow rates, as well as the ability to direct blood flow accurately to tissues that need it the most. In organisms with a closed circulatory system, a lymphatic system returns excess fluid that leaks from the capillaries back to the circulation.

In mammals and birds, a four-chambered heart pumps blood into two circuits, which separately serve the lungs and the rest of the body. Although the heart rate is controlled by electrical signals that originate in the heart itself, heart rate, cardiac output, and contraction of both arterioles and veins are constantly modified by signals from the sympathetic nervous system and the adrenal glands. These signals act to preserve blood flow to vital organs in response to falling blood pressure or situations that are perceived as threatening.

You should be able to describe the circulatory systems expected to be found in terrestrial animals with high versus low activity rates, and in terrestrial versus aquatic animals.

at www.masteringbio.com

The Human Heart

QUESTIONS

Test Your Knowledge

1. O_2 will diffuse from blood to tissue faster in response to which of the following conditions?
   a. an increase in the P_{O_2} of the tissue
   b. a decrease in the P_{O_2} of the tissue
   c. an increase in the thickness of the capillary wall
   d. a decrease in the surface area of the capillary

2. Which of the following does blood *not* do?
   a. transport O_2 and CO_2
   b. distribute body heat
   c. produce red blood cells and other formed elements
   d. buffer against pH changes

3. How is most carbon dioxide transported in the blood?
   a. as a gas in the plasma
   b. bound to amino groups on hemoglobin
   c. bound to the heme group of hemoglobin
   d. as bicarbonate ions (HCO_3^-) in plasma

4. Which of the following is *not* an advantage of breathing air over breathing water?
   a. Air is less dense than water, so it takes less energy to move during ventilation.
   b. Oxygen diffuses faster through air than it does through water.
   c. The oxygen content of air is greater than that of an equal volume of water.
   d. Air breathing leads to high evaporation rates from the respiratory surface.

5. Which of the following promotes oxygen release from hemoglobin?
   a. a decrease in temperature
   b. a decrease in CO_2 levels
   c. a decrease in pH
   d. a decrease in carbonic anhydrase

6. An open circulatory system is less efficient than a closed circulatory system in what respect?
   a. It is harder to deliver O_2 to specific tissues based on need.
   b. Hemolymph does not contain oxygen-binding molecules.
   c. There is no heart to pump the blood.
   d. In closed systems, body movements cannot help circulate the blood.

Test Your Knowledge answers: 1. b; 2. c; 3. d; 4. d; 5. c; 6. a

Test Your Understanding

Answers are available at www.masteringbio.com

1. Compare and contrast the respiratory and circulatory systems of an insect and a human. What are some advantages and disadvantages of each system?

2. Why is the sigmoidal shape of the oxygen-hemoglobin equilibrium curve significant? Describe the changes in oxygen delivery that occur as a person changes from a resting state to intense exercise. How does the Bohr shift affect O_2 delivery to tissues?

3. Explain how carbon dioxide is transported in the blood. In humans, why doesn't intense exercise and rapid production of CO_2 lead to a rapid reduction in blood pH?

4. Draw a diagram summarizing how bird lungs are ventilated. Why is this pattern of ventilation considered much more efficient than the respiratory system of humans and other mammals?

5. Review Fick's law of diffusion. Explain how each parameter in the equation is reflected in the structure of ventilatory organs.

6. Explain why surface tension helps make the lungs of mammals elastic. Why does the large surface area in alveoli increase elasticity?

Applying Concepts to New Situations

Answers are available at www.masteringbio.com

1. Researchers who compared the total amount of ventilatory surface area in the gills of various species of fish found that species that do a lot of high-speed swimming have a larger total surface area than do slow-moving species of the same size. Interpret this pattern in light of the theory of evolution by natural selection.

2. At high elevations, certain cells in humans increase the production of 2,3-diphosphoglycerate (DPG). In blood, DPG shifts the oxygen-hemoglobin equilibrium curve to the right. How does this shift affect O_2 unloading and loading?

3. Carp are fish that thrive in stagnant-water habitats with low-oxygen partial pressures. Compared with the hemoglobin of many other fish species, carp hemoglobin has an extremely high affinity for O_2. Is this trait adaptive? If so, why?

4. The carbon monoxide (CO) found in furnace and engine exhaust and in cigarette smoke binds to the heme groups in hemoglobin 210 times more tightly than does O_2. Explain why exposure to large doses of CO can lead to suffocation.

www.masteringbio.com is also your resource for • Answers to text, table, and figure caption questions and exercises • Answers to *Check Your Understanding* boxes • Online study guides and quizzes • Additional study tools including the *E-Book for Biological Science* 3rd ed., textbook art, animations, and videos.

45 Electrical Signals in Animals

KEY CONCEPTS

- Neurons are cells that transmit electrical signals used in communication. Neurons have a membrane potential, which is an electric potential difference, or voltage, across the plasma membrane. A voltage exists due to differences in the concentrations of ions on the two sides of the plasma membrane.
- Electrical signals are transmitted as graded or all-or-none changes in membrane voltage. All-or-none changes are called action potentials. During an action potential, an inflow of sodium ions is followed by an outflow of potassium ions.
- At synapses, an electrical signal from one neuron is converted to a chemical signal in the form of a neurotransmitter. When it arrives at an adjacent neuron, the neurotransmitter causes a change in that cell's membrane potential.
- Animals have a central nervous system (CNS) and a peripheral nervous system (PNS). PNS neurons receive sensory information and transmit it to the CNS for processing. The CNS then sends signals to muscles, glands, or other target tissues via other PNS neurons.

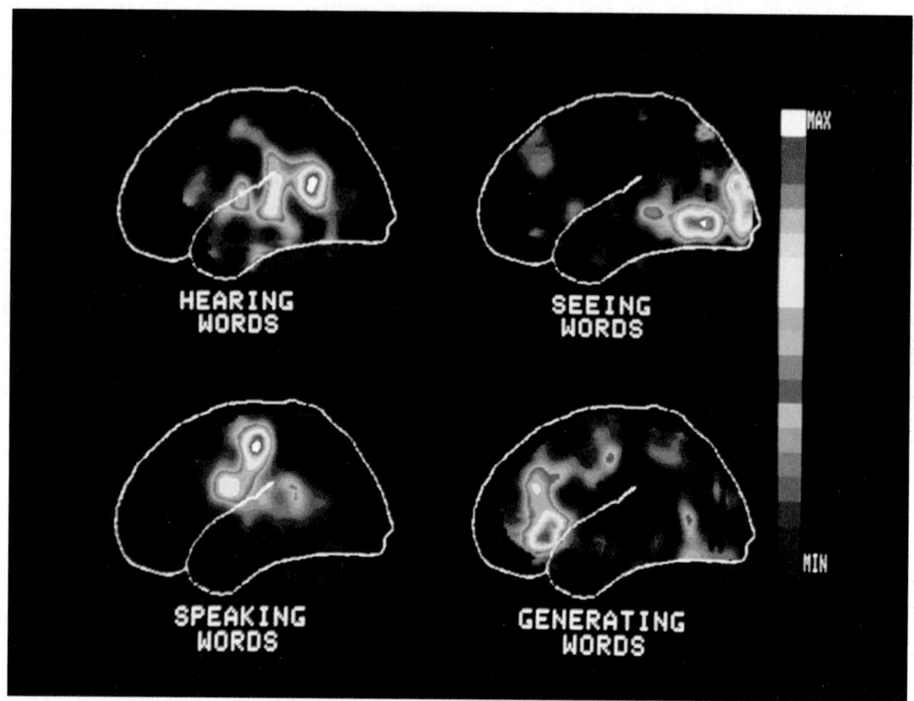

Positron-emission tomography (PET) scans showing changes in the activity of brain cells during different tasks, such as remembering words and speaking them aloud. Different brain areas are specialized for performing different tasks.

If you are unfortunate enough to touch a hot stove burner, your hand will jerk away before your brain perceives pain. When a trout is about to be attacked by an osprey, the trout's tail thrashes so rapidly that the movement is a blur. These and other animal movements are triggered by electrical impulses. Sensory cells in your finger and in the trout's eye send information to the brain and spinal cord, which send electrical signals to muscles in your arm and the trout's tail. Electrical signals are fast and precise. They are conducted from point to point by nerve cells, called **neurons**, at speeds of up to 200 m/sec (450 mph). The result is movement that saves your finger from extensive tissue damage—or saves the trout's life.

The goal of this chapter is to explore how electrical signaling occurs. Section 45.1 introduces the general anatomy of a nerve cell, discusses why a voltage exists across the cell's plasma membrane, and investigates how electrical currents are produced in cells. The next three sections focus on how neurons generate and propagate electrical signals, how information is transferred from cell to cell, and how neurons work in groups to integrate information from different sources. One of this chapter's most fundamental messages is that complex processes such as seeing, moving, and thinking are based on seemingly simple events: flows of ions across plasma membranes.

Key Concept Important Information Practice It

45.1 Principles of Electrical Signaling

Nerve cells, or neurons, transmit information. In many cases, the information they carry originates in sensory cells that are located on the periphery of the body. Sensory receptors inside the skin, eyes, ears, and nose transmit streams of data about the external environment in the form of electrical signals. Similarly, sensory cells inside the body monitor conditions that are important in homeostasis, such as blood pH and oxygen levels. In this way, sensory cells monitor conditions both outside and inside the body.

As **Figure 45.1a** shows, a sensory receptor cell transmits the information it receives by means of a nerve cell known as a **sensory neuron**. (In many cases, the receptor cell itself also acts as a sensory neuron.) In vertebrates, the sensory neuron sends the information to neurons in the brain or spinal cord via nerves; in other animals nerves send the information to the brain. **Nerves** are long, tough strands of nervous tissue. They contain thousands of projections from neurons and carry information to and from the brain and spinal cord. Together, the brain and spinal cord form the **central nervous system**, or **CNS**. One function of the CNS is to integrate information from many sensory neurons. Cells in the CNS called **interneurons** (literally, "between-neurons") perform this integration. Interneurons make connections between sensory neurons and **motor neurons**, which are nerve cells that send signals to effector cells in glands or muscles. Recall from Chapter 41 that effectors are structures that respond to stimuli. All neurons and other components of the nervous system that are outside the CNS are considered part of the **peripheral nervous system**, or **PNS**. (Section 45.4 describes the structure and function of the PNS in more detail.)

Typically, sensory information from receptors in the PNS is sent to the CNS, where it is processed. Then a response is transmitted back to appropriate parts of the body. When you stub your toe, pain receptors in your toe relay sensory information rapidly to the brain, which then modifies your movements to avoid further injury. You might limp to avoid further contact with that toe. In some cases, however, information from a sensory neuron never reaches the brain. This is what happens when you touch a hot burner and jerk your hand back (**Figure 45.1b**). When pain receptors are stimulated, sensory neurons activate interneurons in the spinal cord that directly activate motor neurons serving muscles in the arm. The signals from these motor neurons stimulate muscles that pull the lower part of your arm toward the body and inhibit muscles that extend the lower arm. The movement is a **reflex**—a direct response to a signal without integration by the brain. It is fast and automatic. Later, when your brain integrates information from pain receptors in your hand and from light receptors as you look at the burn, other motor neurons might direct your hand as it plunges into cold water. What do the cells that carry and process this information look like?

The Anatomy of a Neuron

Compared with studying cells in the kidney, digestive system, or lungs, documenting the anatomy of neurons was difficult. Neurons are small, transparent, and morphologically complex. So

(a) The brain integrates sensory information and sends signals to effector cells.

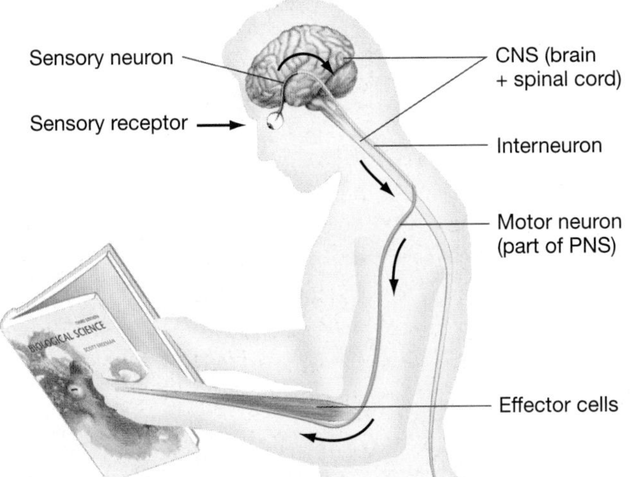

(b) When reflexes occur, sensory information bypasses the brain.

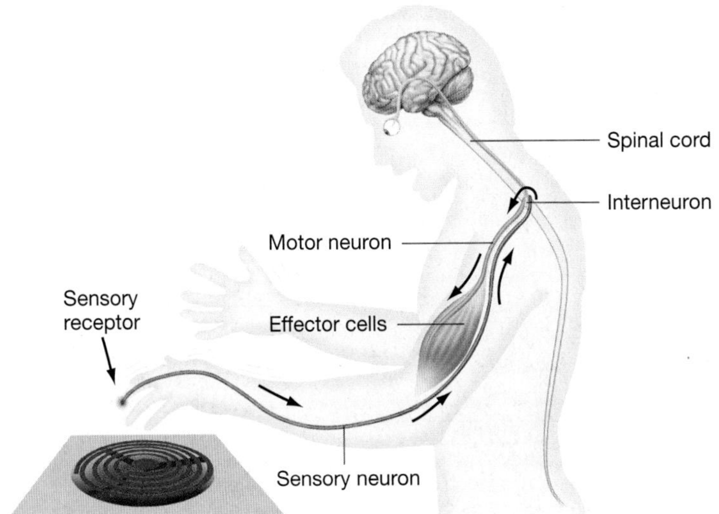

FIGURE 45.1 How Does Information Flow through the Nervous System? (a) In most cases, sensory neurons send information to the brain. There, the electrical signals are integrated with information from other sources. Once integration is complete, a response is sent to effector cells through motor neurons. **(b)** In a spinal reflex, sensory information triggers a direct response by effector cells—without integration by the brain.

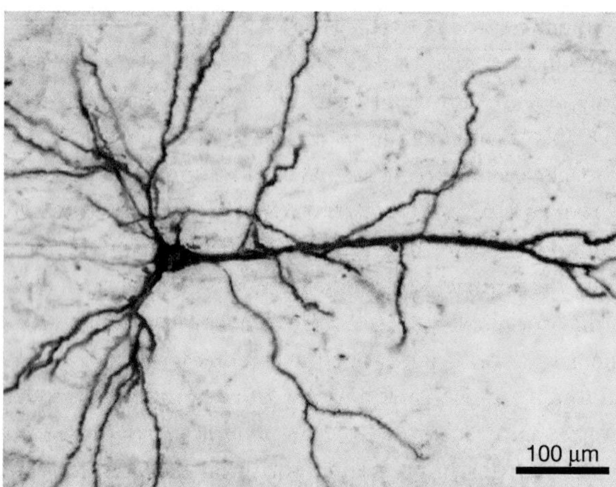

FIGURE 45.2 **Early Methods for Visualizing Neurons.** The discovery of stains that made transparent neurons visible launched the study of electrical signaling in animals.

when Camillo Golgi discovered that some neurons became visible when samples of preserved tissue were treated with a solution containing silver nitrate, his finding was a major advance (**Figure 45.2**).

Golgi published his initial observations in 1898. Through the early decades of the twentieth century, his work and that of Santiago Ramón y Cajal revealed several important points about the anatomy of neurons. Most neurons have the same three parts, shown in **Figure 45.3a**: (1) a cell body, (2) a highly branched group of short projections called dendrites, and (3) one or more long projections called axons. Dendrites are rarely more than 2 mm long, but axons can be over a meter in length. The number of dendrites and their arrangement vary greatly from cell to cell. Further, many brain cells have only dendrites and lack axons.

What do dendrites and axons do? Long before the anatomists began their work, it had been established that neurons carry electrical signals. Based on observations of stained cells, researchers suggested that a **dendrite** receives an electrical signal from the axons of adjacent cells and that a neuron's **axon** then sends the signal to the dendrites and cell bodies of other neurons (**Figure 45.3b**). Dendrites collect electrical signals; axons pass them on. The **cell body**, or **soma**, which includes the nucleus, also receives signals from axons. The cell body is where incoming signals are integrated and an outgoing signal is sent to the axon.

Ramón y Cajal maintained that the plasma membrane of each neuron is distinct and that the membranes of axons and dendrites meet at junctions called synapses. This hypothesis was confirmed in the 1950s, when images from electron microscopes showed that only a small subset of neurons make direct, cytoplasm-to-cytoplasm connections; most interactions occur where the plasma membranes of two neurons meet at synapses.

The basic message of early anatomical studies was that neurons are highly branched and that each neuron makes many connections with other neurons. The focus of research shifted to a new question: If these cells transmit electrical signals, how do they create these signals?

(a) Information flow through neurons

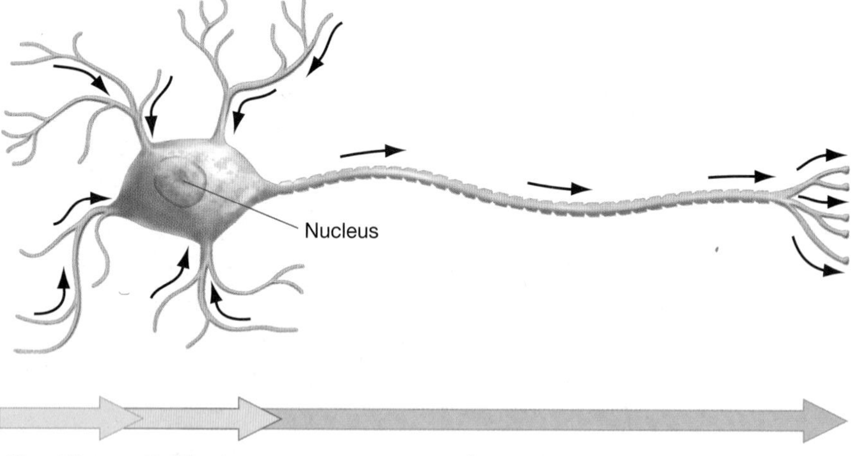

(b) Neurons form networks for information flow

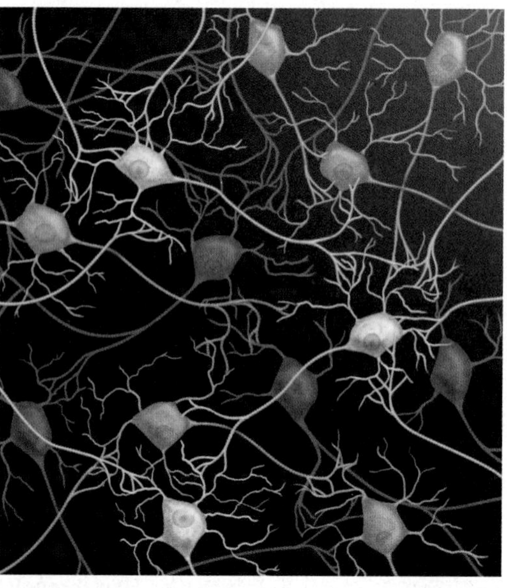

FIGURE 45.3 **How Does Information Flow in a Neuron? (a)** The structure of a generalized neuron. (**b**) Most neurons receive inputs from several different neurons and send projections to several different neurons.

EXERCISE Choose a neuron in part (b). Label the cell body and an axon. Circle points where the cell makes contact with other neurons.

An Introduction to Membrane Potentials

Ions carry an electric charge. In virtually all cells, the cytoplasm and extracellular fluids adjacent to the plasma membrane contain unequal distributions of ions. As a result, cells are inherently electrical in nature. A difference of electrical charge between any two points creates a difference in **electrical potential**, or a **voltage**. For example, if the positive and negative charges on ions that exist on the two sides of a plasma membrane do not balance each other, the membrane will have an electrical potential. When an electrical potential exists on either side of a plasma membrane, the separation of charges is called a **membrane potential**. If there is a large separation of charges across the membrane, then the membrane potential is large. It is important to remember that membrane potentials refer only to a separation of charge immediately adjacent to the plasma membrane, on either side of the membrane. Even if there is a large membrane potential, there may be no charge separation slightly farther from the membrane.

Membrane potentials are measured in units called millivolts. A **millivolt (mV)** is 1/1000 of a volt; a **volt** is the standard unit of electrical potential. To help you get a better understanding of how large a millivolt is, an AA battery that you buy in the store has an electrical potential of 1500 mV between its + and − terminals. In neurons, membrane potentials are typically about 70–80 mV.

Membrane potentials are a form of potential energy. When a membrane potential exists, the ions on both sides of the membrane have potential energy. To convince yourself that this is true, consider that if the membrane were removed, ions would spontaneously move from the area of like charge to the area of unlike charge—causing a flow of charge. A flow of charge is called an **electric current**. This movement would occur because like charges repel and unlike charges attract.

As Chapter 6 emphasized, however, ions move across membranes in response to concentration gradients as well as charge gradients. Therefore, a membrane potential also includes energy stored as the concentration gradient of charged ions on the two sides of the membrane. Recall that the combination of an electric gradient and a concentration gradient is called an **electrochemical gradient**. What do all of these facts have to do with neurons?

How Is the Resting Potential Maintained?

When a neuron is not transmitting an electrical signal but is merely sitting in extracellular fluid at rest, the membrane potential across its membrane is called the **resting potential**. To understand why the resting potential exists, consider the distribution of the various ions and other charged molecules on the two sides of the neuron's plasma membrane. Most of the large molecules inside cells are acids, which tend to release a proton. Proteins, for example, are made up of amino acids. When acids release a proton, they acquire a negative charge. Intracellular proteins that tend to be negatively charged are referred to as *organic anions*. Because of their large size and charge, organic anions do not move across membranes. The interior of a neuron also contains low concentrations of Na^+ and Cl^-. Inside a neuron, the dominant positively charged ion is the potassium ion (K^+). In the extracellular fluid, however, the sodium ion (Na^+) and chloride ion (Cl^-) predominate. K^+, Na^+, and Cl^- can cross the membrane if appropriate channels are present.

Figure 45.4 shows the relative concentrations of organic anions, K^+, Na^+, and Cl^- inside and outside a resting neuron. If each type of ion diffused across the membrane in accordance with its concentration gradient, organic anions and K^+ would leave the cell while Na^+ and Cl^- would enter. But ions cannot cross phospholipid bilayers readily. They cross plasma membranes efficiently in only three ways: (1) flowing along their electrochemical gradient through an ion channel; (2) carried, via a membrane cotransporter, with an ion that experiences a strong electrochemical gradient; or (3) pumped against an electrochemical gradient by a membrane protein that hydrolyzes ATP.

In each case of ion movement across the plasma membrane, a membrane channel or pump is involved. Recall from Chapter 6 that **ion channels** are pores in the membrane that allow only specific ions to pass through. When a neuron is not actively involved in transmitting an electrical signal, the most common

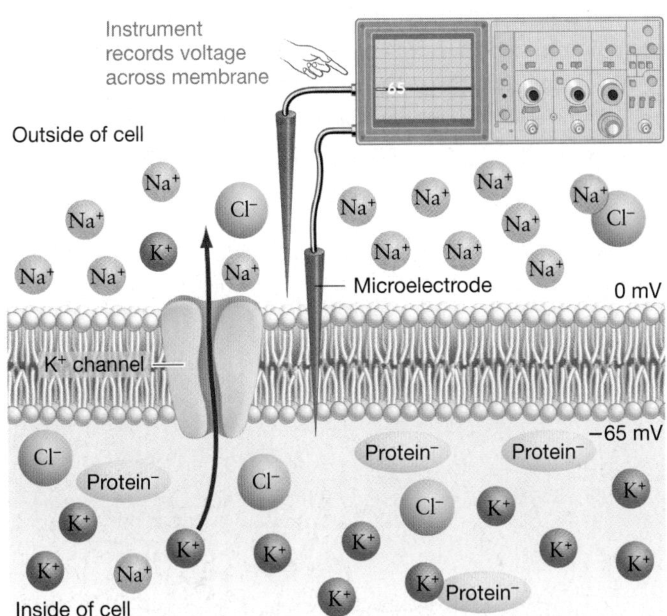

FIGURE 45.4 Neurons Have a Resting Potential. In resting neurons, the membrane is selectively permeable to K^+. As K^+ leaves the cell along its concentration gradient, the inside becomes negatively charged relative to the outside. To measure a neuron's membrane potential, researchers insert a microelectrode into the cell and compare that reading with the reading outside the cell.

QUESTION Will K^+ continue to leave indefinitely? Explain why or why not.

type of channel that is open is one that admits potassium ions. Therefore, resting neurons are most permeable to K^+ ions, which cross the membrane easily along their concentration gradient. To capture this point, biologists say that the membrane is **selectively permeable**—it permits a much greater flow of K^+ ions compared with Cl^- or Na^+ ions. The potassium channels involved are sometimes called **leak channels**, because they allow K^+ to leak out of the cell.

As K^+ moves from the interior of the cell to the exterior through potassium channels, the inside of the cell becomes more and more negatively charged relative to the outside. This buildup of negative charge inside the cell begins to attract K^+ and counteract the concentration gradient that had favored the movement of K^+ out. As a result, the membrane reaches a voltage at which there is an equilibrium between the concentration gradient that moves K^+ out and the electrical gradient that moves K^+ in. At this voltage, there is no longer a net movement of K^+. This voltage is called the **equilibrium potential** for K^+. Although Cl^- and Na^+ cross the plasma membrane much less readily than does K^+, some movement of these ions also occurs through a small number of open ion channels that are selective for each ion. As a result, each type of ion has an equilibrium potential. The membrane as a whole has a membrane potential that combines the effects of the individual ions.

Besides the passive movement of ions that takes place when neurons are at rest, Na^+/K^+-ATPase actively pumps Na^+ out of the cell and K^+ into the cell. More specifically, the sodium-potassium pump uses the energy released from the breakdown of ATP to remove three Na^+ ions that had leaked into the cell and simultaneously recover two K^+ ions that had leaked out of the cell (**Figure 45.5**). In this way, active transport via Na^+/K^+-ATPase ensures that eventually the concentration of K^+ is much higher inside the cell than outside, while the concentration of Na^+ is lower inside than outside.

As a result of these concentration differences on the two sides of the membrane, Na^+ ions would readily enter the cell and K^+ ions would rapidly exit the cell if they could cross the plasma membrane easily. In addition, the inside of the neuron is negatively charged with respect to the extracellular environment. Thus, the neuron has a negative resting membrane potential. The resting potential represents energy stored as concentration gradients in a series of ions. ● If you understand this concept, you should be able to add an Na^+/K^+-ATPase to Figure 45.4 and explain why the Na^+/K^+-ATPase and the K^+ "leak" channels are vital to establishing the resting potential.

Box 45.1 discusses how equilibrium potentials for individual ions are calculated, and how data on the relative concentrations and permeabilities of several ions are used to predict the resting potential. How do researchers actually go about measuring the size of this voltage in actual cells?

MB Web Animation at www.masteringbio.com
Membrane Potentials

Using Microelectr odes to Measure Membrane Potentials

During the 1930s and 1940s, A. L. Hodgkin and Andrew Huxley helped pioneer the study of electrical signaling in animals. Their work focused on what has become a classic model

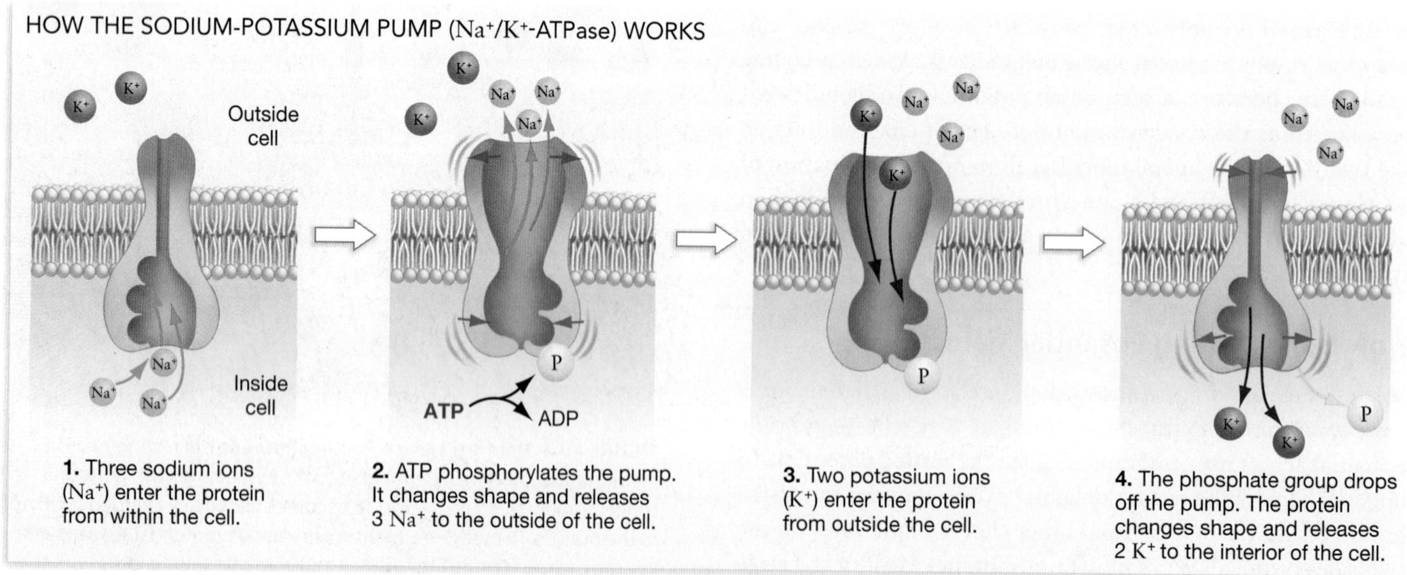

FIGURE 45.5 In Neurons, Na^+/K^+-ATPase Imports Potassium Ions and Exports Sodium Ions. A model of how the pump works. By following radioactive Na^+ and K^+, biologists found that the pump transports 3 Na^+ out of the cell for every 2 K^+ brought in. Notice that this pump operates via a conformational change.

BOX 45.1 Using the Nernst and Goldman Equations

To calculate a membrane voltage created by several ions, biologists begin by calculating the equilibrium potential for each ion involved.

The equilibrium potential for an ion is the voltage that develops across the membrane when the rate at which an ion moves across the membrane down its concentration gradient is equal to the rate at which it moves across the membrane, in the opposite direction, down its electrical gradient. Stated another way, the equilibrium potential is the voltage at which the concentration and electrical gradients acting on an ion balance out. An equilibrium potential exists if there is a concentration gradient for an ion and if the membrane is permeable to that ion.

The ion concentration gradients that contribute to equilibrium potentials are due to the action of Na^+/K^+-ATPase. In this way, chemical energy in the form of ATP is converted to electrical energy in the form of a membrane potential.

To calculate the amount of energy involved, biologists use the Nernst equation—a formula that converts the energy stored in a concentration gradient to the energy stored as an electrical potential. The concentration gradient is symbolized $[\text{ion}]_o/[\text{ion}]_i$, where $[\text{ion}]_o$ and $[\text{ion}]_i$ are the concentrations of the ion outside and inside the cell, respectively. The electrical potential for that ion is symbolized E_{ion}. The Nernst equation specifies the equilibrium potential for a type of ion as

$$E_{\text{ion}} = 2.3\frac{RT}{zF}\log\frac{[\text{ion}]_o}{[\text{ion}]_i}$$

In this expression, z is the valence of the ion (for instance, $+1$ for potassium) and the expression RT/F is known as the thermodynamic potential. The three terms in the thermodynamic potential are the gas constant (R), which acts as a constant of proportionality; the absolute temperature (T), measured in kelvins; and the Faraday constant (F), which specifies the amount of charge carried by one mole of an ion with a valence of $+1$ or -1. Note that the Nernst equation is based the base-10 logarithm of the ion concentration ratio, $[\text{ion}]_o/[\text{ion}]_i$. Thus, the thermodynamic potential specifies the voltage required to balance a tenfold concentration ratio across the membrane. (For more on using logarithms, see **BioSkills 5**.)

It is important to remember that the Nernst equation applies to only a single ion. For example, suppose that the potassium concentration inside the axon of a squid neuron has been measured as 400 mM, yet the outside concentration for this ion is only 20 mM (**Table 45.1**). Potassium has a charge of $+1$, so at 20°C the RT/zF part of the equation yields $+25$ mV. In this case, the equilibrium potential for potassium becomes

$$E_K = 2.3 \times 25\text{ mV} \times \log\frac{20\text{ mM}}{400\text{ mM}}$$

$$E_K = 58\text{ mV} \times \log 0.05 = -75\text{ mV}$$

The minus sign indicates that the interior of the axon is negatively charged with respect to the exterior. Repeating this process for Na^+ ions and Cl^- ions yields the following results for the squid axon:

$$E_{Na} = 58\text{ mV} \times \log\frac{440\text{ mM}}{50\text{ mM}} = +54.8\text{ mV}$$

$$E_{Cl} = -58\text{ mV} \times \log\frac{560\text{ mM}}{51\text{ mM}} = -60\text{ mV}$$

Again, notice that the equilibrium potential given by the Nernst equation is calculated independently for each ion.

How do individual equilibrium potentials relate to the overall resting membrane potential? The Nernst equation cannot be used to calculate resting membrane potentials, because no cell is permeable to only a single ion. David Goldman described a way to calculate the membrane potential when K^+, Na^+, and Cl^- are able to cross the membrane. The Goldman equation calculates a membrane potential by combining the effects of key individual ions. If the permeabilities to these ions are symbolized as P_K, P_{Na}, and P_{Cl}, respectively, then the membrane potential is given by the following expression:

$$V_m = 58\text{ mV} \times \log\frac{[K^+]_o + \frac{P_{Na}}{P_K}[Na^+]_o + \frac{P_{Cl}}{P_K}[Cl^-]_i}{[K^+]_i + \frac{P_{Na}}{P_K}[Na^+]_i + \frac{P_{Cl}}{P_K}[Cl^-]_o}$$

In a squid's neuron, the relative permeabilities for K^+, Na^+, and Cl^- are described by the ratio 1.0 : 0.03 : 0.1. Given these ratios and the observed concentration of each ion inside and outside the axon, the resting potential of the neuron can be calculated as follows:

$$V_m = 58\text{ mV} \times \log\frac{20 + (0.03)440 + (0.1)51}{400 + (0.03)50 + (0.1)560} = -62.5\text{ mV}$$

This is close to the observed value. The Nernst and Goldman equations are remarkably efficient ways to capture the dyamics of equilibrium potentials and membrane potentials.

TABLE **45.1** **Concentration of Important Ions across a Squid Neuron's Plasma Membrane at Rest**

Ion	Cytoplasm Concentration	Extracellular Concentration	Equilibrium Potential
Na^+	50 mM	440 mM	+54.8 mV
K^+	400 mM	20 mM	−75 mV
Cl^-	51 mM	560 mM	−60 mV
Organic ions$^-$	385 mM	—	—

QUESTION Why isn't there an equilibrium potential for organic ions?

system in the study of electrical signaling: the axons of squid. Squid live in the ocean and are preyed on by fish and whales. When a squid is threatened, electrical signals travel down the axon to muscle cells. When these muscles contract, water is expelled from a cavity in the squid's body. As a result, the squid lurches away from danger by jet propulsion (see Chapter 31). Electrical signaling is an adaptation that helps squid avoid predation.

Hodgkin and Huxley decided to study the squid's axon simply because it is so large. Many of the axons found in humans are a mere 2 μm in diameter, but the squid axon is about 500 μm in diameter. The squid axon was large enough that the researchers could record membrane potentials by inserting a wire down its length. That was the best approach available at the time for measuring voltages in cells and required an axon with an exceptionally large diameter.

By measuring the voltage difference between the wire in the interior of the cell and an electrode on the outside of the cell, Hodgkin and Huxley could estimate the voltage that existed across the neuron's plasma membrane and observe how it changed through time. Later, other researchers developed glass microelectrodes that were small enough to be inserted into smaller nerve cells to record membrane voltage.

With their relatively simple early equipment, Hodgkin and Huxley were able to record the axon's resting potential and document that it can be disrupted by an event called the action potential.

What Is an Action Potential?

An **action potential** is a rapid, temporary change in a membrane potential. **Figure 45.6** shows the form of the action potential that Hodgkin and Huxley recorded from the squid's giant axon—the signal that allows these animals to jet away from predators. The action potential has three distinct phases:

1. The initial event is a rapid **depolarization** of the membrane. (A membrane is said to be polarized if the charges on the two sides are different. Depolarization means that the membrane becomes less polarized than before.) For an action potential to begin, the membrane potential must shift from its resting potential of −65 mV to about −55 mV. If the membrane depolarizes less than that, an action potential does not occur. But if this **threshold potential** is reached, certain channels in the axon membrane open and ions rush into the axon, following their electrochemical gradients. The current flow causes further depolarization. More specifically, the inside of the membrane becomes less negative and then positive with respect to the outside.

2. When the membrane potential reaches about +40 mV, an abrupt change occurs. This change is triggered by the closing of certain ion channels and the opening of other ion channels in the membrane. During this second phase of the action potential, the membrane experiences a rapid **repolarization** as ions flow out of the axon.

3. The repolarization event actually results in the membrane briefly becoming more negative than the resting potential, a state called **hyperpolarization.** For this reason, biologists call the final part of the action potential the hyperpolarization phase.

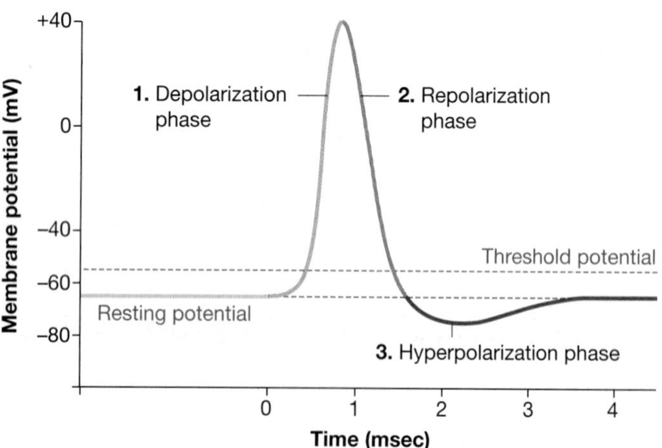

FIGURE 45.6 The Action Potential Is a Stereotyped Change in Membrane Potential. Most action potentials have the general shape shown here.

Taken together, all three phases of an action potential occur in a few milliseconds.

Hodgkin and Huxley made other important observations about the action potential. In addition to being fast, it is an all-or-none event. There is no such thing as a partial action potential, and all action potentials for a given neuron are identical in magnitude and duration. The biologists also found that action potentials could be triggered by artificially depolarizing the membrane with an injection of electrical current through a microelectrode. Finally, action potentials are propagated down the length of the axon. When an impulse was recorded at a particular point on a squid axon, an action potential that was identical in shape and size would be observed farther down the same axon soon afterward. Neurons are said to have **excitable membranes**, because they are capable of generating action potentials that propagate rapidly along the length of their axons.

Taken together, these observations suggested a mechanism for electrical signaling. In the nervous system, information is coded in the form of action potentials that travel along axons. The frequency of action potentials—not their size—is the meaningful signal. In the squid's giant axon, action potentials signal muscles to contract. As a result, the animal escapes from danger.

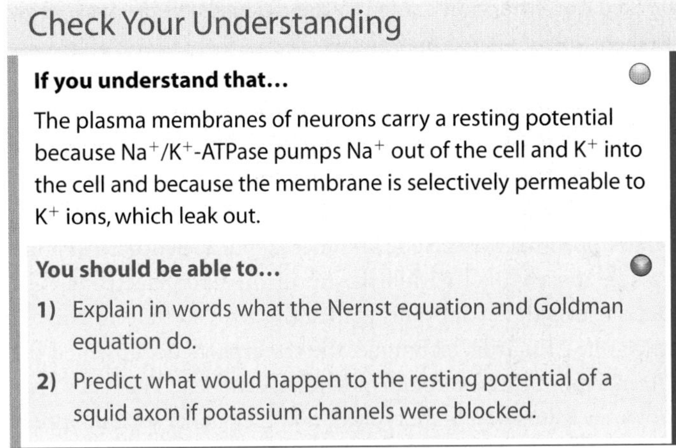

Check Your Understanding

If you understand that...

The plasma membranes of neurons carry a resting potential because Na^+/K^+-ATPase pumps Na^+ out of the cell and K^+ into the cell and because the membrane is selectively permeable to K^+ ions, which leak out.

You should be able to...

1) Explain in words what the Nernst equation and Goldman equation do.
2) Predict what would happen to the resting potential of a squid axon if potassium channels were blocked.

45.2 Dissecting the Action Potential

Once it became possible to record rapid changes in membrane potential accurately, an entirely new field of research opened up. The most urgent question was to understand which ions are involved in the currents that form the action potential. Is Na^+, Cl^-, or K^+ responsible for the depolarization and repolarization phases of the event? One early hypothesis was that the action potential resulted from a temporary breakdown in the membrane's selective permeability. The idea was that all channels opened and allowed a free flow of Na^+ and Cl^- as well as K^+. But the all-channels-open hypothesis ran into trouble when researchers were able to measure the concentrations of each ion on the two sides of the squid's axon membrane. Given the observed concentrations of ions when the neuron was at rest, the all-channels-open hypothesis predicted that the peak of the action potential should occur at about 0 mV, not the +40 mV observed.

In contrast, Hodgkin noted that +40mV was close to the equilibrium potential for Na^+. This insight led to the hypothesis that the action potential's depolarization phase resulted from an influx of sodium ions. How could this idea be tested?

Distinct Ion Currents Are Responsible for Depolarization and Repolarization

To understand the currents responsible for the action potential, Hodgkin and Huxley recorded electrical activity in a squid axon bathed in seawater. Then they replaced the seawater with an isotonic solution containing an uncharged and metabolically inert sugar called dextrose. **Figure 45.7** shows the result of this experiment. Washing Na^+ out of the solution surrounding the axon abolished action potentials. But when the biologists replaced the dextrose solution with seawater, action potentials were restored. Further, when they used solutions with various concentrations of Na^+ instead of seawater, Hodgkin and Huxley found that the peak of the action potential tracked the equilibrium potential of Na^+ as it varied with Na^+ concentration outside the cell. Taken together, the data furnished strong support for the hypothesis

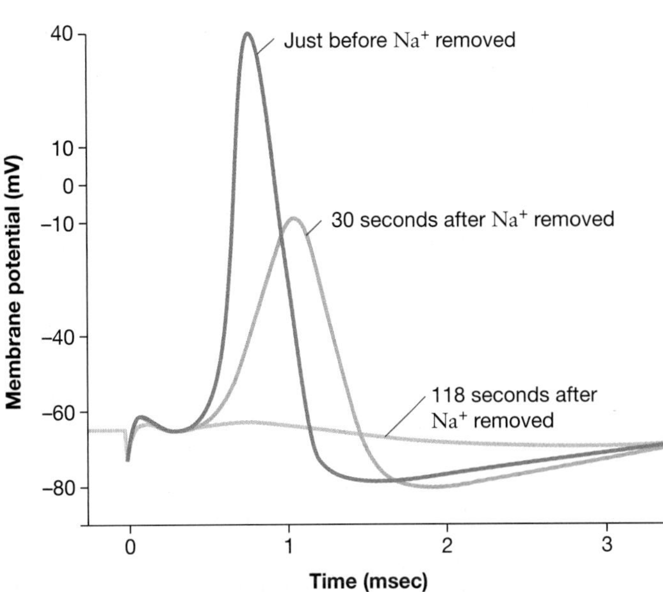

FIGURE 45.7 Action Potentials Diminish when Na^+ Is Removed from the Solution Surrounding a Neuron. Action potentials recorded at intervals after a solution containing Na^+ was flushed away from a squid axon with a solution containing isotonic dextrose.

QUESTION Why was it important that the researchers used an isotonic solution of dextrose?

that the action potential begins when Na^+ flows into the neuron. Sodium ions are responsible for the depolarization phase.

What flow of ions is responsible for the repolarization phase of an action potential? Using radioactive K^+, Hodgkin and Huxley showed that there was a strong flow of potassium ions out of the cell during the repolarization phase. Thus, the action potential consists of a strong inward flow of sodium ions followed by a strong outward flow of potassium ions. If you understand this concept, you should be able to add labels that read, "Sodium channels open—Na^+ enters" and "Potassium channels open—K^+ leaves" to Figure 45.6. Which ion channels are responsible for these currents, and how do they work?

How Do Voltage-Gated Channels Work?

When ion channels were introduced in Chapter 6, they were presented as simple, static proteins that facilitate a constant stream of ions. But research on the action potential suggests that certain ion channels open and close in response to changes in membrane voltage. These types of proteins are called **voltage-gated channels**. The conformation of voltage-gated channels changes in response to the charges present at the membrane surface. These conformational changes lead to the opening or closing of the channels.

To study the relationship between membrane voltage and channel activity, Hodgkin and Huxley used a technique called voltage clamping. **Voltage clamping** allows researchers to hold an axon at any voltage and record the electrical currents generated by flows of Na^+ and K^+ across the membrane.

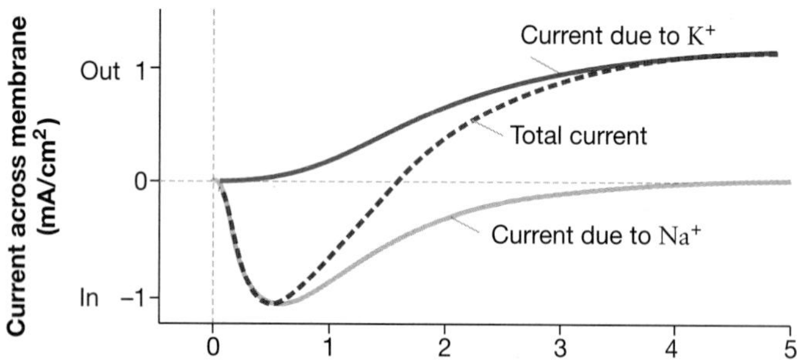

FIGURE 45.8 Voltage Clamping Allows Researchers to Record Current Flows across Membranes. When Hodgkin and Huxley voltage-clamped a squid axon at −9 mV, they recorded the flow of current shown in the dashed record. Current initially flowed into the axon and then back out. As the orange and purple records show, this overall current is due to a rapid flow of Na^+ into the cell and a slower flow of K^+ out of the cell.

Figure 45.8 shows the currents that Hodgkin and Huxley recorded when they voltage-clamped a squid axon at −9 mV. The dashed record shows the current that resulted from the depolarization. Notice that this record indicates a rapid inward-directed current followed by a slower outward-directed current. The orange and purple records break this overall current down into the components due to an inward flow of Na^+ and outward flow of K^+, respectively.

This experiment, and others like it, confirmed that Na^+ and K^+ channels open in response to changes in voltage. When the plasma membrane in the region of the axon is depolarized, voltage-gated Na^+ channels open for less than a millisecond. Sodium ions rush into the cell through these open channels, because the Na^+ concentration is much higher outside the cell and because Na^+ ions are attracted to the negatively charged interior. The result is that the inside of the membrane becomes more positively charged—it goes from a resting potential of −70 mV to −50 mV. This change in electrical potential triggers many additional Na^+ channels to open. As a result, Na^+ ions flood the interior, generating an action potential. Shortly after the action potential is under way, the voltage-gated Na^+ channels close and voltage-gated K^+ channels open.

Hodgkin and Huxley's data also answered an important question: If Na^+ flows in and K^+ flows out, why don't the two currents cancel each other out? The key observation is that the potassium current is delayed. The Na^+ channels close as the membrane potential approaches +54 mV and as the voltage-gated K^+ channels open. At this point, the interior of the membrane contains excess positive charge and favors the outward flow of K^+ through the open K^+ channels. As K^+exits the cell, the membrane potential drops back toward resting levels.

To summarize, early work on electrical signaling showed that action potentials result from the staggered activity of voltage-gated Na^+ and K^+ channels. ● If you understand this concept, you should be able to explain (1) why the opening of Na^+ channels depolarizes a membrane, and (2) why flows of K^+ repolarize the membrane. Once this sequence of events was worked out, attention turned to the channels themselves. How do these voltage-gated proteins work?

Patch Clamping and Studies of Single Channels Studying individual ion channels became possible when Erwin Neher and Bert Sakmann perfected a technique known as **patch clamping**. As **Figure 45.9a** shows, the researchers touched a membrane with a fine-tipped microelectrode and then applied suction to capture a single ion channel within the tip of the microelectrode. The suction seals the membrane against the glass so that no current leaks out. Using this technique, the researchers documented the currents that flowed through individual channels. The recordings shown in **Figure 45.9b** are typical, and they make several important points about the nature of voltage-gated channels:

- *Voltage-gated channels are either open or closed.* There is no gradation in channel behavior. This conclusion is based on the shape of the current records.
- *Sodium channels open quickly after depolarization.* They stay open for about a millisecond and then close. As Na^+ rushes into the axon, the inside becomes electrically positive. Shortly after the initial depolarization, the sodium channels close and remain inactive for 1 to 2 milliseconds. That explains why the cell can repolarize: Once the sodium channels close, they have a lag before they can open again. **Figure 45.9c** shows a simplified model of how voltage-gated sodium channels open.
- *Potassium channels open with a delay after depolarization.* They continue to flip open and closed until the membrane repolarizes. Once the membrane returns to the resting potential, these channels remain closed.

Observations of the behavior of Na^+ channels also helped explain why the action potential is an all-or-none event. The key observation is that Na^+ channels are more likely to open as a membrane depolarizes. An initial depolarization thus leads to the opening of more Na^+ channels, which depolarizes the membrane further and leads to the opening of additional Na^+ channels. In this way, the opening of Na^+ channels exhibits **positive feedback**. This type of control mechanism takes place when the occurrence of an event makes the same event more likely to recur. When a fuse is lit, for example, the heat generated by the oxidation reaction accelerates the reaction itself, which generates still more heat and leads to additional oxidation reactions and keeps the fuse burning. Positive feedback is rare in organisms because it often leads to uncontrolled events. The opening of Na^+ channels during an action potential is one of the few examples known.

● If you understand how sodium and potassium channels work, you should be able to explain (1) why positive feedback occurs in the opening of Na^+ channels, (2) why Na^+ stops flowing across a membrane during an action potential, and (3) why K^+ channels start opening.

Voltage clamping and patch clamping were not the only technical advances that helped researchers explore the dynamics of voltage-gated channels. The discovery of poisons that target neurons, from sources as diverse as venomous snakes and foxglove plants, also provided important experimental tools.

(a) Patch clamping isolates a single ion channel.

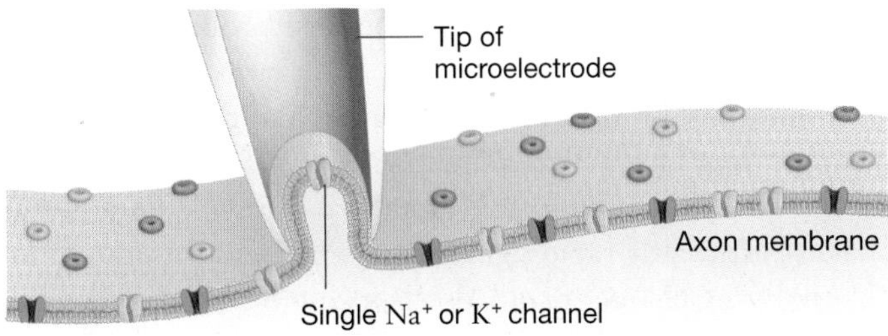

(b) Currents through isolated channels can be measured during an action potential.

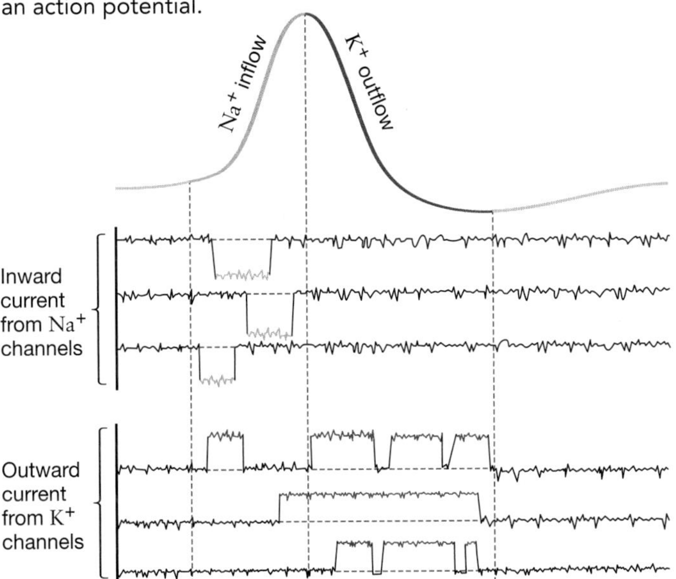

(c) How voltage-gated channels work

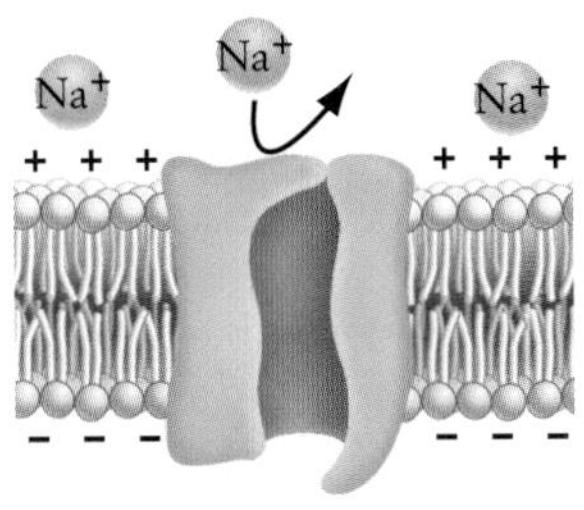

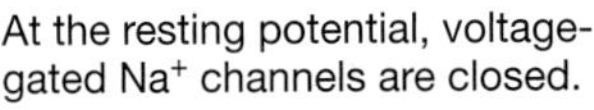

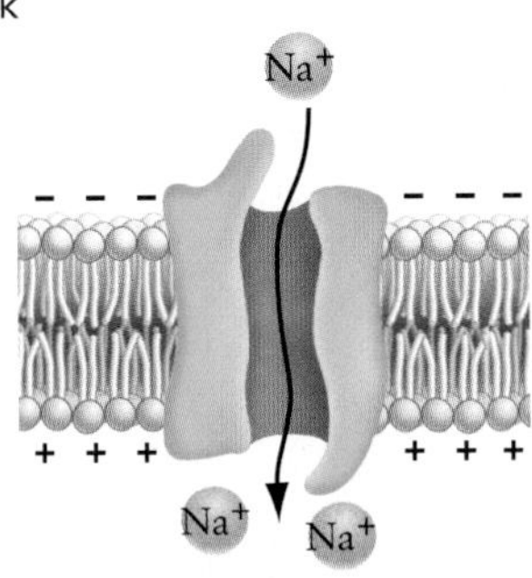

FIGURE 45.9 Patch Clamping Allows Researchers to Record from Individual Channels. **(a)** Patch clamping depends on the use of extremely fine-tipped microelectrodes. The goal is to isolate one channel and record from it. **(b)** Current records from three voltage-gated sodium channels and three voltage-gated potassium channels. No current flows through either type of channel at the resting potential. **(c)** Changes in the conformation of voltage-gated channels are responsible for changes in a neuron's permeability to Na^+ and K^+.

Using Neurotoxins to Identify Channels and Dissect Currents Many of the toxins produced by poisonous animals and plants cause convulsions, paralysis, or unconsciousness when they are ingested. Based on these symptoms, physicians and researchers suspected that at least some poisons affect neuron function; such poisons are known as **neurotoxins.** This hypothesis was supported when biologists used the tetrodotoxin found in puffer fish to treat giant axons from lobsters. Although the resting potential in treated neurons was normal, action potentials were abolished. More specifically, the outward-directed K^+ current was normal but the inward-directed Na^+ flow was wiped out. Based on this result, researchers concluded that puffer-fish toxin specifically blocks the voltage-gated Na^+ channel. The hypothesis is that the toxin binds to a specific site on the channel protein, preventing the channel from functioning. Thus, researchers can use neurotoxins to block only certain channels.

In contrast, when researchers treated neurons with the venom of black mamba snakes, preliminary data hinted that K^+ channels were blocked. To test this finding more rigorously, researchers bathed neurons in a solution containing black-mamba poison and studied the K^+ currents that resulted. As predicted, the poison had a dramatic effect on membrane permeability to K^+. However, the venom did not stop the current completely. Instead, the data suggest that a specific K^+ channel was poisoned but that other K^+ channels still allowed an outflow of current (**Figure 45.10**). The existence of multiple K^+ channels in axons was confirmed by patch-clamping studies that documented K^+ flowing through several types of K^+ channels, each of which had distinct properties. Experiments with neurotoxins helped establish that most neurons contain just one type of sodium channel but several types of potassium channels.

The experiments reviewed thus far haven't considered a fundamental question about electrical signaling, however: How do action potentials move down the axon?

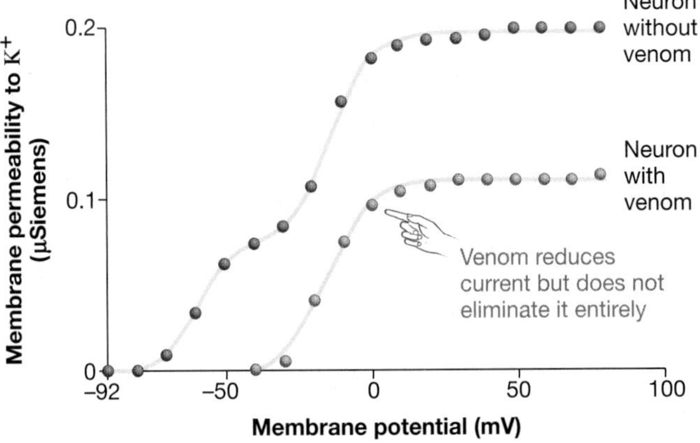

FIGURE 45.10 Experiments with Poisons Showed that There Is More than One Potassium Channel. Neuron permeability to K^+ as a function of membrane voltage (μSiemens are a unit of conductance). Each data point was generated by measuring membrane permeability while the membrane was voltage-clamped at a particular potential.

How Is the Action Potential Propagated?

To explain how action potentials propagate down an axon, Hodgkin and Huxley hypothesized that the influx of Na^+ causes charge to spread away from sodium channels. As **Figure 45.11a** shows, positive charges inside the cell are repulsed by the influx of Na^+, and negative charges are attracted to the Na^+. As these positive charges are pushed farther from the initial sodium channel, they create a local current that causes new sections of the membrane, close to the initial site of an action potential, to become depolarized. Voltage-gated Na^+ channels that are nearby pop open in response, positive feedback occurs, and a full-fledged action potential results. In this way, an action potential is continuously regenerated as it moves down the axon (**Figure 45.11b**). The signal does not diminish as it moves, because the response is all or none.

Why don't action potentials propagate back up the axon? The answer is that Na^+ channels are **refractory**—that is, once they have opened and closed, they are less likely to open again for a short period. Action potentials are propagated in one direction only, because sodium channels "downstream" of the site are not in the refractory state. The hyperpolarization phase, in which the membrane is more negative than the resting potential—also keeps the charge that spreads "upstream" from triggering an action potential in that direction.

Understanding how the action potential propagates helped researchers explain why the squid's axons are so large. The same amount of charge spreads farther in an axon with a large diameter versus a small diameter, because proportionally less charge spreads to the side and is "lost" in terms of conducting the signal. As a result, the squid's giant axon and other large neurons transmit action potentials much faster than small axons can. The squid axon's large size is an adaptation that makes particularly rapid signaling possible.

Analyzing charge spread also helped biologists explain the phenomenon called myelination. Relatively few vertebrates have giant axons. Instead, in vertebrates—and some invertebrates—the membranes of specialized accessory cells wrap around the axons of neurons. In the central nervous system, these cells are **oligodendrocytes**. In the peripheral nervous system, described in Section 45.4, the cells are **Schwann cells** (**Figure 45.12a**). Oligodendrocytes and Schwann cells are two of several types of nervous system cells that support neurons. Collectively, these accessory cells are called **glia**. When oligodendrocytes or Schwann cells wrap around an axon, they form a **myelin sheath**, which acts as a type of electrical

(a) PROPAGATION OF ACTION POTENTIAL

Neuron

Axon

1. Na^+ enters axon.

2. Charge spreads; membrane "downstream" depolarizes.

Depolarization at next ion channel

3. Voltage-gated channel opens in response to depolarization.

(b) Action potential spreads as a wave of depolarization.

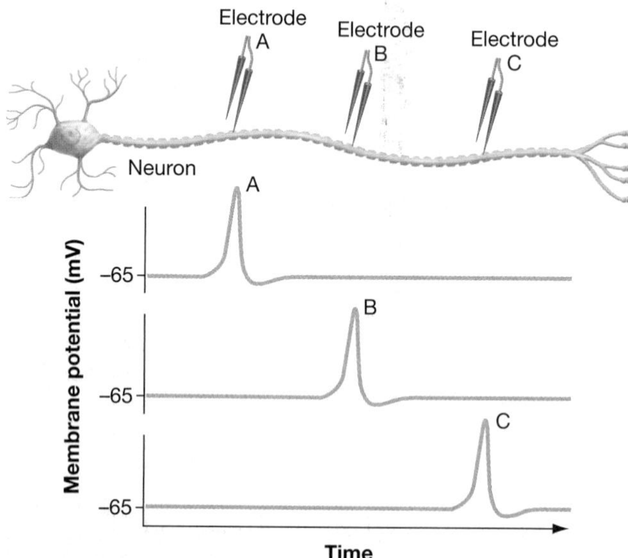

FIGURE 45.11 Action Potentials Propagate because Charge Spreads down the Membrane. (a) An action potential starts with an inflow of Na^+. The influx of positive charge attracts negative charges inside the cell and repels positive charges. As a result, positive charge spreads away from the channel where the Na^+ enters, and depolarizes nearby regions of the neuron. Voltage-gated Na^+ channels open in response. **(b)** The action potential spreads down the axon as a wave of depolarization, but there is no loss of signal because the all-or-none action potential regenerates itself as it travels.

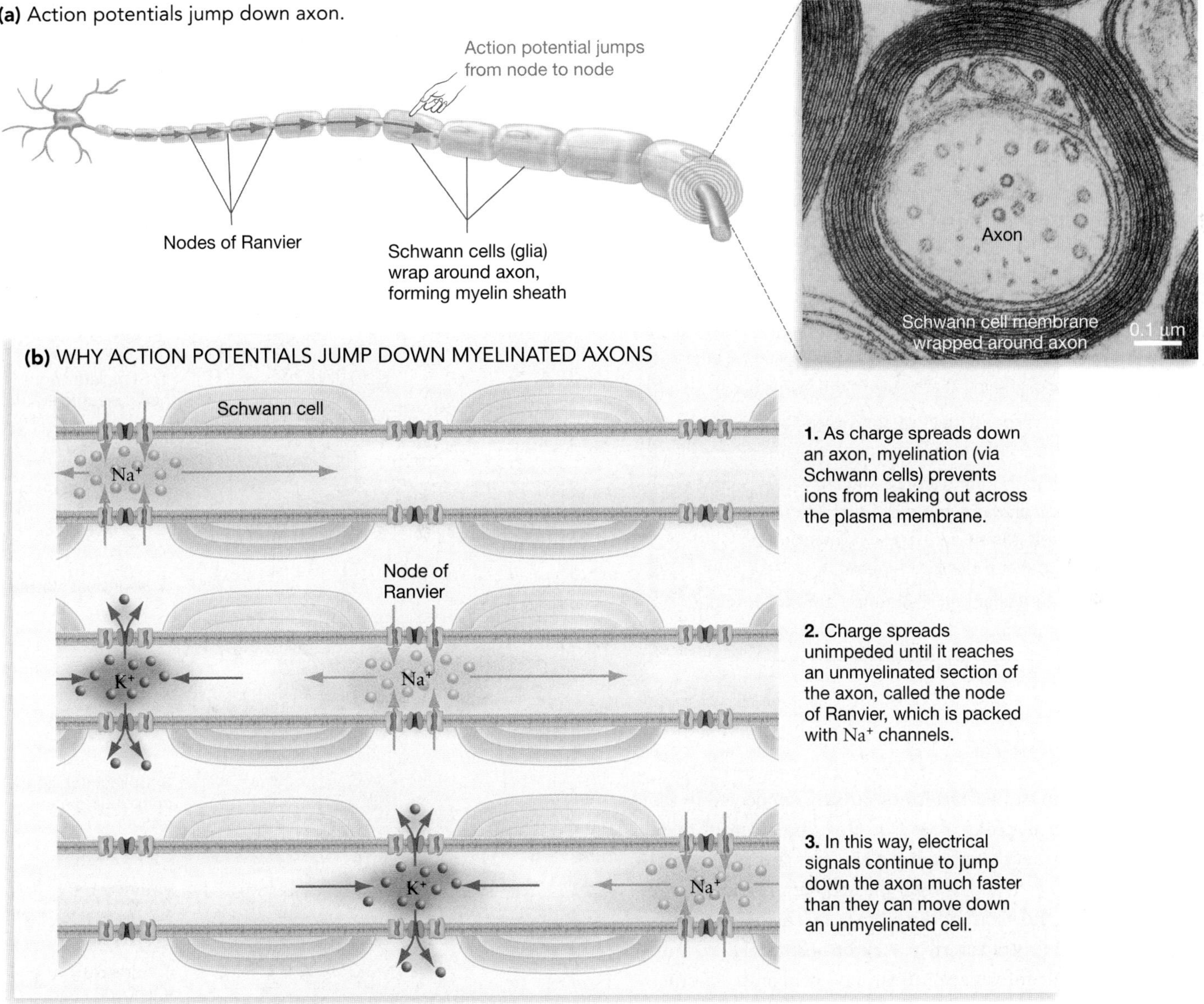

FIGURE 45.12 Action Potentials Propagate Quickly in Myelinated Axons.

insulation. As charge spreads down an axon, the presence of the myelin sheath prevents charge in the form of ions from leaking out across the plasma membrane of the neuron. Consequently, the influx of charge that results from an action potential is able to spread unimpeded until it hits an unmyelinated section of the axon, called a **node of Ranvier** (**Figure 45.12b**). The node has a dense concentration of voltage-gated Na^+ channels, so action potentials can occur. In this way, electrical signals jump down the myelinated axon much faster than they can move down an unmyelinated axon. In an unmyelinated axon, sodium and potassium channels are found in all locations and action potentials occur continuously down its length. ● If you understand this concept, you should be able to draw a diagram like Figure 45.12, showing how action potentials are slowly propagated down the length of unmyelinated axons. (Hint: Consult Figure 45.11a.) Myelination is interpreted as an adaptation that makes rapid transmission of electrical signals possible in axons that have a small diameter.

To drive the importance of myelination home, consider what happens when it is decreased. If myelin degenerates, the transmission of electrical signals slows considerably. The disease **multiple sclerosis** (**MS**) develops as damage to myelin increases and electrical signaling is impaired, causing the muscles to weaken and coordination to lessen. The symptoms of MS are highly variable; in severe cases the disease progresses and can be crippling.

What happens once an action potential has traveled the length of the axon? Ramón y Cajal proposed, and electron microscopy confirmed, that in most neurons the membrane at the end of the axon approaches the membrane of another neuron's dendrite. The surfaces of the two membranes are separated by a tiny gap. What happens when an action potential arrives at this interface between cells?

Check Your Understanding

If you understand that...

- During an action potential, membrane voltage undergoes rapid changes due to an influx of sodium ions, followed by an outflow of potassium ions.
- Action potentials propagate down an axon because inrushing sodium ions depolarize adjacent portions of the membrane.

You should be able to...

1) Diagram the changes in membrane voltage that occur during the action potential, and add notes explaining why—at the molecular level—each step in the signal occurs.
2) Make a labeled diagram that explains why an action potential spreads down an axon.
3) Explain how voltage gating works, why the action potential is an all-or-none phenomenon, why the refractory nature of sodium channels is important, and why myelination increases the speed of electrical signals.
4) Predict what would happen if sodium channels continued to open once membrane depolarization was complete.

45.3 The Synapse

The cytoplasm of most neurons is not directly connected to the cytoplasm of other neurons. Based on this observation, there must be some indirect mechanism that transmits electrical signals from cell to cell, across their plasma membranes. In the 1920s Otto Loewi showed that the indirect mechanism involves molecules called neurotransmitters. **Neurotransmitters** are molecules that transmit information from one neuron to another neuron or from a neuron to a target cell in a muscle or gland.

Loewi was interested in how the nervous system affects the heart rate, and he knew that signals from the vagus nerve—a nerve that originates in the brain stem and extends into the abdomen—slow the heart. He performed the experiment diagrammed in **Figure 45.13** to test his hypothesis that the signal from nerve to muscle is delivered by a chemical. First, he isolated the vagus nerve and heart of a frog. As predicted, the heart rate slowed when he stimulated the nerve electrically. But when he took the solution that bathed the first heart and applied it to another, isolated heart, the second heart rate slowed as well. This result provided strong evidence for the chemical transmission of electrical signals. The vagus nerve had released a neurotransmitter into the bath.

How are neurotransmitters delivered? When transmission electron microscopy became available in the 1950s, biologists finally understood the physical nature of the interface between neurons. This interface is called a **synapse**. As **Figure 45.14** shows, the membranes of axons and dendrites juxtapose closely. The figure also shows that the ends of axons contain numerous

Experiment

Question: How is information transferred from one neuron to another?

Hypothesis: Molecules called neurotransmitters carry information from one neuron to the next.

Null hypothesis: Information is not transferred between neurons in the form of molecules.

Experimental setup:

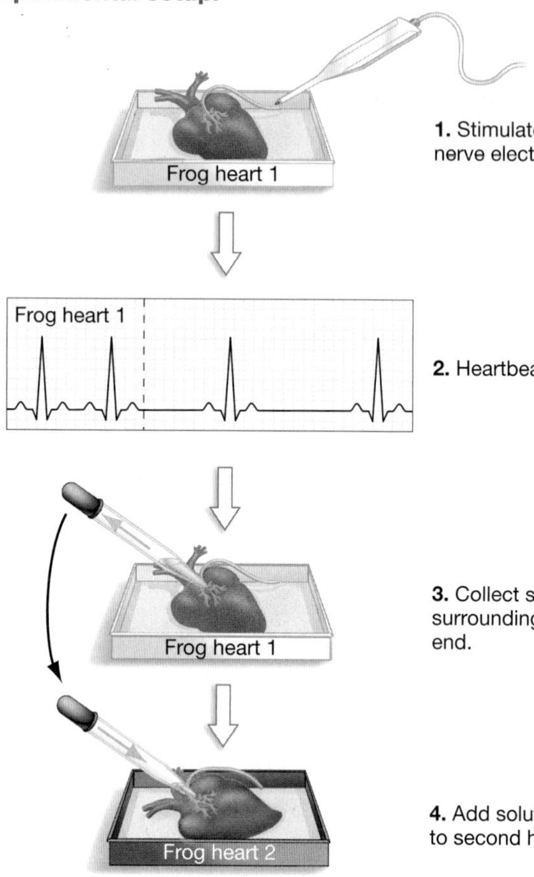

1. Stimulate vagus nerve electrically.

2. Heartbeat slows.

3. Collect solution surrounding nerve end.

4. Add solution to second heart.

Prediction: The heartbeat will slow.

Prediction of null hypothesis: There will be no change in heartbeat.

Results:

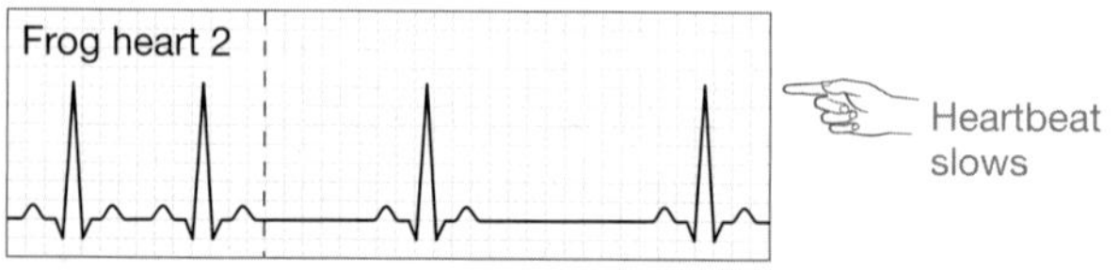

Conclusion: The vagus nerve releases molecules that slow heartbeat. Neurotransmitters carry information.

FIGURE 45.13 Experimental Evidence for the Existence of Neurotransmitters.

QUESTION Prior to this experiment, some researchers contended that most neurons communicated with each other or target cells electrically, not chemically. Why does this experiment fail to support the hypothesis of electrical transmission? (Note: It turns out that a few neurons do make electrical connections with one another.)

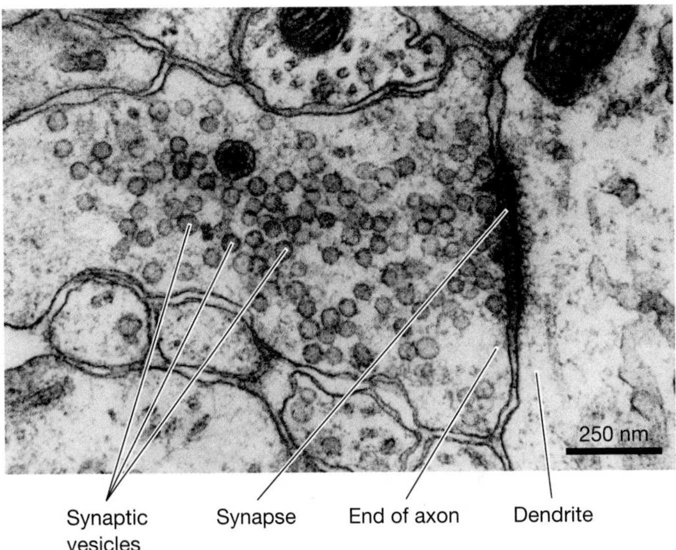

FIGURE 45.14 Synaptic Vesicles Cluster Near Synapses. A cross section of the site where an axon meets a dendrite.

sac-like structures called **synaptic vesicles**. These structures were hypothesized to be storage sites for neurotransmitters.

Anatomical observations such as these, combined with chemical studies of the synapse, led to the model of synaptic transmission illustrated in **Figure 45.15**. Notice that the "sending" cell is called the **presynaptic neuron** and the "receiving" cell is called the **postsynaptic neuron**. The sequence of events begins when the depolarization created by an action potential opens voltage-gated calcium channels located near the synapse, in the presynaptic membrane. The electrochemical gradient for Ca^{2+} results in the inflow of calcium ions through the open channels. In response to the increase in calcium concentration inside the axon, synaptic vesicles fuse with the membrane and release a neurotransmitter into the gap between the cells. This gap is called the **synaptic cleft**. The delivery of neurotransmitters into the cleft is an example of exocytosis, a process introduced in Chapter 7.

What happens next? Biologists hypothesized that neurotransmitters bind to receptors in the membrane of the next cell in the circuit. As step 3 in Figure 45.15 shows, the idea was that interactions between neurotransmitters and receptors on the postsynaptic cell would cause a change in the membrane potential of the postsynaptic cell and possibly trigger the start of an action potential there. Is this part of the model correct?

What Do Neurotransmitters Do?

Once Loewi had established that neurotransmitters exist, investigators were able to isolate and purify them from a variety of neurons. They accomplished this by stimulating a neuron, collecting the neurotransmitter that was released, and analyzing it chemically. To find the receptor for a particular neurotransmitter, researchers could add a radioactive label to the neurotransmitter and add the labeled transmitter to neurons. Once the labeled transmitter had bound to its receptor, the receptor protein could be identified, isolated, and analyzed.

Using techniques such as these, biologists have discovered and characterized a wide array of neurotransmitters and receptors, some of which are listed in **Table 45.2**. To establish that a molecule functions as a neurotransmitter, researchers have to provide evidence for the following three criteria: (1) The neurotransmitter is present at the synapse and released in response to an action potential, (2) it binds to a receptor on a postsynaptic cell, and (3) it is taken up or degraded.

By patch-clamping receptors, biologists confirmed that many neurotransmitters function as ligands. A **ligand** is a molecule

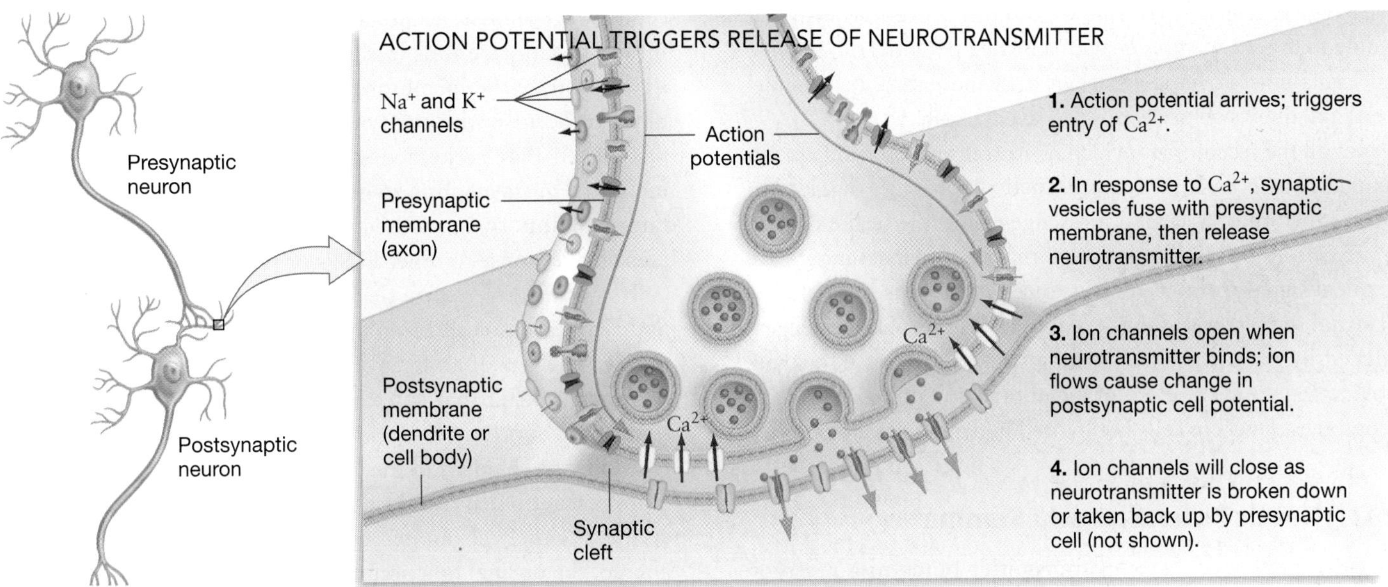

FIGURE 45.15 Neurons Meet and Transfer Information at Synapses. The sequence of events that occurs when an action potential arrives at a synapse.

TABLE **45.2 Categories of Neurotransmitters**

Excitatory neurotransmitters make action potentials more likely in postsynaptic cells. Inhibitory neurotransmitters make them less likely; modulatory neurotransmitters modify the response at other synapses. Drugs that prevent re-uptake of neurotransmitters increase their activity.

Neurotransmitter	Site of Action	Action	Drugs That Interfere
Acetylcholine	Neuromuscular junction, some CNS pathways	Excitatory (inhibitory in some parasympathetic neurons)	• Botulism toxin blocks release • Black widow spider venom increases, then eliminates, release • α-bungarotoxin (in some snake venoms) binds to receptor
Monoamines			
Norepinephrine	Sympathetic neurons, some CNS pathways	Excitatory or inhibitory	• Ritalin (used for ADHD) increases release • Some antidepressants prevent re-uptake
Dopamine	Many CNS pathways	Primarily excitatory	• Cocaine prevents re-uptake • Amphetamines prevent re-uptake
Serotonin	Many CNS pathways	Inhibitory or modulatory	• MDMA (ecstasy) causes increased release • Prozac, Paxil, Zoloft prevent re-uptake
Amino Acids			
Glutamate	Many CNS pathways	Excitatory	• PCP blocks receptor
Gamma-aminobutyric acid (GABA)	Some CNS pathways	Inhibitory	• Ethanol mimics response
Peptides			
Endorphins, enkephalins, substance P	Used in sensory pathways (pain)	Inhibitory	

that binds to a specific site on a receptor molecule. Many neurotransmitters are ligands that bind to receptors called **ligand-gated channels**. These are channel proteins that open in response to binding by a specific ligand. When a neurotransmitter binds to a ligand-gated ion channel in the postsynaptic membrane, the channel opens and admits a flow of ions. In this way, the neurotransmitter's chemical signal is transduced to an electrical signal—a change in the membrane potential of the postsynaptic cell. If you understand this concept, you should be able to draw a membrane with a resting potential of −65 mV and explain what happens when a ligand-gated ion channel opens and allows chloride ions to leave the cell.

Not all the receptors to which neurotransmitters bind are ion channels, however. Some receptors activate enzymes that lead to the production of a second messenger molecule in the postsynaptic cell. Recall from Chapter 8 that **second messengers** are chemical signals produced inside a cell in response to a chemical signal that arrives at the cell surface. The second messengers induced by neurotransmitters may trigger changes in enzyme activity, gene transcription, or membrane potential. The role of second messengers in cells is explored in detail in Chapter 47.

Postsynaptic Potentials and Summation

What happens when a neurotransmitter binds to a receptor and opens an ion channel? As the scanning electron micrograph in **Figure 45.16a** indicates, numerous synapses can be found in the same region of a dendrite or cell body. And as **Figure 45.16b** illustrates, these synapses can be one of two general types. If the receptors at the synapse admit sodium ions in response to the arrival of the neurotransmitter, then the postsynaptic membrane depolarizes in response to an action potential in the presynaptic cell. In most cases, depolarization makes an action potential in the postsynaptic cell more likely. Changes in the postsynaptic cell that make action potentials more likely are called **excitatory postsynaptic potentials** (**EPSPs**). Other receptors, in contrast, lead to an outflow of potassium ions or an inflow of chloride ions in the postsynaptic cell. These events hyperpolarize the membrane and make action potentials less likely to occur in the postsynaptic cell. Changes in the postsynaptic cell that make action potentials less likely are called **inhibitory postsynaptic potentials** (**IPSPs**). Synapses can be excitatory, inhibitory, or modulatory—meaning that they modify a neuron's response to other EPSPs or IPSPs.

It is critical to realize that, unlike action potentials, EPSPs and IPSPs are not all-or-none events. Instead, they are graded in size. The size of an EPSP or IPSP depends on the amount of neurotransmitter that is released at the synapse. Both types of signal are short lived because neurotransmitters do not bind irreversibly to channels in the postsynaptic cell. Instead, they are quickly inactivated or taken up by the presynaptic cell and recycled. If either the amount or life span of neurotransmitters is altered, the normal functioning of neurons is altered. The street drugs cocaine and amphetamine, for example, exert their effects by inhibiting the uptake and recycling of particular neurotransmitters. (see Table 45.2)

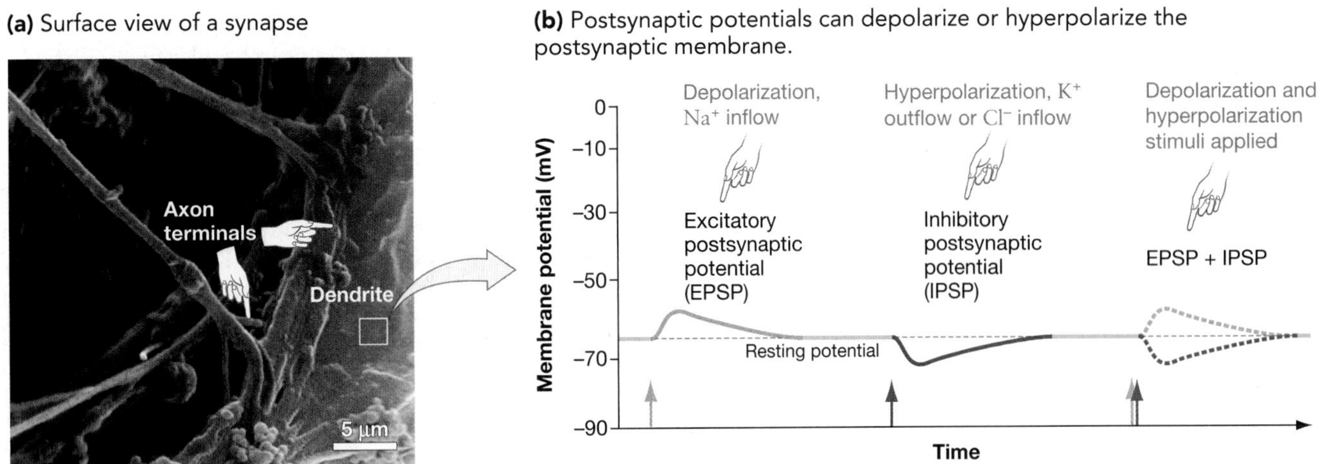

FIGURE 45.16 There Are Two Major Types of Synapses, Distinguished by Whether They Lead to Depolarization or Hyperpolarization of the Membrane. (a) Scanning electron micrograph showing axons from one neuron synapsing with the dendrite of a second neuron. **(b)** Records showing that the arrival of an action potential causes a short-lived depolarization or hyperpolarization of the postsynaptic plasma membrane. The change in postsynaptic potential can be excitatory or inhibitory, depending on whether it makes action potentials more or less likely. If both excitatory and inhibitory signals arrive at the same time, they cancel one another and the cell remains at resting level.

What effect do EPSPs and IPSPs have on the postsynaptic cell? As **Figure 45.17a** shows, the many dendrites and the cell body of a neuron typically make hundreds or thousands of synapses with other cells. At any instant, the EPSPs and IPSPs that occur at each of these synapses lead to short-lived surges of charge in the dendrites and cell body of the postsynaptic cell. If an IPSP and EPSP occur close together in space or time, the changes in membrane potential tend to cancel each other out (see Figure 45.16b). But if several EPSPs occur close together in space or time, they sum and make the neuron likely to fire an action potential (**Figure 45.17b**). The additive nature of postsynaptic potentials is termed **summation.**

The sodium channels that trigger action potentials in the postsynaptic cell are located near the start of the axon at a site called the **axon hillock** (see Figure 45.17a). As IPSPs and EPSPs are received and interact throughout the dendrites and cell body, charge spreads to the axon hillock. If the membrane at the axon hillock depolarizes past the threshold potential, enough sodium channels open to trigger positive feedback and an action potential. Once an action potential starts at the axon hillock, it propagates down the axon to the next synapse.

The summation that occurs prior to the axon hillock is a crucial phenomenon, because it determines whether an action potential begins in the postsynaptic cell. It also explains why the chemical transmission of electrical signals and synapses exists in the first place. If the plasma membranes and cytoplasm of adjacent neurons were continuous, then action potentials would

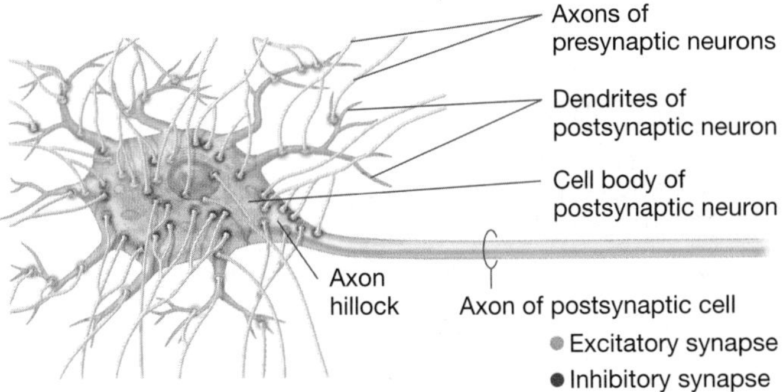

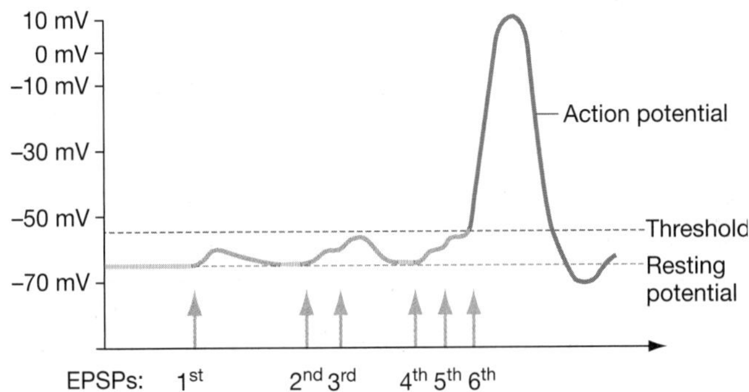

FIGURE 45.17 Neurons Integrate Information from Many Synapses. (a) The dendrites and cell body of a neuron typically receive signals from hundreds or thousands of other neurons. **(b)** When action potentials arrive close together in time from the same axon or from different axons with synapses close to one another, the postsynaptic potentials sum. In this example, the first excitatory signal is insufficient to generate an action potential. Two excitatory signals arriving close together cause summation but do not reach the threshold for generating an action potential. Three excitatory signals arriving closely spaced sum to exceed the threshold. If excitatory postsynaptic potentials depolarize the axon hillock past threshold, enough Na^+ channels open to trigger an action potential.

simply propagate from one cell to another without changing. But because neurons receive input from many synapses and because IPSPs and EPSPs sum, information in the form of electrical signals is modified at the synapse before being passed along. In addition, it turns out that synapses are dynamic—their behavior can change through time in response to events. This "plasticity" in the structure and function of synapses underlies many types of processes, including learning and memory.

The general message is that the postsynaptic cell integrates information from hundreds or thousands of other neurons. The information arrives in the form of action potentials that produce EPSPs and IPSPs. If the combination of postsynaptic potentials depolarizes the membrane at the axon hillock sufficiently, the postsynaptic cell fires an action potential in response. But if the events lead to a depolarization that is below threshold, the rate of firing decreases.

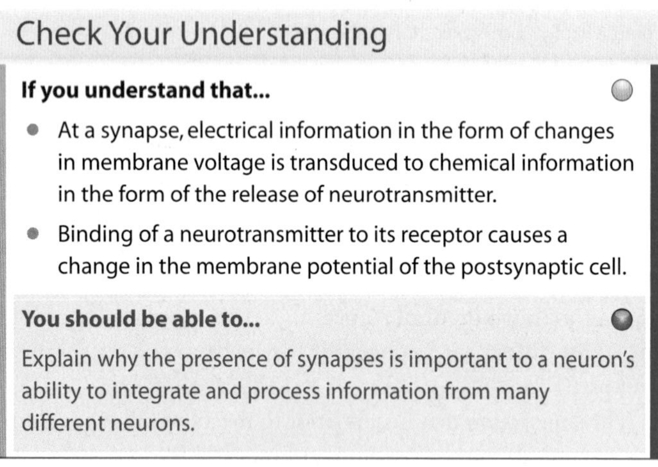

(MB) Web Animation *BioFlix* at www.masteringbio.com
Action Potentials; How Neurons Work

45.4 The Vertebrate Nervous System

The first three sections of this chapter examined electrical signaling at the level of molecules, membranes, and individual cells. The goal of this section is to discuss electrical signaling at the level of tissues, organs, and systems.

To begin, let's consider the overall anatomy of the vertebrate nervous system. Then we can ask how researchers explore the function of the most complex organ known: the human brain. The chapter concludes by returning to the molecular level and introducing recent work on learning and memory.

What Does the Peripheral Nervous System Do?

Biologists routinely distinguish the central nervous system, or CNS, from the peripheral nervous system, or PNS. Recall from Section 45.1 that the CNS is made up of the brain and spinal cord and is concerned primarily with integrating information, and that the PNS is made up of neurons outside the CNS.

What specific functions do the cells of the PNS control? Anatomical and functional studies indicate that the PNS consists of two systems with distinct functions: (1) an **afferent division**, which transmits sensory information to the CNS, and (2) an **efferent division**, which carries commands from the CNS to the body. Neurons in the afferent division monitor conditions inside and outside the body. Once information from afferent neurons has been processed in the CNS, neurons in the efferent division carry signals that allow the body to respond to changed conditions in an appropriate way. The afferent and efferent divisions carry out sensory and motor functions, respectively.

As **Figure 45.18** shows, the afferent and efferent divisions are part of a larger hierarchy of PNS functions. The efferent division is further divided into a **somatic nervous system**, which controls the skeletal muscles, and an **autonomic nervous system**, which controls internal processes such as digestion and heart rate. In effect, the somatic system responds to external stimuli and results in movement, while the autonomic system responds to internal stimuli and controls the activity of internal organs and glands. Many organs and glands are served by two functionally distinct types of autonomic nerves. Those in the **parasympathetic nervous system** promote functions that conserve or restore energy. For example, the parasympathetic nerves that innervate the heart slow it down, while those that serve the digestive tract stimulate its activity. Nerves in the **sympathetic nervous system**, in contrast, typically prepare organs for stressful situations. Sympathetic nerves speed up the heart rate, stimulate the release of glucose from the liver, and inhibit action by digestive organs. **Figure 45.19** summarizes the effects of signals from parasympathetic and sympathetic nerves.

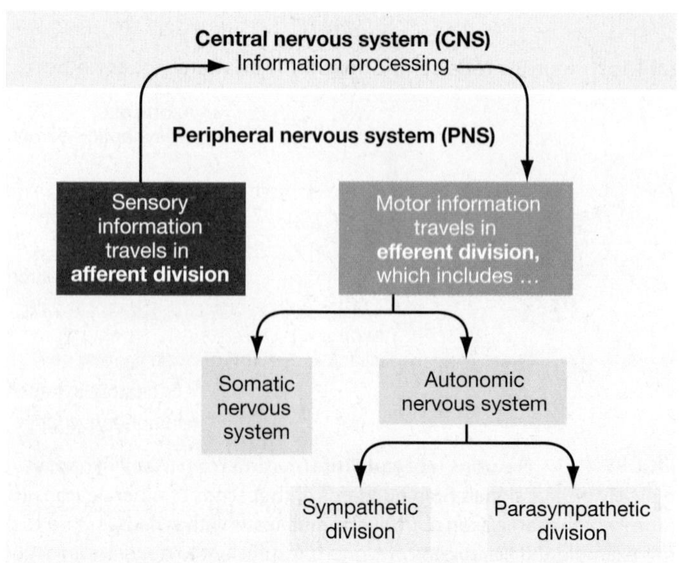

FIGURE 45.18 The Functions of the PNS Form a Hierarchy.

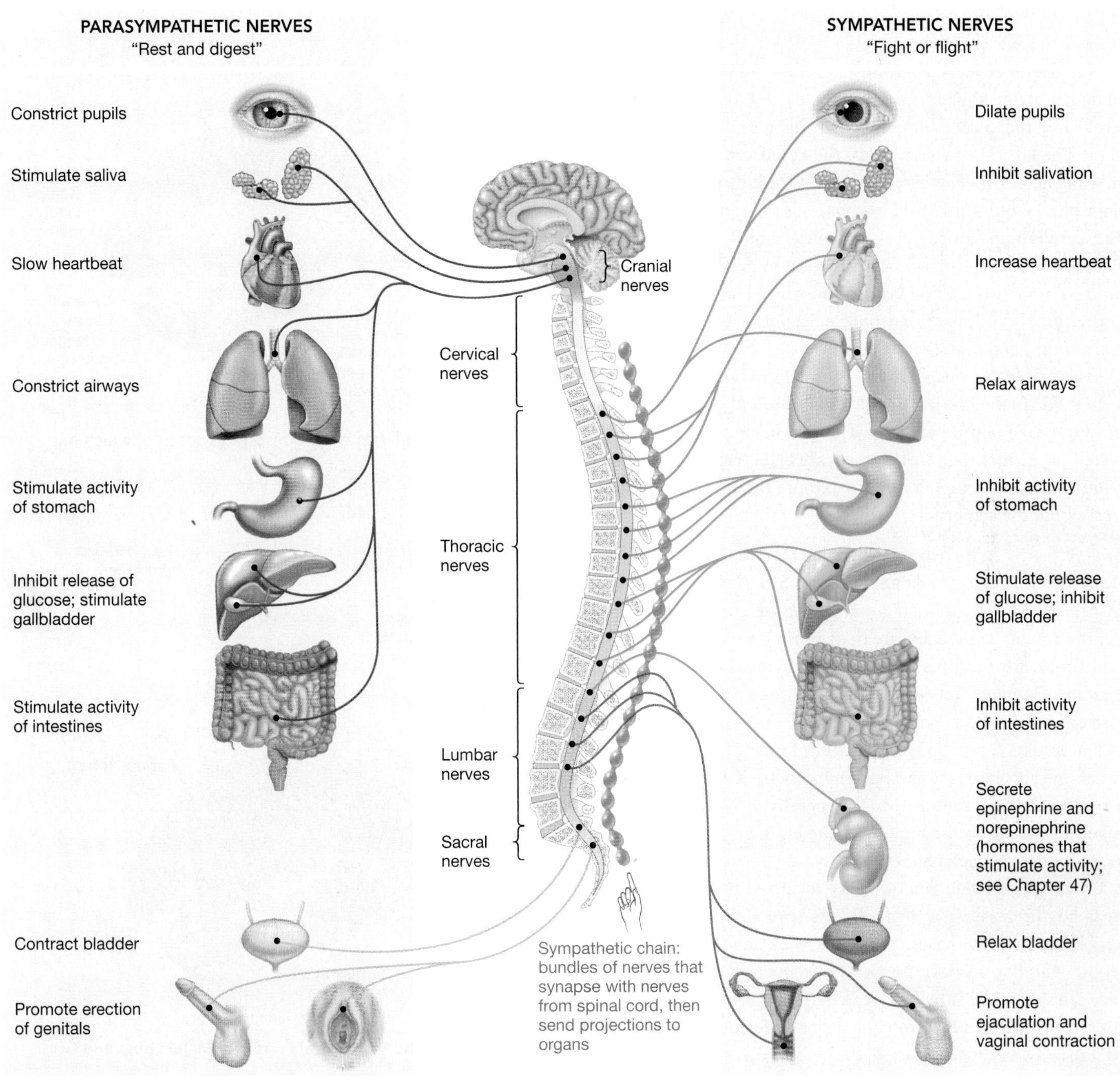

FIGURE 45.19 The Autonomic Nervous System Controls Internal Processes.

QUESTION Many parasympathetic nerves promote digestion and lack of movement; many sympathetic nerves support energy use and muscle activity. How do each of the responses listed here support these functions?

Functional Anatomy of the CNS

Parasympathetic nerves originate at the base of the brain or the base of the spinal cord. Most sympathetic nerves also originate in the spinal cord, but they emerge along the middle of its length. Similarly, most sensory and motor neurons in the somatic nervous system project to or from the spinal cord. In effect, then, the spinal cord serves as an information conduit. It collects and transmits information from throughout the body. With a few exceptions, such as the spinal reflexes illustrated in Figure 45.1, virtually all the information that travels to or from the spinal cord is sent to the brain for processing.

The brain is far and away the most complex organ found in animals. Researchers estimate that the human brain has 100 billion neurons, each making thousands of synaptic connections with other neurons. How do biologists even begin to study how such a fantastically complex structure functions? Researchers begin with general anatomy.

Nineteenth-century anatomists established that the brain is made up of the four structures labeled in **Figure 45.20a**: the cerebrum, the cerebellum, the diencephalon, and the brain stem. The largest of these areas, the **cerebrum**, makes up the bulk of the brain. The cerebrum is divided into left and right hemispheres and is involved in conscious thought and memory. The **cerebellum** coordinates complex motor patterns. The **diencephalon** relays sensory information to the cerebellum and controls homeostasis. The **brain stem**, which connects the brain to the spinal cord, is the autonomic center for regulating the heart, lungs, and digestive system.

Each cerebral hemisphere has four major areas, or lobes: the frontal lobe, the parietal lobe, the occipital lobe, and the temporal lobe (**Figure 45.20b**). The two hemispheres are connected by a thick band of axons called the **corpus callosum**. What tools do researchers use to explore the function of each area within the cerebrum?

(a) The brain is made up of four distinct structures.

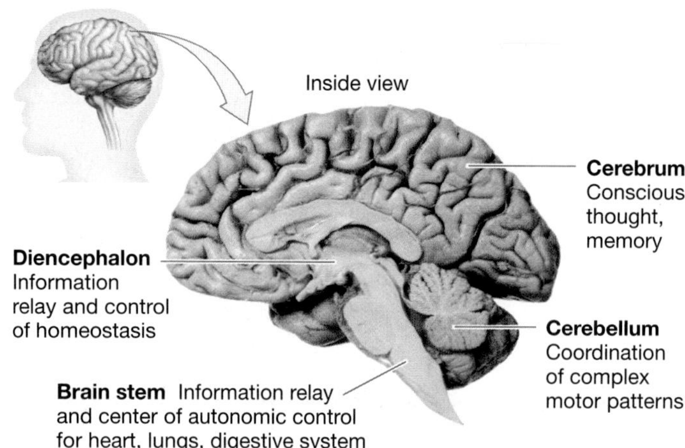

(b) The cerebrum has two hemispheres, each of which has four lobes.

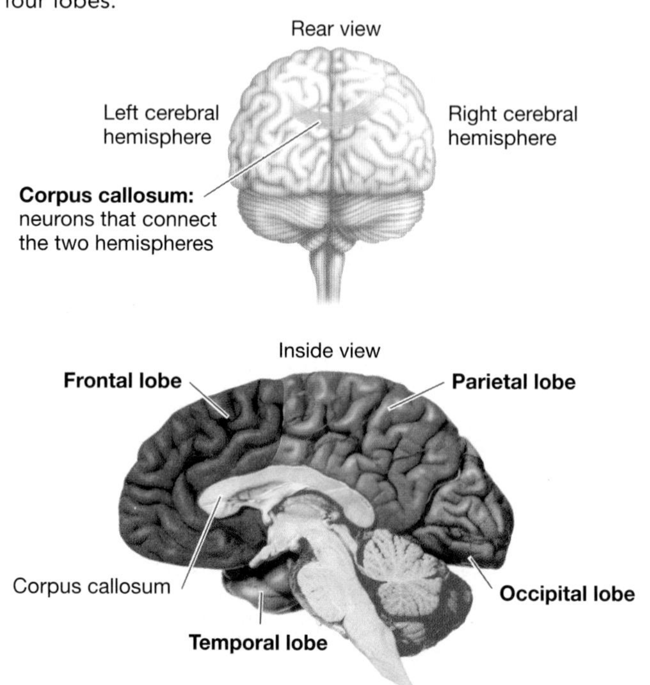

FIGURE 45.20 The Human Brain Has Four Major Lobes and Two Hemispheres.

EXERCISE On your own head, point to each of the labeled areas.

Mapping Functional Areas: Lesion Studies Early work on brain function focused on studying people with specific mental deficits caused by areas of brain damage, or lesions. Paul Broca, for example, studied an individual who could understand language but could not speak. After the individual's death in 1861, Broca examined the patient's brain and discovered a damaged area in the left frontal lobe of the cerebrum. Based on this observation, Broca hypothesized that this region is responsible for speech. More generally, he formulated the hypothesis that specific regions of the brain are specialized for coordinating particular functions.

Broca's claim that functions are localized to specific brain areas has been verified through extensive efforts to map the cerebrum. In some cases, advances were made by studying people who had to have portions of their brains removed. In 1953, for example, surgeons treated a 27-year-old man for life-threatening seizures by removing a small portion of his temporal lobe. The man recovered (he is still alive, in fact), has normal intelligence, and vividly remembers his childhood, but he has no short-term memory. Brenda Milner, who studied this individual for over 40 years, had to introduce herself to him every time they met; he cannot recognize a recent picture of himself. Based on case histories and data from studies of memory in laboratory animals, a consensus has emerged that several aspects of memory localize to interior sections of the temporal lobe.

Electrical Stimulation of Conscious Patients Studying the mental abilities of people who have suffered accidental or surgical brain damage has been extraordinarily fruitful. A different approach to studying brain function, however, was pioneered by Wilder Penfield, who worked with severe epileptics—people suffering from seizures. These individuals were scheduled to have seizure-prone areas of their brains surgically removed. While the patients were awake and under a local anesthetic, Penfield electrically stimulated portions of their cerebrums. His goal was to map areas that are essential and should be spared from removal if possible. When he stimulated specific areas, patients reported sensations or movement in particular regions of the body. Penfield was able to map the sensory regions of the cerebrum shown in **Figure 45.21**, as well as the adjacent motor regions. The technique is

(a) Top view of cerebrum

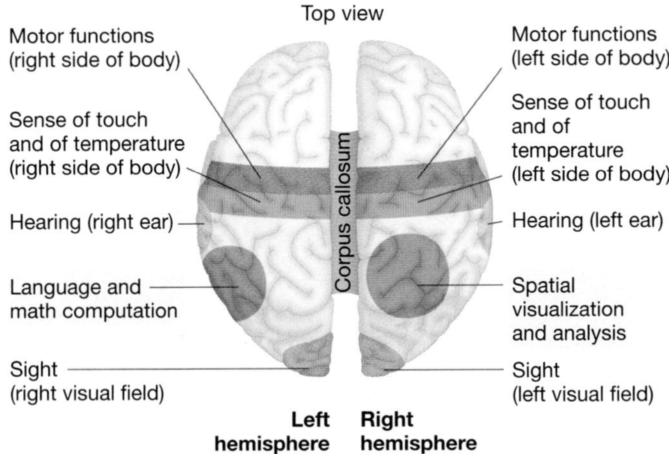

(b) Cross section through area responsible for sense of touch and of temperature

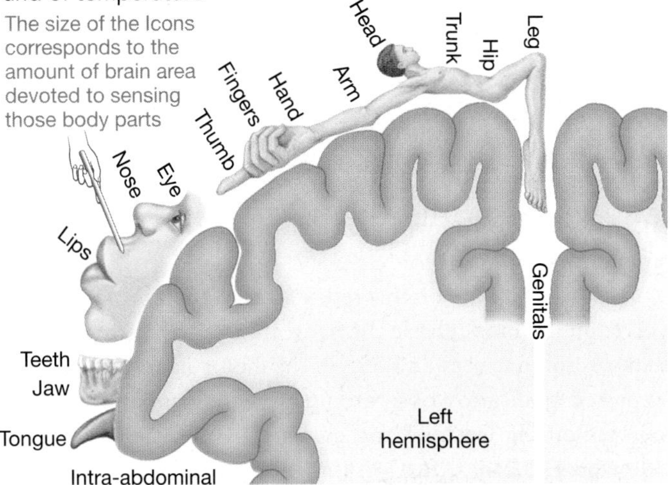

FIGURE 45.21 Specific Brain Areas Have Specific Functions. (a) Map of the brain, in top view (as if the person is looking at the top of the page), showing the functions of some major regions. The map was compiled from studies of people with damaged brain areas or with brain regions that were removed surgically. **(b)** Researchers mapped the area responsible for the sense of touch and of temperature by stimulating neurons in the brains of patients who were awake.

● **QUESTION** Is there a correlation between the size of the brain area devoted to sensing a particular body part and that body part's size? Explain.

still used by brain surgeons to map critical areas near tumors and seizure-prone areas.

Perhaps the most striking of Penfield's findings was that, on occasion, patients would respond to stimulation of their temporal lobe by having what appeared to be flashbacks. After one region was stimulated, a woman said, "I hear voices. It is late at night around the carnival somewhere—some sort of traveling circus. . . . I just saw lots of big wagons that they used to haul animals in."

Was this a memory, stored in a small set of neurons that Penfield happened to simulate? The hypothesis that memories are stored in specialized cells is intensely controversial. As critics have pointed out, Penfield's results are difficult to interpret, because he was working with people who suffered from severe brain dysfunction. In addition, Penfield's patients sometimes described the same memory when other cells were stimulated after the original area had been surgically removed. Have other approaches to studying memory been productive?

How Does Memory Work?

Learning is an enduring, usually adaptive change in behavior that results from a specific experience in an individual's life. **Memory** is the retention of learned information. Learning and memory are thus closely related and are often studied in tandem. Research on these phenomena has been extensive, and progress has been particularly rapid recently. As an introduction to how researchers explore learning and memory, let's first explore work that is focused at the level of whole neurons and then review research at the molecular level.

Recording from Single Neurons during Memory Tasks

One approach to studying learning and memory is to record a sequence of action potentials from individual neurons during learning and memory tasks. How do the action potentials generated by a cell change as learning and memory take place? Researchers have attempted to answer this question by recording from individual neurons in the temporal lobes of humans. Before operating on patients who were still awake and about to undergo surgery to remove seizure-prone areas of their brains, physicians have projected words or names of objects on a screen and asked the individuals to read them silently, read them aloud, and/or remember them and repeat them later. The data shown in **Figure 45.22** are informative. In this case, the neuron being recorded was relatively quiet while the patient first identified objects but extremely active when the individual was remembering the objects and repeating their names aloud.

What do such data mean? Neurons in the temporal lobe are most active during memory tasks. So how could action potentials from particular cells make memory possible?

Documenting Changes in Synapses Research on the molecular basis of memory is based on two fundamental ideas. First, learning and memory must involve some type of short-term or long-term change in the neurons responsible for these processes. This change could be structural or chemical in nature. Structural changes might include modifications in the number of synapses that a particular neuron makes. Chemical (molecular) changes might involve alterations in the amount of neurotransmitter released at certain synapses and changes in

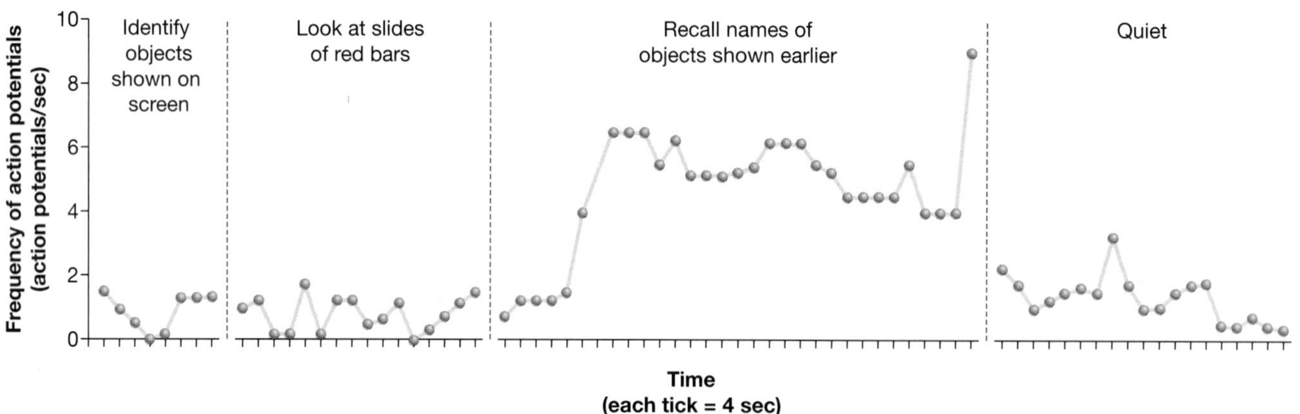

FIGURE 45.22 The Frequency of Action Potentials from Brain Neurons Varies with Activity. Data showing the frequency of action potentials from a single brain neuron from the temporal lobe as a patient identified a series of objects shown on slides, looked at a series of red bars shown on slides, recalled the names of the objects shown previously, and looked at a blank screen while doing nothing.

QUESTION Are the data here consistent or inconsistent with the data from PET scans shown in the chapter opening photo? Explain your answer.

the number of receptors present in postsynaptic cells. Second, it will be much easier to understand what these changes are if an extremely simple system of neurons can be studied.

To explore the molecular basis of learning and memory, Eric Kandel's group has focused on the sea slug *Aplysia californica* (**Figure 45.23a**). Much of their work has explored the reflex diagrammed in **Figure 45.23b**. When a structure in *Aplysia* called the siphon is touched, the individual responds by withdrawing its gill. The reflex involves a sensory neuron that is activated by touch and a motor neuron that projects to a gill muscle. Retracting the gill protects it from predators.

Early work established that this simple reflex is modified by learning. For example, *Aplysia* also withdraw their siphons when their tails are given an electrical shock. If shocks to the tail are combined with a very light touch to the siphon—too light to normally get a response on its own—an *Aplysia* will learn to withdraw its gills in response to a light siphon touch alone.

Follow-up studies on this reflex showed that the neurons involved in learning release the neurotransmitter **serotonin** and that serotonin causes an EPSP in the motor neuron to the gill. Repeated application of serotonin mimics what happens at the synapse during learning and changes the serotonin-secreting cell (**Figure 45.24a**). After learning or increased serotonin application, EPSPs are higher and the motor neuron is more likely to generate action potentials. These results implied that in

(a) Sea slug *Aplysia californica*

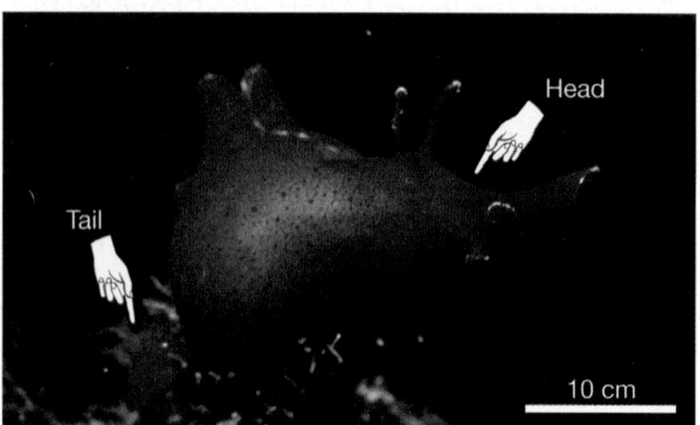

(b) Gill-withdrawal reflex protects the gills during an attack.

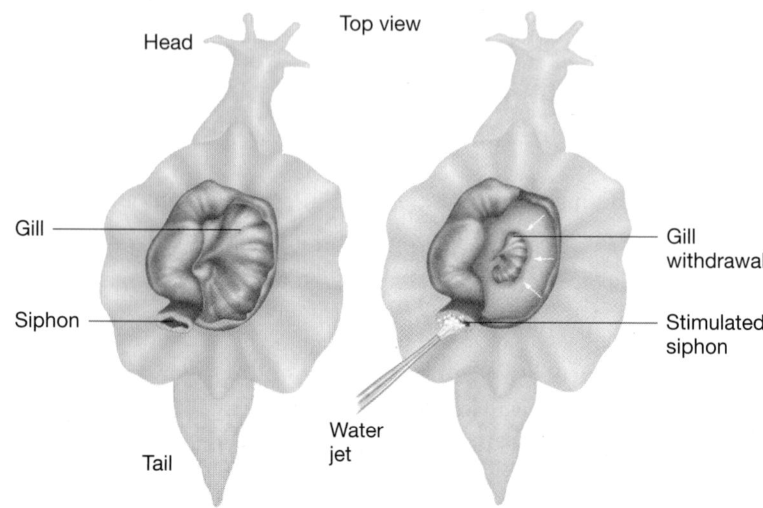

FIGURE 45.23 The Gill-Withdrawal Reflex in *Aplysia* Is a Model System in Learning and Memory.

(a) Postsynaptic potentials change as a result of learning.

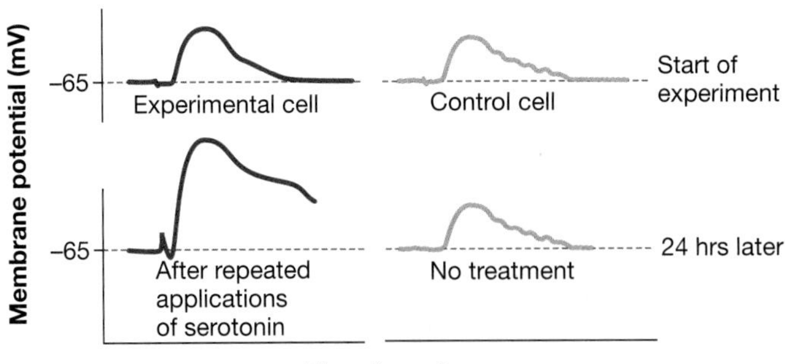

(b) Apply neurotransmitters to neurons in culture.

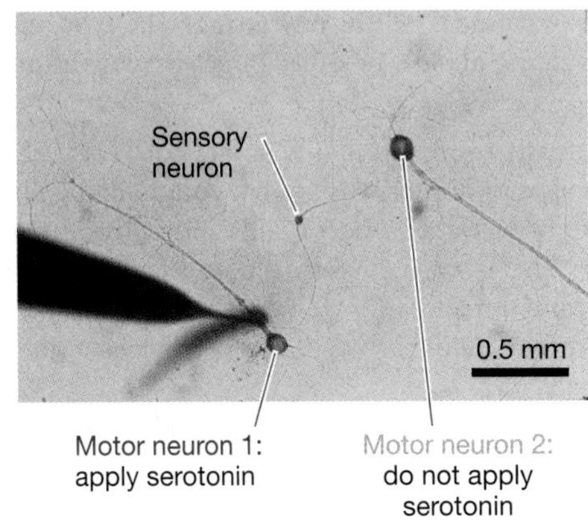

(c) Repeating *Aplysia* experiment in culture

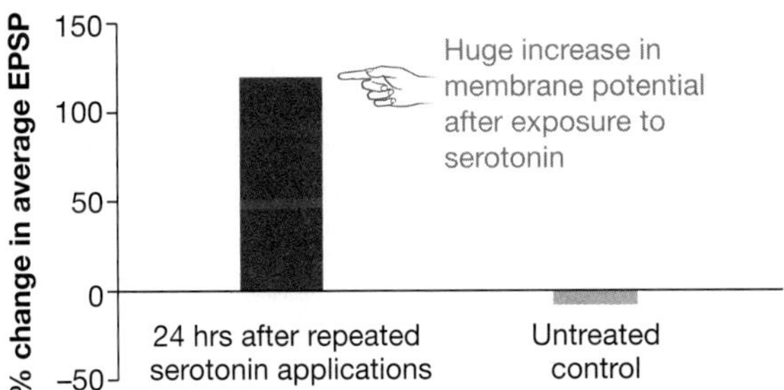

FIGURE 45.24 Learning and Memory Involve Changes in Synapses. **(a)** Experiment showing that repeated application of serotonin changes the behavior of postsynaptic neurons. **(b)** An in-culture study of interactions between *Aplysia* motor neurons and sensory neurons. **(c)** Histograms documenting the percentage increase in postsynaptic potentials that occurs after repeated application of serotonin, as in part (a).

Aplysia, changes in the nature of the synapse form the molecular basis of learning and memory. This process is termed **synaptic plasticity.**

Kandel's team was recently able to replicate these results with sensory and motor neurons growing in culture. **Figure 45.24b** shows two *Aplysia* motor neurons on a culture plate. Each motor neuron is receiving synapses from a sensory neuron. To mimic the learning process, the investigators applied serotonin to one synapse five times over a short period. When they stimulated the sensory neuron a day later, they found a huge increase in EPSPs (**Figure 45.24c**). The cells had also established additional synapses. Neurons that had not received the repeated stimulation with serotonin had normal postsynaptic responses and numbers of synapses.

Results like these reinforce a growing consensus that learning and memory involve both molecular and structural changes in synapses. Further, most researchers now agree that at least some aspects of long-term memory involve changes in gene expression. Chapter 47 explores how the chemical messengers called hormones cause changes in gene expression in target cells. But before investigating how hormones work, let's focus on the electrical signals involved in vision, hearing, taste, and movement—the subject of Chapter 46.

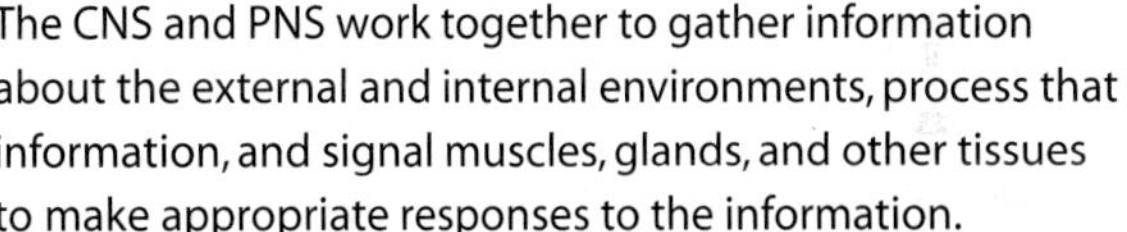

Check Your Understanding

If you understand that...

The CNS and PNS work together to gather information about the external and internal environments, process that information, and signal muscles, glands, and other tissues to make appropriate responses to the information.

You should be able to...

1) Diagram the hierarchical structure of the PNS.
2) Describe the types of research strategies that allowed biologists to localize particular functions to specific regions in the brain.
3) Predict whether the degree of synaptic plasticity is greater in the sensory neurons of sea slugs or in the brain neurons of humans, and why.

Chapter Review

SUMMARY OF KEY CONCEPTS

Neurons are cells that transmit electrical signals used in communication. Neurons have a membrane potential, which is an electric potential difference, or voltage, across the plasma membrane. A voltage exists due to differences in the concentrations of ions on the two sides of the plasma membrane.

Although neurons are highly variable in size and shape, all have a cell body and multiple short dendrites that receive electrical signals from other cells. Most neurons have axons that transmit electrical signals to other neurons or to effector cells in glands or muscles.

Studies of the squid giant axon neurons established that neurons have a resting potential due to differences in the concentrations of ions on the two sides of the membrane and the selective permeability of the membrane to ions. The difference in concentration of ions is created, in part, by the sodium-potassium pump. When Na^+/K^+-ATPase hydrolyzes ATP, it transports 3 Na^+ out of the cell and 2 K^+ in.

You should be able to diagram the plasma membrane of a neuron. Add symbols to show the relative concentrations of Na^+, K^+, and Cl^-. Add labels indicating the role of K^+ leak channels and the Na^+/K^+-ATPase.

MB Web Animation at www.masteringbio.com
Membrane Potentials

Electrical signals are transmitted as graded or all-or-none changes in membrane voltage. All-or-none changes are called action potentials. During an action potential, an inflow of sodium ions is followed by an outflow of potassium ions.

Studies of the squid axon established that the action potential is a rapid, all-or-none change in membrane potential. An action potential begins with an inflow of Na^+ that depolarizes the membrane. An outflow of K^+ follows and repolarizes the membrane. As charge spreads from the site of an action potential, the nearby membrane is depolarized enough to trigger additional Na^+ inflows and propagate the signal. Both Na^+ and K^+ flow through voltage-gated channels. Propagation takes place most rapidly in large axons or myelinated axons.

You should be able to explain why the action potential never stops once a membrane potential depolarizes past threshold, and why every action potential in a neuron is identical.

MB Web Animation *BioFlix* at www.masteringbio.com
Action Potentials; How Neurons Work

At synapses, an electrical signal from one neuron is converted to a chemical signal in the form of a neurotransmitter. When it arrives at an adjacent neuron, the neurotransmitter causes a change in that cell's membrane potential.

When action potentials arrive at a synapse, synaptic vesicles fuse with the axon's membrane and deliver neurotransmitters that bind to receptors on the membrane of a postsynaptic cell. One class of receptors functions as ligand-gated channels. In response to binding by a neurotransmitter, the channels open and admit ions that depolarize or hyperpolarize the postsynaptic cell's membrane. Postsynaptic potentials from nearby synapses sum. If the membrane at the axon hillock depolarizes to a threshold value, then an action potential is triggered.

You should be able to predict the effect of a drug that binds to K^+ channels in the postsynaptic membranes of certain brain neurons.

Animals have a central nervous system (CNS) and a peripheral nervous system (PNS). PNS neurons receive sensory information and transmit it to the CNS for processing. The CNS then sends signals to muscles, glands, or other target tissues via other PNS neurons.

The CNS consists of the brain and spinal cord (in vertebrates); the PNS consists of all nervous system components outside the CNS. In vertebrates, the PNS contains somatic and autonomic components. The somatic PNS is responsible for sensing external stimuli and effecting movement; the autonomic system monitors internal conditions and effects changes in the activity of organs.

Although the CNS is enormously complex, researchers have succeeded in mapping the functions of brain structures. Early mapping studies depended on analyzing deficits in individuals with brain lesions or on the direct stimulation of certain regions of the cerebral cortex. Efforts to understand higher brain functions such as learning and memory form the current focus of research on the CNS. To date, research has established that learning and memory are based on modifications in synapses. After learning takes place, certain neurons release more or less neurotransmitter, or make additional synapses, in response to stimulation.

You should be able to describe what an animal would be like if synapses were "fixed" early in development and unchangeable.

QUESTIONS

Test Your Knowledge

1. Why does the resting potential exist?
   a. because cells contain several types of ions
   b. because ion concentrations differ on the two sides of the membrane and because the membrane is selectively permeable to K^+
   c. because ion concentrations differ on the two sides of the membrane and because the membrane is selectively permeable to Na^+
   d. because ion concentrations differ on the two sides of the membrane and because the membrane is selectively permeable to Cl^-

2. Why did the squid axon become a model system for studying electrical signaling in animals?
   a. Its action potentials are particularly large and frequent.
   b. It is the tissue from which researchers initially isolated Na^+/K^+-ATPase.
   c. Squids are abundant and easy to obtain.
   d. It was large enough to support intracellular recording by the first microelectrodes.

3. How does myelination affect the propagation of an action potential?
   a. It speeds propagation by increasing the density of voltage-gated channels.
   b. It speeds propagation by increasing electrochemical gradients favoring Na^+ entry.
   c. It speeds propagation because charge does not leak out of the membrane as it spreads down the axon.
   d. It slows down propagation because Na^+ channels exist only at unmyelinated nodes (nodes of Ranvier).

4. In a neuron, what creates the electrochemical gradient favoring the outflow of K^+?
   a. Na^+/K^+-ATPase
   b. voltage-gated K^+ channels
   c. voltage-gated Na^+ channels
   d. ligand-gated Na^+/K^+ channels

5. Why do biologists say that positive feedback occurs during an action potential?
   a. The action potential is an all-or-none event, meaning that once it starts, it goes to completion.
   b. The opening of potassium channels repolarizes the membrane, making it less likely that sodium channels will open and depolarize the membrane.
   c. Once sodium channels open and begin to depolarize the membrane, they become more likely to open and cause further depolarization.
   d. Sodium channels are refractory—once they have opened, they are less likely to open again for a few milliseconds.

6. Why is memory thought to involve changes in particular synapses?
   a. In some systems, an increased release of neurotransmitters occurs after learning takes place.
   b. In some systems, the type of neurotransmitter released at the synapse changes after learning takes place.
   c. When researchers stimulated certain neurons electrically, individuals replayed memories.
   d. People who lack short-term memory have specific deficits in synapses within the brain regions responsible for memory.

Test Your Knowledge answers: 1. b; 2. d; 3. c; 4. a; 5. c; 6. a

Test Your Understanding

Answers are available at www.masteringbio.com

1. Explain why the resting potential exists. Be sure to differentiate between the roles of K^+ channels and that of Na^+/K^+-ATPase in the membrane's selective permeability and ion concentration gradients.
2. Draw a graph of an action potential and label the axes. Label the parts of the graph and explain which ion flow or flows are responsible for each part.
3. Draw a diagram of a synapse. Label the parts. Then make a series of diagrams showing what happens when an action potential arrives at a synapse. Explain the events that occur at both the presynaptic cell and the postsynaptic cell.
4. Why do summation and integration occur in postsynaptic cells? Why is it important that synapses exist, instead of neurons having a continuous plasma membrane and direct cytoplasmic connections?
5. Compare and contrast the somatic and autonomic components of the PNS.
6. Compare and contrast the sympathetic and parasympathetic components of the autonomic nervous system.

Applying Concepts to New Situations

Answers are available at www.masteringbio.com

1. Study the circuitry in the nervous system diagrammed in Figure 45.1 and the roles of the parasympathetic nerves described in Figure 45.19. Discuss how the nervous system functions as a mechanism for achieving homeostasis.
2. Discuss the pros and cons of lesion studies and electrical stimulation of conscious patients in determining the functions of particular brain structures.
3. During an epileptic seizure, muscles convulse spasmodically and the person, although awake, is unable to think, see, or hear. The cause of epilepsy is unknown. Speculate on what is happening at the level of the synapse in seizure-prone areas of the brain. How would you test your hypothesis?
4. In some species, researchers are able to identify individual neurons in the brain and record the action potentials they produce. Using these data, how can they infer the function of particular neurons?

www.masteringbio.com is also your resource for • Answers to text, table, and figure caption questions and exercises • Answers to *Check Your Understanding* boxes • Online study guides and quizzes • Additional study tools including the *E-Book for Biological Science* 3rd ed., textbook art, animations, and videos.

46 Animal Sensory Systems and Movement

KEY CONCEPTS

- Sensory receptor cells transduce sound, light, or other stimuli to a change in membrane potential. Receptor cells or sensory neurons send action potentials to the brain, where the signals are processed and integrated.
- Hearing is based on sensory receptor cells that move in response to sound waves of a particular frequency. The movement opens ion channels in the sensory cells and results in a change in membrane potential.
- Vision is based on sensory receptor cells that contain a light-absorbing pigment bound to a protein. The pigment changes conformation when it absorbs light and triggers a series of events that changes the cell's membrane potential.
- Taste and smell sensations are registered by membrane proteins that act as receptors or channels for particular molecules and that trigger changes in membrane potential.
- In many cases, animals respond to sensory stimuli by moving. Movement is based on antagonistic muscle groups that act on a stiff skeleton containing joints that flex or extend in response to electrical signals.

In many species of moth, males have much larger antennae than females do. Receptor cells on the males' feathery antennae detect airborne chemical signals that are produced by sexually mature females. As a result, males can locate females in total darkness.

Adult moths are active at night when it is difficult or impossible to see. It is logical, then, to observe that sexually mature female and male moths do not go looking for each other. Instead, females release a chemical attractant called a pheromone into the air. A **pheromone** is a small molecule that acts as a signal between individuals and that usually causes a change in physiology or behavior. Male moths of the same species that are flying in the area can detect even a single molecule of the pheromone, due to the receptor cells located on their large, feathery antennae. In response to an airborne gradient of pheromone molecules, males fly toward a female.

As they patrol in search of these airborne pheromones, however, male moths are hunted by bats. Like moths, bats are active almost exclusively at night. Instead of hunting by sight, like a falcon or a cheetah, bats hunt with the aid of sonar. Bats emit a train of high-pitched sounds as they fly and then listen for echoes that indicate the direction and shape of objects in their path. If the object is a moth, the bat flies toward it, catches the individual in its mouth, and eats it. Some moth species can hear bat calls, however. When moths detect sounds from an onrushing bat, they tumble out of the sky in chaotic escape flights (see Chapter 50).

If you were out at night as these dramas unfolded, at best you might be dimly aware that bats and moths were flying about. Humans cannot smell moth pheromones or hear the sounds that bats emit when flying. It took decades of careful experimentation for biologists to understand how moths and bats sense the world around them and how they move in response to the information they receive.

Sensing changes in the environment and moving in response to this information is fundamental to how animals work. To explore the mechanisms involved, Section 46.1 introduces how animals receive information from the external or internal environment and respond to it. The next three sections delve into the molecular processes involved in hearing, vision, and taste and smell. The chapter concludes by investigating the cellular and molecular mechanisms responsible for movement.

46.1 How Do Sensory Organs Convey Information to the Brain?

As a moth flies through the night, its brain receives streams of signals from an array of sensory organs. Antennae provide information about the concentration of pheromones; ears located on various parts of the body send data on the presence of high-pitched sounds; detectors for balance and gravity transmit signals about the body's orientation in space. Each type of sensory information is detected by a sensory neuron or by a specialized receptor cell that makes a synapse with a sensory neuron. As **Figure 46.1** shows, the moth's nervous system integrates the sensory input—information from sensory neurons—and responds with motor output, via electrical signals, to specific muscle groups (effectors).

The ability to sense a change in the environment depends on four processes: (1) **transduction**, or the conversion of an external stimulus to an internal signal in the form of an action potential; (2) amplification of the signal; (3) transmission to the central nervous system (CNS); and (4) **integration**, or processing with other incoming signals.

The first step in the sequence requires a sensory receptor cell to convert light, sound, touch, or some other signal into an electrical signal. Sensory receptors are located throughout the body and are categorized by the type of stimulus: **Nociceptors** sense harmful stimuli; **thermoreceptors** detect changes in temperature; **mechanoreceptors** respond to distortion caused by touch or pressure; **chemoreceptors** perceive specific molecules or classes of molecules; **photoreceptors** respond to particular wavelengths of light; and **electroreceptors**, if present, detect electric fields. With such a broad range of sensory detectors available, it is no wonder that animals can monitor and respond to a wide array of changes in their environments (see **Box 46.1** on page 1034).

Now, how do all these sensory cells receive information from the environment and report it to the brain, so an appropriate response can occur?

Sensory Transduction

Chapter 45 analyzed how electrical signals are generated and propagated in nerve cells. Electrical signaling is based on changes in membrane voltage. Recall that an **electrical potential**, or voltage, is created whenever a plasma membrane separates charges in the form of different concentrations of ions. When most sensory cells are in the resting state, the inside of the plasma membrane is more negative than the exterior.

Researchers can record changes in a cell's membrane potential with the aid of microelectrodes. If ion flows cause the interior to become more positive (less negative), the membrane is said to be **depolarized.** If changes in ion channels cause the cell interior to become more negative than the resting potential, the membrane is said to be **hyperpolarized.**

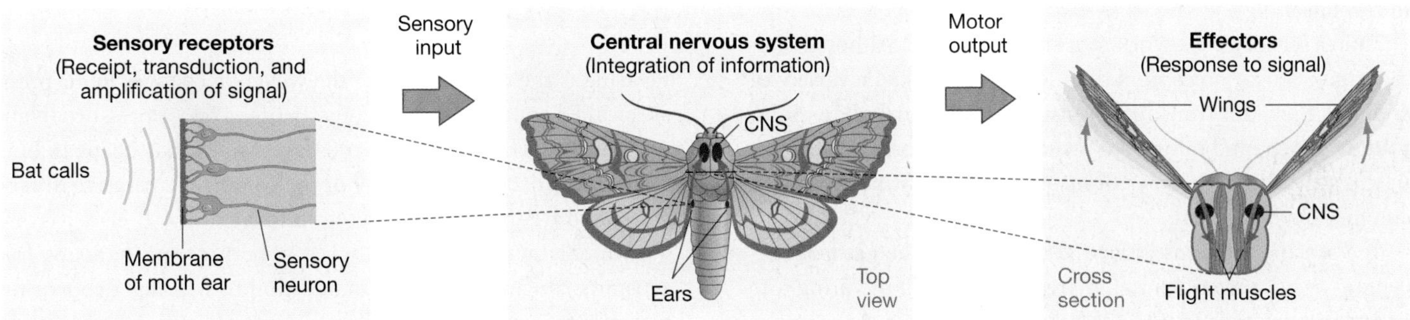

FIGURE 46.1 Sensory Systems, the CNS, and Effectors Such as Muscles Are Linked. Sensory neurons relay information about conditions inside and outside an animal to the central nervous system. After integrating information from many sensory neurons, the CNS sends signals to muscles.

QUESTION Why is it appropriate to call these moth organs "ears"? Why is it inaccurate or misleading? (Hint: Insects and vertebrates made the water-to-land transition independently—see Chapters 33 and 34.)

(a) Sound-receptor cells depolarize in response to sound.

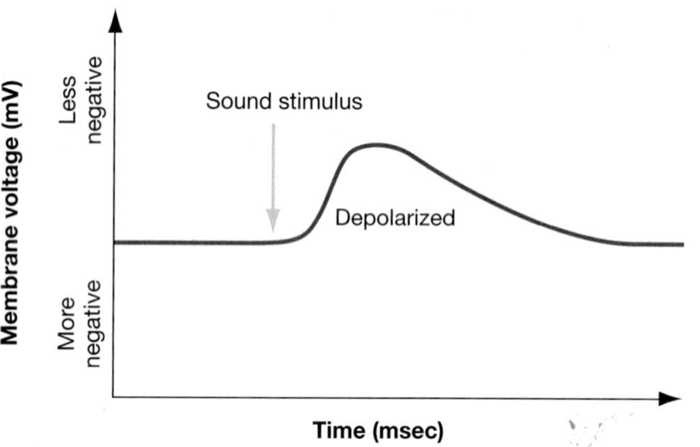

(b) Sound-receptor cells respond more strongly to louder sounds.

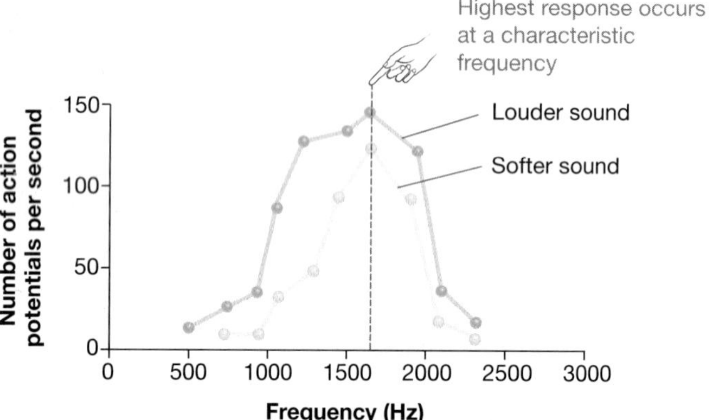

FIGURE 46.2 Sensory Inputs Change the Membrane Potential of Receptor Cells. (a) In response to sensory stimuli, ions flow across the membranes of receptor cells and either depolarize or hyperpolarize the membrane. **(b)** The frequency of action potentials from a receptor transmits information about the nature and intensity of the sensory stimulus.

What do these principles have to do with sensory systems? To answer this question, examine the voltage recording from a sound-receptor cell in **Figure 46.2a**. Notice that when the experimenter played a sound, the sound-receptor cell depolarized in response.

Other sensory cells work in a similar way. Although animals have sensory receptors that detect a remarkable variety of stimuli, they all transduce sensory input—including light, sounds, touch, and odors—to a change in membrane potential. In this way, different types of information are transduced to a common type of signal.

If a sensory stimulus induces a large change in a sensory receptor's membrane potential, there is a change in the firing rate of action potentials sent to the brain. For example, the amount of depolarization that occurs in a sound-receptor cell is proportional to the loudness of the sound. If the depolarization passes threshold, enough voltage-gated sodium channels open to trigger action potentials that are sent to the brain. Recall from Chapter 45 that all action potentials are identical in size and shape. As **Figure 46.2b** shows, however, louder sounds induce a higher frequency of action potentials than do softer sounds. In this way, receptor cells provide information about the intensity of a stimulus.

The universal nature of sensory transduction raises an important question: If all types of external stimuli are converted to electrical signals in the form of action potentials, and if all action potentials are alike, then how is it possible for the brain to interpret the information properly?

Transmitting Information to the Brain

There are two keys to understanding how the brain interprets sensory information. First, receptor cells tend to be highly specific. For example, each receptor cell in a human ear responds best to certain frequencies of sound. Some receptors are more sensitive to low-pitched sounds at a frequency of 1000 Hz (hertz, or cycles per second); others respond best to high-pitched sounds at 8000 Hz. The receptor-cell response depicted in Figure 46.2b is strongest for sounds at about 1650 Hz. At this frequency, the maximum number of action potentials per second emerge from the receptor. In this way, the pattern of action potentials from a cell contains information about the frequency of sound that is being received, its intensity, and how long the stimulus lasts.

Second, each type of sensory neuron sends its signal to a specific portion of the brain. Axons from sensory neurons in the human ear project to a particular area at the side of the brain, but axons from sensory receptors in the eye deliver action potentials to a specific area at the back of the brain. Different regions of the brain are specialized for interpreting different types of stimuli.

Now that the basic principles of sensory reception and transduction have been introduced, let's delve into the details of four sensory systems that are particularly well understood: hearing, vision, taste, and smell.

46.2 Hearing

Hearing is the ability to sense the wavelike changes in air pressure called sound. A sound consists of waves of pressure in air or in water. The number of pressure waves that occur in one second is called the **frequency** of the sound. We perceive different sound frequencies as different **pitches**.

Animals actually have a wide variety of mechanisms for sensing changes in pressure in addition to hearing airborne or waterborne sound waves. Crabs, for example, have a fluid-filled organ that helps them sense the pressure created by gravity. The organ, known as a **statocyst**, is lined with pressure-receptor cells and contains a small calcium-rich structure. The calcium-rich particle normally rests on the bottom of the organ. But if the crab is tipped or flipped over, this structure presses against receptors that are *not* on the bottom of the organ. When the brain receives action potentials from these cells, it responds by activating muscles that restore the animal to its normal posture. It's also common for animals to have cells that are responsible for detecting direct physical pressure on skin, as well as pressure-receptor cells that monitor how far muscles or blood vessels are stretched.

Virtually all the pressure-sensing systems found in animals are based on the same mechanism, however. Let's briefly examine the general nature of a mechanoreceptor cell that responds to pressure, then investigate the specific structures involved in vertebrate hearing.

How Do Sensory Cells Respond to Sound Waves and Other Forms of Pressure?

The mechanoreceptors responsible for sensing sound and vibrations in the environment are relatively simple in design. In every case, direct physical pressure on a plasma membrane or distortion by bending changes the conformation of ion channels in the membrane and causes the channels to open or close. In response to a change in ion flow, the membrane depolarizes or hyperpolarizes. The result is a new pattern of action potentials from a sensory neuron.

In the pressure-sensing organs found in vertebrates, ion channels that respond to pressure are found in hair cells similar to the one shown in **Figure 46.3a**. **Hair cells** are pressure-receptor cells named for their appearance, which they owe to a set of stiff outgrowths called **stereocilia** (singular: **stereocilium**). The "hairy-looking" stereocilia are microvilli that are reinforced by actin filaments. Many hair cells also have a single **kinocilium**; the kinocilium is a true cilium that contains a 9+2 arrangement of microtubules introduced in Chapter 7. Hair cells are found in the ears of land-dwelling vertebrates and the lateral line system observed in many species of fish.

As Figure 46.3a shows, the stereocilia in hair cells are arranged in order of increasing height; if a kinocilium is present, it is the tallest of all the projections. All these structures extend into a fluid-filled chamber. If the stereocilia are bent in the direction of the kinocilium in response to pressure (as illustrated in **Figure 46.3b**), the membranes of the structure are distorted. The distortion causes potassium ion (K^+) channels in the stereocilia to open. This is the common theme in pressure-sensing cells: Bending opens ion channels.

Recall from Chapter 45 that the opening of K^+ channels causes an outflow of K^+ that hyperpolarizes neurons. Hair-cell plasma membranes respond differently, because they are bathed by extracellular fluid with an extraordinarily high K^+ concentration. As a result, the equilibrium potential for K^+ in hair cells is 0 mV instead of the −85 mV in a typical neuron. The resting potential of the hair-cell plasma membrane is −70 mV, so K^+ rushes in and causes a depolarization of approximately 20 mV.

(a) Hair cells have many stereocilia and one kinocilium.

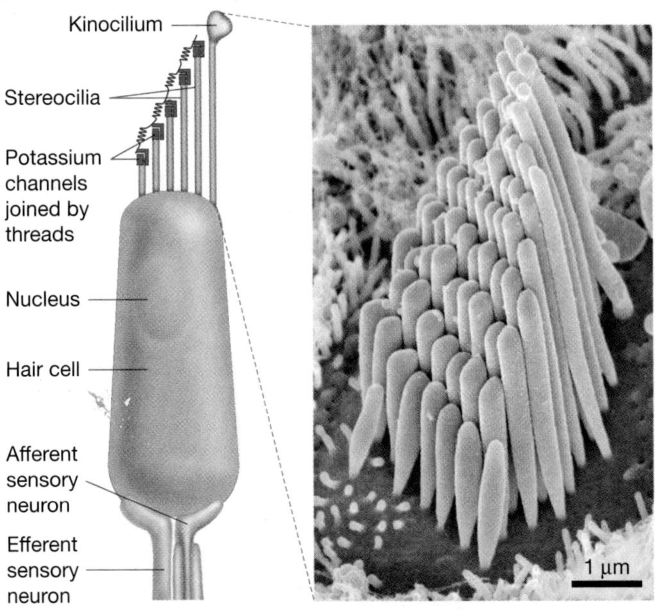

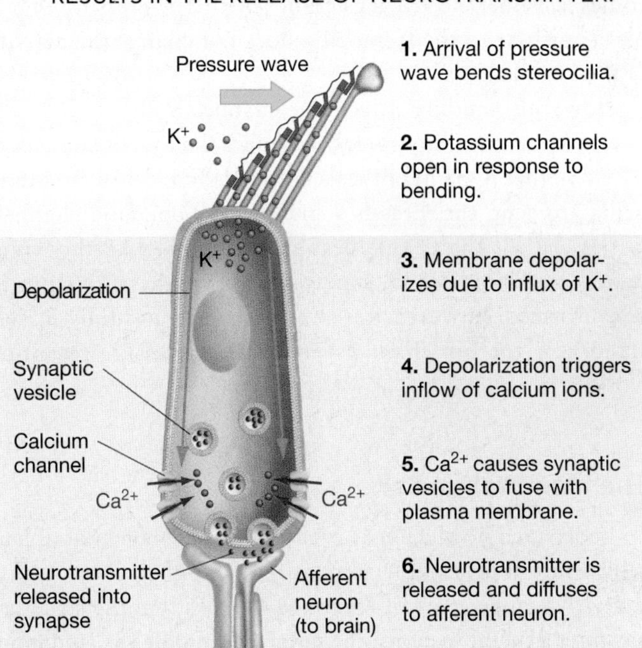

FIGURE 46.3 Hair Cells Transduce Sound Waves to Electrical Signals. **(a)** Hair cells are named for the hairlike stereocilia that project from one end. **(b)** If potassium channels open when stereocilia bend, then the hair cell's membrane depolarizes—resulting in an influx of calcium ions and the release of neurotransmitter to an afferent neuron.

In hair cells, depolarization causes an inflow of calcium ions, which triggers an increase in the amount of neurotransmitter released at the synapse between the hair cell and a sensory neuron. The end result is excitation of the postsynaptic cell, meaning that it becomes more likely to fire an action potential to the brain, via an afferent sensory neuron.

BOX 46.1 Senses That Humans Don't Have

This chapter introduces animals that can sense wavelengths of light, frequencies of sound, and odors that humans cannot perceive. But in addition to having eyes, ears, noses, and taste buds that are more acute than ours, many species have completely different sensory capabilities. Some aquatic predators are so sensitive to electric fields that they can detect electrical activity in the muscles of passing prey. Sharks become badly disoriented in captivity by electric fields generated by water pumps near their tanks. Similarly, many birds, sea turtles, and other animals can detect magnetic fields and use Earth's magnetic field as an aid in navigation. Homing pigeons tend to become disoriented if small magnets are glued to their heads. Some species of birds can also sense changes in barometric pressure. Biologists who work on species such as these are faced with the challenge of studying stimuli that they themselves cannot sense. In addition, it is possible that animals have senses that are yet to be discovered.

If sound pressure waves bend stereocilia the other way, however, then the K^+ channels close and the cell hyperpolarizes by 5 mV. Hyperpolarization of the hair cell results in a decrease in neurotransmitter released at the synapse and the inhibition of the sensory neuron, meaning that the neuron becomes less likely to trigger action potentials. Pressure that is perpendicular to the stereocilia does not change the activity of the ion channels.

How can bending affect ion channels? Electron micrographs show that tiny threads connect the tips of stereocilia to each other. One hypothesis contends that when the stereocilia are bent, the threads somehow pull open ion channels in the wall of the next tallest stereocilium—like tiny trapdoors (see Figure 46.3b, step 2). This hypothesis remains to be confirmed, however. Researchers still do not fully understand how the ion channels involved in pressure reception work.

The Mammalian Ear

To understand how changes in the membrane potential of hair cells result in hearing, let's focus on the human ear as a case study. The upper part of **Figure 46.4** shows that the structure has three general sections: the **outer ear**, **middle ear**, and **inner ear**. Each of these three sections is separated from the others by a membrane.

To trace the path of sound through the three sections of the ear, study the lower part of Figure 46.4, beginning with sound waves that are just reaching your head. Pressure waves are collected by the outer ear, which projects from the head, and are funneled into a tube known as the ear canal. At the end of the ear canal, the waves strike the **tympanic membrane**, or eardrum, which separates the outer ear from the middle ear. The repeated cycles of air compression cause the tympanic membrane to vibrate back and forth with the same frequency as the sound wave. The vibrations are passed to three tiny bones called the **ear ossicles**. In response, these bones vibrate against one another. The last ossicle, called the **stapes** (pronounced *STAY-peez*), vibrates against a membrane. That membrane, called the **oval window**, separates the middle ear from

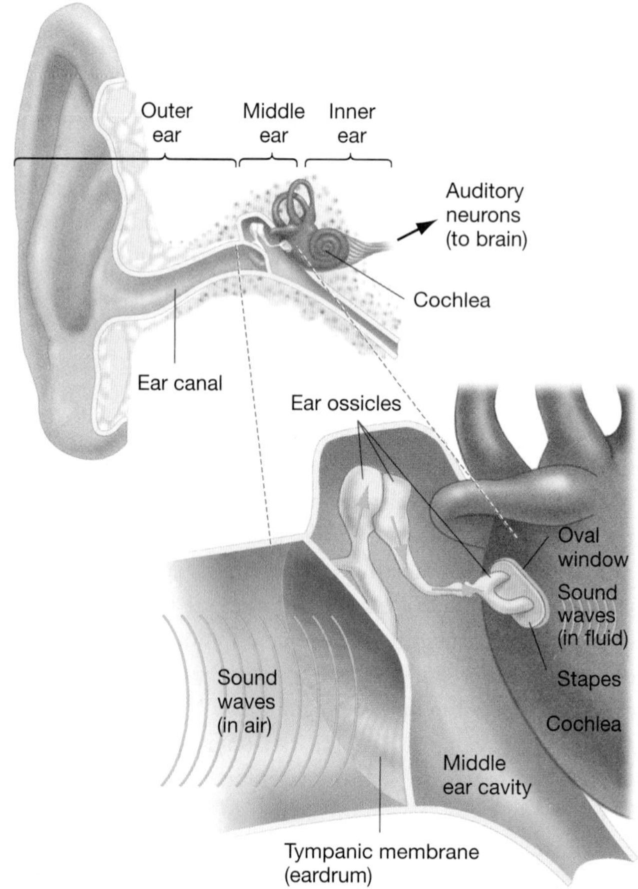

FIGURE 46.4 Mammals Have an Outer Ear, a Middle Ear, and an Inner Ear. The middle ear starts with the tympanic membrane and ends in the oval window of the cochlea.

the inner ear. The oval window oscillates in response. It generates waves in the fluid inside a chamber known as the **cochlea** (pronounced *KOK-lee-ah*). These pressure waves are sensed by hair cells in the cochlea.

In effect, the ear translates airborne waves to waterborne waves. The system seems extraordinarily complex, though, for such a simple result. Why doesn't the outer ear canal lead directly to the oval window? Why have a middle ear at all?

The Middle Ear Amplifies Sounds Biologists began to understand the function of the middle ear when they recognized two key aspects of its structure. First, the size difference between the tympanic membrane and the oval window is important. Because the tympanic membrane is about 15 times larger than the oval window, the amount of vibration induced by sound waves is increased by a factor of 15 by the time it reaches the oval window. This phenomenon is similar to using the same amount of force to bang on a very large door versus a very small door. In addition, the three ossicles act as levers that further amplify the vibrations at the tympanic membrane. In reptiles and birds, there is just one ear ossicle instead of three, and this levering action is not possible. The overall effect in the mammalian middle ear is to amplify sound by a factor of 22. This is important because it means that soft sounds are amplified enough that the hair cells lining the cochlea are stimulated. Biologists interpret the mammalian middle ear as an adaptation for increasing sensitivity to sound.

To summarize, the mammalian outer ear transmits sound waves from the environment to the middle ear; the middle ear amplifies these waves enough to stimulate the hair cells within the cochlea of the inner ear. How can hair cells distinguish different frequencies of sound? If all hair cells responded equally to all frequencies of sound, we would be able to perceive only one pitch. Everyone's voice—indeed, every sound—would sound the same.

The Cochlea Detects the Frequency of Sounds Careful anatomical studies and experimental work have revealed why different hair cells respond to specific frequencies of sound. As the cross section in **Figure 46.5a** shows, the cochlea has a set of internal membranes that divide it into three chambers. Hair cells form rows in the middle chamber. As **Figure 46.5b** indicates, the bottom of each hair cell connects to a structure called the **basilar membrane**. In addition, the hair cells' stereocilia touch yet another, smaller surface called the **tectorial membrane**. (The kinocilium is not present in a mature cochlear hair cell.) In effect, hair cells are sandwiched between membranes.

Researchers struggled for decades to understand how these membranes affect hair-cell function. It is virtually impossible to study cochleas in living organisms, because the cochleas are tiny, complex, coiled, and buried deep inside the skull. During the 1920s and 1930s, however, Georg von Békésy was able to perform experiments on cochleas that he had dissected from human cadavers. Once a cochlea was isolated, von Békésy was able to vibrate the oval window and record how the cochlea's internal membranes moved in response. He found that when a pressure wave traveled down the fluid in the upper and lower chambers, the basilar membrane vibrated in response. His key finding, though, was that sounds of different frequencies caused the basilar membrane to vibrate maximally at specific

(a) The middle chamber of the fluid-filled cochlea contains hair cells.

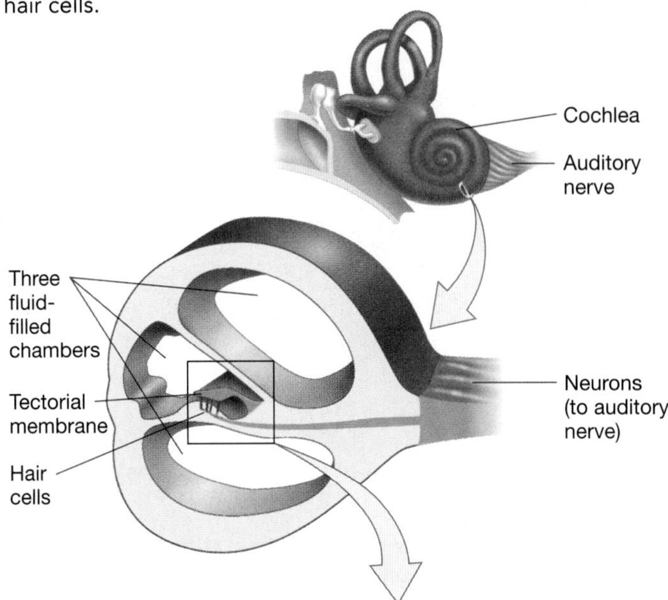

(b) Hair cells are sandwiched between membranes.

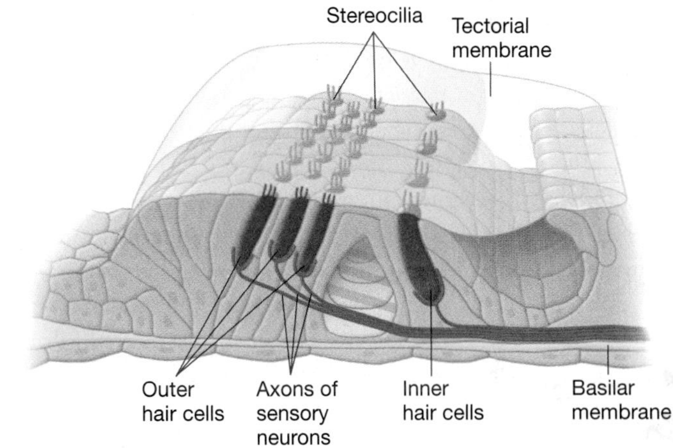

FIGURE 46.5 In the Human Cochlea, Hair Cells Are Sandwiched between Membranes. (a) The cochlea contains fluid-filled chambers separated by membranes. Hair cells are located in the middle chamber and **(b)** are sandwiched between the basilar membrane and the tectorial membrane.

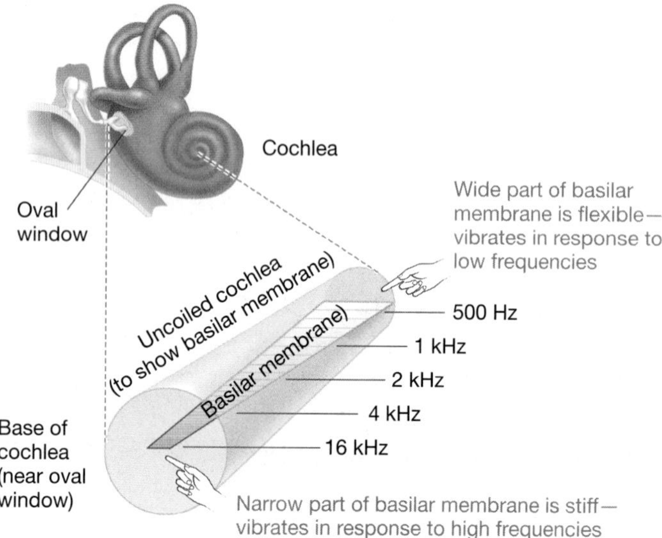

FIGURE 46.6 The Basilar Membrane Varies in Stiffness. High-frequency sounds are detected at the narrow, stiff end of the cochlea; low-frequency sounds are detected at the wider end near the oval window.

spots along its length (**Figure 46.6**). When the basilar membrane vibrated in a particular location, the stereocilia of the hair cells there were bent one way and then the other by the tectorial membrane.

Another observation allowed von Békésy to understand cochlea function: He noted that the basilar membrane varies in stiffness. It is very stiff near the oval window and very flexible at the other end. Thus, each segment of the membrane vibrates in response to a different frequency of sound. Just as a stiff drumhead produces a high-pitched sound and a loose drumhead yields a low-pitched sound, high-frequency sounds cause the stiff part of the basilar membrane to vibrate, and low-frequency sounds cause the more flexible part to vibrate.

To summarize, certain portions of the basilar membrane vibrate in response to specific frequencies and result in the bending of hair-cell stereocilia. In this way, hair cells in a particular place on the membrane respond to sounds of a certain frequency. The brain, in turn, receives action potentials from the neurons associated with each hair cell and interprets them as a particular pitch—meaning a specific frequency of sound. Complex sounds contain a wide variety of frequencies and trigger particular combinations of hair cells. Through experience, the brain learns which combinations of frequencies represent music, a fire alarm, or a parent's voice.

Sensory Worlds: What Do Other Animals Hear?

Compared with the hearing of many mammals, human hearing is not particularly acute. This point was driven home to Katherine Payne as she was observing elephants at a zoo in the mid-1980s. She noticed a throbbing sensation in the air, coming from the elephants. Payne knew that infrasound, or sound frequencies that are too low for humans to hear, can produce such sensations. To test the hypothesis that the elephants were producing infrasonic vocalizations, she returned to the zoo with microphones that were capable of picking up sounds at extremely low frequency. At normal speed, the tape Payne made was silent. But when she raised the pitch of the sounds by speeding up the tape, she heard a chorus of cow-like noises. The elephants were calling to each other, using low-frequency sounds.

According to follow-up research, elephants have the best infrasonic hearing of any land mammal. Biologists note that infrasound can travel exceptionally long distances, and they hypothesize that infrasonic calls allow wild elephants to coordinate their movements when they are miles apart. Recent research suggests that in addition to detecting infrasound via their large ears, elephants use their feet to detect a by-product of infrasound: seismic vibrations traveling in the ground.

The bats introduced at the start of the chapter furnish another example of an animal perceiving sounds that humans cannot. But in this case, the sounds are ultrasonic—that is, they have frequencies *above* the range of human hearing. Humans can hear sounds between 20 Hz and 20,000 Hz (20,000 Hz is equal to 20 kHz, or kilohertz). Ultrasonic hearing in bats was discovered in the late 1930s, when Donald Griffin borrowed the only ultrasonic apparatus then in existence from Robert Galambos, a fellow graduate student. Griffin used the machine to demonstrate that flying bats constantly emit ultrasounds. In follow-up experiments, he documented that a bat with cotton in its ears, or with its mouth taped shut, crashed into walls when released in a room. Blindfolded bats, in contrast, never crashed.

Griffin and Galambos concluded that bats use sound echoes (sonar) to navigate. This concept, termed **echolocation**, was an outlandish idea at the time. When Galambos described the use of sonar by bats at a meeting in 1940, another scientist shook him by the shoulders and said, "You can't really mean that!" But continued research showed that dolphins, shrews, and certain other animals besides bats use sonar. In fact, it is likely that at least some of these species perceive shapes with their ears better than they do with their eyes.

Although the use of sonar is much more extensive than originally believed, most animals rely on vision to detect the size, shape, and location of the objects around them. In fact, it turned out that Griffin and Galambos were fortunate to have started out studying echolocation in the lineage of bats called Microchiroptera—also known as whispering bats. Virtually all members of the other major bat lineage, the Megachiroptera, or Old World flying foxes, do not echolocate. Instead of hunting insects at night by echolocation, flying foxes feed on fruit or nectar and rely on vision to find food. How do megachiropterans and other animals see the world?

Check Your Understanding

If you understand that...

Hearing is a type of pressure detection that begins when the stereocilia on hair cells bend in response to changes in pressure. The bending movement opens ion channels and results in a change in membrane potential.

You should be able to...

1) Make a labeled diagram of the mammalian ear and describe the function of each part.
2) Predict the type of hearing loss that results from each of the following: a punctured eardrum, a severe mutation in a gene affecting stereocilium length, and an age-related loss of flexibility in the basilar membrane.

46.3 Vision

Most animals have a way to sense light. The organs involved range from simple light-sensitive eyespots in flatworms to the sophisticated, image-forming eyes of vertebrates, cephalopod mollusks, and arthropods. Insects, for example, have a **compound eye**, which is composed of hundreds or thousands of light-sensing columns called **ommatidia** (**Figure 46.7**). Each ommatidium contributes information about one small piece of the visual field, not unlike a single pixel on a computer monitor. Thus, the more ommatidia in a compound eye, the better the resolution—meaning the resolving power, or ability to distinguish objects. In addition, the presence of many light-sensing columns makes species with compound eyes particularly good at detecting movement. Vertebrates and cephalopods, in contrast, have a **camera eye**. This is a structure that focuses incoming light onto a layer of receptor cells. Let's examine this structure more closely.

The Vertebrate Eye

Figure 46.8a shows a typical vertebrate eye. The outermost layer of the structure is a tough rind of white tissue called the sclera. This is the "white of the eye." The front of the sclera forms the **cornea**, a transparent sheet of connective tissue. Just inside the cornea is a colored, round muscle called the **iris**. The iris can contract or expand to control the amount of light entering the eye. The hole in the center of the iris is the **pupil**. Light enters the eye through the cornea and passes through the pupil and then through a curved, clear **lens**. Together, the cornea and lens focus incoming light onto the retina in the back of the eye. The **retina** contains a thin layer of light-sensitive cells and several layers of neurons.

Figure 46.8b provides a closer look at the retina. The structure is attached to the rest of the eye by a single layer of pigmented epithelial cells. The sensory cells that respond to light, the photoreceptors, are held in place by the pigmented epithelium and form a layer at the back of the retina. Photoreceptors synapse with an intermediate layer of connecting neurons. Cells in the intermediate neuron layer connect with one another and with the neurons called **ganglion cells**, which form the innermost layer of the retina. The axons of the ganglion cells project to the brain via the **optic nerve**. This is an unusual arrangement of neurons; **Box 46.2** discusses the consequences. Here, though, let's explore how photoreceptors work.

What Do Photoreceptor Cells Do? Early anatomists established that the photoreceptors in vertebrate eyes are small rod-shaped or cone-shaped cells called **rods** and **cones**

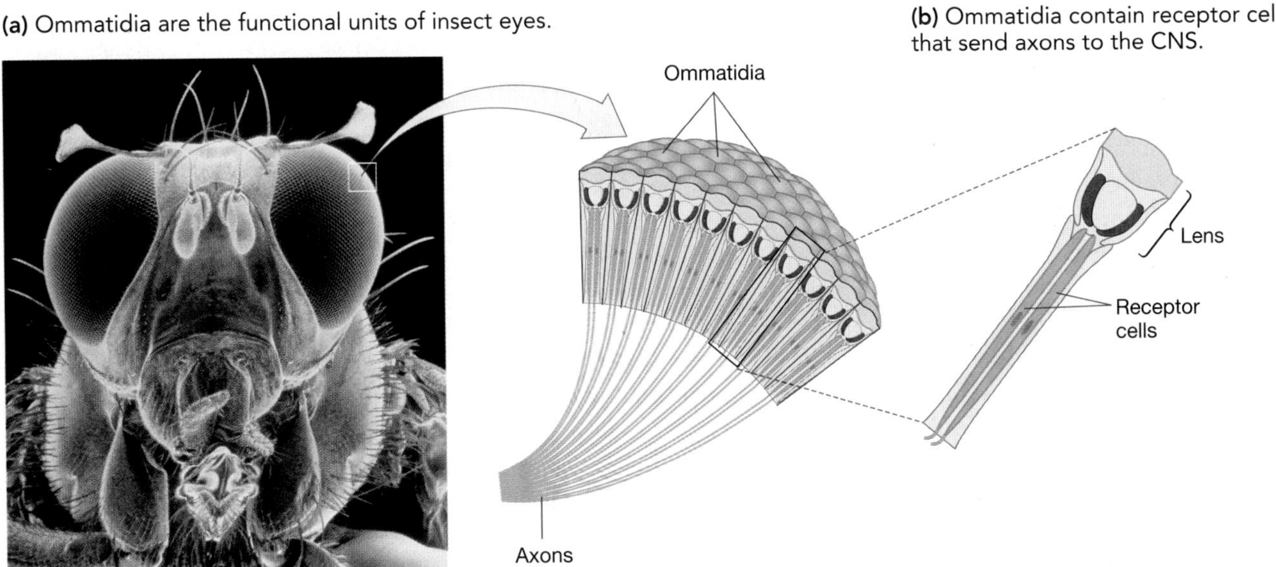

FIGURE 46.7 In the Compound Eyes of Insects, Each Ommatidium Sees Part of the World.

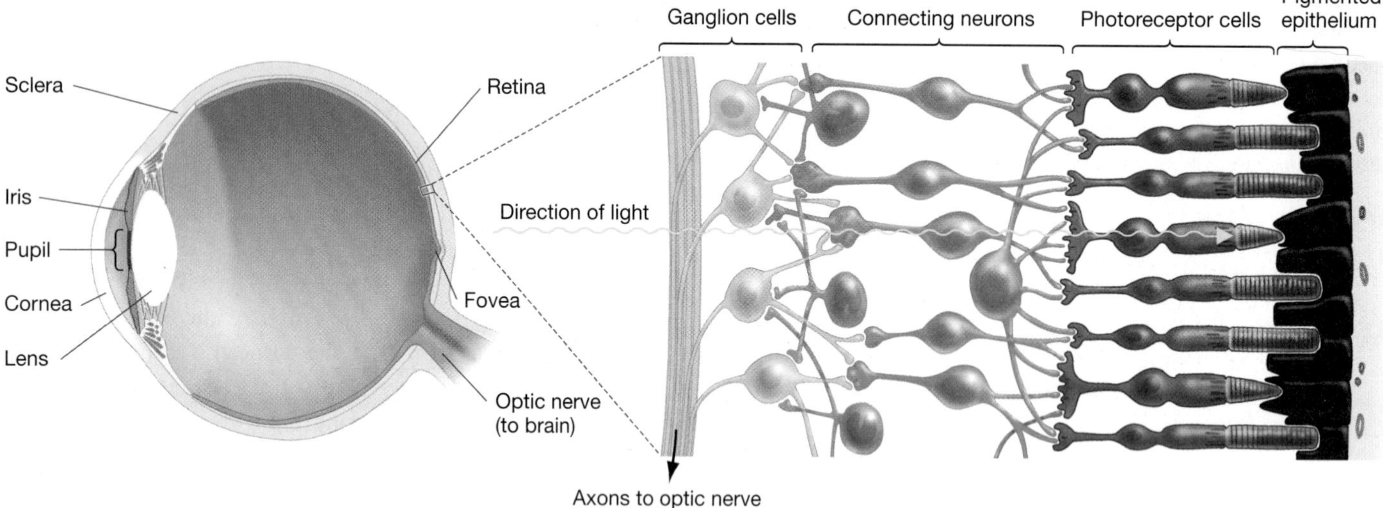

FIGURE 46.8 Camera Eyes Have a Single Lens That Focuses Incoming Light on Receptor Cells. **(a)** Light passes through the pupil of the eye and is focused onto the retina by the cornea and lens. **(b)** The photoreceptor cells that respond to light are in the "outermost" layer of the retina, furthest from the light source.

QUESTION Are any photoreceptor cells located in the spot where the optic nerve leaves the retina?

EXERCISE In part (b), add an arrow indicating the path of an electrical signal from this photoreceptor to the optic nerve.

BOX 46.2 Vertebrate versus Cephalopod Eyes

The photoreceptor cells in the vertebrate eye are located in the outermost layer of the retina—farthest from the incoming light. As a result, light must pass through several layers of neurons before reaching the photoreceptors. This arrangement has an important consequence: The axons of the ganglion cells in the inner layer must block the photoreceptor cells in the outermost layer in order to exit the eye at the optic nerve. This disruption creates a blind spot—that is, a portion of the retina where there are no photoreceptors.

To convince yourself that this blind spot exists, point a finger up and hold it out at arm's length. Close your left eye, and focus your right eye on a location just beyond the tip of your finger. Now slowly move your pointed finger to the right, while continuing to focus on the original spot. Can you find a place where the tip of your finger seems to vanish? Light from this region is falling on the blind spot of your right eye. If an arrow or other projectile approached from this direction, you would not detect the danger.

Squid, octopus, and other cephalopods have camera eyes analogous to those found in vertebrates, but their photoreceptors are located in the innermost layer of the retina (**Figure 46.9**). As a result, light strikes the photoreceptors directly, and the eye has no blind spot. Why do vertebrates have what appears to be an inferior eye design? Recall from Chapter 24 that adaptations are not perfect. To explain the existence of blind spots, biologists hypothesize that the ancestor of vertebrates had an eyelike organ with photoreceptors located in the outermost cell layer. As a camera eye evolved in the descendants of this species, the ancestral arrangement was retained. If this hypothesis is correct, our blind spots furnish another example of historical constraints on adaptation.

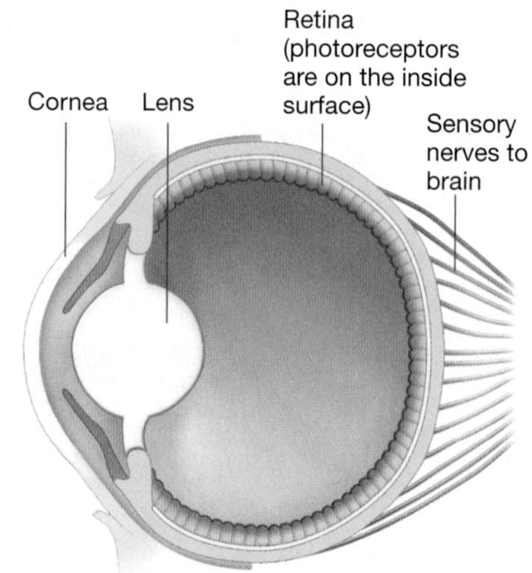

FIGURE 46.9 Cephalopods Have a Camera Eye, but No Blind Spot.

EXERCISE In Figure 46.8a, circle the structure that creates the blind spot in the vertebrate eye.

(**Figure 46.10a**). When technical advances allowed changes in the membrane potentials of these cells to be recorded, it became clear that rods and cones differ in function as well as structure. Rods are sensitive to dim light but not to color. Cones, in contrast, are much less sensitive to faint light but are stimulated by different wavelengths (that is, colors). These discoveries explained why night vision is largely black and white—at night, the rods do most of the work.

Early descriptions of the human retina also established that rods dominate most of the structure. There is one small spot in the center of the retina, however, that has only cones. This is the **fovea** (see Figure 46.8a). Researchers noticed that when people focus on an object, their eyes move so that the image falls on the fovea of each eye. Based on these observations, biologists concluded that the high density of cones in the fovea maximizes the resolution of the image.

How Do Rods and Cones Detect Light? The effort to understand how rods and cones work required a combination of careful anatomical and biochemical studies. As Figure 46.10a shows, rods and cones have segments that are packed with membrane-rich disks. The membranes of these disks contain large quantities of a transmembrane protein called **opsin**. Each opsin molecule is associated with one much smaller molecule, the pigment **retinal**. In rod cells, the two-molecule complex is called rhodopsin (**Figure 46.10b**).

Experiments with isolated retinal, opsin, and rhodopsin molecules confirmed that retinal changes shape when it absorbs light. More specifically, the number-11 carbon in the retinal molecule changes from the *cis* conformation to the *trans* conformation (**Figure 46.10c**). Documenting this conformational change helped confirm that rhodopsin is the light receptor.

The shape change that occurs in retinal triggers a series of events that culminate in a different stream of action potentials being sent to the brain. The sequence of events is unusual, though, because the receipt of a light stimulus does not open ion channels or trigger the release of a neurotransmitter to a

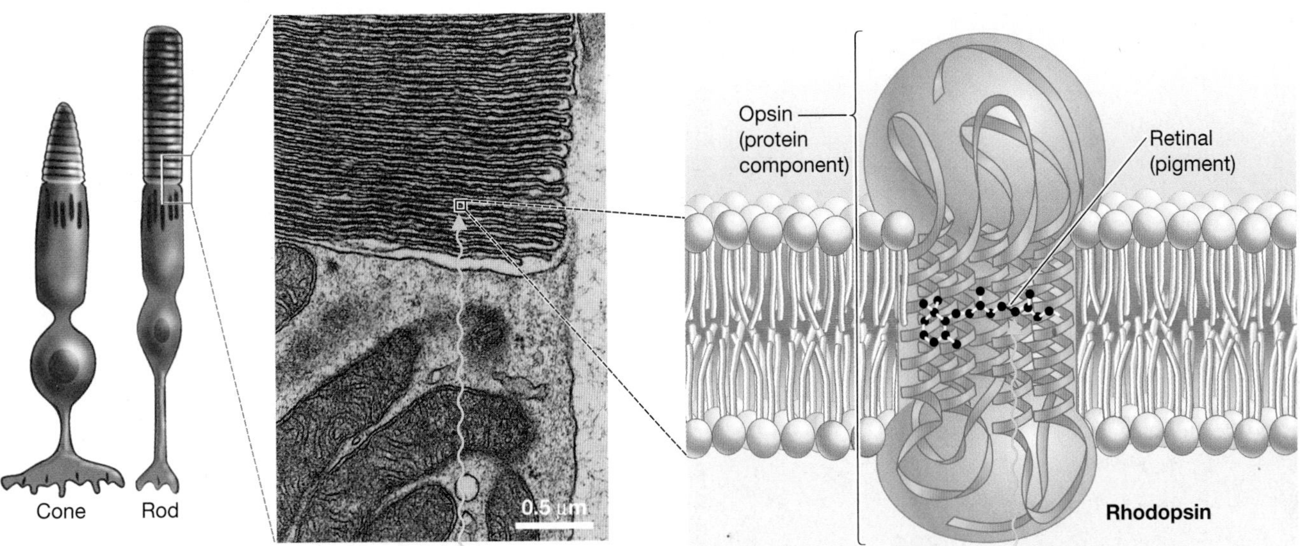

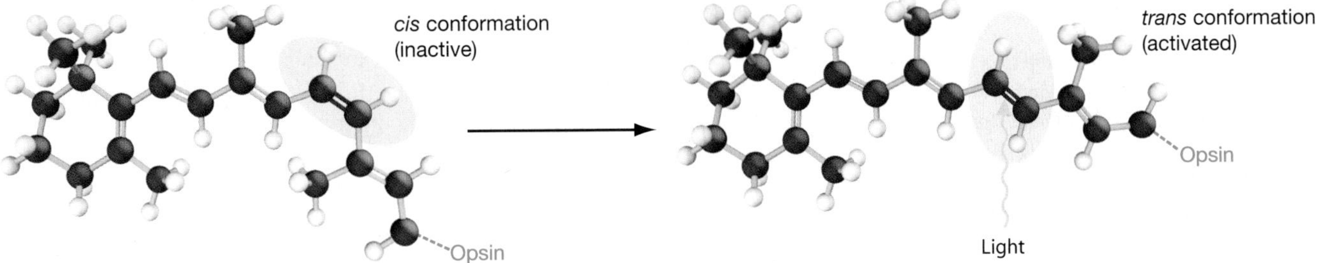

FIGURE 46.10 Rods and Cones Are Packed with Transmembrane Proteins That Contain the Pigment Retinal.
(a) Rods and cones have membranous disks containing thousands of opsin molecules. **(b)** Each opsin holds one retinal molecule. **(c)** Retinal changes conformation when it absorbs light. In response, opsin also changes shape.

QUESTION Explain why the change from *cis* to *trans* shown in part (c) would affect opsin's shape.

sensory neuron. In vertebrates, the molecular basis of vision is a shape change in retinal that shuts down an existing ion channel and causes a decrease in the amount of neurotransmitter being released to the sensory neuron. In rod cells, electrical activity across the membrane, as well as neurotransmitter release, are maximized in the dark. Exposure to light transmits information by shutting down both processes.

As **Figure 46.11** shows, the sequence of events begins when light causes retinal to change shape from the *cis* to *trans* conformation, activating rhodopsin. Rhodopsin activation causes a membrane-bound molecule called transducin to activate the enzyme phosphodiesterase (PDE). PDE breaks down an important nucleotide called cyclic guanosine monophosphate (cGMP), to guanosine monophosphates (GMP). As cGMP levels decline, cGMP-gated sodium channels in the plasma membrane of the rod cell close. This is a critical event, because it causes Na^+ entry to decrease. As a result, the membrane hyperpolarizes. If you understand this sequence of events, you should be able to explain why cGMP acts as both a second messenger and a ligand in this system, and how transducin compares to the G proteins introduced in Chapter 8. You should also be able to explain why shutting Na^+ channels results in a hyperpolarization.

In response to the ensuing change in membrane potential, smaller quantities of the neurotransmitter glutamate are discharged at the synapse. The decrease in neurotransmitter indicates to the postsynaptic cell, called a **bipolar cell**, that the rod absorbed light. As a result, a new pattern of action potentials is sent to the brain. This system is exquisitely sensitive: Biologists have recorded a measurable change in the membrane potentials of rod cells in response to a single photon of light.

(a) The disk of a photoreceptor cell (a rod) before stimulation

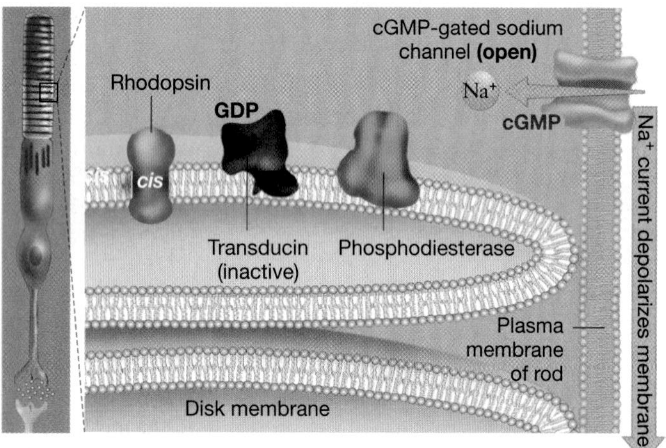

(b) The same disk after stimulation (light)

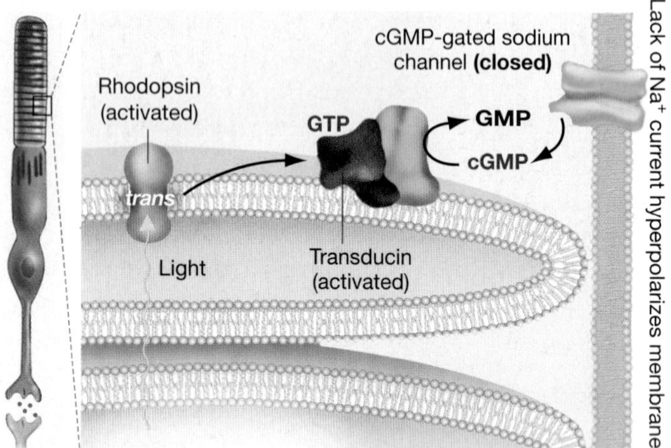

FIGURE 46.11 A Signal Transduction Pathway Connects Light Absorption with Changes in Membrane Potential. (a) An unstimulated photoreceptor. Notice that sodium ions flow into the cell when light is *not* being received. **(b)** A stimulated photoreceptor. When rhodopsin is activated, it leads to a reduction in cGMP concentration in the photoreceptor. With less cGMP available, cGMP-gated sodium channels close and the membrane hyperpolarizes.

QUESTION The inflow of sodium ions into a photoreceptor cell is called "the dark current." Why?

EXERCISE Circle the synapses in the rod cell drawings on the left of each part. Is the rod cell releasing more neurotransmitters into the synapse in the dark or in the light?

Color Vision: The Puzzle of Dalton's Eye Studies of rhodopsin revealed the molecular mechanism responsible for light perception. But it remained to be determined how humans and other animals can respond to specific wavelengths of light. How do we perceive color?

To answer this question, consider the research program initiated by John Dalton[1] in the late eighteenth century. At the age of 26, Dalton realized that he and his brother saw colors differently than did other people. To them, red sealing wax and green laurel leaves appeared to be the same color, and a rainbow exhibited only two hues. Dalton and his brother could not differentiate the colors red and green. This condition is called red-green color deficiency (**Figure 46.12**).

In a lecture delivered in 1794, Dalton explained his perceptions by hypothesizing that red wavelengths failed to reach his retinas. Further, he hypothesized that because a normal eyeball is filled with clear fluid, and because blue fluids absorb red light, his defective vision resulted from the presence of bluish fluid rather than clear fluid in his eyes. To test this hypothesis, Dalton left explicit instructions that his eyes should be removed after his death and examined to see if the fluid inside was blue. When he died 50 years later, an assistant dutifully removed the eyes from Dalton's corpse and examined them. The fluid inside the eye was not blue at all, however, but slightly yellow—the normal color for an older person. Further, when

[1]Dalton was an accomplished physicist. He was the first proponent of the atomic theory and formulated Dalton's law on the partial pressures of gases (introduced in Chapter 44). Red-green color blindness is sometimes called Daltonism in his honor.

FIGURE 46.12 People with Color Deficiencies See Colors Differently than Do People without Deficiencies. These images show what a person with red-green color deficiency would see, as compared to a person with intact color receptors.

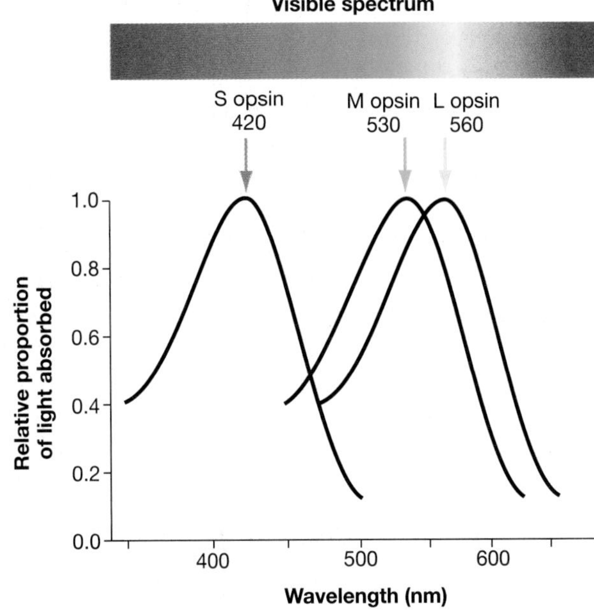

FIGURE 46.13 Color Vision Is Possible because Different Opsins Absorb Different Wavelengths of Light. Each human cone cell contains one of three different types of opsin. Each of the opsins has a different range of absorbed wavelengths.

QUESTION The retinal molecules in S, M, and L opsins are identical. How is it possible that they respond to different wavelengths of light?

the back was cut off of one eye and colored objects were viewed through the lens, the objects looked perfectly normal. Dalton's hypothesis was incorrect.

What caused Dalton's color blindness? The key to answering this question was the hypothesis that the human retina contains just three types of color-sensitive photoreceptors: blue, green, and red cones, named for the colors that they best perceive. Researchers who analyzed the activity of these three types of cone cells found that each absorbed short, medium, or long wavelengths of light, respectively. To follow up on this result, biologists analyzed opsin molecules from the three cell types and found that each had a distinct amino acid sequence. The three proteins are called the blue, green, and red opsins (or S, M, and L, for short, medium, and long wavelengths, respectively). Although retinal is the light-absorbing molecule in all photoreceptor cells, various types of opsins respond only to distinct wavelengths of light.

Based on these results, biologists hypothesized that the brain distinguishes colors by combining signals initiated by the three classes of opsins. For example, notice in **Figure 46.13** that a wavelength of 560 nm stimulates L cones strongly, M cones to an intermediate degree, and S cones hardly at all. In response to the corresponding signals from these cells, the brain perceives the color yellow.

Does this hypothesis explain Dalton's color blindness? According to the data in Figure 46.13, wavelengths from green to red do not stimulate S opsin at all. It is thus unlikely that S opsin is involved in red-green color blindness. Did Dalton fail to distinguish red and green because his M or his L cones were defective? Research has shown that color-blind people lack either functional M or L cones, or both. Was the same true of Dalton? Answering this question became possible in the 1990s, when the genes for the M and L opsins were identified and sequenced. Researchers realized that if they could get a sample of Dalton's DNA, they could analyze his M opsin genes and L opsin genes directly. Remarkably, Dalton's eyes had been preserved. The research group managed to extract DNA from the 150-year-old tissue and found that Dalton had a normal *L* allele but lacked a functional *M* allele. As a result, he did not have green-sensitive cones. The puzzle of Dalton's color vision was solved.

(MB) **Web Animation** at www.masteringbio.com
The Vertebrate Eye

Sensory Worlds: Do Other Animals See Color?

What about other animals—do they see color the way humans do? The short answer to this question is probably not. Animals that are active at night have relatively few cone cells and many rods, giving them high sensitivity to light but poor color vision. And recent research has shown that different species with good color vision have opsins with different peak light sensitivities. A marine fish called the coelacanth (pronounced *SEE-luh-kanth*), which lives in water 200 meters deep, has two opsins that respond to the blue region of the spectrum (wavelengths of 478 nm and 485 nm). Based on this observation, it is likely that coelocanths perceive several distinct hues of blue that we would perceive as a single color. The existence of these opsins is logical, because wavelengths in the yellow and red parts of the spectrum do not penetrate well into deep water. Because only blue light exists in the coelacanth's habitat, biologists interpret the opsins of this fish as an adaptation to life in deep water.

Correlations between opsin structure and visual function are extensive. For example, in humans and other primates that eat fruit, two of the three opsins are sensitive to wavelengths around 550 nm. The presence of these opsins allows individuals to make fine distinctions between the greens, yellows, and reds of unripe and ripe fruits.

In addition, numerous vertebrate and invertebrate species have four or more types of opsins and probably perceive a world of colors that is much richer than ours. Birds, insects, and a few mammals can also see ultraviolet light, which has shorter wavelengths than humans can see. This ability is adaptive. As Figure 40.6 showed, certain flowers have ultraviolet patterns that serve as signals for insect pollinators. In addition, many birds have strong ultraviolet patterns in their plumage that are invisible to us. Recent experiments have confirmed that in some species, females use these ultraviolet patches as a criterion for selecting mates. Beyond the other end of the visible spectrum, rattlesnakes and other pit vipers can sense infrared wavelengths by using specialized infrared-sensing organs on their snouts (**Figure 46.14a**). Infrared light has longer wavelengths than humans can see. Because endothermic animals radiate large amounts of heat in the form of infrared radiation, pit vipers can detect and strike at prey in complete darkness (**Figure 46.14b**).

Check Your Understanding

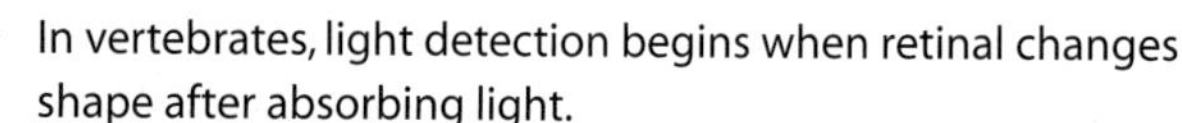

If you understand that...

- In vertebrates, light detection begins when retinal changes shape after absorbing light.
- If opsin changes conformation in response to movement by retinal, a series of events results in the closing of sodium channels in the membrane of a photoreceptor cell, and thus a change in membrane voltage.

You should be able to...

1) Make a labeled diagram of the vertebrate eye and describe the function of each part.
2) Predict the type of vision loss that results from each of the following: a tear in the fovea, a severe mutation in the gene for S opsin, and an age-related clouding and stiffening of the lens.

46.4 Taste and Smell

The senses of taste (**gustation**) and smell (**olfaction**) originate in chemoreceptors. Chemoreceptors can detect the presence of particular molecules because they undergo a change in membrane potential when a specific compound is present. In this way, information about the presence of a particular chemical is transduced to an electrical signal in the body.

Until recently, taste and smell were poorly understood in comparison with vision and hearing. It is easy to see why. Eyes and ears respond to relatively simple stimuli—light and sound waves—that can be reproduced in the lab. The tongue and nose, in contrast, respond to thousands of different chemicals. Consequently, taste and smell were difficult to study until

(a) Pit vipers can detect infrared radiation.

(b) Warm animals emit much more infrared radiation than their surroundings do.

What we see in the light (the top compartment in the white box contains a lightbulb wrapped in dark cloth)

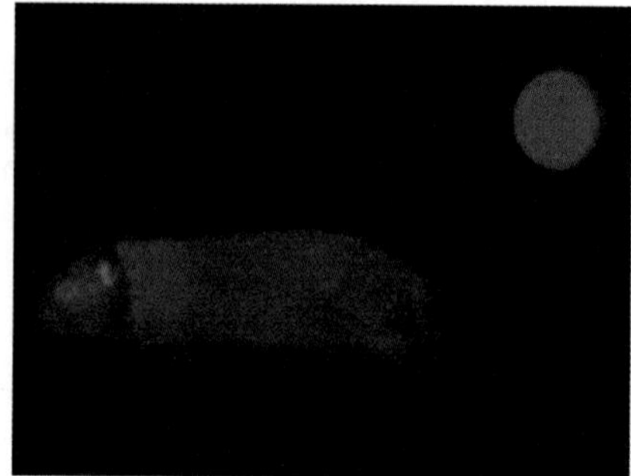

What pit vipers "see" in the dark

FIGURE 46.14 Pit Vipers Detect Infrared Radiation. (a) Pit vipers are named for the "pits" on their heads. These pits contain receptor cells that detect infrared radiation. **(b)** Pit vipers are able to "see" a fuzzy image of warm objects. As a result, pit vipers and other rattlesnakes can strike accurately at a rat or mouse even in darkness.

QUESTION Predict what happens if the viper's pits are covered experimentally.

techniques became available for identifying how particular molecules bind to certain receptors. But in the past 15 years, research on the chemosenses has exploded.

Understanding taste and smell is important because animals use these senses to find food, assess mates, and avoid danger. For researchers in the food and cosmetic industries, understanding particular aspects of chemoreception, such as the sensation of sweetness and the activity of perfumes, is critical to developing new products.

Taste: Detecting Molecules in the Mouth

The chemoreceptor cells that sense taste are clustered in structures known as **taste buds**. Although humans have taste buds scattered around the mouth and throat, most taste buds are located on the tongue (**Figure 46.15**). A taste bud contains about 100 spindle-shaped taste cells, which are taste receptors that synapse onto sensory neurons. How do these receptors work on a molecular level, and how do they produce the sensation of taste?

Early taste research focused almost exclusively on the hypothesis that four "basic tastes" existed: salty, sour, sweet, and bitter. Researchers who analyzed the membrane proteins in taste cells found strong evidence that salt and sour sensations result from the activity of ion channels. Specifically, the sensation of saltiness is due primarily to sodium ions (Na^+) dissolved in food. These ions flow into certain taste cells through open Na^+ channels and depolarize the cells' membranes. Similarly, sourness is due to the presence of protons (H^+), which flow directly into certain taste cells through H^+ channels. The sour taste of grapefruit and other citrus fruits, for example, results from the release of protons by citric acid. In general, the lower the pH of a food, the more it depolarizes a taste cell's plasma membrane and the more sour the food tastes.

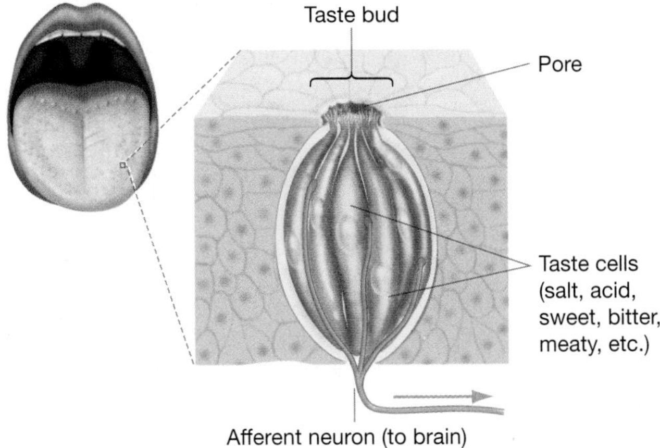

FIGURE 46.15 Taste Buds Contain Many Types of Receptor Cells. The human tongue has a high density of taste cells located in structures called taste buds.

The molecular mechanisms responsible for the sensations of bitterness and sweetness have been much more difficult to identify, however. Researchers have only recently been able to document that certain food molecules actually bind to specific receptors on taste cells to cause bitter and sweet tastes.

Why Do Many Different Foods Taste Bitter? Bitterness has been a particularly difficult taste to understand. The problem is that molecules with very different structures are perceived as bitter. How is this possible? An answer began to emerge after researchers confirmed that some humans genetically lack the ability to taste certain bitter substances. This phenomenon came to light in 1931, when Arthur Fox was synthesizing phenylthiocarbamide (PTC) and accidentally blew some of it into the air. A nearby colleague complained of a bitter taste in his mouth. Fox could not taste anything, however. Follow-up research confirmed that the ability to taste PTC is inherited and polymorphic. About 25 percent of United States citizens cannot sense the molecule. To find the gene responsible for this trait, biologists compared the distribution of genetic markers observed in "tasters" and "nontasters." The mapping effort recently narrowed down this gene's location to several candidate chromosomal segments. (Chapter 19 introduced how this type of gene hunt is done.)

At about the same time, researchers who were working on taste reception in mice found that some individuals are better than others at perceiving bitter substances and avoiding them. Gene mapping narrowed down the location of the bitter receptor gene to a specific region of a mouse chromosome.

These independent lines of research converged when different research teams searched a database of known genes to look for sequences similar to those from the candidate regions of the human and mouse chromosomes. Independently, the two teams identified a family of 40 to 80 genes that encode transmembrane receptor proteins. Follow-up work documented that each protein in the family binds to just one particular type of bitter molecule. A taste cell, however, can have many different receptor proteins from this family. As a result, many different kinds of molecules can depolarize the same cell and cause the taste of bitterness. These results explained why so many different molecules give rise to the same sensation.

Why do animals have such extensive molecular machinery to detect bitterness? To answer this question, biologists note that many of the molecules that bind to bitter receptors are found in toxic plants. Also, most animals react to bitter foods by spitting them out and avoiding them in the future. Based on these observations, researchers hypothesize that bitter receptors evolved in response to selection for avoiding toxic molecules. In essence, bitterness indicates "this food might be dangerous; don't swallow it." The proliferation of genes responsible for detecting bitterness hints that the trait is extremely important—meaning that individuals with the capacity to detect a wide array of bitter compounds produce more offspring than do individuals that lack this ability.

What Is the Molecular Basis of Sweetness and Other Tastes? Inspired by the rapid recent progress on bitter receptors, research teams are using the same techniques to search for sweetness and other types of receptors. Several proteins have been identified as possible sweetness detectors, some of which are closely related to recently discovered membrane receptors for glutamate and other amino acids. The glutamate receptor is responsible for the sensation called **umami**, which is the meaty taste of the molecule monosodium glutamate (MSG). Researchers are still struggling to understand how the sweetness receptors work, however, and exactly how many different membrane proteins might be involved in detecting the presence of sugars in food.

In addition to searching for other types of taste receptors, several teams are studying how the different taste sensations are conveyed to the brain and interpreted. Although taste is beginning to reveal its secrets, the complete story will probably not be known for many more years.

Olfaction: Detecting Molecules in the Air

Taste allows animals to assess the quality of their food before swallowing it. Smell, in contrast, allows animals to monitor airborne molecules that convey information. Wolves and domestic dogs, for example, can distinguish millions of different odors at vanishingly small concentrations. The molecules that cause odor contain information about the movements and activities of prey and other members of their own species.

How does the sense of smell work? When odor molecules, or odorants, reach the nose, they diffuse into a mucus layer in the roof of the nose (**Figure 46.16**). There, they activate olfactory receptor neurons via membrane-bound receptor proteins. Axons from these neurons project up to the olfactory bulb. The **olfactory bulb** is the part of the brain where olfactory signals are processed and interpreted.

Understanding the anatomy of the odor recognition system was a relatively simple task. Understanding how receptor neurons distinguish one molecule from another turned out to be much more difficult. Initially, investigators hypothesized that receptors respond to a small set of "basic odors," such as musky, floral, minty, and so on. The idea was that each basic odor would be detected by its own type of receptor.

In 1991, however, Linda Buck and Richard Axel demolished the basic-odor hypothesis with an astonishing result: Their studies of odor recognition in mice culminated in the discovery of a gene family—containing hundreds of distinct coding regions—that encodes receptor proteins on the surface of odor-receptor neurons. Follow-up experiments confirmed that each receptor protein binds to a small set of molecules. Further work established that most, if not all, vertebrates have this family of genes. Mammals typically have between 500 and 1000 of these genes, meaning mammals can produce 500 to 1000 different odor-receptor proteins. In humans, however, about half of these genes have mutations that probably render them nonfunctional. This observation may explain why the sense of smell is so poor in humans compared with that of other mammals.

Buck and Axel's announcement inspired a series of questions. How many different receptors occur in the membrane of each neuron involved in odor reception? How does the brain make sense of the input from so many different receptors?

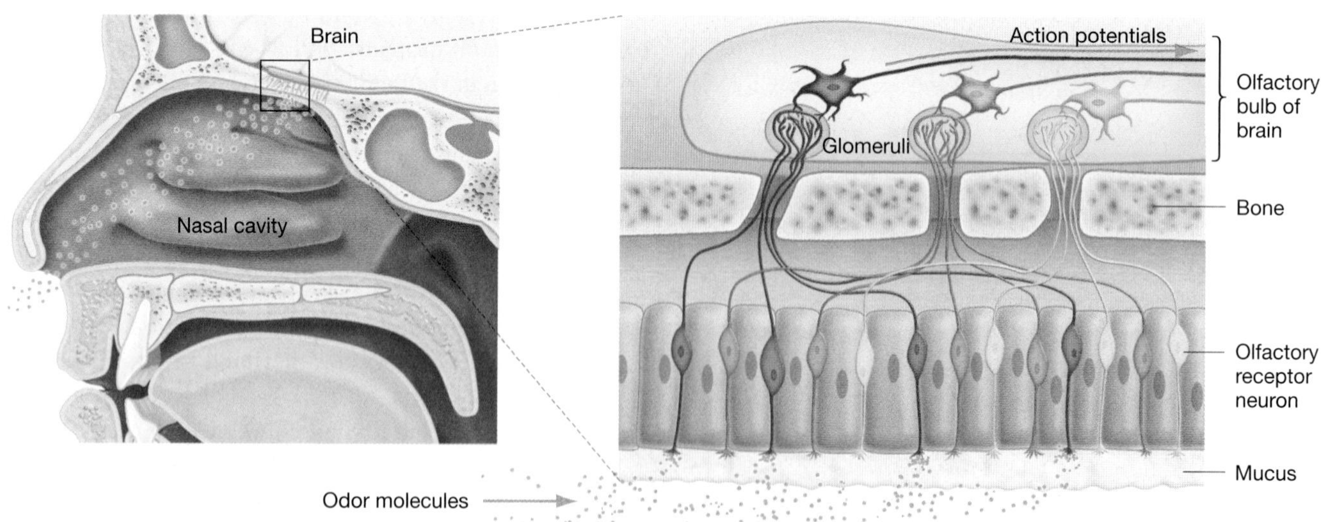

FIGURE 46.16 In Mammals, Receptor Cells in the Nose Respond to Specific Odorants. Odor molecules are detected in the nose by chemoreceptor cells. Each of these sensory neurons carries one type of odor-receptor protein on its dendrites. Sensory neurons with the same receptor project to the same glomerulus, or section within the olfactory bulb of the brain.

Recently, Buck determined that each olfactory neuron has only one type of receptor and that neurons with the same type of receptor are linked to distinct regions in the olfactory bulb of the brain. These regions are called **glomeruli** (meaning "little balls"). Follow-up work from Axel's lab indicated that particular smells in mice are associated with the activation of a certain subset of the 2000 glomeruli in the brain. For example, the activation of clumps 130, 256, and 1502 might be perceived as the smell "cinnamon." In essence, then, the sense of smell is similar to the visual system's use of three cones to perceive many colors, but on a much larger scale. Research on this complex and impressive sense continues at a furious pace.

Check Your Understanding

If you understand that...

In most cases, chemoreception occurs when a specific molecule in air or food binds to a specific receptor in the nose or mouth and the binding event is transduced to a change in membrane voltage.

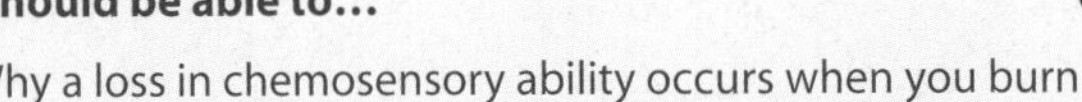

You should be able to...

1) Why a loss in chemosensory ability occurs when you burn your tongue with extremely hot food.
2) Why dogs have a better sense of smell than people do.
3) Why some people cannot smell the compound PTC.

46.5 Movement

The first four sections of this chapter focused on how animals sense different aspects of their environment. Acquiring information is only a first step, however. Information about changes in the environment is useless to an animal unless the individual can respond to it in an appropriate way—usually by moving. Bats and moths not only hear ultrasonic frequencies but also alter their flight paths and activities in response—in a way that allows them to catch food or avoid being eaten. Flight and other types of motion are based on muscle contractions in conjunction with a skeleton.

Recall from Chapters 32–34 that diversification in methods of moving has been extremely important during the evolution of animals. But it's also important to remember that **locomotion**, or the movement of an entire animal, is only one type of movement. Sessile organisms, such as barnacles and sea anemones, may not undergo locomotion, but they rely on movement of the heart muscle or gills to accomplish circulation and respiration.

To explore locomotion in detail, let's briefly examine the skeleton's role as a scaffold for muscle attachment. The analysis of animal skeletons is followed by a description of the diversity of muscle cells and tissues found in vertebrates and how they work.

Skeletons

Skeletons provide attachment sites for muscles and a support system for the body's soft tissues. As Chapters 32–34 indicated, three types of skeletons are found in animals. **Exoskeletons** are hard, hollow structures that envelop the body; **hydrostatic skeletons** use the pressure of enclosed body fluids to support the body; and **endoskeletons** are hard structures inside the body. Because those chapters introduced the design of exoskeletons and hydrostatic skeletons in some detail, our focus here is on the structure and function of endoskeletons.

Endoskeletons are composed of connective tissues called cartilage and bone. You might recall from Figure 41.3b that **cartilage** is made up of cells scattered in a gelatinous matrix of polysaccharides and protein fibers. In the skeletal system, cartilage provides padding between bones. **Bone** is made up of cells in a hard extracellular matrix of calcium phosphate ($CaPO_4$) with small amounts of calcium carbonate ($CaCO_3$) and protein fibers. Bones meet and interact at locations called **articulations**, or **joints**. Bones articulate in ways that allow limbs to swivel, hinge, or pivot (**Figure 46.17**).

How do skeletal muscles attach to and move elements of an endoskeleton? In vertebrates, most skeletal muscles have ends that are attached to different bones by tendons. **Tendons** are bands of tough, fibrous connective tissue. Muscles can exert force only by contracting, so pairs of muscles must work together to move a bone back and forth. In the case of a limb, one muscle pulls the limb in one direction; the other muscle pulls it in the opposite direction. This pairing of muscles is called an antagonistic muscle group. The muscle that swings

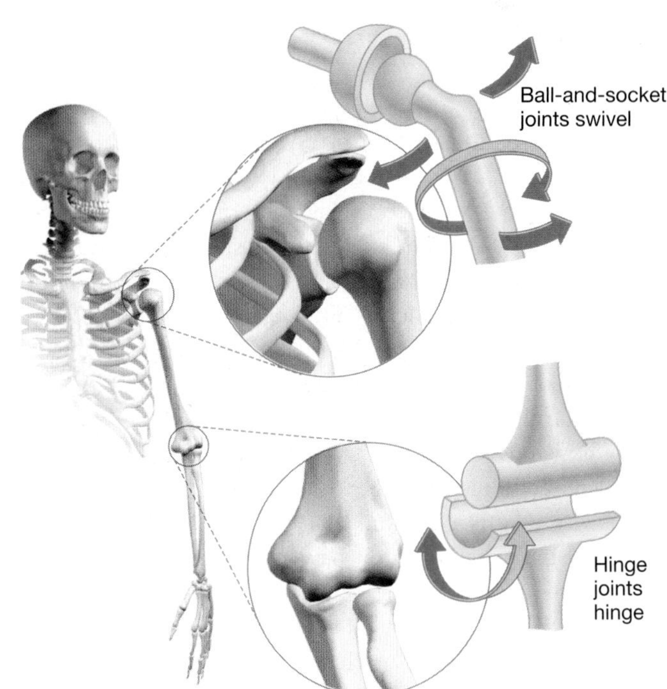

FIGURE 46.17 Bones Articulate at Joints. Bones articulate in ways that make specific types of movement possible.

two long bones in an arc toward each other is called a **flexor**; the muscle that straightens them out is called an **extensor** (**Figure 46.18a**). For example, the hamstring muscle in the back of your thigh flexes your lower leg; the quadriceps muscle in the front of your thigh extends it. When the hamstring contracts, the quadriceps relaxes; when the quadriceps contracts, the hamstring relaxes. The hamstring and quadriceps together make up an antagonistic muscle group.

Locomotion in animals with exoskeletons and hydrostatic skeletons is also based on antagonistic muscle groups. Animals with exoskeletons have paired flexor-extensor muscles inside their leg joints (**Figure 46.18b**). Animals with hydrostatic skeletons have both circumferential muscles and longitudinal muscles, and these muscles work in concert to shorten and lengthen parts of the body (**Figure 46.18c**).

In all vertebrate animals, the movements of paired muscles are coordinated by motor neurons that originate in the brain or spinal cord. These motor neurons project from processing centers that receive input from sensory systems. Motor neuron activity changes in response to information about balance, smells, sights, and sounds. In this way, an animal's movements are directly tied to the functioning of its sensory systems. The interplay of sensory input and motor output results in the coordinated behaviors we call running, eating, swimming, and flying.

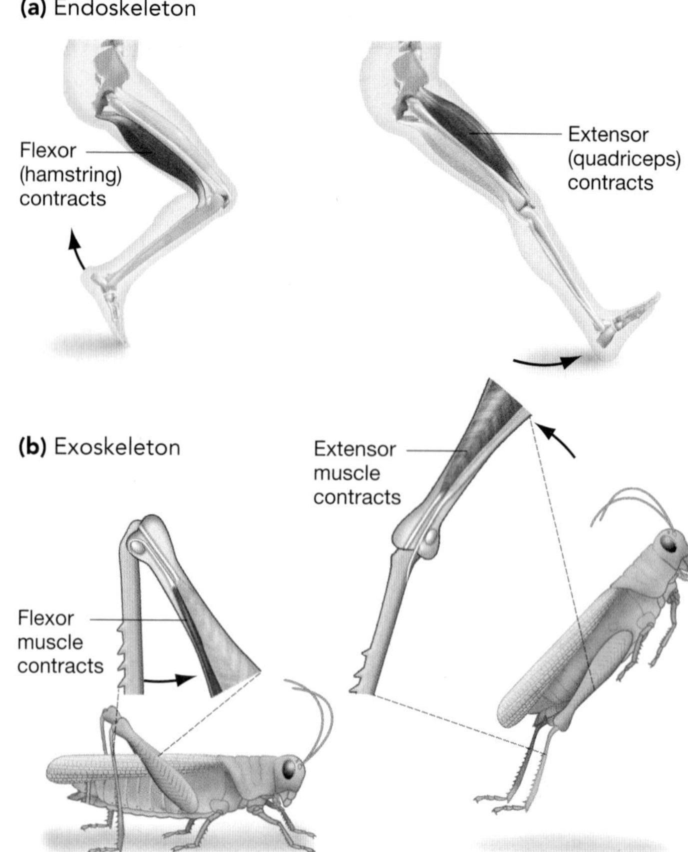

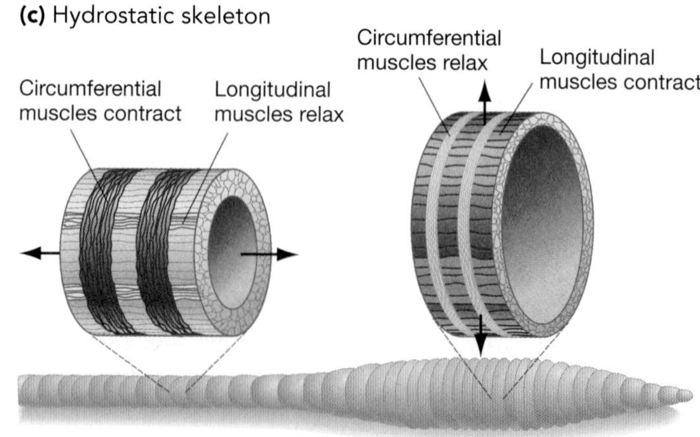

FIGURE 46.18 Antagonistic Muscle Groups Flex or Extend Skeletal Elements.

How Do Muscles Contract?

Early microscopists established that the muscle tissue found in vertebrate limbs and hearts is composed of slender fibers. A **muscle fiber** is a long, thin muscle cell. Within each of the muscle cells from limbs are many small strands called **myofibrils.** How do these strands work?

The Sliding-Filament Model As **Figure 46.19** shows, the myofibrils inside skeletal muscle cells look striped when viewed with the electron microscope. The pattern is caused by alternating light-dark units called **sarcomeres,** which repeat down the length of a myofibril. The sarcomeres in a myofibril shorten as the cell contracts. They lengthen when the cell is relaxed and an external force—exerted by the antagonistic muscle of the pair—stretches the muscle.

Based on these observations, it became clear that the question of how muscles contract simplifies to the question of how sarcomeres contract. Hugh Huxley and Jean Hanson proposed an answer to this question in 1954. These biologists had noticed that the relationship between different parts of the sarcomere—specifically, the points labeled A–D in the photographs in Figure 46.19—change during contraction. Points A and B move closer to each other during contraction, as do points C and D. But the distance from point A to point C does not change, and the distance from point B to point D does not change.

To explain these observations, Huxley and Hanson hypothesized that the banding patterns in the sarcomere are actually caused by two types of long filaments—**thick filaments** and **thin filaments**—and that these filaments slide past one another during contraction. **Figure 46.20** illustrates the model. Notice that thin filaments extend from A to C and thick filaments extend from B to D. According to Huxley and Hanson, the thick and thin filaments slide past one another during a contraction. This explanation became known as the **sliding-filament model**.

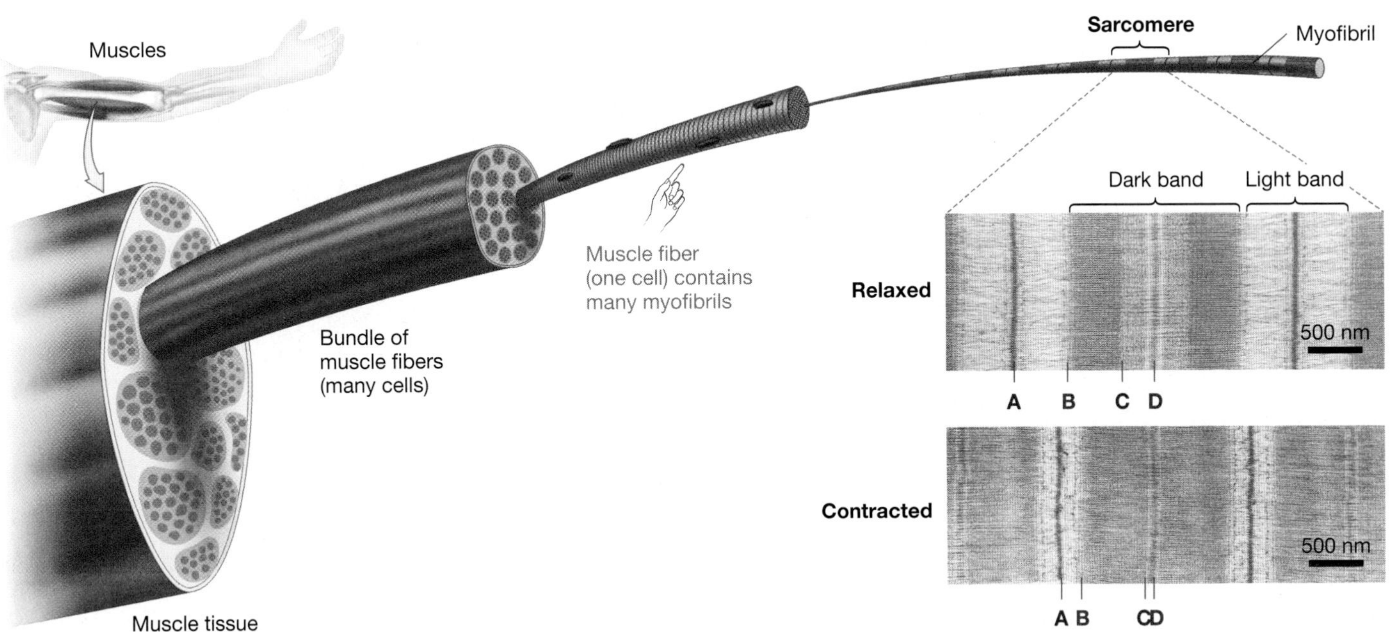

FIGURE 46.19 Muscle Cells Contain Many Myofibrils, Which Contain Many Sarcomeres. Skeletal muscle cells (fibers) have a striped appearance due to repeating sarcomeres, which are units of alternating light-dark bands. When a sarcomere contracts, the distances between the points labeled A, B, C, and D change.

● **QUESTION** Measure the distance from A to B, A to C, A to D, B to C, B to D, and C to D. Which points get closer together during contraction? Which points stay the same distance from each other?

Follow-up research has shown that the Huxley-Hanson model is correct in almost every detail. Thin filaments are composed of two coiled chains of the globular protein **actin**. (Recall from Chapter 7 that actin is the building block of the microfilaments found in the cytoskeleton. The structure of microfilaments is similar to the structure of thin filaments.) One end of each thin filament is bound to a structure called the **Z disk**, which forms the wall of the sarcomere. Z disks contain a protein that binds tightly to actin, anchoring the thin filament. The other end of a thin filament is free to interact with thick filaments. Thick filaments are composed of multiple strands of a long protein called **myosin** and are anchored to the middle of the sarcomere; at both ends they are free to interact with thin filaments. Indeed, researchers who

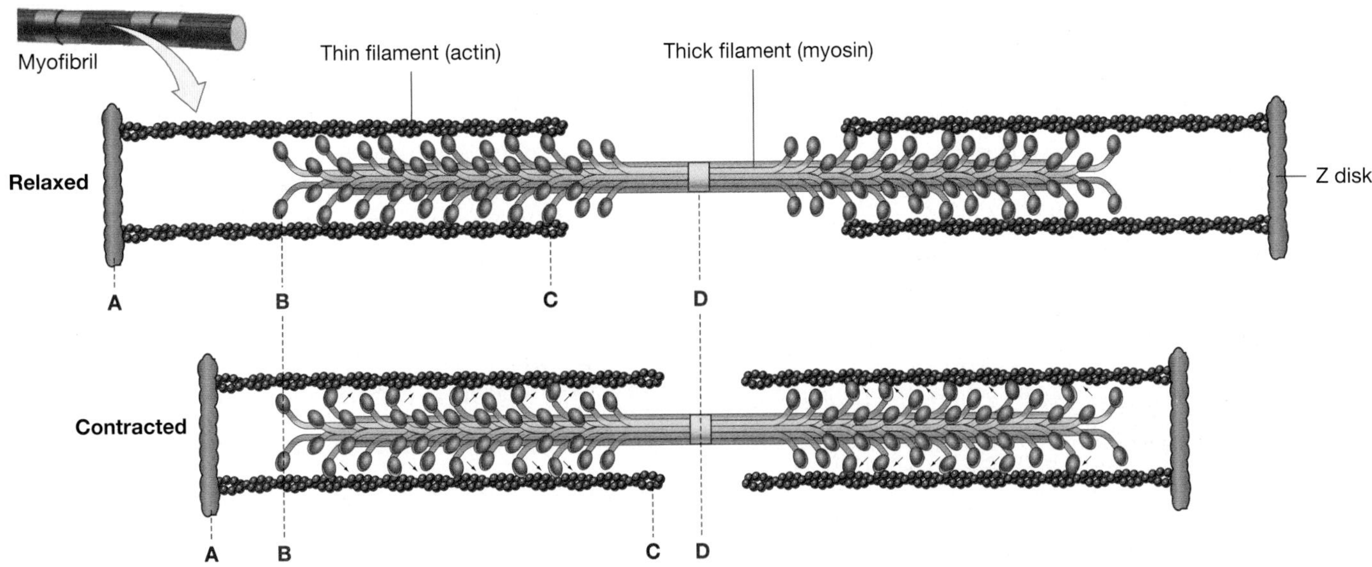

FIGURE 46.20 The Sliding-Filament Model of Sarcomere Contraction. Points A, B, C, and D are the same as those labeled on the sarcomere photographs in Figure 46.19.

● **EXERCISE** Label the Z disks in Figure 46.19. Label the "Dark band" and "Light band" regions in Figure 46.20.

● **QUESTION** Does this model explain the changes in distances that you observed in Figure 46.19? Explain.

isolated actin filaments and myosin molecules and mixed them together on a glass slide in the presence of ATP then observed actin sliding past myosin. As predicted, thick and thin filaments interact in a way that allows them to slide past one another.

To appreciate how the sliding-filament model works, consider the following analogy: Two large trucks are parked 50 m apart, facing each other. Each has a long rope attached to the front bumper. Teams of burly weightlifters grab onto each rope and pull, hand over hand, so that the two trucks roll toward one another. ● If you understand the model, you should be able to explain which elements in the analogy represent the Z disks, the thick filaments, and the thin filaments.

How Do Actin and Myosin Interact? Once the sliding-filament model was confirmed, researchers focused on the question of how the sliding action occurs at the molecular level. Early work on the three-dimensional structure of myosin revealed that each molecule has a head region that projects from the main body of the thick filament. Follow-up experiments established that the myosin head can bind to actin and that the head region can catalyze the hydrolysis of ATP into ADP and a phosphate ion. These results suggested that myosin—not actin—was the site of active movement.

In addition, electron microscopy revealed that myosin and actin are locked together shortly after an animal dies and its muscles enter the stiff state known as rigor mortis. Because ATP is unavailable in dead tissue, the data suggested that ATP is involved in getting myosin to release from actin once the two molecules have bound to each other.

The next major advance in understanding how actin and myosin interact to produce movement was achieved when Ivan Rayment and colleagues solved the detailed three-dimensional structure of the myosin head (**Figure 46.21**), using the X-ray crystallographic techniques introduced in **BioSkills 8**. Rayment's group determined the location of the actin binding site, and they examined how the protein's structure changed when ATP or ADP was bound to it. Because phosphorylation or dephosphorylation can lead to dramatic changes in the shape of a protein, the investigators were not surprised to find significant changes in the protein's conformation when bound to ATP versus ADP.

Based on these data, Rayment and co-workers proposed the model for actin-myosin interaction illustrated in **Figure 46.22**. The cycle of interactions begins at step 1, when the myosin head of a thick filament is attached to ATP but not to actin in a thin filament. (This is the conformation that occurs when a muscle is relaxed.) When ATP is subsequently hydrolyzed to ADP and inorganic phosphate in step 2, the conformation of the protein changes. Specifically, the neck of the myosin straightens out, and the head pivots. The myosin head then binds to a new actin subunit farther down the thin filament. The third step occurs when inorganic phosphate is released and the protein's conformation changes again. Specifically, the neck bends back into its original position. This bending is called the power stroke, because it moves the entire thin filament. In step 4, ADP is released. A new ATP molecule then binds to the myosin, causing myosin to release from actin. The cycle can then start anew. (The transition from step 4 to step 1 cannot occur after death. Rigor mortis sets in as the ATP supplies run out.)

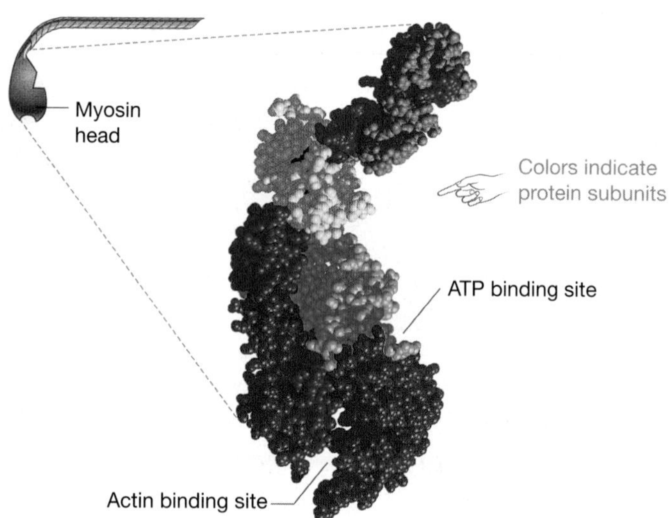

FIGURE 46.21 Myosin's "Head" Binds ATP and Actin. The head region of the myosin protein contains both an ATP binding site and an actin-binding site.

● **EXERCISE** Label the myosin heads in Figure 46.20.

As ATP binding, hydrolysis, and release continue, the two ends of the sarcomere are pulled closer together. The same basic mechanism is observed in cardiac and smooth muscle cells. Most animal movement, in fact, is powered by similar interactions between actin and myosin. These two proteins are also responsible for the amoeboid movement observed in amoebae and slime molds (see Chapter 29) and the streaming of cytoplasm observed in algae and land plants. The message of these observations is that actin and myosin are evolutionarily ancient proteins. Because they make movement possible in the absence of cilia and flagella, they have played a critical role in the diversification of eukaryotes.

How Does Relaxation Occur? The results summarized in Figures 46.20 and 46.22 were obtained by studying myosin and actin in isolation. Although elegant and powerful, experiments on isolated fibers indicate only what is happening in individual fibers. In vitro, thick filaments from muscle continue to crawl along thin filaments as long as ATP and a length of

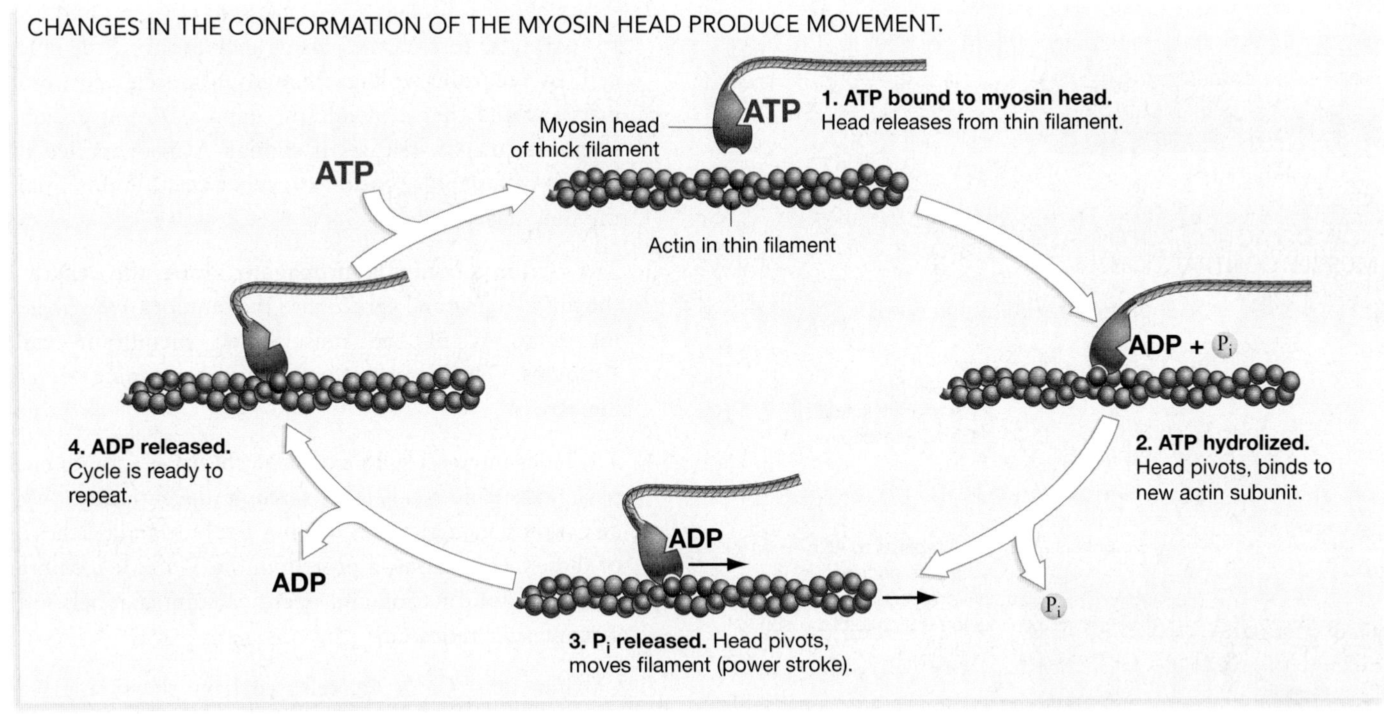

FIGURE 46.22 Myosin and Actin Interact during Muscle Contraction. Summary of the current model of how myosin and actin interact as a sarcomere contracts. The four steps repeat rapidly.

(a) Tropomyosin and troponin work together to block the myosin binding sites on actin.

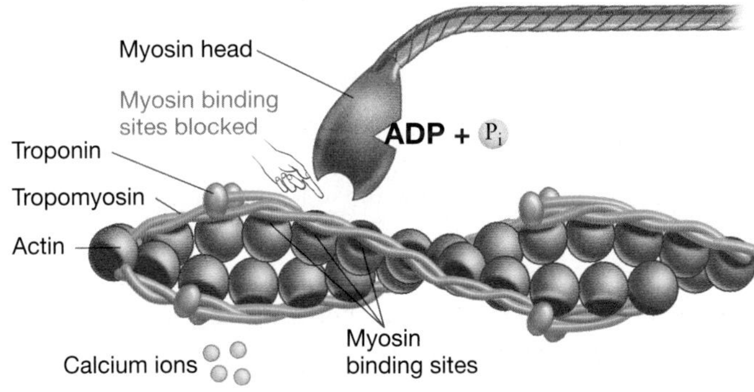

(b) When a calcium ion binds to troponin, the troponin-tropomyosin complex moves, exposing myosin binding sites.

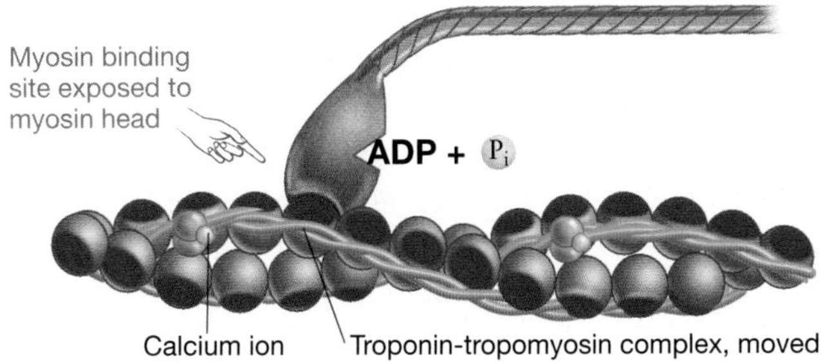

FIGURE 46.23 Troponin and Tropomyosin Regulate Muscle Activity.

thin filament are available. ATP is almost always available in living muscles. How do our muscles ever stop contracting and relax?

Researchers were able to answer this question when they confirmed that, in addition to containing actin, thin filaments contain two key proteins called **tropomyosin** and **troponin** (**Figure 46.23a**). Tropomyosin and troponin work together to block the myosin binding sites on actin. When tropomyosin and troponin are in the conformation shown at the top in **Figure 46.23b**, the myosin-actin interaction cannot occur, and thick and thin filaments cannot slide past each other. But when calcium ions bind to troponin, the troponin-tropomyosin complex moves in a way that exposes the myosin binding sites on actin.

In vivo, how are calcium ions released so that contraction can begin? The answer turned out to have several steps, beginning with the arrival of an electrical signal from a motor neuron.

An Overview of Events at the Neuromuscular Junction

Figure 46.24 summarizes what happens when an action potential from a motor neuron arrives at a muscle cell and initiates contraction:

1. Action potentials trigger the release of the neurotransmitter called **acetylcholine** (**ACh**) from the motor neuron into the synaptic cleft between the motor neuron and the muscle cell.

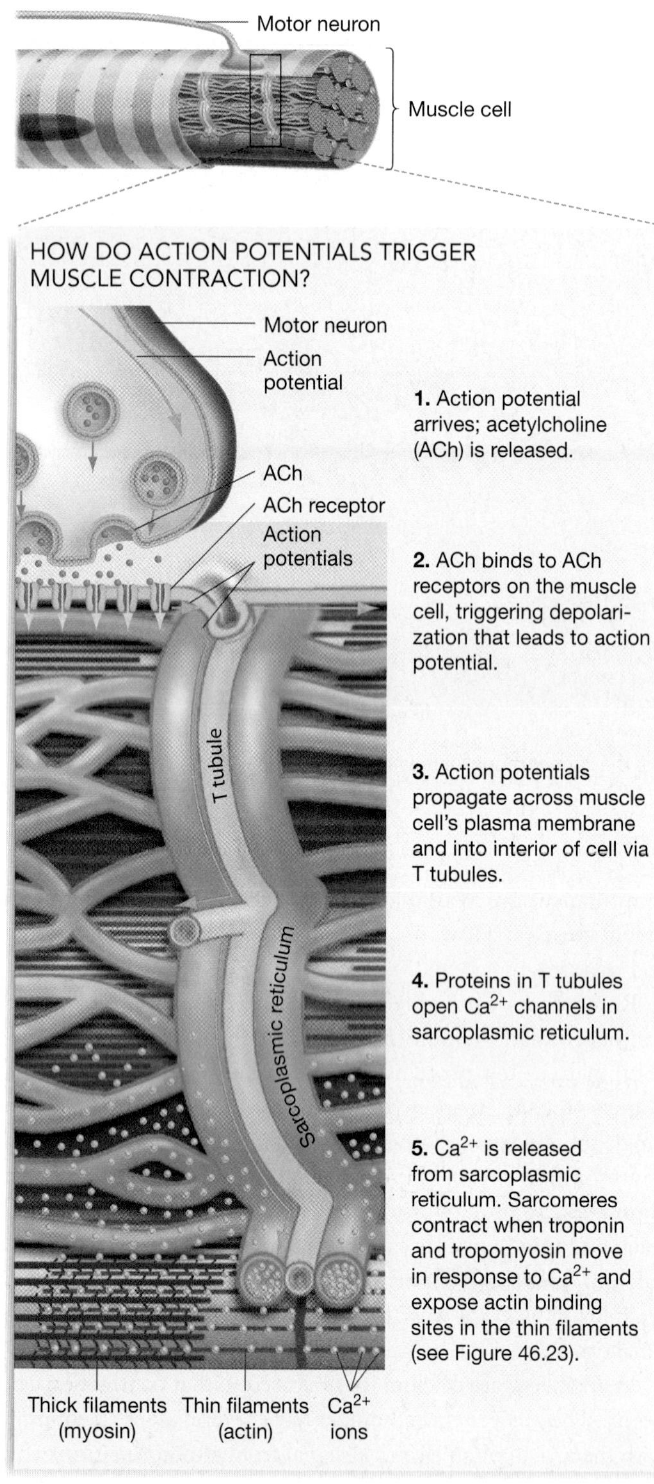

FIGURE 46.24 Action Potentials at the Neuromuscular Junction Trigger the Release of Ca^{2+}, Which Binds to Troponin-Tropomyosin and Allows Myosin to Form a Cross-Bridge with Actin. The arrival of action potentials from a motor neuron triggers a series of events in a muscle cell. The sequence ends with changes in troponin-tropomyosin proteins that allow contraction to begin.

2. Acetylcholine diffuses across the synaptic cleft and binds to ACh receptors on the plasma membrane of the muscle cell. By recording voltage changes in muscle cells, biologists showed that a membrane depolarization occurs in response to ACh release. If enough ACh is applied to a muscle cell, depolarization triggers an action potential in the fiber itself.

3. The action potentials propagate along the length of the muscle fiber and spread into the interior of the fiber via invaginations of the muscle cell membrane called **T tubules.** (The *T* stands for *transverse*, meaning "extending across.")

4. T tubules intersect with extensive sheets of smooth endoplasmic reticulum called the **sarcoplasmic reticulum.** When an action potential passes down a T tubule and reaches one of these intersections, a protein in the T tubule membrane changes conformation and opens calcium channels in the sarcoplasmic reticulum.

5. Calcium ions (Ca^{2+}) are released from the sarcoplasmic reticulum into the cytoplasm. The outward flow occurs because a tremendous concentration gradient exists—the concentration of calcium ions inside the sarcoplasmic reticulum is about 40,000 times that of the cytosol. Contact with Ca^{2+} causes a conformational change in troponin. The conformational change in troponin moves tropomyosin away from the myosin binding sites on actin. Myosin binds to actin.

To summarize, the arrival of an action potential causes a release of calcium ions that bind to troponin, causing tropomyosin to move and allowing contraction to begin. In the absence of an action potential, calcium is not released and the troponin-tropomyosin complex blocks further contraction. The grand result is that an arm or a leg or a wing moves; an animal grabs or runs or flies; and a piece of ripe fruit is picked, a sprinter breaks the tape, or a moth ducks under the gaping jaws of a bat.

MB Web Animation *BioFlix* at www.masteringbio.com
Muscle Contraction

Muscle Types

Early anatomical studies confirmed that vertebrates have three distinct types of muscle tissue: (1) skeletal muscle, (2) cardiac muscle, and (3) smooth muscle. These muscle types were introduced briefly in Chapter 41; **Table 46.1** provides a more detailed comparison. Although all three cell types contract in response to electrical stimulation and actin-myosin interactions, they differ in overall morphology as follows:

SUMMARY TABLE **46.1 Types of Muscle in Vertebrates**

	Skeletal Muscle	Cardiac Muscle	Smooth Muscle
	50 μm	50 μm	50 μm
Location	Attached to bones	Heart	Intestines, arteries, other
Function	Move skeleton	Pump blood	Move food, help regulate blood pressure, etc.
Cell characteristics	Nuclei	Intercalated discs	
	Multinucleate	1 or 2 nuclei	Single nucleus
	Unbranched	Branched; intercalated discs form direct cytoplasmic connection end to end	Unbranched
	Contains myofibrils	Contains myofibrils	No myofibrils
	Activity is "voluntary," meaning that signal from motor neuron is required	Activity is "nonvoluntary," meaning that signal from motor neuron is not required	Activity is "nonvoluntary," meaning that signal from motor neuron is not required

1. **Skeletal muscle** consists of unbranched multinucleate cells—they contain multiple nuclei. This tissue is also known as striated muscle, because it appears to be striated, or striped.

2. **Cardiac muscle** contains branched cells whose ends are connected via specialized regions called intercalated discs. In an **intercalated disc**, protein-lined openings called gap junctions provide a direct cytoplasmic connection between adjacent muscle cells. As Chapter 45 noted, intercalated discs are critical to the flow of electrical signals from cell to cell and to the coordination of the heartbeat.

3. **Smooth muscle** is unbranched, lacks myofibrils, and is often organized into thin sheets. It is essential to the function of the lungs, blood vessels, digestive system, urinary bladder, and reproductive system.

In combination, the three muscle types keep your blood flowing, the food you've eaten moving through your digestive system, and your body moving where you want it to go. Understanding movement is fundamental to understanding how animals work.

Check Your Understanding

If you understand that...

- Antagonistic muscle groups work in conjunction with skeletons to produce movement.
- Many types of movement are based on interactions between actin and myosin.

You should be able to...

1) Make a labeled diagram of a skeletal muscle fiber, showing the relationship between the multinucleate cell, myofibrils, and sarcomeres.
2) Describe the sliding-filament model.
3) Explain how binding by ATP and hydrolysis of ATP affect the interaction between actin and myosin.
4) Predict the effect on muscle function of drugs that have the following actions: increase ACh release at the neuromuscular junction, prevent conformational changes in troponin, and block uptake of calcium ions into the sarcoplasmic reticulum.

Chapter Review

SUMMARY OF KEY CONCEPTS

Sensory receptor cells transduce sound, light, or other stimuli to a change in membrane potential. Receptor cells or sensory neurons send action potentials to the brain, where the signals are processed and integrated.

If the membrane potential of a sensory cell is altered substantially enough in response to a stimulus, the pattern of action potentials that it sends to the brain changes. In this way, sensory stimuli as different as sound and light are transduced to electrical signals. The brain is able to distinguish different types of stimuli because axons from different types of sensory neurons project to different regions of the brain.

You should be able to suggest a hypothesis to explain why people who have had limbs amputated experience "phantom pain"—the perception that their missing tissue hurts.

Hearing is based on sensory receptor cells that move in response to sound waves of a particular frequency. The movement opens ion channels in the sensory cells and results in a change in membrane potential.

Pressure receptors detect direct physical stimulation, including stimulation from sound. Hair cells, the major sensory detectors in the vertebrate ear, undergo a change in membrane potential in response to bending of their stereocilia. Hair cells contact a basilar membrane, and experiments with isolated cochleas showed that sound waves of a certain frequency cause a certain part of the basilar membrane to vibrate. Because of this specificity in hair-cell response, mammals can discriminate different pitches.

You should be able to predict how a hair-cell–before receptor could be involved in gravity sensing in plants, in response to pressure exerted by the dense, starch-filled organelles called amyloplasts.

Vision is based on sensory receptor cells that contain a light-absorbing pigment bound to a protein. The pigment changes conformation when it absorbs light and triggers a series of events that change the cell's membrane potential.

In the vertebrate eye, photoreceptors are located in rods and cones. Although these two cell types differ in structure and function, both contain rhodopsin molecules that consist of retinal paired with opsin. The rhodopsin found in rods is stimulated by even the faintest light. Color vision is possible because cones contain opsins that respond to specific wavelengths of light absorbed by retinal. Humans distinguish colors based on the pattern of stimulation of three types of opsins found in cones. People who lack one of these cone opsins are color blind, meaning they cannot distinguish as many colors as people with all three opsins can.

You should be able to explain why different animal species are able to see different colors.

at www.masteringbio.com
The Vertebrate Eye

Taste and smell are based on membrane proteins that act as receptors or channels for particular molecules and that trigger changes in membrane potential.

Chemoreceptors detect the presence of certain foodborne or airborne molecules. Taste buds, for example, contain taste cells with membrane proteins that respond to toxins, salt, acid, and other types of molecules in food. Sodium ions and protons enter taste cells via channels and depolarize the membrane directly. Toxic compounds bind to membrane receptors and trigger action potentials that are interpreted by the brain as bitterness. Smell, or olfaction, is used to scan molecules from the outside environment. Airborne chemicals are detected by hundreds of different odor-receptor proteins located in the membranes of receptor cells in the nose.

You should be able to explain why the brain can perceive so many different flavors based on inputs from just four or five basic types of taste receptors, and why people with nasal congestion complain that food tastes bland.

In many cases, animals respond to sensory stimuli by moving. Movement is based on antagonistic muscle groups that act on a stiff skeleton containing joints that flex or extend in response to electrical signals.

All animal muscles use the same basic mechanism for contraction. Muscles shorten when thick filaments comprised of myosin slide past thin filaments comprised of actin, in a series of binding events mediated by the hydrolysis of ATP. Calcium ions play an essential role in muscle contraction, by making the actin in thin filaments available for binding by myosin. In animals with exoskeletons or endoskeletons, muscles are usually arranged in opposing pairs of flexors and extensors. Animals with hydrostatic skeletons have muscles arranged in opposing pairs of circular and longitudinal bands. In all cases, muscles work in concert with skeletons.

You should be able to predict the primary symptom of botulism, which occurs when a toxin prevents release of ACh from the neuromuscular junction.

at www.masteringbio.com
Muscle Contraction

QUESTIONS

Test Your Knowledge

1. What is the major function of the mammalian middle ear?
   a. It amplifies the sound energy transmitted to the cochlea.
   b. It transmits sound vibrations to the tympanic membrane.
   c. It circulates fluid through the cochlea.
   d. It contains hair cells that detect specific frequencies of sound.

2. In the human ear, why do different hair cells respond to different frequencies of sound?
   a. Waves of pressure move through the fluid in the cochlea.
   b. Hair cells are "sandwiched" between membranes.
   c. Receptors in the stereocilia of each hair cell are different; each receptor protein responds to a certain range of frequencies.
   d. Because the basilar membrane varies in stiffness, it vibrates in certain places in response to certain frequencies.
3. Which of the following comparisons of rods and cones is *false*?
   a. The eyes of most humans have one type of rod and three types of cones.
   b. Rods are more sensitive to dim light than cones are.
   c. There are more rods than cones in the fovea.
   d. Both rods and cones use retinal and opsins to detect light.
4. Which of the following statements about taste is *true*?
   a. Sweetness is a measure of the concentration of hydrogen ions in food.
   b. Sodium ions from foods can directly depolarize certain taste cells.
   c. All bitter-tasting compounds have a similar chemical structure.
   d. Membrane receptors are involved in detecting acids.
5. In muscle cells, myosin molecules continue moving along actin molecules as long as
   a. ATP is present and troponin is not bound to Ca^{2+}.
   b. ADP is present and tropomyosin is released from intracellular stores.
   c. ADP is present and intracellular ACh is high.
   d. ATP is present and intracellular Ca^{2+} is high.
6. Which of the following is critical to the function of exoskeletons, endoskeletons, and hydrostatic skeletons?
   a. Muscles interact with the skeleton in antagonistic groups.
   b. Muscles attach to each of these types of skeleton via tendons.
   c. Muscles extend joints by pushing them.
   d. Segments of the body or limbs are extended when paired muscles relax in unison.

Test Your Knowledge answers: 1. a; 2. d; 3. c; 4. b; 5. d; 6. a

Test Your Understanding

Answers are available at www.masteringbio.com

1. Describe how a sound, a light, and an odor are transduced into a change in the pattern of action potentials from a sensory cell. How does the brain know which sense is which when the action potentials reach the brain?
2. Give three examples of how the sensory abilities of an animal correlate with its habitat or method of finding food and mates.
3. How did the discovery of odor-receptor genes affect our understanding of how the sense of smell works?
4. Draw a hair cell and explain how bending of stereocilia results in a change in membrane potential.
5. Scientists generally think that a "good hypothesis" is one that is reasonable, is testable, and inspires further research into the field. Using these criteria, was Dalton's hypothesis about color vision a good hypothesis? Was it correct?
6. How did observations of sarcomere structure inspire the sliding-filament model of muscle contraction? Explain why the observation that muscle cells contain a large number of mitochondria and extensive smooth endoplasmic reticulum turned out to be logical once the molecular mechanism of muscular contraction was understood.

Applying Concepts to New Situations

Answers are available at www.masteringbio.com

1. Myasthenia gravis is a disease that develops in humans when the immune system produces proteins that bind to the acetylcholine (ACh) receptors in muscles. The primary symptom of myasthenia gravis is muscle weakness. Why?
2. Houseflies have about 800 ommatidia in each of their compound eyes. Dragonflies, in contrast, have up to 10,000 ommatidia per eye. Houseflies feed by lapping up watery material from piles of excrement or rotting carcasses, which they locate by scent. Dragonflies are aerial predators and hunt by sight. Suggest a hypothesis to explain the difference in the structure of their eyes. How would you test your hypothesis?
3. Skeletal muscles contain two general types of cells. "Slow-twitch fibers" produce ATP via cellular respiration. They are reddish because they have a high concentration of myoglobin—a protein that delivers oxygen to electron transport chains during cellular respiration. Slow-twitch fibers support endurance exercise. Fast-twitch fibers, in contrast, produce ATP primarily via fermentation. They do not have large quantities of myoglobin and are light in color. They support extremely fast contraction but tire easily. Chickens, turkeys, and other ground-dwelling birds can run long distances but escape from predators by flying in short, extremely fast bursts. Based on these observations, explain the distribution of "dark meat" and "white meat" in these birds. Most other birds do not have white meat. Why?
4. When looking at faint stars through a telescope, astronomers will focus their eyes just to the side of the object. Instead of landing on the fovea, then, the star's image falls next to the fovea. With this technique, faint objects pop into view. They vanish if looked at directly, however. Explain what is going on.

www.masteringbio.com is also your resource for • Answers to text, table, and figure caption questions and exercises • Answers to *Check Your Understanding* boxes • Online study guides and quizzes • Additional study tools including the *E-Book for Biological Science* 3rd ed., textbook art, animations, and videos.

47 Chemical Signals in Animals

KEY CONCEPTS

- Animals use at least six major types of chemical signals. Hormones are those chemical signals that are present in tiny concentrations and travel throughout the body to affect target cells.
- The information in hormonal signals helps animals respond to environmental change, develop as embryos, undergo sexual maturation, and achieve homeostasis.
- The production of a hormone is tightly regulated by input from the nervous system and by other hormones.
- Some hormones bind to receptors inside target cells and change gene expression. Other hormones bind to receptors at the cell surface and lead to changes in protein activation.

The spectacular transformation that occurs during insect metamorphosis is triggered by the chemical signals called hormones.

In response to sights, sounds, and other sensory stimuli, an animal's nervous system sends rapid messages to precise locations in the body. In many cases, these messages result in muscle contractions and movement.

As conditions inside and outside an animal change, however, cells in the central nervous system (CNS) or cells in specialized organs that make up the **endocrine system** may trigger the release of molecules that produce longer-term responses in a broad range of tissues and organs. A chemical signal that circulates through body fluids and affects distant target cells is called a **hormone**. As you read this, a large suite of hormones is coursing through your circulatory system. These molecules are regulating the maturation of sperm or eggs by your reproductive system, changing the composition of the urine forming in your kidneys, and controlling the release of digestive enzymes in your gastrointestinal tract. Earlier in your life, changes in hormone concentrations led to the dramatic physical changes associated with puberty.

Animal hormones are present in extremely low concentrations but have enormous effects on their target cells. Unlike action potentials, which are electrical impulses that have a short-term effect on a single cell or on a small population of adjacent cells, the messages that hormones carry are received by many different cells and have a relatively long-lasting effect. In combination, electrical and chemical signals allow animals to coordinate the activities of cells throughout the body. They are the mechanism responsible for maintaining trillions of cells as the integrated unit we call an individual.

Key Concept Important Information Practice It

The goal of this chapter is to explore how hormones and other types of chemical signals work in animals. Let's begin with an overview of their chemical signaling systems, and then plunge into analyzing how hormones regulate the activity of target cells.

47.1 Cell-to-Cell Signaling: An Overview

For over a century, biologists have been exploring how chemical messengers function in animals. One of the most basic conclusions from this work is that the chemical signals found in animals have diverse structures and functions. **Figure 47.1** summarizes how biologists go about organizing the diversity of chemical signals, based on where the molecules originate and where they act. Notice that the names for most of the categories use the Greek word root *crin*, meaning "separated." Its use captures something essential about how chemical signals act: They are released from cells and thus are separated from them. There are six major categories of chemical signals:

1. **Autocrine** (literally, "same-separated") signals act on the same cell that secretes them. Perhaps the best-studied autocrine signals are **cytokines** ("cell-movers"). The function of most cytokines is to amplify the response of a cell to a stimulus. An example is interleukin 2, which is synthesized and released by a type of white blood cell called a T cell in the course of fighting an infection. Interleukin 2 activates T cells to help eliminate the infection. It also causes the cells to divide repeatedly, producing more activated T cells for host defense.

2. **Paracrine** ("beside-separated") signals diffuse locally and act on neighboring cells. Cytokines may act as paracrine signals as well as autocrine signals, because they can trigger responses by other cells of the immune system. It is common, in fact, to observe that a single chemical messenger acts in a variety of ways. As an example, consider molecules that are produced by three distinct populations of cells within a region of the pancreas called the islets of Langerhans. The molecules named **insulin**, **glucagon**, and **somatostatin** act on nearby pancreatic cells as paracrine signals and ensure a smooth, steady response to changing blood-glucose levels. But they also act as hormones—meaning they target distant cells—in controlling the concentration of glucose in the blood.

3. **Endocrine** ("inside-separated") signals are hormones—molecules that are carried between distant cells by blood or other body fluids. The cells that produce endocrine signals may be organized into discrete organs called **glands** or may be interspersed among the cells of other organs.

4. **Neural** signals, introduced in Chapter 45, are the chemical messengers called **neurotransmitters**. Neural signaling can be very fast, because neurotransmitters have to diffuse only a short distance—the tiny gap between two neurons, called the synaptic cleft.

5. **Neuroendocrine** ("nerve-inside-separated") signals are considered hormones because they act on distant cells after being released from neurons, instead of acting at the adjacent synapse. Antidiuretic hormone (ADH; also called vasopressin), introduced in Chapter 42, is a good example. ADH is produced by neurons that originate in the hypothalamus of the brain, but the molecule acts on cells in the collecting duct of the kidney to help regulate water excretion.

6. **Pheromones** are signals that are released into the environment and act on a different individual. In sea urchins, insects, and many other species, pheromones help coordinate reproduction in males and females or function in attracting mates.

Autocrine signals act on the same cell that secretes them.

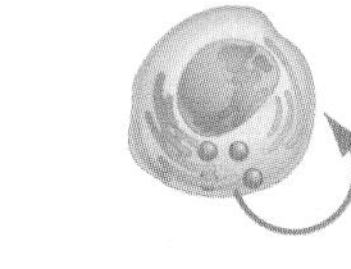

Paracrine signals diffuse locally and act on neighboring cells.

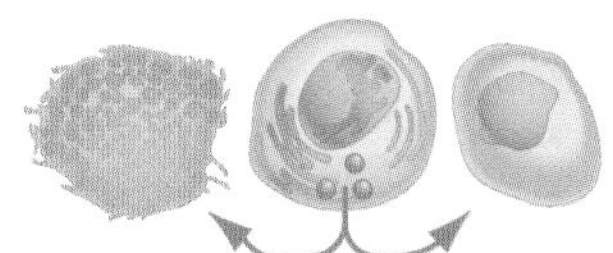

Endocrine signals are hormones carried between cells by blood or other body fluids.

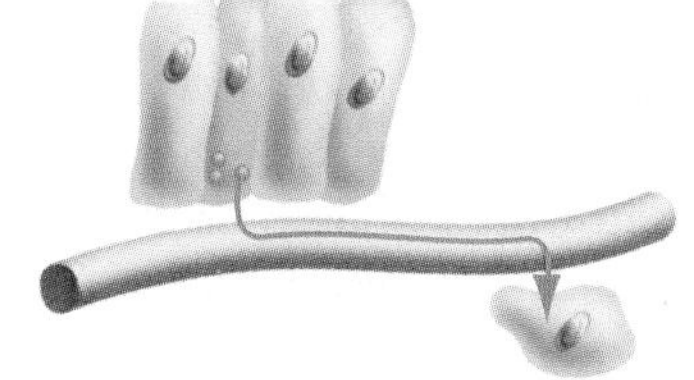

Neural signals diffuse a short distance between neurons.

Neuroendocrine signals are released from neurons but are carried by blood or other body fluids and act on distant cells.

Pheromones are released into the environment and act on a different individual.

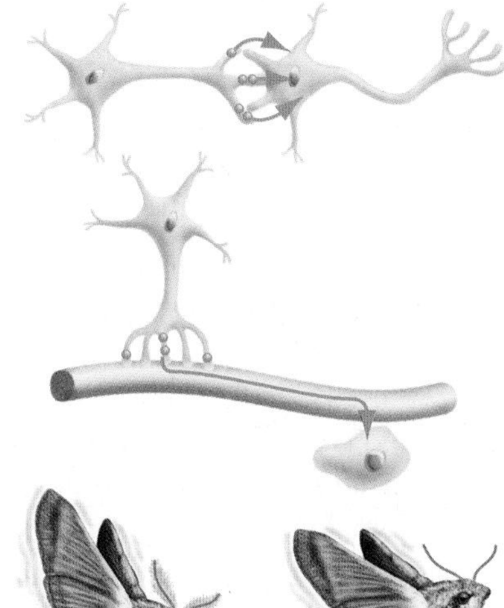

FIGURE 47.1 There Are Six Categories of Chemical Signals in Animals. Chemical signals can be categorized by how they move from the cells that produce them to the target cells that respond to them.

(a) Endocrine pathway

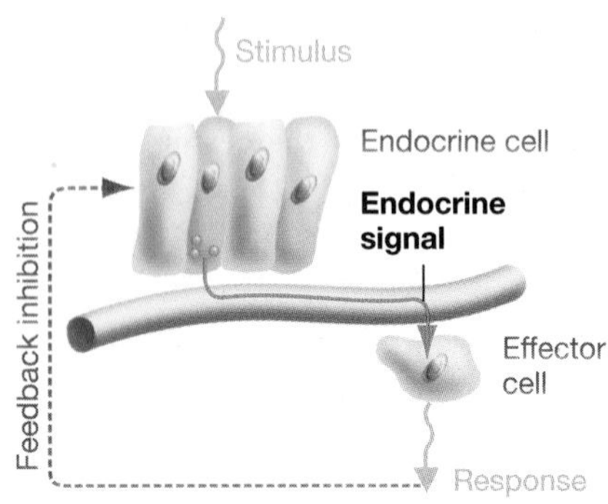

(b) Neuroendocrine pathway

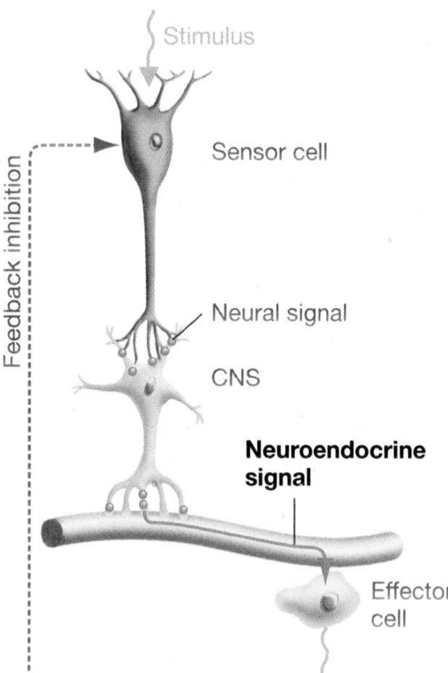

(c) Neuroendocrine-to-endocrine pathway

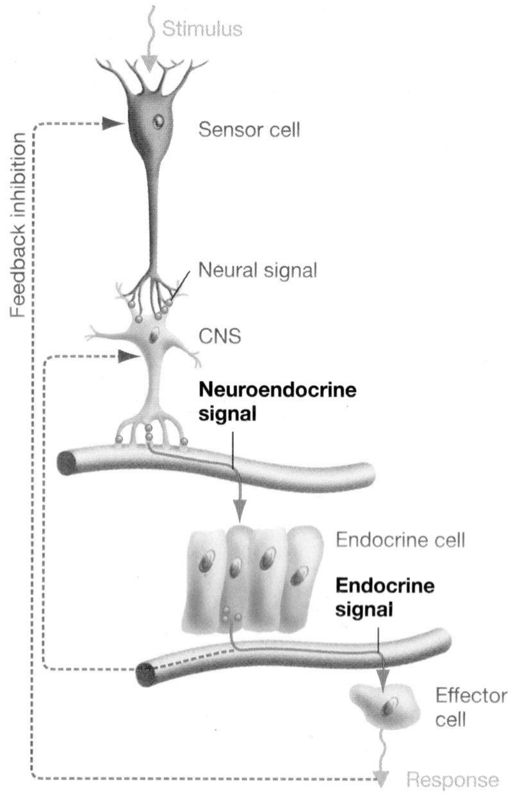

FIGURE 47.2 Hormones Act via Three Pathways and Are Regulated by Negative Feedback.

It's important to note that these six classes of chemical messenger do not coincide with six structurally distinct classes of molecules. For example, the endocrine signals found in a particular organism routinely belong to several families of chemical compounds, ranging from amino acid derivatives to glycoproteins. And a particular family of molecules—say, peptides or the lipids called steroids—may function as endocrine, autocrine, and paracrine signals in the same individual.

The diversity of chemical signal types observed in animals is impressive. This chapter focuses primarily on hormones, however, because they are well studied and because they are particularly important to understanding how animals work. Let's begin with an introduction to how these long-distance signals are produced and received.

Hormone Signaling Pathways

In plants, sensory cells perceive changes in the environment and broadcast a hormonal signal that triggers an appropriate response from effector cells. Some animal hormones are also sent directly from endocrine cells to effector cells, in response to a stimulus (**Figure 47.2a**). But more frequently, hormonal signaling in animals involves additional steps.

In many or most cases, information about external or internal conditions is gathered by receptors and then integrated by neurons in the central nervous system (CNS) prior to the production of a hormonal signal. Neurons in the CNS respond by synthesizing and releasing neuroendocrine signals that either act on effector cells directly (**Figure 47.2b**) or—more commonly—stimulate cells in the endocrine system. The endocrine cells then respond by producing a hormone: an endocrine signal that acts on effector cells and triggers an appropriate response (**Figure 47.2c**).

All three types of signaling pathway—direct from an endocrine cell, direct from CNS, or CNS-to-endocrine system—are regulated by **negative feedback,** or **feedback inhibition**. When feedback inhibition occurs, the product of a process inhibits its production. In the case of the endocrine pathway, the response by the effector cells feeds back on endocrine cells, lowering production of the hormone and down-regulating the response (Figure 47.2a). An effector's response also feeds back to cells that initiate the neuroendocrine and neuroendocrine-to-endocrine pathways. A change in input from these cells then lowers production of the signal and reduces the response (see Figures 47.2b and 47.2c). Neuroendocrine-to-endocrine signaling pathways have an additional layer of regulation, because the endocrine signal usually inhibits production of the neuroendocrine signal (Figure 47.2c).

Feedback inhibition in the endocrine system is analogous to temperature control by a heat-sensitive thermostat. If the temperature is too high, the thermostat sends a signal that turns the furnace off; if the temperature is too low, the thermostat sends a signal that turns the furnace on. The result is a constant air temperature. With hormonal signals, feedback inhibition reduces production of the hormone. ● If you understand this concept, you should be able to explain which parts of the thermostat analogy correspond to the sensory input, CNS, cell-to-cell signal, and effector in a hormone signaling pathway. You should also be able to predict what happens when feedback inhibition fails in a hormone signaling pathway.

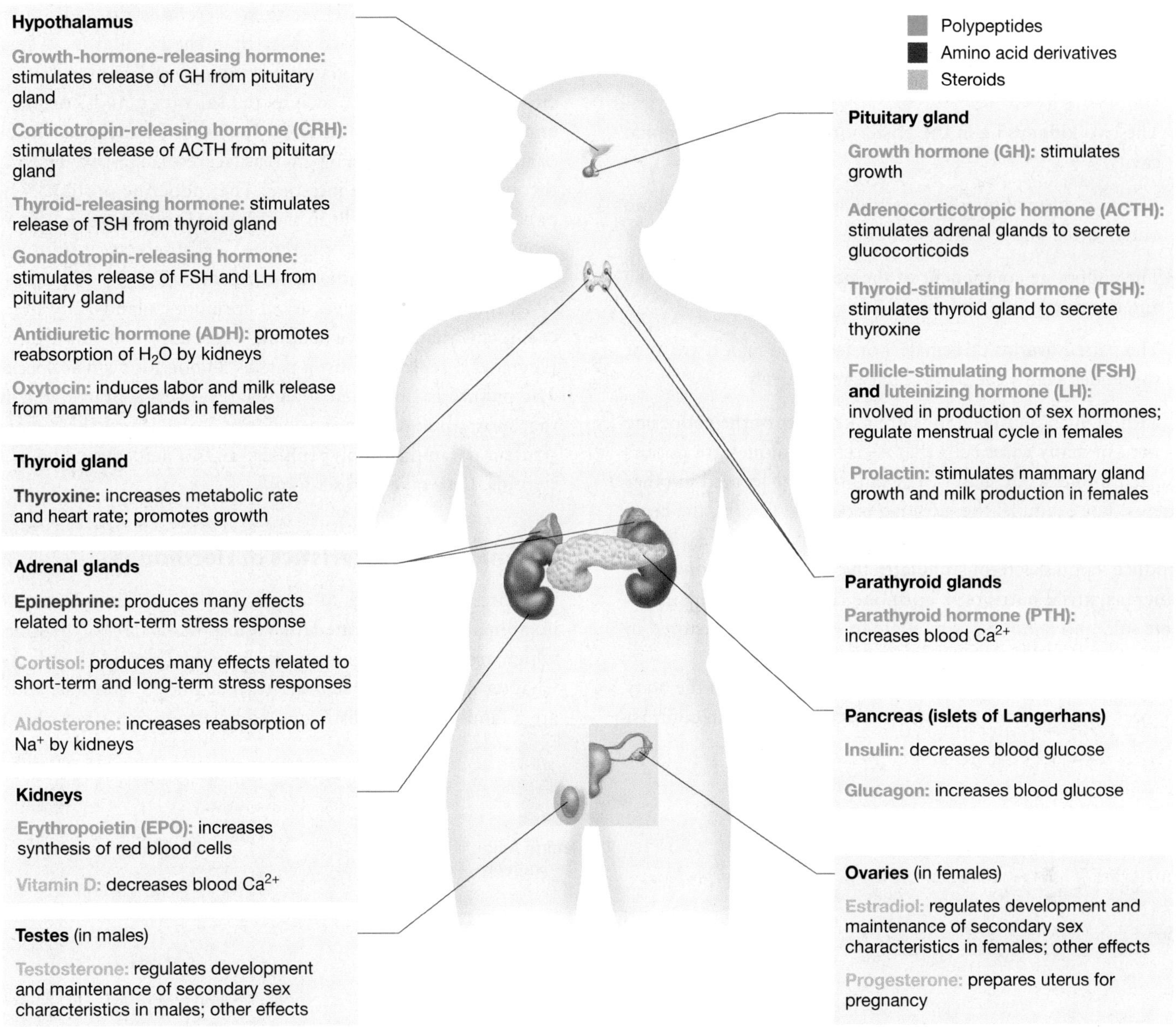

Hypothalamus

Growth-hormone-releasing hormone: stimulates release of GH from pituitary gland

Corticotropin-releasing hormone (CRH): stimulates release of ACTH from pituitary gland

Thyroid-releasing hormone: stimulates release of TSH from thyroid gland

Gonadotropin-releasing hormone: stimulates release of FSH and LH from pituitary gland

Antidiuretic hormone (ADH): promotes reabsorption of H_2O by kidneys

Oxytocin: induces labor and milk release from mammary glands in females

Thyroid gland

Thyroxine: increases metabolic rate and heart rate; promotes growth

Adrenal glands

Epinephrine: produces many effects related to short-term stress response

Cortisol: produces many effects related to short-term and long-term stress responses

Aldosterone: increases reabsorption of Na^+ by kidneys

Kidneys

Erythropoietin (EPO): increases synthesis of red blood cells

Vitamin D: decreases blood Ca^{2+}

Testes (in males)

Testosterone: regulates development and maintenance of secondary sex characteristics in males; other effects

Polypeptides
Amino acid derivatives
Steroids

Pituitary gland

Growth hormone (GH): stimulates growth

Adrenocorticotropic hormone (ACTH): stimulates adrenal glands to secrete glucocorticoids

Thyroid-stimulating hormone (TSH): stimulates thyroid gland to secrete thyroxine

Follicle-stimulating hormone (FSH) and luteinizing hormone (LH): involved in production of sex hormones; regulate menstrual cycle in females

Prolactin: stimulates mammary gland growth and milk production in females

Parathyroid glands

Parathyroid hormone (PTH): increases blood Ca^{2+}

Pancreas (islets of Langerhans)

Insulin: decreases blood glucose

Glucagon: increases blood glucose

Ovaries (in females)

Estradiol: regulates development and maintenance of secondary sex characteristics in females; other effects

Progesterone: prepares uterus for pregnancy

FIGURE 47.3 An Overview of the Human Endocrine System. A partial list of the endocrine glands and hormones found in humans. The heart, gastrointestinal tract, adipose tissue, and many other hormone-producing organs and cells could also be added to this list.

EXERCISE Write the meanings of the following word roots next to the appropriate hormone: *tocus* ("birth"), *insula* ("island"), *pro* ("before"), *lacto* ("milk"), *epi* ("above"), *nephro* ("kidney"), *cortic* ("outer shell," or "cortex"), *erythryo* ("red"), *poiesis* ("make"), *luti* ("yellowing").

What Are the Components of the Endocrine System?

The endocrine system is the collection of cells, tissues, and organs responsible for hormone production and secretion. Organs that secrete a hormone into the bloodstream are called **endocrine glands**.

The tissues and organs that make up the endocrine system vary widely among animals. For example, neurons that manufacture and secrete hormones are particularly important in insects, where they regulate molting, metamorphosis, and other processes. Salmon have an unusual gland that secretes a hormone responsible for regulating calcium ion concentration. And within a species, the diversity of endocrine system components can be impressive. For example, consider the major glands that have endocrine functions in humans (**Figure 47.3**).

- The **hypothalamus** is a region deep within the brain.
- The **pituitary gland** sits just below the hypothalamus and has distinct anterior and posterior regions.

- The **thyroid gland** is situated in the neck.
- The four **parathyroid glands** are embedded in the thyroid gland.
- The two **kidneys** lie in the posterior part of the abdominal cavity.
- The two **adrenal glands** sit atop the kidneys and have an outer cortex and a central medulla.
- The endocrine component of the **pancreas** is located in the anterior part of the abdominal cavity.
- The paired **ovaries** (in females) or **testes** (in males) are in or suspended below the pelvic cavity, respectively.

Although endocrine glands are a key part of the endocrine system, in many cases cells that secrete hormones are not organized into discrete glands. Instead, they are located in other organs. For example, the intestine produces secretin, the heart produces atrial natriuretic hormone, and the cells of fat tissue produce leptin. Secretin stimulates the exocrine portion of the pancreas, atrial natriuretic hormone causes the kidney to excrete salt, and **leptin** helps regulate the amount of fat stored in the body.

Finally, it's important to note that not all glands in the body are part of the endocrine system. **Exocrine glands**, in contrast to endocrine glands, deliver their secretions through outlets called ducts into a space other than the circulatory system. Most of the digestive glands introduced in Chapter 43 were either exocrine glands, such as the salivary glands, or mixed endocrine and exocrine glands, such as the pancreas. The exocrine portions of the pancreas secrete digestive enzymes through ducts into the intestine. The endocrine portion of the pancreas consists of cells in the **islets of Langerhans**. In this region of the pancreas, clusters of cells secrete insulin and glucagon directly into the bloodstream.

At first glance, the diversity of hormones, glands, and effects can seem almost overwhelming, especially considering that Figure 47.3 represents just a partial catalog for a single species. The picture is simplified somewhat, however, by recognizing that most animal hormones belong to one of just three major structural families: polypeptides, amino acid derivatives, or steroids. Let's take a closer look.

Chemical Characteristics of Hormones

The three major classes of chemicals that can act as hormones in animals are illustrated in **Figure 47.4**: (1) polypeptides, which are chains of amino acids linked by peptide bonds (see Chapter 3); (2) amino acid derivatives; and (3) steroids, which are a family of lipids distinguished by a four-ring structure (see

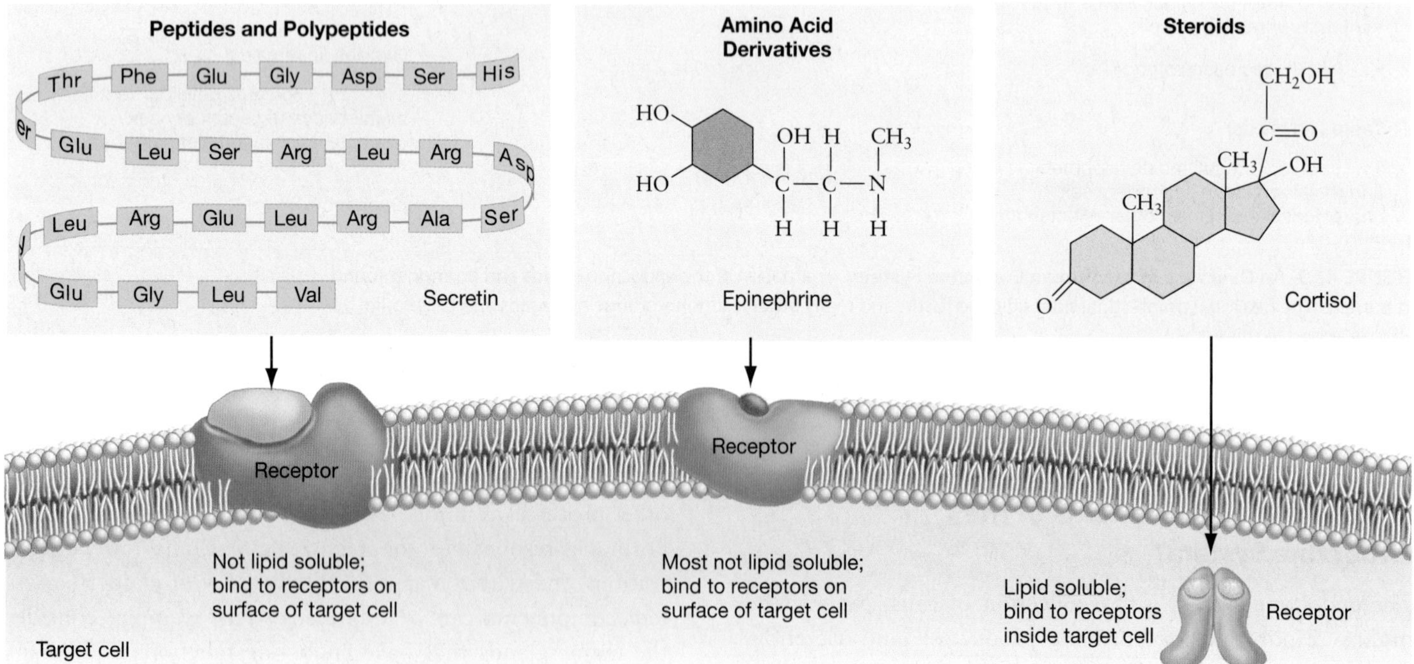

FIGURE 47.4 Most Animal Hormones Belong to One of Three Chemical Families.

QUESTION Secretin and epinephrine are not lipid soluble and do not cross plasma membranes readily. Cortisol, in contrast, is lipid soluble and diffuses through plasma membranes efficiently. How do the size and charge of these molecules explain the difference in lipid solubility?

Chapter 6). The secretin produced in the small intestine is a polypeptide; epinephrine is synthesized in the adrenal medulla from the amino acid tyrosine; and cortisol is synthesized in the adrenal cortex from the steroid cholesterol.

Polypeptide hormones, amino-acid–derived hormones, and steroid hormones have important similarities, as well as important differences. All animal hormones share several defining characteristics:

- All hormones are organic compounds.
- All hormones are secreted from the cells where they are synthesized.
- All hormones act on target cells remote from their point of origin.
- All hormones are present in extremely small concentrations yet have large effects.

To drive this last point home, consider work on **growth hormone** (**GH**), also known as somatotropin. Several researchers had noted that rats and other laboratory animals stopped growing when their pituitary glands were removed. Based on this observation, it was widely suspected that the pituitary produces a chemical signal that promotes cell division and other aspects of growth. A research group was able to purify a polypeptide from cow pituitary glands, inject the polypeptide into lab rats, and document rapidly accelerated growth. When the researchers injected 0.01 mg of the molecule each day for nine days into rats that lacked pituitary glands, the width of the growth plates in the leg bones increased by 50 percent, and the individuals gained an average of 10 g compared with rats that lacked a pituitary and did not receive the hormone treatment. Stated another way, the addition of a total of 0.09 mg of hormone led to a weight gain of 10,000 mg. Further, 1 kg of cow pituitary tissue yielded a mere 0.04 g of growth hormone. By weight, the hormone makes up just four one-thousandths of 1 percent of the cow pituitary.

Given that small amounts of polypeptide, amino-acid–derived and steroid hormones have large effects, how do the three types of hormones differ? The major difference is that steroids are lipid soluble, while polypeptides and most amino acid derivatives are not (see Figure 47.4). An important exception to this rule is thyroxine, which is produced by the thyroid gland. Thyroxine is derived from the amino acid tyrosine but is lipid soluble.

Differences in solubility are important because steroids and thyroxine cross plasma membranes much more readily than do other types of hormones. To affect a target cell, all polypeptides and most amino acid derivatives bind to a receptor on the cell surface. Lipid-soluble hormones, in contrast, can diffuse through the plasma membrane and bind to receptors inside the cell. Before exploring the consequences of this difference in more detail, let's consider how researchers discovered the array of hormones introduced here.

How Do Researchers Identify a Hormone?

Research on animal hormones began in earnest in the early 1900s with the discovery of **secretin**. You might recall from Chapter 43 that researchers who were investigating how secretions from the pancreas are regulated had cut all the nerves connected to the upper part of the small intestine and pancreas of a dog, to address the hypothesis that secretion was stimulated by electrical signals. The biologists then added a small amount of dilute hydrochloric acid (HCl) to the small intestine, mimicking the arrival of acidic material from the stomach. They found that the pancreas secreted compounds that buffered, or neutralized, the acid in the small intestine (**Figure 47.5**).

How could stimulating the small intestine lead to a response by the pancreas? The response could not have been mediated by the nervous system, since all the nerves to the pancreas had been cut. Instead, the biologists proposed that a chemical had traveled from the small intestine to the pancreas by way of the bloodstream to signal the arrival of acid. To test this hypothesis,

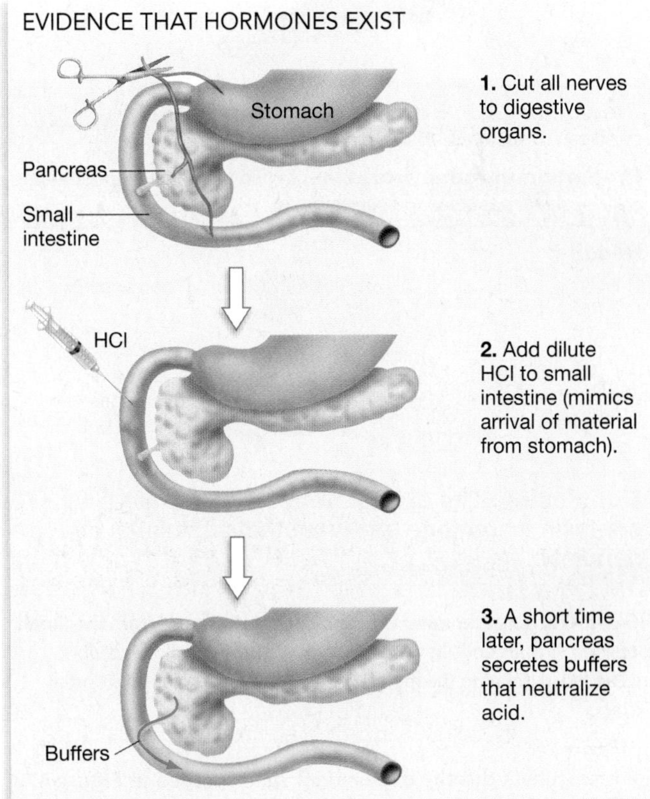

FIGURE 47.5 Experimental Evidence that a Hormone from the Small Intestine Signals Cells in the Pancreas, via Blood. The key assumption here was that a blood-borne signal was the only way for the small intestine to communicate with the pancreas.

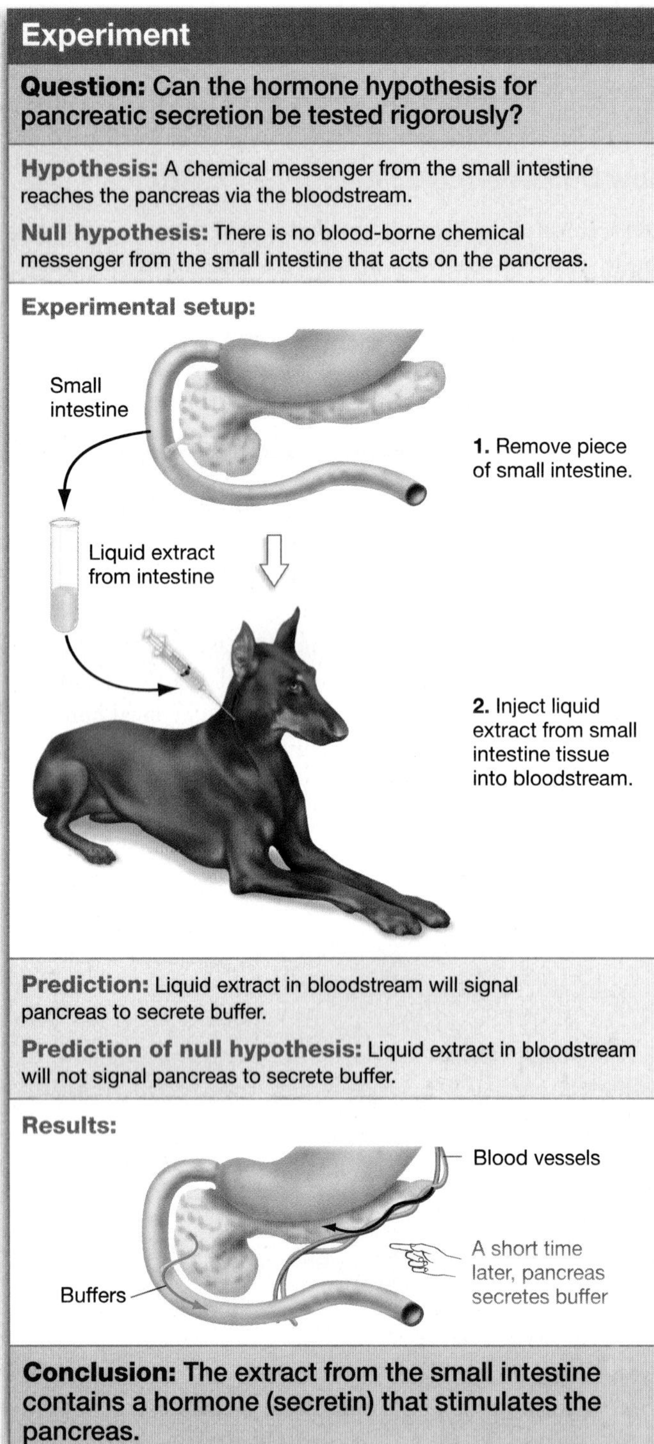

FIGURE 47.6 Experimental Evidence that a Hormone from the Small Intestine Signals Cells in the Pancreas, via Blood. Follow-up work succeeded in isolating the molecule secretin from a soluble intestinal extract.

the researchers did the experiment diagrammed in **Figure 47.6**. An extract from the small intestine was injected into blood vessels in a dog's neck. A short time later, the biologists observed that the pancreas secreted an alkaline solution. This was strong evidence that the extract from the small intestine contained a hormone. The molecule was later purified and named secretin.

In addition to using liquid extracts to identify hormones, biologists have focused on analyzing the results of experimental or natural destruction of organs. For example, the experimental removal of the adrenal glands in animals or damage to the adrenal glands in humans by hemorrhage, tumor, or infection rapidly leads to death associated with very low levels of blood sugar and blood pressure. The injection of adrenal extracts into the blood of animals lacking adrenal glands corrects these abnormalities. Results like these provided strong evidence that hormones secreted by the adrenal glands help regulate blood sugar concentrations and blood pressure.

Documenting an association between a particular gland or hormone and an effect in the body is just the first step in understanding hormone action, however. Researchers gained deeper insights into the role of chemical signaling when they were able to step back and consider how the array of hormones they were cataloging help animals stay alive and produce offspring.

47.2 What Do Hormones Do?

At the beginning of this chapter, you read that hormones are chemical messengers. If so, what do hormones "say"? A first step in answering this question is to recognize that even a single type of hormone can exert a variety of effects. Consider thyroxine, for example. In humans, thyroxine stimulates metabolism and thus oxygen consumption throughout the body. But it also promotes growth, increases the heart rate, and stimulates the synthesis of many important macromolecules. (**Box 47.1** explores the medical implications of abnormalities in the production of thyroxine.) A second step in grasping what hormones do is recognizing that several different types of hormones may affect the same aspect of physiology. For example, the blood-glucose concentration is regulated by insulin, glucagon, epinephrine, and cortisol.

Some hormones have extremely diverse effects, whereas functions of other hormones appear to overlap. These observations begin to make sense when hormone action is viewed in the context of the whole organism. Hormones coordinate the activities of cells in response to three general situations: (1) environmental challenges; (2) growth, development, and reproduction; and (3) homeostasis. Let's analyze each in turn.

How Do Hormones Coordinate Responses to Environmental Change?

The challenges or stimuli that hormones respond to can be simple or complex. Digestive hormones are a good example of how hormones function in simple stimulus-and-response circuits. When acidic food material passes from the stomach to the upper reaches of the small intestine, the food triggers the release of secretin and cholecystokinin by intestinal cells into the bloodstream. Recall that **secretin** induces the pancreas to secrete an alkaline solution that neutralizes acid. **Cholecystokinin** is a hormone that causes the pancreas to secrete digestive enzymes into the small intestine. Cholecystokinin also causes the gallbladder to eject bile salts into the intestine to emulsify fats. In this way, digestive hormones signal the arrival of food and regulate the release of molecules that aid digestion. But what about more complex environmental stimuli?

Short-Term Responses to Stress When a person is thrust into a dangerous or unpredictable situation, hormones are involved in both the short-term and long-term responses. The short-term reaction, called the **fight-or-flight response**, occurs in conjunction with the activation of the sympathetic nervous system, which was described in Chapter 45. Sympathetic nerves prepare organs for stressful situations. If you were being chased by a grizzly bear, action potentials from your sympathetic nerves would stimulate your adrenal medulla and lead to the release of **epinephrine**, also known as **adrenaline**. (The Greek word roots *epi* and *nephron* mean "top-kidney"; the Latin word roots *ad* and *renal* also mean "top-kidney.")

To understand the responses that occur, consider the experiment reported in **Figure 47.7**. To determine how epinephrine affects the body, human volunteers received injections of a saline solution or epinephrine. The graphs in the "Results" section of the figure document dramatic increases in the concentrations of free fatty acids and glucose in the blood following the injection of epinephrine. The data in the table confirm significant increases in pulse rate, blood pressure, and oxygen consumption by the brain. In addition, the volunteers reported strong subjective feelings of anxiety and excitement. Other experiments showed that epinephrine leads to dramatic changes in the distribution of blood. Specifically, the fight-or-flight response includes a redirection of blood toward the heart, brain, and muscles and away from the skin and digestive system. Epinephrine also relaxes smooth muscle and thereby opens blood vessels—increasing blood delivery to targeted tissues.

Taken together, the responses to epinephrine lead to a state of heightened alertness and increased energy use that prepares the body for rapid, intense action such as fighting or fleeing. By coordinating the activities of cells in many organs and systems throughout the body, epinephrine prepares an individual to cope with a life-threatening situation.

Experiment

Question: How does epinephrine affect the body?

Hypothesis: Epinephrine causes changes involved in the fight-or-flight response.

Null hypothesis: Epinephrine does not affect the fight-or-flight response.

Experimental setup:

Control (Saline solution)

Epinephrine

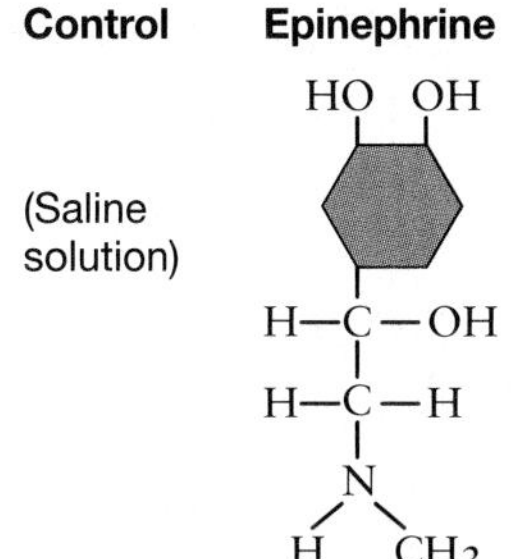

1. Inject human volunteers with saline solution or epinephrine.

2. Document changes in fatty-acid and glucose concentrations in blood, pulse rate, blood pressure, and oxygen consumption in brain.

Prediction: Epinephrine increases fatty-acid and glucose concentrations in blood, pulse rate, blood pressure, and brain oxygen consumption relative to controls.

Prediction of null hypothesis: No differences in physiological state of individuals based on molecule injected.

Results:

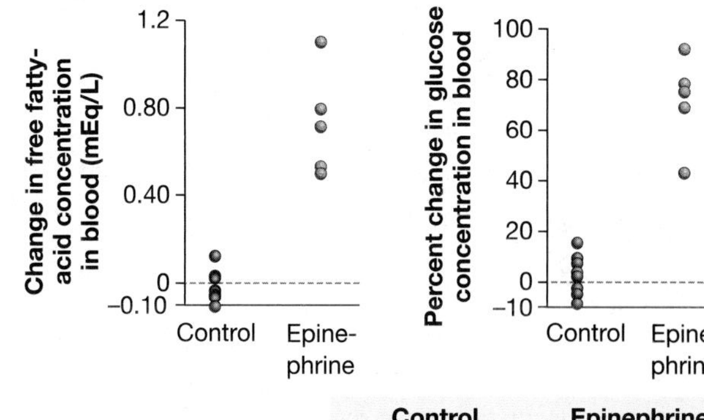

	Control	Epinephrine
Pulse rate (beats/min)	78.3	89.6
Blood pressure (average, mm Hg)	90.9	108.7
O_2 consumption in brain (cc O_2/100 gm blood/min)	3.41	4.16

Conclusion: Epinephrine causes an array of changes associated with the fight-or-flight response.

FIGURE 47.7 What Effect Does Epinephrine Have on the Body? Each data point on the graphs represents a human volunteer. The data in the table are average values from seven volunteers.

Long-Term Responses to Stress If you have ever been in an acutely stressful situation and experienced the fight-or-flight response, you may recall that the state is short lived. Once an epinephrine "rush" has worn off, most people feel exhausted and want to rest and eat.

What happens if the stress continues and turns into a long-term condition? In the course of a lifetime, it is not unusual for a person to experience periods of starvation or fasting, prolonged emotional distress, or chronic illness. How do hormones help humans and other animals cope with extended stress?

Early studies of long-term stress in human subjects suggested a role for the hormone **cortisol**, which is produced in the adrenal cortex. Increased levels of cortisol were found in

BOX 47.1 A Closer Look at Thyroxine and the Thyroid Gland

The thyroid hormone **thyroxine** is an amino-acid–derived hormone that is synthesized from tyrosine. Thyroxine is an unusual molecule, however, because it contains four iodine atoms. This feature inspired its alternate name, T_4. A closely related thyroid hormone called triiodothyronine, or T_3, contains three iodine atoms.

In mammals, T_3 is the more active of the two hormones. Although the thyroid releases much more T_4 than it does T_3, cells in the liver and elsewhere convert T_4 to T_3.

The primary effect of the thyroid hormones on target cells is to trigger increased cellular metabolism. People who produce inadequate amounts of thyroid hormones are lethargic and unable to tolerate cold. In contrast, people who produce excessive amounts of T_3 and T_4 are restless, excitable, and prone to mood swings but have limited energy reserves and tend to tire easily. Other medical conditions have been traced to problems with the iodine atoms incorporated into T_3 and T_4:

- When radioactive iodine atoms are released into the environment, such as during the accident at the Chernobyl nuclear power plant in the Ukraine, people living in the area become highly susceptible to thyroid cancers. This pattern occurs because radioactive iodine molecules are sequestered in the thyroid and because radiation can damage DNA in a way that makes cells cancerous.
- If iodine is deficient in the diet, the thyroid gland cannot produce enough T_3 and T_4. As a result, it gets overstimulated with the hormone that controls release of T_3 and T_4 and may swell to huge proportions in response. The swelling produces the condition called endemic **goiter**. Endemic goiter is now rare in many parts of the world because iodine-containing salt is available for consumption.

airplane pilots and crew members during long flights, athletes who were training for intense contests, the parents of children undergoing treatment for cancer, and college students who were preparing for final exams. What do these signals do?

The primary role of cortisol during prolonged stress in humans is to ensure the continued availability of glucose for use by the brain. This hormone maintains glucose production by inducing the synthesis of the liver enzymes that make glucose from amino acids and other chemical precursors. It also causes adipose tissue —fat tissue—and resting muscles to become resistant to the effects of insulin. When the body is in an unstressed state, insulin stimulates **adipocytes**, or fat cells, and resting muscle cells to remove glucose from the bloodstream. When cortisol makes these cells resistant to insulin, the effect is to reserve glucose for use by the brain and exercising muscles, neither of which depend on insulin to stimulate glucose uptake. The hormone also promotes the release of the body's major fuel molecules, fatty acids, from adipose tissue for use by the heart and muscles. Because of its importance in regulating blood glucose, cortisol is referred to as a **glucocorticoid**.

The long-term stress response comes at a high price, however, as any victim of a serious injury or illness knows. Glucocorticoids make amino acids available for glucose synthesis by promoting the degradation of contractile proteins in muscle. The resulting loss of muscle mass may cause severe weakness. Glucocorticoids also conserve glucose by impairing wound healing and suppressing immune and inflammatory responses. Although these processes are costly in terms of energy use, suppressing them makes the body more susceptible to infection. The overall concept here is that the long-term stress response is a compromise—a fitness trade-off (see Chapter 41). The fuel requirements of the brain are met at the expense of other tissues and organs.

It is instructive to note that glucocorticoids may mediate the response to stressful environmental challenges in species other than humans. Recall from Chapter 42 that when moving from freshwater to saltwater, a salmon gains sodium ions via diffusion and loses water across its gills by osmosis. The salmon counteracts these stresses by actively pumping sodium and chloride ions out of its gills. The chloride cells that perform this pumping are induced to proliferate by a burst of cortisol. Cortisol release occurs as the salmon moves downstream toward the ocean. If this release fails to occur, salmon die shortly after reaching salt water.

In summary, hormones help regulate the body's response to environmental change, as in the human body's response to short- and long-term stresses. Epinephrine and the glucocortocoids are examples of molecules that alert cells and tissues to a particular environmental stimulus or challenge. Other hormones have dramatically different effects, however.

How Do Hormones Direct Developmental Processes?

In animals, as in plants, hormones play a key role in regulating growth and development. Growth hormones and sex hormones, for example, play crucial roles in promoting cell division, increasing overall body size, and promoting sexual differentiation as an individual matures. In addition, certain hormones direct the development of particular cells and tissues at critical junctures in an individual's life. Which hormones are involved in regulating development, and what specific effects do they have?

Let's answer this question by exploring two of the most dramatic examples of hormonal control of development—

metamorphosis in amphibians and insects—and then surveying other developmental processes that are affected by hormone action.

T_3's Role in Amphibian Metamorphosis Frogs, toads, and salamanders are called amphibians ("double-lives") because in most species juveniles live in water while adults live on land. The process of changing from an immature, aquatic tadpole to a sexually mature, terrestrial frog or toad or salamander is an example of **metamorphosis** ("change-form"; **Figure 47.8**).

Two sets of complementary experiments, published in 1912 and 1916, established that frog metamorphosis depends on thyroid hormones. Researchers could induce frog tadpoles to undergo metamorphosis by feeding them ground-up thyroid glands from horses; they could prevent metamorphosis by surgically removing the tadpoles' thyroid glands. Follow-up work showed that the thyroid hormone **triiodothyronine**, or T_3, is responsible for most of the changes observed in metamorphosis, and that T_3 is produced in response to signals from the brain—in the form of thyroid-stimulating hormone produced in the pituitary gland.

In juvenile amphibians, cells respond to increased levels of T_3 in one of three ways:

1. They may begin growing and form new structures, such as legs.
2. They may die, as in cells that form the tadpole's tail.
3. Or, they may change structure and function. For example, changes in existing cells are responsible for the switch from a long intestine specialized for digesting plant material to a short intestine specialized for digesting insects and prey. In the liver, cells respond to T_3 by manufacturing the enzymes required to excrete urea instead of ammonia.

● You should be able to suggest a hypothesis explaining why different cells can respond to the same hormone in different ways.

Some observers would argue that T_3's role in amphibian metamorphosis is the most spectacular example of how hormones affect development and how a tiny concentration of a cell-to-cell signal can trigger dramatic changes in animals. Others would claim instead that frog T_3 pales in comparison to the role that ecdysone plays in insect metamorphosis.

Hormone Interactions Regulate Insect Metamorphosis
Chapter 32 introduced the two major patterns of juvenile-to-adult transition in insects, called hemimetabolous and holometabolous metamorphosis. You might recall that in hemimetabolous metamorphosis, juveniles are called nymphs. They look like miniature adults, live in similar habitats, and eat similar food. But in species that undergo holometabolous metamorphosis, juveniles are called larvae. They look completely different from adults, live in different habitats, and eat different food.

Larvae undergo a series of molts in which they shed their old exoskeleton, expand their bodies, and produce a new exoskeleton. After a specific number of these juvenile molts, however, they secrete a tough case called a pupa. Inside the pupal case, distinctive populations of larval cells give rise to a

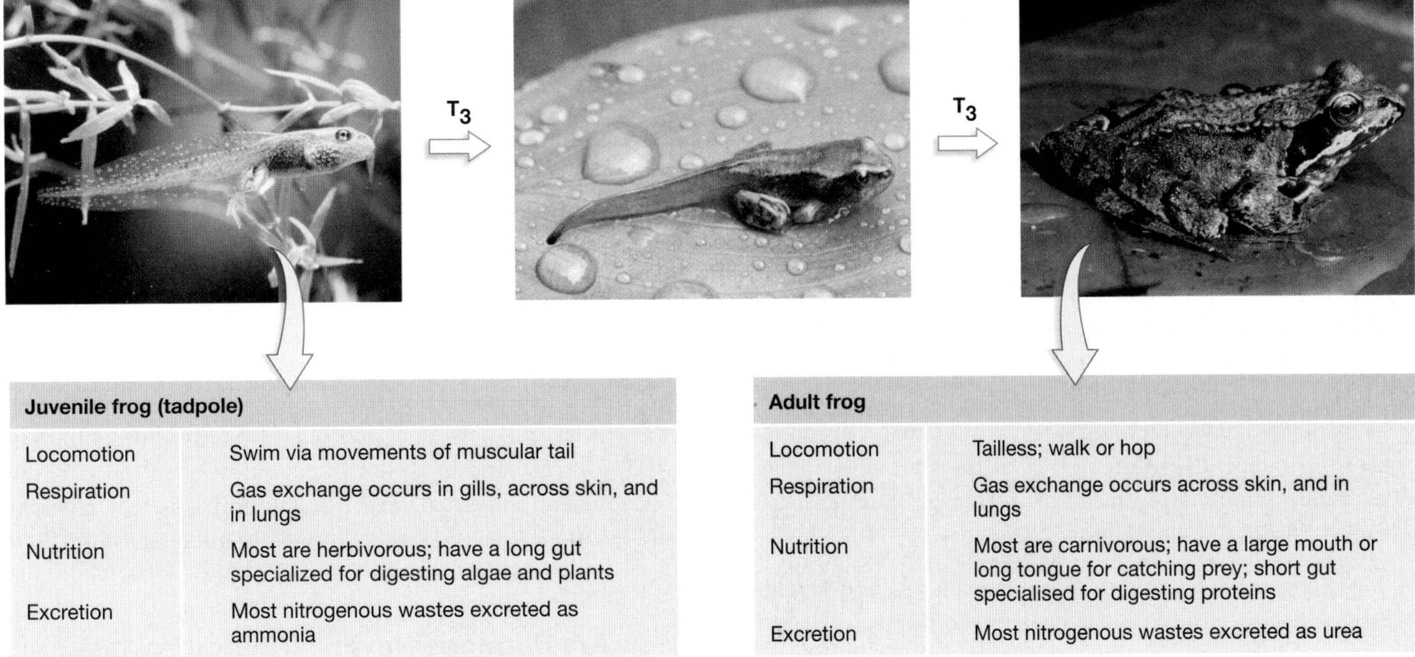

Juvenile frog (tadpole)	
Locomotion	Swim via movements of muscular tail
Respiration	Gas exchange occurs in gills, across skin, and in lungs
Nutrition	Most are herbivorous; have a long gut specialized for digesting algae and plants
Excretion	Most nitrogenous wastes excreted as ammonia

Adult frog	
Locomotion	Tailless; walk or hop
Respiration	Gas exchange occurs across skin, and in lungs
Nutrition	Most are carnivorous; have a large mouth or long tongue for catching prey; short gut specialised for digesting proteins
Excretion	Most nitrogenous wastes excreted as urea

FIGURE 47.8 Amphibian Metamorphosis Is a Continuous Process. When metamorphosis begins in a frog, toad, or salamander, the individual stays active and feeding. The transition from juvenile to adult is continuous and gradual.

FIGURE 47.9 Insect Metamorphosis Occurs during a Resting Stage. When metamorphosis begins in a holometabolous insect, the individual enters a resting stage called the pupa.

completely new adult body. The rest of the larval body is torn down (**Figure 47.9**).

Like amphibian metamorphosis, insect metamorphosis is triggered by a hormone from the brain. In insects metamorphosis depends on interactions between two hormones. If **juvenile hormone** (**JH**) is present at a high concentration in the larva, then surges of the hormone 20-hydroxyecdysone, usually called **ecdysone**, induce the growth of a juvenile insect via molting. But if JH levels are low, ecdysone triggers metamorphosis and the transition to adulthood and sexual maturity.

Other Examples of How Hormones Regulate Development Amphibian and frog metamorphosis provide particularly spectacular examples of how hormones direct animal development. But long distance cell-to-cell signals play an array of key roles as embryos develop—frequently directing changes that occur throughout life.

- *Early embryonic development* Events early in development dictate whether the sex organs, or **gonads**, of an embryo become male (testes) or female (ovaries). This process is called primary sex determination. In mammals it is not dependent on hormone action. Once testes or ovaries develop, though, they begin producing male-specific or female-specific hormones. In humans, the early testes produce two hormones: A steroid hormone called **testosterone** induces early development of the male reproductive tract, and a polypeptide hormone called **Müllerian inhibitory substance** inhibits development of the female reproductive tract. Development of the female reproductive tract depends on the action of the steroid hormone **estradiol**, which is produced in the ovaries and is in the family of molecules called **estrogens**.
- *Juvenile-to-adult transition* When humans reach early adolescence, surges of sex hormones lead to the physical and emotional changes associated with **puberty**. These developmental changes create the adult phenotype and the ability to produce offspring. In boys, surges of sex hormones lead to changes that include enlargement of the penis and testes and growth of facial and body hair. In girls, increased concentrations of estradiol lead to the enlargement of breasts, the onset of menstruation, and other changes.
- *Attaining full adult stature* In humans and other mammals, the long bones in the limbs and the vertebrae of the spinal column must grow for full adult height to be achieved. Growth hormone produced in the pituitary gland regulates growth factors, which are signals that control the cell cycle (see Chapter 11). Puberty is associated with a growth spurt because the effect of growth hormone on the skeleton is enhanced by the action of sex hormones that are produced in increased amounts during adolescence. Even though growth hormone and sex hormones are produced in low levels throughout life, bone growth stops when sex hormone concentrations fall at the end of puberty.
- *Seasonal or cyclical sexual activity* Most long-lived animals reproduce seasonally. In many species, environmental cues such as increasing day length, warmth, or the onset of seasonal rains trigger the release of sex hormones. Chapter 51 details how this flush of testosterone or estrogen induces the development of seasonal traits, such as sexual receptivity in female lizards. Even though humans do not breed seasonally, sex hormones are instrumental in regulating sperm production and the menstrual cycle. Chapter 48 explores these processes in detail.

To summarize, hormones play key roles in growth and development. Growth hormones support overall size increases in juveniles. Sex hormones coordinate the activities of diverse groups of cells and tissues at three critical life stages: (1) the development of sexual structures in embryos, (2) physical and emotional changes at the juvenile-to-adult transition, and (3) the regulation of sexual activity in adults.

How Are Hormones Involved in Homeostasis?

Chapter 41 introduced the concept of homeostasis, the maintenance of relatively constant physical and chemical conditions inside the body. Recall that homeostatic systems depend on

(1) a sensory receptor that monitors conditions relative to a preferred value, or set point; (2) an integrator that processes information from the sensor; and (3) effector cells that return conditions to the set point. (Glance back at Figure 41.14.) In homeostatic systems, messages often travel from integrators to effectors in the form of hormones.

Several hormones that act as messengers in homeostatic systems have already been introduced. Recall from Chapter 42 that when an individual is dehydrated, **antidiuretic hormone** (**ADH**, also called vasopressin) is synthesized in the hypothalamus and released from the pituitary gland. ADH increases the permeability of the kidney's collecting ducts to water. As a result, water is reabsorbed from the urine and saved. In this way, ADH is instrumental in achieving homeostasis with respect to water balance. If ADH's action is inhibited, homeostasis fails and illness can occur. For example, the ethanol in alcoholic beverages inhibits the release of ADH from the pituitary. People who imbibe large quantities of these beverages produce large quantities of dilute urine. The resulting water loss can lead to dehydration and nausea—symptoms associated with an alcoholic hangover.

Chapter 42 also noted that aldosterone is released from the adrenal cortex when ion concentrations in body fluids are low. Because aldosterone increases reabsorption of in the distal tubules of the kidney, it plays a key role in homeostasis with respect to electrolyte concentrations and the overall volume of body fluids. Adrenal hormones with this effect are called **mineralocorticoids**.

Several of the hormones introduced in Figure 47.3 are involved in homeostatic systems. For example, **erythropoietin** (**EPO**) is released from the kidneys and other tissues when blood oxygen levels are low. EPO acts to achieve homeostasis with respect to oxygen by stimulating the production of red blood cells. The more red blood cells, the higher the oxygen-carrying capacity of blood is. In this way, EPO is a crucial element in the homeostatic system for blood oxygen levels.

Figure 47.10 provides a more detailed example by diagramming how the homeostatic system for blood calcium concentration works. The hormonally active form of Vitamin D is synthesized in the kidneys, and **parathyroid hormone** (**PTH**) is produced by the parathyroid glands. These two hormones work in tandem to keep Ca^{2+} levels in the blood close to the set

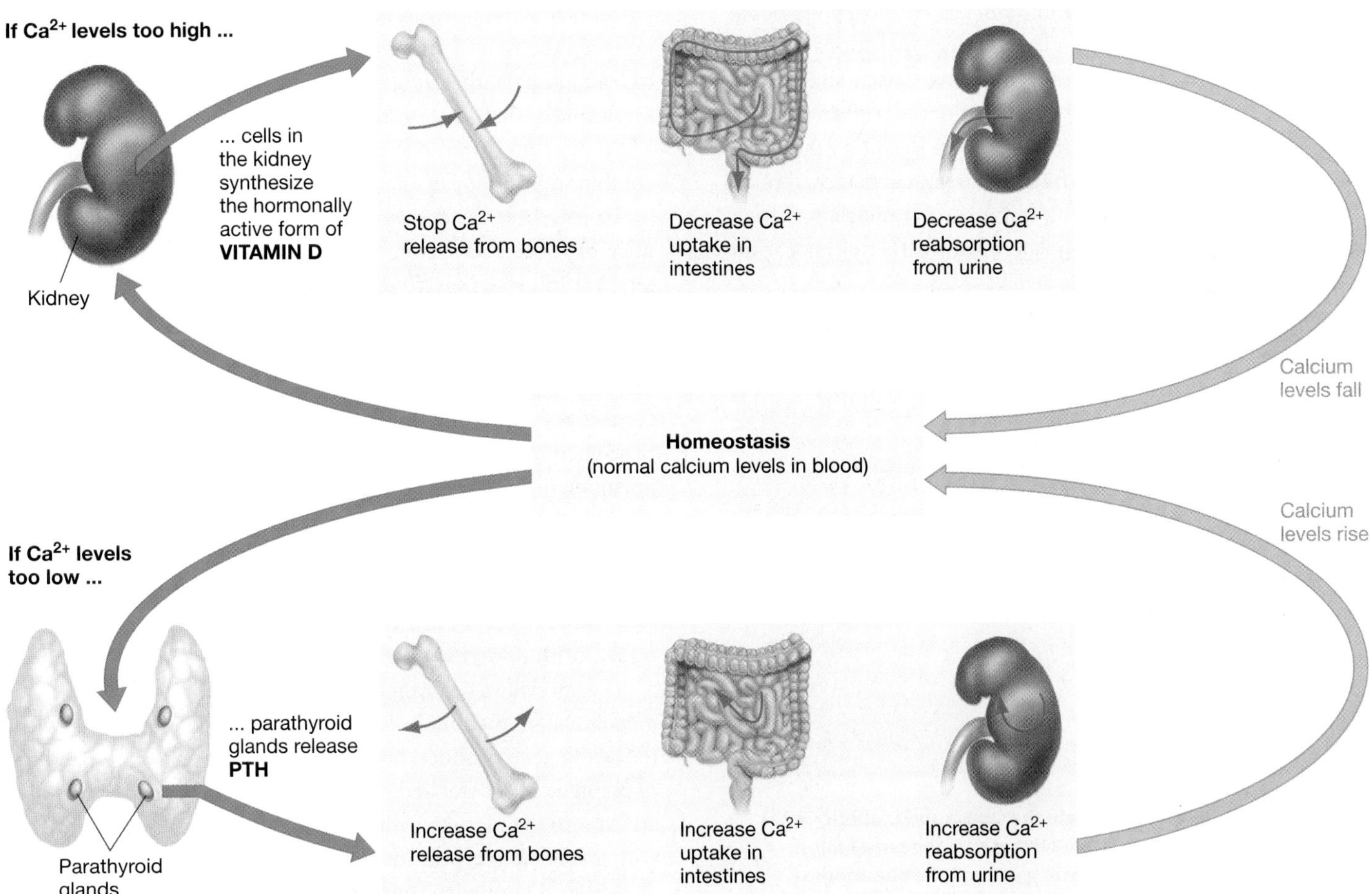

FIGURE 47.10 Hormones Act to Maintain Homeostasis. Parafollicular cells of the thyroid gland and the cells of the four adjacent parathyroid glands sense Ca^{2+} levels and secrete hormones in response to changes. Notice that this homeostatic system does not involve the CNS.

EXERCISE Draw a similar diagram to illustrate the action of EPO, ADH, aldosterone, or glucagon and insulin. (Hint: Look back at Figure 47.3.)

point, much as insulin and glucagon interact to maintain blood glucose concentrations at preferred levels (see Chapter 43).

To drive home how important hormones are in homeostatic processes, consider recent results on homeostasis in energy reserves. This research led to the discovery of new hormones and inspired research on the causes of obesity in humans.

Homeostasis of the Body's Energy Reserve Healthy animals keep energy in reserve for use during periods of decreased food availability. This energy reserve is typically in the form of the lipid called triglyceride. **Triglyceride** is an effective energy-storage molecule, because large amounts of ATP can be generated when its three fatty-acid subunits are oxidized (see Chapter 6). Although some triglyceride is present in muscle cells, most is stored in adipocytes. These cells make up the fat bodies found in insects and other species and the adipose tissue of mammals. Even a lean 70-kg human male stores enough energy in adipose tissue to survive for over 30 days without eating.

For decades researchers thought that total adipose tissue mass varied only in response to food intake—meaning it was not subject to homeostatic regulation. This view started to change in the 1970s, when biologists began studying mutations in mouse genes called *obese* and *diabetic*. Homozygous *obese* (*ob/ob*) and *diabetic* (*db/db*) mice eat large quantities of food and move much less than heterozygous or wild-type siblings do. The homozygotes also become extremely obese (**Figure 47.11**).

To test the hypothesis that the *obese* and *diabetic* gene products are involved in endocrine function, researchers turned to an experimental technique called parabiosis. **Parabiosis** has been used for more than 100 years to determine whether physiological

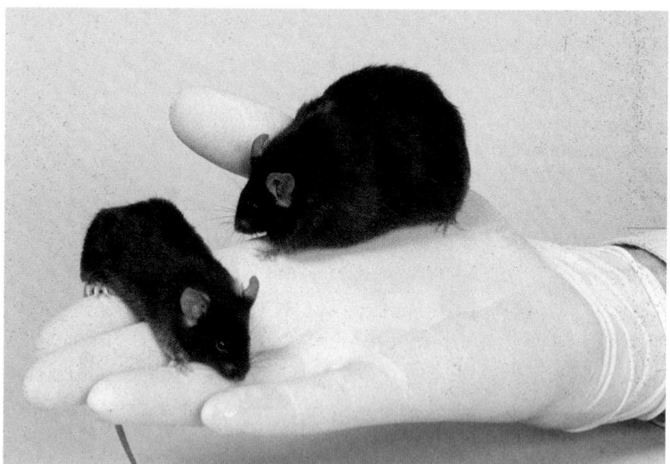

FIGURE 47.11 In Mice, Mutations in the Obese and Diabetic Genes Can Cause Obesity. These mice are siblings from the same litter. At the *ob* gene, the lean mouse is heterozygous and the obese mouse is homozygous. The phenotypes of *ob/ob* and *db/db* mice are indistinguishable.

● **QUESTION** Suppose a mouse were doubly homozygous *(ob/ob db/db)*. Would its phenotype be the same as or different from singly homozygous individuals?

phenomena are hormone mediated. In this technique, two closely related animals are surgically united by suturing the pelvis, shoulders, and abdominal walls together. The skin of the two animals is then joined over the surgical connection, and the newly created twin animal is allowed to recover from the procedure. Within a short time, capillaries form between the two parabiotic partners, and they develop a shared circulatory system. This circulatory system permits the passage of long-lived molecules, such as certain hormones, between the two partners. In contrast, it does not allow the exchange of molecules that are rapidly metabolized, such as glucose and fatty acids. Another key observation is that no new nerves grow between the two animals. As a result, they can influence one another through endocrine signals but not through electrical signals. If an experimental manipulation of one partner produces an effect in the other partner, biologists conclude that the effect is mediated by a hormone.

Researchers performed parabiosis between *db/db*, *ob/ob*, and lean mice as shown in **Figure 47.12**. When a *db/db* animal was joined to either a lean animal or an *ob/ob* animal, the *db/db* mouse continued to eat and grow normally, but its partner stopped eating, lost weight, and eventually died of apparent starvation. But when two *db/db* animals were joined, both partners ate and grew as expected, and there was no difference between the two in survival. When *ob/ob* animals were joined to lean animals, the *ob/ob* mice ate less food and gained weight less rapidly than did *ob/ob* mice joined to other *ob/ob* mice. There was no difference in survival of pairs representing any combination of *ob/ob* and lean mice.

To interpret these results, the biologists hypothesized that mice produce a satiation, or "stop-eating," hormone. The interpretation was that *db/db* mice lack the receptor required for the hormone to affect target cells. As *db/db* mice got fatter and fatter, they would produce more and more hormone—to no avail. This model explained why the signal from *db/db* mice greatly reduced food intake in their parabiotic partners but continued to have no effect on the *db/db* mouse. The satiation hormone itself was postulated to be encoded by the *obese* gene. As a result, *ob/ob* mice do not produce the hormone. They respond to the signal if it is available from a normal partner, however. A key idea here is that the two genotypes produce the same phenotype because they disable different parts of the same hormone-signaling system.

This model was a dramatic break with traditional views of fat metabolism in mammals. It was not confirmed until 1994, when the *obese* gene product, now called leptin, was isolated and the *obese* gene characterized. **Leptin** is a polypeptide hormone secreted into the blood by adipocytes. Leptin signals by interacting with a specific receptor located in many tissues, including areas of the brain known to control feeding behavior. As the model in Figure 47.12 predicts, the administration of leptin corrects the obesity of *ob/ob* (leptin-deficient) mice but not of *db/db* (leptin receptor–deficient) mice. But as **Box 47.2** explains, leptin injections are not helpful for the vast majority of obese humans.

Experiment

Question: Are the *diabetic* (*db*) and *obese* (*ob*) gene products involved in endocrine function?

Hypothesis: The gene products are involved in hormonal signaling that affects appetite and activity level.

Null hypothesis: The gene products are not involved in hormonal signaling.

Experimental setup:

Perform parabiosis surgery on closely related mice with different genotypes, so that blood-borne products will pass between the mice in each pair.

Prediction: Phenotypes of at least some of the parabiotic mice will change.

Prediction of null hypothesis: No changes in phenotypes of parabiotic mice will occur.

Results:

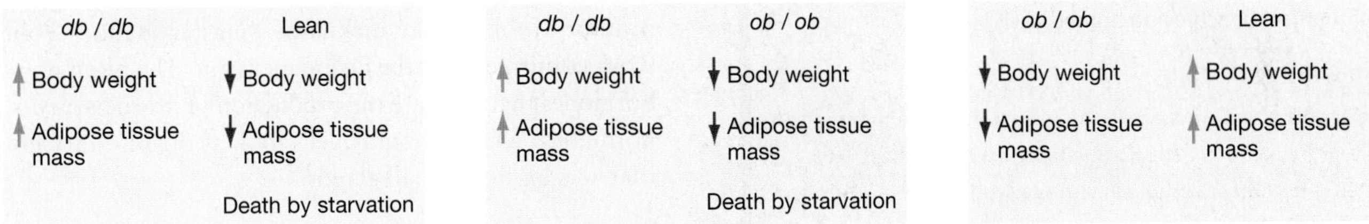

Conclusion: The *diabetic* and *obese* gene products are involved in hormonal signaling that affects appetite and activity level.

Interpretation: The *obese* gene codes for a hormone that reduces appetite and increases activity in response to increases in fatness. The *diabetic* gene codes for the receptor for this hormone.

FIGURE 47.12 Parabiosis of Genetically Obese and Lean Mice Provides Evidence of a "Satiation Hormone."

BOX 47.2 Is Human Obesity Caused by Leptin Deficiency?

Immediately after leptin was discovered, researchers began testing the hypothesis that it could be used as a drug to treat human obesity. Unfortunately, it quickly became apparent that almost all obese individuals already have high leptin levels—consistent with their high adipose tissue mass. Administering more leptin to these individuals did not cause meaningful weight loss.

Worldwide, fewer than 20 humans with true genetic leptin deficiency have been identified. Just like *ob/ob* mice, these individuals eat vast quantities of food and are severely obese. Because these rare individuals lack the hormone, their obesity can be cured by administering leptin.

To explain why the vast majority of obese humans do not respond to their high leptin levels, researchers hypothesize that the hormone's effect on the brain is diminished for some reason. Identifying the factor or factors involved in diminishing the normal response to leptin is currently the subject of intense research. Does the response drop when specific types of food are ingested, or when meal portions get large, or when exercise is inadequate? One intriguing finding is that mice placed on a high-fat diet respond much less strongly to experimental leptin administration than do mice fed a normal diet. This result supports the hypothesis that the increased ingestion of fatty foods disrupts the normal human homeostatic system in some way. If so, then increased fat content in the diet not only increases energy intake but also blocks the normal appetite-diminishing response, leading to overeating and obesity.

Follow-up work has shown that leptin levels in the blood vary in proportion to total adipose tissue mass. When adipose mass falls below a set point, the leptin level in the blood also falls. The decrease in leptin level is sensed by the brain, which generates both an increase in appetite and a decrease in energy expenditure. These responses promote eating and restore energy balance. When sufficient food intake has occurred to restore triglyceride stores, leptin levels rise. The result is diminished appetite, increased energy expenditure, and the stabilization of adipose tissue mass. This is a striking example of homeostasis achieved by the process called feedback inhibition.

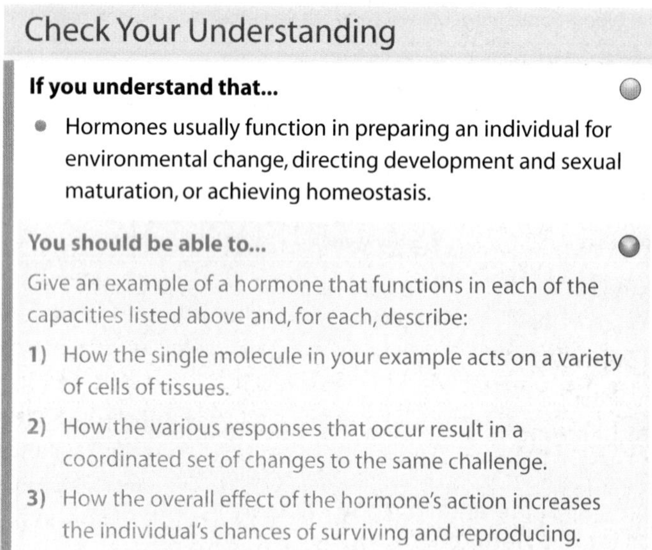

Check Your Understanding

If you understand that...

- Hormones usually function in preparing an individual for environmental change, directing development and sexual maturation, or achieving homeostasis.

You should be able to...

Give an example of a hormone that functions in each of the capacities listed above and, for each, describe:

1) How the single molecule in your example acts on a variety of cells of tissues.
2) How the various responses that occur result in a coordinated set of changes to the same challenge.
3) How the overall effect of the hormone's action increases the individual's chances of surviving and reproducing.

47.3 How Is the Production of Hormones Regulated?

Most hormones are released in response to either an environmental cue or a message from an integrator in a homeostatic system. In both cases, the nervous system may be closely involved. For example, environmental cues that signal the onset of the breeding season or the presence of a predator are received by sensory receptors and interpreted by the brain. Similarly, integration in most homeostatic systems occurs in groups of neurons in the central nervous system (CNS)—that is, the brain and spinal cord (see Chapter 45).

Based on these observations, the short answer to the question posed in the title of this section is simple: In many cases, hormone production is directly or indirectly controlled by the nervous system. How, then, does the nervous system direct the endocrine system? To explore this question, let's focus on how cortisol and epinephrine are regulated by the hypothalamus and pituitary gland.

The Hypothalamus and Pituitary Gland

Researchers who were interested in how hormone production is controlled quickly focused on the pituitary gland. As **Figure 47.13** shows, the pituitary is located at the base of the brain and is directly connected to the brain region called the hypothalamus. This physical link between the hypothalamus and pituitary is the basis of the connection between the CNS and the endocrine system. The pituitary has two distinct segments: the **anterior pituitary** and the **posterior pituitary**. In 1930 a biologist showed that laboratory rats exhibit a variety of debilitating symptoms when their pituitary is removed. The animals stop growing and cannot maintain a normal body temperature. In addition, their genitals, thyroid glands, and adrenal cortexes atrophy (shrink). Not surprisingly, their life span is also dramatically shortened.

These experiments suggested that, in addition to secreting growth hormone, the pituitary secretes substances affecting the gonads, thyroid, and adrenals. Based on this observation, the pituitary took on the nickname "master gland." Follow-up work confirmed that the nickname is apt. The pituitary secretes hormones that regulate the production of a wide array of other hormones. As a case study, let's look at the pituitary hormone that acts on the adrenal glands.

Controlling the Release of Glucocorticoids Early work on rats suggested that a molecule from the pituitary gland affects the adrenal gland. This molecule came to be called **adrenocorticotropic hormone**, or **ACTH**. (*Adreno* refers to the

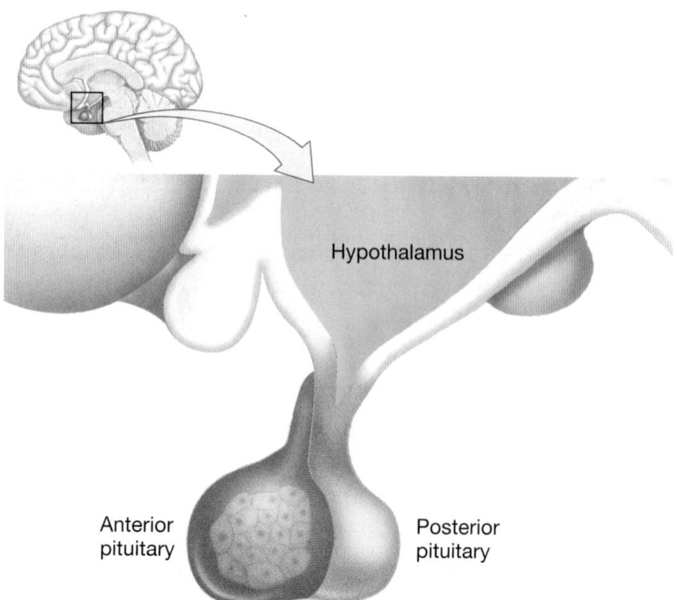

FIGURE 47.13 The Hypothalamus and the Pituitary Gland Are Physically Connected. The pituitary has two distinct segments, designated the anterior and posterior pituitary.

(a) Results of injecting ACTH into human volunteers

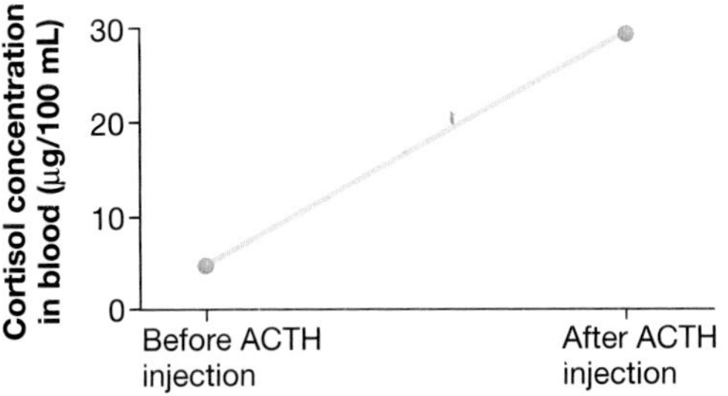

(b) Results of injecting cortisol into human volunteers

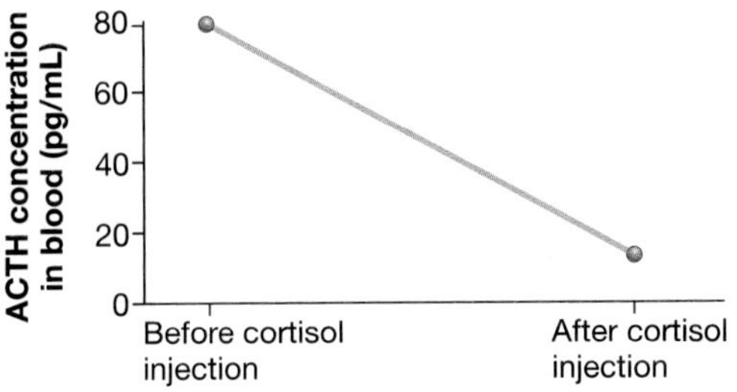

(c) Feedback inhibition by cortisol on ACTH

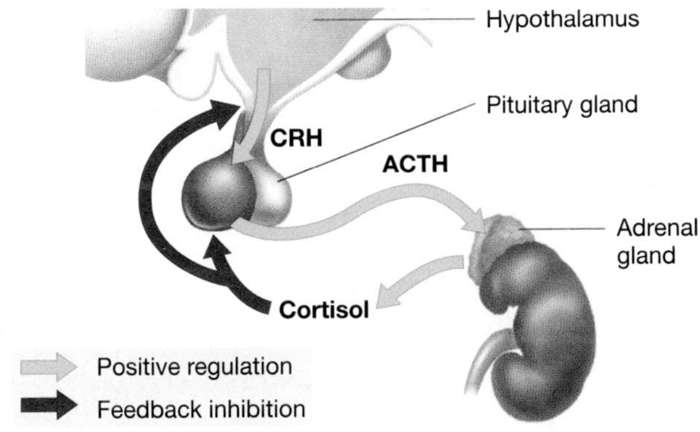

FIGURE 47.14 Feedback Inhibition by Cortisol on ACTH Release. **(a)** ACTH stimulates release of cortisol. **(b)** Cortisol reduces ACTH. **(c)** The interaction between cortisol, ACTH, and CRH is an example of feedback inhibition.

EXERCISE How would you use the data in part (b) to devise a test for adrenal failure in humans?

adrenal glands; *cortico* refers to the outer portion, or cortex, of the gland; and *tropic* means "affecting the activity of.") In 1943 ACTH was purified and characterized. The biologists who did the work confirmed that glucocorticoids are secreted from the adrenal cortex in response to ACTH released from the pituitary. As **Figure 47.14a** shows, when human volunteers are injected with ACTH, cortisol levels in their blood rise. Thus, ACTH is a regulatory hormone.

Not long after ACTH was isolated, biologists from two laboratories simultaneously and independently showed that it is released in response to a molecule produced by the hypothalamus. Researchers initially called this hormone corticotropin-releasing factor, or CRF. After years of effort, a different team of researchers finally succeeded in purifying CRF. It is a peptide—just 41 amino acids long—that has been renamed **corticotropin-releasing hormone (CRH)**. When CRH is released from the hypothalamus, cells in the anterior pituitary are stimulated to secrete ACTH into the bloodstream.

Why is glucocorticoid secretion controlled by the two hormones CRH and ACTH? The key to answering this question is to recognize that glucocorticoids themselves are able to suppress ACTH production by the pituitary gland. As the data in **Figure 47.14b** show, when human volunteers are injected with cortisol, ACTH levels in their bloodstream drop dramatically. Cortisol also inhibits release of CRH from the hypothalamus. Thus, if glucocorticoid levels become too high, then ACTH levels fall. The result is feedback inhibition: glucocorticoid production decreases and levels of glucocorticoid in the blood return to normal. But if glucocorticoid levels become too low, then ACTH levels rise and drive a compensatory increase in glucocorticoid production.

Figure 47.14c summarizes the relationships among CRH, ACTH, and the glucocorticoid cortisol. The importance of feedback inhibition in their interactions can best be appreciated by observing what happens when it fails. Certain pituitary tumors diminish the ability of cortisol to suppress ACTH production. This results in a medical condition called **Cushing's disease**, which is characterized by persistently high blood ACTH and cortisol levels. In essence, Cushing's disease is an unrelenting stress response that results in progressive depletion of the body's protein reserves. It is fatal if not treated.

If the feedback relationship between cortisol and ACTH stabilizes blood cortisol levels, what is the role of CRH? In essence, CRH changes cortisol's ability to suppress ACTH production in the pituitary. Increasing the production of CRH by the hypothalamus is equivalent to raising the temperature setting on a thermostat—it makes things hotter. Under ordinary circumstances, the production of CRH by the hypothalamus displays a daily rhythm, with the highest level in the early morning hours. This drives a corresponding daily rhythm in ACTH production and blood cortisol level, as shown in **Figure 47.15.** The peak in blood cortisol levels that occurs in the morning ordinarily coincides with arousal and initiation of the day's activities—saving glucose for use by the brain. As an aside, the unpleasant symptoms of jet lag are due to the daily cortisol rhythm being out of synchrony with local time for a period of several days after you arrive in a new time zone.

Based on this observation, it should make sense to learn that the long-term stress response is initiated when stimuli that produce pain or anxiety are processed by the brain and result in a

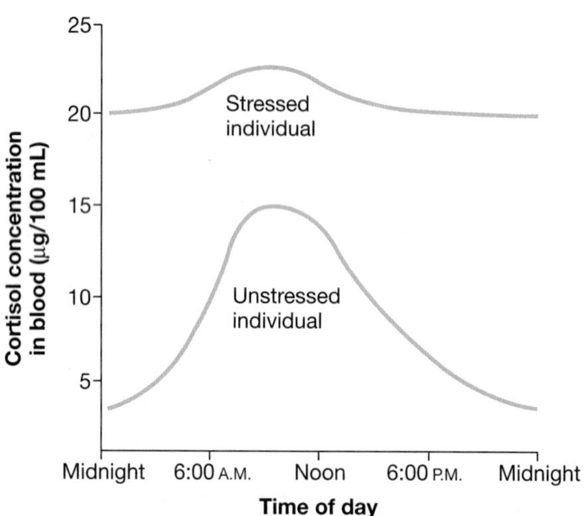

FIGURE 47.15 Blood Cortisol Levels Have a Daily Rhythm. Blood cortisol levels are usually highest in the early morning and drop in the late afternoon and evening. Under stressful circumstances, blood cortisol levels are elevated throughout the day.

sustained increase in CRH production. Increased CRH production causes blood ACTH and cortisol levels to remain higher throughout the day than they are in the unstressed state (see Figure 47.15).

The Hypothalamic-Pituitary Axis—an Overview Six decades of research have revealed that the CRH-ACTH-glucocorticoid relationship is just one example in an array of hormone systems based on interactions among the hypothalamus, pituitary, and target glands or cells. As **Figure 47.16** illustrates, the **hypothalamic-pituitary axis** actually forms two anatomically distinct systems. The anterior pituitary develops from cells in an embryo's mouth and throat lining, while the posterior pituitary is an extension of the brain.

The posterior and anterior sections of the pituitary gland are each influenced by different populations of neurons in the hypothalamus. These populations of hypothalamic neurons are called **neurosecretory cells**, because they are neurons that

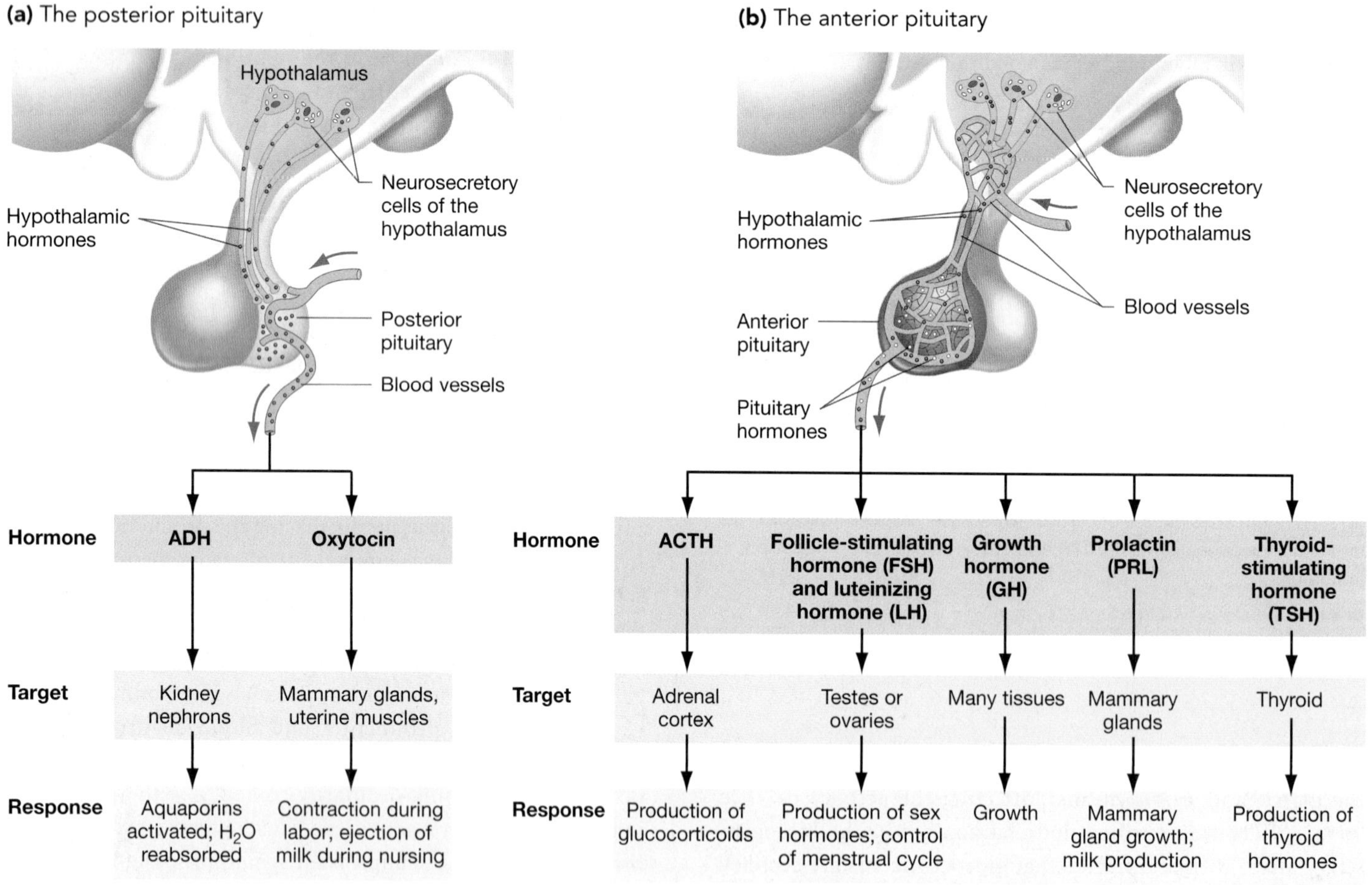

(a) The posterior pituitary

Hypothalamus

Hypothalamic hormones

Neurosecretory cells of the hypothalamus

Posterior pituitary

Blood vessels

Hormone	**ADH**	**Oxytocin**
Target	Kidney nephrons	Mammary glands, uterine muscles
Response	Aquaporins activated; H_2O reabsorbed	Contraction during labor; ejection of milk during nursing

(b) The anterior pituitary

Hypothalamic hormones

Neurosecretory cells of the hypothalamus

Blood vessels

Anterior pituitary

Pituitary hormones

Hormone	**ACTH**	**Follicle-stimulating hormone (FSH) and luteinizing hormone (LH)**	**Growth hormone (GH)**	**Prolactin (PRL)**	**Thyroid-stimulating hormone (TSH)**
Target	Adrenal cortex	Testes or ovaries	Many tissues	Mammary glands	Thyroid
Response	Production of glucocorticoids	Production of sex hormones; control of menstrual cycle	Growth	Mammary gland growth; milk production	Production of thyroid hormones

FIGURE 47.16 The Hypothalamus and Pituitary Interact Closely. (a) Developmentally and anatomically, the posterior pituitary is an extension of the hypothalamus. Neurosecretory cells in the hypothalamus extend directly into the posterior pituitary and secrete ADH (vasopressin) and oxytocin. **(b)** The hypothalamus and the anterior pituitary communicate indirectly, via blood vessels. Hormones produced by neurosecretory cells in the hypothalamus travel to the anterior pituitary in blood, where they control the release of pituitary hormones.

QUESTION Which of these hormones act via the neuroendocrine pathway? Which act via the neuroendocrine-to-endocrine pathway?

synthesize and release hormones. The release of hormones by cells in the hypothalamus is under the control of brain regions responsible for integrating information about the external or internal environment. For example, information about the approach of an important exam or athletic contest might trigger action potentials that lead to the release of CRH from neurosecretory cells in the hypothalamus.

It's important to recognize that the anterior pituitary and posterior pituitary function in different ways. As Figure 47.16a indicates, the posterior portion of the pituitary is actually an extension of the hypothalamus itself. Neurosecretory cells that project from the hypothalamus produce the hormones ADH and oxytocin, which are then stored in the posterior pituitary. From there, ADH and oxytocin are released into the bloodstream. This is an example of the neuroendocrine pathway of hormone action. Recall that ADH aids in the reabsorption of water by the kidneys. **Oxytocin** helps induce labor and milk production in females.

Neurosecretory cells in the hypothalamus do not project into the anterior portion of the pituitary, however. As Figure 47.16b illustrates, the hypothalamus and anterior pituitary are connected indirectly, by blood vessels. What happens when neurosecretory cells from the hypothalamus secrete hormones into these blood vessels? In response to the arrival of stimulatory releasing hormones from the hypothalamus, including CRH and others listed in Figure 47.3, the anterior pituitary secretes hormones that enter the bloodstream and act on target tissues or glands. This is an example of the neuroendocrine-to-endocrine pathway of hormone action.

The anterior pituitary hormones include ACTH; **follicle-stimulating hormone** (**FSH**) and **luteinizing hormone** (**LH**), which are involved in the production of sex hormones and which regulate the menstrual cycle; GH; **prolactin**, which stimulates mammary gland growth and milk production; and **thyroid-stimulating hormone** (**TSH**), which triggers the production of thyroid hormones. Hypothalamic inhibitory hormones, in contrast, stop the secretion of certain pituitary hormones. These pituitary hormones include somatostatin (also produced by the pancreas), which inhibits GH production, and **dopamine**, or prolactin-inhibiting hormone (PIH), which limits prolactin production.

In several cases, the production of a pituitary hormone is controlled by both a stimulatory and an inhibitory hypothalamic hormone. The realization that many of the hormones produced by the anterior pituitary stimulate the production of other hormones justified the structure's designation as the master gland.

MB Web Animation at www.masteringbio.com
Endocrine System Anatomy

To summarize, interactions between the hypothalamus and pituitary gland furnish a key mechanism for the control of hormone action by the CNS. The hypothalamic-pituitary axis consists of two anatomically and functionally distinct systems. The posterior pituitary serves as a storage and release site for hormones produced by neurosecretory cells in the hypothalamus; the anterior pituitary serves as a hormone production and release site that is controlled by hypothalamic hormones.

Control of Epinephrine by Sympathetic Nerves

When biologists analyze how the nervous system and endocrine system interact to control the release of epinephrine, the distinction between the two systems begins to blur. Section 47.2 introduced how epinephrine acts as a hormone. During the fight-or-flight response, sympathetic nerves trigger the release of epinephrine from the adrenal medulla into the bloodstream. But in addition, some sympathetic nerves release the related molecule, **norepinephrine**, directly onto target cells. In effect, the nervous system delivers a chemical messenger directly to particular cells while the endocrine system broadcasts a similar messenger by secreting it into the bloodstream. Epinephrine and norepinephrine, which differ from one another only by the presence of an additional methyl group on epinephrine, are members of the family of molecules called **catecholamines**.

Catecholamines function as neurotransmitters as well as hormones. The close functional similarity between hormones and neurotransmitters does not end there, however. In some cases, the mode of action of hormones and neurotransmitters is similar. To drive this point home, recall from Chapter 45 that neurotransmitters have two major types of effects on target cells. All neurotransmitters trigger a postsynaptic potential, which can make the postsynaptic neuron more or less likely to deliver an action potential. In addition, certain neurotransmitters may initiate changes in gene expression in neurons. Chapter 45 illustrated the effect of neurotransmitters on transcription by reviewing experiments on the sea slug *Aplysia*. Repeated application of serotonin led to gene activation and changes in the behavior of the synapse. By altering synapses in this way, neurotransmitters play a central role in learning and memory. Similarly, many hormones exert their effects by activating particular genes in target cells. Understanding how gene activation occurs in response to neurotransmitters and hormones is currently the subject of intense research at laboratories around the world.

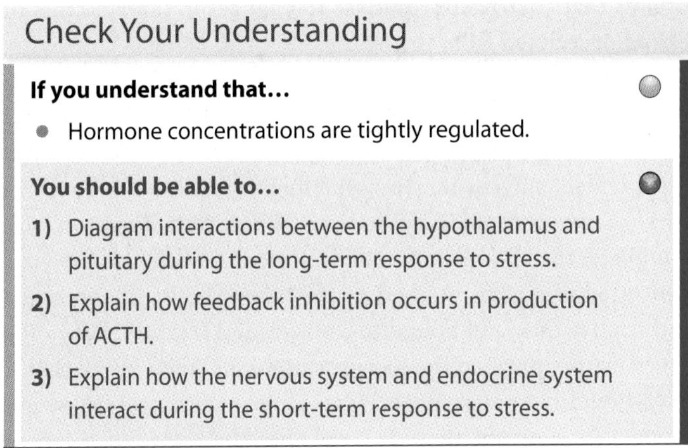
Check Your Understanding

If you understand that...

- Hormone concentrations are tightly regulated.

You should be able to...

1) Diagram interactions between the hypothalamus and pituitary during the long-term response to stress.
2) Explain how feedback inhibition occurs in production of ACTH.
3) Explain how the nervous system and endocrine system interact during the short-term response to stress.

47.4 How Do Hormones Act on Target Cells?

The key to understanding how hormones act on target cells is recognizing that some animal hormones are lipid soluble and cross plasma membranes readily, while others are not (see Figure 47.4). More specifically, steroid hormones are small lipids that enter cells without difficulty. But the peptide and polypeptide hormones and most amino acid derivatives do not cross plasma membranes easily, because of their large size and electrical charge.

Differences in the lipid solubility of hormones are important because they influence where a target cell receives the chemical message. Steroids often act inside the cell, while most amino acid derivatives and all polypeptides act at the cell surface. To explore these two distinct paths of hormone action, let's consider how estrogen and epinephrine affect target cells. As a steroid and a nonsteroid, they serve as model systems for hormone action.

Steroid Hormones and Intracellular Receptors

Estrogens are steroids that direct the development of female secondary sex characteristics in many animal species. In humans and other mammals, the most important estrogen is the molecule estradiol (formally, 17 β-estradiol). Because of its importance in reproduction by humans and domesticated animals, estradiol's mode of action has been the topic of intense investigation for over 50 years. How do target cells receive the signal carried by estradiol?

Identifying the Estrogen Receptor A key advance in research on how hormones affect target cells occurred in 1964, when biologists succeeded in isolating the estradiol receptor in laboratory rats. As **Figure 47.17** indicates, the researchers began by labeling purified estradiol with a radioactive atom and injecting a small amount into five adult female rats. Twenty minutes later, the biologists removed the uterus from each female and homogenized the tissue. The uterus, a muscular sac where embryonic development takes place, is one of the organs that changes in response to changes in the amounts of estradiol secreted from the ovaries.

After centrifugation, the soluble fraction of the homogenized cells was removed from the pellet and placed in another centrifuge tube that contained a sucrose gradient. As Chapter 7 explained, sucrose density centrifugation allows researchers to separate molecules by density. When centrifugation was complete, the biologists found that the radioactivity was concentrated in a narrow band in the tube. The band contained radioactive estradiol bound to the estradiol receptor. When the investigators purified the receptor molecule, they found that it could readily be destroyed by treatment with proteinase enzymes. As a result, they inferred that the estradiol receptor was a protein.

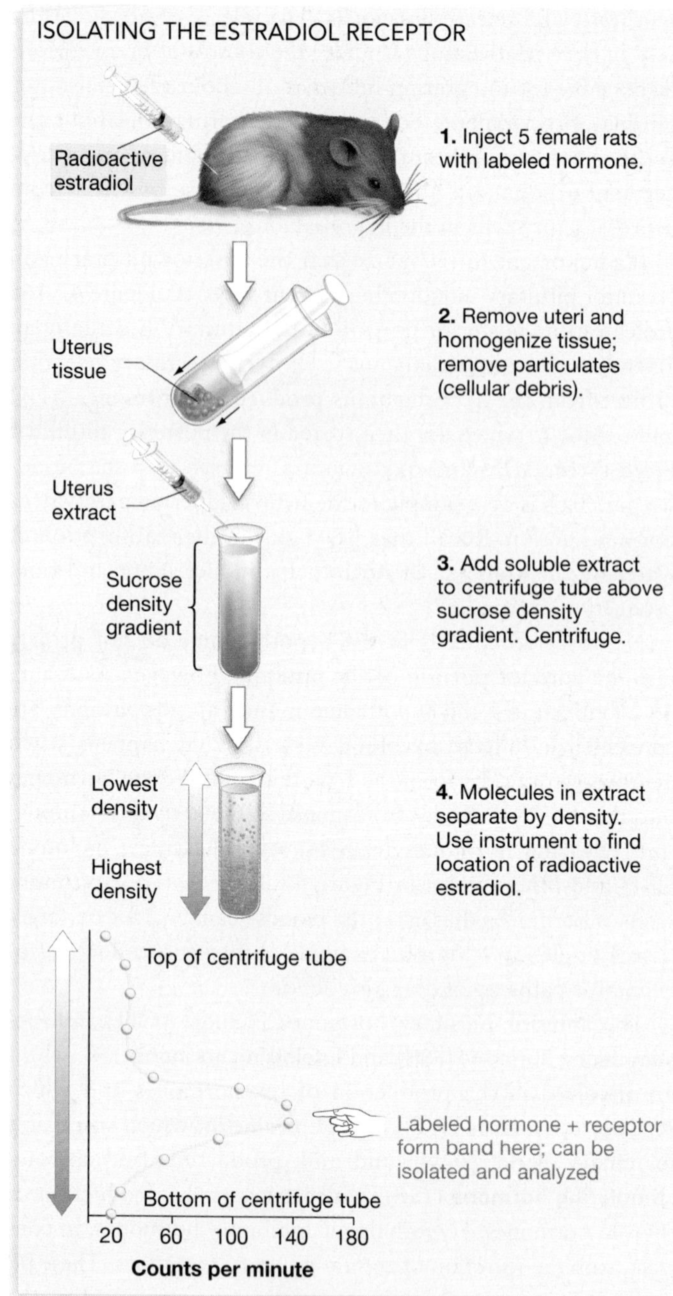

FIGURE 47.17 Discovery of the Estradiol Receptor. If radioactive estradiol bound to its receptor in the uterus, then the hormone-receptor complex should form a distinct band of radioactivity when molecules from uterine cells were separated by centrifugation.

Follow-up experiments established that the estradiol receptor is located in the nucleus but is not associated with the nuclear envelope. Further, the receptor is found only in estradiol target tissues, including the uterus, hypothalamus, and mammary glands. The latter finding was particularly exciting, because it clarified how hormones act in a tissue-specific way. This is a crucial point: Hormones are broadcast throughout the body via the bloodstream, but they act only on cells that express the appropriate receptor. Target cells respond to particular hormones because they contain a receptor for that hormone.

Later, biologists found that the gene for the estradiol receptor is similar to the genes that encode receptors for the glucocorticoids, testosterone, and other steroids. This result suggested that all steroid receptors are related by descent from an ancestral receptor molecule and that the binding of any steroid hormone to its receptor might affect the target cell in a similar way.

Biologists have answered some of the most basic questions about the nature and location of the estradiol receptor. What happens once the hormone-receptor complex forms inside the nucleus of a target cell?

Documenting Changes in Gene Expression During the 1970s and 1980s, work in several laboratories suggested that estradiol and other steroid hormones affect gene transcription after they bind to their receptors. For example, researchers were able to inject laboratory animals with estradiol or other steroid hormones and document changes in the mRNAs and proteins produced in target cells. These data showed that steroid hormones can cause dramatic changes in the amount or timing of mRNA production by a large number of genes.

An important insight into the molecular mechanism of hormone action came when researchers found a distinctive DNA-binding region in the **steroid-hormone receptor**. The region contains the DNA-binding domain called a zinc finger. Follow-up work showed that zinc fingers are found in all proteins in the steroid-hormone receptor family. To understand why this observation is significant, recall from Chapter 18 that DNA-binding domains are protein regions that make physical contact with DNA. The presence of a zinc finger in the estradiol receptor suggested that once estradiol had bound to it, it might affect gene expression by binding directly to DNA.

Follow-up work confirmed that steroid hormone-receptor complexes bind to specific sites in DNA, which came to be called **hormone-response elements**. Hormone-response elements are located just "upstream" (in the 5′ direction) from the start of target genes. Gene expression changes when a regulatory protein such as a steroid hormone-receptor complex binds to the hormone-response element for that gene.

The model in **Figure 47.18** summarizes the current state of thinking about how steroid hormones affect target cells. When estradiol or another steroid hormone enters a target cell, the hormone binds to its receptor (steps 1 and 2 of the figure). The binding event causes a conformational change in the receptor. In response, the hormone-receptor complex binds to DNA along with other proteins required for transcription initiation (step 3), and the expression of specific genes begins. Because each hormone-receptor complex leads to the production of many copies of the gene product, the signal from the hormone is amplified (steps 4 and 5). In this way, a small number of hormone molecules produces a large change in the activity of target cells and tissues.

Currently, research on steroid hormone action is focused on (1) the details of how the hormone-receptor protein interacts with the other proteins required to initiate gene transcription, and (2) how interactions between the hormone-receptor complex and other regulatory proteins alter the timing and the degree of expression of target genes. Researchers are also following up on the relatively recent discovery that at least some mammals, including humans, have two distinct receptors for estradiol. As experiments continue, it will be interesting to learn whether or not these two receptors trigger different responses to the same hormone.

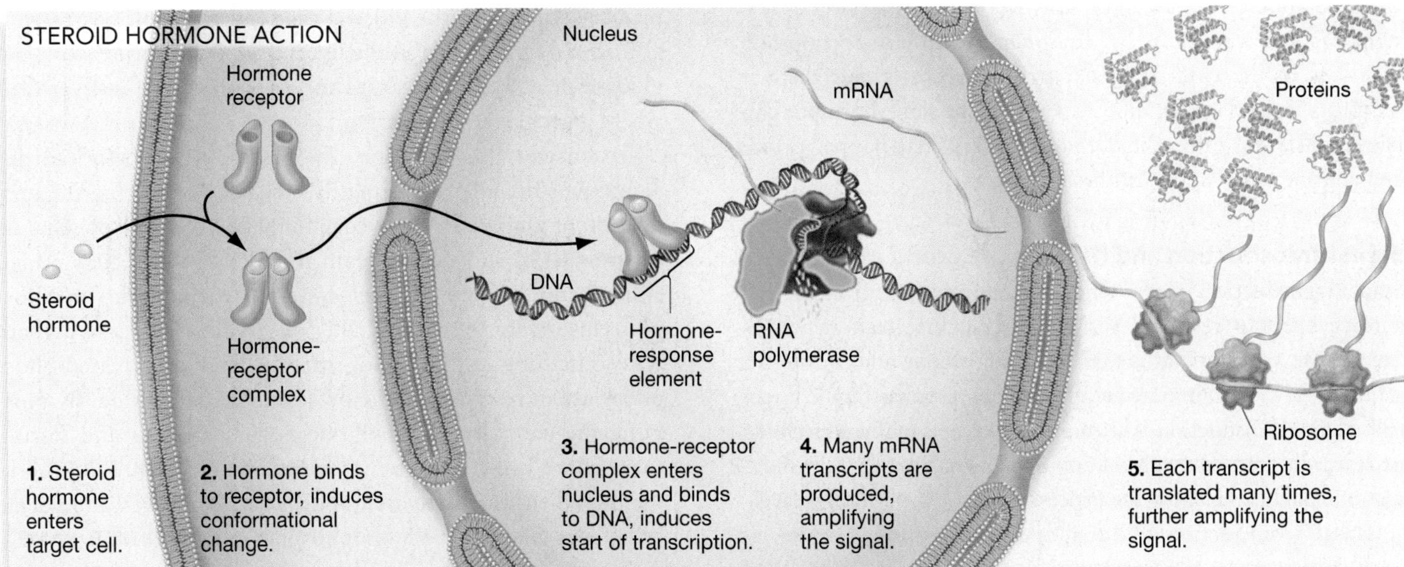

FIGURE 47.18 Steroid Hormones Bind to Receptors inside Cells and Change Gene Expression.

EXERCISE Tamoxifen is a drug that blocks estrogen receptors in breast tissue cells. (Estrogen stimulates growth of breast cells, so tamoxifen is often prescribed as a treatment for breast cancer.) Add symbols and labels to this diagram illustrating tamoxifen's mode of action and its consequences.

Hormones That Bind to Cell-Surface Receptors

Epinephrine, norepinephrine, and the polypeptide hormones are not lipid soluble. For these molecules to affect a cell, they have to bind to receptors on the cell surface. The messenger never enters the target cell, so its message must be transduced—changed into a form that is active inside the cell. Recall from Chapter 8 that this phenomenon is known as **signal transduction.**

To explore how signal transduction occurs, let's first examine the nature of the hormone receptors residing in the plasma membrane. Then we can explore research on the molecules that are responsible for processing the message inside the cell. In both cases, we'll focus on epinephrine as a model system.

Identifying the Epinephrine Receptor In 1948 a biologist published the results of an exhaustive set of studies on how epinephrine affects dogs, cats, rats, and rabbits. The responses observed fell into two distinct categories, depending on the tissue being considered. To explain this observation, the researcher suggested that epinephrine binds two distinct types of receptor. He called these hypothetical proteins the alpha receptor and the beta receptor.

Follow-up work with molecules that block epinephrine receptors confirmed this hypothesis. In fact, experiments documented that there are two types of alpha receptor and two types of beta receptor. In short, there are four distinct epinephrine receptors. Each receptor type is found on a distinct tissue type, and each induces a different response from the cell.

The discovery of four epinephrine receptors reinforces the concept of tissue specificity observed in experiments on the estradiol receptor. Hormones are transmitted throughout the body, not unlike a radio or television signal that is broadcast through the atmosphere. Their message is received only by cells with the appropriate receptor, however—just as a radio or TV signal is received only by equipment with an appropriate antenna. In the case of epinephrine, there are four distinct receptors instead of just one. As a result, the same hormone can trigger different effects in different cells. What happens once epinephrine binds to one of these receptors?

Signal Transduction and the Role of Second Messengers Signal transduction occurs when a chemical message at the cell surface elicits a response inside the cell. In effect, receptors "read" the message that is carried by a hormone and, through a sequence of events, initiate an appropriate response by the target cell. Signal transduction is currently an exceptionally active area of research, and research on epinephrine has produced fundamental insights into how the process works at a molecular level.

Recall from Section 47.2 that one of epinephrine's major effects is to increase glucose levels in the blood. To understand how this happens, biologists focused on phosphorylase. **Phosphorylase** is an enzyme that was discovered in the mid-1940s by biologists who were studying how cells cleave glucose molecules

(a) Phosphorylase catalyzes the production of glucose from glycogen.

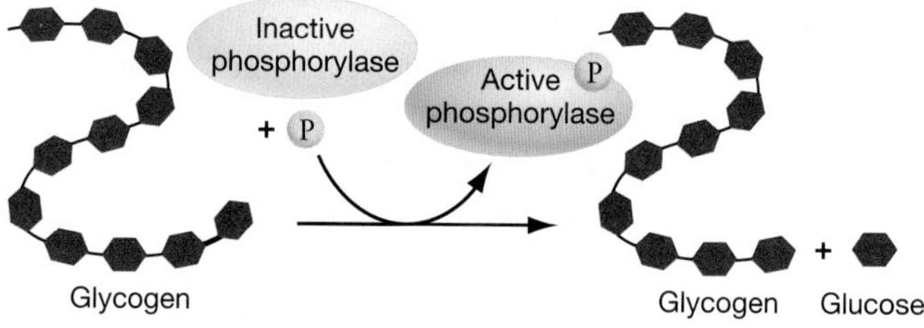

(b) Phosphorylase is activated in response to epinephrine.

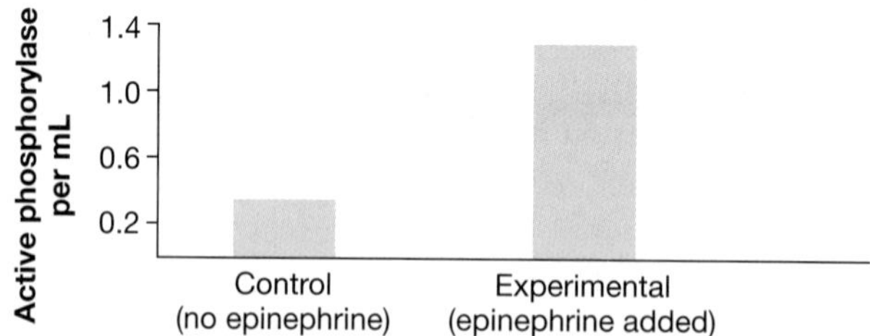

FIGURE 47.19 Epinephrine Activates the Enzyme That Catalyzes the Formation of Glucose from Glycogen. (a) Phosphorylase is activated when an enzyme adds a phosphate group to it. **(b)** When epinephrine is added to cell-free extracts from liver tissue, the amount of activated phosphorylase increases dramatically.

off glycogen. As **Figure 47.19a** shows, phosphorylase catalyzes this reaction. The researchers found that phosphorylase exists in active and inactive forms, and that the enzyme switches between these states when it is phosphorylated or dephosphorylated by another enzyme. Phosphorylase was the first of many enzymes shown to be activated by the addition of a phosphate group.

Investigators in a different laboratory followed up on these results by exploring whether phosphorylase is activated in response to epinephrine. During the fight-or-flight response, the primary source for the glucose that enters the bloodstream is glycogen that is stored in the liver. Does epinephrine stimulate glucose production by activating phosphorylase in liver cells where glycogen is stored?

To answer this question, the research team worked out techniques for homogenizing liver cells and studying the production of glucose from glycogen in vitro—outside of cells. As **Figure 47.19b** shows, the team found that when they added epinephrine to their cell-free system, large amounts of phosphorylase were activated. This observation suggested there was something in the homogenized cells that activated phosphorylase in response to the presence of epinephrine. By purifying the components of the homogenized cells and testing them in the cell-free system, the researchers eventually found the ingredient that activated phosphorylase. It was a molecule called cyclic adenosine monophosphate, or **cyclic AMP** (**cAMP**).

The role of cAMP in signal transduction was confirmed when researchers studied the effects of epinephrine on the activity of the rat heart. During the fight-or-flight response, both the heart rate and the contractile force of the heart rise dramatically—

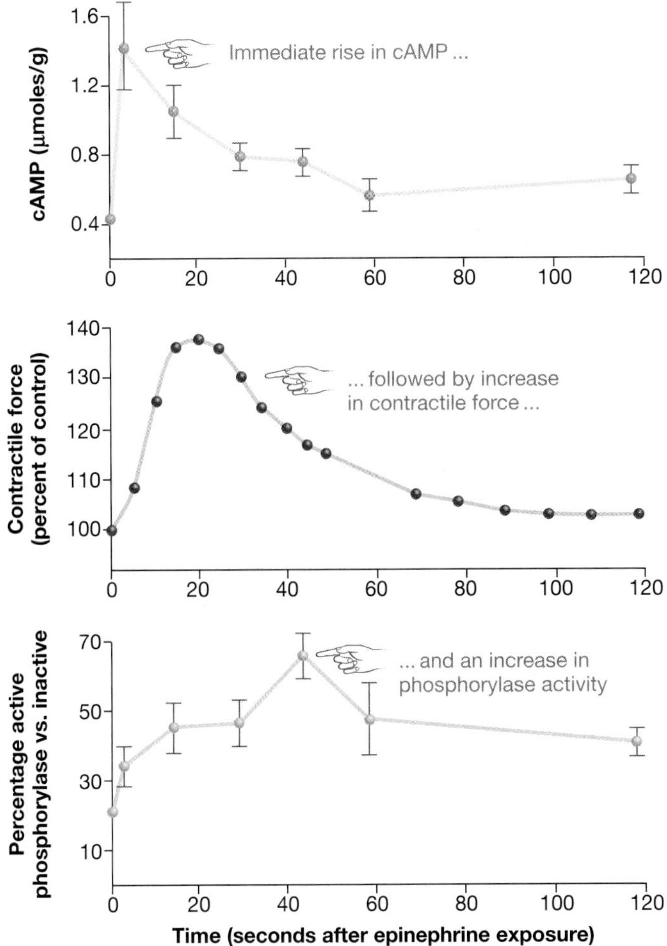

FIGURE 47.20 Chemistry and Activity of Heart-Muscle Cells Change in Response to Epinephrine. When epinephrine is added to heart-muscle cells, there is a rapid and dramatic rise in intracellular cAMP concentration. The contractile force of muscle fibers shows a sharp increase soon thereafter; slightly later, phosphorylase activity increases significantly.

EXERCISE In part (a), draw a line showing what the expected cAMP concentration would be if it had no role in activating phosphorylase.

increasing cardiac output. Cardiac output is a measure of the efficiency of the heart's pumping. As the graphs in **Figure 47.20** show, there is a striking increase in cAMP levels inside heart-muscle cells soon after the cells are treated with epinephrine (at time 0). Investigators confirmed that the contractile force of the heart-muscle cells increased along with phosphorylase activity in these cells.

To capture the importance of cAMP in triggering these effects, biologists referred to it as a second messenger. Recall from Chapter 8 that a **second messenger** is a nonprotein signaling molecule that increases in concentration inside a cell in response to a received signal—a molecule that binds at the surface. What does cAMP do? Follow-up work showed that cAMP binds to an enzyme called cAMP-dependent protein kinase A. The enzyme responds by phosphorylating the enzyme phosphorylase kinase, which then phosphorylates phosphorylase. This chain of events, called a signal transduction cascade, is initially triggered by the synthesis of cAMP, which is produced from ATP in a reaction that is catalyzed by the enzyme adenylyl cyclase. Adenylyl cyclase is activated by a G protein (see Chapter 8), which is activated when epinephrine binds to its receptor.

To pull all these details and reactions together, researchers proposed the model for epinephrine action diagrammed in **Figure 47.21**. In studying this model, it is crucial to recognize two points: First, the second messenger cAMP transmits the signal from the cell surface to the signaling cascade. Second, the other events in the sequence amplify that signal. To drive the latter point home, consider that, in response to stimulation by the hormone-receptor complex, adenylyl cyclase is thought to catalyze the formation of at least 100 molecules of cAMP. In turn, each of these cAMP molecules then activates many molecules of cAMP-dependent protein kinase A. Subsequently, each protein kinase molecule activates many molecules of phosphorylase kinase, and so on. In this way, the binding of just a single molecule of epinephrine may trigger the release of millions

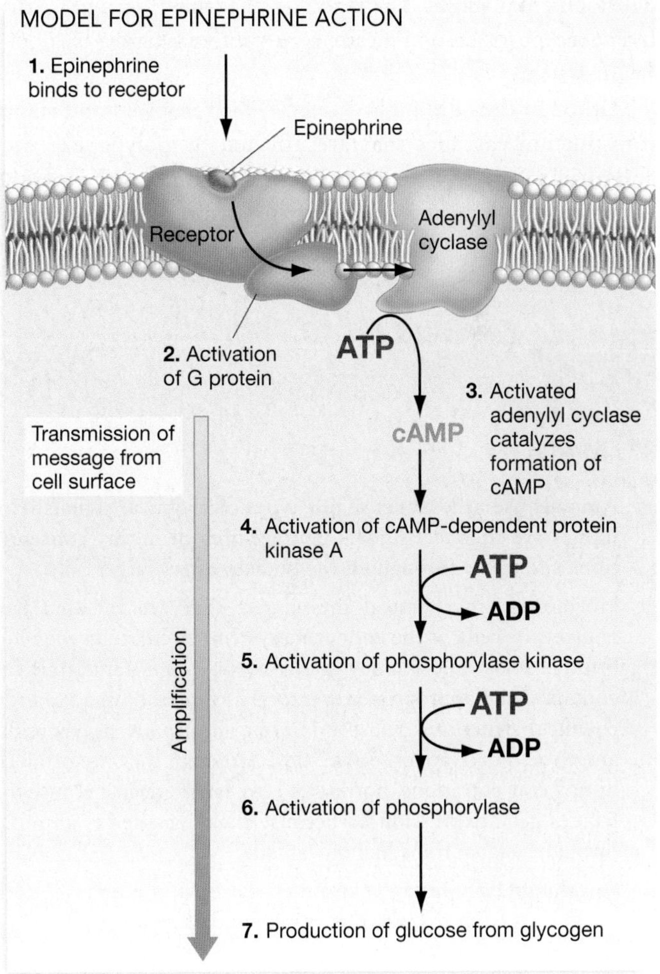

FIGURE 47.21 Epinephrine Triggers a Signal Transduction Cascade.

EXERCISE Using information in the text, label the approximate amount of amplification that occurs in each of steps 3–7.

or even billions of glucose molecules. Amplification through a signal transduction cascade explains why tiny amounts of hormones can have such huge effects on an individual.

The model in Figure 47.21 was inspired by experiments on the epinephrine receptor called the beta-1 receptor. But other groups showed that when epinephrine binds to an alpha-1 receptor, a completely different signal transduction event occurs. In this and many other receptor systems, calcium ions (Ca^{2+}) serve as the second messenger in conjunction with a second-messenger molecule called IP_3. Diacylglycerol (DAG) and 3′, 5′ cyclic GMP (cGMP) are also common second messengers in hormone response systems.

Researchers are increasingly impressed with the diversity and complexity of signal transduction cascades. For example, target cells that have the same receptor protein may have different second messengers or enzyme systems. As a result, the same hormone and receptor can give rise to different responses in different target cells. This observation helps explain one of the most fundamental observations about hormones: that the same chemical messenger can trigger different responses in cells from different organs or in cells at different developmental stages. The reason is that the cells contain different receptors, second messengers, protein kinases, or amplification steps.

Although they are highly diverse, the receptors and signal transduction cascades that are affected by polypeptide and amino-acid–derived hormones lead to a common response: the activation of a specific protein or set of proteins, usually by phosphorylation. In contrast, steroid hormones tend to exert their effects through changes in gene expression. Steroid hormones activate transcription factors that lead to the production of new proteins; nonsteroid hormones trigger signal transduction cascades that activate existing proteins.

Check Your Understanding

If you understand that... 

- Hormones act on target cells by binding to receptors.
- The response to a hormone is tissue specific because only certain cells contain receptors for particular hormones.
- In response to binding, hormone-receptor complexes either bind to DNA and induce changes in gene expression or induce the production of second messengers and the activation of signal transduction cascades that result in the phosphorylation of existing proteins.

You should be able to... 

1) Draw diagrams that summarize both sequences of events.
2) Predict how the systems you diagrammed would respond to a change in the DNA sequence of the hormone-response element or the enzyme responsible for production of the second messenger.

 **Web Animation** at www.masteringbio.com
Hormone Actions on Target Cells

Chapter Review

SUMMARY OF KEY CONCEPTS

Animals use at least six major types of chemical signals. Hormones are chemical signals that are present in tiny concentrations and travel throughout the body to affect target cells.

Hormones are chemical messengers that are released from neurons or cells of the endocrine system, circulate in the blood or other body fluids, and trigger a response in distant target cells containing an appropriate receptor. Hormones have a variety of chemical structures. Most animal hormones are polypeptides, amino acid derivatives, or steroids. Although they are produced in tiny concentrations, hormones have large effects because they trigger gene expression or because their message is amplified through a signal transduction cascade.

You should be able to explain the relationship between electrical signals from the nervous system and chemical signals from the nervous system and endocrine system, as they work in combination to coordinate the body's response to environmental change.

 **Web Animation** at www.masteringbio.com
Endocrine System Anatomy

The information in hormonal signals helps animals respond to environmental change, develop as embryos, undergo sexual maturation, and achieve homeostasis.

With the nervous system, hormones coordinate the activities of diverse cells and tissues. A single hormone may affect a wide array of cells and tissues and induce a variety of responses.

Epinephrine and cortisol are examples of hormones that help individuals cope with environmental changes. These hormones activate the short-term and long-term responses to stress, triggering the fight-or-flight response or inducing changes that conserve glucose for use by the brain.

Estradiol is an example of a hormone that regulates development and sexual maturation. Estradiol stimulates the formation of female sex characteristics in human embryos and the maturation of these tissues in adolescence.

Hormones are involved in a wide array of homeostatic interactions. For example, hormones are involved in directing cells that modify the concentrations of glucose, Na^+, Ca^{2+}, O_2, water, and other molecules in the blood and interstitial fluid. Homeostasis with respect to triglyceride stores is also under endocrine control.

You should be able to explain why electrical signals from the nervous system are not primarily responsible for regulating responses to environmental change, embryonic and sexual development, and homeostasis.

The production of a hormone is tightly regulated by input from the nervous system and by other hormones.

In many cases, the release of a hormone is regulated by chemical messengers from the anterior pituitary. The pituitary gland is located near the base of the brain in humans and other mammals. Hormone-secreting cells in the anterior pituitary are regulated by hormones released by the hypothalamus region of the brain. For example, the brain responds to long-term stress by triggering the release of the hypothalamic hormone CRH. CRH activates the release of ACTH by the pituitary gland, which stimulates the production of cortisol by cells in the adrenal cortex. Because cortisol inhibits the production of ACTH and CRH, the chain of events is regulated by feedback inhibition.

You should be able to predict the consequences of negative feedback failing to occur after the release of CRH.

Some hormones bind to receptors inside target cells and change gene expression. Other hormones bind to receptors at the cell surface and lead to changes in protein activation.

Animal hormones have two basic modes of action. Steroid hormones are lipid soluble, cross plasma membranes readily, and often bind to receptors inside cells. Most polypeptide and amino-acid–derived hormones are not lipid soluble and bind to receptors located in the membranes of target cells. In both cases, the response to a hormone is tissue specific because only certain cells express certain receptors.

Most steroid hormones act by inducing a change in gene expression. Steroid-hormone receptors are closely related structurally and have a distinctive DNA-binding domain. The hormone-receptor complex binds to hormone-response elements near the starting point of one or more genes. In response, the transcription of specific genes is activated or repressed.

Polypeptide and amino-acid–derived hormones trigger a complex sequence of events when they bind to a receptor on the plasma membrane. These signal transduction cascades induce a change in the concentration of a second messenger, such as cAMP, inside the target cell. In many cases, the endpoint of the process is the activation of a target protein by phosphorylation and a dramatic change in the cell's activity.

You should be able to predict whether a cell can respond to more than one hormone at a time, and explain why or why not.

Web Animation at www.masteringbio.com
Hormone Actions on Target Cells

QUESTIONS

Test Your Knowledge

1. Both epinephrine and cortisol are involved in the body's response to stress. How do the two molecules differ?
   a. Cortisol is an amino acid derivative; epinephrine is a steroid.
   b. Cortisol binds to receptors on the plasma membranes of target cells; epinephrine binds to receptors in the interior of target cells.
   c. Epinephrine mediates the short-term response; cortisol mediates the short- and long-term responses.
   d. Cortisol controls the release of epinephrine from the adrenal glands.

2. How do steroid hormones differ from the polypeptide hormones and most amino-acid–derived hormones?
   a. Steroids are lipid soluble and cross plasma membranes readily.
   b. Polypeptide and amino-acid–derived hormones are longer lived in the bloodstream and thus amplify signals more highly.
   c. Polypeptide hormones are the most structurally complex and induce permanent changes in target cells.
   d. Only polypeptide and steroid hormones bind to receptors in the plasma membrane.

3. What is signal transduction?
   a. the binding of a steroid hormone-receptor complex to DNA
   b. the release of a hormone from the anterior pituitary in response to a hypothalamic hormone
   c. the release of a hormone from the posterior pituitary in response to action potentials from the hypothalamus
   d. the production of a second chemical messenger inside a cell in response to the binding of a hormone at the cell surface

4. Which of the following developmental processes is *not* controlled by hormones?
   a. the initial development of male and female gonads, soon after fertilization
   b. overall growth
   c. molting in insects and other invertebrate animals
   d. metamorphosis in insects and other invertebrate animals

5. What is a hormone-response element?
   a. a receptor for a steroid hormone
   b. a receptor for a polypeptide hormone
   c. a segment of DNA where a hormone-receptor complex binds
   d. an enzyme that is activated in response to hormone binding and produces a second messenger

6. In hormone systems, when does feedback inhibition occur?
   a. when the presence of a hormone inhibits its release
   b. when the presence of a hormone stimulates its release
   c. when a second messenger triggers the phosphorylation of an inhibitory protein
   d. when a hormone from the hypothalamus inhibits the release of a hormone from the anterior pituitary

Test Your Knowledge answers: 1. c; 2. a; 3. d; 4. a; 5. c; 6. a

Test Your Understanding

Answers are available at www.masteringbio.com

1. Compare and contrast the functions of electrical signaling and chemical signaling in animals. How do the nervous system and endocrine system interact during the fight-or-flight response? What role does the nervous system play during hormonal control of homeostasis?
2. Compare and contrast the modes of action of steroid and nonsteroid hormones.
3. The pituitary is often referred to as the "master gland." Why?
4. Why is the observation that one hormone may bind to more than one type of receptor important?
5. Hormones are present in tiny concentrations yet have large effects on target cells and on the individual as a whole. How is this possible?
6. Compare and contrast the short-term and long-term responses to stress in humans. Which hormones are involved? Why are their effects logical, in terms of maximizing an individual's ability to survive a potentially life-threatening situation?

Applying Concepts to New Situations

Answers are available at www.masteringbio.com

1. Suppose that during a detailed anatomical study of a marine invertebrate, you found a small, previously undescribed structure. How would you test the hypothesis that the structure is a gland that releases one or more hormones?
2. Cortisone is a glucocorticoid that suppresses inflammation and other aspects of wound healing. Cortisone was once widely used to treat athletes with joint injuries and people with arthritis. In the short term, cortisone is extremely effective in reducing swelling and pain. Over time, however, physicians found that repeated large doses of cortisone had damaging side effects. Predict what these side effects are. Explain your logic.
3. You are a physician supervising a patient's recovery from the surgical removal of the posterior pituitary. Name a hormone that you will have to administer to this patient artificially. Which symptom(s) will you monitor to assess whether the dosage and timing of your injections are having the desired effect?
4. Suppose that a researcher announces the discovery of a hormone that affects the metabolism of fats in lab rats. Preliminary data indicate that the hormone is a polypeptide, about 50 amino acids long. How would you go about isolating the receptor for this hormone? Using fat cells growing in vitro, how could you test the hypothesis that hormone binding causes a change in a second messenger?

www.masteringbio.com is also your resource for • Answers to text, table, and figure caption questions and exercises • Answers to *Check Your Understanding* boxes • Online study guides and quizzes • Additional study tools including the *E-Book for Biological Science* 3rd ed., textbook art, animations, and videos.

Animal Reproduction

48

The swollen, red rump of this female Hamadryas baboon indicates that she is about to produce an egg and will accept courtship and possibly copulation from males. She will probably mate with several males before the egg is fertilized. This chapter discusses an array of questions about animal reproduction, including how hormones control the female reproductive cycle and the consequences of multiple mating.

KEY CONCEPTS

- The reproductive systems of animals are highly variable. Many animals can switch between asexual and sexual reproduction. When sexual reproduction occurs, fertilization may be external or internal and egg development may take place inside or outside the mother's body, depending on the species.
- In humans, the male reproductive system is composed of structures specialized for the production and storage of sperm, the synthesis of important molecules found in semen, or the transport and delivery of semen. The female reproductive system is composed of structures specialized for producing eggs, receiving sperm, and nourishing offspring during early development.
- In humans, the menstrual cycle is regulated by hormones from the pituitary gland and female reproductive organs. These hormones may interact with each other via positive or negative feedback. Pregnancy is maintained by hormonal signals from the embryo and from the mother's reproductive organs.

All the cells, tissues, organs, and systems introduced in Unit 8 exist for one reason: They allow animals to survive long enough and gather enough resources to reproduce. The shark rectal gland, the fish gill, the squid giant axon, the infrared sensors of pit vipers, and the human adrenal glands are all adaptations that help individuals survive in particular environments, acquire energy, and produce offspring. Stated another way, producing offspring is the reason that adaptations exist. Reproduction is the unconscious goal of virtually everything that an animal does.

Evolution by natural selection explains why animals reproduce; the goal of this chapter is to explore *how* reproduction occurs. Section 48.1 begins by introducing research on animals that cycle between asexual and sexual modes of reproduction and then explains how both modes occur at the cellular level. Section 48.2 surveys the diversity of ways in which animals accomplish fertilization once gametes have formed. The final three sections focus on mammalian reproduction, using humans as the primary model organism. The discussion begins by analyzing male and female reproductive organs, explores how hormones control changes in those organs, and concludes with a description of human pregnancy and birth.

In the course of investigating research on the molecular and cellular mechanisms of reproduction, we'll also consider topics of urgent practical interest. These topics include the dangers of steroid hormone use by athletes and bodybuilders, and the recent and dramatic declines in the death rate of mothers during childbirth. Understanding and manipulating animal reproductive systems is an important issue in the working lives of physicians, veterinarians, farmers, zookeepers, and many others in biology-related professions.

48.1 Asexual and Sexual Reproduction

Several earlier chapters have explored how asexual and sexual reproduction differ. **Asexual reproduction** occurs without fusion of gametes. It is usually based on mitosis and results in offspring that are genetically identical to their parent. **Sexual reproduction**, in contrast, is based on meiosis and fusion of gametes. Due to genetic recombination during meiosis and the fusion of haploid gametes—usually from different parents—during fertilization, sexual reproduction results in offspring that are genetically different from each other and from their parents. Let's first consider how animals make genetically identical copies of themselves, then review the mechanisms responsible for sexual reproduction.

How Does Asexual Reproduction Occur?

In thousands of animal species, individuals can **clone** themselves—that is, produce large numbers of identical copies of themselves asexually. Which animals produce offspring asexually, and how is asexual reproduction accomplished? **Figure 48.1** addresses both questions. The hydra in Figure 48.1a is generating offspring by budding. When **budding** occurs, an offspring begins to form within or on a parent. Budding is completed when the offspring—a miniature version of the parent—breaks free and begins to grow on its own. Figure 48.1b illustrates a flatworm undergoing **fission**, in which an individual simply splits into two or more descendants. In some populations of the western whiptail lizard (Figure 48.1c), reproduction occurs solely by **parthenogenesis** (literally, "virgin-origin"), in which female offspring develop from unfertilized eggs. In most cases, parthenogenetic eggs are produced either by mitosis or by meiosis that occurs without crossing over and recombination, after chromosome number has doubled. In both cases, offspring are genetically identical to the mother. If a species reproduces exclusively via parthenogenesis, no males exist. Parthenogenesis occurs in a wide diversity of lineages, including certain guppies, crustaceans, and rotifers as well as lizards.

In addition, many animal species regularly switch between reproducing asexually and reproducing sexually. Why?

(a) Budding in hydra

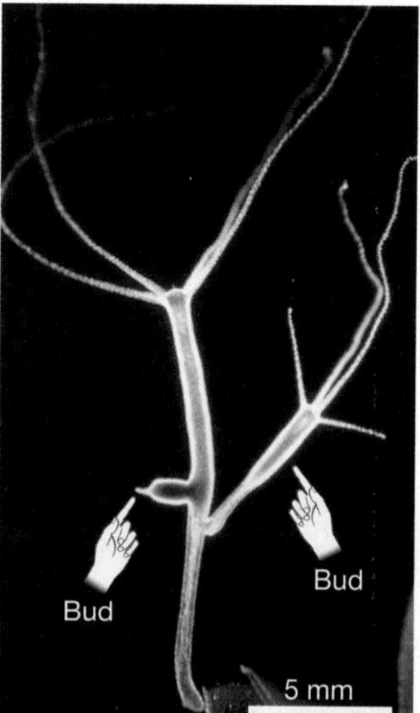

(b) Fission in flatworms

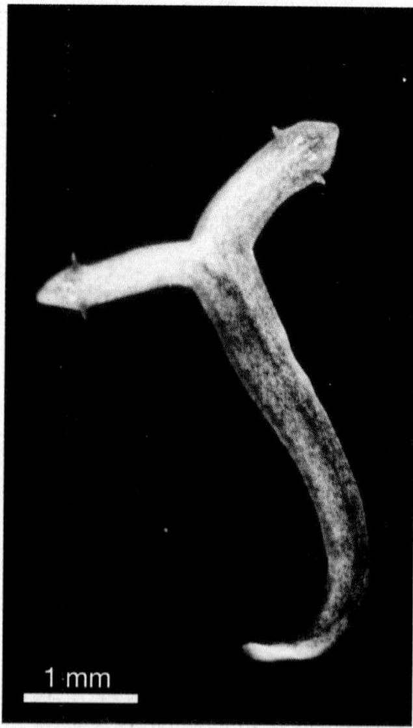

(c) Parthenogenesis in lizards

FIGURE 48.1 Mechanisms of Asexual Reproduction in Animals Are Diverse.

Switching Reproductive Modes: A Case History

Daphnia are crustaceans that live in freshwater habitats throughout the world. In a typical year, *Daphnia* produce only diploid female offspring throughout the spring and summer, via parthenogenesis. As **Figure 48.2a** shows, the eggs produced by parthenogenesis develop in a structure called a brood pouch. They are released when the female molts her exoskeleton.

In late summer or early fall, however, many *Daphnia* females begin producing unfertilized, asexually produced eggs that develop into males as well as females. Once the males have matured, sexual reproduction ensues: Haploid **sperm** (male gametes) produced by meiosis in the males fertilize haploid **eggs** (female gametes) that females produce via meiosis. **Fertilization** is the fusion of sperm and egg. The fertilized eggs are released into a durable case that falls to the bottom of the pond or lake

(a) *Daphnia* produce diploid eggs asexually.

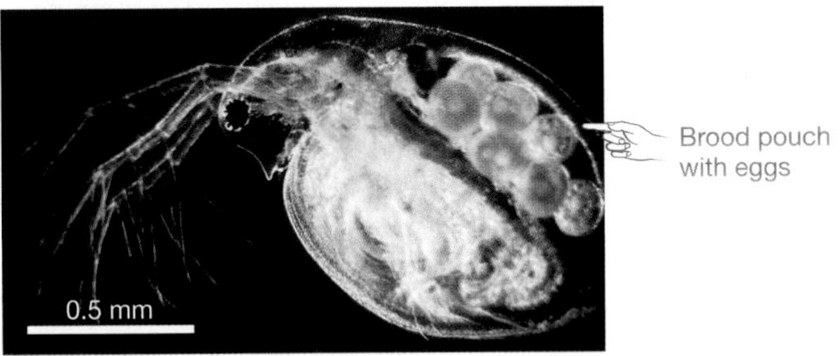

(b) Sexual reproduction is more common in crowded populations than in sparse populations.

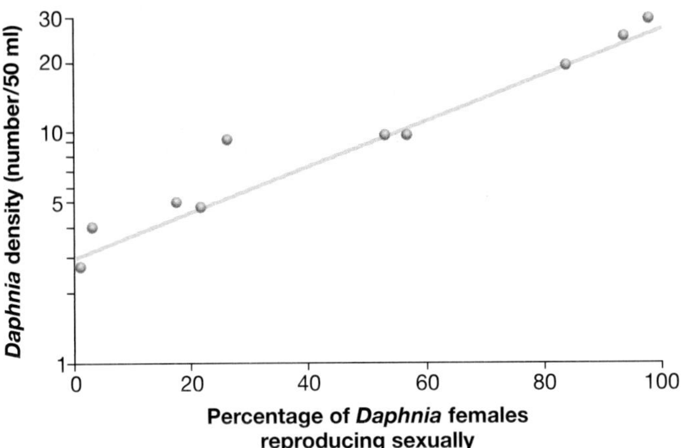

(c) Which environmental cues trigger the switch to sexual reproduction?

Water quality	Food concentration	Day length	% Sexual broods
Clean	Low	Short	0
Crowded	Low	Short	44
Clean	Low	Long	0
Crowded	Low	Long	0
Clean	High	Short	0
Crowded	High	Short	0
Clean	High	Long	0
Crowded	High	Long	0

FIGURE 48.2 In *Daphnia*, Environmental Cues Signal the Switch from Asexual to Sexual Reproduction. (a) The diploid eggs shown here will hatch into offspring that are genetically identical to the mother. **(b)** Graph plotting the percentage of *Daphnia* females that reproduce sexually versus *Daphnia* density (log scale). **(c)** Environmental conditions were varied experimentally for *Daphnia*. "Crowded" water was taken from tanks containing dense populations.

● **QUESTION** How would you go about determining which molecule or molecules in crowded water serve as a signal that triggers sexual reproduction?

for the winter. In spring, the sexually produced offspring hatch and begin reproducing asexually.

Biologists try to explain observations like this at two levels (see Chapters 12 and 51). At the ultimate level, the issue is how the behavior affects fitness. Researchers hypothesize that *Daphnia* offspring do better the following year, when conditions are likely to be different from those the year before, if they are genetically unlike their parent. But at the proximate level, the issue is explaining the mechanism responsible for the observation. For example, what cues trigger the change in *Daphnia*'s modes of reproduction?

For decades, most researchers contended that day length triggered the asexual-sexual switch. The idea was that the shortening days of late summer or fall induced the production of males and haploid eggs. Then in 1965 biologists showed that high population densities are also a factor in causing the switch. Researchers who brought *Daphnia* populations into the lab and kept day length constant found a strong correlation between population density and the percentage of females reproducing sexually (**Figure 48.2b**).

In 1992 another group of investigators extended this result by pinpointing the specific aspects of crowding that were affecting the animals. To do this work, the biologists brought *Daphnia* into the laboratory and altered day length, the amount of food available to individuals, and the quality of the water they occupied. (To vary water quality, the investigators used either clean water or water taken from tanks where *Daphnia* were being maintained at high density.) As **Figure 48.2c** shows, individuals in the study population switched to sexual reproduction only if they were exposed to water from crowded populations, low food availability, and short day lengths. In short, *Daphnia* need three different cues from the environment to switch to sexual reproduction.

These experiments don't address the question of why *Daphnia* make the switch, however. Why is it adaptive to reproduce sexually under the adverse conditions dictated by high population density, nutrient stress, and shortened days? The leading hypothesis is that because sexually produced offspring are genetically diverse, they are more capable of thriving when they hatch the following spring than are asexually produced offspring, which are genetically identical to each other and to their parents. This hypothesis is consistent with research introduced in Chapter 12. You might recall that sexual reproduction in a snail species is associated with high rates of parasite infection. The general idea is that genetically variable offspring have higher fitness in environments with rapidly evolving parasites, deteriorating physical conditions, or other types of rapid environmental change. ● If you understand this concept, you should be able to predict the conditions under which asexual reproduction would be found as the only method of producing offspring. You should also be able to discuss the costs and benefits of propagating crop plants like bananas solely through asexual reproduction.

To date, however, the variable-environment hypothesis for the evolution of sexual reproduction has not been tested rigorously in *Daphnia*. Research on the adaptive significance of sex in this species and other organisms continues.

Mechanisms of Sexual Reproduction: Gametogenesis

The mitotic cell divisions, meiotic cell divisions, and developmental events resulting in the production of male and female gametes, or sperm and eggs, are collectively called **gametogenesis** (see Chapter 22). **Spermatogenesis** is the formation of sperm, and **oogenesis** is the formation of eggs. In the vast majority of animals, gametogenesis occurs in a sex organ, or **gonad**. Male gonads are called **testes**; female gonads are called **ovaries**. Early in development, reproductive cells known as *germ cells* enter the testes and ovaries and give rise to diploid cells that will undergo gametogenesis.

Figure 48.3 summarizes the events that take place during gametogenesis in humans. It's important to note that in the male and female gonads, diploid cells called **spermatogonia** (singular: **spermatogonium**) and **oogonia** (singular: **oogonium**), respectively, divide by mitosis. Some of these daughter cells continue to function as spermatogonia or oogonia; others change to form specialized cells that are committed to producing gametes. In males, these specialized cells are called **primary spermatocytes**. They undergo meiosis I and produce two **secondary spermatocytes**, which then undergo meiosis II. The result is four haploid cells called **spermatids**. Each haploid spermatid matures into a sperm. The production of spermatogonia and primary spermatocytes occurs continuously throughout adult life.

Gametogenesis is markedly different in human females. The specialized cells produced by an oogonium are called **primary oocytes**. When these cells undergo meiosis, only one of the four haploid products, known as an **ovum**, matures into an egg. The other cells produced by meiosis in females have a tiny amount of cytoplasm and do not mature into eggs. Because the distribution of cytoplasm is so unequal during each meiotic division in females, the smaller cells are called **polar bodies**. (Recall that *polar* refers to inequality or opposites.) Further, the production of primary oocytes stops early in development in many species; in humans, it stops before birth. And in humans and many

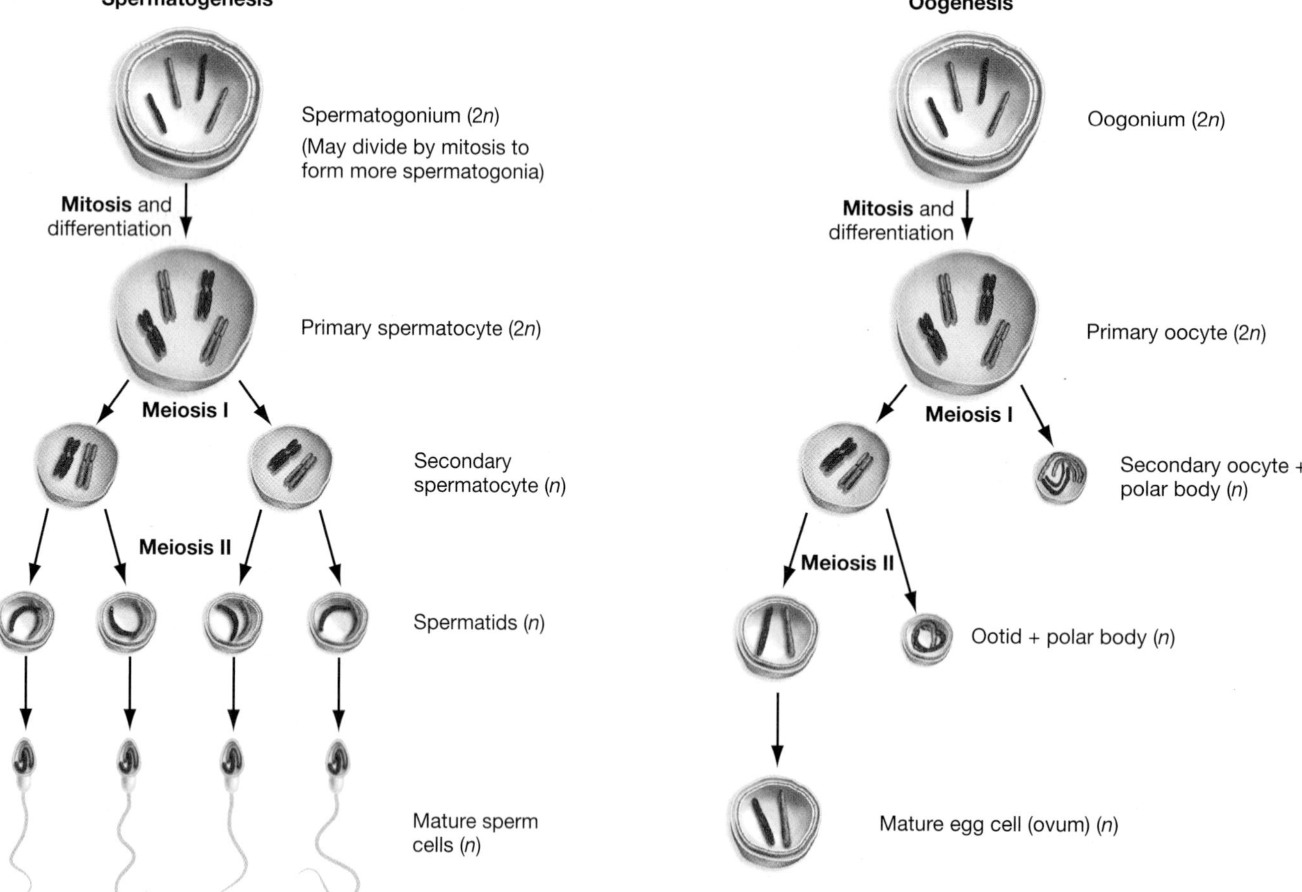

FIGURE 48.3 Gametogenesis in Humans. During gametogenesis, spermatogonia produce sperm and oogonia produce eggs.

● **EXERCISE** In each diagram, label the haploid cells and the diploid cells.

other mammals, the primary oocytes enter prophase of meiosis I during embryogenesis but then stop developing for a period of months or years.

(MB) Web Animation at www.masteringbio.com
Human Gametogenesis

48.2 Fertilization and Egg Development

Fertilization is the joining of a sperm and an egg to form a diploid **zygote**. Chapter 22 introduced this process by describing how a sperm makes contact with an egg and penetrates the egg's membrane. That discussion focused on the molecular mechanisms of fertilization, using the sea urchin as a model organism. The objectives of this section are to introduce the variety of fertilization mechanisms observed among animals and to explore what happens to fertilized eggs. Why are eggs laid into the environment in some species but retained to develop inside the mother's body in other species? An equally striking contrast occurs in modes of fertilization. In many species, individuals release their gametes into their environment and external fertilization occurs. In other animals, males deposit sperm into the reproductive tracts of females and internal fertilization occurs.

External Fertilization

Most animals that rely on external fertilization live in aquatic environments. The correlation between external fertilization and aquatic environments is logical, because gametes and embryos must be protected from drying. If external fertilization occurred in a terrestrial environment, either the gametes or the resulting zygote would likely die of desiccation.

Another general observation is that species with external fertilization tend to produce exceptionally large numbers of gametes. For example, a female sea star *Asterias amurensis* typically releases 100,000,000 eggs into the surrounding seawater during spawning. Males release many times that number of sperm. The leading hypothesis to explain this pattern is that the probability of a sperm and egg meeting in an ocean or lake is extremely small unless extremely large numbers of gametes are present.

If sperm and eggs from different individuals must be released into the environment synchronously (at the same time) for external fertilization to work, how is gamete release coordinated? The answer has two parts. Most research to date indicates that gametogenesis occurs in response to environmental cues such as lengthening days and warmer water temperatures. These results are logical, because the cues indicate a favorable season for breeding. But what triggers the actual release of gametes? In fishes and other aquatic animals with well-developed eyes and external fertilization, spawning is often the culmination of an elaborate courtship ritual between a male and female. In contrast, courtship behavior appears to be much less important—or even absent—in species such as clams, sea urchins, and sea cucumbers. How do animals that cannot see their mates time the release of their gametes?

Researchers have long hypothesized that the chemical messengers called **pheromones** might be involved in synchronizing gamete release. The idea is that when it is ready to mate, a sea star will release a pheromone, which then triggers gamete release in other individuals. Data to back this claim have begun to emerge only recently, however. For example, in one recent test of the pheromone hypothesis for spawning synchronization, biologists maintained two groups of sea cucumbers under natural conditions of light and temperature for 15 months. In one treatment, individuals were kept in isolated tanks; in another, they were kept in tanks with other sea cucumbers. The researchers found that all the individuals that were maintained in groups released gametes during the normal spawning period. In contrast, only about 10 percent of the individuals maintained in isolation released gametes.

Although these data suggest that pheromones might be involved in coordinating external fertilization in sea stars, they are not definitive. Most biologists will not be convinced that the pheromone hypothesis is valid until a chemical messenger is purified and can be shown to induce spawning in isolated individuals.

Internal Fertilization and Sperm Competition

Internal fertilization occurs in the vast majority of terrestrial animals as well as in a significant number of aquatic animals. Internal fertilization occurs after copulation—in which males deposit sperm directly into the female reproductive tract with the aid of a copulatory organ, usually called a **penis**. Alternatively, males may package their sperm into a structure called a **spermatophore**, which is then placed into the female's reproductive tract by the male or female. In some salamander species, for example, the male places the spermatophore on the ground within its territory. Later the female picks it up with her **cloaca**, a chamber used by both the reproductive and excretory systems that opens to the environment. In this case, the result is internal fertilization without any direct physical contact between the sexes.

Although internal fertilization appears to be straightforward, research on the topic has produced some unexpected results. In terms of understanding animal behavior, perhaps the most important insight about internal fertilization originated with Geoff Parker in the 1970s. Parker realized that in many animal species, females mate with more than one male before fertilization occurs. As a result, sperm from different males compete with each other to fertilize the eggs.

In 1970 Parker published the results of experiments on dung flies that confirmed the existence of sperm competition—competition between sperm from different males to fertilize the eggs of the same female. Parker's experiments consisted of a series of carefully controlled matings between one female and two males. In each experiment, the two males were selected in such a way that Parker could distinguish their offspring. As predicted by the sperm competition hypothesis, the proportion of offspring fathered by each male was not 50:50. Instead, whichever male was last to copulate fathered an average of 85 percent of the offspring produced.

Follow-up research has confirmed that this **second-male advantage** is widespread, although not quite universal, in insects and some other animal groups. How does it occur? To answer this question, biologists turned to the fruit fly *Drosophila melanogaster*. A research group was able to introduce a gene into male fruit flies that resulted in the production of sperm with green tails (**Figure 48.4**). When a mating by a green-spermed male was followed with a mating by a male having normal-colored sperm, most of the green-tailed sperm disappeared from the female's sperm-storage area. To interpret this finding, the biologists suggested that the sperm of the second male physically dislodged the first male's gametes and inserted themselves in their place. The researchers also showed that the fluid that accompanies sperm during fertilization is able to displace stored sperm from competing males. These two mechanisms resulted in the second male's sperm fertilizing most of the eggs laid.

Besides documenting the mechanisms responsible for second-male advantage, research on sperm competition has recently contributed another major result. In species in which multiple mating is common, males have extraordinarily large testes for their size and produce proportionately larger numbers of sperm. Although this pattern was first documented in primates, it has now been observed in a wide variety of animal groups. For example, males from species of fruit bats that live in large social groups produce more sperm, on average, than do male fruit bats from species that live in small groups. The leading hypothesis to explain this observation is that fertilization is similar to a lottery in which each sperm represents a ticket. According to the lottery hypothesis, the more tickets a male enters in the competition, the higher his chance of "winning" fertilizations and passing his alleles on to the next generation. As a result, it is logical to observe exceptionally large testes in species in which sperm competition occurs.

The lottery model has recently been challenged, however, by evidence that in some species if not most, females store sperm and exert control over which sperm are successful in fertilization. In other words, females do not always accept the results of sperm competition passively. Females of some species actively choose which male performs the last copulation before fertilization takes place. In other species, females physically eject sperm from undesirable males. This phenomenon has been dubbed *cryptic female choice*. The name is appropriate because the selection of sperm by females is hidden from males.

Experiment

Question: How does the "second-male advantage" occur in sperm competition?

Hypothesis: Sperm from the second male displaces sperm from the first male.

Null hypothesis: The mechanism does not involve sperm displacement.

Experimental setup:

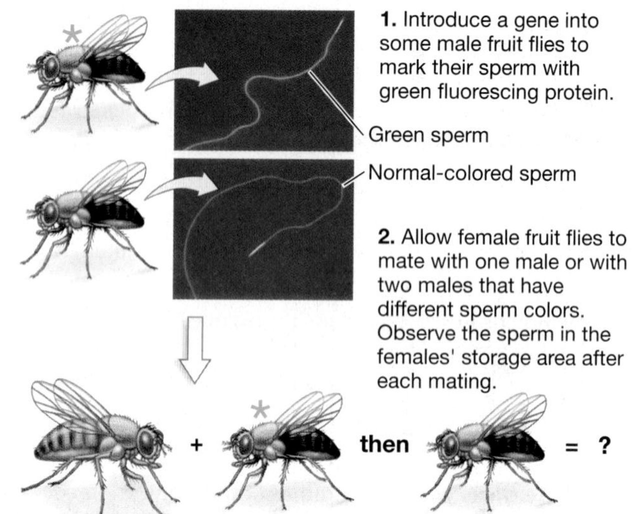

Prediction: When females mate twice, few sperm from the first male remain in storage.

Prediction of null hypothesis: When females mate twice, all of the sperm deposited by the first male are still present.

Results:

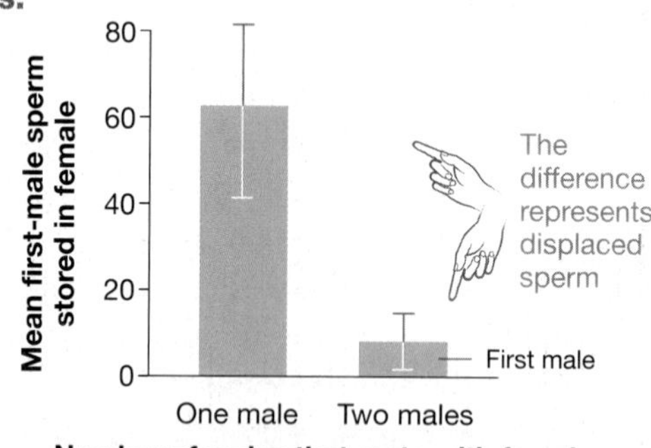

Conclusion: The sperm from the second male displaces the sperm from the first male.

FIGURE 48.4 Experimental Evidence for Second-Male Advantage during Sperm Competition in *Drosophila*. Researchers created fruit-fly males with green-tailed sperm to explore why the second male to mate tends to father more offspring than does the first male in many species. The graph shows the results when males with green sperm were the only male to mate or the first of two males to mate. When sperm competition occurred, the first male to mate fathered a tiny percentage of offspring, probably because most of its sperm disappeared from the sperm-storage organ in females.

Why Do Some Females Lay Eggs while Others Give Birth?

Once fertilization is accomplished, the embryo is either laid as an egg outside the mother's body or retained so that it can develop inside. In **oviparous** ("egg-bearing") animals, the embryo develops in the external environment. In some oviparous species such as sea stars, sea urchins, and most insects, neither the male nor the female provides any further care; the eggs and embryos are left to fend for themselves. But a substantial number of oviparous species continue to care for their young after eggs have emerged from the mother's body. For example, birds incubate their eggs and feed the young after hatching; fish may guard their eggs from predators and fan the clutches to oxygenate them.

In **viviparous** ("live-bearing") species, embryonic development takes place within the mother's body. More specifically, the embryo attaches to the reproductive tract of the mother and receives nutrition directly from the mother—via diffusion from her circulatory system. When **ovoviviparity** occurs, offspring develop inside the mother's body but are nourished by nutrient-rich yolk stored in the egg.

Why does oviparity exist in some groups and viviparity or ovoviviparity in others? Biologists tackled this question by studying the lizard genus *Sceloporus* (**Figure 48.5a**). Some of the populations they studied are oviparous while others are ovoviviparous; all live in the highlands of the southwestern United States and central Mexico, where temperatures can be cool. To understand how oviparity and ovoviviparity evolved, the biologists analyzed a phylogenetic tree—based on molecular and morphological data—of many *Sceloporus* species (**Figure 48.5b**; see **BioSkills 2** for help with interpreting phylogenetic trees). One of the researchers' first conclusions was that because most *Sceloporus* populations are oviparous, egg laying probably represents the original or ancestral condition. As the red branches on the tree show, however, ovoviviparity evolved independently in two groups. Using a similar research strategy, biologists have found that ovoviviparous populations of sea stars have evolved from oviparous populations on several occasions.

Why did natural selection favor these changes between egg laying and live birth? Ovoviviparity or viviparity should evolve when it leads to higher numbers of surviving young. Researchers have long hypothesized that natural selection favoring live birth should be especially strong in cold habitats. Low temperatures slow the development of embryos, so in cold habitats, it might be advantageous for females to retain eggs inside their bodies so that offspring can develop at a more favorable temperature. In support of this hypothesis, a researcher has shown that in at least one oviparous species of lizard, lowering nest temperatures leads to a higher percentage of deformities in offspring. In addition, offspring that developed in colder nests run more slowly than do individuals of the same age that developed at higher temperatures. Research into why ovoviviparity and viviparity has evolved in certain populations continues.

(a) Some *Sceloporus* lizards lay eggs; other members of the same genus bear live young.

(b) Phylogeny of *Sceloporus* from central Mexico

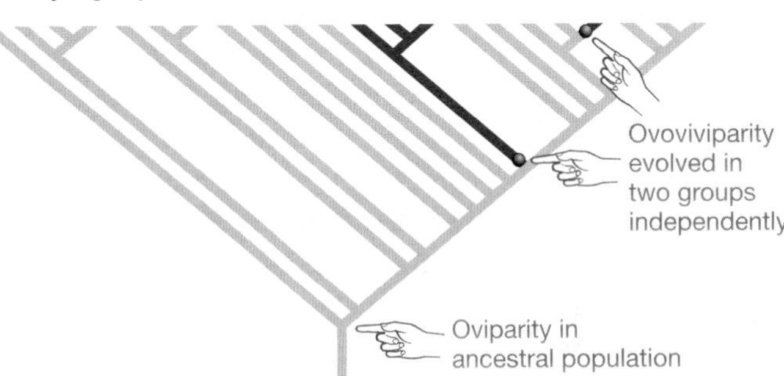

FIGURE 48.5 Ovoviviparity Has Evolved Multiple Times in *Sceloporus* Lizards. (a) *Sceloporus* is a genus of lizards from desert areas in North America. **(b)** Each twig on this phylogenetic tree represents a *Sceloporus* species or population from North America. According to these data, ovoviviparity evolved twice as the lineage diversified.

Check Your Understanding

If you understand that...

Some of the most basic aspects of reproduction are variable among animal populations or species: whether reproduction is sexual or asexual, whether fertilization is external or internal, and whether fertilized eggs develop in the environment or in the mother's body.

You should be able to...

1) Describe the advantages and disadvantages of each of these modes of reproduction.
2) Design experiments to test the following hypotheses:
   - A particular molecule is a gamete-releasing pheromone in sea stars.
   - A particular molecule is a toxin active in sperm competition.
   - In ovoviviparous and oviparous populations of the same species, ovoviviparous populations have higher reproductive success in cold environments.

48.3 Reproductive Structures and Their Functions

The first two sections of this chapter considered the broad contrast between asexual and sexual reproduction and examined general patterns in fertilization and egg care. Now let's narrow the focus and explore the mechanics of sexual reproduction in more detail. The first task is to understand the anatomy of the male and female reproductive systems.

The Male Reproductive System

In humans, the external anatomy of the male reproductive system is simple: It consists of the scrotum and the penis. The saclike **scrotum** holds the testes; the penis functions as the organ of copulation prior to internal fertilization.

Biologists who have compared these structures among animal species have been struck by their variability and complexity. For example, a scrotum occurs only in certain mammal species. Whales, elephants, hedgehogs, moles, and many other mammal groups lack the structure entirely and instead have testes that are located well within the abdominal cavity. Among species that do have a scrotum, the structure's size and shape vary. In numerous species of primates, the scrotum is brightly colored and appears to function in sexual display behaviors.

How Does External Anatomy Affect Sperm Competition?

Diversity in scrotal morphology pales in comparison with variation in the structure of the penis or other types of male **genitalia.** (*Genitalia* is the term used when the copulatory organ has distinct parts.) In many groups of insects and spiders, examining the animals' genitalia is the only way that biologists can distinguish species. **Figure 48.6a** shows one elaborate structure typically observed in spiders. Even among closely related species, the sizes and shapes of these organs tend to be strikingly different. Why are insect genitalia so diverse?

Researchers recently offered data that might help answer this question, based on the hypothesis that certain genitalia shapes are advantageous when sperm competition occurs. To test this idea, biologists studied how much second-male advantage varied among males with different-shaped genitalia when water-strider females mated with two males. As the data in **Figure 48.6b** show, there is substantial variation in the size and shape of two important parts of the male genitalia in this species. When sperm competition occurred, the researchers found that males with certain types of genitalia—those with genitalia shapes plotted at the lower left of Figure 48.6b—fathered a higher percentage of offspring than did others. In this and perhaps many other species, variation in the size and shape of male genitalia has a strong impact on male reproductive success. The exact mechanism responsible for this pattern is not known.

(a) Male genitalia in the spider *Zosis geniculatus*

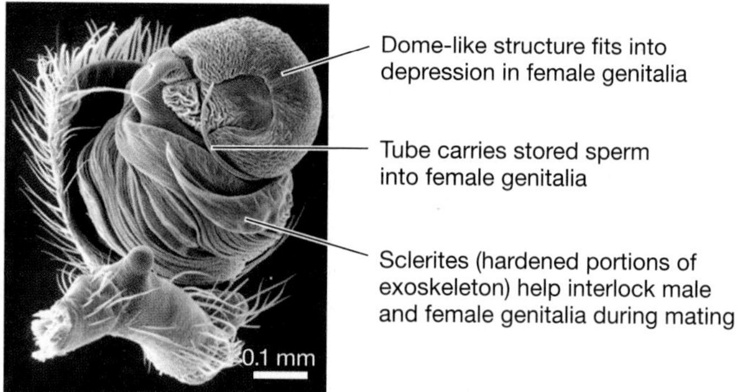

(b) In a population of water striders, male genitalia shape is highly variable.

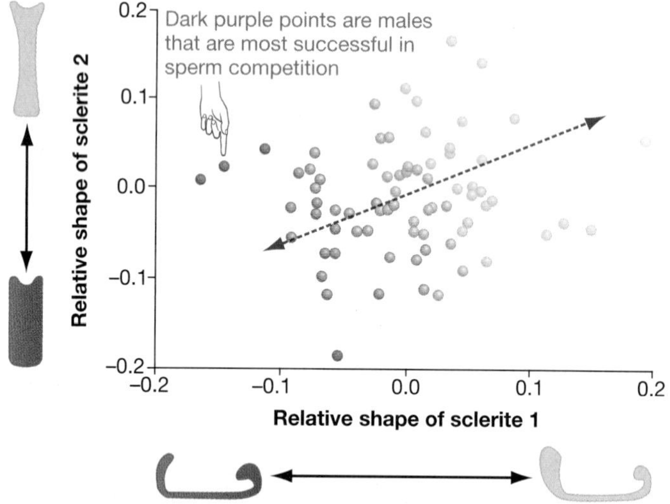

FIGURE 48.6 In Insects, Male Genitalia May Vary among Species and Individuals and May Affect Reproductive Success. (a) An elaborate male reproductive structure, used to transfer sperm to the female, in a species of spider. In insects and spiders, species that look identical often have distinctive genitalia. **(b)** Plot of the relationship between the size and shape of two parts of male genitalia in water striders.

External and Internal Anatomy of Human Males

Figure 48.7 furnishes side and front views of the internal reproductive structures in a human male. Although numerous structures are involved, notice that the reproductive system in males has just three basic functional components:

1. *Spermatogenesis and sperm storage* Sperm are produced in the testes and stored in the **epididymis** nearby.
2. *Production of additional fluids* Complex solutions form in the **seminal vesicles**, **prostate gland**, and **bulbourethral gland**. These solutions, called accessory fluids, are added to sperm prior to **ejaculation**, or expulsion from the body. **Table 48.1** lists some components of these accessory fluids. **Semen** is the combination of sperm and accessory fluids.

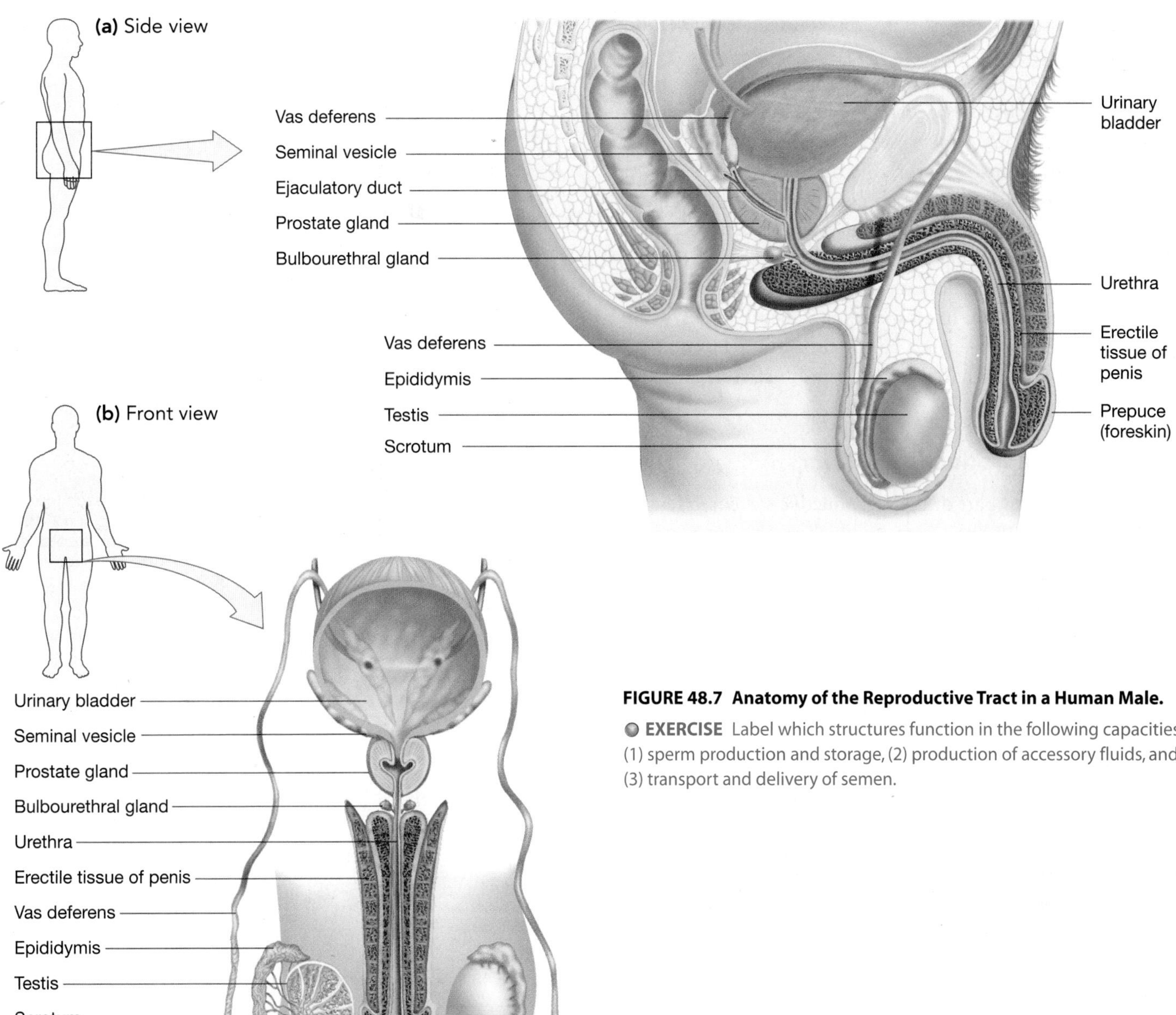

FIGURE 48.7 Anatomy of the Reproductive Tract in a Human Male.

EXERCISE Label which structures function in the following capacities: (1) sperm production and storage, (2) production of accessory fluids, and (3) transport and delivery of semen.

3. *Transport and delivery* The **vas deferens** are a pair of muscular tubes that store sperm and transport semen from the epididymis to the short **ejaculatory duct**. The semen then enters the **urethra**, a longer tube that passes through the penis and services both the reproductive and urinary systems in males. The semen is expelled during ejaculation.

How do these structures vary among animals? Variation in relative testis size among mammal species has already been discussed. In addition, the composition of the accessory fluids varies widely. In many insects, spiders, and vertebrates, molecules in the accessory fluids congeal after they arrive in the female reproductive tract and plug it. Experiments have shown that these copulatory plugs can serve as an effective deterrent

TABLE **48.1 Accessory Fluids in Human Semen**

Source	Content	Function
Seminal vesicles	Fructose (a sugar)	Source of chemical energy for sperm movement
	Prostaglandins	Stimulate smooth-muscle contractions in uterus
Prostate gland	Antibiotic compound	Prevent urinary tract infections in males?
	Citric acid	Nutrient used by sperm
Bulbourethral gland	Alkaline mucus	Lubricates tip of penis; neutralizes acids in urethra

to future matings; in some species, though, females or second males actively remove them.

Another diverse aspect of male internal anatomy is a bone inside the penis called the **baculum**. Some mammal species, including humans, lack this feature. But among rodents, the shape of the baculum is so variable that it can be used as a characteristic that distinguishes species. And in seals, baculum size appears to correlate with mating system. In seal species in which females routinely mate with several males before becoming pregnant, males have not only large testes for their size, but a large baculum as well. In all species in which it appears, the baculum helps stiffen the penis during copulation. **Box 48.1** introduces other usual aspects of male reproductive organs and fertilization.

The Female Reproductive System

The most important part of the reproductive system in female animals is the ovary, the organ in which meiosis occurs and mature egg cells, or ova, are produced. In the vast majority of species, the mature egg cell is a membrane-bound structure consisting of a haploid nucleus, a full complement of other organelles, and a large supply of nutrients in the form of yolk (see Chapter 22).

Earlier chapters have analyzed variations on the basic theme of egg structure. Chapter 22 introduced the complex outer membranes of sea urchin eggs, which play a key role in binding sperm from the same species and initiating fertilization. The amniotic egg was featured in Chapter 34 as an innovation that allowed tetrapods to reproduce efficiently in dry environments. Chapter 41 pointed out that the amount of yolk present in an egg correlates strongly with the size of offspring at hatching. That chapter also presented data indicating that females face a trade-off between egg size and the total number of eggs they can produce.

To probe variation in egg structure and female reproductive systems further, let's consider two highly specialized examples: The reproductive systems of female birds and humans. Birds are an example of a vertebrate that lays an amniotic egg protected by a hard shell; humans are an example of a viviparous vertebrate.

The Reproductive Tract of Female Birds **Figure 48.8** diagrams the bird reproductive system. Starting with the ovary, the labels on the drawing indicate the sequence of events that takes place as the egg moves down the reproductive tract. The result is a hard-shelled egg—like the familiar chicken eggs you buy in the store—that can be laid into the environment and incubated. From the time the egg is released from the ovary until the zygote undergoes mitosis and the embryo begins to develop, a bird egg is a single cell. The ostrich egg, which can be over 15 cm (6 in) in diameter, contains one of the largest single cells known in animals. The structure contains enough nutrients and water to sustain development until hatching.

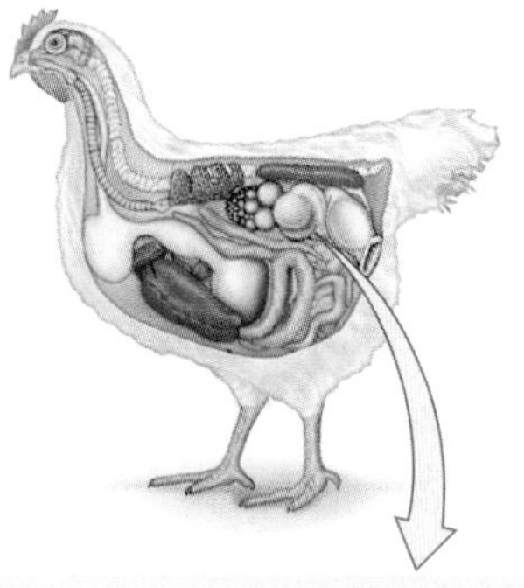

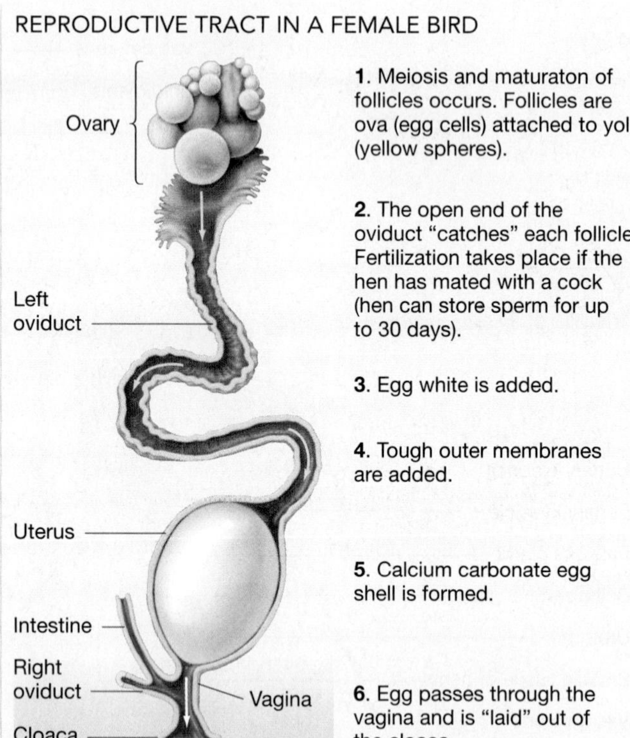

FIGURE 48.8 Anatomy of the Reproductive Tract in a Female Bird.

EXERCISE Two branches of the female reproductive tract empty into the cloaca, but only one is functional. Circle the vestigial branch.

Note that although male birds have two testes, females have just one ovary. The presence of a single ovary is thought to be an adaptation that reduces weight and makes flight more efficient.

External and Internal Anatomy of Human Females **Figure 48.9** shows side and front views of the anatomy of the human female reproductive system. The external anatomy features the **labia minora** (singular: **labium minus**) and the **labia majora** (singular: **labium majus**), the opening of the urethra, and the opening of the vagina. The labia are folds of skin that cover the urethral and vaginal openings. The **clitoris** is a sensitive organ that develops from the same population of embryonic cells that give rise to the penis in males. It likewise becomes erect during sexual stimulation and is covered with a protective sheath called the *prepuce*, homolo-

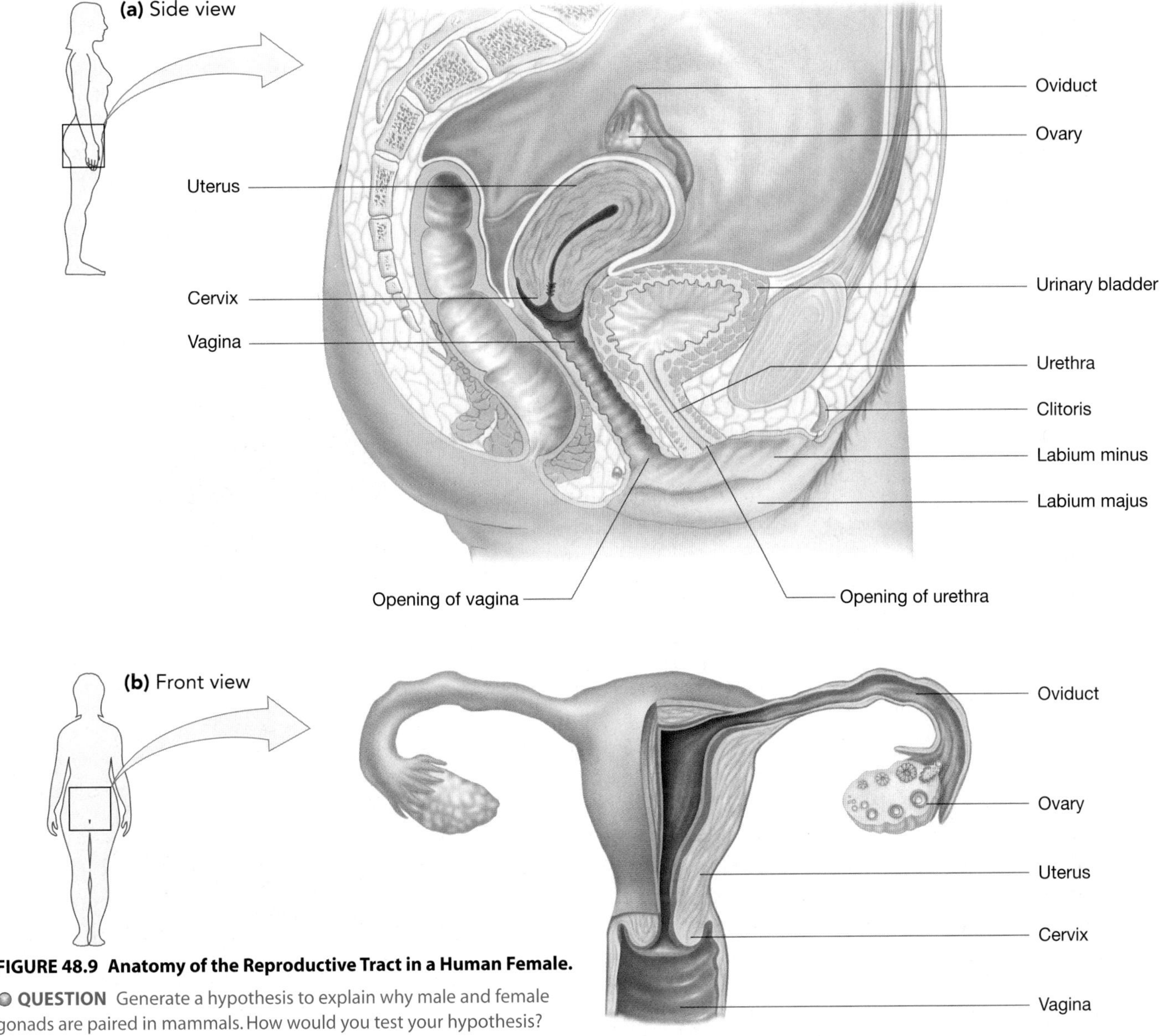

FIGURE 48.9 Anatomy of the Reproductive Tract in a Human Female.

QUESTION Generate a hypothesis to explain why male and female gonads are paired in mammals. How would you test your hypothesis?

gous with the prepuce (foreskin) that covers the end of the penis. In females, however, the urethral opening, where urine is expelled, is separate from the reproductive structures. The **vagina**, or birth canal, is the chamber where semen is deposited during sexual intercourse and where the baby is delivered during childbirth.

The internal anatomy of the female reproductive system in humans and other mammals is dominated by structures with two functions:

1. *Production and transport of eggs* Eggs are produced in the paired ovaries. During **ovulation**, a developing oocyte (egg) is expelled from the ovary and enters the **oviduct**, also known as the **fallopian tube**, where fertilization may take place. As Section 48.5 will detail, fertilized eggs are then transported from the oviduct to the muscular sac called the **uterus**.

2. *Development of offspring* The uterus is where embryonic development takes place. During childbirth, the developed embryo (now called a fetus) passes through the opening of the uterus, known as the **cervix**, and into the vagina.

To appreciate how the ovaries and uterus work in humans, it's essential to explore the monthly menstrual cycle and pregnancy in much more detail. It is not possible to understand either event, though, without investigating the hormones that regulate them. Let's turn now to research investigating the sex hormones of mammals.

BOX 48.1 Unusual Aspects of Fertilization

Studies on animal sexual reproduction have documented some of the most remarkable behaviors and structures observed in nature. The list that follows is by no means exhaustive. It is a sample, offered simply because the diversity of mating arrangements in animals is fascinating to biologists and laypeople alike.

- *Femme fatales* When Australian redback spiders mate, the male does a somersault after inserting his penis-like organ. The somersault behavior places his dorsal surface in front of the female's mouthparts. The female then eats the male. (Biologists refer to females who cannibalize males as femme fatales—a phrase used to describe murderous human females in movies.) Experiments have shown that males that are eaten copulate longer and fertilize more eggs than do males that survive copulation, probably because sperm transfer continues until the meal is over.
- *Giant sperm* In the fruit fly *Drosophila bifurca*, males average 1.5 mm in total body length. Their sperm, however, are each 6 cm long. Recent research suggests that the coiled sperm fill the female's sperm storage area and prevent fertilization by competing males—the bulk of the sperm makes it impossible for another male's sperm to enter.
- *False penises* Although very few bird species have a penis, the red-billed buffalo weaver has a false penis that becomes erect during copulation (**Figure 48.10**). It does not function in sperm transfer, but appears to stimulate the female's reproductive tract and trigger an orgasm-like state in males. Female buffalo weavers have a similar, but smaller, organ of unknown function.
- *Infidelity* Researchers who use genetic markers to assess paternity in birds that appear to be monogamous find that up to 60 percent of nests contain at least one offspring fathered by a male that is not mated to the resident female. In most cases, females actively solicit copulations from males holding nearby territories.
- *Love darts* Many snails and slugs are **hermaphroditic**, meaning an individual has both male and female gonads. During mating, two individuals simultaneously receive and deposit sperm to fertilize each other's eggs. In some species, individuals fire mucus-covered "love darts" from their genitalia into the mating partner. Recent experiments have shown that a substance in the mucus increases the chance that sperm will successfully fertilize eggs. In other species, the penis frequently becomes stuck in the female reproductive tract and is bitten off by one partner.
- *Hypodermic insemination* Some bedbugs have hypodermic penises, meaning that they function like a hypodermic needle. Males force the organ through the female's abdominal wall and deposit sperm directly into her body cavity.

Did the process of sexual selection, introduced in Chapter 25, lead to the evolution of these structures and behaviors? If so, how and why? In many or even most cases, the answers to these questions are not known.

FIGURE 48.10 A False Penis in a Weaverbird. The organ shown here, in the erect state, is called a false penis because it is not inserted in the female's reproductive tract and does not transmit semen.

48.4 The Role of Sex Hormones in Mammalian Reproduction

Chapter 47 introduced the male sex hormone **testosterone** and the female sex hormone **estradiol**, which belongs to a class of hormones known as **estrogens**. Recall that both testosterone and estradiol are steroids, that both bind to receptors inside the nucleus of target cells, and that the resulting hormone-receptor complexes bind to DNA and trigger changes in gene expression. Both are also considered gonadal hormones because they are produced in the gonads. Testosterone is synthesized in specialized cells inside the testes; estradiol and other estrogens are synthesized in the ovaries. More specifically, the female sex hormones are produced by cells that surround each developing egg. These surrounding cells form a structure called a follicle around the egg.

As Chapter 47 pointed out, the human sex hormones play a key role in three events: (1) the development of the reproduc-

TABLE 48.2 **Changes That Take Place during Puberty**

Pediatricians use the five stages of puberty listed here to diagnose the developmental stage of their patients.

	Boys	Girls
Stage 1 (Prepubertal)	No sexual development	No sexual development
Stage 2	Testes enlarge Body odor	Breast budding First pubic hair Body odor Height spurt
Stage 3	Penis enlarges First pubic hair Ejaculation (wet dreams)	Breasts enlarge Pubic hair darkens, becomes curlier Vaginal discharge
Stage 4	Continued enlargement of testes and penis Penis and scrotum deepen in color Pubic hair curlier and coarser Height spurt Male breast development (temporary)	Onset of menstruation Nipple is distinct from surrounding areola Pelvis begins to widen
Stage 5 (Fully mature adult)	Pubic hair extends to inner thighs Increases in height slow, then stop Increased muscle mass	Pubic hair extends to inner thighs Increases in height slow, then stop Deposition of fat in hips, buttocks, thighs, and breasts

tive tract in embryos, (2) the maturation of the reproductive tract during the transition from childhood to adulthood, and (3) the regulation of spermatogenesis and oogenesis in adults. To begin exploring the action of sex hormones in more detail, let's take a closer look at their role in the transition from juveniles to adults.

Which Hormones Control Puberty in Mammals?

Table 48.2 lists some of the changes that occur in boys and girls as they undergo **puberty**—the process that leads to sexual maturation in humans. Chapter 47 pointed out that in amphibians, the juvenile-to-adult transition is triggered by the hormone T_3; in insects the transition to adulthood occurs in response to ecdysone. But in humans, physicians and researchers have long known that the transition occurs in response to increased levels of gonadal hormones—testosterone in boys and estradiol in girls. A group of physicians recently offered dramatic evidence for testosterone's role. They were presented with a 2-year-old boy who showed signs of entering puberty. His symptoms included an increased penis size, the development of pubic hair, and facial acne. Through careful interviewing, the doctors determined the cause of his condition. The child's father was a bodybuilder who was smearing a testosterone cream on his own shoulders and arms in an attempt to build muscle mass. The child was apparently absorbing enough testosterone from being carried by his father to trigger puberty-like symptoms. Fortunately for the 2-year-old, all his symptoms except for penile enlargement subsided once his father discontinued the use of testosterone creams. **Box 48.2** introduces other unfortunate side effects of steroid use by athletes and bodybuilders.

Although the roles of testosterone and estradiol in puberty are well documented, understanding how the process is initiated proved to be a stubborn problem. Some researchers hypothesized that the sex hormones are regulated by the hypothalamic-pituitary axis, introduced in Chapter 47. Recall that chemical signals from the hypothalamus region of the brain lead to the release of regulatory hormones from the pituitary gland, which then cause the release of hormones from other glands. Are the hypothalamus and pituitary involved in regulating the release of sex hormones?

Two advances made it possible to answer this question definitively and affirmatively. First, researchers isolated a hormone called **gonadotropin-releasing hormone** (**GnRH**) from the hypothalamus. Second, investigators noted that boys and girls who were entering puberty experienced distinctive pulses in the concentration of two pituitary hormones, called **luteinizing hormone** (**LH**) and **follicle-stimulating hormone** (**FSH**). Researchers hypothesized that pulses of LH and FSH occurred in response to the release of GnRH and were responsible for increases in testosterone and estradiol. To test the hypothesis of hypothalamus and pituitary control that is diagrammed in **Figure 48.11**, researchers administered pulses of GnRH to boys and girls who had deficits in their hypothalamus and were

BOX 48.2 Abuse of Synthetic Steroids

Among bodybuilders and athletes, the use of synthetic testosterone-like molecules, commonly called **androgens**, has recently exploded in popularity. Both men and women use these drugs because they support the development of increased muscle mass during weight training, just as testosterone supports increased muscle mass during puberty.

Research on bodybuilders has confirmed a wide variety of damaging side effects, however. Women who use synthetic androgens routinely experience voice deepening and growth of facial hair. Both male and female users may develop acne, the emotional changes dubbed 'roid rage (for steroid rage), fluid retention, and liver ailments. There is some evidence that males may also experience fertility problems. The leading hypothesis to explain infertility is that the synthetic androgens mimic the negative feedback that testosterone exerts on luteinizing hormone (LH) production from the pituitary. LH is required to support spermatogenesis, so high doses of androgens may reduce LH concentrations to a level that precludes sperm production.

Unfortunately, opinion polls have shown that most bodybuilders and athletes are willing to tolerate these side effects in hopes of increasing their performance in competitions. In response, some regulatory agencies actively test for synthetic androgens in the urine of competitors and ban individuals who are using them.

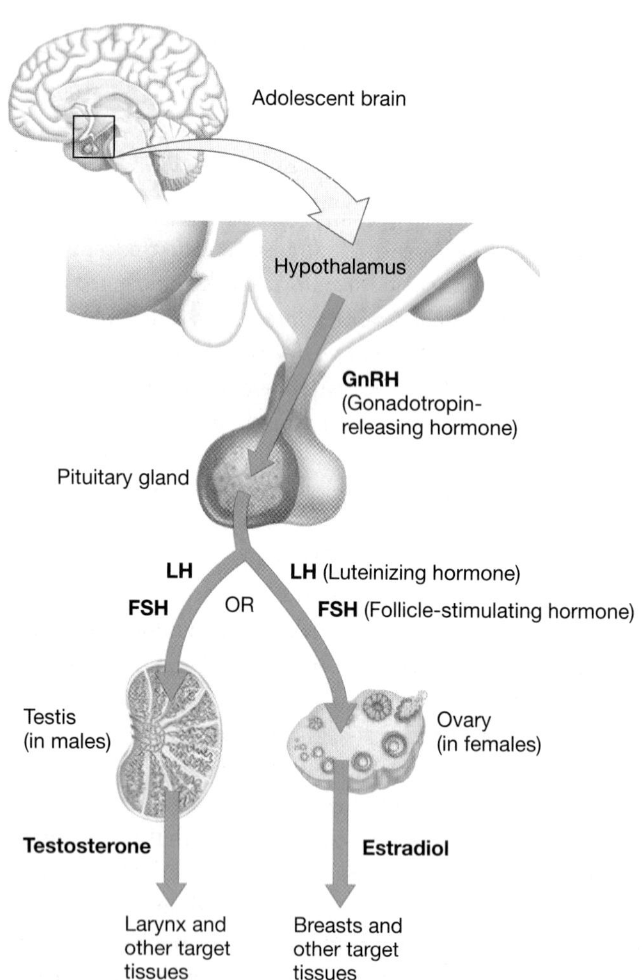

FIGURE 48.11 In Humans, Puberty Is Triggered by Hormones from the Hypothalamus and Pituitary.

● **QUESTION** How does control of testosterone and estradiol production compare with control of cortisol release by the adrenal gland? (See Figure 47.14c.)

experiencing delays in the onset of puberty. As predicted, the GnRH treatment induced surges in LH and FSH that triggered the developmental events listed in Table 48.2. This result provided strong support for the hypothesis that a hormone from the hypothalamus is responsible for initiating puberty.

What triggers GnRH increases at the appropriate age? This is the most pressing question facing researchers and physicians interested in hormonal control of sexual development. Although the question remains unanswered, there is some evidence that nutritional state is involved. For example, girls with large fat stores tend to enter puberty earlier than do girls who are thin. In addition, the current average age for the onset of menstruation in females in the United States is slightly over 12 years. This is much earlier than the average of 17 years of age that had been observed in the United States during the eighteenth and nineteenth centuries, when the general nutritional state of the population was poorer. Research continues on how age, nutritional condition, and perhaps other factors interact to promote the release of GnRH.

If you recall the discussion in Chapter 47 of how the adrenal hormone cortisol is controlled, however, you might suspect that the model of sex-hormone regulation diagrammed in Figure 48.11 is probably simplified. Chapter 47 emphasized that many hormones participate in negative feedback—also called feedback inhibition—meaning that the presence of a hormone inhibits the factor that triggers its release. Do sex hormones participate in negative feedback? The short answer to this question is yes. To appreciate the details, let's investigate hormonal control of the human menstrual cycle.

Which Hormones Control the Menstrual Cycle in Mammals?

Figure 48.12 illustrates the sequence of events in the human **menstrual cycle**, a monthly reproductive cycle that occurs in the uterus. Although the length of this cycle varies among

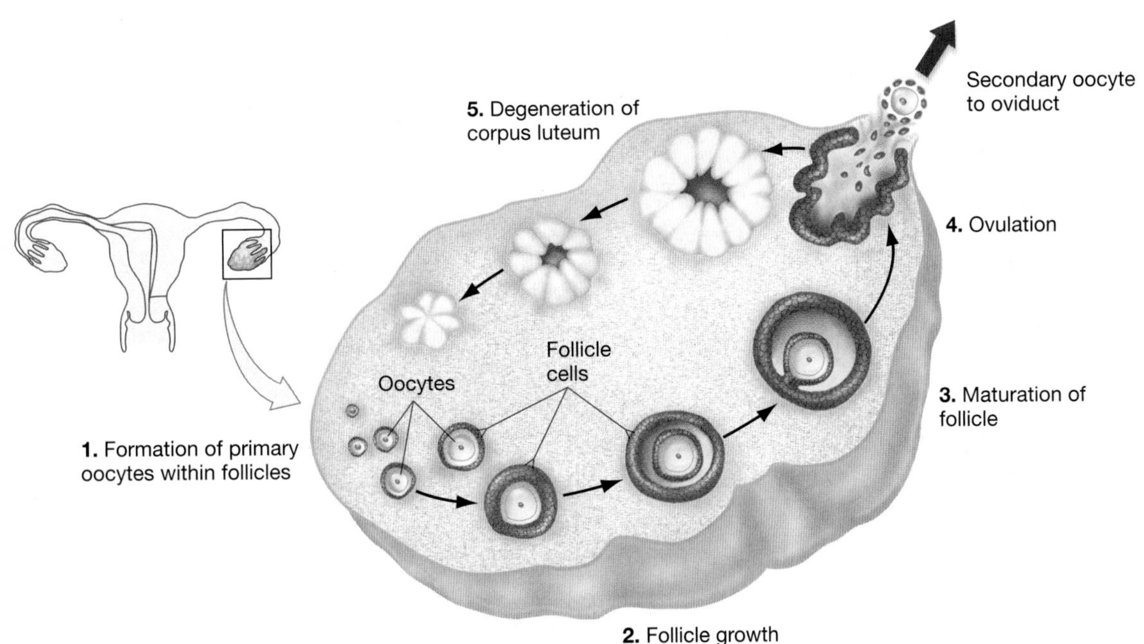

FIGURE 48.12 The Ovary Changes over the Course of a Menstrual Cycle. As a follicle matures, primary oocytes complete meiosis I. The secondary oocyte that results is expelled from the ovary and then travels down the oviduct. (Follicles don't actually move in a circle during the menstrual cycle; they are shown that way for illustrative purposes only.)

QUESTION Which events illustrated here occur during the follicular phase of the menstrual cycle? Which occur during the luteal phase?

women, 28 days is about average. In conjunction with changes in the ovary illustrated in the figure, the lining of the uterus undergoes a dramatic thickening and regression; ultimately, part of the lining sloughs off and is expelled through the vagina. The onset of the combined ovarian and uterine cycles is the start of **menstruation**—the expulsion of the uterine lining. This event is designated as day 0 in the menstrual cycle.

The cycle itself has two distinct phases. A follicle matures during the **follicular phase**, which lasts an average of 14 days. Primary oocytes complete meiosis I during this phase (illustrated in Figure 48.3). Ovulation occurs when the follicle is mature and releases its secondary oocyte into the oviduct. (Researchers are still unsure what controls which ovary—right or left—will release an egg during ovulation.) Although the timing of ovulation is highly variable, the subsequent **luteal phase** also averages 14 days in length. Its name was inspired by the formation and subsequent degeneration of a structure called the **corpus luteum** ("yellowish-body") from the ruptured follicle. Meiosis II occurs only if the ovulated cell is actually fertilized (see Section 48.5).

The regular occurrence of a menstrual cycle throughout the year makes human females extremely unusual among mammals. Although some mammals ovulate multiple times during the year, most ovulate only during a prescribed breeding season—often in response to environmental cues; more infrequently in response to cues from males. In addition, only humans and other great apes menstruate. In the vast majority of mammals, the lining of the uterus is reabsorbed if pregnancy does not occur. Females that do not menstruate have an estrous cycle and are sexually receptive only during estrus—also known as being "in heat." But whether an estrous or menstrual cycle occurs, the basic sequence of events, with a follicular phase preceding ovulation and a luteal phase following ovulation, is shared among mammals. As it turns out, hormonal control of the estrous and menstrual cycles is also similar.

How Do Pituitary and Ovarian Hormones Change during a Menstrual Cycle? Signals from hormones are responsible for the changes that occur in the uterus and ovaries during a menstrual cycle. For biologists who wanted to probe how hypothalamic, pituitary, and ovarian hormones interact to regulate the menstrual cycle, the first task was to document how each hormone changed over time. By monitoring hormone concentrations in the blood or urine of a large number of women, researchers were able to document dramatic changes in the concentrations of estradiol and several other hormones. LH and FSH are produced in the pituitary gland during each cycle in response to GnRH; the steroid hormone **progesterone** is produced along with estrogens, including estradiol, in the ovaries.

Several observations jump out of the data in **Figure 48.13**. LH levels are fairly constant except for a spike that begins just prior to ovulation, suggesting that LH might be the trigger

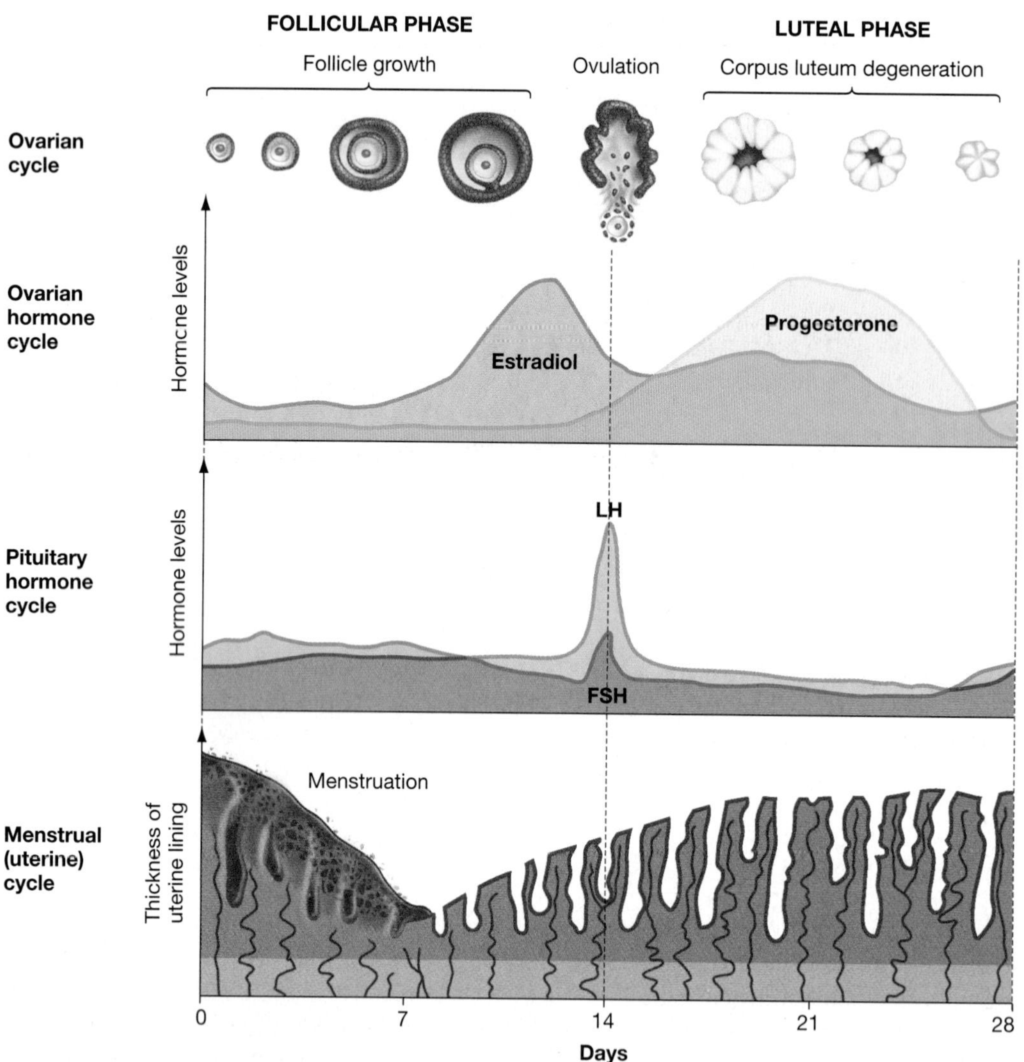

FIGURE 48.13 Changes That Occur during a Menstrual Cycle in a Human Female. During a menstrual cycle, changes occur in the concentrations of the pituitary hormones LH and FSH, the ovary, concentrations of the ovarian hormones estradiol and progesterone, and the lining of the uterus. At the beginning of the cycle, the thickened lining of the uterus sloughs off and is expelled through the vagina.

for this event. FSH concentrations, in contrast, are relatively high during the follicular phase and low during the luteal phase, though they also make a small spike prior to ovulation. Progesterone is present at very low levels during the follicular phase but at high levels during the luteal phase. This observation suggested that progesterone might support the maturation of the thickened uterine lining. Estradiol concentrations change in a much more complex way, highlighted by a peak late in the follicular phase. In general, estradiol surges during the follicular phase while progesterone surges during the luteal phase. Similar patterns are observed in other mammals, ranging from marsupials to mice. How are these changes regulated?

Web Animation at www.masteringbio.com
Human Reproduction

How Do the Pituitary and Ovarian Hormones Interact?

Experiments helped establish that changes in the concentration of estradiol and progesterone affect the release of the pituitary hormones LH and FSH. Researchers worked with three volunteers whose ovaries had been removed because of cancerous growths or other problems. The women were receiving low, maintenance-level doses of estradiol, which appeared to exert negative feedback on LH and FSH. But when the investigators injected the women with larger doses of estradiol or with progesterone, the situation changed dramatically. For example, a large increase in estradiol stimulated a dramatic spike in LH levels. This suggested that positive feedback was occurring—that is, high levels of estradiol increased the release of its regulatory hormone, even though low doses of estradiol suppressed it. Injections of progesterone, in contrast, appeared to inhibit both FSH and LH. Progesterone exerts only negative feedback on the pituitary hormones.

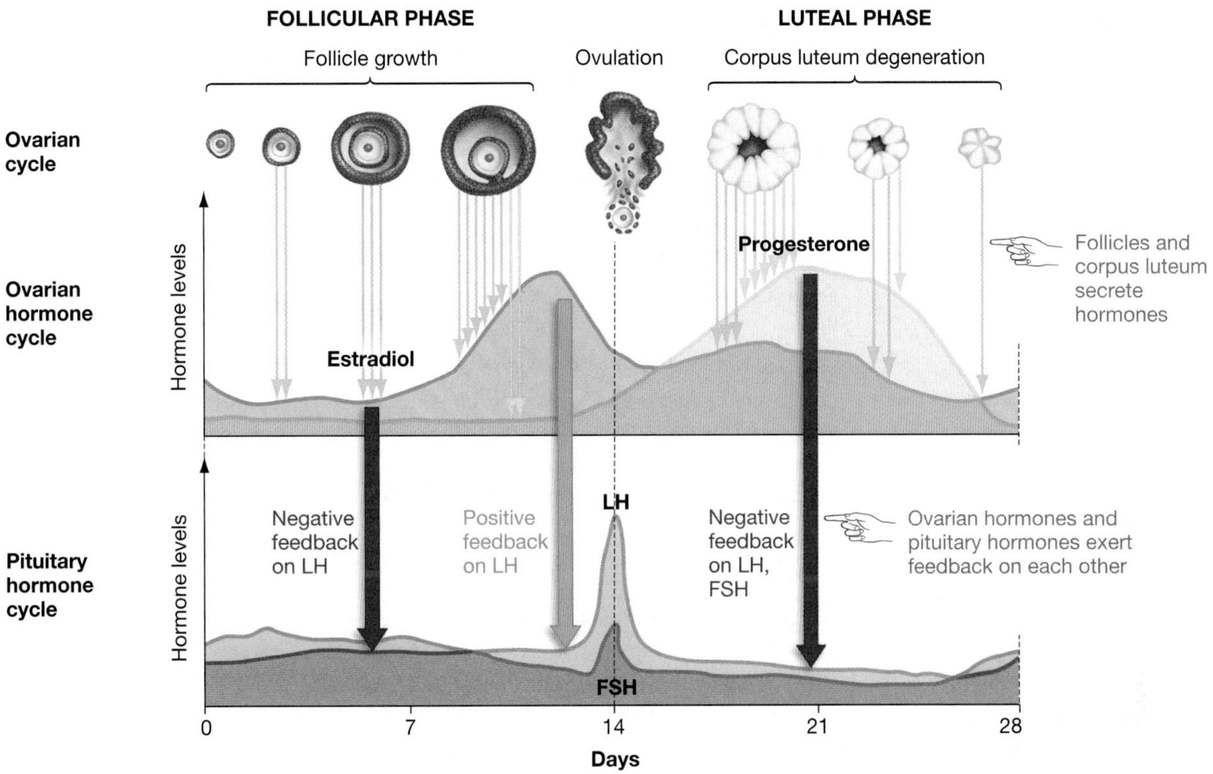

FIGURE 48.14 The Menstrual Cycle Is Regulated by Complex Interactions among Hormones. The pituitary hormones FSH and LH control the release of estradiol and progesterone from reproductive tissues. Estradiol and progesterone may exert negative feedback, positive feedback, or both—depending on their concentration—on the pituitary hormones.

Thanks to dozens of experiments like these, conducted in a wide variety of labs over the past several decades, researchers now have a fairly good grasp of how the various hormones interact. To summarize the interplay between LH, FSH, estradiol, and progesterone, let's start at day 0 and follow key events as the cycle progresses.

As a cycle begins, the uterus is shedding much of its lining during menstruation. In the ovary, a follicle is being stimulated to develop under the influence of FSH. As the follicle grows, its production of estradiol gradually begins to increase. The increase in estradiol stimulates mitosis and an increase in cell number in the uterine lining. The follicle also produces a small amount of progesterone. While estradiol levels are still relatively low, that hormone suppresses LH secretion through negative feedback inhibition. Once the follicle has grown enough to produce large quantities of estradiol, however, the hormone begins to exert positive feedback on LH secretion. As Figure 48.13 shows, this positive feedback results in a spike in LH levels just after estradiol concentrations peak. The LH surge triggers ovulation and ends the follicular phase. If you understand how estradiol's effect on LH changes from negative regulation to positive regulation, you should be able to predict the effect of (1) injecting a female volunteer with high estradiol concentrations early in a cycle, and (2) using a drug to keep estradiol production low throughout the cycle.

As the corpus luteum develops from the remains of the ruptured follicle, it secretes large amounts of progesterone and small quantities of estradiol in response to LH. This rise in progesterone lowers production of LH and FSH and activates the thickened uterine lining, creating a spongy tissue with a well-developed blood supply. In this way, progesterone creates an environment that supports embryonic development if fertilization occurs.

If fertilization does not occur, however, the corpus luteum degenerates approximately 12 days after ovulation. Progesterone levels fall as a result, which causes the thickened lining of the uterus to degenerate. Once progesterone levels have declined, the hypothalamus and pituitary are released from the inhibitory control progesterone exerts. LH and FSH levels rise, and a new cycle begins. The interactions between FSH, LH, estradiol, and progesterone are summarized in **Figure 48.14**, and the interplay between ovarian and pituitary hormones is similar in other mammal species that have been studied to date.

Manipulating Hormone Levels to Prevent Pregnancy

Data on hormonal control of the menstrual cycle opened new avenues in birth control research. Specifically, researchers have found that manipulating levels of progesterone and estrogen can prevent ovulation and serve as a safe and effective method of preventing unwanted pregnancies.

TABLE **48.3 Comparing Methods of Contraception**

See text for explanation of "Percent Effectiveness."

Type	Name	Mode of action	Percent Effectiveness
Hormone-based methods	The Pill, the Patch, the Ring, the Shot, the Implant	Provides continuous or cyclical delivery of progesterone, or progesterone plus estradiol	92 to 99.9*
Barrier methods	Condom	Covers penis and prevents sperm from entering vagina	85
	Female condom	Covers labia, vagina, and cervix and prevents sperm from entering uterus	79
	Diaphragm	Covers cervix and prevents sperm from entering uterus	84
	Sponge	Covers cervix and prevents sperm from entering uterus; also contains a molecule that immobilizes sperm	84
	Spermicide	Foam or jelly covers cervix and prevents sperm from entering uterus; contains a molecule that immobilizes sperm	71
Behavioral methods	Rhythm method	Couple refrains from vaginal intercourse around time of ovulation	80
	Withdrawal	Man withdraws penis prior to ejaculation	73
Prevent implantation	Intrauterine device (IUD)	Small T-shaped structure inserted into uterus; mode of action not known	99
	Emergency contraception	Delivers progesterone or progesterone and estradiol after unprotected vaginal intercourse	92
	RU-486	Blocks progesterone receptors so menstruation occurs even after fertilization and implantation	92

*Depends on delivery system used.

Hormone-based birth-control methods deliver synthetic versions of progesterone or progesterone and estradiol. These hormones suppress the release of GnRH and FSH through negative feedback inhibition. Because an LH spike does not occur during the follicular phase of the cycle, the follicle does not mature and ovulation does not occur.

If a woman uses the hormones in pill form, they are taken in a cyclic fashion. Hormone-containing pills are taken for three weeks and then stopped for one week to allow menstruation to occur. A newer method of inhibiting ovulation involves inserting a small implant of synthetic hormone under the skin; these implants are extremely effective and last for up to five years. In the United States, birth control pills are the most widely used contraceptive method.

As **Table 48.3** indicates, however, hormone-based methods are just one of several distinct approaches to preventing pregnancy. Other types of methods work by preventing sperm from contacting the oocyte, or by interfering with implantation of the embryo.

The "Percent Effectiveness" column on the far right of the table indicates the average percentage of women who do not become pregnant during one year of typical use of a method. In many cases percent effectiveness increases dramatically if couples are able to use a method "perfectly"—meaning, exactly as specified for optimal effectiveness, during every episode of sexual intercourse.

When sperm and egg do unite successfully, the menstrual cycle is interrupted. The corpus luteum does not degenerate, and progesterone and estradiol levels stay high. Menstruation does not occur. Instead, the mother is now pregnant.

Check Your Understanding

If you understand that...

An array of hormones interacts in complex ways to regulate the human menstrual cycle.

You should be able to...

1) Describe the function of FSH, LH, estradiol, and progesterone during one cycle.
2) Explain the consequences of positive feedback between estradiol and LH and negative feedback between progesterone and LH and FSH.
3) Predict the consequences of a drug that inhibits the release of any one of these four hormones during a cycle.

48.5 Pregnancy and Birth in Mammals

Recall from Section 48.2 that viviparity allows the mother to provide a warm, protected environment for offspring during early development. Oviparous species that guard or incubate their eggs are also providing warm, safe surroundings for young offspring. The investments that parents make in each offspring comes at a cost, however: The more a mother invests in each offspring, the lower the number of offspring she can produce. This fitness trade-off between the quantity and quality of offspring was explored in detail in Chapter 41.

Pregnancy and **lactation**—providing milk that nourishes offspring after birth—represent some of the most extreme forms of parental care known in animals. And in some mammal species, parental care continues long after lactation ends. Humans, for example, are largely or completely dependent on their parents for protection and nutrition until puberty or young adulthood.

Let's examine how eutherian mammals make this investment, starting with fertilization and ending with birth, again using humans as a well-studied model organism.

Major Events during Pregnancy

Once it is released from the human ovary, a secondary oocyte is viable for less than 24 hours. Human sperm, in contrast, remain capable of fertilizing an egg for up to five days. Therefore, sexual intercourse in humans has to occur less than five days prior to ovulation, or immediately after ovulation, for pregnancy to result.

Although an ejaculate may contain hundreds of millions of sperm, most die as they travel through the uterus. Only 100 to 300 actually succeed in reaching the oviduct, where fertilization takes place.

Chapter 22 detailed how sperm and oocyte (egg) interact when they meet. Recall that contact between sperm and oocyte triggers the **acrosomal reaction**, in which enzymes released from the head of the sperm chew a path through the material surrounding the oocyte membrane. Once the membranes of the oocyte and sperm have fused, the oocyte nucleus completes meiosis II. The two nuclei then unite to form a diploid zygote.

Smooth-muscle contractions in the oviduct gradually move the zygote toward the uterus. As it travels, the cell begins to divide by mitosis. By the time it reaches the lining of the uterus, the embryo consists of a hollow ball of cells. It then undergoes **implantation,** becoming embedded in the thickened, vascularized wall of the uterus. It will stay in the uterus for the full period of development, or **gestation**, of approximately 270 days (9 months) in humans. **Box 48.3** contrasts the long gestation periods observed in humans and other eutherian mammals with the pattern observed in marsupials.

BOX 48.3 How Do Gestation and Early Development Occur in Marsupials?

You might recall from Chapter 34 that mammals comprise three major monophyletic groups: the monotremes, the eutherians, and the marsupials. Although all mammals are endothermic, have fur, and lactate, the three lineages are distinguished by their mode of reproduction. **Monotremes** lay eggs and incubate them until hatching. In the **eutherians**, mothers carry offspring for relatively long periods of development during gestation and nourish them via a placenta. But in **marsupials**, the corpus luteum is not maintained as it is in eutherians, and no placenta forms. Instead, young are ejected from the mother's body at the end of the estrous cycle, when they are far less developed than eutherian mammal offspring completing gestation.

Figure 48.15a shows a newly born kangaroo. The jaws, gut, lungs, and forelimbs are relatively well developed at birth. As a result, the offspring is able to climb from its mother's vagina to a nipple, which is usually enclosed in a pouch created by a flap of skin. The offspring clamps on and continues to develop, fed by the mother's milk (**Figure 48.15b**). Even after growing large enough to leave the pouch and begin moving and feeding on its own, offspring will return to the pouch for protection (**Figure 48.15c**).

(a) Red kangaroo at birth

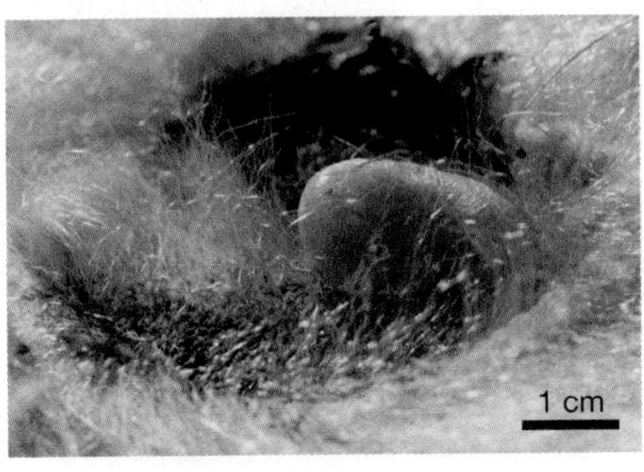

(b) 4 weeks after birth

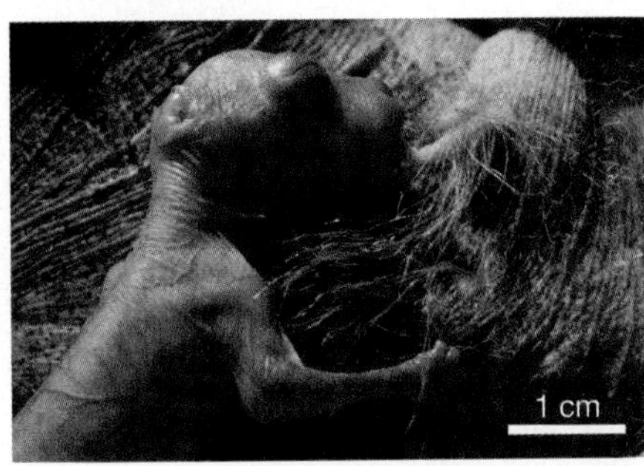

(c) 4 months after birth

FIGURE 48.15 Marsupials Trade a Long Gestation Period for a Long Lactation Period. Eutherian offspring spend a long time inside the mother being fed and protected and a relatively short time being fed milk after birth. Marsupial offspring spend a short time inside the mother and a relatively long time being fed milk after birth and being protected in a pouch.

Once the embryo has become implanted in the uterine lining, its cells begin synthesizing and secreting a hormone called **human chorionic gonadotropin (hCG)**. The hormone hCG is a chemical messenger that prevents the corpus luteum from degenerating. When hCG is present, the ovary continues secreting progesterone and the menstrual cycle is arrested. Enough hCG is produced by the embryo and excreted in the mother's urine to be used as an indicator in pregnancy tests.

The First Trimester In humans, gestation is divided into 3-month stages called trimesters. Not long after implantation is complete, mass movements of cells result in the formation of the three major embryonic tissues called ectoderm, endoderm, and mesoderm (see Chapter 22). By 8 weeks of age, these tissues have differentiated into the various organs and systems of the body. Also by this time, the heart has begun pumping blood through a circulatory system. The embryo at this stage is called a **fetus** (**Figure 48.16a**).

Also in the first trimester, the embryonic ectoderm contributes to several important membranes. One of these membranes, called the **amnion**, completely surrounds the embryo. The amnion eventually fills with amniotic fluid, which provides the embryo with a protective cushion.

The other key event in the first trimester is the formation of the **placenta**. This organ starts to form on the uterine wall a few weeks after implantation. The placenta is composed of tissues from both mother and embryo. Because it contains a dense supply of blood vessels from the mother, it is the primary source of nutrition for the growing fetus. Arteries transport blood from the circulatory system of the fetus, through the **umbilical cord**, to an extensive capillary bed in the placenta. In this way, the placenta provides a large surface area for the exchange of gases, nutrients, and wastes between maternal and fetal blood, even though maternal and fetal blood do not commingle. The placenta also secretes a variety of hormones. For example, by the end of the first trimester, placental cells synthesize and secrete large amounts of progesterone and estrogens—more than replacing the hormones produced by the corpus luteum, which has degenerated by this time. These hormones suppress the release of GnRH, LH, and FSH through negative feedback, and thus prevent the maturation and ovulation of additional follicles. ● If you understand how hormones influence pregnancy, you should be able to explain (1) why women who produce low levels of progesterone from the corpus luteum are prone to miscarriage, and (2) why progesterone therapy to help these women can be safely discontinued after the first trimester.

The Second and Third Trimesters After the fetal organs and placenta form during the first trimester, the remainder of development focuses on growth (**Figures 48.16b** and **48.16c**). During the last weeks of pregnancy, the brain and lungs undergo particularly dramatic growth and development. If a baby is born prematurely, the primary challenge to caregivers is to maintain an adequate oxygen supply. In many cases, dramatic intervention is required to keep the baby alive until the lungs can complete their development (see Box 44.1).

The machinery and level of hospital care required by premature infants emphasize just how superbly adapted mothers are to nourishing a growing fetus in the uterus. Let's take a closer look at this critical aspect of pregnancy.

(a) 1st trimester

5 mm

(b) 2nd trimester

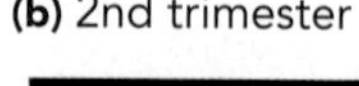

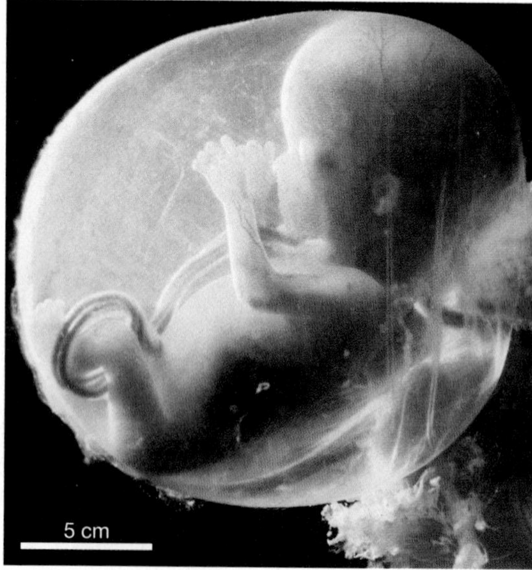

(c) 3rd trimester

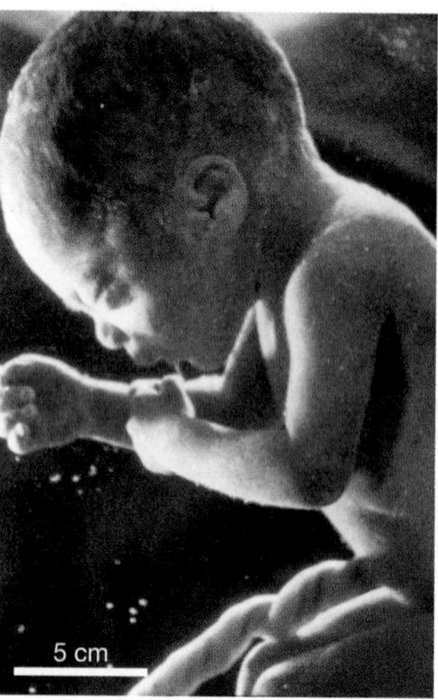

FIGURE 48.16 Development of the Human Fetus.

● **EXERCISE** On each photograph, label the hands, eye, and umbilical cord. Estimate the ratio of head length to total body length in each photograph.

How Does the Mother Nourish the Fetus?

In oviparous species, mothers provide their eggs with all of the nutrients and fluids they need to develop until hatching. But in some viviparous species, eggs are relatively small and contain almost no nutrient supply. In these species, the developing embryo depends on the mother's body for oxygen, chemical energy in the form of sugars, amino acids and other raw materials for growth, and waste removal. What physiological changes occur in human mothers to accommodate these demands?

To answer this question, researchers have focused on the body's basic transport and delivery systems: the heart, lungs, and blood vessels. During pregnancy, a mother's respiratory and circulatory systems change in ways that increase the efficiency of nutrient transfer and gas exchange with the fetus. For example, a woman's total blood volume expands by as much as 50 percent during pregnancy. To accommodate this increase, maternal blood vessels dilate (widen) and blood pressure drops. In addition, dramatic increases occur in the pumping capacity, or stroke volume, of the heart as well as in the heart rate. The mother's heart enlarges and begins to beat faster. The result is that a mother's total cardiac output increases by almost 50 percent over the course of a pregnancy.

The mother's breathing rate and breathing volume also increase to accommodate the fetus's demand for oxygen and its production of carbon dioxide. Important adaptations heighten the efficiency of gas exchange between the mother and the embryo as well. As **Figure 48.17a** shows, maternal and fetal blood flows in the placenta occur in a countercurrent fashion in many species. As Chapter 41 explained, countercurrent flows maintain a concentration gradient that increases the efficiency of diffusion or other types of exchange. The data for partial pressures of oxygen given in Figure 48.17a are from experiments on sheep.

Countercurrent flow does not occur in the human placenta, so fetal blood does not become as highly oxygenated as sheep blood. Still, several traits make oxygen exchange between mother and fetus efficient. For example, maternal arteries in humans empty into a space at the junction of the maternal and fetal portions of the placenta. This space is packed with small projections called villi, which contain the fetal blood vessels. Thus, a large surface area from the fetus is bathed with highly oxygenated maternal blood. The fetal villi are analogous to the alveoli of the lungs (Chapter 44), which provide a large surface area for gas exchange.

Figure 48.17b illustrates another adaptation that increases the rate of oxygen delivery to the fetus. Notice that the axes on this graph give the partial pressure of oxygen in blood or tissue versus the percentage of hemoglobin that holds oxygen at that pressure. (Recall from Chapter 44 that this type of graph is called an **oxygen-hemoglobin equilibrium curve**.) The

(a) Countercurrent blood flow in the placenta

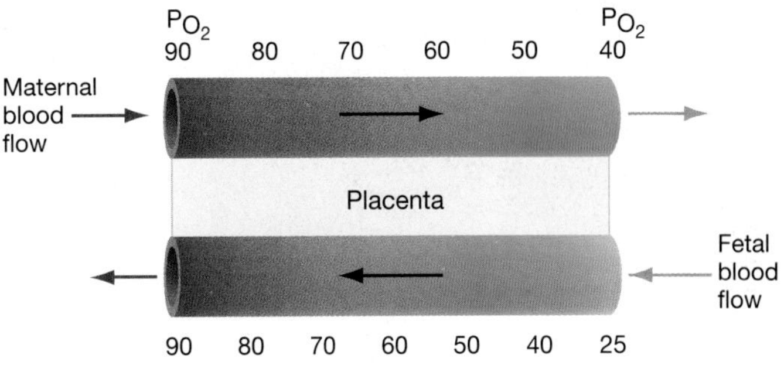

(b) Oxygen-hemoglobin saturation curves in the fetus and mother

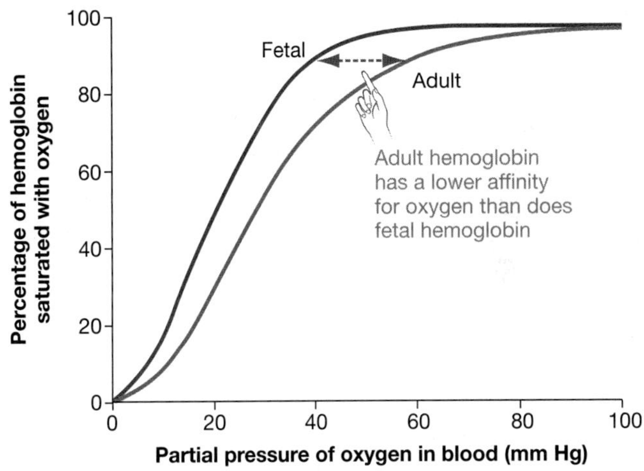

FIGURE 48.17 Adaptations That Increase Delivery of Oxygen to the Fetus.

● **EXERCISE** In part (a), draw oxygen arrows passing from maternal blood to fetal blood.

key point here is that the data for fetal hemoglobin are shifted to the left of the data for adult hemoglobin. As a result, the fetus's blood always has a higher affinity for oxygen than does the mother's blood. Biologists interpret this pattern as an adaptation. The high oxygen affinity of fetal hemoglobin ensures that the fetus is always able to acquire oxygen from the mother.

Before leaving the topic of pregnancy, it is important to acknowledge that mothers and embryos exchange more than nutrients and wastes. For example, consider the thalidomide tragedy. During the 1950s, hundreds of children were affected by their mothers' consumption of the tranquilizer thalidomide, which diffused into the fetal bloodstream and caused birth defects—often a dramatic shortening of the arms. Although thalidomide is now banned for use by pregnant women, alcohol use continues to take a similar toll. Children of mothers who drink alcohol are at high risk of experiencing hyperactivity, severe learning disabilities, and

(a) Brain sections from newborn rats injected with saline versus ethanol

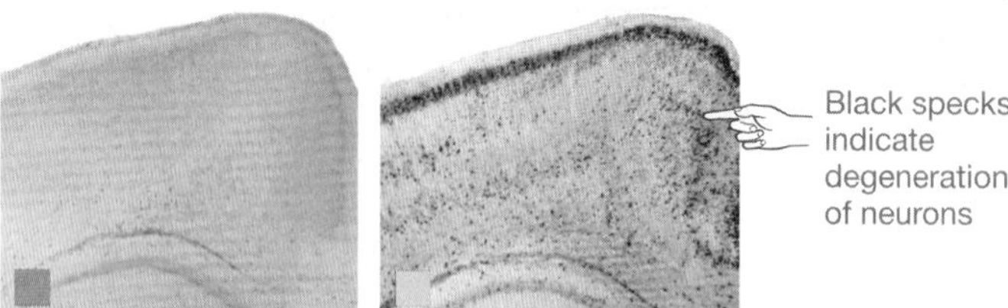

(b) Weights of brain tissue from newborn rats injected with saline versus ethanol

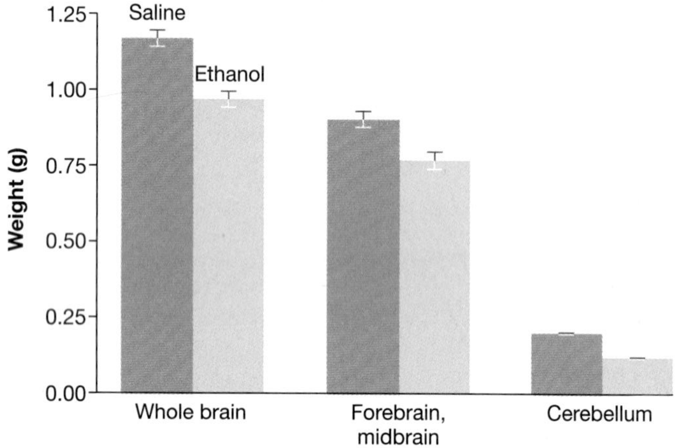

FIGURE 48.18 Ethanol Kills Brain Cells in Newborn Rats.

● **QUESTION** This study was designed to test the effects of a single drinking event. How would you modify the study to test the effects of long-term exposure to alcohol during pregnancy?

depression. Collectively, these symptoms are termed **fetal alcohol syndrome** (**FAS**).

To understand why FAS exists, recall that the brain undergoes especially rapid growth in the third trimester of development, and then consider the data shown in **Figure 48.18**. Figure 48.18a shows two slices from the brains of newborn rats. A day before these photographs were taken, one group of pregnant rats was given two injections of a harmless salt solution at a dosage of 2.5 g per kg body weight; the other group was given the same injections, except that the salt solution contained 20 percent ethanol. For a female human of average weight, this would be equivalent to drinking 2.5 cans of beer or glasses of wine. The black specks in the brains of ethanol-treated rats indicate degenerating neurons. The histograms in Figure 48.18b illustrate one consequence: The ethanol-treated rats had smaller brains than the saline-treated individuals did. The messages of these data are that ethanol destroys growing neurons and that even a single, "moderate" dose of alcohol can have a devastating effect. Based on these results, public health officials strongly advise pregnant women against drinking *any* amount of alcohol.

Birth

Although the mechanisms responsible for triggering the birthing process are not completely understood, the pituitary hormone **oxytocin** is clearly important in stimulating smooth-muscle cells in the uterine wall to begin contractions. The contractions that expel the fetus from the uterus and out of the mother's body constitute **labor**.

During the first stage of birth, the uterus contracts at relatively low frequency. As labor progresses, the cervix at the base of the uterus begins to open, or dilate (**Figure 48.19**). Once the cervix is fully dilated, uterine contractions become more

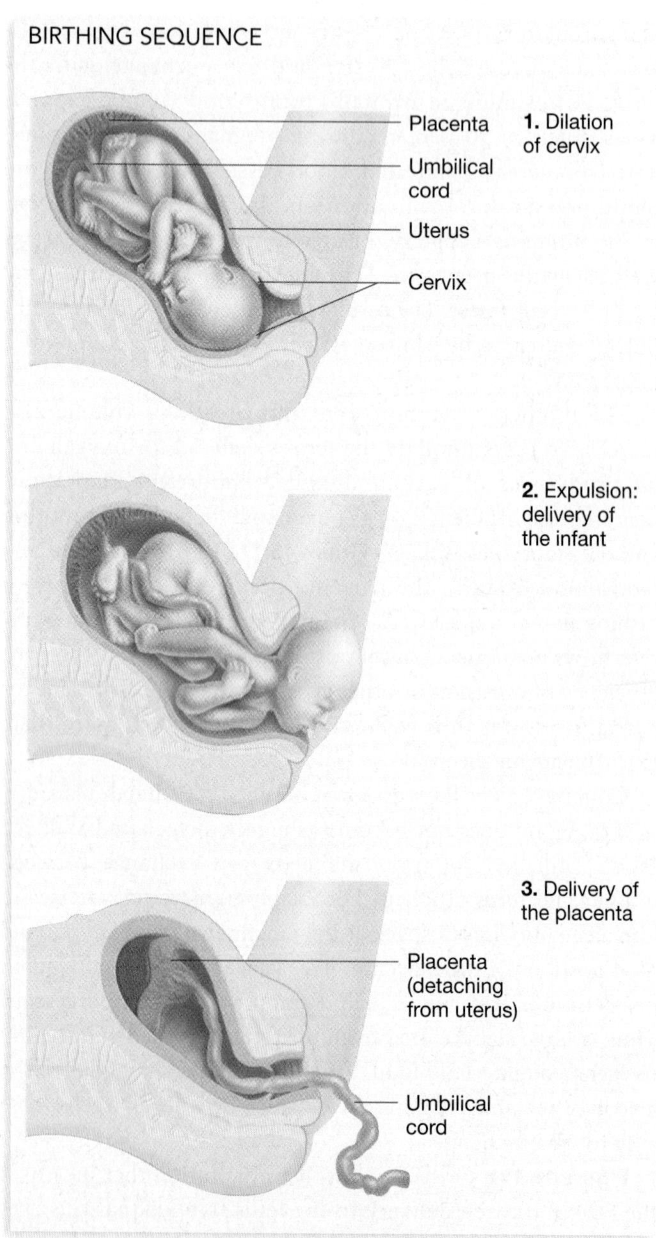

FIGURE 48.19 The Birthing Sequence in Humans.

● **QUESTION** Human babies are normally born face down, as shown. Predict the consequences of a baby positioned face up, or feet first, in the birth canal.

forceful, longer, and more frequent. Eventually, the fetus is expelled through the cervix and into the vagina.

After the baby is delivered, the placenta remains attached to the uterine wall. At this point, caregivers clamp and then cut the umbilical cord that connects the child and the placenta. By gently tugging on the cord, the caregiver helps the mother deliver the placenta and accompanying membranes, and birth is complete.

Although this description sounds straightforward, in reality a large number of complications are possible. Consider the data in **Figure 48.20**. The graph shows the number of Swedish mothers who died, per 100,000 live births, in five-year intervals between 1760 and 1980. In 1760, approximately 1.4 mothers died for every 100 infants successfully delivered. In most cases, the cause of death was blood loss or infection following delivery. Note that the vertical axis is logarithmic, and that the steepest drop in mortality rate occurred with the introduction of hand-washing in the late 1870s. As a result of sterile techniques, antibiotics, and blood transfusion technology, Sweden's mortality rate has now declined to less than 0.007 percent. The advent of cesarean sections, better care for premature infants, and improved treatments for Rh disease have also resulted in steep declines in infant mortality rates in many countries. Although it is generally unheralded, this huge reduction in the rate of death associated with childbirth qualifies as one of the great triumphs of modern medicine. Unfortunately, because many developing nations lack sterile facilities and appropriate antibiotics, the mortality of mothers and infants remains high in those countries.

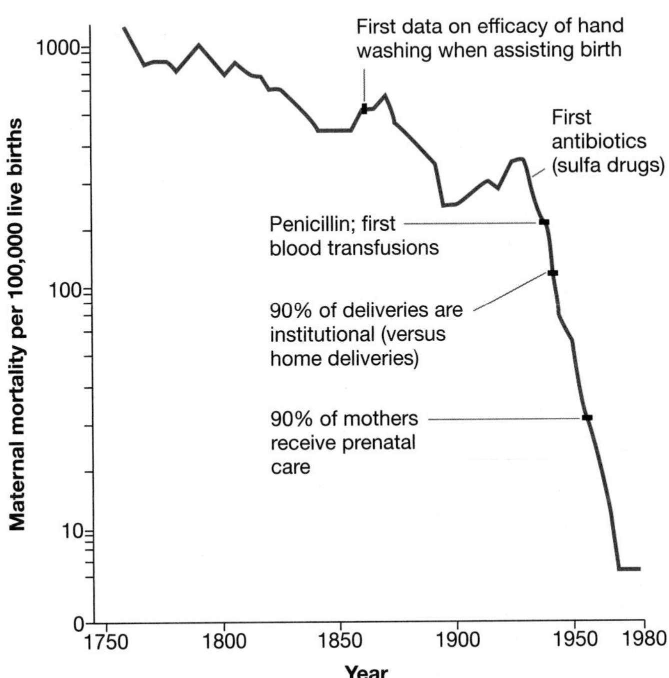

FIGURE 48.20 Maternal Mortality in Childbirth Has Decreased Dramatically. These data are from Sweden; the scale on the vertical axis is logarithmic. Note: In 1861 a researcher published convincing data that hand washing dramatically reduced maternal mortality during childbirth, but doctors resisted the idea until well after Pasteur had published the germ theory of disease in 1878.

QUESTION If a woman living in Sweden in 1760 had 10 children over the course of her lifetime, what was her overall chance of dying in childbirth?

Chapter Review

SUMMARY OF KEY CONCEPTS

The reproductive systems of animals are highly variable. Many animals can switch between asexual and sexual reproduction. When sexual reproduction occurs, fertilization may be external or internal and egg development may take place inside or outside the mother's body, depending on the species.

Only a few animal species reproduce exclusively by asexual means, but many alternate between asexual and sexual reproduction. Asexual reproduction results in offspring that are genetically identical to the parent. In contrast, every sexually produced offspring is genetically unique because of recombination during meiosis and the fusion of haploid gametes from different parents during fertilization.

In humans, the production of sperm and eggs differs in several respects. Spermatogenesis is continuous throughout the adult life of males, but in females all of the cells that divide mitotically to begin oogenesis are formed early in development. Meiosis in females is also arrested for long periods of time. Finally, cell division during meiosis is so unequal in females that just one egg (instead of four) is produced from each primary oocyte.

Fertilization is external in many aquatic animals but internal in almost all terrestrial species. Release of gametes during external fertilization may be synchronized by courtship displays or possibly by pheromones. Research on internal fertilization has documented that multiple mating and sperm competition are widespread. Two important generalizations have emerged from studies of sperm competition. First, the last male to mate usually fathers a disproportionately large number of offspring. Second, males of species that engage in sperm competition have relatively large testes for their overall size.

Once eggs are fertilized, females may lay the eggs or retain them and give birth to live offspring. By estimating the evolutionary relationships among closely related species with viviparity and oviparity, researchers have determined that viviparity has evolved many times independently. In some groups at least, viviparity may be an adaptation that increases the survival of young in cold habitats.

You should be able to explain why fish that care for their young are expected to produce fewer eggs per year than fish that do not care for their young, whether sperm competition can occur in fish, and when internal fertilization would be expected to occur in fish.

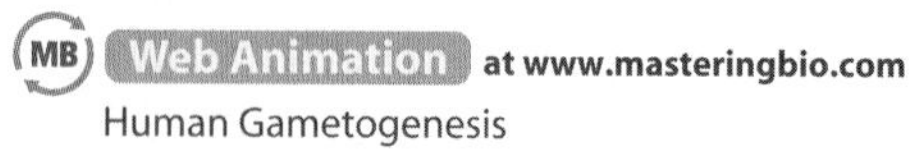

at www.masteringbio.com
Human Gametogenesis

In humans, the male reproductive system is composed of structures specialized for the production and storage of sperm, the synthesis of important molecules found in semen, or the transport and delivery of semen. The female reproductive system is composed of structures specialized for producing eggs, receiving sperm, and nourishing offspring during early development.

In humans, the external anatomy of the male reproductive tract consists of the penis and scrotum. Biologists are unsure why the scrotum evolved in mammals but have documented that sexual selection has produced striking variation in the genital morphology of male insects and spiders. The internal anatomy of the human male reproductive tract includes structures devoted to sperm production and storage, synthesis of accessory fluids, and transport and delivery of semen. The external anatomy of the human female reproductive tract consists of the vaginal opening and clitoris. The internal anatomy of the human female reproductive tract is devoted to egg production and transport or to the development of offspring.

You should be able to predict how surgeons alter the male and female reproductive tracts of human volunteers to sterilize them.

In humans, the menstrual cycle is regulated by hormones from the pituitary gland and female reproductive organs. These hormones may interact with each other via positive or negative feedback. Pregnancy is maintained by hormonal signals from the embryo and from the mother's reproductive organs.

The development and activity of male and female reproductive structures are under hormonal control. In mammals, experiments confirmed that GnRH from the hypothalamus triggers the release of FSH and LH from the pituitary gland. The pituitary hormones, in turn, regulate the production of the gonadal hormones, testosterone and estradiol, in the testes and ovaries, respectively. Research on the human menstrual cycle has confirmed that estradiol and progesterone exert negative and positive feedback on the production of FSH and LH. Interactions between the pituitary and ovarian hormones are responsible for regulating cyclical changes in the ovaries and uterus.

If fertilization occurs, the developing embryo secretes the hormone hCG, which arrests the menstrual cycle and allows pregnancy to continue. During the first trimester, the embryo becomes implanted in the thickened uterine wall, organs develop, and the nutritive organ called the placenta forms. To make rapid growth possible during the second and third trimesters, the mother's heart rate and pumping volume increase. Nutrients and gases are exchanged efficiently in the placenta.

You should be able to explain why administration of progesterone is an effective method of birth control in humans.

at www.masteringbio.com
Human Reproduction

QUESTIONS

Test Your Knowledge

1. What term describes the mode of asexual reproduction in which offspring develop from unfertilized eggs?
   a. parthenogenesis
   b. budding
   c. regeneration
   d. fission

2. In sperm competition, what is "second-male advantage"?
   a. the observation that when females mate with two males, each male fertilizes the same number of eggs
   b. the observation that when females mate with two males, the second male fertilizes most of the eggs
   c. the observation that females routinely mate with at least two males before laying eggs or becoming pregnant
   d. the observation that accessory fluids prevent matings by second males—for example, by forming copulatory plugs

3. How are the human penis and clitoris similar?
   a. They develop from the same population of embryonic cells.
   b. Both develop during the earliest stages of puberty.
   c. Both contain the urethra.
   d. Both produce accessory fluids required during sexual intercourse.

4. In hormone production, how do the follicle and corpus luteum compare?
   a. Both produce primarily estradiol.
   b. Both produce primarily progesterone.
   c. The follicle produces more estradiol than progesterone; the corpus luteum produces more progesterone than estradiol.
   d. The follicle produces mostly progesterone; the corpus luteum produces estradiol and progesterone.

5. What pituitary hormones are involved in regulating the human menstrual cycle?
   a. GnRH and LH
   b. estradiol and progesterone
   c. oxytocin and LH
   d. FSH and LH

6. The corpus luteum is retained upon fertilization due to the presence of which hormone?
   a. LH
   b. estradiol
   c. progesterone
   d. human chorionic gonadotropin (hCG)

Test Your Knowledge answers: 1. a; 2. b; 3. a; 4. c; 5. d; 6. d

Test Your Understanding

Answers are available at *www.masteringbio.com*

1. Summarize the experimental evidence that *Daphnia* require three cues to trigger sexual reproduction. Discuss what these cues indicate about the environment. Generate a hypothesis for why sexual reproduction is adaptive for these animals.
2. Compare and contrast spermatogenesis with oogenesis. How do these two processes differ with respect to numbers of daughter cells produced, gamete size, and timing of the second meiotic division?
3. The drug RU-486 blocks progesterone receptors in the uterus. Why does this drug end pregnancy in humans, even if implantation has occurred?
4. Explain the difference between negative feedback and positive feedback in hormonal control of the human menstrual cycle. How is it possible that estrogen inhibits LH release at one point of the reproductive cycle but stimulates its release at another point?
5. Why are pregnant mothers advised to refrain from smoking, and from drinking alcoholic or caffeinated beverages?
6. Suppose that females of a certain insect species routinely mate with two males and choose their mates based on courtship displays or other traits. Predict whether females should be choosier about the first male or the second male that they mate with, or equally choosy about both males. Explain the logic behind your prediction.

Applying Concepts to New Situations

Answers are available at *www.masteringbio.com*

1. Researchers have recently developed methods for cloning mammals. In effect, biologists can now induce asexual reproduction in species that do not normally reproduce asexually. Suppose this practice becomes so widespread in the future that most of the sheep in the world become genetically identical. Discuss some possible consequences of this development.
2. The text claims that species with external fertilization produce extraordinarily large numbers of gametes. How would you test this hypothesis rigorously? In answering, assume that you have data on the average number of gametes produced by different species, along with data on their average body size.
3. Suppose that you've been given a chance to study populations of *Sceloporus* lizards—some of which are viviparous, and others of which are oviparous. Which populations would you expect to produce especially large eggs, relative to their overall body size? How would you go about testing this prediction?
4. When marsupial eggs are developing, a shell membrane appears for a short time and then disintegrates. State a hypothesis to explain this observation.

www.masteringbio.com is also your resource for • Answers to text, table, and figure caption questions and exercises • Answers to *Check Your Understanding* boxes • Online study guides and quizzes • Additional study tools including the *E-Book for Biological Science* 3rd ed., textbook art, animations, and videos.

49 The Immune System in Animals

KEY CONCEPTS

- The innate immune response to infection is mounted by leukocytes that respond in a nonspecific way to pathogens. The acquired immune response is mounted by leukocytes called lymphocytes. Each individual lymphocyte responds to a specific pathogen.
- The acquired immune response begins when cell-surface proteins on lymphocytes bind to an antigen—usually a protein from a disease-causing agent.
- The humoral response tags pathogens for destruction with antibodies that are produced by lymphocytes called B cells.
- The cell-mediated response destroys infected cells via lymphocytes called cytotoxic T cells.

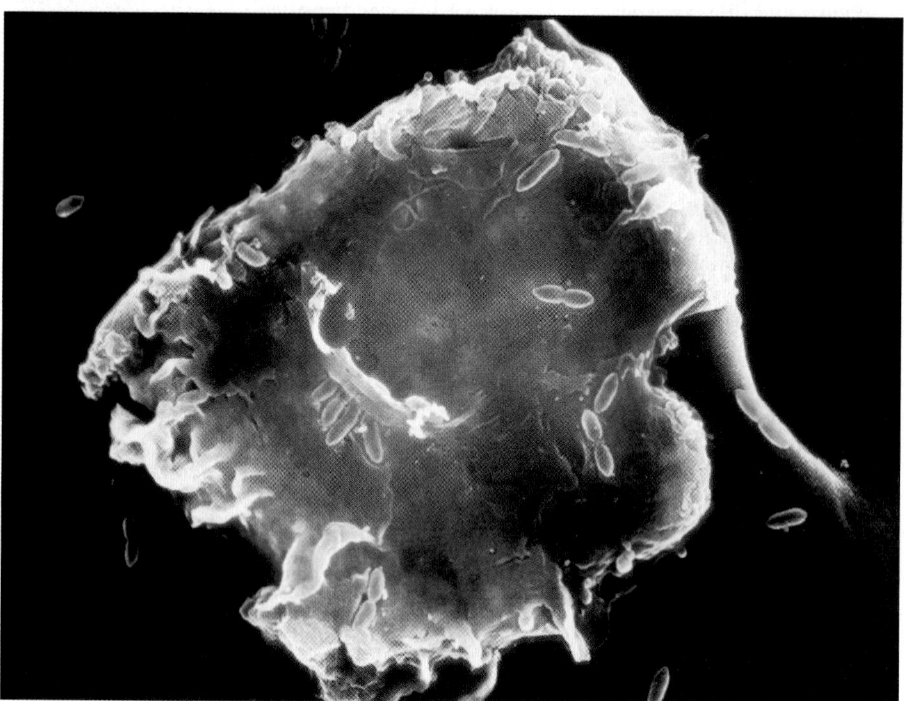

Scanning electron micrograph of an immune system cell engulfing and destroying bacteria (the small green cells). This chapter explores how immune system cells are able to recognize bacteria and viruses as foreign and eliminate them.

Disease threatens every animal. Humans alone are victimized by hundreds of different disease-causing bacteria and viruses and a wide array of parasitic worms, fungi, and protists. Given the ability of these pathogens to cause illness and death, it is remarkable that so many animals stay healthy for most of their lives.

To understand how humans and other animals are protected against disease, biologists have focused on investigating three observations. First, wounds usually heal, even if they become infected. Second, most people who contract a bacterial or viral illness eventually recover, even without help from medications. Third, people who acquire bacterial or viral infections and recover are frequently immune (literally, "safe") to that disease—that is, they do not contract the same disease in the future. **Immunity** is a resistance to or protection from disease-causing pathogens.

This last observation is particularly intriguing. Why do people who get measles or chickenpox fail to get these diseases again later, after a second exposure to the same disease-causing pathogen? Thucydides commented on this pattern in 430 B.C. He wrote that when plague struck Athens, only people who had recovered from the illness could nurse the sick, because they would not become ill a second time. In the middle ages, Chinese and Turkish practitioners protected people from smallpox by intentionally exposing individuals to the dried crusts of smallpox pustules taken from infected people. These are examples of **immunization**—the conferring of immunity to a particular disease.

In the late 1700s Edward Jenner had a key insight that allowed him to refine this immunization technique. In Jenner's day, milkmaids were considered pretty because their faces were not pockmarked with scars from smallpox infections. Jenner knew that cows suffered from a smallpox-like disease called cowpox, and he reasoned that milkmaids were immune to smallpox because they had been exposed to cowpox while milking cows. To test this hypothesis, he injected a boy with fluid taken from a cowpox pustule, then later injected the same child with fluid from a smallpox pustule. That is, he inoculated the boy with cowpox pathogens and then with smallpox pathogens. As predicted, the boy did not contract smallpox. Jenner's technique was quickly adopted throughout Europe and was later named **vaccination**. (The Latin root *vacca* means "cow.")

Why did Jenner's technique work? More precisely, what is the molecular basis of a person's ability to fight off a bacterial or viral infection and achieve immunity? To answer these questions, we need to explore how the human immune system recognizes and eliminates pathogens. We'll consider the body's response to two events: a bacterial infection and an assault by a virus. Both infections could afflict anyone on any given day. For example, suppose that several days ago you were on a crowded bus next to someone with a persistent cough and runny nose, and 30 minutes later you stumbled while sprinting across campus and skinned your elbow. Now you realize that the wound on your arm has gotten sore and red. Worse yet, you are developing a sore throat, runny nose, and fever. Your elbow has become infected with bacteria, and your upper respiratory tract is supporting a growing population of influenza, or "flu," virus. Let's start by analyzing what is happening in the wound at your elbow.

49.1 Innate Immunity

When biologists first began analyzing the immune system, they realized that certain immune system cells are ready to respond to foreign invaders at all times, while other components must be activated first. Cells that are always ready confer **innate immunity**; cells that are selectively activated to eliminate a specific pathogen confer **acquired immunity**. The key to understanding the two types of immunity is to recognize that the cells involved provide different responses to antigens. An **antigen** is any foreign molecule that can initiate an immune system response. Most antigens are proteins or glycoproteins from bacteria or viruses or other invaders, but foreign carbohydrates and lipids can also function as antigens.

The cells involved in innate immunity are nonspecific in their response to antigens. Stated another way, the innate immune system responds in the same way to all antigens. The innate immune response to a wound in your elbow, for example, is the same no matter what species of bacteria or viruses or fungi enter the tissue. In contrast, cells involved in acquired immunity respond in an extremely specific way to each particular strain of bacterium, virus, or fungus that enters the body. In combination, the innate and acquired immune responses form an extremely powerful system for protecting individuals against a formidable and ever-changing array of parasitic bacteria, viruses, and eukaryotes.

To launch this investigation into immune system function, let's focus on how the body prevents entry by foreign invaders. Then we can consider what happens once some do get in.

Barriers to Entry

The most effective way to avoid getting sick is to avoid contact with pathogens. In humans and other animals, the most important barrier to pathogen entry is the skin. In addition to providing a physical barrier that is thick and tough, human skin offers a chemical deterrent. Skin cells are covered by fatty acids, which lower the pH of the surface to between 3 and 5. The dry, acidic environment prevents the growth of most bacterial species that might be a threat to the body. Many bacteria do live on the skin, but most of them are adapted to the acidic conditions and are **commensal**—meaning they do not harm the host organism. The bodies of insects and other animals with exoskeletons are difficult for pathogens to enter because they are covered with the tough layer called cuticle, introduced in Chapter 42.

An animal's body has gaps in these barriers, however, where the digestive tract, reproductive tract, gas-exchange surfaces, and sensory organs make contact with the environment. As **Figure 49.1** shows, these gaps have a protective physical barrier in the form of mucus or other features that discourage pathogen

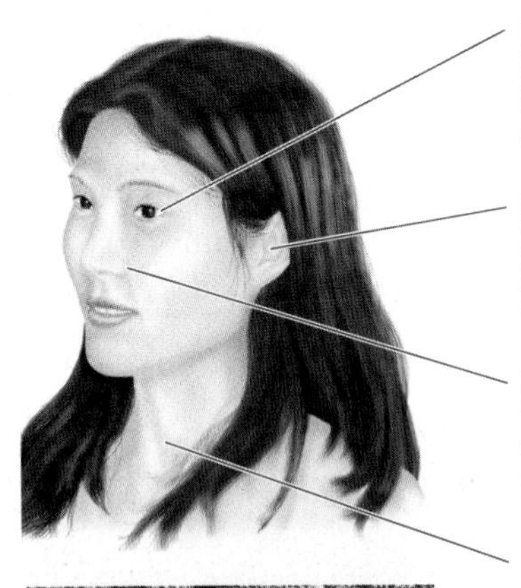
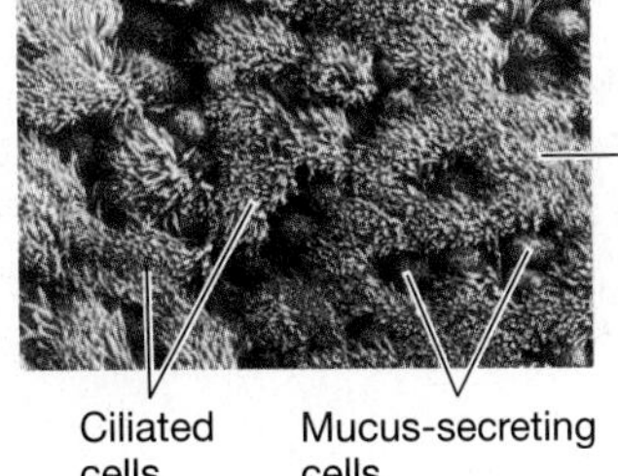

FIGURE 49.1 How Does the Body Keep Pathogens Out?

QUESTION Why is a combination of hair and sticky secretions effective at trapping pathogens?

entry. **Mucus** is a solution secreted by cells within the epithelium. It is rich in proteoglycans—large polysaccharides bonded to proteins. The bodies of slugs, snails, earthworms, and other soft-bodied animals are covered with a layer of mucus that shields their epithelial cells from pathogens in the environment. Mucus is equally important in vertebrates. For example, many of the pathogens that you breathe in or ingest while eating or drinking stick to the mucus that lines your airways and gastrointestinal tract. Pathogens that are stuck in mucus cannot come in contact with the plasma membranes of epithelial tissues. In mammals, these pathogens are coughed out or swallowed and killed in the acidic environment of the stomach. Gaps in the body that are not covered with mucus layers, such as the eyes, are often protected by other types of secretions. Your ears are protected by waxy secretions, and your eyes by tears that contain the enzyme **lysozyme**. Lysozyme acts as an antibiotic by digesting bacterial cell walls.

With regularity, though, preventive measures fail and some pathogens gain entry to tissues beneath the skin. Flu viruses, for example, have an enzyme on their surface that disrupts the mucous lining of the respiratory tract. When the outer surface of the virus makes contact with a host cell beneath the mucous layer, the virus is able to enter the cell and begin an infection. When a fall or other physical trauma breaks your skin, bacteria and other pathogens enter and gain direct access to the tissues inside. To viruses, bacteria, and fungi, your body is a tropical paradise, brimming with resources. Within minutes, they can begin growing. What happens then?

The Innate Immune Response

When bacteria enter your body at the site of a wound, the cells pictured in **Figure 49.2** implement the **innate immune response**—the body's nonspecific response to pathogens. As a group, these cells are called **leukocytes** ("white-cells"); all of them reside in the blood. Leukocytes are sometimes referred to as white blood cells. They include macrophages, neutrophils, and other cells involved in the innate response as well as the lymphocytes responsible for acquired immunity. Like the red blood cells that carry oxygen in the bloodstream, leukocytes are produced in the bone marrow.

The leukocytes involved in the innate response are alerted to the presence of foreign invaders by specific molecules that are found on the surfaces of bacterial or other types of invading cells, but not on host cells. Some of the cells in the innate system have proteins on their plasma membranes that bind to these bacteria-specific compounds. These proteins are known as **pattern-recognition receptors**. When they are activated by antigens, the cells respond.

Figure 49.3 summarizes the major steps in the **inflammatory** ("in-flames") **response**, which is a type of innate immune response observed in mammals. As step 1 shows, a break in the skin allows bacteria to enter the body. If capillaries and other small blood vessels are broken by the wound, blood leaves. Almost immediately (step 2), blood components called **platelets** release proteins that form clots and lessen bleeding. Other clotting proteins in the blood form cross-linked structures that help wall off the wound and reduce blood loss.

Step 3 shows leukocytes called **macrophages** at the wound site, in the process of secreting signaling molecules called **chemokines** ("chemical-movers"). Chemokines are also secreted by wounded tissues. Chemokines are important because they form a gradient that marks a path to the wound site.

Soon after macrophages begin their response, leukocytes called **mast cells** release chemical messengers that cause blood vessels near the wound to constrict. As a result, blood flow and thus blood loss at the wound decreases. Step 4 indicates that mast cells also release **histamine** and other signaling molecules that induce blood vessels slightly farther from the wound to dilate and become more permeable. As these vessels get bigger and increasingly permeable, the delivery of leukocytes and other blood components to the general region increases. Thus,

(a) Mast cells secrete signals that increase blood flow.

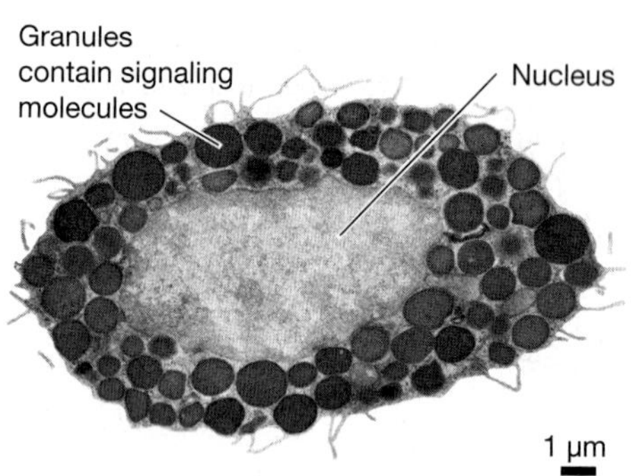

(b) Neutrophils ingest and kill pathogens.

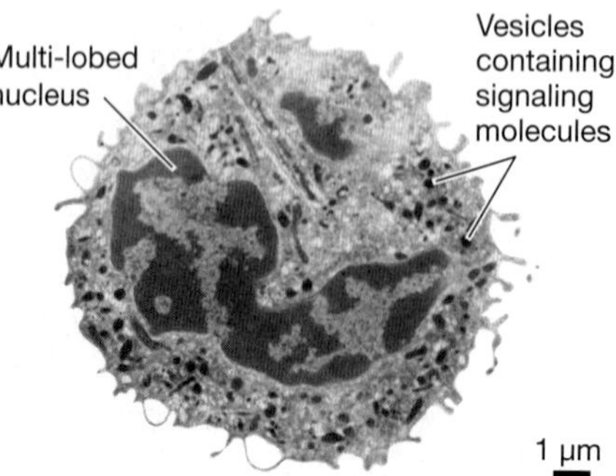

(c) Macrophages recruit other cells and ingest and kill pathogens.

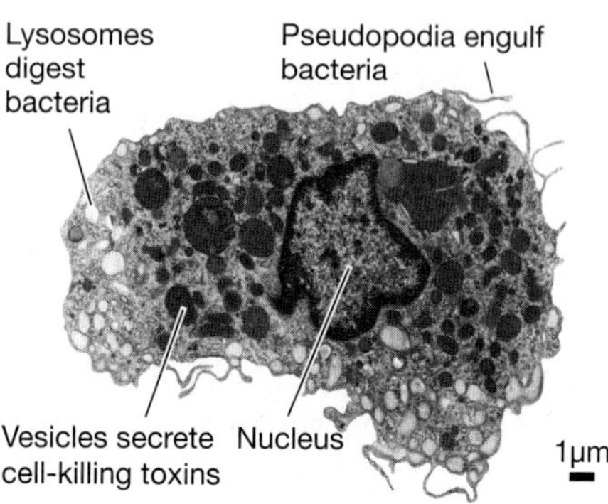

FIGURE 49.2 Several Cells Are Involved in the Innate Immune Response. Transmission electron micrographs illustrating a few of the leukocytes responsible for innate immunity.

QUESTION How does the structure of these cells correlate with their function?

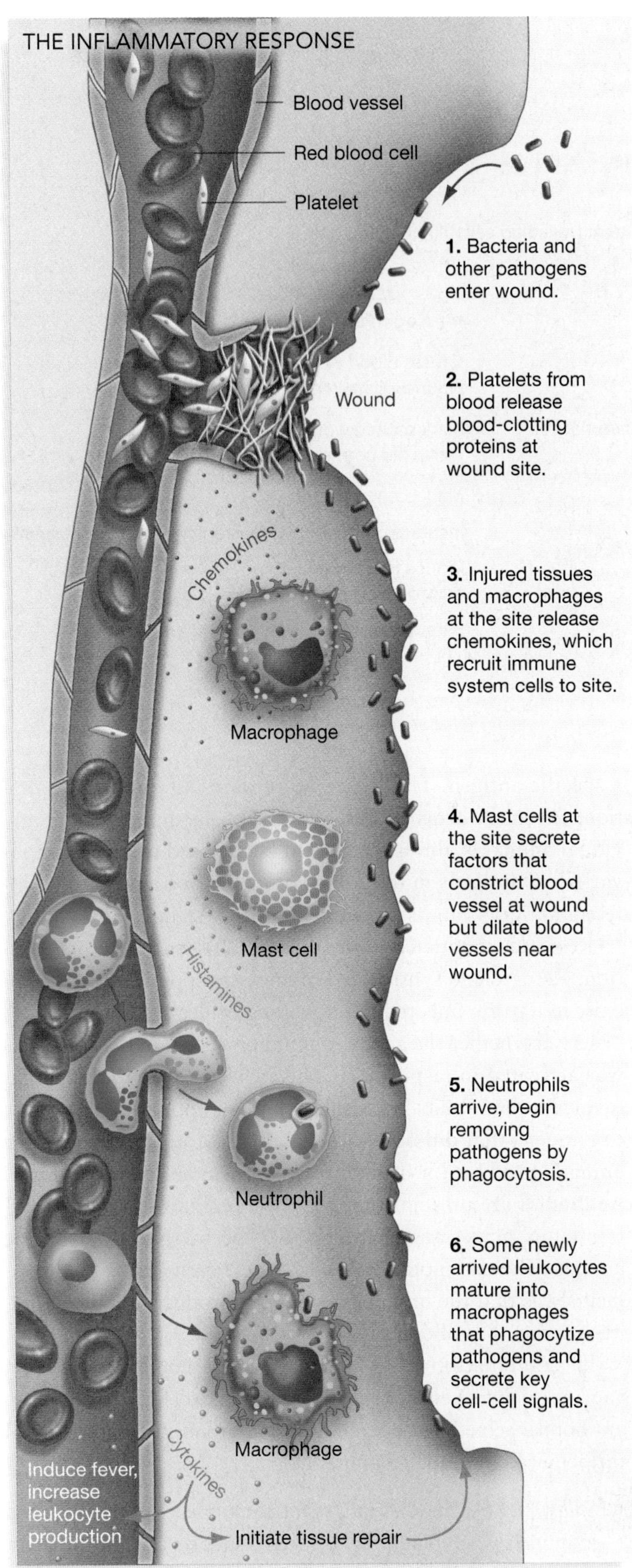

FIGURE 49.3 The Inflammatory Response of Innate Immunity Has Many Elements.

QUESTION Only one blood vessel is shown, for simplicity. If you had space to draw in other vessels in the wound area, where would they be constricted and where would they be dilated?

signals from mast cells stop blood loss at the wound site itself but increase blood flow to surrounding tissues.

The combination of dilated blood vessels and a chemokine gradient is like a 911 call that provides specific directions to the scene of a fire. Step 5 shows a key part of the response: leukocytes called **neutrophils** moving out of dilated blood vessels and migrating to the site of the infection. Neutrophils are major players in the innate response. They destroy invading cells by **phagocytosis** ("eating-cell")—meaning that they engulf them. Once invading cells are inside neutrophils, they are killed with a complex array of toxic compounds. These molecules include lysozyme—which degrades bacterial cell walls—free radicals, nitric oxide (NO), and molecules called **reactive oxygen intermediates** (**ROIs**), including hydrogen peroxide (H_2O_2).

After neutrophils begin arriving in large numbers, cells that will mature into macrophages arrive. In addition to phagocytizing bacteria at the wound, the growing population of macrophages secretes chemical messengers called **cytokines** ("cell-movers"), which have an array of effects (step 6). Cytokines from macrophages attract other immune system cells to the site, stimulate bone marrow to make and release additional neutrophils and macrophages, induce fever—an elevated body temperature that aids in healing—and activate cells involved in tissue repair and wound healing.

The site of inflammation often becomes swollen due to increased numbers of cells and fluids in the area—red and warm due to increased blood flow, and painful due to signals from pain receptors. The inflammatory response continues until all foreign material is eliminated and the wound is repaired. **Table 49.1a** summarizes some of the key cells in the response, while **Table 49.1b** summarizes some of the key molecules involved.

Humans and other vertebrates are not the only animals that have sophisticated innate immune responses. Innate responses make up the entire immune system observed in millions of species of invertebrates. The spectacular abundance and diversity of invertebrate animals supports the hypothesis that their innate immune systems provide efficient protection against invading bacteria, viruses, and fungi. For example, if a pathogen succeeds in entering the main body cavity of an insect through a wound or other break in the barriers to entry, cells respond by synthesizing and secreting peptides with potent antibacterial or antifungal properties. Sea stars have specialized cells that are similar in function to neutrophils and macrophages—they secrete cytokines or engulf and destroy pathogens by phagocytosis. But biologists still have a great deal to learn about defense systems in invertebrates.

What happens when the innate immune system of a vertebrate fails to contain and eliminate an infection that starts at a wound? And what happens to the viruses that blast through the mucous membrane of the human respiratory tract and enter the cells underneath?

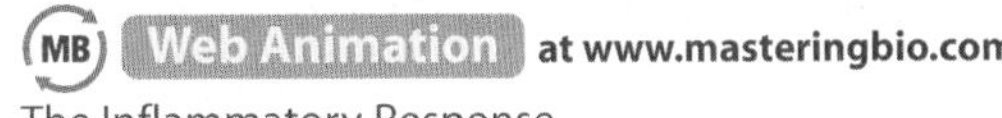

at www.masteringbio.com

The Inflammatory Response

SUMMARY TABLE **49.1 The Innate Immune System**

(a) Key Cells

Name	**Primary function**
Mast cells	Release signals that increase blood flow to wound site
Neutrophils	Kill invading cells via phagocytosis
Macrophages	Release cytokines that recruit other cells to wound site; kill invading cells via phagocytosis

(b) Key Signaling Molecules

Name	**Produced by:**	**Received by:**	**Message/Function:**
Histamine	Mast cells	Blood vessels	Causes blood vessels to constrict next to wound; dilate in nearby areas
Chemokines*	Injured tissues and macrophages in tissues	Neutrophils and macrophages	Mark path to wound; promote dilation and increased permeability of blood vessels
Cytokines other than chemokines	Macrophages	Leukocytes	Mark path to wound
		Bone marrow	Increase production of macrophages and neutrophils
		CNS	Induce fever by increasing set point for control of body temperature
		Local tissues	Stimulate cells involved in wound repair

* Note that chemokines are a subset of cytokines.

Check Your Understanding

If you understand that...

- The innate immune response occurs when macrophages and mast cells that reside in tissues, and neutrophils that circulate in the blood, react in a nonspecific way to signals from invading pathogens.

You should be able to...

Describe how events in the innate response relate to the following steps, which are implemented by first-aid workers when treating a wound:

1) Apply direct pressure to close blood vessels.
2) Apply bandages containing compounds that recruit platelets to the site.
3) Clean the wound once bleeding has stopped.
4) Apply antibiotics.

49.2 The Acquired Immune Response: Recognition

Recall from Chapter 35 that viruses actually enter cells to infect them. Once a virus succeeds in entering a cell, the virus is hidden from the host's innate immune response. In vertebrates, the acquired immune system is responsible for limiting the spread of infections at wound sites and responding to invading viruses and other pathogens. The **acquired immune response** is based on interactions between specific immune system cells and a specific antigen. This observation raises an important question: Given the array of different pathogens that exist, an individual animal is almost certain to be exposed to an enormous variety of antigens in the course of its lifetime. How many different antigens can its acquired immune system respond to?

Research conducted in the early 1920s answered this question. Researchers synthesized organic compounds that do not exist in nature, injected the novel molecules into rabbits, and observed whether the acquired immune system in these animals was activated in response. To the amazement of the scientists involved, the rabbits were able to mount an immune response to antigens that did not exist in nature. More specifically, the animals produced a different antibody against each antigen. **Antibodies** are proteins that are produced and secreted by certain lymphocytes and that bind to a specific part of a specific antigen. The take-home message of these early injection experiments was that the immune system can produce an almost limitless array of antibodies.

In combination with insights from Jenner's work on vaccines—reviewed in the preceding section—early work on antibodies helped focus attention on four key characteristics of the acquired immune response:

- *Specificity* Antibodies and other components of the acquired immune system bind only to specific sites on specific antigens.
- *Diversity* The acquired response recognizes an almost limitless array of antigens.
- *Memory* The acquired response can be reactivated quickly if it recognizes antigens from a previous infection.

- *Self-nonself recognition* Molecules that are produced by an individual do not act as antigens, meaning that the acquired immune system can distinguish between self and nonself. Nonself molecules are antigens; self molecules are not.

To explore these characteristics in detail, let's begin by reviewing the cells and organs that are responsible for the acquired immune response.

An Introduction to Lymphocytes and the Immune System

The cells that carry out the major features of the acquired immune response are called **lymphocytes**. Lymphocytes originate in the immune system and are subsequently activated and transported inside it (**Figure 49.4**):

1. *Lymphocyte origin* All lymphocytes and blood cells are produced in **bone marrow**, tissue that fills the internal cavities in bones.

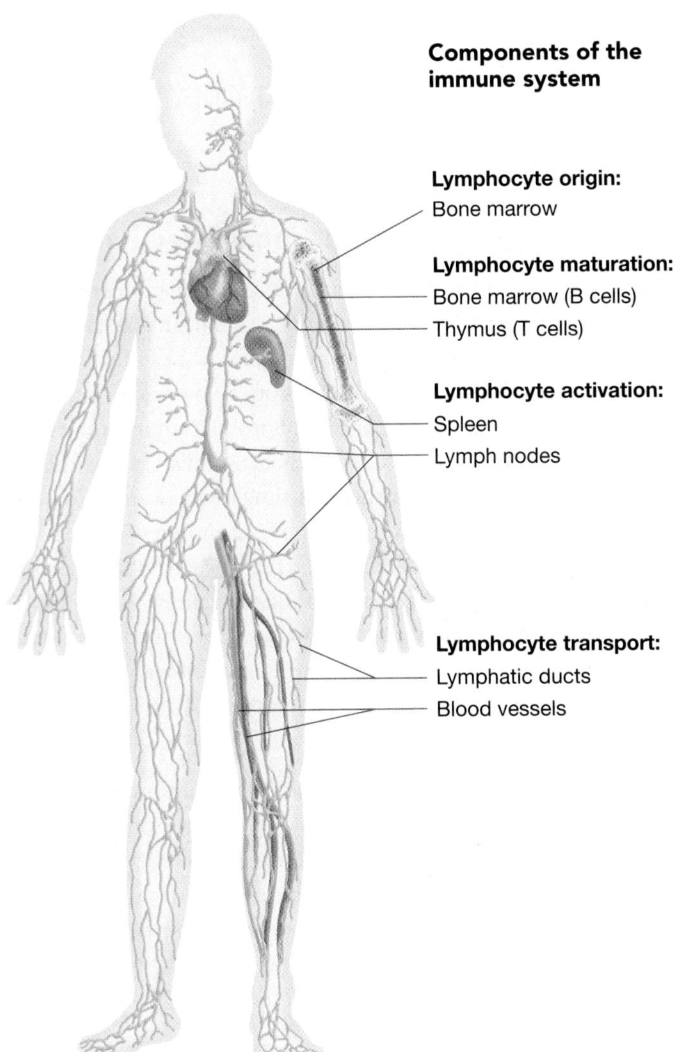

FIGURE 49.4 Lymphocytes Are Created, Activated, and Transported in the Immune System.

2. *Lymphocyte maturation* Certain lymphocytes (called B cells) mature in the bone marrow where they originate. Other lymphocytes (called T cells) mature in the **thymus**, an organ located in the upper part of the chest of vertebrates (just behind the breastbone in humans).
3. *Lymphocyte activation* Lymphocytes recognize antigens and become activated in the spleen and lymph nodes. The **spleen** is a lymphatic organ located in the abdominal cavity. In addition to its function in the immune response, the spleen destroys old blood cells and stores iron from red blood cells. **Lymph nodes** are small, oval organs that are located all around the body. Lymph nodes filter the lymph passing through them. Recall from Chapter 44 that **lymph** is a mixture of fluid and lymphocytes. The liquid portion of lymph originates in fluid that is forced out of capillaries by blood pressure.
4. *Lymphocyte transport* Lymphocytes circulate through the blood and the secondary organs of the immune system—lymph nodes, spleen, and lymphatic ducts. Lymphatic ducts are thin-walled, branching tubules that transport lymph throughout the body in the **lymphatic system**.

In addition to being found inside the structures shown in Figure 49.4, large numbers of leukocytes and lymphocytes are associated with skin cells and with epithelial tissues that secrete mucus—primarily in the digestive tract and respiratory tract. Collectively, the immune system cells found in gut and respiratory organs are called the **mucosal-associated lymphoid tissue** (**MALT**). Leukocytes in the skin and MALT are important because they guard points of entry for pathogens.

Lymphocytes are normally in a resting, or inactive, state as they circulate through the bloodstream and lymphatic system or reside in skin or MALT. As **Figure 49.5a** shows, inactive lymphocytes have a large nucleus, little cytoplasm, few mitochondria, and a ruffled membrane. Over the course of a day, an inactive lymphocyte might migrate through the spleen, enter the blood, cross over into the lymphatic vessels, migrate through a lymph node, return to the blood, and so on.

If an inactive lymphocyte does not encounter the antigen to which it is programmed to respond, then the cell eventually dies. But a lymphocyte that encounters an appropriate antigen becomes activated. **Figure 49.5b** shows an activated lymphocyte that has a massive amount of rough ER. It also has a large number of mitochondria. Recall from Chapter 7 that many of the proteins synthesized in a cell's rough ER are inserted into the plasma membrane or secreted from the cell. The increased rough ER in this activated lymphocyte suggests that it is manufacturing and possibly secreting proteins.

Activation clearly produces dramatic changes in lymphocytes. Now the question is, what *are* these cells? What types of cells execute the acquired immune response?

(a) Inactive lymphocyte

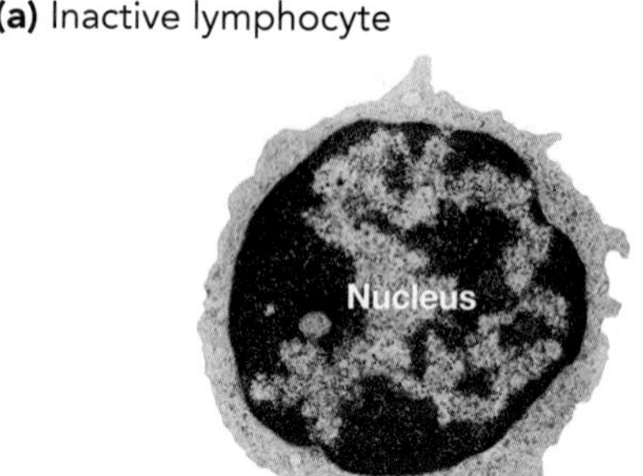

(b) Activated lymphocyte

Large amount of rough ER

Nucleus

1 µm

FIGURE 49.5 Lymphocytes Exist in Two States: Inactive and Activated. (a) This inactive lymphocyte has a small amount of cytoplasm and few organelles. **(b)** This activated lymphocyte has extensive rough endoplasmic reticulum. This observation suggests that a great deal of protein synthesis is taking place.

The Discovery of B Cells and T Cells

In 1956 a group of researchers provided an important insight into the acquired immune system, quite by accident. The biologists were investigating the immune system's response to the bacterium *Salmonella typhimurium*, which is a common cause of food poisoning in humans. To do this work, they needed to produce and isolate antibodies to a particular antigen from *S. typhimurium*—an antigen that is toxic. Their plan was to inject a large number of chickens with the antigen and collect the antibodies that the treated animals produced in response.

In addition to injecting many normal chickens, though, the biologists happened to include some chickens that had undergone experimental removal of an organ called the bursa. Six of the nine chickens that lacked a bursa and that were injected with the antigen died. The other three birds without a bursa survived but failed to produce antibodies to the antigen. In contrast, chickens with the bursa intact produced large quantities of antibodies and survived. To make sense of these results, the researchers proposed that the bursa is critical for antibody production and that antibodies are important in neutralizing antigens.

Not long after this observation was published, three groups of scientists independently conducted a related experiment. To explore the function of the thymus in mammals, these groups removed the organ from newborn mice. Mice lacking a thymus developed pronounced defects in their immune systems. For example, when pieces of skin from other mice were grafted onto the experimental individuals, their immune systems did not recognize the tissue as foreign. In contrast, individuals with an intact thymus quickly mounted an immune response to the foreign skin cells and killed them.

The results of these and follow-up experiments showed that lymphocytes from the bursa and thymus have different functions. The two types of lymphocytes became known as bursa-dependent and thymus-dependent lymphocytes, or B cells and T cells, respectively. **B cells** produce antibodies. (The lymphocytes in Figure 49.5 are B cells.) **T cells** are involved in an array of functions, including recognizing and killing host cells that are being infected with a virus. Later work showed that in humans and other species that lack a bursa, B cells mature in bone marrow.

Now let's turn to one of the most fundamental of all questions in immunology: How are B cells and T cells able to recognize so many different antigens?

Antigen Recognition and Clonal Selection

By the 1950s biologists had made a series of important general observations about the immune system. It was clear that antibodies to a seemingly limitless number of antigens could be produced and that each antibody is specific to an antigen. By studying the time course of antibody production in rabbits and other experimental animals, researchers established that the immune response intensified through time after an infection had begun. In addition, the response is "remembered"—meaning individuals do not get sick at all, or recover extremely quickly, if they are exposed to the same pathogen again in the future.

To explain these patterns, researchers developed the **clonal-selection theory** of immune system function. This theory made three central claims about how the acquired immune system works:

1. Each lymphocyte formed in the bone marrow or thymus has thousands of copies of a unique receptor on its surface. The receptor is a membrane protein and recognizes one antigen. Stated another way, there is a one-to-one correspondence between the receptors found on B cells and T cells and the antigens to which they respond.

2. When the receptor on a lymphocyte binds to an antigen, the lymphocyte is activated. An activated lymphocyte divides and makes many identical copies of itself. In this way, specific cells are selected and cloned in response to an infection.

3. Some of the cloned cells descended from an activated lymphocyte persist long after the pathogen is eliminated. As a result, the cloned cells are able to respond quickly and effectively if the infection recurs.

The clonal-selection theory provided a coherent explanation for most major attributes of the acquired immune system. Just as important, it made a number of specific predictions that could be tested. Which aspects of the theory have been supported, and which rejected?

The Discovery of B-Cell Receptors and T-Cell Receptors

To test the prediction that lymphocytes have unique receptors on their surfaces, researchers injected experimental animals with radioactively labeled antigens. The presence of these labeled antigens allowed researchers to isolate and purify a protein on the surface of B cells that bound to the antigens. Chemical analysis of this **B-cell receptor** (**BCR**) showed that the protein has the same structure as the antibodies in the blood produced by B cells. This was a crucial result and led to the realization that both the BCR and the antibodies produced by B cells belong to a family of proteins called the gamma globulins. Any one gamma globulin is referred to as an **immunoglobulin** (**Ig**). These proteins are the key to the acquired immune response.

As **Figure 49.6a** shows, the BCR has three distinct components. The first is a protein called the **light chain**. The second component is roughly twice the size of the light chain and is called the **heavy chain**. Each BCR has two copies of the light chain and two copies of the heavy chain. Within the heavy chain, transmembrane domains anchor the protein in the plasma membrane of the B cell. B-cell antibodies are identical in structure to the BCR, except that they lack the transmembrane domains. Instead of being inserted into the plasma membrane, antibodies are secreted from the cell.

Table 49.2 shows the five classes of immunoglobulin proteins that act as B-cell receptors or antibodies. The five types are symbolized IgG, IgD, IgE, IgA, and IgM. Each class is distinguished by unique amino acid sequences in the heavy-chain region, and each has a distinct function in the immune response.

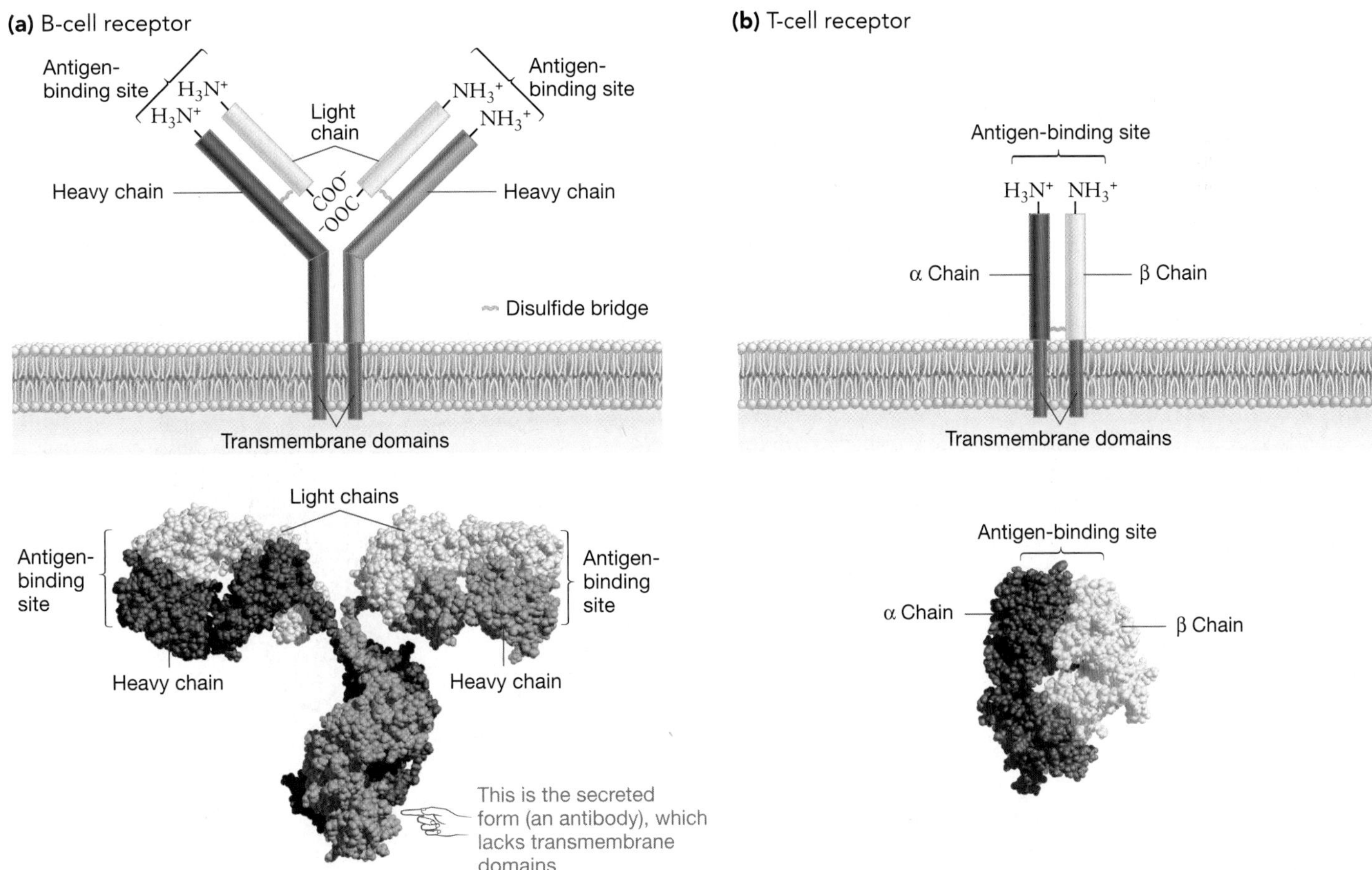

FIGURE 49.6 B-Cell Receptors and T-Cell Receptors Have Transmembrane Domains and Antigen-Binding Sites.
(a) Schematic model of the B-cell receptor, which is shaped like a Y. The antibodies and receptors produced by each B cell are identical, except that antibodies, shown below as a space-filling model, lack the transmembrane domain and are secreted. **(b)** The shape of the T-cell receptor resembles one "arm" of the Y-shaped B-cell receptor.

TABLE 49.2 **Five Classes of Immunoglobulins**

Name	Structure (secreted form)	Function
IgG	Monomer	The most abundant type of secreted antibody. Circulates in blood and interstitial fluid. Protects against bacteria, viruses, and toxins.
IgD	Monomer	Present on membranes of immature B cells; rarely secreted. Function is unknown.
IgE	Monomer	Secreted in minute amounts. Involved in hypersensitive reaction that produces allergies.
IgA	Dimer	Most common antibody in breast milk, tears, saliva, and the mucus lining the respiratory and digestive tracts. Prevents bacteria and viruses from attaching to mucous membranes; helps immunize breastfed newborns.
IgM	Pentamer	First type of secreted antibody to appear during an infection. Binds many antigens at once; effective at clumping viruses and bacteria so that they can be destroyed.

It took much longer for researchers to isolate and characterize the **T-cell receptor** (**TCR**). It turns out that the TCR only binds to antigens that have been modified by other cells and then presented, or displayed, on the plasma membranes of these cells, a complex process called **antigen presentation**. This is a fundamentally important distinction: B cells can bind to antigens directly; T cells bind only to antigens that are displayed by other cells. Other data showed that the TCR is composed of two protein chains: an alpha (α) chain and a beta (β) chain. The overall shape of the TCR is very similar to the "arm" of an antibody or BCR molecule (**Figure 49.6b**).

Antibodies, BCRs, and TCRs do not bind to the entire antigen but to a selected region of the antigen called an **epitope**. To understand the relationship between an antigen and an epitope, consider that every bacterium, virus, fungus, and protist is made up of a large number of different molecules. Each of these molecules is an antigen, because each is foreign to your cells. And each antigen may have many different epitopes, where binding by antibodies and lymphocyte receptors actually takes place. **Figure 49.7** illustrates a protein called hemagglutinin, which is found on the surface of the influenza virus. This antigen has six epitopes, identified by the blue and red lines in the figure. Each of these epitopes is recognized by a particular antibody, BCR, or TCR. It is not unusual for an antigen to have between 10 and 100 different epitopes.

How are the immunoglobulins able to recognize specific epitopes? The answer to this question came through detailed studies of the BCR's heavy and light chains.

What Is the Molecular Basis of Antibody Specificity and Diversity? In the 1950s biologists developed an important model system for studying BCR and antibody production. The cells involved were B-cell tumors, or myelomas, that could be grown in laboratory culture. Like other cancers, myelomas grow in an uncontrolled fashion. Further, each type of

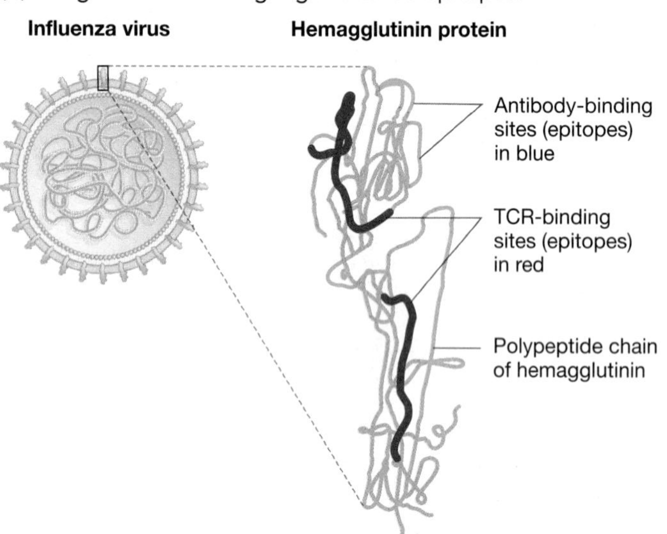

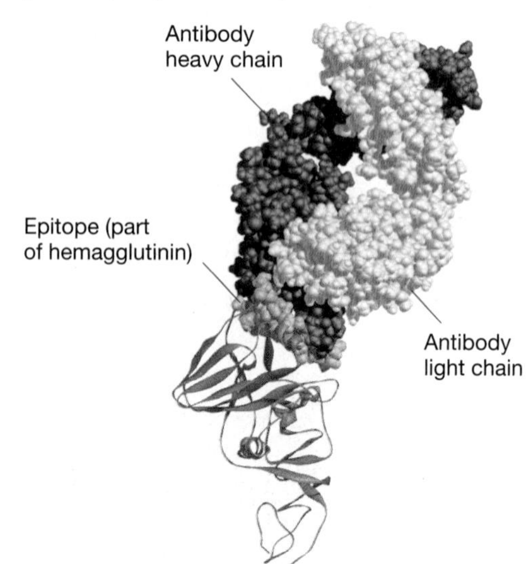

FIGURE 49.7 Most Antigens Have Multiple Binding Sites for B-Cell Receptors, Antibodies, and T-Cell Receptors.
(a) The envelope of the influenza virus includes the protein hemagglutinin. This version of hemagglutinin has four different sites where antibodies bind and two distinct places where T-cell receptors bind. **(b)** The space-filling model shows how closely one of hemagglutinin's epitopes fits into an antibody.

(a) The amino acid sequences of antibodies vary.

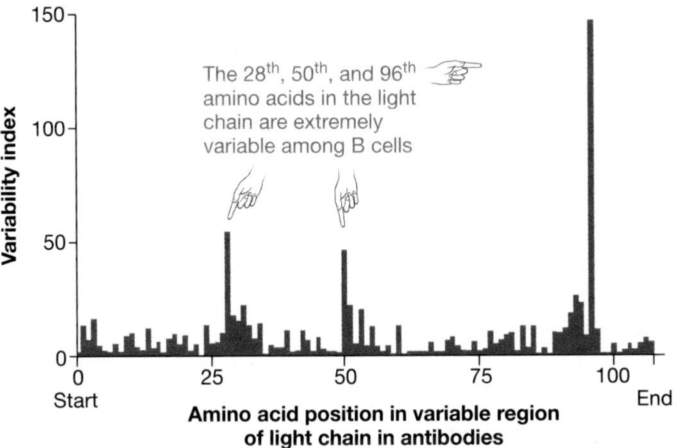

(b) B-cell receptors and T-cell receptors have variable regions and constant regions.

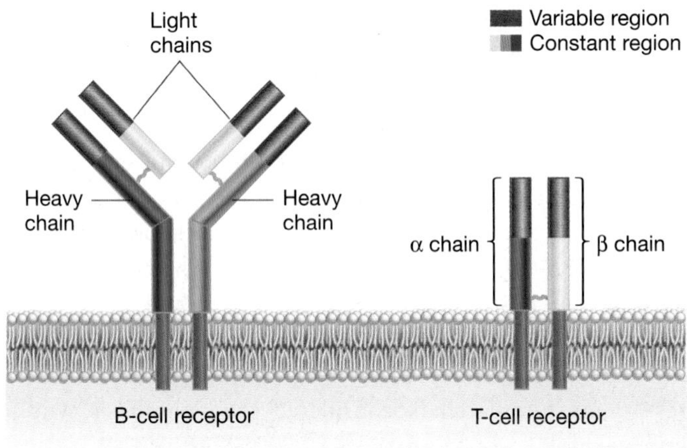

FIGURE 49.8 In B-Cell Receptors and T-Cell Receptors, the Amino Acid Sequence of the Light Chain Is Highly Variable. (a) Data indicating which amino acids in the immunoglobulin light chain are variable when 77 different antibody molecules are compared. "Variability index" is the number of antibodies that had a different amino acid at a position in the light chain, divided by the frequency of the most common amino acid at that position. The data for all 110 amino acids in the light chain are shown. The equivalent graph for the constant chain shows almost no variation. **(b)** The variable regions of a B-cell receptor and T-cell receptor face away from the plasma membrane.

EXERCISE Label the transmembrane domains of the receptors shown in part (b). Circle the antigen-binding sites on both molecules. Label the amino and carboxy terminus of each polypeptide.

myeloma produces a single type of antibody. By isolating and comparing the light chains produced by different myelomas, researchers were able to (1) document that each light chain is different, and (2) infer why different antibodies bind to different epitopes on antigens.

The initial studies exploring the composition of BCRs and antibodies used enzymes to chop the light chains from different B cells into pieces. Researchers used the technique called electrophoresis to separate the pieces and noticed that light chains from different B cells appeared to have one segment in common. In addition, there seemed to be light-chain regions that were unique to each B cell. These light-chain segments came to be known as the **constant (C) regions** and **variable (V) regions**, respectively.

When techniques to determine the amino acid sequence of proteins became available, the resulting data confirmed that both the light chain and the heavy chain consist of a variable (V) region at the amino terminus and a constant (C) region at the carboxy terminus. For example, when biologists compared the amino acid sequences of light chains from 77 different antibodies, they found that certain amino acids were extremely variable among B cells (**Figure 49.8a**). These portions of the light chain are within the V region. In contrast, the amino acid sequences of the C regions were almost identical. Similar patterns are found in the V and C regions of the heavy chain.

Why is this observation important? First, the presence of unique amino acids at certain positions in the light chain and heavy chain explains why the BCR and antibodies bind to unique epitopes: Receptors with different amino acid sequences bind to different epitopes. Second, the binding of epitopes occurs at the variable regions on the "end" (amino terminus) of the BCR molecule's and antibody's arms (**Figure 49.8b**). Analogous studies showed that the TCR also has a variable region near its amino terminus of both the α and β chains.

This observation solved one of the most fundamental questions about the acquired immune system: Your body can respond to an almost limitless number of antigens because there are a virtually limitless number of different BCRs, antibodies, and TCRs. If you understand this concept, you should be able to explain why variation in the composition of the variable region is responsible for the acquired immune system's ability to recognize so many different antigens. You should also be able to explain the molecular basis of specificity in the acquired immune system—meaning, why each BCR, antibody, and TCR binds to one epitope.

Analyses of the variable region raised a paradox, however. The acquired immune system can produce an almost limitless number of antibodies and receptors, but the genome of animals is not limitless. How does the genome code for so many different immune system proteins?

The Discovery of Gene Recombination In 1965 W. J. Dryer and J. Claude Bennett proposed a controversial explanation for how immunoglobulin genes code for so many different variable regions and thus so many different proteins. Dryer and Bennett suggested that a single gene codes for the C region of the light chain and that a separate gene codes for the V region. In addition, they hypothesized that early in the life of a lymphocyte, a segment from the *V* gene is combined with a segment from the *C* gene to form a new "*V* + *C* gene" for the light-chain protein. In effect, Dryer and Bennett hypothesized that as a B cell is maturing, the gene encoding the BCR is recombined in a way that results in the production of a sequence unique to that particular cell. The type of DNA cutting and pasting they were proposing had never been observed, however. (The introns and alternative splicing introduced in Chapter 16 had not been discovered yet.) As a result, most researchers considered the hypothesis wildly implausible.

In 1976, however, Nobumichi Hozumi and Susumu Tonegawa published data supporting the recombination hypothesis. Hozumi and Tonegawa reasoned that if light-chain diversity was produced by DNA recombination, then the variable and constant regions of immature B cells, which are not producing antibodies, should be longer than the same region in mature, activated B cells, which are producing BCR and antibodies. The amount of DNA in the *V* + *C* region should be shorter in mature cells than it is in embryonic cells, because it contains only a portion of the complete *V* and *C* gene regions (**Figure 49.9**). When they examined the *V* + *C* regions from immature B cells versus mature B cells, this is exactly what they found.

Hozumi and Tonegawa's result inspired a flurry of studies on the mechanism of DNA recombination. One central finding is that mammals have two genes that code for the light chain and one for the heavy chain. Each of these genes contains distinctive segments, however. For example, one of the light-chain genes in humans includes about 40 different variable (*V*) segments, 5 joining (*J*) segments, and a single constant (*C*) segment. The heavy-chain locus includes about 51 *V* segments, 27 diversity (*D*) segments, and 6 *J* segments along with the constant (*C*) segment. A light-chain protein is made up of portions encoded by a *V*, a *J*, and a *C* segment; a heavy-chain protein is made up of portions encoded by a *V*, a *D*, a *J*, and a *C* segment. The genes that encode the α and β chains of the TCR have a similar arrangement of distinct segments, each with multiple versions.

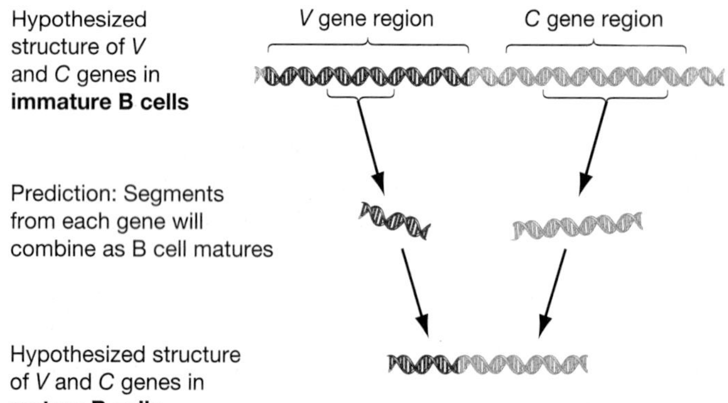

FIGURE 49.9 The Gene Recombination Hypothesis for Antibody Formation in B Cells.

Figure 49.10 summarizes the current model for how these gene segments are processed as a B cell matures. The structure of the *V*, *J*, and *C* regions of light-chain and heavy-chain DNA are illustrated in Figure 49.10a. This is what the genes look like in an immature B cell and all non-B cells. Figure 49.10b walks through the processing steps as a B cell matures, using one of the immunoglobulin light-chain genes as an example (step 1). Early in lymphocyte development, the gene segments are brought closer together by DNA recombination (step 2). For example, in B cells any one of the 40 *V* light-chain segments can recombine with any one of the 5 *J* segments. Thus, $40 \times 5 = 200$ different light chains can be produced by recombination. In the heavy chain, any one of the 51 *V* segments, 27 *D* segments, and 6 *J* segments can recombine, giving a total of $51 \times 27 \times 6 = 8262$ possible heavy chains. Because light-chain and heavy-chain rearrangements occur independently, DNA recombination can produce $200 \times 8262 = 1{,}652{,}400 = 1.65 \times 10^6$ different antigen-specific BCRs. And this number does not include BCRs that result from recombination in the other light-chain gene. The functional gene is then transcribed, and the resulting mRNA is processed and translated into a light-chain protein (steps 3–5). ● If you understand these events, you should be able devise a system for generating a high diversity of lottery numbers that is analogous to gene recombination, but uses only the digits 0–9.

The story does not end here. Gene segments are not always joined precisely during DNA recombination. Some variation occurs where the *V* and *D* segments join and where the *D* and *J* segments join. This junctional diversity results in many additional novel sequences. Thus, by recombining small DNA segments in the two light-chain genes, an estimated 10^{10} to 10^{14} different BCRs can be created. A similar number pertains for the TCRs, which undergo the same type of recombination process. These results revealed why each B cell and T cell in the body can have a different receptor on its surface. Due to gene recombination, each receptor has a unique amino acid sequence. As a result, it has the ability to bind a unique epitope on an antigen. Gene recombination is the molecular mechanism responsible for the acquired immune system's combination of specificity and diversity.

How Does the Immune System Distinguish Self from Nonself?

As B cells and T cells mature and begin producing receptors for specific antigens, an important problem arises: How does the immune system ensure that the receptors on these cells don't respond to molecules that are part of normal host cells? If a

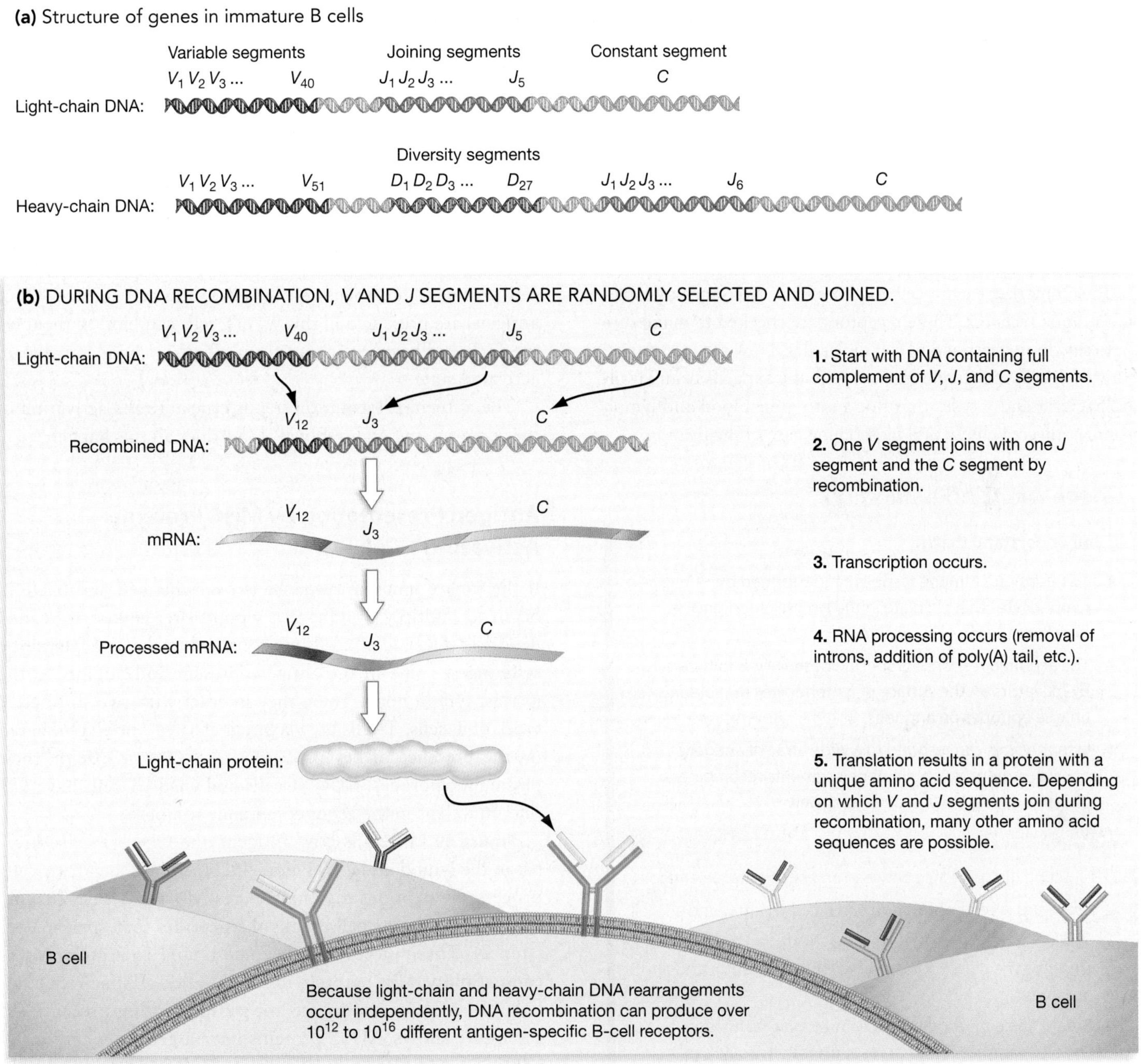

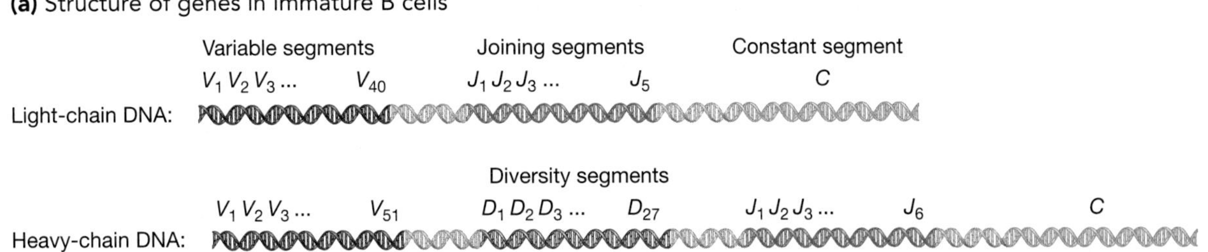

(b) DURING DNA RECOMBINATION, *V* AND *J* SEGMENTS ARE RANDOMLY SELECTED AND JOINED.

$V_1\,V_2\,V_3$ ... V_{40} $J_1\,J_2\,J_3$ ... J_5 C

Light-chain DNA:

1. Start with DNA containing full complement of *V*, *J*, and *C* segments.

V_{12} J_3 C

Recombined DNA:

2. One *V* segment joins with one *J* segment and the *C* segment by recombination.

V_{12} J_3 C

mRNA:

3. Transcription occurs.

V_{12} J_3 C

Processed mRNA:

4. RNA processing occurs (removal of introns, addition of poly(A) tail, etc.).

Light-chain protein:

5. Translation results in a protein with a unique amino acid sequence. Depending on which *V* and *J* segments join during recombination, many other amino acid sequences are possible.

B cell

Because light-chain and heavy-chain DNA rearrangements occur independently, DNA recombination can produce over 10^{12} to 10^{16} different antigen-specific B-cell receptors.

B cell

FIGURE 49.10 DNA Recombination Results in Antibodies with Unique Amino Acid Sequences. Diagram summarizing current data on how light-chain proteins are produced as a B cell matures.

B-cell receptor or T-cell receptor responded to a **self** molecule—that is, a molecule belonging to the host—the receptor would trigger an immune response. An anti-self reaction such as this is known as **autoimmunity** and can lead to immune system cells destroying parts of the host's own body.

Biologists predicted that there must be some mechanism for eliminating young B cells and T cells that have receptors for self molecules. This prediction was confirmed when researchers injected B cells and T cells with anti-self receptors into mice and found that these experimental lymphocytes were eliminated. Follow-up work showed that if B cells and T cells that are maturing in the bone marrow and thymus have anti-self receptors, the cells are destroyed before they leave these organs.

A great deal of research is currently focused on understanding the molecular mechanisms responsible for eliminating lymphocytes with anti-self receptors. In addition to the intrinsic interest it offers, this research has important implications for human health. If the elimination process breaks down, lymphocytes are produced that react to self proteins, and illness results.

For example, multiple sclerosis (MS), introduced in Chapter 45, results from the production of anti-self T cells. The T cells attack the myelin sheath of nerve fibers. Because damage to myelin reduces the efficiency of nerve signaling, muscular and coordination problems result. Other autoimmune diseases include rheumatoid arthritis, type 1 diabetes mellitus, myasthenia gravis, and lupus (systemic lupus erythematosus). In the vast majority of cases, however, cells in the bone marrow and thymus correctly eliminate maturing B cells and T cells that have receptors for self proteins.

To summarize, each B cell and T cell has thousands of copies of a unique receptor. These receptors are checked to make sure that they do not respond to self molecules. Now the question is, How do B cells and T cells interact with the antigens from invading bacteria and viruses that flood into your blood and lymph from an infected elbow or a blossoming case of the flu?

Check Your Understanding

If you understand that...

- The acquired immune response is performed by lymphocytes that circulate in the lymph, blood, and organs of the immune system.
- The acquired response is specific because it is initiated by receptors on the surface of lymphocytes that bind to unique epitopes on antigens.
- Lymphocytes can respond to a wide array of antigens because immunoglobulin gene segments recombine to produce a receptor unique to each cell.

You should be able to...

1) Describe the difference between an epitope and an antigen.
2) Draw and label the structure of a B-cell receptor and a T-cell receptor.
3) Describe the difference between a B-cell receptor and an antibody.
4) Predict the consequences of a mutation that disrupts selection against cells that respond to self molecules.

49.3 The Acquired Immune Response: Activation

When the cell-surface receptor of a B cell or T cell binds to an antigen, the event is like pushing the button that launches a "smart bomb"—a missile that is programmed to destroy a specific target. The binding of a B cell or T cell to an antigen unleashes a series of events that results in the destruction of the antigen. The lethal power of activated B cells and T cells is nothing short of awesome.

Perhaps because lymphocytes are so destructive, the activation of the relevant B cells and T cells is a carefully controlled, stepwise process. The mechanism is reminiscent of the precautions that nations with powerful missiles take to avoid accidental deployment. For the most dangerous weapons, the signal to launch is checked and cross-checked, using a series of codes and signals. In the immune system, the checking and cross-checking occurs through protein-protein interactions on the surfaces of cells, and the release and receipt of cytokines and other signaling molecules.

For T lymphocytes, the sequence that leads to activation begins when antigens are taken up by a specific type of leukocyte or an infected cell, cut into pieces, packaged with specific cell proteins, and then transferred to the cell surface. Once antigens are presented in this way, T cells can bind to them via the TCR and begin their transformation from an inactive to an activated state.

The antigen-presentation step is crucial to the activation of the acquired immune response. How does all this happen?

Antigen Presentation by MHC Proteins: Activating T Cells

If the innate immune response is overwhelmed and bacteria begin to multiply rapidly at a wound site, leukocytes called dendritic ("tree-like") cells are recruited to the area. **Dendritic cells** ingest some of the antigens present and migrate to the nearest lymph node. There they interact with two distinctive types of T cells. T cells are classified as $CD4^+$ or $CD8^+$, based on the presence of key proteins called **CD4** or **CD8** on their plasma membranes. $CD4^+$ T cells and $CD8^+$ T cells have distinct functions in the acquired immune response.

Figure 49.11 details how antigens and $CD8^+$ T cells interact in the lymph node. As Figure 49.11a shows, dendritic cells process the peptides they have ingested (step 1). An enzyme complex inside the cell breaks the peptides into small pieces (step 2). These pieces then become bound to a **major histocompatibility (MHC) protein** inside the endoplasmic reticulum (step 3). MHC proteins are antigen-presenting proteins. As the figure shows, MHC proteins have a groove where small peptide fragments, typically from 8 to 20 amino acids in length, bind. MHC proteins come in two types, called **Class I** and **Class II MHC proteins**. In dendritic cells, peptide fragments attach to Class I proteins and are transported to the cell surface (step 4). When the MHC protein-peptide complex is displayed on the cell surface, antigen presentation is complete (step 5).

Each individual has several genes encoding Class I and Class II MHC proteins, so each person can produce several distinct proteins of each type. Further, the MHC genes are among the most polymorphic of any genes known—meaning that many different alleles exist in the population (see Chapter 13). Because so many distinct alleles exist, most individuals are heterozygous for the Class I and Class II MHC genes. Heterozygous individuals produce an even greater diversity of Class I and Class II MHC proteins. Having a wide array of

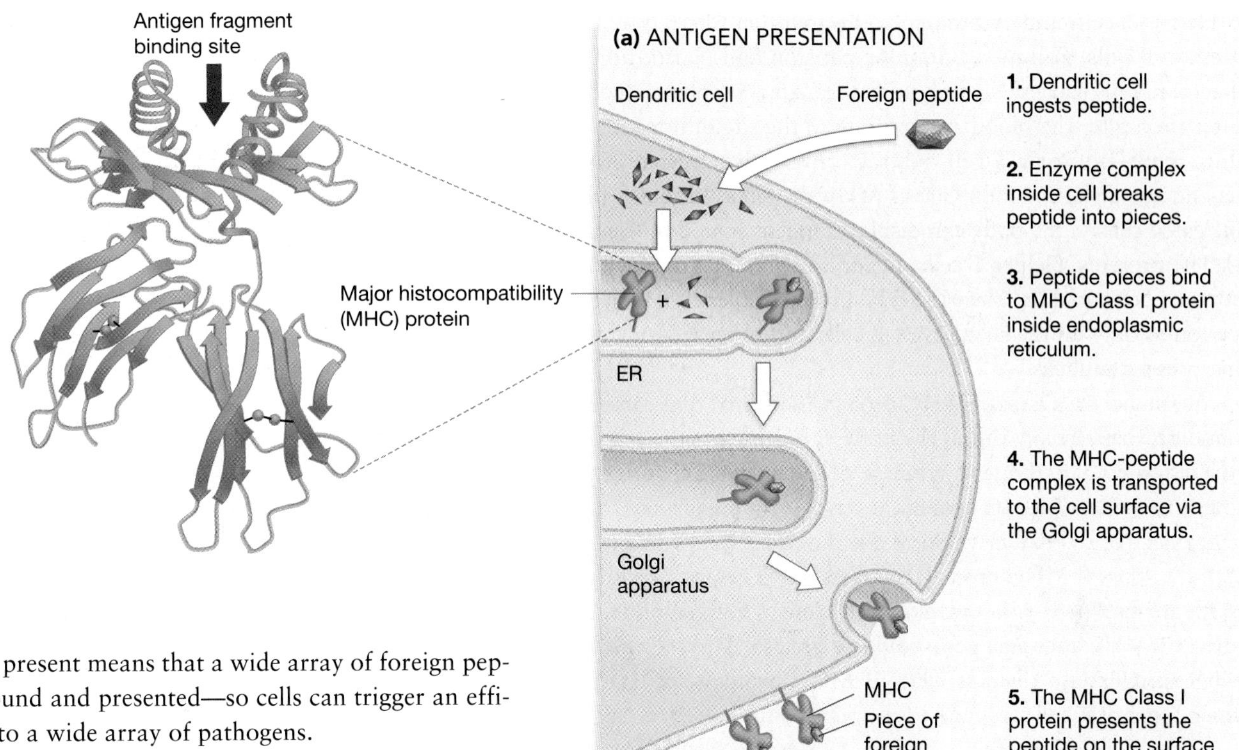

MHC proteins present means that a wide array of foreign peptides can be bound and presented—so cells can trigger an efficient response to a wide array of pathogens.

Figure 49.11b illustrates what happens once a dendritic cell displays an MHC protein-foreign peptide complex. As step 1 shows, the antigen-displaying dendritic cell interacts with a $CD8^+$ T cell that is moving through the lymph organs. To this T cell, the dendritic cell displaying an MHC protein and a bound peptide carries the message "self cell has found antigen—response required." The T cell becomes activated in response to interactions with proteins on the dendritic cell. (In some cases, activation of $CD8^+$ T cells requires interactions with activated $CD4^+$ T cells as well.) As step 2 in the figure shows, the activated T cell begins to divide to produce a series of genetically identical daughter cells. This event is called **clonal expansion.** Clonal expansion is a crucial step in the acquired immune response. It leads to a large population of lymphocytes capable of responding specifically to the antigen that has entered the body. If you understand why clonal expansion is important, you should be able to explain why cancer chemotherapy—which destroys rapidly dividing cells—suppresses the ability of the patient's immune system to respond to infections.

When activated $CD8^+$ T cells undergo clonal expansion, the daughter cells develop into **cytotoxic** ("cell-poison") **T cells,** also known as cytotoxic T lymphocytes (CTLs) or killer T cells. As we'll see, the adjectives cytotoxic and killer are appropriate: $CD8^+$ T cells kill cells that are infected with a virus or other pathogen in their cytoplasm.

The daughter cells of activated $CD4^+$ lymphocytes, in contrast, differentiate into **helper T cells.** As we'll see, the adjective helper is also appropriate—helper T cells assist with the activation of other lymphocytes. There are two types of helper T cells, designated T_H1 and T_H2, and they have distinct functions: T_H1 cells help activate cytotoxic T cells, while T_H2 cells help activate B cells.

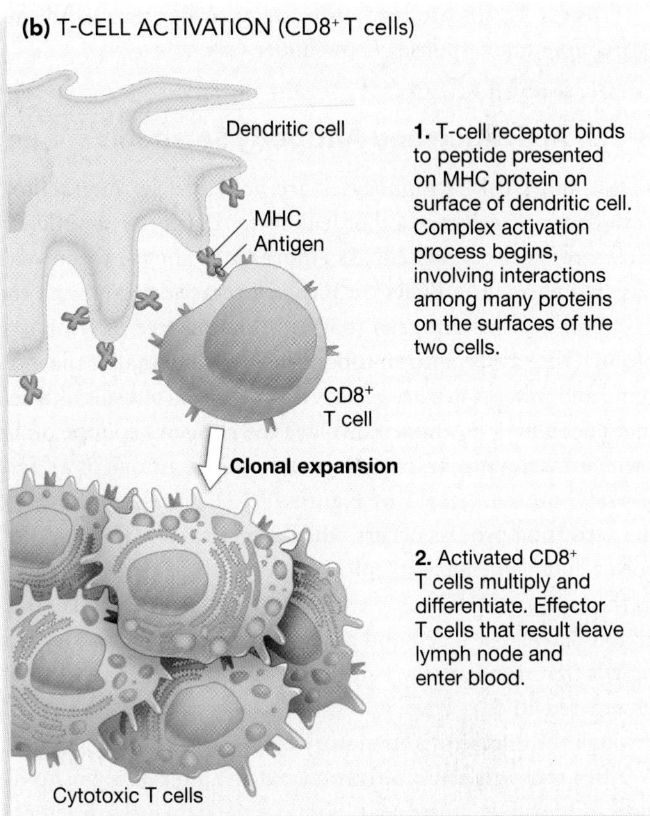

FIGURE 49.11 T Cells Are Activated after Interacting with Antigens Presented by Dendritic Cells. (a) Dendritic cells take in antigens, break them into pieces, and then present the fragments in the groove of an MHC protein. **(b)** If a $CD8^+$ T cell has a receptor that is complementary to the MHC-antigen complex presented on the surface of a dendritic cell, the T cell will begin activation.

Helper T cells and cytotoxic T cells are often referred to as effector T cells. Following clonal expansion and maturation, effector T cells leave the lymph node through a lymphatic duct. Later they enter the blood and migrate to the site of infection. Once there, cytotoxic T cells interact only with cells that display antigens presented on Class I MHC proteins. Virtually all nucleated cells in the body can display antigens bound to Class I MHC proteins. Helper T cells, in contrast, interact only with antigens presented on Class II MHC protein molecules, which are found only on the surfaces of B cells and other leukocytes that present antigens.

An antigen on a Class I MHC protein is a signal that can be found on virtually any cell of the body. It means that the cell is infected with a pathogen. When a cell displays an antigen bound to a Class I MHC protein, it sends a simple message to cytotoxic T cells: "Kill me." But if a leukocyte displays an antigen on a Class II MHC protein, the message to helper T cells is "Activate me." ● If you understand the role of antigen presentation by MHC proteins, you should be able to (1) state which cells express Class I versus Class II MHC proteins, (2) state which types of T cells respond to peptides bound to Class I versus Class II MHC proteins, and (3) summarize the message provided by MHC-peptide complexes on a dendritic cell, an infected cell, and a leukocyte.

Effector T cells are ready for action. A key part of the acquired immune response is now under way.

B-Cell Activation and Antibody Secretion

$CD8^+$ and $CD4^+$ lymphocytes are activated by interactions with dendritic cells and other leukocytes that present antigens. How are B cells activated? As **Figure 49.12** shows, the answer has two parts. The BCRs on B cells interact directly with the bacterial or viral antigens that are floating free in lymph or blood. Once a free antigen is bound, B cells internalize the molecule and process it with a Class II protein. As a result, a B cell that encounters its antigen displays the antigen's epitope on its own surface, with the antigen cradled in the groove of a Class II MHC protein (step 1 of Figure 49.12). The second part of the activation process occurs when an activated $CD4^+$ T_H2 cell with a complementary receptor arrives. This helper T cell binds to the antigen-MHC complex on the B cell (step 2). The interaction between a B cell and a helper T cell supplies activation signals that stimulate the helper T cell. Once the helper T cell is stimulated in this way, it responds by activating the B cell through the release of stimulatory cytokines.

After receiving these activation signals, a B cell begins to divide and form daughter cells. Some of the daughters differentiate into activated B lymphocytes called **plasma cells**. Plasma cells produce and secrete large quantities of antibodies (step 3). Recall that antibodies are identical to B-cell receptors, except that they lack a transmembrane domain and are secreted instead of being found in the plasma membrane.

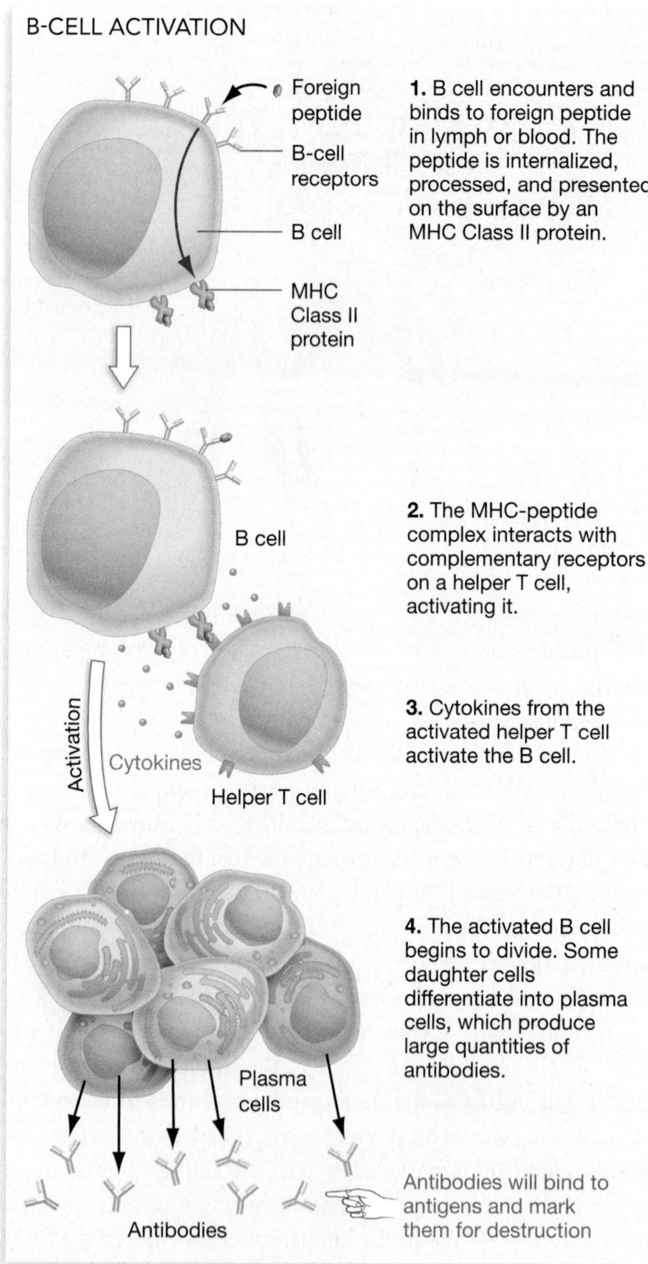

FIGURE 49.12 B Cells Are Activated after Interacting with Helper T Cells. A series of steps are involved in activating B cells. Once receptors on a helper T cell have bound to MHC-antigen complexes on a B cell, the helper T cell releases cytokines that activate the B cell. (Other proteins on the surface of a B cell and T cell must also interact for activation to occur.) Activation eventually leads to the production of antibodies by plasma cells.

Once B cells are activated, the acquired immune response gains an important dimension. Antibodies specific to the invading bacterium or virus begin to circulate in the blood.

Table 49.3 summarizes how B cells and T cells are activated. (The memory cells mentioned in the table are introduced in Section 49.4.) Once these lymphocytes have been activated, most of the major elements in the acquired immune response are in place.

SUMMARY TABLE 49.3 **Activation of the Acquired Immune System**

Type of Lymphocyte	Method of Activation	Cells That Result from Activation and Clonal Expansion
B cell	Receptor binds to free antigen, then interacts with T_H2 cell.	Plasma cells Memory B cells
$CD4^+$ T cell	Receptor binds to antigen-MHC Class II protein complex on dendritic cell or other antigen-presenting cell.	T_H1 (helper T cell) T_H2 (helper T cell) Memory T cells
$CD8^+$ T cell	Receptor binds to antigen-MHC Class I protein complex on infected cell; may also interact with T_H1 cells.	Cytotoxic T cells Memory cells

Check Your Understanding

If you understand that...

- T cells are activated when their receptor binds to an antigen presented by the MHC proteins on a dendritic cell or an infected cell.
- B cells are activated when their receptor binds to an antigen, the antigen is presented by an MHC protein on their surface, and they are stimulated by a helper T cell activated by the same antigen.
- Activated lymphocytes undergo dramatic changes in morphology and divide rapidly.

You should be able to...

Generate a hypothesis to explain the observation that humans who are heterozygous for the genes encoding MHC proteins tend to be healthier than individuals who are homozygous for these genes.

49.4 The Acquired Immune Response: Culmination

Activated B cells, helper T cells, and cytotoxic T cells present a formidable response to invading pathogens. In combination with the leukocytes involved in the innate immune system, the cells of the acquired immune system are almost always successful in eliminating threats from bacteria, parasites, fungi, and viruses. To drive this point home, consider what happens when the acquired immune system does *not* work. For example, children who are born with severe combined immunodeficiency (SCID) have a genetic defect in one of the enzymes responsible for DNA recombination in maturing lymphocytes. As a result, they are unable to generate normal T-cell and B-cell receptors and lack a normal immune system (see Chapter 19). In particular, it is their acquired immune response that is badly impaired. Afflicted individuals suffer debilitating illness from infections that other children fight off easily, and they die before they are two years old. Similarly, people who are infected with the **human immunodeficiency virus** (**HIV**) suffer a progressive loss of $CD4^+$ T cells, because HIV kills $CD4^+$ T cells and macrophages. Eventually, HIV-infected people develop **AIDS—acquired immune deficiency syndrome**. They succumb to illnesses that physicians almost never see in people with healthy immune systems.

To understand how the acquired response actually kills pathogens, let's again return to a bacterial infection in a wounded elbow and an upper respiratory tract infection caused by the flu virus. How do activated B cells and T cells eliminate these invaders?

How Are Bacteria and Other Foreign Cells Killed?

During the innate immune response to bacteria and other pathogenic cells that enter a wound, macrophages at the site phagocytize some of the invaders. In addition to killing the foreign cells, these leukocytes process and present antigens via the MHC Class II proteins. As a result, macrophages at the site of infection display epitopes on their surfaces that can be recognized by helper T cells activated during an acquired immune response. If an activated T_H1 cell binds to these antigen-laden macrophages, two things happen: First, the phagocytic activity of the macrophages is enhanced. Second, the T_H1 cells secrete cytokines that recruit additional phagocytic cells to the site, and increase the inflammatory response.

Bacteria, fungi, and other foreign cells at the infection site also begin to be coated with antibodies from plasma cells. In many cases, antibody action causes **agglutination**, or clumping of cells. Each antibody has at least two binding sites, so a single antibody can bind epitopes on foreign cells and link them, forming a clump. Clumped or single cells that are tagged with antibodies are readily destroyed by macrophages by phagocytosis. Antibodies that are bound to antigens also stimulate a group of proteins called the **complement system**. Complement proteins circulate in the bloodstream and assemble at antigen-antibody complexes. When complement proteins activate, they punch lethal holes in the plasma membranes of invading cells. Within a few days, this combination of killing mechanisms usually eliminates all foreign cells.

How Are Viruses Destroyed?

Just as the acquired immune system has an array of mechanisms to dispose of bacteria, it has two major ways to eliminate viruses. One route involves cytotoxic T cells (activated $CD8^+$ cells) and is called the **cell-mediated response**; the other route involves antibodies and is called the **humoral response**. (The Latin root *humor* means "fluid." The humoral response takes place in blood and lymph; the cell-mediated response takes place at the surface of cells.) Let's take a closer look at both.

Cell-Mediated Response Infected cells respond to the arrival of a virus by processing antigens from the invader. MHC Class I proteins attach to the viral antigens and present the antigens on the surface of the infected cell. Every nucleated cell in the body expresses MHC Class I proteins and has the ability to signal that it is infected. Cells that display viral antigens bound to Class I MHC molecules are effectively waving a flag that says, "I'm infected. If you destroy me, you'll destroy them."

As $CD8^+$ cells migrate into the area, they recognize and bind to the antigen's epitopes and the MHC protein displayed on infected cells. Once binding occurs, the $CD8^+$ cell secretes molecules that assemble on the surface of the infected cell's plasma membrane and produce pores. Granules from the cytotoxic T cell then enter the cell and activate a self-destruct response (**Figure 49.13a**).

Once the infected cell is destroyed, the cytotoxic T cell releases and seeks out another infected cell to kill. Over time, all virus-infected cells are eliminated. Because viruses can reproduce only inside host cells, the cell-mediated response limits the spread of the infection by preventing new generations of virus particles from maturing.

Humoral Response Even as cytotoxic T lymphocytes swing into action, plasma cells are producing antibodies to viral proteins. In most cases, the most effective antibodies are those that bind to epitopes on the surface of the virus. As **Figure 49.13b** shows, two things happen when antibodies bind to the outside of a virus: First, the virus is blocked from making contact with plasma membranes and entering new host cells. Second, macrophages and other phagocytic cells recognize the antibody-coated particles and phagocytize them. Because of this humoral response, the number of virus particles floating free in lymph and blood is eventually reduced to zero.

Why Does the Immune System Reject Foreign Tissues and Organs?

Antibodies and cytotoxic T cells have devastating effects on invading bacteria and viruses. Unfortunately, they are equally deadly in response to tissues or organs that are introduced into a patient in the hopes of healing a wound or curing a disease. To understand why, consider the problems that can arise

(a) CELL-MEDIATED RESPONSE

Granules

Cytotoxic T cell

1. Cytotoxic T cell makes contact with virus-infected cell and releases granules (black dots).

Virus-infected host cell

Virus particle

2. Molecules in the granules induce infected cell to self-destruct, killing the viral particles inside.

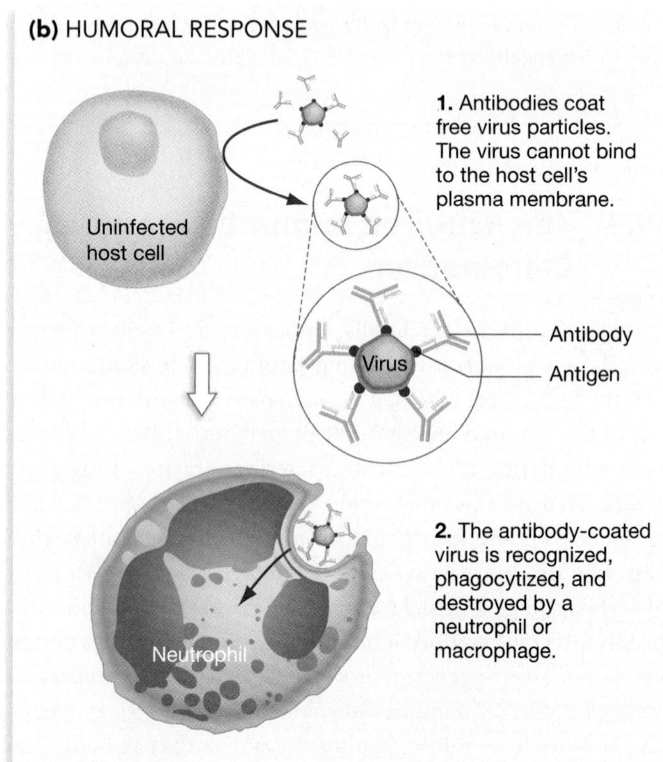

FIGURE 49.13 Lymphocytes Kill Viruses via the Cell-Mediated Response and the Humoral Response.

with blood transfusions. Certain individuals have red blood cells with membrane proteins called A and B. These proteins act as antigens if they are introduced into a person whose own blood cells lack those proteins. For example, if you have type A blood, it means that your blood cells have the A antigen. If your blood is introduced to a person who lacks the A antigen—meaning a person who has type B or type O blood—the recipient's immune system will recognize the A antigen as foreign and mount a devastating response against it. For a blood transfusion to be successful, a recipient has to receive blood that lacks the A and B proteins entirely or that contains the same antigens found in his or her own blood.

Similar problems arise in organ transplants, except that the antigenic molecules in foreign organs are the MHC proteins found on the surfaces of their cells. To prevent strong immune reactions to a transplanted kidney, heart, or liver, physicians do two things: (1) obtain the organ to be transplanted from a sibling or other donor whose MHC proteins are extremely similar in structure to those of the recipient and (2) treat the recipient with drugs that suppress the immune response. Thanks to steady improvements in drug development and in systems for matching MHC types between donors and recipients, the success rate for organ transplants has improved dramatically in recent years.

As the blood transfusion and organ transplant examples show, the immune system rejects foreign tissues because they contain nonself proteins—antigens. To your T cells and B cells, a blood transfusion or an organ transplant is indistinguishable from a massive influx of bacteria, viruses, or other foreign invaders.

Responding to Future Infections: Immunological Memory

When B cells and T cells are activated and undergo clonal expansion, their daughter cells differentiate into plasma cells, helper T cells, and cytotoxic T lymphocytes. **Table 49.4** summarizes what these cells do. But in addition to producing the cells that implement the humoral and cell-mediated responses, activated B cells and T cells produce specialized daughter cells called memory cells. **Memory cells** do not participate in the initial acquired response, or **primary immune response**. Instead, they provide a surveillance service after the original infection has been cleared. Memory cells remain in the spleen and lymph nodes for years or decades, ready to provide an extremely rapid response should an infection with the same antigen recur. The production of memory cells is a hallmark of the vertebrate immune response. It occurs only to a limited degree, if at all, in invertebrates.

If the same antigen enters the body a second time, memory cells bind the antigen and trigger a second acquired response, or **secondary immune response**. The launching of a secondary immune response by means of memory cells is known as **immunological memory**. The secondary response is faster and more efficient than the primary response. It is faster because the presence of memory T and B cells increase the likelihood that lymphocytes with the correct antigen-specific receptors will find the antigen and activate quickly. It is more efficient because some of the memory B cells that respond to the returning antigen migrate to a specialized area in the lymph node called the germinal center. There the DNA sequences that code for the variable region of the immunoglobulin gene begin to undergo rapid mutation. Mutations in the DNA of the variable region modify the receptors produced by the memory cell. Memory B cells with receptors that bind best to the antigen's epitope live and produce daughter cells; those that bind to antigens less well die.

In effect, this process of somatic hypermutation leads to a fine-tuning of the immune response. The antibodies that result from somatic hypermutation bind to the antigen more tightly than the antibodies produced by plasma cells during the primary immune response. As the secondary immune response proceeds and somatic hypermutation continues, better-fitting antibodies are produced.

SUMMARY TABLE **49.4 Function of Acquired Immune System Components**

Type of Lymphocyte	Cells That Result from Activation and Clonal Expansion	Function of Resulting Cells
B cell	Plasma cells	Secrete antibodies
	Memory cells	Participate in secondary response (secrete antibodies)
$CD4^+$ T cell	T_H1 (helper T cell)	Activates cytotoxic T cells, regulates inflammatory response
	T_H2 (helper T cell)	Activates B cells
	Memory cells	Participate in secondary response
$CD8^+$ T cell	Cytotoxic T cells	Kill infected host cells
	Memory cells	Participate in secondary response

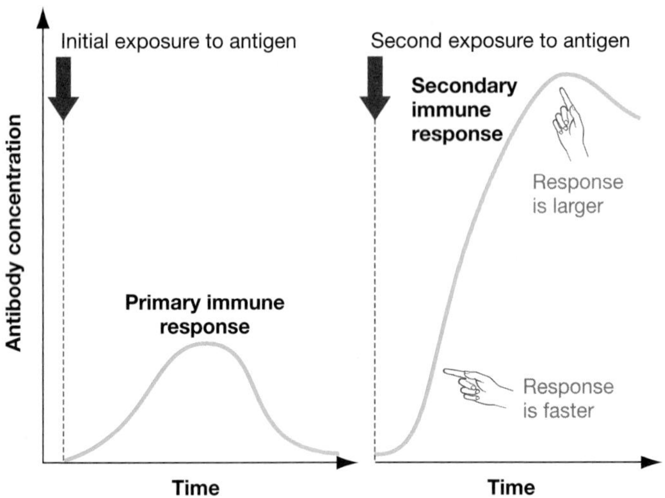

FIGURE 49.14 The Secondary Immune Response Is Faster and Stronger than the Primary Response. The data shown here summarize what happens when biologists inject a mouse with an antigen, document changes in antibody concentration over time, inject the same individual with the same antigen later, and measure the response in antibody concentration.

EXERCISE The primary response takes 10–17 days; the secondary response takes 3–5 days. Add appropriate labels to scale the *x*-axes.

Figure 49.14 underlines the effectiveness of the secondary immune response by comparing the rate of antibody production during the first and second exposures to a virus. The graphs are based on data that researchers collect by inoculating a laboratory mouse with influenza virus, collecting blood from the mouse every other day, and then measuring the amount of antiviral antibodies present in the fluid phase of the blood. (**Box 49.1** explains how researchers measure antibody concentrations in blood.)

The data in Figure 49.14 also explain why vaccination is an effective defense against certain viruses. A **vaccine** contains epitopes from a pathogen or a killed or weakened version of the pathogen itself. After vaccination (inoculation with a vaccine), the body mounts a primary immune response that results in the production of memory cells. If an actual infection occurs later, these memory cells respond quickly and eliminate the threat before illness occurs. Edward Jenner's vaccination strategy worked because the antigens presented by cowpox are extremely similar to the antigens presented by smallpox. As a result, exposing people to cowpox virus elicited the production of memory cells that effectively thwarted future infections by smallpox virus.

BOX 49.1 The ELISA Test

One of the common strategies used to measure protein concentrations in blood is called an ELISA (enzyme-linked immunosorbent assay). The standard screening test for HIV infection, for example, is an ELISA for antibodies that the immune system produces in response to this virus. Pregnancy tests are ELISAs that test for the presence of human chorionic gonadotropin in urine. There are several distinct ELISA tests; **Figure 49.15** illustrates how an indirect ELISA might be used to determine the amount of antibody present in a mouse that has been infected with influenza.

To begin an ELISA test, each well in a plastic plate containing many wells is coated with antigen (step 1 of Figure 49.15). In our example, the antigen is purified hemagglutinin protein from the influenza virus. Next, a defined amount of plasma from the patient is added to the plate. (Recall from Chapter 44 that plasma is the cell-free, fluid fraction of the blood.) If antibodies are present, they bind to the antigen (step 2). To continue the assay, a secondary antibody is added to the mixture of antigen and plasma (step 3). The secondary antibody is an antibody that can attach to the constant region of antibodies in the plasma added to the well. In ELISA tests, the secondary antibody is attached to an enzyme such as alkaline phosphatase (shown in the figure as AP). Alkaline phosphatase is useful because it catalyzes a reaction with a substrate to generate a product that changes the color of the solution. As a result, the presence of color in a well indicates a positive reaction—meaning the antibody to hemagglutinin is present (step 4). A positive result confirms that the patient is infected with influenza virus.

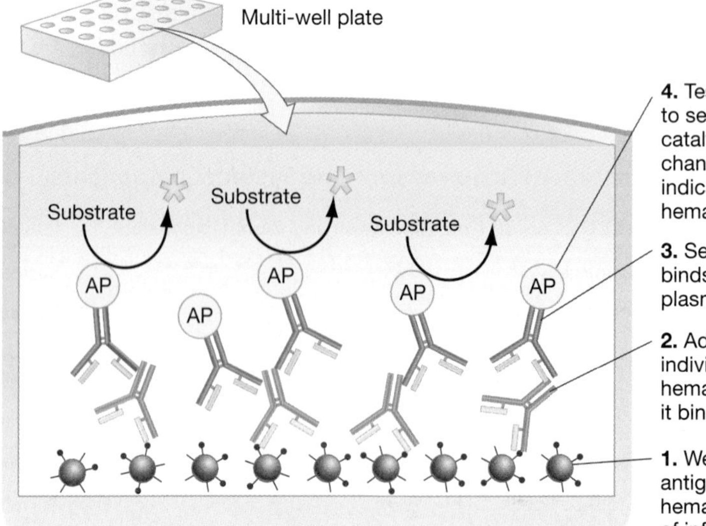

FIGURE 49.15 The ELISA Test Measures Antibody Concentrations in Blood.

Unfortunately, viruses such as influenza and HIV mutate so rapidly that they present the immune system with a constantly changing array of epitopes. Memory cells that were effective during a previous infection by these viruses are unlikely to bind to the changed epitopes and trigger a response to a later infection. As a result, it is extremely difficult for biologists to design an effective vaccine against these agents. Currently, the only "cure" for HIV infection is prevention; to keep up with rapidly mutating flu populations, flu vaccines must be redesigned and administered every year. The high mutation rates observed in these viruses are thought to be an adaptation that helps them escape detection by the immune system.

 Web Animation at www.masteringbio.com
The Acquired Immune Response

Check Your Understanding

If you understand that... 

- The antibodies produced by activated B cells mark invading pathogens for destruction.
- Activated cytotoxic T cells recognize epitopes displayed by infected cells and kill them before the pathogens inside can replicate.
- Memory cells produced during B-cell and T-cell activation remain in the body and provide secondary immunity against future infections.

You should be able to... 

1) Explain how vaccination works.
2) Propose a strategy for developing a vaccine for HIV or the cold virus.

Chapter Review

SUMMARY OF KEY CONCEPTS

The innate immune response to infection is mounted by leukocytes that respond in a nonspecific way to pathogens. The acquired immune response is mounted by leukocytes called lymphocytes. Each individual lymphocyte responds to a specific pathogen.

The innate response occurs in the same way no matter which pathogens have entered the body, but the acquired response is specific to the pathogen present. During the innate immune system's inflammatory response, mast cells release chemical messengers that result in increased blood flow to wounds or areas of tissue damage. After neutrophils migrate to the site, they respond to bacteria that stimulate their pattern-recognition receptors by phagocytizing the invading cells and destroying them. Macrophages also phagocytize pathogens and release chemical messengers that activate other leukocytes and raise body temperature.

You should be able to explain what would happen if blood vessels near a wound did not dilate and become more permeable in response to chemical signals from leukocytes and damaged tissues.

 Web Animation at www.masteringbio.com
The Inflammatory Response

The acquired immune response begins when cell-surface proteins on lymphocytes bind to an antigen—usually a protein from a disease-causing agent.

Lymphocytes are activated when T-cell receptors and B-cell receptors recognize an epitope on an antigen. The receptor protein on the surface of a T cell or B cell is generated through DNA recombination, a rearrangement of gene segments. Because every receptor that results from this process is slightly different, the immune system is able to recognize and respond to an almost limitless array of antigens.

You should be able to explain how a drug that bound to a particular epitope of an antigen would affect the activation of T cells with receptors specific to that antigen, the activation of B cells with receptors specific to that antigen, and the cell-mediated response to that antigen.

The humoral response tags pathogens for destruction with antibodies that are produced by lymphocytes called B cells.

During the humoral response, activated B cells produce antibodies to specific epitopes on the pathogen that has invaded the body. Viruses that become coated with antibodies are unable to enter host cells and are destroyed by macrophages. Bacteria that are tagged with antibodies are destroyed by complement proteins or by macrophages.

You should be able to explain why antibodies to a specific pathogen would act as an effective drug.

The cell-mediated response destroys infected cells via lymphocytes called T cells.

During the cell-mediated response, host cells that are infected with a virus display antigens on their surface and are destroyed by cytotoxic T cells. As a result, the virus particles inside are destroyed and cannot contribute further to the infection.

Because memory cells are created as activated B cells and T cells undergo clonal expansion, the immune system is able to respond rapidly and effectively to future infections by the same antigen. A vaccine triggers the production of memory cells because it contains epitopes from antigens.

You should be able to explain why the boy that Edward Jenner inoculated with cowpox became immune to smallpox.

 Web Animation at www.masteringbio.com
The Acquired Immune Response

QUESTIONS

Test Your Knowledge

1. What is the primary difference between the innate and acquired responses?
   a. The innate response is modified over time; the acquired response occurs in the same way throughout life.
   b. Only the acquired response is triggered by antigens.
   c. The acquired response is specific and is "remembered"; the innate response is nonspecific.
   d. There is no nonself recognition component in the innate response.
2. All of the following events are involved in the inflammatory response *except*:
   a. Cytotoxic T cells kill infected host cells.
   b. Neutrophils phagocytize pathogens that stimulate their pattern-specific receptors.
   c. Mast cells secrete chemical messengers that lead to increased blood flow.
   d. Macrophages release chemical messengers that lead to increased body temperature.
3. What is the difference between an epitope and an antigen?
   a. An epitope is any foreign substance; an antigen is a foreign protein.
   b. An epitope is the part of an antigen where an antibody or lymphocyte receptor binds.
   c. An antigen is the part of an epitope where an antibody or lymphocyte receptor binds.
   d. Antigens are recognized by B cells and antibodies; epitopes are recognized by T cells.
4. How do B-cell receptors and antibodies differ?
   a. B-cell receptors are made up of heavy chains; antibodies are made up of light chains.
   b. B-cell receptors are made up of light chains; antibodies are made up of heavy chains.
   c. Only antibodies include a variable region.
   d. Antibodies lack a transmembrane domain and are secreted.
5. How do memory cells become activated?
   a. They undergo somatic hypermutation.
   b. They encounter an antigen complexed with an MHC protein.
   c. They undergo clonal expansion.
   d. They are stimulated by histamines.
6. In terms of morphology, T cells come in two major types. How are they distinguished?
   a. They have a CD4 or CD8 protein on their surfaces.
   b. They have extensive ER and many mitochondria.
   c. Only one type has pattern-recognition receptors.
   d. One type functions in the primary response; the other in the secondary response.

Test Your Knowledge answers: 1. c; 2. a; 3. b; 4. d; 5. b; 6. a

Test Your Understanding

Answers are available at www.masteringbio.com

1. To a physician, the classical signs of the inflammatory response are rubor (reddening), calor (heat), dolor (pain), and tumor (swelling). Explain why each symptom occurs. Mention at least two of the leukocytes and two of the chemical messengers that are involved in the inflammatory response.
2. Compare and contrast the general structure of a B-cell receptor and a T-cell receptor (make a sketch of each). Explain how each interacts with an antigen.
3. What do vaccines need to contain in order to be effective? Why don't we have vaccines for HIV? Why do people need to get a new flu shot every year?
4. Summarize the clonal-selection theory of acquired immune system function. How was the theory tested and verified?
5. Compare and contrast the interaction between (a) pathogens and the pattern-recognition receptors on leukocytes versus (b) antigens and BCRs or TCRs.
6. Explain how DNA recombination leads to the production of almost limitless numbers of different B-cell receptors, T-cell receptors, and antibodies.

Applying Concepts to New Situations

Answers are available at www.masteringbio.com

1. If you were being treated for a badly skinned knee from a bicycle accident, health care workers would irrigate the wound with large volumes of sterile water, scrub it as thoroughly as possible with warm, soapy water, and then apply antibiotics. Explain why these measures are effective, in the context of how the innate and acquired immune responses operate.
2. Suppose you had an antibody to insulin. Explain how you would use this antibody to study the presence and concentration of insulin in the blood under different conditions (e.g., before and after a meal) and at different times in the development of an individual.
3. It seems astonishing that the immune system can produce antibodies to compounds that were recently synthesized for the first time in the lab. But because viruses and other pathogens are constantly mutating and producing new antigens over time, this observation should not be surprising at all. Explain.
4. During World War II, physicians discovered that if they removed undamaged skin from a patient and grafted it onto the site of a burn on the same patient, the tissue healed well. But if the grafted tissue came from a different individual, the skin graft was rejected—the cells were killed by the immune system. Explain why.

An Introduction to Ecology

50

In addition to providing habitat for a frog, the puddle of water inside this plant provides an environment for an array of microscopic species. The plant itself is growing in the branch of a rain forest tree, and is just one of hundreds of plant, fungal, bacterial, and insect species that live in association with the tree.

KEY CONCEPTS

- Ecology focuses on how organisms interact with their environment. Because its goal is to understand the distribution and abundance of organisms, ecology provides a scientific foundation for the conservation of species and natural areas.
- Physical structure—particularly water depth—is the primary factor that limits the distribution and abundance of aquatic species. Climate—specifically, both the average value and annual variation in temperature and in moisture—is the primary factor that limits the distribution and abundance of terrestrial species.
- Climate varies with latitude, elevation, and other factors. Climate is changing rapidly around the globe.
- In addition to abiotic aspects of the environment such as physical structure and climate, species distribution is constrained by historical and biotic factors.

E**cology** is the study of how organisms interact with their environment. Except for the episode 65 million years ago, when a mountain-sized asteroid struck Earth, environments are changing more rapidly right now than they have at any time in the past 3.5 billion years. As human impacts on the planet accelerate, ecology is becoming an increasingly prominent field in biological science. Efforts to maintain human health and welfare depend on our ability to understand and predict the consequences of the rapid environmental changes occurring all around us.

One of ecology's central goals is to understand the distribution and abundance of organisms. In many cases an ecologist's job is to identify factors that dictate why certain species live where they do, and how many individuals can live there. Some biologists ask why orangutans are restricted to forests in Borneo and why their numbers are declining so rapidly. Other biologists create mathematical models to predict how quickly the human immunodeficiency virus will increase in India, or how long it will take for the current human population of 6.6 billion to double. The distribution and abundance of these species is dictated by their ecology.

This chapter has two goals: to explore how biologists go about studying ecology, and to analyze the major types of environments that organisms occupy. It is a springboard to subsequent chapters in this unit, which analyze how organisms interact with their environment and the consequences of environmental change. In terms of understanding the biological problems that humans currently face, this may be the most important unit in the book.

50.1 Areas of Ecological Study

To understand why organisms live where they do and in what numbers, biologists break ecology into several levels of analysis. This is a common strategy in biological science. Cell biologists study how cells work at increasingly complex levels of organization—from individual molecules to complex multicellular organisms—while physiologists analyze processes at the level of ions and molecules as well as whole cells, tissues and organs, or a complete system. In ecology, researchers work at four main levels: (1) organisms, (2) populations, (3) communities, and (4) ecosystems. Let's examine each.

Organismal Ecology

At the finest level of organization, ecology focuses on how individuals interact with their environment. Researchers who study organismal ecology explore the morphological, physiological, and behavioral adaptations that allow individuals to live successfully in a particular area. The study of behavior is an important aspect of organismal ecology; it focuses on how organisms respond to particular stimuli from their environment (see Chapter 51). The environmental stimuli that trigger behavior may consist of changes in temperature or moisture, an escape response from prey, or a rival challenging for a mate. Organismal ecology also focuses on the physiological adaptations that allow individuals to thrive in heat, drought, cold, or other demanding physical conditions.

As an example, consider the sockeye salmon. After spending four or five years feeding and growing in the ocean, these salmon travel hundreds or thousands of kilometers to return to the stream where they hatched (**Figure 50.1a**). Females create nests in the gravel stream bottom and lay eggs. Nearby males compete for the chance to fertilize eggs as they are laid. When breeding is finished, all of the adults die.

At the level of organismal ecology, biologists want to know how these individuals interact with their physical surroundings and with other organisms in and around the stream. Which females get the best nesting sites and lay the most eggs, and which males are most successful in fertilizing eggs? How do individuals cope with the transition from living in salt water to living in freshwater?

Population Ecology

A **population** is a group of individuals of the same species that lives in the same area at the same time. When biologists study population ecology, they focus on how the numbers of individuals in a population change over time. Some of the tools that

(a) Organismal ecology

How do individuals interact with each other and their physical environment?

Salmon migrate from saltwater to freshwater environments to breed

(b) Population ecology

How and why does population size change over time?

Each female salmon produces thousands of eggs. Only a few will survive to adulthood. On average, only two will return to the stream of their birth to breed

(c) Community ecology

How do species interact, and what are the consequences?

Salmon are prey as well as predators

(d) Ecosystem ecology

How do energy and nutrients cycle through the environment?

Salmon die and then decompose, releasing nutrients that are used by bacteria, archaea, plants, protists, young salmon, and other organisms

FIGURE 50.1 Biologists Study Ecology at Four Main Levels.

population ecologists have developed to analyze and predict changes in population size are now being used to evaluate the fate of endangered species. For example, mathematical models of population growth have been used to predict the future of particular salmon populations (**Figure 50.1b**). Many salmon populations have declined precipitously as their habitats have become dammed or polluted, and salmon are an important source of food for both people and wildlife. If the factors that affect population size can be described accurately enough, mathematical models can assess the impact of proposed dams, changes in weather patterns, altered harvest levels, or specific types of protection efforts.

Community Ecology

A biological **community** consists of the species that interact with each other within a particular area. Researchers who study community ecology ask questions about the nature of the interactions between species and the consequences of those interactions. The work might concentrate on predation, parasitism, or competition. At this level of organization in ecology, biologists also analyze how groups of species respond to disturbances such as fires, floods, and volcanic eruptions. Because human activities are driving species to extinction and disturbing communities on a massive scale, community ecologists are being called on to generate hypotheses and data on how human impacts can be lessened.

As an example of the types of questions asked at the level of community ecology, consider the interactions among salmon and other species in the marine and stream communities where they live. When they are at sea, salmon eat smaller fish and are themselves hunted and eaten by orcas, sea lions, humans, and other mammals; when they return to freshwater to breed, they are preyed on by bears and bald eagles (**Figure 50.1c**). In both marine and freshwater habitats, salmon are subject to parasitism and disease. They are also heavily affected by disturbances—particularly changes in their food supply due to overfishing and the damming and degradation of breeding streams.

Ecosystem Ecology

Ecosystem ecology is an extension of community ecology. An **ecosystem** consists of all the organisms in a particular region along with nonliving components. These physical, or **abiotic** (literally, "not-living"), components include air, water, and soil. At the ecosystem level, biologists study how nutrients and energy move among organisms and between organisms and the surrounding atmosphere and soil or water. Because humans are adding massive amounts of nutrients to ecosystems all over the world and affecting energy flows and climate through global warming, this work has direct public policy implications. Ecosystem ecologists are responsible for assessing the impact of pollution and increased temperature on the distribution and abundance of species—in fact, on Earth's ability to support life.

Salmon are interesting to study at the ecosystem level because they form a link between marine and freshwater ecosystems. They harvest nutrients in the ocean and then, when they die and decompose, transport those molecules to streams (**Figure 50.1d**). In this way, salmon transport chemical energy and nutrients from one habitat to another. Because salmon are sensitive to pollution and to changes in water temperature, human-induced changes in marine and freshwater ecosystems have a large impact on their populations.

How Do Ecology and Conservation Efforts Interact?

The four levels of ecological study are synthesized and applied in conservation biology. **Conservation biology** is the effort to study, preserve, and restore threatened populations, communities, and ecosystems. Ecologists study how interactions between organisms and their environments result in a particular species being found in a particular area at a particular population size; conservation biologists apply these data to preserve species and restore environments. Conservation biologists are like physicians, except that instead of drawing on results from research in molecular biology and physiology, they draw on research in ecology and evolution. Instead of prescribing drugs for sick people, they prescribe remedies for threatened species and manage land to produce a diversity of species, clean air, pure water, and productive soils.

50.2 Types of Aquatic Ecosystems

If ecology is the study of how species interact with their environment, what makes up the environment? The short answer is that an organism's environment has both physical and biological components. The abiotic or physical components include temperature, precipitation, sunlight, and wind. The **biotic** ("living") components consist of other members of the organism's own species as well as individuals of other species.

Chapters 51–53 focus on biological interactions—how individuals interact with offspring, mates, competitors, predators, prey, and parasites. This chapter focuses on how the physical environment affects organisms. The discussion offers two payoffs: (1) Once you understand key aspects of the physical environment, you'll be able to think intelligently about the consequences of altering those conditions through global warming; and (2) knowledge of the physical environment will help you make sense of a fundamental observation—that no species lives everywhere. Organisms have a restricted set of physical conditions in which they can survive and thrive. Why?

Let's answer this question by analyzing the physical attributes of aquatic ecosystems in this section and focusing on the physical structure of terrestrial ecosystems in Section 50.3. Section 50.4 considers how global warming is affecting both types of environments.

What Physical Factors Play a Key Role in Aquatic Ecosystems?

In aquatic ecosystems, water depth and the rate of water movement qualify as the key physical factors that shape the environment. Water depth dictates how much light reaches the organisms that live in a particular region. Water movement presents a physical challenge: It can literally sweep organisms off their feet.

Water absorbs and scatters light, so the amount and types of wavelengths available to organisms change dramatically as water depth increases. As **Figure 50.2a** shows, ocean water specifically removes light in the blue and red regions of the visible spectrum. This is important because wavelengths in the blue and red regions are required for photosynthesis in many species (see Chapter 10). The total amount of light available to organisms also diminishes rapidly with increasing depth. In pure seawater, the total amount of light available at a depth of 10 m is less than 40 percent of what it is at the surface, and virtually no light reaches depths greater than 40 m in pure seawater (**Figure 50.2b**). In seawater that contains organisms or debris, light penetration is dramatically less than that observed in pure seawater. Light has a major influence on **productivity**—the total amount of carbon fixed by photosynthesis per unit area per year. Different species are found at different depths in aquatic environments, simply because the physical environment changes so radically.

The type and amount of water movement or flow is the other major influence in aquatic environments. Organisms that live in fast-flowing streams have to cope with the physical force of the water, which constantly threatens to move them downstream. Marine organisms that live in intertidal regions are exposed to the air periodically each day, as well as to violent wave action during storms.

Different water depths and types of water movement define the array of aquatic environments that are available to organisms. The boxes that follow summarize the types of freshwater and marine environments that organisms occupy.

(a) Only certain wavelengths of light are available under water.

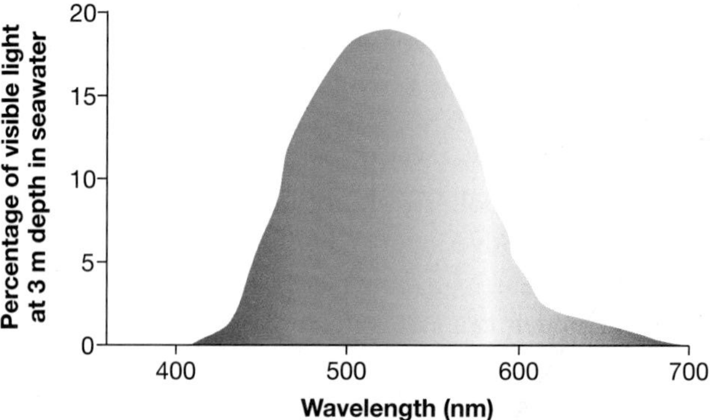

(b) Intensity of light declines with water depth.

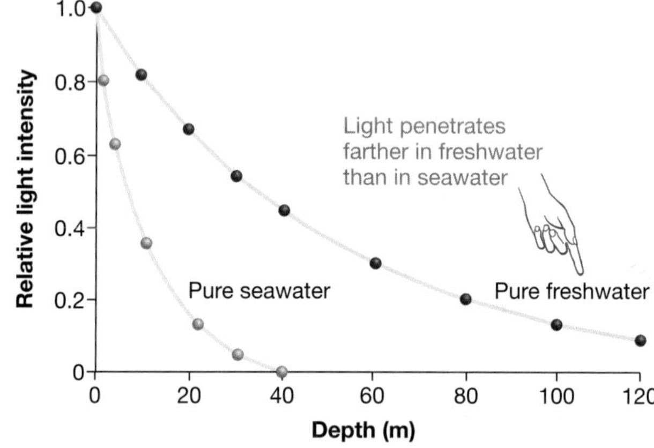

FIGURE 50.2 Availability of Light Changes Dramatically with Increasing Water Depth. (a) Graph showing the wavelengths of light available at a depth of 3 meters in seawater near an ocean coastline. **(b)** Graphs showing how rapidly the amount of light declines with depth of pure freshwater and pure seawater. When water contains organisms and organic debris, light availability declines even faster.

EXERCISE In green algae, photosynthesis is driven most efficiently by wavelengths of about 425 nm and 680 nm. Mark and label these wavelengths in part (a).

QUESTION What is the maximum depth you'd expect to find any photosynthetic organisms in freshwater? In salt water?

Freshwater Environments > Lakes and Ponds

Lakes and ponds are distinguished from each other by size. Ponds are small; lakes are large enough that the water in them can be mixed by wind and wave action. Most lakes and ponds occur in northern latitudes—they formed in depressions that were created by the scouring action of glaciers thousands of years ago. In the tropics, most lakes consist of old river channels. Elsewhere, many of the lakes and ponds that exist now were dug recently by people.

Water Depth Biologists describe the structure of lakes and ponds by naming five zones (**Figure 50.3**).

- The **littoral** ("seashore") **zone** consists of the shallow waters along the shore, where flowering plants are rooted.
- The **limnetic** ("lake") **zone** is offshore and comprises water that receives enough light to support photosynthesis.
- The **benthic** ("depths") **zone** is made up of the substrate.
- Regions of the littoral, limnetic, and benthic zones that receive sunlight are part of the **photic zone**.
- Portions of a lake or pond that do not receive sunlight make up the **aphotic zone**.

(Continued on next page)

Water Flow Water movement in lakes and ponds is driven by wind and temperature. The littoral and limnetic zones are typically much warmer and better oxygenated than the benthic zone, simply because they receive so much more solar radiation and are in contact with oxygen in the atmosphere. The benthic zone, in contrast, is relatively nutrient rich because dead and decomposing bodies sink and accumulate there. But as the temperature of the surface changes throughout the year, water from different depths can mix (see **Box 50.1**). Mixing driven by wind and changes in temperature allows well-oxygenated water from the surface to reach the benthic zone and nutrient-rich water from the benthic zone to enter the littoral and limnetic zones.

Organisms Cyanobacteria, algae, and other microscopic organisms, collectively called **plankton**, live in the photic zone, as do the fish that eat them. Animals (invertebrates and fish) that consume dead organic matter, or **detritus**, are common in the benthic zone.

● You should be able to predict which zone of lakes and ponds produces the most grams of organic material per m^2 per year, and explain your logic.

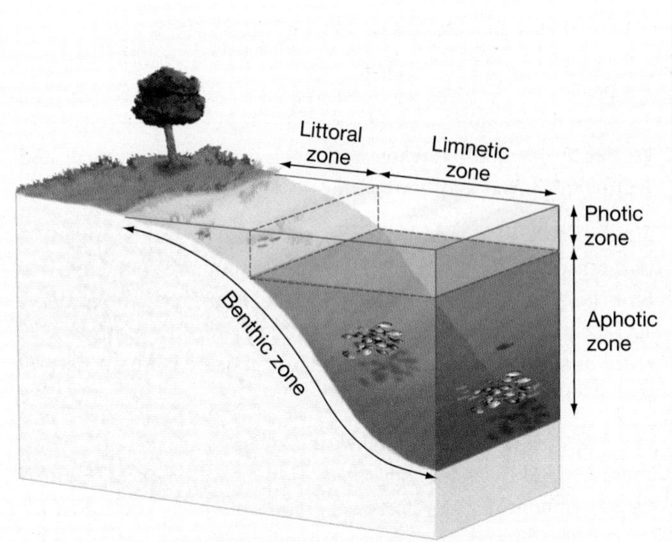

FIGURE 50.3 Lakes Have Distinctive Zones Defined by Water Depth and Distance from Shore.

BOX 50.1 Thermoclines and Lake Turnover

The northern and temperate regions of the world host many lakes. Each year, these bodies of water undergo remarkable changes known as the spring and fall **turnovers**.

The spring and fall lake turnovers occur in response to changes in air temperature. To see how this happens, examine the temperature profiles in **Figure 50.4**. In winter, water at the surface is locked up in ice at a temperature of 0°C. The water just under the ice is slightly warmer and is relatively oxygen rich, because it was exposed to the atmosphere as winter set in. The water at the bottom of the lake, in contrast, is at 4–5°C. A gradient in temperature such as this is called a **thermocline** ("heat-slope"); thermal stratification is said to occur. The water at the bottom is also oxygen poor, because organisms that decompose falling organic material use up available oxygen.

When spring arrives, the ice begins to melt. The temperature of the water rises until it reaches 4°C. This is important, because the density of liquid water is highest at 4°C. The water at the surface of the lake is now heavier than the water below it. As a result, it sinks. The water at the bottom of the lake is displaced and comes to the surface, completing the spring turnover. During the spring turnover, water at the bottom of the lake carries sediments and nutrients from the benthic zone up to the limnetic zone. This flush of nutrients triggers a rapid increase in the growth of algae and bacteria that biologists call the spring bloom.

When temperatures cool in the fall, the water at the surface reaches a temperature of 4°C and sinks, displacing water at the bottom and creating the fall turnover. The fall turnover is important because it brings oxygen-rich water from the surface down to the benthic layer, and because it again brings nutrients from the benthic zone up to the limnetic zone.

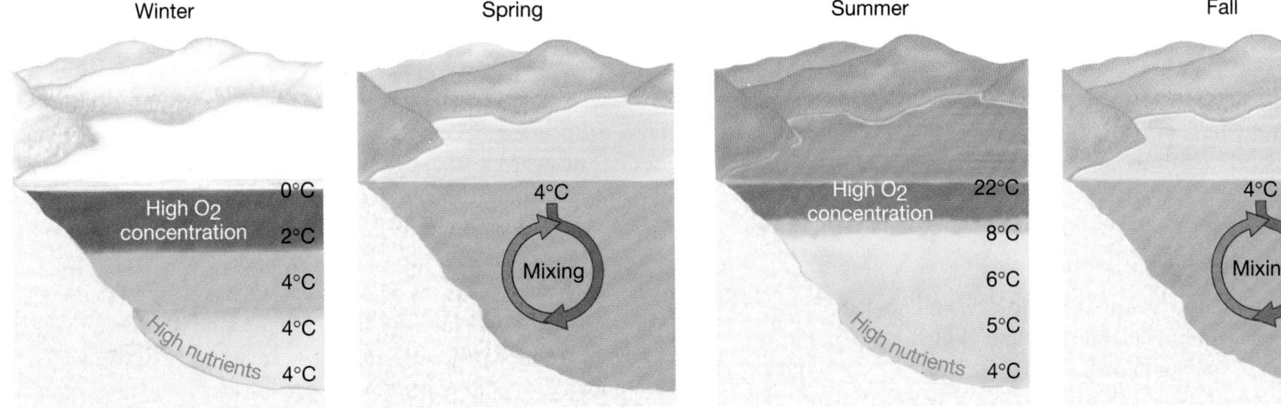

FIGURE 50.4 In Temperate Regions, Lakes Turn Over Each Spring and Fall.

● **QUESTION** In lakes like the one pictured here, there is a dramatic burst of photosynthesis each spring and fall. Why?

Freshwater Environments > Wetlands

Wetlands are shallow-water habitats where the soil is saturated with water for at least part of the year.

Water Depth Wetlands are distinct from lakes and ponds for two reasons: They have only shallow water, and they have **emergent vegetation**—meaning plants that grow above the surface of the water. All or most of the water in wetlands receives sunlight, and emergent plants capture sunlight before it strikes the water.

Water Flow Freshwater marshes and swamps are wetland types characterized by a slow but steady flow of water. **Bogs** (**Figure 50.5a**), in contrast, develop in depressions where water flow is low or nonexistent. If water is stagnant, oxygen is used up during the decomposition of dead organic matter faster than it enters via diffusion from the atmosphere. As a result, bog water is oxygen poor or even anoxic. Once the oxygen in the water is depleted, decomposition slows. Organic acids and other acids build up, lowering the pH of the water. At low pH, nitrogen becomes unavailable to plants (see Chapter 38).

Organisms The combination of acidity, lack of available nitrogen, and anoxic conditions makes bogs extremely unproductive habitats. Marshes and swamps, in contrast, offer ample supplies of oxygenated water and sunlight and are extraordinarily productive. **Marshes** lack trees and typically feature grasses (**Figure 50.5b**); **swamps** are dominated by trees and shrubs (**Figure 50.5c**). Because their physical environments are so different, there is little overlap in the types of species found in bogs, marshes, and swamps.

● You should be able to explain why carnivorous plants, which capture and digest insects, are relatively common in bogs but rare in marshes and swamps.

(a) Bogs are stagnant and acidic.

(b) Marshes have nonwoody plants.

(c) Swamps have trees and shrubs.

FIGURE 50.5 Wetland Types Are Distinguished by Water Flow and Vegetation.

Freshwater Environments > Streams

Streams are bodies of water that move constantly in one direction. Creeks are small streams; rivers are large. In terms of the environment that is available to organisms, the major physical variables in streams are the speed of the current and the availability of oxygen and nutrients.

Water Depth Most streams are shallow enough that sunlight reaches the bottom. Availability of sunlight is usually not a limiting factor for organisms.

Water Flow The structure of a typical stream varies along its length (**Figure 50.6**). Where it originates at a mountain glacier, lake, or spring, a stream tends to be cold, narrow, and fast. As it descends toward a lake, ocean, or larger river, a stream accepts water from tributaries and becomes larger, warmer, and slower.

Oxygen levels tend to be high in fast-moving streams because water droplets are exposed to the atmosphere when moving water splashes over rocks or other obstacles. Oxygen from the atmosphere diffuses into the droplets. In contrast, slow-moving streams that lack riffles or rapids tend to become relatively oxygen poor. Also, cold water holds more oxygen than warm water does (see Chapter 44).

(Continued on next page)

Freshwater Environments > Streams *continued*

Organisms It is rare to find photosynthetic organisms in small, fast-moving streams; nutrient levels tend to be low and most of the organic matter present consists of leaves and other materials that fall into the water from outside the stream. Fish, insect larvae, molluscs, and other animals have adaptations that allow them to maintain their positions in the fast-moving portions of streams. As streams widen and slow down, conditions become more favorable for the growth of algae and plants, and the amount of organic matter and nutrients increases. As a result, the same stream often contains completely different types of organisms near its source and near its end, or mouth.

● You should be able to explain why, in terms of their ability to perform cellular respiration, fish species found in cold, fast-moving streams tend to be much more active than fish species found in warm, slow-moving streams.

FIGURE 50.6 Distinctive Environments Appear along a Stream's Length.

Freshwater/Marine Environments > Estuaries

Estuaries form where rivers meet the ocean—meaning that freshwater mixes with salt water (**Figure 50.7**). In essence, an estuary includes slightly saline marshes as well as the body of water that moves in and out of these environments. Salinity varies with changes in river flows and with proximity to the ocean. Salinity has dramatic effects on osmosis and water balance (see Chapters 37 and 42); species that live in estuaries have adaptations that allow them to cope with variations in salinity.

Water Depth Most estuaries are shallow enough that sunlight reaches the substrate. Water depth may fluctuate dramatically, however, in response to tides, storms, and floods.

Water Flow Water flow in estuaries fluctuates daily and seasonally due to tides, storms, and floods. The fluctuation is important because it alters salinity, which in turn affects which types of organisms are present.

Organisms Because the water is shallow and sunlit, and because nutrients are constantly replenished by incoming river water, estuaries are among the most productive environments on Earth. They are often packed with young fish, which feed on abundant vegetation and plankton while hiding from predators.

● You should be able to compare and contrast the characteristics of estuaries and marshes, and explain why few of the same species are found in both.

FIGURE 50.7 Estuaries Are Highly Productive Environments. This photo was taken at low tide.

● **EXERCISE** Label freshwater environments, saltwater environments, and areas where freshwater and salt water are mixing.

Marine Environments > The Ocean

The world's oceans form a continuous body of salt water and are remarkably uniform in chemical composition. Regions within an ocean vary markedly in their physical characteristics, however, with profound effects on the organisms found there.

Water Depth Biologists describe the structure of an ocean by naming six regions (**Figure 50.8**).

- The **intertidal** ("between tides") **zone** consists of a rocky, sandy, or muddy beach that is exposed to the air at low tide but submerged at high tide.
- The **neritic zone** extends from the intertidal zone to depths of about 200 m. Its outermost edge is defined by the end of the **continental shelf**—the gently sloping, submerged portion of a continental plate.
- The **oceanic zone** is the "open ocean"—the deepwater region beyond the continental shelf.
- The bottom of the ocean is the **benthic zone**.
- The intertidal and sunlit regions of the neritic, oceanic, and benthic zones make up a **photic zone**.
- Areas that do not receive sunlight are in an **aphotic zone**.

Water Flow Water movement in the ocean is dominated by different processes at different depths. In the intertidal zone, tides and wave action are the major influences. In the neritic zone, currents that bring nutrient-rich water from the benthic zone of the deep ocean toward shore have a heavy impact. More specifically, nutrient-rich water is carried toward the surface when it hits the steep slope of the continental plate. Throughout the ocean, large-scale currents circulate water in the oceanic zone in response to prevailing winds and the Earth's rotation.

Organisms Each zone in the ocean is populated by distinct species that are adapted to the physical conditions present. Organisms that live in the intertidal zone must be able to withstand physical pounding from waves and desiccation at low tide. Productivity is high, however, due to the availability of sunlight and nutrients contributed by estuaries as well as by currents that sweep in nutrient-laden sediments from offshore areas.

Productivity is also high on the outer edge of the neritic zone, due to nutrients contributed by upwellings at the edge of the continental plate. Almost all of the world's major marine fisheries exploit organisms that live in the neritic zone. In the tropics, shallow portions of the neritic zone may host **coral reefs**. Because the water is warm and sunlight penetrates to the ocean floor in these habitats, coral reefs are among the most productive environments in the world (see Chapter 54).

If coral reefs are the rain forests of the ocean, then the oceanic zone is the desert. Sunlight is abundant in the photic zone of the open ocean, but nutrients are extremely scarce. When the photosynthetic organisms and the animals that feed on them die, their bodies drift downward out of the photic zone and are lost. In the open ocean, there is no mechanism for bringing nutrients back up from the bottom, as there is in the neritic zone or in most lakes (see Box 50.1). The aphotic zone of the open ocean is also unproductive because light is absent and photosynthesis is impossible. Most organisms present in the aphotic zone survive on the rain of dead bodies from the photic zone.

● You should be able to predict the following characteristics of fish that live in the aphotic zone: what they eat, whether they have eyes, whether they are capable of swimming, and whether they are rare or abundant.

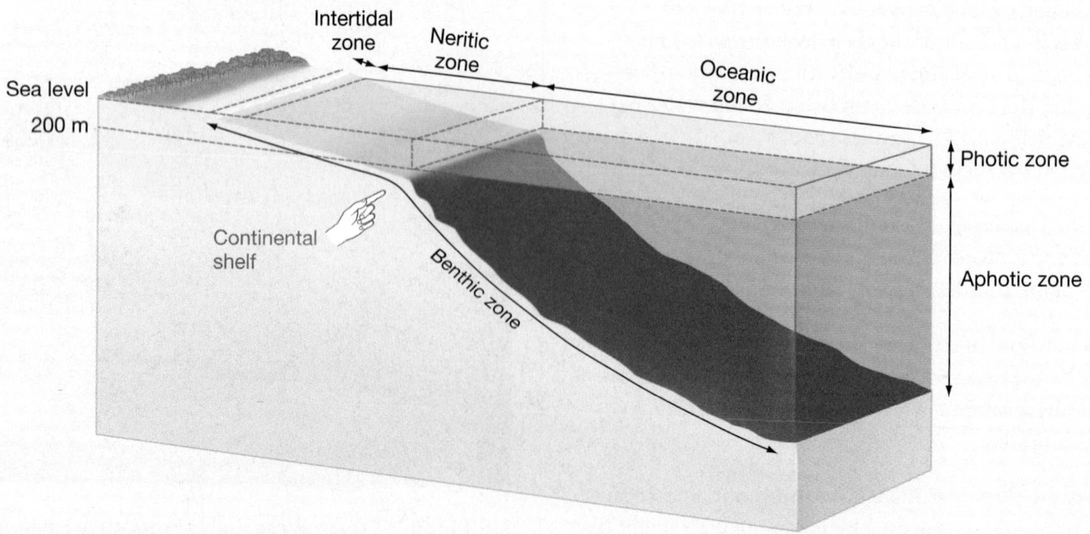

FIGURE 50.8 Oceans Have Distinctive Zones Defined by Water Depth and Distance from Shore.

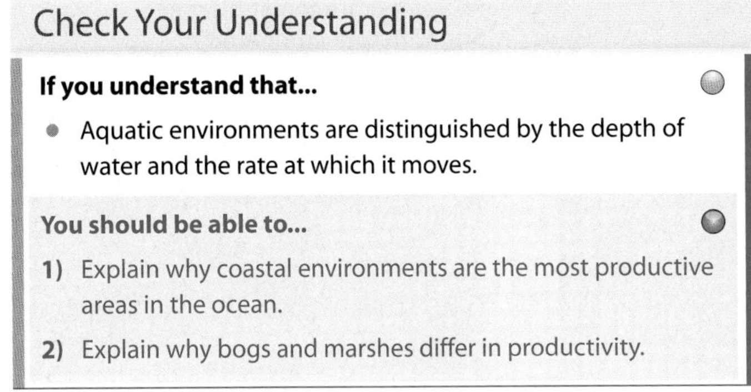

Check Your Understanding

If you understand that...

- Aquatic environments are distinguished by the depth of water and the rate at which it moves.

You should be able to...

1) Explain why coastal environments are the most productive areas in the ocean.
2) Explain why bogs and marshes differ in productivity.

50.3 Types of Terrestrial Ecosystems

If you could walk from the equator in South America to the North Pole, you would notice startling changes in the organisms around you. Lush tropical forests with broad-leaved evergreen trees would give way to seasonally dry forests and then to deserts. The deserts would yield to the vast grasslands of central North America, which terminate at the boreal forests of the subarctic. If you pressed on, you would reach the end of the trees and the beginning of the most northerly community—the arctic tundra.

Broad-leaved evergreen forests, deserts, and grasslands are **biomes**: major groupings of plant and animal communities defined by a dominant vegetation type. **Figure 50.9** shows the global distribution of the most common types of biomes. Many distinct communities exist within each of these broad regions. For example, the Finger Lakes region of the northeastern United States is part of the temperate forest biome. However, the cool, wet, north-facing slopes in the area host plant communities dominated by eastern hemlock trees; warmer and drier south-facing slopes feature communities with red oak trees; and mild slopes and flat areas are covered with forests consisting of sugar maples and white oaks (**Figure 50.10**). The characteristics of these forest communities are similar enough to place them in the same biome, even though the specific species present vary with local conditions.

Each of the biomes found around the world is associated with a distinctive set of abiotic conditions. Just as water depth and movement have an overriding influence on aquatic ecosystems, the type of biome present in a terrestrial region depends on **climate**—the prevailing, long-term weather conditions found in an area. **Weather** consists of the specific short-term atmospheric conditions of temperature, moisture, sunlight, and wind.

- Temperature is critical because the enzymes that make life possible work at optimal efficiency only in a narrow range of temperatures (see Chapter 3). Temperature also affects the availability of moisture: Water freezes at low temperatures and evaporates rapidly at high temperatures.
- Moisture is significant because it is required for life, and because terrestrial organisms constantly lose water to the environment through evaporation or transpiration. To stay alive, they must reduce water loss and replace lost water.
- Sunlight is essential because it is required for photosynthesis.
- Wind is important because it exacerbates the effects of temperature and moisture. Wind increases heat loss due to evaporation and convection, and it increases water loss due to evaporation and transpiration. It also has a direct physical impact on organisms such as birds, flying insects, and plants by pushing them around.

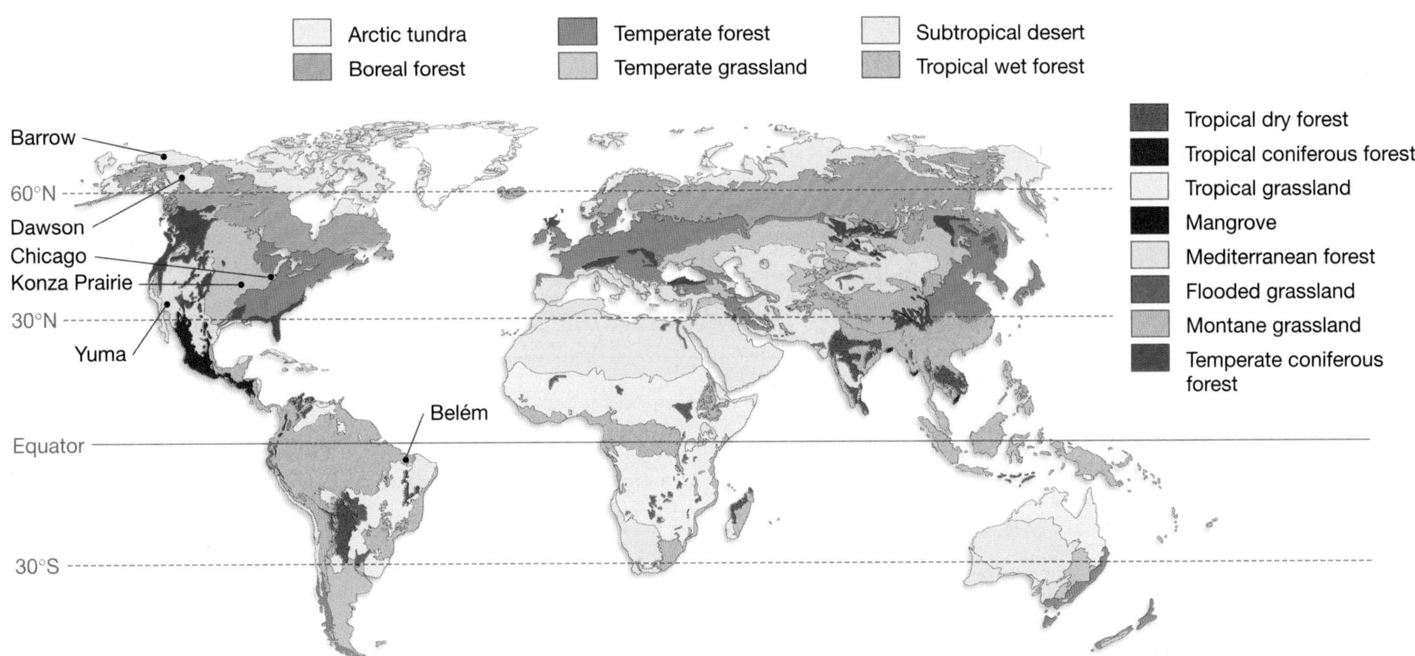

FIGURE 50.9 Distinct Biomes Are Found throughout the World. A recent classification identified the 14 major biomes shown here.

(a) Temperate forest biome:
Eastern hemlock community

(b) Temperate forest biome:
Red oak community

(c) Temperate forest biome:
Sugar maple community

FIGURE 50.10 Biomes Contain Distinct Communities. Within even a small area, the same biome can contain several distinct plant and animal communities.

QUESTION These communities are found in the northeastern United States and southeast Canada. How would temperate forest communities in New Zealand or northern France compare?

Of the four components of climate, temperature and moisture are far and away the most important to plants. More specifically, the nature of the biome that develops in a particular region is governed by (1) average annual temperature and precipitation, and (2) annual variation in temperature and precipitation. Each biome contains species that are adapted to a particular temperature and moisture regime.

The amount and variability of heat and precipitation structure terrestrial environments, much as water depth and flow rate structure aquatic environments. On land, photosynthesis and plant growth are maximized when temperatures are warm and conditions are wet; conversely, photosynthesis cannot occur efficiently at low temperatures or under drought stress. Biologists are particularly concerned with **net primary productivity (NPP)**, which is the total amount of carbon that is fixed per year minus the amount of fixed carbon oxidized during cellular respiration. Fixed carbon that is consumed in cellular respiration provides energy for the organism but is not used for growth—that is, production of **biomass.** NPP is key because it represents the organic matter that is available as food for other organisms. In terrestrial environments NPP is often estimated by measuring **aboveground biomass**—the total mass of living plants, excluding roots.

To help you understand how temperature and precipitation influence biomes, let's take a detailed look at the six biomes that represent Earth's most extensive vegetation types, ranging from the wet tropics to the arctic. In each case, you'll be analyzing temperature and precipitation data from a specific location, typical of the biome in question.

Check Your Understanding

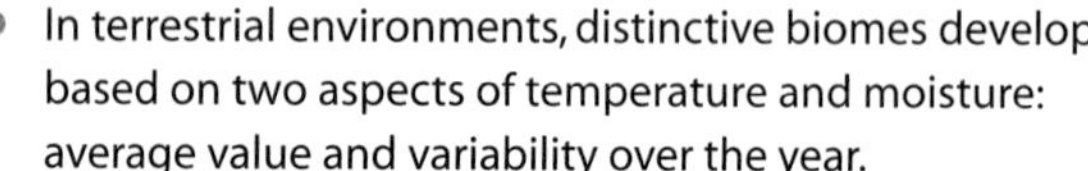

If you understand that...

- In terrestrial environments, distinctive biomes develop based on two aspects of temperature and moisture: average value and variability over the year.

You should be able to...

Predict whether tropical dry forests, which have high year-round temperatures but a distinct dry season, are more or less productive than tropical wet forests.

Terrestrial Biomes > Tropical wet forest

Tropical wet forests—also called tropical rain forests—are found in equatorial regions around the world. Plants in this biome have broad leaves as opposed to narrow, needle-like leaves, and are evergreen. Older leaves are shed throughout the year, but there is no complete, seasonal loss of leaves.

Temperature The data in **Figure 50.11** are from Belém, Brazil, where mean monthly temperatures never drop below 25°C and never exceed 30°C. Compared to other biomes, tropical wet forests show almost no seasonal variation in temperature. This is important because temperatures are high enough to support growth throughout the year.

Precipitation Even in the driest month of the year, November, this region receives over 5 cm (2 in.) of rainfall—considerably more than the *annual* rainfall of many deserts.

Vegetation Favorable year-round growing conditions produce riotous growth, leading to extremely high productivity and aboveground biomass (**Figure 50.12**). Tropical wet forests are also

(Continued on next page)

Terrestrial Biomes > Tropical wet forest *continued*

renowned for their species diversity. It is not unusual to find over 200 tree species in a single 10 m × 100 m study plot. And based on counts of the insects and spiders collected from single trees, some biologists contend that the world's tropical wet forests may hold up to 30 million species of arthropods alone.

The diversity of plant sizes and growth forms in wet forest communities produces extraordinary structural diversity. In tropical wet forests, a few extremely large trees tower over a layer of large trees that form a distinctive **canopy** (the uppermost layers of branches). From the canopy to the ground, there is a complex assortment of vines, **epiphytes** (plants that grow entirely on other plants), small trees, shrubs, and herbs. This diversity of growth forms presents a wide array of habitat types for animals.

● You should be able to explain which aspects of structural diversity in a rain forest are lacking in the biome where you live.

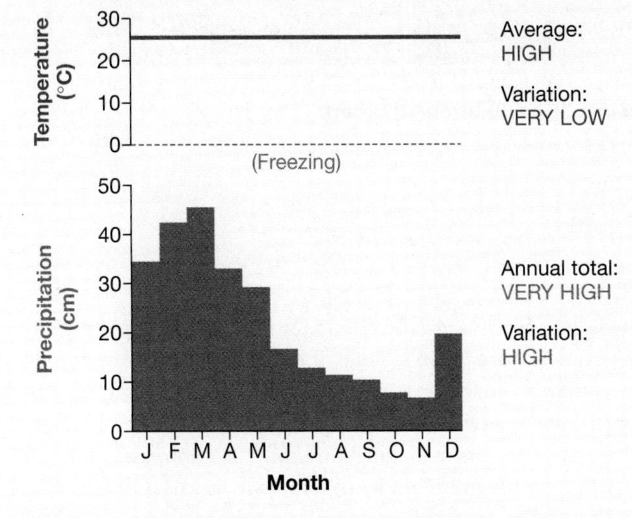

FIGURE 50.11 Belém, Brazil.

FIGURE 50.12 Tropical Wet Forest.

Terrestrial Biomes > Subtropical deserts

Subtropical deserts are found throughout the world in two distinctive locations: 30 degrees latitude, or distance from the equator, both north and south. Most of the world's great deserts—including the Sahara, Gobi, Sonoran, and Australian outback—lie on or about 30°N or 30°S latitude.

Temperature The data in **Figure 50.13** are from Yuma, Arizona, in the Sonoran Desert of southwestern North America. Mean monthly temperatures in subtropical deserts vary more than in tropical wet climates, but in Yuma temperatures still never fall below freezing. (Freezing nighttime temperatures are common elsewhere in the Sonora and in other subtropical deserts.)

Precipitation The most striking feature of the subtropical desert climate is low precipitation. The average annual precipitation in Yuma is just 7.5 cm (3 in.).

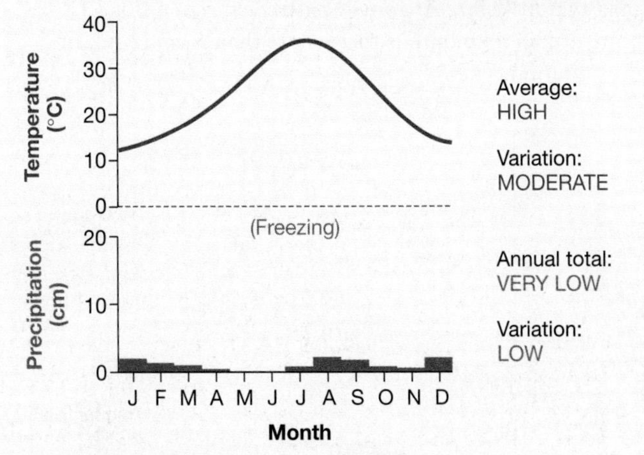

FIGURE 50.13 Sonoran Desert.

Vegetation The scarcity of water in deserts has profound implications. Conditions are rarely adequate to support photosynthesis, so the productivity of desert communities is a tiny fraction of average values for tropical forest communities. Further, as the photo in **Figure 50.14** shows, individual plants are widely spaced—a pattern that may reflect intense competition for water.

(Continued on next page)

Desert species adapt to the extreme temperatures and aridity in one of two ways: Growing at a low rate year-round, or breaking dormancy and growing rapidly in response to any rainfall. Cacti can grow year-round because they possess adaptations to cope with hot, dry conditions: small leaves or no leaves; a thick, waxy coating on leaves and stems; and the CAM pathway for photosynthesis (Chapter 10). The desert shrub called ocotillo, in contrast, is usually dormant. Individuals sprout leaves within days of a rainfall and drop them a week or two later when soils dry out again. The seeds of annual plants can also lie dormant for many years, then germinate after a rain.

● You should be able to be able to state hypotheses explaining why desert species are not found in rain forests, and why rain forest species are not found in deserts.

FIGURE 50.14 Subtropical Desert.

Terrestrial Biomes > Temperate grasslands

Temperate grassland communities are found throughout central North America and the heartland of Eurasia (see Figure 50.9). In North America they are commonly called prairies; in central Eurasia they are known as steppes.

Temperature A region of the world is called **temperate** if it has pronounced annual fluctuations in temperature—typically hot summers and cold winters—but not the temperature extremes recorded in the tropics or arctic. Temperature variation is important because it dictates a well-defined growing season. In the temperate zone, plant growth is possible only in spring, summer, and fall months when moisture and warmth are adequate. **Figure 50.15** presents climate data for a typical grassland community—in this case, the Konza Prairie near Manhattan, Kansas.

Precipitation Precipitation at Konza is over four times greater than that in Yuma, Arizona. Nonetheless, conditions are still quite dry; in no month is there more than 5 cm (2 in.) of precipitation.

Vegetation Grasses are the dominant life-form in temperate grasslands **Figure 50.16** for one of two reasons: (1) Conditions are too dry to enable tree growth, or (2) encroaching trees are burned out by fires. Prairie fires are ignited by lightning strikes or by native people managing land for game animals.

Although the productivity of temperate grasslands is generally lower than that of forest communities, grassland soils are often highly fertile. The subsurface is packed with roots and rhizomes, which add organic material to the soil as they die and decay. Further, grassland soils retain nutrients, because rainfall is low enough to keep key ions from dissolving and leaching out of the soil. It is no accident, then, that the grasslands of North America and Eurasia are the breadbaskets of those continents. The conditions that give rise to natural grasslands are ideal for growing wheat, corn, and other cultivated grasses.

● You should be able to explain why fires are much more common in grasslands than in deserts, even though deserts are drier.

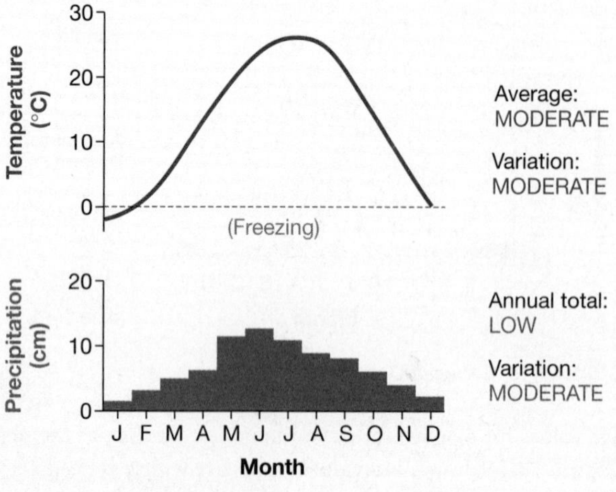

Temperature (°C)
30
20
10
0
(Freezing)
Average: MODERATE
Variation: MODERATE

Precipitation (cm)
20
10
0
J F M A M J J A S O N D
Month
Annual total: LOW
Variation: MODERATE

FIGURE 50.15 Konza Prairie.

FIGURE 50.16 Temperate Grassland.

In temperate areas with relatively high precipitation, grasslands give way to forests. Temperate forests are the most common biome found in eastern North America, western Europe, east Asia, Chile, and New Zealand.

Temperature As **Figure 50.17** shows, temperate forests experience a period in which mean monthly temperatures fall below freezing and plant growth stops. The data in this graph are from Chicago, Illinois, which was forested before being settled.

Precipitation Compared to grassland climates, precipitation in temperate forests is moderately high and relatively constant throughout the year. Chicago, for example, has an annual precipitation of 85 cm (34 in.); during most months, precipitation exceeds 5 cm (2 in.).

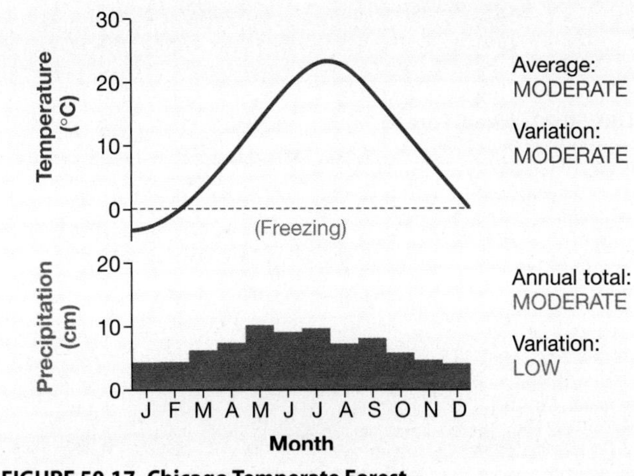

FIGURE 50.17 Chicago Temperate Forest.

FIGURE 50.18 Temperate Forest.

Vegetation In North America and Europe, temperate forests are dominated by deciduous species, which drop their leaves in autumn and grow new leaves in spring (**Figure 50.18**). Needle-leaved evergreens are also common. But in the temperate forests of New Zealand and Chile , broad-leaved evergreens predominate.

Most temperate forests have productivity levels that are lower than those of tropical forests yet higher than those of deserts or grasslands. The level of diversity is also moderate. A temperate forest in southeastern North America may have more than 20 tree species; similar forests to the north may have fewer than 10.

● You should be able to predict the type of biome that is found in regions with climate characteristics intermediate between those of grasslands and temperate forests.

The boreal forest, or **taiga**, stretches across most of Canada, Alaska, Russia, and northern Europe. Because these regions are just south of the Arctic Circle, they are referred to as subarctic.

Temperature The data in **Figure 50.19** are from Dawson, in the Yukon Territory of Canada. The region is characterized by very cold winters and cool, short summers. Temperature variation is extreme; in the course of a year, subarctic areas may be subject to temperature ranges of more than 70°C.

Precipitation Annual precipitation in boreal forests is low, but temperatures are so cold that evaporation is minimal; as a result, moisture is usually abundant enough to support tree growth.

Vegetation Boreal forests are dominated by highly cold-tolerant conifers, including pines, spruce, fir, and larch trees. Except for larches, these species are evergreen. Two hypotheses have been offered to explain why evergreens predominate in cold environ-

(Continued on next page)

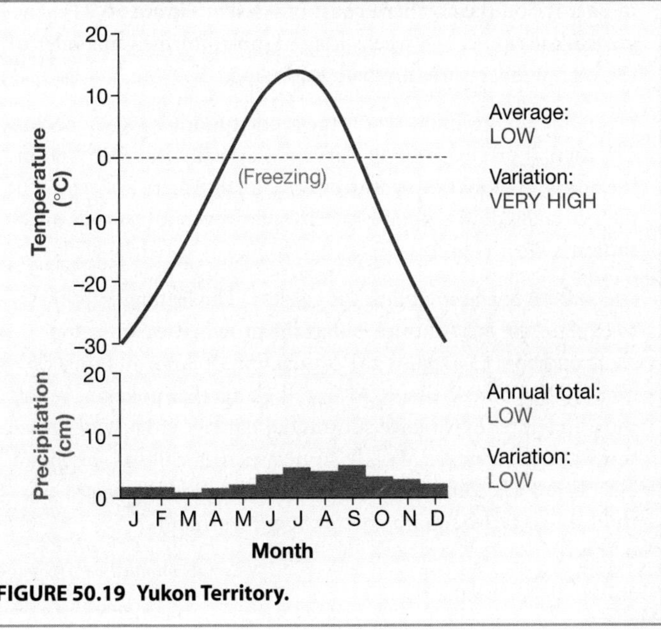

FIGURE 50.19 Yukon Territory.

Terrestrial Biomes > Boreal forests *continued*

ments, even though they do not photosynthesize in winter (**Figure 50.20**). The first is that evergreens can begin photosynthesizing early in the spring, even before the snow melts, when sunshine is intense enough to warm their needles. The second hypothesis is based on the observation that boreal forest soils tend to be acidic and contain little available nitrogen. Because leaves are nitrogen rich, species that must produce an entirely new set of leaves each year might be at a disadvantage. To date, however, these hypotheses have not been tested rigorously.

Based on these observations, it is not surprising that the productivity of boreal forests is low. Aboveground biomass is high, however, because slow-growing tree species may be long lived and gradually accumulate large standing biomass. Boreal forests also have exceptionally low species diversity. The boreal forests of Alaska, for example, typically contain seven or fewer tree species.

● You should be able to predict how the global distribution of boreal forests will change in response to global warming.

FIGURE 50.20 Boreal Forest.

Terrestrial Biomes > Arctic tundra

The arctic **tundra** biome, which lies poleward from the subarctic, is found throughout the arctic regions of the Northern Hemisphere and in regions of Antarctica that are not covered in ice.

Temperature Tundra develops in climatic conditions like those in Barrow, on the northern coast of Alaska (**Figure 50.21**). The growing season is 6–8 weeks long at most; for the remainder of the year, temperatures are below freezing.

Precipitation Precipitation on the arctic tundra is extremely low. The annual precipitation in Barrow is actually less than that in the Sonoran Desert of southwestern North America. Because of the extremely low evaporation rates, however, arctic soils are saturated year-round.

Vegetation The arctic tundra is treeless. The leading hypothesis to explain the lack of trees is that the growing season is too short and cool to support the production of large amounts of non-photosynthetic tissue. Also, tall plants that poke above the snow in winter experience substantial damage from wind-driven snow and ice crystals. Woody shrubs such as willows, birch, and blueberries are common, but they rarely exceed the height of a child. Most arctic tundra species hug the ground.

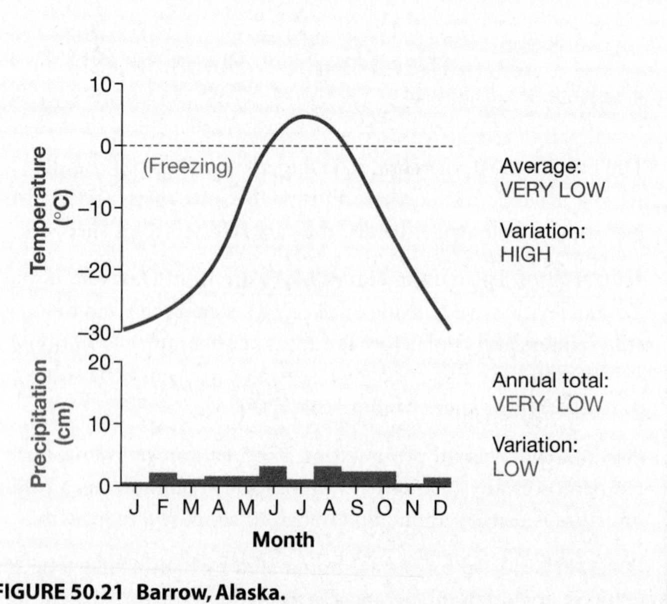

FIGURE 50.21 Barrow, Alaska.

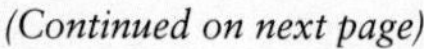
(Continued on next page)

Terrestrial Biomes > Boreal forests *continued*

Arctic tundra has low species diversity, low productivity, and low aboveground biomass. Most tundra soils are in the perennially frozen state known as **permafrost**. The low temperatures inhibit both the release of nutrients from decaying organic matter and the uptake of nutrients into live roots. Unlike desert biomes, however, the ground surface in tundra communities is completely covered with plants or lichens (**Figure 50.22**). Animal diversity also tends to be low in the arctic tundra, although insect abundance—particularly of biting flies—can be staggeringly high.

● You should be able to predict how arctic tundras compare and contrast to alpine tundras, which are found at high elevation.

FIGURE 50.22 Arctic Tundra.

50.4 The Role of Climate and the Consequences of Climate Change

Each type of aquatic environment and terrestrial biome hosts species that are adapted to the abiotic conditions present at that location. Global warming—due to rising concentrations of atmospheric CO_2 (see Chapter 54)—is having a profound impact on these abiotic factors. Runoff from melting glaciers and pole ice is changing water depths along coasts, and warming ocean waters are disturbing long-established patterns in the direction and intensity of ocean currents. On land, higher air temperatures are altering growing seasons and water availability in biomes around the globe.

To understand how global warming is affecting the distribution and abundance of organisms, biologists begin by studying why climate varies in predictable ways around the planet. Then they ask how changes in these climate patterns are affecting the organisms present, with the aim of predicting the consequences of continued change. Let's do the same.

Global Patterns in Climate

Simple questions often have fascinating answers. This turns out to be true for questions about why climate varies around the globe. For example, why are some parts of the world warmer and wetter than others? Why do seasons exist?

Why Are the Tropics Warm and the Poles Cold? In general, areas of the world are warm if they receive a large amount of sunlight per unit area; they are cold if they receive a small amount of sunlight per unit area. Over the course of a year, regions at or near the equator receive much more sunlight per unit area—and thus much more energy in the form of heat—than regions that are closer to the poles. At the equator, the Sun is often directly overhead. As a result, sunlight strikes Earth there at or close to a 90° angle. At these angles, Earth receives a maximum amount of solar radiation per unit area (**Figure 50.23**). But because Earth's surface slopes away from the equator, the Sun strikes the surface at lower and lower angles moving toward the poles. When sunlight arrives at a low angle, much less energy is received per unit area. The pattern of decreasing average temperature with increasing latitude is a result of Earth's spherical shape.

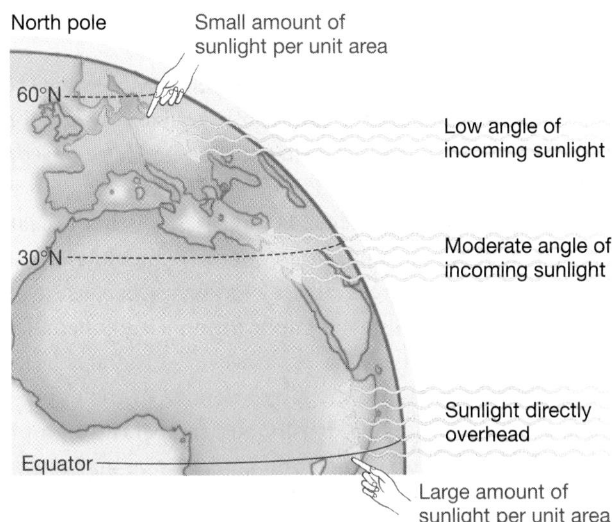

FIGURE 50.23 Solar Radiation per Unit Area Declines with Increasing Latitude. Over the course of a year, the Sun is frequently almost directly overhead at the equator. As a result, equatorial regions receive a large amount of solar radiation per unit area. At latitudes greater than 23.5°, the Sun is never directly overhead. As a result, high-latitude regions receive less solar radiation per unit area than do low-latitude regions.

(a) Circulation cells exist at the equator ...

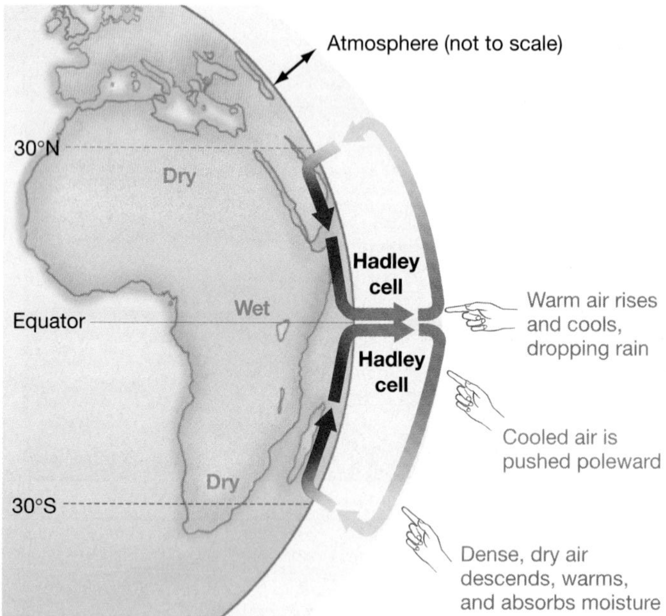

(b) ... and at higher latitudes.

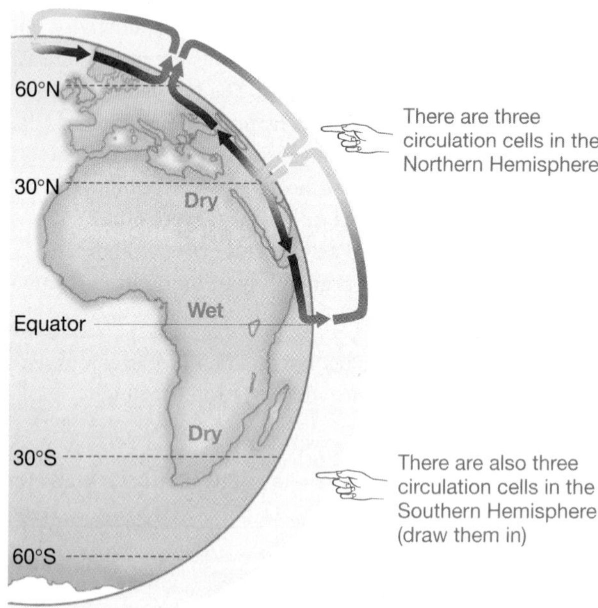

FIGURE 50.24 Global Air Circulation Patterns Affect Rainfall. Hadley cells explain why the tropics are much wetter than regions near 30° latitude north and south. Regions with rising air, such as those at the equator, tend to be wetter than regions with descending air.

● **QUESTION** Is air at the North Pole wet or dry? Is your answer consistent with the data in Figure 50.21?

Why Are the Tropics Wet? One of the most striking weather patterns on Earth involves precipitation. When average annual rainfall is mapped for regions around the globe, it is clear that areas along the equator receive the most moisture, while locations about 30 degrees latitude north and south of the equator are among the driest on the planet. Why do these patterns occur?

A major cycle in global air circulation, called a **Hadley cell**, is responsible for making the Amazon River basin wet and the Sahara Desert dry. Hadley cells are named after George Hadley, who in 1735 conceived of the idea of enormous air circulation patterns. As **Figure 50.24a** indicates, air that is heated by the strong sunlight along the equator expands and rises. Warm air can hold a great deal of moisture, because warm water molecules tend to stay in vapor form instead of condensing into droplets. As the air rises, however, it radiates heat to space. It also expands into the larger volume of the upper atmosphere, which lowers its density and temperature—a phenomenon known as adiabatic cooling. As rising air cools, its ability to hold water declines. When water vapor cools, water condenses. The result? High levels of precipitation occur along the equator.

As more air is heated along the equator, the cooler, "older" air above Earth's surface is pushed poleward. When the air mass has cooled enough, its density increases and it begins to sink. As it sinks, it absorbs more and more solar radiation reflected from Earth's surface and begins to warm. As the air warms, it also gains water-holding capacity. Thus, the air approaching the Earth "holds on" to its water and little rain occurs where it returns to the surface; the area is bathed in warm, dry air. The result is a band of deserts in the vicinity 30° latitude north and south. As this air moves toward each pole, it continues to pick up more moisture.

Similar air circulation cells occur between 30° and 60° of latitude, and between 60° latitude and the poles (**Figure 50.24b**). ● If you understand how Hadley cells work, you should be able to add diagrams to Figure 50.24b showing the three cells that occur in the Southern Hemisphere.

What Causes Seasonality in Weather? The striking global patterns in temperature and moisture that we've just reviewed are complicated by the phenomenon of seasons: regular, annual fluctuations in temperature, precipitation, or both. As **Figure 50.25** shows, seasonality occurs because Earth is tilted on its axis by 23.5 degrees. As a result of this incline, the Northern Hemisphere is tilted toward the Sun in June and faces the Sun most directly. Thus the Northern Hemisphere presents its least-acute angle to the Sun and receives the largest amount of solar radiation per unit area in June. The Southern Hemisphere, in contrast, is tilted away from the Sun in June, presents a steep angle for incoming sunlight, and receives its smallest quantity of solar radiation per unit area. As a result, in June it is summer in the Northern Hemisphere, and winter in the Southern Hemisphere. Conversely, in De-

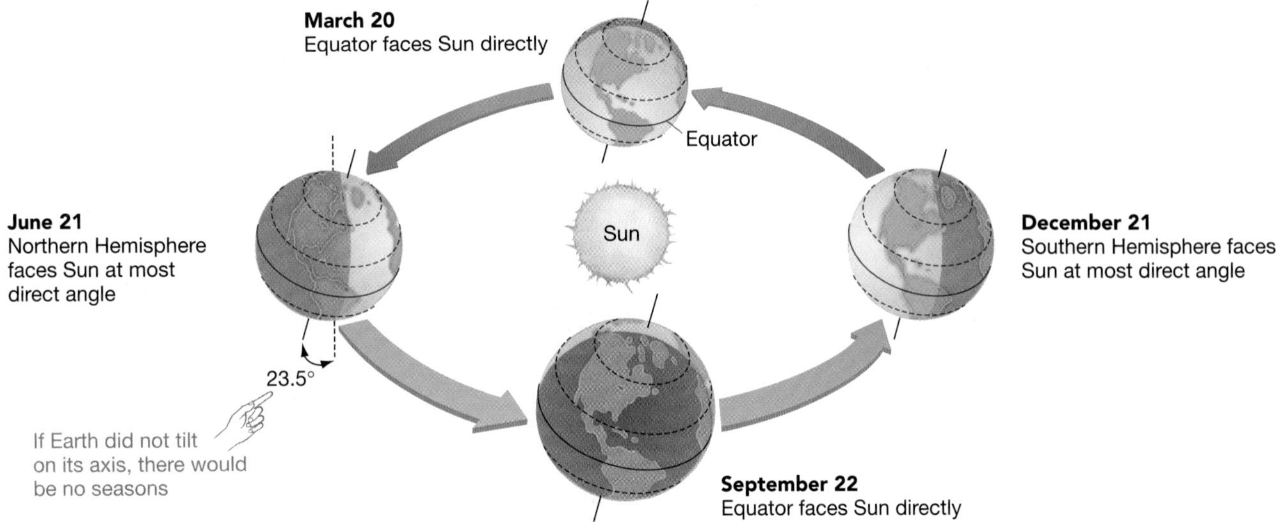

FIGURE 50.25 Earth's Orbit and Tilt Create Seasons at High Latitudes. Seasons occur due to annual variation in the amount of solar radiation that different parts of the Earth receive.

cember it is summer in the Southern Hemisphere, and winter in the Northern Hemisphere. In March and September the equator faces the Sun most directly, so the tropics receive the most solar radiation then. If Earth did not tilt on its axis, there would be no seasons.

Mountains and Oceans: Regional Effects on Climate

The broad patterns of climate that are dictated by global heating patterns, Hadley cells, and seasonality are overlain by regional effects. The most important of these are due to the presence of mountain ranges and proximity to an ocean.

Figure 50.26 shows how mountain ranges affect regional climate, using the Cascade Mountain range along the Pacific Coast of North America as an example. In this area of the world, the prevailing winds are from the west. These winds bring moisture-laden air from the Pacific Ocean onto the continent. As the air masses begin to rise over the mountains, the air cools and releases large volumes of water as rain. As a result, the area along the Pacific Coast from northern California to southeast Alaska hosts some of the only high-latitude rain forests in the world. One area along the Pacific Coast of Washington State gets an average of 340 cm (134 in.) of precipitation each year. (Mountain ranges also occur along the Pacific Coast of Southern California and Mexico; but these areas are arid because they are dominated by dry air that is part of the Hadley circulation.)

Once cooled air has passed the crest of a mountain range, the air is relatively dry because much of its moisture content has already been released. Areas that receive this dry air are said to be in a **rain shadow**. One of the only high-latitude deserts in the world exists to the east of the Cascade Mountains. It is created by a rain shadow and averages less than 25 cm (10 in.) of moisture annually. Similar rain shadows create the Great Basin desert of western North America, the Atacama Desert of South America, and the dry conditions of Asia's Tibetan Plateau.

While the presence of mountain ranges tends to produce extremes in precipitation, the presence of an ocean has a moderating influence on temperature. To understand why this is so, recall from Chapter 2 that water has an extremely high **specific heat**, meaning that it has a large capacity for storing heat energy. Because water molecules form hydrogen bonds with each other, it takes a great deal of heat energy to boil water or melt ice—in fact, to change the temperature of water at all. The hydrogen bonds have to be broken before the water molecules

FIGURE 50.26 Mountain Ranges Create Rain Shadows. When moisture-laden air rises over a mountain range, the air cools enough to condense water vapor. As a result, moisture falls to the ground as precipitation. On the other side of the mountain, little water is left to provide precipitation.

QUESTION What conditions would create the wettest spots on Earth? The driest spots?

themselves can begin to absorb heat and move faster, which we measure as increased temperature.

Because water has a high specific heat, it can absorb a great deal of heat from the atmosphere in summer, when the water temperature is cooler than the air temperature. As a result, the ocean moderates summer temperatures on nearby landmasses. Similarly, the ocean releases heat to the atmosphere in winter, when the water temperature is warmer than the air temperature. As a result, islands and coastal areas have much more moderate climates than do nearby areas inland.

at www.masteringbio.com
Tropical Atmospheric Circulation

FIGURE 50.27 Experimental Increases in Temperature Change Species Composition in Artic Tundra. The transparent, open-topped chambers shown here are an effective way to increase average annual temperature. Inside the chambers, species composition changes relative to otherwise similar sites nearby.

QUESTION Why does a clear chamber like this increase average temperature inside?

How Will Global Warming Affect Ecosystems?

Climate—particularly the amount and variability of heat and precipitation—has a dramatic effect on terrestrial ecosystems. Climate also has a significant impact on aquatic ecosystems. Heat and precipitation influence ocean currents and lake turnover, as well as water levels in lakes, streams, and oceans.

It is now well established that CO_2 pollution is causing climate change (see Chapter 54). Climate has changed frequently and radically throughout the history of life, but recent increases in average temperatures around the globe, combined with a projected increase of up to 5.8°C over the next 100 years, represents one of the most rapid periods of climate change since life began.

Biologists use three tools to predict how global warming will affect aquatic and terrestrial ecosystems.

1. *Simulation studies* are based on computer models of weather patterns in local regions. By increasing average temperatures in these models, researchers can predict how wind and rainfall patterns, storm frequency and intensity, and other aspects of weather and climate will change. In effect, simulation studies predict how temperature and precipitation profiles, like those in Figures 50.11, 13, 15, 17, 19, and 21, will change.
2. *Observational studies* are based on long-term monitoring at fixed sites around the globe. Researchers are documenting changes in key physical variables—water depth, water flow, temperature, and precipitation—as well as changes in the distribution and abundance of organisms. By extrapolation, they can predict how change may continue at these sites.
3. *Experiments* are designed to simulate changed climate conditions, and to record responses by the organisms present.

Experiments That Manipulate Temperature Even though the scientific community did not become alert to the reality of global warming until the 1980s, a large number of multiyear experiments have already been done to test how increased temperature affects organisms in nature. Because arctic regions are projected to experience particularly dramatic changes in temperature and precipitation, many of these experiments have been done in the tundra biome.

Figure 50.27 shows how the most successful of these experiments are done. Transparent, open-topped chambers are placed at random locations around a study site, and the characteristics of the vegetation inside the chambers are recorded annually. The same measurements are made in randomly assigned study plots that lack chambers.

Across experiments, the average annual temperature inside the open-topped chambers increases by 1–3°C. This increase is consistent with conservative projections for how temperatures will change over the next century. In a recent analysis of data from 11 such study sites scattered throughout the arctic in North America and Eurasia, several strong patterns emerged: Overall, species diversity decreases. Compared with control plots, grasses and shrubs increase inside the chambers, and mosses and lichens decrease. These results support simulation and observational studies predicting that arctic tundra environments are giving way to boreal forest. Biomes are changing right before our eyes.

Experiments That Manipulate Precipitation Recent simulation and observational studies have converged on a remarkable conclusion: Increases in average global temperature are increasing variability in temperature and precipitation. Stated another way, global warming is making climates more extreme. For example, the amount of precipitation and the wind speeds associated with monsoon rains, cyclones, hurricanes, and other types of storm events is increasing, even though their frequency may be unchanged.

Experiment

Question: How does increasing variation in rainfall—without changing total amount—affect grassland biomes?

Hypothesis: Increasing variability in rainfall affects soil moisture, net primary productivity (NPP), and species diversity.

Null hypothesis: Increasing variability in rainfall has no effect on grassland biomes.

Experimental setup:

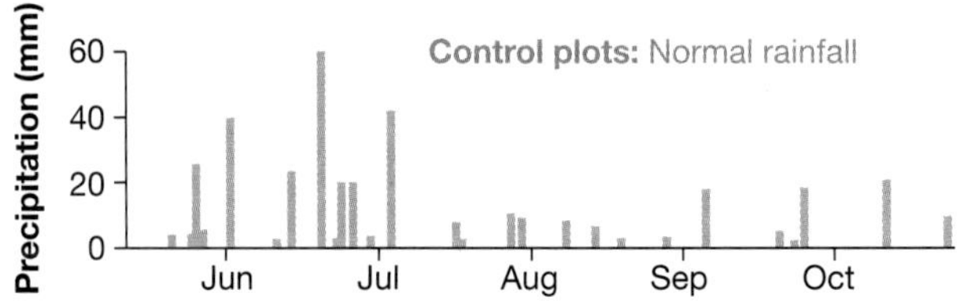

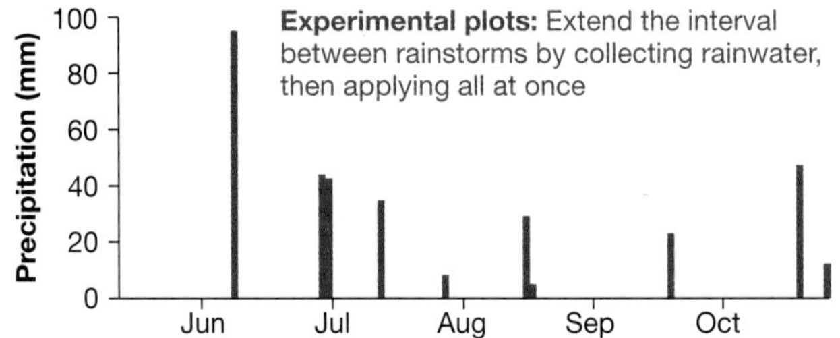

Record soil moisture, aboveground biomass (as an index of NPP), and species diversity, over a 4-year period.

Prediction: Soil moisture, NPP, and/or species diversity will differ between experimental plots and control plots.

Prediction of null hypothesis: Soil moisture, NPP, and species diversity will not differ between experimental plots and control plots.

Results:

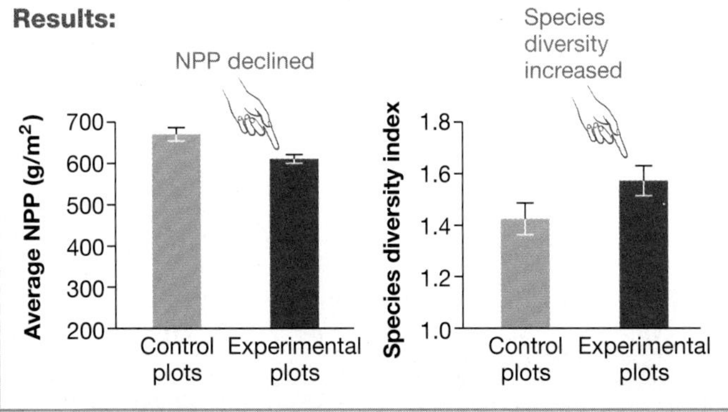

Conclusion: Altering variability in rainfall has dramatic effects on soil moisture (data not shown), NPP, and species diversity.

FIGURE 50.28 Experimental Increases in Rainfall Variability Change Species Composition in Temperate Grasslands.

To test how increased variation in climate will affect a temperate grassland ecosystem, researchers set up the experiment described in **Figure 50.28**. The goal was to keep the average annual rainfall unchanged during the growing season, but increase variability. The biologists increased the severity of storms by capturing rain that fell over study plots during some storms and then applying that water to the same plots during later storms.

The results from the first four years of the experiment are striking: Average soil moisture and overall NPP declined in the experimental plots, compared with similar plots where rainfall was not manipulated. Overall species diversity is increasing in the high-variation plots, however, because the grass that produces most of the biomass at this site is doing poorly in the experimental plots—opening up room for other species to grow. The experiment is ongoing, however, because it is possible that long-term changes will differ from short-term changes, due to continued alterations in the species present.

Experiments like this are documenting that climate change causes dramatic alterations in the composition and distribution of biomes. Your world is changing.

Check Your Understanding

If you understand that...

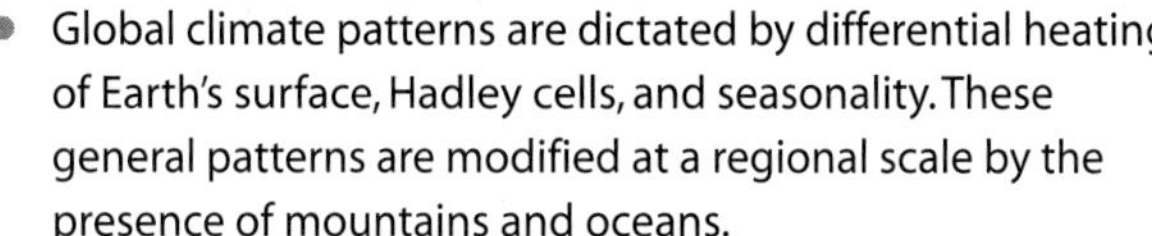

- Global climate patterns are dictated by differential heating of Earth's surface, Hadley cells, and seasonality. These general patterns are modified at a regional scale by the presence of mountains and oceans.

You should be able to...

Predict how the biome where you live will change over the course of your lifetime.

50.5 Biogeography: Why Are Organisms Found Where They Are?

Understanding why organisms are found where they are is one of ecology's most basic tasks. It is a fundamental aspect of predicting the consequences of climate change, and is required for designing nature preserves in a rational and effective way.

The study of how organisms are distributed geographically is called **biogeography**. You were introduced to this subject in Chapter 26, which analyzed how continental drift has contributed to the current distributions of ostriches and their relatives. Biogeography will come up repeatedly in this unit as well: Chapter 53 analyzes general patterns in how species numbers vary around the world, and Chapter 54 presents data showing that the geographic ranges of many species are changing rapidly in response to global warming. Here, let's focus on how interactions with the abiotic and biotic environments affect where a particular species is able to live.

The most fundamental observation about the **range**, or geographic distribution, of species is simple: No one species can survive the full range of environmental conditions present on Earth. *Thermus aquaticus* cells thrive in hot springs with temperatures above 70°C. These archaea can live in this environment because they have enzymes that do not denature at near-boiling temperatures. The same cells would die instantly, however, if they were transplanted to the frigid waters near polar ice. But vast numbers of other archaea are present in cold seawater. These cells have enzymes that can work efficiently at near-freezing temperatures. No organism thrives in both hot springs and cold ocean water, however. This observation is explained by the phenomenon called fitness trade-offs, introduced in Chapters 24 and 41: No enzyme can function well in both extremely high temperatures and extremely low temperatures.

Because of fitness trade-offs, organisms tend to be adapted to a limited set of physical conditions. Although some species have much broader geographic ranges than other species do—humans and the bacteria that live in and on them may have the widest distribution of any species at present—no organism can live everywhere.

To understand a species' distribution thoroughly, though, it is essential to examine historical and biotic factors in addition to the physical conditions present. For example, the temperature and moisture regimes of large areas of southern California, in North America, are almost identical to those found in the Mediterranean region of Europe and north Africa. Both regions are considered part of the same biome. But even though the physical conditions at these sites are extremely similar, the species present are almost completely different. Why?

The Role of History

To explain why certain species occur in a particular region while others don't, the first factor to consider is history—specifically, the history of dispersal. **Dispersal** refers to the movement of an individual from the place of its birth, hatching, or origin to the location where it lives and breeds as an adult. If a particular species is missing from an area, a physical barrier to dispersal—such as a river, a glacier, a mountain range, or an ocean—may be present. For example, the plant species found along North America's southwest coast are different from those found in the Mediterranean because the Pacific and Atlantic Oceans are effective barriers against the dispersal of seeds.

In many cases, human activity circumvents physical barriers to dispersal. For example, public health officials closely monitor annual flu outbreaks in south China—where most new strains originate—because they recognize that new strains disperse all over the planet in a matter of weeks or even days, via the respiratory passages of airplane passengers. Prior to the invention of air travel, disease epidemics in humans were much more likely to be confined to relatively small geographic areas.

Similarly, humans have transported thousands of seeds, birds, insects, and other species across physical barriers to new locations—sometimes purposefully and sometimes by accident. One accidental introduction has had disastrous consequences for the arid shrublands and grasslands of the western United States. In 1889 a native plant of Eurasia called cheatgrass was accidentally introduced to western North America in a shipment of crop seeds. Cheatgrass is an annual plant with seeds that germinate over winter, so that plants grow and set seed in early spring. As a result, cheatgrass seedlings get a "jump" on many North American plant species, which initiate growth later. Early growth allows cheatgrass to use more of the available water and nutrients. Within 30 years, cheatgrass had taken over tens of thousands of square miles of habitat (**Figure 50.29**).

If an **exotic** species —one that is not native—is introduced into a new area, spreads rapidly, and eliminates native species, it is said to be an **invasive species**. In North America alone, dozens of invasive species have had devastating effects on native plants and animals. Examples include kudzu, purple loosestrife, reed canary grass, garlic mustard, Russian thistle (also known as tumbleweed), European starlings, African honeybees, and zebra mussels. Similar lists could be compiled for other regions of the world.

Regarding the distribution of organisms, the story of invasive species is straightforward: They and other introduced species did not exist in certain parts of the world until recently, simply because they had never been dispersed there. But data show that only about 10 percent of the species introduced to an area actually become common, and a smaller percentage than that increases in population enough to be considered invasive. Why is it that some species introduced into an area do not thrive? Conversely, why are invasives so successful?

Biotic Factors

The distribution of a species is often limited by biotic factors—meaning interactions with other organisms. As an example, recall the data presented in Chapter 26 on the current and former

FIGURE 50.29 Cheatgrass Has Invaded Extensive Areas of North America.

distributions of hermit warblers and Townsend's warblers along the Pacific coast of North America. Both hermit warblers and Townsend's warblers live in evergreen forests. Experiments have shown that male Townsend's warblers directly attack male hermit warblers and evict them from breeding territories, and historical data indicate that the geographic range of Townsend's warblers has been expanding steadily at the expense of hermit warblers. The results support the hypothesis that a biotic factor—competition with another species—is limiting the range of hermit warblers.

Competition is not the only biotic factor to consider, however. In Africa, the range of domestic cattle is limited by the distribution of tsetse (pronounced *TSEE-tsee*) flies. Tsetse flies transmit a parasite that causes the disease trypanosomiasis, which is fatal in cattle (**Figure 50.30a**). Similarly, most of the songbirds native to Hawaii are limited to alpine habitats—above the range of the mosquitoes that transmit avian malaria. And the moth pictured in **Figure 50.30b** lays its eggs only in the flowers of yucca plants. As a result, this species does not exist outside the range of yucca plants.

(a) Distribution of cattle is limited by distribution of tsetse flies.

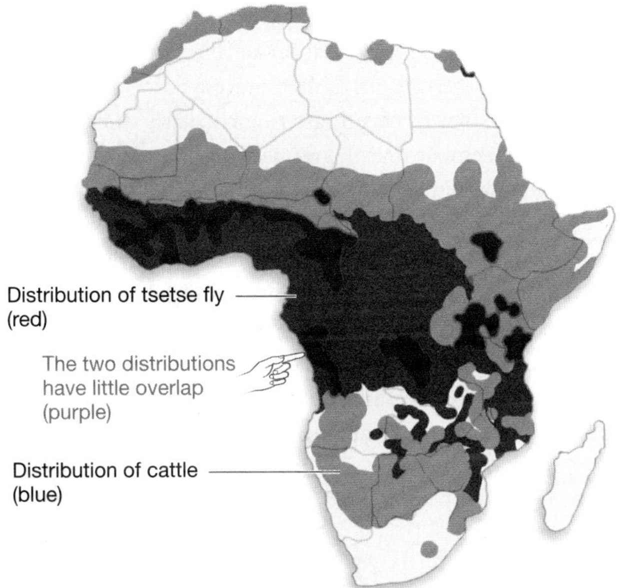

(b) Yucca moth distribution is limited by yucca plant distribution.

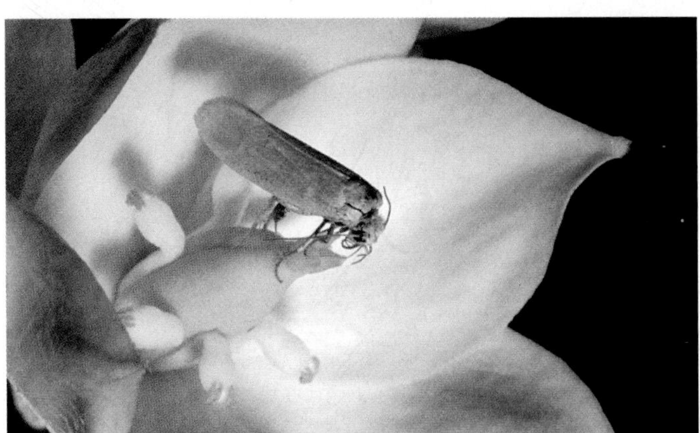

FIGURE 50.30 Distributions Can Be Limited by Biotic Factors.
(a) Because tsetse flies carry a disease that is fatal to cattle, tsetse flies and cattle rarely coexist. **(b)** Because yucca moths lay their eggs only in the flowers of yucca plants, the moths cannot live where yucca plants are absent.

Abiotic Factors

In many cases, introduced species may not become established in a region because of abiotic factors—particularly temperature and moisture. An area may simply be too cold or too dry for an organism to survive. For example, even though the invasive plant kudzu grows prolifically throughout the southeastern United States, it is killed when the soil freezes to a depth of a foot or more. As a result, it has not expanded its range farther north and west.

It is often difficult to separate the effects of biotic and abiotic factors on a species' range, however. To drive this point home, consider how cheatgrass became established in North America. Cheatgrass does not grow in wet grasslands because it does not compete well against the tall species that thrive there. But it has been able to invade two important types of biomes in North America: dry temperate grasslands and the arid, shrub-dominated habitats known as sage-steppe. In each case, the reasons for the success of cheatgrass were different.

Cheatgrass has been able to invade sage-steppe habitats because it is not affected by fire. Fire is an abiotic factor that can kill the shrubs that dominate sage-steppe (**Figure 50.31**). In fact, because cheatgrass grows in dense beds and dies back in

FIGURE 50.31 Sage-Steppe Is an Arid Biome Dominated by Shrubs. Large areas of western North America are covered with sage-steppe vegetation.

QUESTION If cheatgrass were not present, what would happen if a fire started here?

early spring, it actually leads to an increase in the frequency and intensity of wildfires. Normally, fires are rare in sage-steppe biomes, because plants are widely spaced. This distribution usually prevents fires from spreading and affecting a large area. But when this biome begins to dry in late summer, beds of dead cheatgrass offer a continuous, extremely flammable source of fuel. If a fire does come through, it does not affect the cheatgrass for two reasons: (1) Cheatgrass is an annual, so there is no living tissue exposed once the growing season is over, and (2) cheatgrass seeds sprout readily in soils that have been depleted of organic matter by fire. Because sagebrush and other shrubs are perennials, their tissues are heavily damaged or killed by hot fires. By promoting the frequency and intensity of fire in sage-steppe biomes, the presence of cheatgrass creates an environment that favors the spread of cheatgrass.

How has this species been able to invade dry grassland biomes? Prior to the arrival of cheatgrass, grasslands in arid areas of the American West were dominated by the types of bunchgrasses shown in **Figure 50.32**. Bunchgrasses do not form a continuous sod; instead, they grow in compact bunches. The spaces between these bunches are occupied by what biologists call "black-soil crust." This material is a community of soil-dwelling cyanobacteria, archaea, and other microorganisms that grow when moisture is available. Many of these species also fix nitrogen (see Chapter 38). The nitrogen becomes available to the surrounding plants when the microbial cells die and decompose.

When European settlers arrived and began grazing cattle in this biome, however, both the black soils and bunchgrasses were affected. Cattle grazed on the bunchgrasses until the grasses died, and their hooves compacted and disrupted the black soils. Neither component of the ecosystem was allowed to recover, in part because intensive grazing continued and in part because cheatgrass arrived. Cheatgrass shades out black-soil crust organisms. When black-soil crust is gone, the amount of nitrogen available to bunchgrasses is reduced. Cheatgrass also competes successfully with bunchgrasses for water and nutrients, because it completes its growth early in the spring—before the native bunchgrasses have broken dormancy and begun to grow. And as in sage-steppe habitats, the expansion of cheatgrass increased the frequency and severity of fires, which further degraded the integrity of black-soil crust communities.

To summarize, changes in abiotic and biotic factors allowed cheatgrass to invade native biomes in North America and thrive as an invasive species. Currently, managing cheatgrass is the major problem facing conservation biologists and rangeland managers in the western United States.

Although the story of cheatgrass is particularly dramatic, it is important to remember that the range of every species on Earth is limited by a combination of historical, abiotic, and biotic factors. The range of a particular species depends on dispersal ability, the capacity to survive climatic conditions, and the ability to find food and avoid being eaten. Climate and other abiotic factors were the focus of this chapter; the next several chapters focus on the biotic aspects of ecology. How do interactions among organisms of the same species and interactions among different species affect geographic range and population size?

FIGURE 50.32 Arid Grasslands Feature Bunchgrasses and Black-Soil Crust. As their name implies, bunchgrasses grow in compact bunches instead of forming a sod. The black-soil crust that forms between bunchgrasses is composed of cyanobacteria, archaea, and other microscopic organisms that add organic matter and nutrients to the soil.

● **QUESTION** Suggest a hypothesis to explain why grasses with the bunching growth habitat tend to be found in dry habitats, as compared to grasses that form sods.

Chapter Review

SUMMARY OF KEY CONCEPTS

Ecology focuses on how organisms interact with their environment. Because its goal is to understand the distribution and abundance of organisms, ecology provides a scientific foundation for the conservation of species and natural areas.

Biologists study ecology at four main levels: (1) organisms, (2) populations, (3) communities, and (4) ecosystems. The goal of organismal biology is to understand how individuals respond to stimuli from the environment, including members of their own species and other species. Population ecology focuses on how and why populations grow or decline. Community ecology is the study of how two or more species interact, and ecosystem ecology analyzes interactions between communities of organisms and their abiotic environment—particularly the flow of energy and nutrients. In efforts to preserve endangered species and communities, conservation biologists apply analytical methods and results from all four levels.

You should be able to explain how data from ecology's four levels would inform efforts to preserve an endangered species such as polar bears, pandas, tigers, or blue whales.

Physical structure—particularly water depth—is the primary factor that limits the distribution and abundance of aquatic species. Climate—specifically, both the average value and annual variation in temperature and in moisture—is the primary factor that limits the distribution and abundance of terrestrial species.

Water depth is a crucial factor in aquatic ecosystems because the wavelengths of sunlight that are required for photosynthesis do not penetrate water efficiently. Thus, only certain zones in a lake or ocean can support photosynthesis, and organisms in other aquatic zones must depend on food that rains out of the sunlit areas near the surface.

Distinct climatic regimes are associated with distinct types of terrestrial vegetation, or biomes. Biomes represent the major types of environments available to terrestrial organisms. Because photosynthesis is most efficient when temperatures are warm and water supplies are ample, the productivity and degree of seasonality in biomes varies with temperature and moisture.

You should be able to design a man-made stream and lake system that would maximize the diversity of aquatic ecosystems present. You should also be able to comment on how increased temperatures alone (not considering changes in precipitation) might affect the biomes described in Figures 50.11 through 50.16.

Climate varies with latitude, elevation, and other factors. Climate is changing rapidly around the globe.

An array of abiotic factors affect organisms, including the chemistry of water in aquatic habitats; the nature of soils in terrestrial habitats; and climate, which is a combination of temperature, moisture, sunlight, and wind. Climate varies around the globe because sunlight is distributed asymmetrically, with equatorial regions receiving on average more solar radiation than do regions toward the poles. Hadley cells create bands of wet and dry habitats, and the tilt of Earth's axis causes seasonality in the amount of sunlight that non-equatorial regions receive. The presence of mountains can create local areas of wet or dry habitats, and proximity to an ocean moderates temperatures in nearby terrestrial habitats. Global warming is causing temperature profiles and precipitation patterns to change in ecosystems throughout the planet.

You should be able to explain how climate would be expected to vary around the Earth if the planet were cylindrical and spun on an axis that was at a right angle to the Sun.

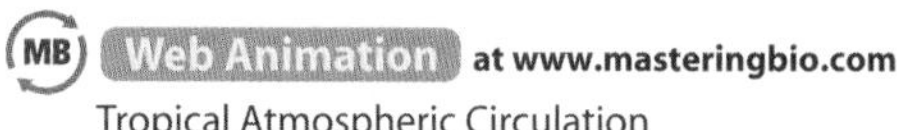

Tropical Atmospheric Circulation

In addition to abiotic aspects of the environment such as physical structure and climate, a species' distribution is constrained by historical and biotic factors.

The range, or geographic distribution, of a particular species depends on dispersal ability, the capacity to survive climatic conditions, and the ability to find food and avoid being eaten. For each species, a unique combination of historical, abiotic factors, and biotic factors dictates where individuals live and the size of the population.

You should be able to describe conditions that would maximize the likelihood that an introduced species becomes invasive.

QUESTIONS

Test Your Knowledge

1. What is a rain shadow?
   a. the part of a mountain that receives prevailing winds and heavy rain
   b. the region beyond a mountain range that receives dry air masses
   c. the region along the equator where precipitation is abundant
   d. the region near 30°N and 30°S latitude that receives hot, dry air masses

2. A region receives less than 5 cm (2 in.) of precipitation annually and has temperatures that never drop below freezing. Which type of biome is present?
   a. subtropical desert
   b. temperate grassland
   c. boreal forest
   d. arctic tundra

3. What is the predominant type of vegetation in a tropical wet forest?
   a. shrubs and bunchgrasses
   b. herbs, grasses, and vines
   c. broad-leaved deciduous trees
   d. broad-leaved evergreen trees

4. The littoral zone of a lake is most similar to which of the following environments?
   a. oceanic zone
   b. intertidal zone
   c. neritic zone
   d. marsh

5. Typically, where are oxygen levels highest and nutrient levels lowest in a stream?
   a. near its source
   b. near its mouth, or end
   c. where it flows through a swamp or marsh
   d. where it forms an estuary

6. Which of the following is most important in the success of an invasive species?
   a. historical factors
   b. abiotic factors
   c. biotic factors
   d. Historical, abiotic, and biotic factors are equally important in limiting invasive species.

Test Your Knowledge answers: 1. b; 2. a; 3. d; 4. b; 5. a; 6. c

Test Your Understanding

Answers are available at www.masteringbio.com

1. Name the four main levels of study in ecology. Write a question about the ecology of humans, trout, or oak trees that is relevant to each level.
2. In June, does the Northern Hemisphere receive more or less solar radiation than the equator does? Explain your answer.
3. Diagram the Hadley cell that exists between the equator and 30° latitude south. Indicate the direction of air flow, regions where air is warming or cooling, and regions where air is gaining or losing moisture. Use your diagram to explain why the Australian Outback is a subtropical desert.
4. Contrast the productivities of the intertidal, neritic, and oceanic zones of marine environments. Explain why large differences in productivity exist.
5. Compare and contrast the biomes found at increasing elevation on a mountain with the biomes found at increasing latitude in Figure 50.9.
6. Explain why productivity is much lower in bogs than it is in marshes or lakes that receive the same amount of solar radiation.

Applying Concepts to New Situations

Answers are available at www.masteringbio.com

1. Mars has an even more pronounced tilt than Earth does. Does Mars experience seasons? Why or why not?
2. Mountaintops are closer to the Sun than the surrounding area is. Why are mountaintops cold?
3. The southwest corners of some Caribbean islands are extremely dry, even though other areas of the islands receive substantial rainfall. Explain this observation. Base your answer on a hypothesis (one that you propose) about the direction of the prevailing winds in this region of the world.
4. Edinburgh, Scotland, and Moscow, Russia, are at the same latitude, yet these two cities have very different climates—Moscow has much colder winters. Explain why.

www.masteringbio.com is also your resource for • Answers to text, table, and figure caption questions and exercises • Answers to *Check Your Understanding* boxes • Online study guides and quizzes • Additional study tools including the *E-Book for Biological Science* 3rd ed., textbook art, animations, and videos.

Behavior

51

Peregrine falcons hunt ducks and other large birds.

KEY CONCEPTS

- After describing a behavior, biologists seek to explain both its proximate and ultimate causes—meaning, how it happens at the genetic and physiological levels and how it affects the individual's fitness.
- In a single species, behavior may range from highly stereotyped, invariable responses to highly flexible, conditional responses and from unlearned to learned responses.
- The types of learning that individuals do, the way they communicate, and the way they orient and navigate all correlate closely with their habitat and with the challenges they face in trying to survive and reproduce.
- When individuals behave altruistically, they are usually helping close relatives or individuals that help them in return.

To biologists, **behavior** is action—specifically, the response to a stimulus. Pond-dwelling bacteria swim toward drops of blood that fall into the water. Time-lapse movies of growing sunflowers document the steady movement of their flowering heads throughout the day, as they turn to face the shifting Sun. Filaments of the fungus *Arthrobotrys* form loops, then release a molecule that attracts roundworms. When a roundworm touches one of these loops, the fungal filaments swell—ensnaring the worm. *Arthrobotrys* cells then proceed to grow into the worm's body and digest it.

Among animals, action may be frequent and even spectacular. Peregrine falcons reach speeds of up to 320 km/hr (200 mph) when they dive in pursuit of ducks or other flying prey. Some Arctic terns migrate over 32,000 km (20,000 miles) each year. Deep inside a hive, in complete darkness, honeybees communicate the position and nature of food sources a mile away.

Ecology is the study of how organisms interact with their physical and biological environments, and behavioral biology is the study of how organisms respond to particular stimuli from those environments. What does an *Anolis* lizard do when it gets too hot? How does that same individual respond when a member of the same species approaches it? Or when a house cat approaches?

Although all organisms respond in some way to signals from their environment, most behavioral research is performed on vertebrates, arthropods, or molluscs. With their sophisticated nervous systems and skeletal-muscular systems, these animals can sense, process, and respond rapidly to a wide array of environmental stimuli.

Although most behavioral studies start by describing what animals do in response to a particular stimulus, the eventual goal is to explain the proximate and ultimate causes of behavior.

Key Concept Important Information Practice It

Proximate, or mechanistic, **causation** explains *how* actions occur in terms of the neurological, hormonal, and skeletal-muscular mechanisms involved. **Ultimate**, or evolutionary, **causation** explains *why* actions occur—based on their evolutionary consequences and history. Is a particular behavior adaptive, meaning that it increases an individual's fitness? If so, how does it help individuals produce offspring in a particular environment?

To illustrate the proximate and ultimate levels of causation, consider the spiny lobster shown in **Figure 51.1**. These animals spend the day hiding in cracks or holes in coral reefs. At night they emerge and wander away from the reef in search of clams, mussels, crabs, or other sources of food. Before dawn, they return to one of several dens they use on a regular basis. How do they find their way back? This question is proximate in nature—it asks how an individual does what it does. Although research is continuing, recent experiments have provided strong support for the hypothesis that spiny lobsters have receptors in their brains that detect changes in Earth's magnetic field, and that they navigate by using information on how the orientation and strength of the field changes as they move about. At the ultimate level of causation, biologists want to know why the behavior occurs. The leading hypothesis to explain homing behavior is that the ability to navigate allows spiny lobsters to search for food over a wide area under cover of darkness, then return to a safe refuge before sharks and other large predators that hunt by sight can find them.

It is important to recognize that efforts to explain behavior at the proximate and ultimate levels are complementary. To understand what an organism is doing, biologists want to know how the behavior happens and why. Accomplishing this task requires them to draw on techniques and data from genetics, physiology, and evolutionary biology. But even though behavioral biologists work at the interface of many fields, their work is essentially ecological in nature. Questions about behavior are almost always aimed at understanding how individuals cope with changing physical conditions or how they interact with individuals of their own or other species—in short, how they interact with their environment.

FIGURE 51.1 Spiny Lobsters Can Find Their Way Back to Their Home Burrow. Spiny lobsters wander widely in search of food at night, then return to their hiding places in coral reefs before dawn.

QUESTION Compared to other lobsters, spiny lobsters have extremely long antennae for their body size. Generate a hypothesis to explain why. (Your hypothesis will be at the ultimate level of causation.)

The goals of this chapter are to explore some key topics in behavioral biology and illustrate the spirit of the field. To begin, let's consider one way to categorize and organize the enormous variety of behavioral responses observed in animals.

51.1 Types of Behavior: An Overview

Some types of behavior are performed nearly the same way every time. When people yawn, the action is so similar that every instance takes about 6 seconds. Other types of behavior are highly flexible. When people smile, the action is variable. The same person may smile in dozens of ways, and different people tend to have their own individual style and frequency of smiling. In addition, some types of behavior are readily modified by learning, while other types of behavior are not. Formally, **learning** is defined as a change in behavior that results from a specific experience in the life of an individual. Human infants learn their native language through imitation and trial and error, but they begin to yawn and smile before they have a chance to learn these actions from their parents.

Figure 51.2 offers one way of summarizing these general observations about behavior. Notice that the graph has two axes.

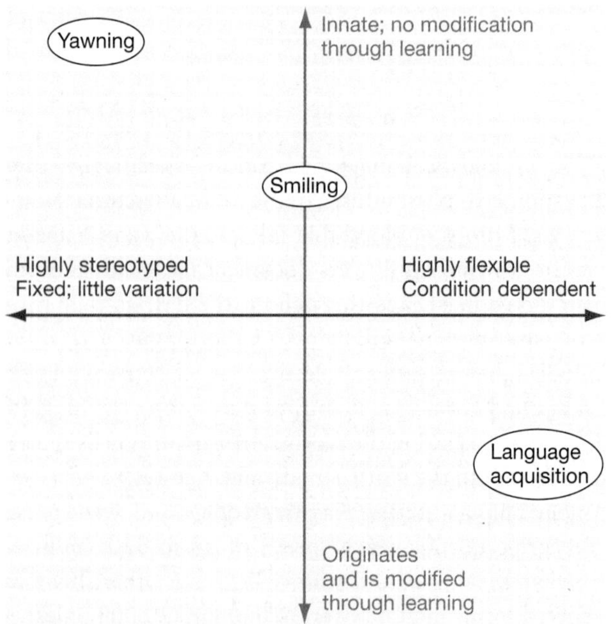

FIGURE 51.2 Behavioral Traits Vary in Flexibility and Dependence on Learning. These axes represent one way to organize the diversity of behavioral traits. Each type of behavior observed in a species could be plotted at a particular point on the graph.

The horizontal axis maps how flexible a behavior is in terms of its expression within and between individuals of the same species. This axis forms a continuum ranging from fixed, invariant, behaviors (often referred to as stereotyped behaviors) to highly flexible and variable responses. The vertical axis plots how much learning is involved in the nature of the response. This axis ranges from behaviors that require little or no learning for normal expression to responses that are highly dependent on learning. In humans, yawning would map in the upper left-hand corner of this graph. Language acquisition would be plotted in the lower right-hand corner.

When researchers first began to explore animal behavior experimentally, they focused on responses that are highly stereotyped and largely unlearned. This general research strategy is common in biological science. When biologists start to analyze a question—ranging from how DNA is transcribed to how plasma membranes work—they start by studying the simplest possible system. Once simple systems are understood, investigators can delve into more complex situations and questions. Behavioral biology started the same way. Early work focused on inflexible behavioral responses—those that map to the upper left-hand corner of the graph in Figure 51.2. Let's look at some conclusions from these pioneering studies of stereotyped responses, then consider examples of work on more flexible types of behavior.

Innate Behavior

Highly inflexible behavior patterns are called **fixed action patterns**, or **FAPs**. FAPs are stereotyped—meaning that they are performed in the same way every time—and are usually triggered by simple stimuli called **releasers** or **sign stimuli**. As an example, consider what happens in a darkened environment when a kangaroo rat hears the sound of a rattlesnake's rattle, or when a person hears a piercing scream. In both cases, a simple sign stimulus releases a FAP—a rapid jumping movement backward, away from the direction of the stimulus. In both cases, the FAP may result in survival. But kangaroo rats and humans don't learn to do this jump-back response, and it is performed in virtually the same way in response to the same stimulus. FAPs are examples of what biologists call **innate behavior**—behavior that is inherited and shows little variation based on learning or the individual's condition. (**Box 51.1** explores the role of genetics in behavior in more detail.)

It is common to observe innate behavior in response to (1) situations that have a high impact on fitness and demand a

BOX 51.1 A Closer Look at Behavior Genetics

Biologists explore the genetic bases of behavior to understand the proximate mechanisms responsible for specific behavioral traits. Are certain alleles associated with certain types of behavior? If so, how do the products of these alleles change what an individual does in response to certain stimuli?

To explore how biologists are answering these questions, consider the first gene shown to affect feeding behavior. When Marla Sokolowski was working as an undergraduate research assistant, she noticed that some of the fruit-fly larvae she was studying tended to move after feeding at a particular location, while others tended to stay put. By rearing these "rovers" and "sitters" to adulthood, breeding them, and studying the offspring, she was able to determine that the trait is inherited. Using information from the *Drosophila melanogaster* genome sequence (see Chapter 20) and genetic mapping techniques introduced in Chapter 19, Sokolowski found and cloned the gene responsible for the rover-sitter difference. She named it *foraging* (*for*).

Follow-up work showed that the protein product of *for* is involved in a signal transduction cascade (see Chapters 8 and 47)—meaning it is involved in the response to a signal. Rovers and sitters behave differently because they have different alleles of the *for* gene. The difference in behavior is apparent only in the presence of food, however.

By altering population density experimentally and documenting the reproductive success of rovers and sitters, Sokolowski has shown that the rover allele is favored when population density is high, but that the sitter allele reaches high frequency in low-density populations. These results are exciting because they link a proximate mechanism with an ultimate outcome—in this case, the presence of certain alleles is responsible for a difference in fitness in specific types of habitats. Genes that are homologous to *for* have now been found in honeybees and the roundworm *Caenorhabditis elegans*. Currently, researchers are focused on finding out which signaling pathway uses the *for* gene product and why different alleles lead to different types of feeding behavior.

Although the *for* gene presents a compelling story, biologists recognize that it is rare for behavioral traits to be so heavily affected by a single gene product. In reality, most aspects of behavior are influenced by the action of many genes. (Think of the number of genes that are likely to be involved in forming the human brain and thus in aspects of human personality and decision making.) To use a term that was introduced in Chapter 13, most behavioral traits are polygenic. Instead of representing discrete states such as roving and sitting, polygenic traits exhibit quantitative variation—meaning that individuals differ by degree and that many different phenotypes are observed when the population is considered as a whole.

reflex-like, unlearned response, or (2) situations where learning is not possible. Web-weaving in spiders and nest-building in birds are largely innate, because offspring have little or no opportunity to watch adults perform these behaviors and learn from them. In addition, the quality of a web or a nest has a high impact on fitness in spiders and birds—individuals need to get it right the first time.

Conditional Strategies and Decision Making

Although all species studied to date show some degree of innate behavior, it is much more common for an individual's behavior to change in response to learning and to show flexibility in response to changing environmental conditions. Animals are not stimulus-response machines. Instead, they take in information from the environment and, based on that information, make decisions about what to do. Animals make choices.

It is generally assumed that animals do not make conscious decisions. In most cases, though, biologists know relatively little about the proximate mechanisms involved in flexible responses and decision making. Fortunately, flexibility in behavior is well understood at the ultimate level. The guiding principle here is what biologists call cost-benefit analysis. Animals appear to take in information about their environment and weigh the costs and benefits of responding in various ways. Costs and benefits are measured in terms of their impact on fitness—the ability to produce offspring. As the following three examples show, individuals usually choose the option that maximizes their fitness.

What Decisions Do White-Fronted Bee-Eaters Make When Foraging? Given a choice, what should an animal eat? Biologists answered this question by assuming that individuals should forage in a way that maximizes the amount of usable energy they take in, given the costs of finding and ingesting their food and the risk of being eaten while they're at it. This claim—that animals should maximize their feeding efficiency—is called **optimal foraging.**

To test whether optimal foraging actually occurs, a researcher began studying feeding behavior in a bird called the white-fronted bee-eater (**Figure 51.3a**). Mated pairs of this species dig tunnels in riverbanks and raise their young inside. Appropriate riverbanks are rare, so many pairs build tunnels in the same location, forming a colony. To feed their young, pairs have to fly away from the colony, capture insects, and bring the prey back to the nest. Each pair defends a specific area where only they find food.

The key observation here is that some individuals forage a few meters from the colony, while some have to look for food hundreds of meters away. It takes only a few seconds to make a round trip to the closest feeding territories versus several minutes to the farthest territories. Thus, the fitness cost of each feeding trip varies widely among individuals.

(a) White-fronted bee-eaters are native to East Africa.

(b) Foraging behavior depends on distance traveled.

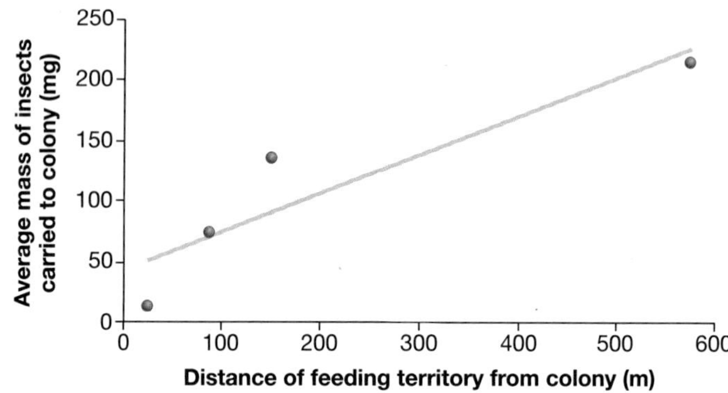

FIGURE 51.3 White-Fronted Bee-Eaters Vary Their Foraging Behavior Depending on Conditions.

If optimal foraging occurs, how do individuals maximize their benefits, given the cost they have to pay in time and energy? Individuals that have to find food far from the colony stay away longer on each trip. And as the data in **Figure 51.3b** show, individuals foraging far from the colony bring back a much larger mass of insects on each trip, on average, than do individuals that fly a short distance each time.

White-fronted bee-eaters appear to make decisions that maximize the energy they deliver to their offspring, given their costs in finding food. This behavior is highly flexible and condition dependent. If the same individual had a territory close to the colony instead of far away, it would make different decisions. Its behavior would change, but in each case it would be acting in a way that maximizes its fitness.

How Do Female Barn Swallows Choose Mates? Given a choice, who should an individual choose for a mate? Chapter 25 presented data indicating that females are usually the gender that is pickiest about mate choice, and that females choose males that contribute good alleles and/or resources to their offspring.

To see this type of decision making in action, consider an experiment on mate choice by female barn swallows native to Denmark. Although both male and female barn swallows help to build the nest and feed the young, the species exhibits a significant amount of sexual dimorphism—males are slightly

FIGURE 51.4 Barn Swallows Are Sexually Dimorphic.

larger and more brightly colored, and their outer tail feathers are about 15 percent longer than the same feathers in females (**Figure 51.4**). Research on other bird species has shown that individuals with particularly long tails, bright colors, and energetic courtship displays are usually the healthiest and best fed in the population (see Chapter 25). Recent work has shown that barn swallows with particularly long tails are more efficient in flight and thus more successful in finding food. If a female barn swallow chooses to mate with a long-tailed male, then she should be picking an individual that will pass high-fitness alleles on to her offspring *and* be able to help her rear them efficiently.

A researcher tested this prediction by altering tail length experimentally—in this case, by cutting off tail feathers and re-gluing them. The operation is painless and does not appear to affect flying ability. As the "Experimental Setup" section of **Figure 51.5** shows, the biologist caught a large series of males and randomly assigned them to one of four groups: shortened tails, "sham-operated" unchanged tails, untouched tails, and elongated tails. When he documented how long it took males in each group to find a mate, a strong trend emerged: Long-tailed males mated earliest—meaning that females preferred

Experiment

Question: Does tail length in barn swallows affect female choice of mates?

Hypothesis: Females prefer to mate with the longest-tailed males.

Null hypothesis: Tail length in males has no influence on female choice of mates.

Experimental setup:

Capture many males at the same site and randomly assign them to 1 of 4 groups:

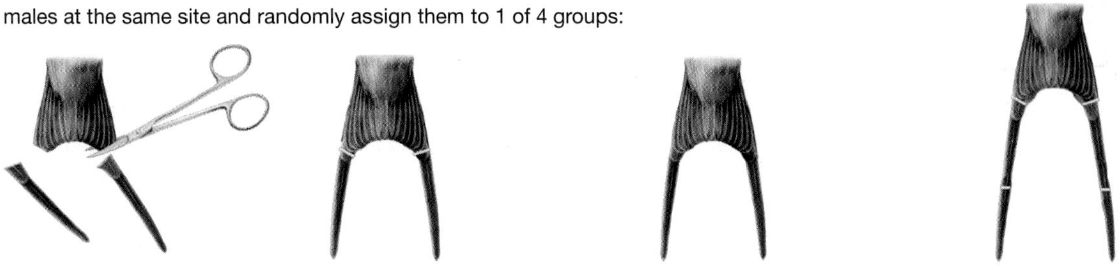

Short tail: Outermost tail feathers cut

Control 1: Outermost tail feathers cut and glued back in place

Control 2: Males caught and handled, but no change in tail feathers

Elongated tail: Outermost feathers cut and elongated with feathers from other males, glued in place

Release males and document which males are first to mate, have a second nest that season, and raise the most offspring.

Results:

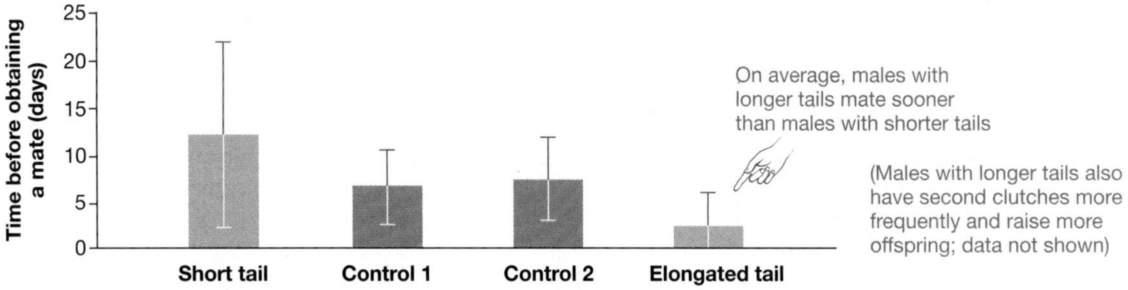

Time before obtaining a mate (days)

0, 5, 10, 15, 20, 25

Short tail, Control 1, Control 2, Elongated tail

On average, males with longer tails mate sooner than males with shorter tails

(Males with longer tails also have second clutches more frequently and raise more offspring; data not shown)

Conclusion: Females prefer to mate with long-tailed males.

FIGURE 51.5 Barn Swallow Females Prefer to Mate with Long-Tailed Males.

QUESTION Why did the investigator bother to artificially lengthen and shorten tails? Why didn't he just record natural variation in tail length and observe mating success?

QUESTION Why did he employ two controls?

them. This finding is important because pairs that nest early have time to complete a second nest later in the season and thus produce the most offspring.

Data on mate choice in barn swallows reinforce a central theme of this chapter: Animals make decisions in a way that maximizes their fitness.

Why Do Some Bluehead Wrasses Undergo a Sex Change? Many species of coral-reef fish have a distinctive mating system. Males defend territories that contain nesting sites and feeding areas. A group of females live inside the boundaries of this territory. When these females lay eggs, the male fertilizes the eggs. Thus, a single male monopolizes all the matings in that territory. Invariably, this male is the largest individual in the group. This is logical because the male guards the territory, and because fights between fish are usually won by the biggest contestant.

When the territory-holding male dies, however, something unusual happens. The largest female in the group changes sex. Her reproductive organs change from egg production to sperm production, and her coloration changes (**Figure 51.6**). She becomes the dominant male and begins fertilizing all of the eggs laid in the territory.

Why does she do this? An idea called the size-advantage hypothesis states that if a group of fish are living in a territory dominated by a single male, and if that male dies, then females should switch from female to male if their body size is very large. The hypothesis is based on the observation that fish have indeterminate growth—meaning they continue growing throughout their lives.

To understand the logic behind the size-advantage hypothesis, suppose that a small female fish can lay 10 eggs a year but a large female of the same species can lay 20 eggs a year. If six small females and two large females live in a harem, the male that owns the territory fertilizes 100 eggs each year. If the male dies, the largest female can increase the number of offspring she produces each year from 20 to 80 by changing sex and taking over the role of dominant male. The switch is costly in terms of time and energy, but the benefit is large. There is no fitness advantage for smaller females to switch sex, though, because they would be defeated in fights and have 0 offspring per year instead of 10.

Like foraging in white-fronted bee-eaters, sex change in bluehead wrasses is flexible and condition dependent. If conditions don't change, females remain female. If conditions change, the same individual will stay female if she is small but become a male if she is large. In many cases animals have alleles that make a wide range of behavior possible; what an individual actually does is based on decisions that change, depending on conditions.

What other types of factors lead to flexible behavior? In many species, learning is crucial. Chapter 45 showed that at the proximate level, learning occurs because individual neurons and connections between neurons are modified in response to experience. For example, recall that in the sea slug *Aplysia*, the production of the neurotransmitter serotonin changes in response to experience. Changes in serotonin release lead to changes in the behavior of sea slugs in response to attacks by predators. Learning is usually based on changes in neurons.

Learning is a fundamentally important phenomenon in animal behavior. Let's take a closer look at learning from both proximate and ultimate perspectives. How does the process occur, and why?

FIGURE 51.6 Many Species of Coral-Reef Fish Can Change Sex, Depending on Conditions. Bluehead wrasses: a group of females surround a male. Bluehead wrasses start life as females but can change sex and become male.

Check Your Understanding

If you understand that...

- Within a given species, behavioral traits may range from highly stereotyped to highly flexible and from innate to learned.
- Fixed action patterns (FAPs) are highly stereotyped, innate responses released by simple stimuli.
- Foraging, mate choice, and sex change are examples of flexible decision-making. Animals usually make decisions that maximize their fitness, given their current environmental conditions.

You should be able to...

1) Think of an example of a FAP and a condition-dependent behavior in a species that you are familiar with.
2) Pose hypotheses to explain why each type of behavior you just listed is adaptive.

51.2 Learning

Learning occurs when behavior changes in response to specific life experiences. Learning is particularly important in species that have large brains and a lifestyle dominated by complex social interactions—including humans, dolphins, chimps, and crows. In species such as these, FAPs and other types of inflexible, stereotyped behaviors are relatively rare. Instead, each individual is capable of a wide range of behavior. What an individual actually does is condition dependent or learned.

An enormous amount of research has been done on learning. This section touches on only a few highlights, to provide examples of what is known and illustrate how biologists go about studying it.

Simple Types of Learning: Classical Conditioning and Imprinting

Learning can be simple. For example, dogs salivate in response to the sight or smell of food. Ivan Pavlov discovered that if he rang a bell each time food was presented to a dog, the dog would learn to associate the sound of the bell with food. When stimulated by a bell in the absence of food, the dogs Pavlov was training would begin to salivate. This type of learning is called classical conditioning. In **classical conditioning**, individuals are trained by experience to give the same response to more than one stimulus—even a stimulus that has nothing to do with the normal response.

Another type of simple learning takes place in newly hatched ducks, geese, and certain other species of birds. Upon hatching, ducklings and goslings adopt as their mother the first moving thing they see. This type of learning is called **imprinting**. In nature, the "thing" that creates the imprint is the mother duck or goose. But Konrad Lorenz found that if he incubated eggs artificially until they hatched, he could be the first thing that the offspring saw. In response, young greylag geese would imprint on whatever boots he was wearing at the time (**Figure 51.7**). In contrast, mallard ducklings would imprint on him only if he crawled on all fours and quacked continuously.

Why does imprinting occur? At the ultimate level of causation, the leading hypothesis is that offspring must quickly learn to recognize and respond to their mother in order to survive. This explanation is most relevant to species that nest on the ground, such as ducks and geese. Ducklings and goslings are easy prey for foxes, raccoons, and other predators, so mother ducks and geese lead their offspring away from the nest to the safety of nearby water almost immediately. In these bird species, the ability to recognize a moving object and follow it immediately should increase the fitness of offspring.

Lorenz was also able to establish that imprinting occurs only in the early life of the animal, during a short interval called the **critical period** or **sensitive period**. In addition, his research showed that imprinting lasts for life and may establish not only the identity of the offspring's mother but also its species identity.

FIGURE 51.7 Imprinting Occurs in Some Species of Birds and Lasts for Life. Goslings imprint on the first moving object they see—in this case, researcher Konrad Lorenz's boots.

The key characteristics of imprinting—that it is fast and irreversible and occurs during a critical period—are not typical of most types of learning. Yet recent research has shown that imprinting and human language acquisition share some common characteristics. You may have personal experience with how quickly children learn a foreign language relative to adults. This observation suggests that there might be something like a critical period for language learning. The commonplace observation that babies learn the language of their parents is also important. The babbling noises that babies make are thought to be imitations of words and sounds uttered by their parents. Children who are born deaf cannot hear these sounds. Learning a spoken language may be almost impossible for these individuals, even with intensive training later in life. In contrast, people who become deaf as adults experience relatively slight changes in their speech. These data suggest that auditory feedback is important during the learning process and that language acquisition should be presented as shifted to the left in the graph depicted in Figure 51.2. Let's explore the topic of language learning further by considering research on how birds learn to sing.

More Complex Types of Learning: Birdsong

Birds sing to attract mates and to mark territories where they find food and raise offspring. In most cases, bird songs are unique to each species. Do young birds learn their species-specific song or suite of songs from their parents, or are songs innate? To answer these questions, biologists perform experiments to control the sound stimuli that birds hear over the course of their lives. The experimental protocols involve taking eggs from the nest, hatching them, and rearing individuals in the laboratory in a controlled auditory environment.

The results of these experiments show that, depending on the species, song-learning behavior falls at various locations on the learning continuum in Figure 51.2. For example, if young chickens or phoebes—birds that have simple vocalizations—are raised in isolation from other members of their species and never hear their species-specific calls, they still produce the correct vocalizations as adults. Song-learning behavior is innate in these species and may be highly stereotyped.

In contrast, white-crowned sparrows that are raised in isolation do not sing a normal song unless they hear a tape recording of their species-specific song during the early months of life. In this species—which sings a complex song—singing is learned during a critical period. Yet if researchers play the song of a song sparrow or other closely related species during the critical period for language acquisition, white-crowned sparrow chicks do not learn it. Instead they learn an abnormal song differing from that of their own species and from that of the alien introduced species. In this species, the ability to learn songs is restricted to songs of their own species (**Figure 51.8**).

In addition, the critical period for song learning in white-crowned sparrows is augmented by practice that occurs when the individual is older. White-crowned sparrows don't begin to sing until they are almost one year old, during the spring. Beginning singers produce a disorganized warbling called sub-song. With practice, this sub-song becomes more and more organized and progressively closer to adult song. Eventually an adult song develops and is "crystallized," meaning that it remains unchanged for the rest of the individual's life. For this improvement to occur, though, the individual must hear itself sing. Individuals that are prevented from hearing their own practice never sing a normal song. Yet if the same individual is deafened after it has developed adult song, it continues to sing a perfectly normal song. These data suggest that individuals have two critical periods for song learning during their lives: one as a nestling, when they must hear the song of their own species, and a second as a 1-year-old, when they must hear themselves practice.

To summarize, singing in white-crowned sparrows is heavily influenced by learning. But in this species, learning is constrained to certain periods of life and occurs only in response to certain types of stimuli.

If chickens are at one end of the learning continuum and if white-crowned sparrows are in the middle, then species such as mynahs, mockingbirds, lyrebirds, and some types of parrots occupy the other end. Individuals of these species continue to learn new songs throughout their lives by imitating sounds in their environment. They have relatively few age-related or species-specific limitations on song learning. "Polly-wanna-cracker" is not a species-specific call in parrots, and the ability to use this call does not depend on hearing it during a critical period early in life.

One of the great challenges that remains for research on song learning is to explain why so much variation exists in song-learning ability and in the size of song repertoires. Why is it adaptive for chickens to make a few innate calls while mynahs sing dozens of learned songs? Research continues.

Can Animals Think?

Research on classical conditioning, imprinting, and song learning has some important take-home messages: Several types of learning exist, and the ability to learn varies widely among species. It is clear that simple types of learning are widespread in animals and that at least some species can learn a great deal by imitation and practice. But can organisms other than humans think?

Cognition is defined as the recognition and manipulation of facts about the world, combined with the ability to form concepts and gain insights. Unlike the results of classical conditioning, imprinting, and song learning, the results of cognition

FIGURE 51.8 White-Crowned Sparrows Have a Critical Period for Song Learning Early in Life. To sing a normal song, white-crowned sparrows have to hear the song of their own species while they are chicks. They also must hear themselves practice their own song when they begin singing at age 1 year.

cannot be observed directly. Instead, to infer that animals are thinking, researchers must design experimental situations that require animals to manipulate facts or information and demonstrate an ability to form novel associations or insights. As an example of work on animal cognition, consider recent experiments on tool use by New Caledonian crows.

New Caledonian crows routinely make and use tools to find food in the wild. Each of these tools is held in the beak and used to remove insect larvae (grubs) or other types of food from plant stems or from crevices (**Figure 51.9a**). Biologists have documented that the crows manufacture the following general types of tools:

- Straight or curved sticks are broken off and cleaned before use. Several other bird species—including one of the Galápagos finches introduced in Chapter 41—are known to modify sticks or plant stems to "fish" for insects. Chimpanzees also make these types of tools. Recent laboratory experiments have shown that when New Caledonian crows are presented with an array of straight sticks along with a piece of food in a clear plastic tube, the crows select a stick whose width or length is appropriate for the tube's width and the food's distance from the end of the tube. The crows then insert the stick into the tube to fish out the food.

- Hooks are made by breaking complex leaves into pieces (**Figure 51.9b**).

- Spearing tools are constructed from a complex series of cuts and rips, made with the beak, along the edge of leaves from pandanus plants (**Figure 51.9c**).

(a) Straight sticks are used to fish or pry food from crevices.

(b) Hooks are made by breaking parts off leaves.

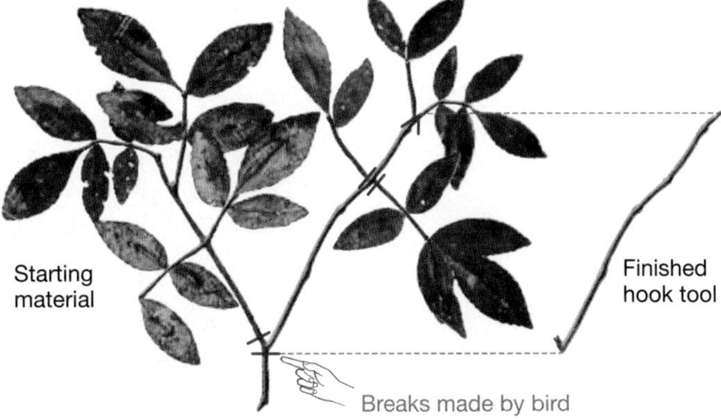

(c) Several cuts and tears are used to make spearing tools.

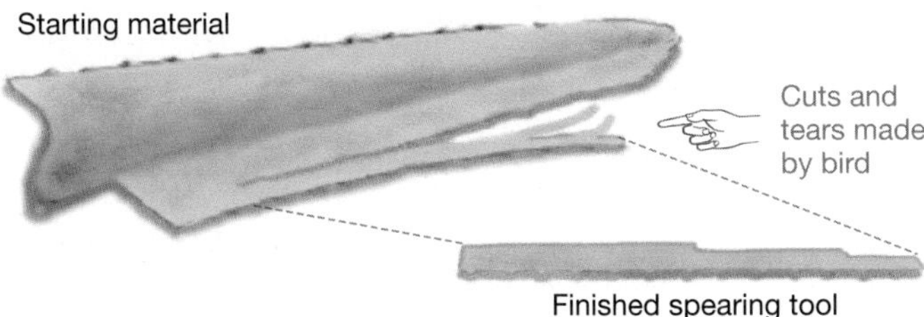

FIGURE 51.9 New Caledonian Crows Make and Use Several Types of Tools.

These observations support the hypothesis that crows can think. More specifically, crows appear to understand facts about the size and shape of raw materials and the location of food. Crows recognize that if they choose or manipulate materials in a certain way, they can use the resulting structure as a tool to acquire the food. The idea is that natural selection has favored the evolution of cognition in this species, because crows with the ability to manufacture and manipulate tools have higher fitness than do crows that lack this ability.

As **Figure 51.10** shows, the crows-can-think hypothesis was supported by recent experiments with two New Caledonian crows in captivity. The experimental setup consisted of a small bucket, containing food, in the bottom of a clear tube that was open at the top, with the handle of the bucket pointing up. In an initial experiment, a male crow and a female crow were given a choice of straight or bent wires to use in retrieving the food from the inside of the tube. The goal of the experiment was to test the hypothesis that the crows can distinguish the two shapes and associate the bent shape with the ability to lift the bucket and acquire the food. During the fifth trial with the apparatus and the choice of two wires, the male happened to remove the hooked wire from the area. Apparently in response, the female picked up the straight wire in her beak, bent it, and successfully used it to lift the bucket and obtain the food.

To confirm the validity of this observation, the researchers gave the pair a straight wire and let them use it until they either successfully retrieved the food or dropped the wire into the tube and were unable to retrieve it. Over the course of 17 trials, the male or female dropped the wire into the tube and was unable to retrieve it 7 times. Once the male was able to lift the bucket out and obtain the food using only the unmodified, straight wire. But 9 times, the female bent the straight wire and used it successfully to obtain the food.

These observations are nothing short of astonishing. They imply that the female was able to associate the shape of a wire

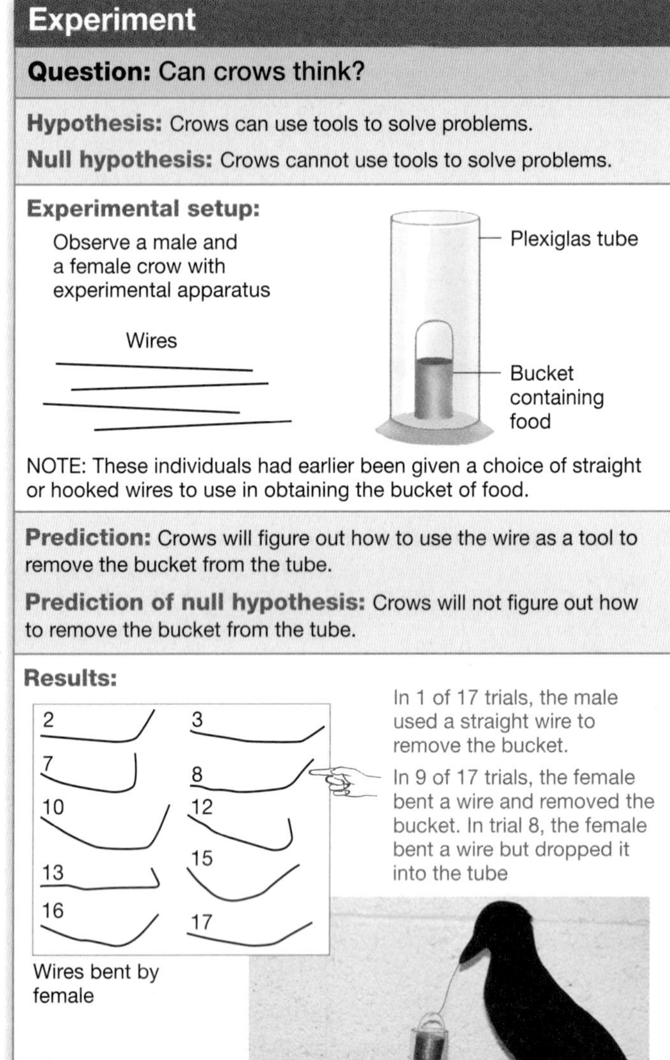

FIGURE 51.10 Experimental Evidence that New Caledonian Crows Understand How to Make Tools. Prior to these experiments, the crows involved had never seen the experimental apparatus or wires of the type available to them. Bending of materials to form hooks has not been observed in the wild.

with the ability to lift the bucket successfully, and that she also had the insight to understand that exerting a force on the wire would bend it into the shape required to be successful. In more colloquial terms, she appeared to understand what sort of tool she needed and how to make one. She had no model to imitate when she made her hooks, however, and she had had no prior experience with bending wires. Instead, she appeared to be thinking.

What Is the Adaptive Significance of Learning?

Biologists view learning as an adaptation that allows individuals to change their behavior in response to a changing and unpredictable environment. Under this hypothesis, the ability to learn varies among species because some species live in environments that are much more unpredictable than others.

In addition, the type of learning that occurs in a given species is correlated with the type of environmental unpredictability it encounters. Norway rats, for example are extremely adept at learning to navigate mazes (**Figure 51.11**). These animals live in sewers, the walls of homes, or other locations where the ability to learn travel routes is essential for survival. Norway rats are also exceptionally good at learning to avoid foods that contain poisons. In nature they feed on a wide variety of fruits, seeds, and insects. The availability of these foods changes with the season, and many potential food items are actually toxic. Thus, it is logical to observe that rats will taste only small quantities of novel foods and that they quickly learn to avoid any foods that induce illness.

Scrub jays, in contrast, are not particularly adept at negotiating mazes or learning to avoid foods. Instead they are proficient at remembering where they have cached (stored) seeds. In this species, spatial learning and a good memory are essential for survival. In addition, scrub jays learn how to prevent theft. This species lives in social groups, and it is not unusual for some members of the group to steal food that was cached by other jays in their group. Experiments have shown that if an individual that has stolen food in the past caches food while being watched by other jays, the individual will come back later, by itself, and move the stored food to a new location.

FIGURE 51.11 Different Species Are Adept at Different Types of Learning. There is often a strong correlation between the type of learning that occurs in a particular species and the problems the species must solve to survive and reproduce.

QUESTION Domestic dogs have been trained by people for thousands of years. Domestic dogs are adept at learning via classical conditioning. What is the relationship between these observations?

To summarize, the ability to learn and the type of learning that occurs vary within and among species. Learning is an adaptation that helps organisms cope with challenges from their environment.

Check Your Understanding

If you understand that...

- The same species of animal may learn in a variety of ways, ranging from simple forms such as imprinting and classical conditioning to imitation to cognition.
- It is common to observe that an individual is capable of learning only certain types of things—such as its species-specific song—only at certain times in its life.

You should be able to...

1) Define learning and give examples of at least two distinct types of learning.
2) Discuss the adaptive value of cognition.
3) Offer a hypothesis to explain why cognition appears to be rare among animals.

51.3 How Animals Act: Hormonal and Neural Control

Most behavior is modified at least slightly by some type of learning, and most animals have behavioral responses that range from highly stereotyped to highly flexible. In many cases, the adaptive value of learning various types of behavior has been documented. But we have yet to explore a question about proximate causation in detail. To understand how biologists go about addressing questions at the proximate level, consider the following: When behavior is flexible and condition dependent, it means that an individual has the potential to behave in a variety of ways. How are changes in behavior implemented and controlled? To answer this question, let's explore how lizards reproduce and how moths escape from bats.

Sexual Activity in *Anolis* Lizards

Anolis carolinensis lives in the woodlands of the southeastern United States. After spending the winter under a log or rock, males emerge in January and establish breeding territories (**Figure 51.12a**). Females become active a month later, and the breeding season begins in April. By May, females are laying an egg every 10–14 days. By the time the breeding season is complete three months later, the eggs produced by a female will total twice her body mass. **Figure 51.12b** maps this series of events, along with the corresponding changes observed in the male and female reproductive systems.

What causes these dramatic seasonal changes in behavior? The proximate answer is sex hormones—testosterone in males and estradiol in females. Testosterone is produced in the testes of males, and estradiol in the ovaries of females. The evidence for effects of these hormones is direct. Testosterone injections induce sexual behavior in castrated males, while estradiol injections induce sexual activity in females whose ovaries have been removed.

This result leaves some key questions unanswered, however. *Anolis* lizards are most successful if they reproduce early in the spring. In springtime their food supplies are increasing, and snakes and other predators are not yet hunting lizards to feed their own young. If testosterone and estradiol levels induced sexual activity in *Anolis* lizards at the wrong time of year, it would be a disaster for the individuals involved. In addition, it is critical for all *Anolis* individuals in a population to start their sexual activity at the same time, so that females can find males that are ready to breed, and vice versa. What environmental cues trigger the production of sex hormones in early spring? How do male and female *Anolis* synchronize their sexual behavior so that they are ready to produce gametes at the same time?

(a) Courtship display of a male *Anolis* lizard

(b) Changes in sexual organs through the year

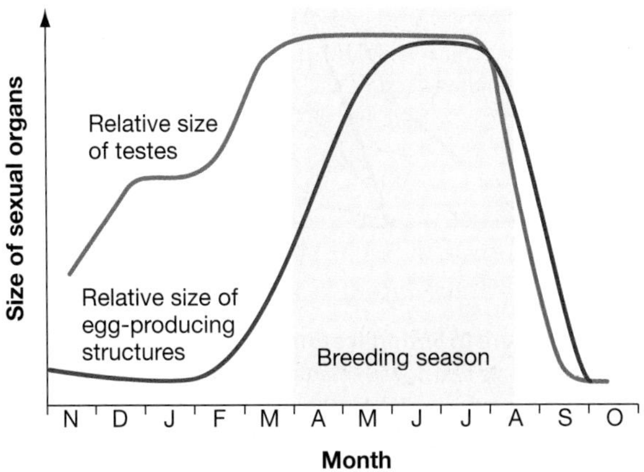

FIGURE 51.12 Sexual Behavior in *Anolis* Lizards.

To answer these questions, a biologist brought a large group of sexually inactive adult lizards into the laboratory during the winter and divided them into five treatment groups. The physical environment was exactly the same in all treatments. Each individual received identical food, and in all treatments high "daytime" temperatures were followed by slightly lower "nighttime" settings. The biologist also continued to monitor the condition of lizards that remained in natural habitats nearby.

To test the hypothesis that changes in day length signal the arrival of spring and trigger initial changes in sex hormones, the biologist exposed the five treatment groups in the laboratory to artificial lighting that simulated the long days and short nights of spring. To test the hypothesis that social interactions among individuals are responsible for synchronizing sexual behavior, the social setting was varied among treatment groups: (1) single isolated females; (2) groups of females only; (3) pairs of lizards: single females, each with a single male; (4) single females, each with a group of castrated (nonbreeding) males; or (5) single females, each with a group of uncastrated (breeding) males.

Each week, the researcher examined the ovaries of females in each group. He also monitored the ovaries of females in nearby natural habitats, since those females were not exposed to springlike conditions. As **Figure 51.13** shows, the differences in the animals' reproductive systems were dramatic. Females that were exposed to springlike conditions began producing eggs; females in the field that were not exposed to springlike conditions did not. But in addition, females that were exposed to breeding males began producing eggs much earlier than did the females placed in the other treatment groups. These results support the hypothesis that two types of stimulation are necessary to produce the hormonal changes that lead to sexual behavior. Females need to experience springlike light and temperatures *and* exposure to breeding males.

What aspect of male behavior causes the difference in female egg production? When males court females, they bob up and down and extend a brightly colored patch of skin called a dewlap. To test the hypothesis that this visual stimulation triggers changes in estradiol production, the investigator repeated the previous experiment but added a twist: He placed some females with males that had intact dewlaps, and other females with males whose dewlaps had been surgically removed. The result? Females grouped with dewlap-less males were slow to produce eggs—just as slow, in fact, as the females in the first experiment that had been grouped with castrated males. These latter females had not been courted at all. The result suggests that the dewlap is a key visual signal. The experiments succeeded in identifying the environmental cues that trigger hormone production and the onset of sexual behavior.

Escape Behavior in Noctuid Moths

Many bats hunt night-flying insects. To find insects in the dark, the bats emit high-pitched pulses of sound and listen for the echo. This process, called **echolocation**, is remarkably acute. If you toss a small stone into the air, a foraging bat will pursue it but then turn away at the last instant. Bats are able to tell the difference between the echos from a stone and an insect.

As **Figure 51.14** shows, bats emit a relatively slow series of pulsed sound when they are searching for prey, switch to a faster series when they are approaching a flying insect, and finish with an extremely rapid sequence as they close in. In this way they get progressively detailed information about the prey's location. The "shouting" involved in echolocation is

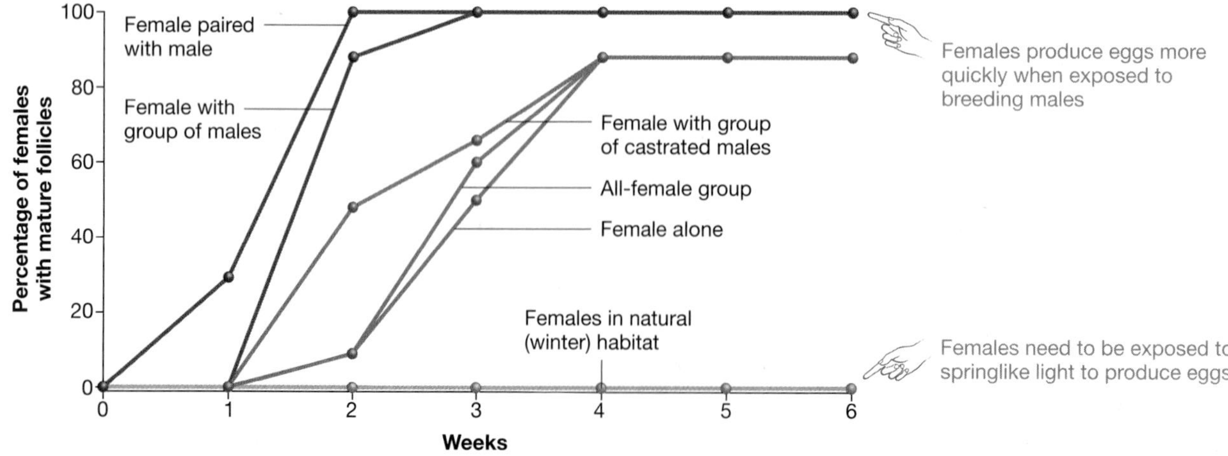

FIGURE 51.13 Exposure to Springlike Conditions and Breeding Males Stimulates Female Lizards to Produce Eggs. The percentage of female lizards with mature follicles, plotted over time. Five treatment groups were exposed to springlike light and temperature; data are also plotted for females left in a natural (winter) habitat. Each data point represents the average value for a group of 6–10 females.

QUESTION Some critics contend that the females that remained outside do not represent a legitimate control in this experiment. What conditions would represent a better control?

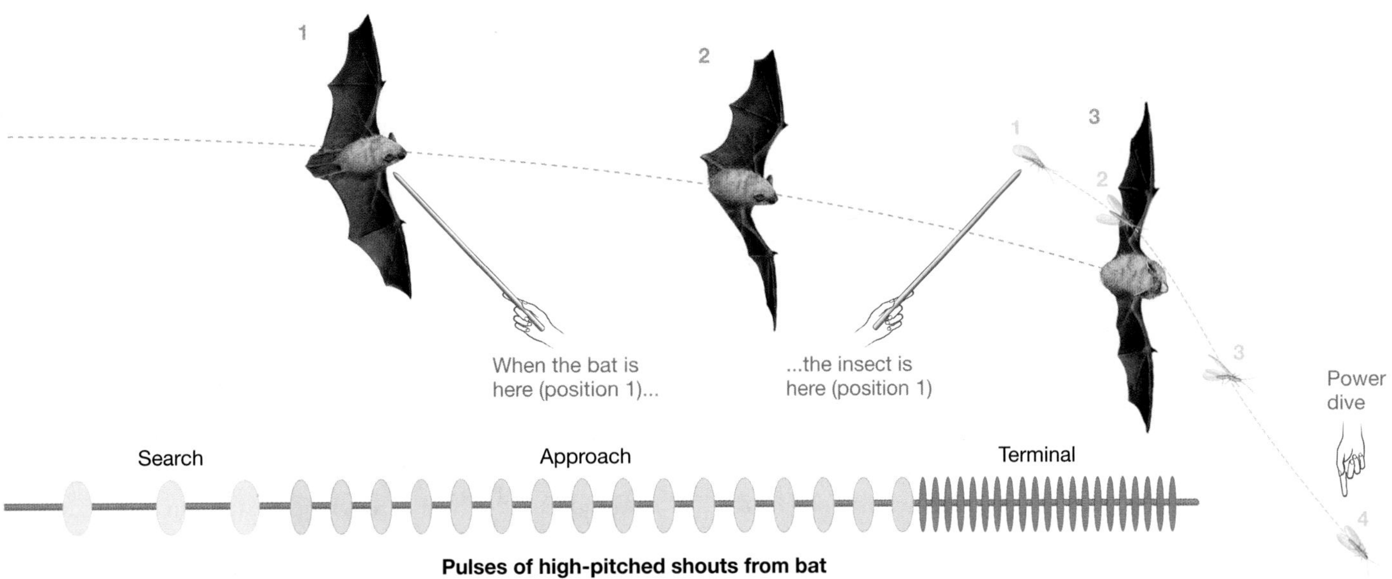

FIGURE 51.14 Bats Hunt By Echolocation. This drawing shows how the pulses of high-pitched sounds change as a bat locates and closes in on a flying insect. To read this diagram, put one finger on the bat at its position 1 and another finger on the insect's position 1. Then change your fingers to the bat's and insect's position 2, then position 3, noting how the bat's calls change as the positions change.

QUESTION Did the bat catch the insect?

energetically expensive, so it is reasonable to observe that bats switch to rapid, high-cost sound production only when it is likely to pay off in a kill.

Humans can't hear the sounds that bats make, but moths and other insects that are preyed on by bats can. When a noctuid moth hears a bat in the distance, it turns and flies away. But if the bat is very close, the moth makes a power dive—it drops from the air like a stone. The adaptive significance of the power-dive behavior is obvious. But at the proximate level, how do moths detect and avoid bats?

Kenneth Roeder did pioneering work on this question. He showed that on either side of their bodies, moths have ears consisting of a tympanic membrane attached to two sensory neurons. When the membrane vibrates in response to bat sounds, the sensory neurons generate action potentials (see Chapter 46). One of the neurons is sensitive to high-pitched sounds of low to moderate loudness, while the other sends action potentials only in response to high-pitched sounds of high intensity. Roeder hypothesized that when the sensory neurons send an extremely rapid sequence of action potentials to the moth brain in response to increasingly loud sounds from an onrushing bat, the moth brain responds by shutting down the neurons that control the flight muscles. The result is chaotic flight or a power dive—desperate behavior that might save the moth's life. In this case, researchers have identified and characterized the neurons responsible for a particularly dramatic example of behavior.

Although a great deal remains to be learned about the hormonal and neural basis for behavior, biologists are making good progress. As techniques advance for measuring hormone concentrations as well as for finding and recording action potentials from individual neurons, the proximate basis for complex behaviors should become increasingly well understood.

51.4 Communication

The song of a bird, the bobbing dewlap of a male *Anolis carolinensis*, and the sentences in this text all have the same overall goal: communication. In biology, **communication** is defined as any process in which a signal from one individual modifies the behavior of a recipient individual. A **signal** is any information-containing behavior. Communication is a crucial component of animal behavior, because it creates a stimulus that elicits a response.

By definition, communication is a social process. For communication to occur, it is not enough that a signal is sent; the signal must be received and acted on. A lizard's bobbing dewlap does not qualify as communication unless another individual sees it and responds to the message. In addition, some biologists maintain that for an event to qualify as communication, the signal that is sent must be intended as a signal. According to these investigators, predicting a person's behavior by watching their body language does not qualify as communication. Similarly, if students unintentionally fall asleep in class and the instructor changes his behavior to wake them up, the exchange does not qualify as communication.

Modes of Communication

The information that organisms communicate may be encoded and delivered in a variety of ways. Communication can be acoustic, as in the song of crickets or birds. It can be visual, as in the color patterns of birds or fish. It can be olfactory, as in scent marking by mice and wolves. It can be tactile, as in communication in some spiders.

One of the most general observations about communication is that the type of signal used by an organism correlates with its habitat. For example, light is quickly dispelled in aquatic habitats, but sound is not. Based on this observation, it is logical to observe that humpback whales rely on songs for long-distance communication. In some cases, humpback whale songs can travel hundreds of kilometers. Groups of whales use acoustic communication to keep together as they move from summer feeding areas to winter breeding grounds, and individual males sing to attract mates and warn rivals. Correlations between habitat and mode of communication are also observed in visual, olfactory, and tactile communication. Animals that are active during the day and that live in open or treeless habitats tend to rely on visual communication during courtship and territorial displays. Bats, wolves, and other animals that are active at night communicate via sound or scent; ants and termites that live underground rely on olfactory and tactile communication.

It is also common to observe several modes of communication being used in conjunction. For example, male red-winged blackbirds establish a breeding territory in the spring by giving a display that combines auditory and visual signals. The birds' display is based on revealing the bright red patches of feathers near their shoulders and giving a loud call. To test the hypothesis that both components of the display are important, researchers have manipulated both acoustic and visual elements. In one experiment, biologists used a speaker to play recordings of red-winged blackbird songs in existing territories (**Figure 51.15a**). In response, the territory-owning males flew toward the speaker and appeared to hunt for it. In a second experiment, researchers made a crude model of a male red-winged blackbird by sewing red patches on a black sock stuffed with rags. When the model was placed in existing territories, territory owners approached the model and inspected it (**Figure 51.15b**). But if the model was presented along with a recorded blackbird song, the resident males attacked the model (**Figure 51.15c**). These experiments support the hypothesis that both auditory and visual information are important in territorial displays of red-winged blackbirds. More specifically, the data indicate that while each type of stimulus alone induces a response, the combination of both types of stimuli provokes a much more powerful change in behavior.

Each mode of communication has advantages and disadvantages. Although acoustic communication can be extremely effective in some habitats, songs and calls are short lived. A red-winged blackbird's red feathers remain in place for at least 6 months, but each call lasts less than 3 seconds. Thus, calls and songs have to be repeated to be effective. Frequent repetition of calls and songs requires a large expenditure of time and energy. In addition, acoustic communication has been shown to attract predators. It is no surprise that when a hawk or falcon approaches a marsh inhabited by red-winged blackbirds, things get very quiet. Communication systems have been honed by natural selection to maximize their benefits and minimize their costs.

(a) Red-winged blackbird call only

(b) Model of male only

(c) Call and model together

FIGURE 51.15 Experimental Evidence that Visual and Auditory Stimuli Are Important in Territorial Behavior. Territory-owning red-winged blackbirds react differently to **(a)** auditory stimuli, **(b)** visual stimuli, and **(c)** auditory plus visual stimuli. Territory owners inspect each stimulus presented in (a) and (b) but attack the combination of the two in (c).

EXERCISE Researchers have clipped the red feathers from male red-winged blackbirds and released the birds into territories. Predict how territory owners respond to these individuals.

Research on animal communication has gone far beyond describing types of communication and documenting correlations between modes of communication and habitat characteristics, however. To appreciate the array of questions that biologists ask about communication and how they go about answering them, let's consider two research programs on animal signaling—one classical and one more recent. The classical work revealed a remarkable form of communication in honeybees; the more recent work has focused on the question of how and why signalers lie.

A Case History: The Honeybee Dance

Honeybees are highly social animals that live in hives. Inside the hive, a queen bee lays eggs that are cared for by individuals called workers. Besides caring for young and building and maintaining the hive, workers obtain food for themselves and other members of the colony by gathering nectar and pollen from flowering plants.

Biologists noticed that bees appear to recruit to food sources, meaning that if a new source is discovered by one or a few individuals, many more bees begin showing up over time. This observation inspired the hypothesis that successful food-finders have some way of communicating the location of food to other individuals. Karl von Frisch suspected that successful food-finders communicate information about food sources when they interact with other workers inside the hive. Because beehives are completely dark, von Frisch hypothesized that communication was tactile in nature.

In the 1930s Von Frisch began studying bee communication by observing bees that built hives inside the glass-walled chambers he had constructed. He found that if he placed a feeder containing sugar water near one of these observation hives, a few of the workers began moving in a circular pattern on the vertical, interior walls of the hive. Von Frisch called these movements the "round dance" (**Figure 51.16a**). Other bee workers appeared to follow the progress of the dance by touching the displaying individual as it danced, and to respond by flying away from the hive in search of the food source.

To investigate the function of these movements further, von Frisch placed feeders containing sugar water at progressively greater distances from the hive. Using this technique, he was able to get bees to visit feeders at a distance of several kilometers from the hive. By catching bees at the feeders and dabbing them with paint, he could individually mark successful food-finders. Follow-up observations at the hive confirmed that marked foragers danced when they returned to the hive. Follow-up observations at the feeders confirmed that marked foragers returned with greater and greater numbers of unmarked bees. To explain these data, von Frisch proposed that the round dance contained information about the location of food and that workers got information from the dance by touching the dancer and following the dancer's movements.

When von Frisch placed feeders at longer distances from the hive, however, he found that successful food-finders no longer did the round dance. Instead, they performed a new type of display. He named these movements the "waggle dance," because they combined circular movements like those of the round dance with short, straight runs (**Figure 51.16b**). During these runs, the dancer vigorously moved her abdomen from side to side.

These observations supported the hypothesis that both the round dance and waggle dance communicated information about food sources. But a key observation allowed von Frisch to push this result further. He noticed that the orientation of the waggle part of the dance varied and that the variation correlated with the direction of the food source from the hive. In addition, he observed that the length of the straight, "waggling" run was

(a) The round dance

(b) The waggle dance

FIGURE 51.16 Honeybees Perform Two Types of Dances. (a) During the round dance, successful food-finders move in a circle. **(b)** During the waggle dance, successful foragers move in a circle but then make straight runs through the circle. During the straight part of the dance, the dancer waggles her abdomen.

proportional to the distance the foragers had to fly to reach the feeder. By varying the location of the food source and observing the orientation of the waggle dance given by marked workers, von Frisch was able to confirm that dancing bees were communicating the position of the food relative to the current position of the Sun. For example, if he placed food directly away from the Sun's current position, marked bees would give the waggle portion of their dance directly downward (**Figure 51.17a**). But if the food was 90 degrees to the right of the Sun, bees would waggle 90 degrees to the right of vertical (**Figure 51.17b**).

Based on these experiments, von Frisch concluded that the difference between the round dance and the waggle dance was a matter of distance. The round dance is used to indicate the presence of food within 80 to 100 m of the hive. The waggle dance is used to indicate the direction and distance to food that is over 100 m from the hive.

These results are nothing short of astonishing. Honeybees do not have large brains, yet they are capable of symbolic language. What's more, they are able to interpret the angle of the waggle dance performed on a vertical surface and to respond by flying horizontally along the corresponding angle. Further work has confirmed that the dance language of bees includes several modes of communication. In addition to the tactile information in the movements themselves, bees make sounds during the dance and give off scents that indicate the nature of the food source.

When Is Communication Honest or Deceitful?

As far as is known, honeybees are always honest. Stated another way, honeybee dances consistently "tell the truth" about the location of food sources. This observation is logical, because the honeybees that occupy a hive are closely related and cooperate closely in the rearing of offspring. As a result, it is advantageous for an individual to convey information accurately. If a food-finder provided inaccurate or misleading information to its hivemates, fewer offspring would be reared and thus the food-finder's fitness would be reduced.

In many cases, however, natural selection has favored the evolution of deceitful communication, or what is commonly called lying. Recent research on deceitful communication has highlighted just how complex interactions between signalers and receivers can be. A few examples will help drive this point home.

Deceiving Individuals of Another Species

Figure 51.18 illustrates a few of the hundreds of examples of deceitful communication that have been documented among members of different species. Many of the best-studied instances involve predation:

- The anglerfish in Figure 51.18a has an appendage that dangles near its mouth and looks remarkably like a minnow. If another fish approaches this "lure" and attempts to eat it, the anglerfish attacks.

(a) Straight runs down the wall of the hive indicate that food is opposite the direction of the Sun.

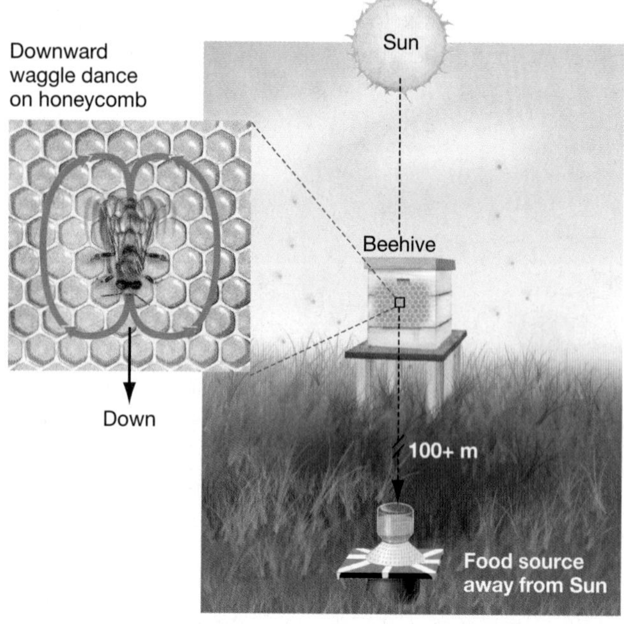

(b) Straight runs to the right indicate that food is 90° to the right of the Sun.

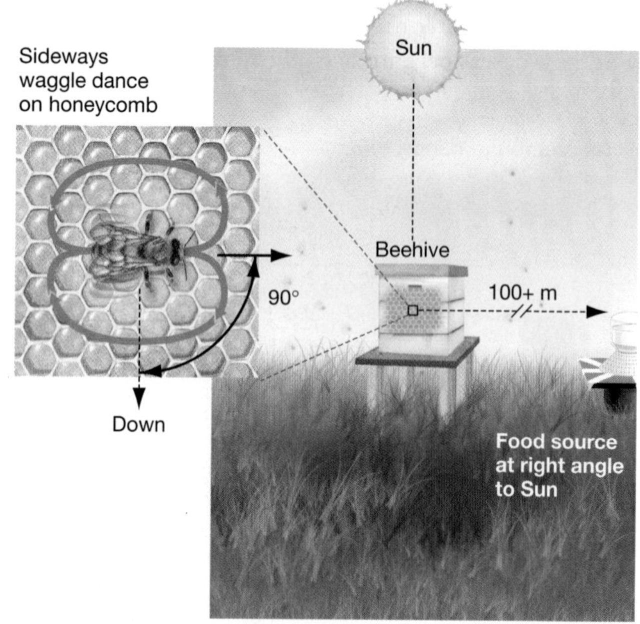

FIGURE 51.17 The Direction of the Waggle Dance Indicates the Location of Food Relative to the Sun. Waggle dances are done only when the food is at least 100 m from the hive.

● **EXERCISE** The length of the straight run in the waggle dance indicates the relative distance of the food. Diagram a waggle dance that indicates food in the direction of the Sun and twice as far away as the food sources indicated by the dances drawn here.

(a) Anglerfish use a "lure" to attract prey.

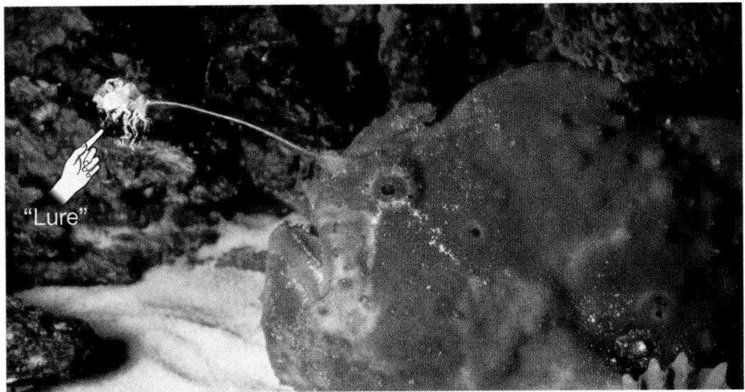

(b) Female *Photuris* fireflies flash the courtship signal of another species, then eat males that respond.

(c) This butterfly looks like a bad-tasting species but actually tastes good.

FIGURE 51.18 Deceitful Communication Is Common in Nature.

- Male and female fireflies flash a species-specific signal to each other during courtship. Predatory *Photuris* fireflies can mimic the pattern of flashes given by females of several other species. A *Photuris* female attracts a male of different species with the appropriate set of flashes, and then attacks and eats him (Figure 51.18b).
- The butterfly shown in Figure 51.18c is highly palatable to birds and other predators. It is rarely attacked, though, because it strongly resembles butterfly species that contain toxins and are highly unpalatable to predators. This phenomenon, known as mimicry, is discussed in more detail in Chapter 53.

In each of these examples, individuals increase their fitness by providing inaccurate or misleading information to members of a different species. Deceitful communication is also known in plants, where it functions in pollination. For example, several orchid species have flowers that look and smell like female wasps and that accomplish pollination by "fooling" male wasps into attempting copulation.

Deceiving Individuals of the Same Species In some cases, natural selection has favored the evolution of traits or actions that deceive members of an organism's own species. When mantis shrimps molt, for example, they lack any external covering and cannot use their large claws. Thus, they are unable to defend the cavities where they live. But if another mantis shrimp approaches with the intent of evicting the individual and taking over the cavity, the molting individual bluffs. It raises its claws in the normal aggressive display and may even lunge at the intruder.

Perhaps the best-studied type of deceit in nonhuman animals, however, involves the mating system of bluegill sunfish. Male bluegills set up nesting territories in the shallow water along lake edges, fertilize the eggs laid in their nests, care for the developing embryos by fanning them with oxygen-rich water, and protect newly hatched offspring from large predators. Some males cheat on this system, however, by mimicking females. To understand how this happens, examine the female, normal male, and female-mimic male in **Figure 51.19**. Female mimic males look like females but have well-developed testes

FIGURE 51.19 In Bluegill Sunfish, Some Males Look and Act Like Females. Female-mimic males have fully functional testes and produce large numbers of sperm.

and produce large volumes of sperm. Mimics also act like females during courtship movements with territory-owning males. They even adopt the usual egg-laying posture. Then when normal females approach the nest and begin courtship, the female-mimic males join in. The territory-owning male tolerates the mimic, apparently thinking that he is successfully courting two females at the same time. But when the actual female begins to lay eggs, the mimic responds by releasing sperm and fertilizing some of them, and then darting away. In this way, the female-mimic male fathers offspring but does not help care for them.

When Does Deception Work? In analyzing deceitful communication in bluegill sunfish and other species, researchers point out that in most cases, lying works only when it is relatively rare. The logic behind this hypothesis is that if deceit becomes extremely common, then natural selection will strongly favor individuals that can detect and avoid or punish liars. But if liars are rare, then natural selection will favor individuals that are occasionally fooled but are more commonly rewarded by responding to signals in a normal way. In mantis shrimps, for example, molting occurs only once or twice a year. Thus, the vast majority of contests over hiding places involve individuals that have an intact exoskeleton and claws that qualify as a deadly weapon. As a result, intruders usually benefit by interpreting threat displays as honest and responding to the stimulus by retreating. If molting were frequent and bluffing behavior common, then intruders would benefit by challenging territory owners, which are likely to be defenseless. Research on deceit and honesty in communication has been extraordinarily productive and continues at a rapid pace.

Check Your Understanding

If you understand that...

- Communication is an exchange of information between individuals.
- In most instances, the mode of communication that animals use maximizes the probability that the information will be transferred efficiently.
- Communication can be honest or deceitful, depending on the intent of the signaler.

You should be able to...

1) Explain the adaptive significance of different modes of communication (e.g. auditory, tactile, visual) and give examples of each.
2) Explain why natural selection favors individuals that can detect deceitful communication and either avoid liars or punish them.

51.5 Migration and Navigation

In response to many stimuli, organisms move. Ants follow odor trails; bees fly toward food sources. A movement that results in a change of position, like the flight paths of hunting bats, escaping moths, and foraging bees, is called **orientation**. The simplest type of orientation is termed **taxis**, and it involves positioning the body, or part of the body, toward or away from a stimulus. A moth attracted to a porch light is an example of a positive **phototaxis**—orientation toward light. A female cricket approaching a calling male is an example of a positive **phonotaxis**—orientation toward sound. A person who retreats from the sound of a siren blaring is an example of negative phonotaxis.

Orienting movements cover fairly short distances. But in some cases, bats and other species travel thousands of miles in search of places to feed and breed. Why?

Why Do Animals Move with a Change of Seasons?

In ecology, **migration** is defined as the long-distance movement of a population associated with a change of seasons. A few examples will bring home the simple point that migratory movements can be spectacular in their extent and the navigation challenge they pose:

- Arctic terns nest along the Atlantic coast of North America in warm months, fly south along the coast of Africa to wintering grounds off Antarctica, and then fly back north along the eastern coast of South America. Chicks follow their parents south but make the return trip on their own. The journey totals over 32,000 km (20,000 miles).
- Many of the monarch butterflies native to North America spend the winter in the mountains of central Mexico or southwest California. Tagged individuals are known to have flown over 3000 km (1870 miles) to get to a wintering area. In spring, monarchs begin the return trip north. They do not live to complete the trip, however. Instead, mating takes place along the way, and females lay eggs en route. Although adults die on the journey, offspring continue to head north. After several generational cycles of reproduction and death in the original habitats in northern North America have passed, individuals again migrate south to overwinter. Instead of a single generation making the entire migratory cycle, then, the round trip takes several generations.
- Salmon that hatch in rivers along the Pacific Coast of North America and northern Asia migrate to the ocean when they are a few months to several years old, depending on the species. After spending several years feeding and growing in the North Pacific Ocean, they return to the stream where they hatched. There they mate and die.

In most cases, it is relatively straightforward to generate hypotheses for why migration exists at the ultimate level. Arctic terns feed on fish that are available in different parts of the world at different seasons. Thus, individuals that migrate achieve higher reproductive success than do individuals that do not migrate. Monarch butterflies may achieve higher reproductive success by migrating to wintering areas and new breeding areas than by trying to overwinter in northern North America. Salmon eggs and young are safer and thrive better in freshwater habitats than they do in the ocean, but adult salmon can find much more food in saltwater habitats than in freshwater.

At the proximate level, however, explaining migratory movements is often extremely difficult. How do all of these animals find their way? What cues guide them on these immense journeys? The surprising answer is that in many or even most cases, biologists don't really know.

How Do Animals Find Their Way?

To organize research into how animals find their way during migratory movements, biologists distinguish three categories of navigation: (1) **Piloting** is the use of familiar landmarks, (2) **compass orientation** is movement that is oriented in a specific direction, and (3) **true navigation** is the ability to locate a specific place on Earth's surface. A thought experiment will help you understand these categories. Suppose you grew up on the shores of Lake Superior and were transported to the middle of this enormous lake. Because you were blindfolded and made to spin in place en route, you are disoriented and have no idea where you are. But you are given a magnetic compass. The compass tells you where north, south, east, and west are. Unfortunately this information will not help you find home, because you have no idea where you are in relation to home. You are stuck because you have no mechanism of true navigation. Now suppose that a helicopter flying overhead drops you a map. The map has an X marking your current position and an H marking your home. The map also has a scale and a compass symbol. From this information, you determine that home is 100 miles to the west. The map has solved your navigational problem. The compass has solved your orientation problem. Now that you know you live to the west, you can use the compass to find west and paddle home. As you approach the shore, you recognize familiar landmarks. Piloting gets you home.

Organisms other than humans do not have magnetic compasses or printed maps. How do they navigate? Although biologists know little about the nature of a map sense in animals, piloting and compass orientation are increasingly well understood.

Piloting A substantial amount of data suggests that at least some migratory animals use piloting to find their way. In some species of migratory birds and mammals, offspring follow their parents south in the fall and north in the spring. Young appear to memorize the route. Piloting even plays a role in species with more sophisticated navigational abilities. Homing pigeons, for example, can find their way home even when they are released in strange terrain a long distance away. But if these birds are equipped with frosted spectacles so that they see the world as a foggy haze, they return to within a mile or so of their home but do not actually find it. Based on these experiments, researchers have concluded that homing pigeons use piloting to navigate the final stages of a journey home.

Compass Orientation How do animals perform compass orientation? To date, most research on compass orientation has been done on migratory birds. To determine where north is, these animals appear to use the Sun during the day and the stars at night. Let's consider each type of compass orientation separately, because they work differently.

The Sun is difficult to use as a compass reference, because its position changes during the day. It rises in the east, is due south at noon (in the Northern Hemisphere), and sets in the west. To use the Sun as a compass reference, then, an animal must have an internal clock that defines morning, noon, and evening. Fortunately, most animals have such a clock. The **circadian clock** that exists in organisms maintains a 24-hour rhythm of chemical activity. The clock is set by the light–dark transitions of day and night. It tells individuals enough about the time of day that they can use the Sun's position to find magnetic north.

The situation is actually simpler on clear nights, because migratory birds in the Northern Hemisphere can use the North Star to find magnetic north and select a direction for migration. But what if the weather is cloudy and neither the Sun nor stars are visible? Under these conditions, migratory birds appear to orient using Earth's magnetic field. Exactly how they detect magnetism is under debate. One hypothesis contends that the birds can detect magnetism by their visual system, through an unknown molecular mechanism. An alternative hypothesis maintains that individuals have small particles of magnetic iron—the mineral called magnetite—in their bodies. Changes in the positions of magnetic particles, in response to Earth's magnetic field, could then be detected and provide reliable information for compass orientation.

Although research on mechanisms of compass orientation continues, one important point is clear: Birds and perhaps other organisms have multiple mechanisms of finding a compass direction. At least some species can use a Sun compass, a star compass, and a magnetic compass. Which system they use depends on the weather and other circumstances.

(MB) Web Animation at www.masteringbio.com
Homing Behavior in Digger Wasps

51.6 The Evolution of Self-Sacrificing Behavior

The types of behavior reviewed thus far—including FAPs, learned behaviors, honest and deceitful communication, and migration—all have a key common element: They help individuals respond to environmental stimuli in a way that increases their fitness. There is a type of behavior that appears to contradict this pattern, however: altruism. **Altruism** is behavior that has a fitness cost to the individual exhibiting the behavior and a fitness benefit to the recipient of the behavior. It is the formal term for self-sacrificing behavior. Altruism decreases an individual's ability to produce offspring but helps others produce more offspring.

The existence of altruistic behavior appears to be paradoxical, because if certain alleles make an individual more likely to be altruistic, those alleles should be selected against. Why does it occur?

Kin Selection

Even though it is not possible for behavior to evolve for the good of the species, self-sacrificing behavior does occur in nature. For example, black-tailed prairie dogs perform a behavior called alarm calling (**Figure 51.20**). These burrowing mammals live in large communities, called towns, throughout the Great Plains region of North America. When a badger, coyote, hawk, or other predator approaches a town, some prairie dogs give alarm calls that alert other prairie dogs to run to mounds and scan for the threat. Giving these calls is risky. In several species of ground squirrels and prairie dogs, researchers have shown that alarm-callers draw attention to themselves by calling and are in much greater danger of being attacked than non-callers are.

How can natural selection favor the evolution of self-sacrificing behavior? William Hamilton answered this question by creating a mathematical model to assess how an allele that contributes to altruistic behavior could increase in frequency in a population. To model the fate of altruistic alleles, Hamilton represented the fitness cost of the altruistic act to the actor as C and the fitness benefit to the recipient as B. Both C and B are measured in units of offspring produced. His model showed that the allele could spread if

$$Br > C$$

where r is the coefficient of relatedness. The **coefficient of relatedness** is a measure of how closely the actor and beneficiary are related. Specifically, r measures the fraction of alleles in the actor and beneficiary that are identical by descent—that is, inherited from the same ancestor (see **Box 51.2**).

This result is called **Hamilton's rule.** It is important because it shows that individuals can pass their alleles on to the next generation not only by having their own offspring but also by helping close relatives produce more offspring. Stated another way, an altruist might produce slightly fewer offspring because of the sacrifices the individual makes to help others. But if those sacrifices help close relatives produce many more offspring than the relatives could produce on their own, then copies of the altruist's alleles will increase in frequency.

According to Hamilton's rule, if the fitness benefits of altruistic behavior are high for the recipients, if the recipients are close relatives, and if the fitness costs to the altruist are low, then alleles associated with altruistic behavior will be favored by natural selection and will spread throughout the population. The keys are that (1) close relatives are very likely to have copies of the altruistic allele, and (2) help from close relatives allows these relatives to produce more offspring than they could produce without help. Biologists use the term **kin selection** to refer to natural selection that acts through benefits to relatives. ● If you understand this concept, you should be able to give an example of a species for which kin selection should be important, and explain why. You should also be able to explain whether kin selection can occur in plants.

Does Hamilton's rule work? Do animals really favor relatives when they act altruistically? To test the kin-selection hypothesis, a researcher studied which of the inhabitants of a black-tailed prairie dog town were most likely to give alarm calls. Within a large prairie dog town, individuals live in small groups called coteries that share the same underground burrow. Members of each coterie defend a territory inside the town. By tagging offspring that were born over several generations, the researcher identified the genetic relationships among individuals in the town. More specifically, the researcher determined, within each coterie, to which of the following three categories each prairie dog belonged: (1) an individual with no close genetic relatives in its coterie; (2) an individual with no offspring in the coterie but at least one sibling, cousin, uncle, aunt, niece, or nephew; and (3) an individual with at least one offspring or grandoffspring in the coterie.

FIGURE 51.20 Black-Tailed Prairie Dogs Are Highly Social. Prairie dogs live with their immediate and extended family within large groups called towns. The individual with the upright posture has spotted an intruder and may give an alarm call.

BOX 51.2 Calculating the Coefficient of Relatedness

The coefficient of relatedness, *r*, varies between 0.0 and 1.0. If two individuals have no identical alleles that were inherited from the same ancestor, then their *r* value is 0.0. Because every allele in pairs of identical twins is identical, their coefficient of relatedness is 1.0.

What about other relationships? **Figure 51.21a** shows how *r* is calculated between half-siblings. (To review what the boxes, circles, and lines in a pedigree mean, see Figure 13.21.) Half-siblings share one parent. Thus, *r* represents the probability that half-siblings share alleles as a result of inheriting alleles from their common parent. It is critical to realize that in each parent-to-offspring link of descent, the probability of any particular allele being transmitted is 1/2. This is so because meiosis distributes alleles from the parent's diploid genome to their haploid gametes randomly. Thus, half the gametes produced by a parent get one of the alleles present at each gene, and half the gametes produced get the other allele. Half-siblings are connected by two such parent-to-offspring links. The overall probability of two half-siblings sharing the same allele by descent is $1/2 \times 1/2 = 1/4$ (To review rules for combining probabilities, see **BioSkills 9**.)

To think about this calculation in another way, focus on the red arrows in Figure 51.21a. The left arrow represents the probability that the mother transmits a particular allele to her son. The right arrow represents the probability that the mother transmits the same allele to her daughter. Both probabilities are 1/2. Thus, the probability that the mother transmitted the same allele to both her son and daughter is $1/2 \times 1/2 = 1/4$.

Figure 51.21b shows how *r* is calculated between full siblings. The challenge here is to calculate the probability of two individuals sharing the same allele as a result of inheriting it through their mother or through their father. The probability that full siblings share alleles as a result of inheriting them from one parent is 1/4. Thus, the probability that full siblings share alleles inherited from either their mother or their father is $1/4 + 1/4 = 1/2$. (To review rules for combining probabilities, see BioSkills 9.) ● If you understand how to calculate *r*, you should be able to draw a pedigree with 3 generations and calculate *r* between any 2 individuals on the diagram.

(a) What is the *r* between half-siblings?

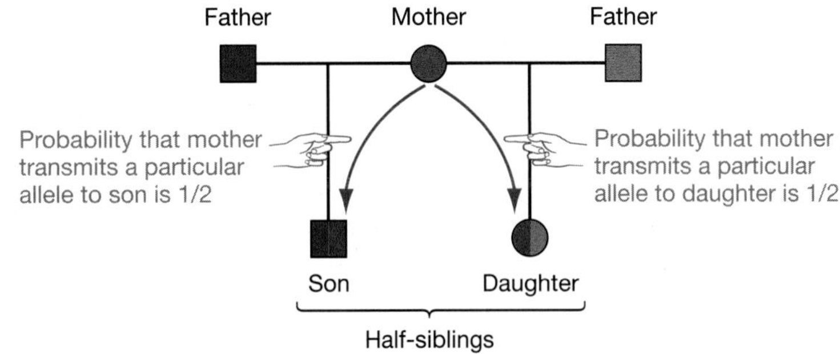

What is the probability that half-siblings inherit the same allele from their common parent?

Answer: *r* between half-siblings $= 1/2 \times 1/2 = 1/4$

(b) What is the *r* between full siblings?

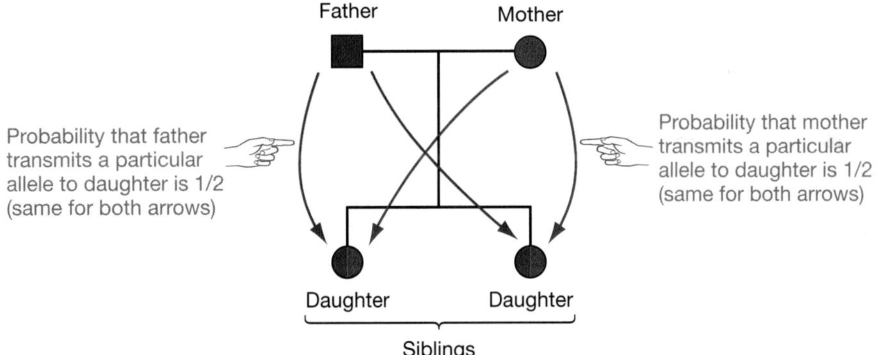

What is the probability that full siblings inherit the same allele from their father or their mother?

Answer: Probability that they inherit same allele from father $= 1/2 \times 1/2 = 1/4$
Probability that they inherit same allele from mother $= 1/2 \times 1/2 = 1/4$
Overall probability that they inherit the same allele $= 1/4 + 1/4 = 1/2$
r between full siblings $= 1/2$

FIGURE 51.21 The Coefficient of Relatedness Is Calculated from Information in Pedigrees.

● **QUESTION** According to Hamilton's rule, who should help each other more: half-siblings or full siblings?

● **QUESTION** What is the *r* between first cousins?

The kin-selection hypothesis predicts that individuals who do not have close genetic relatives nearby will rarely give an alarm call. To evaluate this prediction, the biologist recorded the identity of callers during 698 experiments. In these studies, a stuffed badger was dragged through the colony on a sled. Were prairie dogs with close relatives nearby more likely to call, or did kinship have nothing to do with the probability of alarm calling? The data in **Figure 51.22** illustrate that black tailed prairie dogs are much more likely to call if they live in a coterie that includes close relatives. This same pattern—of preferentially dispensing help to kin—has been observed in many other species of social mammals and birds. Most cases of self-sacrificing behavior that have been analyzed to date are consistent with Hamilton's rule and are hypothesized to be the result of kin selection.

Experiment

Question: Do black-tailed prairie dogs prefer to help relatives when they give an alarm call?

Kin-selection hypothesis: Individuals give an alarm call only when close relatives are near.

Null hypothesis: The presence of relatives has no influence on the probability of alarm calling.

Experimental setup:

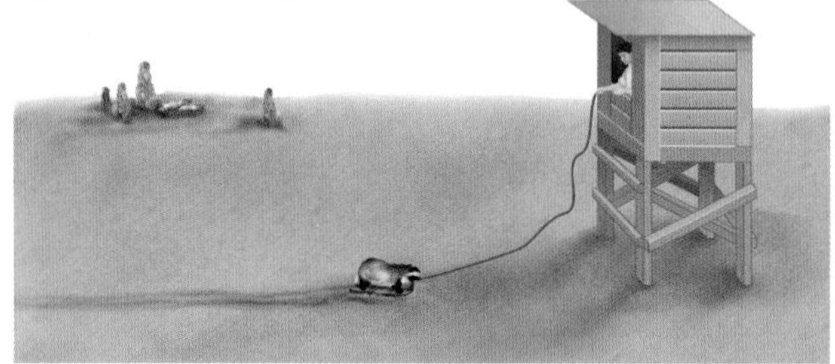

1. Determine relationships among individuals in prairie dog coterie.
2. Drag stuffed badger across territory of coterie.
3. From observation tower, record which members of coterie give an alarm call.
4. Repeat experiment 698 times. Each individual prairie dog coterie is tested 6-9 times over 3-year period.

Prediction of kin-selection hypothesis: Individuals in coteries that contain a close genetic relative are more likely to give an alarm call than are individuals in coteries that do not contain a close genetic relative.

Prediction of null hypothesis: The presence of relatives in coteries will not influence the probability of alarm calling.

Results:

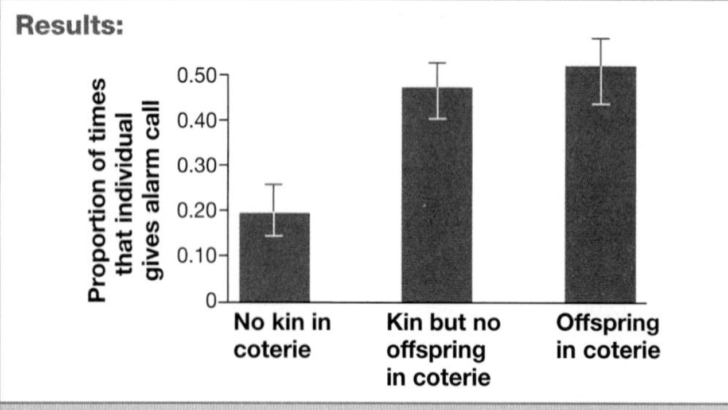

Conclusion: Alarm calling usually benefits relatives.

FIGURE 51.22 Experimental Evidence that Black-Tailed Prairie Dogs Are More Likely to Give Alarm Calls If Relatives Are Nearby.

Reciprocal Altruism

During long-term studies of highly social animals, such as lions, chimpanzees, and vampire bats, biologists have observed nonrelatives helping each other. Chimps and other primates, for example, occasionally spend considerable time grooming unrelated members of their social group—cleaning their fur and removing ticks and other parasites from their skin. In vampire bats, individuals that have been successful in finding food are known to regurgitate blood meals to non-kin that have not been successful and that are in danger of starving.

How can self-sacrificing behavior like this evolve if kin selection is not acting? The leading hypothesis to explain altruism among nonrelatives is called **reciprocal altruism**, which is an exchange of fitness benefits that are separated in time. Data that have been collected so far support the reciprocal-altruism hypothesis in at least some instances. Among vervet monkeys, for example, individuals are most likely to groom unrelated individuals that have groomed or helped them in the past. Similarly, vampire bats are most likely to donate blood meals to non-kin that have previously shared food with them. Reciprocal altruism is also widely invoked as an explanation for the helpful and cooperative behavior commonly observed among unrelated humans.

To summarize, altruistic behavior increases the fitness of individuals by favoring kin or by increasing the likelihood of receiving help in the future from non-kin. Altruism is a flexible, condition-dependent behavior that occurs in a surprisingly wide array of species that live in social groups.

Check Your Understanding

If you understand that...

- Hamilton's rule states that alleles for altruistic behavior increase in frequency if the fitness cost of the behavior for the actor is low, the fitness benefit to the recipient is high, and the actor and recipient are closely related.
- Reciprocal altruism is based on an exchange of fitness benefits, separated in time.

You should be able to...

1) Use Hamilton's rule to describe situations in which a black-tailed prairie dog is likely and unlikely to give an alarm call.
2) Explain why reciprocal altruism has been observed only in species where individuals have good memories and live in small, long-lived social groups.

Chapter Review

SUMMARY OF KEY CONCEPTS

After describing a behavior, biologists seek to explain both its proximate and ultimate causes—meaning how it happens at the genetic and physiological levels and how it affects the individual's fitness.

At the proximate level, experiments and observations focus on understanding how specific gene products, neuron activity, and hormonal signals cause behavior. At the ultimate level, researchers seek to understand the adaptive significance of behavior, or how it enables individuals to survive and reproduce. By combining proximate and ultimate viewpoints and studying the genetic basis of behavior, biologists can seek a comprehensive understanding of how and why animals do what they do.

You should be able to give an explanation at the proximate and ultimate levels for biological phenomena other than behavior—for example, cellular respiration, meiosis, water transport in plants, or the action potential in animal neurons.

In a single species, behavior may range from highly stereotyped, invariable responses to highly flexible, conditional responses and from unlearned to learned responses.

FAPs are examples of highly stereotyped, unlearned behaviors that are released by extremely simple stimuli. Language acquisition and tool making are examples of behavior that are highly dependent on learning, cognition, and the environmental conditions experienced by an individual. Most of the behavior observed in animals is condition dependent, meaning an individual has the genetic, neuronal, and hormonal mechanisms in place for behaving in a variety of ways, and the actual behavior observed depends on environmental conditions. Foraging in white-fronted bee-eaters, mate choice in barn swallow, sex change in coral-reef fish, and the onset of sexual behavior in *Anolis* lizards are examples of flexible, condition-dependent behaviors triggered by changing environmental stimuli.

You should be able to explain why migration in some birds is considered condition dependent, based on the observation that some species do not migrate if people supply food in feeders.

The types of learning that individuals do, the way they communicate, and the way they orient and navigate all correlate closely with their habitat and with the challenges they face in trying to survive and reproduce.

Animals usually behave in a way that increases their ability to survive and reproduce in their current environment. Deceitful communication can be adaptive—as when anglerfish fool prey into attacking a "lure" attached to the anglerfish's face or when molting mantis shrimps try to bluff their way into retaining ownership of a hiding place.

You should be able to predict the consequences for deceitful communication in bluegill sunfish if most large, territory-owning males were fished out of a lake.

MB Web Animation at www.masteringbio.com
Homing Behavior in Digger Wasps

When individuals behave altruistically, they are usually helping close relatives or individuals that help them in return.

Although it is common to observe self-sacrificing behavior in animals, most altruistic behavior is directed toward close relatives. When this is the case, alleles that lead to self-sacrificing can increase in frequency due to kin selection. In addition, some animals that live in close-knit social groups engage in reciprocal altruism—meaning they exchange help over time.

You should be able to describe the characteristics of a species for which altruistic behavior is expected to be extremely common.

QUESTIONS

Test Your Knowledge

1. What do proximate explanations of behavior focus on?
   a. how displays and other types of behavior have changed through time, or evolved
   b. the functional aspect of a behavior, or its "adaptive significance"
   c. genetic, neurological, and hormonal mechanisms of behavior
   d. psychological interpretations of behavior—especially motivation
2. What evidence suggests that there is a critical period for song learning in some bird species?
   a. In mynahs and mockingbirds, individuals continue to sing new songs throughout their lives.
   b. Individuals do not learn to sing normally unless they are allowed to hear themselves practice.
   c. Individuals that never hear their species-specific song can still sing normally as adults.
   d. Birds that are deafened soon after hatching never sing normally, but birds that are deafened several months after hatching sing normally.
3. What is the difference between orienting behavior and piloting?
   a. Orienting movements are fast responses to visual or auditory stimuli; piloting is finding the way using landmarks.
   b. Orienting is the ability to follow a compass during navigation; piloting is the ability to use map information during navigation.
   c. Orienting is used during long-distance movements such as migration; piloting is used in short-distance movements involved in homing.
   d. Orienting is the ability to use a Sun or star compass; piloting is the ability to use a magnetic compass.
4. Which of the following statements about the waggle dance of the honeybee is not correct?
   a. The length of a waggling run is proportional to the distance from the hive to a food source.
   b. Sounds and scents produced by the dancer provide information about the nature of the food source.

c. The dancer uses no elements of the round dance.
d. The orientation of the waggling run provides information about the direction of the food from the hive, relative to the Sun's position.

5. Why are biologists convinced that the sex hormone testosterone is required for normal sexual activity in male *Anolis* lizards?
a. Male *Anolis* lizards with larger testes court females more vigorously than do males with smaller testes.
b. The testosterone molecule is not found in female *Anolis* lizards.
c. Male *Anolis* lizards whose gonads had been removed did not develop dewlaps.
d. Male *Anolis* lizards whose gonads had been removed did not court females.

6. What does Hamilton's rule specify?
a. why animals do things "for the good of the species"
b. why reciprocal altruism can lead to fitness gains for unrelated individuals
c. how alleles that favor self-sacrificing acts increase in frequency via kin selection
d. the conditions under which more complex behaviors evolve from simpler behaviors

Test Your Knowledge answers: 1. c; 2. d; 3. a; 4. c; 5. d; 6. c

Test Your Understanding

Answers are available at www.masteringbio.com

1. Make a graph like that in Figure 51.2, with axes that plot the degree to which a particular behavior is modified by learning and the degree to which behavior is stereotyped and inflexible. Plot the following behaviors on the graph: tool use in New Caledonian crows; exploration of mazes by rats; singing by chickens, by white-crowned sparrows, and by mynahs; and blushing in humans.
2. Discuss the proximate and ultimate causes of the following behaviors: (a) homing behavior by spiny lobsters, (b) sexual behavior by *Anolis* lizards, and (c) the "jump-back" response of kangaroo rats.
3. Propose a hypothesis to explain the adaptive significance of tool use in New Caledonian crows. Recall that a female crow bent an unfamiliar material into a hooked tool. Explain why this observation supports the hypothesis that individuals of this species can think.
4. What environmental stimuli cause changes in hormone levels that lead to egg laying in *Anolis* lizards? Based on the data presented on sexual behavior in *Anolis*, propose a hypothesis to explain the proximate basis of sex switching in coral-reef fish.
5. For an animal to navigate, it must have a "map" and a "compass." Explain why both types of information are needed. Describe three types of compasses that have been identified in migratory or homing species.
6. Compare and contrast kin selection and reciprocal altruism. Be sure to identify the conditions under which self-sacrificing behavior is expected to evolve under kin selection versus reciprocal altruism.

Applying Concepts to New Situations

Answers are available at www.masteringbio.com

1. Mated pairs of cranes give territorial displays called unison calls. Cranes that are raised in isolation from other cranes give unison calls normally, and the display is performed the same way every time. Your friend argues that the unison call of cranes is a FAP, because it is stereotyped and not influenced by learning. Another friend argues that it is a condition-dependent behavior, because only mated pairs do it and because the tendency to give unison calls varies among pairs and with time of year. (Some pairs are more aggressive than others and give unison calls more frequently in territorial contexts.) Who's right?
2. Most tropical habitats are highly seasonal. But instead of alternating warm and cold seasons, there are alternating wet and dry seasons. Most animal species breed during the wet months. If you were studying a species of *Anolis* native to the tropics, what environmental cue would you simulate in the lab to bring them into breeding condition? How would you simulate this cue? Further, think about the sensory organs that lizards use to receive this cue. Are they the same as or different than the receptors that *Anolis carolinensis* uses to recognize that spring has arrived in the southeastern United States?
3. What is the significance of the observation that the product of the *for* gene in fruit flies is involved in a signal transduction pathway? Propose a hypothesis to explain why fruit-fly larvae with the rover genotype and phenotype have higher fitness when population density is high and individuals are crowded.
4. A biologist once remarked that he'd be willing to lay down his life to save two brothers or eight cousins. Explain what he meant. Based on the theory of kin selection and reciprocal altruism, predict the conditions under which people are expected to be nice to one another.

www.masteringbio.com is also your resource for • Answers to text, table, and figure caption questions and exercises • Answers to *Check Your Understanding* boxes • Online study guides and quizzes • Additional study tools including the *E-Book for Biological Science* 3rd ed., textbook art, animations, and videos.

Population Ecology

52

A flock of white pelicans swimming in a channel of the Mississippi River. This chapter explores how and why growth rates in populations change through time.

KEY CONCEPTS

- Life tables summarize how likely it is that individuals of each age class in a population will survive and reproduce.
- The growth rate of a population can be calculated from life-table data or from the direct observation of changes in population size over time.
- Researchers observe a wide variety of patterns when they track changes in population size over time, ranging from no growth, to regular cycles, to continued growth independent of population size.
- Data from population ecology studies help biologists evaluate prospects for endangered species and design effective management strategies.

If you asked a biologist to name two of the most pressing global issues facing your generation, she might say global warming and extinction of species. If you were asked to name two of the most pressing issues facing the region where you live, you might say traffic and the price of housing. All four of these problems have a common cause: recent and dramatic increases in the size of the human population.

A **population** is a group of individuals of the same species that live in the same area at the same time. In both ecology and evolutionary biology, populations are the basic unit of analysis. Evolutionary biologists study how the characteristics of populations change through time; ecologists study how populations interact with their environment. Increasingly, ecologists are studying how species are being affected by growing human populations.

Population ecology is the study of how and why the number of individuals in a population changes over time. Biologists also analyze changes in the ages of individuals in a population, the proportion of males to females, and geographic distribution. With the explosion of human populations across the globe, the massive destruction of natural habitats, and the resulting threats to species throughout the tree of life, population ecology has become a vital field in biological science. The mathematical and analytical tools introduced in this chapter help biologists predict changes in population size and design management strategies to save threatened species.

Biologists ask a wide array of questions in population ecology. How are individuals distributed in space? How old are they, and how likely are they to reproduce or die? Is the population growing, declining, or staying the same through time, and why?

To answer these questions, let's consider some of the basic tools that biologists use to study populations, then follow up with examples of how biologists study changes in the population size of humans and other species over time. The chapter concludes by asking how all of these elements fit together in efforts to limit human population growth and save endangered species.

52.1 Demography

The number of individuals that are present in a population depends on four processes: birth, death, immigration, and emigration. Populations grow due to births—here meaning any form of reproduction—and **immigration**, which occurs when individuals enter a population by moving from another population. Populations decline due to deaths and **emigration**, which occurs when individuals leave a population to join another population. Analyzing birth rates, death rates, immigration rates, and emigration rates is fundamental to **demography**: the study of factors that determine the size and structure of populations through time.

To make detailed predictions about the future of a population, however, biologists must understand the makeup of the population in more detail. If a population consists primarily of young individuals with a high survival rate and reproductive rate, the population size should increase over time. But if a population comprises chiefly old individuals with low reproductive rates and low survival rates, then it is almost certain to decline over time. To predict the future of a population, biologists need to know how many individuals of each age are alive, how likely individuals of different ages are to survive to the following year, how many offspring are produced by females of different ages, and how many individuals of different ages immigrate and emigrate each **generation**—the average time between a mother's first offspring and her daughter's first offspring.

Demographic data provide an important tool for biologists charged with designing management programs for endangered species. To understand the nature of these data and how they are used in conservation programs, let's examine a classical tool for describing the demography of a population.

Life Tables

Formal demographic analyses of populations are based on a type of data set called a life table. A **life table** summarizes the probability that an individual will survive and reproduce in any given year over the course of its lifetime. Life tables were invented almost 2000 years ago; in ancient Rome they were used to predict food needs. In modern times, life tables have been the domain of life-insurance companies, which have a strong financial interest in predicting the likelihood of a person dying at a given age. More recently, biologists have employed life tables to study the demographics of endangered species and other nonhuman populations.

To understand how researchers use life tables, consider the lizard *Lacerta vivipara* (**Figure 52.1**). *L. vivipara* is a common resident of open, grassy habitats in western Europe.

FIGURE 52.1 *Lacerta vivipara* Are Native to Europe. Translated literally, *vivipara* means "live birth." Females in most populations bear live young, though females in *L. vivipara* populations from northern Spain and southwest France lay eggs.

As their name suggests, most populations of the species are ovoviviparous (see Chapter 32) and give birth to live young. Researchers set out to estimate the life table of a low-elevation population in the Netherlands, with the goal of comparing the results to data that other researchers had collected from *L. vivipara* populations in the mountains of Austria and France and in lowland habitats in Britain and Belgium. The researchers wanted to know whether populations that live in different environments vary in basic demographic features.

To complete the study, the researchers visited their study site daily during the seven months that these lizards are active during the year. Each day the researchers captured and marked as many individuals as possible. Because this program of daily monitoring continued for seven years, the biologists were able to document the number of young produced by each female in each year of her life. If a marked individual was not recaptured in a subsequent year, they assumed that it had died sometime during the previous year. These data allowed researchers to calculate the number of individuals that survived each year in each particular age group and how many offspring each female produced. What did the data reveal?

Survivorship

Survivorship is a key component of a life table and is defined as the proportion of offspring produced that survive, on average, to a particular age. For example, suppose 1000 *L. vivipara* are born in a particular year. These individuals represent a **cohort**—a group of the same age that can be followed through time. How many individuals would survive to age 1, age 2, age 3, and so on?

As **Table 52.1** shows, the biologists calculated that in the population living in the Netherlands, survivorship from birth to age 1 was 0.424. If 1000 females were born in a particular year in this population, on average 424 would still be alive one year later. Survivorship from birth to age 2 was 0.308, meaning an average of 308 female lizards would survive for two years.

TABLE 52.1 **Life Table for *Lacerta vivipara* in the Netherlands**

Year	Number Alive	Survivorship	Fecundity	Survivorship × Fecundity = Average Number of Offspring Produced per Female Born
0	1000	1.000	0.00	0.00
1	424	0.424	0.08	0.03
2	308	0.308	2.94	0.91
3	158	0.158	4.13	0.65
4	57	0.057	4.88	0.28
5	10	0.010	6.50	0.07
6	7	0.007	6.50	0.05
7	2	0.002	6.50	0.01

Data are from H. Strijbosch and R.C.M. Creemers, 1988. Comparative population demography of sympatric populations of *Lacerta vivipara* and *Lacerta agilis*. *Oecologia* 76: 20–26.

To recognize general patterns in survivorship and make comparisons among populations or species, biologists plot the logarithm of the number of survivors versus age. The resulting graph is called a **survivorship curve**. Studies on a wide variety of species indicate that three general types of survivorship curves exist (**Figure 52.2a**). Humans have what biologists call a type I survivorship curve. In this pattern, survivorship throughout life is high, and most individuals approach the species' maximum life span. Songbirds, in contrast, experience relatively constant mortality throughout their lives once they have left the nest, resulting in a type II survivorship curve. Many plants have type III curves due to extremely high death rates for seeds and seedlings. **Figure 52.2b** provides a graph for you to plot survivorship of *L. vivipara*.

Survivorship curves are important in conservation work, because they pinpoint the stage of life when endangered species have particularly low survivorship. Biologists can then put measures into place to protect individuals during that period. For example, when they are very young, trees or herbaceous plants can be protected from drought by germinating them in a nursery and from predation by putting a wire cage over seedlings.

Fecundity The number of female offspring produced by each female in the population is termed **fecundity**. (In most cases, biologists keep track only of females when calculating life-table data, because only females produce offspring.) Researchers documented the reproductive output of the same *L. vivipara* lizard females year after year and thus were able to calculate a quantity called **age-specific fecundity**, which is defined as the average number of female offspring produced by a female in age class x. An **age class** is a group of individuals of a specific age—for instance, all female lizards between 4 and 5 years old. As **Box 52.1** shows, data on survivorship and fecundity allow researchers to calculate the growth rate of a population. How do data on survivorship and fecundity in the Netherlands populations compare with populations in different types of habitats?

(a) Three general types of survivorship curves

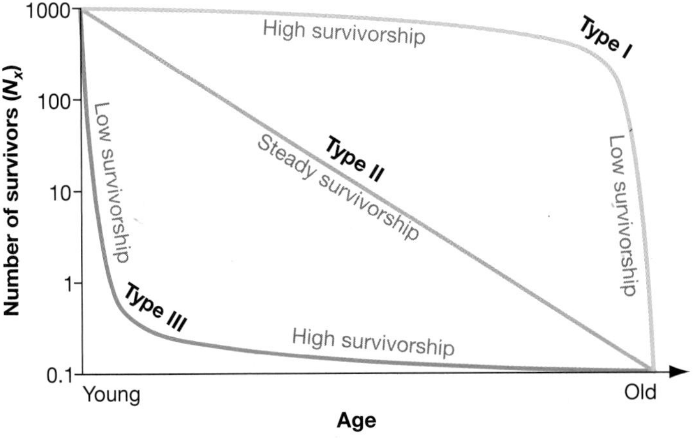

(b) Exercise: Survivorship curve for *Lacerta vivipara*

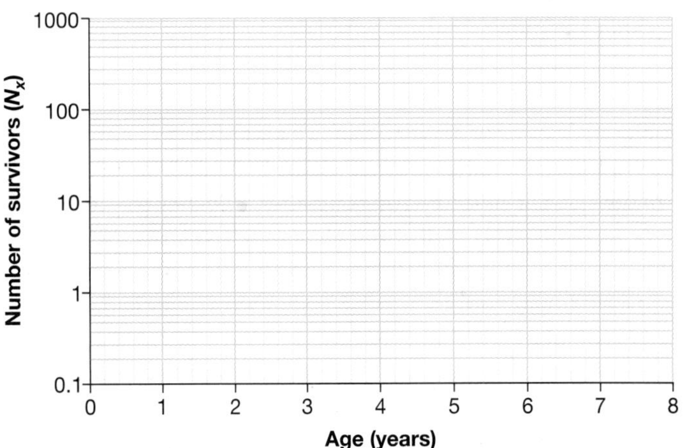

FIGURE 52.2 Survivorship Curves Identify When Mortality Rates Are Low, Steady, or High.

● **EXERCISE** Fill in the graph in part **(b)** with data on survivorship of *Lacerta vivipara* given in Table 52.1. Compare the shape of the curve to the generalized graphs in part (a).

BOX 52.1 Using Life Tables to Calculate Population Growth Rates

If immigration and emigration are ignored, the data in a life table can be used to calculate a population's growth rate. This is logical, because survivorship and fecundity are ways to express death rates and birth rates—the other two factors influencing population size. To see how a population's growth rate can be estimated from life-table data, let's look at each component in a life table more carefully.

Survivorship is symbolized as l_x, where x represents the age class being considered. Survivorship for age class x is calculated by dividing the number of individuals in that age class (N_x) by the number of individuals that existed as offspring (N_0):

$$l_x = \frac{N_x}{N_0} \qquad \text{(Eq. 52.1)}$$

Age-specific fecundity is symbolized m_x, where x again represents the age class being considered. Age-specific fecundity is calculated as the total number of offspring produced by females of a particular age, divided by the total number of females of that age class present. It represents the average number of offspring produced by a female of age x.

Documenting age-specific survivorship and fecundity allows researchers to calculate the net reproductive rate, R_0, of a population:

$$R_0 = \sum_{i=0}^{x} l_x m_x \qquad \text{(Eq. 52.2)}$$

The **net reproductive rate** represents the growth rate of a population per generation. The logic behind the equation for R_0 is that the growth rate of a population per generation equals the average number of female offspring that each female produces over the course of her lifetime. A female's average lifetime reproduction, in turn, is a function of survivorship and fecundity at each age class. In Table 52.1, R_0 is the sum of the survivorship $\times$ fecundity values in the right column.

If R_0 is greater than 1, then the population is increasing in size. If R_0 is less than 1, then the population is declining. ● If you understand these concepts, you should be able to use the data in Table 52.1 to calculate R_0 in the Netherlands population of *L. vivipara*;[1] state how many female offspring an average *L. vivipara* female produces over the course of her lifetime,[2] and describe whether the population is growing, stable, or declining.[3] Answers are given below.

[1]$R_0 = 2.0$; [2]2.00; [3]The population is growing rapidly.

The Role of Life History

The life-table data in Table 52.1 are interesting because they contrast with results from other populations of *L. vivipara*. For example, notice that almost no 1-year-old female *L. vivipara* lizards in the Netherlands reproduces. But in Brittany, France, 50 percent of 1-year-old female lizards reproduce. Both situations contrast sharply with data on females in the mountains of Austria. In populations that live at high elevation, females do not begin breeding until they are 4 years old. In addition, most females in the Austrian population live much longer than do individuals in either lowland population—in the Netherlands or France. When the three sites are compared, it is clear that fecundity is high but survivorship is low in Brittany. In contrast, fecundity is low but survivorship is high in Austria. The population in the Netherlands is intermediate in both characteristics.

Why isn't it possible for *L. vivipara* females to have both high fecundity and high survival? Biologists answer this question using the concept of **fitness trade-offs** (see Chapters 24 and 41). Fitness trade-offs occur because every individual has a restricted amount of time and energy at its disposal—meaning that its resources are limited. If a female lizard devotes a great deal of energy to feeding a large number of offspring as they develop, it is not possible for her to devote that same energy to her immune system, growth, nutrient stores, or other traits that increase survival. (Ask yourself whether a woman who has 10 children would be expected to live a longer or shorter time than a woman who has 2.) A female can maximize fecundity, maximize survival, or strike a balance between the two. An organism's **life history** consists of how an individual allocates resources to growth, reproduction, and activities or structures that are related to survival.

Life history trade-offs are universal. To drive this point home, **Figure 52.3** graphs the average number of eggs laid—a measure of fecundity—in 10 species of birds versus the probability that a female of that species survives to the following year. There are no points in the upper right corner of the graph because it is not possible to have high fecundity and high survivorship. There are no points in the lower left corner of the graph because species with low survivorship and low fecundity have low population growth and go extinct.

In almost all cases, biologists find that life history is shaped by natural selection in a way that maximizes an individual's fitness in its environment. For example, Chapter 41 provided experimental evidence that there is a fitness trade-off between egg size and egg number in side-blotched lizards, and that females in the study population strike a balance in egg size and number that is optimum for their habitat.

To make sense of the life-history data from *L. vivipara* populations, biologist contend that females who live a long time but mature late and have few offspring each year have high fitness in cold, high-elevation habitats. In contrast, females who have short lives but mature early and have large numbers of offspring each year do better in warm, low-elevation habitats. ● If you understand the concept of fitness trade-offs, you

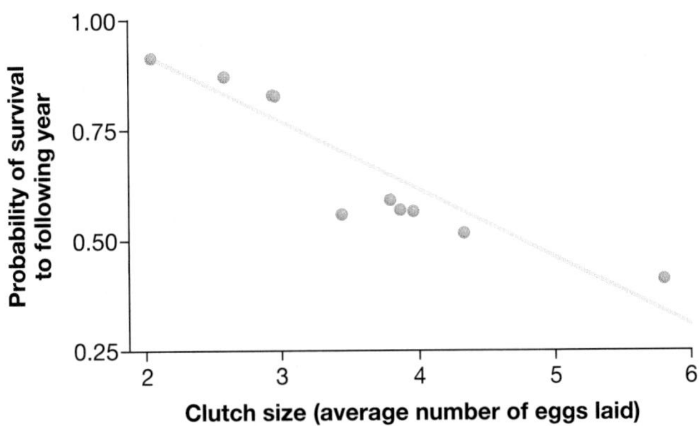

FIGURE 52.3 There Is a Trade-off between Survival and Reproduction. Graph of survivorship versus fecundity. Each data point represents a different species of bird. A statistical analysis called regression was used to create a "line of best fit" to these data points.

QUESTION To study survival-reproduction trade-offs within species of birds, researchers randomly add or subtract eggs to clutches. Predict what happens to survivorship in females that rear additional versus a reduced number of young.

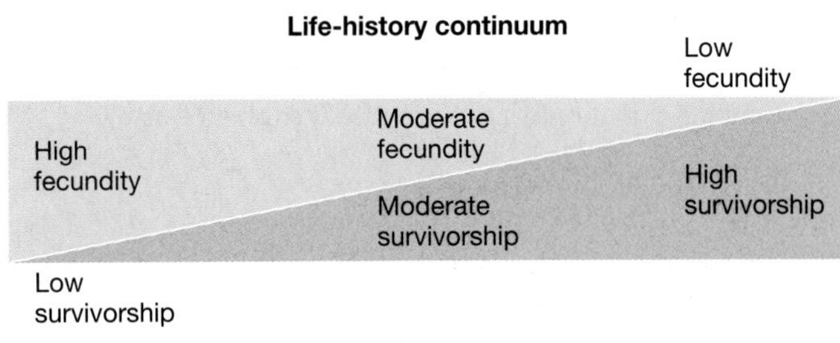

FIGURE 52.4 Life-History Traits Form a Continuum. Every organism can be placed somewhere on this life-history continuum. The placement of a species is most meaningful when it is considered relative to closely related species.

EXERCISE Place *Arabidopsis thaliana* and the coconut palm on the continuum. For plants, list the types of traits you would expect to see on the left, middle, and right of the continuum. Include growth habit (herbaceous, shrub, or tree) as well as relative disease- and predator-fighting ability, seed size, seed number, body size, and investment in roots versus shoots.

should be able to (1) predict how survivorship and fecundity should compare in *L. vivipara* in the warmest part of their range (southwest France and northern Spain), and (2) comment on why females in these populations lay many eggs instead of giving birth to a few live young.

In general, organisms with high fecundity tend to grow quickly, reach sexual maturity at a young age, and produce many small eggs or seeds. The mustard plant *Arabidopsis thaliana*, for example, germinates and grows to sexual maturity in just 4 to 6 weeks. In this species, individuals usually live only a few months but may produce as many as 10,000 tiny seeds. In contrast, organisms with high survivorship tend to grow slowly and invest resources in traits that reduce damage from enemies and increase their own ability to compete for water, sunlight, or food. A coconut palm, for example, may take a decade to reach sexual maturity but live 60–70 years and produce offspring each year. Coconut palms invest resources in making stout stems that allow them to grow tall, a relatively extensive root system, large fruits (coconuts), molecules that make herbivores sick, and enzymes that reduce infections from disease-causing fungi. These traits increase the survivorship of both seedlings and adults but decrease fecundity.

An *Arabidopsis thaliana* plant and a coconut palm represent two ends of a broad continuum of life-history characteristics (**Figure 52.4**). Even within the species *L. vivipara*, analyses of life tables have shown that there is considerable variation in life-history traits.

Research has also documented that within populations, life-history traits can change if conditions change. For example, suppose that a series of wet years produced an abundance of food for lizards. Females might survive better in response, or they might begin breeding at an earlier age or produce more offspring at a particular age. Such changes might directly affect a population's growth rate. How do biologists analyze growth rates in populations? And what factors make growth rates increase or decrease?

52.2 Population Growth

The most fundamental questions that biologists ask about populations involve growth or decline in numbers of individuals. For conservationists, analyzing and predicting changes in population size is fundamental to managing threatened species.

Recall that four processes affect a population's size: Births and immigration add individuals to the population; deaths and emigration remove them. It follows that a population's overall growth rate is a function of birth rates, death rates, immigration rates, and emigration rates. In this section we consider only the impact of births and deaths on population growth; Section 52.4 includes an analysis of the movement of individuals into and out of populations.

A population's growth rate is the change in the number of individuals in the population (ΔN) per unit time (Δt). If no immigration or emigration is occurring, then the growth rate is equal to the number of individuals (N) in the population times the difference between the birth rate per individual (b) and death rate per individual (d). The difference between the birth rate and death rate per individual is called the **per-capita rate of increase** and is symbolized r. (*Per capita* means "for each individual.") If the per-capita birth rate is greater than the per-capita death rate, then r is positive and the population is growing. But if the per-capita death rate begins to exceed the per-capita birth rate, then r becomes negative and the population declines. Within populations, r varies through time and can be positive, negative, or 0.

When conditions are optimal for a particular species—meaning birth rates per individual are as high as possible and death rates per individual are as low as possible—then r reaches a maximal value called the **intrinsic rate of increase,** r_{max}. When this occurs, a population's growth rate is expressed as

$$\Delta N/\Delta t = r_{max}N$$

In species such as *Arabidopsis* and fruit flies, which breed at a young age and produce many offspring each year, r_{max} is high. In contrast, r_{max} is low in species such as giant pandas and coconut palms, which take years to mature and produce few offspring each year. Stated another way, r_{max} is a function of a species' life-history traits.

A species with a certain life history has a characteristic r_{max} that does not change. But at any specific time, each population of that species has an instantaneous growth rate, or per-capita rate of increase, symbolized by r. The instantaneous growth rate of a population at a particular time is likely to be much lower than r_{max}. A population's r is also likely to be different from r values of other populations of the same species, and to change over time.

Exponential Growth

The graphs in **Figure 52.5** plot changes in population size for various values of r, under the condition known as exponential growth. **Exponential population growth** occurs when r does not change over time. The key point about exponential growth is that the growth rate does not depend on the number of individuals in the population. Biologists say that this type of population growth is **density independent.**

It's important to emphasize that exponential growth adds an increasing number of individuals as the total number of individuals, N, gets larger. As an extreme example, an r of 0.02 per year in a population of 1 billion adds over 20 million individuals per year. The same growth rate in a population of 100 adds just over 2 individuals per year. Even if r is constant, the number of individuals added to a population is a function of N.

In nature, exponential growth is observed when a few individuals found a new population in a new habitat or when a population has been devastated by a storm or some other type of catastrophe and then begins to recover, starting with a few surviving individuals. But it is not possible for exponential growth to continue indefinitely. If *Arabidopsis* populations grew exponentially for a long period of time, they would eventually fill all available breeding habitat. When **population density**—the number of individuals per unit area—gets very high, we would expect the population's per-capita birth rate to decrease and the per-capita death rate to increase, causing r to decline. Stated another way, growth in the natural world is most often **density dependent.**

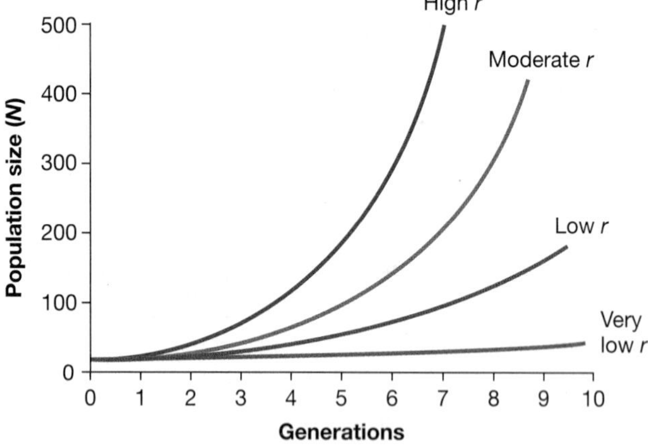

FIGURE 52.5 Exponential Growth Is Independent of Population Size. When the per-capita growth rate r does not change over time, exponential growth occurs. Population size may increase slowly or rapidly, depending on the size of r.

QUESTION Consider a species on the left-hand side of the life-history continuum in Figure 52.4 (high fecundity, low survivorship) versus a species on the right-hand side (low fecundity, high survivorship). Under ideal conditions, which species has higher r? Explain your reasoning.

Logistic Growth

To analyze what happens when growth is density dependent, biologists use a parameter called carrying capacity. **Carrying capacity, *K*,** is defined as the maximum number of individuals in a population that can be supported in a particular habitat over a sustained period of time. The carrying capacity of a habitat depends on a large number of factors: food, space, water, soil quality, resting or nesting sites, and the intensity of disease and predation. Carrying capacity can change from year to year, depending on conditions.

If a population of size N is below the carrying capacity K, then the population should continue to grow. More specifically, a population's growth rate should be proportional to $(K - N)/K$:

$$\frac{\Delta N}{\Delta t} = r_{max}N\left(\frac{K - N}{K}\right)$$

This expression is called the logistic growth equation. In the expression $(K - N)/K$, the numerator defines the number of additional individuals that can be accommodated in a habitat with carrying capacity K; dividing $K - N$ by K turns this number of individuals into a proportion. Thus, $(K - N)/K$ describes the proportion of "unused resources and space" in the habitat. It can be thought of as the environment's resistance to growth. When N is small, then $(K - N)/K$ is large and the growth rate should be high. But as N gets larger, $(K - N)/K$ gets smaller. When N is at carrying capacity, then $(K - N)/K$ is equal to 0 and growth stops. Thus, as a population approaches a habitat's carrying capacity, its growth rate should slow.

The logistic growth equation describes **logistic population growth,** or changes in growth rate that occur as a function of population size. Just as exponential growth is density independent, logistic growth is density dependent. **Box 52.2** explores population growth models and how they are applied in more detail.

BOX 52.2 Developing and Applying Population Growth Equations: A Closer Look

To explore how biologists model changes in population size in more detail, let's consider data on whooping cranes, which are large, wetland-breeding birds native to North America (**Figure 52.6**). Whooping cranes may live more than 20 years in the wild, but it is common for females to be six or seven years old before they breed for the first time. Female cranes lay just two eggs per year and usually rear just one chick. Based on these life-history characteristics, we would expect the growth rate of whooping crane populations to be extremely low.

Conservationists monitor this species closely, because hunting and habitat destruction reduced the total number of individuals in the world to about 20 in the mid-1940s. Since then, intensive conservation efforts have resulted in a current total population of about 518. That number includes a group of 237 individuals that breeds in Wood Buffalo National Park in the Northwest Territories of Canada.

In addition to the Wood Buffalo population, two new populations of cranes have been established recently by releasing offspring raised in captivity. One of these groups lives near the Atlantic coast of Florida year-round; the other group migrates from breeding areas in northern Wisconsin to wintering areas along Florida's Gulf coast.

FIGURE 52.6 Whooping Cranes Have Been the Focus of Intensive Conservation Efforts.

According to the biologists who are managing whooping crane recovery, the species will no longer be considered endangered when the two newly established populations have at least 25 breeding pairs—meaning there would be about 125 individuals in each—and when neither population needs to be supplemented with captive-bred young in order to be self-sustaining. How long will this take?

Discrete Growth

Whooping cranes breed once per year, so the simplest way to express a crane population's growth rate is to compare the number of individuals at the start of one breeding season to the number at the start of the following year's breeding season.

When populations breed during discrete seasons, their growth rate can be calculated as for whooping cranes. To create a general expression for how populations grow over a discrete time interval, biologists use N to symbolize population size. N_0 is the population size at time zero (the starting point), and N_1 the population size one breeding interval later. In equation form, the growth rate is given as

$$N_1 = N_0\lambda \qquad \textbf{(Eq. 52.3)}$$

The parameter λ (lambda) is called the **finite rate of increase**. (In mathematics, a *finite rate* refers to an observed rate over a given period of time. A *parameter* is a variable or constant term that affects the shape of a function but does not affect its general nature.) The current total at Wood Buffalo is 237. Suppose that biologists count 248 cranes on the breeding grounds next year. The population growth rate could be calculated as $248/237 = 1.046 = \lambda$. Stated another way, the population will have grown at the rate of 4.6 percent per year. Rearranging the expression in Equation 52.3 gives

$$N_1 = N_0\lambda \qquad \textbf{(Eq. 52.4)}$$

Stated more generally, the size of the population at the end of time t will be given by

$$N_t = N_0\lambda^t \qquad \textbf{(Eq. 52.5)}$$

This equation summarizes how populations grow when breeding takes place seasonally. The size of the population at time t is equal to the starting size, times the finite rate of increase multiplied by itself t times. In a sense, λ works like the interest rate at a bank. For species that breed once per year, the "interest" on the population is compounded annually. A savings account with a 5 percent annual interest rate increases by a factor of 1.05 per year. If a population is growing, then its λ is greater than one. The population is stable when λ is 1.0 and declining when λ is less than 1.0.

If a population's age structure is stable, meaning that the proportion of females in each age class is not changing over time, then its finite rate of increase also has a simple relationship to its net reproductive rate: $\lambda = R_0/g$, where g is the generation time. In essence, dividing the net reproductive rate by generation time transforms it into a discrete rate.

Continuous Growth

The parameters λ and r—the finite rate of increase and the per-capita growth rate—have a simple mathematical relationship. The best way to understand their relationship is to recall that λ expresses a population's growth rate over a discrete interval of time. In contrast, r gives the population's per-capita growth rate at any particular instant. This is why r is also called the instantaneous rate of increase. The relationship between the two parameters is given by

$$\lambda = e^r \qquad \textbf{(Eq. 52.6)}$$

where e is the base of the natural logarithm, or about 2.72. (For help with using logarithms, see **BioSkills 5**. Also, note that the relationship between any finite rate and any instantaneous rate is given by finite rate $= e^{\text{instantaneous rate}}$.)

(Continued on next page)

(continued)

Substituting Equation 52.6 into Equation 52.5 gives

$$N_t = N_0 e^{rt} \qquad \textbf{(Eq. 52.7)}$$

This expression summarizes how populations grow when they breed continuously, as do humans and bacteria, instead of at defined intervals. For species that breed continuously, the "interest" on the population is compounded continuously. When the growth rates λ and r are equivalent, however, the differences between discrete and continuous growth are negligible. Because r represents the growth rate at any given time, and because r and λ are so closely related, biologists routinely calculate r for species that breed seasonally. In the whooping crane example, $\lambda = 1.046 = e^r$. To solve for r, take the natural logarithm of both sides. (**BioSkills 5** explains how to do this on your calculator.) In this case, $r = 0.045$.

The instantaneous rate of increase, r, is also directly related to the net reproductive rate, R_0, introduced in Box 52.1. In most cases, r is calculated as $\ln R_0/g$, where g is the generation time. Thus, r can be calculated from life-table data. It is a more useful measure of growth rate than R_0, because r is independent of generation time.

To summarize, biologists have developed several ways of calculating and expressing a population's growth rate. Growth rate expressed as λ has the advantage of being easy to understand, and R_0 has the advantage of being calculated directly from life-table data. Although r is slightly more difficult conceptually, it is the most useful expression for growth rate, because it is independent of generation time and is relevant for species that breed either seasonally or continuously.

Applying the Models

To get a better feel for r and for Equation 52.7, consider the following series of questions about whooping cranes. The key to answering these questions is to realize that Equation 52.7 has just four parameters. Given three of these parameters, you can calculate the fourth. ● If you understand this concept, you should be able to solve the following four problems:

1. If 20 individuals were alive in 1941 and 518 existed in 2007, what is r? Here $N_t = 518$, $N_0 = 20$, and $t = 66$ years. Substitute these values into Equation 52.7 and solve for r. Then check your answer at the end of this box.[1]

2. In the most recent report issued by the biologists working on the Wood Buffalo crane recovery program, it is estimated that the flock should be able to sustain an r of 0.046 for the foreseeable future. If the flock currently contains 237 individuals, how long will it take that population to double? Here $N_0 = 237$ and $N_t = 2 \times 237 = 474$. In this case, you solve Equation 52.7 for t. Then check your answer.[2]

3. In 2002 a pair of birds in the flock that lives in Florida year-round successfully raised offspring (nine years after the first cranes were introduced there). In 2003 another pair of cranes in this population bred successfully. If the number of breeding pairs continues to double each year, how long will it take to reach the goal of 25 breeding pairs? (Note that $\lambda = 2.0$ if a population is doubling each year.) Given that $N_0 = 2$ and $N_t = 25$, solve for t again, and check your answer.[3]

4. The whooping crane flock that migrates between Wisconsin and Florida was founded in 2001. If its development is like that of the resident flock in Florida, the first successful breeding attempt will occur in 2010—nine years after the initial introduction. Suppose that the instantaneous growth rate for the number of breeding pairs in this population will be 0.05. In what year will the breeding population reach 25 pairs? This is the year that whooping cranes should come off the endangered species list. Here $N_0 = 1$, $N_t = 25$, and $r = 0.05$. Solve for t, and add this number of years to 2010; then check your answer.[4]

[1] $r = 0.049$; [2] $t = 15$ years; [3] $t = 3.64$ years; [4] 2074

Figure 52.7a illustrates density-dependent growth for a hypothetical population. Notice that the graph has three sections. Initially, growth is exponential—meaning that r is constant. With time, N increases to the point where competition for resources or other density-dependent factors begins to occur. As a result, the growth rate begins to decline and eventually reaches 0. When the population is at the habitat's carrying capacity, the graph of population size versus time is flat.

Figure 52.7b shows data on logistic growth from laboratory populations of two species of protists in the group called ciliates (see Chapter 29): *Paramecium aurelia* and *P. caudatum*. In this experiment, an investigator placed 20 individuals from one of the *Paramecium* species into 5 mL of a solution. He created many replicates of these 5-mL environments for each species separately. He kept conditions as constant as possible in each replicate by adding the same number of bacterial cells every day for food, washing the solution every second day to remove wastes, and maintaining the pH at 8.0. As the graphs show, both species exhibited logistic growth in this environment. The carrying capacity differed in the two species, however. The maximum density of *P. aurelia* averaged 448 individuals per mL, but that for *P. caudatum* averaged just 128 individuals per mL.

In both of the *Paramecium* species in this experiment, exponential growth could be sustained for only about 5 days. Exponential growth is almost always short lived. When population size increases, a population's growth rate has to decline. Why? What factors cause growth rates to change?

(a) Density dependence: Growth rate slows at high density.

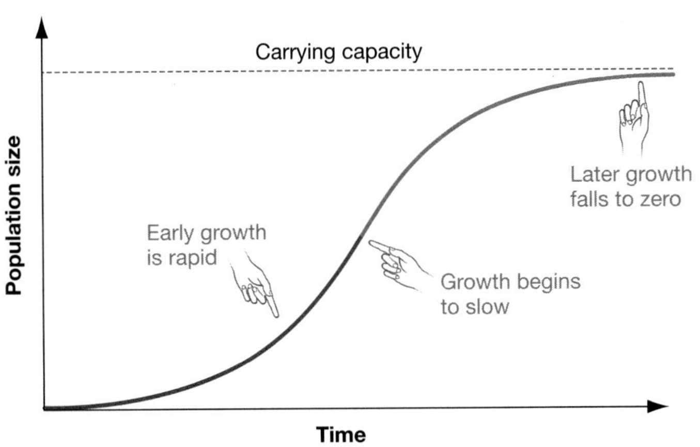

(b) Logistic growth in ciliates

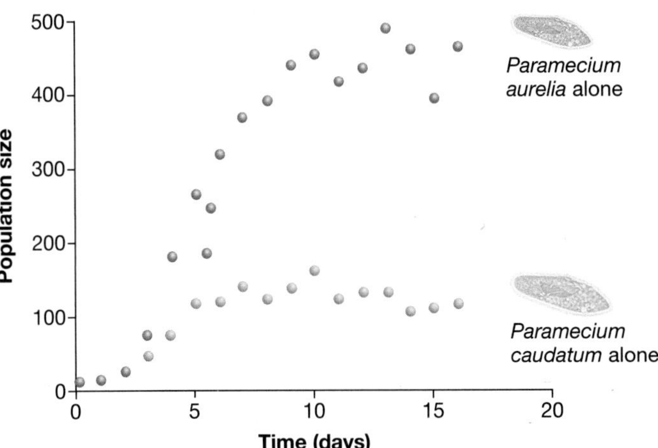

FIGURE 52.7 Logistic Growth Is Dependent on Population Size. (a) Hypothetical curve illustrating the pattern predicted by the logistic growth equation. This pattern often occurs when a small number of individuals colonizes an unoccupied habitat. Initially, *r* is high because competition for resources is low to nonexistent. Carrying capacity depends on the quality of the habitat and can vary over time. **(b)** Data from laboratory experiments with two species of *Paramecium*.

EXERCISE In part (b), draw dashed horizontal lines indicating the carrying capacity for each species.

What Limits Growth Rates and Population Sizes?

Population sizes change as a result of two general types of factors: (1) density-independent and (2) density-dependent factors. Density-independent factors alter birth rates and death rates irrespective of the number of individuals in the population. Density-independent changes are usually triggered by changes in the abiotic environment—variation in weather patterns or catastrophic events such as cold snaps, hurricanes, volcanic eruptions, or drought. In contrast, density-dependent factors are usually biotic and change in intensity as a function of population size. For example, predation rates on deer may increase when deer population sizes are high, or competition for food and starvation may become more common. When trees are crowded, they have less water, nutrients, and sunlight at their disposal and make fewer seeds. Death rates may increase and birth rates decrease when populations are at high density.

To get a better understanding of how biologists study the density-dependent factors that affect population size, consider the data in **Figure 52.8**:

- The graph in Figure 52.8a presents results of an experimental study of a coral-reef fish called the bridled goby. Each data point represents an identical artificial reef, constructed by a researcher from the rubble of real coral reefs. The initial density, plotted along the horizontal axis, is the number of marked bridled gobies that were released on each artificial reef at the start of the experiment. The proportion surviving, plotted on the vertical axis, is the number of these introduced individuals that were still living at that reef 2.5 months later. This graph shows a strong density-dependent relationship in survivorship.
- The graph in Figure 52.8b is from a long-term study of song sparrows on Mandarte Island, British Columbia. Each data point represents a different year. The density of females, plotted along the horizontal axis, is the number of females that bred on the island; the clutch size, plotted on the vertical axis, is the average number of eggs laid by each female. This graph indicates a strong density-dependent relationship in fecundity.

Density-dependent changes in survivorship and fecundity cause logistic population growth. In this way, density-dependent factors define a particular habitat's carrying capacity. If gobies get crowded, they die or emigrate. If song sparrows are crowded, the average number of eggs and offspring that they produce declines.

It's important to recognize that *K* varies among species and populations, however, and that this variation affects both growth rates and population sizes. *K* varies because for any particular species, some habitats are better than other habitats due to differences in food availability, predator abundance, and other density-dependent factors. Stated another way, *K* varies in space. It also varies with time, as conditions in some years are better than in others. In addition, the same region may have a very different carrying capacity for different species. The same area will tend to support many more individuals of a small-bodied species than of a large-bodied species, simply because large individuals demand more space and resources. These simple observations help explain the variation in total population size that exists among species and among populations of the same species.

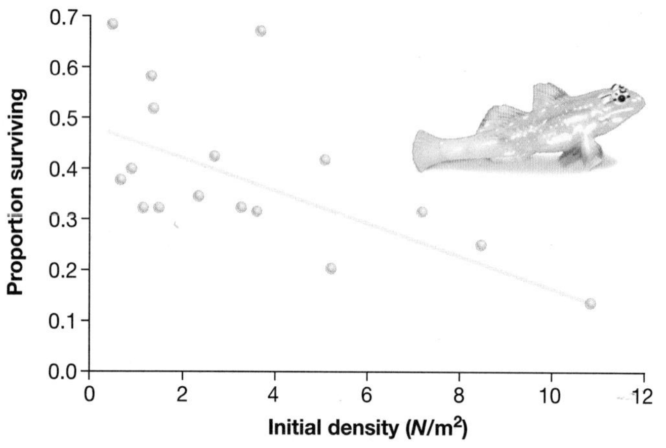

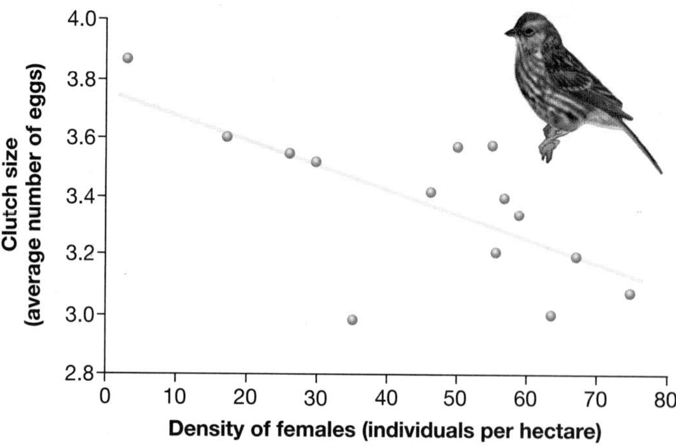

FIGURE 52.8 Density-Dependent Growth Results from Changes in Survivorship and Fecundity. (a) When bridled gobies are introduced at different population densities to artificial reefs, survivorship is highest at low density. **(b)** In the song sparrow population on Mandarte Island, there is a strong negative correlation between population density in a particular year and the average number of eggs (clutch size) produced by females.

QUESTION When song sparrow populations are at high density and extra food is provided to females experimentally, average clutch size is much higher than expected from the data in part (b). Based on these data, state a hypothesis to explain why average clutch size declines at high density.

Check Your Understanding

If you understand that...

- Populations grow exponentially unless slowed by density dependence.
- Density-dependent factors that influence population growth rates include predation, disease, and competition for resources such as food and sunlight.

You should be able to...

1) Propose a density-dependent factor that limits growth in *Paramecium aurelia* and design an experiment to test this hypothesis.
2) Propose a hypothesis to explain why the carrying capacity for *P. caudatum* in Figure 52.7b is so much lower than the carrying capacity for *P. aurelia* and design an experiment to test this hypothesis.

MB Web Animation at www.masteringbio.com
Modeling Population Growth

52.3 Population Dynamics

The tools introduced in the previous two sections provide a foundation for exploring how biologists study **population dynamics**—changes in populations through time. Research on population dynamics has uncovered a wide array of patterns in natural populations in addition to exponential and logistic growth.

As an initial example of how population sizes change through time, consider data on the longest-running experiment in the history of biological science: the Park Grass study in Rothamsted, United Kingdom. In 1856 researchers established a series of 0.1–0.2-ha (about 1/3-acre) plots in a hay meadow that had uniform soil characteristics and vegetation. Since then, some of the plots have received regular fertilizer treatments or lime (a mix of calcium and magnesium oxides), and some have been left untreated as controls. Each year researchers recorded which plant species were present in each plot, as an index of population size.

The weather was the same in all plots, the fertilizing and liming regimes were constant in treated plots, and presumably all plots were exposed to the same predators and diseases. Even so, several dramatic changes in population size have taken place. When the data from 1920 to 1979 were analyzed for the control plots, four major patterns emerged among the 43 species present. Ten of the species showed a clear spike in abundance, followed by a decline over the course of the 60-year period (**Figure 52.9a**). Seven species increased in abundance and then reached a plateau (**Figure 52.9b**). Five species declined steadily, and 21 showed no discernible change (**Figure 52.9c**).

In several cases, patterns of change over time appeared to correlate with life-history traits. For example, species that maintained a constant population size throughout the 60-year interval have longer life spans and a lower r_{max} than do other species at the site. To interpret this observation, biologists suggest that these species allocate more energy and resources to competitive ability than to reproduction. As a result, these species are able to maintain abundance over a longer period. In contrast, species that spiked and declined or that increased and maintained high abundance had a higher r_{max}, an earlier age at first reproduction, and shorter life spans than did species

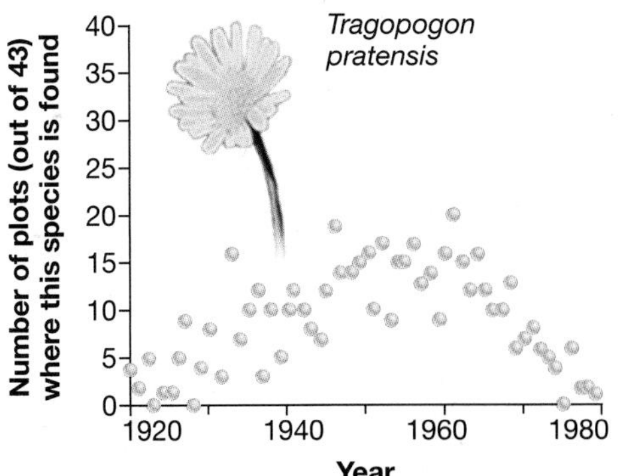

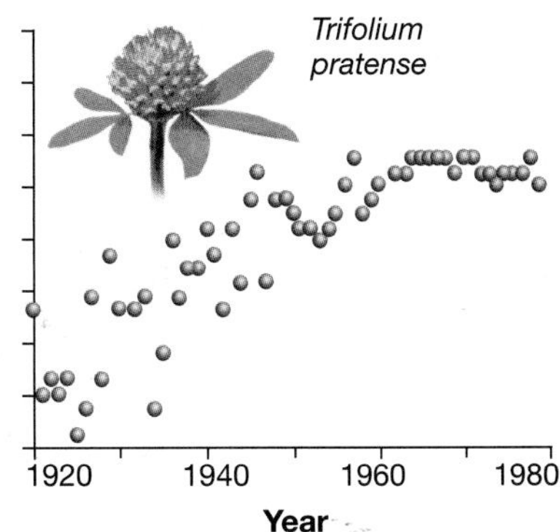

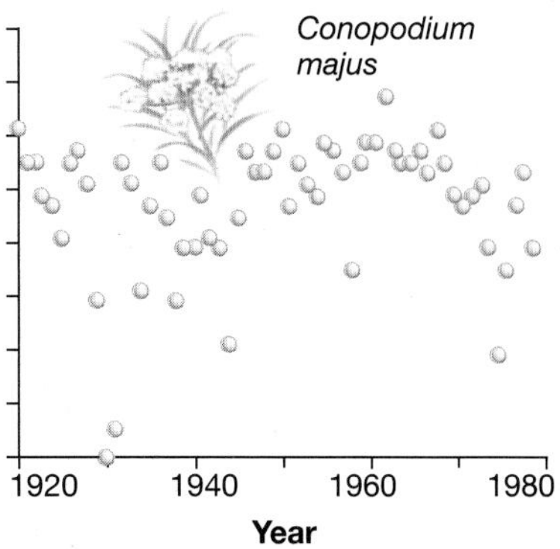

FIGURE 52.9 Distinct Patterns of Long-Term Changes in the Populations of Meadow Species. Graphs showing some of the distinct patterns in population dynamics from 1920 to 1979 among species in the Park Grass experiment, which began in 1856.

that maintained a constant population size. Researchers are still working on the question of why some of these species remained constant over time and why others declined.

How Do Metapopulations Change through Time?

If you browse through an identification guide for trees or birds or butterflies, you'll likely find range maps indicating that most species occupy a broad area. In reality, however, the habitat preferences of many species are restricted, and individuals occupy only isolated patches within that broad area. The meadows occupied by Glanville fritillaries—an endangered species of butterfly native to the Åland islands off the coast of Finland (**Figure 52.10**)—are a well-studied example of this pattern.

If individuals from a species occupy many small patches of habitat, so that they form many independent populations, they are said to represent a **metapopulation**, or population of populations. Glanville fritillaries exist naturally as metapopulations. But because humans are reducing large, contiguous areas of forest and grasslands to isolated patches or reserves, more and more species are being forced into a metapopulation structure.

Recent research on Glanville fritillaries by Ilkka Hanski and colleagues illustrates the consequences of metapopulation structure for endangered species. The work began with a survey of meadow habitats on the islands. Because fritillary caterpillars feed on just two types of host plant, *Plantago lanceolata* and *Veronica spicata*, the team was able to pinpoint potential butterfly habitats. They estimated the fritillary population size within each patch by counting the number of larval webs in each. Of the 1502 meadows that contained the host plants, 536 had Glanville fritillaries. The patches ranged from 6 m^2 to 3 ha in area. Most had only a single breeding pair of adults and one larval group, but the largest population contained hundreds of pairs of breeding adults and 3450 larvae.

Glanville fritillaries are an endangered species

They live on patches of certain host plants within meadows

FIGURE 52.10 Glanville Fritillaries Live in Patches of Meadow. The Glanville fritillary is extinct in much of its former range; it is now found only on the Åland islands near Finland.

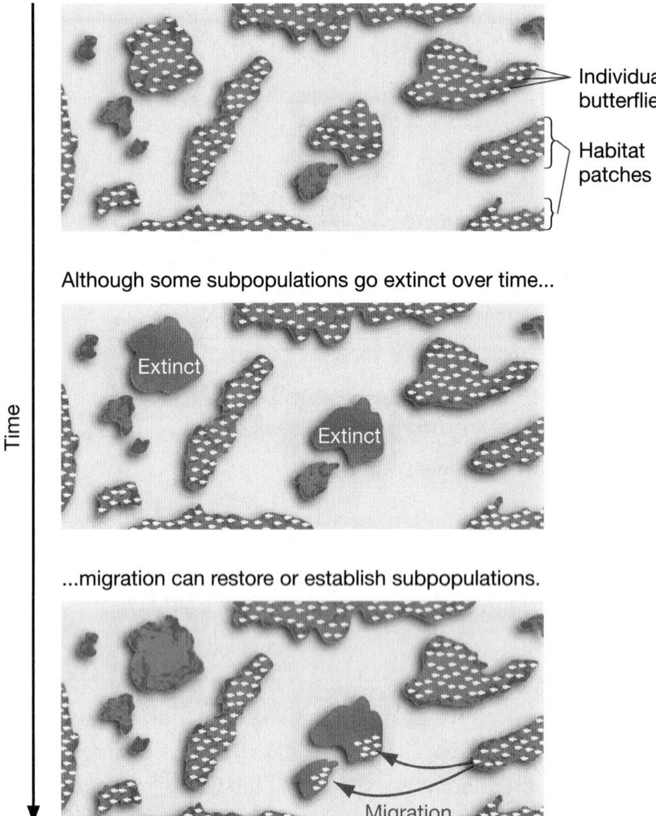

FIGURE 52.11 Metapopulation Dynamics Depend on Extinction and Recolonization. The overall population size of a metapopulation stays relatively stable even if subpopulations go extinct. These populations may be restored by migration, or unoccupied habitats might be colonized.

Figure 52.11 illustrates how a metapopulation like this is expected to change over the years. Given enough time, each population within the larger metapopulation is expected to go extinct. The cause could be a catastrophe, such as a storm or an oil spill; it could also be a disease outbreak or a sudden influx of predators. Migration from nearby populations can reestablish populations in these empty habitat fragments, however. In this way, there is a balance between extinction and recolonization. Even though subpopulations blink on and off over time, the overall population is maintained at a stable number of individuals.

Does this metapopulation model correctly describe the population dynamics of fritillaries? To answer this question, Hanski's group conducted a mark-recapture study. (**Box 52.3** on page 1188 provides details on how a mark-recapture study is done.) Of the 1731 butterflies that the biologists marked and released, 741 were recaptured over the course of the summer study period. Of the recaptured individuals, 9 percent were found in a previously unoccupied patch. This migration rate is high enough to suggest that patches where a population has gone extinct will eventually be recolonized. Hanski and co-workers repeated the survey two years after their initial census. Just as the metapopulation model predicts, some populations had gone extinct and others had been created. On average, butterflies were lost from 200 patches each year; 114 unoccupied patches were colonized and newly occupied each year. The overall population size was relatively stable even though constituent populations came and went.

To summarize, the history and future of a metapopulation is driven by the birth and death of populations, just as the dynamics of a single population are driven by the birth and death of individuals. As the final section in this chapter will show, these dynamics have important implications for saving endangered species.

Why Do Some Populations Cycle?

One of the most striking patterns in population dynamics has also been one of the most difficult to understand. **Population cycles** are regular fluctuations in size that some animal populations exhibit. **Figure 52.12** shows a classic case of population cycling: snowshoe hare and lynx populations in northern Canada. Snowshoe hares are herbivores and are the primary prey of lynx.

Most hypotheses to explain population cycles hinge on some sort of density-dependent factor. The idea is that predation, disease, or food shortages intensify dramatically at high population density and cause population numbers to crash. It has been extremely difficult, however, to document a causal

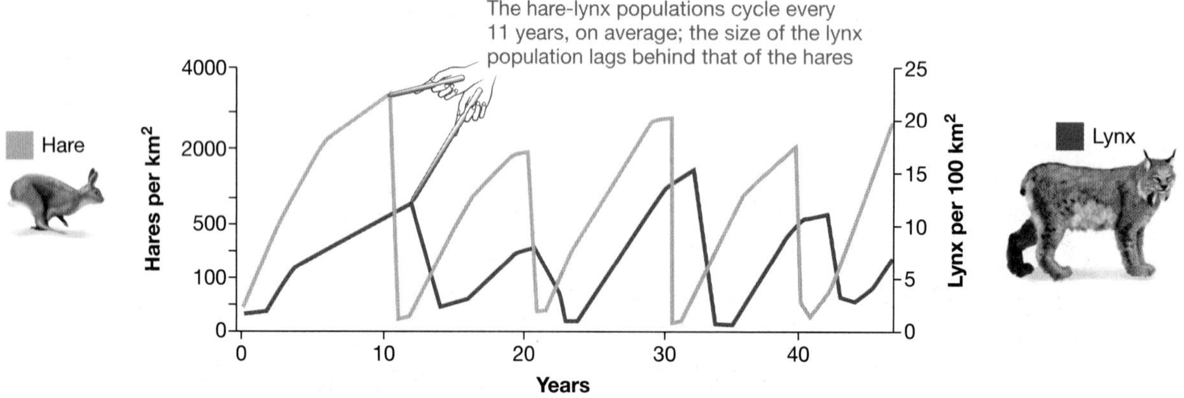

FIGURE 52.12 Hare and Lynx Populations Cycle in Synchrony. To make it easier to read, the left-hand vertical axis, which indicates hare density, is not linear.

relationship between population cycles and factors such as predation intensity, disease outbreaks, or food availability.

Although the hare-lynx cycle was documented in the early 1900s, it was not well understood. Two hypotheses were proposed to explain these population dynamics:

1. Hares use up all their food when their populations reach high density and starve; in response, lynx also starve.
2. Lynx populations reach high density in response to increases in hare density. At high density, lynx eat so many hares that the prey population crashes.

Stated another way, either hares control lynx population size or lynx control hare population size.

To test these hypotheses rigorously, researchers set up a series of 1-km^2 study plots in boreal forest habitats that were as identical as possible (**Figure 52.13**). Three plots were left as unmanipulated controls. One plot was ringed with an electrified fence with a mesh that excluded lynx but allowed hares to pass freely. Two plots received additional food for hares year-round. One plot had a predator-exclusion fence and was also supplemented with food for hares year-round. The biologists then monitored the size of the hare and lynx populations over an 11-year period, or enough time for a complete cycle in the two populations.

As the data in Figure 52.13 show, plots with predators excluded showed higher hare populations at the peak of the cycle than did control plots. This result supports the hypothesis that

Experiment

Question: What factors control the hare-lynx population cycle?

Hypothesis: Predation, food availability, or a combination of those two factors controls the hare-lynx cycle.

Null hypothesis: The hare-lynx cycle isn't driven by predation, food availability, or a combination of those two factors.

Experimental setup:

Document hare population in seven study plots (similar boreal forest habitats, each 1 km^2) over 11 years (duration of one cycle).

3 plots: Unmanipulated controls

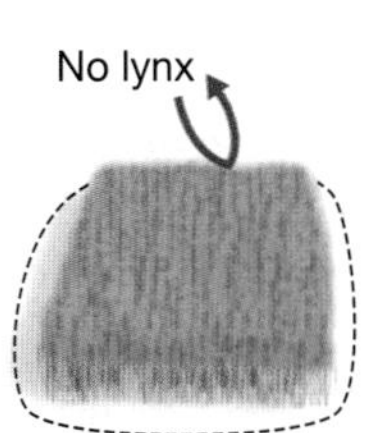

1 plot: Electrified fence excludes lynx but allows free access by hares.

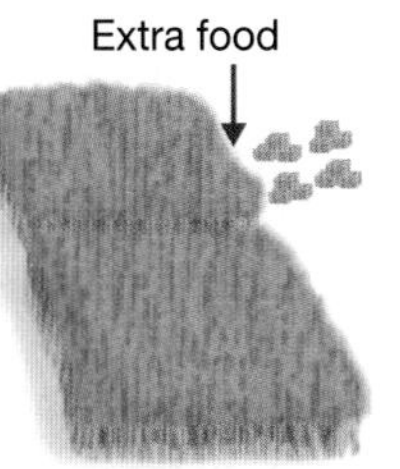

2 plots: Supply extra food for hares.

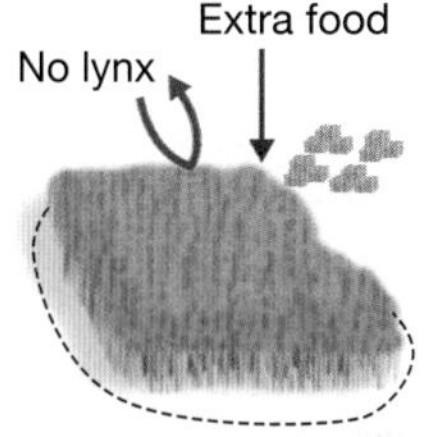

1 plot: Electrified fence excludes lynx but allows free access by hares; supply extra food for hares.

Prediction: Hare populations in at least one type of manipulated plot will be higher than the average population in control plots.

Prediction of null hypothesis: Hare populations in all of the plots will be the same.

Results:

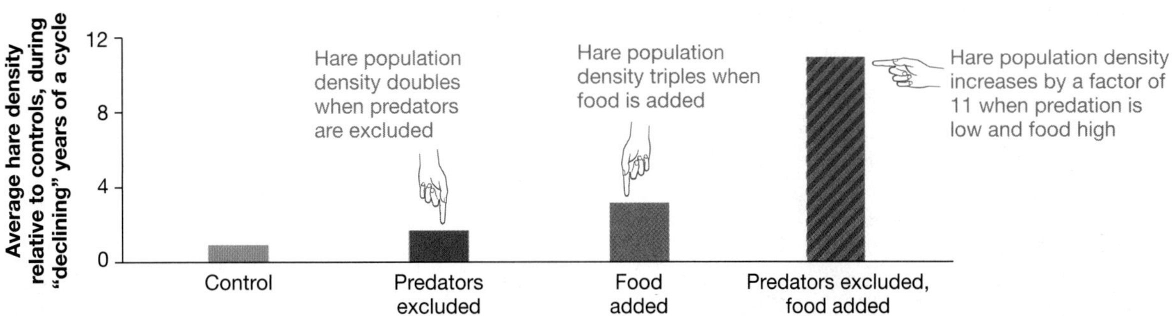

Conclusion: Hare populations are limited by both predation and food availability. When predation and food limitation occur together, they have a greater effect than either factor does independently.

FIGURE 52.13 Experimental Evidence that Predation and Food Availability Drive the Hare-Lynx Cycle.

QUESTION State a hypothesis to explain why lack of predation and increased food, in combination, have such a large effect on hare population size.

predation by lynx reduces hare populations. Plots with supplemental food had even higher peak populations, however, and the plot with supplemental food and predators excluded had a hare population over 16 times as large as the average in control plots.

These data support the hypotheses that hare populations are limited by availability of food as well as by predation, and that food availability and predation intensity interact—meaning the combined effect of food and predation is much larger than their impact in isolation. Research continues to figure out why this is so. In the meantime, the study illustrates the power of a well-designed experiment to solve long-standing controversies in biology.

How Does Age Structure Affect Population Growth?

As the life-table data presented in Section 52.1 showed, age has a dramatic effect on the probability that an individual will reproduce and survive to the following year. But the previous analysis of life-table data left out a critical point about the ages of individuals in a population: A population's **age structure**—meaning the proportion of individuals that are at each possible age—has a dramatic influence on the population's growth over time. To see how changes in age structure can affect population dynamics, let's consider two case histories for which biologists have documented changes in age structure—one example concerns a flowering plant, and the other focuses on our own species.

Age Structure in a Woodland Herb The common primrose, pictured in **Figure 52.14a**, grows in the woodlands of Western Europe. Primroses grow on the forest floor and can germinate, mature, and produce offspring only in the relatively high-light environment created when a tall tree falls and opens a gap in the forest canopy. These sunny spaces are short lived, however, because mature trees and saplings in and around the gap grow and fill the space. As they do, light levels in the gap decline and the environment becomes increasingly unsuitable for herbs such as primrose.

Biologists recently set out to document how primrose populations respond to this ephemeral woodland environment. The researchers hypothesized that populations in new gaps and populations in large gaps—meaning those with the highest light availability—would experience high rates of reproduction and an age structure characterized by large numbers of juveniles. The researchers also hypothesized that as light levels declined in a particular gap due to the growth of surrounding trees, primrose reproduction would decline. As a result, the age structure of populations in older gaps should be characterized by a large proportion of individuals in adult stages, with fewer juveniles.

To test these hypotheses, the investigators selected eight common primrose populations growing in a range of light levels, from high to low. The team established 1-m^2 study plots inside each gap and studied approximately 350 individuals per population. Each individual was assigned to one of five ages: seedlings, juveniles, or adults in one of three size classes. As **Figure 52.14b** shows, the data supported the hypothesis that in gaps with high light levels, populations have a larger proportion of juveniles than do gaps with low light levels. In addition, the proportion of juvenile individuals declined over time in all gaps as light levels declined. These data have three important messages about the dynamics of primrose populations:

1. Populations that are dominated by juveniles should experience rapid growth. Population size should then decline with time, however, due to a density-independent factor: shading by trees.

(a) Common primroses live in sunlit gaps in forests.

(b) Age structure of primrose population varies with age of gap.

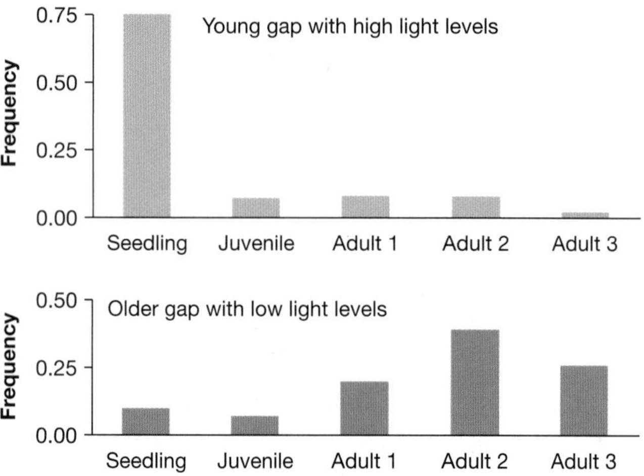

FIGURE 52.14 The Age Structure of Common Primrose Populations Changes over Time. In the graphs, "Adult 1," "Adult 2," and "Adult 3" individuals are of increasingly large size and thus age.

EXERCISE Create a third graph showing the age distribution of a gap that has completely filled in and closed, meaning that light levels are now as low as they were before the gap existed.

2. The long-term trajectory of the overall primrose population in an area may depend primarily on the frequency and severity of windstorms that knock down trees and create sunlit gaps. If so, it means that the dynamics of primrose populations are governed by an abiotic, density-independent factor.

3. In a large tract of forest, the primrose population will consist of a large number of subpopulations, each found in canopy gaps. Some of these subpopulations are likely to be very small, and each subpopulation will arise, grow, and then go extinct over time. Recall that biologists say that a group of small, isolated populations such as this represents a metapopulation, a population of populations. Primroses have metapopulation structure.

Age Structure in Human Populations As another example of how age structure affects population dynamics, consider humans. In countries where industrial and technological development occurred many decades ago, survivorship has been high and fecundity low for several decades. As a result, these populations are not growing quickly. When researchers create a graph with horizontal bars representing the numbers of males and females of each age group for such a population, an age pyramid like that in **Figure 52.15a** results. The most striking pattern in these data is that there are similar numbers of people in most age classes. The evenness occurs because the same number of infants are being born each year and because most survive to old age. In this way, analyzing an age pyramid can give biologists important information about a population's history.

Studying age distributions can also help researchers predict a population's future. For example, the populations of developed nations are not expected to grow especially quickly, because only modest numbers of individuals will reach reproductive age in the near future. The white lines and lighter bars in Figure 52.15a show the projected age structure of a developed nation (Sweden) in 2050. Modest changes in the age distribution have occurred since 2000 because survivorship has increased while fecundity has remained the same. The projections highlight a major public policy concern in countries such as this: how to care for an increasingly aged population.

In contrast, the age distribution is bottom-heavy in less-developed nations. As **Figure 52.15b** shows, these populations are dominated by the very young. This type of age distribution occurs when populations have undergone rapid growth—with more children being born than exist in older age classes. Due to dramatic improvements in health care, most people in the younger age classes now survive to reproductive age.

For Figure 52.15b, the projected age distribution in 2050 in Honduras is based on continued high survivorship and a pattern that has occurred repeatedly around the world: When infant mortality drops in a country, fecundity rates begin to decline within a generation or two. The white lines and light bars in Figure 52.15b are relatively even for ages 0–55—meaning that rapid growth has stopped—because fecundity is expected to be relatively low and survivorship high. The projected data also illustrate a major public policy concern in less-developed countries—providing education and jobs for an enormous group of young people who will be reaching adulthood during your lifetime.

The data in Figure 52.15b make another important point: Because of recent and rapid growth in developing countries, overall population size will increase dramatically in these nations over the course of your lifetime. A large part of the increase will be due to increased survivorship. But the number

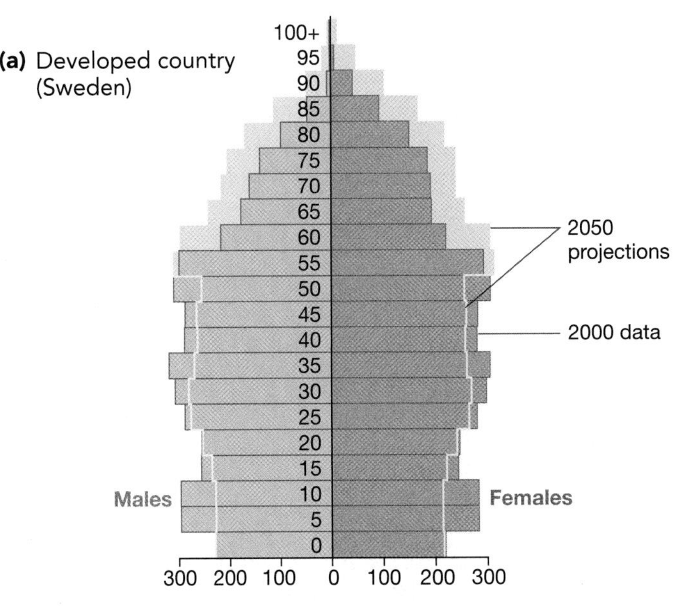

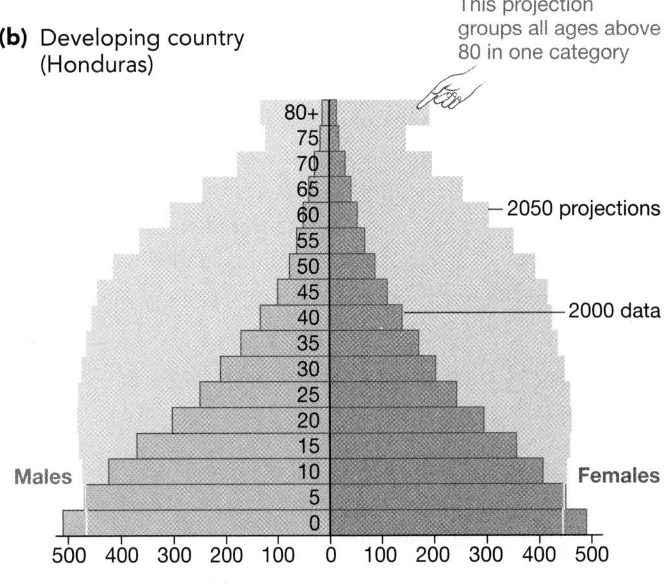

FIGURE 52.15 The Age Structure of Human Populations Varies Dramatically. Age distributions showing the number of males and females in 5-year age increments from 0 to 100.

BOX 52.3 Mark-Recapture Studies

To estimate the population size of sedentary (largely immobile) organisms, researchers sample appropriate habitats by counting the individuals that occur along transects or inside rectangular plots—called quadrats—set up at random locations in the habitat. These counts can be extrapolated to the entire habitat area to estimate the total population size. In addition, they can be compared to later censuses to document trends over time.

In contrast, estimating the total population size of organisms that are mobile and that do not congregate into herds or flocks or schools is much more of a challenge. In species such as the Glanville fritillary, the population inside sample quadrats or along transects changes constantly as individuals move in and out. Further, it can be difficult to track whether a particular individual has already been counted.

If individuals can be captured and then tagged in some way, however, the total population size of a mobile species can be estimated by using an approach called mark-recapture. To begin a mark-recapture study, researchers catch individuals and mark them with leg bands, ear tags, or some other method of identification. After the marked individuals are released, they are allowed to mix with the unmarked animals in the population for a period of time. Then a second trapping effort is conducted, and the percentage of marked individuals that were captured is recorded.

To estimate the total population from these data, researchers make a key assumption: The percentage of marked and recaptured individuals is equal to the percentage of marked individuals in the entire population. This assumption should be valid when no bias exists regarding which individuals are caught in each sample attempt. It is important that individuals do not learn to avoid traps after being caught once and that they do not emigrate or die as a result of being trapped.

The relationship between marked and unmarked individuals can be concisely expressed algebraically:

$$\frac{m_2}{n_2} = \frac{n_1}{N}$$

In this equation, m_2 is the number of marked animals in the second sample (the recapture), n_2 is the total number of animals (marked and unmarked) in the second sample, n_1 is the number of marked animals in the first sample, and N is the total population size. Having measured m_2, n_2, and n_1, the researcher can estimate N. ● If you understand how to estimate population size in a mark-recapture study, you should be able to do so in this example: Suppose researchers marked 255 animals and later were able to trap a total of 162 individuals in the population, of which 78 were marked. What is the estimate for total population size? Solve for N and check your answer.[1]

[1] $N = 530$

of offspring being born each year is also expected to stay high, even though fecundity is predicted to decline significantly. This result seems paradoxical but is based on a key observation: There are now so many young women in these populations that the overall number of births will stay high even though the average number of children per female is much less than it was a generation ago. To capture this point, biologists say that these populations have momentum or inertia. Combined with high survivorship, population pyramids like these make continuing increases in overall population size inevitable.

As these examples show, understanding age structure is a key component of analyzing population dynamics in humans and other species. Now let's take a more detailed look at recent and projected changes in the world population of humans.

Analyzing Change in the Growth Rate of Human Populations

Studies on changes in whooping crane and the Park Grass populations over 60-year periods have been extraordinarily valuable. But six decades turns out to be short in comparison to data available on human population dynamics.

Archeological and anthropological data have been used to estimate the size of human populations over the past 250 years (**Figure 52.16**). Although the shape of the curve is superficially similar to the examples of exponential growth in Figure 52.5, the growth rate for humans has actually increased over time since 1750, leading to a steeply rising curve.

It is difficult to overemphasize just how dramatically the human population has increased over the past 250 years. **Table 52.2** presents data that will allow you to calculate how long it has taken this population to add each billion people, and to determine how many times the population has doubled—from a starting point in the year A.D. 1420—and how many years the

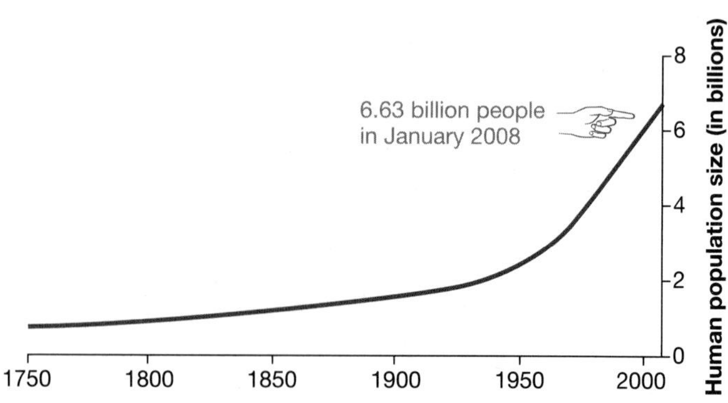

FIGURE 52.16 The Human Population Has Been Growing Rapidly. Graph based on estimates from historical and census data.

TABLE 52.2 **Milestones in Human Population Growth**

Date	Human Population (in hundreds of millions)
1420	375
1720	750
1800	1000
1875	1500
1927	2000
1961	3000
1974	4000
1987	5000
1999	6000
Number of years required to add 1 billion people:	
Number of years required to double population:	

EXERCISE Using the data provided, fill in the correct number of years in the blank cells in the table. To calculate the time required for the human population to reach 1 billion, assume that *Homo sapiens* originated 150,000 years ago.

population took to double each time. To put your calculations on doubling time in perspective, consider that until your grandparents' generation, no individual had lived long enough to see the human population double. But many members of the generation born in the early 1920s have lived to see the population *triple*. And to drive home the impact of continued population doubling, consider the following: If you were given a penny on January 1, then \$0.02 on January 2, \$0.04 on January 3, and so on, you would be handed \$10,737,418 on January 31.

As this book goes to press, the world population is estimated at 6.6 billion. About 77 million additional people—equivalent to the current population of Egypt, or more than double the state of California—are being added each year.

It is not possible to overstate the consequences of recent and current increases in human population. In addition to being the primary cause of the habitat losses and species extinctions analyzed in Chapter 55, overpopulation is linked by researchers to declines in living standards, mass movements of people, political instability, and acute shortages of water, fuel, and other basic resources in many parts of the world.

The one encouraging trend in the data is that the growth rate of the human population has already peaked and begun to decline. The highest growth rates occurred between 1965 and 1970, when populations increased at an average of 2.04 percent per year. Between 1990 and 1995, the overall growth rate in human populations averaged 1.46 percent per year; the current growth rate is 1.2 percent annually. In humans, r may be undergoing the first long-term decline in history. The question is: Will the human population stabilize or begin to decline in time to prevent global—and potentially irreversible—damage?

Will Human Population Size Peak in Your Lifetime? The United Nations Population Division makes regular projections for how human population size will change between now and 2050. For most readers of this book, these projections describe what the world will look like as you reach your early 60s.

The UN projections are based on four scenarios, which hinge on different values for fertility rates—the average number of children that each woman has during her lifetime. Currently, the worldwide average is 2.7. This represents an enormous reduction from the fertility rates during the 1950s, which averaged 5.0 children per woman. **Figure 52.17** shows how total population size is expected to change by 2050 if average fertility continues to decline and averages 2.5 (high), 2.1 (medium), or 1.7 (low) children per woman. The middle number, 2.1, **replacement rate**, where each woman produces exactly enough offspring to replace herself and her offspring's father. When this fertility rate is sustained for a generation, $r = 0$ and there is **zero population growth** (**ZPG**).

A glance at Figure 52.17 should convince you that the four scenarios are starkly different. If average fertility around the world stays at its present level, world population will be closing in on 12 billion about the time you might consider retiring. This is close to double the current population. To realize what this might mean, imagine what traffic or home prices in your city would be like if double the number of people were using the roads and needing housing.

The "high-fertility" projection assumes that fertility rates continue to drop but average 2.5 children per woman over this interval. This model predicts that world population in 2050 will reach nearly 11 billion—a 75 percent increase over today's population—with no signs of peaking. The low-fertility projection, in contrast, predicts that the total human population in 2050 will be about 7.4 billion and will already have peaked.

The graphs make an important point: The future of the human population hinges on fertility rates—on how many children each of the women living today decides to have. Those decisions, in turn, depend on a wide array of factors,

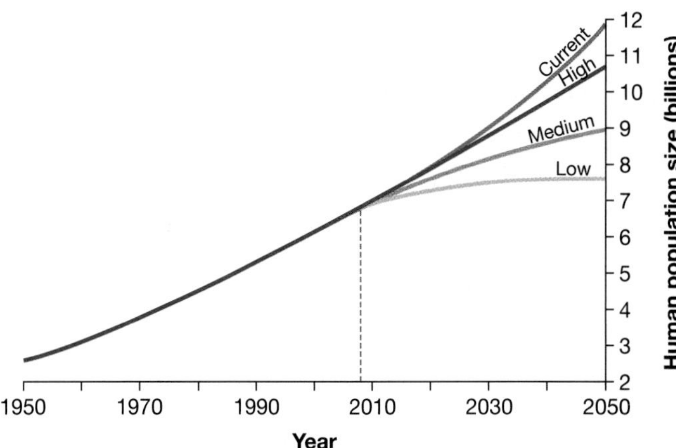

FIGURE 52.17 Projections for Human Population Growth to 2050. The "Current," "High," "Medium," and "Low" labels refer to fertility rates.

including how free women are to choose their family size and how much access women have to education. When women are allowed to become educated, they tend to delay when they start having children and have a smaller overall family size. Access to education and reliable birth control methods (see Chapter 48), in addition to overall economic development, access to quality health care, and other issues—including the course of the AIDS epidemic (see Chapter 35)—will all play a large role in determining how world population changes over your lifetime.

To summarize, humans may be approaching the end of a period of rapid growth that lasted well over 500 years. How quickly growth rates decline and how large the population can eventually become will be decided primarily by changes in fertility rates.

Check Your Understanding

If you understand that...

- Biologists use observational data—such as the life history of species in the Park Grass experiment—or experiments—such as manipulating food and predation in snowshoe hare populations—to text hypotheses about changes in the size of populations over time.
- In many cases, understanding population dynamics depends on understanding changes in a population's age structure.

You should be able to...

1) Explain why the age structure of human populations differs between developed and developing nations.
2) Explain why changes in fertility rates have such a dramatic impact on projections for the total human population in 2050.

at www.masteringbio.com

Human Population Growth and Regulation

52.4 How Can Population Ecology Help Endangered Species?

Conservationists draw heavily on concepts and techniques from population ecology when designing programs to save species threatened with extinction. As habitat destruction and invasive species push species throughout the world into decline, the study of population growth rates and population dynamics has taken on an increasingly applied tone.

To illustrate one of the ways that biologists apply the theory and results introduced in this chapter, let's analyze how an understanding of life-table data and geographic structure can help direct conservation action.

Using Life-Table Data to Make Population Projections

Demographic data such as age-specific survivorship and fecundity are important for saving endangered species and for other applied problems. To understand why, suppose that you were in charge of reintroducing a population of lizards to a nature reserve, and that research conducted when this species occupied the site previously documented the survivorship and fecundity data. Your initial plan is to take 1000 newly hatched females from a captive breeding center and release them into the habitat. Is this population likely to become established and grow, or will you need to keep introducing offspring that have been raised in captivity?

To answer this question, you need to use the life-table data in **Figure 52.18a** and calculate (1) how many adults will survive to each age class each year and (2) how many offspring will be produced by each adult age class each year over the course of several years. **Figure 52.18b** starts these calculations for you. The fate of the original 1000 females is indicated in red. In the second year, just 330 of these individuals have survived; 40 are left as 3-year-olds.

How many offspring did this generation produce? As just stated, 330 of the original 1000 females remain after one year. As the first purple number in Figure 52.18b indicates, these survivors have an average of 3.0 female offspring apiece, so they contribute 990 new juveniles to the population. In the second year, the 200 females that are left have an average of 4.0 female offspring each and contribute 800 juveniles. In their third year, the 40 surviving females average 5.0 offspring and contribute 200 juveniles.

Figure 52.18c extends the calculations by showing what happens as the offspring of the original females begin to breed. Their offspring are shown in green. By adding subsequent generations and continuing the analysis, you should be able to predict whether the population will stay the same, decline, or increase over time. In this way, life-table data can be used to predict the future of populations.

Part of the value of a population projection based on life-table data—like the one begun in Figure 52.18—is that it allows biologists to alter values for survivorship and fecundity at particular ages and assess the consequences. For example, suppose that a predatory snake began preying on juvenile lizards. According to the model in Figure 52.18, what would be the impact of a change in juvenile mortality rate? Analyses like this allow biologists to determine which aspects of survivorship and fecundity are especially sensitive for particular species. The studies done to date support some general conclusions:

- Whooping cranes, sea turtles, spotted owls, and many other endangered species have high juvenile mortality, low adult mortality, and low fecundity. In these species, the fate of a population is extremely sensitive to increases in adult mortality. Based on this insight, conservationists have

(a) Life table

Age (x)	Survivorship (l_x)	Fecundity (m_x)
0 (birth)	----	0.0
1	0.33	3.0
2	0.2	4.0
3	0.04	5.0

(b) Fate of first-generation females

Year	0 (newborns)	1-year-olds	2-year-olds	3-year-olds	Total population size (*N*)
1st	1000 (just introduced)				1000
2nd	990 (= 330 × 3.0)	330 (= 1000 × 0.33)			1320 (= 990 + 330)
3rd	800 (= 200 × 4.0)		200 (= 1000 × 0.20)		
4th	200 (= 40 × 5.0)			40 (= 1000 × 0.04)	
5th					

(c) Fate of first- and second-generation females

Year	0 (newborns)	1-year-olds	2-year-olds	3-year-olds	Total population size (sum across all rows)
1st	1000				1000
2nd	990	330			1320
3rd	800 + 981 (981 = 327 × 3.0)	327 (= 990 × 0.33)	200		2308 (= 800 + 981 + 327 + 200)
4th	200 + 792 (792 = 198 × 4.0)		198 (= 990 × 0.20)	40	
5th	195 (195 = 39 × 5.0)			39 (= 990 × 0.04)	

FIGURE 52.18 Life-Table Data Can Be Used to Project the Future of a Population. (a) Life table providing age-specific survivorship and fecundity for a hypothetical population of lizards. **(b)** Predicted fate of 1000 one-year-old females introduced into a habitat just before the breeding season. The number of individuals in this cohort is shown in red; the number of offspring they produce each year is indicated in purple. **(c)** Extension of the data in part (b), indicating how many of the offspring produced by the original females in their first year survived in subsequent years, shown in purple, and how many offspring they produced in each subsequent year, shown in green.

EXERCISE Assume that all 4-year-old females die after producing three young. Fill in the 4th and 5th years in the table.

recently begun an intensive campaign to reduce the loss of adult female sea turtles in fishing nets. Previously, most conservation action had focused on protecting eggs and nesting sites.

- In humans and other species with high survivorship in most age classes, rates of population growth are extremely sensitive to changes in age-specific fecundity. Because of this, programs to control human population growth focus on two issues: lowering fertility rates through the use of birth control, and delaying the age of first reproduction by improving women's access to education.

In some or even most cases, however, the population projections made from life-table data may be too simplistic to be useful. For example, conservationists may need to expand the basic demographic models to account for occasional disturbances such as fires or storms or disease outbreaks. And what about species where overall population dynamics are dictated by a metapopulation structure?

Preserving Metapopulations

Decades of work have convinced biologists that the conclusions about metapopulation dynamics introduced in the previous section are correct, and that an increasing number of species are being forced into a metapopulation structure. Habitat destruction caused by suburbanization and other human activities leaves small populations isolated in pockets of intact habitat.

Work on Glanville fritillaries and other species has shown that a small, isolated population—even one within a nature preserve—is unlikely to survive over the long term. In Glanville fritillaries, data collected by Hanski's group have identified the attributes of subpopulations that are most likely to persist: They are large, occupy larger geographical areas, and are closer to neighboring populations (and hence more likely to be colonized). The data also indicate that fritillary populations with low genetic diversity—probably due to inbreeding—are more likely to go extinct than are populations of the same size that have higher genetic diversity.

Results like these have important messages for conservation biologists:

- Areas that are being protected for threatened species should be substantial enough in area to maintain large populations that are unlikely to go extinct in the near future.
- When it is not possible to preserve large tracts of land, an alternative is to establish systems of smaller tracts that are connected by corridors of habitat, so that migration between patches is possible.
- If the species that is threatened exists as a metapopulation, it is crucial to preserve at least some patches of unoccupied habitat to provide future homes for immigrants. If a population is lost from a preserve, the habitat should continue to be protected so it can be colonized in the future.

These results also have a more general message: Although traditional population growth models such as the expressions for exponential and logistic growth are simple and elegant, the factors they ignore—immigration and emigration—are crucial to understanding the dynamics of most populations. Because metapopulation structure is common, biologists must use more sophisticated models to predict the fates of populations.

Population Viability Analysis

A **population viability analysis**, or **PVA**, is a model that estimates the likelihood that a population will avoid extinction for a given time period. In most cases, a PVA attempts to combine basic demographic models for a species with data on geographic structure and the rate and severity of habitat disturbance. Typically, a population is considered viable if the analysis predicts that it has a 95 percent probability of surviving for at least 100 years. Natural resource managers are currently using PVA to assess the effects of logging, suburbanization, and other land management practices on sensitive species and to evaluate the merits of alternative recovery plans for endangered species.

A recent study—conducted on an endangered marsupial called Leadbeater's possum—illustrates the value of a carefully constructed PVA. Leadbeater's possum (**Figure 52.19a**) inhabits old-growth forests in southeastern Australia and relies on dead trees for nest sites. The goal of the PVA was to assess the effects that logging these habitats might have on the viability of the species. The question was, if logging reduced the ability of possums to migrate to new habitats, what would the impact on the population be?

(a) Leadbeater's possum

(b) Population viability analysis (PVA)

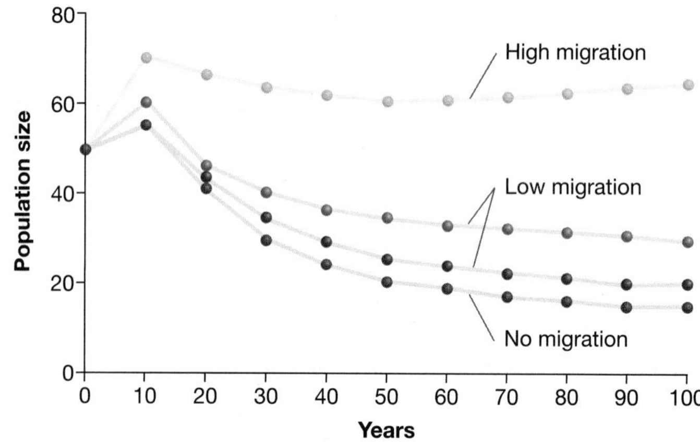

FIGURE 52.19 A PVA Can Predict the Consequences of Habitat Loss and Fragmentation. (a) Leadbeater's possum, a marsupial that breeds in old-growth forests of southeast Australia. **(b)** Results of a PVA: the fate of possum populations for four different rates of migration among habitat patches. The rates range from a case in which patches are isolated and no migration occurs to a case in which 5 percent of the individuals migrate each year. The no-migration scenario simulates what might happen if extensive deforestation occurred and created isolated patches of old-growth forest.

Life-table data are difficult to obtain for Leadbeater's possum, because this species is rare, lives in trees, and is active at night. Enough data were available from field studies, however, to allow for an estimate of age-specific survival and average fecundity per female per year. Biologists used these data to make population projections while varying the spatial configuration of habitat patches. The analysis allowed the researchers to assess the impact of migration and to simulate the effects of logging, fires, storms, and other disturbances.

Figure 52.19b illustrates the researchers' results. Each data point represents a population in a particular year. The four curves describe the changes in population size predicted to occur over time, based on different assumptions about the migration rates between groups of possums in patches of old-growth forest. When migration is high, the overall population size is predicted to stabilize at approximately 65 individuals. When no migration among fragments occurs, the final population size is predicted to be fewer than 20 individuals. Based on these results, the biologists concluded that extensive timber harvesting would pose a serious threat to the species. The PVA showed that reducing the size and number of remaining old-growth fragments would reduce the possibility for migration and lead to rapid population declines.

Like all analyses based on simulations, though, a PVA makes many assumptions about future events and is only as accurate as the data entered into it. In particular, the usefulness of PVAs has been challenged because data on age-specific fecundity and survivorship and other basic demographic information are lacking or poorly documented in many endangered species.

A group of biologists recently defended the approach, however, by analyzing 21 long-term studies that have been completed on threatened populations of animals. To do their analysis, the researchers split each of the 21 data sets in two. Data from the first half of each study was used to run a PVA on each of the 21 species. After comparing the predictions of the PVAs to the actual data from the second half of each study, the researchers found that the correspondence was extremely close. This result strengthens confidence in PVA as a good predictive method for land managers.

Chapter Review

SUMMARY OF KEY CONCEPTS

A population is a group of individuals of the same species that occupy the same area at the same time. Population ecology is the study of how and why the number of individuals in a population changes over time. Population size changes in response to changes in birth, death, immigration, and emigration rates. Biologists use a variety of mathematical and analytical tools to study population ecology.

Life tables summarize how likely it is that individuals of each age class in a population will survive and reproduce.

Life tables are a basic tool in demography—the study of patterns in births and deaths in populations. When life tables of different species or populations are compared, it is clear that individuals have distinct ways of allocating energy and resources to activities that promote survival versus activities that promote reproduction. Because the resources available to an individual are always limited, any increase in allocation of resources to survival and competitive ability necessitates a decrease in the resources allocated to reproduction. This trade-off between survival and reproduction is the most fundamental aspect of a species' life history.

You should be able to explain why a life table is relevant only to a particular population at a particular time, and predict how the first life tables—constructed in ancient Rome—differ from the current life table for the same population.

The growth rate of a population can be calculated from life-table data or from directly observing changes in population size over time.

One of the most basic characteristics of a population is its growth rate. Exponential growth occurs when the per-capita growth rate, r, does not change over time. Eventually, however, growing populations approach the carrying capacity of their environment. When this happens, logistic growth occurs.

Laboratory and field studies have confirmed the existence of exponential and logistic growth in species ranging from whooping cranes to *Paramecium*. Field research has documented that as population density increases, survivorship and fecundity may decrease—leading to increased death rates and lowered birth rates and thus a reduction in growth rate. Density dependence in population growth is due to competition for resources, disease, predation, or other factors that increase in intensity when population size is high.

You should be able to draw a logistic growth curve for a hypothetical human population, describe how r changes along its length, and suggest factors responsible for each change in r.

at www.masteringbio.com

Modeling Population Growth

A wide variety of patterns are observed when researchers track changes in population size over time, ranging from no growth, to regular cycles, to continued growth independent of population size.

Experiments have shown that density-dependent factors drive the regular population cycles observed in certain species. In the case of snowshoe hare and lynx populations, both predator-prey interactions and food availability affect these species' synchronized population cycles.

Changes in population size through time may occur due to changes in the age structure of populations—specifically the number of individuals in various age classes. A population with few juveniles and many adults past reproductive age, like the human populations of developed nations, may be declining or stable in size. In contrast, a population with a large proportion of juveniles is likely to increase rapidly in size. Human populations in the developing nations currently have this type of age distribution.

The total human population has been increasing rapidly since about 1750 and is currently over 6.6 billion. In countries where survivorship has increased due to medical advances, fecundity has decreased. Based on various scenarios for average female fertility, the total human population in 2050 is expected to be between 7 and 11 billion.

You should be able to explain why human populations are projected to continue rapid growth, even though fertility rates are declining.

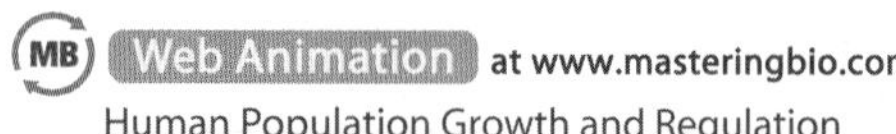

Web Animation at www.masteringbio.com
Human Population Growth and Regulation

Data from population ecology studies help biologists evaluate prospects for endangered species and design effective management strategies.

Human activities are isolating populations into metapopulations occupying small, fragmented habitats. The dynamics of a metapopulation are driven by the birth and death of populations, just as the dynamics of a population are driven by the birth and death of individuals. Migration among habitat patches is essential for the stability of a metapopulation, so conservationists are trying to preserve unoccupied patches of habitat and establish corridors that link habitat fragments.

Demographic data are the basis of population projections that are fundamental to population viability analysis, or PVA. Most PVAs attempt to model the effects that different management strategies might have on populations of endangered species. A PVA estimates the probability that a population will persist for a certain number of years under a prescribed set of demographic and habitat conditions.

You should be able to explain why a small, isolated population is virtually doomed to extinction, but why population size may be stable in a species consisting of a large number of small, isolated populations.

QUESTIONS

Test Your Knowledge

1. What is the defining feature of exponential growth?
   a. The population is growing very quickly.
   b. The growth rate is constant.
   c. The growth rate increases rapidly over time.
   d. The growth rate is very high.

2. What four factors define population growth?
   a. age-specific birth rates, age-specific death rates, age structure, and metapopulation structure
   b. survivorship, age-specific mortality, fecundity, death rate
   c. mark-recapture, census, quadrat sampling, transects
   d. births, deaths, immigration, emigration

3. In what populations does exponential growth tend to occur?
   a. in populations that colonize new habitats
   b. in populations that experience intense competition
   c. in populations that experience high rates of predation
   d. in metapopulations

4. Which of the following is *not* a reason that population growth declines as population size approaches the carrying capacity?
   a. Climate becomes unfavorable.
   b. Competition for resources increases.
   c. Predation rates increase.
   d. Disease rates increase.

5. If most individuals in a population are young, why is the population likely to grow rapidly in the future?
   a. Death rates will be low.
   b. The population has a skewed age distribution.
   c. Immigration and emigration can be ignored.
   d. Many individuals will begin to reproduce soon.

6. Why have population biologists become particularly interested in the dynamics of metapopulations?
   a. because humans exist as a metapopulation
   b. because whooping cranes exist as a metapopulation
   c. because many populations are becoming restricted to small islands of habitat
   d. because metapopulations explain why populations occupying large, contiguous areas are vulnerable to extinction

Test Your Knowledge answers: 1. b; 2. d; 3. a; 4. a; 5. d; 6. c

Test Your Understanding

Answers are available at www.masteringbio.com

1. Explain Equations 52.3 and 52.5 in words.
2. Draw type I, II, and III survivorship curves on a graph with labeled axes. Explain why the growth rate of species with a type I survivorship curve depends primarily on fertility rates. Explain why the growth rate of species with a type III survivorship curve is extremely sensitive to changes in adult survivorship.
3. Offer a hypothesis to explain why humans have undergone near-exponential growth for over 500 years. Why can't exponential growth continue indefinitely? Describe two documented examples of density-dependent factors that influence population growth in natural populations.
4. Compare and contrast the dynamics of a population that resides in a large contiguous habitat to the dynamics of a metapopulation. Be sure to consider density-dependent and density-independent factors on birth rates and death rates, and the importance of immigration and emigration. Assume that the total amount of area occupied is the same in both populations.
5. Make a rough sketch of the age distribution in developing versus developed countries, and explain the significance of the differences. How is AIDS, which is a sexually transmitted disease, affecting the age distribution in countries hard hit by the epidemic?
6. Compare the pros and cons of using R_0, λ, and r to express growth rates. What is the difference between r (the per-capita rate of increase) and r_{max} (the maximum or intrinsic growth rate)?

Applying Concepts to New Situations

Answers are available at www.masteringbio.com

1. When wild plant and animal populations are logged, fished, or hunted, only the oldest or largest individuals tend to be taken. What impact does harvesting have on a population's age structure? How might harvesting affect the population's life-table and growth rate?
2. Design a system of nature preserves for an endangered species of beetle whose larvae feed on only one species of sunflower. The sunflowers tend to be found in small patches that are scattered throughout dry grassland habitats. Explain the rationale behind your proposal.
3. The population of the United States is projected to increase dramatically over the next 50 years, even though fertility rates are only slightly above replacement level and the age distribution is largely stable. How is this possible?
4. In most species the sex ratio is at or near 1.00, meaning that there is an approximately equal number of males and females. In China, however, there is a strong preference for male children. According to the 2000 census there, the sex ratio of newborns is almost 1.17, meaning that close to 117 boys are born for every 100 girls. Based on these data, researchers project that there will soon be about 50 million more men than women of marriageable age in China. Discuss how this skewed sex ratio might affect the population growth rate in China.

www.masteringbio.com is also your resource for • Answers to text, table, and figure caption questions and exercises • Answers to *Check Your Understanding* boxes • Online study guides and quizzes • Additional study tools including the *E-Book for Biological Science* 3rd ed., textbook art, animations, and videos.

53 Community Ecology

KEY CONCEPTS

- Interactions among species, such as competition and consumption, have two main outcomes: (1) They affect the distribution and abundance of the interacting species, and (2) they are agents of natural selection and thus affect the evolution of the interacting species. The nature of interactions between species frequently changes over time.
- The assemblage of species found in a biological community changes over time and is primarily a function of climate and chance historical events.
- Species diversity is high in the tropics and lower toward the poles. The mechanism responsible for this pattern is still being investigated.

This chapter explores how species that live in coral reefs and other communities interact with each other, and how communities change over time.

Chapter 52 focused on the dynamics of populations—how and why they grow or decline and how they change over time and through space. That chapter considered each population as an isolated entity. But in reality, populations of different species form complex assemblages called communities. A biological **community** consists of interacting species, usually living within a defined area.

The concept of a biological community was introduced in Chapter 50. Each of the biomes analyzed in that chapter represents a broad grouping of plant communities. The goal of that chapter was to describe the general characteristics of selected biomes and analyze broad correlations between a region's climate and the type of biome present.

The task in this chapter is to analyze the dynamics of biological communities—how they develop and change over time. Biologists want to know how communities work. How do species inside communities interact with each other, and what are the long-term consequences? What happens to communities when they are disturbed by fires or flood, and why is the number of species higher in some areas than others? Let's delve in.

53.1 Species Interactions

The species in a community interact almost constantly. Members of different species eat one another, pollinate each other, exchange nutrients, compete for resources, and provide habitats for each other. As a result, the fate of a particular population may be tightly linked to the other species that share its habitat.

To study species interactions, biologists focus on analyzing the effects on the fitness of the individuals involved. Recall from Chapter 24 that **fitness** is defined as the ability to survive and produce offspring. Does the relationship between two species provide a fitness benefit to members of one species (a + interaction) but hurt members of the other species (a − interaction)? Or does the association have no effect on the fitness of a participant (a "0" interaction)? Three broad categories of interaction are analyzed in this chapter: the −/− relationship known as competition, the +/− interactions called consumption and parasitism, and the +/+ association termed mutualism. A fourth category of interaction, **commensalism**, is defined as a +/0 association. An example is the birds that follow moving army ants in the tropics. As the ants march along the forest floor, they hunt insects and small vertebrates. As they do, birds follow and pick off prey species that fly or jump up out of the way of the ants (**Figure 53.1**). The birds are commensals that benefit from the association (+) but have no measurable impact on the ants (0).

In addition to introducing the array of tools that biologists use to study competition, consumption, and mutualism, this section focuses on three key themes:

1. Species interactions may affect the distribution and abundance of a particular species. Recall from Chapter 52 that most of the density-dependent factors that produce logistic growth are based on species interactions—predation, disease, or competition for space and resources. In addition, changes in species interactions often explain short-term changes in population size and distribution.

2. Species act as agents of natural selection when they interact. Deer are fast and agile in response to natural selection exerted by their major predators, wolves and cougars. The speed and agility of deer, in turn, promote natural selection that favors wolves and cougars that are fast and that have superior eyesight and senses of smell. To capture this point, biologists say that species interactions resemble an arms race. In humans, an arms race is said to occur when one nation develops a new weapon, which prompts a rival country to develop a defensive weapon, which pushes the original country to manufacture an even more powerful weapon, and so on. In biology, a coevolutionary arms race occurs between predators and prey, parasites and hosts, and other types of interacting species. "Coevolutionary" indicates that species influence each others' evolution, leading to reciprocal adaptation. In this way, changes in species interactions lead to long-term changes in the characteristics of populations, a phenomenon called **coevolution**, in addition to having short-term impacts on population size.

3. The outcome of interactions among species is dynamic and conditional. Consider the relationship between army ants and birds that follow them, which is usually commensal. If bird attacks start to force other insects into the path of the ants, then both ants and birds benefit and the relationship becomes mutualistic. But if birds begin to steal prey that would otherwise be taken by ants, then the relationship becomes parasitic. The outcome of the interaction may depend on the number and types of prey, birds, and ants present and may change over time.

FIGURE 53.1 Commensals Gain a Fitness Advantage but Don't Affect the Species They Depend On. Birds that associate with army ants are commensals. They have no measurable fitness effect on the ants but gain from the association by capturing insects that try to fly or climb out of the way of the ants.

Competition

Competition is a −/− interaction that occurs when different individuals use the same resources and when those resources are limiting—meaning that lack of access to those resources prevents individuals from surviving better and having more offspring.

Ever since ecological studies began, researchers have focused on competition as an important interaction within and between species. The attention is justified in part by the central place that competition holds in the theory of evolution by natural selection. Darwin pointed out that individuals within a population compete for the resources that are required for growth and reproduction. Further, some individuals are more successful in this competition and leave more offspring than others do. If the traits that lead to success are heritable, then the frequency of alleles in the population changes and evolution by natural selection occurs.

Competition that occurs between members of the same species is called **intraspecific** (literally, "within species") **competition**. In addition to creating opportunities for natural selection to occur, intraspecific competition for space, sunlight, food, and other resources intensifies as a population's density increases. As a result, intraspecific competition is a major cause of density-dependent growth (see Chapter 52).

(a) Consumptive competition

These trees are competing for water and nutrients.

(b) Preemptive competition

Space preempted by these barnacles is unavailable to competitors.

(c) Overgrowth competition

The large fern has overgrown other individuals and is shading them.

(d) Chemical competition

Few plants are growing under these *Salvia* shrubs.

(e) Territorial competition

Grizzly bears drive off black bears.

(f) Encounter competition

Spotted hyenas and vultures fight over a kill.

FIGURE 53.2 Many Types of Competition Exist.

Interspecific ("between species") **competition** occurs when individuals from different species use the same limiting resources. As **Figure 53.2** shows, the mechanisms involved vary widely.

1. *Consumptive competition* occurs when two species consume the same resources (Figure 53.2a).
2. *Preemptive competition* exists when one species makes space unavailable to other species (Figure 53.2b).
3. *Overgrowth competition* happens when one species grows above another (Figure 53.2c).
4. *Chemical competition* takes place when one species produces toxins that negatively affect another species (Figure 53.2d).
5. *Territorial competition* arises when a mobile species protects its feeding or breeding territory against other species (Figure 53.2e).
6. *Encounter competition* occurs when two species interfere directly for access to specific resources (Figure 53.2f).

Using the Niche Concept to Analyze Competition Early work on competition focused on the concept of the niche. A **niche** can be thought of as the range of resources that the species is able to use or the range of conditions it can tolerate. G. Evelyn Hutchinson proposed that a species' niche could be envisioned by plotting these habitat requirements along a series of axes. **Figure 53.3a**, for example, represents one niche axis for a hypothetical species. In this case, the habitat requirement plotted is the size of seeds eaten by members of this population, which might be a function of mouth or tooth size. Other niche axes could represent other types of foods used or the temperatures, humidity, and other environmental conditions tolerated by the species.

Interspecific competition occurs when the niches of two species overlap. The two species plotted in **Figure 53.3b**, for example, compete for seeds of intermediate size. When competition occurs, each individual will get fewer seeds on average and fitness will decline. This is why competition is considered a −/− interaction.

What Happens When One Species Is a Better Competitor? G. F. Gause claimed it is not possible for species with the same niche to coexist. This hypothesis is called the **competitive exclusion principle**, and it was inspired by a series of experiments Gause did with similar species of the unicellular pond-dweller *Paramecium*. When Gause placed small populations of *P. caudatum* and *P. aurelia* in separate laboratory cultures, both species exhibited logistic growth (see Chapter 52). But Gause showed that when the two species are put in the same

(a) One species eats seeds of a certain size range.

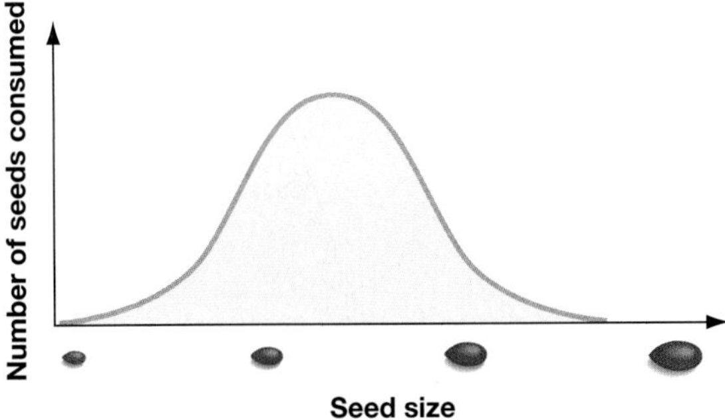

(b) Partial niche overlap: competition for seeds of intermediate size

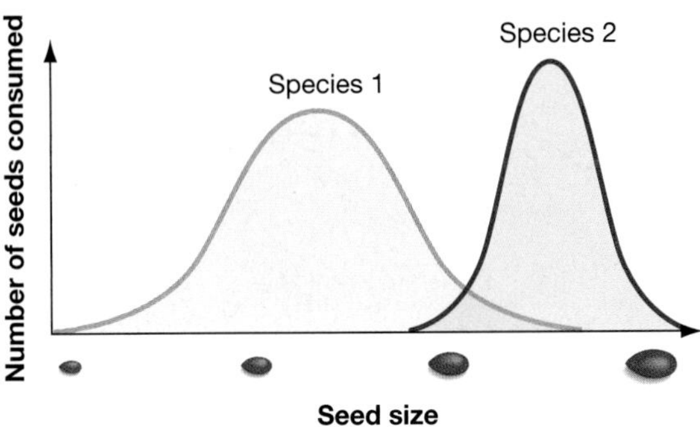

FIGURE 53.3 Niche Overlap Leads to Competition. (a) Graph describing one aspect of a species' fundamental niche, meaning the range of resources that it can use or range of conditions it can tolerate. **(b)** Competition occurs when the niches of different species overlap. In this case, both species use seeds of intermediate size.

● **EXERCISE** Mark the horizontal axis to indicate the range of seed sizes where competition occurs.

(a) Competitive exclusion in two species of *Paramecium*

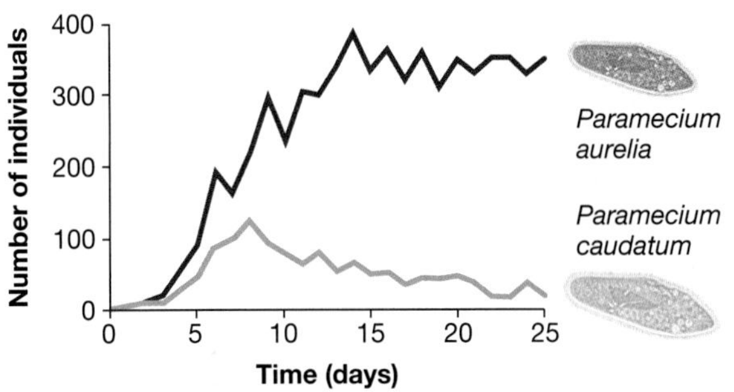

(b) Competitive exclusion occurs when competition is asymmetric ...

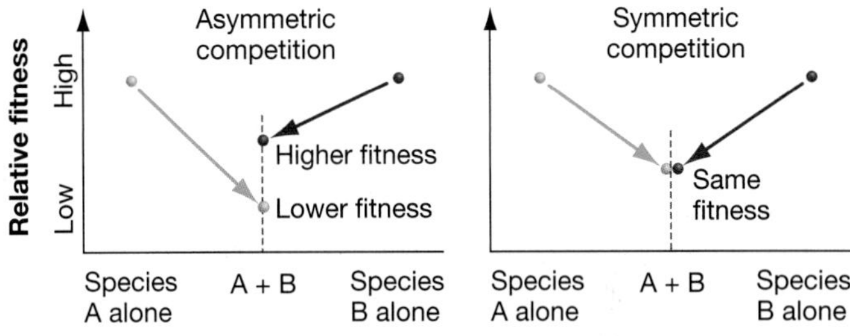

(c) ... and niches overlap completely.

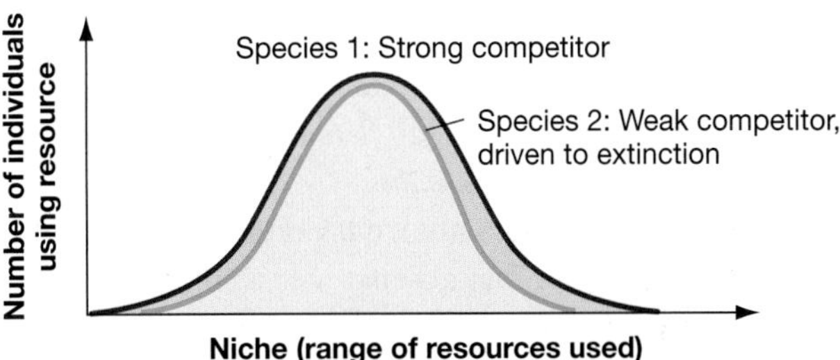

(d) When competition is asymmetric and niches do not overlap completely, weaker competitors use nonoverlapping resources.

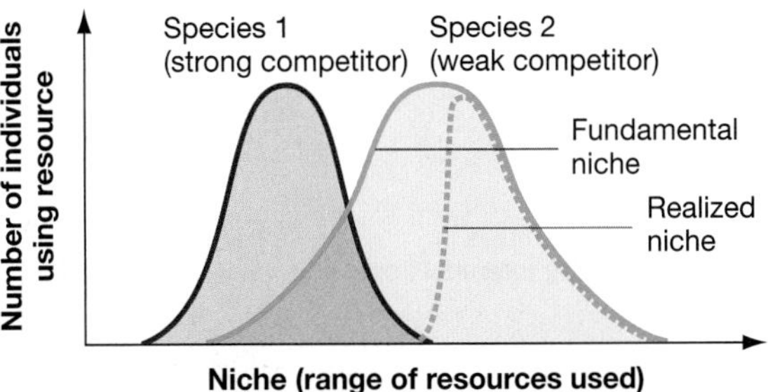

FIGURE 53.4 Niche Overlap Leads to Competitive Exclusion or Restricted Habitat Use. (a–c) If two species have completely overlapping niches, there is no refuge for the weaker competitor. It may be driven to extinction. **(d)** A realized niche is a subset of a species' fundamental niche.

culture together, only the *P. aurelia* population exhibits a logistic growth pattern. *P. caudatum*, in contrast, is driven to extinction (**Figure 53.4a**).

Gause's result is a product of asymmetric competition. When **asymmetric competition** occurs, one species suffers a much greater fitness decline than the other species does (**Figure 53.4b**). Under **symmetric competition**, however, each of the interacting species experiences a roughly equal decrease in fitness.

If asymmetric competition occurs and the two species have completely overlapping niches, as diagrammed in **Figure 53.4c**, then the stronger competitor is likely to drive the weaker competitor to extinction. But if the niches do not overlap completely, then the species that is the weaker competitor should be able to retreat to an area of non-overlap. In cases like this, an important distinction arises between a species' **fundamental niche**, which is the combination of resources or areas used or conditions tolerated in the absence of competitors, and its **realized niche**, which is the portion of resources or areas used or conditions tolerated when competition occurs (**Figure 53.4d**).

In Gause's laboratory cultures, competition between *P. aurelia* and *P. caudatum* was asymmetric and competitive exclusion occurred. How can researchers study competition in the field, under natural conditions?

Experimental Studies of Competition Joseph Connell began a classic study of competition in the late 1950s, after observing an interesting pattern in an intertidal rocky shore in Scotland. He noticed that there were two species of barnacles with distinctive distributions. Barnacle larvae are mobile, but adults live attached to rocks. The adults of one species, *Chthamalus stellatus*, occurred in an upper intertidal zone, while the adults of the other species, *Semibalanus balanoides* (formerly named *Balanus balanoides*), were restricted to a lower intertidal zone (**Figure 53.5**). The upper intertidal zone is a more severe environment for barnacles, because it is exposed to the air for longer periods at low tide each day. The young of both species were found together in the lower intertidal zone, however.

To explain these observations, Connell hypothesized that adult *Chthamalus* were competitively excluded from the lower intertidal zone. The alternative hypothesis is that adult *Chthamalus* are absent from the lower intertidal zone because they do not thrive in the physical conditions there.

Connell tested these hypotheses by performing the experiment shown in **Figure 53.6.** He began by removing a number of rocks that had been colonized by *Chthamalus* from the upper intertidal zone and transplanting them into the lower intertidal zone. He screwed the rocks into place and allowed *Semibalanus* larvae to colonize them. Once the spring colonization period was over, Connell divided each rock into two groups. In one half, he removed all *Semibalanus* that were in contact with or next to a *Chthamalus*.

This experimental design allowed Connell to document *Chthamalus* survival in the absence of competition with *Semibalanus* and compare it with survival during competition. This is a common experimental strategy in competition studies: One of the competitors is removed, and the response by the remaining species is observed.

Connell's results support the hypothesis of competitive exclusion. In the unmodified areas, *Semibalanus* killed many of the young *Chthamalus* by growing against them and lifting them off the substrate. As the graph in Figure 53.6 shows, *Chthamalus* survival was much higher when all of the *Semibalanus* were removed.

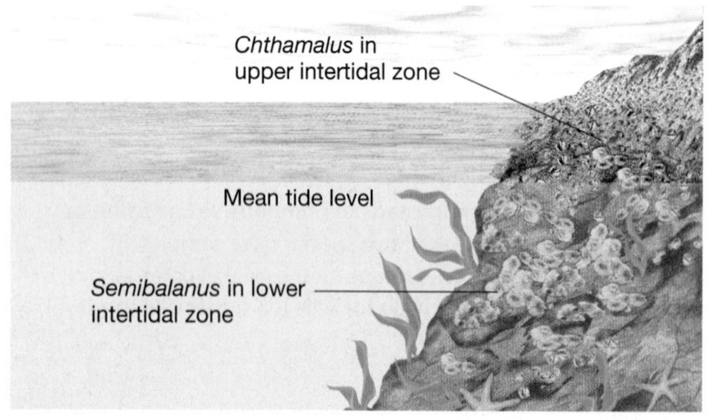

FIGURE 53.5 Barnacle Species Are Distributed in Distinct Zones. In natural habitats, adult *Chthamalus* and *Semibalanus* barnacles do not coexist.

Experiment

Question: Why is the distribution of adult *Chthamalus* restricted to the upper intertidal zone?

Hypothesis: Adult *Chthamalus* are competitively excluded from the lower intertidal zone.

Alternative hypothesis: Adult *Chthamalus* do not thrive in the physical conditions of the lower intertidal zone.

Experimental setup:

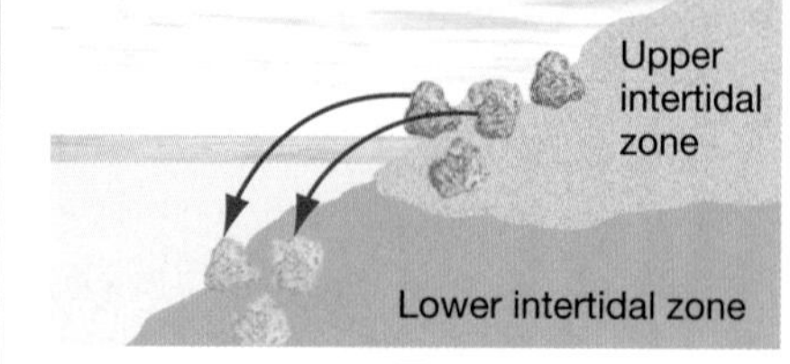

1. Transplant rocks containing young *Chthamalus* to lower intertidal zone.

2. Let *Semibalanus* colonize the rocks.

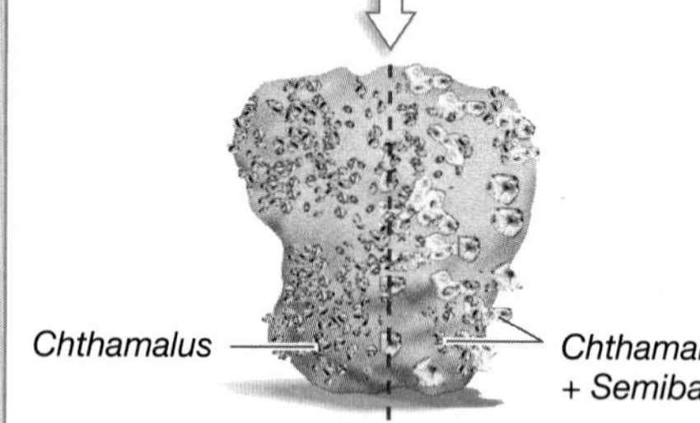

3. Remove *Semibalanus* from half of each rock. Monitor survival of *Chthamalus* on both sides.

Prediction: *Chthamalus* will survive better in the absence of *Semibalanus*.

Prediction of alternative hypothesis: *Chthamalus* survival will be low and the same in the presence or absence of *Semibalanus*.

Results:

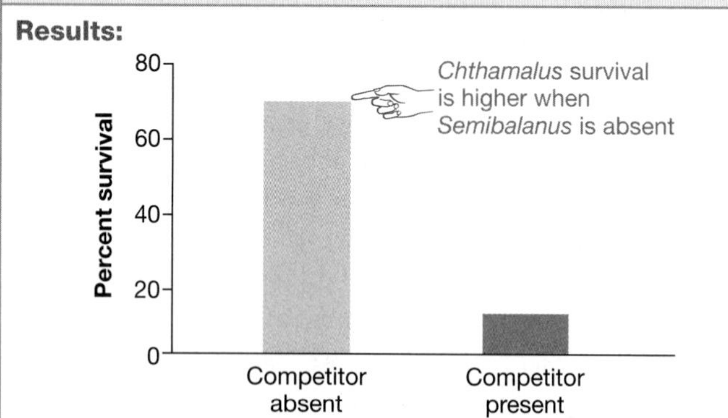

Conclusion: *Semibalanus* is competitively excluding *Chthamalus* from the lower intertidal zone.

FIGURE 53.6 Experimental Evidence for Competitive Exclusion.

● QUESTION Why was it important to carry out both treatments on the same rock? Why not use separate rocks?

● QUESTION Connell also did the reciprocal removal experiment—removing *Chthamalus* from experimental rocks. Predict the outcome of this experiment.

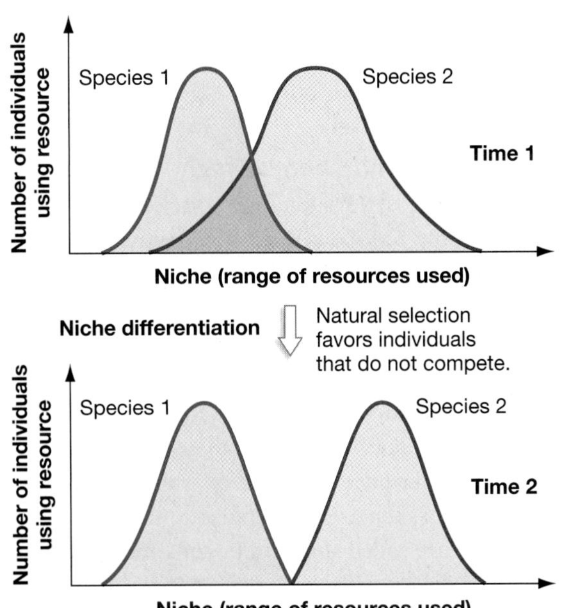

FIGURE 53.7 Over Time, Competition Can Lead to Niche Differentiation.

● QUESTION Biologists call the outcome of this process resource partitioning. Explain why.

Mechanisms of Coexistence: Fitness Trade-offs and Niche Differentiation Why haven't *Semibalanus* and other superior competitors taken over the world? Biologists answer this question by invoking the concept of fitness trade-offs (see Chapter 24). The key here is that the ability to compete for a particular resource—like space on rocks or edible seeds of a certain size—is only one aspect of an organism's niche. If individuals are extremely good at competing for a particular resource, then they are probably less good at enduring drought conditions, warding off disease, or preventing predation.

In the case of *Semibalanus* and *Chthalamus* growing in the intertidal, the fitness trade-off is rapid growth and success in competing for space versus the ability to endure the harsh physical conditions of the upper intertidal. *Semibalanus* are fast-growing and large; *Chthalamus* grow slowly but can survive long exposures to the air and to intense sun and heat. Neither species can do both things well. Fitness trade-offs limit the ability of superior competitors to spread.

It's also important to realize that because competition is a −/− interaction, there is strong natural selection on both species to avoid it. **Figure 53.7** shows the predicted outcome: An evolutionary change in traits reduces the amount of niche overlap, and thus the amount of competition. This change in resource use is called **niche differentiation** or resource partitioning; the change in species' traits is called **character displacement.** The fundamental idea is that competition exerts natural selection, and that the characteristics of species change in a way that reduces competition.

Peter and Rosemary Grant recently documented character displacement occurring in Galápagos finches. You might recall, from Chapter 24, that the Grants observed dramatic increases in average beak size and body size of a *Geospiza fortis* population during a drought in 1977. The changes occurred because the major food source available during the drought was fruits from a plant called *Tribulus cistoides*, and because only the largest-beaked individuals were able to crack these fruits and eat the seeds inside. In 1982, however, individuals from a species called *Geospiza magnirostris* arrived on the island and began breeding. *G. magnirostris* are about twice the size of *G. fortis* and use *T. cistoides* fruits as their primary food source (**Figure 53.8**).

A severe drought recurred in 2003 and 2004; most finches died of starvation. When the Grants measured the surviving *G. fortis* they made a remarkable observation: In contrast to the 1977 drought, only the *smallest*-beaked individuals survived. Data on feeding behavior indicated that *G. magnirostris* were outcompeting *G. fortis* for *T. cistoides* fruits. Only *G. fortis* that could eat extremely small seeds efficiently could survive.

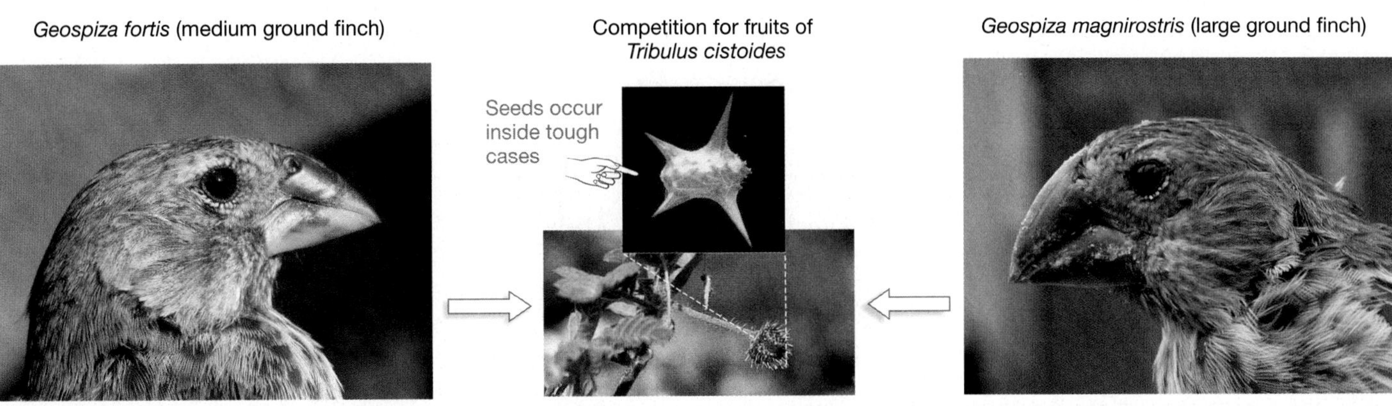

FIGURE 53.8 A Natural Experiment in Niche Differentiation. During drought years, *Geospiza fortis* have to compete for *Tribulus cistoides* fruits with *Geospiza magnirostris* if they are present.

Under similar environmental conditions, then, the presence of competition altered the course of evolution. This is a dramatic example of character displacement. Large ground finches and medium ground finches are partitioning a resource to avoid competition.

Consumption

Consumption occurs when one organism eats another. There are three major types of consumption:

1. **Herbivory** takes place when **herbivores** ("plant-eaters") consume plant tissues. Bark beetles mine cambium; cicadas suck sap; caterpillars chew leaves.
2. **Parasitism** occurs when a **parasite** consumes relatively small amounts of tissue or nutrients from another individual, called the **host**. Parasitism often occurs over a long period of time. It is not necessarily fatal, and parasites are usually small relative to their host. An array of worms and unicellular protists parasitizes humans; ticks parasitize cattle and other large mammals. Not all parasitism involves consumption, however—social parasites in birds and insects lay their eggs in other species' nests and induce them to raise the young.
3. **Predation** occurs when a predator kills and consumes all or most of another individual. The consumed individual is called the prey. Woodpeckers eat bark beetles; finches eat seeds; ladybird beetles devour aphids; wasps kill caterpillars.

To illustrate how biologists analyze the impact of consumption, let's consider a series of questions about predators, herbivores, parasites, and their victims.

How Do Prey Defend Themselves? With respect to fitness, predation is costly for the prey species and beneficial for the predator. Prey individuals do not passively give up their lives to increase the fitness of their predators, however. Instead, prey may hide or run, fly, or swim away when they sense the presence of a predator. Analyses have shown that many species find safety in numbers—schooling and flocking behavior is an effective way to reduce the risk of predation, in part because predators become confused when they dive or swim into a mass of prey. Other prey species sequester or spray toxins or employ weaponry such as sharp spines or kicking hooves. The traits just listed are called **standing** or **constitutive defenses**, because they are always present (**Figure 53.9a**).

Some of the best-studied constitutive defenses involve the phenomenon called mimicry. **Mimicry** occurs when one species closely resembles another species. **Figure 53.9b** illustrates two major types of mimicry. The wasp on the far left is brightly colored and dangerous, because of its stinger. It looks like other dangerous insects—particularly other wasps and bees. When harmful prey species resemble each other, **Müllerian mimicry** is said to occur. To explain the existence of **Müllerian** mimics, biologists propose that the existence of similar-looking dangerous prey in the same habitat increases the likelihood that predators will learn to avoid them. In this way, Müllerian mimicry

(a) Constitutive defenses of animals vary.

Camouflage: blending into the background

Schooling: safety in numbers

Weaponry: fighting back

(b) Mimicry can protect both dangerous and harmless species.

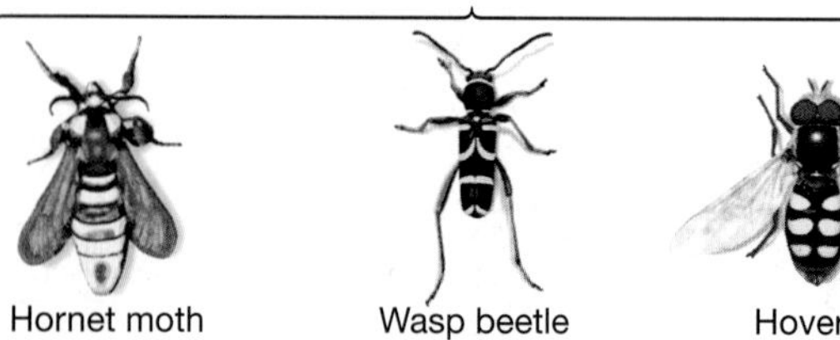

FIGURE 53.9 Constitutive Defenses Are Always Present. (a) Prey have an array of adaptations to reduce the likelihood of predation. **(b)** Müllerian mimicry occurs among dangerous prey species; Batesian mimicry occurs between a dangerous prey species and harmless prey species.

(a) Prey and predator

Blue mussels

Crabs

(b) Correlation between predation rate and prey defense

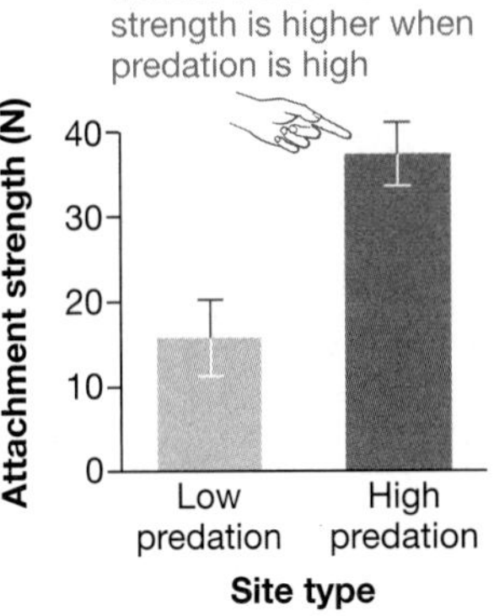

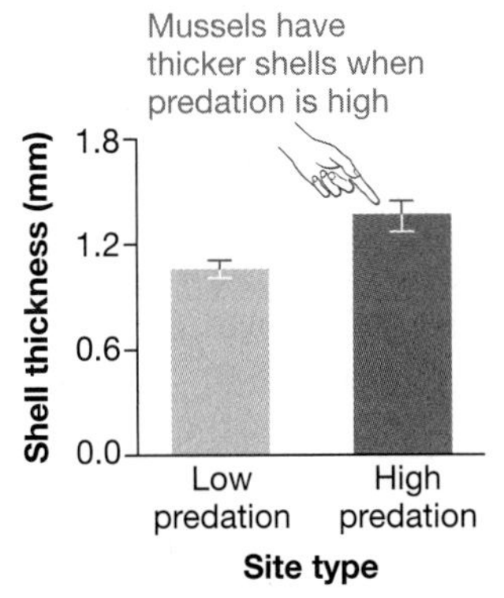

FIGURE 53.10 Inducible Defenses Are Produced Only when Prey Are Threatened. (a) In shallow-water environments in Maine, blue mussels (left) are preyed on by crabs (right). **(b)** To measure how strongly mussels are attached to the substrate, researchers drilled a hole in their shells and measured the force, in newtons (N), required to pull them off. They also measured how large mussel shells were relative to an individual's size by dividing shell weight by soft-tissue weight.

should reduce the likelihood of dangerous individuals being attacked. But wasps also act as a model for harmless species of moth, beetle, and fly that resemble wasps. To explain these mimics, biologists propose that predators avoid the harmless mimics because they mistake them for a dangerous wasp. This resemblance is known as **Batesian mimicry**.

The key point here is that prey have adaptations that reduce their likelihood of becoming victims. These adaptations are responses to natural selection exerted by predators. Constitutive defenses are expensive, however, in terms of the energy and resources that must be devoted to producing and maintaining them. Based on this observation, it should not be surprising to learn that many prey species have **inducible defenses**—meaning defensive traits that are produced only in response to the presence of a predator. Induced defenses then decline if predators leave the habitat. Inducible defenses are efficient energetically, but they are slow—it takes time to produce them.

To see how inducible defenses work, consider recent research on the blue mussels that live in an estuary along the coast of Maine (**Figure 53.10a**). Biologists had documented that predation on mussels by crabs was high in an area of the estuary with relatively slow tidal currents (a "low-flow" area) but low in an area of the estuary with relatively rapid tidal currents (a "high-flow" area). The researchers hypothesized that if blue mussels possess inducible defenses, then heavily defended prey individuals should occur in the low-flow area, where predation pressure is higher, but not in the high-flow area, where water movement reduces the number of crabs present.

To evaluate this hypothesis, the biologists measured mussel shell characteristics in the two areas. They found that mussels in the high-predation area were more strongly attached to their base or substrate and had thicker shells than did mussels in the low-predation area (**Figure 53.10b**). These traits make the mussels more difficult to remove from the substrate and harder to crush and so function as effective antipredator defenses.

The data in Figure 53.10b are correlational in nature, however, so they are open to interpretation. A critic of the inducible defense explanation could offer a reasonable alternative hypothesis—for example, that only constitutive defenses exist and that crabs have eliminated weakly attached mussels with thin shells from the low-flow areas. The observed differences could also be due to differences in light, temperature, or other abiotic factors that might affect mussel traits but have nothing to do with predation.

To test the inducible-defense hypothesis more rigorously, biologists carried out the experiment diagrammed in **Figure 53.11**. The tank on the left allowed the researchers to measure shell growth in mussels that were "downstream" from crabs. As predicted by the induced-defense hypothesis, the mussels exposed to a crab in this way developed significantly tougher shells than did mussels that were not exposed to a crab. These results suggest that, even without direct contact, mussels can sense the presence of crabs and increase their investment in defenses. In a similar experiment, the investigators compared mussels that were exposed to water running through broken mussel shells versus intact but empty mussel shells. They recorded a significant increase in shell thickness in the tank downstream from the broken shells. This result supports the hypothesis that mussels can detect the presence of predators from molecules released by broken shells.

Results like these underscore several themes of this section: Species interactions are dynamic and result in coevolution—in this case, adaptations and counteradaptations in an evolutionary arms race between predators and prey.

Are Animal Predators Efficient Enough to Reduce Prey Populations? Research on animal defense systems supports the hypothesis that species interactions have a strong impact on the evolution of predator and prey populations. Can predators also affect the short-term distribution and abundance of prey populations? Prey are typically smaller than predators, have larger litter or clutch sizes, and tend to begin reproduction at a younger age. As a result, they have a much larger intrinsic growth rate, r_{max}—that is, the maximum growth rate that a population can achieve under ideal

Experiment

Question: Are mussel defenses induced by the presence of crabs?

Hypothesis: Mussels increase investment in defense in the presence of crabs.

Null hypothesis: Mussels do not increase investment in defense in the presence of crabs.

Experimental setup:

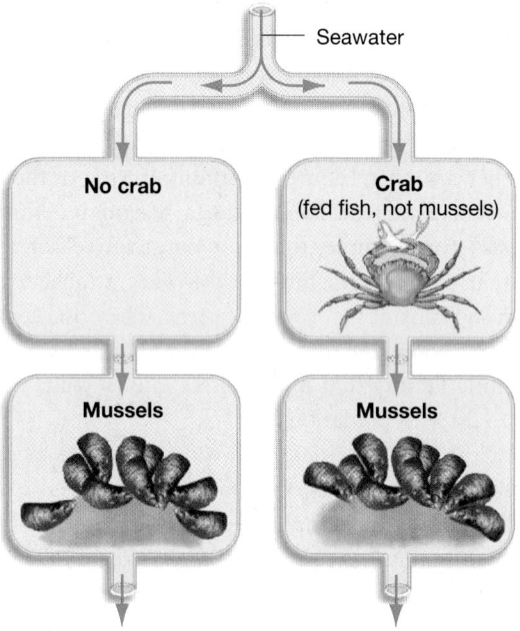

Seawater
No crab
Crab (fed fish, not mussels)
Mussels
Mussels

Prediction: Mussels downstream of the crab tank will have thicker shells than mussels downstream of the empty tank.

Prediction of null hypothesis: Mussels in the two tanks will have shells of equal thickness.

Results:

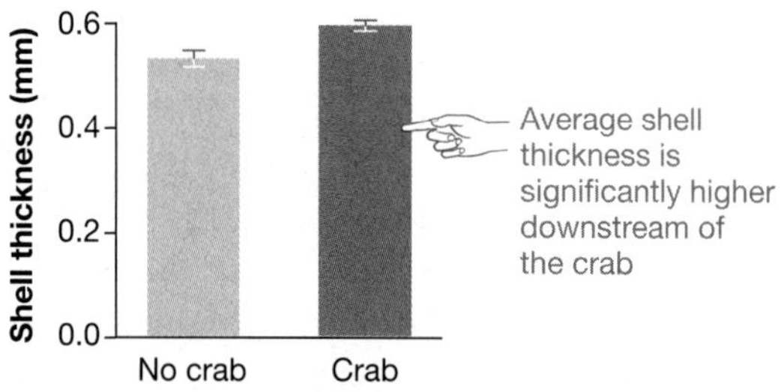

Shell thickness (mm)
0.6
0.4
0.2
0.0
No crab
Crab
Average shell thickness is significantly higher downstream of the crab

Conclusion: Mussels increase investment in defense when they detect crabs. Shell thickness is an inducible defense.

FIGURE 53.11 Experimental Evidence for Inducible Defenses.

QUESTION Why did the researchers feed the crabs fish instead of mussels?

conditions (see Chapter 52). If prey reproduce rapidly and are also well defended, it is not clear whether predators should be able to kill enough of them to reduce the prey population significantly—particularly if predators tend to take old or sick members of a population.

In several cases, data from predator-removal programs—in which wolves, cougars, coyotes, or other predators are actively killed by human hunters—support the hypothesis that predators actually do reduce the size of prey populations. For example, a wolf control program in Alaska during the 1970s decreased predator abundance to 55–80 percent below pre-control density. Concurrently, the population of moose, on which wolves prey, tripled. This observation suggests that wolf predators had reduced this moose population far below the number that could be supported by the available space and food.

Other types of experiments have also supported the hypothesis that predators play a role in density-dependent growth of prey populations. Recall from Chapter 52 that some populations go through regular cycles and that the regular population cycles of snowshoe hare and lynx appears to be driven, at least in part, by density-dependent increases in predation. Taken together, the data available to date indicate that in many instances, predators are efficient enough to reduce prey populations.

Why Don't Herbivores Eat Everything—Why Is the World Green? If predators affect the size of populations in prey that can run or fly or swim away, then consumers should have a devastating impact on plants and on mussels, anemones, sponges, and other sessile (nonmoving) animals.

In some cases, this prediction turns out to be correct. For example, consider the results of a recent **meta-analysis**—a study of studies, meaning an analysis of a large number of data sets on a particular question. Biologists who compiled the results of more than 100 studies on herbivory found that the median percentage of mass removed from aquatic algae by herbivores was 79 percent. Herbivores eat the vast majority of algal food available in aquatic biomes. The figure dropped to just 30 percent for aquatic plants, however, and only 18 percent for terrestrial plants.

Why don't herbivores don't eat more of the food available on land? Stated another way, why is the world green? Biologists routinely consider three possible answers to this question. Herbivores could be kept in check by predation or disease; plant tissues could offer poor or incomplete nutrition; or plants could defend themselves effectively against attack:

1. The **top-down control hypothesis** states that herbivore populations are limited by predation and disease. The "top-down" name is inspired by the food chains introduced in Chapter 30 and explored in detail in Chapter 54. It's appropriate because predators and parasites remove herbivores that eat plants. In a recent test of this idea, researchers monitored herbivory on islands created by a 1986 dam project in Venezuela. On some small islands in the new lake, predators disappeared. There are now many more herbivores on these islands than on similar sites nearby where predators were present, and a much higher percentage of primary production is being eaten. Predator-free islands, for example, had

just 25 percent of the small trees found on similar islands that contained predators.

2. The **poor-nutrition hypothesis** contends that plants are a poor food source in terms of the nutrients they provide for herbivores. More specifically, plant tissues have less than 10 percent of the nitrogen found in animal tissues, by weight. If the growth and reproduction of herbivores are limited by the availability of nitrogen, then their populations will be low and the impact of herbivory relatively slight. Herbivores could eat more plant material to gain nitrogen, but at a cost—they would be exposed to predation and expend energy processing the food. To evaluate the poor-nutrition hypothesis, a recent meta-analysis examined 185 studies of how insect herbivores had responded when the nitrogen concentration of plants was increased experimentally via fertilization. In over half of the cases, herbivores showed a significant increase in growth rate or reproduction when the plants that they feed on were fertilized. Based on this result, the researchers concluded that nitrogen limitation is an important factor in a large fraction of plant-herbivore interactions.

3. The **plant-defense hypothesis** holds that plants defend themselves effectively enough to limit herbivory. Most plant tissues are defended by weapons such as thorns, prickles, or hairs, or by potent poisons such as nicotine, caffeine, and cocaine. In addition, no animal species can, without help from protists or bacteria, digest cellulose or lignin, which are components of wood. Manufacturing defensive compounds seems like the perfect solution to the problem of herbivory. In practice, however, plants face a complex challenge in defending themselves. Consider data on interactions between cottonwood trees and two of their herbivores: beavers and leaf beetles. Cottonwoods resprout after they are cut down by a beaver (**Figure 53.12a**), and resprouted trees contain high concentrations of a defensive compound that deters further attack by beavers (**Figure 53.12b**). This is an induced defense. But larvae of a particular leaf beetle species eat this defensive compound readily. In fact, the larvae store enough of it in their bodies to act as a defensive compound against their own major predator—ants. The data in **Figure 53.12c** show that larvae that grow on resprouted cottonwood trees survive longer than do larvae that grow on normal cottonwoods when the larvae are placed on an ant mound. This is an example of an indirect effect in species interactions. The response by cottonwoods to herbivory by beavers benefits another herbivore—leaf beetles. The net result is that there is no perfect, one-size-fits-all defensive strategy. Natural selection should favor plants that evolve an ever-changing suite of compounds to deter the ever-changing array of herbivores they face.

Taken together, the data reviewed here suggest that there is no single answer to the question of why herbivores don't eat a greater fraction of the available plant food. All three of the hypotheses we have examined are correct. Top-down control, nitrogen limitation, and effective defense are all important factors in limiting the impact of herbivory, although the particular mix of factors will vary from plant species to plant species and from habitat to habitat.

Adaptation and Arms Races Over the long term, how do species that interact via consumption affect each other's evolution? When predators and prey or herbivores and plants interact over time, coevolutionary arms races result: Consumers evolve traits that increase their efficiency; in response, prey

(a) Cottonwood tree felled by beavers

(b) Resprouted trees have more defensive compounds.

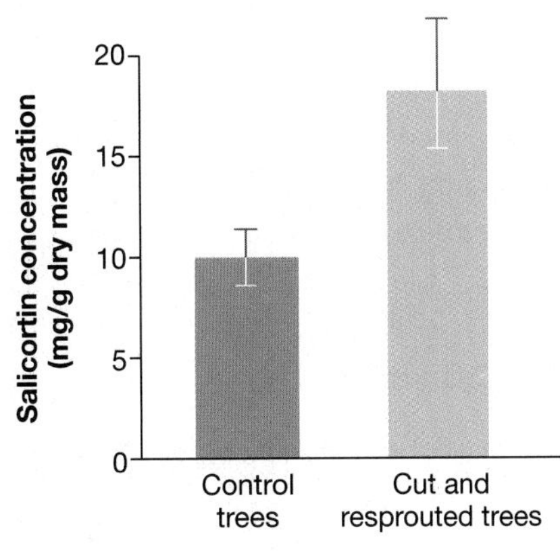

(c) Survival of beetle larvae placed on ant mound

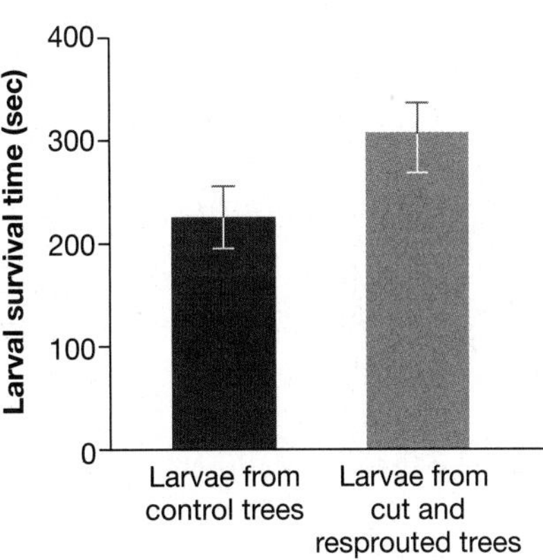

FIGURE 53.12 Defensive Compounds in Cottonwood Trees Discourage Beavers but Favor Beetles.

QUESTION Beetle larvae survive better on resprouted versus control trees, so they should damage resprouted trees more. If natural selection has maximized cottonwood fitness, do beavers or beetle larvae present the greater overall threat to them?

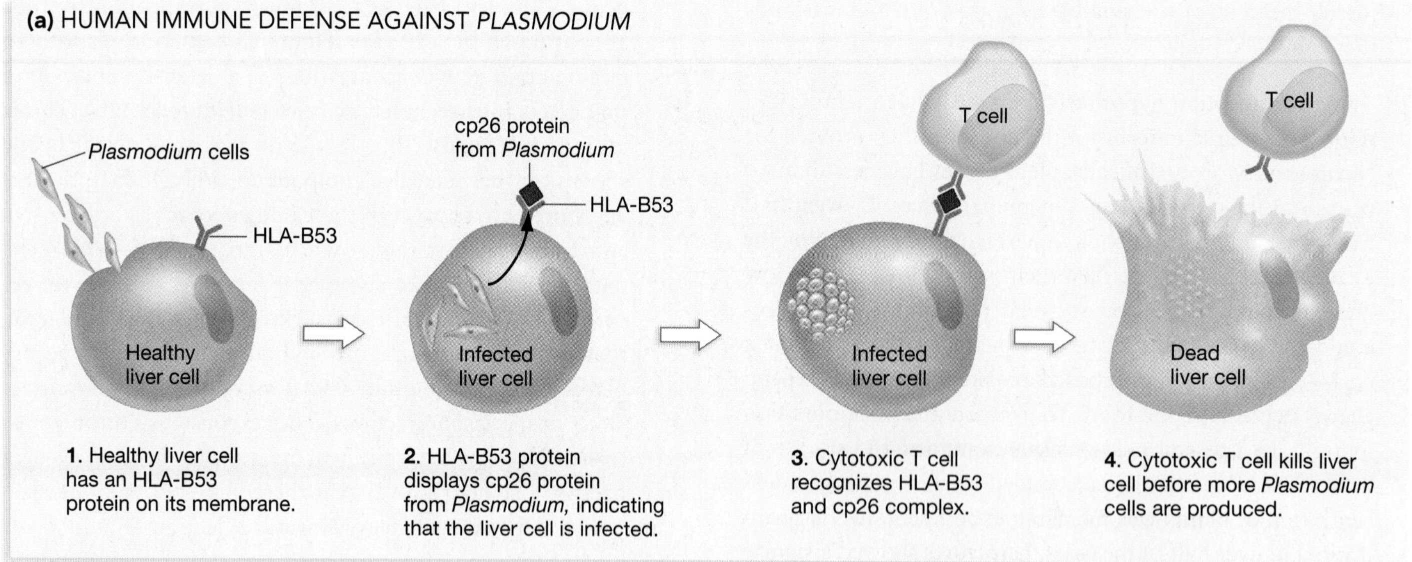

evolve traits that make them unpalatable or elusive, which leads to selection on consumers for traits that counter the prey adaptation, and so on.

To see a coevolutionary arms race in action, consider interactions between humans and the most serious of all human parasites—species in the genus *Plasmodium*. Recall from Chapter 29 that *Plasmodium* are unicellular protists that cause malaria in a wide array of vertebrates. Malaria kills at least a million people a year, most of them preschool-age children. Recent data suggest that humans and the causative agent of malaria—four species in the genus *Plasmodium*—are locked in a coevolutionary arms race.

Plasmodium has several distinct cell types, each of which infects liver, red blood cell, or other types of cells in mosquitoes or humans (see Chapter 29). Natural selection favors alleles that allow *Plasmodium* cells to infect their hosts efficiently and multiply inside. Cells in the human immune system do not sit passively by as *Plasmodium* destroys liver cells and red blood cells. Instead, immune system cells (introduced in Chapter 49) seek out and destroy *Plasmodium* cells. Natural selection favors alleles that allow mosquitoes and humans to resist infection.

In West Africa, for example, there is a strong association between an allele called *HLA-B53* and protection against malaria. In liver cells that are infected by *Plasmodium*, HLA-B53 proteins on the surface of the liver cell display a parasite protein called cp26 (**Figure 53.13a**). The display is a signal that immune system cells can read. It means "I'm an infected cell. Kill me before they kill all of us." Immune system cells destroy the liver cell before the parasite cells inside can multiply. In this way, people who have at least one copy of the *HLA-B53* allele are better able to beat back malarial infections. People who have the *HLA-B53* allele appear to be winning the arms race against malaria.

(b) *Plasmodium* strains have different versions of the cp protein.

Plasmodium strain	Infection rate	Interpretation
cp26	Low	HLA-B53 binds to these proteins. Immune response is effective.
cp29	Low	
cp26 and cp29 strains together	High	Immune response fails when these strains infect the same person.
cp27	High	HLA-B53 does not bind to these proteins. Immune response is not as effective.
cp28	Average	

FIGURE 53.13 Interactions between the Human Immune System and Plasmodium. (a) If HLA-B53 binds to a particular *Plasmodium* protein, then infected cells are recognized and destroyed. **(b)** Some strains of *Plasmodium* appear to avoid detection by the immune system better than others do. Also, certain strains defeat the immune response if they infect the same person at the same time.

Follow-up research has shown that the arms race is far from won, however. *Plasmodium* populations in West Africa now have a variety of alleles for the protein recognized by HLA-B53. Some of these variants bind to HLA-B53 and trigger an immune response in the host, but others escape detection. Furthermore, many people in West Africa are infected with several different strains of *Plasmodium*, some of which are better at evading the immune system response. In some cases, the recognition step by HLA-B53 breaks down when certain strains are found together (**Figure 53.13b**). To make sense of these observations, researchers suggest that natural selection has favored the evolution of *Plasmodium* strains with weapons that counter HLA-B53.

An arms race continues to rage between *Plasmodium* and humans. Certain human proteins act as antimalarial weapons. But as predicted by coevolutionary theory, *Plasmodium* has evolved effective responses and continues to evolve.

MB Web Animation at www.masteringbio.com
Life Cycle of a Malaria Parasite

Can Parasites Manipulate Their Hosts? To thrive, parasites do not just have to invade tissues and grow while evading defensive responses by their host. They also have to be transmitted to new hosts. To a parasite, an uninfected host represents uncolonized habitat, teeming with resources. What have biologists learned about how parasites are transmitted to new hosts?

To answer this question, consider species of land snails that are parasitized by flatworms—specifically, by flukes in the genus *Leucochloridium*. Researchers who studied this association discovered something unusual. When the flukes have matured and are ready to be transmitted to their next host, a bird, they burrow into the snail's tentacles and wriggle. In addition, infected snails become attracted to light, even though uninfected snails avoid sunlit areas and prefer dark, shady environments (**Figure 53.14**). When infected snails move out of the shade into the open and glide about with wriggling tentacles, they are more easily spotted and consumed by birds. To interpret these observations, biologists suggest that the worms manipulate the behavior of the snail, and that the change in snail behavior makes the parasite more likely to be transmitted to a new host.

Studies of how parasites are transmitted to new hosts reinforce a general message: Extensive coevolution occurs among species that interact via consumption. Experiments on mutualistic interactions carry the same message.

FIGURE 53.14 A Parasite That Manipulates Host Behavior. The behavior of snails that are infected with flukes in the genus *Leucochloridium* is dramatically different from the behavior of uninfected snails.

EXERCISE Design an experiment to test the hypothesis that infected snails are more likely than uninfected snails to be eaten by birds.

Mutualism

Mutualisms are +/+ interactions that involve a wide variety of organisms and rewards. Many species of bees, for example, visit flowers to harvest nectar and pollen. Bees benefit because nectar is used as a food source for adult bees and the pollen is fed to larvae. Flowering plants also benefit because, in the process of visiting flowers, foraging bees carry pollen from one plant to another and accomplish pollination. Chapters 30 and 40 detailed some of the adaptations found in flowering plants that increase the efficiency of pollination. Other chapters highlighted an array of other mutualisms:

- One of the most important of all mutualisms occurs between fungi and plant roots. Chapter 31 reviewed experimental evidence indicating that fungi receive sugars and other carbon-containing compounds in exchange for nitrogen or phosphorus needed by the plant partner.
- Arguably the most critical of all mutualisms involves bacteria that fix nitrogen and certain species of plants. As Chapters 28 and 38 pointed out, the partnership is based on host plants providing sugars and protection and the bacteria supplying nitrogen in return.
- Chapter 32 cited mutualisms between (1) rancher ants and the aphids that the ants protect in exchange for sugar-rich honeydew, and (2) farmer ants and the fungi that they cultivate for food.

Figure 53.15 illustrates some other interesting mutualisms. Figure 53.15a shows ants in the genus *Crematogaster*, which live in acacia trees native to Africa and the New World tropics. The ants live in bulbs at the base of acacia thorns and feed on small structures that grow from tree branches. These ants protect the tree by attacking and biting herbivores and by cutting vegetation from the ground below the host tree. Figure 53.15b illustrates cleaner shrimp in action. These shrimp pick external parasites from the jaws and gills of fish. In this mutualism, one species receives dinner while the other obtains medical attention.

As these examples show, the rewards from mutualistic interactions range from the transportation of gametes to food, housing, medical help, and protection. It is important to note, however, that even though mutualisms benefit both species, the interaction does not involve individuals from different species being altruistic or "nice" to each other. Expanding on a point that Charles Darwin introduced in 1862, Judith Bronstein described mutualisms as "a kind of reciprocal parasitism; that is, each partner is out to do the best it can by obtaining what it needs from its mutualist at the lowest possible cost to itself." Her point is that the benefits received in a mutualism are a by-product of each individual pursuing its own self-interest, by maximizing its ability to survive and reproduce.

(a) Mutualism between ants and acacia trees

(b) Mutualism between cleaner shrimp and fish

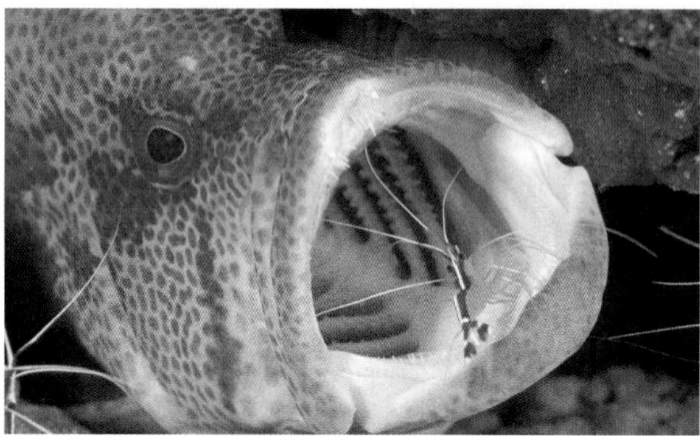

FIGURE 53.15 Mutualisms Take Many Forms. (a) In certain species of acacia tree, ants in the genus *Crematogaster* live in large bulbs at the base of spines and attack herbivores that threaten the tree. The ants eat nutrient-rich tissue produced at the tips of leaves or on branches. **(b)** Cleaner shrimp remove and eat parasites that take up residence on the gills of fish.

QUESTION What is the cost of these associations to the acacia tree, the *Crematogaster* the cleaner shrimp, and the host fish?

In this light, it is not surprising that some species "cheat" on mutualistic systems. For example, deceit pollination occurs when certain species of plants produce a showy flower but no nectar reward. Pollinators have to be deceived to make a visit and carry out pollination. Evolutionary studies show that deceit pollinators evolved from ancestral species that did provide a reward. Over time, a +/+ interaction evolved into a +/− interaction.

A recent experimental study of mutualism provides another good example of the dynamic nature of these interactions. This study focused on ants and treehoppers. Ants are insects that live in colonies; treehoppers are small, herbivorous insects that feed by sucking sugar out of the phloem of plants. Treehoppers excrete the sugary solution honeydew from their posteriors. The honeydew, in turn, is harvested for food by ants.

It is clear that ants benefit from this association. But do the treehoppers? Biologists hypothesized that the ants might protect the treehoppers from their major predator, jumping spiders. These spiders feed heavily on juvenile treehoppers.

To test the hypothesis that ants protect treehoppers, the researchers studied ant-treehopper interactions over a three-year period. As **Figure 53.16** shows, the researchers marked out a 1000-m^2 study plot. Each year they removed the ants from one group of the treehopper host plants inside the plot but left the others alone to serve as a control. Then they compared the growth and survival of treehoppers on plants with and without ants. Recall that this is a common research strategy for studying species interactions. To assess the fitness costs or benefits of the interaction, researchers remove one of the participants experimentally and document the effect on the other

Experiment

Question: Is the relationship between ants and treehoppers mutualistic?

Hypothesis: Ants harvest food from treehoppers. In return, they protect treehoppers from jumping spiders.

Null hypothesis: Ants harvest food from treehoppers but are not beneficial to treehopper survival.

Experimental setup:

Ant
Treehoppers
Plants with ants
Study plot, 1000 m^2
Treehoppers
Plants with ants removed

Prediction: Treehopper reproduction will be higher when ants are present than when ants are absent.

Prediction of null hypothesis: There will be no difference in the number of young treehoppers on the plants.

Results (Year 1):

Average number of young treehoppers per plant
100
80
60
40
20
0
20 25 30 5 10 15
July August
Plants with ants
Plants without ants

Conclusion: Treehoppers benefit from the interaction with ants.

FIGURE 53.16 Experimental Evidence that the Treehopper-Ant Interaction Is Mutualistic.

SUMMARY TABLE 53.1 **Species Interactions**

Type of Interaction	Fitness Effects	Short-Term Impact: Distribution and Abundance	Long-Term Impact: Coevolution
Competition	−/−	Reduces population size of both species; if competition is asymmetric, competitive exclusion reduces range of one species.	Niche differentiation via selection to reduce competition
Consumption	+/−	Impact on prey population depends on prey density and effectiveness of defenses.	Strong selection on prey for effective defense; strong selection on consumer for traits that overcome defenses
Parasitism	+/−	Impact on host population depends on parasite density and effectiveness of defenses.	Strong selection on host for effective defense; strong selection on parasite for traits that overcome defenses
Mutualism	+/+	Population size and range of both species are dependent on each other.	Strong selection on both species to maximize fitness benefits and minimize fitness costs of relationship
Commensalism	+/0	Population size and range of commensal may depend on size and distribution of host.	Strong selection on commensal to increase fitness benefits in relationship; no selection on host

participant's survival and reproduction, compared with the survival and reproduction of control individuals that experience a normal interaction.

In both the first and third years of the study, the number of treehopper young on host plants increased in the treatment with ants but showed a significant decline in the treatment with no ants. This result supports the hypothesis that treehoppers benefit from the interaction with ants because the ants protect the treehoppers from predation by jumping spiders.

In the second year of the study, however, the researchers found a very different pattern. There was no difference in offspring survival, adult survival, or overall population size between treehopper populations with ants and those without ants. Why? The researchers were able to answer this question because they also measured the abundance of spiders that prey on treehoppers in each of the three years. Their census data showed that in the second year of the study, spider populations were very low.

Based on these results, the investigators concluded that the benefits of the ant-treehopper interaction depend entirely on predator abundance. Treehoppers benefit from their interaction with ants in years when predators are abundant but are unaffected in years when predators are scarce. If producing honeydew is costly to treehoppers, then the +/+ mutualism changes to a +/− interaction when spiders are rare.

Mutualism is like parasitism, competition, and other types of species interactions in an important respect: The outcome of the interaction depends on current conditions. Because the costs and benefits of species interactions are fluid, an interaction between the same two species may range from parasitism to mutualism to competition. **Table 53.1** summarizes the fitness effects, short-term impacts on population size, and long-term evolutionary aspects of species interactions.

Check Your Understanding

If you understand that... 

- Natural selection favors the evolution of traits that decrease competition, resulting in changes in a species' niche over time.
- Consumers may reduce the population size of the species on which they feed and often exert strong natural selection for effective defense mechanisms.
- Mutualisms benefit the species involved and can lead to highly coevolved associations, such as mycorrhizal fungi and symbiotic nitrogen-fixing bacteria.

You should be able to... 

1) Explain what a coevolutionary arms race is and give an example.
2) Give specific examples of how competition, consumption, and mutualism each affect population size and long-term evolution.

53.2 Community Structure

Biologists have made important progress in understanding the nature of species interactions and their consequences. In terms of understanding the structure and function of biological communities, however, research on species interactions has a limitation: It usually focuses on just two species at a time. But biological communities contain many thousands of species. To understand how communities work, biologists broaden the scope of research and explore how combinations of many species interact.

The question that biologists initially asked about communities concerned structure: Do biological communities have a tightly prescribed organization and composition, or are they merely loose assemblages of species? If communities are highly structured entities, then their makeup should be predictable. For example, if a community is destroyed by a disturbance and then allowed to recover, the diversity and abundance of species at that site should be identical when recovery is complete. But if communities can be made up of many different combinations of species, depending on which arrive earlier or later, then community composition will be difficult to predict. The diversity and abundance of species found in a region would vary substantially before a disturbance and after recovery.

How Predictable Are Communities?

Beginning with a paper published in 1936, Frederick Clements promoted the view that biological communities are stable, integrated, and orderly entities with a highly predictable composition. His hypothesis was that species interactions are so extensive and coevolution is so important that the groups of species called communities have become highly integrated and interdependent units in nature. Stated another way, the species within a community cannot live without each other.

To drive his idea home, Clements likened the development of a plant community to the development of an individual organism. He argued that communities develop by passing through a series of predictable stages dictated by extensive interactions among species and that this development culminates in a stable final stage known as a **climax community**. According to Clements, the nature of the climax community is determined by the area's climate and does not change over time. Further, he held that if a fire or other disturbance destroys the climax community, it will reconstitute itself by repeating its predictable developmental stages.

Henry Gleason, in contrast, contended that the community found in a particular area is neither stable nor predictable. He claimed that plant and animal communities are ephemeral associations of species that just happen to share similar climatic requirements. According to Gleason, it is largely a matter of chance whether a similar community develops in the same area after a disturbance occurs. Gleason downplayed the role of biotic factors, such as species interactions, in structuring communities. To him, abiotic factors and history—for example, which seeds and juvenile animals happened to arrive after a disturbance—were the key elements in determining which species are found at a particular location.

Which viewpoint is more accurate? Let's consider observational and experimental data.

Mapping Current and Past Species' Distributions If communities are predictable assemblages, then the ranges of species that make up a particular community should be congruent. Stated another way, the same group of species should almost always be found growing together.

When biologists began documenting the ranges of tree species along elevational gradients, however, they found that species came and went independently of each other. As you go up a mountain, for example, you might find white oak trees growing with hickories and chestnuts. As you continue to gain elevation, chestnuts might disappear but white oaks and hickories remain. Farther upslope, hickories might drop out of the species mix while white pines and red pines start to appear.

Data on the historical composition of plant communities supported these observations. Studies of fossil pollen documented that the distribution of plant species and communities at specific locations throughout North America has changed radically since the end of the last ice age about 11,000 years ago. An important pattern emerged from these data: Species do not come and go in the fossil record in tightly integrated units. Instead, the ranges of individual species tended to change independently of one another. For example, at a study site in Mirror Lake, New Hampshire, the percentage of pollen present from pines and oaks rose and fell in tandem over the past 14,000 years—just as predicted for a tightly integrated, pine-oak forest community. In contrast, the percentage of pollen from spruce, fir, cedar, beech, and most other tree species changed independently of each other. In general, studies of fossil pollen suggest that the composition of most plant communities has been dynamic and contingent on historical events, rather than static.

Experimental Tests To explore the predictability of community structure experimentally, biologists constructed 12 identical ponds (**Figure 53.17**). They filled the ponds at the same time with water that contained enough chlorine to kill any preexisting organisms. If community structure is predictable, then each pond should develop the same community of species once the chlorine vaporized and made the water habitable. If community structure is unpredictable, then each pond should develop a different community.

To test these predictions, the researchers sampled water from the ponds repeatedly for one year. They measured temperature, chemical makeup, and other physical characteristics of the water and recorded the diversity and abundance of each planktonic species by examining the samples under the microscope. They found a total of 61 species in all of the ponds but discovered that individual ponds each had only 31 to 39 species. This observation is important. Each pond contained just half to two-thirds of the total number of species that lived in the experimental area and that were available for colonization. A number of species occurred in most or all of the 12 ponds, but each pond had a unique species assemblage. Why?

To explain their results, the researchers contended that some species are particularly good at dispersing and are likely to colonize all or most of the available habitats. Other species disperse more slowly and tend to reach only one or a few of the

Experiment

Question: Are communities predictable or unpredictable?

Communities-are-predictable hypothesis: The group of species present at a particular site is highly predictable.

Communities-are-unpredictable hypothesis: The group of species present at a particular site is highly unpredictable.

Experimental setup:

1. Construct 12 identical ponds. Fill at the same time and sterilize water so that there are no preexisting organisms.

2. Examine water samples from each pond. Identify each plankton species present in each sample.

Communities-are-predictable prediction: Identical plankton communities will develop in all 12 ponds.

Communities-are-unpredictable prediction: Different plankton communities will develop in different ponds.

Results (after 1 year):

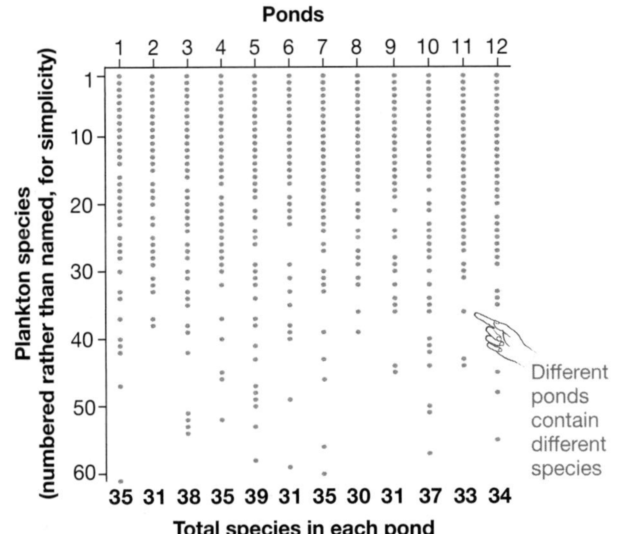

Conclusion: Although about half of the species present appear in all or most ponds, each pond has a unique composition. Both hypotheses are partially correct.

FIGURE 53.17 Experimental Evidence that Identical Communities Do Not Develop in Identical Habitats.

● QUESTION How many species are found in all 12 ponds?

available habitats. Further, the investigators proposed that the arrival of certain competitors or predators early in the colonization process greatly affects which species are able to invade successfully later. As a result, the specific details of community assembly and composition are somewhat contingent and difficult to predict. At least to some degree, communities are a product of chance and history.

The overall message of research on community structure is that Clements's position was too extreme and that Gleason's view is closer to being correct. ● Although both biotic interactions and climate are important in determining which species exist at a certain site, chance and history also play a large role.

How Do Keystone Species Structure Communities?

Even though communities are not predictable assemblages dictated by obligatory species interactions, the presence of certain consumers can have an enormous impact on the species present. Experiments have shown that, in some cases, the structure of an entire community can change dramatically if a single species of predator or herbivore is removed from a community or added to it.

As an example of this research, consider an experiment that Robert Paine conducted in intertidal habitats of the Pacific Northwest of North America. In this environment, the sea star *Pisaster ochraceous* is an important predator (**Figure 53.18a**). When Paine removed *Pisaster* from experimental areas, what had been diverse communities of algae and invertebrates became overgrown with solid stands of the California mussel *Mytilus californianus* (**Figure 53.18b**). Although *M. californianus* is a dominant competitor, its populations had been held in

(a) Predator: *Pisaster ochraceous*

(b) Prey: *Mytilus californianus*

FIGURE 53.18 A Keystone Predator and Its Prey

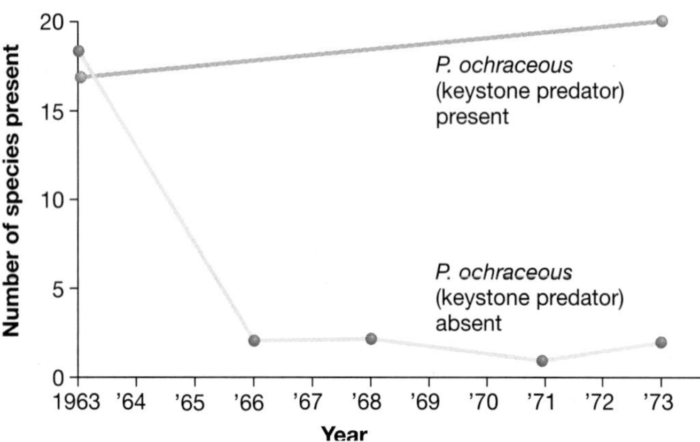

FIGURE 53.19 Keystone Predation Alters Community Structure in a Rocky Intertidal Habitat. Each data point represents the average number of species present in several study plots.

check by sea-star predation. When this predator was gone, the species richness and structural complexity of the habitat changed radically (**Figure 53.19**).

To capture the effect that a predator such as *Pisaster* can have on a community, Paine coined the term "keystone species." A **keystone species** has a much greater impact on the surrounding species than its abundance and total biomass would suggest. For example, wolves were introduced to the Greater Yellowstone Ecosystem of western North America in 1995. Although their population has grown to only about 150 individuals, their presence has led to far-reaching changes in plant and animal communities. Wolves feed on elk and do not tolerate the presence of coyotes. As elk declined, populations of favorite elk foods like aspen, willow, and cottonwood have increased, triggering an increase in beavers, which compete with elk for these plants. As coyotes declined, mouse populations increased and led to larger populations of hawks.

Sea stars, wolves, and other keystone species vary in abundance in space or over time. As a result, the composition and structure of their communities vary. Communities are ever-changing entities whose composition is difficult to predict.

53.3 Community Dynamics

Once biologists had a basic understanding of how species interact and how communities are structured, they turned to questions about how communities change through time. Like cells, individuals, and species, communities can be described in one word: dynamic.

Disturbance and Change in Ecological Communities

Community composition and structure may change radically in response to changes in abiotic and biotic conditions. Biologists have become particularly interested in how communities respond to disturbance. A **disturbance** is any event that removes some individuals or biomass from a community. (Recall from Chapter 50 that biomass is the total mass of living organisms.)

Forest fires, windstorms, floods, the fall of a large canopy tree, disease epidemics, and short-term explosions in herbivore numbers all qualify as disturbances. These events are important because they alter light levels, nutrients, unoccupied space, or some other aspect of resource availability.

Biologists have come to realize that the impact of disturbance is a function of three factors: (1) the type of disturbance, (2) its frequency, and (3) its severity—for example, the speed and duration of a flood, or the intensity of heat during a fire. Most communities experience a characteristic type of disturbance, and in most cases, disturbances occur with a predictable frequency and severity. To capture this point, biologists refer to a community's **disturbance regime.** For example, fires kill all or most of the existing trees in a boreal forest every 100 to 300 years, on average. In contrast, numerous small-scale tree falls, usually caused by windstorms, occur in temperate and tropical forests every few years.

How Do Researchers Determine a Community's Disturbance Regime? Ecologists use two strategies to determine a community's natural pattern of disturbance. The first approach is based on inferring long-term patterns from data obtained in a short-term analysis. For example, an observational study might document that 1 percent of all boreal forest on Earth burns in a given year. Assuming that fires occur randomly, researchers project that any particular piece of boreal forest has a 1 in 100 chance of burning each year. According to this reasoning, fires will recur in that particular area every 100 years, on average.

This extrapolation approach is straightforward to implement, but it has important drawbacks. In boreal forests, for example, fires do not occur randomly in either space or time. They are more likely in some areas than in others, and they tend to occur in particularly dry years. Unless sampling is extensive, it is difficult to avoid errors caused by extrapolating from particularly disturbance-prone or disturbance-free areas or years.

The second approach to determining disturbance regimes is based on reconstructing the history of a particular site. Flooding frequency, for example, can be estimated by analyzing sediments, because floods deposit distinctive groups of sediment particles. Researchers estimate the frequency and impact of storms by finding wind-killed trees and determining their date of death. (Investigators do so by comparing patterns in the growth rings of the dead trees with those of living individuals nearby; see Chapter 36.) The disturbances that have been most extensively studied using historical techniques, however, are forest fires.

Forest fires often leave a layer of burned organic matter and charcoal on the surface of the ground. As a result, researchers can dig a soil pit, find charcoal layers, and use radioisotope dating to establish when the fires occurred. It is also possible to date the death of trees killed by fire by comparing their growth rings with those of living trees. Further, trees that are not killed by fire are often scarred. When a fire burns close enough to kill a patch of cambium tissue, a scar forms. These fire scars occur most often at the tree's base, where dead leaves and twigs accumulate and furnish fuel. Fire scars can be dated by analyzing growth rings.

Why Is It Important to Understand Disturbance Regimes?

To appreciate why biologists are so interested in understanding disturbance regimes, consider a recent study on the fire history of giant sequoia groves in California (**Figure 53.20a**). Giant sequoias grow in small, isolated groves on the western side of the Sierra Nevada range. Individuals live more than a thousand years, and many have been scarred repeatedly by fires. A biologist obtained samples of cross sections through the bases of 90 giant sequoias in five different groves. As **Figure 53.20b** shows, the cross sections contained numerous rings that had been scarred by fire. To determine the date of each disturbance, the researcher counted tree rings back from the present. He found that, in most of the groves, 10 to 53 fires had occurred each century for the past 1530 to 2000 years (**Figure 53.20c**). The data indicated that each tree had been burned an average of 64 times.

This study established that fires are extremely frequent in the community examined. Because not enough time would pass for large amounts of fuel to accumulate between fires, they were probably of low severity. Partly because of this work, the biologists responsible for managing sequoia groves now set controlled fires or let low-intensity natural fires burn instead of suppressing them immediately. Similarly, studies of disturbance regimes along the Colorado River in southwestern North America inspired land managers to release a huge pulse of water from the reservoirs behind dams on the waterway recently. The flood that resulted was designed to mimic a natural disturbance. According to follow-up studies, the artificial flood appears to have benefited the plant and animal communities downstream.

Succession: The Development of Communities after Disturbance

Severe disturbances remove all or most of the organisms from an area. The recovery that follows is called **succession**. The name was inspired by the observation that certain species succeed others over time. **Primary succession** occurs when a disturbance removes the soil and its organisms as well as organisms that live above the surface. Glaciers, floods, volcanic eruptions, and landslides often initiate primary succession. **Secondary succession** occurs when a disturbance removes some or all of the organisms from an area but leaves the soil intact. Fire and logging are examples of disturbances that initiate secondary succession.

(a) Giant sequoias after a fire

(b) Fire scars in the growth rings

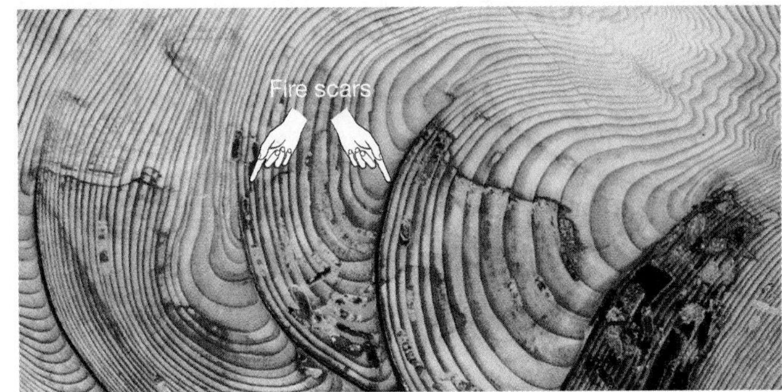

(c) Reconstructing history from fire scars

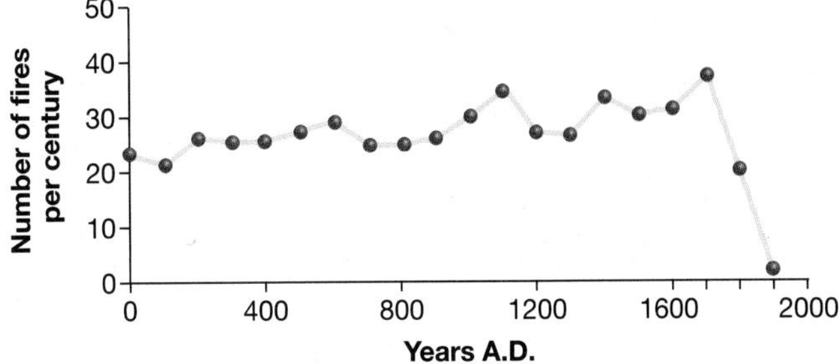

FIGURE 53.20 History of Disturbance in a Fire-Prone Community. Because trees form one ring (light band/dark band) every year, researchers can count the rings to determine how often fires have occurred during the last 2000 years.

QUESTION About how many fires occurred during the last century?

As **Figure 53.21** shows, a sequence of plant communities develops as succession proceeds. Early successional communities are dominated by species that are short lived and small and that disperse their seeds over long distances. Late successional communities are dominated by species that tend to be long lived, large, and good competitors for resources such as light and nutrients. The specific sequence of species that appears over time is called a successional pathway. What determines the pattern and rate of species replacement during succession at a particular time and place?

Theoretical Considerations Biologists focus on three factors to predict the outcome of succession in a community: (1) the particular traits of the species involved, (2) how species interact, and (3) historical and environmental circumstances such as the size of the area involved and weather conditions. Before going on to consider a detailed case history, we need to explore these factors in more detail.

Species traits, such as dispersal capability and the ability to withstand extreme dryness, are particularly important early in succession. As common sense would predict, recently disturbed sites tend to be colonized by plants and animals with good dispersal ability. When these organisms arrive, however, they often have to endure harsh environmental conditions. These **pioneering species** tend to have "weedy" life histories. (A **weed** is a plant that is adapted for growth in disturbed soils.) Early successional species devote most of their energy to reproduction and little to competitive ability. They have small seeds, rapid growth, and a short life span, and they begin reproducing at an early age. As a result, they have a high reproductive rate. But in addition, they can tolerate severe abiotic conditions, such as high light levels, poor nutrient availability, and drying.

Once colonization is under way, the course of succession tends to depend less on how species cope with aspects of the abiotic environment and more on how they interact with other species. This change occurs because plants that grow early in succession change abiotic conditions in a way that makes the conditions less severe. Because plants provide shade, they reduce temperatures and increase humidity. Their dead bodies also add organic material and nutrients to the soil. As abiotic conditions improve, biotic interactions become more important.

During succession, existing species can have one of three effects on subsequent species: (1) facilitation, (2) tolerance, or (3) inhibition. **Facilitation** takes place when the presence of an early arriving species makes conditions more favorable for the arrival of certain later species by providing shade or nutrients. **Tolerance** means that existing species do not affect the probability that subsequent species will become established. **Inhibition** occurs when the presence of one species inhibits the establishment of another. For example, a plant species that requires high light levels to germinate may be inhibited late in

Old field

Disturbance (plowing) ends, site is invaded by short-lived weedy species.

Pioneering species

Weedy species are replaced by longer-lived herbaceous species and grasses.

Early successional community

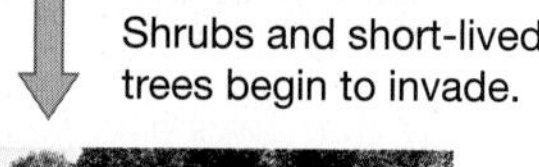

Mid-successional community

Short-lived tree species mature; long-lived trees begin to invade.

Late-successional community

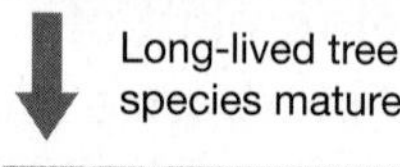

Climax community

FIGURE 53.21 Succession in Midlatitude Temperate Forests. Sequence of photos showing how succession leads to the development of a temperate forest from a disturbed state (in this case, an abandoned agricultural field).

succession by the presence of mature trees that prevent sunlight from reaching the forest floor.

In addition to species traits and species interactions, the pattern and rate of succession depend on the historical and environmental context in which they occur. For example, researchers found that the communities that developed after forest fires disturbed Yellowstone National Park in 1988 depended on the size of the burned patch and how hot the fire had been at that location. Succession is also affected by the particular weather or climate conditions that occur during the process. Variation in weather and climate causes different successional pathways to occur in the same place at different times.

Analyzing species traits, species interactions, and the historical/environmental context provides a useful structure for understanding why particular successional pathways occur. To see this theoretical framework in action, let's examine data on the course of primary succession that has occurred in Glacier Bay, Alaska.

A Case History: Glacier Bay, Alaska An extraordinarily rapid and extensive glacial recession is occurring at Glacier Bay (**Figure 53.22**, bottom). In just 200 years, glaciers that once filled the bay have retreated approximately 100 km, exposing extensive tracts of barren glacial sediments to colonization. Because of this event, Glacier Bay has become an important site for studying succession.

The top of Figure 53.22 shows the plant communities found in the area. The oldest sites are dense forests of Sitka spruce and western hemlock. Areas that have been deglaciated for about 100 years are inhabited by scattered spruce trees and dense thickets of a shrub called Sitka alder. Sites that have been deglaciated for 45 to 80 years are also covered with dense alder thickets, but the emergent trees are primarily cottonwood. Locations that have been ice free for 20 years or less do not have a continuous plant cover. Instead, they host scattered individuals of willow and a small shrub called *Dryas*.

These observations inspired a hypothesis for the pattern of succession in Glacier Bay: With time, the youngest communities of *Dryas* and willow succeed to alder thickets, which subsequently become dense spruce-hemlock forests. Stated another way, there is a single successional pathway throughout the bay. A recent study has challenged this hypothesis, however. Researchers who reconstructed the history of each community by studying tree rings found that three distinct successional pathways have occurred (**Figure 53.23**): (1) In the lower part of the bay, soon after the ice retreated, Sitka spruce began growing and quickly formed dense forests. Western hemlock arrived after spruce and is now common in the understory. (2) At middle-aged sites in the upper part of the bay, alder thickets were dominant for several decades, and spruce is just beginning to become common. These forests will probably never be as dense as the ones in the lower bay, however, and there is no sign that western hemlock has begun to establish itself. (3) In contrast, the youngest sites in the uppermost part of the bay may be following a third pathway. Alder thickets became dominant fairly early, but spruce trees are scarce. Instead, cottonwood trees are abundant.

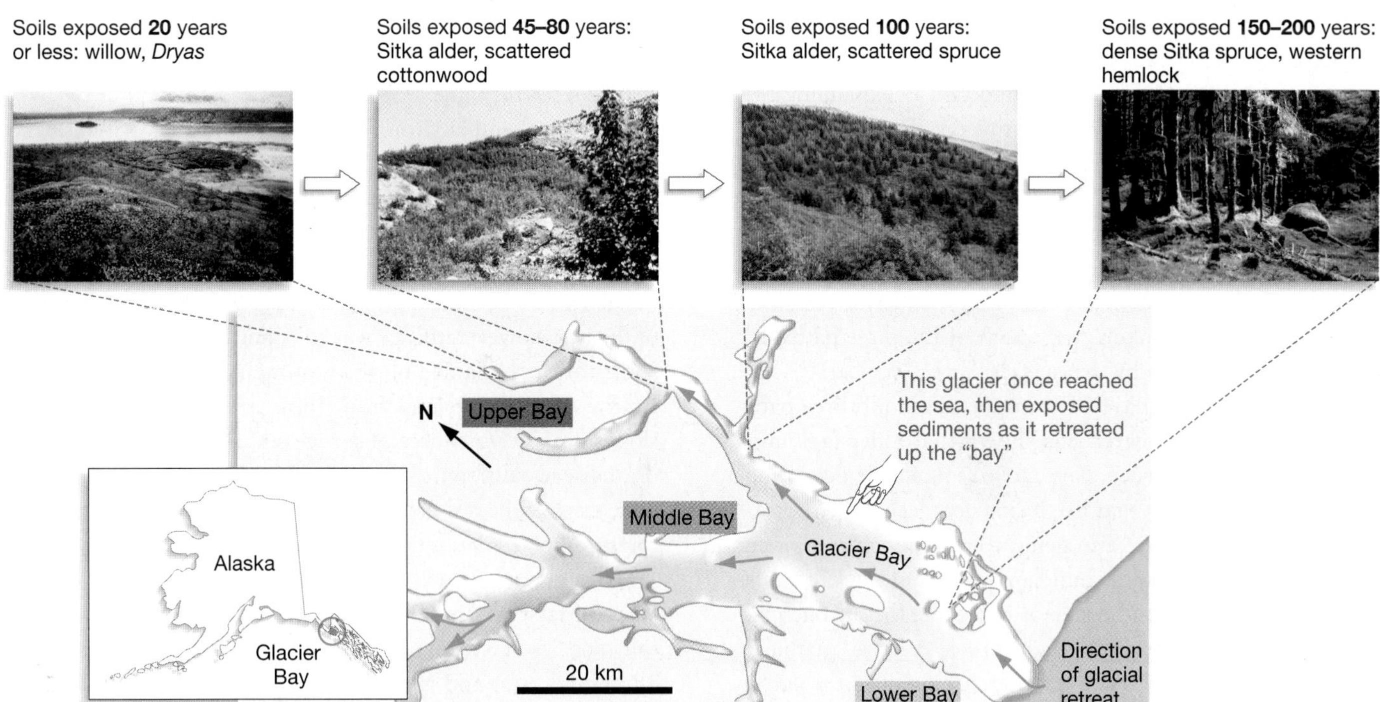

FIGURE 53.22 Evidence That a Single Successional Pathway Occurs in Glacier Bay.

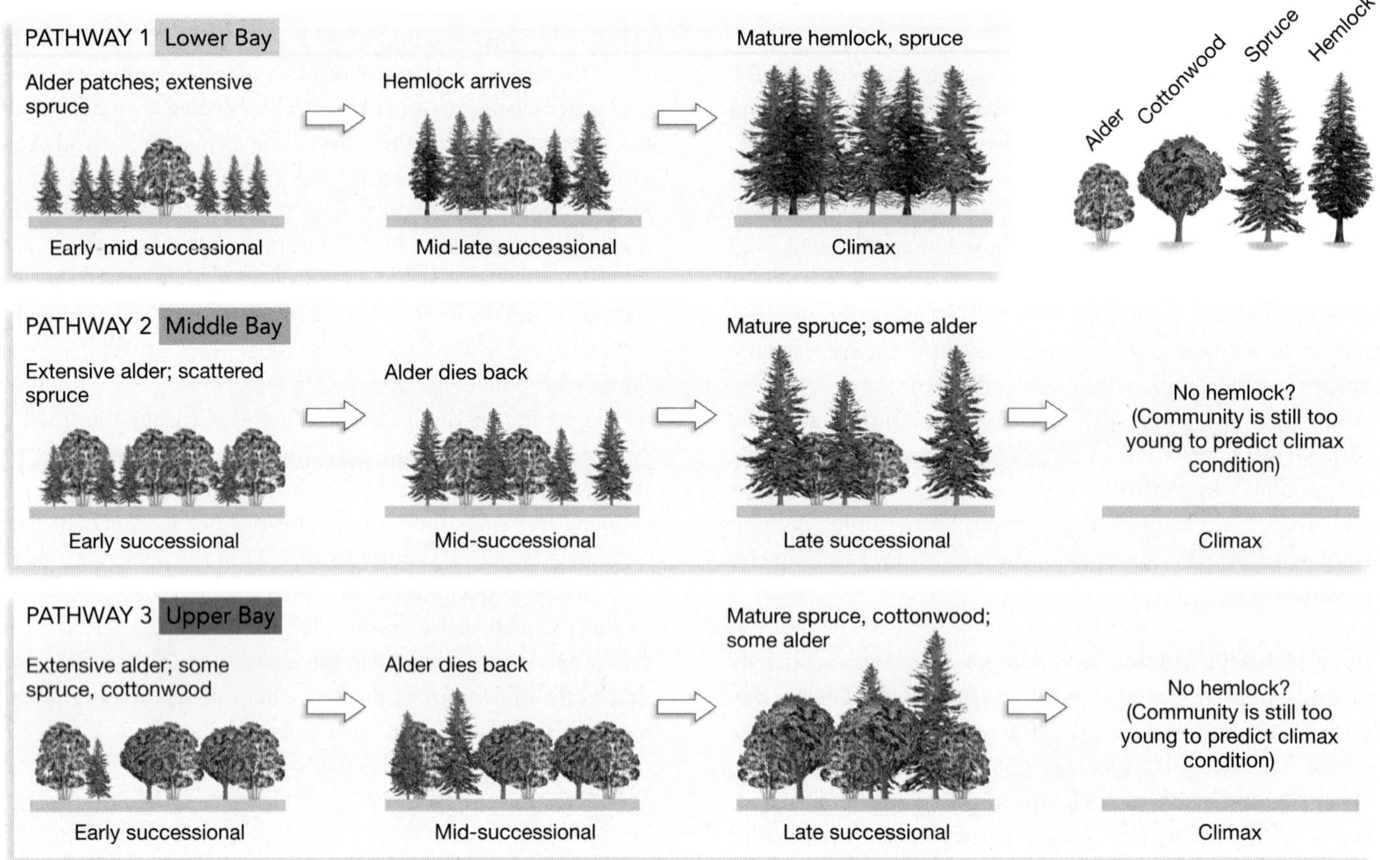

FIGURE 53.23 Evidence for Multiple Successional Pathways in Glacier Bay.

QUESTION Suppose that a forest fire wiped out all of the plant communities throughout Glacier Bay. Would you predict that the successional pathways described here would reoccur?

These data present a challenge: How do species traits, species interactions, and dispersal patterns interact to generate the three observed successional pathways?

Species traits may be especially important in explaining certain details about the successional pathways. Western hemlock, for example, is abundant at older sites but largely absent from young ones. This is logical, because its seeds germinate and grow only in soils containing a substantial amount of organic matter and because the young trees can tolerate deep shade but not bright sunlight. Western hemlock's intolerance of early successional conditions explains why none of the three pathways began with colonization by this species.

Species interactions have been important in all three pathways. For example, research has shown that alder facilitates the growth of Sitka spruce. The facilitating effect occurs because symbiotic bacteria that live inside nodules on the roots of alder convert atmospheric nitrogen (N_2) to nitrogen-containing molecules that alder use to build proteins and nucleic acids. When alder leaves fall and decay or roots die, the nitrogen becomes available to spruce. Although spruce trees are capable of invading and growing without the presence of alder, they grow faster when alder stands have added nitrogen to the soil.

Competition is another important species interaction. For example, shading by alder reduces the growth of spruce until spruce trees are tall enough to protrude above the alder thicket. Once the spruce trees breach the alder canopy, however, alder dies out, because it is unable to compete with spruce trees for light.

Historical and environmental context also clearly influences succession at Glacier Bay. For example, geologists have found evidence that the ice was more than 1100 m thick in the upper part of Glacier Bay during the mid-1700s. Because forests grow to an elevation of only 700 m or 800 m in this part of Alaska, the glacier eliminated all of the existing forests. In the lower part of the bay, however, the ice was substantially thinner. As a result, some forests remained on the mountain slopes beyond the ice. As the glacier retreated from these areas, the forests on the slopes provided a source of spruce and hemlock seeds and set a dramatically different successional pathway in motion. In this way, environmental context—in this case, distance to existing forests—helped determine how the community developed.

To summarize, successional pathways are determined by an array of factors. These factors include the adaptations that certain species have to their abiotic environment, interactions among species, and the history of the site. Species traits and species interactions tend to make succession predictable, while history and chance events contribute a degree of unpredictability to succession.

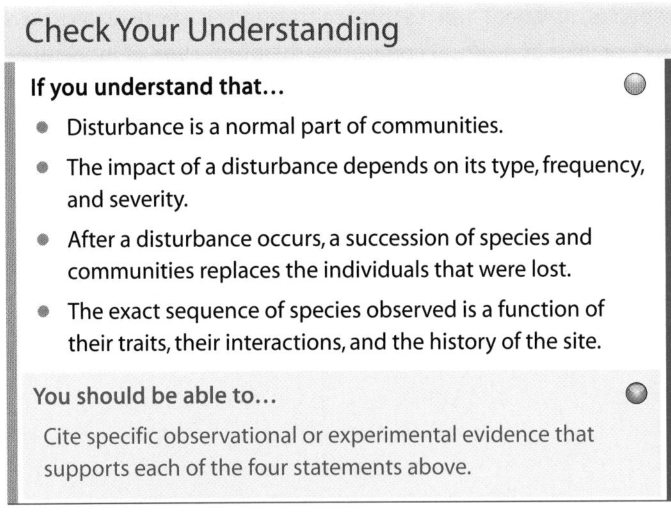

Check Your Understanding

If you understand that...

- Disturbance is a normal part of communities.
- The impact of a disturbance depends on its type, frequency, and severity.
- After a disturbance occurs, a succession of species and communities replaces the individuals that were lost.
- The exact sequence of species observed is a function of their traits, their interactions, and the history of the site.

You should be able to...

Cite specific observational or experimental evidence that supports each of the four statements above.

(MB) Web Animation at www.masteringbio.com
Primary Succession

53.4 Species Richness in Ecological Communities

The diversity of species present is a key feature of biological communities, and it can be quantified in two ways. **Species richness** is a simple count of how many species are present in a given community. **Species diversity**, in contrast, is a weighted measure that incorporates a species' relative abundance as well as its presence or absence (see **Box 53.1**). But at scales above relatively small study plots, it is rare to have data on relative abundance. As a result, ecologists sometimes use species richness and species diversity interchangeably. To introduce research on richness and diversity, let's focus on a simple question. Why are some communities more species rich than others?

Predicting Species Richness: The Theory of Island Biogeography

When researchers first began counting how many species are present in various areas, a strong pattern emerged: Larger patches of habitat contain more species than do smaller patches of habitat. The observation is logical because large areas should contain more types of niches and thus support higher numbers of species. But early work on species richness highlighted another pattern—one that was harder to explain. Islands in the ocean have smaller numbers of species than do areas of the same size on continents.

Robert MacArthur and Edward O. Wilson tackled this question by assuming that speciation occurs so slowly that the number of species present on an island is a product of just two events: immigration and extinction. The rates of both of these processes, they contended, should vary with the number of species present on an island. Immigration rates should decline as the number of species on the island increases, because individuals that arrive are more likely to represent a species that is already present. But extinction rates should increase as species richness increases, because niche overlap and competition for resources will be more intense. The result is an equilibrium—a balance between the arrival of new species and the extinction of existing ones (**Figure 53.24a**). If species richness changes due to a hurricane or fire or some other disturbance, then continued immigration and extinction should restore the equilibrium value.

MacArthur and Wilson also realized that immigration and extinction rates should vary as a function of island size and how far the island was to a continent or other source of immigrants. Immigration rates should be higher on large islands that are close to mainlands, because immigrants are more likely to find large islands that are close to shore than small ones that are far

(a) Species richness depends on the number of existing species.

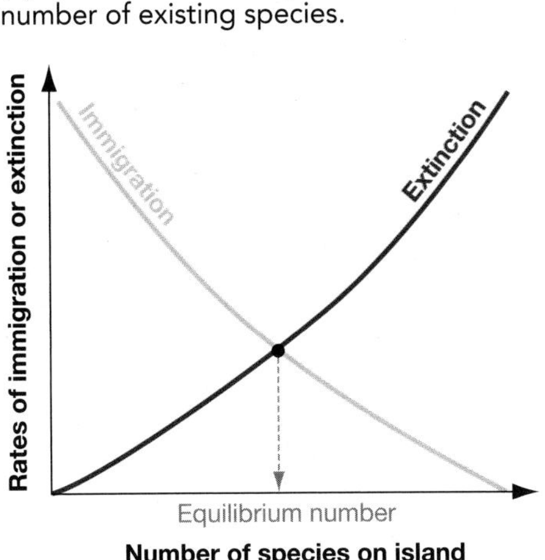

(b) Species richness depends on island size.

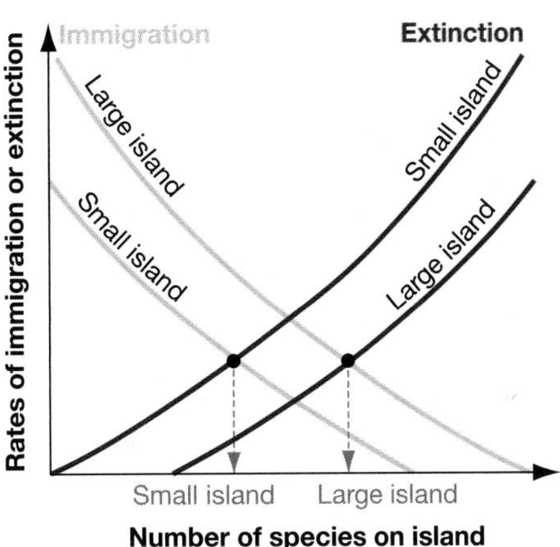

(c) Species richness depends on remoteness of the island.

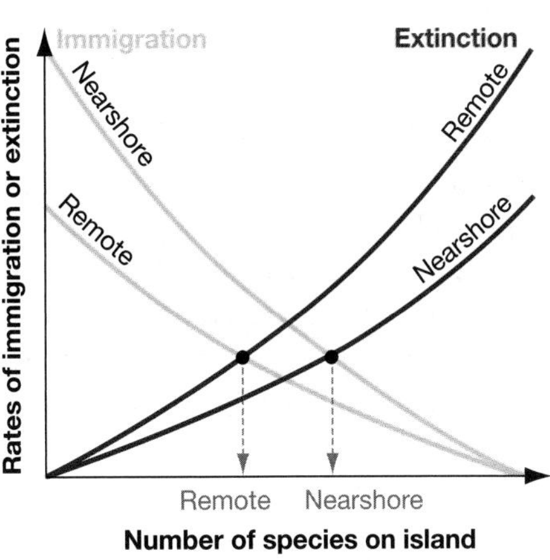

FIGURE 53.24 Species Richness Varies as a Function of Island Characteristics. If species richness on islands is dictated only by immigration and extinction rates, then species richness will vary as a function of (a) number of existing species, (b) island size, and (c) remoteness of the island from large landmasses.

QUESTION Suppose that the mainland habitats closest to an island were wiped out by suburbanization. How would this affect the curves in part (c)?

BOX 53.1 Measuring Species Diversity

To measure species diversity, a biologist could simply count the number of species present in a community. The problem is that simple counts provide an incomplete picture of diversity. The relative abundance of species is also an important component of diversity. To grasp this point, consider the composition of the three hypothetical communities shown in **Figure 53.25.** These communities are nearly identical in species composition but differ greatly in the relative abundance of each species.

These data can be used to compare two measures of species diversity. Species richness is simply the number of species found in a community. In this case, communities 1 and 2 have equal species richness and community 3 is lower in richness by one species. It is important to note, however, that communities 2 and 3 have similar relative abundances of each species, or what biologists call high *evenness*. Community 1, in contrast, is highly uneven. Fifty-five percent of the individuals in community 1 belong to species A, and other species are relatively rare. An uneven community has lower effective diversity than its species richness would indicate.

To take evenness into account, other diversity indices have been developed. A simple example is the Shannon index, given by the following equation:

$$H' = -\sum_{i=1}^{S} p_i \ln p_i$$

In this equation, p_i is the proportion of individuals in the community that belong to species i. The index is summed over all of the species in the study. The Shannon index for the three hypothetical communities is shown at the bottom of Figure 53.25.

In each case, the index is calculated by (1) computing the proportion of individuals in each community that belong to each species, (2) taking the natural logarithm of each of these proportions (see **BioSkills 5** for an introduction to logarithms), (3) multiplying each natural logarithm times the proportion for each species, and (4) summing the total across the six species in the community. ● If you understand the equation, you should be able to do these calculations and get the same result given in the figure. Notice that while communities 1 and 2 have the same species richness, community 2 has higher diversity because of its greater evenness. Community 3 has lower species richness than community 1 but higher diversity.

● If you understand how to use and interpret the Shannon index, you should be able to calculate species richness and the Shannon diversity index for 3 communities that "double" Communities 1–3 in Figure 53.25. For example, one of the three new communities should be identical to Community 1, except that there are 2 species with 10 individuals present like species "A", 2 species with 1 individual present like species "B", 2 species with 1 individual present like species "C", 2 species with 3 individuals present like species "D", and so on. There are a total of 12 species in each community. ● You should also be able to compare and contrast your results with the richness and diversity values given in the figure.

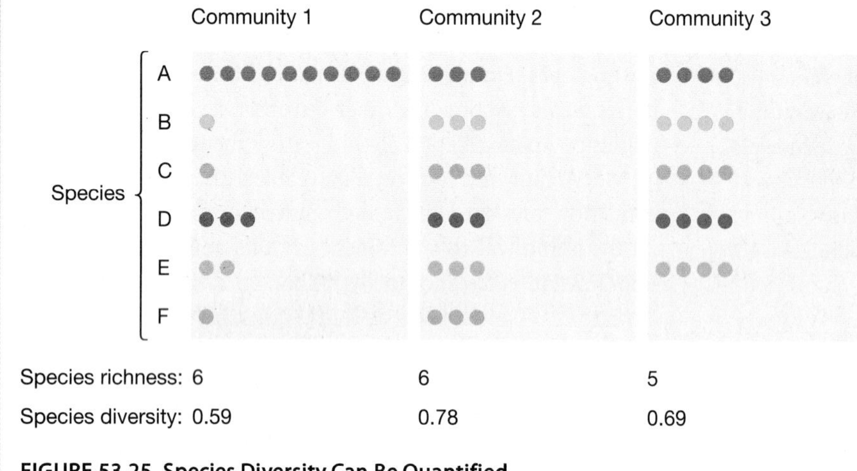

FIGURE 53.25 Species Diversity Can Be Quantified.

from shore. Extinction rates should be highest on small islands that are far from shore, because fewer resources are available to support large populations and because fewer individuals arrive to keep the population going. As a result, species richness should be higher on larger islands than smaller islands (**Figure 53.24b**), and on nearshore islands versus remote islands (**Figure 53.24c**).

This model, called the theory of island biogeography, is important for several reasons:

- It is relevant to a wide variety of island-like habitats such as alpine meadows, lakes and ponds, and caves.
- It made specific predictions that could be tested. For example, researchers have measured species richness on tiny islands, removed all of the species present, and then measured whether the same number of species recolonized the island and reached an equilibrium number. Predictions about immigration and extinction rates have also been measured by observing islands over time.
- It is relevant to the metapopulation dynamics introduced in Chapter 52 and can help inform decisions about the design of natural preserves. In general, the most-species rich reserves should be ones that are (1) relatively large, and (2) located close to other relatively large habitat areas.

Global Patterns in Species Richness

Biologists have long understood that large habitat areas tend to be species-rich, and the theory of island biography has been

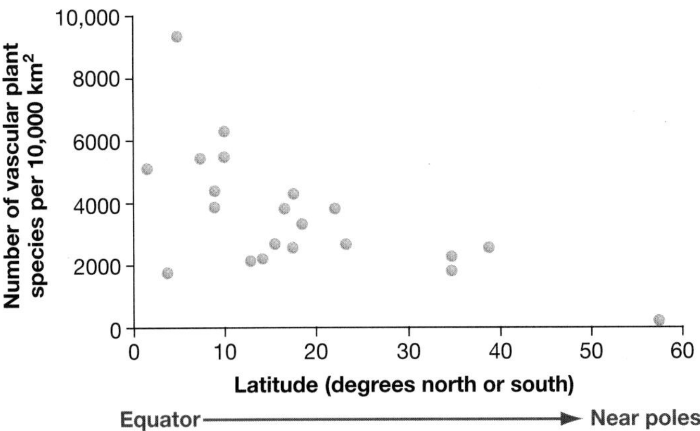

FIGURE 53.26 There Is a Strong Latitudinal Gradient in Species Diversity. One example of a latitudinal gradient in species diversity.

successful in framing thinking about how species richness should vary among island-like habitats. But researchers have had a much more difficult time explaining what may be the most striking pattern in species richness.

Biologists who began cataloguing the flora and fauna of the tropics in the mid-1800s quickly recognized that these habitats contain many more species than do temperate or subarctic environments. Data compiled in the intervening years have confirmed the existence of a strong latitudinal gradient in species diversity—for communities as a whole as well as for many taxonomic groups. In birds, mammals, fish, reptiles, many aquatic and terrestrial invertebrates, orchids, and trees, for example, species diversity declines as latitude increases (**Figure 53.26**). Although this pattern is not universal—a number of marine groups, as well as shorebirds, show a positive relationship between latitude and diversity—it is widespread. Why does it occur?

To explain why species diversity might decline with increasing latitude, biologists have to consider two fundamental principles. First, the causal mechanism must be abiotic, because latitude is a physical phenomenon produced by Earth's shape. The explanation must be a physical factor that varies predictably with latitude and that could produce changes in species diversity. Second, the species diversity of a particular area is the sum of four processes: speciation, extinction, immigration (colonization), and emigration (dispersal). Thus, the latitudinal gradient must be caused by an abiotic factor that affects the rate of speciation, extinction, immigration, or emigration in a way that would lead to more species in the tropics and fewer near the poles.

Over 30 hypotheses have been proposed to explain the latitudinal gradient. One of these maintains that high productivity in the tropics promotes high diversity by increasing speciation rates and decreasing extinction rates. (Recall from Chapter 50 that productivity is the total amount of photosynthesis per unit area per year.) The idea is that increased biomass production supports more herbivores and thus more predators and parasites and scavengers. In addition, speciation rates should increase when niche differentiation occurs within populations of herbivores, predators, parasites, and scavengers. Although this high-productivity hypothesis is supported by the global-scale correlation between productivity and diversity, experimental studies challenge it. For example, researchers who add fertilizer to aquatic or terrestrial communities routinely observe significant increases in productivity but decreases in diversity.

In light of results such as these, researchers have concluded that productivity alone is probably not a sufficient explanation for the higher species diversity in the tropics. Research continues, however, and investigators have recently begun to focus on an abiotic factor that is correlated with productivity: temperature. Biologists who analyzed data on gastropods and other marine invertebrates have documented a strong correlation between the temperature of marine waters and species diversity. This observation has inspired the energy hypothesis: High temperatures increase species diversity by increasing productivity and the likelihood that organisms can tolerate the physical conditions in a region.

A third hypothesis is that tropical regions have had more time for speciation to occur than other regions have. Temperate and arctic latitudes were repeatedly scoured by ice sheets over the last 2 million years, but tropical regions were not. Recent data suggest, however, that tropical forests were dramatically reduced in size by widespread drying trends during the ice ages. Existing forests may be much younger than originally thought. If so, then the contrast in the age of northern and southern habitats may not be enough to explain the dramatic difference in species diversity.

A fourth hypothesis was inspired by the observation that species diversity is much higher in mid-successional communities than in pioneer or mature communities. The **intermediate disturbance hypothesis** holds that regions with a moderate type, frequency, and severity of disturbance should have high species richness and diversity. The logic here is that, with intermediate levels of disturbance, communities will contain pioneering species as well as species better adapted to late-successional conditions. For example, recent studies have confirmed that tree falls and canopy gaps occur regularly in tropical forests and that fires occur in these biomes occasionally. As yet, however, there are no convincing data showing that intermediate levels of disturbance are more likely to occur in the tropics than they are at higher latitudes.

Each of the factors discussed here may influence diversity, but no single one offers a convincing explanation for the global diversity gradient. Although the high temperatures present in the tropics seem to have an especially important role in promoting species richness, there is no simple answer to the question of why the tropics are so much more species rich than temperate and high-latitude habitats.

Chapter Review

SUMMARY OF KEY CONCEPTS

Interactions among species, such as competition and consumption, have two main outcomes: (1) They affect the distribution and abundance of the interacting species, and (2) they are agents of natural selection and thus affect the evolution of the interacting species. The nature of interactions between species frequently changes over time.

A community is an assemblage of interacting species.

To categorize the different types of interactions that occur among species, biologists consider whether each participant experiences a net fitness cost or benefit from the interaction. These costs and benefits depend on the conditions that prevail at a particular time and place and may change through time.

Competition is a −/− interaction that occurs when the niches of two species overlap—meaning they use the same resources. Competition may result in the complete exclusion of one species. It may also result in niche differentiation, in which competing species evolve traits that allow them to exploit different resources or live in different areas.

Consumption is a +/− interaction that occurs when consumers eat prey, which resist through standing defenses or inducible defenses. Predators are efficient enough to reduce the size of many prey populations. Levels of herbivory are relatively low in terrestrial ecosystems, however, because predation and disease limit herbivore populations, because plants provide little nitrogen, and because many plants contain toxic compounds or other types of defenses. Parasites generally spend all or part of their life cycle in or on their host (or hosts) and usually have traits that allow them to escape host defenses. In turn, hosts have evolved counteradaptations that help fight off parasites.

Mutualism is a +/+ interaction that provides participating individuals with food, shelter, transport of gametes, or defense against predators. For each species involved, the costs and benefits of a mutualism may vary over time and from place to place.

You should be able to give an example of how competition can evolve into commensalism, and how a mutualistic relationship can evolve into a parasitic one.

MB Web Animation at www.masteringbio.com
Life Cycle of a Malaria Parasite

The assemblage of species found in a biological community changes over time and is primarily a function of climate and chance historical events.

Ecologists have debated whether communities are fixed, predictable entities or simply places where the distributions of various species overlap. Historical and experimental evidence support the view that communities are dynamic rather than static and that their composition is neither entirely predictable nor stable over time.

In addition to climate, the composition of a community is influenced by disturbance. In extreme cases, disturbance may remove all organisms and all soil from a large area. Each community has a characteristic disturbance regime—meaning a type, severity, and frequency of disturbance that it experiences. Three types of factors influence the pattern of succession that occurs after a disturbance. First, the historical and environmental context of the site affects which species are available to join the resulting communities. The dispersal ability of different species also affects their availability. Second, the physiological traits of any given species influence the kinds of abiotic environmental conditions it can tolerate and dictate when it can successfully join a community. Third, interactions among species influence if and when a species appears during succession.

You should be able to to explain why climate makes the three successional pathways documented in Glacier Bay similar, and why chance historical events make them different.

MB Web Animation at www.masteringbio.com
Primary Succession

Species diversity is high in the tropics and lower toward the poles. The mechanism responsible for this pattern is still being investigated.

One of the most widely studied patterns in community ecology is the latitudinal gradient in species diversity. Within most specific taxonomic groups and within communities as a whole, species diversity declines from the equator to the poles. Recent research suggests that the high species richness observed in the tropics results from a combination of factors—the most important of which may be high temperatures that increase productivity and provide relatively benign abiotic conditions.

You should be able to suggest a hypothesis to explain an exception to the latitudinal gradient rule—specifically, that most species of shorebirds live at high latitudes, not the tropics. (Note that many shorebird species nest near the abundant lakes found in arctic tundra).

QUESTIONS

Test Your Knowledge

1. What is competitive exclusion?
   a. the evolution of traits that reduce niche overlap and competition
   b. interactions that allow species to occupy their fundamental niche
   c. the degree to which the niches of two species overlap
   d. the claim that species with the same niche cannot coexist

2. What is niche differentiation?
   a. the evolution of traits that reduce niche overlap and competition
   b. interactions that allow species to occupy their fundamental niche
   c. the degree to which the niches of two species overlap
   d. the claim that species with the same niche cannot coexist

3. Why is the phrase "coevolutionary arms race" an appropriate way to characterize the long-term effects of species interactions?
   a. Both plants and animals have evolved weapons for defense that are so effective that many plants are not eaten and predators cannot reduce prey populations to extinction.
   b. Adaptations that give one species a fitness advantage in an interaction are likely to be countered by adaptations in the other species that eliminate this advantage.
   c. In all species interactions except for mutualism, at least one species loses (suffers decreased fitness).
   d. Even mutualistic interactions can become parasitic if conditions change. As a result, interacting species are always "at war."
4. Why are inducible defenses advantageous?
   a. They are always present—thus, an individual is always able to defend itself.
   b. They make it impossible for a consumer to launch surprise attacks.
   c. They result from a coevolutionary arms race.
   d. They make efficient use of resources, because they are produced only when needed.
5. Which of the following is *not* correlated with species diversity?
   a. latitude
   b. productivity
   c. longitude
   d. island size
6. What is net primary productivity?
   a. an individual's lifetime reproductive success (lifetime fitness)
   b. an individual's average annual reproductive success
   c. the total amount of photosynthesis that occurs in an area of a given size per year
   d. the amount of energy that is stored in standing biomass per year

Test Your Knowledge answers: 1. d; 2. a; 3. b; 4. d; 5. c; 6. d

Test Your Understanding

Answers are available at www.masteringbio.com

1. The text claims that species interactions are conditional and dynamic. Do you agree with this statement? Why or why not? Cite specific examples to support your answer.
2. State three hypotheses that have been proposed to explain the low level of herbivory in terrestrial plant communities. Are these hypotheses mutually exclusive? (In other words, can more than one be correct?) Explain why or why not.
3. Biologists have tested the hypotheses that communities are highly predictable versus highly unpredictable. State the predictions that these hypotheses make with respect to (a) the presence and impact of keystone species, (b) changes in the distribution of the species in a particular community over time, and (c) the communities that should develop at sites where abiotic conditions are identical. Which hypothesis appears to be more accurate?
4. What is a disturbance? List five examples of disturbance. Compare and contrast their effects. For each type of disturbance, compare and contrast the consequences of high-frequency and low-frequency disturbances and high and low severity of disturbances.
5. Summarize the life-history attributes of early successional versus late successional species. Why are these attributes considered adaptations?
6. Describe the latitudinal gradient in species diversity that exists for most taxonomic groups. Discuss the pros and cons of one hypothesis to explain this pattern.

Applying Concepts to New Situations

Answers are available at www.masteringbio.com

1. Some insects harvest nectar by chewing through the wall of the structure that holds the nectar As a result, they obtain a nectar reward, but pollination does not occur. Suppose that you observed a certain bee species obtaining nectar in this way from a particular orchid species. Over time, how might you expect the characteristics of the orchid population to change in response to this bee behavior?
2. Using this chapter's information on fire regimes in giant sequoia groves, propose a management plan for Sequoia National Park. Explain the logic behind your plan.
3. Suppose that a two-acre lawn on your college's campus is allowed to undergo succession. Describe how species traits, species interactions, and the site's history might affect the community that develops.
4. Design an experiment to test the hypothesis that increasing species richness increases a community's productivity and ability to resist disturbance and recover from disturbance.

www.masteringbio.com is also your resource for • Answers to text, table, and figure caption questions and exercises • Answers to *Check Your Understanding* boxes • Online study guides and quizzes • Additional study tools including the *E-Book for Biological Science* 3rd ed., textbook art, animations, and videos.

54 Ecosystems

KEY CONCEPTS

- An ecosystem has four components: (1) the abiotic environment, (2) primary producers, (3) consumers, and (4) decomposers. These components are linked by the movement of energy and nutrients.
- The productivity of terrestrial ecosystems is limited by warmth and moisture, while nutrient availability is the key constraint in aquatic ecosystems. As energy flows from producers to consumers and decomposers, much of it is lost.
- To analyze nutrient cycles, biologists focus on the nature of the reservoirs where elements reside and on how quickly elements move between reservoirs.
- Humans are causing large, global changes in the abiotic environment. The burning of fossil fuels has led to rapid global warming. Extensive fertilization is increasing productivity and causing pollution.

The Qori Kalis glacier in the Andes Mountains of Peru in the year 2006. In 1978 the glacier extended past the edge of the lake in the lower left-hand corner of this photograph. Global warming is causing dramatic reductions in the size of glaciers and ice fields in many locations throughout the world.

An **ecosystem** consists of the organisms that live in an area together with their physical, or abiotic, environment. With the explosive growth of human populations, dramatic changes are occurring in both the biotic and abiotic components of ecosystems around the globe. Extinctions and other effects on the biotic elements of ecosystems are explored in Chapter 55; this chapter analyzes changes in the chemical and physical characteristics of the environment. Human impacts on the abiotic environment include global warming, acid rain, a hole in the atmosphere's ozone layer, and nitrate pollution (see Chapter 28). Ecosystem studies are taking a prominent place in biological science.

How does an ecosystem differ from a community? In most cases, ecosystems are composed of multiple communities along with their chemical and physical environments. For example, biologists who study lakes recognize a number of communities within a lake. There are distinct communities of interacting species along the lake bottom, at the surface, and in the shallow water of the littoral zone. Those different communities are studied as a unit called the lake ecosystem, because energy and matter flow among them.

Adjacent ecosystems—such as a river, a lake, and the surrounding forest—can be distinguished because their energy and nutrient flows are self-contained. Occasionally, though, energy and nutrients can also be transferred between ecosystems. This happens when nutrients dissolved in rainwater run off of agricultural ecosystems into nearby streams, or when an eagle kills a fish and carries it into a forest ecosystem.

This chapter introduces ecosystem studies with a look at how energy flows among the components of an ecosystem; it continues by exploring how carbon, nitrogen, and other key elements cycle through organisms, sediments, the oceans, and the atmosphere; and it ends by analyzing how humans are affecting the abiotic environment. Ecosystem ecology has taken on an increasingly applied focus, because humans are adding massive amounts of energy to ecosystems, disrupting nutrient cycles, and changing the chemistry of lakes, oceans, and the atmosphere. Understanding ecosystem ecology is fundamental to managing the future of our planet.

54.1 Energy Flow and Trophic Structure

If you imagine that an ecosystem is like an economy, then energy is its currency. Ecosystems have four components: (1) the abiotic environment, (2) primary producers, (3) consumers, and (4) decomposers. The four components are linked by a flow of energy (**Figure 54.1**).

A **primary producer** is an **autotroph** (literally, "self-feeder")—meaning, an organism that can synthesize its own food from inorganic sources. In most ecosystems, primary producers use solar energy to manufacture their own food via photosynthesis. But in some ecosystems, such as deep-sea hydrothermal vents and iron-rich rocks deep below Earth's surface, bacteria use the chemical energy contained in inorganic compounds such as hydrogen (H_2), methane (CH_4), or hydrogen sulfide (H_2S) to make food (see Chapter 28).

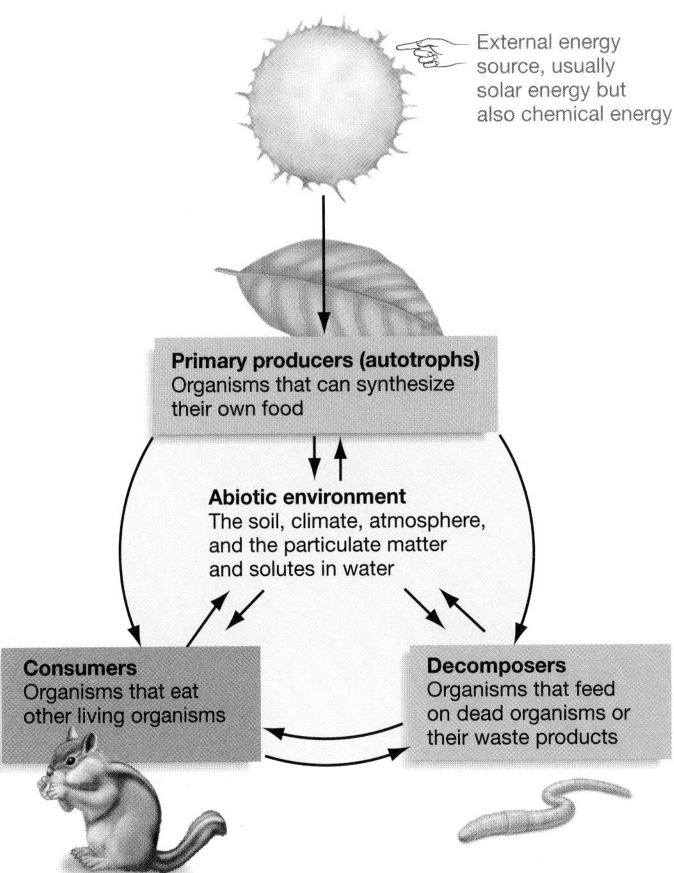

FIGURE 54.1 The Four Components of an Ecosystem Interact. The arrows represent energy. A similar diagram could be drawn to represent the flow of nutrients.

Primary producers form the basis of ecosystems because they transform the energy in sunlight or inorganic compounds into the chemical energy stored in sugars. Primary producers use this chemical energy in two ways: Most supports maintenance or respiratory costs, but some makes growth and reproduction possible. The energy that is invested in new tissue is called **net primary productivity (NPP)**.

Net primary productivity represents the amount of energy that is available to the second and third components of an ecosystem: consumers and decomposers. **Consumers** eat living organisms. **Herbivores** are consumers that eat plants; **carnivores** are consumers that eat animals. **Decomposers**, or **detritivores**, obtain energy by feeding on the dead remains of other organisms or waste products.

The fourth and final component of an ecosystem is the abiotic environment, which includes soil, climate, the atmosphere, the Sun, and the particulate matter and solutes in water. The four components are linked because energy moves from the Sun or inorganic compounds to consumers, decomposers, and the abiotic environment.

Understanding NPP is a primary focus of ecosystem ecology because it dictates the amount of energy available to consumers and decomposers, and because global warming is altering NPP worldwide. Let's consider two of the most fundamental questions about net primary productivity: How does it vary among the world's ecosystems, and what happens to it?

Global Patterns in Productivity

Figure 54.2 summarizes data on NPP from around the globe. A quick look at the color key should convince you that the terrestrial ecosystems with highest productivity are located in the wet tropics. Notice that, with the exception of the world's major deserts, NPP on land declines from the equator toward the poles. Productivity patterns in marine ecosystems are different, however. Marine productivity is highest along coastlines, and it can be as high near the poles as it is in tropics. The oceanic zones, introduced in Chapter 50, have extremely low NPP. Typically, a square meter of open ocean produces a maximum of 35 g (1.2 oz) of organic matter each year. In terms of productivity per m^2, the open ocean is a desert.

Figure 54.3 presents the NPP data a different way—organized by biome instead of by geography. Figure 54.3a provides data on average NPP per square meter per year for each

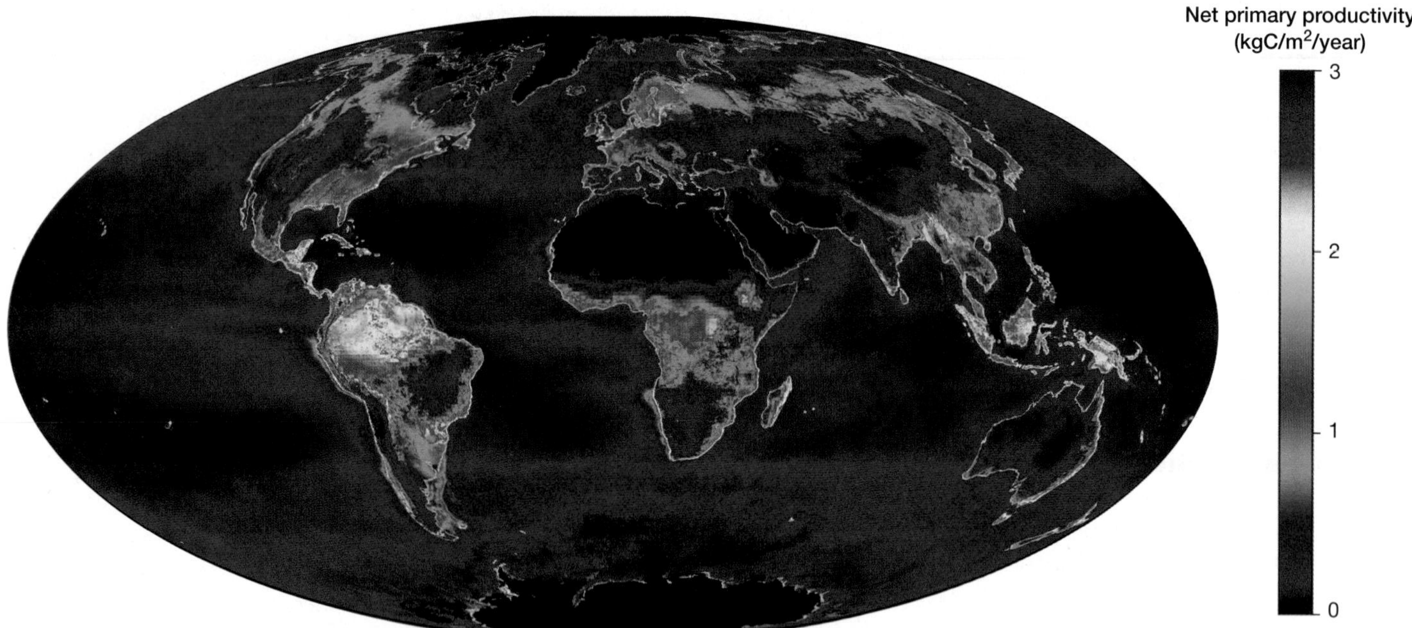

FIGURE 54.2 Net Primary Productivity Varies among Regions. The terrestrial ecosystems with the highest primary productivity are found in the tropics, where warm temperatures and high moisture encourage high rates of photosynthesis. Tundras and deserts have the lowest productivity. The highest productivity in the oceans occurs in nutrient-rich coastal areas.

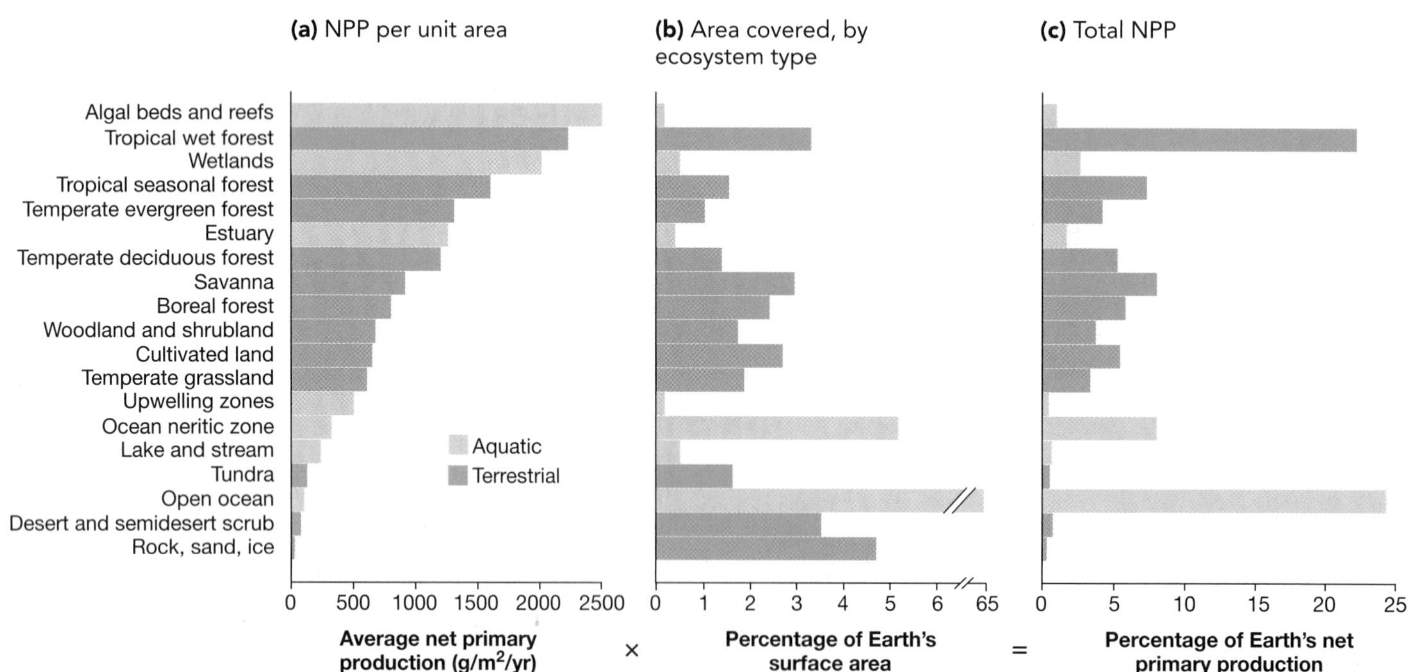

FIGURE 54.3 Net Primary Productivity Varies among Ecosystems. (a) Among biomes, average annual NPP per square meter varies by over three orders of magnitude. **(b)** Most of Earth's surface is covered by open ocean. The most common terrestrial habitat consists of unvegetated rock, sand, or ice. **(c)** Even though it has a low NPP per square meter, the open ocean is so vast that it is responsible for over 25 percent of Earth's total NPP.

EXERCISE Using information in Chapter 50, explain the patterns in part (a).

biome; Figure 54.3b documents the total area that is covered by each type of ecosystem; and Figure 54.3c presents the percentage of the world's total productivity—the result of multiplying the data in part (a) by the data in part (b). Note that tropical wet forests and tropical seasonal forests, which have a dry season, cover less than 5 percent of Earth's surface but together account for over 30 percent of total NPP. Among aquatic ecosystems, the most productive habitats are algal beds and coral reefs, wetlands, and estuaries. Most of the total NPP provided by aquatic ecosystems derives from the open ocean, however. Even though NPP per square meter is low in these regions, the open ocean is so extensive that its total production is high.

These patterns raise an interesting question: What limits NPP in terrestrial and marine ecosystems? The short answer is that productivity is limited by any factor that limits the rate of photosynthesis—specifically, temperature and the availabilities of water, sunlight, and nutrients. Different limiting factors prevail in different environments, however. Let's take a closer look.

What Limits Productivity? The data in Figure 54.2 and Figure 54.3 document that terrestrial productivity is lowest in deserts and arctic regions. This observation suggests that the overall productivity of terrestrial ecosystems is limited by a combination of temperature and availability of water and sunlight.

To explain why the productivity of marine habitats is so much higher along coastlines than in deepwater regions, biologists focus on nutrient limitation. As Chapter 50 pointed out, the neritic and intertidal zones along coasts receive nutrients from two major sources: (1) rivers that carry and deposit nutrients from terrestrial ecosystems, and (2) nearshore ocean currents that bring nutrients from the cold, deep water of the oceanic zone back up to the surface. Both of these sources are absent in the surface waters of the open ocean. In addition, nutrients found in organisms near the surface of the open ocean—where sunlight is abundant—constantly fall to dark, deeper waters in the form of dead cells and are lost.

Analyses have confirmed that trace elements such as zinc, iron, and magnesium are particularly rare in the open ocean. These atoms are important because they are required as enzyme cofactors. Iron, for example, is essential to the proteins that are involved in electron transport chains (see Chapters 9 and 10). On the basis of these observations, biologists have proposed that the productivity of open-ocean ecosystems could be increased dramatically by fertilizing them with iron. The results of one iron-fertilization experiment are shown in **Figure 54.4**. The data show large increases in the concentration of chlorophyll *a* in surface waters over a two-week interval. Consistent with this result, recent research on ocean waters that are naturally enriched by nearby iron-containing rocks indicates that these regions have exceptionally high NPP. Results like these provide strong support for the hypothesis that NPP in marine ecosystems is limited primarily by the availability of nutrients and that iron is particularly important in the open ocean.

Experiment

Question: Is net primary productivity (NPP) in the open ocean limited by nutrients?

Hypothesis: NPP in the open ocean is limited by availability of iron.

Null hypothesis: NPP in the open ocean is not limited by availability of iron.

Experimental setup:

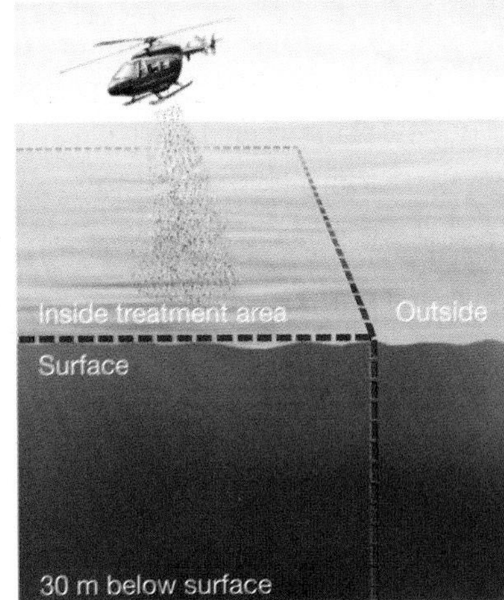

1. Add 350 kg iron (as $FeSO_4$) to a treatment area—a patch of ocean 8 km × 10 km.

2. Take water samples for a two-week period outside and inside the treatment area, at surface and at a depth of 30 m, and record amount of chlorophyll *a* present (as indicator of NPP).

Prediction: Amount of chlorophyll *a* near the surface inside the treatment area will increase relative to amounts outside the treatment area or at 30 m below the surface.

Prediction of null hypothesis: Amount of chlorophyll *a* will be the same in all measurements.

Results:

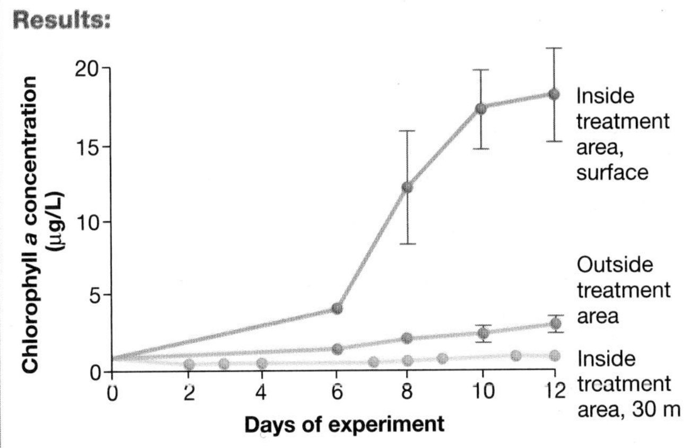

Conclusion: NPP in the open ocean is limited by the scarcity of nutrients—specifically, iron.

FIGURE 54.4 Fertilization with Iron Increases NPP in the Open Ocean.

QUESTION Most of the increased chlorophyll *a* was present in a single species of diatom (see Chapter 29). What happens to the biomass present in diatoms after they die?

How Does Energy Flow through an Ecosystem?

What happens to NPP? Net primary productivity results in **biomass**—organic material that non-photosynthetic organisms can eat. In every terrestrial and marine environment in the world, the chemical energy in primary producers eventually moves to one of two types of organisms: primary consumers or primary decomposers.

A **primary consumer** is an herbivore—an organism that eats plants or algae or other photosynthetic cells. As Chapter 53 pointed out, however, only a small percentage of all plant tissue is consumed by herbivores. Tissues that are not consumed eventually die. In forest ecosystems, for example, dead animals and dead plant tissues ("plant litter") are collectively known as **detritus**. Detritus is consumed by a variety of **primary decomposers**, including bacteria, archaea, and fungi.

Trophic Levels, Food Chains, and Food Webs Biomass represents chemical energy. Every time an herbivore eats a leaf or a fungus absorbs molecules from decaying wood, chemical energy flows from primary producers to primary consumers and decomposers. To describe these energy flows, biologists identify distinct feeding levels in an ecosystem. Organisms that obtain their energy from the same type of source are said to occupy the same **trophic** ("feeding") **level**.

A **food chain** connects the trophic levels in a particular ecosystem. In doing so, it describes how energy moves from one trophic level to another. As **Figure 54.5** shows, primary consumers are a key link in what biologists call the **grazing food chain**: the collection of organisms that eat plants, along with the organisms that eat herbivores. Consumers that eat herbivores are called **secondary consumers**. Organisms at the top trophic level in a food chain are not killed and eaten by any other organisms but enter the **decomposer food chain** when they die—along with biomass from dead primary producers and other consumers. A decomposer food chain starts with primary decomposers and includes primary consumers that usually specialize in eating primary decomposers. At higher trophic levels, though, the grazing and decomposer food chains merge. For example, the robin in Figure 54.5 functions as a secondary consumer in the grazing food chain and a tertiary consumer in the decomposer food chain.

In addition to feeding at several trophic levels, most organisms eat more than one type of food. As a result, food chains are almost always embedded in more complex **food webs**. The food web shown in **Figure 54.6** is a more complete description of the trophic relationships among the organisms in an ecosystem.

Food webs are a compact way of summarizing energy flows and documenting the complex trophic interactions that occur in communities. The keystone species introduced in Chapter 53, for example, are usually at the top of food chains and food webs. Removing or adding them has dramatic effects on a wide array of species at lower trophic levels. ● If you understand this concept, you should be able to predict how removing the top predator in Figure 54.5 would affect each of the other species in the food web.

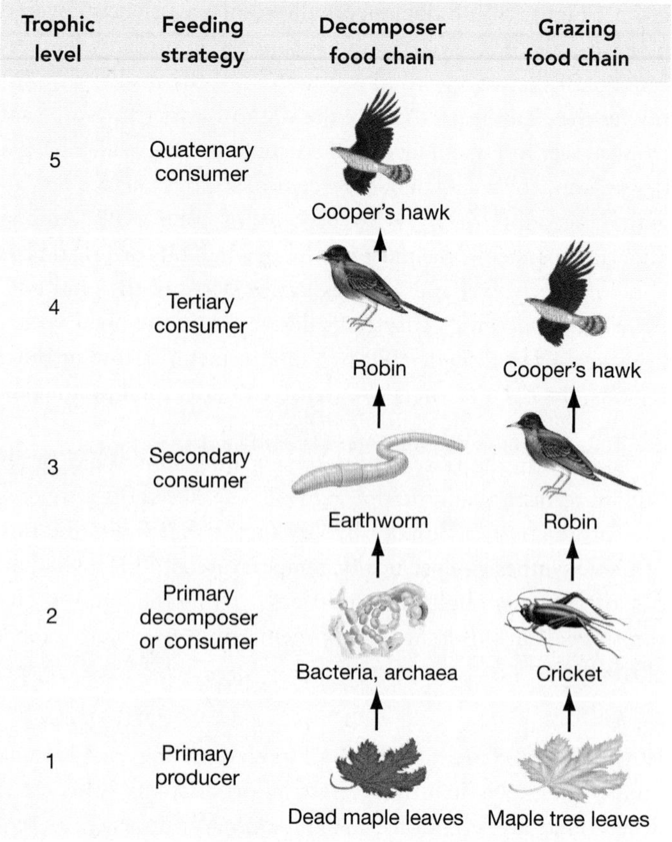

Trophic level	Feeding strategy	Decomposer food chain	Grazing food chain
5	Quaternary consumer	Cooper's hawk	
4	Tertiary consumer	Robin	Cooper's hawk
3	Secondary consumer	Earthworm	Robin
2	Primary decomposer or consumer	Bacteria, archaea	Cricket
1	Primary producer	Dead maple leaves	Maple tree leaves

FIGURE 54.5 Trophic Levels Identify Steps in Energy Transfer. Examples from a temperate-forest ecosystem are shown; many other species exist at each level in this ecosystem.

● **EXERCISE** Label the "top predator" in these food chains—a species that is not eaten by any other species.

Why Is Energy Lost at Each Trophic Level? All ecosystems share a characteristic pattern: The total biomass produced each year declines from lower trophic levels up to the higher levels. Although biomass production at each trophic level varies widely among ecosystems, the data in **Figure 54.7** are representative. The most striking observation is that biomass production at upper trophic levels is often only 10 percent of the production at the next lowest level.

To understand why this pattern occurs, consider the interface between primary producers and primary consumers. Much of the NPP produced each year is unavailable to herbivores, because it resides in indigestible substances such as the lignin in wood or because it is protected by noxious defensive compounds. NPP that is not eaten by an herbivore enters the decomposer food web. Even if the material is ingested by an herbivore,

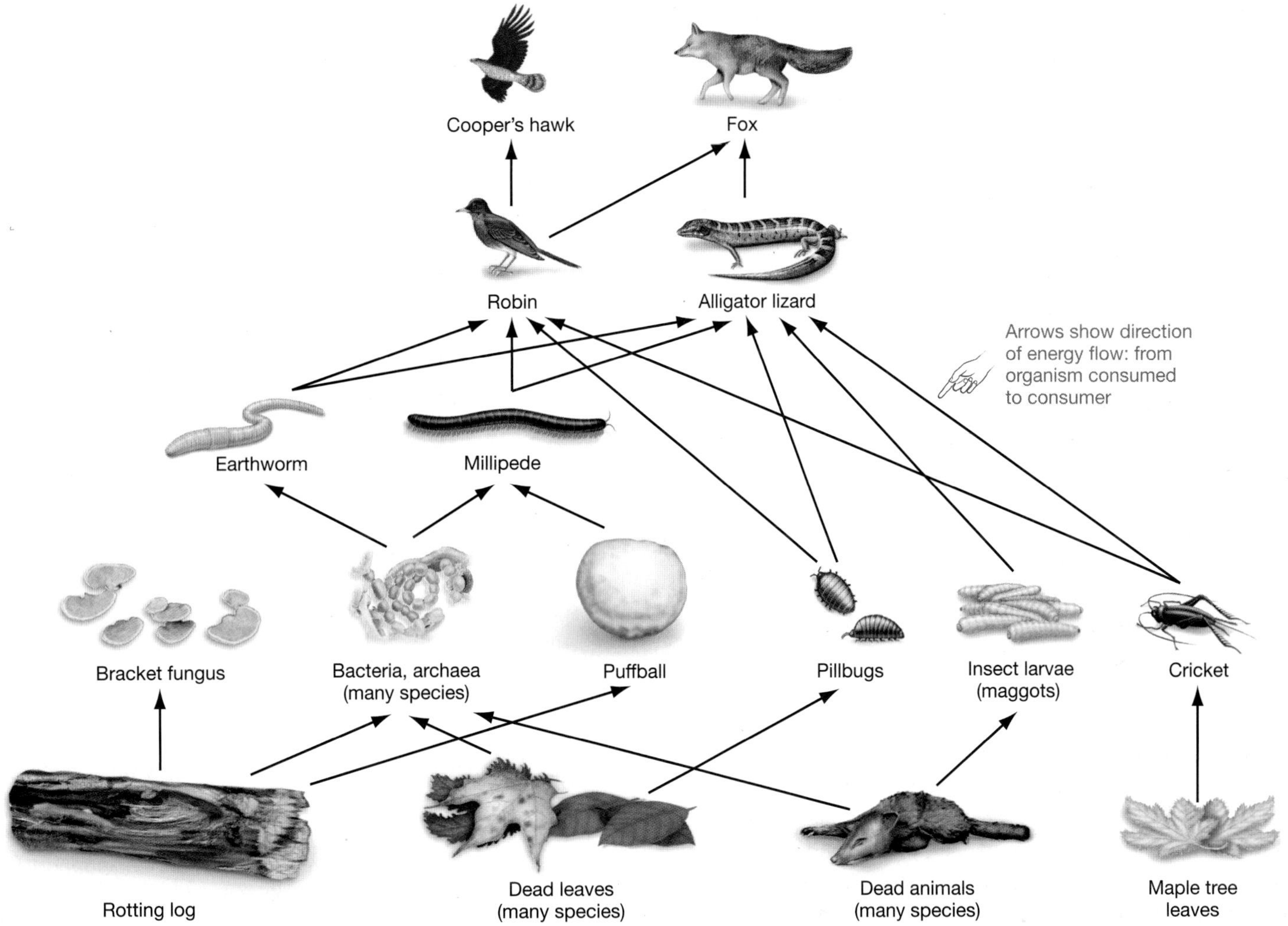

Cooper's hawk
Fox
Robin
Alligator lizard
Arrows show direction of energy flow: from organism consumed to consumer
Earthworm
Millipede
Bracket fungus
Bacteria, archaea (many species)
Puffball
Pillbugs
Insect larvae (maggots)
Cricket
Rotting log
Dead leaves (many species)
Dead animals (many species)
Maple tree leaves

FIGURE 54.6 Food Webs Describe Trophic Relationships in an Ecosystem. Food webs offer a more comprehensive analysis of feeding relationships than food chains do. This food web shows only a fraction of the total feeding relationships and species of a temperate deciduous forest. More complete food webs include species in the grazing food chain, along with data on the average amount of energy transferred between each link in the web.

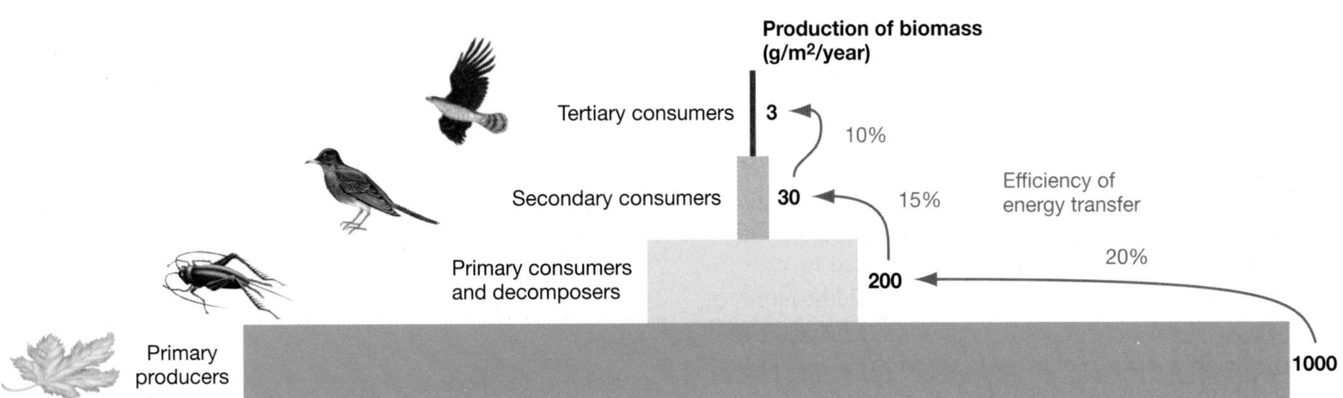

FIGURE 54.7 Productivity Declines at Higher Trophic Levels. In all ecosystems, annual production of biomass is highest at the lowest (bottom) trophic level and declines at higher levels. This pattern is the pyramid of productivity.

most of the energy stored in the chemical bonds of carbon compounds is lost as heat as it is metabolized or used to keep the consumer alive—not to grow or reproduce. At the next higher trophic level, many herbivores are never consumed by secondary consumers, because these herbivores defend themselves effectively and die of other causes. Of the energy that is successfully ingested by secondary consumers, some is again lost as heat or used up in efforts to stay alive and capture prey.

The general point here is simple: The amount of biomass produced at the second trophic level must be less than productivity at the first trophic level; productivity at the third trophic level must be less than that at the second. This pattern holds true for the entire food chain and produces a pyramid of productivity. Production of biomass is highest at the lowest trophic level. ● If you understand this concept, you should be able to explain why keystone species have relatively low biomass. You should also be able to explain why biologists consider vegetarian diets in humans much more efficient than diets that include large quantities of meat.

What Limits the Length of Food Chains? It is interesting to note that none of the food chains, food webs, or trophic structures presented in this chapter thus far have more than four or five trophic levels. When a biologist reviewed the characteristics of food chains that had been documented by researchers working in a wide variety of ecosystems, he found that the maximum number of links ranged from 1 to 6. Each link joins two trophic levels. In this study, terrestrial and lake ecosystems had 3.7 links per food chain, on average, while streams had an average of 3.2 links. Why don't ecosystems have 8 or 9 or 10 trophic levels? Why does the overall average number of levels seem to be about 3.5? Several competing hypotheses have been offered to explain this observation. Let's consider three:

- *Hypothesis 1: Energy Transfer* As energy is transferred up a food chain, a large fraction of that energy is lost. By the time energy reaches the top trophic level, there may not be enough left to support an additional suite of consumers. Suppose that the efficiency of energy transfer between each pair of trophic levels is 10 percent. If the initial trophic level in a hypothetical ecosystem produces 10,000 kcal per day, then the second, third, and fourth levels will produce 1000, 100, and 10 kcal per day, respectively. Can any organism obtain enough energy to survive at a fifth trophic level?

  The hypothesis that food-chain length is limited by productivity leads to a strong prediction: There should be more trophic levels in ecosystems with higher productivity or higher efficiency of energy transfer than in ecosystems with lower productivity or lower efficiency of energy transfer. When a researcher analyzed data on the productivity of four aquatic and four terrestrial ecosystems, however, he found that low-productivity ecosystems were as likely to contain four trophic levels as high-productivity ones were. To date, research has not supported the prediction that the most productive ecosystems have the longest food chains.

- *Hypothesis 2: Stability* Stuart Pimm proposed an alternative hypothesis to the energy-transfer explanation for why food chains are short. His idea was that long food chains are easily disrupted by droughts, floods, or other disturbances and thus tend to be eliminated. Using mathematical models, he demonstrated that long food chains took longer to return to their previous state following a disturbance than did short food chains. He then proposed that long food chains are unlikely to persist in a variable environment.

  Pimm's hypothesis also predicts that the length of food chains should increase with the stability of the environment. A biologist tested this prediction by comparing the animal communities that develop inside tree holes in Australia and Great Britain. After water accumulates in depressions or in the holes in tree branches or roots, leaf litter from the trees falls in and forms the basis of a food web. The researcher found that tree-hole communities in Australia and Great Britain were similar in many respects, except that annual leaf fall was much more variable in the British communities than in the Australian ones. As predicted by the stability hypothesis, the British ecosystems had only two trophic levels while Australian habitats supported three.

  Other researchers have challenged the assumptions behind Pimm's theoretical analysis, however, and support for the stability hypothesis remains tentative. Experimental and observational tests of this hypothesis are continuing.

- *Hypothesis 3: Environmental Complexity* Biologists have also hypothesized that food-chain length is a function of an ecosystem's physical structure. Specifically, researchers have proposed that tundra, grasslands, lake and sea bottoms, streambeds, and intertidal zones offer a largely flat, or two-dimensional, surface to the organisms living there. In contrast, forests and open-water environments in lakes and rivers offer three-dimensional volumes. This hypothesis predicts that three-dimensional ecosystems should have longer food chains than two-dimensional sites do.

  To test this hypothesis, researchers examined 113 publications that described food chains in a wide variety of aquatic and terrestrial environments. These data supported the prediction that food webs are significantly longer in three-dimensional ecosystems. As a result, the investigators concluded that dimensionality does influence food-web structure. As yet, however, no one knows why. The mechanism for this pattern remains to be determined.

To summarize, there is unlikely to be a single, simple answer to the question of what limits food-chain length. Inefficiencies of energy transfer, environmental stability, and environmental complexity may all influence the number of trophic levels that can be supported in a given ecosystem. Multiple causation is not unusual in biological science. In genetics, for example, gene expression is rarely triggered by a single regulatory site in DNA and just one regulatory protein. Instead, the amount and timing of gene expression result from interactions among dozens of regulatory DNA sequences and proteins. Recognizing multiple causation is productive, because it can suggest exciting new experiments and analyses. For example, if food-chain length does have multiple causes, then habitats that are particularly productive, stable, and highly complex in physical structure should have the longest food chains. Research continues on this fundamental question in ecosystem ecology.

Analyzing Energy Flow: A Case History

Research on energy flow in ecosystems has uncovered a series of important general principles. In both terrestrial and marine ecosystems, NPP varies with region and biome. On land, the most important constraints on productivity are warmth and moisture; in aquatic habitats, nutrient availability is key. The chemical energy in biomass flows into consumer or decomposer food webs, where it is passed up a series of trophic levels. In terrestrial ecosystems, the total biomass declines with each trophic level.

To see these general principles in action, let's analyze energy flow in a specific ecosystem. For almost five decades, a team of researchers has been studying how energy and nutrients flow through a temperate forest ecosystem in the northeastern United States: the Hubbard Brook Experimental Forest in New Hampshire. Hubbard Brook may qualify as the most intensively studied ecosystem in the world.

As **Figure 54.8** shows, energy flow in this ecosystem begins when plants capture the energy in solar radiation via photosynthesis. At Hubbard Brook, the amount of energy entering the ecosystem from sunlight varies throughout the year and, to a lesser extent, from year to year. From June 1, 1969, to May 31, 1970, which was a typical year, 1,254,000 kilocalories (kcal) of solar radiation per square meter (m^2) reached the forest. If this amount of energy were available in the form of electricity, it would easily power two 75-watt lightbulbs that burned continuously for one year.

By documenting rates of photosynthesis in a large sample of forest plants, the biologists calculated that the plants used 10,400 kcal/m^2 of energy in photosynthesis. This value represents **gross primary productivity**, which is the total amount of photosynthesis in a given area and time period. The team also calculated **gross photosynthetic efficiency**, or the efficiency with which plants use the total amount of energy available to them, as the ratio of gross photosynthesis to solar radiation in kcal/m^2. At Hubbard Brook, efficiency was $10{,}400 \div 1{,}254{,}000 = 0.8$ percent. This value is typical of other ecosystems as well.

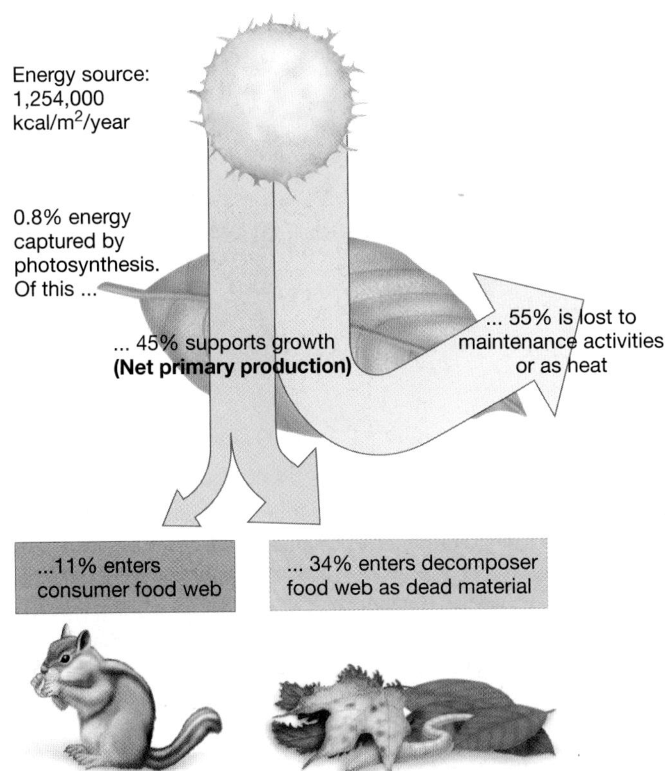

FIGURE 54.8 Energy Flow through the Hubbard Brook Forest Ecosystem. Energy from sunlight is transformed to chemical energy by photosynthesis. Photosynthetic products fuel new plant growth, which eventually enters either the consumer or decomposer food web.

EXERCISE Calculate the amount of energy transferred to primary consumers and decomposers each year, in kcal/m^2.

QUESTION If NPP in this ecosystem increases due to global warming, predict the impact on primary consumers and decomposers.

Why is efficiency so low? The answer has several components. Plants in temperate biomes cannot photosynthesize at all in winter, although sunlight is available. If conditions get dry during the summer, stomata close to conserve water and photosynthesis stops due to a lack of CO_2. Even when conditions are ideal, the pigments that drive photosynthesis can respond to only a fraction of the wavelengths available and thus a fraction of the total energy received.

Given that only a tiny fraction of incoming sunlight is converted to chemical energy and that only a fraction of this gross primary productivity is used to build biomass, let's consider the next question: What happens to the NPP at Hubbard Brook?

How Does Energy Flow through the Hubbard Brook Ecosystem? Between 1969 and 1971 the amount of energy that entered the consumer food web at Hubbard Brook varied from about 1 percent of net primary productivity per year to more than 40 percent. Leaves, seeds, and fruits were the most commonly consumed plant parts; wood was rarely eaten.

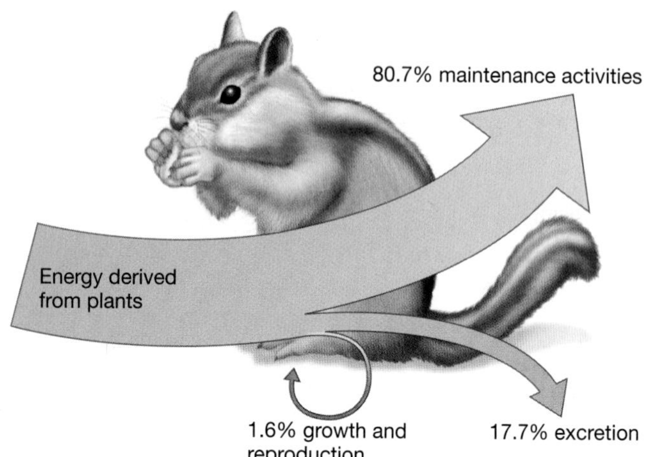

FIGURE 54.9 How Do Consumers Use Primary Production? A small fraction of the energy consumed by chipmunks, which are primary consumers (herbivores), is used for secondary production. Most of the energy is used for cellular respiration.

QUESTION Aphids are ectothermic, largely sedentary, and feed on high-quality foods containing almost no indigestible molecules. Describe how the percentages given here would compare in aphids.

Figure 54.9 illustrates what happened to the energy obtained by consumers, using chipmunks as an example consumer. Chipmunks are small rodents that are seed predators. On average, these primary consumers harvest 31 kcal/m^2 of energy each year. Of that total, 17.7 percent is unused and excreted and 80.7 percent is lost to respiration and other maintenance processes. Just 1.6 percent of the yearly 31 kcal/m^2 goes into the production of new chipmunk tissue by growth and reproduction. The production of new tissue by primary consumers is called **secondary production.**

Compared to endotherms such as chipmunks, secondary production is much higher in ectothermic consumers. Caterpillars, for example, transform about 5.4 percent of the energy they ingest into new tissue. Because ectotherms rely principally on heat gained from the environment and do not oxidize sugars to keep warm, they devote much less energy to cellular respiration than endotherms do. Even in the case of the ectotherms, though, it is clear that only a tiny fraction of the available solar radiation is involved in secondary production.

What about biomass that enters the decomposer food web? At Hubbard Brook, about 75 percent of total net primary productivity is not eaten while alive. This energy enters the decomposer food web and is passed up subsequent trophic levels, just as it is in the consumer food web. But in addition, the researchers at Hubbard Brook found that large amounts of energy leave the decomposer food web in the form of detritus that washes into streams. This transfer of energy from the forest ecosystem to the aquatic ecosystem is important, because it is a major source of energy for aquatic organisms. At Hubbard Brook, photosynthesis by aquatic algae and plants introduced only about 10 kcal/m^2/yr of energy. In contrast, each year 6039 kilocalories washed into each square meter of streambed from the surrounding forest.

Long-term studies have documented the flow of energy into, through, and out of Hubbard Brook and other types of ecosystems. Although the specific numbers found at Hubbard Brook are unique to that site and the time interval of that study, the general patterns have turned out to be typical of ecosystems around the globe.

Check Your Understanding

If you understand that... 

- In an ecosystem, energy flows from sunlight or inorganic compounds with high potential energy to producers, and from there to consumers and decomposers.
- In most ecosystems, NPP is limited by conditions that limit the rate of photosynthesis: temperature and/or the availability of sunlight, water, and nutrients.
- Productivity diminishes at each subsequent trophic level in the ecosystem.

You should be able to... 

1) Summarize variation in annual NPP among ecosystems and briefly explain why some ecosystems are particularly high or low.
2) Give two reasons that the amount of biomass present in terrestrial ecosystems diminishes with increasing trophic levels.
3) Evaluate at least one hypothesis to explain why most food chains have only two or three links.
4) Predict what will happen to consumer and decomposer food chains in ecosystems where global warming leads to increases or decreases in NPP.

54.2 Biogeochemical Cycles

Energy is not the only quantity that is transferred when one organism eats another. The organisms that are eaten also contain carbon (C), nitrogen (N), phosphorus (P), calcium (Ca), and other elements that act as nutrients. Atoms are constantly reused as they cycle through trophic levels and air, water, and soil. The path that an element takes as it moves from abiotic systems through organisms and back again is referred to as its **biogeochemical** ("life-Earth-chemical") **cycle.**

Because humans are now disturbing biogeochemical cycles on a massive scale, research on this aspect of ecosystem ecology is exploding. To get a basic understanding of how a particular biogeochemical cycle works, researchers focus on three fundamental questions:

1. What are the nature and size of the **reservoirs**, or areas where elements are stored for a period of time? In the case of carbon, the biomass of living organisms is an important reservoir, as are sediments and soils. Another significant carbon reservoir is buried in the form of coal and oil.

2. How fast does the element move between reservoirs, and what factors influence these rates? The global photosynthetic rate, for example, measures the rate of carbon flow from CO_2 in the atmosphere to living biomass. When fossil fuels burn, carbon that was buried in coal or petroleum moves into the atmosphere as carbon dioxide (CO_2).

3. How does a biogeochemical cycle interact with other cycles? For example, researchers are trying to understand how changes in the nitrogen cycle affect carbon.

Before taking up these issues, though, let's begin with a general overview of what nutrient cycling is.

Biogeochemical Cycles in Ecosystems

Figure 54.10 shows a simplified version of a generalized terrestrial nutrient cycle. In this case, the cycle starts when nutrients are taken up from the soil by plants and assimilated into plant tissue. If the plant is eaten, the nutrients pass to the animal members of the ecosystem's consumer food web; if the plant dies, the nutrients enter the decomposer food web. Once consumed by an animal, nutrients are excreted in fecal matter or urine, taken up by a parasite or predator, or added to the dead biomass reservoir when the animal dies.

The nutrients in plant litter, animal excretions, and dead animal bodies are used by bacteria, archaea, roundworms, fungi, and other primary decomposers. The combination of microscopic decomposers and the ions and molecules they release combine to form the soil organic matter. Soil organic matter is a complex mixture of partially and completely decomposed detritus. Completely decayed organic material is called **humus**, because it is rich in a family of carbon-containing molecules called humic acids. (Chapter 38 described other components of the soil.) Eventually, the nutrients in soil organic matter are converted to an inorganic form. For example, cellular respiration by soil-dwelling bacteria and archaea converts the nitrogen present in amino acids that are found in detritus to ammonium (NH_4^+) or nitrate (NO_3^-) ions. Once this step is accomplished, the nutrients are available for uptake by plants.

A key feature of this process is that nutrients are reused. Reuse is not complete, however. Nutrients leave the ecosystem whenever plant or animal biomass leaves. For example, plant biomass is removed if an herbivore enters the ecosystem, eats a plant, and migrates out of the ecosystem before excreting nutrients or dying. Nutrients also leave ecosystems when flowing water or wind removes particles or inorganic ions and deposits them somewhere else. Soil erosion has a huge impact on nutrient cycles: It removes nutrients rapidly and in potentially large quantities.

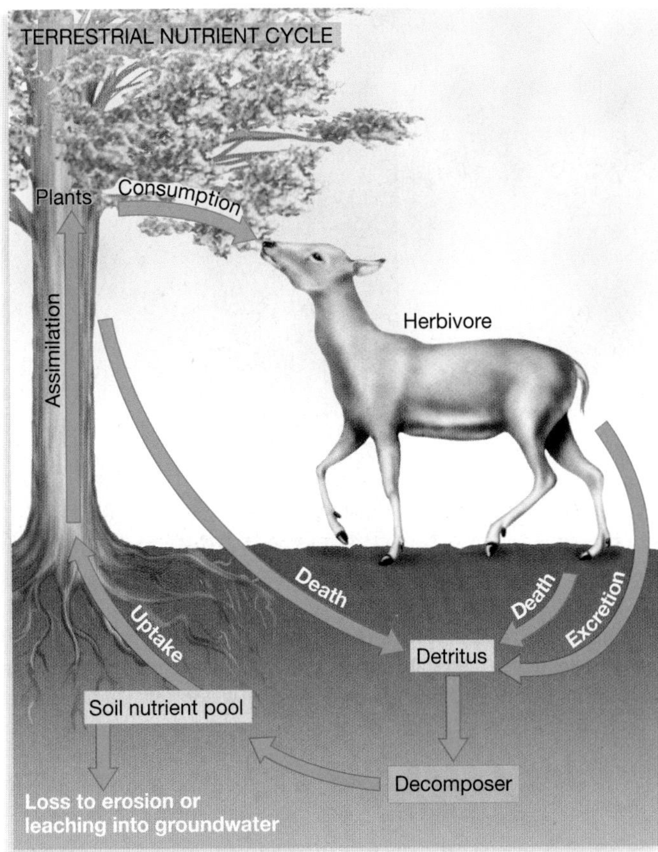

FIGURE 54.10 Generalized Terrestrial Nutrient Cycle. Nutrients cycle from organism to organism in an ecosystem via assimilation, consumption, and decomposition. Nutrients are exported from ecosystems when water or organisms leave the area.

What Factors Control the Rate of Nutrient Cycling? Of the many links in a nutrient cycle, the decomposition of detritus most often limits the overall rate at which nutrients move through an ecosystem. Until decomposition occurs, nutrients stay tied up in intact tissues. The decomposition rate, in turn, is influenced by two types of factors: abiotic conditions such as temperature and precipitation, and the quality of the detritus as a nutrient source for the fungi, bacteria, and archaea that accomplish decomposition.

To appreciate the importance of abiotic conditions on the decomposition rate, consider the difference in detritus accumulation between boreal forests and tropical wet forests. Chapter 50 indicated that boreal forests are found in areas where temperature is low. As a result, soils in these ecosystems are cold and wet. Tropical wet forests, in contrast, occur in areas where temperatures and rainfall are high. Soils tend to remain moist and warm all year long.

(a) Boreal forest: Accumulation of detritus and organic matter

(b) Tropical wet forest: Almost no organic accumulation

FIGURE 54.11 Temperature and Moisture Affect Decomposition Rates. (a) In boreal forests, decomposition rates are limited by cold soil temperatures. Organic matter builds up, because the input of detritus into the soil exceeds the decomposition rate. **(b)** In tropical wet forests, warm temperatures allow decomposition to proceed rapidly, so organic matter does not build up.

Figure 54.11 illustrates typical soils from boreal forests and tropical wet forests. Notice that the uppermost part of the soil in a boreal forest consists of partially decomposed detritus and organic matter. There is no such layer at the top of the soil in a tropical forest. The contrast occurs because the cold and wet conditions in boreal forests limit the metabolic rates of decomposers. As a result, decomposition fails to keep up with the input of detritus, and organic matter accumulates there. In the tropics, conditions are so favorable for fungi, bacteria, and archaea that decomposition keeps pace with detrital inputs. Nutrients cycle slowly through boreal forests but rapidly through wet tropical forests.

The quality of detritus also exerts a powerful influence on the decomposition rate and thus on nutrient availability. For example, decomposers can be hampered by the presence of large compounds in detritus that are difficult to digest, such as lignin. The presence of lignin, which is a primary constituent of wood, is one reason that wood takes much longer to decompose than leaves do. As Chapter 31 pointed out, only basidiomycete fungi, ascomycete fungi, and a few bacteria have the enzymes required to completely degrade lignin. The growth of decomposers is also inhibited if detritus is low in nitrogen.

What Factors Influence the Rate of Nutrient Loss?

Nutrient availability has a profound effect on productivity, so the rate of nutrient loss is an important characteristic of an ecosystem. Several of the major impacts that humans have on ecosystems—such as farming, logging, burning, and soil erosion—accelerate nutrient loss.

To test the effect of vegetation removal on nutrient export, researchers initiated a large-scale experiment at the Hubbard Brook Experimental Forest. They chose two similar **watersheds**—areas drained by a single stream—for study (**Figure 54.12**). They then cut all vegetation, including the trees, from the forests in one of the two watersheds. In the following three years, this clear-cut area was treated with an herbicide to prevent vegetation from regrowing. As a result, the experimental watershed was devegetated. An untreated watershed served as a control.

Before removing the vegetation, the researchers had documented that 90 percent of the nutrients in the ecosystems was in soil organic matter, and an additional 9.5 percent was in plant biomass. After the vegetation was removed, the team monitored the concentrations of nutrients in the streams exiting the two watersheds. The graph in Figure 54.12 documents the amount of dissolved substances that subsequently washed out of the stream in each watershed over the course of four years. Losses from the deforested site were typically 10 times as high as they were from the control site.

Based on these data, the researchers concluded that devegetation has a huge impact on nutrient export. Instead of being held in the ecosystem and recycled, nutrients wash out in the absence of vegetation. The loss occurs because nutrients in the soil are either dissolved in water or attached to small particles of sand or clay (see Chapter 38). If plant roots no longer hold the soil particles in place and if they no longer take up and recycle aqueous nutrients, then the molecules and ions wash out of the soil and are lost to the ecosystem. Consequently, productivity may decline over time if the area is kept

(a) Hubbard Brook Experimental Forest

Experiment

(b) Question: How does the presence of vegetation affect the rate of nutrient export in a temperate-forest ecosystem?

Hypothesis: Presence of vegetation lowers the rate of nutrient export because it increases soil stability and recycling of nutrients.

Null hypothesis: Presence of vegetation has no effect on the rate of nutrient export.

Experimental setup:

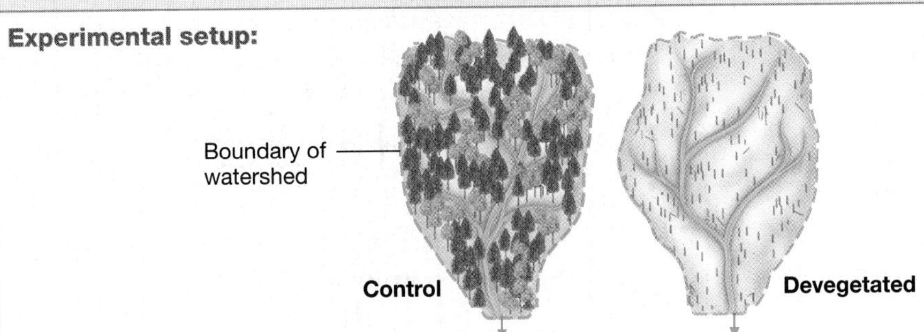

1. Choose two similar watersheds. Document nutrient levels in soil organic matter, plants, and streams.

2. Devegetate one watershed, and leave the other intact.

3. Monitor the amount of dissolved substances in streams.

Prediction: Amount of dissolved substances in stream in devegetated watershed will be much higher than amount of dissolved substances in stream in control watershed.

Prediction of null hypothesis: No difference will be observed in amount of dissolved substances in the two streams.

Results:

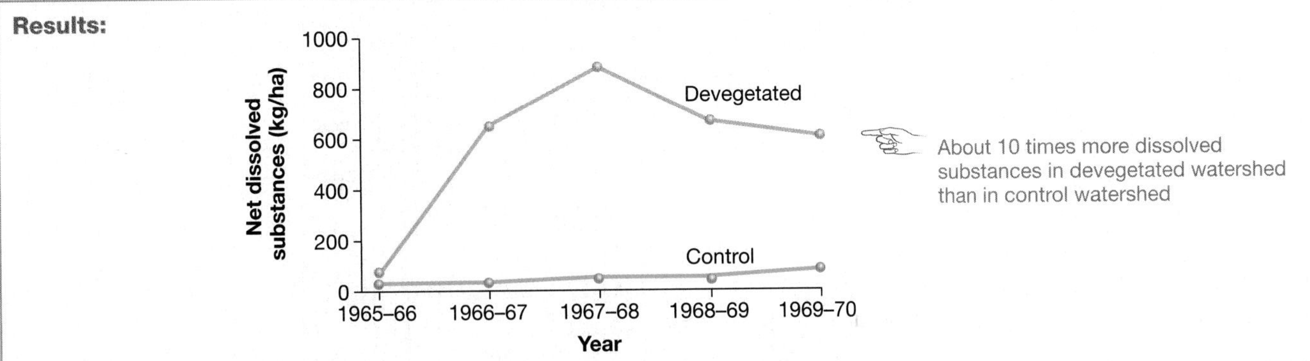

Conclusion: Presence of vegetation limits nutrient loss. Removing vegetation leads to large increases in nutrient export.

FIGURE 54.12 Experimental Evidence that Deforestation Increases the Rate of Nutrient Loss from Ecosystems.

QUESTION In effect, this experiment removed one of the arrows in Figure 54.10. Which one?

in a devegetated state. Long-term devegetation of this type has occurred in formerly forested areas of the Middle East, North Africa, and elsewhere due to intensive farming and grazing. As a result, the productivity of these regions is a tiny fraction of what it once was.

Global Biogeochemical Cycles

When nutrients leave one ecosystem, they enter another. In this way, the movement of ions and molecules among ecosystems links local biogeochemical cycles into one massive global system. Local and global cycles interact when water, organisms, or wind move nutrients. As an introduction to how these global biogeochemical cycles work, let's consider the global water, carbon, and nitrogen cycles. These cycles have recently been heavily modified by human activities—with serious ecological consequences.

The Global Water Cycle A simplified version of the **global water cycle** appears in **Figure 54.13**. The diagram shows the estimated amount of water that moves between major components of the cycle over the course of a year.

To analyze this cycle, begin with evaporation of water out of the ocean and the subsequent precipitation of water back into the water. For the marine component of the cycle, evaporation exceeds precipitation—meaning that, over the oceans, there is a net gain of water to the atmosphere. When this water vapor moves over the continents, it is joined by a small amount of water that evaporates from lakes and streams and a large volume of water that is transpired by plants. The total volume of water in the atmosphere over land is balanced by the amount of rain and other forms of precipitation that occurs on the continents. The cycle is completed by the water that moves from the land to the oceans via streams and **groundwater**—water that is found in soil.

Humans are affecting the water cycle in complex ways. Perhaps the simplest and most direct impacts concern rates of groundwater replenishment. Asphalt and concrete surfaces added through suburbanization reduce the amount of precipitation that percolates from the surface to enter deep soil layers. The conversion of grasslands and forests into agricultural fields also increases the amount of water that runs off Earth's surface into streams instead of penetrating to groundwater layers. Intact fields and forests lose less water than croplands do because extensive root systems that hold water are in place year-round. Most importantly, the dramatic increases in irrigated agriculture over the past three decades have resulted in massive quantities of water being removed from groundwater storage and brought to the surface.

In combination, these impacts have resulted in alarming drops in groundwater storage. The **water table** is the upper limit below the surface that the ground is saturated with stored water, and it is dropping on every continent. Between 1986 and 2006, the water table north of Beijing, China, experienced drops averaging almost 61 m (200 ft). Throughout India, water tables are falling at a rate of between 1 m and 3 m per year. Similar rates are being documented in Yemen, parts of Mexico, and states in the southern Great Plains region of the United States. Lack of water is exacerbating political tensions in several areas of the world, including the Middle East. Humans are mining water in many parts of the world—meaning that populations have already grown beyond carrying capacity for water.

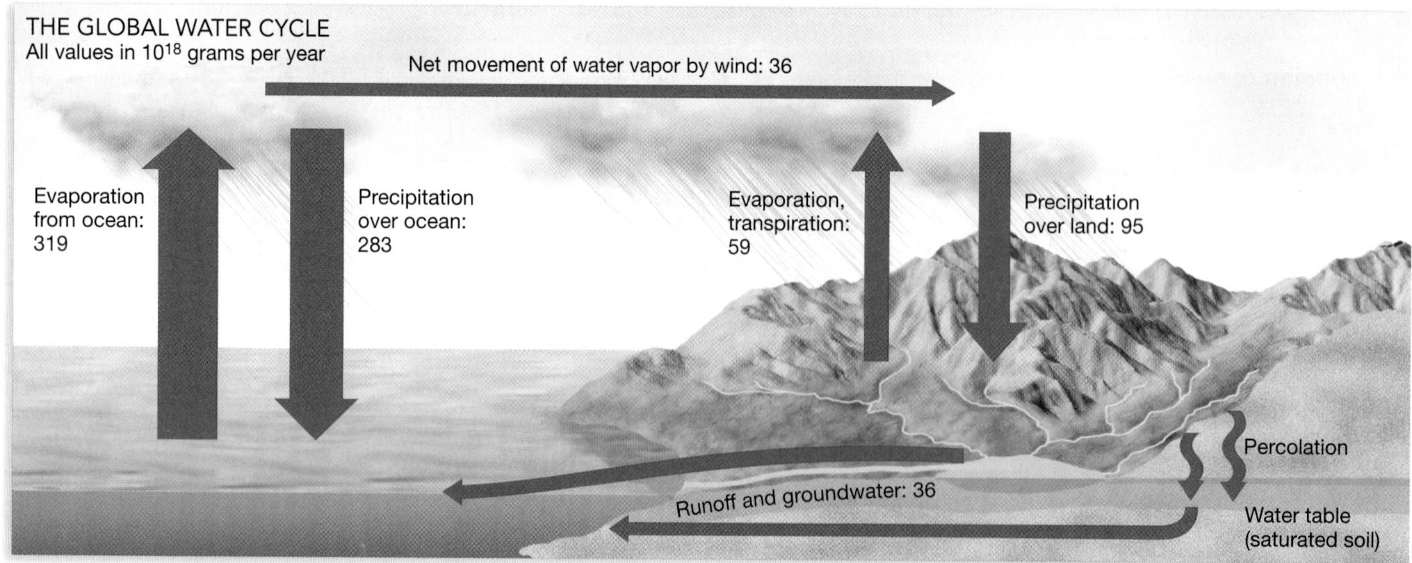

FIGURE 54.13 The Global Water Cycle.

QUESTION Predict how the amount of water evaporated from the ocean is changing in response to global warming. Discuss one possible consequence of this change.

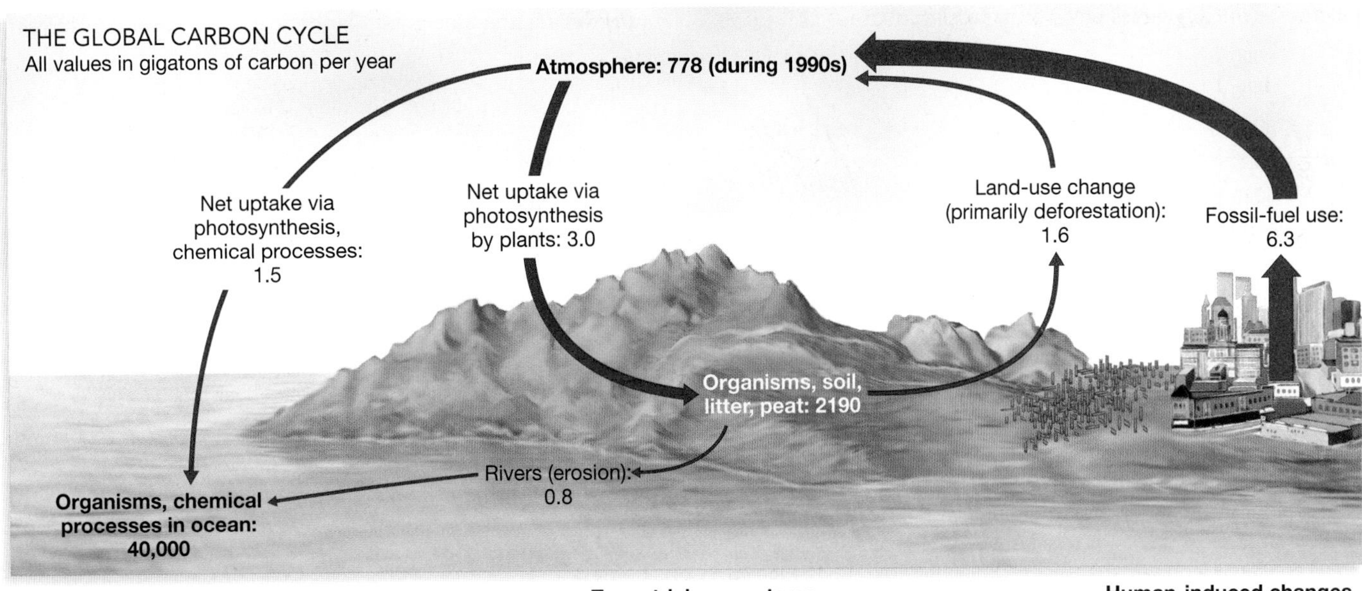

FIGURE 54.14 The Global Carbon Cycle. The arrows indicate how carbon moves into and out of ecosystems. Deforestation and the use of fossil fuels are adding 7.9 gigatons of carbon to the atmosphere each year. Of that 7.9 gigatons of carbon produced by human activities, 2 gigatons are fixed by photosynthesis in terrestrial ecosystems and 2 gigatons are fixed by physical and chemical processes in the oceans. The remainder—3.9 gigatons—is added to the atmosphere.

The Global Carbon Cycle As **Figure 54.14** shows, the **global carbon cycle** documents the movement of carbon among terrestrial ecosystems, the oceans, and the atmosphere. Of these three reservoirs, the ocean is by far the largest. The atmospheric reservoir is also important despite its relatively small size, because carbon moves into and out of it rapidly.

The arrows in Figure 54.14 emphasize that carbon frequently moves into and out of the atmospheric pool through organisms. In both terrestrial and aquatic ecosystems, for example, photosynthesis is responsible for taking carbon out of the atmosphere and incorporating it into tissue. Cellular respiration, in contrast, releases carbon that has been incorporated into living organisms to the atmosphere, in the form of carbon dioxide.

How have humans changed the carbon cycle? The fossil fuels found on Earth, which are derived from carbon-rich sediments, are estimated to contain a total of 5000–10,000 gigatons of carbon. (A gigaton, symbolized Gt, is a billion tons). This is one-eighth to one-fourth the size of the oceanic reservoir and 2.5 to 5 times the size of the terrestrial reservoir. In effect, burning fossil fuels moves carbon from an inactive geological reservoir to an active reservoir—the atmosphere. When you burn gasoline, you are releasing carbon atoms that have been locked up in coal or petroleum reservoirs for hundreds of millions of years.

Land-use changes have also altered the global carbon cycle. Deforestation, for example, reduces an area's net primary productivity. It also releases CO_2 when fire is used for clear-cutting or when dead limbs, twigs, and stumps are left to decompose. Researchers who used data on the amount of land cleared for agriculture and forestry to estimate the amount of carbon released found that, at a global scale, a net movement of carbon to the atmosphere has been occurring from terrestrial ecosystems for at least the past 100 years. The change from net carbon storage to net carbon release is the result of expanding human populations (see Chapter 52).

Figure 54.15a shows the amount of carbon released from fossil-fuel burning over the past century; **Figure 54.15b** shows the consequences of this accelerated carbon release. In just 45 years, CO_2 in the atmosphere at Mauna Loa Observatory on the island of Hawaii has increased from about 315 to over 375 parts per million (ppm)—meaning milligrams of CO_2 per kilogram of air. The same trend has been observed at sites around the globe. As the data in **Figure 54.15c** show, CO_2 concentrations in the atmosphere have now risen to levels far above the 280 ppm that was typical prior to 1860.

These changes in the global carbon cycle are important because carbon dioxide functions as a **greenhouse gas:** It traps heat that has been radiated from Earth and keeps it from being lost to space, similar to the way the glass of a greenhouse traps heat. More specifically, carbon dioxide is one of several gases in the atmosphere that absorb and reflect the infrared wavelengths radiating from Earth's surface. Increases in amounts of greenhouse gases are warming Earth's climate by increasing the atmosphere's heat-trapping potential.

at www.masteringbio.com
The Global Carbon Cycle

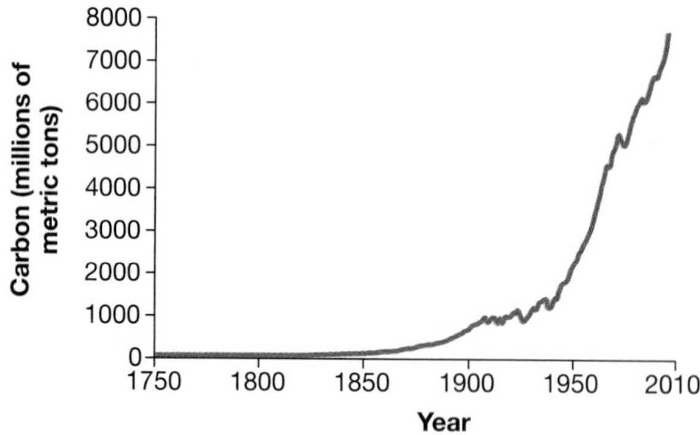

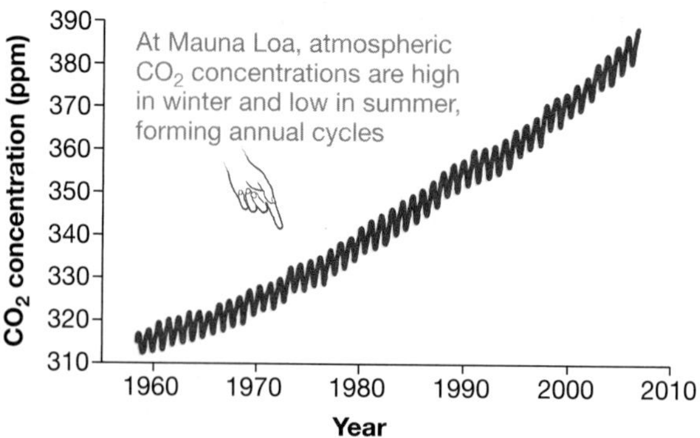

FIGURE 54.15 Humans Are Causing Increases in Atmospheric Carbon Dioxide. **(a)** Rates of carbon flow from fossil-fuel burning have increased as human populations have increased (see Chapter 52). **(b)** Because the Mauna Loa Observatory in Hawaii is far from large-scale human influences, it should accurately represent the average condition of the atmosphere in the Northern Hemisphere. **(c)** For centuries, average CO_2 concentrations in the atmosphere were about 280 parts per million (ppm).

QUESTION Why are atmospheric CO_2 concentrations low in the Northern Hemisphere in summer and high in winter? What pattern would you expect in the Southern Hemisphere?

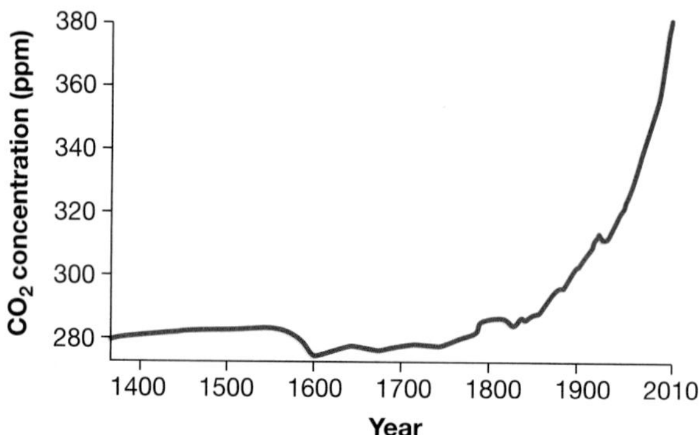

The Global Nitrogen Cycle **Figure 54.16** illustrates the **global nitrogen cycle**. A key aspect of this biogeochemical cycle is that plants are able to use nitrogen only in the form of ammonium or nitrate ions (NH_4^+ or NO_3^-). As a result, the vast pool of molecular nitrogen (N_2) that is in the air blanketing Earth—N_2 makes up 78 percent of the atmosphere—is unavailable to plants. Nitrogen is added to ecosystems in a usable form only when it is reduced, or "fixed," meaning when it is converted from N_2 to NH_3. Nitrogen fixation results from lightning-driven reactions in the atmosphere and from enzyme-catalyzed reactions in bacteria that live in the soil and oceans. (Chapter 38 explained how bacteria fix nitrogen.)

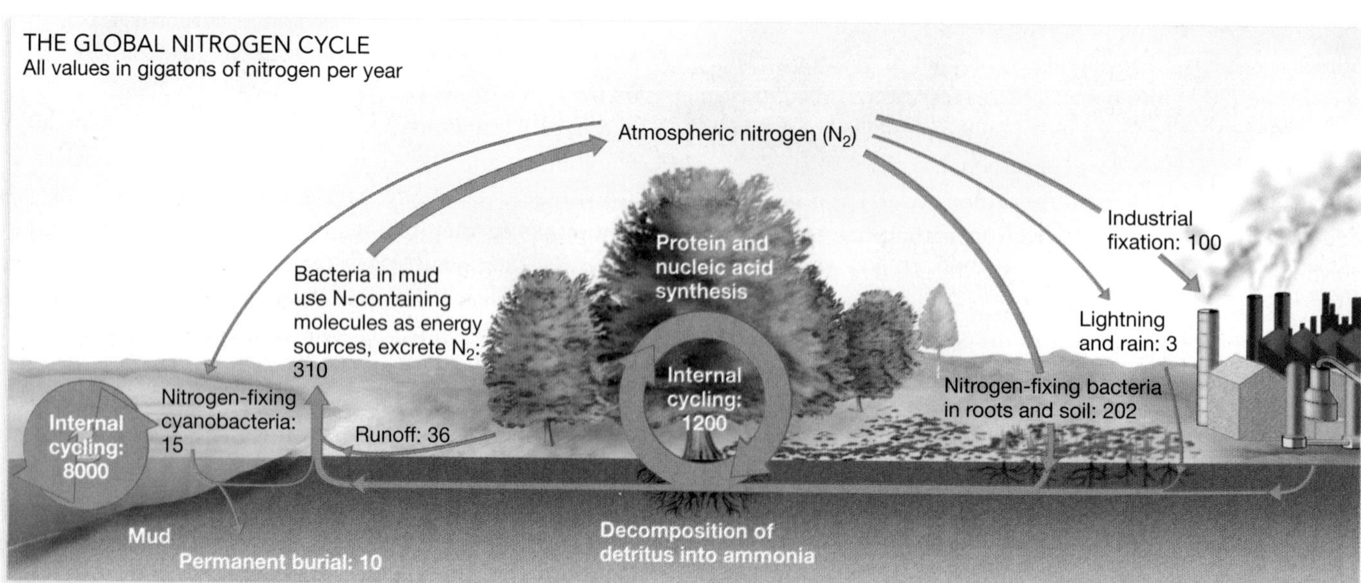

FIGURE 54.16 The Global Nitrogen Cycle. Nitrogen enters ecosystems as ammonia or nitrate via fixation from atmospheric nitrogen. It is exported in runoff and as nitrogen gas given off by bacteria that use nitrogen-containing compounds as an electron acceptor.

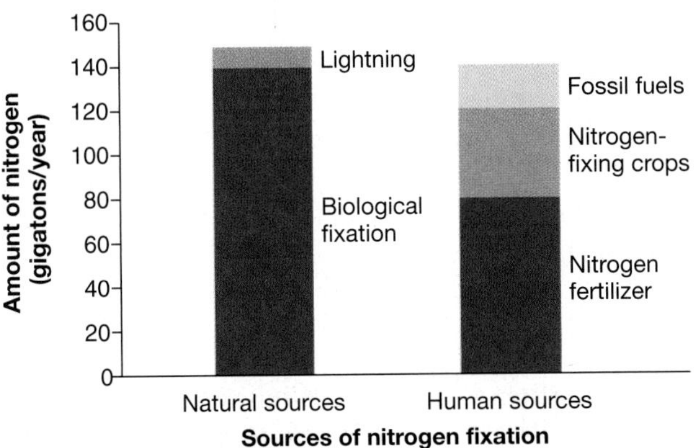

FIGURE 54.17 Humans Are Adding Large Amounts of Nitrogen to Ecosystems. Human activities now fix almost as much nitrogen each year as natural sources do. Thus, human activities have almost doubled the total amount of nitrogen available to organisms.

The nitrogen cycle has been profoundly altered by human activities. The amount of nitrogen fixation from human sources is now approximately equal to the amount of nitrogen fixation from natural sources (**Figure 54.17**). There are three major sources of this human-fixed nitrogen: (1) industrially produced fertilizers, (2) the cultivation of crops, such as soybeans and peas that harbor nitrogen-fixing bacteria, and (3) the release of nitric oxide during the combustion of fossil fuels.

Adding nitrogen to terrestrial ecosystems usually increases productivity. In some cases, then, a massive increase in nitrogen availability is beneficial. But in other situations, it is not. Chapter 28, for example, detailed how excessive applications of nitrogen-containing fertilizers on farmlands have produced nitrogen-laced runoff, which has had disastrous impacts—the formation of oxygen-free "dead zones"—in numerous aquatic ecosystems. Researchers have also documented that nitrogen inputs can lead to a significant loss of biodiversity in terrestrial ecosystems. For example, when biologists added nitrogen to plots of native grasslands in midwestern North America, a few competitively dominant species tended to take over. As those species grew rapidly, they displaced other species that did not respond to nitrogen inputs as strongly. In the grassland ecosystem, increased nitrogen boosted productivity but decreased species diversity, by altering the balance of competitive interactions. The same result has occurred in study plots within the Park Grass experiment, introduced in Chapter 52. The plots that have been fertilized with nitrogen since 1856 contain many fewer species than unfertilized plots do.

As the data in this section make clear, several of the most pressing environmental problems facing our species result from recent and massive alterations in biogeochemical cycles. Even local changes in biogeochemical cycles tend to have large-scale consequences, because nutrients are transported among ecosystems. If you are in your late teens or early twenties as you read this text, you are part of a generation that is expected to experience the most traumatic episode of environmental change in human history. The trauma has two sources: the massive loss of species documented in Chapter 55 and the profound changes in global biogeochemistry recorded here. Let's take a closer look at the consequences of altering biogeochemical cycles.

BOX 54.1 What Is Your "Ecological Footprint"?

The Ecological Footprint is an online tool that helps individuals evaluate their resource consumption. Once you've completed a questionnaire, the program calculates your resource-use "footprint"—meaning, the amount of productive land and water area required to support your lifestyle. The calculation is based on information about four aspects of resource use:

- *Food consumption* Because so much energy is lost between trophic levels, people who eat only primary producers have much smaller footprints than do individuals who eat a great deal of meat and dairy products. Eating foods that are transported long distances also inflates a footprint.
- *Consumer goods* Purchasing manufactured goods and generating large amounts of solid waste increases a footprint; recycling reduces it.
- *Shelter* Footprints are maximized by people who live alone in large homes and do not limit their use of electricity and running water.
- *Transportation* As you might predict, footprints increase when individuals drive gas-guzzling cars long distances by themselves, and when people fly a great deal. Footprints are reduced by fuel-efficient vehicles, car-pooling, bicycling, walking, and public transportation.

Many students (and professors!) who live in industrialized nations are shocked to find that their lifestyle demands over 20 acres of productive land—even though the world average is 2.2 acres/person.

In addition to raising personal awareness, the footprint makes another important point: It is not possible for Earth's resources to support a "westernized" lifestyle for more than a tiny fraction of the total human population. Most researchers agree that the human population has already overshot carrying capacity for high-resource-use lifestyles.

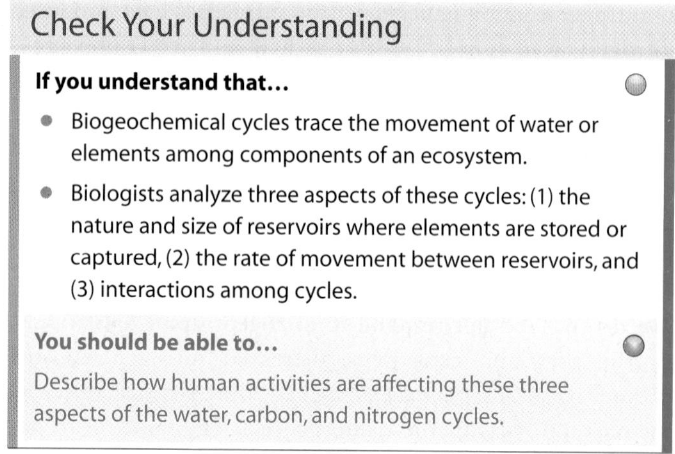

Check Your Understanding

If you understand that...

- Biogeochemical cycles trace the movement of water or elements among components of an ecosystem.
- Biologists analyze three aspects of these cycles: (1) the nature and size of reservoirs where elements are stored or captured, (2) the rate of movement between reservoirs, and (3) interactions among cycles.

You should be able to...

Describe how human activities are affecting these three aspects of the water, carbon, and nitrogen cycles.

54.3 Human Impacts on Ecosystems

Two factors are responsible for the human impacts on ecosystems being documented by biologists. The first is the increase in human population size, analyzed in Chapter 52. The second is an increase in resources used by people. Figure 54.15a documented increases in fossil-fuel use over about the past 150 years; **Figure 54.18** provides data from 2006 on the average annual oil consumption per person around the world. The key point is that residents of industrialized countries, though relatively few in number, have a disproportionately large impact on biogeochemical cycles because they use so much energy, water, food, and other resources (**Box 54.1** on page 1237).

Chapter 55 will assess how changes in human population size and resource use are affecting species extinctions. Here let's focus on the consequences for biogeochemical cycles. Explosive human population growth and intensive resource consumption are causing the average global temperature to rise and, in many biomes, NPP to increase. How are ecosystems responding?

Global Warming

Recall that carbon dioxide concentration in the atmosphere has been increasing throughout the twentieth century and that most analyses indicate that the increase is due to the burning of fossil fuels and clear-cutting of land, particularly forests, for homes and agriculture. Whether the increase in CO_2 is producing global warming—an increase in Earth's surface temperature, averaged over the globe, has been controversial until recently, however.

In 1988 an international group of scientists, called the Intergovernmental Panel on Climate Change (IPCC), was formed to evaluate the consequences of rising CO_2. The group has since produced a series of reports summarizing the state of scientific knowledge on the issue. The 1998 IPCC report concluded that current evidence suggests a "discernible human influence on climate." This was the IPCC's first statement supporting the hypothesis that rising CO_2 concentrations due to human activities are having a measurable impact on climate. The IPCC's next major report, released in 2001, stated that "There is new

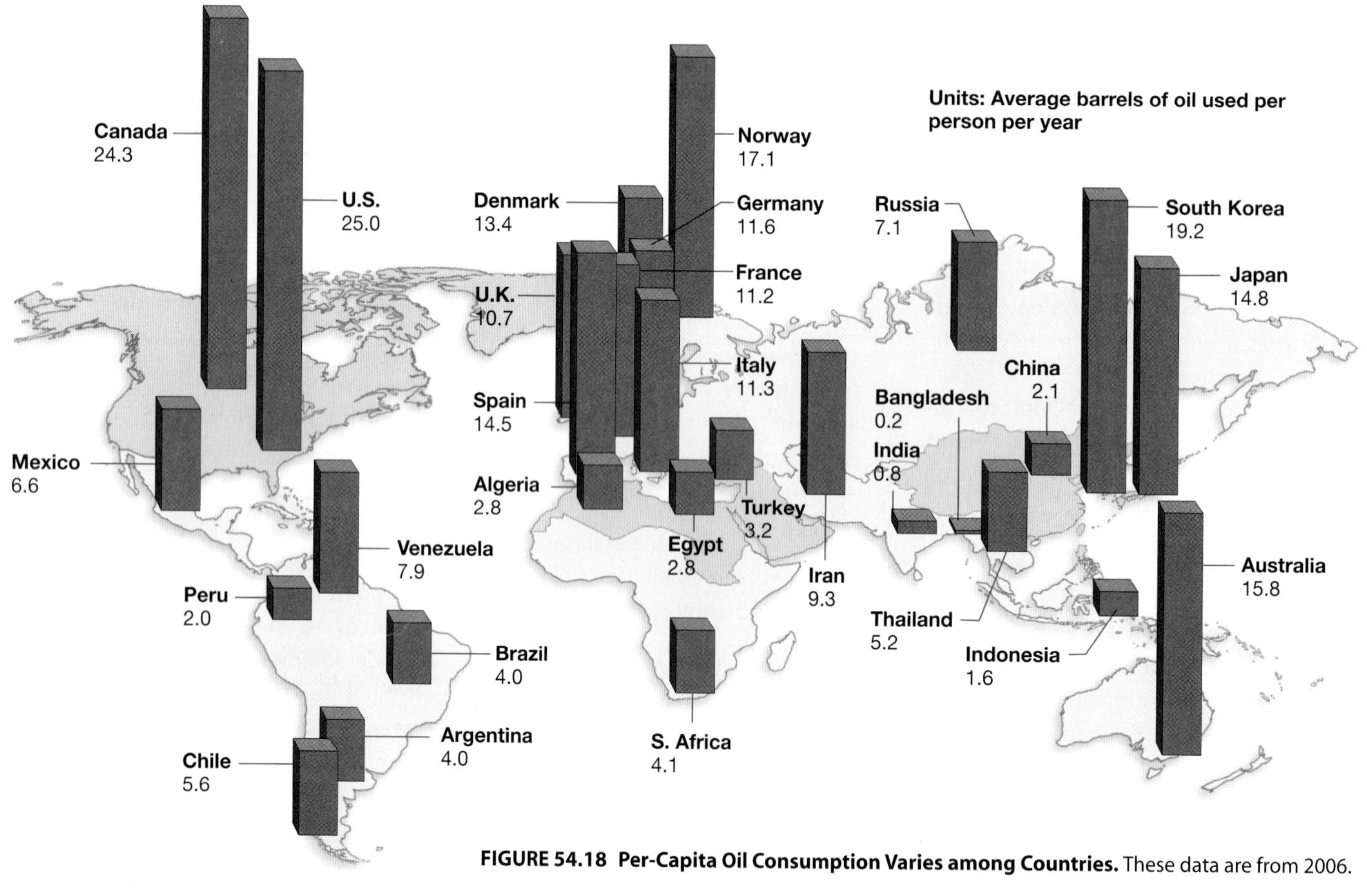

FIGURE 54.18 Per-Capita Oil Consumption Varies among Countries. These data are from 2006.

and stronger evidence that most of the warming observed over the last 50 years is attributable to human activities." In their most recent report, issued in 2007, the IPCC declared that evidence for global warming is unequivocal and that it is "very likely" due to human-induced changes in greenhouse gases.

How much will average temperatures rise in our lifetimes? Predicting the future state of a system as complex and variable as Earth's climate is extremely difficult. Global climate models are the primary tools that scientists use to make these projections. A global climate model is based on a large series of equations that describe how the concentrations of various gases in the atmosphere, solar radiation, transpiration rates, and other parameters interact to affect climate. The models currently being used suggest that average global temperature will undergo additional increases of 1.1–6.4°C (2.0–11.5°F) by the year 2100. The low number is based on models that assume no further increase in greenhouses gases over present levels; the high number is based on models that assume continued intensive use of fossil fuels and increases in greenhouse gases.

How will ecosystems respond to this increase? Answering this question is equally difficult, because ecosystems respond to warming in ways that increase or decrease CO_2 concentrations and thus exacerbate or mitigate warming. Positive feedbacks, for example, occur when warmer and drier climate conditions lead to more fires, which in turn release more CO_2, which leads to even more warming. Researchers recently documented that a form of positive feedback is already occurring in arctic tundras. Traditionally, tundras sequester carbon in the form of soil organic matter, because decomposition rates are extremely low there. During a series of warm summers in the 1980s, however, researchers found that decomposition rates increased sufficiently to release carbon from stored soil organic matter and to produce a net flow of carbon to the atmosphere.

Negative feedbacks, in contrast, arise when warmer conditions lead to increased rates of photosynthesis and hence to an increase in the uptake of CO_2. For example, experiments have shown that the growth rates of several tree species and some agricultural crops increase in direct response to increasing atmospheric CO_2. Because CO_2 is required for photosynthesis, it can act as a fertilizer.

Will ecosystem responses increase or decrease global warming? Currently, it is not clear whether positive or negative feedbacks will predominate. Researchers are working to answer this question by using computer models, experiments like the two highlighted in Chapter 50, and analyses of how ecosystems responded to past climate changes.

Even though global temperatures have risen only slightly in comparison with projections for the next 50–100 years, biologists have already documented dramatic impacts on organisms.

- The geographic ranges of many organisms—including the malaria parasite—are changing. Figure 54.19a shows recent data on the ranges of small crustaceans, called copepods, in the North Atlantic. Copepods are important predators in marine plankton and are eaten by many fish. The maps show changes that occurred in a 40-year period, beginning in 1960, in the number of copepod species typical of marine waters off southern Europe versus the number of copepod species typical of northern European waters. The key observation is that southern species have moved north, while the range of cold-water species has declined.
- The timing of events is changing in many seasonal environments. The graph in **Figure 54.19b** shows changes in the date of first flowering for a plant native to the Midwestern United States. Over the past 60 years, the first day of the year when flowers are produced by *Baptisia leucantha* plants has moved up by about 10 days. To interpret this observation, biologists hypothesize that the climate has warmed enough to promote growth and flowering earlier in the year. Similarly, recent studies have confirmed that migratory

(a) Cold-water copepods are declining in the North Atlantic.

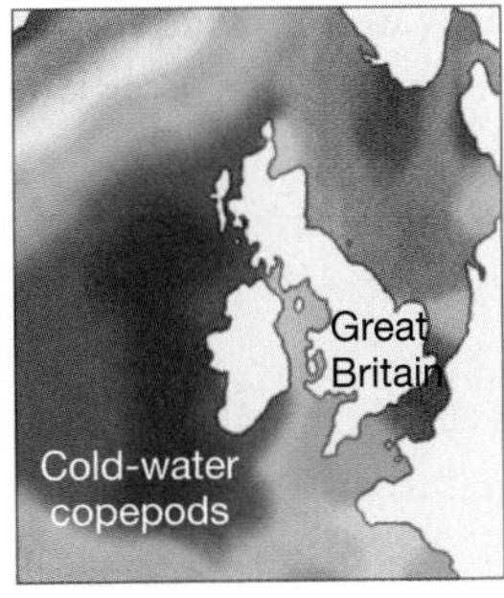

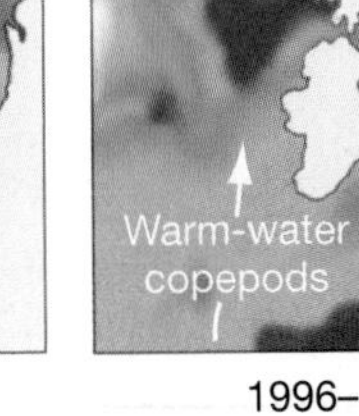

(b) Flowering times for some species in midwestern North America are earlier in the year.

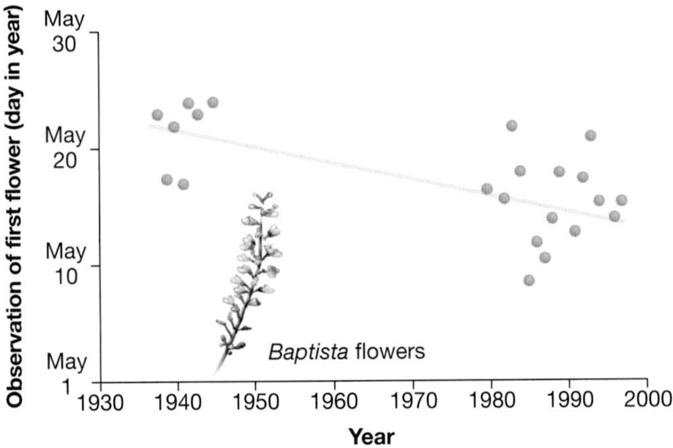

FIGURE 54.19 Global Warming Is Causing Changes in Geographic Ranges and Behavior. (a) Most species of copepod are restricted to waters of a certain temperature. **(b)** In a data set from Wisconsin, 19 of the 55 events tracked over a 61-year period occurred significantly earlier. Only one event occurred later; the remaining 35 showed no change.

bird species are arriving on their breeding grounds in Europe earlier and that fungi are fruiting earlier in some seasonal environments.

- There is strong evidence that most of the 122 frog species that have gone extinct recently succumbed to a parasitic fungus. Data published in 2006 suggest that global warming has increased the frequency and severity of these fungal infections.

- Coral reefs are among the most productive and species-rich ecosystems in the world. Their integrity is threatened, however, by increases in water temperature that cause reef-building corals to expel their photosynthetic algae. When this "coral bleaching" continues due to sustained exposure to warm water, corals begin to die of starvation.

- In some populations, changing temperatures are already causing allele frequencies to change—meaning that some species are evolving in response to global warming. In the fruit fly *Drosophila subobscura*, for example, alleles that increase fitness in hot habitats have increased in frequency in Europe, South America, and North America independently.

Global warming has already caused significant changes in geographic ranges, behavior, and allele frequencies. If the climate models published to date are correct, it will cause many more changes over the course of your lifetime.

Productivity Changes

Several of the changes that humans are inducing in biogeochemical cycles have the same effect: They alter NPP. For example, experiments have documented that warming temperatures, the addition of nitrogen and other nutrients, and rising CO_2 levels increase NPP in some terrestrial ecosystems.

Figure 54.20 shows how average annual NPP has changed globally, in (1) terrestrial environments between 1982 and 1999 (Figure 54.20a) and (2) marine environments from 1999 to 2004 (Figure 54.20b). These are the most recent intervals for which data are available. On land, global NPP increased by over 6 percent during the 17-year study interval. In the oceans, though, the 1999–2004 data show a convincing correlation between increased surface water temperatures and *decreased* productivity.

The overall increase in terrestrial productivity is thought to be due to rising temperatures, increased rainfall in the tropics, and CO_2 fertilization—all factors that increase the rate of photosynthesis in plants. The leading explanation for the drop in marine productivity is more involved. You might recall from Chapter 50 that lakes can become stratified because water is most dense at 4°C and much less dense at higher temperatures. Seawater also becomes stratified. For example, water in the benthic zone is at 4°C year-round in large regions of the ocean (**Figure 54.21a**). When the temperature of surface water rises due to global warming, the water at the surface becomes even less dense than benthic water. This is important because water in the benthic zone is nutrient rich, due to the rain of decomposing organic material from the surface. When surface water becomes lighter, water currents are much less likely to be strong enough to overcome the density difference and bring nutrient-rich, 4°C water all the way up to the surface, where nutrients can spur the growth of photosynthetic bacteria and algae (**Figure 54.21b**). If currents don't become stronger, then global warming causes surface waters to become more nutrient poor.

It's important to note, though, that these global changes are underlain by considerable local variation. On land, productivity during the 1982–1999 interval tended to increase near the equator and at 60°N latitude but to decrease in the arctic and an array of more localized areas. In the ocean, productivity during the 1999–2004 interval dropped dramatically in large areas of the pelagic zone but increased in a number of other regions.

What are the consequences of these local and global changes in NPP? The short answer is, "It depends."

In local areas of the ocean, the consequences of dramatically increased productivity have usually been negative. For example, Chapter 28 detailed how nitrate ions—derived from fertilizers applied to cornfields in midwestern North America—are washing into the Gulf of Mexico. Increased nitrate concentrations have stimulated the growth of planktonic organisms, whose subsequent decomposition has used up available oxygen and triggered the formation of an anoxic "dead zone." Biologists also hypothesize that increases in nitrate and other fertilizers of human origin have contributed to the frequency or intensity of harmful algal blooms in the ocean—large populations of dinoflagellates that release toxins into the surrounding water (see Chapter 29).

Biologists are concerned that if overall NPP in the ocean continues to decline in response to global warming, positive feedback will occur: Atmospheric CO_2 will rise due to less atmospheric CO_2 being fixed by photosynthesis. In contrast, increased NPP on land provides a negative feedback on CO_2 concentrations and global warming. In addition, lower NPP in the ocean means that the productivity of the world's fisheries may decline, but biomass increases in terrestrial environments should boost the numbers of primary and higher-level consumers.

Overall, it is not clear whether changes in productivity will be beneficial or detrimental to ecosystems. The answer should be forthcoming, however. Because global temperatures continue to rise and large-scale additions of nitrogen, phosphorus, and carbon dioxide are ongoing, humans are implementing a global-scale experiment on the effects of altering NPP. Some of the most urgent problems facing your generation are rooted in ecosystem ecology.

(a) Average annual change in terrestrial NPP, 1982–1999

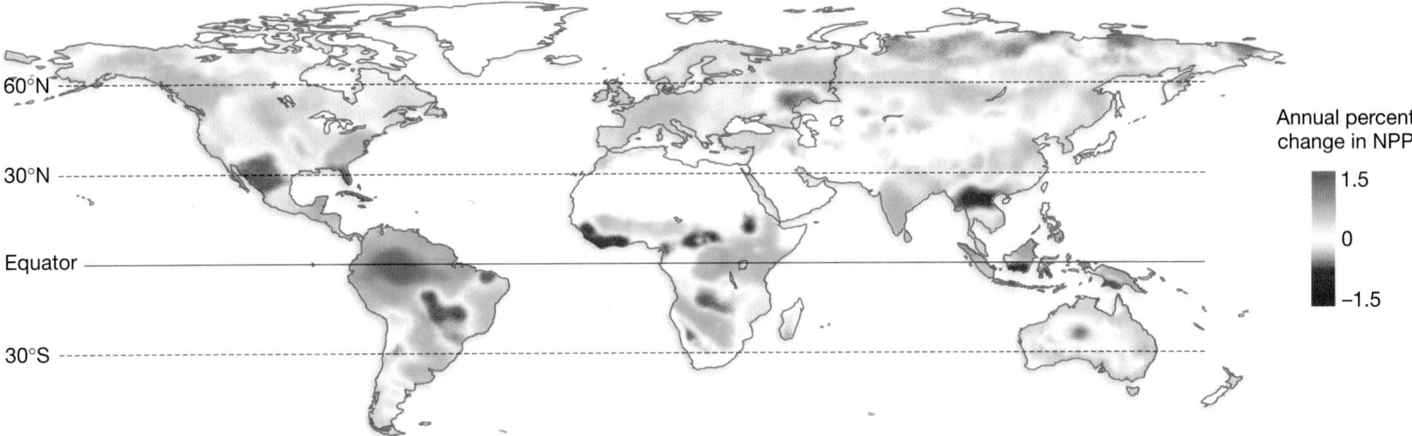

(b) Average annual change in marine NPP, 1999–2004

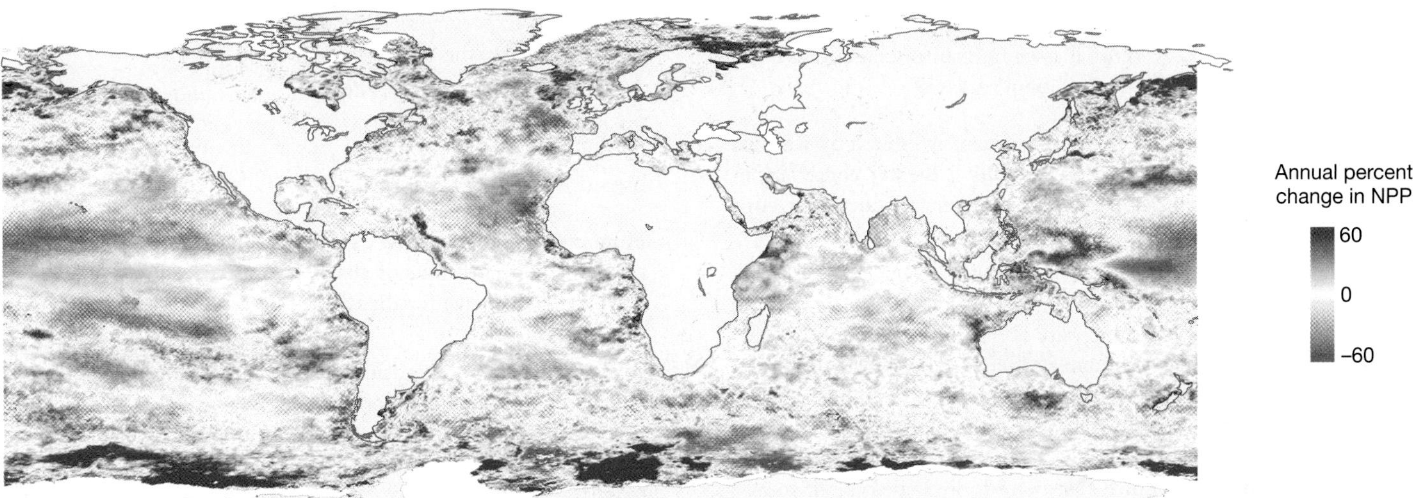

FIGURE 54.20 Consequences of Global Warming: Recent Changes in Terrestrial and Marine NPP.

(a) Much of the ocean is stratified by density and temperature.

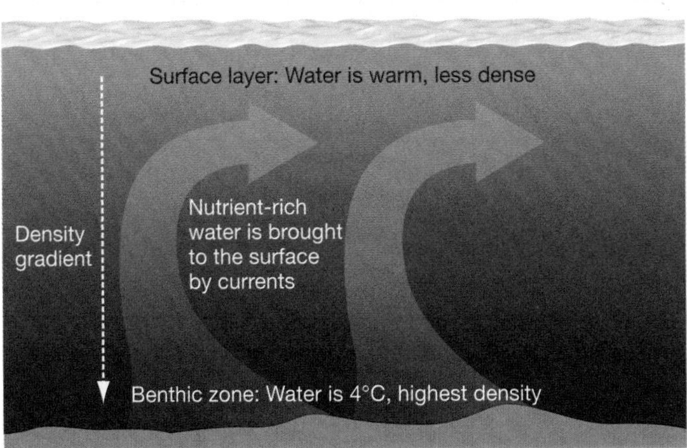

(b) Global warming increases the density gradient, making it less likely for layers to mix.

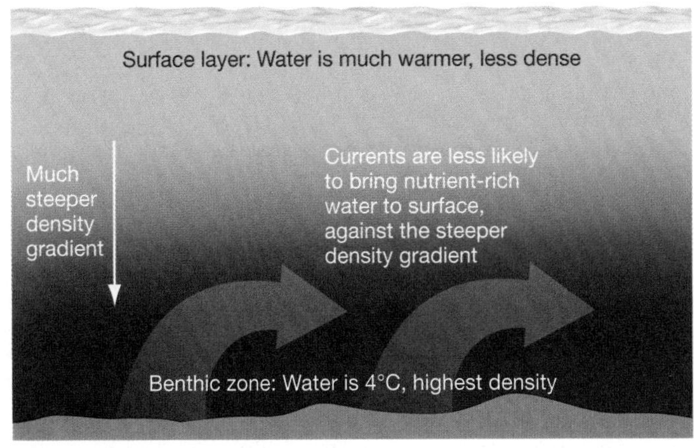

FIGURE 54.21 Temperature Stratification in the Ocean May Affect Productivity.

Chapter Review

SUMMARY OF KEY CONCEPTS

An ecosystem has four components: (1) the abiotic environment, (2) primary producers, (3) consumers, and (4) decomposers. These components are linked by the movement of energy and nutrients.

An ecosystem consists of one or more communities of interacting species and their abiotic environment. As energy flows through ecosystems and as nutrients cycle through them, energy and nutrients are exchanged between biotic and abiotic components of the ecosystem. Energy flows into ecosystems as a result of photosynthesis or the respiration of inorganic molecules with high potential energy. Chemical energy from producers enters food webs via primary consumers or decomposers.

You should be able to draw the relationships among the four components of an ecosystem where no photosynthesis occurs—the outside source of energy is iron- and sulfur-containing compounds released from magma (liquid rock).

The productivity of terrestrial ecosystems is limited by warmth and moisture, while nutrient availability is the key constraint in aquatic ecosystems. As energy flows from producers to consumers and decomposers, much of it is lost.

Among terrestrial ecosystems, net primary productivity (NPP) is highest in tropical wet forests and tropical dry forests. Among aquatic ecosystems, productivity is highest in coral reefs, wetlands, and estuaries. Although productivity is extremely low in the oceanic zone, the area covered by this ecosystem is so extensive that the open ocean accounts for the highest percentage of Earth's overall productivity.

Organisms that acquire energy from the same type of source are said to occupy the same trophic level. Most ecosystems have at least three trophic levels: primary producers, herbivores or primary decomposers, and carnivores. Because energy transfer from one trophic level to the next is inefficient, ecosystems have a pyramid of productivity: Biomass production is highest at the lowest trophic level and lower at each higher trophic level.

The feeding relationships among species in a particular ecosystem are described by a food chain or food web. Food chains rarely exceed five or six trophic levels. The energy-flow, dynamic-stability, and environmental-complexity hypotheses have been proposed to explain why food webs and food chains are not longer.

You should be able to describe the nature of a food chain where a minimum of energy would be lost.

To analyze nutrient cycles, biologists focus on the nature of the reservoirs where elements reside and on how quickly elements move between reservoirs.

Nutrients move through ecosystems in biogeochemical cycles. The rate of nutrient cycling is strongly affected by the rate of decomposition of detritus. The decomposition rate, in turn, is affected by abiotic environmental conditions such as temperature and by the quality of the detritus. Nutrients are also lost from ecosystems. Experiments have shown that the loss of vegetation greatly increases the rate of nutrient loss.

You should be able to describe conditions under which decomposition rates in terrestrial and marine environments are extremely low—creating a carbon sink due to a buildup of organic matter.

Humans are causing large, global changes in the abiotic environment. The burning of fossil fuels has led to rapid global warming. Extensive fertilization is increasing productivity and causing pollution.

Average global temperatures are increasing rapidly because land-use changes and burning of fossil fuels have increased the flow of carbon in the form of CO_2 into the atmosphere, and because carbon dioxide acts as a greenhouse gas. Nitrogen fixation from fertilizer production and from the planting of nitrogen-fixing crop species is now approximately equal to the amount of nitrogen fixation from natural sources. These increases have led to increased productivity, but also to pollution and to loss of biodiversity.

You should be able to propose three steps to reduce global warming and explain the logic behind each.

Web Animation at www.masteringbio.com
The Global Carbon Cycle

QUESTIONS

Test Your Knowledge

1. What is the difference between a community or group of communities and an ecosystem?
   a. An ecosystem comprises a community and the abiotic environment.
   b. An ecosystem is a type of community.
   c. A biome includes only the plant community or communities present in an environment.
   d. An ecosystem includes only the abiotic aspects of a particular environment.

2. Which of the following ecosystems would you expect to have the highest primary production?
   a. subtropical desert
   b. temperate grassland
   c. boreal forest
   d. tropical dry forest

3. Most of the net primary productivity that is consumed is used for what purpose?
   a. respiration by primary consumers
   b. respiration by secondary consumers
   c. growth by primary consumers
   d. growth by secondary consumers
4. According to the dynamic-stability hypothesis for food-chain length, food chains will be shorter in which type of environment?
   a. cold
   b. constant
   c. variable
   d. low in nutrient availability
5. Which of the following is normally the longest-lived reservoir for carbon?
   a. atmosphere (CO_2)
   b. marine plankton (primary producers *and* consumers)
   c. petroleum
   d. wood
6. Devegetation has what effect on ecosystem dynamics?
   a. It increases belowground biomass.
   b. It increases nutrient export.
   c. It increases rates of groundwater recharge (penetration of precipitation to the water table).
   d. It increases the pool of soil organic matter.

Test Your Knowledge answers: 1. a; 2. d; 3. a; 4. c; 5. c; 6. b

Test Your Understanding

Answers are available at www.masteringbio.com

1. Draw a pyramid of productivity for a temperate-forest ecosystem, and explain its shape.
2. Explain the difference between gross primary productivity and net primary productivity. Which is larger, and why?
3. Explain why decomposition rates are higher in some ecosystems than in others, and give examples. How does the decomposer food web regulate nutrient availability in an ecosystem?
4. Compare and contrast the energy-flow, dynamic-stability, and environmental-complexity hypotheses for food-chain length.
5. Draw a diagram of the global water, carbon, or nitrogen cycle. Label major reservoirs and flows. Compare and contrast the life span of water or an element in each reservoir, and evaluate factors that affect the rate of movement between reservoirs.
6. Why are the open oceans nutrient poor? Why are neritic zones and intertidal habitats relatively nutrient rich?

Applying Concepts to New Situations

Answers are available at www.masteringbio.com

1. Suppose you had a small set of experimental ponds at your disposal and an array of pond-dwelling algae, plants, and animals. How could you use radioactive isotopes of carbon or phosphorus to study energy flows or nutrient cycling in these experimental ecosystems?
2. Some researchers are concerned that fertilizing the open oceans with iron would lead to overfertilization of neritic zones, because larger amounts of nutrients would be carried into coastal regions by ocean currents. Outline an experiment and a computer simulation study to test this hypothesis. Evaluate which approach—the experiment or the computer simulation—would likely be more effective in addressing the hypothesis.
3. Suppose that herbivores were removed from a temperate deciduous forest ecosystem. Predict what would happen to the rate of nitrogen cycling. Explain the logic behind your prediction.
4. Explain why human-caused changes to the global carbon cycle are affecting Earth's climate. State why you think that these changes are beneficial or detrimental.

www.masteringbio.com is also your resource for • Answers to text, table, and figure caption questions and exercises • Answers to *Check Your Understanding boxes* • Online study guides and quizzes • Additional study tools including the *E-Book for Biological Science* 3rd ed., textbook art, animations, and videos.

55 Biodiversity and Conservation Biology

KEY CONCEPTS

- Biodiversity can be analyzed at the genetic, species, and ecosystem levels.
- If recent rates of extinction due to human population growth and habitat destruction continue, a mass extinction will occur.
- Humans depend on biodiversity for the products that wild species provide and for ecosystem services that protect the quality of the abiotic environment.

Wangari Maathai, 2004 Nobel Peace Prize Winner, founded the Green Belt Movement that has sponsored the planting of over 1 billion trees, beginning in Kenya and expanding internationally. She is shown here planting a tree.

Most biologists choose their profession for two reasons: They love organisms, and they love answering questions about organisms. But even as the extent of our knowledge about life explodes and the science of biology expands, the number of species decreases. For people who have devoted their lives to the study of life, the irony is cruel. In response, many students cite a third reason for becoming a biologist: Learning how to preserve biodiversity. That is the focus of this chapter.

Let's start by getting a better understanding of what biodiversity is, and how biologists quantify and analyze the diversity of life. Section 55.2 follows up by exploring data on where biodiversity is highest.

The second half of the chapter focuses on efforts to preserve biodiversity. Section 55.3 delves into data on the extent and causes of extinction, and Section 52.4 focuses on efforts to prevent extinction—beginning with the question of why we should bother. As biologists, what reasons can we provide for preserving species? Are these reasons compelling enough to influence real estate developers, farmers, fishermen, loggers, miners, and others whose lives are directly affected by conservation efforts?

Biologists who are concerned about conservation realize that to be effective, they have to do more than collect and analyze data on threatened species and communities. They must also be comfortable in the economic and public policy arenas. A conservation biologist is similar to a physician managing the recovery of an acutely ill patient. Success demands more than the clinical skills of making a correct diagnosis and prescribing medication; it requires an empathy with the person affected, the ability to manage the cost and implementation of care, and a devotion to collaborating with other professionals to achieve a cure. Conservation biology is a demanding and synthetic field.

55.1 What Is Biodiversity?

Perhaps the simplest way to think about biodiversity is in terms of the tree of life—the phylogenetic tree of all organisms, introduced in Chapter 1 and explored in detail in Chapters 28 through 35. If biologists are eventually able to estimate the complete tree of life, using the techniques reviewed in Chapter 27, the branches would represent all of the lineages of organisms living today and the tips would represent all of the species.

When biodiversity increases, branches and tips are added and the tree of life gets fuller and bushier. When extinctions occur, tips and perhaps branches are removed; the tree of life becomes thinner and sparser.

Coordinated, multinational efforts are under way to estimate the phylogenetic tree of major branches on the tree of life, including the fungi, brown algae, and land plants. These and other efforts to estimate phylogenies are producing a better and better picture of the relationships among lineages and species, and a clearer understanding of how extinction is affecting biodiversity. But even though the tree of life is a compelling way to document biodiversity, it does not tell the entire story.

Biodiversity Can Be Measured and Analyzed at Several Levels

To get a complete understanding of the diversity of life, biologists recognize and analyze biodiversity on three levels:

1. **Genetic diversity** is the total genetic information contained within all individuals of a species and is measured as the number and relative frequency of all alleles present in a species. Because no two members of the same species are genetically identical, each species is the repository of an immense array of alleles. At this level, biodiversity is everywhere around you, from the variety of apples at the grocery store to the variation in coloration and singing ability of sparrows in your backyard. **Box 55.1** details recent technical innovations that are allowing biologists to document genetic diversity at a large scale.

2. **Species diversity** is based on the variety of species on Earth. Recall from Chapter 53 that, in practice, species diversity is measured by quantifying the number and relative frequency of species in a particular region. There is an additional aspect of species diversity, however, that biologists call taxonomic diversity and document by estimating phylogenies. Besides understanding the number and relative frequency of species present, researchers also want to know their

BOX 55.1 Environmental Sequencing: A New Approach to Quantifying Genetic Diversity

When the first techniques for analyzing genes and proteins became available in the 1960s, biologists immediately began quantifying the level of genetic diversity within species. The basic approach was to collect tissue samples from many individuals within a population or species, isolate proteins or DNA from the samples, and document how many different alleles existed at a particular gene or suite of genes. With some notable exceptions, the overall message of this work is that most species contain a remarkably high level of allelic diversity. This was an important result, because it indicated that changes in natural selection or other evolutionary processes could cause rapid changes in the frequencies of existing alleles and thus a rapid evolutionary response to environmental change. It also demonstrated that losing a species to extinction means that a large number of unique alleles are lost.

Recent work on genetic diversity has focused on studying entire communities or ecosystems rather than individual populations or species. Research on bacteria from the Sargasso Sea furnishes a good example of this approach. The Sargasso Sea is located in the Caribbean region, near the islands of the Bahamas. Because it is exceptionally nutrient poor and species poor, it is considered one of the ocean's great deserts in terms of biodiversity. The Sargasso Sea was an attractive ecosystem to study for just this reason: Biologists routinely start by studying simple systems before progressing to more complex situations.

To inventory the complete array of bacterial genes present, a research team collected cells from different water depths and locations. The team isolated DNA from the samples and sequenced it using techniques introduced in Chapter 20. They called the work **environmental sequencing**, because their goal was to analyze all of the genes present in an ecosystem. After analyzing over 1 billion base pairs, the team concluded that at least 1800 bacterial species were present, of which 148 were previously undiscovered. They also identified more than 1.2 million alleles that had never before been characterized.

The results suggest that an enormous amount of genetic diversity exists even in Earth's simplest ecosystems. To grasp why this is important, consider that the researchers discovered more than 780 new bacterial alleles that code for proteins similar to the rhodopsin molecules found in your eyes. Do these proteins function as light receptors, like rhodopsin does? Or are these molecules absorbing light energy and using it to pump protons out of the cells, establishing a proton gradient that can drive synthesis of ATP and keep the organisms alive? No one knows—yet. New knowledge inspires new questions.

evolutionary relationships. For example, some lineages on the tree of life are extremely species rich. Prominent examples include the African cichlid fish, introduced in Chapter 27, and the 35,000 species of orchids. Other lineages, in contrast, are represented by a single species (**Figure 55.1**). Some biologists argue that it is particularly important to preserve populations from species-poor lineages, because they are the last living representatives of their lineages. If those populations go extinct, an entire lineage is lost forever.

(a) Red panda

(b) Yangtze river dolphin

FIGURE 55.1 Phylogenetically Distinct Species May Be High-Priority Targets for Conservation. The red panda and the Yangtze river dolphin have few close relatives and represent distinct branches on the tree of life. If conservation measures attempt to preserve taxonomic diversity, these species would be a high priority for preservation. Unfortunately, preservation efforts for the Yangtze river dolphin are too little too late—this species is now functionally extinct (too few individuals remain to breed and sustain the population).

3. **Ecosystem diversity** is the variety of biotic communities in a region along with abiotic components, such as soil, water, and nutrients. Ecosystem diversity is more difficult to define and measure than genetic diversity or species diversity, because ecosystems do not have sharp boundaries. Recall from Chapter 54 that ecosystems are complex and dynamic assemblages of organisms that interact with each other and their nonliving environment. Attempts to measure ecosystem diversity focus on capturing the array of biotic communities in a region, along with variation in the physical conditions present. Areas around estuaries, for example, tend to have high ecosystem diversity due to the combination of stream, wetland, intertidal, neritic, and upland habitats (see Chapter 50). When an estuary is dredged or filled, biodiversity is affected at all three levels: genetic and species diversity change due to the different numbers and types of individuals present, and ecosystem diversity is altered due to the change in abiotic conditions.

Biodiversity can be recognized and quantified on several distinct levels, but it is also dynamic. Mutations that create new alleles increase genetic diversity; natural selection, genetic drift, and gene flow may eliminate certain alleles or change their frequency, leading to an increase or decrease in overall genetic diversity. Speciation increases species diversity; extinction decreases it. Changes in climate or other physical conditions can result in the formation of new ecosystems, as can the evolution of new species that interact in novel ways. Disturbances such as volcanic eruptions, human activities, and glaciation can destroy ecosystems.

Biodiversity is not static. It has been changing since life on Earth began.

How Many Species Are Living Today?

One of the simplest questions about biodiversity is also one of the most difficult to answer. How many species are there on Earth? The answer is not known. Given the massive effort that it would take to document every form of life on Earth, the answer will probably never be known.

Biologists are well aware, though, that the approximately 1.5 million species cataloged to date represent a tiny fraction of the number actually present. Chapter 31, for example, noted that an average of 6 fungal species live on each well-studied plant species. Because over 300,000 plant species have been described, a conservative estimate based on the 6:1 fungi:plant ratio suggests that 1.65 million species of fungi exist, even though only 80,000 have been described thus far. The issue is even more acute in poorly studied lineages like bacteria and archaea, where researchers routinely discover dozens or hundreds of previously unknown species with each new direct sequencing or environmental sequencing study (see Chapter 28 and Box 55.1). For example, a recent direct sequencing study estimated that over 500 species of bacteria live in the human mouth, although only about half have been described and named. Even among well-studied groups like birds and mammals, new species are discovered almost every year.

Given that only a fraction of the organisms alive have been discovered to date, how can biologists go about estimating the total number of species on Earth? Two general approaches have been used. One is based on intensive surveys of species-rich groups at small sites. A second is based on attempts to identify all of the species present in a particular region.

Taxon-Specific Surveys Terry Erwin and J. C. Scott published a classic study of species diversity in a single taxon. They began by estimating the number of insect species that live in the canopy of a single tropical tree (**Figure 55.2**). After using an insecticidal fog to knock down insects from the top of a *Luehea seemannii* tree, they identified over 900 species of beetles among the individuals that fell. Most of these species were new to science.

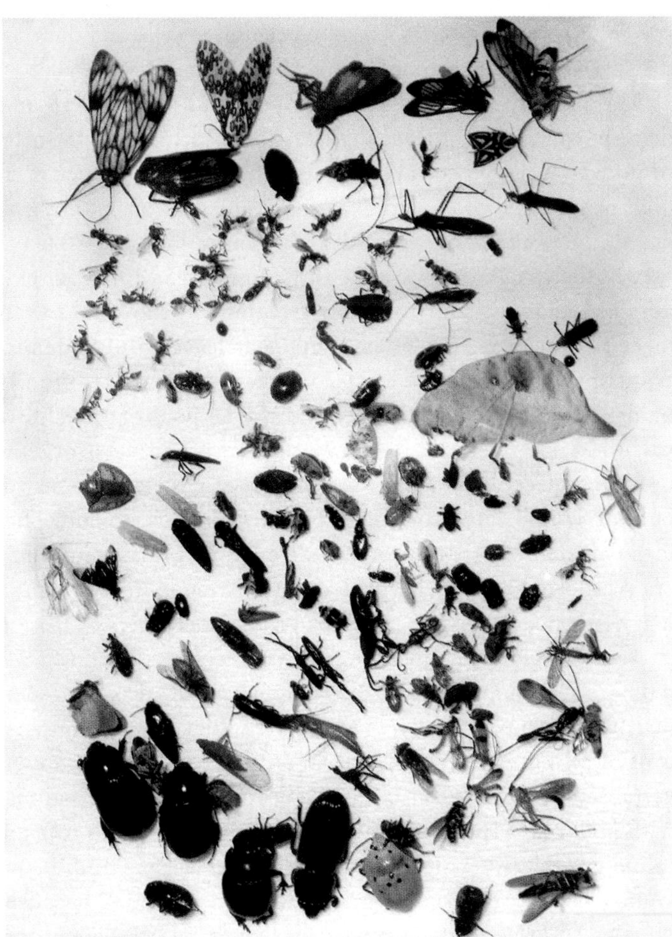

FIGURE 55.2 Estimating Species Richness via Intensive Local Sampling. A small sample of the insects collected from a single tree in the Amazonian rain forest.

To use these data as an indicator of global arthropod species diversity, Erwin and Scott used the following train of logic: Based on earlier work with insects on this tree species, they estimated that 160 of the 900 beetle species live only on *L. seemannii*. Worldwide, beetles represent about 40 percent of all known arthropods. Thus, it was reasonable to suggest that 400 species of arthropods live only in the canopy of *L. seemannii*. By adding an estimate of arthropods specializing on the trunk and roots of this tree, Erwin projected that it is host to 600 specialist arthropods. If each of the 50,000 species of tropical tree harbors the same number of arthropod specialists, then the world total of arthropod species exceeds 30 million species. Based on such studies, biologists estimate that at least 10 million, and possibly as many as 100 million, species of all types exist today.

More recent single-taxon surveys have sampled larger geographic areas. For example, Philippe Bouchet and colleagues conducted a massive survey of marine molluscs in coral-reef habitats along the west coast of New Caledonia, a tropical island in the southwest Pacific Ocean. The team spent more than a year collecting molluscs at 42 sites over a total area of almost 300 square kilometers. The survey represented the most thorough sampling effort ever made to determine the species diversity of molluscs, and produced more than 127,000 individuals representing 2738 species. These numbers far exceed the mollusc diversity recorded for any comparable-sized area. The species total was 2 to 3 times the total number of mollusc species that had been reported for similar habitats in the region.

In reporting their findings, the biologists emphasized that 20 percent of the species found were represented by a single specimen. This observation suggests that many species are exceedingly rare and thus likely to be missed by less-intensive sampling efforts. And when the investigators compared their data with a survey in progress at a second site, different in reef structure but only 200 kilometers away from the original site, they found that only 36 percent of species were shared between the two sites. These results support the hypothesis that the global biodiversity of molluscs—currently thought to be about 93,000 species—is an underestimate.

All-Taxa Surveys The first effort to find and catalog all of the species present in a large area is now under way. The location is the Great Smoky Mountains National Park in the southeastern United States (**Figure 55.3**). A consortium of biologists, volunteers, and research organizations initiated this all-taxa survey in 1999. To date, the survey has discovered over 650 species that are new to science and over 4650 species that had never before been found in the park. When the inventory is complete, in 2015, biologists will have a much better database to use in estimating the extent of global biodiversity.

(a) Great Smoky Mountains National Park

(b) All-taxon survey workers assess organisms in a sample.

FIGURE 55.3 The First All-Taxon Survey Is Now Under Way. The first attempt to document every species living in a prescribed area is under way at Great Smoky Mountains National Park, along the border of Tennessee and North Carolina.

Check Your Understanding

If you understand that...

- Biologists document biodiversity at the genetic, species, and ecosystem levels.

You should be able to...

Outline a study that would quantify levels of genetic, species, and ecosystem diversity on your campus.

55.2 Where Is Biodiversity Highest?

If documenting the extent of genetic, species, and ecosystem diversity is the most fundamental goal of research on biodiversity, then the second-most important aim is to understand its geographic distribution. Chapter 53 introduced the most prominent geographic pattern in the distribution of biodiversity: In most taxonomic groups, species richness is highest in the tropics and declines toward the poles. Tropical rain forests are particularly species rich. Even though they represent just 7 percent of Earth's land area, they are thought to contain at least 50 percent of all species present. As Chapter 53 noted, understanding why tropical forests are so species rich and why a latitudinal gradient occurs is one of the most active areas of research in community ecology and biodiversity.

Biologists have also been working to understand the distribution of species richness at finer spatial scales, however. For example, the map in **Figure 55.4a** was constructed by dividing the world's landmasses into a grid of cells 1° latitude by 1° longitude. In the tropics, this translates to rectangles approximately 111 km by 111 km (69 mi by 69 mi); because lines of longitude converge at the poles, the rectangles are about 111 km by 55 km at 60° latitude. Using published data, researchers then plotted how many species of birds breed in each cell in the grid. These data are consistent with the latitudinal gradient in species richness, but indicate that some areas of the tropics are much more species rich than others. Biologists use the term **hotspot** to capture this point. In terms of bird species richness, the Andes mountains, the Amazon River basin, portions of East Africa, and southwest China are important hotspots.

Researchers have also been keenly interested in understanding which regions of the world have a high proportion of **endemic species**—meaning, species that are found in an area and nowhere else. **Figure 55.4b** maps the location of endemic species of breeding birds, based on the same cells and data as the species richness map in Figure 55.4a.

Mapping species richness hotspots and centers of endemism can inspire interesting questions—principally, efforts to understand *why* certain regions contain many species or a high proportion of endemic taxa. But in addition, biologists are studying the geographic distribution of biodiversity as a way of focusing conservation efforts. In the year 2000, for example, a team set out to identify regions of the world that meet two criteria: (1) They contain at least 1500 endemic plant species, and (2) at least 70 percent of their traditional or primary vegetation has been lost. The goal of the study was to identify regions of the world that are in most urgent need of conservation action—areas where efforts to preserve habitat would have the highest return on investment. The idea was that efforts to protect a diversity of primary producers would guarantee the protection of a wide array of primary, secondary, tertiary, and quaternary consumers.

The 25 regions that met the two "conservation hotspot" criteria are shown in **Figure 55.4c**. Although these areas represent just 1.4 percent of the Earth's land area, they contain 44 percent of all known plant species and 35 percent of all known vertebrate species. It's also important to note that key regions with high plant species diversity—specifically the rain forests found in the Amazon, the Congo River basin, and New Guinea—are not included as conservation hotspots, because they have not yet lost 70 percent of their primary vegetation. Effectively protecting these areas along with the hotspots defined on the map would provide protection for 60 percent of all land plants in just 5 percent of the Earth's land surface.

(a) Hotspots in terms of species richness of birds

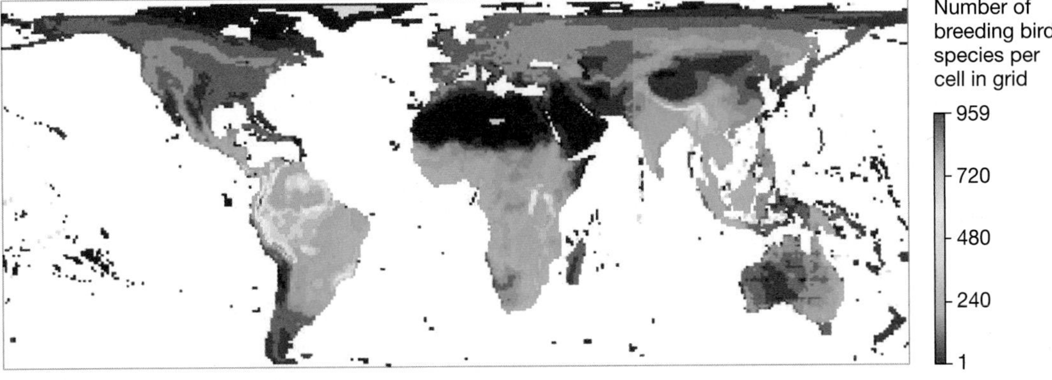

(b) Hotspots in terms of endemic species of birds

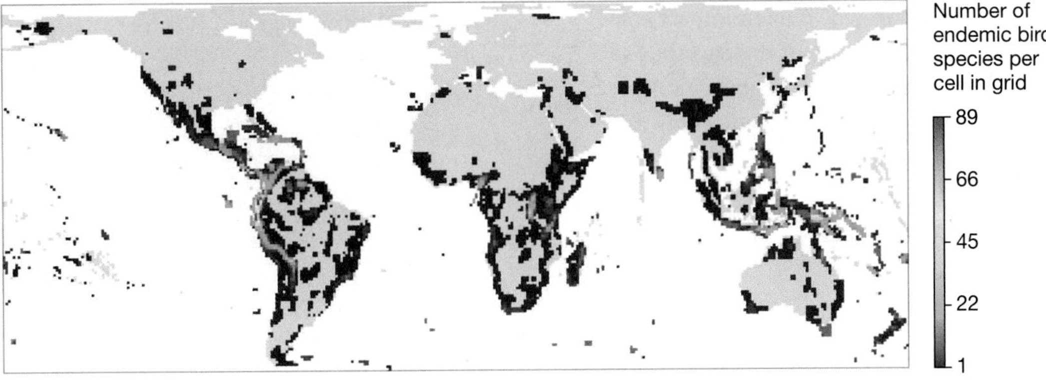

(c) Hotspots in terms of high proportion of endemic plants *and* high threat

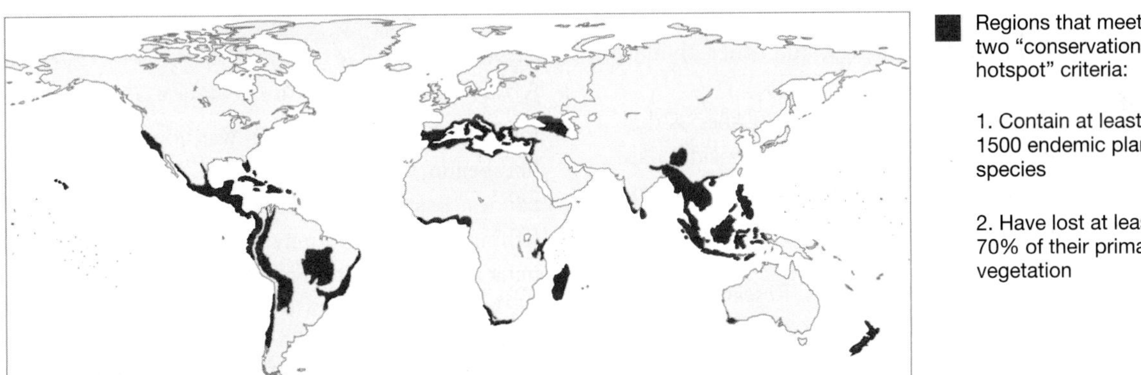

FIGURE 55.4 Conservation "Hotspots" Can Be Defined in Different Ways. The maps in **(a)** and **(b)** are based on a grid that is 1° longitude by 1° latitude. The areas shown in red in **(c)** have lost 70 percent of their original vegetation and contain a high diversity of vascular plant species.

QUESTION Suppose you were going to set up natural areas focused on preserving bird species. Which areas on the three maps are in agreement in terms of conservation priorities? Which are in conflict?

This analysis serves as your introduction to the applied aspects of biodiversity studies—the field known as conservation biology. To understand the importance of the data in Figure 55.4 thoroughly, we need to explore two things in depth: why so many species are threatened with annihilation, and why we should do something about it. Let's consider each issue in turn.

55.3 Threats to Biodiversity

No species lasts forever. Climate change, disease, competition from newly arrived species, and habitat alteration are all natural processes. They have been happening since life began. Extinction, like death, is a fact of life.

If extinction is natural, why are biologists so concerned about habitat and species conservation? The answer is rate. Today species are vanishing faster than at virtually any other time in Earth's history. Modern rates of extinction are 100 to 1000 times greater than the average, or "background," rate recorded in the fossil record over the past 550 million years.

Either directly or indirectly, current extinctions are being caused by the demands of a rapidly growing human population, which is currently increasing by about 77 million people per year. If present trends in human population growth and species extinctions continue, a mass extinction on the scale of events described in Chapter 27 will occur. Based on the data in hand, the vast majority of biologists agree with the claim that the sixth mass extinction in the history of multicellular life is now under way. Between the time you are reading this and the time your great-grandchildren are grown, human impacts on the planet promise to equal or exceed those of the gigantic asteroid that smashed into Earth 65 million years ago. It's important to note, however, that human impacts on biodiversity are not a new phenomenon.

Humans Have Affected Biodiversity throughout History

Historically, humans have a poor record of conserving resources and protecting species. For example, recent research on fossil birds found on islands in the South Pacific suggests that about 2000 bird species were wiped out as people colonized this area about A.D. 1200. Many of these extinctions occurred due to predation by humans or rats, pigs, and other animals introduced by humans.

Humans even have a poor record of preserving species they depend on for survival. To drive this point home, consider data on the easternmost island in the South Pacific—Easter Island (**Figure 55.5**). When European explorers arrived on Easter Island in 1722, about 1000 people lived there. The island was treeless and dotted with gigantic stone statues. Researchers who analyzed buried pollen samples taken from swamps on the island discovered that it had once been covered with lush forest dominated by palm trees. A rapid decline in tree pollen coincided with the arrival of the first human settlers.

Fossil digs confirmed that the fauna of the island underwent drastic changes at the same time. In the oldest human garbage piles excavated by biologists, bones from dolphins, seabirds, and land birds are abundant. But these species dropped out of the fossil record by around A.D. 1300—about the time that loss of forests, or deforestation, was complete.

Jared Diamond interpreted these data as evidence of an ecological disaster. In his view, people arrived to find a lush tropical island brimming with natural resources. The human population flourished and people had the leisure time and resources to carve the gigantic statues and roll them into place on beds of palm logs. But after deforestation was complete and local extinctions had begun, the system collapsed. Without palm trunks to make canoes, Easter Island natives did less dolphin hunting and less fishing. Soil erosion may have cut into the productivity of banana and sweet potato plantations. The population crashed, and the great statues fell.

FIGURE 55.5 On Easter Island, Social Decline Coincided with the Exhaustion of Natural Resources. Easter Island was one of the last Pacific Islands to be colonized by humans. The prosperity that supported the artwork shown in the inset was based on resources that were overexploited.

Current Threats to Biodiversity

Most extinctions that have occurred over the past 1000 years took place on islands and were caused by overhunting or the introduction of **exotic species**—nonnative competitors, diseases, or predators. Starting in the twentieth century, though, the situation began to change. **Endangered species**, which are almost certain to go extinct unless effective conservation programs are put in place, are now more likely to live on continents than islands. In addition, habitat destruction has replaced overhunting and species introductions as the primary threat to such species.

Figure 55.6 shows the results of a recent analysis on causes of endangerment for 488 endangered species native to Canada. Several patterns jump out of the data:

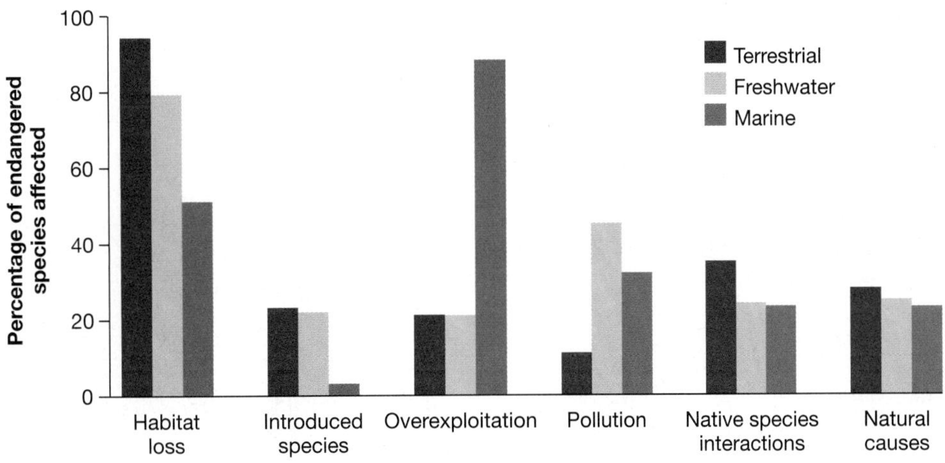

FIGURE 55.6 Endangerment Is Caused by Multiple Factors. These data are from Canada.

QUESTION Why do the data for terrestrial, freshwater, and marine species each add up to more than 100 percent?

- Virtually all of the endangered species are affected by more than one factor. This is important, because it means that conservation biologists may have to solve more than one problem for any given species to recover.
- Habitat loss is the single most important factor in the decline of these species. It is a significant issue for over 90 percent of endangered species in terrestrial environments.
- Overharvesting is the dominant problem for marine species, while pollution plays a large role for freshwater species.
- Factors beyond human control can be important. These include predation or competition with native species, natural disturbances such as fires or droughts, or the fact that some species have narrow niches and have historically been rare. Stated another way, background extinctions will continue to occur.

Do these conclusions hold up for other parts of the world? Although invasive species are a more significant problem in some regions (**Figure 55.7**; also see Chapter 50), most analyses completed to date are broadly consistent with the Canadian data. For example, overexploitation is a grave concern for marine fisheries worldwide. Two-thirds of harvestable species are now considered fully exploited or depleted, and a recent study of the global fishing industry concluded that 90 percent of large bottom-dwelling and open-water fish have been removed from the world's oceans. The species affected include tuna, swordfish, marlin, cod, halibut, and flounder. Overhunting has also emerged recently as a dire threat to

(a) Invasive species increase **competition**.

Purple loosestrife is crowding out native organisms in North American marshes.

(b) Invasive species introduce **disease**.

An introduced fungus has virtually wiped out the American chestnut.

(c) Invasive species increase **predation**.

The brown tree snake has extinguished dozens of bird species on Guam.

FIGURE 55.7 Invasive Species Are Destructive. In many parts of the world, invasive species are a leading cause of endangerment. [Photo (b) © Gary Braasch.]

FIGURE 55.8 Overhunting Continues to Be a Problem in Some Regions. The recent growth of the "bushmeat" trade has devastated many populations of primates and other mammals in Africa. Overharvesting is also a major cause of endangerment in marine species worldwide (see Figure 55.6).

FIGURE 55.9 Suburbanization Is a Major Cause of Habitat Loss in North America.

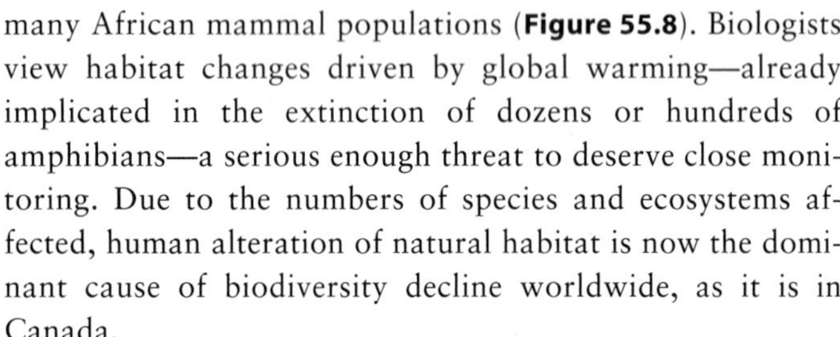

many African mammal populations (**Figure 55.8**). Biologists view habitat changes driven by global warming—already implicated in the extinction of dozens or hundreds of amphibians—a serious enough threat to deserve close monitoring. Due to the numbers of species and ecosystems affected, human alteration of natural habitat is now the dominant cause of biodiversity decline worldwide, as it is in Canada.

Habitat Destruction Humans cause **habitat destruction** by logging and burning forests, damming rivers, dredging or filling estuaries and wetlands, plowing prairies, grazing livestock, excavating minerals, and building housing developments, golf courses, shopping centers, office complexes, airports, and roads (**Figure 55.9**).

On a global scale, one of the most important types of habitat destruction is deforestation—especially the conversion of primary forests to agricultural fields and human settlements. To appreciate the extent of deforestation in some areas, consider satellite images of wet tropical rain forest in Rondônia, Brazil, that were made in 1975 and in 2001 (**Figure 55.10**). Analyses of satellite photos such as these have shown that as many as 3 million hectares, an area about 10 percent larger than the state of Maryland, was deforested in the Amazon each year during the 1990s. (A hectare, abbreviated ha, is 100 meters by 100 meters, approximately the size of two football fields.)

More recent analyses suggest that the global rate of deforestation has now slowed compared to peak rates in the 1990s. The latest report from the United Nations Food and Agriculture Organization, released in 2005, states that the world's

(a) The devastation of deforestation

(b) Satellite view of deforestation in Rondônia, Brazil

FIGURE 55.10 Rates of Deforestation in Tropical Wet Forests Are Approaching 1 Percent per Year. These satellite photos show the extent of deforestation in Rondônia, Brazil, in less than 30 years. The dark green areas are intact forest; the lighter areas have been burned or logged and converted to agricultural fields and pastures.

forests experienced a net loss of about 7.3 million ha/year between 2000 and 2005. This is down from the average net loss of 8.9 million ha/year during the previous decade. Although most conservationists greeted this report as good news, they also urged caution. Net losses dropped because of extensive forest planting and regrowth in China, North America, and Europe, even though over 8 million ha/year of primary forest were lost in South America and Africa each year from 2000 to 2005. Forest loss in South America and Africa is particularly important because it affects biodiversity hotspots.

Habitat Fragmentation In addition to destroying natural areas outright, human activities fragment large, contiguous areas of natural habitats into small, isolated fragments. **Habitat fragmentation** concerns biologists for several reasons.

1. It can reduce habitats to a size that is too small to support some species. This is especially true for keystone predators such as mountain lions, grizzly bears, and bluefin tuna, which need vast natural spaces in which to feed, find mates, and reproduce successfully.
2. By creating islands of habitat in a sea of human-dominated landscapes, fragmentation reduces the ability of individuals to disperse from one habitat to another. In effect, habitat fragmentation is forcing many species into the metapopulation structure introduced in Chapter 52. The small, isolated populations that make up a metapopulation are much more likely than large populations to be wiped out by catastrophic events such as storms, disease outbreaks, or fires. They can also suffer from inbreeding depression and random loss of alleles due to genetic drift (see **Box 55.2**).
3. Fragmentation creates large amounts of "edge" habitat. The edges of intact habitat are subject to invasion by weedy species and are exposed to more intense sunlight and wind, creating difficult conditions for plants.

The decline in habitat quality caused by fragmentation is being documented in a long-term experiment in a tropical wet forest. In an area near Manaus, Brazil, that was slated for clear-cutting, a research group set up 66 square, 1-hectare experimental plots that remained uncut. Thirty-nine of these study plots were located in fragments designed to contain 1, 10, or 100 hectares of intact forest. Twenty-seven of the plots were set up nearby, in continuous wet forest. As the "Experimental setup" section of **Figure 55.11** shows, the distribution of the study plots allowed the research team to monitor changes inside forest fragments of different sizes and to compare these changes with conditions in unfragmented forest.

When the research group surveyed the plots at least 10 years after the initial cut, they recorded two predominant effects: a rapid loss of species diversity, especially from the smaller fragments, and a startling drop in **biomass**, or

Experiment

Question: How does fragmentation affect the quality of tropical wet forest habitats?

Hypothesis: Fragmentation reduces the quality of wet forest habitats.

Null hypothesis: Fragmentation has no effect on the quality of wet forest habitats.

Experimental setup:

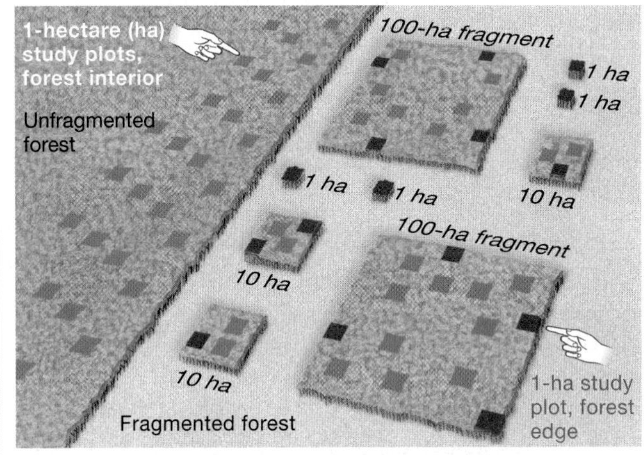

Prediction: Species diversity and biomass will decline in forest fragments compared with those of the forest interior, particularly along edges of fragments.

Prediction of null hypothesis: Species diversity and biomass will be the same inside forest fragments and along edges as in the forest interior.

Results:

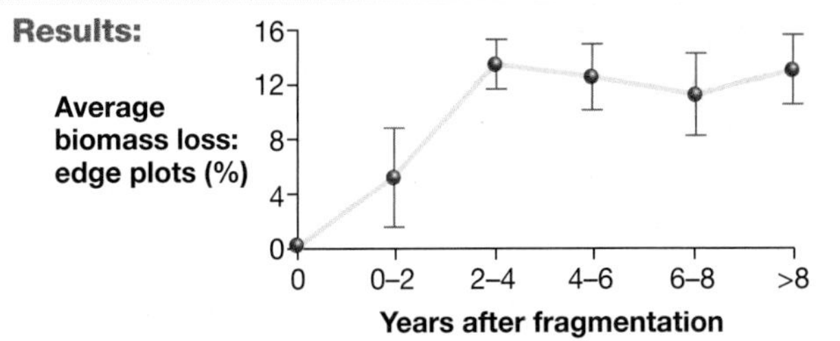

Conclusion: Biomass declines sharply along edges of forest fragments. To interpret this result, hypothesize that large trees near forest edges tend to die.

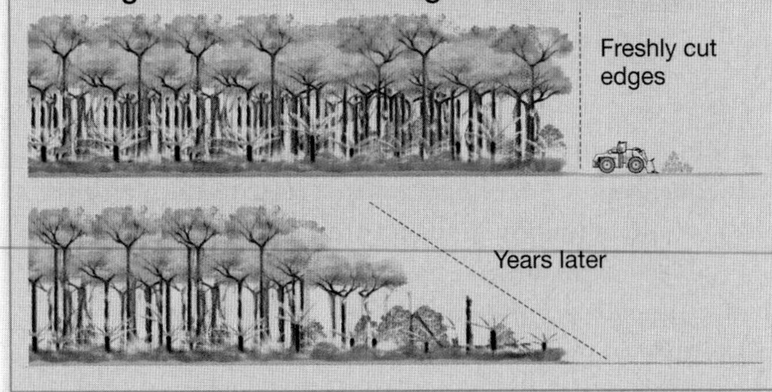

FIGURE 55.11 Experimental Evidence for Edge Effects in Fragmented Forests. Researchers tracked 66 study plots among four 1-hectare fragments, three 10-hectare fragments, and two 100-hectare fragments.

BOX 55.2 An Introduction to Conservation Genetics

Analyses presented in Chapters 24 and 25 highlighted the four evolutionary forces—natural selection, genetic drift, gene flow, and mutation—that affect allele frequencies and cause evolution. In addition, inbreeding and other forms of nonrandom mating change genotype frequencies. Conservation biologists have been keenly interested in how these processes are affecting small, endangered populations.

Several decades of observational and experimental work on conservation genetics supports the following conclusions:

- When populations are fragmented into a metapopulation structure, gene flow between isolated groups is reduced or eliminated. You might recall from Chapter 25 that the major impact of gene flow between populations is to homogenize them genetically. When gene flow is reduced, then, isolated populations begin to differentiate.
- Genetic drift is much more pronounced in small populations than large populations. Chapter 25 presented theory and experiments on the major impact of drift: Because it leads to the random loss or fixation of alleles, drift reduces genetic variation in populations. Loss of genetic variation has now been documented in a wide array of endangered species and other small populations. This is a concern because if the environment is changed by the evolution of a new disease or through global warming, the population may lack alleles that confer high fitness in the new environment.
- Small populations become inbred, and inbreeding often leads to lowered fitness—the phenomenon known as inbreeding depression.

In general, then, small populations that are isolated from each other are expected to be much less healthy, genetically, than robust populations that occupy large, contiguous areas.

When biologists document fitness declines in small, isolated populations, one option is to increase gene flow experimentally by importing individuals of the same species from a different population. For example, **Figure 55.12** shows data on annual population growth rate (λ; see Chapter 52) in a small population of bighorn sheep isolated on a refuge in northwest Montana. From the time the population was founded in 1922 until gene flow was experimentally increased, the average population size was about 42 individuals and average growth rate declined. After introducing 5 new individuals in 1985 and 10 additional sheep in 1990–1994, though, λ has stabilized and population size has increased. Introducing new alleles in this way can counteract the effects of genetic drift and inbreeding.

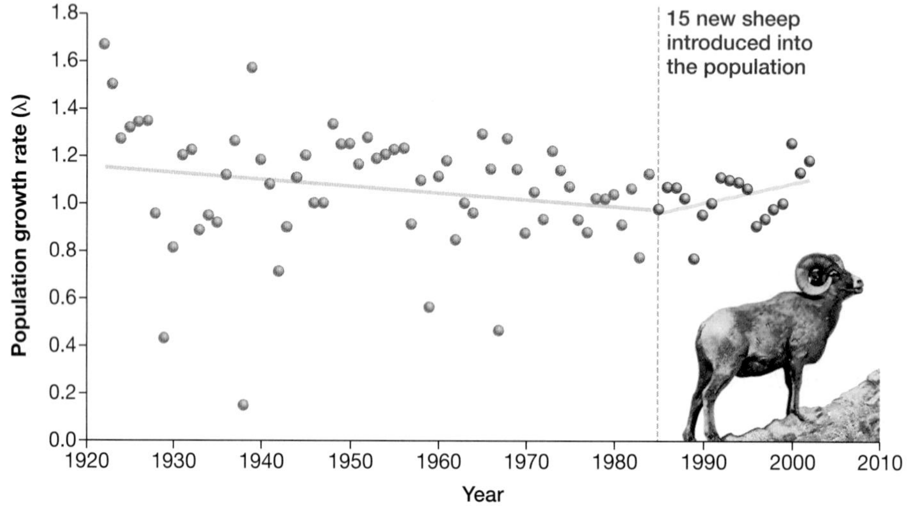

FIGURE 55.12 Experimental Evidence that Gene Flow Can Alleviate Genetic Problems in Small Populations. The y-axis plots the finite rate of increase (λ) in an isolated population of bighorn sheep, before and after unrelated individuals were introduced.

EXERCISE Draw a horizontal line indicating the value of λ where population size is constant (no growth).

the total amount of fixed carbon, in the study plots located near the edges of logged fragments. In tropical wet forests, most of the biomass is concentrated in large trees. The edges of the experimental plots contained many downed and dying trees. Based on this observation, the researchers inferred that the decrease in biomass occurred because large trees near the edges of fragments died from exposure to high winds and dry conditions. Follow-up work showed that in edge habitats and in the fragmented patches, large-seeded, slow-growing trees typical of undisturbed forests are being replaced by early successional, "weedy" species (see Chapter 53). A large number of bird and understory plant species have disappeared as well.

The take-home message of this experiment and other research is clear: When habitats are fragmented, the quality *and* quantity of habitat decline drastically. In addition to losing over 8 million hectares of primary forest in South America and Africa each year, we are losing large amounts of high-quality habitat to fragmentation. Just how many extinctions are all of these problems projected to cause?

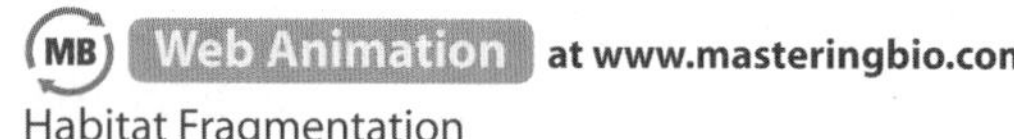

at www.masteringbio.com

Habitat Fragmentation

How Can Biologists Predict Future Extinction Rates?

If the most fundamental question about biodiversity is how many species are alive, then the most basic question in conservation biology is how rapidly species are going extinct. Both questions have been difficult to answer.

Biologists use two approaches to estimate current extinction rates and predict how they might change in the near future. The first is based on direct counts of species that are known to have gone extinct recently or are in imminent danger of going extinct. The second is based on predicting the consequences of habitat loss, using existing data on the relationship between the size of habitats and the number of species present.

Estimates Based on Direct Counts The best information on current extinction rates comes from studies on the best-studied of all major lineages on the tree of life: birds. Based on data in the fossil record, background extinctions in lineages like birds are estimated to occur at the rate of 1 extinction per million species alive per year. Of the approximately 10,000 bird species known, then, one species should go extinct every 100 years due to normal or background events.

Recent analyses have focused on species that were wiped out over the past 1000 years, coincident with the expansion of Polynesian people throughout the Pacific and Europeans throughout the world. These data suggest that birds have been going extinct at a rate 100 times background levels—meaning that a bird species has gone extinct every year, on average. Although the success of conservation programs targeted at birds has lowered this rate in the last few decades, estimates based on the numbers of species that are currently endangered suggest that the rate could reach 1000–1500 extinctions per million species alive per year by the end of this century. If this prediction holds, you may live long enough to see 10 or more species of bird go extinct each year.

Similar trends are occurring in listings of other well-studied groups. For example, a research group looked at endemic species of birds, mammals, reptiles, amphibians, and conifers, and determined that 794 of the species are in imminent danger of extinction. This is three times the number of all species from these groups known to have gone extinct since the year 1500. Instead of a handful of species going extinct each year, we may be losing dozens or hundreds. The message from these analyses is that the current rate of extinction is already many times higher than background or normal levels, and it is poised to increase dramatically in the near future.

Species–Area Relationships A second strategy for predicting the future of Earth's biodiversity focuses on the consequences of the most urgent problem: habitat destruction. Given reasonable projections of habitat loss, biologists can estimate rates of extinction based on well-documented **species–area relationships.**

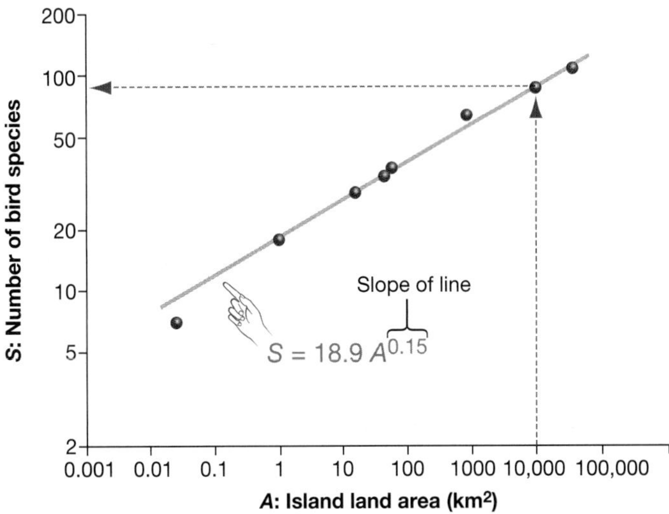

FIGURE 55.13 Species–Area Plots Quantify the Relationship between Species Richness and Habitat Area. Graph plotting the sizes of islands in the South Pacific (horizontal axis) versus the number of bird species present on each island (vertical axis). Notice the log-log scale. The dashed arrows show the number of bird species that are expected to live on an island with an area of 10,000 km^2.

To understand how this work is done, consider that recent estimates of habitat destruction in the world's tropical wet forests suggest that about 10 percent of the original area covered by these biomes is being lost each decade. It is not unreasonable, then, to project that 90 percent of the world's tropical forest habitat will be lost over the next century.

To predict how many species will be driven to extinction as a result, biologists use species–area curves like the one shown in **Figure 55.13.** This graph was generated by a biologist who analyzed the number of bird species found on islands in the Bismarck Archipelago near New Guinea, in the South Pacific. The graph shows the number of species on islands of various sizes. Notice that both axes on the graph are logarithmic. The solid line drawn through the points is described by the function $S = (18.9)A^{0.15}$, where S is the number of species and A is the area. In this island group, the function indicates that each tenfold increase in island area increases the number of bird species present by about 40 percent; a tenfold decline in area reduces bird species by about 30 percent.

These data turn out to be typical for other habitats and taxonomic groups as well. When biologists have plotted species–area relationships for plants, butterflies, mammals, or birds from islands or continental habitats around the globe, the relationship is consistently described by a function of the form $S = cA^z$. The c term is a constant that scales the data. Its value is high in species-rich areas, such as coral reefs or tropical wet forests, and low in species-poor areas, such as arctic tundras. The exponent z represents the slope of the line on a log-log plot. Thus, z describes how rapidly species numbers change with area. Typically, z is about 0.25.

To understand how biologists use these analyses to project extinction rates, study Figure 55.13 again. Ask yourself: If 90 percent of the habitats were destroyed—for example, if *A* were reduced from 10,000 km^2 to 1000 km^2—what percentage of species would disappear? According to the graph, the number of bird species in the Bismarck Archipelago would drop from about 75 to about 53. Thus, the answer is roughly 30 percent. This is an important conclusion. If z is higher than 0.25 for taxa that inhabit tropical wet forests and other threatened habitats, the prediction is that more than 30 percent of all species will be wiped out in the next 100 years.

Check Your Understanding

If you understand that...

- Until recently, most human-induced extinctions have occurred on islands and were due to overhunting or the impact of introduced species.
- Although invasive species remain a major threat to biodiversity, the majority of current problems are on continents and are due to habitat destruction and fragmentation.
- Current and projected extinction rates can be estimated from direct counts and from an analysis of species–area relationships.

You should be able to...

1) Explain why fragmentation reduces habitat quality and leads to genetic problems.
2) Use the species–area relationship to compare the proportion of species lost when 90 percent of a 100,000 km^2 habitat is destroyed in a region where $c = 0.19$ and $z = 0.20$ versus the same-sized region where $c = 0.19$ and $z = 0.25$.

55.4 Preserving Biodiversity

The conservation of biodiversity is among the most urgent but least-recognized issues facing us today. Unlike ozone holes, acid rain, phosphorus pollution, global warming, and other environmental problems that humans have recently begun to address, extinction is irreversible. The only solution to the biodiversity crisis is to prevent the loss of alleles, species, and ecosystems.

Solving any global problem requires a common goal to be defined. In the case of preserving biodiversity, the objective is to sustain diverse communities in natural landscapes while supporting the extraction of resources required to maintain the health and well-being of the human population.

In almost every case, the underlying causes of the biodiversity crisis are political and economic pressures that encourage short-term overexploitation of land and other resources and discourage **sustainability**—the managed use of resources at a rate only as fast as the rate at which they are replaced. What reasons can biologists give for preserving biodiversity, in addition to the direct economic benefits listed in **Box 55.3**? From a biological point of view, why does biodiversity matter?

Why Is Biodiversity Important?

The benefits of biodiversity extend beyond the direct use of diverse genes and species by humans to include **ecosystem services**—processes that increase the quality of the abiotic environment. Recall from Chapter 10 that green plants and other photosynthetic organisms produce the oxygen we breathe, and from Chapter 30 that the presence of land plants builds soil, reduces soil erosion, moderates local temperature and wind conditions, and increases the volume of water retained in lakes, streams, and soils. Species from throughout the tree of life are involved in cycling nitrogen, carbon, and other nutrients through ecosystems (see Chapters 28–31 and 54). But do ecosystems improve the abiotic environment more when biodiversity is high?

The effort to answer this question began in the mid-1980s and has blossomed into one of the most active research frontiers in community ecology and conservation biology.

Biodiversity Increases Productivity **Figure 55.14** summarizes a classic experiment on how species richness affects one of the most basic aspects of an ecosystem function: the production of biomass. As the "Experimental Setup" portion of the figure shows, David Tilman and colleagues classified 32 grassland plant species into five functional categories. The functional groups differ by the timing of their growing season and by whether they allocate most of their resources to manufacturing woody stems or seeds. The researchers planted plots measuring 13 m × 13 m with a mixture of between 0 and 32 randomly chosen species, representing from 0 to 5 functional categories. After 2 years of plant growth, the researchers harvested and weighed the aboveground tissues from the plots. The data provided an index of **net primary productivity** (**NPP**)—the total amount of photosynthesis per unit area per year that ends up in biomass.

Tilman's group compared the productivity of plots with different numbers of species and functional groups, and found that both the number and type of species present had important effects. Plots with more species and with a wider diversity of functional groups were more productive. Total biomass leveled off as species richness and functional diversity increased, however. This observation suggests that increasing species diversity improves ecosystem function only up to a point. The data support the hypothesis that at least some species in ecosystems are redundant.

Follow-up experiments in an array of ecosystems not only supported the conclusion that species richness has a positive impact on NPP, but showed that several causal mechanisms may be at work.

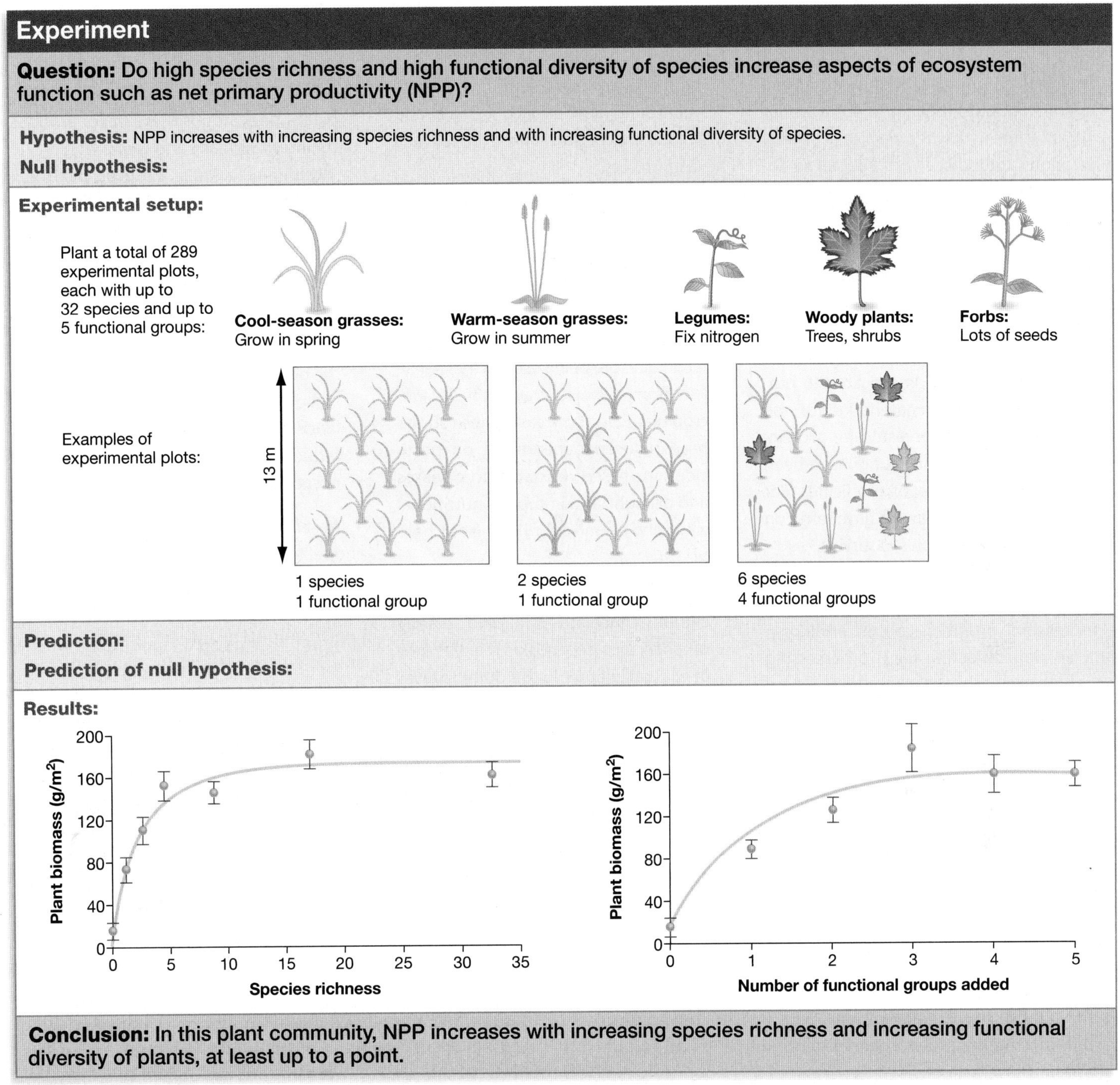

Experiment

Question: Do high species richness and high functional diversity of species increase aspects of ecosystem function such as net primary productivity (NPP)?

Hypothesis: NPP increases with increasing species richness and with increasing functional diversity of species.

Null hypothesis:

Experimental setup:

Plant a total of 289 experimental plots, each with up to 32 species and up to 5 functional groups:

Cool-season grasses: Grow in spring

Warm-season grasses: Grow in summer

Legumes: Fix nitrogen

Woody plants: Trees, shrubs

Forbs: Lots of seeds

Examples of experimental plots:

13 m

1 species
1 functional group

2 species
1 functional group

6 species
4 functional groups

Prediction:

Prediction of null hypothesis:

Results:

Plant biomass (g/m²)
200
160
120
80
40
0
0 5 10 15 20 25 30 35
Species richness

Plant biomass (g/m²)
200
160
120
80
40
0
0 1 2 3 4 5
Number of functional groups added

Conclusion: In this plant community, NPP increases with increasing species richness and increasing functional diversity of plants, at least up to a point.

FIGURE 55.14 Evidence that the Productivity of Ecosystems Depends on the Number and Types of Species Present.

EXERCISE Write in the null hypothesis and both sets of predictions.

- *Resource use efficiency* Diverse assemblages of plant species make more efficient use of the sunlight, water, and other resources available and thus lead to greater overall productivity. For example, some prairie plants extract water near the soil surface, while others use water available a meter or more below the surface. When species diversity is high, more overall water is used and more photosynthesis can occur.

- *Facilitation* Certain species or functional groups facilitate the growth of other species by providing them with nutrients, partial shade, or other benefits. In prairies, the presence of onion-family plants may discourage herbivores, or decaying roots from nitrogen-fixing species may fertilize other species.

- *Sampling effects* In many habitats it is common to observe that one or two species are extremely productive. If the number of species in a study plot is low, it is likely that the "big producer" will be missing and NPP will be low. But if the number of species in a study plot is high, then it is likely

BOX 55.3 Economic Benefits of Biodiversity

The first large-scale, highly organized human societies began to form about 10,000 years ago, when people at several locations around the world began to domesticate wild plants and farm their food. Since that time, wild species have provided the raw material to fuel the continued development of human societies. Vast tracts of forest have been felled for building materials and fuel; wild plants have been selectively bred to yield food and fibers; animals have been domesticated to provide food, labor, and material goods; the oceans have been harvested for protein; and plants, animals, and fungi have been processed as sources of medicines. For thousands of years, humans have relied on a diversity of wild species to survive.

The direct use of biodiversity continues today:

- Research programs collectively known as bioprospecting focus on assessing bacteria, archaea, plants, fungi, and frogs as novel sources of drugs or ingredients in consumer products. Bioprospecting has benefited from the recent explosion of genetic and phylogenetic information, because biologists can now search genomes from a wide array of species to find alleles with desired functions. Recent advances in biotechnology facilitated the development of a new painkiller from the paralyzing sting of tropical cone snails, and a blood anticoagulant from the saliva of vampire bats.
- Agricultural scientists are preserving diverse strains of crop plants in seed banks and continue to use wild relatives of domesticated species in breeding programs aimed at improving crop traits. In addition, genetic engineering techniques are being used to transfer alleles from a diverse array of species into crop plants. In some cases these efforts have reduced dependence on pesticides, increased resistance to diseases and drought, and improved nutritional value and overall crop yields (see Chapter 19).
- The production of almonds, apples, cherries, chocolate, alfalfa, and an array of other crops depends on the presence of wild pollinators. In the United States alone, insect-pollinated crops produce $40 billion worth of products annually.
- Strategies for cleaning up oil spills, abandoned mines, and contaminated industrial sites are incorporating bioremediation—the use of bacteria, archaea, and plants to metabolize pollutants and render them harmless (see Chapters 28 and 30).
- Recreation based on visiting wild places, or ecotourism, is a major industry internationally and is growing rapidly. In South Africa, for example, the number of tourists visiting wildlife preserves increased from 454,428 in 1986 to over 6 million in 1999. In 2004, ecotourism grew three times faster than the tourism industry as a whole.
- High ecosystem diversity—particularly the presence of forests or grasslands on steep slopes and the presence of wetlands in low-lying areas—dramatically reduces flood damage and the danger posed by mudslides.

An array of biologists, philosophers, and religious leaders argue that, in addition to making money from biodiversity, humans have an ethical obligation to preserve species and ecosystems. Their position is that organisms have intrinsic worth, and that humans diminish the world by extinguishing species and destroying ecosystems. If this is so, then extinction is a moral issue as well as a biological and economic problem. One of the most important reasons to preserve biodiversity may simply be that it is the right thing to do.

the big producer is present and NPP will be high. Simply due to sampling, then, high-species plots will tend to outproduce low-species plots.

The three mechanisms listed are not mutually exclusive—several can operate at the same time. Further, long-term experiments like the one diagrammed in Figure 55.14 have shown that the reason for the pattern can change over time. As plants mature, resource use efficiency and facilitation may become more important than sampling effects.

Does Biodiversity Lead to Stability? When biologists refer to the stability of a community, they mean its ability to (1) withstand a disturbance without changing, (2) recover to former levels of productivity or species richness after a disturbance, and (3) maintain productivity and other aspects of ecosystem function as conditions change over time. **Resistance** is a measure of how much a community is affected by a disturbance. **Resilience** is a measure of how quickly a community recovers following a disturbance.

To test the hypothesis that diversity increases resistance and resilience, the team that did the work highlighted in Figure 55.14 followed up on a natural experiment. In 1987–1988 a severe drought hit their study site. When the drought ended, the team asked whether species richness affected the response to the disturbance. The results are shown in **Figure 55.15**. The graph documents the change in total biomass that occurred from the year prior to the drought until the height of the drought. This quantity reflects how resistant the community is to disturbance. A completely resistant community would show no change. As predicted, drought resistance appeared to be higher in more diverse communities than in less diverse ones.

Experiment

Question: Are more species-rich communities more stable than less-diverse communities?

Hypothesis: Resistance to disturbance should increase with increasing species richness.

Null hypothesis: There is no relationship between species richness and resistance.

Experimental setup:

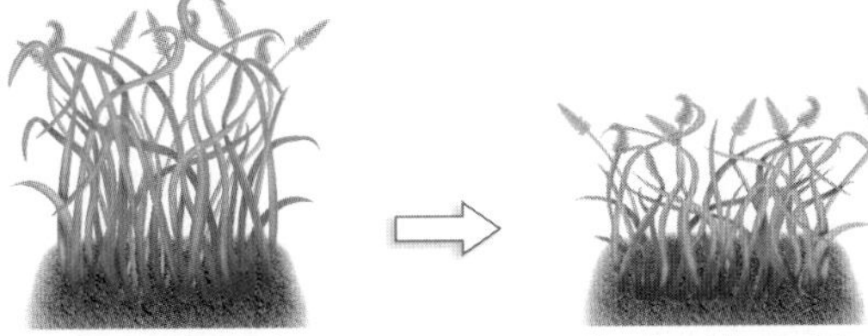

Before drought **During drought**

Compare biomass of experimental plots before drought and at peak of drought. (This was a natural experiment—severe drought just happened to occur during study.)

Prediction: Plots that were more species rich before the drought will be more resistant to change.

Prediction of null hypothesis: All the plots will have similar resistance regardless of species richness before the drought.

Results:

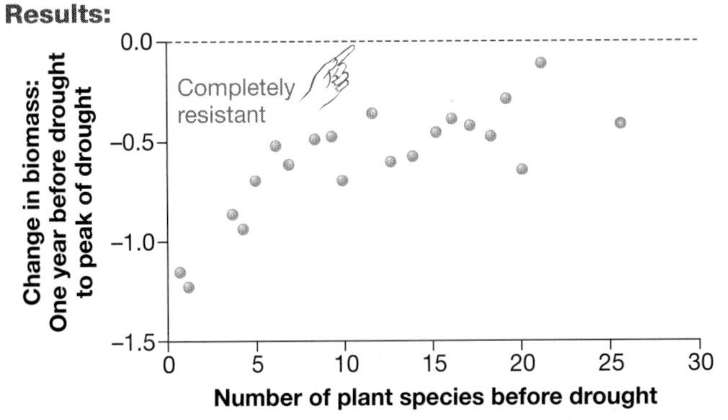

Conclusion: Resistance to disturbance increases with increasing species richness.

FIGURE 55.15 Evidence that Species-Rich Communities Resist Disturbance. Data support the hypothesis that diverse communities are more stable, meaning they change less during a disturbance, than less-diverse communities are.

● **QUESTION** The 0.0 values on the vertical graph axes are labeled "Completely resistant." Why?

In addition, the group analyzed the change in biomass in each plot four years after the drought versus biomass prior to the drought. This analysis focused on how resilient the community is. A completely resilient community would recover quickly from the disturbance and have the same biomass at both times. The data indicated that most plots containing five or fewer species showed a significant lowering of biomass after the disturbance, indicating that they had not recovered. But in all of the plots that contained more than five species, biomass after the drought was the same as biomass prior to the disturbance.

More recent experiments on California grasslands suggests that species-rich ecosystems may also be less prone to invasion by exotic species. Taken together, these results lead biologists to be increasingly confident that species richness has a strongly positive effect on how ecosystems function. In North American grasslands, at least, communities that are more diverse appear to be more productive, more resistant to disturbance and invasion, and more resilient than communities that are less diverse. Increased richness increases the services provided by ecosystems. By implication, biologists can infer that if ecosystems are simplified by extinctions, then productivity and other attributes might decrease.

Designing Effective Protected Areas

When biologists first recognized the growing threats to biodiversity worldwide, they joined with government agencies, economists, community leaders, private landowners, and others to set aside protected areas. To assess the progress of this effort, the International Union for the Conservation of Nature (IUCN) sponsors the World Parks Congress every decade. In 1992 the Congress met in Caracas, Venezuela, and established a goal of setting aside 10 percent of Earth's land surface in protected areas. In 2003 the Congress met in Durban, South Africa, and announced that this goal had been surpassed: Protected areas covered 11.5 percent of Earth's terrestrial surface.

How efficiently is the existing network of protected areas protecting biodiversity? Researchers are attempting to answer this question via a geographic approach called the **Gap Analysis Program** (**GAP**). A GAP analysis tries to identify gaps between geographic areas that are particularly rich in biodiversity and areas that are actually managed for the preservation of biodiversity. One GAP analysis combined data sets on the distribution of mammals, birds, amphibians, and freshwater turtles with a map of world protected areas. The analysis revealed that many species' ranges occur completely outside any protected areas. It also pinpointed regions in Mexico, Madagascar, and elsewhere where the gap between species' ranges and protected areas is particularly high.

To date, most GAP analyses suggest that, because relatively few species richness hotspots are included in existing protected areas, the 11.5 percent of Earth's surface area that is now being managed for biodiversity will not be enough to conserve many species. Efforts to fill these gaps are now focused on preserving species richness hotspots and centers of endemism like those illustrated in Figure 55.4.

In addition to working on where reserves are set up, biologists are focusing more attention on how reserves are designed. For example, it is often impossible to preserve large areas of high-quality habitat. In most cases, biologists have to accept the fact that habitats are fragmented and species exist as metapopulations. Given this reality, what is the best way to prevent small, isolated populations from going extinct?

The leading hypothesis in reserve design was to make sure that strips of undeveloped habitat, called **wildlife corridors**, connected populations that would otherwise be isolated. In some cases wildlife corridors were as simple as a walkway under a major highway, so that animals could move from one side to the other without being killed. By facilitating the movement of individuals, the goals of corridors were to (1) allow areas to be recolonized if a species was lost, and (2) introduce new alleles that would counteract the deleterious effects of genetic drift and inbreeding, detailed in Box 55.2.

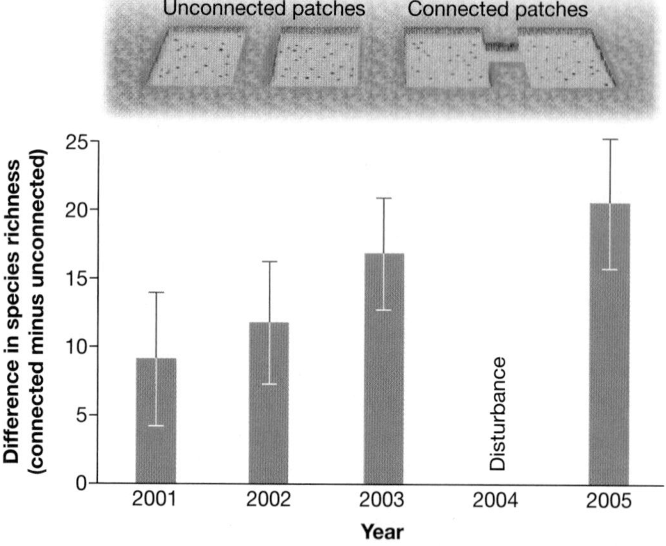

FIGURE 55.16 Experimental Evidence that Wildlife Corridors Work. Species richness could not be assessed in 2004 because the study plots were burned as part of a program to restore natural habitats.

EXERCISE Draw and label bars showing what this graph would look like if corridors had no effect on species richness.

Do corridors work? A recent experiment suggests that the answer is yes. A research team established a series of restored natural areas in the middle of a large, species-poor pine plantation. This was the equivalent of restoring patches of grassland habitat in the middle of an enormous cornfield. Some of the restoration sites were connected by corridors, while others were isolated. By monitoring the composition of species inside each habitat patch over time, the group was able to show that connected patches are steadily gaining more species over time compared to unconnected patches (**Figure 55.16**). This is strong evidence that wildlife corridors can increase overall species richness in a metapopulation.

Experiments like these bring us to the forefront of work on biodiversity and conservation biology. Your generation is facing the most serious global environmental crisis in the history of our species. As someone with a background in biology, you have the intellectual tools to help solve it. In addition to studying life, biologists have to save it.

Check Your Understanding

If you understand that...

- Experimental plots with higher species richness tend to have higher productivity and show increased stability in response to disturbance or changed abiotic conditions.
- Wildlife corridors are an effective way to minimize the loss of subpopulations from a metapopulation due to (1) genetic problems, and (2) catastrophic events.

You should be able to...

Design an experiment to test the hypothesis that species-rich areas are more effective at building soil, retaining soil nutrients, and minimizing soil erosion than are species-poor areas.

Chapter Review

SUMMARY OF KEY CONCEPTS

Biodiversity can be analyzed at the genetic, species, and ecosystem levels.

Genetic diversity is well characterized within some species, but biologists are only beginning to use genome sequencing techniques to explore the extent of genetic diversity in ecosystems. Research on the total number of species alive today is also at a preliminary stage, with estimates ranging from 10 to 100 million. To date, the message from efforts to characterize biodiversity is that it is much more extensive than expected and that a great deal remains to be learned.

You should be able to describe conditions that maximize genetic, species, and ecosystem diversity.

If recent rates of extinction due to human population growth and habitat destruction continue, a mass extinction will occur.

Historically, most human-caused extinctions have occurred on islands because of direct exploitation or the introduction of exotic herbivores and predators. Habitat destruction is currently the leading cause of extinctions, however. Experiments in the Brazilian Amazon have shown that habitat loss leads not only to a rapid decline in biodiversity but also to a decline in the quality of the remaining habitats due to fragmentation.

To estimate how many species will go extinct in the near future, biologists combine data on current rates of habitat loss—usually estimated from satellite images taken over time—with data on the average number of species found in habitats of a given size. These species–area analyses suggest that if 90 percent of habitats are destroyed as expected during the next century, then over 30 percent of all species will become extinct. According to data on the rates of extinction in birds and other well-studied groups, it is likely that 60 percent of all species will be wiped out within 500 years.

You should be able to explain why species that have a metapopulation structure, due to habitat fragmentation, may be at higher risk of extinction than a species of similar abundance that exists in a large, contiguous habitat.

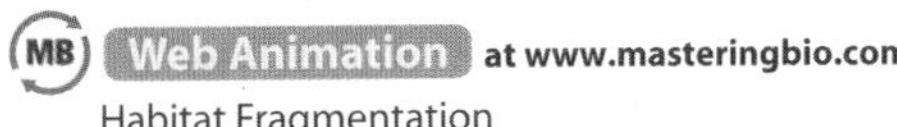

Habitat Fragmentation

Humans depend on biodiversity for the products that wild species provide and for ecosystem services that protect the quality of the abiotic environment.

Species diversity is important for maintaining the productivity of natural ecosystems and their ability to build and hold soil, moderate local climates, retain and cycle nutrients, retain surface water and recharge groundwater, prevent flooding, and produce oxygen. At the ecosystem level, experiments have shown that high species richness increases aspects of ecosystem function such as productivity, resistance to disturbance, and ability to recover from disturbance. Humans also gain direct economic benefits from fishing, forestry, agriculture, tourism, and other activities that depend on biodiversity.

You should be able to outline a plan for restoring an abandoned airstrip that would maximize its ability to deliver ecosystem services.

QUESTIONS

Test Your Knowledge

1. What does a species–area plot show?
   a. The overall distribution, or area, occupied by a species.
   b. The relationship between the body size of a species and the amount of territory or home range it requires.
   c. The number of species found, on average, in tropical versus northern areas.
   d. The number of species found, on average, in a habitat of a given size.

2. What does a GAP analysis do?
   a. It compares the current distributions of species with the locations of preserved habitats.
   b. It quantifies gross *a*boveground *p*roductivity.
   c. It uses data on the rates at which lists of threatened species are growing to project the rate of future extinctions.
   d. It uses genome sequencing techniques to quantify genetic (allelic) diversity in an ecosystem.

3. What is the difference between species richness and species diversity?
   a. Species diversity incorporates data on species interactions.
   b. Species diversity takes relative abundance into account.
   c. Species diversity is weighted by taxonomic (phylogenetic) diversity.
   d. Species diversity is adjusted for conservation priority (endangered status of species present).

4. What is a biodiversity "hotspot?"
   a. an area where an all-taxon survey is underway
   b. an area where an environmental sequencing study has been completed
   c. a habitat with high NPP
   d. an area with high species richness

5. Why do small populations become inbred?
   a. They are often part of a metapopulation structure.
   b. Genetic drift becomes a prominent evolutionary force.
   c. Over time, all individuals become related.
   d. Natural selection does not operate efficiently in small populations.

6. What is the primary cause of endangerment in marine environments?
   a. overexploitation
   b. pollution
   c. global warming
   d. invasive species

Test Your Knowledge answers: 1. d; 2. a; 3. b; 4. d; 5. c; 6. a

Test Your Understanding

Answers are available at www.masteringbio.com

1. The primary cause of endangerment on continents has changed from direct exploitation to habitat loss. Explain why this change occurred.

2. Compare and contrast biodiversity at the genetic, species, and ecosystem levels. How do biologists analyze the extent of biodiversity at each of these levels?

3. Biologists claim that the all-taxa survey now under way at the Great Smoky Mountains National Park in the United States will improve their ability to estimate the total number of species living today. Discuss the benefits and limitations that this data set will provide in understanding the extent of global biodiversity.

4. How are species–area curves used to relate rates of habitat destruction to projected extinction rates?

5. Discuss evidence to support the hypotheses that species richness increases ecosystem functions such as productivity, resistance to disturbance, and resilience. Describe the logic behind the resource use efficiency, facilitation, and sampling effect hypotheses to explain why species richness increases productivity, and thus the ecosystem services that benefit humans.

6. Explain why the fragmentation of habitats reduces their ability to support biodiversity. Explain why the construction of wildlife corridors can help maintain biodiversity in a fragmented landscape.

Applying Concepts to New Situations

Answers are available at www.masteringbio.com

1. Projections for an impending mass extinction are contingent on the continuation of present trends in human population growth and in habitat destruction. Do you believe that these trends will continue? In general, are you optimistic or pessimistic about the future of biodiversity? Explain your logic.

2. You are helping design a series of reserves in a tropical country. List the steps you would recommend for gathering data and creating a plan that would protect a large number of species in a small amount of land.

3. The maps that follow chronicle the loss of old-growth forest (>200 years old) that occurred in Warwickshire, England, and in the United States. In your opinion, under what conditions is it ethical for conservationists who live in these countries to lobby government officials in Brazil, Indonesia, and other tropical countries to slow the rate of loss of old-growth forest?

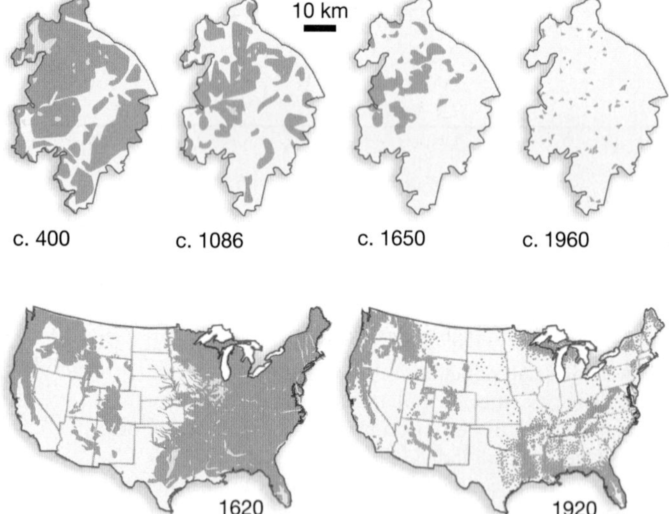

4. Make a list of characteristics that would render a species particularly vulnerable to extinction by humans. Make a list of characteristics that would render a species particularly resistant to pressure from humans. Try to think of an example of each type of species.

www.masteringbio.com is also your resource for • Answers to text, table, and figure caption questions and exercises • Answers to *Check Your Understanding* boxes • Online study guides and quizzes • Additional study tools including the *E-Book for Biological Science* 3rd ed., textbook art, animations, and videos.

Reading Graphs

Graphs are the most common way to report data, for a simple reason. Compared to reading raw numerical values in the form of a table or list, a graph makes it much easier to understand what the data mean.

Learning how to read and interpret graphs is one of the most basic skills you'll need to acquire as a biology student. As when learning piano or soccer or anything else, you need to understand a few key ideas to get started and then have a chance to practice—a *lot*—with some guidance and feedback.

Getting Started

To start reading a graph, ask yourself three questions:

1. *What do the axes represent?* As **Figure BS1.1a** shows, the horizontal axis of a graph is also called the *x*-axis or the abscissa. The vertical axis of a graph is also called the *y*-axis or the ordinate. Each axis represents a quantity that varies over a range of values. These values are indicated by the ticks and labels on the axis. In our example, the *x*-axis represents the width of broccoli stalks, in centimeters, while the *y*-axis represents the number of individuals in the sample. Note that the units being represented on each axis should *always* be clearly labeled or implied.

   To create a graph, researchers plot the independent variable on the *x*-axis and the dependent variable on the *y*-axis. The *independent* and *dependent* terms are appropriate because the values on the *y*-axis depend on the *x*-axis values. In our example, the researchers wanted to show the distribution of flowering stalk sizes in a large sample of individuals. Thus, the number of individuals present depended on the size of the flowering stalk. (This is similar to the way you would plot a graph showing the heights of students in your class.)

   In many graphs in biology, the independent variable is either time or the various treatments used in an experiment. In these cases, the *y*-axis records how some quantity changes as a function of time or as the outcome of the treatments applied to the experimental cells or organisms. The value on the *y*-axis depends on the value on the *x*-axis, but not vice versa.

(a) Read the axes—what's being plotted?

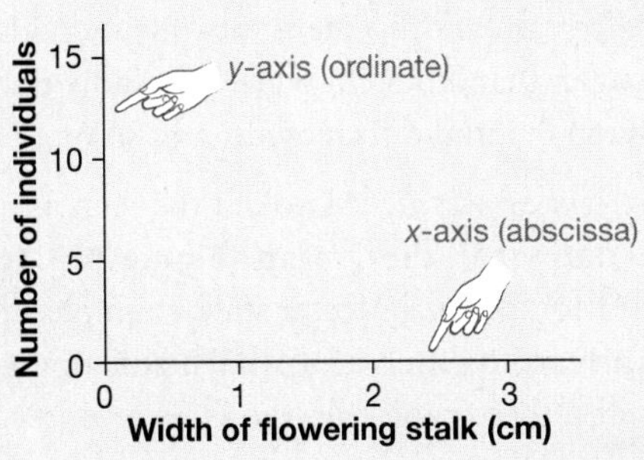

(b) Look at the bars or data points—what do they represent?

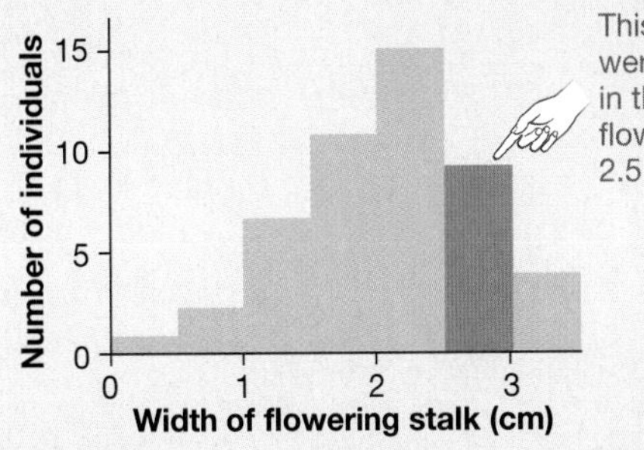

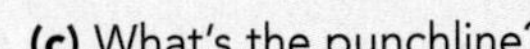

(c) What's the punchline?

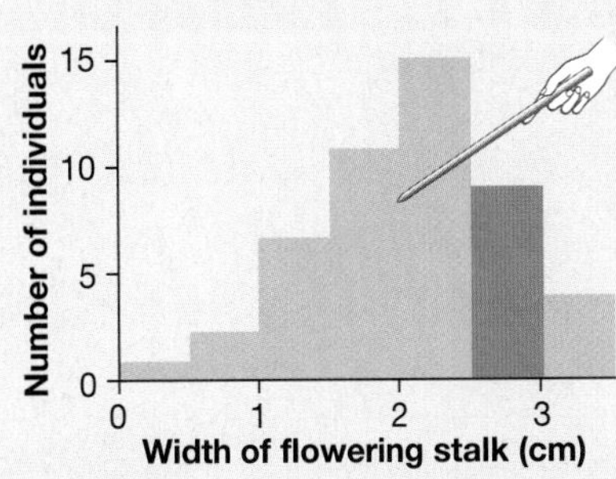

FIGURE BS1.1 Histograms Are a Common Way to Graph Data.

EXERCISE Draw what this histogram would look like if almost all the individuals in the sample had flowering stalks between 3.0 and 3.5 cm wide.

EXERCISE Suppose you measured the width of flowering stalks in a large sample of individuals over time—measuring the same individuals every week for 10 weeks. Draw a scatter plot predicting how the average width of flowering stalks might change over time.

2. *What do the bars or the data points represent?* Most of the graphs in this text are histograms like the one shown here, or bar charts—which are similar to histograms, but plot data that have discrete values instead of a continuous range of values—or scatter plots, where individual data points are plotted. Once you've read the axes, you need to figure out what each bar in the bar chart or each data point in the scatter plot represents and how the researchers determined it. In our broccoli stalk example, the bars in **Figure BS1.1b** represent the number of individuals in the sample with flowering stalks of a certain width. There was one individual with a stalk between 0 and 0.5 cm wide, two individuals with a stalk between 0.5 and 1.0 cm wide, and so on.
3. *What is the overall trend or message?* Look at the data as a whole, and figure out what they mean. **Figure BS1.1c** suggests an interpretation of the broccoli stalk example. If a bar chart plots values from different treatments in an experiment, ask yourself if these values are the same or different. If a scatter plot shows how some quantity changes over time, ask yourself if that quantity is increasing, decreasing, or staying the same.

Getting Practice

Working with this text will give you lots of practice with reading graphs—they appear in almost every chapter. In many cases we've put a little pointing hand, like your instructor's hand at the whiteboard, with a label that suggests an interpretation or draws your attention to an important point on the graph. In other cases, you should be able to figure out what the data mean on your own or with the help of other students or your instructor.

Make an effort to develop this skill. One of the most common complaints from professors who teach upper-level courses is, "My students can't read graphs!" When you become competent and confident at interpreting graphs, your performance will improve in this course, in subsequent courses, and on professional and graduate school admission tests.

Reading a Phylogenetic Tree

Phylogenetic trees show the evolutionary relationships among species, just as a genealogy shows the relationships among people in your family. They are unusual diagrams, however, and it can take practice to interpret them correctly.

To understand how evolutionary trees work, consider **Figure BS2.1**. Notice that a phylogenetic tree consists of branches, nodes, and tips. Branches represent populations through time. Nodes (also called forks) occur where an ancestral group splits into two or more descendant groups (see point B in Figure BS2.1). If more than two descendant groups emerge from a node, the node is called a polytomy (see node C). Tips (also called terminal nodes) are the tree's endpoints, which represent groups living today or a dead end—a branch ending in extinction. The names at the tips can represent species or larger groups such as mammals or conifers. Recall from Chapter 1 that a taxon (plural: *taxa*) is any named group of organisms. A taxon could be a single species, such as *Homo sapiens*, or a large group of species, such as Primates. Groups that occupy adjacent branches on the tree are called sister taxa.

The phylogenetic trees used in this text are all rooted—meaning the bottom, or most basal, node on the tree is the most ancient. To determine where the root on a tree occurs, biologists include one or more out-group species when they are collecting data to estimate a particular phylogeny. An out-group is a taxonomic group that is known to have diverged prior to the rest of the taxa in the study. In Figure BS2.1, "Taxon 1" is an out-group to the monophyletic group consisting of taxa 2–6. A monophyletic group consists of an ancestral species and all of its descendants. The root of a tree is placed between the out-group and the monophyletic group being studied. This position in Figure BS2.1 is node A.

Understanding monophyletic groups is fundamental to reading and estimating phylogenetic trees. Monophyletic groups may also be called lineages or clades and can be identified using the "one-snip test": If you cut any branch on a phylogenetic tree, all of the branches and tips that fall off represent a monophyletic group. Using the one-snip test, you should be able to convince yourself that the monophyletic groups on a tree are nested. In Figure BS2.1, for example, the monophyletic group comprising node A and taxa 1–6 contains a monophyletic group consisting of node B and taxa 2–6, which includes the monophyletic group represented by node C and taxa 4–6.

To put all these new terms and concepts to work, consider the phylogenetic tree in **Figure BS2.2**, which shows the relationships

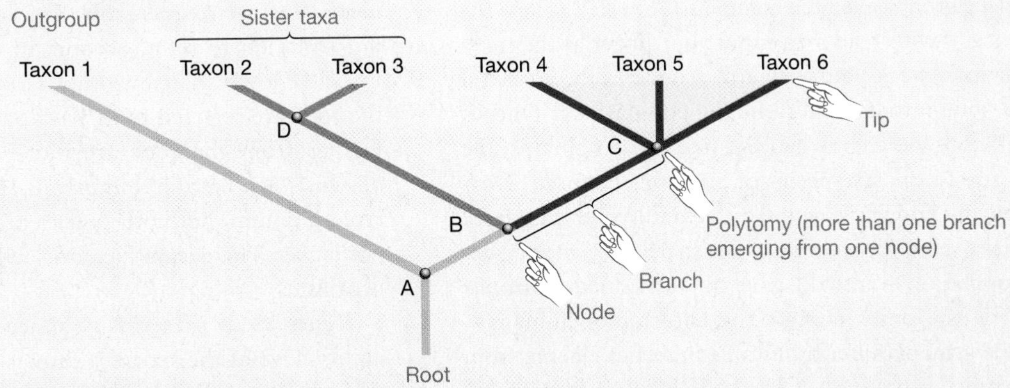

FIGURE BS2.1 Phylogenetic Trees Have Roots, Branches, Nodes, and Tips.
● **EXERCISE** Circle all nine monophyletic groups present.

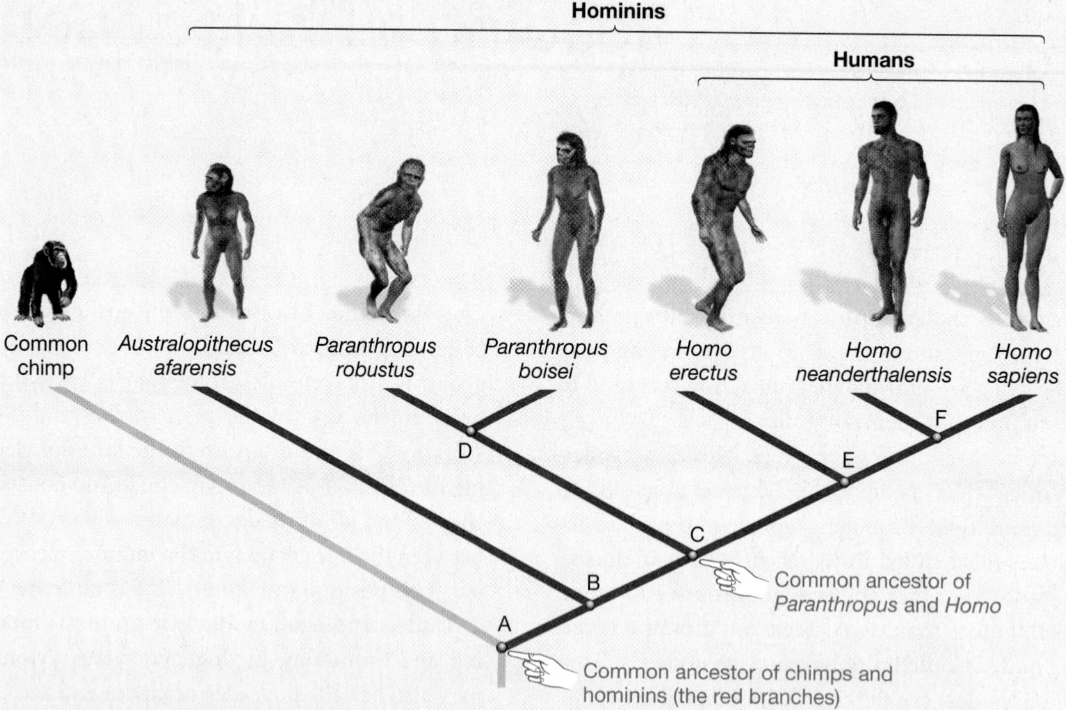

FIGURE BS2.2 An Example of a Phylogenetic Tree. A phylogenetic tree showing the relationships of species in the monophyletic group called hominins.

EXERCISE All of the hominins walked on two legs—unlike chimps and ali of the other primates. Add a mark on the phylogeny to show where upright posture evolved, and label it "origin of walking on two legs." Circle and label a pair of sister species. Circle the monophyletic group called hominins. Label an out-group to the monophyletic group called humans (species in the genus *Homo*).

between common chimpanzees and six human and humanlike species that lived over the past 5–6 million years. Chimps functioned as an out-group in the analysis that led to this tree, so the root was placed at node A. The branches marked in red identify a monophyletic group called the hominins.

To practice how to read a tree, put your finger at the tree's root and begin to work your way up. At node A, the ancestral population split into two descendant populations. One of these populations eventually evolved into today's chimps; the other gave rise to the six species of hominins pictured. Now continue moving your finger up the tree until you hit node C. It should make sense to you that at this splitting event, one descendant population eventually gave rise to two *Paranthropus* species, while the other became the ancestor of humans—species in the genus *Homo*. If multiple branches emerge from a node, creating a polytomy, it would be because the populations involved split from one another so quickly that it is not possible to tell which split off earlier or later.

As you study Figure BS2.2, you should consider a couple of important points. First, there are many equivalent ways of drawing this tree. For example, this version shows *Homo sapiens* on the far right. But the tree would be identical if the two branches emerging from node E were rotated 180°, so that the species appeared in the order *Homo sapiens*, *Homo neanderthalensis*, *Homo erectus*. Trees are read from root to tips, not from left to right. Second, no species on any tree is any higher or lower than any other. Groups that branch off close to the root are referred to as basal, and groups that branch off further from the root are more derived. *Australopithecus afarensis* is a basal hominin and *Homo erectus* is a more derived hominin, but both species are at tips—*Homo erectus* is not higher. There is, in fact, no such thing as a higher or lower organism.

Figure BS2.3 presents a chance to test your tree-reading ability. Five of the six trees shown in this diagram are identical in terms of the evolutionary relationships they represent. One differs. You should be able to answer the question posed in the figure's caption and then, on tree (a), circle three nested monophyletic groups and label a node, a branch, a tip, and the out-group.

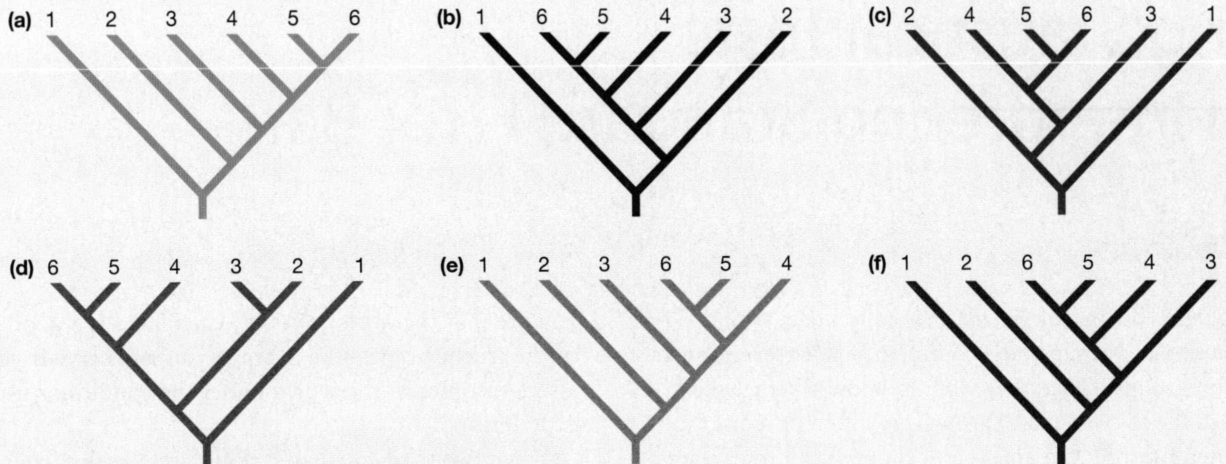

FIGURE BS2.3 Alternative Ways of Drawing the Same Tree.

QUESTION Five of these six trees describe exactly the same relationships among taxa 1 through 6. Identify the tree that is different from the other five.

Answer: The unlike tree is (d).

BioSkills 3

Using Statistical Tests and Interpreting Standard Error Bars

When biologists do an experiment, they collect data on individuals in a treatment group and a control group (or several such comparison groups). Then they want to know whether the individuals in the two (or more) groups are different. For example, Chapter 1 introduced an experiment in which researchers fed three types of fruits to two different types of fruit-eating organisms—cactus mice and birds called thrashers—and measured how much of each type of fruit the animals ate.

Figure BS3.1 graphs the data for cactus mice and thrashers. The treatments are plotted on the *x*-axes as hackberry (H), non-pungent chilies (NP), and pungent chilies (P). The heights of the bars on the graphs indicate the average percentage of each type of fruit consumed by the 5 mice and 10 thrashers tested. The thin "I-beams" on each bar indicate the standard error of each average. The standard error is a quantity that indicates the uncertainty in the calculation of an average. For example, if two of the mice ate all of the hackberry offered, two ate none, and one ate half, then your estimate of how much an average mouse consumes would be 50 percent. The standard error of that average would be very large, though, because the amounts that the mice ate ranged from 0 to 100 percent. In contrast, if two of the mice ate 49 percent of the hackberry, two ate 51 percent, and one ate half, the average would still be 50 percent but the standard error would be very small. In effect, the standard error quantifies how confident you are that the average you've calculated is a good estimate of the true average—the average you'd observe if you tested all cactus mice in the world under the conditions used in the experiment.

Once they had calculated these averages and standard errors, the biologists wanted to know the answers to two questions: Do cactus mice eat all three types of fruit equally? And, do thrashers eat all three types of fruit equally?

After looking at the data, you might conclude that cactus mice ate different amounts of the fruits offered—specifically, that they ate much more hackberry than non-pungent chilies or pungent chilies. You might be less sure about suggesting that thrashers ate less hackberry and more of the non-pungent and pungent chilies. But how could you come to conclusions like this rigorously, instead of subjectively?

The answer is to use a statistical test. The first step in a statistical test is to specify the null hypothesis, which is that there is no difference among the groups. The second step is to calculate a test statistic, which is a number that characterizes the size of the difference among the groups. In this case, the test statistic compares the actual differences in amount of fruit consumed to the differences predicted by the null hypothesis. The third step is to determine the probability of getting a test statistic as large as the one calculated just by chance. The answer comes from a reference distribution—a mathematical function

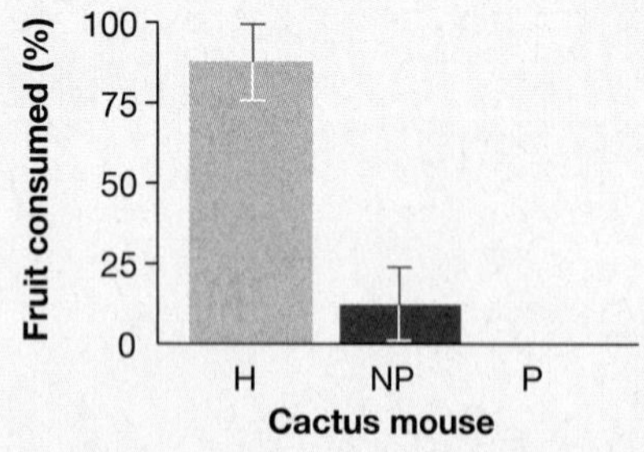

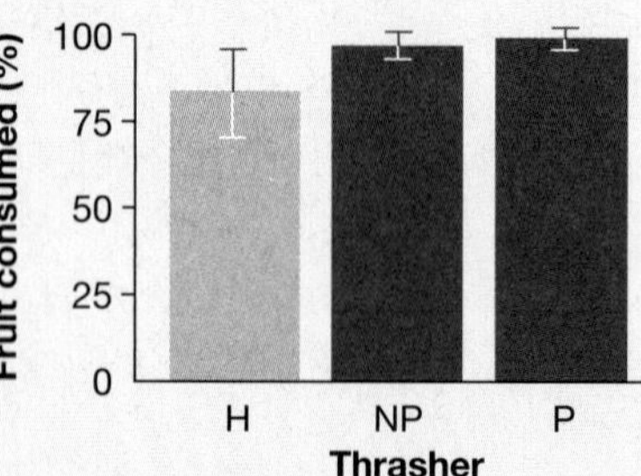

FIGURE BS3.1 Standard Error Bars Indicate the Uncertainty in an Average.

that specifies the probability of getting various values of the test statistic if the null hypothesis is correct. (If you take a statistics course, you'll learn which reference distributions are relevant to different types of data.)

You are very likely to see small differences among treatment groups just by chance—even if no differences actually exist. If you flipped a coin 10 times, for example, you are unlikely to get exactly five heads and five tails, even if the coin is fair. A reference distribution tells you how likely you are to get each of the possible outcomes of the 10 flips if the coin is fair, just by chance.

In this case, the reference distribution indicated that if the null hypothesis of no actual fruit preference is correct, you would see differences as large as those observed for mice only 0.8 percent of the time just by chance. For thrashers, though, you would expect to see differences as large as those observed in the experiment 22 percent of the time, just by chance, if the null hypothesis is correct. By convention, biologists consider a difference among treatment groups to be statistically significant if you have less than a 5 percent probability of observing it just by chance. Based on this convention, the researchers were able to claim that the null hypothesis is not correct for cactus mice—they really do prefer hackberries over chilies. But the null hypothesis is correct for thrashers—the data indicate that they show no preference among the three fruits.

It is likely that you'll be doing actual statistical tests early in your undergraduate career. To use this text, though, you only need to be aware of what statistical testing does. And you should take care to inspect the standard error bars on graphs in this book. As a *very* rough rule of thumb, averages often turn out to be significantly different, according to an appropriate statistical test, if two times the standard errors do not overlap.

BioSkills 4

Reading Chemical Structures

If you haven't had much chemistry yet, learning basic biological chemistry can be a challenge. One of the stumbling blocks is simply being able to read chemical structures efficiently and understand what they mean. This skill will come much easier once you have a little notation under your belt and you understand some basic symbols.

Atoms are the basic building blocks of everything in the universe, just as cells are the basic building blocks of your body. Every atom has a 1- or 2-letter symbol. The following table shows the symbols for most of the atoms you'll encounter in this book. You should memorize these. The table also offers details on how the atoms form bonds as well as how they are represented in visual models.

When atoms attach to each other by covalent bonding, a molecule forms. Biologists have a couple of different ways of representing molecules—you'll see each of these in the book and in class. A molecular formula like those in **Figure BS4.1a** simply lists the atoms present in a molecule, with subscripts indicating how many of each atom are present. If the formula has no subscript, only one of that type of atom is present. A methane (natural gas) molecule, for example, can be written as CH_4 because it consists of one carbon atom and four hydrogen atoms. Structural formulas like those in **Figure BS4.1b** show which atoms in the molecule are bonded to each other, with each bond indicated by a dash. The structural formula for methane indicates that each of the four hydrogen atoms forms one covalent bond with carbon, and that carbon makes a total of four covalent bonds. Note that single covalent bonds are symbolized by a single dash; double bonds are indicated by two dashes.

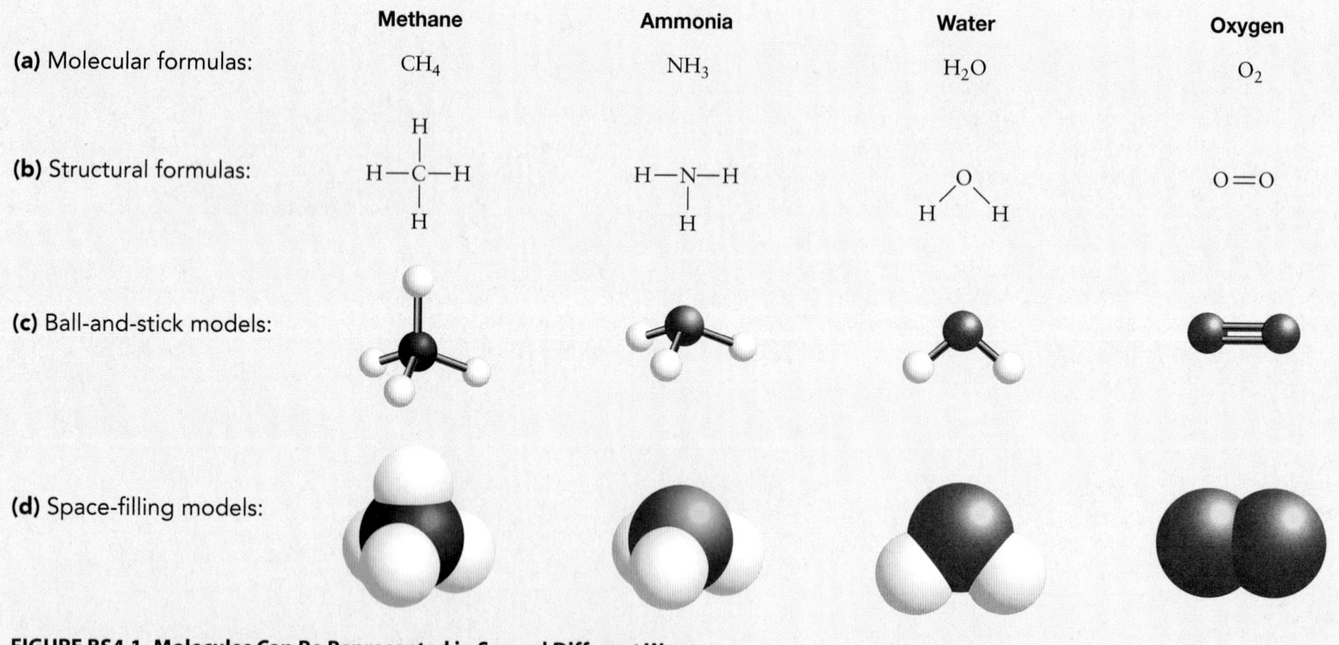

	Methane	Ammonia	Water	Oxygen
(a) Molecular formulas:	CH_4	NH_3	H_2O	O_2
(b) Structural formulas:	H—C—H (with H above and below C)	H—N—H (with H below N)	H—O—H (bent)	O=O
(c) Ball-and-stick models:				
(d) Space-filling models:				

FIGURE BS4.1 Molecules Can Be Represented in Several Different Ways.

EXERCISE Carbon dioxide consists of a carbon atom that forms a double bond with each of two oxygen atoms, for a total of four bonds. It is a linear molecule. Write carbon dioxide's molecular formula, then draw its structural formula, a ball-and-stick model, and a space-filling model.

Even simple molecules have distinctive shapes, because different atoms make covalent bonds at different angles. Ball-and-stick and space-filling models show the geometry of the bonds accurately. In a ball-and-stick model, a stick is used to represent each covalent bond (see **Figure BS4.1c**). In space-filling models, the atoms are simply stuck onto each other in their proper places (see **Figure BS4.1d**).

Atom	Symbol	Number of Bonds It Can Form	Standard Color Code*
Hydrogen	H	1	white
Carbon	C	4	black
Nitrogen	N	3	blue
Oxygen	O	2	red
Sodium	Na	1	(not used in this text)
Magnesium	Mg	2	(not used in this text)
Phosphorus	P	5	orange or purple
Sulfur	S	2	yellow
Calcium	Ca	2	(not used in this text)

*In ball-and-stick or space-filling models.

To learn more about a molecule when you look at a chemical structure, ask yourself three questions:

1. *Is the molecule polar—meaning that some parts are more negatively or positively charged than others?* Molecules that contain nitrogen or oxygen atoms are often polar, because these atoms have such high electronegativity (see Chapter 2). This trait is important because polar molecules dissolve in water.
2. *Does the structural formula show atoms that might participate in chemical reactions?* For example, are there charged atoms or amino ($—NH_2$) or carboxyl (—COOH) groups that might act as a base or an acid?
3. *In ball-and-stick and especially space-filling models of large molecules, are there interesting aspects of overall shape?* For example, is there a groove where a protein might bind to DNA, or a cleft where a substrate might undergo a reaction in an enzyme?

BioSkills 5

Using Logarithms

You have probably been introduced to logarithms and logarithmic notation in algebra courses, and you will encounter logarithms at several points in this course. Logarithms are a way of working with powers—meaning, numbers that are multiplied by themselves one or more times. Scientists use exponential notation to represent powers. For example,

$$a^x = y$$

means that if you multiply a by itself x times, you get y. In exponential notation, a is called the base and x is called the exponent. The entire expression is called an exponential function.

What if you know y and a, and you want to know x? This is where logarithms come in.

$$x = \log_a y$$

This equation reads, x is equal to the logarithm of y to the base a. Logarithms are a way of solving exponential functions. They are important because so many processes in biology (and chemistry and physics, for that matter) are exponential in nature. To understand what's going on, you have to describe the process with an exponential function and then use logarithms to work with that function.

Although a base can be any number, most scientists use just two bases when they employ logarithmic notation: 10 and e. Logarithms to the base 10 are so common that they are usually symbolized in the form $\log y$ instead of $\log_{10} y$. A logarithm to the base e is called a natural logarithm and is symbolized ln (pronounced *EL-EN*) instead of log. You write "the natural logarithm of y" as $\ln y$. The base e is an irrational number (like π) that is approximately equal to 2.718. Like 10, e is just a number. But both 10 and e have qualities that make them convenient to use in biology (and chemistry, and physics).

Most scientific calculators have keys that allow you to solve problems involving base 10 and base e. For example, if you know y, they'll tell you what $\log y$ or $\ln y$ are—meaning that they'll solve for x in our example above. They'll also allow you to find a number when you know its logarithm to base 10 or base e. Stated another way, they'll tell you what y is if you know x, and y is equal to e^x or 10^x. This is called taking an antilog. In most cases, you'll use the inverse or second function button on your calculator to find an antilog (above the log or ln key).

To get some practice with your calculator, consider the equation

$$10^2 = 100$$

If you enter 100 in your calculator and then press the log key, the screen should say 2. The logarithm tells you what the exponent is. Now press the antilog key while 2 is on the screen. The calculator screen should return to 100. The antilog solves the exponential function, given the base and the exponent.

If your background in algebra isn't strong, you'll want to get more practice working with logarithms—you'll see them frequently during your undergraduate career. Remember that once you understand the basic notation, there's nothing mysterious about logarithms. They are simply a way of solving exponential functions, which describe what happens when something is multiplied by itself a number of times—like cells that divide and then divide again and then again.

Using logarithms will also come up when you are studying something that can have a huge range of values, like the concentration of hydrogen ions in a solution or the intensity of sound that the human ear can detect. In cases like this, it's convenient to express the numbers involved as exponents. Using exponents makes a large range of numbers smaller and more tractable. For example, instead of saying that hydrogen ion concentration in a solution can range from 1 to 10^{-14}, the pH scale allows you to simply say that it ranges from 1 to 14. Instead of giving the actual value, you're expressing it as an exponent. It just simplifies things.

BioSkills 6

Making Concept Maps

A concept map is a graphical device for organizing and expressing what you know about a topic. It has two main elements: (1) concepts that are identified by words or short phrases and placed in a box or circle, and (2) labeled arrows that physically link two concepts and explain the relationship between them. The concepts are arranged hierarchically on a page, with the most general concepts at the top and the most specific ideas at the bottom.

The combination of a concept, a linking word, and a second concept is called a proposition. Good concept maps also have cross-links—meaning, labeled arrows that connect different elements in the hierarchy, as you read down the page.

Concept maps were initially developed by Joseph Novak in 1972 and have proven to be an effective studying and learning tool. They can be particularly valuable if constructed by a group, or when different individuals exchange and critique concept maps they have created independently. Although concept maps vary widely in quality and can be graded using objective criteria, there are many equally valid ways of making a high-quality concept map on a particular topic.

When you are asked to make a concept map in this text, you will usually be given at least a partial list of concepts to use. As an example, suppose you were asked to create a concept map on experimental design and were given the following concepts: results, predictions, control treatment, experimental treatment, controlled (identical) conditions, conclusions, experiment, hypothesis to be tested, null hypothesis. One possible concept map is shown in **Figure BS6.1**.

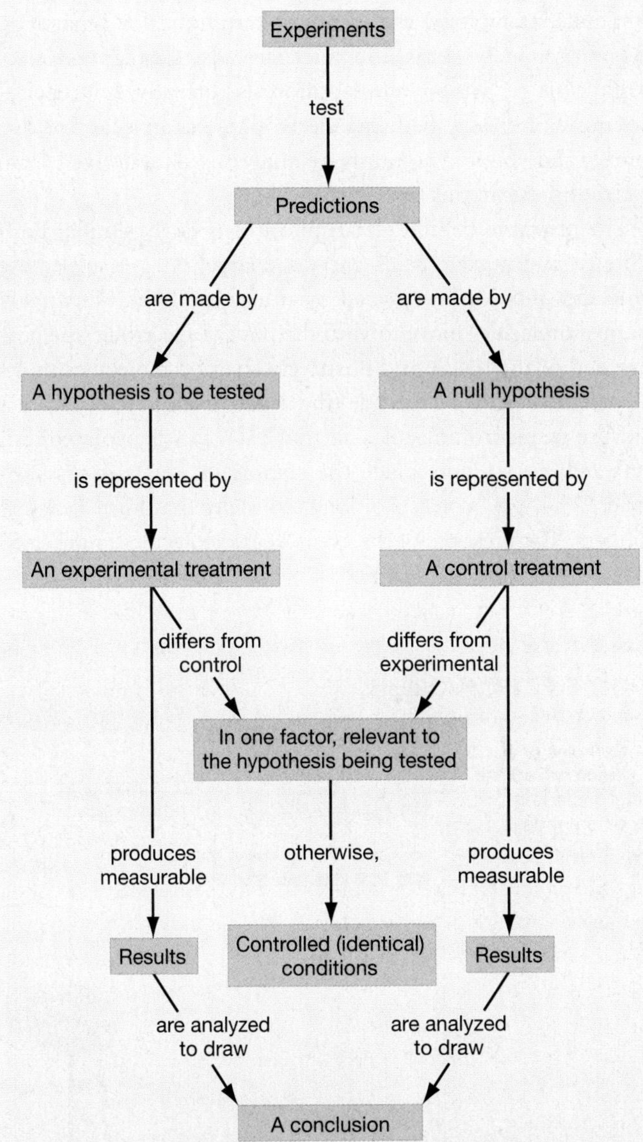

FIGURE BS6.1 A Concept Map on Principles of Experimental Design.

Dr. Doug Luckie, who uses concept maps extensively in his teaching at Michigan State University, points out that good concept maps have four qualities.

- They exhibit an organized hierarchy, indicating how each concept on the map relates to larger and smaller concepts.
- The concept words are specific—not vague.
- The propositions are accurate.
- There is cross-linking between different elements in the hierarchy of concepts.

As you practice making concept maps, go through these criteria and use them to evaluate your own work, as well as the work of fellow students. For more about concept maps, see the website for this text. It has additional information as well as links to concept-mapping software developed by Dr. Luckie.

BioSkills 7

Using Electrophoresis to Separate Molecules

In molecular biology, the standard technique for separating and analyzing proteins and nucleic acids is called gel electrophoresis or, simply, electrophoresis (literally "electricity-moving"). You may be using electrophoresis in a lab for this course, and you will certainly be analyzing data derived from electrophoresis in this text.

The principle behind electrophoresis is fairly simple. Both proteins and nucleic acids carry a charge. As a result, these molecules move when placed in an electric field. Negatively charged molecules move toward the positive electrode (the positive end of the field), and positively charged molecules move toward the negative electrode (the negative end). To separate a mixture of macromolecules so that each can be isolated and analyzed, researchers place the sample in a gelatinous substance. The "gel" consists of long molecules that form a matrix of fibers. The presence of the fibers keeps molecules in the sample from moving around randomly, but the gelatinous matrix also has pores through which the molecules can pass. When an electrical field is applied across the gel, the molecules in the well move through the gel toward an electrode. Molecules that are smaller or more highly charged for their size move faster than do larger or less highly charged molecules. As they move, the molecules separate by size and by charge.

Figure BS7.1 shows the electrophoresis setup used in an experiment investigating how RNA molecules polymerize, described in Chapter 4. Step 1 shows how investigators loaded samples of macromolecules, taken on different days during the experiment, into cavities or "wells" at the top of the gel slab. In this and many other cases, the researchers also filled a well with a sample containing fragments of known size, called a size standard or "ladder." In step 2 the researchers immersed the gel in a solution that conducts electricity and applied a voltage

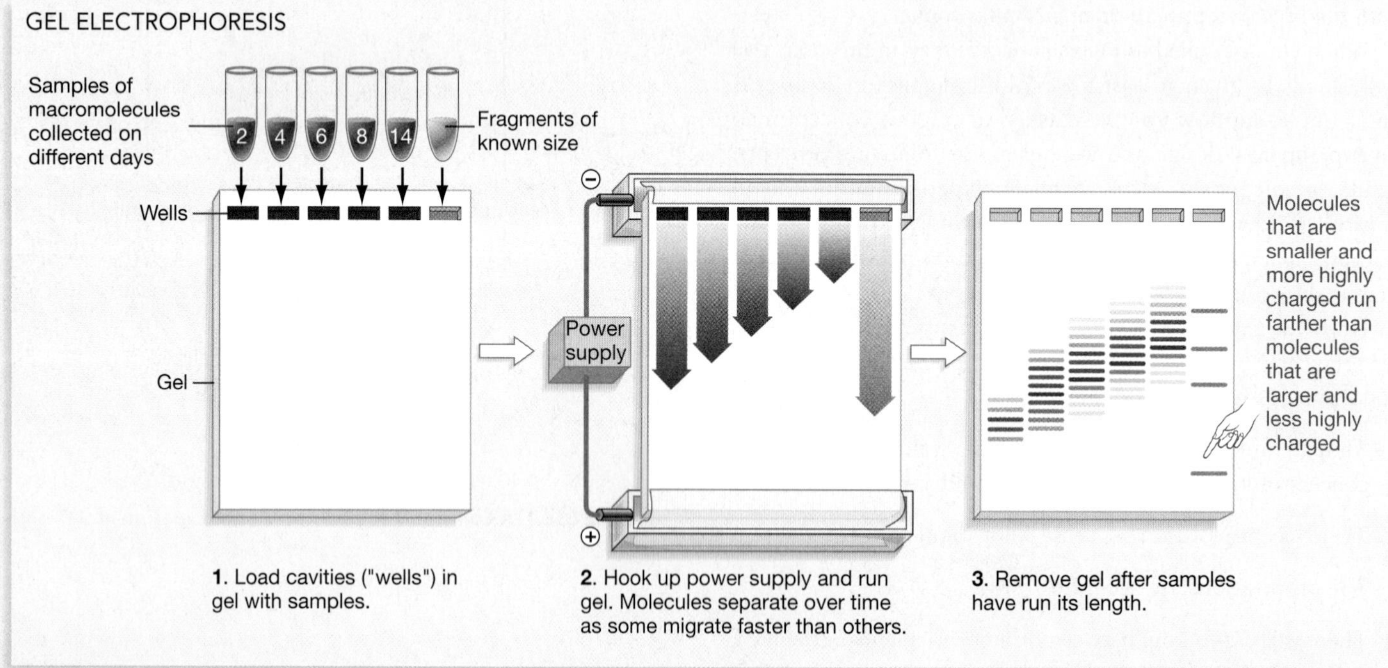

FIGURE BS7.1 Macromolecules Can Be Separated via Gel Electrophoresis.

QUESTION DNA and RNA run toward the positive electrode. Why are these molecules negatively charged?

across the gel. After the samples had run down the gel for some time (step 3), they removed the electric field. By then, molecules of different size and charge had separated from one another. In this case, small RNA molecules had reached the bottom of the gel. Above them were larger RNA molecules, which had run more slowly.

Once molecules have been separated in this way, they have to be detected. Often, proteins or nucleic acids can be stained or dyed. In this case, however, the researchers had attached a radioactive atom to the monomers used in the experiment, so they could visualize the resulting polymers by laying X-ray film over the gel. Because radioactive emissions expose film, a black dot appears wherever a radioactive atom is located in the gel. This technique for visualizing macromolecules is called autoradiography. If the samples are loaded into a rectangular well, as is commonly done, molecules of a particular size and charge form a band in an autoradiograph.

The autoradiograph that resulted from the polymerization experiment is shown in **Figure BS7.2**. The samples, taken on days 2, 4, 6, 8, and 14 of the experiment, are labeled along the bottom. The far right lane contains macromolecules of known size; this lane is used to estimate the size of the molecules in the experimental samples. The bands that appear in each sample lane represent the different polymers that had formed. Darker bands contain more radioactive marker, indicating the presence of many radioactive molecules. Lighter bands contain fewer molecules.

To read a gel, you look for (1) the presence or absence of bands in some lanes—meaning, some experimental samples—versus others, and (2) contrasts in the darkness of the bands present—meaning, differences in the amount of molecule present. For example, several conclusions can be drawn from the data in Figure BS7.2. First, a variety of polymers formed at each stage. After the second day, for example, polymers from 12 to 18 monomers long had formed on the clay particles used in this experiment. Second, the overall length of polymers produced increased with time. At the end of the fourteenth day, most of the RNA molecules were between 20 and 40 monomers long.

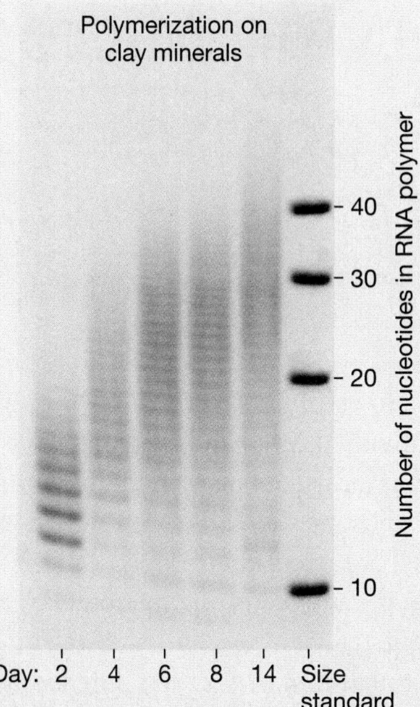

FIGURE BS7.2 Autoradiography Is a Technique for Visualizing Macromolecules. The molecules in a gel can be visualized in a number of ways. In this case, the RNA molecules in the gel exposed an X-ray film because they had radioactive atoms attached. When developed, the film is called an autoradiograph.

EXERCISE Add labels to the photograph that read: "Top of gel—large molecules" and "Bottom of gel—small molecules."

BioSkills 8

Observing Microscopic Structures and Processes

With the unaided eye, it is not possible to see a lot of biology. Biologists have to use microscopes to study small multicellular organisms, individual cells, and the contents of cells. And to understand what individual macromolecules or multimolecular machines like ribosomes look like, researchers use data from a technique called X-ray crystallography.

You'll probably use dissecting microscopes and compound light microscopes to view specimens during your labs for this course, and throughout this text you'll be seeing images generated from other types of microscopy and from X-ray crystallographic data. One of the fundamental skills you'll be acquiring as an introductory student, then, is a basic understanding of how these techniques work. The key is to recognize that each approach for visualizing microscopic structures has strengths and weaknesses. As a result, each technique is appropriate for studying certain types or aspects of cells or molecules.

Trends in Microscopy: Increasing Magnification and Clarity

If you use a dissecting microscope during labs, you'll recognize that it works by magnifying light that bounces off a whole specimen—often a live organism. You'll be able to view the specimen in three dimensions, which is why these instruments are sometimes called stereomicroscopes, but the maximum magnification is only about 20 to 40 times normal size (20× to 40×).

To view smaller objects, you'll probably use a compound microscope. Compound microscopes magnify light that is passed through a specimen. The instruments used in introductory labs are usually capable of 400× magnifications; the most sophisticated compound microscopes available can achieve magnifications of about 2000×. This is enough to view individual bacterial or eukaryotic cells and see large structures inside cells, like condensed chromosomes (see Chapter 11). To prepare a specimen for viewing under a compound light microscope, the tissues or cells are usually sliced to create a thin enough section for light to pass through efficiently. The section is then dyed to increase contrast and make structures visible. In many cases, different types of dyes are used to highlight different types of structures.

Until the 1950s, the compound microscope was the biologist's only tool for viewing cells directly. But the invention of the electron microscope provided a new way to view specimens. Two basic types of electron microscopy are now available: one that allows researchers to examine cross sections of cells at extremely high magnification, and one that offers a view of surfaces at somewhat lower magnification.

Transmission Electron Microscopy (TEM)

The transmission electron microscope is an extraordinarily effective tool for viewing cell structure at high magnification. TEM forms an image from electrons that pass through a specimen, just as a light microscope forms an image from light rays that pass through a specimen.

Biologists who want to view a cell under a transmission electron microscope begin by "fixing" the cell, meaning that they treat it with a chemical agent that stabilizes the cell's structure and contents while disturbing them as little as possible. Then they permeate the cell with an epoxy plastic that stiffens the structure. Once this epoxy hardens, the cell can be cut into extremely thin sections with a glass or diamond knife. Finally, the sectioned specimens are impregnated with a metal—often lead. (The reason for this last step is explained shortly.)

Figure BS8.1a outlines how the transmission electron microscope works. A beam of electrons is produced by a tungsten filament at the top of a column and directed downward. (All of the air is pumped out of the column, so that the electron beam isn't scattered by collisions with air molecules.) The electron beam passes through a series of lenses and through the specimen. The lenses are actually electromagnets, which alter the path of the beam much like a glass lens in a dissecting or compound microscope bends light. The lenses magnify and focus the image on a screen at the bottom of the column. There the electrons strike a coating of fluorescent crystals, which emit visible light in response—just like a television screen. When the microscopist moves the screen out of the way and allows the electrons to expose a sheet of black-and-white film, the result is a micrograph—a photograph of an image produced by microscopy.

The image itself is created by electrons that pass through the specimen. If no specimen were in place, all the electrons would pass through and the screen (and micrograph) would be uniformly bright. Unfortunately, cell materials by themselves would also appear fairly uniform and bright. This is because an atom's ability to deflect an electron depends on its mass. In

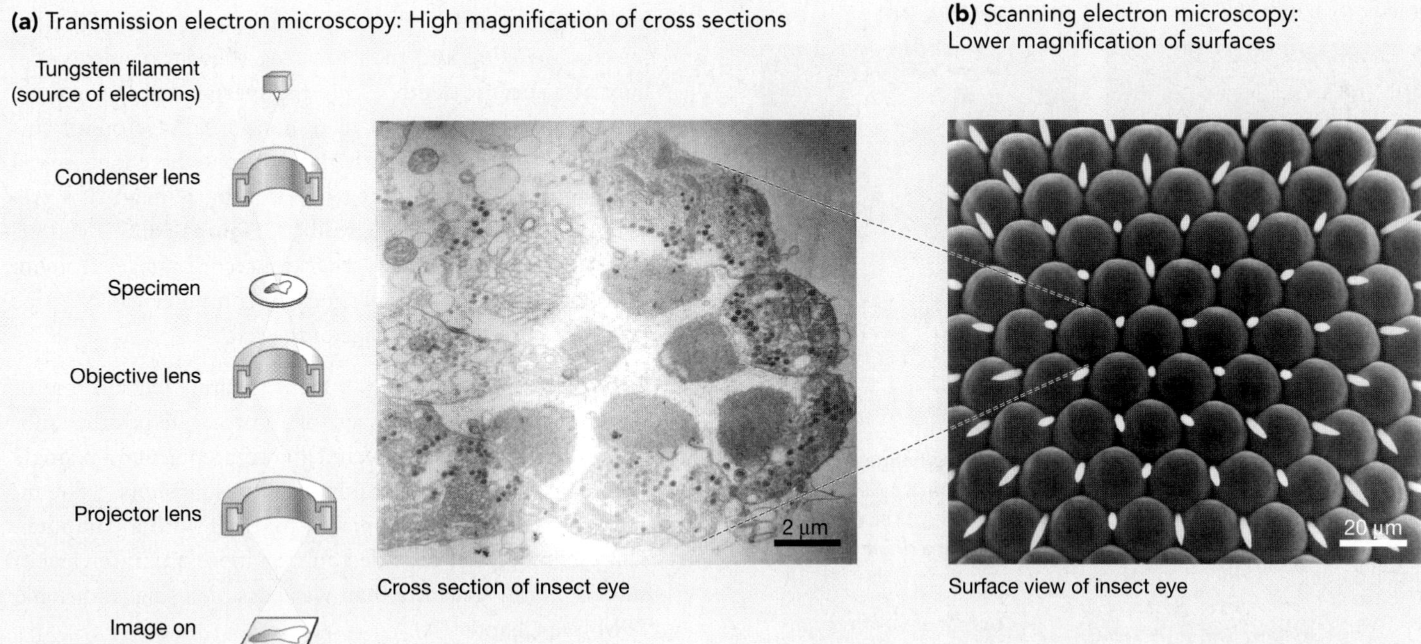

FIGURE BS8.1 There Are Two Basic Types of Electron Microscopy.

turn, an atom's mass is a function of its atomic number. The hydrogen, carbon, oxygen, and nitrogen atoms that dominate biological molecules have low atomic numbers. This is why cell biologists must saturate cell sections with lead solutions. Lead has a high atomic number and scatters electrons effectively. Different macromolecules take up lead atoms in different amounts, so the metal acts as a "stain" that produces contrast. With TEM, areas of dense metal scatter the electron beam most, producing dark areas in micrographs.

The advantage of TEM is that it can magnify objects up to 250,000×—meaning that intracellular structures are clearly visible. The downsides are that researchers are restricted to observing dead, sectioned material, and they must take care that the preparation process does not distort the specimen.

Scanning Electron Microscopy (SEM)

The scanning electron microscope is the most useful tool biologists have for looking at the surfaces of cells. Materials are prepared for scanning electron microscopy by coating their surfaces with a layer of metal atoms. To create an image of this surface, the microscope scans the surface with a narrow beam of electrons. Electrons that are reflected back from the surface or that are emitted by the metal atoms in response to the beam then strike a detector. The signal from the detector controls a second electron beam, which scans a TV-like screen and forms an image magnified up to 50,000 times the object's size. Because SEM records shadows and highlights, it provides images with a three-dimensional appearance (**Figure BS8.1b**). It cannot magnify objects nearly as much as TEM can, however.

Studying Live Cells and Real-Time Processes

Until the 1960s, it was not possible for biologists to get clear, high-magnification images of living cells. But a series of innovations over the past several decades has made it possible to observe organelles and subcellular structures in action.

The development of video microscopy, where the image from a light microscope is captured by a video camera instead of by an eye or a film camera, proved revolutionary. It allowed specimens to be viewed at higher magnification, because video cameras are more sensitive to small differences in contrast than are the human eye or still cameras. It also made it easier to keep live specimens functioning normally, because the increased light sensitivity of video cameras allows them to be used with low illumination, so specimens don't overheat. And when it became possible to digitize video images, researchers began using computers to remove out-of-focus background material and increase image clarity.

Another important innovation was the use of a fluorescent molecule called green fluorescent protein, or GFP, which allows researchers to tag specific molecules or structures and follow their movement over time. GFP is naturally synthesized in jellyfish that fluoresce, or emit light. By affixing GFP molecules

(a) Conventional fluorescence image of single cell

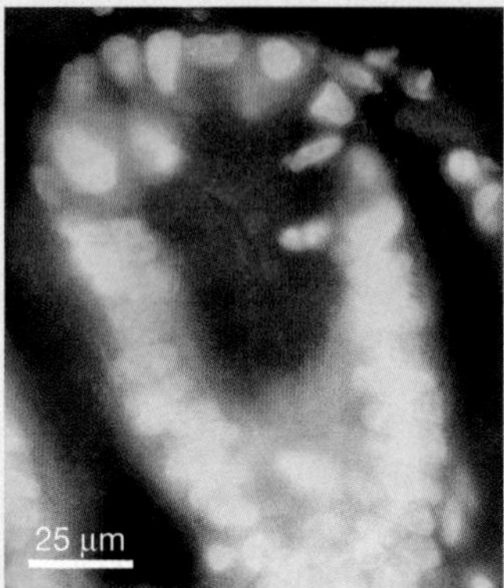

(b) Confocal fluorescence image of same cell

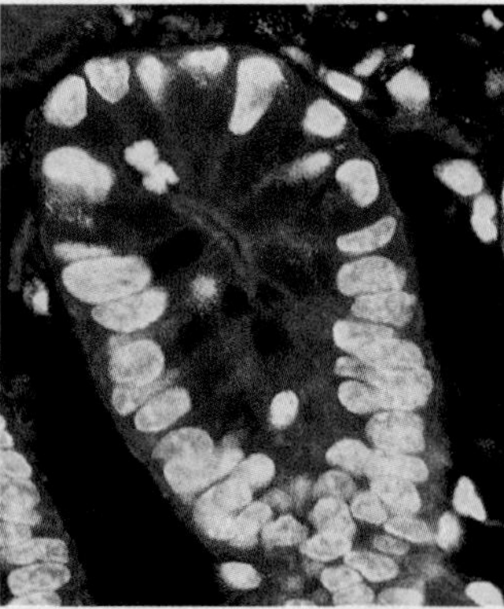

FIGURE BS8.2 Confocal Microscopy Provides Sharp Images of Living Cells. (a) The conventional image of this mouse intestinal cell is blurred, because it results from light emitted by the entire cell. **(b)** The confocal image is sharp, because it results from light emitted at a single plane inside the cell.

to another protein and then inserting it into a cell, investigators can follow the protein's fate over time and even videotape its movement. For example, researchers have videotaped GFP-tagged proteins being transported from the rough ER through the Golgi apparatus and out to the plasma membrane. This is cell biology: the movie.

Visualizing Structures in 3-D

The world is three-dimensional. To understand how microscopic structures and macromolecules work, it is essential to understand their shape and spatial relationships. Consider three techniques currently being used to reconstruct the 3-D structure of cells, organelles, and macromolecules.

- Confocal microscopy is carried out by mounting cells that have been treated with one or more fluorescing tags on a microscope slide and then focusing a beam of ultraviolet light at a specific depth within the specimen. The fluorescing tag emits visible light in response. A detector for this light is then set up at exactly the position where the emitted light comes into focus. The result is a sharp image of a precise plane in the cell being studied (**Figure BS8.2**). By altering the focal plane, a researcher can record images from an array of depths in the specimen; a computer can then be used to generate a 3-D image of the cell.
- Electron tomography uses a transmission electron microscope to generate a 3-D image of an organelle or other subcellular structure. The specimen is rotated around a single axis, with the researcher taking many "snapshots." The individual images are then pieced together with a computer. This technique has provided a much more accurate view of mitochondrial structure than was possible using traditional TEM (see Chapter 7).
- X-ray crystallography, or X-ray diffraction analysis, is the most widely used technique for reconstructing the 3-D structure of molecules. As its name implies, the procedure is based on bombarding crystals of a molecule with X-rays. X-rays are scattered in precise ways when they interact with the electrons surrounding the atoms in a crystal, producing a diffraction pattern that can be recorded on X-ray film or other types of detectors (**Figure BS8.3**). By varying the orientation of the X-ray beam as it strikes a crystal and documenting the diffraction patterns that result, researchers can construct a map representing the density of electrons in the crystal. By relating these electron-density maps to information about the primary structure of the nucleic acid or protein, a 3-D model of the molecule can be built. Virtually all of the molecular models used in this book were built from X-ray crystallographic data.

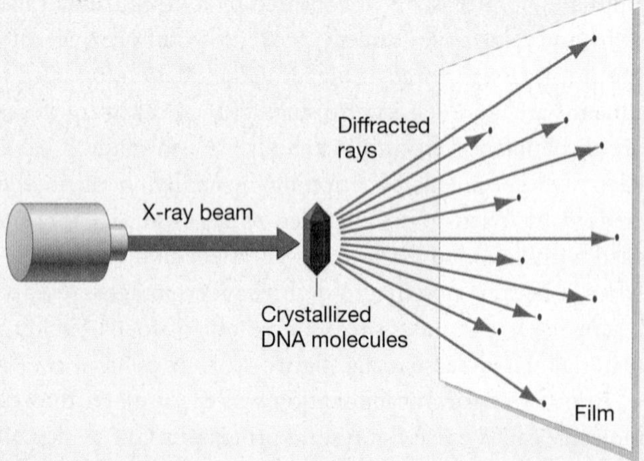

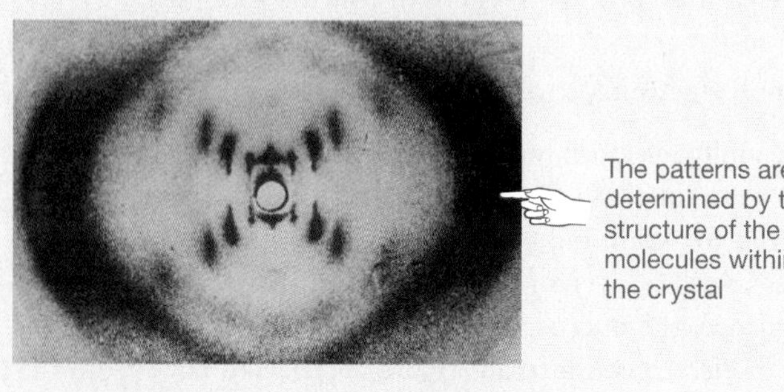

FIGURE BS8.3 X-Ray Crystallography. When crystallized molecules are bombarded with X-rays, the radiation is scattered in distinctive patterns. The photograph at the right shows an X-ray film that recorded the pattern of scattered radiation from DNA molecules.

BioSkills 9

Combining Probabilities

In several cases in this text, you'll need to combine probabilities from different events in order to solve a problem. One of the most common applications is in genetics problems. For example, Punnett squares work because they are based on two fundamental rules of probability, called the "both-and rule" and the "either-or" rule. Each rule pertains to a distinct situation.

The both-and rule applies when you want to know the probability that two or more independent events occur together. Let's use the rolling of two dice as an example. What is the probability of rolling two sixes? These two events are independent, because the probability of rolling a six on one die has no effect on the probability of rolling a six on the other die. (In the same way, the probability of getting a gamete with allele *R* from one parent has no effect on the probability of getting a gamete with allele *R* from the other parent. Gametes fuse randomly.) The probability of rolling a six on the first die is 1/6. The probability of rolling a six on the second die is also 1/6. The probability of rolling a six on both die, then, is $1/6 \times 1/6 = 1/36$. In other words, if you rolled two die 36 times, on average you would expect to roll two sixes once. It should make sense that the both-and rule is also called the multiplication rule or product rule. In the case of a cross between two parents heterozygous at the *R* gene, the probability of getting allele *R* from the father is 1/2 and the probability of getting *R* from the mother is 1/2. Thus, the probability of getting both alleles and creating an offspring with genotype *RR* is $1/2 \times 1/2 = 1/4$.

The either-or rule, in contrast, applies when you want to know the probability of an event happening when there are several ways for the event to occur. In this case, the probability that the event will occur is the sum of the probabilities of each way that it can occur. For example, suppose you wanted to know the probability of rolling either a one or a six when you toss a die. The probability of drawing each is 1/6, so the probability of getting one or the other is $1/6 + 1/6 = 1/3$. (The either-or rule is also called the addition rule or sum rule.) If you rolled a die three times, on average you'd expect to get a one or a six once. In the case of a cross between two parents heterozygous at the *R* gene, the probability of getting an *R* allele from the father and an *r* allele from the mother is $1/2 \times 1/2 = 1/4$. Similarly, the probability of getting an *r* allele from the father and an *R* allele from the mother is $1/2 \times 1/2 = 1/4$. Thus, the combined probability of getting the *Rr* genotype in either of the two ways is $1/4 + 1/4 = 1/2$.

Questions

1. Suppose that four students each toss a coin. What is the probability of four "tails"?
2. After a single roll of a die, what is the probability of getting either a two, a three, or a six?

Answers: 1. $1/2 \times 1/2 \times 1/2 \times 1/2 = 1/16$; 2. $1/6 + 1/6 + 1/6 = 1/2$

Glossary

abiotic Not alive (e.g., air, water, and soil). Compare with **biotic.**

ABO blood types Genetically determined classes of human blood that are distinguished by the presence or absence of specific glycolipids on the surface of red blood cells. Also called *ABO blood groups.*

aboveground biomass The total mass of living plants in an area, excluding roots.

abscisic acid (ABA) A plant hormone that inhibits cell elongation and stimulates leaf shedding and dormancy.

abscission In plants, the normal (often seasonal) shedding of leaves, fruits, or flowers.

abscission zone The region at the base of a petiole that thins and breaks during dropping of leaves.

absorption In animals, the uptake of ions and small molecules derived from food across the lining of the intestine and into the bloodstream.

absorption spectrum The amount of light of different wavelengths absorbed by a pigment. Usually depicted as a graph of light absorbed versus wavelength. Compare with **action spectrum.**

acclimation, acclimatization Gradual physiological adjustment of an organism to new environmental conditions that occur naturally or as part of a laboratory experiment.

acetylation Addition of an acetyl group (CH_3COO^-) to a molecule.

acetylcholine (ACh) A neurotransmitter, released by nerve cells at neuromuscular junctions, that triggers contraction of muscle cells. Also used as a neurotransmitter between neurons.

acetyl CoA A molecule produced by oxidation of pyruvate (the final product of glycolysis) in a reaction catalyzed by pyruvate dehydrogenase. Can enter the Krebs cycle and also is used as a carbon source in the synthesis of fatty acids, steroids, and other compounds.

acid Any compound that gives up protons or accepts electrons during a chemical reaction or that releases hydrogen ions when dissolved in water.

acid-growth hypothesis The hypothesis that auxin triggers elongation of plant cells by inducing the synthesis of proton pumps whose activity makes the cell wall more acidic, leading to expansion of the cell wall and an influx of water.

acoelomate An animal that lacks an internal body cavity (coelom). Compare with **coelomate** and **pseudocoelomate.**

acquired immune deficiency syndrome (AIDS) A human disease characterized by death of immune system cells (in particular helper T cells and macrophages) and subsequent vulnerability to other infections. Caused by the human immunodeficiency virus (HIV).

acquired immune response See **acquired immunity.**

acquired immunity Immunity to a particular pathogen or other antigen conferred by antibodies and activated B and T cells following exposure to the antigen. Is characterized by specificity, diversity, memory, and self-nonself recognition. Compare with **innate immunity.**

acrosomal reaction A set of events occurring in a sperm cell upon encountering an egg cell, including release of acrosomal enzymes and formation of the acrosomal process, which helps the sperm cell reach the egg.

acrosome A caplike structure, located on the head of a sperm cell, that contains enzymes capable of dissolving the outer coverings of an egg.

ACTH See **adrenocorticotropic hormone.**

actin A globular protein that can be polymerized to form filaments. Actin filaments are part of the cytoskeleton and constitute the thin filaments in skeletal muscle cells.

actin filament A long fiber, about 7 nm in diameter, composed of two intertwined strands of polymerized actin protein; one of the three types of cytoskeletal fibers. Involved in cell movement. Also called a *microfilament.* Compare with **intermediate filament** and **microtubule.**

action potential A rapid, temporary change in electrical potential across a membrane, from negative to positive and back to negative. Occurs in cells, such as neurons and muscle cells, that have an excitable membrane.

action spectrum The relative effectiveness of different wavelengths of light in driving a light-dependent process such as photosynthesis. Usually depicted as a graph of some measure of the process versus wavelength. Compare with **absorption spectrum.**

activation energy The amount of energy required to initiate a chemical reaction; specifically, the energy required to reach the transition state.

active site The portion of an enzyme molecule where substrates (reactant molecules) bind and react.

active transport The movement of ions or molecules across a plasma membrane or organelle membrane against an electrochemical gradient. Requires energy (e.g., from hydrolysis of ATP) and assistance of a transport protein (e.g., pump).

adaptation Any heritable trait that increases the fitness of an individual with that trait, compared with individuals without that trait, in a particular environment.

adaptive radiation Rapid evolutionary diversification within one lineage, producing numerous descendant species with a wide range of adaptive forms.

adenosine diphosphate (ADP) A molecule consisting of adenine, a sugar, and two phosphate groups. Addition of a third phosphate group produces adenosine triphosphate (ATP).

adenosine triphosphate (ATP) A molecule consisting of adenine, a sugar, and three phosphate groups that can be hydrolyzed to release energy. Universally used by cells to store and transfer energy.

adenylyl cyclase An enzyme that can catalyze the formation of cyclic AMP (cAMP) from ATP. Involved in controlling transcription of various operons in prokaryotes and in some eukaryotic signal-transduction pathways.

adhesion The tendency of certain dissimilar molecules to cling together due to attractive forces. Compare with **cohesion.**

adipocyte A fat cell.

adipose tissue A type of connective tissue whose cells store fats.

ADP See **adenosine diphosphate.**

adrenal glands Two small endocrine glands that sit above each kidney. The outer portion (cortex) secretes several steroid hormones; the inner portion (medulla) secretes epinephrine and norepinephrine.

adrenaline See **epinephrine.**

adrenocorticotropic hormone (ACTH) A peptide hormone, produced and secreted by the anterior pituitary, that stimulates release of steroid hormones (e.g., cortisol, aldosterone) from the adrenal cortex.

adventitious root A root that develops from a plant's shoot system instead of from the plant's root system.

aerobic Referring to any metabolic process, cell, or organism that uses oxygen as an electron acceptor. Compare with **anaerobic.**

afferent division The part of the nervous system, consisting mainly of sensory neurons, that transmits information about the internal and external environment to the central nervous system. Compare with **efferent division.**

agar A gelatinous mix of polysaccharides that is commonly used to grow cells on solid media.

age class All the individuals of a specific age in a population.

age structure The proportion of individuals in a population that are of each possible age.

age-specific fecundity The average number of female offspring produced by a female in a certain age class.

agglutination Clumping together of cells, typically caused by antibodies.

aggregate fruit A fruit (e.g., raspberry) that develops from a single flower that has many separate carpels. Compare with **multiple** and **simple fruit.**

AIDS See **acquired immune deficiency syndrome.**

albumen A solution of water and protein (particularly albumins), found in amniotic eggs, that nourishes the growing embryo. Also called *egg white.*

albumin A class of large proteins found in plants and animals, particularly in the albumen of eggs and in blood plasma.

alcohol fermentation Catabolic pathway in which pyruvate produced by glycolysis is converted to ethanol in the absence of oxygen.

aldosterone A hormone produced in the adrenal cortex that stimulates the kidney to conserve salt and water and promotes retention of sodium.

alimentary canal See **digestive tract.**

allele A particular version of a gene.

allopatric speciation The divergence of populations into different species by physical isolation of populations in different geographic areas. Compare with **sympatric speciation.**

allopatry Condition in which two or more populations live in different geographic areas. Compare with **sympatry.**

allopolyploidy (adjective: allopolyploid) The state of having more than two full sets of chromosomes (polyploidy) due to hybridization between different species. Compare with **autopolyploidy**.

allosteric regulation Regulation of a protein's function by binding of a regulatory molecule, usually to a specific site distinct from the active site, causing a change in the protein's shape.

alpha(α)-helix A protein secondary structure in which the polypeptide backbone coils into a spiral shape stabilized by hydrogen bonds between atoms.

alternation of generations A life cycle involving alternation of a multicellular haploid stage (gametophyte) with a multicellular diploid stage (sporophyte). Occurs in most plants and some protists.

alternative splicing In eukaryotes, the splicing of primary RNA transcripts from a single gene in different ways to produce different mature mRNAs and thus different polypeptides.

altruism Any behavior that has a cost to the individual (such as lowered survival or reproduction) and a benefit to the recipient. See **reciprocal altruism.**

alveolus (plural: alveoli) One of the tiny air-filled sacs of a mammalian lung.

ambisense virus A virus whose genome contains both positive-sense and negative-sense sequences.

amino acid A small organic molecule with a central carbon atom bonded to an amino group ($-NH_3$), a carboxyl group ($-COOH$), a hydrogen atom, and a side group. Proteins are polymers of 20 common amino acids.

aminoacyl tRNA A transfer RNA molecule that is covalently bound to an amino acid.

aminoacyl tRNA synthetase An enzyme that catalyzes the addition of a particular amino acid to its corresponding tRNA molecule.

ammonia (NH_3) A small molecule, produced by the breakdown of proteins and nucleic acids, that is very toxic to cells. Is a strong base that gains a proton to form the ammonium ion (NH_4^+).

amnion The membrane within an amniotic egg that surrounds the embryo and encloses it in a protective pool of fluid (amniotic fluid).

amniotes A major lineage of vertebrates (Amniota) that reproduce with amniotic eggs. Includes all reptiles (including birds) and mammals.

amniotic egg An egg that has a watertight shell or case enclosing a membrane-bound water supply (the amnion), food supply (yolk sac), and waste sac (allantois).

amoeba Any unicellular protist that lacks a cell wall, is extremely flexible in shape, and moves by means of pseudopodia.

amoeboid motion See **cell crawling.**

amphibians A lineage of vertebrates many of whom breathe through their skin and feed on land but lay their eggs in water; represent the earliest tetrapods. Include frogs, salamanders, and caecilians.

amphipathic Containing hydrophilic and hydrophobic elements.

amylase Any enzyme that can break down starch by catalyzing hydrolysis of the glycosidic linkages between the glucose residues.

amyloplasts Dense, starch-storing organelles that settle to the bottom of plant cells and that may be used as gravity detectors.

anabolic pathway Any set of chemical reactions that synthesizes larger molecules from smaller ones. Generally requires an input of energy. Compare with **catabolic pathway.**

anadromous Having a life cycle in which adults live in the ocean (or large lakes) but migrate up freshwater streams to breed and lay eggs.

anaerobic Referring to any metabolic process, cell, or organism that uses an electron acceptor other than oxygen, such as nitrate or sulfate. Compare with **aerobic.**

anaphase A stage in mitosis or meiosis during which chromosomes are moved to opposite ends of the cell.

anatomy The study of the physical structure of organisms.

androgens A class of steroid hormones that generally promote male-like traits (although females have some androgens, too). Secreted primarily by the gonads and adrenal glands.

aneuploidy (adjective: aneuploid) The state of having an abnormal number of copies of a certain chromosome.

angiosperm A flowering vascular plant that produces seeds within mature ovaries (fruits). The angiosperms form a single lineage. Compare with **gymnosperm.**

animal A member of a major lineage of eukaryotes (Animalia) whose members typically have a complex, large, multicellular body, eat other organisms, and are mobile.

animal model Any disease that occurs in a nonhuman animal and has many parallels to a similar disease of humans. Studied by medical researchers in hope that findings may apply to human disease.

anion A negatively charged ion.

annelids Members of the phylum Annelida (segmented worms). Distinguished by a segmented body and a coelom that functions as a hydrostatic skeleton. Annelids belong to the lophotrochozoan branch of the protostomes.

annual Referring to a plant whose life cycle normally lasts only one growing season—less than one year. Compare with **perennial.**

anoxygenic Referring to any process or reaction that does not produce oxygen. Photosynthesis in purple sulfur and purple nonsulfur bacteria, which does not involve photosystem II, is anoxygenic. Compare with **oxygenic.**

antenna (plural: antennae) A long appendage that is used to touch or smell.

antenna complex Part of a photosystem, containing an array of chlorophyll molecules and accessory pigments, that receives energy from light and directs the energy to a central reaction center during photosynthesis.

anterior Toward an animal's head and away from its tail. The opposite of posterior.

anterior pituitary The part of the pituitary gland containing endocrine cells that produce and release a variety of peptide hormones in response to other hormones from the hypothalamus. Compare with **posterior pituitary.**

anther The pollen-producing structure at the end of a stamen in flowering plants (angiosperms).

antheridium (plural: antheridia) The sperm-producing structure in most land plants except angiosperms.

anthropoids One of the two major lineages of primates, including apes, humans, and all monkeys. Compare with **prosimians.**

antibiotic Any substance, such as penicillin, that can kill or inhibit the growth of bacteria.

antibody An immunoglobulin protein, produced by B cells, that can bind to a specific part of an antigen, tagging it for attack by the immune system. All antibody molecules have a similar Y-shaped structure and, in their monomer form, consist of two identical light chains and two identical heavy chains.

anticodon The sequence of three bases (triplet) in a transfer RNA molecule that can bind to a mRNA codon with a complementary sequence.

antidiuretic hormone (ADH) A peptide hormone, secreted from the posterior pituitary gland, that stimulates water retention by the kidney. Also called *vasopressin.*

antigen Any foreign molecule, often a protein, that can stimulate a specific response by the immune system.

antigen presentation Process by which small peptides, derived from ingested particulate antigens (e.g., bacteria) or intracellular antigens (e.g., viruses in infected cell) are complexed with MHC proteins and transported to the cell surface where they are displayed and can be recognized by T cells.

antiparallel Describing the opposite orientation of the strands in a DNA double helix with one strand running in the $5' \rightarrow 3'$ direction and the other in the $3' \rightarrow 5'$ direction.

antiporter A carrier protein that allows an ion to diffuse down an electrochemical gradient, using the energy of that process to transport a different substance in the opposite direction *against* its concentration gradient. Compare with **symporter.**

antiviral Any drug or other agent that can kill or inhibit the transmission or replication of viruses.

anus In a multicellular animal, the end of the digestive tract where wastes are expelled.

aorta In terrestrial vertebrates, the major artery carrying oxygenated blood away from the heart.

aphotic zone Deep water receiving no sunlight. Compare with **photic zone.**

apical Toward the top. In plants, at the tip of a branch. In animals, on the side of an epithelial layer that faces the environment and not other body tissues. Compare with **basal.**

apical bud A bud at the tip of a stem, where growth occurs to lengthen the stem.

apical dominance Inhibition of lateral bud growth by the apical meristem at the tip of a plant branch.

apical meristem A group of undifferentiated plant cells, at the tip of a stem or root, that is responsible for primary growth. Compare with **lateral meristem.**

apomixis The formation of mature seeds without fertilization occurring; a type of asexual reproduction.

apoplast In plant roots, a continuous pathway through which water can flow, consisting of the porous cell walls of adjacent cells and the intervening extracellular space. Compare with **symplast.**

apoptosis Series of genetically controlled changes that lead to death of a cell. Occurs frequently during embryological development and later may occur in response to infections or cell damage. Also called *programmed cell death.*

aquaporin A type of channel protein through which water can move by osmosis across a plasma membrane.

arbuscular mycorrhizal fungi (AMF) Fungi whose hyphae enter the root cells of their host plants.

Archaea One of the three taxonomic domains of life consisting of unicellular prokaryotes distinguished by cell walls made of certain polysaccharides not found in bacterial or eukaryotic cell walls, plasma membranes composed of unique isoprene-containing phospholipids, and ribosomes and RNA polymerase similar to those of eukaryotes. Compare with **Bacteria** and **Eukarya.**

archegonium (plural: archegonia) The egg-producing structure in most land plants except angiosperms.

arteriole One of the many tiny vessels that carry blood from arteries to capillaries.

artery Any thick-walled blood vessel that carries blood (oxygenated or not) under relatively high pressure away from the heart to organs throughout the body. Compare with **vein.**

arthropods Members of the phylum Arthropoda. Distinguished by a segmented body; a hard, jointed exoskeleton; paired appendages; and an extensive body cavity called a hemocoel. Arthropods belong to the ecdysozoan branch of the protostomes.

articulation A movable point of contact between two bones of a skeleton. See **joint.**

artificial selection Deliberate manipulation by humans, as in animal and plant breeding, of the genetic composition of a population by allowing only individuals with desirable traits to reproduce.

ascocarp A large, cup-shaped reproductive structure produced by some ascomycete fungi. Contains many microscopic asci, which produce spores.

ascomycete See **sac fungus.**

ascus (plural: asci) Specialized spore-producing cell found at the ends of hyphae in sac fungi (ascomycetes).

asexual reproduction Any form of reproduction resulting in offspring that are genetically identical to the parent. Includes binary fission, budding, and parthenogenesis. Compare with **sexual reproduction.**

asymmetric competition Ecological competition between two species in which one species suffers a much greater fitness decline than the other. Compare with **symmetric competition.**

atomic mass unit (amu) A unit of mass equal to 1/12 the mass of one carbon-12 atom; about the mass of 1 proton or 1 neutron. Also called *dalton.*

atomic number The number of protons in the nucleus of an atom, giving the atom its identity as a particular chemical element.

ATP See **adenosine triphosphate.**

ATP synthase A large membrane-bound protein complex in chloroplasts, mitochondria, and some bacteria that uses the energy of protons flowing through it to synthesize ATP. Also called F_oF_1 *complex.*

atrioventricular (AV) node A region of the heart between the right atrium and right ventricle where electrical signals from the atrium are slowed briefly before spreading to the ventricle. This delay allows the ventricle to fill with blood before contracting. Compare with **sinoatrial (SA) node.**

atrium (plural: atria) A thin-walled chamber of the heart that receives blood from veins and pumps it to a neighboring chamber (the ventricle).

autocrine Relating to a chemical signal that affects the same cell that produced and released it.

autoimmunity A pathological condition in which the immune system attacks self cells or tissues of an individual's own body.

autonomic nervous system The part of the peripheral nervous system that controls internal organs and involuntary processes, such as stomach contraction, hormone release, and heart rate. Includes parasympathetic and sympathetic nerves. Compare with **somatic nervous system.**

autophagy The process by which damaged organelles are surrounded by a membrane and delivered to a lysosome to be destroyed.

autopolyploidy (adjective: autopolyploid) The state of having more than two full sets of chromosomes (polyploidy) due to a mutation that doubled the chromosome number.

autoradiography A technique for detecting radioactively labeled molecules separated by gel electrophoresis by placing an unexposed film over the gel. A black dot appears on the film wherever a radioactive atom is present in the gel. Also can be used to locate labeled molecules in fixed tissue samples.

autosomal inheritance The inheritance patterns that occur when genes are located on autosomes rather than on sex chromosomes.

autosome Any chromosome that does not carry genes involved in determining the sex of an individual.

autotroph Any organism that can synthesize reduced organic compounds from simple inorganic sources such as CO_2 or CH_4. Most plants and some bacteria and archaea are autotrophs. Also called *primary producer.* Compare with **heterotroph.**

auxin Indoleacetic acid, a plant hormone that stimulates phototropism and some other responses.

avirulence (*avr*) loci Genes in pathogens encoding proteins that trigger a defense response in plants. Compare with **resistance (R) loci.**

axillary bud A bud that forms in the angle between a leaf and a stem and may develop into a lateral (side) branch. Also called *lateral bud.*

axon A long projection of a neuron that can propagate an action potential and transmit it to another neuron.

axon hillock The site in a neuron where an axon joins the cell body and where action potentials are first triggered.

axoneme A structure found in eukaryotic cilia and flagella and responsible for their motion; composed of two central microtubules surrounded by nine doublet microtubules (9 + 2 arrangement).

background extinction The average rate of low-level extinction that has occurred continuously throughout much of evolutionary history. Compare with **mass extinction.**

Bacteria One of the three taxonomic domains of life consisting of unicellular prokaryotes distinguished by cell walls composed largely of peptidoglycan, plasma membranes similar to those of eukaryotic cells, and ribosomes and RNA polymerase that differ from those in archaeans or eukaryotes. Compare with **Archaea** and **Eukarya.**

bacterial artificial chromosome (BAC) An artificial version of a bacterial chromosome that can be used as a cloning vector to produce many copies of large DNA fragments.

bacteriophage Any virus that infects bacteria.

baculum A bone inside the penis usually present in mammals with a penis that lacks erectile tissue.

bark The protective outer layer of woody plants, composed of cork cells, cork cambium, and secondary phloem.

baroreceptors Specialized nerve cells in the walls of the heart and certain major arteries that detect changes in blood pressure and trigger appropriate responses by the brain.

basal Toward the base. In plants, at the base of a branch where it joins the stem. In animals, on the side of an epithelial layer that abuts underlying body tissues. Compare with **apical.**

basal body A structure of nine pairs of microtubules arranged in a circle at the base of eukaryotic cilia and flagella where they attach to the cell. Structurally similar to a centriole.

basal lamina A thick, collagen-rich extracellular matrix that underlies most epithelial tissues (e.g., skin) in animals.

basal metabolic rate (BMR) The total energy consumption by an organism at rest in a comfortable environment. For aerobes, often measured as the amount of oxygen consumed per hour.

basal transcription complex A large multiprotein structure that assembles near the promoter of eukaryotic genes and initiates transcription. Composed of basal transcription factors, TATA-binding protein, coactivators, and RNA polymerase.

basal transcription factor General term for proteins, present in all cell types, that bind to eukaryotic promoters and help initiate transcription. Compare with **regulatory transcription factor.**

base Any compound that acquires protons or gives up electrons during a chemical reaction or accepts hydrogen ions when dissolved in water.

basidiomycete See **club fungus.**

basidium (plural: basidia) Specialized spore-producing cell at the ends of hyphae in club fungi.

basilar membrane The membrane in the vertebrate cochlea on which the bottom portion of hair cells sit.

basolateral Toward the bottom and sides. In animals, the side of an epithelial layer that faces other body tissues and not the environment.

Batesian mimicry A type of mimicry in which a harmless or palatable species resembles a dangerous or poisonous species. Compare with **Müllerian mimicry.**

B cell A type of leukocyte that matures in the bone marrow and, with T cells, is responsible for acquired immunity. Produces antibodies and also functions in antigen presentation. Also called *B lymphocyte.*

B-cell receptor (BCR) An immunoglobulin protein (antibody) embedded in the plasma membrane of mature B cells and to which antigens bind.

behavior Any action by an organism.

benign tumor A mass of abnormal tissue that grows slowly or not at all, does not disrupt surrounding tissues, and does not spread to other organs. Benign tumors are not cancers. Compare with **malignant tumor.**

benthic Living at the bottom of an aquatic environment.

benthic zone The area along the bottom of an aquatic environment.

beta(β)-pleated sheet A protein secondary structure in which the polypeptide backbone folds into a sheetlike shape stabilized by hydrogen bonding.

bilateral symmetry An animal body pattern in which there is one plane of symmetry dividing the body into a left side and a right side. Typically, the body is long and narrow, with a distinct head end and tail end. Compare with **radial symmetry.**

bilaterian A member of a major lineage of animals (Bilateria) that are bilaterally symmetrical at some point in their life cycle, have three embryonic germ layers, and have a coelom. All protostomes and deuterostomes are bilaterians.

bile A complex solution produced by the liver, stored in the gall bladder, and secreted into the intestine. Contains steroid derivatives called bile salts that are responsible for emulsification of fats during digestion.

binary fission See **fission** (1).

binomial nomenclature A system of naming species by using two-part Latinized names composed of a genus name and a species name (e.g., *Homo sapiens,* humans; *Canis familiaris,* dogs). Always italicized, with genus name capitalized.

biodiversity The diversity of life considered at three levels: genetic diversity (variety of alleles in a population, species, or group of species); species diversity (variety and relative abundance of species present in a certain area); and ecosystem diversity (variety of communities and abiotic components in a region).

biogeochemical cycle The pattern of circulation of an element or molecule among living organisms and the environment.

biogeography The study of how species and populations are distributed geographically.

bioinformatics The field of study concerned with managing, analyzing, and interpreting biological information, particularly DNA sequences.

biological species concept The definition of a species as a population or group of populations that are reproductively isolated from other groups. Members of a species have the potential to interbreed in nature to produce viable, fertile offspring but cannot produce viable, fertile hybrid offspring with members of other species. Compare with **morphological** and **phylogenetic species concept.**

bioluminescence The emission of light by a living organism.

biomass The total mass of all organisms in a given population or geographical area; usually expressed as total dry weight.

biome A large terrestrial ecosystem characterized by a distinct type of vegetation and climate.

bioremediation The use of living organisms, usually bacteria or archaea, to degrade environmental pollutants.

biotechnology The application of biological techniques and discoveries to medicine, industry, and agriculture.

biotic Living, or produced by a living organism. Compare with **abiotic.**

bipedal Walking primarily on two legs.

bipolar cell A cell in the vertebrate retina that receives information from one or more photoreceptors and passes it to other bipolar cells or ganglion cells.

bivalves A lineage of molluscs that have two shells, such as clams and mussels.

bladder A mammalian organ that holds urine until it can be excreted.

blade The wide, flat part of a plant leaf.

blastocoel Fluid-filled cavity in the blastula of many animal species.

blastocyst Specialized type of blastula in mammals. A spherical structure composed of trophoblast cells on the exterior and a cluster of cells (the inner cell mass), which fills part of the interior space.

blastomere A small cell created by cleavage divisions in early animal embryos.

blastopore A small opening (pore) in the surface of an early vertebrate embryo, through which cells move during gastrulation.

blastula In vertebrate development, a hollow ball of cells (blastomere cells) that is formed by cleavage of a zygote and immediately undergoes gastrulation. See **blastocyst.**

blood A type of connective tissue consisting of red blood cells and leukocytes suspended in a fluid portion called plasma.

blood pressure See **diastolic blood pressure** and **systolic blood pressure.**

body mass index (BMI) A mathematical relationship of weight and height used to assess obesity in humans. Calculated as weight (kg) divided by the square of height (m^2).

body plan The basic architecture of an animal's body, including the number and arrangement of limbs, body segments, and major tissue layers.

bog A wetland that has no or almost no water flow, resulting in very low oxygen levels and acidic conditions.

Bohr shift The rightward shift of the oxygen-hemoglobin dissociation curve that occurs with decreasing pH. Results in hemoglobin being more likely to release oxygen in the acidic environment of exercising muscle.

bone A type of vertebrate connective tissue consisting of living cells and blood vessels within a hard extracellular matrix composed of calcium phosphate ($CaPO_4$) and small amounts of calcium carbonate ($CaCO_3$) and protein fibers.

bone marrow The soft tissue filling the inside of long bones containing stem cells that develop into red blood cells and leukocytes throughout life.

Bowman's capsule The hollow, double-walled cup-shaped portion of a nephron that surrounds a glomerulus in the vertebrate kidney.

braincase See **cranium.**

brain stem The most posterior portion of the vertebrate brain, connecting to the spinal cord and responsible for autonomic body functions such as heart rate, respiration, and digestion.

branch (1) A part of a phylogenetic tree that represents populations through time. (2) Any extension of a plant's shoot system.

bronchiole One of the small tubes in mammalian lungs that carry air from the bronchi to the alveoli.

bronchus (plural: bronchi) In mammals, one of a pair of large tubes that lead from the trachea to each lung.

brown adipose tissue A specialized form of adipose tissue whose cells contain a high density of mitochondria as well as stored fats and that can produce extra body heat. Found in some mammals.

bryophytes Members of several phyla of green plants that lack vascular tissue including liverworts, hornworts, and mosses. Also called *nonvascular plants.*

budding Asexual reproduction via outgrowth from the parent that eventually break free as an independent individual; occurs in yeasts and some invertebrates.

buffer A substance that, in solution, acts to minimize changes in the pH of that solution when acid or base is added.

bulbourethral glands In male mammals, small paired glands at the base of the urethra that secrete an alkaline mucus (part of semen), which lubricates the tip of the penis and neutralizes acids in the urethra during copulation. In humans, also called *Cowper's glands.*

bulk flow The directional movement of a substantial volume of fluid due to pressure differences, such as movement of water through plant phloem and movement of blood in animals.

bundle-sheath cell A type of cell found around the vascular tissue (veins) of plant leaves.

C_3 photosynthesis The most common form of photosynthesis in which atmospheric CO_2 is used to form 3-phosphoglycerate, a three-carbon sugar.

C_4 photosynthesis A variant type of photosynthesis in which atmospheric CO_2 is first fixed into four-carbon sugars, rather than the three-carbon sugars of classic C_3 photosynthesis. Enhances photosynthetic efficiency in hot, dry environments, by reducing loss of oxygen due to photorespiration.

cadherin Any of a class of cell-surface proteins involved in cell adhesion and important for coordinating movements of cells during embryological development.

callus In plants, a mass of undifferentiated cells that can generate roots and other tissues necessary to create a mature plant.

Calorie A unit of energy often used to measure the energy content of food. Also called *kilocalorie.*

Calvin cycle In photosynthesis, the set of light-independent reactions that use NADPH and ATP formed in the light-dependent reactions to drive the fixation of atmospheric CO_2 and reduction of the fixed carbon, ultimately producing sugars. Also called *carbon fixation* and *light-independent reactions.*

calyx All of the sepals of a flower.

CAM See **crassulacean acid metabolism.**

cambium (plural: cambia) See **lateral meristem.**

Cambrian explosion The rapid diversification of animal body types that began about 543 million years ago and continued for approximately 40 million years.

camera eye The type of eye in vertebrates and cephalopods, consisting of a hollow chamber with a hole at one end (through which light enters) and a sheet of light-sensitive cells against the opposite wall.

cAMP See **cyclic AMP.**

cancer General term for any tumor whose cells grow in an uncontrolled fashion, invade nearby tissues, and spread to other sites in the body.

canopy The uppermost layers of branches in a forest (i.e., those fully exposed to the Sun).

5′ cap A chemical grouping, consisting of 7-methylguanylate and three phosphate groups, that is added to the 5′ end of newly transcribed messenger RNA molecules.

CAP binding site A DNA sequence upstream of certain prokaryotic operons to which catabolite activator protein can bind, increasing gene transcription.

capillarity The tendency of water to move up a narrow tube due to surface tension, adhesion, and cohesion.

capillary One of the numerous small, thin-walled blood vessels that permeate all tissues and organs, and allow exchange of gases and other molecules between blood and body cells.

capillary bed A thick network of capillaries.

capsid A shell of protein enclosing the genome of a virus particle.

carapace In crustaceans, a large platelike section of the exoskeleton that covers and protects the cephalothorax (e.g., a crab's "shell").

carbohydrate Any of a class of molecules that contain a carbonyl group, several hydroxyl groups, and several to many carbon-hydrogen bonds. See **monosaccharide** and **polysaccharide.**

carbon cycle, global The worldwide movement of carbon among terrestrial ecosystems, the oceans, and the atmosphere.

carbon fixation See **Calvin cycle.**

carbonic anhydrase An enzyme that catalyzes the formation of carbonic acid (H_2CO_3) from carbon dioxide and water.

cardiac cycle One complete heartbeat cycle, including systole and diastole.

cardiac muscle The muscle tissue of the vertebrate heart. Consists of long branched fibers that are electrically connected and that initiate their own contractions; not under voluntary control. Compare with **skeletal** and **smooth muscle.**

carnivore (adjective: carnivorous) An animal whose diet consists predominantly of meat. Most members of the mammalian taxon Carnivora are carnivores. Some plants are carnivorous, trapping and killing small animals, then absorbing nutrients from the prey's body. Compare with **herbivore** and **omnivore.**

carotenoid Any of a class of accessory pigments, found in chloroplasts, that absorb wavelengths of light not absorbed by chlorophyll; typically appear yellow, orange, or red. Includes carotenes and xanthophylls.

carpel The female reproductive organ in a flower. Consists of the stigma, to which pollen grains adhere; the style, through which pollen grains move; and the ovary, which houses the ovule. Compare with **stamen.**

carrier A heterozygous individual carrying a normal allele and a recessive allele for an inherited trait; does not display the phenotype of the trait but can pass the recessive gene to offspring.

carrier protein A membrane protein that facilitates diffusion of a small molecule (e.g., glucose) across the plasma membrane by a process involving a reversible change in the shape of the protein. Also called *carrier* or *transporter.*

carrying capacity (K) The maximum population size of a certain species that a given habitat can support.

cartilage A type of vertebrate connective tissue that consists of relatively few cells scattered in a stiff matrix of polysaccharides and protein fibers.

Casparian strip In plant roots, a waxy layer containing suberin, a water-repellent substance, that prevents movement of water through the walls of endodermal cells, thus blocking the apoplastic pathway.

catabolic pathway Any set of chemical reactions that breaks down larger, complex molecules into smaller ones, releasing energy in the process. Compare with **anabolic pathway.**

catabolite activator protein (CAP) A protein that can bind to the CAP binding site upstream of certain prokaryotic operons, facilitating binding of RNA polymerase and stimulating gene expression.

catabolite repression A type of positive transcriptional control in which the end product of a catabolic pathway inhibits further transcription of the gene encoding an enzyme early in the pathway.

catalysis (verb: catalyze) Acceleration of the rate of a chemical reaction due to a decrease in the free energy of the transition state, called the activation energy.

catalyst Any substance that increases the rate of a chemical reaction without itself undergoing any permanent chemical change.

catecholamines A class of small compounds, derived from the amino acid tyrosine, that are used as hormones or neurotransmitters. Include epinephrine, norepinephrine, and dopamine.

cation A positively charged ion.

cation exchange In botany, the release (displacement) of cations, such as magnesium and calcium from soil particles, by protons in acidic soil water. The released cations are available for uptake by plants.

CD4 A membrane protein on the surface of some T cells in humans. $CD4^+$ T cells can give rise to helper T cells.

CD8 A membrane protein on the surface of some T cells in humans. $CD8^+$ T cells can give rise to cytotoxic T cells.

Cdk See **cyclin-dependent kinase.**

cDNA See **complementary DNA.**

cDNA library A set of cDNAs from a particular cell type or stage of development. Each cDNA is carried by a plasmid or other cloning vector and can be separated from other cDNAs. Compare with **genomic library.**

cecum A blind sac between the small intestine and the colon. Is enlarged in some species (e.g., rabbits) that use it as a fermentation vat for digestion of cellulose.

cell A highly organized compartment bounded by a thin, flexible structure (plasma membrane) and containing concentrated chemicals in an aqueous (watery) solution. The basic structural and functional unit of all organisms.

cell body The part of a neuron that contains the nucleus and where incoming signals are integrated. Also called the *soma.*

cell crawling A form of cellular movement involving actin filaments in which the cell produces bulges (pseudopodia) that stick to the substrate and pull the cell forward. Also called *amoeboid motion.*

cell cycle Ordered sequence of events in which a eukaryotic cell replicates its chromosomes, evenly partitions the chromosomes to two daughter cells, and then undergoes division of the cytoplasm.

cell-cycle checkpoint Any of several points in the cell cycle at which progression of a cell through the cycle can be regulated.

cell division Creation of new cells by division of pre-existing cells.

cell-mediated (immune) response The type of immune response that involves generation of cytotoxic T cells from $CD8^+$ T cells. Defends against pathogen-infected cells, cancer cells, and transplanted cells. Compare with **humoral (immune) response.**

cell membrane See **plasma membrane.**

cell plate A double layer of new plasma membrane that appears in the middle of a dividing plant cell; ultimately divides the cytoplasm into two separate cells.

cell sap An aqueous solution found in the vacuoles of plant cells.

cell theory The theory that all organisms are made of cells and that all cells come from preexisting cells.

cellular respiration A common pathway for production of ATP, involving transfer of electrons from compounds with high potential energy (often NADH and $FADH_2$) to an electron transport chain and ultimately to an electron acceptor (often oxygen).

cellulose A structural polysaccharide composed of β-glucose monomers joined by β-1,4-glycosidic linkages. Found in the cell wall of algae, plants, bacteria, fungi, and some other groups.

cell wall A protective layer located outside the plasma membrane and usually composed of polysaccharides. Found in algae, plants, bacteria, fungi, and some other groups.

Cenozoic era The most recent period of geologic time, beginning 65 million years ago. during which mammals became the dominant vertebrates and angiosperms became the dominant plants. Also called *Age of Mammals.*

central dogma The long-accepted hypothesis that information in cells flows in one direction: DNA codes for RNA, which codes for proteins. Exceptions are now known (e.g., retroviruses).

central nervous system (CNS) The brain and spinal cord of vertebrate animals. Compare with **peripheral nervous system (PNS).**

centriole One of two small cylindrical structures, structurally similar to a basal body, found together within the centrosome near the nucleus of a eukaryotic cell.

centromere Constricted region of a replicated chromosome where the two sister chromatids are joined and the kinetochore is located.

centrosome Structure in animal and fungal cells, containing two centrioles, that serves as a microtubule-organizing center for the cell's cytoskeleton and for the mitotic spindle during cell division.

cephalization The formation of a distinct anterior region (the head) where sense organs and a mouth are clustered.

cephalochordates One of the three major chordate lineages (Cephalochordata), comprising small, mobile organisms that live in marine sands; also called *lancelets* or *amphioxi.* Compare with **urochordates** and **vertebrates.**

cephalopods A lineage of molluscs including the squid, octopuses, and nautiluses. Distinguished by large brains, excellent vision, tentacles, and a reduced or absent shell.

cerebellum Posterior section of the vertebrate brain that is involved in coordination of complex muscle movements, such as those required for locomotion and maintaining balance.

cerebrum The most anterior section of the vertebrate brain. Divided into left and right hemispheres and four lobes: parietal lobe, involved in complex decision making (in humans); occipital lobe, receives and interprets visual information; parietal lobe, involved in integrating sensory and motor functions; and temporal lobe, functions in memory, speech (in humans), and interpreting auditory information.

cervix The narrow passageway between the vagina and the uterus of female mammals.

chaetae (singular: chaeta) Bristle-like extensions found in some annelids.

channel protein A membrane protein that facilitates diffusion of a specific ion or small molecule across a plasma membrane through a central pore in the protein. Includes water channels (aquaporins) and ion channels. Also called simply *channel*.

character displacement The tendency for the traits of similar species that occupy overlapping ranges to change in a way that reduces interspecific competition.

chelicerae A pair of clawlike appendages found around the mouth of certain arthropods called chelicerates (spiders, mites, and allies).

chemical bond An attractive force binding two atoms together. Covalent bonds, ionic bonds, and hydrogen bonds are types of chemical bonds.

chemical carcinogen Any chemical that can cause cancer.

chemical energy The potential energy stored in covalent bonds between atoms.

chemical equilibrium A dynamic but stable state of a reversible chemical reaction in which the forward reaction and reverse reactions proceed at the same rate, so that the concentrations of reactants and products remain constant.

chemical evolution The theory that simple chemical compounds in the ancient atmosphere and ocean combined by spontaneous chemical reactions to form larger, more complex substances, eventually leading to the origin of life and the start of biological evolution.

chemical reaction Any process in which one compound or element is combined with others or is broken down; involves the making and/or breaking of chemical bonds.

chemiosmosis An energetic coupling mechanism whereby energy stored in an electrochemical proton gradient (proton-motive force) is used to drive an energy-requiring process such as production of ATP.

chemokine Any of several chemical signals that attract leukocytes to a site of tissue injury or infection.

chemolithotroph An organism that produces ATP by oxidizing inorganic molecules with high potential energy such as ammonia (NH_3) or methane (CH_4). Also called *lithotroph*. Compare with **chemoorganotroph.**

chemoorganotroph An organism that produces ATP by oxidizing organic molecules with high potential energy such as sugars. Also called *organotroph*. Compare with **chemolithotroph.**

chemoreceptor A sensory cell or organ specialized for detection of specific molecules or classes of molecules.

chemotaxis Movement toward or away from a certain chemical.

chiasma (plural: chiasmata) The X-shaped structure formed during meiosis by crossing over between non-sister chromatids in a pair of homologous chromosomes.

chitin A structural polysaccharide composed of *N*-acetylglucosamine monomers joined end to end by β-1,4-glycosidic linkages. Found in cell walls of fungi and many algae, and in external skeletons of insects and crustaceans.

chitons A lineage of marine mollusc that have a protective shell formed of eight calcium carbonate plates.

chlorophyll Any of several closely related green pigments, found in chloroplasts and photosynthetic protists, that absorb light during photosynthesis.

chloroplast A chlorophyll-containing organelle, bounded by a double membrane, in which photosynthesis occurs; found in plants and photosynthetic protists. Also the location of amino acid, fatty acid, purine, and pyrimidine synthesis.

choanocyte A specialized flagellated feeding cell found in choanoflagellates (protists that are the closest living relatives of animals) and sponges (the oldest animal phylum).

cholecystokinin A peptide hormone secreted by cells in the lining of the small intestine. Stimulates the secretion of digestive enzymes from the pancreas and of bile from the liver and gallbladder.

chordates Members of the phylum Chordata, deuterostomes distinguished by a dorsal hollow nerve cord, pharyngeal gill slits, a notochord, a dorsal hollow nerve cord, and a post-anal tail. Include vertebrates, cephalochordata, and urochordata.

chromatid One of the two identical strands composing a replicated chromosome that is connected at the centromere to the other strand.

chromatin The complex of DNA and proteins, mainly histones, that compose eukaryotic chromosomes. Can be highly compact (heterochromatin) or loosely coiled (euchromatin).

chromatin remodeling The process by which the DNA in chromatin is unwound from its associated proteins to allow transcription or replication. May involve chemical modification of histone proteins or reshaping of the chromatin by large multi-protein complexes in an ATP-requiring process.

chromosome Gene-carrying structure consisting of a single long molecule of DNA and associated proteins (e.g., histones). Most prokaryotic cells contain a single, circular chromosome; eukaryotic cells contain multiple noncircular (linear) chromosomes located in the nucleus.

chromosome inversion See **inversion.**

chromosome painting A technique for producing high-resolution karyotypes by "painting" chromosomes with fluorescent tags that bind to particular regions of certain chromosomes. Also called *spectral karyotyping*.

chromosome theory of inheritance The principle that genes are located on chromosomes and that patterns of inheritance are determined by the behavior of chromosomes during meiosis.

chromosome translocation See **translocation** (2).

chylomicron A ball of protein-coated lipids used to transport the lipids through the bloodstream.

chytrid A member of a paraphyletic group of fungi (Chytridiomycota) that produce motile spores and gametes (both have flagella similar to those in sperm cells). Also called *chytridiomycete*.

cilium (plural: cilia) One of many short, filamentous projections of some eukaryotic cells containing a core of microtubules. Used to move the cell and/or to move fluid or particles along a stationary cell. See **axoneme.**

circadian clock An internal mechanism found in most organisms that regulates many body processes (sleep-wake cycles, hormonal patterns, etc.) in a roughly 24-hour cycle.

circulatory system The system in animals responsible for moving oxygen, carbon dioxide, and other materials (hormones, nutrients, wastes) around the body.

cisternae (singular: cisterna) Flattened, membrane-bound compartments that make up the Golgi apparatus.

clade See **monophyletic group.**

cladistic approach A method for constructing a phylogenetic tree that is based on identifying the unique traits of each monophyletic group. Compare with **phenetic approach.**

class In Linnaeus' system, a taxonomic category above the order level and below the phylum level.

Class I MHC protein An MHC protein that is present on the plasma membrane of virtually all nucleated cells and functions in presenting antigen to $CD8^+$ T cells.

Class II MHC protein An MHC protein that is present only on the plasma membrane of dendritic cells, macrophages, and B cells and functions in presenting antigen to $CD4^+$ T cells.

classical conditioning A type of learning in which an animal learns to associate two stimuli, so that a response originally given to just one stimulus can be evoked by the second stimulus as well.

cleavage In animal development, the series of rapid mitotic cell divisions, with little cell growth, that produces successively smaller cells and transforms a zygote into a multicellular blastula, or blastocyst in mammals.

cleavage furrow A pinching-in of the plasma membrane that occurs as cytokinesis begins in animal cells and deepens until the cytoplasm is divided.

climate The prevailing long-term weather conditions in a particular region.

climax community The stable, final community that develops from ecological succession.

clitoris A small rod of erectile tissue in the external genitalia of female mammals. Is formed from the same embryonic tissue as the male penis and has a similar function in sexual arousal.

cloaca An opening to the outside used by the excretory and reproductive systems in many nonmammalian vertebrate but few vertebrates.

clonal expansion Rapid cell division by a T cell or B cell in response to its binding the specific antigen recognized by receptors on the cell surface. Produces a large population of descendant cells all specific for that particular antigen.

clonal-selection theory The dominant explanation of the development of acquired immunity in vertebrates. According to the theory, the immune system retains a vast pool of inactive lymphocytes, each with a unique receptor for a unique antigen. Lymphocytes that encounter their antigens are stimulated to divide (selected and cloned), producing daughter cells that combat infection and confer immunity.

clone (1) An individual that is genetically identical to another individual. (2) A lineage of genetically identical individuals or cells. (3) As a verb, to make one or more genetic replicas of a cell or individual.

cloning vector A plasmid or other agent used to transfer recombinant genes into cultured host cells. Also called simply *vector.*

closed circulatory system A circulatory system in which the circulating fluid (blood) is confined to blood vessels and flows in a continuous circuit. Compare with **open circulatory system.**

club fungus A member of a lineage of fungi (Basidiomycota) that produce complex reproductive structures (e.g., mushrooms, puffballs) containing specialized spore-forming cells (basidia). Also called *basidiomycete.*

cnidocyte A specialized stinging cell found in cnidarians (e.g., jellyfish, corals, and anemones) that is used in capturing prey.

coactivator Any regulatory protein that help initiate transcription by bringing together the necessary transcription factors but does not itself bind to DNA.

cochlea The organ of hearing in the inner ear of mammals, birds, and crocodilians. A coiled, fluid-filled tube containing specialized pressure-sensing neurons (hair cells) that detect sounds of different pitches.

coding strand See **non-template strand.**

codominance An inheritance pattern in which heterozygotes exhibit both of the traits seen in either kind of homozygous individual.

codon A sequence of three nucleotides in DNA or RNA that codes for a certain amino acid or that initiates or terminates protein synthesis.

coefficient of relatedness (*r*) A measure of how closely two individuals are related. Calculated as the probability that an allele in two individuals is inherited from the same ancestor.

coelom An internal, usually fluid-filled, body cavity that is lined with mesoderm.

coelomate An animal that has a true coelom. Compare with **acoelomate** and **pseudocoelomate.**

coenocytic Containing many nuclei and a continuous cytoplasm through a filamentous body, without the body being divided into distinct cells. Some fungi are coenocytic.

coenzyme A small organic molecule that is a required cofactor for an enzyme-catalyzed reaction. Often donates or receives electrons or functional groups during the reaction.

coenzyme A (CoA) A nonprotein molecule that is required for many cellular reactions involving transfer of acetyl groups ($-COCH_3$).

coenzyme Q A nonprotein molecule that shuttles electrons between membrane-bound complexes in the mitochondrial electron transport chain. Also called *ubiquinone* or *Q.*

coevolution A pattern of evolution in which two interacting species reciprocally influence each other's adaptations over time.

cofactor A metal ion or small organic compound that is required for an enzyme to function normally. May be bound tightly to an enzyme or associate with it transiently during catalysis.

cognition The mental processes involved in recognition and manipulation of facts about the world, particularly to form novel associations or insights.

cohesion The tendency of certain like molecules (e.g., water molecules) to cling together due to attractive forces. Compare with **adhesion.**

cohesion-tension theory The theory that water movement upward through plant vascular tissues is due to loss of water from leaves (transpiration), which pulls a cohesive column of water upward.

cohort A group of individuals that are the same age and can be followed through time.

coleoptile A modified leaf that covers and protects the stems and leaves of young grasses.

collagen A fibrous, pliable, cable-like glycoprotein that is a major component of the extracellular matrix of animal cells. Various subtypes differ in their tissue distribution.

collecting duct In the vertebrate kidney, a large straight tube that receives filtrate from the distal tubules of several nephrons. Involved in the regulated reabsorption of water.

collenchyma cell In plants, an elongated cell with cell walls thickened at the corners that provides support to growing plant parts; usually found in strands along leaf veins and stalks. Compare with **parenchyma cell** and **sclerenchyma cell.**

colon The portion of the large intestine where feces are formed by compaction of wastes and reabsorption of water.

colony An assemblage of individuals. May refer to an assemblage of semi-independent cells or to a breeding population of multicellular organisms.

commensalism (adjective: commensal) A symbiotic relationship in which one organism (the commensal) benefits and the other (the host) is not harmed. Compare with **mutualism** and **parasitism.**

communication In ecology, any process in which a signal from one individual modifies the behavior of another individual.

community All of the species that interact with each other in a certain area.

companion cell In plants, a cell in the phloem that is connected via numerous plasmodesmata to adjacent sieve-tube members. Companion cells provide materials to maintain sieve-tube members and function in the loading and unloading of sugars into sieve-tube members.

compass orientation A type of navigation in which movement occurs in a specific direction.

competition In ecology, the interaction of two species or two individuals trying to use the same limited resource (e.g., water, food, living space). May occur between individuals of the same species (intraspecific competition) or different species (interspecific competition).

competitive exclusion principle The principle that two species cannot coexist in the same ecological niche in the same area because one species will out-compete the other.

competitive inhibition Inhibition of an enzyme's ability to catalyze a chemical reaction via a nonreactant molecule that competes with the substrate(s) for access to the active site.

complementary base pairing The association between specific nitrogenous bases of nucleic acids stabilized by hydrogen bonding. Adenine pairs only with thymine (in DNA) or uracil (in RNA), and guanine pairs only with cytosine.

complementary DNA (cDNA) DNA produced in the laboratory using an RNA transcript as a template and reverse transcriptase; corresponds to a gene but lacks introns. Also produced naturally by retroviruses.

complementary strand A newly synthesized strand of RNA or DNA that has a base sequence complementary to that of the template strand.

complement system A set of proteins that circulate in the bloodstream and can form holes in the plasma membrane of bacteria, leading to their destruction.

complete metamorphosis See **holometabolous metamorphosis.**

compound eye An eye formed of many independent light-sensing columns (ommatidia); occurs in arthropods. Compare with **simple eye.**

concentration gradient Difference across space (e.g., across a membrane) in the concentration of a dissolved substance.

condensation reaction A chemical reaction in which two molecules are joined covalently with the removal of an $-OH$ from one and an $-H$ from another to form water. Also called a *dehydration reaction.* Compare with **hydrolysis.**

conduction (1) Direct transfer of heat between two objects that are in physical contact. Compare with **convection.** (2) Transmission of an electrical impulse along the axon of a nerve cell.

cone In conifers (e.g., pines, cedars, and spruces), the reproductive structure in which microspores and macrospores are formed.

cone cell A photoreceptor cell with a cone-shaped outer portion that is particularly sensitive to bright light of a certain color. Also called simply *cone.* Compare with **rod cell.**

conjugation The process by which DNA is exchanged between unicellular individuals. Occurs in bacteria, archaea, and some protists.

connective tissue An animal tissue consisting of scattered cells in a liquid, jellylike, or solid extracellular matrix. Includes bone, cartilage, tendons, ligaments, and blood.

conservation biology The effort to study, preserve, and restore threatened populations, communities, and ecosystems.

constant (C) region The portion of an antibody's light chains or heavy chains that has the same amino acid sequence in the antibodies produced by every B cell of an individual. Compare with **variable (V) region.**

constitutive Always occurring; always present. Commonly used to describe enzymes and other proteins that are synthesized continuously or mutants in which one or more genetic loci are constantly expressed due to defects in gene control.

constitutive defense A defensive trait that is always manifested even in the absence of a predator or pathogen. Also called *standing defense.* Compare with **inducible defense.**

consumer See **heterotroph.**

continuous strand See **leading strand.**

control In a scientific experiment, a group of organisms or samples that do not receive the experimental treatment but are otherwise identical to the group that does.

convection Transfer of heat by movement of large volumes of a gas or liquid. Compare with **conduction.**

convergent evolution The independent evolution of analogous traits in distantly related organisms due to adaptation to similar environments and a similar way of life.

cooperative binding The tendency of the protein subunits of hemoglobin to affect each other's oxygen binding such that each bound oxygen molecule increases the likelihood of further oxygen binding.

copulation The act of transferring sperm from a male directly into a female's reproductive tract.

coral reef A large assemblage of colonial marine corals that usually serves as shallow-water, sunlit habitat for many other species as well.

co-receptor Any membrane protein that acts with some other membrane protein in a cell interaction or cell response.

cork cambium One of two types of lateral meristem, consisting of a ring of undifferentiated plant cells found just under the cork layer of woody plants; produces new cork cells on its outer side. Compare with **vascular cambium.**

cork cell A waxy cell in the protective outermost layer of a woody plant.

corm A rounded, thick underground stem that can produce new plants via asexual reproduction.

cornea The transparent sheet of connective tissue at the very front of the eye in vertebrates and some other animals. Protects the eye and helps focus light.

corolla All of the petals of a flower.

corona The cluster of cilia at the anterior end of a rotifer.

corpus callosum A thick band of neurons that connects the two hemispheres of the cerebrum in the mammalian brain.

corpus luteum A yellowish structure in an ovary that secretes progesterone. Is formed from a follicle that has recently ovulated.

cortex (1) The outermost region of an organ, such as the kidney or adrenal gland. (2) In plants, a layer of ground tissue found outside the vascular bundles and pith of a plant stem.

cortical granules Small enzyme-filled vesicles in the cortex of an egg cell. Involved in formation of the fertilization envelope after fertilization.

corticosterone A steroid hormone, produced and secreted by the adrenal cortex, that increases blood glucose and prepares the body for stress. The major glucocorticoid hormone in most reptiles, birds, and many mammals.

corticotropin-releasing hormone (CRH) A peptide hormone, produced and secreted by the hypothalamus, that stimulates the anterior pituitary to release ACTH.

cortisol A steroid hormone, produced and secreted by the adrenal cortex, that increases blood glucose and prepares the body for stress. The major glucocorticoid hormone in some mammals. Also called *hydrocortisone.*

cotransporter A transmembrane protein that facilitates diffusion of an ion down its previously established electrochemical gradient and uses the energy of that process to transport some other substance, in the same or opposite direction, *against* its concentration gradient. See **antiporter** and **symporter.**

cotyledon The first leaf, or seed leaf, of a plant embryo. Used for storing and digesting nutrients and/or for early photosynthesis.

countercurrent exchanger In animals, any anatomical arrangement that allows the maximum transfer of heat or a soluble substance from one fluid to another. The two fluids must be flowing in opposite directions and have a heat or concentration gradient between them.

covalent bond A type of chemical bond in which two atoms share one or more pairs of electrons. Compare with **hydrogen bond** and **ionic bond.**

cranium A bony, cartilaginous, or fibrous case that encloses and protects the brain of vertebrates. Forms part of the skull. Also called *braincase.*

crassulacean acid metabolism (CAM) A variant type of photosynthesis in which CO_2 is stored in organic acids at night when stomata are open and then released to feed the Calvin cycle during the day when stomata are closed. Helps reduce water loss and oxygen loss by photorespiration in hot, dry environments.

cristae (singular: crista) Sac-like invaginations of the inner membrane of a mitochondrion. Location of the electron transport chain and ATP synthase.

critical period See **sensitive period.**

Cro-Magnon A prehistoric European population of modern humans (*Homo sapiens*) known from fossils, paintings, sculptures, and other artifacts.

crossing over The exchange of segments of non-sister chromatids between a pair of homologous chromosomes that occurs during meiosis I.

cross-pollination Pollination of a flower by pollen from another individual, rather than by self-fertilization. Also called *crossing.*

crustaceans A lineage of arthropods that includes shrimp, lobster, and crabs. Many have a carapace (a platelike portion of the exoskeleton) and mandibles for biting or chewing.

culture In cell biology, a collection of cells or a tissue growing under controlled conditions, usually in suspension or on the surface of a dish on solid growth medium.

cup fungus See **sac fungus.**

Cushing's disease A human endocrine disorder caused by loss of feedback inhibition of cortisol on ACTH secretion. Characterized by high ACTH and cortisol levels and wasting of body protein reserves.

cuticle A protective coating secreted by the outermost layer of cells of an animal or a plant.

cyanobacteria A lineage of photosynthetic bacteria formerly known as blue-green algae. Likely the first life-forms to carry out oxygenic photosynthesis.

cyclic AMP (cAMP) Cyclic adenosine monophosphate; a small molecule, derived from ATP, that is widely used by cells in signal transduction and transcriptional control.

cyclic photophosphorylation Path of electron flow during the light-dependent reactions of photosynthesis in which photosystem I transfers excited electrons back to the electron transport chain of photosystem II, rather than to $NADP^+$. Also called *cyclic electron flow.* Compare with **Z scheme.**

cyclin One of several regulatory proteins whose concentrations fluctuate cyclically throughout the cell cycle.

cyclin-dependent kinase (Cdk) Any of several related protein kinases that are active only when bound to a cyclin. Involved in control of the cell cycle.

cytochrome *c* (cyt *c*) A soluble iron-containing protein that shuttles electrons between membrane-bound complexes in the mitochondrial electron transport chain.

cytokines A diverse group of autocrine signaling proteins, secreted largely by cells of the immune system, whose effects include stimulating leukocyte production, tissue repair, and fever. Generally function to regulate the intensity and duration of an immune response.

cytokinesis Division of the cytoplasm to form two daughter cells. Typically occurs immediately after division of the nucleus by mitosis or meiosis.

cytokinins A class of plant hormones that stimulate cell division and retard aging.

cytoplasm All of the contents of a cell, excluding the nucleus, and bounded by the plasma membrane.

cytoplasmic determinant A regulatory transcription factor or signaling molecule that is distributed unevenly in the cytoplasm of the egg cells of many animals and that directs early pattern formation in an embryo.

cytoplasmic streaming The directed flow of cytosol and organelles that facilitates distribution of materials within some large plant and fungal cells. Occurs along actin filaments and is powered by myosin.

cytoskeleton In eukaryotic cells, a network of protein fibers in the cytoplasm that are involved in cell shape, support, locomotion, and transport of materials within the cell. Prokaryotic cells have a similar but much less extensive network of fibers.

cytosol The fluid portion of the cytoplasm.

cytotoxic T cell An effector T cell that destroys infected cells and cancer cells. Is descended from an activated $CD8^+$ T cell that has interacted with antigen on an infected cell or cancer cell. Also called *cytotoxic T lymphocyte (CTL)* and *killer T cell.* Compare with **helper T cell.**

dalton (Da) See **atomic mass unit.**

Darwinian fitness See **fitness.**

day-neutral plant A plant whose flowering time is not affected by the relative length of day and night (the photoperiod). Compare with **long-day** and **short-day plant.**

dead space Portions of the air passages that are not involved in gas exchange with the blood, such as the trachea and bronchi.

deciduous Describing a plant that sheds leaves or other structures at regular intervals (e.g., each fall)

decomposer See **detritivore.**

decomposer food chain An ecological network of detritus, decomposers that eat detritus, and predators and parasites of the decomposers.

definitive host The host species in which a parasite reproduces sexually. Compare with **intermediate host.**

dehydration reaction See **condensation reaction.**

deleterious In genetics, referring to any mutation, allele, or trait that reduces an individual's fitness.

demography The study of factors that determine the size and structure of populations through time.

denaturation (verb: denature) For a macromolecule, loss of its three-dimensional structure and biological activity due to breakage of hydrogen bonds and disulfide bonds, usually caused by treatment with excess heat or extreme pH conditions.

dendrite A short extension from a neuron's cell body that receives signals from other neurons.

dendritic cell A type of leukocyte that ingests and digests foreign antigens, moves to a lymph node, and presents the antigens displayed on its membrane to $CD4^+$ T cells.

density dependent In population ecology, referring to any characteristic that varies depending on population density.

density independent In population ecology, referring to any characteristic that does not vary with population density.

deoxyribonucleic acid (DNA) A nucleic acid composed of deoxyribonucleotides that carries the genetic information of a cell. Generally occurs as two intertwined strands, but these can be separated. See **double helix.**

deoxyribonucleoside triphosphate (dNTP) A monomer that can be polymerized to form DNA. Consists of deoxyribose, a base (A, T, G, or C), and three phosphate groups; similar to a nucleotide, but with two more phosphate groups.

deoxyribonucleotide See **nucleotide.**

depolarization Change in membrane potential from its resting negative state to a less negative or a positive state; a normal phase in an action potential. Compare with **hyperpolarization.**

deposit feeder An animal that eats its way through a food-containing substrate.

dermal tissue system The tissue forming the outer layer of an organism. In plants, also called *epidermis;* in animals, forms two distinct layers: *dermis* and *epidermis.*

descent with modification The phrase used by Darwin to describe his hypothesis of evolution by natural selection.

desmosome A type of cell-cell attachment structure, consisting of cadherin proteins, that bind the cytoskeletons of adjacent animal cells together. Found where cells are strongly attached to each other. Compare with **gap junction** and **tight junction.**

determination In embryogenesis, progressive changes in a cell that commit it to a particular cell fate. Once a cell is fully determined, it can differentiate only into a particular cell type (e.g., liver cell, brain cell).

detritivore An organism whose diet consists mainly of dead organic matter (detritus). Various bacteria, fungi, and protists are detritivores. Also called *decomposer.*

detritus A layer of dead organic matter that accumulates at ground level or on seafloors and lake bottoms.

deuterostomes A major lineage of animals that share a pattern of embryological development, including radial cleavage, formation of the anus earlier than the mouth, and formation of the coelom by pinching off of layers of mesoderm from the gut. Includes echinoderms and chordates. Compare with **protostomes.**

developmental homology A similarity in embryonic form, or in the fate of embryonic tissues, that is due to inheritance from a common ancestor.

diabetes insipidus A human disease caused by defects in the kidney's system for conserving water. Characterized by production of large amounts of dilute urine.

diabetes mellitus A human disease caused by defects in insulin production (type I) or the response of cells to insulin (type II). Characterized by abnormally high blood glucose levels and large amounts of glucose-containing urine.

diaphragm An elastic, sheetlike structure. In mammals, the muscular sheet of tissue that separates the chest and abdominal cavities. Contracts and moves downward during inhalation, expanding the chest cavity.

diastole The portion of the heartbeat cycle during which the atria or ventricles of the heart are relaxed. Compare with **systole.**

diastolic blood pressure The force exerted by blood against artery walls during relaxation of the heart's left ventricle. Compare with **systolic blood pressure.**

dicot Any plant that has two cotyledons (embryonic leaves) upon germination. The dicots do not form a monophyletic group. Also called *dicotyledonous plant.* Compare with **eudicot** and **monocot.**

dideoxy sequencing A laboratory technique for determining the exact nucleotide sequence of DNA. Relies on the use of dideoxynucleotide triphosphates (ddNTPs), which terminate DNA replication.

diencephalon The part of the mammalian brain that relays sensory information to the cerebellum and functions in maintaining homeostasis.

differential centrifugation Procedure for separating cellular components according to their size and density by spinning a cell homogenate in a series of centrifuge runs. After each run, the supernatant is removed from the deposited material (pellet) and spun again at progressively higher speeds.

differential gene expression Expression of different sets of genes in cells with the same genome. Responsible for creating different cell types.

differentiation The process by which a relatively unspecialized cell becomes a distinct specialized cell type (e.g., liver cell, brain cell) usually by changes in gene expression. Also call *cell differentiation.*

diffusion Spontaneous movement of a substance from a region of high concentration to one of low concentration (i.e., down a concentration gradient).

digestion The physical and chemical breakdown of food into molecules that can be absorbed into the body of an animal.

digestive tract The long tube that begins at the mouth and ends at the anus. Also called *alimentary canal, gastrointestinal(GI) tract,* or the *gut.*

dihybrid cross A mating between two parents that are heterozygous for both of the two genes being studied.

dimer An association of two molecules, which may be identical (homodimer) or different (heterodimer).

dioecious Describing an angiosperm species that has male and female reproductive structures on separate plants. Compare with **monoecious.**

diploblast (adjective: diploblastic) An animal whose body develops from two basic embryonic cell layers—ectoderm and endoderm. Compare with **triploblast.**

diploid (1) Having two sets of chromosome s ($2n$). (2) A cell or an individual organism with two sets of chromosomes, one set inherited from the maternal parent and one set from the paternal parent. Compare with **haploid.**

directional selection A pattern of natural selection that favors one extreme phenotype with the result that the average phenotype of a population changes in one direction. Generally reduces overall genetic variation in a population.

direct sequencing A technique for identifying and studying microorganisms that cannot be grown in culture. Involves detecting and amplifying copies of certain specific genes in their DNA, sequencing these genes, and then comparing the sequences with the known sequences from other organisms.

disaccharide A carbohydrate consisting of two monosaccharides (sugar residues) linked together.

discontinuous strand See **lagging strand.**

discrete trait An inherited trait that exhibits distinct phenotypic forms rather than the continuous variation characteristic of a quantitative traits such as body height.

dispersal The movement of individuals from their place of origin (birth, hatching) to a new location.

disruptive selection A pattern of natural selection that favors extreme phenotypes at both ends of the range of phenotypic variation. Maintains overall genetic variation in a population. Compare with **stabilizing selection.**

distal tubule In the vertebrate kidney, the convoluted portion of a nephron into which filtrate moves from the loop of Henle. Involved in the regulated reabsorption of sodium and water. Compare with **proximal tubule.**

disturbance In ecology, any event that disrupts a community, usually causing loss of some individuals or biomass from it.

disturbance regime The characteristic disturbances that affect a given ecological community.

disulfide bond A covalent bond between two sulfur atoms, typically in the side groups of some amino acids (e.g., cysteine). Often contributes to tertiary structure of proteins.

DNA See **deoxyribonucleic acid.**

DNA cloning Any of several techniques for producing many identical copies of a particular gene or other DNA sequence.

DNA fingerprinting Any of several methods for identifying individuals by unique features of their genomes. Commonly involves using PCR to produce many copies of certain simple sequence repeats (microsatellites) and then analyzing their lengths.

DNA footprinting A technique used to identify stretches of DNA that are bound by particular regulatory proteins.

DNA library See **cDNA library** and **genomic library.**

DNA ligase An enzyme that joins pieces of DNA by catalyzing formation of a phosphodiester bond between the pieces.

DNA microarray A set of single-stranded DNA fragments, representing thousands of different genes, that are permanently fixed to a small glass slide. Can be used to determine which genes are expressed in different cell types, under different conditions, or at different developmental stages.

DNA polymerase Any enzyme that catalyzes synthesis of DNA from deoxyribonucleotides.

domain (1) A section of a protein that has a distinctive tertiary structure and function. (2) A taxonomic category, based on similarities in basic cellular biochemistry, above the kingdom level. The three recognized domains are Bacteria, Archaea, and Eukarya.

dominant Referring to an allele that determines the phenotype of a heterozygous individual. Compare with **recessive.**

dopamine A catecholamine neurotransmitter that functions mainly in a part of the mammalian brain involved with muscle control. Also functions as a hypothalamic inhibitory hormone that inhibits release of prolactin from the anterior pituitary; also called *prolactin-inhibiting hormone (PIH).*

dormancy A temporary state of greatly reduced, or no, metabolic activity and growth in plants or plant parts (e.g., seeds, spores, bulbs, and buds).

dorsal Toward an animal's back and away from its belly. The opposite of ventral.

double fertilization An unusual form of reproduction seen in flowering plants, in which one sperm nucleus fuses with an egg to form a zygote and the other sperm nucleus fuses with two polar nuclei to form the triploid endosperm.

double helix The secondary structure of DNA, consisting of two antiparallel DNA strands wound around each other.

downstream In genetics, the direction in which RNA polymerase moves along a DNA strand. Compare with **upstream.**

Down syndrome A human developmental disorder caused by trisomy of chromosome 21.

dyad symmetry A type of symmetry in which an object can be superimposed on itself if rotated 180°. Occurs in some regulatory sequences of DNA. Also called *two-fold rotational symmetry.*

dynein Any one of a class of motor proteins that use the chemical energy of ATP to "walk" along an adjacent microtubule. Dyneins are responsible for bending of cilia and flagella, play a role in chromosome movement during mitosis, and can transport certain organelles.

early endosome A small membrane-bound vesicle, formed by endocytosis, that is an early stage in the formation of a lysosome.

ecdysone An insect hormone that triggers either molting (to a larger larval form) or metamorphosis (to the adult form), depending on the level of juvenile hormone.

ecdysozoans A major lineage of protostomes (Ecdysozoa) that grow by shedding their external skeletons (molting) and expanding their bodies. Includes arthropods, insects, crustaceans, nematodes, and centipedes. Compare with **lophotrochozoans.**

echinoderms A major lineage of deuterostomes (Echinodermata) distinguished by adult bodies with five-sided radial symmetry, a water vascular system, and tube feet. Includes sea urchins, sand dollars, and sea stars.

echolocation The use of echoes from vocalizations to obtain information about locations of objects in the environment.

ecology The study of how organisms interact with each other and with their surrounding environment.

ecosystem All the organisms that live in a geographic area, together with the nonliving (abiotic) components that affect or exchange materials with the organisms; a community and its physical environment.

ecosystem diversity The variety of biotic components in a region along with abiotic components, such as soil, water, and nutrients.

ecosystem services Alterations of the physical components of an ecosystem by living organisms, especially beneficial changes in the quality of the atmosphere, soil, water, etc.

ectoderm The outermost of the three basic cell layers in most animal embryos; gives rise to the outer covering and nervous system. Compare with **endoderm** and **mesoderm.**

ectomycorrhizal fungi (EMF) Fungi whose hyphae form a dense network that covers their host plant's roots but do not enter the root cells.

ectoparasite A parasite that lives on the outer surface of the host's body.

ectotherm An animal that does not use internally generated heat to regulate its body temperature. Compare with **endotherm.**

effector Any cell, organ, or structure with which an animal can respond to external or internal stimuli. Usually functions, along with a sensor and integrator, as part of a homeostatic system.

efferent division The part of the nervous system, consisting primarily of motor neurons, that carries commands from the central nervous system to the body.

egg A mature female gamete and any associated external layers (such as a shell). Larger and less mobile than the male gamete. In animals, also called *ovum.*

ejaculation The release of semen from the copulatory organ of a male animal.

ejaculatory duct A short duct connecting the vas deferens to the urethra, through which sperm move during ejaculation.

elastic Referring to a structure (e.g., lungs) with the ability to stretch and then spring back to its original shape.

electrical potential Potential energy created by a separation of electric charges between two points. Also called *voltage.*

electric current A flow of electrical charge past a point. Also called *current.*

electrocardiogram (EKG) A recording of the electrical activity of the heart, as measured through electrodes on the skin.

electrochemical gradient The combined effect of an ion's concentration gradient and electrical (charge) gradient across a membrane that affects the diffusion of ions across the membrane.

electrolyte Any compound that dissociates into ions when dissolved in water. In nutrition, refers to the major ions necessary for normal cell function.

electromagnetic spectrum The entire range of wavelengths of radiation extending from short wavelengths (high energy) to long wavelengths (low energy). Includes gamma rays, x-rays, ultraviolet, visible light, infrared, microwaves, and radio waves (from short to long wavelengths).

electron acceptor A reactant that gains an electron and is reduced in a reduction-oxidation reaction.

electron carrier Any molecule that readily accepts electrons from and donates electrons to other molecules.

electron donor A reactant that loses an electron and is oxidized in a reduction-oxidation reaction.

electronegativity A measure of the ability of an atom to attract electrons toward itself from an atom to which it is bonded.

electron microscope See **scanning electron microscope** and **transmission electron microscope.**

electron shell A group of orbitals of electrons with similar energies. Electron shells are arranged in roughly concentric layers around the nucleus of an atom, with electrons in outer shells having more energy than those in inner shells. Electrons in the outermost shell, the valence shell, often are involved in chemical bonding.

electron transport chain (ETC) Any set of membrane-bound protein complexes and smaller soluble electron carriers involved in a coordinated series of redox reactions in which the potential energy of electrons transferred from reduced donors is successively decreased and used to pump protons from one side of a membrane to the other.

electroreceptor A sensory cell or organ specialized to detect electric fields.

element A substance, consisting of atoms with a specific number of protons, that cannot be separated into or broken down to any other substance. Elements preserve their identity in chemical reactions.

elongation (1) The process by which messenger RNA lengthens during transcription. (2) The process by which a polypeptide chain lengthens during translation.

elongation factors Proteins involved in the elongation phase of translation, assisting ribosomes in the synthesis of the growing peptide chain.

embryo A young developing organism; the stage after fertilization and zygote formation.

embryo sac The female gametophyte in flowering plants that exhibit alternation of generations.

embryogenesis The process by which a single-celled zygote becomes a multicellular embryo.

embryophyte A plant that nourishes its embryos inside its own body. All land plants are embryophytes.

emergent vegetation Any plants in an aquatic habitat that extend above the surface of the water.

emerging disease Any infectious disease, often a viral disease, that suddenly afflicts significant numbers of humans for the first time; often due to changes in the host species for a pathogen or host population movements.

emigration The migration of individuals away from one population to other populations. Compare with **immigration.**

emulsification (verb: emulsify) The dispersion of fat into an aqueous solution. Usually requires the aid of an amphipathic substance such as a detergent or bile salts, which can break large fat globules into microscopic fat droplets.

endangered species A species whose numbers have decreased so much that it is in danger of extinction throughout all or part of its range.

endemic species A species that lives in one geographic area and nowhere else.

endergonic Referring to a chemical reaction that requires an input of energy to occur and for which the Gibbs free-energy change (ΔG) > 0. Compare with **exergonic.**

endocrine Relating to a chemical signal (hormone) that is released into the bloodstream by a producing cell and acts on a distant target cell.

endocrine gland A gland that secretes hormones directly into the bloodstream or interstitial fluid instead of into ducts. Compare with **exocrine gland.**

endocrine system All of the glands and tissues that produce and secrete hormones into the bloodstream.

endocytosis General term for any pinching off of the plasma membrane that results in the uptake of material from outside the cell. Includes phagocytosis, pinocytosis, and receptor-mediated endocytosis. Compare with **exocytosis.**

endoderm The innermost of the three basic cell layers in most animal embryos; gives rise to the digestive tract and organs that connect to it (liver, lungs, etc.). Compare with **ectoderm** and **mesoderm.**

endodermis In plant roots, a cylindrical layer of cells that separates the cortex from the vascular tissue.

endomembrane system A system of organelles in eukaryotic cells that performs most protein and lipid synthesis. Includes the endoplasmic reticulum (ER), Golgi apparatus, and lysosomes.

endoparasite A parasite that lives inside the host's body.

endophyte (adjective: endophytic) A fungus that lives inside the aboveground parts of a plant in a symbiotic relationship. Compare with **epiphyte.**

endoplasmic reticulum (ER) A network of interconnected membranous sacs and tubules found inside eukaryotic cells. See **rough** and **smooth endoplasmic reticulum.**

endoskeleton Bony and/or cartilaginous structures within the body that provide support. Examples are the spicules of sponges, the plates in echinoderms, and the bony skeleton of vertebrates. Compare with **exoskeleton.**

endosome See **early** and **late endosome.**

endosperm A triploid ($3n$) tissue in the seed of a flowering plant (angiosperm) that serves as food for the plant embryo. Functionally analogous to the yolk in some animal eggs.

endosymbiont An organism that lives in a symbiotic relationship inside the body of its host.

endosymbiosis theory The theory that mitochondria and chloroplasts evolved from prokaryotes that were engulfed by host cells and took up a symbiotic existence within those cells, a process termed primary endosymbiosis. In some eukaryotes, chloroplasts originated by secondary endosymbiosis, that is, by engulfing a chloroplast-containing protist and retaining its chloroplasts.

endotherm An animal whose primary source of body heat is internally generated heat. Compare with **ectotherm.**

endothermic Referring to a chemical reaction that absorbs heat. Compare with **exothermic.**

energetic coupling In cellular metabolism, the mechanism by which energy released from an exergonic reaction (commonly, hydrolysis of ATP) is used to drive an endergonic reaction.

energy The capacity to do work or to supply heat. May be stored (potential energy) or available in the form of motion (kinetic energy).

enhancer A regulatory sequence in eukaryotic DNA that may be located far from the gene it controls or within introns of the gene. Binding of specific proteins to an enhancer enhances the transcription of certain genes.

enrichment culture A method of detecting and obtaining cells with specific characteristics by placing a sample, containing many types of cells, under a specific set of conditions (e.g., temperature, salt concentration, available nutrients) and isolating those cells that grow rapidly in response.

entropy (*S*) A quantitative measure of the amount of disorder of any system, such as a group of molecules.

envelope, viral A membrane-like covering that encloses some viruses and their capsid coats, shielding them from attack by the host's immune system.

environmental sequencing The inventory of all the genes in a community or ecosystem by sequencing, analyzing, and comparing the genomes of the component organisms.

enzyme A protein catalyst used by living organisms to speed up and control biological reactions.

epicotyl In some embryonic plants, a portion of the embryonic stem that extends above the cotyledons.

epidemic The spread of an infectious disease throughout a population in a short time period. Compare with **pandemic.**

epidermis The outermost layer of cells of any multicellular organism.

epididymis A coiled tube wrapped around the testis in reptiles, birds, and mammals. The site of the final stages of sperm maturation and storage.

epigenetic inheritance Pattern of inheritance involving differences in phenotype that are not due to changes in the nucleotide sequence of genes.

epinephrine A catecholamine hormone, produced and secreted by the adrenal medulla, that triggers rapid responses relating to the fight-or-flight response. Also called *adrenaline.*

epiphyte (adjective: epiphytic) A nonparasitic plant that grows on trees or other solid objects and is not rooted in soil.

epithelium (plural: epithelia) An animal tissue consisting of sheet-like layers of tightly packed cells that lines an organ, a duct, or a body surface. Also called *epithelial tissue.*

epitope A small region of a particular antigen to which an antibody, B-cell receptor, or T-cell receptor binds.

equilibrium curve See **oxygen-hemoglobin equilibrium curve.**

equilibrium potential The membrane potential at which there is no net movement of a particular ion into or out of a cell.

ER signal sequence A short amino acid sequence that marks a polypeptide for transport to the endoplasmic reticulum where synthesis of the polypeptide chain is completed and the signal sequence removed. See **signal recognition particle.**

erythrocyte See **red blood cell.**

erythropoietin (EPO) A peptide hormone, released by the kidney in response to low blood oxygen levels, that stimulates the bone marrow to produce more red blood cells.

esophagus The muscular tube that connects the mouth to the stomach.

essential amino acid An amino acid that an animal cannot synthesize and must obtain from the diet. May refer specifically to one of the eight essential amino acids of adult humans: isoleucine, leucine, lysine, methionine, phenylalanine, threonine, tryptophan, and valine.

essential nutrient Any chemical element, ion, or compound that is required for normal growth, reproduction, and maintenance of a living organism and that cannot be synthesized by the organism.

ester linkage The covalent bond formed by a condensation reaction between a carboxyl group ($-COOH$) and a hydroxyl group ($-OH$). Ester linkages join fatty acids to glycerol to form a fat or phospholipid.

estradiol The major estrogen produced by the ovaries of female mammals. Stimulates development of the female reproductive tract, growth of ovarian follicles, and growth of breast tissue.

estrogens A class of steroid hormones, including estradiol, estrone, and estriol, that generally promote female-like traits. Secreted by the gonads, fat tissue, and some other organs.

estrous cycle A female reproductive cycle, seen in all mammals except Old World monkeys and apes (including humans), in which the uterine lining is reabsorbed rather than shed in the absence of pregnancy, and the female is sexually receptive only briefly during mid-cycle (estrus). Compare with **menstrual cycle.**

estuary An environment of brackish (partly salty) water where a river meets the ocean.

ethylene A gaseous plant hormone that induces fruits to ripen, flowers to fade, and leaves to drop.

eudicot A member of a monophyletic group (lineage) of angiosperms that includes complex flowering plants and trees (e.g., roses, daisies, maples). All eudicots have two cotyledons, but not all dicots are members of this lineage. Compare with **dicot** and **monocot.**

Eukarya One of the three taxonomic domains of life consisting of unicellular organisms (most protists, yeast) and multicellular organisms (fungi, plants, animals) distinguished by a membrane-bound cell nucleus, numerous organelles, and an extensive cytoskeleton. Compare with **Archaea** and **Bacteria.**

eukaryote A member of the domain Eukarya; an organism whose cells contain a nucleus, numerous membrane-bound organelles, and an extensive cytoskeleton. May be unicellular or multicellular. Compare with **prokaryote.**

eutherians A lineage of mammals (Eutheria) whose young develop in the uterus and are not housed in an abdominal pouch. Also called *placental mammals.*

evaporation The energy-absorbing phase change from a liquid state to a gaseous state. Many organisms evaporate water as a means of heat loss.

evo-devo Research field focused on how changes in developmentally important genes have led to the evolution of new phenotypes.

evolution (1) The theory that all organisms on Earth are related by common ancestry and that they have changed over time, predominantly via natural selection. (2) Any change in the genetic characteristics of a population over time, especially, a change in allele frequencies.

excitable membrane A plasma membrane that is capable of generating an action potential. Neurons, muscle cells, and some other cells have excitable membranes.

excitatory postsynaptic potential (EPSP) A change in membrane potential, usually depolarization, at a neuron dendrite that makes an action potential more likely.

exergonic Referring to a chemical reaction that can occur spontaneously, releasing heat and/or increasing entropy, and for which the Gibbs free-energy change (ΔG) < 0. Compare with **endergonic.**

exocrine gland A gland that secretes some substance through a duct into a space other than the circulatory system, such as the digestive tract or the skin sufrace. Compare with **endocrine gland.**

exocytosis Secretion of intracellular molecules (e.g., hormones, collagen), contained within membrane-bounded vesicles, to the outside of the cell by fusion of vesicles to the plasma membrane. Compare with **endocytosis.**

exon A region of a eukaryotic gene that is translated into a peptide or protein. Compare with **intron.**

exoskeleton A hard covering secreted on the outside of the body, used for body support, protection, and muscle attachment. Examples are the shell of molluscs and the outer covering (cuticle) of arthropods. Compare with **endoskeleton.**

exothermic Referring to a chemical reaction that releases heat. Compare with **endothermic.**

exotic species A nonnative species that is introduced into a new area. Exotic species often are competitors, pathogens, or predators of native species.

expansins A class of plant proteins that actively increase the length of the cell wall when the pH of the wall falls below 4.5.

exponential population growth The accelerating increase in the size of a population that occurs when the growth rate is constant and density independent. Compare with **logistic population growth.**

extensor A muscle that pulls two bones farther apart from each other, as in the extension of a limb or the spine. Compare with **flexor.**

extinct Said of a species that has died out.

extracellular digestion Digestion that takes place outside of an organism, as occurs in many fungi that make and secrete digestive enzymes.

extracellular matrix (ECM) A complex meshwork of proteins (e.g., collagen, fibronectin) and polysaccharides secreted by animal cells and in which they are embedded.

extremophile A bacterium or archaean that thrives in an "extreme" environment (e.g., high-salt, high-temperature, low-temperature, or low-pressure).

F_1 generation First filial generation. The first generation of offspring produced from a mating (i.e., the offspring of the parental generation).

facilitated diffusion Movement of a substance across a plasma membrane down its concentration gradient with the assistance of transmembrane carrier proteins or channel proteins.

facilitation In ecological succession, the phenomenon in which early-arriving species make conditions more favorable for later-arriving species. Compare with **inhibition** and **tolerance.**

facultative aerobe Any organism that can perform aerobic respiration when oxygen is available to serve as an electron acceptor but can switch to fermentation when it is not.

FAD/FADH$_2$ Oxidized and reduced forms, respectively, of flavin adenine dinucleotide. A nonprotein electron carrier that functions in the Krebs cycle and oxidative phosphorylation.

fallopian tube A narrow tube connecting the uterus to the ovary in humans, through which the egg travels after ovulation. Site of fertilization and cleavage. In nonhuman animals, called *oviduct.*

family In Linnaeus' system, a taxonomic category above genus and below order. In animals, family names usually end in the suffix *-idae*; Canidae (dogs, wolves and foxes) and Felidae (cats) are both in the order Carnivora.

fat A lipid consisting of three fatty acid molecules joined by ester linkages to a glycerol molecule. Also called *triacylglycerol* or *triglyceride.*

fatty acid A lipid consisting of a hydrocarbon chain bonded to a carboxyl group ($-$COOH) at one end. Used by many organisms to store chemical energy; a major component of animal and plant fats.

fauna All the animals characteristic of a particular region, period, or environment.

feather A specialized skin outgrowth, composed of β-keratin, present in all birds and only in birds. Used for flight, insulation, display, and other purposes.

feces The waste products of digestion.

fecundity The average number of female offspring produced by a single female in the course of her lifetime.

feedback inhibition A type of metabolic control in which high concentrations of the product of a metabolic pathway inhibit one of the enzymes early in the pathway. A form of negative feedback.

fermentation Any of several metabolic pathways that make ATP by transferring electrons from a reduced compound such as glucose to a final electron acceptor other than oxygen. Allows glycolysis to proceed in the absence of oxygen.

ferredoxin In photosynthetic organisms, an iron- and sulfur-containing protein in the electron transport chain of photosystem I. Can transfer electrons to the enzyme NADP$^+$ reductase, which catalyzes formation of NADPH.

fertilization Fusion of the nuclei of two haploid gametes to form a zygote with a diploid nucleus.

fertilization envelope A physical barrier that forms around a fertilized egg in amphibians and some other animals. Formed by an influx of water under the vitelline membrane.

fetal alcohol syndrome A condition, marked by hyperactivity, severe learning disabilities, and depression, thought to be caused by exposure of an individual to high blood alcohol concentrations during embryonic development.

fetus In live-bearing animals, the unborn offspring after the embryonic stage, which usually are developed sufficiently to be recognizable as belonging to a certain species. In humans, from 9 weeks after fertilization until birth.

fiber In plants, a type of elongated sclerenchyma cell that provides support to vascular tissue. Compare with **sclereid.**

fibronectin An abundant protein in the extracellular matrix that binds to other ECM components and to integrins in plasma membranes; helps anchor cells in place. Numerous subtypes are found in different tissues.

Fick's law of diffusion A mathematical relationship that describes the rates of gas exchange in animal respiratory systems.

fight-or-flight response Rapid physiological changes that prepare the body for emergencies. Includes increased heart rate, increased blood pressure, and decreased digestion.

filament Any thin, threadlike structure, particularly (1) the threadlike extensions of a fish's gills or (2) the slender stalk that bears the anthers in a flower.

filter feeder See **suspension feeder.**

filtrate Any fluid produced by filtration, in particular the fluid (pre-urine) in the nephrons of vertebrate kidneys.

filtration A process of removing large components from a fluid by forcing it through a filter. Occurs in a renal corpuscle of the vertebrate kidney, allowing water and small solutes to pass from the blood into the nephron.

finite rate of increase (λ) The rate of increase of a population over a given period of time. Calculated as the ending population size divided by the starting population size. Compare with **intrinsic rate of increase.**

first law of thermodynamics The principle of physics that energy is conserved in any process. Energy can be transferred and converted into different forms, but it cannot be created or destroyed.

fission (1) A form of asexual reproduction in which a prokaryotic cell divides to produce two genetically similar daughter cells by a process similar to mitosis of eukaryotic cells. Also called *binary fission.* (2) A form of asexual reproduction in which an animal splits into two or more individuals of approximately equal size; common among invertebrates.

fitness The relative ability of an individual to produce viable offspring compared with other individuals in the same population. Also called *Darwinian fitness.*

fitness trade-off See **trade-off.**

fixed action pattern (FAP) Highly stereotyped behavior pattern that occurs in a certain invariant way in a certain species. A form of innate behavior.

flaccid Limp as a result of low internal pressure (e.g., a wilted plant leaf). Compare with **turgid.**

flagellum (plural: flagella) A long, cellular projection that undulates (in eukaryotes) or rotates (in prokaryotes) to move the cell through an aqueous environment. See **axoneme.**

flatworms Members of the phylum Platyhelminthes. Distinguished by a broad, flat, unsegmented body that lacks a coelom. Flatworms belong to the lophotrochozoan branch of the protostomes.

flavin adenine dinucleotide See **FAD/FADH$_2$.**

flexor A muscle that pulls two bones closer together, as in the flexing of a limb or the spine. Compare with **extensor.**

floral meristem A group of undifferentiated plant cells that can give rise to the four organs making up a flower.

florigen In plants, a protein hormone that is synthesized in leaves and transported to the shoot apical meristem where it stimulates flowering.

flower In angiosperms, the part of a plant that contains reproductive structures. Typically includes a calyx, a corolla, and one or more stamens and/or carpels. See **perfect** and **inperfect flower.**

fluid feeder An animal that feeds by sucking or mopping up liquids such as nectar, plant sap, or blood.

fluid-mosaic model The widely accepted hypothesis that the plasma membrane and organelle membranes consist of proteins embedded in a fluid phospholipid bilayer.

fluorescence The spontaneous emission of light from an excited electron falling back to its normal (ground) state.

follicle An egg cell and its surrounding ring of supportive cells in a mammalian ovary.

follicle-stimulating hormone (FSH) A peptide hormone, produced and secreted by the anterior pituitary, that stimulates (in females) growth of eggs and follicles in the ovaries or (in males) sperm production in the testes.

follicular phase The first major phase of a menstrual cycle during which follicles grow and estrogen levels increase; ends with ovulation.

food Any nutrient-containing material that can be consumed and digested by animals.

food chain A relatively simple pathway of energy flow through a few species, each at a different trophic level, in an ecosystem. Might include, for example, a primary producer, a primary consumer, a secondary consumer, and a decomposer. Compare with **food web.**

food vacuole A membrane-bound organelle containing food ingested by a cell via phagocytosis.

food web Any complex pathway along which energy moves among many species at different trophic levels of an ecosystem.

foot One of the three main parts of the mollusc body; a muscular appendage, used for movement and/or burrowing into sediment.

forebrain One of the three main regions of the vertebrate brain; includes the cerebrum, thalamus, and hypothalamus. Compare with **hindbrain** and **midbrain.**

fossil Any trace of an organism that existed in the past. Includes tracks, burrows, fossilized bones, casts, etc.

fossil record All of the fossils that have been found anywhere on Earth and that have been formally described in the scientific literature.

founder effect A change in allele frequencies that often occurs when a new population is established from a small group of individuals (founder event) due to sampling error (i.e., the small group is not a representative sample of the source population).

free energy See **Gibbs free-energy change.**

free radical Any substance containing one or more atoms with an unpaired electron. Unstable and highly reactive.

frequency The number of wave crests per second traveling past a stationary point. Determines the pitch of sound and the color of light.

fronds The large leaves of ferns.

fruit In flowering plants (angiosperms), a mature, ripened plant ovary (or group of ovaries), along with the seeds it contains and any adjacent fused parts. See **aggregate, multiple,** and **simple fruit.**

fruiting body A structure formed in some prokaryotes, fungi, and protists for spore dispersal; usually consists of a base, a stalk, and a mass of spores at the top.

functional genomics The study of how a genome works, that is, when and where specific genes are expressed and how their products interact to produce a functional organism.

functional group A small group of atoms bonded together in a precise configuration and exhibiting particular chemical properties that it imparts to any organic molecule in which it occurs.

fundamental niche The ecological space that a species occupies in its habitat in the absence of competitors. Compare with **realized niche.**

fungi A lineage of eukaryotes that typically have a filamentous body (mycelium) and obtain nutrients by absorption.

fungicide Any substance that can kill fungi or slow their growth.

G_1 phase The phase of the cell cycle that constitutes the first part of interphase before DNA synthesis (S phase).

G_2 phase The phase of the cell cycle between synthesis of DNA (S phase) and mitosis (M phase); the last part of interphase.

gall A tumorlike growth that forms on plants infected with certain bacteria or parasites.

gallbladder A small pouch that stores bile from the liver and releases it as needed into the small intestine during digestion of fats.

gametangium (plural: gametangia) (1) The gamete-forming structure found in all land plants except angiosperms. Contains a sperm-producing antheridium and an egg-producing archegonium. (2) The gamete-forming structure of some chytrid fungi.

gamete A haploid reproductive cell that can fuse with another haploid cell to form a zygote. Most multicellular eukaryotes have two distinct forms of gametes: egg cells (ova) and sperm cells.

gametogenesis The production of gametes (eggs or sperm).

gametophyte In organisms undergoing alternation of generations, the multicellular haploid form that arises from a single haploid spore and produces gametes. A female gametophyte is commonly called an *embryo sac;* a male gametophyte, a *pollen grain.* Compare with **sporophyte.**

ganglion cell A neuron in the vertebrate retina that collects visual information from one or several bipolar cells and sends it to the brain via the optic nerve.

gap junction A type of cell-cell attachment structure that directly connects the cytoplasms of adjacent animal cells, allowing passage of water, ions, and small molecules between the cells. Compare with **desmosome** and **tight junction.**

gastrin A hormone produced by cells in the stomach lining in response to the arrival of food or to a neural signal from the brain. Stimulates other stomach cells to release hydrochloric acid.

gastrointestinal (GI) tract See **digestive tract.**

gastropods A lineage of mollusc distinguished by a large muscular foot and a unique feeding structure, the radula. Include slugs and snails.

gastrulation The process by which some cells on the outside of a young embryo move to the interior of the embryo, resulting in the three distinct germ layers (endoderm, mesoderm, and ectoderm).

gated channel A channel protein that opens and closes in response to a specific stimulus, such as the binding of a particular molecule or a change in the electrical charge on the outside of the membrane.

gel electrophoresis A technique for separating molecules on the basis of size and electric charge, which affect their differing rates of movement through a gelatinous substance in an electric field.

gemma (plural: gemmae) A small reproductive structure that is produced in some liverworts during the gametophyte phase and can grow into mature gametophyte.

gene A section of DNA (or RNA, for some viruses) that encodes information for building one or more related polypeptides or functional RNA molecules along with the regulatory sequences required for its transcription.

gene duplication The formation of an additional copy of a gene, typically by misalignment of chromosomes during crossing over. Thought to be an important evolutionary process in creating new genes.

gene expression Overall process by which the information encoded in genes is converted into an active product, most commonly a protein. Includes transcription and translation of a gene and in some cases protein activation.

gene family A set of genetic loci whose DNA sequences are extremely similar. Thought to have arisen by duplication of a single ancestral gene and subsequent mutations in the duplicated sequences.

gene flow The movement of alleles between populations; occurs when individuals leave one population, join another, and breed.

gene-for-gene hypothesis The hypothesis that there is a one-to-one correspondence between the resistance (*R*) loci of plants and the avirulence (*avr*) loci of pathogenic fungi; particularly, that *R* genes produce receptors and *avr* genes produce molecules that bind to those receptors.

gene pool All of the alleles of all of the genes in a certain population.

generation The average time between a mother's first offspring and her daughter's first offspring.

gene therapy The treatment of an inherited disease by introducing normal alleles.

genetic bottleneck A reduction in allelic diversity resulting from a sudden reduction in the size of a large population (population bottleneck) due to a random event.

genetic code The set of all 64 codons and the particular amino acids that each specifies.

genetic correlation A type of evolutionary constraint in which selection on one trait causes a change in another trait as well; may occur when the same gene(s) affect both traits.

genetic diversity The diversity of alleles in a population, species, or group of species.

genetic drift Any change in allele frequencies due to random events. Causes allele frequencies to drift up and down randomly over time, and eventually can lead to the fixation or loss of alleles.

genetic homology Similarities in DNA sequences or amino acid sequences that are due to inheritance from a common ancestor.

genetic map An ordered list of genes on a chromosome that indicates their relative distances from each other. Also called a *linkage map* or *meiotic map*. Compare with **physical map.**

genetic marker A genetic locus that can be identified and traced in populations by laboratory techniques or by a distinctive visible phenotype.

genetic model A set of hypotheses that explain how a certain trait is inherited.

genetic recombination A change in the combination of genes or alleles on a given chromosome or in a given individual. Also called *recombination.*

genetics The field of study concerned with the inheritance of traits.

genetic screen Any of several techniques for identifying individuals with a particular type of mutation. Also called a *screen.*

genetic variation (1) The number and relative frequency of alleles present in a particular population. (2) The proportion of phenotypic variation in a trait that is due to genetic rather than environmental influences in a certain population in a certain environment.

genitalia External copulatory organs.

genome All of the hereditary information in an organism, including not only genes but also other non-gene stretches of DNA.

genomic library A set of DNA segments representing the entire genome of a particular organism. Each segment is carried by a plasmid or other cloning vector and can be separated from other segments. Compare with **cDNA library.**

genomics The field of study concerned with sequencing, interpreting, and comparing whole genomes from different organisms.

genotype All of the alleles of every gene present in a given individual. May refer specifically to the alleles of a particular set of genes under study. Compare with **phenotype.**

genus (plural: genera) In Linnaeus' system, a taxonomic category of closely related species. Always italicized and capitalized to indicate that it is a recognized scientific genus.

geologic time scale The sequence of eons, epochs, and periods used to describe the geologic history of Earth.

geometric isomer A molecule that shares the same molecular formula as another molecule but differs in the arrangement of atoms or groups on either side of a double bond or ring structure. Compare with **optical isomer** and **structural isomer.**

germ cell In animals, any cell that can potentially give rise to gametes. Also called *germ-line cells.*

germination The process by which a seed becomes a young plant.

germ layer In animals, one of the three basic types of tissue formed during gastrulation; gives rise to all other tissues. See **endoderm, mesoderm,** and **ectoderm.**

germ theory of disease The theory that infectious diseases are caused by bacteria, viruses, and other microorganisms.

gestation The duration of embryonic development from fertilization to birth in those species that have live birth.

gibberellins A class of hormones found in plants and fungi that stimulate growth. Gibberellic acid is one of the major gibberellins.

Gibbs free-energy change (ΔG) A measure of the change in potential energy and entropy that occurs in a given chemical reaction. $\Delta G < 0$ for spontaneous reactions and >0 for nonspontaneous reactions.

gill Any organ in aquatic animals that exchanges gases and other dissolved substances between the blood and the surrounding water. Typically, a filamentous outgrowth of a body surface.

gill arch In aquatic vertebrates, curved region of tissue between the gills. Gills are suspended from the gill arches.

gill filament In fish, one of the many long, thin structures that extend from gill arches into the water and across which gas exchange occurs.

gill lamella (plural: gill lamellae) One of hundreds to thousands of sheetlike structures, each containing a capillary bed, that makes up a gill filament.

gland An organ whose primary function is to secrete some substance, either into the blood (endocrine gland) or into some other space such as the gut or skin (exocrine gland).

glia Collective term for several types of cells in nervous tissue that are not neurons and do not conduct electrical signals but provide support, nourishment, and electrical insulation and perform other functions. Also called *glial cells.*

global warming A sustained increase in Earth's average surface temperature.

glomerulus (plural: glomeruli) (1) In the vertebrate kidney, a ball-like cluster of capillaries, surrounded by Bowman's capsule, at the beginning of a nephron. (2) In the brain, a ball-shaped cluster of neurons in the olfactory bulb.

glucagon A peptide hormone produced by the pancreas in response to low blood glucose. Raises blood glucose by triggering breakdown of glycogen and stimulating gluconeogenesis. Compare with **insulin.**

glucocorticoids A class of steroid hormones, produced and secreted by the adrenal cortex, that increase blood glucose and prepare the body for stress. Include cortisol and corticosterone. Compare with **mineralocorticoids.**

gluconeogenesis Synthesis of glucose from non-carbohydrate sources (e.g., proteins and fatty acids). Occurs in the liver in response to low insulin levels and high glucagon levels.

glucose Six-carbon monosaccharide whose oxidation in cellular respiration is the major source of ATP in animal cells.

glyceraldehyde-3-phosphate (G3P) The phosphorylated three-carbon compound formed as the result of carbon fixation in the first step of the Calvin cycle.

glycerol A three-carbon molecule that forms the "backbone" of phospholipids and most fats.

glycogen A highly branched storage polysaccharide composed of α-glucose monomers joined by 1,4- and 1,6-glycosidic linkages. The major form of stored carbohydrate in animals.

glycolipid Any lipid molecule that is covalently bonded to a carbohydrate group.

glycolysis A series of 10 chemical reactions that oxidize glucose to produce pyruvate and ATP. Used by all organisms as part of fermentation or cellular respiration.

glycoprotein Any protein with one or more covalently bonded carbohydrate groups.

glycosidic linkage The covalent bond formed by a condensation reaction between two sugar monomers; joins the residues of a polysaccharide.

glycosylation Addition of a carbohydrate group to a molecule.

glyoxisome Specialized type of peroxisome found in plant cells and packed with enzymes for processing the products of photosynthesis.

goblet cell A cell in the stomach lining that secretes mucus.

goiter A pronounced swelling of the thyroid gland; usually caused by a deficiency of iodine in the diet.

Golgi apparatus A eukaryotic organelle, consisting of stacks of flattened membranous sacs (cisternae), that functions in processing and sorting proteins and lipids destined to be secreted or directed to other organelles. Also called *Golgi complex.*

gonad An organ that produces reproductive cells, such as a testis or an ovary.

gonadotropin-releasing hormone (GnRH) A peptide hormone, produced and secreted by the hypothalamus, that stimulates release of FSH and LH from the anterior pituitary.

G protein Any of various peripheral membrane proteins that bind GTP and function in signal transduction. Binding of a signal to its receptor triggers activation of the G protein, leading to production of a second messenger or initiation of a phosphorylation cascade.

grade In taxonomy, a group of species that share a position in an inferred evolutionary sequence of lineages but that are not a monophyletic group. Also called a *paraphyletic group.*

Gram-negative Describing bacteria that look pink when treated with a Gram stain. These bacteria have a cell wall composed of a thin layer of peptidoglycan and an outer phospholipid layer.

Gram-positive Describing bacteria that look purple when treated with a Gram stain. These bacteria have cell walls composed of a thick layer of peptidoglycan.

Gram stain A dye that distinguishes the two general types of cell walls found in bacteria. Used to routinely classify bacteria as Gram-negative or Gram-positive.

granum (plural: grana) In chloroplasts, a stack of flattened, membrane-bound vesicles (thylakoids) where the light reactions of photosynthesis occur.

gravitropism The growth or movement of a plant in a particular direction in response to gravity.

grazing food chain The ecological network of herbivores and the predators and parasites that consume them.

great apes See **hominids.**

green algae A paraphyletic group of photosynthetic organisms that contain chloroplasts similar to those in green plants. Often classified as protists, green algae are the closest living relatives of land plants and form a monophyletic group with them.

greenhouse gas An atmospheric gas that absorbs and reflects infrared radiation, so that heat radiated from Earth is retained in the atmosphere instead of being lost to space.

green plant A member of a lineage of eukaryotes that includes green algae and land plants.

gross photosynthetic productivity The efficiency with which all the plants in a given area use the light energy available to them to produce sugars.

gross primary productivity In an ecosystem, the total amount of carbon fixed by photosynthesis, including that used for cellular respiration, over a given time period. Compare with **net primary productivity.**

ground meristem The middle layer of a young plant embryo. Gives rise to the ground tissue system.

ground tissue system In plants, all the tissues beneath the outer protective layers of epidermis and cork except for vascular tissue. Also called simply *ground tissue.*

groundwater Any water below the land surface.

growth factor Any of a large number of signaling molecules that are secreted by certain cells and that stimulate other cells to divide or to differentiate.

growth hormone (GH) A peptide hormone, produced and secreted by the mammalian anterior pituitary, that promotes lengthening of the long bones in children and muscle growth, tissue repair, and lactation in adults. Also called *somatotropin.*

GTP See **guanosine triphosphate.**

guanosine triphosphate (GTP) A molecule consisting of guanine, a sugar, and three phosphate groups. Can be hydrolyzed to release free energy. Commonly used in many cellular reactions; also functions in signal transduction in association with G proteins.

guard cell One of two specialized, crescent-shaped cells forming the border of a plant stoma. Guard cells can change shape to open or close the stoma. See also **stoma.**

gustation The perception of taste.

guttation Excretion of water droplets from plant leaves in the early morning, caused by root pressure.

gymnosperm A vascular plant that makes seeds but does not produce flowers. The gymnosperms include four lineages of green plants (cycads, ginkgoes, conifers, and gnetophytes). Compare with **angiosperm.**

habitat destruction Human-caused destruction of a natural habitat with replacement by an urban, suburban, or agricultural landscape.

habitat fragmentation The breakup of a large region of a habitat into many smaller regions, separated from others by a different type of habitat.

Hadley cell An atmospheric cycle of large-scale air movement in which warm equatorial air rises, moves north or south, and then descends at approximately 30°N or 30°S latitude.

hair cell A pressure-detecting sensory cell, found in the cochlea, that has tiny "hairs" (stereocilia) jutting from its surface.

hairpin A secondary structure in RNA consisting of a stable loop formed by hydrogen bonding between purine and pyrimidine bases on the same strand.

halophile A bacterium or archaean that thrives in high-salt environments.

halophyte A plant that thrives in salty habitats.

Hamilton's rule The proposition that an allele for altruistic behavior will be favored by natural selection only if $Br > C$, where B = the fitness benefit to the recipient, C = the fitness cost to the actor, and r = the coefficient of relatedness between recipient and actor.

haploid (1) Having one set of chromosomes ($1n$). (2) A cell or an individual organism with one set of chromosomes. Compare with **diploid.**

haploid number The number of distinct chromosome sets in a cell. Symbolized as n.

haplotype The set of alleles found on a single chromosome.

Hardy-Weinberg principle A principle of population genetics stating that genotype frequencies in a large population do not change from generation to generation in the absence of evolutionary processes (e.g., mutation, migration, genetic drift, random mating, and selection).

heart A muscular pump that circulates blood throughout the body.

heart murmur A distinctive sound caused by backflow of blood through a defective heart valve.

heartwood The older xylem in the center of an older stem or root, containing protective compounds and no longer functioning in water transport.

heat Thermal energy that is transferred from an object at higher temperature to one at lower temperature.

heat of vaporization The energy required to vaporize 1 gram of a liquid into a gas.

heavy chain The larger of the two types of polypeptide chains in an antibody molecule; composed of a variable (V) region, which contributes to the antigen-binding site, and a constant (C) region. Differences in heavy-chain constant regions determine the different classes of immunoglobulins (IgA, IgE, etc.). Compare with **light chain.**

helicase An enzyme that catalyzes the breaking of hydrogen bonds between nucleotides of DNA, "unzipping" a double-stranded DNA molecule.

helix-turn-helix motif A motif seen in many repressor proteins in prokaryotes, consisting of two α-helices connected by a short stretch of amino acids that form a turn.

helper T cell An effector T cell that secretes cytokines and in other ways promotes the activation of other lymphocytes. Is descended from an activated $CD4^+$ T cell that has interacted with antigen presented by dendritic cells, macrophages, or B cells.

hemimetabolous metamorphosis A type of metamorphosis in which the animal increases in size from one stage to the next, but does not dramatically change its body form. Also called *incomplete metamorphosis.*

hemocoel A body cavity, present in arthropods and some molluscs, containing a pool of circulatory fluid (hemolymph) bathing the internal organs.

hemoglobin An oxygen-binding protein consisting of four polypeptide subunits, each containing an oxygen-binding heme group. The major oxygen carrier in mammalian blood.

hemolymph The circulatory fluid of animals with open circulatory systems (e.g., insects) in which the fluid is not confined to blood vessels.

hemophilia A human disease, caused by an X-linked recessive allele, that is characterized by defects in the blood-clotting system.

herbaceous Referring to a plant that is not woody.

herbivore (adjective: herbivorous) An animal that eats primarily plants and rarely or never eats meat. Compare with **carnivore** and **omnivore.**

herbivory The practice of eating plant tissues.

heredity The transmission of traits from parents to offspring via genetic information.

heritable Referring to traits that can be transmitted from one generation to the next.

hermaphrodite An organism that produces both male and female gametes.

heterokaryotic Describing a cell or fungal mycelium containing two or more nuclei that are genetically distinct.

heterospory (adjective: heterosporous) In seed plants, the production of two distinct types of spore-producing structures and thus two distinct types of spores: microspores, which become the male gametophyte, and megaspores, which become the female gametophyte. Compare with **homospory.**

heterotherm An animal whose body temperature varies markedly with environmental conditions. Compare with **homeotherm.**

heterotroph Any organism that cannot synthesize reduced organic compounds from inorganic sources and that must obtain them by eating other organisms. Some bacteria, some archaea, and virtually all fungi and animals are heterotrophs. Also called *consumer.* Compare with **autotroph.**

heterozygote advantage A pattern of natural selection that favors heterozygous individuals compared with homozygotes. Tends to maintain genetic variation in a population. Also called *heterozygote superiority.*

heterozygous Having two different alleles of a certain gene.

hexose A monosaccharide (simple sugar) containing six carbon atoms.

hibernation An energy-conserving physiological state, marked by a decrease in metabolic rate, body temperature, and activity, that lasts for a prolonged period (weeks to months). Occurs in some animals in response to winter cold and scarcity of food. Compare with **torpor.**

hindbrain One of the three main regions of the vertebrate brain; includes the cerebellum and medulla oblongata. Compare with **forebrain** and **midbrain.**

histamine A molecule released from mast cells during an inflammatory response that causes blood vessels to dilate and become more permeable.

histone One of several positively charged (basic) proteins associated with DNA in the chromatin of eukaryotic cells.

histone acetyl transferase (HAT) In eukaryotes, one of a class of enzymes that loosen chromatin structure by adding acetyl groups to histone proteins.

histone deacetylase (HDAC) In eukaryotes, one of a class of enzymes that recondense chromatin by removing acetyl groups from histone proteins.

HIV See human immunodeficiency virus (HIV).

holoenzyme A multipart enzyme consisting of a core enzyme (containing the active site for catalysis) along with other required proteins.

holometabolous metamorphosis A type of metamorphosis in which the animal completely changes its form. Also called *complete metamorphosis.*

homeosis Replacement of one body part by another normally found elsewhere in the body as the result of mutation in certain developmentally important genes (homeotic genes).

homeostasis (adjective: homeostatic) The array of relatively stable chemical and physical conditions in an animal's cells, tissues, and organs. May be achieved by the body's passively matching the conditions of a stable external environment (conformational homeostasis) or by active physiological processes (regulatory homeostasis) triggered by variations in the external or internal environment.

homeotherm An animal that has a constant or relatively constant body temperature. Compare with **heterotherm.**

homeotic gene Any gene that specifies a particular location within an embryo, leading to the development of structures appropriate for that location. Mutations in homeotic genes cause the development of extra body parts or body parts in the wrong places.

hominids Members of the family Hominidae, which includes humans and extinct related forms; chimpanzees, gorillas, and orangutans. Distinguished by large body size, no tail, and an exceptionally large brain. Also called *great apes.*

hominins Humans and extinct related forms; species in the lineage that branched off from chimpanzees and eventually led to humans.

homologous chromosomes In a diploid organism, chromosomes that are similar in size, shape, and gene content. Also called *homologs.*

homology (adjective: homologous) Similarity among organisms of different species due to their inheritance from a common ancestor. Features that exhibit such similarity (e.g., DNA sequences, proteins, body parts) are said to be homologous. Compare with **homoplasy.**

homoplasy Similarity among organisms of different species due to convergent evolution. Compare with **homology.**

homospory (adjective: homosporous) In seedless vascular plants, the production of just one type of spore. Compare with **heterospory.**

homozygous Having two identical alleles of a certain gene.

hormone Any of numerous different signaling molecules that circulate throughout the body in blood or other body fluids and can trigger characteristic responses in distant target cells at very low concentrations.

hormone-response element A specific sequence in DNA to which a steroid hormone-receptor complex can bind and affect gene transcription.

host An individual or a species in or on which a parasite lives.

host cell A cell that has been invaded by an organism such as a parasite or a virus.

***Hox* genes** A class of homeotic genes found in several animal phyla, including vertebrates, that are expressed in a distinctive pattern along the anterior-posterior axis in early embryos and control formation of segment-specific structures.

human Any member of the genus *Homo,* which includes modern humans (*Homo sapiens*) and several extinct species.

human chorionic gonadotropin (hCG) A glycoprotein hormone produced by the human placenta from about week 3 to week 14 of pregnancy. Maintains the corpus luteum, which produces hormones that preserve the uterine lining.

Human Genome Project The multinational research project that sequenced the human genome.

human immunodeficiency virus (HIV) A retrovirus that causes AIDS (acquired immune deficiency syndrome) in humans.

humoral (immune) response The type of immune response that involves generation of antibody-secreting plasma cells from activated B cells. Defends against extracellular pathogens. Compare with **cell-mediated (immune) response.**

humus The completely decayed organic matter in soils.

Huntington's disease A degenerative brain disease of humans caused by an autosomal dominant allele.

hybrid The offspring of parents from two different strains, populations, or species.

hybrid zone A geographic area where interbreeding occurs between two species, sometimes producing fertile hybrid offspring.

hydrocarbon An organic molecule that contains only hydrogen and carbon atoms.

hydrogen bond A weak interaction between two molecules or different parts of the same molecule resulting from the attraction between a hydrogen atom with a partial positive charge and another atom (usually O or N) with a partial negative charge. Compare with **covalent bond** and **ionic bond.**

hydrogen ion (H^+) A single proton with a charge of 1+; typically, one that is dissolved in solution or that is being transferred from one atom to another in a chemical reaction.

hydrolysis A chemical reaction in which a molecule is split into smaller molecules by reacting with water. In biology, most hydrolysis reactions involve the splitting of polymers into monomers. Compare with **condensation reaction.**

hydrophilic Interacting readily with water. Hydrophilic compounds are typically polar compounds containing charged or electronegative atoms. Compare with **hydrophobic.**

hydrophobic Not interacting readily with water. Hydrophobic compounds are typically nonpolar compounds that lack charged or electronegative atoms and often contain many C–C and C–H bonds. Compare with **hydrophilic.**

hydroponic growth Growth of plants in liquid cultures instead of soil.

hydrostatic skeleton A system of body support involving fluid-filled compartments that can change in shape but cannot easily be compressed.

hydroxide ion (OH^-) An oxygen atom and a hydrogen atom joined by a single covalent bond and carrying a negative charge; formed by dissociation of water.

hyperpolarization Change in membrane potential from its resting negative state to an even more negative state; a normal phase in an action potential. Compare with **depolarization.**

hypersensitive response In plants, the rapid death of a cell that has been infected by a pathogen, thereby reducing the potential for infection to spread throughout a plant. Compare with **systemic acquired resistance.**

hypertension Abnormally high blood pressure.

hypertonic Comparative term designating a solution that has a greater solute concentration, and therefore a lower water concentration, than another solution. Compare with **hypotonic** and **isotonic.**

hypha (plural: hyphae) One of the strands of a fungal mycelium (the meshlike body of a fungus). Also found in some protists.

hypocotyl The stem of a very young plant; the region between the cotyledon (embryonic leaf) and the radicle (embryonic root).

hypothalamic-pituitary axis The functional interaction of the hypothalamus and the pituitary gland, which are anatomically distinct but work together to regulate most of the other endocrine glands in the body.

hypothalamus A part of the brain that functions in maintaining the body's internal physiological state by regulating the autonomic nervous system, endocrine system, body temperature, water balance, and appetite.

hypothesis A proposed explanation for a phenomenon or for a set of observations.

hypotonic Comparative term designating a solution that has a lower solute concentration, and therefore a higher water concentration, than another solution. Compare with **hypertonic** and **isotonic.**

immigration The migration of individuals into a particular population from other populations. Compare with **emigration.**

immune system In vertebrates, the system whose primary function is to defend the body against pathogens. Includes several types of cells (e.g., lymphocytes and macrophages) and several organs where they develop or reside (e.g., lymph nodes and thymus).

immunity (adjective: immune) State of being protected against infection by disease-causing pathogens either by relatively nonspecific mechanisms (innate immunity) or by specific mechanisms triggered by exposure to a particular antigen (acquired immunity).

immunization The conferring of immunity to a particular disease by artificial means.

immunoglobulin (Ig) Any of the class of proteins that function as antibodies.

immunological memory The ability of the immune system to "remember" an antigen and mount a rapid, effective response to a pathogen encountered years or decades earlier.

imperfect flower A flower that contains male parts (stamens) *or* female parts (carpels) but not both. Compare with **perfect flower.**

implantation The process by which an embryo buries itself in the uterine wall and forms a placenta. Occurs in mammals and a few other vertebrates.

imprinting A type of rapid, irreversible learning in which a young animal learns to recognize the individual caring for it. Occurs in birds and mammals.

inbreeding Mating between closely related individuals. Increases homozygosity of a population and often leads to a decline in the average fitness (inbreeding depression).

incomplete dominance An inheritance pattern in which the heterozygote phenotype is a blend or combination of both homozygote phenotypes.

incomplete metamorphosis See **hemimetabolous metamorphosis.**

independent assortment, principle of The concept that each pair of hereditary elements (alleles of the same gene) behaves independently of other genes during meiosis. One of Mendel's two principles of genetics.

indeterminate growth A pattern of growth in which an individual continues to increase its overall body size throughout its life.

indicator plate A laboratory technique for detecting mutant cells by growing them on agar plates containing a compound that when metabolized by wild-type cells yields a colored product.

induced fit Change in the shape of the active site of an enzyme, as the result of the initial weak binding of a substrate, so that it binds substrate more tightly.

inducer A small molecule that triggers transcription of a specific gene, often by binding to and inactivating a repressor protein.

inducible defense A defensive trait that is manifested only in response to the presence of a predator or pathogen. Compare with **constitutive defense.**

induction (1) The process by which one embryonic cell, or group of cells, alters the differentiation of neighboring cells. (2) Positive control of gene expression by a regulatory protein that binds to DNA and triggers transcription of a specific gene(s).

infection thread An invagination of the membrane of a root hair through which beneficial nitrogen-fixing bacteria enter the roots of their host plants (legumes).

inflammatory response An aspect of the innate immune response, seen in most cases of infection or tissue injury, in which the affected tissue becomes swollen, red, warm, and painful.

inhibition In ecological succession, the phenomenon in which early-arriving species make conditions less favorable for the establishment of certain later-arriving species. Compare with **facilitation** and **tolerance.**

inhibitory postsynaptic potential (IPSP) A change in membrane potential, usually hyperpolarization, at a neuron dendrite that makes an action potential less likely.

initiation (1) In an enzyme-catalyzed reaction, the stage during which enzymes orient reactants precisely as they bind at specific locations within the enzyme's active site. (2) In DNA transcription, the stage during which RNA polymerase and other proteins assemble at the promoter sequence. (3) In RNA translation, the stage during which a complex consisting of a ribosome, a mRNA molecule, and an aminoacyl tRNA corresponding to the start codon is formed.

initiation factors A class of proteins that assist ribosomes in binding to a messenger RNA molecule to begin translation.

innate behavior Behavior that is inherited genetically, does not have to be learned, and is typical of a species.

innate immune response See **innate immunity.**

innate immunity A set of nonspecific defenses against pathogens that exist before exposure to an antigen and involves mast cells, neutrophils, and macrophages; typically results in an inflammatory response. Compare with **acquired immunity.**

inner cell mass (ICM) A cluster of cells in the interior of a mammalian blastocyst that undergo gastrulation and eventually develop into the embryo.

inner ear The innermost portion of the mammalian ear, consisting of a fluid-filled system of tubes that includes the cochlea (which receives sound vibrations from the middle ear) and the semicircular canals (which function in balance).

insects The largest lineage of arthropods distinguished by having three body regions (a head, thorax, and abdomen), walking legs, and in most species one or two pairs of wings.

in situ hybridization A technique for detecting specific DNAs and mRNAs in cells and tissues by use of labeled probes. Can be used to determine where and when particular genes are expressed in embryos.

insulin A peptide hormone produced by the pancreas in response to high levels of glucose (or amino acids) in blood. Enables cells to absorb glucose and coordinates synthesis of fats, proteins, and glycogen. Compare with **glucagon.**

integral membrane protein Any membrane protein that spans the entire lipid bilayer. Also called *transmembrane protein*. Compare with **peripheral membrane protein.**

integration In the nervous system, processing of information from many sources.

integrator A component of an animal's nervous system that functions as part of a homeostatic system by evaluating sensory information and triggering appropriate responses. See **effector** and **sensor.**

integrin Any of a class of cell-surface proteins that bind to fibronectins and other proteins in the extracellular matrix, thus holding cells in place.

intercalated disc A specialized junction between adjacent heart muscle cells that contains gap junctions, allowing electrical signals to pass between the cells.

intermediate disturbance hypothesis The hypothesis that moderate ecological disturbance is associated with higher species diversity than either low or high disturbance.

intermediate filament A long fiber, about 10 nm in diameter, composed of one of various proteins (e.g., keratins, lamins); one of the three types of cytoskeletal fibers. Form networks that help maintain cell shape and hold the nucleus in place. Compare with **actin filament** and **microtubule.**

intermediate host The host species in which a parasite reproduces asexually. Compare with **definitive host.**

interneuron A neuron that passes signals from one neuron to another. Compare with **motor neuron** and **sensory neuron.**

internode The section of a plant stem between two nodes (sites where leaves attach).

interphase The portion of the cell cycle between one mitotic (M) phase and the next. Includes the G_1 phase, S phase, and G_2 phase.

interspecific competition Competition between members of different species for the same limited resource. Compare with **intraspecific competition.**

interstitial fluid The plasma-like fluid found in the region (interstitial space) between cells.

intertidal zone The region between the low-tide and high-tide marks on a seashore.

intraspecific competition Competition between members of the same species for the same limited resource. Compare with **interspecific competition.**

intrinsic rate of increase (r_{max}) The rate at which a population will grow under optimal conditions (i.e., when birthrates are as high as possible and death rates are as low as possible). Compare with **finite rate of increase.**

intron A region of a eukaryotic gene that is transcribed into RNA but is later removed, so it is not translated into a peptide or protein. Compare with **exon.**

invasive species An exotic (nonnative) species that, upon introduction to a new area, spreads rapidly and competes successfully with native species.

inversion A mutation in which a segment of a chromosome breaks from the rest of the chromosome, flips, and rejoins with the opposite orientation as before.

invertebrates A paraphyletic group composed of animals without a backbone; includes about 95 percent of all animal species. Compare with **vertebrates.**

ion An atom or a molecule that has lost or gained electrons and thus carries an electric charge, either positive (cation) or negative (anion), respectively.

ion channel A type of channel protein that allows certain ions to diffuse across a plasma membrane down an electochemical gradient.

ionic bond A chemical bond that is formed when an electron is completely transferred from one atom to another so that the atoms remain associated due to their opposite electric charges. Compare with **covalent bond** and **hydrogen bond.**

iris A ring of pigmented muscle just below the cornea in the vertebrate eye that contracts or expands to control the amount of light entering the eye through the pupil.

islets of Langerhans Clusters of cells in the pancreas that secrete insulin and glucagon directly into the blood.

isomer A molecule that has the same molecular formula as another molecule but differs from it in three-dimensional structure.

isotonic Comparative term designating a solution that has the same solute concentration and water concentration than another solution. Compare with **hypertonic** and **hypotonic.**

isotope Any of several forms of an element that have the same number of protons but differ in the number of neutrons.

joint A place where two components (bones, cartilages, etc.) of a skeleton meet. May be movable (an articulated joint) or immovable (e.g., skull sutures).

juvenile hormone An insect hormone that prevents larvae from metamorphosing into adults.

karyogamy Fusion of two haploid nuclei to form a diploid nucleus. Occurs in many fungi, and in animals and plants during fertilization of gametes.

karyotype The distinctive appearance of all of the chromosomes in an individual, including the number of chromosomes and their length and banding patterns (after staining with dyes).

keystone species A species that has an exceptionally great impact on the other species in its ecosystem relative to its abundance.

kidney In terrestrial vertebrates, one of a paired organ situated at the back of the abdominal cavity that filters the blood, produces urine, and secretes several hormones.

kilocalorie (kcal) A unit of energy often used to measure the energy content of food. Also called *Calorie.*

kinesin Any one of a class of motor proteins that use the chemical energy of ATP to transport vesicles, particles, or chromosomes along microtubules.

kinetic energy The energy of motion. Compare with **potential energy.**

kinetochore A protein structure at the centromere where spindle fibers attach to the sister chromatids of a replicated chromosome. Contains motor proteins that move a chromosome along a microtubule.

kingdom In Linnaeus' system, a taxonomic category above the phylum level and below the domain level.

kinocilium (plural: kinocilia) A single cilium that juts from the surface of many hair cells and functions in detection of sound or pressure.

kin selection A form of natural selection that favors traits that increase survival or reproduction of an individual's kin at the expense of the individual.

Klinefelter syndrome A syndrome seen in humans who have an XXY karyotype. People with this syndrome have male sex organs, may have some female traits, and are sterile.

knock-out mutant A mutant allele that does not function at all, or an organism homozygous for such a mutation. Also called *null mutant* or *loss-of-function mutant.*

Koch's postulates Four criteria used to determine whether a suspected infectious agent causes a particular disease.

Krebs cycle A series of chemical reactions, occurring in mitochondria, in which acetyl CoA is oxidized to CO_2, producing ATP and reduced compounds for the electron transport chain. Also called *citric acid cycle.*

labia major (plural: labium majus) One of two outer folds of skin that protect the labia minora, clitoris, and vaginal opening of female mammals.

labia minora (plural: labium minus) One of two inner folds of skin that protect the opening of the urethra and vagina.

labor The strong muscular contractions of the uterus that expel the fetus during birth.

lactation (verb: lactate) Production of milk from mammary glands of mammals.

lacteal A small lymphatic vessel extending into the center of a villus in the small intestine. Receives chylomicrons containing fat absorbed from food.

lactic acid fermentation Catabolic pathway in which pyruvate produced by glycolysis is converted to lactic acid in the absence of oxygen.

lagging strand In DNA replication, the strand of new DNA that is synthesized discontinuously in a series of short pieces that are later joined together. Also called *discontinuous strand.* Compare with **leading strand.**

lamellae (singular: lamella) Any set of parallel platelike structures (e.g., the crescent-shaped flaps on the gill filaments of fish gills that serve to increase surface area for gas exchange).

large intestine The distal portion of the digestive tract consisting of the cecum, colon, and rectum. Its primary function is to compact the wastes delivered from the small intestine and absorb enough water to form feces.

larva (plural: larvae) An immature stage of a species in which the immature and adult stages have different body forms.

late endosome A membrane-bound vesicle that arises from an early endosome and develops into a lysosome.

lateral bud A bud that forms in the angle between a leaf and a stem and may develop into a lateral (side) branch. Also called *axillary bud.*

lateral gene transfer Transfer of DNA between two different species, especially distantly related species. Commonly occurs among bacteria and archaea via plasmid exchange; also can occur in eukaryotes via viruses and some other mechanisms.

lateral meristem A layer of undifferentiated plant cells found in older stems and roots that is responsible for secondary growth. Also called *cambium* or *secondary meristem.* Compare with **apical meristem.**

lateral root A plant root extending from another, older root.

leaching Loss of nutrients from soil via percolating water.

leading strand In DNA replication, the strand of new DNA that is synthesized in one continuous piece, with nucleotides added to the 3′ end of the growing molecule. Also called *continuous strand.* Compare with **lagging strand.**

leaf The main photosynthetic organ of vascular plants.

leak channel Potassium channel that allows potassium ions to leak out of a neuron in its resting state.

learning An enduring change in an individual's behavior that results from specific experience(s).

leghemoglobin An iron-containing protein similar to hemoglobin. Found in root nodules of legume plants where it binds oxygen, preventing it from poisoning a bacterial enzyme needed for nitrogen fixation.

legumes Members of the pea plant family that form symbiotic associations with nitrogen-fixing bacteria in their roots.

lens A transparent, crystalline structure that focuses incoming light onto a retina or other light-sensing apparatus of an eye.

leptin A hormone produced and secreted by fat cells (adipocytes) that acts to stabilize fat tissue mass in part by inhibiting appetite and increasing energy expenditure.

leukocytes Several types of blood cells, including neutrophils, macrophages, and lymphocytes, that circulate in blood and lymph and function in defense against pathogens. Also called *white blood cells.*

lichen A symbiotic association of a fungus and a photosynthetic alga.

life cycle The sequence of developmental events and phases that occurs during the life span of an organism, from fertilization to offspring production.

life history The sequence of events in an individual's life from birth to reproduction to death, including how an individual allocates resources to growth, reproduction, and activities or structures that are related to survival.

life table A data set that summarizes the probability that an individual in a certain population will survive and reproduce in any given year over the course of its lifetime.

ligand Any molecule that binds to a specific site on a receptor molecule.

ligand-gated channel An ion channel that opens or closes in response to binding by a certain molecule. Compare with **voltage-gated channel.**

light chain The smaller of the two types of polypeptide chains in an antibody molecule; composed of a variable (V) region, which contributes to the antigen-binding site, and a constant (C) region. Compare with **heavy chain.**

light-dependent reactions In photosynthesis, the set of reactions, occurring in photosystem I and II, that use the energy of sunlight to split water, producing ATP, NADPH, and oxygen.

light-independent reactions See **Calvin cycle.**

lignin A substance found in the secondary cell walls of some plants that is exceptionally stiff and strong. Most abundant in woody plant parts.

limiting nutrient Any essential nutrient whose scarcity in the environment significantly reduces growth and reproduction of organisms.

limnetic zone Open water (not near shore) that receives enough light to support photosynthesis.

lineage See **monophyletic group.**

LINEs (long interspersed nuclear elements) The most abundant class of transposable elements in human genomes; can create copies of itself and insert them elsewhere in the genome. Compare with **SINEs.**

linkage In genetics, a physical association between two genes because they are on the same chromosome; the inheritance patterns resulting from this association.

linkage map See **genetic map.**

lipase Any enzyme that can break down fat molecules into fatty acids and monoglycerides.

lipid Any organic subtance that does not dissolve in water, but dissolves well in nonpolar organic solvents. Lipids include fats, oils, phospholipids, and waxes.

lipid bilayer The basic structural element of all cellular membranes consisting of a two-layer sheet of phospholipid molecules whose hydrophobic tails are oriented toward the inside and hydrophilic heads, toward the outside. Also called *phospholipid bilayer.*

littoral zone Shallow water near shore that receives enough sunlight to support photosynthesis. May be marine or freshwater; often flowering plants are present.

liver A large, complex organ of vertebrates that performs many functions including storage of glycogen, processing and conversion of food and wastes, and production of bile.

locomotion Movement of an organism under its own power.

locus (plural: loci) A gene's physical location on a chromosome.

logistic population growth The density-dependent decrease in growth rate as population size approaches the carrying capacity. Compare with **exponential population growth.**

long-day plant A plant that blooms in response to short nights (usually in late spring or early summer in the northern hemisphere). Compare with **day-neutral** and **short-day plant.**

long interspersed nuclear elements See **LINEs.**

loop of Henle In the vertebrate kidney, a long U-shaped loop in a nephron that extends into the medulla. Functions as a countercurrent exchanger to set up an osmotic gradient that allows reabsorption of water from a subsequent portion of the nephron.

loose connective tissue A type of connective tissue consisting of fibrous proteins in a soft matrix. Often functions as padding for organs.

lophophore A specialized feeding structure found in some lophotrochozoans and used in filter feeding.

lophotrochozoans A major lineage of protostomes (Lophotrochozoa) that grow by extending the size of their skeletons rather than by molting. Many phyla have a specialized feeding structure (lophophore) and/or ciliated larvae (trochophore). Includes rotifers, flatworms, segmented worms, and molluscs. Compare with **ecdysozoans.**

loss-of-function mutant See **knock-out mutant.**

lumen The interior space of any hollow structure (e.g., the rough ER) or organ (e.g., the stomach).

lung Any respiratory organ used for gas exchange between blood and air.

luteal phase The second major phase of a menstrual cycle, after ovulation, when the progesterone levels are high and the body is preparing for a possible pregnancy.

luteinizing hormone (LH) A peptide hormone, produced and secreted by the anterior pituitary, that stimulates estrogen production, ovulation, and formation of the corpus luteum in females and testosterone production in males.

lymph The mixture of fluid and white blood cells that circulates through the ducts and lymph nodes of the lymphatic system in vertebrates.

lymphatic system In vertebrates, a body-wide network of thin-walled ducts (or vessels) and lymph nodes, separate from the circulatory system. Collects excess fluid from body tissues and returns it to the blood; also functions as part of the immune system.

lymph node One of numerous small oval structures through which lymph moves in the lymphatic system. Filter the lymph and screen it for pathogens and other antigens. Major sites of lymphocyte activation.

lymphocytes Two types of leukocyte—B cells and T cells—that circulate through the bloodstream and lymphatic system and that are responsible for the development of acquired immunity.

lysogenic cycle A type of viral replication in which a viral genome enters a host cell, is inserted into the host's chromosome, and is replicated whenever the host cell divides. When activated, the viral DNA enters the lytic cycle, leading to production of new virus particles. Also called *lysogeny* or *latent growth*. Compare with **lytic cycle.**

lysosome A small organelle in an animal cell containing acids and enzymes that catalyze hydrolysis reactions and can digest large molecules. Compare with **vacuole.**

lysozyme An enzyme that functions in innate immunity by digesting bacterial cell walls. Occurs in saliva, tears, mucus, and egg white.

lytic cycle A type of viral replication in which a viral genome enters a host cell, new virus particles (virions) are made using host enzymes and eventually burst out of the cell, killing it. Also called *replicative growth*. Compare with **lysogenic cycle.**

macromolecule Any very large organic molecule, usually made up of smaller molecules (monomers) joined together into a polymer. The main biological macromolecules are proteins, nucleic acids, and polysaccharides.

macronutrient Any element (e.g., carbon, oxygen, nitrogen) that is required in large quantities for normal growth, reproduction, and maintenance of a living organism. Compare with **micronutrient.**

macrophage A type of leukocyte, capable of moving through body tissues, that engulfs and digests pathogens and other foreign particles; also secretes cytokines and presents foreign antigens to $CD4^+$ T cells.

MADS box A DNA sequence that codes for a DNA-binding motif in proteins; present in floral organ identity genes in plants. Functionally similar sequences are found in some fungal and animal genes.

major histocompatibility protein See **MHC protein.**

malaria A human disease caused by four species of the protist *Plasmodium* and passed to humans by mosquitoes.

malignant tumor A tumor that is actively growing and disrupting local tissues and/or is spreading to other organs. Cancer consists of one or more malignant tumors. Compare with **benign tumor.**

Malpighian tubules A major excretory organ of insects, consisting of blind-ended tubes that extend from the gut into the hemocoel. Filter hemolymph to form pre-urine and then send it to the hindgut for further processing.

mammals One of the two lineages of amniotes (vertebrates that produce amniotic eggs) distinguished by hair (or fur) and mammary glands. Includes the monotremes (platypuses), marsupials, and eutherians (placental mammals).

mammary glands Specialized exocrine glands that produce and secrete milk for nursing offspring. A diagnostic feature of mammals.

mandibles Any mouthpart used in chewing. In vertebrates, the lower jaw. In insects, crustaceans, and myriapods, the first pair of mouthparts.

mantle One of the three main parts of the mollusc body; the thick outer tissue that protects the visceral mass and may secrete a calcium carbonate shell.

Marfan syndrome A human syndrome involving increased height, long limbs and fingers, an abnormally shaped chest, and heart disorders. Probably caused by mutation in one pleiotropic gene.

marsh A wetland that lacks trees and usually has a slow but steady rate of water flow.

marsupials A lineage of mammals (Marsupiala) that nourish their young in an abdominal pouch after a very short period of development in the uterus.

mass extinction The extinction of a large number of diverse evolutionary groups during a relatively short period of geologic time (about 1 million years). May occur due to sudden and extraordinary environmental changes. Compare with **background extinction.**

mass feeder An animal that takes chunks of food into its mouth.

mass number The total number of protons and neutrons in an atom.

mast cell A type of leukocyte that is stationary (embedded in tissue) and helps trigger the inflammatory response to infection or injury, including secretion of histamine. Particularly important in allergic responses and defense against parasites.

maternal chromosome A chromosome inherited from the mother.

mechanoreceptor A sensory cell or organ specialized for detecting distortions caused by touch or pressure. One example is hair cells in the cochlea.

medium A liquid or solid in which cells can grow in vitro.

medulla The innermost part of an organ (e.g., kidney or adrenal gland).

medulla oblongata In vertebrates, a region of the brain stem that along with the cerebellum forms the hindbrain.

medusa (plural: medusae) The free-floating stage in the life cycle of some cnidarians (e.g., jellyfish). Compare with **polyp.**

megapascal (MPa) A unit of pressure (force per unit area), equivalent to 1 million pascals (Pa).

megaspore In seed plants, a haploid (n) spore that is produced in a megasporangium by meiosis of a diploid ($2n$) megasporocyte; develops into a female gametophyte. Compare with **microspore.**

meiosis In sexually reproducing organisms, a special two-stage type of cell division in which one diploid ($2n$) parent cell produces four haploid (n) reproductive cells (gametes); results in halving of the chromosome number. Also called *reduction division*.

meiosis I The first cell division of meiosis, in which synapsis and crossing over occur, and homologous chromosomes are separated from each other, producing daughter cells with half as many chromosomes (each composed of two sister chromatids) as the parent cell.

meiosis II The second cell division of meiosis, in which sister chromatids are separated from each other. Similar to mitosis.

meiotic map See **genetic map.**

membrane potential A difference in electric charge across a cell membrane; a form of potential energy. Also called *membrane voltage*.

memory Retention of learned information.

memory cells A type of lymphocyte responsible for maintenance of immunity for years or decades after an infection. Descended from a B cell or T cell activated during a previous infection.

meniscus (plural: menisci) The concave boundary layer formed at most air-water interfaces due to surface tension.

menstrual cycle A female reproductive cycle seen in Old World monkeys and apes (including humans) in which the uterine lining is shed (menstruation) if no pregnancy occurs. Compare with **estrous cycle.**

menstruation The periodic shedding of the uterine lining through the vagina that occurs in females of Old World monkeys and apes, including humans.

meristem (adjective: meristematic) In plants, a group of undifferentiated cells that can develop into various adult tissues throughout the life of a plant.

mesoderm The middle of the three basic cell layers in most animal embryos; gives rise to muscles, bones, blood, and some internal organs (kidney, spleen, etc.). Compare with **ectoderm** and **endoderm.**

mesoglea A gelatinous material, containing scattered ectodermal cells, that is located between the ectoderm and endoderm of cnidarians (e.g., jellyfish, corals, and anemones).

mesophyll cell A type of cell, found near the surfaces of plant leaves, that is specialized for the light-dependent reactions of photosynthesis.

Mesozoic era The period of geologic time, from 250 million to 65 million years ago, during which gymnosperms were the dominant plants and dinosaurs the dominant vertebrates. Ended with extinction of the dinosaurs. Also called *Age of Reptiles.*

messenger RNA (mRNA) An RNA molecule that carries encoded information, transcribed from DNA, that specifies the amino acid sequence of a polypeptide.

meta-analysis A comparative analysis of the results of many smaller, previously published studies.

metabolic pathway An ordered series of chemical reactions that build up or break down a particular molecule. Often, each reaction is catalyzed by a different enzyme.

metabolic rate The total energy use by all the cells of an individual. For aerobic organisms, often measured as the amount of oxygen consumed per hour.

metabolic water The water that is produced as a by-product of cellular respiration.

metabolism All the chemical reactions occurring in a living cell or organism.

metallothioneins Small plant proteins that bind to and prevent excess metal ions from acting as toxins.

metamorphosis Transition from one developmental stage to another, such as from the larval to the adult form of an animal.

metaphase A stage in mitosis or meiosis during which chromosomes line up in the middle of the cell.

metaphase plate The plane along which chromosomes line up during metaphase of mitosis or meiosis; not an actual structure.

metastasis The spread of cancerous cells from their site of origin to distant sites in the body where they may establish additional tumors.

methanogen A prokaryote that produces methane (CH_4) as a by-product of cellular respiration.

methanotroph An organism that uses methane (CH_4) as its primary electron donor and source of carbon.

methylation The addition of a methyl ($-CH_3$) group to a molecule.

MHC protein One of a large set of mammalian cell-surface glycoproteins involved in marking cells as self and in antigen presentation to T cells. Also called *MHC molecule.* See **Class I** and **Class II MHC protein.**

microbe Any microscopic organism, including bacteria, archaea, and various tiny eukaryotes.

microbiology The field of study concerned with microscopic organisms.

microfilament See **actin filament.**

micrograph A photograph of an image produced by a microscope.

micronutrient Any element (e.g., iron, molybdenum, magnesium) that is required in very small quantities for normal growth, reproduction, and maintenance of a living organism. Compare with **macronutrient.**

micropyle The tiny pore in a plant ovule through which the pollen tube reaches the embryo sac.

microRNA (miRNA) A small, single-stranded RNA associated with proteins in an RNA-induced silencing complex. Can bind to complementary sequences in mRNA molecules, allowing the associated proteins to degrade the bound mRNA or inhibit its translation. See **RNA interference.**

microsatellite A noncoding stretch of eukaryotic DNA consisting of a repeating sequence 1- to 5-base pair long. Also called *simple sequence repeat.*

microspore In seed plants, a haploid (n) spore that is produced in a microsporangium by meiosis of a diploid ($2n$) microsporocyte; develops into a male gametophyte. Compare with **megaspore.**

microtubule A long, tubular fiber, about 25 nm in diameter, formed by polymerization of tubulin protein dimers; one of the three types of cytoskeletal fibers. Involved in cell movement and transport of materials within the cell. Compare with **actin filament** and **intermediate filament.**

microtubule-organizing center (MTOC) General term for any structure (e.g., centrosome and basal body) that organizes microtubules in cells.

microvilli (singular:microvillus) Tiny protrusions from the surface of an epithelial cell that increase the surface area for absorption of substances.

midbrain One of the three main regions of the vertebrate brain; includes sensory integrating and relay centers. Compare with **forebrain** and **hindbrain.**

middle ear The air-filled middle portion of the mammalian ear, which contains three small bones (ossicles) that transmit and amplify sound from the tympanic membrane to the inner ear. Is connected to the throat via the eustachian tube.

middle lamella The layer of gelatinous pectins between the primary cell walls of adjacent plant cells. Helps hold the cells together.

migration (1) In ecology, a cyclical movement of large numbers of organisms from one geographic location or habitat to another. (2) In population genetics, movement of individuals from one population to another.

millivolt (mV) A unit of voltage equal to 1/1000 of a volt.

mimicry A phenomenon in which one species has evolved (or learns) to look or sound like another species. See **Batesian mimicry** and **Müllerian mimicry.**

mineralocorticoids A class of steroid hormones, produced and secreted by the adrenal cortex, that regulate electrolyte levels and the overall volume of body fluids. Aldosterone is the principal one in humans. Compare with **glucocorticoids.**

minisatellite A noncoding stretch of eukaryotic DNA consisting of a repeating sequence that is 6 to 500 base pairs long. Also called *variable number tandem repeat (VNTR).*

mismatch repair The process by which mismatched base pairs in DNA are fixed.

missense mutation A point mutation (change in a single base pair) that causes a change in the amino acid sequence of a protein. Also called *replacement mutation.*

mitochondrial matrix Central compartment of a mitochondrion, which is lined by the inner membrane; contains the enzymes and substrates of the Krebs cycle and mitochondrial DNA.

mitochondrion (plural: mitochondria) A eukaryotic organelle that is bounded by a double membrane and is the site of aerobic respiration.

mitosis In eukaryotic cells, the process of nuclear division that results in two daughter nuclei genetically identical to the parent nucleus. Subsequent cytokinesis (division of the cytoplasm) yields two daughter cells.

mitosis-promoting factor (MPF) A complex of a cyclin and cyclin-dependent kinase that phosphorylates a number of specific proteins needed to initiate mitosis in eukaryotic cells.

mitotic (M) phase The phase of the cell cycle during which cell division occurs. Includes mitosis and cytokinesis.

mitotic spindle Temporary structure, composed largely of microtubules, that is involved in the movement of chromosomes to the equatorial plate and then to opposite sides of the cell during mitosis and meiosis.

model organism An organism selected for intensive scientific study based on features that make it easy to work with (e.g., body size, life span), in the hope that findings will apply to other species.

molarity A common unit of solute concentration equal to the number of moles of a dissolved solute in 1 liter of solution.

mole The amount of a substance that contains 6.022×10^{23} of its elemental entities (e.g., atoms, ions, or molecules). This number of molecules of a compound will have a mass equal to the molecular weight of that compound expressed in grams.

molecular chaperone A protein that facilitates the three-dimensional folding of newly synthesized proteins, usually by an ATP-dependent mechanism.

molecular clock The hypothesis that certain types of mutations tend to reach fixation in populations at a steady rate over large spans of time. As a result, comparisons of DNA sequences can be used to infer the timing of evolutionary divergences.

molecular formula A notation that indicates only the numbers and types of atoms in a molecule, such as H_2O for the water molecule. Compare with **structural formula.**

molecular weight The sum of the mass numbers of all of the atoms in a molecule; roughly, the total number of protons and neutrons in the molecule.

molecule A combination of two or more atoms held together by covalent bonds.

molluscs Members of the phylum Mollusca. Distinguished by a body plan with three main parts: a muscular foot, a visceral mass, and a mantle. Include bivalves (clams, oysters), gastropods (snails, slugs), chitons, and cephalopods (squid, octopuses). Molluscs belong to the lophotrochozoan branch of the protostomes.

molting A method of body growth, used by ecdysozoans, that involves the shedding of an external protective cuticle or skeleton, expansion of the soft body, and growth of a new external layer.

monocot Any plant that has a single cotyledon (embryonic leaf) upon germination. Monocots form a monophyletic group. Also called a monocotyledonous plant. Compare with **dicot.**

monoecious Describing an angiosperm species that has both male and female reproductive structures on each plant. Compare with **dioecious.**

monohybrid cross A mating between two parents that are both heterozygous for a given gene.

monomer A small molecule that can covalently bind to other similar molecules to form a larger macromolecule. Compare with **polymer.**

monophyletic group An evolutionary unit that includes an ancestral population and all of its descendants but no others. Also called a *clade* or *lineage*. Compare with **paraphyletic group.**

monosaccharide A small carbohydrate, such as glucose, that has the molecular formula $(CH_2O)_n$ and cannot be hydrolyzed to form any smaller carbohydrates. Also called *simple sugar.* Compare with **disaccharide** and **polysaccharide.**

monosomy Having only one copy of a particular type of chromosome.

monotremes A lineage of mammals (Monotremata) that lay eggs and then nourish the young with milk. Includes just three living species: the platypus and two species of echidna.

morphological species concept The definition of a species as a population or group of populations that have measurably different anatomical features from other groups. Also called *morphospecies concept.* Compare with **biological** and **phylogenetic species concept.**

morphology The shape and appearance of an organism's body and its component parts.

motif In molecular biology, a domain (a section of a protein with a distinctive tertiary structure) found in many different proteins and often having specific functional properties.

motor neuron A nerve cell that carries signals from the central nervous system (brain and spinal cord) to an effector, such as a muscle or gland. Compare with **interneuron** and **sensory neuron.**

motor protein A class of proteins whose major function is to convert the chemical energy of ATP into motion. Includes dynein, kinesin, and myosin.

MPF See **mitosis-promoting factor.**

mRNA. See **messenger RNA.**

mucigel A slimy substance secreted by plant root caps that eases passage of the growing root through the soil.

mucosal-associated lymphoid tissue (MALT) Collective term for lymphocytes and other leukocytes associated with skin cells and with mucus-secreting epithelial tissues in the gut and respiratory tract. Plays important role in preventing entry of pathogens into the body.

mucus (adjective: mucous) A slimy mixture of glycoproteins (called mucins) and water that is secreted in many animal organs for lubrication.

Müllerian inhibitory substance A peptide hormone secreted by the embryonic testis that causes regression (withering away) of the female reproductive ducts.

Müllerian mimicry A type of mimicry in which two (or more) harmful species resemble each other. Compare with **Batesian mimicry.**

multicellularity The state of being composed of many cells that adhere to each other and do not all express the same genes with the result that some cells have specialized functions.

multiple fruit A fruit (e.g., pineapple) that develops from many separate flowers and thus many carpels. Compare with **aggregate** and **simple fruit.**

multiple sclerosis (MS) A human autoimmune disease caused by the immune system attacking the myelin sheaths that insulate nerve axons.

muscle fiber A single muscle cell.

muscle tissue An animal tissue consisting of bundles of long, thin contractile cells (muscle fibers).

mutagen Any physical or chemical agent that increases the rate of mutation.

mutant An individual that carries a mutation, particularly a new or rare mutation.

mutation Any change in the hereditary material of an organism (DNA in most organisms, RNA in some viruses).

mutualism (adjective: mutualistic) A symbiotic relationship between two organisms (mutualists) that benefits both. Compare with **commensalism** and **parasitism.**

mycelium (plural: mycelia) A mass of underground filaments (hyphae) that form the body of a fungus. Also found in some protists and bacteria.

mycorrhiza (plural: mycorrhizae) A mutualistic association between certain fungi and most vascular plants, sometimes visible as nodules or nets in or around plant roots.

myelin sheath Multiple layers of myelin, derived from the cell membranes of certain glial cells, that is wrapped around the axon of a neuron, providing electrical insulation.

myofibril Long, slender structure composed of contractile proteins organized into repeating units (sarcomeres) in vertebrate heart muscle and striated muscle.

myosin Any one of a class of motor proteins that use the chemical energy of ATP to move along actin filaments in muscle contraction, cytokinesis, and vesicle transport.

myriapods A lineage of arthropods with long segmented trunks, each segment bearing one or two pairs of legs. Includes millipedes and centipedes.

NAD^+/NADH Oxidized and reduced forms, respectively, of nicotinamide adenine dinucleotide. A nonprotein electron carrier that functions in many of the redox reactions of metabolism.

$NADP^+$/NADPH Oxidized and reduced forms, respectively, of nicotinamide adenine dinucleotide phosphate. A nonprotein electron carrier that is reduced during the light-dependent reactions in photosynthesis and extensively used in biosynthetic reactions.

natural experiment A situation in which groups to be compared are created by an unplanned, natural change in conditions rather than by manipulation of conditions by researchers.

natural selection The process by which individuals with certain heritable traits tend to produce more surviving offspring than do individuals without those traits, often leading to a change in the genetic makeup of the population. A major mechanism of evolution.

nauplius A distinct planktonic larval stage seen in many crustaceans.

Neanderthal A recently extinct European species of hominid, *Homo neanderthalensis*, closely related to but distinct from modern humans.

nectar The sugary fluid produced by flowers to attract and reward pollinating animals.

negative feedback A self-limiting, corrective response in which a deviation in some variable (e.g., body temperature, blood pH, concentration of some compound) triggers responses aimed at returning the variable to normal. Compare with **positive feedback.**

negative pressure ventilation Ventilation of the lungs that is accomplished by "pulling" air into the lungs by expansion of the rib cage. Compare with **positive pressure ventilation.**

negative-sense virus A virus whose genome contains sequences complementary to those in the mRNA required to produce viral proteins. Compare with **ambisense virus** and **positive-sense virus.**

nematodes. See **roundworms.**

nephron One of the tiny tubes within the vertebrate kidney that filter blood and concentrate salts to produce urine. Also called *renal tubule.*

neritic zone Shallow marine waters beyond the intertidal zone, extending down to about 200 meters, where the continental shelf ends.

nerve A long, tough strand of nervous tissue typically containing thousands of axons wrapped in connective tissue; carries impulses between the central nervous system and some other part of the body.

nerve cord A bundle of nerves extending from the brain along the dorsal (back) side of a chordate animal, with cerebrospinal fluid inside a hollow central channel. One of the defining features of chordates.

nervous tissue An animal tissue consisting of nerve cells (neurons) and various supporting cells.

net primary productivity (NPP) In an ecosystem, the total amount of carbon fixed by photosynthesis over a given time period minus the amount oxidized during cellular respiration. Compare with **gross primary productivity.**

net reproductive rate (R_0) The growth rate of a population per generation; equivalent to the average number of female offspring that each female produces over her lifetime.

neural Relating to nerve cells (neurons) and the nervous system.

neural tube A folded tube of ectoderm that forms along the dorsal side of a young vertebrate embryo and that will give rise to the brain and spinal cord.

neuroendocrine Referring to nerve cells (neurons) that release hormones into the blood or to such hormones themselves.

neuron A cell that is specialized for the transmission of nerve impulses. Typically has dendrites, a cell body, and a long axon that forms synapses with other neurons. Also called *nerve cell.*

neurosecretory cell A nerve cell (neuron) that produces and secretes hormones into the bloodstream. Principally found in the hypothalamus. Also called *neuroendocrine cell.*

neurotoxin Any substance that specifically destroys or blocks the normal functioning of neurons.

neurotransmitter A molecule that transmits electrical signals from one neuron to another or from a neuron to a muscle or gland. Examples are acetylcholine, dopamine, serotonin, and norepinephrine.

neutral In genetics, referring to any mutation or mutant allele that has no effect on an individual's fitness.

neutrophil A type of leukocyte, capable of moving through body tissues, that engulfs and digests pathogens and other foreign particles; also secretes various compounds that attack bacteria and fungi.

niche The particular set of habitat requirements of a certain species and the role that species plays in its ecosystem.

niche differentiation The change in resource use by competing species that occurs as the result of character displacement.

nicotinamide adenine dinucleotide See **NAD^+/NADH.**

nicotinamide adenine dinucleotide phosphate See **$NADP^+$/NADPH.**

nitrogen cycle, global The movement of nitrogen among terrestrial ecosystems, the oceans, and the atmosphere.

nitrogen fixation The incorporation of atmospheric nitrogen (N_2) into forms such as ammonia (NH_3) or nitrate (NO_3^-), which can be used to make many organic compounds. Occurs in only a few lineages of bacteria and archaea.

nociceptor A sensory cell or organ specialized to detect tissue damage, usually producing the sensation of pain.

node (1) In animals, any small thickening (e.g., a lymph node). (2) In plants, the part of a stem where leaves or leaf buds are attached. (3) In a phylogenetic tree, the point where two branches diverge, representing the point in time when an ancestral group split into two or more descendant groups. Also called *fork.*

node of Ranvier One of the periodic unmyelinated sections of a neuron's axon at which an action potential can be regenerated.

Nod factors Molecules produced by nitrogen-fixing bacteria that help them recognize and bind to roots of legumes.

nodule Lumplike structure on roots of legume plants that contain symbiotic nitrogen-fixing bacteria.

noncyclic electron flow See **Z scheme.**

nondisjunction An error that can occur during meiosis or mitosis in which one daughter cell receives two copies of a particular chromosome, and the other daughter cell receives none.

nonpolar covalent bond A covalent bond in which electrons are equally shared between two atoms of the same or similar electronegativity. Compare with **polar covalent bond.**

non-sister chromatids The chromatids of a particular type of chromosome (after replication) with respect to the chromatids of its homologous chromosome. Crossing over occurs between non-sister chromatids. Compare with **sister chromatids.**

non-template strand The strand of DNA that is not transcribed during synthesis of RNA. Its sequence corresponds to that of the mRNA produced from the other strand. Also called *coding strand.*

nonvascular plants See **bryophytes.**

norepinephrine A catecholamine used as a neurotransmitter in the sympathetic nervous system. Also is produced by the adrenal medulla and functions as a hormone that triggers rapid responses relating to the fight-or-flight response.

Northern blotting A technique for identifying specific RNAs separated by gel electrophoresis by transferring them to filter paper and hybridizing with a labeled DNA probe complementary to the RNA of interest. Compare with **Southern blotting** and **Western blotting.**

notochord A long, gelatinous, supportive rod down the back of a chordate embryo, below the developing spinal cord. Replaced by vertebrae in most adult vertebrates. A defining feature of chordates.

nuclear envelope The double-layered membrane enclosing the nucleus of a eukaryotic cell.

nuclear lamina A lattice-like sheet of fibrous nuclear lamins, which are one type of intermediate filaments. Lines the inner membrane of the nuclear envelope, stiffening the envelope and helping organize the chromosomes.

nuclear localization signal (NLS) A short amino acid sequence that marks a protein for delivery to the nucleus.

nuclear pore An opening in the nuclear envelope that connects the inside of the nucleus with the cytoplasm and through which molecules such as mRNA and some proteins can pass.

nuclear pore complex A large complex of dozens of proteins lining a nuclear pore, defining its shape and transporting substances through the pore.

nuclease Any enzyme that can break down RNA or DNA molecules.

nucleic acid A macromolecule composed of nucleotide monomers. Generally used by cells to store or transmit hereditary information. Includes ribonucleic acid and deoxyribonucleic acid.

nucleic acid hybridization Base pairing between a single-stranded nucleic acid and a complementary sequence in a different nucleic acid (e.g., a labeled probe). Is used experimentally in Southern blotting and Northern blotting.

nucleoid In prokaryotic cells, a dense, centrally located region that contains DNA but is not surrounded by a membrane.

nucleolus In eukaryotic cells, specialized structure in the nucleus where ribosomal RNA processing occurs and ribosomal subunits are assembled.

nucleosome A repeating, bead-like unit of eukaryotic chromatin, consisting of about 200 nucleotides of DNA wrapped twice around eight histone proteins.

nucleotide A molecule consisting of a five-carbon sugar (ribose or deoxyribose), a phosphate group, and one of several nitrogen-containing bases. DNA and RNA are polymers of nucleotides containing deoxyribose (deoxyribonucleotides) and ribose (ribonucleotides), respectively. Equivalent to a nucleoside plus one phosphate group.

nucleotide excision repair The process of removing a damaged region in one strand of DNA and correctly replacing it using the undamaged strand as a template.

nucleus (1) The center of an atom, containing protons and neutrons. (2) In eukaryotic cells, the large organelle containing the chromosomes and surrounded by a double membrane. (3) A discrete clump of neuron cell bodies in the brain, usually sharing a distinct function.

null hypothesis A hypothesis that specifies what the results of an experiment will be if the main hypothesis being tested is wrong. Often states that there will be no difference between experimental groups.

null mutant See **knock-out mutant.**

nutrient A substance that an organism requires for normal growth, maintenance, or reproduction.

nutritional balance A state in which an organism is taking in enough nutrients to maintain normal health and activity.

nymph The juvenile form of an animal that undergoes hemimetabolous (incomplete) metamorphosis; resembles a miniature version of the adult form.

oceanic zone The waters of the open ocean beyond the continental shelf.

oil A fat that is liquid at room temperature.

Okazaki fragment Short segment of DNA produced during replication of the 5′ to 3′ template strand. Many Okazaki fragments make up the lagging strand in newly synthesized DNA.

olfaction The perception of odors.

olfactory bulb A bulb-shaped projection of the brain just above the nose. Receives and interprets odor information from the nose.

oligodendrocyte A type of glial cell that wraps around axons of some neurons in the central nervous system, forming a myelin sheath that provides electrical insulation. Compare with **Schwann cell.**

oligopeptide A chain composed of fewer than 50 amino acids linked together by peptide bonds. Often referred to simply as *peptide.*

ommatidium (plural: ommatidia) A light-sensing column in an arthropod's compound eye.

omnivore (adjective: omnivorous) An animal whose diet regularly includes both meat and plants. Compare with **carnivore** and **herbivore.**

oncogene Any gene whose protein product stimulates cell division at all times and thus promotes cancer development. Often is a mutated form of a gene involved in regulating the cell cycle. See **proto-oncogene.**

one-gene, one-enzyme hypothesis The hypothesis that each gene is responsible for making one (and only one) protein, in most cases an enzyme that catalyzes a specific reaction. Many exceptions to this hypothesis are now known.

oocyte A cell in the ovary that can undergoes meiosis to produce an ovum.

oogenesis The production of egg cells (ova).

oogonia (singular: oogonia) The diploid cells in an ovary that can divide by mitosis to create more oogonia and primary oocytes, which can undergo meiosis.

open circulatory system A circulatory system in which the circulating fluid (hemolymph) is not confined to blood vessels. Compare with **closed circulatory system.**

open reading frame (ORF) Any DNA sequence, ranging in length from several hundred to thousands of base pairs long, that is flanked by a start codon and a stop codon. ORFs identified by computer analysis of DNA may be functional genes, especially if they have other features characteristic of genes (e.g., promoter sequence).

operator In prokaryotic DNA, a binding site for a repressor protein; located near the start of an operon.

operculum The stiff flap of tissue that covers the gills of teleost fishes.

operon A region of prokaryotic DNA that codes for a series of functionally related genes and is transcribed from a single promoter into a polycistronic mRNA.

opsin A transmembrane protein that is covalently linked to retinal, the light-detecting pigment in rod and cone cells.

optical isomer A molecule that shares the same molecular formula as another molecule but differs in the arrangement of atoms or groups around a carbon atom; left-handed or right-handed form of a molecule. Compare with **geometric** and **structural isomer.**

optimal foraging The concept that animals forage in a way that maximizes the amount of usable energy they take in, given the costs of finding and ingesting their food and the risk of being eaten while they're at it.

orbital The spherical region around an atomic nucleus in which an electron is present most of the time.

order In Linnaeus' system, a taxonomic category above the family level and below the class level.

ORF See **open reading frame.**

organ A group of tissues organized into a functional and structural unit.

organelle Any discrete, membrane-bound structure within a cell (e.g., mitochondrion) that has a characteristic structure and functions.

organic For a compound, containing carbon and hydrogen and usually containing carbon-carbon bonds. Organic compounds are widely used by living organisms.

organism Any living entity that contains one or more cells.

organogenesis A stage of embryonic development, just after gastrulation in vertebrate embryos, during which major organs develop from the three embryonic germ layers.

orientation A deliberate movement that results in a change in position relative to some external cue, such as toward the Sun or away from a sound.

origin of replication The site on a chromosome at which DNA replication begins.

osmoconformer An animal that does not actively regulate the osmolarity of its tissues but conforms to the osmolarity of the surrounding environment.

osmolarity The concentration of dissolved substances in a solution, measured in moles per liter.

osmoregulation The process by which a living organism controls the concentration of water and salts in its body.

osmoregulator An animal that actively regulates the osmolarity of its tissues.

osmosis Diffusion of water across a selectively permeable membrane from a region of high water concentration (low solute concentration) to a region of low water concentration (high solute concentration).

osmotic potential See **solute potential.**

ossicle One of three small bones, in the middle ear of mammals, that transmit and amplify sound from the tympanic membrane to the inner ear.

ouabain A plant toxin that poisons the sodium-potassium pumps of animals.

outcrossing Reproduction by fusion of the gametes of different individuals, rather than self-fertilization. Typically refers to plants.

outer ear The outermost portion of the mammalian ear, consisting of the pinna (ear flap) and the ear canal. Funnels sound to the tympanic membrane.

outgroup A taxon that is closely related to a particular monophyletic group but is not part of it.

out-of-Africa hypothesis The hypothesis that modern humans (*Homo sapiens*) evolved in Africa and spread to other continents, replacing other *Homo* species without interbreeding with them.

oval window A membrane separating the fluid-filled cochlea from the air-filled middle ear through which sound vibrations pass from the middle ear to the inner ear in mammals.

ovary The egg-producing organ of a female animal, or the seed-producing structure in the female part of a flower.

oviduct See **fallopian tube.**

oviparous Producing eggs that are laid outside the body where they develop and hatch. Compare with **ovoviviparous** and **viviparous.**

ovoviviparous Producing eggs that are retained inside the body until they are ready to hatch. Compare with **oviparous** and **viviparous.**

ovulation The release of an ovum from an ovary of a female vertebrate. In humans, an ovarian follicle releases an egg at the end of the follicular phase of the menstrual cycle.

ovule In flowering plants, the structure inside an ovary that contains the female gametophyte and eventually (if fertilized) becomes a seed.

ovum (plural: ova) See **egg.**

oxidation The loss of electrons from an atom during a redox reaction, either by donation of an electron to another atom or by the shared electrons in covalent bonds moving farther from the atomic nucleus.

oxidative phosphorylation Production of ATP molecules from the redox reactions of an electron transport chain.

oxygen-hemoglobin equilibrium curve The graphical depiction of the percentage of hemoglobin in the blood that will bind to oxygen at various partial pressures of oxygen.

oxygenic Referring to any process or reaction that produces oxygen. Photosynthesis in plants, algae, and cyanobacteria, which involves photosystem II, is oxygenic. Compare with **anoxygenic.**

oxytocin A peptide hormone, secreted by the posterior pituitary, that triggers labor and milk production in females and that stimulates pair bonding, parental care, and affiliative behavior in both sexes.

p53 A tumor-suppressor protein (molecular weight of 53 kilodaltons) that responds to DNA damage by stopping the cell cycle and/or triggering apoptosis. Encoded by the *p53* gene.

pacemaker cell A specialized cardiac muscle cell in the sinoatrial (SA) node of the vertebrate heart that has an inherent rhythm and can generate an electrical impulse that spreads to other heart cells.

paleontology The study of organisms that lived in the distant past.

Paleozoic era The period of geologic time, from 543 million to 250 million years ago, during which fungi, land plants, and animals first appeared and diversified. Began with the Cambrian explosion and ended with the extinction of many invertebrates and vertebrates.

pancreas A large gland in vertebrates that has both exocrine and endocrine functions. Secretes digestive enzymes into a duct connected to the intestine and several hormones (notably, insulin and glucagon) into the bloodstream.

pandemic The spread of an infectious disease in a short time period over a wide geographic area and affecting a very high proportion of the population. Compare with **epidemic.**

parabiosis An experimental technique for determining whether a certain physiological phenomenon is regulated by a hormone, by surgically uniting two individuals so that hormones can pass between them.

parabronchus (plural: parabronchi) One of the many tiny parallel air tubes that run through a bird's lung where gas exchange occurs.

paracrine Relating to a chemical signal that is released by one cell and affects neighboring cells.

paraphyletic group An evolutionary unit that includes an ancestral population and *some* but not all of its descendants. Paraphyletic groups are not meaningful units in evolution. Compare with **monophyletic group.**

parapodia (singular: parapodium) Appendages found in some annelids from which bristle-like structures (chaetae) extend.

parasite An organism that lives on or in a host species and that damages its host.

parasitism (adjective: parasitic) A symbiotic term relationship between two organisms that is beneficial to one organism (the parasite) but detrimental to the other (the host). Compare with **commensalism** and **mutualism.**

parasitoid An organism that has a parasitic larval stage and a free-living adult stage. Most parasitoids are insects that lay eggs in the bodies of other insects.

parasympathetic nervous system The part of the autonomic nervous system that stimulates functions for conserving or restoring energy, such as reduced heart rate and increased digestion. Compare with **sympathetic nervous system.**

parathyroid glands Four small glands, located near or embedded in the thyroid gland of vertebrates, that secrete parathyroid hormone.

parathyroid hormone (PTH) A peptide hormone, secreted from the parathyroid glands, that increases blood calcium by promoting Ca^{2+} release from bones, Ca^{2+} uptake in the intestines, and Ca^{2+} reabsorption in the kidneys.

parenchyma cell In plants, a general type of cell with a relatively thin primary cell wall. These cells, found in leaves, the centers of stems and roots, and fruits, are involved in photosynthesis, starch storage, and new growth. Compare with **collenchyma cell** and **sclerenchyma cell.**

parental care Any action by which an animal expends energy or assumes risks to benefit its offspring (e.g., nest-building, feeding of young, defense).

parental generation The adult organisms used in the first experimental cross in a formal breeding experiment.

parietal cell A cell in the stomach lining that secretes hydrochloric acid.

parsimony The logical principle that the most likely explanation of a phenomenon is the most economical or simplest. When applied to comparison of alternative phylogenetic trees, it suggests that the one requiring the fewest evolutionary changes is most likely to be correct.

parthenogenesis Development of offspring from unfertilized eggs; a type of asexual reproduction.

partial pressure The pressure of one particular gas in a mixture; the contribution of that gas to the overall pressure.

pascal (Pa) A unit of pressure (force per unit area).

passive transport Diffusion of a substance across a plasma membrane or organelle membrane. When this occurs with the assistance of membrane proteins, it is called facilitated diffusion.

patch clamping A technique for studying the electrical currents that flow through individual ion channels by sucking a tiny patch of membrane to the hollow tip of a microelectrode.

paternal chromosome A chromosome inherited from the father.

pathogen (adjective: pathogenic) Any entity capable of causing disease, such as a microbe, virus, or prion.

pattern formation The series of events that determines the spatial organization of an embryo, including alignment of the major body axes and orientation of the limbs.

pattern-recognition receptor One of a class of membrane proteins on leukocytes that bind to molecules on the surface of many bacteria. Part of the innate immune response.

PCR See **polymerase chain reaction.**

peat Semidecayed organic matter that accumulates in moist, low-oxygen environments such as bogs.

pectin A gelatinous polysaccharide found in the primary cell wall and middle lamella of plant cells. Attracts and holds water, forming a gel that helps keep the cell wall moist.

pedigree A family tree of parents and offspring, showing inheritance of particular traits of interest.

pellet Any solid material that collects at the bottom of a test tube below a layer of liquid (the supernatant) during centrifugation.

penis The copulatory organ of male mammals, used to insert sperm into a female.

pentose A monosaccharide (simple sugar) containing five carbon atoms.

PEP carboxylase An enzyme that catalyzes addition of CO_2 to phosphoenol pyruvate, a three-carbon compound, forming a four-carbon organic acid. Found in mesophyll cells of plants that perform C_4 photosynthesis.

pepsin A protein-digesting enzyme present in the stomach.

pepsinogen The precursor of the digestive enzyme pepsin. Is secreted from cells in the stomach lining and converted to pepsin by the acidic environment of the stomach lumen.

peptide See **oligopeptide.**

peptide bond The covalent bond (C–N) formed by a condensation reaction between two amino acids; links the residues in peptides and proteins.

peptidoglycan A complex structural polysaccharide found in bacterial cell walls.

per capita rate of increase (*r*) The growth rate of a population, expressed per individual. Calculated as the per capita birthrate minus the per capita death rate. Also called *per capita growth rate.*

perennial Describing a plant whose life cycle normally lasts for **more than one year.** Compare with **annual.**

perfect flower A flower that contains both male parts (stamens) and female parts (carpels). Compare with **imperfect flower.**

perforation In plants, a small hole in the primary and secondary cell walls of vessel elements that allow passage of water.

pericarp The part of a fruit, formed from the ovary wall, that surrounds the seeds and protects them. Corresponds to the flesh of most edible fruits and the hard shells of most nuts.

pericycle In plant roots, a layer of cells that give rise to lateral roots.

periderm The outermost portion of bark consisting of cork cambium, cork cells, and a third small layer of cells (phelloderm).

peripheral membrane protein Any membrane protein that does not span the entire lipid bilayer and associates with only one side of the bilayer. Compare with **integral membrane protein.**

peripheral nervous system (PNS) All the components of the nervous system that are outside the central nervous system (the brain and spinal cord). Includes the somatic nervous system and the autonomic nervous system.

peristalsis Rhythmic waves of muscular contraction that push food along the digestive tract.

permafrost A permanently frozen layer of icy soil found in most tundra and some taiga.

permeability The tendency of a structure, such as a membrane, to allow a given substance to diffuse across it.

peroxisome An organelle found in most eukaryotic cells that contains enzymes for oxidizing fatty acids and other compounds including many toxins, rendering them harmless. See **glyoxisome.**

petal One of the leaflike organs arranged around the reproductive organs of a flower. Often colored and scented to attract pollinators.

petiole The stalk of a leaf.

pH A measure of the concentration of protons in a solution and thus of acidity or alkalinity. Defined as the negative of the base-10 logarithm of the proton concentration: $pH = -\log[H^+]$.

phagocytosis Uptake by a cell of small particles or cells by pinching off the plasma membrane to form small membrane-bound vesicles; one type of endocytosis.

pharyngeal gill slits A set of parallel openings from the throat through the neck to the outside. A diagnostic trait of chordates.

phelloderm In the stems of woody plants, a thin layer of cells located between the outer cork cells and inner cork cambium.

pharyngeal jaw A secondary jaw in the back of the mouth, found in some fishes. Derived from modified gill arches.

phenetic approach A method for constructing a phylogenetic tree by computing a statistic that summarizes the overall similarity among populations, based on the available data. Compare with **cladistic approach.**

phenotype The detectable physical and physiological traits of an individual, which are determined its genetic makeup. Also the specific trait associated with a particular allele. Compare with **genotype.**

phenotypic plasticity Within-species variation in phenotype that is due to differences in environmental conditions. Occurs more commonly in plants than animals.

pheophytin In photosystem II, a molecule that accepts excited electrons from a reaction center chlorophyll and passes them to an electron transport chain.

pheromone A chemical signal, released by one individual into the external environment, that can trigger responses in a different individual.

phloem A plant vascular tissue that conducts sugars; contains sieve-tube members and companion cells. Primary phloem develops from the procambium of apical meristems; secondary phloem, from the vascular cambium of lateral meristems. Compare with **xylem.**

phonotaxis Orientation toward or away from sound.

phosphodiester linkage Chemical linkage between adjacent **nucleotide residues** in DNA and RNA. Forms when the phosphate group of one nucleotide condenses with the hydroxyl group on the sugar of another nucleotide. Also known as *phosphodiester bond.*

phosphofructokinase The enzyme that catalyzes synthesis of fructose-1,6-bisphosphate from fructose-6-phosphate, a key reaction (step 3) in glycolysis.

phospholipid A class of lipid having a hydrophilic head (a phosphate group) and a hydrophobic tail (one or more fatty acids). Major components of the plasma membrane and organelle membranes.

phosphorylase An enzyme that breaks down glycogen by catalyzing hydrolysis of the α-glycosidic linkages between the glucose residues.

phosphorylation (verb: phosphorylate) The addition of a phosphate group to a molecule.

phosphorylation cascade A series of enzyme-catalyzed phosphorylation reactions commonly used in signal transduction pathways to amplify and convey a signal inward from the plasma membrane.

photic zone In an aquatic habitat, water that is shallow enough to receive some sunlight (whether or not it is enough to support photosynthesis). Compare with **aphotic zone.**

photon A discrete packet of light energy; a particle of light.

photoperiodism Any response by an organism to the relative lengths of day and night (i.e., photoperiod).

photophosporylation Production of ATP molecules using the energy released as light-excited electrons flow through an electron transport chain during photosynthesis. Involves generation of a proton-motive force during electron transport and its use to drive ATP synthesis.

photoreceptor A molecule, a cell, or an organ that is specialized to detect light.

photorespiration A series of light-driven chemical reactions that consumes oxygen and releases carbon dioxide, basically reversing photosynthesis. Usually occurs when there are high O_2 and low CO_2 concentrations inside plant cells, often in bright, hot, dry environments when stomata must be kept closed.

photoreversibility A change in conformation that occurs in certain plant pigments when they are exposed to their preferred wavelengths of light and that triggers responses by the plant.

photosynthesis The complex biological process that converts the energy of light into chemical energy stored in glucose and other organic molecules. Occurs in plants, algae, and some bacteria.

photosystem One of two types of units, consisting of a central reaction center surrounded by antenna complexes, that is responsible for the light-dependent reactions of photosynthesis.

photosystem I A photosystem that contains a pair of P700 chlorophyll molecules and uses absorbed light energy to produce NADPH.

photosystem II A photosystem that contains a pair of P680 chlorophyll molecules and uses absorbed light energy to split water into protons and oxygen and to produce ATP.

phototaxis Orientation toward or away from light.

phototroph An organism that produces ATP through photosynthesis.

phototropins A class of plant photoreceptors that detect blue light and initiate phototropic responses.

phototropism Growth or movement of an organism in a particular direction in response to light.

phylogenetic species concept The definition of a species as the smallest monophyletic group in a phylogenetic tree. Compare with **biological** and **morphological species concept.**

phylogenetic tree A diagram that depicts the evolutionary history of a group of species and the relationships among them.

phylogeny The evolutionary history of a group of organisms.

phylum (plural: phyla) In Linnaeus' system, a taxonomic category above the class level and below the kingdom level. In plants, sometimes called a *division.*

physical map A map of a chromosome that shows the number of base pairs between various genetic markers. Compare with **genetic map.**

physiology The study of how an organism's body functions.

phytoalexin Any small compound produced by a plant to combat an infection (usually a fungal infection).

phytochrome A specialized plant photoreceptor that exists in two shapes depending on the ratio of red to far-red light and is involved in the timing certain physiological processes, such as flowering, stem elongation, and germination.

phytoplankton Small drifting aquatic organisms (plankton) that are photosynthetic.

phytoremediation The use of plants to clean contaminated soils.

pigment Any molecule that absorbs certain wavelengths of visible light and reflects or transmits other wavelengths.

piloting A type of navigation in which animals use familiar landmarks to find their way.

pinocytosis Uptake by a cell of extracellular fluid by pinching off the plasma membrane to form small membrane-bound vesicles; one type of endocytosis.

pioneering species Those species that appear first in recently disturbed areas.

pit In plants, a small hole in the secondary cell walls of tracheids that allow passage of water.

pitch The sensation produced by a particular frequency of sound. Low frequencies are perceived as low pitches; high frequencies, as high pitches.

pituitary gland A small gland directly under the brain that is physically and functionally connected to the hypothalamus. Produces and secretes an array of hormones that affect many other glands and organs.

placenta A structure that forms in the pregnant uterus from maternal and fetal tissues. Exchanges nutrients and wastes between mother and fetus, anchors the fetus to the uterine wall, and produces some hormones. Occurs in most mammals and in a few other vertebrates.

placental mammals See **eutherians.**

plankton Any small organism that drifts near the surface of oceans or lakes and swims little if at all.

Plantae The monophyletic group that includes red, green, and glaucophyte algae and land plants.

plant-defense hypothesis The hypothesis that rates of herbivory are limited by plant defenses such as toxins and spines.

plantlet A small plant, particularly one that forms on a parent plant via asexual reproduction and drops, becoming an independent individual.

plasma cell An effector B cell, which produces large quantities of antibodies. Is descended from an activated B cell that has interacted with antigen.

plasma membrane A membrane that surrounds a cell, separating it from the external environment and selectively regulating passage of molecules and ions into and out of the cell. Also called *cell membrane.*

plasmid A small, usually circular, supercoiled DNA molecule independent of the cell's main chromosome(s) in prokaryotes and some eukaryotes.

plasmodesmata (singular: plasmodesma) Physical connection between two plant cells, consisting of gaps in the cell walls through which the two cells' plasma membranes, cytoplasm, and smooth ER can connect directly. Functionally similar to gap junctions in animal cells.

plasmogamy Fusion of the cytoplasm of two individuals. Occurs in many fungi.

plastid One of a family of plant organelles, bounded by a double membrane, that includes chloroplasts; chromoplasts, which house pigment-containing vacuoles; and leucoplasts, which store oils, starch, or proteins.

plastocyanin A small protein that shuttles electrons from photosystem II to photosystem I during photosynthesis.

plastoquinone (PQ) A nonprotein electron carrier in the chloroplast electron transport chain. Receives excited electrons from pheophytin and passes them to more electronegative molecules in the chain. Also carries protons to the lumen side of the thylakoid membrane, generating a proton-motive force.

platelet A small membrane-bound cell fragment in vertebrate blood that functions in blood clotting. Derived from large cells in the bone marrow.

pleiotropy (adjective: pleiotropic) The ability of a single gene to affect more than one phenotypic trait.

ploidy The number of complete chromosome sets present. *Haploid* refers to a ploidy of 1; *diploid,* a ploidy of 2; *triploid,* a ploidy of 3; and *tetraploid,* a ploidy of 4.

podium (plural: podia) See **tube foot.**

point mutation A mutation that results in a change in a single nucleotide pair in a DNA molecule.

polar (1) Asymmetrical or unidirectional. (2) Carrying a partial positive charge on one side of a molecule and a partial negative charge on the other. Polar molecules are generally hydrophilic.

polar bodies The tiny, nonfunctional cells produced during meiosis of a primary oocyte, due to most of the cytoplasm going to the ovum.

polar covalent bond A covalent bond in which electrons are shared unequally between atoms differing in electronegativity, resulting in the more electronegative atom having a partial negative charge and the other atom, a partial positive charge. Compare with **nonpolar covalent bond.**

polar nuclei In flowering plants, the nuclei in the female gametophyte that fuse with one sperm nucleus to produce the endosperm. Most species have two.

pollen grain In seed plants, a male gametophyte enclosed within a protective coat.

pollen tube In flowering plants, a structure that grows out of a pollen grain after it reaches the stigma, extends down the style, and through which two sperm cells are delivered to the ovule.

pollination The process by which pollen reaches the carpel of a flower (in flowering plants) or reaches the ovule directly (in conifers and their relatives).

poly(A) tail In eukaryotes, a sequence of 100–250 adenine nucleotides added to the 3′ end of newly transcribed messenger RNA molecules.

polycistronic mRNA An mRNA molecule that contains more than one protein-coding segment, each with its own start and stop codons and each coding for a different protein. Common in prokaryotes.

polygenic inheritance The inheritance patterns that result when many genes influence one trait.

polymer Any long molecule composed of small repeating units (monomers) bonded together. The main biological polymers are proteins, nucleic acids, and polysaccharides.

polymerase chain reaction (PCR) A laboratory technique for rapidly generating millions of identical copies of a specific stretch of DNA by incubating the original DNA sequence of interest with primers, nucleotides, and DNA polymerase.

polymerization (verb: polymerize) The process by which many identical or similar small molecules (monomers) are covalently bonded to form a large molecule (polymer).

polymorphism (adjective: polymorphic) (1) The occurrence of more than one allele at a certain genetic locus in a population. (2) The occurrence of more than two distinct phenotypes of a trait in a population.

polyp The immotile (sessile) stage in the life cycle of some cnidarians (e.g., jellyfish). Compare with **medusa.**

polypeptide A chain of 50 or more amino acids linked together by peptide bonds. Compare with **oligopeptide** and **protein.**

polyploidy (adjective: polyploid) The state of having more than two full sets of chromosomes.

polyribosome A structure consisting of one messenger RNA molecule along with many attached ribosomes and their growing peptide strands.

polysaccharide A linear or branched polymer consisting of many monosaccharides joined by glycosidic linkages. Carbohydrate polymers with relatively few residues often are called *oligosaccharides.*

polyspermy Fertilization of an egg by multiple sperm.

poor-nutrition hypothesis The hypothesis that herbivore populations are limited by the poor nutritional content of plants, especially low nitrogen.

population A group of individuals of the same species living in the same geographic area at the same time.

population cycle A regular fluctuation in size exhibited by certain populations.

population density The number of individuals of a population per unit area.

population dynamics Changes in the size and other characteristics of populations through time.

population ecology The study of how and why the number of individuals in a population changes over time.

population viability analysis (PVA) A method of estimating the likelihood that a population will avoid extinction for a given time period.

positive feedback A physiological mechanism in which a change in some variable stimulates a response that increases the change. Relatively rare in organisms but is important in generation of the action potential. Compare with **negative feedback.**

positive pressure ventilation Ventilation of the lungs that is accomplished by "pushing" air into the lungs by positive pressure in the mouth. Compare with **negative pressure ventilation.**

positive-sense virus A virus whose genome contains the same sequences as the mRNA required to produce viral proteins. Compare with **ambisense virus** and **negative-sense virus.**

posterior Toward an animal's tail and away from its head. The opposite of anterior.

posterior pituitary The part of the pituitary gland that contains the ends of hypothalamic neurosecretory cells and from which oxytocin and antidiuretic hormone are secreted. Compare with **anterior pituitary.**

postsynaptic neuron A neuron that receives signals, usually via neurotransmitters, from another neuron at a synapse. Muscle and gland cells also may receive signals from presynaptic neurons.

post-translational control Regulation of gene expression by modification of proteins (e.g., addition of a phosphate group or sugar residues) after translation.

postzygotic isolation Reproductive isolation resulting from mechanisms that operate after mating of individuals of two different species occurs. The most common mechanisms are the death of hybrid embryos or reduced fitness of hybrids.

potential energy Energy stored in matter as a result of its position or molecular arrangement. Compare with **kinetic energy.**

prebiotic soup A hypothetical solution of sugars, amino acids, nitrogenous bases, and other building blocks of larger molecules that may have formed in shallow waters or deep-ocean vents of ancient Earth and given rise to larger biological molecules.

Precambrian era The interval between the formation of the Earth, about 4.6 billion years ago, and the appearance of most animal groups about 543 million years ago. Unicellular organisms were dominant for most of this era, and oxygen was virtually absent for the first 2 billion years.

predation The killing and eating of one organism (the prey) by another (the predator).

predator Any organism that kills other organisms for food.

prediction A measurable or observable result of an experiment based on a particular hypothesis. A correct prediction provides support for the hypothesis being tested.

pressure-flow hypothesis The hypothesis that sugar movement through phloem tissue is due to differences in the turgor pressure of phloem sap.

pressure potential (ψ_P) A component of the potential energy of water caused by physical pressures on a solution. In plant cells, it equals the wall pressure plus turgor pressure. Compare with **solute potential (ψ_S).**

presynaptic neuron A neuron that transmits signals, usually by releasing neurotransmitters, to another neuron or to an effector cell at a synapse.

pre-urine See **filtrate.**

prezygotic isolation Reproductive isolation resulting from any one of several mechanisms that prevent individuals of two different species from mating.

primary cell wall The outermost layer of a plant cell wall, made of cellulose fibers and gelatinous polysaccharides, that defines the shape of the cell and withstands the turgor pressure of the plasma membrane.

primary consumer An herbivore; an organism that eats plants, algae, or other primary producers. Compare with **secondary consumer.**

primary decomposer A decomposer (detritivore) that consumes detritus from plants.

primary growth In plants, an increase in the length of stems and roots due to the activity of apical meristems. Compare with **secondary growth.**

primary immune response An acquired immune response to a pathogen that the immune system has not encountered before. Compare with **secondary immune response.**

primary oocyte The large diploid cell in an ovarian follicle that can initiate meiosis to produce a haploid ovum.

primary producer Any organism that creates its own food by photosynthesis or from reduced inorganic compounds and that is a food source for other species in its ecosystem. Also called *autotroph.*

primary RNA transcript In eukaryotes, a newly transcribed messenger RNA molecule that has not yet been processed (i.e., it has not received a 5′ cap or poly(A) tail, and still contains introns).

primary spermatocyte A diploid cell in the testis that can initiate meiosis I to produce two secondary spermatocytes.

primary structure The sequence of amino acids in a peptide or protein; also the sequence of nucleotides in a nucleic acid. Compare with **secondary, tertiary,** and **quaternary structure.**

primary succession The gradual colonization of a habitat of bare rock or gravel, usually after an environmental disturbance that removes all soil and previous organisms. Compare with **secondary succession.**

primase An enzyme that synthesizes a short stretch of RNA to use as a primer during DNA replication.

primates The lineage of mammals that includes prosimians (lemurs, lorises, etc.), monkeys, and great apes (including humans).

primer A short, single-stranded RNA molecule that base pairs with the 5′ end of a DNA template strand and is elongated by DNA polymerase during DNA replication.

prion An infectious form of a protein that is thought to cause disease by inducing the normal form to assume an abnormal three-dimensional structure. Cause of spongiform encephalopathies, such as mad cow disease.

probe A radioactively or chemically labeled single-stranded fragment of a known DNA or RNA sequence that can bind to and thus detect its complementary sequence in a sample containing many different sequences.

proboscis A long, narrow feeding appendage through which food can be obtained.

procambium A group of cells in the center of a young plant embryo that gives rise to the vascular tissue.

product Any of the final materials formed in a chemical reaction.

productivity The total amount of carbon fixed by photosynthesis per unit area per year.

progesterone A steroid hormone produced in the ovaries and secreted by the corpus luteum after ovulation; causes the uterine lining to thicken.

programmed cell death See **apoptosis.**

prokaryote A member of the domain Bacteria or Archaea; a unicellular organism lacking a nucleus and containing relatively few organelles or cytoskeletal components. Compare with **eukaryote.**

prolactin A peptide hormone, produced and secreted by the anterior pituitary, that promotes milk production in female mammals and has a variety of effects on parental behavior and seasonal reproduction in other vertebrates.

prometaphase A stage in mitosis or meiosis during which the nuclear envelope breaks down and spindle fibers attach to chromatids.

promoter A short nucleotide sequence in DNA that binds RNA polymerase, enabling transcription to begin. In prokaryotic DNA, a single promoter often is associated with several contiguous genes. In eukaryotic DNA, each gene generally has its own promoter.

promoter-proximal elements In eukaryotes, regulatory sequences in DNA that are close to a promoter and that can bind regulatory transcription factors.

proofreading The process by which a DNA polymerase recognizes and removes a wrong base added during DNA replication and then continues synthesis.

prophase The first stage in mitosis or meiosis during which chromosomes become visible and the mitotic spindle forms. Synapsis and crossing over occur during prophase of meiosis I.

prosimians One of the two major lineages of primates, including lemurs, tarsiers, pottos, and lorises. Compare with **anthropoids.**

prostate gland A gland in male mammals that surrounds the base of the urethra and secretes a fluid that is a component of semen.

protease An enzyme that can degrade proteins by cleaving the peptide bonds between amino acid residues.

protein A macromolecule consisting of one or more polypeptide chains composed of 50 or more amino acids linked together. Each protein has a unique sequence of amino acids and, in its native state, a characteristic three-dimensional shape.

protein kinase An enzyme that catalyzes the addition of a phosphate group to another protein, typically activating or inactivating the substrate protein.

proteome The complete set of proteins produced by a particular cell type.

proteomics The systematic study of the interactions, localization, functions, regulation, and other features of the full protein set (proteome) in a particular cell type.

protist Any eukaryote that is not a green plant, animal, or fungus. Protists are a diverse paraphyletic group. Most are unicellular, but some are multicellular or form aggregations called colonies.

protoderm The exterior layer of a young plant embryo that gives rise to the epidermis.

proton-motive force The combined effect of a proton gradient and an electric potential gradient across a membrane, which can drive protons across the membrane. Used by mitochondria and chloroplasts to power ATP synthesis via the mechanism of chemiosmosis.

proton pump A membrane protein that can hydrolyze ATP to power active transport of protons (H^+ ions) across a plasma membrane against an electrochemical gradient. Also called *H^+ ATPase.*

proto-oncogene Any gene that normally encourages cell division in a regulated manner, typically by triggering specific phases in the cell cycle. Mutation may convert it into an oncogene.

protostomes A major lineage of animals that share a pattern of embryological development, including spiral cleavage, formation of the mouth earlier than the anus, and formation of the coelom by splitting of a block of mesoderm. Includes arthropods, mollusks, and annelids. Compare with **deuterostomes.**

proximal tubule In the vertebrate kidney, the convoluted section of a nephron into which filtrate moves from Bowman's capsule. Involved in the largely unregulated reabsorption of electrolytes, nutrients, and water. Compare with **distal tubule.**

proximate causation In biology, the immediate, mechanistic cause of a phenomenon (how it happens), as opposed to why it evolved. Also called *proximate explanation.* Compare with **ultimate causation.**

pseudocoelomate An animal with an internal fluid-filled body cavity (coelom) that is lined with endoderm and mesoderm layers.

pseudogene A DNA sequence that closely resembles a functional gene but is not transcribed. Thought to have arisen by duplication of the functional gene followed by inactivation due to a mutation.

pseudopodium (plural: pseudopodia) A temporary bulge-like extension of certain cells used in cell crawling and ingestion of food.

puberty The various physical and emotional changes that an immature animal undergoes leading to reproductive maturity. Also the period when such changes occur.

pulmonary artery A short, thick-walled artery that carries oxygen-poor blood from the heart to the lungs.

pulmonary circulation The part of the circulatory system that sends oxygen-poor blood to the lungs. Is separate from the rest of the circulatory system (the systemic circulation) in mammals and birds.

pulmonary vein A short, thin-walled vein that carries oxygen-rich blood from the lungs to the heart.

pulse-chase experiment A type of experiment in which a population of cells or molecules at a particular moment in time is marked by means of a labeled molecule and then their fate is followed over time.

pump Any membrane protein that can hydrolyze ATP to power active transport of a specific ion or small molecule across a plasma membrane against its electrochemical gradient. See **proton pump.**

Punnett square A diagram that depicts the genotypes and phenotypes that should appear in offspring of a certain cross.

pupa (plural: pupae) A metamorphosing insect that is enclosed in a protective case.

pupation A developmental stage of many insects, in which the body metamorphoses from the larval form to the adult form while enclosed in a protective case.

pupil The hole in the center of the iris through which light enters a vertebrate or cephalopod eye.

pure line In animal or plant breeding, a strain of individuals that produce offspring identical to themselves when self-pollinated or crossed to another member of the same population. Pure lines are homozygous for most, if not all, genetic loci.

purines A class of small, nitrogen-containing, double-ringed bases (guanine, adenine) found in nucleotides. Compare with **pyrimidines.**

pyrimidines A class of small, nitrogen-containing, single-ringed bases (cytosine, uracil, thymine) found in nucleotides. Compare with **purines.**

pyruvate dehydrogenase A large enzyme complex, located in the inner mitochondrial membrane, that is responsible for conversion of pyruvate to acetyl CoA during cellular respiration.

quantitative trait A heritable feature that exhibits phenotypic variation along a smooth, continuous scale of measurement (e.g., human height), rather than the distinct forms characteristic of discrete traits.

quaternary structure The overall three-dimensional shape of a protein containing two or more polypeptide chains (subunits); determined by the number, relative positions, and interactions of the subunits. Compare with **primary, secondary,** and **tertiary structure.**

radial cleavage The pattern of embryonic cleavage seen in protostomes, in which cells divide at right angles to each other to form tiers. Compare with **spiral cleavage.**

radial symmetry An animal body pattern in which there are least two planes of symmetry. Typically, the body is in the form of a cylinder or disk, with body parts radiating from a central hub. Compare with **bilateral symmetry.**

radiation Transfer of heat between two bodies that are not in direct physical contact. More generally, the emission of electromagnetic energy of any wavelength.

radicle The root of a plant embryo.

radula A rasping feeding appendage in gastropods (snails, slugs).

rain shadow The dry region on the side of a mountain range away from the prevailing wind.

range The geographic distribution of a species.

Ras protein A type of G protein that is activated by binding of signaling molecules to receptor tyrosine kinases and then initiates a phosphorylation cascade, culminating in a cell response.

Rb protein A tumor-suppressor protein that helps regulate progression of a cell from the G_1 phase to the S phase of the cell cycle. Defects in Rb protein are found in many types of cancer.

reactant Any of the starting materials in a chemical reaction.

reaction center Centrally located component of a photosystem containing proteins and a pair of specialized chlorophyll molecules. Is surrounded by antenna complexes and receives excited electrons from them.

reactive oxygen intermediates (ROIs) Highly reactive oxygen-containing compounds that are used in plant and animal cells to kill infected cells and for other purposes.

reading frame The division of a sequence of DNA or RNA into a particular series of three-nucleotide codons. There are three possible reading frames for any sequence.

realized niche The ecological niche that a species occupies in the presence of competitors. Compare with **fundamental niche.**

receptor-mediated endocytosis Uptake by a cell of certain extracellular macromolecules, bound to specific receptors in the plasma membrane, by pinching off the membrane to form small membrane-bound vesicles.

receptor tyrosine kinase (RTK) Any of a class of cell-surface signal receptors that undergo phosphorylation after binding a signaling molecule. The activated, phosphorylated receptor then triggers a signal-transduction pathway inside the cell.

recessive Referring to an allele whose phenotypic effect is observed only in homozygous individuals. Compare with **dominant.**

reciprocal altruism Altruistic behavior that is exchanged between a pair of individuals at different points in time (i.e., sometimes individual A helps individual B, and sometimes B helps A).

reciprocal cross A breeding experiment in which the mother's and father's phenotypes are the reverse of that examined in a previous breeding experiment.

recombinant Possessing a new combination of alleles. May refer to a single chromosome or DNA molecule, or to an entire organism.

recombinant DNA technology A variety of techniques for isolating specific DNA fragments and introducing them into different regions of DNA and/or a different host organism.

recombination See **genetic recombination.**

rectal gland A salt-excreting gland in the digestive system of sharks, skates, and rays.

rectum The last portion of the digestive tract where feces are held until they are expelled.

red blood cell A hemoglobin-containing cell that circulates in the blood and delivers oxygen from the lungs to the tissues. Also called *erythrocyte.*

redox reaction Any chemical reaction that involves the transfer of one or more electrons from one reactant to another. Also called *reduction-oxidation reaction.*

reduction An atom's gain of electrons during a redox reaction, either by acceptance of an electron from another atom or by the electrons in covalent bonds moving closer to the atomic nucleus.

reduction-oxidation reaction See **redox reaction.**

reflex An involuntary response to environmental stimulation. May involve the brain (e.g., conditioned reflex) or not (e.g., spinal reflex).

refractory No longer responding to stimuli that previously elicited a response. For example, the tendency of voltage-gated sodium channels to remain closed immediately after an action potential.

regeneration Growth of a new body part to replace a lost body part.

regulatory cascade In embryonic development, a progressive series of interactions among genes and/or cytoplasmic determinants that organizes the body plan of the embryo.

regulatory sequence, DNA Any segment of DNA that is involved in controlling transcription of a specific gene by binding certain proteins.

regulatory site A site on an enzyme to which a regulatory molecule can bind and affect the enzyme's activity; separate from the active site where catalysis occurs.

regulatory transcription factor General term for proteins that bind to DNA regulatory sequences (eukaryotic enhancers, silencers, and promoter-proximal elements), but not to the promoter itself, leading to an increase or decrease in transcription of specific genes. Compare with **basal transcription factor.**

reinforcement In evolutionary biology, the natural selection for traits that prevent interbreeding between recently diverged species.

release factors Proteins that can trigger termination of RNA translation when a ribosome reaches a stop codon.

releaser See **sign stimulus.**

renal corpuscle In the vertebrate kidney, the ball-like structure at the beginning of a nephron, consisting of a glomerulus and the surrounding Bowman's capsule. Acts as a filtration device.

replacement mutation See **missense mutation.**

replacement rate The number of offspring each female must produce over her entire life to "replace" herself and her mate, resulting in zero population growth. The actual number is slightly more than 2 because some offspring die before reproducing.

replica plating A method of identifying bacterial colonies that have certain mutations by transferring cells from each colony on a master plate to a second (replica) plate and observing their growth when exposed to different conditions.

replication fork The Y-shaped site at which a double-stranded molecule of DNA is separated into two single strands for replication.

repolarization Return to a normal membrane potential after it has changed; a normal phase in an action potential.

repressor Any regulatory protein that inhibits transcription.

reproduction The ability of an organism to make an exact or nearly exact copy of itself.

reptiles One of the two lineages of amniotes (vertebrates that produce amniotic eggs) distinguished by adaptations for reproduction on land. Includes turtles, snakes and lizards, crocodiles and alligators, and birds. Except for birds, all are ectotherms.

reservoir In biogeochemical cycles, a location in the environment where elements are stored for a time.

residues In a polymer, the individual units derived from the monomers that covalently bind to form the polymer. Proteins contain amino acid residues; nucleic acids, nucleotide residues; and polysaccharides, sugar residues.

resilience, community A measure of how quickly a community recovers following a disturbance.

resistance, community A measure of how much a community is affected by a disturbance.

resistance (*R*) loci Genes in plants encoding proteins involved in sensing the presence of pathogens and mounting a defensive response. Compare with **avirulence (*avr*) loci.**

respiration See **cellular respiration.**

respiratory distress syndrome A syndrome in which premature infants can suffocate due to insufficient surfactant in their lungs.

respiratory system The collection of cells, tissues, and organs responsible for gas exchange between an animal and its environment.

resting potential The membrane potential of a cell in its resting, or normal, state.

restriction endonucleases Bacterial enzymes that cut DNA at a specific base-pair sequence (restriction site). Also called *restriction enzymes.*

restriction site A specific short sequence in a DNA molecule that is recognized and cleaved by a particular restriction endonuclease. Also called *restriction endonuclease recognition site.*

retina A thin layer of light-sensitive cells (rods and cones) and neurons at the back of a camera-type eye, such as that of cephalopods and vertebrates.

retinal A light-absorbing pigment, derived from vitamin A, that is linked to the protein opsin in rods and cones of the vertebrate eye.

retrovirus A virus with an RNA genome that reproduces by transcribing its RNA into a DNA sequence and then inserting that DNA into the host's genome for replication.

reverse transcriptase A enzyme of retroviruses (RNA viruses) that can synthesize double-stranded DNA from a single-stranded RNA template.

Rh factor A protein present on the surface of red blood cells in some but not all humans. Clumping may occur when blood from an Rh^+ individual is mixed with that from an Rh^- individual.

rhizobia (singular: rhizobium) Members of the bacterial genus *Rhizobia;* nitrogen-fixing bacteria that live in root nodules of members of the pea family (legumes).

rhizoid The hairlike structure that anchors a bryophyte (nonvascular plant) to the substrate.

rhizome A modified stem that runs horizontally underground and produces new plants at the nodes (a form of asexual reproduction). Compare with **stolon.**

rhodopsin A transmembrane complex that is instrumental in detection of light by rods and cones of the vertebrate eye. Is composed of the transmembrane protein opsin covalently linked to retinal, a light-absorbing pigment.

ribonucleic acid (RNA) A nucleic acid composed of ribonucleotides that usually is single stranded and functions as structural components of ribosomes (rRNA), transporters of amino acids (tRNA), and translators of the message of the DNA code (mRNA).

ribonucleotide See **nucleotide.**

ribosomal RNA (rRNA) A RNA molecule that forms part of the structure of a ribosome.

ribosome A large complex structure that synthesizes proteins by using the genetic information encoded in messenger RNA strands. Consists of two subunits, each composed of ribosomal RNA and proteins.

ribosome binding site In a bacterial mRNA molecule, the sequence just upstream of the start codon to which a ribosome binds to initiate translation. Also called the *Shine-Dalgarno sequence.*

ribozyme Any RNA molecule that can act as a catalyst, that is, speed up a chemical reaction.

ribulose bisphosphate (RuBP) A five-carbon compound that combines with CO_2 in the first step of the Calvin cycle during photosynthesis.

RNA See **ribonucleic acid.**

RNA interference (RNAi) Degradation of an mRNA molecule or inhibition of its translation following its binding by a short RNA (microRNA) whose sequence is complementary to a portion of the mRNA.

RNA polymerase One of a class of enzymes that catalyze synthesis of RNA from ribonucleotides using a DNA template. Also called *RNA pol.*

RNA processing In eukaryotes, the changes that a primary RNA transcript undergoes in the nucleus to become a mature mRNA molecule, which is exported to the cytoplasm. Includes the addition of a 5′ cap and poly(A) tail and splicing to remove introns.

RNA replicase A viral enzyme that can synthesize RNA from an RNA template.

rod cell A photoreceptor cell with a rod-shaped outer portion that is particularly sensitive to dim light, but not used to distinguish colors. Also called simply *rod.* Compare with **cone cell.**

root (1) An underground part of a plant that anchors the plant and absorbs water and nutrients. (2) In a phylogenetic tree, the bottom, most ancient node.

root apical meristem (RAM) A group of undifferentiated plant cells at the tip of a plant root that can differentiate into mature root tissue.

root cap A small group of cells that covers and protects the tip of a plant root. Senses gravity and determines the direction of root growth.

root hair A long, thin outgrowth of the epidermal cells of plant roots, providing increased surface area for absorption of water and nutrients.

root pressure Positive (upward) pressure of xylem sap in the vascular tissue of roots. Is generated during the night as a result of the accumulation of ions from the soil and subsequent osmotic movement of water into the xylem.

root system The belowground part of a plant.

rotifers Members of the phylum Rotifera. Distinguished by a cluster of cilia (corona) at the anterior end and a pseudocoelom. Rotifers belong to the lophotrochozoan branch of the protostomes.

rough endoplasmic reticulum (rough ER) The portion of the endoplasmic reticulum that is dotted with ribosomes. Involved in synthesis of plasma membrane proteins, secreted proteins, and proteins localized to the ER, Golgi apparatus, and lysosomes. Compare with **smooth endoplasmic reticulum.**

roundworms Members of the phylum Nematoda. Distinguished by an unsegmented body with a pseudocoelom and no appendages. Roundworms belong to the ecdysozoan branch of the protostomes. Also called *nematodes.*

rRNA See **ribosomal RNA.**

rubisco The enzyme that catalyzes the first step of the Calvin cycle during photosynthesis: the addition of a molecule of CO_2 to ribulose bisphosphate. Also called *ribulose 1,5-bisphosphate carboxylase/oxygenase.*

ruminants A group of hoofed mammals (e.g., cattle, sheep, deer) that have a four-chambered stomach specialized for digestion of plant cellulose. Ruminants regurgitate the cud, a mixture of partially digested food and cellulose-digesting bacteria, from the largest chamber (the rumen) for further chewing.

sac fungus A member of a monophyletic lineage of fungi (Ascomycota) that produce large, often cup-shaped reproductive structures that contain asci. Also called *cup fungus* and *ascomycete.*

salicylic acid A compound produced by plants that may play a role in systemic acquired resistance (SAR) against pathogens; a component of aspirin.

salivary glands Vertebrate glands that secrete saliva (a mixture of water, mucus-forming glycoproteins, and digestive enzymes) into the mouth.

sampling error The accidental selection of a nonrepresentative sample from some larger population, due to chance.

saprophyte An organism that feeds primarily on dead plant material.

sapwood The younger xylem in the outer layer of wood of a stem or root, functioning primarily in water transport.

sarcomere The repeating contractile unit of a skeletal muscle cell; the portion of a myofibril located between adjacent Z disks.

sarcoplasmic reticulum Sheets of smooth endoplasmic reticulum in a muscle cell. Contains high concentrations of calcium, which can be released into the cytoplasm to trigger contraction.

saturated Referring to fats and fatty acids in which all the carbon-carbon bonds are single bonds. Such fats have relatively high melting points. Compare with **unsaturated.**

scanning electron microscope (SEM) A microscope that produces images of the surfaces of objects by reflecting electrons from a specimen coated with a layer of metal atoms. Compare with **transmission electron microscope.**

scarify To scrape, rasp, cut, or otherwise damage the coat of a seed. Necessary in some species to trigger germination.

Schwann cell A type of glial cell that wraps around axons of some neurons outside the brain and spinal cord, forming a myelin sheath that provides electrical insulation. Compare with **oligodendrocyte.**

sclereid In plants, a type of sclerenchyma cell that usually functions in protection, such as in seed coats and nutshells. Compare with **fiber.**

sclerenchyma cell In plants, a cell that has a thick secondary cell wall and provides support; typically contains the tough structural polymer lignin and usually is dead at maturity. Includes fibers and sclereids. Compare with **collenchyma cell** and **parenchyma cell.**

screen See **genetic screen.**

scrotum A sac of skin, containing the testes, suspended just outside the abdominal body cavity of many male mammals.

secondary cell wall The inner layer of a plant cell wall formed by certain cells as they mature. Provides support or protection.

secondary consumer A carnivore; an organism that eats herbivores. Compare with **primary consumer.**

secondary growth In plants, an increase in the width of stems and roots due to the activity of lateral meristems. Compare with **primary growth.**

secondary immune response The acquired immune response to a pathogen that the immune system has encountered before. Compare with **primary immune response.**

secondary production The total amount of new body tissue produced by animals that eat plants. May involve growth and/or reproduction.

secondary spermatocyte A cell produced by meiosis I of a primary spermatocyte in the testis. Can undergo meiosis II to produce spermatids.

secondary structure In proteins, localized folding of a polypeptide chain into regular structures (e.g., α-helix and β-pleated sheet) stabilized by hydrogen bonding between atoms of the backbone. In nucleic acids, elements of structure (e.g., helices and hairpins) stabilized by hydrogen bonding and other interactions between complementary bases. Compare with **primary, tertiary,** and **quaternary structure.**

secondary succession Gradual colonization of a habitat after an environmental disturbance (e.g., fire, windstorm, logging) that removes some or all previous organisms but leaves the soil intact. Compare with **primary succession.**

second law of thermodynamics The principle of physics that the entropy of the universe or any closed system increases during any spontaneous process.

second-male advantage The reproductive advantage of a male who mates with a female last, after other males have mated with her.

second messenger A nonprotein signaling molecule produced or activated inside a cell in response to stimulation at the cell surface. Commonly used to relay the message of a hormone or other extracellular signaling molecule.

secretin A peptide hormone produced by cells in the small intestine in response to the arrival of food from the stomach. Stimulates secretion of bicarbonate (HCO_3^-) from the pancreas.

sedimentary rock A type of rock formed by gradual accumulation of sediment, as in riverbeds and on the ocean floor. Most fossils are found in sedimentary rocks.

seed A plant reproductive structure consisting of an embryo, associated nutritive tissue (endosperm), and an outer protective layer (seed coat). In angiosperms, develops from the fertilized ovule of a flower.

seed coat A protective layer around a seed that encases both the embryo and the endosperm.

seed plants Members of several phyla of green plants that have vascular tissue and make seeds. Include **angiosperms** and **gymnosperms.**

seedless vascular plants Members of several phyla of green plants that have vascular tissue but do not make seeds. Include horsetails, ferns, lycophytes, and whisk ferns.

segment A well-defined region of the body along the anterior-posterior body axis, containing similar structures as other, nearby segments.

segmentation Division of the body or a part of it into a series of similar structures; exemplified by the body segments of insects and worms and by the somites of vertebrates.

segmentation genes A group of genes that affect body segmentation in embryonic development. Includes gap genes, pair-rule genes, and segment polarity genes.

segmented worms See **annelids.**

segregation, principle of The concept that each pair of hereditary elements (alleles of the same gene) separate from each other during the formation of offspring (i.e., during meiosis). One of Mendel's two principles of genetics.

selective adhesion The tendency of cells of one tissue type to adhere to other cells of the same type.

selectively permeable membrane Any membrane across which some solutes can move more readily than others.

selective permeability The property of a membrane that allows some substances to diffuse across it much more readily than other substances.

self Property of a molecule or cell such that immune system cells do not attack it, due to certain molecular similarities to other body cells.

self-fertilization In plants, the fusion of two gametes from the same individual to form a diploid offspring. Also called *selfing.*

self-incompatible Incapable of self-fertilization.

semen The combination of sperm and accessory fluids that is released by male mammals and reptiles during ejaculation.

semiconservative replication The mechanism of replication used by cells to copy DNA. Results in each daughter DNA molecule containing one old strand and one new strand.

seminal vesicles In male mammals, paired reproductive glands that secrete a sugar-containing fluid into semen, which provides energy for sperm movement. In other vertebrates and invertebrates, often stores sperm.

senescence The process of aging.

sensitive period A short time period in a young animal's life during which learning of certain critical behaviors can occur. Also called the *critical period.*

sensor Any cell, organ, or structure with which an animal can sense some aspect of the external or internal environment. Usually functions, along with an integrator and effector, as part of a homeostatic system.

sensory neuron A nerve cell that carries signals from sensory receptors to the central nervous system. Compare with **interneuron** and **motor neuron.**

sepal One of the protective leaflike organs enclosing a flower bud and later supporting the blooming flower.

septum (plural: septa) Any wall-like structure. In fungi, septa divide the filaments (hyphae) of mycelia into cell-like compartments.

serotonin A neurotransmitter involved in many brain functions, including sleep, pleasure, and mood.

serum The liquid that remains when cells and clot material are removed from clotted blood. Contains water, dissolved gases, growth factors, nutrients, and other soluble substances. Compare with **plasma.**

sessile Permanently attached to a substrate; not capable of moving to another location.

set point A normal or target value for a regulated internal variable, such as body heat or blood pH.

severe combined immunodeficiency disease (SCID) A human disease characterized by an extremely high vulnerability to infectious disease. Caused by a genetic defect in the immune system.

sex chromosome Any chromosome carrying genes involved in determining the sex of an individual. Compare with **autosome.**

sex-linked inheritance Inheritance patterns observed in genes carried on sex chromosomes, so females and males have different numbers of alleles of a gene and may pass its trait only to one sex of offspring. Also called *sex-linkage.*

sexual dimorphism Any trait that differs between males and females.

sexual reproduction Any form of reproduction in which genes from two parents are combined via fusion of gametes, producing offspring that are genetically distinct from both parents. Compare with **asexual reproduction.**

sexual selection A pattern of natural selection that favors individuals with traits that increase their ability to obtain mates. Acts more strongly on males than females.

shell A hard protective outer structure. In protists, also called a *test.*

Shine-Dalgarno sequence See **ribosome binding sequence.**

shoot apical meristem (SAM) A group of undifferentiated plant cells at the tip of a plant stem that can differentiate into mature shoot tissues.

shoot system The aboveground part of a plant comprising stems, leaves, and flowers (in angiosperms).

short-day plant A plant that blooms in response to long nights (usually in late summer or fall in the northern hemisphere). Compare with **day-neutral** and **long-day plant.**

short interspersed nuclear elements See **SINES.**

shotgun sequencing A method of sequencing genomes that is based on breaking the genome into small pieces, sequencing each piece separately, and then figuring out how the pieces are connected.

sieve plate In plants, a pore-containing structure at one end of a sieve-tube member in phloem.

sieve-tube member In plants, an elongated sugar-conducting cell in phloem that has sieve plates at both ends, allowing sap to flow to adjacent cells.

signal In behavioral ecology, any information-containing behavior.

signal receptor Any cellular protein that binds to a particular signaling molecule (e.g., a hormone or neurotransmitter) and triggers a response by the cell. Receptors for water-soluble signals are transmembrane proteins in the plasma membrane; those for many lipid-soluble signals (e.g., steroid hormones) are located inside the cell.

signal recognition particle (SRP) A RNA-protein complex that binds to the ER signal sequence in a polypeptide as it emerges from a ribosome and transports the ribosome-polypeptide complex to the ER membrane where synthesis of the polypeptide is completed.

signal transducers and activators of transcription (STATs) In mammals, a group of regulatory transcription factors that, upon phosphorylation, can activate transcription of certain genes.

signal transduction The process by which a stimulus (e.g., a hormone, a neurotransmitter, or sensory information) outside a cell is amplified and converted into a response by the cell. Usually involves a specific sequence of molecular events, or signal transduction pathway.

signal transduction cascade See **phosphorylation cascade.**

sign stimulus A simple stimulus that elicits an invariant, stereotyped behavioral response (fixed action pattern) from an animal. Also called a *releaser.*

silencer A regulatory sequence in eukaryotic DNA to which repressor proteins can bind, inhibiting transcription of certain genes.

silent mutation A mutation that does not detectably affect the phenotype of the organism.

simple eye An eye with only one light-collecting apparatus (e.g., one lens), as in vertebrates. Compare with **compound eye.**

simple fruit A fruit (e.g., apricot) that develops from a single flower that has a single carpel or several fused carpels. Compare with **aggregate** and **multiple fruit.**

simple sequence repeat See **microsatellite.**

SINEs (short interspersed nuclear elements) The second most abundant class of transposable elements in human genomes; can create copies of itself and insert them elsewhere in the genome. Compare with **LINEs.**

single nucleotide polymorphism (SNP) A site on a chromosome where individuals in a population have different nucleotides. Can be used as a genetic marker to help track the inheritance of nearby genes.

single-strand DNA-binding proteins (SSBPs) A class of proteins that attach to separated strands of DNA during replication or transcription, preventing them from re-forming a double helix.

sink Any tissue, site, or location where an element or a molecule is consumed or taken out of circulation (e.g., in plants, a tissue where sugar exits the phloem). Compare with **source.**

sinoatrial (SA) node A cluster of cardiac muscle cells, in the right atrium of the vertebrate heart, that initiates the heartbeat and determines the heart rate. Compare with **atrioventricular (AV) node.**

siphon A tubelike appendage of many molluscs, that is often used for feeding or propulsion.

sister chromatids The paired strands of a recently replicated chromosome, which are connected at the centromere and eventually separate during anaphase of mitosis and meiosis II. Compare with **non-sister chromatids.**

sister groups Closely related taxa, which occupy adjacent branches in a phylogenetic tree. Also called *sister taxa.*

skeletal muscle The muscle tissue attached to the bones of the vertebrate skeleton. Consists of long, unbranched muscle fibers with a characteristic striped (striated) appearance; controlled voluntarily. Also called *striated muscle.* Compare with **cardiac** and **smooth muscle.**

sliding-filament model The hypothesis that thin (actin) filaments and thick (myosin) filaments slide past each other, thereby shortening the sarcomere. Shortening of all the sarcomeres in a myofibril results in contraction of the entire myofibril.

slug (1) A member of a certain lineage of terrestrial gastropods, closely related to snails but lacking a shell. (2) A mobile aggregation of cells of a cellular slime mold.

small intestine The portion of the digestive tract between the stomach and the large intestine. The site of the final stages of digestion and of most nutrient absorption.

small nuclear ribonucleoproteins See **snRNPs.**

smooth endoplasmic reticulum (smooth ER) The portion of the endoplasmic reticulum that does not have ribosomes attached to it. Involved in synthesis and secretion of lipids. Compare with **rough endoplasmic reticulum.**

smooth muscle The unstriated muscle tissue that lines the intestine, blood vessels, and some other organs. Consists of tapered, unbranched cells that can sustain long contractions. Not voluntarily controlled. Compare with **cardiac** and **skeletal muscle.**

snRNPs (small nuclear ribonucleoproteins) Complexes of proteins and small RNA molecules that function in splicing (removal of introns from primary RNA transcripts) as components of spliceosomes.

sodium-potassium pump A transmembrane protein that uses the energy of ATP to move sodium ions out of the cell and potassium ions in. Also called Na^+/K^+*-ATPase.*

solute Any substance that is dissolved in a liquid.

solute potential (ψ_S) A component of the potential energy of water caused by a difference in solute concentrations at two locations. Also called *osmotic potential*. Compare with **pressure potential** (ψ_P).

solution A liquid containing one or more dissolved solids or gases in a homogeneous mixture.

solvent Any liquid in which one or more solids or gases can dissolve.

soma See **cell body.**

somatic cell Any type of cell in a multicellular organism except eggs, sperm, and their precursor cells. Also called *body cells.*

somatic nervous system The part of the peripheral nervous system (outside the brain and spinal cord) that controls skeletal muscles and is under voluntary control. Compare with **autonomic nervous system.**

somatostatin A hormone secreted by the pancreas and hypothalamus that inhibits the release of several other hormones.

somites Paired blocks of mesoderm on both sides of the developing spinal cord in a vertebrate embryo. Give rise to muscle tissue, vertebrae, ribs, limbs, etc.

soredium (plural: soredia) In lichens, a small reproductive structure that consists of fungal hyphae surrounding a cluster of green algae.

source Any tissue, site, or location where a substance is produced or enters circulation (e.g., in plants, the tissue where sugar enters the phloem). Compare with **sink.**

Southern blotting A technique for identifying specific DNA fragments separated by gel electrophoresis by transferring them to filter paper, separating the strands, and hybridizing with a labeled probe complementary to the fragment of interest. Compare with **Northern** and **Western blotting.**

speciation The evolution of two or more distinct species from a single ancestral species.

species A distinct, identifiable group of populations that is thought to be evolutionarily independent of other populations and whose members can interbreed. Generally distinct from other species in appearance, behavior, habitat, ecology, genetic characteristics, etc.

species–area relationship The mathematical relationship between the area of a certain habitat and the number of species that it can support.

species diversity The variety and relative abundance of the species present in a given ecological community.

species richness The number of species present in a given ecological community.

specific heat The amount of energy required to raise the temperature of 1 gram of a substance by 1°C; a measure of the capacity of a substance to absorb energy.

spectral karyotyping See **chromosome painting.**

spectrophotometer An instrument used to measure the wavelengths of light that are absorbed by a particular pigment.

sperm A mature male gamete; smaller and more mobile than the female gamete.

spermatid An immature sperm cell.

spermatogenesis The production of sperm. Occurs continuously in a testis.

spermatogonia (singular: spermatogonium) The diploid cells in a testis that can give rise to primary spermatocytes.

spermatophore A gelatinous package of sperm cells that is produced by males of species that have internal fertilization without copulation.

sphincter A muscular valve that can close off a tube, as in a blood vessel or a part of the digestive tract.

spicule Stiff spike of silica or calcium carbonate found in the body of many sponges.

spindle fibers Groups of microtubules that attach to chromosomes and push and pull them during mitosis and meiosis.

spiracle In insects, a small opening that connects air-filled tracheae to the external environment, allowing for gas exchange.

spiral cleavage The pattern of embryonic cleavage seen in deuterostomes, in which cells divide at oblique angles to form a spiral coil of cells. Compare with **radial cleavage.**

spleen A dark red organ, found near the stomach of most vertebrates, that filters blood, stores extra red blood cells in case of emergency, and plays a role in immunity.

spliceosome In eukaryotes, a large, complex assembly of snRNPs (small nuclear ribonucleoproteins) that catalyzes removal of introns from primary RNA transcripts.

splicing The process by which introns are removed from primary RNA transcripts and the remaining exons are connected together.

sporangium (plural: sporangia) A spore-producing structure found in seed plants, some protists, and some fungi (e.g., chytrids).

spore (1) In bacteria, a dormant form that generally is resistant to extreme conditions. (2) In eukaryotes, a single cell produced by mitosis or meiosis (not by fusion of gametes) that is capable of developing into an adult organism.

sporophyte In organisms undergoing alternation of generations, the multicellular diploid form that arises from two fused gametes and produces haploid spores. Compare with **gametophyte.**

sporopollenin A watertight material that encases spores and pollen of modern land plants.

stabilizing selection A pattern of natural selection that favors phenotypes near the middle of the range of phenotypic variation. Reduces overall genetic variation in a population. Compare with **disruptive selection.**

stamen The male reproductive structure of a flower. Consists of an anther, in which pollen grains are produced, and a filament, which supports the anther. Compare with **carpel.**

standing defense See **constitutive defense.**

stapes The last of three small bones (ossicles) in the middle ear of vertebrates. Receives vibrations from the tympanic membrane and by vibrating against the oval window passes them to the cochlea.

starch A mixture of two storage polysaccharides, amylose and amylopectin, both formed from α-glucose monomers. Amylopectin is branched, and amylose is unbranched. The major form of stored carbohydrate in plants.

start codon The AUG triplet in mRNA at which protein synthesis begins; codes for the amino acid methionine.

statocyst A sensory organ of many arthropods that detects the animal's orientation in space (i.e., whether the animal is flipped upside down).

statolith A tiny stone or dense particle found in specialized gravity-sensing organs in some animals such as lobsters.

statolith hypothesis The hypothesis that amyloplasts (dense, starch-storing plant organelles) serve as statoliths in gravity detection by plants.

STATs See **signal transducers and activators of transcription.**

stem cell Any relatively undifferentiated cell that can divide to produce daughter cells identical to itself or more specialized daughter cells, which differentiate further into specific cell types.

stereocilium (plural: stereocilia) One of many stiff outgrowths from the surface of a hair cell that are involved in detection of sound by terrestrial vertebrates or of waterborne vibrations by fishes.

steroid A class of lipid with a characteristic four-ring structure.

steroid-hormone receptor One of a family of intracellular receptors that bind to various steroid hormones, forming a hormone-receptor complex that acts as a regulatory transcription factor and activates transcription of specific target genes.

sticky end The short, single-stranded ends of a DNA molecule cut by a restriction endonuclease. Tend to form hydrogen bonds with other sticky ends that have complementary sequences.

stigma The moist tip at the end of a flower carpel to which pollen grains adhere.

stolon A modified stem that runs horizontally over the soil surface and produces new plants at the nodes (a form of asexual reproduction). Compare with **rhizome.**

stoma (plural: stomata) Generally, a pore or opening. In plants, a microscopic pore on the surface of a leaf or stem through which gas exchange occurs.

stomach A tough, muscular pouch in the vertebrate digestive tract between the esophagus and small intestine. Physically breaks up food and begins digestion of proteins.

stop codon One of three mRNA triplets (UAG, UGA, or UAA) that cause termination of protein synthesis. Also called a *termination codon.*

strain A population of genetically similar or identical individuals.

stream A body of water that moves constantly in one direction.

striated muscle See **skeletal muscle.**

stroma The fluid matrix of a chloroplast in which the thylakoids are embedded. Site where the Calvin cycle reactions occur.

structural formula A two-dimensional notation in which the chemical symbols for the constituent atoms are joined by straight lines representing single (—), double (=), or triple (≡) covalent bonds. Compare with **molecular formula.**

structural gene A stretch of DNA that codes for a functional protein or functional RNA molecule, not including any regulatory sequences (e.g., a promoter, enhancer).

structural homology Similarities in organismal structures (e.g., limbs, shells, flowers) that are due to inheritance from a common ancestor.

structural isomer A molecule that shares the same molecular formula as another molecule but differs in the order in which covalently bonded atoms are attached. Compare with **geometric** and **optical isomer.**

style The slender stalk of a flower carpel connecting the stigma and the ovary.

subspecies A population that has distinctive traits and some genetic differences relative to other populations of the same species but that is not distinct enough to be classified as a separate species.

substrate (1) A reactant that interacts with an enzyme in a chemical reaction. (2) A surface on which a cell or organism sits.

substrate-level phosphorylation Production of ATP by transfer of a phosphate group from an intermediate substrate directly to ADP. Occurs in glycolysis and in the Krebs cycle.

succession In ecology, the gradual colonization of a habitat after an environmental disturbance (e.g., fire, flood), usually by a series of species. See **primary** and **secondary succession.**

sucrose A disaccharide formed from glucose and fructose. One of the two main products of photosynthesis.

sulfate reducer A prokaryote that produces hydrogen sulfide (H_2S) as a by-product of cellular respiration.

summation The additive effect of different postsynaptic potentials at a nerve or muscle cell, such that several subthreshold stimulations can cause an action potential.

supernatant The liquid above a layer of solid particles (the pellet) in a tube after centrifugation.

surface tension The cohesive force that causes molecules at the surface of a liquid to stick together, thereby resisting deformation of the liquid's surface and minimizing its surface area.

surfactant A mixture of phospholipids and proteins produced by lung cells that reduces surface tension, allowing the lungs to expand more.

survivorship curve A graph depicting the percentage of a population that survives to different ages.

suspension culture A population of cells grown in a flask containing a liquid nutrient medium. The flask is rotated continuously to keep the cells suspended and the nutrients mixed.

suspension feeder Any organism that obtains food by filtering small particles or small organisms out of water or air. Also called *filter feeder.*

sustainability The planned use of environmental resources at a rate no faster than the rate at which they are naturally replaced.

sustainable agriculture Agricultural techniques that are designed to maintain long-term soil quality and productivity.

swim bladder A gas-filled organ of many ray-finned fishes that regulates buoyancy.

swamp A wetland that has a steady rate of water flow and is dominated by trees and shrubs.

symbiosis (adjective: symbiotic) Any close and prolonged physical relationship between individuals of two different species. See **commensalism, mutualism,** and **parasitism.**

symmetric competition Ecological competition between two species in which both suffer similar declines in fitness. Compare with **asymmetric competition.**

sympathetic nervous system The part of the autonomic nervous system that stimulates fight-or-flight responses, such as increased heart rate, increased blood pressure, and decreased digestion. Compare with **parasympathetic nervous system.**

sympatric speciation The divergence of populations living within the same geographic area into different species as the result of their genetic (not physical) isolation. Compare with **allopatric speciation.**

sympatry Condition in which two or more populations live in the same geographic area, or close enough to permit interbreeding. Compare with **allopatry.**

symplast In plant roots, a continuous pathway through which water can flow through the cytoplasm of adjacent cells that are connected by plasmodesmata. Compare with **apoplast.**

symporter A carrier protein that allows an ion to diffuse down an electrochemical gradient, using the energy of that process to transport a different substance in the same direction *against* its concentration gradient. Compare with **antiporter.**

synapomorphy A shared, derived trait found in two or more taxa that is present in their most recent common ancestor but is missing in more distant ancestors. Useful for inferring evolutionary relationships.

synapse The interface between two neurons or between a neuron and an effector cell.

synapsis The physical pairing of two homologous chromosomes during prophase I of meiosis. Crossing over occurs during synapsis.

synaptic cleft The space between two communicating nerve cells (or between a neuron and effector cell) at a synapse, across which neurotransmitters diffuse.

synaptic plasticity Long-term changes in the responsiveness or physical structure of a synapse that can occur after particular stimulation patterns. Thought to be the basis of learning and memory.

synaptic vesicle A small neurotransmitter-containing vesicle at the end of an axon that releases neurotransmitter into the synaptic cleft by exocytosis.

synaptonemal complex A network of proteins that holds non-sister chromatids together during synapsis in meiosis I.

synthesis (S) phase The phase of the cell cycle during which DNA is synthesized and chromosomes are replicated.

system In biology, a more complex organization resulting from the combination of various components, such as a group of organs that work together to perform a physiological function.

systemic acquired resistance (SAR) A slow, widespread response of plants to a localized infection that protects healthy tissue from invasion by pathogens. Compare with **hypersensitive response.**

systemic circulation The part of the circulatory system that sends oxygen-rich blood from the lungs out to the rest of the body. Is separate from the pulmonary circulation in mammals and birds.

systemin A peptide hormone, produced by plant cells damaged by herbivores, that initiates a protective response in undamaged cells.

systole The portion of the heartbeat cycle during which the heart muscles are contracting. Compare with **diastole.**

systolic blood pressure The force exerted by blood against artery walls during contraction of the heart's left ventricle. Compare with **diastolic blood pressure.**

taiga A vast forest biome throughout subarctic regions, consisting primarily of short conifer trees. Characterized by intensely cold winters, short summers, and high annual variation in temperature.

taproot A large vertical main root of a plant.

taste bud Sensory structure, found chiefly in the mammalian tongue, containing spindle-shaped cells that respond to chemical stimuli.

TATA-binding protein (TBP) A protein that binds to the TATA box in eukaryotic promoters and is a component of the basal transcription complex.

TATA box A short DNA sequence in many eukaryotic promoters about 30 base pairs upstream from the transcription start site.

taxis Movement toward or away from some external cue.

taxon (plural: taxa) Any named group of organisms at any level of a classification system.

taxonomy The branch of biology concerned with the classification and naming of organisms.

TBP See **TATA-binding protein.**

T cell A type of leukocyte that matures in the thymus and, with B cells, is responsible for acquired immunity. Involved in activation of B cells ($CD4^+$ helper T cells) and destruction of infected cells ($CD8^+$ cytotoxic T cells). Also called *T lymphocytes.*

T-cell receptor (TCR) A transmembrane protein found on T cells that can bind to antigens displayed on the surfaces of other cells. Composed of two polypeptides called the alpha chain and beta chain. See **antigen presentation.**

tectorial membrane A membrane in the vertebrate cochlea that takes part in the transduction of sound by bending the stereocilia of hair cells in response to sonic vibrations.

telomerase An enzyme that replicates the ends of chromosome (telomeres) by catalyzing DNA synthesis from an RNA template that is part of the enzyme.

telomere The region at the end of a linear chromosome.

telophase The final stage in mitosis or meiosis during which sister chromatids (replicated chromosomes in meiosis I) separate and new nuclear envelopes begin to form around each set of daughter chromosomes.

temperate Having a climate with pronounced annual fluctuations in temperature (i.e., warm summers and cold winters) but typically neither as hot as the tropics nor as cold as the poles.

temperature A measurement of thermal energy present in an object or substance, reflecting how much the constituent molecules are moving.

template strand (1) The strand of DNA that is transcribed by RNA polymerase to create RNA. (2) An original strand of RNA used to make a complementary strand of RNA.

tendon A band of tough, fibrous connective tissue that connects a muscle to a bone.

tentacle A long, thin, muscular appendage of gastropod molluscs.

termination (1) In enzyme-catalyzed reactions, the final stage in which the enzyme returns to its original conformation and products are released. (2) In DNA transcription, the dissociation of RNA polymerase from DNA when it reaches a termination signal sequence. (3) In RNA translation, the dissociation of a ribosome from mRNA when it reaches a stop codon.

territory An area that is actively defended by an animal from others of its species.

tertiary structure The overall three-dimensional shape of a single polypeptide chain, resulting from multiple interactions among the amino acid side chains and the peptide backbone. Compare with **primary, secondary,** and **quaternary structure.**

test A hard protective outer structure seen in some protists. Also called a *shell.*

testcross The breeding of an individual of unknown genotype with an individual having only recessive alleles for the traits of interest in order to infer the unknown genotype from the phenotypic ratios seen in offspring.

testis (plural: testes) The sperm-producing organ of a male animal.

testosterone A steroid hormone, produced and secreted by the testes, that stimulates sperm production and various male traits and reproductive behaviors.

tetrapod Any member of the taxon Tetrapoda, which includes all vertebrates with two pairs of limbs (amphibians, mammals, birds, and other reptiles).

theory A proposed explanation for a broad class of phenomena or observations.

thermal energy The kinetic energy of molecular motion.

thermocline A gradient (cline) in environmental temperature across a large geographic area.

thermophile A bacterium or archaean that thrives in very hot environments.

thermoreceptor A sensory cell or an organ specialized for detection of changes in temperature.

thermoregulation Regulation of body temperature.

thick filament A filament composed of bundles of the motor protein myosin; anchored to the center of the sarcomere. Compare with **thin filament.**

thigmotropism Growth or movement of an organism in response to contact with a solid object.

thin filament A filament composed of two coiled chains of actin and associated regulatory proteins; anchored at the Z disk of the sarcomere. Compare with **thick filament.**

thorn A modified plant stem shaped as a sharp protective structure. Helps protect a plant against feeding by herbivores.

threshold potential The membrane potential that will trigger an action potential in a neuron or other excitable cell. Also called simply *threshold.*

thylakoid A flattened, membrane-bound vesicle inside a plant chloroplast that functions in converting light energy to chemical energy. A stack of thylakoids is a granum.

thymus An organ, located in the anterior chest or neck of vertebrates, in which immature T cells generated in the bone marrow undergo maturation.

thyroid gland A gland in the neck that releases thyroid hormone (which increases metabolic rate) and calcitonin (which lowers blood calcium).

thyroid-stimulating hormone (TSH) A peptide hormone, produced and secreted by the anterior pituitary, that stimulates release of thyroid hormones from the thyroid gland.

thyroxine (T_4) A peptide hormone containing four iodine atoms that is produced and secreted by the thyroid gland. Acts primarily to increase cellular metabolism. In mammals, T_4 is converted to the more active hormone triiodothyronine (T_3) in the liver.

tight junction A type of cell-cell attachment structure that links the plasma membranes of adjacent animal cells, forming a barrier that restricts movement of substances in the space between the cells. Most abundant in epithelia (e.g., the intestinal lining). Compare with **desmosome** and **gap junction.**

tip The end of a branch on a phylogenetic tree. Represents a specific species or larger taxon that has not (yet) produced descendants—either a group living today or a group that ended in extinction. Also called *terminal node.*

Ti plasmid A plasmid carried by *Agrobacterium* (a bacterium that infects plants) that can integrate into a plant cell's chromosomes and induce formation of a gall.

tissue A group of similar cells that function as a unit, such as muscle tissue or epithelial tissue.

tolerance In ecological succession, the phenomenon in which early-arriving species do not affect the probability that subsequent species will become established. Compare with **facilitation** and **inhibition.**

tonoplast The membrane surrounding a plant vacuole.

top-down control hypothesis The hypothesis that the size of herbivore populations is limited by predation or disease rather than by limited or toxic nutritional resources.

topoisomerase An enzyme that cuts and rejoins DNA downstream of the replication fork, to ease the twisting that would otherwise occur as the DNA "unzips."

torpor An energy-conserving physiological state, marked by a decrease in metabolic rate, body temperature, and activity, that lasts for a short period (overnight to a few days or weeks). Occurs in some small mammals when the ambient temperature drops significantly. Compare with **hibernation.**

totipotent Capable of dividing and developing to form a complete, mature organism.

trachea (plural: tracheae) (1) In insects, one of the small air-filled tubes that extend throughout the body and function in gas exchange. (2) In terrestrial vertebrates, the airway connecting the larynx to the bronchi. Also called *windpipe.*

tracheid In vascular plants, a long, thin water-conducting cell that has gaps in its secondary cell wall, allowing water movement between adjacent cells. Compare with **vessel element.**

trade-off In evolutionary biology, an inescapable compromise between two traits that cannot be optimized simultaneously. Also called *fitness trade-off.*

trait Any heritable characteristic of an individual.

transcription The process by which RNA is made from a DNA template.

transcriptional control Regulation of gene expression by various mechanisms that change the rate at which genes are transcribed to form messenger RNA. In negative control, binding of a regulatory protein to DNA represses transcription; in positive control binding of a regulatory protein to DNA promotes transcription.

transcriptome The complete set of genes transcribed in a particular cell.

transduction Conversion of information from one mode to another. For example, the process by which a stimulus outside a cell is converted into a response by the cell.

transfer cell In land plants, a cell that transfers nutrients from a parent plant to a developing plant seed.

transfer RNA (tRNA) One of a class of RNA molecules that have an anticodon at one end and an amino acid binding site at the other. Each tRNA picks up a specific amino acid and binds to the corresponding codon in messenger RNA during translation.

transformation (1) Incorporation of external DNA into the genome. Occurs naturally in some bacteria; can be induced in the laboratory by certain processes. (2) Conversion of a normal cell to a cancerous one.

transgenic Referring to an individual plant or animal whose genome contains DNA introduced from another individual, either from the same or a different species.

transitional form A fossil species or population with traits that are intermediate between older and younger species.

transition state A high-energy intermediate state of the reactants during a chemical reaction that must be achieved for the reaction to proceed. Compare with **activation energy.**

transition state facilitation The second stage (after initiation) in enzyme-catalyzed reactions, in which the enzyme enables formation of the transition state.

translation The process by which proteins and peptides are synthesized from messenger RNA.

translational control Regulation of gene expression by various mechanisms that alter the life span of messenger RNA or the efficiency of translation.

translocation (1) In plants, the movement of sugars and other organic nutrients through the phloem by bulk flow. (2) A type of mutation in which a piece of a chromosome moves to a nonhomologous chromosome. (3) The process by which a ribosome moves down a messenger RNA molecule during translation.

transmembrane protein Any membrane protein that spans the entire lipid bilayer. Also called *integral membrane protein.*

transmission The passage or transfer (1) of a disease from one individual to another or (2) of electrical impulses from one neuron to another.

transmission electron microscope (TEM) A microscope that forms an image from electrons that pass through a specimen. Compare with **scanning electron microscope.**

transpiration Water loss from aboveground plant parts. Occurs primarily through stomata.

transporter See **carrier protein.**

transport protein Collective term for any membrane protein that enables a specific ion or small molecule to cross a plasma membrane. Includes carrier proteins and channel proteins, which carry out passive transport (facilitated diffusion), and pumps, which carry out active transport.

transposable elements Any of several kinds of DNA sequences that are capable of moving themselves, or copies of themselves, to other locations in the genome. Include LINEs and SINEs.

tree of life A diagram depicting the genealogical relationships of all living organisms on Earth, with a single ancestral species at the base.

triacylglycerol See **fat.**

trichome A hairlike appendage that grows from epidermal cells of some plants. Trichomes exhibit a variety of shapes, sizes, and functions depending on species.

triglyceride See **fat.**

triiodothyronine (T_3) A peptide hormone containing three iodine atoms that is produced and secreted by the thyroid gland. Acts primarily to increase cellular metabolism. In mammals, T_3 has a stronger effect than does the related hormone thyroxine (T_4).

triose A monosaccharide (simple sugar) containing three carbon atoms.

triplet code A code in which a "word" of three letters encodes one piece of information. The genetic code is a triplet code because a codon is three nucleotides long and encodes one amino acid.

triploblast (adjective: triploblastic) An animal whose body develops from three basic embryonic cell layers: ectoderm, mesoderm, and endoderm. Compare with **diploblast.**

trisomy The state of having three copies of one particular type of chromosome.

tRNA See **transfer RNA.**

trochophore A larva with a ring of cilia around its middle that is found in some lophotrochozoans.

trophic level A feeding level in an ecosystem.

tropomyosin A regulatory protein present in thin (actin) filaments that blocks the myosin-binding sites on these filaments, thereby preventing muscle contraction.

troponin A regulatory protein, present in thin (actin) filaments, that can move tropomyosin off the myosin-binding sites on these filaments, thereby triggering muscle contraction. Activated by high intracellular calcium.

true navigation The type of navigation by which an animal can reach a specific point on Earth's surface.

trypsin A protein-digesting enzyme present in the small intestine that activates several other protein-digesting enzymes.

trypsinogen The precursor of protein-digesting enzyme trypsin. Secreted by the pancreas and activated by the intestinal enzyme enterokinase.

T tubules Membranous tubes that extend into the interior of muscle cells. Propagate action potentials throughout a muscle cell and trigger release of calcium from the sarcoplasmic reticulum.

tube foot One of the many small, mobile, fluid-filled extensions of the water vascular system of echinoderms; the part extending outside the body is called a podium. Used in locomotion and feeding.

tuber A modified plant rhizome that functions in storage of carbohydrates.

tumor A mass of cells formed by uncontrolled cell division. Can be benign or malignant.

tumor suppressor A gene (e.g., *p53* and *Rb*) or the protein it encodes that prevents cell division, particularly when the cell has DNA damage. Mutated forms are associated with cancer.

tundra The treeless biome in polar and alpine regions, characterized by short, slow-growing vegetation, permafrost, and a climate of long, intensely cold winters and very short summers.

turgid Swollen and firm as a result of high internal pressure (e.g., a plant cell containing enough water for the cytoplasm to press against the cell wall). Compare with **flaccid.**

turgor pressure The outward pressure exerted by the fluid contents of a plant cell against its cell wall.

Turner syndrome A human genetic disorder caused by the presence of only one X chromosome and no Y chromosome ("XO"). Individuals with this condition are female but sterile.

turnover In lake ecology, the complete mixing of upper and lower layers of water that occurs each spring and fall in temperate-zone lakes.

twofold rotational symmetry See **dyad symmetry.**

tympanic membrane The membrane separating the middle ear from the outer ear in terrestrial vertebrates, or similar structures in insects. Also called the *eardrum.*

ubiquinone See **coenzyme Q.**

ultimate causation In biology, the reason that a trait or phenomenon is thought to have evolved; the adaptive advantage of that trait. Also called *ultimate explanation.* Compare with **proximate causation.**

umami The taste of glutamate, responsible for the "meaty" taste of most proteins and of monosodium glutamate.

umbilical cord The cord that connects a developing mammalian embryo or fetus to the placenta and through which the embryo or fetus receives oxygen and nutrients.

unequal crossover An error in crossing over during meiosis I in which the two non-sister chromatids match up at different sites. Results in gene duplication in one chromatid and gene loss in the other.

unsaturated Referring to fats and fatty acids in which at least one carbon-carbon bond is a double bond. Double bonds produce kinks in the fatty acid chains and decrease the compound's melting point. Compare with **saturated.**

upstream In genetics, opposite to the direction in which RNA polymerase moves along a DNA strand. Compare with **downstream.**

urea A water-soluble excretory product of mammals and sharks. Used to remove from the body excess nitrogen derived from the breakdown of amino acids. Compare with **uric acid.**

ureter In vertebrates, a tube that transports urine from one kidney to the bladder.

urethra The tube that drains urine from the bladder to the outside environment. In male vertebrates, also used for passage of sperm during ejaculation.

uric acid A whitish excretory product of birds, reptiles, and terrestrial arthropods. Used to remove from the body excess nitrogen derived from the breakdown of amino acids. Compare with **urea.**

urochordates One of the three major chordate lineages (Urochordata), comprising sessile, filter-feeding animals that have a polysaccharide exoskeleton (tunic) and two siphons through which water enters and leaves; also called tunicates or sea squirts. Compare with **cephalochordates** and **vertebrates.**

uterus The organ in which developing embryos are housed in those vertebrates that give live birth. Common in most mammals and in some lizards, sharks, and other vertebrates.

vaccination The introduction into an individual of weakened, killed, or altered pathogens to stimulate development of acquired immunity against those pathogens.

vaccine A preparation designed to stimulate an immune response against a particular pathogen without causing illness. Vaccines consist of inactivated (killed) pathogens, live but weakened (attenuated) pathogens, or portions of a viral capsid (subunit vaccine).

vacuole A large organelle in plant and fungal cells that usually is used for bulk storage of water, pigments, oils, or other substances. Some vacuoles contain enzymes and have a digestive function similar to lysosomes in animal cells.

vagina The birth canal of female mammals; a muscular tube that extends from the uterus through the pelvis to the exterior.

valence The number of unpaired electrons in the outermost electron shell of an atom; determines how many covalent bonds the atom can form.

valence electron An electron in the outermost electron shell, the valence shell, of an atom. Valence electrons tend to be involved in chemical bonding.

valves In circulatory systems, flaps of tissue that prevent backward flow of blood, particularly in veins and between the chambers of the heart.

van der Waals interactions A weak electrical attraction between two hydrophobic side chains. Often contributes to tertiary structure in proteins.

variable number tandem repeat See **minisatellite.**

variable (V) region The portion of an antibody's light chains or heavy chains that has a highly variable amino acid sequence and forms part of the antigen-binding site. Compare with **constant (C) region.**

vasa recta In the vertebrate kidney, a network of blood vessels that runs alongside the loop of Henle of a nephron. Functions in reabsorption of water and solutes from the filtrate.

vascular bundle A cluster of xylem and phloem strands in a plant stem.

vascular cambium One of two types of lateral meristem, consisting of a ring of undifferentiated plant cells inside the cork cambium of woody plants; produces secondary xylem (wood) and secondary phloem. Compare with **cork cambium.**

vascular tissue system In plants, any tissue that is involved in conducting water or solutes from one part of a plant to another. Also called simply *vascular tissue.* See **phloem** and **xylem.**

vas deferens A pair of muscular tubes that store and transport semen from the epididymis to the ejaculatory duct. In nonhuman animals, called the *ductus deferens.*

vector A biting insect or other organism that transfers pathogens from one species to another. See also **cloning vector.**

vegetative organs The nonreproductive parts of a plant including roots, leaves, and stems.

vein Any blood vessel that carries blood (oxygenated or not) under relatively low pressure from the tissues toward the heart. Compare with **artery.**

vena cava (plural: vena cavae) A large vein that returns oxygen-poor blood to the heart.

ventral Toward an animal's belly and away from its back. The opposite of dorsal.

ventricle (1) A thick-walled chamber of the heart that receives blood from an atrium and pumps it to the body or to the lungs. (2) One of several small fluid-filled chambers in the vertebrate brain.

vertebra (plural: vertebrae) One of the cartilaginous or bony elements that form the spine of vertebrate animals.

vertebrates One of the three major chordate lineages (Vertebrata), comprising animals with a dorsal column of cartilaginous or bony structures (vertebrae) and a skull enclosing the brain. Includes fishes, amphibians, mammals, reptiles, and birds. Compare with **cephalochordates** and **urochordates.**

vessel element In vascular plants, a short, wide water-conducting cell that has gaps through both the primary and secondary cell walls, allowing unimpeded passage of water between adjacent cells. Compare with **tracheid.**

vestigial trait Any rudimentary structure of unknown or minimal function that is homologous to functioning structures in other species. Vestigial traits are thought to reflect evolutionary history.

vicariance The physical splitting of a population into smaller, isolated populations by a geographic barrier.

villi (singular: villus) Small, fingerlike projections (1) of the lining of the small intestine or (2) of the fetal portion of the placenta adjacent to maternal arteries. Function to increase the surface area available for absorption of nutrients and gas exchange (in the placenta).

virion A single mature virus particle.

virulence The ability of a pathogen or parasite to cause disease and death.

virulent Referring to pathogens that can cause severe disease in susceptible hosts.

virus A tiny intracellular parasite that uses host cell enzymes to replicate; consists of a DNA or RNA genome enclosed within a protein shell (capsid). In enveloped viruses, the capsid is surrounded by a phospholipid bilayer derived from the host cell plasma membrane, whereas nonenveloped viruses lack this protective covering.

visceral mass One of the three main parts of the mollusc body; contains most of the internal organs and external gill.

visible light The range of wavelengths of electromagnetic radiation that humans can see, from about 400 to 700 nanometers.

vitamin An organic micronutrient that usually functions as a coenzyme.

vitelline envelope A fibrous sheet of glycoproteins that surrounds mature egg cells in many vertebrates. Surrounded by a thick gelatinous matrix (the jelly layer) in some species. In mammals, called the *zona pellucida.*

viviparous Producing live young (instead of eggs) that develop within the body of the mother before birth. Compare with **oviparous** and **ovoviviparous.**

volt (V) A unit of electrical potential (voltage).

voltage Potential energy created by a separation of electric charges between two points. Also called *electrical potential.*

voltage clamping A technique for imposing a constant membrane potential on a cell. Widely used to investigate ion channels.

voltage-gated channel An ion channel that opens or closes in response to changes in membrane voltage. Compare with **ligand-gated channel.**

wall pressure The inward pressure exerted by a cell wall against the fluid contents of a plant cell.

water cycle, global The movement of water among terrestrial ecosystems, the oceans, and the atmosphere.

water potential (ψ) The potential energy of water in a certain environment compared with the potential energy of pure water at room temperature and atmospheric pressure. In living organisms, ψ equals the solute potential (ψ_S) plus the pressure potential (ψ_P).

water potential gradient A difference in water potential in one region compared with that in another region. Determines the direction that water moves, always from regions of higher water potential to regions of lower water potential.

watershed The area drained by a single stream or river.

water table The upper limit of the underground layer of soil that is saturated with water.

water vascular system In echinoderms, a system of fluid-filled tubes and chambers that functions as a hydrostatic skeleton.

Watson-Crick pairing See **complementary base-pairing.**

wavelength The distance between two successive crests in any regular wave, such as light waves, sound waves, or waves in water.

wax A class of lipid with extremely long hydrocarbon tails, usually combinations of long-chain alcohols with fatty acids. Harder and less greasy than fats.

weather The specific short-term atmospheric conditions of temperature, moisture, sunlight, and wind in a certain area.

weathering The gradual wearing down of large rocks by rain, running water, and wind; one of the processes that transform rocks into soil.

weed Any plant that is adapted for growth in disturbed soils.

Western blotting A technique for identifying specific proteins separated by gel electrophoresis by transferring them to filter paper and exposing them to a labeled antibody that binds to the protein of interest. Compare with **Northern blotting** and **Southern blotting.**

wetland A shallow-water habitat where the soil is saturated with water for at least part of the year.

white blood cells See **leukocytes.**

wild type The most common phenotype seen in a population; especially the most common phenotype in wild populations compared with inbred strains of the same species.

wildlife corridor Strips of wildlife habitat connecting populations that otherwise would be isolated by man-made development.

wilt To lose turgor pressure in a plant tissue.

wobble hypothesis The hypothesis that some tRNA molecules can pair with more than one mRNA codon, tolerating some variation in the third base, as long as the first and second bases are correctly matched.

wood Xylem resulting from secondary growth. Also called *secondary xylem.*

xeroderma pigmentosum A human disease characterized by extreme sensitivity to ultraviolet light. Caused by an autosomal recessive allele that results in a defective DNA repair system.

X-linked inheritance Inheritance patterns for genes located on the mammalian X chromosome. Also called *X-linkage.*

X-ray crystallography A technique for determining the three-dimensional structure of large molecules, including proteins and nucleic acids, by analysis of the diffraction patterns produced by X-rays beamed at crystals of the molecule.

xylem A plant vascular tissue that conducts water and ions; contains tracheids and/or vessel elements. Primary xylem develops from the procambium of apical meristems; secondary xylem, or wood, from the vascular cambium of lateral meristems. Compare with **phloem.**

xylem sap The watery fluid found in the xylem of plants.

yeast Any fungus growing as a single-celled form. Also, a specific lineage of ascomycetes.

Y-linked inheritance Inheritance patterns for genes located on the mammalian Y chromosome. Also called *Y-linkage.*

yolk The nutrient-rich cytoplasm inside an egg cell; used as food for the growing embryo.

Z disk The structure that forms each end of a sarcomere. Contains a protein that binds tightly to actin, thereby anchoring thin filaments.

zero population growth (ZPG) A state of stable population size due to fertility staying at the replacement rate for at least one generation.

zona pellucida The gelatinous layer around a mammalian egg cell. In other vertebrates, called the *vitelline envelope.*

zone of (cellular) division In plant roots, a group of apical meristematic cells just behind the root cap where cells are actively dividing.

zone of (cellular) elongation In plant roots, a group of young cells, located behind the apical meristem, that are increasing in length.

zone of (cellular) maturation In plant roots, a group of plant cells, located several centimeters behind the root cap, that are differentiating into mature tissues.

Z scheme Path of electron flow in which electrons pass from photosystem II to photosystem I and ultimately to $NADP^+$ during the light-dependent reactions of photosynthesis. Also called *noncyclic electron flow.*

zygomycete A member of a paraphyletic group of fungi (Zygomycota) characterized by a durable, thick-walled zygosporangium that forms from yoked hyphae of different mating types.

zygosporangium (plural: zygosporangia) The spore-producing structure in fungi that are members of the Zygomycota.

zygote The diploid cell formed by the union of two haploid gametes; a fertilized egg. Capable of undergoing embryological development to form an adult.

Credits

IMAGE CREDITS

Frontmatter

viT Natalie B. Fobes Photography **viB** David Quillin **x** Jeff Rotman/Nature Picture Library **xv** Anthony Bannister/Gallo Images/Corbis; **xix** Lee W. Wilcox **xxiii** Robert Fried/robertfriedphotography.com

Chapter 1

Opener Jeff Rotman/Nature Picture Library. **1.1a** Burndy Library/Omikron/Photo Researchers, Inc. **1.1bM** Walker/Photo Researchers, Inc. **1.3a** Kelly Buono/Dr. Richard Amasino. **1.3c** Bruce Forster/Getty Images Inc. **1.4aL** University of Florida. **1.4aR** Michael Lustbader/Photo Researchers, Inc. **1.4bLR** Scott P. Carroll. **1.4cL** Tom Murray. **1.4cR** Andrew A. Forbes. **1.6a** Samuel F. Conti and Thomas D. Brock. **1.6b** Kwangshin Kim/Photo Researchers, Inc. **1.7/1** Dr. David Phillips/Visuals Unlimited. **1.7/2** Dennis Kunkel/Dennis Kunkel Microscopy, Inc. **1.7/3** Kolar, Richard/Animals Animals/Earth Scenes **1.7/4** Biophoto Associates/Photo Researchers, Inc. **1.7/5** Darwin Dale/Photo Researchers, Inc. **1.10b** Michael Hughes/Aurora & Quanta Productions Inc. **1.11a** Joshua J. Tewksbury/Joshua J. Tewksbury. **1.11b** William Weber/Visuals Unlimited. **1.11c** Robert Dobbs/Joshua J. Tewksbury

Chapter 2

Opener Colin Monteath/Hedgehog House/Minden Pictures. **2.1** Mitchell Layton/PCN Photography. **2.6c** Albert Copley/Visuals Unlimited. **2.11** Beth Plowes–Proteapix. **2.14** Dietmar Nill/Picture Press/Photolibrary.com. **2.15c** Geostock/Getty Images Inc. **2.19bL** David Glick/Getty Images Inc. Stone Allstock. **2.19bR** Sergio Bartelsman/eStock

Chapter 3

Opener The Scripps Research Institute **3.4b** Martin Bough/Fundamental Photographs, NYC **3.10a-c** Clare Sansom. **3.11a** Microworks/Phototake NYC. **3.11b** Walter Reinhart/Phototake NYC. **3.13BLMR** Clare Sansom. **3.14a-b** Clare Sansom. **3.20aLR** Thomas A. Steitz. **3.23aLR, 3.23bLR** Thomas A. Steitz

Chapter 4

4.5 A. Barrington Brown/Science Source Photo Researchers, Inc. **4.11** Reprinted with permission from Science 292: 1319-1325 Fig 4B (2001) by Wendy K. Johnston, Peter J. Unrau, Michael S. Lawrence, Margaret E. Glasner, David P. Bartel "RNA-Catalyzed RNA Polymerization: Accurate and General RNA-Templated Primer Extension." Copyright 2004 American Association for the Advancement of Science

Chapter 5

Opener Dr. Jeremy Burgess Photo Researchers, Inc. **5.6a** Biophoto Associates/Photo Researchers, Inc. **5.6b** Dr. Jacob S. Ishay. **5.6c** Dr. Manfred Jericho Terry J. Beveridge.

Chapter 6

Opener Kit Pogliano . **6.1a** Alec D. Bangham, M.D., F.R.S. Alec D. Bangham, M.D., F.R.S. **6.1b** Fred Hossler Visuals Unlimited. **6.7aL** James J. Cheetham James J. Cheetham. **6.11a** Dorling Kindersley Dorling Kindersley Media Library. **6.11b** Clive Streeter Dorling Kindersley Media Library. **6.11c** Phil Degginger Color-Pic, Inc. **6.24b** Timothy A. Cross. **6.26L** Andrew Syred Getty Images Inc.-Stone Allstock. **6.26M** Dr. David Phillips Visuals Unlimited. **6.26R** Joseph F. Hoffman Joseph F. Hoffman/Yale University School of Medicine

Chapter 7

Opener Albert Tousson Phototake NYC. **7.1** Dr. T.J. Beveridge Visuals Unlimited. **7.2** Stanley C. Holt/Biological Photo Service **7.3** Gopal Murti/ Visuals Unlimited **7.5** Wanner/Eye of Science Photo Researchers, Inc. **7.7** Fawcett Photo Researchers, Inc. **7.8** Omikron Photo Researchers, Inc. **7.9** Dr. Don Fawcett Photo Researchers, Inc. **7.10** Biophoto Associates Photo Researchers, Inc. **7.12** Fawcett/Friend Photo Researchers, Inc. **7.12** Dr. Don Fawcett Photo Researchers, Inc. **7.13** Dr. Gopal Murti Visuals Unlimited. **7.16** E.H. Newcomb & W.P. Wergin/Biological Photo Service Biological Photo Service. **7.17** T. Kanaseki & Donald Fawcett Visuals Unlimited. **7.18** E.H. Newcomb & W.P. Wergin Biological Photo Service. **7.19** E. H. Newcomb & S.E. Frederick Biological Photo Service. **7.20a** Don Fawcett/S. Ito & A. Like/Photo Researchers, Inc. **7.20b** Don W. Fawcett/Photo Researchers, Inc. **7.20c** Biophoto Associates Photo Researchers, Inc. **7.20d** Dr. Dennis Kunkel Visuals Unlimited. **7.23** Don W. Fawcett Photo Researchers, Inc. **7.24a** Don W. Fawcett Photo Researchers, Inc. **7.27a** James D. Jamieson, M.D. James D. Jamieson, M.D. **7.27b** James D. Jamieson, M.D. **7.27c** James D. Jamieson, M.D. **7.31a-c** Dr. Victor Small. **7.33** Conly L. Rieder Biological Photo Service. **7.34a-b** American Society for Cell Biology. **7.35a** John E. Heuser, M.D. **7.36L** Dennis Kunkel Phototake NYC. **7.36R** Dennis Kunkel Phototake NYC. **7.37a** Don W. Faucett Photo Researchers, Inc. **7.38** Charles J. Brokaw The Rockefeller University Press

Chapter 8

Opener E.H. Newcomb & W.P. Wergin Biological Photo Service. **8.2** Henry M. Walker, Samuel R. & Marie-Louise Rosenthal, Profs **8.3** Biophoto Associates Photo Researchers, Inc. **8.4** Barry F. King Biological Photo Service. **8.6** C. T. Huang, Karen Xu, Gordon McFeters, and Philip S. Stewart Philip S. Stewart. **8.7** SPL Photo Researchers, Inc. **8.9** Photo Researchers, Inc. Photo Researchers, Inc. **8.10aL** Don W. Fawcett Photo Researchers, Inc. **8.10aR** Dr. Don Fawcett Photo Researchers, Inc. **8.11a** Dr. Don Fawcett/Gida Matoltsy Photo Researchers, Inc. **8.14a** E. H. Newcomb & W.P. Wergin Biological Photo Service. **8.14b** Dr. Don Fawcett Photo Researchers, Inc.

Chapter 9

Opener Richard Megna Fundamental Photographs, NYC. **9.6b** Darren McCollester/Reuters Corbis/Bettmann. **9.6TL** Peter Anderson Dorling Kindersley Media Library. **9.6TR** Dorling Kindersley Dorling Kindersley Media Library. **9.16** Terry Frey. **9.25a** Dr. Yasuo Kagawa

Chapter 10

Opener Murray Cooper/NPL Minden Pictures. **10.2aM** John Durham/Science Photo Library Photo Researchers, Inc. **10.2aR(detail)** John Durham/Science Photo Library Photo Researchers, Inc. **10.2bR** Dr. J. Burgess Photo Researchers, Inc. **10.3bL** Biophoto Associates Photo Researchers, Inc. **10.3bM** Dr. George Chapman Visuals Unlimited **10.3bR** Richard Green Photo Researchers, Inc. **10.3T** E. H. Newcomb & P. K. Hepler Biological Photo Service **10.5b** Sinclair Stammers/Science Photo Library Photo Researchers, Inc. **10.10** David Newman Visuals Unlimited. **10.18L** James A. Bassham James A. Bassham. **10.18R** James A. Bassham. **10.21a** Dr. Jeremy Burgess/Science Photo Library Photo Researchers, Inc. **10.24a** Adam Hart-Davis Photo Researchers, Inc. **10.24b** David Muench Corbis/Bettmann

Chapter 11

Opener Collection CNRI Phototake NYC. **11.2a** Conly L. Rieder, Ph.D. **11.2b** Photo Researchers, Inc. **11.3a** From: Paulson, J.R. and Laemmli, U.K. Cell 12 (1977) 817-828. **11.9a** David M. Phillips/Visuals Unlimited **11.9bR** Calentine/Visuals Unlimited. **11.10** Micrographs by Conly L. Rieder, Division of Molecular Medicine, Wadsworth Center, Albany, New York 12201-0509

Chapter 12

Opener David Phillips/The Population Council Photo Researchers, Inc. **12.2a** Hesed M. Padilla-Nash. **12.6c** Wessex Reg. Genetics Centre Wellcome Trust Medical Photographic Library. **12.8a-d** David A. Jones. **12.12** Doug Sokell Visuals Unlimited

Chapter 13

Opener Brian Johnston Brian Johnston. **13.9a** Robert Calentine Visuals Unlimited. **13.9bLR** Carolina Biological Supply Company Phototake NYC. **13.17a** Robert Calentine Visuals Unlimited

Chapter 14

Opener Dr. Gopal Murti/Science Photo Library Photo Researchers, Inc. **14.1** Belanger et al., 2004. Pyruvate oxidase is a determinant of Avery's rough morphology. J. Bacteriology 186:8164-8174. American Society of Microbiology. **14.4b** Oliver Meckes/Max-Planck-Institut-Tubingen/Photo Researchers, Inc. **14.10a** Dr. Gopal Murti/Science Photo Library Photo Researchers, Inc.

Chapter 15

Opener Halaska, Jacob Photolibrary.com

Chapter 16

Opener Joseph G. Gall. **16.05** Fred Stevens/Mykoweb. **16.06** Bert W. O'-Malley, M.D. Bert W. O'Malley, M.D. **16.09** Oscar Miller/Science Photo Library/Photo Researchers, Inc. **16.15a** Venkitaraman Ramakrishnan. **16.16** E.V. Kiseleva/Donald Fawcett/Visuals Unlimited. **16.19b** Bill Longcore /Photo Researchers, Inc. **16.21b** Bill Longcore /Photo Researchers, Inc.

Chapter 17

Opener EM Unit, VLA/Science Photo Library/Photo Researchers, Inc. **applying concepts** Michael Gabridge/Visuals Unlimited

Chapter 18

Opener Dr. Timothy J. Richmond Nature Magazine DC. **18.02a** Ada Olins/Don Fawcett/Photo Researchers, Inc. **18.04** Victoria E. Foe **18.04** Barbara Hamkalo

Chapter 19

Opener Tsien Laboratory **19.14** Baylor Coll. of Medicine/Peter Arnold, Inc. **19.17a** Brad Mogen/Visuals Unlimited

Chapter 20

Opener Sanger Institute Welcome Trust Medical Photographic Library. **20.07** and **20.07b** Reprinted by Permission from Macmillan Publishers Ltd. From Fig 4 in "Nature Medicine", 11 October 1, 2005 "Genetic Fingerprinting" by Alec J. Jeffreys pp1035-1039. The first application of DNA fingerprinting—an immigration case. **20.12** Patrick O. Brown, M.D.

Chapter 21

Opener Dr. Yorgos Nikas/Photo Researchers, Inc. **21.01** Chin-Sang. **21.02a-b** Elsevier Science Ltd. **21.04** Roslin Institute Empics. **21.05a** Richard Hutchings/Photo Researchers, Inc. **21.05b** Institute for Laboratory Animal Research ILAR/The National Academies. **21.06a** F. Rudolf Turner. **21.06b** Gary Grumbling. **21.07** Prof. Dr. Christiane Nusslein-Volhard. **21.09** Wolfgang Driever. **21.10a** Stephen Paddock. **21.10b** Stephen J. Small. **21.10c** Stephen Paddock. **21.12L** David Scharf/Photo Researchers, Inc. **21.12M** David Scharf/Peter Arnold, Inc. **21.12R** (c)Science VU/Dr. F.R. Turner/Visuals Unlimited. **21.14** Anthony Bannister/NHPA/Photo Researchers, Inc. **21.15a-b** Natural History Magazine

Chapter 22

Opener Yorgos Nikas/Photo Researchers, Inc. **22.02** Holger Jastrow. **22.04.** Gregory Ochocki/Photo Researchers, Inc. **22.07a** Michael Whitaker/Science Photo Library/Photo Researchers, Inc. **22.07bRML** Victor D. Vacquier. **22.14a** Kathryn W. Tosney. **22.14b** Gary C. Schoenwolf

Chapter 23

Opener Dr. John Runions/Oxford Brookes University **23.01** Holt Studios International/Photo Researchers, Inc. **23.03a** Ray Evert. **23.03c** Cabisco/Visuals Unlimited **23.04b** Biodisc/Visuals Unlimited **23.08b** Ken Wagner/Phototake NYC **23.10a-c** Neelima Sinha **23.12** John L.

Chapter 24

Opener Kim Taylor/NPL/Minden Pictures **24.02a** Bettmann/Corbis **24.02b** National Portrait Gallery, London **24.03a** Ken Lucas/Visuals Unlimited **24.03b** T. A. Wiewandt/DRK Photo **24.03c** Knut Finstermeier/Dr. Svante Paabo **24.04aR** Robert Lubeck/Animals Animals Earth Scenes **24.05aL** CMCD/Getty Images Inc. **24.05aR** Vincent Zuber/Custom Medical Stock Photo, Inc. **24.05bL** Mary Beth Angelo/Photo Researchers, Inc. **24.05bR** Custom Medical Stock Photo, Inc. **24.06aBL** George D. Lepp/Photo Researchers, Inc. **24.06aBR** Mickey Gibson Animals Animals/Earth Scenes **24.06aTL** Marie Read Animals Animals/Earth Scenes **24.06aTR** Tui De Roy Bruce Coleman Inc. **24.08L** Michael K. Richardson Springer-Verlag GmbH & Co KG **24.08M** Ronan O'Rahilly Springer-Verlag GmbH & Co KG **24.08R** National Museum of Health and Medicine **24.12** Frans Lanting/Minden Pictures **24.17aTL** From Fig 1C in Abzanhov et al. 2004. Science 305: 1462-1465 **24.17aTR** From Fig 1C in Abzanhov et al. 2004. Science 305: 1462-1465 **24.17aBL,BR** The Royal Society of London Reproduced by permission from K. Petren et al., A phylogeny of Darwin's finches based on microsatellite DNA length variation. Proceedings of the Royal Society of London, Series B, 266:321-329 (1999), p. 327, fig. 3. Copyright (c) 1999 Royal Society of London. Image courtesy of Kenneth Petren, University of Cincinnati. **24.17bL** From Figure 3D in Abzanhov et al. 2004. Science 305: 1462-1465 **24.17bR** From Fig 3E in Abzhanov et al. 2004. Science 305: 1462-1465

Chapter 25

Opener Phil Savoie/Minden Pictures **25.07a** Max Westby **25.08a** James Hughes/Visuals Unlimited **25.12a** Cyril Laubscher/Dorling Kindersley Media Library **25.13** Otorohanga Zoological Society Inc. **25.14a** Marc Moritsch/National Geographic Image Collection **25.15aT** Linda Mitchell Photography **25.15bB** R. & A. Simpson VIREO/The Academy of Natural Sciences **25.15bT** B. Schorre VIREO/The Academy of Natural Sciences **25.15cB** Van Os, Joseph/Getty Images Inc. **25.15cT** Jeremy Woodhouse/Getty Images, Inc

Chapter 26

Opener Wayne Lankinen/DRK Photo **26.09** Jordan Rofkar **26.12LMR** Jason Rick Loren H. Rieseberg

Chapter 27

Opener Pete Oxford/Minden Pictures **27.06a** William L. Crepet/American Journal of Botany **27.06b** Martin Land/Science Photo Library/Photo Researchers, Inc. **27.06c** Monte Hieb & Harrison Hieb/Monte Hieb/Geocraft **27.06d** John Gerlach/DRK Photo **27.09b** Nature Magazine **27.09b** Shuhai Xiao/Nature Magazine **27.09cB** Ken Lucas/Visuals Unlimited **27.09cT** Simon Conway Morris Simon Conway Morris **27.09dB** Ed Reschke/Peter Arnold, Inc. **27.09dT** B.J. Miller Biological Photo Service **27.12a** left, middle Jonathan B. Losos and Kevin de Quieroz/Prof. Jonathan B. Losos **27.13a** Colin Keates/Dorling Kindersley Media Library **27.13b** Tim Fitzharris/Minden Pictures **27.13c** Don P. Northup www.africancichlidphotos.com **27.13d** Joe Tucciarone/Photo Researchers, Inc. **27.15** David Hardy/Photo Researchers, Inc. **27.16bLR** Glen A. Izett/U.S. Geological Survey, Denver **27.16cR** Peter H. Schultz

Chapter 28

Opener Beth Donidow/Visuals Unlimited **28.03** Science VU/Visuals Unlimited **28.04** J. Robert Waaland/Biological Photo Service **28.06** Richard L. Carlton/Visuals Unlimited **28.08** Yul Roh American Association for the Advancement of Science **28.10** Kwangshin Kim/Photo Researchers, Inc. **28.10b** Todd Bannor/Custom Medical Stock Photo, Inc. **28.10c** Carolina Biological/Visuals Unlimited **28.13aL** CNRI/Science Photo Library/Photo Researchers, Inc. **28.13aR** Dr. Heide Schulz American Association for the Advancement of Science **28.13bL** Gary Gaugler/Visuals Unlimited **28.13bR** David M. Phillips/Visuals Unlimited **28.13cL** Linda Stannard University of Cape Town/Science Photo Library/Photo Researchers, Inc. **28.13c** Richard W. Castenholz **28.14a** R. Hubert, Microbiology Program, Dept. of Animal Science, Iowa State University **28.18** J. C. Revy/Phototake NYC **28.19** Centers for Disease Control & Prevention **28.20** S. Amano, S. Miyadoh and T. Shomura (The Society for Actinomycetes Japan) **28.21** R. Calentine/Visuals Unlimited **28.22** National Cancer Institute/SPL/Photo Researchers, Inc. **28.23a** Yves. V. Brun **28.23b** Prof. Hans Reichenbach **28.26** Kenneth M. Stedman **28.26** Corale Brierley/Visuals Unlimited **28.27** Marli Miller **28.27** Bonnie K. Baxter **28.29** Roland Birke/Tierbild Okapia/Photo Researchers, Inc.

Chapter 29

Opener Steve Gschmeissner/Photo Researchers, Inc. **29.02a** Norman T. Nicoll/Natural Visions **29.02b** Gregory Ochocki/Photo Researchers, Inc. **29.02c** Tom and Therisa Stack/Tom Stack & Associates, Inc. **29.02r** John Anderson/Animals Animals Earth Scenes **29.03a** Michael Abbey/Photo Researchers, Inc. **29.03b** Wim van Egmond/Visuals Unlimited **29.03c** Cabisco/Visuals Unlimited **29.03d** Sherman Thomson/Visuals Unlimited **29.03e** National Institute for Environmental Studies, Japan http://www.nies.go.jp/biology/mcc/home.htm **29.03f** Dr. Sc. Yuuji Tsuki **29.04** Sanford Berry/Visuals Unlimited **29.04** J. Robert Waaland/Biological Photo Service **29.07** Biophoto Associates/Photo Researchers, Inc. **29.13a** Andrew Syred/Science Photo Library/Photo Researchers, Inc. **29.13b** David M. Phillips/Visuals Unlimited **29.13c** Andrew Syred/Science Photo Library/Photo Researchers, Inc. **29.14a** Biophoto Associates/Photo Researchers, Inc. **29.14b** Bruce Coleman Inc. **29.15** M.I. Walker/Photo Researchers, Inc. **29.18** D. Fleetham/Animals Animals Earth Scenes **29.23** M. Abbey/Photo Researchers, Inc. **29.24** Carolina Biological/Visuals Unlimited **29.26** Michael Abbey/Visuals Unlimited **29.28** Dr. Philip J. Spagnuolo **29.30** Wim van Egmond/Visuals Unlimited **29.32** Phototake NYC **29.33** Andrew Syred/SPL/Photo Researchers, Inc. **29.34** Joyce Photographics/Photo Researchers, Inc. **29.36** Peter Parks/Image Quest 3-D **29.38** Robert De Goursey/Visuals Unlimited **29.40** Scott Camazine/Photo Researchers, Inc.

Chapter 30

Opener Darrell Gulin/DRK Photo **30.01a** Tom Myers/Photo Researchers, Inc. **30.01b** D. P. Burnside/Photo Researchers, Inc. **30.03a** Runk/Schoenberger/Grant Heilman Photography, Inc. **30.05** Holt Studios Int./Photo

Researchers, Inc. **30.06L** Linda Graham **30.06R** Lee W. Wilcox **30.07a/1** K.G. Vock/Okapia/Photo Researchers, Inc. **30.07a/2-3** Lee W. Wilcox **30.07b-1** Alvin E. Staffan/Photo Researchers, Inc. **30.07b/2** Biophoto Associates/Photo Researchers, Inc. **30.07b/3** Milton Rand/Tom Stack & Associates, Inc. **30.07b/4** Rod Planck/Photo Researchers, Inc. **30.07c-bM** Larry Lefever/Grant Heilman Photography, Inc. **30.07cBR** Stephen J. Krasemann/DRK Photo **30.07cTL** Walter H. Hodge/Peter Arnold, Inc. **30.07cTM** Degginger, E R/Animals Animals Earth Scenes **30.07cTR** ©ENLIGHTENED IMAGES/Animals Animals Earth Scenes **30.08/1** Paul K. Strother **30.08/2** Robert & Linda Mitchell Photography **30.08/3** Thomas A. Wiewandt/DRK Photo **30.08/4** David L. Dilcher **30.10** Lee W. Wilcox **30.13a-b** Lee W. Wilcox **30.14** David T. Webb **30.18a-bLR** Lee W. Wilcox **30.23a** James L. Castner **30.23b** Gerald C. Kelley/Photo Researchers, Inc. **30.23c** Gerry Ellis/GLOBIO **30.25a** John Cancalosi/Peter Arnold, Inc. **30.25b** Holt Studios Int.(Nigel Cattlin)/Photo Researchers, Inc. **30.27b/1** Robert & Linda Mitchell Photography **30.27b/2** Runk/Schoenberger/Grant Heilman Photography, Inc. **30.27b/3** Don Farrall/Getty Images, Inc. **30.27b/4** Kim Heacox Photography/DRK Photo **30.27T/2** Ed Reschke/Peter Arnold, Inc. **30.27T/3** Pat O'Hara/Corbis-Bettmann **30.27T/4** Robert & Linda Mitchell Photography **30.29a** Carolina Biological Supply Company/Phototake NYC **30.20bBL** C.C. Lockwood/Animals Animals Earth Scenes **30.20bTL** Runk/Schoenberger/Grant Heilman Photography, Inc. **30.20bR** Bill Banaszewski/Visuals Unlimited **30.29a-c** Lee W. Wilcox **30.31** Wim van Egmond **30.32** Andrew J. Martinez/Photo Researchers, Inc. **30.33** John Clayton/NIWA National Institute of Water and Atmospheric Research **30.35a-b** Robert and Beth Plowes Photograhy/Proteapix **30.36** Ed Reschke/Peter Arnold, Inc. **30.37** Visuals Unlimited **30.39** Lee W. Wilcox **30.40** Karen S. Renzaglia **30.41** Michael Clayton **30.42aL** David T. Webb **30.42aR** David Cavagnaro/DRK Photo **30.42b** Lee W. Wilcox **30.44** David T. Webb **30.45a** Biophoto Associates/Photo Researchers, Inc. **30.45b** Martin Land/Science Photo Library/Photo Researchers, Inc. **30.46** NICK GREEN/Photolibrary.com **30.47a-b** Lee W. Wilcox **30.49a** Brian Johnston **30.49bLR** Lee W. Wilcox. 30.48 Peter Chadwick/Dorling Kindersley Media Library

Chapter 31

Opener Steve Austin/CORBIS **31.01L** Mycorrhizal Applications, Inc. **31.01R** Jim Deacon Institute of Cell and Molecular Biology, The University of Edinburgh **31.03** John Cang Photography **31.04a** David M. Dennis /Tom Stack & Associates, Inc. **31.04b** Poehlmann/Phototake NYC **31.05a** George Musil/Visuals Unlimited **31.05b** Tony Brain/Science Photo Library/Photo Researchers, Inc. **31.06b** Biophoto Associates/Photo Researchers, Inc. **31.11a** Stan Flegler Visuals Unlimited **31.11b** David T. Webb **31.12** Michael P. Gadomski/Photo Researchers, Inc. **31.15** James A. Wubah **31.16** J. I. Ronny Larsson, Professor of Zoology **31.17** Biophoto Associates/Photo Researchers, Inc. **31.18** Jim Deacon Institute of Cell and Molecular Biology, The University of Edinburgh **31.19** Jeff Lepore/Photo Researchers, Inc. **31.20b** Pat O'Hara/DRK Photo **31.21** Stephen Sharnoff/Visuals Unlimited

Chapter 32

Opener David B. Fleetham/SeaPics.com **32.01** Perennou/Nuridsa /Photo Researchers, Inc. **32.01b** Mark W. Moffett/Minden Pictures **32.02a** Harry Rogers/Photo Researchers, Inc. **32.02b** Glenn Oliver/Visuals Unlimited **32.02cCM** Fallows/SeaPics.com **32.03a** Walker England/Photo Researchers, Inc. **32.03b** Ken Lucas/Visuals Unlimited **32.03c** David J. Wrobel/Visuals Unlimited **32.04** Roland Birke /Tierbild Okapia/Photo Researchers, Inc. **32.05a** Jeffrey L. Rotman/CORBIS **32.05b** Pitchal Frederic/Corbis Sygma **32.05c** J. Alex Halderman **32.09bB** L. Newman & A. Flowers/Photo Researchers, Inc. **32.09bT** Roger Steene/Image Quest 3-D **32.13a** J.P. Ferrero/Jacana/Photo Researchers, Inc. **32.12a** Flip Nicklin/Minden Pictures **32.12a** Flip Nicklin/Minden Pictures **32.12b** inset Marc Chamberlain/SeaPics.com **32.12b** Hiroya Minakuchi/Minden Pictures **32.13b** Steve Earley/Animals Animals Earth Scenes **32.14a** Pete Oxford/Minden Pictures **32.14b** Heidi & Hans-Jurgen Koch/Minden Pictures **32.15a** Daniel Sambraus/Photo Researchers, Inc. **32.15b** Anthony Bannister/NHPA/Photo Researchers, Inc. **32.15b** Josef Ramsauer & Prof. Dr. Robert Patzner **32.16a** Satoshi Kuribayashi/DRK Photo **32.16a** Satoshi Kuribayashi/OSF/DRK Photo **32.16b** Jim Brandenburg/Minden Pictures **32.17a** Andrew Syred/Photo Researchers, Inc. **32.17b** Eye of Science/Photo Researchers, Inc. **32.18a** Peter Parks/Image Quest 3-D **32.19a** Roland Seitre/Peter Arnold, Inc. **32.19b** Jeff Foott/DRK Photo **32.19c** Chamberlain, MC/DRK Photo **32.19d** Jim Greenfield/Image Quest 3-D **32.18b** Armin Maywald/Foto Natura/Minden Pictures **32.20bLBMBR** Proceedings of the National Academy of Sciences **32.22a** Doug Perrine/DRK Photo **32.22b** Dan Suzio/Photo Researchers, Inc. **32.21a** Karl-Heinz Marschner Holger de Groot/Biologis **32.21b** D. Parer & E. Parer-Cook/Auscape/Minden Pictures **32.23aTM** Phototake NYC **32.23aB** Sinclair Stammers/Science Photo Library/Photo Researchers, Inc. **32.23bT** Jose B. Ruiz/Nature Picture Library **32.B** Jose B. Ruiz/Nature Picture Library **32.26** Andrew J. Martinez/Photo Researchers, Inc. **32.27a-b** Holger de Groot/Biologis **32.28** Andrew J. Martinez/Photo Researchers, Inc. **32.29** Matthew D. Hooge

Chapter 33

Opener Barbara Strnadova/Photo Researchers, Inc. **33.01a** Jean Claude Revy/Phototake NYC **33.01b** Sinclair Stammers/Photo Researchers, Inc. **33.04a** Peter Parks/Image Quest 3-D **33.05** Bill Beatty/Visuals Unlimited **33.04b** Carmel McDougall **33.09a** Mark W. Moffett/Minden Pictures **33.09b** Phototake NYC **33.09c** Dennis Kunkel Phototake NYC **33.10b** M.C. Chamberlain/DRK Photo **33.10c** Stephen Dalton/Photo Researchers, Inc. **33.12** Jan Van Arkel/Minden Pictures **33.13a** Newman & Flowers/Photo Researchers, Inc. **33.13b** Manfred Kage/Peter Arnold, Inc. **33.13c** Oliver Meckes/Photo Researchers, Inc. **33.14a** Peter Batson/Image Quest 3-D **33.14b** Alexa Bely **33.14c** Martin Dohrn/Photo Researchers, Inc. **33.15a** Degginger, E R/Animals Animals Earth Scenes **32.16b** Norbert Wu/Peter Arnold, Inc. **33.16a** Roger Steen/Image Quest Marine **33.17** Kjell B. Sandved/Photo Researchers, Inc. **33.18** Fred Bavendam/Peter Arnold, Inc. **33.20a** Michael Fogden/DRK Photo **33.20b** Diane Nelson/Visuals Unlimited **33.21** Luis M. de la Maza/Phototake NYC. **33.22** Mark Smith/Photo Researchers, Inc. **33.23a** Gerard Blondeau/Phototake NYC **T33.1/1** Jay Cossey **T33.1/2** Gustav W. Verderber **T33.1/3** Joachim Lippi **T33.1/4** Trounce Wikimedia Commons **T33.1/5** J. Gall/Photo Researchers, Inc. **T33.1/6** Nature's Images/Photo Researchers, Inc. **T33.1/7** Dr. J. K. Lindsey **T33.1/8** Scott Camazine/Photo Researchers, Inc. **33.24a** E. R. Degginger/Photo Researchers, Inc. **33.24b** M C Chamberlain/DRK Photo

Chapter 34

Opener Paul Nicklen/National Geographic Image Collection **34.02b** Thurston Lacalli **34.02a** Kaj R. Svensson/Photo Researchers, Inc. **34.03b** John D. Cunningham/Visuals Unlimited **34.09a** Gerald & Buff Corsi/Visuals Unlimited **34.09b** Greg Rouse/Scripps Institution of Oceanography Explorations **34.11** Rudie Kuiter/OceanwideImages.com **34.18a** James D. Watt/SeaPics.com **34.18b** S. Boon Fu/Peter Arnold, Inc. **34.20** James D. Watt/Image Quest 3-D **34.21R** Tom Stack/Tom Stack & Associates, Inc. **34.21** Southestern Regional Taxonomic Center (SERTC)/South Carolina Department of Natural Resources **34.23a** Sue Daly/Nature Picture Library **34.23b** David Wrobel/SeaPics.com **34.25a** Tom McHugh/Photo Researchers, Inc. **34.25b** Tom Stack/Tom Stack & Associates, Inc. **34.24** Heather Angel/Natural Visions **34.26a** Fred McConnaughey/Photo Researchers, Inc. **34.26b** Jeff Jaskolski/SeaPics.com **34.27** Amar & Isabelle Guillen/SeaPics.com **34.28** Peter Scoones/Getty Images Inc **34.29a** Michael Fogden/DRK Photo **34.29b** R.Andrew Odum/Peter Arnold, Inc. **34.31** Tom McHugh/Photo Researchers, Inc. **34.32** Tom Vezo/Peter Arnold, Inc. **34.33** Gerry Ellis/Minden Pictures **34.35** Gerald and Buff Corsi/Visuals Unlimited **34.36** George Grall/National Geographic Image Collection **34.37** Doug Perrine/DRK Photo **34.38** Gerald and Buff Corsi/Visuals Unlimited **34.39** M.S. Mayilvahnan **34.39b** Tom McHugh/Photo Researchers, Inc.

Chapter 35

Opener Jed Fuhrman **35.02** NIBSC/Science Photo Library/Photo Researchers, Inc. **35.06a** Omikron/Photo Researchers, Inc. **35.06b** Biophoto Associates/Photo Researchers, Inc. **35.06c** K.G. Murti/Visuals Unlimited **35.06d** Oliver Meckes/E.O.S./Max-Planck-Institut-Tubingen/Photo Researchers, Inc. **35.13a-b** Abbott Laboratories **35.15aR** NIBSC/Science Photo Library/Photo Researchers, Inc. **35.15bR** James L. Van Etten **35.17** Hans Gelderblom/Eye of Science/Meckes/Ottawa/Photo Researchers, Inc. **35.18** Philip Leder **35.19** Nigel Cattlin/Holt Studios/Photo Researchers, Inc. **35.20** Lowell Georgia/Photo Researchers, Inc. **35.21** David Parker/SPL/Photo Researchers, Inc.

Chapter 36

Opener Tom McHugh/Photo Researchers, Inc. **36.04a** Tim Hauf Photography/Visuals Unlimited **36.04bLMR** John E. Weaver University of Nebraska-Lincoln Libraries **36.05** Phil Stoffer U.S. Geological Survey **36.06a** Wally Eberhart/Visuals Unlimited **36.06b** Wolfgang Poelzer/Water Frame/Peter Arnold, Inc. **36.08a-b** Gerald D. Carr **36.08c** Frans Lanting/Minden Pictures **36.09R** Alan Majchrowicz/Peter Arnold, Inc. **36.10e** Will Cook/Charles W. Cook **36.10a** Ken Wagner/Phototake NYC **36.10b** Kenneth W. Fink/Photo Researchers, Inc. **36.10c** Lee W. Wilcox **36.10d** Bill Beatty/Visuals Unlimited **36.12a-d** Lee W. Wilcox **36.14a** Lee W. Wilcox **36.14b** M. Harvey/DRK Photo **36.14d-f** Lee W. Wilcox **36.14c** Doug Wechsler/Animals Animals Earth Scenes **36.14e** TORSTEN BREHM/Nature Picture Library **36.15b** Walker/Photo Researchers, Inc. **36.15a** Ed Reschke/Peter Arnold, Inc. **36.18b** Lee W. Wilcox **36.18b** Ed Reschke/Peter Arnold,

Inc. **36.20RL** Lee W. Wilcox **36.22b** Lee W. Wilcox **36.21** Andrew Syred/Science Photo Library/Photo Researchers, Inc. **36.22a** Dr. Brad Mogen/Visuals Unlimited/Getty Images **36.23LR** Lee W. Wilcox **36-24a-c** Lee W. Wilcox **36.25a** Biophoto Associates/Science Source/Photo Researchers, Inc. **36.25b** Lee W. Wilcox **36.26a** Bruce Iverson/Bruce Iverson, Photomicrography **36.26b** G. Shih and R. Kessel/Visuals Unlimited **36.26c** Richared Kessel & Dr. Gene Shih Visuals Unlimited **36.27LMR** Lee W. Wilcox **36.28a-b** Michael Clayton **36.29c** Adam Hart-Davis/Science Photo Library/Photo Researchers, Inc. **36.29b** James W. Richardson/Visuals Unlimited **36.29a** Lee W. Wilcox;

Chapter 37

Opener David Nunuk/Science Photo Library/Photo Researchers, Inc. **37.02** Dr. Ulf Mehlig/Wikipedia Commons **37.02R** Alan Watson "Peter Arnold, Inc." **37.05a-b** David T. Webb **37.07LR** Lee W. Wilcox **37.08a** Lee W. Wilcox **37.09** Larry Lefever/Grant Heilman Photography, Inc. **37.11M** G. Shih and R. Kessel/Visuals Unlimited **37.11B** Lee W. Wilcox **37.11T** Ken Wagner/Phototake NYC **37.16a** John D. Cunningham/Visuals Unlimited **37.16a** Lee W. Wilcox **37.16b** Lee W. Wilcox **37.19R** Biophoto Associates/Photo Researchers, Inc. **36.18L** Jean Claude Revy/Phototake NYC **37.21** Martin H. Zimmerman/ Harvard Forest **37.24** Michael R. Sussman/American Society of Plant Biologists

Chapter 38

Opener Angelo Cavalli/Getty Images **38.03** Emanuel Epstein **38.03a** Photo Researchers, Inc. **38.02b** Nigel Cattlin/Photo Researchers, Inc. **38.02c** Photo Researchers, Inc. **38.06a** The Institute of Texan Cultures **38.06b** Thony Belizaire/Getty Images, Inc. **38.09R** Dennis Drenner/Visuals Unlimited **38.13b** Eduardo Blumwald **38.14** Hugh Spencer/Photo Researchers, Inc. **38.14** Andrew Syred/SPL/Photo Researchers, Inc. **38.15** E.H. Newcomb & S.R. Tandon/Biological Photo Service **38.17a** Richard Thom/Visuals Unlimited **38.17b** Gerry Ellis/GLOBIO **38.19** Frank Greenaway/Dorling Kindersley Media Library **38.18** Carlos Munoz-Yague Eurelios/Photographic Press Agency **38.20** Ken W. Davis /Tom Stack & Associates, Inc.

Chapter 39

Opener Lee W. Wilcox **39.04a-b** Malcolm B. Wilkins **39.12LR** Malcolm B. Wilkins **39.16a-b** Runk/Schoenberger/Grant Heilman Photography, Inc. **39.17a** American Society of Plant Biologists **39.20** Thomas Bjorkman **39.22L** Carolina Biological Supply Company/Phototake NYC **39.21** Donald Specker/Animals Animals Earth Scenes **39.23a-b** Lee W. Wilcox **39.27** Malcolm B. Wilkins **39.30b** American Society of Plant Biologists **39.32** Adel Kader **39.39** Nigel Cattlin/Holt Studios International /Photo Researchers, Inc.

Chapter 40

Opener Brian Johnston **40.01aL** Wayne P. Armstrong **40.01aR** Mitsuhiko Imamori Nature Production **40.01bL** Jerome Wexler/Photo Researchers, Inc. **40.01bR** Danny Ellinger Minden Pictures **40.03a** Dan Suzio/Photo Researchers, Inc. **40.03b-c** Jerome Wexler/Photo Researchers, Inc. **40.04bB** Rod Planck/Photo Researchers, Inc. **40.04bTL** John Gerlach/DRK Photo **40.04bTR** Tom & Therisa Stack/Tom Stack & Associates, Inc. **40.07a** D. Cavagnaro/DRK Photo **40.06a-b** Leonard Lessin/Photo Researchers, Inc. **40.07bTB** Lee W. Wilcox **40.13** "Candace Galen, University of Missouri. American Journal of Botany. 2003; 90:724-729. Sunny-side up: flower heliotropism as a source of parental environmental effects on pollen quality and performance in the snow buttercup, Ranunculus adoneus (Ranunculaceae) by Candace Galen, and Maureen L. Stanton. Reproduced by permission of the publisher. **40.18** Lee W. Wilcox **40.17TL** R.J. Erwin "Photo Researchers, Inc." **40.17TM** Mark Stouffer Animals Animals/Earth Scenes **40.17TR** Gerard Lacz/Animals Animals Earth Scenes **40.17ML** Tom Edwards/Visuals Unlimited **40.17MM** David Stuckel/Visuals Unlimited **40.17MR** Gregory K. Scott/Photo Researchers, Inc **40.17BL** Fritz Prenzel/Peter Arnold, Inc. **40.17BM** David M. Schleser/Nature's Images, Inc/Photo Researchers, Inc. **40.17BR** Sylvan Wittwer/Visuals Unlimited

Chapter 41

Opener Rob Nunnington/Minden Pictures **41.02** The Royal Society of London **41.03a** Educational Images/Custom Medical Stock Photo **41.03bT** Carolina Biological Supply Company/Phototake NYC **41.03bB** Carolina Biological Supply Company/Phototake NYC **41.03c** Carolina Biological Supply Company/Phototake NYC **41.04** Ed Reschke/Peter Arnold, Inc. **41.05a** Innerspace Imaging/Photo Researchers, Inc. **41.05b** Ed Reschke/Peter Arnold, Inc. **41.05c** G.W. Willis/Animals Animals Earth Scenes **41.06a** Nina Zanetti, Pearson Benjamin Cummings **41.10** Tom Stewart/CORBIS **41.11** Natalie Fobes/Getty Images Inc. **41.13a** Fred Hossler/Visuals Unlimited **41.13b** Oliver Meckes & Nicole Ottawa/Photo Researchers, Inc. **41.13c** P.M. Motta, A. Caggiati, G. Macchiarelli/Science Photo Library/Photo Researchers, Inc. **41.18a** Kent, Breck P./Animals Animals Earth Scenes **41.18b** Mendez, Raymond A/Animals Animals Earth Scenes **41.18c** Wayne Lynch/DRK Photo **41.19b** Editrice Kurtis S.r.l. **41.19a** Ed Reschke/Peter Arnold, Inc.;

Chapter 42

Opener Frans Lanting/Minden Pictures **42.13a** Fred Hossler/Visuals Unlimited

Chapter 43

Opener Jonathan Blair/Corbis Bettmann **43.03** Mark Smith/Photo Researchers, Inc. **43.04bLMR** Karel F. Liem **43.09a** Carolina Biological Supply Company/Phototake NYC **43.11a** Ed Reschke Peter Arnold, Inc.

Chapter 44

Opener Tony Freeman/PhotoEdit Inc. **44.06a** Walter E. Harvey/Photo Researchers, Inc. **44.06b** John D. Cunningham/Visuals Unlimited **44.22a** Lennart Nilsson Albert Bonniers Forlag AB **44.22b** Carolina Biological Supply/Visuals Unlimited

Chapter 45

Opener Marcus E. Raichle **45.02** Ronald F. Mervis, PhD, Neurostructural Research Labs and The Center for Aging and Brain Repair, University of South Florida College of Medicine **45.12a** C. Raines/Visuals Unlimited **45.14** Dennis Kunkel/Visuals Unlimited **45.16a** Oliver Meckes & Nicole Ottawa/Photo Researchers, Inc. **45.23a** Norbert Wu/Peter Arnold, Inc. **45.24b** Elsevier Science Ltd.

Chapter 46

Opener Stanley Breeden/DRK Photo **46.03a** Carole M. Hackney **46.07a** David Scharf/Peter Arnold, Inc. **46.10aR** Don Fawcett/T. Kwwabara/Photo Researchers, Inc. **46.13RL** David Quillin **46.14a** Michael and Patricia Fogden **46.14bLR** John A. Pearce **46.19B** James E. Dennis/Phototake NYC **46.19T** James E. Dennis/Phototake NYC **T46.01L** Brian Eyden/Science Photo Library/Photo Researchers, Inc. **T46.01M** Innerspace Imaging/Science Photo Library/Photo Researchers, Inc. **T46.01R** Brian Eyden/Science Photo Library/Photo Researchers, Inc.

Chapter 47

Opener Ralph A. Clevenger/Corbis **47.08R** Bernard Castelein/Nature Picture Library **47.08M** Duncan McEwan/Nature Picture Library **47.08L** Stephen Dalton/Photo Researchers, Inc. **47.09R** Perennou Nuridsany/Photo Researchers, Inc. **47.09M** Hans Pfletschinger/Peter Arnold Inc. **47.09L** BIOS/Heras Joël/Peter Arnold Inc. **47.11** Jordan Rehm/D. Scott Weigle

Chapter 48

Opener Leszczynski, Zigmund/Animals Animals/Earth Scenes **48.01a** Nova Scientific Corp/Oxford Scientific Films Ltd. **48.01b** Tom Adams/Visuals Unlimited **48.01c** Charles J. Cole **48.02a** Oxford Scientific Films/Animals Animals Earth Scenes **48.04TB** Reprinted by permission from Catherine S.C. Price et al., Sperm competition between Drosophila males involves both development and incapacitation. Nature 400:449-452 (1999), figs. 2 and 3. Copyright (c) 1999 Macmillan Magazines Limited. Image courtesy of Jerry A. Coyne, University of Chicago/Nature Magazine **48.05a** G. & C. Merker/B. Tomberlin/Visuals Unlimited **48.06a** Brent D. Opell Brent D. Opell **48.10B** From Nature "A stimulatory phalloid organ in a weaver bird" M. Winterbottom, T. Burke & T.R. Birkhead/Nature/ 6 May 1999, 399: 28 Fig 1b. Photo courtesy T.R. Birkhead **48.10T** From Nature "A stimulatory phalloid organ in a weaver bird" M. Winterbottom, T. Burke & T.R. Birkhead/Nature/6 May 1999, 399: 28 Fig 1a. Photo courtesy T.R. Birkhead **48.15a-c** Mitsuaki Iwago/Minden Pictures **48.16a** Claude Edelmann /Petit Format/Photo Researchers, Inc. **48.16b** Lennart Nilsson Albert Bonniers Forlag AB **48.16c** Petit Format/Nestle/Science Source/Photo Researchers, Inc. **48.18aLR** American Association for the Advancement of Science

Chapter 49

Opener Dennis Kunkel/CNRI Phototake NYC **49.01B** Fred Hossler/Visuals Unlimited **49.02a** Photo Researchers Inc **49.02b-c** Steve Gschmeissner/Photo Researchers, Inc. **49.05b** Steve Gschmeissner/SPL/ Photo Researchers, Inc. **49.05a** David M. Phillips/Visuals Unlimited

Chapter 50

Opener Stephen J. Krasemann/DRK Photo **50.01a-d** Natalie B. Fobes Photography **50.05b** Gerry Ellis/Minden Pictures **50.05c** Tim Fitzharris/Minden Pictures **50.05a** Gerry Ellis/Minden Pictures **50.06** George Gerster/Photo Researchers, Inc. **50.07** Michael Collier **50.10a-c** Edward Kinsman/Photo Researchers, Inc. **50.12** Hans Reinhard/OKAPIA/Photo Researchers, Inc. **50.14** C.K. Lorenz/Photo Researchers, Inc. **50.16** © Judd Patterson **50.18** Tom Edwards/Animals Animals/Earth Scenes **50.20** Francis Lepine/Animals Animals Earth Scenes **50.22** Tom Bean/DRK Photo **50.27** Sandy Scheltema/Courtesy of The Age **50.29 inset** William H. Mullins/Photo Researchers, Inc. **50.29** Raymond Gehman/Corbis Bettmann **50.30b** Michael Fogden/DRK Photo **50.31** Ric Ergenbright/Corbis **50.32** Mark Newman/Photo Researchers, Inc. **50.32** Scott T. Smith Corbis Bettmann

Chapter 51

Opener Thomas Mangelsen/Minden Pictures **51.01** A. Flowers & L. Newman/Photo Researchers, Inc. **51.03a** © Callie de Wet **51.04** © Gary Woodburn/http://pbase.com/woody **51.06** Fred McConnaughey/Photo Researchers, Inc. **51.07** Nina Leen/Getty Images/Time Life Pictures **51.09a** Gavin Hunt **51.10** Photo courtesy of the Behavioural Ecology Research Group, Department of Zoology, Oxford University, Alexander Weir **51.11** Matt Meadows/Peter Arnold, Inc. **51.12a** Manuel Leal, Department of Biological Sciences, Union College **51.18a** Ken Lucas/Visuals Unlimited **51.18b** Paul A. Zahl/Photo Researchers, Inc. **51.18c** Ken Preston-Mafham/Premaphotos Wildlife **51.19** Bryan D. Neff **51.20** Tom Vezo/Peter Arnold, Inc.

Chapter 52

Opener National Geographic Image Collection **52.01** Robert Maier Animals Animals/Earth Scenes **52.06** Mark Trabue/USDA/NRCS **52.10LR** Marko Nieminen Ilkka Hanski **52.14a** Paal Hermansen /NHPA/Photo Researchers, Inc. **52.19a** BIOS (F. Mercay) Peter Arnold, Inc.

Chapter 53

Opener Ergenbright Scott Tuason/Image Quest Marine **53.01** PREMAPHOTOS/Nature Picture Library **53-01** Christian Ziegler **53.02a** G.R.Dick Roberts Photo Library/The Natural Sciences Image Library (NSIL) **53.02b** Fred Bavendam/Minden Pictures **53.02c** Tom Till/DRK Photo **53.02d** D. Cavagnaro/Visuals Unlimited **53.02e** Stephen J. Krasemann/DRK Photo **53.02f** Hal Beral/Visuals Unlimited **53.08LR** Jeff Podos **53.08MB** B.R. Grant **53.08MT** From Fig 1D in Science 14 July 2006 Vol. 313. no. 5784, pp. 224 – 226. Evolution of Character Displacement in Darwin's Finches Peter R. Grant* and B. Rosemary Grant. [Photos are by the authors] **53.09aL** Dave B. Fleetham/Visuals Unlimited **53.09aM** Chris Newbert/Minden Pictures **53.09aR** J. Sneesby/B. Wilkins/Getty Images Inc **53.10aL** Andrew J. Martinez Photo Researchers, Inc. **53.10aR** Zig Leszczynski/Animals Animals Earth Scenes **53.12a** Thomas G. Whitham/Northern Arizona University **53.15a** Mark Moffett/Minden Pictures **53.15b** Doug Perrine/SeaPics.com **53.18b** Thomas Kitchin/Tom Stack & Associates, Inc. **53.18a** Nancy Sefton/Photo Researchers, Inc. **53.20a** David Kjaer/Nature Picture Library **53.20b** Tony C. Caprio/Tony C. Caprio **53.21/1** Breck P. Kent/Animals Animals Earth Scenes **53.21/2** G. Carleton Ray/Photo Researchers, Inc. **53.21/3** Michael P. Gadomski/Animals Animals/Earth Scenes **53.21/4** James P. Jackson/Photo Researchers, Inc. **53.21/5** Bruce Heinemann/Getty Images, Inc. **53.21/6** Michael P. Gadomski/Photo Researchers, Inc. **53.22/1-4** Glacier Bay National Park Photo Glacier Bay National Park and Preserve **53.22/3** Christopher L. Fastie

Chapter 54

Opener Lonnie G. Thompson Byrd Polar Research Center The Ohio State University **54.02** MODIS NASA **54.11b** Randall J. Schaetzl, Michigan State University **54.11a** Richard Hartnup/Wikipedia Comons **54.12a** U.S. Forest Service, Northern Research Station

Chapter 55

Opener Micheline Pelletier/Corbis Bettmann **55.01b** Institute of Hydrobiology, Wuhan, China/The New York Times **55.01a** Tom & Pat Lesson/DRK Photo **55.02** Edward S. Ross, California Academy of Sciences **55.03a** Carr Clifton/Minden Pictures **55.03b** Kevin W. Fitz All Species Photography & Sound **55.04b** from: Fig1C in Nature Vol 436 August 18, 2005 page 1016 "Global hotspots of species richness are not congruent with endemism or threat". C. David L. Orne et al. **55.04a** from Fig 1A in Nature Vol 436 August 18, 2005 page 1016 "Global hotspots of species richness are not congruent with endemism or threat". C. David L. Orne et al. **55.05B** Fred Bruemmer/DRK Photo **55.07a** John Mitchell/Photo Researchers, Inc. **55.07b** Gary Braasch Gary Braasch Photography **55.07c** David Dennis/Animals Animals Earth Scenes **55.09** Jim Wark/Airphoto **55.08** Martin Harvey/Alamy **55.10a** Jacques Jangoux/Peter Arnold, Inc. **55.10bLR** NASA/Goddard Space Flight Center

Backmatter

BS7.02 Reproduced by permission from J.P. Ferris et al., Synthesis of long prebiotic oligomers on mineral surfaces. Nature 381:59-61 (1996), Fig. 2. Copyright (c) 1996 Macmillan Magazines Limited. Image courtesy of James P. Ferris, Rensselaer Polytechnic Institute **BS8.01aL** Photo courtesy of Peter M. O'Day, Juan Bacigalupo, Joan E. Haab, and Cecilia Vergara. The Journal of Neuroscience, October 1, 2000. 20(19): 7193-7198, Fig. 1C. (c) 2004 by the Society for Neuroscience **BS8-01bR** Dennis Kunkel/Phototake NYC **BS8.02aTL** Michael W. Davidson/Florida State University/Molecular Expressions **BS8.02bTR** Michael W. Davidson/Florida State University/Molecular Expressions; **BS8.03B** Rosalind Franklin/Photo Researchers, Inc.

ILLUSTRATION CREDITS

1.12 Data from J.J Tewksbury and G.P. Nabhan. 2001. *Nature* 412:403–404, Fig. 1a.

3.25a Data from N.N Nawani and B.P. Kapadnis. 2001. *Journal of Applied Microbiology* 90:803–808, Fig. 3. Data also from N.N. Nawani et al. 2002. *Journal of Applied Microbiology* 93:965–975, Fig. 7. **3.25b** Data from T. Hansen et al. 2002. *FEMS Microbiology Letters* 216:249–253, Fig. 1.

6.25a PDB Document Object Identifier (DOI):10.2210/pdb1J4N/pdb. **6.25b** PDB DOI:10.2210/pdb1ORS/pdb and PDB DOI:10.2210/pdb1ORQ/pdb.

7.4 PDB DOI:10.2210/pdb2AW4/pdb and PDB DOI:10.2210/pdb2AVY/pdb.

8.10b After *Molecular Biology of the Cell*, 4e, Fig. 19.5. © 2002 by Bruce Alberts, Alexander Johnson, Julian Lewis, Martin Raff, Keith Roberts, and Peter Walter. Adapted by permission of Garland Science/Taylor & Francis Books, Inc.

9.2 PDB DOI:10.2210/pdb1ERK/pdb and PDB DOI:10.2210/pdb2ERK/pdb. **9.15** PDB DOI:10.2210/pdb1PFK/pdb. **9.25** Modified from H. Wang and G. Oster. 1998. *Nature* 396:279–282, Fig. 1.

11.16 American Cancer Society's *Cancer Facts and Figures–2003*. Reprinted with permission.

16.15 PDB DOI:10.2210/pdb2J00/pdb and PDB DOI:10.2210/pdb2J01/pdb.

17.18b PDB DOI:10.2210/pdb2ORI/pdb.

18.2b After B. Dorigo et al. 2004. *Science* 306:1571–1573, Fig. 1. **18.15** PDB DOI:10.2210/pdb1TSR/pdb.

20.1 Data from "Public collections of DNA and RNA reach 100 gigabases." 2005. EMBL-EBI Press Release, Hixton. **20.10** After Lodish et al., *Molecular Cell Biology,* 5e, Fig. 9.34. © 2004 W.H. Freeman and Company. Adapted with permission.

21.13 After S.B. Carroll. 1995. Nature 376:479–485 (Review).

24.4b *Pakicetus* based on J.G.M. Thewissen et al. 2001. *Nature* 413:277–281, Fig. 2. *Rhodocetus* based on P.D. Gingerich et al. 2001. *Science* 293:2239–2242, Fig. 3. **24.13** Data from P.T. Boag and P.R. Grant. 1981. *Science* 214:82-85. **24.14** Modified from P.R. Grant and B.R. Grant. 2001. *Science* 296:707–711, Fig. 1.

25.6 Data from S. Freeman and J. Herron, *Evolutionary Analysis,* 3e, Figs. 6.15a and 6.15c. © 2003 by Prentice-Hall, Inc. **25.7b** Data from W.E. Kerr and S. Wright. 1954. *Evolution* 8:172–177. **25.11** Data from M.O. Johnston. 1992. *Evolution* 46:688–702.

28.2 After G.L. Armstrong et al. 1999. *JAMA* 281:61–66, Fig. 1. © American Medical Association. All rights reserved. **28.5** Data from M.T. Madigan and J.M. Martinko. 2006. *Brock Biology of Microorganisms,* 11e, Fig. 5.9.

34.12 After E.B. Daeschler et al. 2006. *Nature* 440:757–763, Fig. 6. Also after N.H. Shubin et al. 2006. *Nature* 440:764–771, Fig. 4.

43.17 Data from L.O. Schulz et al. 2006. *Diabetes Care* 29:1866–1871.

48.4 Data from C.S.C. Price et al. 1999. *Nature* 400:449–452, Fig. 3c. © Macmillan Magazines Ltd.

49.6a PDB DOI:10.2210/pdb1IGT/pdb. **49.6b** PDB DOI:10.2210/pdb1TCR/pdb. **49.7** PDB DOI:10.2210/pdb2VIR/pdb.

50.28 Data from A.K. Knapp et al. 2002. *Science* 298:2202–2205, Figs. 1, 2a, and 3b.

52.3 Modified from C.K. Ghalambor and T.E. Martin. 2001. *Science* 292:494–497, Fig. 1c. **52.7b** Data from G.F. Gause. 1934. *The Struggle for Existence.* New York: Hafner Press. **52.9** Data from M. Dodd et al. 1995. *Journal of Ecology* 83:277–285. **52.14b** Data from T. Valverde and J. Silvertown. 1998. *Journal of Ecology* 86:545–562.

53.4 Data from G.F. Gause. 1934. *The Struggle for Existence.* New York: Hafner Press. **53.19** Data from R.T. Paine. 1974. *Oecologia* 15:93–120.

54.4 Data from A. Tsuda et al. 2003. *Science* 300:958–961. **54.18** Data from *BP Statistical Review of World Energy 2007,* pp. 6–21. **54.19a** Data from G. Beaugrand et al. 2002. *Science* 296:1692–1694. **54.19b** Data from N.L. Bradley et al. 1999. *Proceedings of the National Academy of Science* 96:9701–9704.

55.4c N. Myers et al. 2000. *Nature* 403:853_858, Fig. 1. **55.6** O. Venter et al. 2006. *BioScience* 56:903_910, Fig. 2 **55.12** Data from J.T. Hogg et al. 2006. *Proceedings of the Royal Society B: Biological Sciences* 273:1491_1499, Fig.1. Additional unpublished from J.T. Hogg. **55.16** Data from E.I. Damschen et al. 2006. *Science* 313:1284–1286, Fig. 2b.

Index

Boldface page numbers indicate a glossary entry; page numbers followed by an *f* indicate a figure; those followed by *t* indicate a table.

D

S

SUMMARY TABLES

BIOSKILLS

Dear Student:

As the author of *Biological Science*, **Third Edition,** I set out to write a book that would be an effective learning tool for you. I would appreciate hearing about your experience with this textbook and its multimedia support, and I invite your comments and suggestions for improvements!

Many thanks,

Scott Freeman

Grading System **A**=excellent/very useful, **B**=good/useful, **C**=moderately helpful, **D**=poor/very little help, **F**=not helpful

1. Please grade the following study aids in *Biological Science*, Third Edition:

(circle one)

Gold highlighting of Key Concepts and important information within the narrative — Grade: **A B C D F**

Comments ______________________________

____

"If you understand... you should be able to..." in blue typeface within the narrative — Grade: **A B C D F**

Comments ______________________________

____

Check Your Understanding boxes — Grade: **A B C D F**

Comments ______________________________

____

Figure caption questions and exercises — Grade: **A B C D F**

Comments ______________________________

____

Summary tables — Grade: **A B C D F**

Comments ______________________________

____

Overall Comments ______________________________

2. Did you use your *MasteringBiology*™ subscription? Yes [ ] No [ ] If yes, please grade the following online resources:

Web Animations — Grade: **A B C D F**

Comments ______________________________

____

BioFlix animations and tutorials — Grade: **A B C D F**

Comments ______________________________

____

Instructor-assigned tutorials — Grade: **A B C D F**

Comments ______________________________

____

Chapter quizzes — Grade: **A B C D F**

Comments ______________________________

____

Answer keys for textbook questions — Grade: **A B C D F**

Comments ______________________________

____

3. Please list three things you liked most about the textbook and/or online resources:

Comments ______________________________

4. Please describe two or three topics that were difficult for you to understand, and please include page numbers.

Comments ______________________________

5. Do you have any other suggestions for improving future editions of *Biological Science*?
Please provide page numbers where appropriate.

Comments ______________________________

School: ______________________________

Your Name (Optional): ______________________________

Email: ______________________________

Date: ______________________________

May Benjamin Cummings have permission to quote your comments in promotions for ***Biological Science***? ☐ Yes ☐ No

NO POSTAGE NECESSARY IF MAILED IN THE UNITED STATES

BUSINESS REPLY MAIL

FIRST-CLASS MAIL PERMIT NO. 275 SAN FRANCISCO CA

POSTAGE WILL BE PAID BY ADDRESSEE

BENJAMIN CUMMINGS
PEARSON EDUCATION
1301 SANSOME STREET
SAN FRANCISCO CA 94111-9328

SUMMARY TABLES

BIOSKILLS